BIOLOGY

SIXTH EDITION

Neil A. Campbell

University of California, Riverside

Jane B. Reece

Palo Alto, California

Contributors and Advisors

Charles J. Krebs, University of British Columbia

Mark Ridley, Oxford University

Peter V. Minorsky, Vassar College

Mark A. Chappell, University of California, Riverside

Linda E. Graham, University of Wisconsin, Madison

M. Deric Bownds, University of Wisconsin, Madison

Mary Jane Niles, University of San Francisco

Jeff Hardin, University of Wisconsin, Madison

Garland E. Allen, Washington University

San Francisco Boston New York
Cape Town Hong Kong London Madrid Mexico City
Montreal Munich Paris Singapore Sydney Tokyo Toronto

WORLD STUDENT SERIES

Senior Editor: *Beth Wilbur*

Developmental Manager, Text and Media: *Pat Burner*

Senior Project Manager: *Ginnie Simione Jutson*

Project Editor: *Evelyn Dahlgren*

Senior Producer, Art and Media: *Russell Chun*

Developmental Artist: *Carla Simmons*

Senior Art Editor: *Donna Kalal*

Marketing Manager: *Josh Frost*

Manufacturing Supervisor: *Vivian McDougal*

Media Assistant Editor: *Aaron Gass*

Publishing Assistants: *David De Rouen, Krystina Sibley*

Senior Production Editor: *Jamie Sue Brooks*

Illustrations: *Precision Graphics*

Text Designer: *Detta Penna*

Cover Designer: *Roy Neuhaus*

Photo Researchers: *Travis Amos, Stephen Forsling*

Copy Editor: *Janet Greenblatt*

Compositor and Prepress: *GTS Graphics*

Cover Printer: *The Lehigh Press, Inc.*

Printer: *Von Hoffmann Press, Inc.*

On the cover: Photograph of an agave plant, 1991,
by Brett Weston. Courtesy of the Brett Weston Archive.
Special thanks to Dennis High, Center for Photographic Art,
for his advice and assistance with cover research.

Credits continue following the appendices.

ISBN 0-201-75054-6

4 5 6 7 8 9 10—VHP—06 05 04 03

www.aw.com/bc

ABOUT THE AUTHORS

Neil A. Campbell combines the investigative nature of a research scientist with the heart of an experienced and caring teacher. He earned his M.A. in Zoology from UCLA and his Ph.D. in Plant Biology from the University of California, Riverside, where he received the Distinguished Alumnus Award in 2001. Dr. Campbell has published numerous research articles on how certain desert and coastal plants thrive in salty soil and how the sensitive plant (*Mimosa*) and other legumes move their leaves. His 30 years of teaching in diverse environments include general biology courses at Cornell University, Pomona College, and San Bernardino Valley College, where he received the college's first Outstanding Professor Award in 1986. Dr. Campbell is currently a visiting scholar in the Department of Botany and Plant Sciences at the University of California, Riverside. In addition to his authorship of this book, he coauthors *Biology: Concepts and Connections* and *Essential Biology* with Jane Reece. Each year, over 500,000 students worldwide use Campbell/Reece biology textbooks.

Jane B. Reece has worked in biology publishing since 1978, when she joined the editorial staff of Benjamin Cummings. Her education includes an A.B. in Biology from Harvard University, an M.S. in Microbiology from Rutgers University, and a Ph.D. in Bacteriology from the University of California, Berkeley. At UC Berkeley, and later as a postdoctoral fellow in genetics at Stanford University, her research focused on genetic recombination in bacteria. She taught biology at Middlesex County College (New Jersey) and Queensborough Community College (New York). As an editor at Benjamin Cummings, Dr. Reece played major roles in a number of successful textbooks, including *Molecular Biology of the Gene*, Fourth Edition, by J. D. Watson et al. In addition to being a coauthor with Neil Campbell on *BIOLOGY, Biology: Concepts and Connections* and *Essential Biology,* she coauthored *The World of the Cell*, Third Edition, with W. M. Becker and M. F. Poenie.

PREFACE

Students come to biology during its golden age. The exhilarating progress in our understanding of life at all levels, from molecules and cells to ecosystems and the biosphere, is bound to inspire the inquisitive minds of undergraduates. And life becomes more fascinating even as it becomes less mysterious, for headway on one question leads to a dozen others that will captivate curiosity for decades to come. Today's students will also find biology at the center of current culture, more visible and more important than ever. Biology news has moved to the front page.

Reflecting the changing landscape of the subject it surveys, this Sixth Edition of *BIOLOGY* is our most sweeping revision ever. But throughout the revision process, we have kept in mind the two core goals of all earlier editions: to explain the key concepts of biology clearly and accurately within a context of unifying themes, and to help students develop positive and realistic impressions of science as a process of inquiry. These two teaching values evolved in the classroom, and we are obviously gratified that the book's dual emphases on concept-building and the process of science have appealed to the educators and students who have made *BIOLOGY* the most widely used college science textbook. But with this privilege of sharing biology with so many students comes the responsibility to continue improving the book to serve the biology community even better. Thus, throughout the entire planning and revision process, we visited dozens of campuses to listen to what students and their professors had to say about their biology courses and textbooks. These conversations with faculty and students led to the many improvements you'll find in this Sixth Edition.

An Even Greater Emphasis on the Process of Science

"How we know" and "what we do not yet know" are as important for students to appreciate as "what we know." The process of science has always been one of *BIOLOGY*'s [unifying] themes, but we have increased its presence in the [book]. First, we revised the introduction to science in [Chapter 1] by [setting] up the 215 examples of scientific inquiry embedded [throughout] the book. These case studies are announced [with a special heading, "]The Process of Science." An example is the [greater] prairie chicken and the extinction

vortex in Chapter 55 (Conservation Biology). Also new to this edition, a Process of Science question at the end of each chapter encourages student inquiry. On the CD-ROM and website that accompany *BIOLOGY*, students will find Case Studies in the Process of Science that give them additional practice with science skills, such as forming and testing hypotheses, collecting and analyzing data, and critically evaluating the evidence bearing on current debates in biology. And eight new interviews with influential researchers, which introduce the eight units of the book, humanize science and portray it as a social activity of creative men and women.

An Evolution Theme More Pervasive Than Ever

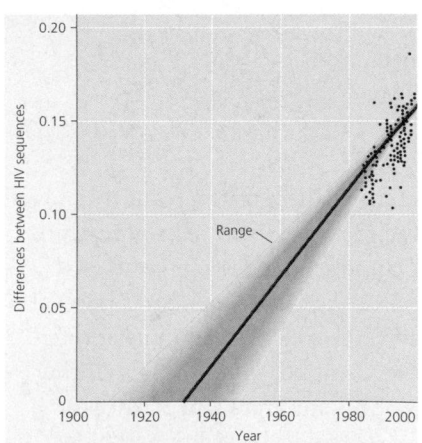

Dating the origin of HIV-1 M with a molecular clock.

The use of ten unifying themes, introduced in Chapter 1 and serving as touchstones for students throughout the book, continues to distinguish *BIOLOGY* from a topical, encyclopedic approach. And as the one central theme that unifies all of biology by accounting for both the unity and diversity of life, evolution is woven into the fabric of every chapter of our book. The power of this integrating theme becomes even more essential as the discovery explosion in biology threatens to suffocate students under an avalanche of what would otherwise seem to be unrelated information. Thus, we have made the evolution theme more pervasive than ever in this new edition of *BIOLOGY*. For example, a new section in Chapter 25 (Phylogeny and Systematics) explains how researchers used a molecular clock to date the origin of HIV. A new Evolution Connection question at the end of each chapter will help students fit what they have learned in the chapter into a broader biological context.

New Guided Tour Diagrams and Other Major Improvements of *BIOLOGY*'s Graphics

Biology is a visual science, and many of our students are visual learners. As authors, we take as much care in creating new ways to illustrate biology as we do in writing the narrative. With each edition, we have found ways to improve the teaching effectiveness of the book's diagrams, and many students have taken the time to let us know how much the illustrations help them construct their understanding of key concepts. Still, in planning the Sixth Edition, we thought it was time for our most ambitious revision of the art program ever. To illustrate new examples and reinforce the most difficult concepts, we have increased the total number of illustrations by more than 150. In addition, new Guided Tour Diagrams merge art and text to walk students through complex biological structures and processes. You can see an example below.

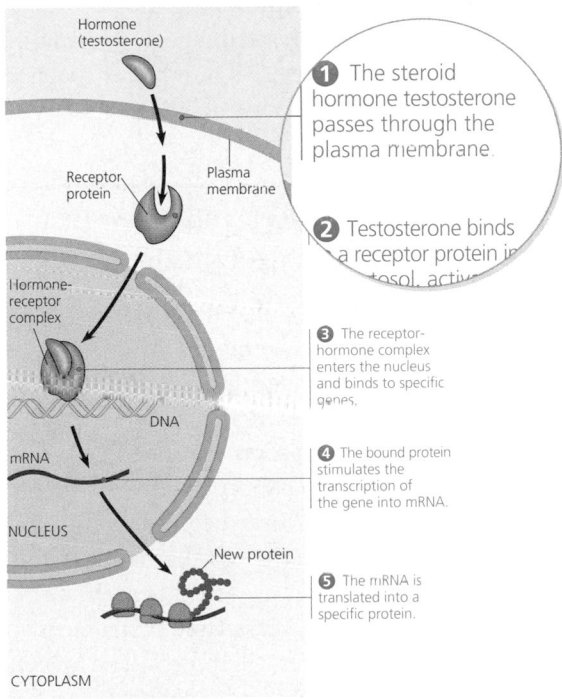

Enriched Chapter Reviews

We have learned from students how much they value chapter reviews, especially the practice questions. Therefore, we have countered the trend of reducing the learning tools available to students at the ends of chapters. In fact, the chapter reviews in this Sixth Edition of *BIOLOGY* are more robust than ever. In addition to the chapter summary and multiple-choice questions of earlier editions, the chapter reviews now include short-answer questions, as requested by many students and instructors. And three categories of essay questions—Evolution Connection; The Process of Science; and Science, Technology, and Society—give students an opportunity to analyze and write about biology and its applications.

The New Media Package That Instructors and Students Have Asked For

Just one example of *BIOLOGY*'s expanded media package is the new Campbell Image Presentation Library for instructors, a chapter-by-chapter digital archive that includes over 1600 photos (the text's photo program plus hundreds of additional selections), all of the book's art in a variety of formats, tables, over 100 animations, and over 80 video clips. The diagrams, photos, and tables are also provided as PowerPoint slides.

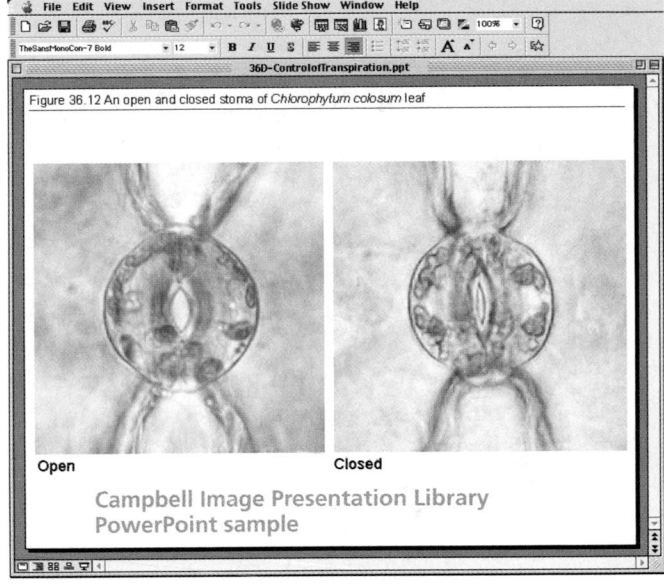

Campbell Image Presentation Library PowerPoint sample

For the student, the CD-ROM and website that support the book emphasize active learning, including 55 Case Studies in the Process of Science, 230 Activities, interactive Chapter Reviews, word roots, key terms, pre-tests, activities quizzes, chapter quizzes, essay questions, web links, news links, news articles, further readings, art from the book, over 80 videos, the Campbell Biology Interviews from all editions, and a glossary with audio pronunciations.

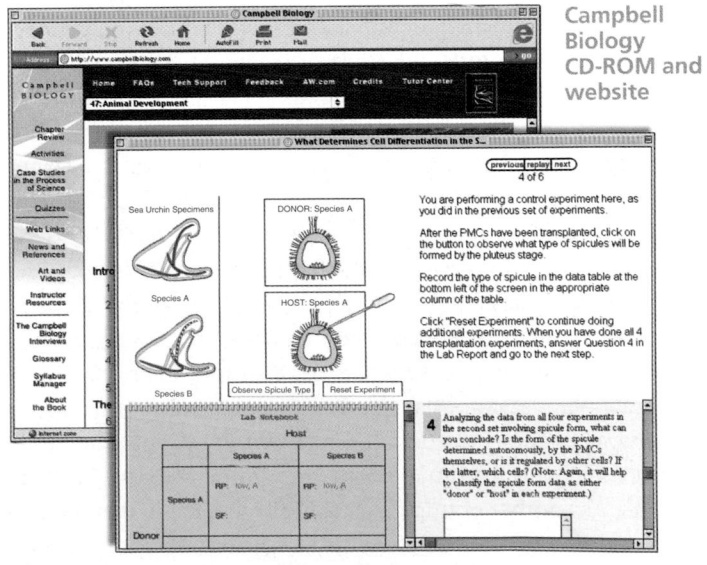

Campbell Biology CD-ROM and website

Concepts and Themes That Work Together

BIOLOGY's key concepts and unifying themes work together to help students develop a coherent view of life. For example, "Feedback mechanisms control cellular respiration" is a concept specific to the topic of Chapter 9. However, one of the book's overarching themes, the theme of regulation, helps students fit the concept of controlling respiration into a broader context that applies to many other biological processes. The key concepts give form to each chapter; the integrating themes connect the concepts and give form to the whole book.

The Breadth and Depth to Support a Diversity of Courses and to Serve Students Throughout Their Biology Education

Even by limiting our scope to each chapter's key concepts, *BIOLOGY* spans more biological territory than most introductory courses could or should attempt to explore. But given the great diversity of general biology syllabi, we have opted for a survey broad enough and deep enough to support each instructor's special emphases. Students also seem to appreciate *BIOLOGY*'s balance of breadth and depth—more than 70% of students who have used *BIOLOGY* have kept it after completing their introductory course, apparently to have a more basic context to complement their upper division studies. In fact, we receive numerous letters and emails from upper division and graduate students, including medical students, expressing their appreciation for the long-term value of *BIOLOGY* as a general resource for their continuing education.

A Versatile Organization

Just as we recognize that few courses will cover all 55 chapters of the textbook, *BIOLOGY* also makes no pretense that there is one "correct" sequence of topics for a general biology course. Though a biology textbook's table of contents must be linear, biology itself is more like a web of related ideas without a single starting point or a prescribed path. Diverse courses can navigate this network of concepts starting with molecules and cells, or with evolution and the diversity of organisms, or with the big-picture ideas of ecology. We have built *BIOLOGY* to be versatile enough to serve these different syllabi. The eight units of the book are largely self-contained, and most of the chapters within each unit can be assigned in a different sequence without substantial loss of coherence. For example, instructors who integrate plant and animal physiology into a systems approach can merge chapters from Unit Six (Plant Form and Function) and Unit Seven (Animal Form and Function) to fit their courses. As another example, instructors who begin their course with ecology and continue with this "top-down" approach can

assign Unit Eight (Ecology) right after Chapter 1 (Ten Themes in the Study of Life). It is the themes introduced in Chapter 1 that make the book so versatile in organization by providing students with a strong overall context no matter what the topic order of the course syllabus.

The Added Expertise of Leading Scientists

Given the rapid progress across biology's many fields, our challenge for this new edition was to *update* without simply *adding-on*. What, we asked, are the most important breakthroughs in each field that should be included in the Sixth Edition? And, even harder to decide, what can be left out? In adapting to the growth of biology, we did not want the book to move away from its thematic, conceptual tradition to become encyclopedic. The goal was to identify the most important questions in each field of biology and make the research advances on those questions accessible to students without compromising the book's reputation for scientific accuracy. We achieved this goal by recruiting a team of stellar specialists to help revise a number of chapters. These scientists are listed on the title page.

An Overview of *BIOLOGY* and a Few Examples of What's New in the Sixth Edition

Unit One: The Chemistry of Life Many students struggle in general biology courses because they are uncomfortable with basic chemistry. Chapters 2–4 help such students by developing just the chemical concepts 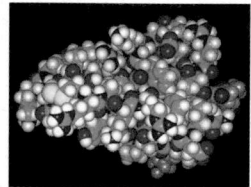 that are essential to success in a beginning course. We designed the chapters of Unit One so that students of diverse educational backgrounds can use them for self-study, reducing the amount of valuable class time instructors need for reviewing chemistry. However, Chapter 5 (The Structure and Function of Macromolecules) and Chapter 6 (An Introduction to Metabolism) provide important orientation even for students with solid chemistry backgrounds. Just one example of the many improvements in Unit One is an updated discussion of chaperonins in Chapter 5.

Unit Two: The Cell Chapters 7–12 build the study of cells upon the theme of correlation between structure and function. For example, we emphasize the role of mem- 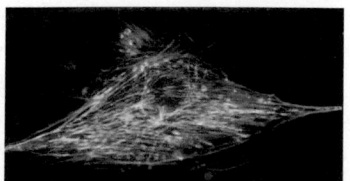 branes in ordering cell functions throughout the unit. In Chapter 7 (A Tour of the Cell), the improvements in this edition include completely new overview figures of animal and plant cells. Chapter 8 now includes the facilitated diffusion of water across plasma membranes by aquaporins. We have improved Chapter 11 (Cell Communication) in response to suggestions from instructors and students. This chapter synthesizes our current understanding of the basic mechanisms of chemical signaling at the cellular level, giving students a conceptual foundation for a variety of topics that arise in later chapters. For this edition, we have enhanced the treatment of intracellular (steroid) receptors, while maintaining coverage of this topic in Chapter 45 (Chemical Signals in Animals).

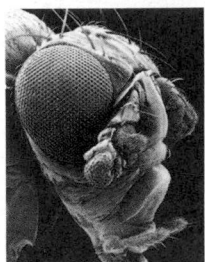

Unit Three: Genetics Chapters 13–21 trace the history of genetics, from Mendel to DNA technology, with the process of science as a major theme. We integrate extensive coverage of human genetics throughout the unit. We have thoroughly updated all of the molecular biology chapters—Chapters 16–21—to reflect recent advances in this fast-moving area. Recent progress in genomics has affected Chapter 20 (DNA Technology and Genomics) most extensively; new figures include a diagram outlining the two main strategies used in genome sequencing. This unit has been further strengthened by the redesign of many existing figures (for instance, the series of diagrams explaining DNA replication). Concluding Unit Three is Chapter 21 (The Genetic Basis of Development). This chapter, updated with the guidance of Jeff Hardin (University of Wisconsin, Madison), builds on molecular, cellular, and genetic principles to introduce the basic concepts of development that apply to both animals and plants. *Drosophila*, *C. elegans*, and *Arabidopsis* provide the main case studies here. (More extensive presentations of vertebrate development and plant development appear in later units.)

Unit Four: Mechanisms of Evolution As the central theme of *BIOLOGY*, evolution unifies the entire book. Chapters 22–25 focus specifically on *how* life evolves and how biologists study evolution and test evolutionary hypotheses. Oxford University biologist Mark Ridley helped us support current concepts throughout the unit with many fresh case studies from the research literature. Here are just a few examples of the improvements: Chapter 22 (Descent with Modification: A Darwinian View of Life) now includes HIV as a case study of natural selection in action; Chapter 23 (The Evolution of Populations) has a new section on the evolutionary biology of sex; in Chapter 24 (The Origin of Species), we have examined the different definitions of species in greater depth in this edition, and we have thoroughly revised the section on "evo-devo," the relationship of evolution to development; and Chapter 25 (Phylogeny and Systematics) uses new examples to walk students through the steps of cladistic analysis, including the construction of cladograms based on DNA-sequence data.

Unit Five: The Evolutionary History of Biological Diversity Renewed interest in the fossil record is complementing molecular systematics and cladistic analysis in arousing reevaluation of the history and classification of life. We have tried to capture the excitement of this scientific revolution. Unit Five's introductory chapter (Chapter 26) now features a chronology of major events in the history of life as an overview of the whole unit. In Chapter 27, the survey of prokaryotic diversity is based on the major clades of archaea and bacteria. Chapter 28 (The Origins of Eukaryotic Diversity) includes the hypothesis that all three domains of life had their origins in communities of gene-swapping prokaryotes. The chapter also relates the diversity of algae to the evolutionary history of plastids. In the chapters on plant diversity (29 and 30), Linda Graham of the University of Wisconsin, Madison, helped us explain the evolution of plants from charophycean algae and reorganized our survey of plant phyla to fit recent progress in plant systematics. In Chapter 31 on the fungi, the diagrams of life cycles have been thoroughly revised to emphasize the key reproductive adaptations of each phylum. The chapters on animal diversity (Chapters 32–34), like all others in the unit, present phylogenetic hypotheses and taxonomy as works in progress, subject to revision in light of new evidence. Throughout our extensive revision of Unit Five, our goal was to help students make sense of the current taxonomic turmoil while still providing a lucid survey of life's diversity. We would like students to understand why such issues as the relationship of annelids to arthropods remain unsettled, but we have not let our discussion of such debates obscure the biology of annelids, arthropods, and other major groups of organisms.

Unit Six: Plant Form and Function Plant biology is in the midst of a renaissance. The announcement in 2000 of the complete DNA sequence for the genome of *Arabidopsis*, the laboratory "white mouse" of modern plant research, commenced what promises to be an even livelier era in the study of plants. Peter Minorsky, science writer for the journal *Plant Physiology*, helped us reshape the plant unit to emphasize how research at the molecular and cellular levels is increasing our understanding of plant morphology, physiology, and development. We have updated the sections on the cellular basis of plant development and moved them to Chapter 35 to integrate development with our discussions of plant structure and growth. For example, Chapter 35 (Plant Structure and Growth) now features a research case study on the development of root hairs. The role of signal-transduction pathways is now an organizing concept throughout Chapter 39 (Plant Responses to Internal and External Signals).

Unit Seven: Animal Form and Function Chapters 40–49 explore adaptations that have evolved in diverse animal groups. Mark Chappell, University of California, Riverside, helped us give the themes of bioenergetics and regulation a stronger presence throughout the unit. For example, the introductory chapter (Chapter 40) now features a new section on the energy budgets of diverse animals. Mary Jane Niles, University of San Francisco, helped us update the immunology chapter (Chapter 43), with special attention to the immune systems of invertebrates, the vertebrate MHC, autoimmune diseases, and AIDS. Chapter 47 (Animal Development) is now both clearer and current, thanks to input from Jeff Hardin (University of Wisconsin, Madison) and Lisa Urry (Mills College). With the help of Deric Bownds, University of Wisconsin, Madison, we thoroughly updated the neuroscience in Chapters 48 and 49. For example, Chapter 48 features recent breakthroughs in our understanding of how the brain works. There are also new sections on how axons find their way during development and on the discovery and medical significance of stem cells in the brain.

Unit Eight: Ecology Chapters 50–55 emphasize the connections between ecology and evolution. The unit also makes the case for basic ecological research in an age when human activities are threatening the biosphere. We targeted the ecology chapters for especially extensive revision in this edition, and so we asked Charles Krebs, University of British Columbia, to contribute his expertise to the project. One of our main goals was to increase the emphasis on research case studies throughout the chapters, including analysis of many field experiments. The result amounts to a new ecology unit for *BIOLOGY*. Here are just a few examples of the improvements: Chapter 52 provides case studies of experiments ecologists perform to determine the factors that limit population growth; Chapter 53 contrasts the rivet model with the redundancy model in a discussion of community structure, and also features research on the bottom-up and top-down modes in the control of community structure; Chapter 54 on ecosystems includes a survey of experiments that test which nutrients limit productivity in various aquatic ecosystems; and Chapter 55 on conservation biology contrasts the small-population approach with the declining-population approach as strategies in slowing losses of biodiversity.

■   ■   ■

The real test for any textbook is how well it helps instructors teach and students learn. We welcome comments from the students and professors who use *BIOLOGY*. Please address your suggestions directly to one of the authors:

Neil Campbell
Department of Botany and Plant Sciences
University of California
Riverside, California 92521

or

Jane Reece
c/o Benjamin Cummings Publishing Company
1301 Sansome St.
San Francisco, California 94111

Acknowledgments

This Sixth Edition of *BIOLOGY* reflects the talents and hard work of many people, and the authors wish to express their deepest thanks to the numerous instructors, researchers, students, publishing professionals, and artists who have contributed to this and previous editions.

We are particularly grateful to the biologists who are listed as Contributors and Advisors on the title page. These nine experts guided our revision of material in several critical areas of biology. Distinguished ecologist Charles Krebs was an important advisor on the revision of the ecology unit, and contributed new material to Chapters 50 and 52–55. Evolutionary biologist Mark Ridley did the same for Chapters 22–26, 34, and 51 (and also wrote a number of challenging Evolution Connection questions). Plant biologists Linda Graham and Peter Minorsky lent their considerable expertise and good judgment to Chapters 29–30 and Chapters 35–39, respectively. Animal physiologist Mark Chappell provided important new insights and material for the revisions of Chapters 40–42 and 44, as did neuroscientist Deric Bownds for Chapters 48–49. Once again for this edition, immunologist Mary Jane Niles was the mastermind behind the revision of Chapter 43, and developmental biologist Jeff Hardin was our main guide for the updating of Chapters 21 and 47. Discussions with science historian Gar Allen inspired our efforts to give the process of science enhanced prominence in this edition. We also wish to acknowledge the contributions of Larry Mitchell to previous editions of this book.

Thanks also to the instructors who contributed new questions for the chapter reviews. In addition to Mark Ridley, these include James Costa (Western Carolina University), William Fixsen (Harvard University), Martha Taylor (Cornell University), and Lisa Urry (Mills College). It's not easy to write good questions, and we appreciate the time and effort these dedicated educators put into their creation of excellent ones.

Further helping us improve *BIOLOGY*'s scientific accuracy and pedagogy, 47 biologists, cited on page xi, provided detailed reviews of one or more chapters for this edition. Among these reviewers, we would like to single out Lisa Urry (Mills College) for contributions beyond the call of duty; her detailed suggestions for the genetics and development chapters were immensely helpful. Numerous other professors and students offered suggestions by writing directly to the authors. Those correspondents include: Scott Poethig (University of Pennsylvania), Michael Marcotrigiano (University of Massachusetts), Cyril Thong (Simon Fraser University), D. Reid Wiseman (College of Charleston), Paul Broady (University of Canterbury, NZ), Allan Markezich (Black Hawk College), Barbara Demming-Adams (University of Colorado), Jon Havenhand (Flinders University), Joseph Frankel (University of Iowa), Jim Grier (North Dakota State University), Lee Sola (Glendale Community College), Barbara Beitch (Hamden Hall Country Day School), Susan Riechert (University of Tennessee, Knoxville), Christa Schwintzer (University of Maine), Tony Hulbert (University of Wollongong), Tim Nelson (Seattle Pacific University), Sandra Baldauf (University of York), Philip Bishop (Otago University), Alison Cree (Otago University), Raymond White (City College of San Francisco), and Grahame Kelley (Queensland University of Technology).

Of course, the authors alone bear the responsibility for any errors that remain in the text, but the dedication of our contributors, reviewers, and correspondents makes us especially confident in the accuracy of this edition.

Many scientists have also helped shape this Sixth Edition by discussing their research fields and ideas about biology education. Neil Campbell thanks many UC Riverside colleagues, including Ring Carde, Richard Cardullo, Mark Chappell, Darleen DeMason, Norman Ellstrand, Alan Fix, Kimberly Hammond, Anthony Huang, Bradley Hyman, Tracy Kahn, Elizabeth Lord, Carol Lovatt, Eugene Nothnagel, John Oross, Timothy Paine, David Reznick, Rodolfo Ruibal, Clay Sassaman, Mark Springer, Nelson Thompson, William Thomson, John Trumble, and John Moore (whose "Science as a Way of Knowing" essays have had such an important influence on the evolution of *BIOLOGY*). Jane Reece thanks Mark Guyer of the National Human Genome Research Institute for continuing to serve as an invaluable resource for Chapter 20.

The value of *BIOLOGY* as a teaching tool is greatly enhanced by the supplementary materials that have been created for instructors and students. We wish to thank Ed Zalisko and David Reid of Blackburn College and Eric Simon of Fordham University for their work on the Instructor's Guide and Steve Norton (East Carolina University) for the PowerPoint Lectures. Special thanks are due to Bill Barstow (University of Georgia) and his team of contributors (Michael Dini, Eugene Fenster, Conrad Firling, Kurt Redborg, Marshall Sundberg, Catherine Wilcoxson Ueckert, and Robert Yost) for their extensive revision of the Test Bank. Martha Taylor has once again written a superb student Study Guide. And we are grateful to the many people—biology instructors, editors, artists, production experts, and narrators— who are listed in the credits for the very impressive electronic media that accompany the book.

Interviews with prominent scientists have been a hallmark of *BIOLOGY* since its inception, and conducting these interviews was one of the great pleasures of revising the text. To open the eight units of this Sixth Edition, we are proud to include interviews with Thomas Eisner, George Langford, Nancy Hopkins, Peter and Rosemary Grant, Paul Sereno, Joanne Chory, Flossie Wong-Staal, and David Schindler.

BIOLOGY, Sixth Edition, results from an unusually strong synergism between a team of scientists and a team of publishing professionals. A new design, the sweeping revision of the art as well as the text, and a rich media package combined to create unprecedented challenges for the publishing team.

The members of the core book team at Benjamin Cummings brought extraordinary talents and extraordinary hard work to this project. Developmental Manager Pat Burner, our colleague of many years, again worked closely with the text and art from manuscript stage through production and directed and helped create the elaborate media program that accompanies the book. For this Sixth Edition, however, Pat also amazingly found time to research and write up-to-date sections on nerve cell development and neural stem cells for Chapter 48! So, in addition to our thanks for Pat's exceptional editorial talents and dedication, we are now delighted to express our gratitude for her biological scholarship.

Pat's partner in directing the media program, Senior Producer Russell Chun, brought to the project not only his formidable skills in biological illustration and computer animation but an innate pedagogical sense that enhanced everything he touched. Also serving as art director for the book, Russell was the originator of the Guided Tour concept used for many of the more complex illustrations in this edition, and an important technical expert on the extended media package.

The other senior members of the core book team brought complementary talents to the project. Senior Project Manager Ginnie Simione Jutson miraculously managed to keep the whole team, including the authors, on track; somehow she managed to be both tough and encouraging. She was also the project leader on the complex new Campbell Image Presentation Library. Project Editor Evelyn Dahlgren made numerous contributions during manuscript development and production; this book has benefited greatly from her sharp-eyed and patient attention to text, art, and layout. In addition, Evelyn managed the supplements package and was a key member of the Image Presentation Library team. Senior Production Editor Jamie Sue Brooks was responsible for the complex design and production process and brought a wonderful design sensibility to the project.

We are also grateful to production professionals Diane Southworth, Steven Anderson, and Ariel Sosna, and to Manufacturing Supervisor Vivian McDougal for their expertise and hard work. For their tireless efforts in producing the student media package, we would like to thank the in-house media production team: Lauren Fogel, Andrew Corbett, Jim Hufford, Marlene Dabis, Michael Walsh, Roman Tobe, and Phaedra Schroeder. Little would have gotten done, of course, without the support of our bright and talented publishing assistants, including Aaron Gass (who went on to work as Assistant Editor coordinating the student media), Nina Lewallen, David De Rouen, and Krystina Sibley, all of whom performed numerous tasks relating to both the text and media package.

And now we come to our incomparable editor, Senior Editor Beth Wilbur. Joining the company just as this complex project was getting under way, Beth has faced a series of daunting challenges. But with every new challenge, she has risen to the task with creative solutions and fresh enthusiasm, and she has earned our heartfelt thanks for pulling it all together. We are proud to have Beth represent our book in the academic community, and we are proud to have her as our colleague and friend.

In addition to the in-house editorial and production staff, a number of talented freelancers made valuable contributions to this book. We are grateful for the help of developmental editors Robin Fox, Moira Lerner, and Richard Morel; copyeditors Janet Greenblatt (who also helped develop the Ecology unit), Jan McDearmon, and Eleanor Renner Brown; text designer Detta Penna; cover designer Roy Neuhaus; layout artists Carolyn Deacy, Andrew Ogus, Leigh McLellan, and Jennifer Dunn; proofreaders Martha Ghent, Anita Wagner, Yonie Overton, Joseph Curran, Joan Saunders, Sue Rudolph, Maria Paras, Anna Trabucco, and Arlene Larson (our biologist proofreader); indexer Charlotte Shane, SAIS; and production editor Sharon Page Ritchie. Stephanie Kuhns and Kathi Townes at TechArts keyboarded and proofread final manuscript. Also, we thank Shelley Parlante for her work on the early stages of development and for taking time off from her current high-tech job to help check dummy. We are grateful for the production expertise and extra commitment of Larry Lazopoulos, Kim Johnson, Suzanne Olivier, and especially for the devotion of John Burner, who worked on review of dummy and other late-production tasks.

The illustrations in *BIOLOGY* are key elements of its teaching effectiveness. Russell Chun personally developed much of the art. His partner in that endeavor was our longtime colleague Carla Simmons, a gifted freelance biological artist. This book has benefited from Carla's artistic and pedagogical talents since its first edition. We were again fortunate to have the indomitable Senior Art Editor Donna Kalal on the team; she shepherded all of our art through production with scrupulous attention to thousands of details and unswerving commitment to quality. We wish to thank all the artists who worked on the new and revised figures for this edition, including Russell, Carla, Caitlin Duckwall, and the artists of Precision Graphics. Photo researchers Stephen Forsling and Travis Amos searched for just the right photos to reinforce key concepts. Travis also carried out the herculean task of collecting a large number of excellent photos for the Campbell Image Presentation Library, and was an important advisor, contributor, and colleague on the development of that important new supplement. Robin Heyden worked with Travis on the Image Library, as did Graham Kent (Smith College), who also contributed to the photo program for the text.

We would also like to thank the hard-working employees of GTS Graphics, especially Sherrill Redd, and Von Hoffman Press.

Linda Davis, President of Benjamin Cummings Publishing, has shared our commitment to excellence and provided strong support for two editions now, and new Editorial Director Frank Ruggirello has added his own unique brand of support and enthusiasm in the final stages of preparation of this edition. We thank our former editor Erin Mulligan for her contributions to the early stages of the project.

Both before and after publication, we are fortunate to have the support of the Benjamin Cummings marketing professionals. We gratefully acknowledge the contributions of Marketing Manager Josh Frost and Marketing VP Stacy Treco and her team. We also thank Lillian Carr for designing an elegant brochure and Jessica McFadden for the e-brochure.

The Benjamin Cummings field staff, which represents *BIOLOGY* on campus, is our living link to the students and professors who use the text. The field representatives tell us what you like and don't like about the book, and they provide prompt service to biology departments. The field reps are good allies in science education, and we thank them for their professionalism in communicating the features of our book.

Finally, we wish to thank our families and friends for their encouragement and for enduring our obsession with *BIOLOGY*.

Neil Campbell
Jane Reece

Reviewers of the Sixth Edition

Estry Ang, *University of Pittsburgh at Greensburg*
Howard J. Arnott, *University of Texas at Arlington*
Andrew R. Blaustein, *Oregon State University*
Deric Bownds, *University of Wisconsin, Madison*
Robert Boyd, *Auburn University*
Paul Broady, *University of Canterbury*
David Byres, *Florida Community College, Jacksonville*
Robert E. Cannon, *University of North Carolina at Greensboro*
Joseph P. Chinnici, *Virginia Commonwealth University*
Ross C. Clark, *Eastern Kentucky University*
James T. Costa, *Western Carolina University*
Joe W. Crim, *University of Georgia*
W. Marshall Darley, *University of Georgia*
Michael Dini, *Texas Tech University*
William D. Eldred, *Boston University*
Margaret T. Erskine, *Lansing Community College*
Peter Fajer, *Florida State University*
Eugene Fenster, *Longview Community College*
Robert G. Fowler, *San Jose State University*
Bill Freedman, *Dalhousie University*
Frank S. Gilliam, *Marshall University*
Wayne Goodey, *University of British Columbia*
Jeff Hardin, *University of Wisconsin, Madison*
Mark Ikeda, *San Bernardino Valley College*
John C. Jahoda, *Bridgewater State College*
Randall Johnson, *University of California, San Diego*
Thomas C. Kane, *University of Cincinnati*
Harvey Liftin, *Broward Community College*
Margaret A. Lynch, *Tufts University*
Jeffrey D. May, *Marshall University*
Phillip Meneely, *Haverford College*
Darrel L. Murray, *University of Illinois at Chicago*
Peter Nonacs, *University of California, Los Angeles*
Steve Norton, *East Carolina University*
Gary P. Olivetti, *University of Vermont*
John Olsen, *Rhodes College*
Shelley Penrod, *North Harris College*
Kurt Redborg, *Coe College*
David Reid, *Blackburn College*
Gary W. Saunders, *University of New Brunswick*
Frederick W. Spiegel, *University of Arkansas*
Marshall D. Sundberg, *Emporia State University*
Martha R. Taylor, *Cornell University*
Lisa A. Urry, *Mills College*
Thomas J. Volk, *University of Wisconsin, La Crosse*
David J. Westenberg, *University of Missouri, Rolla*
Edward Zalisko, *Blackburn College*

Reviewers of Previous Editions

Kenneth Able (State University of New York, Albany), Martin Adamson (University of British Columbia), John Alcock (Arizona State University), Richard Almon (State University of New York, Buffalo), Katherine Anderson (University of California, Berkeley), Richard J. Andren (Montgomery County Community College), J. David Archibald (Yale University), Robert Atherton (University of Wyoming), Leigh Auleb (San Francisco State University), P. Stephen Baenziger (University of Nebraska), Katherine Baker (Millersville University), William Barklow (Framingham State College), Steven Barnhart (Santa Rosa Junior College), Ron Basmajian (Merced College), Tom Beatty (University of British Columbia), Wayne Becker (University of Wisconsin, Madison), Jane Beiswenger (University of Wyoming), Anne Bekoff (University of Colorado, Boulder), Marc Bekoff (University of Colorado, Boulder), Tania Beliz (College of San Mateo), Adrianne Bendich (Hoffman-La Roche, Inc.), Barbara Bentley (State University of New York, Stony Brook), Darwin Berg (University of California, San Diego), Werner Bergen (Michigan State University), Gerald Bergstrom (University of Wisconsin, Milwaukee), Anna W. Berkovitz (Purdue University), Dorothy Berner (Temple University), Annalisa Berta (San Diego State University), Paulette Bierzychudek (Pomona College), Charles Biggers (Memphis State University), Judy Bluemer (Morton College), Robert Blystone (Trinity University), Robert Boley (University of Texas, Arlington), Eric Bonde (University of Colorado, Boulder), Richard Boohar (University of Nebraska, Omaha), Carey L. Booth (Reed College), James L. Botsford (New Mexico State University), J. Michael Bowes (Humboldt State University), Richard Bowker (Alma College), Barry Bowman (University of California, Santa Cruz), Deric Bownds (University of Wisconsin, Madison), Jerry Brand (University of Texas, Austin), Theodore A. Bremner (Howard University), James Brenneman (University of Evansville), Charles H. Brenner (Berkeley, California), Donald P. Briskin (University of Illinois, Urbana), Danny Brower (University of Arizona), Carole Browne (Wake Forest University), Mark Browning (Purdue University), Herbert Bruneau (Oklahoma State University), Gary Brusca (Humboldt State University), Alan H. Brush (University of Connecticut, Storrs), Meg Burke (University of North Dakota), Edwin Burling (De Anza College), William Busa (Johns Hopkins University), John Bushnell (University of Colorado), Linda Butler (University of Texas, Austin), Iain Campbell (University of Pittsburgh), Deborah Canington (University of California, Davis), Gregory Capelli (College of William and Mary), Richard Cardullo (University of California, Riverside), Nina Caris (Texas A & M University), Bruce Chase (University of Nebraska, Omaha), Doug Cheeseman (De Anza College), Shepley Chen (University of Illinois, Chicago), Henry Claman (University of Colorado Health Science Center), Lynwood Clemens (Michigan State University), William P. Coffman (University of Pittsburgh), J. John Cohen (University of Colorado Health Science Center), David Cone (Saint Mary's University), John Corliss (University of Maryland), Stuart J. Coward (University of Georgia), Charles Creutz (University of Toledo), Bruce Criley (Illinois Wesleyan University), Norma Criley (Illinois Wesleyan University), Richard Cyr (Pennsylvania State University), Marianne Dauwalder (University of Texas, Austin), Bonnie J. Davis (San Francisco State University), Jerry Davis (University of Wisconsin, La Crosse), Thomas Davis (University of New Hampshire), John Dearn (University of Canberra), James Dekloe (University of California, Santa Cruz), T. Delevoryas (University of Texas, Austin), Diane C. DeNagel (Northwestern University), Jean DeSaix (University of North Carolina), Michael Dini (Texas Tech University), Andrew Dobson (Princeton University), John Drees (Temple University School of Medicine), Charles Drewes (Iowa State University), Marvin Druger (Syracuse University), Susan Dunford (University of Cincinnati), Betsey Dyer (Wheaton College), Robert Eaton (University of Colorado), Robert S. Edgar (University of California, Santa Cruz), Betty J. Eidemiller (Lamar University), David Evans (University of Florida), Robert C. Evans (Rutgers University, Camden), Sharon Eversman (Montana State University), Lincoln Fairchild (Ohio State University), Bruce Fall (University of Minnesota), Lynn Fancher (College of DuPage), Larry Farrell (Idaho State University), Jerry F. Feldman (University of California, Santa Cruz), Russell Fernald (University of Oregon), Milton Fingerman (Tulane University), Barbara Finney (Regis College), David Fisher (University of Hawaii, Manoa), William Fixsen (Harvard University), Abraham Flexer (Manuscript Consultant, Boulder, Colorado), Kerry Foresman (University of Montana), Norma Fowler (University of Texas, Austin), David Fox (University of Tennessee, Knoxville), Carl Frankel (Pennsylvania State University, Hazleton), Otto Friesen (University of Virginia), Virginia Fry (Monterey Peninsula College), Alice Fulton (University of Iowa), Sara Fultz (Stanford University), Berdell Funke (North Dakota State University), Anne Funkhouser (University of the Pacific), Arthur W. Galston (Yale University), Carl Gans (University of Michigan), John Gapter (University of Northern Colorado), Reginald Garrett (University of Virginia), Patricia Gensel (University of North Carolina), Chris George (California Polytechnic State University, San Luis Obispo), Robert George (University of Wyoming), Frank Gilliam (Marshall University), Simon Gilroy (Pennsylvania State University), Todd Gleeson (University of Colorado), David Glenn-Lewin (Wichita State University), William Glider (University of Nebraska), Elizabeth A. Godrick (Boston University), Lynda Goff (University of California, Santa Cruz), Elliott Goldstein (Arizona State University), Paul Goldstein (University of Texas, El Paso), Anne Good (University of California, Berkeley), Judith Goodenough (University of Massachusetts, Amherst), Ester Goudsmit (Oakland University), Linda Graham (University of Wisconsin, Madison), Robert Grammer (Belmont University), Joseph Graves (Arizona State University), A. J. F. Griffiths (University of British Columbia), William Grimes (University of Arizona), Mark Gromko (Bowling Green State University), Serine Gropper (Auburn University), Katherine L. Gross (Ohio State University), Gary Gussin (University of Iowa), Mark Guyer (National Human Genome Research Institute), Ruth Levy Guyer (Bethesda, Maryland), R. Wayne Habermehl (Montgomery County Community College), Mac Hadley (University of Arizona), Jack P. Hailman (University of Wisconsin), Leah Haimo (University of California, Riverside), Rebecca Halyard (Clayton State College), Penny Hanchey-Bauer (Colorado State University), Laszlo Hanzely (Northern Illinois University), Jeff Hardin (University of Wisconsin, Madison), Richard Harrison (Cornell University), H. D. Heath (California State University, Hayward), George Hechtel (State University of New York, Stony Brook), Jean Heitz-Johnson (University of Wisconsin, Madison), Colin Henderson (University of Montana), Caroll Henry (Chicago State University), Frank Heppner (University of Rhode Island), Ira Herskowitz (University of California, San Francisco), Paul E. Hertz (Barnard College), R. James Hickey (Miami University), Ralph Hinegardner (University of California, Santa Cruz), William Hines (Foothill College), Helmut Hirsch (State University of New York, Albany), Tuan-hua David Ho (Washington University), Carl Hoagstrom (Ohio Northern University), James Hoffman (University of Vermont), James Holland (Indiana State University, Bloomington), Charles Holliday (Lafayette College), Laura Hoopes (Occidental College), Nancy Hopkins (Massachusetts Institute of Technology), Kathy Hornberger (Widener University), Pius F. Horner (San Bernardino Valley College), Margaret Houk (Ripon College), Ronald R. Hoy (Cornell University), Donald Humphrey (Emory University School of Medicine), Robert J. Huskey (University of Virginia), Steven Hutcheson (University of Maryland, College Park), Bradley Hyman (University of California, Riverside), Alice Jacklet (State University of New York, Albany), John Jackson (North Hennepin Community College), Dan Johnson (East Tennessee State University), Wayne Johnson (Ohio State University), Kenneth C. Jones (California State University, Northridge), Russell Jones (University of California, Berkeley), Alan Journet (Southeast Missouri State University), Thomas Kane (University of Cincinnati), E. L. Karlstrom (University of Puget Sound), George Khoury (National Cancer Institute), Robert Kitchin (University of Wyoming), Attila O. Klein (Brandeis University), Greg Kopf (University of Pennsylvania School of Medicine), Thomas Koppenheffer (Trinity University), Janis Kuby (San Francisco State University), J. A. Lackey (State University of New York, Oswego), Lynn Lamoreux (Texas A & M University), Carmine A. Lanciani (University of Florida), Kenneth Lang (Humboldt State University), Allan Larson (Washington University), Diane K. Lavett (State University of New York, Cortland, and Emory University), Charles Leavell (Fullerton College), C. S. Lee (University of Texas), Robert Leonard (University of California, Riverside), Joseph Levine (Boston College), Bill Lewis (Shoreline Community College), John Lewis (Loma Linda University), Lorraine Lica (California State University, Hayward), Harvey Lillywhite (University of Florida, Gainesville), Sam Loker (University of New Mexico), Jane Lubchenco (Oregon State University), James MacMahon (Utah State University), Charles Mallery (University of Miami), Lynn Margulis (Boston University), Edith Marsh (Angelo State University), Karl Mattox (Miami University of Ohio), Joyce Maxwell (California State University, Northridge), Richard McCracken (Purdue University), Jacqueline McLaughlin (Pennsylvania State University, Lehigh Valley), Paul Melchior (North Hennepin Community College), John Merrill (University of Washington), Brian Metscher (University of California, Irvine), Ralph Meyer (University of Cincinnati), Roger Milkman (University of Iowa), Helen Miller (Oklahoma State University), John Miller (University of California, Berkeley), Kenneth R. Miller (Brown University), John E. Minnich (University of Wisconsin, Milwaukee), Michael Misamore (Louisiana State University), Kenneth Mitchell (Tulane University School of Medicine), Russell Monson (University of Colorado, Boulder), Frank Moore (Oregon State University), Randy Moore (Wright State University), William Moore (Wayne State University), Carl Moos (Veterans Administration Hospital, Albany, New York), Michael Mote (Temple University), Deborah Mowshowitz (Columbia University), John Mutchmor (Iowa State University), Elliot Myerowitz (California Institute of Technology), Gavin Naylor (Iowa State University), John Neess (University of Wisconsin, Madison), Raymond Neubauer (University of Texas, Austin), Todd Newbury (University of California, Santa Cruz), Harvey Nichols (University of Colorado, Boulder), Deborah Nickerson (University of South Florida), Bette Nicotri (University of Washington), Caroline Niederman (Tomball College), Maria Nieto (California State University, Hayward), Charles R. Noback (College of Physicians and Surgeons, Columbia University), Mary C. Nolan (Irvine Valley College), David O. Norris (University of Colorado, Boulder), Cynthia Norton (University of Maine, Augusta), Bette H. Nybakken (Hartnell College), Brian O'Conner (University of Massachusetts, Amherst), Gerard O'Donovan (University of North Texas), Eugene Odum (University of Georgia), Patricia O'Hern (Emory University), John Olsen (Rhodes College), Sharman O'Neill (University of California, Davis), Wan Ooi (Houston Community College), Gay Ostarello (Diablo Valley College), Barry Palevitz (University of Georgia), Peter Pappas (County College of Morris), Bulah Parker (North Carolina State University), Stanton Parmeter (Chemeketa Community College), Robert Patterson (San Francisco State University), Crellin Pauling (San Francisco State University), Kay Pauling (Foothill Community College), Daniel Pavuk (Bowling Green State University), Debra Pearce (Northern Kentucky University), Patricia Pearson (Western Kentucky University), Bob Pittman (Michigan State University), James Platt (University of Denver), Martin Poenie (University of Texas, Austin), Scott Poethig (University of Pennsylvania), Jeffrey Pommerville (Texas A & M University), Warren Porter (University of Wisconsin), Donald Potts (University of California, Santa Cruz), David Pratt (University of California, Davis), Halina Presley (University of Illinois, Chicago), Rebecca Pyles (East Tennessee State University), Scott Quackenbush (Florida International University), Ralph Quatrano (Oregon State University), Deanna Raineri (University of Illinois, Champaign-Urbana), Charles Ralph (Colorado State University), Brian Reeder (Morehead State University), C. Gary Reiness (Lewis & Clark College), Charles Remington (Yale University), David Reznick (University of California, Riverside), Fred Rhoades (Western Washington State University), Christopher Riegle (Irvine Valley College), Donna Ritch (Pennsylvania State University), Thomas Rodella (Merced College), Rodney Rogers (Drake University), Wayne Rosing (Middle Tennessee State University), Thomas Rost (University of California, Davis), Stephen I. Rothstein (University of California, Santa Barbara), John Ruben (Oregon State University), Albert Ruesink (Indiana University), Don Sakaguchi (Iowa State University), Walter Sakai (Santa Monica College), Mark F. Sanders (University of California, Davis), Ted Sargent (University of Massachusetts, Amherst), Gary Saunders (University of New Brunswick), Carl Schaefer (University of Connecticut), Lisa Shimeld (Crafton Hills College), David Schimpf (University of Minnesota, Duluth), William H. Schlesinger (Duke University), Erik P. Scully (Towson State University), Edna Seaman (Northeastern University), Elaine Shea (Loyola College, Maryland), Stephen Sheckler (Virginia Polytechnic Institute and State University), James Shinkle (Trinity University), Barbara Shipes (Hampton University), Peter Shugarman (University of Southern California), Alice Shuttey (DeKalb Community College), James Sidie (Ursinus College), Daniel Simberloff (Florida State University), Susan Singer (Carleton College), John Smarrelli (Loyola University), Andrew T. Smith (Arizona State University), John Smol (Queen's University), Andrew J. Snope (Essex Community College), Mitchell Sogin (Woods Hole Marine Biological Laboratory), Susan Sovonick-Dunford (University of Cincinnati), Karen Steudel (University of Wisconsin), Barbara Stewart (Swarthmore College), Cecil Still (Rutgers University, New Brunswick), John Stolz (California Institute of Technology), Richard D. Storey (Colorado College), Stephen Strand (University of California, Los Angeles), Eric Strauss (University of Massachusetts, Boston), Russell Stullken (Augusta College), John Sullivan (Southern Oregon State University), Gerald Summers (University of Missouri), Marshall Sundberg (Louisiana State University), Daryl Sweeney (University of Illinois, Champaign-Urbana), Samuel S. Sweet (University of California, Santa Barbara), Lincoln Taiz (University of California, Santa Cruz), Samuel Tarsitano (Southwest Texas State University), David Tauck (Santa Clara University), James Taylor (University of New Hampshire), Roger Thibault (Bowling Green State University), William Thomas (Colby-Sawyer College), John Thornton (Oklahoma State University), Robert Thornton (University of California, Davis), James Traniello (Boston University), Robert Tuveson (University of Illinois, Urbana), Maura G. Tyrrell (Stonehill College), Gordon Uno (University of Oklahoma), James W. Valentine (University of California, Santa Barbara), Joseph Vanable (Purdue University), Theodore Van Bruggen (University of South Dakota), Kathryn VandenBosch (Texas A & M University), Frank Visco (Orange Coast College), Laurie Vitt (University of California, Los Angeles), Susan D. Waaland (University of Washington), William Wade (Dartmouth Medical College), John Waggoner (Loyola Marymount University), Dan Walker (San Jose State University), Robert L. Wallace (Ripon College), Jeffrey Walters (North Carolina State University), Margaret Waterman (University of Pittsburgh), Charles Webber (Loyola University of Chicago), Peter Webster (University of Massachusetts, Amherst), Terry Webster (University of Connecticut, Storrs), Peter Wejksnora (University of Wisconsin, Milwaukee), Kentwood Wells (University of Connecticut), Stephen Williams (Glendale Community College), Christopher Wills (University of California, San Diego), Fred Wilt (University of California, Berkeley), Robert T. Woodland (University of Massachusetts Medical School), Joseph Woodring (Louisiana State University), Patrick Woolley (East Central College), Philip Yant (University of Michigan), Hideo Yonenaka (San Francisco State University), John Zimmerman (Kansas State University), Uko Zylstra (Calvin College).

Supplements for the Instructor

New! **Campbell Image Presentation Library** (0-8053-6632-6)
The new Campbell Image Presentation Library is a chapter-by-chapter visual archive that includes over 1600 photos from the text plus additional sources, all text art with and without labels in several convenient formats (pdf and jpeg or gif), selected figures layered for step-by-step presentation, all text tables, over 100 animations, over 80 video clips, and PowerPoint slides. All of the diverse images—art, tables, animations, photos, and videos—are organized by chapter. The art, photos, and tables are also provided as PowerPoint slides.

New! **Instructor's Guide to Text and Media** (0-8053-6635-0)
This comprehensive guide to *BIOLOGY*, Sixth Edition, includes chapter-by-chapter references to all the media resources. Thumbnail-sized images provide easy viewing of the Campbell Image Presentation Library.

Transparency Acetates (0-8053-6636-9) Approximately 1,000 acetates include all illustrations and tables in full color, many of which incorporate photographs from the text. New to this edition are selected figures broken down into a series of images for step-by-step lecture presentation. Brightened colors provide excellent viewing in the classroom or lecture hall.

Test Bank (0-8053-6637-7)
Computerized Test Bank (0-8053-6748-9)
Edited by William Barstow, University of Georgia Thoroughly revised and updated, the Sixth Edition test bank also includes new media activity test questions and the self-quizzes from the text. Available in print and on a cross-platform CD-ROM.

New! **PowerPoint Lectures** (0-8053-6756-X)
Steve Norton, East Carolina University Prepared PowerPoint lectures integrate the art, photos, tables, and lecture outline from each chapter. The PowerPoint Lectures can be used as is, or customized for your course with your own images and text and/or additional images from the Campbell Image Presentation Library.

Investigating Biology, Annotated Instructor's Edition, Fourth Edition (0-8053-7366-7) Teaching information, added to the original Student Edition text, includes margin notes with hints on lab procedures, additional art, and answers to in-text and end-of-chapter questions from the Student Edition. Also featured is a detailed Teaching Plan at the end of each lab with specific suggestions for organizing labs, including estimated time allotments and suggestions for encouraging independent thinking and collaborative discussion. Also available: **Preparation Guide for Investigating Biology**, **Fourth Edition** (0-8053-7367-5) Guides lab coordinators in ordering materials as well as in planning, setting up, and running labs.

New! **Symbiosis Book Building Kit—Customized Lab Manuals** (0-201-72142-2) Build a customized lab manual, choosing the labs you want, importing artwork from our graphics library, and even adding your own material, and get a made-to-order black and white lab manual. Visit **http://www.pearsoncustom.com/database/symbiosis.html** for more information.

New! **Course Management Systems** The content from the Campbell Biology website is also available in these popular course management systems: CourseCompass™, Blackboard, and WebCT. Visit **http://cms.aw.com** for more information.

Supplements for the Student

New! **Campbell Biology CD-ROM and Website**
www.campbellbiology.com The CD-ROM and website included with each book contain over 230 activities, over 80 videos, one or more Case Studies in the Process of Science for each chapter, interactive chapter reviews, objectives, word roots, key terms, several forms of assessment for each chapter (Pre-Test, Activities Quiz, Chapter Quiz, Essay Questions), a glossary with audio pronunciations, and the Campbell Biology Interviews from previous editions. In addition, the website provides access to an e-book, the Biology Tutor Center, web links, news, further readings, and Syllabus Manager.

Student Study Guide (0-8053-6634-2)
Martha R. Taylor This printed learning aid provides a concept map of each chapter, chapter summaries, word roots, chapter tests, and a variety of interactive questions, including multiple choice, short-answer essay, labeling art, and interpreting graphs.

Biology Tutor Center
www.aw.com/tutorcenter This center provides one-to-one tutoring for college students four ways—phone, fax, email, and the Internet—during evening hours and on weekends. Qualified college instructors are available to answer questions and provide instruction regarding self-quizzes and other content found in *BIOLOGY*, **Sixth Edition**. Visit the web site for more information.

Investigating Biology, Fourth Edition (0-8053-7365-9)
Judith Giles Morgan, Emory University, and M. Eloise Brown Carter, Oxford College of Emory University With its distinctive investigative approach to learning, this laboratory manual encourages students to practice science. Students are invited to pose hypotheses, make predictions, conduct open-ended experiments, collect data, and then apply the results to new problems.

Biology Labs On-Line
www.biologylabsonline.com Enables students to expand their scientific horizons beyond the traditional wet lab setting and perform potentially dangerous, lengthy, or expensive experiments in a safe electronic environment. Log on to Biology Labs On-Line and perform 12 different experiments perfectly suited to introductory biology courses. Each experiment can be repeated as often as necessary, employing a unique set of variables each time.

New! **Understanding the Human Genome Project** (0-8053-6774-8)
Michael Palladino, Monmouth University A brief booklet that explains in accessible language what students need to know about the Human Genome Project, presenting the background, the findings, and the social and ethical implications.

The Chemistry of Life CD-ROM (0-8053-8150-3)
Robert M. Thornton, University of California, Davis Teaches the essentials of chemistry to biology students by presenting them with animations, encouraging interaction, and then testing their grasp of the material.

An Introduction to Chemistry for Biology Students, Seventh Edition (0-8053-3075-5)
George I. Sackheim, University of Illinois, Chicago

THE INTERVIEWS

xv

BRIEF CONTENTS

DETAILED CONTENTS

UNIT TWO

THE CELL 106

CHAPTER 1

INTRODUCTION: TEN THEMES IN THE STUDY OF LIFE

EXPLORING LIFE ON ITS MANY LEVELS

- Each level of biological organization has emergent properties
- Cells are an organism's basic units of structure and function
- The continuity of life is based on heritable information in the form of DNA
- Structure and function are correlated at all levels of biological organization
- Organisms are open systems that interact continuously with their environments
- Regulatory mechanisms ensure a dynamic balance in living systems

EVOLUTION, UNITY, AND DIVERSITY

- Diversity and unity are the dual faces of life on Earth
- Evolution is the core theme of biology

THE PROCESS OF SCIENCE

- Science is a process of inquiry that includes repeatable observations and testable hypotheses
- Science and technology are functions of society

REVIEW: USING THEMES TO CONNECT THE CONCEPTS OF BIOLOGY

Biology, the study of life, is rooted in the human spirit. People keep pets, nurture houseplants, invite avian visitors with backyard birdhouses, and visit zoos and nature preserves. Biology is the scientific extension of this human tendency to feel connected to and curious about all forms of life. It is a science for adventurous minds. It takes us, personally or vicariously, into jungles, deserts, seas, and other environments, where a variety of living forms and their physical surroundings are interwoven into complex webs called ecosystems. Studying life leads us into laboratories to examine more closely how living things, called organisms, work. Biology draws us into the microscopic world of the fundamental units of life known as cells and into the submicroscopic realm of the molecules that make up those cells.

Our intellectual journey also takes us back in time, for biology encompasses not only contemporary life, but also a history of ancestral forms stretching nearly 4 billion years into the past. The scope of biology is immense. The purpose of this book is to introduce you to this multifaceted science.

You are becoming involved with biology during its most exciting era. The largest and best-equipped community of scientists in history is beginning to solve biological puzzles that once seemed unsolvable. We are moving ever closer to understanding how a single cell becomes a plant or animal; how the human mind works; how plants convert solar energy to the chemical energy of food; how organisms network in biological communities such as forests and coral reefs; and how the great diversity of life on Earth evolved from the first microbes.

Modern biology is as important as it is inspiring. Genetics and cell biology are revolutionizing medicine and agriculture. Molecular biology is providing new tools for anthropology, helping us trace the origin and dispersal of early humans. Ecology is helping us evaluate environmental issues, such as the causes and consequences of global warming. Neuroscience and evolutionary biology are reshaping psychology and sociology. These are just a few examples of how biology is weaving into the fabric of our culture as never before.

These are also the most challenging times to learn biology. The same discovery explosion that makes modern biology so exhilarating can also be intimidating, even to professional biologists. How, then, can beginning biology students hope to keep their heads above water in this deluge of data and discovery? The key is to recognize unifying themes that pervade all of biology—themes that will still apply decades from now, when much of the specific information presented in any textbook will be obsolete. This chapter introduces some broad, enduring themes in the study of life. The bulleted list on this page previews these ten themes.

1

EXPLORING LIFE ON ITS MANY LEVELS

Biologists study life on many different levels, from the molecular to the global (FIGURE 1.1). This first set of themes will help you integrate the many biological levels into a cohesive view of life.

Each level of biological organization has emergent properties

A basic characteristic of life is a high degree of order. You can see it in the intricate pattern of veins throughout a leaf or in the colorful pattern of a bird's feathers. Biological order exists at all levels, even those invisible to the unaided eye.

A Hierarchy of Organization

Biological organization is based on a hierarchy of structural levels, each level building on the levels below it (FIGURE 1.2). Starting at the lowest level, atoms, the chemical building blocks of all matter, are ordered into complex biological molecules. Many of the molecules of life are arranged into minute structures called organelles, which are in turn the components of cells.

Cells are subunits of organisms, and organisms are the units of life. Some organisms, such as amoebas, consist of single cells, but others are multicellular aggregates of many specialized types of cells. What an amoeba accomplishes with a single cell—the uptake and processing of nutrients, excretion of wastes, response to environmental stimuli, reproduction, and other functions—a human or other multicellular organism accomplishes with a division of labor among specialized cells. Unlike the amoeba, none of your cells could live for long on its own. The organism we recognize as an animal or plant is not a random collection of individual cells, but a multicellular cooperative.

Multicellular organisms exhibit three major structural levels above the cell: Similar cells are grouped into tissues; specific arrangements of different tissues form organs; and organs are grouped into organ systems. For example, the signals (nerve impulses) that coordinate your movements are transmitted along specialized cells called neurons. The nervous

FIGURE 1.1 Biologists study life on many different scales of size and time. These are a few of the biologists you will meet in the interviews that introduce this book's eight units of chapters.

(a) Paul Sereno, paleontologist (fossil specialist)

(b) Flossie Wong-Staal, HIV researcher

(c) George Langford, cell biologist

(d) Joanne Chory, plant biologist

4 **Tissue.** In multicellular organisms, cells are usually organized into tissues, groups of similar cells forming a functional unit. The leaf in this artificially-colored micrograph has been cut obliquely, revealing two different specialized tissues. The honeycomb-like tissue (upper half) consists of photosynthetic cells within the leaf. The dark green tissue with the small pores (lower half) is the epidermis, the "skin" of the plant. The pores in the epidermis allow carbon dioxide, a raw material converted to sugar by photosynthesis, to enter the leaf.

5 **Organ.** The maple leaf, a plant organ, has a specific organization of many different tissues, including photosynthetic tissue, epidermis, and the vascular tissue that transports water from the roots to the leaves.

6 **Organism.** A maple tree is a member of a biological community that includes many other species of organisms.

Cell

3 **Cell.** Many organelles cooperate in the functioning of the living unit we call a cell. Chloroplasts are evident in these leaf cells.

50 μm

10 μm

1 μm

2 **Organelle.** The process of photosynthesis requires the participation of many other molecules organized within the cellular organelle called the chloroplast (the large structure in this micrograph, a photograph taken with a microscope).

Atoms

1 **Molecule.** Chlorophyll, represented here by a computer graphic model, is a molecule built from many atoms. This molecule in the leaves of plants absorbs sunlight as a source of energy for driving photosynthesis, the manufacture of food in the leaf.

FIGURE 1.2 The hierarchy of biological organization. This sequence of images takes us all the way from atoms to a biological community of many interacting species.

tissue within your brain has billions of neurons organized into a communications network of spectacular complexity. The brain, however, is not pure nervous tissue; it is an organ built of many different tissues, including a type called connective tissue that forms the protective covering of the brain. The brain is itself part of the nervous system, which also includes the spinal cord and the many nerves that transmit messages between the spinal cord and other parts of the body. The nervous system is only one of several organ systems characteristic of humans and other complex animals.

In the hierarchy of biological organization, there are tiers beyond the individual organism. A population is a localized group of organisms belonging to the same species; populations of different species living in the same area make up a biological community; and community interactions that include nonliving features of the environment, such as soil and water, form an ecosystem.

Investigating biological organization at its many levels is fundamental to the study of life. This text essentially follows such an organization, beginning with the chemistry of life and ending with the study of ecosystems and the biosphere, the sum of all Earth's ecosystems. However, we will also see that biological processes often involve several levels of biological organization. For example, when a rattlesnake explodes from its coiled posture and strikes a mouse, the snake's coordinated movements result from complex interactions at the molecular, cellular, tissue, and organ levels within its body. This behavior also affects the biological community in which the snake and its prey live. Such episodes of predation can have an important cumulative impact on the sizes of both the mouse and the rattlesnake populations. Most biologists specialize in the study of life at a particular level, but they gain broader perspective when they integrate their discoveries with processes occurring at lower and higher levels.

Emergent Properties

With each step upward in the hierarchy of biological order, novel properties emerge that were not present at the simpler levels of organization. These emergent properties result from interactions between components. A molecule such as a protein has attributes not exhibited by any of its component atoms, and a cell is certainly much more than a bag of molecules. If the intricate organization of the human brain is disrupted by a head injury, that organ will cease to function properly, even though all its parts may still be present. And an organism is a living whole greater than the sum of its parts.

This theme of emergent properties accents the importance of structural arrangement and applies to inanimate material as well as to life. By itself, neither the head nor the handle of a hammer is very useful for driving nails; but put these parts together in a certain way, and the functional properties of a hammer emerge. Diamonds and graphite are both made of carbon, but they have different properties because their carbon atoms are arranged differently. The emergent properties of life are not supernatural, but simply reflect a hierarchy of structural organization without counterpart among inanimate objects.

Life resists a simple, one-sentence definition because it is associated with numerous emergent properties. Yet almost any child perceives that a dog or a bug or a tree is alive and a rock is not. We recognize life by what living things do. FIGURE 1.3 illustrates and describes some of the properties and processes we associate with the state of being alive.

Reductionism in Biology

Because the properties of life emerge from complex organization, scientists seeking to understand biological processes confront a dilemma. One horn of the dilemma is that we cannot fully explain a higher level of order by breaking it down into its parts. A dissected animal no longer functions; a cell reduced to its chemical ingredients is no longer a cell. Disrupting a living system interferes with the meaningful explanation of its processes. The other horn of the dilemma is the futility of trying to analyze something as complex as an organism or a cell without taking it apart. Reductionism—reducing complex systems to simpler components that are more manageable to study—is a powerful strategy in biology. For example, by studying the molecular structure of a substance called DNA that had been extracted from cells, James Watson and Francis Crick deduced, in 1953, how this molecule could serve as the chemical basis of inheritance. The central role of DNA was better understood, however, when it was possible to study its interactions with other substances in the cell. Biology balances the reductionist strategy with the longer-range objective of understanding emergent properties—how the parts of cells, organisms, and higher levels of order, such as ecosystems, are functionally integrated.

Cells are an organism's basic units of structure and function

The cell is the lowest level of structure capable of performing *all* the activities of life. All organisms are composed of cells, the basic units of structure and function (see FIGURE 1.2).

The Cell Theory

Robert Hooke, an English scientist, first described and named cells in 1665, when he observed a slice of cork (bark from an oak tree) with a microscope that magnified 30 times (30×). Apparently believing that the tiny boxes, or "cells," that he saw were unique to cork, Hooke never realized the significance of his discovery. His contemporary, a Dutchman named Anton van Leeuwenhoek, discovered organisms we now know to be single-celled. Using grains of sand that he had polished into magnifying glasses as powerful as 300×, Leeuwenhoek discovered a microbial world in droplets of pond water and also observed the blood cells and sperm cells of animals. In 1839, nearly two centuries after the discoveries of Hooke and Leeuwenhoek, cells were finally acknowledged as the universal units of life by Matthias Schleiden and Theodor Schwann, two German biologists. In a classic case of inductive reasoning—reaching a generalization based on many concurring observations—Schleiden and Schwann summarized their own microscopic studies and those of others by concluding that all living things consist of cells. This generalization forms the basis of what is known as the cell theory. This theory was later expanded to include the idea that all cells come from other cells. The ability of cells to divide to form new cells is the basis for all reproduction and for the growth and repair of multicellular organisms, including humans.

The Two Main Cell Types

All cells are enclosed by a membrane that regulates the passage of materials between the cell and its surroundings. And every cell, at some stage in its life, contains DNA, the heritable material that directs the cell's many activities.

Two major kinds of cells—prokaryotic cells and eukaryotic cells—can be distinguished by their structural organization. The cells of the microorganisms called bacteria and archaea are prokaryotic. All other forms of life are composed of eukaryotic cells. Much more complex than the prokaryotic cell,

FIGURE 1.3 Some properties of life.

(a) Order. All other characteristics of life emerge from an organism's highly ordered structure, which is apparent in this closeup of a sunflower.

(b) Reproduction. Organisms reproduce their own kind. Life comes only from life, an axiom known as biogenesis. Here, a Japanese macaque protects its offspring.

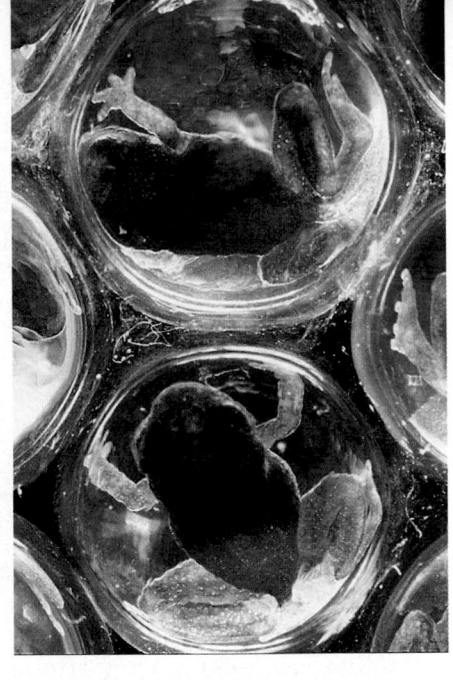

(c) Growth and development. Heritable programs in the form of DNA direct the pattern of growth and development, producing an organism that is characteristic of its species. Shown here are embryos of a Costa Rican species of frog.

(d) Energy utilization. Organisms take in energy and transform it to do many kinds of work. This bat obtains fuel in the form of nectar from the saguaro cactus. The bat will use energy stored in the molecules of its food to power flight and other work.

(e) Response to the environment. This soon-to-be-digested cricket "tripped" the Venus flytrap when it stimulated hair cells on the surface of the modified leaves that make up the trap. The plant responded to this environmental stimulus with a rapid closure of the trap.

(g) Evolutionary adaptation. Life evolves as a result of the interaction between organisms and their environments. One consequence of evolution is the adaptation of organisms to their environment. The white feathers of the white-tailed ptarmigan in winter plumage make the bird nearly invisible against its snowy surroundings.

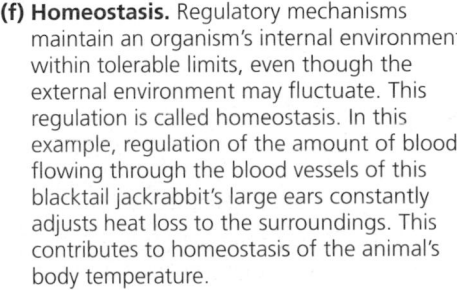

(f) Homeostasis. Regulatory mechanisms maintain an organism's internal environment within tolerable limits, even though the external environment may fluctuate. This regulation is called homeostasis. In this example, regulation of the amount of blood flowing through the blood vessels of this blacktail jackrabbit's large ears constantly adjusts heat loss to the surroundings. This contributes to homeostasis of the animal's body temperature.

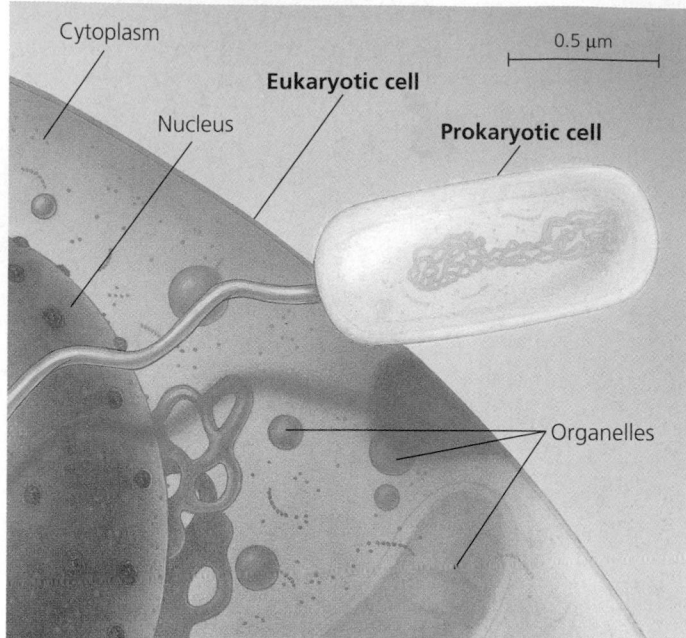

Cytoplasm

Eukaryotic cell

Nucleus

0.5 μm

Prokaryotic cell

Organelles

FIGURE 1.4 Structural organization of eukaryotic and prokaryotic cells. The eukaryotic cell, found in plants, animals, and all other organisms except bacteria and archaea, is characterized by an extensive subdivision into many different functional compartments called organelles. The prokaryotic cell, unique to bacteria and archaea, is much simpler, lacking the organelles found in eukaryotic cells. Most prokaryotic cells are also much smaller than eukaryotic cells.

the eukaryotic cell is subdivided by internal membranes into many different functional compartments—membrane-enclosed organelles (FIGURE 1.4). In eukaryotic cells, the DNA is organized along with certain proteins into structures called chromosomes contained within a nucleus, the largest organelle of most eukaryotic cells. Surrounding the nucleus is the cytoplasm, consisting of a thick fluid (cytosol) in which are suspended the various organelles that perform most of the cell's functions. Some eukaryotic cells, including those of plants, have tough walls external to their membranes. Animal cells lack walls.

In the much simpler prokaryotic cell, the DNA is not separated from the rest of the cell in a nucleus. Prokaryotic cells also lack the cytoplasmic organelles typical of eukaryotic cells. Almost all prokaryotic cells have tough external cell walls.

Although prokaryotic and eukaryotic cells contrast sharply in complexity, we will see that they have some key similarities. Cells vary widely in size, shape, and specific structural features, but all are highly ordered structures that carry out complicated processes necessary for maintaining life.

The continuity of life is based on heritable information in the form of DNA

Order implies information; instructions are required to arrange parts or processes in an organized way. Biological instructions are encoded in the molecule known as DNA (deoxyribonucleic acid). DNA is the substance of genes, the units of inheritance that transmit information from parents to offspring (FIGURE 1.5).

Each DNA molecule is made up of two long chains arranged into what is called a double helix. Each chain is composed of four kinds of chemical building blocks called nucleotides. The way DNA conveys information is analogous to the way we arrange the letters of the alphabet into precise sequences with specific meanings. The word *rat*, for example, conjures up an image of a rodent; *tar* and *art*, which contain the same letters, mean quite different things. Libraries are filled with books containing information encoded in varying sequences of only 26 letters. We can think of nucleotides as the alphabet of inheritance. Specific sequential arrangements of these four chemical letters encode the precise information in a gene, which is typically hundreds or thousands of nucleotides long.

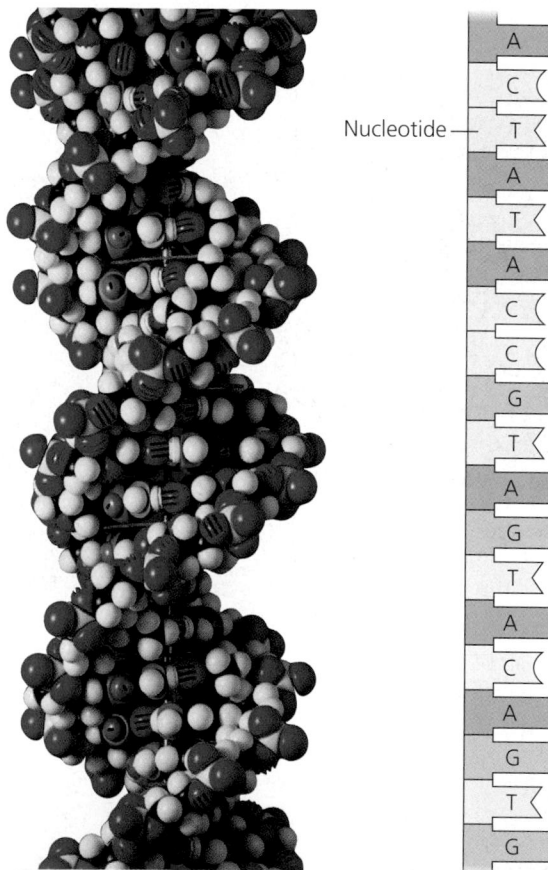

Nucleotide

A
C
T
A
T
A
C
C
G
T
A
G
T
A
C
A
G
T
G

(a) DNA double helix. This model shows each atom in a segment of DNA. Made up of two long chains of building blocks called nucleotides, a DNA molecule takes the three-dimensional form of a double helix.

(b) Single strand of DNA. These geometric shapes and letters represent the nucleotides in a small section of one chain of a DNA molecule. Genetic information is encoded in specific sequences of the four types of nucleotides.

FIGURE 1.5 The genetic material: DNA. DNA molecules carry biological information from one generation to the next.

All forms of life employ essentially the same genetic code. A particular sequence of nucleotides says the same thing to one organism as it does to another; differences between organisms reflect differences between their nucleotide sequences. The diverse forms of life are different expressions of a common language for programming biological order.

Inheritance itself depends on a mechanism for copying DNA and passing its sequence of chemical letters on to offspring. As a cell prepares to divide to form two cells, it copies its DNA. A mechanical system that moves chromosomes then distributes the duplicated DNA equally to the two "daughter" cells. In species that reproduce sexually, offspring inherit copies of DNA from the parents' sperm and egg cells. The continuity of life over the generations and over the eons has its molecular basis in the replication of DNA.

The entire "library" of genetic instructions that an organism inherits is called its genome. The nucleus of each human cell packs a genome that is 3 billion chemical letters long. If this sequence of nucleotides were written in letters the size of those you are now reading, the information would fill more than a hundred books the size of this one. In 2001, scientists tabulating the sequence of these letters published a "rough draft" of the human genome. The press and world leaders acclaimed this international achievement as the

greatest scientific triumph ever. But unlike past cultural zeniths, such as the landing of Apollo astronauts on the moon, the sequencing of the human genome is more a commencement than a climax. As the quest continues, biologists will learn the functions of thousands of genes and how their activities are coordinated in the development of an organism. It is a striking example of human curiosity about life at its many levels.

Structure and function are correlated at all levels of biological organization

Given a choice of tools, you would not loosen a screw with a hammer or pound a nail with a screwdriver. How a device works is correlated with its structure: Form fits function. Applied to biology, this theme is a guide to the anatomy of life at its many structural levels, from molecules to organisms. Analyzing a biological structure gives us clues about what it does and how it works. Conversely, knowing the function of a structure provides insight about its construction.

An example of this structure-function theme is the aerodynamically efficient shape of a bird's wing (FIGURE 1.6). The skeleton of the bird also has structural qualities that contribute to flight, with bones that have a strong but light

(a) A bird's build makes flight possible. The correlation between structure and function can apply to the shape of an entire organism, as you can see from this gull in flight.

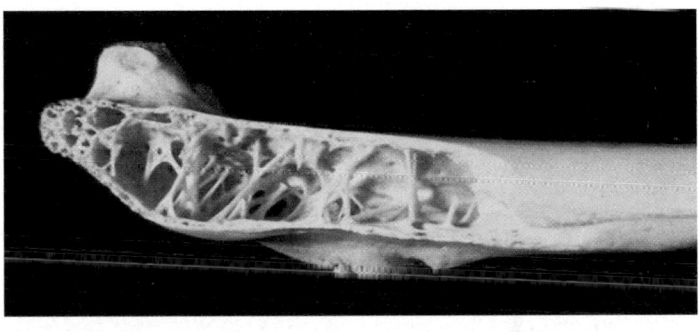

(b) The structure-function theme also applies to organs and tissues. For example, the honeycombed construction of a bird's bones provides a lightweight skeleton of great strength.

100 μm

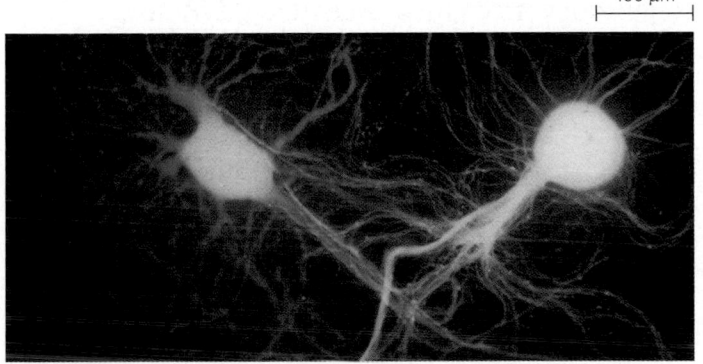

(c) The form of a cell fits its specialized function. Nerve cells, or neurons, have long extensions (processes) that transmit nervous impulses.

FIGURE 1.6 Form fits function.

Mitochondrion   Infoldings of membrane   0.5 μm

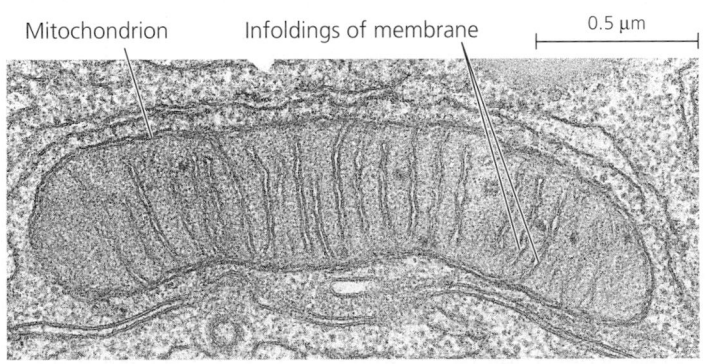

(d) Functional beauty is also apparent at the subcellular level. This organelle, called a mitochondrion, has an inner membrane that is extensively folded, a structural solution to the problem of packing a relatively large amount of this membrane into a very small container.

honeycombed internal structure. The flight muscles of a bird are controlled by neurons (nerve cells), which transmit impulses. With their long extensions, neurons are especially well structured for communication. The flight muscles need plenty of energy, which they obtain from organelles called mitochondria. These organelles are the sites of cellular respiration, the chemical process that powers the cell by using oxygen to help tap the energy stored in sugar and other food molecules. A mitochondrion has an inner membrane with many infoldings. Molecules embedded in the inner membrane carry out many of the steps in cellular respiration, and the infoldings pack a large amount of this membrane into a minute container (see FIGURE 1.6d). In exploring life on its different structural levels, we discover functional beauty at every turn.

Organisms are open systems that interact continuously with their environments

Life does not exist in a vacuum. An organism is an example of what scientists call an open system, an entity that exchanges materials and energy with its surroundings. Each organism interacts continuously with its environment, which includes other organisms as well as nonliving factors. The roots of a tree, for example, absorb water and minerals from the soil, and the leaves take in carbon dioxide from the air. Solar energy absorbed by chlorophyll, the green pigment of leaves, drives photosynthesis, which converts water and carbon dioxide to sugar and oxygen. The tree releases oxygen to the air, and its roots change the soil by breaking up rocks into smaller particles, secreting acid, and absorbing minerals. Both organism and environment are affected by the interaction between them. The tree also interacts with other life, including soil microorganisms associated with its roots and animals that eat its leaves and fruit.

Ecosystem Dynamics

The many interactions between organisms and their environment are interwoven to form the fabric of an ecosystem. The dynamics of any ecosystem include two major processes. One is the cycling of nutrients. For example, minerals acquired by plants will eventually be returned to the soil by microorganisms that decompose leaf litter, dead roots, and other organic debris. The second major process in an ecosystem is the flow of energy from sunlight to photosynthetic life (producers) to organisms that feed on plants (consumers) (FIGURE 1.7).

Energy Conversion

The exchange of energy between an organism and its surroundings involves the transformation of one form of energy to another. For example, when a leaf produces sugar, it con-

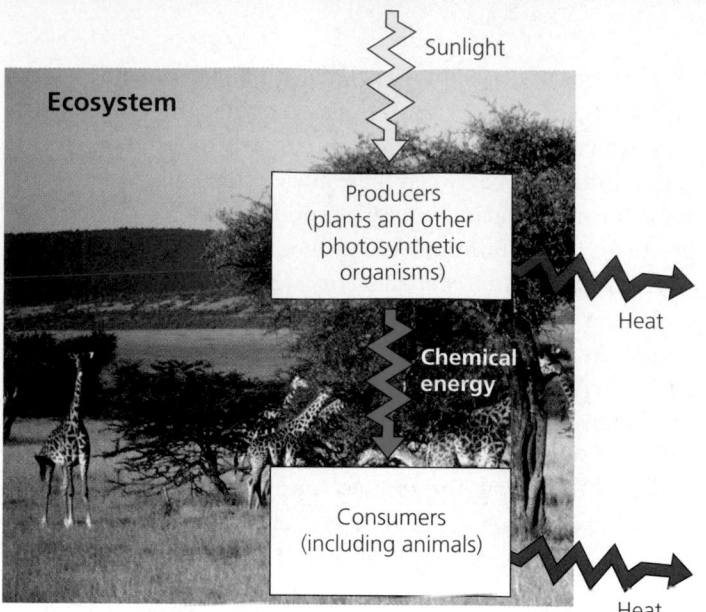

FIGURE 1.7 An introduction to energy flow and energy transformation in an ecosystem. Living is work, and work requires that organisms obtain and use energy. Most ecosystems are solar powered. Plants and other photosynthetic organisms convert light energy to the chemical energy stored in sugar and other complex molecules. By breaking these fuel molecules down to simpler molecules, organisms can harvest the stored energy and put it to work. Animals and other consumers acquire their energy in chemical form by eating plants, by eating animals that ate plants, or by decomposing organic refuse, such as leaf litter and dead animals. The energy that enters an ecosystem in the form of sunlight exits in the form of heat, which all organisms dissipate to their surroundings.

verts solar energy to chemical energy in sugar molecules. When an animal's muscle cells use sugar as fuel to power movements, they convert chemical energy into kinetic energy, the energy of motion. All of the work of cells involves the transformation of chemical energy (which is ordered) into heat, which is the unordered energy of random molecular motion. Life requires the continual uptake of ordered energy and the release of some unordered energy to the surroundings.

Regulatory mechanisms ensure a dynamic balance in living systems

If you strike a match, it undergoes a chemical reaction in which the chemical energy in its molecules is transformed into heat and light. Burning is unregulated energy transformation, obviously not a suitable way for living cells to transform energy. Organisms are able to obtain useful energy from fuel molecules such as sugar because cells break the molecules down in a series of closely regulated chemical reactions.

Regulation of the chemical reactions within cells centers on protein molecules called enzymes. Produced by the cells in which they function, enzymes are catalysts, substances that speed up chemical reactions. When your muscle cells need a lot of energy during exercise, enzymes catalyze the rapid breakdown of sugar (glucose) molecules, releasing energy that can be put to work. In contrast, when you rest, other enzymes catalyze the chemical conversion of glucose to storage substances that may be added to the body's fuel reserves. With one group of enzymes catalyzing the breakdown of glucose and another group catalyzing its formation, how is order maintained in the cell? A big part of the answer is that regulatory mechanisms determine precisely when, where, and how fast certain reactions occur in a cell.

Many biological processes are self-regulating, operating by a mechanism called feedback, in which an output or product of a process regulates that process. Negative feedback, also called feedback inhibition, slows or stops processes; positive feedback speeds a process up (FIGURE 1.8).

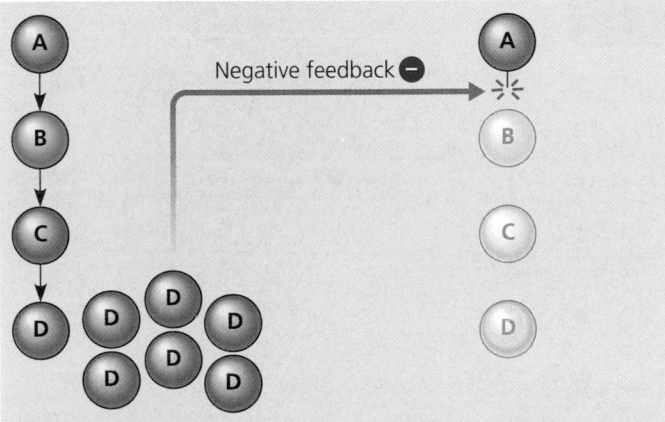

(a) Negative feedback. This simple model illustrates the principle of negative feedback, or feedback inhibition, in the regulation of a chemical reaction sequence in a cell. The sequence involves four molecules (A–D). The black arrows represent three different enzymes catalyzing the conversion of one molecule to the next. The final product (D) inhibits the first enzyme in the sequence; when the concentration of D rises to a certain point, the reaction shuts itself down.

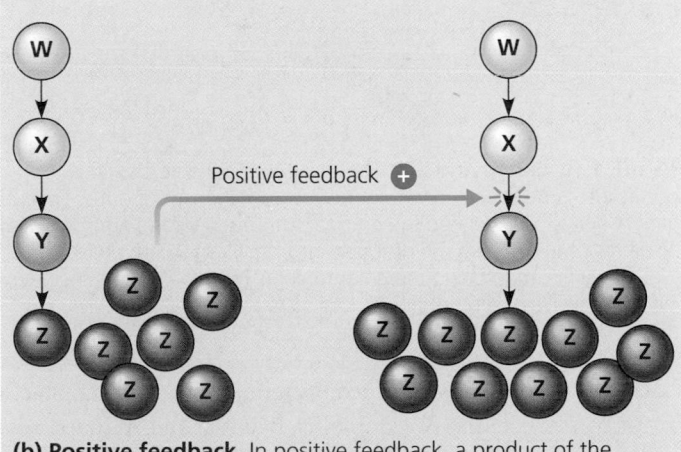

(b) Positive feedback. In positive feedback, a product of the reaction sequence enhances the action of one of the enzymes, increasing the rate of production of the product. Positive feedback is less common than negative feedback in living systems.

FIGURE 1.8 Regulation by feedback mechanisms.

Mammals and birds have a negative-feedback system that keeps body temperature within a narrow range, despite wide fluctuations in the animal's surroundings. A "thermostat" in the brain controls processes that hold the temperature of the blood close to a set point (about 37°C in mammals). For instance, when the human body starts getting hot, signals from the brain's control center increase the activity of sweat glands and the diameter of blood vessels in the skin. Evaporative cooling results as sweating increases, and heat radiates from the blood vessels as they fill with warm blood. Negative feedback occurs as soon as the blood cools back to the set point, causing the control center to stop sending signals to the skin. If the blood temperature drops below the set point, the brain's control center inactivates the sweat glands and constricts the skin's blood vessels. This shunts blood to deeper tissues, reducing heat loss. When the blood warms back to the set point, negative feedback occurs again, and signals from the control center cease. This type of steady-state regulation, which keeps an internal factor such as body temperature within narrow tolerance limits, is called homeostasis (see FIGURE 1.3f).

The clotting of blood is an example of positive feedback. When a blood vessel is injured, structures in the blood called platelets start accumulating at the site. Positive feedback occurs as chemicals released by the platelets attract more platelets, and the platelet cluster initiates a complex sequence of chemical reactions that seals the wound with a clot.

Regulation by positive and negative feedback is a pervasive theme in biology, and we will see numerous examples throughout this text.

EVOLUTION, UNITY, AND DIVERSITY

We can think of biology's enormous scope as having two dimensions. The "vertical" dimension is the size scale that reaches all the way from molecules to the biosphere, as you learned in the preceding section. But there is also a "horizontal" dimension stretching across the great diversity of life, now and throughout life's history. Evolution is the key to understanding this biological diversity. Evolutionary change has been a central feature of life since it arose about 4 billion years ago. The evolutionary connections among all organisms explain the unity and diversity of life. This section introduces two related themes that will help you see these evolutionary connections throughout your study of life.

Diversity and unity are the dual faces of life on Earth

Diversity is a hallmark of life. Biologists have identified and named about 1.5 million species, including over 280,000 plants, almost 50,000 vertebrates (animals with backbones),

FIGURE 1.9 A small sample of biological diversity.
Shown here are just some of the tens of thousands of
species in the butterfly and moth collection at the
National Museum of Natural History in Washington,
D.C. As diverse as the species are, they are all
variations on a common anatomical theme. One of
biology's major goals is to explain how such diversity
arises while also accounting for characteristics
common to different species.

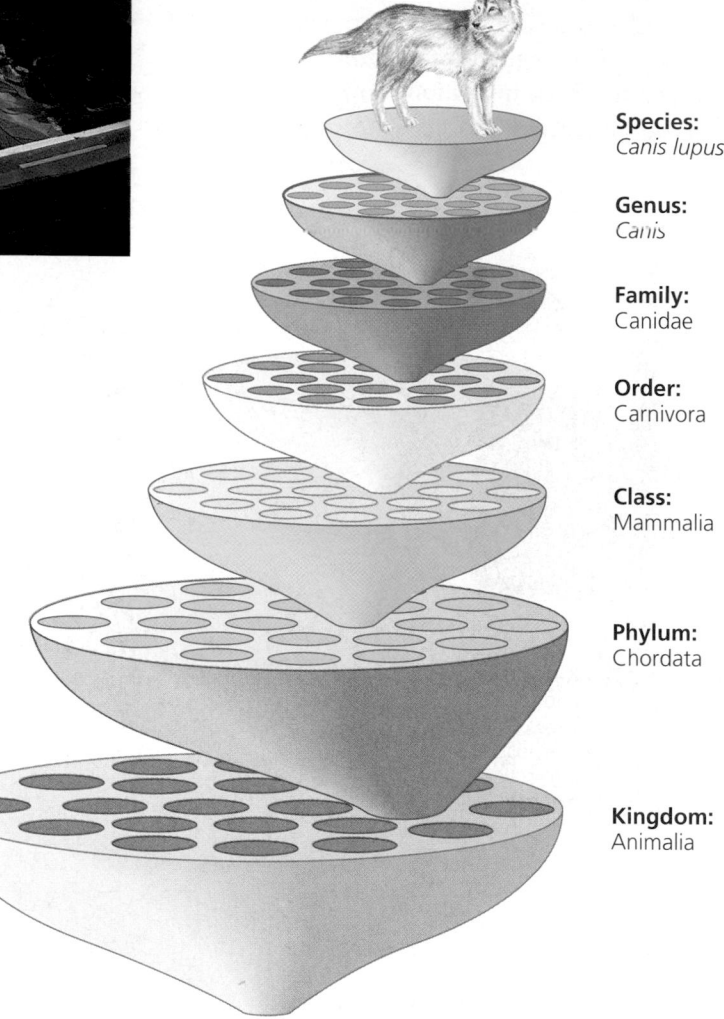

Species:
Canis lupus

Genus:
Canis

Family:
Canidae

Order:
Carnivora

Class:
Mammalia

Phylum:
Chordata

Kingdom:
Animalia

and more than 750,000 insects. Thousands of newly identified species are added to the list each year. Estimates of the
total diversity of life range from about 5 million to over 30
million species.

Grouping Species: The Basic Concept

Biological diversity is something to relish and preserve, but it
can also be a bit overwhelming (FIGURE 1.9). Confronted with
complexity, humans are inclined to categorize diverse items
into a smaller number of groups. Grouping species that are
similar is natural for us. We may speak of squirrels and butterflies, though we recognize that many different species belong to each group. We may even sort groups into broader
categories, such as rodents (which include squirrels) and
insects (which include butterflies). Taxonomy, the branch of
biology that names and classifies species, formalizes this hierarchical ordering (FIGURE 1.10). You will learn more about
this taxonomic scheme in Chapter 25. For now, we focus on
kingdoms and domains, the broadest units of classification.

The Three Domains of Life

Until the past decade, most biologists divided the diversity of
life into five main groups, or kingdoms. (The most familiar
two are the plant and animal kingdoms.) But new methods,
such as comparisons of DNA among organisms, have led to an
ongoing reassessment of the number and boundaries of kingdoms. Various classification schemes are now based on six,
eight, or more kingdoms. But as the debate continues on the
kingdom level, there is broader consensus that the kingdoms
of life can now be assigned to three even higher levels of classification called domains.

FIGURE 1.10 Classifying life. The taxonomic scheme classifies
species into groups nested within more comprehensive groups. Species
that are very closely related (green circles in this diagram) are placed in
the same genus, genera are grouped into families, and so on, each
level of classification being more comprehensive than those it includes.
This example classifies the species *Canis lupus,* the wolf.

The three domains are named Bacteria, Archaea, and Eukarya
(FIGURE 1.11). The first two domains, Bacteria and Archaea, recognize two very different groups of organisms that both have
prokaryotic cells. In the five-kingdom system, these prokaryotes
were combined in a single kingdom. But newer evidence suggests that the organisms known as archaea may be more closely
related to eukaryotes than they are to bacteria.

DOMAIN BACTERIA

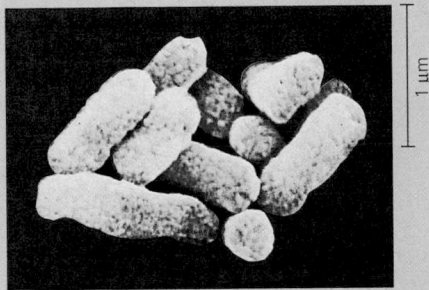

(a) Members of **Domain Bacteria** are the most diverse and widespread prokaryotes.

DOMAIN ARCHAEA

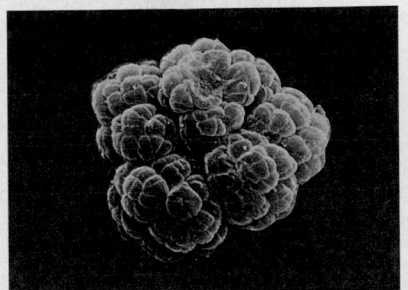

(b) Most prokaryotes in **Domain Archaea** live in Earth's extreme environments, such as salty lakes and boiling hot springs. Molecular evidence indicates that archaea have at least as much in common with eukaryotes as they do with members of the domain Bacteria.

DOMAIN EUKARYA

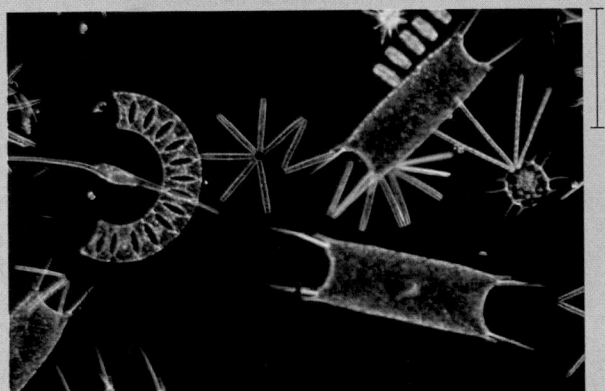

(c) Kingdom Protista consists of unicellular eukaryotes and their relatively simple multicellular relatives. Pictured here is an assortment of protists inhabiting pond water. Scientists are currently debating how to split the protists into several kingdoms that better represent evolution and diversity.

(d) Kingdom Plantae consists of multicellular eukaryotes, such as these tulips, that carry out photosynthesis.

(e) Kingdom Fungi is defined in part by the nutritional mode of its members, such as these mushrooms, which absorb nutrients after decomposing organic material.

(f) Kingdom Animalia consists of multicellular eukaryotes that ingest other organisms.

FIGURE 1.11 Three domains of life. The domains Bacteria, Archaea, and Eukarya represent three fundamentally different kinds of organisms. The domains Bacteria and Archaea consist of organisms, mostly unicellular, with prokaryotic cells (see FIGURE 1.4). The more traditional five-kingdom system of classification recognized all prokaryotes as members of one kingdom, Monera, plus the four kingdoms in the domain Eukarya (organisms with eukaryotic cells) shown here.

All the eukaryotes (organisms with eukaryotic cells) are grouped into at least four kingdoms in the domain Eukarya. Kingdom Protista consists of eukaryotic organisms that are generally single-celled—for example, the microscopic protozoans, such as the amoebas. Many biologists extend the boundaries of the kingdom Protista to include certain multicellular forms, such as seaweeds, that seem to be closely related to the unicellular protists. Other biologists split the protists into multiple kingdoms. The remaining three eukaryotic kingdoms—Plantae, Fungi, and Animalia—consist of multicellular eukaryotes. These three kingdoms are distinguished partly by their modes of nutrition. Plants produce their own sugars and other foods by photosynthesis. Fungi are mostly decomposers that absorb nutrients by breaking down dead organisms and organic wastes, such as leaf litter and animal feces. Animals obtain food by ingestion, which is the eating and digesting of other organisms. It is, of course, the kingdom to which we belong.

Unity in the Diversity of Life

If life is so diverse, how can biology have any unifying themes at all? What, for instance, can a tree, a mushroom, and a human possibly have in common? As it turns out, a great deal! Underlying the diversity of life is a striking unity, especially at the lower levels of biological organization. We have already seen one example: the universal genetic language of DNA. That ubiquitous feature connects all kingdoms of life, even uniting prokaryotes such as bacteria with eukaryotes such as humans. And among eukaryotes, unity is evident in many of the details of cell structure (FIGURE 1.12). Above the cellular level, however, organisms are so variously adapted to their ways of life that describing and classifying biological diversity remains an important goal of biology. It is the process called evolution that accounts for this combination of unity and diversity in life.

Evolution is the core theme of biology

The history of life, as documented by fossils and other evidence, is a saga of a restless Earth billions of years old, inhabited by a changing cast of living forms (FIGURE 1.13). Life evolves. Just as an individual has a family history, each species is one twig of a branching tree of life extending back in time through ancestral species more and more remote. Species that are very similar, such as the brown bear and the polar bear, share a common ancestor that represents a relatively recent branch point on the tree of life. But through an ancestor that lived much farther back in time, all bears are also related to squirrels, humans, and all other mammals. Hair and milk-producing mammary glands are just two of a long list of uniquely mammalian traits. It is what we would expect if all

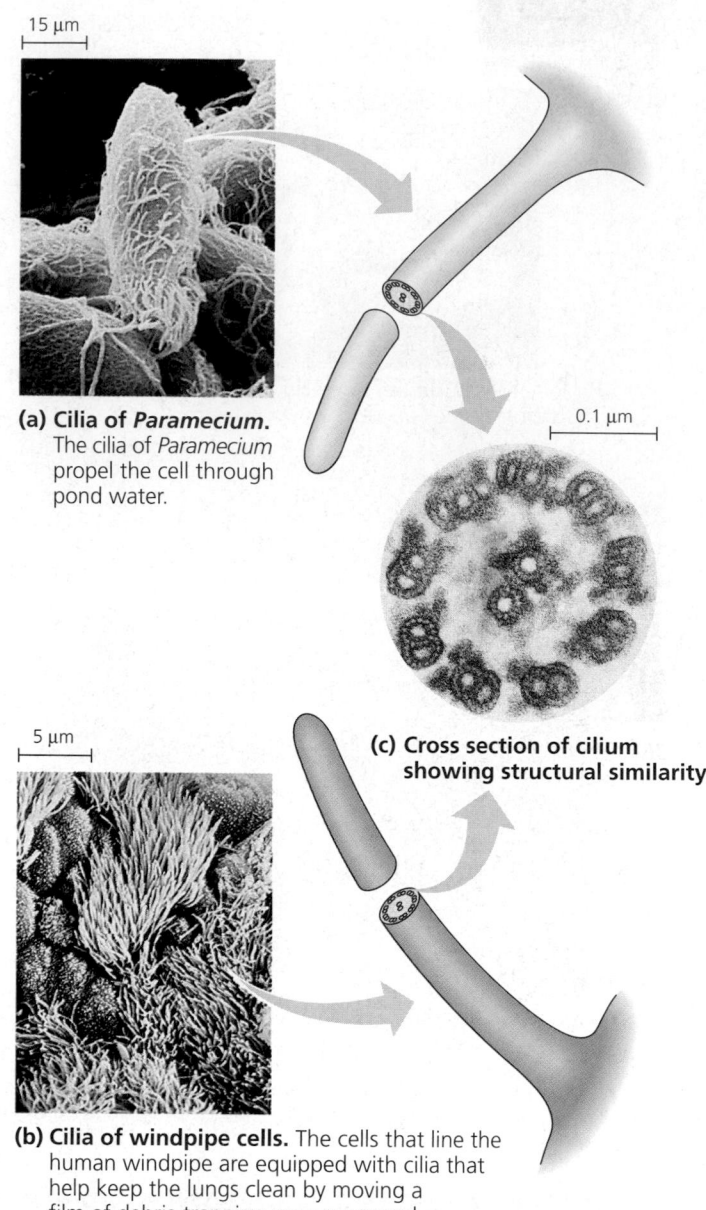

(a) Cilia of *Paramecium*. The cilia of *Paramecium* propel the cell through pond water.

(c) Cross section of cilium showing structural similarity

(b) Cilia of windpipe cells. The cells that line the human windpipe are equipped with cilia that help keep the lungs clean by moving a film of debris-trapping mucus upward.

FIGURE 1.12 An example of unity underlying the diversity of life: the architecture of eukaryotic cilia. Eukaryotic organisms as diverse as the single-celled *Paramecium* and animals possess cilia, locomotor "hairs" that extend from cells. Comparing cross sections of cilia from diverse eukaryotes reveals a common structural organization. Such striking similarity in complex components contributes to the evidence that organisms as different as *Paramecium* and humans are, to some degree, related.

mammals descended from a common ancestor, a prototypical mammal. And mammals, birds, reptiles, and all other vertebrates share a common ancestor even more ancient. Evidence of a still broader relationship can be found in such similarities as the matching machinery of all eukaryotic cilia (see FIGURE 1.12). Trace life back far enough, and there are only fossils of the primeval prokaryotes that inhabited Earth over 3.5 billion years ago. We can recognize some of their vestiges in our own cells—in the universal genetic code, for example. All of life is

FIGURE 1.13 Digging into the past. Paleontologist Paul Sereno excavates the leg bones of a dinosaur named *Jobaria*. See the interview on pages 508–509 for the story of Paul Sereno's expedition to Niger in Africa, where this giant plant-eating dinosaur was unearthed. The fossil record supports other evidence that life has changed dramatically over Earth's long history.

FIGURE 1.14 Charles Darwin (1809–1882). Darwin and his son William posed for this photograph in 1842. The author of numerous books and monographs on topics as diverse as barnacles, plant movements, and island geology, Darwin would be remembered as one of the greatest naturalists of the 19th century even if he had never published on the topic of evolution. But it was *The Origin of Species* that established Darwin's place as the most influential scientist in the development of modern biology. He is buried next to Isaac Newton in London's Westminster Abbey.

connected. And the basis for this kinship is evolution, the process that has transformed life on Earth from its earliest beginnings to the extensive diversity we see today. Evolution is the one theme that ties together all others and unifies biology.

Darwin and Natural Selection

Charles Darwin (FIGURE 1.14) brought biology into focus in 1859 when he published *The Origin of Species*. Darwin's book presented two main concepts. First, Darwin argued convincingly from several lines of evidence that contemporary species arose from a succession of ancestors through a process of "descent with modification," his phrase for evolution. (The evidence for evolution is discussed in detail in Chapter 22.) Darwin's second concept in *The Origin of Species* was his theory for *how* life evolves. This proposed mechanism of evolution is called natural selection.

Darwin synthesized the concept of natural selection from observations that by themselves were neither new nor profound. Others had the pieces of the puzzle, but Darwin saw how they fit together. He inferred natural selection by connecting two observations:

OBSERVATION #1: *Individual variation.* Individuals in a population of any species vary in many heritable traits.

OBSERVATION #2: *Struggle for existence.* Any population

of a species has the potential to produce far more offspring than the environment can possibly support with food, space, and other resources. This overproduction makes a struggle for existence among the variant members of a population inevitable.

INFERENCE: *Differential reproductive success.* Those individuals with traits best suited to the local environment generally leave a disproportionately large number of surviving, fertile offspring. This differential reproductive success of some individuals over others means that certain heritable traits (those carried by the best-suited individuals) are more likely to appear in each new generation. Darwin called differential reproductive success natural selection, and he envisioned it as the cause of evolution (FIGURE 1.15, p. 14).

We see the products of natural selection in the exquisite adaptations of organisms to the special problems posed by their environments (FIGURE 1.16, p. 14). Notice, however, that natural selection does not *create* adaptations; rather, it screens the heritable variations in each generation, increasing the frequencies of some variations and decreasing the frequencies of others over the generations. Natural selection is an editing process, with heritable variations exposed to environmental factors that favor the reproductive success of some individuals over others. The camouflage of the sea horse in FIGURE 1.16 did

❶ Populations with varied inherited traits. This imaginary beetle population has colonized a locale where the soil has been blackened by a recent brush fire. Initially, the population varies extensively in the coloration of the individuals, from very light gray to charcoal.

❷ Elimination of individuals with certain traits. For hungry birds that prey on the beetles, it is easiest to spot the beetles that are lightest in color.

❸ Reproduction of survivors. The selective predation favors the survival and reproductive success of the darker beetles. Thus, genes for dark color are passed along to the next generations in greater frequency than genes for light color.

❹ Increasing frequency of traits that enhance survival and reproductive success. Generation after generation, the beetle population adapts to its environment through natural selection.

FIGURE 1.15 Natural selection.

FIGURE 1.16 Evolutionary adaptation is a product of natural selection. This sea horse lives among kelp (seaweed). The fish looks so much like a seaweed that it lures prey into the seeming safety of the kelp forest and then eats them. The camouflage also helps prevent the sea horse itself from becoming prey.

not result from individuals changing during their lifetimes to look more like their backgrounds and then passing that improvement on to offspring. The adaptation evolved over many generations by the greater reproductive success in each generation of individuals that were innately better camouflaged than the average member of the sea horse population.

Natural Selection and the Diversity of Life

Darwin proposed that natural selection, by its cumulative effects over vast spans of time, could produce new species from ancestral species. This would occur, for example, if a population fragmented into several populations isolated in different environments. In these various arenas of natural selection, what began as one species could gradually diversify into many species as the geographically isolated populations adapted over many generations to different sets of environmental problems (FIGURE 1.17).

Descent with modification accounts for both the unity and the diversity we observe in life. In many cases, features shared by two species are due to their descent from common ancestors, and differences between the species are due to natural selection modifying the ancestral equipment in different environmental contexts. Evolution is the core theme of biology—a unifying thread that ties every chapter of this text together.

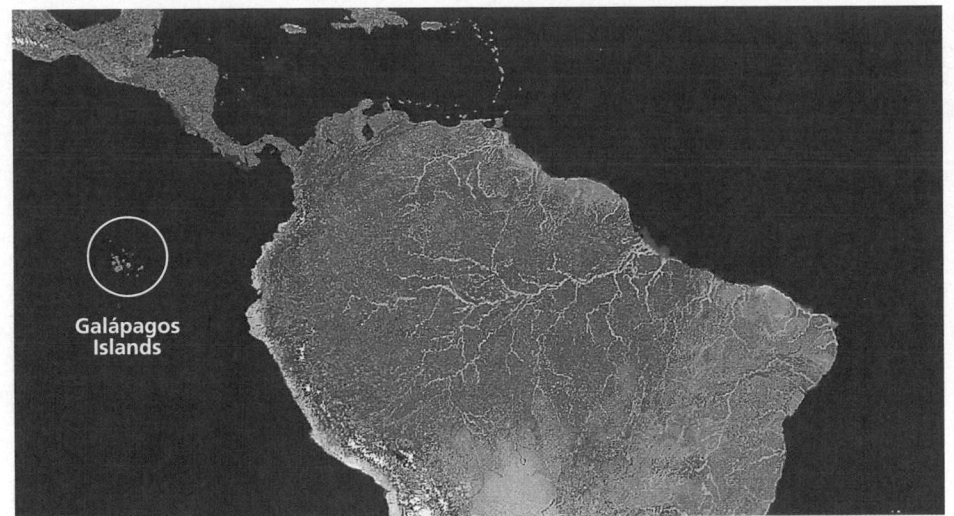

(a) The Galápagos Islands off the coast of South America

FIGURE 1.17 Diversification of finches on the Galápagos Islands. (a) The Galápagos, which Darwin visited in 1835, are a relatively young cluster of volcanic islands about 900 kilometers (540 miles) off the Pacific coast of South America. The islands are home to many species of plants and animals found nowhere else in the world, though Galápagos life-forms are clearly related to species inhabiting the South American mainland. **(b)** Fourteen species of finches classified in three different genera diversified on the various islands of the Galápagos. These birds probably descended from a common ancestor that managed to reach the Galápagos from the South American mainland several million years ago after volcanism built the islands. Note the specialization of beaks, which are adapted to various food sources on the different islands.

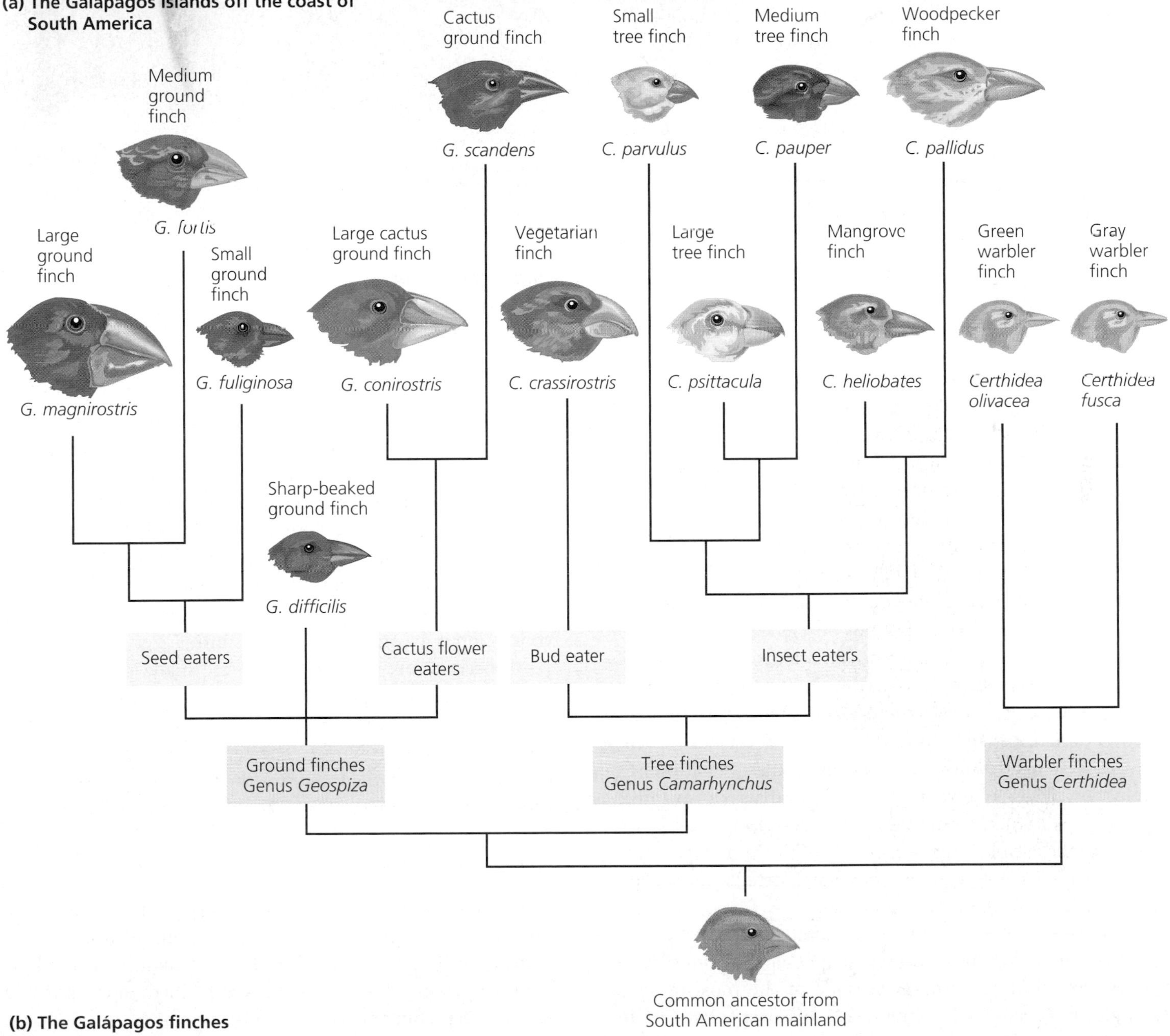

Medium ground finch

G. fortis

Cactus ground finch

G. scandens

Small tree finch

C. parvulus

Medium tree finch

C. pauper

Woodpecker finch

C. pallidus

Large ground finch

G. magnirostris

Small ground finch

G. fuliginosa

Large cactus ground finch

G. conirostris

Vegetarian finch

C. crassirostris

Large tree finch

C. psittacula

Mangrove finch

C. heliobates

Green warbler finch

Certhidea olivacea

Gray warbler finch

Certhidea fusca

Sharp-beaked ground finch

G. difficilis

Seed eaters

Cactus flower eaters

Bud eater

Insect eaters

Ground finches Genus *Geospiza*

Tree finches Genus *Camarhynchus*

Warbler finches Genus *Certhidea*

Common ancestor from South American mainland

(b) The Galápagos finches

THE PROCESS OF SCIENCE

Darwin helped develop biology as a science by seeking natural rather than supernatural causes for the unity and diversity of life. But what *is* science? And how do we tell the difference between science and other ways we try to make sense of nature?

Science is a process of inquiry that includes repeatable observations and testable hypotheses

The word *science* is derived from a Latin verb meaning "to know." Science is a way of knowing. It emerges from our curiosity about ourselves, the world, and the universe. Striving to understand seems to be one of our basic drives. At the heart of science are people asking questions about nature and believing that those questions are answerable. Scientists tend to be quite passionate in their quest for discovery. Max Perutz, a Nobel Prize–winning biochemist, put it this way: "A discovery is like falling in love and reaching the top of a mountain after a hard climb all in one, an ecstasy induced not by drugs but by the revelation of a face of nature that no one has seen before."

The process of science blends two main types of exploration: discovery science and hypothetico-deductive science. Most scientists practice a combination of these two forms of inquiry.

Discovery Science and Induction

Science seeks natural causes for natural phenomena. This limits the scope of science to the study of structures and processes that we can observe and measure, either directly or indirectly with the help of tools such as microscopes that extend our senses. This dependence on observations that other people can confirm demystifies nature and distinguishes science from supernatural explanations. Science can neither prove nor disprove that angels, ghosts, deities, or spirits, both benevolent and evil, cause storms, rainbows, illnesses, and cures, for such explanations are outside the bounds of science.

Verifiable observations and measurements are the data of discovery science (FIGURE 1.18). In our quest to describe nature accurately, we discover its structure and behavior. In biology, discovery science enables us to describe life at its many levels, from ecosystems down to cells and molecules. A naturalist's careful description of diverse plants and animals is an example of discovery science, sometimes called descriptive science. A more recent example is the sequencing of the human genome; it's not really a set of experiments, but a detailed dissection and description of the genetic material.

In contrast to the carefully structured description of the human genome, curious and observant people sometimes make totally serendipitous discoveries. One of the most famous examples is Alexander Fleming's 1928 discovery that certain

FIGURE 1.18 Careful observation and measurement provide the raw data for science. Cornell University's Eloy Rodriguez collects a sample of chemicals extracted from a plant. His analysis will determine the types and amounts of chemicals in the plant.

fungi produce chemicals that kill bacteria. Fleming, a Scottish physician and microbiologist, was culturing bacteria in glass dishes in his laboratory. After finding that a mold (fungus) had contaminated some of his bacterial cultures, he was in the process of discarding the cultures when he noticed that no bacteria were growing in the vicinity of the mold. The fungus turned out to be a common bread mold named *Penicillium*, which produces an antibacterial chemical that was later named penicillin. Quite by accident, Fleming had made a discovery that led to the use of penicillin and other antibiotics to cure syphilis, meningitis, and many other diseases caused by bacteria.

Discovery science can lead to important conclusions based on a type of logic called inductive reasoning. An inductive conclusion is a generalization that summarizes many concurrent observations. "All organisms are made of cells" is an example. That induction was based on two centuries of biologists discovering cells in every biological specimen they observed with microscopes. The careful observations of discovery science and the inductive conclusions they sometimes produce are fundamental to our understanding of nature.

Hypothetico-Deductive Science

The observations of discovery science engage inquisitive minds to ask questions and seek explanations. Ideally, such investigation consists of what is called the scientific method. As a formal process of inquiry, the scientific method consists of a series of steps, but few scientists adhere rigidly to this prescrip-

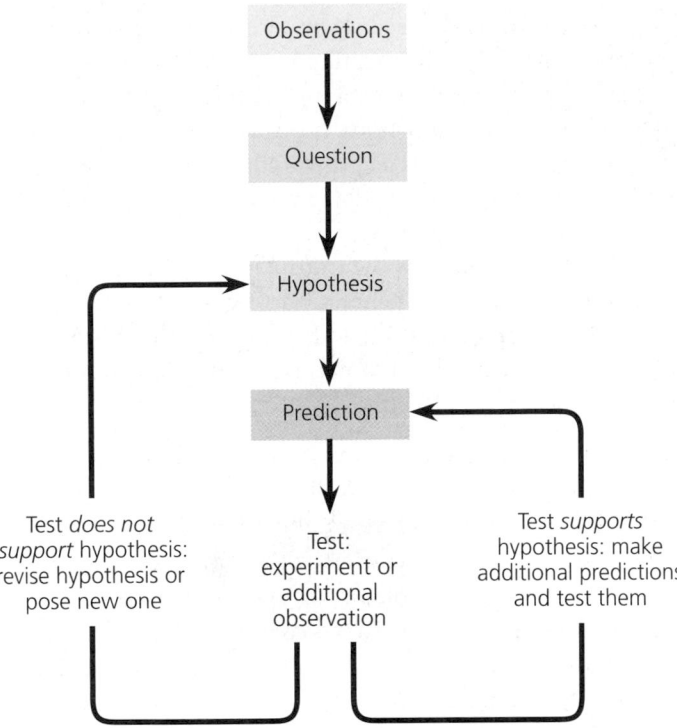

FIGURE 1.19 Idealized version of the scientific method. Although science rarely conforms exactly to this protocol of steps, inquiry usually involves posing and testing hypotheses.

tion (FIGURE 1.19). Science is a less structured process than most people realize. Although it would be misleading to reduce science to a stereotyped method, we *can* identify the key element of the method. It is called hypothetico-deductive reasoning. The "hypothetico" refers to what scientists call hypotheses.

A hypothesis is a tentative answer to some question—an explanation on trial. It is usually an educated guess. We all use hypotheses in solving everyday problems. Let's say, for example, that your flashlight fails during a camp-out. That's an observation. The question is obvious: Why doesn't the flashlight work? A reasonable hypothesis based on past experience is that the batteries in the flashlight are dead.

"If . . . then" Logic. The *deductive* in hypothetico-deductive reasoning refers to the use of deductive logic to test hypotheses. Deduction differs from induction, which, remember, is reasoning from a set of specific observations to reach a general conclusion. In deduction, the reasoning flows in the opposite direction, from the general to the specific. From general premises, we extrapolate to the specific results we should expect if the premises are true. If all organisms are made of cells (premise 1) and humans are organisms (premise 2), then humans are composed of cells (deductive prediction about a specific case).

In the process of science, the deduction usually takes the form of predictions about what outcomes of experiments or observations we should expect if a particular hypothesis (premise) is correct. We then test the hypothesis by performing the ex-

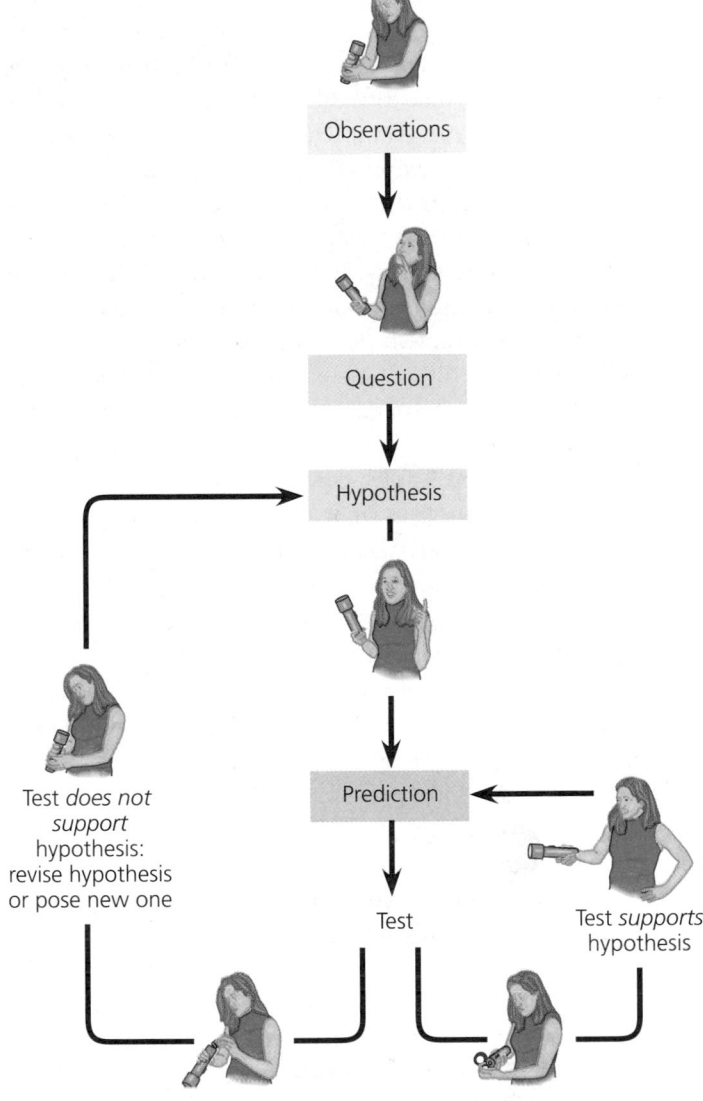

FIGURE 1.20 Applying hypothetico-deductive reasoning to a campground problem.

periment to see whether or not the results are as predicted. This deductive testing takes the form of "*If . . . then*" logic:

OBSERVATION:	My flashlight doesn't work.
QUESTION:	What's wrong with my flashlight?
HYPOTHESIS:	The flashlight's batteries are dead.
PREDICTION:	*If* this hypothesis is correct
EXPERIMENT:	and I replace the batteries with new ones,
PREDICTED RESULT:	*then* the flashlight should work.

Let's say the flashlight *still* doesn't work. We can test an alternative hypothesis if new flashlight bulbs are available (FIGURE 1.20). We could also blame the dead flashlight on campground ghosts playing tricks, but that hypothesis is untestable and therefore outside the realm of science.

A Case Study. You should now be able to recognize the hypothetico-deductive process in an elegant study reported in the scientific literature. For many years, David Reznick of the University of California, Riverside, and John Endler of the University of California, Santa Barbara, have been investigating differences between populations of guppies in Trinidad, a Caribbean island. Guppies *(Poecilia reticulata)* are small freshwater fishes you probably recognize as common aquarium pets. In the Aripo River system of Trinidad, guppies live in small pools as populations that are relatively isolated from one another. In some cases, two populations inhabiting the same stream live less than 100 m (meters) apart, but they are separated by a waterfall that impedes the migration of guppies between the two pools.

When Reznick and Endler compared guppy populations, they observed differences in what are called *life history characteristics*. These characteristics included the average age and size of guppies when they reach sexual maturity and begin to reproduce. The researchers were able to correlate variations in these life history characteristics with the types of predators present in different locations. In some pools, the main predator is a small fish called a killifish, which preys predominantly on small, juvenile guppies. In other locations, a larger predator called a pike-cichlid preys more intensely on guppies and mainly eats relatively large, sexually mature individuals. Guppies in populations exposed to these pike-cichlids reproduce at a younger age and are smaller at maturity, on average, than guppies that coexist with killifish.

What causes these life history differences between the guppy populations? Correlation with the type of predator present is suggestive, but a correlation does not necessarily imply a cause-and-effect relationship. The type of predator present and the life history characteristics of the guppy populations in a particular location may be independent consequences of some third factor. In fact, Reznick and Endler tested the hypothesis that the life history variations were due to differences in water temperature or other features of the physical environment. Notice the *"If . . . then"* logic characteristic of the hypothetico-deductive approach.

> HYPOTHESIS #1: **If** differences in physical environments cause variations in the life histories of guppy populations,
> EXPERIMENT: and samples from different wild guppy populations are collected and maintained for several generations in identical environments in predator-free aquaria,
> PREDICTED RESULT: **then** the laboratory populations should become more similar in their life history characteristics.

When the researchers performed this experiment, the differences persisted for many generations. This result eliminates hypothesis #1 and also indicates that the life history differences in guppy populations are inherited. Based on the assumption that natural selection can lead to genetic differences in populations, Reznick and Endler tested the following explanation:

> HYPOTHESIS #2: **If** the feeding preferences of different predators caused contrasting life histories in different guppy populations to evolve by natural selection,
> EXPERIMENT: and guppies are transplanted from locations with pike-cichlids (predators of mature guppies) to guppy-free sites inhabited by killifish (predators of juvenile guppies),
> PREDICTED RESULT: **then** the transplanted guppy populations should show a generation-to-generation trend toward later maturation and larger size—life history characteristics typical of natural guppy populations that coexist with killifish.

Reznick and Endler introduced guppies from locations with pike-cichlids to sites that had killifish but no guppies (FIGURE 1.21). Over a period of 11 years, the scientists compared the transplanted guppy populations with populations left in the pike-cichlid pools. The biologists measured such life history as age and size at maturity. To be certain that only heritable differences were counted, the scientists made their measurements after samples from the transplanted and non-transplanted guppy populations had been reared for two generations in identical aquarium environments. After 11 years, or 30 to 60 generations, the transplanted guppies had an average weight at maturity that was about 14% greater than the guppies that were not transplanted to killifish pools. Other life history traits also changed in the direction predicted by the hypothesis that the type of predator affects the evolution of these traits in guppy populations (bar graphs in FIGURE 1.21).

The experimental design in our case study is an example of what scientists call a *controlled experiment*. In a controlled experiment, comparisons are made between two sets of subjects—guppy populations, in this case. The set that receives the experimental treatment is called the *experimental group*. In our example, the experimental group consisted of the guppies that Reznick and Endler transplanted from pike-cichlid pools to killifish pools. In contrast, the *control group*—the guppies that remained in the pike-cichlid pools, in this case—does not receive the experimental treatment. Ideally, the experimental treatment is the only difference between the experimental group and the control group. Such a controlled experiment enables researchers to focus on responses to a single variable—a change of predator in the guppy experiments.

Without a control group for comparison, there would be no way to tell whether it was the killifish or some *other* factor that caused the transplanted guppy populations to change. But because control sites and experimental sites were often nearby pools of the same stream, the main variable was probably the different predators. And the researchers observed similar results when guppies were reared in artificial streams that were identical except for the type of predator.

Of the several hypotheses Reznick and Endler have tested (we examined only two), they are left with natural selection due to differential predation on larger versus smaller guppies as the

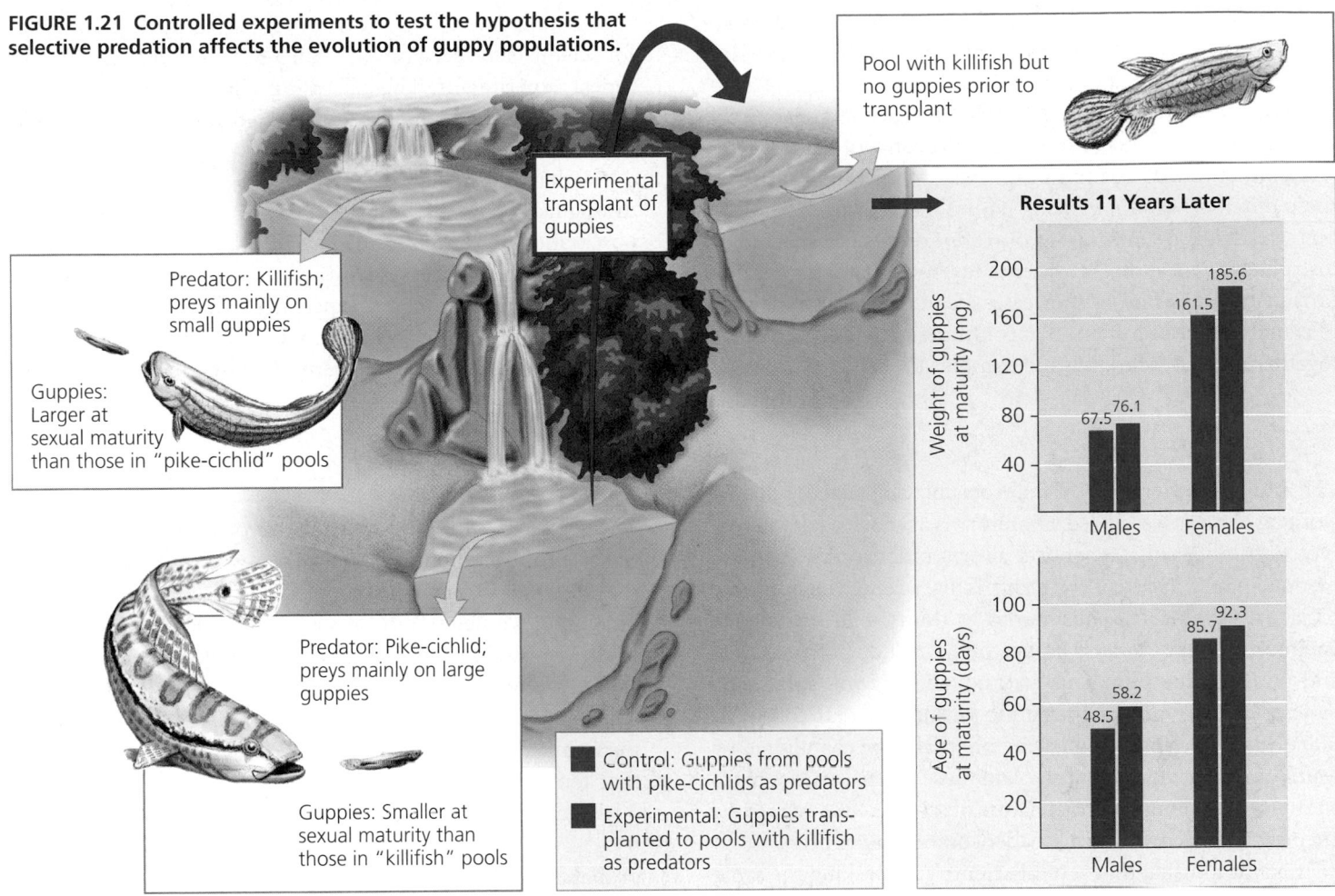

FIGURE 1.21 Controlled experiments to test the hypothesis that selective predation affects the evolution of guppy populations.

Pool with killifish but no guppies prior to transplant

Experimental transplant of guppies

Predator: Killifish; preys mainly on small guppies

Guppies: Larger at sexual maturity than those in "pike-cichlid" pools

Predator: Pike-cichlid; preys mainly on large guppies

Guppies: Smaller at sexual maturity than those in "killifish" pools

■ Control: Guppies from pools with pike-cichlids as predators
■ Experimental: Guppies transplanted to pools with killifish as predators

Results 11 Years Later

Weight of guppies at maturity (mg)

Males: 67.5, 76.1
Females: 161.5, 185.6

Age of guppies at maturity (days)

Males: 48.5, 58.2
Females: 85.7, 92.3

---

most likely explanation for the observed differences between guppy populations. Apparently, when predators such as pike-cichlids prey mainly on reproductively mature adults, the chance that a guppy will survive to reproduce several times is relatively low. The guppies with greatest reproductive success should then be the individuals that mature at a young age and small size and produce at least one brood before growing to a size preferred by the local predator. The controlled experiments of Reznick and Endler documented evolution in a natural setting over a relatively short time (within only 11 years in the study we have followed).

Our case study reinforces the important point that scientific hypotheses must be *testable*. Without such testing, ideas about nature, such as speculations on size differences in guppies, are "just so" stories. And explaining that something is true because "it's just so" is not very convincing.

Theories in Science

Many people associate facts with science, but accumulating facts is not what science is primarily about. A telephone book is an impressive catalog of factual information, but it has little to do with science. It is true that facts, in the form of verifiable observations and repeatable experimental results, are the prerequisites of science. What really advances science, however, is some new theory that ties together a number of observations and experimental results that previously seemed unrelated. The cornerstones of science are the explanations that apply to the greatest variety of phenomena. People like Newton, Darwin, and Einstein stand out in the history of science not because they discovered a great many facts, but because their theories had such broad explanatory power.

What is a scientific theory, and how is it different from a hypothesis? Compared with a hypothesis, a theory is much broader in scope. This is a hypothesis: "Feeding preferences of different predators cause contrasting life histories in different guppy populations to evolve by natural selection." But *this* is a theory: "Populations can evolve by natural selection because of feeding preferences of predators."

Because theories are so comprehensive, they only become widely accepted in science if they are supported by an accumulation of extensive and varied evidence. This use of the term *theory* in science for a comprehensive explanation supported by abundant evidence contrasts with our everyday usage, which equates theories more with speculations or hypotheses. Natural selection qualifies as a scientific theory because of its broad application and because it has been validated by a continuum of observations and experiments. (You will find many examples in Units Four and Eight of this textbook.)

Scientific theories are not the only way of "knowing nature," of course. A comparative religion course would be an excellent opportunity to learn about the diverse legends that tell of a supernatural creation of Earth and its life. Science and religion are two very different ways of trying to make sense of nature. Art is still another way. A broad education should include exposure to these different ways of viewing nature. Each of us synthesizes our worldview by integrating our life experiences and multidisciplinary education. As a science textbook that is part of that broad education, *Biology* showcases life in the scientific context of evolution, the one theme that continues to hold all of biology together no matter how big and complex the subject becomes.

Science as a Social Process

The Hollywood image of the socially challenged scientist working alone in a secluded laboratory is as much a false stereotype as the idea that scientists always practice the scientific method in rigid lockstep. The reality is that science is an intensely social activity. Most scientists work in teams, and in academic environments the research groups often include both graduate and undergraduate students (FIGURE 1.22). If you survey the table of contents of some scientific journals in your college library, you'll find that the majority of articles are coauthored by teams of two or more scientists. And those journal articles make another point about the social nature of science: Research results are inert until shared with a broader community of peers.

It is not unusual that several scientists are asking the same questions. Such convergence contributes to the progressive and self-correcting qualities of science. Scientists build on what has been learned from earlier research, and they pay close attention to contemporary scientists working in the same field. Scientists share information through publications, seminars, meetings, and personal communication. The Internet has added a new medium for this exchange of ideas and data.

Both cooperation and competition characterize the scientific culture. Scientists working in the same research field subject one another's work to careful scrutiny. It is common for scientists to check on each other's claims by attempting to repeat experiments. This obsession with evidence and confirmation helps characterize the scientific style of inquiry. Scientists are generally skeptics.

We have seen that science has two related features that distinguish it from other styles of inquiry: (1) a dependence on observations and measurements that others can verify and (2) the requirement that ideas (hypotheses and theories) are testable by observations and experiments that others can repeat.

Throughout this book, you will find many examples of science in action, each marked by the icon that is shown on the right.

THE PROCESS OF SCIENCE

The Cultural Context of Science

The sociology of biology extends beyond the boundaries of the scientific community; science as a whole is embedded in the culture of its times. Just one example is the increasing proportion of women in biology and the impact they are having on the research being performed. A few decades ago, for instance, biologists who studied the mating behavior of animals focused most of their attention on competition among males for access to females. More recent research is revealing the role that females play in choosing mates. For example, in many bird species, females prefer the bright coloration that "advertises" a male's vigorous health, a behavior that enhances the probability of having healthy offspring.

Some philosophers of science argue that scientists are so influenced by cultural and political values that science is no more objective than other ways of "knowing nature." At the other extreme are people who speak of scientific theories as though they were natural laws instead of human interpretations of nature. The reality of science is probably somewhere in between. The cultural milieu affects scientific fashion, but the criterion of repeatability in observation and hypothesis testing still defines science as fundamentally different from other strategies of knowing nature. This distinction also contrasts the provisional status of science-based knowledge with unconditional beliefs based on faith. There are many examples of cherished hypotheses and even theories in science that were eventually remodeled or rejected in light of compelling new data. A hundred years ago, for instance, most biologists agreed that fungi and bacteria were types of plants. And there are also cases of entrenched theories that went unchallenged for too long because they fit so comfortably into the current cultural context or because there was a tendency to reject evidence that "rocked the boat." The theory of a geocentric (Earth-centered) universe is an example. But such historical examples of theories that persisted in science in spite of their incompatibility with an accumulation of repeatable measurements and experiments are relatively rare. If there is "truth" in science, it is based on a preponderance of the available evidence.

FIGURE 1.22 Science as a social process. Here, New York University plant biologist Gloria Coruzzi (left) mentors one of her students in the methods of molecular biology.

(a) Producing hepatitis B vaccine. This employee of a biotechnology company is monitoring a tank for growing yeast cells that have been engineered to carry genes of the virus that causes hepatitis B. The genetically engineered cells produce large amounts of a protein molecule that is found on the surface of the virus. The protein is used to make a vaccine against the virus.

(b) Solving crimes. Forensic technicians can use traces of DNA extracted from a blood sample or other body fluid to produce a molecular "fingerprint." The stained bands visible in this photograph represent fragments of DNA. The pattern of bands varies from one person to another. The legal applications of DNA technology have become very public in recent years. You will learn more about DNA technology in Chapter 20.

FIGURE 1.23 Two examples of DNA technology.

Science and technology are functions of society

Science and technology are associated. In many cases, technology results from scientific discoveries applied to the development of goods and services. Watson and Crick discovered the structure of DNA through the process of science. This breakthrough sparked an explosion of scientific activity that led to a better understanding of DNA chemistry and the genetic code. These discoveries eventually made it possible to manipulate DNA, enabling genetic technologists to transplant foreign genes into microorganisms and mass-produce such valuable products as human insulin. The new biotechnology is revolutionizing the pharmaceutical industry, and DNA technology has also had an enormous impact in other areas, including agriculture and the legal profession (FIGURE 1.23). Perhaps Watson and Crick envisioned that their discovery would someday have technological applications, but that was probably not what motivated their research; nor could they have predicted exactly what the applications would be.

Not all technology can be described as applied science. In fact, technology predates science, driven by inventive humans who built tools, crafted pots, mixed paints, designed musical instruments, and made clothing—all without necessarily understanding why their inventions worked. Science catalyzes certain technologies by complementing trial and error with more informed design. But the direction technology takes depends less on science than it does on the needs of humans and the values of society.

Technology has improved our standard of living in many ways, but not without introducing some new problems. Technology, especially technology that keeps people healthier, has enabled the human population to grow more than tenfold in the past three centuries. The environmental consequences are enormous. Acid rain, deforestation, global warming, nuclear accidents, ozone holes, toxic wastes, landscapes degraded by exploration for oil and other natural resources, and extinction of species are just a few of the repercussions from more and more people wielding more and more technology. Science can help us identify problems and provide insight about what course of action may prevent further damage. But solutions to these problems have as much to do with politics, economics, culture, and values as with science and technology.

Now that both science and technology have become such powerful functions of society, it is more important than ever to distinguish "what we would like to understand" from "what we would like to build." Scientists should not distance themselves from technology but instead try to influence how scientific discoveries are applied. And scientists have a responsibility to help educate politicians, bureaucrats, corporate leaders, and voters about how science works and about the potential benefits and hazards of specific technologies. The crucial relationship between science, technology, and society is a theme that increases the significance of any biology course.

REVIEW: USING THEMES TO CONNECT THE CONCEPTS OF BIOLOGY

In some ways, biology is the most demanding of all sciences, partly because living systems are so complex and partly because biology is a multidisciplinary science that requires a knowledge of chemistry, physics, and mathematics. Modern biology is the decathlon of natural science. And of all the sciences, biology is the most connected to the humanities and social sciences. If you are a biology major or a preprofessional student, you have an opportunity to become a versatile scientist. If you are a physical science major or an engineering student, you will discover in the study of life many applications for what you have learned in your other science courses. If you are a nonscience student enrolled in biology as part of a liberal arts education, you have selected a course in which you can sample many scientific disciplines.

No matter what brings you to biology, you will find the study of life to be challenging and uplifting. Do not let the de-

Table 1.1 Review of Ten Unifying Themes in Biology

Theme	Description	Web/CD Activity*
1. Emergent properties	The living world has a hierarchical organization, extending from molecules to the biosphere. With each step upward in organizational level, novel properties emerge as a result of interactions among components at the lower levels.	Activity 1A: *Emergent Properties: The Levels of Life Card Game*
2. The cell	Cells are every organism's basic units of structure and function. The two main types of cells are prokaryotic cells (in bacteria and archaea) and eukaryotic cells (in protists, plants, fungi, and animals).	Activity 1B: *Comparing Prokaryotic and Eukaryotic Cells*
3. Heritable information	The continuity of life depends on the inheritance of biological information in the form of DNA molecules. This genetic information is encoded in the nucleotide sequences of the DNA.	Activity 1C: *Heritable Information: DNA*
4. Structure/function	Form and function are correlated at all levels of biological organization.	Activity 1D: *Correlating Structure and Function of Cells*
5. Interaction with the environment	Organisms are open systems that exchange materials and energy with their surroundings. An organism's environment includes other organisms as well as nonliving factors.	Activity 1E: *Energy Flow and Chemical Cycling*

tails of biology spoil a good time. The complexity of life is inspiring, but it can be overwhelming. To help you from "getting lost in the forest because of all the trees," each chapter of this book is constructed from a manageable number of key concepts. The concepts are listed at the beginning of the chapter, are displayed throughout the chapter, and then reappear in the summary at the end of the chapter. The details of a chapter enrich your understanding of the concepts and how they fit together. This introductory chapter is the exception; instead of presenting the key concepts of a particular area of biology, this chapter has introduced themes that cut across *all* biological fields—ways of thinking about biology (TABLE 1.1). These ten themes, along with the key concepts in each chapter, will provide you with a framework for fitting together the many things you will learn in your multidisciplinary exploration of life—and will encourage you to begin asking important questions of your own.

Table 1.1 Review of Ten Unifying Themes in Biology *(continued)*

Theme	Description	Web/CD Activity*
6. Regulation	Feedback mechanisms regulate biological systems. In some cases, the regulation maintains homeostasis, a relatively steady state for internal factors such as body temperature.	Activity 1F: *Regulation: Negative and Positive Feedback*
7. Unity and diversity	Biologists group the diversity of life into three domains: Bacteria, Archaea, and Eukarya. As diverse as life is, we can also find unity, such as a universal genetic code. The more closely related two species are, the more characteristics they share.	Activity 1G: *Unity and Diversity: Classification Schemes*
8. Evolution	Evolution, biology's core theme, explains both the unity and diversity of life. The Darwinian theory of natural selection accounts for adaptation of populations to their environment through the differential reproductive success of varying individuals.	Activity 1H: *Evolution: Sea Horse Camouflage Video* Case Study in the Process of Science: *How Do Environmental Changes Affect a Population?*
9. Scientific inquiry Observations Question Hypothesis Prediction Test	The process of science includes observation-based discovery and the testing of explanations through the hypothetico-deductive method. Scientific credibility depends on the repeatability of observations and experiments.	Case Study in the Process of Science: *How Does Acid Precipitation Affect Trees?*
10. Science, technology, and society	Many technologies are goal-oriented applications of science. The relationships of science and technology to society are now more crucial to understand than ever before.	Activity 1I: *Science, Technology, and Society: DDT*

*Go to the Campbell Biology CD-ROM or website (www.campbellbiology.com) to explore an interactive Chapter Review, Activities, Case Studies in the Process of Science, Self-Quizzes, and more.

THE CHEMISTRY OF LIFE

The work of Thomas Eisner highlights the central role of chemistry in the living world. Dr. Eisner is a pioneer of chemical ecology, the study of the chemical language of nature. Insects are his "beasts of choice," and his research on their chemical interactions with each other and with plants has yielded insights into animal behavior, ecology, and evolution. Among his many distinctions, Dr. Eisner is a member of the National Academy of Sciences and a recipient of the National Medal of Science and the Tyler Prize for Environmental Achievement. He is also an accomplished musician and nature photographer. We met at Cornell University, where Dr. Eisner is a professor and the director of the Cornell Institute of Chemical Ecology.

Interview
THOMAS EISNER

When did you first become interested in science?

According to my parents, I got interested in insects when I could walk—and in chemistry at about the same time. My father was a chemist, and I share with him a sensitive nose and a memory for scents. I would know that my grandmother had visited the previous evening because I recognized the residual scent of her coat in the closet. So from an early age I've paid attention to the chemical images in the world around me.

A few years later, I actually did some experiments with insects—though without knowing what an experiment was. While I was growing up in Uruguay, we stored food in an ice box, and to keep the ants out, we would put little containers of turpentine around the legs of the box. My father had told me that turpentine was made from pine resin, and when I was about 10, it occurred to me that the resin might be a defense of the pine tree against insects. I was struck by the idea that plants might use chemical weaponry. To test my idea about pine resin, I streaked some across an ant trail—and found that the ants wouldn't cross it. But I didn't consider this science; to me it was just fooling around.

What happened after high school, when you came to the United States?

Soon after arriving, I volunteered to work at the American Museum of Natural History for a summer. The scientist I assisted,

Charles Michener, encouraged me and gave me a list of books to read about insects. Then I had a setback: I was turned down by every college I applied to (including Cornell). I went to secretarial school for a few months, then got into Champlain College, a two-year college in Plattsburgh, New York. There I took a chemistry course that I liked a lot. But I also took a course in comparative anatomy, which was an eye opener. It taught me about evolution—for example, that you could relate the bones in the inner ear of a mammal to the jaws of a fish.

From Champlain College I transferred to Harvard, where after considering chemistry and medicine, I finally committed myself to biology. My entomology course as a senior and then meeting fellow student Ed Wilson [E. O. Wilson] during my first year of graduate school won me over completely. I remember thinking, how could I ever have considered being anything but a biologist!

What led you to chemical ecology?

My Ph.D. research was on the stomach of ants—how ants manage to save food for themselves and also for fellow ants. On the side, though, I was starting to collect and study insects that gave off all sorts of fascinating scents and stinks.

At that time, new discoveries in biology were laying the foundation for chemical ecology as a science. It was the heyday of insect hormones. The work on hormones, chem-

icals that act as signals *within* organisms, suggested to me that there might be chemicals that act as signals *between* organisms. So when I arrived at Cornell for my first job, I went looking for a chemist I could work with who would be able to isolate and characterize the signal chemicals. I soon met Jerry Meinwald and started a collaboration that's lasted over 40 years.

It turns out that chemical signaling is ubiquitous. In fact, I've stopped thinking of air as air; I think of it as a carrier of messages. When I see a meadow filled with insects and other animals browsing on the vegetation, I think of the perfumes that attract pollinators to the flowers as well as the repellent chemicals that plants produce that discourage a butterfly from laying her eggs or a caterpillar from feeding on leaves. These repellent chemicals are a defensive strategy for the plant; the attractive floral scents are a reproductive strategy. Meanwhile, the insects are fighting with each other. The ants are repelled by substances produced by beetles. The beetles produce many of these chemicals because they've got a problem: They can't take to flight right away like a fly does; they have to unfold their wings first. They buy safety during that period with chemical weapons. Another insect, a moth, procures defensive substances from the plants it eats as a caterpillar; later, as an adult, it bestows the chemicals on its own offspring. And these and many other insects use chemicals to attract mates. It's all chemical!

Does chemical communication go on in humans, too?

I think there *is* chemical signaling in humans, though not necessarily where you might expect it. Humans don't have a chemical that attracts males to ovulating females. On the contrary, the human female seems to be programmed to hide ovulation, chemically as well as anatomically. In this way, she induces the male—who can't be sure when she is fertile—to remain in attendance for long periods. (So what we call love has a subtle biological basis!) But there are clearly some kinds of chemical signaling going on between men and women. For instance, there are chemicals in the armpit of a male that when detected by a female can regularize her ovulatory cycling.

How do you choose subjects for research?

I'm an opportunist. I go out into the field and look around for something that catches my fancy. For example, I got interested in the bombardier beetle when I found some under a rock and picked one up. It made a popping sound, and a little puff of mist came out of its rear!

Is it true that you once put a bombardier beetle in your mouth?

Yes, I used to put insects in my mouth as a way of sensing their chemicals. With the bombardier beetle, I also sensed heat. I later learned that Charles Darwin may have had the same experience. In his autobiography, Darwin mentions something that happened while he was collecting beetles as a student at Cambridge University. One day he saw two beetles he wanted and grabbed one with each hand. But then he spotted a third interesting beetle. So, having only two hands like the rest of us, he popped one of the beetles into his mouth! Startled by a hot, irritating sensation, he spat out the one in his mouth and dropped the other two. I can't be certain that the one in his mouth was a bombardier beetle, but when I visited Darwin's home, I did see a bombardier in his insect collection, which goes back to his Cambridge days.

Tell us more about the bombardier beetle.

The bombardier beetle has an amazing defense system. When attacked, this animal generates a chemical explosion in less than 80 milliseconds and shoots chemicals out at 100°C with an uncanny aim. If an ant is biting part of the beetle's leg, for example, the beetle hits that part of its leg, no matter what its position [see p. 26].

It took the contributions of a number of scientists to work out all the details. When I was first studying this beetle, I guessed from the smell that its spray contained chemicals called quinones. I wanted a way to make the spray pattern visible, and my father was able to tell me how to make a solution that would turn dark in the presence of quinones. I soaked a piece of filter paper with the solution and positioned the beetle on it. I then pinched one of the beetle's legs with forceps, heard the popping sound, and saw on the paper a clear pattern

of the ejection. Soon after, a researcher in Germany identified the quinones in the beetle's spray and showed that the explosion is produced when the beetle mixes chemicals stored in two compartments of glands in its abdomen.

I then followed up on the physical characteristics of the spray in collaboration with several other scientists. One was Dan Aneshansley, an engineering student. Together we consulted physical chemists at Cornell, who calculated that the reaction of the chemicals in the beetle's glands should produce enough heat to raise the temperature of the spray to 100°C. We confirmed the prediction by rigging up an electronic setup to measure the temperature. Then we found evidence that the spray was pulsed, like the spray from a Water Pik but at a rate of 500 to 1,000 pulses per second. With the help of Harold Edgerton of MIT, the inventor of the electronic flash, we were able to photograph the beetle in action. We used his high-speed movie camera, shooting at 4,000 frames per second.

You've done a lot of work on behalf of the environment. What is your main focus?

Preservation of biodiversity—stopping the ongoing encroachment on nature. To me, science is like a three-cornered hat. The first corner is the discovery aspect, exploration. The second is the attempt to explain the findings and to relate them to other findings—to put together a view of the world. And the third corner is conservation.

You can argue that conservation is the ultimate purpose of it all. When you're doing science, you're trying to explain how you fit into nature and how nature operates around you. I feel that scientists—who all study nature in one way or another—

have an obligation to be conservationists. I'm somewhat impatient with students who think of a scientific career that's totally insulated from activism.

Can you tell us one of your environmental success stories?

While speaking at colleges in Texas as a Sigma Xi lecturer, I learned that the Big Thicket, a wonderful wilderness area, was being chopped down for lumber. A graduate student who was with me, Jim Carroll, and I created an organization to save it—complete with a fancy title and, initially, a membership of two. Shortly after, I was scheduled to give a keynote address at a scientific meeting where I knew I would have a big audience. So we got together a group of students and prepared a petition and little yellow ribbons saying, "Save the Big Thicket." Everyone who came to the lecture was given a ribbon. (It was 1970, and people wore slogans then.) I mentioned the Big Thicket problem in my lecture and boldly said I would hold a press conference the next day. Well, the *New York Times* picked up the story, and suddenly this issue came to public attention. The eventual result was the establishment of the Big Thicket National Preserve, which includes almost 100,000 acres.

Why is biodiversity important?

The value of biodiversity is partly aesthetic, because we're nothing without the green surrounds. And there's a practical aspect, the treasury of nature—countless potentially useful chemicals waiting to be discovered. But I prefer to emphasize the value of nature for its own sake. Its ecosystems provide clean water and air and everything else we need for our survival. So we love nature, we can use it, and we need it.

CHAPTER 2

THE CHEMICAL CONTEXT OF LIFE

CHEMICAL ELEMENTS AND COMPOUNDS

- Matter consists of chemical elements in pure form and in combinations called compounds
- Life requires about 25 chemical elements

ATOMS AND MOLECULES

- Atomic structure determines the behavior of an element
- Atoms combine by chemical bonding to form molecules
- Weak chemical bonds play important roles in the chemistry of life
- A molecule's biological function is related to its shape
- Chemical reactions make and break chemical bonds

This unit of chapters introduces key concepts of chemistry that will apply throughout our study of life. We will make many connections to the themes introduced in Chapter 1. One of those themes is the organization of life into a hierarchy of structural levels (FIGURE 2.1), with additional properties emerging at each successive level. In this unit, we will see how the theme of emergent properties applies to the lowest levels of biological organization— to the ordering of atoms into molecules and to the interactions of those molecules within cells. Somewhere in the transition from molecules to cells, we will cross the blurry boundary between nonlife and life.

*L*ike other animals, *beetles have evolved structures and mechanisms that defend them from attack. As chemical ecologist Thomas Eisner described in the interview opening this unit, the soil-dwelling bombardier beetle has a particularly effective mechanism for dealing with the ants that plague it. Upon detecting an ant on its body, this beetle ejects a spray of boiling hot liquid from glands in its abdomen, aiming the spray directly at the ant. (In the photo on this page, the beetle aims its spray at Dr. Eisner's forceps.) The spray contains irritating chemicals that are generated at the moment of ejection by the explosive reaction of two sets of chemicals stored separately in the glands. The reaction produces heat and an audible pop.*

Research on the bombardier beetle has involved chemistry, physics, and engineering, as well as biology. This is not surprising, for unlike a college catalog of courses, nature is not neatly packaged into the individual natural sciences. Biologists specialize in the study of life, but organisms and the world they live in are natural systems to which basic concepts of chemistry and physics apply. Biology is a multidisciplinary science.

CHEMICAL ELEMENTS AND COMPOUNDS

Matter consists of chemical elements in pure form and in combinations called compounds

Organisms are composed of **matter,** which is anything that takes up space and has mass.* Matter exists in many diverse forms, each with its own characteristics. Rocks, metals, oils, gases, and you and I are just a few examples of what seems an endless assortment of matter.

---

* Sometimes we substitute the term *weight* for *mass,* although the two are not identical. Mass is the amount of matter in an object, whereas the weight of an object is how strongly that mass is pulled by gravity. An astronaut in orbit in a space shuttle is weightless, but the astronaut's mass is the same as it would be on Earth. However, as long as we are earthbound, the weight of an object is a measure of its mass; so we can use the terms interchangeably.

Some of the ancient Greek philosophers believed that the great variety of matter arises from four basic ingredients, or elements. They believed these elements to be air, water, fire, and earth—supposedly pure substances that could not be decomposed to other forms of matter. All other substances were thought to be formed by blending various proportions of two or more of the elements. Although these ancient philosophers proposed the wrong elements, their basic idea was correct.

An **element** is a substance that cannot be broken down to other substances by chemical reactions. Today, chemists recognize 92 elements occurring in nature; gold, copper, carbon, and oxygen are examples. Each element has a symbol, usually the first letter or two of its name. Some of the symbols are derived from Latin or German names; for instance, the symbol for sodium is Na, from the Latin word *natrium.*

A **compound** is a substance consisting of two or more elements combined in a fixed ratio. Table salt, for example, is sodium chloride (NaCl), a compound composed of the elements sodium (Na) and chlorine (Cl) in a 1:1 ratio. Pure sodium is a metal and pure chlorine is a poisonous gas. Chemically combined, however, sodium and chlorine form an edible compound. This is a simple example of organized matter having emergent properties: A compound has characteristics beyond those of its combined elements (FIGURE 2.2).

Life requires about 25 chemical elements

About 25 of the 92 natural elements are known to be essential to life. Just four of these—carbon (C), oxygen (O), hydrogen (H), and nitrogen (N)—make up 96% of living matter. Phosphorus (P), sulfur (S), calcium (Ca), potassium (K), and a few other elements account for most of the remaining 4% of an organism's weight. TABLE 2.1 (p. 28) lists by percentage the elements that make up the human body; the percentages for other organisms are similar. FIGURE 2.3 (p. 28) illustrates the effect of nitrogen deficiency in a plant.

Trace elements are those required by an organism in only minute quantities. Some trace elements, such as iron (Fe), are

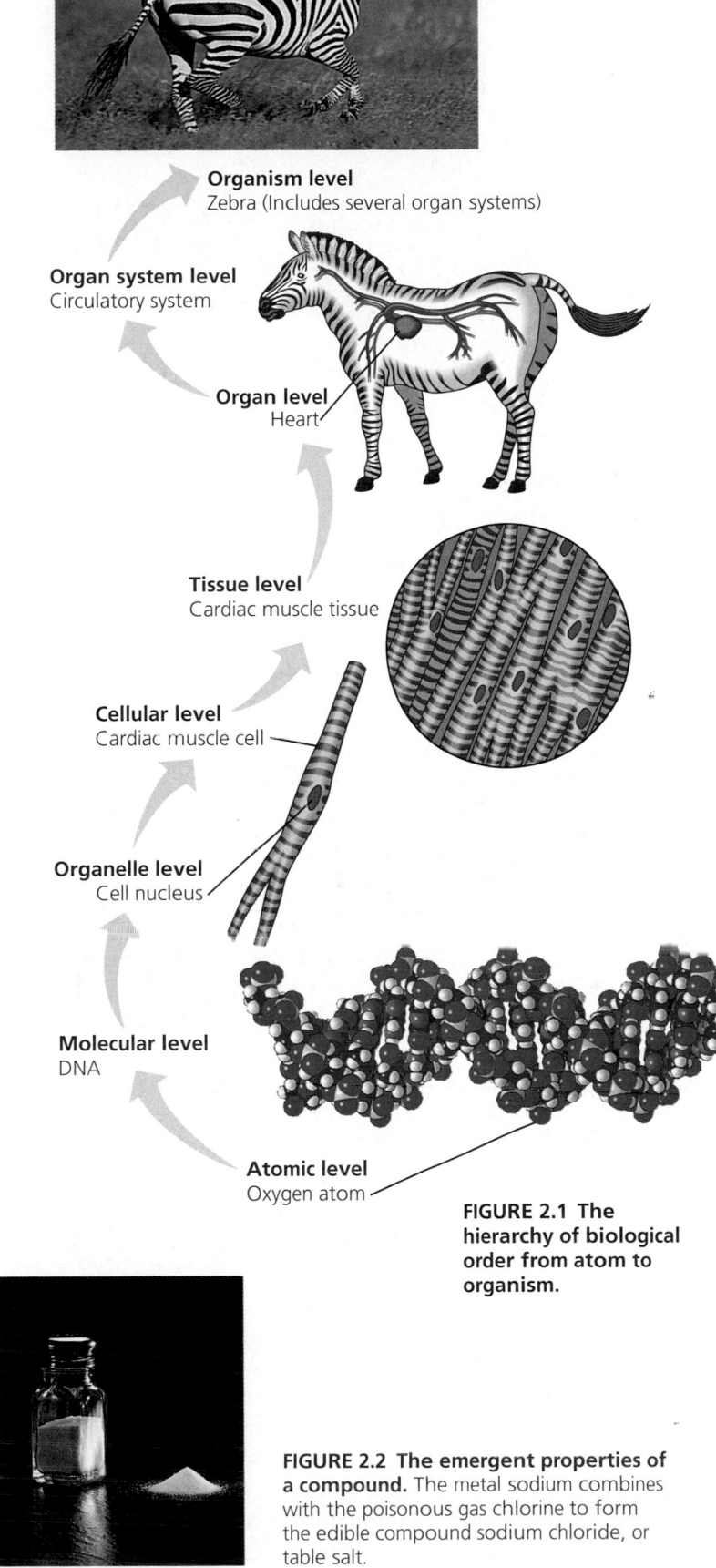

Organism level
Zebra (Includes several organ systems)

Organ system level
Circulatory system

Organ level
Heart

Tissue level
Cardiac muscle tissue

Cellular level
Cardiac muscle cell

Organelle level
Cell nucleus

Molecular level
DNA

Atomic level
Oxygen atom

FIGURE 2.1 The hierarchy of biological order from atom to organism.

Sodium

Chlorine

Sodium chloride

FIGURE 2.2 The emergent properties of a compound. The metal sodium combines with the poisonous gas chlorine to form the edible compound sodium chloride, or table salt.

Table 2.1 Naturally Occurring Elements in the Human Body

Symbol	Element	Atomic Number (See p. 29)	Percentage of Human Body Weight
O	Oxygen	8	65.0
C	Carbon	6	18.5
H	Hydrogen	1	9.5
N	Nitrogen	7	3.3
Ca	Calcium	20	1.5
P	Phosphorus	15	1.0
K	Potassium	19	0.4
S	Sulfur	16	0.3
Na	Sodium	11	0.2
Cl	Chlorine	17	0.2
Mg	Magnesium	12	0.1

Trace elements (less than 0.01%): boron (B), chromium (Cr), cobalt (Co), copper (Cu), fluorine (F), iodine (I), iron (Fe), manganese (Mn), molybdenum (Mo), selenium (Se), silicon (Si), tin (Sn), vanadium (V), and zinc (Zn).

FIGURE 2.4 Goiter. The enlarged thyroid gland of this Malaysian woman is due to an iodine deficiency.

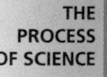

FIGURE 2.3 The effect of nitrogen deficiency in corn. In this controlled experiment, the plants on the left are growing in soil that was fertilized with compounds containing nitrogen, an essential element. The soil on the right is deficient in nitrogen. If the poorly nourished crop growing in this deficient soil is harvested, it will yield less food—and less nutritious food—than the crop on the left.

needed by all forms of life; others are required only by certain species. For example, in vertebrates (animals with backbones), the element iodine (I) is an essential ingredient of a hormone produced by the thyroid gland. A daily intake of only 0.15 milligram (mg) of iodine is adequate for normal activity of the human thyroid. An iodine deficiency in the diet causes the thyroid gland to grow to abnormal size, a condition called goiter (FIGURE 2.4). Where it is available, iodized salt has reduced the incidence of goiter.

ATOMS AND MOLECULES

The properties of chemical elements and of the compounds they form, including the compounds crucial to life, ultimately result from the structure of atoms.

Atomic structure determines the behavior of an element

Each element consists of a certain kind of atom that is different from the atoms of any other element. An **atom** is the smallest unit of matter that still retains the properties of an element. Atoms are so small that it would take about a million of them to stretch across the period printed at the end of this sentence. We symbolize atoms with the same abbreviation used for the element made up of those atoms; thus, C stands for both the element carbon and a single carbon atom.

Subatomic Particles

Although the atom is the smallest unit having the properties of its element, these tiny bits of matter are composed of even smaller parts, called subatomic particles. Physicists have split the atom into more than a hundred types of particles, but only three kinds of particles are stable enough to be of relevance here: **neutrons, protons,** and **electrons.** Neutrons and protons are packed together tightly to form a dense core, or **atomic nucleus,** at the center of the atom. The electrons, moving at nearly the speed of light, form a cloud around the nucleus (FIGURE 2.5).

Electrons and protons are electrically charged. Each electron has one unit of negative charge, and each proton has one

THE PROCESS OF SCIENCE

unit of positive charge. A neutron, as its name implies, is electrically neutral. Protons give the nucleus a positive charge, and it is the attraction between opposite charges that keeps the rapidly moving electrons in the vicinity of the nucleus.

The neutron and proton are almost identical in mass, each about 1.7×10^{-24} gram (g). Grams and other conventional units are not very useful for describing the mass of objects so minuscule. Thus, for atoms and subatomic particles (and for molecules, as well), we use a unit of measurement called the **dalton,** in honor of John Dalton, the British scientist who helped develop atomic theory around 1800. (The dalton is the same as the *atomic mass unit,* or *amu,* a unit you may have encountered elsewhere.) Neutrons and protons have masses close to 1 dalton. Because the mass of an electron is only about $\frac{1}{2,000}$ that of a neutron or proton, we can ignore electrons when computing the total mass of an atom.

Atomic Number and Atomic Weight

Atoms of the various elements differ in their number of subatomic particles. All atoms of a particular element have the same number of protons in their nuclei. This number of protons, which is unique to that element, is referred to as the **atomic number** and is written as a subscript to the left of the symbol for the element. The abbreviation $_2$He, for example, tells us that an atom of the element helium has 2 protons in its nucleus. Unless otherwise indicated, an atom is neutral in electrical charge, which means that its protons must be balanced by an equal number of electrons. Therefore, the atomic number tells us the number of protons and also the number of electrons in an electrically neutral atom.

We can deduce the number of neutrons from a second quantity, the **mass number,** which is the sum of protons plus

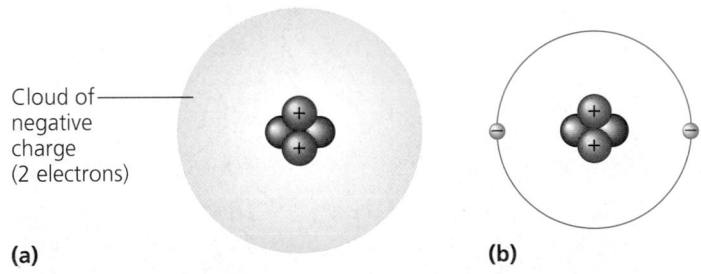

Cloud of negative charge (2 electrons)

(a)   **(b)**

FIGURE 2.5 Two simplified models of a helium (He) atom. The helium nucleus consists of 2 neutrons (gray) and 2 protons (magenta). Two electrons move rapidly around the nucleus. In model **(a),** the electrons are represented as a cloud of negative charge. The cloud is not to scale; it is actually *much* bigger relative to the nucleus. In **(b),** the electrons are shown as small blue spheres; the circle indicates that their average distance from the nucleus is about two-thirds of the radius of the electron cloud. (Our model of an atom will be refined as the chapter progresses.)

neutrons in the nucleus of an atom. The mass number is written as a superscript to the left of an element's symbol. For example, we can use this shorthand to write an atom of helium as $_2^4$He. Because the atomic number indicates how many protons there are, we can determine the number of neutrons by subtracting the atomic number from the mass number: A $_2^4$He atom has 2 neutrons. An atom of sodium, $_{11}^{23}$Na, has 11 protons, 11 electrons, and 12 neutrons. The simplest atom is hydrogen $_1^1$H, which has no neutrons; it consists of a lone proton with a single electron moving around it.

Almost all of an atom's mass is concentrated in its nucleus, because the contribution of electrons to mass is negligible. Because neutrons and protons each have a mass very close to 1 dalton, the mass number is an approximation of the total mass of an atom, usually called its **atomic weight.** So, we might say that the atomic weight of helium ($_2^4$He) is 4 daltons, although more precisely it is 4.003 daltons.

Isotopes

All atoms of a given element have the same number of protons, but some atoms have more neutrons than other atoms of the same element and therefore weigh more. These different atomic forms are referred to as **isotopes** of the element. In nature, an element occurs as a mixture of its isotopes. For example, consider the three isotopes of the element carbon, which has the atomic number 6. The most common isotope is carbon-12, $_6^{12}$C, which accounts for about 99% of the carbon in nature. It has 6 neutrons. Most of the remaining 1% of carbon consists of atoms of the isotope $_6^{13}$C, with 7 neutrons. A third, even rarer isotope, $_6^{14}$C, has 8 neutrons. Notice that all three isotopes of carbon have 6 protons—otherwise, they would not be carbon. Although isotopes of an element have slightly different masses, they behave identically in chemical reactions. (The number usually given as the atomic weight of an element is actually an average of the atomic weights of all the element's naturally occurring isotopes.)

Both ^{12}C and ^{13}C are stable isotopes, meaning that their nuclei do not have a tendency to lose particles. The isotope ^{14}C, however, is unstable, or radioactive. A **radioactive isotope** is one in which the nucleus decays spontaneously, giving off particles and energy. When the decay leads to a change in the number of protons, it transforms the atom to an atom of a different element. For example, radioactive carbon decays to form nitrogen.

Radioactive isotopes have many useful applications in biology. In Chapter 25, you will learn how researchers use measurements of radioactivity in fossils to date those relics of past life. Radioactive isotopes are also useful as tracers to follow atoms through metabolism, the chemical processes of an organism. Cells use the radioactive atoms as they would nonradioactive isotopes of the same element, but the radioactive

FIGURE 2.6 Using radioactive isotopes to study cell chemistry. Scientists use radioactive isotopes to label certain chemical substances in order to follow a metabolic process or locate the substance within a cell or an organism. **(a)** This researcher is performing an experiment to determine how temperature affects the rate at which certain cells make new copies of their DNA. First, the cells are placed in a medium that includes the ingredients for making DNA. One ingredient is labeled with ^3H, a radioactive isotope of hydrogen. Any new DNA the cells make will incorporate ^3H. Separate samples of the cells in their radioactive medium are incubated at different temperatures. After an allotted amount of time, samples are taken from each culture and their DNA is tested for radioactivity in a device called a scintillation counter (shown here). The DNA samples are placed in vials containing a fluid that emits flashes of light whenever radiation from the decay of the ^3H tracer in the DNA excites chemicals in the fluid. The frequency of flashes, proportional to the amount of radioactive DNA present, is measured in counts per minute. When the counts for the various DNA samples are plotted against the temperature at which the cells were grown, as in the graph here, it is apparent that temperature dramatically affects the rate of DNA synthesis. This result might be useful to researchers studying the details of DNA synthesis. **(b)** A technique known as autoradiography can be used to determine the cellular localization of the radioactively labeled DNA synthesized by the cells described above in Figure 2.6(a). Thin slices of the cells are placed on glass slides and kept for some time in the dark covered by a layer of photographic emulsion. Radiation from the radioactive tracer present in any new DNA exposes the photographic emulsion, creating a pattern of black dots. In the photo, radioactive DNA in the cell on the left is clearly located in the cell's nucleus.

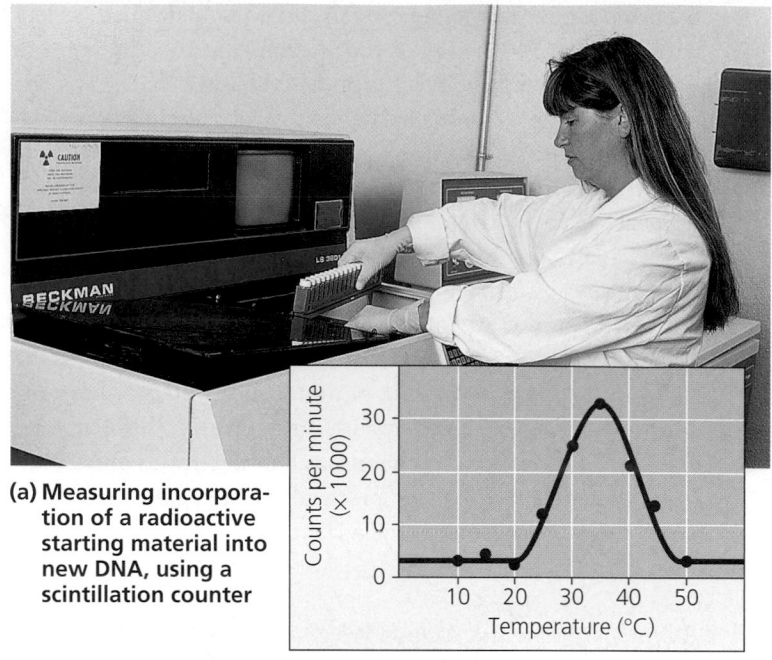

(a) Measuring incorporation of a radioactive starting material into new DNA, using a scintillation counter

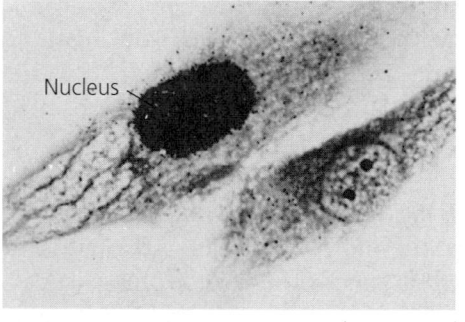

(b) Autoradiography   25 µm

tracers can be readily detected. FIGURE 2.6 shows two methods biologists use to monitor radioactive tracers.

Radioactive tracers are important diagnostic tools in medicine. For example, certain kidney disorders can be diagnosed by injecting small doses of substances containing radioactive isotopes into the blood and then measuring the amount of tracer excreted in the urine. Radioactive tracers are also used in combination with sophisticated imaging instruments, such as PET scanners, which can monitor chemical processes as they actually occur in the body (FIGURE 2.7).

Although radioactive isotopes are very useful in biological research and medicine, radiation from decaying isotopes also poses a hazard to life by damaging cellular molecules. The severity of this damage depends on the type and amount of radiation an organism absorbs. One of the most serious environmental threats is radioactive fallout from nuclear accidents (FIGURE 2.8).

FIGURE 2.7 A PET scan, a medical use for radioactive isotopes. PET, an acronym for positron-emission tomography, detects locations of intense chemical activity in the body. The patient is first injected with a nutrient such as glucose labeled with a radioactive isotope that emits subatomic particles called positrons. The positrons collide with electrons made available by chemical reactions in the body. A PET scanner detects the energy released in these collisions and maps as "hot spots" the regions of an organ that are most chemically active at the time. The color of the image varies with the amount of the isotope present in an area. In this image the bright yellow color identifies cancerous bone tissue in the patient's spine and at the bottom of one shoulder blade.

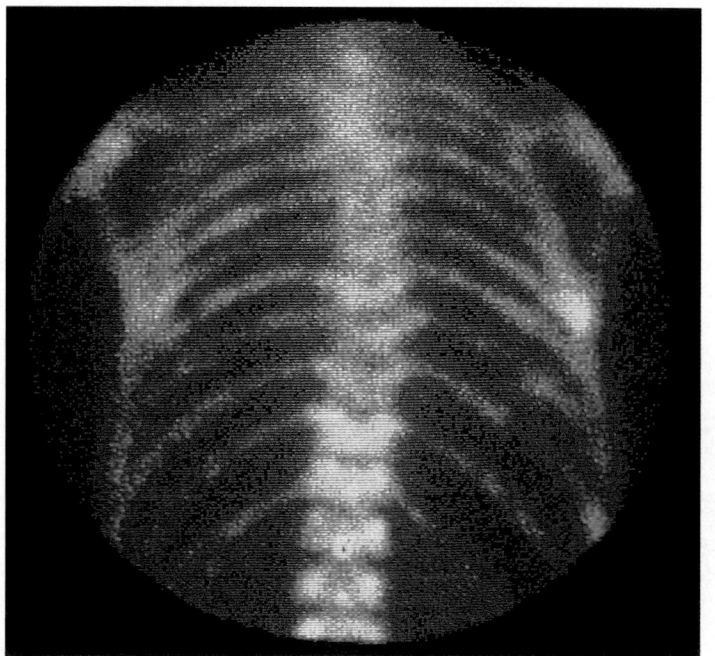

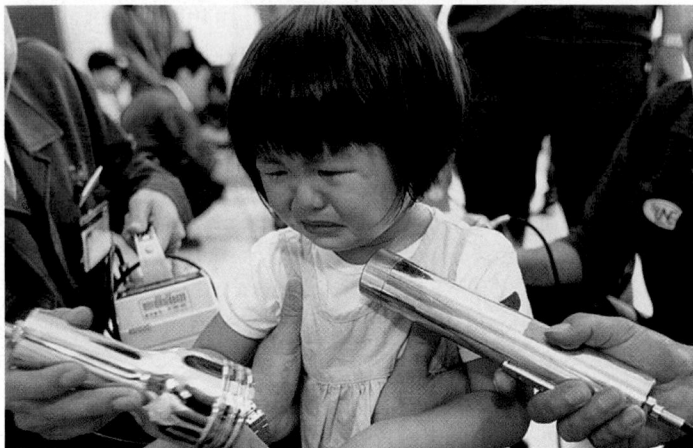

FIGURE 2.8 The Tokaimura nuclear accident. In 1999, workers at a nuclear power plant in Tokaimura, Japan, accidentally triggered a chain reaction that leaked large amounts of radioactivity into the surroundings. The accident exposed 46 workers in the plant to dangerous radiation doses. A mile from the plant, radiation levels peaked at 15,000 times higher than normal. About 300,000 people in nearby towns were checked for radioactive contamination, as shown here, and will continue to be monitored for possible long-term damage, such as cancers.

The Energy Levels of Electrons

The simplified models of the atom in FIGURE 2.5 greatly exaggerate the size of the nucleus relative to the volume of the whole atom. If the nucleus were the size of a golf ball, the electrons would be moving about the nucleus at an average distance of approximately 1 kilometer (km). Atoms are mostly empty space.

When two atoms approach each other during a chemical reaction, their nuclei do not come close enough to interact. Of the three kinds of subatomic particles we have discussed, only electrons are directly involved in the chemical reactions between atoms.

An atom's electrons vary in the amount of energy they possess. **Energy** is defined as the ability to do work. **Potential energy** is the energy that matter stores because of its position or location. For example, because of its altitude, water in a reservoir on a hill has potential energy. When the gates of the reservoir's dam are opened and the water runs downhill, the energy comes out of storage to do work, such as turning generators. Because potential energy has been expended, the water stores less energy at the bottom of the hill than it did in the reservoir. Matter has a natural tendency to move to the lowest possible state of potential energy; in this example, water runs downhill. To restore the potential energy of a reservoir, work must be done to elevate the water against gravity.

The electrons of an atom also have potential energy because of their position in relation to the nucleus. The negatively charged electrons are attracted to the positively charged nucleus; the more distant the electrons are from the nucleus, the greater their potential energy. Unlike the continuous flow

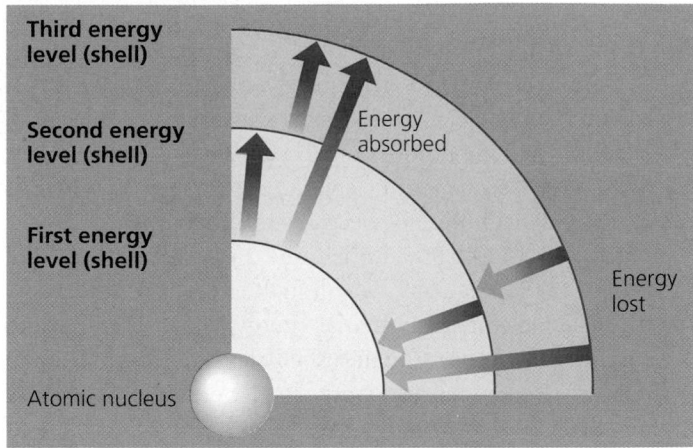

Figure 2.9 Energy levels of an atom's electrons. Electrons exist only at fixed levels of potential energy, which are also called electron shells. An electron can move from one level to another only if the energy it gains or loses is exactly equal to the difference in energy between the two levels. Arrows indicate some of the stepwise changes in potential energy that are possible.

of water downhill, changes in the potential energy of electrons can occur only in steps of fixed amounts. An electron having a certain discrete amount of energy is something like a ball on a staircase. The ball can have different amounts of potential energy, depending on which step it is on, but it cannot spend much time between the steps. An electron cannot exist in between its fixed states of potential energy.

The different states of potential energy that electrons have in an atom are called **energy levels,** or **electron shells** (FIGURE 2.9). The first shell is closest to the nucleus, and electrons in this shell have the lowest energy. Electrons in the second shell have more energy, electrons in the third shell more energy still, and so on. An electron can change its shell, but only by absorbing or losing an amount of energy equal to the difference in potential energy between the old shell and the new shell. To move to a shell farther out from the nucleus, the electron must absorb energy. For example, light energy can excite an electron to a higher energy level. (Indeed, this is the first step when plants harness the energy of sunlight for photosynthesis, the process that produces food from carbon dioxide and water.) To move to a shell closer to the nucleus, an electron must lose energy, which is usually released to the environment in the form of heat. For example, when sunlight excites electrons in the paint of a dark car roof to higher energy levels, the roof heats up as the electrons fall back to their original levels.

Electron Configuration and Chemical Properties

The chemical behavior of an atom is determined by its electron configuration—that is, the distribution of electrons in the atom's electron shells. Beginning with hydrogen, the simplest atom, we can imagine building the atoms of the other elements by adding 1 proton and 1 electron at a time (along with an appropriate number of neutrons). FIGURE 2.10, an abbreviated version of what is called a periodic table, shows this distribution of electrons for the first 18 elements, from hydrogen ($_1$H) to argon ($_{18}$Ar). The elements are arranged in three rows, or periods, corresponding to the number of electron shells in their atoms. The left-to-right sequence of elements in each row corresponds to the sequential addition of electrons (and protons).

Hydrogen's 1 electron and helium's 2 electrons are located in the first shell. Electrons, like all matter, tend to exist in the lowest available state of potential energy, which they have in the first shell. However, the first shell can hold no more than 2 electrons. An atom with more than 2 electrons must use higher shells because the first shell is full. The next element, lithium, has 3 electrons. Two of these electrons fill the first shell while the third electron occupies the second shell. The second shell holds a maximum of 8 electrons. Neon, at the end of the second row, has 8 electrons in the second shell, giving it a total of 10 electrons.

The chemical behavior of an atom depends mostly on the number of electrons in its *outermost* shell. We refer to those outer electrons as **valence electrons** and to the outermost electron shell as the **valence shell.** In the case of lithium, there is only 1 valence electron, and the second shell is the valence shell.

Atoms with the same number of electrons in their valence shells exhibit similar chemical behavior. For example, fluorine (F) and chlorine (Cl) both have 7 valence electrons, and both combine with the element sodium to form compounds (see FIGURE 2.2). An atom with a completed valence shell is unreactive; that is, it will not interact readily with other atoms it encounters. At the far right of the periodic table are helium, neon, and argon, the only three elements shown in FIGURE 2.10 that have full valence shells. These elements are said to be inert, meaning chemically unreactive. All the other atoms in FIGURE 2.10 are chemically reactive because they have incomplete valence shells.

Electron Orbitals

Early in the 20th century, the electron shells of an atom were visualized as concentric paths of electrons orbiting the nucleus, somewhat like planets orbiting the sun. It is still convenient to use concentric-circle diagrams to symbolize electron shells, as in FIGURE 2.10. The atom is not this simple, however. In fact, we can never know the exact path of an electron. What we can do instead is describe the space in which an electron spends most of its time. The three-dimensional space where an electron is found 90% of the time is called an **orbital.** Each

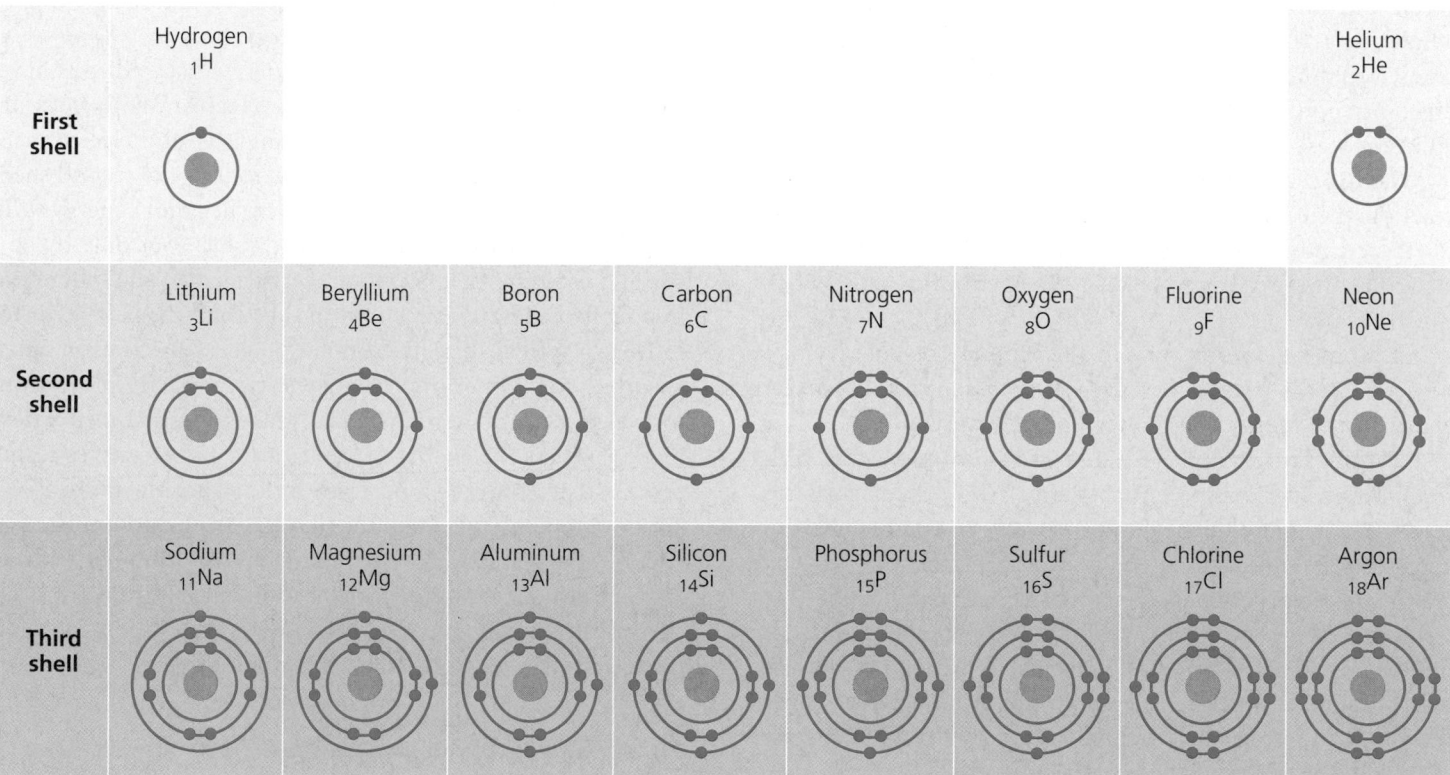

FIGURE 2.10 Electron configurations of the first 18 elements. In these diagrams, electrons are shown as blue dots and energy levels (electron shells) as concentric rings. The elements are arranged in rows, each representing the filling of an electron shell.

As electrons are added, they occupy the lowest available shell. Hydrogen's 1 electron and helium's 2 electrons are located in the first shell. The next element, lithium, has 3 electrons. Two electrons fill the first shell while the third electron occupies the second shell

(the lowest available shell). The outermost energy level occupied by electrons is called the valence shell. Elements with the same number of valence electrons—fluorine and chlorine, for instance—have similar chemical properties.

electron shell consists of a specific number of orbitals of specific shapes (FIGURE 2.11).

No more than 2 electrons can occupy the same orbital. The first electron shell has a single orbital and can thereby accommodate a maximum of 2 electrons. This orbital, which is spherical in shape, is designated the 1s orbital. The lone electron of a hydrogen atom occupies the 1s orbital, as do the 2 electrons of a helium atom. The second electron shell has four orbitals and therefore can hold 8 electrons. Electrons in each of the four orbitals have nearly the same energy, but they move in different volumes of space. There is a 2s orbital, spherical in shape like the 1s orbital, but with a greater diameter. The other three orbitals, called 2p orbitals, are dumbbell-shaped. Each 2p orbital is oriented at right angles to the other two. (The third and higher electron shells also have s and p orbitals, as well as orbitals of more complex shapes.)

The reactivity of atoms arises from the presence of unpaired electrons in one or more orbitals of their valence shells. Recall that the electronic configurations in FIGURE 2.10 build up with the addition of 1 electron at a time. Like strangers getting on a bus with seats for two, each additional electron occupies a separate orbital until no empty orbitals remain in a shell. Only then do the orbitals begin to receive a second electron. The paired dots in FIGURE 2.10 represent paired electrons

sharing an orbital. When atoms interact to complete their valence shells, it is the *unpaired* electrons that are involved.

Atoms combine by chemical bonding to form molecules

Now that we have looked at the structure of atoms, we can move up in the hierarchy of organization and see how atoms combine to form molecules. Atoms with incomplete valence shells interact with certain other atoms in such a way that each partner completes its valence shell. Atoms do this by either sharing or transferring valence electrons. These interactions usually result in atoms staying close together, held by attractions called **chemical bonds.** The strongest kinds of chemical bonds are covalent bonds and ionic bonds.

Covalent Bonds

A **covalent bond** is the sharing of a pair of valence electrons by two atoms. For example, let's consider what happens when two hydrogen atoms approach each other. Recall that hydrogen has 1 valence electron in the first shell, but the shell's capacity is 2 electrons. When the two hydrogen atoms come

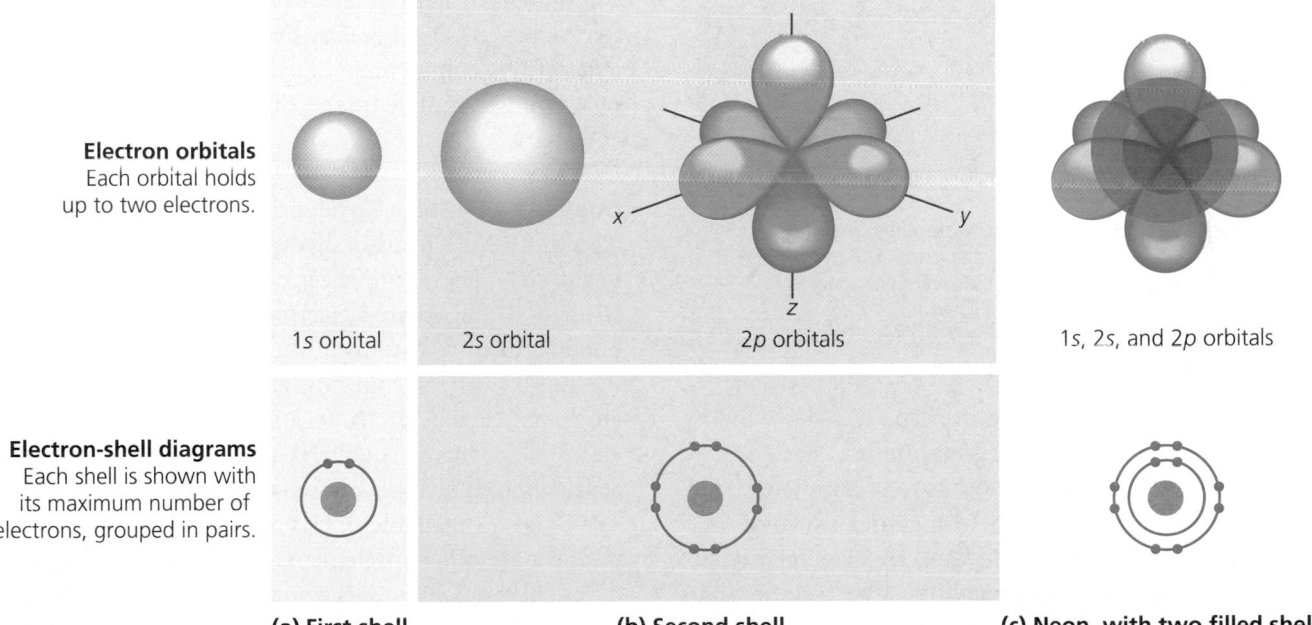

Electron orbitals
Each orbital holds up to two electrons.

1s orbital   2s orbital   2p orbitals   1s, 2s, and 2p orbitals

Electron-shell diagrams
Each shell is shown with its maximum number of electrons, grouped in pairs.

(a) First shell   **(b) Second shell**   **(c) Neon, with two filled shells**

FIGURE 2.11 Electron orbitals. The three-dimensional shapes in the top half of this figure represent electron orbitals—the volumes of space where the electrons of an atom are most likely to be found. Each orbital holds a maximum of 2 electrons. The bottom half of the figure shows the equivalent electron-shell diagrams. **(a)** The first electron shell has one spherical (s) orbital, designated 1s. Only this orbital is occupied in hydrogen, which has 1 electron, and helium, which has 2 electrons. **(b)** The second and all higher shells each have one larger s orbital (designated 2s for the second shell) plus three dumbbell-shaped orbitals called p orbitals (2p for the second shell). The three 2p orbitals lie at right angles to one another along imaginary x, y, and z axes of the atom. (The third and higher electron shells can hold additional electrons in orbitals of more complex shapes.) **(c)** To symbolize the electron orbitals of the element neon, which has a total of 10 electrons, we superimpose the 1s orbital of the first shell and the 2s and three 2p orbitals of the second shell. (Each orbital, remember, can hold 2 electrons.)

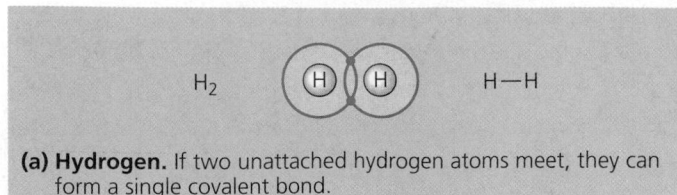

(a) Hydrogen. If two unattached hydrogen atoms meet, they can form a single covalent bond.

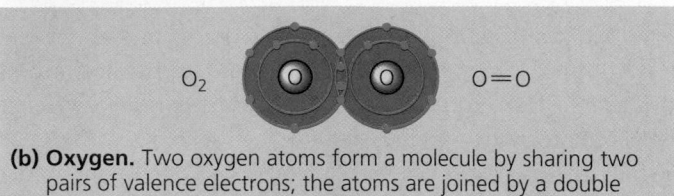

(b) Oxygen. Two oxygen atoms form a molecule by sharing two pairs of valence electrons; the atoms are joined by a double covalent bond.

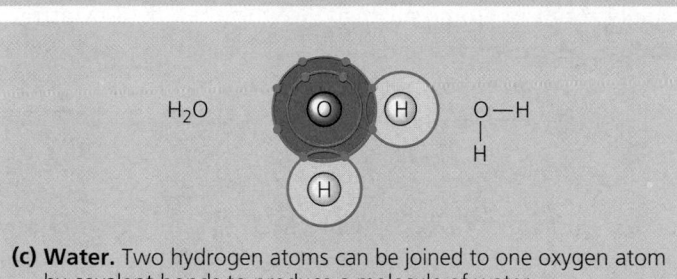

(c) Water. Two hydrogen atoms can be joined to one oxygen atom by covalent bonds to produce a molecule of water.

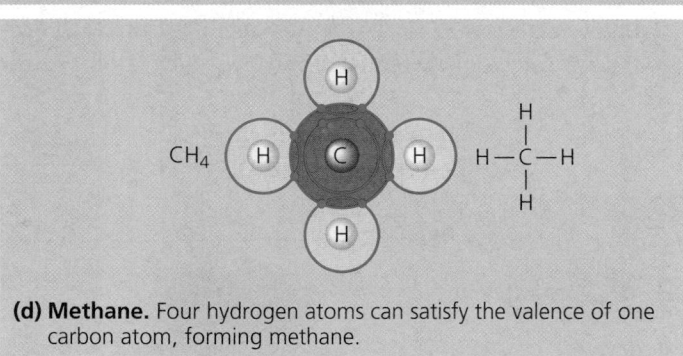

(d) Methane. Four hydrogen atoms can satisfy the valence of one carbon atom, forming methane.

FIGURE 2.12 Covalent bonding in four molecules. A single covalent bond consists of a pair of shared electrons. The number of electrons required to complete an atom's valence shell generally determines how many bonds that atom will form.

close enough for their 1s orbitals to overlap, they share their electrons (FIGURE 2.12a). Each hydrogen atom now has 2 electrons associated with it in what amounts to a completed valence shell. Two or more atoms held together by covalent bonds constitute a **molecule.** In this case, we have formed a hydrogen molecule. We can abbreviate the structure of this molecule as H—H, where the line represents a covalent bond—that is, a pair of shared electrons. This notation, which represents both atoms and bonding, is called a **structural formula.** We can abbreviate even further by writing H$_2$, a **molecular formula** indicating simply that the molecule consists of two atoms of hydrogen.

With 6 electrons in its second electron shell, oxygen needs 2 more electrons to complete its valence shell. Two oxygen atoms form a molecule by sharing *two* pairs of valence elec-

trons (FIGURE 2.12b). The atoms are thus joined by what is called a **double covalent bond.**

Each atom that can share valence electrons has a bonding capacity corresponding to the number of covalent bonds the atom can form. When the bonds form, they give the atom a full complement of valence electrons. The bonding capacity of oxygen, for example, is 2. This bonding capacity is called the atom's **valence** and generally equals the number of unpaired electrons in the atom's outermost (valence) shell. You can readily determine the valences of life's most abundant elements from the electron configurations in FIGURE 2.10: The valence of hydrogen is 1; oxygen, 2; nitrogen, 3; and carbon, 4. One more complicated case is phosphorus (P), another element important to life. Phosphorus can have a valence of 3, as we would predict from its 3 unpaired electrons. In biologically important molecules, however, it generally has a valence of 5, forming three single bonds and one double bond.

The molecules H$_2$ and O$_2$ are pure elements, not compounds. (Recall that a compound is a combination of two or more *different* elements.) An example of a molecule that is a compound is water, with the molecular formula H$_2$O. It takes two atoms of hydrogen to satisfy the valence of one oxygen atom. FIGURE 2.12c shows the structure of a water molecule. Water is so important to life that Chapter 3 is devoted entirely to its structure and behavior.

Another molecule that is a compound is methane, the main component of natural gas, with the molecular formula CH$_4$ (FIGURE 2.12d). It takes four hydrogen atoms, each with a valence of 1, to complement one atom of carbon, with its valence of 4. We will look at many other compounds of carbon in Chapter 4.

Nonpolar and Polar Covalent Bonds. The attraction of an atom for the electrons of a covalent bond is called its **electronegativity.** The more electronegative an atom, the more strongly it pulls shared electrons toward itself. In a covalent bond between two atoms of the same element, the outcome of the tug-of-war for common electrons is a standoff; the two atoms are equally electronegative. Such a bond, in which the electrons are shared equally, is a **nonpolar covalent bond.** The covalent bond of H$_2$ is nonpolar, as is the double bond of O$_2$. The bonds of methane (CH$_4$) are also nonpolar; although the partners are different elements, carbon and hydrogen do not differ substantially in electronegativity. In other compounds, however, where one atom is bonded to a more electronegative atom, the electrons of the bond will not be shared equally. This sort of bond is called a **polar covalent bond.** For example, the bonds between the oxygen and hydrogen atoms of a water molecule are polar (FIGURE 2.13). Oxygen is one of the most electronegative of the 92 elements, attracting shared electrons much more strongly than hydrogen does. In a covalent bond between oxygen and hydrogen, the electrons spend more time near the oxygen nucleus than they do near the hydrogen nucleus. Because

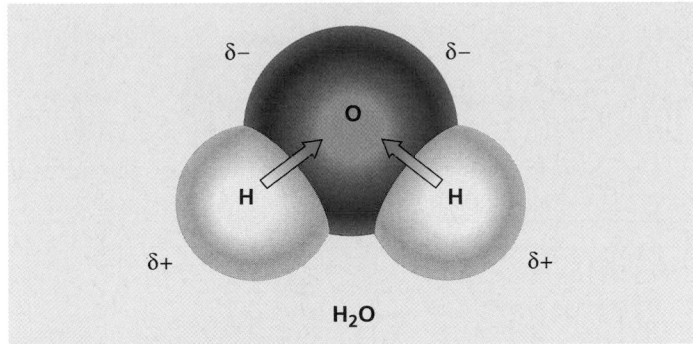

δ−   δ−

O

H   H

δ+   δ+

H₂O

FIGURE 2.13 Polar covalent bonds in a water molecule. Oxygen, being much more electronegative than hydrogen, pulls the shared electrons of the covalent bonds toward itself, as indicated by the arrows. This unequal sharing of electrons gives the oxygen a partial negative charge and the hydrogens a partial positive charge. The Greek symbol delta (δ) indicates that the charges are less than full units. This "space-filling" model approximates the true shape of H₂O.

electrons have a negative charge, the unequal sharing of electrons in water causes the oxygen atom to have a partial negative charge and each hydrogen atom a partial positive charge.

Ionic Bonds

In some cases, two atoms are so unequal in their attraction for valence electrons that the more electronegative atom strips an electron completely away from its partner. This is what happens when an atom of sodium ($_{11}$Na) encounters an atom of chlorine ($_{17}$Cl) (FIGURE 2.14). A sodium atom has a total of 11 electrons, with its single valence electron in the third electron shell. A chlorine atom has a total of 17 electrons, with 7 electrons in its valence shell. When these two atoms meet, the lone valence electron of sodium is transferred to the chlorine atom, and both atoms end up with their valence shells complete. (Because sodium no longer has an electron in the third shell, the second shell is now the valence shell.)

The electron transfer between the two atoms moves one unit of negative charge from sodium to chlorine. Sodium, now with 11 protons but only 10 electrons, has a net electrical charge of +1. A charged atom (or molecule) is called an **ion.** When the charge is positive, the ion is specifically called a **cation;** the sodium atom has become a cation. Conversely, the chlorine atom, having gained an extra electron, now has 17 protons and 18 electrons, giving it a net electrical charge of −1. It has become a chloride ion—an **anion,** or negatively charged ion. Because of their opposite charges, cations and anions attract each other in what is called an **ionic bond.** Any two ions of opposite charge can form an ionic bond. The ions need not have acquired their charge by an electron transfer with each other.

Compounds formed by ionic bonds are called **ionic compounds,** or **salts.** We know the ionic compound sodium chloride (NaCl) as table salt (FIGURE 2.15). Salts are often found in nature as crystals of various sizes and shapes, each an aggregate of vast numbers of cations and anions bonded by their electrical attraction and arranged in a three-dimensional lattice. A salt crystal does not consist of molecules in the sense that a covalent compound does, because a covalently bonded molecule

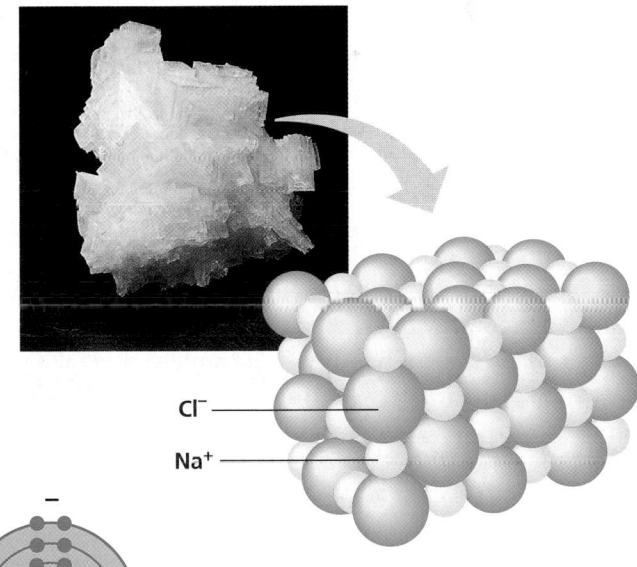

Cl⁻

Na⁺

FIGURE 2.15 A sodium chloride crystal. The sodium ions (Na⁺) and chloride ions (Cl⁻) are held together by ionic bonds. The formula NaCl tells us that the ratio of Na⁺ to Cl⁻ is 1:1.

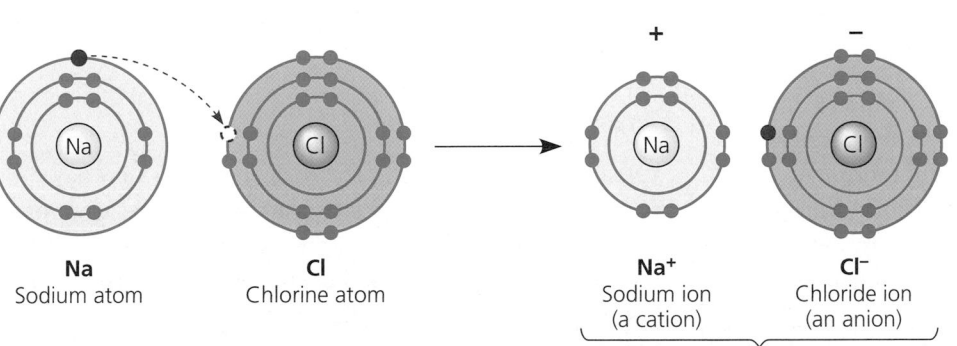

+   −

Na	**Cl**		**Na⁺**	**Cl⁻**
Sodium atom	Chlorine atom		Sodium ion (a cation)	Chloride ion (an anion)

Sodium chloride (NaCl)

FIGURE 2.14 Electron transfer and ionic bonding. A valence electron can be transferred from sodium (Na) to chlorine (Cl), giving both atoms completed valence shells. The electron transfer leaves the sodium atom with a net charge of +1 (cation) and the chlorine atom with a net charge of −1 (anion). The attraction between the oppositely charged atoms, or ions, is an ionic bond. Ions can bond not only to atoms with which they have reacted, but to any other ions of opposite charge.

has a definite size and number of atoms. The formula for an ionic compound, such as NaCl, indicates only the ratio of elements in a crystal of the salt. "NaCl" is not a molecule.

Not all salts have equal numbers of cations and anions. For example, the ionic compound magnesium chloride ($MgCl_2$) has two chloride ions for each magnesium ion. Magnesium ($_{12}Mg$) must lose 2 outer electrons if the atom is to have a complete valence shell, so it tends to become a cation with a net charge of $+2$ (Mg^{2+}). One magnesium cation can therefore form ionic bonds with two chloride anions.

The term *ion* also applies to entire molecules that are electrically charged. In the salt ammonium chloride (NH_4Cl), for instance, the anion is a single chloride ion (Cl^-), but the cation is ammonium (NH_4^+), a nitrogen atom with four covalently bonded hydrogen atoms. The whole ammonium ion has an electrical charge of $+1$ because it is 1 electron short.

Environment affects the strength of ionic bonds. In a dry salt crystal, the bonds are so strong that it takes a hammer and chisel to break enough of them to crack the crystal in two. Place the same salt crystal in water, however, and the salt dissolves as the attractions between its ions decrease. In the next chapter, you will learn how water dissolves salts.

Weak chemical bonds play important roles in the chemistry of life

In living organisms, most of the strongest chemical bonds are covalent ones, which link atoms to form a cell's molecules. But bonding *between* molecules is also indispensable in the cell, where the properties of life emerge from molecular interactions. When two molecules in the cell make contact, they may adhere temporarily by types of chemical bonds that are much weaker than covalent bonds. The advantage of weak bonding is that the contact between the molecules can be brief; the molecules come together, respond to one another in some way, and then separate.

The importance of weak bonding can be seen in the example of chemical signaling in the brain. One brain cell signals another by releasing molecules that use weak bonds to dock onto receptor molecules on the nearby surface of a receiving cell. The bonds last just long enough to trigger a momentary response by the receiving cell. If the signal molecules attached by stronger bonds, the receiving cell would continue to respond long after the transmitting cell ceased dispatching the message, with perhaps disastrous consequences. (Imagine what it would be like, for instance, if your brain continued to perceive the ringing sound of a bell for hours after nerve cells transmitted the information from the ears to the brain.)

Several types of weak chemical bonds are important in living organisms. One is the ionic bond, which is relatively weak in the presence of water. Another type of weak bond, crucial to life, is known as a hydrogen bond.

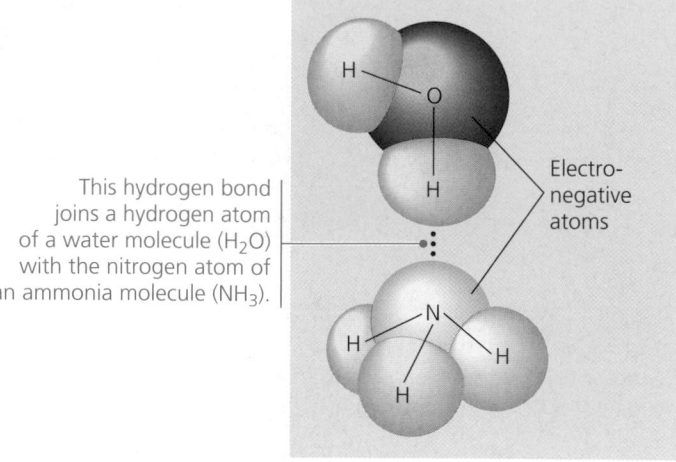

This hydrogen bond joins a hydrogen atom of a water molecule (H_2O) with the nitrogen atom of an ammonia molecule (NH_3).

Electro-negative atoms

FIGURE 2.16 A hydrogen bond. Through a weak electrical attraction, a hydrogen atom covalently bonded to one electronegative atom weakly bonds to another electronegative atom.

Hydrogen Bonds

Among the various kinds of weak chemical bonds, hydrogen bonds are so important in the chemistry of life that they deserve special attention. A **hydrogen bond** forms when a hydrogen atom covalently bonded to one electronegative atom is also attracted to another electronegative atom. In living cells, the electronegative partners involved are usually oxygen or nitrogen atoms.

Let's examine the simple case of hydrogen bonding between water (H_2O) and ammonia (NH_3) (FIGURE 2.16). You have seen how the polar covalent bonds of water result in the oxygen atom having a partial negative charge and the hydrogen atoms having a partial positive charge. A similar situation arises in the ammonia molecule, where an electronegative nitrogen atom has a partial negative charge because of its pull on the electrons it shares covalently with hydrogen. If a water molecule and an ammonia molecule are close together, a weak attraction will occur between the negatively charged nitrogen atom and a positively charged hydrogen atom of the adjacent water molecule. This attraction is a hydrogen bond.

Van der Waals Interactions

Even a molecule with nonpolar covalent bonds may have positively and negatively charged regions. Because electrons are in constant motion, they are not always symmetrically distributed in the molecule; at any instant, they may accumulate by chance in one part of the molecule or another. The results are ever-changing "hot spots" of positive and negative charge that enable all atoms and molecules to stick to one another. These **van der Waals interactions** are weak and occur only when atoms and molecules are very close together.

Van der Waals interactions, hydrogen bonds, ionic bonds, and other weak bonds may form not only between molecules but also between different regions of a single large molecule, such as a protein. Although these bonds are individually weak, their cumulative effect is to reinforce the three-dimensional shape of a large molecule. You will learn more about the biological roles of weak bonds in Chapter 5.

A molecule's biological function is related to its shape

A molecule has a characteristic size and shape. The precise shape of a molecule is usually very important to its function in the living cell.

A molecule consisting of two atoms, such as H_2 or O_2, is always linear, but molecules with more than two atoms have more complicated shapes. These shapes are determined by the positions of the atoms' orbitals. When an atom forms covalent bonds, the orbitals in its valence shell rearrange. For atoms with valence electrons in both s and p orbitals (see FIGURE 2.11), the single s and three p orbitals hybridize to form four new orbitals shaped like identical teardrops extending from the region of the atomic nucleus (FIGURE 2.17a). If we connect the larger ends of the teardrops with lines, we have the outline of a pyramidal shape called a tetrahedron.

For the water molecule (H_2O), two of the hybrid orbitals in the oxygen atom's valence shell are shared with hydrogen atoms (FIGURE 2.17b). The result is a molecule shaped roughly like a V, its two covalent bonds spread apart at an angle of 104.5°.

The methane molecule (CH_4) has the shape of a completed tetrahedron, because all four hybrid orbitals of carbon are shared (FIGURE 2.17c). The nucleus of the carbon is at the center, with its four covalent bonds radiating to hydrogen nuclei at the corners of the tetrahedron. Larger molecules containing multiple carbon atoms, including many of the molecules that make up living matter, have more complex overall shapes. However, the tetrahedral shape of a carbon atom bonded to four other atoms is often a repeating motif within such molecules.

Molecular shape is crucial in biology because it determines how most biological molecules recognize and respond to one another. For instance, in the chemical signaling example mentioned earlier, the signal molecules released by the transmitting brain cell have a unique shape that specifically fits together with the shape of the receptor molecules on the surface of the receiving cell, much as a key fits into a lock (FIGURE 2.18). This complementarity of shape aids the formation of weak bonds

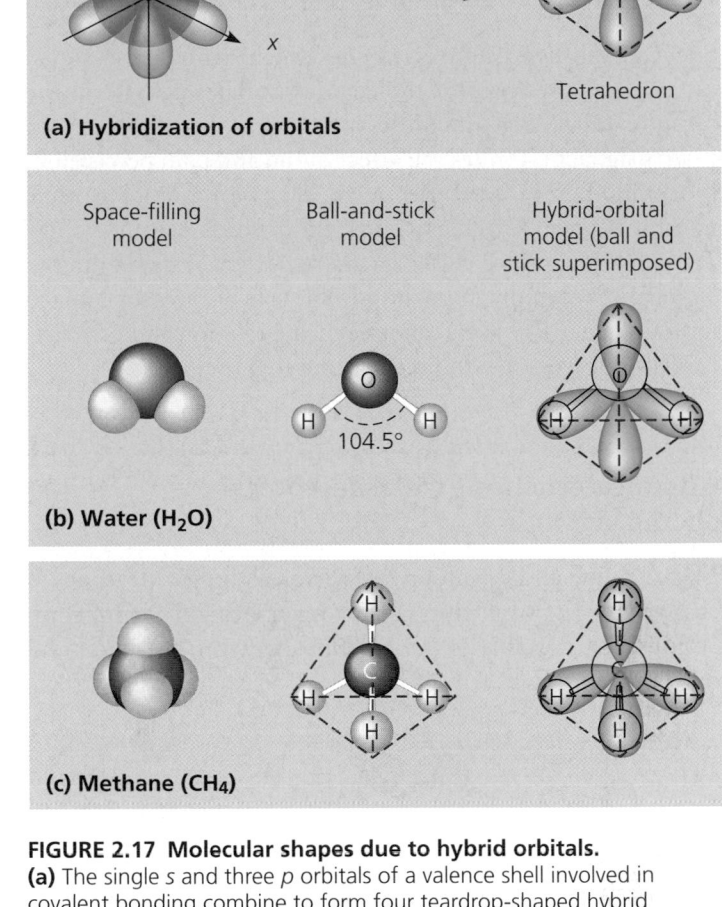

(a) Hybridization of orbitals

Tetrahedron

Space-filling model   Ball-and-stick model   Hybrid-orbital model (ball and stick superimposed)

(b) Water (H₂O)

104.5°

(c) Methane (CH₄)

FIGURE 2.17 Molecular shapes due to hybrid orbitals. (a) The single s and three p orbitals of a valence shell involved in covalent bonding combine to form four teardrop-shaped hybrid orbitals. These orbitals extend to the four corners of an imaginary tetrahedron (outlined in magenta). **(b)** Because of the positions of the hybrid orbitals, the two covalent bonds of water are angled at 104.5°. This is seen most clearly in the ball-and-stick model. The space-filling model comes closer to the actual shape of the molecule. **(c)** The hydrogens of methane occupy all four corners of the tetrahedron, giving methane a tetrahedral shape.

FIGURE 2.18 Molecular shape and brain chemistry. One nerve cell in the brain signals another by releasing signal molecules into the gap between the cells. The shape of the signal molecules is complementary to the shape of receptor molecules located on the surface of the receiving cell. The actual molecules have much more complex shapes than represented here.

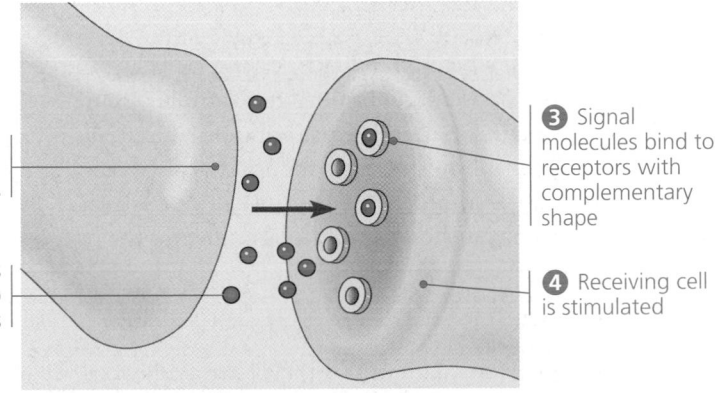

❶ Transmitting cell releases signal molecules

❷ Signal molecules pass across gap between nerve cells

❸ Signal molecules bind to receptors with complementary shape

❹ Receiving cell is stimulated

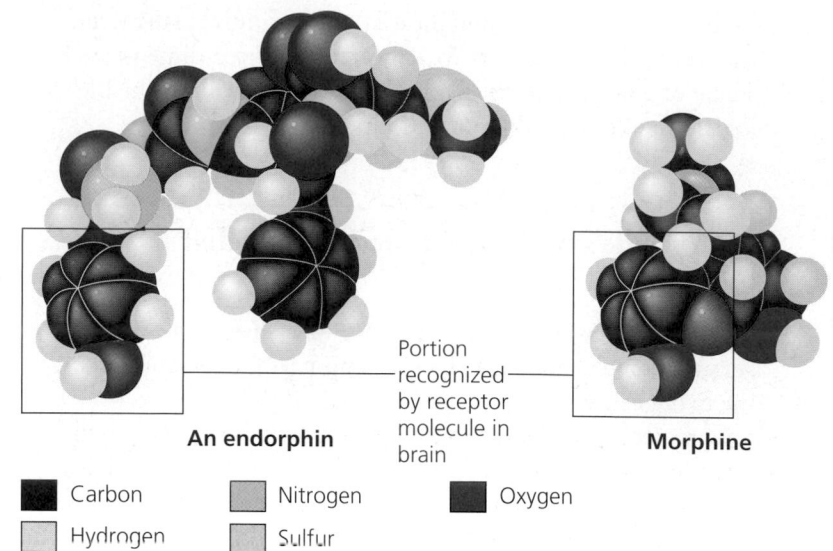

FIGURE 2.19 A molecular mimic. Morphine affects pain perception and emotional state by mimicking the brain's natural endorphins. The boxed portion of the endorphin molecule is recognized by receptor molecules on target cells in the brain. The boxed portion of the morphine molecule, an opiate drug, is a close match.

An endorphin Portion recognized by receptor molecule in brain **Morphine**

■ Carbon ▨ Nitrogen ▨ Oxygen
▨ Hydrogen ▨ Sulfur

between the two molecules. The attachment of the signal molecule to the receptor molecule stimulates activity in the receptor cell. Molecules with shapes similar to those of the brain's signal molecules can affect mood and pain perception. Morphine, heroin, and other opiate drugs, for example, mimic natural signal molecules called endorphins (FIGURE 2.19). These drugs produce euphoria and relieve pain by binding to endorphin receptors in the brain. The role of molecular shape in brain chemistry is an example of the relationship of structure to function, one of biology's unifying themes.

Chemical reactions make and break chemical bonds

The making and breaking of chemical bonds, leading to changes in the composition of matter, are called **chemical reactions.** An example is the reaction between hydrogen and oxygen to form water:

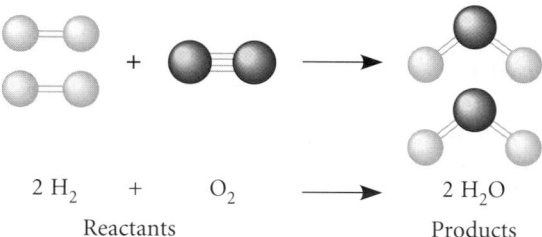

$$2\,H_2 \quad + \quad O_2 \quad \longrightarrow \quad 2\,H_2O$$
Reactants   Products

This reaction breaks the covalent bonds of H_2 and O_2 and forms the new bonds of H_2O. When we write a chemical reaction, we use an arrow to indicate the conversion of the starting materials, called **reactants,** to the **products.** The coefficients indicate the number of molecules involved; for example, the 2 in front of the H_2 means that the reaction starts with two molecules of hydrogen. Notice that all atoms of the reactants must be accounted for in the products. Matter is conserved in a chemical reaction: Reactions cannot create or destroy matter but can only rearrange it.

Photosynthesis, which takes place within the cells of green plant tissues, is a particularly important example of chemical reactions rearranging matter. Humans and other animals ultimately depend on photosynthesis for food and oxygen, and this process is at the foundation of almost all ecosystems. Following is the chemical shorthand that summarizes the process of photosynthesis:

$$6\,CO_2 + 6\,H_2O \longrightarrow C_6H_{12}O_6 + 6\,O_2$$

The raw materials of photosynthesis are carbon dioxide

(CO_2), which is taken from the air, and water (H_2O), which is absorbed from the soil. Within the plant cells, sunlight powers the conversion of these ingredients to a sugar called glucose ($C_6H_{12}O_6$) and oxygen molecules (O_2), a by-product that the plant releases into the surroundings (FIGURE 2.20). Although photosynthesis is actually a sequence of many chemical reactions, we still end up with the same number and kinds of atoms we had when we started. Matter has simply been rearranged.

Some chemical reactions go to completion; that is, all the reactants are converted to products. But most reactions are reversible, the products of the forward reaction becoming the reactants for the reverse reaction. For example, hydrogen and nitrogen molecules can combine to form ammonia, but ammonia can also decompose to regenerate hydrogen and nitrogen:

$$3\,H_2 + N_2 \rightleftharpoons 2\,NH_3$$

The opposite-headed arrows indicate that the reaction is reversible.

One of the factors affecting the rate of a reaction is the concentration of reactants. The greater the concentration of reactant molecules, the more frequently they collide with one

FIGURE 2.20 Photosynthesis: a solar-powered rearrangement of matter. This *Elodea,* a freshwater plant, produces sugar by rearranging the atoms of carbon dioxide and water in the chemical process known as photosynthesis. Sunlight powers this chemical transformation. Much of the sugar is then converted to other food molecules. Oxygen gas (O_2) is a by-product of photosynthesis; notice the bubbles of oxygen escaping from the leaves in the photo.

another and have an opportunity to react to form products. The same holds true for the products. As products accumulate, collisions resulting in the reverse reaction become increasingly frequent. Eventually, the forward and reverse reactions occur at the same rate, and the relative concentrations of products and reactants stop changing. The point at which the reactions offset one another exactly is called **chemical equilibrium.** This is a dynamic equilibrium; reactions are still going on, but with no net effect on the concentrations of reactants and products. Equilibrium does *not* mean that the reactants and products are equal in concentration, but only that their concentrations have stabilized. The reaction involving ammonia reaches equilibrium when ammonia decomposes as rapidly as it forms. In this case, there is far more ammonia than hydrogen and nitrogen at equilibrium.

■   ■   ■

We will return to the subject of chemical reactions after more detailed study of the various types of molecules that are important to life. In the next chapter, we focus on water, the substance in which all the chemical processes of living organisms occur.

CHAPTER 2 REVIEW

Go to the Campbell Biology website (www.campbellbiology.com) to explore an interactive version of the Chapter Review.

Summary of Key Concepts

CHEMICAL ELEMENTS AND COMPOUNDS

■ **Matter consists of chemical elements in pure form and in combinations called compounds (pp. 26–27, FIGURE 2.2)** Elements cannot be broken down to other substances. A compound contains two or more elements in a fixed ratio.
  Web/CD Activity 2A: *The Levels of Life Card Game*

■ **Life requires about 25 chemical elements (pp. 27–28, TABLE 2.1)** Carbon, oxygen, hydrogen, and nitrogen make up approximately 96% of living matter.
  Web/CD Case Study in the Process of Science: *How Are Space Rocks Analyzed for Signs of Life?*

ATOMS AND MOLECULES

■ **Atomic structure determines the behavior of an element (pp. 28–33, FIGURE 2.10)** An atom is the smallest unit of an element. An atom has a nucleus made up of positively charged protons and uncharged neutrons, as well as a surrounding cloud of negatively charged electrons. The number of electrons in an electrically neutral atom equals the number of protons. Most elements have two or more isotopes, different in neutron number and therefore mass. Some isotopes are unstable and give off particles and energy as radioactivity.
  Electron configuration determines the chemical behavior of an atom. Electrons occupy specific energy levels, or shells, of the atom. Chemical behavior depends on the number of valence electrons, those in the outermost shell. An atom with an incomplete valence shell is reactive. Electrons move within orbitals, three-dimensional spaces with specific shapes located within successive shells.
  Web/CD Activity 2B: *Structure of the Atomic Nucleus*
  Web/CD Activity 2C: *Electron Arrangement*
  Web/CD Activity 2D: *Build an Atom*

■ **Atoms combine by chemical bonding to form molecules (pp. 33–36, FIGURES 2.12 and 2.14)** Chemical bonds form when atoms interact and complete their valence shells. A covalent bond is the sharing of a pair of valence electrons by two atoms. Molecules consist of two or more covalently bonded atoms. Electrons of a polar covalent bond are pulled closer to the more electronegative atom. A covalent bond is nonpolar if both atoms are equally electronegative.
  Two atoms may differ so much in electronegativity that one or more electrons are actually transferred from one atom to the other. The result is a negatively charged ion (anion) and a positively charged ion (cation). The attraction between two ions of opposite charge is called an ionic bond.
  Web/CD Activity 2E: *Covalent Bonds*
  Web/CD Activity 2F: *Nonpolar and Polar Molecules*
  Web/CD Activity 2G: *Ionic Bonds*

■ **Weak chemical bonds play important roles in the chemistry of life (pp. 36–37, FIGURE 2.16)** A hydrogen bond is a weak attraction between one electronegative atom and a hydrogen atom that is covalently linked to another electronegative atom. Van der Waals interactions occur when transiently positive and negative regions of molecules attract each other. Weak bonds reinforce the shapes of large molecules and help molecules adhere to each other.
  Web/CD Activity 2H: *Hydrogen Bonds*

■ **A molecule's biological function is related to its shape (pp. 37–38, FIGURE 2.17)** A molecule's shape is determined by the positions of its atoms' valence orbitals. When covalent bonds form, the *s* and *p* orbitals in the valence shell of an atom may combine to form four hybrid orbitals that extend to the corners of a tetrahedron; such orbitals are responsible for the shapes of H_2O, CH_4, and many more complex biological molecules. Shape is usually the basis of the recognition of one biological molecule by another.

■ **Chemical reactions make and break chemical bonds (pp. 38–39)** Chemical reactions change reactants into products while conserving matter. Most chemical reactions are reversible. Chemical equilibrium is reached when the forward and reverse reaction rates are equal.

Self-Quiz

1. An element is to a (an) _____ as a tissue is to a (an) _____.
   a. atom; organism
   b. compound; organ
   c. molecule; cell
   d. atom; organ
   e. compound; organelle

2. In the term *trace element,* the modifier *trace* means
   a. the element is required in very small amounts.
   b. the element can be used as a label to trace atoms through an organism's metabolism.
   c. the element is very rare on Earth.
   d. the element enhances health but is not essential for the organism's long-term survival.
   e. the element passes rapidly through the organism.

3. Compared to ^{31}P, the radioactive isotope ^{32}P has
   a. a different atomic number.   d. one more electron.
   b. one more neutron.   e. a different charge.
   c. one more proton.

4. Atoms can be represented by simply listing the number of protons, neutrons, and electrons—for example, $2p^+$; $2n^0$; $2e^-$ for helium. Which atom represents the ^{18}O isotope of oxygen?
   a. $6p^+$; $8n^0$; $6e^-$ d. $7p^+$; $2n^0$; $9e^-$
   b. $8p^+$; $10n^0$; $8e^-$ e. $10p^+$; $8n^0$; $10e^-$
   c. $9p^+$; $9n^0$; $9e^-$

5. The atomic number of sulfur is 16. Sulfur combines with hydrogen by covalent bonding to form a compound, hydrogen sulfide. Based on the electron configuration of sulfur, we can predict that the molecular formula of the compound will be (explain your answer)
   a. HS   b. HS_2 c. H_2S d. H_3S_2 e. H_4S

6. Review the valences of carbon, oxygen, hydrogen, and nitrogen, and then determine which of the following molecules is most likely to exist.

   a. O=C—H

   c.
   $$H-\underset{\underset{H}{|}}{\overset{\overset{H}{|}}{C}}-H-\overset{\overset{H}{|}}{C}=O$$

   b.
   $$H-O-\underset{\underset{H}{|}}{\overset{\overset{H}{|}}{C}}-C=O$$

   d.
   $$H-\overset{\overset{O}{|}}{N}=H$$

7. The reactivity of an atom arises from
   a. the average distance of the outermost electron shell from the nucleus.
   b. the existence of unpaired electrons in the valence shell.
   c. the sum of the potential energies of all the electron shells.
   d. the potential energy of the valence shell.
   e. the energy difference between the *s* and *p* orbitals.

8. Which of these statements is true of all anionic atoms?
   a. The atom has more electrons than protons.
   b. The atom has more protons than electrons.
   c. The atom has fewer protons than does a neutral atom of the same element.
   d. The atom has more neutrons than protons.
   e. The net charge is −1.

9. What coefficients must be placed in the blanks to balance this chemical reaction?

   $$C_6H_{12}O_6 \longrightarrow \_\_C_2H_6O + \_\_CO_2$$

   a. 1; 2   b. 2; 2   c. 1; 3   d. 1; 1   e. 3; 1

10. Which of the following statements correctly describes any chemical reaction that has reached equilibrium?
    a. The concentration of products equals the concentration of reactants.
    b. The rate of the forward reaction equals the rate of the reverse reaction.
    c. Both forward and reverse reactions have halted.
    d. The reaction is now irreversible.
    e. No reactants remain.

11. What four chemical elements are most abundant in living matter?

12. Why are DNA, water, and sodium chloride all classified as chemical compounds?

13. A nitrogen atom has 7 protons, and the most common isotope of nitrogen has 7 neutrons. A radioactive isotope of nitrogen has 8 neutrons. What is the atomic number and mass number of this radioactive nitrogen?

14. Magnesium has an atomic number of 12. How many electron shells does a magnesium atom have, and how many electrons are in its valence shell?

15. Explain what holds together the atoms in a crystal of magnesium chloride ($MgCl_2$).

Go to the website or CD-ROM for more quiz questions.

Evolution Connection

The text states that the percentages of naturally occurring elements making up the human body (see TABLE 2.1) are similar to the percentages of these elements found in other organisms. How could you account for this similarity among organisms?

The Process of Science

Female silkworm moths *(Bombyx mori)* attract males by emitting chemical signals that spread through the air. A male hundreds of meters away can detect these molecules and fly toward their source. The sensory organs responsible for this behavior are the comblike antennae visible in the photograph here. Each filament of an antenna is equipped with thousands of receptor cells that detect the sex attractant. Based on what you learned in this chapter, propose a hypothesis to account for the ability of the male moth to detect a specific molecule in the presence of many other molecules in the air. What predictions does your hypothesis make? Design an experiment to test one of these predictions.

Analyze space rocks for signs of life in the Case Study in the Process of Science, available on the website and CD-ROM.

Science, Technology, and Society

While waiting at an airport, Neil Campbell once overhead this claim: "It's paranoid and ignorant to worry about industry or agriculture contaminating the environment with their chemical wastes. After all, this stuff is just made of the same atoms that were already present in our environment." How would you counter this argument?

CHAPTER 3

WATER AND THE FITNESS OF THE ENVIRONMENT

As astronomers study *newly discovered planets orbiting distant stars, they hope to find evidence of water on these far-off worlds, for water is the substance that makes possible life as we know it here on Earth. All organisms familiar to us are made mostly of water and live in an environment dominated by water. Water is the biological medium here on Earth, and possibly on other planets as well.*

Life on Earth began in water and evolved there for 3 billion years before spreading onto land. Modern life, even terrestrial (land-dwelling) life, remains tied to water. Most cells are surrounded by water, and cells themselves are about 70–95% water. Three-quarters of Earth's surface is submerged in water. Although most of this water is in liquid form, water is also present on Earth as ice and vapor. Water is the only common substance to exist in the natural environment in all three physical states of matter: solid, liquid, and gas. These three states of water are visible in the photograph on this page, a view of Earth from space.

The abundance of water is a major reason Earth is habitable. In a classic book called The Fitness of the Environment, *Lawrence Henderson highlights the importance of water to life. While acknowledging that life adapts to its environment through natural selection, Henderson emphasizes that for life to exist at*

all, the environment must first be a suitable abode. Your objective in this chapter is to develop a conceptual understanding of how water contributes to the fitness of Earth for life.

THE EFFECTS OF WATER'S POLARITY

Water is so common that it is easy to overlook the fact that it is an exceptional substance with many extraordinary qualities. Following the theme of emergent properties, we can trace water's unique behavior to the structure and interactions of its molecules.

The polarity of water molecules results in hydrogen bonding

Studied in isolation, the water molecule is deceptively simple. Its two hydrogen atoms are joined to the oxygen atom by single covalent bonds. Because oxygen is more electronegative than hydrogen, the electrons of the polar bonds spend more time closer to the oxygen atom. In other words, the bonds that hold together the atoms in a water molecule are polar covalent bonds, with the oxygen region of the molecule having a partial negative charge and the hydrogens having a partial positive charge. The water molecule, shaped something like a wide V, is a **polar molecule,** meaning that opposite ends of the molecule have opposite charges (see FIGURE 2.13).

The anomalous properties of water arise from attractions between these polar molecules. The attraction is electrical; a slightly positive hydrogen of one molecule is attracted to the slightly negative oxygen of a nearby molecule. The two molecules

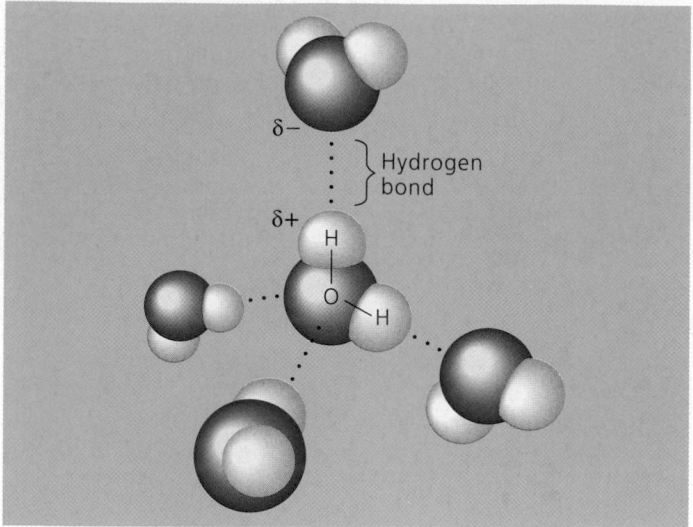

FIGURE 3.1 Hydrogen bonds between water molecules. The charged regions of a polar water molecule are attracted to oppositely charged parts of neighboring molecules. (The oxygen has a slight negative charge; the hydrogens have a slight positive charge.) Each molecule can hydrogen-bond to a maximum of four partners. At any instant in liquid water at 37°C (human body temperature), about 15% of the molecules are bonded to four partners in short-lived clusters.

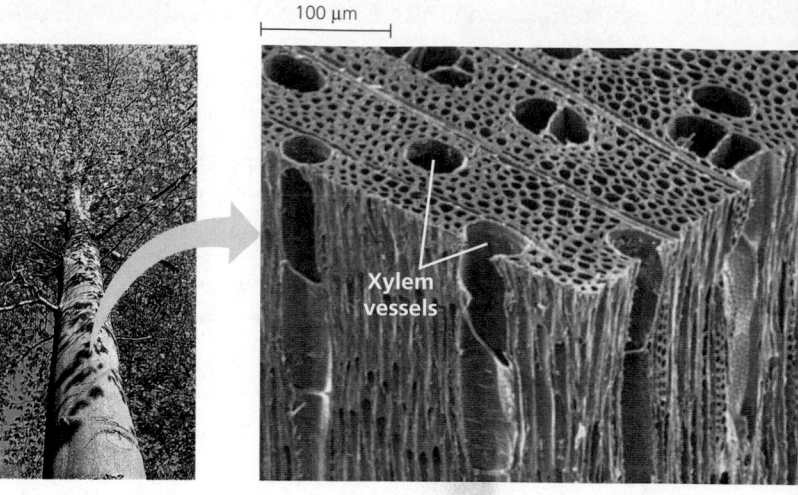

FIGURE 3.2 Water transport in plants. Evaporation from leaves pulls water upward from the roots through microscopic tubes called xylem vessels, in this case located in the trunk of a tree. Cohesion due to hydrogen bonding helps hold together the column of water within a vessel. Adhesion of the water to the vessel wall also helps in resisting the downward pull of gravity.

are thus held together by a hydrogen bond (FIGURE 3.1). Each water molecule can form hydrogen bonds to a maximum of four neighbors, and at any given moment, many of the molecules in a sample of liquid water are linked in this way. The extraordinary qualities of water are emergent properties resulting from the hydrogen bonding that orders molecules into a higher level of structural organization.

We will examine four of water's properties that contribute to the fitness of Earth as an environment for life: water's cohesive behavior, its ability to stabilize temperature, its expansion upon freezing, and its versatility as a solvent.

Organisms depend on the cohesion of water molecules

Water molecules stick to each other as a result of hydrogen bonding. When water is in its liquid form, its hydrogen bonds are very fragile, about one-twentieth as strong as covalent bonds. They form, break, and re-form with great frequency. Each hydrogen bond lasts only a few trillionths of a second, but the molecules are constantly forming new bonds with a succession of partners. Thus, at any instant, a substantial percentage of all the water molecules are bonded to their neighbors, making water more structured than most other liquids. Collectively, the hydrogen bonds hold the substance together, a phenomenon called **cohesion.**

Cohesion due to hydrogen bonding contributes to the transport of water against gravity in plants. Water reaches the

leaves through microscopic vessels that extend upward from the roots (FIGURE 3.2). Water that evaporates from a leaf is replaced by water from the vessels in the veins of the leaf. Hydrogen bonds cause water molecules leaving the veins to tug on molecules farther down in the vessel, and the upward pull is transmitted along the vessel all the way down to the roots. **Adhesion,** the clinging of one substance to another, also plays a role. Adhesion of water to the walls of the vessels helps counter the downward pull of gravity.

Related to cohesion is **surface tension,** a measure of how difficult it is to stretch or break the surface of a liquid. Water has a greater surface tension than most other liquids. At the interface between water and air is an ordered arrangement of water molecules, hydrogen-bonded to one another and to the water below. This makes the water behave as though coated with an invisible film. We can observe the surface tension of water by slightly overfilling a drinking glass; the water will stand above the rim. Water's surface tension also allows us to skip rocks on a pond. In a more biological example, some animals can stand, walk, or run on water without breaking the surface (FIGURE 3.3).

Water moderates temperatures on Earth

Water stabilizes air temperatures by absorbing heat from air that is warmer and releasing the stored heat to air that is cooler. Water is effective as a heat bank because it can absorb

FIGURE 3.3 Walking on water. The high surface tension of water, resulting from the collective strength of its hydrogen bonds, allows the water strider to walk on a pond without breaking the surface.

or release a relatively large amount of heat with only a slight change in its own temperature. To understand this quality of water, we must first look briefly at heat and temperature.

Heat and Temperature

Anything that moves has **kinetic energy,** the energy of motion. Atoms and molecules have kinetic energy because they are always moving, although not necessarily in any particular direction. The faster a molecule moves, the greater its kinetic energy. **Heat** is a measure of the *total* quantity of kinetic energy due to molecular motion in a body of matter. **Temperature** measures the intensity of heat due to the *average* kinetic energy of the molecules. When the average speed of the molecules increases, a thermometer records this as a rise in temperature. Heat and temperature are related, but they are not the same. A swimmer crossing the English Channel has a higher temperature than the water, but the ocean contains far more heat because of its volume.

Whenever two objects of different temperature are brought together, heat passes from the warmer to the cooler body until the two are the same temperature. Molecules in the cooler object speed up at the expense of the kinetic energy of the warmer object. An ice cube cools a drink not by adding coldness to the liquid, but by absorbing heat as the ice melts.

Throughout this book, we will use the **Celsius scale** to indicate temperature (Celsius degrees are abbreviated as °C). At sea level, water freezes at 0°C and boils at 100°C. The temperature of the human body averages 37°C, and comfortable room temperature is about 20–25°C.

One convenient unit of heat used in this book is the **calorie (cal).** A calorie is the amount of heat energy it takes to raise the temperature of 1 g of water by 1°C. Conversely, a calorie is also the amount of heat that 1 g of water releases when it cools by 1°C. A **kilocalorie (kcal),** 1,000 cal, is the quantity of heat required to raise the temperature of 1 kilogram (kg) of water by 1°C. (The "calories" on food packages are actually kilocalories.) Another energy unit used in this book is the **joule (J).** One joule equals 0.239 cal; a calorie equals 4.184 J.

Water's High Specific Heat

The ability of water to stabilize temperature stems from its relatively high specific heat. The **specific heat** of a substance is defined as the amount of heat that must be absorbed or lost for 1 g of that substance to change its temperature by 1°C. We already know water's specific heat because we have defined a calorie as the amount of heat that causes 1 g of water to change its temperature by 1°C. Therefore, the specific heat of water is 1 calorie per gram per degree Celsius, abbreviated as 1 cal/g/°C. Compared with most other substances, water has an unusually high specific heat. For example, ethyl alcohol, the type of alcohol in alcoholic beverages, has a specific heat of 0.6 cal/g/°C.

Because of the high specific heat of water relative to other materials, water will change its temperature less when it absorbs or loses a given amount of heat. The reason you can burn your fingers by touching the metal handle of a pot on the stove when the water in the pot is still lukewarm is that the specific heat of water is ten times greater than that of iron. In other words, it will take only 0.1 cal to raise the temperature of 1 g of iron 1°C. Specific heat can be thought of as a measure of how well a substance resists changing its temperature when it absorbs or releases heat. Water resists changing its temperature; when it does change its temperature, it absorbs or loses a relatively large quantity of heat for each degree of change.

We can trace water's high specific heat, like many of its other properties, to hydrogen bonding. Heat must be absorbed in order to break hydrogen bonds, and heat is released when hydrogen bonds form. A calorie of heat causes a relatively small change in the temperature of water because much of the heat energy is used to disrupt hydrogen bonds before the water molecules can begin moving faster. And when the temperature of water drops slightly, many additional hydrogen bonds form, releasing a considerable amount of energy in the form of heat.

What is the relevance of water's high specific heat to life on Earth? A large body of water can absorb and store a huge amount of heat from the sun in the daytime and during summer, while warming up only a few degrees. At night and during winter, the gradually cooling water can warm the air. This is the reason coastal areas generally have milder climates than inland regions. The high specific heat of water also tends to stabilize ocean temperatures, creating a favorable environment for marine life. Thus, because of its high specific heat, the water that covers most of Earth keeps temperature fluctuations on land and in water within limits that permit life. Also, because organisms are made primarily of water, they are more able to resist changes in their own temperatures than if they were made of a liquid with a lower specific heat.

Evaporative Cooling

Molecules of any liquid stay close together because they are attracted to one another. Molecules moving fast enough to overcome these attractions can depart the liquid and enter the air as gas. This transformation from a liquid to a gas is called vaporization, or evaporation. Recall that the speed of molecular movement varies and that temperature is the *average* kinetic energy of molecules. Even at low temperatures, the speediest molecules can escape into the air. Some evaporation occurs at any temperature; a glass of water, for example, will eventually evaporate at room temperature. If a liquid is heated, the average kinetic energy of molecules increases and the liquid evaporates more rapidly.

Heat of vaporization is the quantity of heat a liquid must absorb for 1 g of it to be converted from the liquid to the gaseous state. Compared with most other liquids, water has a high heat of vaporization. To evaporate one gram of water at 25°C, about 580 cal of heat is needed—nearly double the amount needed to vaporize a gram of alcohol or ammonia. Water's high heat of vaporization is another emergent property caused by hydrogen bonds, which must be broken before the molecules can make their exodus from the liquid.

Water's high heat of vaporization helps moderate Earth's climate. A considerable amount of solar heat absorbed by tropical seas is consumed during the evaporation of surface water. Then, as moist tropical air circulates poleward, it releases heat as it condenses to form rain.

As a liquid evaporates, the surface of the liquid that remains behind cools down. This **evaporative cooling** occurs because the "hottest" molecules, those with the greatest kinetic energy, are the most likely to leave as gas. It is as if the 100 fastest runners at a college transferred to another school; the average speed of the remaining students would decline.

Evaporative cooling of water contributes to the stability of temperature in lakes and ponds and also provides a mechanism that prevents terrestrial organisms from overheating. For example, evaporation of water from the leaves of a plant helps keep the tissues in the leaves from becoming too warm in the sunlight. Evaporation of sweat from human skin dissipates body heat and helps prevent overheating on a hot day or when excess heat is generated by strenuous activity (FIGURE 3.4). High humidity on a hot day increases discomfort because the high concentration of water vapor in the air inhibits the evaporation of sweat from the body.

FIGURE 3.4 Evaporative cooling. Because of water's high heat of vaporization, evaporation of sweat significantly cools the body surface.

Oceans and lakes don't freeze solid because ice floats

Water is one of the few substances that are less dense as a solid than as a liquid. In other words, ice floats. While other materials contract when they solidify, water expands. The cause of this exotic behavior is, once again, hydrogen bonding. At temperatures above 4°C, water behaves like other liquids, expanding as it warms and contracting as it cools. Water begins to freeze when its molecules are no longer moving vigorously enough to break their hydrogen bonds. As the temperature reaches 0°C, the water becomes locked into a crystalline lattice, each water molecule bonded to the maximum of four partners (FIGURE 3.5). The hydrogen bonds keep the molecules at "arm's length," far enough apart to make ice about 10% less dense (10% fewer molecules for the same volume) than liquid water at 4°C. When ice absorbs enough heat for its temperature to rise above 0°C, hydrogen bonds between molecules are disrupted. As the crystal collapses, the ice melts, and molecules are free to slip closer together. Water reaches its greatest density at 4°C and then begins to expand as the molecules move faster. Keep in mind, however, that even in liquid water, many of the molecules are connected by hydrogen bonds, though only transiently: The hydrogen bonds are constantly breaking and re-forming.

The ability of ice to float because of the expansion of water as it solidifies is an important factor in the fitness of the environment. If ice sank, then eventually all ponds, lakes, and even oceans would freeze solid, making life as we know it impossible on Earth. During summer, only the upper few inches of the ocean would thaw. Instead, when a deep body of water cools, the floating ice insulates the liquid water below, preventing it from freezing and allowing life to exist under the frozen surface (FIGURE 3.6).

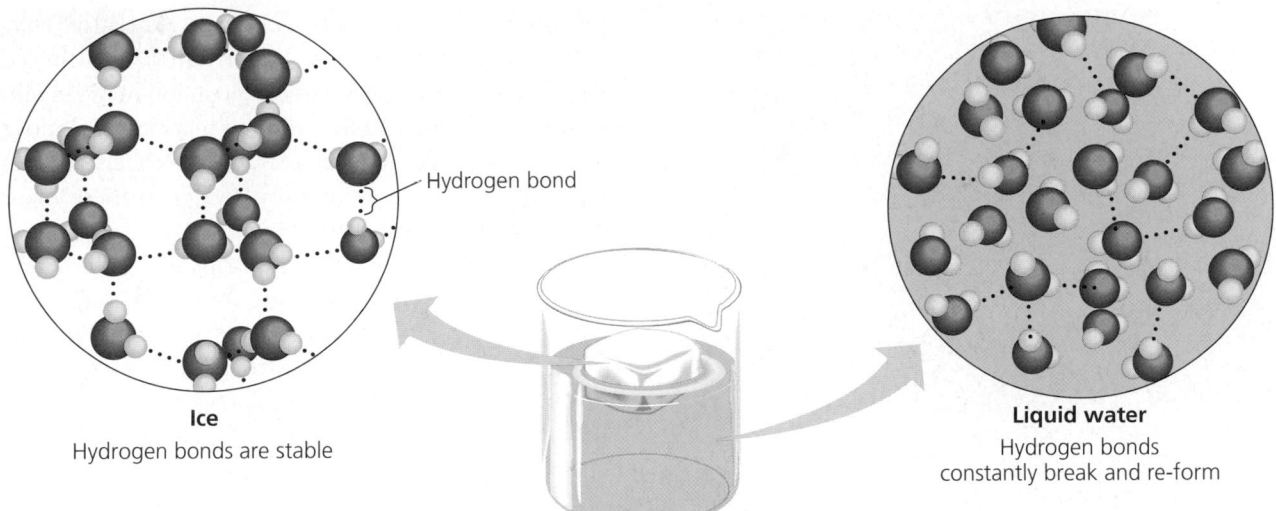

Ice
Hydrogen bonds are stable

Liquid water
Hydrogen bonds
constantly break and re-form

Hydrogen bond

FIGURE 3.5 The structure of ice. In ice, each molecule is hydrogen-bonded to four neighbors in a three-dimensional crystal. Because the crystal is spacious, ice has fewer molecules than an equal volume of liquid water. In other words, ice is less dense than liquid water.

FIGURE 3.6 Floating ice and the fitness of the environment. Floating ice becomes a barrier that protects the liquid water below from the colder air. The invertebrates shown here are called krill; they were photographed beneath the antarctic ice.

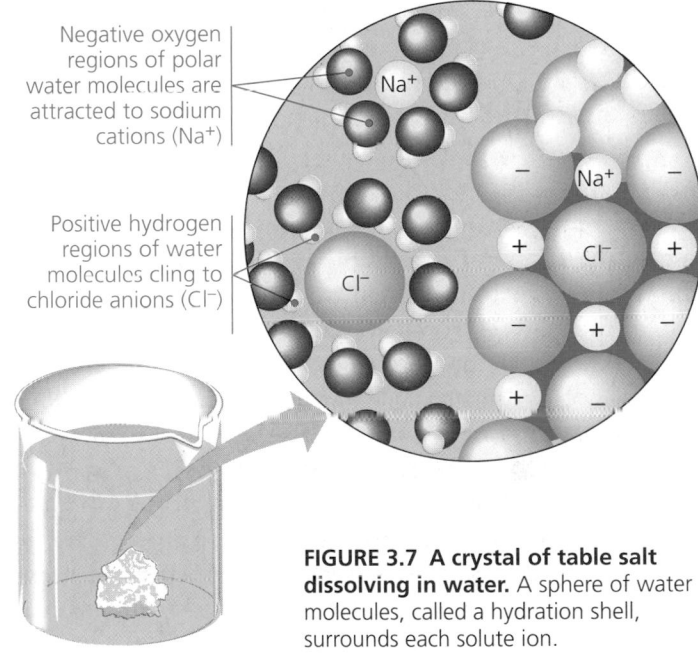

Negative oxygen regions of polar water molecules are attracted to sodium cations (Na^+)

Positive hydrogen regions of water molecules cling to chloride anions (Cl^-)

FIGURE 3.7 A crystal of table salt dissolving in water. A sphere of water molecules, called a hydration shell, surrounds each solute ion.

Water is the solvent of life

A sugar cube placed in a glass of water will dissolve. The glass will then contain a uniform mixture of sugar and water; the concentration of dissolved sugar will be the same everywhere in the mixture. A liquid that is a completely homogeneous mixture of two or more substances is called a **solution.** The dissolving agent of a solution is the **solvent,** and the substance that is dissolved is the **solute.** In this case, water is the solvent and sugar is the solute. An **aqueous solution** is one in which water is the solvent.

The medieval alchemists tried to find a universal solvent, one that would dissolve anything. They learned that nothing works better than water. However, water is not a universal solvent; if it were, it could not be stored in any container, including our cells. But water is a very versatile solvent, a quality we can trace to the polarity of the water molecule.

Suppose, for example, that a crystal of the ionic compound sodium chloride is placed in water (FIGURE 3.7). At the surface of the crystal, the sodium and chloride ions are exposed to the solvent. These ions and the water molecules have a mutual affinity through electrical attraction. The oxygen regions of the water molecules are negatively charged and cling to sodium cations. The hydrogen regions of the water molecules are positively charged and are attracted to chloride anions. As a result, water molecules surround the individual sodium and

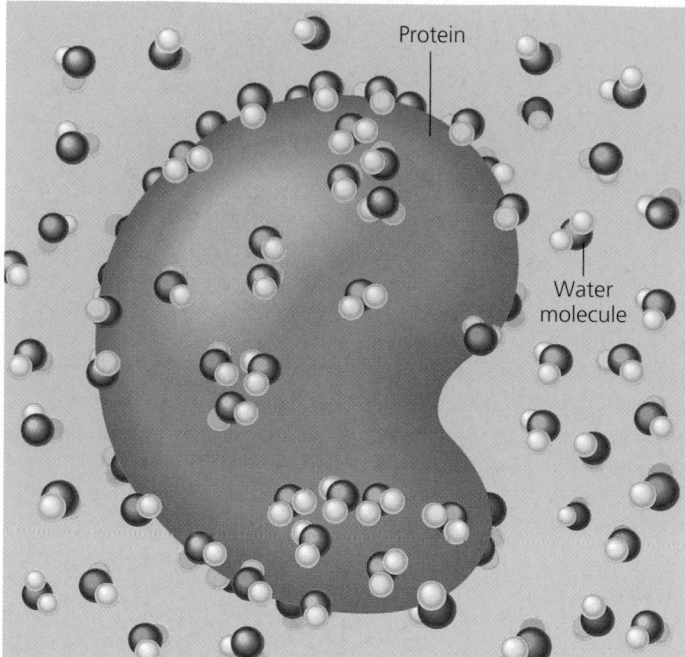

Protein

Water molecule

FIGURE 3.8 A water-soluble protein. Even a molecule as large as a protein can dissolve in water if it has enough ionic and polar regions on its surface. The mass of purple here represents a single such protein molecule, which is being surrounded by water molecules.

chloride ions, separating and shielding them from one another. The sphere of water molecules around each dissolved ion is called a **hydration shell.** Working inward from the surface of the salt crystal, water eventually dissolves all the ions. The result is a solution of two solutes, sodium and chloride, homogeneously mixed with water, the solvent. Other ionic compounds also dissolve in water. Seawater, for instance, contains a great variety of dissolved ions, as do living cells.

A compound does not need to be ionic to dissolve in water; compounds made up of polar molecules, such as sugars, are also water-soluble. Such compounds dissolve when water molecules surround each of the solute molecules. Even molecules as large as proteins can dissolve in water if they have ionic and polar regions on their surface (FIGURE 3.8). Many different kinds of polar compounds are dissolved (along with ions) in the water of such biological fluids as blood, the sap of plants, and the liquid within all cells. Water is the solvent of life.

Hydrophilic and Hydrophobic Substances

Whether ionic or polar, any substance that has an affinity for water is said to be **hydrophilic** (from the Greek *hydro,* water, and *philios,* loving). This term is used even if the substance does not dissolve—because the molecules are too large, for instance. Cotton, a plant product, is an example of a hydrophilic substance that absorbs water without dissolving. Cotton consists of giant molecules of cellulose, a compound with numer-

ous regions of partial positive and partial negative charges associated with polar bonds. Water adheres to the cellulose fibers. Thus, a cotton towel does a great job of drying the body, yet does not dissolve in the washing machine. Cellulose is also present in the walls of water-conducting vessels in a plant; you read earlier how the adhesion of water to these hydrophilic walls functions in water transport.

There are, of course, substances that do not have an affinity for water. Substances that are non-ionic and nonpolar actually seem to repel water; these substances are termed **hydrophobic** (from the Greek *phobos,* fearing). An example from the kitchen is vegetable oil, which, as you know, does not mix stably with watery substances such as vinegar. The hydrophobic behavior of the oil molecules results from a prevalence of nonpolar bonds, in this case bonds between carbon and hydrogen, which share electrons almost equally. Hydrophobic molecules related to oils are major ingredients of cell membranes. (Imagine what would happen to a cell if its membrane dissolved.)

Solute Concentration in Aqueous Solutions

Biological chemistry is "wet" chemistry. Most of the chemical reactions that occur in organisms involve solutes dissolved in water. To carry out experiments on the chemistry of life, it is important to learn how to calculate the concentrations of solutes dissolved in aqueous solutions.

To understand chemical reactions, we need to know how many atoms and molecules are involved. Suppose we wanted to prepare an aqueous solution of table sugar having a specified concentration of sugar molecules (a certain number of solute molecules in a certain volume of solution). Because counting or weighing individual molecules is not practical, we usually measure substances in units called moles. A **mole (mol)** is equal in number to the molecular weight of a substance, but upscaled from daltons to units of grams. Imagine weighing out 1 mol of table sugar (sucrose), which has the molecular formula $C_{12}H_{22}O_{11}$. In round numbers, a carbon atom weighs 12 daltons, a hydrogen atom weighs 1 dalton, and an oxygen atom weighs 16 daltons. **Molecular weight** is the sum of the weights of all the atoms in a molecule; thus, the molecular weight of sucrose is 342 daltons. To obtain 1 mol of sucrose, we weigh out 342 g, the molecular weight of sucrose expressed in grams.

The practical advantage of measuring a quantity of chemicals in moles is that a mole of one substance has exactly the same number of molecules as a mole of any other substance. If substance A has a molecular weight of 10 daltons and substance B has a molecular weight of 100 daltons, then 10 g of A will have the same number of molecules as 100 g of B. The number of molecules in a mole, called Avogadro's number, is 6.02×10^{23}. A mole of table sugar contains 6.02×10^{23} sucrose molecules and weighs 342 g. A mole of ethyl alcohol (C_2H_6O) also contains 6.02×10^{23} molecules, but it weighs only 46 g because the molecules are smaller than those of sucrose. Measuring in moles makes it convenient for scientists working in the laboratory to combine substances in fixed ratios of molecules.

How would we make a liter (L) of solution consisting of 1 mol of sucrose dissolved in water? We would weigh out 342 g of sucrose and then gradually add water, while stirring, until the sugar was completely dissolved. We would then add enough water to bring the total volume of the solution up to 1 L. At that point, we would have a 1-molar (1 M) solution of sucrose. **Molarity**—the number of moles of solute per liter of solution—is the unit of concentration most often used by biologists for aqueous solutions.

THE DISSOCIATION OF WATER MOLECULES

Occasionally, a hydrogen atom shared by two water molecules in a hydrogen bond shifts from one molecule to the other. When this happens, the hydrogen atom leaves its electron behind, and what is actually transferred is a **hydrogen ion,** a single proton with a charge of +1. The water molecule that lost a proton is now a **hydroxide ion** (OH^-), which has a charge of −1. The proton binds to the other water molecule, making that molecule a hydronium ion (H_3O^+). We can picture the chemical reaction this way:

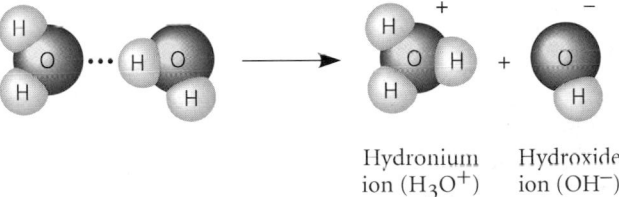

Hydronium   Hydroxide
ion (H_3O^+)   ion (OH^-)

Although this is what actually happens, we can think of the process in a simplified way, as the dissociation (separation) of a water molecule into a hydrogen ion and a hydroxide ion:

$$H_2O \rightleftharpoons H^+ + OH^-$$

Hydrogen   Hydroxide
ion   ion

As the double arrows indicate, this is a reversible reaction that will reach a state of dynamic equilibrium when water dissociates at the same rate that it is being re-formed from H^+ and OH^-. At this equilibrium point, the concentration of water molecules greatly exceeds the concentrations of H^+ and OH^-. In fact, in pure water, only one water molecule in every 554 million is dissociated. The concentration of each ion in pure water is $10^{-7}\ M$ (at 25°C). This means that there is only one ten-millionth of a mole of hydrogen ions per liter of pure water and an equal number of hydroxide ions.

Although the dissociation of water is reversible and statistically rare, it is exceedingly important in the chemistry of life.

Hydrogen and hydroxide ions are very reactive. Changes in their concentrations can drastically affect a cell's proteins and other complex molecules. As we have seen, the concentrations of H^+ and OH^- are equal in pure water, but adding certain kinds of solutes, called acids and bases, disrupts this balance. Biologists use something called the pH scale to describe how acidic or basic (the opposite of acidic) a solution is. In the remainder of this chapter, you will learn about acids, bases, and pH and why changes in pH can adversely affect organisms.

Organisms are sensitive to changes in pH

Before discussing the pH scale, let's see what acids and bases are and how they interact with water.

Acids and Bases

What would cause an aqueous solution to have an imbalance in its H^+ and OH^- concentrations? When the substances called acids dissolve in water, they donate additional H^+ to the solution. An **acid,** according to the definition most often used by biologists, is a substance that increases the hydrogen ion concentration of a solution. For example, when hydrochloric acid (HCl) is added to water, hydrogen ions dissociate from chloride ions:

$$HCl \rightarrow H^+ + Cl^-$$

This additional source of H^+ (dissociation of water is the other source) results in the solution having more H^+ than OH^-. Such a solution is known as an acidic solution.

A substance that reduces the hydrogen ion concentration of a solution is called a **base.** Some bases reduce the H^+ concentration directly by accepting hydrogen ions. Ammonia (NH_3), for instance, acts as a base when the unshared electron pair in nitrogen's valence shell attracts a hydrogen ion from the solution, resulting in an ammonium ion (NH_4^+):

$$NH_3 + H^+ \rightleftharpoons NH_4^+$$

Other bases reduce the H^+ concentration indirectly by dissociating to form hydroxide ions, which then combine with hydrogen ions in the solution to form water. One base that acts this way is sodium hydroxide (NaOH), which in water dissociates into its ions:

$$NaOH \rightarrow Na^+ + OH^-$$

In either case, the base reduces the H^+ concentration. Solutions with a higher concentration of OH^- than H^+ are known as basic solutions. A solution in which the H^+ and OH^- concentrations are equal is said to be neutral.

Notice that single arrows were used in the reactions for HCl and NaOH. These compounds dissociate completely when mixed with water. Hydrochloric acid is called a strong acid and sodium hydroxide a strong base because they dissociate completely. In contrast, ammonia is a relatively weak base. The double arrows in the reaction for ammonia indicate that the binding and release of the hydrogen ion are reversible, although at equilibrium there will be a fixed ratio of NH_4^+ to NH_3.

There are also weak acids, which reversibly release and reaccept hydrogen ions. An example is carbonic acid, which has essential functions in many organisms:

$$H_2CO_3 \rightleftharpoons HCO_3^- + H^+$$

| Carbonic acid | Bicarbonate ion | Hydrogen ion |

Here the equilibrium so favors the reaction in the left direction that when carbonic acid is added to water, only 1% of the molecules are dissociated at any particular time. Still, that is enough to shift the balance of H^+ and OH^- from neutrality.

The pH Scale

In any aqueous solution, the *product* of the H^+ and OH^- concentrations is constant at 10^{-14}. This can be written

$$[H^+][OH^-] = 10^{-14}$$

In such an equation, brackets indicate molar concentration for the substance enclosed within them. In a neutral solution at room temperature (25°C), $[H^+] = 10^{-7}$ and $[OH^-] = 10^{-7}$, so in this case, 10^{-14} is the product of $10^{-7} \times 10^{-7}$. If enough acid is added to a solution to increase $[H^+]$ to 10^{-5} M, then $[OH^-]$ will decline by an equivalent amount to 10^{-9} M (note that $10^{-5} \times 10^{-9} = 10^{-14}$). This constant relationship expresses the behavior of acids and bases in a solution. An acid not only adds hydrogen ions to a solution, but also removes hydroxide ions because of the tendency for H^+ to combine with OH^- to form water. A base has the opposite effect, increasing OH^- concentration but also reducing H^+ concentration by the formation of water. If enough of a base is added to raise the OH^- concentration to 10^{-4} M, the H^+ concentration will drop to 10^{-10} M. Whenever we know the concentration of either H^+ or OH^- in a solution, we can deduce the concentration of the other ion.

Because the H^+ and OH^- concentrations of solutions can vary by a factor of 100 trillion or more, scientists have developed a way to express this variation more conveniently than in moles per liter. The pH scale, which ranges from 0 to 14 (FIGURE 3.9), compresses the range of H^+ and OH^- concentrations by employing logarithms. The **pH** of a solution is defined as the negative logarithm (base 10) of the hydrogen ion concentration:

$$pH = -\log [H^+]$$

For a neutral solution, $[H^+]$ is 10^{-7} M, giving us

$$-\log 10^{-7} = -(-7) = 7$$

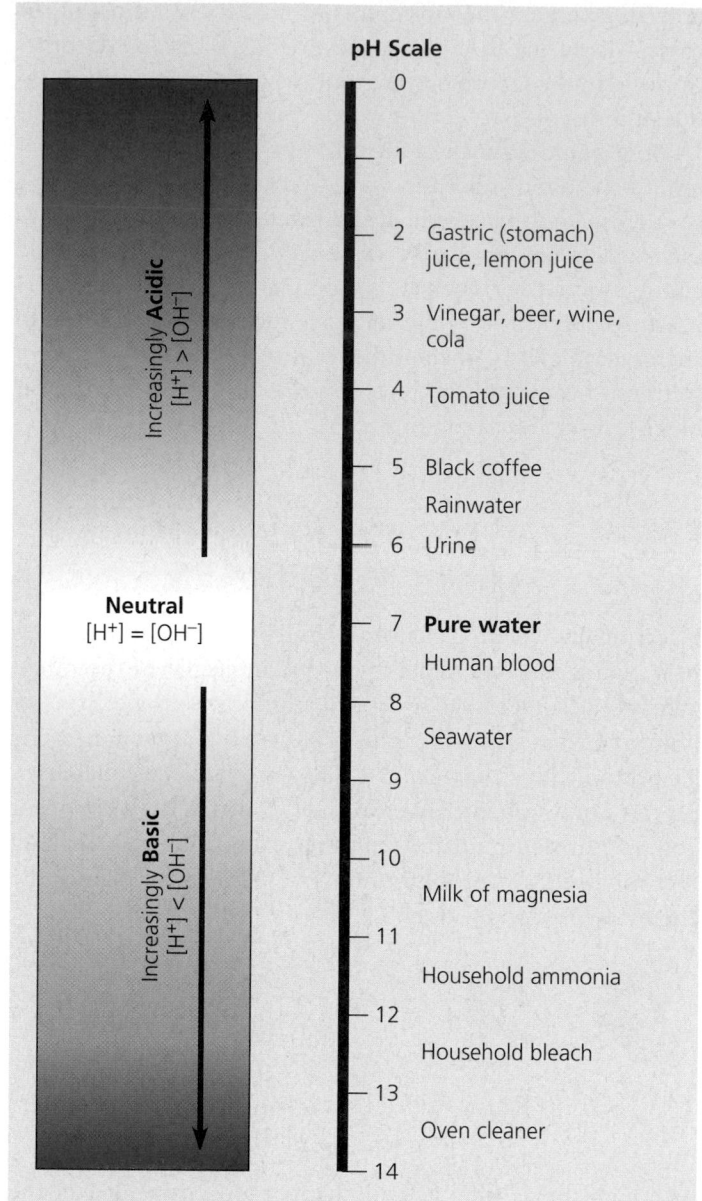

FIGURE 3.9 The pH of some aqueous solutions.

Notice that pH *declines* as H^+ concentration *increases*. Notice, too, that although the pH scale is based on H^+ concentration, it also implies OH^- concentration. A solution of pH 10 has a hydrogen ion concentration of 10^{-10} M and a hydroxide ion concentration of 10^{-4} M.

The pH of a neutral solution is 7, the midpoint of the scale. A pH value less than 7 denotes an acidic solution; the lower the number, the more acidic the solution. The pH for basic solutions is above 7. Most biological fluids are within the range pH 6 to pH 8. There are a few exceptions, however, including the strongly acidic digestive juice of the human stomach, which has a pH of about 2.

It is important to remember that each pH unit represents a tenfold difference in H^+ and OH^- concentrations. It is this mathematical feature that makes the pH scale so compact. A solution of pH 3 is not twice as acidic as a solution of pH 6, but a thousand times more acidic. When the pH of a solution changes slightly, the actual concentrations of H^+ and OH^- in the solution change substantially.

Buffers

The internal pH of most living cells is close to 7. Even a slight change in pH can be harmful, because the chemical processes of the cell are very sensitive to the concentrations of hydrogen and hydroxide ions.

Thanks to the presence of buffers, biological fluids resist changes to their own pH when acids or bases are introduced. **Buffers** are substances that minimize changes in the concentrations of H^+ and OH^- in a solution. For example, buffers normally maintain the pH of human blood very close to 7.4. A person cannot survive for more than a few minutes if the blood pH drops to 7 or rises to 7.8. Under normal circumstances, the buffering capacity of the blood prevents such swings in pH.

A buffer works by accepting hydrogen ions from the solution when they are in excess and donating hydrogen ions to the solution when they have been depleted. Most buffer solutions contain a weak acid and its corresponding base, which combine reversibly with hydrogen ions. One of the buffers that contribute to pH stability in human blood and many other biological solutions is carbonic acid (H_2CO_3), which, as already mentioned, dissociates to yield a bicarbonate ion (HCO_3^-) and a hydrogen ion (H^+):

$$\underset{\substack{H^+ \text{ donor} \\ (\text{acid})}}{H_2CO_3} \underset{\substack{\text{Response to} \\ \text{a drop in pH}}}{\overset{\substack{\text{Response to} \\ \text{a rise in pH}}}{\rightleftharpoons}} \underset{\substack{H^+ \text{ acceptor} \\ (\text{base})}}{HCO_3^-} + \underset{\substack{\text{Hydrogen} \\ \text{ion}}}{H^+}$$

The chemical equilibrium between carbonic acid and bicarbonate acts as a pH regulator, the reaction shifting left or right as other processes in the solution add or remove hydrogen ions. If the H^+ concentration in blood begins to fall (that is, if pH rises), more carbonic acid dissociates, replenishing hydrogen ions. But when H^+ concentration in blood begins to rise (pH drops), the bicarbonate ion acts as a base and removes the excess hydrogen ions from the solution. Thus, the carbonic acid–bicarbonate buffering system consists of an acid and a base in equilibrium with each other. Most other buffers are also acid–base pairs.

Acid precipitation threatens the fitness of the environment

Considering the dependence of all life on water, contamination of rivers, lakes, and seas is a dire environmental problem. One of the most serious assaults on water quality is acid precipitation.

Uncontaminated rain has a pH of about 5.6, slightly acidic, owing to the formation of carbonic acid from carbon dioxide and water. **Acid precipitation** refers to rain, snow, or fog that is more acidic than pH 5.6. What causes acid precipitation, and what are its effects on the fitness of the environment?

Acid precipitation is caused primarily by the presence in the atmosphere of sulfur oxides and nitrogen oxides, gaseous compounds that react with water in the air to form strong acids, which fall to earth with rain or snow. A major source of these oxides is the burning of fossil fuels (coal, oil, and gas) in factories and automobiles. Electrical power plants that burn coal produce more of these pollutants than any other single source. Ironically, the tall smokestacks built to reduce local pollution by dispersing factory exhaust help spread airborne pollutants. Winds carry the pollutants away, and acid rain may fall thousands of miles away from industrial centers. In the Adirondack Mountains of upstate New York, the pH of rainfall averages 4.2, about 25 times more acidic than normal rain. Acid precipitation falls on many other regions, including the Cascade Mountains of the Pacific Northwest and certain parts of Europe and Asia. One West Virginia storm dropped rain having a pH of 1.5, as acidic as the digestive juices in our stomachs!

The effect of acids in lakes and streams is most pronounced in the spring, as snow begins to melt. The surface snow melts first, drains down, and sends much of the acid that has accumulated over the winter into lakes and streams all at once. Early meltwater often has a pH as low as 3, and this acid surge hits when fish and other forms of aquatic life are producing eggs and young, which are especially vulnerable to acidic conditions. Strong acidity can alter the structure of biological molecules and prevent them from carrying out the essential chemical processes of life.

Although acid precipitation can clearly damage life in lakes and streams, its direct effects on forests and other terrestrial life are controversial. However, recent research indicates that acid rain and snow can bring about profound changes in soils by affecting the solubility of soil minerals. Acid precipitation falling on land washes away certain mineral ions, such as calcium and magnesium ions, that ordinarily help buffer the soil solution and are essential nutrients for plant growth. At the same time, other minerals, such as aluminum, reach toxic concentrations when acidification increases their solubility. The effects of acid precipitation on soil chemistry have contributed to the decline of European forests and are

THE PROCESS OF SCIENCE

FIGURE 3.10 The effects of acid precipitation on a forest. Acid fog and acid rain are thought to be responsible for directly or indirectly killing many of the trees on Mount Mitchell, in North Carolina.

taking a toll on some North American forests (FIGURE 3.10). Nevertheless, studies indicate that the majority of North American forests are not currently suffering substantially from acid precipitation.

If there is reason for optimism about the future quality of water resources, it is our progress in reducing certain kinds of pollution. For example, in the United States, Canada, and Europe, emissions of sulfur oxides have declined markedly in recent decades, causing a decrease in acid precipitation. Continued progress can come only from the actions of people who are concerned about environmental quality. An essential part of their education is to understand the crucial role that water plays in the environment's fitness for continued life on Earth.

CHAPTER 3 REVIEW

Go to the Campbell Biology website (www.campbellbiology.com) to explore an interactive version of the Chapter Review.

Summary of Key Concepts

THE EFFECTS OF WATER'S POLARITY

■ **The polarity of water molecules results in hydrogen bonding (pp. 41–42, FIGURE 3.1)** A hydrogen bond forms when the oxygen of one water molecule is electrically attracted to the hydrogen of a nearby molecule. Hydrogen bonding between water molecules is the basis for water's unusual properties.
Web/CD Activity 3A: *The Polarity of Water*

■ **Organisms depend on the cohesion of water molecules (p. 42)** Hydrogen bonding makes water molecules stick to each other, and this cohesion helps pull water upward in the microscopic vessels of plants. Hydrogen bonding is also responsible for water's surface tension.
Web/CD Activity 3B: *Cohesion of Water*

■ **Water moderates temperatures on Earth (pp. 42–44)** Hydrogen bonding gives water a high specific heat. Heat is absorbed when hydrogen bonds break and is released when hydrogen bonds form, helping minimize temperature fluctuations to within limits that permit life. Evaporative cooling is based on water's high heat of vaporization. Water molecules must have a relatively high kinetic energy to break hydrogen bonds. The evaporative loss of these energetic water molecules cools a surface.

■ **Oceans and lakes don't freeze solid because ice floats (pp. 44–45, FIGURE 3.6)** Ice is less dense than liquid water because its more or-

ganized hydrogen bonding causes expansion into a crystal formation. Floating ice allows life to exist under the frozen surfaces of lakes and polar seas.

■ **Water is the solvent of life (pp. 45–47, FIGURE 3.7)** Water is an unusually versatile solvent because its polar molecules are attracted to charged and polar substances. When ions or polar substances are surrounded by water molecules, they dissolve and are called solutes. Hydrophilic substances have an affinity for water. Hydrophobic substances do not; they seem to repel water. Biologists usually use molarity, the number of moles of solute per liter of solution, as a measure of solute concentration in solutions. A mole is the number of grams of a substance equal to its molecular weight.

THE DISSOCIATION OF WATER MOLECULES

■ **Organisms are sensitive to changes in pH (pp. 47–49, FIGURE 3.9)** Water can dissociate into H^+ and OH^-. The concentration of H^+ is expressed as pH, where $pH = -\log [H^+]$. Acids donate additional H^+ in aqueous solutions; bases donate OH^- or accept H^+. In a neutral solution, $[H^+] = [OH^-] = 10^{-7}$, and pH = 7. In an acidic solution, $[H^+]$ is greater than $[OH^-]$, and the pH is less than 7. In a basic solution, $[H^+]$ is less than $[OH^-]$, and the pH is greater than 7. Buffers in biological fluids resist changes in pH. A buffer consists of an acid-base pair that combines reversibly with hydrogen ions.
Web/CD Activity 3C: *Dissociation of Water Molecules*
Web/CD Activity 3D: *Acids, Bases, and pH*

■ **Acid precipitation threatens the fitness of the environment (pp. 49–50)** Acid precipitation is rain, snow, or fog with a pH below 5.6. It often results from a reaction in the air between water vapor and sulfur oxides and nitrogen oxides produced by the combustion of fossil fuels.
Web/CD Case Study in the Process of Science: *How Does Acid Precipitation Affect Trees?*

1. What is the main thesis of Lawrence Henderson's *The Fitness of the Environment*?
   a. Earth's environment is constant.
   b. It is the physical environment, not life, that has evolved.
   c. The environment of Earth has adapted to life.
   d. Life as we know it depends on certain environmental qualities on Earth.
   e. Water and other aspects of Earth's environment exist because they make the planet more suitable for life.

2. Air temperature often increases slightly as clouds begin to drop rain or snow. Which behavior of water is *most directly* responsible for this phenomenon?
   a. water's change in density when it condenses
   b. water's reactions with other atmospheric compounds
   c. release of heat by the formation of hydrogen bonds
   d. release of heat by the breaking of hydrogen bonds
   e. water's high surface tension

3. For two bodies of matter in contact, heat always flows from
   a. the body with greater heat to the one with less heat.
   b. the body of higher temperature to the one of lower temperature.
   c. the denser body to the less dense body.
   d. the body with more water to the one with less water.
   e. the larger body to the smaller body.

4. A slice of pizza has 500 kcal. If we could burn the pizza and use all the heat to warm a 50-L container of cold water, what would be the approximate increase in the temperature of the water? (*Note:* A liter of cold water weighs about 1 kg.)
   a. 50°C   d. 100°C
   b. 5°C   e. 1°C
   c. 10°C

5. The bonds that are broken when water vaporizes are
   a. ionic bonds
   b. bonds between water molecules
   c. bonds between atoms within individual water molecules
   d. polar covalent bonds
   e. nonpolar covalent bonds

6. Which of the following is an example of a hydrophobic material?
   a. paper   d. sugar
   b. table salt   e. pasta
   c. wax

7. We can be sure that a mole of table sugar and a mole of vitamin C are equal in their
   a. weight in daltons   d. number of atoms
   b. weight in grams   e. volume
   c. number of molecules

8. How many grams of acetic acid ($C_2H_4O_2$) would you use to make 10 L of a 0.1 M aqueous solution of acetic acid? (*Note:* The atomic weights, in daltons, are approximately 12 for carbon, 1 for hydrogen, and 16 for oxygen.)
   a. 10 g   d. 60
   b. 0.1 g   e. 0.6 g
   c. 6 g

9. Acid precipitation has lowered the pH of a particular lake to 4.0. What is the hydrogen ion concentration of the lake?
   a. 4.0 M d. $10^4 M$
   b. $10^{-10} M$ e. 4%
   c. $10^{-4} M$

10. What is the *hydroxide* ion concentration of the lake described in question 9?
    a. $10^{-7} M$ d. $10^{-14} M$
    b. $10^{-4} M$ e. 10 M
    c. $10^{-10} M$

11. Why is it unlikely that two neighboring water molecules would be arranged like this?

    $$\begin{array}{ccc} & \text{H H} & \\ \text{O} & & \text{O} \\ & \text{H H} & \end{array}$$

12. In a tall tree, water in thin tubes within the trunk is pulled upward by evaporation from the leaves. What keeps the water molecules at the bottom of the tree moving?

13. Explain the popular adage "It's not the heat, it's the humidity."

14. Explain how the freezing of water can crack boulders.

15. Compared to a basic solution at pH 9, the same volume of an acidic solution at pH 4 has _____ × more hydrogen ions (H^+).

Go to the website or CD-ROM for more quiz questions.

Evolution Connection

The surface of the planet Mars has many landscape features reminiscent of those formed by flowing water on Earth, including what appear to be meandering channels and outwash areas. Thus far, probes sent to Mars have revealed only trace quantities of water, in the form of atmospheric water vapor, but some scientists suspect a great deal more water may be present beneath the Martian surface. Why has there been so much interest in the presence of water on Mars? Would the presence of water make it more likely that life had evolved there? What other physical factors might also be important?

The Process of Science

1. Design a controlled experiment to test the hypothesis that acid precipitation inhibits the growth of *Elodea*, a common freshwater plant.

2. In agricultural areas, farmers pay close attention to the weather forecast. Right before a predicted overnight freeze, farmers spray water on crops to protect the plants. Use the properties of water to explain how this works. Be sure to mention why hydrogen bonds are responsible for this phenomenon.

Design experiments to test the effects of acid precipitation on trees in the Case Study in the Process of Science, available on the website and CD-ROM.

Science, Technology, and Society

Agriculture, industry, and the growing populations of cities all compete, through political influence, for water. If you were in charge of water resources in an arid region, what would your priorities be for allocating the limited water supply for various uses? How would you try to build consensus among the different special-interest groups?

Answers: **1.** d; **2.** c; **3.** b; **4.** c; **5.** b; **6.** c; **7.** c; **8.** d; **9.** c; **10.** c; **11.** The hydrogen atoms of one molecule, with their partial positive charges, would repel the hydrogen atoms of the adjacent molecule. **12.** Hydrogen bonds hold neighboring water molecules together ; this cohesion helps the molecules move against the downward pull of gravity. **13.** High humidity hampers cooling by resisting the evaporation of sweat. **14.** Water expands as it freezes, because the water molecules move farther apart in forming ice crystals. When there is water in a crevice of a boulder, expansion of the water due to freezing may crack the rock. **15.** 100,000.

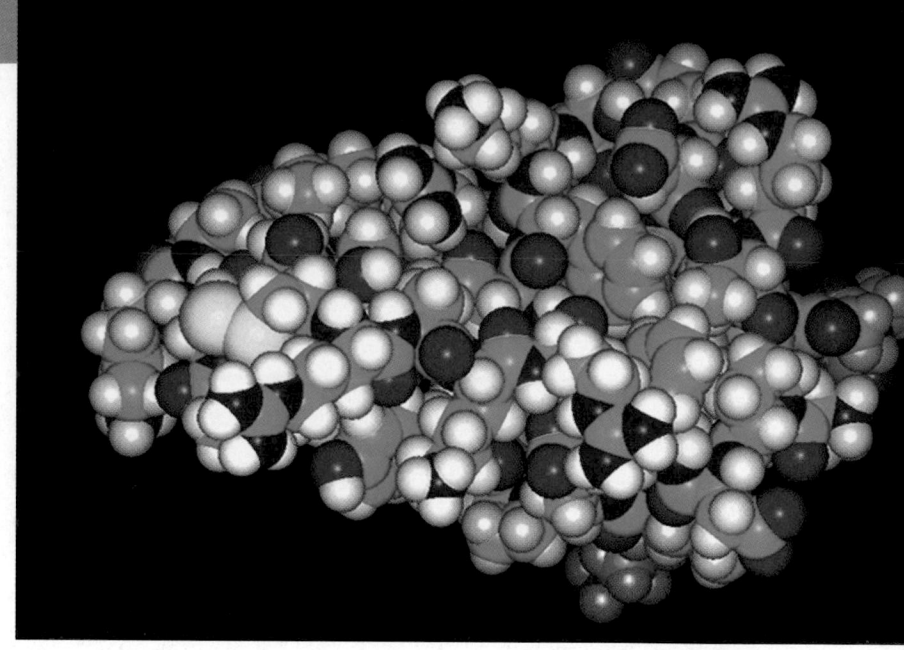

CHAPTER 4

CARBON AND THE MOLECULAR DIVERSITY OF LIFE

Although water *is the universal medium for life on Earth, most of the chemicals that make up living organisms are based on the element carbon. Of all chemical elements, carbon is unparalleled in its ability to form molecules that are large, complex, and diverse, and this molecular diversity has made possible the diversity of organisms that have evolved on Earth. The protein shown in the computer graphic image above is an example of a large, complex molecule based on carbon (the green atoms). Proteins are a major topic of Chapter 5. In this chapter, we focus on smaller molecules, using them to illustrate a few concepts of molecular architecture that highlight carbon's importance to life and the theme that emergent properties arise from the organization of the matter of living organisms.*

THE IMPORTANCE OF CARBON

Although a cell is composed of 70–95% water, the rest consists mostly of carbon-based compounds. Proteins, DNA, carbohydrates, and other molecules that distinguish living matter from inanimate material are all composed of carbon atoms bonded to one another and to atoms of other elements. Hydrogen (H), oxygen (O), nitrogen (N), sulfur (S), and phosphorus (P) are other common ingredients of these compounds, but it is carbon (C) that accounts for the large diversity of biological molecules.

Organic chemistry is the study of carbon compounds

Compounds containing carbon are said to be organic, and the branch of chemistry that specializes in the study of carbon compounds is called **organic chemistry.** Once thought to come only from living things, organic compounds range from simple molecules, such as carbon dioxide (CO_2) and methane (CH_4), to colossal ones, such as proteins, with thousands of atoms and molecular weights in excess of 100,000 daltons. Most organic compounds contain hydrogen atoms.

The overall percentages of the major elements of life—C, H, O, N, S, and P—are quite uniform from one organism to another. Because of carbon's versatility, however, this limited assortment of atomic building blocks, taken in roughly the same proportions, can be used to build an inexhaustible variety of organic molecules. Different species of organisms, and different individuals within a species, are distinguished by variations in their organic molecules.

Since the dawn of human history, people have used other organisms as sources of valued substances—from foods to medicines and fabrics. The science of organic chemistry originated in attempts to purify and improve the yield of such products. By the early 19th century, chemists had learned to make many simple compounds in the laboratory by combining elements under the right conditions. Artificial synthesis of the complex molecules extracted from living matter seemed impossible, however. It was at that time that the Swedish chemist Jöns Jakob Berzelius first made the distinc-

THE PROCESS OF SCIENCE

tion between organic compounds, those that seemingly could arise only within living organisms, and inorganic compounds, those that were found in the nonliving world. The new discipline of organic chemistry was first built on a foundation of *vitalism*, the belief in a life force outside the jurisdiction of physical and chemical laws.

Chemists began to chip away at the foundation of vitalism when they learned to synthesize organic compounds in their laboratories. In 1828, Friedrich Wöhler, a German chemist who had studied with Berzelius, attempted to make an inorganic salt, ammonium cyanate, by mixing solutions of ammonium (NH_4^+) and cyanate (CNO^-) ions. Wöhler was astonished to find that instead of the expected product, he had made urea, an organic compound present in the urine of animals. Wöhler challenged the vitalists when he wrote, "I must tell you that I can prepare urea without requiring a kidney or an animal, either man or dog." However, one of the ingredients used in the synthesis, the cyanate, had been extracted from animal blood, and the vitalists were not swayed by Wöhler's discovery. However, a few years later, Hermann Kolbe, a student of Wöhler's, made the organic compound acetic acid from inorganic substances that could themselves be prepared directly from pure elements.

The foundation of vitalism finally crumbled after several more decades of laboratory synthesis of increasingly complex organic compounds. In 1953, Stanley Miller, then a graduate student at the University of Chicago, helped bring this abiotic (nonliving) synthesis of organic compounds into the context of evolution. Miller used a laboratory simulation of chemical conditions on the primitive Earth to demonstrate that the spontaneous synthesis of organic compounds could have been an early stage in the origin of life (FIGURE 4.1).

The pioneers of organic chemistry helped shift the mainstream of biological thought from vitalism to *mechanism,* the belief that all natural phenomena, including the processes of life, are governed by physical and chemical laws. Organic chemistry was redefined as the study of carbon compounds, regardless of their origin. Most naturally occurring organic compounds are produced by organisms, and these molecules represent a diversity and range of complexity unrivaled by inorganic compounds. However, the same rules of chemistry apply to inorganic and organic molecules alike. The foundation of organic chemistry is not some intangible life force, but the unique chemical versatility of the element carbon.

Carbon atoms are the most versatile building blocks of molecules

The key to the chemical characteristics of an atom, as you learned in Chapter 2, is in its configuration of electrons, because electron configuration determines the kinds and number of bonds an atom will form with other atoms. Carbon has

FIGURE 4.1 Abiotic synthesis of organic compounds under "early Earth" conditions. Here Stanley Miller re-creates his 1953 experiment, a laboratory simulation demonstrating that environmental conditions on the lifeless, primordial Earth allowed the spontaneous synthesis of some organic molecules. Miller used electrical discharges (simulated lightning) to trigger reactions in a primitive "atmosphere" of H_2O, H_2, NH_3 (ammonia), and CH_4 (methane)—some of the gases released by volcanoes. From these ingredients, Miller's apparatus made a variety of organic compounds that play key roles in living cells. Similar chemical reactions may have set the stage for the origin of life on Earth, a hypothesis we will explore in more detail in Chapter 26.

THE PROCESS OF SCIENCE

a total of 6 electrons, with 2 in the first electron shell and 4 in the second shell. Having 4 valence electrons in a shell that holds 8, carbon has little tendency to gain or lose electrons and form ionic bonds; it would have to donate or accept 4 electrons to do so. Instead, a carbon atom usually completes its valence shell by sharing electrons with other atoms in four covalent bonds. Each carbon atom thus acts as an intersection point from which a molecule can branch off in up to four directions. This *tetravalence* is one facet of carbon's versatility that makes large, complex molecules possible.

In Chapter 2, you also learned that when a carbon atom forms single covalent bonds, the arrangement of its four hybrid orbitals causes the bonds to angle toward the corners of an imaginary tetrahedron (see FIGURE 2.15c). The bond angles in methane (CH_4) are 109° (FIGURE 4.2a, p. 54), and they are approximately the same in any group of atoms where carbon has four single bonds. For example, ethane (C_2H_6) is shaped like two tetrahedrons joined at their apexes (FIGURE 4.2b). In molecules with still more carbons, every grouping of a carbon bonded to four other atoms has a tetrahedral shape. But when two carbon atoms are joined by a double bond, all bonds around those carbons are in the same plane. For example, ethene is a flat molecule; its atoms all lie in the same plane (FIGURE 4.2c). It is convenient to write all structural formulas

FIGURE 4.2 The shapes of three simple organic molecules.

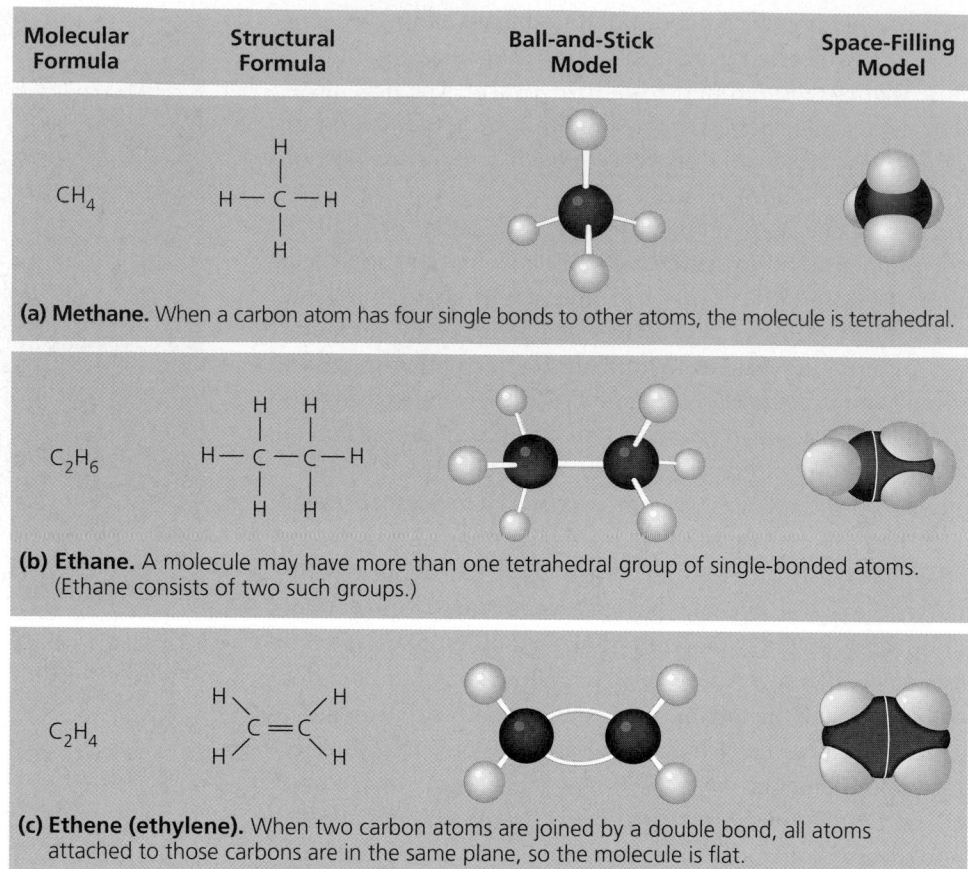

Molecular Formula	Structural Formula	Ball-and-Stick Model	Space-Filling Model

(a) Methane. When a carbon atom has four single bonds to other atoms, the molecule is tetrahedral.

(b) Ethane. A molecule may have more than one tetrahedral group of single-bonded atoms. (Ethane consists of two such groups.)

(c) Ethene (ethylene). When two carbon atoms are joined by a double bond, all atoms attached to those carbons are in the same plane, so the molecule is flat.

as though the molecules represented were flat, but it is important to remember that molecules are three-dimensional and that the shape of a molecule often determines its function.

The electron configuration of carbon gives it covalent compatibility with many different elements. FIGURE 4.3 reviews the valences of the four major atomic components of organic molecules: carbon and its most frequent partners—oxygen, hydrogen, and nitrogen. We can think of these valences as the rules of covalent bonding in organic chemistry—the building code that governs the architecture of organic molecules.

A couple of additional examples will show how the rules of covalent bonding apply to carbon atoms with partners other than hydrogen. In the carbon dioxide molecule (CO_2), a single carbon atom is joined to two atoms of oxygen by double covalent bonds. The structural formula for CO_2 is O=C=O.

Each line (bond) in a structural formula represents a pair of shared electrons. Notice that the carbon atom in CO_2 is involved in four covalent bonds, two with each oxygen atom. The arrangement completes the valence shells of all atoms in the molecule. Because carbon dioxide is a very simple molecule and lacks hydrogen, it is often considered inorganic, even though it contains carbon. Whether we call CO_2 organic or inorganic is an arbitrary distinction, but there is no ambiguity about its importance to the living world. Taken from the air by plants and incorporated into sugar and other foods during photosynthesis, CO_2 is the source of carbon for all the organic molecules found in organisms.

Another relatively simple molecule is urea, $CO(NH_2)_2$. This is the organic compound found in urine that Wöhler learned to synthesize in the early 19th century. The structural formula for urea is shown on the following page.

FIGURE 4.3 Valences for the major elements of organic molecules. Valence is the number of covalent bonds an atom will usually form. It is generally equal to the number of electrons required to complete the atom's outermost (valence) electron shell.

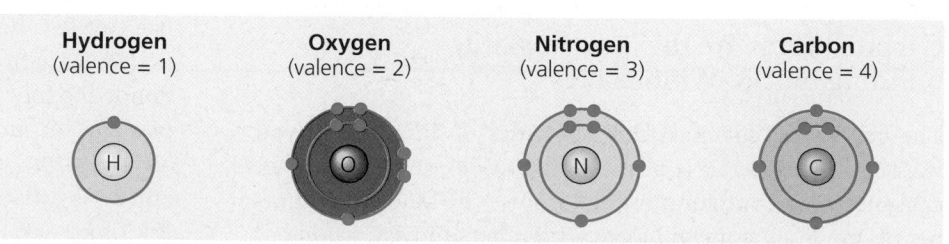

Hydrogen (valence = 1)	Oxygen (valence = 2)	Nitrogen (valence = 3)	Carbon (valence = 4)

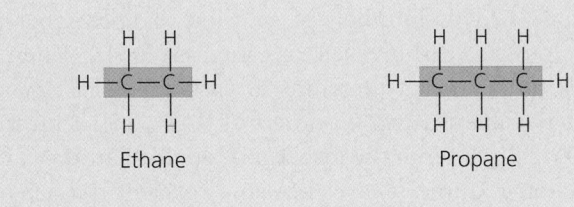

(a) Length. Carbon skeletons vary in length.

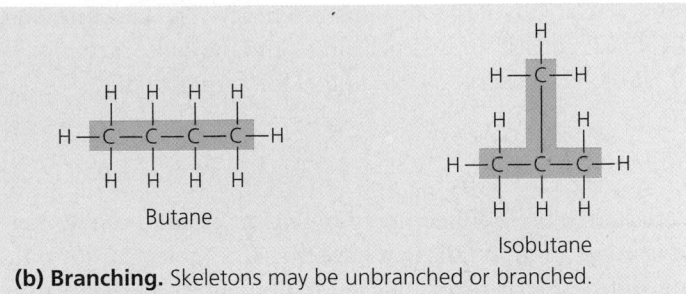

(b) Branching. Skeletons may be unbranched or branched.

1-Butene   2-Butene

(c) Double bonds. The skeleton may have double bonds, which can vary in location.

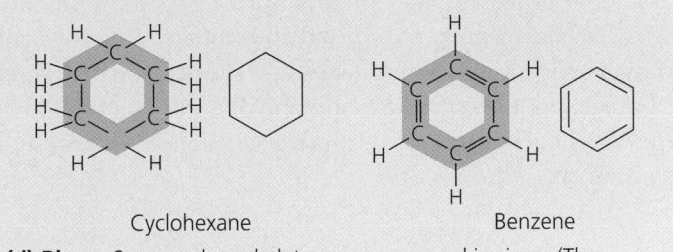

Cyclohexane   Benzene

(d) Rings. Some carbon skeletons are arranged in rings. (The abbreviated structural formulas omit the corner carbons and the hydrogens attached to them.)

FIGURE 4.4 Variations in carbon skeletons. Hydrocarbons, organic molecules consisting only of carbon and hydrogen, illustrate the diversity of the carbon skeletons of organic molecules.

Urea

Again, each atom has the required number of covalent bonds. In this case, one carbon atom is involved in both single and double bonds.

Both urea and carbon dioxide are molecules with only one carbon atom. But as FIGURE 4.2 shows, a carbon atom can also use one or more of its valence electrons to form covalent bonds to other carbon atoms, making it possible to link the atoms into chains of seemingly infinite variety.

Variation in carbon skeletons contributes to the diversity of organic molecules

Carbon chains form the skeletons of most organic molecules. The skeletons vary in length and may be straight, branched, or arranged in closed rings (FIGURE 4.4). Some carbon skeletons have double bonds, which vary in number and location. Such variation in carbon skeletons is one important source of the molecular complexity and diversity that characterize living matter. In addition, atoms of other elements can be bonded to the skeletons at available sites.

All the molecules shown in FIGURES 4.2 and 4.4 are **hydrocarbons,** organic molecules consisting only of carbon and hydrogen. Atoms of hydrogen are attached to the carbon skeleton wherever electrons are available for covalent bonding.

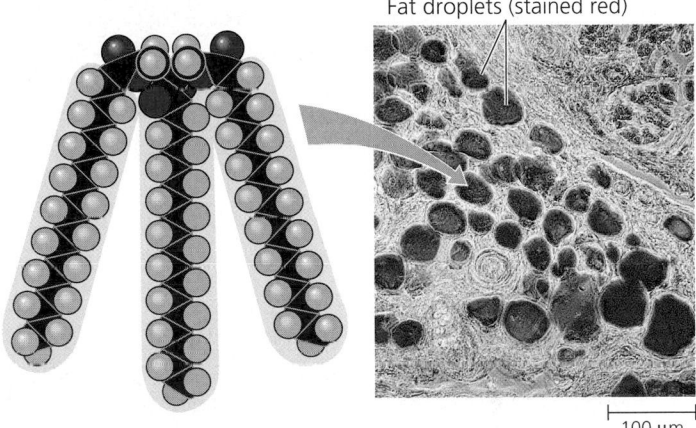

Fat droplets (stained red)

100 μm

(a) A fat molecule   **(b) Mammalian adipose cells**

FIGURE 4.5 The role of hydrocarbons in fats. (a) A fat molecule consists of a headpiece and three hydrocarbon tails. The tails store energy and account for the hydrophobic behavior of fats. (Black = carbon; gray = hydrogen; red = oxygen)
(b) Mammalian adipose cells stockpile fat molecules as a fuel reserve. Each adipose cell in this micrograph is almost filled by a large fat droplet, which stockpiles a huge number of fat molecules.

Hydrocarbons are the major components of petroleum, which is called a fossil fuel because it consists of the partially decomposed remains of organisms that lived millions of years ago.

Although hydrocarbons are not prevalent in living organisms, many of a cell's organic molecules have regions consisting of only carbon and hydrogen. For example, the molecules known as fats have long hydrocarbon tails attached to a nonhydrocarbon component (FIGURE 4.5). Neither petroleum nor fat mixes with water; both are hydrophobic compounds because the bonds between the carbon and hydrogen atoms are nonpolar. Another characteristic of hydrocarbons is that they store a relatively large amount of energy. The gasoline that fuels a car consists of hydrocarbons, and the hydrocarbon tails of fat molecules serve as stored fuel for animal bodies.

Isomers

Variation in the architecture of organic molecules can be seen in **isomers,** compounds that have the same molecular formula but different structures and hence different properties. Compare, for example, the two butanes in FIGURE 4.4b. Both have the molecular formula C_4H_{10}, but they differ in the covalent arrangement of their carbon skeletons. The skeleton is straight in butane, but branched in isobutane. We will examine three types of isomers: structural isomers, geometric isomers, and enantiomers (FIGURE 4.6).

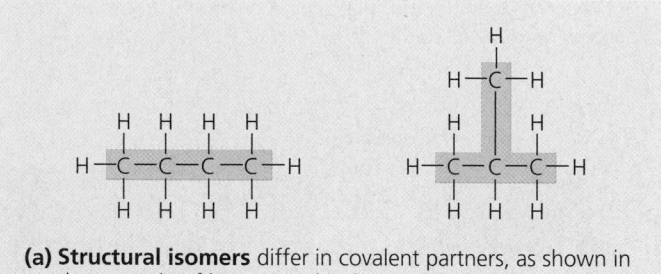

(a) Structural isomers differ in covalent partners, as shown in the example of butane and isobutane.

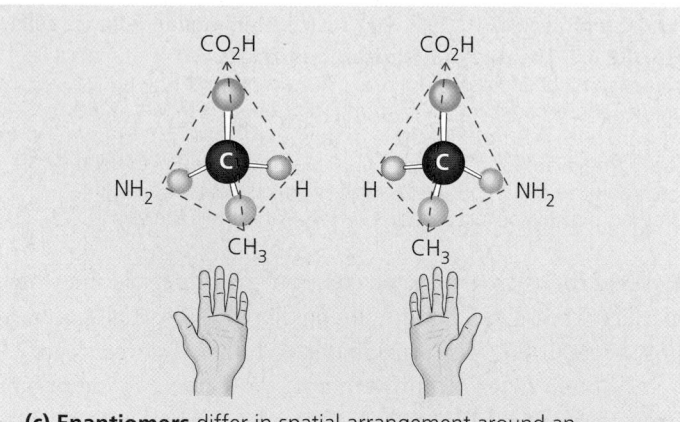

(b) Geometric isomers differ in arrangement about a double bond. (In these diagrams, X represents an atom or group of atoms attached to a double-bonded carbon.)

(c) Enantiomers differ in spatial arrangement around an asymmetric carbon, resulting in molecules that are mirror images, like left and right hands. Enantiomers cannot be superimposed on each other.

FIGURE 4.6 Three types of isomers. Compounds with the same molecular formula but different structures, isomers are a source of diversity in organic molecules.

Structural isomers differ in the covalent arrangements of their atoms. The number of possible isomers increases tremendously as carbon skeletons increase in size. There are only two butanes, but there are 18 variations of C_8H_{18} and 366,319 possible structural isomers of $C_{20}H_{42}$. Structural isomers may also differ in the location of double bonds.

Geometric isomers have the same covalent partnerships, but they differ in their spatial arrangements. Geometric isomers arise from the inflexibility of double bonds, which, unlike single bonds, will not allow the atoms they join to rotate freely about the bond axis. The subtle difference in shape between geometric isomers can dramatically affect the biological activities of organic molecules. For example, the biochemistry of vision involves a light-induced change of rhodopsin, a chemical compound in the eye, from one geometric isomer to another.

Enantiomers are molecules that are mirror images of each other. In the ball-and-stick models shown in FIGURE 4.6C, the middle carbon is called an *asymmetric carbon* because it is attached to four different atoms or groups of atoms. The four groups can be arranged in space about the asymmetric carbon in two different ways that are mirror images. They are, in a way, left-handed and right-handed versions of the molecule. A cell can distinguish these isomers based on their different shapes. Usually, one isomer is biologically active and the other is inactive.

The concept of enantiomers is important in the pharmaceutical industry because the two enantiomers of a drug may not be equally effective (FIGURE 4.7). In some cases, one of the isomers may even produce harmful effects. This was the case with thalidomide, a drug prescribed for thousands of pregnant women in the late 1950s and early 1960s. The drug was a mixture of two enantiomers. One enantiomer reduced morning sickness, the desired effect, but the other caused severe birth defects. (And unfortunately, even if the "good" thalidomide enantiomer is used in purified form, some of it soon converts to the "bad" enantiomer in the patient's body.) The

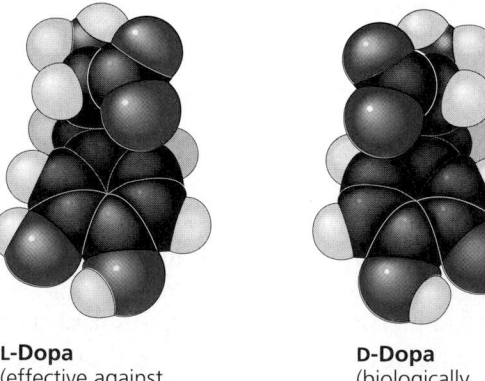

L-Dopa
(effective against Parkinson's disease)

D-Dopa
(biologically inactive)

FIGURE 4.7 The pharmacological importance of enantiomers. L-Dopa is a drug used to treat Parkinson's disease, a disorder of the central nervous system. The drug's enantiomer, the mirror-image molecule designated D-Dopa, has no effect on patients.

differing effects of enantiomers in the body demonstrate that organisms are sensitive to even the most subtle variations in molecular architecture. Once again, we see that molecules have emergent properties that depend on the specific arrangement of their atoms.

FUNCTIONAL GROUPS

The distinctive properties of an organic molecule depend not only on the arrangement of its carbon skeleton, but also on the molecular components attached to that skeleton. We will now examine certain groups of atoms that are frequently attached to the skeletons of organic molecules.

Functional groups contribute to the molecular diversity of life

The components of organic molecules that are most commonly involved in chemical reactions are known as **functional groups.** If we think of hydrocarbons as the simplest organic molecules, we can view functional groups as attachments that replace one or more of the hydrogens bonded to the carbon skeleton of the hydrocarbon. (However, some functional groups include atoms of the carbon skeleton, as we will see.)

Each functional group behaves consistently from one organic molecule to another, and the number and arrangement of the groups help give each molecule its unique properties. Consider the differences between testosterone and estradiol (a type of estrogen). These compounds are male and female

sex hormones, respectively, in humans and other vertebrates (FIGURE 4.8). Both are steroids, organic molecules with a common carbon skeleton in the form of four fused rings. These sex hormones differ mainly in the functional groups attached to the rings. The different actions of these two molecules on many targets throughout the body help produce the contrasting features of females and males. Thus, even our sexuality has its biological basis in variations of molecular architecture.

The six functional groups most important in the chemistry of life are the hydroxyl, carbonyl, carboxyl, amino, sulfhydryl, and phosphate groups (TABLE 4.1, p. 58). All are hydrophilic and thus increase the solubility of organic compounds in water.

The Hydroxyl Group

In a **hydroxyl group,** a hydrogen atom is bonded to an oxygen atom, which in turn is bonded to the carbon skeleton of the organic molecule. Organic compounds containing hydroxyl groups are called **alcohols,** and their specific names usually end in *-ol,* as in ethanol, the drug present in alcoholic beverages. In a structural formula, the hydroxyl group is usually abbreviated by omission of the covalent bond between the oxygen and hydrogen and is written as —OH or HO—. (Do not confuse this functional group with the hydroxide ion, OH^-, formed by the dissociation of bases such as sodium hydroxide.) The hydroxyl group is polar as a result of the electronegative oxygen atom drawing electrons toward itself. Consequently, water molecules are attracted to the hydroxyl group, and this helps dissolve organic compounds containing such groups. Sugars, for example, owe their solubility in water to the presence of multiple hydroxyl groups (see FIGURE 5.3).

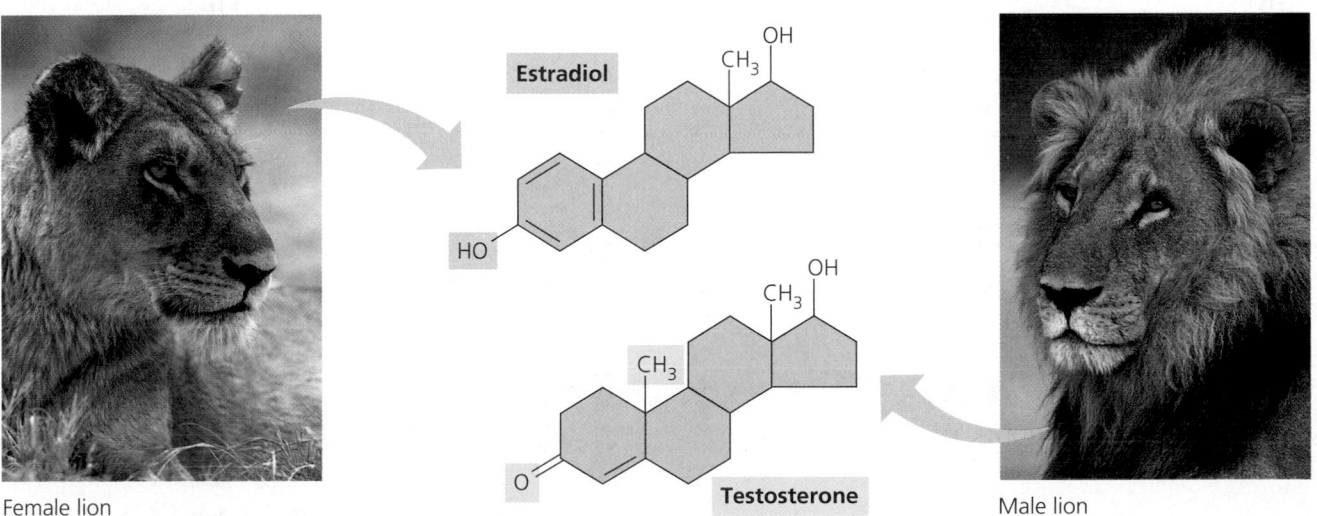

Female lion

Male lion

FIGURE 4.8 A comparison of functional groups of female (estradiol) and male (testosterone) sex hormones. The two molecules differ mainly in the attachment of functional groups to a common carbon skeleton of four fused rings. (The carbon skeleton has been simplified here by omitting the carbons in the rings, as well as their hydrogens.) These subtle variations in molecular architecture influence the development of the anatomical and physiological differences between female and male vertebrates.

Table 4.1 Functional Groups of Organic Compounds

Functional Group	Formula	Name of Compounds	Example
Hydroxyl	—OH	Alcohols	Ethanol (the drug of alcoholic beverages)
Carbonyl	(aldehyde carbonyl)	Aldehydes	Propanal
	(ketone carbonyl)	Ketones	Acetone
Carboxyl	(non-ionized) (ionized)	Carboxylic acids	Acetic acid* (the acid of vinegar)
Amino	(non-ionized) (ionized)	Amines	Glycine* (an amino acid)
Sulfhydryl	—SH	Thiols	Ethanethiol
Phosphate	(phosphate group)	Organic phosphates	Glycerol phosphate

*The ionized forms of the carboxyl and amino groups prevail in cells. However, acetic acid and glycine are represented here in their non-ionized forms.

The Carbonyl Group

The **carbonyl group** ($>$CO) consists of a carbon atom joined to an oxygen atom by a double bond. If the carbonyl group is on the end of a carbon skeleton, the organic compound is called an **aldehyde;** otherwise the compound is called a **ketone.**

The simplest ketone is acetone, which is three carbons long (see TABLE 4.1). Acetone has different properties from propanal, a three-carbon aldehyde. (Acetone and propanal are structural isomers.) Thus, variation in locations of functional groups along carbon skeletons is a major source of molecular diversity.

The Carboxyl Group

When an oxygen atom is double-bonded to a carbon atom that is also bonded to a hydroxyl group, the entire assembly of atoms is called a **carboxyl group** (—COOH). Compounds containing carboxyl groups are known as **carboxylic acids,** or organic acids. The simplest is the one-carbon compound called formic acid (HCOOH), the substance some ants inject when they sting. Acetic acid, which has two carbons, gives vinegar its sour taste. (In general, acids taste sour.)

Why does a carboxyl group have acidic properties? A carboxyl group is a source of hydrogen ions. The covalent bond between the oxygen and the hydrogen is so polar that the hydrogen tends to dissociate reversibly from the molecule as an ion (H^+). In the case of acetic acid, we have

Acetic acid   Acetate ion   Hydrogen ion

Dissociation occurs as a result of the two electronegative oxygen atoms of the carboxyl group pulling shared electrons away from hydrogen. If the double-bonded oxygen and the hydroxyl group were attached to *separate* carbon atoms, there would be less tendency for the hydrogen of the hydroxyl group to dissociate because the second oxygen would be farther away. Here is another example of how emergent properties result from a specific arrangement of building components.

The Amino Group

The **amino group** (—NH$_2$) consists of a nitrogen atom bonded to two hydrogen atoms and to the carbon skeleton. Organic compounds with this functional group are called **amines.** An example is glycine, illustrated in TABLE 4.1. Because glycine also has a carboxyl group, it is both an amine and a carboxylic acid. Most of the cell's organic compounds have two or more different functional groups. Glycine and similar compounds having both amino and carboxyl groups are called amino acids; these are the molecular building blocks of proteins.

The amino group acts as a base. You learned in Chapter 3 that ammonia (NH_3) can pick up a proton from the surrounding solution. Amino groups of organic compounds can do the same:

This process gives the amino group a charge of $+1$, its most common state within the cell.

The Sulfhydryl Group

Sulfur is directly below oxygen in the periodic table; both have 6 valence electrons and form two covalent bonds. The organic functional group known as the **sulfhydryl group** (—SH), which consists of a sulfur atom bonded to an atom of hydrogen, resembles a hydroxyl group in shape (see TABLE 4.1). Organic compounds containing sulfhydryls are called **thiols.** In Chapter 5, you will learn how sulfhydryl groups can interact to help stabilize the intricate structure of a protein.

The Phosphate Group

Phosphate is an anion formed by dissociation of an inorganic acid called phosphoric acid (H_3PO_4). The loss of hydrogen ions by dissociation leaves the phosphate with two negative charges. Organic compounds containing a **phosphate group** (—OPO_3^{2-}) have a phosphate ion covalently attached by one of its oxygen atoms to the carbon skeleton (see TABLE 4.1). One function of phosphate groups is the transfer of energy between organic molecules. In Chapter 6, you will learn how cells harness the transfer of phosphate groups to perform work, such as the contraction of muscle cells.

The chemical elements of life: *a review*

Living matter, as you have learned, consists mainly of carbon, oxygen, hydrogen, and nitrogen, with smaller amounts of sulfur and phosphorus. These elements share the characteristic of forming strong covalent bonds, a quality that is essential in the architecture of complex organic molecules. Of all these elements, carbon is the virtuoso of the covalent bond. The chemical behavior of carbon makes it exceptionally versatile as a building block in molecular architecture: It can form four covalent bonds, link together into intricate molecular skeletons, and join with several other elements. The versatility of carbon makes possible the great diversity of organic molecules, each with special properties that emerge from the unique arrangement of its carbon skeleton and the functional groups appended to that skeleton. At the foundation of all biological diversity lies this variation at the molecular level.

■   ■   ■

Now that we have examined the basic architectural principles of organic compounds, we can move on to the next chapter, where we will explore the specific structures and functions of the large and complex molecules made by living cells: carbohydrates, lipids, proteins, and nucleic acids.

CHAPTER 4 REVIEW

Go to the Campbell Biology website (www.campbellbiology.com) to explore an interactive version of the Chapter Review.

Summary of Key Concepts

THE IMPORTANCE OF CARBON

- **Organic chemistry is the study of carbon compounds (pp. 52–53)** Organic compounds were once thought to arise only within living organisms, but this idea (vitalism) was disproved when chemists were able to synthesize organic compounds in the laboratory.

- **Carbon atoms are the most versatile building blocks of molecules (pp. 53–55, FIGURE 4.2)** A covalent-bonding capacity of four contributes to carbon's ability to form diverse molecules. Carbon can bond to a variety of atoms, including O, H, and N. Carbon atoms can also bond to other carbons, forming the carbon skeletons of organic compounds.

- **Variation in carbon skeletons contributes to the diversity of organic molecules (pp. 55–57, FIGURE 4.4)** The carbon skeletons of organic molecules vary in length and shape and have bonding sites for atoms of other elements. Hydrocarbons consist only of carbon and hydrogen. Carbon's versatile bonding is the basis for isomers, molecules with the same molecular formula but different structures and thus different properties. Three types of isomers are structural isomers, geometric isomers, and enantiomers.
  Web/CD Activity 4A: *Diversity of Carbon-Based Molecules*
  Web/CD Activity 4B: *Isomers*
  Web/CD Case Study in the Process of Science: *What Factors Determine the Effectiveness of Drugs?*

FUNCTIONAL GROUPS

- **Functional groups contribute to the molecular diversity of life (pp. 57–59, TABLE 4.1)** Functional groups are specific chemically reactive groups of atoms within organic molecules that give the overall molecule distinctive chemical properties. The hydroxyl group (—OH), found in alcohols, has a polar covalent bond, which helps alcohols dissolve in water. The carbonyl group ($>$CO) can be either at the end of a carbon skeleton (aldehyde) or within the skeleton (ketone). The carboxyl group (—COOH) is found in carboxylic acids. The hydrogen of this group can dissociate, making the molecule a weak acid. The amino group (—NH$_2$) can accept a proton (H$^+$), thereby acting as a base. The sulfhydryl group (—SH) helps stabilize the structure of some proteins. The phosphate group (—OPO$_3^{2-}$) has an important role in the transfer of cellular energy.
  Web/CD Activity 4C: *Functional Groups*

- **The chemical elements of life: *a review* (p. 59)** Living matter is made mostly of carbon, oxygen, hydrogen, and nitrogen, with some sulfur and phosphorus. Biological diversity has its molecular basis in carbon's ability to form a huge number of molecules with particular shapes and chemical properties.

Self-Quiz

1. Organic chemistry is currently defined as
   a. the study of compounds that can be made only by living cells.
   b. the study of carbon compounds.
   c. the study of vital forces.
   d. the study of natural (as opposed to synthetic) compounds.
   e. the study of hydrocarbons.

2. Choose the pair of terms that correctly completes this sentence: Hydroxyl is to _____ as _____ is to aldehyde.
   a. carbonyl; ketone
   b. oxygen; carbon
   c. alcohol; carbonyl
   d. amine; carboxyl
   e. alcohol; ketone

3. Which of the following hydrocarbons has a double bond in its carbon skeleton?
   a. C$_3$H$_8$
   b. C$_2$H$_6$
   c. CH$_4$
   d. C$_2$H$_4$
   e. C$_2$H$_2$

4. The gasoline consumed by an automobile is a fossil fuel consisting mostly of
   a. aldehydes.
   b. amino acids.
   c. alcohols.
   d. hydrocarbons.
   e. thiols.

5. Choose the term that correctly describes the relationship between these two sugar molecules:

   a. structural isomers
   b. geometric isomers
   c. enantiomers
   d. isotopes

6. Identify the asymmetric carbon in this molecule:

7. Which functional group is *not* present in this molecule?

   a. carboxyl
   b. sulfhydryl
   c. hydroxyl
   d. amino

60 UNIT ONE THE CHEMISTRY OF LIFE

8. Which action could produce a carbonyl group?
   a. the replacement of the hydroxyl of a carboxyl group with hydrogen
   b. the addition of a thiol to a hydroxyl
   c. the addition of a hydroxyl to a phosphate
   d. the replacement of the nitrogen of an amine with oxygen
   e. the addition of a sulfhydryl to a carboxyl

9. Which functional group is most likely to be responsible for an organic molecule behaving as a base?
   a. hydroxyl
   b. carbonyl
   c. carboxyl
   d. amino
   e. phosphate

10. Which of the following molecules would be the strongest acid? Explain your answer.

11. Draw a structural formula for C_2H_4.
12. Why is the following compound called an amino acid?

13. Draw three structural isomers of the hydrocarbon pentane (C_5H_{12}).
14. What is the chemical similarity between gasoline and fat?
15. What is a sulfhydryl group?

Go to the website or CD-ROM for more quiz questions.

Evolution Connection

Some scientists believe that life elsewhere in the universe might be based on the element silicon, rather than on carbon, as on Earth. What properties does silicon share with carbon that would make silicon-based life more likely than, say, neon-based life or aluminum-based life? (see FIGURE 2.30, p. 32)

The Process of Science

In 1918, an epidemic of sleeping sickness caused an unusual rigid paralysis in some survivors, similiar to symptoms of advanced Parkinson's disease. Years later, L-Dopa, a chemical used to treat Parkinson's disease (see FIGURE 4.7), was given to some of these patients, as dramatized in the movie *Awakenings,* L-Dopa was remarkably effective at eliminating the paralysis, at least temporarily. However, its enantiomer, D-Dopa, was subsequently shown to have no effect at all, as is the case for Parkinson's disease. Suggest a hypothesis to explain why, for *both* diseases, one enantiomer is effective and the other is not.

Investigate the effect of enantiomers on drug development in the Case Study in the Process of Science, available on the CD-ROM and website.

Science, Technology, and Society

Thalidomide achieved notoriety 40 years ago because of a wave of birth defects among children born to women who took thalidomide during pregnancy as a treatment for morning sickness. However, in 1998 the U.S. Food and Drug Administration (FDA) approved this drug for the treatment of certain conditions associated with Hansen's disease (leprosy). In clinical trials, thalidomide also shows promise for use in treating patients suffering from AIDS, tuberculosis, and a number of other diseases, including some types of cancer. Do you think approval of this drug is appropriate? If so, under what conditions? What criteria do you think the FDA should use in weighing a drug's benefits against its dangers?

Answers: 1. b; 2. c; 3. d; 4. d; 5. a; 6. b; 7. b; 8. a; 9. d; 10. b;
11.

12. It has both an amino group ($-NH_2$) and a carboxyl group ($-COOH$), which also makes it a carboxylic acid.
13.

14. Both consist largely of hydrocarbon chains. 15. The functional group $-SH$.

CHAPTER 5

THE STRUCTURE AND FUNCTION OF MACROMOLECULES

POLYMER PRINCIPLES

- Most macromolecules are polymers
- An immense variety of polymers can be built from a small set of monomers

CARBOHYDRATES—FUEL AND BUILDING MATERIAL

- Sugars, the smallest carbohydrates, serve as fuel and carbon sources
- Polysaccharides, the polymers of sugars, have storage and structural roles

LIPIDS—DIVERSE HYDROPHOBIC MOLECULES

- Fats store large amounts of energy
- Phospholipids are major components of cell membranes
- Steroids include cholesterol and certain hormones

PROTEINS—MANY STRUCTURES, MANY FUNCTIONS

- A polypeptide is a polymer of amino acids connected in a specific sequence
- A protein's function depends on its specific conformation

NUCLEIC ACIDS—INFORMATIONAL POLYMERS

- Nucleic acids store and transmit hereditary information
- A nucleic acid strand is a polymer of nucleotides
- Inheritance is based on replication of the DNA double helix
- We can use DNA and proteins as tape measures of evolution

We have applied *the concept of emergent properties to our study of water and relatively simple organic molecules. These substances are central to life, each one having unique behavior arising from the orderly arrangement of its atoms. Another level in the hierarchy of biological organization is reached when cells join small organic molecules together to form larger molecules. The four main classes of large biological molecules are carbohydrates, lipids, proteins, and nucleic acids. Many of these cellular molecules are, on the molecular scale, huge. For example, a protein may consist of thousands of covalently connected atoms*

that form a molecular colossus weighing over 100,000 daltons. Biologists use the term **macromolecule** *for such giant molecules.*

*Considering the size and complexity of macromolecules, it is remarkable that biochemists have determined the detailed structures of so many of them (*FIGURE 5.1*). The architecture of a macromolecule helps explain how that molecule works. For example, the pleated structure of the silk protein from which the orb spider in the photo on this page weaves its web gives the fibers their strength and springiness. Proteins and the other large molecules of life are the main subject of this chapter. For these molecules, as at all levels in the biological hierarchy, form and function are inseparable.*

POLYMER PRINCIPLES

In examining the relationship between the structure and function of life's macromolecules, we begin with a key generalization about how cells build these large molecules from smaller ones.

Most macromolecules are polymers

The large molecules in three of the four classes of life's organic compounds—carbohydrates, proteins, and nucleic acids—are chainlike molecules called polymers (from the Greek *polys*, many, and *meris*, part). A **polymer** is a long molecule consisting of many similar or identical building blocks linked by covalent bonds, much as a train consists of a chain of cars. The repeating units that serve as the building blocks of a polymer are small molecules called **monomers.** Some of the molecules that serve as monomers also have other functions of their own.

(a)

(b)

FIGURE 5.1 Building models to study the structure of macromolecules.
(a) Linus Pauling (1901–1994) with a model of part of a protein. In the 1950s, Pauling discovered several of the basic structural features of proteins, which he demonstrated by building physical models. **(b)** Today, scientists use computers to help build molecular models, but the goal remains the same: to correlate the structure of macromolecules with their functions.

The classes of polymeric macromolecules differ in the nature of their monomers, but the chemical mechanisms that cells use to make and break polymers are basically the same in all cases (FIGURE 5.2). Monomers are connected by a reaction in which two molecules are covalently bonded to each other through loss of a water molecule; this is called a **condensation reaction,** specifically a **dehydration reaction,** because the molecule lost is water (FIGURE 5.2a). When a bond forms between two monomers, each monomer contributes part of the water molecule that is lost: One molecule provides a hydroxyl group (—OH), while the other provides a hydrogen (—H). To make a polymer, this reaction is repeated as monomers are added to the chain one by one. The cell must expend energy to carry out these dehydration reactions, and the process occurs only with the help of enzymes, specialized proteins that speed up chemical reactions in cells.

Polymers are disassembled to monomers by **hydrolysis,** a process that is essentially the reverse of the dehydration reaction (FIGURE 5.2b). Hydrolysis means to break with water (from the Greek *hydro,* water, and *lysis,* break). Bonds between monomers are broken by the addition of water molecules, a hydrogen from the water attaching to one monomer and a hydroxyl attaching to the adjacent monomer. An example of hydrolysis working in our bodies is the process of digestion. The bulk of the organic material in our food is in the form of polymers that are much too large to enter our cells. Within the digestive tract, various enzymes attack the polymers, speeding up hydrolysis. The released monomers are then absorbed into the bloodstream for distribution to all body cells. Those cells can then use dehydration reactions to assemble the monomers into new polymers that differ from the ones that were digested.

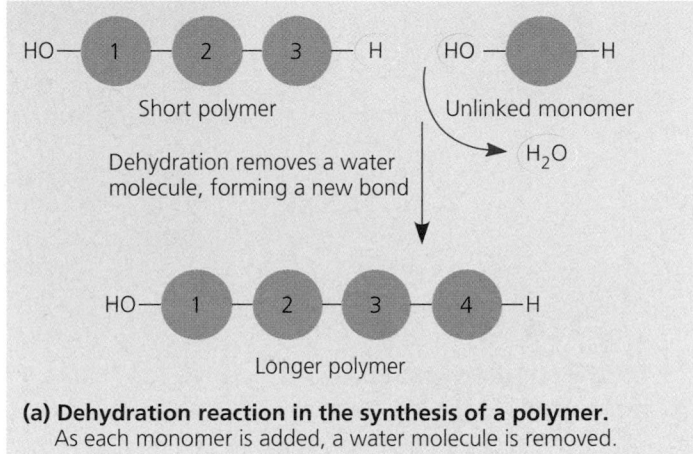

(a) Dehydration reaction in the synthesis of a polymer. As each monomer is added, a water molecule is removed.

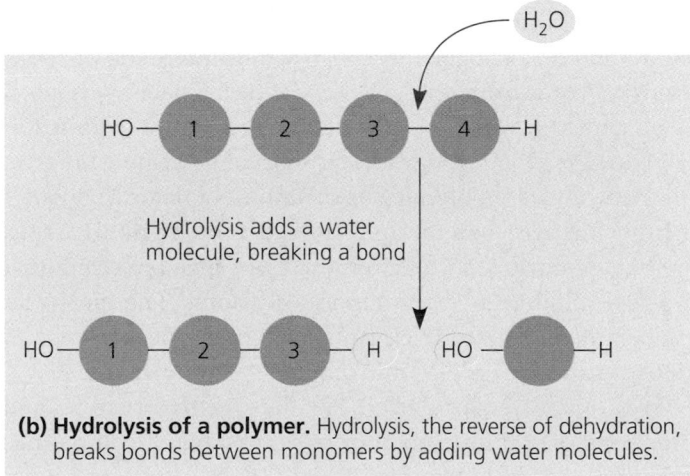

(b) Hydrolysis of a polymer. Hydrolysis, the reverse of dehydration, breaks bonds between monomers by adding water molecules.

FIGURE 5.2 The synthesis and breakdown of polymers.

An immense variety of polymers can be built from a small set of monomers

Each cell has thousands of different kinds of macromolecules; the collection varies from one type of cell to another in the same organism. The inherent differences between human siblings reflect variations in polymers, particularly DNA and proteins. Molecular differences between unrelated individuals are more extensive—and between species greater still. The diversity of macromolecules in the living world is vast, and the potential variety is effectively limitless.

What is the basis for such diversity in life's polymers? These molecules are constructed from only 40 to 50 common monomers and some others that occur rarely. Building an

FIGURE 5.3 The structure and classification of some monosaccharides.

Sugars may be aldoses (aldehyde sugars) or ketoses (ketone sugars), depending on the location of the carbonyl group (pink). Sugars are also classified according to the length of their carbon skeletons. A third point of variation is the spatial arrangement around asymmetric carbons (compare, for example, the gray portions of glucose and galactose).

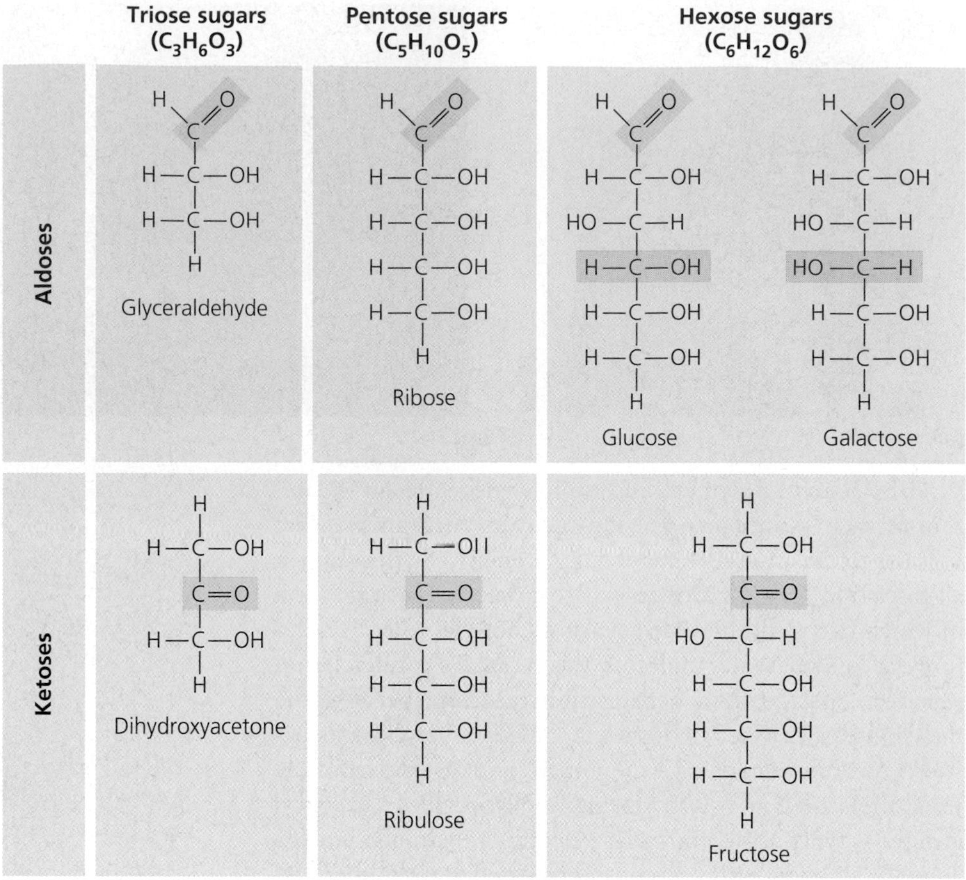

Triose sugars ($C_3H_6O_3$) Pentose sugars ($C_5H_{10}O_5$) Hexose sugars ($C_6H_{12}O_6$)

Aldoses: Glyceraldehyde, Ribose, Glucose, Galactose

Ketoses: Dihydroxyacetone, Ribulose, Fructose

---

enormous variety of polymers from such a limited list of monomers is analogous to constructing hundreds of thousands of words from only 26 letters of the alphabet. The key is arrangement—variation in the linear sequence the units follow. However, this analogy falls far short of describing the great diversity of macromolecules, because most biological polymers are much longer than the longest word. Proteins, for example, are built from 20 kinds of amino acids arranged in chains that are typically hundreds of amino acids long. The molecular logic of life is simple but elegant: Small molecules common to all organisms are ordered into unique macromolecules.

We are now ready to investigate the specific structures and functions of the four major classes of organic compounds found in cells. For each class, we will see that the large molecules have emergent properties not found in their individual building blocks.

CARBOHYDRATES—FUEL AND BUILDING MATERIAL

Carbohydrates include both sugars and their polymers. The simplest carbohydrates are the monosaccharides, or single sugars, also known as simple sugars. Disaccharides are double sugars, consisting of two monosaccharides joined by conden-

sation. The carbohydrates that are macromolecules are polysaccharides, polymers of many sugars.

Sugars, the smallest carbohydrates, serve as fuel and carbon sources

Monosaccharides (from the Greek *monos*, single, and *sacchar*, sugar) generally have molecular formulas that are some multiple of CH_2O (FIGURE 5.3). Glucose ($C_6H_{12}O_6$), the most common monosaccharide, is of central importance in the chemistry of life. In the structure of glucose, we can see the trademarks of a sugar: The molecule has a carbonyl group ($>C=O$) and multiple hydroxyl groups. Depending on the location of the carbonyl group, a sugar is either an aldose (aldehyde sugar) or a ketose (ketone sugar). Glucose, for example, is an aldose; fructose, a structural isomer of glucose, is a ketose. (Most names for sugars end in *-ose*.) Another criterion for classifying sugars is the size of the carbon skeleton, which ranges from three to seven carbons long. Glucose, fructose, and other sugars that have six carbons are called hexoses. Trioses (three-carbon sugars) and pentoses (five-carbon sugars) are also common.

Still another source of diversity for simple sugars is in the spatial arrangement of their parts around asymmetric car-

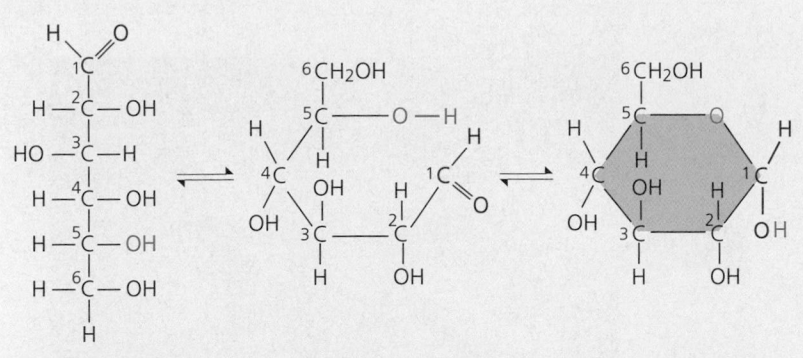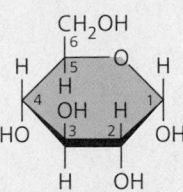

(a) Linear and ring forms. Chemical equilibrium between the linear and ring structures greatly favors the formation of rings. To form the glucose ring, carbon 1 bonds to the oxygen attached to carbon 5.

(b) Abbreviated ring structure. The carbons in the ring are omitted. The ring's thicker edge indicates that you are looking at the ring edge-on; the components attached to the ring lie above or below the plane of the ring.

FIGURE 5.4 Linear and ring forms of glucose.

bons. (Recall from Chapter 4 that an asymmetric carbon is one attached to four different kinds of covalent partners.) Glucose and galactose, for example, differ only in the placement of parts around one asymmetric carbon (see the gray boxes in FIGURE 5.3). What may seem at first a small difference is significant enough to give the two sugars distinctive shapes and behaviors.

Although it is convenient to draw glucose with a linear carbon skeleton, this representation is not accurate. In aqueous solutions, glucose molecules, as well as most other sugars, form rings (FIGURE 5.4).

Monosaccharides, particularly glucose, are major nutrients for cells. In the process known as cellular respiration, cells extract the energy stored in glucose molecules. Not only are simple sugar molecules a major fuel for cellular work, but their carbon skeletons serve as raw material for the synthesis

of other types of small organic molecules, such as amino acids and fatty acids. Sugar molecules that are not immediately used in these ways are generally incorporated as monomers into disaccharides or polysaccharides.

A **disaccharide** consists of two monosaccharides joined by a **glycosidic linkage,** a covalent bond formed between two monosaccharides by a dehydration reaction. For example, maltose is a disaccharide formed by the linking of two molecules of glucose (FIGURE 5.5a). Also known as malt sugar, maltose is an ingredient for brewing beer. The most prevalent disaccharide is sucrose, which is table sugar. Its two monomers are glucose and fructose (FIGURE 5.5b). Plants generally transport carbohydrates from leaves to roots and other nonphotosynthetic organs in the form of sucrose. Lactose, the sugar present in milk, is another disaccharide, consisting of a glucose molecule joined to a galactose molecule.

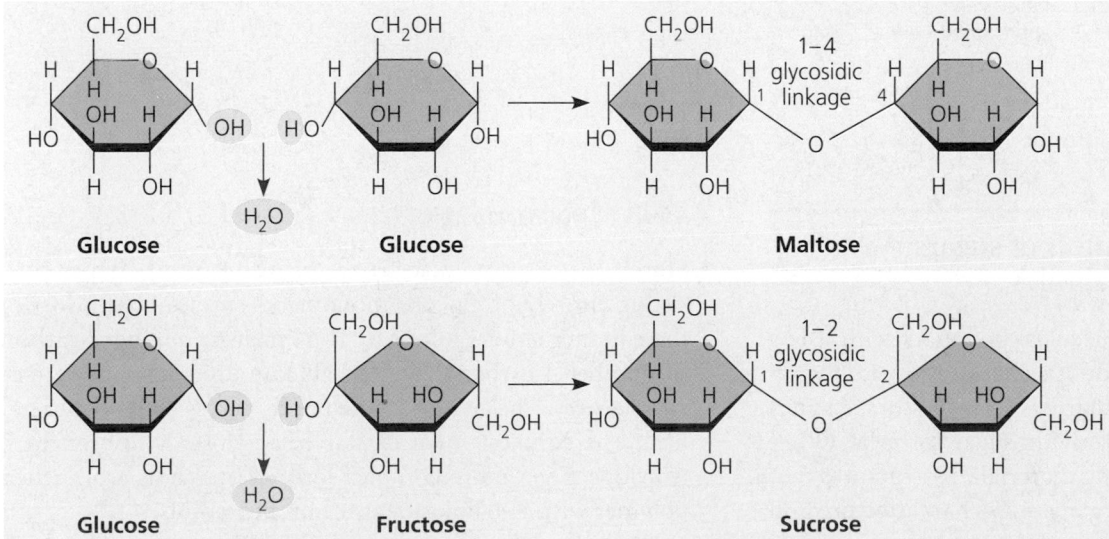

(a) Dehydration synthesis of maltose. The bonding of two glucose units forms maltose. The glycosidic link joins the number 1 carbon of one glucose to the number 4 carbon of the second glucose. Joining the glucose monomers in a different way would result in a different disaccharide.

(b) Dehydration synthesis of sucrose. Sucrose is a disaccharide formed from glucose and fructose. Notice that fructose, though a hexose like glucose, forms a five-sided ring.

FIGURE 5.5 Examples of disaccharide synthesis.

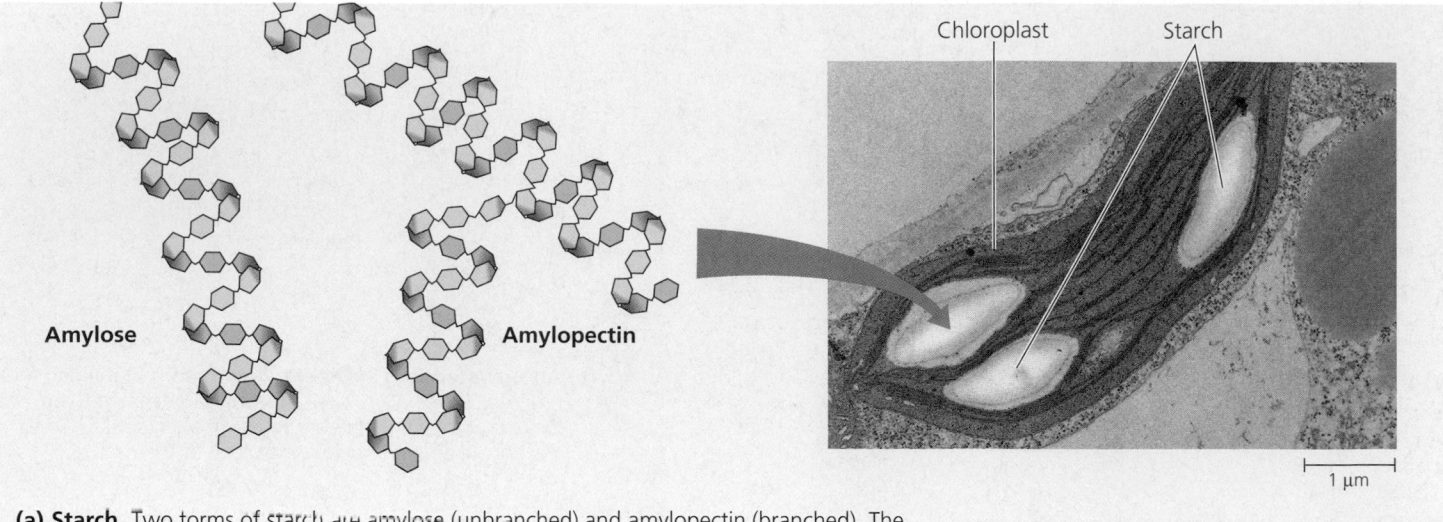

(a) Starch. Two forms of starch are amylose (unbranched) and amylopectin (branched). The light ovals in the micrograph are granules of starch within a chloroplast of a plant cell.

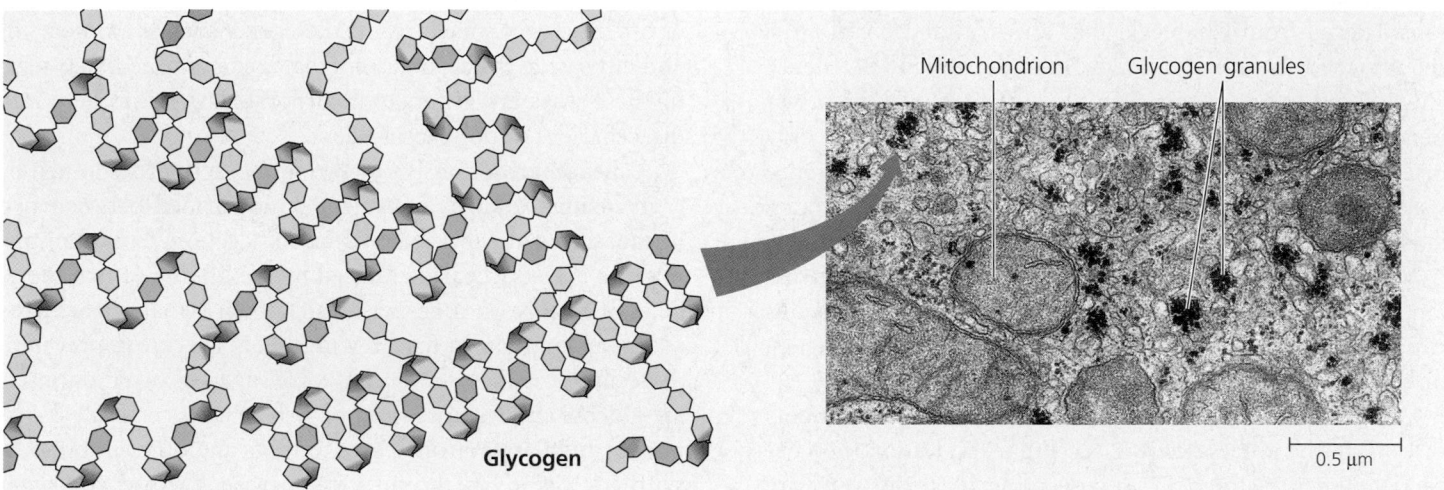

(b) Glycogen. Glycogen is more extensively branched than amylopectin. Animal cells stockpile glycogen as dense clusters of granules within liver and muscle cells. Hydrolysis frees the glucose from storage. (The micrograph shows part of a liver cell.)

FIGURE 5.6 Storage polysaccharides. These examples, starch and glycogen, are composed entirely of glucose monomers, abbreviated here as hexagons. The polymer chains tend to spiral to form helices.

Polysaccharides, the polymers of sugars, have storage and structural roles

Polysaccharides are macromolecules, polymers with a few hundred to a few thousand monosaccharides joined by glycosidic linkages. Some polysaccharides serve as storage material, hydrolyzed as needed to provide sugar for cells. Other polysaccharides serve as building material for structures that protect the cell or the whole organism. The architecture and function of a polysaccharide are determined by its sugar monomers and by the positions of its glycosidic linkages.

Storage Polysaccharides

Starch, a storage polysaccharide of plants, is a polymer consisting entirely of glucose monomers (FIGURE 5.6a). Most of these monomers are joined by 1–4 linkages (number 1 carbon to number 4 carbon), like the glucose units in maltose (see FIGURE 5.5a). The angle of these bonds makes the polymer helical. The simplest form of starch, amylose, is unbranched. Amylopectin, a more complex form of starch, is a branched polymer with 1–6 linkages at the branch points.

Plants store starch as granules within cellular structures called plastids, including chloroplasts (see FIGURE 5.6a). By

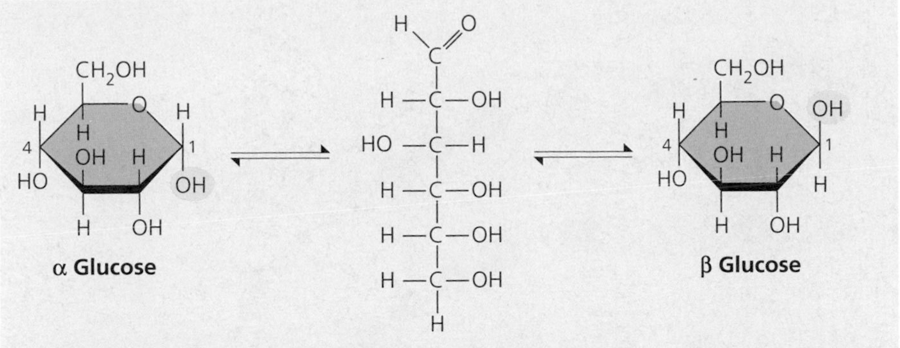

(a) α and β glucose ring structures. These two interconvertible forms of glucose differ in the placement of the hydroxyl group attached to the number 1 carbon.

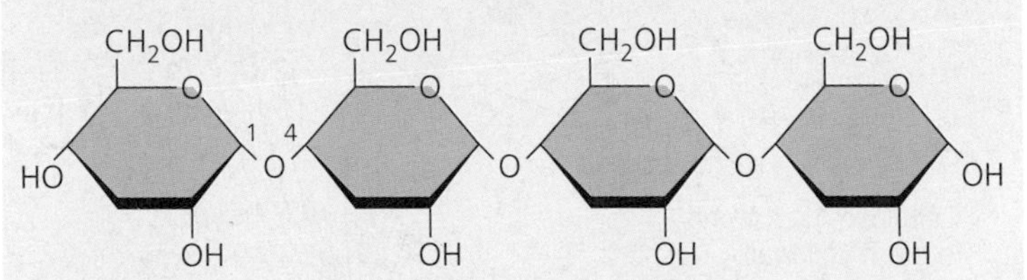

(b) Starch: 1–4 linkage of α glucose monomers.

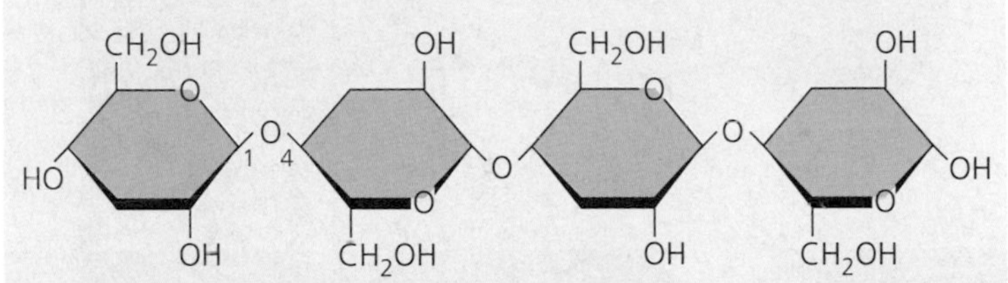

(c) Cellulose: 1–4 linkage of β glucose monomers. The angles of the bonds that link the rings make every other glucose monomer "upside down."

FIGURE 5.7 Starch and cellulose structures.

synthesizing starch, the plant can stockpile surplus glucose. Because glucose is a major cellular fuel, starch represents stored energy. The sugar can later be withdrawn from this carbohydrate bank by hydrolysis, which breaks the bonds between the glucose monomers. Most animals, including humans, also have enzymes that can hydrolyze plant starch, making glucose available as a nutrient for cells. Potato tubers and grains—the fruits of wheat, corn, rice, and other grasses—are the major sources of starch in the human diet.

Animals store a polysaccharide called **glycogen,** a polymer of glucose that is like amylopectin but more extensively branched (see FIGURE 5.6b). Humans and other vertebrates store glycogen mainly in liver and muscle cells. Hydrolysis of glycogen in these cells releases glucose when the demand for sugar increases. This stored fuel cannot sustain an animal for long, however. In humans, for example, the glycogen bank is depleted in about a day unless it is replenished by consumption of food.

Structural Polysaccharides

Organisms build strong materials from structural polysaccharides. For example, the polysaccharide called **cellulose** is a major component of the tough walls that enclose plant cells. On a global scale, plants produce almost 10^{11} (100 billion) tons of cellulose per year; it is the most abundant organic compound on Earth. Like starch, cellulose is a polymer of glucose, but the glycosidic linkages in these two polymers differ. The difference is based on the fact that there are actually two slightly different ring structures for glucose (FIGURE 5.7a). When glucose forms a ring, the hydroxyl group attached to the number 1 carbon is locked into one of two alternative positions: either below or above the plane of the ring. These two ring forms for glucose are called alpha (α) and beta (β), respectively. In starch, all the glucose monomers are in the α configuration (FIGURES 5.7b), the arrangement we saw in FIGURES 5.4 and 5.5. In contrast, the glucose monomers of cellulose are all in the β configuration, making every other glucose monomer upside down with respect to the others (FIGURE 5.7c).

The differing glycosidic links in starch and cellulose give the two molecules distinct three-dimensional shapes. Whereas a starch molecule is mostly helical, a cellulose molecule is straight (and never branched), and its hydroxyl groups are free to hydrogen-bond with the hydroxyls of other cellulose molecules lying parallel to it. In plant cell walls, parallel cellulose molecules held together in this way are grouped into units called microfibrils (FIGURE 5.8, p. 68). These cables are a strong building material for plants—as well as for humans, who use wood, which is rich in cellulose, for lumber.

Enzymes that digest starch by hydrolyzing its α linkages are unable to hydrolyze the β linkages of cellulose. In fact, few organisms possess enzymes that can digest cellulose. Humans do not; the cellulose fibrils in our food pass through the digestive tract and are eliminated with the feces. Along the way, the fibrils abrade the wall of the digestive tract and stimulate the lining to secrete mucus, which aids in the smooth passage of food through the tract. Thus, although cellulose is not a nutrient for humans, it is an important part of a healthful diet. Most fresh fruits, vegetables, and grains are rich in cellulose. On food packages, "insoluble fiber" refers mainly to cellulose.

Some microbes can digest cellulose, breaking it down to glucose monomers. A cow harbors cellulose-digesting bacteria in the rumen, the first compartment in its stomach. The bacteria hydrolyze the cellulose of hay and grass and convert the glucose to other nutrients that nourish the cow. Similarly, a termite, which is unable to digest cellulose for itself, has microbes living in its gut that can make a meal of wood. Some fungi can also digest cellulose, thereby helping recycle chemical elements within Earth's ecosystems.

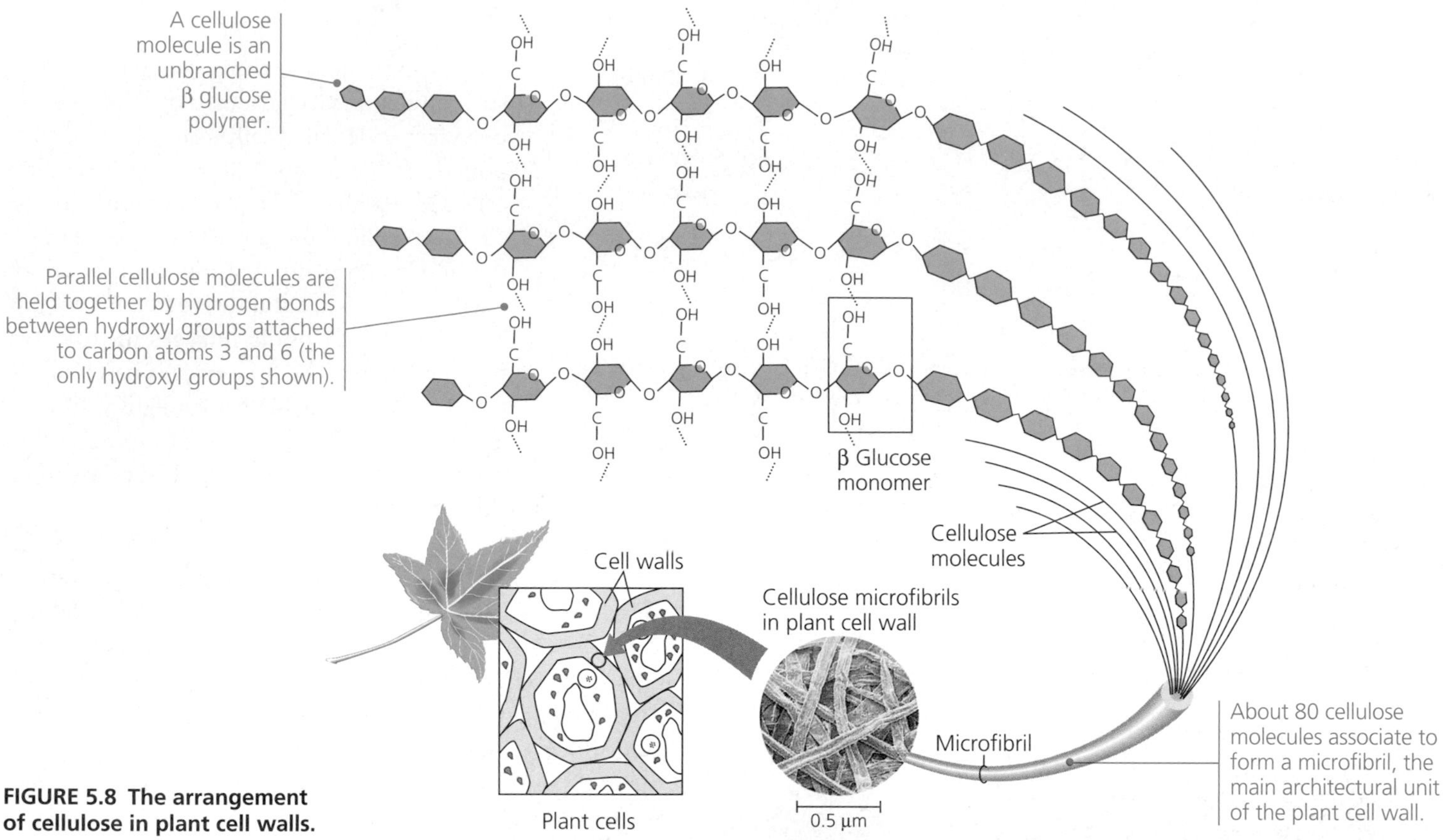

A cellulose molecule is an unbranched β glucose polymer.

Parallel cellulose molecules are held together by hydrogen bonds between hydroxyl groups attached to carbon atoms 3 and 6 (the only hydroxyl groups shown).

β Glucose monomer

Cellulose molecules

Cell walls

Cellulose microfibrils in plant cell wall

Microfibril

About 80 cellulose molecules associate to form a microfibril, the main architectural unit of the plant cell wall.

Plant cells

0.5 μm

FIGURE 5.8 The arrangement of cellulose in plant cell walls.

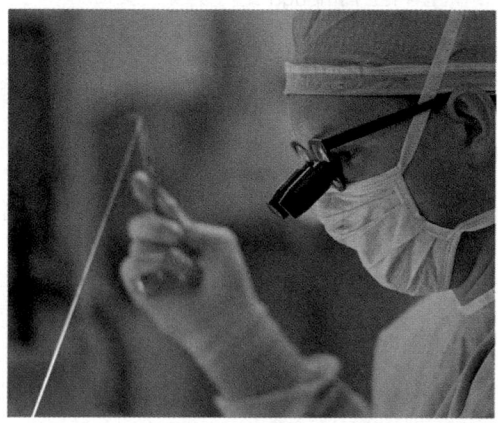

(a)

(b)

FIGURE 5.9 Chitin, a structural polysaccharide. (a) Chitin forms the exoskeleton of arthropods. This cicada is molting, shedding its old exoskeleton and emerging in adult form. **(b)** Chitin is used to make a strong and flexible surgical thread that decomposes after the wound or incision heals.

Another important structural polysaccharide is **chitin,** the carbohydrate used by arthropods (insects, spiders, crustaceans, and related animals) to build their exoskeletons (FIGURE 5.9). An exoskeleton is a hard case that surrounds the soft parts of an animal. Pure chitin is leathery, but it becomes hardened when encrusted with calcium carbonate, a salt. Chitin is also found in many fungi, which use this polysaccharide rather than cellulose as the building material for their cell walls. Chitin is similar to cellulose, except that the glucose monomer of chitin has a nitrogen-containing appendage:

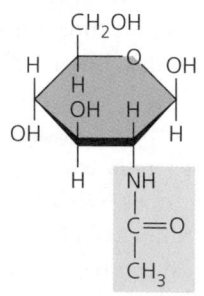

LIPIDS—DIVERSE HYDROPHOBIC MOLECULES

Lipids are the one class of large biological molecules that does not include polymers. The compounds called **lipids** are grouped together because they share one important trait: They have little or no affinity for water. The hydrophobic behavior of lipids is based on their molecular structure. Although they may have some polar bonds associated with oxygen, lipids consist mostly of hydrocarbons. Smaller than true (polymeric) macromolecules, lipids are a highly varied group in both form and function. Lipids include waxes and certain pigments, but we will focus on the most important families of lipids: the fats, phospholipids, and steroids.

Glycerol

(a) Dehydration synthesis

Ester linkage

(b) Fat molecule (triacylglycerol)

FIGURE 5.10 The synthesis and structure of a fat, or triacylglycerol. The molecular building blocks of a fat are one molecule of glycerol and three molecules of fatty acids. **(a)** One water molecule is removed for each fatty acid joined to the glycerol. **(b)** The result is a fat. Although the fat shown here has three identical fatty acid units, other fats have two or even three different kinds of fatty acids. The carbons of the fatty acid chains are arranged zig-zag to suggest the actual orientations of the four single bonds extending from each carbon (see FIGURE 4.2).

Fatty acid
(Palmitic acid)

Fats store large amounts of energy

Although fats are not polymers, they are large molecules, and they are assembled from smaller molecules by dehydration reactions. A **fat** is constructed from two kinds of smaller molecules: glycerol and fatty acids (FIGURE 5.10). Glycerol is an alcohol with three carbons, each bearing a hydroxyl group. A **fatty acid** has a long carbon skeleton, usually 16 or 18 carbon atoms in length. At one end of the fatty acid is a carboxyl group, the functional group that gives these molecules the name fatty *acids*. Attached to the carboxyl group is a long hydrocarbon chain. The nonpolar C—H bonds in the hydrocarbon chains of fatty acids are the reason fats are hydrophobic. Fats separate from water because the water molecules hydrogen-bond to one another and exclude the fats. A common example of this phenomenon is the separation of vegetable oil (a liquid fat) from the aqueous vinegar solution in a bottle of salad dressing.

In making a fat, three fatty acids each join to glycerol by an ester linkage, a bond between a hydroxyl group and a carboxyl group. The resulting fat, also called a **triacylglycerol**, thus consists of three fatty acids ("tails") linked to one glycerol molecule (the "head"). (Still another name for a fat is triglyceride, a word often found in the list of ingredients on packaged foods.) The fatty acids in a fat can be the same, as in FIGURE 5.10b, or they can be of two or three different kinds.

Fatty acids vary in length and in the number and locations of double bonds. The terms *saturated fats* and *unsaturated fats* are commonly used in the context of nutrition (FIGURE 5.11). These terms refer to the structure of the hydrocarbon chains

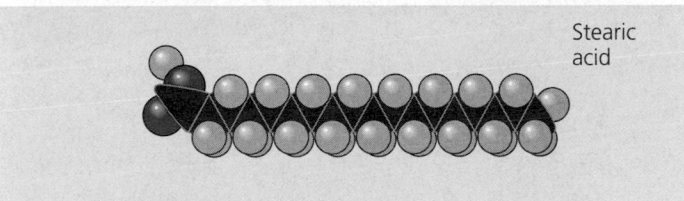

Stearic acid

(a) Saturated fat and fatty acid. At room temperature, the molecules of a saturated fat are packed closely together, forming a solid.

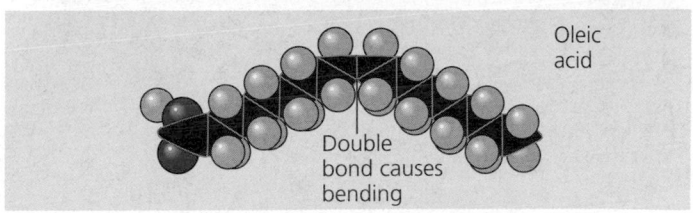

Oleic acid

Double bond causes bending

(b) Unsaturated fat and fatty acid. At room temperature, the molecules of an unsaturated fat cannot pack together closely enough to solidify because of the kinks in their fatty acid tails.

FIGURE 5.11 Examples of saturated and unsaturated fats and fatty acids.

of the fatty acids. If there are no double bonds between the carbon atoms composing the chain, then as many hydrogen atoms as possible are bonded to the carbon skeleton. Such a structure is described as being *saturated* with hydrogen, so the resulting fatty acid is called a **saturated fatty acid** (see FIGURE 5.11a). An **unsaturated fatty acid** has one or more double bonds, formed by the removal of hydrogen atoms from the carbon skeleton. The fatty acid will have a kink in its tail wherever a double bond occurs (FIGURE 5.11b).

A fat made from saturated fatty acids is called a saturated fat. Most animal fats are saturated: The "tails" of their fatty acids lack double bonds. Saturated animal fats—such as lard and butter—are solid at room temperature. In contrast, the fats of plants and fishes are generally unsaturated, meaning that they are built of one or more types of unsaturated fatty acids. Usually liquid at room temperature, plant and fish fats are referred to as oils—for instance, corn oil and cod liver oil. The kinks where the double bonds are located prevent the molecules from packing together closely enough to solidify at room temperature. The phrase "hydrogenated vegetable oils" on food labels means that unsaturated fats have been synthetically converted to saturated fats by adding hydrogen. Peanut butter, margarine, and many other products are hydrogenated to prevent lipids from separating out in liquid (oil) form.

A diet rich in saturated fats is one of several factors that may contribute to the cardiovascular disease known as atherosclerosis. In this condition, deposits called plaques develop on the internal lining of blood vessels, impeding blood flow and reducing the resilience of the vessels.

Fat has come to have such a negative connotation in our culture that you might wonder whether fats serve any useful purpose. The major function of fats is energy storage. The hydrocarbon chains of fats are similar to gasoline molecules and are just as rich in energy. A gram of fat stores more than twice as much energy as a gram of a polysaccharide, such as starch. Because plants are relatively immobile, they can function with bulky energy storage in the form of starch. (Vegetable oils are generally obtained from seeds, where more compact storage is an asset to the plant.) Animals, on the other hand, must carry their energy stores with them, so there is an advantage to having a more compact reservoir of fuel—fat. Humans and other mammals stock their long-term food reserves in adipose cells (see FIGURE 4.5), which swell and shrink as fat is deposited and withdrawn from storage. In addition to storing energy, adipose tissue also cushions such vital organs as the kidneys, and a layer of fat beneath the skin insulates the body. This subcutaneous layer is especially thick in whales, seals, and most other marine mammals.

Phospholipids are major components of cell membranes

Phospholipids are similar to fats, but they have only two fatty acid tails rather than three. The third hydroxyl group of glycerol is joined to a phosphate group, which is negative in electrical charge. Additional small molecules, usually charged or polar, can be linked to the phosphate group to form a variety of phospholipids (FIGURE 5.12).

Phospholipids show ambivalent behavior toward water. Their tails, which consist of hydrocarbons, are hydrophobic and are excluded from water. However, the phosphate group and its attachments form a hydrophilic head that has an affinity for water.

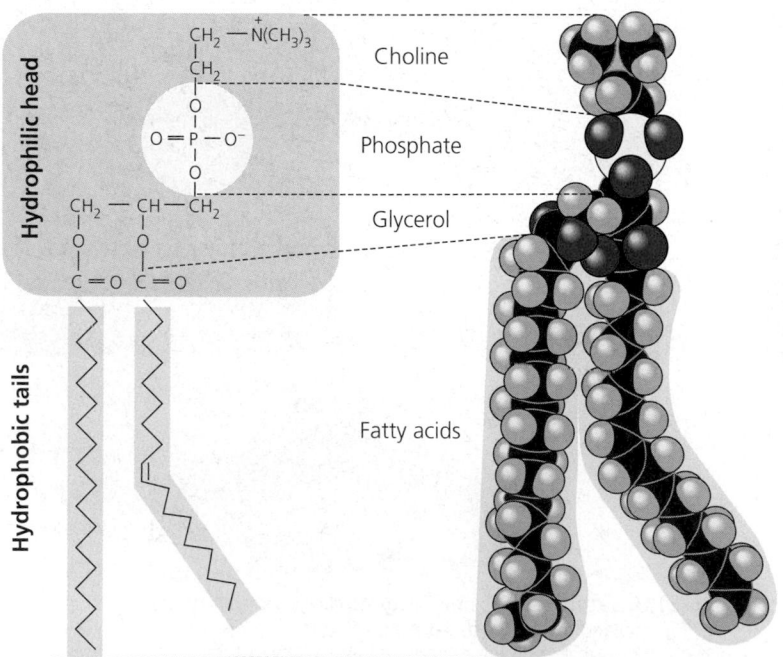

(a) Structural formula
(b) Space-filling model

FIGURE 5.12 The structure of a phospholipid. A phospholipid has a hydrophilic (polar) head and two hydrophobic tails. Phospholipid diversity is based on differences in the two fatty acid tails and in the groups attached to the phosphate group of the head. This particular phospholipid, called a phosphatidylcholine, has an attached choline group. The kink in one of its tails is due to a double bond. **(a)** The structural formula follows a common chemical convention of omitting the carbons and attached hydrogens of the hydrocarbon tails. **(b)** In the space-filling model, black = carbon, gray = hydrogen, red = oxygen, yellow = phosphorus, and blue = nitrogen. **(c)** This symbol for a phospholipid will appear throughout the book.

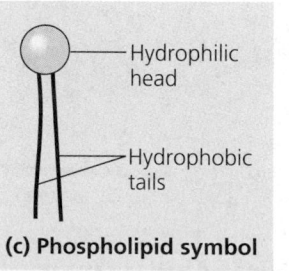

(c) Phospholipid symbol

When phospholipids are added to water, they self-assemble into aggregates that shield their hydrophobic portions from water. One kind of cluster is a micelle, a phospholipid droplet with the phosphate heads on the outside, in contact with water. The hydrocarbon tails are restricted to the water-free interior of the micelle (FIGURE 5.13a).

At the surface of a cell, phospholipids are arranged in a bilayer, or double layer (FIGURE 5.13b). The hydrophilic heads of the molecules are on the outside of the bilayer, in contact with the aqueous solutions inside and outside the cell. The hydrophobic tails point toward the interior of the membrane, away from the water. The phospholipid bilayer forms a boundary between the cell and its external environment; in fact, phospholipids are major components of cell membranes. This behavior provides another example of how form fits function at the molecular level.

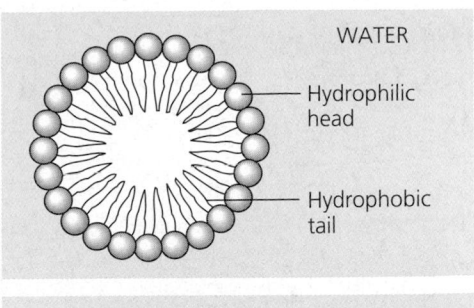

(a) Micelle

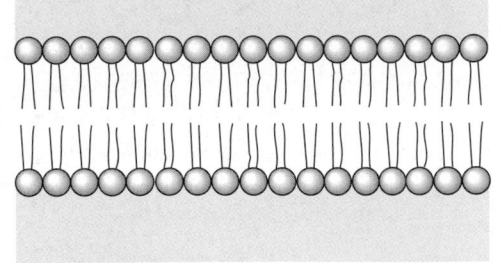

(b) Phospholipid bilayer

FIGURE 5.13 Two structures formed by self-assembly of phospholipids in aqueous environments. The hydrophilic heads of the phospholipids are in contact with water, whereas the hydrophobic tails are in contact with each other and remote from water. **(a)** A micelle, in cross section. **(b)** A cross section of a phospholipid bilayer between two aqueous compartments. Such bilayers are the main fabric of biological membranes.

Steroids include cholesterol and certain hormones

Steroids are lipids characterized by a carbon skeleton consisting of four fused rings (FIGURE 5.14). Different steroids vary in the functional groups attached to this ensemble of rings. One steroid, **cholesterol,** is a common component of animal cell membranes and is also the precursor from which other steroids are synthesized. Many hormones, including vertebrate sex hormones, are steroids produced from cholesterol (see FIGURE 4.8). Thus, cholesterol is a crucial molecule in animals, although a high level of it in the blood may contribute to atherosclerosis.

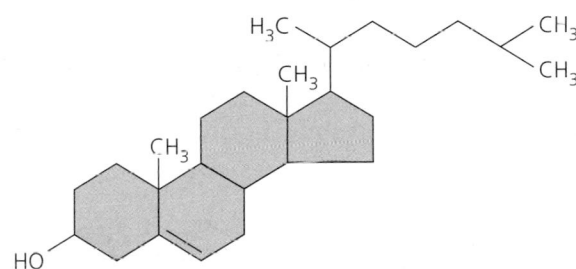

FIGURE 5.14 Cholesterol, a steroid. Cholesterol is the molecule from which other steroids, including the sex hormones, are synthesized. Steroids vary in the functional groups attached to their four interconnected rings (shown in gold).

PROTEINS—MANY STRUCTURES, MANY FUNCTIONS

The importance of proteins is implied by their name, which comes from the Greek word *proteios*, meaning "first place." Proteins account for more than 50% of the dry weight of most cells, and they are instrumental in almost everything organisms do (TABLE 5.1, p. 72). Proteins are used for structural support, storage, transport of other substances, signaling from one part of the organism to another, movement, and defense against foreign substances. In addition, as enzymes, proteins regulate metabolism by selectively accelerating chemical reactions in the cell. A human has tens of thousands of different proteins, each with a specific structure and function.

Proteins are the most structurally sophisticated molecules known. Consistent with their diverse functions, they vary extensively in structure, each type of protein having a unique three-dimensional shape, or conformation. Diverse though proteins may be, they are all polymers constructed from the same set of 20 amino acids. Polymers of amino acids are called polypeptides. A **protein** consists of one or more polypeptides folded and coiled into specific conformations.

A polypeptide is a polymer of amino acids connected in a specific sequence

As mentioned in Chapter 4, **amino acids** are organic molecules possessing both carboxyl and amino groups. The figure at the right shows the general formula for an amino acid. At the center of the amino acid is an asymmetric carbon atom called the *alpha (α) carbon.* Its four different partners are

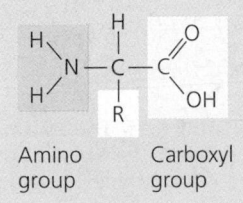

Amino group
Carboxyl group

Table 5.1 An Overview of Protein Functions

Type of Protein	Function	Examples
Structural proteins	Support	Insects and spiders use silk fibers to make their cocoons and webs, respectively. Collagen and elastin provide a fibrous framework in animal connective tissues. Keratin is the protein of hair, horns, feathers, and other skin appendages.
Storage proteins	Storage of amino acids	Ovalbumin is the protein of egg white, used as an amino acid source for the developing embryo. Casein, the protein of milk, is the major source of amino acids for baby mammals. Plants have storage proteins in their seeds.
Transport proteins	Transport of other substances	Hemoglobin, the iron-containing protein of vertebrate blood, transports oxygen from the lungs to other parts of the body. Other proteins transport molecules across cell membranes.
Hormonal proteins	Coordination of an organism's activities	Insulin, a hormone secreted by the pancreas, helps regulate the concentration of sugar in the blood of vertebrates.
Receptor proteins	Response of cell to chemical stimuli	Receptors built into the membrane of a nerve cell detect chemical signals released by other nerve cells.
Contractile proteins	Movement	Actin and myosin are responsible for the movement of muscles. Other proteins are responsible for the undulations of the organelles called cilia and flagella.
Defensive proteins	Protection against disease	Antibodies combat bacteria and viruses.
Enzymatic proteins	Selective acceleration of chemical reactions	Digestive enzymes catalyze the hydrolysis of the polymers in food.

an amino group, a carboxyl group, a hydrogen atom, and a variable group symbolized by R. The R group, also called the side chain, differs with the amino acid. FIGURE 5.15 shows the 20 amino acids that cells use to build their thousands of proteins. Here the amino and carboxyl groups are all depicted in ionized form, as they usually exist at the neutral pH inside a cell. The R group may be as simple as a hydrogen atom, as in the amino acid glycine (the one amino acid lacking an asymmetric carbon, since two of its α carbon's partners are hydrogen atoms), or it may be a carbon skeleton with various functional groups attached, as in glutamine. (Organisms have some other amino acids, not shown in FIGURE 5.15, but these are not incorporated into proteins.)

The physical and chemical properties of the side chain determine the unique characteristics of a particular amino acid. In

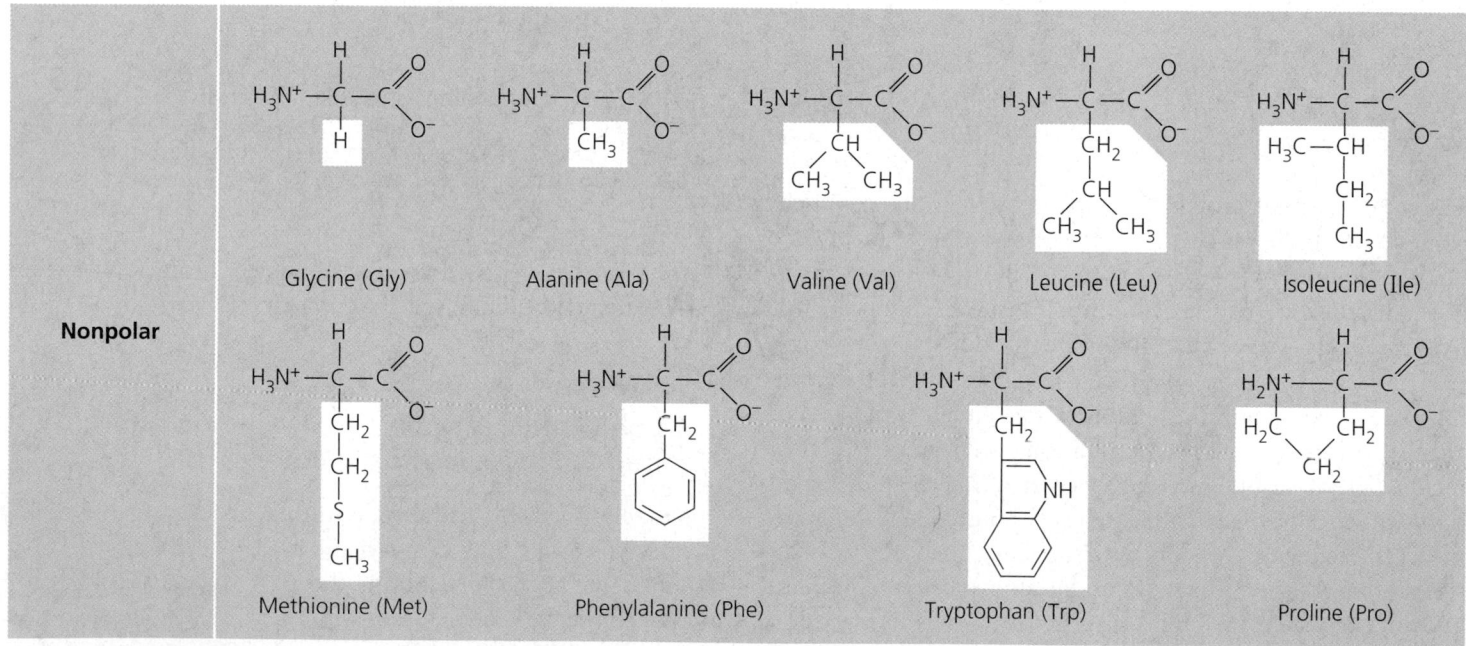

FIGURE 5.15 The 20 amino acids of proteins. The amino acids are grouped here according to the properties of their side chains (R groups), highlighted in white. (One group is shown above; the other groups are on page 73.) The amino acids are shown in their prevailing ionic forms at pH 7, the approximate pH within a cell. In parentheses are the three-letter abbreviations for the amino acids. All the amino acids used in proteins are the same enantiomer, called the L form (see FIGURE 4.6).

FIGURE 5.16 Making a polypeptide chain. (a) Peptide bonds formed by dehydration reactions link the carboxyl group of one amino acid to the amino group of the next. **(b)** The peptide bonds are formed one at a time, starting with the amino acid at the amino end (N-terminus). The polypeptide has a repetitive backbone (purple) to which the amino acid side chains are attached.

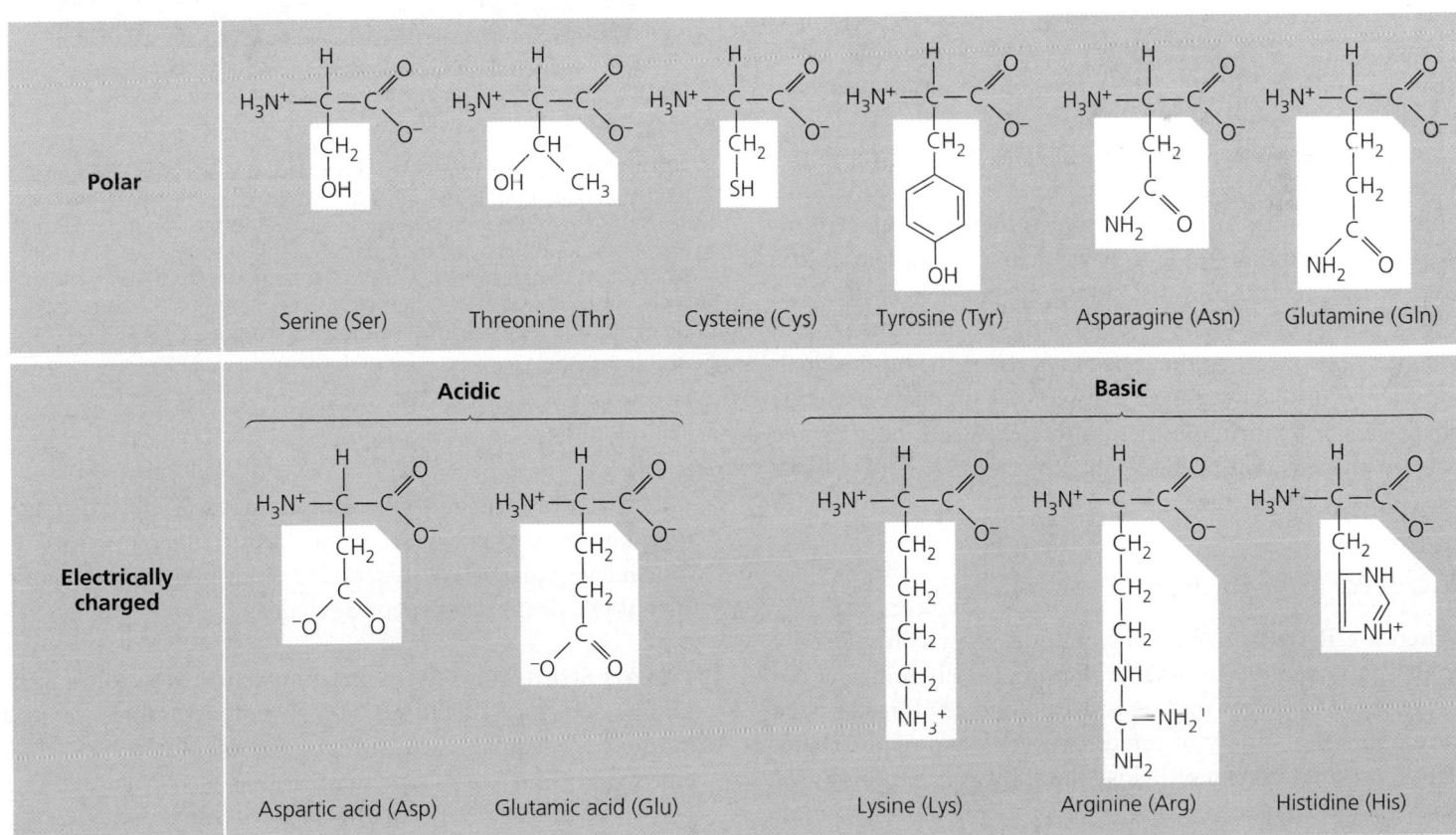

FIGURE 5.15, the amino acids are grouped according to the properties of their side chains. One group consists of amino acids with nonpolar side chains, which are hydrophobic. Another group consists of amino acids with polar side chains, which are hydrophilic. Acidic amino acids are those with side chains that are generally negative in charge owing to the presence of a carboxyl group, which is usually dissociated (ionized) at cellular pH. Basic amino acids have amino groups in their side chains that are generally positive in charge. (Notice that *all* amino acids have carboxyl groups and amino groups; the terms *acidic* and *basic* in this context refer only to the side chains.) Because they are charged, acidic and basic side chains are also hydrophilic.

Now that we have examined amino acids, let's see how they are linked to form polymers (FIGURE 5.16). When two amino acids are positioned so that the carboxyl group of one is adjacent to the amino group of the other, an enzyme can cause them to join by catalyzing a dehydration reaction. The resulting covalent bond is called a **peptide bond.** Repeated over and

FIGURE 5.15 The 20 amino acids of proteins *(continued).*

over, this process yields a polypeptide, a polymer of many amino acids linked by peptide bonds (FIGURE 5.16). At one end of the polypeptide chain is a free amino group, and at the opposite end is a free carboxyl group. Thus, the chain has an amino end (N-terminus) and a carboxyl end (C-terminus). The repeating sequence of atoms highlighted in purple in FIGURE 5.16b is called the polypeptide backbone. Attached to this backbone are different kinds of appendages, the side chains of the amino acids. Polypeptides range in length from a few monomers to a thousand or more. Each specific polypeptide has a unique linear sequence of amino acids. The immense variety of polypeptides in nature illustrates an important concept introduced earlier—that cells can make many different polymers by linking a limited set of monomers into diverse sequences.

A protein's function depends on its specific conformation

The term *polypeptide* is not quite synonymous with *protein*. The relationship is somewhat analogous to that between a long strand of yarn and a sweater of particular size and shape that one can knit from the yarn. A functional protein is not *just* a polypeptide chain, but one or more polypeptides precisely twisted, folded, and coiled into a molecule of unique shape (FIGURE 5.17). It is the amino acid sequence of a polypeptide that determines what three-dimensional conformation the protein will take. Many proteins are globular (roughly spherical), while others are fibrous in shape. However, within these broad categories, countless variations are possible.

A protein's specific conformation determines how it works. In almost every case, the function of a protein depends on its ability to recognize and bind to some other molecule. For instance, an antibody binds to a particular foreign substance that has invaded the body, and an enzyme recognizes and binds to its substrate, the substance the enzyme works on. In Chapter 2, you learned that one nerve cell in the brain signals another by dispatching specific molecules that have a unique shape. The receptor molecules on the surface of the receiving cell are proteins that fit the signal molecules something like a lock and key (see FIGURE 2.18).

Four Levels of Protein Structure

When a cell synthesizes a polypeptide, the chain generally folds spontaneously to assume the functional conformation for that protein. This folding is driven and reinforced by the formation of a variety of bonds between parts of the chain. Thus, the function of a protein—the ability of a receptor protein to identify and associate with a particular chemical messenger, for instance—is an emergent property resulting from exquisite molecular order. In the complex architecture of a

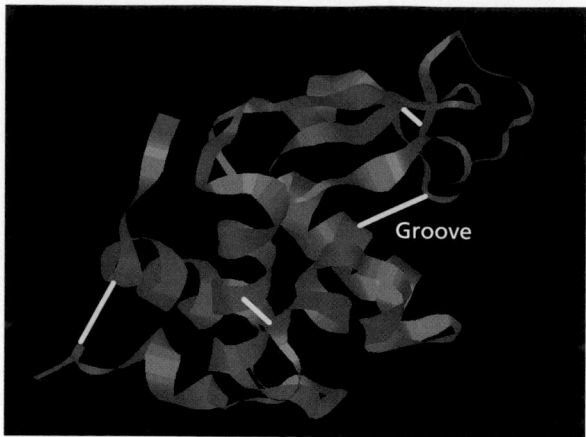

(a) A **ribbon model** shows how the single polypeptide chain folds and coils to form the functional protein.

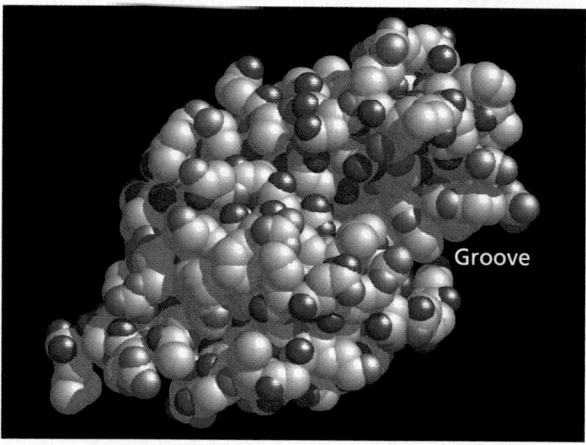

(b) A **space-filling model** shows more clearly the globular shape seen in many proteins, as well as the specific conformation unique to lysozyme.

FIGURE 5.17 Conformation of a protein, the enzyme lysozyme. Present in our sweat, tears, and saliva, lysozyme is an enzyme that helps prevent infection by binding to and destroying specific molecules on the surface of many kinds of bacteria. The groove is the part of the protein that recognizes and binds to the target molecules on bacterial walls. (In the ribbon model, the yellow lines represent one type of chemical bond that stabilizes the protein's shape.)

protein, we can recognize three superimposed levels of structure, known as primary, secondary, and tertiary structure. A fourth level, quaternary structure, arises when a protein consists of two or more polypeptide chains.

Primary Structure. The **primary structure** of a protein is its unique sequence of amino acids. As an example, we will examine the primary structure of lysozyme, the antibacterial enzyme illustrated in its three-dimensional form in FIGURE 5.17. Lysozyme is a relatively small protein, its single polypeptide chain only 129 amino acids long. In FIGURE 5.18, the polypeptide chain is unraveled for a closer look at its primary structure.

FIGURE 5.18 The primary structure of a protein. This is the unique amino acid sequence, or primary structure, of the enzyme lysozyme. The names of the amino acids are given as their three-letter abbreviations. The chain was drawn in this serpentine fashion to make the entire sequence readily visible. The actual shape of lysozyme is shown in FIGURE 5.17.

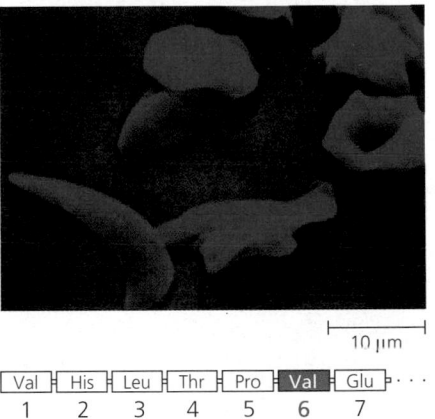

10 μm

| Val | His | Leu | Thr | Pro | Glu | Glu | · · · |
| 1 | 2 | 3 | 4 | 5 | 6 | 7 | |

(a) Normal red blood cells and the primary structure of normal hemoglobin. Normal human red blood cells are disk-shaped (with center depression), as seen in this micrograph. The first seven amino acids of one of the polypeptides of normal hemoglobin are shown below the micrograph; this polypeptide has a total of 146 amino acids.

10 μm

| Val | His | Leu | Thr | Pro | **Val** | Glu | · · · |
| 1 | 2 | 3 | 4 | 5 | 6 | 7 | |

(b) Sickled red blood cells and the primary structure of sickle-cell hemoglobin. A slight change in the primary structure of hemoglobin (the polypeptide shown in part a)—an inherited substitution at amino acid number 6—causes sickle-cell disease.

FIGURE 5.19 A single amino acid substitution in a protein causes sickle-cell disease.

A specific one of the 20 amino acids occupies each of the 129 positions along the chain. The primary structure is like the order of letters in a very long word. If left to chance, there would be 20^{129} different ways of arranging amino acids into a polypeptide chain of this length. However, the precise primary structure of a protein is determined not by the random linking of amino acids, but by inherited genetic information.

Even a slight change in primary structure can affect a protein's conformation and ability to function. For instance, the substitution of one amino acid for another at a particular position in the primary structure of hemoglobin, the protein that carries oxygen in red blood cells, causes sickle-cell disease, an inherited blood disorder. Normal red blood cells are disk-shaped, but in sickle-cell disease, the abnormal hemoglobin

molecules tend to crystallize, deforming some of the cells into a sickle shape (FIGURE 5.19). The life of someone with the disease is punctuated by "sickle-cell crises," which occur when the angular cells clog tiny blood vessels, impeding blood flow.

The pioneer in determining the primary structure of proteins was Frederick Sanger, who, with his colleagues at Cambridge University in England, worked out the amino acid sequence of the hormone insulin in the late 1940s and early 1950s. His approach was to use protein-digesting enzymes and other catalysts that break polypeptides at specific places rather than completely hydrolyzing the chains. Treatment with one of these agents cleaves a polypeptide into fragments that can be separated by a technique called chromatography. Hydrolysis with another agent breaks the polypeptide at different sites,

THE PROCESS OF SCIENCE

yielding a second group of fragments. Sanger used chemical methods to determine the sequence of amino acids in these small fragments. Then he searched for overlapping regions among the pieces obtained by hydrolyzing with the different agents. Consider, for instance, two fragments with the following sequences:

Cys-Ser-Leu-Tyr-Gln-Leu

Tyr-Gln-Leu-Glu-Asn

We can deduce from the overlapping regions that the intact polypeptide contains in its primary structure the following segment:

Cys-Ser-Leu-Tyr-Gln-Leu-Glu-Asn

Just as we could reconstruct this sentence from a collection of fragments with overlapping sequences of letters, Sanger and his co-workers were able, after years of effort, to reconstruct the complete primary structure of insulin. Since then, most of the steps involved in sequencing a polypeptide have been automated. However, it was Sanger's analysis of insulin that first demonstrated what is now a fundamental axiom of molecular biology: A protein has a unique primary structure, a precise sequence of amino acids.

Secondary Structure. Most proteins have segments of their polypeptide chain repeatedly coiled or folded in patterns that contribute to the protein's overall conformation. These coils and folds, collectively referred to as **secondary structure,** are the result of hydrogen bonds at regular intervals along the polypeptide backbone (FIGURE 5.20). Only the atoms of the backbone are involved, not the amino acid side chains. Both the oxygen and the nitrogen atoms of the backbone are electronegative, with partial negative charges (see Chapter 2). The weakly positive hydrogen atom attached to the nitrogen atom has an affinity for the oxygen atom of a nearby peptide bond. Individually, these hydrogen bonds are weak, but because they are repeated many times over a relatively long region of the polypeptide chain, they can support a particular shape for that part of the protein. One such secondary structure is the **α helix,** a delicate coil held together by hydrogen bonding between every fourth amino acid. The regions of α helix in the enzyme lysozyme are evident in FIGURE 5.20, where one α helix is enlarged to show the hydrogen bonds. Lysozyme is fairly typical of a globular protein in that it has a few stretches of α helix separated by nonhelical regions. In contrast, some fibrous proteins, such as α-keratin, the structural protein of hair, have the α-helix formation over most of their length.

The other main type of secondary structure is the **β pleated sheet,** in which two or more regions of the polypeptide chain lie parallel to each other. Hydrogen bonds between the parts of the backbone in the parallel regions hold the structure together. Pleated sheets make up the core of many globular proteins, and we can see one such region in lysozyme in FIGURE 5.20. Also, pleated sheets dominate some fibrous proteins,

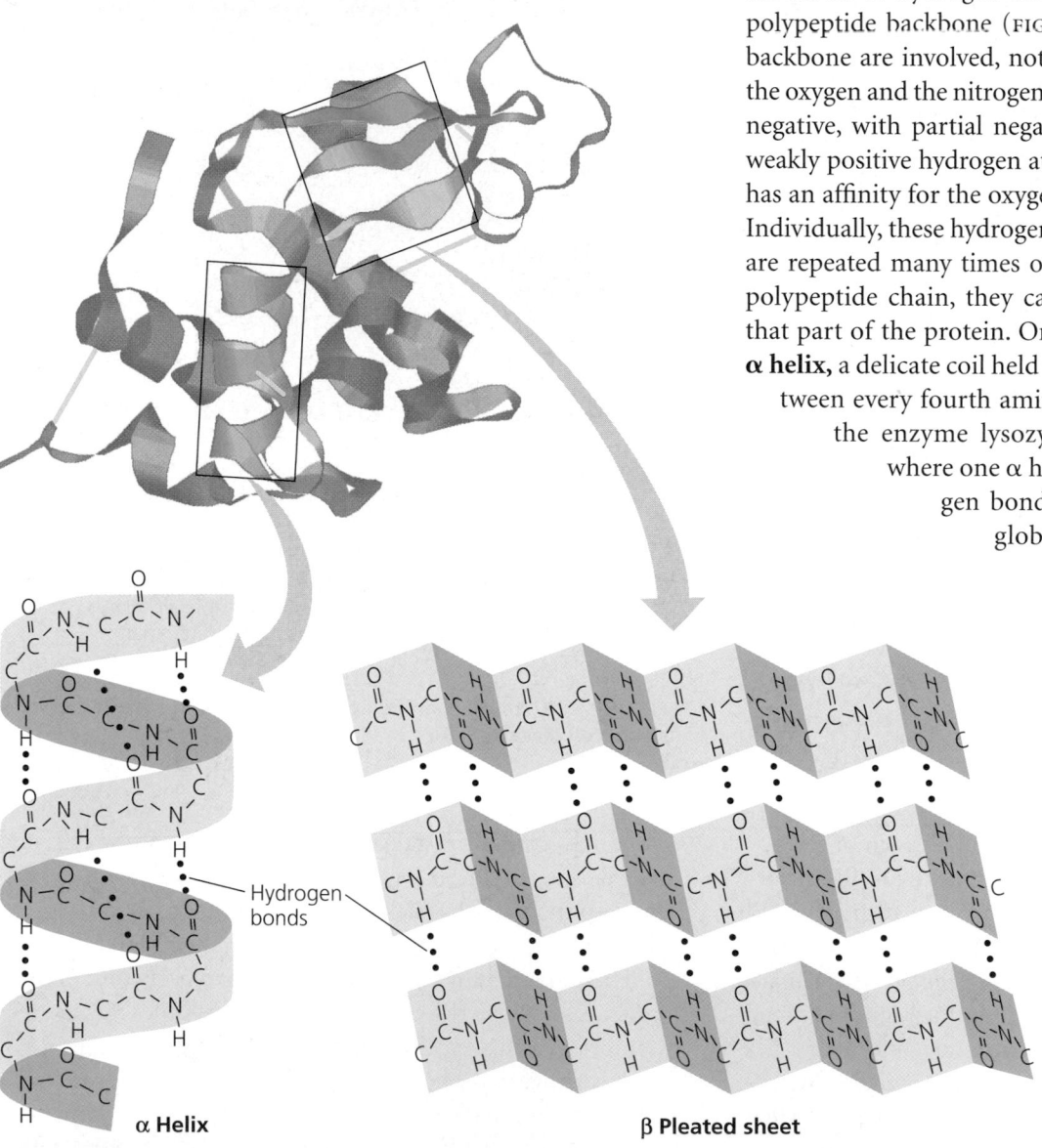

Hydrogen bonds

α Helix

β Pleated sheet

FIGURE 5.20 The secondary structure of a protein. Two types of secondary structure, the α helix and the β pleated sheet, can both be found in the protein lysozyme. Both patterns depend on hydrogen bonding between $>C=O$ and $>N-H$ groups along the polypeptide backbone. The R groups of the amino acids are omitted in these diagrams, as are the hydrogen atoms not involved in hydrogen bonding.

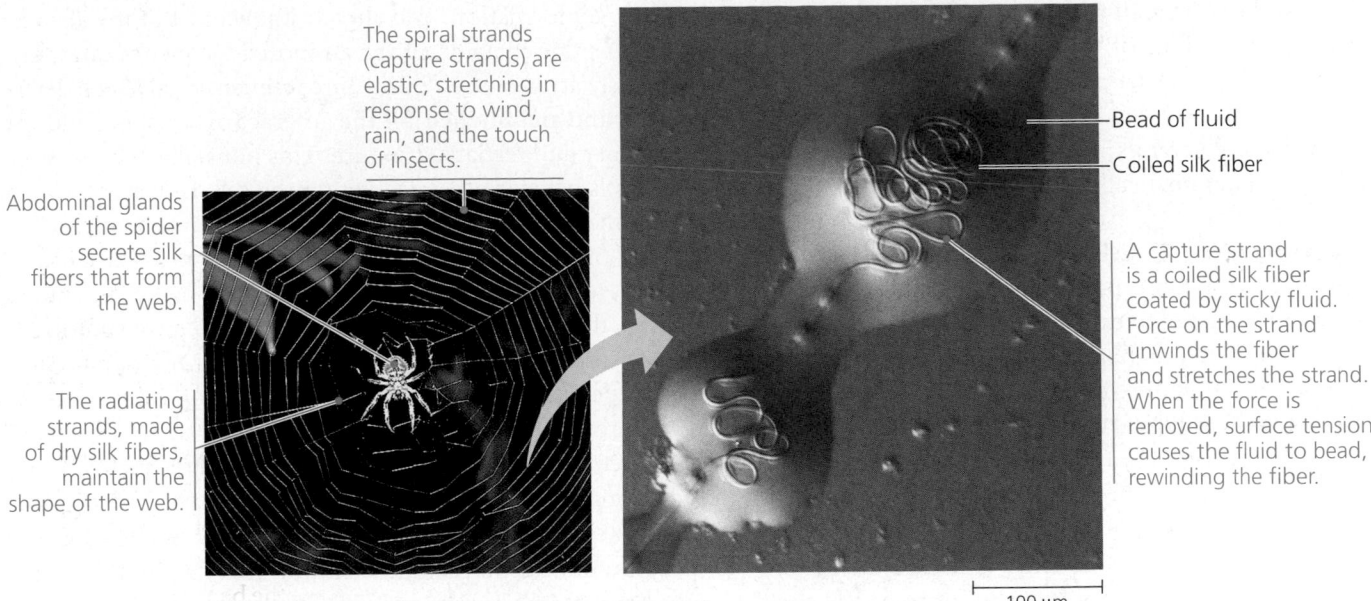

The spiral strands (capture strands) are elastic, stretching in response to wind, rain, and the touch of insects.

Abdominal glands of the spider secrete silk fibers that form the web.

The radiating strands, made of dry silk fibers, maintain the shape of the web.

Bead of fluid

Coiled silk fiber

A capture strand is a coiled silk fiber coated by sticky fluid. Force on the strand unwinds the fiber and stretches the strand. When the force is removed, surface tension causes the fluid to bead, rewinding the fiber.

100 μm

FIGURE 5.21 Spider silk: a structural protein. The silk protein owes its strength and resilience largely to its secondary structure: It contains many regions of β pleated sheets. When an insect is unfortunate enough to fly into the web, the silk fibers first uncoil and stretch, absorbing the shock, and then rewind, trapping the prey.

including the silk produced by many insects and spiders (FIG-URE 5.21). The silk protein of a spider's web contains many regions of β pleated sheet. The teamwork of so many hydrogen bonds makes each silk fiber stronger than steel.

Tertiary Structure.   Superimposed on the patterns of secondary structure is a protein's **tertiary structure,** consisting of irregular contortions from interactions between side chains (R groups) of the various amino acids (FIGURE 5.22). One of the types of interaction that contributes to tertiary structure is—somewhat misleadingly—called a **hydrophobic interaction.** As a polypeptide folds into its functional conformation, amino acids with hydrophobic (nonpolar) side chains usually end up in clusters at the core of the protein, out of contact with water. Thus, what we call a hydrophobic interaction is actually caused by the action of water molecules, which exclude nonpolar substances as they hydrogen-bond to one another and to hydrophilic parts of the protein. Once nonpolar amino acid side chains are close together, van der Waals interactions help hold them together. Meanwhile, hydrogen bonds between polar side chains and ionic bonds between positively and negatively charged side chains also help stabilize tertiary structure. These are all weak interactions, but their cumulative effect helps give the protein a specific shape.

The conformation of a protein may be reinforced further by strong, covalent bonds called **disulfide bridges.** Disulfide bridges form where two cysteine monomers, amino acids with sulfhydryl groups (—SH) on their side chains, are brought

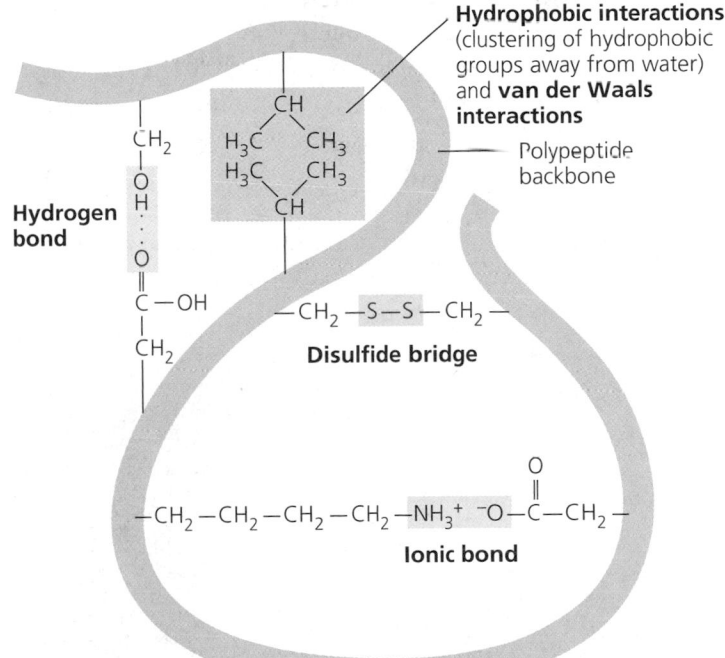

FIGURE 5.22 Examples of interactions contributing to the tertiary structure of a protein. Hydrophobic side chains usually end up in the interior of a protein, away from water. Along with this clustering of hydrophobic groups, misleadingly called hydrophobic interactions, hydrogen bonds, ionic bonds, and van der Waals interactions are all weak interactions (weak bonds) between side chains that collectively hold the protein in a specific conformation. Much stronger are the disulfide bridges, covalent bonds between the side chains of two cysteine amino acids. This diagram shows only one small part of a hypothetical protein.

close together by the folding of the protein. The sulfur of one cysteine bonds to the sulfur of the second, and the disulfide bridge (—S—S—) rivets parts of the protein together. (The yellow lines in FIGURES 5.17a and 5.20 represent disulfide bridges.) All of these different kinds of bonds can occur in one protein, as shown diagramatically in FIGURE 5.22.

Quaternary Structure.   As mentioned previously, some proteins consist of two or more polypeptide chains aggregated into one functional macromolecule. **Quaternary structure** is the overall protein structure that results from the aggregation of these polypeptide subunits. For example, collagen is a fibrous protein that has helical subunits supercoiled into a larger triple helix (FIGURE 5.23a). This supercoiled organization of collagen, which is similar to the construction of a rope, gives the long fibers great strength. This suits collagen fibers to their function as the girders of connective tissue in skin, bone, tendons, ligaments, and other body parts. Hemoglobin, the oxygen-binding protein of red blood cells, is an example of a globular protein with quaternary structure (FIGURE 5.23b). It consists of two kinds of polypeptide chains, with two of each kind per hemoglobin molecule.

We have taken the reductionist approach in dissecting proteins to their four levels of structural organization. However, it is the overall product, a macromolecule with a unique shape, that works in a cell. The specific function of a protein is an emergent property that arises from the architecture of the molecule. FIGURE 5.24 reviews the levels of protein structure.

What Determines Protein Conformation?

You've learned that unique conformation endows each protein with a specific function, but what are the key factors deter-mining conformation? You already know most of the answer: A polypeptide chain of a given amino acid sequence can spontaneously arrange itself into a three-dimensional shape determined and maintained by the interactions responsible for secondary and tertiary structure. This normally occurs as the protein is being synthesized within the cell. However, protein conformation also depends on the physical and chemical conditions of the protein's environment. If the pH, salt concentration, temperature, or other aspects of its environment are altered, the protein may unravel and lose its native conformation, a change called **denaturation** (FIGURE 5.25). Because it is misshapen, the denatured protein is biologically inactive. Most proteins become denatured if they are transferred from an aqueous environment to an organic solvent, such as ether or chloroform; the protein turns inside out, its hydrophobic regions changing places with its hydrophilic portions. Other denaturation agents include chemicals that disrupt the hydrogen bonds, ionic bonds, and disulfide bridges that maintain a protein's shape. Denaturation can also result from excessive heat, which agitates the polypeptide chain enough to overpower the weak interactions that stabilize conformation. The white of an egg becomes opaque during cooking because the denatured proteins are insoluble and solidify.

When a protein in a test-tube solution has been denatured by heat or chemicals, it will often return to its functional shape when the denaturing agent is removed. We can conclude that the information for building specific shape is intrinsic in the protein's primary structure. The sequence of amino acids determines conformation—where an α helix can form, where β pleated sheets can occur, where disulfide bridges are located, where ionic bonds can form, and so on. However, in the crowded environment inside a cell, correct folding may be more of a problem than it is in a test tube.

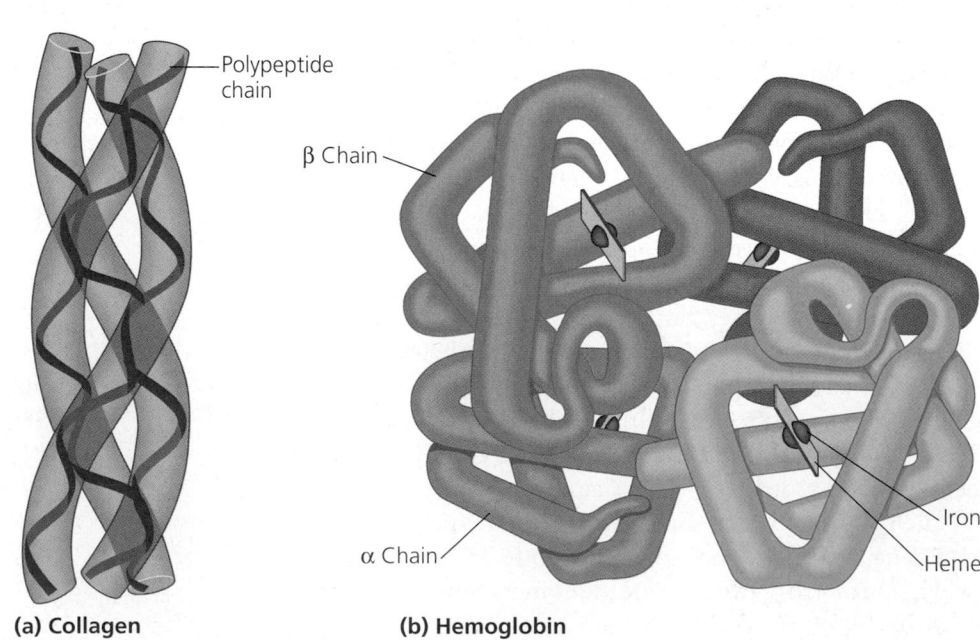

FIGURE 5.23 The quaternary structure of proteins. At this level of structure, two or more polypeptide subunits associate to form a functional protein. **(a)** Collagen is a fibrous protein consisting of three helical polypeptides that are supercoiled to form a ropelike structure of great strength. Accounting for 40% of the protein in the human body, collagen strengthens connective tissue throughout the body. **(b)** Hemoglobin is a globular protein with four polypeptide subunits, two of one kind (α chains) and two of another kind (β chains). Both α and β subunits consist primarily of α-helical secondary structure, represented by the thicker cylindrical sections of the polypeptides in this model. (Each subunit has a nonpolypeptide component, called heme, with an iron atom that binds oxygen.)

Polypeptide chain

β Chain

α Chain

Iron

Heme

(a) Collagen

(b) Hemoglobin

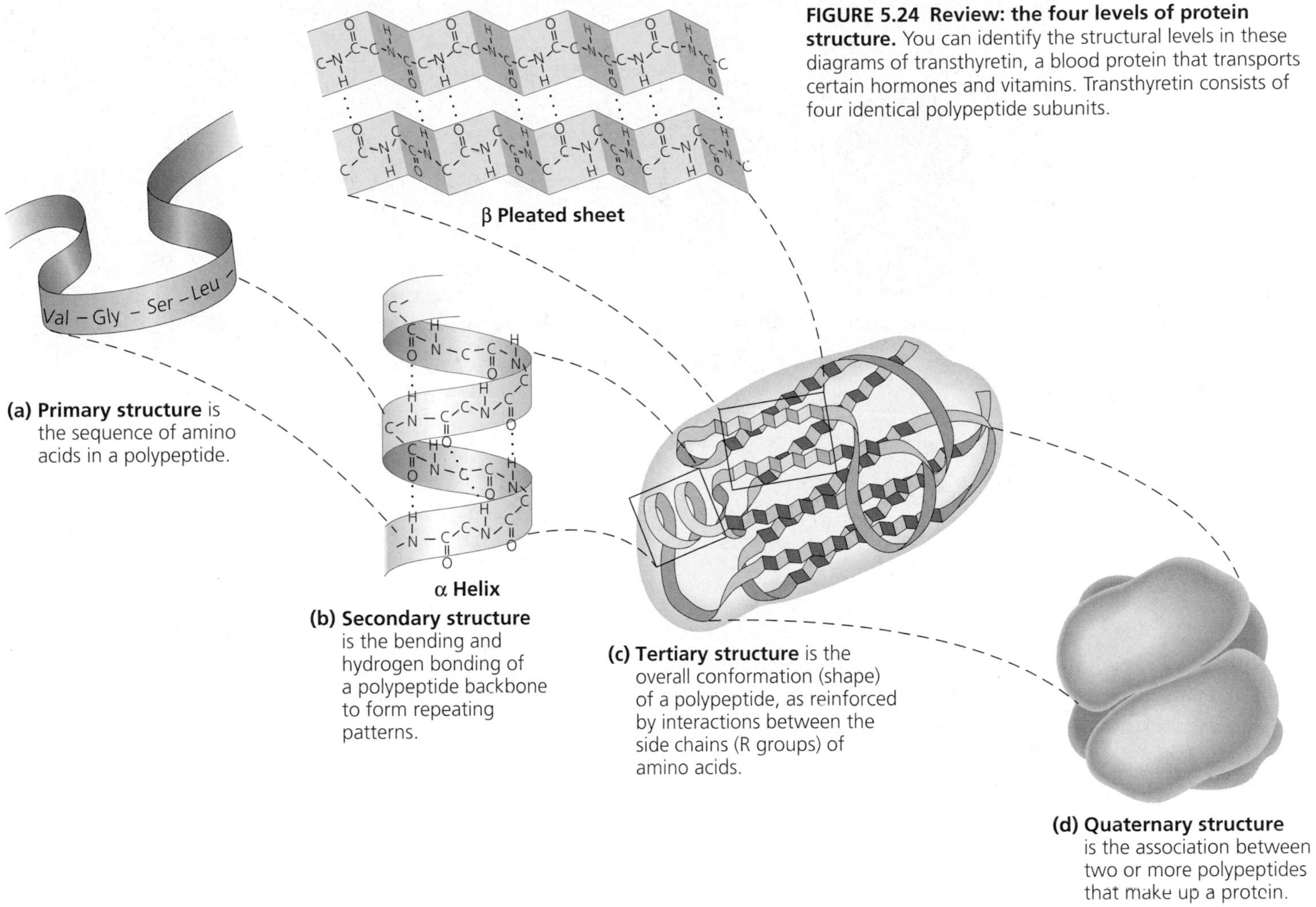

FIGURE 5.24 Review: the four levels of protein structure. You can identify the structural levels in these diagrams of transthyretin, a blood protein that transports certain hormones and vitamins. Transthyretin consists of four identical polypeptide subunits.

β Pleated sheet

α Helix

(a) Primary structure is the sequence of amino acids in a polypeptide.

(b) Secondary structure is the bending and hydrogen bonding of a polypeptide backbone to form repeating patterns.

(c) Tertiary structure is the overall conformation (shape) of a polypeptide, as reinforced by interactions between the side chains (R groups) of amino acids.

(d) Quaternary structure is the association between two or more polypeptides that make up a protein.

The Protein-Folding Problem

Biochemists now know the amino acid sequences of more than 100,000 proteins and the three-dimensional shapes of about 10,000. One would think that by correlating the primary structures of many proteins with their conformations, it would be relatively easy to discover the rules of protein folding. Unfortunately, the protein-folding problem is not that simple. Most proteins probably go through several intermediate states on their way to a stable conformation, and looking at the "mature" conformation does not reveal the stages of folding that are required to achieve that form. However, biochemists have developed methods for tracking a protein through its intermediate stages of folding. Researchers have also discovered **chaperonins** (also called chaperone proteins), protein molecules that assist the proper folding of other proteins. Chaperonins do not actually specify the correct final

Denaturation

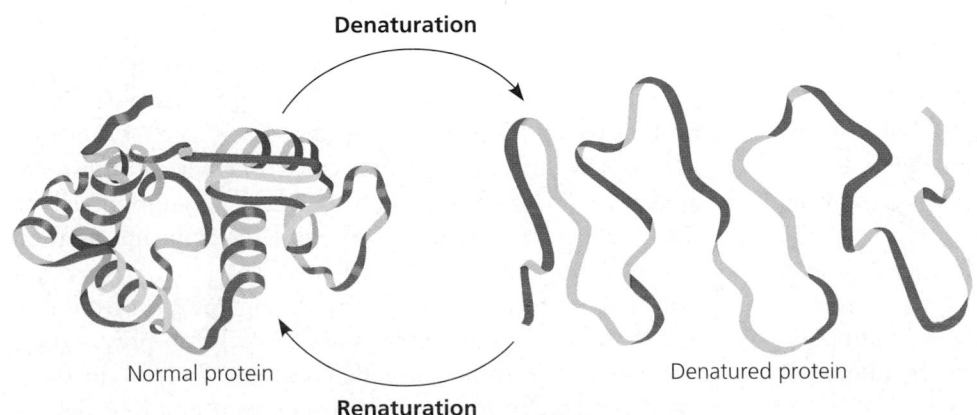

Normal protein

Denatured protein

Renaturation

FIGURE 5.25 Denaturation and renaturation of a protein. High temperatures or various chemical treatments will denature a protein, causing it to lose its conformation and hence its ability to function. If the denatured protein remains dissolved, it can often renature when the chemical and physical aspects of its environment are restored to normal.

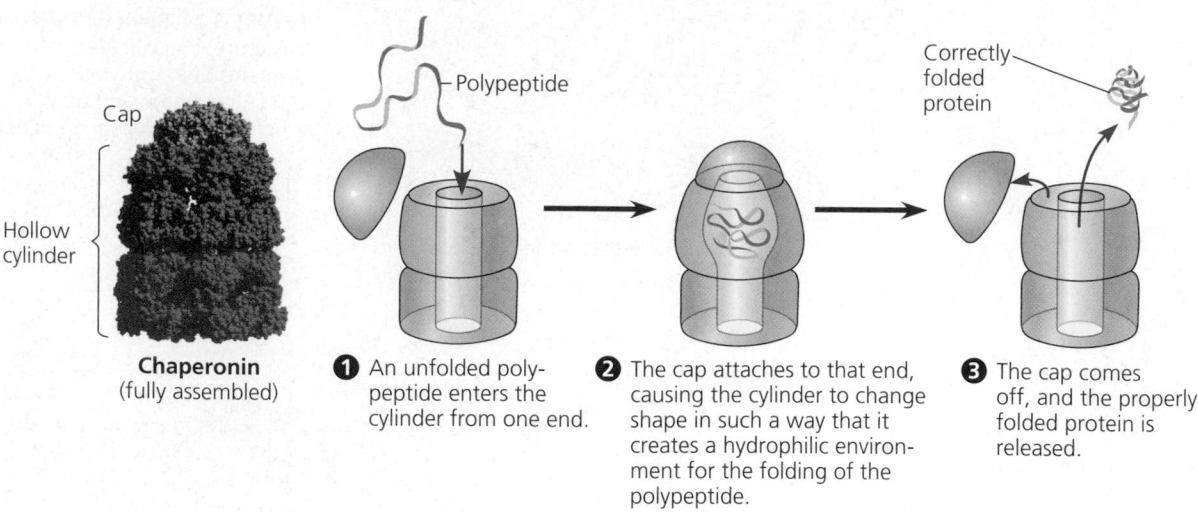

FIGURE 5.26
A chaperonin in action. The computer graphic shows a large chaperonin protein "complex" with an interior space that provides a shelter for the proper folding of newly made polypeptides. The complex consists of two proteins, with a total of 21 polypeptide subunits and a mass of almost 900,000 daltons! One protein is a hollow cylinder; the other is a cap that can fit on either end.

Cap

Hollow cylinder

Chaperonin (fully assembled)

Polypeptide

❶ An unfolded polypeptide enters the cylinder from one end.

❷ The cap attaches to that end, causing the cylinder to change shape in such a way that it creates a hydrophilic environment for the folding of the polypeptide.

Correctly folded protein

❸ The cap comes off, and the properly folded protein is released.

structure of a polypeptide. Instead they work by keeping the new polypeptide segregated from "bad influences" in the cytoplasmic environment while it folds spontaneously. One well-studied chaperonin, from the bacterium *E. coli,* is a giant multiprotein complex shaped like a hollow cylinder. The cavity provides a shelter for folding polypeptides of various types (FIGURE 5.26).

Beyond the reach of today's computers, the simulation of protein folding is the goal of a five-year project at the IBM Corporation. Scientists at IBM aim to develop an extraordinarily powerful supercomputer, to be called Blue Gene, able to generate the three-dimensional structure of any protein starting from its amino acid sequence (or from the gene sequence that encodes the amino acid sequence). Among the practical benefits of this project may be the discovery of protein-folding principles that can be used to design custom proteins for medical and other uses.

Determining the Structure of a Protein

THE PROCESS OF SCIENCE

Even when scientists have an actual protein in hand, determining its exact three-dimensional structure is not simple, for a single protein molecule is built of thousands of atoms. FIGURE 5.27 describes **X-ray crystallography,** the main method used for the task. X-ray crystallography depends on the diffraction (deflection) of an X-ray beam by the individual atoms in a crystal of the protein. After the spatial coordinates of the atoms are determined in this way, a model of the protein is built. Linus Pauling and other pioneers of molecular biology built models from wood, wire, and plastic model sets. Computers have made it possible to build models much more quickly.

NUCLEIC ACIDS—INFORMATIONAL POLYMERS

If the primary structure of polypeptides determines the conformation of a protein, what determines primary structure? The amino acid sequence of a polypeptide is programmed by a unit of inheritance known as a **gene.** Genes consist of DNA, which is a polymer belonging to the class of compounds known as **nucleic acids.**

Nucleic acids store and transmit hereditary information

There are two types of nucleic acids: **deoxyribonucleic acid (DNA)** and **ribonucleic acid (RNA).** These are the molecules that enable living organisms to reproduce their complex components from one generation to the next. Unique among molecules, DNA provides directions for its own replication. DNA also directs RNA synthesis and, through RNA, controls protein synthesis.

DNA is the genetic material that organisms inherit from their parents. A DNA molecule is very long and usually consists of hundreds or thousands of genes. When a cell reproduces itself by dividing, its DNA molecules (one per chromosome) are copied and passed along from one generation of cells to the next. Encoded in the structure of DNA is the information that programs all the cell's activities. The DNA, however, is not directly involved in running the operations of the cell, any more than computer software by itself can print a bank statement or read the bar code on a box of cereal. Just as a printer is needed to print out a statement and a scanner is

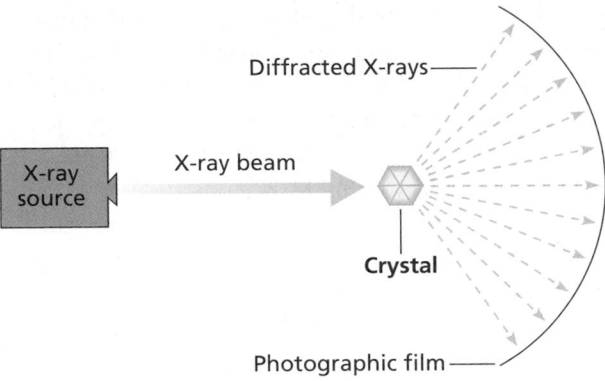

❶ X-ray crystallography. An instrument aims an X-ray beam through the protein crystal. The regularly spaced atoms of the crystal diffract (deflect) the X-rays into an orderly array.

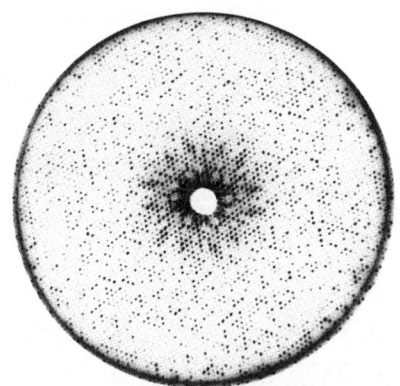

❷ X-ray diffraction pattern from the crystal of a protein. The diffracted X-rays expose photographic film, producing a pattern of spots.

❸ Electron density map. From such diffraction patterns, computers generate electron density maps of successive, cross-sectional slices through the protein. By combining the information from electron density maps with the primary structure of the protein, as determined by chemical methods, it is possible to plot the three-dimensional (x, y, z) coordinates of each atom.

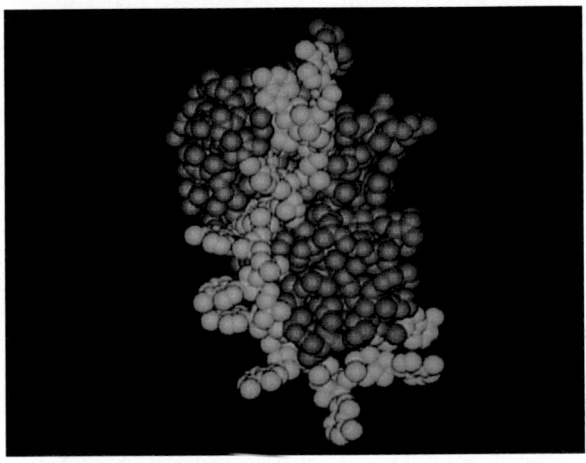

❹ A computer graphic model of the protein ribonuclease (purple) bound to a short strand of nucleic acid (green). Finally, graphics software enables the computer to create a picture showing the position of each atom in the molecule. The scientist can use the software to view the molecule's appearance from various angles.

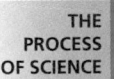

THE PROCESS OF SCIENCE

FIGURE 5.27 X-ray crystallography. This series of illustrations (from the Department of Biochemistry at the University of California, Riverside) shows how scientists determine the three-dimensional structure of a protein using X-ray crystallography. The protein here is an enzyme called ribonuclease, whose function involves binding to a nucleic acid molecule. Before X-ray crystallography can be carried out, the protein must be crystallized, in this case in combination with a short strand of nucleic acid. Computer analysis of the results of X-ray crystallography result in a map of all the atoms of the molecules in three-dimensional space. Finally, the scientists use other computer software to generate a three-dimensional model of the enzyme and nucleic acid.

needed to read a bar code, proteins are required to implement genetic programs. The molecular hardware of the cell—the tools for most biological functions—consists of proteins. For example, it is the protein hemoglobin that carries oxygen in the blood, not the DNA that specifies the structure of hemoglobin.

How does RNA, the other type of nucleic acid, fit into the flow of genetic information from DNA to proteins? Each gene along the length of a DNA molecule directs the synthesis of a type of RNA called messenger RNA (mRNA). The mRNA molecule then interacts with the cell's protein-synthesizing machinery to direct the production of a polypeptide. We can summarize the flow of genetic information as DNA → RNA → protein (FIGURE 5.28, p. 82). The actual sites of protein synthesis are cellular structures called ribosomes. In a eukaryotic cell, ribosomes are located in the cytoplasm, but DNA resides in the nucleus. Messenger RNA conveys the genetic instructions for building proteins from the nucleus to the cytoplasm. Prokaryotic cells lack nuclei, but they still use RNA to send a message from the DNA to the ribosomes and other equipment of the cell that translate the coded information into amino acid sequences.

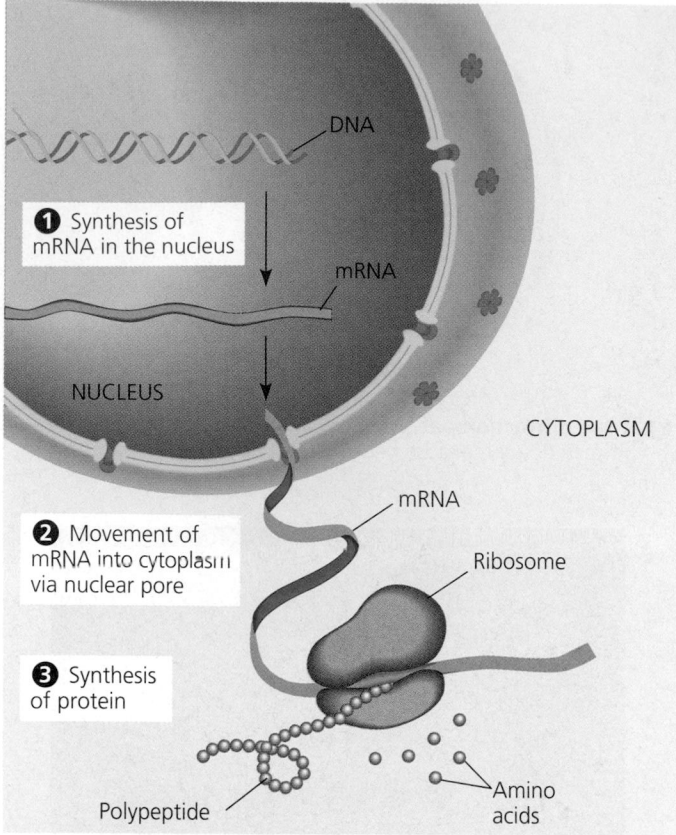

① Synthesis of mRNA in the nucleus

DNA

mRNA

NUCLEUS

CYTOPLASM

② Movement of mRNA into cytoplasm via nuclear pore

mRNA

Ribosome

③ Synthesis of protein

Polypeptide

Amino acids

FIGURE 5.28 DNA ⟶ RNA ⟶ protein: a diagrammatic overview of information flow in a cell. In a eukaryotic cell, DNA in the nucleus programs protein production in the cytoplasm by dictating the synthesis of messenger RNA (mRNA), which travels to the cytoplasm and binds to ribosomes. As a ribosome (greatly enlarged in this drawing) moves along the mRNA, the genetic message is translated into a polypeptide of specific amino acid sequence.

A nucleic acid strand is a polymer of nucleotides

Nucleic acids are polymers of monomers called **nucleotides.** Each nucleotide is itself composed of three parts: an organic molecule called a nitrogenous base, a pentose (five-carbon sugar), and a phosphate group (FIGURE 5.29a).

There are two families of nitrogenous bases: pyrimidines and purines. A **pyrimidine** has a six-membered ring of carbon and nitrogen atoms. (The nitrogen atoms tend to take up H⁺ from solution, which explains the term *nitrogenous base.*) The members of the pyrimidine family are cytosine (C), thymine (T), and uracil (U). **Purines** are larger, with the six-membered ring fused to a five-membered ring. The purines are adenine (A) and guanine (G). The specific pyrimidines and purines differ in the functional groups attached to the rings. Adenine, guanine, and cytosine are found in both types of nucleic acid. Thymine is found only in DNA and uracil only in RNA.

The pentose connected to the nitrogenous base is **ribose** in the nucleotides of RNA and **deoxyribose** in DNA. The only difference between these two sugars is that deoxyribose lacks an oxygen atom on its number 2 carbon—hence its name.

So far, we have built a nucleoside, which is a nitrogenous base joined to a sugar. To complete the construction of a nucleotide, we attach a phosphate group to the number 5 carbon of the sugar (FIGURE 5.29b). The molecule is now a nucleoside monophosphate, better known as a nucleotide.

In a nucleic acid polymer, or **polynucleotide,** nucleotides are joined by covalent bonds called phosphodiester linkages between the phosphate of one nucleotide and the sugar of the next. This bonding results in a backbone with a repeating pattern of sugar-phosphate units (FIGURE 5.29c). All along this sugar-phosphate backbone are appendages consisting of the nitrogenous bases.

The sequence of bases along a DNA (or mRNA) polymer is unique for each gene. Because genes are hundreds to thousands of nucleotides long, the number of possible base sequences is effectively limitless. A gene's meaning to the cell is encoded in its specific sequence of the four DNA bases. For example, the sequence AGGTAACTT means one thing, whereas the sequence CGCTTTAAC has a different translation. (Real genes, of course, are much longer.) The linear order of bases in a gene specifies the amino acid sequence—the primary structure—of a protein, which in turn specifies that protein's three-dimensional conformation and function in the cell.

Inheritance is based on replication of the DNA double helix

The RNA molecules of cells consist of a single polynucleotide chain like the one shown in FIGURE 5.29. In contrast, cellular DNA molecules have two polynucleotides that spiral around an imaginary axis to form a **double helix** (FIGURE 5.30, p. 84). James Watson and Francis Crick, working at Cambridge University, first proposed the double helix as the three-dimensional structure of DNA in 1953. The two sugar-phosphate backbones are on the outside of the helix, and the nitrogenous bases are paired in the interior of the helix. The two polynucleotides, or strands, as they are called, are held together by hydrogen bonds between the paired bases and by van der Waals interactions between the stacked bases. Most DNA molecules are very long, with thousands or even millions of base pairs connecting the two chains. One long DNA double helix includes many genes, each one a particular segment of the molecule.

Only certain bases in the double helix are compatible with each other. Adenine (A) always pairs with thymine (T), and guanine (G) always pairs with cytosine (C). If we were to read

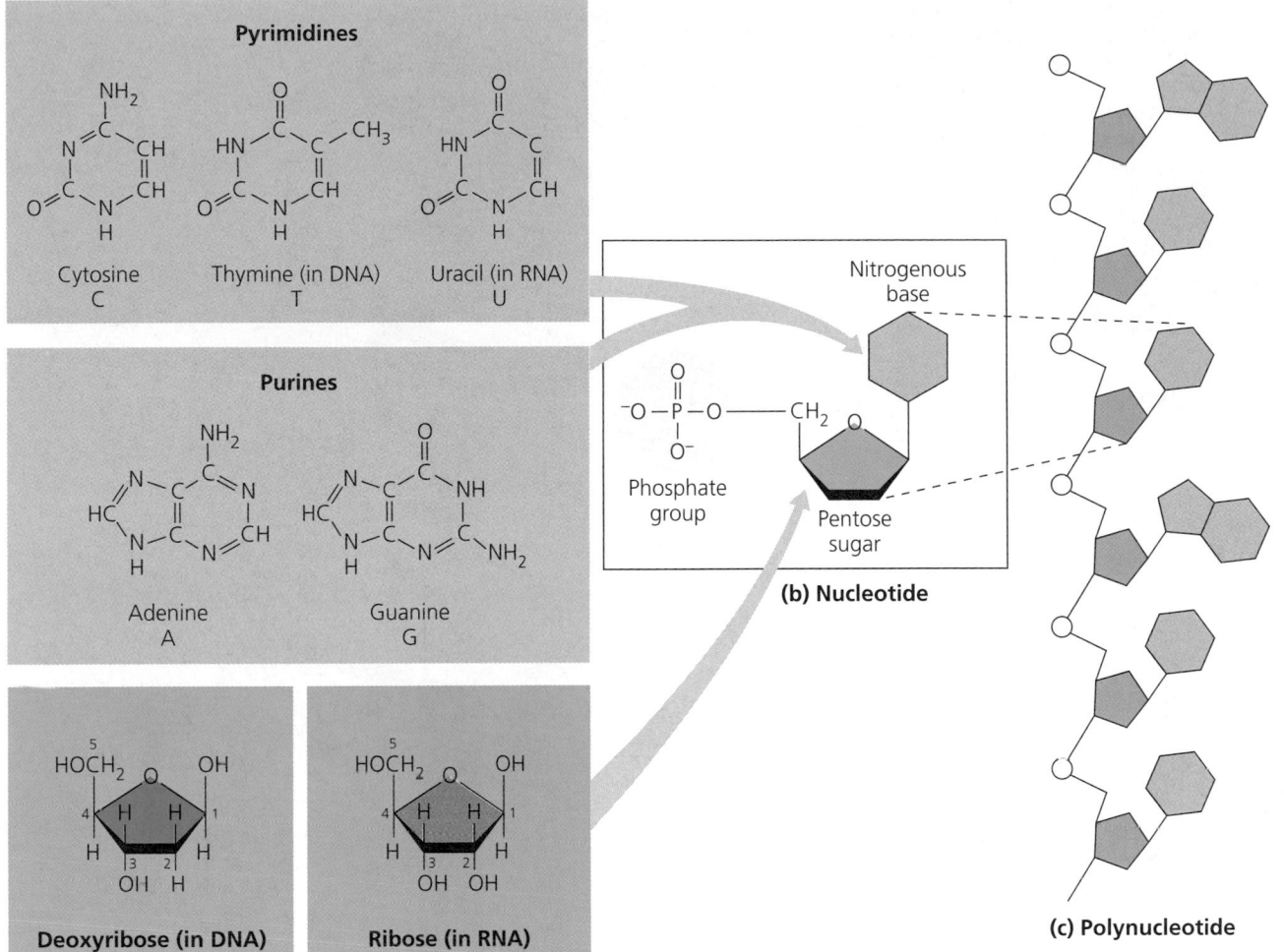

Pyrimidines

Cytosine
C

Thymine (in DNA)
T

Uracil (in RNA)
U

Purines

Adenine
A

Guanine
G

Deoxyribose (in DNA)

Ribose (in RNA)

(a) Nucleotide components

Phosphate group

Nitrogenous base

Pentose sugar

(b) Nucleotide

(c) Polynucleotide

FIGURE 5.29 **The components of nucleic acids. (a)** Nucleotides, the monomers of nucleic acids, are themselves composed of three smaller molecular building blocks: a nitrogenous base (either a purine or a pyrimidine), a pentose sugar, (either deoxyribose or ribose), and a phosphate group. RNA has ribose as its sugar, and DNA has deoxyribose. Also, uracil is found only in RNA and thymine is found only in DNA. **(b)** The three components of a nucleotide are linked together as shown here. **(c)** In polynucleotides, each nucleotide monomer has its phosphate group bonded to the sugar of the next nucleotide. The polymer has a regular sugar-phosphate backbone with variable appendages, the four kinds of nitrogenous bases. RNA usually exists in the form of a single polynucleotide, like the one shown here.

the sequence of bases along one strand as we traveled the length of the double helix, we would know the sequence of bases along the other strand. If a stretch of one strand has the base sequence AGGTCCG, then the base-pairing rules tell us that the same stretch of the other strand must have the sequence TCCAGGC. The two strands of the double helix are *complementary,* each the predictable counterpart of the other. It is this feature of DNA that makes possible the precise copying of

genes that is responsible for inheritance (FIGURE 5.30, p. 84). In preparation for cell division, each of the two strands of a DNA molecule serves as a template to order nucleotides into a new complementary strand. The result is two identical copies of the original double-stranded DNA molecule, which are then distributed to the two daughter cells. Thus, the structure of DNA accounts for its function in transmitting genetic information whenever a cell reproduces.

FIGURE 5.30 The DNA double helix and its replication. The DNA molecule is usually double-stranded, with the sugar-phosphate backbone of the polynucleotides (symbolized here by blue ribbons) on the outside of the helix. In the interior are pairs of nitrogenous bases, holding the two strands together by hydrogen bonds. Hydrogen bonding between the bases is specific. As illustrated here with symbolic shapes for the bases, adenine (A) can pair only with thymine (T), and guanine (G) can pair only with cytosine (C). As a cell prepares to divide, the two strands of the double helix separate, and each serves as a template for the precise ordering of nucleotides into new complementary strands (orange). Each DNA strand in this figure is the structural equivalent of the polynucleotide diagrammed in FIGURE 5.29c.

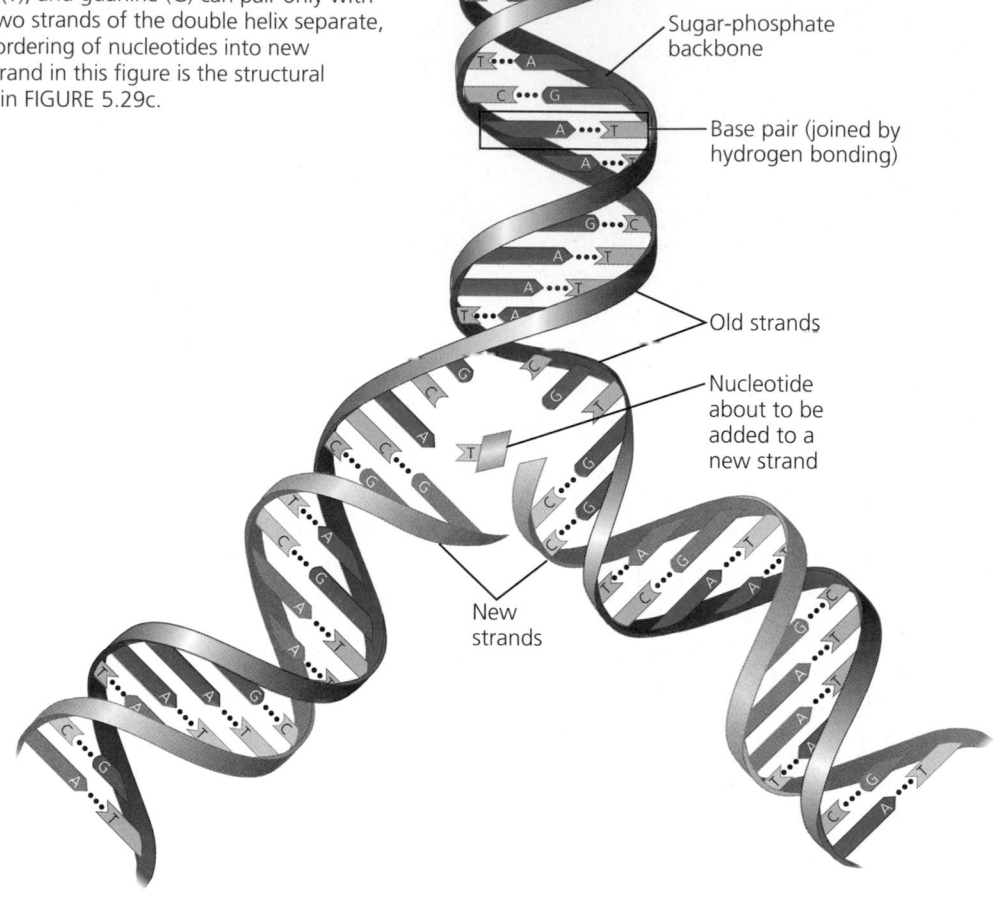

Sugar-phosphate backbone

Base pair (joined by hydrogen bonding)

Old strands

Nucleotide about to be added to a new strand

New strands

We can use DNA and proteins as tape measures of evolution

Genes (DNA) and their products (proteins) document the hereditary background of an organism. The linear sequences of nucleotides in DNA molecules are passed from parents to offspring, and these DNA sequences determine the amino acid sequences of proteins. Siblings have greater similarity in their DNA and proteins than do unrelated individuals of the same species. If the evolutionary view of life is valid, we should be able to extend this concept of "molecular genealogy" to relationships *between* species: We should expect two species that appear to be closely related based on fossil and anatomical evidence to also share a greater proportion of their DNA and protein sequences than do more distantly related species. That is the case. For example, TABLE 5.2 compares a polypeptide chain of human hemoglobin with the corresponding hemoglobin polypeptide in five other vertebrates. In this chain of 146 amino acids, humans and gorillas differ in just one amino acid. More distantly related species have chains that are less similar. Molecular biology has added a new tape measure to the toolkit biologists use to assess evolutionary kinship.

◼   ◼   ◼

We have concluded our survey of macromolecules, but not our study of the chemistry of life. Applying the reductionist strategy, we have examined the architecture of molecules, but we have yet to explore the dynamic interactions between molecules that result in the biochemical changes collectively called cellular metabolism. Chapter 6, the final chapter of this unit, will take us another step up the hierarchy of biological order by introducing the fundamental principles of metabolism.

Table 5.2 Polypeptide Sequence as Evidence for Evolutionary Relationships	
Species	**Number of Amino Acid Differences in the β Chain of Hemoglobin, Compared to Human Hemoglobin** (total chain length = 146 amino acids)
Human	0
Gorilla	1
Gibbon	2
Rhesus monkey	8
Mouse	27
Frog	67

CHAPTER 5 REVIEW

Go to the Campbell Biology website (www.campbellbiology.com) to explore an interactive version of the Chapter Review.

Summary of Key Concepts

POLYMER PRINCIPLES

- **Most macromolecules are polymers (pp. 62–63, FIGURE 5.2)** Carbohydrates, lipids, proteins, and nucleic acids are the four major classes of organic compounds in cells. Some of these compounds are very large and are called macromolecules. Most macromolecules are polymers, chains of identical or similar building blocks called monomers. Monomers form larger molecules by condensation reactions in which water molecules are released (dehydration). Polymers can disassemble by the reverse process, hydrolysis.
  Web/CD Activity 5A: *Making and Breaking Polymers*

- **An immense variety of polymers can be built from a small set of monomers (pp. 63–64)** Each class of polymer is formed from a specific set of monomers. Although organisms share the same limited number of monomer types, each organism is unique because of the specific arrangement of monomers into polymers.

CARBOHYDRATES—FUEL AND BUILDING MATERIAL

- **Sugars, the smallest carbohydrates, serve as fuel and carbon sources (pp. 64–65, FIGURES 5.3–5.5)** Monosaccharides are the simplest carbohydrates. They are used directly for fuel, converted to other types of organic molecules, or used as monomers for polymers. Disaccharides consist of two monosaccharides connected by a glycosidic linkage.
  Web/CD Activity 5B: *Models of Glucose*

- **Polysaccharides, the polymers of sugars, have storage and structural roles (pp. 66–68, FIGURES 5.6–5.9)** The monosaccharide monomers of polysaccharides are connected by glycosidic linkages. Starch in plants and glycogen in animals are both storage polymers of glucose. Cellulose is an important structural polymer of glucose in plant cell walls. Starch, glycogen, and cellulose differ in the positions and orientations of their glycosidic linkages.
  Web/CD Activity 5C: *Carbohydrates*

LIPIDS—DIVERSE HYDROPHOBIC MOLECULES

- **Fats store large amounts of energy (pp. 68–70, FIGURES 5.10–5.11)** Fats, also known as triacylglycerols, are constructed by the joining of a glycerol molecule to three fatty acids by dehydration reactions. Saturated fatty acids have the maximum number of hydrogen atoms. Unsaturated fatty acids (present in oils) have one or more double bonds in their hydrocarbon chains.

- **Phospholipids are major components of cell membranes (pp. 70–71, FIGURE 5.12–5.13)** Where fats have a third fatty acid linked to glycerol, phospholipids have a negatively charged phosphate group, which may be joined, in turn, to another small hydrophilic molecule. Thus, the "head" of a phospholipid is hydrophilic.

- **Steroids include cholesterol and certain hormones (p. 71, FIGURE 5.14)** Steroids have a basic structure of four fused rings of carbon atoms.
  Web/CD Activity 5D: *Lipids*

PROTEINS—MANY STRUCTURES, MANY FUNCTIONS

- A protein consists of one or more polypeptide chains folded into a specific three-dimensional conformation (p. 71).

- **A polypeptide is a polymer of amino acids connected in a specific sequence (pp. 71–74, FIGURES 5.15–5.16, TABLE 5.1)** Polypeptides are constructed from 20 different amino acids, each with a characteristic side chain (R group). The carboxyl and amino groups of adjacent amino acids link together in peptide bonds.

- **A protein's function depends on its specific conformation (pp. 74–80, FIGURES 5.17–5.27)** The primary structure of a protein is its unique sequence of amino acids. Secondary structure is the folding or coiling of the polypeptide into repeating configurations, mainly the α helix and the β pleated sheet, which result from hydrogen bonding between parts of the polypeptide backbone. Tertiary structure is the overall three-dimensional shape of a polypeptide and results from interactions between amino acid side chains. Proteins made of more than one polypeptide chain (subunits) have a quaternary level of structure. The structure and function of a protein are sensitive to physical and chemical conditions. Protein shape is ultimately determined by its primary structure, but in the cell, proteins called chaperonins may help the folding process.
  Web/CD Activity 5E: *Protein Functions*
  Web/CD Activity 5F: *Protein Structure*
  Biology Labs On-Line: *HemoglobinLab*

NUCLEIC ACIDS—INFORMATIONAL POLYMERS

- **Nucleic acids store and transmit hereditary information (pp. 80–81, FIGURE 5.28)** DNA stores information for the synthesis of specific proteins. RNA (specifically, mRNA) carries this genetic information to the protein-synthesizing machinery.

- **A nucleic acid strand is a polymer of nucleotides (p. 82, FIGURE 5.29)** Each nucleotide monomer consists of a pentose covalently bonded to a phosphate group and to one of four different nitrogenous bases (A, G, C, and T or U). RNA has ribose as its pentose; DNA has deoxyribose. RNA has U and DNA, T. In making a polynucleotide, nucleotides join to form a sugar-phosphate backbone from which the nitrogenous bases project. The sequence of bases along a gene specifies the amino acid sequence of a particular protein.

- **Inheritance is based on replication of the DNA double helix (pp. 82–83, FIGURE 5.30)** DNA is a helical, double-stranded macromolecule with bases projecting into the interior of the molecule. Because A always hydrogen-bonds to T, and C to G, the nucleotide sequences of the two strands are complementary. One strand can serve as a template for the formation of the other. This unique feature of DNA provides a mechanism for the continuity of life.

- **We can use DNA and proteins as tape measures of evolution (p. 84, TABLE 5.2)** Molecular comparisons help biologists sort out the evolutionary connections among species.
  Web/CD Activity 5G: *Nucleic Acid Functions*
  Web/CD Activity 5H: *Nucleic Acid Structure*

Self-Quiz

1. Which of the following terms includes all others in the list?
   a. monosaccharide
   b. disaccharide
   c. starch
   d. carbohydrate
   e. polysaccharide

2. The molecular formula for glucose is $C_6H_{12}O_6$. What would be the molecular formula for a polymer made by linking ten glucose molecules together by dehydration reactions? Explain your answer.
   a. $C_{60}H_{120}O_{60}$
   b. $C_6H_{12}O_6$
   c. $C_{60}H_{102}O_{51}$
   d. $C_{60}H_{100}O_{50}$
   e. $C_{60}H_{111}O_{51}$

3. The two ring forms of glucose (α and β)
   a. are made from different structural isomers of glucose.
   b. arise from different linear (nonring) glucose molecules.
   c. arise when different carbons of the linear structure join to form the rings.
   d. arise because the hydroxyl group at the point of ring closure can be trapped in either one of two possible positions.
   e. include an aldose and a ketose.

4. Choose the pair of terms that correctly completes this sentence: Nucleotides are to _____ as _____ are to proteins.
   a. nucleic acids; amino acids
   b. amino acids; polypeptides
   c. glycosidic linkages; polypeptide linkages
   d. genes; enzymes
   e. polymers; polypeptides

5. Which of the following statements concerning *unsaturated* fats is correct?
   a. They are more common in animals than in plants.
   b. They have double bonds in the carbon chains of their fatty acids.
   c. They generally solidify at room temperature.
   d. They contain more hydrogen than saturated fats having the same number of carbon atoms.
   e. They have fewer fatty acid molecules per fat molecule.

6. The structural level of a protein least affected by a disruption in hydrogen bonding is the
   a. primary level.
   b. secondary level.
   c. tertiary level.
   d. quaternary level.
   e. All structural levels are equally affected.

7. To convert a nucleoside to a nucleotide, it would be necessary to
   a. combine two nucleosides using dehydration synthesis.
   b. remove the pentose from the nucleoside.
   c. replace the purine with a pyrimidine.
   d. add phosphate to the nucleoside.
   e. replace ribose with deoxyribose.

8. Which of the following is *not* a protein?
   a. hemoglobin
   b. cholesterol
   c. an antibody
   d. an enzyme
   e. insulin

9. Compare and contrast starch and cellulose, two plant polysaccharides.

10. Human sex hormones belong to what family of lipids?

11. Why does a denatured protein no longer function normally?

12. How many molecules of water are needed to completely hydrolyze a polymer that is 100 monomers long?

13. A genetic mutation can change the primary structure of a protein. How can this destroy the protein's function?

14. In a DNA double helix, a region along one DNA strand has this sequence of nitrogenous bases: TAGGCCT. What is the base sequence along the other strand of the molecule?

Go to the website or CD-ROM for more quiz questions.

Evolution Connection

Comparisons of the amino acid sequences of proteins or the nucleotide sequences of genes can shed light on the evolutionary divergence of related organisms. Would you expect all the proteins or genes of a given set of organisms living on Earth today to show the same degree of divergence? Why or why not?

The Process of Science

1. A particular small polypeptide is nine amino acids long. Using three different enzymes to hydrolyze the polypeptide at various sites, we obtain the following five fragments (N denotes the amino end of the chain): Ala-Leu-Asp-Tyr-Val-Leu; Tyr-Val-Leu; N-Gly-Pro-Leu; Asp-Tyr-Val-Leu; N-Gly-Pro-Leu-Ala-Leu. Determine the primary structure of this polypeptide.

2. You are studying a cellular enzyme involved in breaking down fatty acids for energy. Looking at the R groups of the amino acids in FIGURE 5.15, what amino acids would you predict to occur in the parts of the enzyme that interact with the fatty acids? Why might this region of the enzyme need to be sequestered in a pocket, rather than on the enzyme's surface?

From the Campbell Biology website you can link to the Hemoglobin-Lab of Biology Labs On-Line, where you can explore the biochemistry of the protein hemoglobin.

Science, Technology, and Society

Some amateur and professional athletes take anabolic steroids to help them "bulk up" or build strength. The health risks of this practice are extensively documented. Apart from health considerations, how do you feel about the use of chemicals to enhance athletic performance? Is an athlete who takes anabolic steroids cheating, or is his or her use of such chemicals just part of the preparation required to succeed in a competitive sport? Defend your answer.

AN INTRODUCTION TO METABOLISM

METABOLISM, ENERGY, AND LIFE

- The chemistry of life is organized into metabolic pathways
- Organisms transform energy
- The energy transformations of life are subject to two laws of thermodynamics
- Organisms live at the expense of free energy
- ATP powers cellular work by coupling exergonic reactions to endergonic reactions

ENZYMES

- Enzymes speed up metabolic reactions by lowering energy barriers
- Enzymes are substrate specific
- The active site is an enzyme's catalytic center
- A cell's physical and chemical environment affects enzyme activity

THE CONTROL OF METABOLISM

- Metabolic control often depends on allosteric regulation
- The localization of enzymes within a cell helps order metabolism
- The theme of emergent properties is manifest in the chemistry of life: *a review*

*T*he living cell *is a chemical industry in miniature, where thousands of reactions occur within a microscopic space. Sugars are converted to amino acids, and vice versa. Small molecules are assembled into polymers, which may be hydrolyzed later as the needs of the cell change. In plants and animals, many cells export chemical products that are used in other parts of the organism. The process known as cellular respiration drives the cellular economy by extracting the energy stored in sugars and other fuels. Cells apply this energy to perform various types of work, such as the synthesis of the macromolecules featured in Chapter 5. In a more exotic example, cells of the fungus in the photo above convert the energy stored in certain organic molecules to light, a process called bioluminescence. (The glow may attract insects that benefit the fungus by dispersing its spores.) Bioluminescence and all other metabolic activities carried out by a cell are precisely coordinated and controlled. In its complexity, its efficiency, its integration, and its responsiveness to subtle changes, the cell is peerless as a chemical institution. The concepts of metabolism you learn in this chapter will help you further understand the connections between chemistry and life.*

METABOLISM, ENERGY, AND LIFE

The totality of an organism's chemical reactions is called **metabolism** (from the Greek *metabole*, change). Metabolism is an emergent property of life that arises from interactions between molecules within the orderly environment of the cell.

The chemistry of life is organized into metabolic pathways

We can think of a cell's metabolism as an elaborate road map of the thousands of chemical reactions that occur in that cell (FIGURE 6.1, p. 88). These reactions are arranged in intricately branched metabolic pathways that alter molecules by a series of steps. Enzymes route matter through the metabolic pathways by selectively accelerating each step. Analogous to the red, yellow, and green lights that control the flow of traffic and prevent snarls, mechanisms that regulate enzymes balance metabolic supply and demand, averting deficits and surpluses of chemicals.

As a whole, metabolism is concerned with managing the material and energy resources of the cell. Some metabolic

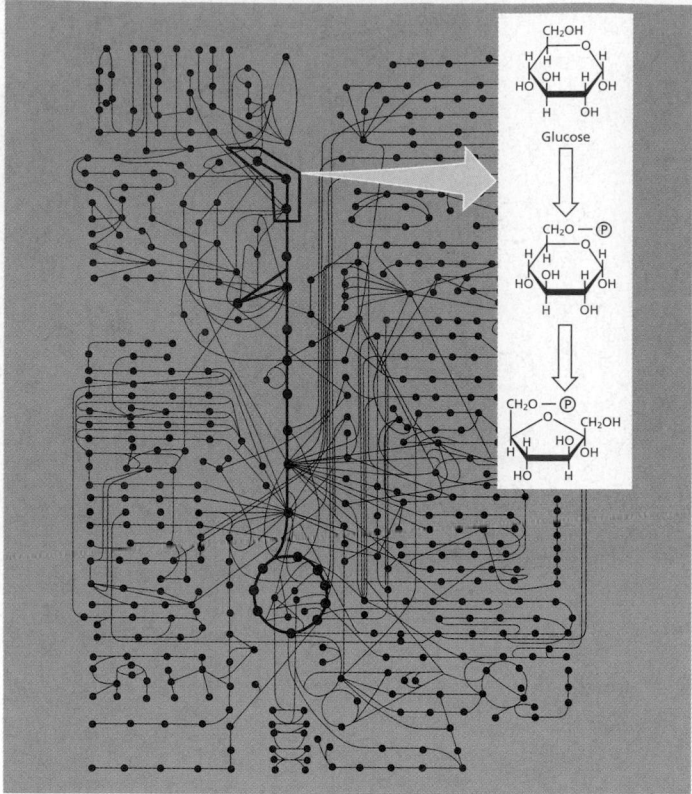

FIGURE 6.1 The complexity of metabolism. This diagram traces only a few hundred of the thousands of metabolic reactions that occur in a cell. The dots represent molecules, and the lines represent the chemical reactions that transform them. The reactions proceed in stepwise sequences called metabolic pathways, in which each step is catalyzed by a specific enzyme. The inset shows the first two steps in the catabolic pathway that breaks down glucose. (You'll learn more about this pathway in Chapter 9.)

Sliding down converts potential energy to kinetic energy.

FIGURE 6.2 Transformations between kinetic and potential energy. The children have more potential energy at the top of the slide (because of the effect of gravity) than they do at the bottom.

pathways release energy by breaking down complex molecules to simpler compounds. These degradative processes are called **catabolic pathways.** A major thoroughfare of catabolism is cellular respiration, in which the sugar glucose and other organic fuels are broken down to carbon dioxide and water. Energy that was stored in the organic molecules becomes available to do the work of the cell. **Anabolic pathways,** in contrast, consume energy to build complicated molecules from simpler ones. An example of anabolism is the synthesis of a protein from amino acids. Catabolic and anabolic pathways are the downhill and uphill avenues of the metabolic map. The metabolic pathways intersect in such a way that energy released from the "downhill" reactions of catabolism can be used to drive the "uphill" reactions of the anabolic pathways. This transfer of energy from catabolism to anabolism is called energy coupling.

In this chapter, we will focus on the mechanisms common to metabolic pathways. Because energy is fundamental to all metabolic processes, a basic knowledge of energy is necessary to understand how the living cell works. Although we will use some nonliving examples to study energy, keep in mind that the concepts demonstrated by these examples also apply to **bioenergetics,** the study of how organisms manage their energy resources. An understanding of energy is as important for students of biology as it is for students of physics, chemistry, and engineering.

Organisms transform energy

Energy is the capacity to do work—that is, to move matter against opposing forces, such as gravity and friction. Put another way, energy is the ability to rearrange a collection of matter. For example, you expend energy to turn the pages of this book. Energy exists in various forms, and the work of life depends on the ability of cells to transform energy from one type into another.

Anything that moves possesses a form of energy called **kinetic energy,** the energy of motion. Moving objects perform work by imparting motion to other matter: A pool player uses the motion of the cue stick to push the cue ball, which in turn moves the other balls; water gushing through a dam turns turbines; electrons flowing along a wire run household appliances; the contraction of leg muscles pushes bicycle pedals. Light is also a type of kinetic energy that can be harnessed to perform work, such as powering photosynthesis in green

(b) Second law of thermodynamics: The entropy (disorder) of the universe is increasing.

(a) First law of thermodynamics: Energy can be transformed but not created or destroyed.

plants. Heat, or thermal energy, is kinetic energy that results from the random movement of molecules.

A resting object not presently at work may still possess energy, which, remember, is the *capacity* to do work. Stored energy, or **potential energy,** is energy that matter possesses because of its location or structure. Water behind a dam, for instance, stores energy because of its altitude. **Chemical energy,** a form of potential energy especially important to biologists, is stored in molecules as a result of the arrangement of the atoms in those molecules.

How is energy converted from one form to another? Consider, for example, the playground scene in FIGURE 6.2. The girl at the bottom of the slide transformed kinetic energy to potential energy when she climbed the ladder up to the slide. This stored energy was converted back to kinetic energy as she slid down. It was another source of potential energy—the chemical energy in the food she ate for breakfast—that enabled the girl to climb the ladder in the first place.

Chemical energy can be tapped when chemical reactions rearrange the atoms of molecules in such a way that potential energy stored in the molecules is converted to kinetic energy. This transformation occurs, for example, in the engine of an automobile when the hydrocarbons of gasoline react explosively with oxygen, releasing the energy that pushes the pistons. Similarly, chemical energy fuels organisms. Cellular respiration and other catabolic pathways unleash energy stored in sugar and other complex molecules and make that energy available for cellular work. Each child who climbed the ladder in FIGURE 6.2 transformed some of the chemical energy that was stored in the organic molecules of food to the kinetic energy of his or her movements. The chemical energy stored in these fuel molecules had itself been derived from light energy by plants during photosynthesis. Organisms are energy transformers.

The energy transformations of life are subject to two laws of thermodynamics

The study of the energy transformations that occur in a collection of matter is called **thermodynamics.** Scientists use the word *system* to denote the matter under study and refer to the rest of the universe—everything outside the system— as the *surroundings.* A *closed system,* such as that approximated by liquid in a thermos bottle, is isolated from its surroundings. In an *open system,* energy (and often matter) can be transferred between the system and its surroundings. Organisms are open systems. They absorb energy—for instance, light energy or chemical energy in the form of organic molecules— and release heat and metabolic waste products, such as carbon dioxide, to the surroundings. Two laws of thermodynamics govern energy transformations in organisms and all other collections of matter (FIGURE 6.3).

The First Law of Thermodynamics

According to the **first law of thermodynamics,** the energy of the universe is constant. *Energy can be transferred and transformed, but it cannot be created or destroyed.* The first law is also known as the *principle of conservation of energy.* The electric company does not manufacture energy, but merely converts it to a form that is convenient to use. By converting light to chemical energy, a green plant acts as an energy transformer, not an energy producer.

The car in FIGURE 6.3a will convert the chemical energy of gasoline fuels to kinetic and other forms of energy as it is driven away from the gas station. What happens to this energy after it has performed work in a machine or an organism? If energy cannot be destroyed, what prevents organisms from behaving like closed systems and recycling their energy? The second law answers these questions.

The Second Law of Thermodynamics

The **second law of thermodynamics** can be stated many ways. Let's begin with the following statement of the second law: Every energy transfer or transformation makes the universe more disordered. Scientists use a quantity called **entropy** as a measure of disorder, or randomness. The more random a collection of matter is, the greater its entropy. We can therefore restate the second law as follows: *Every energy transfer or transformation increases the entropy of the universe.* Although order can increase locally, there is an unstoppable trend toward randomization of the universe as a whole.

In many cases, increased entropy is evident in the physical disintegration of a system's organized structure. For example, you can observe increasing entropy in the gradual decay of an unmaintained building. Much of the increasing entropy of the universe is less apparent, however, because it takes the form of an increasing amount of heat, which is the energy of random molecular motion. As the car in FIGURE 6.3b converts chemical energy to kinetic energy, it is also increasing the disorder of its surroundings in the form of heat and the small molecules that are the breakdown products of gasoline.

In most energy transformations, ordered forms of energy are at least partly converted to heat. Only about 25% of the chemical energy stored in the fuel tank of an automobile is transformed into the motion of the car; the remaining 75% is lost from the engine as heat, which dissipates rapidly through the surroundings. Similarly, the children in FIGURE 6.2 convert only a fraction of the energy stored in their food to the kinetic energy of ladder climbing and other play. In performing various kinds of work, living cells unavoidably convert organized forms of energy to heat. (This can make a room crowded with people uncomfortably warm.)

In machines and organisms, even energy that performs useful work is eventually converted to heat. The organized energy of an automobile's forward movement becomes heat when the friction of the brakes and tires stops the car. Conversion to heat is the fate of *all* the chemical energy a child uses to climb a slide: Metabolic breakdown of food generates heat during the climb, and the fraction of energy temporarily stored as gravitational potential energy is converted to heat on the way down, as friction warms the surrounding air.

Conversion of other forms of energy to heat does not violate the first law of thermodynamics. Energy has been conserved, because heat is a form of energy, though energy in its most random state. By combining the first and second laws of thermodynamics, we can conclude that the *quantity* of energy in the universe is constant, but its *quality* is not. In a sense, heat is the lowest grade of energy, as uncoordinated movement of molecules. A system can put heat to work only when

there is a temperature difference that results in the heat flowing from a warmer location to a cooler one. If temperature is uniform, as it is in a living cell, then the only use for heat energy is to warm a body of matter, such as an organism.

How can we reconcile the second law of thermodynamics—the unstoppable increase in the entropy of the universe—with the orderliness of life (FIGURE 6.4), which is one of this book's themes? The key is to remember another theme: Organisms are open systems that exchange energy and materials with their surroundings. It is true that cells create ordered structures from less organized starting materials. For example, amino acids are ordered into the specific sequences of polypeptide chains. But an organism also takes in organized forms of matter and energy from the surroundings and replaces them with less ordered forms. For example, an animal obtains starch, proteins, and other complex molecules from the food it eats. As catabolic pathways break these molecules down, the animal releases carbon dioxide and water—small, simple molecules that store less chemical energy than the food did. The depletion of chemical energy is accounted for by heat generated during metabolism. On a larger scale, energy flows into an ecosystem in the form of light and leaves in the form of heat. Living systems increase the entropy of their surroundings, as predicted by thermodynamic law.

During the early history of life, complex organisms evolved from simpler ancestors. For example, we can trace the ancestry of the plant kingdom to much simpler organisms called green algae. However, this increase in organization over time in no way violates the second law. The entropy of a particular system, such as an organism, may actually decrease, so long as the total entropy of the *universe*—the system plus its surroundings—increases. Thus, organisms are islands of low entropy in an increasingly random universe. The evolution of biological order is perfectly consistent with the laws of thermodynamics.

50 μm

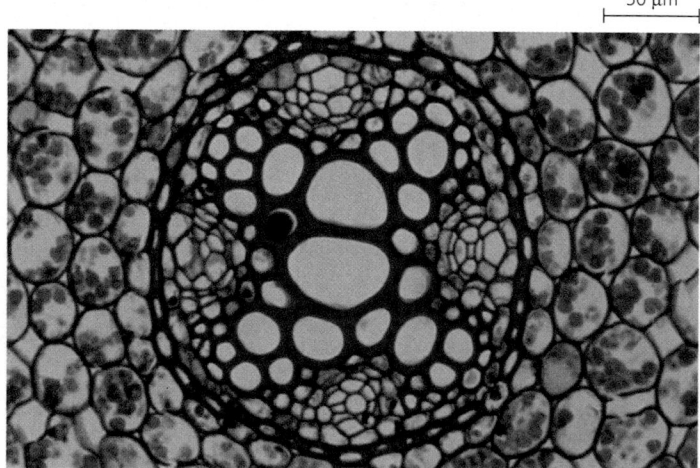

FIGURE 6.4 Order as a characteristic of life. Order is evident in the detailed anatomy of this root tissue from a buttercup plant (the micrograph shows a cross section). As open systems, organisms can increase their order at the expense of the order of their surroundings.

Organisms live at the expense of free energy

How can we predict what can and cannot occur in nature? How can we distinguish the possible from the impossible? We know from experience that certain events occur spontaneously and others do not. For instance, we know that water flows downhill, that objects of opposite charge move toward each other, that an ice cube melts at room temperature, and that a sugar cube dissolves in water. Explaining *why* these processes occur spontaneously is tricky.

Let's begin by defining a spontaneous process as a change that can occur without outside help. A spontaneous change can be harnessed to perform work. The downhill flow of water can be used to turn a turbine in a power plant, for example. A process that cannot occur on its own is said to be nonspontaneous; it will happen only if energy is added to the system. Water moves uphill only when a windmill or some other machine pumps the water against gravity, and a cell must expend energy to synthesize a protein from amino acids.

When a spontaneous process occurs in a system, the stability of that system increases. Unstable systems tend to change in such a way that they become more stable. A body of elevated water, such as a reservoir, is less stable than the same water at sea level. A system of charged particles is less stable when opposite charges are apart than when they are together. A compressed spring is less stable than a relaxed one. In all of these

examples, the system moves toward greater stability when nothing prevents such a change: The water falls, the opposite charges come together, the spring relaxes. In situations less familiar to us, how can we predict which changes lead to greater stability in a system—that is, which changes are spontaneous? You have already learned that a process can occur spontaneously only if it increases the disorder (entropy) of the universe. This principle is helpful in theory, but it does not give us a practical criterion to apply to biological systems because it requires that we measure changes in both the system and the surroundings. We need some standard for spontaneity that is based on the system alone. That criterion is called free energy.

Free Energy: A Criterion for Spontaneous Change

The concept of free energy is not easy to grasp, but the effort is worthwhile because we can apply the idea to many biological problems. **Free energy** *is the portion of a system's energy that can perform work when temperature is uniform throughout the system,* as in a living cell. It is called *free* energy because it is available for work, not because it can be spent without cost to the universe. In fact, you will soon understand that organisms can live only at the expense of free energy acquired from the surroundings. FIGURE 6.5 illustrates the relationship of free energy to spontaneity, stability, and work.

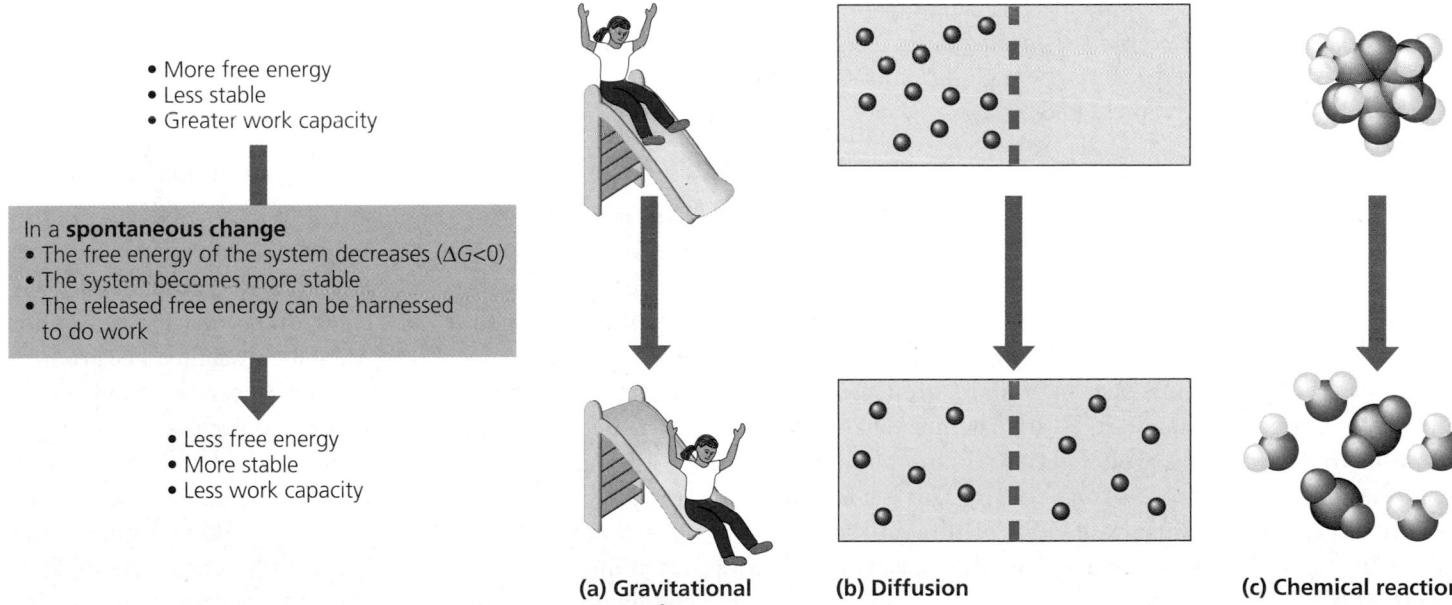

- More free energy
- Less stable
- Greater work capacity

In a **spontaneous change**
- The free energy of the system decreases ($\Delta G < 0$)
- The system becomes more stable
- The released free energy can be harnessed to do work

- Less free energy
- More stable
- Less work capacity

(a) Gravitational motion

(b) Diffusion

(c) Chemical reaction

FIGURE 6.5 The relationship of free energy to stability, work capacity, and spontaneous change. Unstable systems (top diagrams) are rich in free energy. They have a tendency to change spontaneously to a more stable state (bottom), and it is possible to harness this "downhill" change to perform work. **(a)** In this case, free energy is proportional to the girl's altitude.

(b) The free-energy concept also applies on the molecular scale, in this case to the movement of molecules known as diffusion. Here, molecules of a solute are distributed unequally across a membrane separating two aqueous compartments. This ordered state is unstable; it is rich in free energy. If the solute molecules can cross the membrane, there will be a net movement (diffusion) of the molecules until

they are equally concentrated in the two compartments. **(c)** Chemical reactions also involve free energy. The sugar molecule on top is less stable than the simpler molecules below. When catabolic pathways break down complex organic molecules, a cell can perform work using the free energy that was stored in the more complex molecules.

A system's quantity of free energy is symbolized by the letter G. There are two components to G: the system's total energy (symbolized by H) and its entropy (symbolized by S). Free energy is related to these factors in the following way:

$$G = H - TS$$

where T stands for absolute temperature in Kelvin (K) units (K = °C + 273; see Appendix 1). Notice that temperature amplifies the entropy term of the equation. This makes sense if you remember that temperature measures the intensity of random molecular motion (heat), which tends to disrupt order. What does this equation tell us about free energy? Not all the energy stored in a system (H) is available for work. The system's disorder, the entropy factor, is subtracted from total energy in computing the maximum capacity of the system to perform useful work. We are then left with free energy, which is somewhat less than the system's total energy.

How does the concept of free energy help us determine whether a particular process can occur spontaneously? Think of free energy G as a measure of a system's instability—its tendency to change to a more stable state. Systems that are rich in energy, such as compressed springs or separated charges, are unstable; so are highly ordered systems, such as complex molecules. In other words, systems that tend to change spontaneously to a more stable state have high energy, low entropy, or both. The free-energy equation weighs these two factors, which are consolidated in the system's G content. Now we can state a versatile criterion for spontaneous change: *In any spontaneous process, the free energy of a system decreases.*

The change in free energy as a system goes from a starting state to a different state is represented by ΔG:

$$\Delta G = G_{\text{final state}} - G_{\text{starting state}}$$

Or, put another way:

$$\Delta G = \Delta H - T\Delta S$$

For a process to occur spontaneously, the system must either give up energy (a decrease in H), give up order (an increase in S), or both. When these changes in H and S are tallied, ΔG must have a negative value ($\Delta G < 0$). The greater this decrease in free energy, the greater the maximum amount of work the spontaneous process can perform. This is a formal, mathematical way of stating the obvious: Nature runs "downhill" in the sense of a loss in useful energy, the capacity to perform work.

Free Energy and Equilibrium

Another term for a state of maximum stability is *equilibrium*, which you learned about in Chapter 2 in connection with chemical reactions. There is an important relationship between free energy and equilibrium, including chemical equilibrium. Recall that most chemical reactions are reversible and

proceed until the forward and backward reactions occur at the same rate. The reaction is then said to be at chemical equilibrium, and there is no further change in the concentration of products or reactants. As a reaction proceeds toward equilibrium, the free energy of the mixture of reactants and products decreases. Free energy increases when a reaction is somehow pushed away from equilibrium. For a reaction at equilibrium, $\Delta G = 0$, because there is no net change in the system. We can think of equilibrium as an energy valley. A chemical reaction or physical process at equilibrium performs no work. A process is spontaneous and can perform work when sliding toward equilibrium. Movement away from equilibrium is nonspontaneous; it can occur only with the help of an outside energy source. We can now apply the free-energy concept more specifically to the chemistry of life.

Free Energy and Metabolism

Exergonic and Endergonic Reactions in Metabolism. Based on their free-energy changes, chemical reactions can be classified as either exergonic ("energy outward") or endergonic ("energy inward"). An **exergonic reaction** proceeds with a net release of free energy (FIGURE 6.6a). Because the chemical mixture loses free energy, ΔG is negative for an exergonic reaction. In other words, exergonic reactions are those that occur spontaneously. The magnitude of ΔG for an exergonic reaction is the maximum amount of work the reaction can perform. We can use the overall reaction for cellular respiration as an example:

$$C_6H_{12}O_6 + 6\,O_2 \longrightarrow 6\,CO_2 + 6\,H_2O$$
$$\Delta G = -686 \text{ kcal/mol (or } -2,870 \text{ kJ/mol)}$$

For each mole (180 g) of glucose broken down by respiration, 686 kcal (or 2,870 kJ) of energy are made available for work (under what scientists call standard conditions). Because energy must be conserved, the chemical products of respiration store 686 kcal less free energy than the reactants. The products are, in a sense, the spent exhaust of a process that tapped most of the free energy stored in the sugar molecules.

An **endergonic reaction** is one that absorbs free energy from its surroundings (FIGURE 6.6b). Because this kind of reaction *stores* free energy in molecules, ΔG is positive. Such reactions are nonspontaneous, and the magnitude of ΔG is the quantity of energy required to drive the reaction. If a chemical process is exergonic (downhill) in one direction, then the reverse process must be endergonic (uphill). A reversible process cannot be downhill in both directions. If $\Delta G = -686$ kcal/mol for respiration, then for photosynthesis to produce sugar from carbon dioxide and water, $\Delta G = +686$ kcal/mol. Sugar production in the leaf cells of a plant is steeply endergonic, an uphill process powered by the absorption of light energy.

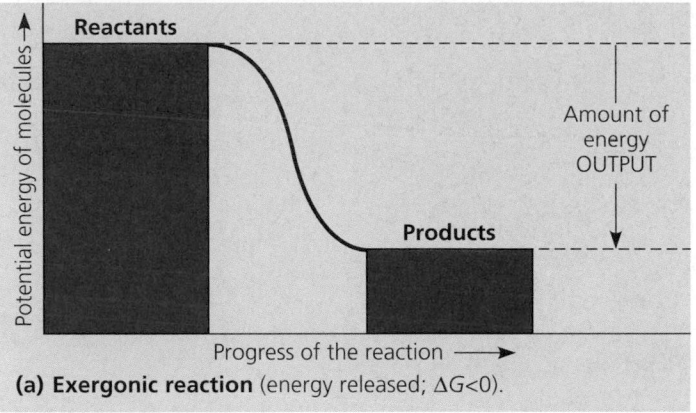

(a) Exergonic reaction (energy released; $\Delta G < 0$).

(b) Endergonic reaction (energy required; $\Delta G > 0$).

FIGURE 6.6 Energy changes in exergonic and endergonic reactions.

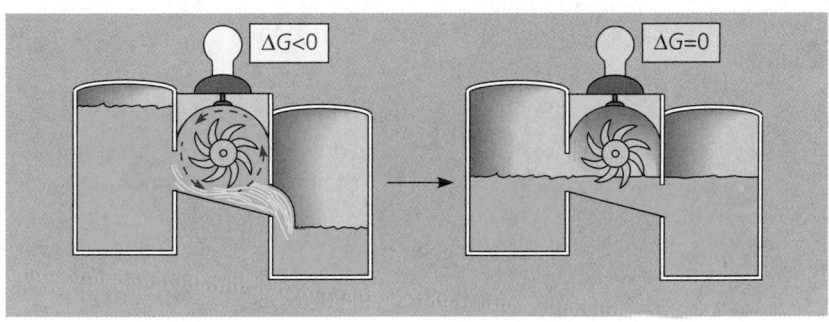

(a) A closed hydroelectric system. Flowing water drives the generator only until the system reaches equilibrium.

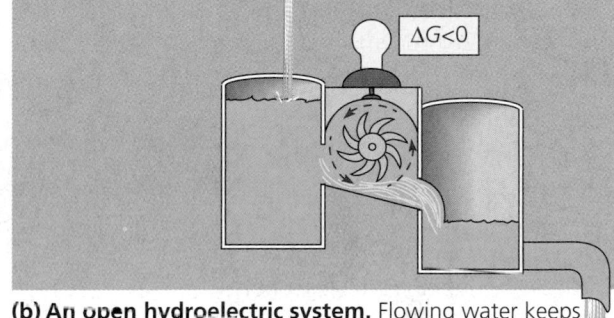

(b) An open hydroelectric system. Flowing water keeps driving the generator because intake and outflow of water keep the system from reaching equilibrium.

(c) A multistep open hydroelectric system. In these simple hydroelectric systems, water flowing downhill turns a turbine that provides electricity to a light bulb. Cellular respiration is analogous to part c: Glucose is broken down in a series of exergonic reactions that power the work of the cell. The product of each reaction becomes the reactant for the next, so no reaction reaches equilibrium.

FIGURE 6.7 Disequilibrium and work in closed and open systems.

Metabolic Disequilibrium.   Reactions in a closed system eventually reach equilibrium and can do no work, as illustrated by the closed hydroelectric system in FIGURE 6.7a. The chemical reactions of metabolism are reversible, and they, too, would reach equilibrium if they occurred in the isolation of a test tube. Because systems at equilibrium have a ΔG of zero and can do no work, a cell that has reached metabolic equilibrium is dead! In fact, metabolic disequilibrium is one of the defining features of life.

A cell can maintain disequilibrium because it is an open system. The constant flow of materials in and out of the cell keeps the metabolic pathways from ever reaching equilibrium,

and the cell continues to do work throughout its life. This principle is illustrated by the open (and more realistic) hydroelectric system in FIGURE 6.7b. However, unlike this simple single-step system, a catabolic pathway in a cell releases free energy in a series of reactions. An example is cellular respiration, illustrated by analogy in FIGURE 6.7c. Some of the reversible reactions of respiration are constantly "pulled" in one direction—that is, they are kept out of equilibrium. The key to sustaining this disequilibrium is that the product of one reaction does not accumulate, but instead becomes a reactant in the next step. The overall sequence of reactions is kept going by the huge free-energy difference between glucose at the "uphill" end of the pathway and carbon dioxide and water at the "downhill"end. As long as the cell has a steady supply of glucose or other fuels and is able to expel waste products to the surroundings, its metabolic pathways never reach equilibrium and continue to do the work of life.

We see once again how important it is to think of organisms as open systems. Sunlight provides a daily source of free energy for an ecosystem's plants and other photosynthetic organisms. Animals and other nonphotosynthetic organisms in an ecosystem depend on free-energy transfusions in the form of the organic products of photosynthesis.

Now that we have applied the free-energy concept to metabolism, we are ready to see how a cell actually performs the work of life. A key feature of bioenergetics is **energy coupling,** the use of an exergonic process to drive an endergonic one. A molecule called ATP is responsible for mediating most energy coupling in cells.

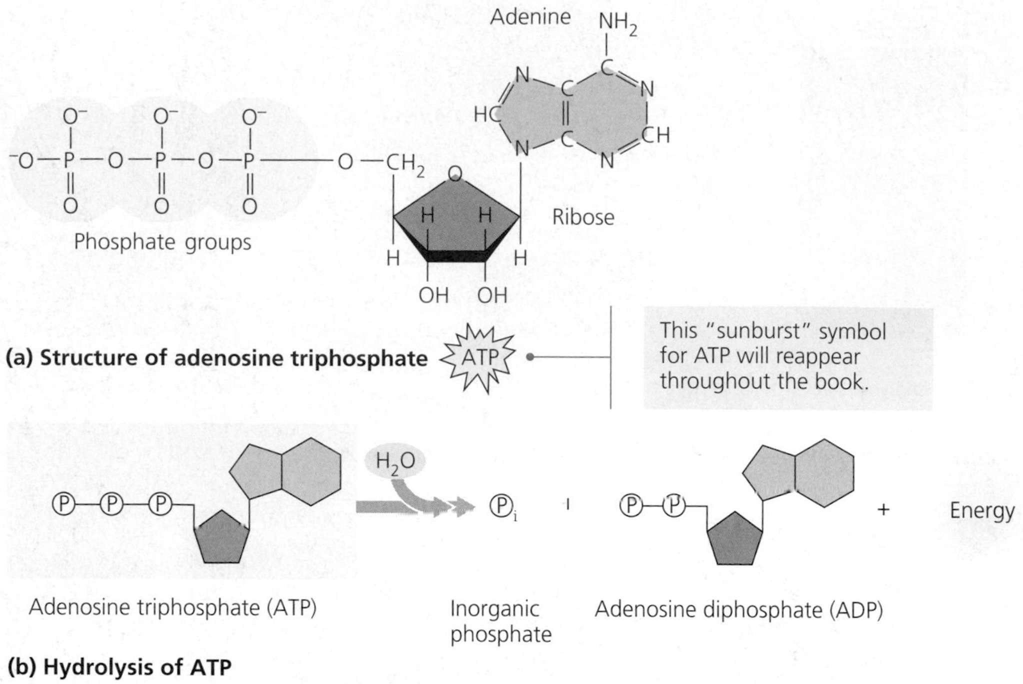

(a) Structure of adenosine triphosphate

This "sunburst" symbol for ATP will reappear throughout the book.

(b) Hydrolysis of ATP

FIGURE 6.8 The structure and hydrolysis of ATP. The hydrolysis of ATP yields inorganic phosphate and ADP. In the cell, most hydroxyl groups of phosphates are ionized ($—O^-$).

ATP powers cellular work by coupling exergonic reactions to endergonic reactions

A cell does three main kinds of work:

1. *Mechanical work,* such as the beating of cilia, the contraction of muscle cells, and the movement of chromosomes during cellular reproduction
2. *Transport work,* the pumping of substances across membranes against the direction of spontaneous movement
3. *Chemical work,* the pushing of endergonic reactions that would not occur spontaneously, such as the synthesis of polymers from monomers

In most cases, the immediate source of energy that powers cellular work is ATP.

The Structure and Hydrolysis of ATP

ATP (adenosine triphosphate) is closely related to one type of nucleotide found in nucleic acids. ATP has the nitrogenous base adenine bonded to ribose, as in an adenine nucleotide of RNA. In RNA, however, only *one* phosphate group is attached

to the ribose (see FIGURE 5.29b), whereas adenosine *tri*phosphate has a chain of *three* phosphate groups attached to the ribose (FIGURE 6.8a).

The bonds between the phosphate groups of ATP's tail can be broken by hydrolysis. When the terminal phosphate bond is broken, a molecule of inorganic phosphate (abbreviated $\mathcal{P}_i$ throughout this book) leaves the ATP, which becomes adenosine diphosphate, or ADP (FIGURE 6.8b). The reaction is exergonic and under laboratory conditions releases 7.3 kcal of energy per mole of ATP hydrolyzed:

$$\text{ATP} + \text{H}_2\text{O} \longrightarrow \text{ADP} + \mathcal{P}_i$$

$$\Delta G = -7.3 \text{ kcal/mol (or } -31 \text{ kJ/mol)}$$

This is the free-energy change measured under what are called standard conditions. However, the chemical and physical conditions in the cell do not conform to standard conditions. When the reaction occurs in the cellular environment rather than in a test tube, the actual ΔG is about -13 kcal/mol, 78% greater than the energy released by ATP hydrolysis under standard conditions.

Because their hydrolysis releases energy, the phosphate bonds of ATP are sometimes referred to as high-energy phosphate bonds, but the term is misleading. The phosphate bonds of ATP are not unusually strong bonds, as "high-energy" may imply. In fact, compared to most bonds in organic molecules, these bonds are relatively weak, and it is *because* they are somewhat unstable that their hydrolysis yields energy. The products of hydrolysis (ADP and $\circledP_i$) are more stable than ATP. When a system changes in the direction of greater stability—when a compressed spring relaxes, for instance—the change is exergonic. Thus, the release of energy during the hydrolysis of ATP comes from the chemical change to a more stable condition, not from the phosphate bonds themselves.

Why are the phosphate bonds so fragile? If we reexamine the ATP molecule in FIGURE 6.8a, we can see that all three phosphate groups are negatively charged. These like charges are crowded together, and their repulsion contributes to the instability of this region of the ATP molecule. The triphosphate tail of ATP is the chemical equivalent of a loaded spring.

How ATP Performs Work

When ATP is hydrolyzed in a test tube, the release of free energy merely heats the surrounding water. In the cell, that would be an inefficient and dangerous use of a valuable energy resource. With the help of specific enzymes, the cell is able to couple the energy of ATP hydrolysis directly to endergonic processes by transferring a phosphate group from ATP to some other molecule. The recipient of the phosphate group is then said to be **phosphorylated.** The key to the coupling of exergonic and endergonic reactions is the formation of this phosphorylated intermediate, which is more reactive (less stable) than the original molecule. FIGURE 6.9 shows a cellular process that exemplifies this mechanism: the synthesis of the amino acid glutamine from glutamic acid (another amino acid) and ammonia.

Nearly all cellular work depends on ATP's energizing of other molecules by transferring phosphate groups. For instance, when ATP powers the movement of muscles, it transfers phosphate groups to contractile proteins.

The Regeneration of ATP

An organism at work uses ATP continuously, but ATP is a renewable resource that can be regenerated by the addition of phosphate to ADP (FIGURE 6.10). The free energy required to phosphorylate ADP comes from breakdown reactions (catabolism) in the cell. This shuttling of inorganic phosphate and energy is called the ATP cycle, and it couples the cell's energy-yielding processes to the energy-producing ones. The ATP cycle moves at an astonishing pace. For example, a working muscle cell recycles its entire pool of ATP about once each minute. That turnover represents 10 million molecules of ATP consumed and regenerated per second per cell. If ATP could not be regenerated by the phosphorylation of ADP, humans would consume nearly their body weight in ATP each day.

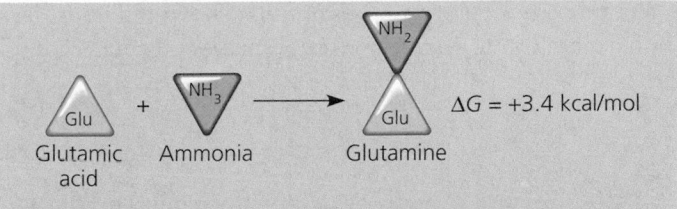

(a) Without ATP. The conversion is nonspontaneous (ΔG is positive) without the help of ATP.

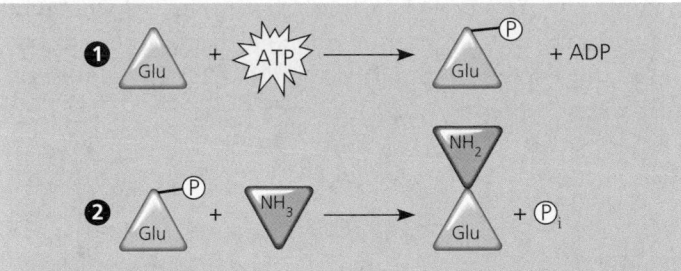

(b) With ATP. As it actually occurs in the cell, the synthesis of glutamine is a two-step reaction driven by ATP. The formation of a phosphorylated intermediate couples the two steps. ❶ ATP phosphorylates glutamic acid, making the amino acid less stable. ❷ Ammonia displaces the phosphate group, forming glutamine.

$$Glu + NH_3 \longrightarrow Glu-NH_2 \qquad \Delta G = +3.4 \text{ kcal/mol}$$
$$ATP \longrightarrow ADP + \circledP_i \qquad \Delta G = -7.3 \text{ kcal/mol}$$
$$\text{Net } \Delta G = -3.9 \text{ kcal/mol}$$

(c) Free energy change with ATP. Adding the ΔG for the amino acid conversion to the ΔG for the ATP hydrolysis gives the free-energy change for the overall reaction. Because the overall process is exergonic (has a negative ΔG), it occurs spontaneously.

FIGURE 6.9 Energy coupling by phosphate transfer. In this example, ATP hydrolysis drives the conversion of the amino acid glutamic acid (Glu) to another amino acid, glutamine (Glu—NH_2).

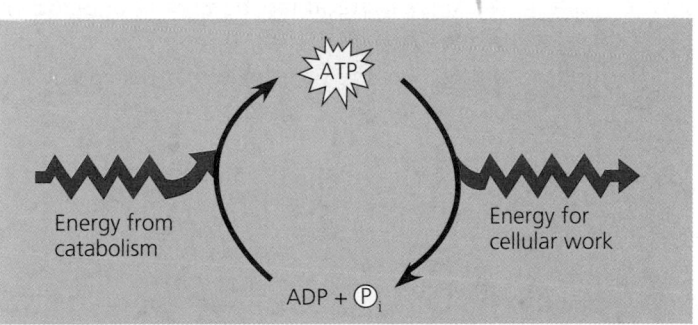

FIGURE 6.10 The ATP cycle. Energy released by breakdown reactions (catabolism) in the cell is used to phosphorylate ADP, regenerating ATP. Energy stored in ATP drives most cellular work. Thus, ATP couples the cell's energy-yielding processes to the energy-consuming ones.

Because a reversible process cannot go downhill both ways, the regeneration of ATP from ADP is necessarily endergonic:

$$ADP + Ⓟi \longrightarrow ATP + H_2O$$

$$\Delta G = +7.3 \text{ kcal/mol (standard conditions)}$$

Catabolic (exergonic) pathways, especially cellular respiration, provide the energy for the endergonic process of making ATP. Plants also use light energy to produce ATP.

Thus, the ATP cycle is a turnstile through which energy passes during its transfer from catabolic to anabolic pathways. In fact, the energy temporarily stored in ATP drives most cellular work.

ENZYMES

The laws of thermodynamics tell us what can and cannot happen under given conditions but say nothing about the speed of these processes. A spontaneous chemical reaction may occur so slowly that it is imperceptible. For example, the hydrolysis of sucrose (table sugar) to glucose and fructose (FIGURE 6.11) is exergonic, occurring spontaneously with a release of free energy ($\Delta G = -7$ kcal/mol). Yet a solution of sucrose dissolved in sterile water will sit for years at room temperature with no appreciable hydrolysis. But if we add a small amount of the enzyme sucrase to the solution, then all the sucrose may be hydrolyzed within seconds. How does an enzyme do this?

Enzymes speed up metabolic reactions by lowering energy barriers

A **catalyst** is a chemical agent that changes the rate of a reaction without being consumed by the reaction; an **enzyme** is a catalytic protein. (Another class of biological catalysts, ribozymes, made of RNA, is discussed in Chapters 17 and 26.) In the absence of enzymes, chemical traffic through the pathways of metabolism would become hopelessly congested. In the next two sections, we will see what impedes a spontaneous reaction and how an enzyme changes the situation.

The Activation Energy Barrier

Every chemical reaction between molecules involves both bond breaking and bond forming. For example, the hydrolysis of sucrose involves first breaking the bond between glucose and fructose and then forming new bonds with a hydrogen and a hydroxyl group from water (see FIGURE 6.11). Whenever a reaction rearranges the atoms of molecules, existing bonds in the reactants must be broken and the new bonds of the products formed. The reactant molecules must absorb energy from their surroundings for their bonds to break, and energy is released when the new bonds of the product molecules are formed.

The initial investment of energy for starting a reaction—the energy required to break bonds in the reactant molecules—is known as the **free energy of activation,** or **activation energy,** abbreviated E_A in this book. It is usually provided in the form of heat that the reactant molecules absorb from the surroundings. The bonds of the reactants break only when the molecules have absorbed enough energy to become unstable. (Recall that systems rich in free energy are unstable, and unstable systems are reactive.) The absorption of thermal energy increases the speed of the reactant molecules, so they collide more often and more forcefully. Also, thermal agitation of the atoms in the molecules makes the bonds more likely to break. As the molecules settle into their new, more stable bonding arrangements, energy is released to the surroundings. If the reaction is exergonic, E_A will be repaid with dividends, as the formation of new bonds releases more energy than was invested in the breaking of old bonds.

We can think of activation energy as the amount of energy needed to push the reactants over an energy barrier, or hill, so that the "downhill" part of the reaction can begin. FIGURE 6.12 graphs the energy changes for a hypothetical exergonic reaction that swaps portions of two reactant molecules:

$$AB + CD \longrightarrow AC + BD$$

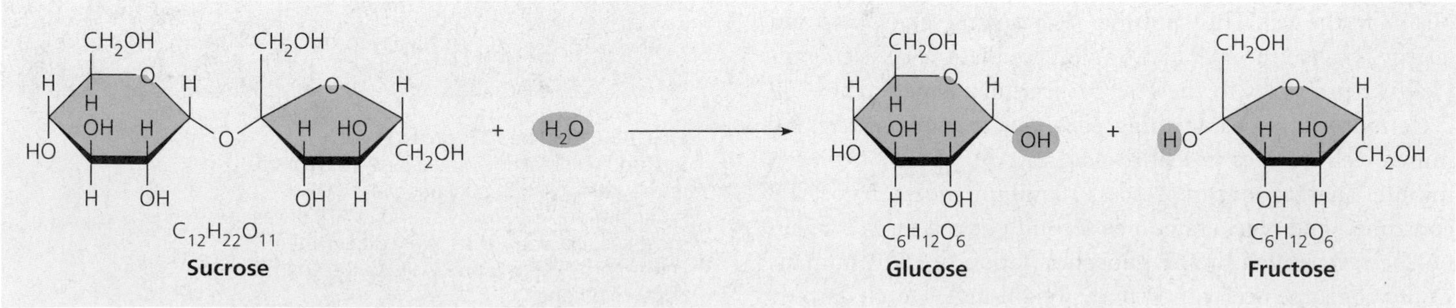

FIGURE 6.11 Example of an enzyme-catalyzed reaction: Hydrolysis of sucrose.

The energizing, or activation, of the reactants is represented by the uphill portion of the graph, with the free-energy content of the reactant molecules increasing. At the summit, the reactants are in an unstable condition known as the transition

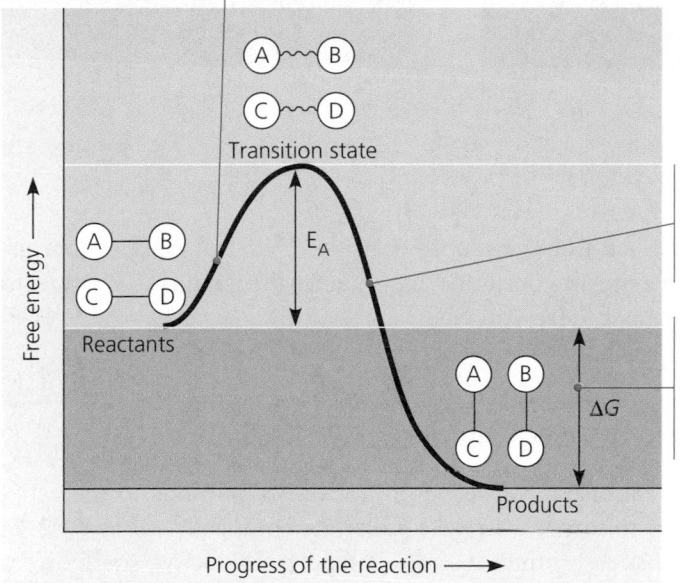

The reactants AB and CD must absorb enough energy from the surroundings to surmount the hill of activation energy (E_A) and reach the unstable transition state.

Transition state

E_A

Reactants

Bonds break, and new bonds form. In the process, energy is released to the surroundings.

This is an exergonic reaction, which has a negative ΔG; the products have less free energy than the reactants.

ΔG

Products

Free energy

Progress of the reaction ⟶

FIGURE 6.12 Energy profile of an exergonic reaction. The "molecules" are hypothetical, with A, B, C, and D representing *parts* of actual molecules. Using the reaction in FIGURE 6.11 as an example, you can imagine that AB = sucrose (glucose monomer—fructose monomer), CD = water (with C = hydroxyl group and D = hydrogen), AC = glucose, and BD = fructose.

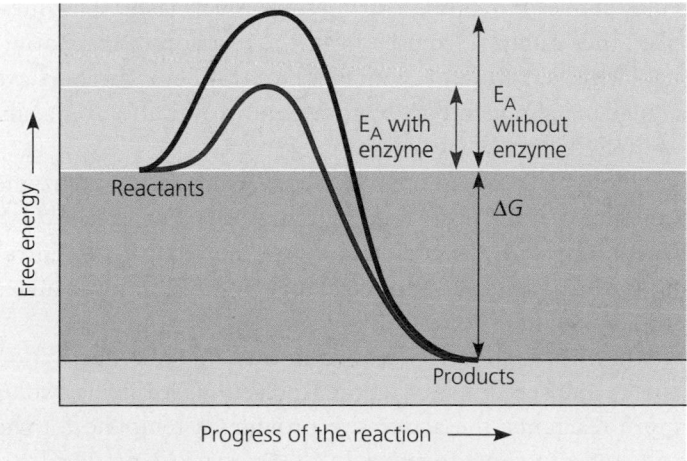

Reactants

E_A with enzyme

E_A without enzyme

ΔG

Products

Free energy

Progress of the reaction ⟶

FIGURE 6.13 Enzymes lower the barrier of activation energy. Without affecting the free-energy change (ΔG) for the reaction, an enzyme speeds the reaction by reducing the uphill climb to the transition state. The black curve shows the course of the reaction without an enzyme; the red curve shows the course of the reaction with an enzyme.

state: They are activated, and the breaking and making of bonds can occur. The bond-forming phase of the reaction corresponds to the downhill part of the curve, which shows the loss of free energy by the molecules. The difference between the free energy of the products and the free energy of the reactants is ΔG for the overall reaction. Because this is an exergonic reaction, ΔG is negative.

As FIGURE 6.12 shows, even for an exergonic reaction, which is energetically downhill overall, the barrier of activation energy must be scaled before the reaction can occur. For some reactions, E_A is modest enough that even at room temperature there is sufficient thermal energy for many of the reactants to reach the transition state. In most cases, however, the E_A barrier is loftier, and the reaction will occur at a noticeable rate only if the reactants are heated. The spark plugs in an automobile engine energize the gasoline-oxygen mixture so that the molecules reach the transition state and react; only then can there be the explosive release of energy that pushes the pistons. Without a spark, the hydrocarbons of gasoline are too stable to react with oxygen.

Enzymes and Activation Energy

The barrier of activation energy is essential to life. Proteins, DNA, and other complex molecules of the cell are rich in free energy and have the potential to decompose spontaneously; that is, the laws of thermodynamics favor their breakdown. These molecules exist only because at temperatures typical for cells, few molecules can make it over the hump of activation energy. Occasionally, however, the barrier for selected reactions must be surmounted, or else the cell would be metabolically stagnant. Heat speeds a reaction, but high temperature denatures proteins and kills cells. Organisms must therefore use an alternative: a catalyst.

An enzyme speeds a reaction by lowering the E_A barrier (FIGURE 6.13), so that the precipice of the transition state is within reach even at moderate temperatures. An enzyme cannot change the ΔG for a reaction; it cannot make an endergonic reaction exergonic. Enzymes can only hasten reactions that would occur eventually anyway, but this function makes it possible for the cell to have a dynamic metabolism. And because enzymes are very selective in the reactions they catalyze, they determine which chemical processes will be going on in the cell at any particular time.

Enzymes are substrate specific

The reactant an enzyme acts on is referred to as the enzyme's **substrate.** The enzyme binds to its substrate (or substrates,

FIGURE 6.14 The induced fit between an enzyme and its substrate. (a) The active site of this enzyme (hexokinase) can be seen in this computer graphic model as a groove on the surface of the protein (blue). **(b)** On entering the active site, the substrate, which is glucose (red), induces a change in the shape of the protein that causes the active site to embrace the substrate.

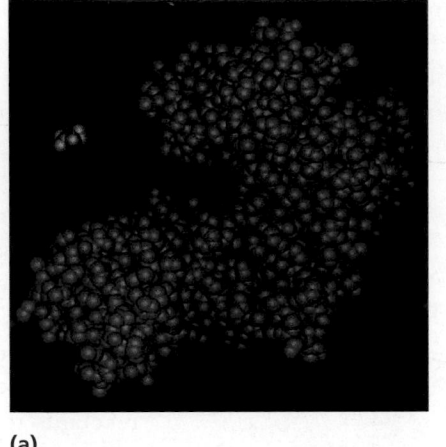

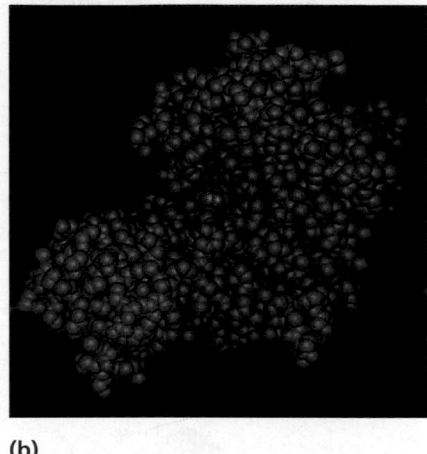

(a)   (b)

when there are two or more reactants). While enzyme and substrate are joined, the catalytic action of the enzyme converts the substrate to the product (or products) of the reaction. The overall process can be summarized as follows, with the name of the enzyme written above the reaction arrow:

$$\text{Substrate(s)} \xrightarrow{\text{Enzyme}} \text{Product(s)}$$

For example, the enzyme sucrase (most enzyme names end in *-ase*) breaks the disaccharide sucrose into its two monosaccharides, glucose and fructose (see FIGURE 6.11):

$$\text{Sucrose} + \text{H}_2\text{O} \xrightarrow{\text{Sucrase}} \text{Glucose} + \text{Fructose}$$

An enzyme can distinguish its substrate from even closely related compounds, such as isomers, so that each type of enzyme catalyzes a particular reaction. For instance, sucrase will act only on sucrose and will reject other disaccharides, such as maltose. What accounts for this molecular recognition? Recall that enzymes are proteins, and proteins are macromolecules with unique three-dimensional conformations. The specificity of an enzyme results from its shape.

Only a restricted region of the enzyme molecule actually binds to the substrate. This region, called the **active site,** is typically a pocket or groove on the surface of the protein (FIGURE 6.14a). Usually, the active site is formed by only a few of the enzyme's amino acids, with the rest of the protein molecule providing a framework that reinforces the configuration of the active site. The specificity of an enzyme is attributed to a compatible fit between the shape of its active site and the shape of the substrate. The active site, however, is not a rigid receptacle for the substrate. As the substrate enters the active site, it induces the enzyme to change its shape slightly so that the active site fits even more snugly around the substrate (FIGURE 6.14b). This **induced fit** is like a clasp-

ing handshake. Induced fit brings chemical groups of the active site into positions that enhance their ability to catalyze the chemical reaction.

The active site is an enzyme's catalytic center

In an enzymatic reaction, the substrate binds to the active site to form an enzyme-substrate complex (FIGURE 6.15). In most cases, the substrate is held in the active site by weak interactions, such as hydrogen bonds and ionic bonds. Side chains (R groups) of a few of the amino acids that make up the active site catalyze the conversion of substrate to product, and the product departs from the active site. The enzyme is then free to take another substrate molecule into its active site. The entire cycle happens so fast that a single enzyme molecule typically acts on about a thousand substrate molecules per second. Some enzymes are much faster. Enzymes, like other catalysts, emerge from the reaction in their original form. Therefore, very small amounts of enzyme can have a huge metabolic impact by functioning over and over again in catalytic cycles.

Most metabolic reactions are reversible, and an enzyme can catalyze both the forward and the reverse reactions. Which reaction prevails depends mainly on the relative concentrations of reactants and products; that is, the enzyme catalyzes the reaction in the direction of equilibrium.

Enzymes use a variety of mechanisms that lower activation energy and speed up a reaction. In reactions involving two or more reactants, the active site provides a template for the substrates to come together in the proper orientation for a reaction to occur between them. As the active site clutches the substrates with an induced fit, the enzyme may stress the substrate molecules, stretching and bending critical chemical bonds that must be broken during the reaction. Because E_A is proportional to the difficulty of breaking bonds, distorting

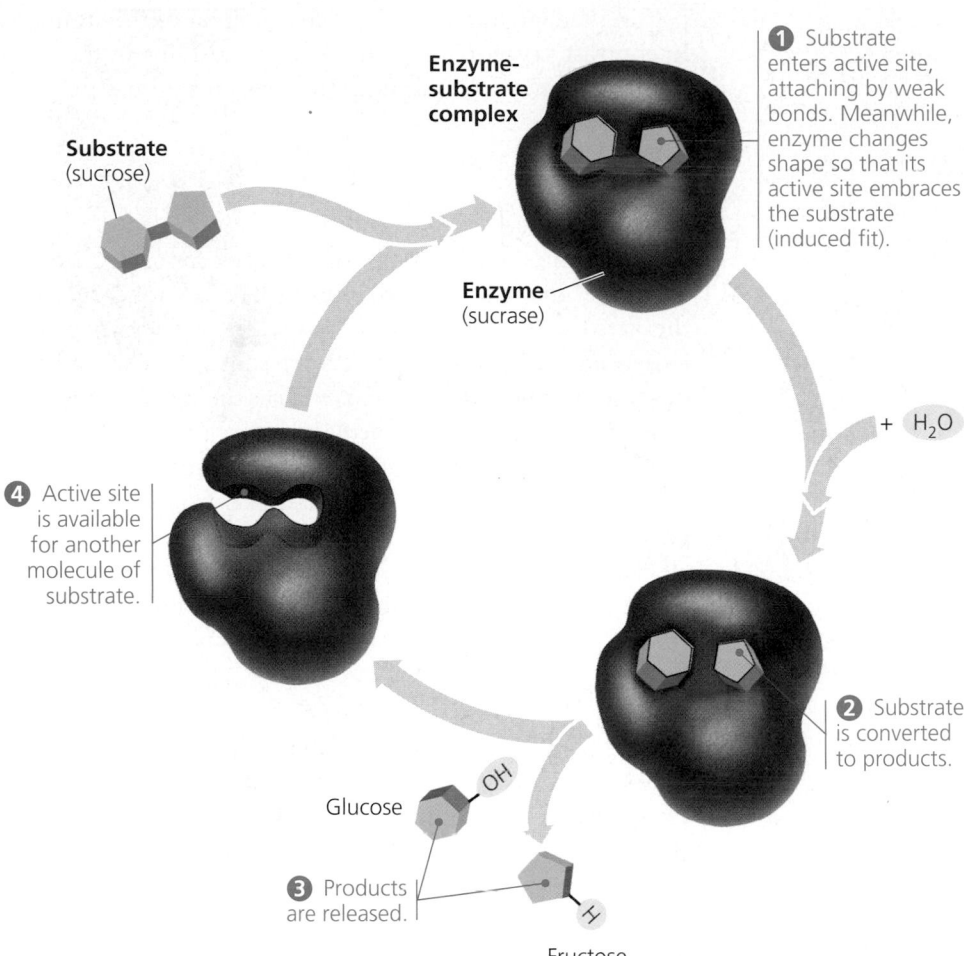

Substrate (sucrose)

Enzyme-substrate complex

Enzyme (sucrase)

① Substrate enters active site, attaching by weak bonds. Meanwhile, enzyme changes shape so that its active site embraces the substrate (induced fit).

+ H_2O

② Substrate is converted to products.

④ Active site is available for another molecule of substrate.

Glucose

OH

③ Products are released.

H

Fructose

FIGURE 6.15 The catalytic cycle of an enzyme. In this example, the enzyme sucrase catalyzes the hydrolysis of sucrose to glucose and fructose.

the substrate reduces the amount of thermal energy that must be absorbed to achieve a transition state.

The active site may also provide a microenvironment that is conducive to a particular type of reaction. For example, if the active site has amino acids with acidic side chains (R groups), the active site may be a pocket of low pH in an otherwise neutral cell. In such cases, an acidic amino acid may facilitate H^+ transfer to the substrate as a key step in catalyzing the reaction. Still another mechanism of catalysis is the direct participation of the active site in the chemical reaction. Sometimes this process even involves brief covalent bonding between the substrate and a side chain of an amino acid of the enzyme. Subsequent steps of the reaction restore the side chains to their original states, so the active site is the same after the reaction as it was before.

The rate at which a given amount of enzyme converts substrate to product is partly a function of the initial concentration of the substrate: The more substrate molecules available, the more frequently they access the active sites of the enzyme molecules. However, there is a limit to how fast the reaction can be pushed by adding more substrate to a fixed concentration of enzyme. At some point, the concentration of substrate will be high enough that all enzyme molecules have their active sites engaged. As soon as the product exits an active site, another substrate molecule enters. At this substrate concentration, the enzyme is said to be saturated, and the rate of the reaction is determined by the speed at which the active site can convert substrate to product. When an enzyme population is saturated, the only way to increase productivity is to add more enzyme. Cells sometimes do this by making more enzyme molecules.

A cell's physical and chemical environment affects enzyme activity

The activity of an enzyme is affected by general environmental factors, such as temperature and pH, and also by particular chemicals that specifically influence that enzyme.

Effects of Temperature and pH

Recall from Chapter 5 that the three-dimensional structures of proteins are sensitive to their environment. As a protein, an

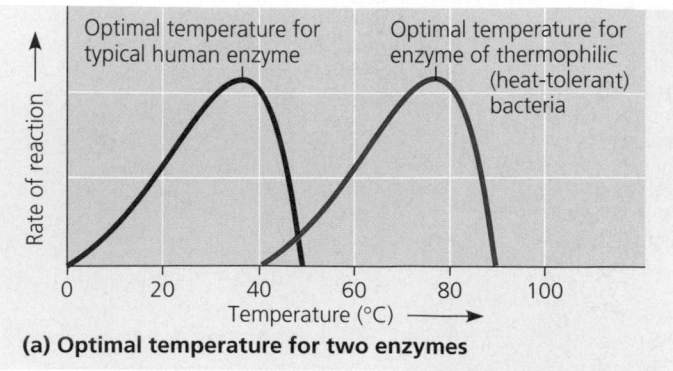

(a) Optimal temperature for two enzymes

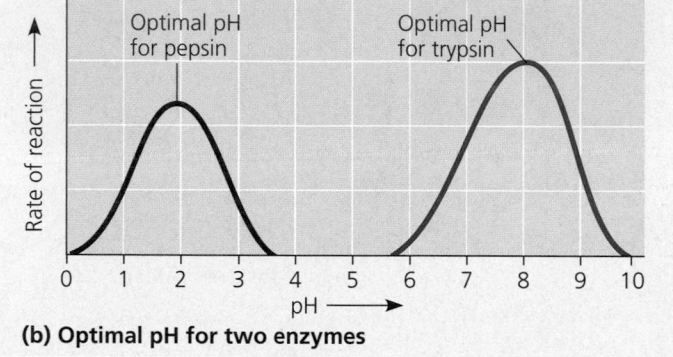

(b) Optimal pH for two enzymes

FIGURE 6.16 Environmental factors affecting enzyme activity. Each enzyme has an optimal **(a)** temperature and **(b)** pH that favor the active conformation of the protein molecule.

enzyme has conditions under which it works optimally, because that environment favors the most active conformation for the enzyme molecule.

Temperature and pH are environmental factors important in the activity of an enzyme. Up to a point, the velocity of an enzymatic reaction increases with increasing temperature, partly because substrates collide with active sites more frequently when the molecules move rapidly. Beyond that temperature, however, the speed of the enzymatic reaction drops sharply. The thermal agitation of the enzyme molecule disrupts the hydrogen bonds, ionic bonds, and other weak interactions that stabilize the active conformation, and the protein molecule denatures. Each enzyme has an optimal temperature at which its reaction rate is fastest. This temperature allows the greatest number of molecular collisions without denaturing the enzyme. Most human enzymes have optimal temperatures of about 35–40°C (close to human body temperature). Bacteria that live in hot springs contain enzymes with optimal temperatures of 70°C or higher (FIGURE 6.16a).

Just as each enzyme has an optimal temperature, it also has a pH at which it is most active. The optimal pH values for most enzymes fall in the range of pH 6–8, but there are exceptions. For example, pepsin, a digestive enzyme in the stomach, works best at pH 2. Such an acidic environment denatures most enzymes, but the active conformation of

pepsin is adapted to the acidic environment of the stomach. In contrast, trypsin, a digestive enzyme residing in the alkaline environment of the intestine, has an optimal pH of 8 (FIGURE 6.16b).

Cofactors

Many enzymes require nonprotein helpers for catalytic activity. These adjuncts, called **cofactors,** may be bound tightly to the active site as permanent residents, or they may bind loosely and reversibly along with the substrate. The cofactors of some enzymes are inorganic, such as the metal atoms zinc, iron, and copper in ionic form. If the cofactor is an organic molecule, it is more specifically called a **coenzyme.** Most vitamins are coenzymes or raw materials from which coenzymes are made. Cofactors function in various ways, but in all cases they are necessary for catalysis to take place. You'll encounter examples of cofactors later in the book.

Enzyme Inhibitors

Certain chemicals selectively inhibit the action of specific enzymes (FIGURE 6.17). If the inhibitor attaches to the enzyme by covalent bonds, inhibition is usually irreversible. It is reversible, however, if the inhibitor binds by weak bonds.

Some reversible inhibitors resemble the normal substrate molecule and compete for admission into the active site. These mimics, called **competitive inhibitors,** reduce the productivity of enzymes by blocking substrates from entering active sites. This kind of inhibition can be overcome by increasing the concentration of substrate so that as active sites become available, more substrate molecules than inhibitor molecules are around to gain entry to the sites.

So-called **noncompetitive inhibitors** do not directly compete with the substrate at the active site. Instead, they impede enzymatic reactions by binding to another part of the enzyme. This interaction causes the enzyme molecule to change its shape, rendering the active site unreceptive to substrate or leaving the enzyme less effective at catalyzing the conversion of substrate to product.

Some poisons absorbed from an organism's environment act by inhibiting enzymes. For example, the pesticides DDT and parathion are inhibitors of key enzymes in the nervous system. Many antibiotics are inhibitors of specific enzymes in bacteria. For instance, penicillin blocks the active site of an enzyme that many bacteria use to make their cell walls.

Mentioning enzyme inhibitors that are metabolic poisons may give the impression that enzyme inhibition is generally abnormal and harmful. In fact, selective inhibition and activation of enzymes by molecules naturally present in the cell are essential mechanisms in metabolic control, as we discuss next.

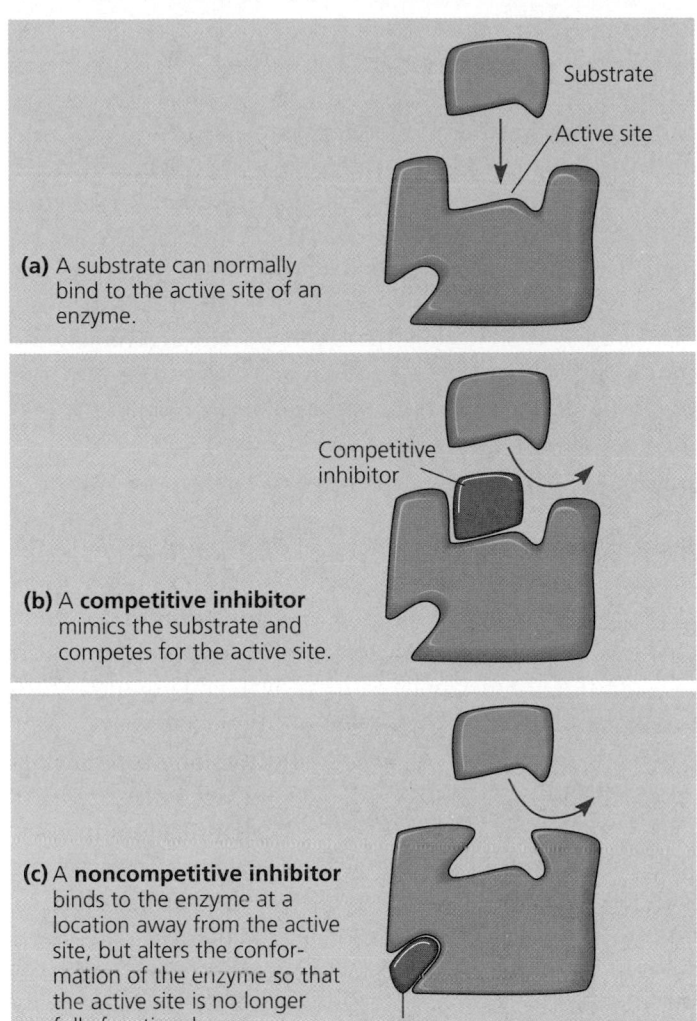

(a) A substrate can normally bind to the active site of an enzyme.

(b) A **competitive inhibitor** mimics the substrate and competes for the active site.

(c) A **noncompetitive inhibitor** binds to the enzyme at a location away from the active site, but alters the conformation of the enzyme so that the active site is no longer fully functional.

FIGURE 6.17 Inhibition of enzyme activity.

Chemical chaos would result if all of a cell's metabolic pathways were open simultaneously. Imagine, for example, a substance synthesized by one pathway and immediately broken down by another. The cell would be spinning its metabolic wheels. Actually, a cell tightly regulates its metabolic pathways by controlling when and where its various enzymes are active. It does this either by switching on and off the genes that encode specific enzymes (as we will discuss in Unit Three) or, as we discuss here, by regulating the activity of enzymes once they are made.

Metabolic control often depends on allosteric regulation

In many cases, the molecules that naturally regulate enzyme activity in a cell behave something like reversible noncompetitive inhibitors (see FIGURE 6.17c): These regulatory molecules change an enzyme's shape and function by binding weakly to an **allosteric site,** a specific receptor site on some part of the enzyme molecule remote from the active site. The effect of this allosteric regulation may be either inhibition or stimulation of the enzyme's activity.

Allosteric Regulation

Most allosterically regulated enzymes are constructed from two or more polypeptide chains, or subunits (FIGURE 6.18). Each subunit has its own active site, and allosteric sites are

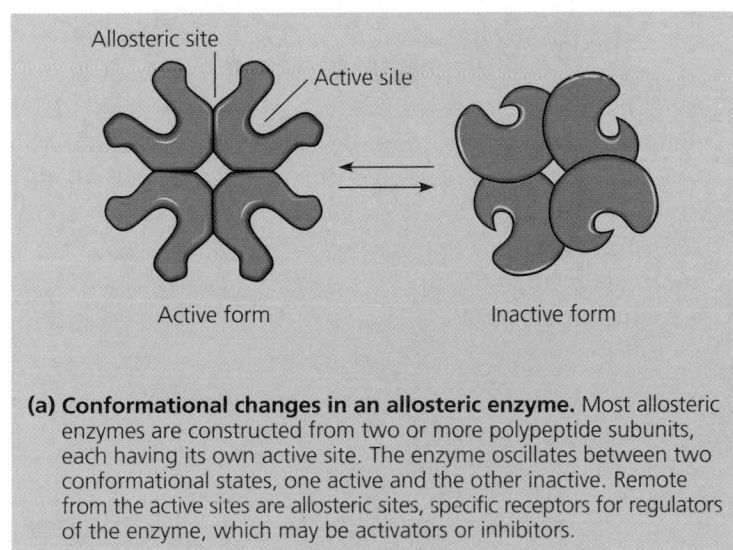

(a) **Conformational changes in an allosteric enzyme.** Most allosteric enzymes are constructed from two or more polypeptide subunits, each having its own active site. The enzyme oscillates between two conformational states, one active and the other inactive. Remote from the active sites are allosteric sites, specific receptors for regulators of the enzyme, which may be activators or inhibitors.

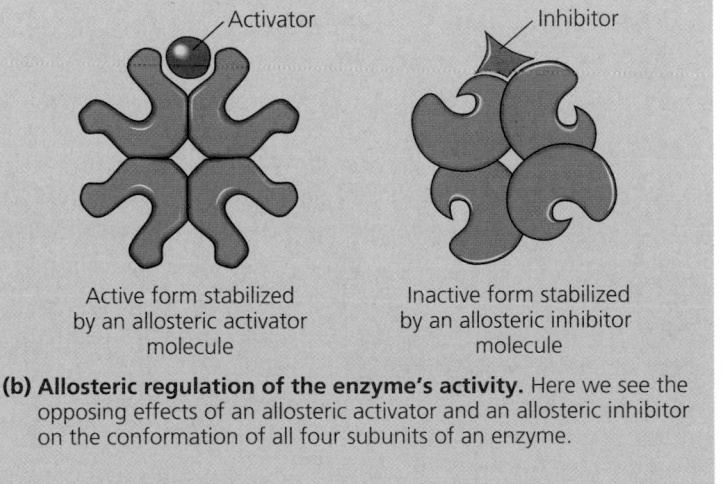

(b) **Allosteric regulation of the enzyme's activity.** Here we see the opposing effects of an allosteric activator and an allosteric inhibitor on the conformation of all four subunits of an enzyme.

FIGURE 6.18 Allosteric regulation of enzyme activity.

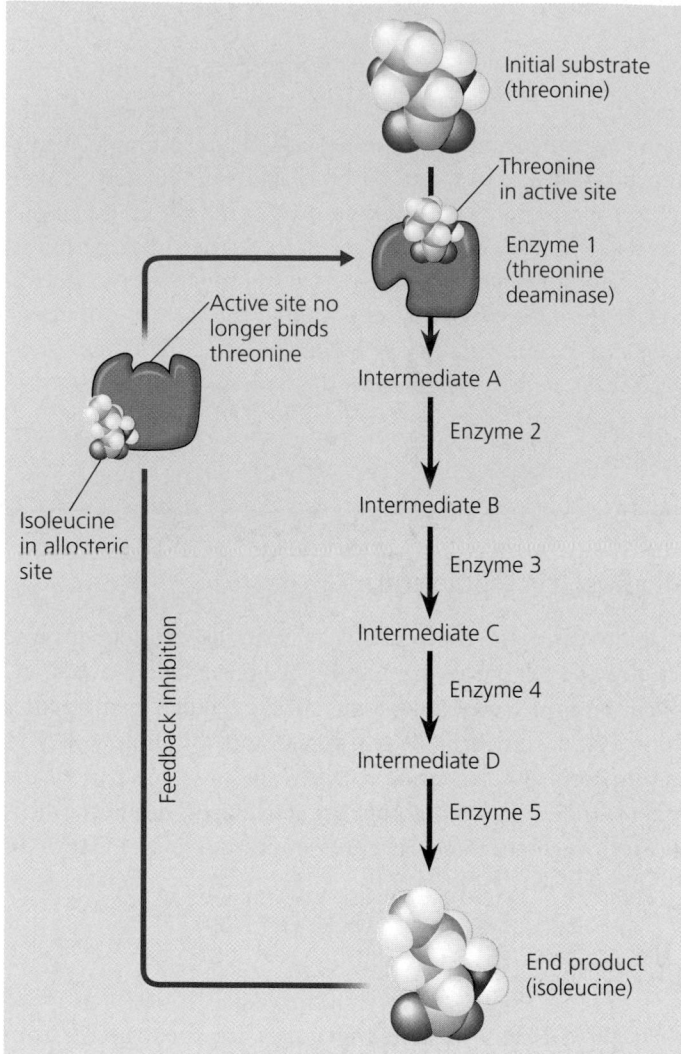

Initial substrate
(threonine)

Threonine
in active site

Enzyme 1
(threonine
deaminase)

Active site no
longer binds
threonine

Intermediate A

Enzyme 2

Intermediate B

Enzyme 3

Intermediate C

Enzyme 4

Intermediate D

Enzyme 5

Isoleucine
in allosteric
site

Feedback inhibition

End product
(isoleucine)

FIGURE 6.19 Feedback inhibition. Many metabolic pathways are switched off by an end product, which acts as an allosteric inhibitor of an enzyme earlier in the pathway. This example is the pathway for synthesizing the amino acid isoleucine. (Enzyme 1 is actually a tetramer of four identical subunits, like the hypothetical enzyme in FIGURE 6.18.)

often located where subunits are joined. The entire complex oscillates between two conformational states, one catalytically active and the other inactive. The binding of an activator to an allosteric site stabilizes the conformation that has a functional active site, whereas the binding of an allosteric inhibitor stabilizes the inactive form of the enzyme. The areas of contact between the subunits of an allosteric enzyme fit together in such a way that a conformational change in one subunit is transmitted to all others. Through this interaction of subunits, a single activator or inhibitor molecule that binds to one allosteric site will affect the active sites of all subunits.

The activity of an allosteric enzyme changes in response to fluctuating concentrations of the regulators. In some cases, an

inhibitor and an activator are similar enough in shape to compete for the same allosteric site. For example, some enzymes of catabolic pathways may have an allosteric site that fits both ATP and AMP (adenosine monophosphate), which the cell routinely derives from ADP. Such enzymes are inhibited by ATP and activated by AMP. This is logical because a major function of catabolism is to regenerate ATP. If ATP production lags behind its use, AMP accumulates and activates key enzymes that speed up catabolism. If the supply of ATP exceeds demand, then catabolism slows down as ATP molecules accumulate and compete for allosteric sites. In this way, allosteric enzymes control the rates of key reactions in metabolic pathways.

Feedback Inhibition

The inhibition of an ATP-generating catabolic pathway by the allosteric binding of ATP to an enzyme in the pathway is an example of feedback inhibition, one of the most common methods of metabolic control. **Feedback inhibition** is the switching off of a metabolic pathway by its end product, which acts as an inhibitor of an enzyme within the pathway. FIGURE 6.19 shows an example of this control mechanism operating on an anabolic pathway. Some cells use this pathway of five steps to synthesize the amino acid isoleucine from threonine, another amino acid. As isoleucine, the end product of the pathway, accumulates, it slows down its own synthesis by allosterically inhibiting the enzyme for the very first step of the pathway. Feedback inhibition thereby prevents the cell from wasting chemical resources to synthesize more isoleucine than is necessary.

Cooperativity

By a mechanism that resembles allosteric activation, substrate molecules may stimulate the catalytic powers of an enzyme (FIGURE 6.20). Recall that the binding of a substrate to an enzyme can induce a favorable change in the shape of the enzyme's active site (induced fit). If an enzyme has two or more subunits, such an interaction with one substrate molecule can trigger the same favorable conformational change in all other subunits of the enzyme. Called **cooperativity,** this mechanism amplifies the response of enzymes to substrates: One substrate molecule primes an enzyme to accept additional substrate molecules more readily.

The localization of enzymes within a cell helps order metabolism

The cell is not just a bag of chemicals with thousands of different kinds of enzymes and substrates wandering about randomly. Structures within the cell help bring order to metabolic

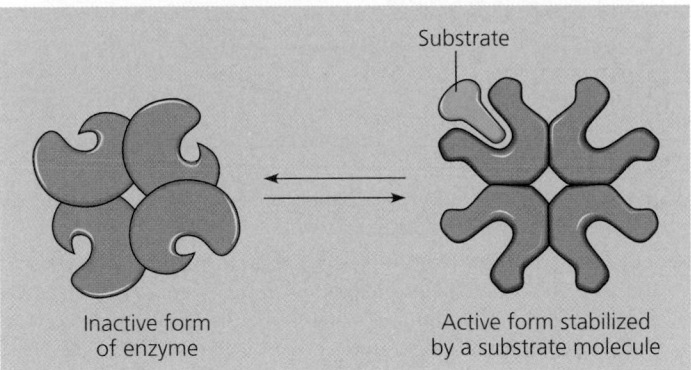

FIGURE 6.20 Cooperativity. In an enzyme molecule with multiple subunits, the binding of one substrate molecule to the active site of one subunit causes all the subunits to assume their active conformation.

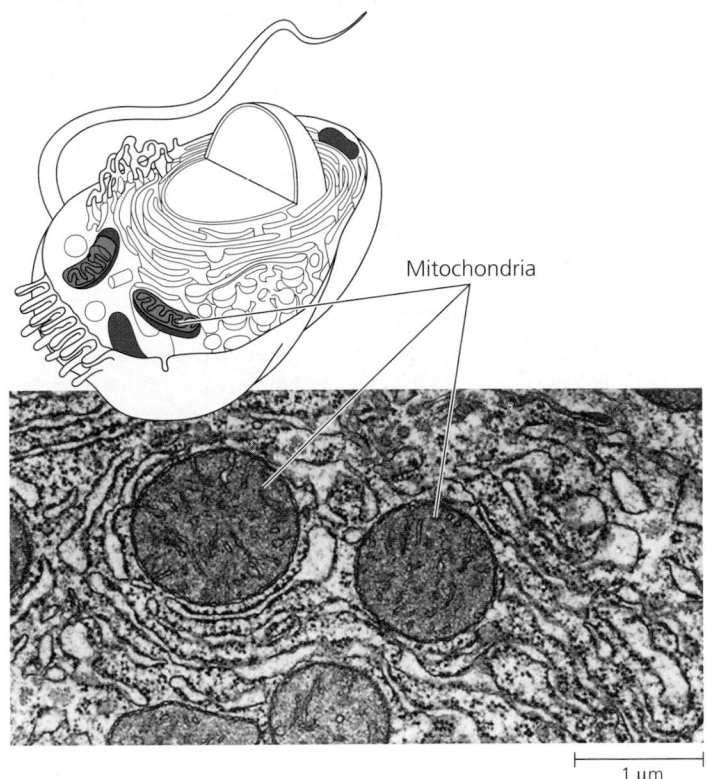

FIGURE 6.21 Organelles and structural order in metabolism. Membranes partition a eukaryotic cell into various metabolic compartments, or organelles, each with a corps of enzymes that carry out specific functions. The organelles shown in this micrograph are mitochondria, the sites of cellular respiration.

pathways. In some cases, a team of enzymes for several steps of a metabolic pathway is assembled together as a multienzyme complex. The arrangement controls the sequence of reactions, as the product from the first enzyme becomes substrate for the adjacent enzyme in the complex, and so on, until the end product is released. Some enzymes and enzyme complexes have fixed locations within the cell as structural components of particular membranes. Others are in solution within specific membrane-enclosed eukaryotic organelles, each with its own internal chemical environment. For example, in eukaryotic cells the enzymes for cellular respiration reside within mitochondria (FIGURE 6.21). If the cell had the same number of enzyme molecules for respiration but they were diluted throughout the entire volume of the cell, respiration would be very inefficient.

The structural basis of metabolic order brings us back to the theme with which this unit of chapters began.

The theme of emergent properties is manifest in the chemistry of life: *a review*

Recall that life is organized along a hierarchy of structural levels. With each increase in the level of order, new properties emerge in addition to those of the component parts. In Chapters 2–6, we have dissected the chemistry of life using the strategy of the reductionist. But we have also begun to develop a more integrated view of life as we have seen how properties emerge with increasing order.

We have seen that the unusual behavior of water, so essential to life on Earth, results from interactions of the water molecules, themselves an ordered arrangement of hydrogen and oxygen atoms. We reduced the great complexity and diversity of organic compounds to the chemical characteristics of carbon, but we also saw that the unique properties of organic compounds are related to the specific structural arrangements of carbon skeletons and their appended functional groups. We learned that small organic molecules are often assembled into giant molecules, but we also discovered that a macromolecule does not behave like a simple composite of its monomers. For example, the unique form and function of a protein are consequences of a hierarchy of primary, secondary, and tertiary structures. And in this chapter we have learned that metabolism, that orderly chemistry that characterizes life, is a concerted interplay of thousands of different kinds of molecules in an organized cell.

■   ■   ■

By completing our overview of metabolism with an introduction to its structural basis in the compartmentalized cell, we have built a bridge to Unit Two, where we will study the cell's structure and function in more depth. We will maintain our balance between the need to reduce life to a conglomerate of simpler processes and the ultimate satisfaction of viewing those processes in their integrated context.

CHAPTER 6 REVIEW

Go to the Campbell Biology website (www.campbellbiology.com) to explore an interactive version of the Chapter Review.

Summary of Key Concepts

METABOLISM, ENERGY, AND LIFE

- **The chemistry of life is organized into metabolic pathways (pp. 87–88)** Metabolism is the collection of chemical reactions that occur in an organism. Aided by enzymes, it follows intersecting pathways, which may be catabolic (breaking down molecules, releasing energy) or anabolic (building molecules, consuming energy).

- **Organisms transform energy (pp. 88–89)** Energy is the capacity to do work by moving matter. A moving object has kinetic energy. Potential energy is stored in the location or structure of matter and includes chemical energy stored in molecular structure. Energy can change form, governed by the laws of thermodynamics.
  Web/CD Activity 6A: *Energy Transformations*

- **The energy transformations of life are subject to two laws of thermodynamics (pp. 89–90, FIGURES 6.3 and 6.4)** The first, conservation of energy, states that energy cannot be created or destroyed. The second states that when energy changes form, entropy (S), or the disorder of the universe, increases. Matter can become more ordered only if the surroundings become more disordered.

- **Organisms live at the expense of free energy (pp. 91–94, FIGURES 6.5–6.7)** A living system's free energy is energy that can do work under cellular conditions. Free energy (G) is related directly to total energy (H) and to entropy (S): $\Delta G = \Delta H - T\Delta S$. Spontaneous changes involve a decrease in free energy ($-\Delta G$). In an exergonic (spontaneous) chemical reaction, the products have less free energy than the reactants ($-\Delta G$). Endergonic (nonspontaneous) reactions require an input of energy ($+\Delta G$). In cellular metabolism, exergonic reactions power endergonic reactions (energy coupling). The addition of starting materials and the removal of end products prevent metabolism from reaching equilibrium.

- **ATP powers cellular work by coupling exergonic reactions to endergonic reactions (pp. 94–96, FIGURES 6.8–6.10)** ATP is the cell's energy shuttle. Release of its terminal phosphate group produces ADP, inorganic phosphate, and free energy. ATP drives endergonic reactions by transfer of the phosphate to specific reactants, making them more reactive. In this way, cells can carry out work, such as movement and anabolism. Catabolic pathways drive the regeneration of ATP from ADP and phosphate.
  Web/CD Activity 6B: *The Structure of ATP*
  Web/CD Activity 6C: *Chemical Reactions and ATP*

ENZYMES

- **Enzymes speed up metabolic reactions by lowering energy barriers (pp. 96–97, FIGURES 6.12 and 6.13)** Enzymes, which are proteins, are biological catalysts. They speed up reactions by lowering activation energy (E_A), allowing bonds to break at moderate temperatures.

- **Enzymes are substrate specific (pp. 97–98, FIGURE 6.14)** Each type of enzyme has a unique active site that combines specifically with its substrate, the reactant molecule on which it acts. The enzyme changes shape slightly when it binds the substrate (induced fit).
  Web/CD Activity 6D: *How Enzymes Work*

- **The active site is an enzyme's catalytic center (pp. 98–99, FIGURE 6.15)** The active site can lower activation energy by orienting substrates correctly, straining their bonds, and providing a microenvironment that favors the reaction.

- **A cell's physical and chemical environment affects enzyme activity (pp. 99–101, FIGURES 6.16 and 6.17)** As proteins, enzymes are sensitive to conditions that influence their three-dimensional structure. Each has an optimal temperature and pH. Cofactors are metal ions or molecules required for some enzymes to function. Coenzymes are organic cofactors. Inhibitors reduce enzyme function. A competitive inhibitor binds to the active site, while a noncompetitive inhibitor binds to a different site on the enzyme.
  Web/CD Case Study in the Process of Science: *How Is the Rate of Enzyme Catalysis Measured?*
  Biology Labs On-Line: *EnzymeLab*

THE CONTROL OF METABOLISM

- **Metabolic control often depends on allosteric regulation (pp. 101–102, FIGURES 6.18–6.20)** Many enzymes change shape when regulatory molecules, either activators or inhibitors, bind to specific allosteric sites. In feedback inhibition, the end product of a metabolic pathway allosterically inhibits the enzyme for an earlier step in the pathway. In cooperativity, a substrate molecule binding to one active site of a multi-subunit enzyme activates the other subunits.

- **The localization of enzymes within a cell helps order metabolism (pp. 102–103, FIGURE 6.21)** Some enzymes are grouped into complexes, some are incorporated into membranes, and others are contained inside organelles.

- **The theme of emergent properties is manifest in the chemistry of life: *a review* (p. 103)** Higher levels of organization result in the emergence of new properties. Organization is the key to the chemistry of life.

Self-Quiz

1. Choose the pair of terms that correctly completes this sentence: Catabolism is to anabolism as _____ is to _____ .
   a. exergonic; spontaneous
   b. exergonic; endergonic
   c. free energy; entropy
   d. work; energy
   e. entropy; order

2. Most cells cannot harness heat to perform work because
   a. heat is not a form of energy.
   b. cells do not have much heat; they are relatively cool.
   c. temperature is usually uniform throughout a cell.
   d. heat cannot be used to do work.
   e. heat denatures enzymes.

3. According to the first law of thermodynamics,
   a. matter can be neither created nor destroyed.
   b. energy is conserved in all processes.
   c. all processes increase the order of the universe.
   d. systems rich in energy are intrinsically stable.
   e. the universe constantly loses energy because of friction.

4. Which of the following metabolic processes can occur without a net influx of energy from some other process?
   a. $ADP + \text{\textcircled{P}}_i \longrightarrow ATP + H_2O$
   b. $C_6H_{12}O_6 + 6\,O_2 \longrightarrow 6\,CO_2 + 6\,H_2O$

c. $6 CO_2 + 6 H_2O \longrightarrow C_6H_{12}O_6 + 6 O_2$

d. amino acids $\longrightarrow$ protein

e. glucose + fructose $\longrightarrow$ sucrose

5. If an enzyme has been noncompetitively inhibited,

   a. the ΔG for the reaction it catalyzes will always be negative.

   b. the active site will be occupied by the inhibitor molecule.

   c. increasing the substrate concentration will increase the inhibition.

   d. a higher activation energy will be necessary to initiate the reaction.

   e. the inhibitor molecule may be chemically unrelated to the substrate.

6. If an enzyme solution is saturated with substrate, the most effective way to obtain an even faster yield of products is to

   a. add more of the enzyme.

   b. heat the solution to 90°C.

   c. add more substrate.

   d. add an allosteric inhibitor.

   e. add a noncompetitive inhibitor.

7. An enzyme accelerates a metabolic reaction by

   a. altering the overall free-energy change for the reaction.

   b. making an endergonic reaction occur spontaneously.

   c. lowering the activation energy.

   d. pushing the reaction away from equilibrium.

   e. making the substrate molecule more stable.

8. Some bacteria are metabolically active in hot springs because

   a. they are able to maintain an internal temperature much cooler than that of the surrounding water.

   b. the high temperatures facilitate active metabolism without the need of catalysis.

   c. their enzymes have high optimal temperatures.

   d. their enzymes are insensitive to temperature.

   e. they use molecules other than proteins as their main catalysts.

9. Which of the following characteristics is not associated with allosteric regulation of an enzyme's activity?

   a. A molecule mimics the substrate and competes for the active site.

   b. A naturally occurring molecule stabilizes a catalytically active conformation.

   c. Regulatory molecules bind to a site remote from the active site.

   d. Inhibitor and activator molecules may compete with one another.

   e. The enzyme usually has a quaternary structure.

10. In the following branched metabolic pathway, a dotted arrow with a minus sign symbolizes inhibition of a metabolic step by an end product:

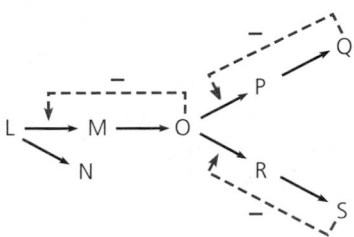

Which reaction would prevail if both Q and S were present in the cell in high concentrations?

   a. L $\longrightarrow$ M   d. O $\longrightarrow$ P

   b. M $\longrightarrow$ O   e. R $\longrightarrow$ S

   c. L $\longrightarrow$ N

11. How can an object at rest have energy?

12. Cellular respiration is an exergonic process. Remembering that energy must be conserved, what becomes of the energy released from glucose during cellular respiration?

13. In general, how does ATP transfer energy from exergonic to endergonic processes in the cell?

14. What is meant by "induced fit"?

15. A competitive inhibitor of the enzyme sucrase (see FIGURE 6.15) slows the production of glucose and fructose in a test-tube reaction. How could you overcome the effect of the inhibitor?

Go to the website or CD-ROM for more quiz questions.

Evolution Connection

A recent revival of the anti-evolutionary "argument from design" holds that biochemical pathways are too complex to have evolved, because all intermediate steps in a given pathway must be present to produce the final product. Critique this argument. How could you use the existing diversity of metabolic pathways that produce the same or similar products to support your case?

The Process of Science

A researcher has developed an assay to measure the activity of an important liver enzyme present in liver cells which are being cultured in a lab. She adds the substrate for the enzymatic reaction to the dish of cells, then measures the appearance of the reaction products. The results are graphed as the amount of product on the y axis versus time on the x axis. The researcher noted four sections of the graph. For a short period of time, no products appeared (section A). In section B, the reaction rate was quite rapid (the slope of the line was steep). After some time, the reaction slowed down considerably (section C), although products continued to appear (the line was not flat). Still later, the reaction resumed its original rapid rate (section D). Draw the graph, and propose a model to explain the molecular events underlying this interesting reaction profile.

Go to the Case Studies in the Process of Science on the CD-ROM or website to learn how to measure the rate of enzyme catalysis. Also link to Biology Labs On-Line to investigate the biochemistry of enzymes, including the effects of pH and temperature.

Science, Technology, and Society

As mandated by the 1996 Food Quality Protection Act, the EPA has announced its intention to evaluate the safety of the most commonly used organophosphate insecticides (organic compounds containing phosphate groups). In agriculture, the most frequently used organophosphates account for half of the 58 million pounds applied annually nationwide. Organophosphates typically interfere with nerve transmission by inhibiting the enzymes that degrade the transmitter molecules that diffuse from one neuron to another. Noxious insects are not uniquely susceptible; humans and other vertebrates can be affected as well. Thus, the use of organophosphate pesticides creates some health risks. As a consumer, what levels of risk are you willing to accept in exchange for an abundant and affordable food supply? Is it prudent to expect "a reasonable certainty that no harm will result from pesticide exposure"? What other facts would you like to know about this situation before you defend your opinion?

Answers: **1.** b; **2.** c; **3.** b; **4.** b; **5.** e; **6.** a; **7.** c; **8.** c; **9.** a; **10.** c; **11.** It can have potential energy due to its location. **12.** Some of it is stored in ATP molecules; the rest is released at heat. **13.** By phosphorylation, the addition of phosphate groups: Exergonic processes phosphorylate ADP to form ATP, which transfers the energy to endergonic processes by phosphorylating other molecules. **14.** Induced fit is the slight change in shape of the active site of an enzyme as it embraces its substrate. In its new shape, the active site catalyzes the reaction. **15.** Add a lot more sucrose.

THE CELL

Cell biologist George M. Langford is the Ernest Everett Just Professor of Biology at Dartmouth College. He grew up in rural North Carolina and received a B.S. degree from Fayetteville State University. He earned a Ph.D. from the Illinois Institute of Technology and did postdoctoral training at the University of Pennsylvania. Before arriving at Dartmouth, Dr. Langford held positions at the University of Massachusetts at Boston and at the medical schools of Howard University and the University of North Carolina. He has served as program director for the cell biology program of the National Science Foundation (NSF) and is a member of the National Science Board. He has made important discoveries about how organelles move inside cells; he also works to combat the underrepresentation of minorities in science. We met at the Marine Biological Laboratory (MBL) in Woods Hole, Massachusetts, where Dr. Langford spends his summers.

Interview
GEORGE M. LANGFORD

How did you choose science as a career?

It was actually by a process of elimination. I've always loved the arts, especially music, but in high school, I realized that I just didn't have the training to pursue a career in music. Luckily, I became interested in science and had some wonderful teachers who nurtured my interest. My high school didn't have good science facilities—North Carolina was still segregated. But perhaps because there weren't a lot of opportunities for blacks, many talented people became teachers. And the science teachers tried to encourage promising students. When I got to college, I decided to major in biology.

What drew you to cell biology?

In high school math, I had been particularly good at geometry, and it was shapes and structures that first attracted me to cell biology. Even as an undergraduate I loved microscopy, and I was good at visualizing the three-dimensional structures represented by two-dimensional images. This was the era when the electron microscope was first revealing the fine details of cell structure.

Who were some of your mentors during your graduate and postgraduate training?

I was fortunate to be able to do my Ph.D. research under Bill Danforth, who ran one of the strongest biology research groups at my university and was a supportive mentor. My research was on the biochemistry of glucose metabolism in *Euglena*—not microscopy. But while I was there, I was a teaching assistant for a famous embryologist, Jean Clark Dan. She urged me to do postdoctoral research with Shinya Inoue, a reknowned microscopist at Penn. Shinya taught me the importance of choosing the best experimental system for the question being asked. At his suggestion, I studied the movement of an exotic organelle, the axostyle, that certain protozoa use for swimming. It's made of microtubules, tiny protein tubes just like the ones in cilia [see FIGURE 1.12]. But unlike cilia, the axostyle is entirely inside the cell and causes the entire protozoan surface to undulate.

You went on to be a medical school professor. What led you to go to Dartmouth?

Dartmouth had just established a professorship in honor of Ernest Everett Just, a black alumnus who had graduated in 1907. I was invited by Dartmouth to apply for that position, and I discovered that I liked the community there very much. Also, I missed undergraduate teaching. So I left Chapel Hill for the cold North.

Who was Ernest Everett Just?

He was a developmental biologist and a pioneer who worked hard to break down barriers for black scientists in this country. His mother, a schoolteacher, had managed to enroll him in a prep school that sent most of its graduates to Dartmouth. Just was a stellar student there and at Dartmouth. After graduating and earning a Ph.D. at the University of Chicago, he did important research on fertilization and early development in animals, working mainly with marine invertebrates. At the time, genetics was an exciting new field in biology, and many biologists thought that embryonic development was completely controlled by the nucleus. It was Just who established the crucial role of the cytoplasm in fertilization and the early events of development. He showed that there were important signals going from the cytoplasm to the nucleus, as well as the other way around.

Did he achieve recognition in his day?

No, he experienced racial discrimination that eventually led him to emigrate to Europe; he returned to the United States only when the Nazis took over Germany. He was very dispirited, and he became ill and died at 59. I think his heart was broken by his failure to get a position at a major American research university. His graduate training had been with F. R. Lillie—for whom this MBL building is named—and Lillie was a father figure for him. But when he asked Lillie to recommend him for a position at a white university, Lillie was unwilling. By the way, in the summer Just worked with Lillie right here at the MBL.

What exactly is the MBL?

The MBL is an independent research institution owned and governed by the scientists who work here. Founded in 1888, it has had an enormous impact on biology in America because so many biologists have come here to do research or take a course. It's a major gathering place for biologists in the summer, and there are now year-round programs, too.

Is research limited to marine organisms?

No, lots of scientists here work on mice or other nonmarine animals. But a number of us do come here to work with marine organisms. I use the nervous system of the squid in my research, primarily because of the squid's giant axon.

What is a giant axon, and how do you use it in your research?

An axon is a long, thin extension of a nerve cell that carries electrical signals from the cell body (the vicinity of the nucleus) to another cell, which can be far away. The speed of electrical conduction along an axon is proportional to its thickness. Squids benefit from this principle by having an unusually thick axon that connects their brain area to muscles in their mantle, which they use for locomotion. When the squid sees a predator, it can signal its muscles very quickly and escape. The so-called giant axon results from the fusion of about a thousand ordinary axons. The length of the giant axon we use is about 25 cm.

The question we're investigating is the mechanism by which molecular "motors" move cargo around in the cell, using protein filaments as tracks. The filaments I'm concerned with are microtubules (just like the ones in cilia and axostyles) and thinner filaments called actin filaments, which are two of the elements of a structural network called the cytoskeleton. Molecular motors are proteins that utilize chemical energy from ATP to do mechanical work—such as moving an object along a track.

Now let's return to nerve cells and their axons. The cell body of a nerve cell produces a lot of materials that need to be transported to the end of the axon. For example, neurotransmitter molecules are synthesized and packaged in vesicles (small sacs) for eventual release at synapses [see FIGURE 2.18]. Molecular motors are responsible for transporting these vesicles from the cell body to the axon terminals, and they do so using filaments of the cytoskeleton as tracks. We use the squid giant axon for studying this transport not only because of its size but also because you can strip away the plasma membrane and leave the cytoplasm intact with transport still going on. We can then watch and record the filaments and vesicles by high-resolution video microscopy, which cannot be done with other cells. And we can easily test the effects of different chemicals on transport.

We also use the methods of molecular biology, including recombinant DNA technology. For instance, we've cloned the gene for the motor protein we're working on, called myosin-5, and we can use bacteria to produce the different parts of the motor separately. We can then add different combinations of these parts to our giant-axon preparations to determine what the various parts do.

What's the most exciting discovery you've made using squid giant axons?

When we first started looking at vesicle transport in giant axons, it was thought that the only filaments involved were microtubules. But one day, as we were observing vesicles moving under the microscope, we suddenly noticed that there were vesicles moving in areas of the preparation lacking microtubules. And we said, aha, something interesting is going on here—there have to be *other* kinds of filaments supporting the transport, ones not visible with our microscope. Actin filaments were a prime candidate. To make them visible, we labeled them with a fluorescent dye. Sure enough, fluorescing images of actin filaments then showed up in the "empty" regions where the vesicle transport was taking place.

Based on those observations, we proposed a *dual transport mechanism*, by which transport vesicles use microtubules as expressways and actin filaments as local streets. The vesicles move down the axon along microtubules, but once at the axon terminal, they transfer onto actin filaments to get to the plasma membrane.

What questions are you addressing now?

We're trying to learn how a vesicle transfers from a microtubule to an actin filament. We know that there are two different kinds of motors: The motor molecule that

moves a vesicle along a microtubule will not work on an actin filament. So there has to be a switch from one motor to the other. We've learned that the two motors interact directly with each other, and we're now trying to understand this interaction.

Do you also study other kinds of cells?

I'm using mammalian kidney and lung cells in collaborating with a group at Dartmouth that studies cystic fibrosis (CF), a devastating genetic disease. The CF defect is in a protein called CFTR, which normally is located in the plasma membrane and helps transport chloride ions across. We're studying how CFTR gets from its site of synthesis to the membrane, a process that requires molecular motors. In most CF patients, the CFTR protein has a defect that prevents it from reaching the membrane. If we can figure out exactly how CFTR is transported, we may be able to develop a treatment to help patients.

Tell us about some of your efforts on behalf of minorities in science.

A legacy of the racism encountered by Ernest Just continues to discourage minorities. While at the NSF, I attacked this problem at the postdoc level by starting a program to help minority Ph.D.s get the advanced training and mentoring they need to develop research careers. Later, at Dartmouth, I set up the E. E. Just Program, with the goal of raising the number of minority science majors. We have a forum that brings students together with professors, who talk about what excites them in science and serve as resources and mentors for the students. We also provide internships enabling students to do research in faculty labs. We're now seeing an increase in the number of minority science majors, though there is still more to be done.

A TOUR
OF THE CELL

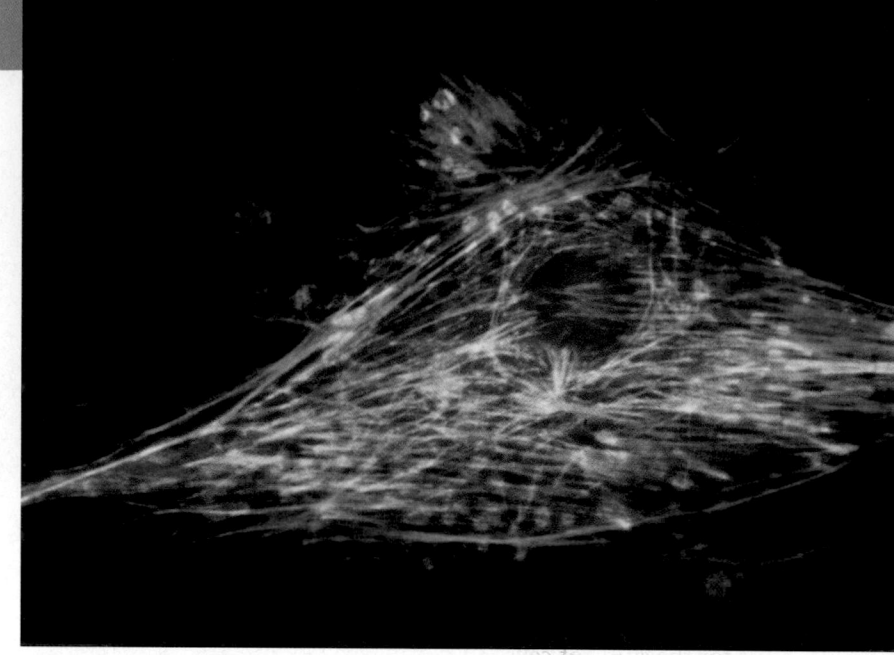

HOW WE STUDY CELLS

- Microscopes provide windows to the world of the cell
- Cell biologists can isolate organelles to study their functions

A PANORAMIC VIEW OF THE CELL

- Prokaryotic and eukaryotic cells differ in size and complexity
- Internal membranes compartmentalize the functions of a eukaryotic cell

THE NUCLEUS AND RIBOSOMES

- The nucleus contains a eukaryotic cell's genetic library
- Ribosomes build a cell's proteins

THE ENDOMEMBRANE SYSTEM

- The endoplasmic reticulum manufactures membranes and performs many other biosynthetic functions
- The Golgi apparatus finishes, sorts, and ships cell products
- Lysosomes are digestive compartments
- Vacuoles have diverse functions in cell maintenance

OTHER MEMBRANOUS ORGANELLES

- Mitochondria and chloroplasts are the main energy transformers of cells
- Peroxisomes generate and degrade H_2O_2 in performing various metabolic functions

THE CYTOSKELETON

- Providing structural support to the cell, the cytoskeleton also functions in cell motility and regulation

CELL SURFACES AND JUNCTIONS

- Plant cells are encased by cell walls
- The extracellular matrix (ECM) of animal cells functions in support, adhesion, movement, and regulation
- Intercellular junctions help integrate cells into higher levels of structure and function
- The cell is a living unit greater than the sum of its parts

The cell is as fundamental to biology *as the atom is to chemistry: All organisms are made of cells. In the hierarchy of biological organization, the cell is the simplest collection of matter that can live. Indeed, there are diverse forms of life existing as single-celled organisms. More complex organisms, including plants and animals, are multicellular; their bodies are cooperatives of many kinds of specialized cells that could not survive for long on their own. However, even when they are arranged into higher levels of organization, such as tissues and organs, cells can be singled out as the organism's basic units of structure and function. The contraction of muscle cells moves your eyes as you read this sentence; when you decide to turn the next page, nerve cells will transmit that decision from your brain to the muscle cells of your hand. Everything an organism does occurs fundamentally at the cellular level. This chapter introduces the microscopic world of the cell.*

This book takes a thematic approach to the study of life, and the cell is a microcosm that demonstrates most of the themes introduced in Chapter 1. We will see that life at the cellular level arises from structural order, reinforcing the themes of emergent properties and the correlation between structure and function. For example, the movement of an animal cell depends on an intricate interplay of structures that make up a cellular skeleton (green in the micrograph above). Another recurring theme in biology is the interaction of organisms with their environment. Cells sense and respond to environmental fluctuations. As open systems, they continuously exchange both materials and energy with their surroundings. And keep in mind the one biological theme that unifies all others: evolution. All cells are related by their descent from earlier cells, but they have been modified in various ways during the long evolutionary history of life on Earth.

HOW WE STUDY CELLS

It can be difficult to understand how a cell, usually too small to be seen by the unaided eye, can be so complex. How can cell biologists possibly investigate the inner workings of such tiny entities? Before we actually tour the cell, it will be helpful to learn how cells are studied.

Microscopes provide windows to the world of the cell

THE PROCESS OF SCIENCE

The evolution of a science often parallels the invention of instruments that extend human senses to new limits. The discovery and early study of cells progressed with the invention and improvement of microscopes in the 17th century. Microscopes of various types are still indispensable tools for the study of cells.

The microscopes first used by Renaissance scientists, as well as the microscopes you are likely to use in the laboratory, are all **light microscopes (LMs).** Visible light is passed through the specimen and then through glass lenses. The lenses refract (bend) the light in such a way that the image of the specimen is magnified as it is projected into the eye, onto photograph film, or onto a video screen. (See the appendix at the back of the book that diagrams microscope structure.)

Two important parameters in microscopy are magnification and resolving power, or resolution. Magnification in microscopy is the ratio of an object's image to its real size. **Resolving power** is a measure of the clarity of the image; it is the minimum distance two points can be separated and still be distinguished as two separate points. For example, what appears to the unaided eye as one star in the sky may be resolved as twin stars with a telescope.

Just as the resolving power of the human eye is limited, the resolving power of telescopes and microscopes is limited. Microscopes can be designed to magnify objects as much as desired, but the light microscope can never resolve detail finer than about 0.2 micrometer (μm), the size of a small bacterium (FIGURE 7.1). This resolution is limited by the shortest wavelength of light used to illuminate the specimen. Light microscopes can magnify effectively to about 1,000 times the size of the actual specimen; at greater magnifications, the image becomes increasingly blurry. Most of the improvements in light microscopy since the beginning of the 20th century have involved new methods for enhancing contrast, which clarifies the details that can be resolved (TABLE 7.1, p. 110). In addition, scientists have developed methods for labeling particular cell components so that they stand out visually.

Although cells were discovered by Robert Hooke in 1665, the geography of the cell was largely uncharted until the 1950s. Most subcellular structures, or **organelles,** are too small to be resolved by the light microscope. Cell biology advanced rapidly

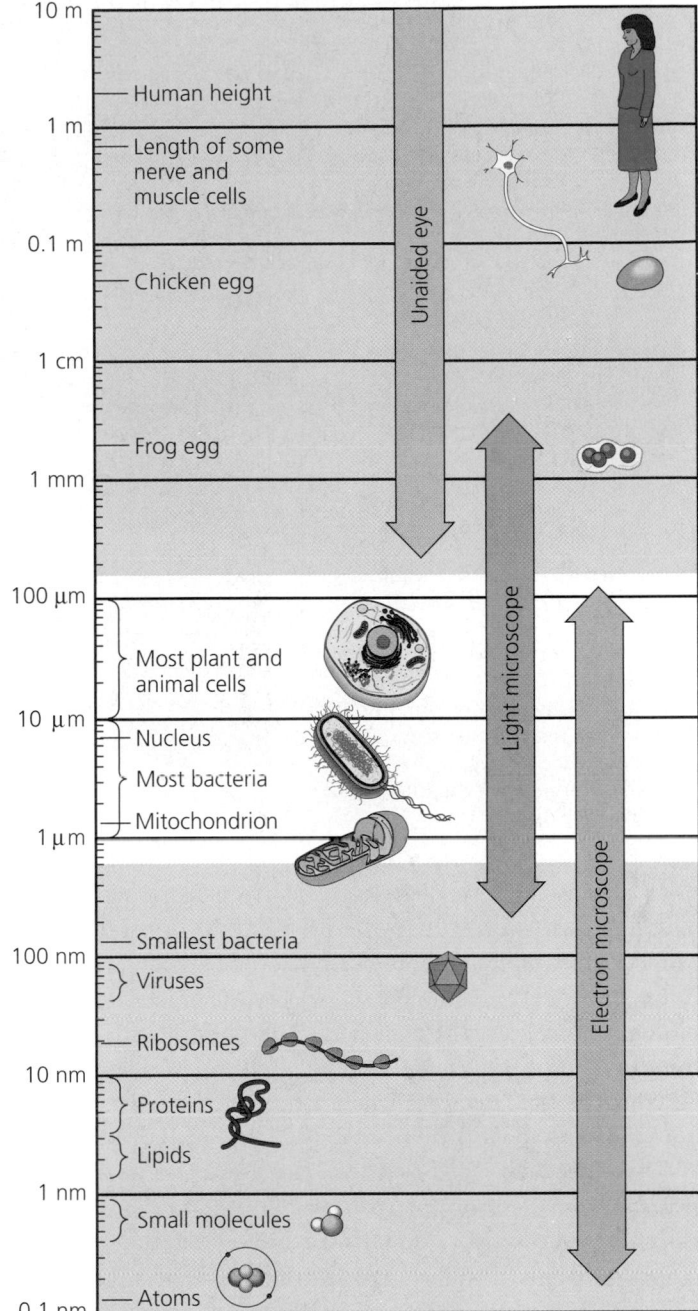

Measurements
1 centimeter (cm) = 10^{-2} meter (m) = 0.4 inch
1 millimeter (mm) = 10^{-3} m
1 micrometer (μm) = 10^{-3} mm = 10^{-6} m
1 nanometer (nm) = 10^{-3} μm = 10^{-9} m

FIGURE 7.1 The size range of cells. Most cells (yellow region of chart) are between 1 and 100 μm in diameter and are therefore visible only under a microscope. Notice that the scale along the left side is logarithmic to accommodate the range of sizes shown. Starting at the top of the scale with 10 m and going down, each reference measurement marks a tenfold decrease in diameter or length.

in the 1950s with the introduction of the electron microscope. Instead of using light, the **electron microscope (EM)** focuses a beam of electrons through the specimen or onto its surface. Resolving power is inversely related to the wavelength

Table 7.1 Different Types of Light Microscopy: A Comparison

Type of Microscopy	Light Micrographs of Human Cheek Epithelial Cells		Type of Microscopy
Brightfield (unstained specimen). Passes light directly through specimen; unless cell is naturally pigmented or artificially stained, image has little contrast.	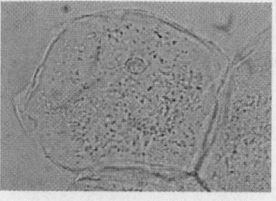	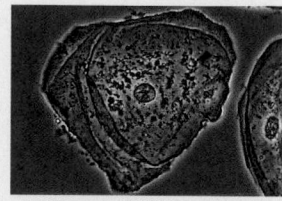	**Phase-contrast.** Enhances contrast in unstained cells by amplifying variations in density within specimen; especially useful for examining living, unpigmented cells.
Brightfield (stained specimen). Staining with various dyes enhances contrast, but most staining procedures require that cells be fixed (preserved).	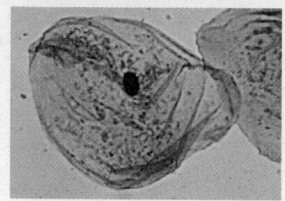	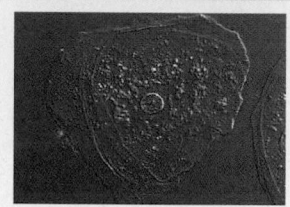	**Differential-interference-contrast (Nomarski).** Like phase-contrast microscopy, it uses optical modifications to exaggerate differences in density.
Fluorescence. Shows the locations of specific molecules in the cell. Fluorescent substances absorb short-wavelength, ultraviolet radiation and emit longer-wavelength, visible light. The fluorescing molecules may occur naturally in the specimen but more often are made by tagging the molecules of interest with fluorescent dyes or antibodies.	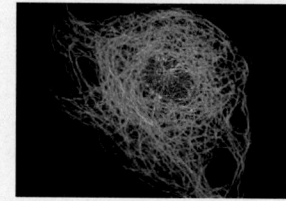	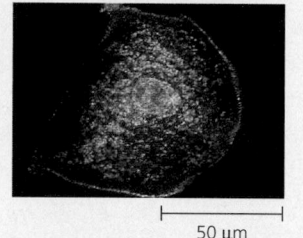 50 µm	**Confocal.** Uses lasers and special optics for "optical sectioning." Only those regions within a narrow depth of focus are imaged. Regions above and below the selected plane of view appear black rather than blurry. This microscope is typically used with fluorescently stained specimens, as in the example here.

of radiation a microscope uses, and electron beams have wavelengths much shorter than the wavelengths of visible light. Modern electron microscopes can theoretically achieve a resolution of about 0.1 nanometer (nm), but the practical limit for biological structures is generally only about 2 nm—still a hundredfold improvement over the light microscope. Biologists use the term *cell ultrastructure* to refer to a cell's anatomy as revealed by an electron microscope.

There are two basic types of electron microscopes: the **transmission electron microscope (TEM)** and the **scanning electron microscope (SEM).** Cell biologists use the TEM mainly to study the internal ultrastructure of cells. The TEM aims an electron beam through a thin section of the specimen, similar to the way a light microscope transmits light through a slide. However, instead of using glass lenses, the TEM uses electromagnets as lenses to focus and magnify the image by bending the paths of the electrons. The image is ultimately focused onto a screen for viewing or onto photographic film. To enhance contrast in the image, very thin sections of preserved cells are stained with atoms of heavy metals, which attach to certain cellular structures (FIGURE 7.2a).

The SEM is especially useful for detailed study of the surface of the specimen (FIGURE 7.2b). The electron beam scans the surface of the sample, which is usually coated with a thin

FIGURE 7.2 Electron micrographs. (a) This micrograph, taken with a transmission electron microscope (TEM), profiles a thin section of part of a cell from a rabbit trachea (windpipe), revealing its ultrastructure. **(b)** A scanning electron microscope (SEM) produced this three-dimensional image of the surface of the same type of cell. Both micrographs show motile organelles called cilia. Beating of the cilia that line the windpipe helps move inhaled debris upward toward the pharynx (throat).

From this point on in the book, micrographs are identified by the type of microscopy: LM for a light micrograph, TEM for a transmission electron micrograph, and SEM for a scanning electron micrograph.

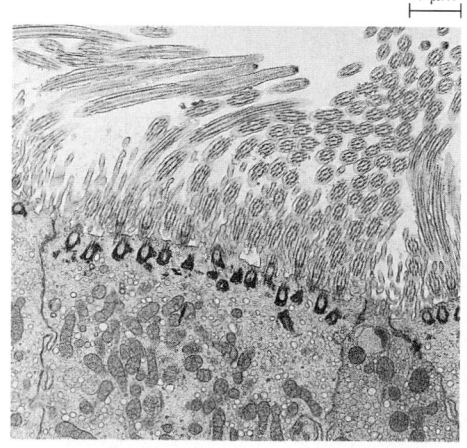
1 µm

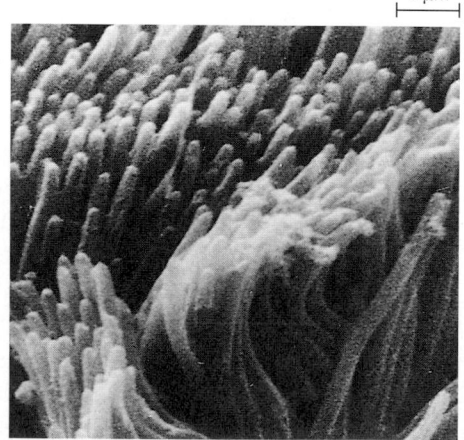
1 µm

(a) Transmission electron micrograph (TEM)   **(b) Scanning electron micrograph (SEM)**

film of gold. The beam excites electrons on the sample's surface, and these secondary electrons are collected and focused onto a screen. The result is an image of the topography of the specimen. The SEM has great depth of field, which results in an image that appears three-dimensional.

Electron microscopes reveal many organelles that are impossible to resolve with the light microscope. But the light microscope offers advantages, especially for the study of live cells. A disadvantage of electron microscopy is that the methods used to prepare the specimen kill the cells. Also, these methods may introduce artifacts, structural features seen in micrographs that do not exist in the living cell. (Artifacts can occur in light microscopy, too.)

Microscopes are the most important tools of *cytology,* the study of cell structure. But simply describing the diverse organelles within the cell reveals little about their function. Modern cell biology developed from an integration of cytology with *biochemistry,* the study of the molecules and chemical processes of metabolism. A biochemical approach called cell fractionation has been particularly important in cell biology.

Cell biologists can isolate organelles to study their functions

The goal of **cell fractionation** is to take cells apart, separating the major organelles so that their functions can be studied (FIGURE 7.3). The instrument used to fractionate cells is the centrifuge, a merry-go-round for test tubes that can spin at various speeds. The most powerful machines, called **ultracentrifuges,** can spin as fast as 130,000 revolutions per minute (rpm) and apply forces on particles of more than 1 million times the force of gravity (1,000,000 *g*).

Fractionation begins with homogenization, the disruption of cells. The objective is to break the cells apart without damaging their organelles. Spinning the soupy homogenate in a centrifuge at low speed separates the parts of the cell into two fractions: the pellet, consisting of the larger, heavier structures that become packed at the bottom of the test tube; and the supernatant, consisting of the smaller, lighter parts of the cell suspended in the liquid above the pellet. The supernatant is decanted into another tube and centrifuged again at a higher speed. The process is repeated, increasing the speed with each step, collecting smaller and smaller components of the homogenized cells in successive pellets.

Cell fractionation enables the researcher to prepare specific components of cells in bulk quantity in order to study their composition and functions. By following this approach, biologists have been able to assign various functions of the cell to the different organelles, a task that would be far more difficult with intact cells. For example, one cellular fraction collected by centrifugation has enzymes that function in the metabolic process known as cellular respiration. The electron microscope reveals this fraction to be very rich in the organelles called mitochondria. This evidence helped cell biologists determine that mitochondria are the sites of cellular respiration. Cytology and biochemistry complement each other in correlating cellular structure and function.

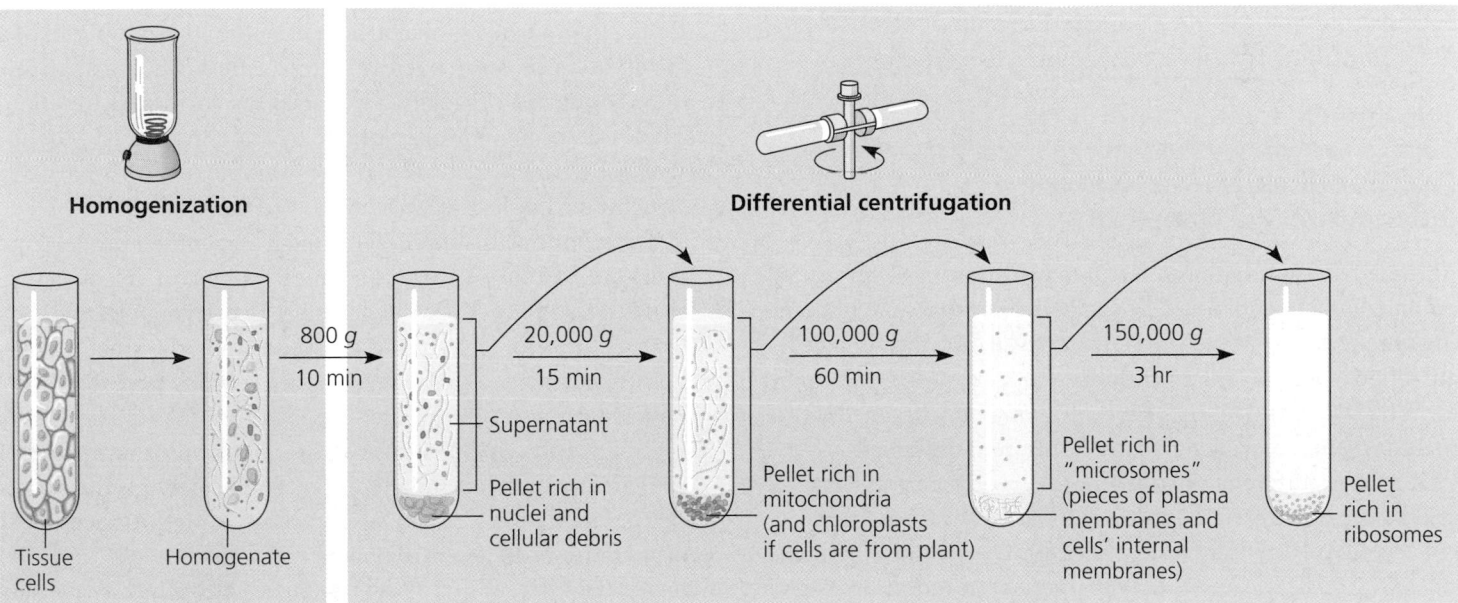

FIGURE 7.3 Cell fractionation. Disrupted cells are centrifuged at various speeds and durations to isolate (fractionate) components of different sizes. By determining which cell fractions are associated with particular metabolic processes, researchers can tie those functions to certain organelles.

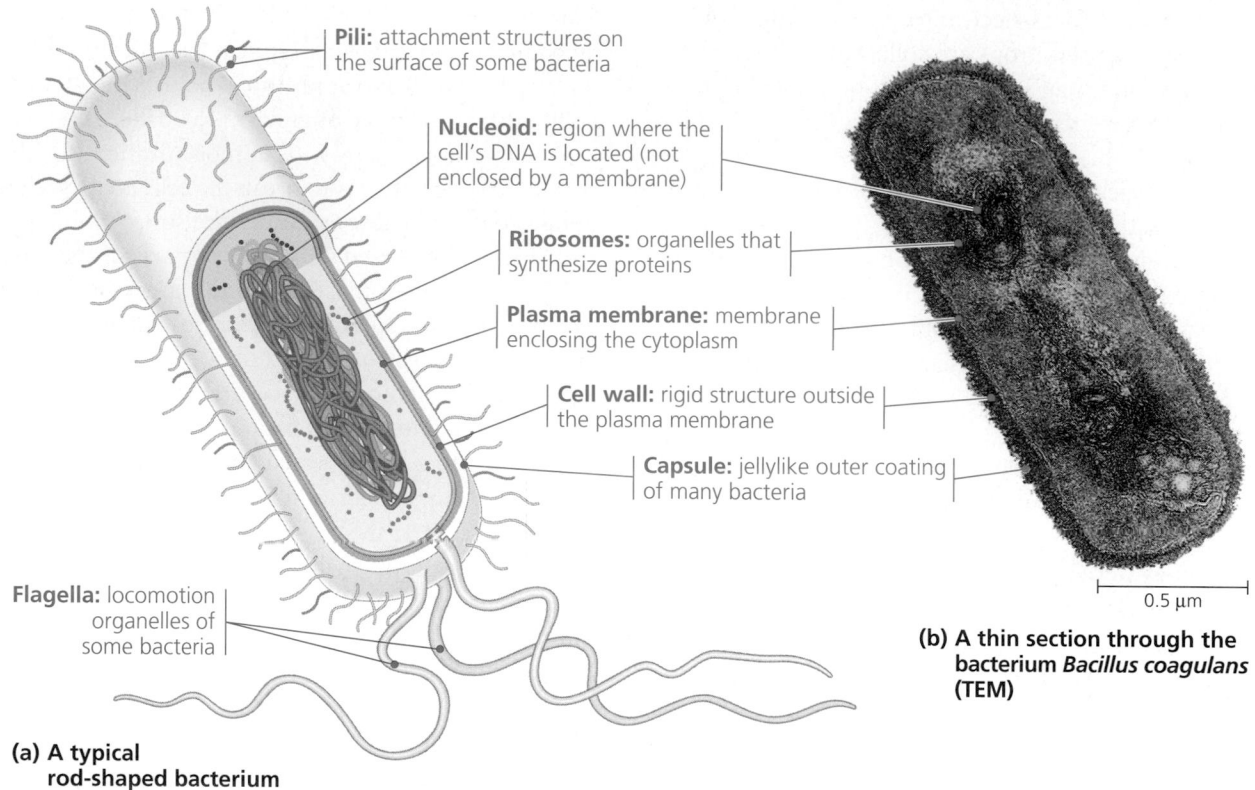

Pili: attachment structures on the surface of some bacteria

Nucleoid: region where the cell's DNA is located (not enclosed by a membrane)

Ribosomes: organelles that synthesize proteins

Plasma membrane: membrane enclosing the cytoplasm

Cell wall: rigid structure outside the plasma membrane

Capsule: jellylike outer coating of many bacteria

Flagella: locomotion organelles of some bacteria

0.5 µm

(b) A thin section through the bacterium *Bacillus coagulans* (TEM)

(a) A typical rod-shaped bacterium

FIGURE 7.4 A prokaryotic cell. Lacking a true nucleus and the other membrane-enclosed organelles of the eukaryotic cell, the prokaryotic cell is much simpler in structure. Only organisms of the domains Bacteria and Archaea have prokaryotic cells.

A PANORAMIC VIEW OF THE CELL

Every organism is composed of one of two structurally different types of cells: prokaryotic cells or eukaryotic cells. Only the bacteria and archaea have prokaryotic cells. Protists, plants, fungi, and animals all have eukaryotic cells.

Prokaryotic and eukaryotic cells differ in size and complexity

All cells have several basic features in common: They are all bounded by a membrane, called a *plasma membrane.* Within the membrane is a semifluid substance, **cytosol,** in which organelles are found. All cells contain chromosomes, carrying genes in the form of DNA. And all cells have *ribosomes,* tiny organelles that make proteins according to instructions from the genes.

A major difference between prokaryotic and eukaryotic cells, indicated by their names, is that the chromosomes of a eukaryotic cell are located in a membrane-enclosed organelle called the *nucleus.* The word *prokaryotic* is from the Greek *pro,* before, and *karyon,* kernel, referring here to the nucleus. In a **prokaryotic cell** (FIGURE 7.4), the DNA is concentrated in a region called the **nucleoid,** but no membrane separates this region from the rest of the cell. In contrast, the eukaryotic cell

(Greek *eu,* true, and *karyon*) has a true nucleus, bounded by a membranous nuclear envelope (see FIGURES 7.7 and 7.8, pp. 114–115). The entire region between the nucleus and the plasma membrane is called the **cytoplasm,** a term also used for the interior of a prokaryotic cell. Within the cytoplasm of a eukaryotic cell, suspended in cytosol, are a variety of membrane-bounded organelles of specialized form and function. These are absent in prokaryotic cells. Thus, the presence or absence of a true nucleus is just one example of the disparity in structural complexity between the two types of cells.

Eukaryotic cells are generally much bigger than prokaryotic cells (see FIGURE 7.1). Size is a general aspect of cell structure that relates to function. The logistics of carrying out metabolism set limits on cell size. At the lower limit, the smallest cells known are bacteria called mycoplasmas, which have diameters between 0.1 and 1.0 µm. These are perhaps the smallest packages with enough DNA to program metabolism and enough enzymes and other cellular equipment to carry out the activities necessary for a cell to sustain itself and reproduce. Most bacteria are 1–10 µm in diameter, a dimension about ten times bigger than that of mycoplasmas. Eukaryotic cells are typically 10–100 µm in diameter, ten times bigger than bacteria.

Metabolic requirements also impose upper limits on the size that is practical for a single cell. As an object of a particular shape increases in size, its volume grows proportionately

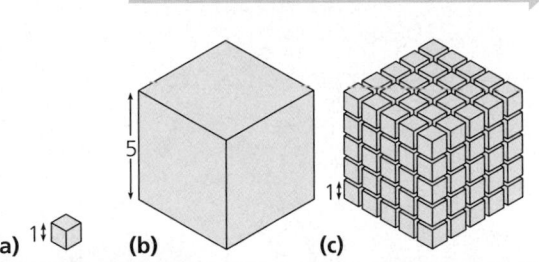

Surface area increases while total volume remains constant

FIGURE 7.5 Geometric relationships explain why most cells are microscopic. In this diagram, cells are represented as boxes. Using arbitrary units of length, we can calculate the cell's surface area (in square units), volume (in cubic units), and surface-to-volume ratio. A high surface-to-volume ratio facilitates the exchange of materials between a cell and its environment.

	(a)	(b)	(c)
Total surface area (height × width × number of sides × number of boxes)	6	150	750
Total volume (height × width × length × number of boxes)	1	125	125
Surface-to-volume ratio (area ÷ volume)	6	1.2	6

more than its surface area. (Area is proportional to a linear dimension squared, whereas volume is proportional to the linear dimension cubed.) Thus, the smaller the object, the greater its ratio of surface area to volume (FIGURE 7.5).

At the boundary of every cell, the **plasma membrane** functions as a selective barrier that allows sufficient passage of oxygen, nutrients, and wastes to service the entire volume of the cell (FIGURE 7.6). For each square micrometer of membrane, only so much of a particular substance can cross per second. Rates of chemical exchange with the extracellular environ-

ment might be inadequate to maintain a cell with a very large cytoplasm. The need for a surface sufficiently large to accommodate the volume helps explain the microscopic size of most cells. Larger organisms do not generally have *larger* cells than smaller organisms—simply *more* cells.

Prokaryotic cells will be described in detail in Chapters 18 and 27, and the possible evolutionary relationships between prokaryotic and eukaryotic cells will be discussed in Chapter 28. Most of the discussion of cell structure that follows in this chapter applies to eukaryotic cells.

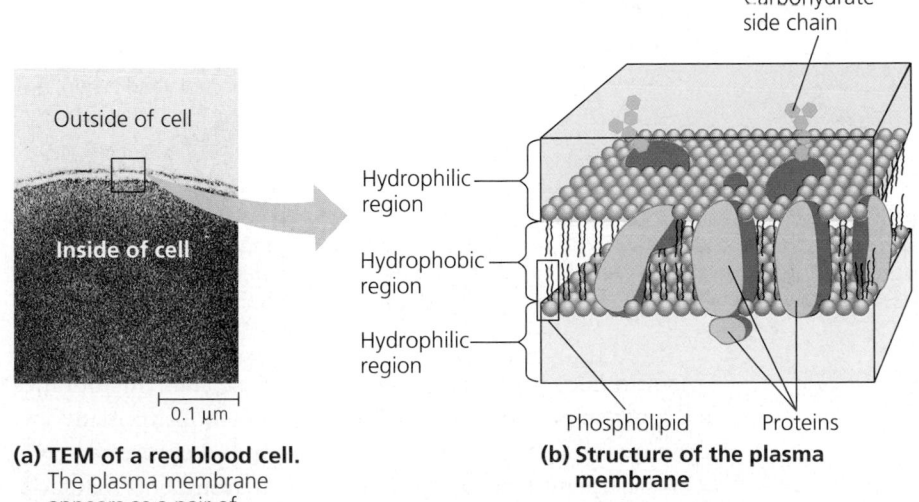

Outside of cell

Inside of cell

0.1 μm

(a) TEM of a red blood cell. The plasma membrane appears as a pair of dark bands separated by a light band.

Carbohydrate side chain

Hydrophilic region

Hydrophobic region

Hydrophilic region

Phospholipid   Proteins

(b) Structure of the plasma membrane

FIGURE 7.6 The plasma membrane. The plasma membrane and the membranes of organelles consist of a double layer (bilayer) of phospholipids with various proteins attached to or embedded in it. The phospholipid tails in the interior of a membrane are hydrophobic; the phospholipid heads, the exterior proteins and parts of proteins, and any carbohydrate side chains are hydrophilic and in contact with the aqueous solution on either side of the membrane. Carbohydrate side chains are found only on the outer surface of the plasma membrane. The specific functions of a membrane depend on the kinds of phospholipids and proteins present.

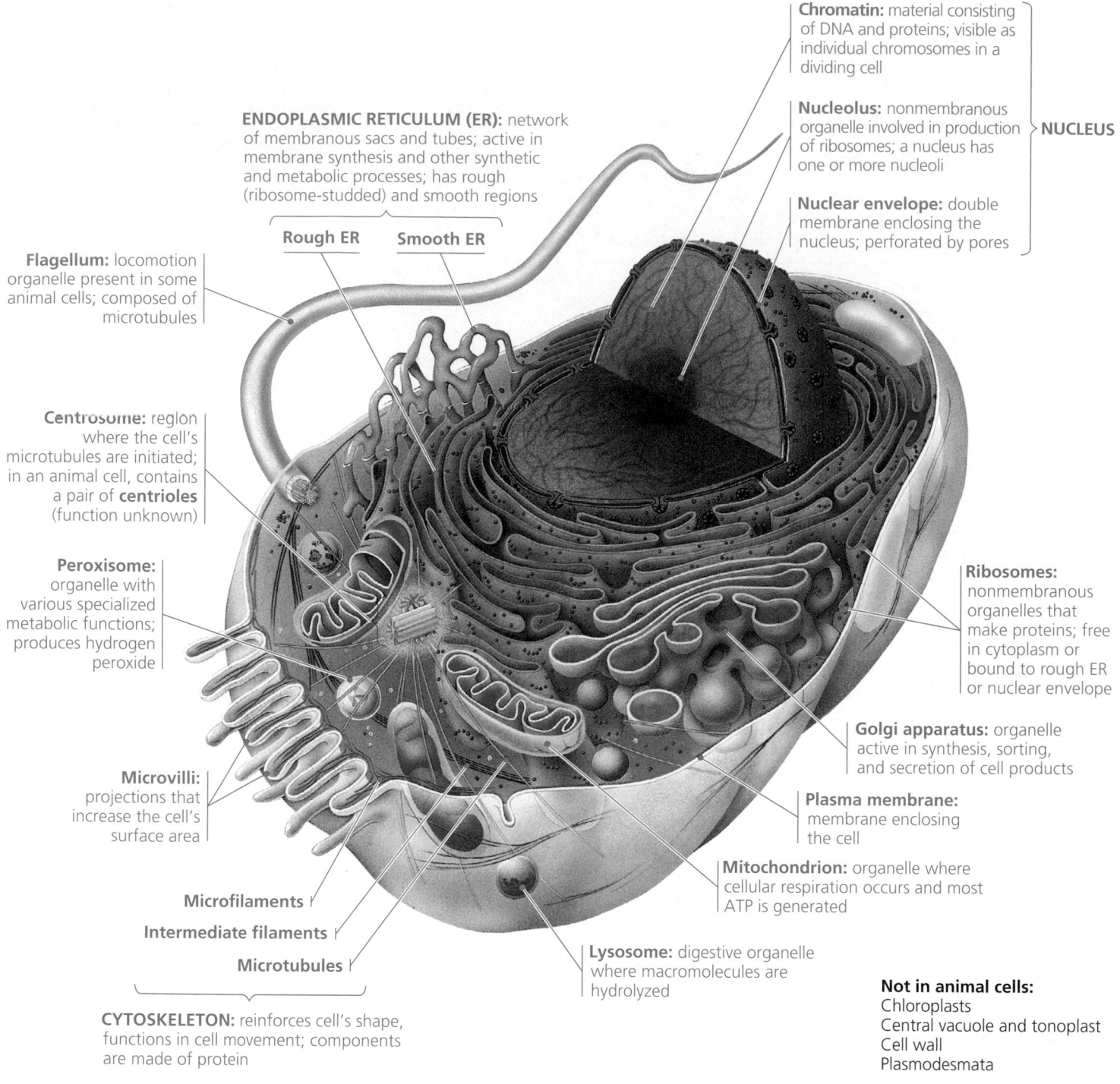

Chromatin: material consisting of DNA and proteins; visible as individual chromosomes in a dividing cell

Nucleolus: nonmembranous organelle involved in production of ribosomes; a nucleus has one or more nucleoli ⎱ **NUCLEUS**

Nuclear envelope: double membrane enclosing the nucleus; perforated by pores

ENDOPLASMIC RETICULUM (ER): network of membranous sacs and tubes; active in membrane synthesis and other synthetic and metabolic processes; has rough (ribosome-studded) and smooth regions

Rough ER | Smooth ER

Flagellum: locomotion organelle present in some animal cells; composed of microtubules

Centrosome: region where the cell's microtubules are initiated; in an animal cell, contains a pair of **centrioles** (function unknown)

Peroxisome: organelle with various specialized metabolic functions; produces hydrogen peroxide

Microvilli: projections that increase the cell's surface area

Microfilaments
Intermediate filaments
Microtubules

CYTOSKELETON: reinforces cell's shape, functions in cell movement; components are made of protein

Ribosomes: nonmembranous organelles that make proteins; free in cytoplasm or bound to rough ER or nuclear envelope

Golgi apparatus: organelle active in synthesis, sorting, and secretion of cell products

Plasma membrane: membrane enclosing the cell

Mitochondrion: organelle where cellular respiration occurs and most ATP is generated

Lysosome: digestive organelle where macromolecules are hydrolyzed

Not in animal cells:
Chloroplasts
Central vacuole and tonoplast
Cell wall
Plasmodesmata

FIGURE 7.7 Overview of an animal cell. This drawing of an animal cell incorporates the most common structures of animal cells (no cell actually looks just like this). The cell has a variety of organelles ("little organs"), many of which are bounded by membranes. The most prominent organelle in an animal cell is usually the nucleus. Most of the cell's metabolic activities occur in the cytoplasm, the entire region between the nucleus and the plasma membrane. The cytoplasm contains many organelles suspended in a semifluid medium, the cytosol. Pervading much of the cytoplasm is a labyrinth of membranes called the endoplasmic reticulum (ER).

Internal membranes compartmentalize the functions of a eukaryotic cell

In addition to the plasma membrane at its outer surface, a eukaryotic cell has extensive and elaborately arranged internal membranes, which partition the cell into compartments—the membranous organelles mentioned earlier. These membranes also participate directly in the cell's metabolism; many enzymes are built right into the membranes. Because the cell's compartments provide different local environments that facilitate specific metabolic functions, incompatible processes can go on simultaneously inside the same cell.

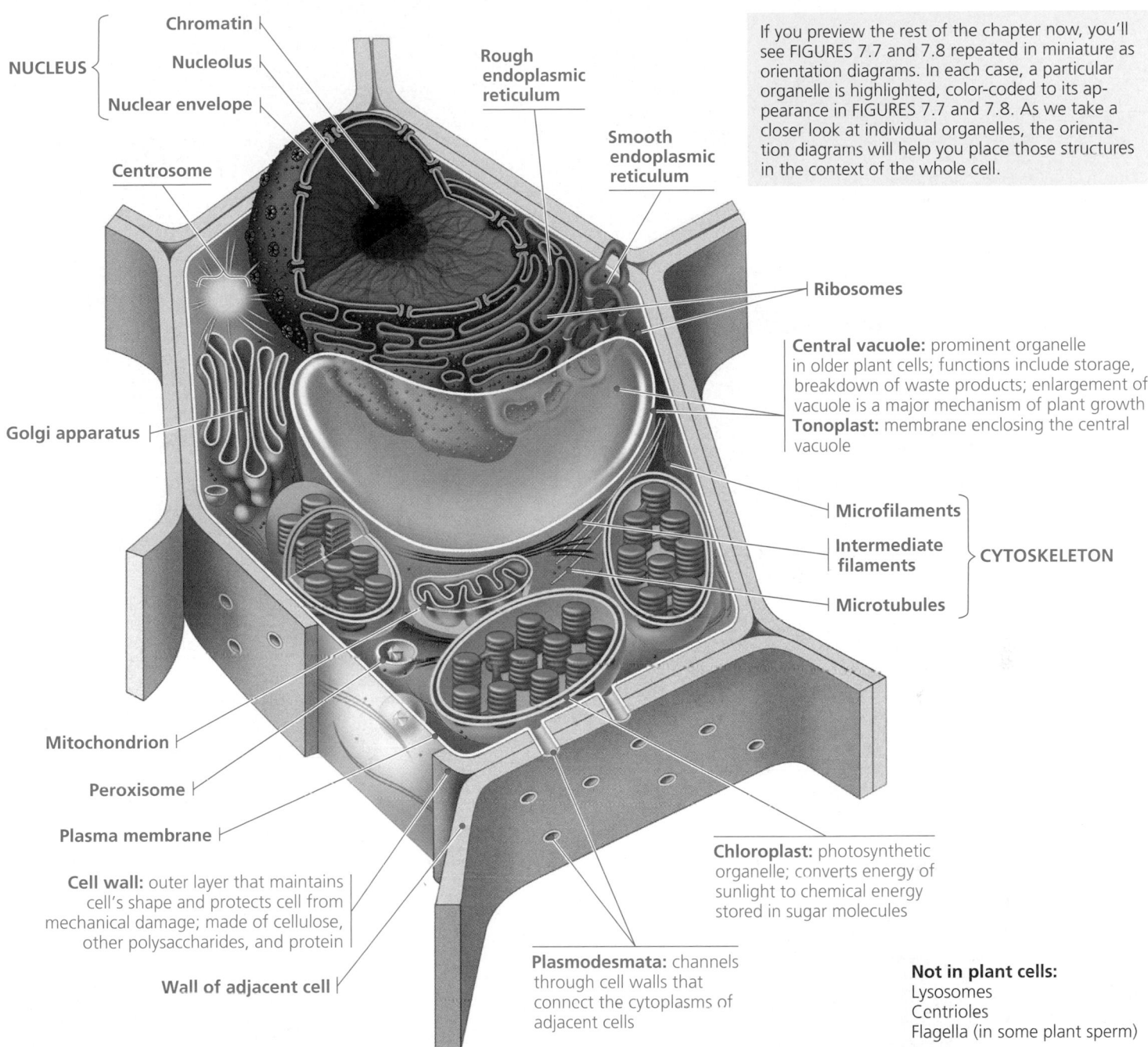

NUCLEUS
- Chromatin
- Nucleolus
- Nuclear envelope

Centrosome

Rough endoplasmic reticulum

Smooth endoplasmic reticulum

If you preview the rest of the chapter now, you'll see FIGURES 7.7 and 7.8 repeated in miniature as orientation diagrams. In each case, a particular organelle is highlighted, color-coded to its appearance in FIGURES 7.7 and 7.8. As we take a closer look at individual organelles, the orientation diagrams will help you place those structures in the context of the whole cell.

Ribosomes

Central vacuole: prominent organelle in older plant cells; functions include storage, breakdown of waste products; enlargement of vacuole is a major mechanism of plant growth
Tonoplast: membrane enclosing the central vacuole

Golgi apparatus

Microfilaments
Intermediate filaments
Microtubules
CYTOSKELETON

Mitochondrion

Peroxisome

Plasma membrane

Cell wall: outer layer that maintains cell's shape and protects cell from mechanical damage; made of cellulose, other polysaccharides, and protein

Wall of adjacent cell

Plasmodesmata: channels through cell walls that connect the cytoplasms of adjacent cells

Chloroplast: photosynthetic organelle; converts energy of sunlight to chemical energy stored in sugar molecules

Not in plant cells:
Lysosomes
Centrioles
Flagella (in some plant sperm)

FIGURE 7.8 Overview of a plant cell. This drawing of a generalized plant cell reveals the similarities and differences between an animal cell and a plant cell. In addition to most of the features seen in an animal cell, a plant cell has membrane-enclosed organelles called plastids. The most important type of plastid is the chloroplast, which carries out photosynthesis. Many plant cells have a large central vacuole. Outside a plant cell's plasma membrane is a thick cell wall, perforated by channels called plasmodesmata.

Membranes of various kinds are fundamental to the organization of the cell. In general, biological membranes consist of a double layer of phospholipids and other lipids. Embedded in this lipid bilayer or attached to its surfaces are diverse proteins (see FIGURE 7.6). However, each type of membrane has a unique composition of lipids and proteins suited to that membrane's specific functions. For example, enzymes embedded in the membranes of the organelles called mitochondria function in cellular respiration.

Before continuing with this chapter, examine the overviews of eukaryotic cells in FIGURES 7.7 and 7.8 on these two pages. These figures introduce the various organelles and provide a map of the cell for the detailed tour upon which we will now embark. FIGURES 7.7 and 7.8 also contrast animal and plant cells. As eukaryotic cells, they have much more in common than either has with any prokaryote. As you will see, however, there are important differences between plant and animal cells.

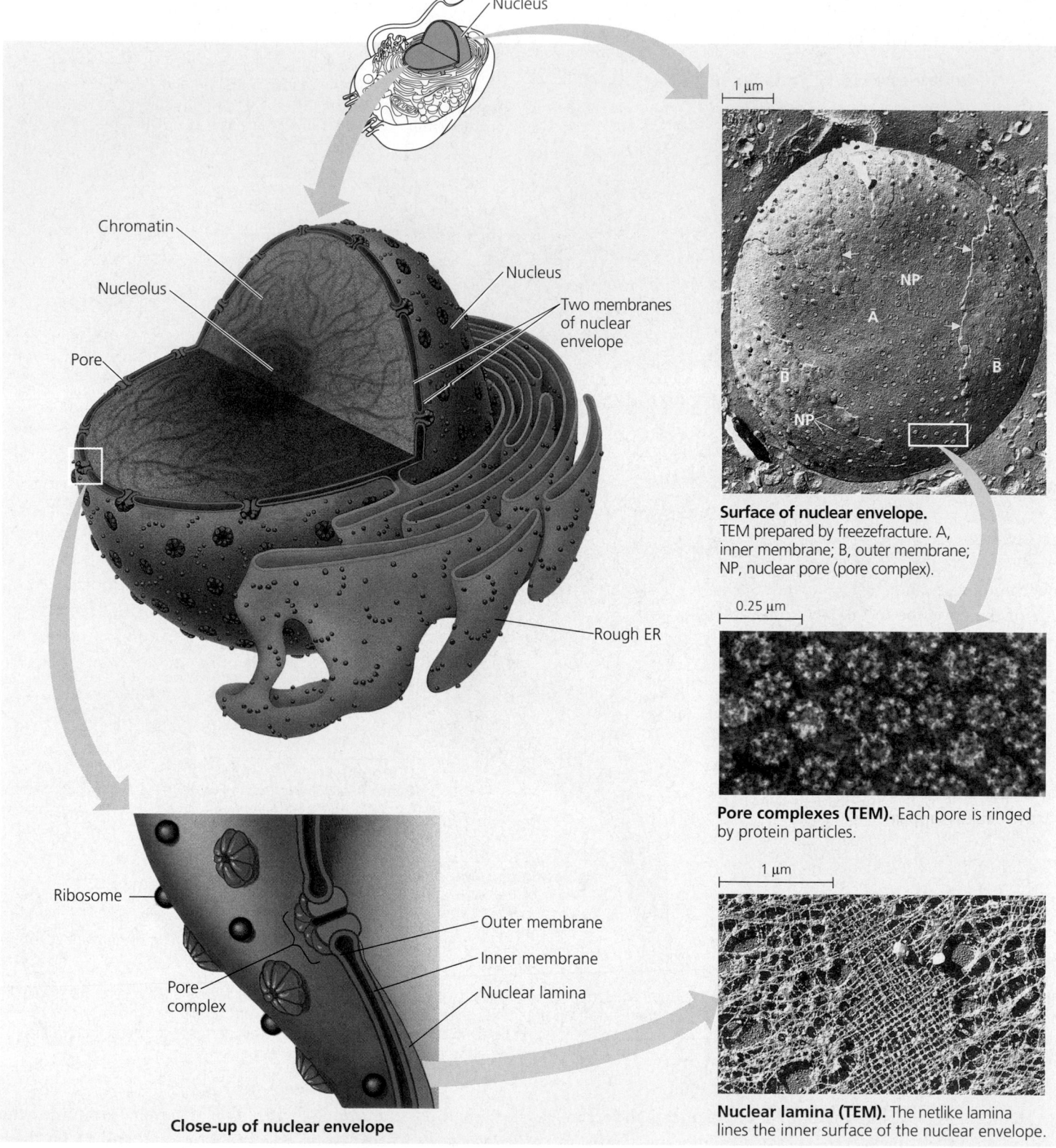

Chromatin

Nucleus

Nucleolus

Nucleus

Two membranes
of nuclear
envelope

Pore

Rough ER

Surface of nuclear envelope.
TEM prepared by freezefracture. A,
inner membrane; B, outer membrane;
NP, nuclear pore (pore complex).

1 µm

0.25 µm

Pore complexes (TEM). Each pore is ringed
by protein particles.

1 µm

Nuclear lamina (TEM). The netlike lamina
lines the inner surface of the nuclear envelope.

Ribosome

Pore
complex

Outer membrane

Inner membrane

Nuclear lamina

Close-up of nuclear envelope

FIGURE 7.9 The nucleus and its envelope. Within the nucleus is chromatin, consisting
of DNA and proteins. When a cell prepares to divide, individual chromosomes become
visible as the chromatin condenses. The nucleolus functions in ribosome synthesis. The
nuclear envelope, which consists of two membranes separated by a narrow space, is
perforated with pores and lined by a nuclear lamina.

Upper TEM from L. Orci and A. Perrelet, *Freeze-Etch Histology* (Heidelberg: Springer-Verlag, 1975). ©1975 Springer-Verlag.
Middle TEM from A. C. Faberge, *Cell Tissue Res.* 151(1974):403. ©1974 Springer-Verlag.

THE NUCLEUS AND RIBOSOMES

At the first stop of our detailed tour of the cell are two organelles involved in the genetic control of the cell: the nucleus, which houses most of the cell's DNA, and the ribosomes, which use information from the DNA to make proteins.

The nucleus contains a eukaryotic cell's genetic library

The **nucleus** contains most of the genes in the eukaryotic cell (some genes are located in mitochondria and chloroplasts). It is generally the most conspicuous organelle in a eukaryotic cell, averaging about 5 μm in diameter. The nuclear envelope encloses the nucleus (FIGURE 7.9), separating its contents from the cytoplasm.

The nuclear envelope is a *double* membrane. The two membranes, each a lipid bilayer with associated proteins, are separated by a space of about 20–40 nm. The envelope is perforated by pores that are about 100 nm in diameter. At the lip of each pore, the inner and outer membranes of the nuclear envelope are fused. An intricate protein structure called a pore complex lines each pore and regulates the entry and exit of certain large macromolecules and particles. Except at the pores, the nuclear side of the envelope is lined by the **nuclear lamina,** a netlike array of protein filaments (intermediate filaments) that maintains the shape of the nucleus. There is also much evidence for a *nuclear matrix,* a framework of fibers extending throughout the nuclear interior. (We will examine possible functions of the nuclear lamina and matrix in Chapter 19.)

Within the nucleus, the DNA is organized along with proteins into a fibrous material called **chromatin.** Stained chromatin usually appears through both light microscopes and electron microscopes as a diffuse mass. As a cell prepares to divide (reproduce), however, the thin chromatin fibers coil up (condense), becoming thick enough to be discerned as separate structures called **chromosomes.** Each eukaryotic species has a characteristic number of chromosomes. A typical human cell, for example, has 46 chromosomes in its nucleus; the exceptions are the sex cells (eggs and sperm), which have only 23 chromosomes in humans.

A prominent structure within the nondividing nucleus is the **nucleolus,** which appears through the electron microscope as a mass of densely stained granules and fibers adjoining part of the chromatin. Here a special type of RNA called ribosomal RNA is synthesized and assembled with proteins imported from the cytoplasm into the main components of ribosomes, called ribosomal subunits. These subunits then pass through the nuclear pores to the cytoplasm, where they can combine to form ribosomes. Sometimes there are two or more nucleoli; the number depends on the species and the stage in the cell's reproductive cycle.

As we saw in FIGURE 5.28, the nucleus directs protein synthesis by synthesizing messenger RNA (mRNA) and sending it to the cytoplasm via the nuclear pores. The mRNA is made according to instructions provided by the DNA (as is ribosomal RNA). Once an mRNA molecule reaches the cytoplasm, ribosomes translate its genetic message into the primary structure of a specific polypeptide. This process of translating genetic information is described in detail in Chapter 17.

Ribosomes build a cell's proteins

Ribosomes, particles made of ribosomal RNA and protein, are the organelles that carry out protein synthesis; each is composed of two subunits (FIGURE 7.10). Cells that have high

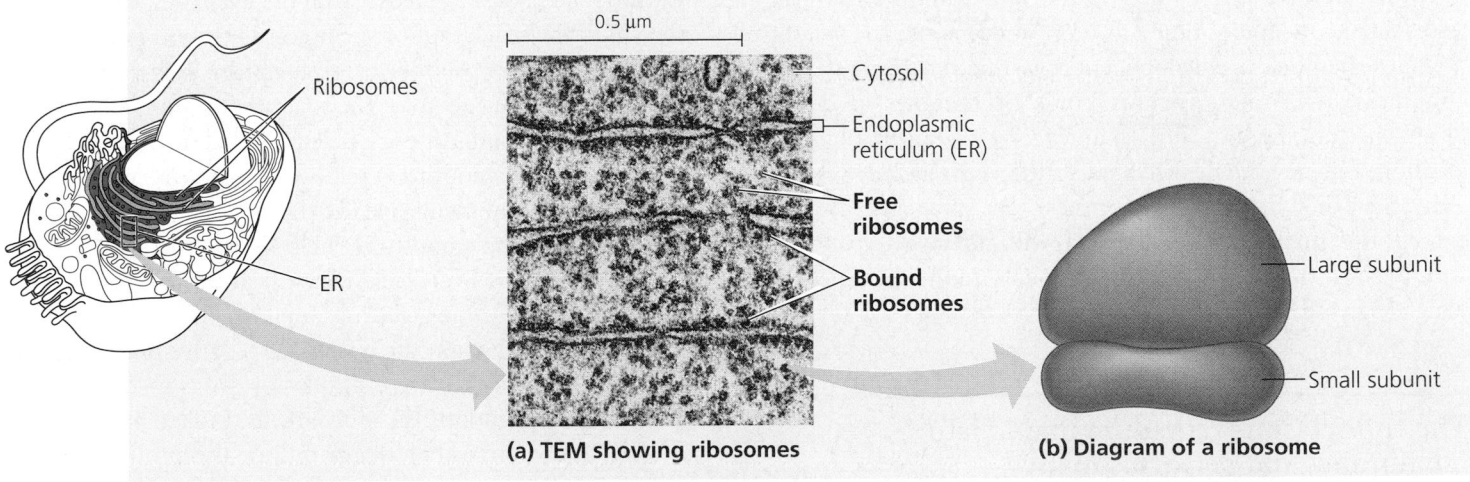

(a) TEM showing ribosomes

(b) Diagram of a ribosome

FIGURE 7.10 Ribosomes. (a) This electron micrograph of part of a pancreas cell shows many ribosomes, both free (in the cytosol) and bound (to the endoplasmic reticulum). The bound ribosomes of pancreas cells make a number of secretory proteins, including the hormone insulin and digestive enzymes. Bound ribosomes also make proteins destined for insertion into membranes or the interiors of other organelles. Free ribosomes mainly make proteins that remain dissolved in the cytosol. Bound and free ribosomes are identical and can alternate between these two roles. **(b)** This simplified diagram of a ribosome shows its two subunits.

rates of protein synthesis have a particularly large number of ribosomes. For example, a human pancreas cell has a few million ribosomes. Not surprisingly, cells active in protein synthesis also have prominent nucleoli. (Keep in mind that both nucleoli and ribosomes, unlike most other organelles, are not enclosed in membrane.)

Ribosomes build proteins in two cytoplasmic locales (see FIGURE 7.10). *Free* ribosomes are suspended in the cytosol, while *bound* ribosomes are attached to the outside of the endoplasmic reticulum or nuclear envelope. Most of the proteins made by free ribosomes will function within the cytosol; examples are enzymes that catalyze the first steps of sugar breakdown. Bound ribosomes generally make proteins that are destined either for insertion into membranes, for packaging within certain organelles such as lysosomes, or for export from the cell (secretion). Cells that specialize in protein secretion—for instance, the cells of the pancreas and other glands that secrete digestive enzymes—frequently have a high proportion of bound ribosomes. Bound and free ribosomes are structurally identical and can alternate between the two roles; the cell adjusts the relative numbers of each as its metabolism changes. You will learn more about ribosome structure and function in Chapter 17.

THE ENDOMEMBRANE SYSTEM

Many of the different membranes of the eukaryotic cell are part of an **endomembrane system.** These membranes are related either through direct physical continuity or by the transfer of membrane segments as tiny **vesicles** (sacs made of membrane). Despite these relationships, the various membranes are not identical in structure and function. Moreover, the thickness, molecular composition, and metabolic behavior of a membrane are not fixed, but may be modified several times during the membrane's life. The endomembrane system includes the nuclear envelope, endoplasmic reticulum, Golgi apparatus, lysosomes, various kinds of vacuoles, and the plasma membrane (not actually an *endo*membrane in physical location, but nevertheless related to the endoplasmic reticulum and other internal membranes). We have already discussed the nuclear envelope and will now focus on the endoplasmic reticulum and the other endomembranes to which it gives rise.

The endoplasmic reticulum manufactures membranes and performs many other biosynthetic functions

The **endoplasmic reticulum (ER)** is a membranous labyrinth so extensive that it accounts for more than half the total membrane in many eukaryotic cells. (The word *endoplasmic* means "within the cytoplasm," and *reticulum* is Latin for "little net.") The ER consists of a network of membranous tubules and sacs called cisternae (Latin, *cisterna,* a reservoir for a liquid). The ER membrane separates the internal compartment of the ER, called the cisternal space, from the cytosol. And because the ER membrane is continuous with the nuclear envelope, the space between the two membranes of the envelope is continuous with the cisternal space of the ER (FIGURE 7.11).

There are two distinct, though connected, regions of ER that differ in structure and function: smooth ER and rough ER. **Smooth ER** is so named because its cytoplasmic surface lacks ribosomes. **Rough ER** appears rough through the electron microscope because ribosomes stud the cytoplasmic surface of the membrane. As already mentioned, ribosomes are also attached to the cytoplasmic side of the nuclear envelope's outer membrane, which is confluent with rough ER.

Functions of Smooth ER

The smooth ER of various cell types functions in diverse metabolic processes, including synthesis of lipids, metabolism of carbohydrates, and detoxification of drugs and poisons.

Enzymes of the smooth ER are important to the synthesis of lipids, including oils, phospholipids, and steroids. Among the steroids produced by the smooth ER in animal cells are the sex hormones of vertebrates and the various steroid hormones secreted by the adrenal glands. The cells that actually synthesize and secrete these hormones—in the testes and ovaries, for example—are rich in smooth ER, a structural feature that fits the function of these cells.

Liver cells provide one example of the role of smooth ER in carbohydrate metabolism. Liver cells store carbohydrate in the form of glycogen, a polysaccharide. The hydrolysis of glycogen leads to the release of glucose from the liver cells, which is important in the regulation of sugar concentration in the blood. However, the first product of glycogen hydrolysis is glucose phosphate, an ionic form of the sugar that cannot exit the cell and enter the blood. An enzyme embedded in the membrane of the liver cell's smooth ER removes the phosphate from the glucose, which can then leave the cell.

Enzymes of the smooth ER help detoxify drugs and poisons, especially in liver cells. Detoxification usually involves adding hydroxyl groups to drugs, making them more soluble and easier to flush from the body. The sedative phenobarbital and other barbiturates are examples of drugs metabolized in this manner by smooth ER in liver cells. In fact, barbiturates, alcohol, and many other drugs induce the proliferation of smooth ER and its associated detoxification enzymes. This, in turn, increases tolerance to the drugs, meaning that higher doses are required to achieve a particular effect, such as sedation. Also, because some of the detoxification enzymes have

cytosol into the cisternal space. When a muscle cell is stimulated by a nerve impulse, calcium rushes back across the ER membrane into the cytosol and triggers contraction of the muscle cell.

Rough ER and the Synthesis of Secretory Proteins

Many types of specialized cells secrete proteins produced by ribosomes attached to rough ER. For example, certain cells in the pancreas secrete the protein insulin, a hormone, into the bloodstream (see FIGURE 7.10a). As a polypeptide chain grows from a bound ribosome, it is threaded into the cisternal space through a pore formed by a protein in the ER membrane. As it enters the cisternal space, the new protein folds into its native conformation. Most secretory proteins are **glycoproteins,** proteins that are covalently bonded to carbohydrates. The carbohydrate is attached to the protein in the ER by specialized molecules built into the ER membrane. The carbohydrate appendage of a glycoprotein is an oligosaccharide, the term for a relatively small polymer of sugar units.

Once secretory proteins are formed, the ER membrane keeps them separate from the proteins, produced by free ribosomes, that will remain in the cytosol. Secretory proteins depart from the ER wrapped in the membranes of vesicles that bud like bubbles from a specialized region called transitional ER. Such vesicles in transit from one part of the cell to another are called **transport vesicles,** and we will soon learn their fate.

Rough ER and Membrane Production

In addition to making secretory proteins, rough ER is a membrane factory that grows in place by adding proteins and phospholipids. As polypeptides destined to be membrane proteins grow from the ribosomes, they are inserted into the ER membrane itself and are anchored there by their hydrophobic portions. The rough ER also makes its own membrane phospholipids; enzymes built into the ER membrane assemble phospholipids from precursors in the cytosol. The ER membrane expands and can be transferred in the form of transport vesicles to other components of the endomembrane system.

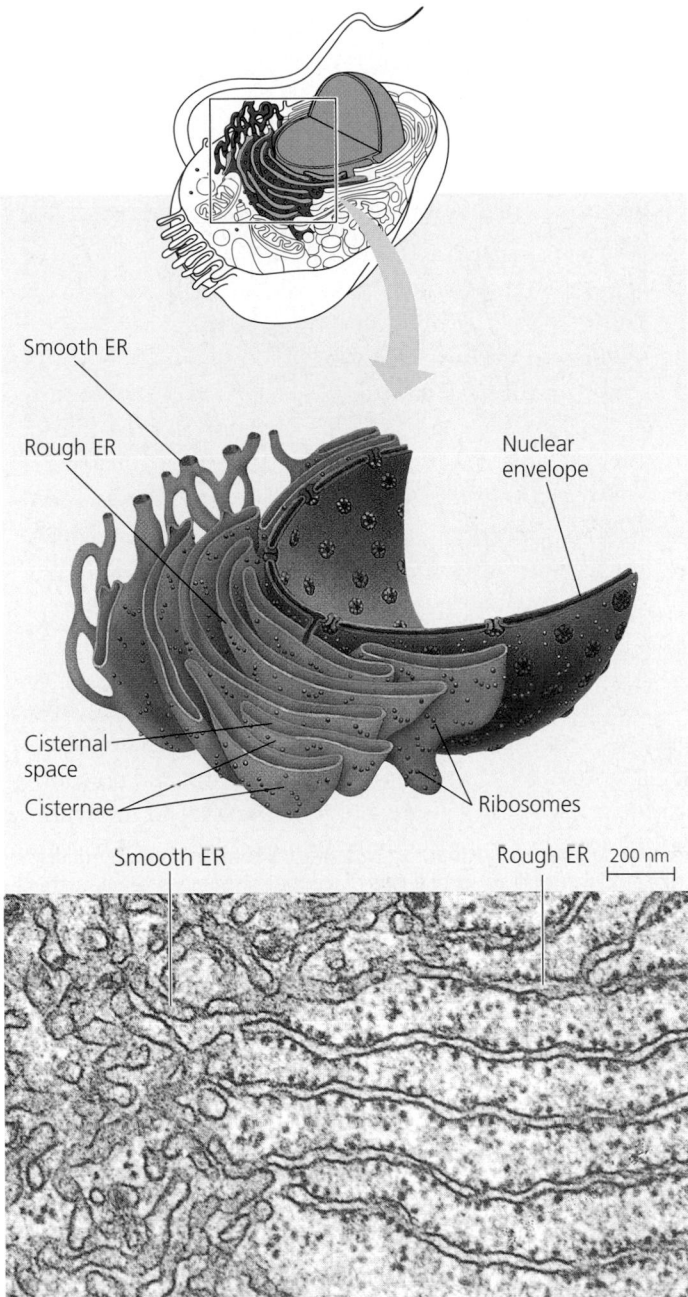

FIGURE 7.11 Endoplasmic reticulum (ER). A membranous system of interconnected tubules and flattened sacs called cisternae, the ER is also continuous with the nuclear envelope. (The drawing is a cutaway view.) The membrane of the ER encloses a compartment called the cisternal space. Rough ER, which is studded on its outer surface with ribosomes, can be distinguished from smooth ER in the electron micrograph (TEM).

relatively broad action, the proliferation of smooth ER in response to one drug can increase tolerance to other drugs as well. Barbiturate abuse, for example, may decrease the effectiveness of certain antibiotics and other useful drugs.

Muscle cells exhibit still another specialized function of smooth ER. The ER membrane pumps calcium ions from the

The Golgi apparatus finishes, sorts, and ships cell products

After leaving the ER, many transport vesicles travel to the **Golgi apparatus.** We can think of the Golgi as a center of manufacturing, warehousing, sorting, and shipping. Here, products of the ER are modified and stored and then sent to other destinations. Not surprisingly, the Golgi apparatus is especially extensive in cells specialized for secretion.

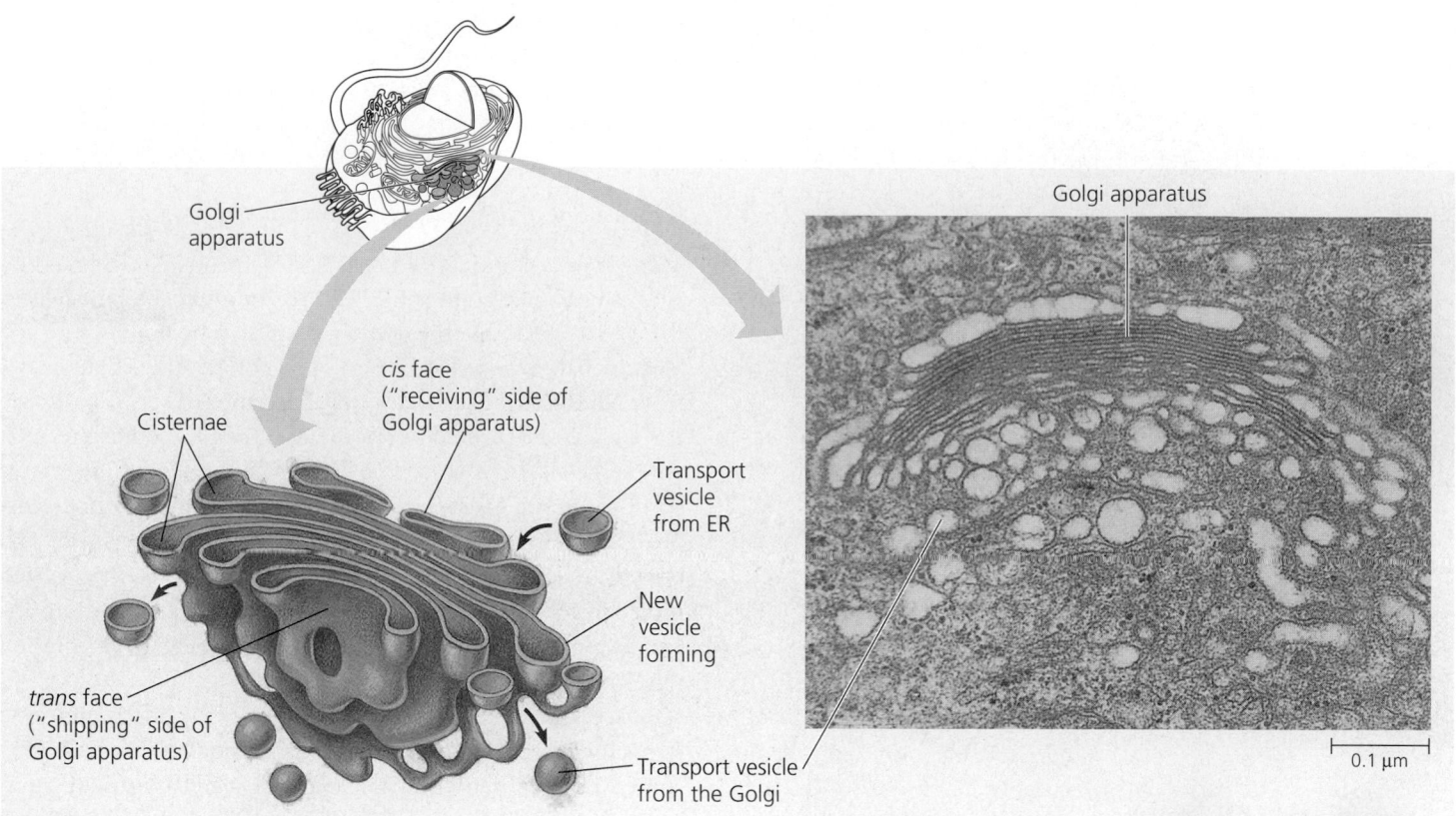

Golgi
apparatus

Golgi apparatus

Cisternae

cis face
("receiving" side of
Golgi apparatus)

Transport
vesicle
from ER

New
vesicle
forming

trans face
("shipping" side of
Golgi apparatus)

Transport vesicle
from the Golgi

0.1 µm

FIGURE 7.12 The Golgi apparatus. The Golgi apparatus consists of stacks of flattened sacs, or cisternae, which, unlike ER cisternae, are not physically connected. (The drawing is a cutaway view.) A Golgi stack receives and dispatches transport vesicles and the products they contain. Materials received from the ER are modified and stored in the Golgi and eventually shipped to the cell surface or other destinations. Note the vesicles joining and leaving the cisternae. A Golgi stack has a structural and functional polarity, with a *cis* face that receives vesicles containing ER products and a *trans* face that dispatches vesicles (at right, TEM).

The Golgi apparatus consists of flattened membranous sacs—cisternae—looking like a stack of pita bread (FIGURE 7.12). A cell may have several or even hundreds of these stacks. The membrane of each cisterna in a stack separates its internal space from the cytosol. Vesicles concentrated in the vicinity of the Golgi apparatus are engaged in the transfer of material between the Golgi and other structures.

A Golgi stack has a distinct polarity, with the membranes of cisternae at opposite ends of the stack differing in thickness and molecular composition. The two poles of a Golgi stack are referred to as the *cis* face and the *trans* face; these act, respectively, as the receiving and shipping departments of the Golgi apparatus. The *cis* face is usually located near the ER. Transport vesicles move material from the ER to the Golgi. A vesicle that buds from the ER will add its membrane and the contents of its lumen (cavity) to the *cis* face by fusing with a Golgi membrane. The *trans* face gives rise to vesicles, which pinch off and travel to other sites.

Products of the ER are usually modified during their transit from the *cis* pole to the *trans* pole of the Golgi. Proteins and phospholipids of membranes may be altered. For example, various Golgi enzymes modify the oligosaccharide portions of glycoproteins. Oligosaccharides are first added to proteins in the rough ER, often during the process of polypeptide synthesis. The resulting glycoprotein is then modified as it passes through the rest of the ER and the Golgi. The Golgi removes some sugar monomers and substitutes others, producing a large variety of oligosaccharides.

In addition to its finishing work, the Golgi apparatus manufactures certain macromolecules by itself. Many polysaccharides secreted by cells are Golgi products, including pectins and certain other noncellulose polysaccharides made by plant cells and incorporated along with cellulose into their cell walls. (Cellulose is made by enzymes located within the plasma membrane, which directly deposit this polysaccharide on the outside surface.) Golgi products that will be secreted depart from the *trans* face of the Golgi inside transport vesicles that eventually fuse with the plasma membrane.

The Golgi manufactures and refines its products in stages, with different cisternae between the *cis* and *trans* ends containing unique teams of enzymes. Products in various stages of processing seem to be transferred from one cisterna to the next by vesicles.

Before a Golgi stack dispatches its products by budding vesicles from the *trans* face, it sorts these products and targets them for various parts of the cell. Molecular identification tags, such as phosphate groups that have been added to the Golgi products, aid in sorting. Finally, transport vesicles budded from the Golgi may have external molecules on their membranes that recognize "docking sites" on the surface of specific organelles or on the plasma membrane.

Lysosomes are digestive compartments

A **lysosome** is a membrane-bounded sac of hydrolytic enzymes that the cell uses to digest macromolecules (FIGURE 7.13). There are lysosomal enzymes that can hydrolyze proteins, polysaccharides, fats, and nucleic acids—all the major classes of macromolecules. These enzymes work best in an acidic environment, at about pH 5. The lysosomal membrane maintains this low internal pH by pumping hydrogen ions from the cytosol into the lumen of the lysosome. If a lysosome breaks open or leaks its contents, the released enzymes are not very active, because the cytosol has a neutral pH. However, excessive leakage from a large number of lysosomes can destroy a cell by autodigestion. From this example we can see once again how important compartmental organization is to the functions of the cell: The lysosome provides a space where the cell can digest macromolecules safely, without the general destruction that would occur if hydrolytic enzymes roamed at large.

Hydrolytic enzymes and lysosomal membrane are made by rough ER and then transferred to the Golgi apparatus for further processing. At least some lysosomes probably arise by budding from the *trans* face of the Golgi apparatus (FIGURE 7.14, p. 122). Proteins of the inner surface of the lysosomal membrane and the digestive enzymes themselves are probably spared from destruction by having three-dimensional conformations that protect vulnerable bonds from enzymatic attack.

Lysosomes carry out intracellular digestion in a variety of circumstances. Amoebas and many other protists eat by engulfing smaller organisms or other food particles, a process called **phagocytosis** (from the Greek *phagein*, to eat, and *kytos*, vessel, referring here to the cell). The *food vacuole* formed in this way then fuses with a lysosome, whose enzymes digest the food (see FIGURE 7.14). Digestion products, including simple sugars, amino acids, and other monomers, pass into the cytosol and become nutrients for the cell. Some human cells also carry out phagocytosis. Among them are macrophages, cells that help defend the body by destroying bacteria and other invaders (see FIGURE 7.13a).

Lysosomes also use their hydrolytic enzymes to recycle the cell's own organic material, a process called *autophagy*. This occurs when a lysosome engulfs another organelle or a small amount of cytosol (see FIGURE 7.13b). The lysosomal enzymes dismantle the ingested material, and the organic monomers are returned to the cytosol for reuse. With the help of lysosomes, the cell continually renews itself. A human liver cell, for example, recycles half of its macromolecules each week.

Programmed destruction of cells by their own lysosomal enzymes is important in the development of many multicellular organisms. During the transforming of a tadpole into a

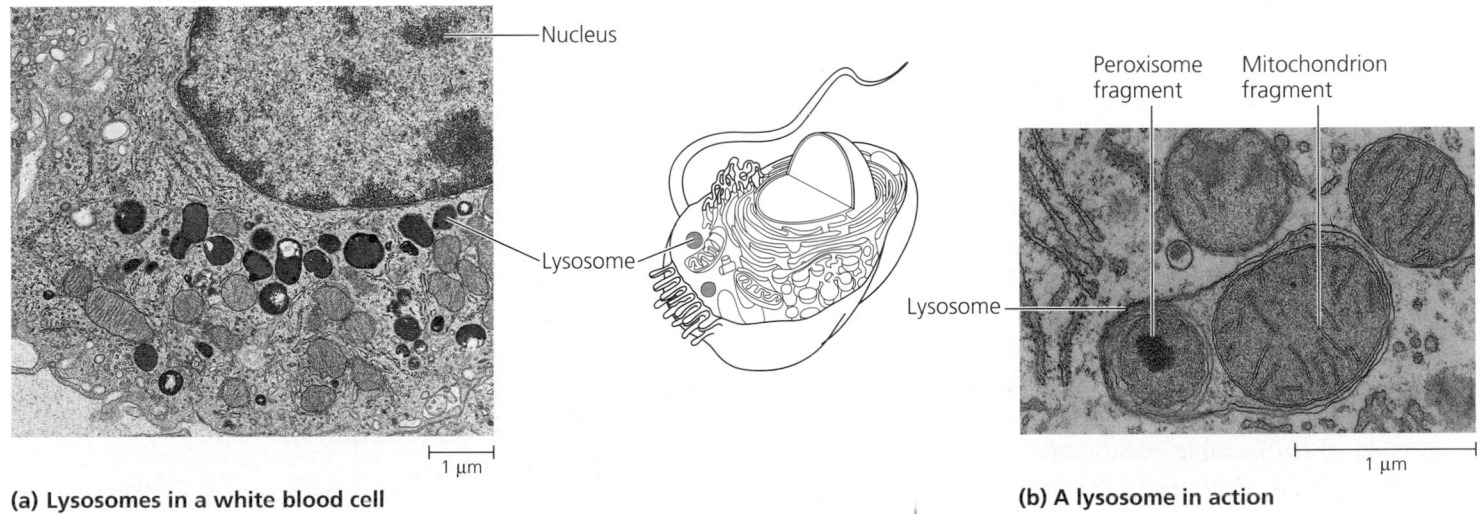

(a) **Lysosomes in a white blood cell**

(b) **A lysosome in action**

FIGURE 7.13 Lysosomes. (a) In this white blood cell from a rat, the lysosomes are very dark because of a specific stain that reacts with one of the products of digestion within the lysosome. This type of white blood cell ingests bacteria and viruses and destroys them in the lysosomes (TEM). **(b)** In the cytoplasm of this rat liver cell, an autophagic lysosome has engulfed two disabled organelles, a mitochondrion and a peroxisome (TEM).

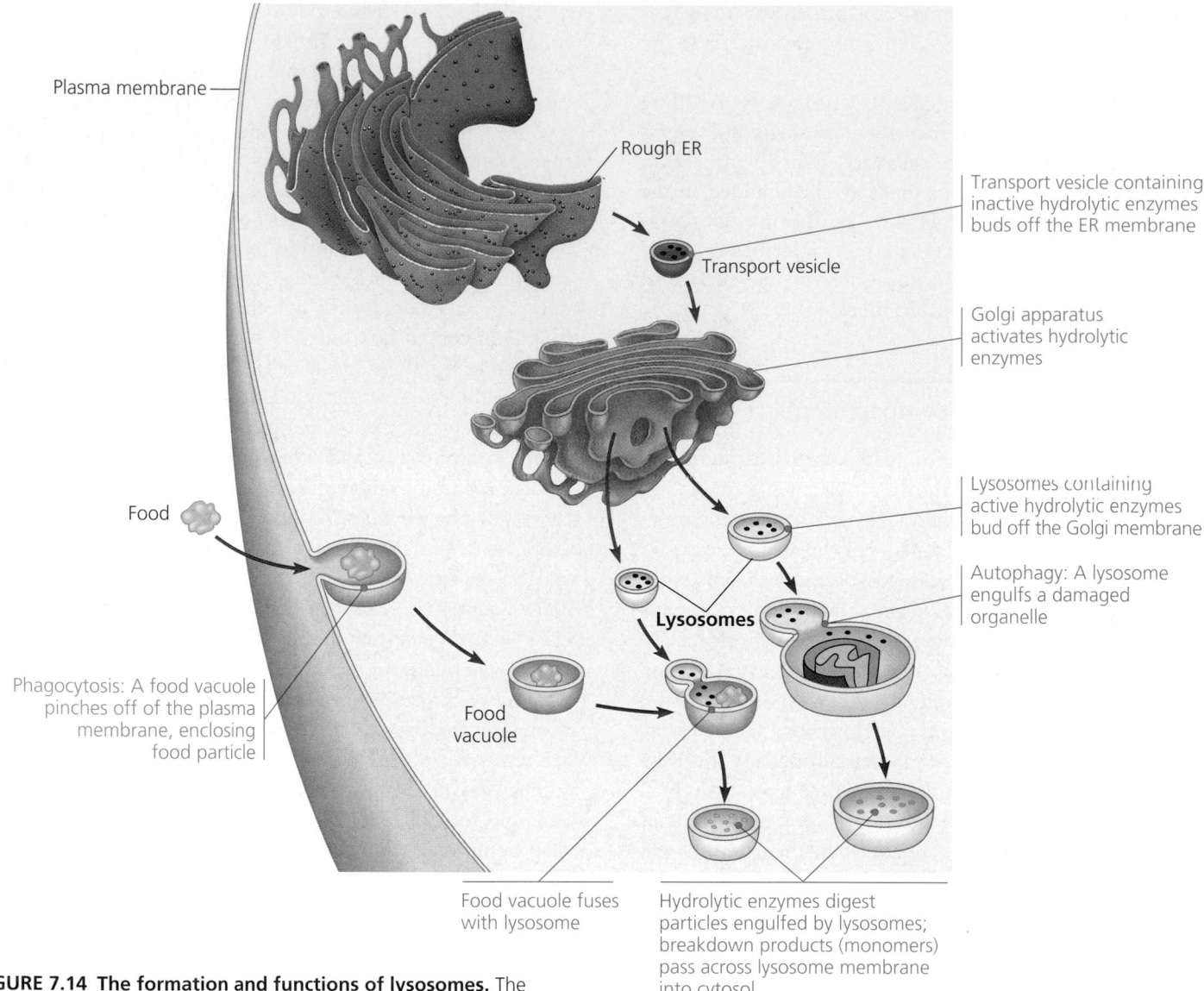

Plasma membrane

Rough ER

Transport vesicle containing inactive hydrolytic enzymes buds off the ER membrane

Transport vesicle

Golgi apparatus activates hydrolytic enzymes

Food

Lysosomes containing active hydrolytic enzymes bud off the Golgi membrane

Autophagy: A lysosome engulfs a damaged organelle

Lysosomes

Phagocytosis: A food vacuole pinches off of the plasma membrane, enclosing food particle

Food vacuole

Food vacuole fuses with lysosome

Hydrolytic enzymes digest particles engulfed by lysosomes; breakdown products (monomers) pass across lysosome membrane into cytosol.

FIGURE 7.14 The formation and functions of lysosomes. The ER and Golgi apparatus generally cooperate in the production of lysosomes containing active hydrolytic enzymes. Lysosomes digest (hydrolyze) materials taken into the cell and recycle materials from intracellular refuse. This figure shows one lysosome fusing with a food vacuole and another engulfing a damaged mitochondrion.

frog, for instance, lysosomes in the cells of the tail destroy these cells. And the hands of human embryos are webbed until lysosomes digest the tissue between the fingers.

A variety of inherited disorders called lysosomal storage diseases affect lysosomal metabolism. A person afflicted with a storage disease lacks a functioning version of a hydrolytic enzyme normally present in lysosomes. The lysosomes become engorged with indigestible substrates, which begin to interfere with other cellular activities. In Pompe's disease, for example, the liver is damaged by an accumulation of glycogen due to the absence of a lysosomal enzyme needed to break down that polysaccharide. In Tay-Sachs disease, a lipid-digesting enzyme is missing or inactive, and the brain becomes impaired by an

accumulation of lipids in the cells. Fortunately, storage diseases are rare in the general population. In the future, it may be possible to cure such a disorder by inserting genes (DNA) for the missing enzyme into the appropriate cells (see Chapter 20).

Vacuoles have diverse functions in cell maintenance

Vacuoles and vesicles are both membrane-bounded sacs within the cell, but vacuoles are larger than vesicles. Vacuoles have various functions. **Food vacuoles,** formed by phagocytosis, have already been mentioned (see FIGURE 7.14). Many freshwater protists have **contractile vacuoles** that pump excess water out of the cell (see FIGURE 8.13). Mature plant cells generally contain a large **central vacuole** (FIGURE 7.15) enclosed by a membrane called the **tonoplast,** which is part of their endomembrane system. The central vacuole develops by the coalescence of smaller vacuoles, themselves derived from the endoplasmic reticulum and Golgi apparatus. The vacuole is in this way an integral part of the endomembrane system. Like all cellular membranes, the tonoplast is selective in transporting solutes; therefore, the solution inside the vacuole, called cell sap, differs in composition from the cytosol.

The plant cell's central vacuole is a versatile compartment. It can hold reserves of important organic compounds, such as the proteins stockpiled in the vacuoles of storage cells in seeds. It is also the plant cell's main repository of inorganic ions, such as potassium and chloride. Many plant cells use their vacuoles as disposal sites for metabolic by-products that would endanger the cell if they accumulated in the cytosol. Some vacuoles contain pigments that color the cells, such as the red and blue pigments of petals that help attract pollinating insects to flowers. Vacuoles may also help protect the plant against predators by containing compounds that are poisonous or unpalatable to animals. The vacuole has a major role in the growth of plant cells, which elongate as their vacuoles absorb water, enabling the cell to become larger with a minimal investment in new cytoplasm. And because the cytosol often occupies only a thin layer between the plasma membrane and the tonoplast, the ratio of membrane surface to cytosolic volume is great, even for a large plant cell.

FIGURE 7.16 reviews the endomembrane system, showing the flow of membranes through the various organelles. As the membrane moves from the ER to the Golgi and then elsewhere, its molecular composition and metabolic functions are modified. The endomembrane system is a complex and dynamic player in the cell's compartmental organization.

We'll continue our tour of the cell with some membranous organelles that are *not* closely related to the endomembrane system.

OTHER MEMBRANOUS ORGANELLES

Mitochondria and chloroplasts are the main energy transformers of cells

Organisms are open systems that transform the energy they acquire from their surroundings. In eukaryotic cells, mitochondria and chloroplasts are the organelles that convert energy to forms that

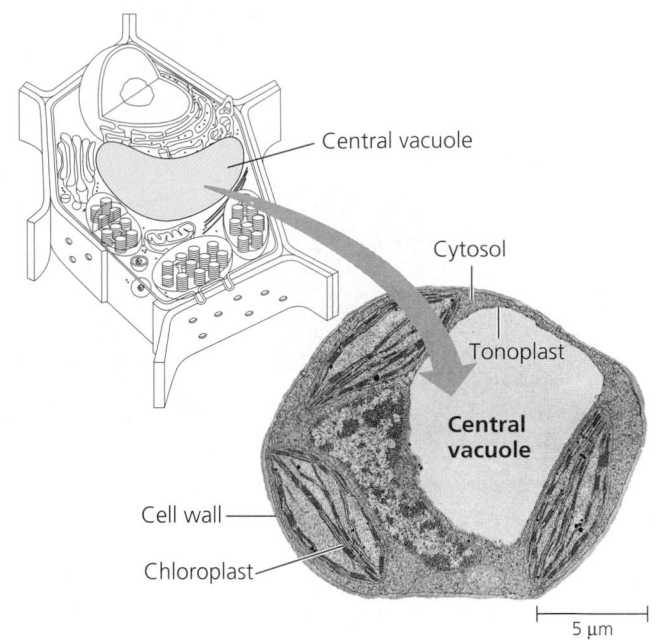

FIGURE 7.15 The plant cell vacuole. The central vacuole is usually the largest compartment in a plant cell, comprising 80% or more of a mature cell. The rest of the cytoplasm is generally confined to a narrow zone between the vacuolar membrane (tonoplast) and the plasma membrane. Functions of the vacuole include storage, waste disposal, protection, and growth (TEM).

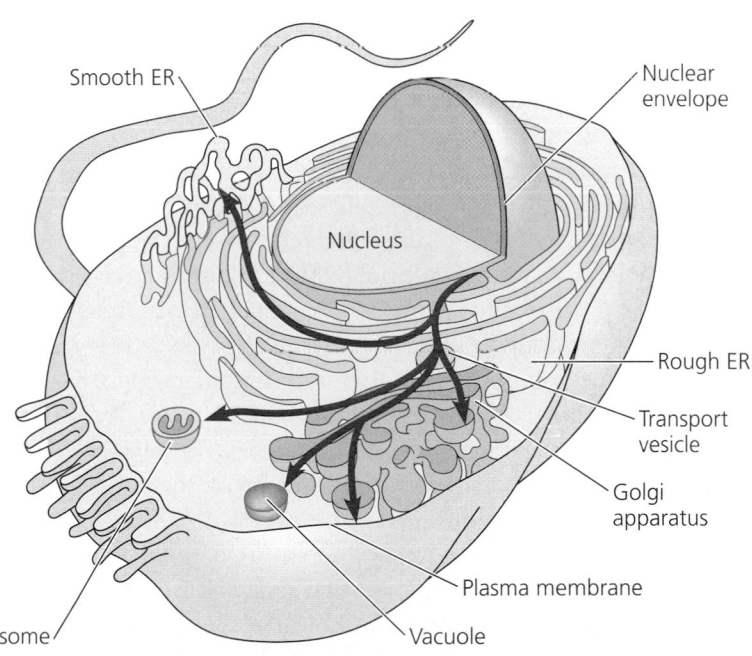

FIGURE 7.16 Review: relationships among organelles of the endomembrane system. The red arrows show some of the pathways of membrane migration. The nuclear envelope is connected to the rough ER, which is also confluent with smooth ER. Membrane produced by the ER flows in the form of transport vesicles to the Golgi, which in turn pinches off vesicles that give rise to lysosomes and vacuoles. Even the plasma membrane expands by the fusion of vesicles born in the ER and Golgi. Coalescence of vesicles with the plasma membrane also releases secretory proteins and other products to the outside of the cell.

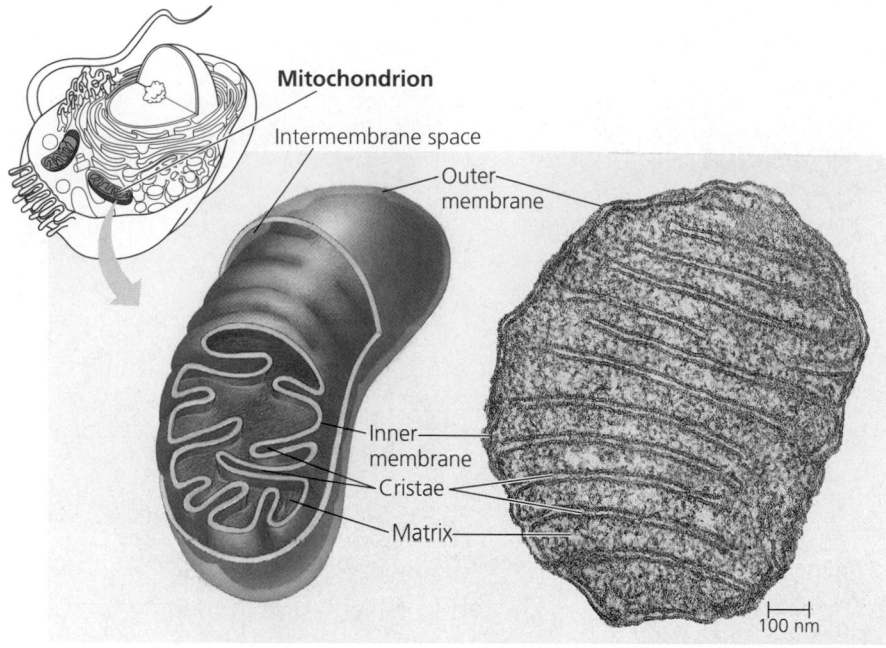

Mitochondrion

Intermembrane space

Outer
membrane

Inner
membrane

Cristae

Matrix

100 nm

FIGURE 7.17 The mitochondrion, site of cellular respiration. The two membranes of the mitochondrion are evident in the drawing and micrograph (TEM). The cristae are infoldings of the inner membrane. The cutaway drawing shows the two compartments bounded by the membranes: the intermembrane space and the mitochondrial matrix.

cells can use for work. **Mitochondria** (singular, *mitochondrion*) are the sites of cellular respiration, the catabolic process that generates ATP by extracting energy from sugars, fats, and other fuels with the help of oxygen. **Chloroplasts,** found only in plants and algae, are the sites of photosynthesis. They convert solar energy to chemical energy by absorbing sunlight and using it to drive the synthesis of organic compounds from carbon dioxide and water.

Although mitochondria and chloroplasts are enclosed by membranes, they are not part of the endomembrane system. Their membrane proteins are made not by the ER, but by free ribosomes in the cytosol and by ribosomes contained within the mitochondria and chloroplasts themselves. Not only do these organelles have ribosomes, but they also contain a small amount of DNA. It is this DNA that programs the synthesis of the proteins made on the organelle's own ribosomes. (Proteins imported from the cytosol—constituting most of the organelle's proteins—are programmed by nuclear DNA.) Mitochondria and chloroplasts are semiautonomous organelles that grow and reproduce within the cell. In Chapters 9 and 10, we will focus on how mitochondria and chloroplasts function. We will consider the evolution of these organelles in Chapter 28. Here we are concerned mainly with the structure of these energy transformers.

Mitochondria

Mitochondria are found in nearly all eukaryotic cells, including those of plants, animals, fungi, and protists. Some cells

have a single large mitochondrion, but more often a cell has hundreds or even thousands of mitochondria; the number is correlated with the cell's level of metabolic activity. Mitochondria are about 1–10 μm long. Time-lapse films of living cells reveal mitochondria moving around, changing their shapes, and dividing in two, unlike the static cylinders seen in electron micrographs of dead cells.

The mitochondrion is enclosed by two membranes, each a phospholipid bilayer with a unique collection of embedded proteins (FIGURE 7.17). The outer membrane is smooth, but the inner membrane is convoluted, with infoldings called **cristae.** The inner membrane divides the mitochondrion into two internal compartments. The first is the intermembrane space, the narrow region between the inner and outer membranes. The second compartment, the **mitochondrial matrix,** is enclosed by the inner membrane. The matrix contains many different enzymes as well as the mitochondrial DNA and ribosomes. Some of the metabolic steps of cellular respiration are catalyzed by enzymes in the matrix. Other proteins that function in respiration, including the enzyme that makes ATP, are built into the inner membrane. The cristae give the inner mitochondrial membrane a large surface area that enhances the productivity of cellular respiration, another example of structure fitting function.

Chloroplasts

The chloroplast is a specialized member of a family of closely related plant organelles called **plastids.** Amyloplasts are colorless plastids that store starch (amylose), particularly in roots and tubers. Chromoplasts have pigments that give fruits and flowers their orange and yellow hues. Chloroplasts contain the green pigment chlorophyll, along with enzymes and other molecules that function in the photosynthetic production of sugar. These lens-shaped organelles, measuring about 2 μm by 5 μm, are found in leaves and other green organs of plants and in algae (FIGURE 7.18).

The contents of a chloroplast are partitioned from the cytosol by an envelope consisting of two membranes separated by a very narrow intermembrane space. Inside the chloroplast is another membranous system in the form of flattened sacs called **thylakoids.** In some regions, thylakoids are stacked like poker chips; each stack is called a **granum** (plural, *grana*). The fluid outside the thylakoids is the **stroma,** which contains the chloroplast DNA and ribosomes as well as many enzymes. Note that the thylakoid membrane divides the interior of the chloroplast into two compartments: the thylakoid space and

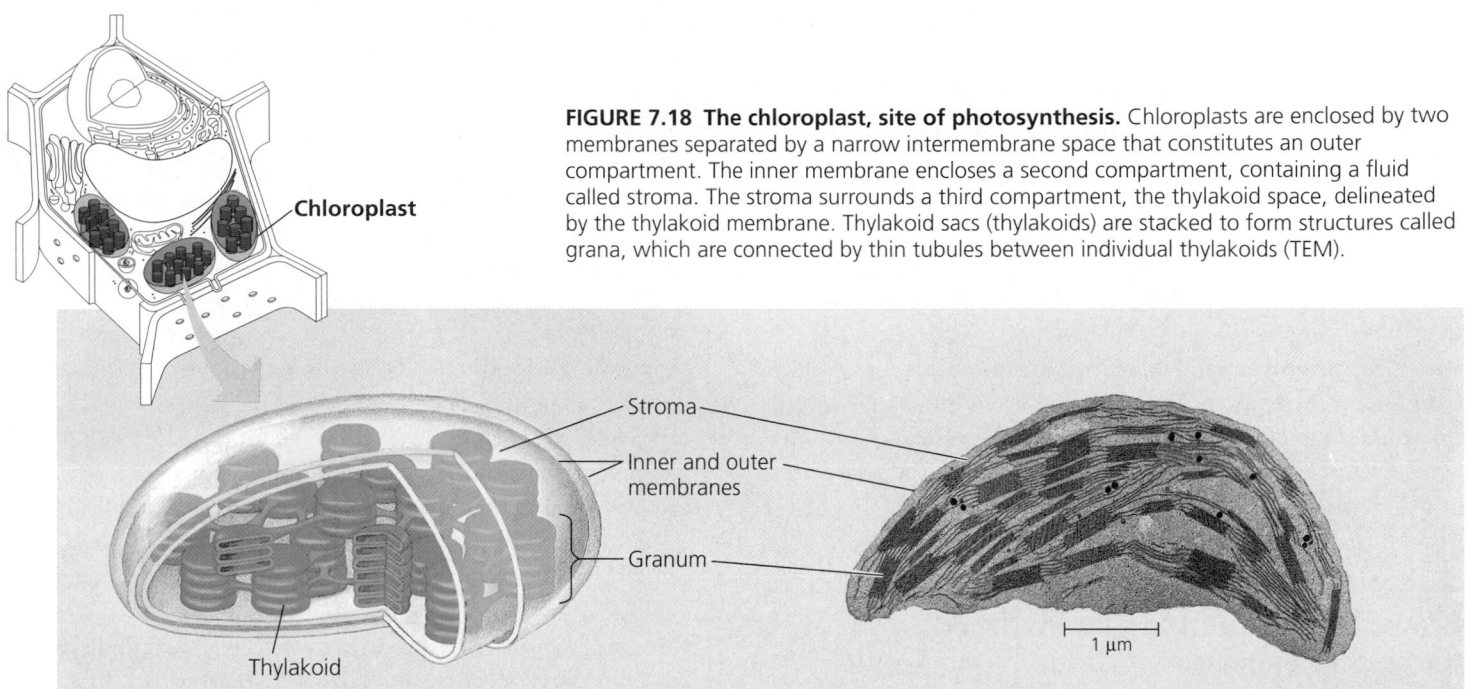

Chloroplast

FIGURE 7.18 The chloroplast, site of photosynthesis. Chloroplasts are enclosed by two membranes separated by a narrow intermembrane space that constitutes an outer compartment. The inner membrane encloses a second compartment, containing a fluid called stroma. The stroma surrounds a third compartment, the thylakoid space, delineated by the thylakoid membrane. Thylakoid sacs (thylakoids) are stacked to form structures called grana, which are connected by thin tubules between individual thylakoids (TEM).

Stroma

Inner and outer membranes

Granum

Thylakoid

1 μm

the stroma. In Chapter 10, you will learn how this compartmental organization enables the chloroplast to convert light energy to chemical energy during photosynthesis.

As with mitochondria, the static and rigid appearance of chloroplasts in micrographs is not true to their dynamic behavior in the living cell. Their shapes are plastic, and they grow and occasionally pinch in two, reproducing themselves. They are mobile and move around the cell with mitochondria and other organelles along tracks of the cytoskeleton.

Peroxisomes generate and degrade H$_2$O$_2$ in performing various metabolic functions

The **peroxisome** is a specialized metabolic compartment bounded by a single membrane (FIGURE 7.19). Peroxisomes contain enzymes that transfer hydrogen from various substrates to oxygen, producing hydrogen peroxide (H$_2$O$_2$) as a by-product, from which the organelle derives its name. These reactions may have many different functions. Some peroxisomes use oxygen to break fatty acids down into smaller molecules that can then be transported to mitochondria as fuel for cellular respiration. Peroxisomes in the liver detoxify alcohol and other harmful compounds by transferring hydrogen from the poisons to oxygen. The H$_2$O$_2$ formed by peroxisome metabolism is itself toxic, but the organelle contains an enzyme that converts the H$_2$O$_2$ to water. Enclosing in the same space both the enzymes that produce hydrogen peroxide and those that dispose of this toxic compound is another example of how the cell's compartmental structure is crucial to its functions.

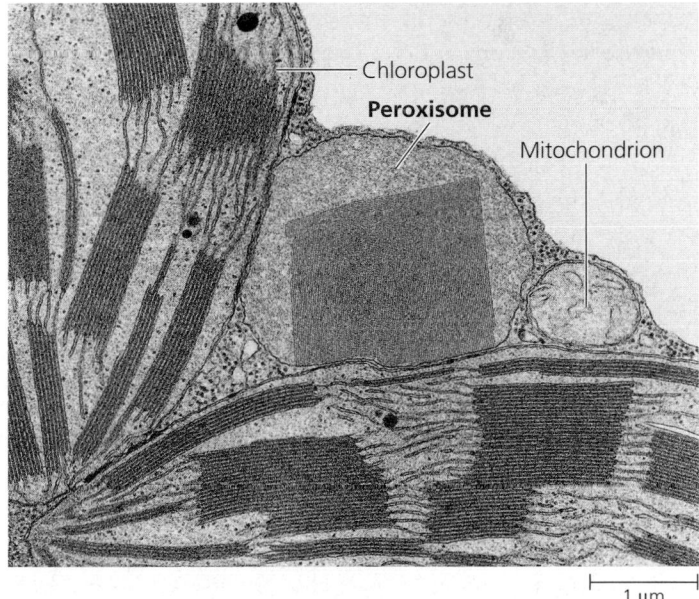

Chloroplast

Peroxisome

Mitochondrion

1 μm

FIGURE 7.19 Peroxisomes. Peroxisomes are roughly spherical and often have a granular or crystalline core that is probably a dense collection of enzyme molecules. This peroxisome is in a leaf cell. Notice its proximity to two chloroplasts and a mitochondrion. These organelles cooperate with peroxisomes in certain metabolic functions (TEM).

Specialized peroxisomes called glyoxysomes are found in the fat-storing tissues of plant seeds. These organelles contain enzymes that initiate the conversion of fatty acids to sugar, which the emerging seedling can use as an energy and carbon source until it is able to produce its own sugar by photosynthesis.

Unlike lysosomes, peroxisomes do not bud from the endomembrane system. They grow by incorporating proteins and lipids made in the cytosol, and they increase in number by splitting in two when they reach a certain size.

THE CYTOSKELETON

In the early days of electron microscopy, biologists thought that the organelles of a eukaryotic cell floated freely in the cytosol. But improvements in both light microscopy and electron microscopy have revealed the **cytoskeleton,** a network of fibers extending throughout the cytoplasm (FIGURE 7.20). The cytoskeleton plays a major role in organizing the structures and activities of the cell.

Providing structural support to the cell, the cytoskeleton also functions in cell motility and regulation

The most obvious function of the cytoskeleton is to give mechanical support to the cell and maintain its shape. This is especially important for animal cells, which lack walls. The remarkable strength and resilience of the cytoskeleton as a whole is based on its architecture. Like a geodesic dome, the cytoskeleton is stabilized by a balance between opposing forces exerted by its elements. And just as the skeleton of an animal helps fix the positions of other body parts, the cytoskeleton provides anchorage for many organelles and even cytosolic enzyme molecules. The cytoskeleton is more dynamic than an animal skeleton, however. It can be quickly dismantled in one part of the cell and reassembled in a new location, changing the shape of the cell.

The cytoskeleton is also involved in several types of cell motility (movement). The term *cell motility* encompasses both changes in cell location and more limited movements of parts of the cell. Cell motility generally requires the interaction of the cytoskeleton with proteins called motor molecules (FIGURE 7.21). Examples of such cell motility abound. Motor molecules of the cytoskeleton bring about the movements of cilia and flagella by allowing components of the cytoskeleton to slide past each other. The same mechanism causes muscle cells to contract. Vesicles often travel to their destinations in the cell along "monorails" provided by the cytoskeleton. For example, as discussed in the interview with George Langford (pp. 106–107),

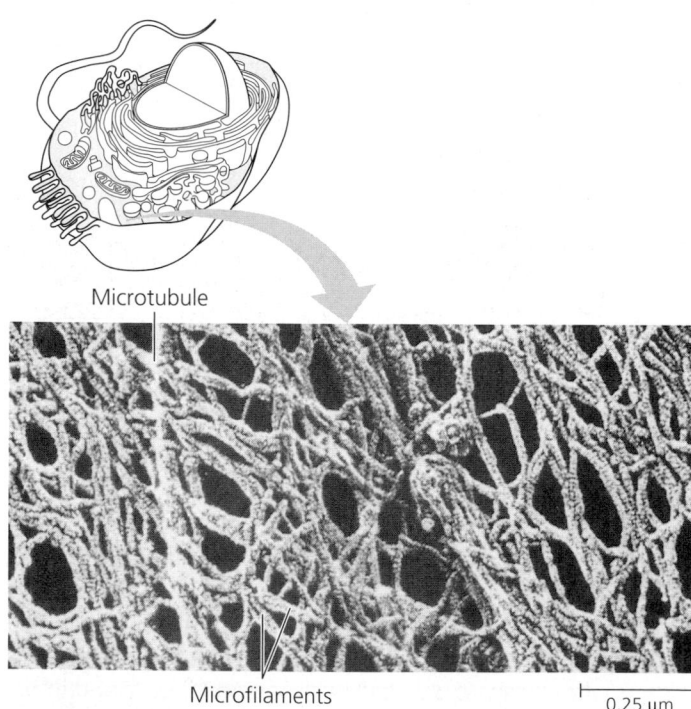

Microtubule

Microfilaments

0.25 μm

FIGURE 7.20 The cytoskeleton. In this TEM, prepared by a method known as deep-etching, microtubules and microfilaments are visible. A third component of the cytoskeleton, intermediate filaments, is not evident here.

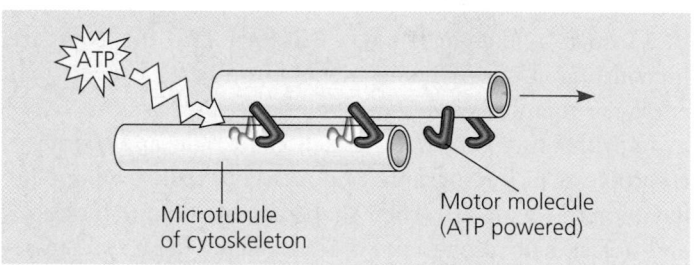

Microtubule of cytoskeleton

Motor molecule (ATP powered)

(a) Motor molecules attached to one microtubule (or microfilament) slide it past another. Sliding of neighboring microtubules moves cilia and flagella. In the contraction of muscle cells, motor molecules slide microfilaments rather than microtubules.

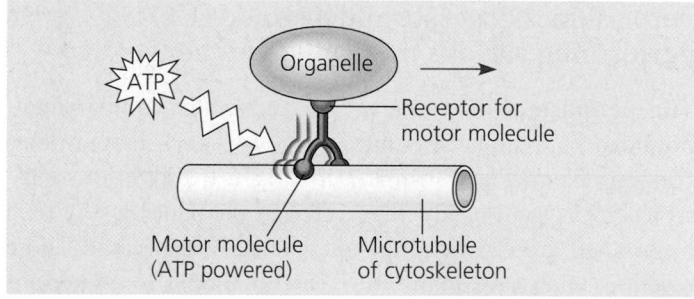

Organelle

Receptor for motor molecule

Motor molecule (ATP powered)

Microtubule of cytoskeleton

(b) Motor molecules that attach to receptors on organelles can "walk" the organelles along microtubules or, in some cases, microfilaments. For example, this is how vesicles containing neurotransmitters migrate to the tips of nerve cell axons.

FIGURE 7.21 Motor molecules and the cytoskeleton. The microtubules and microfilaments of the cytoskeleton function in motility by interacting with proteins called motor molecules. Motor molecules work by changing their shapes, moving back and forth something like microscopic legs. ATP powers these conformational changes. With each cycle of shape change, the motor molecule releases at its free end and then grips at a site farther along a microtubule or microfilament.

Table 7.2 The Structure and Function of the Cytoskeleton

Property	Microtubules	Microfilaments (Actin Filaments)	Intermediate Filaments
Structure	Hollow tubes; wall consists of 13 columns of tubulin molecules	Two intertwined strands of actin	Fibrous proteins supercoiled into thicker cables
Diameter	25 nm with 15-nm lumen	7 nm	8–12 nm
Protein subunits	Tubulin, consisting of α-tubulin and β-tubulin	Actin	One of several different proteins of the keratin family, depending on cell type
Main functions	Maintenance of cell shape (compression-resisting "girders") Cell motility (as in cilia or flagella) Chromosome movements in cell division Organelle movements	Maintenance of cell shape (tension-bearing elements) Changes in cell shape Muscle contraction Cytoplasmic streaming Cell motility (as in pseudopodia) Cell division (cleavage furrow formation)	Maintenance of cell shape (tension-bearing elements) Anchorage of nucleus and certain other organelles Formation of nuclear lamina

SOURCE: Adapted from W. M. Becker, L. J. Kleinsmith, and J. Hardin, *The World of the Cell,* 4th ed. (San Francisco, CA: Benjamin Cummings, 2000), p. 753.

this is how vesicles containing neurotransmitter molecules migrate to the tips of axons, the long extensions of nerve cells that release these molecules as chemical signals to adjacent nerve cells. It is the cytoskeleton that manipulates the plasma membrane to form food vacuoles during phagocytosis. The streaming of cytoplasm that circulates materials within many large plant cells is yet another kind of cellular movement brought about by components of the cytoskeleton.

The most recent addition to the list of possible cytoskeletal functions is the regulation of biochemical activities in the cell. Mounting evidence suggests that the cytoskeleton can transmit mechanical forces from the surface of the cell to its interior—and even, via other fibers, into the nucleus. In one experiment, investigators used a micromanipulation device to pull on certain plasma membrane proteins attached to the cytoskeleton. A video microscope captured almost instantaneous rearrangements of nucleoli and other structures in the nucleus. In this way, the transmission of naturally occurring mechanical signals by the cytoskeleton may help regulate cell function.

Now let's look more closely at the three main types of fibers that make up the cytoskeleton (TABLE 7.2). **Microtubules** are the thickest of the three types; **microfilaments** (also called actin filaments) are the thinnest; and **intermediate filaments** are fibers with diameters in a middle range.

Microtubules

Microtubules are found in the cytoplasm of all eukaryotic cells. They are hollow rods measuring about 25 nm in diameter and from 200 nm to 25 μm in length. The wall of the hollow tube is constructed from a globular protein called tubulin.

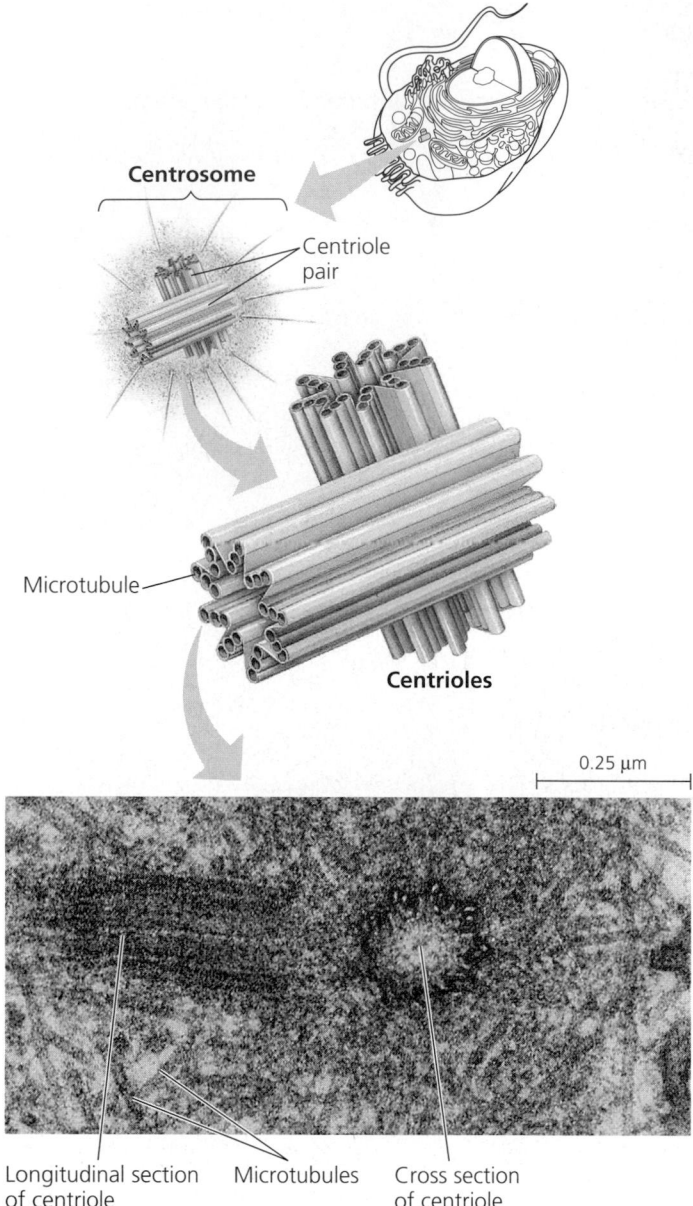

Centrosome

Centriole
pair

Microtubule

Centrioles

0.25 µm

Longitudinal section   Microtubules   Cross section
of centriole   of centriole

FIGURE 7.22 Centrosome containing a pair of centrioles. An animal cell has a pair of centrioles within its centrosome, the region near the nucleus where the cell's microtubules are initiated. The centrioles, each about 250 nm (0.25 µm) in diameter, are arranged at right angles to each other, and each is made up of nine sets of three microtubules. The blue portions of the drawing represent nontubulin proteins that connect the microtubule triplets (TEM).

Each tubulin molecule is a dimer consisting of two slightly different polypeptide subunits, α-tubulin and β-tubulin. A microtubule grows in length by adding tubulin dimers to its ends. Microtubules can be disassembled and their tubulin used to build microtubules elsewhere in the cell.

Microtubules shape and support the cell and also serve as tracks along which organelles equipped with motor molecules can move (see FIGURE 7.21). For example, microtubules guide secretory vesicles from the Golgi apparatus to the plasma membrane. Microtubules are also responsible for the separation of chromosomes during cell division (see Chapter 12).

Centrosomes and Centrioles. In many cells, microtubules grow out from a **centrosome**, a region often located near the nucleus. These microtubules function as compression-resisting girders of the cytoskeleton. Within the centrosome of an animal cell are a pair of **centrioles**, each composed of nine sets of triplet microtubules arranged in a ring (FIGURE 7.22). Before a cell divides, the centrioles replicate. Although centrioles may help organize microtubule assembly, they are not essential for this function in all eukaryotes; centrosomes of most plants lack centrioles altogether.

Cilia and Flagella. In eukaryotes, a specialized arrangement of microtubules is responsible for the beating of **flagella** and **cilia,** locomotor appendages that protrude from some cells. Many unicellular eukaryotic organisms are propelled through water by cilia or flagella, and the sperm of animals, algae, and some plants are flagellated. If cilia or flagella extend from cells that are held in place as part of a tissue layer, they function to move fluid over the surface of the tissue. For example, the ciliated lining of the windpipe sweeps mucus containing trapped debris out of the lungs (see FIGURE 7.2).

Cilia usually occur in large numbers on the cell surface. They are about 0.25 µm in diameter and about 2–20 µm in length. Flagella are the same diameter but longer than cilia, measuring 10–200 µm in length. Also, flagella are usually limited to just one or a few per cell.

Flagella and cilia also differ in their beating patterns (FIGURE 7.23). A flagellum has an undulating motion that generates force in the same direction as the flagellum's axis. In contrast, cilia work more like oars, with alternating power and recovery strokes generating force in a direction perpendicular to the cilium's axis.

Though different in length, number per cell, and beating pattern, cilia and flagella share a common ultrastructure. A cilium or flagellum has a core of microtubules sheathed in an extension of the plasma membrane (FIGURE 7.24). Nine doublets of microtubules, the members of each pair sharing part of their walls, are arranged in a ring. In the center of the ring are two single microtubules. This arrangement, referred to as the "9 + 2" pattern, is found in nearly all eukaryotic flagella and cilia. (The flagella of motile prokaryotes, which will be discussed in Chapter 27, are entirely different.) Flexible "wheels" of proteins, evenly spaced along the length of the cilium or flagellum, connect the outer doublets to each other and to the two central microtubules (note the radial spokes in FIGURE 7.24c). Each outer doublet also has pairs of side-arms spaced along its length and reaching toward the neighboring doublet; these are motor molecules. The microtubule assembly of a cilium or flagellum is anchored in the cell by a **basal body,** which is structurally identical to a centriole. In fact, in many animals

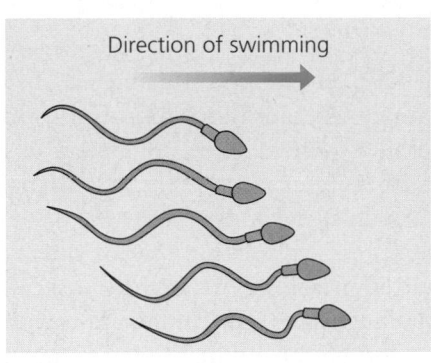

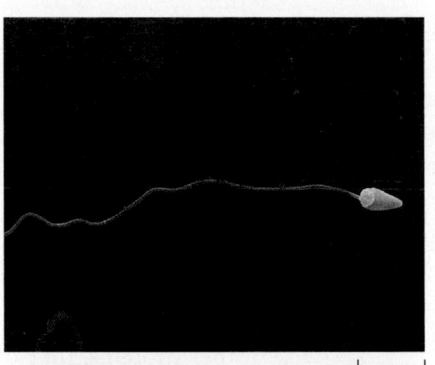

(a) Motion of flagella. A flagellum usually undulates, its snakelike motion driving a cell in the same direction as the axis of the flagellum. Propulsion of a sperm cell is an example of flagellate locomotion (SEM).

1 μm

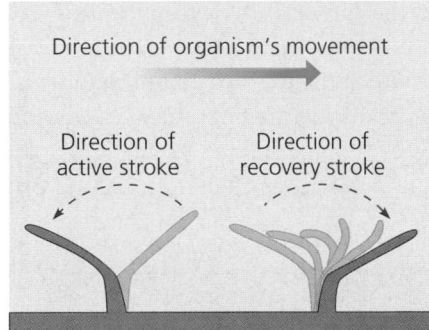

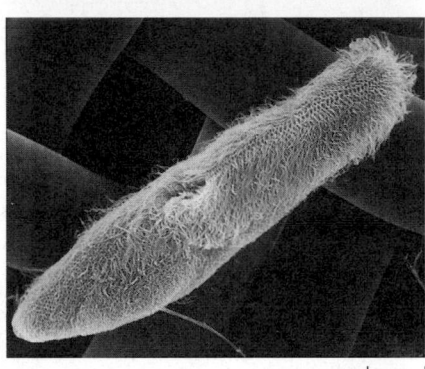

(b) Motion of cilia. Cilia have a back-and-forth motion that move the cell in a direction perpendicular to the axis of the cilium. A dense nap of cilia, beating at a rate of about 40 to 60 strokes a second, covers this *Paramecium*, a motile protist (SEM).

25 μm

FIGURE 7.23 A comparison of the beating of flagella and cilia.

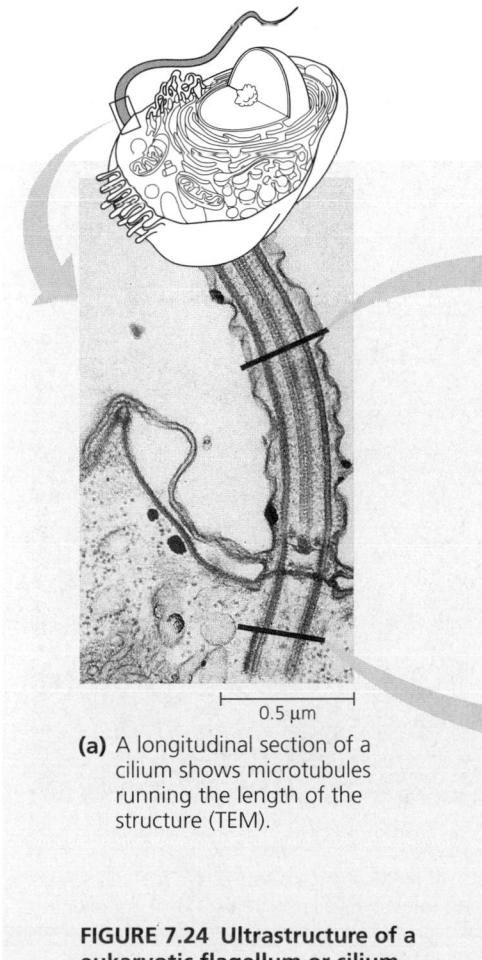

0.5 μm

(a) A longitudinal section of a cilium shows microtubules running the length of the structure (TEM).

FIGURE 7.24 Ultrastructure of a eukaryotic flagellum or cilium.

Outer microtubule doublet

Dynein arms

Central microtubule

Radial spoke

0.1 μm

(b) A cross section through the cilium shows the "9 + 2" arrangement of microtubules (TEM).

Triplets

0.1 μm

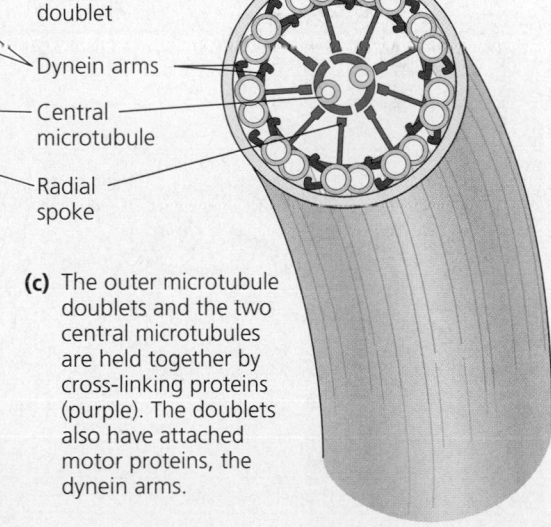

Plasma membrane

(c) The outer microtubule doublets and the two central microtubules are held together by cross-linking proteins (purple). The doublets also have attached motor proteins, the dynein arms.

(d) Basal body: The nine outer doublets of a cilium or flagellum extend into the basal body, where each doublet joins another microtubule to form a ring of nine triplets. The two central microtubules terminate above the basal body (TEM).

(including humans), the basal body of the fertilizing sperm's flagellum enters the egg and becomes a centriole.

The motor molecules extending from each microtubule doublet to the next are made of a large protein called **dynein.** These dynein arms are responsible for the bending movements of cilia and flagella. A dynein arm performs a complex cycle of movements caused by changes in the conformation of the protein, with ATP providing the energy for these changes.

The mechanics of dynein "walking" are reminiscent of a cat climbing a tree by attaching its claws, moving its legs, releasing its front claws, and grabbing again farther up the tree. Similarly, the dynein arms of one doublet attach to an adjacent doublet and pull so that the doublets slide past each other in opposite directions. The arms then release from the other doublet and reattach a little farther along its length. Without any restraints on the movement of the microtubule doublets, one doublet would continue to "walk" along the surface of the other, elongating the cilium or flagellum rather than bending it (see FIGURE 7.21a). For lateral movement of a cilium or flagellum, the dynein "walking" must have something to pull against, as when the muscles in your leg pull against your bones to move your knee. In cilia and flagella, the microtubule doublets seem to be held in place by the protein cross-links between the doublets, the radial spokes, and the other structural elements. Thus, neighboring doublets cannot slide past each other very far. Instead, the forces exerted by the dynein arms cause the doublets to curve, bending the cilium or flagellum (FIGURE 7.25).

Microfilaments (Actin Filaments)

Microfilaments are solid rods about 7 nm in diameter. They are also called actin filaments, because they are built from molecules of **actin,** a globular protein. A microfilament is a twisted double chain of actin subunits (see TABLE 7.2). Microfilaments seem to be present in all eukaryotic cells.

In contrast to the compression-resisting role of microtubules, the structural role of microfilaments in the cytoskeleton is to bear tension (pulling forces). In combination with other proteins, they often form a three-dimensional network just inside the plasma membrane, helping support the cell's shape. This network gives the cortex (outer cytoplasmic layer) of such a cell the semisolid consistency of a gel, in contrast with the more fluid (sol) state of the interior cytoplasm. In animal cells specialized for transporting materials across the plasma membrane, bundles of microfilaments make up the core of microvilli, delicate projections that increase the cell surface area (FIGURE 7.26).

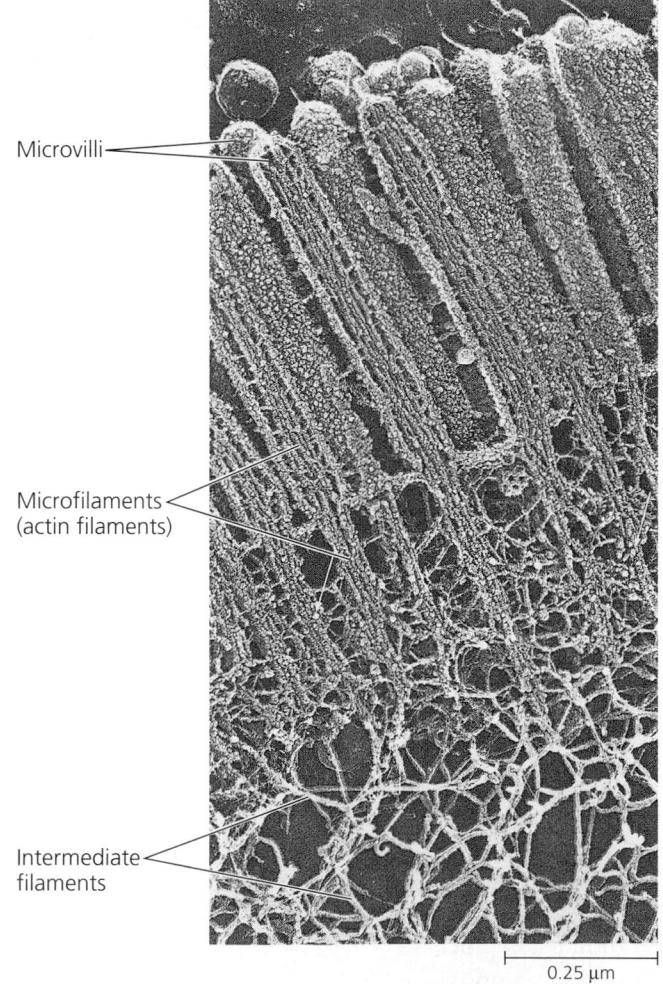

FIGURE 7.26 A structural role of microfilaments. The surface area of this nutrient-absorbing intestinal cell is increased by its many microvilli, cellular extensions reinforced by bundles of microfilaments. These actin filaments are anchored to a network of intermediate filaments (TEM).

Hirokawa et al. 1982. *J. Cell Biol.* 94, pp. 425–443, Fig. 1.

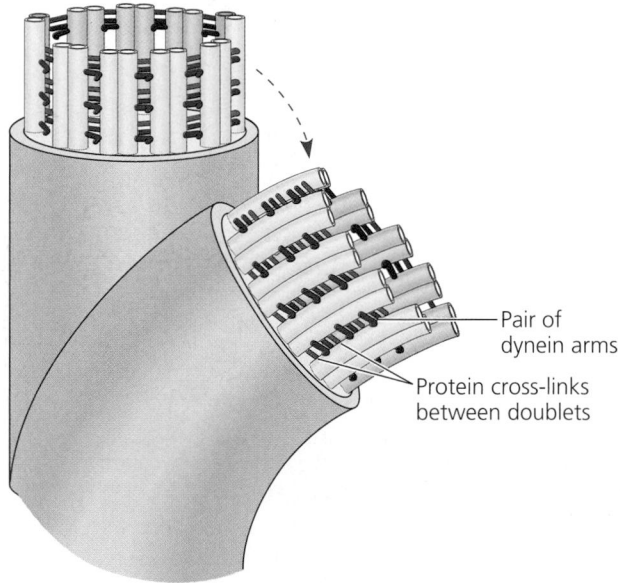

FIGURE 7.25 How dynein "walking" moves cilia and flagella. Powered by ATP, the dynein arms of one microtubule doublet grip the adjacent doublet, pull, release, and then grip again. The doublets cannot slide far because they are physically restrained within the cilium or flagellum, so they bend. For clarity, the two central microtubules and the radial spokes are omitted here.

Microfilaments are well known for their role in cell motility, particularly as part of the contractile apparatus of muscle cells. Thousands of actin filaments are arranged parallel to one another along the length of a muscle cell, interdigitated with thicker filaments made of a protein called **myosin** (FIGURE 7.27a). Myosin acts as a motor molecule by means of projections (arms) that "walk" along the actin filaments. Contraction of the muscle cell results from the actin and myosin filaments sliding past one another in this way, shortening the cell. In other kinds of cells, actin filaments are associated with myosin in miniature and less elaborate versions of the arrangement in muscle cells. These actin-myosin aggregates are responsible for localized contractions of cells. For example, a contracting belt of microfilaments forms a cleavage furrow that pinches a dividing animal cell into two daughter cells.

Localized contraction brought about by actin and myosin also plays a role in amoeboid movement (FIGURE 7.27b), in which a cell crawls along a surface by extending and flowing into cellular extensions called **pseudopodia** (from the Greek *pseudes*, false, and *pod*, foot). Pseudopodia extend and contract through the reversible assembly of actin subunits into microfilaments and of microfilaments into networks that convert cytoplasm from sol to gel. According to a widely accepted model, filaments near the cell's trailing end interact with myosin, causing contraction. Like squeezing on a toothpaste tube, this contraction forces the interior fluid into the pseudopodium, where the actin network has been weakened. The pseudopodium extends until the actin reassembles into a network. Not only do amoebas move by crawling, but so do many cells in the animal body, such as white blood cells.

In plant cells, both actin-myosin interactions and sol-gel transformations brought about by actin may be involved in **cytoplasmic streaming,** a circular flow of cytoplasm within cells (FIGURE 7.27c). This movement, which is especially common in large plant cells, speeds the distribution of materials within the cell.

Intermediate Filaments

Intermediate filaments are named for their diameter, which, at 8–12 nm, is larger than the diameter of microfilaments but smaller than that of microtubules (see TABLE 7.2 and FIGURE 7.26). Specialized for bearing tension (like microfilaments), intermediate filaments are a diverse class of cytoskeletal elements. Each type is constructed from a different molecular subunit belonging to a family of proteins called keratins. Microtubules and microfilaments, in contrast, are consistent in diameter and composition in all eukaryotic cells.

Intermediate filaments are more permanent fixtures of cells than are microfilaments and microtubules, which are often disassembled and reassembled in various parts of a cell. Chemical treatments that remove microfilaments and microtubules from the cytoplasm leave a web of intermediate fila-

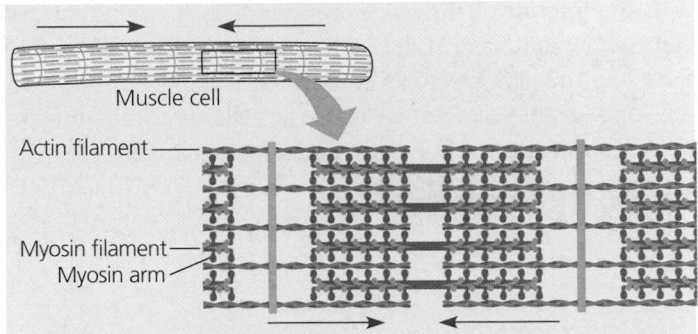

(a) Myosin motors in muscle cell contraction. The "walking" of myosin arms drives the parallel myosin and actin filaments past each other. The teamwork of many such sliding filaments enables the entire cell to shorten.

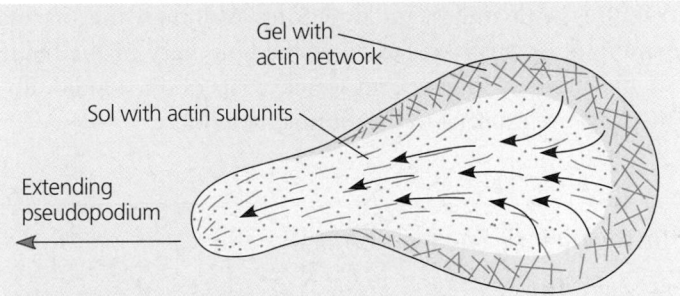

(b) Amoeboid movement. Interaction of actin filaments with myosin near the cell's trailing end squeezes the interior fluid forward into the pseudopodium.

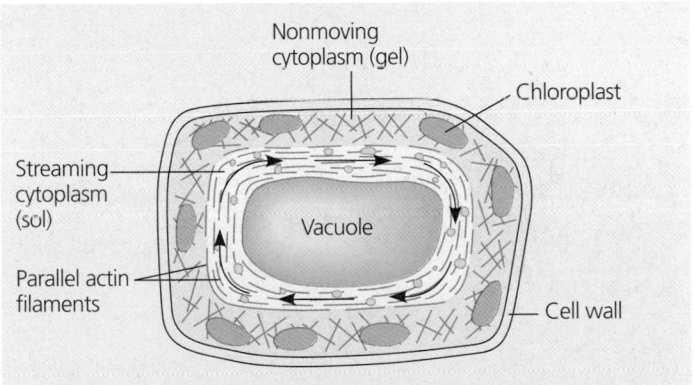

(c) Cytoplasmic streaming in plant cells. A layer of cytoplasm cycles around the cell, moving over a carpet of parallel actin filaments. Myosin motors attached to organelles in the fluid cytosol may drive the streaming by interacting with the actin.

FIGURE 7.27 Microfilaments and motility. In the three examples shown in this figure, cell nuclei and most other organelles have been omitted.

ments that retains its original shape. Such experiments suggest that intermediate filaments are especially important in reinforcing the shape of a cell and fixing the position of certain organelles. For example, the nucleus commonly sits within a cage made of intermediate filaments, fixed in location by branches of the filaments that extend into the cytoplasm. Other intermediate filaments make up the nuclear lamina that lines the interior of the nuclear envelope (see FIGURE 7.9). In cases where the shape of the entire cell is correlated with function, intermediate filaments support that shape. For instance, the long extensions (axons) of nerve cells that transmit impulses are strengthened by one class of intermediate filament. The various kinds of intermediate filaments may function as the framework of the entire cytoskeleton.

CELL SURFACES AND JUNCTIONS

Having crisscrossed the interior of the cell to explore various organelles, we complete our tour of the cell by returning to the surface of this microscopic world, where there are additional structures with important functions. Although the plasma membrane is usually regarded as the boundary of the living cell, most cells synthesize and secrete coats of one kind or another that are external to the plasma membrane.

Plant cells are encased by cell walls

The **cell wall** is one of the features of plant cells that distinguishes them from animal cells. The wall protects the plant cell, maintains its shape, and prevents excessive uptake of water. On the level of the whole plant, the strong walls of special-ized cells hold the plant up against the force of gravity. Prokaryotes, fungi, and some protists also have cell walls, but we will postpone discussion of them until Unit Five.

Plant cell walls are much thicker than the plasma membrane, ranging from 0.1 μm to several micrometers. The exact chemical composition of the wall varies from species to species and from one cell type to another in the same plant, but the basic design of the wall is consistent. Microfibrils made of the polysaccharide cellulose (see FIGURE 5.8) are embedded in a matrix of other polysaccharides and protein. This combination of materials, strong fibers in a "ground substance" (matrix), is the same basic architectural design found in steel-reinforced concrete and in fiberglass.

A young plant cell first secretes a relatively thin and flexible wall called the **primary cell wall** (FIGURE 7.28). Between primary walls of adjacent cells is the **middle lamella,** a thin layer rich in sticky polysaccharides called pectins. The middle lamella glues the cells together (pectin is used as a thickening agent in jams and jellies). When the cell matures and stops growing, it strengthens its wall. Some plant cells do this simply by secreting hardening substances into the primary wall. Other cells add a **secondary cell wall** between the plasma membrane and the primary wall. The secondary wall, often deposited in several laminated layers, has a strong and durable matrix that affords the cell protection and support. Wood, for example, consists mainly of secondary walls.

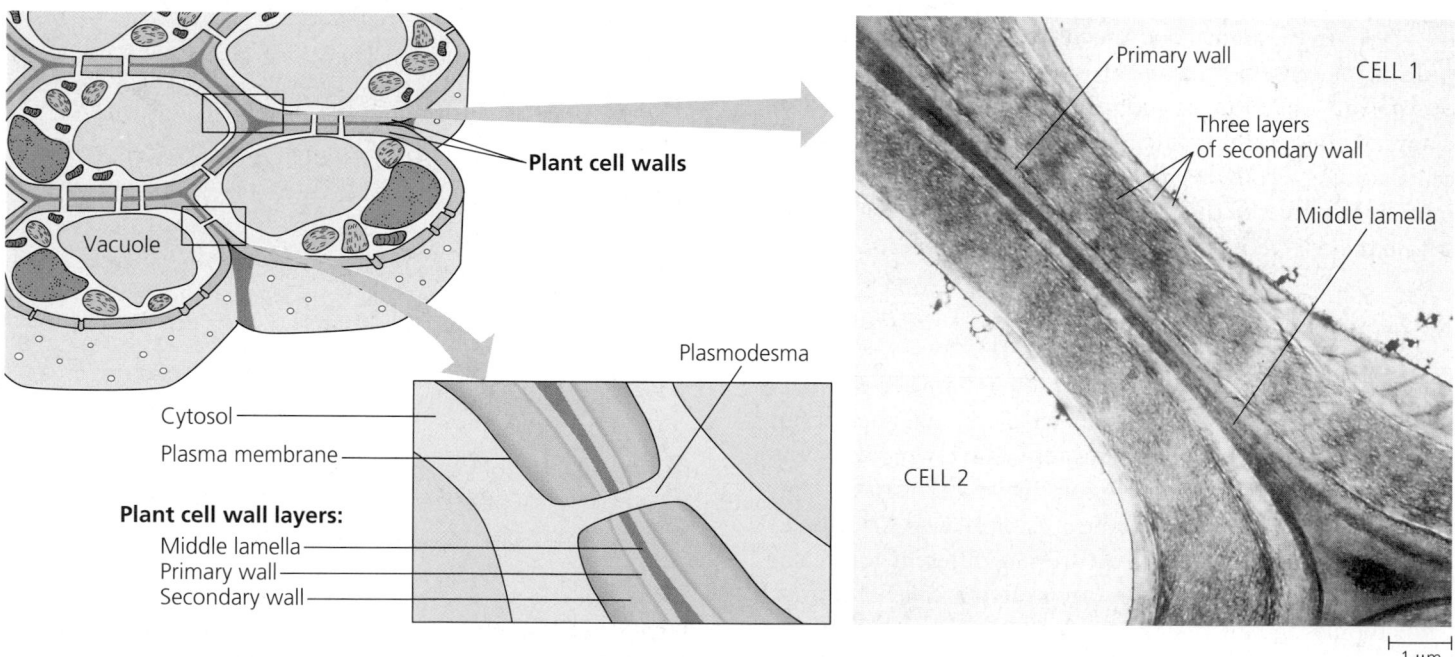

FIGURE 7.28 Plant cell walls. Plant cells first construct thin primary walls, often adding stronger secondary walls to the inside of the primary wall when the cell's growth ceases. A sticky middle lamella cements adjacent cells together. Thus, the multilayered partition between these cells consists of adjoining walls individually secreted by the cells. The walls do not isolate the cells: The cytoplasm of one cell is continuous with the cytoplasm of its neighbors via plasmodesmata, channels through the walls (TEM).

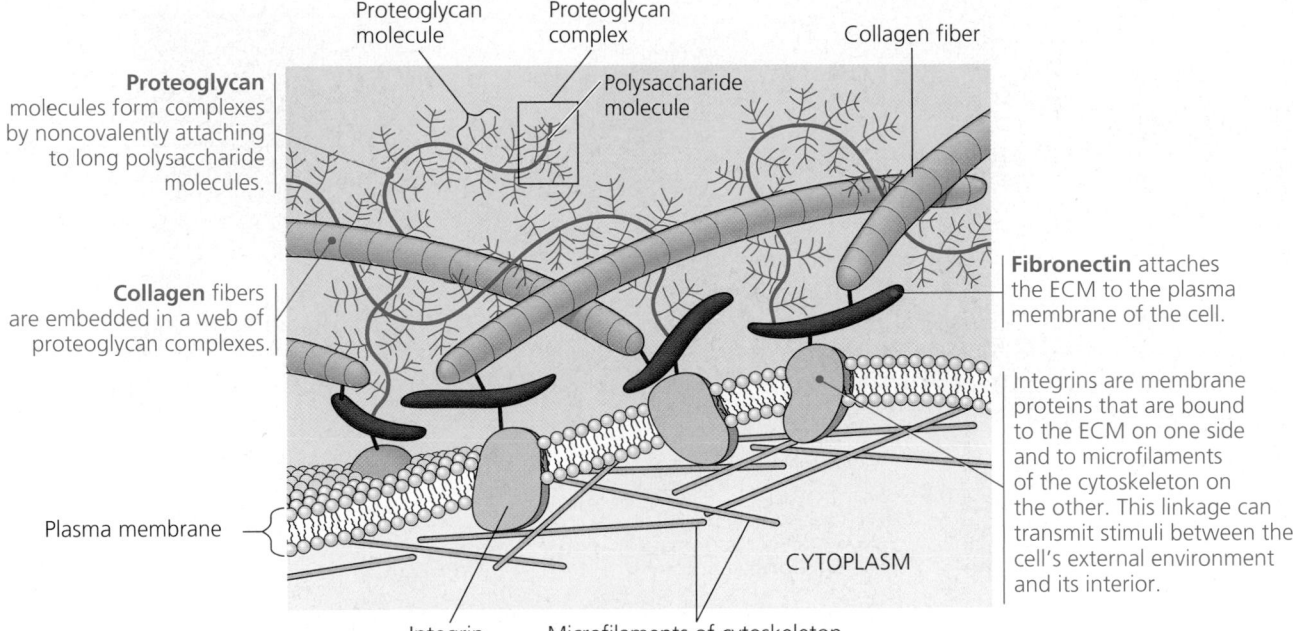

Proteoglycan molecule

Proteoglycan complex

Collagen fiber

Polysaccharide molecule

Proteoglycan molecules form complexes by noncovalently attaching to long polysaccharide molecules.

Fibronectin attaches the ECM to the plasma membrane of the cell.

Collagen fibers are embedded in a web of proteoglycan complexes.

Integrins are membrane proteins that are bound to the ECM on one side and to microfilaments of the cytoskeleton on the other. This linkage can transmit stimuli between the cell's external environment and its interior.

Plasma membrane

CYTOPLASM

Integrin

Microfilaments of cytoskeleton

FIGURE 7.29 Extracellular matrix (ECM) of an animal cell. The molecular composition and structure of the ECM varies from one cell type to another. In this example, three different types of glycoproteins are present: proteoglycan, collagen, and fibronectin. Collagen fibers are embedded in proteoglycan complexes, which consist of proteoglycan molecules extending like little trees from long polysaccharide molecules. Fibronectin molecules are the adhesive that attaches the ECM to the plasma membrane of the cell, by connecting to membrane proteins called integrins.

The extracellular matrix (ECM) of animal cells functions in support, adhesion, movement, and regulation

Although animal cells lack walls akin to those of plant cells, they do have an elaborate **extracellular matrix (ECM)** (FIG-URE 7.29). The main ingredients of the ECM are glycoproteins secreted by the cells. (Recall that glycoproteins are proteins with covalently bonded carbohydrate, usually short chains of sugars.) The most abundant glycoprotein in the ECM of most animal cells is **collagen,** which forms strong fibers outside the cells. In fact, collagen accounts for about half of the total protein in the human body. The collagen fibers are embedded in a network woven from **proteoglycans,** which are glycoproteins of another class. Proteoglycan molecules are especially rich in carbohydrate—up to 95%—and they can form large complexes, as shown in FIGURE 7.29. Some cells are attached to the ECM by still other kinds of glycoproteins, most commonly **fibronectins.** Fibronectins bind to receptor proteins called **integrins** that are built into the plasma membrane. Integrins span the membrane and bind on their cytoplasmic side to microfilaments of the cytoskeleton. Thus, integrins are in a position to transmit changes in the ECM to the cytoskeleton, and vice versa—to integrate changes occurring outside and inside the cell.

Current research on fibronectins and integrins is revealing the influential role of the extracellular matrix in the lives of cells. Communicating with a cell through integrins, the ECM can regulate a cell's behavior. For example, some cells in a developing embryo migrate along specific pathways by matching the orientation of their microfilaments to the "grain" of fibers in the extracellular matrix. Researchers are also learning that the extracellular matrix around a cell can influence the activity of genes in the nucleus. Information about the ECM probably reaches the nucleus by a combination of mechanical and chemical signaling pathways. Mechanical signaling involves fibronectins, integrins, and the cytoskeleton. Changes in the cytoskeleton may in turn trigger chemical signaling pathways inside the cell. In this way, the extracellular matrix of a particular tissue could help coordinate the behavior of all the cells within that tissue. Direct connections between cells also function in this coordination, as we discuss next.

Intercellular junctions help integrate cells into higher levels of structure and function

The many cells of an animal or plant are organized into tissues, organs, and organ systems. Neighboring cells often adhere, interact, and communicate through special patches of direct physical contact.

It might seem that the nonliving cell walls of plants would isolate cells from one another. In fact, as already mentioned, plant cell walls are perforated with channels called **plasmo-desmata** (singular, *plasmodesma*; from the Greek *desmos,* to

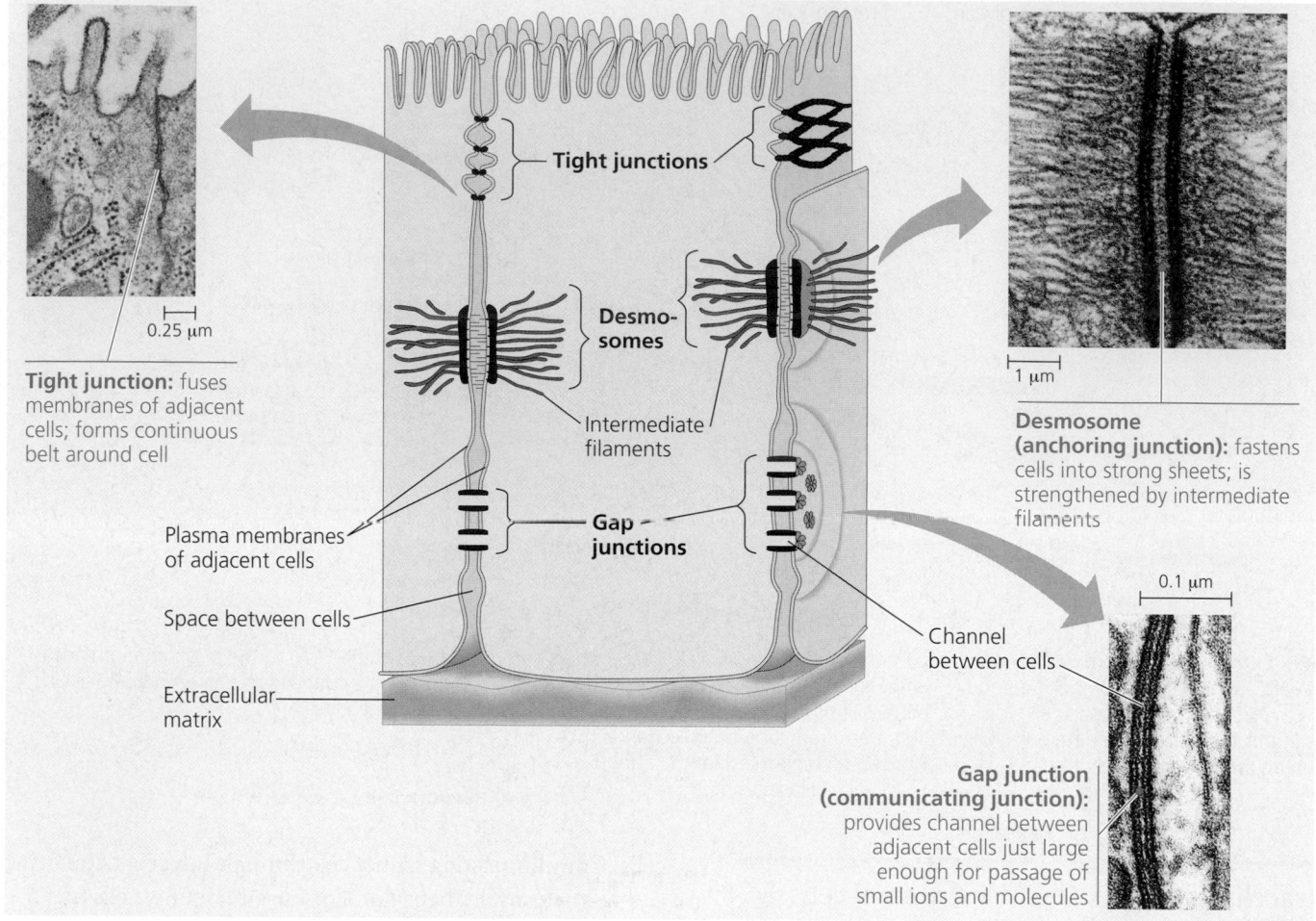

Tight junctions

Tight junction: fuses membranes of adjacent cells; forms continuous belt around cell

0.25 µm

Desmosomes

Intermediate filaments

Plasma membranes of adjacent cells

Space between cells

Extracellular matrix

Gap junctions

Channel between cells

Desmosome (anchoring junction): fastens cells into strong sheets; is strengthened by intermediate filaments

1 µm

0.1 µm

Gap junction (communicating junction): provides channel between adjacent cells just large enough for passage of small ions and molecules

FIGURE 7.30 Intercellular junctions in animal tissues. Epithelial cells lining the intestine illustrate the three main kinds of intercellular junctions (TEMs).

Desmosome TEM from L. Ord and A. Perrelet, *Freeze-Etch Histology* (Heidelberg: Springer-Verlag, 1975). ©1975 Springer-Verlag.

bind). Cytosol passes through the plasmodesmata and connects the living contents of adjacent cells (see FIGURES 7.8 and 7.28). This unifies most of the plant into one living continuum. The plasma membranes of adjacent cells are continuous through a plasmodesma; the membrane lines the channel. Water and small solutes can pass freely from cell to cell, and recent experiments have shown that in certain circumstances, particular proteins and RNA molecules can also do this. The macromolecules to be transported to neighboring cells seem to reach the plasmodesmata by moving along fibers of the cytoskeleton.

In animals, there are three main types of intercellular junctions: tight junctions, desmosomes, and gap junctions. All three types are especially common in epithelial tissue, which lines the internal surfaces of the body. FIGURE 7.30 uses epithelial cells of the intestinal lining to illustrate the junctions. Each kind of junction has a structure well adapted for its function. At **tight junctions,** the membranes of neighboring cells are

actually fused. Forming continuous belts around the cells, these junctions prevent leakage of extracellular fluid across a layer of epithelial cells. In our example, the tight junctions of the intestinal epithelium keep the contents of the intestine separate from the body fluid on the opposite side of the epithelium. **Desmosomes** (also called *anchoring junctions*) function like rivets, fastening cells together into strong sheets. Intermediate filaments made of the sturdy protein keratin reinforce desmosomes. **Gap junctions** (also called *communicating junctions*) provide cytoplasmic channels between adjacent animal cells. Special membrane proteins surround each pore, which is wide enough for salt ions, sugars, amino acids, and other small molecules to pass. In the muscle tissue of the heart, the flow of ions through gap junctions coordinates the contractions of the cells. Gap junctions are especially common in animal embryos, in which chemical communication between cells is essential for development.

The cell is a living unit greater than the sum of its parts

From our panoramic view of the cell's overall compartmental organization to our closeup inspection of each organelle's architecture, this tour of the cell has provided many opportunities to correlate structure with function. (This would be a good time to review cell structure by returning to FIGURES 7.7 and 7.8 on pages 114 and 115.) But even as we dissect the cell, remember that none of its organelles works alone. As an example of cellular integration, consider the microscopic scene in FIGURE 7.31. The large cell is a macrophage. It helps defend the body against infections by ingesting bacteria (the smaller yellow cells). The macrophage crawls along a surface and reaches out to the bacteria with thin pseudopodia ("filopodia"). Actin filaments interact with other elements of the cytoskeleton in these movements. After the macrophage engulfs the bacteria, they are destroyed by lysosomes. The elaborate endomembrane system, which includes the ER and the Golgi apparatus, produces the lysosomes. The digestive enzymes of the lysosomes and the proteins of the cytoskeleton are all made on ribosomes. And the synthesis of these proteins is programmed by genetic messages dispatched from the DNA in the nucleus. All these processes require energy, which mitochondria supply in the form of ATP. Cellular functions arise from cellular order: The cell is a living unit greater than the sum of its parts.

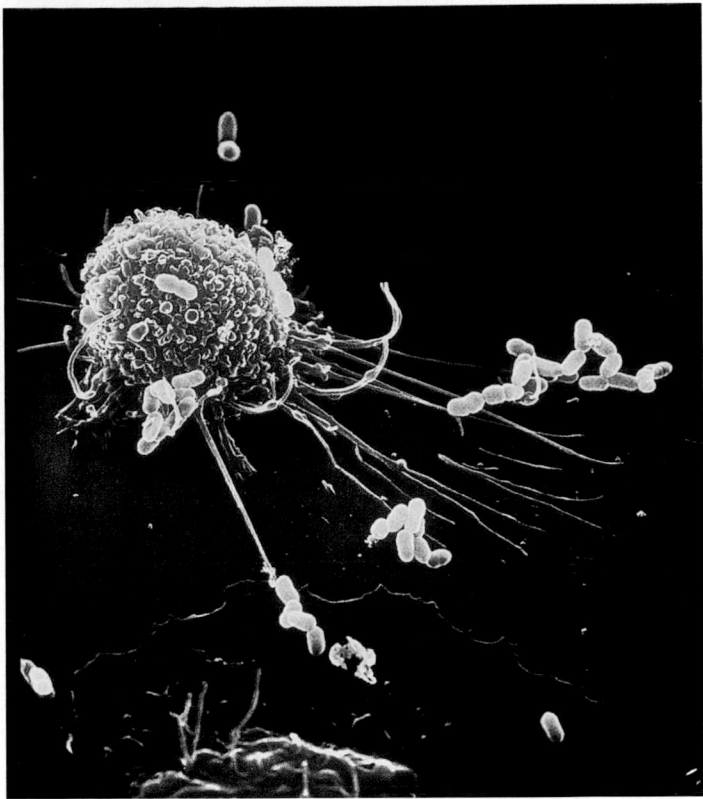

FIGURE 7.31 The emergence of cellular functions from the cooperation of many organelles. The ability of this macrophage (brown) to recognize, apprehend, and destroy bacteria (yellow) is a coordinated activity of the whole cell. Its cytoskeleton, lysosomes, and plasma membrane are among the components that function in phagocytosis. This and other cellular functions are emergent properties that depend on interactions of the cell's parts (colorized SEM).

CHAPTER 7 REVIEW

Go to the Campbell Biology website (www.campbellbiology.com) to explore an interactive version of the Chapter Review.

Summary of Key Concepts

HOW WE STUDY CELLS

- Microscopes provide windows to the world of the cell (pp. 109–111, FIGURES 7.1, 7.2) Improvements in microscopy have catalyzed progress in the study of cell structure.
  Web/CD Activity 7A: *Metric System Review*
  Web/CD Case Study in the Process of Science: *What Is the Size and Scale of Our World?*

- Cell biologists can isolate organelles to study their functions (p. 111, FIGURE 7.3) Cell biologists use the ultracentrifuge to produce pellets enriched in specific organelles.

A PANORAMIC VIEW OF THE CELL

- Prokaryotic and eukaryotic cells differ in size and complexity (pp. 112–113, FIGURES 7.4–7.6) All cells are bounded by a plasma membrane. Bacteria and archaea have prokaryotic cells, without nuclei or other membrane-enclosed organelles. All other organisms have eukaryotic cells, with membrane-enclosed nuclei and other membranous organelles in their cytoplasm. The need for a high surface-to-volume ratio limits cell size.
  Web/CD Activity 7B: *Prokaryotic Cell Structure and Function*
  Web/CD Activity 7C: *Comparing Prokaryotic and Eukaryotic Cells*

- Internal membranes compartmentalize the functions of a eukaryotic cell (pp. 114–115, FIGURES 7.7, 7.8) Plant and animal cells have most of the same organelles.
  Web/CD Activity 7D: *Build an Animal Cell and a Plant Cell*

THE NUCLEUS AND RIBOSOMES

- The nucleus contains a eukaryotic cell's genetic library (pp. 116–117, FIGURE 7.9) DNA is organized with proteins into thin fibers of chromatin, which coil to form thick chromosomes in dividing cells. Associated with the chromatin in nondividing cells are one or more nucleoli, sites of ribosome synthesis. Macromolecules and ribosomal subunits pass between nucleus and cytoplasm through pores in the nuclear envelope.
Web/CD Activity 7E: *Role of the Nucleus and Ribosomes in Protein Synthesis*

- Ribosomes build a cell's proteins (pp. 117–118, FIGURE 7.10) Free ribosomes in the cytosol and bound ribosomes on the outside of ER and the nuclear envelope synthesize proteins.

THE ENDOMEMBRANE SYSTEM

- Many of the eukaryotic cell's membranes are connected either by physical continuity or through transport vesicles made of pinched-off pieces of membrane.

- The endoplasmic reticulum manufactures membranes and performs many other biosynthetic functions (pp. 118–119, FIGURE 7.11) Continuous with the nuclear envelope, the endoplasmic reticulum (ER) is a network of cisternae, membrane-enclosed compartments. Smooth ER lacks ribosomes; it synthesizes steroids, metabolizes carbohydrates, stores calcium in muscle, and detoxifies poisons in liver. Rough ER has bound ribosomes and produces cell membrane and secretory proteins. These products are distributed by transport vesicles budded from the ER.

- The Golgi apparatus finishes, sorts, and ships cell products (pp. 119–121, FIGURE 7.12) Stacks of separate cisternae make up the Golgi. The *cis* face of a Golgi stack receives secretory proteins from the ER in transport vesicles. The proteins are modified, sorted, and released in transport vesicles from the *trans* face. The Golgi also synthesizes some macromolecules on its own.

- Lysosomes are digestive compartments (pp. 121–122, FIGURE 7.14) Lysosomes are membranous sacs of hydrolytic enzymes. They break down cell macromolecules for recycling as well as substances ingested by phagocytosis.

- Vacuoles have diverse functions in cell maintenance (pp. 122–123, FIGURE 7.15) A plant cell's central vacuole functions in storage, waste disposal, cell growth, and protection.
Web/CD Activity 7F: *The Endomembrane System*

OTHER MEMBRANOUS ORGANELLES

- Mitochondria and chloroplasts are the main energy transformers of cells (pp. 123–125, FIGURES 7.17, 7.18) Mitochondria, the sites of cellular respiration in eukaryotes, have an outer membrane and an inner membrane that is folded into cristae. Some reactions of respiration occur in the mitochondrial matrix enclosed by the inner membrane, and others are catalyzed by enzymes built into the inner membrane. Chloroplasts contain chlorophyll and other pigments that function in photosynthesis. In chloroplasts, two membranes surround the fluid stroma, which contains thylakoids stacked into grana.
Web/CD Activity 7G: *Build a Chloroplast and a Mitochondrion*

- Peroxisomes generate and degrade H_2O_2 in performing various metabolic functions (pp. 125–126, FIGURE 7.19) Peroxisomes carry out processes that produce hydrogen peroxide (H_2O_2) as waste, and their enzymes convert the toxic peroxide to water.

THE CYTOSKELETON

- Providing structural support to the cell, the cytoskeleton also functions in cell motility and regulation (pp. 126–132, TABLE 7.2, FIGURES 7.22–7.24) The cytoskeleton is made of microtubules, microfilaments, and intermediate filaments. Microtubules grow out from the centrosome, a region near the nucleus that includes two centrioles in most animal cells. Microtubules shape the cell, guide movement of organelles, and help separate the chromosome copies in dividing cells. Cilia and flagella are motile appendages containing microtubule doublets that are moved past each other by the motor protein dynein. Microfilaments are thin rods built from actin; they function in muscle contraction, amoeboid movement, cytoplasmic streaming, and support for microvilli. Intermediate filaments support cell shape and fix organelles in place.
Web/CD Activity 7H: *Cilia and Flagella*

CELL SURFACES AND JUNCTIONS

- Plant cells are encased by cell walls (p. 132, FIGURE 7.28) Plant cell walls are composed of cellulose fibers embedded in other polysaccharides and protein.

- The extracellular matrix (ECM) of animal cells functions in support, adhesion, movement, and regulation (p. 133, FIGURE 7.29) Animal cells secrete glycoproteins that form the ECM. Important components include collagen, proteoglycan complexes, and fibronectin attached to integrins in the plasma membrane.

- Intercellular junctions help integrate cells into higher levels of structure and function (pp. 133–134, FIGURE 7.30) Plants have plasmodesmata, channels that pass through adjoining cell walls. Animal cell contact is by tight junctions, desmosomes, and gap junctions.
Web/CD Activity 7I: *Cell Junctions*

- The cell is a living unit greater than the sum of its parts (p. 135, FIGURE 7.31)
Web/CD Activity 7J: *Review: Animal Cell Structure and Function*
Web/CD Activity 7K: *Review: Plant Cell Structure and Function*

Self-Quiz

1. The symptoms of a certain inherited disorder in humans include respiratory problems and, in males, sterility. Which of the following is a reasonable hypothesis for the molecular basis of this disorder? (Explain your answer.)
   a. a defective enzyme in the mitochondria
   b. defective actin molecules in cellular microfilaments
   c. defective dynein molecules in cilia and flagella
   d. abnormal hydrolytic enzymes in the lysosomes
   e. a defective secretory protein

2. Choose the statement that correctly characterizes bound ribosomes.
   a. Bound ribosomes are enclosed in their own membrane.
   b. Bound ribosomes are structurally different from free ribosomes.
   c. Bound ribosomes generally synthesize membrane proteins and secretory proteins.
   d. The most common location for bound ribosomes is the cytoplasmic surface of the plasma membrane.
   e. Bound ribosomes are concentrated in the cisternal space of rough ER.

3. Which of the following organelles is least closely associated with the endomembrane system?
   a. nuclear envelope
   b. chloroplast
   c. Golgi apparatus
   d. plasma membrane
   e. ER

4. Cells of the pancreas will incorporate radioactively labeled amino acids into proteins. This "tagging" of newly synthesized proteins enables a researcher to track the location of these proteins in a cell. In this case, we are tracking an enzyme that is eventually secreted by pancreatic cells. Which of the following is the most likely pathway for movement of this protein in the cell?
   a. ER $\longrightarrow$ Golgi $\longrightarrow$ nucleus
   b. Golgi $\longrightarrow$ ER $\longrightarrow$ lysosome
   c. nucleus $\longrightarrow$ ER $\longrightarrow$ Golgi
   d. ER $\longrightarrow$ Golgi $\longrightarrow$ vesicles that fuse with plasma membrane
   e. ER $\longrightarrow$ lysosomes $\longrightarrow$ vesicles that fuse with plasma membrane

5. Which of the following organelles is common to plant *and* animal cells?
   a. chloroplasts
   b. wall made of cellulose
   c. tonoplast
   d. mitochondria
   e. centrioles

6. Which of the following components is present in a prokaryotic cell?
   a. mitochondria
   b. ribosomes
   c. nuclear envelope
   d. chloroplasts
   e. ER

7. Which type of cell would probably provide the best opportunity to study lysosomes? (Explain your answer.)
   a. muscle cell
   b. nerve cell
   c. phagocytic white blood cell
   d. leaf cell of a plant
   e. bacterial cell

8. Which of the following statements is a correct distinction between prokaryotic and eukaryotic cells attributable to the absence of a prokaryotic cytoskeleton?
   a. Compartmentalized organelles are found only in eukaryotic cells.
   b. Cytoplasmic streaming is not observed in prokaryotes.
   c. Only eukaryotic cells are capable of movement.
   d. Prokaryotic cells are usually 10 μm or less in diameter.
   e. Only the eukaryotic cell concentrates its genetic material in a region separate from the rest of the cell.

9. Which of the following structure-function pairs is *mismatched*?
   a. nucleolus; ribosome production
   b. lysosome; intracellular digestion
   c. ribosome; protein synthesis
   d. Golgi; secretion of cell products
   e. microtubules; muscle contraction

10. Cyanide binds with at least one of the molecules involved in the production of ATP. Following exposure of a cell to cyanide, most of the cyanide could be expected to be found within the
    a. mitochondria.
    b. ribosomes.
    c. peroxisomes.
    d. lysosomes.
    e. endoplasmic reticulum.

11. Which type of microscope would you use to study (a) the changes in shape of a living human white blood cell; (b) the details of surface texture of a human hair; (c) the detailed structure of an organelle in the cytoplasm of a human liver cell?

12. After very small viruses infect a plant cell by crossing its membrane, the viruses often spread rapidly throughout the entire plant without crossing additional membranes. Explain how this occurs.

13. Consider the following organelles: mitochondrion, chloroplast, ribosome, lysosome, peroxisome. Which organelle does not belong in the list? Why not?

14. What is the relationship of chromosomes to chromatin?

15. What are three functions of smooth endoplasmic reticulum?

16. What organelles have energy conversion as their primary function?

17. How do transport vesicles integrate the endomembrane system?

18. Which component of the cytoskeleton is most important in (a) holding the nucleus in place; (b) guiding transport vesicles from the Golgi to the plasma membrane; (c) amoeboid movement?

19. How do cilia and flagella bend?

20. How is a plant or animal tissue different from just a collection of similar cells?

Go to the website or CD-ROM for more quiz questions.

Evolution Connection

The similarities among cells reveal the evolutionary unity of life, yet cells can differ dramatically in structure. Which aspects of cell structure best reveal their evolutionary unity? What are some examples of specialized cellular modifications?

The Process of Science

Imagine protein X, destined to go to the plasma membrane of a cell. Assume the mRNA carrying the genetic message for protein X has already been translated on ribosomes in a cell culture. You collect the cells, break them open, and then fractionate them by differential centrifugation as shown in FIGURE 7.3. In the pellet of which fraction would you expect to find protein X? Explain your answer by describing the transit of protein X through the cell.

Take a hyperspace tour from outer space to the hydrogen atom and learn how to estimate size in the Case Study in the Process of Science, available on the website and CD-ROM.

Science, Technology, and Society

Doctors at a California university removed a man's spleen, standard treatment for a type of leukemia. The disease did not recur. Researchers kept some of the spleen cells alive in a nutrient medium. They found that some of the cells produced a blood protein that showed promise for treating cancer and AIDS. The researchers patented the cells. The patient sued, claiming a share in profits from any products derived from his cells. The California Supreme Court ruled against the patient, stating that his suit "threatens to destroy the economic incentive to conduct important medical research." The U.S. Supreme Court agreed. Do you think the patient was treated fairly? What else you would like to know about this case that might help you make up your mind?

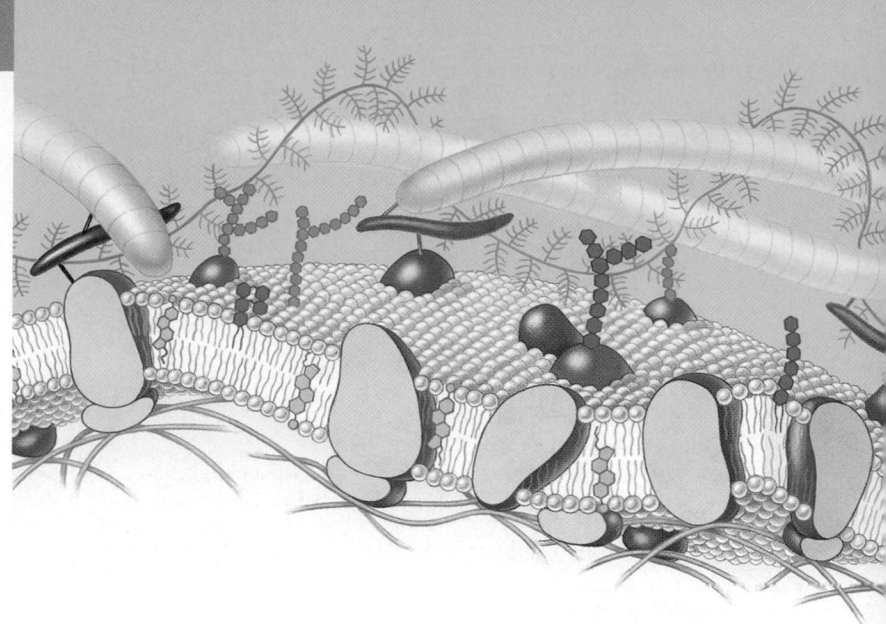

CHAPTER 8

MEMBRANE STRUCTURE AND FUNCTION

MEMBRANE STRUCTURE

- Membrane models have evolved to fit new data
- Membranes are fluid
- Membranes are mosaics of structure and function
- Membrane carbohydrates are important for cell-cell recognition

TRAFFIC ACROSS MEMBRANES

- A membrane's molecular organization results in selective permeability
- Passive transport is diffusion across a membrane
- Osmosis is the passive transport of water
- Cell survival depends on balancing water uptake and loss
- Specific proteins facilitate the passive transport of water and selected solutes: *a closer look*
- Active transport is the pumping of solutes against their gradients
- Some ion pumps generate voltage across membranes
- In cotransport, a membrane protein couples the transport of two solutes
- Exocytosis and endocytosis transport large molecules

The plasma membrane *is the edge of life, the boundary that separates the living cell from its nonliving surroundings. A remarkable film only about 8 nm thick—it would take over 8,000 to equal the thickness of this page—the plasma membrane controls traffic into and out of the cell it surrounds. Like all biological membranes, the plasma membrane exhibits* **selective permeability;** *that is, it allows some substances to cross it more easily than others. One of the earliest episodes in the evolution of life may have been the formation of a membrane that enclosed a solution different from the surrounding solution while still permitting the uptake of nutrients and elimination of waste products. This ability of the cell to discriminate in its chemical exchanges with its environment is fundamental to life, and it is the plasma membrane that makes this selectivity possible.*

In this chapter, you will learn how cellular membranes control the passage of substances. We will concentrate on the plasma

membrane, the outermost membrane of the cell, represented in the drawing on this page. However, the same general principles of membrane traffic also apply to the many varieties of internal membranes that partition the eukaryotic cell. To understand how membranes work, we begin by examining their architecture.

MEMBRANE STRUCTURE

Lipids and proteins are the staple ingredients of membranes, although carbohydrates are also important. The most abundant lipids in most membranes are phospholipids. Their ability to form membranes is built into their molecular structure. A phospholipid is an **amphipathic molecule,** meaning it has both a hydrophilic region and a hydrophobic region (see FIG-URE 5.12). Other types of membrane lipids are also amphipathic. Furthermore, most of the proteins of membranes have both hydrophobic and hydrophilic regions.

How are phospholipids and proteins arranged in the membranes of cells? You encountered the currently accepted model for the arrangement of these molecules in Chapter 7 (see FIG-URE 7.6). In this **fluid mosaic model,** the membrane is a fluid structure with various proteins embedded in or attached to a double layer (bilayer) of phospholipids. We'll discuss this model in detail, starting with the story of how it was developed.

Membrane models have evolved to fit new data

Scientists began building molecular models of the membrane decades before membranes were first seen with the electron microscope in the 1950s. In 1895, Charles Overton hypothesized that membranes are made of lipids, based on his observations that substances that dissolve in lipids enter cells much more rapidly than substances that are

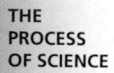

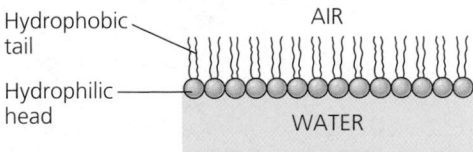

Hydrophobic tail

Hydrophilic head

AIR

WATER

(a) The hydrophilic heads of the phospholipids are immersed in water, and the hydrophobic tails are excluded from water.

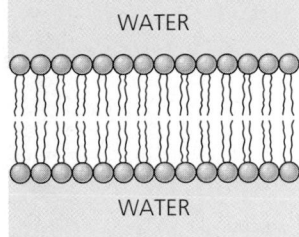

WATER

WATER

(b) A bilayer of phospholipids forms a stable boundary between two aqueous compartments, exposing the hydrophilic parts of the molecules to water and shielding the hydrophobic parts from water.

FIGURE 8.1 Artificial membranes (cross sections).

insoluble in lipids. Twenty years later, membranes isolated from red blood cells were chemically analyzed and found to be composed of lipids and proteins.

In 1917, Irving Langmuir made artificial membranes by adding phospholipids dissolved in benzene (an organic solvent) to water. After the benzene evaporated, the phospholipids remained as a film covering the surface of the water, with only the hydrophilic heads of the phospholipids immersed in the water (FIGURE 8.1a). In 1925, two Dutch scientists, E. Gorter and F. Grendel, reasoned that cell membranes must actually be phospholipid bilayers, two molecules thick. Such a bilayer could exist as a stable boundary between two aqueous compartments because the molecular arrangement shelters the hydrophobic tails of the phospholipids from water while exposing the hydrophilic heads to water (FIGURE 8.1b).

Gorter and Grendel measured the phospholipid content of membranes isolated from red blood cells and found just enough of the lipid to cover the cells with two layers. (Ironically, Gorter and Grendel underestimated both the phospholipid content and the surface area of the cells, but the two errors canceled each other. Thus, what turned out to be a correct conclusion was based on flawed measurements.)

If we assume that a phospholipid bilayer is the main fabric of the membrane, where do we place the proteins? Although the heads of phospholipids are hydrophilic, the surface of an artificial membrane consisting of a phospholipid bilayer adheres less strongly to water than does the surface of an actual biological membrane. This difference could be accounted for if the membrane were coated on both sides with hydrophilic proteins. In 1935, Hugh Davson and James Danielli proposed a sandwich model: a phospholipid bilayer between two layers of globular protein (FIGURE 8.2a).

When researchers first used electron microscopes to study cells in the 1950s, the pictures seemed to support the Davson-Danielli model. In electron micrographs of cells stained with atoms of heavy metals, the plasma membrane is triple-layered, having two dark (stained) bands separated by an unstained layer (see FIGURE 7.6a). Most early electron microscopists assumed that the stain adhered to the proteins and hydrophilic heads of the phospholipids, leaving the hydrophobic core of the membrane unstained. By the 1960s, the Davson-Danielli sandwich had become widely accepted as the structure not only of the plasma membrane, but of all the internal membranes of the cell. However, by the end of that decade, many cell biologists recognized two problems with the model.

First, the generalization that all membranes of the cell are identical was challenged. Not all membranes look alike through the electron microscope. For example, whereas the plasma membrane is 7–8 nm thick and has the three-layered structure, the inner membrane of the mitochondrion is only 6 nm thick and in electron micrographs looks like a row of beads. Mitochondrial membranes also have a substantially greater percentage of proteins than plasma membranes, and there are differences in the specific kinds of phospholipids and other lipids. Membranes with different functions differ in chemical composition and structure. A second, more serious problem with the sandwich model is in the placement of the proteins. Unlike proteins dissolved in the cytosol, membrane proteins are not very soluble in water. Membrane proteins have hydrophobic regions as well as hydrophilic regions (that is, they are amphipathic). If such proteins were layered on the

Protein

Hydrophilic zone

Hydrophobic zone

Hydrophilic zone

Hydrophilic region of protein

Phospholipid bilayer

Hydrophobic region of protein

(a) The Davson-Danielli model. This model, proposed in 1935, sandwiched the phospholipid bilayer between two protein layers. With later modifications, this model was widely accepted until about 1970.

(b) Current fluid mosaic model. This model disperses the proteins and immerses them in the phospholipid bilayer, which is in a fluid state. Shown here in simplified form, this is our present working model of the membrane.

FIGURE 8.2 Two generations of membrane models.

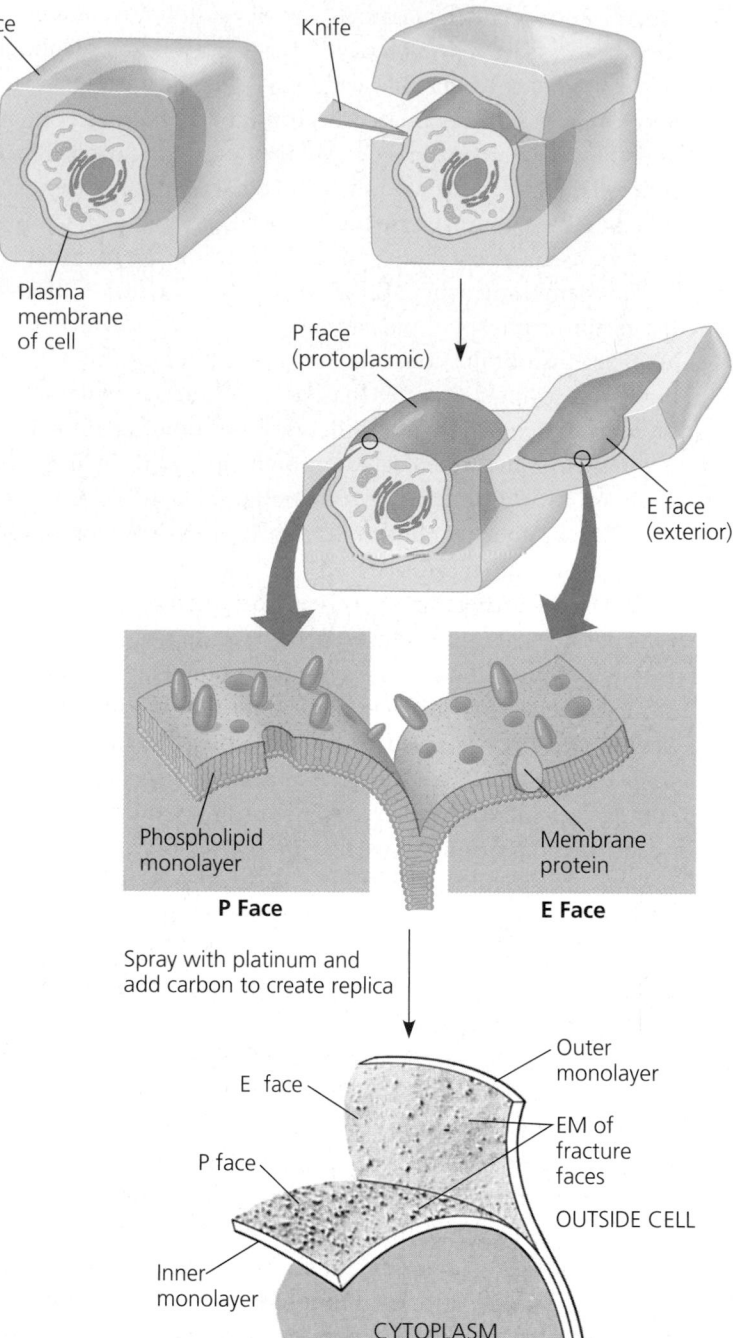

(a) A researcher freezes the specimen at the temperature of liquid nitrogen, then fractures the cells with a cold knife. The knife does not cut cleanly through the frozen cells; instead, it cracks the specimen, with the fracture plane following the path of least resistance.

(b) The fracture plane often follows the hydrophobic interior of a membrane, splitting the lipid bilayer down the middle into a P ("protoplasmic"—that is, cytoplasmic) face and an E (exterior) face. The membrane proteins are not split but go with one or the other of the phospholipid layers. The topography of the fractured surface may be enhanced by etching, the removal of frozen water by direct evaporation (sublimation).

(c) A fine mist of platinum is sprayed from an angle onto the fractured surface of the cell. There will be "shadows" where elevated regions of the fractured cell block the platinum. Adding a film of carbon strengthens the platinum coat.

   The orginal specimen is digested away with bleach, acids, and enzymes, leaving the platinum-carbon film as a replica of the fractured surface. It is this replica, not the membrane itself, that is examined through the electron microscope.

   Electron micrographs have been inserted into this drawing of a delaminated membrane. Notice the protein particles (the "bumps").

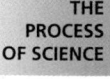

THE PROCESS OF SCIENCE

FIGURE 8.3 Freeze-fracture and freeze-etch. With this method, the two-layer plasma membrane (or other cell membrane) can be split into its two layers. When the separated layers are examined with an electron microscope, the ultrastructure of the membrane's interior is revealed.

surface of the membrane, their hydrophobic parts would be in an aqueous environment.

In 1972, S. J. Singer and G. Nicolson advocated a revised membrane model that placed the proteins in a location compatible with their amphipathic character. They proposed that membrane proteins are dispersed and individually inserted into the phospholipid bilayer, with only their hydrophilic regions protruding far enough from the bilayer to be exposed to water (FIGURE 8.2b). This molecular arrangement would max-

imize contact of hydrophilic regions of proteins and phospholipids with water while providing their hydrophobic parts with a nonaqueous environment. According to this model, the membrane is a mosaic of protein molecules bobbing in a fluid bilayer of phospholipids—hence the term *fluid mosaic model*.

A method of preparing cells for electron microscopy called freeze-fracture has demonstrated visually that proteins are indeed embedded in the phospholipid bilayer of the membrane (FIGURE 8.3). Freeze-fracture splits a membrane along the middle of the phospholipid bilayer. When the halves of the fractured membrane are viewed in the electron microscope, the interior of the bilayer appears cobblestoned, with protein particles interspersed in a smooth matrix, as in the fluid mosaic model. Other kinds of evidence further support this arrangement.

Models are proposed by scientists as hypotheses, ways of organizing and explaining existing information. Replacing one model of membrane structure with another does not imply that the original model was worthless. The acceptance or rejection of a model depends on how well it fits observations and explains experimental results. A good model also makes predictions that shape future research. Models inspire experiments, and few models survive these tests without modification. New findings may make a model obsolete; even then, it may not be totally scrapped, but revised to incorporate the new observations. The fluid mosaic model is continually being refined and may one day undergo major revision.

Now let's take a closer look at membrane structure, beginning with the word *fluid*.

Membranes are fluid

Membranes are not static sheets of molecules locked rigidly in place. A membrane is held together primarily by hydrophobic interactions, which are much weaker than covalent bonds (see Chapter 5). Most of the lipids and some of the proteins can drift about laterally—that is, in the plane of the membrane (FIGURE 8.4a). It is rare, however, for a molecule to flip-flop transversely across the membrane, switching from one phospholipid layer to the other; to do so, the hydrophilic part of the molecule would have to cross the hydrophobic core of the membrane.

The lateral movement of phospholipids within the membrane is rapid, averaging about 2 μm—the length of a typical bacterial cell—per second. Proteins are much larger than lipids and move more slowly, but some membrane proteins do, in fact, drift (FIGURE 8.5). And some membrane proteins seem to move in a highly directed manner, perhaps driven along cytoskeletal fibers by motor proteins connected to the membrane proteins' cytoplasmic regions. However, many other membrane proteins seem to be held virtually immobile by their attachment to the cytoskeleton.

A membrane remains fluid as temperature decreases, until finally the phospholipids settle into a closely packed arrangement and the membrane solidifies, much as bacon grease forms lard when it cools. The temperature at which a membrane solidifies depends on the types of lipids it is made of. The membrane remains fluid to a lower temperature if it is rich in phospholipids with unsaturated hydrocarbon tails (see FIGURES 5.11 and 5.12). Because of kinks in the tails where double

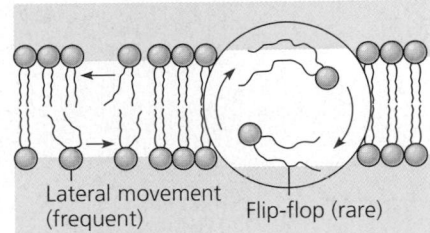

(a) Movement of phospholipids. Lipids move laterally in a membrane, but flip-flopping across the membrane is rare.

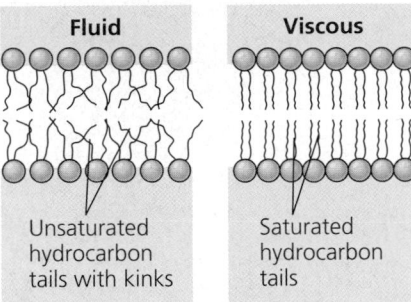

(b) Membrane fluidity. Unsaturated hydrocarbon tails of phospholipids have kinks that keep the molecules from packing together, enhancing membrane fluidity.

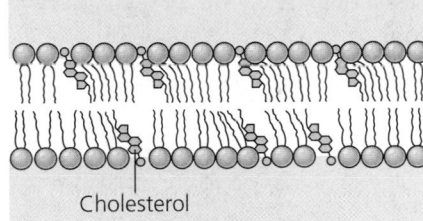

(c) Cholesterol within the membrane. Cholesterol reduces membrane fluidity at moderate temperatures by reducing phospholipid movement, but at low temperatures it hinders solidification by disrupting the regular packing of phospholipids.

FIGURE 8.4 The fluidity of membranes.

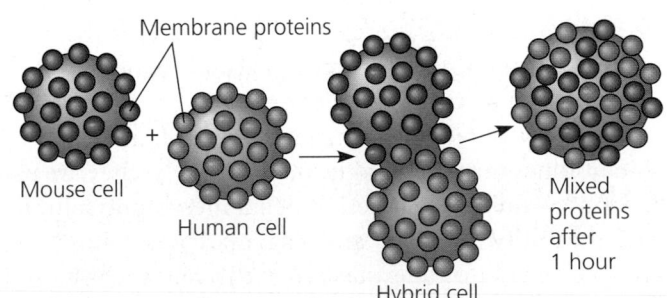

FIGURE 8.5 Evidence for the drifting of membrane proteins. When researchers fuse a human cell with a mouse cell, it takes less than an hour for the membrane proteins of the two species to completely intermingle in the membrane of the hybrid cell.

THE PROCESS OF SCIENCE

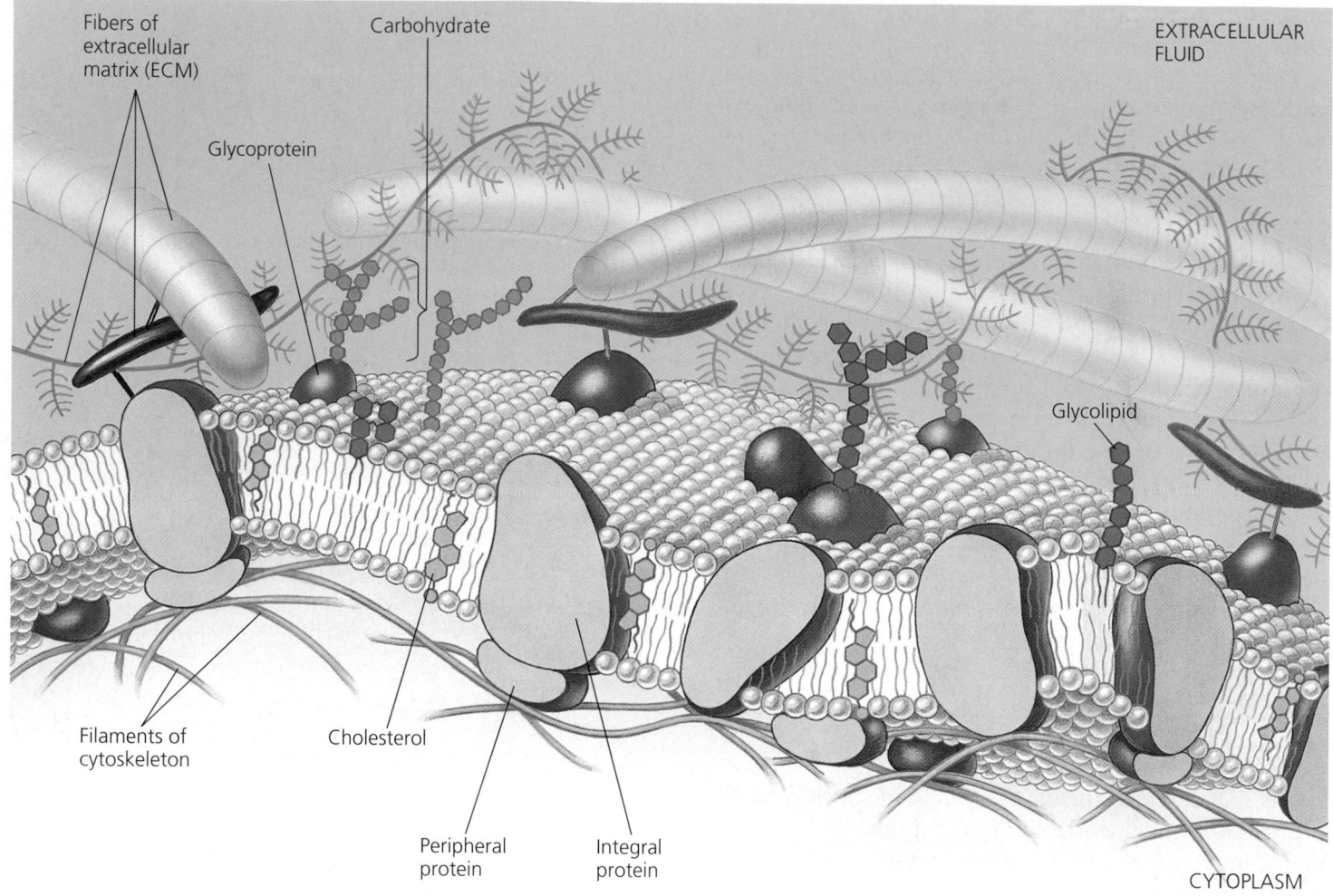

Fibers of
extracellular
matrix (ECM)

Carbohydrate

EXTRACELLULAR
FLUID

Glycoprotein

Glycolipid

Filaments of
cytoskeleton

Cholesterol

Peripheral
protein

Integral
protein

CYTOPLASM

FIGURE 8.6 The detailed structure of an animal cell's plasma membrane, in cross section. See FIGURE 7.29 for details of the ECM.

bonds are located, unsaturated hydrocarbons do not pack together as closely as saturated hydrocarbons (see FIGURE 8.4b).

The steroid cholesterol, which is wedged between phospholipid molecules in the plasma membranes of animal cells, has different effects on membrane fluidity at different temperatures (see FIGURE 8.4c). At relatively warm temperatures—at 37°C, the body temperature of humans, for example—cholesterol makes the membrane less fluid by restraining the movement of phospholipids. However, because cholesterol also hinders the close packing of phospholipids, it lowers the temperature required for the membrane to solidify.

Membranes must be fluid to work properly; they are usually about as fluid as salad oil. When a membrane solidifies, its permeability changes, and enzymatic proteins in the membrane may become inactive. Cells can alter the lipid composition of their membranes as an adjustment to changing temperature. For instance, in many plants that tolerate extreme cold, such as winter wheat, the percentage of unsaturated phospholipids increases in autumn, an adaptation that keeps the membranes from solidifying during winter.

Membranes are mosaics of structure and function

Now we come to the word *mosaic*. A membrane is a collage of different proteins embedded in the fluid matrix of the lipid bilayer (FIGURE 8.6). The lipid bilayer is the main fabric of the membrane, but proteins determine most of the membrane's specific functions. The plasma membrane and the membranes of the various organelles each have unique collections of proteins. More than 50 kinds of proteins have been found so far in the plasma membrane of red blood cells, for example.

Notice in FIGURE 8.6 that there are two major populations of membrane proteins. **Integral proteins** penetrate the hydrophobic core of the lipid bilayer. Many are *transmembrane* proteins, which completely span the membrane. The hydrophobic regions of an integral protein consist of one or more stretches of nonpolar amino acids (see FIGURE 5.15), usually coiled into α helices (FIGURE 8.7). The hydrophilic parts of the molecule are exposed to the aqueous solutions on either side of the membrane. **Peripheral proteins** are not embedded in the lipid

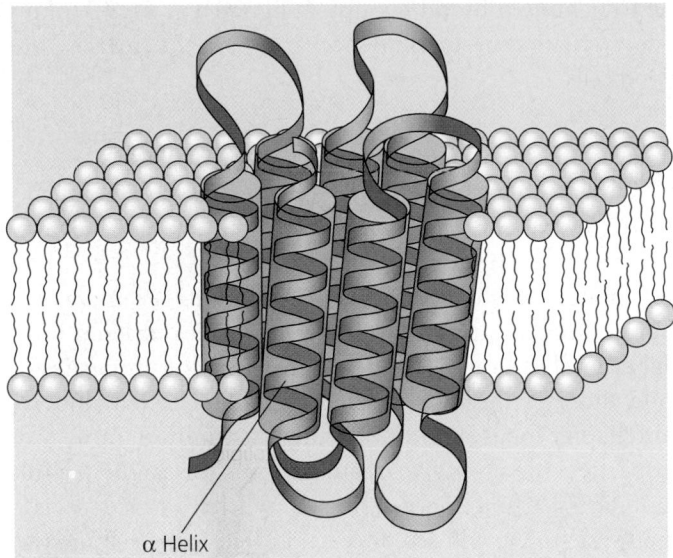

FIGURE 8.7 The structure of a transmembrane protein. This ribbon model highlights the α-helical secondary structure of the hydrophobic parts of the protein, which lie mostly within the hydrophobic core of the membrane. This particular protein, bacteriorhodopsin, has seven transmembrane helices (outlined with cylinders for emphasis). The nonhelical hydrophilic segments of the protein are in contact with the aqueous solutions on either side of the membrane. Bacteriorhodopsin is a specialized transport protein found in certain bacteria.

α Helix

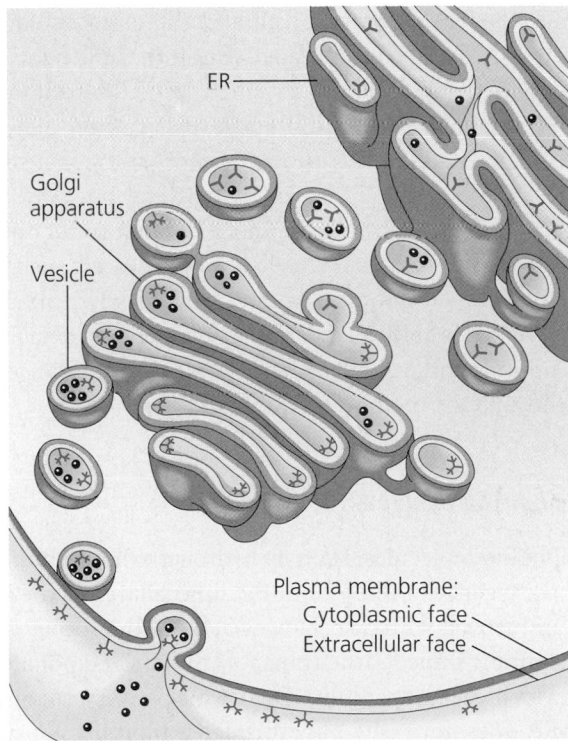

FR

Golgi apparatus

Vesicle

Plasma membrane:
Cytoplasmic face
Extracellular face

FIGURE 8.8 Sidedness of the plasma membrane. The plasma membrane has distinct cytoplasmic and extracellular sides, or faces. The extracellular face is topologically equivalent to the inside face of ER, Golgi, and vesicle membranes. Vesicle fusion with the plasma membrane is responsible for enlarging the membrane and for secretion of cell products (purple). Carbohydrates on the extracellular surface (green) are synthesized in the ER and modified in the Golgi.

bilayer at all; they are appendages loosely bound to the surface of the membrane, often to the exposed parts of integral proteins (see FIGURE 8.6).

On the cytoplasmic side of the plasma membrane, some membrane proteins are held in place by attachment to the cytoskeleton. And on the exterior side, certain membrane proteins are attached to fibers of the extracellular matrix (see FIGURE 7.29; *integrins* are one particular type of integral protein). These attachments combine to give animal cells a stronger external framework than the plasma membrane itself could provide.

Membranes have distinct inside and outside faces. The two lipid layers may differ in specific lipid composition, and each protein has directional orientation in the membrane. The plasma membrane also has carbohydrates, which are restricted to the exterior surface. This asymmetrical distribution of proteins, lipids, and carbohydrates is determined as the membrane is being built by the endoplasmic reticulum. Molecules that start out on the *inside* face of the ER end up on the *outside* face of the plasma membrane (FIGURE 8.8).

FIGURE 8.9 (p. 144) gives an overview of six major functions exhibited by proteins of the plasma membrane. A single cell may have membrane proteins performing several of these functions, and a single protein may have multiple functions. Thus, the membrane is a functional mosaic as well as a structural one.

Membrane carbohydrates are important for cell-cell recognition

Cell-cell recognition, a cell's ability to distinguish one type of neighboring cell from another, is crucial to the functioning of an organism. It is important, for example, in the sorting of cells into tissues and organs in an animal embryo. It is also the basis for the rejection of foreign cells (including those of transplanted organs) by the immune system, an important line of defense in vertebrate animals (see Chapter 43). The way cells recognize other cells is by keying on surface molecules, often carbohydrates, on the plasma membrane.

Membrane carbohydrates are usually branched oligosaccharides with fewer than 15 sugar units. (*Oligo* is Greek for "few"; an oligosaccharide is a short polysaccharide.) Some of these oligosaccharides are covalently bonded to lipids, forming molecules called glycolipids. (Recall that *glyco* refers to the presence of carbohydrate.) Most, however, are covalently bonded to proteins, which are thereby glycoproteins (see FIGURE 8.6).

The oligosaccharides on the external side of the plasma membrane vary from species to species, among individuals of the same species, and even from one cell type to another in a single individual. The diversity of the molecules and their location on the cell's surface enable oligosaccharides to function as markers that distinguish one cell from another. For example,

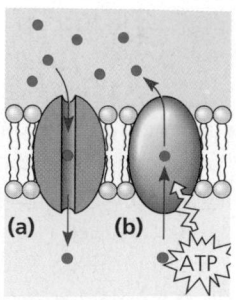

Transport. (a) A protein that spans the membrane may provide a hydrophilic channel across the membrane that is selective for a particular solute. **(b)** Some transport proteins hydrolyze ATP as an energy source to actively pump substances across the membrane.

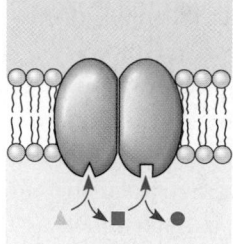

Enzymatic activity. A protein built into the membrane may be an enzyme with its active site exposed to substances in the adjacent solution. In some cases, several enzymes in a membrane are ordered as a team that carries out sequential steps of a metabolic pathway.

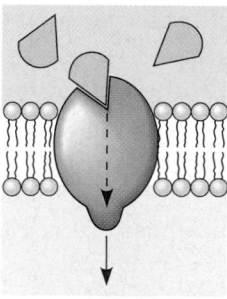

Signal transduction. A membrane protein may have a binding site with a specific shape that fits the shape of a chemical messenger, such as a hormone. The external messenger (signal) may cause a conformational change in the protein that relays the message to the inside of the cell.

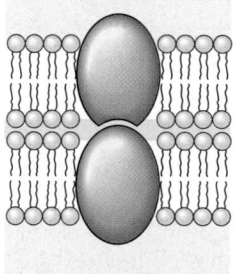

Intercellular joining. Membrane proteins of adjacent cells may be hooked together in various kinds of junctions (see FIGURE 7.30).

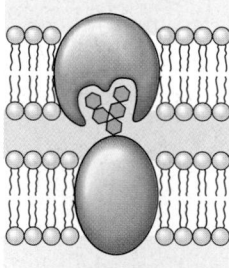

Cell-cell recognition. Some glycoproteins (proteins with short chains of sugars) serve as identification tags that are specifically recognized by other cells.

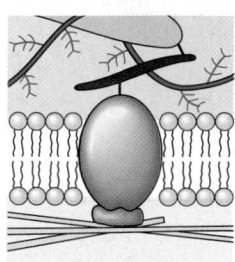

Attachment to the cytoskeleton and extracellular matrix (ECM). Microfilaments or other elements of the cytoskeleton may be bonded to membrane proteins, a function that helps maintain cell shape and fixes the location of certain membrane proteins. Proteins that adhere to the ECM can coordinate extracellular and intracellular changes.

FIGURE 8.9 Some functions of membrane proteins. In many cases, a single protein performs some combination of these tasks.

the four human blood groups designated A, B, AB, and O reflect variation in the oligosaccharides on the surface of red blood cells.

TRAFFIC ACROSS MEMBRANES

The biological membrane is an exquisite example of a supramolecular structure—many molecules ordered into a higher level of organization—with emergent properties beyond those of the individual molecules. The remainder of this chapter focuses on one of the most important of those properties: the ability to regulate transport across cellular boundaries, a function essential to the cell's existence as an open system. We will see once again that form fits function: The fluid mosaic model helps explain how membranes regulate the cell's molecular traffic.

A membrane's molecular organization results in selective permeability

A steady traffic of small molecules and ions moves across the plasma membrane in both directions. Consider the chemical exchanges between a muscle cell and the extracellular fluid that bathes it. Sugars, amino acids, and other nutrients enter the cell, and metabolic waste products leave. The cell takes in oxygen for cellular respiration and expels carbon dioxide. It also regulates its concentrations of inorganic ions, such as Na^+, K^+, Ca^{2+}, and Cl^-, by shuttling them one way or the other across the plasma membrane. Although traffic through the membrane is extensive, cell membranes are selectively permeable, and substances do not cross the barrier indiscriminately. The cell is able to take up many varieties of small molecules and ions and exclude others. Moreover, substances that move through the membrane do so at different rates.

Permeability of the Lipid Bilayer

Hydrophobic molecules, such as hydrocarbons, carbon dioxide, and oxygen, can dissolve in the lipid bilayer of the membrane and cross it with ease. However, the hydrophobic core of the membrane impedes the transport of ions and polar molecules, which are hydrophilic. Even water, a very small polar molecule, does not easily pass through a lipid bilayer, nor do glucose and other sugars, which are considerably larger. Similarly, a charged atom or molecule and its surrounding shell of water also find the hydrophobic layer of the membrane difficult to penetrate. Fortunately, the lipid bilayer is only part of the story of a membrane's selective permeability. Proteins built into the membrane play key roles in regulating transport.

Transport Proteins

Cell membranes *are* permeable to specific ions and polar molecules, including water. These hydrophilic substances avoid contact with the lipid bilayer by passing through **transport proteins** that span the membrane (see FIGURE 8.9, top). Some transport proteins function by having a hydrophilic channel that certain molecules or atomic ions use as a tunnel through the membrane. Other transport proteins hold onto their passengers and physically move them across the membrane. In both cases, the transport protein is specific for the substances it translocates (moves), allowing only a certain substance or class of closely related substances to cross the membrane. For example, glucose carried in blood to the human liver enters liver cells rapidly through specific transport proteins in the plasma membrane. The protein is so selective that it even rejects fructose, a structural isomer of glucose.

Thus, the selective permeability of a membrane depends on both the discriminating barrier of the lipid bilayer and the specific transport proteins built into the membrane. But what determines the *direction* of traffic across a membrane? At a given time, will a particular substance enter the cell or leave? And what mechanisms actually drive molecules across membranes? We will address these questions next as we explore two modes of membrane traffic: passive transport and active transport.

Passive transport is diffusion across a membrane

Molecules have intrinsic kinetic energy called thermal motion (heat). One result of thermal motion is **diffusion,** the tendency for molecules of any substance to spread out into the available space. Each molecule moves randomly, yet diffusion of a *population* of molecules may be directional. For example, imagine a membrane separating pure water from a solution of a dye in water. Assume that this membrane is permeable to the dye molecules (FIGURE 8.10a). Each dye molecule wanders randomly, but there will be a *net* movement of the dye molecules across the membrane to the side that began as pure water. The spreading of the dye across the membrane will continue until both solutions have equal concentrations of the dye. Once that point is reached, there will be a dynamic equilibrium, with as many dye molecules crossing the membrane in one direction as in the other each second.

We can now state a simple rule of diffusion: In the absence of other forces, a substance will diffuse from where it is more concentrated to where it is less concentrated. Put another way, any substance will diffuse down its **concentration gradient.** No work must be done to make this happen; diffusion is a spontaneous process because it decreases free energy (see FIGURE 6.5b). Recall that in any system there is a tendency for entropy, or disorder, to increase. Diffusion of a solute in water increases entropy by producing a more random mixture than exists when there are localized concentrations of the solute. It is important to note that each substance diffuses down its *own* concentration gradient, unaffected by the concentration differences of other substances (FIGURE 8.10b).

Much of the traffic across cell membranes occurs by diffusion. When a substance is more concentrated on one side of a membrane than on the other, there is a tendency for the substance to diffuse across the membrane down its concentration gradient (assuming that the membrane is permeable to that substance). One important example is the uptake of oxygen by a cell performing cellular respiration. Dissolved oxygen diffuses into the cell across the plasma membrane. As long as cellular respiration consumes the O_2 as it enters, diffusion into the cell will continue, because the concentration gradient favors movement in that direction.

The diffusion of a substance across a biological membrane is called **passive**

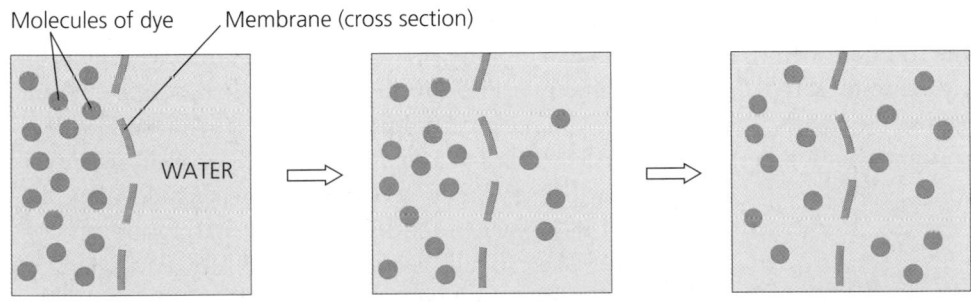

Molecules of dye   Membrane (cross section)

WATER   Equilibrium

(a) Diffusion of one solute. The membrane has pores large enough for molecules of dye to pass through. The dye diffuses from where it is more concentrated to where it is less concentrated (called diffusing down a concentration gradient). This leads to a dynamic equilibrium: The solute molecules continue to cross the membrane, but at equal rates in both directions.

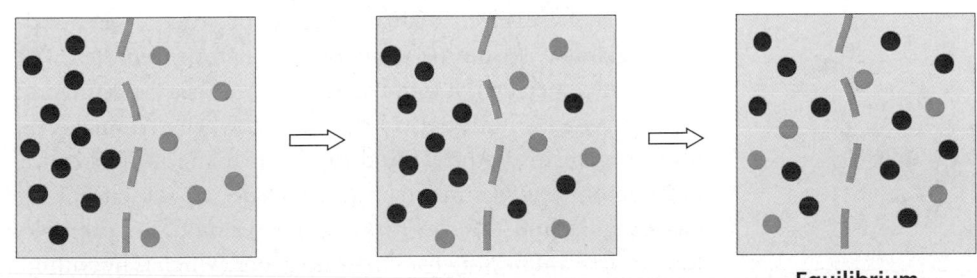

Equilibrium

(b) Diffusion of two solutes. Solutions of two different dyes are separated by a membrane that is permeable to both. Each dye diffuses down its own concentration gradient. There will be a net diffusion of the orange dye toward the left, even though the *total* solute concentration was initially greater on the left side.

FIGURE 8.10 The diffusion of solutes across membranes.

transport because the cell does not have to expend energy to make it happen. The concentration gradient itself represents potential energy and drives diffusion. Remember, however, that membranes are selectively permeable and therefore affect the rates of diffusion of various molecules. Thanks to the widespread presence of the necessary transport protein, one molecule that diffuses freely across most membranes is water, a fact that has important consequences for cells.

Osmosis is the passive transport of water

In comparing two solutions of unequal solute concentration, the solution with a higher concentration of solutes is said to be **hypertonic.** The solution with a lower solute concentration is **hypotonic.** (*Hyper* and *hypo* mean "more" and "less," respectively, referring here to solute concentration.) These are relative terms that are meaningful only in a comparative sense. For example, tap water is hypertonic to distilled water but hypotonic to seawater. In other words, tap water has a higher solute concentration than distilled water, but a lower concentration than seawater. Solutions of equal solute concentration are said to be **isotonic** (*iso* means "the same").

Picture a U-shaped vessel with a selectively permeable membrane separating two sugar solutions of different concentrations (FIGURE 8.11). Pores in this synthetic membrane are too small for sugar molecules to pass but large enough for water molecules to cross the membrane. In effect, the solution with higher solute concentration (hypertonic) has a lower water concentration. Therefore the water will diffuse across the membrane from the hypotonic solution to the hypertonic solution.

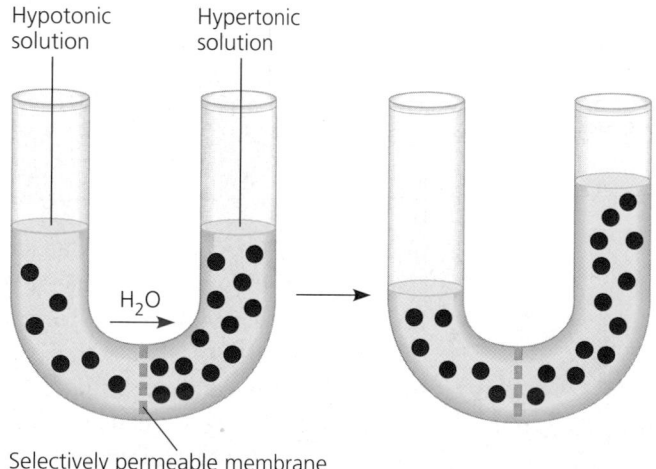

Hypotonic solution   Hypertonic solution

H₂O

Selectively permeable membrane

FIGURE 8.11 Osmosis. Two sugar solutions of different concentrations are separated by a porous membrane that is permeable to the solvent (water) but not to the solute (sugar). Water diffuses from the less concentrated (hypotonic) solution to the more concentrated (hypertonic) solution. This passive transport of water, or osmosis, reduces the difference in sugar concentrations.

This diffusion of water across a selectively permeable membrane is a special case of passive transport called **osmosis.***

The direction of osmosis is determined only by a difference in *total* solute concentration. Water moves from a hypotonic to a hypertonic solution even if the hypotonic solution has more *kinds* of solutes. Seawater, which has a great variety of solutes, will lose water to a very concentrated solution of a single sugar because the total solute concentration of the seawater is lower. If two solutions are isotonic, water moves across a membrane separating the solutions at an equal rate in both directions; that is, there is no net osmosis between isotonic solutions. (In Chapters 36 and 44, we will look at osmosis in a more quantitative way.)

Cell survival depends on balancing water uptake and loss

The movement of water across cell membranes and the balance of water between the cell and its environment are crucial to organisms. Let's now apply to living cells what we have learned about osmosis in artificial systems.

Water Balance of Cells Without Walls

If an animal cell is immersed in an environment that is isotonic to the cell, there will be no net movement of water across the plasma membrane. Water flows across the membrane, but at the same rate in both directions. In an isotonic environment, the volume of an animal cell is stable (FIGURE 8.12). Now let's transfer the cell to a solution that is hypertonic to the cell. The cell will lose water to its environment, shrivel, and probably die. This is one reason why an increase in the salinity (saltiness) of a lake can kill the animals there. However, taking up too much water can be just as hazardous to an animal cell as losing water. If we place the cell in a solution that is hypotonic to the cell, water will enter faster than it leaves, and the cell will swell and lyse (burst) like an overfilled water balloon.

A cell without rigid walls can tolerate neither excessive uptake nor excessive loss of water. This problem of water balance is automatically solved if such a cell lives in isotonic surroundings. Seawater is isotonic to many marine invertebrates. The cells of most terrestrial (land-dwelling) animals are bathed in an extracellular fluid that is isotonic to the cells. Animals and other organisms without rigid cell walls living in hypertonic or hypotonic environments must have special adaptations for **osmoregulation,** the control of water balance. For example, the protist *Paramecium* lives in pond water, which is hypotonic

---

* Actually, at any instant, a fraction of the water molecules in a solution lose their freedom of independent movement by being bound to solute molecules in hydration shells. It is not really a difference in *total* water concentration that causes osmosis, but a difference in the concentration of *unbound* water molecules that are free to cross the membrane.

Hypotonic solution

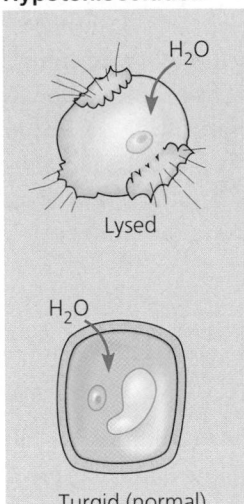

H₂O

Lysed

H₂O

Turgid (normal)

Isotonic solution

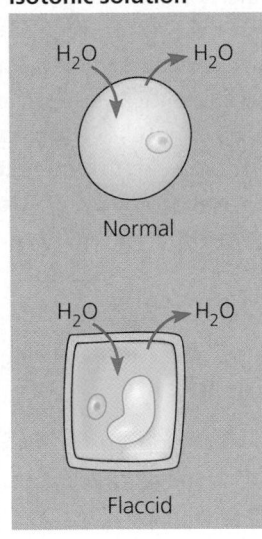

H₂O → H₂O

Normal

H₂O → H₂O

Flaccid

Hypertonic solution

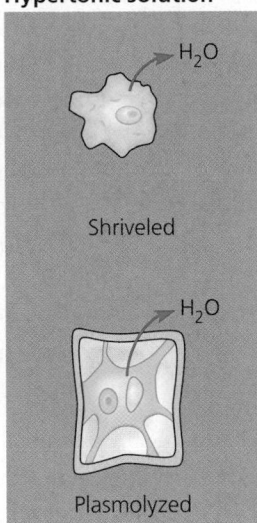

→ H₂O

Shriveled

→ H₂O

Plasmolyzed

FIGURE 8.12 The water balance of living cells. How living cells react to changes in the solute concentrations of their environments depends on whether or not they have cell walls. Animal cells do not have cell walls (top row); plant cells do (bottom row). (Arrows indicate net water movement when the cells are *first* placed in these solutions.)

Animal cell. An animal cell fares best in an isotonic environment, unless it has special adaptations to offset the osmotic uptake or loss of water.

Plant cell. Plant cells are turgid (firm) and generally healthiest in a hypotonic environment, where the tendency for uptake of water is balanced by the elastic wall pushing back on the cell.

Filling vacuole

50 μm

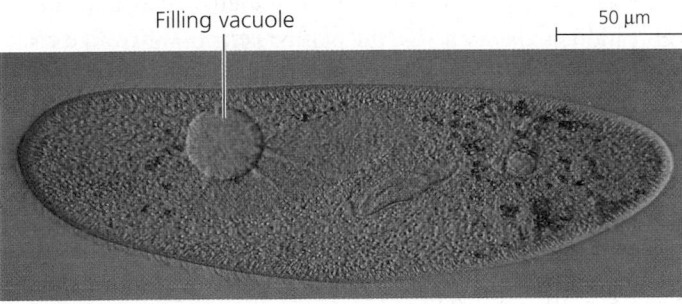

(a) A contractile vacuole fills with fluid that enters from a system of canals radiating throughout the cytoplasm.

Contracting vacuole

50 μm

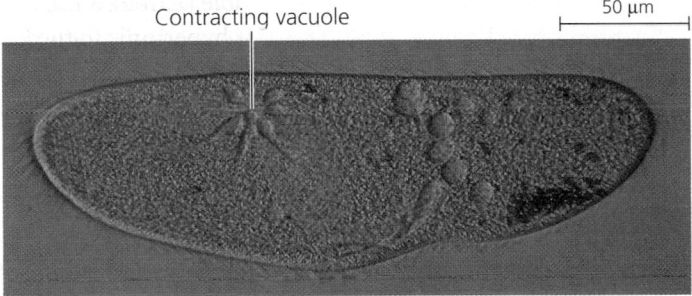

(b) When full, the vacuole and canals contract, expelling fluid from the cell.

FIGURE 8.13 The contractile vacuole of *Paramecium*: an evolutionary adaptation for osmoregulation. The contractile vacuole of this freshwater protist offsets osmosis by bailing water out of the cell.

to the cell. *Paramecium* has a plasma membrane that is much less permeable to water than the membranes of most other cells, but this only slows the uptake of water, which continually enters the cell. Fortunately, *Paramecium* is also equipped with a contractile vacuole, an organelle that functions as a bilge pump to force water out of the cell as fast as it enters by osmosis (FIGURE 8.13). We will examine other evolutionary adaptations for osmoregulation in Chapter 44.

Water Balance of Cells with Walls

The cells of plants, prokaryotes, fungi, and some protists have walls. When such a cell is in a hypotonic solution—when bathed by rainwater, for example—the wall helps maintain the cell's water balance. Consider a plant cell. Like an animal cell, the plant cell swells as water enters by osmosis (see FIGURE 8.12). However, the elastic wall will expand only so much before it exerts a back pressure on the cell that opposes further water uptake. At this point, the cell is **turgid** (very firm), which is the healthy state for most plant cells. Plants that are not woody, such as most house plants, depend for mechanical support on cells kept turgid by a surrounding hypotonic solution. If a plant's cells and their surroundings are isotonic, there is no net tendency for water to enter, and the cells become **flaccid** (limp), causing the plant to wilt.

On the other hand, a wall is of no advantage if the cell is immersed in a hypertonic environment. In this case, a plant cell, like an animal cell, will lose water to its surroundings and shrink. As the plant cell shrivels, its plasma membrane pulls away from the wall. This phenomenon, called **plasmolysis,** is usually lethal. The walled cells of bacteria and fungi also plasmolyze in hypertonic environments.

Specific proteins facilitate the passive transport of water and selected solutes: *a closer look*

Let's look more closely at how water and certain hydrophilic solutes cross a membrane. As mentioned earlier, many polar molecules and ions impeded by the lipid bilayer of the membrane diffuse passively with the help of transport proteins that span the membrane. This phenomenon is called **facilitated diffusion.**

A transport protein has many of the properties of an enzyme. Just as an enzyme is specific for its substrate, a transport protein is specialized for the solute it transports and may even have a specific binding site akin to the active site of an enzyme. Like enzymes, transport proteins can be saturated. There are only so many molecules of each type of transport protein built into the plasma membrane, and when these molecules are translocating passengers as fast as they can, transport is occurring at a maximum rate. Also like enzymes, transport proteins can be inhibited by molecules that resemble the normal "substrate." This occurs when the imposter competes with the normally transported solute by binding to the transport protein. Unlike enzymes, however, transport proteins do not usually catalyze chemical reactions. Their function is to catalyze a *physical process*—the transport of a molecule across a membrane that would otherwise be relatively impermeable to the substance.

Cell biologists are still trying to learn exactly how various transport proteins facilitate diffusion. Many transport proteins simply provide corridors allowing a specific molecule or ion to cross the membrane (FIGURE 8.14a). The hydrophilic passageways provided by such *channel proteins* allow water molecules or small ions to flow very quickly from one side of the membrane to the other. In fact, it is water channel proteins, called **aquaporins,** that facilitate the massive amounts of diffusion illustrated in FIGURE 8.12. Some channel proteins function as **gated channels;** a stimulus causes them to open or close. The stimulus may be electrical or chemical; if chemical, it is a substance other than the one to be transported. For example, stimulation of a nerve cell by neurotransmitter molecules opens gated channels that allow sodium ions into the cell.

Not all transport proteins are channel proteins. Many transport proteins seem to undergo a subtle change in shape that somehow translocates the solute-binding site across the membrane (FIGURE 8.14b). The changes in shape could be triggered by the binding and release of the transported molecule.

In certain inherited diseases, specific transport systems are either defective or missing altogether. An example is cystinuria, a human disease characterized by the absence of a protein that transports cystine and other amino acids across the membranes of kidney cells. Kidney cells normally reabsorb these amino acids from the urine and return them to the blood, but an individual afflicted with cystinuria develops painful stones from amino acids that accumulate and crystallize in the kidneys.

Active transport is the pumping of solutes against their gradients

Despite the help of transport proteins, facilitated diffusion is still considered passive transport because the solute is moving down its concentration gradient. Facilitated diffusion speeds the transport of a solute by providing an efficient passage through the membrane, but it does not alter the direction of transport. Some transport proteins, however, *can* move solutes against their concentration gradients, across the plasma membrane from the side where they are less concentrated to the side where they are more concentrated. This transport is "uphill" and therefore requires work. To pump a molecule across a membrane against its gradient, the cell must expend its own metabolic energy; therefore, this type of membrane traffic is called **active transport.**

Active transport is a major factor in the ability of a cell to maintain internal concentrations of small molecules that differ from concentrations in its environment. For example, compared to its surroundings, an animal cell has a much higher concentration of potassium ions and a much lower concentration of sodium ions. The plasma membrane helps maintain these steep gradients by pumping sodium out of the cell and potassium into the cell.

The work of active transport is performed by specific proteins embedded in membranes. As in other types of cellular

FIGURE 8.14 Two models for facilitated diffusion.

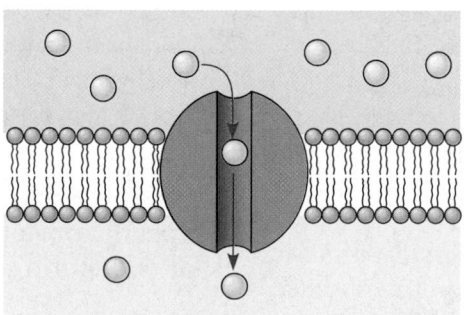

(a) The transport protein (purple) forms a channel through which water molecules or a specific solute can pass.

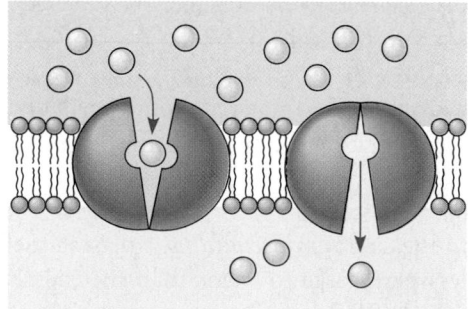

(b) The transport protein alternates between two conformations, moving a solute across the membrane as the shape of the protein changes. The protein can transport the solute in either direction, with the net movement being down the concentration gradient of the solute.

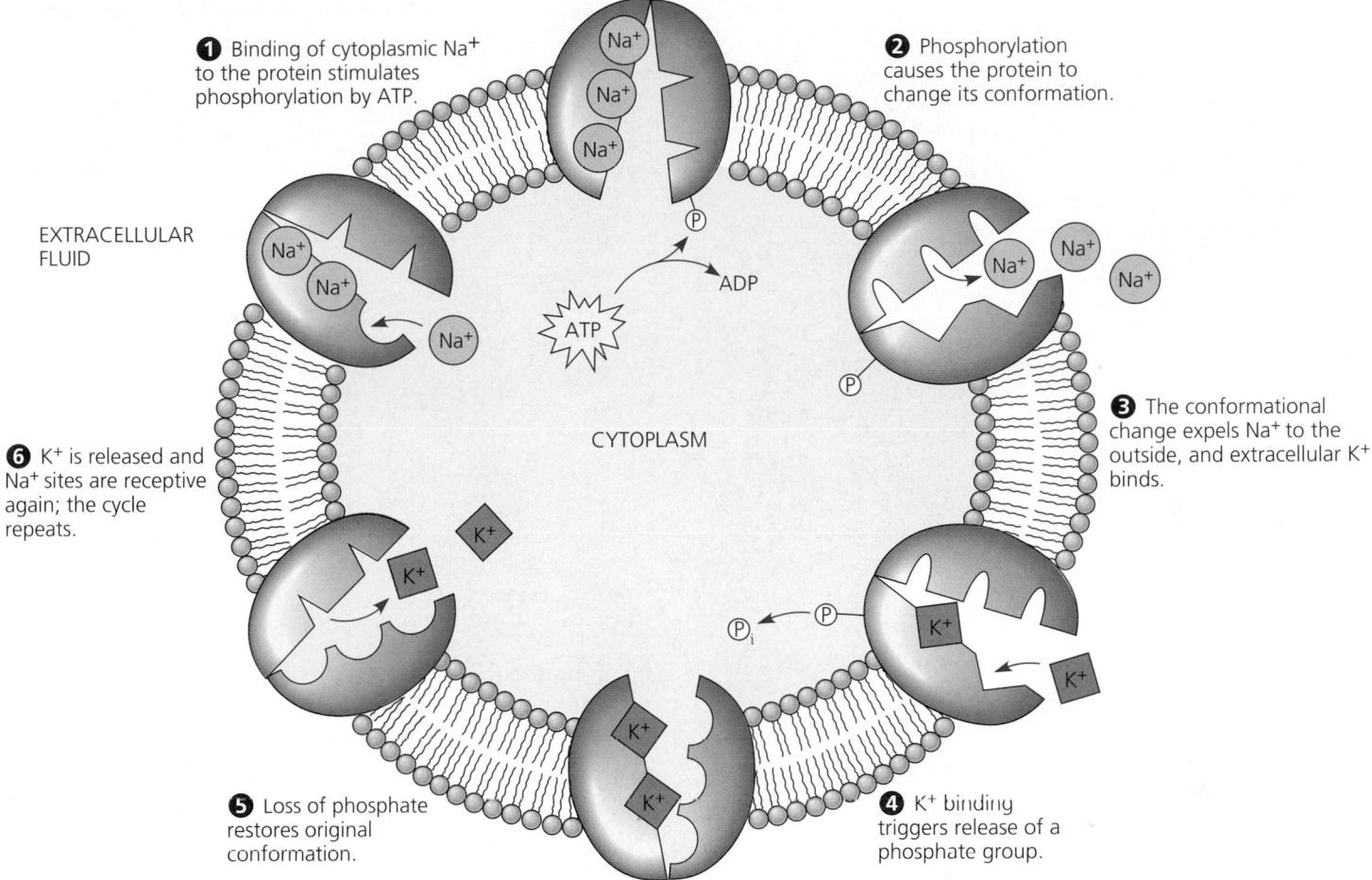

1 Binding of cytoplasmic Na⁺ to the protein stimulates phosphorylation by ATP.

2 Phosphorylation causes the protein to change its conformation.

3 The conformational change expels Na⁺ to the outside, and extracellular K⁺ binds.

4 K⁺ binding triggers release of a phosphate group.

5 Loss of phosphate restores original conformation.

6 K⁺ is released and Na⁺ sites are receptive again; the cycle repeats.

EXTRACELLULAR FLUID

CYTOPLASM

FIGURE 8.15 The sodium-potassium pump: a specific case of active transport. This transport system pumps ions against steep concentration gradients. The pump oscillates between two conformational states in a pumping cycle that translocates three Na⁺ ions out of the cell for every two K⁺ ions pumped into the cell. ATP powers the changes in conformation by phosphorylating the transport protein (that is, by transferring a phosphate group to the protein).

work, ATP supplies the energy for most active transport. One way ATP can power active transport is by transferring its terminal phosphate group directly to the transport protein. This may induce the protein to change its conformation in a manner that translocates a solute bound to the protein across the membrane. One transport system that works this way is the **sodium-potassium pump,** which exchanges sodium (Na⁺) for potassium (K⁺) across the plasma membrane of animal cells (FIGURE 8.15). FIGURE 8.16 (p. 150) reviews the distinction between passive transport and active transport.

Some ion pumps generate voltage across membranes

All cells have voltages across their plasma membranes. Voltage is electrical potential energy—a separation of opposite charges. The cytoplasm of a cell is negative in charge compared to the extracellular fluid because of an unequal distribution of anions and cations on opposite sides of the membrane. The voltage across a membrane, called a **membrane potential,** ranges from about −50 to −200 millivolts. (The minus sign indicates that the inside of the cell is negative compared to the outside.)

The membrane potential acts like a battery, an energy source that affects the traffic of all charged substances across the membrane. Because the inside of the cell is negative compared to the outside, the membrane potential favors the passive transport of cations into the cell and anions out of the cell. Thus, *two* forces drive the diffusion of ions across a membrane: a chemical force (the ion's concentration gradient) and an electrical force (the effect of the membrane potential on the ion's movement). This combination of forces acting on an ion is called the **electrochemical gradient.** In the case of ions, we must refine our concept of passive transport: An ion does not simply diffuse down its *concentration* gradient, but diffuses

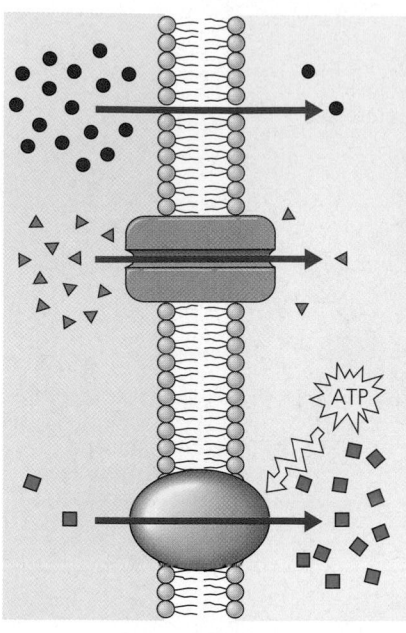

Diffusion. Hydrophobic molecules and (at a slow rate) very small uncharged polar molecules can diffuse through the lipid bilayer.

Facilitated diffusion. Hydrophilic substances, including water molecules, diffuse through membranes with the assistance of transport proteins.

Passive transport. Substances diffuse spontaneously down their concentration gradients, crossing a membrane with no expenditure of energy by the cell.

Active transport. Some transport proteins act as pumps, moving substances across a membrane against their concentration gradients. Energy for this work is usually supplied by ATP.

FIGURE 8.16 Review: passive and active transport compared.

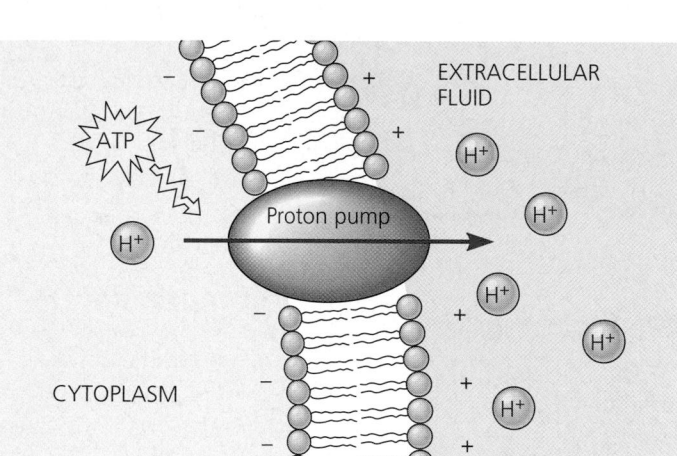

FIGURE 8.17 An electrogenic pump. Proton pumps, the main electrogenic pumps of plants, fungi, and bacteria, are membrane proteins that store energy by generating voltage (charge separation) across membranes. Using ATP for power, a proton pump translocates positive charge in the form of hydrogen ions. The voltage and H+ gradient represent a dual energy source that can drive other processes, such as the uptake of nutrients.

down its *electrochemical* gradient. For example, the concentration of sodium ions (Na^+) inside a resting nerve cell is much lower than outside it. When the cell is stimulated, gated channels that facilitate Na^+ diffusion open. Sodium ions then "fall" down their electrochemical gradient, driven by the concentration gradient of Na^+ and by the attraction of cations to the negative side of the membrane.

Some membrane proteins that actively transport ions contribute to the membrane potential. An example is the sodium-potassium pump. Notice in FIGURE 8.15 that the pump does not translocate Na^+ and K^+ one for one, but actually pumps three sodium ions out of the cell for every two potassium ions

it pumps into the cell. With each crank of the pump, there is a net transfer of one positive charge from the cytoplasm to the extracellular fluid, a process that stores energy in the form of voltage. A transport protein that generates voltage across a membrane is called an **electrogenic pump.** The sodium-potassium pump seems to be the major electrogenic pump of animal cells. The main electrogenic pump of plants, bacteria, and fungi is a **proton pump,** which actively transports hydrogen ions (protons) out of the cell. The pumping of H^+ transfers positive charge from the cytoplasm to the extracellular solution (FIGURE 8.17).

By generating voltage across membranes, electrogenic pumps store energy that can be tapped for cellular work, including a type of membrane traffic called cotransport.

In cotransport, a membrane protein couples the transport of two solutes

A single ATP-powered pump that transports a specific solute can indirectly drive the active transport of several other solutes in a mechanism called **cotransport.** A substance that has been pumped across a membrane can do work as it leaks back by diffusion, analogous to water that has been pumped uphill and performs work as it flows back down. Another specialized transport protein, separate from the pump, can couple the "downhill" diffusion of this substance to the "uphill" transport of a second substance against its own concentration gradient. For example, a plant cell uses the gradient of hydrogen ions generated by its proton pumps to drive the active transport of amino acids, sugars, and several other nutrients into the cell. One specific transport protein couples the return of

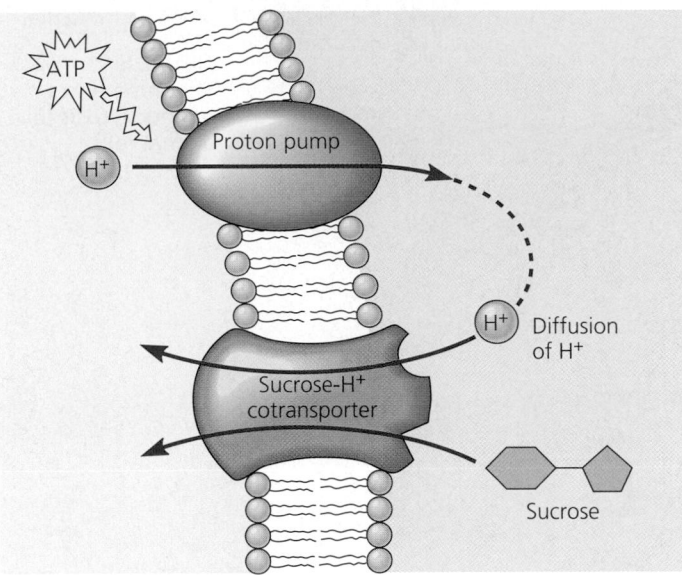

FIGURE 8.18 Cotransport. An ATP-driven pump stores energy by concentrating a substance (here H$^+$) on one side of the membrane. As the substance leaks back across the membrane through specific transport proteins, it escorts other substances (here sucrose) into the cell. In this case, the proton pump of the membrane is indirectly driving sucrose uptake by a plant cell with the help of a protein that cotransports the two solutes.

hydrogen ions to the transport of sucrose into the cell (FIGURE 8.18). The protein can translocate sucrose into the cell against a concentration gradient, but only if the sucrose molecule travels in the company of a hydrogen ion. The sucrose rides on the coattails of the hydrogen ion, which uses the common transport protein as an avenue to diffuse down the concentration gradient maintained by the proton pump. Plants use the mechanism of sucrose-H$^+$ cotransport to load sucrose produced by photosynthesis into specialized cells in the veins of leaves. The sugar can then be distributed by the vascular tissue of the plant to nonphotosynthetic organs, such as roots.

Exocytosis and endocytosis transport large molecules

Water and small solutes enter and leave the cell by passing through the lipid bilayer of the plasma membrane or by being pumped or carried across the membrane by transport proteins. Large molecules, such as proteins and polysaccharides, generally cross the membrane by a different mechanism involving vesicles. As we described in Chapter 7, the cell secretes macromolecules by the fusion of vesicles with the plasma membrane; this is **exocytosis.** A transport vesicle that has budded from the Golgi apparatus moves along fibers of the cytoskeleton to the plasma membrane. When the vesicle membrane and plasma membrane come into contact, the lipid molecules of the two bilayers rearrange themselves so that the

two membranes fuse. The contents of the vesicle then spill to the outside of the cell (see FIGURE 8.8).

Many secretory cells use exocytosis to export their products. For example, certain cells in the pancreas manufacture the hormone insulin and secrete it into the blood by exocytosis. Another example is the neuron, or nerve cell, which uses exocytosis to release chemical signals that stimulate other neurons or muscle cells (see FIGURE 2.18). When plant cells are making walls, exocytosis delivers carbohydrates from Golgi vesicles to the outside of the cell.

In **endocytosis,** the cell takes in macromolecules and particulate matter by forming new vesicles from the plasma membrane. The steps are basically the reverse of exocytosis. A small area of the plasma membrane sinks inward to form a pocket. As the pocket deepens, it pinches in, forming a vesicle containing material that had been outside the cell.

There are three types of endocytosis: phagocytosis ("cellular eating"), pinocytosis ("cellular drinking"), and receptor-mediated endocytosis. In **phagocytosis,** a cell engulfs a particle by wrapping pseudopodia around it and packaging it within a membrane-enclosed sac large enough to be classified as a vacuole (FIGURE 8.19a, p. 152). The particle is digested after the vacuole fuses with a lysosome containing hydrolytic enzymes. In **pinocytosis,** the cell "gulps" droplets of extracellular fluid into tiny vesicles (FIGURE 8.19b). Because any and all solutes dissolved in the droplet are taken into the cell, pinocytosis is unspecific in the substances it transports. In contrast, **receptor-mediated endocytosis** is very specific (FIGURE 8.19c). Embedded in the membrane are proteins with specific receptor sites exposed to the extracellular fluid. The extracellular substances that bind to the receptors are called **ligands,** a general term for any molecule that binds specifically to a receptor site of another molecule (from the Latin *ligare,* to bind). The receptor proteins are usually clustered in regions of the membrane called coated pits, which are lined on their cytoplasmic side by a fuzzy layer of protein. These coat proteins probably help deepen the pit and form the vesicle. After the ingested material is liberated from the vesicle for metabolism, the receptors are recycled to the plasma membrane.

Receptor-mediated endocytosis enables the cell to acquire bulk quantities of specific substances, even though those substances may not be very concentrated in the extracellular fluid. For example, human cells use the process to take in cholesterol for use in the synthesis of membranes and as a precursor for the synthesis of other steroids. Cholesterol travels in the blood in particles called low-density lipoproteins (LDLs), complexes of lipids and proteins. These particles bind to LDL receptors on membranes and then enter the cells by endocytosis. In humans with familial hypercholesterolemia, an inherited disease characterized by a very high level of cholesterol in the blood, the LDL receptor proteins are defective, and the LDL particles cannot enter cells. Cholesterol accumulates in the blood, where it contributes to early atherosclerosis (the buildup of fat deposits on blood vessel linings).

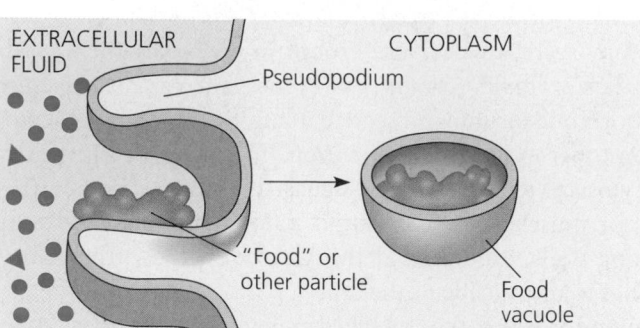
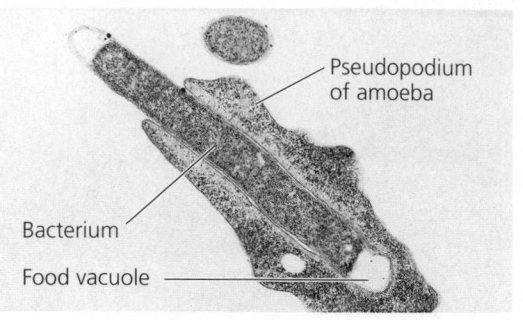

1 μm

FIGURE 8.19
The three types
of endocytosis in
animal cells.

EXTRACELLULAR FLUID — CYTOPLASM

Pseudopodium

"Food" or other particle

Food vacuole

Pseudopodium of amoeba

Bacterium

Food vacuole

(a) Phagocytosis. Pseudopodia engulf a particle and package it in a vacuole. The micrograph shows an amoeba engulfing a bacterium (TEM).

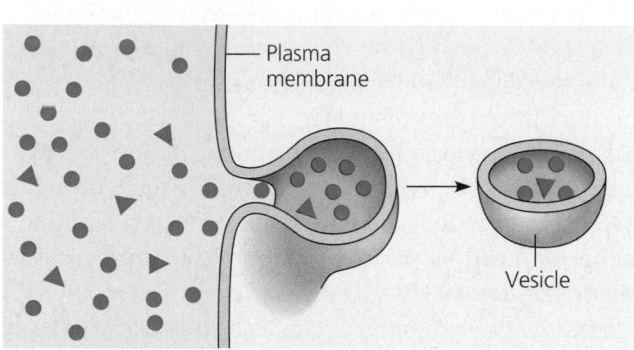
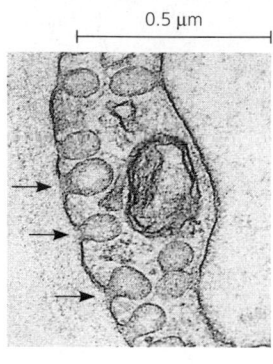

0.5 μm

Plasma membrane

Vesicle

(b) Pinocytosis. Droplets of extracellular fluid are incorporated into the cell in small vesicles. The micrograph shows pinocytotic vesicles forming (arrows) in a cell lining a small blood vessel (TEM).

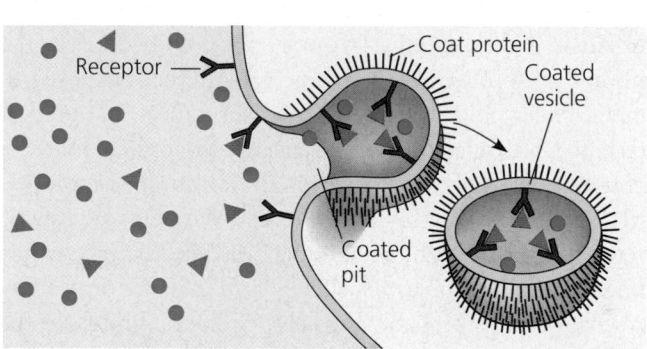
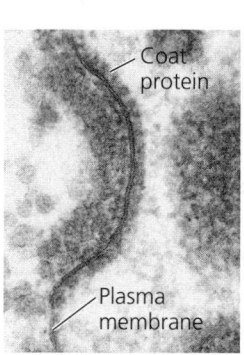
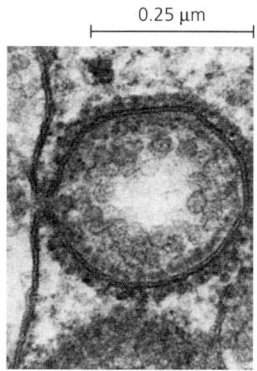

0.25 μm

Receptor

Coat protein

Coated vesicle

Coated pit

Coat protein

Plasma membrane

(c) Receptor-mediated endocytosis. Coated pits form vesicles when specific molecules (ligands) bind to receptors on the cell surface. Notice that there are relatively more bound molecules (purple) inside the vesicles, though other molecules (green) are also present. The micrographs show two progressive stages of receptor-mediated endocytosis (TEMs).

Vesicles not only transport substances between the cell and its surroundings but also provide a mechanism for rejuvenating or remodeling the plasma membrane. Endocytosis and exocytosis occur continually to some extent in most eukaryotic cells, and yet the amount of plasma membrane in a nongrowing cell remains fairly constant over the long run. Apparently, the addition of membrane by one process offsets the loss of membrane by the other.

■   ■   ■

Energy and cellular work have figured prominently in our study of membranes. We have seen, for example, that active transport is powered by ATP. In the next two chapters, you will learn more about how cells acquire chemical energy to do the work of life. We'll call membranes back for an encore in these chapters, as they play a major role in how mitochondria and chloroplasts make energy available to cells.

CHAPTER 8 REVIEW

Go to the Campbell Biology website (www.campbellbiology.com) to explore an interactive version of the Chapter Review.

Summary of Key Concepts

MEMBRANE STRUCTURE

- **Membrane models have evolved to fit new data (pp. 138–141, FIGURES 8.1–8.3)** The Davson-Danielli model, placing layers of proteins on either side of a phospholipid bilayer, has been replaced by the fluid mosaic model.

- **Membranes are fluid (pp. 141–142, FIGURES 8.4, 8.5)** Phospholipids and, to a lesser extent, proteins move laterally within the membrane. Cholesterol and unsaturated hydrocarbon tails in the phospholipids affect membrane fluidity.

- **Membranes are mosaics of structure and function (pp. 142–144, FIGURES 8.6–8.9)** Integral proteins are embedded in the lipid bilayer; peripheral proteins are attached to the surfaces. The inside and outside membrane faces differ in composition. The functions of membrane proteins include transport, enzymatic activity, signal transduction, intercellular joining, cell-cell recognition, and attachment to the cytoskeleton and extracellular matrix.

- **Membrane carbohydrates are important for cell-cell recognition (pp. 143–144)** Short chains of sugars are linked to proteins and lipids on the exterior side of the plasma membrane, where they can interact with the surface molecules of other cells.

  Web/CD Activity 8A: *Membrane Structure*

TRAFFIC ACROSS MEMBRANES

- **A membrane's molecular organization results in selective permeability (pp. 144–145)** A cell must exchange small molecules and ions with its surroundings, a process controlled by the plasma membrane. Hydrophobic substances are soluble in lipid and pass through membranes rapidly. Polar molecules and ions generally require specific transport proteins to help them cross.

  Web/CD Activity 8B: *Selective Permeability of Membranes*

- **Passive transport is diffusion across a membrane (pp. 145–146, FIGURE 8.10)** Diffusion is the spontaneous movement of a substance down its concentration gradient.

  Web/CD Activity 8C: *Diffusion*

- **Osmosis is the passive transport of water (p. 146, FIGURE 8.11)** Water flows across a membrane from the side where solute is less concentrated (hypotonic) to the side where solute is more concentrated (hypertonic). If the concentrations are equal (isotonic), no net osmosis occurs.

- **Cell survival depends on balancing water uptake and loss (pp. 146–147, FIGURE 8.12)** Cells lacking walls (as in animals and some protists) are isotonic with their environments or have adaptations for osmoregulation. Plants, prokaryotes, fungi, and some protists have elastic cell walls, so the cells don't burst in a hypotonic environment.

  Web/CD Activity 8D: *Osmosis and Water Balance in Cells*
  Web/CD Activity 8E: *Plasmolysis Video*
  Web/CD Case Study in the Process of Science: *How Do Salt Concentrations Affect Cells?*

- **Specific proteins facilitate the passive transport of water and selected solutes:** *a closer look* **(pp. 147–148, FIGURE 8.14)** In facilitated diffusion, a transport protein speeds the movement of water or a solute across a membrane down its concentration gradient.

  Web/CD Activity 8E: *Facilitated Diffusion*

- **Active transport is the pumping of solutes against their gradients (pp. 148–149, FIGURE 8.15)** Specific membrane proteins use energy, usually in the form of ATP, to do this work.

  Web/CD Activity 8F: *Active Transport*

- **Some ion pumps generate voltage across membranes (pp. 149–150, FIGURE 8.17)** Ions can have both a concentration (chemical) gradient and an electrical gradient (voltage). These forces combine in the electrochemical gradient, which determines the net direction of ionic diffusion. Electrogenic pumps, such as sodium-potassium pumps and proton pumps, are transport proteins that contribute to electrochemical gradients.

- **In cotransport, a membrane protein couples the transport of two solutes (pp. 150–151, FIGURE 8.18)** One solute's "downhill" diffusion drives the other's "uphill" transport.

- **Exocytosis and endocytosis transport large molecules (pp. 151–152, FIGURE 8.19)** In exocytosis, transport vesicles migrate to the plasma membrane, fuse with it, and release their contents. In endocytosis, large molecules enter cells within vesicles pinched inward from the plasma membrane. The three types of endocytosis are phagocytosis, pinocytosis, and receptor-mediated endocytosis.

  Web/CD Activity 8G: *Exocytosis and Endocytosis*

Self-Quiz

1. In what way do the various membranes of a eukaryotic cell differ?
   a. Phospholipids are found only in certain membranes.
   b. Certain proteins are unique to each membrane.
   c. Only certain membranes of the cell are selectively permeable.
   d. Only certain membranes are constructed from amphipathic molecules.
   e. Some membranes have hydrophobic surfaces exposed to the cytosol, while others have hydrophilic surfaces facing the cytosol.

2. According to the fluid mosaic model of membrane structure, proteins of the membrane are mostly
   a. spread in a continuous layer over the inner and outer surfaces of the membrane.
   b. confined to the hydrophobic core of the membrane.
   c. embedded in a lipid bilayer.
   d. randomly oriented in the membrane, with no fixed inside-outside polarity.
   e. free to depart from the fluid membrane and dissolve in the surrounding solution.

3. Which of the following factors would tend to increase membrane fluidity?
   a. a greater proportion of unsaturated phospholipids
   b. a lower temperature
   c. a relatively high protein content in the membrane
   d. a greater proportion of relatively large glycolipids compared to lipids having smaller molecular weights
   e. a high membrane potential

4. Which of the following processes includes all others?
   a. osmosis
   b. diffusion of a solute across a membrane
   c. facilitated diffusion
   d. passive transport
   e. transport of an ion down its electrochemical gradient

5. Based on the model of sucrose uptake in FIGURE 8.18, which of the following experimental treatments would increase the rate of sucrose transport into the cell?
   a. decreasing extracellular sucrose concentration
   b. decreasing extracellular pH
   c. decreasing cytoplasmic pH
   d. adding an inhibitor that blocks the regeneration of ATP
   e. adding a substance that makes the membrane more permeable to hydrogen ions

6. Why do phospholipids tend to organize into a bilayer in an aqueous environment?

7. The carbohydrates attached to some of the proteins and lipids of the plasma membrane are added as the membrane is refined in the Golgi apparatus; the new membrane then forms transport vesicles that travel to the cell surface. On which side of the vesicle membrane are the carbohydrates?

8. The hormone epinephrine can cause a liver cell to hydrolyze its stored glycogen and release sugar without the hormone ever entering the cell. Explain.

9. How does the second law of thermodynamics help explain the diffusion of a substance across a membrane?

10. Explain why it is not enough just to say that a solution is "hypotonic."

11. In what way is active transport an endergonic process?

12. How do transport proteins contribute to a membrane's selective permeability?

13. How can a protein-secreting cell synthesize and secrete its product without the protein ever having to cross a membrane?

Questions 14–18

An artificial cell consisting of an aqueous solution enclosed in a selectively permeable membrane has just been immersed in a beaker containing a different solution. The membrane is permeable to water and to the simple sugars glucose and fructose but completely impermeable to the disaccharide sucrose.

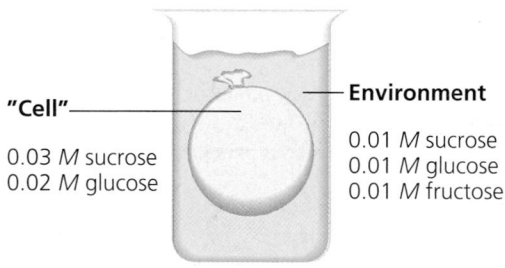

"Cell"
0.03 *M* sucrose
0.02 *M* glucose

Environment
0.01 *M* sucrose
0.01 *M* glucose
0.01 *M* fructose

14. Which solute(s) will exhibit a net diffusion into the cell?

15. Which solute(s) will exhibit a net diffusion out of the cell?

16. Which solution—the cell contents or the environment—is hypertonic to the other?

17. In which direction will there be a net osmotic movement of water?

18. After the cell is placed in the beaker, which of the following changes will occur?
   a. The artificial cell will become more flaccid.
   b. The artificial cell will become more turgid.

   c. The entropy of the system (cell plus surrounding solution) will decrease.
   d. The overall free energy stored in the system will increase.
   e. The membrane potential will decrease.

Go to the website or CD-ROM for more quiz questions.

Go to the website or CD-ROM for more quiz questions.

Evolution Connection

Paramecium and other protists that live in hypotonic environments have cell membrane adaptations that slow osmotic water uptake, while those living in isotonic environments have more permeable cell membranes. What water regulation adaptations would you expect to have evolved in protists living in hypertonic habitats such as Great Salt Lake? How about those living in habitats where salt concentration fluctuates?

The Process of Science

An experiment is designed to study the mechanism of sucrose uptake by plant cells. Cells are immersed in a sucrose solution, and the pH of the solution is monitored with a pH meter. Samples of the cells are taken at intervals, and the sucrose in the sampled cells is measured. The measurements show that sucrose uptake by the cells correlates with a rise in the pH of the surrounding solution. The magnitude of the pH change is proportional to the starting concentration of sucrose in the extracellular solution. A metabolic poison known to block the ability of cells to regenerate ATP is found to inhibit the pH changes in the extracellular solution. Propose a hypothesis accounting for these results. Suggest an additional experiment to test your hypothesis.

Experiment with water balance in cells in the Case Study in the Process of Science, available on the website and CD-ROM.

Science, Technology, and Society

Extensive irrigation in arid regions causes salts to accumulate in the soil. (The water contains low concentrations of salts, but when the water evaporates from the fields, the salts are left behind to concentrate in the soil.) Based on what you have learned about water balance in plant cells, explain why increasing soil salinity (saltiness) has an adverse effect on agriculture. Suggest some ways to minimize this damage. What costs are attached to your solutions?

Answers: 1. b; 2. c; 3. a; 4. d; 5. b; 6. This structure shields the hydrophobic tails of the phospholipids from water, while exposing the hydrophilic heads to water. 7. They are on the inner side of the transport vesicle membrane. 8. Epinephrine binds to a receptor on the liver cell surface, activating a signal transduction pathway inside the cell that leads to sugar release. 9. The second law is the trend toward randomness. Equal concentrations of a substance on both sides of a membrane is a more random distribution than unequal concentrations. Diffusion of a substance to a region where it is initially less concentrated increases entropy, as mandated by the second law. 10. Hypertonic and hypotonic are relative terms: A solution that is hypertonic to tap water could be hypotonic to seawater. You must say what the solution is compared to. 11. Active transport is energetically "uphill" in the sense that it requires a net input of energy (in the form of ATP). 12. Transport proteins are specific for the substances they transport. Thus, the numbers and kinds of different transport proteins embedded in the membrane affect its permeability to various substances. 13. From the time the protein is made by rough ER, it is topologically "outside" the cell, first in the ER interior, then within the Golgi and transport vesicles, and finally outside the plasma membrane as the vesicles release their contents by exocytosis. 14. Fructose. 15. Glucose. 16. Cell contents. 17. Into the cell. 18. b.

CHAPTER 9

CELLULAR RESPIRATION: HARVESTING CHEMICAL ENERGY

Living is work. *Cells organize small organic molecules into polymers, such as proteins and DNA. They pump substances across membranes. Many cells move or change their shapes. They grow and reproduce. Cells must work just to maintain their complex structure, because order is intrinsically unstable. To perform their many tasks, cells require transfusions of energy from outside sources. Energy enters most ecosystems in the form of sunlight, the energy source for plants and other photosynthetic organisms*

(FIGURE 9.1, *p. 156*). *Animals, such as the orangutan in the photograph above, obtain fuel by eating plants, or by eating other organisms that eat plants. In this chapter, you will learn how cells harvest the chemical energy stored in organic molecules and use it to regenerate ATP, the molecule that drives most cellular work.*

THE PRINCIPLES OF ENERGY HARVEST

In harvesting chemical energy, cells usually employ metabolic pathways with many steps. Fortunately, we can organize this complexity with the help of just a handful of general principles. The first part of this chapter develops these principles, which will then enable you to make sense of cellular respiration and related pathways.

Cellular respiration and fermentation are catabolic, energy-yielding pathways

Organic compounds store energy in their arrangement of atoms. With the help of enzymes, a cell systematically degrades complex organic molecules that are rich in potential energy to simpler waste products that have less energy. Some of the energy taken out of chemical storage can be used to do work; the rest is dissipated as heat. As you learned in Chapter 6, metabolic pathways that release stored energy by breaking down complex molecules are called catabolic pathways. One catabolic process, **fermentation,** is a partial degradation of sugars that occurs without the help of oxygen. However, the most prevalent and efficient catabolic pathway is **cellular respiration,** in which oxygen is consumed as a reactant along

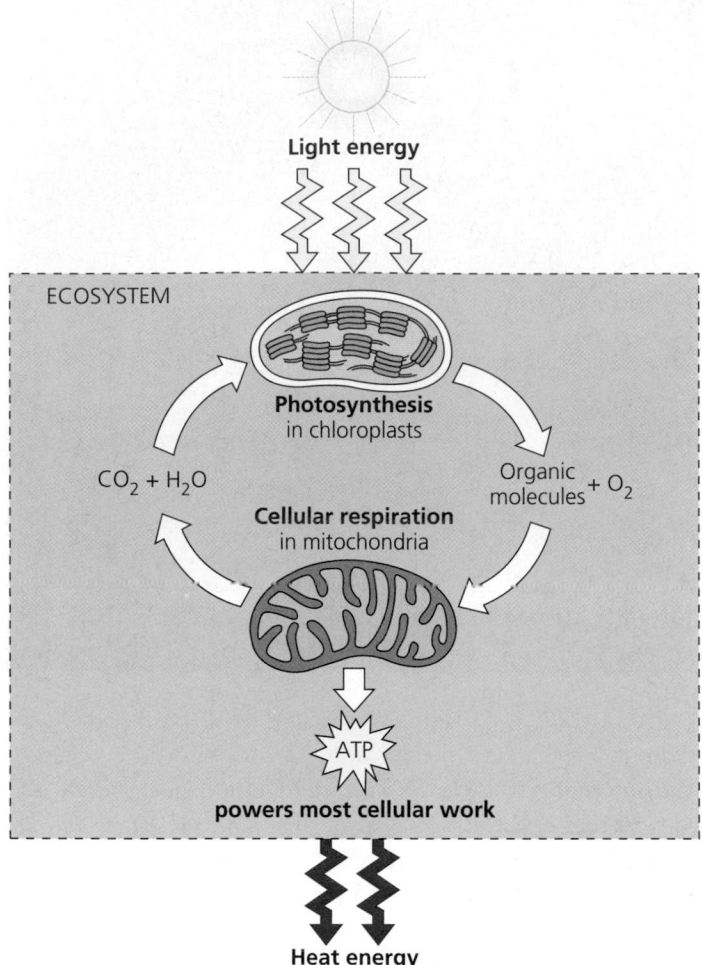

Light energy

ECOSYSTEM

Photosynthesis
in chloroplasts

$CO_2 + H_2O$

Organic molecules $+ O_2$

Cellular respiration
in mitochondria

ATP

powers most cellular work

Heat energy

FIGURE 9.1 Energy flow and chemical recycling in ecosystems. The mitochondria of eukaryotes (including plants and algae) use the organic products of photosynthesis as fuel for cellular respiration, which also consumes the oxygen produced by photosynthesis. Respiration harvests the energy stored in organic molecules to generate ATP, which powers most cellular work. The waste products of respiration, carbon dioxide and water, are the very substances that chloroplasts use as raw materials for photosynthesis. Thus, the chemical elements essential to life are recycled. But energy is not: It flows into an ecosystem as sunlight and leaves as heat.

with the organic fuel. In eukaryotic cells, mitochondria house most of the metabolic equipment for cellular respiration.

Although very different in mechanism, respiration is in principle similar to the combustion of gasoline in an automobile engine after oxygen is mixed with the fuel (hydrocarbons). Food is the fuel for respiration, and the exhaust is carbon dioxide and water. The overall process can be summarized as follows:

$$\text{Organic compounds} + \text{Oxygen} \longrightarrow \text{Carbon dioxide} + \text{Water} + \text{Energy}$$

Although carbohydrates, fats, and proteins can all be processed and consumed as fuel, it is traditional to learn the steps of cellular respiration by tracking the degradation of the sugar glucose ($C_6H_{12}O_6$):

$$C_6H_{12}O_6 + 6\,O_2 \longrightarrow 6\,CO_2 + 6\,H_2O + \text{Energy (ATP + heat)}$$

This breakdown of glucose is exergonic, having a free-energy change of -686 kcal ($-2{,}870$ kJ) per mole of glucose decomposed ($\Delta G = -686$ kcal/mol); recall that a negative ΔG indicates that the products of the chemical process store less energy than the reactants.

Catabolic pathways do not directly move flagella, pump solutes across membranes, polymerize monomers, or perform other cellular work. Catabolism is linked to work by a chemical drive shaft: ATP. The processes of cellular respiration and fermentation are complex and challenging to learn. Therefore, as you study this chapter, it will help you to keep in mind your main objective: discovering how cells use the energy stored in food molecules to make ATP.

Cells recycle the ATP they use for work

The molecule known as ATP, short for adenosine triphosphate, is the central character in bioenergetics. Recall from Chapter 6 that the triphosphate tail of ATP is the chemical equivalent of a loaded spring; the close packing of the three negatively charged phosphate groups is an unstable, energy-storing arrangement (because like charges repel each other). The chemical "spring" tends to "relax" by losing the terminal phosphate (see FIGURE 6.8). The cell taps this energy source by using enzymes to transfer phosphate groups from ATP to other compounds, which are then said to be phosphorylated. Phosphorylation primes a molecule to undergo some kind of change that performs work, and the molecule loses its phosphate group in the process (FIGURE 9.2). The price of most cellular work, then, is the conversion of ATP to ADP and inorganic phosphate (abbreviated $\text{\textcircled{P}}_i$ in this book), products that store less energy than ATP. To keep working, the cell must regenerate its supply of ATP from ADP and inorganic phosphate. A working muscle cell, for example, recycles its ATP at a rate of about 10 million molecules per second. To understand how cellular respiration regenerates ATP, we need to examine the fundamental chemical processes known as oxidation and reduction.

Redox reactions release energy when electrons move closer to electronegative atoms

Just what happens when catabolic pathways decompose glucose and other organic fuels? And why do these metabolic pathways yield energy? The answers are based on the transfer of electrons during the chemical reactions. The relocation of electrons releases the energy stored in food molecules, and this energy is used to synthesize ATP.

In many chemical reactions, there is a transfer of one or more electrons (e^-) from one reactant to another. These electron transfers are called oxidation-reduction reactions, or **redox reactions** for short. In a redox reaction, the loss of

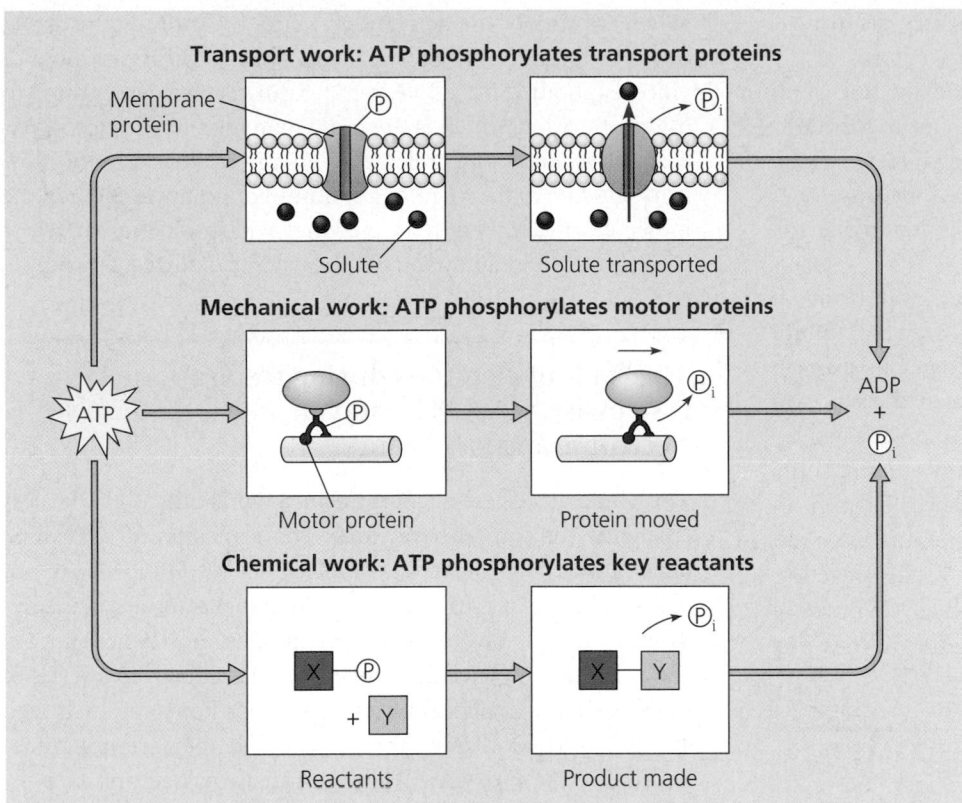

Transport work: ATP phosphorylates transport proteins

Membrane protein

Solute

Solute transported

Mechanical work: ATP phosphorylates motor proteins

ATP

Motor protein

Protein moved

Chemical work: ATP phosphorylates key reactants

X + Y

Reactants

X–Y

Product made

ADP + P_i

FIGURE 9.2 A review of how ATP drives cellular work. Phosphate group transfer is the mechanism responsible for most types of cellular work. Enzymes shift a phosphate group (P) from ATP to some other molecule, and this phosphorylated molecule undergoes a change that performs work. For example, ATP drives active transport by phosphorylating certain membrane proteins; drives mechanical work by phosphorylating motor proteins, such as the ones that move organelles along cytoskeletal "tracks" in the cell; and drives chemical work by phosphorylating key reactants. The phosphorylated molecules lose the phosphate groups as work is performed, leaving ADP and inorganic phosphate (P_i) as products. Cellular respiration replenishes the ATP supply by powering the phosphorylation of ADP.

electrons from one substance is called **oxidation,** and the addition of electrons to another substance is known as **reduction.*** To take a simple, nonbiological example, consider the reaction between sodium and chlorine to form table salt:

$$\text{Na} + \text{Cl} \longrightarrow \text{Na}^+ + \text{Cl}^-$$

with Oxidation spanning Na to Na$^+$ and Reduction spanning Cl to Cl$^-$.

We could generalize a redox reaction this way:

$$\text{X}e^- + \text{Y} \longrightarrow \text{X} + \text{Y}e^-$$

with Oxidation spanning Xe$^-$ to X and Reduction spanning Y to Ye$^-$.

---

* This term defies intuition; *adding* electrons is called *reduction.* The term was derived from the electrical effects of adding electrons: Negatively charged electrons added to a cation *reduce* the amount of positive charge of the cation.

In the generalized reaction, substance X, the electron donor, is called the **reducing agent;** it reduces Y. Substance Y, the electron acceptor, is the **oxidizing agent;** it oxidizes X. Because an electron transfer requires both a donor and an acceptor, oxidation and reduction always go together.

Not all redox reactions involve the complete transfer of electrons from one substance to another; some change the degree of electron sharing in covalent bonds. The reaction between methane and oxygen to form carbon dioxide and water, shown in FIGURE 9.3, is an example. As explained in Chapter 2, the covalent electrons in methane are shared equally between the bonded atoms because carbon and hydrogen have about the same affinity for valence electrons; they are about equally electronegative. But when methane reacts with oxygen to form carbon dioxide, electrons are shifted away from the carbon atoms

FIGURE 9.3 Methane combustion as an energy-yielding redox reaction. During the reaction, covalently shared electrons move away from carbon and hydrogen atoms and closer to oxygen, which is very electronegative. The reaction releases energy to the surroundings, because the electrons lose potential energy as they move closer to electronegative atoms.

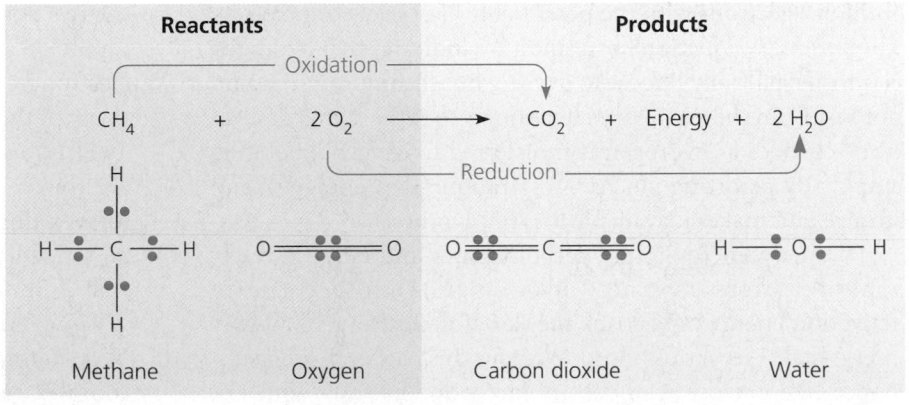

Reactants		Products		
CH_4 +	$2 O_2$	$\longrightarrow$ CO_2 +	Energy +	$2 H_2O$

with Oxidation spanning CH_4 to CO_2 and Reduction spanning $2 O_2$ to $2 H_2O$.

Methane   Oxygen   Carbon dioxide   Water

to their new covalent partner, oxygen, which is very electronegative. Methane has thus been oxidized. The two atoms of the oxygen molecule also share their electrons equally. But when the oxygen reacts with the hydrogen from methane to form water, the electrons of the covalent bonds are drawn closer to the oxygen; the oxygen molecule has been reduced. Because oxygen is so electronegative, it is one of the most potent of all oxidizing agents.

Energy must be added to pull an electron away from an atom, just as energy must be added to push a large ball uphill. The more electronegative the atom (the stronger its pull on electrons), the more energy is required to keep the electron away from it, just as more energy is required to push a ball up a steeper hill. An electron loses potential energy when it shifts from a less electronegative atom (one with a weaker pull on electrons) toward a more electronegative one, just as a ball loses potential energy when it rolls downhill. A redox reaction that relocates electrons closer to oxygen, such as the burning of methane, releases chemical energy that can be put to work.

Electrons "fall" from organic molecules to oxygen during cellular respiration

The oxidation of methane by oxygen is the main combustion reaction that occurs at the burner of a gas stove. The combustion of gasoline in an automobile engine is also a redox reaction; the energy released pushes the pistons. But the energy-yielding redox process of greatest interest here is respiration: the oxidation of glucose and other fuel molecules in food. Examine again the summary equation for cellular respiration, but this time think of it as a redox process:

$$\overbrace{C_6H_{12}O_6 \ + \ 6O_2}^{\text{Oxidation}} \longrightarrow 6CO_2 \ + \ \underbrace{6H_2O}_{\text{Reduction}}$$

As in the combustion of methane or gasoline, the fuel (glucose) is oxidized and oxygen is reduced, and the electrons lose potential energy along the way.

In general, organic molecules that have an abundance of hydrogen are excellent fuels because their bonds are a source of "hilltop" electrons with the potential to "fall" closer to oxygen. The summary equation for respiration indicates that hydrogen is transferred from glucose to oxygen. But the important point, not visible in the summary equation, is that the status of electrons changes as hydrogen is transferred to oxygen, liberating energy. By oxidizing glucose, respiration takes energy out of storage and makes it available for ATP synthesis.

The main energy foods, carbohydrates and fats, are reservoirs of electrons associated with hydrogen. Only the barrier of activation energy holds back the flood of electrons to a lower energy state (see FIGURE 6.12). Without this barrier, a food substance like glucose would combine spontaneously with O_2.

When we supply the activation energy by igniting glucose, it burns in air, releasing 686 kcal (2,870 kJ) of heat per mole of glucose (about 180 g). Body temperature is not high enough to initiate burning, which is the rapid oxidation of fuel accompanied by an enormous release of energy as heat. But swallow some glucose in the form of a spoonful of honey, and when the molecules reach your cells, enzymes will lower the barrier of activation energy, allowing the sugar to be oxidized slowly.

The "fall" of electrons during respiration is stepwise, via NAD^+ and an electron transport chain

The wholesale release of energy from a fuel is difficult to harness efficiently for constructive work: The explosion of a gasoline tank cannot drive a car very far. Cellular respiration does not oxidize glucose in a single explosive step that transfers all the hydrogen from the fuel to the oxygen at one time. Rather, glucose and other organic fuels are broken down gradually in a series of steps, each one catalyzed by an enzyme. At key steps, hydrogen atoms are stripped from the glucose, but they are not transferred directly to oxygen. They are usually passed first to a coenzyme called **NAD^+** (nicotinamide adenine dinucleotide). Thus, NAD^+ functions as an oxidizing agent during respiration.

How does NAD^+ trap electrons from glucose and other fuel molecules? Enzymes called dehydrogenases remove a pair of hydrogen atoms from the substrate, a sugar or some other fuel. We can think of this as the removal of two electrons and two protons (the nuclei of hydrogen atoms). The enzyme delivers the *two* electrons along with *one* proton to its coenzyme, NAD^+ (FIGURE 9.4). The other proton is released as a hydrogen ion (H^+) into the surrounding solution:

$$\text{H}-\overset{|}{\underset{|}{\text{C}}}-\text{OH} + \text{NAD}^+ \xrightarrow{\text{Dehydrogenase}} \overset{|}{\underset{|}{\text{C}}}{=}\text{O} + \text{NADH} + \text{H}^+$$

Though the oxidized form, NAD^+, has a positive charge, the reduced form, NADH, is electrically neutral. By receiving two negatively charged electrons but only one positively charged proton, NAD^+ has its charge neutralized. The name NADH for the reduced form shows the hydrogen that has been received in the reaction. Since NAD^+ gains electrons, it is an electron acceptor (a synonym for oxidizing agent). The most versatile electron acceptor in cellular respiration, NAD^+ functions in many of the redox steps during the breakdown of sugar.

Electrons lose very little of their potential energy when they are transferred from food to NAD^+. Each NADH molecule formed during respiration represents stored energy that can be tapped to make ATP when the electrons complete their "fall" from NADH to oxygen.

How do electrons extracted from food and stored by NADH finally reach oxygen? It will help to compare this complex redox chemistry of cellular respiration to a much

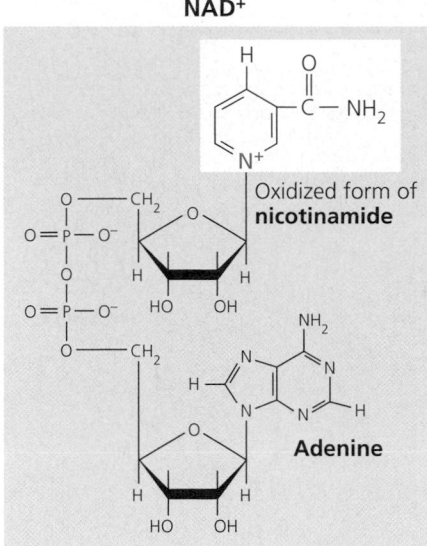

NAD⁺

Oxidized form of **nicotinamide**

Adenine

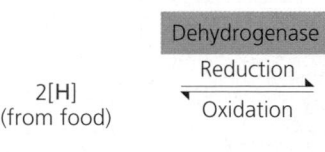

$+ \quad 2[H]$
(from food)

Dehydrogenase
Reduction
Oxidation

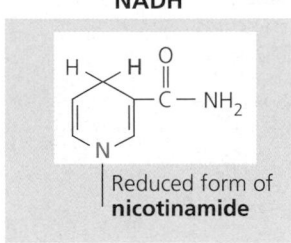

NADH

Reduced form of **nicotinamide**

FIGURE 9.4 NAD⁺ as an electron shuttle. The full name for NAD⁺, nicotinamide adenine dinucleotide, describes its structure; the molecule consists of two nucleotides joined together. (Nicotinamide is a nitrogenous base, although not one that is present in DNA or RNA.) The enzymatic transfer of two electrons and one proton from some organic substrate to NAD⁺ reduces the NAD⁺ to NADH. Most of the electrons removed from food are transferred initially to NAD⁺.

simpler reaction: the reaction between hydrogen and oxygen to form water (FIGURE 9.5a). Mix H₂ and O₂, provide a spark for activation energy, and the gases combine explosively. The explosion represents a release of energy as the electrons of hydrogen fall closer to the electronegative oxygen atoms. Cellular respiration also brings hydrogen and oxygen together to form water, but there are two important differences. First, in cellular respiration, the hydrogen that reacts with oxygen is derived from organic molecules rather than H₂. Second, respiration uses an **electron transport chain** to break the fall of electrons to oxygen into several energy releasing steps instead of one explosive reaction (FIGURE 9.5b). The transport chain consists of a number of molecules, mostly proteins, built into the inner membrane of a mitochondrion. Electrons removed

from food are shuttled by NADH to the "top" end of the chain. At the "bottom" end, oxygen captures these electrons along with hydrogen nuclei (H⁺), forming water.

Electron transfer from NADH to oxygen is an exergonic reaction with a free-energy change of −53 kcal/mol (−222 kJ/mol). Instead of this energy being released and wasted in a single explosive step, electrons cascade down the chain from one carrier molecule to the next, losing a small amount of energy with each step until they finally reach oxygen, the terminal electron acceptor. What keeps the electrons moving is that each carrier is more electronegative than its "uphill" neighbor in the chain. At the bottom of the chain is oxygen, which has a very great affinity for electrons. Thus, electrons removed from food by NAD⁺ fall down the electron transport chain to a far more stable location in the electronegative oxygen atom. Put another way, oxygen pulls electrons down the chain in an energy-yielding tumble analogous to gravity pulling objects downhill.

Thus, during cellular respiration, most electrons travel this "downhill" route: food → NADH → electron transport chain → oxygen. In the next major section of the chapter, you will learn more about how the cell uses the energy released from this exergonic electron fall to regenerate its supply of ATP.

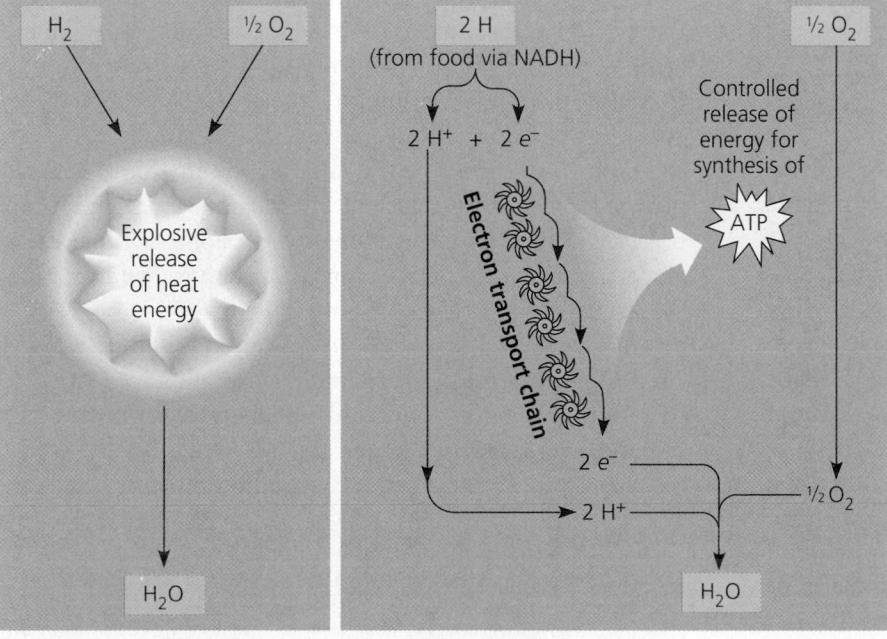

(a) Uncontrolled reaction

(b) Cellular respiration

FIGURE 9.5 An introduction to electron transport chains. (a) The uncontrolled exergonic reaction of hydrogen with oxygen to form water releases a large amount of energy in the form of heat and light: an explosion. **(b)** In cellular respiration, the same reaction occurs in stages: An electron transport chain breaks the "fall" of electrons in this reaction into a series of smaller steps and stores some of the released energy in a form that can be used to make ATP. (The rest of the energy is released as heat.)

FIGURE 9.6 An overview of cellular respiration. In a eukaryotic cell, glycolysis occurs outside the mitochondria in the cytosol. The Krebs cycle and the electron transport chains are located inside the mitochondria. During glycolysis, each glucose molecule is broken down into two molecules of the compound pyruvate. The pyruvate crosses the double membrane of the mitochondrion to enter the matrix, where the Krebs cycle decomposes it to carbon dioxide. NADH or FADH$_2$ (see p. 164) transfers electrons from molecules undergoing glycolysis and the Krebs cycle to electron transport chains, which are built into the inner mitochondrial membrane. The electron transport chain converts the chemical energy to a form that can be used to drive oxidative phosphorylation, which accounts for most of the ATP generated by cellular respiration. A smaller amount of ATP is formed directly during glycolysis and the Krebs cycle by substrate-level phosphorylation.

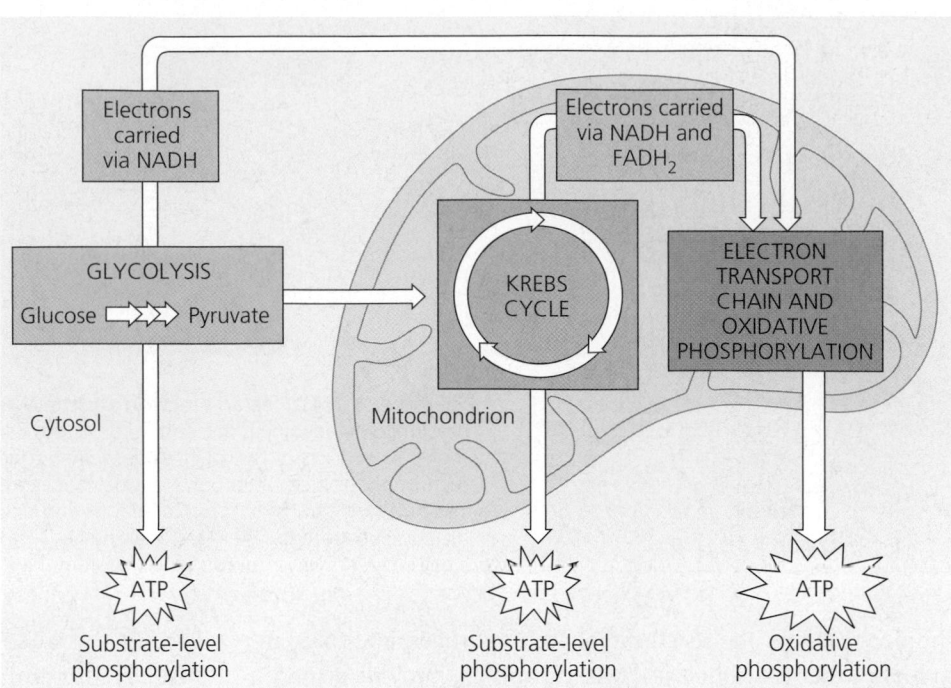

THE PROCESS OF CELLULAR RESPIRATION

Now that we have covered the basic redox mechanisms of respiration, let's look at the entire process.

Respiration involves glycolysis, the Krebs cycle, and electron transport: *an overview*

Respiration is a cumulative function of three metabolic stages, diagrammed in FIGURE 9.6:

1. Glycolysis (color-coded teal throughout the chapter)
2. The Krebs cycle (color-coded salmon)
3. The electron transport chain and oxidative phosphorylation (color-coded violet)

The first two stages, glycolysis and the Krebs cycle, are the catabolic pathways that decompose glucose and other organic fuels. **Glycolysis,** which occurs in the cytosol, begins the degradation by breaking glucose into two molecules of a compound called pyruvate. The **Krebs cycle,** which takes place within the mitochondrial matrix, completes the job by decomposing a derivative of pyruvate to carbon dioxide. Thus, the carbon dioxide produced by respiration represents fragments of oxidized organic molecules.

Some of the steps of glycolysis and the Krebs cycle are redox reactions in which dehydrogenase enzymes transfer electrons from substrates to NAD$^+$, forming NADH. In the third stage of respiration, the electron transport chain accepts electrons from the breakdown products of the first two stages (usually via NADH) and passes these electrons from one molecule to

another. At the end of the chain, the electrons are combined with hydrogen ions and molecular oxygen to form water (see FIGURE 9.5b). The energy released at each step of the chain is stored in a form the mitochondrion can use to make ATP. This mode of ATP synthesis is called **oxidative phosphorylation** because it is powered by the redox reactions that transfer electrons from food to oxygen.

The site of electron transport and oxidative phosphorylation is the inner membrane of the mitochondrion (see FIGURE 7.17). Oxidative phosphorylation accounts for almost 90% of the ATP generated by respiration. A smaller amount of ATP is formed directly in a few reactions of glycolysis and the Krebs cycle by a mechanism called **substrate-level phosphorylation** (FIGURE 9.7). This mode of ATP synthesis occurs when an enzyme transfers a phosphate group from a substrate molecule to ADP. "Substrate molecule" here refers to an organic molecule generated during the catabolism of glucose.

Respiration cashes in the large denomination of energy banked in glucose for the small change of ATP, which is more practical for the cell to spend on its work. For each molecule of glucose degraded to carbon dioxide and water by respiration, the cell makes up to about 38 molecules of ATP.

This overview has introduced how glycolysis, the Krebs cycle, and electron transport fit into the overall process of cellular respiration. We are now ready to take a closer look at each of these three stages of respiration.*

---

* Technically, cellular respiration is defined to include only the processes that require the presence of O$_2$ to operate: the Krebs cycle, the electron transport chain, and the accompanying synthesis of ATP by oxidative phosphorylation. We include glycolysis because most respiring cells deriving energy from glucose do use this process to produce starting material for the Krebs cycle.

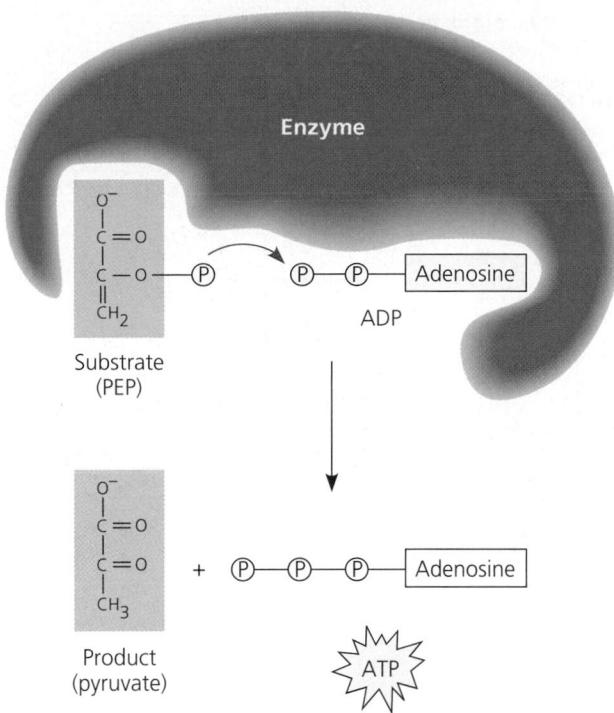

FIGURE 9.7 Substrate-level phosphorylation. Some ATP is made by direct enzymatic transfer of a phosphate group from a substrate to ADP. The phosphate donor in this case is phosphoenolpyruvate (PEP), which is formed from the breakdown of sugar during glycolysis.

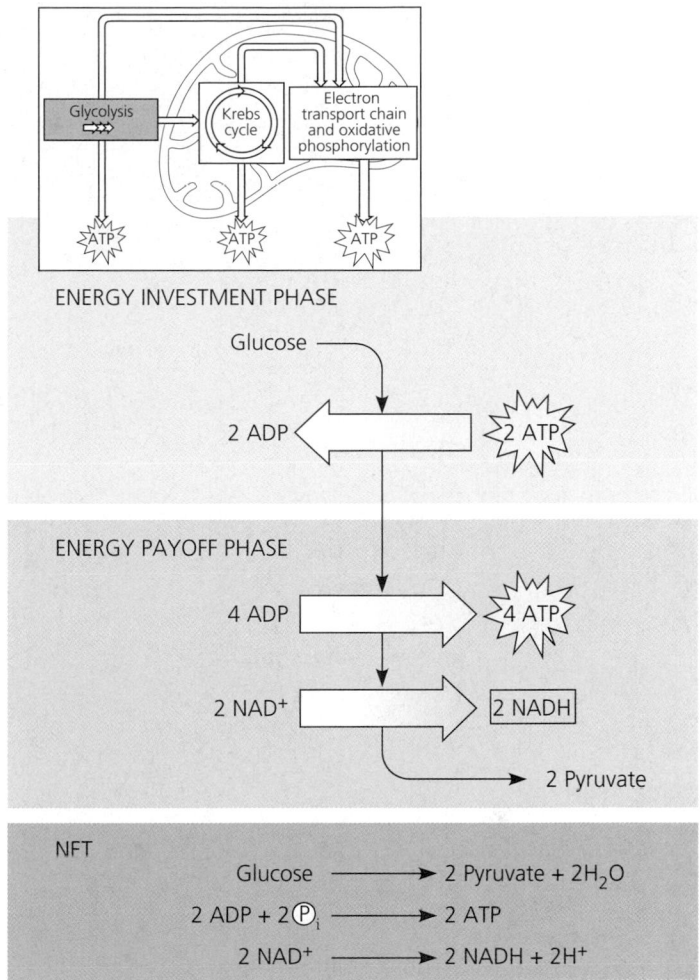

FIGURE 9.8 The energy input and output of glycolysis.

Glycolysis harvests chemical energy by oxidizing glucose to pyruvate: *a closer look*

The word *glycolysis* means "splitting of sugar," and that is exactly what happens during this pathway. Glucose, a six-carbon sugar, is split into two three-carbon sugars. These smaller sugars are then oxidized and their remaining atoms rearranged to form two molecules of pyruvate. (Pyruvate is the ionized form of a three-carbon acid, pyruvic acid.)

As summarized in FIGURE 9.8 and described in detail in FIGURE 9.9 (pp. 162–163) the pathway of glycolysis consists of ten steps, each catalyzed by a specific enzyme. We can divide these ten steps into two phases. The energy investment phase includes the first five steps, and the energy payoff phase includes the next five steps. During the energy investment phase, the cell actually spends ATP to phosphorylate the fuel molecules. This investment is repaid with dividends during the energy payoff phase, when ATP is produced by substrate-level phosphorylation and NAD^+ is reduced to NADH by oxidation of the food. The net energy yield from glycolysis, per glucose molecule, is 2 ATP plus 2 NADH.

Notice in FIGURE 9.9 that all of the carbon originally present in glucose is accounted for in the two molecules of pyruvate; no CO_2 is released during glycolysis. Glycolysis occurs whether or not molecular oxygen (O_2) is present. However, if oxygen *is* present, the energy stored in NADH can be converted to ATP energy by the electron transport chain and oxidative phosphorylation. And in the presence of oxygen, the chemical energy left in pyruvate can be extracted by the Krebs cycle.

The Krebs cycle completes the energy-yielding oxidation of organic molecules: *a closer look*

Glycolysis releases less than a quarter of the chemical energy stored in glucose; most of the energy remains stocked in the two molecules of pyruvate. If molecular oxygen is present, the pyruvate enters the mitochondrion, where the enzymes of the Krebs cycle complete the oxidation of the organic fuel.

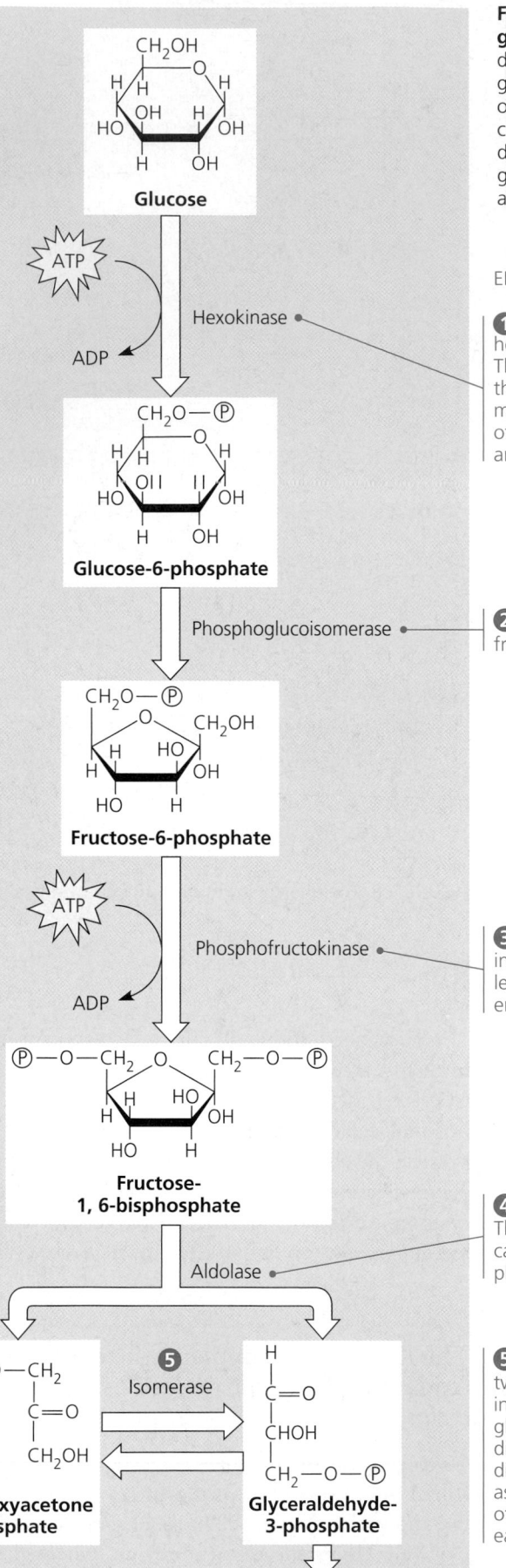

FIGURE 9.9 A closer look at glycolysis. The orientation diagram at the right relates glycolysis to the whole process of respiration. Do not let the chemical detail in the main diagram block your view of glycolysis as a source of ATP and NADH.

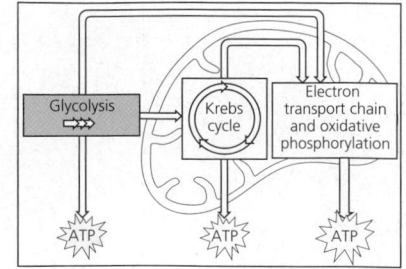

ENERGY INVESTMENT PHASE

1 Glucose enters the cell and is phosphorylated by the enzyme hexokinase, which transfers a phosphate group from ATP to the sugar. The charge of the phosphate group traps the sugar in the cell because the plasma membrane is impermeable to ions. Phosphorylation also makes glucose more chemically reactive. In this diagram, the transfer of a phosphate group or pair of electrons from one reactant to another is indicated by coupled arrows:

2 Glucose-6-phosphate is rearranged to convert it to its isomer, fructose-6-phosphate.

3 This enzyme transfers a phosphate group from ATP to the sugar, investing another molecule of ATP in glycolysis. So far, the ATP ledger shows a debit of 2. With phosphate groups on its opposite ends, the sugar is now ready to be split in half.

4 This is the reaction from which glycolysis gets its name. The enzyme cleaves the sugar molecule into two different three-carbon sugars: glyceraldehyde-3-phosphate and dihydroxyacetone phosphate. These two sugars are isomers of each other.

5 Isomerase catalyzes the reversible conversion between the two three-carbon sugars. This reaction never reaches equilibrium in the cell because the next enzyme in glycolysis uses only glyceraldehyde-3-phosphate as its substrate (and not dihydroxyacetone phosphate). This pulls the equilibrium in the direction of glyceraldehyde-3-phosphate, which is removed as fast as it forms. Thus, the net result of steps 4 and 5 is cleavage of a six-carbon sugar into two molecules of glyceraldehyde-3-phosphate; each will progress through the remaining steps of glycolysis.

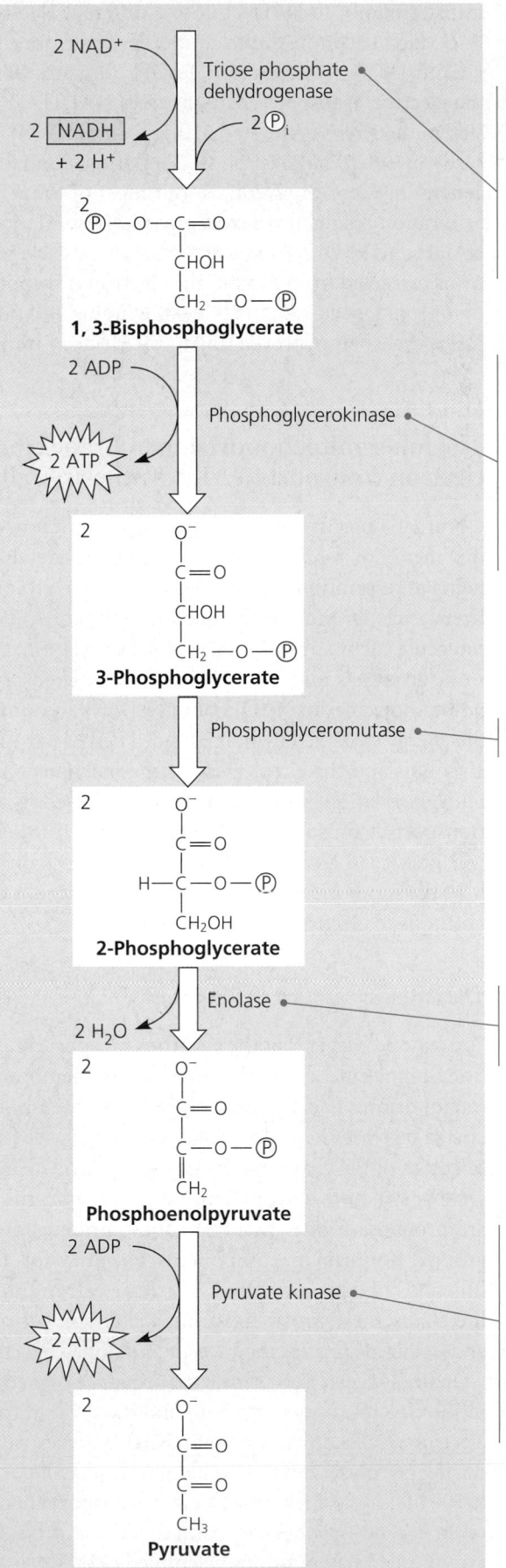

6 This enzyme catalyzes two sequential reactions while it holds glyceraldehyde-3-phosphate in its active site. First, the sugar is oxidized by the transfer of electrons and H$^+$ to NAD$^+$, forming NADH (a redox reaction). This reaction is very exergonic, and the enzyme uses the released energy to attach a phosphate group to the oxidized substrate, making a product of very high potential energy. The source of the phosphates is the pool of inorganic phosphate ions that are always present in the cytosol. Notice that the coefficient 2 precedes all molecules in the energy payoff phase; these steps occur after glucose is split into two three-carbon sugars (step 4).

7 Glycolysis finally produces some ATP. The phosphate group added in the previous step is transferred to ADP in an exergonic reaction. For each glucose molecule that began glycolysis, step 7 produces two molecules of ATP, since every product after the sugar-splitting step (step 4) is doubled. Since two ATPs were invested to get sugar ready for splitting, the ATP ledger now stands at zero. Glucose has been converted to two molecules of 3-phosphoglycerate, which is not a sugar. The carbonyl group that characterizes a sugar has been oxidized to a carboxyl group, the hallmark of an organic acid. The sugar was oxidized in step 6, and now the energy made available by that oxidation has been used to make ATP.

8 Next, this enzyme relocates the remaining phosphate group. This step prepares the substrate for the next reaction.

9 This enzyme causes a double bond to form in the substrate by extracting a water molecule, yielding phosphoenolpyruvate (PEP). The electrons of the substrate are rearranged in such a way that the remaining phosphate bond becomes very unstable, preparing the substrate for the next reaction.

10 The last reaction of glycolysis produces more ATP by transferring the phosphate group from PEP to ADP. Since this step occurs twice for each glucose molecule, the ATP ledger now shows a net gain of two ATPs. A debt of two ATPs was incurred from steps 1 and 3, but steps 7 and 10 each produced two ATPs for a total credit of four. Glycolysis has repaid the ATP investment with 100% interest. Additional energy was stored by step 6 in NADH, which can be used to make ATP by oxidative phosphorylation if oxygen is present. Glucose has been broken down and oxidized to two molecules of pyruvate, the end product of the glycolytic pathway.

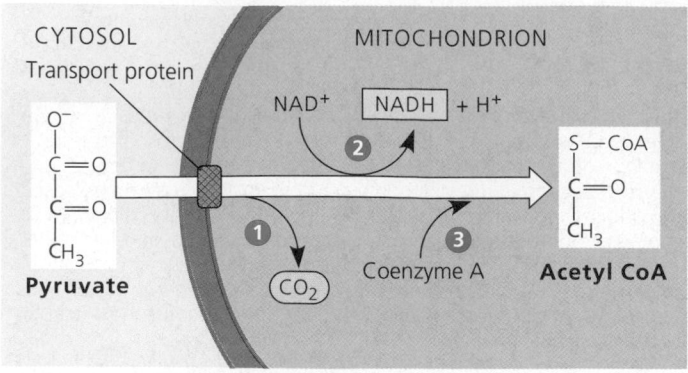

FIGURE 9.10 Conversion of pyruvate to acetyl CoA, the junction between glycolysis and the Krebs cycle. A complex of several enzymes catalyzes the three numbered steps, which are described in the text. The acetyl CoA will enter the Krebs cycle. The CO_2 molecule will diffuse out of the cell.

Upon entering the mitochondrion, pyruvate is first converted to a compound called acetyl coenzyme A, or **acetyl CoA** (FIGURE 9.10). This step, the junction between glycolysis and the Krebs cycle, is accomplished by a multienzyme complex that catalyzes three reactions: ❶ Pyruvate's carboxyl group, which is already fully oxidized and thus has little chemical energy, is removed and given off as a molecule of CO_2. (This is the first step in respiration where CO_2 is released.) ❷ The remaining two-carbon fragment is oxidized to form a compound named acetate (the ionized form of acetic acid). An enzyme transfers the extracted electrons to NAD^+, storing energy in the form of NADH. ❸ Finally, coenzyme A, a sulfur-containing compound derived from a B vitamin, is attached to the acetate by an unstable bond that makes the acetyl group (the attached acetate) very reactive. The product of this chemical grooming, acetyl CoA, is now ready to feed its acetate into the Krebs cycle for further oxidation.

The Krebs cycle is named after Hans Krebs, the German-British scientist who was largely responsible for elucidating the pathway in the 1930s. The cycle has eight steps, each catalyzed by a specific enzyme (FIGURE 9.11). You can see in the diagram that for each turn of the Krebs cycle, two carbons enter in the relatively reduced form of acetate (step 1), and two different carbons leave in the completely oxidized form of CO_2 (steps 3 and 4). The acetate joins the cycle by its enzymatic addition to the compound oxaloacetate, forming citrate. Subsequent steps decompose the citrate back to oxaloacetate, giving off CO_2 as "exhaust." It is this regeneration of oxaloacetate that accounts for the "cycle" in the Krebs cycle. Except for the enzyme that catalyzes step 6, which resides in the inner mitochondrial membrane, all the Krebs cycle enzymes are located in the mitochondrial matrix.

Most of the energy harvested by the oxidative steps of the cycle is conserved in NADH. For each acetate that enters the cycle, three molecules of NAD^+ are reduced to NADH (steps 3, 4, and 8). In one oxidative reaction, step 6, electrons are transferred not to NAD^+, but to a different electron acceptor, FAD (flavin adenine dinucleotide, derived from riboflavin, a B vitamin). The reduced form, $FADH_2$, donates its electrons to the electron transport chain, as does NADH. There is also a step in the Krebs cycle, step 5, that forms an ATP molecule directly by substrate-level phosphorylation, similar to the ATP-generating steps of glycolysis. But most of the ATP output of respiration results from oxidative phosphorylation, when the NADH and $FADH_2$ produced by the Krebs cycle relay the electrons extracted from food to the electron transport chain. Use FIGURE 9.12, page 166, to review the inputs and outputs of the Krebs cycle before proceeding to the electron transport chain.

The inner mitochondrial membrane couples electron transport to ATP synthesis: *a closer look*

Our main objective in this chapter is to learn how cells harvest the energy of food to make ATP. But the metabolic components of respiration we have dissected so far, glycolysis and the Krebs cycle, produce only four molecules of ATP per glucose molecule, all by substrate-level phosphorylation: two net ATPs from glycolysis and two ATPs from the Krebs cycle. At this point, molecules of NADH (and $FADH_2$) account for most of the energy extracted from the food. These electron escorts link glycolysis and the Krebs cycle to the machinery for oxidative phosphorylation, which uses energy released by the electron transport chain to power ATP synthesis. In this section, you will first learn how the electron transport chain works, then how the inner membrane of the mitochondrion couples ATP synthesis to electron flow down the chain.

The Pathway of Electron Transport

You learned earlier that the electron transport chain is a collection of molecules embedded in the inner membrane of the mitochondrion. The folding of the inner membrane to form cristae increases its surface area, providing space for thousands of copies of the chain in each mitochondrion. (Once again, we see that structure fits function.) Most components of the chain are proteins. Tightly bound to these proteins are prosthetic groups, nonprotein components essential for the catalytic functions of certain enzymes. During electron transport along the chain, these prosthetic groups alternate between reduced and oxidized states as they accept and donate electrons.

FIGURE 9.13 (p. 166) shows the sequence of electron carriers in the electron transport chain and the drop in free energy as electrons travel down the chain. Electrons removed from food during glycolysis and the Krebs cycle are transferred by NADH to the first molecule of the electron transport chain. This molecule is a flavoprotein, so named because it has a prosthetic group called flavin mononucleotide (FMN in FIGURE 9.13). In the next redox reaction, the flavoprotein returns to its oxidized form as it passes electrons to an iron-sulfur protein (Fe•S in

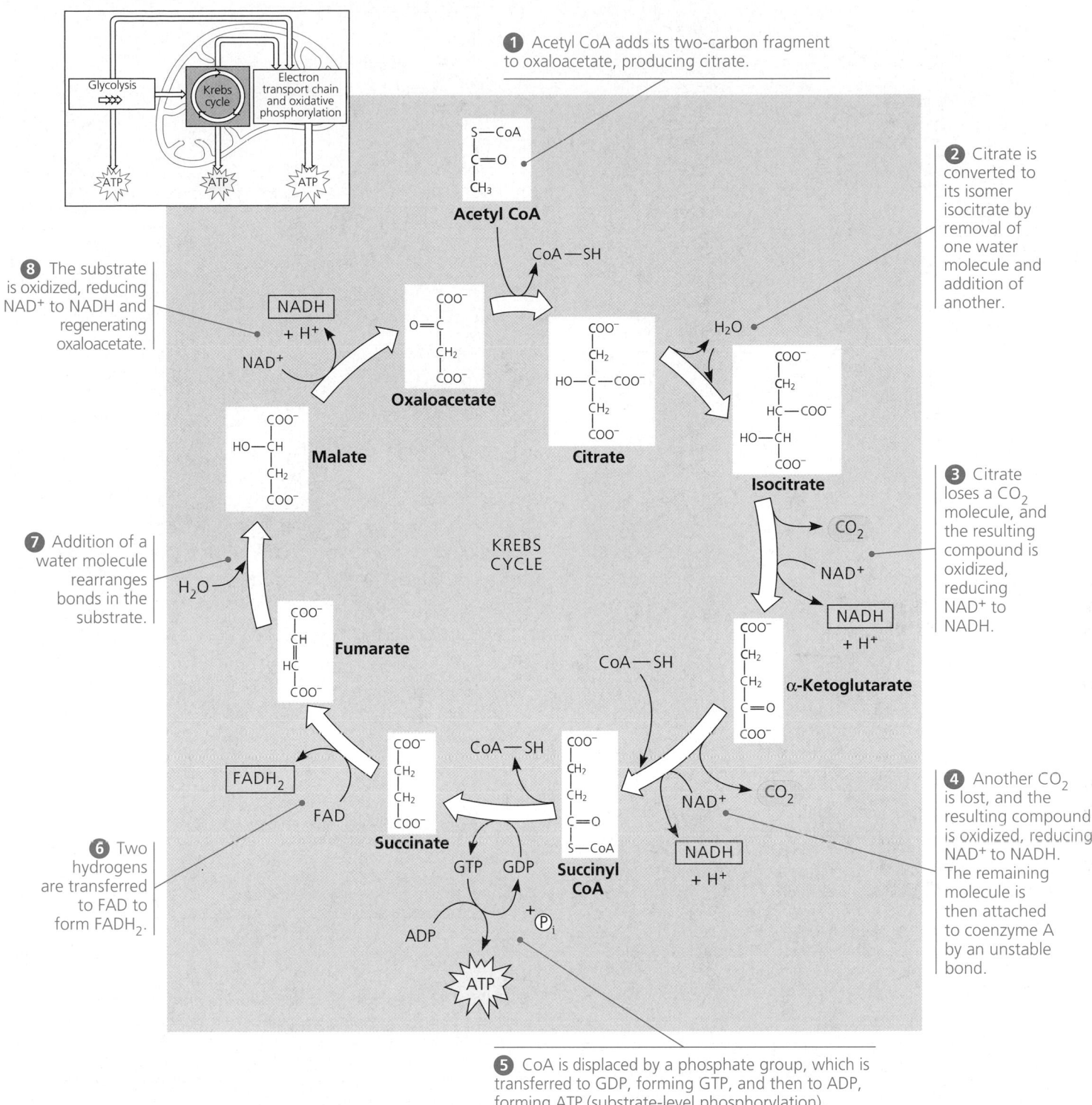

1 Acetyl CoA adds its two-carbon fragment to oxaloacetate, producing citrate.

2 Citrate is converted to its isomer isocitrate by removal of one water molecule and addition of another.

3 Citrate loses a CO_2 molecule, and the resulting compound is oxidized, reducing NAD^+ to NADH.

4 Another CO_2 is lost, and the resulting compound is oxidized, reducing NAD^+ to NADH. The remaining molecule is then attached to coenzyme A by an unstable bond.

5 CoA is displaced by a phosphate group, which is transferred to GDP, forming GTP, and then to ADP, forming ATP (substrate-level phosphorylation).

6 Two hydrogens are transferred to FAD to form $FADH_2$.

7 Addition of a water molecule rearranges bonds in the substrate.

8 The substrate is oxidized, reducing NAD^+ to NADH and regenerating oxaloacetate.

Acetyl CoA

CoA—SH

Oxaloacetate

Malate

Citrate

Isocitrate

H_2O

CO_2

NAD^+

NADH + H^+

α-Ketoglutarate

KREBS CYCLE

Fumarate

$FADH_2$

FAD

Succinate

CoA—SH

CoA—SH

CO_2

NAD^+

NADH + H^+

Succinyl CoA

GTP   GDP

ADP

+ P_i

ATP

NADH + H^+

NAD^+

H_2O

Glycolysis

Krebs cycle

Electron transport chain and oxidative phosphorylation

ATP   ATP   ATP

FIGURE 9.11 A closer look at the Krebs cycle. In the chemical structures, red type traces the fate of the two carbon atoms that enter the cycle via acetyl CoA (step 1), and blue type indicates the two carbons that exit the cycle as CO_2 in steps 3 and 4. Notice that carboxylic acids are represented in their ionized forms, as —COO^-. For example, citrate is the ionized form of citric acid.

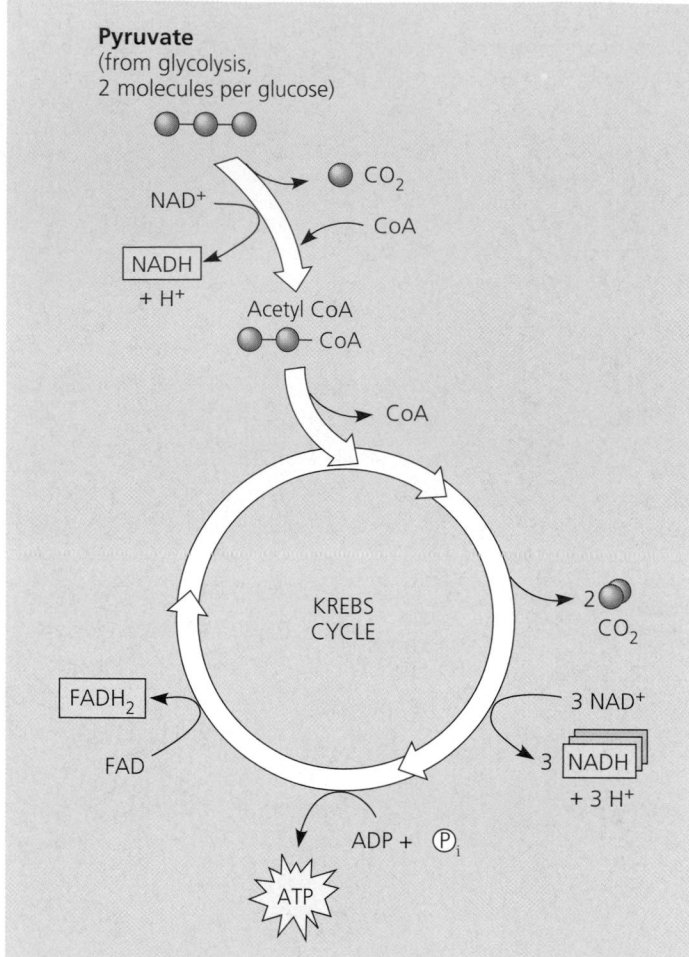

FIGURE 9.12 A summary of the Krebs cycle. The cycle functions as a metabolic furnace that oxidizes organic fuel derived from pyruvate, the product of glycolysis. This diagram summarizes the inputs and outputs as pyruvate is broken down to three molecules of CO_2, and it includes the molecule of CO_2 released during pre–Krebs cycle conversion of pyruvate to acetyl CoA. The cycle generates 1 ATP per turn by substrate phosphorylation, but most of the chemical energy is transferred during the redox reactions to NAD^+ and FAD. The reduced coenzymes, NADH and $FADH_2$, shuttle their cargo of high-energy electrons to the electron transport chain, which uses the energy to synthesize ATP by oxidative phosphorylation. (To calculate the inputs and outputs on a "per-glucose" basis, multiply by 2, because each glucose molecule is split during glycolysis into two pyruvate molecules.)

FIGURE 9.13), one of a family of proteins with both iron and sulfur tightly bound. The iron-sulfur protein then passes the electrons to a compound called ubiquinone (Q in FIGURE 9.13). This electron carrier is a lipid, the only member of the electron transport chain that is not a protein.

Most of the remaining electron carriers between Q and oxygen are proteins called **cytochromes (cyt).** Their prosthetic group, called a heme group, has four organic rings surrounding a single iron atom. It is similar to the iron-containing prosthetic group found in hemoglobin, the red protein of blood that transports oxygen. But the iron of cytochromes transfers electrons, not oxygen. The electron transport chain has several types of cytochromes, each a different protein with a heme

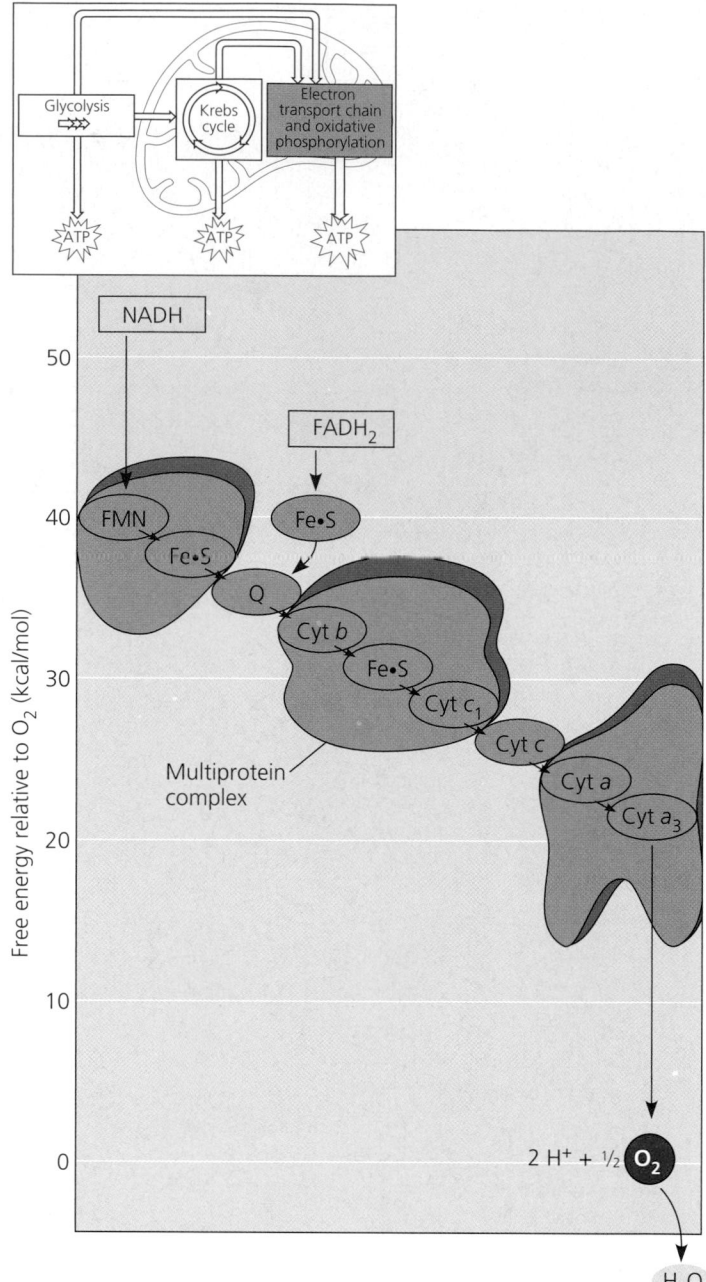

FIGURE 9.13 Free-energy change during electron transport. Each member of the chain oscillates between a reduced state and an oxidized state. A component of the chain becomes reduced when it accepts electrons from its "uphill" neighbor, which has a lower affinity for electrons (is less electronegative). It then returns to its oxidized form as it passes electrons to its "downhill," more electronegative, neighbor. At the bottom of the chain is oxygen, which is *very* electronegative. The overall energy drop for electrons traveling from NADH to oxygen is 53 kcal/mol, but this fall is broken up into a series of smaller steps by the electron transport chain. The various molecules of the electron transport chain, shown as ovals here, are described in the text. Most of these electron carriers are grouped into multiprotein complexes, shown here and in FIGURE 9.15 as irregularly shaped purple blobs.

group. The last cytochrome of the chain, cyt a_3, passes its electrons to oxygen, which also picks up a pair of hydrogen ions from the aqueous solution to form water. (An oxygen atom is represented in FIGURE 9.13 as ½ O_2 to emphasize that the electron transport chain reduces molecular oxygen, O_2, not individual oxygen atoms. For every two NADH molecules, one O_2 molecule is reduced to two molecules of water.)

Another source of electrons for the transport chain is $FADH_2$, the other reduced product of the Krebs cycle. Notice in FIGURE 9.13 that $FADH_2$ adds its electrons to the electron transport chain at a lower energy level than NADH does. Consequently, the electron transport chain provides about one-third less energy for ATP synthesis when the electron donor is $FADH_2$ rather than NADH.

The electron transport chain makes no ATP directly. Its function is to ease the fall of electrons from food to oxygen, breaking a large free-energy drop into a series of smaller steps that release energy in manageable amounts. How does the mitochondrion couple this electron transport and energy release to ATP synthesis? The answer is a mechanism called chemiosmosis.

Chemiosmosis: The Energy-Coupling Mechanism

Populating the inner membrane of the mitochondrion are many copies of a protein complex called **ATP synthase,** the enzyme that actually makes ATP from ADP and inorganic phosphate (FIGURE 9.14). ATP synthase works like an ion pump running in reverse. Recall from Chapter 8 that ion pumps use ATP as an energy source to transport ions against their gradients. In the reverse of that process, ATP synthase uses the energy of an existing ion gradient to power ATP synthesis. The ion gradient that drives oxidative phosphorylation is a proton (hydrogen ion) gradient; that is, the power source for the ATP synthase is a difference in the concentration of H^+ on opposite sides of the inner mitochondrial membrane. (We can also think of this gradient as a difference in pH, since pH is a measure of H^+ concentration.)

How does the mitochondrial membrane generate and maintain an H^+ gradient? That is the function of the electron transport chain, which is shown in its mitochondrial location in FIGURE 9.15 (p. 168). The chain is an energy converter that uses the exergonic flow of electrons to pump H^+ across the membrane, from the matrix into the intermembrane space. The H^+ leaks back across the membrane, diffusing down its gradient. But the ATP synthases are the only patches of the membrane that are freely permeable to H^+. The ions pass through a channel in ATP synthase, and this protein complex uses the exergonic flow of H^+ to drive the oxidative phosphorylation of ADP. Thus, an H^+ gradient across a membrane couples the redox reactions of the electron transport chain to ATP synthesis. This coupling mechanism is called **chemiosmosis** (from the Greek osmos, push), a term that highlights the connection between the chemical reaction that makes ATP

and transport across a membrane. We have previously used the word *osmosis* in discussing water transport, but here we are referring to the flow of H^+ across a membrane.

If you have followed this complex story of chemiosmosis so far, you should have at least two questions. How does the electron transport chain pump hydrogen ions? And how does the ATP synthase use H^+ backflow to make ATP? In answer to the first question, researchers have found that certain members of the electron transport chain accept and release protons (H^+) along with electrons. At certain steps along the chain, electron transfers cause H^+ to be taken up and released back into the surrounding solution. The electron carriers are spatially arranged in the membrane in such a way that H^+ is accepted from the mitochondrial matrix and deposited in the intermembrane space (see FIGURE 9.15 again). The H^+ gradient that results is referred to as a **proton-motive force,** emphasizing the capacity of the gradient to perform work. The force drives H^+ back across the membrane through the specific H^+ channels provided by ATP synthase complexes.

From studying the structure of ATP synthase, scientists have learned how the flow of H^+ through this large enzyme powers ATP generation. ATP synthase is a multisubunit complex with the four main parts illustrated in FIGURE 9.14: a rotor in the inner mitochondrial membrane; a knob that protrudes into the mitochondrial matrix; an internal rod

THE PROCESS OF SCIENCE

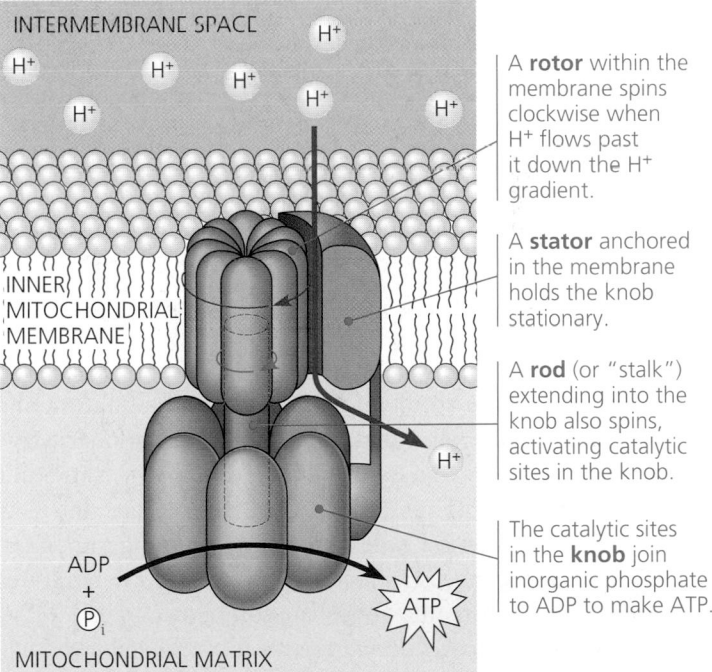

A **rotor** within the membrane spins clockwise when H^+ flows past it down the H^+ gradient.

A **stator** anchored in the membrane holds the knob stationary.

A **rod** (or "stalk") extending into the knob also spins, activating catalytic sites in the knob.

The catalytic sites in the **knob** join inorganic phosphate to ADP to make ATP.

FIGURE 9.14 ATP synthase, a molecular mill. The ATP synthase protein complex functions as a mill, powered by the flow of hydrogen ions. This complex resides in mitochondrial and chloroplast membranes of eukaryotes and in the plasma membranes of prokaryotes. Each of the four parts of ATP synthase consists of a number of polypeptide subunits.

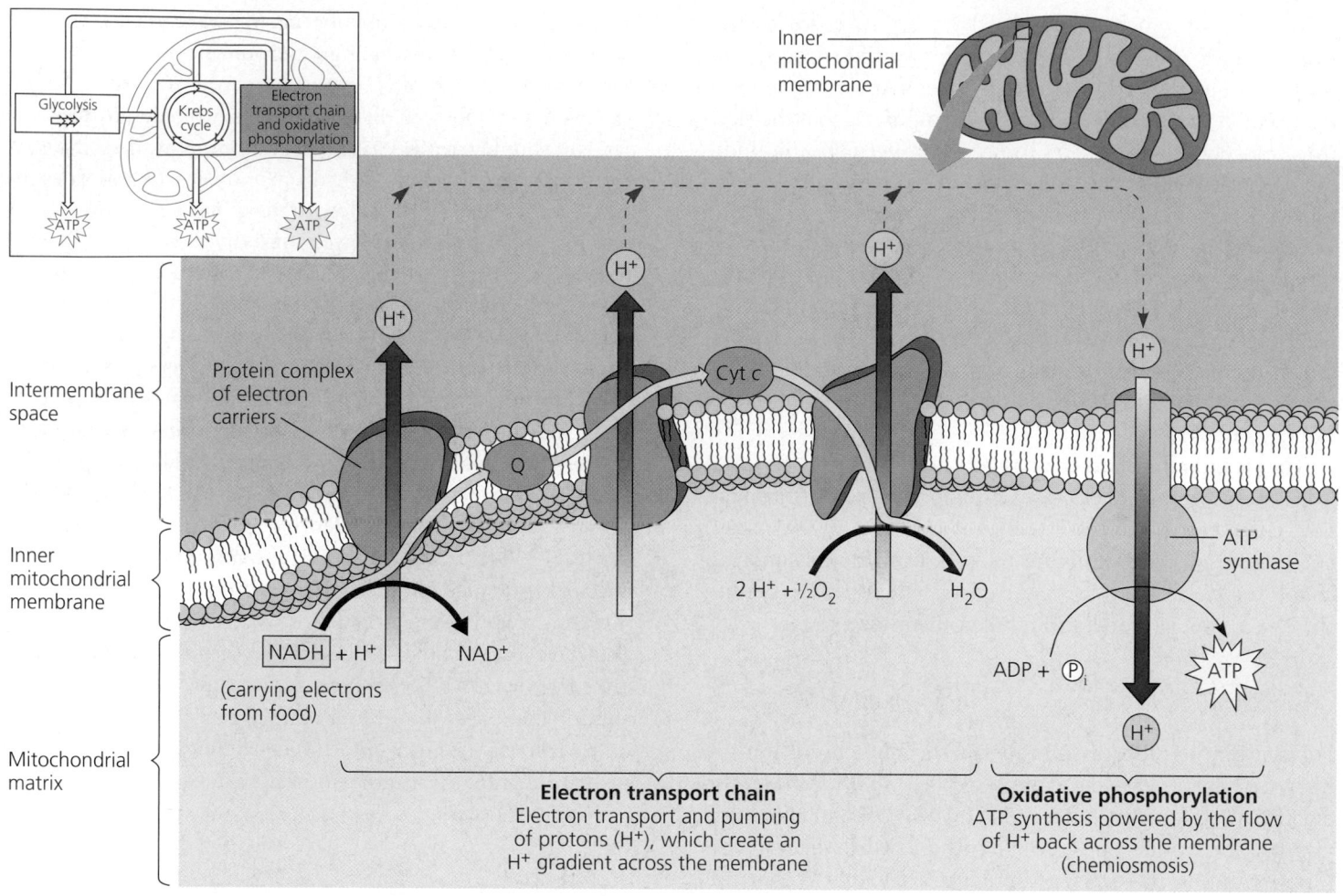

FIGURE 9.15 Chemiosmosis couples the electron transport chain to ATP synthesis. NADH shuttles high-energy electrons extracted from food during glycolysis and the Krebs cycle to an electron transport chain built into the inner mitochondrial membrane. The gold arrow traces the transport of electrons, which finally pass to oxygen at the "downhill" end of the chain to form water. As FIGURE 9.13 showed, most of the electron carriers of the chain are grouped into three complexes, each represented here by a purple blob embedded in the membrane. Two mobile carriers, ubiquinone (Q) and cytochrome c, move rapidly along the membrane, ferrying electrons between the three large complexes. As each complex of the chain accepts and then donates electrons, it pumps hydrogen ions (protons) from the mitochondrial matrix into the intermembrane space. Thus, chemical energy originally harvested from food is transformed into a proton-motive force, a gradient of H^+ across the membrane. The hydrogen ions flow back, down their gradient, through a channel in an ATP synthase, another protein complex built into the membrane. The ATP synthase harnesses the proton-motive force to phosphorylate ADP, forming ATP. (This phosphorylation is called *oxidative* because it is driven by the loss of electrons from food molecules.) The use of an H^+ gradient (proton-motive force) to transfer energy from redox reactions to cellular work (ATP synthesis, in this case) is called chemiosmosis.

extending from the rotor into the knob; and a "stator," anchored next to the rotor, that holds the knob stationary. Hydrogen ions flow down a narrow space between the stator and rotor, causing the rotor and its attached rod to rotate, much as a rushing stream turns a water-wheel. The spinning rod causes conformational changes in the knob, activating catalytic sites where ADP and inorganic phosphate combine to make ATP.

In general terms, *chemiosmosis is an energy-coupling mechanism that uses energy stored in the form of an H^+ gradient across a membrane to drive cellular work.* In mitochondria, the energy for gradient formation comes from exergonic redox reactions, and ATP synthesis is the work performed. But chemiosmosis also occurs elsewhere and in other variations. Chloroplasts use chemiosmosis to generate ATP during photosynthesis; in these organelles, light (rather than chemical energy) drives both electron flow down an electron transport chain and H^+ gradient formation. Prokaryotes, which lack both mitochondria and chloroplasts, generate H^+ gradients across their plasma membranes. They then tap the proton-motive force not only to make ATP but also to pump nutrients and waste products across the membrane and to rotate their flagella.

Experiments with bacteria first led British biochemist Peter Mitchell to propose chemiosmosis as an energy-coupling mechanism in 1961. Nearly two decades later, after many scientists had confirmed the centrality of chemiosmosis in energy conversions within bacteria, mitochondria, and chloroplasts, Mitchell was awarded the Nobel Prize. Chemiosmosis has helped unify the study of bioenergetics.

Cellular respiration generates many ATP molecules for each sugar molecule it oxidizes: *a review*

Now that we have looked more closely at the key processes of cellular respiration, let's return to its overall function: harvesting the energy of food for ATP synthesis.

During respiration, most energy flows in this sequence: Glucose $\rightarrow$ NADH $\rightarrow$ electron transport chain $\rightarrow$ proton-motive force $\rightarrow$ ATP. We can do some bookkeeping to calculate the ATP profit when cellular respiration oxidizes a molecule of glucose to six molecules of carbon dioxide. The three main departments of this metabolic enterprise are glycolysis, the Krebs cycle, and the electron transport chain, which drives oxidative phosphorylation. FIGURE 9.16 gives a detailed accounting of the ATP yield per glucose molecule oxidized. The tally adds the few molecules of ATP produced directly by substrate-level phosphorylation during glycolysis and the Krebs cycle to the many more molecules of ATP generated by oxidative phosphorylation. Each NADH that transfers a pair of electrons from food to the electron transport chain contributes enough to the proton-motive force to generate a maximum of about three ATPs. (The average ATP yield per NADH is probably between two and three; we are rounding off to three here to simplify the bookkeeping.) The Krebs cycle also supplies electrons to the electron transport chain via FADH$_2$, but each molecule of this electron carrier is worth a maximum of only about two molecules of ATP.

In some eukaryotic cells, this lower ATP yield per electron pair also applies to the NADH produced by glycolysis in the cytosol. The mitochondrial inner membrane is impermeable to NADH, so NADH in the cytosol is segregated from the machinery of oxidative phosphorylation. The two electrons of NADH captured in glycolysis must be conveyed into the mitochondrion by one of several electron shuttle systems. Depending on which shuttle is operating, the electrons are passed either to NAD$^+$ or to FAD. If the electrons are passed to FAD, only about 2 ATP can result from each cytosolic NADH$_2$. If passed to mitochondrial NAD$^+$, the yield is closer to 3 ATP.

Assuming that the more energy-efficient type of shuttle is active, we can add a maximum of 34 ATP produced by oxidative phosphorylation to the net of 4 ATP from substrate-level phosphorylation to give a bottom line of 38 ATP. This is only an estimate of the maximum ATP yield from a glucose molecule and is probably somewhat high. One variable that reduces ATP yield is the use of the proton-motive force generated by the redox reactions of respiration to drive other kinds of work. For example, the proton-motive force powers the mitochondrion's uptake of pyruvate from the cytosol.

We can now make a rough estimate of the efficiency of respiration—that is, the percentage of chemical energy stored in glucose that has been restocked in ATP. Recall that the complete

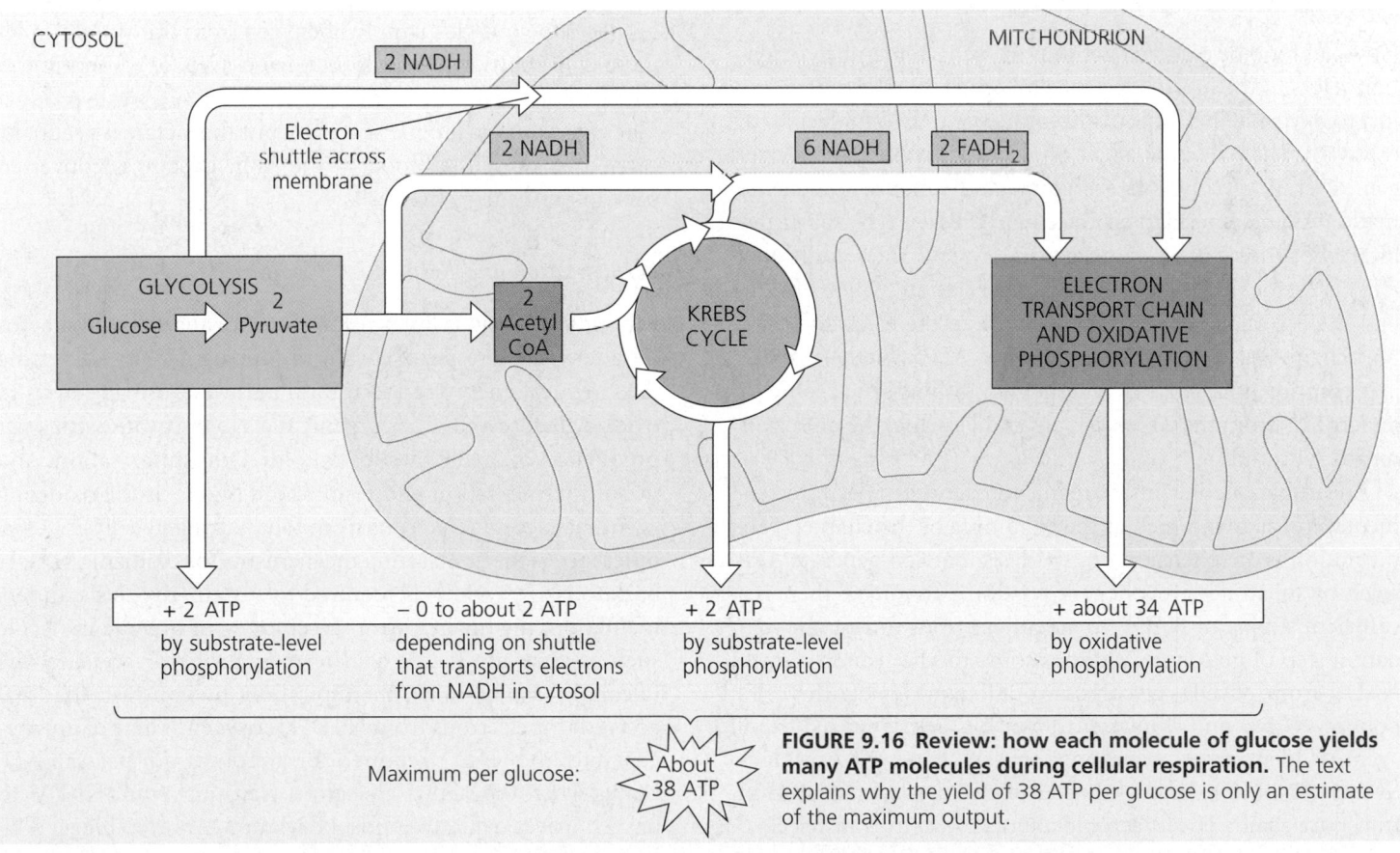

FIGURE 9.16 Review: how each molecule of glucose yields many ATP molecules during cellular respiration. The text explains why the yield of 38 ATP per glucose is only an estimate of the maximum output.

oxidation of a mole of glucose releases 686 kcal of energy ($\Delta G = -686$ kcal/mol). Phosphorylation of ADP to form ATP stores at least 7.3 kcal per mole of ATP (p. 94 explains why this number is probably higher under cellular conditions). Therefore, the efficiency of respiration is 7.3 times 38 (maximum ATP yield per glucose) divided by 686, or about 40%. The rest of the stored energy is lost as heat. We use some of this heat to maintain our relatively high body temperature (37°C), and we dissipate the rest through sweating and other cooling mechanisms. Cellular respiration is remarkably efficient in its energy conversion. By comparison, the most efficient automobile converts only about 25% of the energy stored in gasoline to move the car.

RELATED METABOLIC PROCESSES

Because most of the ATP generated by cellular respiration is the work of oxidative phosphorylation, our estimate of ATP yield from respiration is contingent upon an adequate supply of oxygen to the cell. Without the electronegative oxygen to pull electrons down the transport chain, oxidative phosphorylation ceases. However, fermentation provides a mechanism by which some cells can oxidize organic fuel and generate ATP *without* the help of oxygen.

Fermentation enables some cells to produce ATP without the help of oxygen

How can food be oxidized without oxygen? Remember, oxidation refers to the loss of electrons to *any* electron acceptor, not just to oxygen. Glycolysis oxidizes glucose to two molecules of pyruvate. The oxidizing agent of glycolysis is NAD^+, *not* oxygen. Overall, glycolysis is exergonic, and some of the energy made available is used to produce two ATPs (net) by substrate-level phosphorylation. If oxygen *is* present, then additional ATP is made by oxidative phosphorylation when NADH passes electrons removed from glucose to the electron transport chain. But glycolysis generates two ATPs whether oxygen is present or not—that is, whether conditions are **aerobic** or **anaerobic** (from the Greek *aer*, air, and *bios*, life; the prefix *an-* means "without").

Anaerobic catabolism of organic nutrients can occur by fermentation, as mentioned at the beginning of the chapter. Fermentation is an extension of glycolysis that can generate ATP solely by substrate-level phosphorylation—as long as there is a sufficient supply of NAD^+ to accept electrons during the oxidation step of glycolysis. Without some mechanism to recycle NAD^+ from NADH, glycolysis would soon deplete the cell's pool of NAD^+ and shut itself down for lack of an oxidizing agent. Under aerobic conditions, NAD^+ is recycled productively from NADH by the transfer of electrons to the electron transport chain. The anaerobic alternative is to transfer electrons from NADH to pyruvate, the end product of glycolysis.

Fermentation consists of glycolysis plus reactions that regenerate NAD^+ by transferring electrons from NADH to pyruvate or derivatives of pyruvate. The NAD^+ can then be reused to oxidize sugar by glycolysis, which nets two molecules of ATP by substrate-level phosphorylation. There are many types of fermentation, differing in the waste products formed from pyruvate. Two common types are alcohol fermentation and lactic acid fermentation.

In **alcohol fermentation** (FIGURE 9.17a), pyruvate is converted to ethanol (ethyl alcohol) in two steps. The first step releases carbon dioxide from the pyruvate, which is converted to the two-carbon compound acetaldehyde. In the second step, acetaldehyde is reduced by NADH to ethanol. This regenerates the supply of NAD^+ needed for glycolysis. Alcohol fermentation by yeast, a fungus, is used in brewing and winemaking. Many bacteria also carry out alcohol fermentation under anaerobic conditions.

During **lactic acid fermentation** (FIGURE 9.17b), pyruvate is reduced directly by NADH to form lactate as a waste product, with no release of CO_2. (Lactate is the ionized form of lactic acid.) Lactic acid fermentation by certain fungi and bacteria is used in the dairy industry to make cheese and yogurt. Acetone and methanol (methyl alcohol) are among the by-products of other types of microbial fermentation that are commercially important.

Human muscle cells make ATP by lactic acid fermentation when oxygen is scarce. This occurs during the early stages of strenuous exercise, when sugar catabolism for ATP production outpaces the muscle's supply of oxygen from the blood. Under these conditions, the cells switch from aerobic respiration to fermentation. The lactate that accumulates as a waste product may cause muscle fatigue and pain, but the lactate is gradually carried away by the blood to the liver. Lactate is converted back to pyruvate by liver cells.

Fermentation and Respiration Compared

Fermentation and cellular respiration are anaerobic and aerobic alternatives, respectively, for producing ATP by harvesting the chemical energy of food. Both pathways use glycolysis to oxidize glucose and other organic fuels to pyruvate, with a net production of 2 ATP by substrate-level phosphorylation. And in both fermentation and respiration, NAD^+ is the oxidizing agent that accepts electrons from food during glycolysis. A key difference is the contrasting mechanisms for oxidizing NADH back to NAD^+, which is required to sustain glycolysis. In fermentation, the final electron acceptor is an organic molecule such as pyruvate (lactic acid fermentation) or acetaldehyde (alcohol fermentation). In respiration, by contrast, the final acceptor for electrons from NADH is oxygen. This not only regenerates the NAD^+ required for glycolysis but pays an ATP bonus when the stepwise electron transport from NADH to oxygen drives oxidative phosphorylation. An even bigger ATP payoff comes from the oxidation of pyruvate in the Krebs

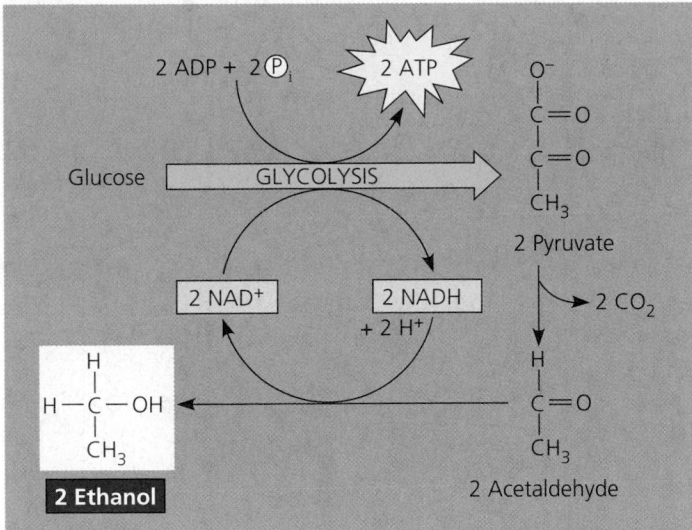

(a) Alcohol fermentation

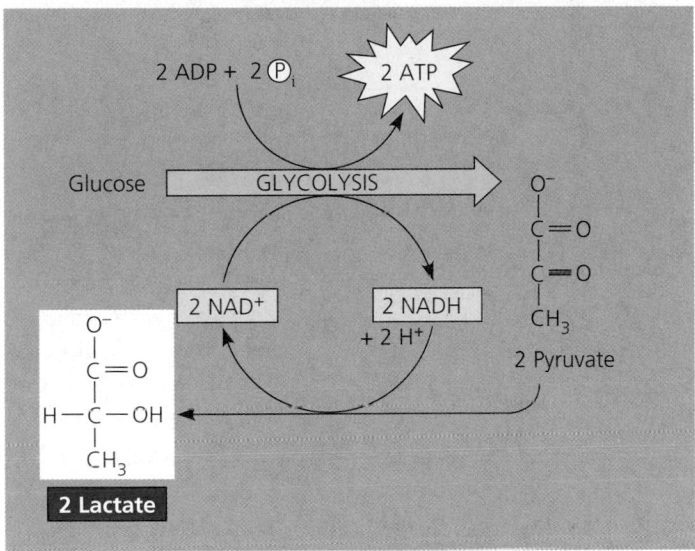

(b) Lactic acid fermentation

FIGURE 9.17 Fermentation. In the absence of oxygen, many cells use fermentation to produce ATP by substrate-level phosphorylation. Pyruvate, the end product of glycolysis, serves as an electron acceptor for oxidizing NADH back to NAD^+, which can then be reused in glycolysis. Two of the common waste products formed from fermentation are **(a)** ethanol and **(b)** lactate, the ionized form of lactic acid.

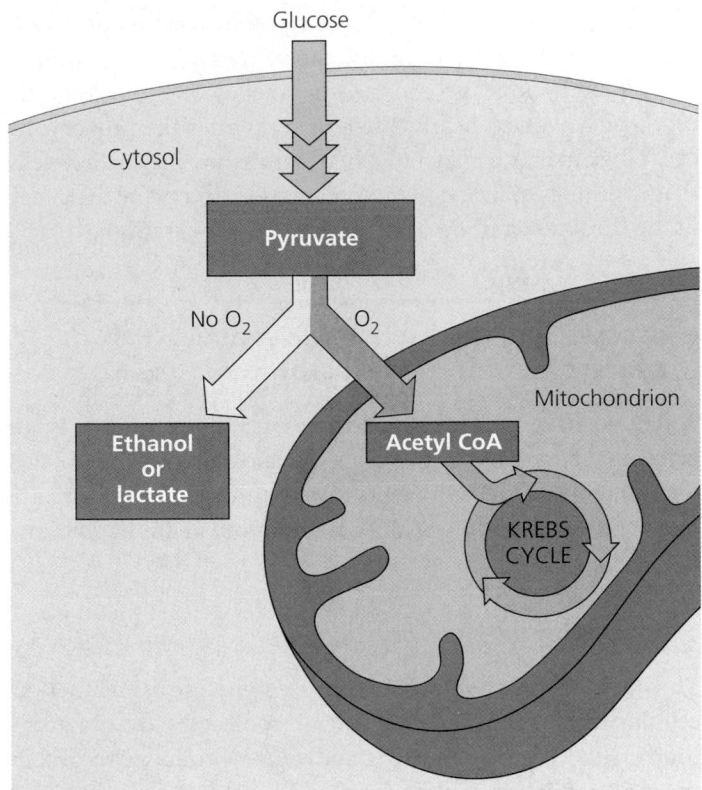

FIGURE 9.18 Pyruvate as a key juncture in catabolism. Glycolysis is common to fermentation and respiration. The end product of glycolysis, pyruvate, represents a fork in the catabolic pathways of glucose oxidation. In a cell capable of both respiration and fermentation, pyruvate is committed to one of those two pathways, usually depending on whether or not oxygen is present.

road that leads to two alternative catabolic routes (FIGURE 9.18). Under aerobic conditions, pyruvate can be converted to acetyl CoA, and oxidation continues in the Krebs cycle. Under anaerobic conditions, pyruvate is diverted from the Krebs cycle, serving instead as an electron acceptor to recycle NAD^+. To make the same amount of ATP, a facultative anaerobe would have to consume sugar at a much faster rate when fermenting than when respiring.

The Evolutionary Significance of Glycolysis

The role of glycolysis in both fermentation and respiration has an evolutionary basis. Ancient prokaryotes probably used glycolysis to make ATP long before oxygen was present in Earth's atmosphere. The oldest known fossils of bacteria date back over 3.5 billion years, but appreciable quantities of oxygen probably did not begin to accumulate in the atmosphere until about 2.7 billion years ago. (According to fossil evidence, the cyanobacteria that produce O_2 as a by-product of photosynthesis had evolved by then.) Therefore, the first prokaryotes may have generated ATP exclusively from glycolysis, which does not require oxygen. In addition, glycolysis is the most widespread metabolic pathway, which suggests that it evolved

cycle, which is unique to respiration. Without oxygen, the energy still stored in pyruvate is unavailable to the cell. Thus, cellular respiration harvests much more energy from each sugar molecule than fermentation can. In fact, respiration yields as much as 19 times more ATP per glucose molecule than does fermentation—38 ATP for respiration, compared to 2 ATP produced by substrate-level phosphorylation in fermentation.

Some organisms, including yeasts and many bacteria, can make enough ATP to survive using either fermentation or respiration. Such species are called **facultative anaerobes.** On the cellular level, our muscle cells behave as facultative anaerobes. In a facultative anaerobe, pyruvate is a fork in the metabolic

very early in the history of life. The cytosolic location of glycolysis also implies great antiquity; the pathway does not require any of the membrane-enclosed organelles of the eukaryotic cell, which evolved nearly 2 billion years after the prokaryotic cell. Glycolysis is a metabolic heirloom from the earliest cells that continues to function in fermentation and as the first stage in the breakdown of organic molecules by respiration.

Glycolysis and the Krebs cycle connect to many other metabolic pathways

So far, we have treated the oxidative breakdown of glucose in isolation from the cell's overall metabolic economy. In this section, you will learn that glycolysis and the Krebs cycle are major intersections of various catabolic and anabolic (biosynthetic) pathways.

The Versatility of Catabolism

Throughout this chapter, we have used glucose as the fuel for cellular respiration. But free glucose molecules are not common in the diets of humans and other animals. We obtain most of our calories in the form of fats, proteins, sucrose and other disaccharides, and starch, a polysaccharide. All these food molecules can be used by cellular respiration to make ATP (FIGURE 9.19).

Glycolysis can accept a wide range of carbohydrates for catabolism. In the digestive tract, starch is hydrolyzed to glucose, which can then be broken down in the cells by glycolysis and the Krebs cycle. Similarly, glycogen, the polysaccharide that humans and many other animals store in their liver and muscle cells, can be hydrolyzed to glucose between meals as fuel for respiration. The digestion of disaccharides, including sucrose, provides glucose and other monosaccharides as fuel for respiration.

Proteins can also be used for fuel, but first they must be digested to their constituent amino acids. Many of the amino acids, of course, are used by the organism to build new proteins. Amino acids present in excess are converted by enzymes to intermediates of glycolysis and the Krebs cycle. Before amino acids can feed into glycolysis or the Krebs cycle, their amino groups must be removed, a process called deamination. The nitrogenous refuse is excreted from the animal in the form of ammonia, urea, or other waste products.

Catabolism can also harvest energy stored in fats obtained either from food or from storage cells in the body. After fats are digested, the glycerol is converted to glyceraldehyde phosphate, an intermediate of glycolysis. Most of the energy of a fat is stored in the fatty acids. A metabolic sequence called **beta oxidation** breaks the fatty acids down to two-carbon fragments, which enter the Krebs cycle as acetyl CoA. Fats make excellent fuel. A gram of fat oxidized by respiration produces more than twice as much ATP as a gram of carbohydrate. Unfortunately, this also means that a dieter must be patient while

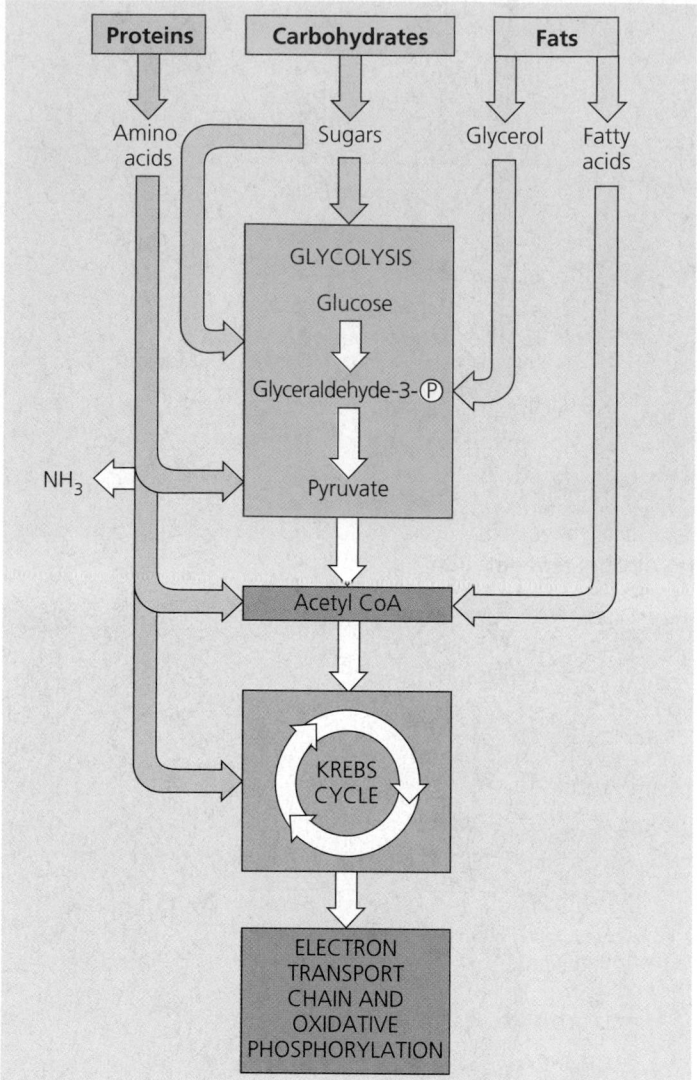

FIGURE 9.19 The catabolism of various food molecules.
Carbohydrates, fats, and proteins can all be used as fuel for cellular respiration. Monomers of these food molecules enter glycolysis or the Krebs cycle at various points. Glycolysis and the Krebs cycle are catabolic funnels through which electrons from all kinds of food molecules flow on their exergonic fall to oxygen.

using fat stored in the body, because so many calories are stockpiled in each gram of fat.

Biosynthesis (Anabolic Pathways)

Cells need substance as well as energy. Not all the organic molecules of food are destined to be oxidized as fuel to make ATP. In addition to calories, food must also provide the carbon skeletons that cells require to make their own molecules. Some organic monomers obtained from digestion can be used directly. For example, as previously mentioned, amino acids from the hydrolysis of proteins in food can be incorporated into the organism's own proteins. Often, however, the body needs specific molecules that are not present as such in food. Compounds formed as intermediates of glycolysis and the

Krebs cycle can be diverted into anabolic pathways as precursors from which the cell can synthesize the molecules it requires. For example, humans can make about half of the 20 amino acids in proteins by modifying compounds siphoned away from the Krebs cycle. Also, glucose can be made from pyruvate, and fatty acids can be synthesized from acetyl CoA. Of course, these anabolic, or biosynthetic, pathways do not generate ATP, but instead consume it.

In addition, glycolysis and the Krebs cycle function as metabolic interchanges that enable our cells to convert some kinds of molecules to others as we need them. For instance, carbohydrates and proteins can be converted to fats through intermediates of glycolysis and the Krebs cycle. If we eat more food than we need, we store fat even if our diet is fat-free. Metabolism is remarkably versatile and adaptable.

Feedback mechanisms control cellular respiration

Basic principles of supply and demand regulate the metabolic economy. The cell does not waste energy making more of a particular substance than it needs. If there is a glut of a certain amino acid, for example, the anabolic pathway that synthesizes that amino acid from an intermediate of the Krebs cycle is switched off. The most common mechanism for this control is feedback inhibition: The end product of the anabolic pathway inhibits the enzyme that catalyzes an early step of the pathway (see FIGURE 6.18). This prevents the needless diversion of key metabolic intermediates from uses that are more urgent.

The cell also controls its catabolism. If the cell is working hard and its ATP concentration begins to drop, respiration speeds up. When there is plenty of ATP to meet demand, respiration slows down, sparing valuable organic molecules for other functions. Again, control is based mainly on regulating the activity of enzymes at strategic points in the catabolic pathway. One important switch is phosphofructokinase, the enzyme that catalyzes step 3 of glycolysis (see FIGURE 9.9). That is the earliest step that commits substrate irreversibly to the glycolytic pathway. By controlling the rate of this step, the cell can speed up or slow down the entire catabolic process; phosphofructokinase is thus the pacemaker of respiration (FIGURE 9.20).

An allosteric enzyme with receptor sites for specific inhibitors and activators, phosphofructokinase is inhibited by ATP and stimulated by AMP, which the cell derives from ADP (see p. 101). As ATP accumulates, inhibition of the enzyme slows down glycolysis. The enzyme becomes active again as cellular work converts ATP to ADP (and AMP) faster than ATP is being regenerated. Phosphofructokinase is also sensitive to citrate, the first product of the Krebs cycle. If citrate accumulates in mitochondria, some of it passes into the cytosol and inhibits phosphofructokinase. This mechanism helps synchronize the rates of glycolysis and the Krebs cycle. As citrate

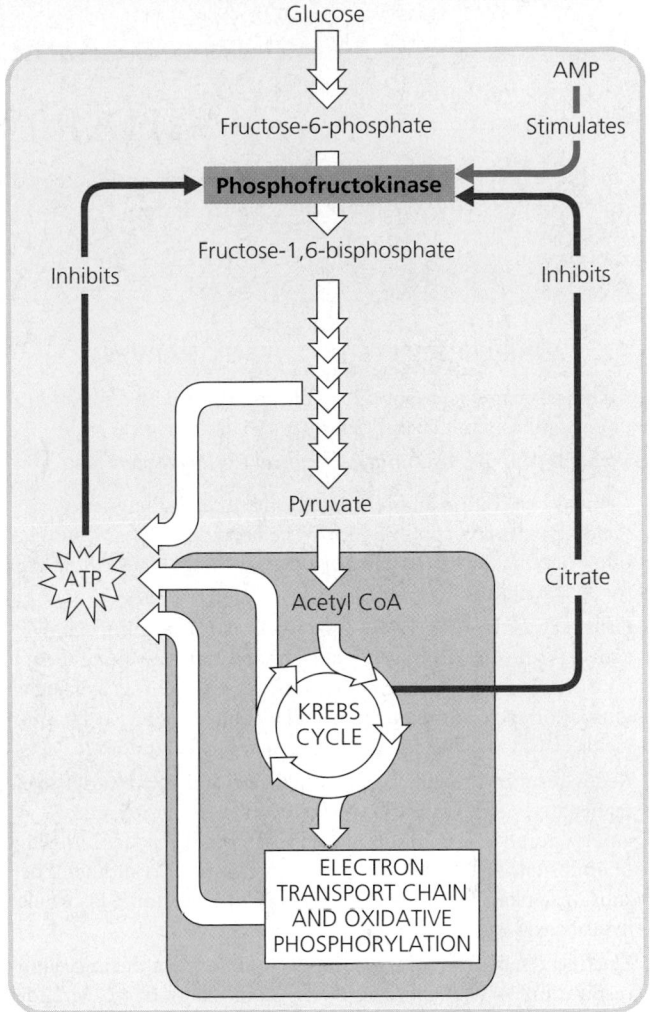

FIGURE 9.20 The control of cellular respiration. Allosteric enzymes at certain points in the respiratory pathway respond to inhibitors and activators that help set the pace of glycolysis and the Krebs cycle. Phosphofructokinase, the enzyme that catalyzes step 3 of glycolysis, is one such enzyme. It is stimulated by AMP (derived from ADP), but it is inhibited by ATP and by citrate. This feedback regulation adjusts the rate of respiration as the cell's catabolic and anabolic demands change.

accumulates, glycolysis slows down, and the supply of acetate to the Krebs cycle decreases. If citrate consumption increases, either because of a demand for more ATP or because anabolic pathways are draining off intermediates of the Krebs cycle, glycolysis accelerates and meets the demand. Metabolic balance is augmented by the control of other enzymes at other key locations in glycolysis and the Krebs cycle. Cells are thrifty, expedient, and responsive in their metabolism.

■　　　■　　　■

Examine FIGURE 9.1 again to put cellular respiration into the broader context of energy flow and chemical cycling in ecosystems. The energy that keeps us alive is *released*, but not *produced*, by cellular respiration. We are tapping energy that was stored in food by photosynthesis. In the next chapter, you will learn how photosynthesis captures light and converts it to chemical energy.

CHAPTER 9 REVIEW

Go to the Campbell Biology website (www.campbellbiology.com) to explore an interactive version of the Chapter Review.

Summary of Key Concepts

THE PRINCIPLES OF ENERGY HARVEST

■ Chemical elements important to life are recycled by respiration and photosynthesis, but energy is not (p. 156, FIGURE 9.1).
Web/CD Activity 9A: *Build a Chemical Cycling System*

■ **Cellular respiration and fermentation are catabolic, energy-yielding pathways (pp. 155–156)** The breakdown of glucose and other organic fuels to simpler products is exergonic, yielding energy for ATP synthesis.

■ **Cells recycle the ATP they use for work (p. 156, FIGURE 9.2)** ATP transfers phosphate groups to various substrates, priming them to do work. To keep working, a cell must regenerate ATP. Starting with glucose or another organic fuel, and using O_2, cellular respiration yields H_2O, CO_2, and energy in the form of ATP and heat.

■ **Redox reactions release energy when electrons move closer to electronegative atoms (pp. 156–158, FIGURE 9.3)** The cell taps the energy stored in food molecules through redox reactions, in which one substance partially or totally shifts electrons to another. The substance receiving electrons is reduced; the substance losing electrons is oxidized.

■ **Electrons "fall" from organic molecules to oxygen during cellular respiration (p. 158)** Glucose ($C_6H_{12}O_6$) is oxidized to CO_2, and O_2 is reduced to H_2O. Electrons lose potential energy during their transfer from organic compounds to oxygen, and this energy drives ATP synthesis.

■ **The "fall" of electrons during respiration is stepwise, via NAD^+ and an electron transport chain (pp. 158–159, FIGURE 9.5)** Electrons from food are usually passed to NAD^+, reducing it to NADH. NADH passes the electrons to an electron transport chain, which conducts them to O_2 in energy-releasing steps. The energy released is used to make ATP by oxidative phosphorylation.

THE PROCESS OF CELLULAR RESPIRATION

■ **Respiration involves glycolysis, the Krebs cycle, and electron transport: an overview (pp. 160–161, FIGURE 9.6)** Glycolysis and the Krebs cycle supply electrons (via NADH) to the transport chain, which drives oxidative phosphorylation. Glycolysis occurs in the cytosol, the Krebs cycle in the mitochondrial matrix. The electron transport chain is built into the inner mitochondrial membrane.
Web/CD Activity 9B: *Overview of Cellular Respiration*

■ **Glycolysis harvests chemical energy by oxidizing glucose to pyruvate: a closer look (p. 161, FIGURES 9.8, 9.9)** Glycolysis nets 2 ATP, produced by substrate-level phosphorylation, and 2 NADH.
Web/CD Activity 9C: *Glycolysis*

■ **The Krebs cycle completes the energy-yielding oxidation of organic molecules: a closer look (pp. 161–166, FIGURES 9.11, 9.12)** The conversion of pyruvate to acetyl CoA links glycolysis to the Krebs cycle. The two-carbon acetate of acetyl CoA joins the four-carbon oxaloacetate to form the six-carbon citrate, which is degraded back

to oxaloacetate. The cycle releases CO_2, forms 1 ATP by substrate-level phosphorylation, and passes electrons to 3 NAD^+ and 1 FAD.
Web/CD Activity 9D: *The Krebs Cycle*

■ **The inner mitochondrial membrane couples electron transport to ATP synthesis: a closer look (pp. 164–168, FIGURE 9.15)** Most of the ATP made in cellular respiration is produced by oxidative phosphorylation when NADH and $FADH_2$ donate electrons to the series of electron carriers in the electron transport chain. At the end of the chain, electrons are passed to O_2, reducing it to H_2O. Electron transport is coupled to ATP synthesis by chemiosmosis. At certain steps along the chain, electron transfer causes electron-carrying protein complexes to move H^+ from the matrix to the intermembrane space, storing energy as a proton-motive force (H^+ gradient). As H^+ diffuses back into the matrix through ATP synthase, its exergonic passage drives the endergonic phosphorylation of ADP.
Web/CD Activity 9E: *Electron Transport*

■ **Cellular respiration generates many ATP molecules for each sugar molecule it oxidizes: a review (pp. 169–170, FIGURE 9.16)** The oxidation of glucose to CO_2 produces a maximum of about 38 ATP.
Biology Labs On-Line: *MitochondriaLab*

Web/CD Case Study in the Process of Science: *How Is the Rate of Cellular Respiration Measured?*

RELATED METABOLIC PROCESSES

■ **Fermentation enables some cells to produce ATP without the help of oxygen (pp. 170–172, FIGURES 9.17, 9.18)** Fermentation is anaerobic catabolism of organic nutrients. It yields ATP from glycolysis. The electrons from NADH made in glycolysis are passed to pyruvate, restoring the NAD^+ required to sustain glycolysis. Yeasts and certain bacteria are facultative anaerobes, capable of making ATP by either aerobic respiration or fermentation. Of the two pathways, respiration is the more efficient in terms of ATP yield per glucose. Glycolysis occurs in nearly all organisms and probably evolved in ancient prokaryotes before there was O_2 in the atmosphere.
Web/CD Activity 9F: *Fermentation*

■ **Glycolysis and the Krebs cycle connect to many other metabolic pathways (pp. 172–173, FIGURE 9.19)** These catabolic pathways combine to funnel electrons from all kinds of food molecules into cellular respiration. Carbon skeletons for anabolism (biosynthesis) come directly from digestion or from intermediates of glycolysis and the Krebs cycle.

■ **Feedback mechanisms control cellular respiration (p. 173, FIGURE 9.20)** Cellular respiration is controlled by allosteric enzymes at key points in glycolysis and the Krebs cycle. This helps the cell strike a moment-to-moment balance between catabolism and anabolism.

Self-Quiz

1. What is the reducing agent in the following reaction?

$$\text{Pyruvate} + \text{NADH} + \text{H}^+ \longrightarrow \text{Lactate} + \text{NAD}^+$$

a. oxygen   b. NADH   c. NAD^+ d. lactate   e. pyruvate

2. The *immediate* energy source that drives ATP synthesis during oxidative phosphorylation is
   a. the oxidation of glucose and other organic compounds.
   b. the flow of electrons down the electron transport chain.
   c. the affinity of oxygen for electrons.
   d. a difference of H^+ concentration on opposite sides of the inner mitochondrial membrane.
   e. the transfer of phosphate from Krebs cycle intermediates to ADP.

3. Which metabolic pathway is common to both fermentation and cellular respiration?
   a. the Krebs cycle
   b. the electron transport chain
   c. glycolysis
   d. synthesis of acetyl CoA from pyruvate
   e. reduction of pyruvate to lactate

4. In mitochondria, exergonic redox reactions
   a. are the source of energy driving prokaryotic ATP synthesis.
   b. are directly coupled to substrate-level phosphorylation.
   c. provide the energy to establish the proton gradient.
   d. reduce carbon atoms to carbon dioxide.
   e. are coupled via phosphorylated intermediates to endergonic processes.

5. The final electron acceptor of the electron transport chain that functions in oxidative phosphorylation is
   a. oxygen   b. water   c. NAD^+ d. pyruvate   e. ADP

6. When electrons flow along the electron transport chains of mitochondria, which of the following changes occurs? (Explain your answer.)
   a. The pH of the matrix increases.
   b. ATP synthase pumps protons by active transport.
   c. The electrons gain free energy.
   d. The cytochromes of the chain phosphorylate ADP to form ATP.
   e. NAD^+ is oxidized.

7. In the presence of a metabolic poison that specifically and completely inhibits the function of mitochondrial ATP synthase, which of the following would you expect? (Explain your answer.)
   a. a decrease in the pH difference across the inner mitochondrial membrane
   b. an increase in the pH difference across the inner mitochondrial membrane
   c. increased synthesis of ATP
   d. oxygen consumption to cease
   e. proton pumping by the electron transport chain to cease

8. Cells do not catabolize carbon dioxide because
   a. its double bonds are too stable to be broken.
   b. CO_2 has fewer bonding electrons than other organic compounds.
   c. the carbon atom is already completely reduced.
   d. most of the available electron energy was released by the time the CO_2 was formed.
   e. the molecule has too few atoms.

9. Which of the following is a true distinction between fermentation and cellular respiration?
   a. Only respiration oxidizes glucose.
   b. NADH is oxidized by the electron transport chain only in respiration.
   c. Fermentation, but not respiration, is an example of a catabolic pathway.
   d. Substrate-level phosphorylation is unique to fermentation.
   e. NAD^+ functions as an oxidizing agent only in respiration.

10. Most CO_2 from catabolism is released during
    a. glycolysis.
    b. the Krebs cycle.
    c. lactate fermentation.
    d. electron transport.
    e. oxidative phosphorylation.

11. What chemical characteristic of oxygen accounts for its function in cellular respiration?

12. In the following redox reaction, which compound is oxidized and which is reduced?

$$C_4H_6O_5 + NAD^+ \longrightarrow C_4H_4O_5 + NADH + H^+$$

13. For each glucose molecule processed, what are the net molecular products of glycolysis?

14. What effect would an absence of O_2 have on the process shown in FIGURE 9.15?

15. A glucose-fed yeast cell is moved from an aerobic environment to an anaerobic one. If the cell continues to generate ATP at the same rate, how will its rate of glucose consumption compare to consumption in the aerobic environment?

Go to the website or CD-ROM for more quiz questions.

Evolution Connection

ATP synthase enzymes are found in the prokaryotic plasma membrane and in mitochondria and chloroplasts. What does this suggest about the evolutionary relationship of these eukaryotic organelles to prokaryotes? How might the amino acid sequences of the ATP synthases from the different sources support or refute your hypothesis?

The Process of Science

In the 1940s, some physicians prescribed low doses of a drug called dinitrophenol (DNP) to help patients lose weight. This unsafe method was abandoned after a few patients died. DNP uncouples the chemiosmotic machinery by making the lipid bilayer of the inner mitochondrial membrane leaky to H^+. Explain how this causes weight loss.

Learn how to use a respirometer in the Case Study in the Process of Science, available on the website and CD-ROM. Also link to *MitochondriaLab* in Biology Labs On-Line to simulate some of the pioneering experiments that led to our current understanding of cellular respiration.

Science, Technology, and Society

Nearly all human societies use fermentation to produce alcoholic drinks such as beer and wine. The practice dates back to the earliest days of agriculture. How do you suppose this use of fermentation was first discovered? Why did wine prove to be a more useful beverage, especially to a preindustrial culture, than the grape juice from which it was made?

Answers: 1. b; **2.** d; **3.** c; **4.** c; **5.** a; **6.** a; **7.** b; **8.** d; **9.** b; **10.** b; **11.** Compared to other elements, oxygen is very electronegative, meaning that it is very powerful in removing electrons from other elements. **12.** $C_4H_6O_5$ is oxidized and NAD^+ is reduced. **13.** Two molecules of pyruvate, two molecules of ATP, and two molecules of NADH. **14.** There would be no production of ATP by oxidative phosphorylation. Without oxygen to "pull" electrons down the electron transport chain, H^+ would not be pumped into the mitochondrion's intermembrane space and chemiosmosis would not occur. **15.** The cell must consume glucose at a rate about $19\times$ the consumption rate in the aerobic environment (2 ATP by fermentation vs. 38 ATP by cellular respiration).

CHAPTER 10

PHOTOSYNTHESIS

Life on Earth *is solar powered. The chloroplasts of plants capture light energy that has traveled 160 million kilometers from the sun and convert it to chemical energy stored in sugar and other organic molecules. The conversion process is called* **photosynthesis.** *In this chapter, you will learn how photosynthesis works. We begin by placing photosynthesis in its ecological context.*

PHOTOSYNTHESIS IN NATURE

Plants and other autotrophs are the producers of the biosphere

Photosynthesis nourishes almost all of the living world directly or indirectly. An organism acquires the organic compounds it uses for energy and carbon skeletons by one of two major modes: autotrophic or heterotrophic nutrition. At first, the term *autotrophic* (from the Greek *autos,* self, and *trophos,* feed) may seem to contradict the principle that organisms are open systems, taking in resources from their environment. **Autotrophs** are not totally self-sufficient, however; they are self-feeders only in the sense that they sustain themselves without eating other organisms or substances derived from other organisms. Autotrophs produce their organic molecules from CO_2 and other inorganic raw materials obtained from the environment and are the ultimate sources of organic compounds for all nonautotrophic organisms. For this reason, biologists refer to autotrophs as the *producers* of the biosphere (the global ecosystem).

Plants are autotrophs; the only nutrients they require are carbon dioxide from the air, and water and minerals from the soil. Specifically, plants are *photo*autotrophs, organisms that use light as a source of energy to synthesize organic substances. Photosynthesis also occurs in algae, including certain protists, and in some prokaryotes (FIGURE 10.1). In this chapter, our emphasis will be on plants. Variations in photosynthesis that occur in algae and bacteria will be discussed in Unit Five. A much rarer form of self-feeding is unique to those bacteria that are *chemo*autotrophs. These organisms produce their organic compounds without the help of light, obtaining their energy by oxidizing inorganic substances, such as sulfur or ammonia. (We will postpone further discussion of this type of autotrophic nutrition until Chapter 27.)

Heterotrophs obtain their organic material by the second major mode of nutrition. Unable to make their own food, they live on compounds produced by other organisms; heterotrophs are the biosphere's *consumers*. The most obvious form of this "other-feeding" (*hetero* means "other, different") occurs when an animal eats plants or other animals.

176

(a) Plants

(b) Multicellular alga

(c) Unicellular protist   5 µm

(d) Cyanobacteria   20 µm

(e) Purple sulfur bacteria   25 µm

FIGURE 10.1 Photoautotrophs. These organisms use light energy to drive the synthesis of organic molecules from carbon dioxide and (in most cases) water. They feed not only themselves, but the entire living world. **(a)** On land, plants are the predominant producers of food. Three major groups of land plants—mosses, ferns, and flowering plants—are represented in this scene. In oceans, ponds, lakes, and other aquatic environments, photosynthetic organisms include **(b)** multicellular algae, such as this kelp; **(c)** some unicellular protists, such as *Euglena;* **(d)** the prokaryotes called cyanobacteria; and **(e)** other photosynthetic prokaryotes, such as these purple sulfur bacteria (c, d, e: LMs).

But heterotrophic nutrition may be more subtle. Some heterotrophs consume the remains of dead organisms, decomposing and feeding on organic litter such as carcasses, feces, and fallen leaves; they are known as decomposers. Most fungi and many types of bacteria get their nourishment this way. Almost all heterotrophs, including humans, are completely dependent on photoautotrophs for food, and also for oxygen, a by-product of photosynthesis. Thus, we can trace the food we eat and the oxygen we breathe to the chloroplast.

FIGURE 10.2 Focusing in on the location of photosynthesis in a plant. Leaves are the major organs of photosynthesis in plants. These pictures take you into a leaf, then into a cell, and finally into a chloroplast, the organelle where photosynthesis occurs. Gas exchange between the leaf's mesophyll tissue and the atmosphere occurs through microscopic pores called stomata. Chloroplasts, found mainly in the mesophyll, are bounded by two membranes that enclose the stroma, a dense fluid. Membranes of the thylakoid system separate the stroma from the thylakoid space. Thylakoids are concentrated in stacks called grana. (Middle right, LM; bottom right, TEM.)

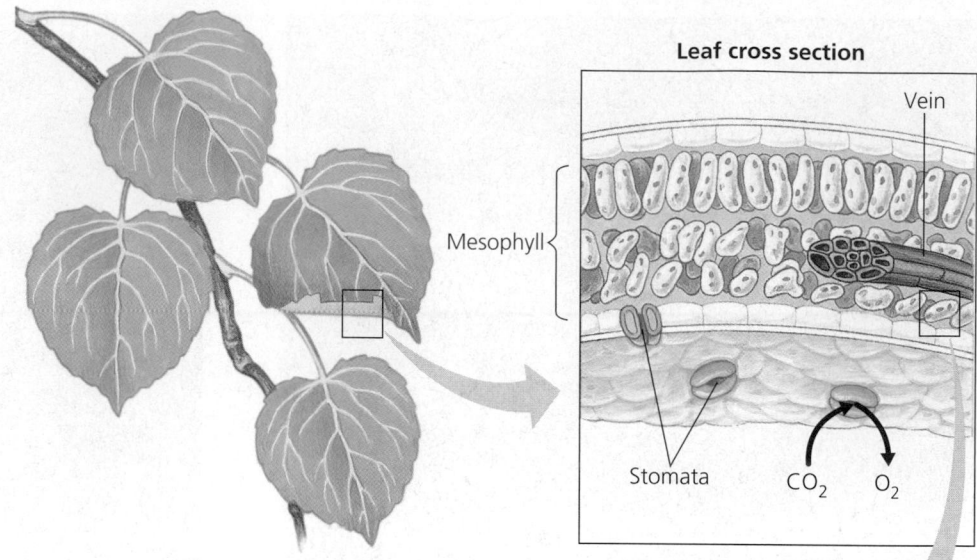

Leaf cross section

Vein

Mesophyll

Stomata

CO₂   O₂

Chloroplasts are the sites of photosynthesis in plants

All green parts of a plant, including green stems and unripened fruit, have chloroplasts, but the leaves are the major sites of photosynthesis in most plants (FIGURE 10.2). There are about half a million chloroplasts per square millimeter of leaf surface. The color of the leaf is from **chlorophyll,** the green pigment located within the chloroplasts. It is the light energy absorbed by chlorophyll that drives the synthesis of food molecules in the chloroplast. Chloroplasts are found mainly in the cells of the **mesophyll,** the tissue in the interior of the leaf. Carbon dioxide enters the leaf, and oxygen exits, by way of microscopic pores called **stomata** (singular, *stoma;* from the Greek, meaning "mouth"). Water absorbed by the roots is delivered to the leaves in veins. Leaves also use veins to export sugar to roots and other nonphotosynthetic parts of the plant.

A typical mesophyll cell has about 30 to 40 chloroplasts, each a watermelon-shaped organelle measuring about 2–4 μm by 4–7 μm. An envelope of two membranes encloses the stroma, the dense fluid within the chloroplast. An elaborate system of interconnected thylakoid membranes segregates the stroma from another compartment, the thylakoid space (or lumen). In some places, thylakoid sacs are stacked in columns called grana. Chlorophyll resides in the thylakoid membranes. (Photosynthetic prokaryotes lack chloroplasts; but as you will see in Chapter 27, they do have membranes that function in a manner similar to the thylakoid membranes of chloroplasts.) Now that we have looked at the sites of photosynthesis in plants, we are ready to see how these organelles convert the light energy absorbed by chlorophyll to chemical energy.

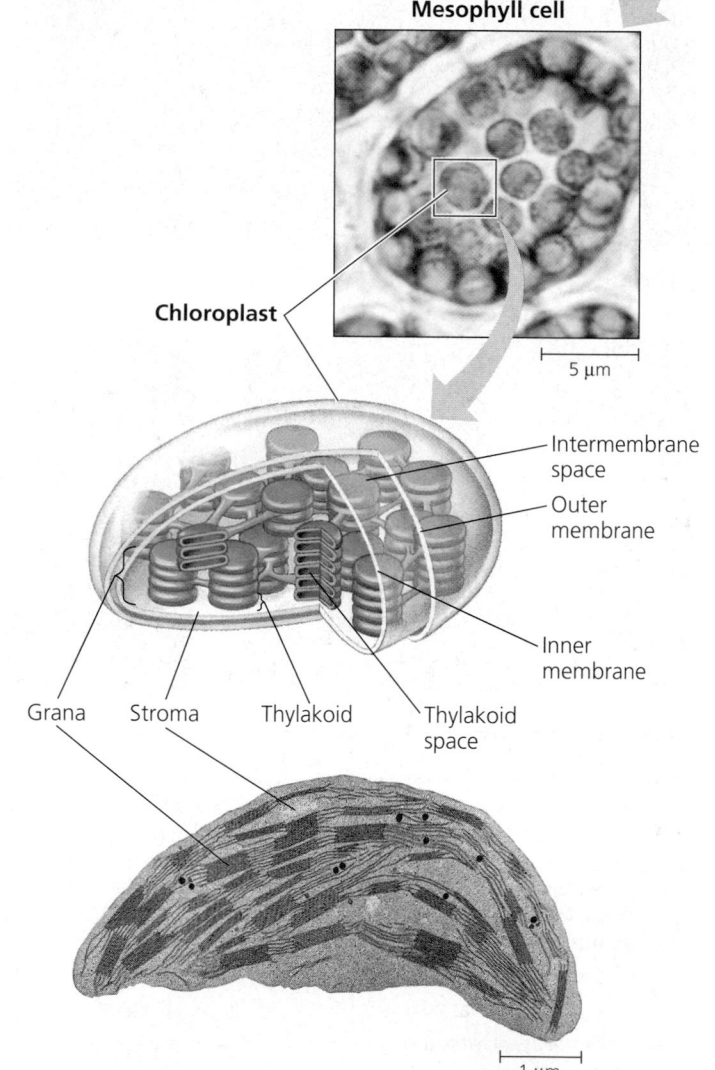

Mesophyll cell

Chloroplast

5 μm

Intermembrane space

Outer membrane

Inner membrane

Thylakoid space

Grana   Stroma   Thylakoid

1 μm

THE PATHWAYS OF PHOTOSYNTHESIS

Evidence that chloroplasts split water molecules enabled researchers to track atoms through photosynthesis

THE PROCESS OF SCIENCE

Scientists have tried for centuries to piece together the process by which plants make food. Although some of the steps are still not completely understood, the overall photosynthetic equation has been known since the 1800s: In the presence of light, the green parts of plants produce organic compounds and oxygen from carbon dioxide and water. Using molecular formulas, we can summarize photosynthesis with this chemical equation:

$$6CO_2 + 12H_2O + \text{Light energy} \longrightarrow C_6H_{12}O_6 + 6O_2 + 6H_2O$$

The carbohydrate $C_6H_{12}O_6$ is glucose.* Water appears on both sides of the equation because 12 molecules are consumed and 6 molecules are newly formed during photosynthesis. We can simplify the equation by indicating only the net consumption of water:

$$6CO_2 + 6H_2O + \text{Light energy} \longrightarrow C_6H_{12}O_6 + 6O_2$$

Writing the equation in this form, we can see that the overall chemical change during photosynthesis is the reverse of the one that occurs during cellular respiration. Both of these metabolic processes occur in plant cells. However, as you will soon learn, plants do not make food by simply reversing the steps of respiration.

Now let's divide the photosynthetic equation by 6 to put it in its simplest possible form:

$$CO_2 + H_2O \longrightarrow CH_2O + O_2$$

Here, CH_2O is not an actual sugar but represents the general formula for a carbohydrate. In other words, we are imagining the synthesis of a sugar molecule one carbon at a time. Six repetitions would produce a glucose molecule. Let's now use this simplified formula to see how researchers tracked the chemical elements (C, H, and O) from the reactants of photosynthesis to the products.

The Splitting of Water

One of the first clues to the mechanism of photosynthesis came from the discovery that the oxygen given off by plants is derived from water and not from carbon dioxide. The chloroplast splits water into hydrogen and oxygen. Before this discovery, the prevailing hypothesis was that photosynthesis split carbon dioxide and then added water to the carbon.

---

* The direct product of photosynthesis is actually a three-carbon sugar. Glucose is used here only to simplify the relationship between photosynthesis and respiration.

Step 1: $CO_2 \longrightarrow C + O_2$

Step 2: $C + H_2O \longrightarrow CH_2O$

This hypothesis predicted that the O_2 released during photosynthesis came from CO_2. This idea was challenged in the 1930s by C. B. van Niel of Stanford University. Van Niel was investigating photosynthesis in bacteria that make their carbohydrate from CO_2 but do not release O_2. Van Niel concluded that, at least in these bacteria, CO_2 is not split into carbon and oxygen. One group of bacteria used hydrogen sulfide (H_2S) rather than water for photosynthesis, forming yellow globules of sulfur as a waste product (these globules are visible in FIGURE 10.1e). Here is the chemical equation:

$$CO_2 + 2H_2S \longrightarrow CH_2O + H_2O + 2S$$

Van Niel reasoned that the bacteria split H_2S and used the hydrogen to make sugar. He then generalized that idea, proposing that all photosynthetic organisms require a hydrogen source, but that the source varies:

General: $CO_2 + 2H_2X \longrightarrow CH_2O + H_2O + 2X$

Sulfur bacteria: $CO_2 + 2H_2S \longrightarrow CH_2O + H_2O + 2S$

Plants: $CO_2 + 2H_2O \longrightarrow CH_2O + H_2O + O_2$

Thus, van Niel hypothesized that plants split water as a source of hydrogen, releasing oxygen as a by-product.

Nearly 20 years later, scientists confirmed van Niel's hypothesis by using oxygen-18 (^{18}O), a heavy isotope, as a tracer to follow the fate of oxygen atoms during photosynthesis. The O_2 that came from plants was labeled with ^{18}O *only* if water was the source of the tracer. If the ^{18}O was introduced to the plant in the form of CO_2, the label did not turn up in the released O_2. In the following summary of these experiments, red denotes labeled atoms of oxygen:

Experiment 1: $CO_2 + 2H_2O \longrightarrow CH_2O + H_2O + O_2$

Experiment 2: $CO_2 + 2H_2O \longrightarrow CH_2O + H_2O + O_2$

The most important result of the shuffling of atoms during photosynthesis is the extraction of hydrogen from water and its incorporation into sugar. The waste product of photosynthesis, O_2, restores the atmospheric oxygen consumed during cellular respiration. FIGURE 10.3 shows the fates of all atoms in photosynthesis.

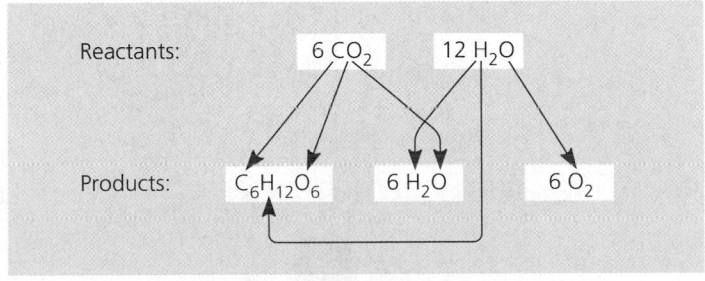

FIGURE 10.3 Tracking atoms through photosynthesis.

Photosynthesis as a Redox Process

Let's briefly contrast photosynthesis with cellular respiration. During respiration, energy is released from sugar when electrons associated with hydrogen are transported by carriers to oxygen, forming water as a by-product. The electrons lose potential energy as they are pulled down the electron transport chain to electronegative oxygen, and the mitochondrion uses the energy to synthesize ATP. Photosynthesis, also a redox process, reverses the direction of electron flow. Water is split, and electrons are transferred along with hydrogen ions from the water to carbon dioxide, reducing it to sugar. The electrons increase in potential energy as they move from water to sugar. The required energy boost is provided by light.

The light reactions and the Calvin cycle cooperate in converting light energy to the chemical energy of food: *an overview*

The equation for photosynthesis is a deceptively simple summary of a very complex process. Actually, photosynthesis is not a single process, but two processes, each with multiple steps. These two stages of photosynthesis are known as the **light reactions** (the *photo* part of photosynthesis) and the **Calvin cycle** (the *synthesis* part) (FIGURE 10.4).

The light reactions are the steps of photosynthesis that convert solar energy to chemical energy. Light absorbed by chlorophyll drives a transfer of electrons and hydrogen from water to an acceptor called **NADP$^+$** (nicotinamide adenine dinucleotide phosphate), which temporarily stores the energized electrons. Water is split in the process, and thus it is the light reactions of photosynthesis that give off O_2 as a by-product. The electron acceptor of the light reactions, NADP$^+$, is first cousin to NAD$^+$, which functions as an electron carrier in cellular respiration; the two molecules differ only by the presence of an extra phosphate group in the NADP$^+$ molecule. The light reactions use solar power to reduce NADP$^+$ to NADPH by adding a pair of electrons along with a hydrogen nucleus, or H$^+$. The light reactions also generate ATP by powering the addition of a phosphate group to ADP, a process called **photophosphorylation.** Thus, light energy is initially converted to chemical energy in the form of two compounds: NADPH, a source of energized electrons ("reducing power"), and ATP, the versatile energy currency of cells. Notice that the light reactions produce no sugar; that happens in the second stage of photosynthesis, the Calvin cycle.

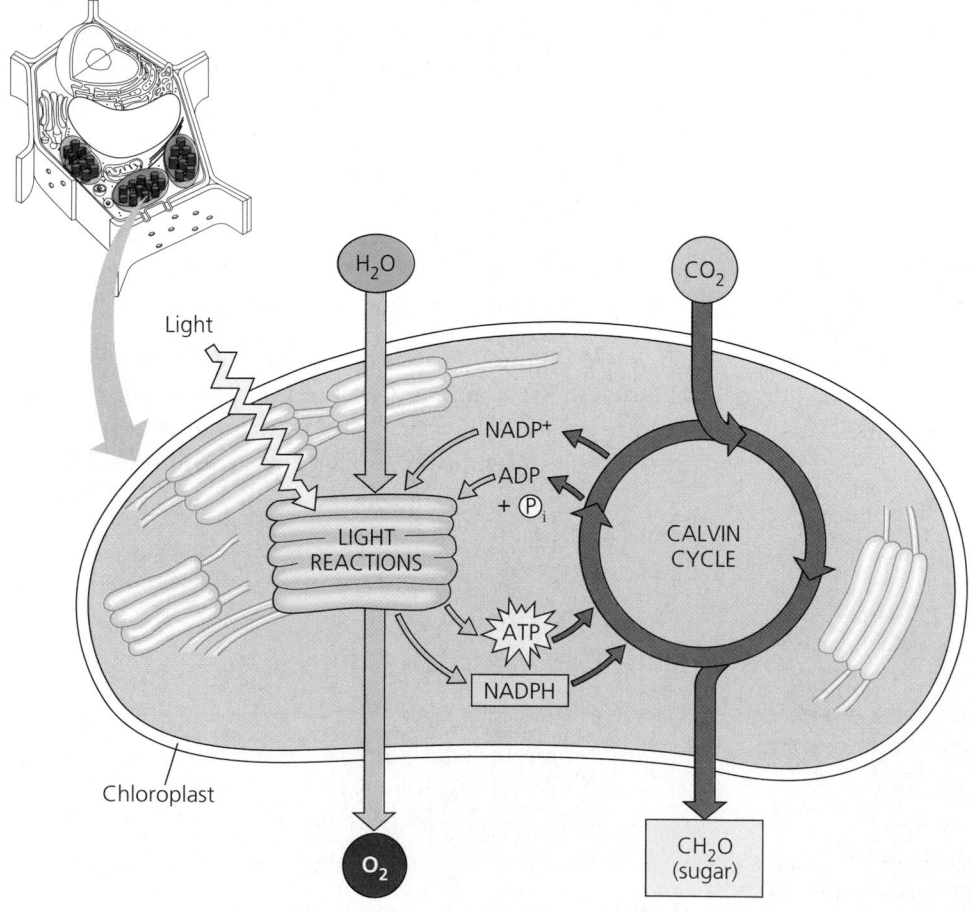

FIGURE 10.4 An overview of photosynthesis: cooperation of the light reactions and the Calvin cycle. In the chloroplast, the thylakoid membranes are the sites of the light reactions, whereas the Calvin cycle occurs in the stroma. The light reactions use solar energy to make ATP and NADPH, which function as chemical energy and reducing power, respectively, in the Calvin cycle. The Calvin cycle incorporates CO_2 into organic molecules, which are converted to sugar. (Recall from Chapter 5 that most simple sugars have formulas that are some multiple of CH_2O.)

A smaller version of this diagram will reappear in several subsequent figures as a reminder of whether the events being described occur in the light reactions or in the Calvin cycle.

The Calvin cycle is named for Melvin Calvin, who, along with his colleagues, began to elucidate its steps in the late 1940s. The cycle begins by incorporating CO_2 from the air into organic molecules already present in the chloroplast. This initial incorporation of carbon into organic compounds is known as **carbon fixation.** The Calvin cycle then reduces the fixed carbon to carbohydrate by the addition of electrons. The reducing power is provided by NADPH, which acquired energized electrons in the light reactions. To convert CO_2 to carbohydrate, the Calvin cycle also requires chemical energy in the form of ATP, which is also generated by the light reactions. Thus, it is the Calvin cycle that makes sugar, but it can do so only with the help of the NADPH and ATP produced by the light reactions. The metabolic steps of the Calvin cycle are sometimes referred to as the dark reactions, or light-independent reactions, because none of the steps requires light *directly.* Nevertheless, the Calvin cycle in most plants occurs during daylight, for only then can the light reactions regenerate the NADPH and ATP spent in the reduction of CO_2 to sugar. In essence, the chloroplast uses light energy to make sugar by coordinating the two stages of photosynthesis.

As FIGURE 10.4 indicates, the thylakoids of the chloroplast are the sites of the light reactions, while the Calvin cycle occurs in the stroma. As molecules of $NADP^+$ and ADP bump into the thylakoid membrane, they pick up electrons and phosphate, respectively, and then transfer their high-energy cargo to the Calvin cycle. The two stages of photosynthesis are treated in this figure as metabolic modules that take in ingredients and crank out products. Our next step toward understanding photosynthesis is to look more closely at how the two stages work, beginning with the light reactions.

The light reactions convert solar energy to the chemical energy of ATP and NADPH: *a closer look*

Chloroplasts are chemical factories powered by the sun. Their thylakoids transform light energy into the chemical energy of ATP and NADPH. To understand this conversion better, it is necessary to know about some important properties of light.

The Nature of Sunlight

Light is a form of energy known as electromagnetic energy, also called electromagnetic radiation. Electromagnetic energy travels in rhythmic waves analogous to those created by dropping a pebble into a puddle of water. Electromagnetic waves, however, are disturbances of electrical and magnetic fields rather than disturbances of a material medium such as water.

The distance between the crests of electromagnetic waves is called the **wavelength.** Wavelengths range from less than a

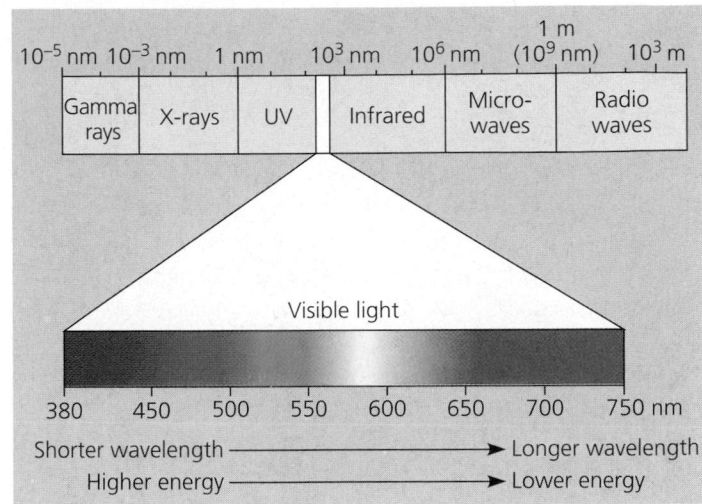

FIGURE 10.5 The electromagnetic spectrum. Visible light and other forms of electromagnetic energy radiate through space as waves of various lengths. We perceive different wavelengths of visible light, which range from about 380 to 750 nm, as different colors. White light is a mixture of all wavelengths of visible light. A prism can sort white light into its component colors by bending light of different wavelengths at different angles. Visible light drives photosynthesis.

nanometer (for gamma rays) to more than a kilometer (for radio waves). This entire range of radiation is known as the **electromagnetic spectrum** (FIGURE 10.5). The segment most important to life is the narrow band that ranges from about 380 to 750 nm in wavelength. This radiation is known as **visible light,** because it is detected as various colors by the human eye.

The model of light as waves explains many of light's properties, but in certain respects light behaves as though it consists of discrete particles, called **photons.** Photons are not tangible objects, but they act like objects in that each of them has a fixed quantity of energy. The amount of energy is inversely related to the wavelength of the light; the shorter the wavelength, the greater the energy of each photon of that light. Thus, a photon of violet light packs nearly twice as much energy as a photon of red light.

Although the sun radiates the full spectrum of electromagnetic energy, the atmosphere acts like a selective window, allowing visible light to pass through while screening out a substantial fraction of other radiation. The part of the spectrum we can see—visible light—is also the radiation that drives photosynthesis.

Photosynthetic Pigments: The Light Receptors

As light meets matter, it may be reflected, transmitted, or absorbed. Substances that absorb visible light are called pigments. Different pigments absorb light of different wavelengths, and the wavelengths that are absorbed disappear. If a pigment is illuminated with white light, the color we see is

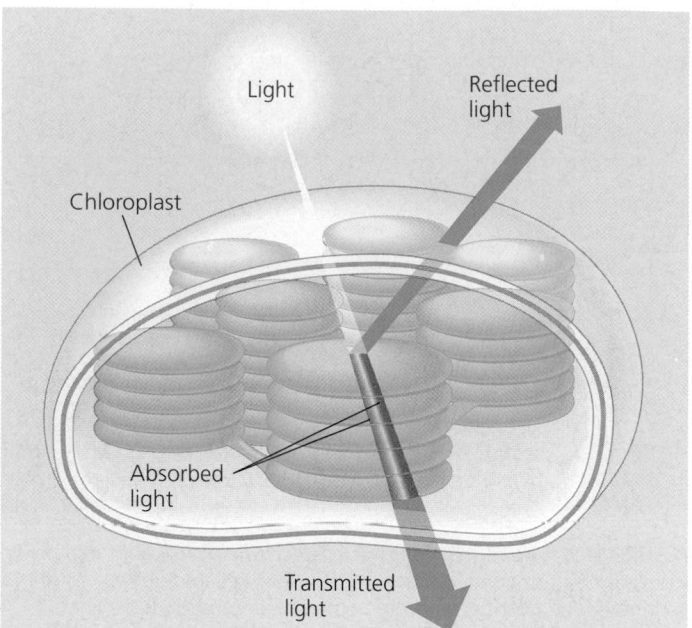

FIGURE 10.6 Why leaves are green: interaction of light with chloroplasts. The pigment molecules of chloroplasts absorb blue and red light and reflect or transmit green light. This is why leaves appear green. It turns out that blue and red are the colors of light most effective in photosynthesis.

the color most reflected or transmitted by the pigment. (If a pigment absorbs all wavelengths, it appears black.) We see green when we look at a leaf because chlorophyll absorbs red and blue light while transmitting and reflecting green light (FIGURE 10.6). The ability of a pigment to absorb various wavelengths of light can be measured with an instrument called a **spectrophotometer.** This machine directs beams of light of different wavelengths through a solution of the pigment and measures the fraction of the light transmitted at each wavelength (FIGURE 10.7). A graph plotting a pigment's

light absorption (the fraction *not* transmitted or reflected) versus wavelength is called an **absorption spectrum.**

The absorption spectra of chloroplast pigments provide clues to the relative effectiveness of different wavelengths for driving photosynthesis, since light can perform work in chloroplasts only if it is absorbed. FIGURE 10.8a shows the absorption spectra of a type of chlorophyll called **chlorophyll *a*** and some other pigments in the chloroplast. If we look first at the absorption spectrum of chlorophyll *a*, it suggests that blue and red light work best for photosynthesis, while green is the

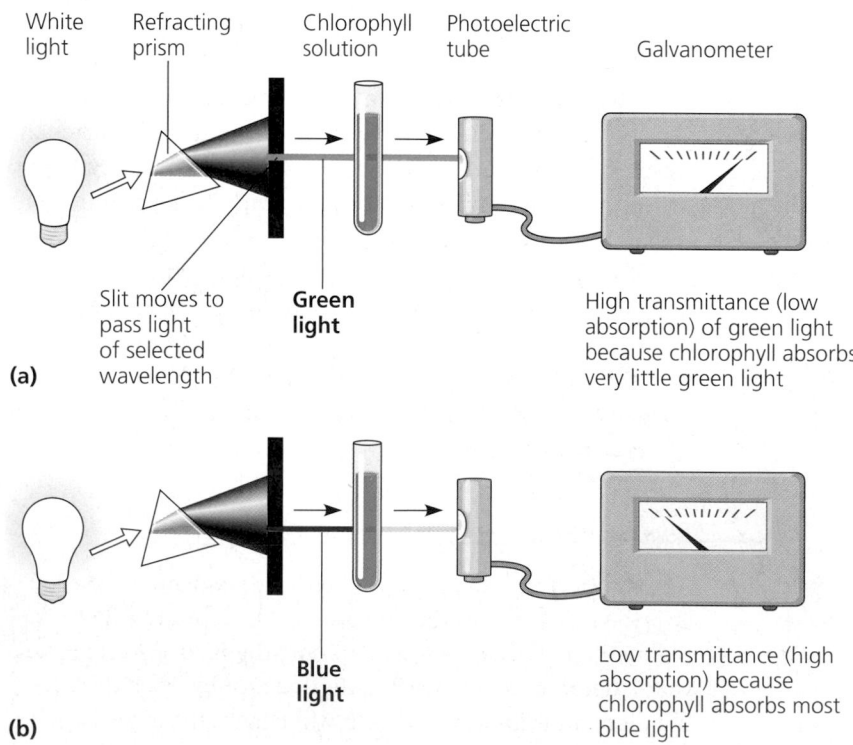

FIGURE 10.7 Determining an absorption spectrum.

THE PROCESS OF SCIENCE

A spectrophotometer measures the relative amounts of light of different wavelengths absorbed and transmitted by a pigment solution. Inside the spectrophotometer, white light is separated into colors (wavelengths) by a prism. Then, one by one, the different colors of light are passed through the sample. The transmitted light strikes a photoelectric tube, which converts the light energy to electricity, and the electrical current is measured by a galvanometer. Each time the wavelength of light is changed, the meter indicates the fraction of light transmitted through the sample (or, conversely, the fraction of light absorbed). This figure shows the transmittance reading on the meter when **(a)** green light and then **(b)** blue light are passed through a chlorophyll solution.

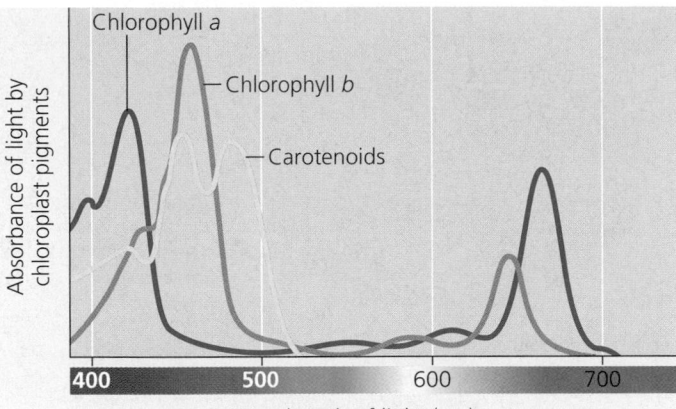

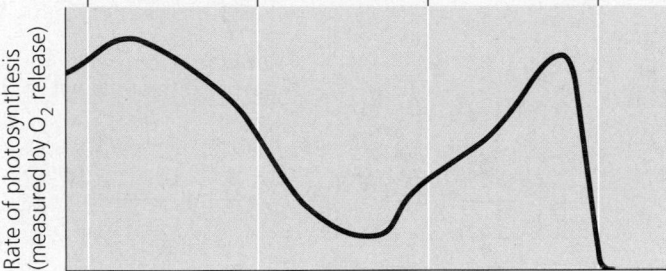

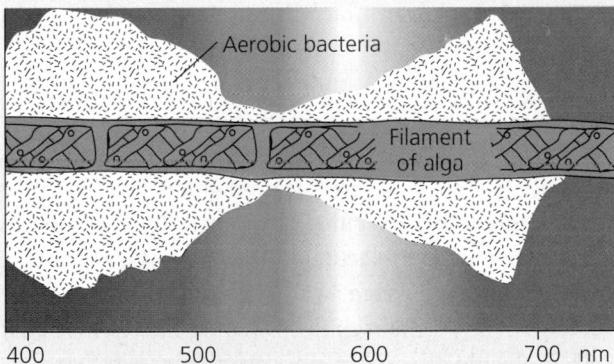

(a) Absorption spectra. The three curves show the wavelengths of light best absorbed by three types of pigments extracted from chloroplasts.

(b) Action spectrum. This graph plots the effectiveness of different wavelengths of light in driving photosynthesis. The peaks in the action spectrum are broader than the peaks in the absorption spectrum for chlorophyll *a* (see part a above), and the valley is narrower and not as deep. This is partly due to the absorption of light by accessory pigments such as chlorophyll *b* and carotenoids, which broaden the spectrum of colors that can be used for photosynthesis.

(c) Engelmann's experiment. Engelmann illuminated a filamentous alga with light that had been passed through a prism, exposing different segments of the alga to different wavelengths. He used aerobic bacteria, which concentrate near an oxygen source, to determine which segments of the alga were releasing the most O_2. Bacteria congregated in greatest numbers around the parts of the alga illuminated with red or blue light. Notice the close match of the bacterial distribution to the action spectrum in part b.

THE PROCESS OF SCIENCE

FIGURE 10.8 Evidence that chloroplast pigments participate in photosynthesis: absorption and action spectra for photosynthesis in an alga.

least effective color. This is confirmed by an **action spectrum** for photosynthesis (FIGURE 10.8b), which profiles the relative performance of the different wavelengths. An action spectrum is prepared by illuminating chloroplasts with different colors of light and then plotting wavelength against some measure of photosynthetic rate, such as carbon dioxide consumption or oxygen release. The action spectrum for photosynthesis was first demonstrated in 1883 in an elegant experiment performed by the German botanist Thomas Engelmann, who used bacteria to measure rates of photosynthesis in filamentous algae (FIGURE 10.8c).

Notice by comparing FIGURES 10.8a and 10.8b that the action spectrum for photosynthesis does not exactly match the absorption spectrum of chlorophyll *a*. The absorption spectrum of chlorophyll *a* alone underestimates the effectiveness of certain wavelengths in driving photosynthesis. This is partly because chlorophyll *a* is not the only photosynthetically important pigment in chloroplasts. Only chlorophyll *a* can participate directly in the light reactions, which convert solar energy to chemical energy. But other pigments in the thylakoid membrane can absorb light and transfer the energy to chlorophyll *a*, which then initiates the light reactions. One of these accessory pigments is another form of chlorophyll, **chlorophyll *b***. Chlorophyll *b* is almost identical to chlorophyll *a* (see FIGURE 10.9 on page 184), but a slight structural difference between them is enough to give the two pigments slightly different absorption spectra and hence different colors. Chlorophyll *a* is blue-green, whereas chlorophyll *b* is yellow-green. If a photon of sunlight is absorbed by chlorophyll *b*, energy is conveyed to chlorophyll *a*, which then behaves just as though it had absorbed the photon. Other accessory pigments include **carotenoids,** hydrocarbons that are various shades of yellow and orange (see FIGURE 10.8a). Carotenoids may broaden the spectrum of colors that can drive photosynthesis. However, a more important function of at least some carotenoids seems to be *photoprotection:* Rather than transmit energy to chlorophyll, these compounds absorb and dissipate excessive light energy that would otherwise damage chlorophyll. (Interestingly, carotenoids similar to the photoprotective ones in chloroplasts have a photoprotective role in the human eye.)

Excitation of Chlorophyll by Light

What exactly happens when chlorophyll and other pigments absorb photons? The colors corresponding to the absorbed wavelengths disappear from the spectrum of the transmitted and reflected light, but energy cannot disappear. When a molecule absorbs a photon, one of the molecule's electrons is elevated to an orbital where it has more potential energy. When the electron is in its normal orbital, the pigment molecule is said to be in its ground state. After absorption of a photon boosts an electron to an orbital of higher energy, the pigment molecule is said to be in an excited state. The only photons

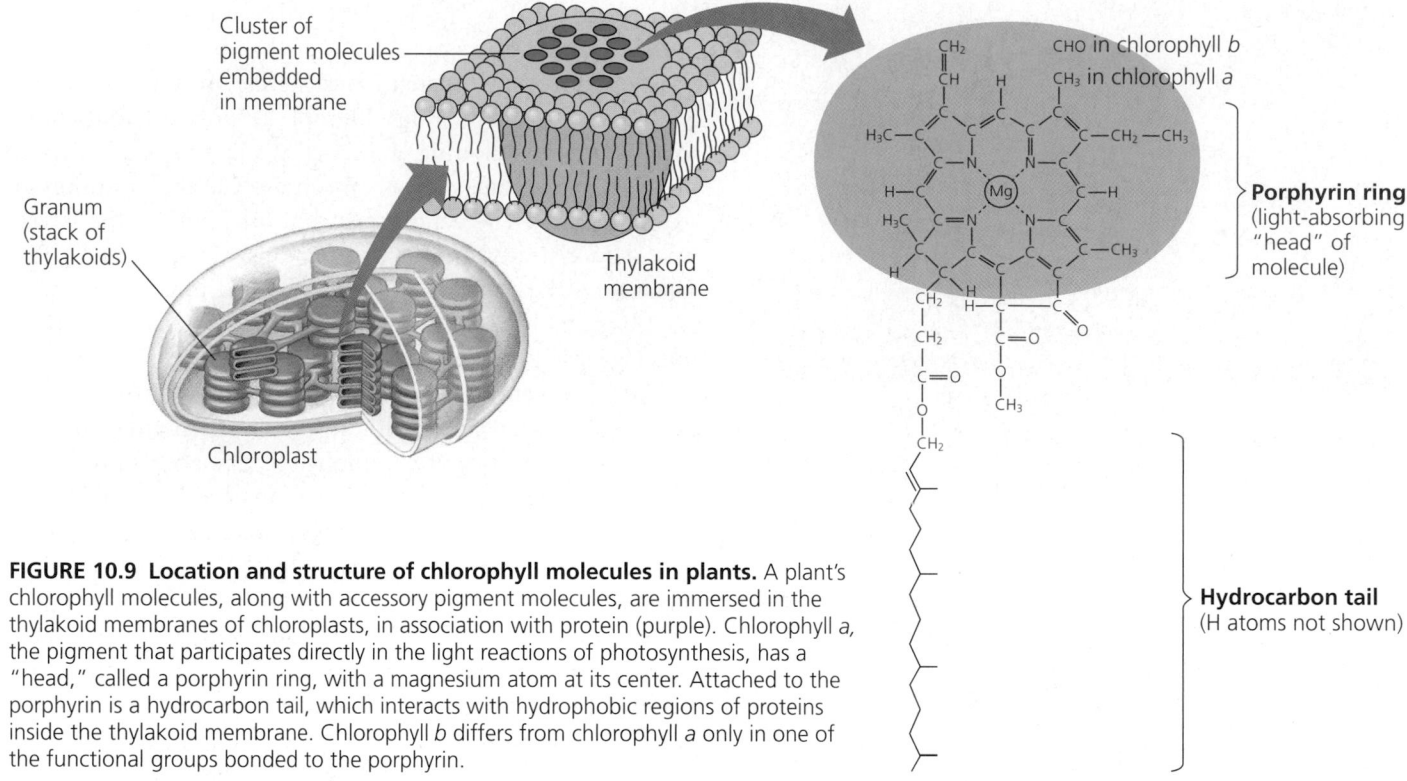

Cluster of pigment molecules embedded in membrane

Granum (stack of thylakoids)

Chloroplast

Thylakoid membrane

CH₂
CH

CHO in chlorophyll b
CH₃ in chlorophyll a

H₃C

CH₂—CH₃

Mg

CH₃

H₃C

CH₃

CH₂

CH₂

C=O

O

CH₂

C=O

O

CH₃

Porphyrin ring (light-absorbing "head" of molecule)

Hydrocarbon tail (H atoms not shown)

FIGURE 10.9 Location and structure of chlorophyll molecules in plants. A plant's chlorophyll molecules, along with accessory pigment molecules, are immersed in the thylakoid membranes of chloroplasts, in association with protein (purple). Chlorophyll *a*, the pigment that participates directly in the light reactions of photosynthesis, has a "head," called a porphyrin ring, with a magnesium atom at its center. Attached to the porphyrin is a hydrocarbon tail, which interacts with hydrophobic regions of proteins inside the thylakoid membrane. Chlorophyll *b* differs from chlorophyll *a* only in one of the functional groups bonded to the porphyrin.

absorbed are those whose energy is exactly equal to the energy difference between the ground state and an excited state, and this energy difference varies from one kind of atom or molecule to another. Thus, a particular compound absorbs only photons corresponding to specific wavelengths, which is why each pigment has a unique absorption spectrum.

Photons are absorbed by clusters of pigment molecules embedded in the thylakoid membrane (FIGURE 10.9). The energy of an absorbed photon is converted to the potential energy of an electron raised from the ground state to an excited state. But the electron cannot remain there long; the excited state, like all high-energy states, is unstable. Generally, when pigments absorb light, their excited electrons drop back down to the ground-state orbital in a billionth of a second, releasing their excess energy as heat. This conversion of light energy to heat is what makes the top of an automobile so hot on a sunny day. (White cars are coolest because their paint reflects all wavelengths of visible light, although it may absorb ultraviolet and other invisible radiation.) Some pigments, including chlorophyll, emit light as well as heat after absorbing photons. The electron jumps to a state of greater energy, and as it falls back to ground state, a photon is given off. This afterglow is called fluorescence. If a solution of chlorophyll isolated from chloroplasts is illuminated, it will fluoresce in the red part of the spectrum and also give off heat (FIGURE 10.10).

Photosystems: Light-Harvesting Complexes of the Thylakoid Membrane

Chlorophyll excited by absorption of light energy produces very different results in an intact chloroplast than it does in isolation. In its native environment of the thylakoid membrane, chlorophyll is organized along with proteins and other kinds of smaller organic molecules into **photosystems.**

A photosystem has a light-gathering "antenna complex" consisting of a cluster of a few hundred chlorophyll *a*, chlorophyll *b*, and carotenoid molecules (FIGURE 10.11). The number and variety of pigment molecules enable a photosystem to harvest light over a larger surface and a larger portion of the spectrum than any single pigment molecule could harvest. When any antenna molecule absorbs a photon, the energy is transmitted from pigment molecule to pigment molecule until it reaches a particular chlorophyll *a*. What is special about *this* chlorophyll *a* molecule is not its molecular structure, but its position. Only this chlorophyll molecule is located in the region of the photosystem called the **reaction center,** where the first light-driven chemical reaction of photosynthesis occurs.

Sharing the reaction center with the chlorophyll *a* molecule is a specialized molecule called the **primary electron acceptor.** In an oxidation-reduction reaction, the chlorophyll *a* molecule at the reaction center loses one of its electrons to the primary electron acceptor. This redox reaction occurs when light excites the

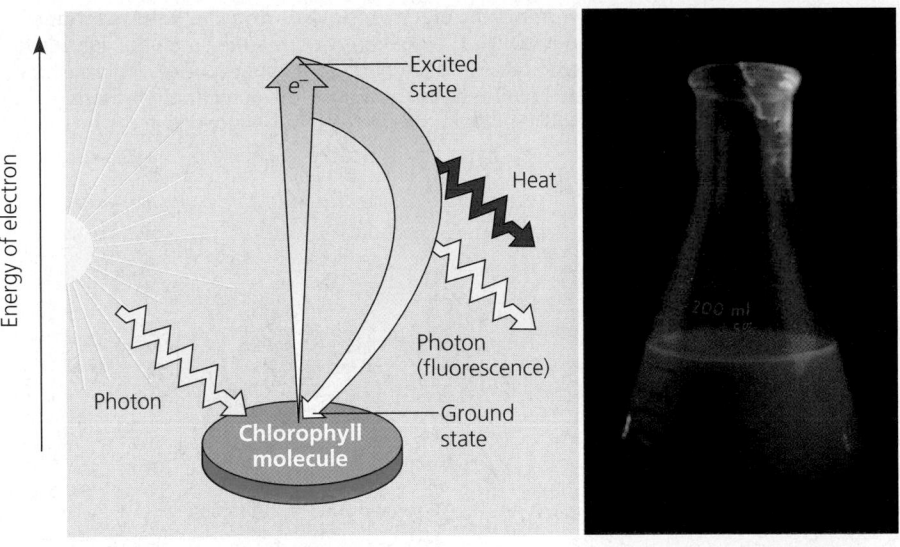

(a) Excitation of isolated chlorophyll molecule

(b) Fluorescence

FIGURE 10.10 Excitation of isolated chlorophyll by light. (a) Absorption of a photon causes a transition of the chlorophyll molecule from its ground state to its excited state. The photon boosts an electron to an orbital where it has more potential energy. If the illuminated molecule exists in isolation, its excited electron immediately drops back down to the ground-state orbital, and its excess energy is given off as heat and fluorescence (light). **(b)** A chlorophyll solution excited with ultraviolet light will fluoresce with a red-orange glow.

electron to a higher energy level in chlorophyll and the electron acceptor traps the high-energy electron before it can return to the ground state in the chlorophyll molecule. Isolated chlorophyll fluoresces because there is no electron acceptor to prevent electrons of photoexcited chlorophyll from dropping right back to the ground state. In a chloroplast, the acceptor molecule functions like a dam that prevents this immediate plunge of high-energy electrons back to the ground state. Thus, each photosystem—reaction-center chlorophyll and primary electron acceptor surrounded by an antenna complex—functions

in the chloroplast as a light-harvesting unit. The solar-powered transfer of electrons from chlorophyll to the primary electron acceptor is the first step of the light reactions.

The thylakoid membrane is populated by two types of photosystems that cooperate in the light reactions of photosynthesis. They are called **photosystem I** and **photosystem II,** in order of their discovery. Each has a characteristic reaction center—a particular kind of primary electron acceptor next to a chlorophyll *a* molecule associated with specific proteins. The reaction-center chlorophyll of photosystem I is known as P700 because this pigment is best at absorbing light having a wavelength of 700 nm (the far-red part of the spectrum). The chlorophyll at the reaction center of photosystem II is called P680 because its absorption spectrum has a peak at 680 nm (also in the red part of the spectrum). These two pigments, P700 and P680, are actually identical chlorophyll *a* molecules. However, their association with different proteins in the thylakoid membrane affects the electron distribution in the chlorophyll molecules and accounts for the slight differences in light-absorbing properties. Now let's see how the two photosystems work together in using light energy to generate ATP and NADPH, the two main products of the light reactions.

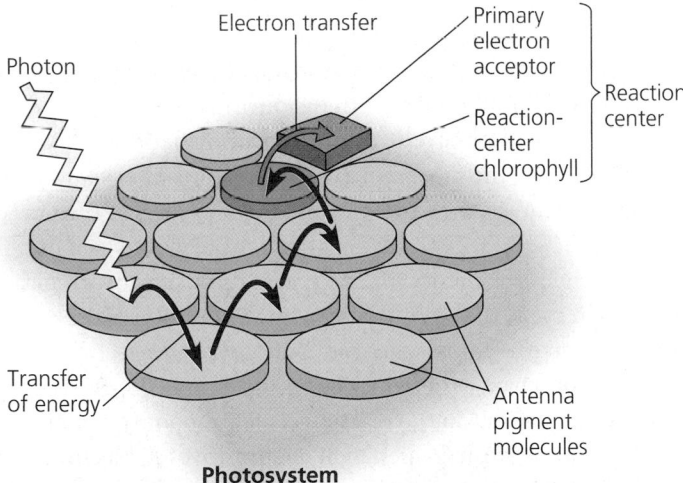

FIGURE 10.11 How a photosystem harvests light. Photosystems are the light-harvesting units of the thylakoid membrane. Each photosystem is a complex of proteins and other kinds of molecules and includes an antenna consisting of a few hundred pigment molecules. When a photon strikes a pigment molecule, the energy is passed from molecule to molecule until it reaches the reaction center. At the reaction center, an excited electron from the reaction-center chlorophyll is captured by a specialized molecule called the primary electron acceptor.

Noncyclic Electron Flow

Light drives the synthesis of NADPH and ATP by energizing the two photosystems embedded in the thylakoid membranes of chloroplasts. The key to this energy transformation is a flow of electrons through the photosystems and other molecular components built into the thylakoid membrane. During the light reactions of photosynthesis, there are two possible routes for electron flow: cyclic and noncyclic. **Noncyclic electron flow,** the predominant route, is shown in FIGURE 10.12 on p. 186. The numbers in the text description correspond to the numbered steps in the figure.

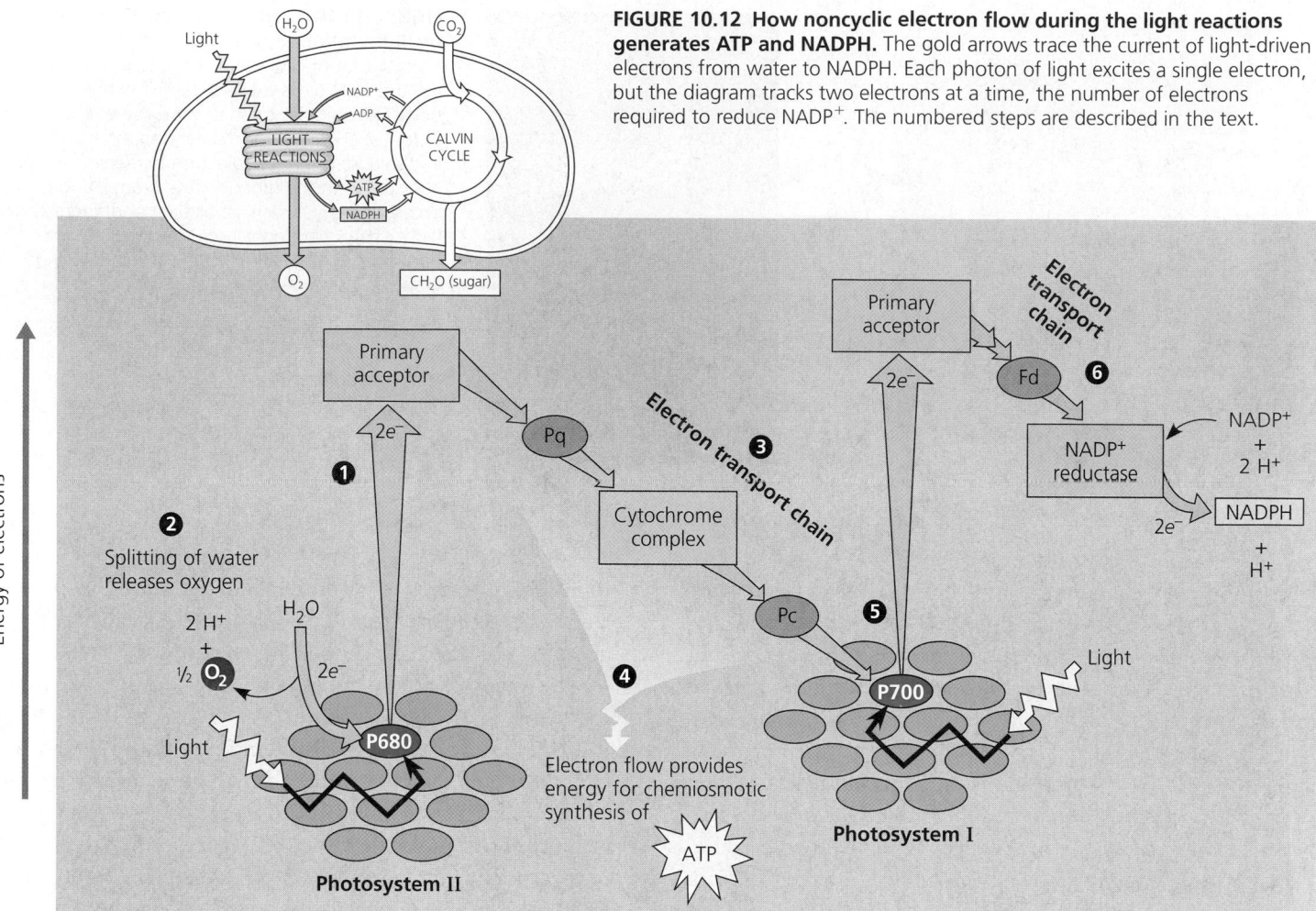

FIGURE 10.12 How noncyclic electron flow during the light reactions generates ATP and NADPH. The gold arrows trace the current of light-driven electrons from water to NADPH. Each photon of light excites a single electron, but the diagram tracks two electrons at a time, the number of electrons required to reduce NADP⁺. The numbered steps are described in the text.

❶ When photosystem II absorbs light, an electron excited to a higher energy level in the reaction-center chlorophyll (P680) is captured by the primary electron acceptor. The oxidized chlorophyll is now a very strong oxidizing agent; its electron "hole" must be filled.

❷ An enzyme extracts electrons from water and supplies them to P680, replacing the electrons that the chlorophyll molecule lost when it absorbed light energy. This reaction splits a water molecule into two hydrogen ions and an oxygen atom, which immediately combines with another oxygen atom to form O_2. This is the water-splitting step of photosynthesis that releases O_2.

❸ Each photoexcited electron passes from the primary electron acceptor of photosystem II to photosystem I via an electron transport chain. This chain is very similar to the one that functions in cellular respiration. The chloroplast version consists of an electron carrier called plastoquinone (Pq), a complex of two cytochromes (closely related to the cytochromes of mitochondria), and a copper-containing protein called plastocyanin (Pc).

❹ As electrons cascade down the chain, their exergonic "fall" to a lower energy level is harnessed by the thylakoid

membrane to produce ATP. This ATP synthesis is called photophosphorylation because it is driven by light energy. Specifically, ATP synthesis during noncyclic electron flow is called **noncyclic photophosphorylation.** (As we will discuss, the mechanism for photophosphorylation is chemiosmosis, the same process that operates in respiration.) This ATP generated by the light reactions will provide chemical energy for the synthesis of sugar during the Calvin cycle, the second major stage of photosynthesis.

❺ When an electron reaches the "bottom" of the electron transport chain, it fills an electron "hole" in P700, the chlorophyll *a* molecule in the reaction center of photosystem I. This hole is created when light energy drives an electron from P700 to the primary acceptor of photosystem I.

❻ The primary electron acceptor of photosystem I passes the photoexcited electrons to a second electron transport chain, which transmits them to ferredoxin (Fd), an iron-containing protein. An enzyme called NADP⁺ reductase then transfers the electrons from Fd to NADP⁺. This is the redox reaction that stores the high-energy electrons in NADPH, the molecule that will provide reducing power for the synthesis of sugar in the Calvin cycle.

The energy changes of electrons as they flow through the light reactions are analogous to the cartoon in FIGURE 10.13. As complicated as the scheme is, do not lose track of its functions: The light reactions use solar power to generate ATP and NADPH, which provide chemical energy and reducing power, respectively, to the sugar-making reactions of the Calvin cycle.

Cyclic Electron Flow

Under certain conditions, photoexcited electrons take an alternative path called **cyclic electron flow,** which uses photosystem I but not photosystem II. You can see in FIGURE 10.14 that cyclic flow is a short circuit: The electrons cycle back from ferredoxin (Fd) to the cytochrome complex and from there continue on to the P700 chlorophyll. There is no production of NADPH and no release of oxygen. Cyclic flow does, however, generate ATP. This is called **cyclic photophosphorylation,** to distinguish it from noncyclic photophosphorylation.

What is the function of cyclic electron flow? Noncyclic electron flow produces ATP and NADPH in roughly equal quantities, but the Calvin cycle consumes more ATP than NADPH. Cyclic electron flow makes up the difference. The concentration of NADPH in the chloroplast may help regulate which pathway, cyclic versus noncyclic, electrons take through the light reactions. If the chloroplast runs low on ATP for the Calvin cycle, NADPH will begin to accumulate as the Calvin cycle slows down. The rise in NADPH may stimulate a temporary shift from noncyclic to cyclic electron flow until ATP supply catches up with demand.

Whether photophosphorylation is driven by noncyclic or cyclic electron flow, the actual mechanism for ATP synthesis is the same. This is a good time to review chemiosmosis, the process that uses membranes to couple redox reactions to ATP production.

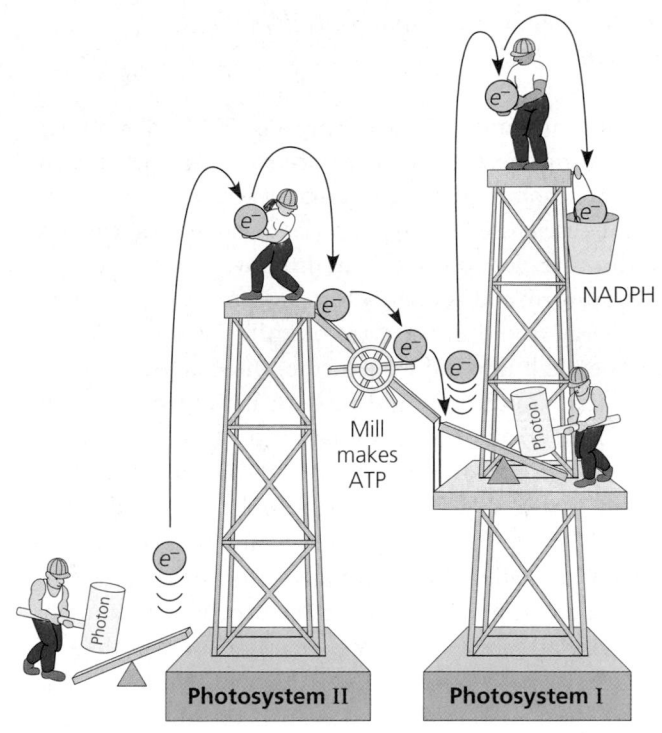

FIGURE 10.13 A mechanical analogy for the light reactions.

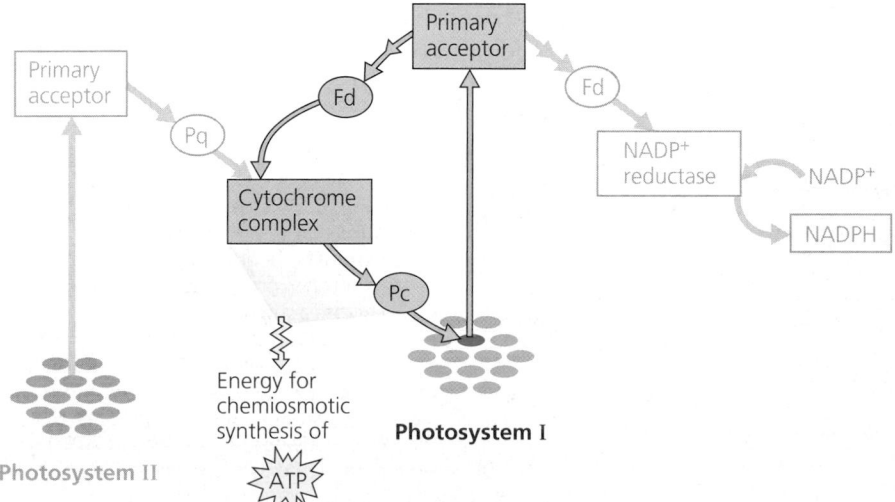

FIGURE 10.14 Cyclic electron flow. Photoexcited electrons from photosystem I are occasionally shunted back from ferredoxin (Fd) to chlorophyll via the cytochrome complex and plastocyanin (Pc). This electron shunt supplements the supply of ATP but produces no NADPH. The "shadow" of noncyclic electron flow is included in the diagram for comparison with the cyclic route. The two ferredoxin molecules shown in this diagram are actually one and the same—the final electron carrier in the electron transport chain of photosystem I.

A Comparison of Chemiosmosis in Chloroplasts and Mitochondria

Chloroplasts and mitochondria generate ATP by the same basic mechanism: chemiosmosis. An electron transport chain assembled in a membrane pumps protons across the membrane as electrons are passed through a series of carriers that are progressively more electronegative. In this way, electron transport chains transform redox energy to a proton-motive force, potential energy stored in the form of an H^+ gradient across a membrane. Built into the same membrane is an ATP synthase complex that couples the diffusion of hydrogen ions down their gradient to the phosphorylation of ADP. Some of the electron carriers, including the iron-containing proteins called cytochromes, are very similar in chloroplasts and mitochondria. The ATP synthase complexes of the two organelles are also very much alike. But there are noteworthy differences between oxidative phosphorylation in mitochondria and photophosphorylation in chloroplasts. In mitochondria, the high-energy electrons dropped down the transport chain are extracted from food molecules (which are thus oxidized). Chloroplasts do not need food to make ATP; their photosystems capture light energy and use it to drive electrons to the top of the transport chain. In other words, mitochondria transfer chemical energy from food molecules to ATP, whereas chloroplasts transform light energy into chemical energy.

The spatial organization of chemiosmosis also differs in chloroplasts and mitochondria (FIGURE 10.15). The inner membrane of the mitochondrion pumps protons from the mitochondrial matrix out to the intermembrane space, which then serves as a reservoir of hydrogen ions that powers the ATP synthase. The thylakoid membrane of the chloroplast pumps protons from the stroma into the thylakoid space, which functions as the H^+ reservoir. The thylakoid membrane makes ATP as the hydrogen ions diffuse from the thylakoid space back to the stroma through ATP synthase complexes, whose catalytic knobs are on the stroma side of the membrane. Thus, ATP forms in the stroma, where it is used to help drive sugar synthesis during the Calvin cycle.

The proton gradient, or pH gradient, across the thylakoid membrane is substantial. When chloroplasts are illuminated, the pH in the thylakoid space drops to about 5, and the pH in the stroma increases to about 8. This gradient of three pH units corresponds to a thousandfold difference in H^+ concentration. If in the laboratory the lights are turned off, the pH gradient is abolished, but it can quickly be restored by turning the lights back on. Such experiments add to the evidence described in Chapter 9 in support of the chemiosmotic model.

Based on studies in several laboratories, FIGURE 10.16 shows a current model for the organization of the light-reaction "machinery" within the thylakoid membrane. Each of the molecules and molecular complexes in the figure is present in numerous copies in each thylakoid. Notice that NADPH, like ATP, is produced on the side of the membrane facing the stroma, where the Calvin cycle reactions take place.

Let's summarize the light reactions. Noncyclic electron flow pushes electrons from water, where they are at a low state of potential energy, to NADPH, where they are stored at a high state of potential energy. The light-driven electron current also generates ATP. Thus, the equipment of the thylakoid

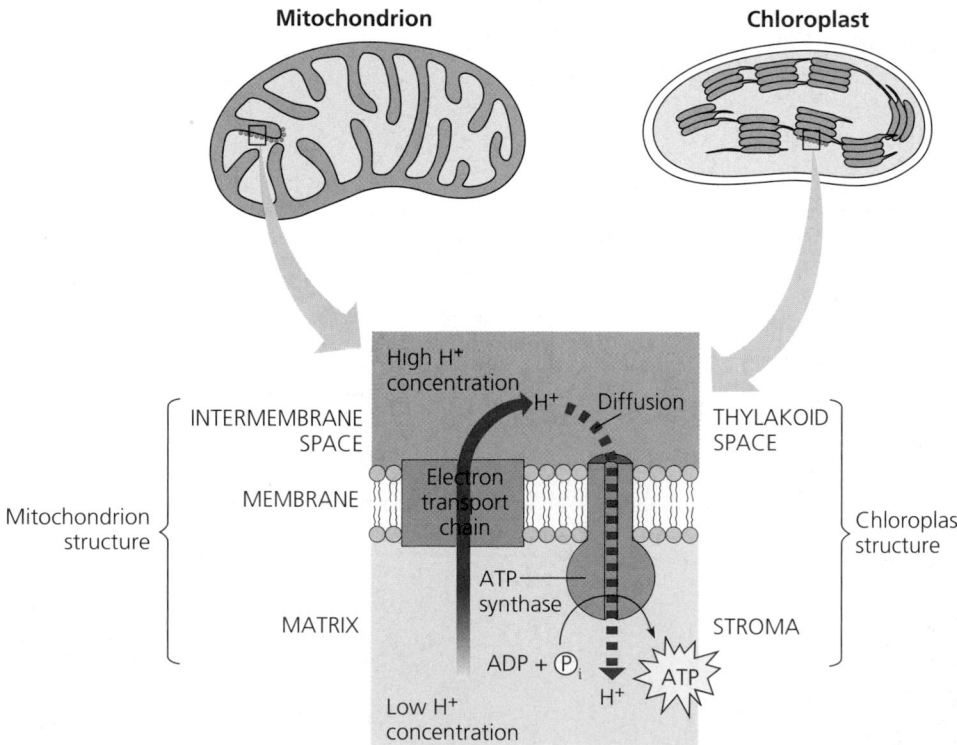

FIGURE 10.15 Comparison of chemiosmosis in mitochondria and chloroplasts. In both kinds of organelles, electron transport chains pump protons (H^+) across a membrane from a region of low H^+ concentration (light brown in this diagram) to one of high H^+ concentration (darker brown). The protons then diffuse back across the membrane through ATP synthase, driving the synthesis of ATP. The diagram identifies the regions of high and low H^+ concentration in the two organelles.

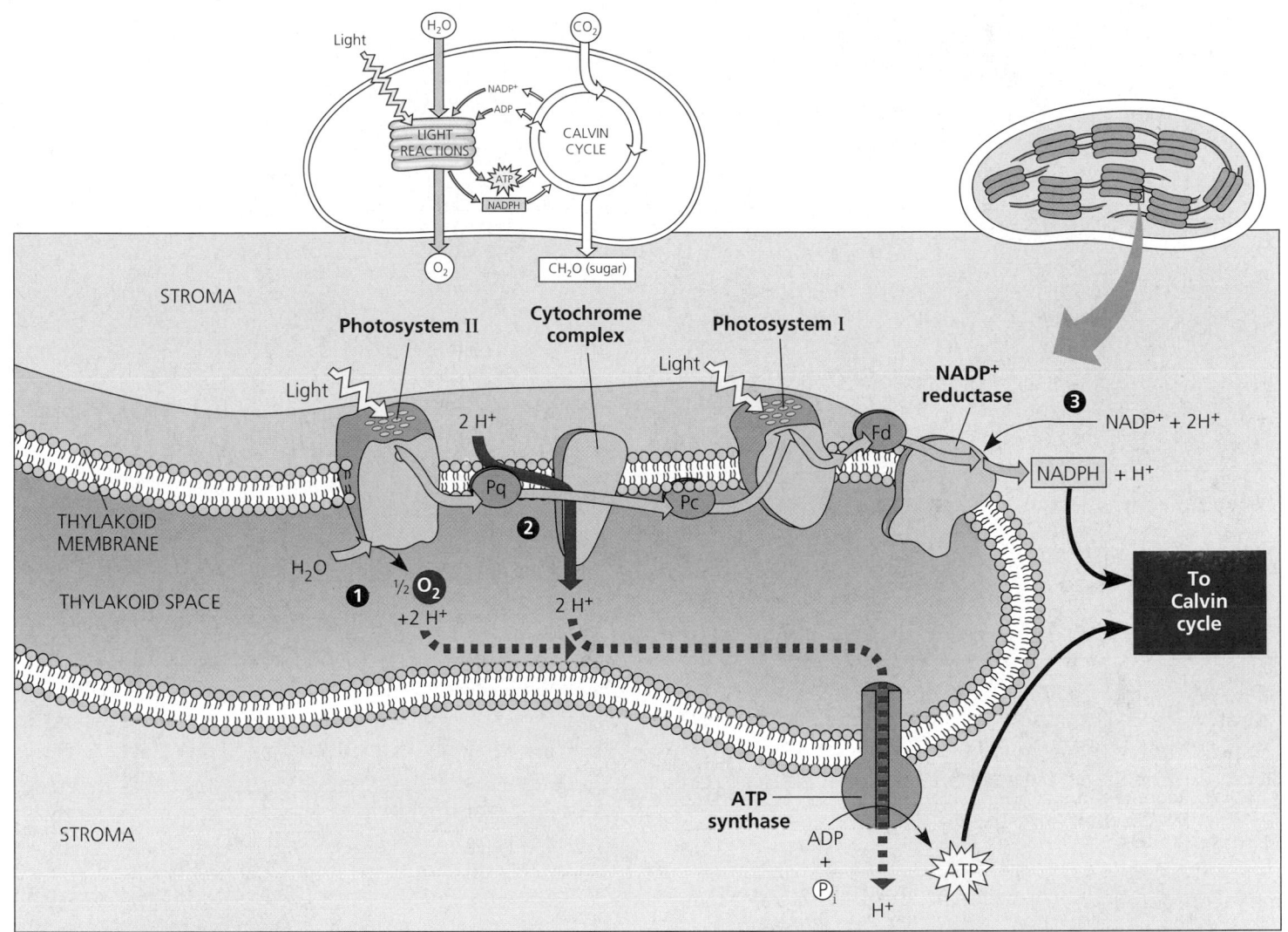

FIGURE 10.16 The light reactions and chemiosmosis: the organization of the thylakoid membrane. This diagram shows a current model for the organization of the thylakoid membrane. The gold arrows track the noncyclic electron flow outlined in FIGURE 10.12. As electrons pass from carrier to carrier in redox reactions, hydrogen ions removed from the stroma are deposited in the thylakoid space, storing energy as a proton-motive force (H^+ gradient). At least three steps in the light reactions contribute to the proton gradient: ❶ Water is split by photosystem II on the side of the membrane facing the thylakoid space; ❷ as plastoquinone (Pq), a mobile carrier, transfers electrons to the cytochrome complex, protons are translocated across the membrane; and ❸ a hydrogen ion is removed from the stroma when it is taken up by $NADP^+$. The diffusion of H^+ from the thylakoid space to the stroma (along the H^+ concentration gradient) powers the ATP synthase. These light-driven reactions store chemical energy in NADPH and ATP, which shuttle the energy to the sugar-producing Calvin cycle.

membrane converts light energy to the chemical energy stored in NADPH and ATP. (Oxygen is a by-product.) Let's now see how the Calvin cycle uses the products of the light reactions to synthesize sugar from CO_2.

The Calvin cycle uses ATP and NADPH to convert CO_2 to sugar: *a closer look*

The Calvin cycle is a metabolic pathway similar to the Krebs cycle in that a starting material is regenerated after molecules enter and leave the cycle. Carbon enters the Calvin cycle in the form of CO_2 and leaves in the form of sugar. The cycle spends ATP as an energy source and consumes NADPH as reducing power for adding high-energy electrons to make the sugar.

The carbohydrate produced directly from the Calvin cycle is actually not glucose, but a three-carbon sugar named **glyceraldehyde-3-phosphate (G3P).** For the net synthesis of one molecule of this sugar, the cycle must take place three times, fixing three molecules of CO_2. (Recall that carbon fixation refers to the initial incorporation of CO_2 into organic material.) As we trace the steps of the cycle, keep in mind that we

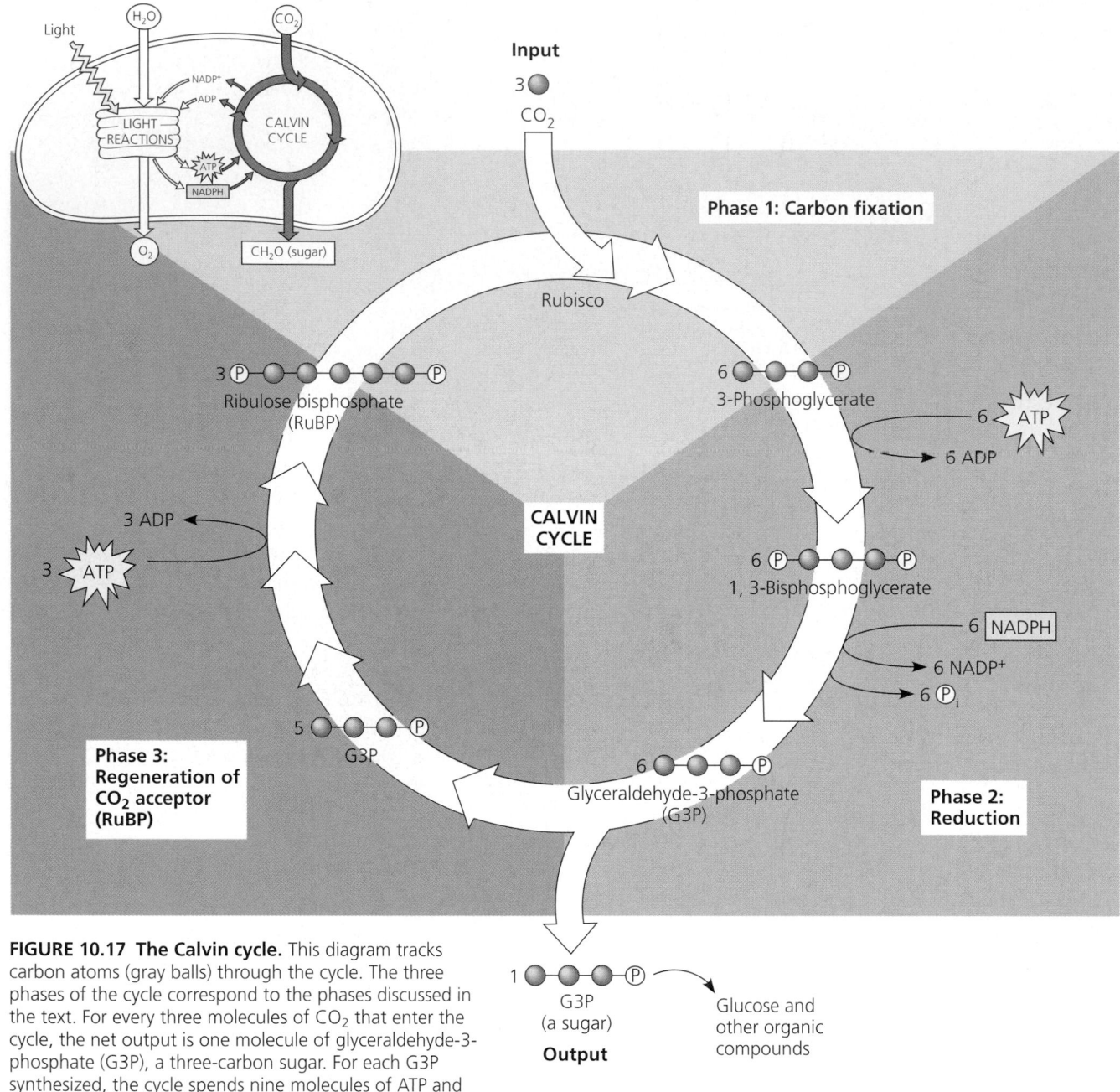

FIGURE 10.17 The Calvin cycle. This diagram tracks carbon atoms (gray balls) through the cycle. The three phases of the cycle correspond to the phases discussed in the text. For every three molecules of CO_2 that enter the cycle, the net output is one molecule of glyceraldehyde-3-phosphate (G3P), a three-carbon sugar. For each G3P synthesized, the cycle spends nine molecules of ATP and six molecules of NADPH. The light reactions sustain the Calvin cycle by regenerating ATP and NADPH.

are following three molecules of CO_2 through the reactions. FIGURE 10.17 divides the Calvin cycle into three phases :

Phase 1: Carbon fixation. The Calvin cycle incorporates each CO_2 molecule by attaching it to a five-carbon sugar named ribulose bisphosphate (abbreviated RuBP). The enzyme that catalyzes this first step is RuBP carboxylase, or **rubisco.** (It is the most abundant protein in chloroplasts and probably the most abundant protein on Earth.) The product of the reaction is a six-carbon intermediate so unstable that it immediately splits in half to form two molecules of 3-phosphoglycerate (for each CO_2).

Phase 2: Reduction. Each molecule of 3-phosphoglycerate

receives an additional phosphate group from ATP, becoming 1,3-bisphosphoglycerate. Next, a pair of electrons donated from NADPH reduces 1,3-bisphosphoglycerate to G3P. Specifically, the electrons from NADPH reduce the carboxyl group of 3-phosphoglycerate to the carbonyl group of G3P, which stores more potential energy. G3P is a sugar—the same three-carbon sugar formed in glycolysis by the splitting of glucose. Notice in FIGURE 10.17 that for every *three* molecules of CO_2, there are *six* molecules of G3P. But only one molecule of this three-carbon sugar can be counted as a net gain of carbohydrate. The cycle began with 15 carbons' worth of carbohydrate in the form of three molecules of the five-carbon sugar RuBP. Now there are 18 carbons' worth of carbohydrate in the form of six molecules of G3P. One molecule exits the cycle to be used by the plant cell, but the other five molecules must be recycled to regenerate the three molecules of RuBP.

Phase 3: Regeneration of CO₂ acceptor (RuBP). In a complex series of reactions, the carbon skeletons of five molecules of G3P are rearranged by the last steps of the Calvin cycle into three molecules of RuBP. To accomplish this, the cycle spends three more molecules of ATP. The RuBP is now prepared to receive CO_2 again, and the cycle continues.

For the net synthesis of one G3P molecule, the Calvin cycle consumes a total of nine molecules of ATP and six molecules of NADPH. The light reactions regenerate the ATP and NADPH. The G3P spun off from the Calvin cycle becomes the starting material for metabolic pathways that synthesize other organic compounds, including glucose and other carbohydrates. Neither the light reactions nor the Calvin cycle alone can make sugar from CO_2. Photosynthesis is an emergent property of the intact chloroplast, which integrates the two stages of photosynthesis.

Alternative mechanisms of carbon fixation have evolved in hot, arid climates

Ever since plants first moved onto land about 425 million years ago, they have been adapting to the problems of terrestrial life, particularly the problem of dehydration. In Chapters 29 and 36, we will consider anatomical adaptations that help plants conserve water. Here we are concerned with metabolic adaptations. The solutions often involve trade-offs. An important example is the compromise between photosynthesis and the prevention of excessive water loss from the plant. The CO_2 required for photosynthesis enters a leaf via stomata, the pores through the leaf surface (see FIGURE 10.2). However, stomata are also the main avenues of transpiration, the evaporative loss of water from leaves. On a hot, dry day, most plants close their stomata, a response that conserves water. This response also reduces photosynthetic yield by limiting access to CO_2. With stomata even partially closed, CO_2 concentrations begin to decrease in the air spaces within the leaf, and the concentration of O_2 released from photosynthesis begins to increase. These conditions within the leaf favor a seemingly wasteful process called photorespiration.

Photorespiration: An Evolutionary Relic?

In most plants, initial fixation of carbon occurs via rubisco, the Calvin cycle enzyme that adds CO_2 to ribulose bisphosphate. Such plants are called **C₃ plants** because the first organic product of carbon fixation is a three-carbon compound,

3-phosphoglycerate (see FIGURE 10.17). Rice, wheat, and soybeans are among the C₃ plants that are important in agriculture. These plants produce less food when their stomata close on hot, dry days. The declining level of CO_2 in the leaf starves the Calvin cycle. Making matters worse, rubisco can accept O_2 in place of CO_2. As O_2 concentrations overtake CO_2 concentrations within the air spaces of the leaf, rubisco adds O_2 to the Calvin cycle instead of CO_2. The product splits, and one piece, a two-carbon compound, is exported from the chloroplast. Mitochondria and peroxisomes then break the two-carbon molecule down to CO_2. The process is called **photorespiration** because it occurs in the light (*photo*) and consumes O_2 (*respiration*). However, unlike normal cellular respiration, photorespiration generates no ATP. And unlike photosynthesis, photorespiration produces no food. In fact, photorespiration *decreases* photosynthetic output by siphoning organic material from the Calvin cycle.

How can we explain the existence of a metabolic process that seems to be counterproductive to the plant? According to one hypothesis, photorespiration is evolutionary baggage—a metabolic relic from a much earlier time, when the atmosphere had less O_2 and more CO_2 than it does today. In the ancient atmosphere present when rubisco first evolved, the inability of the enzyme's active site to exclude O_2 would have made little difference. The hypothesis speculates that modern rubisco retains some of its ancestral affinity for O_2, which is now so concentrated in the atmosphere that a certain amount of photorespiration is inevitable.

It is not known whether photorespiration is beneficial to plants in any way. It *is* known that in many types of plants—including some of agricultural importance, such as soybeans—photorespiration drains away as much as 50% of the carbon fixed by the Calvin cycle. As heterotrophs that depend on carbon fixation in chloroplasts for our food, we naturally view photorespiration as wasteful. Indeed, if photorespiration could be reduced in certain plant species without otherwise affecting photosynthetic productivity, crop yields and food supplies would increase.

The environmental conditions that foster photorespiration are hot, dry, bright days—the conditions that cause stomata to close. In certain plant species, alternate modes of carbon fixation that minimize photorespiration—even in hot, arid climates—have evolved. The two most important of these photosynthetic adaptations are C₄ photosynthesis and CAM.

C₄ Plants

The **C₄ plants** are so named because they preface the Calvin cycle with an alternate mode of carbon fixation that forms a four-carbon compound as its first product. Several thousand species in at least 19 plant families use the C₄ pathway. Among the C₄ plants important to agriculture are sugarcane and corn, members of the grass family.

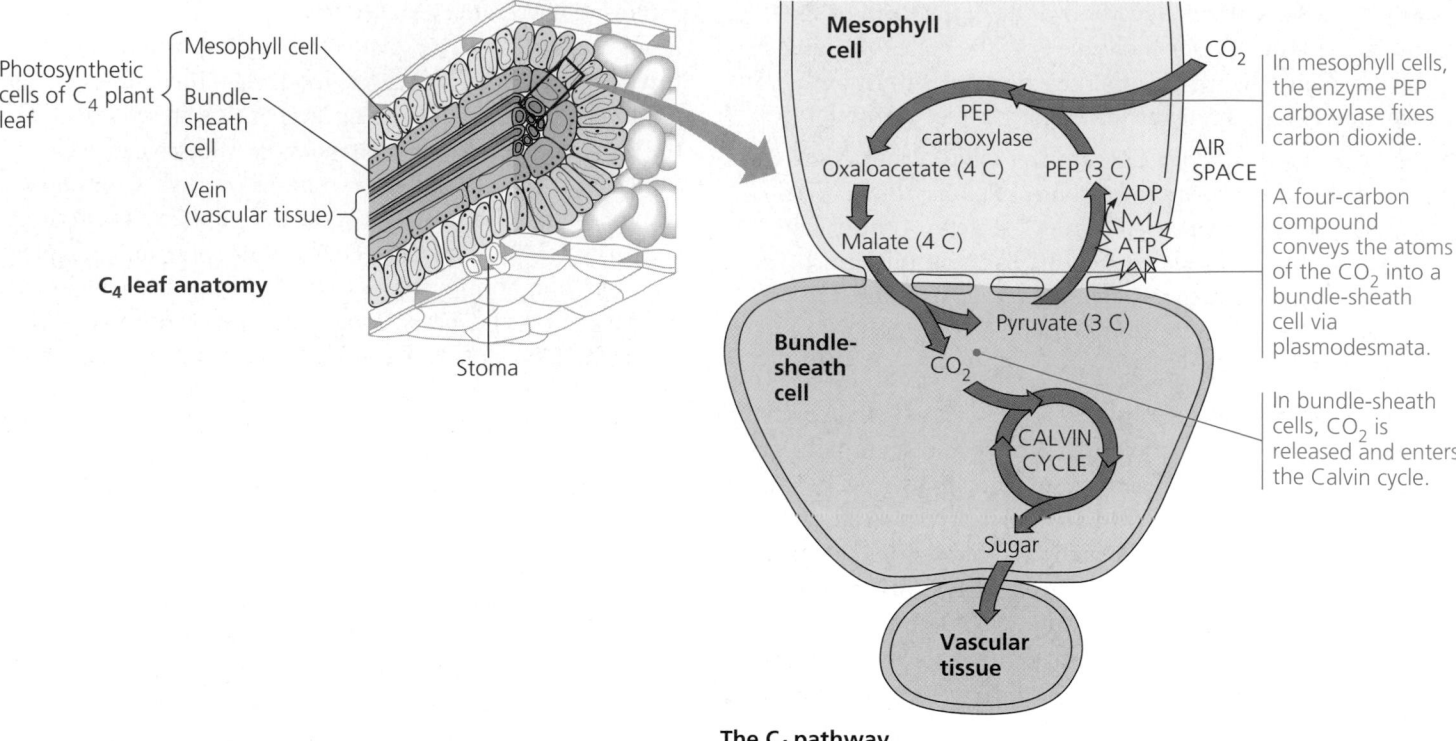

Photosynthetic cells of C_4 plant leaf
- Mesophyll cell
- Bundle-sheath cell
- Vein (vascular tissue)

C_4 leaf anatomy

Stoma

Mesophyll cell

CO_2

In mesophyll cells, the enzyme PEP carboxylase fixes carbon dioxide.

PEP carboxylase

Oxaloacetate (4 C) PEP (3 C)

AIR SPACE

ADP

ATP

A four-carbon compound conveys the atoms of the CO_2 into a bundle-sheath cell via plasmodesmata.

Malate (4 C)

Pyruvate (3 C)

Bundle-sheath cell

CO_2

In bundle-sheath cells, CO_2 is released and enters the Calvin cycle.

CALVIN CYCLE

Sugar

Vascular tissue

The C_4 pathway

FIGURE 10.18 C_4 leaf anatomy and the C_4 pathway. The structure and biochemical functions of the leaves of C_4 plants are an evolutionary adaptation to hot, dry climates. This adaptation maintains a CO_2 concentration in the bundle sheath that favors photosynthesis over photorespiration.

A unique leaf anatomy is correlated with the mechanism of C_4 photosynthesis (FIGURE 10.18; compare with FIGURE 10.2). In C_4 plants, there are two distinct types of photosynthetic cells: bundle-sheath cells and mesophyll cells. **Bundle-sheath cells** are arranged into tightly packed sheaths around the veins of the leaf. Between the bundle sheath and the leaf surface are the more loosely arranged **mesophyll cells.** The Calvin cycle is confined to the chloroplasts of the bundle sheath. However, the cycle is preceded by incorporation of CO_2 into organic compounds in the mesophyll. The first step is the addition of CO_2 to phosphoenolpyruvate (PEP) to form the four-carbon product oxaloacetate. The enzyme **PEP carboxylase** adds CO_2 to PEP. Compared to rubisco, PEP carboxylase has a very high affinity for CO_2. Therefore, PEP carboxylase can fix CO_2 efficiently when rubisco cannot—that is, when it is hot and dry and stomata are partially closed, causing CO_2 concentration in the leaf to fall and O_2 concentration to rise. After the C_4 plant fixes CO_2, the mesophyll cells export their four-carbon products (malate in the example shown in FIGURE 10.18) to bundle-sheath cells through plasmodesmata (see FIGURE 7.28). Within the bundle-sheath cells, the four-carbon compounds release CO_2, which is reassimilated into organic material by rubisco and the Calvin cycle.

In effect, the mesophyll cells of a C_4 plant pump CO_2 into the bundle sheath, keeping the CO_2 concentration in the bundle-sheath cells high enough for rubisco to accept carbon dioxide rather than oxygen. In this way, C_4 photosynthesis minimizes photorespiration and enhances sugar production. This adaptation is especially advantageous in hot regions with intense sunlight, and it is in such environments that C_4 plants evolved and thrive today.

CAM Plants

A second photosynthetic adaptation to arid conditions has evolved in succulent (water-storing) plants (including ice plants), many cacti, pineapples, and representatives of several other plant families. These plants open their stomata during the night and close them during the day, just the reverse of how other plants behave. Closing stomata during the day helps desert plants conserve water, but it also prevents CO_2 from entering the leaves. During the night, when their stomata are open, these plants take up CO_2 and incorporate it into a variety of organic acids. This mode of carbon fixation is called **crassulacean acid metabolism,** or **CAM,** after the plant family Crassulaceae, the succulents in which the process was first discovered. The mesophyll cells of **CAM plants** store the organic acids they make during the night in their vacuoles until morning, when the stomata close. During the day, when the light reactions can supply ATP and NADPH for the Calvin cycle, CO_2 is released from the organic acids made the night before to become incorporated into sugar in the chloroplasts.

FIGURE 10.19 C₄ and CAM photosynthesis compared. Both adaptations are characterized by ❶ preliminary incorporation of CO_2 into organic acids, followed by ❷ transfer of the CO_2 to the Calvin cycle. The C₄ and CAM pathways are two evolutionary solutions to the problem of maintaining photosynthesis with stomata partially or completely closed on hot, dry days.

Sugarcane

Pineapple

C₄

CO_2

Mesophyll cell

Organic acid

❶ CO_2 incorporated into four-carbon organic acids (carbon fixation)

Bundle-sheath cell

CO_2

CALVIN CYCLE

❷ Organic acids release CO_2 to Calvin cycle

Sugar

CAM

CO_2

Organic acid

Night

CO_2

Day

CALVIN CYCLE

Sugar

(a) Spatial separation of steps. In C₄ plants, carbon fixation and the Calvin cycle occur in different types of cells.

(b) Temporal separation of steps. In CAM plants, carbon fixation and the Calvin cycle occur in the same cells at different times.

Notice in FIGURE 10.19 that the CAM pathway is similar to the C₄ pathway in that carbon dioxide is first incorporated into organic intermediates before it enters the Calvin cycle. The difference is that in C₄ plants, the initial steps of carbon fixation are separated structurally from the Calvin cycle, whereas in CAM plants, the two steps occur at separate times. (Keep in mind that CAM, C₄, and C₃ plants all eventually use the Calvin cycle to make sugar from carbon dioxide.)

Photosynthesis is the biosphere's metabolic foundation: *a review*

In this chapter, we have followed photosynthesis from photons to food. The light reactions capture solar energy and use it to make ATP and transfer electrons from water to $NADP^+$. The Calvin cycle uses the ATP and NADPH to produce sugar from carbon dioxide. The energy that enters the chloroplasts as sunlight becomes stored as chemical energy in organic compounds. See FIGURE 10.20 on page 194 for a review of the entire process.

What are the fates of photosynthetic products? The sugar made in the chloroplasts supplies the entire plant with chemical energy and carbon skeletons to synthesize all the major organic molecules of cells. About 50% of the organic material made by photosynthesis is consumed as fuel for cellular respiration in the mitochondria of the plant cells. Sometimes there is a loss of photosynthetic products to photorespiration.

Technically, green cells are the only autotrophic parts of the plant. The rest of the plant depends on organic molecules exported from leaves via veins. In most plants, carbohydrate is transported out of the leaves in the form of sucrose, a disaccharide. After arriving at nonphotosynthetic cells, the sucrose provides raw material for cellular respiration and a multitude of anabolic pathways that synthesize proteins, lipids, and other products. A considerable amount of sugar in the form of glucose is linked together to make the polysaccharide cellulose, especially in plant cells that are still growing and maturing. Cellulose, the main ingredient of cell walls, is the most abundant organic molecule in the plant—and probably on the surface of the planet.

Most plants manage to make more organic material each day than they need to use as respiratory fuel and precursors for biosynthesis. They stockpile the extra sugar by synthesizing starch, storing some in the chloroplasts themselves and

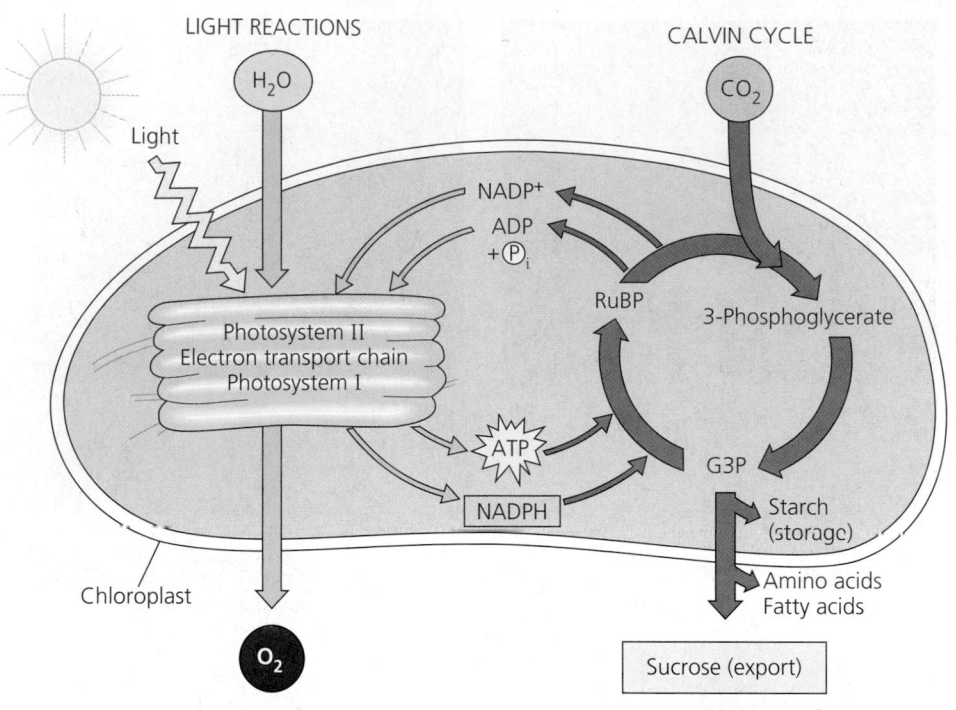

LIGHT REACTIONS

CALVIN CYCLE

FIGURE 10.20 A review of photosynthesis. This diagram outlines the main reactants and products of the light reactions and the Calvin cycle as they occur in the chloroplasts of plant cells. The entire ordered operation depends on the structural integrity of the chloroplast and its membranes. Enzymes in the chloroplast and cytosol convert glyceraldehyde-3-phosphate (G3P), the direct product of the Calvin cycle, into many other organic compounds.

Light reactions:
- Are carried out by molecules in the thylakoid membranes
- Convert light energy to the chemical energy of ATP and NADPH
- Split H_2O and release O_2 to the atmosphere

Calvin cycle reactions:
- Take place in the stroma
- Use ATP and NADPH to convert CO_2 to the sugar G3P
- Return ADP, inorganic phosphate, and $NADP^+$ to the light reactions

some in storage cells of roots, tubers, seeds, and fruits. In accounting for the consumption of the food molecules produced by photosynthesis, let's not forget that most plants lose leaves, roots, stems, fruits, and sometimes their entire bodies to heterotrophs, including humans.

On a global scale, the collective productivity of the minute chloroplasts is prodigious; it is estimated that photosynthesis makes about 160 billion metric tons of carbohydrate per year (a metric ton is 1,000 kg, about 1.1 tons). That's organic matter equivalent to a stack of about 60 trillion copies of this textbook—17 stacks of books reaching from Earth to the sun! No other chemical process on the planet can match the output of photosynthesis. And no process is more important than photosynthesis to the welfare of life on Earth.

CHAPTER 10 REVIEW

Go to the Campbell Biology website (www.campbellbiology.com) to explore an interactive version of the Chapter Review.

Summary of Key Concepts

PHOTOSYNTHESIS IN NATURE

- **Plants and other autotrophs are the producers of the biosphere** (pp. 176–177, FIGURE 10.1) Autotrophs nourish themselves without ingesting organic molecules. Photoautotrophs use the energy of sunlight to synthesize organic molecules from CO_2 and H_2O.

Heterotrophs ingest organic molecules from other organisms to get energy and carbon.

- **Chloroplasts are the sites of photosynthesis in plants** (p. 178, FIGURE 10.2) In autotrophic eukaryotes, photosynthesis occurs in chloroplasts, organelles containing thylakoid membranes that separate the thylakoid space from the chloroplast's stroma. Stacks of thylakoids form grana.
  Web/CD Activity 10A: *The Sites of Photosynthesis*

THE PATHWAYS OF PHOTOSYNTHESIS

■ **Evidence that chloroplasts split water molecules enabled researchers to track atoms through photosynthesis (pp. 179–180, FIGURE 10.3)** Photosynthesis is summarized by the equation

$$6\,CO_2 + 12\,H_2O + \text{Light energy} \longrightarrow C_6H_{12}O_6 + 6\,O_2 + 6\,H_2O$$

Experiments show that the chloroplast splits water into hydrogen and oxygen, incorporating the electrons of hydrogen into the bonds of sugar molecules. Photosynthesis is a redox process: H_2O is oxidized, CO_2 is reduced.

■ **The light reactions and the Calvin cycle cooperate in converting light energy to the chemical energy of food:** *an overview* **(pp. 180–181, FIGURE 10.4)** The light reactions in the grana produce ATP and split water, releasing O_2 and forming NADPH by transferring electrons from water to $NADP^+$. The Calvin cycle in the stroma forms sugar from CO_2, using ATP for energy and NADPH for reducing power.
Web/CD Activity 10B: *Overview of Photosynthesis*

■ **The light reactions convert solar energy to the chemical energy of ATP and NADHP:** *a closer look* **(pp. 181–189, FIGURES 10.12, 10.16)** Light is a form of electromagnetic energy, which travels in waves. The colors we see as visible light are a part of the electromagnetic spectrum. A pigment is a substance that absorbs visible light of specific wavelengths. The action spectrum of photosynthesis does not exactly match the absorption spectrum of chlorophyll *a*, the main photosynthetic pigment in plants, because accessory pigments (chlorophyll *b* and various carotenoids) absorb different wavelengths of light and pass the energy on to chlorophyll *a*.

A pigment goes from a ground state to an excited state when a photon boosts one of its electrons to a higher-energy orbital. The pigments of chloroplasts are built into the thylakoid membrane near molecules called primary electron acceptors, which trap the excited electrons before they return to the ground state. Pigment molecules are clustered in an antenna complex surrounding a chlorophyll *a* molecule at the reaction center. Photons absorbed anywhere in the antenna can pass their energy along to energize this chlorophyll *a*, which then passes an electron to a nearby primary electron acceptor. The antenna complex, the reaction-center chlorophyll, and the primary electron acceptor make up a photosystem, a light-harvesting unit built into the thylakoid membrane. There are two kinds of photosystems. Photosystem I contains P700 chlorophyll *a* molecules at the reaction center; photosystem II contains P680 molecules.

Noncyclic electron flow involves both photosystems and produces NADPH, ATP, and oxygen. Cyclic electron flow employs only photosystem I, producing ATP but no NADPH or O_2. ATP production during the light reactions is called photophosphorylation. The mechanism is chemiosmosis. The redox reactions of the electron transport chain that connects the two photosystems generate an H^+ gradient across the thylakoid membrane. ATP synthase uses this proton-motive force to make ATP.
Web/CD Activity 10C: *Light Energy and Pigments*
Web/CD Case Study in the Process of Science: *How Does Paper Chromatography Separate Plant Pigments?*
Web/CD Activity 10D: *The Light Reactions*

■ **The Calvin cycle uses ATP and NADPH to convert CO_2 to sugar:** *a closer look* **(pp. 189–191, FIGURE 10.17)** The Calvin cycle is a metabolic pathway in the chloroplast stroma. An enzyme (rubisco) combines CO_2 with ribulose bisphosphate (RuBP), a five-carbon sugar. Then, using electrons from NADPH and energy from ATP,

the cycle synthesizes the three-carbon sugar glyceraldehyde-3-phosphate. Most of the G3P is reused in the cycle to reconstitute RuBP, but some exits the cycle and is converted to glucose and other essential organic molecules.
Web/CD Activity 10E: *The Calvin Cycle*
Web/CD Case Study in the Process of Science: *How Is the Rate of Photosynthesis Measured?*
Biology Labs On-Line: *LeafLab*

■ **Alternative mechanisms of carbon fixation have evolved in hot, arid climates (pp. 191–193, FIGURE 10.19)** On dry, hot days, plants close their stomata, conserving water. Oxygen from the light reactions builds up. When O_2 substitutes for CO_2 in the active site of rubisco, the product formed leaves the cycle and is oxidized to CO_2 and H_2O in the peroxisomes and mitochondria. This process, photorespiration, consumes organic fuel without producing ATP. C_4 plants avert photorespiration by incorporating CO_2 into four-carbon compounds in mesophyll cells. These compounds are exported to photosynthetic bundle-sheath cells, where they release carbon dioxide for use in the Calvin cycle. CAM plants open their stomata during the night, incorporating the CO_2 that enters into organic acids, which they store in mesophyll cells. During the day the stomata close, and the CO_2 is released from the organic acids for use in the Calvin cycle.
Web/CD Activity 10F: *Photosynthesis in Dry Climates*

■ **Photosynthesis is the biosphere's metabolic foundation:** *a review* **(pp. 193–194, FIGURE 10.20)** The organic compounds produced by photosynthesis provide the energy and building material for ecosystems.

Self-Quiz

1. The light reactions of photosynthesis supply the Calvin cycle with
   a. light energy.
   b. CO_2 and ATP.
   c. H_2O and NADPH.
   d. ATP and NADPH.
   e. sugar and O_2.

2. Which of the following sequences correctly represents the flow of electrons during photosynthesis?
   a. NADPH $\longrightarrow O_2 \longrightarrow CO_2$
   b. $H_2O \longrightarrow$ NADPH $\longrightarrow$ Calvin cycle
   c. NADPH $\longrightarrow$ chlorophyll $\longrightarrow$ Calvin cycle
   d. $H_2O \longrightarrow$ photosystem I $\longrightarrow$ photosystem II
   e. NADPH $\longrightarrow$ electron transport chain $\longrightarrow O_2$

3. Which of the following conclusions does *not* follow from studying the absorption spectrum for chlorophyll *a* and the action spectrum for photosynthesis?
   a. Not all wavelengths are equally effective for photosynthesis.
   b. There must be accessory pigments that broaden the spectrum of light that contributes energy for photosynthesis.
   c. The red and blue areas of the spectrum are most effective in driving photosynthesis.
   d. Chlorophyll owes its color to the absorption of green light.
   e. Chlorophyll *a* has two absorption peaks.

4. Cooperation of the *two* photosystems of the chloroplast is required for
   a. ATP synthesis.
   b. reduction of $NADP^+$.
   c. cyclic photophosphorylation.
   d. oxidation of the reaction center of photosystem I.
   e. generation of a proton-motive force.

5. In *mechanism*, photophosphorylation is most similar to
   a. substrate-level phosphorylation in glycolysis.
   b. oxidative phosphorylation in cellular respiration.
   c. the Calvin cycle.
   d. carbon fixation.
   e. reduction of $NADP^+$.

6. In what respect are the photosynthetic adaptations of C_4 plants and CAM plants similar?
   a. In both cases, the stomata normally close during the day.
   b. Both types of plants make their sugar without the Calvin cycle.
   c. In both cases, an enzyme other than rubisco carries out the first step in carbon fixation.
   d. Both types of plants make most of their sugar in the dark.
   e. Neither C_4 plants nor CAM plants have grana in their chloroplasts.

7. Which of the following processes is most directly driven by light energy?
   a. creation of a pH gradient by pumping protons across the thylakoid membrane
   b. carbon fixation in the stroma
   c. reduction of NADP molecules
   d. removal of electrons from membrane-bound chlorophyll molecules
   e. ATP synthesis

8. Which of the following statements is a correct distinction between cyclic and noncyclic photophosphorylation?
   a. Only noncyclic photophosphorylation produces ATP.
   b. In addition to ATP, cyclic photophosphorylation also produces O_2 and NADPH.
   c. Only cyclic photophosphorylation utilizes light at 700 nm.
   d. Chemiosmosis is unique to noncyclic photophosphorylation.
   e. Only cyclic photophosphorylation can operate in the absence of photosystem II.

9. Which of the following statements is a correct distinction between autotrophs and heterotrophs?
   a. Only heterotrophs require chemical compounds from the environment.
   b. Cellular respiration is unique to heterotrophs.
   c. Only heterotrophs have mitochondria.
   d. Autotrophs, but not heterotrophs, can nourish themselves beginning with CO_2 and other nutrients that are entirely inorganic.
   e. Only heterotrophs require oxygen.

10. Which of the following processes could still occur in a chloroplast in the presence of an inhibitor that prevents H^+ from passing through ATP synthase complexes? (Explain your answer.)
    a. sugar synthesis
    b. generation of a proton-motive force
    c. photophosphorylation
    d. the Calvin cycle
    e. oxidation of NADPH

11. Chloroplast is to _____ as _____ is to cellular respiration.

12. Which redox process, photosynthesis or cellular respiration, is endergonic?

13. For chloroplasts to produce sugar from carbon dioxide in the dark, they would require an artificial supply of _____ and _____.

14. What color of light is *least* effective in driving photosynthesis?

15. Compared to a solution of isolated chlorophyll, why do intact chloroplasts release less heat and fluorescence when illuminated?

16. What is the advantage of the light reactions producing NADPH and ATP on the stroma side of the thylakoid membrane?

17. To synthesize one glucose molecule, the Calvin cycle uses _____ molecules of CO_2, _____ molecules of ATP, and _____ molecules of NADPH.

18. Explain why a poison that inhibits an enzyme of the Calvin cycle will also inhibit the light reactions.

19. How would you expect the relative abundance of C_3 versus C_4 and CAM species to change in a geographic region whose climate becomes much hotter and drier?

Go to the website or CD-ROM for more quiz questions.

Evolution Connection

In this chapter, you learned that photorespiration can substantially decrease a plant's photosynthetic output. In soybeans, for example, it can eliminate up to 50% of the fixed carbon. Would you expect this figure to be higher or lower in wild relatives of soybeans? Why?

The Process of Science

The diagram below represents an experiment with isolated chloroplasts. The chloroplasts were first made acidic by soaking them in a solution at pH 4. After the thylakoid space reached pH 4, the chloroplasts were transferred to a basic solution at pH 8. The chloroplasts then made ATP in the dark. Explain this result.

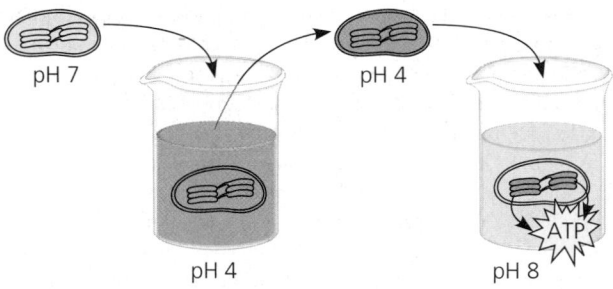

Learn how to separate plant pigments by chromatography and measure the rate of photosynthesis in the Case Studies in the Process of Science, available on the website and CD-ROM. Conduct virtual photosynthesis experiments at the Biology Labs On-Line *LeafLab*.

Science, Technology, and Society

CO_2 in the atmosphere traps heat and warms the air, just as clear glass does in a greenhouse. Most scientists believe that the CO_2 added to the air by the burning of wood and fossil fuels by humans is contributing to a dangerous rise in global temperature. Tropical rain forests are estimated to be responsible for more than 20% of global photosynthesis. It seems reasonable to expect that the lush growth of jungle foliage would reduce global warming by consuming large amounts of CO_2, but many experts now believe that rain forests make little or no *net* contribution to reduction of global warming. Why might this be? (*Hint:* What happens to the food produced by a rain forest tree when it is eaten by animals or the tree dies?)

CHAPTER 11

CELL COMMUNICATION

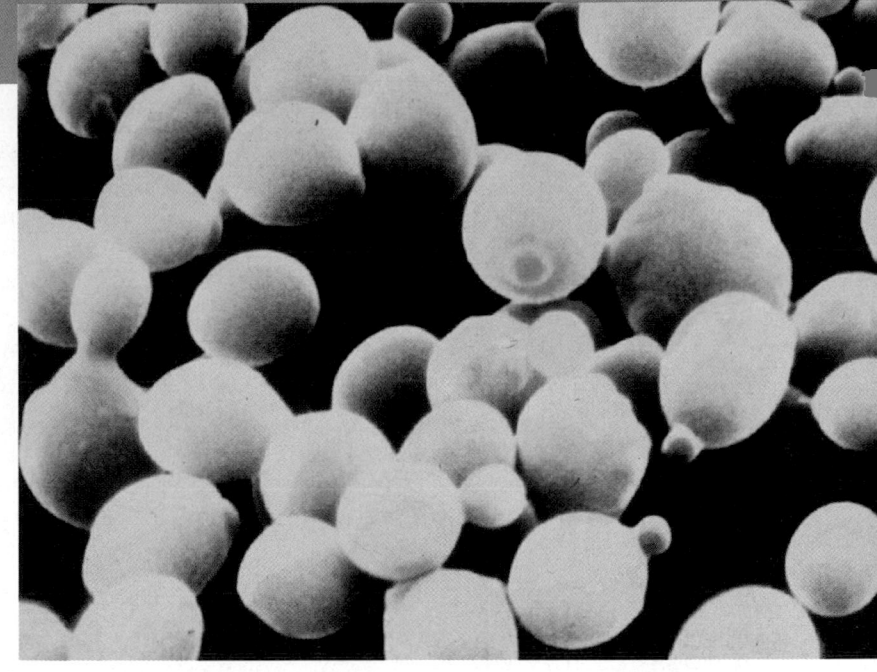

AN OVERVIEW OF CELL SIGNALING

- Cell signaling evolved early in the history of life
- Communicating cells may be close together or far apart
- The three stages of cell signaling are reception, transduction, and response

SIGNAL RECEPTION AND THE INITIATION OF TRANSDUCTION

- A signal molecule binds to a receptor protein, causing the protein to change shape
- Most signal receptors are plasma membrane proteins

SIGNAL-TRANSDUCTION PATHWAYS

- Pathways relay signals from receptors to cellular responses
- Protein phosphorylation, a common mode of regulation in cells, is a major mechanism of signal transduction
- Certain small molecules and ions are key components of signaling pathways (second messengers)

CELLULAR RESPONSES TO SIGNALS

- In response to a signal, a cell may regulate activities in the cytoplasm or transcription in the nucleus
- Elaborate pathways amplify and specify the cell's response to signals

"**W**atch out! There's a car coming!" *We don't really need such a warning to remind us of the importance of communication in our lives. Perhaps less obvious is the critical role of communication in life at the cellular level. Cell-to-cell communication is absolutely essential for multicellular organisms. The billions of cells in a human or an oak tree must communicate in order to coordinate their activities in a way that enables the organism to develop from a fertilized egg and then survive and reproduce in turn. Communication between cells is also important for many unicellular organisms, such as the yeast pictured in the micrograph on this page.*

In studying how cells signal to each other and how they interpret the signals they receive, biologists have discovered some uni-versal mechanisms of cellular regulation, additional evidence for the evolutionary relatedness of all life. The same small set of cell-signaling mechanisms are showing up again and again in many lines of biological research. And to the delight of scientists, studies of cell signaling are helping to answer some of the most important questions in biology and medicine—in areas ranging from embryological development to hormone action to cancer and other kinds of disease.

The signals received by cells, whether originating from other cells or from changes in the physical environment, take various forms. For instance, cells can sense and respond to electromagnetic signals, such as light, and to mechanical signals, such as touch. (In later units of this text, we discuss cellular detection of these and other kinds of physical signals.) However, cells most often communicate with each other by chemical signals. In this chapter, we focus on the main mechanisms by which cells detect, process, and respond to chemical signals sent from other cells.

AN OVERVIEW OF CELL SIGNALING

What do cells talk about? What kinds of things does a "talking" cell say to a "listening" cell, and how does the latter cell respond to the message? Let's approach these questions by first looking at communication among microorganisms, for modern microbes are a window to the role of cell signaling in the evolution of life on Earth.

Cell signaling evolved early in the history of life

One topic of cell "conversation" is sex—at least for the yeast *Saccharomyces cerevisiae*, the fungus people have used for millennia for making bread, wine, and beer. Researchers have

197

FIGURE 11.1 Communication between mating yeast cells.

Cells of the yeast *Saccharomyces cerevisiae* use chemical signaling to identify cells of opposite mating type and initiate the mating process. The two mating types and their corresponding chemical signals, or "mating factors," are called **a** and **α**. The mating factors are peptides about 12 amino acids in length.

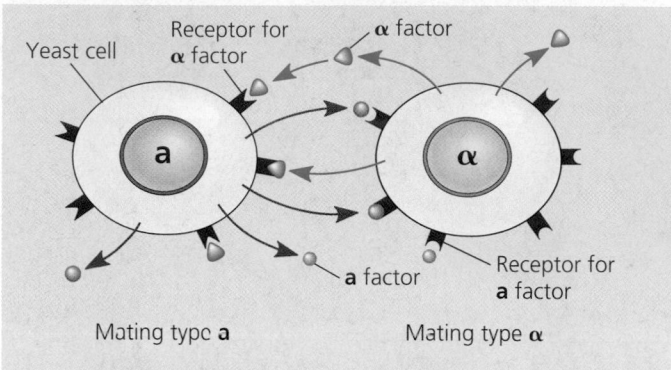

Exchange of mating factors. Each cell secretes its mating factor, which binds to the other cell.

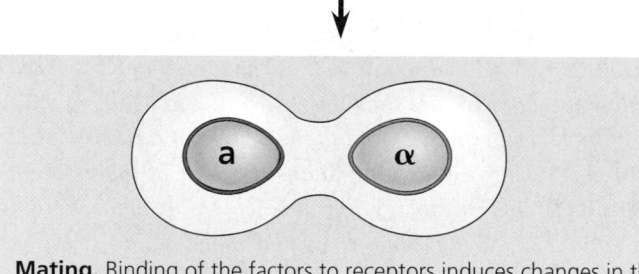

Mating. Binding of the factors to receptors induces changes in the cells that lead to their fusion.

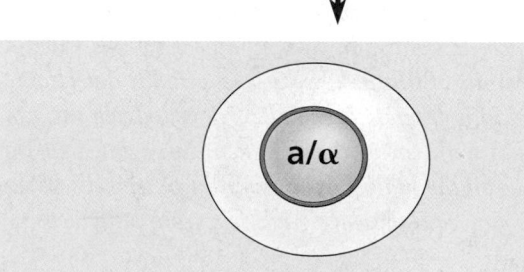

New a/α cell. This cell combines in its nucleus all the genes from the **a** and **α** cells.

learned that cells of this yeast identify their mates by chemical signaling (FIGURE 11.1). There are two sexes, or mating types, called **a** and **α**. Cells of mating type **a** secrete a chemical signal called **a** factor, which can bind to specific receptor proteins on nearby **α** cells. At the same time, **α** cells secrete **α** factor, which binds to receptors on **a** cells. Without actually entering the cells, the receptor-bound molecules of the two mating factors cause the cells to grow toward each other and bring about other cellular changes. The result is the fusion, or mating, of two cells of opposite type. The new **a/α** cell contains all the genes of both original cells, a combination of genetic resources that provides advantages to this cell's descendants.

How is the mating signal at the yeast cell surface transduced, or changed, into a form that brings about the cellular response of mating? The process by which a signal on a cell's surface is converted into a specific cellular response is a series of steps called a **signal-transduction pathway.** Many such pathways have been extensively studied in both yeast and animal cells. Amazingly, the molecular details of signal transduction in yeast and mammals are strikingly similar, even though the last common ancestor of these two groups of organisms lived over a billion years ago. These similarities—and others recently uncovered between signaling systems in bacteria and plants—suggest that early versions of the cell-signaling mechanisms used today evolved well before the first multicellular creatures appeared on Earth. Scientists think that signaling mechanisms evolved first in ancient prokaryotes and single-celled eukaryotes and were then adopted for new uses by their multicellular descendants. Meanwhile, cell signaling has remained important in the microbial world. FIGURE 11.2 shows an example in a modern species of bacterium.

FIGURE 11.2 Communication among bacteria.

Soil-dwelling bacteria called myxobacteria ("slime bacteria") use chemical signaling to share information about nutrient availability. When food is scarce, starving cells secrete a molecule that enters neighboring cells and stimulates them to aggregate. The cells form a structure that produces thick-walled spores capable of surviving until the environment improves. The bacteria shown here are *Myxococcus xanthus*.

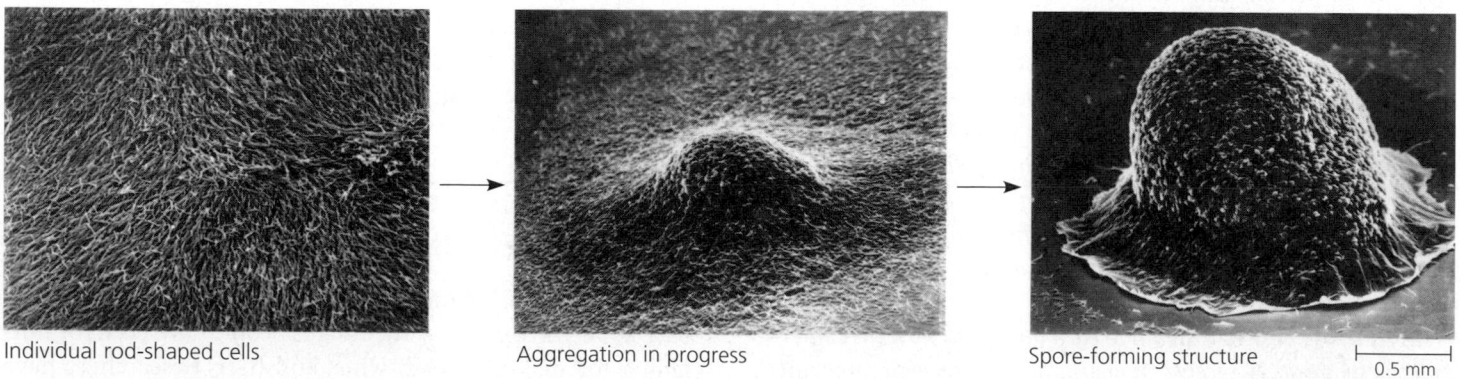

Individual rod-shaped cells   Aggregation in progress   Spore-forming structure   0.5 mm

Communicating cells may be close together or far apart

Like microbes, cells in a multicellular organism usually communicate by releasing chemical messengers targeted for cells that may not be immediately adjacent. Some messengers travel only short distances: The transmitting cell secretes molecules of a **local regulator,** a substance that influences cells in the vicinity (FIGURE 11.3a). One class of local regulators in animals, *growth factors,* are compounds that stimulate nearby target cells to grow and multiply. Numerous cells can simultaneously receive and respond to the molecules of growth factor produced by a single cell in their vicinity. This type of local signaling in animals is called *paracrine signaling.*

Another, more specialized type of local signaling occurs in the animal nervous system. Here a nerve cell produces a chemical signal, a neurotransmitter, that diffuses to a single target cell that is almost touching the first cell. An electrical signal transmitted the length of the nerve cell triggers the secretion of neurotransmitter molecules into the synapse, the narrow space between the nerve cell and its target cell (often another nerve cell). Because specific nerve cells are so close together at synapses, a nerve signal can travel along a series of nerve cells from your brain to your big toe, for example, without causing unwanted responses in other parts of your body.

Local signaling in plants is not as well understood. Because of their cell walls, plants must use mechanisms different from those operating locally in animals.

Both animals and plants use chemicals called **hormones** for signaling at greater distances. In hormonal signaling in animals, also known as endocrine signaling, specialized cells release hormone molecules into vessels of the circulatory system, by which they travel to target cells in other parts of the body (FIGURE 11.3b). In plants, hormones sometimes travel in vessels but more often reach their targets by moving through cells (see FIGURE 39.6) or by diffusion through the air as a gas. Hormones range widely in molecular size and type, as do local regulators. For instance, the plant hormone ethylene, a gas that promotes fruit ripening and helps regulate growth, is a hydrocarbon of only six atoms (C_2H_4). In contrast, the mammalian hormone insulin, which regulates sugar levels in the blood, is a protein with thousands of atoms.

Cells may also communicate by direct contact, as we saw in Chapters 7 and 8. Both animals and plants have cell junctions that, where present, directly connect the cytoplasms of adjacent cells (FIGURE 11.4a on p. 200). In these cases, signaling substances dissolved in the cytosol can pass freely between adjacent cells. Moreover, animal cells may communicate via direct contact between molecules on their surfaces (FIGURE 11.4b on p. 200). This sort of signaling is important in embryonic development and in the operation of the immune system.

What happens when a cell encounters a signal? The signal must be recognized by a specific receptor molecule, and the information it carries must be changed into another form — transduced — inside the cell before the cell can respond. The remainder of the chapter discusses this process, primarily as it occurs in animal cells.

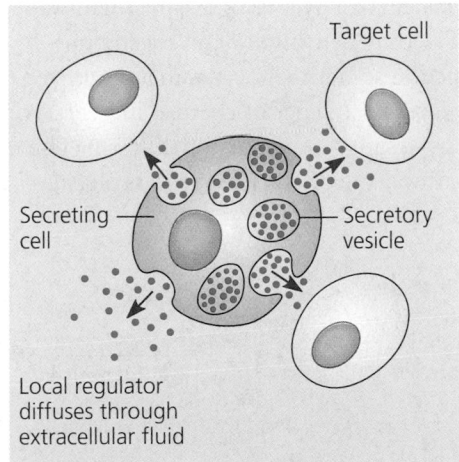

Paracrine signaling. A secreting cell acts on nearby target cells by discharging molecules of a local regulator into the extracellular fluid.

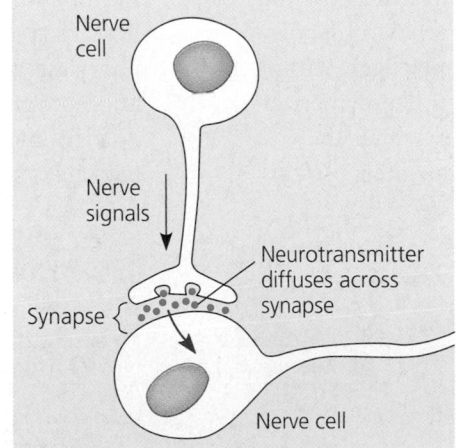

Synaptic signaling. A nerve cell releases neurotransmitter molecules into a synapse, the narrow space between the transmitting cell and the target cell.

(a) Local signaling

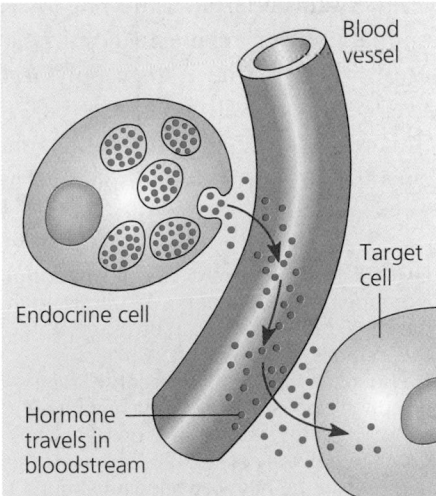

(b) Long distance (hormonal) signaling. Specialized endocrine cells secrete hormones into body fluids, often the blood. Hormones may reach virtually all body cells.

FIGURE 11.3 Local and long-distance cell communication in animals. In both local and long-distance signaling, only specific target cells recognize and respond to a given chemical signal.

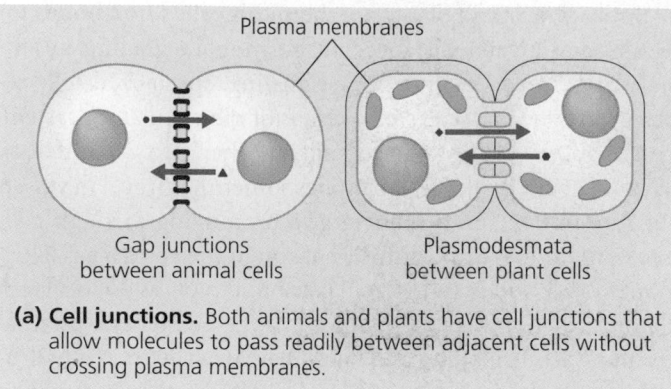

Plasma membranes

Gap junctions
between animal cells

Plasmodesmata
between plant cells

(a) Cell junctions. Both animals and plants have cell junctions that allow molecules to pass readily between adjacent cells without crossing plasma membranes.

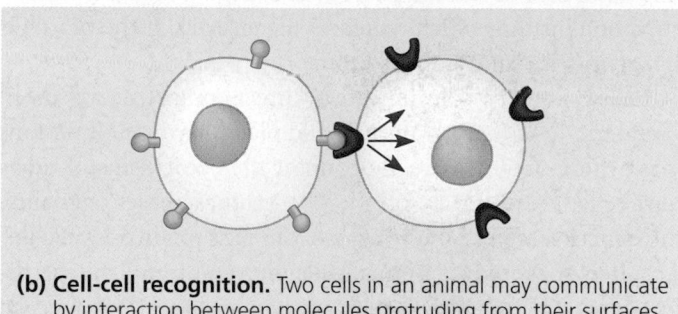

(b) Cell-cell recognition. Two cells in an animal may communicate by interaction between molecules protruding from their surfaces.

FIGURE 11.4 Communication by direct contact between cells.

The three stages of cell signaling are reception, transduction, and response

THE PROCESS OF SCIENCE

Our current understanding of how chemical messengers act via signal-transduction pathways had its origins in the pioneering work of Earl W. Sutherland, whose research led to a Nobel Prize in 1971. Sutherland and his colleagues at Vanderbilt University were investigating how the animal hormone epinephrine stimulates breakdown (depolymerization) of the storage polysaccharide glycogen within liver cells and skeletal muscle cells. Glycogen depolymerization releases the sugar glucose-1-phosphate, which the cell converts to glucose-6-phosphate. The cell can then use this

compound, an early intermediate in glycolysis, for energy production. Alternatively, the compound can be stripped of phosphate and released into the blood as glucose, which can fuel cells throughout the body. Thus, one effect of epinephrine, which is secreted from the adrenal gland during times of physical or mental stress, is the mobilization of fuel reserves.

Sutherland's research team discovered that epinephrine stimulates glycogen breakdown by somehow activating a cytosolic enzyme, glycogen phosphorylase. However, when epinephrine was added to a test-tube mixture containing the phosphorylase and its substrate, glycogen, no depolymerization occurred. Epinephrine could activate glycogen phosphorylase only when the hormone was added to a solution containing *intact* cells. This result told Sutherland two things. First, epinephrine does not interact directly with the enzyme responsible for glycogen breakdown; an intermediate step or series of steps must be occurring inside the cell. Second, the plasma membrane is somehow involved in transmitting the epinephrine signal.

Sutherland's early work suggested that the process going on at the receiving end of a cellular conversation can be dissected into three stages: reception, transduction, and response (FIGURE 11.5):

❶ **Reception.** Reception is the target cell's detection of a signal coming from outside the cell. A chemical signal is "detected" when it binds to a cellular protein, usually at the cell's surface.

❷ **Transduction.** The binding of the signal molecule changes the receptor protein in some way, initiating the process of transduction. The transduction stage converts the signal to a form that can bring about a specific cellular response. In Sutherland's system, the binding of epinephrine to the outside of a receptor protein in a liver cell's plasma membrane leads via a series of steps to activation of glycogen phosphorylase. Transduction sometimes occurs in a single step but more often requires a sequence of changes in a series of different molecules—a signal-transduction *pathway*. The molecules in the pathway are often called relay molecules.

FIGURE 11.5 Overview of cell signaling. From the perspective of the cell receiving the message, cell signaling can be divided into three stages: signal reception, signal transduction, and cellular response. When reception occurs at the plasma membrane, as shown here, the transduction stage is usually a pathway of several steps, with each molecule in the pathway bringing about a change in the next. The last molecule in the pathway triggers the cell's response. The three stages are explained in the text.

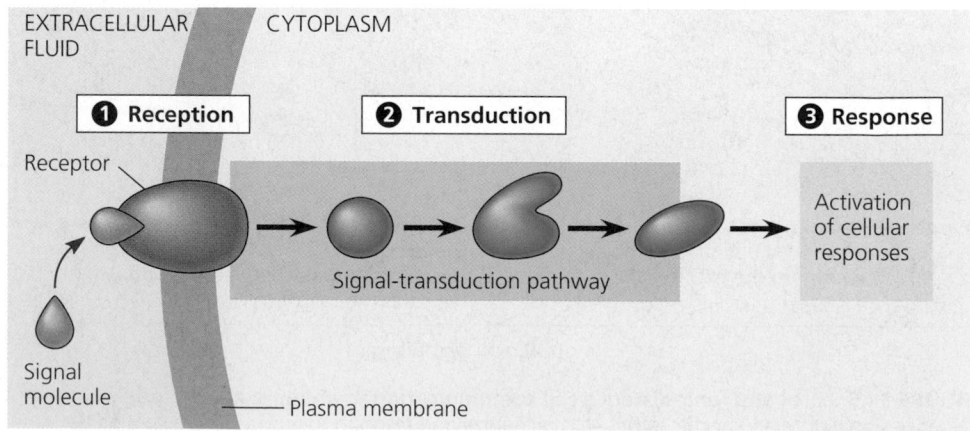

EXTRACELLULAR FLUID   CYTOPLASM

❶ Reception   ❷ Transduction   ❸ Response

Receptor

Signal-transduction pathway

Activation of cellular responses

Signal molecule

Plasma membrane

❸ **Response.** In the third stage of cell signaling, the transduced signal finally triggers a specific cellular response. The response may be almost any imaginable cellular activity—such as catalysis by an enzyme (for example, glycogen phosphorylase), rearrangement of the cytoskeleton, or activation of specific genes in the nucleus. The cell-signaling process helps ensure that crucial activities like these occur in the right cells, at the right time, and in proper coordination with the other cells of the organism.

We'll now explore the mechanisms of cell signaling in more detail.

SIGNAL RECEPTION AND THE INITIATION OF TRANSDUCTION

When we speak to someone, others nearby may hear our message, sometimes with unfortunate consequences. However, errors of this kind rarely occur among cells. The signals emitted by an α yeast cell are "heard" only by its prospective mates, **a** cells. Similarly, although epinephrine encounters many types of cells as it circulates in the blood, only certain target cells detect and react to the hormone. The signal receptor is the identity tag on the target cell.

A signal molecule binds to a receptor protein, causing the protein to change shape

A cell targeted by a particular chemical signal has molecules of a receptor protein that recognize the signal molecule. The signal molecule is complementary in shape to a specific site on the receptor and attaches there, like a key in a lock or a substrate in the catalytic site of an enzyme. The signal molecule behaves as a **ligand,** the term for a small molecule that specifically binds to a larger one. Ligand binding generally causes a receptor protein to undergo a change in conformation—that is, to change shape. For many receptors, this shape change directly activates the receptor so that it can interact with another cellular molecule. For other kinds of receptors, as we'll see shortly, the immediate effect of ligand binding is more limited, mainly causing the aggregation of two or more receptor molecules.

Most signal receptors are plasma membrane proteins

Most signal molecules are water-soluble and too large to pass freely through the plasma membrane. But as Sutherland learned for epinephrine, they can still influence cellular activity in major ways. Like yeast mating factors, most water-soluble signal molecules bind to specific sites on receptor proteins embedded in the cell's plasma membrane. Such a receptor transmits information from the extracellular environment to the inside of the cell by changing shape or aggregating when a specific ligand binds to it.

We'll see how membrane receptors work by looking at three major types: G-protein-linked receptors, tyrosine-kinase receptors, and ion-channel receptors.

G-Protein-Linked Receptors

A **G-protein-linked receptor** is a plasma membrane receptor that works with the help of a protein called a G protein. Many different signal molecules use G-protein-linked receptors, including yeast mating factors, epinephrine and many other hormones, and neurotransmitters. These receptors vary in their binding sites for recognizing signal molecules and for recognizing different G proteins inside the cell. Nevertheless, G-protein-linked receptor proteins are all remarkably similar in structure. They each have seven α helices spanning the membrane, as shown in FIGURE 11.6.

Loosely attached to the cytoplasmic side of the membrane, the **G protein** functions as a switch that is on or off, depending on which of two guanine nucleotides is attached, GDP or GTP. (GTP, or guanosine triphosphate, is similar to ATP.) As indicated in FIGURE 11.7 (p. 202), when GDP is bound, the G protein is inactive; when GTP is bound, the G protein is active.

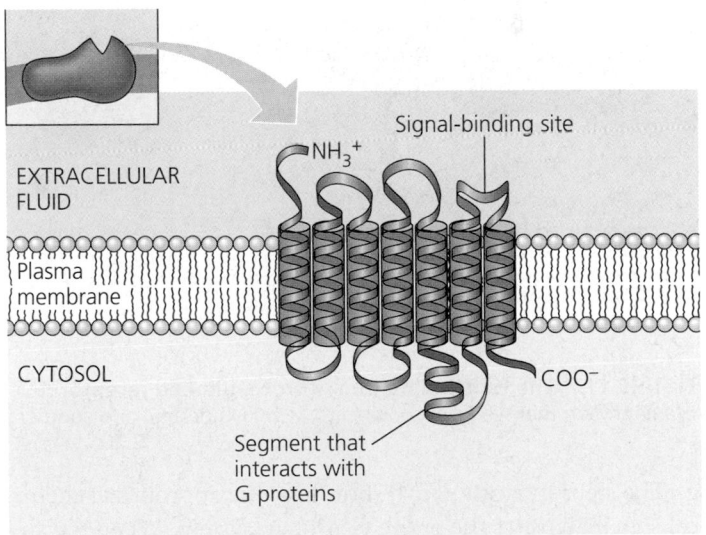

FIGURE 11.6 The structure of a G-protein-linked receptor. A large family of eukaryotic receptor proteins have this secondary structure, where the single polypeptide, represented as a ribbon, has seven transmembrane α helices. For clarity, the α helices are depicted as cylinders and arranged in a straight line. Specific loops correspond to the sites where signal molecules and G-protein molecules bind; the loops labeled here are the binding sites in one particular case.

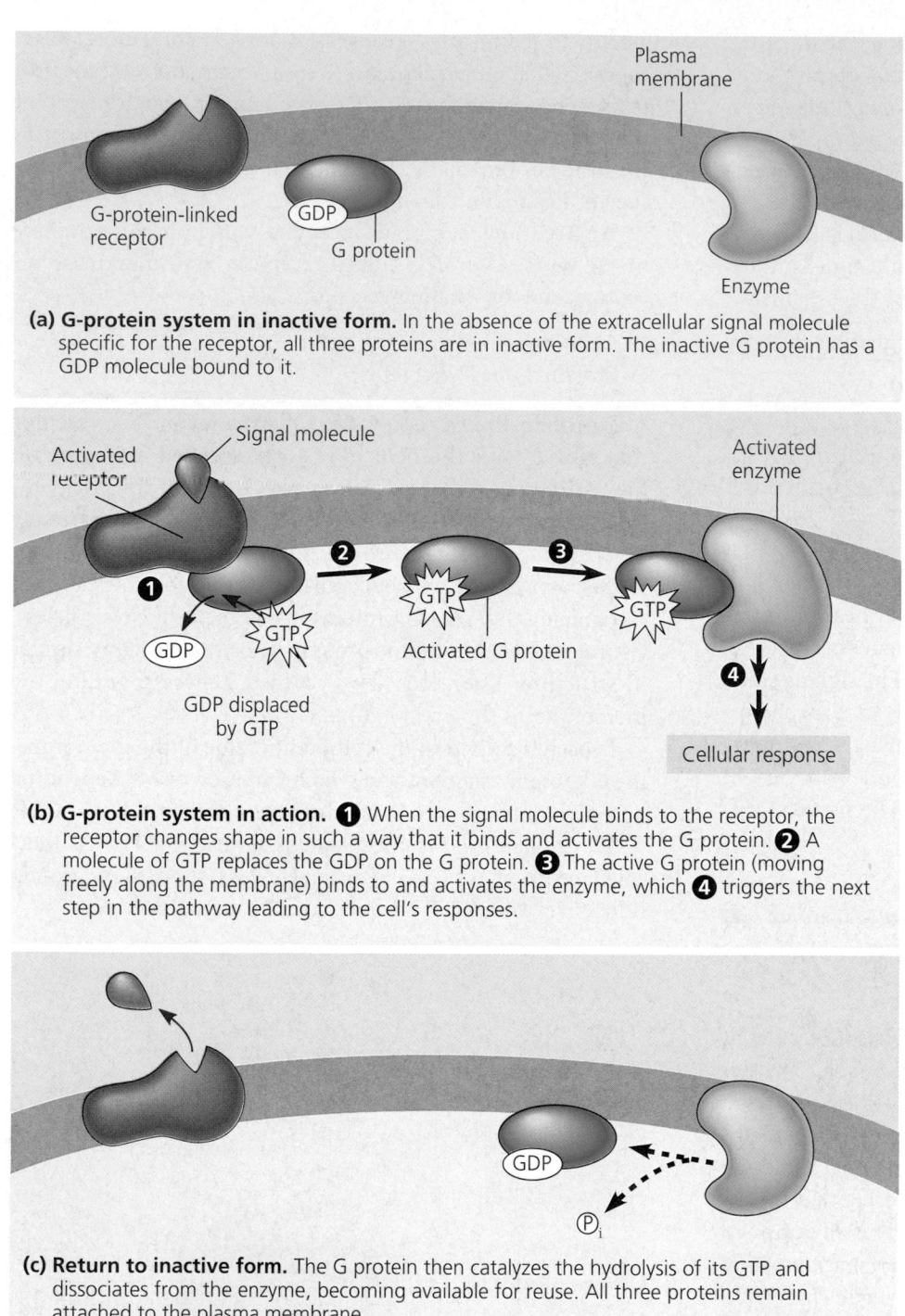

(a) G-protein system in inactive form. In the absence of the extracellular signal molecule specific for the receptor, all three proteins are in inactive form. The inactive G protein has a GDP molecule bound to it.

(b) G-protein system in action. ❶ When the signal molecule binds to the receptor, the receptor changes shape in such a way that it binds and activates the G protein. ❷ A molecule of GTP replaces the GDP on the G protein. ❸ The active G protein (moving freely along the membrane) binds to and activates the enzyme, which ❹ triggers the next step in the pathway leading to the cell's responses.

(c) Return to inactive form. The G protein then catalyzes the hydrolysis of its GTP and dissociates from the enzyme, becoming available for reuse. All three proteins remain attached to the plasma membrane.

FIGURE 11.7 The functioning of a G-protein-linked receptor. This type of receptor is a membrane protein that works in conjunction with a G protein and another protein, usually an enzyme.

zyme is activated, ❹ it can trigger the next step in the pathway.

The changes in the enzyme and G protein are only temporary, for the G protein also functions as a GTPase enzyme and soon hydrolyzes its bound GTP to GDP (FIGURE 11.7c). Now inactive again, the G protein leaves the enzyme, which returns to its original state. The GTPase function of the G protein allows the pathway to shut down rapidly when the extracellular signal molecule is no longer present.

G-protein receptor systems are extremely widespread and diverse in their functions. In addition to the functions already mentioned, they are important in embryonic development, as indicated by genetic studies. For instance, mutant mouse embryos lacking a certain G protein do not develop normal blood vessels and die in utero. Furthermore, G proteins are involved in sensory reception; in humans, for example, both vision and smell depend on such proteins. Similarities in structure among G proteins and G-protein-linked receptors of modern organisms suggest that G proteins and G-protein-linked receptors evolved very early, possibly as sensory receptors of ancient microbes.

G-protein systems are involved in many human diseases, including bacterial infections. The bacteria that cause cholera, pertussis (whooping cough), and botulism, among others, make their victims ill by producing toxins that interfere with G-protein function. Drugs for treating infections and other kinds of diseases have often been discovered by trial and error, but pharmacologists now realize that up to 60% of all medicines used today exert their effects by influencing G-protein pathways.

The steps in FIGURE 11.7b show how the appropriate chemical signal activates the entire G-protein system. When the signal molecule binds as a ligand to the extracellular side of a G-protein-linked receptor, ❶ the receptor is activated, changing conformation in such a way that it, in turn, binds a specific, inactive G protein and ❷ causes a GTP to displace the GDP. This activates the G protein, which then ❸ binds to another protein, usually an enzyme, and alters *its* activity. If the en-

Tyrosine-Kinase Receptors

Among the chemical signals impinging on cells in an animal's body are growth factors, the local regulators that stimulate cells to grow and reproduce. As we'll see in Chapter 12, cell reproduction involves a variety of activities by different parts of the cell, including protein synthesis in the cytoplasm, chromosome duplication in the nucleus, and the rearrangement of elements of the cytoskeleton. Helping the cell regulate and coordinate these activities is a type of receptor specialized for triggering more than one signal-transduction pathway at once.

The receptor for a growth factor is often a tyrosine-kinase receptor, one of a major class of plasma membrane receptors characterized by having enzymatic activity. Part of the receptor protein on the cytoplasmic side of the membrane functions as an enzyme, called **tyrosine kinase,** that catalyzes the transfer of phosphate groups from ATP to the amino acid tyrosine on a substrate protein. Thus **tyrosine-kinase receptors** are membrane receptors that attach phosphates to protein tyrosines.

Many tyrosine-kinase receptors have the structure roughly depicted in FIGURE 11.8. Before the signal molecule binds, the receptors exist as individual polypeptides. Notice that each has an extracellular signal-binding site, a single α helix spanning the membrane, and an intracellular tail containing a number of tyrosines. The binding of a signal molecule to such a receptor does not cause enough of a conformational change to activate the cytoplasmic side of the protein directly. Instead, as shown in FIGURE 11.8b, receptor activation occurs in two steps:

❶ The ligand binding causes two receptor polypeptides to aggregate, forming a dimer (a protein consisting of two polypeptides). ❷ This aggregation activates the tyrosine-kinase parts of both polypeptides, each of which then adds phosphates to the tyrosines on the tail of the other polypeptide. In summary, the effect of the signal molecule on a tyrosine-kinase receptor is polypeptide aggregation and phosphorylation of the receptor.

Now that the receptor protein is fully activated, ❸ it is recognized by specific relay proteins inside the cell. Each such protein binds to a specific phosphorylated tyrosine, undergoing a structural change that activates it (the relay protein may or may not be phosphorylated by the tyrosine kinase). One tyrosine-kinase receptor dimer may activate ten or more different intracellular proteins simultaneously, ❹ triggering as many different transduction pathways and cellular responses. The ability of a single ligand-binding event to trigger so many pathways is a key difference between these receptors and G-protein-linked receptors. Abnormal tyrosine-kinase receptors that aggregate even without ligand cause some kinds of cancer.

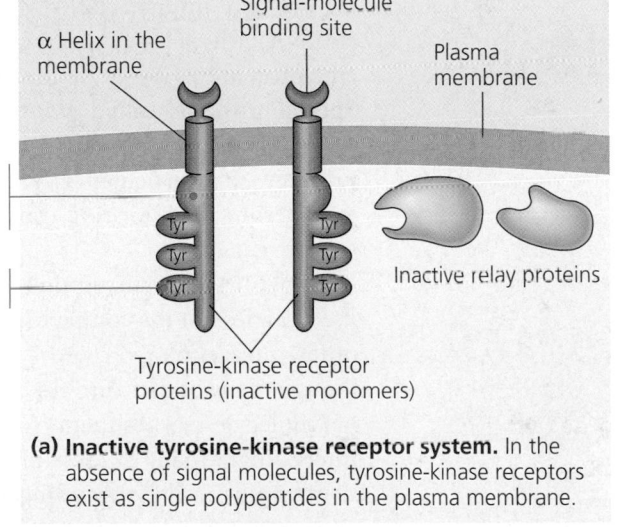

FIGURE 11.8 The structure and function of a tyrosine-kinase receptor.

α Helix in the membrane

Signal-molecule binding site

Plasma membrane

The part of the protein extending into the cytoplasm has the receptor's tyrosine kinase activity.

The amino acids forming the tail include a series of tyrosines.

Inactive relay proteins

Tyrosine-kinase receptor proteins (inactive monomers)

(a) Inactive tyrosine-kinase receptor system. In the absence of signal molecules, tyrosine-kinase receptors exist as single polypeptides in the plasma membrane.

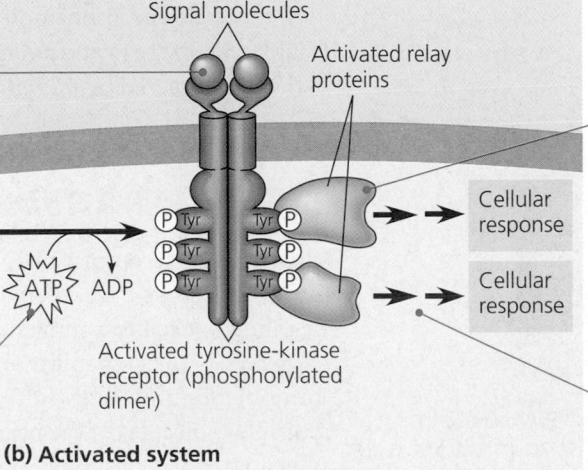

Signal molecules

❶ When signal molecules (such as a growth factor) attach to their binding sites, two polypeptides aggregate, forming a dimer.

Activated relay proteins

❸ Now fully activated, the receptor protein can bind specific intracellular proteins. These relay proteins attach to particular phosphorylated tyrosines and in the process are themselves activated.

Cellular response

Cellular response

❷ Using phosphate groups from ATP, the tyrosine-kinase region of each polypeptide phosphorylates the tyrosines on the other polypeptide.

Activated tyrosine-kinase receptor (phosphorylated dimer)

❹ Each of these proteins can then initiate a signal-transduction pathway leading to a specific cellular response. Tyrosine-kinase receptors often activate several different signal-transduction pathways at once.

(b) Activated system

Ion-Channel Receptors

Some membrane receptors of chemical signals are **ligand-gated ion channels.** These channels are protein pores in the plasma membrane that open or close in response to a chemical signal, allowing or blocking the flow of specific ions, such as Na^+ or Ca^{2+}. Like the other receptors we have discussed, these channel proteins bind a signal molecule as a ligand at a specific site on their extracellular side (FIGURE 11.9). The shape change produced in the channel protein immediately leads to a change in the concentration of a particular ion inside the cell. Often this change directly affects cell functioning in some way. At a synapse between nerve cells, for example, it may trigger an electrical signal that propagates down the length of the receiving cell. Ligand-gated ion channels are very important in the nervous system, as are gated ion channels that are controlled by electrical signals (see Chapter 48).

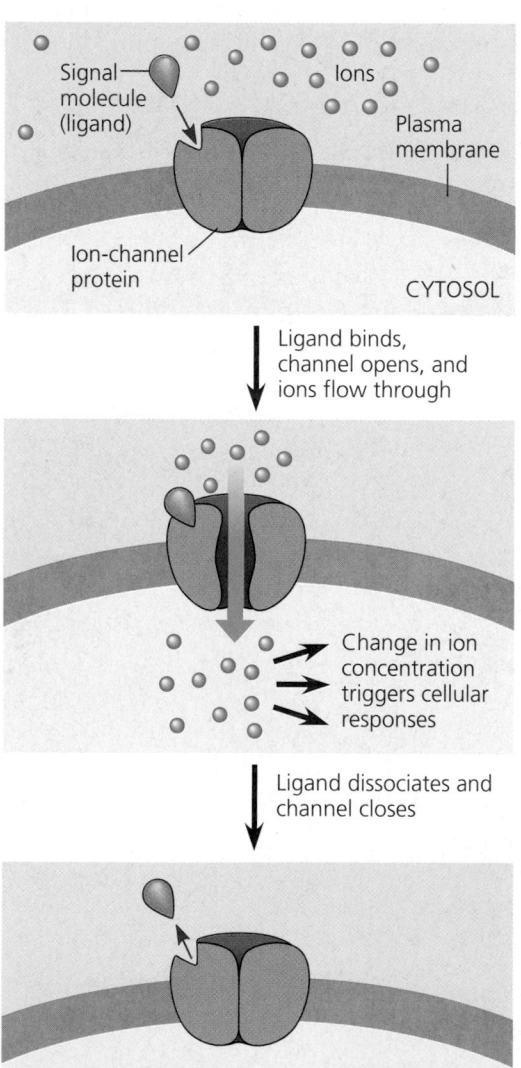

Ligand binds, channel opens, and ions flow through

Change in ion concentration triggers cellular responses

Ligand dissociates and channel closes

FIGURE 11.9 A ligand-gated ion-channel receptor. This receptor is a transmembrane protein in the plasma membrane that opens to allow the flow of a specific kind of ion across the membrane when a specific signal molecule binds to the extracellular side of the protein.

Intracellular Receptors

Not all signal receptors are membrane proteins. Some are proteins dissolved in the cytosol or nucleus of target cells. To reach such a receptor, a chemical messenger must be able to pass through the target cell's plasma membrane. A number of important signaling molecules can do just that because they are hydrophobic enough to cross the phospholipid interior of the membrane. Such hydrophobic chemical messengers include the steroid hormones and thyroid hormones of animals. Another chemical signal with an intracellular receptor is nitric oxide (NO), a gas; its very small molecules readily pass between the membrane phospholipids.

The behavior of testosterone is representative of steroid hormones: Secreted by cells of the testis, the hormone travels through the blood and enters cells all over the body. In target cells, which contain receptor molecules for testosterone in their cytosol, the hormone binds to the receptor protein, activating it (FIGURE 11.10). With the hormone attached, the active form of the receptor protein then enters the nucleus and turns on specific genes that control male sex characteristics.

How does the activated hormone-receptor protein turn on genes? Recall that the genes in a cell's DNA function by being transcribed into an RNA version called messenger RNA (mRNA), which leaves the nucleus and is translated into a specific protein by ribosomes in the cytoplasm (see FIGURE 5.28). Special proteins called *transcription factors* control which genes are turned on—that is, which genes are transcribed into mRNA—in a particular cell at a particular time. The activated testosterone receptor is a transcription factor that regulates specific genes.

By acting as a transcription factor, the testosterone receptor itself carries out the complete transduction of the signal. Most other intracellular receptors function in the same way, although many of them are already in the nucleus before the signal molecule reaches them (for example, estrogen receptors). Interestingly, many of these intracellular receptor proteins are structurally similar, suggesting an evolutionary kinship analogous to that displayed by G-protein-linked receptors. We will look more closely at hormones with intracellular receptors in Chapter 45. In the next section, we focus on signal-transduction pathways triggered by membrane receptors.

SIGNAL-TRANSDUCTION PATHWAYS

When signal receptors are plasma membrane proteins, like most of those we have discussed, the transduction stage of cell signaling is usually a multistep pathway. One benefit of such pathways is the possibility of greatly amplifying a signal. If some of the molecules in a pathway transmit the signal to multiple molecules of the next component in the series, the result can be a large number of activated molecules at the end of the pathway. In other words, a small number of extracellular

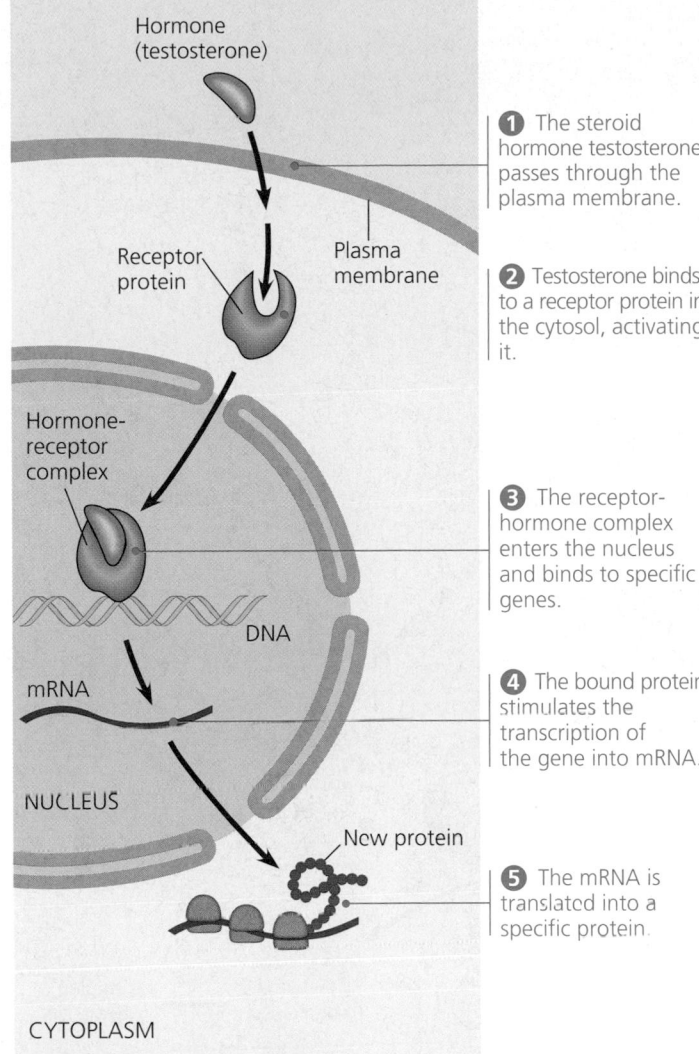

1 The steroid hormone testosterone passes through the plasma membrane.

2 Testosterone binds to a receptor protein in the cytosol, activating it.

3 The receptor-hormone complex enters the nucleus and binds to specific genes.

4 The bound protein stimulates the transcription of the gene into mRNA.

5 The mRNA is translated into a specific protein.

Hormone (testosterone)

Receptor protein

Plasma membrane

Hormone-receptor complex

DNA

mRNA

NUCLEUS

New protein

CYTOPLASM

FIGURE 11.10 Steroid hormone interacting with an intracellular receptor.

signal molecules can produce a large cellular response. Moreover, multistep pathways provide more opportunities for coordination and regulation than simpler systems do, as we'll discuss later.

Pathways relay signals from receptors to cellular responses

The binding of a specific signal molecule to a receptor in the plasma membrane triggers the first step in the chain of molecular interactions—the signal-transduction pathway—that leads to a particular response within the cell. Like falling dominoes, the signal-activated receptor activates another protein, which activates another molecule, and so on, until the protein that produces the final cellular response is activated. The molecules that relay a signal from receptor to response,

sometimes called relay molecules, are mostly proteins. The interaction of proteins is a major theme of cell signaling. Indeed, protein interaction is a unifying theme of all regulation at the cellular level.

Keep in mind that the original signal molecule is not physically passed along a signaling pathway; in most cases, it never even enters the cell. When we say that the signal is relayed along a pathway, we mean that certain information is passed on. At each step the signal is transduced into a different form, commonly a conformational change in a protein. Very often, the conformational change is brought about by phosphorylation.

Protein phosphorylation, a common mode of regulation in cells, is a major mechanism of signal transduction

Previous chapters introduced the concept of activating a protein by adding one or more phosphate groups to it (see FIGURE 9.2). In this chapter, we have already seen how phosphorylation is involved in the activation of tyrosine-kinase receptors. In fact, the phosphorylation of proteins is a widespread cellular mechanism for regulating protein activity. The general name for an enzyme that transfers phosphate groups from ATP to a protein is **protein kinase.** Unlike receptor tyrosine kinases, most cytoplasmic protein kinases act not on themselves, but on other substrate proteins; also, most phosphorylate their substrates on either of two other amino acids, serine or threonine. Such serine/threonine kinases are widely involved in signaling pathways in animals, plants, and fungi.

Many of the relay molecules in signal-transduction pathways are protein kinases, and they often act on each other. FIGURE 11.11 (p. 206) depicts a hypothetical pathway containing three different protein kinases, which create a "phosphorylation cascade." The sequence shown is similar to many known pathways, including those triggered in yeast by mating factors and in animal cells by many growth factors. The signal is transmitted by a cascade of protein phosphorylations, each bringing with it a conformational change. Each shape change results from the interaction of the charged phosphate groups with charged or polar amino acids (see FIGURE 5.15). The addition of phosphates often changes a protein from an inactive form to an active form (although in other cases phosphorylation *decreases* the activity of the protein).

The importance of protein kinases can hardly be overstated. Fully 1% of our own genes are thought to code for protein kinases. A single cell may have hundreds of different kinds, each specific for a different substrate protein. Together, they probably regulate a large proportion of the thousands of proteins in a cell. Among these are most of the proteins that, in turn, regulate cell reproduction. Abnormal activity of such a kinase frequently causes abnormal cell growth and contributes to the development of cancer.

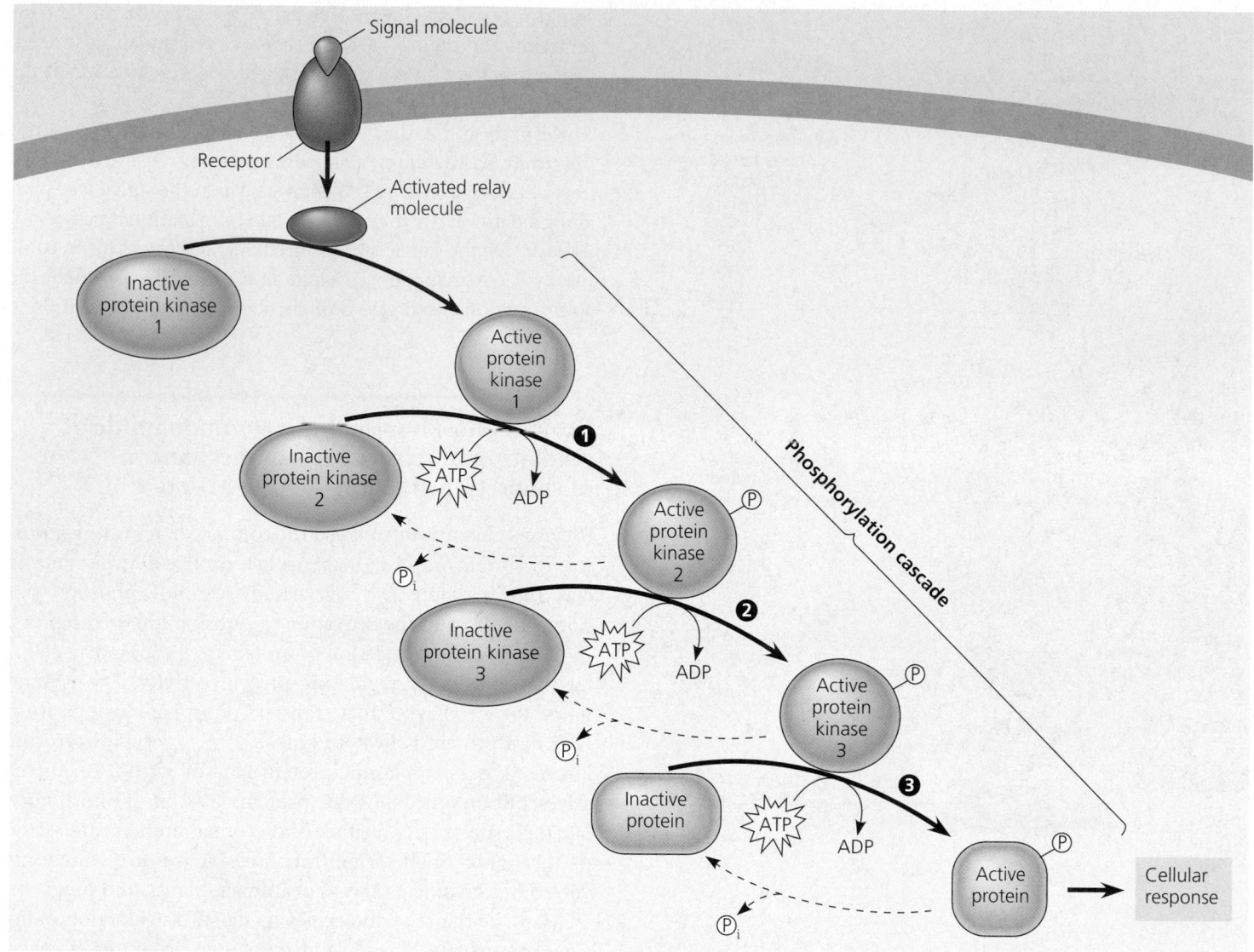

FIGURE 11.11 A phosphorylation cascade.
In a phosphorylation cascade, a series of different molecules in a pathway are phosphorylated in turn, each molecule adding a phosphate group to the next one in line. The phosphorylation cascade shown here begins after a relay molecule activates an enzyme we call protein kinase 1. ❶ Active protein kinase 1 transfers a phosphate from ATP to an inactive molecule of protein kinase 2, thus activating this second kinase. ❷ Active protein kinase 2 then catalyzes the phosphorylation (and activation) of protein kinase 3. ❸ Finally, active protein kinase 3 phosphorylates a protein (pink) that brings about the cell's response to the signal. The dashed arrows represent *inactivation* of the phosphorylated proteins; enzymes called phosphatases catalyze the removal of the phosphate groups from the proteins, making them available for reuse. The active and inactive proteins are represented by different shapes to remind you that activation is usually associated with a change in molecular conformation.

For a cell to respond normally to an extracellular signal, it must have mechanisms for turning off the signal-transduction pathway when the initial signal is no longer present. The effects of protein kinases are rapidly reversed in the cell by **protein phosphatases,** enzymes that remove phosphate groups from proteins. At any given moment, the activity of a protein regulated by phosphorylation depends on the balance in the cell between active kinase molecules and active phosphatase molecules. When the extracellular signal molecule is not present, active phosphatase molecules predominate, and the signaling pathway and cellular response shut down.

Certain small molecules and ions are key components of signaling pathways (second messengers)

Not all components of signal-transduction pathways are proteins. Many signaling pathways also involve small, nonprotein, water-soluble molecules or ions, called **second messengers.** (The extracellular signal molecule that binds to the membrane receptor is a pathway's "first messenger.") Because second messengers are both small and water-soluble, they can readily spread throughout the cell by diffusion. For example,

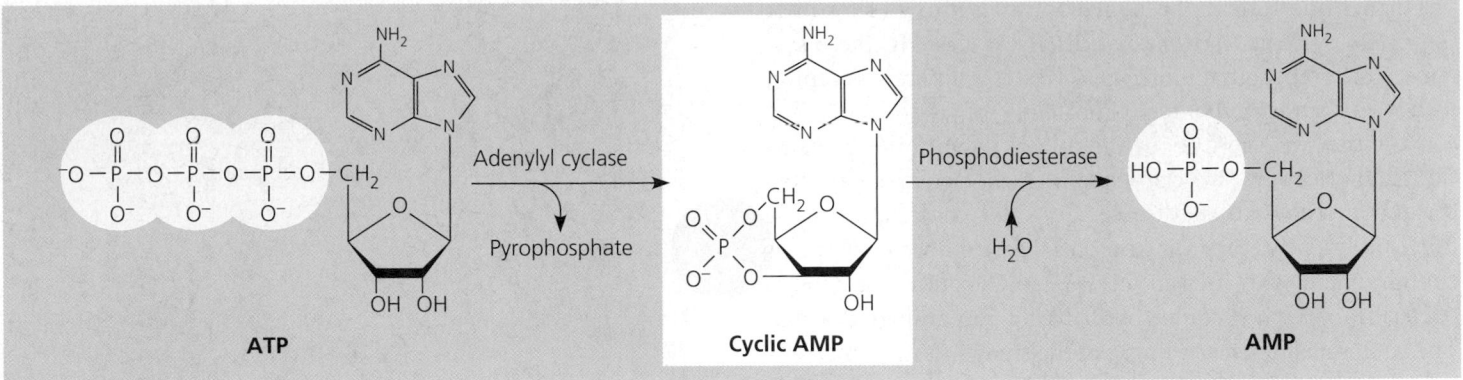

FIGURE 11.12 Cyclic AMP. Cyclic AMP (cAMP) is made from ATP by adenylyl cyclase, an enzyme embedded in the plasma membrane. Cyclic AMP functions as a second messenger that can relay a signal from the membrane to the metabolic machinery of the cytoplasm. Cyclic AMP is inactivated by phosphodiesterase, an enzyme that converts it to AMP.

as we'll see shortly, it is a second messenger called cyclic AMP that carries the signal initiated by epinephrine from the plasma membrane of a liver or muscle cell into the cell's interior, where it brings about glycogen breakdown. Second messengers participate in pathways initiated by both G-protein-linked receptors and tyrosine-kinase receptors. The two most widely used second messengers are cyclic AMP and calcium ions, Ca^{2+}. A large variety of relay proteins are sensitive to the cytosolic concentration of one or the other of these second messengers.

Cyclic AMP

Once Earl Sutherland had established that epinephrine somehow causes glycogen breakdown without passing through the plasma membrane, the search began for the second messenger (he coined the term) that transmits the signal from the plasma membrane to the metabolic machinery in the cytoplasm.

Sutherland found that the binding of epinephrine to the plasma membrane of a liver cell elevates the cytosolic concentration of a compound called cyclic adenosine monophosphate, abbreviated **cyclic AMP** or **cAMP** (FIGURE 11.12). An enzyme built into the plasma membrane, **adenylyl cyclase,** converts ATP to cAMP in response to an extracellular signal—in this case, epinephrine. Adenylyl cyclase becomes active only after epinephrine binds to a specific receptor protein. Thus, the first messenger, the hormone, causes a membrane enzyme to make cAMP, which broadcasts the signal to the cytoplasm. The cAMP does not persist for long in the absence of the hormone, because another enzyme converts the cAMP to an inactive product, AMP. Another surge of epinephrine is needed to boost the cytosolic concentration of cAMP again.

Subsequent research revealed that epinephrine is only one of many hormones and other signal molecules that trigger the

formation of cAMP. It also brought to light the other components of cAMP pathways, including G proteins, G-protein-linked receptors, and protein kinases (FIGURE 11.13). The immediate effect of cAMP is usually the activation of a serine/threonine kinase called *protein kinase A.* The activated kinase then phosphorylates various other proteins, depending on the cell. (The complete pathway for epinephrine's stimulation of glycogen breakdown is shown later, in FIGURE 11.16.)

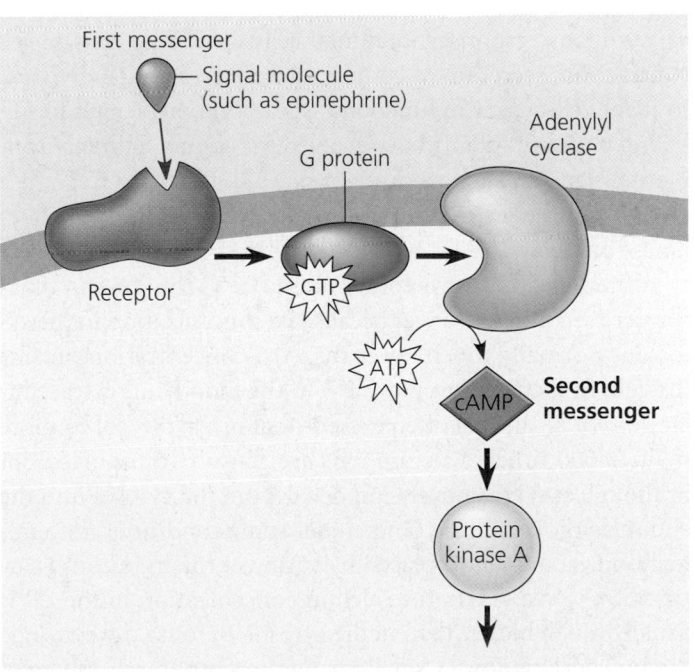

FIGURE 11.13 cAMP as a second messenger. Cyclic AMP is a component of many G-protein-signaling pathways. The signal molecule—the "first messenger"—activates a G-protein-linked receptor, which activates a specific G protein. In turn, the G protein activates adenylyl cyclase, which catalyzes the conversion of ATP to cAMP. The cAMP then activates another protein, usually protein kinase A.

Further fine-tuning of cell metabolism is provided by other G-protein systems that *inhibit* adenylyl cyclase. In these systems, a different signal molecule activates a different receptor, which activates an *inhibitory* G protein.

Now that we know about the role of cAMP in G-protein-signaling pathways, we can explain in molecular detail how certain microbes cause disease. Consider cholera, a disease that is frequently epidemic in places where the water supply is contaminated with human feces. People acquire the cholera bacterium, *Vibrio cholerae*, by drinking contaminated water. The bacteria colonize the lining of the small intestine and produce a toxin, which is an enzyme that chemically modifies a G protein involved in regulating salt and water secretion. Because the modified G protein is unable to hydrolyze GTP to GDP, it remains stuck in its active form, continuously stimulating adenylyl cyclase to make cAMP. The resulting high concentration of cAMP causes the intestinal cells to secrete large amounts of water and salts into the intestines. An infected person quickly develops profuse diarrhea and if left untreated can soon die from the loss of water and salts.

Calcium Ions and Inositol Trisphosphate

Many signal molecules in animals, including neurotransmitters, growth factors, and some hormones, induce responses in their target cells via signal-transduction pathways that increase the cytosolic concentration of calcium ions (Ca^{2+}). Calcium is even more widely used than cAMP as a second messenger. Increasing the cytosolic concentration of Ca^{2+} causes many responses in animal cells, including muscle cell contraction, secretion of certain substances, and cell division. In plant cells, calcium functions as a second messenger in signaling pathways plants have evolved for coping with environmental stresses, such as drought or cold. Cells use Ca^{2+} as a second messenger in both G-protein pathways and tyrosine-kinase pathways.

Although cells always contain some Ca^{2+}, this ion can function as a second messenger because its concentration in the cytosol is normally much lower than the concentration outside the cell. In fact, the level of Ca^{2+} in the blood and extracellular fluid of an animal often exceeds that in the cytosol by more than 10,000 times. Calcium ions are actively transported out of the cell and are actively imported from the cytosol into the endoplasmic reticulum (and, under some conditions, into mitochondria and chloroplasts) by various protein pumps (FIGURE 11.14). As a result, the calcium concentration in the ER is usually much higher than in the cytosol. Because the cytosolic calcium level is low, a small change in absolute numbers of ions represents a relatively large percentage change in calcium concentration.

In response to a signal relayed by a signal-transduction pathway, the cytosolic calcium level may rise, usually by a

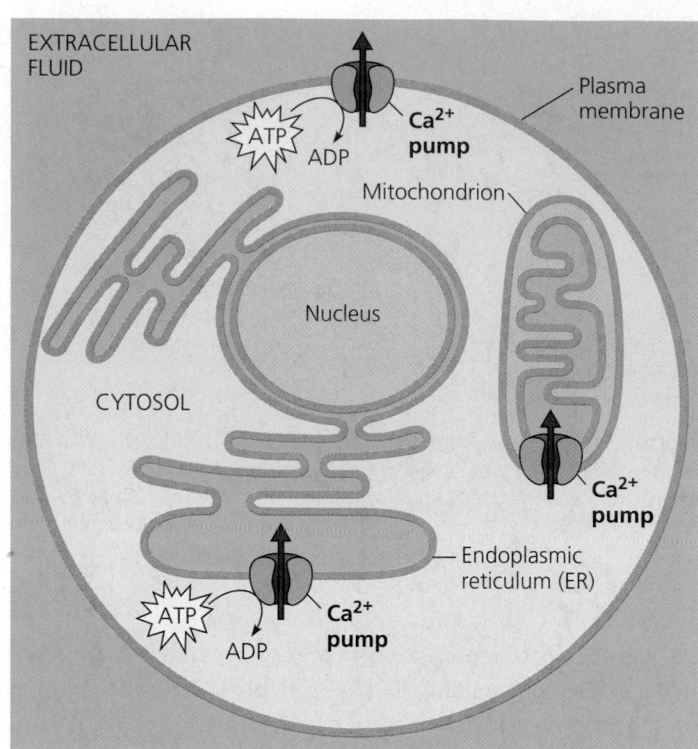

FIGURE 11.14 The maintenance of calcium ion concentrations in an animal cell. The Ca^{2+} concentration in the cytosol is usually much lower (light blue) than in the extracellular fluid and ER (darker blue). Protein pumps in the plasma membrane and the ER membrane move Ca^{2+} from the cytosol into the extracellular fluid and into the lumen of the ER. Mitochondrial pumps, driven by chemiosmosis (see Chapter 9), move Ca^{2+} into mitochondria when the calcium level in the cytosol rises significantly.

mechanism that releases Ca^{2+} from the cell's ER. The pathways leading to calcium release involve still other second messengers, **diacylglycerol (DAG)** and **inositol trisphosphate (IP_3)**. These two messengers are produced by cleavage of a certain kind of phospholipid in the plasma membrane. FIGURE 11.15 shows how this occurs and how IP_3 stimulates the release of calcium from the ER. Because IP_3 acts before calcium in these pathways, calcium could be considered a "*third* messenger." However, scientists use the term *second messenger* for all small, nonprotein components of signal-transduction pathways.

In some cases, calcium ions activate a signal-transduction protein directly, but often they function by means of **calmodulin**, a Ca^{2+}-binding protein present at high levels in eukaryotic cells. (In an animal cell, for example, calmodulin may represent as much as 1% of the total protein.) Calmodulin mediates many calcium-regulated processes in cells. When calcium ions bind to it, calmodulin changes conformation and

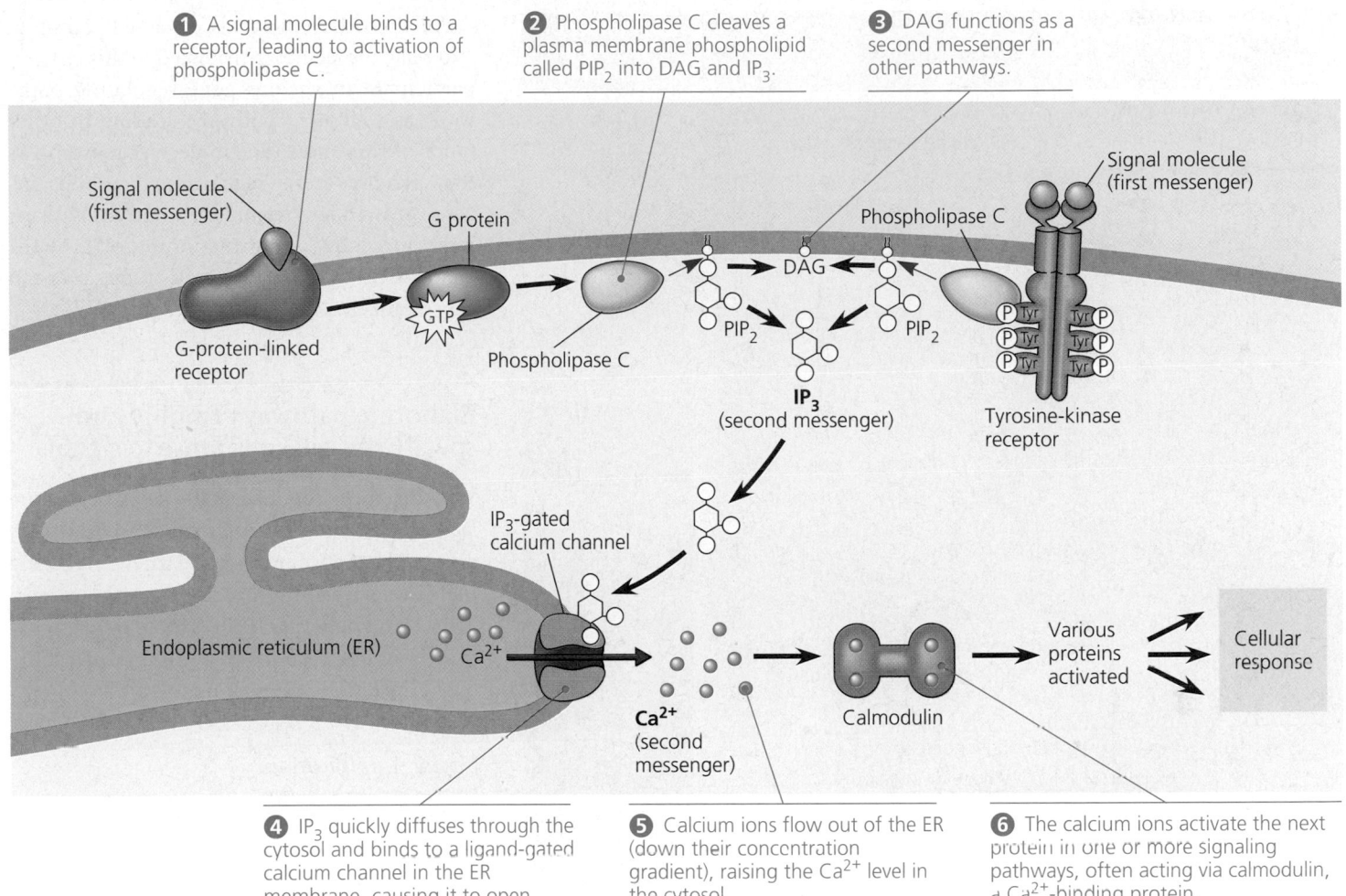

① A signal molecule binds to a receptor, leading to activation of phospholipase C.

② Phospholipase C cleaves a plasma membrane phospholipid called PIP$_2$ into DAG and IP$_3$.

③ DAG functions as a second messenger in other pathways.

Signal molecule (first messenger)

G protein

Phospholipase C

Signal molecule (first messenger)

G-protein-linked receptor

Phospholipase C

DAG

PIP$_2$

PIP$_2$

Tyrosine-kinase receptor

IP$_3$ (second messenger)

IP$_3$-gated calcium channel

Endoplasmic reticulum (ER)

Ca^{2+}

Ca^{2+} (second messenger)

Calmodulin

Various proteins activated

Cellular response

④ IP$_3$ quickly diffuses through the cytosol and binds to a ligand-gated calcium channel in the ER membrane, causing it to open.

⑤ Calcium ions flow out of the ER (down their concentration gradient), raising the Ca^{2+} level in the cytosol.

⑥ The calcium ions activate the next protein in one or more signaling pathways, often acting via calmodulin, a Ca^{2+}-binding protein.

FIGURE 11.15 Calcium and inositol trisphosphate in signaling pathways. Calcium ions (Ca^{2+}) and inositol trisphosphate (IP$_3$) function as second messengers in many signal-transduction pathways. The process is initiated by the binding of a signal molecule to either a G-protein-linked receptor (left) or a tyrosine-kinase receptor (right). The circled numbers trace the former pathway.

then binds to other proteins, activating or inactivating them. The proteins most often regulated by calmodulin are protein kinases and phosphatases—the most common relay proteins in signaling pathways.

CELLULAR RESPONSES TO SIGNALS

We now take a closer look at the cell's eventual response to an extracellular signal—what some researchers call the "output response." What is the nature of the final step in a signaling pathway?

In response to a signal, a cell may regulate activities in the cytoplasm or transcription in the nucleus

Ultimately, a signal-transduction pathway leads to the regulation of one or more cellular activities. In the cytoplasm, a signal may cause, for example, the opening or closing of an ion channel in the plasma membrane or a change in cell metabolism. As we have discussed already, the response of liver cells to signaling by the hormone epinephrine helps regulate cellular energy metabolism. The final step in the signaling pathway activates the enzyme that catalyzes the breakdown of glycogen.

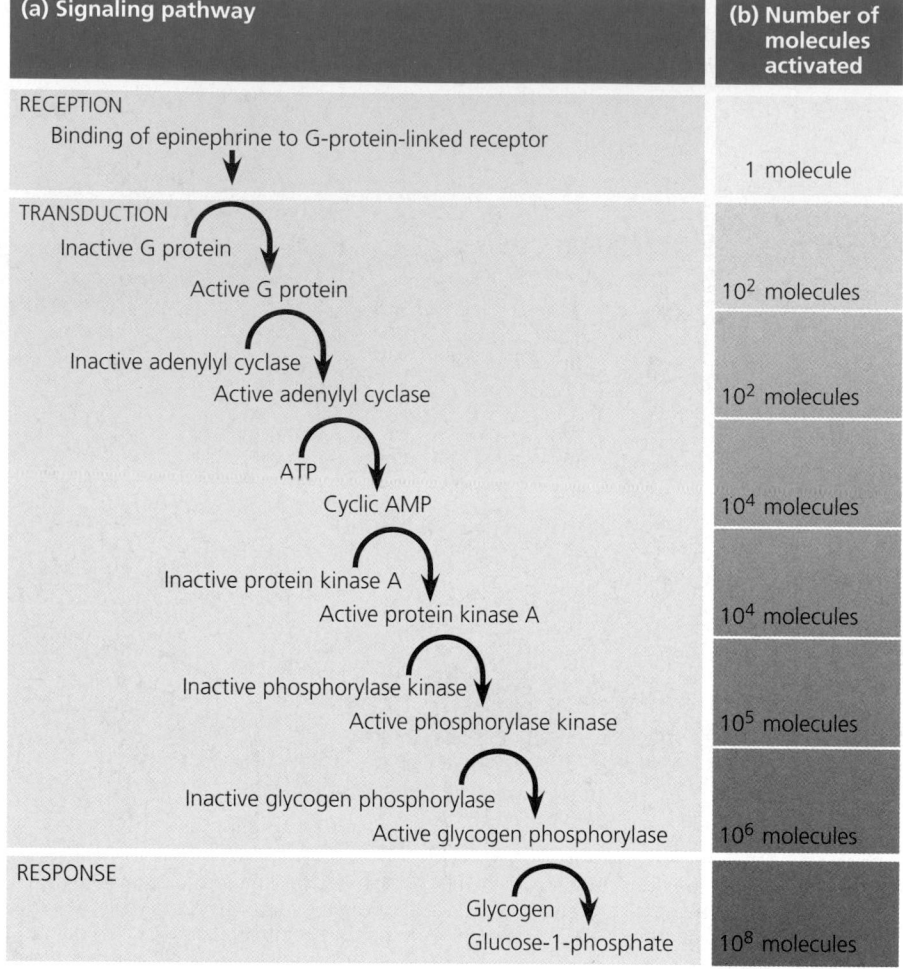

(a) Signaling pathway	(b) Number of molecules activated
RECEPTION Binding of epinephrine to G-protein-linked receptor	1 molecule
TRANSDUCTION Inactive G protein → Active G protein	10^2 molecules
Inactive adenylyl cyclase → Active adenylyl cyclase	10^2 molecules
ATP → Cyclic AMP	10^4 molecules
Inactive protein kinase A → Active protein kinase A	10^4 molecules
Inactive phosphorylase kinase → Active phosphorylase kinase	10^5 molecules
Inactive glycogen phosphorylase → Active glycogen phosphorylase	10^6 molecules
RESPONSE Glycogen → Glucose-1-phosphate	10^8 molecules

FIGURE 11.16 Cytoplasmic response to a signal: the stimulation of glycogen breakdown by epinephrine. (a) In this signaling system, the hormone epinephrine acts through a G-protein-linked receptor to activate a succession of relay molecules, including cAMP and two protein kinases. The final protein to be activated is the cytosolic enzyme glycogen phosphorylase, which releases glucose-1-phosphate units from glycogen. **(b)** As discussed in the next section of the text, this pathway *amplifies* the hormonal signal, because the receptor protein can activate many molecules of G protein, and each enzyme molecule in the pathway can act on many molecules of its substrate, the next molecule in the cascade. The number of activated molecules given for each step is only approximate.

FIGURE 11.16 shows the complete pathway leading to the release of glucose-1-phosphate from glycogen.

Many other signaling pathways ultimately regulate not the *activity* of enzymes but the *synthesis* of enzymes or other proteins, usually by turning specific genes on or off in the nucleus. Like an activated steroid receptor (see FIGURE 11.10), the final activated molecule in a signaling pathway may function as a transcription factor. In FIGURE 11.17, you see an example where a signaling pathway activates a transcription factor that turns a gene on: The response to the growth-factor signal is the synthesis of mRNA that will be translated in the cytoplasm into a specific protein. In other cases, the transcription factor might regulate a gene by turning it off. Often a transcription factor regulates several different genes.

All the different kinds of signal receptors and relay molecules introduced in this chapter participate in various gene-regulating pathways, as well as in pathways leading to other kinds of responses. The molecular messengers that produce gene regulation responses include growth factors and certain plant and animal hormones. Malfunctioning of growth-factor pathways like the one in FIGURE 11.17 can cause cancer, as we will see in Chapter 19.

Elaborate pathways amplify and specify the cell's response to signals

Why are there often so many steps between a signaling event at the cell surface and the cell's response? As mentioned earlier, signaling pathways with a multiplicity of steps have two important benefits: They amplify the signal (and thus the response), and they contribute to the specificity of response.

Signal Amplification

Elaborate enzyme cascades amplify the cell's response to a signal. At each catalytic step in the cascade, the number of activated products is much greater than in the preceding step. For example, in the epinephrine-triggered pathway in FIGURE 11.16, each adenylyl cyclase molecule catalyzes the formation of many cAMP molecules, each molecule of protein kinase A phosphorylates many molecules of the next kinase in the pathway, and so on. The amplification effect stems from the fact that these proteins persist in active form long enough to process numerous molecules of substrate before they become inactive again. As a result of the signal's amplification, a small number of epinephrine molecules binding to receptors on the surface of a liver cell or muscle cell can lead to the release of hundreds of millions of glucose molecules from glycogen.

The Specificity of Cell Signaling

Consider two different cells in your body—a liver cell and a heart muscle cell, for example. Both are in contact with your bloodstream and are therefore constantly exposed to many different hormone molecules, as well as to local regulators secreted by nearby cells. Yet the liver cell responds to some signals but ignores others, and the same is true for the heart cell. And some kinds of signals trigger responses in both cells—but

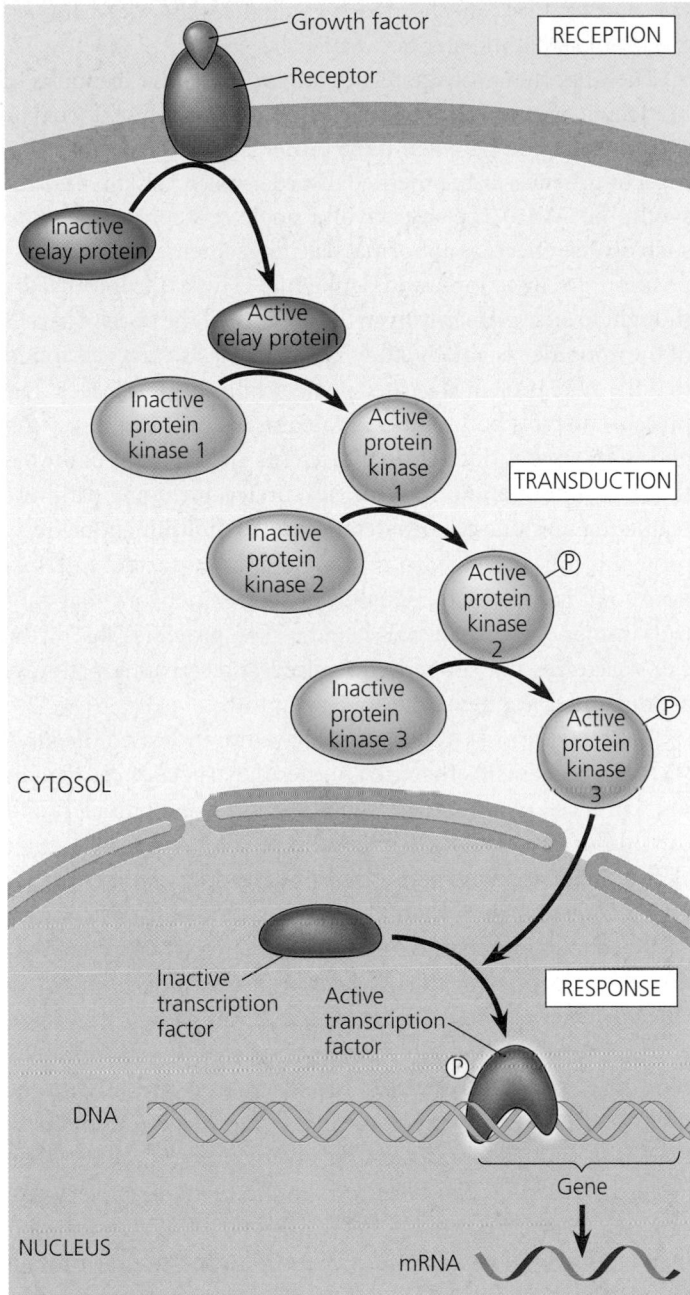

FIGURE 11.17 Nuclear response to a signal: the activation of a specific gene by a growth factor. This diagram is a simplified representation of a typical signaling pathway that leads to the regulation of gene activity in the cell nucleus. The initial signal molecule, a local regulator called a growth factor, triggers a phosphorylation cascade. (The ATP molecules that serve as sources of phosphate are not shown.) The last kinase in the sequence enters the nucleus and there activates a gene-regulating protein, a transcription factor. This protein stimulates a specific gene to be transcribed into mRNA, which then directs the synthesis of a particular protein in the cytoplasm.

different responses. For instance, epinephrine stimulates the liver cell to break down glycogen, but the main response of the heart cell to epinephrine is contraction, leading to a more rapid heartbeat. How do we account for this difference?

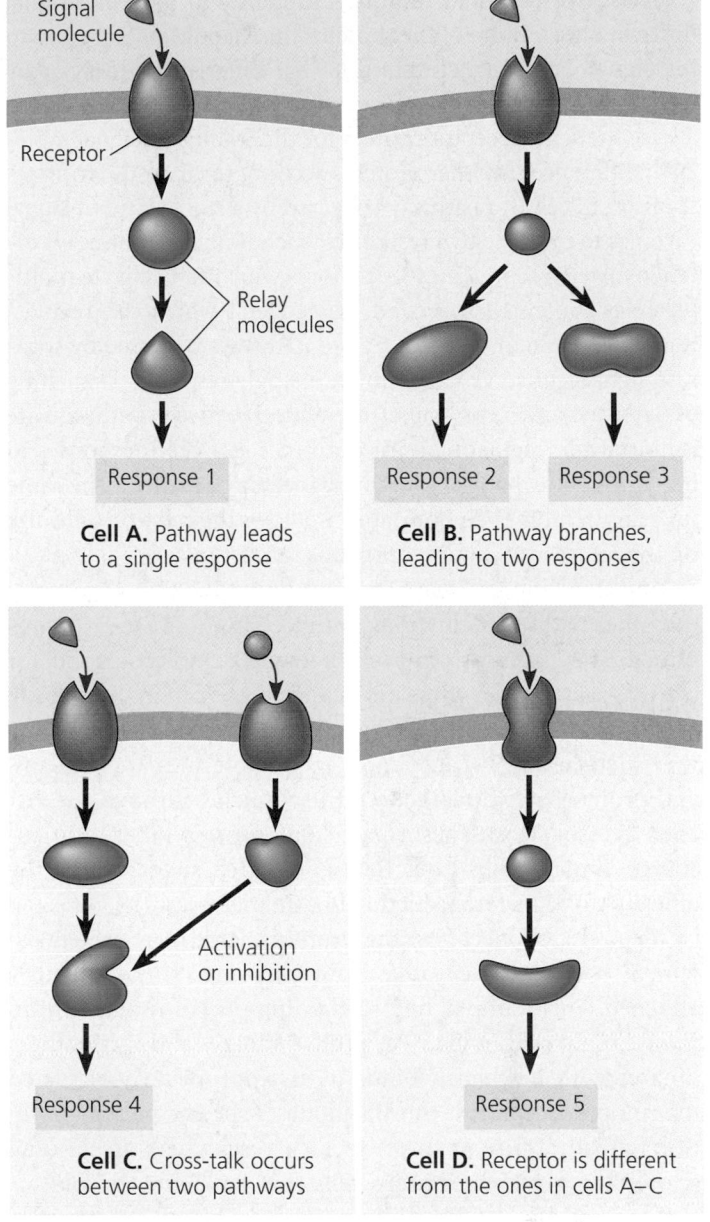

FIGURE 11.18 The specificity of cell signaling. The particular proteins a cell possesses determine what signal molecules it responds to and the nature of the response. The four cells in these diagrams respond to the same signal molecule (orange triangle) in different ways, because each has a different set of proteins (purple and green shapes). Note, however, that the same kinds of molecules can participate in more than one pathway.

The explanation for the specificity exhibited in cellular responses to signals is the same as the basic explanation for virtually all differences between cells: *Different kinds of cells have different collections of proteins* (FIGURE 11.18). The response of a particular cell to a signal depends on its particular collection of signal receptor proteins, relay proteins, and proteins needed to carry out the response. A liver cell, for example, is poised to respond appropriately to epinephrine by having the proteins listed in FIGURE 11.16, as well as those needed to manufacture glycogen.

Thus, two cells that respond differently to the same signal differ in one or more of the proteins that handle and respond to the signal. Notice in FIGURE 11.18 that different pathways may have some molecules in common. For example, cells A, B, and C all use the same receptor protein for the triangular signal molecule; differences in other proteins account for their differing responses. In cell B, a pathway triggered by a single kind of signal diverges to produce two responses; such branched pathways often involve tyrosine-kinase receptors (which can activate multiple relay proteins) or second messengers (which can regulate numerous proteins). In cell C, two pathways triggered by separate signals converge to modulate a single response. Branching of pathways and "cross-talk" (interaction) between pathways are important in regulating and coordinating a cell's responses to incoming information. Moreover, the use of some of the same proteins in more than one pathway allows the cell to economize on the number of different proteins it must make.

The signaling pathways in FIGURE 11.18 (as well as some of the other pathway depictions in this chapter) are greatly simplified. The diagrams show only a few relay molecules and, for clarity's sake, display these molecules spread out in the cytosol. If this were true in the cell, signaling pathways would operate very inefficiently, because most relay molecules are proteins, and proteins are too large to diffuse quickly through the viscous cytosol. How does a particular protein kinase, for instance, find its substrate? Recent research suggests that the efficiency of signal transduction may in many cases be increased by the presence of **scaffolding proteins,** large relay proteins to which several other relay proteins are simultaneously attached. For example, one scaffolding protein isolated from mouse brain cells holds three protein kinases and carries these kinases with it when it binds to an appropriately activated membrane receptor; it thus facilitates a specific phosphorylation cascade (FIGURE 11.19). In fact, researchers are finding scaffolding proteins in brain cells that *permanently* hold together networks of signaling-pathway proteins at synapses

(see FIGURE 2.18). This hardwiring enhances the speed and accuracy of signal transfer between cells.

The importance of the relay proteins that serve as points of branching or intersection in signaling pathways is highlighted by the problems arising when these proteins are defective or missing. For instance, in an inherited disorder called Wiskott-Aldrich syndrome (WAS), the absence of a single relay protein leads to such diverse effects as abnormal bleeding, eczema, and a predisposition to infections and leukemia. These symptoms are thought to arise primarily from the absence of the protein in cells of the immune system. By studying normal cells, scientists found that the WAS protein is located just beneath the cell surface. The protein interacts both with microfilaments of the cytoskeleton and with several different components of signaling pathways that relay information from the cell surface, including pathways regulating immune cell proliferation. This multifunctional relay protein is thus both a branch point and an important intersection point in a complex signal transduction network that controls immune cell behavior. When the WAS protein is absent, the cytoskeleton is not properly organized and signaling pathways are disrupted, leading to the WAS symptoms.

To keep FIGURE 11.18 simple, we have not indicated the *inactivation* mechanisms that are an essential aspect of cell signaling. For a cell of a multicellular organism to remain alert and capable of responding to incoming signals, each molecular change in its signaling pathways must last only a short time. As we saw in the cholera example, if a signaling pathway component becomes locked into one state, whether active or inactive, dire consequences for the organism can result.

Thus, a key to a cell's continuing receptiveness to regulation is the reversibility of the changes that signals produce. The binding of signal molecules to receptors is reversible, with the result that the lower the concentration of signal molecules, the fewer will be bound at any given moment. When they leave the receptor, the receptor reverts to its inactive form. Then, by a variety of means, the relay molecules return to their inactive forms: The GTPase activity intrinsic to a G protein hydrolyzes its bound GTP; the enzyme phosphodiesterase converts cAMP to AMP; protein phosphatases inactivate phosphorylated kinases and other proteins; and so forth. As a result, the cell is soon ready to respond to a fresh signal.

■   ■   ■

This chapter has introduced you to a number of details of cellular signal processing. More important than the details, however, are the general mechanisms of cell communication—mechanisms involving ligand binding, conformational changes in proteins, interactions among proteins, cascades of interactions that transduce and amplify signals, and protein phosphorylations by kinases. As you continue through the text, you will encounter numerous examples of cell signaling. In the very next chapter, in fact, you'll learn about the enormously important role of cell signaling in the regulation of cell reproduction.

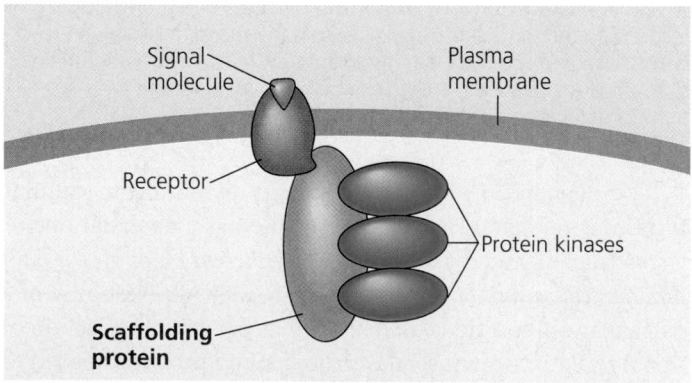

FIGURE 11.19 A scaffolding protein. The scaffolding protein shown here (pink) simultaneously binds to a specific activated membrane receptor and three different protein kinases. This physical arrangement facilitates signal transduction by these molecules.

CHAPTER 11 REVIEW

Go to the Campbell Biology website (www.campbellbiology.com) to explore an interactive version of the Chapter Review.

Summary of Key Concepts

AN OVERVIEW OF CELL SIGNALING

■ **Cell signaling evolved early in the history of life (pp. 197–198, FIGURES 11.1, 11.2)** The evidence is that signaling in microbes has much in common with the process in multicellular organisms.

■ **Communicating cells may be close together or far apart (p. 199, FIGURES 11.3, 11.4)** Animal cells signal with nearby cells by secreting local regulators or, if nerve cells, by secreting neurotransmitters at synapses. Both animal and plant cells use hormones for signaling over long distances. Cells can also communicate by direct contact.
Web/CD Case Study in the Process of Science: *How Do Cells Communicate with Each Other?*

■ **The three stages of cell signaling are reception, transduction, and response (pp. 200–201, FIGURE 11.5)** Earl Sutherland discovered how the hormone epinephrine acts on cells. The signal molecule epinephrine binds to receptors on a cell's surface (reception), leading to a series of changes in the receptor and other molecules inside the cell (transduction) and finally to the activation of an enzyme that breaks down glycogen (response).
Web/CD Activity 11A: *Overview of Cell Signaling*

SIGNAL RECEPTION AND THE INITIATION OF TRANSDUCTION

■ **A signal molecule binds to a receptor protein, causing the protein to change shape (p. 201)** The binding between signal molecule (ligand) and receptor is highly specific. A conformational change in a receptor is often the initial transduction of the signal.

■ **Most signal receptors are plasma membrane proteins (pp. 201–204, FIGURES 11.6, 11.8, 11.9, 11.10)** A G-protein-linked receptor is a membrane receptor that works with the help of a cytoplasmic G protein. Ligand binding activates the receptor, which then activates a specific G protein, which activates yet another protein in a signal-transduction pathway. Epinephrine uses this sort of receptor.

Tyrosine-kinase receptors react to the binding of signal molecules by forming dimers and then adding phosphate groups to tyrosines on the cytoplasmic side of the receptor. Relay proteins in the cell can then be activated by binding to different phosphorylated tyrosines, allowing this receptor to trigger several pathways at once. Growth factors commonly use tyrosine-kinase receptors.

Specific signal molecules cause ligand-gated ion channels in a membrane to open or close, regulating the flow of specific ions.

Intracellular receptors are cytosolic or nuclear proteins. Signal molecules that can readily cross the plasma membrane, such as steroid hormones and nitric oxide, use these receptors.
Web/CD Activity 11B: *Reception*

SIGNAL-TRANSDUCTION PATHWAYS

■ **Pathways relay signals from receptors to cellular responses (pp. 204–205)** At each step in a pathway, the signal is transduced into a different form, commonly a conformational change in a protein.

■ **Protein phosphorylation, a common mode of regulation in cells, is a major mechanism of signal transduction (pp. 205–206, FIGURE 11.11)** Many signal-transduction pathways include phosphorylation cascades, in which a series of protein kinases successively add phosphate groups to the next one in line, activating it. Phosphatase enzymes soon remove the phosphates.

■ **Certain small molecules and ions are key components of signaling pathways (second messengers) (pp. 206–209, FIGURES 11.12, 11.13, 11.14, 11.15)** Second messengers, such as cyclic AMP (cAMP) and Ca^{2+}, diffuse readily through the cytosol and thus help broadcast signals quickly. Many G proteins activate adenylyl cyclase, which makes cAMP from ATP. Although continually present in the fluids of organisms, Ca^{2+} can serve as a messenger because protein pumps usually keep it at low concentrations in the cytosol. Many G proteins and tyrosine-kinase receptors activate an enzyme that splits a plasma membrane phospholipid into two second messengers, one of which is inositol trisphosphate (IP_3). IP_3 is the ligand for a gated calcium channel in the membrane of the ER, which stores Ca^{2+} at high concentrations. When IP_3 binds, Ca^{2+} flows into the cytosol, where it activates proteins of many signaling pathways.
Web/CD Activity 11C: *Signal-Transduction Pathways*

CELLULAR RESPONSES TO SIGNALS

■ **In response to a signal, a cell may regulate activities in the cytoplasm or transcription in the nucleus (pp. 209–210, FIGURES 11.16, 11.17)** For example, signaling pathways regulate enzyme activity and cytoskeleton rearrangement in the cytoplasm. Other pathways regulate genes; they do this by activating transcription factors, proteins that turn specific genes on or off.
Web/CD Activity 11D: *Cellular Responses*
Web/CD Activity 11E: *Build a Signaling Pathway*

■ **Elaborate pathways amplify and specify the cell's response to signals (pp. 210–212, FIGURE 11.18)** Each catalytic protein in a signaling pathway amplifies the signal by activating multiple copies of the next component of the pathway; for long pathways, the total amplification may be a millionfold or more. The particular combination of proteins in a cell gives the cell great specificity in both the signals it detects and the responses it carries out. Scaffolding proteins can increase signal-transduction efficiency by holding multiple components of a pathway together. Pathway branching and cross-talk further help the cell coordinate signals and responses.

Self-Quiz

1. Phosphorylation cascades involving a series of protein kinases are useful for cellular signal transduction because
   a. they are species specific.
   b. they always lead to the same cellular response.
   c. they amplify the original signal many fold.
   d. they counter the harmful effects of phosphatases.
   e. the number of molecules used is small and fixed.

2. Binding of a signal molecule to which type of receptor leads to a change in membrane potential?
   a. tyrosine-kinase receptor
   b. G-protein-linked receptor
   c. phosphorylated tyrosine-kinase dimer
   d. ligand-gated ion channel
   e. intracellular receptor

3. The activation of tyrosine-kinase receptors is characterized by
   a. aggregation and phosphorylation.
   b. IP_3 binding.
   c. calmodulin formation.
   d. GTP hydrolysis.
   e. channel protein conformational change.

4. Cell signaling is believed to have evolved early in the history of life because
   a. it is seen in "primitive" organisms such as bacteria.
   b. yeast cells of different mating types signal one another.
   c. signal-transduction molecules found in distantly related organisms are similar.
   d. signaling can operate over large distances, a function required before the development of multicellular life.
   e. signal molecules typically interact with the outer surface of the plasma membrane.

5. Which observation suggested to Sutherland the involvement of a second messenger in epinephrine's effect on liver cells?
   a. Enzymatic activity was proportional to the amount of calcium added to a cell-free extract.
   b. Receptor studies indicated epinephrine was a ligand.
   c. Glycogen depolymerization was observed only when epinephrine was administered to intact cells.
   d. Glycogen depolymerization was observed when epinephrine and glycogen phosphorylase were combined.
   e. Epinephrine was known to have different effects on different types of cells.

6. Protein phosphorylation is commonly involved with all of the following *except*
   a. regulation of transcription by extracellular signal molecules.
   b. enzyme activation.
   c. activation of G-protein-linked receptors.
   d. activation of tyrosine-kinase receptors.
   e. activation of protein-kinase molecules.

7. Amplification of a chemical signal occurs when
   a. a receptor in the plasma membrane activates several G-protein molecules while a signal molecule is bound to it.
   b. a cAMP molecule activates one protein-kinase molecule before being converted to AMP.
   c. phosphorylase and phosphatase activities are balanced.
   d. numerous calcium ions flow through an open ligand-gated calcium channel.
   e. both a and d occur.

8. Lipid-soluble signal molecules, such as testosterone, cross the membranes of all cells but affect only target cells because
   a. only target cells retain the appropriate DNA segments.
   b. intracellular receptors are present only in target cells.
   c. most cells lack the Y chromosome required.
   d. only target cells possess the cytosolic enzymes that transduce the testosterone.
   e. only in target cells is testosterone able to initiate the phosphorylation cascade leading to activated transcription factor.

9. Signal-transduction pathways benefit cells for all of the following reasons *except*
   a. they help cells respond to signal molecules that are too large or too polar to cross the plasma membrane.
   b. they enable different cells to respond appropriately to the same signal.
   c. they help cells use up phosphate generated by ATP breakdown.
   d. they can amplify a signal.
   e. variations in the signal-transduction pathways can enhance response specificity.

10. Consider this pathway: epinephrine $\longrightarrow$ G-protein-linked receptor $\longrightarrow$ G protein $\longrightarrow$ adenylyl cyclase $\longrightarrow$ cAMP. Identify the "second messenger."
    a. cAMP
    b. G protein
    c. GTP
    d. adenylyl cyclase
    e. G-protein-linked receptor

11. How do the cellular receptors for water-soluble hormones and lipid-soluble hormones differ?

12. The addition of norepinephrine (a water-soluble hormone) to the solution bathing thyroid cells in culture causes an increase in cytosolic Ca^{2+} levels and the release of thyroxine (another hormone) by these cells. What is the likely mechanism of this effect?

13. In Question 12, would injection of norepinephrine into these cells have the same effect? Explain your answer.

14. How can a target cell's response to a hormone be amplified more than a millionfold?

15. When a signal-transduction pathway involves a phosphorylation cascade, how does the cell's response get turned off?

Go to the website or CD-ROM for more quiz questions.

Evolution Connection

You learned in this chapter that cell-cell signaling is thought to have arisen early in the history of life, because the same mechanisms of signaling are found in distantly related organisms. But why hasn't some "better" mechanism arisen? Is it too difficult to evolve wholly new signaling mechanisms, or are existing mechanisms simply adequate and therefore maintained? Put another way, need superior signaling mechanisms necessarily evolve if existing mechanisms are adequate and effective? Why or why not?

The Process of Science

Cell biologists recently reported the discovery of orexin, a signaling molecule that appears to regulate appetite in humans and other mammals. Orexin concentrations were measurably higher in fasting individuals. Using your knowledge of membrane receptors and signal-transduction pathways, suggest ways in which the understanding of orexin function could lead to treatments for both anorexia and obesity.

Determine the chemical nature of the molecule used for cell communication in the cellular slime mold in the Case Study in the Process of Science, available on the website and CD-ROM.

Science, Technology, and Society

The aging process is thought to be initiated at the cellular level. Among the changes that can occur after a certain number of cell divisions is the loss of a cell's ability to respond to growth factors and other chemical signals. Much research into aging is aimed at understanding such losses, with the ultimate goal of significantly extending the human life span. Not everyone, however, agrees that this is a desirable goal. If life expectancy were greatly increased, what might be the social and ecological consequences? How might we cope with these?

Answers to Self-Quiz: 1. c; 2. d; 3. a; 4. c; 5. c; 6. c; 7. a; 8. b; 9. c; 10. a; 11. Receptors for water-soluble hormones are in the plasma membrane; those for lipid-soluble hormones are inside the cell. 12. Signaling via a plasma membrane receptor, phospholipase C, and IP_3 opens ER channels that release Ca^{2+} into the cytosol; the Ca^{2+} triggers thyroxine release. 13. No, because norepinephrine must bind to the extracellular side of its receptor to activate the signal-transduction pathway. 14. By a cascade of sequential activations in which some of the steps activate numerous molecules. 15. Protein phosphatases reverse the effects of the kinases.

CHAPTER 12

THE CELL CYCLE

The ability *of organisms to reproduce their kind is the one characteristic that best distinguishes living things from nonliving matter. This unique capacity to procreate, like all biological functions, has a cellular basis. Rudolf Virchow, a German physician, put it this way in 1855: "Where a cell exists, there must have been a preexisting cell, just as the animal arises only from an animal and the plant only from a plant." He summarized with the Latin axiom, "Omnis cellula e cellula," meaning "Every cell from a cell." The continuity of life is based on the reproduction of cells, or* **cell division.** *The series of images in the micrograph on this page follows an animal cell's chromosomes through one round of cell division, starting at the lower left.*

*In this chapter, you will learn how cells reproduce to form genetically equivalent daughter cells.** This division process is an integral part of the* **cell cycle,** *the life of a cell from its origin in the division of a parent cell until its own division into two.*

THE KEY ROLES OF CELL DIVISION

Before describing the cell cycle or the mechanics of cell division, let's look at the roles that cellular reproduction plays in the lives of organisms.

Cell division functions in reproduction, growth, and repair

When a unicellular organism such as *Amoeba* divides to form duplicate offspring, the division of one cell reproduces an entire organism (FIGURE 12.1a, p. 216), and cell division on a larger scale can produce progeny from some multicellular organisms (such as plants that grow from cuttings). But cell division also enables sexually reproducing organisms to develop from a single cell—the fertilized egg, or zygote (FIGURE 12.1b). And after an organism is fully grown, cell division continues to function in renewal and repair, replacing cells that die from normal wear and tear or accidents. For example, dividing cells in your bone marrow continuously make new blood cells (FIGURE 12.1c).

The reproduction of an ensemble as complex as a cell cannot occur by a mere pinching in half; the cell is not like a soap bubble that simply enlarges and splits in two. Cell division involves the distribution of identical genetic material—DNA—to two daughter cells. What is most remarkable about cell division is the fidelity with which the DNA is passed along, without dilution, from one generation of cells to the next. A dividing cell duplicates its DNA, allocates the two copies to opposite ends of the cell, and only then splits into daughter cells.

---

* Although the terms *daughter cells* and *sister chromatids* (a term you will encounter later in the chapter) are traditional and will be used throughout this book, the structures they refer to have no gender.

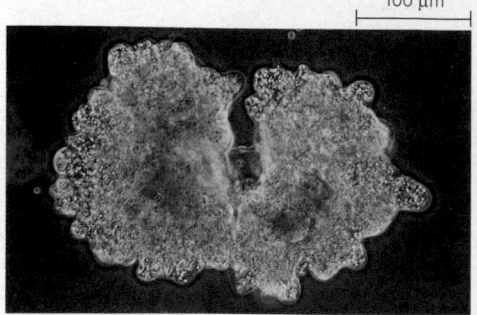

(a) Reproduction. *Amoeba,* a single-celled eukaryote, divides to form two cells, each an individual organism.

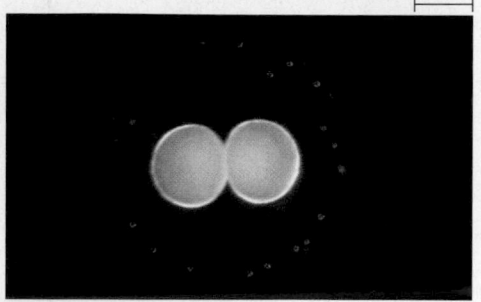

(b) Growth and development. This darkfield micrograph shows a sand dollar embryo shortly after the fertilized egg divided to form two cells.

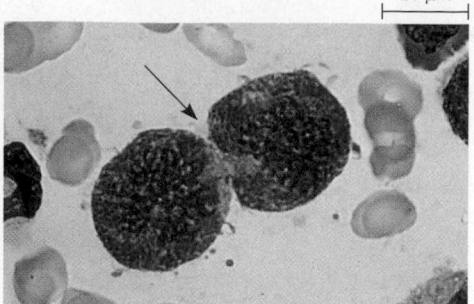

(c) Tissue renewal. These dividing bone marrow cells give rise to new blood cells.

FIGURE 12.1 The functions of cell division (all LMs).

Cell division distributes identical sets of chromosomes to daughter cells

A cell's endowment of DNA, its genetic information, is called its **genome.** Although a prokaryotic genome is often a single long DNA molecule, eukaryotic genomes usually consist of a number of DNA molecules. The overall length of DNA in a eukaryotic cell is enormous. A typical human cell, for example, has about 3 m of DNA—a length about 300,000 times greater than the cell's diameter. Yet before the cell can divide, all of this DNA must be copied and then the two copies separated so that each daughter cell ends up with a complete genome.

The replication and distribution of so much DNA is manageable because the DNA molecules are packaged into **chromosomes,** so named because they take up certain dyes used in microscopy (*chromo,* colored, and *somes,* bodies) (FIGURE 12.2). Every eukaryotic species has a characteristic number of chromosomes in each cell nucleus. For example, the nuclei of human **somatic cells** (all body cells except the reproductive cells) each contain 46 chromosomes. Reproductive cells, or **gametes**—sperm cells and egg cells—have half as many chromosomes as somatic cells, or 23 chromosomes in humans.

Incorporated into each eukaryotic chromosome is one very long, linear DNA molecule representing hundreds or thousands of genes, the units that specify an organism's inherited traits. The DNA is associated with various proteins that maintain the structure of the chromosome and help control the activity of the genes. This DNA-protein complex, called **chromatin,** is organized into a long, thin fiber. After a cell duplicates its DNA in preparation for division, the chromatin condenses: It becomes densely coiled and folded, making the chromosomes much shorter and so thick that we can see them with a light microscope.

Each duplicated chromosome has two **sister chromatids.** The two chromatids, containing identical copies of the chromosome's DNA molecule, are initially attached by proteins all

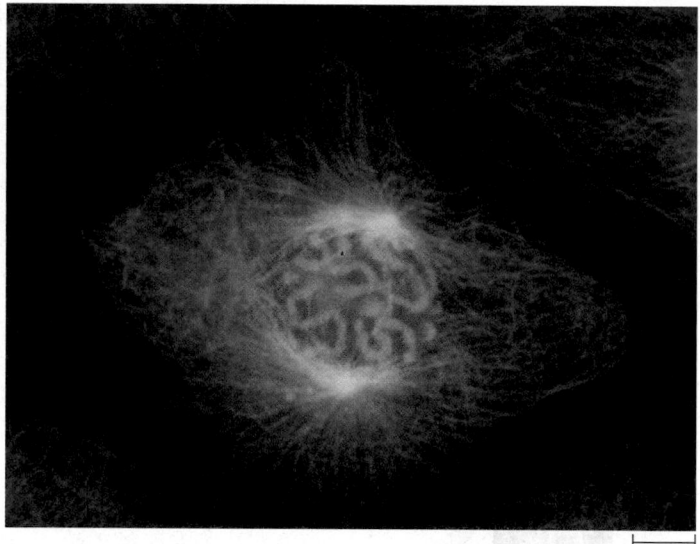

FIGURE 12.2 Eukaryotic chromosomes. A tangle of chromosomes (stained orange) is visible within the nucleus of this kangaroo rat epithelial cell. The cell is preparing to divide (LM).

Photo courtesy of J. M. Murray, University of Pennsylvania Medical School.

along their lengths. In its condensed form, the chromosome has a narrow "waist" at a specialized region called the **centromere** (FIGURE 12.3). Later in the cell division process, the sister chromatids of all the chromosomes are pulled apart and repackaged as complete chromosome sets in two new nuclei, one at each end of the cell. **Mitosis,** the division of the nucleus, is usually followed immediately by **cytokinesis,** the division of the cytoplasm. Where there was one cell, there are now two, each the genetic equivalent of the parent cell.

What happens to chromosome number as we follow the human life cycle through the generations? You inherited 46 chromosomes, one set of 23 from each parent. They were combined in the nucleus of a single cell when a sperm cell from your father united with an egg cell from your mother to form a fertilized egg, or zygote. Mitosis and cytokinesis produced the trillions of somatic cells that now make up your body, and the same processes continue to generate new cells to replace dead and damaged ones. In contrast, you produce gametes—eggs or sperm cells—by a variation of cell division called **meiosis,** which yields daughter cells that have half as many chromosomes as the parent cell. Meiosis occurs only in your gonads (ovaries or testes). In each generation of humans, meiosis reduces the chromosome number from 46 to 23. Fertilization fuses two gametes together and doubles the chromosome number to 46 again, and mitosis conserves that number in every somatic cell nucleus of the new individual. In Chapter 13, we will examine the role of meiosis in reproduction and inheritance in more detail. In the remainder of this chapter, we focus on mitosis and the rest of the mitotic cell cycle.

The mitotic phase alternates with interphase in the cell cycle: *an overview*

Mitosis is just one part of the cell cycle (FIGURE 12.4). In fact, the **mitotic (M) phase,** which includes both mitosis and cytokinesis, is usually the shortest part of the cell cycle. Mitotic cell division alternates with a much longer **interphase,** which often accounts for about 90% of the cycle. It is during interphase that the cell grows and copies its chromosomes in preparation for cell division. Interphase can be divided into subphases: the G_1 **phase** ("first gap"), the **S phase,** and the G_2 **phase** ("second gap"). During all three subphases, the cell grows by producing proteins and cytoplasmic organelles. However, chromosomes are duplicated only during the S phase (S stands for synthesis of DNA). Thus, a cell grows (G_1), continues to grow as it copies its chromosomes (S), grows more as it completes preparations for cell division (G_2), and divides (M). The daughter cells may then repeat the cycle.

Time-lapse films of living, dividing cells reveal the dynamics of mitosis as a continuum of changes. For purposes of description, however, mitosis is conventionally broken down into five subphases: **prophase, prometaphase, metaphase, anaphase,** and **telophase.** FIGURE 12.5, on pages 218–219, describes these stages in an animal cell. Be sure to study this figure thoroughly before progressing to the next section, which examines mitosis more closely.

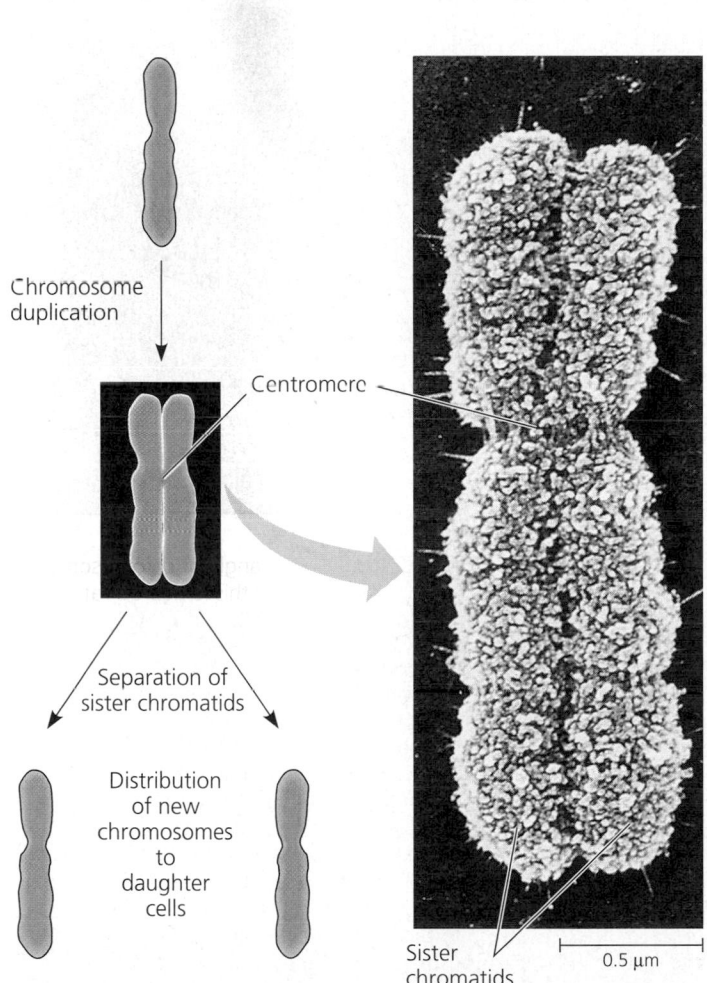

FIGURE 12.3 Chromosome duplication and distribution during mitosis. A eukaryotic cell preparing to divide duplicates each of its multiple chromosomes, one of which is represented here. A duplicated chromosome consists of two sister chromatids, which narrow at their centromeres. The micrograph shows a human chromosome in this state (SEM). Each chromatid consists of one very long chromatin fiber folded and coiled in a compact arrangement. The DNA molecules of the sister chromatids are identical. As mitosis continues, mechanical processes separate the sister chromatids and distribute them to two daughter cells. Chromosomes normally exist in the highly condensed state shown here only during the process of mitosis.

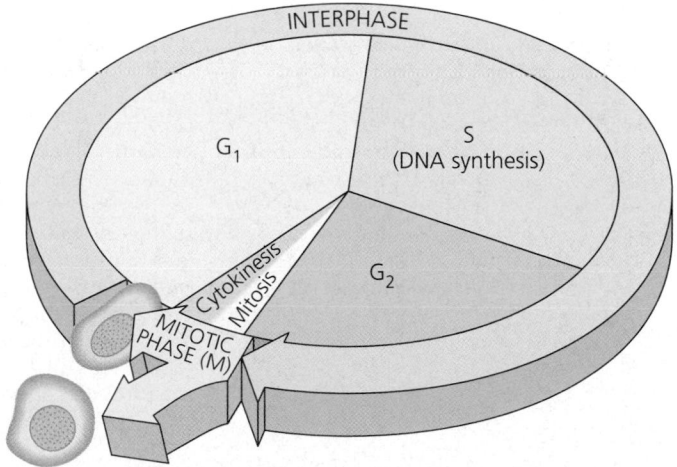

FIGURE 12.4 The cell cycle. In a dividing cell, the mitotic (M) phase alternates with interphase, a growth period. The first part of interphase, called G_1, is followed by the S phase, when the chromosomes replicate; the last part of interphase is called G_2. In the M phase, mitosis divides the nucleus and distributes its chromosomes to the daughter nuclei, and cytokinesis divides the cytoplasm, producing two daughter cells.

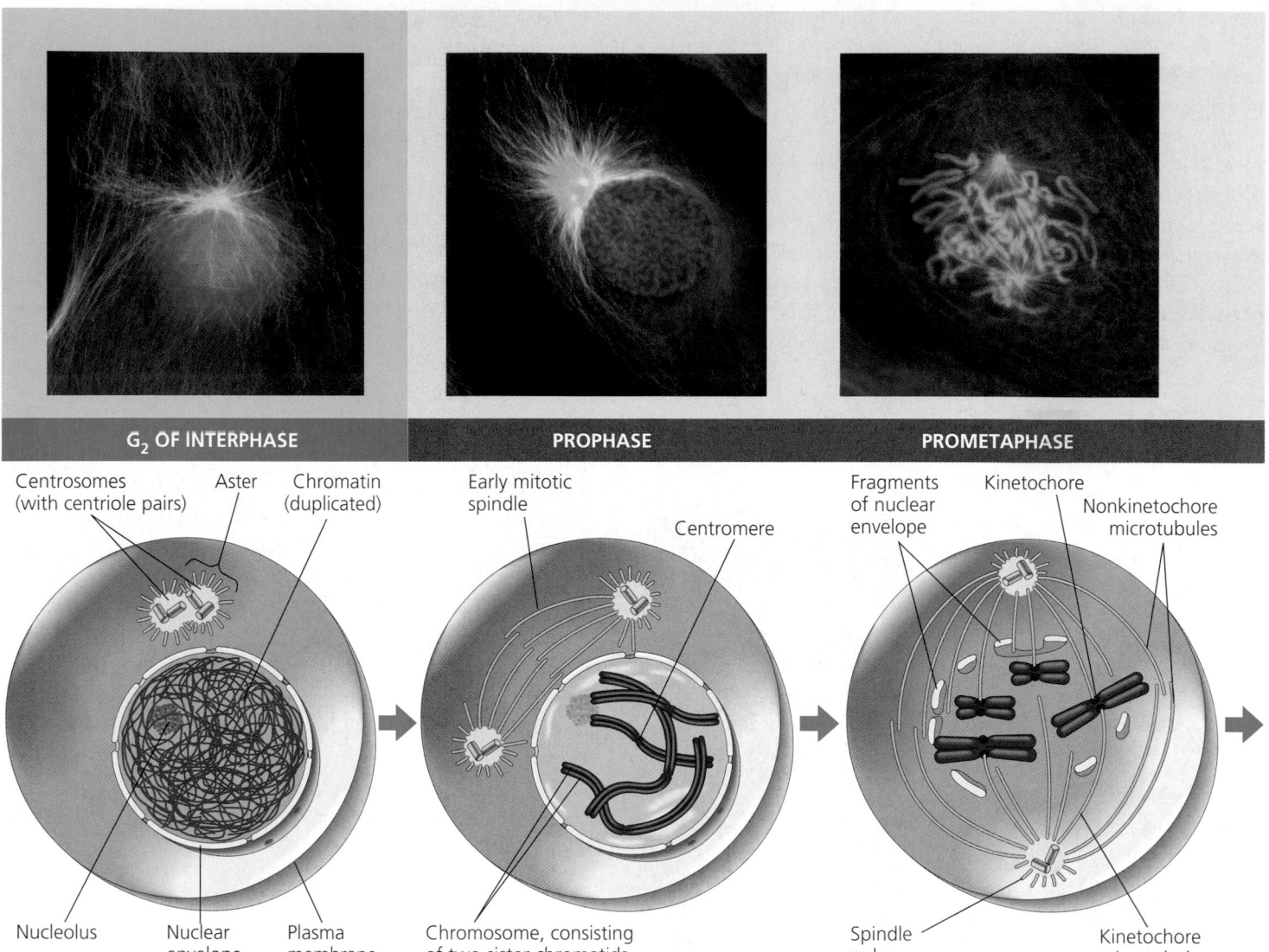

G₂ OF INTERPHASE	PROPHASE	PROMETAPHASE

G₂ OF INTERPHASE

Centrosomes (with centriole pairs) Aster Chromatin (duplicated)

Nucleolus Nuclear envelope Plasma membrane

PROPHASE

Early mitotic spindle Centromere

Chromosome, consisting of two sister chromatids

PROMETAPHASE

Fragments of nuclear envelope Kinetochore Nonkinetochore microtubules

Spindle pole Kinetochore microtubule

During late interphase, the nucleus is well defined and bounded by the nuclear envelope. It contains one or more nucleoli. Just outside the nucleus are two centrosomes, formed earlier by replication of a single centrosome. In animal cells, each centrosome features a pair of centrioles. Microtubules extend from the centrosomes in radial arrays called **asters** ("stars"). The chromosomes have already duplicated (during the S phase), but at this stage they cannot be distinguished individually because they are still in the form of loosely packed chromatin fibers.

During prophase, changes occur in both the nucleus and the cytoplasm. In the nucleus, the chromatin fibers become more tightly coiled, condensing into discrete chromosomes observable with a light microscope. The nucleoli disappear. Each duplicated chromosome appears as two identical sister chromatids joined together. In the cytoplasm, the mitotic spindle begins to form; it is made of microtubules extending from the two centrosomes. The centrosomes move away from each other, apparently propelled along the surface of the nucleus by the lengthening bundles of microtubules between them.

During prometaphase, the nuclear envelope fragments. The microtubules of the spindle can now invade the nuclear area and interact with the chromosomes, which have become even more condensed. Bundles of microtubules extend from each pole toward the middle of the cell. Each of the two chromatids of a chromosome now has a specialized structure called a **kinetochore**, located at the centromere region. Some of the microtubules attach to the kinetochores, causing the chromosomes to begin jerky movements. Nonkinetochore microtubules interact with those from the opposite pole of the cell.

FIGURE 12.5 The stages of mitotic cell division in an animal cell. The light micrographs show dividing lung cells from a newt, which has 22 chromosomes in its somatic cells. The chromosomes appear blue and the microtubules green. (The red fibers are intermediate filaments.) The schematic drawings show details not visible in the micrographs. For the sake of simplicity, only four chromosomes are drawn. (In plant cells, centrioles are lacking and cytokinesis occurs differently.)

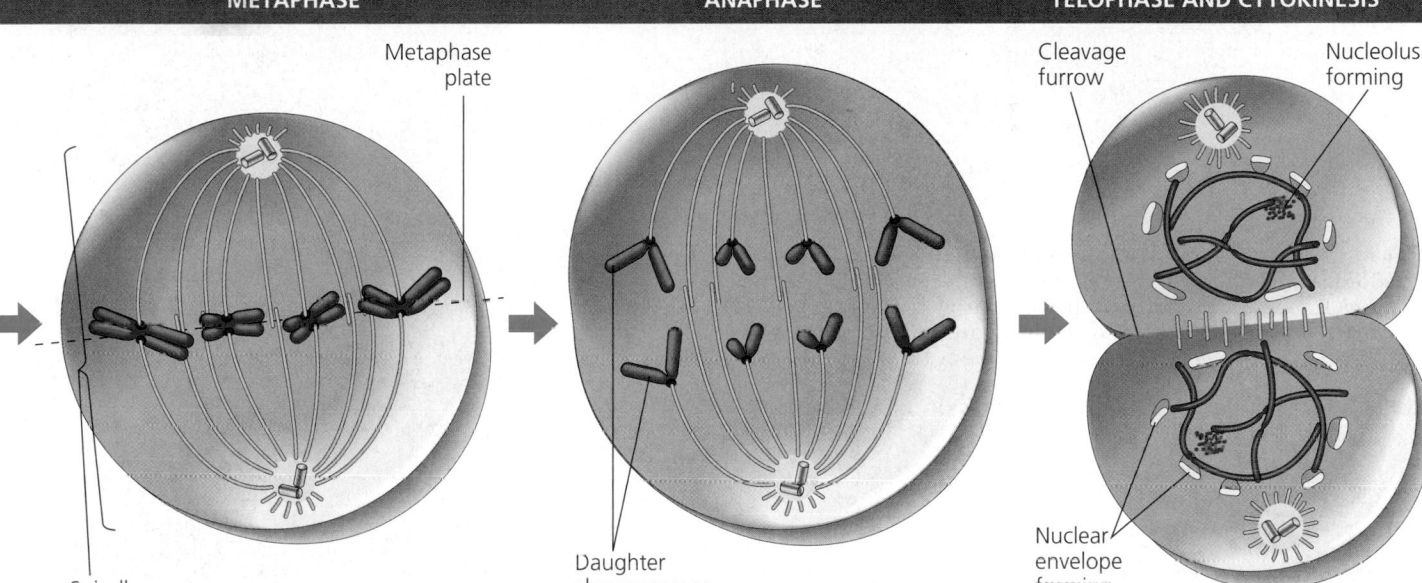

METAPHASE

Metaphase plate

Spindle

ANAPHASE

Daughter chromosomes

TELOPHASE AND CYTOKINESIS

25 µm

Cleavage furrow

Nucleolus forming

Nuclear envelope forming

The centrosomes are now at opposite poles of the cell. The chromosomes convene on the **metaphase plate**, an imaginary plane that is equidistant between the spindle's two poles. The centromeres of the chromosomes are all on the metaphase plate. For each chromosome, the kinetochores of the sister chromatids are attached to microtubules coming from opposite poles of the cell. The entire apparatus of microtubules is called the spindle because of its shape.

Anaphase begins suddenly, when the paired centromeres of each chromosome separate, finally liberating the sister chromatids from each other. The once-joined sisters, each now considered a full-fledged chromosome, begin moving toward opposite poles of the cell, as their kinetochore microtubules shorten. Because these microtubules are attached at the centromere, the chromosomes move centromere first (at about 1 µm/min). At the same time, the poles of the cell move farther apart, as the nonkinetochore microtubules lengthen. By the end of anaphase, the two poles of the cell have equivalent—and complete—collections of chromosomes.

At telophase, the nonkinetochore microtubules elongate the cell still more, and the daughter nuclei form at the two poles of the cell. Nuclear envelopes arise from the fragments of the parent cell's nuclear envelope and other portions of the endomembrane system. In a further reversal of prophase and prometaphase events, the chromatin fiber of each chromosome becomes less tightly coiled. Mitosis, the division of one nucleus into two genetically identical nuclei, is now complete. Cytokinesis, the division of the cytoplasm, is usually well under way by this time, so two daughter cells appear shortly after the end of mitosis. In animal cells, cytokinesis involves the formation of a cleavage furrow, which pinches the cell in two.

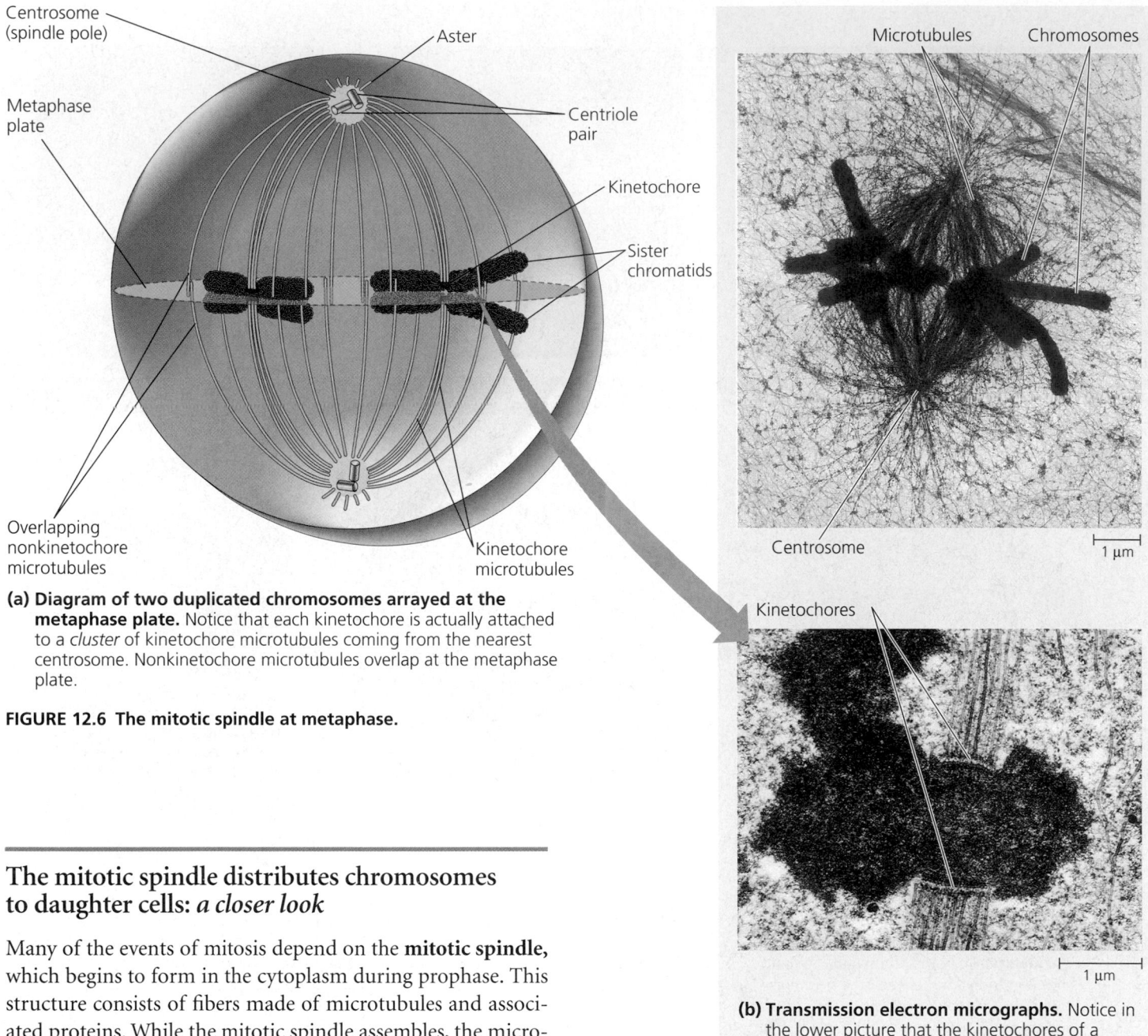

(a) **Diagram of two duplicated chromosomes arrayed at the metaphase plate.** Notice that each kinetochore is actually attached to a *cluster* of kinetochore microtubules coming from the nearest centrosome. Nonkinetochore microtubules overlap at the metaphase plate.

FIGURE 12.6 The mitotic spindle at metaphase.

(b) **Transmission electron micrographs.** Notice in the lower picture that the kinetochores of a chromosome's two sister chromatids face in opposite directions. From Dr. Matthew Schibler, *Photoplasma* 137 (1987):29–44. Reprinted by permission of Springer-Verlag.

The mitotic spindle distributes chromosomes to daughter cells: *a closer look*

Many of the events of mitosis depend on the **mitotic spindle,** which begins to form in the cytoplasm during prophase. This structure consists of fibers made of microtubules and associated proteins. While the mitotic spindle assembles, the microtubules of the cytoskeleton partially disassemble, probably providing the material used to construct the spindle. The spindle microtubules elongate by incorporating more subunits of the protein tubulin (see TABLE 7.2).

The assembly of spindle microtubules starts in the **centrosome,** a nonmembranous organelle that functions throughout the cell cycle to organize the cell's microtubules (it is also called the *microtubule-organizing center*). In animal cells, a pair of centrioles is located at the center of the centrosome, but the centrioles are not essential for cell division. In fact, the centrosomes of most plants lack centrioles, and if a researcher destroys the centrioles of an animal cell with a laser microbeam, a spindle nevertheless forms during mitosis.

During interphase, the single centrosome replicates to form two centrosomes (see FIGURE 12.5). As mitosis starts, the two centrosomes are located near the nucleus; they then move apart from each other during prophase and prometaphase, as spindle microtubules grow out from them. By the end of prometaphase, the two centrosomes, referred to as *spindle poles* in this context, are at opposite poles of the cell.

Each of the two joined chromatids of a chromosome has a **kinetochore,** a structure of proteins and specific sections of chromosomal DNA at the centromere. The chromosome's two kinetochores face in opposite directions. During prometaphase, some of the spindle microtubules attach to the kinetochores. When one of a chromosome's kinetochores is "captured" by microtubules, the chromosome begins to move toward the pole from which those microtubules come. However, this movement is checked as soon as microtubules from the opposite pole attach to the other kinetochore. What happens next is like a tug-of-war that ends in a draw. The chromosome moves first in one direction, then the other, back and forth, finally settling midway between the two poles of the cell (FIGURE 12.6). Meanwhile, microtubules that do not attach to kinetochores interact with nonkinetochore microtubules from the opposite pole of the cell. At metaphase, these microtubules overlap, and the centromeres of all the duplicated chromosomes are on a plane midway between the two poles. This plane is called the **metaphase plate** of the cell. The spindle is now complete.

Let's now see how the structure of the completed spindle correlates with its function during anaphase. Anaphase commences suddenly when proteins holding together the sister chromatids of each chromosome are inactivated. Now that the chromatids are separate, full-fledged chromosomes, they move toward opposite poles of the cell. How do the kinetochore microtubules function in this poleward movement of chromosomes? Experimental evidence supports the hypothesis that kinetochores are equipped with motor proteins that "walk" a chromosome along the attached microtubules toward the nearest pole. Meanwhile, the microtubules shorten by depolymerizing at their kinetochore ends (FIGURE 12.7). (To review how motor proteins move an object along a microtubule, see FIGURE 7.21b.)

What is the function of the *non*kinetochore microtubules? In a dividing animal cell, these microtubules are responsible for elongating the whole cell during anaphase (see FIGURE 12.5). Nonkinetochore microtubules interdigitate across the metaphase plate, and during anaphase ones originating from opposite spindle poles move past each other toward their poles. The mechanism seems to be similar to the one that slides neighboring microtubules in a flagellum: Motor proteins attached to the nonkinetochore microtubules drive them past one another, using energy from ATP (see FIGURE 7.21a). At the same time, the nonkinetochore microtubules lengthen by the addition of tubulin subunits to their ends.

At the end of anaphase, duplicate sets of chromosomes have arrived at opposite poles of the elongated parent cell. Nuclei re-form during telophase. Cytokinesis generally begins during this last stage of mitosis.

Cytokinesis divides the cytoplasm: *a closer look*

In animal cells, cytokinesis occurs by a process known as **cleavage.** The first sign of cleavage is the appearance of a **cleavage furrow,** which begins as a shallow groove in the cell

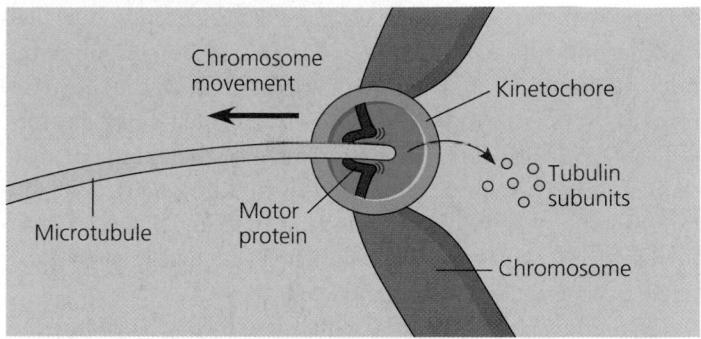

(a) Hypothesis. In this model, a chromosome tracks along a microtubule as the microtubule depolymerizes at its kinetochore end, releasing tubulin subunits.

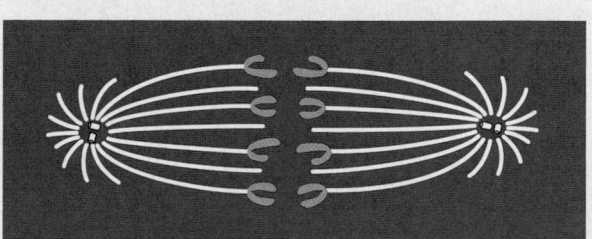

❶ The microtubules of the dividing cell were labeled with a fluorescent dye that glows in the microscope (yellow).

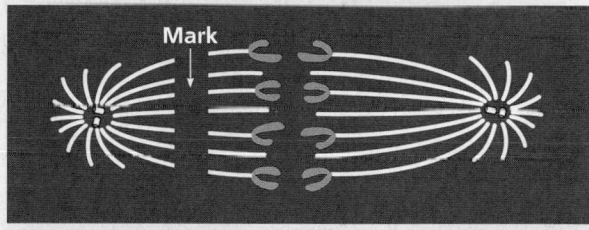

❷ The laser marked the kinetochore microtubules, about midway between one spindle pole and the chromosomes, by eliminating fluorescence in the targeted region.

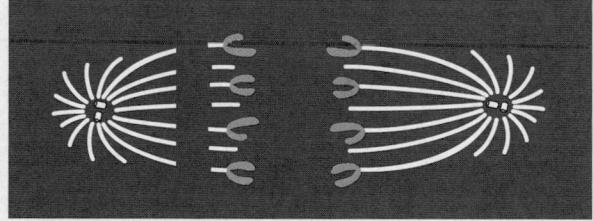

❸ As the chromosomes moved toward the poles, the microtubule segments on the kinetochore side of the laser mark shortened, while those on the centrosome side stayed the same length.

(b) Experiment. In one of the experiments supporting the hypothesis shown in (a), researchers used a laser microbeam to mark the kinetochore microtubules of an early anaphase cell. They then monitored the changes in the lengths of the microtubules on either side of the mark as anaphase proceeded.

FIGURE 12.7 Testing a hypothesis for chromosome migration during anaphase.

THE PROCESS OF SCIENCE

surface near the old metaphase plate (FIGURE 12.8a). On the cytoplasmic side of the furrow is a contractile ring of actin microfilaments associated with molecules of the protein myosin. Actin and myosin are the same proteins responsible for muscle contraction, as well as many other kinds of cell movement. The contraction of the dividing cell's ring of microfilaments is like the pulling of drawstrings. The cleavage furrow deepens until the parent cell is pinched in two, producing two completely separated cells.

Cytokinesis in plant cells, which have walls, is markedly different. There is no cleavage furrow. Instead, during telophase, vesicles derived from the Golgi apparatus move along microtubules to the middle of the cell, where they coalesce, producing a **cell plate** (FIGURE 12.8b). Cell wall materials carried in the vesicles collect in the cell plate as it grows. The cell plate enlarges until its surrounding membrane fuses with the plasma membrane along the perimeter of the cell. Two daughter cells result, each with its own plasma membrane. Meanwhile, a new cell wall arising from the contents of the cell plate has formed between the daughter cells.

FIGURE 12.9 is a series of micrographs of a dividing plant cell. Examining this figure will help you review mitosis and cytokinesis.

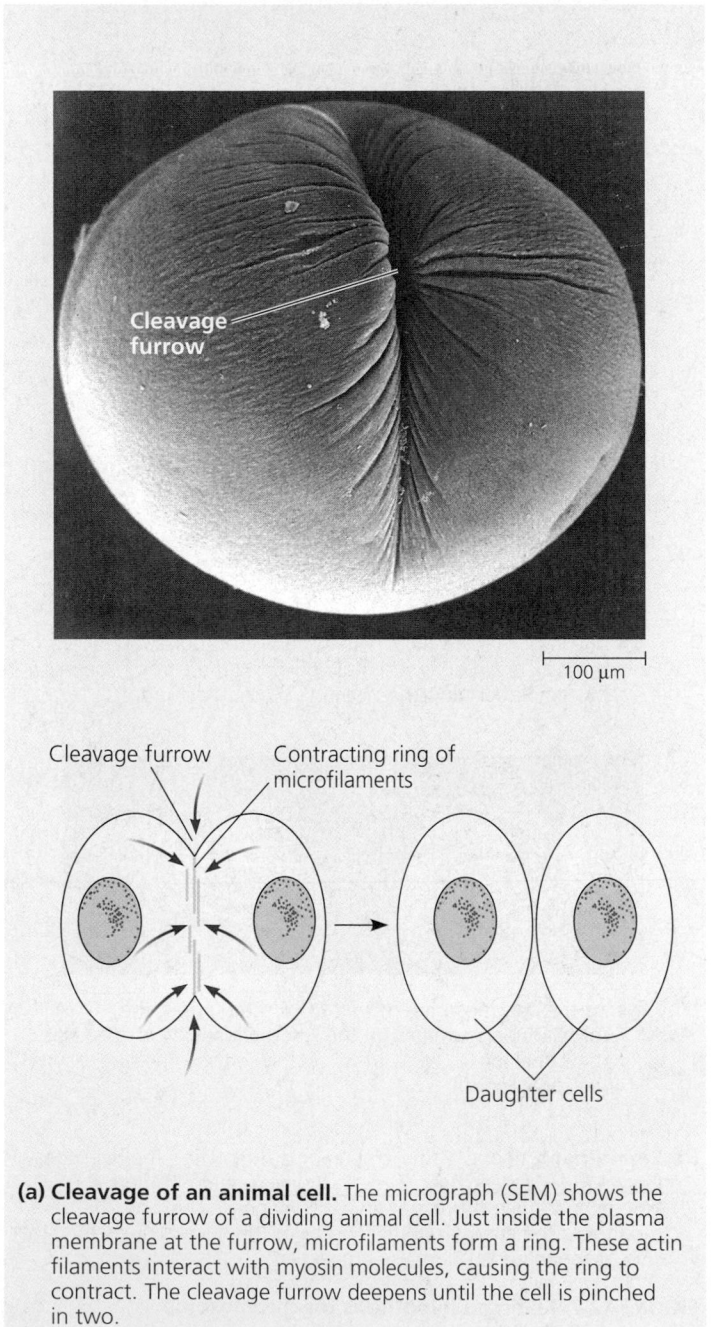

(a) Cleavage of an animal cell. The micrograph (SEM) shows the cleavage furrow of a dividing animal cell. Just inside the plasma membrane at the furrow, microfilaments form a ring. These actin filaments interact with myosin molecules, causing the ring to contract. The cleavage furrow deepens until the cell is pinched in two.

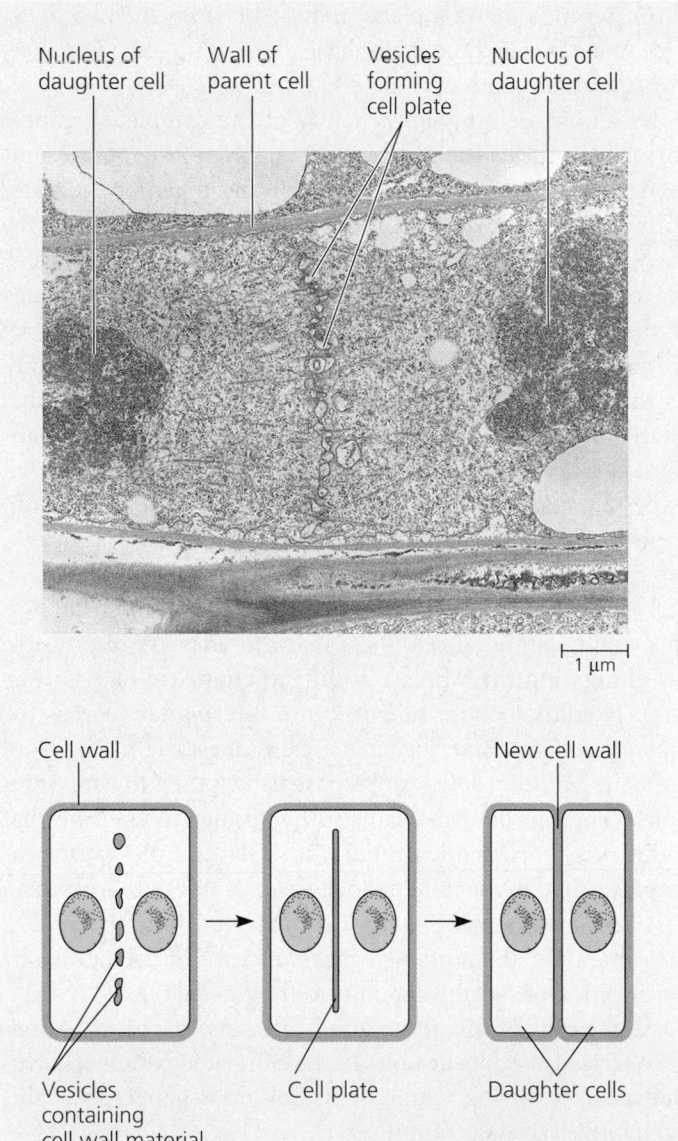

(b) Cell plate formation in a plant cell. In this micrograph (TEM) of a soybean root cell at telophase, you can see the two daughter nuclei and vesicles from the Golgi apparatus that are coming together to form a cell plate at the equatorial plane of the cell. A cell wall between the daughter cells will form from the contents of the cell plate.

FIGURE 12.8 Cytokinesis in animal and plant cells.

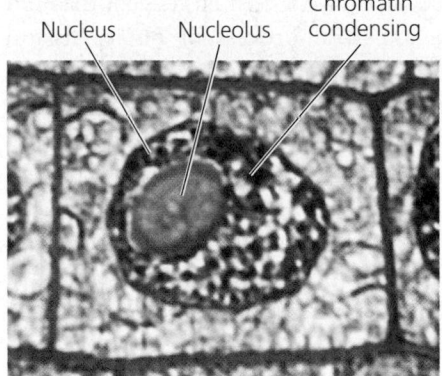

Nucleus   Nucleolus   Chromatin condensing

(a) Prophase. The chromatin is condensing. The nucleolus is still clearly present but will soon begin to disappear. Although not yet visible in the micrograph, the mitotic spindle is starting to form.

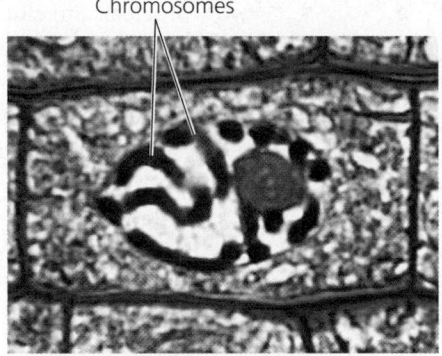

Chromosomes

(b) Prometaphase. We now see discrete chromosomes; each consists of two identical sister chromatids attached all along their lengths. Later in prometaphase, the nuclear envelope will fragment, and spindle microtubules will attach to the kinetochores of the chromosomes.

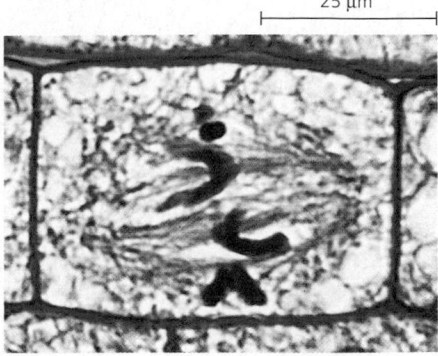

25 µm

(c) Metaphase. The spindle is complete, and the chromosomes, attached to microtubules at their kinetochores, are all at the metaphase plate.

FIGURE 12.9 Mitosis in a plant cell. These light micrographs show mitosis in cells of an onion root.

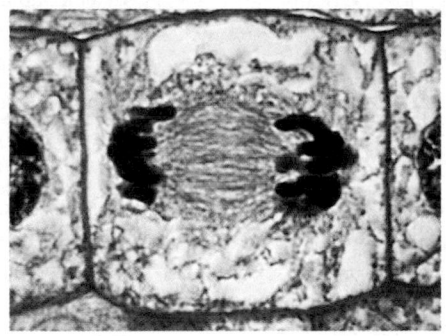

(d) Anaphase. The chromatids of each chromosome have separated, and the daughter chromosomes are moving to the poles of the cell as their kinetochore microtubules shorten.

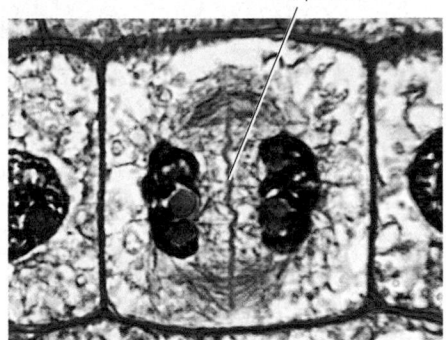

Cell plate

(e) Telophase. Daughter nuclei are forming. Meanwhile, cytokinesis has started: The cell plate, which will divide the cytoplasm in two, is growing toward the perimeter of the parent cell.

Mitosis in eukaryotes may have evolved from binary fission in bacteria

The complex cellular choreography of mitotic cell division solves the problem of correctly distributing copies of eukaryotic genomes to daughter cells. How did mitosis evolve? Given that prokaryotes preceded eukaryotes on Earth by billions of years, we might hypothesize that mitosis had its origins in simpler bacterial mechanisms of cell reproduction.

Prokaryotes (bacteria) reproduce by a type of cell division called **binary fission,** meaning literally "division in half." Most bacterial genes are carried on a single *bacterial chromosome* that consists of a circular DNA molecule and associated proteins. Although bacteria are smaller and simpler than eukaryotic cells, the problem of replicating their genomes in an orderly fashion and distributing the copies equally to two daughter cells is still formidable. The chromosome of the bacterium *Escherichia coli,* for example, when it is fully stretched out, is about 500 times longer than the length of the cell.

Clearly, such a long chromosome must be highly coiled and folded within the cell—and it is.

Prokaryotes do not have mitotic spindles, so what brings about the separation of the two daughter chromosomes in a dividing bacterial cell? A hypothesis proposed in the 1960s suggested that separation of bacterial chromosomes results simply from the growth of new plasma membrane between two sites on the membrane where the chromosome copies are attached. Recent research, however, has challenged this model (FIGURE 12.10, p. 224). Rather than being a passive process, separation of daughter bacterial chromosomes involves active chromosomal movement. Once the DNA of the chromosome begins to replicate, the copies of the first replicated region—called the **origin of replication**—move apart rapidly. Using the techniques of modern DNA technology to tag the origins of replication with molecules that glow green in fluorescence microscopy (see TABLE 7.1), researchers have directly observed the movement of bacterial chromosomes. This movement is reminiscent of the poleward

THE PROCESS OF SCIENCE

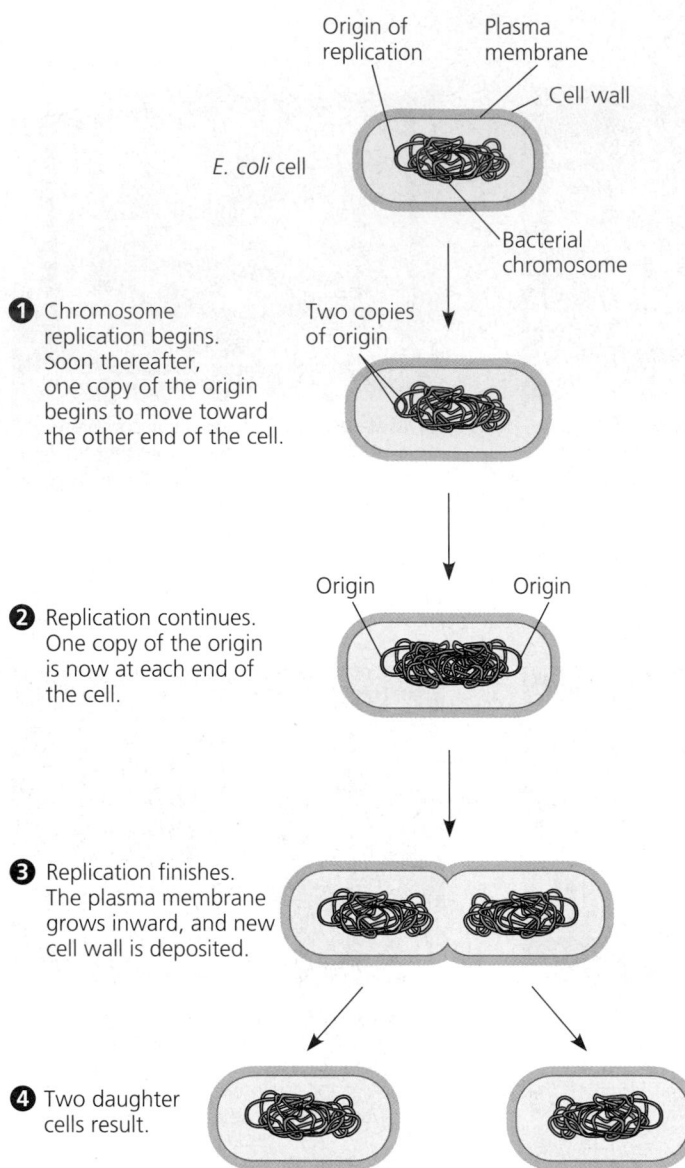

Origin of replication Plasma membrane

Cell wall

E. coli cell

Bacterial chromosome

1 Chromosome replication begins. Soon thereafter, one copy of the origin begins to move toward the other end of the cell.

Two copies of origin

2 Replication continues. One copy of the origin is now at each end of the cell.

Origin   Origin

3 Replication finishes. The plasma membrane grows inward, and new cell wall is deposited.

4 Two daughter cells result.

FIGURE 12.10 Bacterial cell division (binary fission). The single, circular bacterial chromosome replicates, and the two copies move apart by an unknown mechanism. Meanwhile, the cell grows in size. When chromosomal replication is complete, the plasma membrane grows inward to divide the cell in two as a new cell wall is deposited between the daughter cells. The example shown here is the bacterium *Escherichia coli*.

movements of the centromere regions of eukaryotic chromosomes during anaphase of mitosis, even though bacteria don't have mitotic spindles or even microtubules. How bacterial chromosomes move is still a mystery. The idea that prokaryotes might have molecules in any way like the microtubules and motor proteins used in eukaryotic mitosis is surprising and intriguing.

While the bacterial chromosome is replicating, the cell is growing. When replication is complete and the bacterium has reached about twice its initial size, its plasma membrane grows inward, dividing the parent cell into two daughter cells. Each cell inherits a complete genome.

As eukaryotes evolved, along with their larger genomes and nuclear envelopes, the ancestral process of binary fission somehow gave rise to mitosis. FIGURE 12.11 traces a hypothesis for the stepwise evolution of mitosis. Possible intermediate stages are represented by two unusual types of nuclear division found in certain modern unicellular algae. In both types, the nuclear envelope remains intact. In dinoflagellates, replicated chromosomes are attached to the nuclear envelope and separate as it elongates prior to cell division. In diatoms, a spindle within the nucleus separates the chromosomes.

REGULATION OF THE CELL CYCLE

The timing and rate of cell division in different parts of a plant or animal are crucial to normal growth, development, and maintenance. The frequency of cell division varies with the type of cell. For example, human skin cells divide frequently throughout life, whereas liver cells maintain the ability to divide but keep it in reserve until an appropriate need arises—say, to repair a wound. Some of the most specialized cells, such as mature nerve cells and muscle cells, do not divide at all in a mature human. These cell cycle differences result from regulation at the molecular level. The mechanisms of this regulation are of intense interest, not only for understanding the life cycles of normal cells but also for understanding how cancer cells manage to escape normal controls.

A molecular control system drives the cell cycle

What drives the cell cycle? One reasonable hypothesis might be that each event in the cycle triggers the next. According to this hypothesis, for example, the replication of chromosomes in the S phase might cause cell growth during G_2, which might in turn directly trigger the onset of mitosis. However, this apparently logical hypothesis is not in fact correct.

In the early 1970s, a variety of experiments suggested an alternative hypothesis: that the cell cycle is driven by specific chemical signals present in the cytoplasm. Some of the first strong evidence for this hypothesis came from experiments with mammalian cells grown in culture (see FIGURE 12.15). In these experiments, two cells in different phases of the cell cycle were fused to form a single cell with two nuclei. If one of the original cells was in the S phase and the other was in G_1, the G_1 nucleus immediately entered the S phase, as though stimulated by chemicals present in the cytoplasm of the first cell. Similarly, if a cell undergoing mitosis (M phase) was fused with another cell in any stage of its cell cycle, even G_1, the second nucleus immediately entered mitosis, with condensation of the chromatin and formation of a spindle (FIGURE 12.12).

THE PROCESS OF SCIENCE

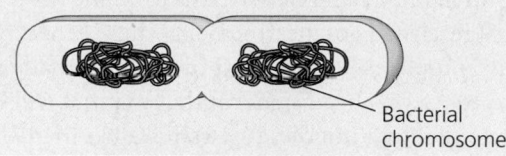

(a) Prokaryotes. During binary fission of bacteria, the daughter chromosomes move apart, ending up at opposite ends of the parent cell. The mechanism is unknown, though attachment of the chromosome to the plasma membrane may be involved.

Bacterial chromosome

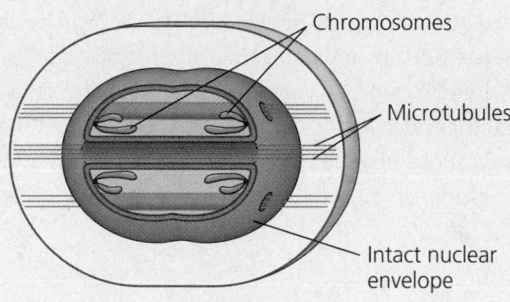

Chromosomes

Microtubules

Intact nuclear envelope

(b) Dinoflagellates. In unicellular algae called dinoflagellates, the nuclear envelope remains intact during cell division, and chromosomes attach to the nuclear envelope. Microtubules pass through the nucleus inside cytoplasmic tunnels, reinforcing the spatial orientation of the nucleus, which then divides in a fission process reminiscent of bacterial reproduction.

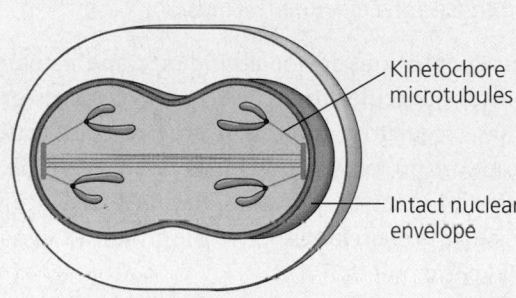

Kinetochore microtubules

Intact nuclear envelope

(c) Diatoms. In another group of unicellular algae, the diatoms, the nuclear envelope also remains intact during cell division. But in these organisms, the microtubules form a spindle *within* the nucleus. The microtubules separate the chromosomes, and the nucleus splits into two daughter nuclei.

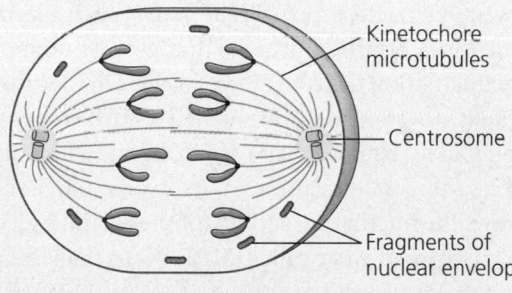

Kinetochore microtubules

Centrosome

Fragments of nuclear envelope

(d) Most eukaryotes. In most other eukaryotes, including plants and animals, the spindle forms outside the nucleus, and the nuclear envelope breaks down during mitosis. Microtubules separate the chromosomes, and the nuclear envelope then re-forms.

THE PROCESS OF SCIENCE

FIGURE 12.11 A hypothesis for the evolution of mitosis. Researchers interested in the evolution of eukaryotic cell division have observed in modern organisms what they believe are mechanisms of division intermediate between the binary fission of bacteria and mitosis as it occurs in most eukaryotes. These schematic diagrams of a proposed evolutionary sequence do not show cell walls.

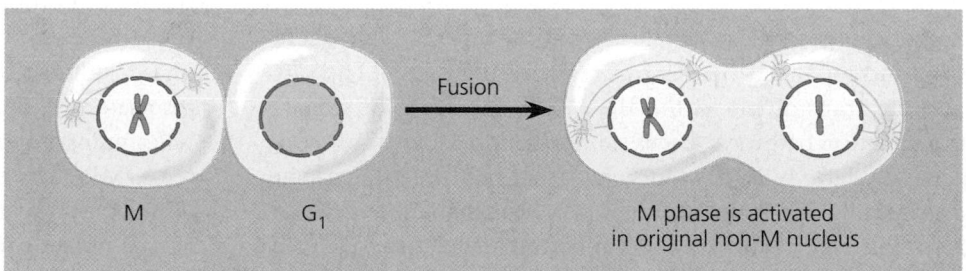

Fusion

M   G₁   M phase is activated in original non-M nucleus

FIGURE 12.12 Evidence for cytoplasmic chemical signals in cell cycle regulation. Cultured mammalian cells can be induced to fuse, forming a single cell with two nuclei. The results of fusing cells at two different phases of the cell cycle suggested that chemicals control the progression of phases. For example, when a cell in M phase was fused with a cell in any other phase, the second nucleus immediately began mitosis. If the second cell was in G₁, as here, the condensed chromosomes that appeared had single chromatids.

THE PROCESS OF SCIENCE

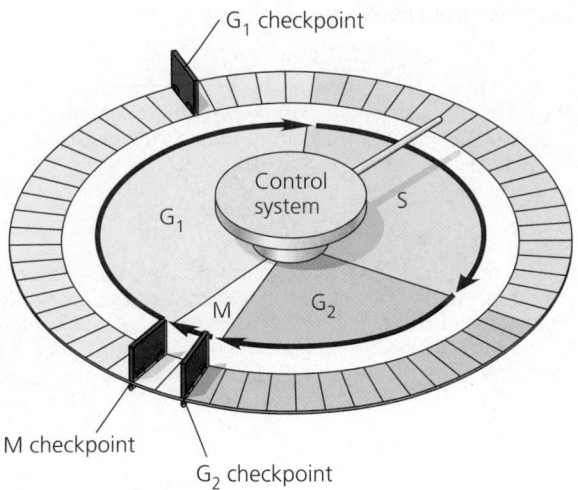

G₁ checkpoint

Control
system

G₁

S

M

G₂

M checkpoint

G₂ checkpoint

FIGURE 12.13 Mechanical analogy for the cell cycle control system. In this diagram of the cell cycle, the flat "stepping stones" around the perimeter represent sequential events. Like the control device of an automatic washer, the cell cycle control system proceeds on its own, driven by a built-in clock. However, the system is subject to regulation at various checkpoints, of which three are shown (red).

These and other experiments demonstrated that the sequential events of the cell cycle are directed by a distinct **cell cycle control system,** a cyclically operating set of molecules in the cell that both triggers and coordinates key events in the cell cycle. The cell cycle control system has been compared to the control device of an automatic washing machine (FIGURE 12.13). Like the washer's device, the cell cycle control system proceeds on its own, driven by a built-in clock. However, just as a washer's cycle is subject to both external adjustment (such as faucets controlling the water supply) and internal control (such as the sensor that detects when the tub is filled with water), the cell cycle is regulated at certain checkpoints by both internal and external controls.

Cell Cycle Checkpoints

A **checkpoint** in the cell cycle is a critical control point where stop and go-ahead signals can regulate the cycle. (The signals are transmitted within the cell by the kinds of signal-transduction pathways discussed in Chapter 11.) Animal cells generally have built-in stop signals that halt the cell cycle at checkpoints until overridden by go-ahead signals. Many signals registered at checkpoints come from cellular surveillance mechanisms; the signals report whether crucial cellular processes up to that point have been completed correctly and thus whether or not the cell cycle should proceed. Checkpoints also register signals from outside the cell, as we will discuss later. Three major checkpoints are found in the G₁, G₂, and M phases.

For many cells, the G₁ checkpoint—dubbed the "restriction point" in mammalian cells—seems to be the most important. If a cell receives a go-ahead signal at the G₁ checkpoint, it will usually complete the cycle and divide. Alternatively, if it does not receive a go-ahead signal at that point, it will exit the cycle, switching into a nondividing state called the **G₀ phase.** Most cells of the human body are actually in the G₀ phase. As mentioned earlier, highly specialized nerve and muscle cells never divide. Other cells, such as liver cells, can be "called back" to the cell cycle by certain environmental cues, such as growth factors released during injury.

To understand how cell cycle checkpoints work, we need to see what kinds of molecules make up the cell cycle control system.

The Cell Cycle Clock:
Cyclins and Cyclin-Dependent Kinases

Rhythmic fluctuations in the abundance and activity of cell cycle control molecules pace the sequential events of the cell cycle. These regulatory molecules are proteins of two main types. Some are protein kinases, enzymes that activate or inactivate other proteins by phosphorylating them (see Chapter 11). Particular protein kinases give the go-ahead signals at the G₁ and G₂ checkpoints.

The kinases that drive the cell cycle are actually present at a constant concentration in the growing cell, but much of the time they are in inactive form. To be active, such a kinase must be attached to a **cyclin,** a protein that gets its name from its cyclically fluctuating concentration in the cell. Because of this requirement, these kinases are called **cyclin-dependent kinases,** or **Cdks.** The activity of a Cdk rises and falls with changes in the concentration of its cyclin partner. FIGURE 12.14a shows the fluctuating activity of the cyclin-Cdk complex that was discovered first, called **MPF.** Note that the peaks of MPF activity correspond to the peaks of cyclin concentration. The cyclin level rises sharply throughout interphase (G₁, S, and G₂), then falls abruptly during mitosis (M).

The initials MPF stand for "maturation-promoting factor," but we can think of MPF as "M-phase-promoting factor" because it triggers the cell's passage past the G₂ checkpoint into M phase (FIGURE 12.14b). When cyclins that accumulate during G₂ associate with Cdk molecules, the resulting MPF complex initiates mitosis, apparently by phosphorylating a variety of proteins. MPF acts both directly and indirectly. For example, it causes the nuclear envelope to fragment by phosphorylating—*and* stimulating still other kinases to phosphorylate—proteins of the nuclear lamina, the lining of the nuclear envelope (see FIGURE 7.9).

Later in the M phase, MPF helps switch itself off by initiating a process that leads to the destruction of its cyclin by a special protein breakdown mechanism you will learn about later (see FIGURE 19.11). Protein breakdown is also involved in driv-

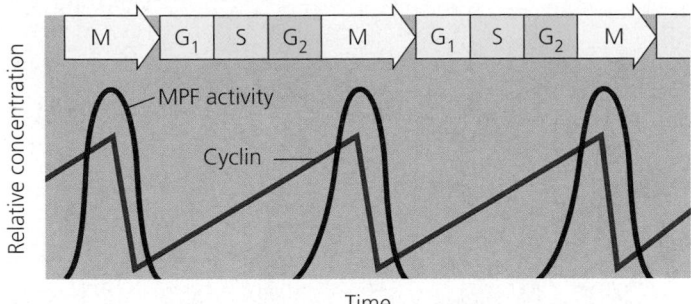

(a) **Fluctuation of MPF activity and cyclin during the cell cycle**

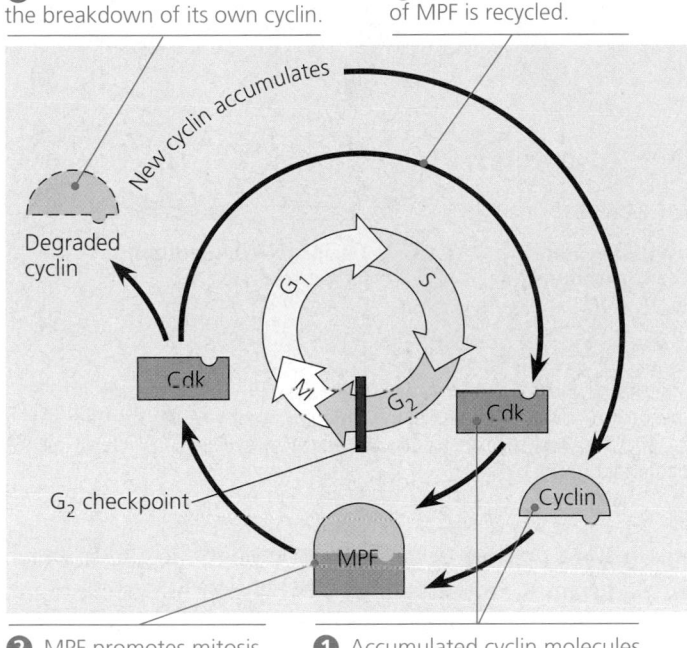

3 One indirect effect of MPF is the breakdown of its own cyclin.

4 The Cdk component of MPF is recycled.

New cyclin accumulates

Degraded cyclin

G_1

S

M

G_2

Cdk

Cdk

G_2 checkpoint

Cyclin

MPF

2 MPF promotes mitosis by phosphorylating various proteins, including other enzymes.

1 Accumulated cyclin molecules combine with Cdk molecules to produce many molecules of MPF by the G_2 checkpoint.

(b) **Molecular mechanisms**

FIGURE 12.14 Molecular control of the cell cycle at the G_2 checkpoint. The steps of the cell cycle are timed by rhythmic fluctuations in the activity of a special class of protein kinases. These enzymes are called cyclin-dependent kinases (Cdks) because they are active only when bound to a cyclin, a protein whose concentration varies cyclically. Here we focus on a Cdk-cyclin complex called MPF, which acts at the G_2 checkpoint to trigger mitosis.

ing the cell cycle past the M phase checkpoint, which controls the onset of anaphase. The noncyclin part of MPF, the Cdk, persists in the cell in inactive form until it associates with new cyclin molecules synthesized during interphase of the next round of the cycle.

What about the G_1 checkpoint? Recent research suggests the involvement of at least three Cdk proteins and several cyclins at this checkpoint. The fluctuating activities of different cyclin-Cdk complexes seem to control all the stages of the cell cycle.

Internal and external cues help regulate the cell cycle

Research scientists are only in the early stages of working out the signaling pathways that link cyclin-dependent kinases to other molecules and events inside and outside the cell. For example, they know that active Cdks function by phosphorylating substrate proteins that affect particular steps in the cell cycle. But identifying the specific substrates of the various Cdks that become active at different phases of the cell cycle has proved difficult. In other words, scientists don't yet know what Cdks actually do in most cases. However, they have identified some steps of the signaling pathways that convey information to the cell cycle machinery. We'll next discuss two examples of such signaling, one originating inside the cell and one outside.

Internal Signals: Messages from the Kinetochores

Anaphase, the separation of sister chromatids, does not begin until all the chromosomes are properly attached to the spindle at the metaphase plate. The gatekeeper is the M phase checkpoint, and it ensures that daughter cells do not end up with missing or extra chromosomes. Researchers have learned that a signal that delays anaphase originates at kinetochores that are not yet attached to spindle microtubules. Certain associated proteins trigger a signaling pathway that keeps an anaphase-promoting complex (APC) in an inactive state. Only when all the kinetochores are attached to the spindle does this "wait" signal cease. The APC then becomes active and indirectly triggers both the breakdown of cyclin and the inactivation of proteins holding the sister chromatids together.

External Signals: Growth Factors

By growing animal cells in culture, researchers have been able to identify many external factors, both chemical and physical, that can influence cell division. For example, cells fail to divide if an essential nutrient is left out of the culture medium. And even if all other conditions are favorable, most types of mammalian cells divide in culture only if the growth medium includes specific growth factors. As mentioned in Chapter 11, a **growth factor** is a protein released by certain body cells that stimulates other cells to divide.

One example of a growth factor is *platelet-derived growth factor (PDGF),* which is made by blood cells called platelets. The experiment illustrated in FIGURE 12.15, p. 228, demonstrates that PDGF is required for the division of fibroblasts in culture. Fibroblasts, a type of connective tissue cell, have PDGF receptors on their plasma membranes. The binding of PDGF molecules to these receptors (which are tyrosine-kinase receptors; see Chapter 11) triggers a signal-transduction pathway that leads to stimulation of cell division. Presumably, the pathway activates one or more components of the cell cycle control system.

THE PROCESS OF SCIENCE

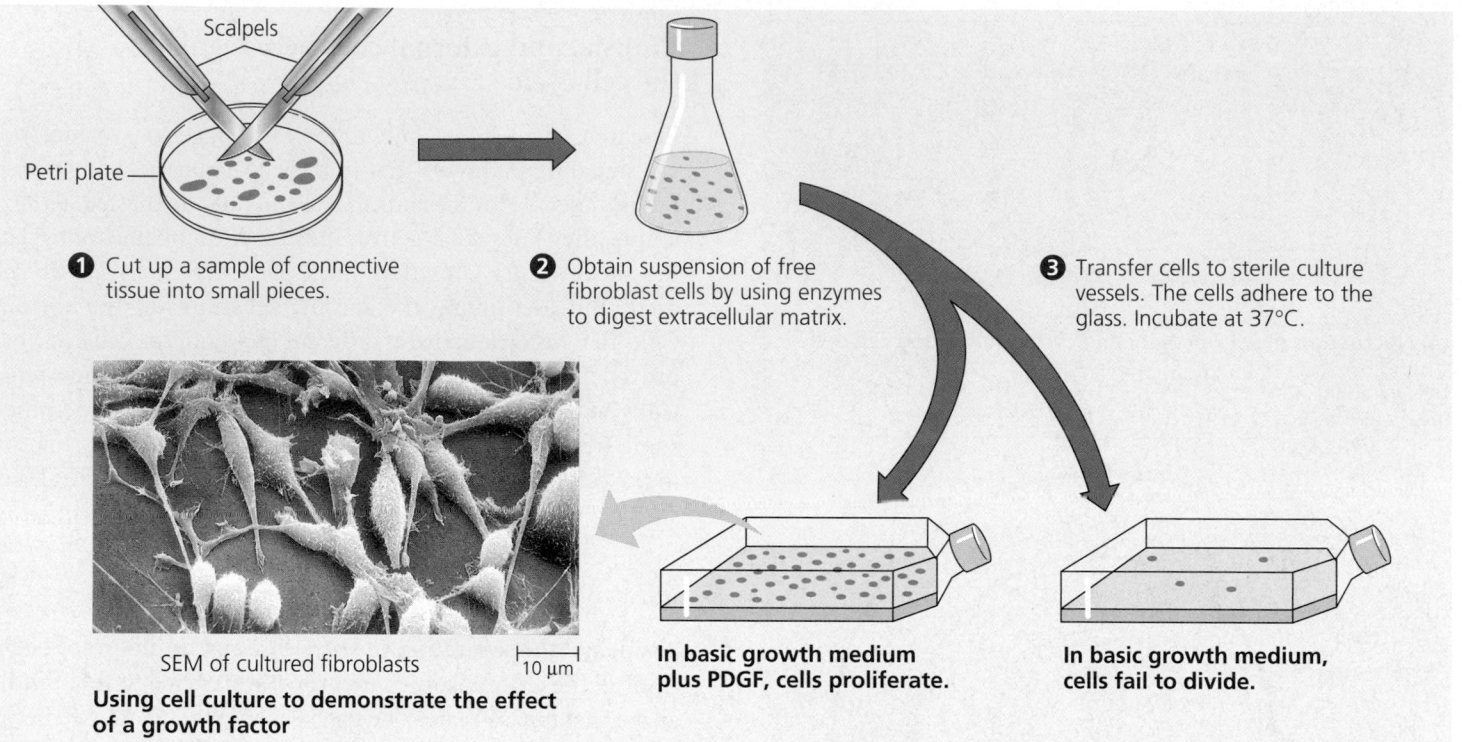

Scalpels

Petri plate

1 Cut up a sample of connective tissue into small pieces.

2 Obtain suspension of free fibroblast cells by using enzymes to digest extracellular matrix.

3 Transfer cells to sterile culture vessels. The cells adhere to the glass. Incubate at 37°C.

SEM of cultured fibroblasts   10 μm

Using cell culture to demonstrate the effect of a growth factor

In basic growth medium plus PDGF, cells proliferate.

In basic growth medium, cells fail to divide.

FIGURE 12.15 The effect of a growth factor on cell division. This experiment confirms that platelet-derived growth factor, PDGF, stimulates the division of human fibroblast cells in culture. The basic growth medium is a complex mixture of glucose, amino acids, salts, and antibiotics (as a precaution against bacterial growth). Some culture vessels (controls) contain only this basic medium while others contain the basic medium plus PDGF.

PDGF stimulates fibroblast division not only in the artificial conditions of cell culture, but in an animal's body as well. When an injury occurs, platelets release PDGF in the vicinity. The resulting proliferation of fibroblasts helps heal the wound. Researchers have discovered a number of different growth factors. Each cell type probably responds specifically to a certain growth factor or combination of growth factors.

The discovery of growth factors provided the key to understanding **density-dependent inhibition** of cell division, a phenomenon in which crowded cells stop dividing (FIGURE 12.16a). As first observed many years ago, cultured cells normally divide until they form a single layer of cells on the inner surface of the culture container, at which point the cells stop dividing. If some cells are removed, those bordering the open space begin dividing again and continue until the vacancy is filled. Apparently, when a cell population reaches a certain density, the amount of required growth factors and nutrients available to each cell becomes insufficient to allow continued cell growth.

Most animal cells also exhibit **anchorage dependence.** To divide, they must be attached to a substratum, such as the inside of a culture jar or the extracellular matrix of a tissue. Experiments suggest that anchorage is signaled to the cell cycle control system via pathways involving plasma membrane proteins and elements of the cytoskeleton linked to them. Density-dependent inhibition and anchorage dependence probably function in the body's tissues as well as in cell culture, checking the growth of cells at some optimal density and location. Cancer cells, which we discuss next, exhibit neither density-dependent inhibition nor anchorage dependence (FIGURE 12.16b).

Cancer cells have escaped from cell cycle controls

Cancer cells do not respond normally to the body's control mechanisms. They divide excessively and invade other tissues. If unchecked, they can kill the organism.

By studying cells growing in culture, researchers have learned that cancer cells do not heed the normal signals that regulate the cell cycle. For example, as FIGURE 12.16b shows, cancer cells do not exhibit density-dependent inhibition when growing in culture; they do not stop dividing when growth factors are depleted. A logical hypothesis to explain this behavior is that cancer cells do not need growth factors in their culture medium in order to grow and divide. They may make a required growth factor themselves or have an abnormality in the signaling pathway that conveys the growth factor's signal to the cell cycle control system—or the cell cycle control system itself may be abnormal. In fact, as you will learn in Chapter 19, these are all possible explanations.

There are other important differences between normal cells and cancer cells that reflect derangements of the cell cycle. If

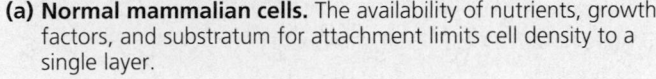

Cells anchor to dish surface and divide (anchorage dependence).

When cells have formed a complete single layer, they stop dividing (density-dependent inhibition).

If some cells are scraped away, the remaining cells divide to fill the gap and then stop (density-dependent inhibition).

(a) Normal mammalian cells. The availability of nutrients, growth factors, and substratum for attachment limits cell density to a single layer.

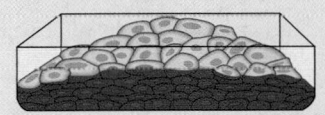

Cancer cells do not exhibit anchorage dependence or density-dependent inhibition.

(b) Cancer cells. Cancer cells usually continue to divide well beyond a single layer, forming a clump of overlapping cells.

FIGURE 12.16 Density-dependent inhibition of cell division. Individual cells are shown disproportionately large.

and when they stop dividing, cancer cells do so at random points in the cycle, rather than at the normal checkpoints. Moreover, in culture, cancer cells can go on dividing indefinitely if they are given a continual supply of nutrients; they are said to be "immortal." A striking example is a cell line that has been reproducing in culture since 1951. Cells of this line are called HeLa cells because their original source was from a tumor removed from a woman named Henrietta Lacks. By contrast, nearly all normal mammalian cells growing in culture divide only about 20 to 50 times before they stop dividing, age, and die. (We'll see a possible

FIGURE 12.17 The growth and metastasis of a malignant breast tumor. The cells of malignant (cancerous) tumors grow in an uncontrolled way and can spread to neighboring tissues and, via the circulatory system, to other parts of the body. The spread of cancer cells beyond their original site is called metastasis.

reason for this phenomenon when we discuss chromosome replication in Chapter 16.)

The abnormal behavior of cancer cells can be catastrophic when it occurs in the body. The problem begins when a single cell in a tissue undergoes **transformation,** the process that converts a normal cell to a cancer cell. The body's immune system normally recognizes a transformed cell as an insurgent and destroys it. However, if the cell evades destruction, it may proliferate to form a **tumor,** a mass of abnormal cells within otherwise normal tissue. If the abnormal cells remain at the original site, the lump is called a **benign tumor.** Most benign tumors do not cause serious problems and can be completely removed by surgery. In contrast, a **malignant tumor** becomes invasive enough to impair the functions of one or more organs (FIGURE 12.17). An individual with a malignant tumor is said to have cancer.

The cells of malignant tumors are abnormal in many ways besides their excessive proliferation. They may have unusual numbers of chromosomes. Their metabolism may be deranged, and they cease to function in any constructive way. Also, owing to abnormal changes on the cells' surfaces, they lose their attachments to neighboring cells and the extracellular matrix and can spread into nearby tissues. Cancer cells may also separate from the original tumor, enter the blood and lymph vessels of the circulatory system, and invade other parts of the body, where they proliferate to form more tumors. This spread of cancer cells to locations distant from their original site is called **metastasis** (see FIGURE 12.17). If a tumor metastasizes, treatments may include high-energy radiation and chemotherapy with toxic drugs that are especially harmful to actively dividing cells.

Researchers are beginning to understand how a normal cell is transformed into a cancer cell. Though the causes of cancer are diverse, cellular transformation always involves the alteration of genes that somehow influence the cell cycle control system. Our knowledge of how changes in the genome lead to the various abnormalities of cancer cells remains rudimentary, however.

Perhaps the reason we have so many unanswered questions about cancer cells is that there is still so much to learn about how normal cells function. The cell, life's basic unit of structure and function, holds enough secrets to engage researchers well into the future.

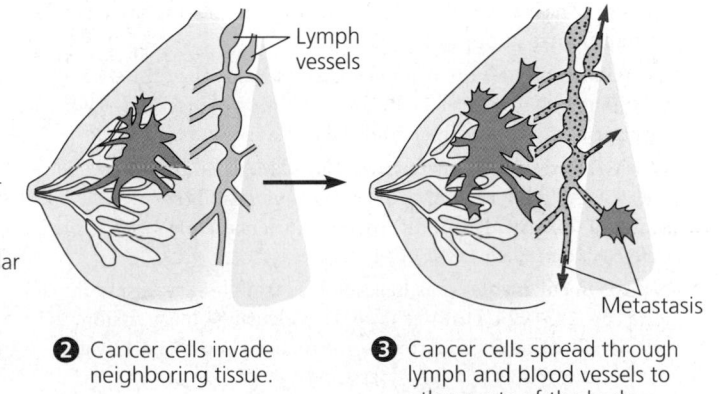

Lymph vessels

Tumor

Glandular tissue

Metastasis

❶ A tumor grows from a single cancer cell.

❷ Cancer cells invade neighboring tissue.

❸ Cancer cells spread through lymph and blood vessels to other parts of the body.

CHAPTER 12 REVIEW

Go to the Campbell Biology website (www.campbellbiology.com) to explore an interactive version of the Chapter Review.

Summary of Key Concepts

THE KEY ROLES OF CELL DIVISION

■ **Cell division functions in reproduction, growth, and repair (p. 215, FIGURE 12.1)** Unicellular organisms reproduce by cell division. Multicellular organisms depend on it for development from a fertilized egg, growth, and repair.

■ **Cell division distributes identical sets of chromosomes to daughter cells (pp. 216–217)** Eukaryotic cell division consists of mitosis (division of the nucleus) and cytokinesis (division of the cytoplasm). DNA is partitioned among chromosomes, making it easier for the eukaryotic cell to replicate and distribute its huge amounts of DNA. Chromosomes consist of chromatin, a complex of DNA and protein that condenses during mitosis. When chromosomes replicate, they form identical sister chromatids. The chromatids separate during mitosis, becoming the chromosomes of the new daughter cells.
  Web/CD Activity 12A: Roles of Cell Division

THE MITOTIC CELL CYCLE

■ **The mitotic phase alternates with interphase in the cell cycle: *an overview* (p. 217, FIGURES 12.4, 12.5)** Mitosis and cytokinesis make up the M (mitotic) phase of the cell cycle. Between divisions, cells are in interphase: the G_1, S, and G_2 phases. The cell grows throughout interphase, but DNA is replicated only during the S (synthesis) phase. Mitosis is a continuous process, often described as occurring in five stages: prophase, prometaphase, metaphase, anaphase, and telophase.
  Web/CD Activity 12B: The Cell Cycle

■ **The mitotic spindle distributes chromosomes to daughter cells: *a closer look* (pp. 220–221, FIGURE 12.6)** The mitotic spindle is an apparatus of microtubules that controls chromosome movement during mitosis. The spindle arises from the centrosomes, organelles near the nucleus that in animal cells include centrioles. Spindle microtubules attach to the kinetochores of chromatids and move the chromosomes to the metaphase plate. In anaphase, sister chromatids separate and move toward opposite poles of the cell. Using motor proteins, each kinetochore moves along shortening microtubules. Meanwhile, nonkinetochore microtubules from opposite poles slide past each other, elongating the cell. In telophase, daughter nuclei form at opposite ends of the cell.

■ **Cytokinesis divides the cytoplasm: *a closer look* (pp. 221–222, FIGURE 12.8)** Mitosis is usually followed by cytokinesis, involving cleavage furrows in animals and cell plates in plants.
  Web/CD Activity 12C: Mitosis and Cytokinesis Animation
  Web/CD Activity 12D: Mitosis and Cytokinesis Video
  Web/CD Case Study in The Process of Science: How Much Time Do Cells Spend in Each Phase of Mitosis?

■ **Mitosis in eukaryotes may have evolved from binary fission in bacteria (pp. 223–224, FIGURES 12.10, 12.11)** During binary fission, the two daughter bacterial chromosomes actively move apart by a mechanism that is not yet understood.

REGULATION OF THE CELL CYCLE

■ **A molecular control system drives the cell cycle (pp. 224–227, FIGURES 12.13–12.14)** Cyclic changes in regulatory proteins work as a mitotic clock. The key molecules are cyclin dependent kinases, complexes of cyclins (whose concentrations build during the cell cycle) and specific protein kinases that are only active when combined with cyclin.

■ **Internal and external cues help regulate the cell cycle (pp. 227–228, FIGURE 12.15)** Cell culture has enabled researchers to study the molecular details of cell division. Both internal signals, such as those emanating from kinetochores not yet attached to the spindle, and external signals, such as growth factors, control the cell cycle checkpoints via signal-transduction pathways. Growth factor depletion explains density-dependent inhibition.

■ **Cancer cells have escaped from cell cycle controls (p. 228–229, FIGURES 12.16, 12.17)** Cancer cells elude normal regulation and divide out of control, forming tumors. Malignant tumors invade surrounding tissues and can metastasize, exporting cancer cells to other parts of the body.
  Web/CD Activity 12E: Causes of Cancer

Self-Quiz

1. During the cell cycle, increases in the enzymatic activity of protein kinases are due to
   a. kinase synthesis by ribosomes.
   b. activation of inactive kinase by binding to cyclin.
   c. conversion of inactive cyclin to the active kinase by means of phosphorylation.
   d. cleavage of the inactive kinase molecules by cytoplasmic proteases.
   e. a decline in external growth factors to a concentration below the inhibitory threshold.

2. Through a microscope, you can see a cell plate beginning to develop across the middle of the cell and nuclei re-forming at opposite poles of the cell. This cell is most likely
   a. an animal cell in the process of cytokinesis.
   b. a plant cell in the process of cytokinesis.
   c. an animal cell in the S phase of the cell cycle.
   d. a bacterial cell dividing.
   e. a plant cell in metaphase.

3. Vinblastine is a standard chemotherapeutic drug used to treat cancer. Since it interferes with the assembly of microtubules, its effectiveness must be related to
   a. disruption of mitotic spindle formation.
   b. inhibition of regulatory protein phosphorylation.
   c. suppression of cyclin production.
   d. myosin denaturation and inhibition of cleavage furrow formation.
   e. inhibition of DNA synthesis.

4. A particular cell has half as much DNA as some of the other cells in a mitotically active tissue. The cell in question is most likely in
   a. G_1.   d. metaphase.
   b. G_2.   e. anaphase.
   c. prophase.

5. One difference between a cancer cell and a normal cell is that
   a. the cancer cell is unable to synthesize DNA.
   b. the cell cycle of the cancer cell is arrested at the S phase.
   c. cancer cells continue to divide even when they are tightly packed together.
   d. cancer cells cannot function properly because they suffer from density-dependent inhibition.
   e. cancer cells are always in the M phase of the cell cycle.

6. The decline of MPF at the end of mitosis is caused by
   a. the destruction of the protein kinase (Cdk).
   b. decreased synthesis of cyclin.
   c. the enzymatic destruction of cyclin.
   d. synthesis of DNA.
   e. an increase in the cell's volume-to-genome ratio.

7. A red blood cell (RBC) has a 120-day life span. If an average adult has 5 L (5,000 cm^3) of blood and each cubic millimeter contains 5 million RBCs, how many new cells must be produced each second to replace the entire RBC population?
   a. 30,000   d. 18,000
   b. 2,400   e. 30,000,000
   c. 2,400,000

8. In function, the plant cell structure that is analogous to an animal cell's cleavage furrow is the
   a. chromosome.   d. centrosome.
   b. cell plate.   e. spindle apparatus.
   c. nucleus.

9. In some organisms, mitosis occurs without cytokinesis occurring. This will result in
   a. cells with more than one nucleus.
   b. cells that are unusually small.
   c. cells lacking nuclei.
   d. destruction of chromosomes.
   e. cell cycles lacking an S phase.

10. Which of the following does *not* occur during mitosis?
   a. packaging of the chromosomes
   b. replication of the DNA
   c. separation of sister chromatids
   d. spindle formation
   e. separation of the centrosomes

11. In the light micrograph below of dividing cells near the tip of an onion root, identify a cell in each of the following stages: interphase, prophase, metaphase, and anaphase. Describe the major events occurring at each stage.

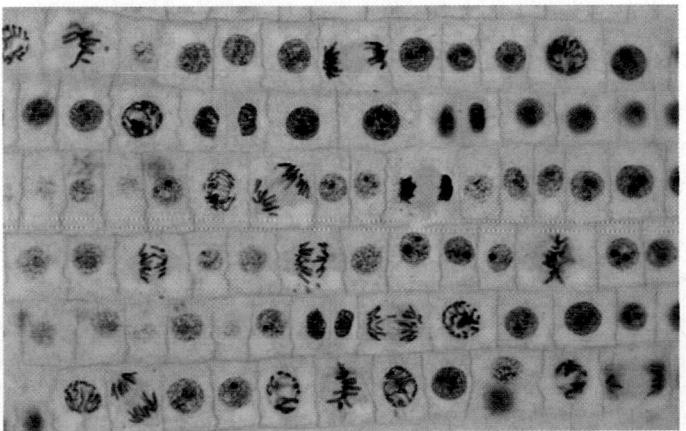

12. Starting with a fertilized egg (zygote), a series of five cell divisions would produce an early embryo with how many cells?

13. Based on what you've read in this chapter, list three similarities between bacterial chromosomes and eukaryotic chromosomes. Consider both chromosome structure and chromosome behavior during cell division.

14. When would a chromosome consist of two identical chromatids?

15. A researcher treats cells with a chemical that prevents DNA synthesis from starting. This treatment traps the cells in which part of the cell cycle?

Go to the website or CD-ROM for more quiz questions.

Evolution Connection

During the mitotic cell cycle, cells double their chromosomes and then return to their original state by mitosis and cell division. The result is that the daughter cells end up with the same number of chromosomes as the parent cell had. Another way to maintain the number of chromosomes would be to carry out cell division first and then replicate the chromosomes in each daughter cell. What would be the problems with this alternative? Or do you think it would be an equally good way of organizing the cell cycle?

The Process of Science

Microtubules are polar structures in that one end (called the "+" end) polymerizes and depolymerizes at a much higher rate than the other end (the "−" end). The experiment shown in FIGURE 12.7 clearly identifies these two ends.
   a. From the results, identify the "+" end and explain your reasoning.
   b. If the opposite end were the "+" end, redraw part 3 of FIGURE 12.7b to show what the result would have been.
   c. Redesign the model in FIGURE 12.7a to reflect these new data.

In the Case Study in the Process of Science on the CD-ROM and website, count cells in different phases of the cell cycle to calculate the percentage of time these cells spend in each phase.

Science, Technology, and Society

Hundreds of millions of dollars are spent each year in the search for effective treatments for cancer; far less money is spent on preventing cancer. Why do you think this is true? What kinds of lifestyle changes could we make to help prevent cancer? What kinds of prevention programs could be initiated or strengthened to encourage these changes? What factors might impede such changes and programs?

Answers: 1. b; 2. b; 3. a; 4. a; 5. c; 6. c; 7. c; 8. b; 9. a; 10. b 11. See Figure 12.5. 12. 32 cells. 13. Each type of chromosome consists of a single molecule of DNA with attached proteins. If stretched out, the molecule of DNA would be many times longer than the cell in which it resides. During cell division, the two copies of a chromosome actively move apart, and one ends up in each of the two daughter cells. 14. After the chromosome replicates during S phase of interphase, throughout G$_2$, and during prophase, prometaphase, and metaphase of mitosis. 15. G$_1$

GENETICS

Nancy Hopkins, the Amgen Professor of Biology at the Massachusetts Institute of Technology, is a molecular biologist, a scientist whose specialty is genetics at the molecular level. She graduated from Radcliffe College and attended graduate school at Yale and Harvard, receiving her Ph.D. in Molecular Biology and Biochemistry from Harvard. After postdoctoral work at the Cold Spring Harbor Laboratory, Dr. Hopkins joined the faculty at MIT. Her research has ranged widely, from gene regulation in bacterial viruses to cancer viruses to embryonic development of the zebrafish. She was also a coauthor of the Fourth Edition of the classic textbook The Molecular Biology of the Gene.

Interview
NANCY HOPKINS

How did you get interested in molecular biology?

When I was a junior in college and trying to decide what to do with my life, I signed up for the introductory biology course. The second lecturer was James D. Watson, and by the end of his first class, that was it—I wanted to be a molecular biologist. I was convinced that out of DNA was going to come the answer to every question in biology.

By that time, spring of 1963, we knew what a gene was, and the genetic code had been figured out. But how were genes regulated? That was the hot question. The French researchers François Jacob and Jacques Monod had recently done a brilliant genetic analysis of gene control in bacteria. They had come up with the hypothesis that the proteins encoded by certain genes could bind to DNA and repress the expression of other genes. But there was no direct evidence for this.

What was your first exposure to research?

I arranged to work in Jim Watson's lab, though I didn't know how to do much of anything. At first I was really just a science groupie. I hung around the laboratory, absorbing the atmosphere. There was a feeling of tremendous excitement in the air. At the same time, I saw that scientists worked 60–70 hours a week, and I worried that a career in science would require me to give up my other interests. So I was drawn to the idea of being a scientist, but a bit repelled by it, too.

However, I really wanted to know how genes were regulated, and I had become obsessed with the question of whether the repressor hypothesis of Jacob and Monod was correct. So I went off to graduate school at Yale hoping I could work on that problem. Unfortunately, I couldn't find a faculty sponsor; everyone said the problem was too hard.

After a year I gave up and went back to Harvard to work as a technician for Mark Ptashne, who was trying to isolate the repressor protein of a bacterial virus called lambda. I didn't care as much about getting a Ph.D. as I did about being involved in isolating a repressor. So I worked in Mark's lab while he isolated the repressor, and we showed that it could bind to DNA—clear evidence that Jacob and Monod were right. That was a day I'll never forget! Having had my fun (as Jim Watson put it), I then enrolled in graduate school at Harvard.

Tell us more about viruses and about the research you did as a graduate student.

Viruses are tiny entities, made chiefly of nucleic acid and protein, that multiply inside cells. Lambda is a type of virus, called a phage, that grows inside bacteria. It has genes that enable it either to replicate and kill its host or to shut down and hide out in the host. Lambda's repressor plays a key role in regulating its genes and determining which pathway is taken.

For my Ph.D., I showed that there were specific sites on lambda DNA where the repressor attached. If one made mutations in those sites—damaged them—the repressor didn't bind, and then the phage didn't behave normally.

What did you do after graduate school?

I was very interested in cancer and wanted to move on to work on animal viruses that were known to cause cancer. For my postdoctoral training, I worked on DNA tumor viruses, and then when I went to MIT I turned to RNA tumor viruses of mice. (These viruses have RNA rather than DNA as their genetic material.) We used genetics to identify genes that made various strains of the viruses differ in their ability to cause cancer, or made them cause different types of cancer.

Are viruses involved in human cancers?

Some viruses can cause cancer in people; for example, papilloma viruses can cause cervical cancer. But viruses aren't thought to be responsible for the major cancers in this country—those of colon, breast, prostate, and lung. Furthermore, in healthy people exposed to cancer viruses, the immune system probably prevents cancers from developing.

After 17 years with cancer viruses, you made a major change in research area. Tell us about that.

In my work with phages and cancer viruses, I'd learned that I loved using genetics to figure out how genes function to produce biological processes. By this time, it seemed possible that, with the new DNA technologies, you might be able to use genetics to dissect one of the most complex processes in biology: the development of a vertebrate animal. For years we had had the idea that molecular genetics held the key to the mysteries of embryonic development—how a heart develops, for instance, or even why you look like your mother; we just didn't believe we would be able to make much progress in our lifetime. But now we were no longer limited to the genes of viruses; we could get our hands on individual genes of eukaryotic cells, of complex animals.

A revolution in developmental biology started with an organism that had been an important object of genetic study since early in the 20th century, the fruit fly *Drosophila*. German researcher Christiane Nüsslein-Volhard and American Eric Wieschaus had shown that one could use a large-scale genetic approach to identify the genes required for the development of a fly embryo. The strategy was to treat flies with a mutagen—something that makes changes (mutations) in DNA—and then look to see how the descendants of the treated flies developed. Essentially, you're damaging one gene or just a few genes at a time and asking, "What happens when I take these genes away?" The amazing success of this *Drosophila* research raised the question of whether you could do the same thing for a vertebrate animal. Many people thought it would be too hard.

Why is this sort of genetic study so hard?

Because animals have so many genes. If you have to screen all the genes one by one, or even a few at a time, it's a lot of work. For the *Drosophila* project, the researchers had to breed about 20,000 families of flies altogether and then examine about a million embryos and larvae, one by one, to find the defective ones. Imagine doing such an experiment with vertebrate animals!

Undaunted, Nüsslein-Volhard was setting up to repeat her fruit fly study with the zebrafish, a vertebrate first proposed for genetic study by a phage geneticist, George Streisinger. So I took a sabbatical in Germany to learn about zebrafish.

What's good about the zebrafish?

The zebrafish is good for genetic analysis because it will breed year-round in the lab, and it's conveniently small and hardy. Moreover, its early development is extremely fast: It develops from a fertilized egg to a free-swimming little fish in only five days. This fish is a juvenile, a larva, but it already has a beating heart, functioning digestive organs, a vision system that can focus on swimming prey (*Paramecium*), and the brain and muscles needed to catch and eat them. And its development is easy to follow, because for the first five days, the animal is transparent!

How does your experimental strategy differ from Nüsslein-Volhard's?

We invented a method to make mutations in fish using a virus rather than a chemical. Some viruses can insert DNA copies of their genes at many places in their host cell's DNA. If the inserted viral DNA lands within a host gene, the gene is damaged—it is mutated. Making mutations by insertion greatly simplifies the task of identifying the mutated genes and cloning them (making many copies for further study), because the inserted DNA serves as a tag. (Ironically, it was a variant of the mouse virus I had studied for 17 years that worked for us!) Having devised a method, we were in a position to find the genes that are needed to make a 5-day-old zebrafish.

What progress have you made?

Our method has really worked. We are now rapidly identifying essential developmental genes and cloning them. It's the most exciting project I've worked on. Many of the genes we've found so far are ones not previously known to be important for development. And these genes turn out to have close relatives in the human genome. This is good news. In fact, we hope that the human versions of some of the genes we find will someday be used for regrowing damaged organs in people.

You spent several years investigating discrimination against women at MIT. How did you get involved?

Reluctantly! I began with a strong belief that science was a merit-based occupation. But, as time went on, I couldn't help observing how women's scientific contributions were valued differently from those of men, how women themselves were valued differently. In part what drove me to action was that I needed more room for my fish!

Trying to get a little more space, I was working my way up through the administrative layers of MIT, and finally I worked myself right up to the president. But before I complained to him, I thought I'd check out my perception of the situation with other women on the faculty. This turned out to be easy to do, because there were then only 15 women among the 212 tenured science faculty at MIT. I was surprised to learn that almost all of these women had reached the same conclusions I had.

So we went as a group to ask the administration to let us study the problem, to let us collect data about the distribution of space and resources. Doing the study was fascinating. We found that subtle discrimination in the individual case is often invisible. But when we looked across all 15 women—and remember these were very outstanding scientists—their experiences formed an undeniable pattern which one could understand was a subtle but damaging type of gender bias.

We wrote up our report, MIT accepted it and corrected all the documented inequities, and everybody was happy as a clam. But it didn't stop there. We soon learned that our results had ramifications far beyond MIT. When a short summary of our study was published in the faculty newsletter, the *New York Times* picked it up, and we soon heard from hundreds of other women, at all sorts of institutions, who had independently come to the same conclusions we had. I was invited to the White House, where the President and First Lady thanked MIT for doing this study. The problem is clearly a longstanding one relating to the role of women in society. Changing that situation will take time. But MIT's recognition of the problem turns out to have been a monumentally important step for women scientists.

CHAPTER 13

MEIOSIS AND SEXUAL LIFE CYCLES

AN INTRODUCTION TO HEREDITY

- Offspring acquire genes from parents by inheriting chromosomes
- Like begets like, more or less: a comparison of asexual and sexual reproduction

THE ROLE OF MEIOSIS IN SEXUAL LIFE CYCLES

- Fertilization and meiosis alternate in sexual life cycles
- Meiosis reduces chromosome number from diploid to haploid: a closer look

ORIGINS OF GENETIC VARIATION

- Sexual life cycles produce genetic variation among offspring
- Evolutionary adaptation depends on a population's genetic variation

*L*iving organisms *are distinguished by their ability to reproduce their kind. Like begets like. Only oak trees produce oaks, and only elephants can make more elephants. Furthermore, offspring resemble their parents more than they do less closely related individuals of the same species. The transmission of traits from one generation to the next is called inheritance, or* **heredity** *(from the Latin* heres, heir)*. Along with inherited similarity, there is also* **variation:** *Offspring differ somewhat in appearance from parents and siblings. These observations have been exploited for the thousands of years that people have bred plants and animals. Curiosity about human similarities and differences is just as ancient. The photo on this page illustrates similarities and differences in Gwyneth Paltrow's family. The mechanisms of heredity and variation, however, eluded biologists until the development of genetics in the twentieth century.*

Genetics, the scientific study of heredity and hereditary variation, is the subject of this unit. Here you will learn about genetics at the levels of organism, cell, and molecule. You will find out how geneticists are helping answer age-old questions about life, including the mystery of how multicellular animals and plants arise from a single cell, the fertilized egg. And you will learn that genetic methods and discoveries are catalyzing progress in every other biological field, including physiology, evolutionary biology, ecology, and even behavior. On the practical side, you will learn how modern genetics is revolutionizing medicine and agriculture. Finally, you will consider some social and ethical questions raised by our ability to manipulate DNA, the genetic material. In this chapter, we begin our study of genetics by examining how chromosomes pass from parents to offspring in sexual reproduction.*

AN INTRODUCTION TO HEREDITY

Offspring acquire genes from parents by inheriting chromosomes

Family friends may tell you that you have your mother's freckles, even though *she* still *has* hers. Parents do not, in any literal sense, give their children freckles, eyes, hair, or any other traits. What, then, actually *is* inherited? Parents endow their offspring with coded information in the form of hereditary units called **genes.** The tens of thousands of genes we inherit from our mothers and fathers constitute our genome. Our genetic link to our parents accounts for family resemblance. Your genome may include a gene for freckles, which you inherited from your mother. Our genes program the specific traits that emerge as we develop from fertilized eggs into adults.

Genes are segments of DNA. You learned in Chapters 1 and 5 that DNA is a polymer of four different kinds of monomers

234

called nucleotides. Inherited information is passed on in the form of each gene's specific sequence of nucleotides, much as printed information is communicated in the form of meaningful sequences of letters. Language is symbolic. The brain translates words and sentences into mental images and ideas; for example, the object you imagine when you read "apple" looks nothing like the word itself. Analogously, cells translate genetic "sentences" into freckles and other features with no resemblance to genes. Most genes program cells to synthesize specific enzymes and other proteins, and it is the cumulative action of these proteins that produces an organism's inherited traits. The programming of these traits in the form of DNA is one of the unifying themes of biology.

The transmission of hereditary traits has its molecular basis in the precise replication of DNA, which produces copies of genes that can be passed along from parents to offspring. In animals and plants, the cellular vehicles that transmit genes from one generation to the next are sperm and ova (unfertilized eggs). After a sperm cell unites with an ovum (a single egg), genes from both parents are present in the nucleus of the fertilized egg.

Except for tiny amounts of DNA in mitochondria and chloroplasts, the DNA of a eukaryotic cell is subdivided into chromosomes within the nucleus. Every living species has a characteristic number of chromosomes. For example, humans have 46 chromosomes in almost all of their cells. Each chromosome consists of a single long DNA molecule elaborately coiled in association with various proteins. One chromosome includes hundreds or thousands of genes, each of which is a specific part of the DNA molecule. A gene's specific location along the length of a chromosome is called the gene's **locus** (plural, *loci*). Our genetic endowment consists of whatever genes happened to be part of the chromosomes we inherited from our parents.

Like begets like, more or less: a comparison of asexual and sexual reproduction

Strictly speaking, "Like begets like" applies only to organisms that reproduce asexually. In **asexual reproduction,** a single individual is the sole parent and passes copies of all its genes to its offspring. For example, single-celled eukaryotic organisms can reproduce asexually by mitotic cell division, in which DNA is copied and allocated equally to two daughter cells. The genomes of the offspring are virtually exact copies of the parent's genome. Some multicellular organisms are also capable of reproducing asexually. *Hydra,* a relative of the jellyfishes, can reproduce by budding (FIGURE 13.1). Because the cells of the bud were derived by mitosis in the parent, the "chip off the old block" is usually genetically identical to its parent. Occasional genetic differences are due to relatively rare changes in

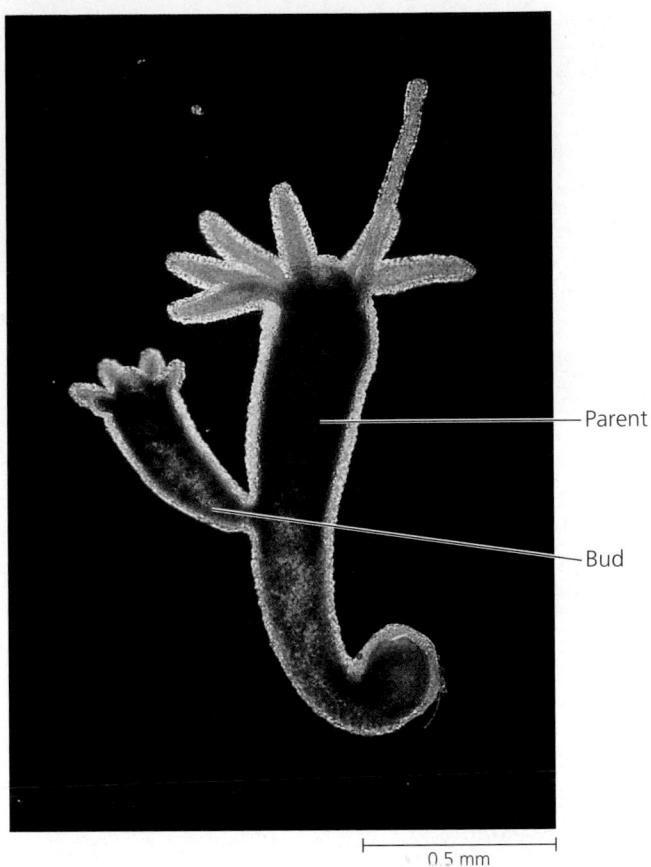

0.5 mm

FIGURE 13.1 The asexual reproduction of a hydra. This relatively simple multicellular animal reproduces by budding. The bud, a localized mass of mitotically dividing cells, develops into a small hydra, which detaches from the parent (LM).

the DNA called mutations, which we will discuss in Chapter 17. An individual that reproduces asexually gives rise to a **clone,** a group of genetically identical individuals.

Sexual reproduction results in greater variation than does asexual reproduction; two parents give rise to offspring that have unique combinations of genes inherited from the two parents. In contrast to a clone, offspring of sexual reproduction vary genetically from their siblings and both parents: "Like begets like" only in the general sense of family resemblance. Genetic variation like that shown in FIGURE 13.2 (p. 236) is an important consequence of sexual reproduction. What mechanisms generate this genetic variation? The key is the behavior of chromosomes during the sexual life cycle.

THE ROLE OF MEIOSIS IN SEXUAL LIFE CYCLES

A **life cycle** is the generation-to-generation sequence of stages in the reproductive history of an organism, from conception to production of its own offspring. In this section, we track the behavior of chromosomes through sexual life cycles.

Couple 1   Couple 2

FIGURE 13.2 Two families. Two sets of parents make up the top row. Each couple has two children represented among the four photos in the bottom row, which are randomly arranged. (All individuals were photographed at about the same age; these are senior pictures from high school yearbooks.) Can you match the offspring with their parents? See the bottom of the page for the answers.*

Fertilization and meiosis alternate in sexual life cycles

We begin with a familiar example—the human life cycle—and use it to introduce some basic terms.

The Human Life Cycle

In humans, each **somatic cell**—any cell other than a sperm or ovum—has 46 chromosomes. With a light microscope, condensed (mitotic) chromosomes can be distinguished from one another by their appearance. The sizes of chromosomes and the positions of their centromeres differ. When chromosomes are stained with certain dyes, each chromosome also has a distinctive pattern of colored bands.

Careful examination of a micrograph of the 46 human chromosomes reveals that there are two of each type. This becomes clear when images of the chromosomes are arranged in pairs, starting with the longest chromosomes. The resulting display is called a **karyotype** (FIGURE 13.3). The chromosomes that make up a pair—that have the same length, centromere position, and staining pattern—are called **homologous chromosomes,** or homologues. The two chromosomes of each pair carry genes controlling the same inherited characters. For example, if a gene for eye color is situated at a particular locus on a certain chromosome, then the homologue of that chromosome will also have a gene specifying eye color at the equivalent locus.

There is an important exception to the rule of homologous chromosomes for human somatic cells: the two distinct chromosomes referred to as X and Y. Human females have a homologous pair of X chromosomes (XX), but males have one X and one Y chromosome (XY). Only small parts of the X and Y are homologous; most of the genes carried on the X chromosome do not have counterparts on the tiny Y, and the Y has genes lacking on the X. Because they determine an individual's sex, the X and Y chromosomes are called **sex chromosomes.** The other chromosomes are called **autosomes.**

The occurrence of homologous pairs of chromosomes in our karyotype is a consequence of our sexual origins. We inherit one chromosome of each pair from each parent. So the 46 chromosomes in our somatic cells are actually two sets of 23 chromosomes—a maternal set (from our mother) and a paternal set (from our father).

Sperm cells and ova are different from somatic cells in chromosome count. Each of these reproductive cells, or **gametes,** has a single set of the 22 autosomes plus a single sex chromosome, either X or Y. A cell with a single chromosome set is called a **haploid cell.** For humans, the haploid number, abbreviated n, is 23 ($n = 23$).

By means of sexual intercourse, a haploid sperm cell from the father reaches and fuses with a haploid ovum of the mother. This union of gametes is called **fertilization,** or **syngamy.** The resulting fertilized egg, or **zygote,** contains the two haploid sets of chromosomes bearing genes representing the maternal and paternal family lines. The zygote and all other cells having two sets of chromosomes are called **diploid cells.** For humans, the diploid number of chromosomes, abbreviated $2n$, is 46 ($2n = 46$).

* Answer: The boy and girl in the center of the bottom row are children of couple 1. The boy and girl on either end of the bottom row are children of couple 2.

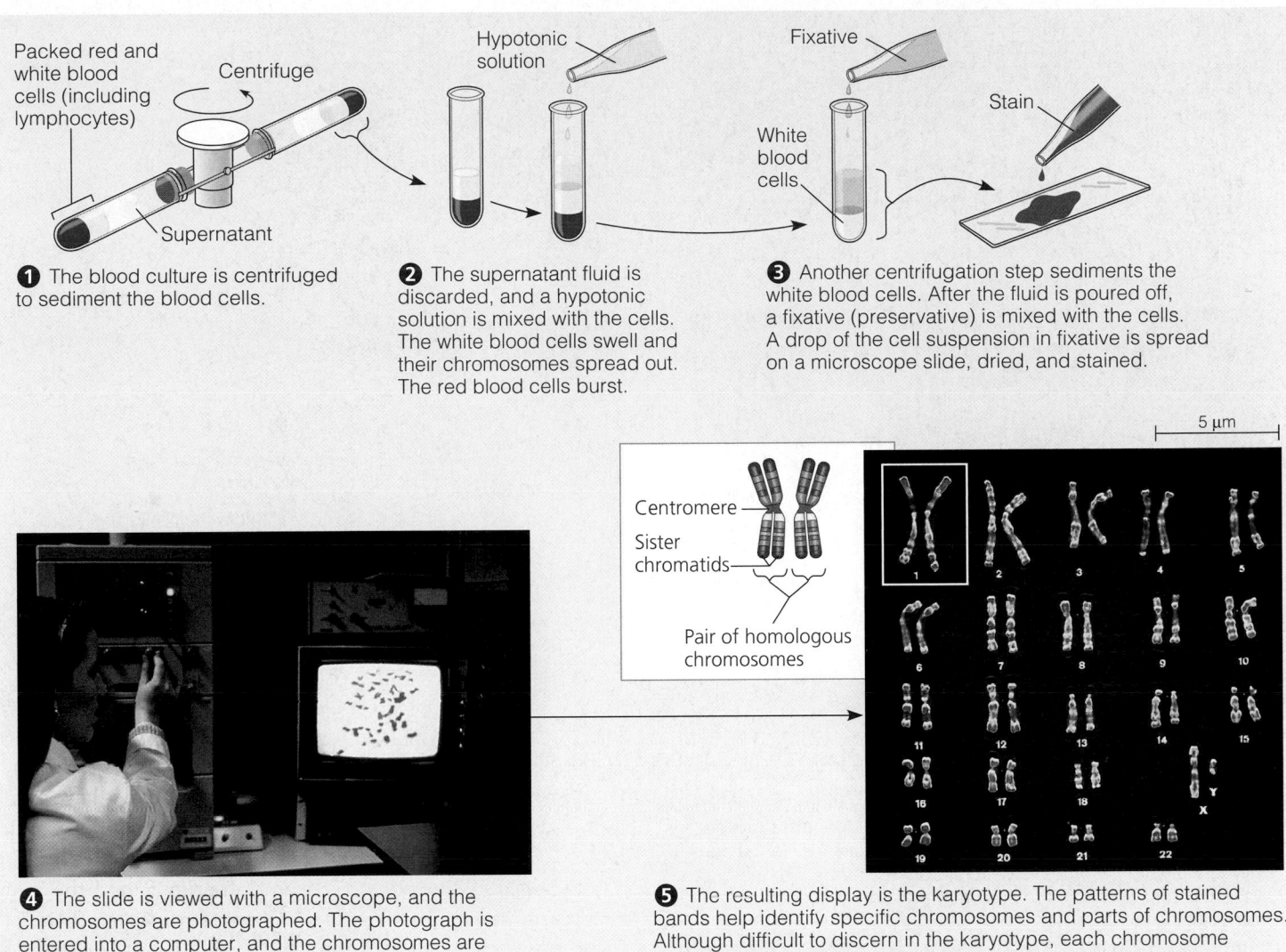

① The blood culture is centrifuged to sediment the blood cells.

② The supernatant fluid is discarded, and a hypotonic solution is mixed with the cells. The white blood cells swell and their chromosomes spread out. The red blood cells burst.

③ Another centrifugation step sediments the white blood cells. After the fluid is poured off, a fixative (preservative) is mixed with the cells. A drop of the cell suspension in fixative is spread on a microscope slide, dried, and stained.

④ The slide is viewed with a microscope, and the chromosomes are photographed. The photograph is entered into a computer, and the chromosomes are electronically rearranged into pairs according to size and shape.

⑤ The resulting display is the karyotype. The patterns of stained bands help identify specific chromosomes and parts of chromosomes. Although difficult to discern in the karyotype, each chromosome consists of two sister chromatids, closely attached all along their lengths (see diagram).

THE PROCESS OF SCIENCE

FIGURE 13.3 Preparation of a human karyotype. Karyotypes, ordered displays of an individual's chromosomes, are often prepared using lymphocytes, a type of white blood cell. The cells are treated with a drug to stimulate mitosis and are then grown in culture for several days. Another drug is then added to arrest mitosis at metaphase, when the chromosomes, each consisting of two joined sister chromatids, are highly condensed and easy to identify in the microscope. The figure above outlines the further steps in the preparation of the karyotype. Karyotyping can be used to screen for abnormal numbers of chromosomes or defective chromosomes associated with certain congenital disorders, such as Down syndrome. (The causes and effects of chromosomal disorders are discussed in Chapter 15.)

As a human develops from a zygote to a sexually mature adult, the zygote's genes are passed on with precision to all somatic cells of the body by the process of mitosis. Thus, somatic cells, like the zygote from which they are derived, are diploid.

The only cells of the human body *not* produced by mitosis are the gametes, which develop in the gonads (ovaries in females and testes in males). Imagine what would happen if human gametes *were* made by mitosis: They would be diploid like the somatic cells. At the next round of fertilization, when two gametes fused, the normal chromosome number of 46 would double to 92, and each subsequent generation would double the number of chromosomes yet again. But sexually reproducing organisms carry out a process that halves the chromosome number in the gametes, compensating for the doubling that occurs at fertilization. This process is a form of cell division called **meiosis,** and in animals it occurs only in the ovaries or testes. While mitosis conserves chromosome number, meiosis reduces the chromosome number by half. As a result, human sperm and ova each have a haploid set of 23 different chromosomes, one from each homologous pair.

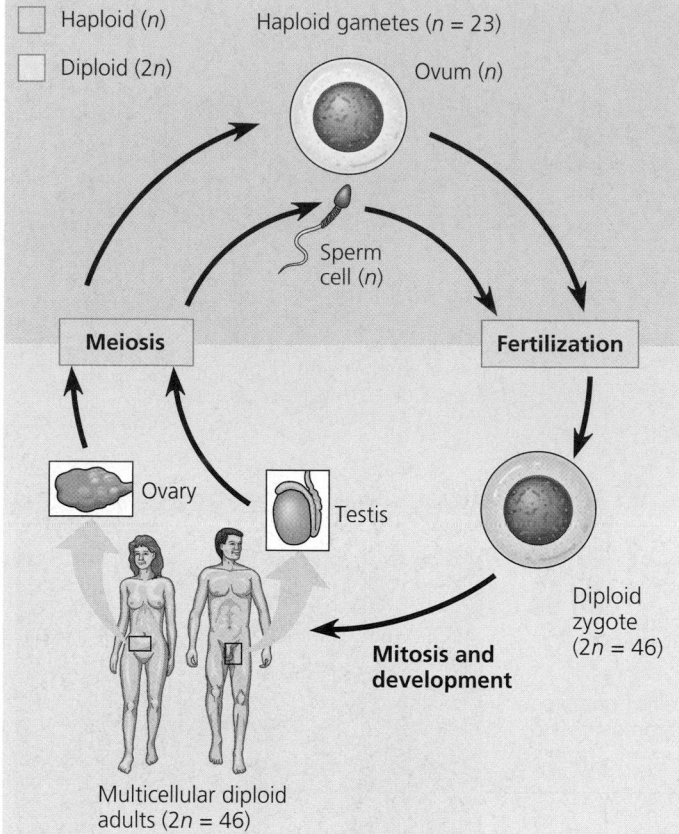

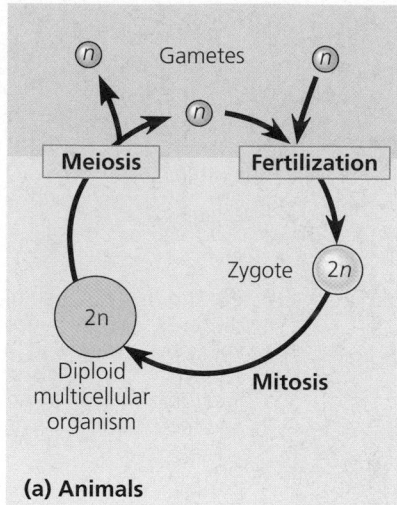

FIGURE 13.4 The human life cycle. In each generation, the doubling of chromosome number that results from fertilization is offset by the halving of chromosome number that results from meiosis. For humans, the number of chromosomes in a haploid cell is 23 ($n = 23$); the number of chromosomes in the diploid zygote and all somatic cells arising from it is 46 ($2n = 46$).

This figure introduces a color code that will be used for other life cycles later in the book. A blue-gray background represents haploid stages of a life cycle, and a tan background represents diploid stages.

(a) Animals

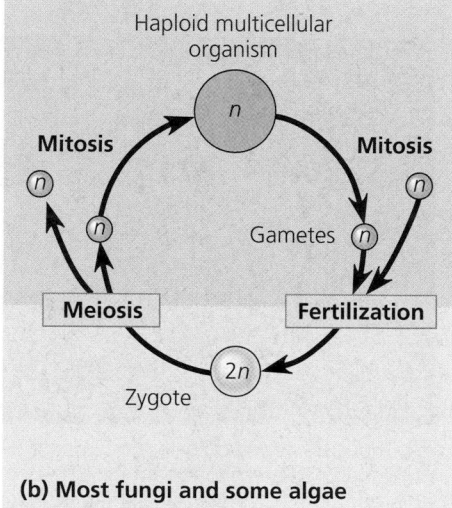

(b) Most fungi and some algae

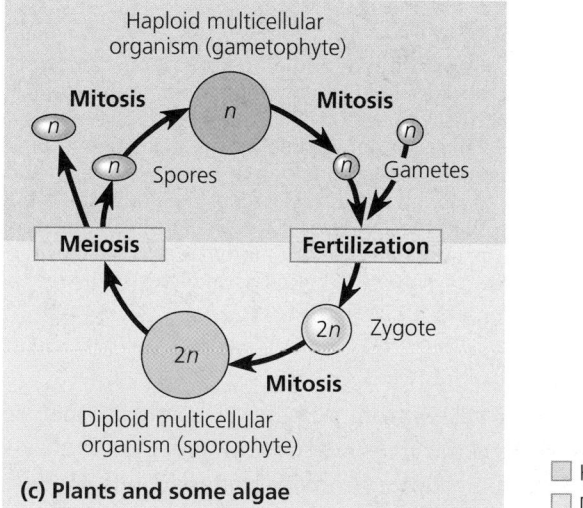

(c) Plants and some algae

☐ Haploid
☐ Diploid

FIGURE 13.5 Three sexual life cycles differing in the timing of meiosis and fertilization (syngamy). The common feature of all three cycles is the alternation of these two key events, which contribute to genetic variation among offspring.

Fertilization restores the diploid condition by combining two haploid sets of chromosomes, and the human life cycle is repeated, generation after generation (FIGURE 13.4).

In general outline, the human life cycle is typical of many animals. Indeed, the processes of fertilization and meiosis are the unique trademarks of sexual reproduction. Fertilization and meiosis alternate in sexual life cycles, offsetting each other's effects on the chromosome number and thus perpetuating a species' chromosome count.

The Variety of Sexual Life Cycles

Although the alternation of meiosis and fertilization is common to all organisms that reproduce sexually, the timing of these two events in the life cycle varies, depending on the species. These variations can be grouped into three main types of life cycles (FIGURE 13.5). The human life cycle is an example of one type, characteristic of most animals. Gametes are the only haploid cells. Meiosis occurs during the production of gametes, which undergo no further cell division prior to fertilization. The diploid zygote divides by mitosis, producing a multicellular organism that is diploid (FIGURE 13.5a).

A second type of life cycle occurs in most fungi and some protists, including some algae. After gametes fuse to form a diploid zygote, meiosis occurs before offspring develop. This meiosis produces not gametes but haploid cells that then divide by mitosis to give rise to a haploid multicellular adult organism. Subsequently, the haploid organism produces gametes by mitosis, rather than by meiosis. The only diploid stage is the zygote (FIGURE 13.5b). (Note that *either* haploid or diploid cells can divide by mitosis, depending on the type of life cycle. Only diploid cells, however, can undergo meiosis.)

Plants and some species of algae exhibit a third type of life cycle called **alternation of generations.** This type of life cycle includes both diploid and haploid multicellular stages. The multicellular diploid stage is called the **sporophyte.** Meiosis in the sporophyte produces haploid cells called **spores.** Unlike a gamete, a spore gives rise to a multicellular individual without fusing with another cell. A spore divides mitotically to generate a multicellular haploid stage called the **gametophyte.** The haploid gametophyte makes gametes by mitosis. Fertilization results in a diploid zygote, which develops into the next sporophyte generation. In this type of life cycle, therefore, the sporophyte and gametophyte generations take turns reproducing each other (FIGURE 13.5c).

Though the three types of sexual life cycles differ in the timing of meiosis and fertilization, they share a fundamental result: Each cycle of chromosome halving and doubling contributes to genetic variation among offspring. A closer look at meiosis will reveal the sources of this variation.

Meiosis reduces chromosome number from diploid to haploid: *a closer look*

Many of the steps of meiosis closely resemble corresponding steps in mitosis. Meiosis, like mitosis, is preceded by the replication of chromosomes. However, this single replication is followed by two consecutive cell divisions, called **meiosis I** and **meiosis II.** These divisions result in four daughter cells (rather than the two daughter cells of mitosis), each with only half as many chromosomes as the parent cell. Study the overview of meiosis in FIGURE 13.6, and be sure you understand the difference between homologous chromosomes and sister chromatids. The two chromosomes of a homologous pair are individual chromosomes that were inherited from different parents. Homologues appear alike in the microscope, but they have different versions of genes at some of their corresponding loci (for example, a gene for freckles on one chromosome and a gene for the absence of freckles at the same locus on the homologue).

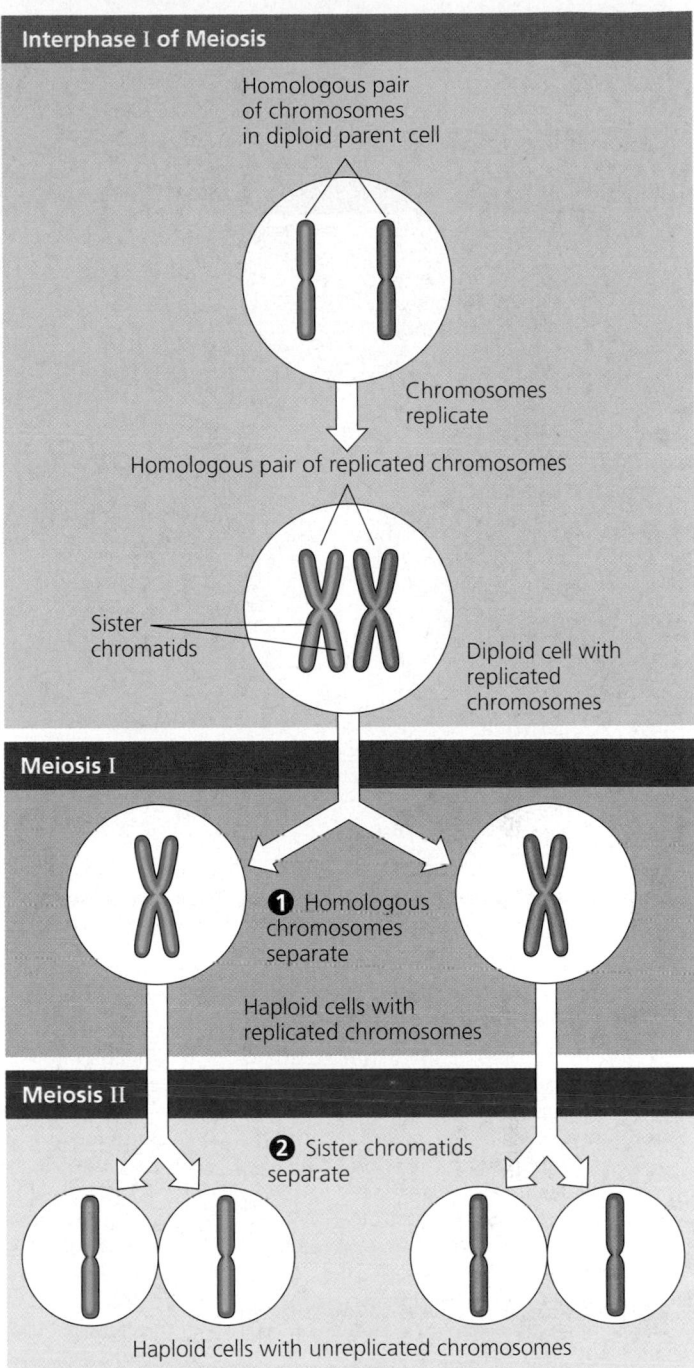

FIGURE 13.6 Overview of meiosis: how meiosis reduces chromosome number. After the chromosomes replicate once, the diploid cell divides *twice,* yielding four haploid daughter cells. This overview tracks just one pair of homologous chromosomes, which for the sake of simplicity are drawn in the condensed state throughout (they would not normally be condensed during interphase). Homologues in this and later figures are colored red and blue to remind you that they carry different versions of some genes.

FIGURE 13.7 on pages 240–241 describes in some detail the two divisions of meiosis for an animal cell whose diploid number is 4. Study FIGURE 13.7 thoroughly before going on to the next section.

INTERPHASE	PROPHASE I	METAPHASE I	ANAPHASE I

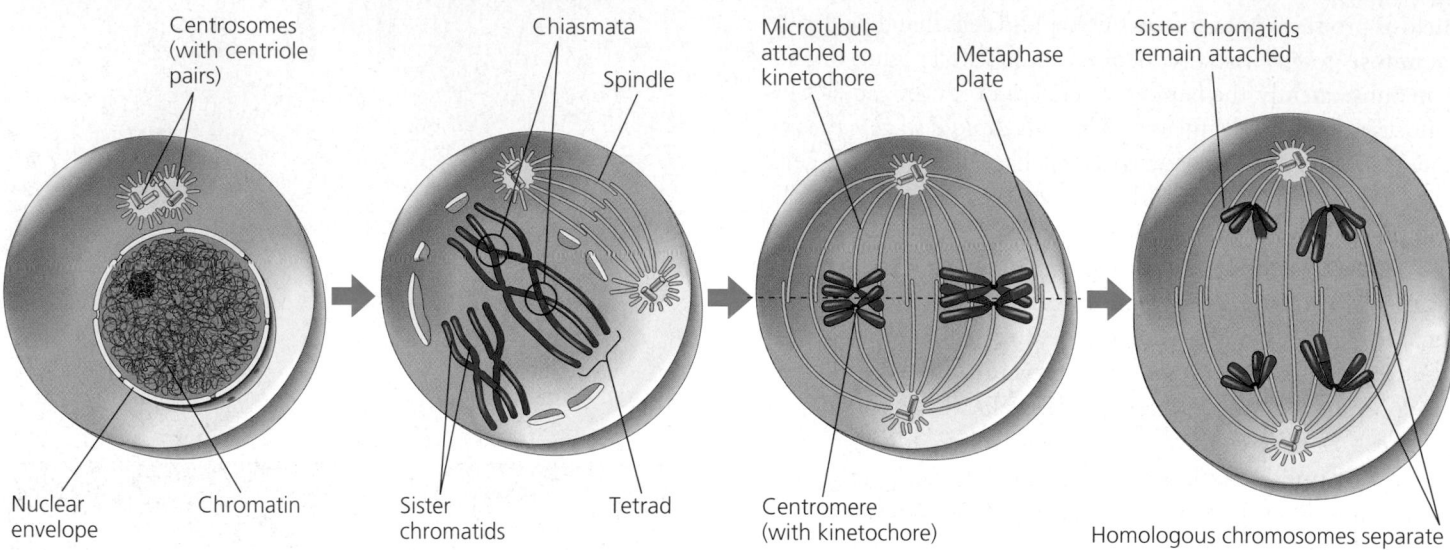

Chromosomes duplicate | **Homologous chromosomes pair and exchange segments** | **Tetrads line up** | **Pairs of homologous chromosomes split up**

INTERPHASE

Meiosis is preceded by an interphase, during which each of the chromosomes replicates. This process is similar to the chromosome replication preceding mitosis. For each chromosome, the result is two genetically identical sister chromatids, which remain attached at their centromeres. The centrosomes also replicate to form the two represented in this drawing.

PROPHASE I

Meiotic prophase I lasts longer and is more complex than prophase in mitosis. The chromosomes begin to condense, and homologues, each consisting of two sister chromatids, pair up. In a process called synapsis, a protein structure—the synaptonemal complex—attaches the homologous chromosomes tightly together all along their lengths. When the synaptonemal complex disappears in late prophase, each chromosome pair becomes visible in the microscope as a tetrad, a cluster of four chromatids. At various places along their length, chromatids of homologous chromosomes are crisscrossed. These crossings are called chiasmata (singular, chiasma). The chiasmata hold the homologous pairs of chromosomes together until anaphase I. Notice that the chromosomes have traded segments at the chiasmata.

Meanwhile, other cellular components prepare for the division of the nucleus in a manner similar to that observed during mitosis. The centrosomes move away from each other, and spindle microtubules form between them. The nuclear envelope and nucleoli disperse. Finally, spindle microtubules capture the kinetochores that form on the chromosomes, and the chromosomes begin moving to the metaphase plate. Prophase I, which can last for days or even longer, typically occupies more than 90% of the time required for meiosis.

METAPHASE I

The chromosomes are now arranged on the metaphase plate, still in homologous pairs. Kinetochore microtubules from one pole of the cell are attached to one chromosome of each pair, while microtubules from the opposite pole are attached to the homologue.

ANAPHASE I

As in mitosis, the spindle apparatus guides the movement of the chromosomes toward the poles. However, sister chromatids remain attached at their centromeres and move as a single unit toward the same pole. The homologous chromosome moves toward the opposite pole. (This contrasts with the behavior of chromosomes during mitosis. In mitosis, chromosomes appear individually on the metaphase plate rather than in pairs, and the sister chromatids of each chromosome separate.)

FIGURE 13.7 The stages of meiotic cell division. These diagrams show meiotic cell division for an animal cell with a diploid number of 4 ($2n = 4$). The behavior of the chromosomes is emphasized by the use of red and blue to differentiate the members of each homologous pair. For a discussion of the spindle and other features common to mitosis and meiosis, see Chapter 12.

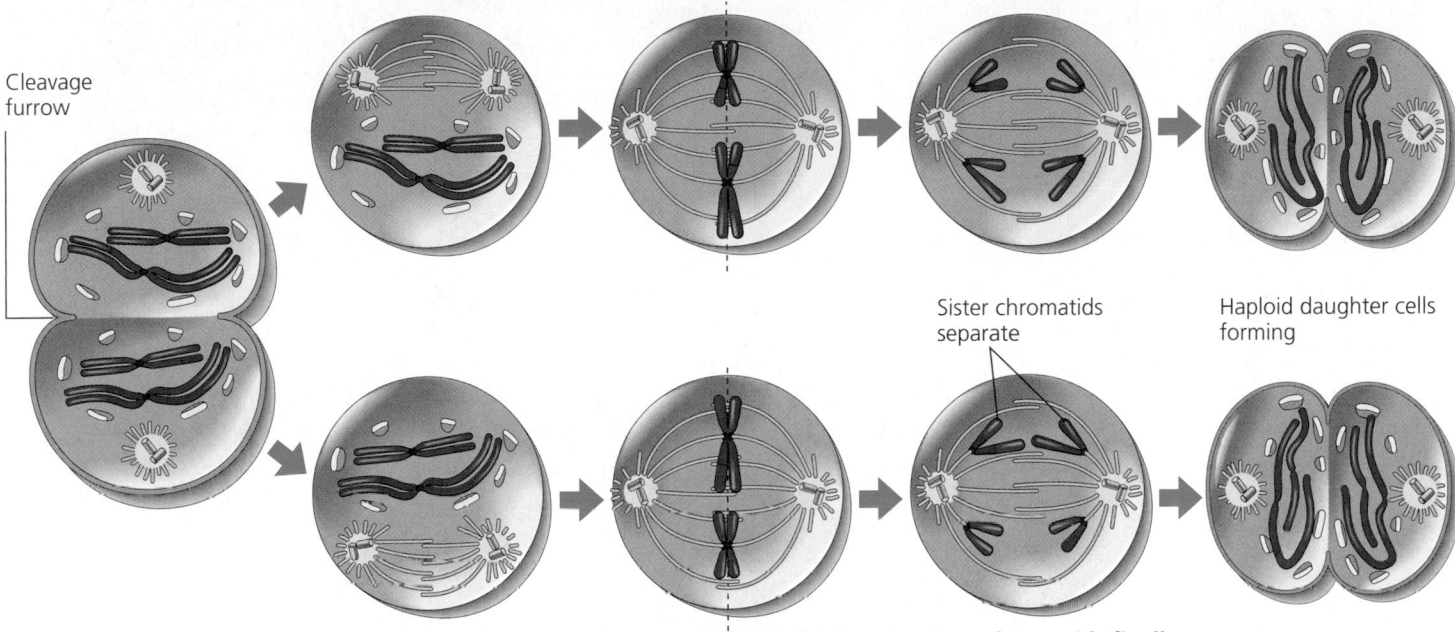

Cleavage furrow

Sister chromatids separate

Haploid daughter cells forming

Two haploid cells form; chromosomes are still double

During another round of cell division, the sister chromatids finally separate; four haploid daughter cells result, containing single chromosomes

TELOPHASE I AND CYTOKINESIS

The members of each pair of homologous chromosomes continue to move apart until they reach the poles of the cell. Each pole now has a haploid chromosome set, but each chromosome still has two sister chromatids. Usually cytokinesis (division of the cytoplasm) occurs simultaneously with telophase I, forming two daughter cells. Cleavage furrows form in animal cells, and cell plates appear in plant cells. In some species, the chromosomes decondense and the nuclear membranes and nucleoli re-form. In no case, however, is there further replication of the genetic material prior to the second division of meiosis.

PROPHASE II

A spindle apparatus forms, and the chromosomes progress toward the metaphase II plate.

METAPHASE II

The chromosomes are positioned on the metaphase plate in mitosis-like fashion, with the kinetochores of sister chromatids of each chromosome pointing toward opposite poles.

ANAPHASE II

The centromeres of sister chromatids finally separate, and the sister chromatids of each pair, now individual chromosomes, move toward opposite poles of the cell.

TELOPHASE II AND CYTOKINESIS

Nuclei form at opposite poles of the cell, and cytokinesis occurs. At the completion of cytokinesis, there are four daughter cells, each with the haploid number of unreplicated chromosomes.

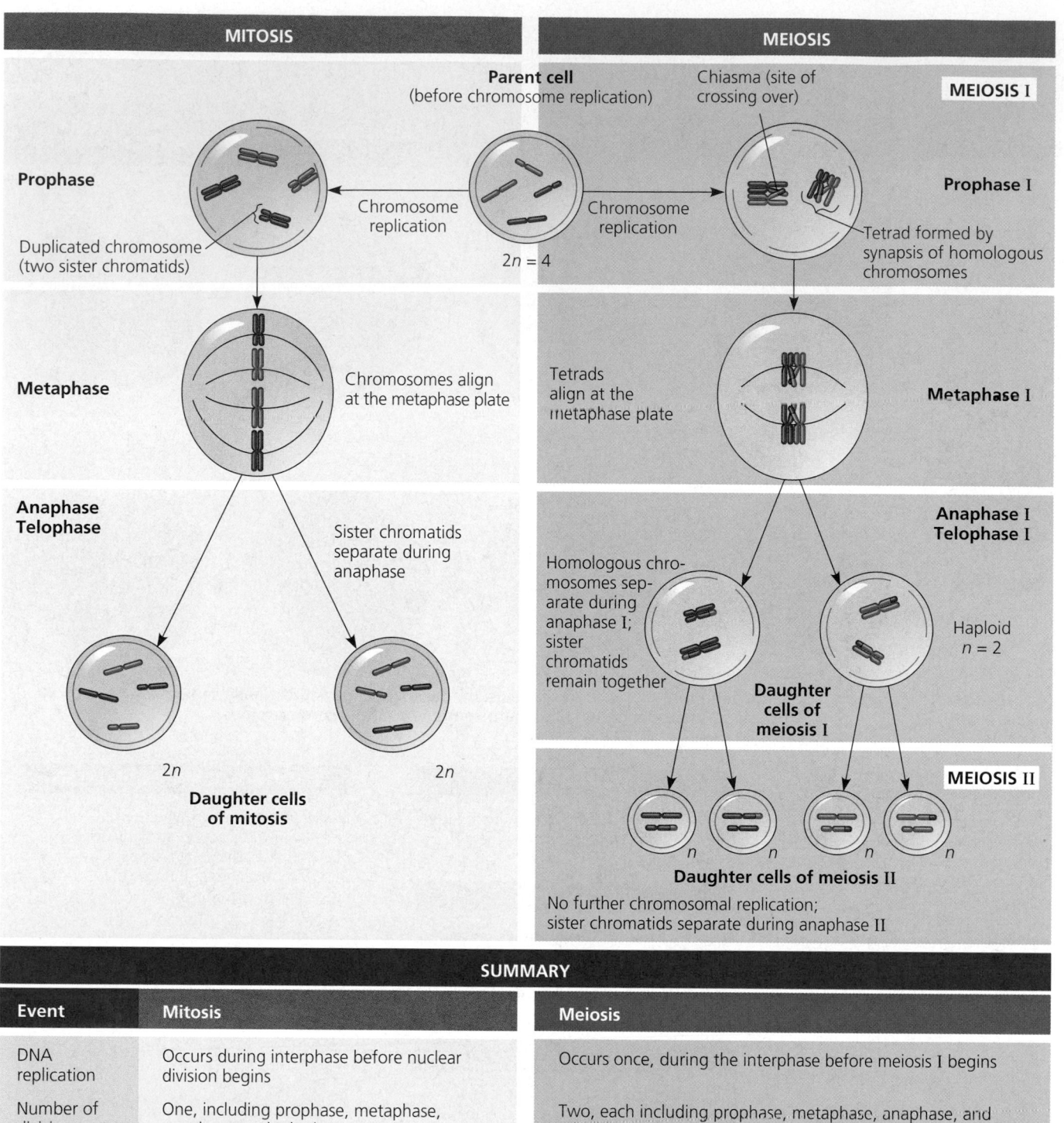

MITOSIS	MEIOSIS

Parent cell (before chromosome replication)

Chiasma (site of crossing over)

MEIOSIS I

Prophase

Chromosome replication

$2n = 4$

Chromosome replication

Prophase I

Duplicated chromosome (two sister chromatids)

Tetrad formed by synapsis of homologous chromosomes

Metaphase

Chromosomes align at the metaphase plate

Tetrads align at the metaphase plate

Metaphase I

Anaphase Telophase

Sister chromatids separate during anaphase

Anaphase I Telophase I

Homologous chromosomes separate during anaphase I; sister chromatids remain together

Haploid $n = 2$

$2n$

$2n$

Daughter cells of mitosis

Daughter cells of meiosis I

MEIOSIS II

n n n n

Daughter cells of meiosis II

No further chromosomal replication; sister chromatids separate during anaphase II

SUMMARY

Event	Mitosis	Meiosis
DNA replication	Occurs during interphase before nuclear division begins	Occurs once, during the interphase before meiosis I begins
Number of divisions	One, including prophase, metaphase, anaphase, and telophase	Two, each including prophase, metaphase, anaphase, and telophase
Synapsis of homologous chromosomes	Does not occur	Synapsis is unique to meiosis: During prophase I, the homologous chromosomes join along their length, forming tetrads (groups of four chromatids); synapsis is associated with crossing over between nonsister chromatids
Number of daughter cells and genetic composition	Two, each diploid ($2n$) and genetically identical to the parent cell	Four, each haploid (n), containing half as many chromosomes as the parent cell; genetically nonidentical to the parent cell and to each other
Role in the animal body	Enables multicellular adult to arise from zygote; produces cells for growth and tissue repair	Produces gametes; reduces chromosome number by half and introduces genetic variability among the gametes

FIGURE 13.8 A comparison of mitosis and meiosis. (In this chapter, we consider prophase to include prometaphase.)

Mitosis and Meiosis Compared

Now that you have followed chromosomes through meiosis in FIGURE 13.7, let's summarize the key differences between meiosis and mitosis. The chromosome number is reduced by half in meiosis but not in mitosis. The genetic consequences of this difference are important. Whereas mitosis produces daughter cells genetically identical to their parent cell and to each other, meiosis produces cells that differ genetically from their parent cell and from each other.

FIGURE 13.8 compares the key steps of mitosis and meiosis. Although meiosis involves two cell divisions, three events that are unique to meiosis all occur during the first of the divisions, meiosis I:

1. During prophase I of meiosis, the duplicated chromosomes pair with their homologues, a process called **synapsis.** For part of prophase I, a protein "zipper"—the *synaptonemal complex*—holds the homologous chromosomes tightly together all along their lengths. When the synaptonemal complex disappears in late prophase, the four closely associated chromatids of a homologous pair are visible in the light microscope as a **tetrad.** Also visible in the light microscope are X-shaped regions called **chiasmata** (singular, *chiasma*). They represent a crossing of *nonsister* chromatids, which are two chromatids belonging to separate but homologous chromosomes. Chiasmata are the physical manifestations of a genetic rearrangement called crossing over, discussed in the next section. Neither synapsis nor chiasma formation occurs during mitosis.

2. At metaphase I of meiosis, homologous pairs of chromosomes, rather than individual chromosomes, align on the metaphase plate.

3. At anaphase I of meiosis, sister chromatids do not separate, as they do in mitosis. Rather, the two sister chromatids of each chromosome remain attached and go to the same pole of the cell. *Meiosis I separates homologous pairs of chromosomes, not sister chromatids of individual chromosomes.*

The second meiotic division, meiosis II, separates sister chromatids and is virtually identical in mechanism to mitosis. However, since the chromosomes do not replicate between meiosis I and meiosis II, the final outcome of meiosis is a halving of the number of chromosomes per cell—a reduction from two haploid sets to one haploid set in each cell.

How do we account for the genetic variation we observed in FIGURE 13.2? We are now ready to address this question.

Sexual life cycles produce genetic variation among offspring

In species that reproduce sexually, the behavior of chromosomes during meiosis and fertilization is responsible for most of the variation that arises each generation. Let's examine three mechanisms that contribute to the genetic variation arising from sexual reproduction: independent assortment of chromosomes, crossing over, and random fertilization.

Independent Assortment of Chromosomes

One way sexual reproduction generates genetic variation is shown in FIGURE 13.9, which features meiosis of a diploid cell with two homologous pairs of chromosomes. The red and blue colors distinguishing the maternal and paternal chromosomes of each homologous pair allow us to track individual chromosomes as meiosis proceeds and they are packaged in gametes. At metaphase I, the homologous pairs of chromosomes, each consisting of one maternal and one paternal chromosome, are situated on the metaphase plate. The orientations of the

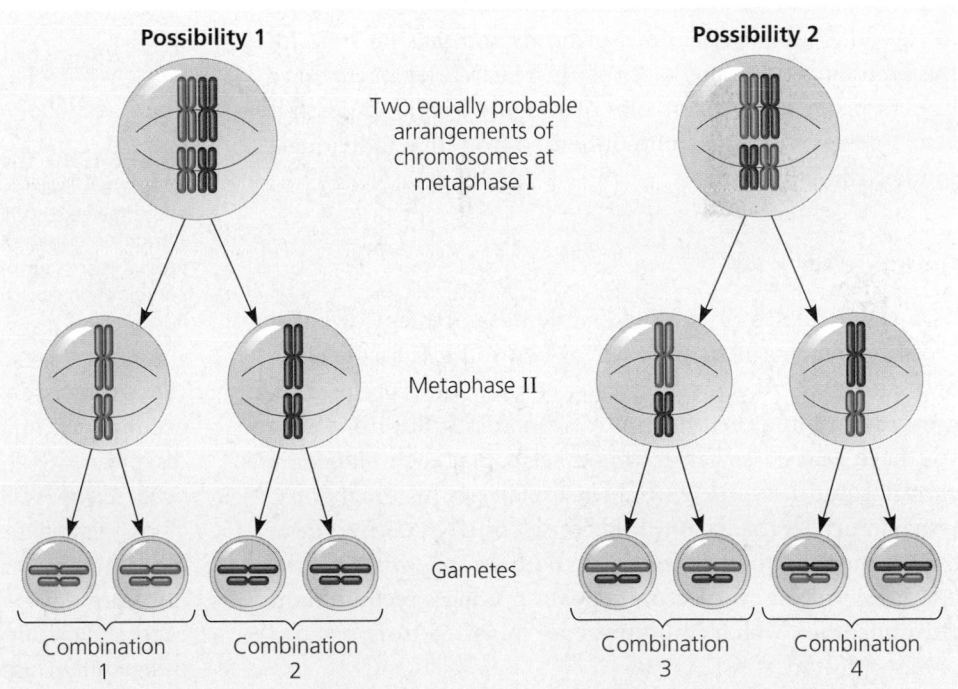

FIGURE 13.9 The results of alternative arrangements of two homologous chromosome pairs on the metaphase plate in meiosis I. In this figure we consider the consequences of meiosis in a hypothetical organism with a diploid chromosome number of 4 ($2n = 4$). The parental origins of the chromosomes are designated with different colors, blue for chromosomes inherited from one parent, red for chromosomes from the other parent. The positioning of each homologous pair of chromosomes at metaphase I is a matter of chance and determines which chromosomes will be packaged together in the haploid daughter cells.

homologous pairs relative to the poles of the cell are random; there are two alternative possibilities for each pair. Thus, there is a fifty-fifty chance that a particular daughter cell of meiosis I will get the maternal chromosome of a certain homologous pair and a fifty-fifty chance that it will receive the paternal chromosome.

Because each homologous pair of chromosomes is positioned independently of the other pairs at metaphase I—its orientation is as random as the flip of a coin—the first meiotic division results in independent assortment of maternal and paternal chromosomes into daughter cells. Each gamete represents one outcome of all possible combinations of maternal and paternal chromosomes. The number of combinations possible for gametes formed by meiosis starting with two homologous pairs of chromosomes ($2n = 4$, $n = 2$) is four, as shown in FIGURE 13.9. (Only two of the four combinations of gametes shown in the figure would result from meiosis of a single diploid cell, but starting with a large number of diploid cells, gametes of all four types would be produced in approximately equal numbers.) In the case of $n = 3$, eight combinations of chromosomes are possible for gametes. More generally, the number of combinations possible when chromosomes assort independently into gametes during meiosis is 2^n, where n is the haploid number of the organism.

In the case of humans, the haploid number (n) in the formula is 23. Thus, the number of possible combinations of maternal and paternal chromosomes in the resulting gametes is 2^{23}, or about 8 million. The variety of gametes is analogous to the different combinations of heads and tails possible for the simultaneous tossing of 23 coins. Thus, each gamete that a human produces contains one of roughly 8 million possible assortments of chromosomes inherited from that individual's mother and father.

Crossing Over

As a consequence of the independent assortment of chromosomes during meiosis, each of us produces a collection of gametes differing greatly in their combinations of the chromosomes we inherited from our two parents. But from what you have learned so far, it would seem that each *individual* chromosome in a gamete would be exclusively maternal or paternal in origin; that is, it would consist of DNA derived from our mother or father, but not from both. In fact, this is *not* the case. The process called **crossing over** produces **recombinant chromosomes,** which combine genes inherited from our two parents (FIGURE 13.10).

Recent research has revealed that crossing over begins very early in prophase I, as homologous chromosomes pair loosely along their lengths and before the synaptonemal complex forms between them. The pairing is precise, the homologues aligning with each other gene by gene. In crossing over, homologous portions of two nonsister chromatids trade places.

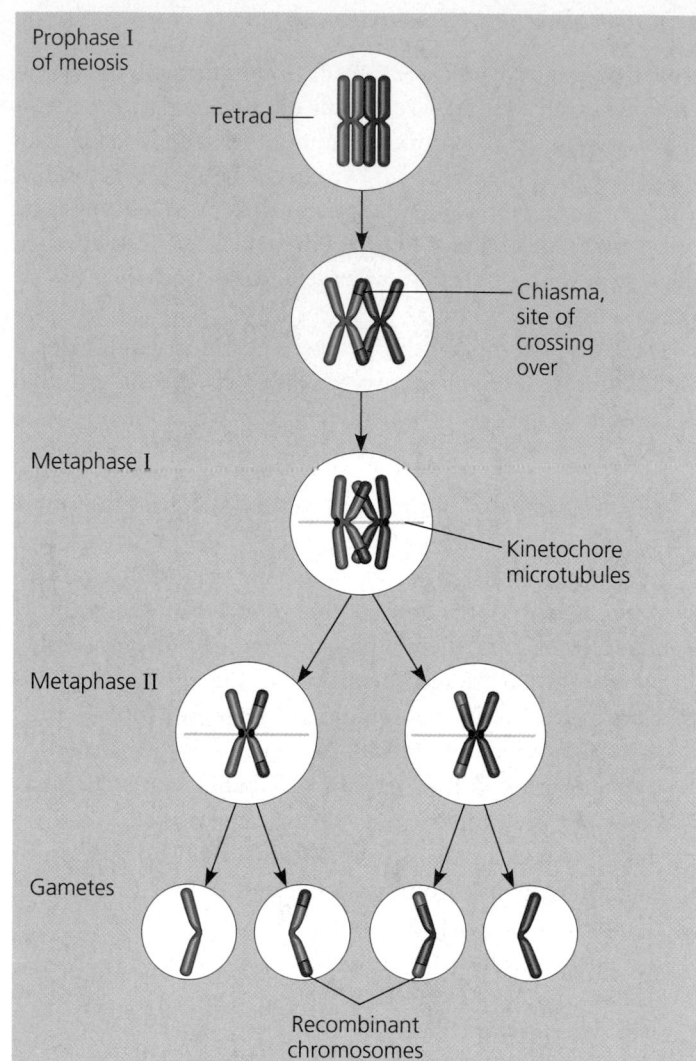

FIGURE 13.10 The results of crossing over during meiosis. During prophase of meiosis I, nonsister chromatids of homologous chromosomes exchange corresponding segments. Following these chromosomes through meiosis, we can see that crossing over gives rise to recombinant chromosomes, individual chromosomes that have some combination of DNA originally derived from two different parents.

(For humans, an average of two or three such crossover events occur per chromosome pair.) After the synaptonemal complex disappears, the locations where these genetic exchanges have occurred are visible as chiasmata.

At metaphase II, chromosomes that contain one or more recombinant chromatids can be oriented in two alternative, nonequivalent ways with respect to other chromosomes, because their sister chromatids are no longer identical twins. The independent assortment of these nonidentical sister chromatids during meiosis II increases still more the number of genetic types of gametes that can result from meiosis.

You will learn more about crossing over in Chapter 15. The important point for now is that crossing over, by combining DNA inherited from two parents into a single chromosome, is an important source of genetic variation in sexual life cycles.

Random Fertilization

The random nature of fertilization adds to the genetic variation arising from meiosis. Consider a zygote resulting from a mating between a woman and a man. A human ovum, representing one of approximately 8 million possible chromosome combinations, is fertilized by a single sperm cell, which represents one of 8 million *different* possibilities. Thus, even without considering crossing over, any two parents will produce a zygote with any of about 64 trillion (8 million × 8 million) diploid combinations. (If you calculate $2^{23} \times 2^{23}$ exactly, you will find that the total is actually over 70 trillion.) Adding in the variation brought about by crossing over, the number of possibilities is truly astronomical. No wonder brothers and sisters can be so different. You really *are* unique.

■   ■   ■

So far, we have seen that there are three sources of genetic variability in a sexually reproducing population of organisms:

- Independent assortment of homologous chromosomes during meiosis I (and of nonidentical sister chromatids during meiosis II)
- Crossing over between homologous chromosomes during prophase of meiosis I
- Random fertilization of an ovum by a sperm

All three mechanisms reshuffle the various genes carried by the individual members of a population. However, as you will learn in subsequent chapters, mutations are what ultimately create a population's diversity of genes.

Evolutionary adaptation depends on a population's genetic variation

Having considered how sexual reproduction contributes to genetic variation in a population, we can relate these concepts to evolution, biology's core theme. Darwin recognized the importance of genetic variation in the evolutionary mechanism he called natural selection. Recall from Chapter 1 that a population evolves through the differential reproductive success of its variant members. On average, those individuals best suited to the local environment leave the most offspring, transmitting their genes in the process. This natural selection results in adaptation, the accumulation of the genetic variations favored by the environment. As the environment changes or a population moves, the population may survive if, in each generation, at least some of its members can cope effectively with the new conditions. Different genetic variations may work better than those that prevailed in the old time or place. Sex and mutations are the two sources of this variation, and we have considered the sexual contribution in this chapter.

Although Darwin realized that heritable variation is what makes evolution possible, he could not explain why offspring resemble—but are not identical to—their parents. Ironically, Gregor Mendel, a contemporary of Darwin, published a theory of inheritance that helps explain genetic variation, but his discoveries had no impact on biologists until 1900, more than 15 years after Darwin (1809–1882) and Mendel (1822–1884) had died. In the next chapter, you will learn how Mendel discovered the basic rules governing the inheritance of specific traits.

CHAPTER 13 REVIEW

Go to the Campbell Biology website (www.campbellbiology.com) to explore an interactive version of the Chapter Review.

Summary of Key Concepts

AN INTRODUCTION TO HEREDITY

- **Offspring acquire genes from parents by inheriting chromosomes (pp. 234–235)** Genetics is the study of heredity and gene-based variation. Each gene in an organism's DNA has a specific locus on a certain chromosome.

- **Like begets like, more or less: a comparison of asexual and sexual reproduction (p. 235, FIGURE 13.2)** In asexual reproduction, one parent produces genetically identical offspring by mitosis. Sexual reproduction combines genes from two different parents to form genetically diverse offspring.

Web/CD Activity 13A: *Asexual and Sexual Life Cycles*

THE ROLE OF MEIOSIS IN SEXUAL LIFE CYCLES

- **Fertilization and meiosis alternate in sexual life cycles (pp. 236–239, FIGURE 13.5)** Normal human somatic cells have 46 chromosomes, half from each parent. Each of the 22 maternal autosomes has a homologous paternal chromosome. The 23rd pair, the sex chromosomes, determines whether the person is female (XX) or male (XY). Single, haploid (*n*) sets of chromosomes in ovum and sperm unite during fertilization to form a diploid (*2n*) single-celled zygote, which develops into a multicellular organism by mitosis. At sexual maturity, ovaries and testes (the gonads) produce haploid gametes by meiosis. Sexual life cycles differ in the timing of meiosis in relation to fertilization. Multicellular organisms may be diploid (as in animals), or haploid (as in most fungi), or may alternate between haploid and diploid generations (as in plants).

- **Meiosis reduces chromosome number from diploid to haploid: *a closer look* (pp. 239–243, FIGURES 13.6–13.8)** The two cell divisions of meiosis, meiosis I and meiosis II, produce four haploid daughter cells. Meiosis is distinguished from mitosis by the events of meiosis I.

In prophase I, replicated homologous chromosomes, each chromosome with two chromatids, undergo synapsis. Nonsister chromatids cross over, exchanging segments (the crossover sites appear as chiasmata). The paired chromosomes (tetrads) align on the metaphase plate, and at anaphase I, the two chromosomes of each homologous pair (not the sister chromatids) move to separate poles. The cell divides, with half the chromosomes going to each daughter cell. Meiosis II separates the sister chromatids, yielding four haploid daughter cells.

Web/CD Activity 13B: *Meiosis Animation*

ORIGINS OF GENETIC VARIATION

■ **Sexual life cycles produce genetic variation among offspring (pp. 243–245, FIGURES 13.9, 13.10)** The events of sexual reproduction that contribute to genetic variation in a population are independent assortment of chromosomes during meiosis I, crossing over between homologous chromosomes during meiosis I, and random fertilization of ova by sperm.

Web/CD Activity 13C: *Origins of Genetic Variation*
Web/CD Case Study in the Process of Science: *How Can the Frequency of Crossing Over Be Estimated?*

■ **Evolutionary adaptation depends on a population's genetic variation (p. 245)** Genetic variation among a population's members is the raw material for evolution by natural selection. Sexual reproduction and mutations generate this variation.

Self-Quiz

1. A human cell containing 22 autosomes and a Y chromosome is
   a. a somatic cell of a male.
   b. a zygote.
   c. a somatic cell of a female.
   d. a sperm cell.
   e. an ovum.

2. Homologous chromosomes move to opposite poles of a dividing cell during
   a. mitosis.
   b. meiosis I.
   c. meiosis II.
   d. fertilization.
   e. binary fission.

3. Meiosis II is similar to mitosis in that
   a. homologous chromosomes synapse.
   b. DNA replicates before the division.
   c. the daughter cells are diploid.
   d. sister chromatids separate during anaphase.
   e. the chromosome number is reduced.

4. The DNA content of a diploid cell in the G_1 phase of the cell cycle is measured (see Chapter 12). If this DNA content is x, then the DNA content of the same cell at metaphase of meiosis I would be
   a. $0.25x$.
   b. $0.5x$.
   c. x.
   d. $2x$.
   e. $4x$.

5. If we continued to follow the cell lineage from question 4, then the DNA content at metaphase of meiosis II would be
   a. $0.25x$.
   b. $0.5x$.
   c. x.
   d. $2x$.
   e. $4x$.

6. How many different combinations of maternal and paternal chromosomes can be packaged in gametes made by an organism with a diploid number of 8 ($2n = 8$)?
   a. 2
   b. 4
   c. 8
   d. 16
   e. 32

7. The immediate product of meiosis in a plant is a
   a. spore.
   b. gamete.
   c. zygote.
   d. sporophyte.
   e. gametophyte.

8. Multicellular haploid organisms
   a. are typically called sporophytes.
   b. produce new cells for growth by meiosis.
   c. produce gametes by mitosis.
   d. are found only in aquatic environments.
   e. are the direct result of syngamy.

9. Crossing over usually contributes to genetic variation by exchanging chromosomal segments between
   a. sister chromatids of a chromosome.
   b. chromatids of nonhomologues.
   c. nonsister chromatids of homologues.
   d. nonhomologous loci of the genome.
   e. autosomes and sex chromosomes.

10. In comparing the typical life cycles of plants and animals, a stage found in plants but not in animals is a
    a. gamete.
    b. zygote.
    c. multicellular diploid.
    d. multicellular haploid.

11. _____ is to somatic cells as haploid is to _____.

12. If a diploid cell with 18 chromosomes undergoes meiosis, the resulting gametes will each have _____ chromosomes.

13. Explain briefly how mitosis conserves chromosome number while meiosis reduces the number in half.

14. Name two events during meiosis that contribute to genetic variety among gametes.

15. How does the karyotype of a human female differ from that of a human male?

Go to the website or CD-ROM for more quiz questions.

Evolution Connection

Many species can reproduce either asexually or sexually. It is often when the environment changes in some way that is unfavorable to an existing population that the organisms begin to reproduce sexually. Speculate about the evolutionary significance of this switch from asexual to sexual reproduction.

The Process of Science

You prepare a karyotype of an animal you are studying and discover that its somatic cells each have *three* homologous sets of chromosomes, a condition called triploidy. You suspect that this condition results from an error in meoisis in one of the parents. What might have happened?

Estimate the frequency of crossing over in a fungus in the Case Study in the Process of Science, available on the website and CD-ROM.

Science, Technology, and Society

It is possible to grow seedlings of pine trees from short pieces of their needles. A few of the straightest, fastest-growing trees are selected for this treatment. By this method, thousands of genetically identical trees can be grown to create a forest that is a superior producer of lumber. What are the short-term and long-term advantages and disadvantages of this approach?

Answers: **1.** d; **2.** b; **3.** d; **4.** d; **5.** c; **6.** d; **7.** a; **8.** c; **9.** c; **10.** d; **11.** Diploid; gametes. **12.** 9. **13.** In mitosis, a single replication of the chromosomes is followed by one division of the cell. In meiosis, a single replication of the chromosomes is followed by two cell divisions. **14.** Crossing over between homologous chromosomes during prophase I and independent orientation of tetrads at metaphase I. **15.** A female has two X chromosomes; a male has an X and a Y.

CHAPTER 14

MENDEL AND THE GENE IDEA

E yes of brown, *blue, green, or gray; hair of black, brown, blond, or red—these are just a few examples of heritable variations that we may observe among individuals in a population. What genetic principles account for the transmission of such traits from parents to offspring?*

One possible explanation of heredity is a "blending" hypothesis, the idea that genetic material contributed by the two parents mixes in a manner analogous to the way blue and yellow paints blend to make green. This hypothesis predicts that over many generations, a freely mating population will give rise to a uniform population of individuals. However, our everyday observations and the results of breeding experiments with animals and plants contradict such a prediction. The blending hypothesis also fails to explain other phenomena of inheritance, such as traits skipping a generation.

An alternative to the blending model is a "particulate" hypothesis of inheritance: the gene idea. According to this model, parents pass on discrete heritable units—genes—that retain their separate identities in offspring. An organism's collection of genes is more like a bucket of marbles than a pail of paint. Like marbles, genes can be sorted and passed along, generation after generation, in undiluted form.

Modern genetics had its genesis in an abbey garden, where a monk named Gregor Mendel documented a particulate mechanism of inheritance. In the painting on this page, Mendel works with his experimental organism, garden peas. Mendel developed his theory of inheritance several decades before the behavior of chromosomes was observed in the microscope and their significance understood. So in this chapter, we digress from the study of chromosomes to recount how Mendel arrived at his theory. We will also see how the Mendelian model applies to the inheritance of human variations.

GREGOR MENDEL'S DISCOVERIES

Mendel discovered the basic principles of heredity by breeding garden peas in carefully planned experiments. As we retrace his work in this and the following sections, we will be able to recognize the key elements of the scientific process that were introduced in Chapter 1.

THE PROCESS OF SCIENCE

Mendel brought an experimental and quantitative approach to genetics

Mendel grew up on his parents' small farm in a region of Austria that is now part of the Czech Republic. At school in this agricultural area, Mendel and the other children received agricultural training along with basic education. Later, Mendel overcame financial hardship and illness to excel in high school and at the Olmutz Philosophical Institute.

In 1843, Mendel entered an Augustinian monastery. After three years of theological studies, he was assigned to a school as a temporary teacher but failed the teacher's examination. An administrator sent Mendel to the University of Vienna, where he studied from 1851 to 1853. These were very important years for Mendel's development as a scientist. Two professors were especially influential. One was the physicist Doppler, who encouraged his students to learn science through experimentation and trained Mendel to use mathematics to help explain natural phenomena. The second was a botanist named Unger, who aroused Mendel's interest in the causes of variation in plants. These influences came together in Mendel's subsequent experiments with garden peas.

After attending the university, Mendel was assigned to teach at the Brünn Modern School, where several teachers shared his enthusiasm for scientific research. At the monastery where Mendel lived, he also found stimulating colleagues, many of them university professors and active researchers. Moreover, there had been a long tradition of interest in the breeding of plants, including peas, at the monastery. Thus, it was probably not extraordinary when, around 1857, Mendel began breeding garden peas in the abbey garden in order to study inheritance. What *was* extraordinary was Mendel's fresh approach to very old questions about heredity.

Mendel probably chose to work with peas because they are available in many varieties. For example, one variety has purple flowers, while another variety has white flowers. Geneticists use the term **character** for a heritable feature, such as flower color, that varies among individuals. Each variant for a character, such as purple or white color for flowers, is called a **trait.**

The use of peas also gave Mendel strict control over which plants mated with which. The sex organs of a pea plant are in its flowers, and each pea flower has both male and female organs—stamens and carpel, respectively. In nature, the plants usually self-fertilize: Pollen grains released from the stamens land on the carpel of the same flower, and sperm from the pollen fertilize ova in the carpel. To achieve cross-pollination (fertilization between different plants), Mendel removed the immature stamens of a plant before they produced pollen and then dusted pollen from another plant onto the emasculated flowers (FIGURE 14.1). Each resulting zygote then developed into a plant embryo encased in a seed (pea). Whether ensuring self-pollination or executing artificial cross-pollination, Mendel could always be sure of the parentage of new seeds.

Mendel chose to track only those characters that varied in an "either-or" rather than a "more-or-less" manner. For example, his plants had either purple flowers or white flowers; there was nothing intermediate between these two varieties. Had Mendel focused instead on characters that vary in a continuum among individuals—seed weight, for example—he would not have discovered the particulate nature of inheritance.

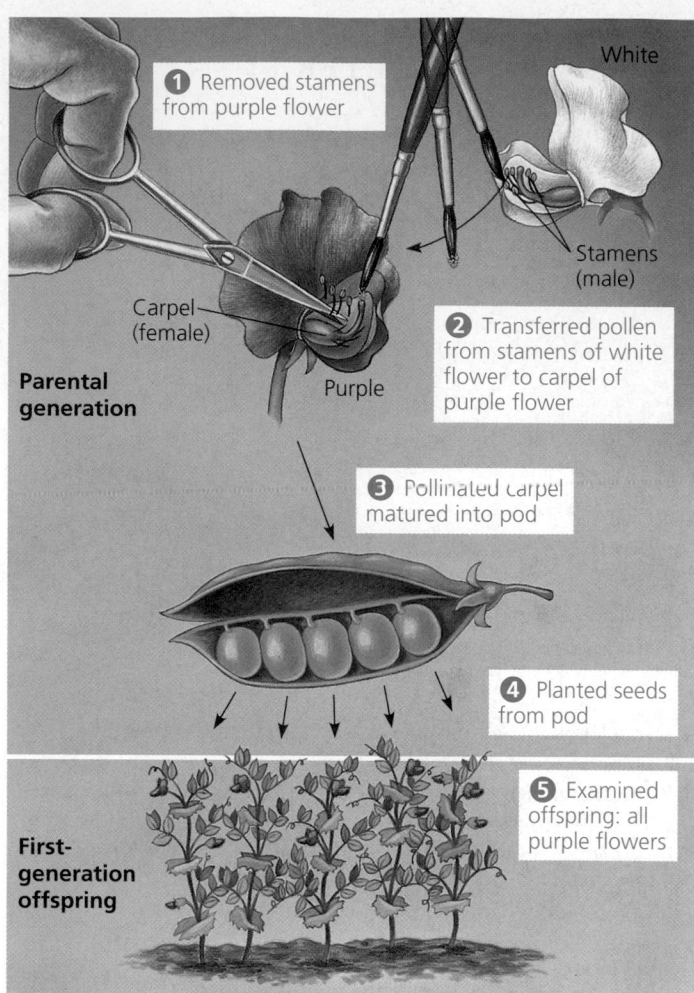

FIGURE 14.1 A genetic cross. To mate (hybridize) two varieties of pea plants, Mendel used an artist's brush to transfer sperm-bearing pollen from one plant to the egg-bearing carpel of another. In this example, the character of interest is flower color. When pollen from a white flower fertilizes ova of a purple flower, the first-generation hybrids all have purple flowers. The result is the same for the reciprocal cross, the transfer of pollen from purple flowers to white flowers.

THE PROCESS OF SCIENCE

Mendel also made sure he started his experiments with varieties that were **true-breeding,** which means that when the plants self-pollinate, all their offspring are of the same variety. For example, a plant with purple flowers is true-breeding if the seeds produced by self-pollination all give rise to plants that also have purple flowers.

In a typical breeding experiment, Mendel would cross-pollinate two contrasting, true-breeding pea varieties—for example, purple-flowered plants and white-flowered plants (see FIGURE 14.1). This mating, or crossing, of two true-breeding varieties is called **hybridization.** The true-breeding parents are referred to as the **P generation** (parental generation), and their hybrid offspring are the F_1 **generation** (first filial generation, the word *filial* from the Latin word for "son"). Allowing these F_1 hybrids to self-pollinate produces an F_2 **generation** (second filial generation). Mendel usually followed traits for at least these three generations: the P, F$_1$, and F$_2$ generations. Had Mendel stopped his experiments with the F$_1$ generation, the basic patterns of inheritance would have eluded him. It was mainly Mendel's quantitative analysis of F$_2$ plants that revealed the two fundamental principles of heredity that are now known as the law of segregation and the law of independent assortment.

By the law of segregation, the two alleles for a character are packaged into separate gametes

If the blending model of inheritance were correct, the F$_1$ hybrids from a cross between purple-flowered and white-flowered pea plants would have pale purple flowers, intermediate between the two varieties of the P generation. Notice in FIGURE 14.1 that the experiment produced a very different result: The F$_1$ offspring all had flowers just as purple as the purple-flowered parents. What happened to the white-flowered plants' genetic contribution to the hybrids? If it were lost, then the F$_1$ plants could produce only purple-flowered offspring in the F$_2$ generation. But when Mendel allowed the F$_1$ plants to self-pollinate and planted their seeds, the white-flower trait reappeared in the F$_2$ generation. Mendel used very large sample sizes and kept accurate records of his results: 705 of the F$_2$ plants had purple flowers, and 224 had white flowers. These data fit a ratio of about three purple to one white (FIGURE 14.2). Mendel reasoned

that the heritable factor for white flowers did not disappear in the F$_1$ plants, but only the purple-flower factor was affecting flower color in these hybrids. In Mendel's terminology, purple flower is a dominant trait and white flower is a recessive trait. The occurrence of white-flowered plants in the F$_2$ generation was evidence that the heritable factor causing that recessive trait had not been diluted in any way by coexisting with the purple-flower factor in the F$_1$ hybrids.

Mendel observed the same pattern of inheritance in six other characters, each represented by two different varieties (TABLE 14.1, p. 250). For example, the parental pea seeds either had a smooth, round shape or were wrinkled. When Mendel crossed his two true-breeding varieties, all the F$_1$ hybrids produced round seeds; this is the dominant trait. In the F$_2$ generation, 75% of the seeds were round and 25% were wrinkled—a 3:1 ratio, as in FIGURE 14.2. How did Mendel explain this pattern, which he consistently observed in his crosses? He developed a hypothesis that we can break down into four related ideas. (We will replace some of Mendel's original terms with modern words; for example, "gene" will be used in place of Mendel's "heritable factor.")

1. *Alternative versions of genes (different alleles) account for variations in inherited characters.* The gene for flower color, for example, exists in two versions, one for purple flowers and the other for white. These alternative versions of a gene are now called **alleles** (FIGURE 14.3). Today, we can relate this concept to chromosomes and DNA. As we mentioned in Chapter 13, each gene resides at a specific locus on a specific chromosome. The DNA at that locus, however, can vary somewhat in its sequence of nucleotides and hence in its information content.

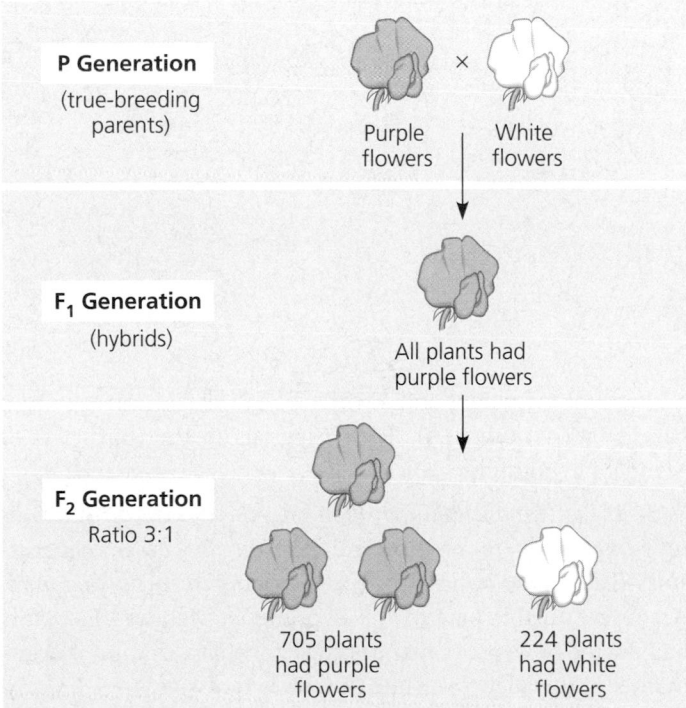

FIGURE 14.2 Mendel tracked heritable characters for three generations. When F$_1$ hybrids were allowed to self-pollinate, or when they were cross-pollinated with other F$_1$ hybrids, a 3:1 ratio of the two varieties occurred in the F$_2$ generation. A genetic cross, or mating, is symbolized by ×.

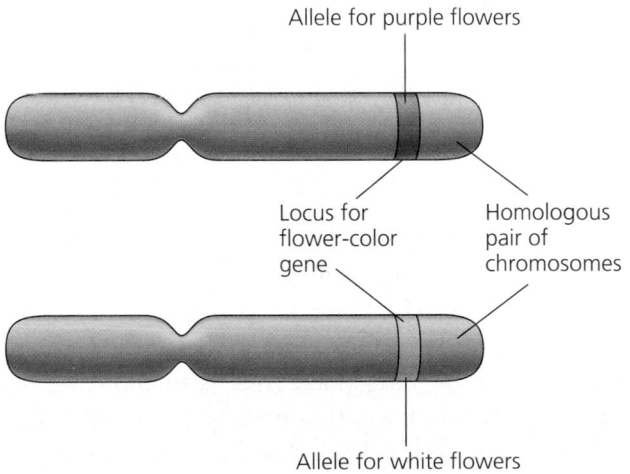

FIGURE 14.3 Alleles, alternative versions of a gene. The gene for a particular inherited character, such as flower color in garden peas, resides at a specific locus (position) on a certain chromosome. Alleles are variants of that gene. In the case of peas, the flower-color gene exists in two versions: the allele for purple flowers and the allele for white flowers. This homologous pair of chromosomes represents an F$_1$ hybrid, which inherited the allele for purple flowers from one parent and the allele for white flowers from the other parent.

Table 14.1 The Results of Mendel's F₁ Crosses for Seven Characters in Pea Plants

Character	Dominant Trait	×	Recessive Trait	F₂ Generation Dominant:Recessive	Ratio
Flower color	Purple	×	White	705:224	3.15:1
Flower position	Axial	×	Terminal	651:207	3.14:1
Seed color	Yellow	×	Green	6022:2001	3.01:1
Seed shape	Round	×	Wrinkled	5474:1850	2.96:1
Pod shape	Inflated	×	Constricted	882:299	2.95:1
Pod color	Green	×	Yellow	428:152	2.82:1
Stem length	Tall	×	Dwarf	787:277	2.84:1

The purple-flower allele and the white-flower allele are two DNA variations possible at the flower-color locus on one of a pea plant's chromosomes.

2. *For each character, an organism inherits two alleles, one from each parent.* Mendel made this deduction without knowing about the role of chromosomes, but what you learned about chromosomes in Chapter 13 will help you understand Mendel's idea. Recall that a diploid organism has homologous pairs of chromosomes, one chromosome of each pair inherited from each parent. Thus, a genetic locus is actually represented twice in a diploid cell. These homologous loci may have identical alleles, as in the true-breeding plants of Mendel's P generation. Or the two alleles may differ, as in the F₁ hybrids. In the flower-color example, the hybrids inherited a purple-flower allele from one parent and a white-flower allele from the other parent (see FIGURE 14.3). This brings us to the third part of Mendel's hypothesis.

3. *If the two alleles differ, then one, the **dominant allele,** is fully expressed in the organism's appearance; the other, the **recessive allele,** has no noticeable effect on the organism's appearance.* According to this part of the hypothesis, Mendel's F₁ plants had purple flowers because the allele for that variation is dominant and the allele for white flowers is recessive.

4. *The two alleles for each character segregate (separate) during gamete production.* Thus, an ovum and a sperm each get only one of the two alleles that are present in the somatic cells of the organism. In terms of chromosomes, this segregation corresponds to the distribution of homologous chromosomes

Each true-breeding plant of the parental generation has matching alleles, *PP* or *pp*.

Gametes (circles) each contain only one allele for the flower-color gene. In this case, every gamete produced by one parent has the same allele.

Union of the parental gametes produces F_1 hybrids having a *Pp* combination. Because the purple-flower allele is dominant, all these hybrids have purple flowers.

When the hybrid plants produce gametes, the two alleles segregate, half the gametes receiving the *P* allele and the other half the *p* allele.

This box, a Punnett square, shows all possible combinations of alleles in offspring. Each square represents an equally probable product of fertilization. For example, the box at the left corner shows the genetic combination resulting from a (*p*) ovum fertilized by a (*P*) sperm.

Random combination of the gametes results in the 3:1 ratio that Mendel observed in the F_2 generation.

P Generation

Appearance: Purple flowers   White flowers
Genetic makeup: *PP*   *pp*

Gametes: *P*   *p*

F_1 Generation

Appearance: Purple flowers
Genetic makeup: *Pp*

Gametes: ½ *P*   ½ *p*

F_2 Generation

F_1 ova *P*   *P* F_1 sperm

p   *PP*

p   *Pp*   *Pp*

pp

3   : 1

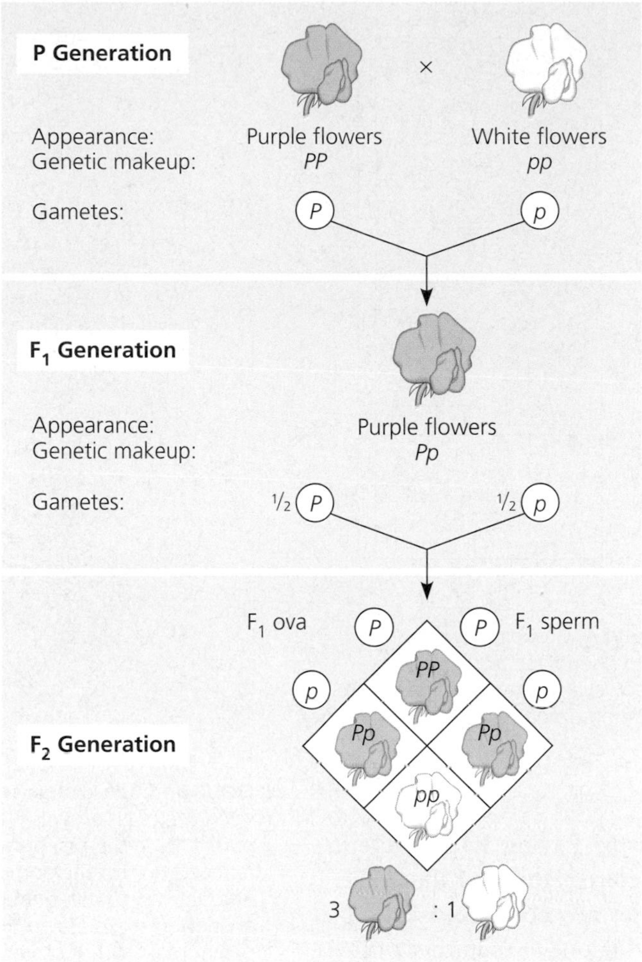

FIGURE 14.4 Mendel's law of segregation. A genetically specific version of FIGURE 14.2, this diagram illustrates Mendel's model for the inheritance of the alleles of a single gene. The purple-flower allele (*P*) is dominant, and the white-flower allele (*p*) is recessive. Each plant has two alleles for the gene controlling flower color, one allele inherited from each parent.

to different gametes in meiosis. Note that if an organism has identical alleles for a particular character—that is, the organism is true-breeding for that character—then that allele exists in a single copy in all gametes. But if different alleles are present, as in the F_1 hybrids, then 50% of the gametes receive the dominant allele, while 50% receive the recessive allele. It is this last part of the hypothesis, the separation of alleles into separate gametes, for which Mendel's **law of segregation** is named.

One test of Mendel's segregation hypothesis is whether or not it can account for the 3:1 ratio he observed in the F_2 generation of his numerous crosses. The hypothesis predicts that the F_1 hybrids will produce two classes of gametes. When alleles segregate, half the gametes receive a purple-flower allele, while the other half get a white-flower allele. During self-pollination, the gametes of these two classes unite randomly. An ovum with a purple-flower allele has an equal chance of being fertilized by a sperm with a purple-flower allele or one with a white-flower allele. Since the same is true for an ovum with a white-flower allele, there are a total of four equally likely combinations of sperm and ovum. FIGURE 14.4 illustrates these combinations using a type of diagram called a **Punnett square,** a handy device for predicting the results of a genetic cross between individuals of known genotype. Notice

that we use a capital letter to symbolize a dominant allele and a lowercase letter for a recessive allele. In our example, *P* is the purple-flower allele, and *p* is the white-flower allele.

What will be the physical appearance of these F_2 offspring? One-fourth of the plants have two alleles specifying purple flowers; clearly, these plants will have purple flowers. But one-half of the F_2 offspring have inherited one allele for purple flowers and one allele for white flowers; like the F_1 plants, these plants will also have purple flowers, the dominant trait. Finally, one-fourth of the F_2 plants have inherited two alleles specifying white flowers and will, in fact, express the recessive trait. Thus, Mendel's model accounts for the 3:1 ratio that he observed in the F_2 generation.

Some Useful Genetic Vocabulary

An organism having a pair of identical alleles for a character is said to be **homozygous** for the gene controlling that character. A pea plant that is true-breeding for purple flowers (*PP*) is an example. Pea plants with white flowers are also homozygous, but for the recessive allele (*pp*). If we cross dominant homozygotes with recessive homozygotes, as in the parental (P generation) cross of FIGURE 14.4, every offspring will have two different alleles—*Pp* in the case of the F_1 hybrids of our

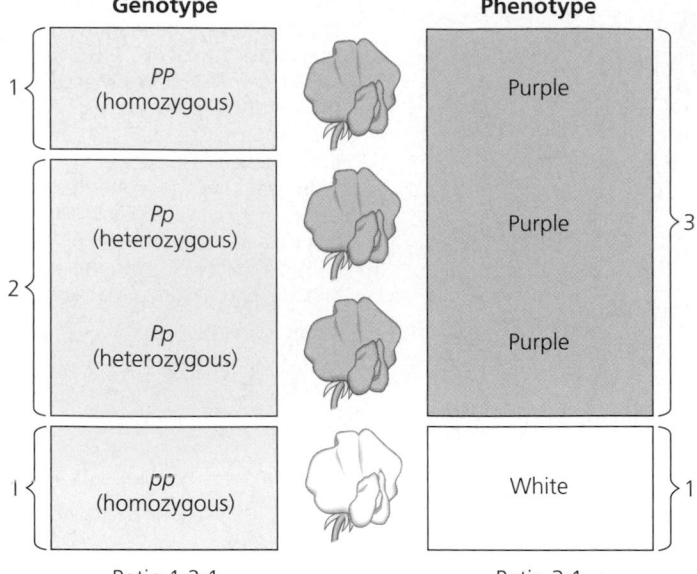

Ratio 1:2:1   Ratio 3:1

FIGURE 14.5 Genotype versus phenotype. Grouping F₂ offspring from a cross for flower color according to phenotype results in the typical 3:1 ratio. In terms of genotype, however, there are actually two categories of purple-flowered plants: *PP* (homozygous) and *Pp* (heterozygous).

flower-color experiment. Organisms having two different alleles for a gene are said to be **heterozygous** for that gene. Unlike homozygotes, heterozygotes are not true-breeding, because they produce gametes having one *or* the other of the different alleles. We have seen that a *Pp* plant of the F₁ generation will produce both purple-flowered and white-flowered offspring when it self-pollinates.

Because of dominance and recessiveness, an organism's traits do not always reveal its genetic composition. Therefore, we have to distinguish between an organism's traits, called its **phenotype,** and its genetic makeup, its **genotype.** In the case of flower color in peas, *PP* and *Pp* plants have the same phenotype (purple) but different genotypes. FIGURE 14.5 reviews these terms. Note that phenotype refers to physiological traits as well as traits relating directly to appearance. For example, there is a pea variety that lacks the normal trait of being able to self-pollinate. This physiological variation is a phenotype.

The Testcross

Suppose we have a pea plant that has purple flowers. We cannot tell from its flower color if this plant is homozygous or heterozygous because the genotypes *PP* and *Pp* result in the same phenotype. But if we cross this pea plant with one having white flowers, the appearance of the offspring will reveal the genotype of the purple-flowered parent (FIGURE 14.6). The genotype of the plant with white flowers is known: Because this is the recessive trait, the plant must be homozygous (*pp*).

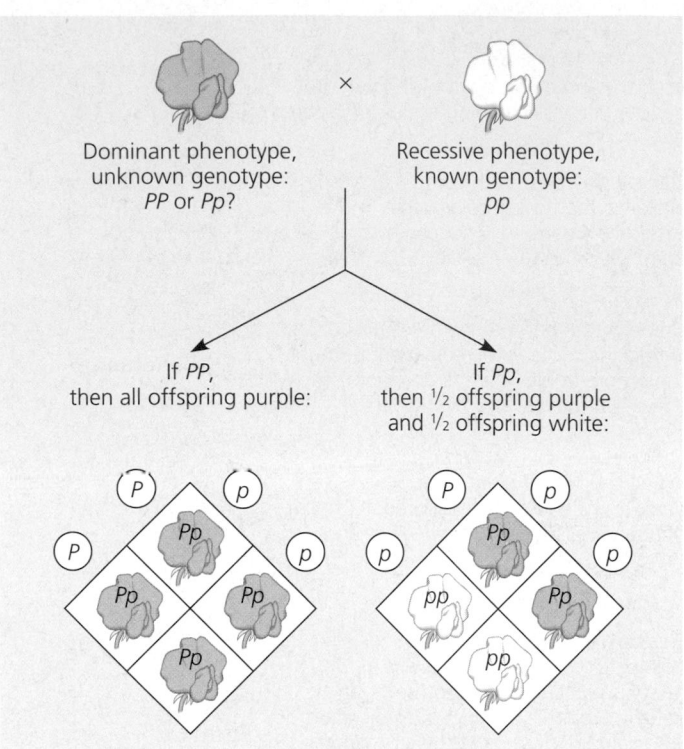

FIGURE 14.6 A testcross. A testcross is designed to reveal the genotype of an organism that exhibits a dominant trait, such as purple flowers in pea plants. Such an organism could be either homozygous for the dominant allele or heterozygous. The most efficient way to determine the genotype is to cross the organism with an individual expressing the recessive trait. Since the white-flowered parent must be homozygous, we can deduce the genotype of the purple-flowered parent by observing the phenotypes of the offspring.

If all the offspring of the cross have purple flowers, then the other parent must be homozygous for the dominant allele; a *PP* × *pp* cross produces nothing but *Pp* offspring. But if both the purple and the white phenotypes appear among the offspring, then the purple-flowered parent must be heterozygous. The offspring of a *Pp* × *pp* cross will have a 1:1 phenotypic ratio. This breeding of a recessive homozygote with an organism of dominant phenotype but unknown genotype is called a **testcross.** It was devised by Mendel and continues to be an important tool of geneticists.

By the law of independent assortment, each pair of alleles segregates into gametes independently

Gregor Mendel derived the law of segregation by performing breeding experiments in which he followed only a *single* character, such as flower color. The F₁ hybrids produced in such crosses are called **monohybrids.** What would happen if he followed *two* characters at the same time? For instance, two of the seven characters Mendel studied were seed color and seed

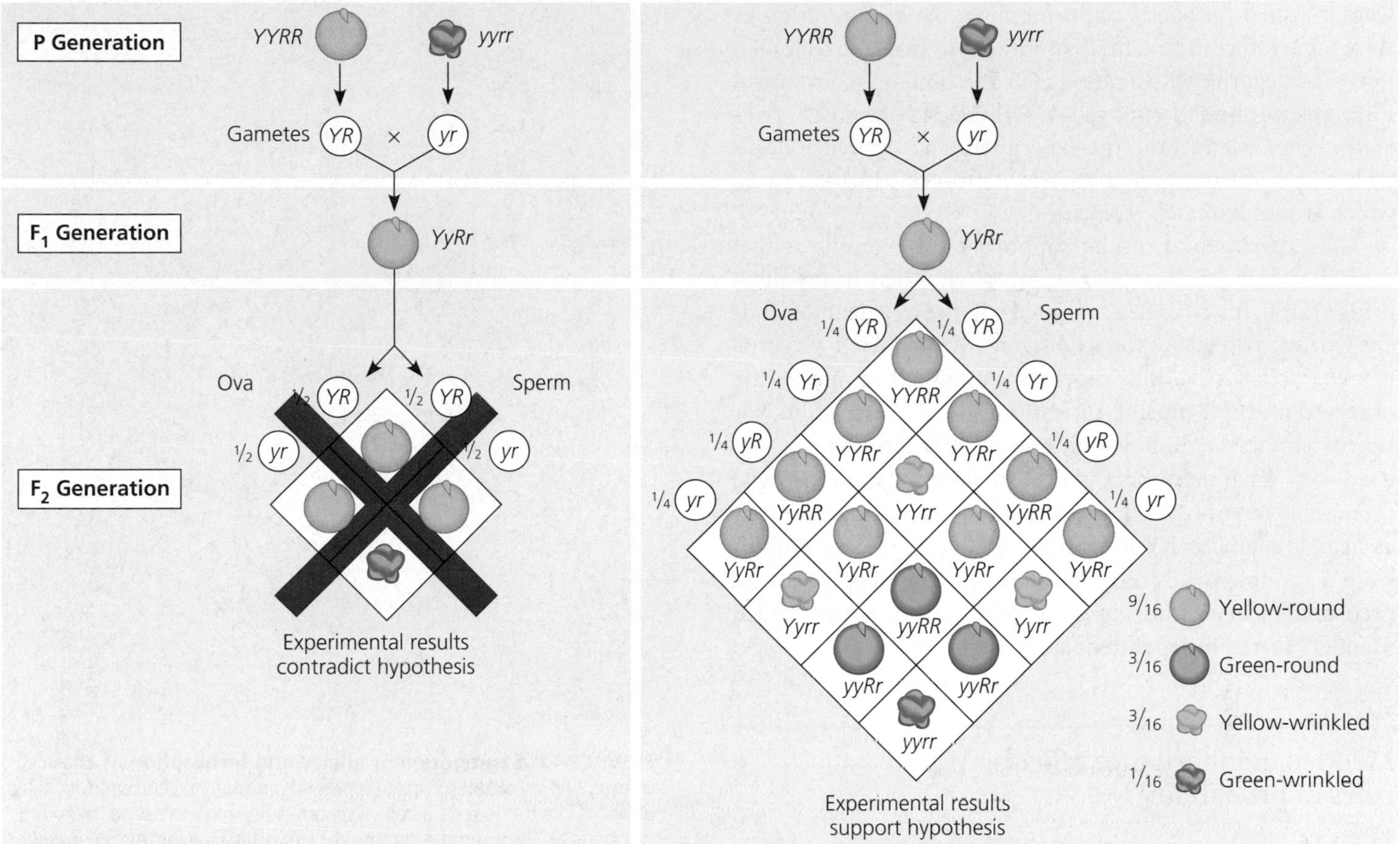

P Generation — YYRR × yyrr

Gametes: YR × yr

F₁ Generation — YyRr

F₂ Generation

(a) Hypothesis: dependent assortment. Ova: ½ YR, ½ yr — Sperm: ½ YR, ½ yr

Experimental results contradict hypothesis

(b) Hypothesis: independent assortment. Ova: ¼ YR, ¼ Yr, ¼ yR, ¼ yr — Sperm: ¼ YR, ¼ Yr, ¼ yR, ¼ yr

YYRR, YYRr, YYRr, YYrr, YyRR, YyRr, YyRr, Yyrr, YyRR, YyRr, YyRr, Yyrr, yyRR, yyRr, yyRr, yyrr

9/16 Yellow-round
3/16 Green-round
3/16 Yellow-wrinkled
1/16 Green-wrinkled

Experimental results support hypothesis

(a) Hypothesis: dependent assortment. If the two characters segregate dependently (together), the F₁ hybrids can only produce the same two classes of gametes that they received from the parents, and the F₂ offspring will show only the parental phenotypes, in a 3:1 phenotypic ratio.

(b) Hypothesis: independent assortment. If the two characters segregate independently, four classes of gametes will be produced by the F₁ generation, and, in the F₂ generation, there will be all possible combinations of traits, in a 9:3:3:1 phenotypic ratio.

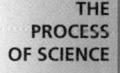

THE PROCESS OF SCIENCE

FIGURE 14.7 Testing two hypotheses for segregation in a dihybrid cross. A cross between true-breeding parent plants that differ in two characters produces F₁ plants that are *dihybrids:* They are heterozygous for both characters. The F₂ phenotypes show that the characters assort (segregate) independently. In this example, the two characters are seed color and seed shape. Yellow color (*Y*) and round shape (*R*) are dominant.

shape. Seeds may be either yellow or green. They also may be either round (smooth) or wrinkled. From single-character crosses, Mendel knew that the allele for yellow seeds is dominant (*Y*) and that the allele for green seeds is recessive (*y*). For the seed-shape character, the allele for round is dominant (*R*), and the allele for wrinkled is recessive (*r*). Imagine crossing two true-breeding pea varieties differing in *both* of these characters—a parental cross between a plant with yellow-round seeds (*YYRR*) and a plant with green-wrinkled seeds (*yyrr*). The F₁ plants will be **dihybrids,** heterozygous for both characters (*YyRr*). But are these two characters, seed color and seed shape, transmitted from parents to offspring as a package? Put another way, will the *Y* and *R* alleles always stay together, generation after generation? Or are seed color and seed shape inherited independently of each other? FIGURE 14.7 illustrates how a *dihybrid cross,* a cross between F₁ dihybrids, can determine which of these two hypotheses is correct.

The F₁ plants, of genotype *YyRr,* exhibit both dominant phenotypes, yellow seeds with round shapes, no matter which hypothesis is correct. The key step in the experiment is to see what happens when F₁ plants self-pollinate to produce F₂ offspring. If the hybrids must transmit their alleles in the same combinations in which they were inherited from the P generation, then there will only be two classes of gametes: *YR* and *yr*. This hypothesis predicts that the phenotypic ratio of the F₂ generation will be 3:1, just as in a cross between monohybrids (see FIGURE 14.7a).

The alternative hypothesis is that the two pairs of alleles segregate independently of each other. In other words, genes are packaged into gametes in all possible allelic combinations, as long as each gamete has one allele for each gene. In our example, four classes of gametes would be produced in equal quantities: *YR, Yr, yR,* and *yr*. If sperm of four classes are mixed with ova of four classes, there will be 16 (4 × 4) equally probable ways in which the alleles can combine in the F₂ generation, as shown in FIGURE 14.7b. These combinations make up four phenotypic categories with a ratio of 9:3:3:1 (nine yellow-round to three green-round to three yellow-wrinkled to one green-wrinkled). When Mendel did the experiment and "scored" (classified) the F₂ offspring, he obtained a ratio of 315:108:101:32, which is approximately 9:3:3:1.

The experimental results supported the hypothesis that each character is independently inherited—that in the dihybrids ($YyRr$), the two alleles for seed color segregate independently of the two alleles for seed shape. Mendel tried his seven pea characters in various dihybrid combinations and always observed a 9:3:3:1 phenotypic ratio in the F₂ generation. Notice in FIGURE 14.7b, however, that, if you consider the two characters separately, there is a 3:1 phenotypic ratio for each: three yellow to one green; three round to one wrinkled. As far as a single character is concerned, the alleles segregate as if this were a monohybrid cross. The independent segregation of each pair of alleles during gamete formation is now called Mendel's **law of independent assortment.**

Mendelian inheritance reflects rules of probability

Mendel's laws of segregation and independent assortment reflect the same rules of probability that apply to tossing coins or rolling dice. A basic understanding of these rules of chance is essential for genetic analysis.

The probability scale ranges from 0 to 1. An event that is certain to occur has a probability of 1, while an event that is certain *not* to occur has a probability of 0. With a two-headed coin, the probability of tossing heads is 1, and the probability of tossing tails is 0. With a normal coin, the chance of tossing heads is ½, and the chance of tossing tails is ½. The probability of rolling the number 3 with a die, which is six-sided, is ⅙. The probabilities of all possible outcomes for an event must add up to 1. With a die, the chance of rolling a number other than 3 is ⅚.

We can learn an important lesson about probability from tossing a coin. For every toss, the probability of heads is ½. The outcome of any particular toss is unaffected by what has happened on previous trials. We refer to phenomena such as successive coin tosses, or simultaneous tosses of several coins, as independent events. The question of whether a tossed coin will come up heads or tails is analogous to the question of whether a gamete produced by a Pp heterozygote will carry allele P or allele p (FIGURE 14.8). In every fertilization involving such gametes, the ovum has a ½ chance of carrying the dominant allele and a ½ chance of carrying the recessive allele. The same odds apply to the sperm cell. Like two separate coin tosses, allele segregation during formation of the sperm and ovum occurs as two independent events.

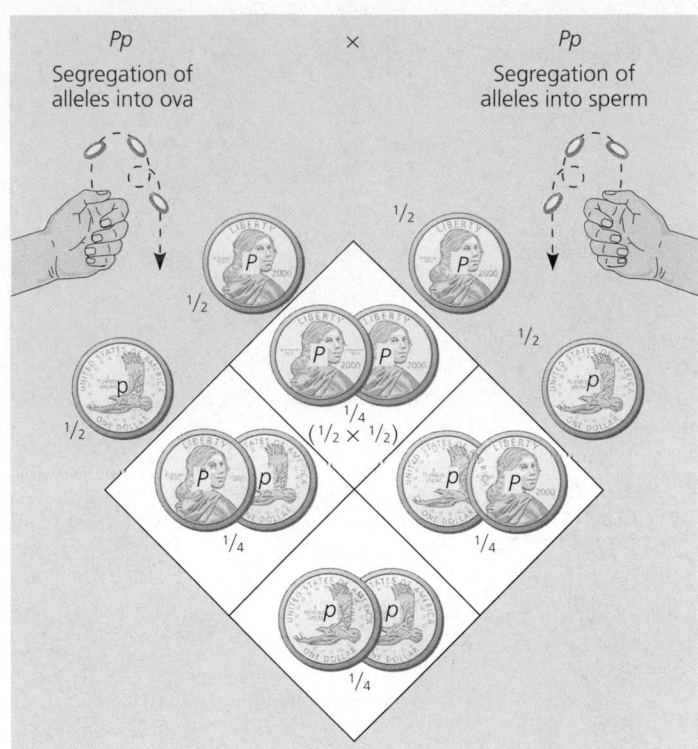

FIGURE 14.8 Segregation of alleles and fertilization as chance events. When a heterozygote (Pp) forms gametes, segregation of alleles is like the toss of a coin. We can determine the probability for any genotype among the offspring of two heterozygotes, as explained in the text.

We can use our understanding of independent events to predict the outcome of genetic crosses. Two basic laws of probability that can help us in games of chance and in solving genetic problems are the rule of multiplication and the rule of addition.

The Rule of Multiplication

What is the chance that two coins tossed simultaneously will land heads up? More generally, how do we determine the chance that two or more independent events will occur together in some specific combination? The solution is in computing the probability for each independent event, then multiplying these individual probabilities to obtain the overall probability of these events occurring together. By this rule of multiplication, the probability that both coins will land heads up is ½ × ½ = ¼. An F₁ monohybrid cross is analogous to this game of chance. With flower color in pea plants as the heritable character, the genotype of a given F₁ plant is Pp. What is the probability that a particular F₂ plant will have white flowers? For this to happen, both the ovum and the sperm must carry the p allele, so we invoke the rule of multiplication. Segregation in the heterozygous plant is like flipping a coin. The probability that an ovum will have the p allele is ½.

The chance that a sperm will have the p allele is $\frac{1}{2}$. Thus, the probability that two p alleles will come together at fertilization is $\frac{1}{2} \times \frac{1}{2} = \frac{1}{4}$, equivalent to the probability that two independently tossed coins will land heads up (see the Punnett square in FIGURE 14.8).

We can also apply the rule of multiplication to dihybrid crosses, such as the one shown in FIGURE 14.7b. For a parent with the genotype $YyRr$, the probability that a gamete will carry the Y and R alleles is $\frac{1}{4}$. We can use the rule of multiplication to determine the probability of specific genotypes in the F_2 generation without having to construct a 16-part Punnett square. For example, the probability of an F_2 plant having the genotype $YYRR$ is $\frac{1}{16}$ ($\frac{1}{4}$ chance for a YR ovum $\times$ $\frac{1}{4}$ chance for a YR sperm). This corresponds to the top box in the Punnett square of FIGURE 14.7b.

The Rule of Addition

What is the probability that an F_2 plant from a monohybrid cross will be heterozygous? Notice in FIGURE 14.8 that there are two ways F_1 gametes can combine to produce a heterozygous result. The dominant allele can come from the ovum and the recessive allele from the sperm, or vice versa. By the rule of addition, the probability of an event that can occur in two or more different ways is the sum of the separate probabilities of those ways. Using this rule, we can calculate the probability of an F_2 heterozygote as $\frac{1}{4} + \frac{1}{4} = \frac{1}{2}$.

Using Rules of Probability to Solve Genetics Problems

We can combine the rules of multiplication and addition to solve complex problems in Mendelian genetics. For instance, imagine a cross of two pea varieties in which we track the inheritance of three characters. Suppose we cross a trihybrid with purple flowers and yellow, round seeds (heterozygous for all three genes) with a plant with purple flowers and green, wrinkled seeds (heterozygous for flower color but homozygous recessive for the other two characters). Using Mendelian symbols, our cross is

$$PpYyRr \times Ppyyrr$$

Let's use the rules of probability to calculate the fraction of offspring predicted to exhibit the recessive phenotypes for *at least two* of the three traits. We can start by listing all genotypes that fulfill this condition: *ppyyRr, ppYyrr, Ppyyrr, PPyyrr,* and *ppyyrr.* (Because the condition is *at least two* recessive traits, this last genotype, which produces all three recessive phenotypes, counts.) Next, we use the rule of multiplication to calculate the probability for each of these genotypes from our $PpYyRr \times Ppyyrr$ cross (that is, we multiply the individual probabilities for the allele pairs). Finally, we use the rule of addition to pool the probabilities for fulfilling the condition of at least two recessive traits as shown at the top of the next column.

ppyyRr	$\frac{1}{4}$ (probability of pp) $\times \frac{1}{2}$ (yy) $\times \frac{1}{2}$ (Rr)	$= \frac{1}{16}$
ppYyrr	$\frac{1}{4} \times \frac{1}{2} \times \frac{1}{2}$	$= \frac{1}{16}$
Ppyyrr	$\frac{1}{2} \times \frac{1}{2} \times \frac{1}{2}$	$= \frac{2}{16}$
PPyyrr	$\frac{1}{4} \times \frac{1}{2} \times \frac{1}{2}$	$= \frac{1}{16}$
ppyyrr	$\frac{1}{4} \times \frac{1}{2} \times \frac{1}{2}$	$= \frac{1}{16}$
Chance of *at least two* recessive traits		$= \frac{6}{16}$ or $\frac{3}{8}$

With practice, you'll be able to apply the rules of probability to solve genetics problems faster than you could by filling in Punnett squares.

Mendel discovered the particulate behavior of genes: *a review*

If we plant a seed from the F_2 generation of FIGURE 14.4, we cannot predict with certainty that the plant will yield white flowers, any more than we can predict with certainty that two tossed coins will both come up heads. What we *can* say is that there is exactly a $\frac{1}{4}$ chance that the plant will have white flowers. Stated another way, among a large sample of F_2 plants, one-fourth (25%) will have white flowers. Usually, the larger the sample size, the closer the results will conform to our predictions. The fact that Mendel counted so many offspring from his crosses suggests that he understood this statistical feature of inheritance and had a keen sense of the rules of chance.

Thus, Mendel's two laws, segregation and independent assortment, explain heritable variations in terms of alternative forms of genes (hereditary "particles") that are passed along, generation after generation, according to simple rules of probability. This particulate theory of inheritance, first discovered in garden peas, is equally valid for figs, flies, fish, birds, and human beings. Mendel's impact endures, not only on genetics, but on all of science, as a case study of the power of scientific reasoning (hypothetico-deductive reasoning; see Chapter 1).

EXTENDING MENDELIAN GENETICS

The relationship between genotype and phenotype is rarely simple

In the 20th century, geneticists have extended Mendelian principles not only to diverse organisms, but also to patterns of inheritance more complex than Mendel actually described. It was brilliant (or lucky) that Mendel chose pea plant characters that turned out to have a relatively simple genetic basis: Each character he studied is determined by one gene, for which there are only two alleles, one completely dominant to the other.* But

---

* There is one exception: Geneticists have found that Mendel's flower-position character is actually determined by two genes.

these conditions are not met by all heritable characters, not even in garden peas. The relationship between genotype and phenotype is rarely so simple. This does not diminish the utility of Mendelian genetics, for the basic principles of segregation and independent assortment apply even to more complex patterns of inheritance. In this section, we will extend Mendelian genetics to hereditary patterns that were not reported by Mendel.

Incomplete Dominance

The F$_1$ offspring of Mendel's classic pea crosses always looked like one of the two parental varieties because of the complete dominance of one allele over another. But for some genes, there is **incomplete dominance,** where the F$_1$ hybrids have an appearance somewhere in between the phenotypes of the two parental varieties. For instance, when red snapdragons are crossed with white snapdragons, all the F$_1$ hybrids have pink flowers (FIGURE 14.9). This third phenotype results from flowers of the heterozygotes having less red pigment than the red

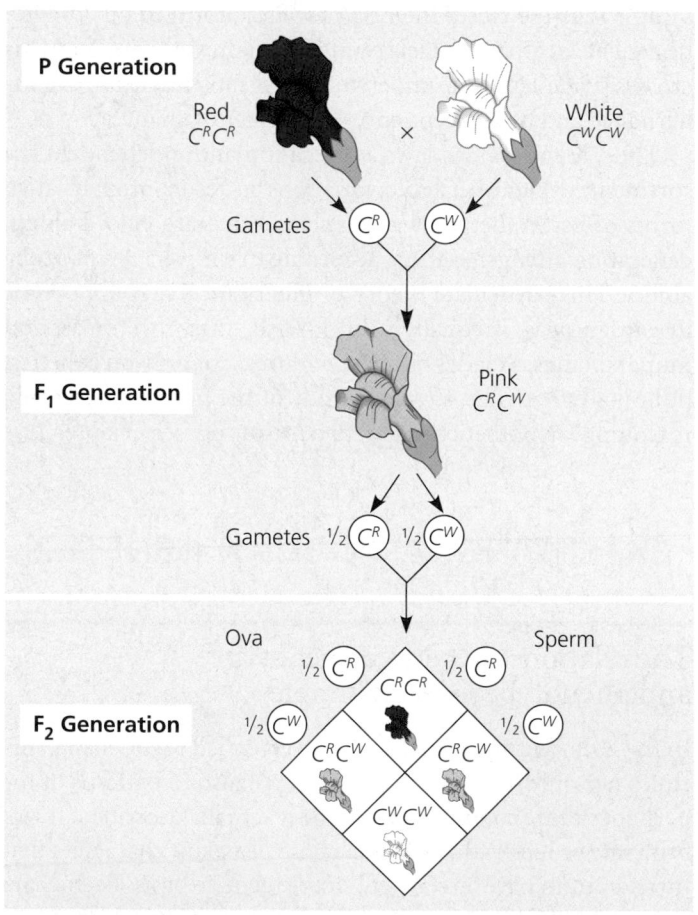

FIGURE 14.9 Incomplete dominance in snapdragon color. When red snapdragons are crossed with white ones, the F$_1$ hybrids have pink flowers. Segregation of alleles into gametes of the F$_1$ plants results in an F$_2$ generation with a 1:2:1 ratio for both genotype and phenotype. C^R = allele for red flower color; C^W = allele for white flower color.

homozygotes (unlike the situation in Mendel's pea plants, where the *Pp* heterozygotes make enough pigment for the flowers to be a purple color indistinguishable from those of *PP* plants). We should not, however, regard incomplete dominance as evidence for the blending hypothesis, which would predict that the red or white traits could never be retrieved from the pink hybrids. In fact, breeding the F$_1$ hybrids produces F$_2$ offspring with a phenotypic ratio of one red to two pink to one white. (Because heterozygotes have a separate phenotype, the genotypic and phenotypic ratios for the F$_2$ generation are the same, 1:2:1.) The segregation of the red-flower and white-flower alleles in the gametes produced by the pink-flowered plants confirms that the alleles for flower color are heritable factors that maintain their identity in the hybrids; that is, inheritance is particulate.

What Is *a Dominant Allele?*

Now that you have learned about incomplete dominance, let's reexamine the meaning of dominance and recessiveness. What *is* a dominant allele? Or, more importantly, what is it *not?*

In **complete dominance,** the situation described by Mendel, the phenotypes of the heterozygote and dominant homozygote are indistinguishable. This represents one extreme of a spectrum in the dominance/recessiveness relationships of alleles. At the other extreme is **codominance,** in which the two alleles affect the phenotype in separate, distinguishable ways. One example is the existence of three different human blood groups called the M, N, and MN blood groups. These groupings are based on two specific molecules located on the surface of red blood cells. People of group M have one of these two types of molecules, and people of group N have the other type. Group MN is characterized by the presence of *both* molecules on red blood cells. What is the genetic basis of these phenotypes? A single gene locus, at which two allelic variations are possible, determines these blood groups. M individuals are homozygous for one allele; N individuals are homozygous for the other allele. A heterozygous condition results in the blood of the MN group.

Note that the MN phenotype is *not* intermediate between the M and N phenotypes; both the M and N phenotypes are individually expressed by the presence of the two types of molecules on red blood cells. In contrast, incomplete dominance is characterized by an intermediate phenotype, as in the pink flowers of snapdragon hybrids. Thus, the range of relative effects of two alleles includes complete dominance, codominance, and different degrees of incomplete dominance. These variations are reflected in the phenotypes of heterozygotes.

For any character, the dominance/recessiveness relationship we observe depends on the level at which we examine phenotype. For example, consider Tay-Sachs disease, an inherited disorder in humans. The brain cells of a baby with Tay-Sachs disease are unable to metabolize gangliosides, a type of lipid, because a crucial enzyme does not work properly. As the lipids accumulate in the brain, the brain cells gradually cease to function normally, leading to death. Only children who inherit two copies of the Tay-Sachs allele (homozygotes) have the disease. Thus, on the *organismal* level of normal versus Tay-Sachs phenotype, the Tay-Sachs allele qualifies as a recessive. At the *biochemical* level, however, we observe an intermediate phenotype characteristic of incomplete dominance: The enzyme deficiency that causes Tay-Sachs disease can be detected in heterozygotes, who have an activity level of the lipid-metabolizing enzyme that is intermediate between individuals homozygous for the normal allele and individuals with Tay-Sachs disease. Heterozygotes lack symptoms of the disease, apparently because half the normal amount of functional enzyme is sufficient to prevent lipid accumulation in the brain. In fact, heterozygous individuals produce equal numbers of normal and dysfunctional enzyme molecules. Thus, at the *molecular* level, the normal allele and the Tay-Sachs allele are codominant. As you can see, dominance/recessiveness relationships are rarely as straightforward as Mendel reported.

It is also important to understand that an allele is not termed *dominant* because it somehow subdues a recessive allele. Recall that alleles are simply variations in a gene's nucleotide sequence. When a dominant allele coexists with a recessive allele in a heterozygous genotype, they do not actually interact at all. It is in the pathway from genotype to phenotype that dominance and recessiveness come into play. We can use one of Mendel's characters—round versus wrinkled pea seed shape—as an example. The dominant allele codes for the synthesis of an enzyme that helps convert sugar to starch in the seed. The recessive allele codes for a defective form of this enzyme. Thus, in a recessive homozygote, sugar accumulates in the seed because it is not converted to starch. As the seed develops, the high sugar concentration causes the osmotic uptake of water, and the seed swells. When the mature seed dries, it develops wrinkles. In contrast, if a dominant allele is present, sugar is converted to starch, and the seeds do not wrinkle when they dry. One dominant allele results in enough of the enzyme to convert sugar to starch, and thus dominant homozygotes and heterozygotes have the same phenotype: round seeds. By exploring the mechanisms responsible for phenotype, we can demystify the concepts of dominance and recessiveness.

There is another important lesson about the meaning of the term *dominance*. Because an allele for a particular character is dominant does not necessarily mean that it is more common in a population than the recessive allele for that character. For example, about one baby out of 400 in the United States is born with extra fingers or toes, a condition known as polydactyly. The allele for polydactyly is *dominant* to the allele for five digits per appendage. In other words, 399 out of every 400 people are recessive homozygotes for this character; the recessive allele is far more prevalent than the dominant allele in the population. In Chapter 23, you will learn how the relative frequencies of alleles in a population are affected by natural selection.

Let's summarize three important points about dominance/recessiveness relationships:

1. They range from complete dominance, through various degrees of incomplete dominance, to codominance.
2. They reflect the mechanisms by which specific alleles are expressed in phenotype and do not involve the ability of one allele to subdue another at the level of the DNA.
3. They do not determine or correlate with the relative abundance of alleles in a population.

Multiple Alleles

Most genes actually exist in populations in more than two allelic forms. The ABO blood groups in humans are one example of multiple alleles of a single gene. There are four possible phenotypes for this character: A person's blood group may be either A, B, AB, or O (FIGURE 14.10a). These letters refer to two carbohydrates—the A substance and the B substance—that may be found on the surface of red blood cells. (These groups are based on blood-cell molecules different from those used for the MN classification discussed earlier.) A person's blood cells may have one substance or the other (type A or B), both (type AB), or neither (type O).

(a) Phenotype (blood group)	(b) Genotypes (see p.258)	(c) Antibodies present in blood serum	(d) Results from adding red blood cells from groups below to serum from groups at left			
			A	B	AB	O
A	$I^A I^A$ or $I^A i$	Anti-B				
B	$I^B I^B$ or $I^B i$	Anti-A				
AB	$I^A I^B$	—				
O	ii	Anti-A Anti-B				

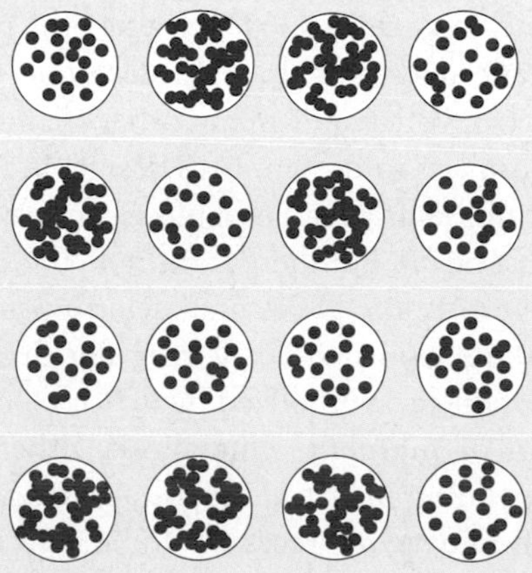

FIGURE 14.10 Multiple alleles for the ABO blood groups.

The four blood groups result from various combinations of three different alleles of one gene, symbolized as I^A (for the A carbohydrate), I^B (for B), and i (giving rise to neither A nor B). Because each person carries two alleles, six genotypes are possible (FIGURE 14.10b). Both the I^A and the I^B alleles are dominant to the i allele. Thus, I^AI^A and I^Ai individuals have type A blood, and I^BI^B and I^Bi individuals have type B. Recessive homozygotes, ii, have type O blood, because neither the A nor the B substance is produced. The I^A and I^B alleles are codominant; both are expressed in the phenotype of the I^AI^B heterozygote, who has type AB blood.

Matching compatible blood groups is critical for blood transfusions, because a person produces specific proteins called antibodies against foreign blood factors (FIGURE 14.10c). If the donor's blood has a factor (A or B) that is foreign to the recipient, antibodies produced by the recipient bind to the foreign molecules and cause the donated blood cells to clump together (FIGURE 14.10d). This agglutination (clumping) can kill the recipient.

Pleiotropy

So far, we have treated Mendelian inheritance as though each gene affects one phenotypic character. Most genes, however, have multiple phenotypic effects. The ability of a gene to affect an organism in many ways is called **pleiotropy** (from the Greek *pleion*, more). For example, pleiotropic alleles responsible for certain hereditary diseases in humans, such as sickle-cell disease, cause multiple symptoms (see FIGURE 14.15). Considering the intricate molecular and cellular interactions responsible for an organism's development, it is not surprising that a gene can affect a number of an organism's characteristics.

Epistasis

Dominance, multiple alleles, and pleiotropy all involve the effects of alleles for single genes. We now turn to situations involving more than one gene. One such situation is **epistasis** (from the Greek for "stopping" or "causing to stand"), in which a gene at one locus alters the phenotypic expression of a gene at a second locus. An example will help clarify this concept. In mice and many other mammals, black coat color is dominant to brown. Let's designate B and b as the two alleles for this character. For a mouse to have brown fur, its genotype must be bb. But there is more to the story. A second gene, said to be epistatic to the first, determines whether or not pigment will be deposited in the hair. For this second gene, the dominant allele, symbolized by C (for color), results in the deposition of pigment. This allows either black or brown color, depending on the genotype at the first locus. But if the mouse is homozygous recessive for the second locus (cc), then the coat is white (albino), regardless of the genotype at the black/brown locus.

What happens if we mate black mice that are heterozygous for both genes ($BbCc$)? Although the two genes affect the same phenotypic character (coat color), they follow the law of independent assortment (the two genes are inherited separately). Thus, our breeding experiment represents an F_1 dihybrid cross, like those that produced a 9:3:3:1 ratio in Mendel's experiments. In the case of coat color, however, the ratio of phenotypes among F_2 offspring is nine black to three brown to four white. FIGURE 14.11 uses a Punnett square to account for this ratio in terms of epistasis. Other types of epistatic interactions produce different ratios.

Polygenic Inheritance

Mendel studied characters that could be classified on an either-or basis, such as purple versus white flower color. For many characters, however, such as human skin color and height, an either-or classification is impossible, because the characters vary in the population along a continuum (in gradations). These are called **quantitative characters.** Quantitative varia-

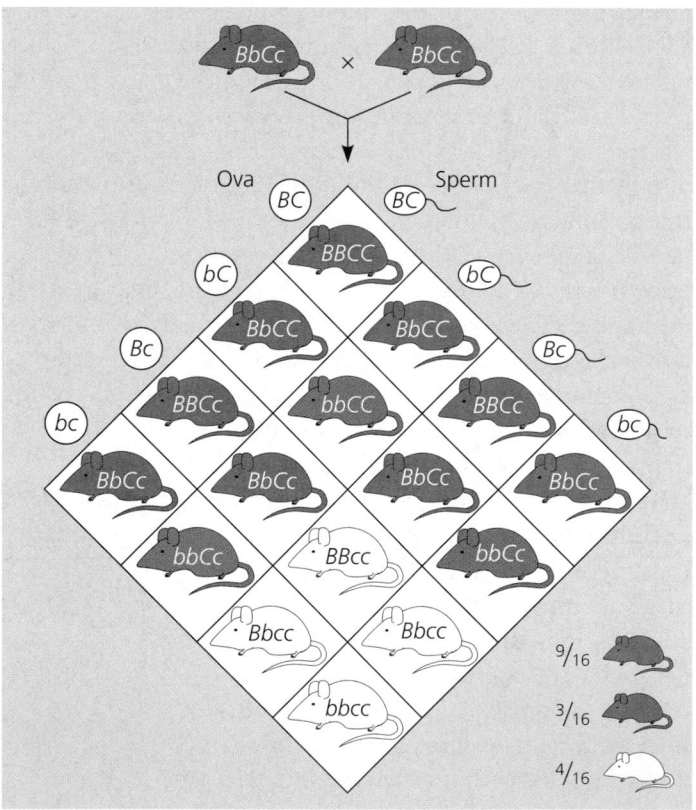

FIGURE 14.11 An example of epistasis. This Punnett square illustrates the genotypes and phenotypes predicted for offspring of matings between two black mice of genotype *BbCc,* where the *C/c* gene is epistatic to the *B/b* gene. One gene determines whether the coat will be black (dominant, *B*) or brown (recessive, *b*). The second gene controls whether or not pigment of any color will be deposited in the hair, with the allele for the presence of color (*C*) dominant to the allele for the absence of color (*c*). The epistatic relationship of the color gene to the black/brown gene results in an F_2 phenotypic ratio of 9 black to 3 brown to 4 white.

tion usually indicates **polygenic inheritance,** an additive effect of two or more genes on a single phenotypic character (the converse of pleiotropy, where a single gene affects several phenotypic characters).

There is evidence, for instance, that skin pigmentation in humans is controlled by at least three separately inherited genes (probably more, but we will simplify). Let's consider three genes, with a dark-skin allele for each gene (*A, B, C*) contributing one "unit" of darkness to the phenotype and being incompletely dominant to the other alleles (*a, b, c*). An *AABBCC* person would be very dark, while an *aabbcc* individual would be very light. An *AaBbCc* person would have skin of an intermediate shade. Because the alleles have a cumulative effect, the genotypes *AaBbCc* and *AABbcc* would make the same genetic contribution (three units) to skin darkness. FIGURE 14.12 shows how this polygenic inheritance could result in a bell-shaped curve, called a normal distribution, for skin darkness among the members of a hypothetical population. (You are probably familiar with the concept of a normal

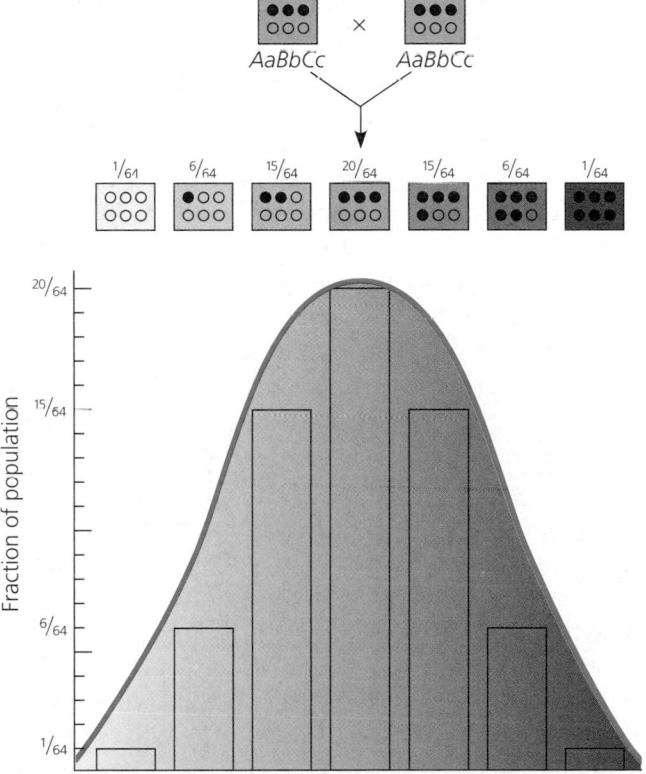

FIGURE 14.12 A simplified model for polygenic inheritance of skin color. According to this model, three separately inherited genes affect the darkness of skin. For each gene, an allele for dark skin (*A, B, C*) is incompletely dominant to an allele for light skin (*a, b, c*). Thus, the heterozygous individuals (*AaBbCc*), represented by the two rectangles at the top of this figure, each carry three "units" of darkness. (Black dots symbolize dark-skin alleles.) Imagine a large number of matings between such heterozygotes. Along the top of the graph are the variations that can occur among offspring. The *y* axis represents the fractions of these variations among offspring of the matings. The resulting histogram is smoothed into a bell-shaped curve by environmental factors that affect skin color.

distribution for class curves of test scores.) Environmental factors, such as exposure to the sun, also affect the skin-color phenotype and help make the graph a smooth curve rather than a stairlike histogram.

Nature and Nurture: The Environmental Impact on Phenotype

Phenotype depends on environment as well as on genes. A single tree, locked into its inherited genotype, has leaves that vary in size, shape, and greenness, depending on exposure to wind and sun. For humans, nutrition influences height, exercise alters build, sun-tanning darkens the skin, and experience improves performance on intelligence tests. Even identical twins, who are genetic equals, accumulate phenotypic differences as a result of their unique experiences.

Whether it is genes or the environment—nature or nurture—that most influences human characteristics is a very old and hotly contested debate that we will not attempt to settle here. We can say, however, that the product of a genotype is generally not a rigidly defined phenotype, but a range of phenotypic possibilities over which there may be variation due to environmental influence. This phenotypic range is called the **norm of reaction** for a genotype (FIGURE 14.13). There are cases where the norm of reaction has no breadth whatsoever; that is, a given genotype mandates a very specific phenotype. An example is the gene locus that determines a person's ABO blood group. In contrast, a person's blood count of red and white cells varies, depending on such factors as the altitude of one's home, the person's customary level of physical activity, and the presence of infectious agents.

Generally, norms of reaction are broadest for polygenic characters. Environment contributes to the quantitative nature of these characters, as we have seen in the continuous variation of skin color. Geneticists refer to such characters as **multifactorial,** meaning that many factors, both genetic and environmental, collectively influence phenotype.

FIGURE 14.13 The effect of environment on phenotype. The outcome of a genotype lies within its norm of reaction, a phenotypic range that depends on the environment in which the genotype is expressed. For example, hydrangea flowers of the same genetic variety range in color from blue-violet to pink, depending on the acidity of the soil.

Integrating a Mendelian View of Heredity and Variation

Over the past several pages, we have broadened our view of Mendelian inheritance by exploring incomplete dominance and other variations in dominance/recessiveness relationships, as well as multiple alleles, pleiotropy, polygenic inheritance, and the phenotypic impact of the environment. How can we integrate these refinements into a comprehensive theory of Mendelian genetics? The key is to make the transition from the reductionist emphasis on single genes and phenotypic characters to the idea of the organism as a whole, one of the themes of this book. In fact, the term *phenotype* does double duty. We have been using the word in the context of specific characters, such as flower color and blood group. But phenotype is also used to describe the organism in its entirety—*all* aspects of its physical appearance, internal anatomy, physiology, and behavior. Similarly, the term *genotype* can also refer to an organism's entire genetic makeup, not just its alleles for a single genetic locus. In most cases, a gene's impact on phenotype is affected by other genes and by the environment. In this integrated view of heredity and variation, an organism's phenotype reflects its overall genotype and unique environmental history.

Considering all that can occur in the pathway from genotype to phenotype, it is indeed impressive that Mendel could simplify the complexities to reveal the fundamental principles governing the transmission of individual genes from parents to offspring. By extending the principles of segregation and independent assortment to help explain such hereditary patterns as epistasis and quantitative characters, we begin to see how broadly Mendelism applies. From Mendel's abbey garden came a theory of particulate inheritance that anchors modern genetics. In the last section of this chapter, we will apply Mendelian genetics to human inheritance, especially the transmission of hereditary diseases.

MENDELIAN INHERITANCE IN HUMANS

Whereas peas are convenient subjects for genetic research, humans are not. The human generation span is about 20 years, and human parents produce relatively few offspring compared to peas and most other species. Furthermore, breeding experiments like the ones Mendel performed are unacceptable with humans. In spite of these difficulties, the study of human genetics continues to advance, powered by the incentive to understand our own inheritance. New techniques in molecular biology have led to many breakthrough discoveries, as we will see in Chapter 20, but basic Mendelism endures as the foundation of human genetics.

Pedigree analysis reveals Mendelian patterns in human inheritance

Unable to manipulate the mating patterns of people, geneticists must analyze the results of matings that have already occurred. As much information as possible is collected about a family's history for a particular trait, and this information is assembled into a family tree describing the interrelationships of parents and children across the generations—the family **pedigree.** A simple example of a pedigree appears in FIGURE 14.14a, which traces the occurrence of widow's peak (a pointed contour of the hairline on the forehead). The trait is due to a dominant allele, which we symbolize as *W*.

We know that all individuals in this family who lack a widow's peak are homozygous recessive, and thus we can fill in their genotypes on the pedigree (*ww*). We also know that both grandparents with widow's peaks are heterozygous (*Ww*); if they were homozygous dominant (*WW*), then all of their offspring would have widow's peaks. The offspring in this second generation who *do* have widow's peaks must also be heterozygous, because they are the products of *Ww* × *ww* matings. The third generation in this pedigree consists of two sisters. The one who has a widow's peak could be either homozygous (*WW*) or heterozygous (*Ww*), given what we know about the genotypes of her parents (both *Ww*).

FIGURE 14.14b is a pedigree of the same family, but this time we focus on a recessive trait, attached earlobes. We'll use *f* for the recessive allele and *F* for the dominant allele, which results in free earlobes. As you work your way through the pedigree, notice once again that you can apply what you have learned about Mendelian inheritance to fill in the genotypes for most individuals. (As you may have already guessed, the ears shown in FIGURE 14.14b do not actually belong to the women shown in FIGURE 14.14a. However, their earlobes would resemble those in the photographs.)

A pedigree not only helps us understand the past; it also helps us predict the future. Suppose that the couple represented in the second generation of FIGURE 14.14 decide to have one more child. What is the probability that the child will have a widow's peak? This is a Mendelian F_1 cross (*Ww* × *Ww*), and thus the probability that a child will exhibit the dominant phenotype (widow's peak) is ¾. What is the probability that the child will have attached earlobes? Again, we can treat this as a monohybrid cross (*Ff* × *Ff*), but this time we want to know the chance that the offspring will be homozygous recessive. The probability is ¼. What is the chance that the child will have a widow's peak *and* attached earlobes? If the two pairs of alleles assort independently in this dihybrid problem (*WwFf* × *WwFf*), then we can use the rule of multiplication: ¾ (chance of widow's peak) × ¼ (chance of attached earlobes) = ³⁄₁₆ (chance of widow's peak and attached earlobes).

FIGURE 14.14 Pedigree analysis. In these family trees, squares symbolize males and circles represent females. A horizontal line connecting a male and female (□—○) indicates a mating, with offspring listed below in their order of birth, from left to right. Shaded symbols stand for individuals with the trait being traced.

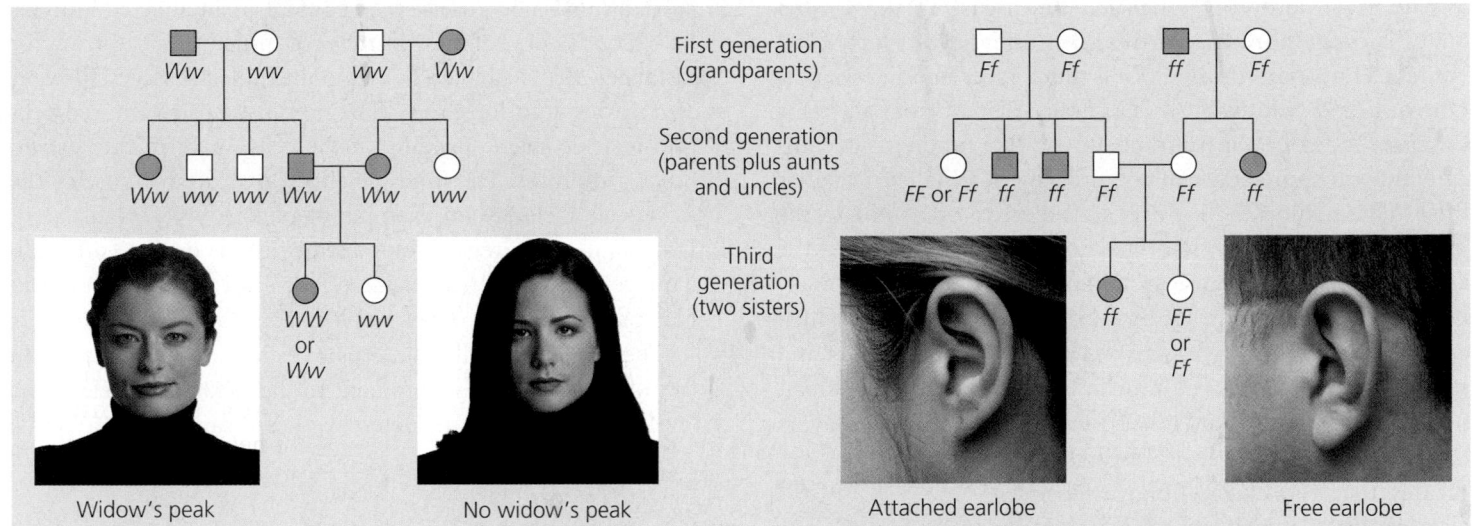

(a) Dominant trait (widow's peak). This pedigree traces the trait called widow's peak through three generations of a family. Notice in the third generation that the second-born daughter lacks a widow's peak, although both of her parents had the trait. Such a pattern of inheritance supports the hypothesis that the trait is due to a dominant allele. If the trait were due to a *recessive* allele, and both parents had the recessive phenotype, then *all* of their offspring would also have the recessive phenotype.

(b) Recessive trait (attached earlobe). This is the same family, but in this case we are tracing the inheritance of a recessive trait, attached earlobes. Notice that the first-born daughter in the third generation has attached earlobes, although both of her parents lack that trait (they have free earlobes). Such a pattern is easily explained if the attached-lobe phenotype is due to a recessive allele. If it were due to a *dominant* allele, then at least one parent would also have had the trait.

Many human disorders follow Mendelian patterns of inheritance

Pedigrees are a more serious matter when the alleles in question cause disabling or deadly hereditary diseases instead of innocuous human variations such as hairline or earlobe configuration. However, for disorders inherited as simple Mendelian traits, the same techniques of pedigree analysis apply.

Recessively Inherited Disorders

Thousands of genetic disorders are known to be inherited as simple recessive traits. These disorders range in severity from relatively mild, such as albinism (lack of pigmentation, which results in susceptibility to skin cancers and vision problems), to life-threatening, such as cystic fibrosis.

How can we account for the recessive behavior of the alleles causing these disorders? Recall that genes code for proteins of specific function. An allele that causes a genetic disorder codes either for a malfunctional protein or for no protein at all. In the case of disorders classified as recessive, heterozygotes are normal in phenotype because one copy of the "normal" allele produces a sufficient amount of the specific protein. Thus, a recessively inherited disorder shows up only

in the homozygous individuals who inherit one recessive allele from each parent. We can symbolize the genotype of such people as *aa,* with individuals lacking the disorder being either *AA* or *Aa.* The heterozygotes (*Aa*), who are phenotypically normal with regard to the disorder, are called **carriers** of the disorder because they may transmit the recessive allele to their offspring.

Most people who have recessive disorders are born to parents of normal phenotype who are both carriers. A mating between two carriers corresponds to a Mendelian F_1 cross (*Aa* × *Aa*), with the zygote having a ¼ chance of inheriting a double dose of the recessive allele. A child of normal phenotype from such a cross has a ⅔ chance of being a carrier. (The genotypic ratio for the offspring is 1*AA*:2*Aa*:1*aa*. Thus, two out of three offspring with *normal* phenotype—*AA* or *Aa*—are predicted to be heterozygous carriers.) Recessive homozygotes could also result from *Aa* × *aa* and *aa* × *aa* matings, but if the disorder is lethal before reproductive age or results in sterility, no *aa* individuals will reproduce. Even if recessive homozygotes are able to reproduce, such individuals will still account for a much smaller percentage of the population than heterozygous carriers, for reasons we will examine in Chapter 23.

In general, a genetic disorder is not evenly distributed among all groups of humans. These disparities result from the different genetic histories of the world's peoples during less technological times, when populations were more geographically (and hence genetically) isolated. We will now examine three examples of recessively inherited disorders.

The most common lethal genetic disease in the United States is **cystic fibrosis,** which strikes one out of every 2,500 whites of European descent but is much rarer in other groups. One out of 25 whites (4%) is a carrier. The normal allele for this gene codes for a membrane protein that functions in chloride ion transport between certain cells and the extracellular fluid. These chloride channels are defective or absent in the plasma membranes of children who have inherited two of the recessive alleles that cause cystic fibrosis. The result is an abnormally high concentration of extracellular chloride, which causes the mucus that coats certain cells to become thicker and stickier than normal. The mucus builds up in the pancreas, lungs, digestive tract, and other organs, a condition that favors bacterial infections. Recent research indicates that the extracellular chloride also contributes to infection by disabling a natural antibiotic made by some body cells. When immune cells come to the rescue, their remains add to the mucus, creating a vicious cycle. Untreated, most children with cystic fibrosis die before their fifth birthday. Gentle pounding on the chest to clear mucus from clogged airways, daily doses of antibiotics to prevent infection, and other preventive treatments can prolong life. In the United States, more than half of the people with cystic fibrosis now survive into their late 20s or beyond.

Another lethal disorder inherited as a recessive allele is **Tay-Sachs disease,** described earlier in the chapter. Recall that the disease is caused by a dysfunctional enzyme that fails to break down brain lipids of a certain class. The symptoms of Tay-Sachs disease usually become manifest a few months after birth. The infant begins to suffer seizures, blindness, and degeneration of motor and mental performance. Inevitably, the child dies within a few years. There is a disproportionately high incidence of Tay-Sachs disease among Ashkenazic Jews, Jewish people whose ancestors lived in central Europe. In that population, the frequency of Tay-Sachs disease is one out of 3,600 births, about 100 times greater than the incidence among non-Jews or Mediterranean (Sephardic) Jews.

The most common inherited disease among blacks is **sickle-cell disease,** which affects one out of 400 African Americans. Sickle-cell disease is caused by the substitution of a single amino acid in the hemoglobin protein of red blood cells (see FIGURE 5.19). When the oxygen content of an affected individual's blood is low (at high altitudes or under physical stress, for instance), the sickle-cell hemoglobin molecules crystallize by aggregating into long rods. The crystals deform the red cells into a sickle shape. Sickling of the cells, in turn, can lead to other symptoms. The multiple effects of a double dose of the sickle-cell allele are an example of pleiotropy (FIGURE 14.15). Doctors now use regular blood transfusions to ward off brain damage in children with sickle-cell disease, and new drugs can help prevent or treat other problems, but there is no cure.

The non-sickle-cell counterpart of the sickle-cell allele is in fact only incompletely dominant to the sickle-cell allele at the level of the organism. Heterozygotes—carriers of a single

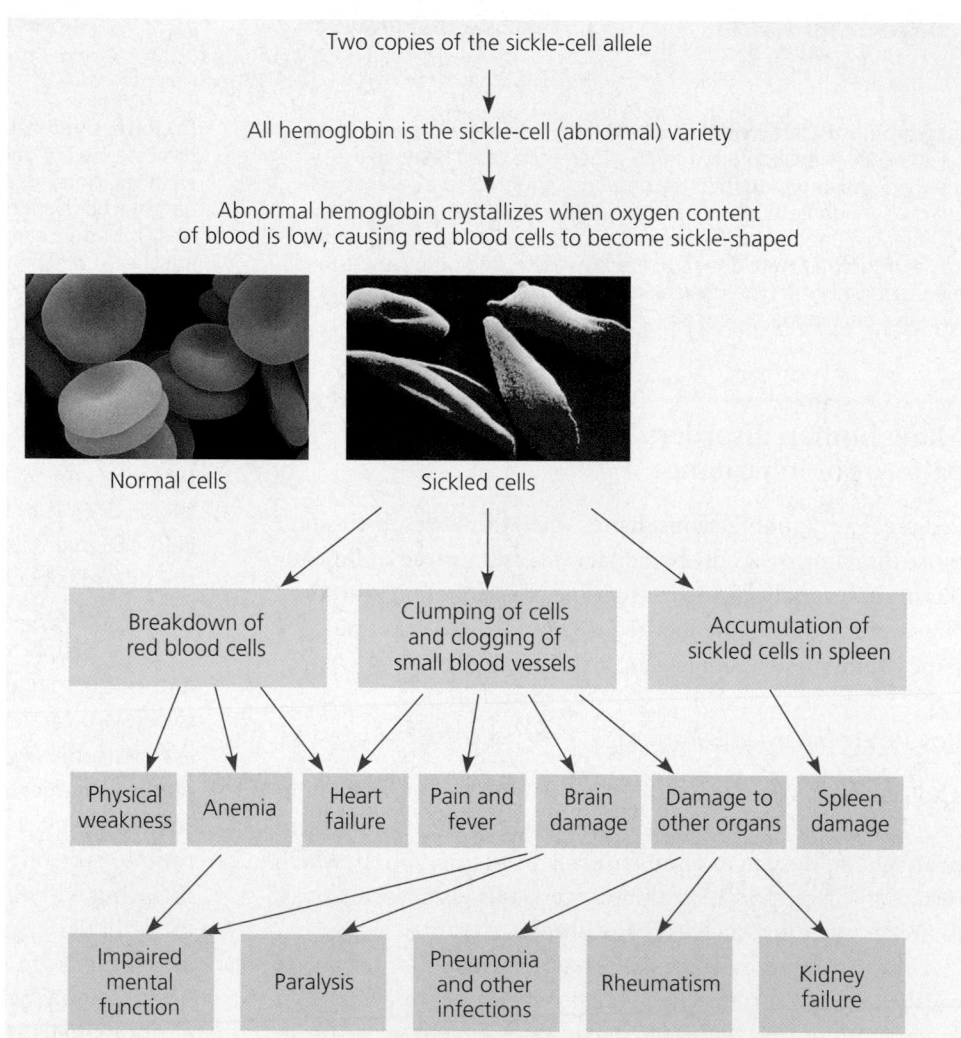

FIGURE 14.15 Pleiotropic effects of the sickle-cell allele in a homozygote. At the molecular level, the recessive allele responsible for sickle-cell disease has a single direct effect: It causes red blood cells to produce an abnormal version of the protein hemoglobin. An individual who inherits a sickle-cell allele from each parent has only the abnormal variety of hemoglobin. The abnormal hemoglobin deforms the red blood cells, starting a cascade of symptoms throughout the body.

sickle-cell allele—are said to have *sickle-cell trait.* Such people are usually healthy, although a fraction suffer some symptoms of sickle-cell disease when there is an extended reduction of blood oxygen. At the molecular level, the two alleles are codominant; both normal and abnormal hemoglobins are made.

About one out of ten African Americans has sickle-cell trait, an unusually high frequency of heterozygotes for an allele with severe detrimental effects in homozygotes. The reason for the prevalence of this allele appears to be that a single copy of the sickle-cell allele, while not usually producing sickle-cell symptoms, benefits the carrier by increasing resistance to malaria. The malaria parasite spends part of its life cycle in red blood cells (see FIGURE 28.13), and the presence of even heterozygous amounts of sickle-cell hemoglobin gives these cells an increased fragility that tends to interrupt the cycle. Thus, in tropical Africa, where malaria is common, the sickle-cell allele is both boon and bane. The relatively high frequency of African Americans with sickle-cell trait is a vestige of their African roots.

Although it is relatively unlikely that two carriers of the same rare harmful allele will meet and mate, the probability increases greatly if the man and woman are close relatives (for example, siblings or first cousins). These are called consanguineous ("same blood") matings, and they are indicated in pedigrees by double lines. Because people with recent common ancestors are more likely to carry the same recessive alleles than are unrelated people, it is more likely that a mating of close relatives will produce offspring homozygous for recessive traits—including harmful ones. Such effects can be observed in many types of domesticated and zoo animals that have become inbred.

There is debate among geneticists about the extent to which human consanguinity increases the risk of inherited diseases. Many deleterious alleles have such severe effects that a homozygous embryo spontaneously aborts long before birth. Most societies and cultures have laws or taboos forbidding marriages between close relatives. These rules may have evolved out of empirical observation that in most populations, stillbirths and birth defects are more common when parents are closely related. But social and economic factors have also influenced the development of customs and laws against consanguineous marriages.

Dominantly Inherited Disorders

Although most harmful alleles are recessive, many human disorders are due to dominant alleles. One example is *achondroplasia,* a form of dwarfism with an incidence of one case among every 10,000 people. Heterozygous individuals have the dwarf phenotype. Therefore, all people who are not achondroplastic dwarfs—99.99% of the population—are homozygous for the recessive allele.

Lethal dominant alleles are much less common than lethal recessives. Presumably, the two kinds of alleles arise by muta-

tion (changes to the DNA) of a sperm or egg equally often. However, if a lethal dominant allele kills an offspring before it is mature and can reproduce, the allele will not be passed on to future generations. This is in contrast to what happens to lethal recessive mutations, which are perpetuated from generation to generation by the reproduction of heterozygous carriers who have normal phenotypes.

A lethal dominant allele can escape elimination if it is late-acting, causing death at a relatively advanced age. By the time the symptoms become evident, the individual may have already transmitted the lethal allele to his or her children. For example, **Huntington's disease,** a degenerative disease of the nervous system, is caused by a lethal dominant allele that has no obvious phenotypic effect until the individual is about 35 to 45 years old. Once the deterioration of the nervous system begins, it is irreversible and inevitably fatal. Any child born to a parent who has the allele for Huntington's disease has a 50% chance of inheriting the allele and the disorder. (The mating can be symbolized as $Aa \times aa$, with A being the dominant allele that causes Huntington's disease.) Until recently, it was impossible to tell before the onset of symptoms if a person at risk for Huntington's disease had actually inherited the allele, but that has changed. Analyzing DNA samples from a large family with a high incidence of the disorder, molecular geneticists have tracked the Huntington's allele to a locus near the tip of chromosome 4 (FIGURE 14.16). It is now possible to test for the presence of the allele in an individual's genome. (The methods that make such tests possible are discussed in Chapter 20.) For those with a family history of Huntington's

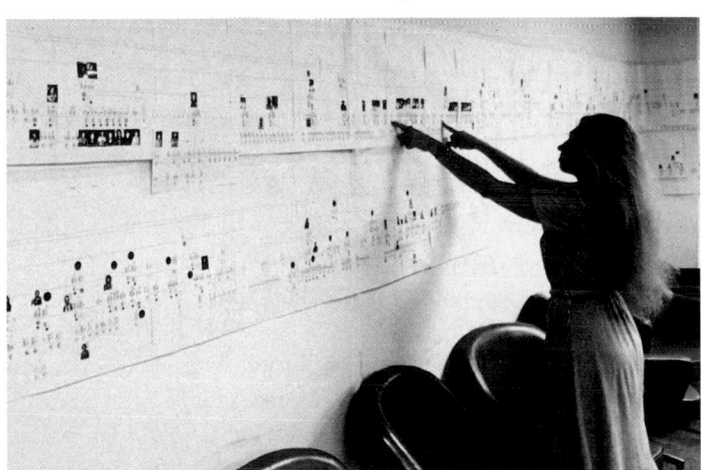

FIGURE 14.16 Large families provide excellent case studies of human genetics. Here, Nancy Wexler, of Columbia University and the Hereditary Disease Foundation, studies a huge pedigree that traces Huntington's disease through several generations of one large family in Venezuela. Classical Mendelian analysis of this family, coupled with the techniques of molecular biology, enabled scientists to develop a test for the presence of the dominant allele that causes Huntington's disease—a test that can be used before symptoms appear. Dr. Wexler is herself at risk for developing Huntington's disease. Her mother died of the disorder, and there is a 50% chance that Dr. Wexler inherited the dominant allele that causes the disease.

disease, the availability of this test poses an agonizing dilemma: Under what circumstances is it beneficial for a presently healthy person to find out whether he or she has inherited a fatal and not yet curable disease?

Multifactorial Disorders

The hereditary diseases we have discussed so far are sometimes described as simple Mendelian disorders because they result from certain alleles at a single genetic locus. Many more people are susceptible to diseases that have a multifactorial basis—a genetic component plus a significant environmental influence. The long list of multifactorial diseases includes heart disease, diabetes, cancer, alcoholism, and certain mental illnesses, such as schizophrenia and manic-depressive disorder. In many cases, the hereditary component is polygenic. For example, many genes affect our cardiovascular health, making some of us more prone than others to heart attacks and strokes. But our lifestyle intervenes tremendously between genotype and phenotype for cardiovascular health and other multifactorial characters. Exercise, a healthful diet, abstinence from smoking, and an ability to put stressful situations in perspective all reduce our risk of heart disease and some types of cancer.

At present, so little is understood about the genetic contributions to most multifactorial diseases that the best public health strategy is to educate people about the importance of environmental factors and to promote healthful behavior.

Technology is providing new tools for genetic testing and counseling

A preventive approach to simple Mendelian disorders is sometimes possible, because the risk that a particular genetic disorder will occur can sometimes be assessed before a child is conceived or during the early stages of the pregnancy. Many hospitals have genetic counselors who can provide information to prospective parents concerned about a family history for a specific disease.

Let's consider the case of a hypothetical couple, John and Carol, who are planning to have their first child and are seeking genetic counseling because of family histories of a lethal disease known to be recessively inherited. John and Carol each had a brother who died of the disorder, so they want to determine the risk of their having a child with the disease. From the information about their brothers, we know that both parents of John and both parents of Carol must have been carriers of the recessive allele. Thus, John and Carol are both products of $Aa \times Aa$ crosses, where a symbolizes the allele that causes this particular disease. We also know that John and Carol are not homozygous recessive (aa), because they do not have the disease. Therefore, their genotypes are either AA or Aa. Given a

genotypic ratio of $1AA:2Aa:1aa$ for offspring of an $Aa \times Aa$ cross, John and Carol each have a $\frac{2}{3}$ chance of being carriers (Aa). Using the rule of multiplication, we can determine that the overall probability of their firstborn having the disorder is $\frac{2}{3}$ (the chance that John is a carrier) multiplied by $\frac{2}{3}$ (the chance that Carol is a carrier) multiplied by $\frac{1}{4}$ (the chance of two carriers having a child with the disease), which equals $\frac{1}{9}$. Suppose that Carol and John decide to take the risk and have a child—after all, there is an $\frac{8}{9}$ chance that their baby will not have the disorder—but their child is born with the disease. We no longer have to guess about John's and Carol's genotypes. We now know that both John and Carol are, in fact, carriers. If the couple decides to have another child, they now know there is a $\frac{1}{4}$ chance that the second child will have the disease.

When we use Mendel's laws to predict possible outcomes of matings, it is important to keep in mind that chance has no memory: Each child represents an independent event in the sense that its genotype is unaffected by the genotypes of older siblings. Suppose that John and Carol have three more children, and *all three* have the hypothetical hereditary disease. This is an unfortunate family, for there is only one chance in 64 ($\frac{1}{4} \times \frac{1}{4} \times \frac{1}{4}$) that such an outcome will occur. But this run of misfortune will in no way affect the result if John and Carol decide to have still another child. There is still a $\frac{1}{4}$ chance that the additional child will have the disease and a $\frac{3}{4}$ chance that it will not. Mendel's laws, remember, are simply rules of probability applied to heredity.

Carrier Recognition

Because most children with recessive disorders are born to parents with normal phenotypes, the key to assessing the genetic risk for a particular disease is determining whether the prospective parents are heterozygous carriers of the recessive trait. For some heritable disorders, there are now tests that can distinguish individuals of normal phenotype who are dominant homozygotes from those who are heterozygotes, and the number of such tests increases each year. Examples are tests that can identify carriers of the alleles for Tay-Sachs disease, sickle-cell disease, and the most common form of cystic fibrosis. On one hand, these tests enable people with family histories of genetic disorders to make informed decisions about having children. On the other hand, these new methods for genetic screening could be abused. If confidentiality is breached, will carriers be stigmatized? Will they be denied health or life insurance, even though they themselves are healthy? Will misinformed employers equate "carrier" with disease? And will sufficient genetic counseling be available to help a large number of individuals understand their test results?

New biotechnology offers possibilities of reducing human suffering, but not before key ethical issues are resolved. The dilemmas posed by human genetics reinforce one of this book's themes: the immense social implications of biology.

Fetal Testing

Suppose a couple learns that they are both Tay-Sachs carriers, but they decide to have a child anyway. Tests performed in conjunction with a technique known as **amniocentesis** can determine, beginning at the 14th to 16th week of pregnancy, whether the developing fetus has Tay-Sachs disease (FIGURE 14.17a). To perform this procedure, a physician inserts a needle into the uterus and extracts about 10 milliliters (mL) of amniotic fluid, the liquid that bathes the fetus. Some genetic disorders can be detected from the presence of certain chemicals in the amniotic fluid itself. Tests for other disorders, including Tay-Sachs disease, are performed on cells grown in the laboratory from the fetal cells that had been sloughed off into the amniotic fluid. These cultured cells can also be used for karyotyping to identify certain chromosomal defects (see FIGURE 13.3).

In an alternative technique called **chorionic villus sampling (CVS)** (FIGURE 14.17b) a physician inserts a narrow tube through the cervix into the uterus and suctions out a tiny sample of fetal tissue from the placenta, the organ that transmits nutrients and fetal wastes between the fetus and the mother. Because the cells of the chorionic villi of the placenta are proliferating rapidly, enough cells are undergoing mitosis to allow karyotyping to be carried out immediately, giving results within 24 hours. This is an advantage over amniocentesis, in which the cells must be cultured for several weeks before karyotyping. Another advantage of CVS is that it can be performed as early as the eighth to tenth week of pregnancy. However, CVS is not suitable for tests requiring amniotic fluid, and it is less widely available than amniocentesis. Recently, medical scientists have developed methods for isolating fetal cells that

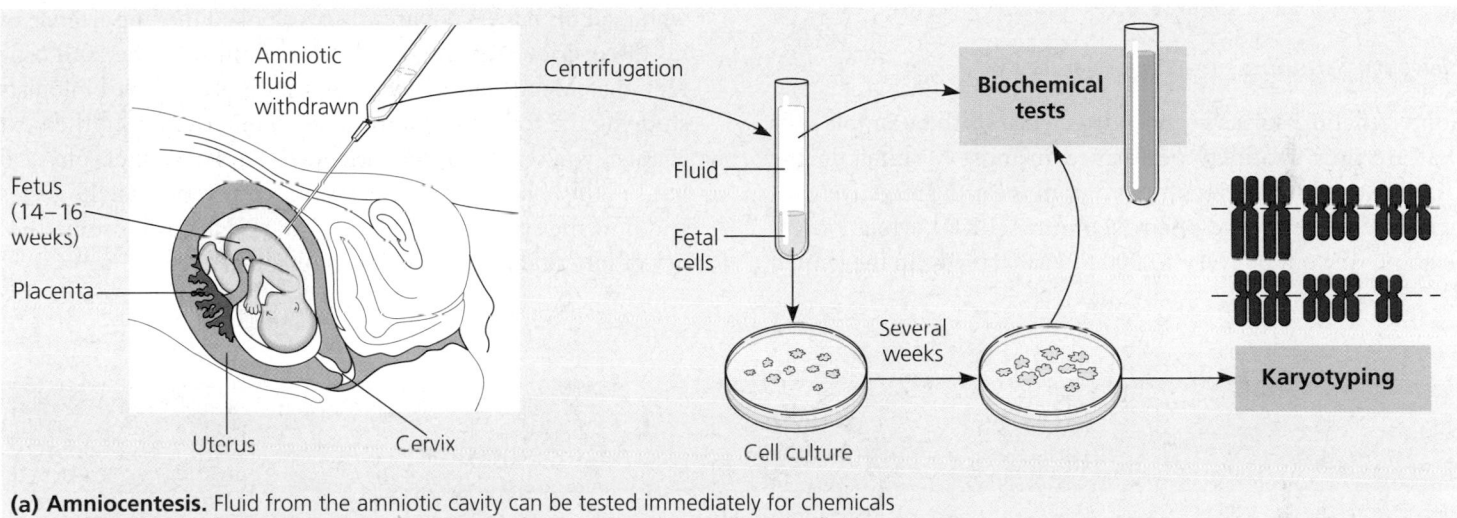

(a) Amniocentesis. Fluid from the amniotic cavity can be tested immediately for chemicals produced by the fetus that reveal the presence of certain disorders. Cells in the fluid must be cultured before they can be tested for other disorders or used for karyotyping. Karyotyping shows whether the chromosomes of the fetus are normal in number and appearance.

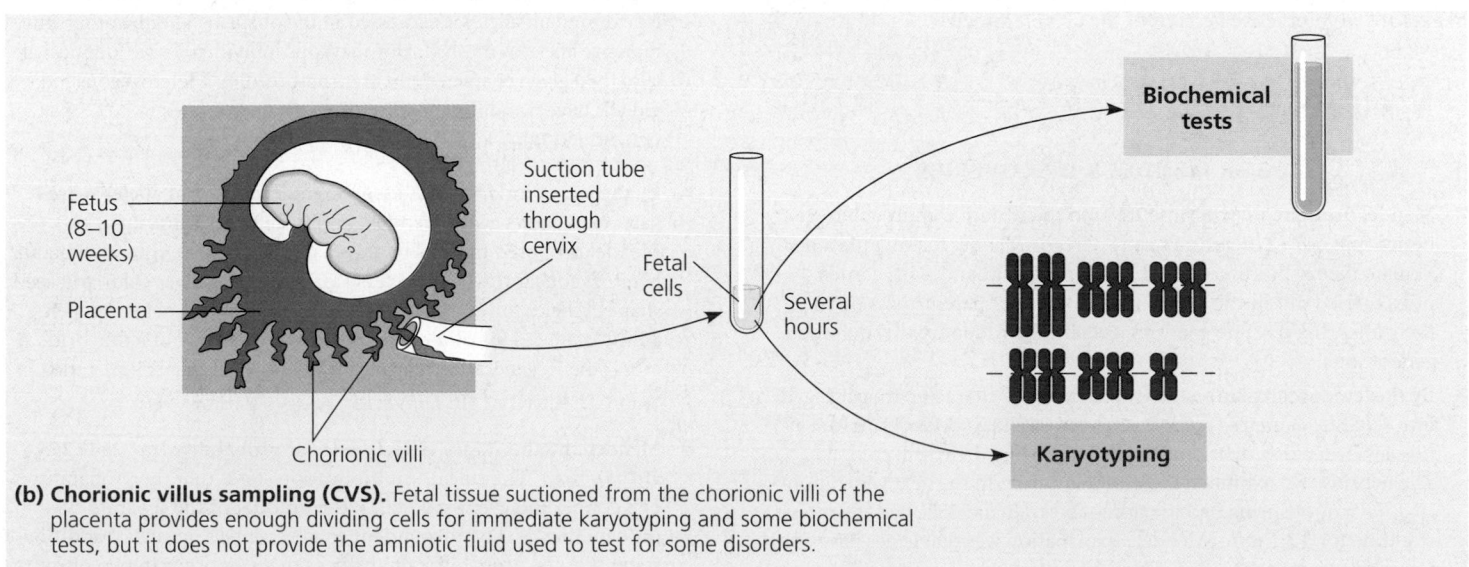

(b) Chorionic villus sampling (CVS). Fetal tissue suctioned from the chorionic villi of the placenta provides enough dividing cells for immediate karyotyping and some biochemical tests, but it does not provide the amniotic fluid used to test for some disorders.

FIGURE 14.17 Testing a fetus for genetic disorders.

have escaped into the mother's blood. Although very few in number, these cells can be cultured and then tested.

Other techniques allow a physician to examine a fetus directly for major anatomical abnormalities. One such technique is *ultrasound,* which uses sound waves to produce an image of the fetus by a simple noninvasive procedure. This procedure, which is also used to locate the fetus during amniocentesis, has no known risk to either fetus or mother. With another technique, *fetoscopy,* a needle-thin tube containing a viewing scope and fiber optics (to transmit light) is inserted into the uterus.

In about 1% of the cases, amniocentesis or fetoscopy causes complications, such as maternal bleeding or fetal death. Thus, these techniques are usually reserved for cases in which the risk of a genetic disorder or other type of birth defect is relatively great. If the fetal tests reveal a serious disorder, the parents face the difficult choice of terminating the pregnancy or preparing to care for a child with a genetic disorder.

Newborn Screening

Some genetic disorders can be detected at birth by simple tests that are now routinely performed in most hospitals in the United States. One screening program is for the recessively inherited disorder called phenylketonuria (PKU), which occurs in about one out of every 10,000 to 15,000 births in the United

States. Children with this disease cannot properly break down the amino acid phenylalanine. This compound and its by-product, phenylpyruvate, can accumulate to toxic levels in the blood, causing mental retardation. However, if the deficiency is detected in the newborn, a special diet low in phenylalanine can usually promote normal development and prevent retardation. Thus, screening newborns for PKU and other treatable disorders can be vitally important. Unfortunately, very few genetic disorders are treatable at the present time.

■   ■   ■

In this chapter, you have learned about the Mendelian model of inheritance and its application to human genetics. We owe the "gene idea," the concept of heritable factors transmitted according to simple rules of chance, to the elegant experiments of Gregor Mendel. Mendel's quantitative approach was foreign to the biology of his era, and even the few biologists who read his papers apparently overlooked the importance of his discoveries. It wasn't until the beginning of the 20th century that Mendelian genetics was rediscovered by biologists studying the role of chromosomes in inheritance. In the next chapter, you will learn how Mendel's laws have their physical basis in the behavior of chromosomes during sexual life cycles and how the synthesis of Mendelism and a chromosome theory of inheritance catalyzed progress in genetics.

CHAPTER 14 REVIEW

Go to the Campbell Biology website (www.campbellbiology.com) to explore an interactive version of the Chapter Review.

Summary of Key Concepts

GREGOR MENDEL'S DISCOVERIES

■ **Mendel brought an experimental and quantitative approach to genetics (pp. 247–249, FIGURE 14.1)** Gregor Mendel formulated a particulate theory of inheritance based on experiments with garden peas, carried out in the 1860s. He showed that parents pass on to their offspring discrete genes that retain their identity through the generations.

■ **By the law of segregation, the two alleles for a character are packaged into separate gametes (pp. 249–252, FIGURE 14.4)** Mendel arrived at this law by making hybrid offspring and letting them self-pollinate. The hybrids (F_1) exhibited the dominant trait. In the next generation (F_2), 75% of offspring had the dominant trait and 25% had the recessive trait, for a 3:1 ratio. Mendel's explanation was that genes have alternative forms (now called alleles) and that each organism inherits one allele for each gene from each parent. These separate (segregate) during gamete formation, so that a sperm or an egg carries only one

allele. After fertilization, if the two alleles of a gene are different, one (the dominant allele) is expressed in the offspring and the other (the recessive allele) is masked. Homozygous individuals have identical alleles for a given character and are true-breeding. Heterozygous individuals have two different alleles for a given character.
Web/CD Activity 14A: *Monohybrid Cross*

■ **By the law of independent assortment, each pair of alleles segregates into gametes independently (pp. 252–254, FIGURE 14.7)** Mendel proposed this law based on dihybrid crosses between plants heterozygous for two characters (for example, flower color and seed shape). Alleles for each character segregate into gametes independently of alleles for other characters. The offspring of a dihybrid cross (the F_2 generation) have four phenotypes in a 9:3:3:1 ratio.
Web/CD Activity 14B: *Dihybrid Cross*

■ **Mendelian inheritance reflects rules of probability (pp. 254–255, FIGURE 14.8)** The rule of multiplication states that the probability of a compound event is equal to the product of the separate probabilities of the independent single events. The rule of addition states that the probability of an event that can occur in two or more independent ways is the sum of the separate probabilities.
Web/CD Activity 14C: *Gregor's Garden*

- **Mendel discovered the particulate behavior of genes: *a review*** (p. 255) Mendel's quantitative analysis of carefully planned experiments exemplifies the process of science.

EXTENDING MENDELIAN GENETICS

- **The relationship between genotype and phenotype is rarely simple** (pp. 255–260, FIGURES 14.9–14.13) In incomplete dominance, a heterozygous individual has a phenotype intermediate between those of the two types of homozygotes. In codominance, a heterozygote exhibits phenotypes for *both* its alleles. Many genes exist in *multiple* (more than two) alleles in a population. Pleiotropy is the ability of a single gene to affect multiple phenotypic traits. In epistasis, one gene affects the expression of another gene. Certain characters are quantitative; they vary continuously, indicating polygenic inheritance, an additive effect of two or more genes on a single phenotypic character. Quantitative characters also influenced by environment are said to be multifactorial.

  Web/CD Activity 14D: *Incomplete Dominance*

MENDELIAN INHERITANCE IN HUMANS

- **Pedigree analysis reveals Mendelian patterns in human inheritance** (p. 260, FIGURE 14.14) Family pedigrees can be used to deduce the possible genotypes of individuals and make predictions about future offspring. Any predictions are usually statistical probabilities rather than certainties.

- **Many human disorders follow Mendelian patterns of inheritance** (pp. 261–264) Certain genetic disorders are inherited as simple recessive traits from phenotypically normal, heterozygous carriers. Some human disorders are due to dominant alleles. Medical researchers are beginning to sort out the genetic and environmental components of multifactorial disorders, such as heart disease and cancer.

  Web/CD Case Study in the Process of Science: *How Do You Diagnose a Genetic Disorder?*

- **Technology is providing new tools for genetic testing and counseling** (pp. 264–266, FIGURE 14.17) Using family histories, genetic counselors help couples determine the odds that their children will have genetic disorders. For certain diseases, tests that identify carriers define the odds more accurately. Once a child is conceived, amniocentesis and chorionic villus sampling can help determine whether a suspected genetic disorder is present.

Genetics Problems

1. A rooster with gray feathers is mated with a hen of the same phenotype. Among their offspring, 15 chicks are gray, 6 are black, and 8 are white. What is the simplest explanation for the inheritance of these colors in chickens? What offspring would you predict from the mating of a gray rooster and a black hen?

2. In some plants, a true-breeding, red-flowered strain gives all pink flowers when crossed with a white-flowered strain: RR (red) $\times$ rr (white) $\longrightarrow$ Rr (pink). If flower position (axial or terminal) is inherited as it is in peas (see TABLE 14.1, p. 250), what will be the ratios of genotypes and phenotypes of the F_1 generation resulting from the following cross: axial-red (true-breeding) $\times$ terminal-white? What will be the ratios in the F_2 generation?

3. Flower position, stem length, and seed shape were three characters that Mendel studied. Each is controlled by an independently assorting gene and has dominant and recessive expression as follows:

Character	Dominant	Recessive
Flower position	Axial (*A*)	Terminal (*a*)
Stem length	Tall (*L*)	Dwarf (*l*)
Seed shape	Round (*R*)	Wrinkled (*r*)

If a plant that is heterozygous for all three characters is allowed to self-fertilize, what proportion of the offspring would you expect to be as follows? (*Note:* Use the rules of probability instead of a huge Punnett square.)
   a. homozygous for the three dominant traits
   b. homozygous for the three recessive traits
   c. heterozygous
   d. homozygous for axial and tall, heterozygous for seed shape

4. A black guinea pig crossed with an albino guinea pig produces 12 black offspring. When the albino is crossed with a second black one, 7 blacks and 5 albinos are obtained. What is the best explanation for this genetic situation? Write genotypes for the parents, gametes, and offspring.

5. In sesame plants, the one-pod condition (*P*) is dominant to the three-pod condition (*p*), and normal leaf (*L*) is dominant to wrinkled leaf (*l*). Pod type and leaf type are inherited independently. Determine the genotypes for the two parents for all possible matings producing the following offspring:
   a. 318 one-pod normal, 98 one-pod wrinkled
   b. 323 three-pod normal, 106 three-pod wrinkled
   c. 401 one-pod normal
   d. 150 one-pod normal, 147 one-pod wrinkled, 51 three-pod normal, 48 three-pod wrinkled
   e. 223 one-pod normal, 72 one-pod wrinkled, 76 three-pod normal, 27 three-pod wrinkled

6. A man with group A blood marries a woman with group B blood. Their child has group O blood. What are the genotypes of these individuals? What other genotypes, and in what frequencies, would you expect in offspring from this marriage?

7. Color pattern in a species of duck is determined by one gene with three alleles. Alleles *H* and *I* are codominant, and allele *i* is recessive to both. How many phenotypes are possible in a flock of ducks that contains all the possible combinations of these three alleles?

8. Phenylketonuria (PKU) is an inherited disease caused by a recessive allele. If a woman and her husband are both carriers, what is the probability of each of the following?
   a. All three of their children will be of normal phenotype.
   b. One or more of the three children will have the disease.
   c. All three children will have the disease.
   d. At least one child will be phenotypically normal.
   (*Note:* Remember that the probabilities of all possible outcomes always add up to 1.)

9. The genotype of F_1 individuals in a tetrahybrid cross is *AaBbCcDd*. Assuming independent assortment of these four genes, what are the probabilities that F_2 offspring will have the following genotypes?
   a. *aabbccdd*
   b. *AaBbCcDd*
   c. *AABBCCDD*
   d. *AaBBccDd*
   e. *AaBBCCdd*

10. What is the probability that each of the following pairs of parents will produce the indicated offspring? (Assume independent assortment of all gene pairs.)
   a. *AABBCC* $\times$ *aabbcc* $\longrightarrow$ *AaBbCc*
   b. *AABbCc* $\times$ *AaBbCc* $\longrightarrow$ *AAbbCC*
   c. *AaBbCc* $\times$ *AaBbCc* $\longrightarrow$ *AaBbCc*
   d. *aaBbCC* $\times$ *AABbcc* $\longrightarrow$ *AaBbCc*

11. Karen and Steve each have a sibling with sickle-cell disease. Neither Karen nor Steve nor any of their parents have the disease, and none of them have been tested to reveal sickle-cell trait. Based on this incomplete information, calculate the probability that if this couple has a child, the child will have sickle-cell disease.

12. In 1981, a stray black cat with unusual rounded, curled-back ears was adopted by a family in California. Hundreds of descendants of the cat have since been born, and cat fanciers hope to develop the "curl" cat into a show breed. Suppose you owned the first curl cat and wanted to develop a true-breeding variety. How would you determine whether the curl allele is dominant or recessive? How would you select for true-breeding cats? How could you be sure they are true-breeding?

13. Imagine that a newly discovered, recessively inherited disease is expressed only in individuals with type O blood, although the disease and blood group are independently inherited. A normal man with type A blood and a normal woman with type B blood have already had one child with the disease. The woman is now pregnant for a second time. What is the probability that the second child will also have the disease? Assume that both parents are heterozygous for the gene that causes the disease.

14. In tigers, a recessive allele causes an absence of fur pigmentation (a "white tiger") and a cross-eyed condition. If two phenotypically normal tigers that are heterozygous at this locus are mated, what percentage of their offspring will be cross-eyed? What percentage will be white?

15. In corn plants, a dominant allele I inhibits kernel color, while the recessive allele i permits color when homozygous. At a different locus, the dominant gene P causes purple kernel color, while the homozygous recessive genotype pp causes red kernels. If plants heterozygous at both loci are crossed, what will be the phenotypic ratio of the F_1 generation?

16. The pedigree below traces the inheritance of alkaptonuria, a biochemical disorder. Affected individuals, indicated here by the filled-in circles and squares, are unable to break down a substance called alkapton, which colors the urine and stains body tissues. Does alkaptonuria appear to be caused by a dominant or recessive allele? Fill in the genotypes of the individuals whose genotypes you know. What genotypes are possible for each of the other individuals?

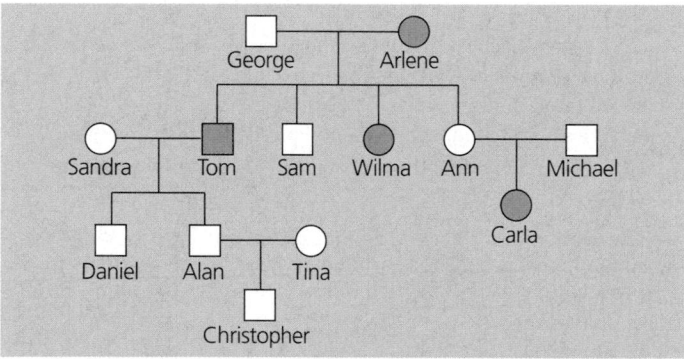

17. A man has six fingers on each hand and six toes on each foot. His wife and their daughter have the normal number of digits. Extra digits is a dominant trait. What fraction of this couple's children would be expected to have extra digits?

18. Imagine that you are a genetic counselor, and a couple planning to start a family come to you for information. Charles was married once before, and he and his first wife had a child with cystic fibrosis. The brother of his current wife Elaine died of cystic fibrosis. What is the probability that Charles and Elaine will have a baby with cystic fibrosis? (Neither Charles nor Elaine has cystic fibrosis.)

19. In mice, black color (B) is dominant to white (b). At a different locus, a dominant allele (A) produces a band of yellow just below the tip of each hair in mice with black fur. This gives a frosted appearance known as agouti. Expression of the recessive allele (a) results in a solid coat color. If mice that are heterozygous at both loci are crossed, what is the expected phenotypic ratio of their offspring?

Go to the website or CD-ROM for multiple-choice quiz questions.

Evolution Connection

Over the past half century, there has been a trend in the United States and other developed countries for people to marry and start families later in life than did their parents and grandparents. Speculate on the effects this trend may have on the incidence (frequency) of late-acting dominant lethal alleles in the population.

The Process of Science

You are handed a "mystery" pea plant with long stems and axial flowers, and asked to determine its genotype as quickly as possible. You know the allele for long stems (L) is dominant to that for dwarf stems (l) and that the allele for axial flowers (A) is dominant to that for terminal flowers (a).

a. What are *all* the possible genotypes for your mystery plant?

b. Describe the *one* cross you would do, out in your garden, to determine the exact genotype of your mystery plant.

c. While you are waiting for the results of your cross, you go back inside and sit at your desk. You make a separate prediction of results for each possible genotype listed in part a. How do you do this?

d. Make these predictions, using the following format: "If the genotype of my mystery plant is _____, the plants resulting from my cross will be _____."

e. If ½ of your offspring plants have long stems with axial flowers and ½ have long stems with terminal flowers, what must be the genotype of your mystery plant?

f. Explain why the activities you performed in parts c and d were not "doing a cross."

Act as a research physician and medical sleuth to investigate a patient's condition in the Case Study in the Process of Science, available on the website and CD-ROM.

Science, Technology, and Society

Imagine that one of your parents had Huntington's disease. What is the probability that you, too, will someday manifest the disease? There is no cure for Huntington's. Would you want to be tested for the Huntington's allele? Why or why not?

Answers: 1. Incomplete dominance, with heterozygotes being gray in color. Mating a gray rooster with a black hen should yield approximately equal numbers of gray and black offspring. **2.** Parental cross is $AARR \times aarr$. Genotype of F_1 is $AaRr$, phenotype is all axial-pink. Genotypes of F_2 are 4 $AaRr$: 2 $AaRR$: 2 $AARr$: 2 $Aarr$: 2 $aaRr$: 1 $AARR$: 1 $aaRR$: 1 $AArr$: 1 $aarr$; phenotypes are 6 axial-pink : 3 axial-red : 3 axial-white : 2 terminal-pink : 1 terminal-white : 1 terminal-red. **3. a.** ¼₆₄ **b.** ¼₆₄ **c.** ⅛ **d.** ¹⁄₃₂ **4.** Albino is a recessive trait; black is dominant. First cross: parents $BB \times bb$; gametes B and b; offspring all Bb. Second cross: parents $bb \times Bb$; gametes b and ½ B, ½ b; offspring ½ Bb, ½ bb. **5. a.** $PPll \times PPLL$, $PPLl$, or $ppLl$ **b.** $ppLl \times ppLl$ **c.** $PPLL$ × any of the 9 possible genotypes or $PPll \times ppLL$ **d.** $PpLl \times PpLl$ **e.** $PpLl \times PpLl$ **6.** Man $I^A i$; woman $I^B i$; child ii. Other genotypes for children are ¼ $I^A I^B$, ¼ $I^A i$, ¼ $I^B i$. **7.** Four **8. a.** ¾ × ¾ × ¾ = ²⁷⁄₆₄ **b.** 1 − ²⁷⁄₆₄ = ³⁷⁄₆₄ **c.** ¼ × ¼ × ¼ = ¹⁄₆₄ **d.** 1 − ¹⁄₆₄ = ⁶³⁄₆₄ **9. a.** ¹⁄₂₅₆ **b.** ¹⁄₁₆ **c.** ¹⁄₂₅₆ **d.** ¼₆₄ **e.** ¹⁄₁₂₈ **10. a.** 1 **b.** ¹⁄₃₂ **c.** ⅛ **d.** ½ **11.** ⅑ **12.** If the "curl" allele is dominant, then the original mutant crossed with noncurl cats will produce both curl and noncurl offspring. If the mutation is recessive, then only curl offspring will result from curl × curl matings. You know that cats are true-breeding when curl × curl matings produce only curl offspring. A pure-bred curl cat is homozygous (as it turns out, for the dominant allele, which causes the curled ears). **13.** ¹⁄₁₆ **14.** Twenty-five percent will be cross-eyed; all of the cross-eyed offspring will also be white. **15.** The dominant allele I is epistatic to the P/p locus, and thus the F_1 generation will be 9 $I\_p\_$ (colorless) : 3 I_pp (colorless) : 3 $iiP\_$ (purple) : 1 $iipp$ (red). Overall, 12 colorless : 3 purple : 1 red. **16.** Recessive; George = Aa, Arlene = aa, Sandra = AA or Aa, Tom = aa, Sam = Aa, Wilma = aa, Ann = Aa, Michael = Aa, Daniel = Aa, Alan = Aa, Tina = AA or Aa, Carla = aa, Christopher = AA or Aa **17.** ½ **18.** ⅙ **19.** 9 $B\_A\_$ (agouti) : 3 B_aa (black) : 3 $bbA\_$ (white) : 1 $bbaa$ (white). Overall, 9 agouti : 3 black : 4 white.

CHAPTER 15

THE CHROMOSOMAL BASIS OF INHERITANCE

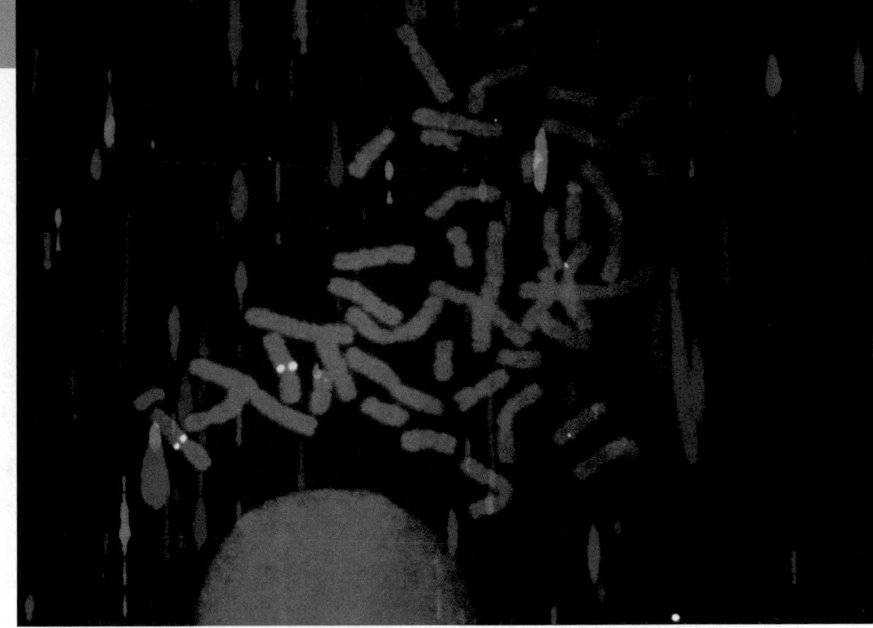

Reprinted with permission from Peter Lichter and David Ward, *Science* 247(1990). Copyright ©1990 American Association for the Advancement of Science.

RELATING MENDELISM TO CHROMOSOMES

- Mendelian inheritance has its physical basis in the behavior of chromosomes during sexual life cycles
- Morgan traced a gene to a specific chromosome
- Linked genes tend to be inherited together because they are located on the same chromosome
- Independent assortment of chromosomes and crossing over produce genetic recombinants
- Geneticists can use recombination data to map a chromosome's genetic loci

SEX CHROMOSOMES

- The chromosomal basis of sex varies with the organism
- Sex-linked genes have unique patterns of inheritance

ERRORS AND EXCEPTIONS IN CHROMOSOMAL INHERITANCE

- Alterations of chromosome number or structure cause some genetic disorders
- The phenotypic effects of some mammalian genes depend on whether they were inherited from the mother or the father (imprinting)
- Extranuclear genes exhibit a non-Mendelian pattern of inheritance

I t was not until the year 1900 *that biology finally caught up with Gregor Mendel. At that time, three botanists, working independently on plant-breeding experiments, reproduced Mendel's results. By searching the scientific literature, the German Karl Correns, the Austrian Erich von Tschermak, and the Dutchman Hugo de Vries all found that Mendel had explained the same results 35 years before. During the intervening years, biology had grown more experimental and quantitative and thus more receptive to Mendelism. Nevertheless, many biologists remained incredulous about Mendel's laws of segregation and independent assortment until evidence had mounted that these principles of heredity had a physical basis in the behavior of chromosomes. Mendel's hereditary factors—genes—are located*

on chromosomes. *For example, the yellow dots (a fluorescent dye) in the light micrograph on this page mark the locus of a specific gene on a homologous pair of human chromosomes (which have already replicated, hence the two dots per chromosome). This chapter integrates and extends what you have learned in the past two chapters by describing the chromosomal basis for the transmission of genes from parents to offspring, along with some important exceptions.*

RELATING MENDELISM TO CHROMOSOMES

Mendelian inheritance has its physical basis in the behavior of chromosomes during sexual life cycles

Using improved microscopy techniques, cytologists worked out the process of mitosis in 1875 and meiosis in the 1890s. Then, around 1900, cytology and genetics converged as biologists began to see parallels between the behavior of chromosomes and the behavior of Mendel's factors. For example, chromosomes and genes are both present in pairs in diploid cells, homologous chromosomes separate and alleles segregate during meiosis, and fertilization restores the paired condition for both chromosomes and genes. Around 1902, Walter S. Sutton, Theodor Boveri, and others independently noted these parallels, and a **chromosome theory of inheritance** began to take form. According to this theory, Mendelian genes have specific loci on chromosomes, and it is the chromosomes that undergo segregation and independent assortment (FIGURE 15.1, p. 270).

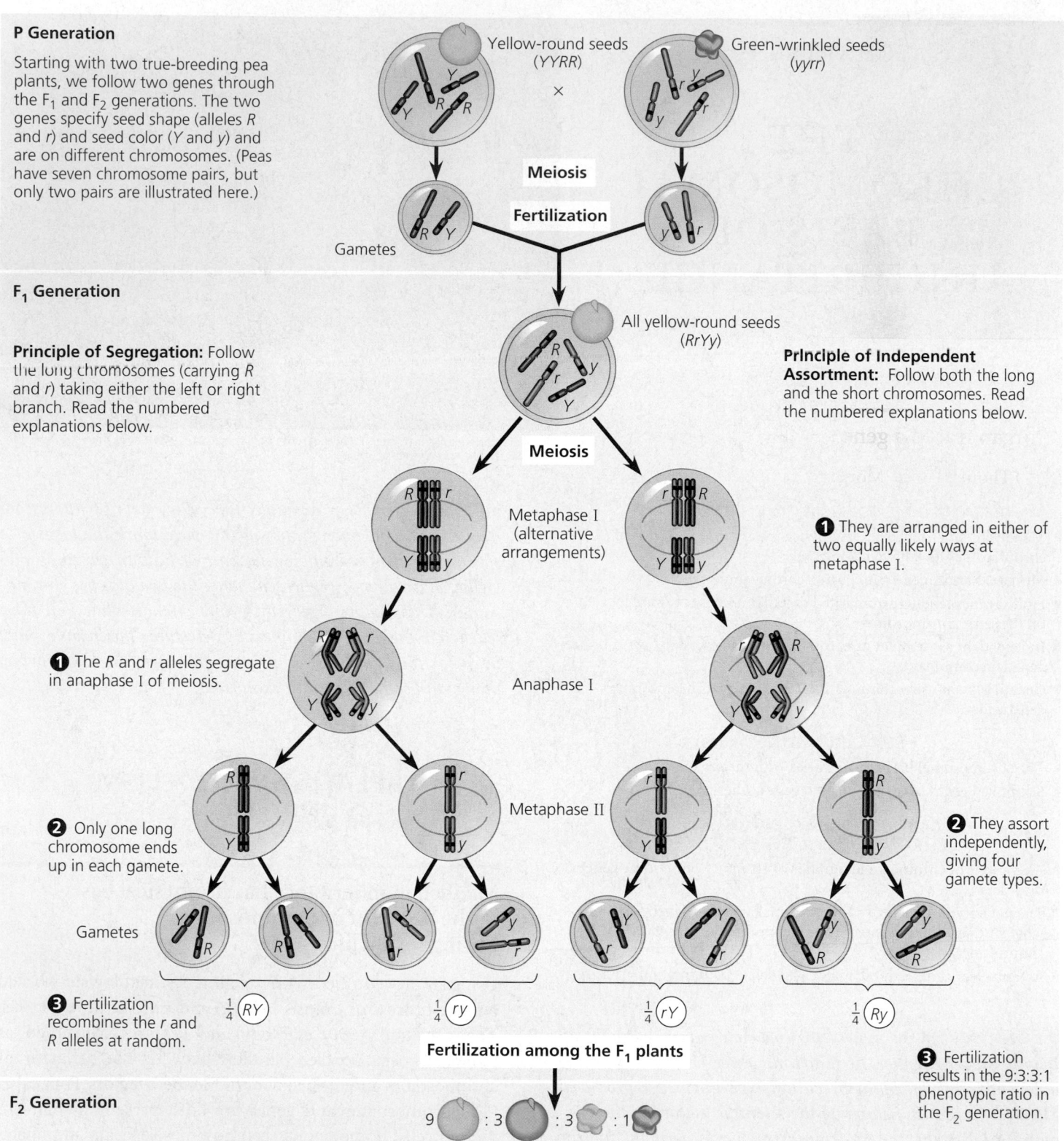

FIGURE 15.1 The chromosomal basis of Mendel's laws. Here we correlate the results of one of Mendel's dihybrid crosses (see FIGURE 14.7b) with the behavior of chromosomes. The arrangement of chromosomes at metaphase I of meiosis and their movement during anaphase I account for the segregation and independent assortment of the alleles for seed color and shape. Because the alternatives at metaphase I are equally likely, F₁ plants produce equal numbers of four kinds of gametes.

FIGURE 15.2 Morgan's first mutant.
Wild-type *Drosophila* flies have red eyes (left). Among his flies, Morgan discovered a mutant male with white eyes (right). This variation made it possible for Morgan to trace a gene for eye color to a specific chromosome (LMs).

THE PROCESS OF SCIENCE

Morgan traced a gene to a specific chromosome

THE PROCESS OF SCIENCE

Thomas Hunt Morgan, an embryologist at Columbia University, was the first to associate a specific gene with a specific chromosome, early in the 20th century. Although Morgan was skeptical about both Mendelism and the chromosome theory, his early experiments provided convincing evidence that chromosomes are indeed the location of Mendel's heritable factors.

Morgan's Choice of Experimental Organism

Many times in the history of biology, important discoveries have come to those insightful enough or lucky enough to choose an experimental organism suitable for the research problem being tackled. Mendel chose the garden pea because a number of distinct varieties were available. For his work, Morgan selected a species of fruit fly, *Drosophila melanogaster*, a common, generally innocuous insect that feeds on the fungi growing on fruit. (Some other kinds of fruit flies, such as "medflies," *Ceratitis capitata*, can seriously damage crops.) Fruit flies are prolific breeders; a single mating will produce hundreds of offspring, and a new generation can be bred every two weeks. These characteristics make the fruit fly a convenient organism for genetic studies. Morgan's laboratory soon became known as "the fly room."

Another advantage of the fruit fly is that it has only four pairs of chromosomes, which are easily distinguishable with a light microscope. There are three pairs of autosomes and one pair of sex chromosomes. Female fruit flies have a homologous pair of X chromosomes, and males have one X chromosome and one Y chromosome.

While Mendel could readily obtain different pea varieties, there were no convenient suppliers of fruit fly varieties for Morgan to employ. Indeed, he was probably the first person to want different varieties of this common insect. After a year of breeding flies and looking for variant individuals, Morgan was rewarded with the discovery of a single male fly with white eyes instead of the usual red. The normal phenotype for a character (the phenotype most common in natural populations), such as red eyes in *Drosophila*, is called the **wild type** (FIGURE 15.2). Traits that are alternatives to the wild type, such as white eyes in *Drosophila*, are called *mutant phenotypes*, because they are due to alleles assumed to have originated as changes, or mutations, in the wild-type allele.

Discovery of Sex Linkage

After Morgan discovered his white-eyed male fly, he mated it with a red-eyed female. All the F_1 offspring had red eyes, suggesting that the wild type was dominant. When Morgan bred the F_1 flies to each other, he observed the classical 3:1 phenotypic ratio among the F_2 offspring. However, there was a surprising additional result: The white-eye trait showed up only in males. All the F_2 females had red eyes, while half the males had red eyes and half had white eyes. Somehow, a fly's eye color was linked to its sex.

From this and other evidence, Morgan deduced that the gene affected in his white-eyed mutant is located exclusively on the X chromosome; there is no corresponding eye-color locus on the Y chromosome (FIGURE 15.3, p. 272). Thus, females (XX) carry two copies of the gene for this character, while males (XY) carry only one. Because the mutant allele is recessive, a female will have white eyes only if she receives that allele on both X chromosomes—an impossibility for the F_2 females in Morgan's experiment. For a male, on the other hand, a single copy of the mutant allele confers white eyes; since a male has only one X chromosome, there can be no wild-type allele present to offset the recessive allele.

Genes located on a sex chromosome are called **sex-linked genes.** Morgan's evidence that a specific gene is carried on the X chromosome added credibility to the chromosome theory of inheritance. Recognizing the importance of this work, many bright students were attracted to Morgan's fly room, and his laboratory dominated genetics research for the next three decades. We will see the influence of Morgan and his colleagues as we consider some other important aspects of the chromosomal basis of inheritance.

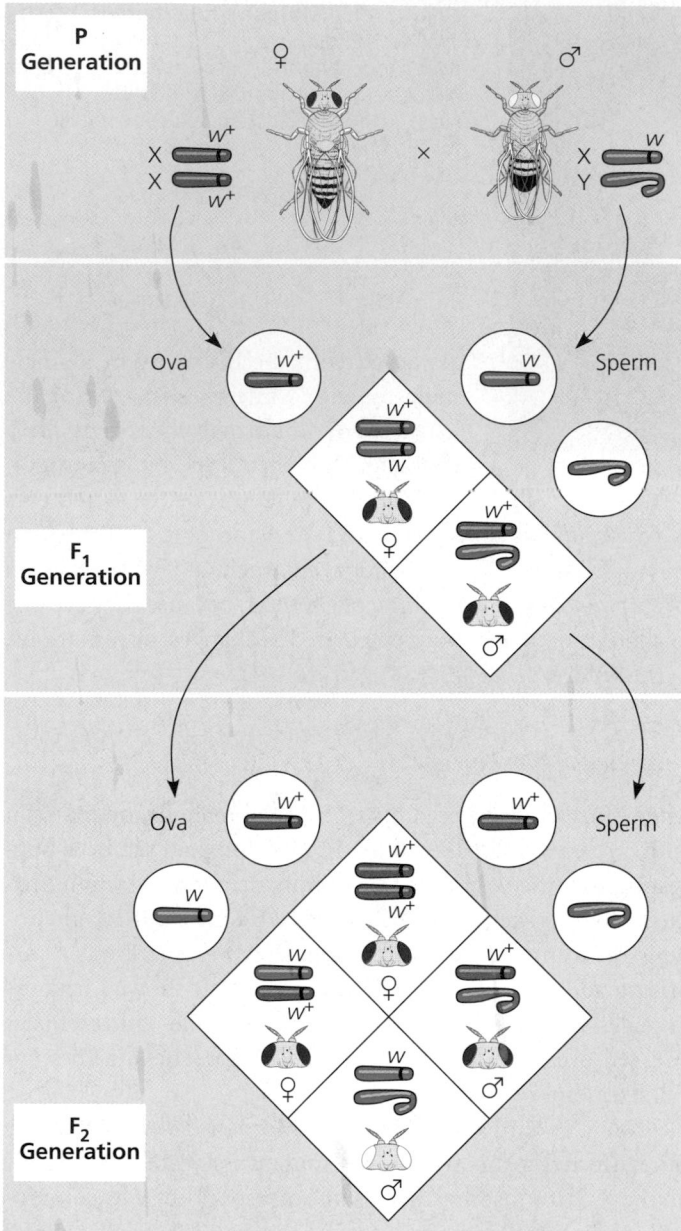

FIGURE 15.3 Sex-linked inheritance. When Morgan bred his mutant male to a wild-type female, all F_1 offspring had red eyes. The F_2 generation showed a typical Mendelian 3:1 ratio of traits, but the recessive trait—white eyes—was linked to sex. All females had red eyes, but half the males had white eyes. Morgan hypothesized that the gene responsible was located on the X chromosome and that there was no corresponding locus on the Y chromosome.

Linked genes tend to be inherited together because they are located on the same chromosome

The number of genes in a cell is far greater than the number of chromosomes; in fact, each chromosome has hundreds or thousands of genes. Genes located on the same chromosome tend to be inherited together in genetic crosses because the chromosome is passed along as a unit. Such genes are said to be **linked genes.** (Note that these genes are linked to each

other. Here the use of the word *linked* is different from its use in the term *sex-linked gene,* where the gene is on a sex chromosome.) When geneticists follow linked genes in breeding experiments, the results deviate from those expected according to the Mendelian principle of independent assortment.

Let's examine another of Morgan's *Drosophila* experiments to see how linkage between genes affects the inheritance of two different characters (FIGURE 15.4). In this case, the characters are body color and wing size. Wild-type flies have gray bodies and normal wings. Morgan had flies with mutant versions of these characters: black body and vestigial wings (much smaller than normal wings). The alleles for these traits are represented by the following symbols: b^+ = gray, b = black; vg^+ = normal wings, vg = vestigial wings.* The mutant alleles are recessive to the wild-type alleles, and neither gene is sex-linked.

Morgan mated true-breeding wild-type flies $(b^+ b^+ vg^+ vg^+)$ with black, vestigial-winged ones $(b b vg vg)$ to produce F_1 dihybrids $(b^+ b vg^+ vg)$, which are wild type in appearance. He then crossed female dihybrids with males of the "double-mutant" phenotype $(b b vg vg)$. Notice that this cross corresponds to a Mendelian testcross. According to Mendel's law of independent assortment, Morgan's *Drosophila* testcross should produce four phenotypic classes of offspring that are about equal in number: 1 gray-normal : 1 black-vestigial : 1 gray-vestigial : 1 black-normal. (To see why, rework FIGURE 14.6 with a dihybrid.) But the actual results were different: There were disproportionately large numbers of wild-type (gray-normal) and double-mutant (black-vestigial) flies among the offspring. These two phenotypes correspond to those of the P generation parents, as well as to those of the parents in the testcross. Morgan reasoned that body color and wing shape are usually inherited together in specific combinations because the genes for these characters are on the same chromosome:

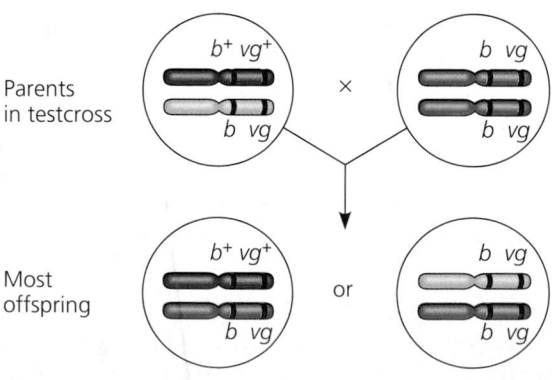

---

* **A note on genetic symbols:** Morgan and his students invented a convention for symbolizing alleles that is now widely used. For a given character, the gene takes its symbol from the first mutant (non-wild type) discovered. For example, the allele for white eyes in *Drosophila* is symbolized by *w.* A superscript + identifies the allele for the wild-type trait—for red eyes, w^+. Gene nomenclature differs somewhat with the type of organism. For example, the symbols for human genes are usually written in all capitals, as in *HD*, the Huntington's disease gene. Depending on the context, a symbol like *HD* or *w* can refer specifically to a mutant allele or to the gene in general.

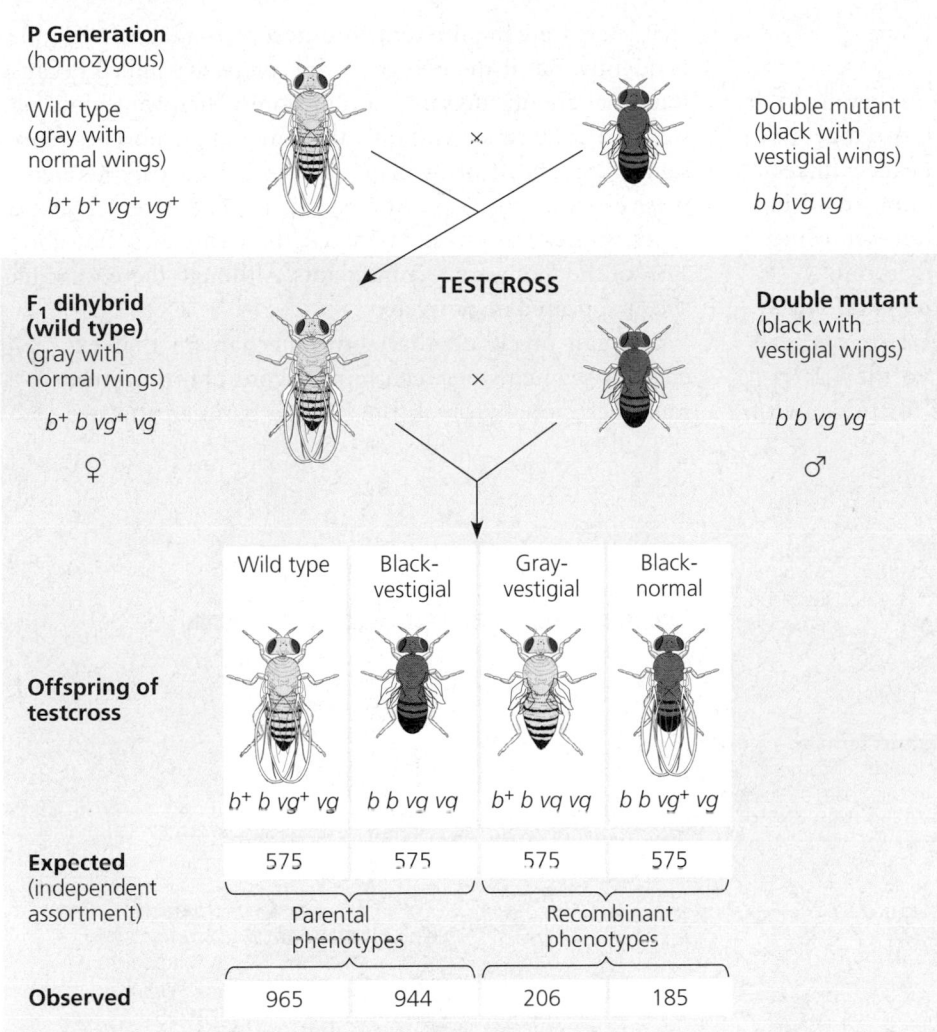

P Generation
(homozygous)

Wild type
(gray with
normal wings)

$b^+ b^+ vg^+ vg^+$

×

Double mutant
(black with
vestigial wings)

$b\ b\ vg\ vg$

TESTCROSS

**F₁ dihybrid
(wild type)**
(gray with
normal wings)

$b^+ b\ vg^+ vg$

♀

×

Double mutant
(black with
vestigial wings)

$b\ b\ vg\ vg$

♂

Offspring of
testcross

Wild type	Black-vestigial	Gray-vestigial	Black-normal
$b^+ b\ vg^+ vg$	$b\ b\ vg\ vg$	$b^+ b\ vg\ vg$	$b\ b\ vg^+ vg$

Expected
(independent
assortment)

575	575	575	575

Parental phenotypes | Recombinant phenotypes

Observed

965	944	206	185

THE
PROCESS
OF SCIENCE

FIGURE 15.4 Evidence for linked genes in *Drosophila*. The tinted area highlights a testcross between a dihybrid female fly, heterozygous for both body color and wing size, and a double-recessive male, homozygous for both recessive alleles. (The male has a "double mutant" phenotype.) When Morgan "scored" (classified according to phenotype) 2,300 offspring from such matings, he observed a much higher proportion of parental phenotypes than would be expected if the two genes assorted independently. Morgan concluded that these genes are usually transmitted together because they are located on the same chromosome. Crossing over accounts for the recombinant phenotypes (combinations of traits different from the combination in either parent).

Although the other two phenotypes (gray-vestigial and black-normal) numbered fewer than expected based on independent assortment, these phenotypes *were* represented among the offspring of Morgan's cross. These new phenotypic variations resulted from crossing over, discussed next.

Independent assortment of chromosomes and crossing over produce genetic recombinants

In Chapter 13, we saw that meiosis and random fertilization generate genetic variation among offspring of sexually reproducing organisms. The general term for the production of offspring with new combinations of traits inherited from two parents is **genetic recombination.** Here we will examine the chromosomal basis of recombination in more detail.

The Recombination of Unlinked Genes: Independent Assortment of Chromosomes

Mendel learned from his study of dihybrid plants that some offspring have combinations of traits that do not match either parent in the parental (P) generation. To see an example, draw a Punnett square for a testcross between a pea plant with yellow-round seeds that is heterozygous for both seed color and seed shape (*YyRr*, the F₁ in FIGURE 15.1) and a plant with green-wrinkled seeds (homozygous for both recessive alleles, *yyrr*). Your Punnett square will predict ratios of genotypes and phenotypes in the offspring similar to what Mendel actually found because the gene loci for seed color and seed shape are on separate chromosomes. In other words, the genes for seed shape and seed color are unlinked, their alleles assorting independently. Notice in your Punnett square that one-half of the offspring of the testcross should inherit a phenotype that matches one of the parental phenotypes—either yellow-round seeds or wrinkled-green seeds. These offspring are called **parental types** (strictly speaking, "parental" here refers to the phenotypes of the P generation plants, which are the grandparents in this case). But other phenotypes are also represented: One-fourth of the offspring have green-round seeds, and one-fourth have yellow-wrinkled seeds. Because these offspring have new combinations of seed shape and color, they are called **recombinants.** When 50% of all offspring are recombinants, as in the example here, geneticists say that there is a 50% frequency of recombination.

A 50% frequency of recombination is observed for any two genes that are located on different chromosomes. The physical basis of recombination between unlinked genes is the random orientation of homologous chromosomes at metaphase I of meiosis, which leads to the independent assortment of alleles (again, see FIGURE 15.1).

The Recombination of Linked Genes: Crossing Over

Linked genes do not assort independently because they are located on the same chromosomes and tend to move together through meiosis and fertilization. We would not expect linked genes to recombine into assortments of alleles not found in the parents. But in fact, recombination between linked genes *does* occur. To see how, let's return to Morgan's fly room.

How can we explain the results of the *Drosophila* cross illustrated in FIGURE 15.4? The offspring of the testcross for body color and wing shape did not conform to the 1:1:1:1 phenotypic ratio we would expect if the genes for these two characters were on different chromosomes and assorted independently. But if the two genes are *completely* linked because their loci are on the same chromosome, then we should observe a 1:1:0:0 ratio, with only the parental phenotypes represented among offspring. The actual results satisfy neither of these expectations. Most of the offspring had parental phenotypes, suggesting linkage between the two genes, but about 17% of the flies were recombinants. Although there was linkage, it appeared incomplete.

Morgan proposed that some mechanism that exchanges segments between homologous chromosomes must occasionally break the linkage between the two

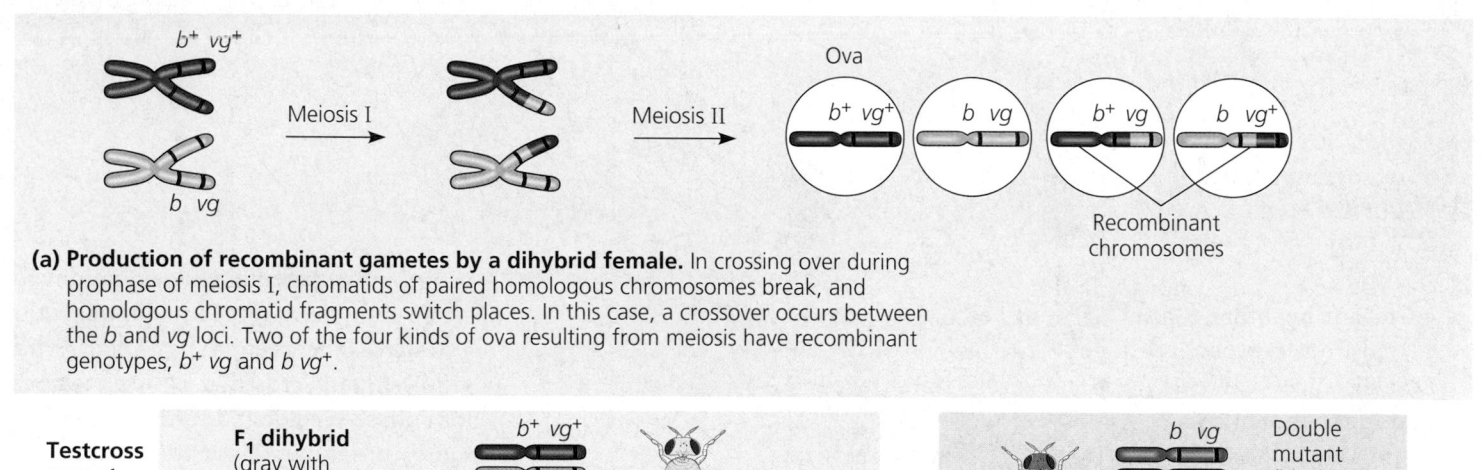

(a) Production of recombinant gametes by a dihybrid female. In crossing over during prophase of meiosis I, chromatids of paired homologous chromosomes break, and homologous chromatid fragments switch places. In this case, a crossover occurs between the *b* and *vg* loci. Two of the four kinds of ova resulting from meiosis have recombinant genotypes, *b⁺ vg* and *b vg⁺*.

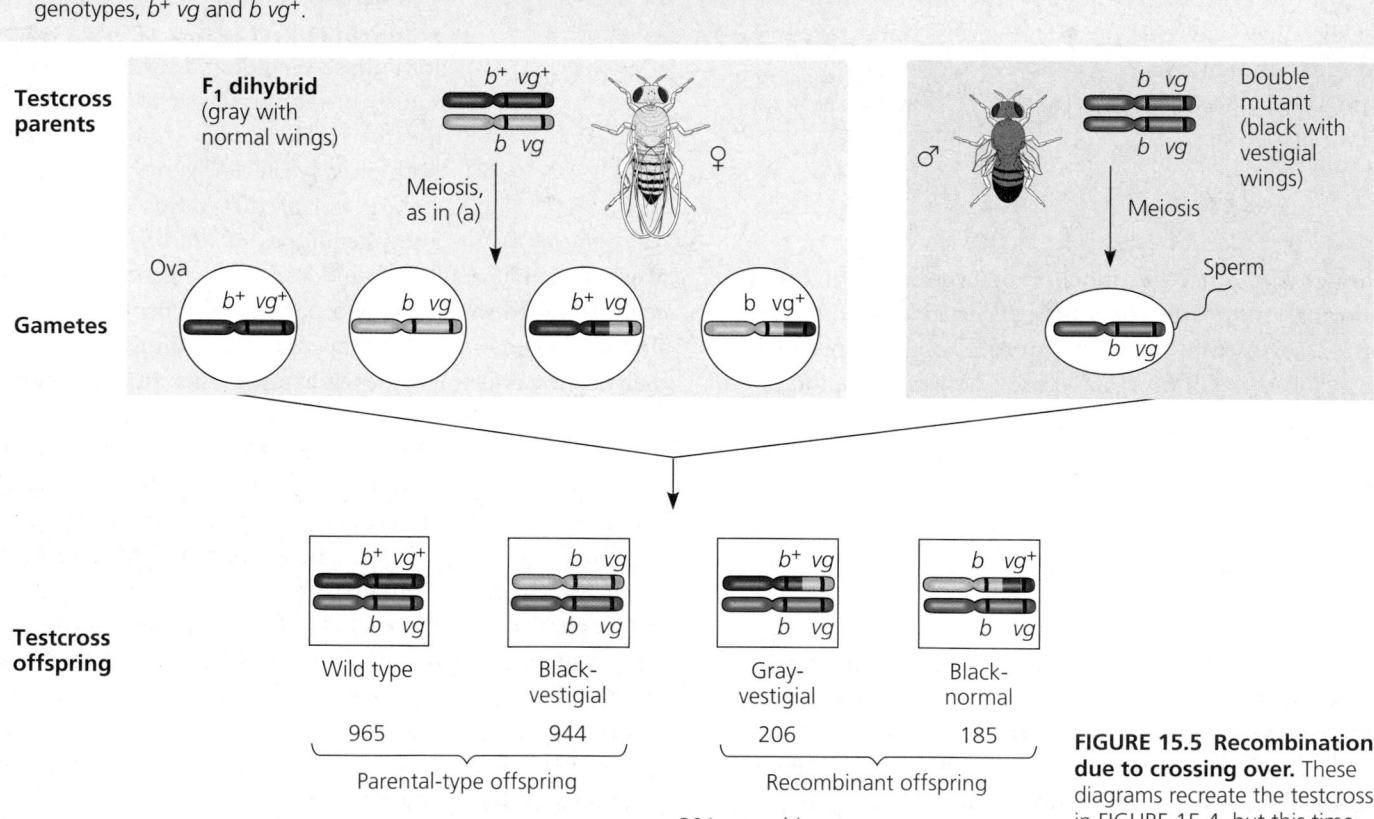

$$\text{Recombination frequency} = \frac{391 \text{ recombinants}}{2,300 \text{ total offspring}} \times 100 = 17\%$$

(b) Production of recombinant offspring. If we follow the recombinant chromosomes through fertilization of the ova by the sperm of genotype *b vg*, we see that they give rise to some recombinant offspring, with genotypes and phenotypes different from either of their parents. The recombinant frequency is the percentage of recombinant flies in the total pool of offspring.

FIGURE 15.5 Recombination due to crossing over. These diagrams recreate the testcross in FIGURE 15.4, but this time we track chromosomes as well as genes. The *b* and *vg* loci are linked; they are on the same chromosome. The maternal chromosomes are color-coded with two shades of red so that we can distinguish one homologue from the other.

genes. Subsequent experiments demonstrated that such an exchange—crossing over—accounts for the recombination of linked genes. While homologous chromosomes are paired during prophase of meiosis I, nonsister chromatids may break at corresponding points and switch fragments; that is, they cross over (see FIGURE 13.9). The recombinant chromosomes resulting from crossing over may bring together alleles in new combinations, and the subsequent events of meiosis distribute the recombinant chromosomes to gametes (FIGURE 15.5a). This sequence of events accounts for the occurrence of recombinant phenotypes in the offspring of Morgan's testcross (FIGURE 15.5b).

Geneticists can use recombination data to map a chromosome's genetic loci

The discovery of linked genes and recombination due to crossing over led one of Morgan's students, Alfred H. Sturtevant, to a method for constructing a **genetic map,** an ordered list of the genetic loci along a particular chromosome.

Sturtevant hypothesized that recombination frequencies calculated from experiments like the one in FIGURE 15.5 reflect the distances between genes on a chromosome. Assuming that the chance of crossing over is approximately equal at all points on a chromosome, he predicted that the farther apart two genes are, the higher the probability that a crossover will occur between them and therefore the higher the recombination frequency. His reasoning was simple: The greater the distance between two genes, the more points there are between them where crossing over can occur. With this idea in mind, Sturtevant began using recombination data from fruit fly crosses to assign to genes relative positions on chromosomes—that is, to *map* genes. A genetic map based on recombination frequencies is specifically called a **linkage map.**

FIGURE 15.6 shows Sturtevant's map of three genes relative to each other, the body-color *(b)* and wing-size *(vg)* genes depicted in FIGURE 15.5 and a third gene on the same chromosome, called cinnabar *(cn)*. Cinnabar is one of many *Drosophila* genes affecting eye color. Cinnabar eyes, a mutant phenotype, are a brighter red than the wild-type color. The recombination frequency between *cn* and *b* is 9%, and the recombination frequency between *cn* and *vg* is 9.5%. In other words, crossovers between *cn* and *b* and between *cn* and *vg* are only about half as frequent as crossovers between *b* and *vg* (17%). Only a map that locates *cn* about midway between *b* and *vg* is consistent with the data (as you can prove to yourself by drawing alternative maps). Sturtevant expressed the distances between genes in **map units,** defining one map unit as equivalent to a 1% recombination frequency. Today, the word *centimorgan* is often used, in honor of Morgan.

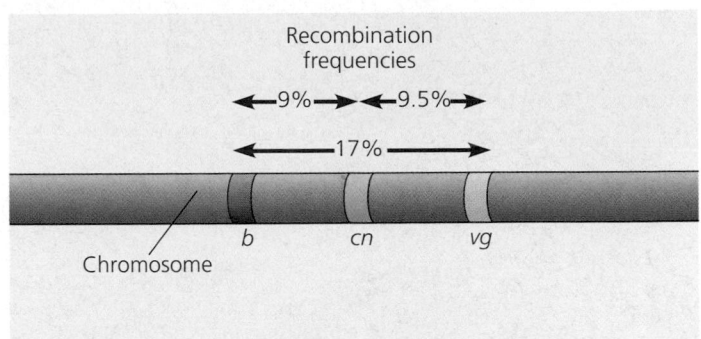

FIGURE 15.6 Using recombination frequencies to construct a genetic map. The method for making this type of map, called a linkage map, is based on the assumption that the probability of a crossover between two genetic loci is proportional to the distance separating the loci. In this example of a linkage map, recombination frequencies have been used to locate three *Drosophila* genes: *b*, *vg*, and *cn*. The data best fit a sequence with *cn* positioned about halfway between the other two genes. The likely reason for the slightly low *b–vg* frequency is given in the text below.

Some genes on a chromosome are so far apart from each other that a crossover between them is virtually certain. The frequency of recombination measured between such genes can have a maximum value of 50%, a result indistinguishable from that for genes on different chromosomes. In fact, the seven characters that Mendel studied in his peas are not all on separate chromosomes, although the pea coincidentally has seven chromosome pairs. Seed color and flower color, for instance, are now known to be on chromosome 1. But they are so far apart on that chromosome that linkage is not observed in genetic crosses. Genes located far apart on a chromosome are mapped by adding the recombination frequencies from crosses involving each of the distant genes and a number of genes lying between them. Only for one pair of the genes Mendel studied, the genes for plant height and pod shape, do modern biologists observe linkage. Although Mendel observed segregation of alleles for each of these characters in monohybrid crosses, he did not report the results of dihybrid crosses for this particular combination of characters.

Even when the distance between genetic loci on a chromosome is not great enough to "unlink" them, it may be great enough to create a significant likelihood of more than one crossover between the loci. A second crossover "cancels out" the first and thus reduces the observed number of recombinant offspring. Multiple crossovers are the main reason for the numerical discrepancy in FIGURE 15.6, where the recombination frequency between *b* and *vg* is less than the sum of the *b–cn* and *cn–vg* frequencies.

Using crossover data, Sturtevant and his colleagues were able to map the other identified *Drosophila* genes in linear arrays. They found that the genes clustered into four groups of linked genes. Because microscopists had found four pairs of chromosomes in *Drosophila* cells, this clustering of genes was additional evidence that genes are located on chromosomes.

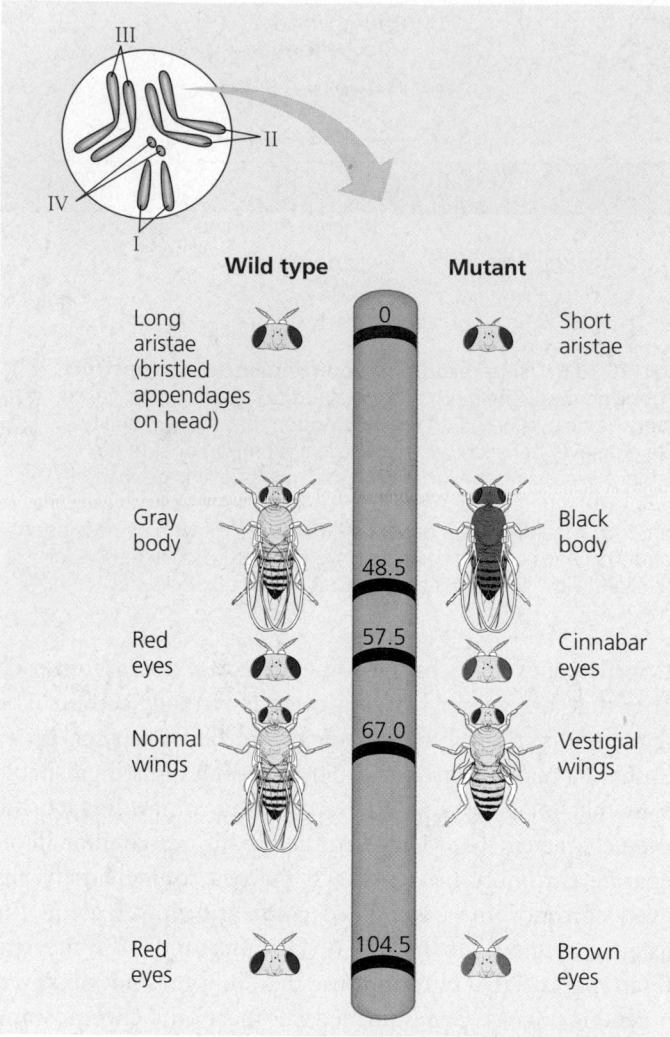

Wild type **Mutant**

Long aristae (bristled appendages on head) — 0 — Short aristae

Gray body — 48.5 — Black body

Red eyes — 57.5 — Cinnabar eyes

Normal wings — 67.0 — Vestigial wings

Red eyes — 104.5 — Brown eyes

FIGURE 15.7 A partial genetic map of a *Drosophila* chromosome. This simplified map shows just a few of the genes that have been mapped on *Drosophila* chromosome II. Notice that more than one gene can affect a given phenotypic characteristic, such as eye color.

Each chromosome has a linear array of specific gene loci (FIGURE 15.7).

Because a linkage map is based on recombination frequencies, it is not really a picture of a chromosome. The frequency of crossing over is not actually uniform over the length of the chromosome, and therefore map units do not have absolute size (in nanometers, for instance). Thus, a linkage map portrays the sequence of genes along a chromosome, but it does not give the precise locations of genes. Other methods enable geneticists to construct **cytological maps** of chromosomes, which locate genes with respect to chromosomal features, such as stained bands, that can be seen in the microscope (see FIGURE 15.14). The ultimate genetic maps, which we'll discuss in Chapter 20, show the distances between gene loci in DNA nucleotides. Comparing a linkage map with this type of map (or even with a cytological map) of the same chromosome, we find that the sequence of genes matches, but the spacing between them does not.

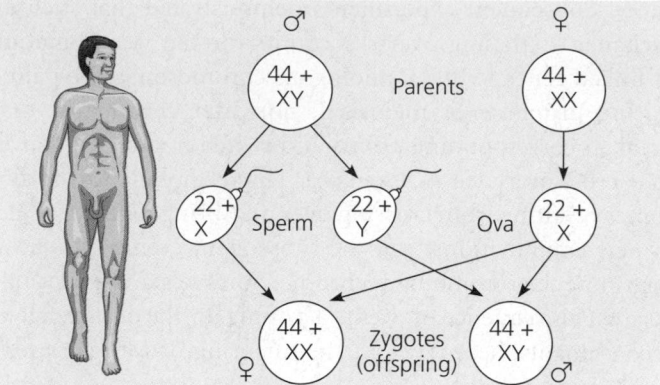

(a) The X-Y system. In mammals, the sex of an offspring depends on whether the sperm cell carried an X chromosome or a Y chromosome.

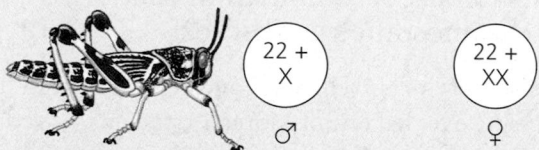

(b) The X-0 system. In grasshoppers, crickets, roaches, and some other insects, there is only one type of sex chromosome, the X. Females are XX; males are X0 (the 0 is a zero; males have only one sex chromosome). Sex of the offspring is determined by whether the sperm cell contains an X chromosome or no sex chromosome.

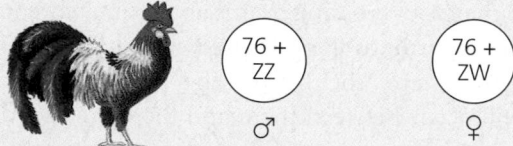

(c) The Z-W system. In birds, some fishes, and some insects, including butterflies and moths, the variable that determines sex is the sex chromosome present in the ovum (not the sperm, as is the case in the X-Y and X-0 systems). The sex chromosomes are designated Z and W to avoid confusions with the X-Y system. Males are ZZ and females are ZW.

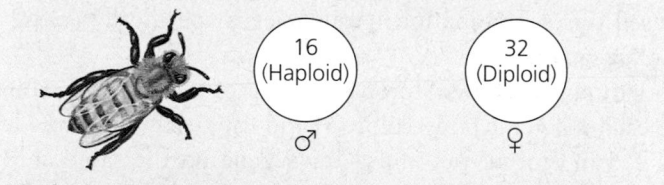

(d) The haplo-diploid system. There are no sex chromosomes in most species of bees and ants. Females develop from fertilized ova and are thus diploid. Males develop from unfertilized eggs and are haploid; they have no fathers.

FIGURE 15.8 Some chromosomal systems of sex determination. Numerals indicate the number of autosomes, the nonsex chromosomes. In *Drosophila*, males are XY, but sex is actually determined by the number of X chromosomes, as in part b.

SEX CHROMOSOMES

You learned earlier that Morgan's discovery of a sex-linked trait (white eyes) was a key episode in the development of a chromosome theory of inheritance. In this section, we consider the role of sex chromosomes in inheritance in more detail. We begin by reviewing the genetics of sex in humans and comparing it with sex determination in some other animals.

The chromosomal basis of sex varies with the organism

Our sex is one of our more obvious phenotypic characters. Although the anatomical and physiological differences between women and men are numerous, the chromosomal basis of sex is rather simple. In humans and other mammals, as in fruit flies, there are two varieties of sex chromosomes, designated X and Y. A person who inherits two X chromosomes, one from each parent, usually develops as a female. A male usually develops from a zygote containing one X chromosome and one Y chromosome (see FIGURE 15.8a and the karyotype in FIGURE 13.3). When meiosis occurs in the testis, the X and Y chromosomes behave like homologous chromosomes, although they are only partially homologous and undergo very little crossing over with each other.

In both testes and ovaries, the two sex chromosomes segregate during meiosis, and each gamete receives one. Each ovum contains one X chromosome. In contrast, sperm fall into two categories: Half the sperm cells a male produces contain an X chromosome, and half contain a Y chromosome. We can trace the sex of each offspring to the moment of conception: If a sperm cell bearing an X chromosome happens to fertilize an ovum, the zygote is XX; if a sperm cell containing a Y chromosome fertilizes an ovum, the zygote is XY. Sex is a matter of chance—a fifty-fifty chance. FIGURE 15.8 compares this X-Y system of mammals to chromosomal systems of sex determination in other animal groups.

In humans, the anatomical signs of sex begin to emerge when the embryo is about two months old. Before then, the rudiments of the gonads are generic—they can develop into either ovaries or testes, depending on hormonal conditions within the embryo. Which of these two possibilities occurs depends on whether or not a Y chromosome is present. In 1990, a British research team identified a gene required for the development of testes. They named the gene *SRY*, for sex-determining region of Y. In the absence of *SRY*, the gonads develop into ovaries. The researchers emphasized that the presence (or absence) of *SRY* is just a trigger. The biochemical, physiological, and anatomical features of sex are complex, and many genes are involved in their development. *SRY* codes for a protein that regulates many other genes. Recently, researchers have identified a number of additional genes on the Y chromosome that are required for normal testis functioning. In the absence of these genes, an XY individual is male but does not produce normal sperm.

Sex-linked genes have unique patterns of inheritance

In addition to their role in determining sex, the sex chromosomes, especially X chromosomes, have genes for many characters unrelated to sex. In humans, the term *sex-linked* usually refers to genes on the X chromosome. These genes follow the same pattern of inheritance that Morgan observed for the white-eye locus in *Drosophila*. Fathers pass sex-linked alleles to all of their daughters but none of their sons (FIGURE 15.9). In contrast, mothers can pass sex-linked alleles to both sons and daughters.

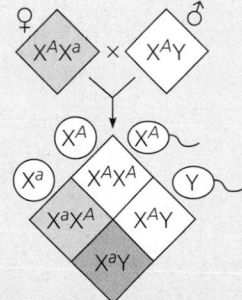

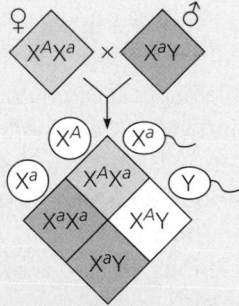

(a) A father with the trait will transmit the mutant allele to all daughters but to no sons. When the mother is a dominant homozygote, the daughters will have the normal phenotype but will be carriers of the mutation.

(b) A carrier who mates with a male of normal phenotype will pass the mutation to half her sons and half her daughters. The sons with the mutation will have the disorder. The daughters who have inherited the mutation in single dose will have the normal phenotype but will be carriers like their mother.

(c) If a carrier mates with a male who has the trait, there is a 50% chance that each child born to them will have the trait, regardless of sex. Daughters who do not have the trait will be carriers, whereas males without the trait will be completely free of the harmful recessive allele.

FIGURE 15.9 The transmission of sex-linked recessive traits. In this diagram, the superscript A represents a dominant allele carried on the X chromosome, and the superscript a represents a recessive allele. Imagine that this recessive allele is a mutation that causes a sex-linked disorder, such as hemophilia. White boxes indicate unaffected individuals, light-colored boxes indicate carriers, and dark-colored boxes indicate individuals with the sex-linked disorder.

If a sex-linked trait is due to a recessive allele, a female will express the phenotype only if she is a homozygote. Because males have only one locus, the terms *homozygous* and *heterozygous* lack meaning for describing their sex-linked genes (the term *hemizygous* is used in such cases). Any male receiving the recessive allele from his mother will express the trait. For this reason, far more males than females have disorders that are inherited as sex-linked recessives. However, even though the chance of a female inheriting a double dose of the mutant allele is much less than the probability of a male inheriting a single dose, there *are* females with sex-linked disorders. For instance, color blindness is a mild disorder inherited as a sex-linked trait. A color-blind daughter may be born to a color-blind father whose mate is a carrier (see FIGURE 15.9c). However, because the sex-linked allele for color blindness is relatively rare, the probability that such a man and woman will mate is low.

Sex-Linked Disorders in Humans

A number of human sex-linked disorders are much more serious than color blindness. An example is **Duchenne muscular dystrophy,** which affects about one out of every 3,500 males born in the United States. People with Duchenne muscular dystrophy rarely live past their early 20s. The disease is characterized by a progressive weakening of the muscles and loss of coordination. Researchers have traced the disorder to the absence of a key muscle protein called dystrophin and have tracked the gene for this protein to a specific locus on the X chromosome. These discoveries may eventually lead to treatments to prevent the disease from progressing.

Hemophilia is a sex-linked recessive trait defined by the absence of one or more of the proteins required for blood clotting. When a person with hemophilia is injured, bleeding is prolonged because a firm clot is slow to form. Small cuts in the skin are usually not a problem, but bleeding in the muscles or joints, which can occur without known cause, can be painful and can lead to serious damage. Today, people with hemophilia are treated as needed with intravenous injections of the missing protein.

Hemophilia has an interesting history. The ancient Hebrews must have had some understanding of its hereditary pattern because sons born to women having a family history of hemophilia were exempted from circumcision. Moreover, a high frequency of sex-linked hemophilia has plagued the royal families of Europe. The first hemophiliac in the royal line seems to have been Leopold, a son of Queen Victoria (1819–1901) of England. The recessive allele for hemophilia was probably introduced to the royal family through a mutation in one of the sex cells of Victoria's mother or father, making Victoria a heterozygote, a carrier of the deadly allele. Leopold survived to father a daughter who was also a carrier, transmitting hemophilia to one of her sons. Hemophilia was eventually brought to the royal families of Prussia, Russia, and Spain through the marriages of two of Victoria's daughters.

X Inactivation in Female Mammals

Although female mammals, including humans, inherit two X chromosomes, one X chromosome in each cell becomes almost completely inactivated during embryonic development. As a result, the cells of females and males have the same effective dose (one copy) of genes with loci on the X chromosome. The inactive X in each cell of a female condenses into a compact object, called a **Barr body,** which lies along the inside of the nuclear envelope. Most of the genes of the X chromosome that forms the Barr body are not expressed, although some remain active. (Barr-body chromosomes are reactivated in the ovary cells that give rise to ova.)

British geneticist Mary Lyon has demonstrated that the selection of which of the two X chromosomes will form the Barr body occurs randomly and independently in each of the embryonic cells present at the time of X inactivation. As a consequence, females consist of a *mosaic* of two types of cells

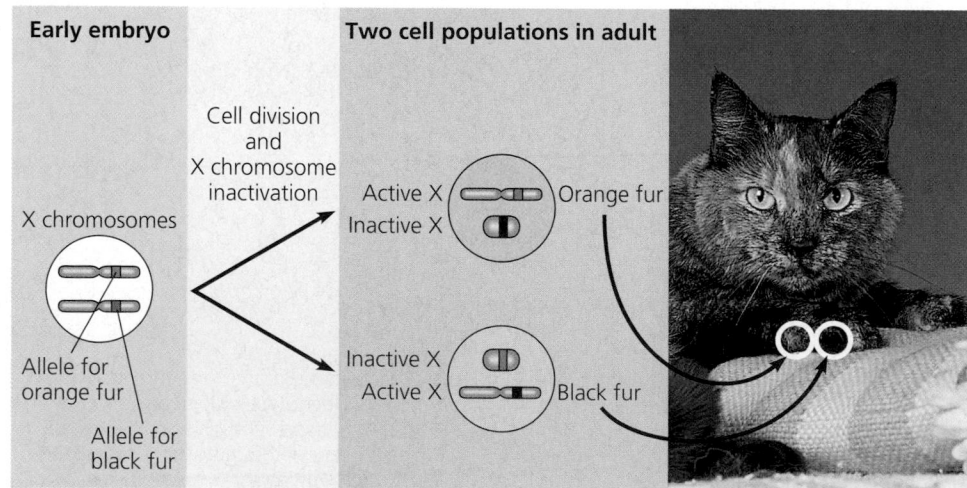

FIGURE 15.10 X inactivation and the tortoiseshell cat. The tortoiseshell gene is on the X chromosome, and the tortoiseshell phenotype requires the presence of two different alleles, one for orange fur and one for nonorange (black) fur. Normally, only females can have both alleles, because only they have two X chromosomes. If a female is heterozygous for the tortoiseshell gene, she is tortoiseshell. Orange patches are formed by populations of cells in which the X chromosome with the orange allele is active; black patches have cells in which the X chromosome with the nonorange allele is active. ("Calico" cats also have white areas, which are determined by yet another gene.)

(FIGURE 15.10): those with the active X derived from the father and those with the active X derived from the mother. After an X chromosome is inactivated in a particular cell, all mitotic descendants of that cell have the same inactive X. Therefore, if the female is heterozygous for a sex-linked trait, approximately half her cells will express one allele, while the others will express the alternate allele. This mosaicism can be seen in the coloration of a tortoiseshell cat, which has patches of orange and black fur (see FIGURE 15.10). In humans, mosaicism can be seen in a recessive X-linked mutation that prevents the development of sweat glands. A woman who is heterozygous for this trait has patches of normal skin and patches of skin lacking sweat glands.

Inactivation of an X chromosome involves attachment of methyl groups ($—CH_3$) to cytosine, one of the nitrogenous bases of DNA nucleotides. (The regulatory role of DNA methylation is discussed in more detail in Chapter 19.) But what determines which of the two X chromosomes is targeted for methylation? Researchers have discovered a gene that is active *only* on the Barr-body chromosome. The gene is called *XIST*, for X-inactive specific transcript. The gene's product, or "specific transcript," is an RNA molecule, and multiple copies of this molecule apparently attach to the X chromosome where they are made, eventually almost covering it. Interaction of this RNA with the chromosome seems to initiate X inactivation. But this finding leads to more questions. What exactly is the connection between *XIST* RNA and DNA methylation? And what determines which of the two X chromosomes in each of a female's cells will have a continually active *XIST* gene and become the Barr body? Our understanding of X inactivation is still rudimentary.

ERRORS AND EXCEPTIONS IN CHROMOSOMAL INHERITANCE

Sex-linked traits are not the only notable deviation from the inheritance patterns observed by Mendel, and the gene mutations that create new alleles are not the only kind of changes to the genome that can affect phenotype. The last section of this chapter discusses major chromosomal aberrations and their consequences, as well as two types of normal inheritance that are exceptions to the standard chromosome theory.

Alterations of chromosome number or structure cause some genetic disorders

Physical and chemical disturbances, as well as errors during meiosis, can damage chromosomes in major ways or alter their number in a cell. Here we survey such chromosomal alterations and see how they relate to some important human disorders.

Alterations of Chromosome Number: Aneuploidy and Polyploidy

Ideally, the meiotic spindle distributes chromosomes to daughter cells without error. But there is an occasional mishap called a **nondisjunction,** in which the members of a pair of homologous chromosomes do not move apart properly during meiosis I or sister chromatids fail to separate during meiosis II. In these cases, one gamete receives two of the same type of chromosome and another gamete receives no copy (FIGURE 15.11). The other chromosomes are usually distributed normally. If either of the aberrant gametes unites with a normal one at fertilization, the offspring will have an abnormal chromosome number, known as **aneuploidy.** If a chromosome is present in triplicate in the fertilized egg (so that the cell has a total of $2n + 1$ chromosomes), the aneuploid cell is said to be **trisomic** for that chromosome. If a chromosome is missing (so that the cell has $2n - 1$ chromosomes), the aneuploid cell is **monosomic** for that chromosome. Mitosis will subsequently transmit the anomaly to all embryonic cells. If the organism survives, it usually has a set of symptoms caused by the abnormal dose of the genes associated with the extra or missing chromosome. Nondisjunction can also occur during mitosis. If such an error takes place

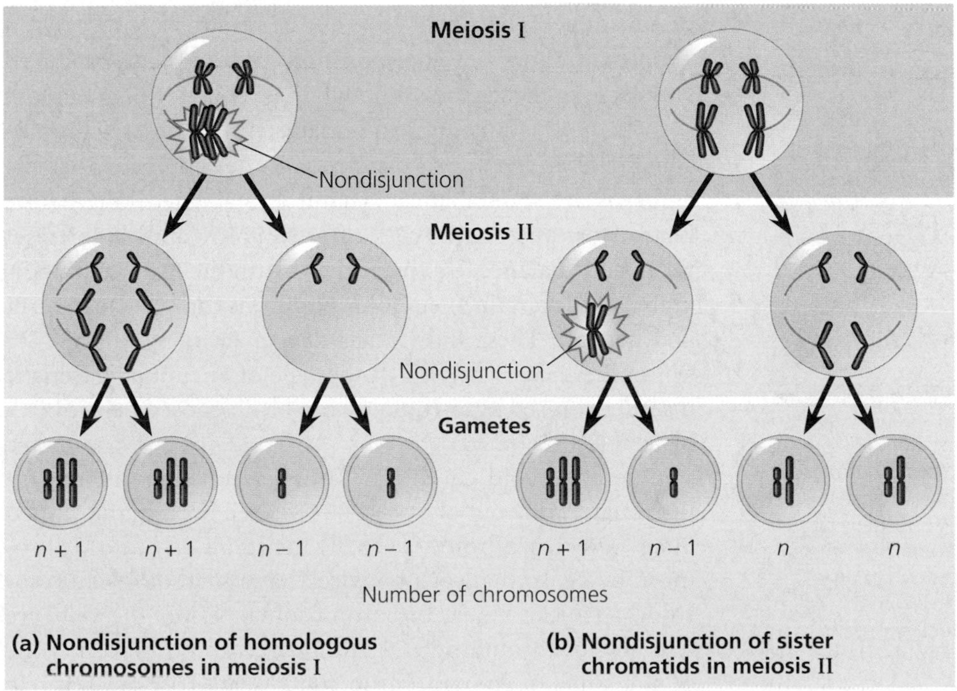

Meiosis I

Nondisjunction

Meiosis II

Nondisjunction

Gametes

$n + 1$ $n + 1$ $n - 1$ $n - 1$ $n + 1$ $n - 1$ n n

Number of chromosomes

(a) Nondisjunction of homologous chromosomes in meiosis I

(b) Nondisjunction of sister chromatids in meiosis II

FIGURE 15.11 Meiotic nondisjunction. Either type of meiotic error will produce gametes with an abnormal chromosome number.

early in embryonic development, then the aneuploid condition is passed along by mitosis to a large number of cells and is likely to have a substantial effect on the organism.

Some organisms have more than two complete chromosome sets. The general term for this chromosomal alteration is **polyploidy,** with the specific terms *triploidy* (3n) and *tetraploidy* (4n) indicating three or four chromosomal sets, respectively. One way a triploid cell may be produced is by the fertilization of an abnormal diploid egg produced by nondisjunction of all its chromosomes. An example of an accident that would result in tetraploidy is the failure of a 2n zygote to divide after replicating its chromosomes. Subsequent mitosis would then produce a 4n embryo.

Polyploidy is relatively common in the plant kingdom, and we will see in Chapter 24 that the spontaneous origin of polyploid individuals plays an important role in the evolution of plants. In the animal kingdom, polyploid species are much less common, although they are known to occur among the fishes and amphibians. Recently, researchers in Chile have identified the first mammalian candidate for polyploidy, a rodent (*Tympanoctomys barrerae*) whose cells seem to be tetraploid (FIGURE 15.12). In general, polyploids are more nearly normal in appearance than aneuploids. One extra (or missing) chromosome apparently disrupts genetic balance more than does an entire extra set of chromosomes.

Alterations of Chromosome Structure

Breakage of a chromosome can lead to four types of changes in chromosome structure. A **deletion** occurs when a chromosomal fragment lacking a centromere is lost during cell division. The chromosome from which the fragment originated will then be missing certain genes. In some cases, if meiosis is in progress, such a fragment may become attached as an extra

FIGURE 15.12 A tetraploid mammal? The somatic cells of this red viscacha rat from Argentina, *Tympanoctomys barrerae*, have about twice as many chromosomes as those of closely related species. (Interestingly, its sperm's head is unusually large, presumably a necessity for holding all that genetic material.) Scientists think that this rat may be a tetraploid species that arose when an ancestor somehow doubled its chromosome number, presumably by errors in mitosis or meiosis within the animal's reproductive organs. Researchers are studying the rat's chromosomes to determine if it actually has four homologous sets.

segment to a sister chromatid, producing a **duplication** in the recipient chromosome. (Alternatively, a detached fragment could produce a duplication by attaching to a homologous chromosome, but in that case, the "duplicated" pieces of chromosome might not be identical.) A chromosomal fragment may also reattach to the original chromosome but in the reverse orientation, producing an **inversion.** A fourth possible result of chromosomal breakage is for the fragment to join a nonhomologous chromosome, a rearrangement called a **translocation.** FIGURE 15.13 illustrates these different types of structural alterations of chromosomes.

Deletions and duplications are especially likely to occur during meiosis. Homologous (nonsister) chromatids sometimes break and rejoin at "incorrect" places, so that one partner gives up more genes than it receives. The products of such a nonreciprocal crossover are one chromosome with a deletion and one chromosome with a duplication.

A diploid embryo that is homozygous for a large deletion (or has a single X chromosome with a large deletion, in a male) is usually missing a number of essential genes, a condition that is ordinarily lethal. Duplications and translocations also tend to have harmful effects. In reciprocal translocations, in which segments are exchanged between nonhomologous chromosomes, and in inversions, the balance of genes is not abnormal—all genes are present in their normal doses. Nevertheless, inversions and translocations can alter phenotype because a gene's expression can be influenced by its location among neighboring genes.

Human Disorders Due to Chromosomal Alterations

Alterations of chromosome number and structure are associated with a number of serious human disorders. When nondisjunction occurs in meiosis, the result is aneuploidy, an abnormal number of chromosomes in the gamete produced and, later, in the zygote. Although the frequency of aneuploid zygotes may be quite high in humans, most of these chromosomal alterations are so disastrous to development that the embryos are spontaneously (naturally) aborted long before birth. However, some types of aneuploidy appear to upset the genetic balance less than others, with the result that individuals with certain aneuploid conditions can survive to birth and beyond. These individuals have a set of symptoms—a syndrome—characteristic of the type of aneuploidy. Genetic disorders caused by aneuploidy can be diagnosed before birth by fetal testing (p. 265).

One aneuploid condition, **Down syndrome,** affects approximately one out of every 700 children born in the United States. Down syndrome is usually the result of an extra chromosome 21, so that each body cell has a total of 47 chromosomes (FIGURE 15.14). In chromosomal terms, the cells are trisomic for chromosome 21. Although chromosome 21 is one of the smallest human chromosomes, its trisomy severely

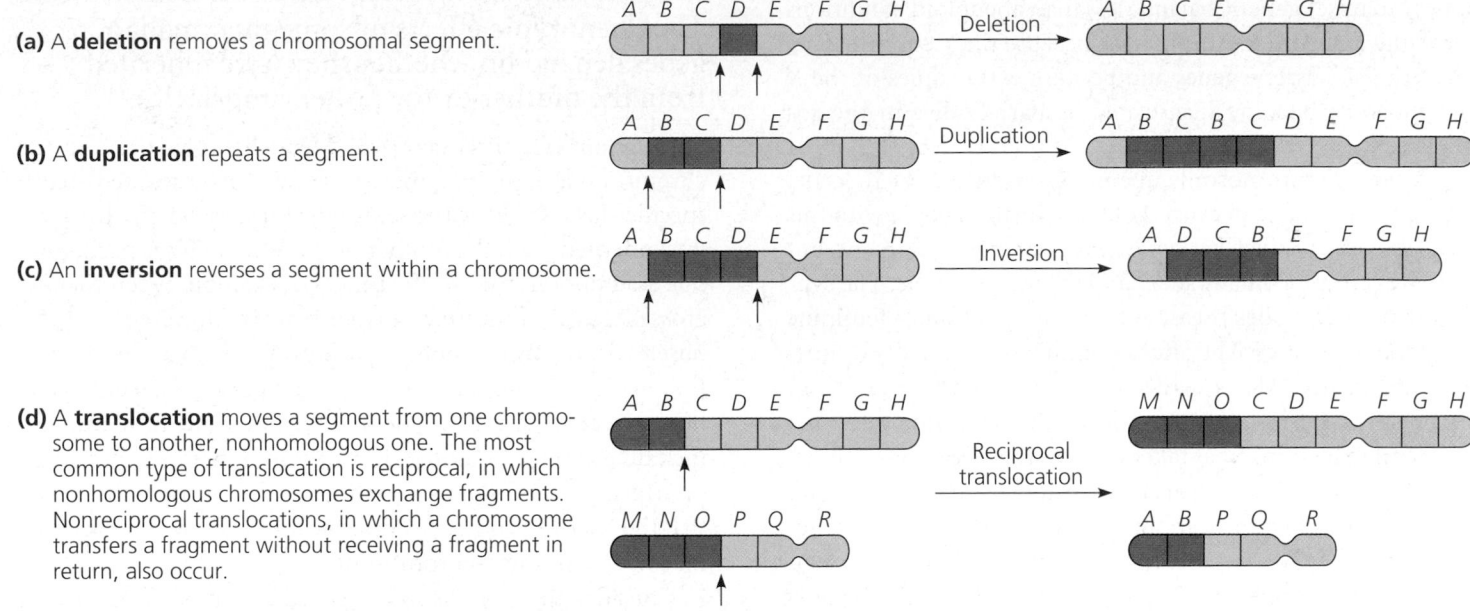

(a) A **deletion** removes a chromosomal segment.

A B C D E F G H → Deletion → A B C E F G H

(b) A **duplication** repeats a segment.

A B C D E F G H → Duplication → A B C B C D E F G H

(c) An **inversion** reverses a segment within a chromosome.

A B C D E F G H → Inversion → A D C B E F G H

(d) A **translocation** moves a segment from one chromosome to another, nonhomologous one. The most common type of translocation is reciprocal, in which nonhomologous chromosomes exchange fragments. Nonreciprocal translocations, in which a chromosome transfers a fragment without receiving a fragment in return, also occur.

A B C D E F G H → Reciprocal translocation → M N O C D E F G H

M N O P Q R → A B P Q R

FIGURE 15.13 Alterations of chromosome structure. Vertical arrows indicate points of chromosome breakage. Dark purple highlights the chromosomal parts affected by the rearrangements.

alters the individual's phenotype. Down syndrome includes characteristic facial features, short stature, heart defects, susceptibility to respiratory infection, and mental retardation. Furthermore, individuals with Down syndrome are prone to developing leukemia and Alzheimer's disease. (It is probably not a coincidence that certain genes associated with the latter two diseases are located on chromosome 21.) Although people with Down syndrome, on average, have a life span shorter than normal, some live to middle age or beyond. Most are sexually underdeveloped and sterile. Most cases of Down syndrome result from nondisjunction during gamete production in one of the parents.

The frequency of Down syndrome correlates with the age of the mother. Down syndrome occurs in 0.04% of children born to women under age 30. The risk climbs to 1.25% for mothers in their early 30s and is even higher for older mothers. Because of this relatively high risk, pregnant women over 35 are candidates for fetal testing to check for trisomy 21 in the embryo. The correlation of Down syndrome with maternal age has not yet been explained. Most cases result from nondisjunction during meiosis I, and recent research suggests the involvement of

some age-dependent abnormality in the spindle checkpoint, which delays anaphase until all the kinetochores are attached to the spindle (like the M-phase checkpoint of the mitotic cell cycle; see Chapter 12). Trisomies of other chromosomes also increase in incidence with maternal age, although infants with these autosomal trisomies rarely survive for long.

Nondisjunction of sex chromosomes produces a variety of aneuploid conditions in humans. Most of these conditions

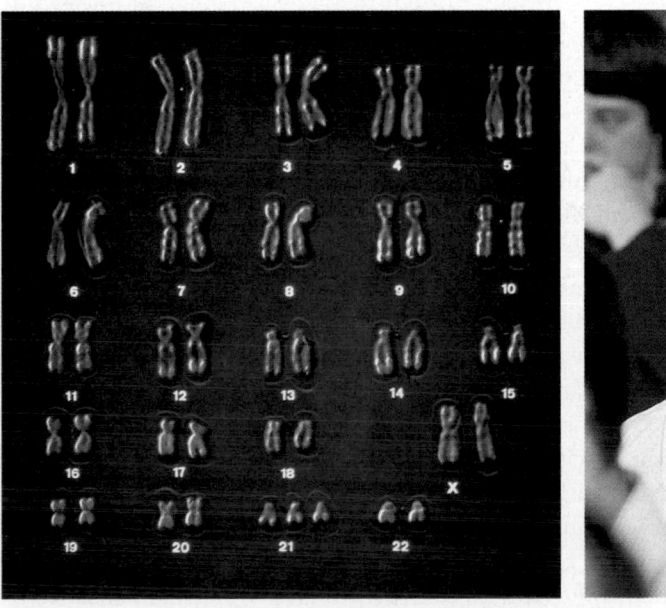

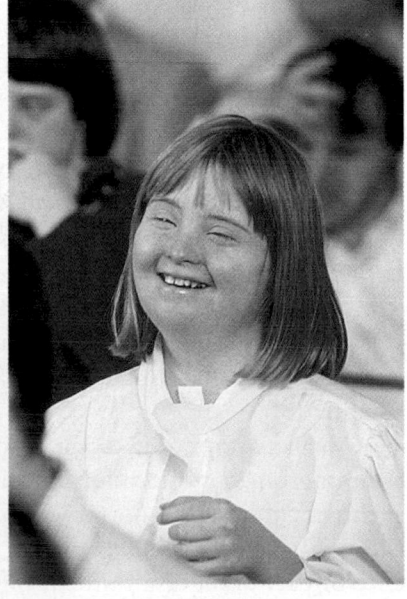

FIGURE 15.14 Down syndrome. The karyotype shows trisomy 21. The child exhibits the facial features characteristic of Down syndrome.

appear to upset genetic balance less than aneuploid conditions involving autosomes. This may be because the Y chromosome carries relatively few genes and because extra copies of the X chromosome become inactivated as Barr bodies in the somatic cells.

An extra X chromosome in a male, producing XXY, occurs approximately once in every 2,000 live births. People with this disorder, called *Klinefelter syndrome,* have male sex organs, but the testes are abnormally small and the man is sterile. The syndrome often includes breast enlargement and other feminine body characteristics. The affected individual is usually of normal intelligence. Males with an extra Y chromosome (XYY) are not characterized by any well-defined syndrome, although they tend to be somewhat taller than average. Females with trisomy X (XXX), which occurs once in approximately 1,000 live births, are healthy and cannot be distinguished from XX females except by karyotype. Monosomy X, called *Turner syndrome,* occurs about once in every 5,000 births and is the only known viable monosomy in humans. Although these X0 individuals are phenotypically female, their sex organs do not mature at adolescence, and they are sterile. However, when provided with estrogen replacement therapy, girls with Turner syndrome do develop secondary sex characteristics. Most have normal intelligence.

Even if chromosome *number* is normal, structural alterations of chromosomes can cause human disorders. Many deletions in human chromosomes (see FIGURE 15.13a), even in a heterozygous state, cause severe physical and mental problems. One such syndrome is known as *cri du chat* ("cry of the cat"). A child born with this specific deletion in chromosome 5 is mentally retarded, has a small head with unusual facial features, and has a cry that sounds like the mewing of a distressed cat. Such individuals usually die in infancy or early childhood.

Another type of chromosomal structural alteration associated with human disorders is chromosomal translocation, the attachment of a fragment from one chromosome to another, nonhomologous chromosome (see FIGURE 15.13d). Chromosomal translocations have been implicated in certain cancers, including *chronic myelogenous leukemia (CML).* Leukemia is a cancer affecting the cells that give rise to white blood cells, and in the cancerous cells of CML patients, a reciprocal translocation has occurred. A portion of chromosome 22 has switched places with a small fragment from a tip of chromosome 9. (How such a switch might cause cancer will be discussed in Chapter 19.)

A small fraction of individuals with Down syndrome have a chromosomal translocation of a different sort. All the cells of such people have the normal number of chromosomes, 46. Close inspection of the karyotype, however, shows the presence of part or all of a third chromosome 21 attached to another chromosome by translocation.

The phenotypic effects of some mammalian genes depend on whether they were inherited from the mother or the father (imprinting)

Throughout our discussions of Mendelian genetics and the chromosomal basis of inheritance, we have assumed that a specific allele will have the same effect regardless of whether it was inherited from the mother or the father. This is probably a safe assumption most of the time. For example, when Mendel crossed purple-flowered peas with white-flowered peas, he observed the same results regardless of whether the purple-flowered parent supplied the ova or pollen. In recent years, however, geneticists have identified some traits in mammals, including certain inherited disorders in humans, that depend on which parent passed along the alleles for those traits. (Note that the issue here is not sex linkage; the genes involved may or may not lie on the X chromosome.)

Consider the two disorders called *Prader-Willi syndrome* and *Angelman syndrome.* The symptoms are different. Prader-Willi syndrome is characterized by mental retardation, obesity, short stature, and unusually small hands and feet. People with Angelman syndrome exhibit spontaneous (uncontrollable) laughter, jerky movements, and other motor and mental symptoms. For both disorders, the genetic cause seems to be the same: deletion of a particular segment of chromosome 15. If a child inherits the abnormal chromosome from the father, the result is Prader-Willi syndrome. If the abnormal chromosome is inherited from the mother, the result is Angelman syndrome. It appears that the genes of the deleted region normally behave differently in offspring, depending on whether they belong to the maternal or the paternal chromosome.

A process called **genomic imprinting** can explain the Prader-Willi/Angelman enigma and some similar phenomena. In this process, a gene on one chromosome is somehow silenced, while its allele on the homologous chromosome is left free to be expressed. In mammals, certain genes are imprinted in some way in each generation, with the imprinting status of a given gene depending on whether the gene resides in a female or in a male (FIGURE 15.15). In other words, the same alleles may have different effects on offspring, depending on whether they arrive in the zygote via the ovum or via the sperm. In the new generation, both maternal and paternal imprints are apparently "erased" in gamete-producing cells, and all the chromosomes are reimprinted according to the sex of the individual in which they now reside.

What exactly is a genomic imprint? In many cases, it seems to consist of methyl ($—CH_3$) groups that are added to cytosine nucleotides of one of the alleles. The hypothesis that this methylation directly silences the allele is consistent with evidence that heavily methylated genes are usually inactive (see p. 363) and suggests that in these cases, the animal uses the allele that is *not* imprinted. But in other

THE PROCESS OF SCIENCE

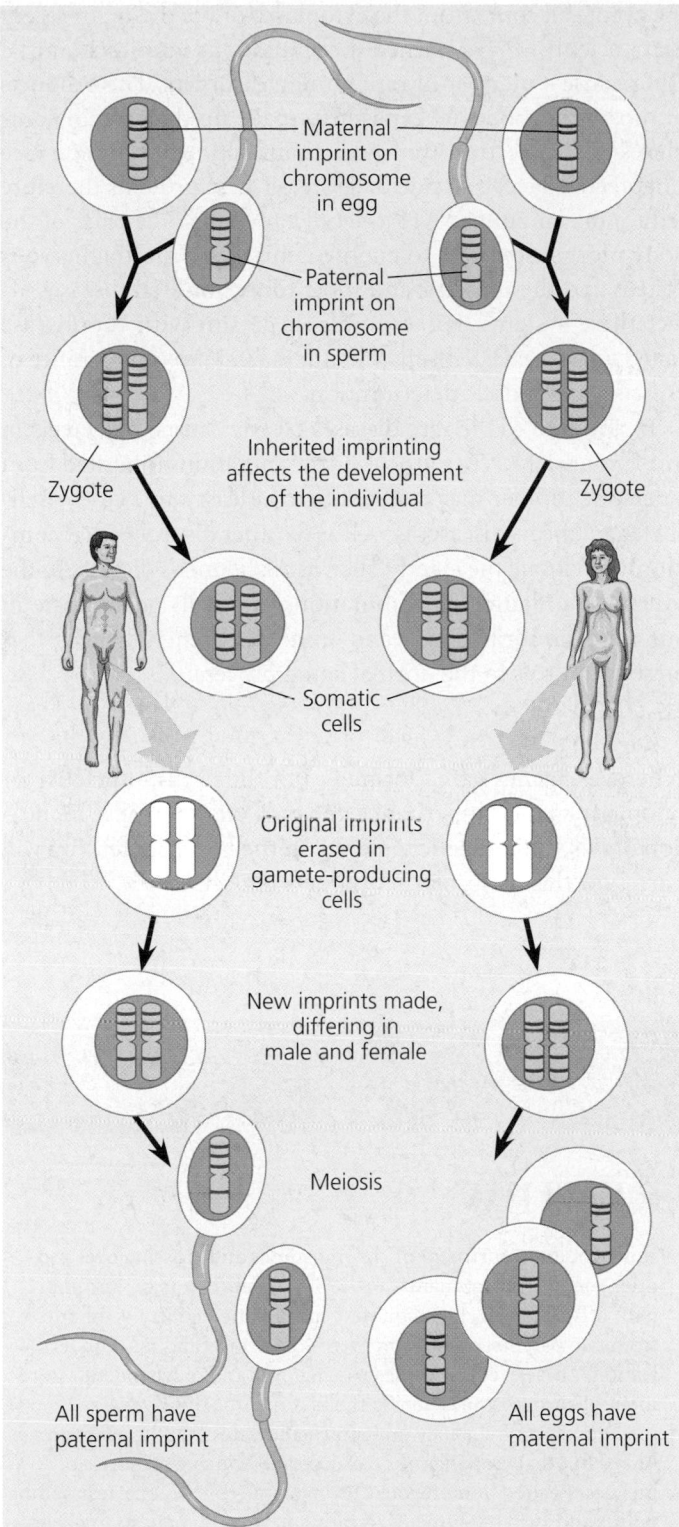

FIGURE 15.15 Genomic imprinting. The sperm and ova of mammals convey chromosomes that are differently imprinted. The phenotypic effect of a particular allele in an offspring may therefore depend on whether the allele came from the mother or the father. In each generation, the old imprints are "erased" when sperm or ova are produced, and all the chromosomes are newly imprinted according to the sex of the individual.

Labels in figure:
Maternal imprint on chromosome in egg
Paternal imprint on chromosome in sperm
Zygote
Inherited imprinting affects the development of the individual
Zygote
Somatic cells
Original imprints erased in gamete-producing cells
New imprints made, differing in male and female
Meiosis
All sperm have paternal imprint
All eggs have maternal imprint

cases of imprinting, the absence of methylation in the vicinity of an allele plays a role in silencing it, whereas methylation does occur near the active allele on the homologous chromosome. Researchers are trying to work out the mechanisms involved.

Researchers have so far identified about 20 mammalian genes subject to imprinting, and there may be a couple of hundred more. Most of the known imprinted genes are critical for embryonic development. In experiments with mice, embryos engineered to inherit both copies of certain chromosomes from the same parent inevitably die before birth, whether that parent is male or female. Normal development apparently requires that certain genes have exactly one active copy—not zero, not two.

In humans, in addition to the Prader-Willi/Angelman case, genomic imprinting may help explain the inheritance pattern of other disorders, including one called **fragile X syndrome.** This disorder is named for the physical appearance of an abnormal X chromosome, the tip of which hangs on to the rest of the chromosome by a thin thread of DNA. Children with fragile X syndrome—about one in every 1,500 males and one in every 2,500 females—are mentally retarded. Of all forms of mental retardation with a genetic basis, fragile X is the most common.

Inheritance of fragile X has a complex pattern, but the syndrome is more common when the abnormal chromosome is inherited from the mother rather than the father. This is consistent with the disorder being more common in males: If a male (XY) inherits a fragile X chromosome, it *has* to be from his mother. The fragile X case is one in which imprinting (methylation) of the abnormal allele by the mother somehow causes (rather than "silences") the syndrome. An unusual molecular characteristic of this abnormal allele will be described in Chapter 19.

Extranuclear genes exhibit a non-Mendelian pattern of inheritance

Although our focus in this chapter has been on the chromosomal basis of inheritance, we end with an important amendment: Not all of a eukaryotic cell's genes are located on nuclear chromosomes, or even in the nucleus. Extranuclear genes are found on small circles of DNA in mitochondria and in plants' plastids, including chloroplasts. Both mitochondria and plastids reproduce themselves and transmit their genes to daughter organelles. These cytoplasmic genes do not display Mendelian inheritance because they are not distributed to offspring according to the same rules that direct the distribution of nuclear chromosomes during meiosis.

Cytoplasmic genes were first observed in plants. In 1909, Karl Correns studied the inheritance of yellow or white patches on the leaves of an otherwise green plant.

FIGURE 15.16 Cytoplasmic inheritance in tomato leaves. Variegated (striped or spotted) leaves result from genes located in the plastids rather than on the nuclear chromosomes of plant cells. Because only the ovum contributes plastids to a plant zygote, all plastid genes are inherited from the maternal parent.

He found that the coloration of the offspring was determined only by the maternal parent (the source of seeds that germinate to give rise to the offspring) and not by the paternal parent (the pollen source). Subsequent research has shown that such coloration patterns are due to genes in the plastids that control pigmentation. In most plants, a zygote receives all its plastids from the cytoplasm of the ovum and none from pollen. Thus, as the zygote of these plants develops, its pattern of leaf coloration depends only on maternal plastid genes (FIGURE 15.16).

Maternal inheritance is also the rule for the mitochondrial genes in mammals, because the mitochondria passed on by the zygote all come from the cytoplasm of the ovum. In recent years, scientists have learned that mutations in mitochondrial DNA cause a number of rare human disorders. The products of most mitochondrial genes help make up the protein complexes of the electron transport chain and ATP synthase (see Chapter 9). Defects in one or more of these proteins therefore reduce the amount of ATP the cell can make. The parts of the body most susceptible to energy deprivation are the nervous system and the muscles, and most mitochondrial diseases affect these systems. For example, a person with the disease called *mitochondrial myopathy* suffers weakness, intolerance of exercise, and muscle deterioration.

In addition to the rare diseases clearly caused by defects in mitochondrial DNA, mitochondrial mutations inherited from a person's mother may contribute to at least some cases of diabetes and heart disease, as well as to other disorders that commonly debilitate the elderly, such as Alzheimer's disease. In the course of a lifetime, new mutations gradually accumulate in our mitochondrial DNA, and some researchers believe that these play a role in the normal aging process.

■   ■   ■

Wherever genes are located in the cell—nucleus or cytoplasm—their inheritance depends on the precise replication of DNA, the genetic material. In the next chapter, you will learn how this molecular reproduction occurs.

CHAPTER 15 REVIEW

Go to the Campbell Biology website (www.campbell biology.com) to explore an interactive version of the Chapter Review.

Summary of Key Concepts

RELATING MENDELISM TO CHROMOSOMES

■ **Mendelian inheritance has its physical basis in the behavior of chromosomes during sexual life cycles (p. 269, FIGURE 15.1)** In the early 1900s, geneticists showed that chromosomal movements in meiosis account for Mendel's laws.

■ **Morgan traced a gene to a specific chromosome (p. 271, FIGURE 15.3)** Morgan's discovery that the X chromosome in *Drosophila* carries a gene for eye color supported the chromosome theory of inheritance.

■ **Linked genes tend to be inherited together because they are located on the same chromosome (pp. 272–273, FIGURE 15.4)** Each chromosome has hundreds or thousands of genes. Linked genes do not assort independently.

■ **Independent assortment of chromosomes and crossing over produce genetic recombinants (pp. 273–275, FIGURE 15.5)** Recombinant offspring, which exhibit new combinations of traits inherited from two parents, result from events of meiosis and random fertilization. These events include crossing over and independent assortment of chromosomes during the first meiotic division. A recombination frequency under 50% indicates that the genes are linked but that crossing over has occurred. During prophase of meiosis I, paired homologous chromosomes break at corresponding points and switch fragments, creating new combinations of alleles that are then passed on to the gametes.

■ **Geneticists can use recombination data to map a chromosome's genetic loci (pp. 275–276, FIGURE 15.6)** One way to map genes is to deduce their order and a rough indication of the relative distances between them from crossover data. The further apart genes are on a chromosome, the more likely they are to be separated during crossing over. Cytological mapping is a technique that pinpoints the physical locus of a gene by associating a mutant phenotype with a chromosomal defect seen in the microscope.

Web/CD Activity 15A: *Linked Genes and Crossing Over*

SEX CHROMOSOMES

■ **The chromosomal basis of sex varies with the organism (pp. 276–277, FIGURE 15.8)** Sex is an inherited phenotypic character usually determined by the presence or absence of special chromosomes; the exact mechanism varies among different species. Humans and other mammals have an X-Y system, as do fruit flies. An XY male gives either an X chromosome or a Y chromosome to the sperm, which combines with an ovum containing an X chromosome from an XX female. The offspring's sex is determined at conception by whether the sperm carries X or Y.

■ **Sex-linked genes have unique patterns of inheritance (pp. 277–279, FIGURE 15.9)** The sex chromosomes carry certain genes for traits that are unrelated to maleness or femaleness. Hemophilia is a sex-linked recessive disorder whose gene is on the X chromosome. In mammalian females, one of the two X chromosomes in each cell is randomly inactivated during early embryonic development.

Web/CD Activity 15B: *Sex-Linked Genes*

Web/CD Case Study in the Process of Science: *What Can Fruit Flies Reveal about Inheritance?*

Biology Labs On-Line: *FlyLab*

Biology Labs On-Line: *PedigreeLab*

ERRORS AND EXCEPTIONS IN CHROMOSOMAL INHERITANCE

■ **Alterations of chromosome number or structure cause some genetic disorders (pp. 279–282, FIGURES 15.11, 15.13)** Errors during meiosis can change the number of chromosomes per cell or the structure of individual chromosomes. Such alterations can affect phenotype. Aneuploidy, an abnormal chromosome number, can arise when a normal gamete unites with one containing two copies or no copies of a particular chromosome as a result of nondisjunction during meiosis. Polyploidy, in which there are more than two complete sets of chromosomes, can result from complete nondisjunction during gamete formation. A variety of rearrangements can result from chromosome breakage. A lost fragment leaves one chromosome with a deletion and may produce a duplication, translocation, or inversion by reattaching to another chromosome. Such alterations cause a variety of human disorders, such as Down syndrome (usually due to trisomy of chromosome 21).

Web/CD Activity 15C: *Polyploid Plants*

■ **The phenotypic effects of some mammalian genes depend on whether they were inherited from the mother or the father (imprinting) (pp. 282–283, FIGURE 15.15)** Individuals imprint certain parts of chromosomes in their gamete-producing cells with either a male or a female "stamp," probably in the form of methylation. This affects the way some genes are expressed in offspring. Genomic imprinting helps explain the inheritance pattern of some hereditary disorders, including fragile X syndrome.

■ **Extranuclear genes exhibit a non-Mendelian pattern of inheritance (pp. 283–284)** Mitochondria and chloroplasts contain some of their own genes. Because the zygote's cytoplasm comes from the ovum, certain features of the offspring's phenotype depend solely on these maternal cytoplasmic genes. Some diseases affecting the nervous and muscular systems are caused by defects in mitochondrial DNA that prevent cells from making enough ATP.

Genetics Problems

1. A man with hemophilia (a recessive, sex-linked condition) has a daughter of normal phenotype. She marries a man who is normal for the trait. What is the probability that a daughter of this mating will be a hemophiliac? That a son will be a hemophiliac? If the couple has four sons, what is the probability that all four will be born with hemophilia?

2. Pseudohypertrophic muscular dystrophy is a disorder that causes gradual deterioration of the muscles. It is seen only in boys born to apparently normal parents and usually results in death in the early teens. Is this disorder caused by a dominant or a recessive allele? Is its inheritance sex-linked or autosomal? How do you know? Explain why this disorder is seen only in boys and never in girls.

3. Red-green color blindness is caused by a sex-linked recessive allele. A color-blind man marries a woman with normal vision whose father was color-blind. What is the probability that they will have a color-blind daughter? What is the probability that their first son will be color-blind? (*Note:* The two questions are worded a bit differently.)

4. A wild-type fruit fly (heterozygous for gray body color and normal wings) is mated with a black fly with vestigial wings. The offspring have the following phenotypic distribution: wild type, 778; black-vestigial, 785; black-normal, 158; gray-vestigial, 162. What is the recombination frequency between these genes for body color and wing type?

5. What pattern of inheritance would lead a geneticist to suspect that an inherited disorder of cell metabolism is due to a defective mitochondrial gene?

6. An aneuploid person is obviously female, but her cells have two Barr bodies. What is the probable complement of sex chromosomes in this individual?

7. Determine the sequence of genes along a chromosome based on the following recombination frequencies: A–B, 8 map units; A–C, 28 map units; A–D, 25 map units; B–C, 20 map units; B–D, 33 map units.

8. About 5% of individuals with Down syndrome are the result of chromosomal translocation in which one copy of chromosome 21 becomes attached to chromosome 14. How does this translocation lead to Down syndrome?

9. More common than completely polyploid animals are mosaic polyploids, animals that are diploid except for patches of polyploid cells. How might a mosaic tetraploid—an animal with some cells containing four sets of chromosomes—arise from an error in *mitosis*?

10. Assume that genes A and B are linked and are 50 map units apart. An animal heterozygous at both loci is crossed with one that is homozygous recessive at both loci. What percentage of the offspring will show phenotypes resulting from crossovers? If you did not know that genes A and B were linked, how would you interpret the results of this cross?

11. In *Drosophila*, the gene for white eyes and the gene that produces "hairy" wings have both been mapped to the same chromosome and have a crossover frequency of 1.5%. A geneticist notices that in a particular stock of flies, these two genes assort independently; that is, they behave as though they were on different chromosomes. What explanation can you offer for this observation?

12. In another cross, a wild-type fruit fly (heterozygous for gray body color and red eyes) is mated with a black fruit fly with purple eyes. The offspring are as follows: wild type, 721; black-purple, 751; gray-purple, 49; black-red, 45. What is the recombination frequency between these genes for body color and eye color? Using information from problem 4, what fruit flies (genotypes and phenotypes) would you mate to determine the sequence of the body-color, wing-shape, and eye-color genes on the chromosome?

13. A space probe discovers a planet inhabited by creatures who reproduce with the same hereditary patterns seen in humans. Three phenotypic characters are height (T = tall, t = dwarf), head appendages (A = antennae, a = no antennae), and nose morphology (S = upturned snout, s = downturned snout). Since the creatures are not "intelligent," Earth scientists are able to do some controlled breeding experiments, using various heterozygotes in testcrosses. For a tall heterozygote with antennae, the offspring are: tall-antennae, 46; dwarf-antennae, 7; dwarf-no antennae, 42; tall-no antennae, 5. For a heterozygote with antennae and an upturned snout, the offspring were: antennae-upturned snout, 47; antennae-downturned snout, 2; no antennae-downturned snout, 48; no antennae-upturned snout, 3. Calculate the recombination frequencies for both experiments.

14. Using the information from problem 13, a further testcross is done using a heterozygote for height and nose morphology. The offspring are: tall-upturned snout, 40; dwarf-upturned snout, 9; dwarf-downturned snout, 42; tall-downturned snout, 9. Calculate the recombination frequency from these data; then use your answer from problem 13 to determine the correct sequence of the three linked genes.

15. The ABO blood type locus has been mapped on chromosome 9. A father who has blood type AB and a mother who has blood type O have a child with trisomy 9 and blood type A. Using this information, can you tell in which parent the nondisjunction occurred? Explain your answer.

16. Two genes of a flower, one controlling blue (B) versus white (b) petals and the other controlling round (R) versus oval (r) stamens, are linked and are 10 map units apart. You cross a homozygous blue-oval plant with a homozygous white-round plant. The resulting F_1 progeny are crossed with homozygous white-oval plants, and 1000 progeny are obtained. How many plants of each of the four phenotypes do you expect?

Go to the website or CD-ROM for multiple-choice quiz questions.

Evolution Connection

You have seen that crossing over, or recombination, is thought to be evolutionarily advantageous because this process continually shuffles genetic alleles into novel combinations. Some organisms, however, have apparently lost the recombination mechanism, while in others certain chromosomes do not recombine. What factors do you think may favor reduced levels of recombination?

The Process of Science

Consider FIGURE 15.4, in which the F_1 dihybrid females resulted from a cross between parental (P) flies with genotypes $b^+ b^+ vg^+ vg^+$ and $b b$ $vg vg$. Now, imagine you make F_1 females by crossing two different P generation flies: $b^+ b^+ vg vg \times b b vg^+ vg^+$.

a. What will be the genotype of your F_1 females? Is this the same as that for the F_1 females in FIGURE 15.4?

b. Draw the chromosomes for the F_1 females, indicating the position of each allele. Are these the same as for the F_1 females in FIGURE 15.4?

c. Knowing that the distance between these two genes is 17 map units, predict the phenotypic ratios you will get from a cross. Will they be the same as in FIGURE 15.4?

d. Draw the chromosomes of the P, F_1, and F_2 generations (as is done in FIGURE 15.5 for the cross in FIGURE 15.4), showing how this arrangement of alleles in the P generation leads via F_1 gametes to the phenotypic ratios seen in the F_2 flies.

Study patterns of inheritance by breeding fruit flies and analyzing the frequency of traits in each generation in the Case Study in the Process of Science, available on the website and CD-ROM. Also link to FlyLab and PedigreeLab at Biology Labs On-Line.

Science, Technology, and Society

Opinions differ about whether children with learning disorders should be tested by karyotyping for the presence of a fragile X chromosome. Some argue that it's always better to know the cause of the problem so that education specialized for that disorder can be prescribed. Others counter that attaching a specific biological cause to a learning disability stigmatizes a child and limits his or her opportunities. What is your evaluation of these arguments?

Answers: 1. 0; $\frac{1}{2}$, $\frac{1}{16}$ 2. Recessive. If the disorder were dominant, it would affect at least one parent of a child born with the disorder. For a girl to have the disorder, she would need to inherit recessive alleles from *both* parents. This would be very rare, especially since males with the allele die in their early teens. 3. $\frac{1}{4}$ for each daughter ($\frac{1}{2}$ chance that child will be female $\times$ $\frac{1}{2}$ chance of a homozygous recessive genotype); $\frac{1}{2}$ for first son. 4. 17% 5. The disorder would always be inherited from the mother. 6. *XXX* 7. *D–A–B–C* 8. In meiosis, the combined 14-21 chromosome will behave as one chromosome. If a gamete receives the combined 14-21 chromosome and a normal copy of chromosome 21, trisomy 21 will result when this gamete combines with a normal gamete. 9. At some point during development, one of the embryo's cells may have failed to carry out mitosis after duplicating its chromosomes. Subsequent normal cell cycles would produce genetic copies of this tetraploid cell. 10. Fifty percent of the offspring would show phenotypes that resulted from crossovers. These results would be the same as those from a cross where *A* and *B* were not linked. Further crosses involving other genes on the same chromosome would reveal the linkage and map distances. 11. One hypothesis is that a translocation has moved one of the genes to a different chromosome. 12. 6%. Wild type (heterozygous for normal wings and red eyes) $\times$ recessive homozygote with vestigial wings and purple eyes. 13. Between T and A, 12%; between A and S, 5%. Sequence of genes is *T-A-S.* 14. Between T and S, 18%. Sequence of genes is *T-A-S.* 15. No; the child can be either $i^A i^A i$ or $i^A i i$. An ovum with the genotype $i^A i^A$ would be the result of nondisjunction during meiosis II in the mother, while a sperm of genotype ii would result from nondisjunction during meiosis I or II in the father. 16. 450 each of blue-oval and white-round (parentals) and 50 each of blue-round and white-oval (recombinants).

THE MOLECULAR BASIS OF INHERITANCE

I n April 1953, *James Watson and Francis Crick shook the scientific world with an elegant double-helical model for the structure of deoxyribonucleic acid, or DNA. The photograph on this page captures Watson and Crick admiring their DNA model. DNA, the substance of inheritance, is the most celebrated molecule of our time. Mendel's heritable factors and Morgan's genes on chromosomes are, in fact, composed of DNA. Chemically speaking, your genetic endowment is the DNA you inherited from your parents.*

Of all nature's molecules, nucleic acids are unique in their ability to direct their own replication. Indeed, the resemblance of offspring to their parents has its basis in the precise replication of DNA and its transmission from one generation to the next. In other words, DNA is the substance behind the adage "Like begets like." Hereditary information is encoded in the chemical language of DNA and reproduced in all the cells of your body. It is this DNA program that directs the development of your biochemical, anatomical, physiological, and, to some extent, behavioral traits. In this chapter, you will learn how biologists deduced that DNA is the genetic material, how Watson and Crick discovered its structure, and how cells replicate and repair their DNA—the molecular basis of inheritance.

DNA AS THE GENETIC MATERIAL

Today, even schoolchildren have heard of DNA, and scientists routinely manipulate DNA in the laboratory and use it to change the heritable characteristics of cells. Early in the 20th century, however, the identification of the molecules of inheritance loomed as a major challenge to biologists.

The search for the genetic material led to DNA

Once T. H. Morgan's group showed that genes are located on chromosomes, the two chemical components of chromosomes—DNA and protein—became the candidates for the genetic material. Until the 1940s, the case for proteins seemed stronger, especially since biochemists had identified them as a class of macromolecules with great heterogeneity and specificity of function, essential requirements for the hereditary material. Moreover, little was known about nucleic acids, whose physical and chemical properties seemed far too uniform to account for the multitude of specific inherited traits exhibited by every organism. This view gradually changed, as experiments with microorganisms yielded unexpected results. As with the work of Mendel and Morgan, a key factor in determining the identity of the genetic material was the choice of appropriate experimental organisms. Bacteria and the viruses that infect them are far simpler than pea plants, fruit flies, or humans, and the role of DNA in heredity was first worked out by studying such microbes. In this section, we will trace the search for the genetic material in some detail as a case study of the scientific process.

THE PROCESS OF SCIENCE

Evidence That DNA Can Transform Bacteria

We can trace the discovery of the genetic role of DNA back to 1928. Frederick Griffith, a British medical officer, was studying *Streptococcus pneumoniae*, a bacterium that causes pneumonia in mammals. Griffith had two strains (varieties) of the bacterium, a pathogenic (disease-causing) one and a variant that was harmless. He was surprised to find that when he killed the pathogenic bacteria with heat and then mixed the cell remains with living bacteria of the harmless strain, some of the living cells were converted to the pathogenic form (FIGURE 16.1). Furthermore, this new trait of pathogenicity was inherited by all the descendants of the transformed bacteria. Clearly, some chemical component of the dead pathogenic cells caused this heritable change, although the identity of the substance was not known. Griffith called the phenomenon **transformation,** now defined as a change in genotype and phenotype due to the assimilation of external DNA by a cell. (This usage of *transformation* should not be confused with the conversion of a normal animal cell to a cancerous one, discussed in Chapter 12.)

Griffith's work set the stage for a 14-year search for the identity of the transforming substance by American bacteriologist Oswald Avery. Avery purified various chemicals from the heat-killed pathogenic bacteria, then tried to transform live nonpathogenic bacteria with each chemical. Only DNA worked. Finally, in 1944, Avery and his colleagues Maclyn McCarty and Colin MacLeod announced that the transforming agent was DNA. Their discovery was greeted with considerable skepticism, in part because of the lingering belief that proteins were better candidates for the genetic material. More-over, many biologists were not convinced that the genes of bacteria would be similar in composition and function to those of more complex organisms. But the major reason for the continued doubt was that so little was known about DNA. No one could imagine how DNA could carry genetic information.

Evidence That Viral DNA Can Program Cells

Additional evidence for DNA as the genetic material came from studies of a virus that infects bacteria. Viruses are much simpler than cells. A virus is little more than DNA (or sometimes RNA) enclosed by a protective coat of protein. To reproduce, a virus must infect a cell and take over the cell's metabolic machinery.

Viruses that infect bacteria are widely used as research tools in molecular genetics. These viruses are called **bacteriophages** (meaning "bacteria eaters"), or just **phages.** In 1952, Alfred Hershey and Martha Chase performed experiments showing that DNA is the genetic material of a phage known as T2. This is one of many phages that infect *Escherichia coli (E. coli)*, a bacterium that normally lives in the intestines of mammals. At that time, biologists already knew that T2, like many other viruses, was composed almost entirely of DNA and protein. They also knew that the phage could quickly turn an *E. coli* cell into a T2-producing factory that released phages when the cell ruptured. Somehow, T2 could reprogram its host cell to produce viruses. But which viral component—protein or DNA—was responsible?

Hershey and Chase answered this question by devising an experiment showing that only one of the two components of

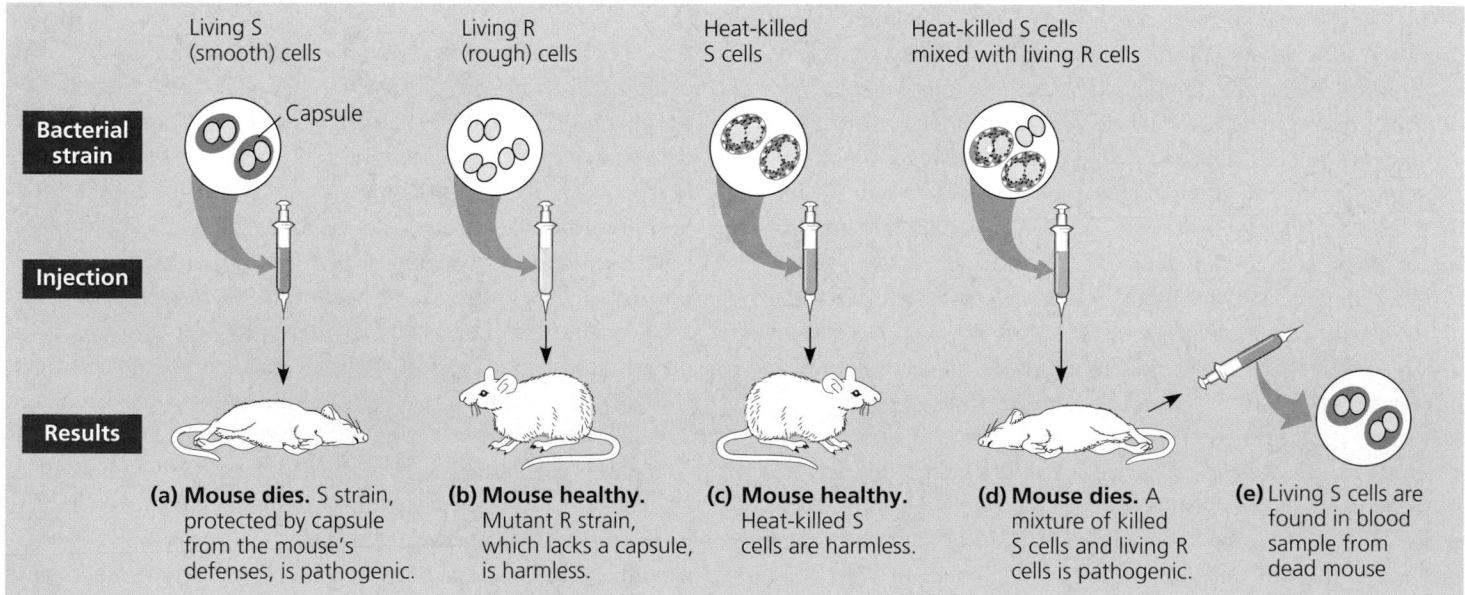

FIGURE 16.1 Transformation of bacteria. The S strain of *Streptococcus pneumoniae* causes pneumonia; the R strain is nonpathogenic. Griffith discovered that a mixture of heat-killed S cells and live R cells killed mice and that live S bacteria could be retrieved from the dead mice. He concluded that molecules from the dead S cells had genetically transformed living R bacteria into S bacteria.

THE PROCESS OF SCIENCE

T2 actually enters the *E. coli* cell during infection (FIGURE 16.2). In preparation for their experiment, they used different radioactive isotopes to tag phage DNA and protein. First, they grew T2 with *E. coli* in the presence of radioactive sulfur. Because protein, but not DNA, contains sulfur, the radioactive

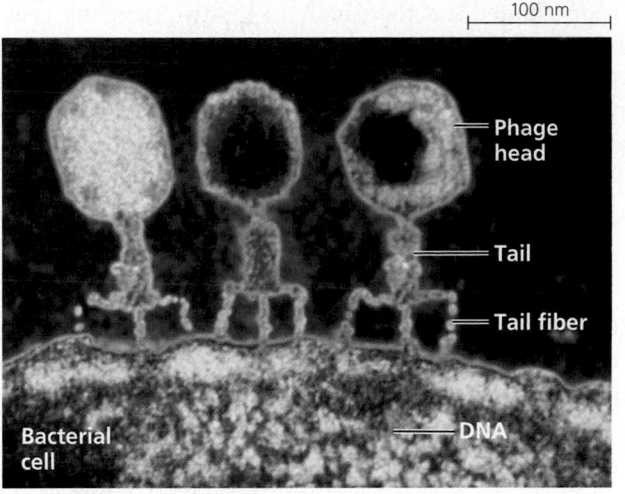

(a) T2 and related phages use their tail pieces to attach to the host cell and inject their genetic material (TEM).

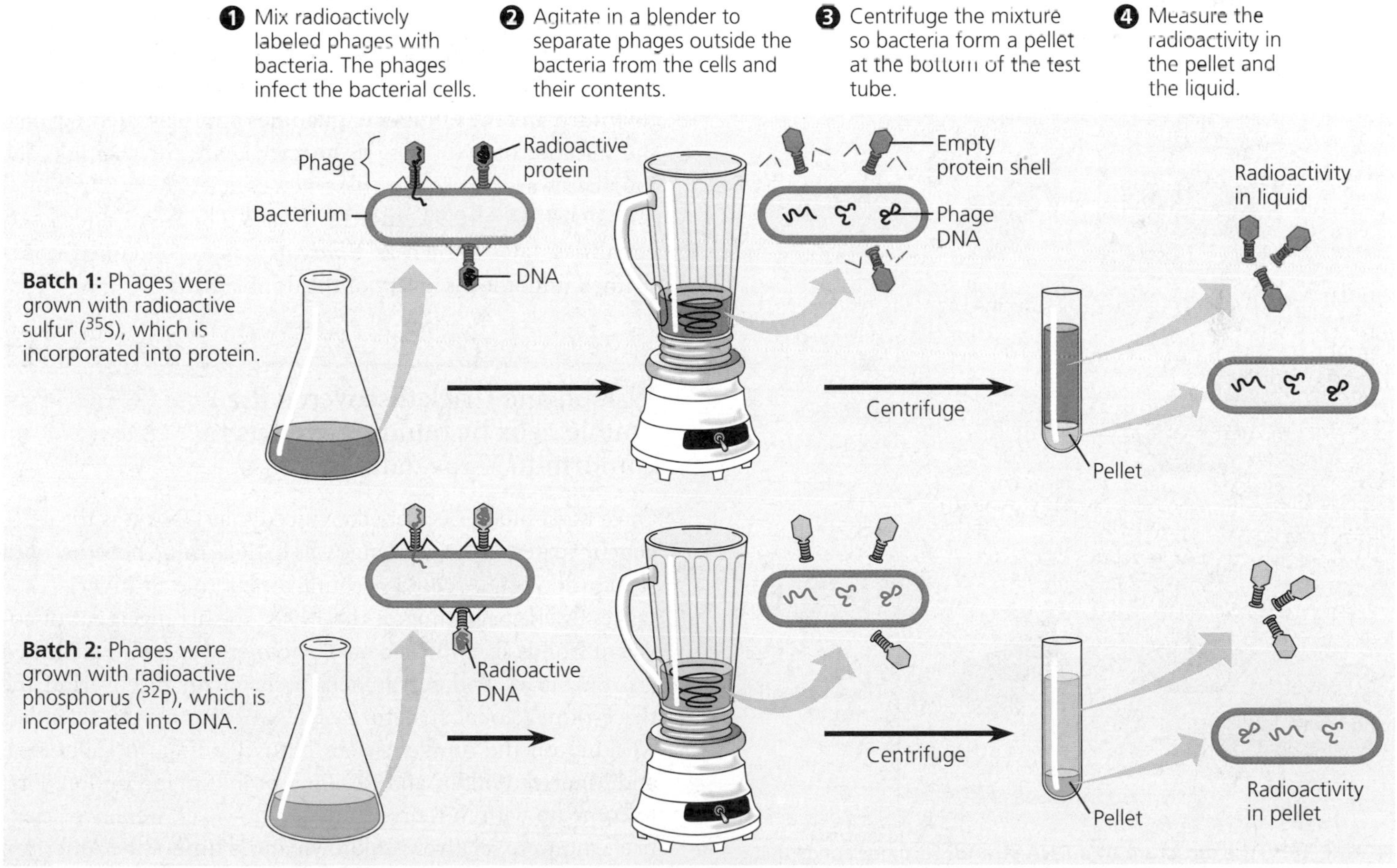

atoms were incorporated only into the protein of the phage. Next, in a similar way, the DNA of a separate batch of phage was labeled with atoms of radioactive phosphorus; because nearly all the phage's phosphorus is in its DNA, this procedure left the phage protein unlabeled. In the experiment, the protein-labeled and DNA-labeled batches of T2 were each allowed to infect separate samples of nonradioactive *E. coli* cells. Shortly after the onset of infection, the cultures were whirled in a kitchen blender to shake loose any parts of the phages that remained outside the bacterial cells. The mixtures were then spun in a centrifuge, forcing the bacterial cells to form a pellet at the bottom of the centrifuge tubes, but allowing free phages and parts of phages, which are lighter, to remain suspended in the liquid, or supernatant. The scientists then measured the radioactivity in the pellet and in the supernatant.

Hershey and Chase found that when the bacteria had been infected with the T2 phage containing radioactively labeled proteins, most of the radioactivity was found in the supernatant,

❶ Mix radioactively labeled phages with bacteria. The phages infect the bacterial cells.

❷ Agitate in a blender to separate phages outside the bacteria from the cells and their contents.

❸ Centrifuge the mixture so bacteria form a pellet at the bottom of the test tube.

❹ Measure the radioactivity in the pellet and the liquid.

Batch 1: Phages were grown with radioactive sulfur (³⁵S), which is incorporated into protein.

Batch 2: Phages were grown with radioactive phosphorus (³²P), which is incorporated into DNA.

(b) The experiment showed that T2 proteins remain outside the host cell during infection, while T2 DNA enters the cell.

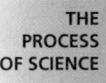

FIGURE 16.2 The Hershey-Chase experiment. In their famous 1952 experiment, Hershey and Chase demonstrated that DNA, not protein, functions as the T2 phage's genetic material.

THE PROCESS OF SCIENCE

which contained phage particles (but not bacteria). This result suggested that the protein of the phage did not enter the host cells. But when the bacteria had been infected with T2 phage whose DNA was tagged with radioactive phosphorus, then the pellet of mainly bacterial material contained most of the radioactivity. Moreover, when these bacteria were returned to a culture medium, the infection ran its course, and the *E. coli* released phages that contained some radioactive phosphorus.

Hershey and Chase concluded that the DNA of the virus is injected into the host cell during infection, leaving the protein outside. The injected DNA provides genetic information that makes the cells produce new viral DNA and proteins, which assemble into new viruses. Thus, the Hershey-Chase experiment provided powerful evidence that nucleic acids, rather than proteins, are the hereditary material, at least for viruses.

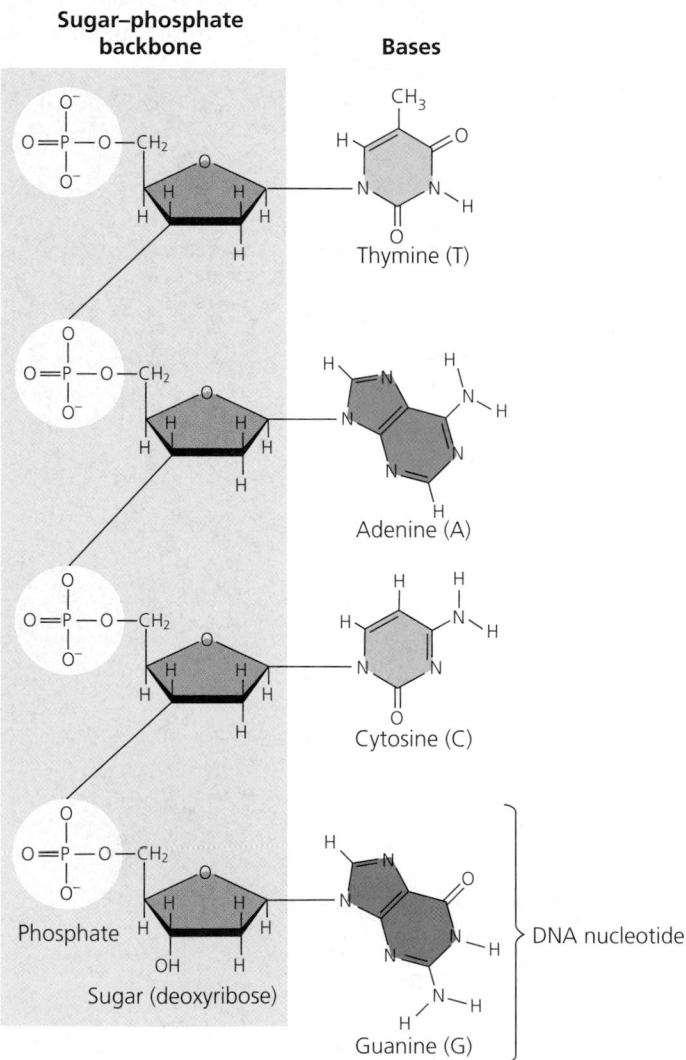

FIGURE 16.3 The structure of a DNA strand. Each nucleotide unit of the polynucleotide chain consists of a nitrogenous base (T, A, C, or G), the sugar deoxyribose, and a phosphate group. The phosphate of one nucleotide is attached to the sugar of the next nucleotide in line. The result is a "backbone" of alternating phosphates and sugars, from which the bases project.

Additional Evidence That DNA Is the Genetic Material of Cells

Additional circumstantial evidence pointed to DNA as the genetic material in eukaryotes. Prior to mitosis, a eukaryotic cell exactly doubles its DNA content, and during mitosis, this DNA is distributed exactly equally to the two daughter cells. Also, diploid sets of chromosomes have twice as much DNA as the haploid sets found in the gametes of the same organism.

Still more evidence came from the laboratory of biochemist Erwin Chargaff. It was already known that DNA is a polymer of nucleotides, each consisting of three components: a nitrogenous (nitrogen-containing) base, a pentose sugar called deoxyribose, and a phosphate group. (See FIGURE 16.3 for review.) The base can be adenine (A), thymine (T), guanine (G), or cytosine (C). Chargaff analyzed the base composition of DNA from a number of different organisms. In 1947, he reported that DNA composition varies from one species to another: In the DNA of any one species, the amounts of the four nitrogenous bases are not all equal but are present in a characteristic ratio. Such evidence of molecular diversity, which had been presumed absent from DNA, made DNA a more credible candidate for the genetic material.

Chargaff also found a peculiar regularity in the ratios of nucleotide bases. In the DNA of each species he studied, the number of adenines approximately equaled the number of thymines, and the number of guanines approximately equaled the number of cytosines. In human DNA, for example, the four bases are present in these percentages: A = 30.9% and T = 29.4%; G = 19.9% and C = 19.8%. The A = T and G = C equalities, later known as *Chargaff's rules,* remained unexplained until the discovery of the double helix.

Watson and Crick discovered the double helix by building models to conform to X-ray data

Once most biologists were convinced that DNA was the genetic material, the challenge was to determine how the structure of DNA could account for its role in inheritance. By the beginning of the 1950s, the arrangement of covalent bonds in a nucleic acid polymer was well established (FIGURE 16.3), and researchers focused on discovering the three-dimensional structure of DNA. Among the scientists working on the problem were Linus Pauling, in California, and Maurice Wilkins and Rosalind Franklin, in London. First to come up with the correct answer, however, were two scientists who were relatively unknown at the time—the American James Watson and the Englishman Francis Crick.

The brief but celebrated partnership that solved the DNA puzzle began soon after the young Watson journeyed to Cambridge University, where Crick was studying protein structure

THE PROCESS OF SCIENCE

with a technique called X-ray crystallography (see FIGURE 5.27). While visiting the laboratory of Maurice Wilkins at King's College in London, Watson saw an X-ray diffraction image of DNA produced by Wilkins's colleague, Rosalind Franklin (FIGURE 16.4a). Images produced by X-ray crystallography are not actually pictures of molecules. The spots and smudges in FIGURE 16.4b were produced by X-rays that were diffracted (deflected) as they passed through aligned fibers of purified DNA. Crystallographers use mathematical equations to translate such patterns of spots into information about the three-dimensional shapes of molecules, and Watson was familiar with the types of patterns that helical molecules produce. Just a glance at Franklin's X-ray diffraction photo of DNA not only told him that DNA was helical in shape but also enabled him to deduce the width of the helix and the spacing of the nitrogenous bases along it. The width of the helix suggested that it was made up of two strands, contrary to a three-stranded model that Linus Pauling had recently proposed. The presence of two strands accounts for the now-familiar term **double helix.**

Using molecular models made of wire, Watson and Crick began building models of a double helix that would conform to the X-ray measurements and what was then known about the chemistry of DNA. After failing to make a satisfactory model that placed the sugar-phosphate chains on the inside of the molecule, Watson tried putting them on the outside and forcing the nitrogenous bases to swivel to the interior of the double helix (FIGURE 16.5). Imagine this double helix as a rope

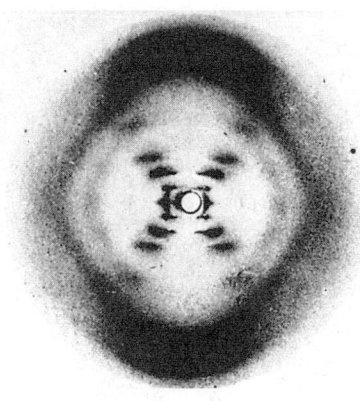

(a) Rosalind Franklin   **(b) Franklin's X-ray diffraction photograph of DNA**

FIGURE 16.4 Rosalind Franklin and her X-ray diffraction photo of DNA. Franklin, an X-ray crystallographer, took the photo that Watson and Crick used in deducing the double-helical structure of DNA. Franklin died of cancer when she was only 38. Her colleague, Maurice Wilkins, received the Nobel Prize in 1962 along with Watson and Crick. Because the Nobel Prize is not awarded posthumously, science historians can only speculate about whether the committee would have recognized Franklin's contribution to the discovery of the double helix.

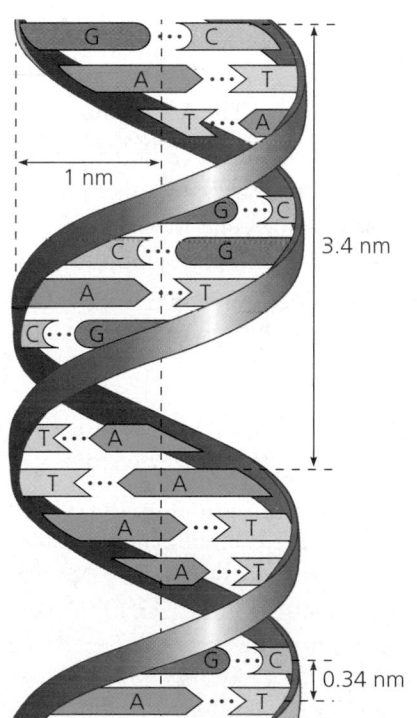

(a) Key features of DNA structure

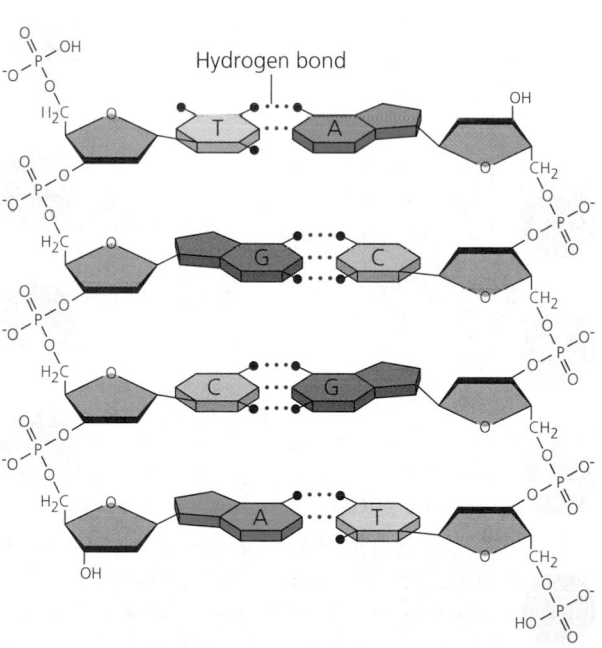

(b) Partial chemical structure

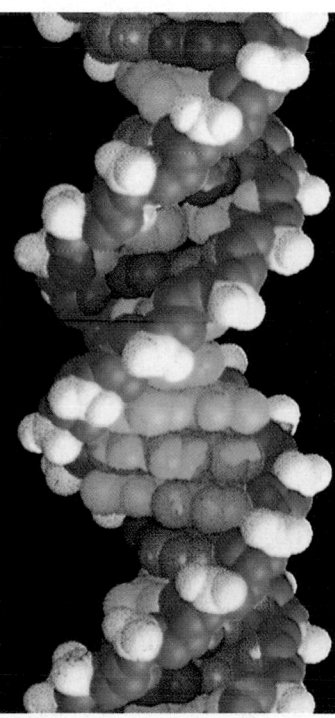

(c) Space-filling model

FIGURE 16.5 The double helix. (a) The "ribbons" in this diagram represent the sugar-phosphate backbones of the two DNA strands. The helix is "right-handed," curving up to the right. The two strands are held together by hydrogen bonds (dotted lines) between the nitrogenous bases, which are paired in the interior of the double helix. **(b)** For clarity, the two DNA strands are shown untwisted in this partial chemical structure. Notice that the strands are oriented in opposite directions. **(c)** The tight stacking of the base pairs is clear in this computer model. Van der Waals attractions between the stacked pairs play a major role in holding the molecule together.

ladder having rigid rungs, with the ladder twisted into a spiral. The side ropes are the equivalent of the sugar-phosphate backbones, and the rungs represent pairs of nitrogenous bases. Franklin's X-ray data indicated that the helix makes one full turn every 3.4 nm along its length. With the bases stacked just 0.34 nm apart, there are ten layers of base pairs, or rungs on the ladder, in each turn of the helix. This arrangement was appealing because it put the relatively hydrophobic nitrogenous bases in the molecule's interior and thus away from the surrounding aqueous medium.

The nitrogenous bases of the double helix are paired in specific combinations: adenine (A) with thymine (T), and guanine (G) with cytosine (C). It was mainly by trial and error that Watson and Crick arrived at this key feature of DNA. At first, Watson imagined that the bases paired like-with-like—for example, A with A and C with C. But this model did not fit the X-ray data, which suggested that the double helix had a uniform diameter. Why is this requirement inconsistent with like-with-like pairing of bases? Adenine and guanine are purines, nitrogenous bases with two organic rings. In contrast, cytosine and thymine belong to the family of nitrogenous bases known as pyrimidines, which have a single ring. Thus, purines (A and G) are about twice as wide as pyrimidines (C and T). A purine-purine pair is too wide and a pyrimidine-pyrimidine pair too narrow to account for the 2-nm diameter of the double helix. The solution is to always pair a purine with a pyrimidine:

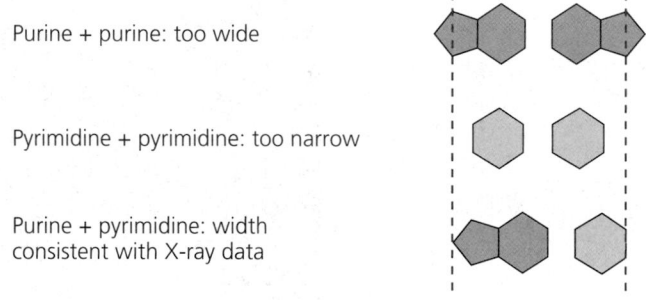

Purine + purine: too wide

Pyrimidine + pyrimidine: too narrow

Purine + pyrimidine: width consistent with X-ray data

Watson and Crick reasoned that there must be additional specificity of pairing dictated by the structure of the bases. Each base has chemical side groups that can form hydrogen bonds with its appropriate partner: Adenine can form two hydrogen bonds with thymine and only thymine; guanine forms three hydrogen bonds with cytosine and only cytosine. In shorthand, A pairs with T, and G pairs with C (FIGURE 16.6).

The Watson-Crick model explained Chargaff's rules. Wherever one strand of a DNA molecule has an A, the partner strand has a T. And a G in one strand is always paired with a C in the complementary strand. Therefore, in the DNA of any organism, the amount of adenine equals the amount of thymine, and the amount of guanine equals the amount of cytosine. Although the base-pairing rules dictate the combinations of nitrogenous bases that form the "rungs" of the double

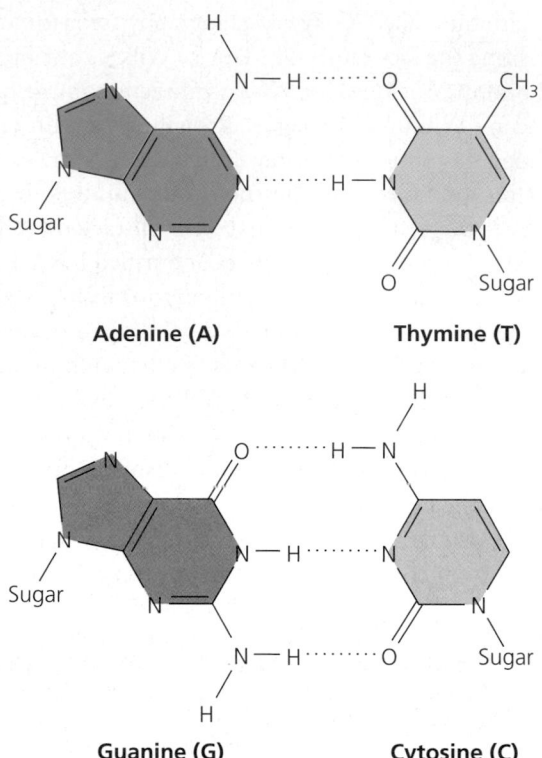

Adenine (A)   Thymine (T)

Guanine (G)   Cytosine (C)

FIGURE 16.6 Base pairing in DNA. The pairs of nitrogenous bases in a DNA double helix are held together by hydrogen bonds as shown here.

helix, they do not restrict the sequence of nucleotides *along* each DNA strand. The linear sequence of the four bases can be varied in countless ways, and each gene has a unique order, or base sequence.

In April 1953, Watson and Crick surprised the scientific world with a succinct, one-page paper in the British journal *Nature.** The paper reported their molecular model for DNA: the double helix, which has since become the symbol of molecular biology. The beauty of the model was that its structure suggested the basic mechanism of DNA replication.

DNA REPLICATION AND REPAIR

The relationship between structure and function, one of the themes of biology, is manifest in the double helix. The idea that there is specific pairing of nitrogenous bases in DNA was the flash of inspiration that led Watson and Crick to the correct double helix. At the same time, they saw the functional significance of the base-pairing rules. They ended their classic paper with this wry statement: "It has not escaped our notice that the specific pairing we have postulated immediately suggests a possible copying mechanism for the genetic material." In the next section, you will learn about this basic mechanism

* J. D. Watson and F. H. C. Crick, "Molecular Structure of Nucleic Acids: A Structure for Deoxynucleic Acids." *Nature* 171 (1953): 738.

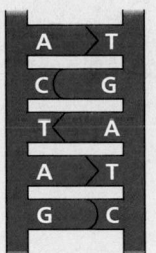
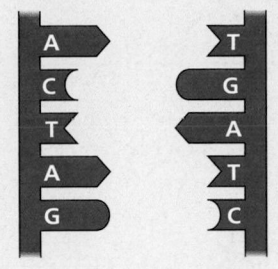
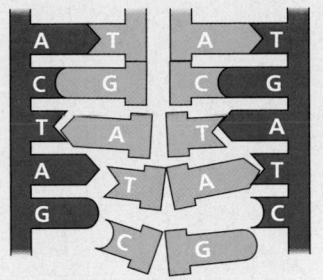
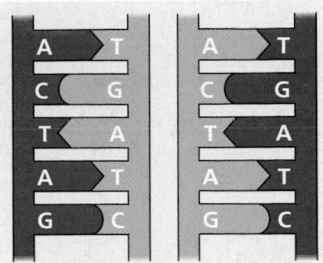

(a) The parent molecule has two complementary strands of DNA. Each base is paired by hydrogen bonding with its specific partner, A with T and G with C.

(b) The first step in replication is separation of the two DNA strands.

(c) Each parental strand now serves as a template that determines the order of nucleotides along a new complementary strand.

(d) The nucleotides are connected to form the sugar-phosphate backbones of the new strands. Each "daughter" DNA molecule consists of one parental strand and one new strand.

FIGURE 16.7 A model for DNA replication: the basic concept. In this simplification, a short segment of DNA has been untwisted into a structure that resembles a ladder. The rails of the ladder are the sugar-phosphate backbones of the two DNA strands; the rungs are the pairs of nitrogenous bases. Simple shapes symbolize the four kinds of bases. Dark blue represents DNA strands originally present in the parent cell; light blue represents newly synthesized DNA.

of DNA replication. Some important details of the process will be presented in the section that follows.

During DNA replication, base pairing enables existing DNA strands to serve as templates for new complementary strands

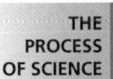

In a second paper that followed their announcement of the double helix, Watson and Crick stated their hypothesis for how DNA replicates:

Now our model for deoxyribonucleic acid is, in effect, a pair of templates, each of which is complementary to the other. We imagine that prior to duplication the hydrogen bonds are broken, and the two chains unwind and separate. Each chain then acts as a template for the formation onto itself of a new companion chain, so that eventually we shall have two pairs of chains, where we only had one before. Moreover, the sequence of the pairs of bases will have been duplicated exactly.*

FIGURE 16.7 illustrates Watson and Crick's basic idea. To make it easier to follow, the diagram shows only a short section of double helix, in untwisted form. Notice that if you cover one of the two DNA strands of FIGURE 16.7a, you can still determine its linear sequence of bases by referring to the unmasked strand and applying the base-pairing rules. The two strands are complementary; each stores the information necessary to reconstruct the other. When a cell copies a DNA molecule, each strand serves as a template (mold) for ordering nucleotides

into a new complementary strand. One at a time, nucleotides line up along the template strand according to the base-pairing rules. The nucleotides are linked to form the new strands. Where there was one double-stranded DNA molecule at the beginning of the process, there are now two, each an exact replica of the "parent" molecule. The copying mechanism is analogous to using a photographic negative to make a positive image, which can in turn be used to make another negative, and so on. (See FIGURE 5.30 for a helical version of FIGURE 16.7.)

This model of gene replication remained untested for several years following publication of the DNA structure. The requisite experiments were simple in concept but difficult to perform. Watson and Crick's model predicts that when a double helix replicates, each of the two daughter molecules will have one old strand, derived from the parent molecule, and one newly made strand. This **semiconservative model** can be distinguished from a conservative model of replication, in which the parent molecule somehow emerges from the replication process intact (that is, it is conserved). In yet a third model, called the dispersive model, all four strands of DNA following replication have a mixture of old and new DNA (FIGURE 16.8, p. 294). Although mechanisms for conservative or dispersive DNA replication are not easy to devise, these models remained possibilities until they could be ruled out. Finally, in the late 1950s, Matthew Meselson and Franklin Stahl devised experiments that tested the three hypotheses. Their experiments supported the semiconservative model, as predicted by Watson and Crick (FIGURE 16.9, p. 294).

The basic principle of DNA replication is elegantly simple. However, the actual process involves complex biochemical gymnastics, as we will now see.

---

* F. H. C. Crick and J. D. Watson, "The Complementary Structure of Deoxyribonucleic Acid." *Proc. Roy. Soc.* (A) 223 (1954): 80.

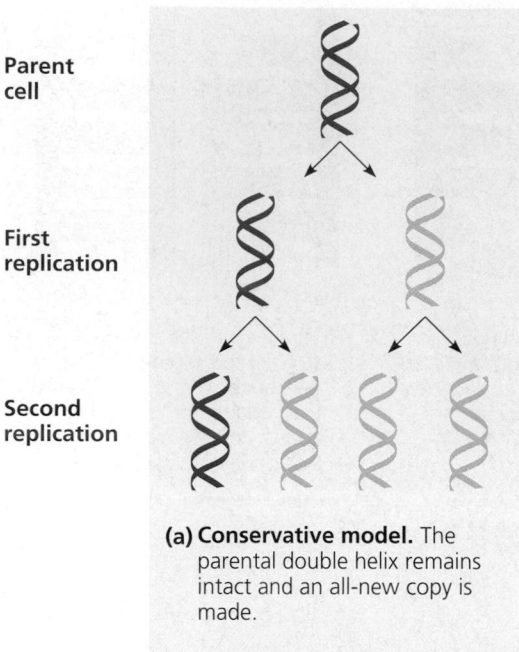
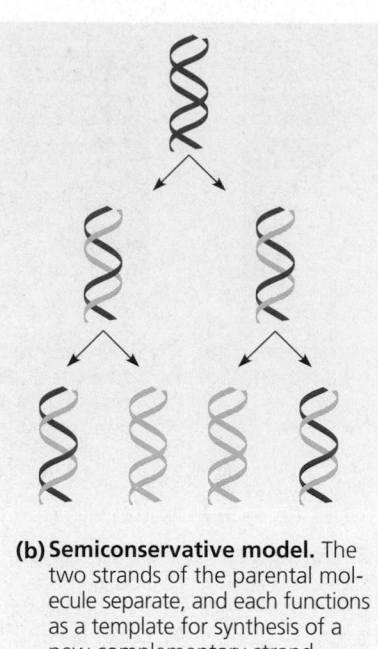
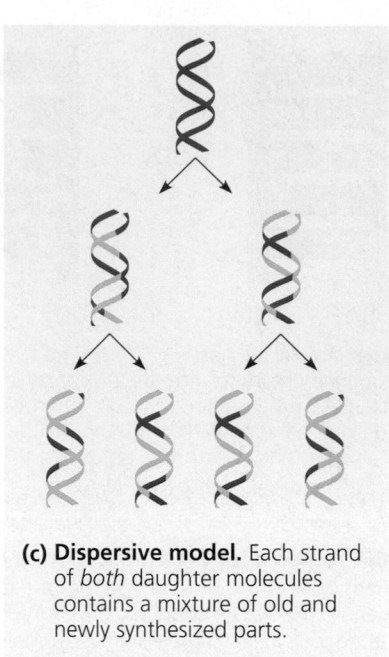

(a) **Conservative model.** The parental double helix remains intact and an all-new copy is made.

(b) **Semiconservative model.** The two strands of the parental molecule separate, and each functions as a template for synthesis of a new complementary strand.

(c) **Dispersive model.** Each strand of *both* daughter molecules contains a mixture of old and newly synthesized parts.

FIGURE 16.8 Three alternative models of DNA replication. The short segments of double helix here symbolize the DNA within a cell. Beginning with a parent cell, we follow the DNA for two generations of cells—two rounds of DNA replication. Newly made DNA is lighter blue.

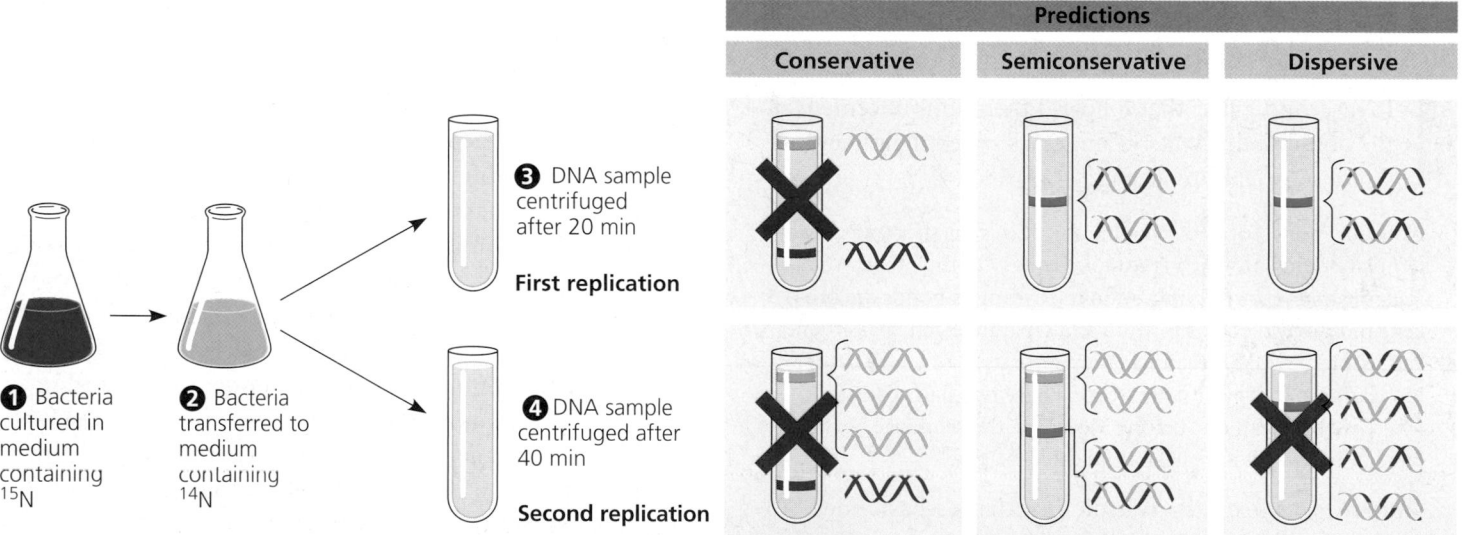

FIGURE 16.9 The Meselson-Stahl experiment tested three models of DNA replication. Meselson and Stahl cultured *E. coli* bacteria for several generations on a medium containing a heavy isotope of nitrogen, ^{15}N. The bacteria incorporated the heavy nitrogen into their nucleotides and then into their DNA. The scientists then transferred the bacteria to a medium containing ^{14}N, the lighter, more common isotope of nitrogen. Any new DNA that the bacteria synthesized would be lighter than the "old" DNA made in the ^{15}N medium. Meselson and Stahl could distinguish DNA of different densities by centrifuging DNA extracted from the bacteria. The centrifuge tubes in this drawing represent the results predicted by each of the three models in FIGURE 16.8. The first replication in the ^{14}N medium produced a band of hybrid (^{15}N-^{14}N) DNA. This result eliminated the conservative model. A second replication produced both light and hybrid DNA, a result that eliminated the dispersive model and supported the semiconservative model.

A large team of enzymes and other proteins carries out DNA replication

The bacterium *E. coli* has a single chromosome of about 5 million base pairs. In a favorable environment, an *E. coli* cell can copy all this DNA and divide to form two genetically identical daughter cells in less than an hour. Each of *your* cells has 46 DNA molecules in its nucleus, one giant molecule per chromosome. In all, that represents about 6 billion base pairs, or over a thousand times more DNA than is found in a bacterial cell. If we were to print the one-letter symbols for these bases (A, G, C, and T) the size of the letters you are now reading, the 6 billion bases of a single human cell would fill about 900 books as thick as this text. Yet it takes a cell just a few hours to copy all this DNA. This replication of an enormous amount of genetic information is achieved with very few errors—only one per billion nucleotides. The copying of DNA is remarkable in its speed and accuracy.

More than a dozen enzymes and other proteins participate in DNA replication. Much more is known about how this "replication machine" works in bacteria than in eukaryotes. However, most of the process seems to be fundamentally similar for prokaryotes and eukaryotes. In this section, we take a closer look at the basic steps.

Getting Started: Origins of Replication

The replication of a DNA molecule begins at special sites called **origins of replication.** The bacterial chromosome, which is circular, has a single origin, a stretch of DNA having a specific sequence of nucleotides. Proteins that initiate DNA replication recognize this sequence and attach to the DNA, separating the two strands and opening up a replication "bubble." Replication of DNA then proceeds in both directions, until the entire molecule is copied (see FIGURE 18.11). In contrast to the bacterial chromosome, a eukaryotic chromosome may have hundreds or even thousands of replication origins. Multiple replication bubbles form and eventually fuse, thus speeding up the copying of the very long DNA molecules (FIGURE 16.10). As in bacteria, DNA replication proceeds in both directions from each origin. At each end of a replication bubble is a **replication fork,** a Y-shaped region where the new strands of DNA are elongating.

Elongating a New DNA Strand

Elongation of new DNA at a replication fork is catalyzed by enzymes called **DNA polymerases.** As nucleotides align with complementary bases along a template strand of DNA, they are added by polymerase, one by one, to the growing end of the new DNA strand. The rate of elongation is about 500 nucleotides per second in bacteria and 50 per second in human cells.

What is the source of energy that drives the polymerization of nucleotides to form new DNA strands? The nucleotides that serve as substrates for DNA polymerase are actually nucleoside triphosphates, which are nucleotides with three phosphate groups (FIGURE 16.11, p. 296). You have already encountered such a molecule—ATP. The only difference between the ATP of energy metabolism and the nucleoside triphosphate that supplies adenine to DNA is the sugar component, which is deoxyribose in the building block of DNA, but ribose in ATP. (As you might guess, ribose-containing ATP is a substrate for *RNA*

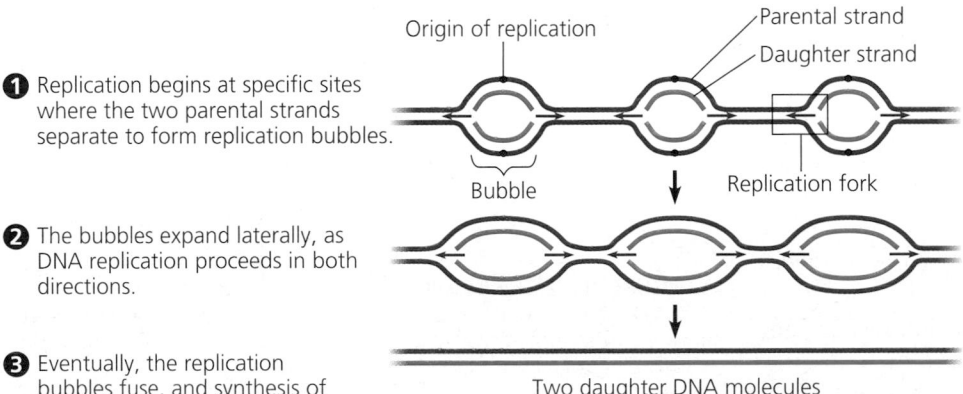

❶ Replication begins at specific sites where the two parental strands separate to form replication bubbles.

❷ The bubbles expand laterally, as DNA replication proceeds in both directions.

❸ Eventually, the replication bubbles fuse, and synthesis of the daughter strands is complete.

Origin of replication

Parental strand

Daughter strand

Bubble

Replication fork

Two daughter DNA molecules

(a) In eukaryotes, DNA replication begins at many sites along the giant DNA molecule of each chromosome.

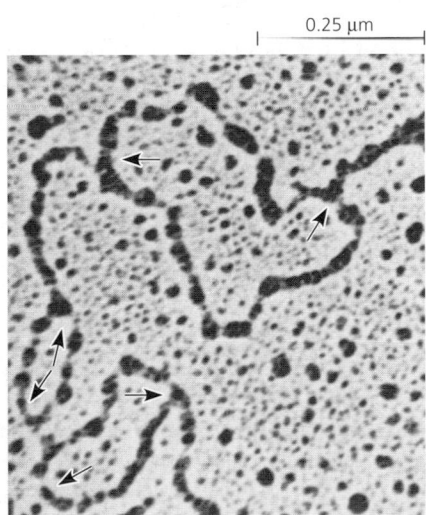

0.25 µm

(b) In this micrograph, three replication bubbles are visible along the DNA of cultured Chinese hamster cells. The arrows indicate the direction of DNA replication at the two ends of each bubble (TEM).

FIGURE 16.10 Origins of replication in eukaryotes.

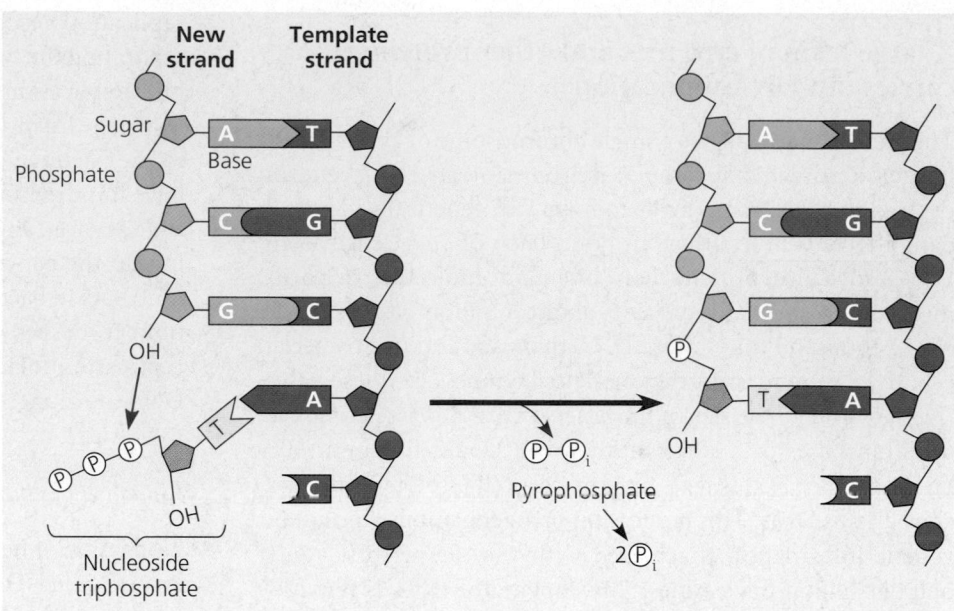

FIGURE 16.11 Incorporation of a nucleotide into a DNA strand. When a nucleoside triphosphate links to the sugar-phosphate backbone of a growing DNA strand, it loses two of its phosphates as a pyrophosphate molecule. The enzyme catalyzing the reaction is a DNA polymerase, and hydrolysis of the bonds between the phosphate groups of the pyrophosphate provides the energy for the reaction.

synthesis.) Like ATP, the triphosphate monomers used for DNA synthesis are chemically reactive, partly because their triphosphate tails have an unstable cluster of negative charge. As each monomer joins the growing end of a DNA strand, it loses two phosphate groups as a molecule of pyrophosphate ($\mathrm{P}-\mathrm{P_i}$). Subsequent hydrolysis of the pyrophosphate to two molecules of inorganic phosphate ($\mathrm{P_i}$) is the exergonic reaction that drives the polymerization reaction.

The Antiparallel Arrangement of the DNA Strands

There is more to the scenario of DNA synthesis at the replication fork. Until now, we have ignored an important feature of the double helix: The two DNA strands are *antiparallel*; that is, their sugar-phosphate backbones run in opposite directions. In FIGURE 16.12, the five carbons of one deoxyribose sugar of each DNA strand are numbered from 1′ to 5′. (The prime sign is used to distinguish the carbon atoms of the sugar from the carbon and nitrogen atoms of the nitrogenous bases.) Notice in FIGURE 16.12 that a nucleotide's phosphate group is attached to the 5′ carbon of deoxyribose. Notice also that the phosphate group of one nucleotide is joined to the 3′ carbon of the adjacent nucleotide. The result is a DNA strand of distinct polarity. At one end, denoted the 3′ end, a hydroxyl group is attached to the 3′ carbon of the terminal deoxyribose. At the opposite end, the 5′ end, the sugar-phosphate backbone terminates with the phosphate group attached to the 5′ carbon of the last nucleotide. In the double helix, the two sugar-phosphate backbones are essentially upside down (antiparallel) relative to each other.

How does the antiparallel structure of the double helix affect replication? DNA polymerases add nucleotides only to the free 3′ end of a growing DNA strand, never to the 5′ end. Thus, a new DNA strand can elongate only in the 5′ ⟶ 3′ direction. With this in mind, let's examine a replication fork (FIGURE 16.13). Along one template strand, DNA polymerase can synthesize a continuous complementary strand by elongating the new DNA in the mandatory 5′ ⟶ 3′ direction. The polymerase simply nestles in the replication fork on that tem-

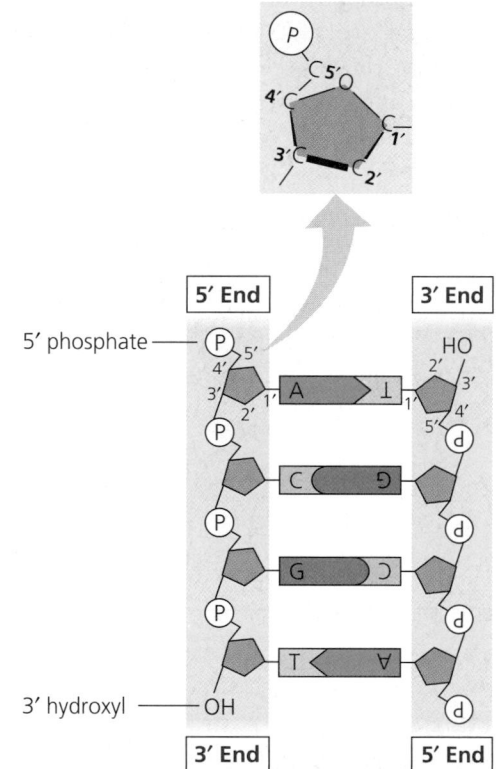

FIGURE 16.12 The two strands of DNA are antiparallel. The 5′ ⟶ 3′ direction of one strand runs counter to the 5′ ⟶ 3′ direction of the other strand. The numbers assigned to the carbon atoms of the deoxyribose units are shown for two of them.

plate strand and continuously adds nucleotides to a complementary strand as the fork progresses. The DNA strand made by this mechanism is called the **leading strand.**

To elongate the other new strand of DNA, polymerase must work along the other template strand in the direction *away*

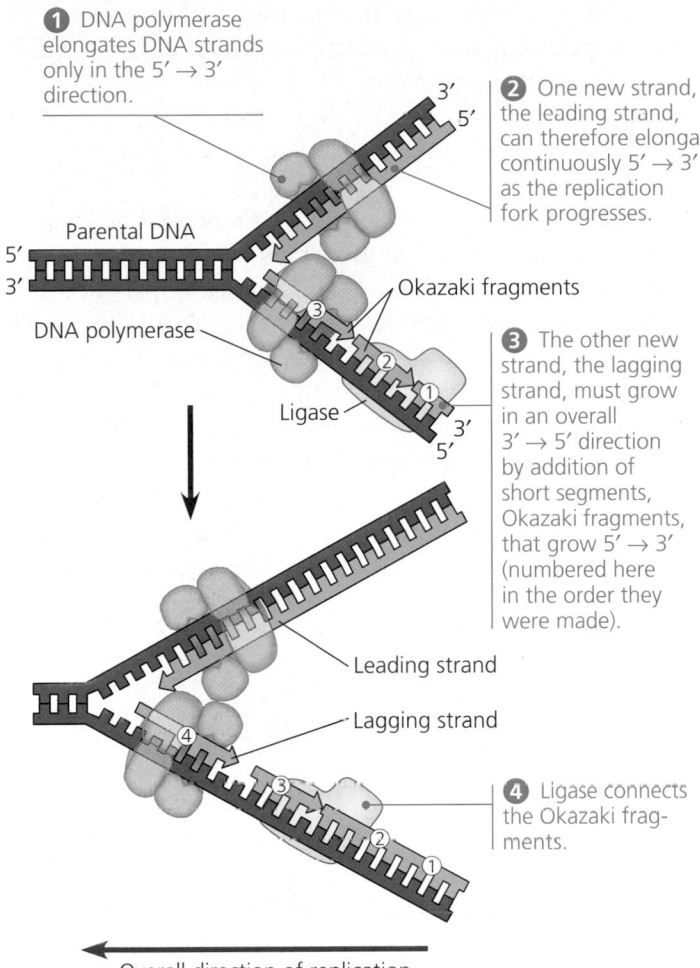

① DNA polymerase elongates DNA strands only in the 5′ → 3′ direction.

Parental DNA

DNA polymerase

Ligase

Okazaki fragments

② One new strand, the leading strand, can therefore elongate continuously 5′ → 3′ as the replication fork progresses.

③ The other new strand, the lagging strand, must grow in an overall 3′ → 5′ direction by addition of short segments, Okazaki fragments, that grow 5′ → 3′ (numbered here in the order they were made).

Leading strand

Lagging strand

④ Ligase connects the Okazaki fragments.

Overall direction of replication

FIGURE 16.13 Synthesis of leading and lagging strands during DNA replication.

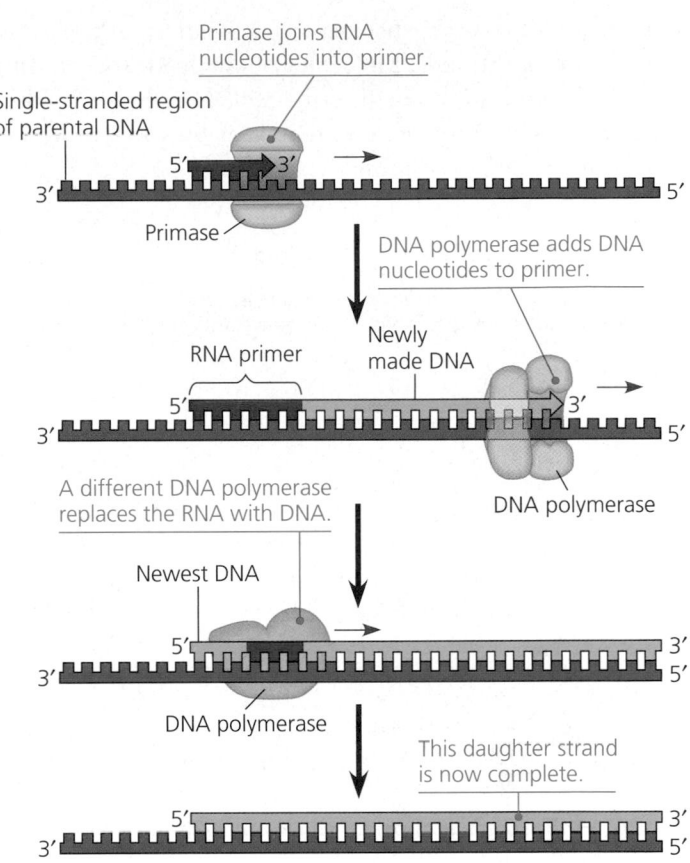

Primase joins RNA nucleotides into primer.

Single-stranded region of parental DNA

Primase

DNA polymerase adds DNA nucleotides to primer.

RNA primer

Newly made DNA

A different DNA polymerase replaces the RNA with DNA.

DNA polymerase

Newest DNA

DNA polymerase

This daughter strand is now complete.

FIGURE 16.14 Priming DNA synthesis with RNA. DNA polymerase cannot initiate a polynucleotide strand; it can only add to the 3′ end of an already-started strand. The primer is a short segment of RNA synthesized by the enzyme primase. Each primer is eventually replaced by DNA.

from the replication fork. The DNA synthesized in this direction is called the **lagging strand.** The process is analogous to a sewing method called backstitching. As a replication bubble opens, a polymerase molecule can work its way away from a replication fork and synthesize a short segment of DNA. As the bubble grows, another short segment of the lagging strand can be made in a similar way. In contrast to the leading strand, which elongates continuously, the lagging strand is first synthesized as a series of segments. These pieces are called Okazaki fragments, after the Japanese scientist who discovered them. The fragments are about 100 to 200 nucleotides long in eukaryotes. Another enzyme, **DNA ligase,** joins (ligates) the sugar-phosphate backbones of the Okazaki fragments to create a single DNA strand.

Priming DNA Synthesis

There is another important restriction for DNA polymerases. None of these enzymes can actually *initiate* synthesis of a polynucleotide; they can only add nucleotides to the end of an already existing chain that is base-paired with the template strand (see FIGURE 16.11). In the replication of cellular DNA, the start of a new chain, its **primer,** is not DNA, but a short stretch of RNA, the other class of nucleic acid. An enzyme called **primase** joins RNA nucleotides to make the primer, which is about 10 nucleotides long in eukaryotes (FIGURE 16.14). (Like all RNA-synthesizing enzymes, primase can start an RNA chain from scratch.) Another DNA polymerase later replaces the RNA nucleotides of the primers with DNA versions. Only one primer is required for a DNA polymerase to begin synthesizing the leading strand of new DNA. For the lagging strand, each Okazaki fragment must be primed; the primers are converted to DNA before ligase joins the fragments together.

Other Proteins Assisting DNA Replication

You have learned about three kinds of proteins that function in DNA synthesis: DNA polymerase, ligase, and primase. Other kinds of proteins also participate; two of these are helicase and single-strand binding protein. A **helicase** is an enzyme that untwists the double helix at the replication fork, separating the two old strands. Molecules of **single-strand binding protein** then line up along the unpaired DNA strands, holding them apart while they serve as templates for the synthesis of new complementary strands.

FIGURE 16.15 summarizes the functions of the main proteins that cooperate in DNA replication. FIGURE 16.16 is a visual summary of DNA replication.

The DNA Replication Machine as a Stationary Complex

It is traditional—and convenient—to represent DNA polymerase molecules as locomotives moving along a DNA "railroad track," but such a model is inaccurate in two important ways. First, the various proteins that participate in DNA replication actually form a single large complex, a DNA replication "machine." Second, this machine is probably stationary during the replication process. In eukaryotic cells, the multiple copies of the machine, perhaps grouped into "factories," may anchor to the nuclear matrix, a framework of fibers extending through the interior of the nucleus. Recent studies support a model in which DNA polymerase molecules "reel in" the parental DNA and "extrude" newly made daughter DNA molecules.

Initiation of replication

Double helix unwinds, providing single-stranded DNA templates	**Helicases** and **single-strand binding proteins**

Synthesis of leading strand

Priming	**Primase**
Elongation	**DNA polymerase**
Replacement of RNA primer by DNA	**DNA polymerase**

Synthesis of lagging strand

Priming for Okazaki fragment	**Primase**
Elongation of fragment	**DNA polymerase**
Replacement of RNA primer by DNA	**DNA polymerase**
Joining of fragments	**Ligase**

FIGURE 16.15 The main proteins of DNA replication and their functions.

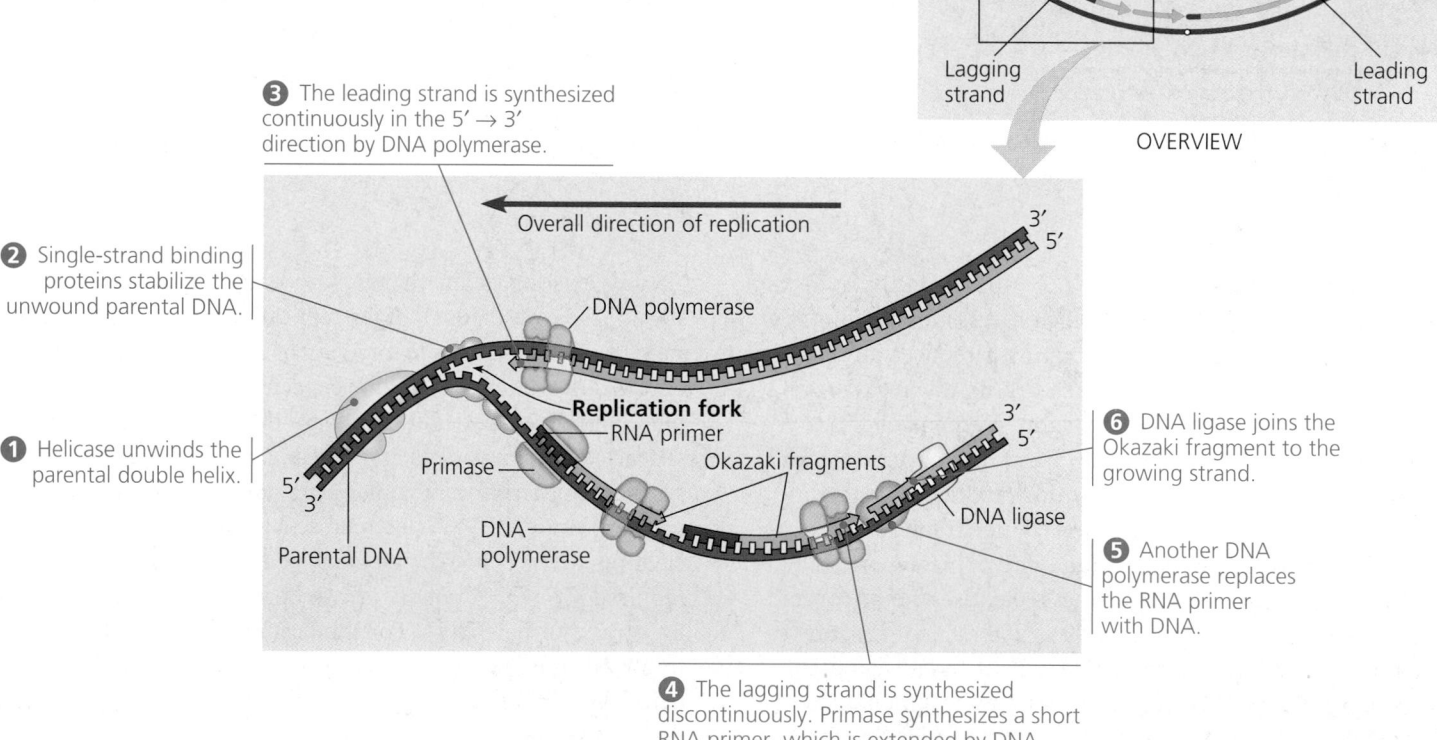

❸ The leading strand is synthesized continuously in the 5′ → 3′ direction by DNA polymerase.

❷ Single-strand binding proteins stabilize the unwound parental DNA.

❶ Helicase unwinds the parental double helix.

Overall direction of replication

DNA polymerase

Replication fork
RNA primer
Primase
Okazaki fragments
DNA polymerase
Parental DNA

DNA ligase

❻ DNA ligase joins the Okazaki fragment to the growing strand.

❺ Another DNA polymerase replaces the RNA primer with DNA.

❹ The lagging strand is synthesized discontinuously. Primase synthesizes a short RNA primer, which is extended by DNA polymerase to form an Okazaki fragment.

Leading strand
Origin of replication
Lagging strand
Lagging strand
Leading strand

OVERVIEW

FIGURE 16.16 A summary of DNA replication. The detailed diagram shows one replication fork, but as indicated in the overview diagram, replication usually occurs simultaneously at two forks, one at either end of a replication bubble. Notice in the overview diagram that a leading strand is initiated by an RNA primer (magenta), as is each Okazaki fragment in a lagging strand. Viewing each daughter strand in its entirety, you can see that half of it is made continuously as a leading strand, while the other half (on the other side of the origin) is synthesized in fragments as a lagging strand.

Enzymes proofread DNA during its replication and repair damage in existing DNA

We cannot attribute the accuracy of DNA replication solely to the specificity of base pairing. Although errors in the completed DNA molecule amount to only one in a billion nucleotides, initial pairing errors between incoming nucleotides and those in the template strand are 100,000 times more common—an error rate of one in 10,000 base pairs. During DNA replication, DNA polymerase itself proofreads each nucleotide against its template as soon as it is added to the growing strand. Upon finding an incorrectly paired nucleotide, the polymerase removes the nucleotide and then resumes synthesis. (This action resembles correcting a word-processing error by using the "delete" key and then entering the correction.)

Mismatched nucleotides sometimes evade proofreading by DNA polymerase or arise after DNA synthesis is completed—by damage to a nucleotide base, for instance. In **mismatch repair,** cells use special enzymes to fix incorrectly paired nucleotides. Researchers spotlighted the importance of such proteins when they found that a hereditary defect in one of them is associated with a form of colon cancer. Apparently, this defect allows cancer-causing errors to accumulate in the DNA.

Maintenance of the genetic information encoded in DNA requires frequent repair of various kinds of damage to existing DNA. DNA molecules are constantly subjected to potentially harmful chemical and physical agents. Reactive chemicals (in the environment and occurring naturally in cells), radioactive emissions, X-rays, and ultraviolet light can change nucleotides in ways that can affect encoded genetic information, usually adversely. In addition, DNA bases often undergo spontaneous chemical changes under normal cellular conditions. Fortunately, changes in DNA are usually corrected before they become self-perpetuating mutations. Each cell continuously monitors and repairs its genetic material. Because repair of damaged DNA is so important to the survival of an organism, it is no surprise that many different DNA repair enzymes have evolved. Almost 100 are known in *E. coli,* and 130 have been identified so far in humans.

Most mechanisms for repairing DNA damage take advantage of the base-paired structure of DNA. Usually, a segment of the strand containing the damage is cut out (excised) by a DNA-cutting enzyme—a **nuclease**—and the resulting gap is filled in with nucleotides properly paired with the nucleotides in the undamaged strand. The enzymes involved in filling the gap are a DNA polymerase and ligase. DNA repair of this type is called **nucleotide excision repair** (FIGURE 16.17).

One function of the DNA repair enzymes in our skin cells is to repair genetic damage caused by the ultraviolet rays of sunlight. One type of damage, the type shown in FIGURE 16.17, is the covalent linking of thymine bases that are adjacent on a DNA strand. Such thymine dimers cause the DNA to buckle

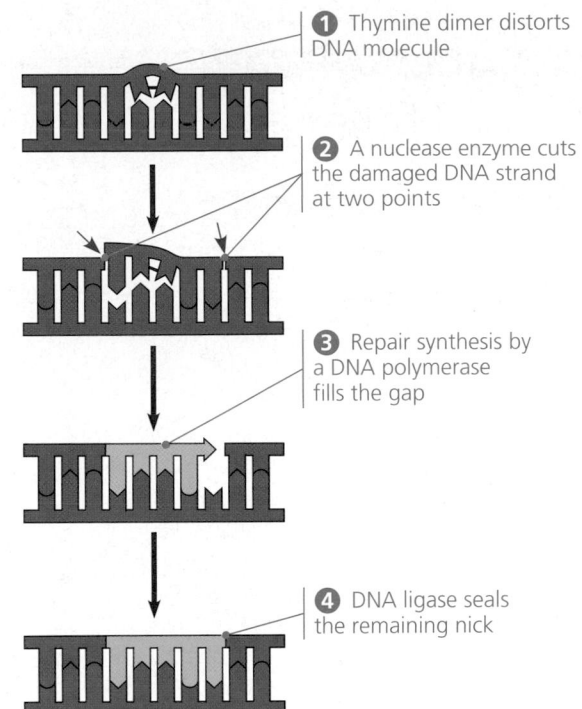

1 Thymine dimer distorts DNA molecule

2 A nuclease enzyme cuts the damaged DNA strand at two points

3 Repair synthesis by a DNA polymerase fills the gap

4 DNA ligase seals the remaining nick

FIGURE 16.17 Nucleotide excision repair of DNA damage. A team of enzymes detects and repairs damaged DNA. This figure shows DNA containing a thymine dimer, a type of damage often caused by ultraviolet radiation. Repair enzymes can excise the damaged region from the DNA and replace it with a normal DNA segment.

and interfere with DNA replication. The importance of repairing this kind of damage is underscored by the disorder xeroderma pigmentosum, which in most cases is caused by an inherited defect in a nucleotide excision repair enzyme. Individuals with this disorder are hypersensitive to sunlight; mutations in their skin cells caused by ultraviolet light are left uncorrected and cause skin cancer.

The ends of DNA molecules are replicated by a special mechanism

Most DNA repair processes involve DNA polymerases, but these enzymes are helpless to fix a "defect" that results from their own limitations. For linear DNA, such as the DNA of eukaryotic chromosomes, the fact that *a DNA polymerase can only add nucleotides to the 3′ end of a preexisting polynucleotide* leads to a potential problem. The usual replication machinery provides no way to complete the 5′ ends of daughter DNA strands; as a result, repeated rounds of replication produce shorter and shorter DNA molecules (FIGURE 16.18, p. 300). If a cell divided enough times, essential genes would be deleted. Clearly, if this trend continued over generations, we would not be here today!

Prokaryotes avoid this problem by having circular DNA molecules (which have no ends), but what about eukaryotes?

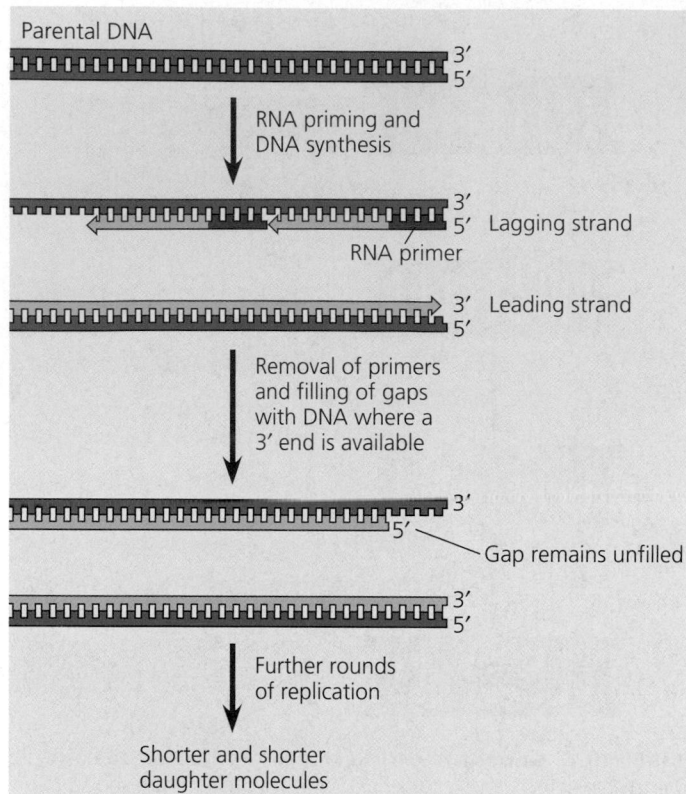

FIGURE 16.18 The end-replication problem. When a linear DNA molecule replicates, a gap is left at the 5' end of each new strand (light blue) because DNA polymerase can only add nucleotides to a 3' end. As a result, with each round of replication, the DNA molecules get slightly shorter. For simplicity we show only one end of a linear DNA molecule.

Eukaryotic chromosomal DNA molecules have special nucleotide sequences called **telomeres** at their ends (FIGURE 16.19). Telomeres do not contain genes; instead, the DNA consists of multiple repetitions of one short nucleotide sequence. The repeated unit in human telomeres, which is typical, is the six-nucleotide sequence TTAGGG. The number of repetitions in a telomere varies between about 100 and 1,000. Telomeric DNA protects the organism's genes from being eroded through successive rounds of DNA replication. In addition, telomeric DNA and special proteins associated with it somehow prevent the ends from activating the cell's systems for monitoring DNA damage. (The end of a DNA molecule that is "seen" as a double-strand break may otherwise trigger signal-transduction pathways leading to cell cycle arrest or cell death.)

In the long term, over the course of generations, eukaryotic organisms need a way of restoring their shortened telomeres. This is provided by **telomerase,** a special enzyme that catalyzes the lengthening of telomeres. But how does telomerase synthesize DNA where the DNA template has been lost? Telomerase is unusual in having a short molecule of RNA along with its protein. The RNA contains a nucleotide sequence that serves as the template for new telomere segments at the 3' end of the

telomere. FIGURE 16.19 shows how telomerase and DNA polymerase work together to lengthen telomeres.

Telomerase is *not* present in most cells of multicellular organisms like ourselves, and the DNA of dividing somatic cells does tend to be shorter in older individuals and in cultured cells that have divided many times. Thus, it is possible that telomeres are a limiting factor in the life span of certain tissues

(a) The bright orange stain marks the telomeres of these mouse chromosomes (LM).

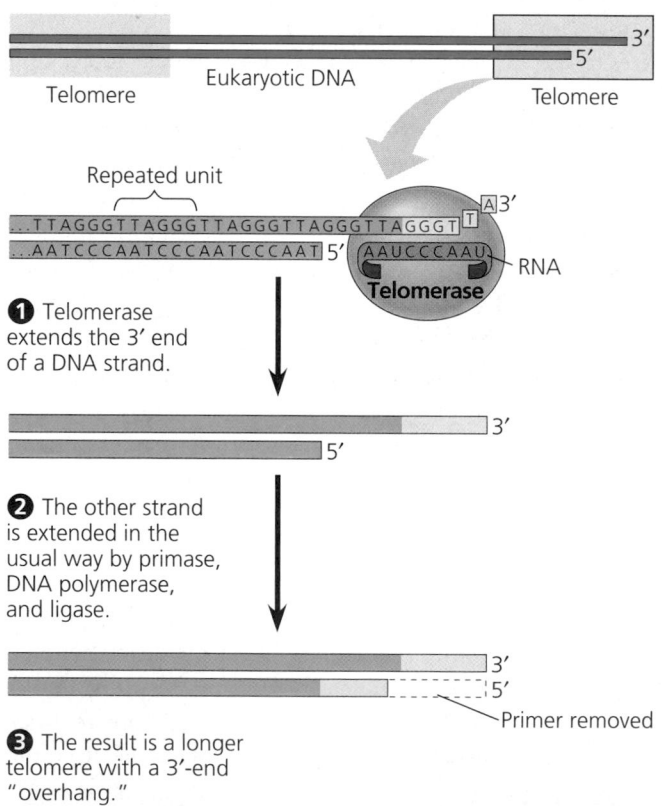

(b) Telomerase has a short molecule of RNA with a sequence that serves as a template for extending the 3' end of the telomere.

FIGURE 16.19 Telomeres and telomerase. Eukaryotes deal with the end-replication issue by having expendable, noncoding sequences called telomeres at the ends of their DNA and the enzyme telomerase in some of their cells.

and even the organism as a whole. In any case, telomerase *is* present in germ-line cells, those that give rise to gametes. The enzyme produces long telomeres in these cells and hence in the newborn.

Intriguingly, researchers have also found telomerase in somatic cells that are cancerous. Cells from large tumors often have unusually short telomeres, as one would expect for cells that have undergone many cell divisions. Progressive shortening would presumably lead eventually to self-destruction of the cancer unless telomerase became available to stabilize telomere length. This is exactly what seems to happen in cancer cells and also in immortal strains of cultured cells (see

Chapter 12). If telomerase is indeed an important factor in many cancers, it may provide a useful target for both cancer diagnosis and chemotherapy.

■   ■   ■

DNA replication provides the copies of genes that parents pass to offspring via gametes. However, it is not enough that genes be copied and transmitted; they must also be expressed. How do genes manifest themselves in phenotypic characters such as eye color? In the next chapter, we will examine the molecular basis of gene expression—how the cell translates genetic information encoded in DNA.

CHAPTER 16 REVIEW

Go to the Campbell Biology website (www.campbellbiology.com) to explore an interactive version of the Chapter Review.

Summary of Key Concepts

DNA AS THE GENETIC MATERIAL

■ **The search for the genetic material led to DNA (pp. 287–290, FIGURES 16.1, 16.2)** Experiments with bacteria and, later, with phages provided the first strong evidence that the genetic material is DNA.
Web/CD Activity 16A: *The Hershey-Chase Experiment*

■ **Watson and Crick discovered the double helix by building models to conform to X-ray data (pp. 290–292, FIGURES 16.3, 16.5, 16.6)** Watson and Crick discovered that DNA is a double helix. Two antiparallel sugar-phosphate chains wind around the outside of the molecule; the nitrogenous bases project into the interior, where they hydrogen-bond in specific pairs, A with T and G with C.
Web/CD Activity 16B: *DNA and RNA Structure*
Web/CD Activity 16C: *3D DNA Double Helix*

DNA REPLICATION AND REPAIR

■ **During DNA replication, base pairing enables existing DNA strands to serve as templates for new complementary strands (pp. 293–294, FIGURE 16.7)** DNA replication is semiconservative: The parent molecule unwinds, and each strand then serves as a template for the synthesis of a new half-molecule according to base-pairing rules.
Web/CD Activity 16D: *DNA Replication: An Overview*
Web/CD Case Study in the Process of Science: *What Is the Correct Model for DNA Replication?*

■ **A large team of enzymes and other proteins carries out DNA replication (pp. 295–298, FIGURES 16.15, 16.16)** Replication begins at ori-

gins of replication. Y-shaped replication forks form at opposite ends of a replication bubble, where the two DNA strands separate. DNA polymerases catalyze the synthesis of new DNA strands, working in the $5' \rightarrow 3'$ direction. DNA synthesis at a replication fork yields a continuous leading strand and short, discontinuous segments of lagging strand. The fragments are then joined together by DNA ligase. DNA synthesis must start on the end of a primer, which is a short segment of RNA.
Web/CD Activity 16E: *DNA Replication: A Closer Look*
Web/CD Activity 16F: *DNA Replication Review*

■ **Enzymes proofread DNA during its replication and repair damage in existing DNA (p. 299, FIGURE 16.17)** DNA polymerase proofreads newly made DNA, replacing any incorrect nucleotides. In mismatch repair of DNA, repair enzymes correct errors in base pairing. In excision repair, enzymes cut out and replace damaged stretches of DNA.

■ **The ends of DNA molecules are replicated by a special mechanism (pp. 299–301, FIGURE 16.19)** The ends of the linear DNA molecules of eukaryotic chromosomes, called telomeres, get shorter with each replication. The enzyme telomerase, present in certain cells, can extend the ends.

Self-Quiz

1. In his work with pneumonia-causing bacteria and mice, Griffith found that
   a. the protein coat from pathogenic cells was able to transform nonpathogenic cells.
   b. heat-killed pathogenic cells caused pneumonia.
   c. some chemical from pathogenic cells was transferred to non-pathogenic cells, making them pathogenic.
   d. the polysaccharide coat of bacteria caused pneumonia.
   e. bacteriophages injected DNA into bacteria.

2. *E. coli* cells grown on ^{15}N medium are transferred to ^{14}N medium and allowed to grow for two generations (two rounds of DNA replication). DNA extracted from these cells is centrifuged. What density distribution of DNA would you expect in this experiment? Explain your answer.

   a. one high-density and one low-density band
   b. one intermediate-density band
   c. one high-density and one intermediate-density band
   d. one low-density and one intermediate-density band
   e. one low-density band

3. A biochemist isolates and purifies various molecules needed for DNA replication. When she adds some DNA, replication occurs, but the DNA molecules formed are defective. Each consists of a normal DNA strand paired with numerous segments of DNA a few hundred nucleotides long. What has she probably left out of the mixture? Explain your answer.

   a. DNA polymerase   d. Okazaki fragments
   b. ligase   e. primers
   c. nucleotides

4. What is the basis for the difference in the synthesis of the leading and lagging strands of DNA molecules?

   a. The origins of replication occur only at the 5′ end.
   b. Helicases and single-strand binding proteins work at the 5′ end.
   c. DNA polymerase can join new nucleotides only to the 3′ end of a growing strand.
   d. DNA ligase works only in the 3′ → 5′ direction.
   e. Polymerase can only work on one strand at a time.

5. In analyzing the number of different bases in a DNA sample, which result would be consistent with the base-pairing rules? Explain your answer.

   a. A = G   d. A = C
   b. A + G = C + T   e. G = T
   c. A + T = G + T

6. The primer that initiates synthesis of a new DNA strand is usually

   a. RNA   d. a structural protein
   b. DNA   e. a thymine dimer
   c. an Okazaki fragment

7. A eukaryotic cell lacking telomerase would

   a. be unable to take up DNA from the surrounding solution.
   b. be unable to identify and correct mismatched nucleotides in its daughter DNA strands.
   c. experience a gradual reduction of chromosome length with each replication cycle.
   d. have a greater potential to become cancerous.
   e. incorporate one extraneous nucleotide for each Okazaki fragment added.

8. The elongation of the *leading* strand during DNA synthesis

   a. progresses away from the replication fork.
   b. occurs in the 3′ → 5′ direction.
   c. produces Okazaki fragments.
   d. depends on the action of DNA polymerase.
   e. does not require a template strand.

9. The spontaneous loss of amino groups from adenine results in hypoxanthine, an unnatural base, opposite thymine. What combination of molecules could the cell use to repair such damage?

   a. nuclease, DNA polymerase, DNA ligase
   b. telomerase, primase, DNA polymerase
   c. telomerase, helicase, single-strand binding protein
   d. DNA ligase, replication fork proteins, adenase
   e. nuclease, telomerase, primase

10. Of the following, the most reasonable inference from the observation that defects in DNA repair enzymes contribute to some forms of cancer is that

    a. cancer is generally inherited.
    b. uncorrected changes in DNA can cause cancer.
    c. cancer cannot occur when DNA repair enzymes work properly.
    d. mutations generally lead to cancer.
    e. cancer is caused by environmental factors that damage DNA repair enzymes.

11. How did Hershey and Chase use radioactively labeled viruses to show that DNA, not protein, is the genetic material?

12. How does complementary base pairing make possible the replication of DNA?

13. Along one strand of a double helix is the nucleotide sequence 5′-GGCATAGGT-3′. What is the corresponding sequence for the other DNA strand?

14. What are three functions of DNA polymerase in DNA replication?

Go to the website or CD-ROM for more quiz questions.

Evolution Connection

Many bacteria may be able to respond to environmental stress by increasing the rate at which mutations occur during cell division. How might this be accomplished, and what might be an evolutionary advantage of this ability?

The Process of Science

Demonstrate your understanding of the Meselson-Stahl experiment by answering the following questions.

a. Describe in your own words exactly what each of the centrifugation bands pictured in FIGURE 16.9 represents.

b. Imagine that the experiment is done as follows: Bacteria are first grown for several generations in a medium containing the *lighter* isotope of nitrogen, ^{14}N, then switched into a medium containing ^{15}N. The rest of the experiment is identical. Redraw FIGURE 16.9 to reflect this experiment, predicting what band positions you would expect after one and after two generations, if each of the three models shown in FIGURE 16.8 were true.

Virtually reenact the Meselson-Stahl experiment in the Case Study in the Process of Science, available on the website and CD-ROM.

Science, Technology, and Society

Cooperation and competition are both common in science. What roles did these two social behaviors play in Watson and Crick's discovery of the double helix? How might competition between scientists accelerate progress in a scientific field? How might it slow progress?

CHAPTER 17

FROM GENE TO PROTEIN

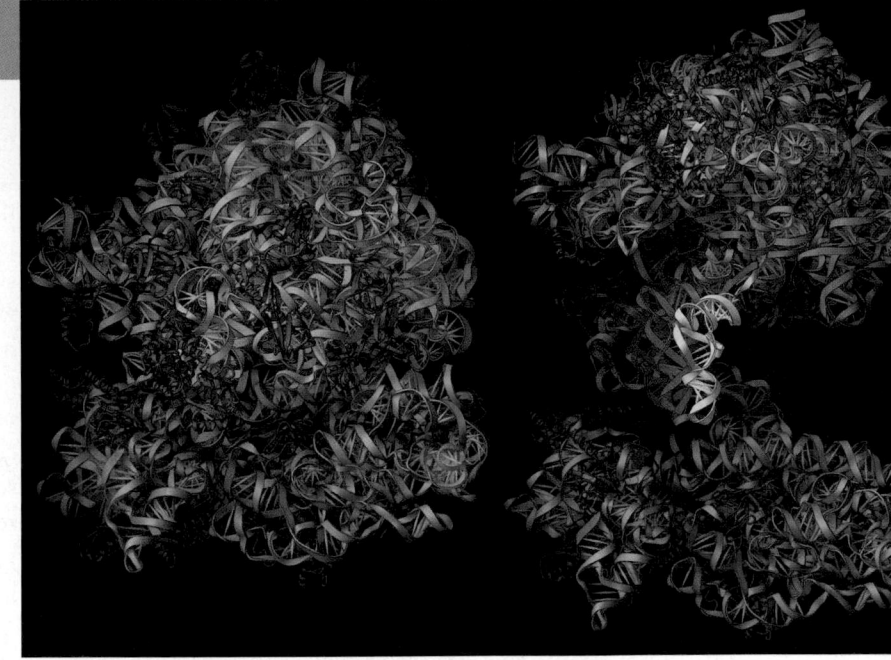

THE CONNECTION BETWEEN GENES AND PROTEINS

- The study of metabolic defects provided evidence that genes specify proteins
- Transcription and translation are the two main processes linking gene to protein: *an overview*
- In the genetic code, nucleotide triplets specify amino acids
- The genetic code must have evolved very early in the history of life

THE SYNTHESIS AND PROCESSING OF RNA

- Transcription is the DNA-directed synthesis of RNA: *a closer look*
- Eukaryotic cells modify RNA after transcription

THE SYNTHESIS OF PROTEIN

- Translation is the RNA-directed synthesis of a polypeptide: *a closer look*
- Signal peptides target some eukaryotic polypeptides to specific destinations in the cell
- RNA plays multiple roles in the cell: *a review*
- Comparing protein synthesis in prokaryotes and eukaryotes: *a review*
- Point mutations can affect protein structure and function
- What is a gene? *revisiting the question*

The information content of DNA, the genetic material, is in the form of specific sequences of nucleotides along the DNA strands. But how is this information related to an organism's inherited traits? Put another way, what does a gene actually say? And how is its message translated by cells into a specific trait, such as brown hair or type A blood?

Consider, once again, Mendel's peas. One of the characters Mendel studied was stem length (see TABLE 14.1*). Variation in a single gene accounts for the difference between the tall and dwarf varieties of pea plants. Mendel did not know the physiological basis of this phenotypic difference, but plant scientists have since worked out the explanation: Dwarf peas lack growth hormones called gibberellins, which stimulate the normal elongation of* stems. *A dwarf plant treated with gibberellins grows to normal height. Why do dwarf peas fail to make their own gibberellins? They are missing a key protein, an enzyme required for gibberellin synthesis. This example illustrates the main point of this chapter: The DNA inherited by an organism leads to specific traits by dictating the synthesis of proteins. Proteins are the links between genotype and phenotype. This chapter explores the flow of information from genes to proteins.*

THE CONNECTION BETWEEN GENES AND PROTEINS

Researchers are now working out the structures and processes of protein synthesis in amazing detail. See, for example, the computer graphic on this page, which shows a ribosome at the molecular level. The exploded view at the right also shows three transfer RNA molecules, in red, orange, and yellow. As you will learn in this chapter, these molecules cooperate with the ribosome in polypeptide synthesis. But before going into the details of how genes direct protein synthesis, let's step back and examine how the fundamental relationship between genes and proteins was discovered.

The study of metabolic defects provided evidence that genes specify proteins

In 1909, British physician Archibald Garrod was the first to suggest that genes dictate phenotypes through enzymes that catalyze specific chemical reactions in the cell. Garrod postulated that the symptoms of an inherited disease reflect a person's inability to make a particular enzyme. He referred to such diseases as "inborn errors of metabolism."

THE PROCESS OF SCIENCE

303

Garrod gave as one example the hereditary condition called alkaptonuria, in which the urine is black because it contains the chemical alkapton, which darkens upon exposure to air. Garrod reasoned that normal individuals have an enzyme that breaks down alkapton, whereas alkaptonuric individuals have inherited an inability to make the enzyme that metabolizes alkapton.

How Genes Control Metabolism: One Gene–One Enzyme

Garrod's idea was ahead of its time, but research conducted several decades later supported his hypothesis that a gene dictates the production of a specific enzyme. Biochemists accumulated much evidence that cells synthesize and degrade most organic molecules via metabolic pathways, in which each chemical reaction in a sequence is catalyzed by a specific enzyme. Such metabolic pathways lead, for instance, to the synthesis of the pigments that give fruit flies (*Drosophila*) their eye color (see FIGURE 15.2). In the 1930s, George Beadle and Boris Ephrussi speculated that each of the various mutations affecting eye color in *Drosophila* blocks pigment synthesis at a specific step by preventing production of the enzyme that catalyzes that step. However, neither the chemical reactions nor the enzymes that catalyze them were known at the time.

A breakthrough in demonstrating the relationship between genes and enzymes came a few years later, when Beadle and Edward Tatum began working with a bread mold, *Neurospora crassa*. They treated *Neurospora* with X-rays and then looked among the survivors for mutants that differed from the wild-type mold in their nutritional needs. Wild-type *Neurospora* has modest food requirements. It can survive in the laboratory on agar (a moist support medium) mixed only with inorganic salts, glucose, and the vitamin biotin. From this *minimal medium,* the mold uses its metabolic pathways to produce all the other molecules it needs. Beadle and Tatum identified mutants that could not survive on minimal medium, apparently because they were unable to synthesize certain essential molecules from the minimal ingredients. However, most such nutritional mutants *can* survive on a *complete growth medium,* minimal medium supplemented with all 20 amino acids and a few other nutrients.

To characterize the metabolic defect in each nutritional mutant, Beadle and Tatum took samples from the mutant growing on complete medium and distributed them to a number of different vials. Each vial contained minimal medium plus a single additional nutrient. The particular supplement that allowed growth indicated the metabolic defect. For example, if the only supplemented vial that supported growth of the mutant was the one fortified with the amino acid arginine, the researchers could conclude that the mutant was defective in the biochemical pathway that wild-type cells use to synthesize arginine.

Beadle and Tatum went on to pin down each mutant's defect more specifically. Their work with arginine-requiring mutants was especially instructive. Using genetic crosses, they determined that their mutants fell into three classes, each mutated in a different gene. The researchers then showed that they could distinguish among the classes of mutants nutritionally by additional tests of their growth requirements (FIGURE 17.1). In the synthetic pathway leading to arginine, they suspected, a precursor nutrient is converted to ornithine, which is converted to citrulline, which is converted to arginine. When they tested their arginine mutants for growth on ornithine and citrulline, they found that one class could grow on either compound (or arginine), the second class only on citrulline (or arginine), and the third on neither—it absolutely required arginine. The three classes of mutants, the researchers reasoned, must be blocked at different steps in the pathway that synthesizes arginine, with each mutant class lacking the enzyme that catalyzes the blocked step.

Because each mutant was defective in a single gene, Beadle and Tatum's results provided strong support for the *one gene–one enzyme hypothesis,* as they dubbed it, which states that the function of a gene is to dictate the production of a specific enzyme. The researchers also showed how a combination of genetics and biochemistry could be used to work out the steps in a metabolic pathway. Further support for the one gene–one enzyme hypothesis came with biochemical experiments that identified the specific enzymes lacking in the mutants.

One Gene–One Polypeptide

As researchers learned more about proteins, they made minor revisions in the one gene–one enzyme hypothesis. Not all proteins are enzymes. Keratin, the structural protein of animal hair, and the hormone insulin are two examples of nonenzyme proteins. Because proteins that are not enzymes are nevertheless gene products, molecular biologists began to think in terms of one gene–one protein. However, many proteins are constructed from two or more different polypeptide chains, and each polypeptide is specified by its own gene. For example, hemoglobin, the oxygen-transporting protein of vertebrate red blood cells, is built from two kinds of polypeptides, and thus two genes code for this protein (see FIGURE 5.23b). We can therefore restate Beadle and Tatum's idea as the **one gene–one polypeptide hypothesis.** Note, however, that it is common to refer to proteins, rather than polypeptides, as the gene products, a practice you will encounter in this book.

Transcription and translation are the two main processes linking gene to protein: *an overview*

Genes provide the instructions for making specific proteins. But a gene does not build a protein directly. The bridge between DNA and protein synthesis is RNA. You learned in Chapter 5

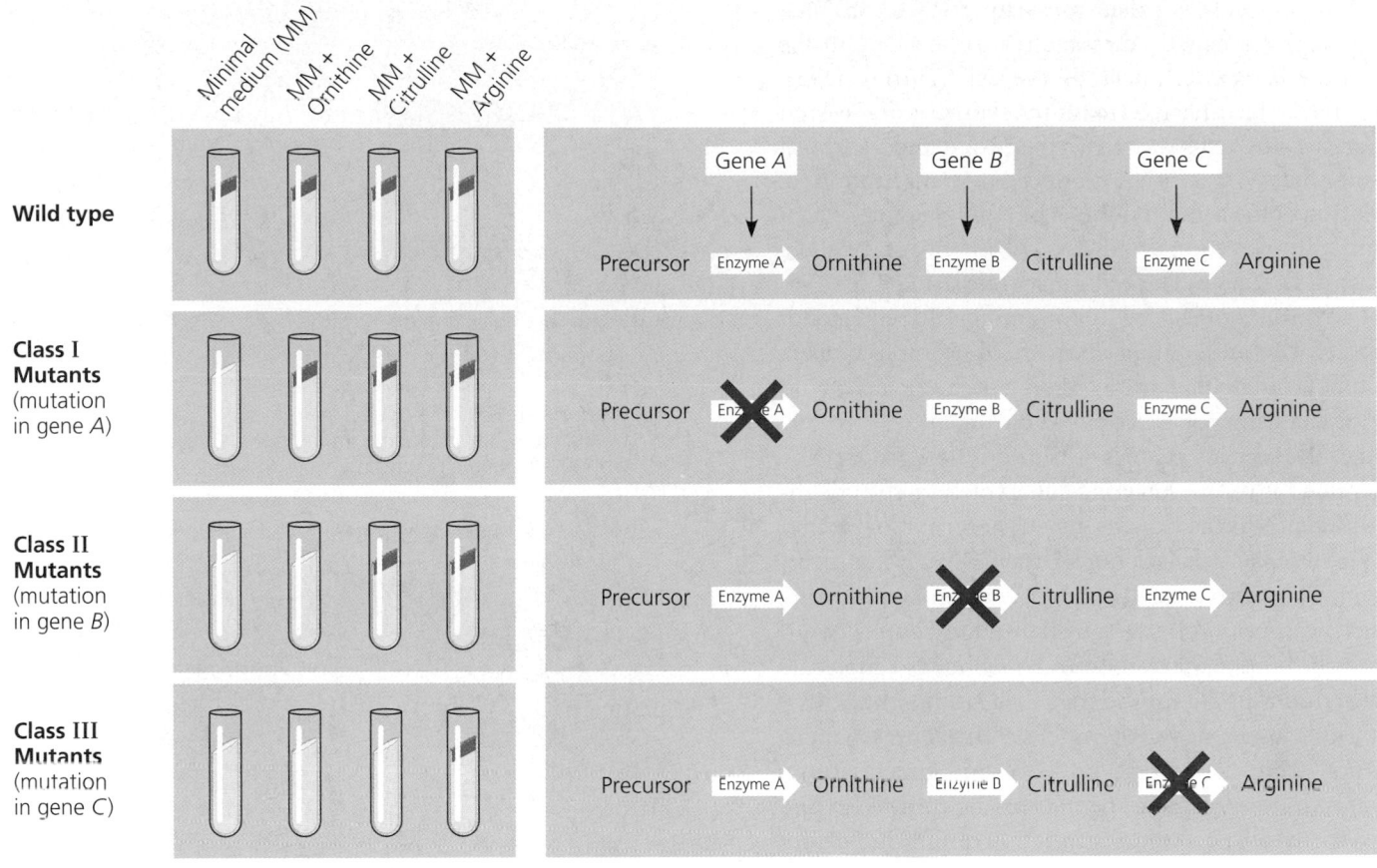

Wild type

Class I Mutants (mutation in gene *A*)

Class II Mutants (mutation in gene *B*)

Class III Mutants (mutation in gene C)

Minimal medium (MM) | MM + Ornithine | MM + Citrulline | MM + Arginine

Gene *A*   Gene *B*   Gene *C*

Precursor   Enzyme A   Ornithine   Enzyme B   Citrulline   Enzyme C   Arginine

(a) Experiment. The researchers tested three classes of arginine mutants for their ability to grow on minimal medium supplemented with ornithine or citruline.

(b) Interpretation. From the growth patterns of the mutants, they deduced that each mutant was unable to carry out one step in the arginine pathway, presumably because it lacked the necessary enzyme.

FIGURE 17.1 Beadle and Tatum's evidence for the one gene–one enzyme hypothesis. In this experiment, the researchers studied three classes of mutants of the mold *Neurospora crassa*, all defective in synthesizing the amino acid arginine. The wild-type strain requires only a minimal nutritional medium for growth; it makes arginine by using a multistep pathway in which ornithine and citrulline are intermediates. The different types of mutants had different growth requirements. Beadle and Tatum concluded that each gene mutated must normally dictate the production of one enzyme: the one gene–one enzyme hypothesis.

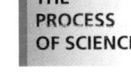

THE PROCESS OF SCIENCE

that RNA is chemically similar to DNA, except that it contains ribose instead of deoxyribose as its sugar and has the nitrogenous base uracil rather than thymine (see FIGURE 5.29). Thus, each nucleotide along a DNA strand has deoxyribose as its sugar and A, G, C, or T as its base; each nucleotide along an RNA strand has ribose as its sugar and A, G, C, or U as its base. An RNA molecule almost always consists of a single strand.

It is customary to describe the flow of information from gene to protein in linguistic terms because both nucleic acids and proteins are polymers with specific sequences of monomers that convey information, much as specific sequences of letters communicate information in a language like English. In DNA or RNA, the monomers are the four types of nucleotides, which differ in their nitrogenous bases. Genes are typically hundreds or thousands of nucleotides long, each gene having a specific sequence of bases. Each polypeptide of a protein also has monomers arranged in a particular linear order (the protein's primary structure), but its monomers are the 20 amino acids. Thus, nucleic acids and proteins contain information written in two different chemical languages. To get from DNA, written in one language, to protein, written in the other, requires two major stages, transcription and translation.

Transcription is the synthesis of RNA under the direction of DNA. Both nucleic acids use the same language, and the information is simply transcribed, or copied, from one molecule to the other. Just as a DNA strand provides a template for the synthesis of a new complementary strand during DNA replication, it provides a template for assembling a sequence of RNA nucleotides. The resulting RNA molecule is a faithful transcript of the gene's protein-building instructions. This type of RNA molecule is called **messenger RNA (mRNA),** because it carries a genetic message from the DNA to the protein-synthesizing machinery of the cell (FIGURE 17.2a). (Transcription is the general term for the synthesis of *any* kind of RNA on a DNA template. Later in this chapter, you will learn about other types of RNA produced by transcription.)

Translation is the actual synthesis of a polypeptide, which occurs under the direction of mRNA. During this stage, there is a change in language: The cell must translate the base sequence of an mRNA molecule into the amino acid sequence of a polypeptide. The sites of translation are ribosomes, complex particles that facilitate the orderly linking of amino acids into polypeptide chains.

Although the basic mechanics of transcription and translation are similar for prokaryotes and eukaryotes, there is an important difference in the flow of genetic information within the cells. Because bacteria lack nuclei, their DNA is not segregated from ribosomes and the other protein-synthesizing equipment. Transcription and translation are coupled, with ribosomes attaching to the leading end of an mRNA molecule while transcription is still in progress (see FIGURE 17.22). In a eukaryotic cell, by contrast, the nuclear envelope separates transcription from translation in space and time (FIGURE 17.2b). Transcription occurs in the nucleus, and mRNA is dispatched to the cytoplasm, where translation occurs. But before they can leave the nucleus, eukaryotic RNA transcripts are modified in various ways to produce the final, functional mRNA. Thus, in a two-step process, the transcription of a eukaryotic gene results in *pre-mRNA,* and **RNA processing** yields the finished mRNA. A more general term for an initial RNA transcript is **primary transcript.**

Let's summarize the main point of our overview of protein synthesis: Genes program protein synthesis via genetic messages in the form of messenger RNA. Put another way, cells are governed by a molecular chain of command: DNA $\rightarrow$ RNA $\rightarrow$ protein. The next section discusses how the instructions for assembling amino acids into a specific order are encoded in nucleic acids.

In the genetic code, nucleotide triplets specify amino acids

When biologists began to suspect that the instructions for protein synthesis were encoded in DNA, they recognized a problem: There are only four nucleotides to specify 20 amino acids. Thus, the genetic code cannot be a language like Chinese, where each written symbol corresponds to a single word. If each nucleotide base were translated into an amino acid, only 4 of the 20 amino acids could be specified. Would a language of two-letter code words suffice? The base sequence AG, for example, could specify one amino acid, and GT could

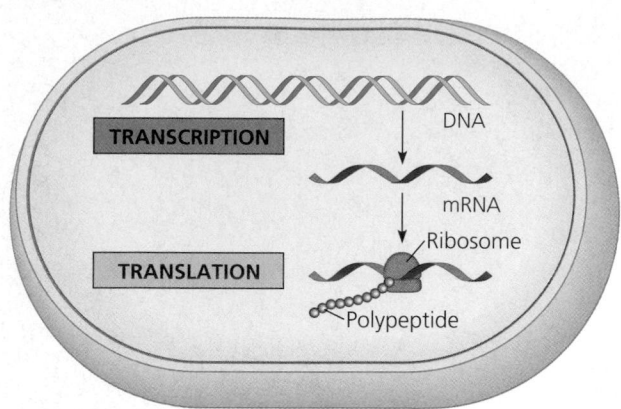

(a) Prokaryotic cell. In a cell lacking a nucleus, mRNA produced by transcription is immediately translated without additional processing

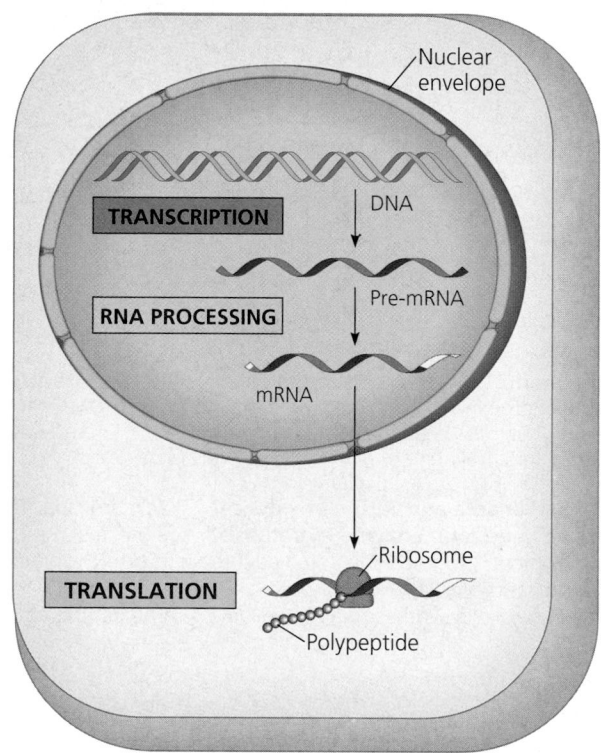

(b) Eukaryotic cell. The nucleus provides a separate compartment for transcription. The original RNA transcript, called pre-mRNA, is processed in various ways before leaving the nucleus as mRNA.

FIGURE 17.2 Overview: the roles of transcription and translation in the flow of genetic information. In a cell, inherited information flows from DNA to RNA to protein. The two main stages of information flow are transcription and translation. In transcription, a gene provides the instructions for synthesizing a messenger RNA (mRNA) molecule. In translation, the information encoded in mRNA determines the order of amino acids that are joined to form a specific polypeptide. The sites of translation are ribosomes.

A miniature version of part b (or sometimes part a) accompanies several figures later in the chapter as an orientation diagram to help you see where a particular figure fits into the overall scheme.

specify another. Since there are four bases, this would give us 16 (that is, 4^2) possible arrangements—still not enough to code for all 20 amino acids.

Triplets of nucleotide bases are the smallest units of uniform length that can code for all the amino acids. If each arrangement of three consecutive bases specifies an amino acid, there can be 64 (that is, 4^3) possible code words—more than enough to specify all the amino acids. Experiments have verified that the flow of information from gene to protein is based on a **triplet code:** The genetic instructions for a polypeptide chain are written in the DNA as a series of three-nucleotide words. For example, the base triplet AGT at a particular position along a DNA strand results in the placement of the amino acid serine at the corresponding position of the polypeptide to be produced.

As you know, a cell does not directly translate a gene into amino acids. The intermediate step is transcription, during which the gene determines the sequence of base triplets along the length of an mRNA molecule. For each gene, only one of the two DNA strands is transcribed (FIGURE 17.3). This strand is called the **template strand,** because it provides the template for ordering the sequence of nucleotides in an RNA transcript. A given DNA strand can be the template strand in some regions of a DNA molecule, while in other regions along the double helix it is the complementary strand that functions as the template for RNA synthesis.

An mRNA molecule is complementary rather than identical to its DNA template because RNA bases are assembled on the template according to base-pairing rules. The pairs are similar to those that form during DNA replication, except that U, the RNA substitute for T, pairs with A. Thus, when a DNA strand is transcribed, the base triplet ACC in DNA provides a template for UGG in the mRNA molecule. The mRNA base triplets are called **codons.** For example, UGG is the codon for the amino acid tryptophan (abbreviated Trp). (The term *codon* is also sometimes used for the complementary DNA base triplet. For example, the DNA codon corresponding to the RNA codon UGG is ACC.)

During translation, the sequence of codons along an mRNA molecule is decoded, or translated, into a sequence of amino acids making up a polypeptide chain. The codons are read in the $5' \rightarrow 3'$ direction along the mRNA. (To review what is meant by the 5' and 3' ends of a nucleic acid chain, see FIGURE 16.12.) Each codon specifies which one of the 20 amino acids will be incorporated at the corresponding position along a polypeptide. Because codons are base triplets, the number of nucleotides making up a genetic message must be three times the number of amino acids making up the protein product. For example, it takes 300 nucleotides along an RNA strand to code for a polypeptide that is 100 amino acids long.

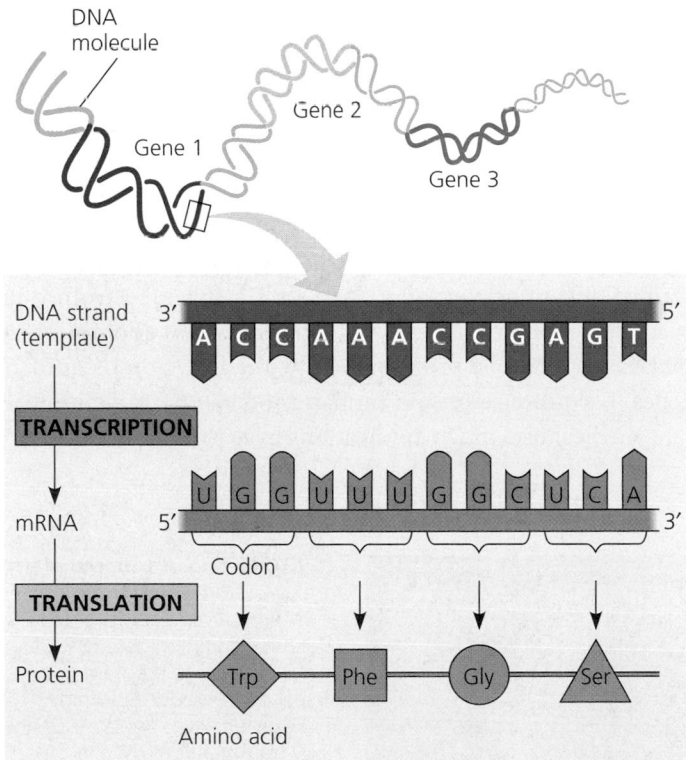

FIGURE 17.3 The triplet code. For each gene, one DNA strand functions as a template for transcription—the synthesis of a complementary mRNA molecule. The base-pairing rules for DNA synthesis also guide transcription, but uracil (U) takes the place of thymine (T) in RNA. During translation, the mRNA is read as a sequence of base triplets, called codons. Each codon specifies an amino acid to be added to the growing polypeptide chain. The mRNA is read in the $5' \rightarrow 3'$ direction.

Cracking the Genetic Code

Molecular biologists cracked the code of life in the early 1960s, when a series of elegant experiments disclosed the amino acid translations of each of the RNA codons. The first codon was deciphered in 1961 by Marshall Nirenberg, of the National Institutes of Health, and his colleagues. Nirenberg synthesized an artificial mRNA by linking identical RNA nucleotides containing uracil as their base. No matter where this message started or stopped, it could contain only one codon in repetition: UUU. Nirenberg added this "poly(U)" to a test-tube mixture containing amino acids, ribosomes, and the other components required for protein synthesis. His artificial system translated the poly(U) into a polypeptide containing a single amino acid, phenylalanine (Phe), strung together as a long polyphenylalanine chain. Thus, Nirenberg determined that the mRNA codon UUU specifies the amino acid phenylalanine. Soon, the amino acids specified by the codons AAA, GGG, and CCC were also determined.

Although more elaborate techniques were required to decode mixed triplets such as AUA and CGA, all 64 codons were

FIGURE 17.4 The dictionary of the genetic code

		Second base			
	U	**C**	**A**	**G**	
U	UUU ⎤ Phe UUC ⎦ UUA ⎤ Leu UUG ⎦	UCU ⎤ UCC UCA Ser UCG ⎦	UAU ⎤ Tyr UAC ⎦ UAA Stop UAG Stop	UGU ⎤ Cys UGC ⎦ UGA Stop UGG Trp	U C A G
C	CUU ⎤ CUC CUA Leu CUG ⎦	CCU ⎤ CCC CCA Pro CCG ⎦	CAU ⎤ His CAC ⎦ CAA ⎤ Gln CAG ⎦	CGU ⎤ CGC CGA Arg CGG ⎦	U C A G
A	AUU ⎤ AUC Ile AUA ⎦ AUG Met or start	ACU ⎤ ACC ACA Thr ACG ⎦	AAU ⎤ Asn AAC ⎦ AAA ⎤ Lys AAG ⎦	AGU ⎤ Ser AGC ⎦ AGA ⎤ Arg AGG ⎦	U C A G
G	GUU ⎤ GUC GUA Val GUG ⎦	GCU ⎤ GCC GCA Ala GCG ⎦	GAU ⎤ Asp GAC ⎦ GAA ⎤ Glu GAG ⎦	GGU ⎤ GGC GGA Gly GGG ⎦	U C A G

First base (5′ end) · Third base (3′ end)

FIGURE 17.4 The dictionary of the genetic code. The three bases of an mRNA codon are designated here as the first, second, and third bases, reading in the 5′ → 3′ direction along the mRNA. (Practice using this dictionary by finding the codons in FIGURE 17.3.) The codon AUG not only stands for the amino acid methionine (Met) but also functions as a "start" signal for ribosomes to begin translating the mRNA at that point. Three of the 64 codons function as "stop" signals. Any one of these termination codons marks the end of a genetic message.

deciphered by the mid-1960s. As FIGURE 17.4 shows, 61 of the 64 triplets code for amino acids. Notice that the codon AUG has a dual function: It not only codes for the amino acid methionine (Met), but also functions as a "start" signal, or initiation codon. Genetic messages begin with the mRNA codon AUG, which signals the protein-synthesizing machinery to begin translating the mRNA at that location. (Because AUG also stands for methionine, polypeptide chains begin with methionine when they are synthesized. However, an enzyme may subsequently remove this starter amino acid from a chain.) The remaining three codons do not designate amino acids. Instead, they are "stop" signals, or termination codons, marking the end of translation.

Notice in FIGURE 17.4 that there is *redundancy* in the genetic code, but no ambiguity. For example, although codons GAA and GAG both specify glutamic acid (redundancy), neither of them ever specifies any other amino acid (no ambiguity). The redundancy in the code is not altogether random. In many cases, codons that are synonyms for a particular amino acid differ only in the third base of the triplet. We will consider a possible benefit for this redundancy later in the chapter.

Our ability to extract the intended message from a written language depends on reading the symbols in the correct groupings—that is, in the correct **reading frame.** Consider this statement: "The red dog ate the cat." Group the letters incorrectly by starting at the wrong point, and the result will probably be gibberish: for example, "her edd oga tet hec at." The reading frame is also important in the molecular language of cells. The short stretch of polypeptide shown in FIGURE 17.3, for instance, will only be made correctly if the mRNA nucleotides are read from left to right (5′ → 3′) in the groups of three shown in the figure: UGG UUU GGC UCA. Although a genetic message is written with no spaces between the codons, the cell's protein-synthesizing machinery reads the message as a series of nonoverlapping three-letter words. The message is *not* read as a series of overlapping words—UGGUUU, and so on—which would convey a very different message.

Let's summarize what we have just covered. Genetic information is encoded as a sequence of nonoverlapping base triplets, or codons, each of which is translated into a specific amino acid during protein synthesis.

The genetic code must have evolved very early in the history of life

The genetic code is nearly universal, shared by organisms from the simplest bacteria to the most complex plants and animals. The RNA codon CCG, for instance, is translated as the amino acid proline in all organisms whose genetic code has been examined. In laboratory experiments, genes can be transcribed and translated after they are transplanted from one species to another (FIGURE 17.5). One important application is that bacteria can be programmed by the insertion of human genes to synthesize certain human proteins that have important medical uses. Such applications have produced many ex-

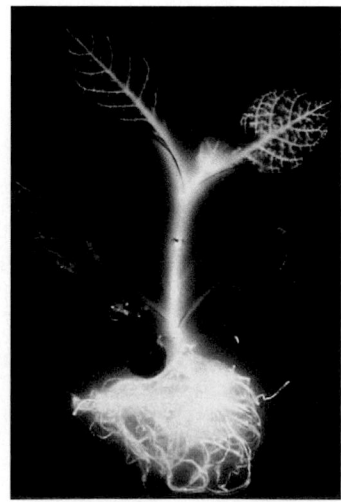

FIGURE 17.5 A tobacco plant expressing a firefly gene. Because diverse forms of life share a common genetic code, it is possible to program one species to produce proteins characteristic of another species by transplanting DNA. In this experiment, researchers were able to incorporate a gene from a firefly into the DNA of a tobacco plant. The gene codes for the firefly enzyme that catalyzes the chemical reaction that releases energy in the form of light.

citing developments in biotechnology, which you will learn about in Chapter 20.

Exceptions to the universality of the genetic code are translation systems where a few codons differ from the standard ones. The main examples are found in certain single-celled eukaryotes, such as *Paramecium,* an organism you may know from the lab. Other examples are found in certain mitochondria and chloroplasts, which transcribe and translate the genes carried by their small amount of DNA. However, the evolutionary significance of the code's *near* universality is clear. A language shared by all living things must have been operating very early in the history of life—early enough to be present in the common ancestors of all modern organisms. A shared genetic vocabulary is a reminder of the kinship that bonds all life on Earth.

Now that we have considered the linguistic logic and evolutionary significance of the genetic code, we are ready to re-examine transcription, translation, and related topics in more detail.

THE SYNTHESIS AND PROCESSING OF RNA

Transcription is the DNA-directed synthesis of RNA: *a closer look*

Messenger RNA, the carrier of information from DNA to the cell's protein-synthesizing machinery, is transcribed from the template strand of a gene. An enzyme called an **RNA polymerase** pries the two strands of DNA apart and hooks together the RNA nucleotides as they base-pair along the DNA template (FIGURE 17.6). Like the DNA polymerases

Promoter — Transcription unit — **Terminator**

DNA of gene

Start point

Termination point

RNA polymerase

❶ Initiation. After RNA polymerase binds to the promoter, the DNA strands unwind, and the enzyme initiates RNA synthesis at the start point on the template strand.

Unwound DNA

RNA transcript

Template strand of DNA

❷ Elongation. The polymerase moves downstream, unwinding the DNA and elongating the RNA transcript 5' → 3'. In the wake of transcription, the DNA strands re-form a double helix.

Rewound DNA

RNA transcript

❸ Termination. Eventually, the polymerase transcribes a terminator sequence, which signals the end of the transcription unit. Shortly thereafter, the RNA transcript is released, and the polymerase detaches from the DNA.

Completed RNA transcript

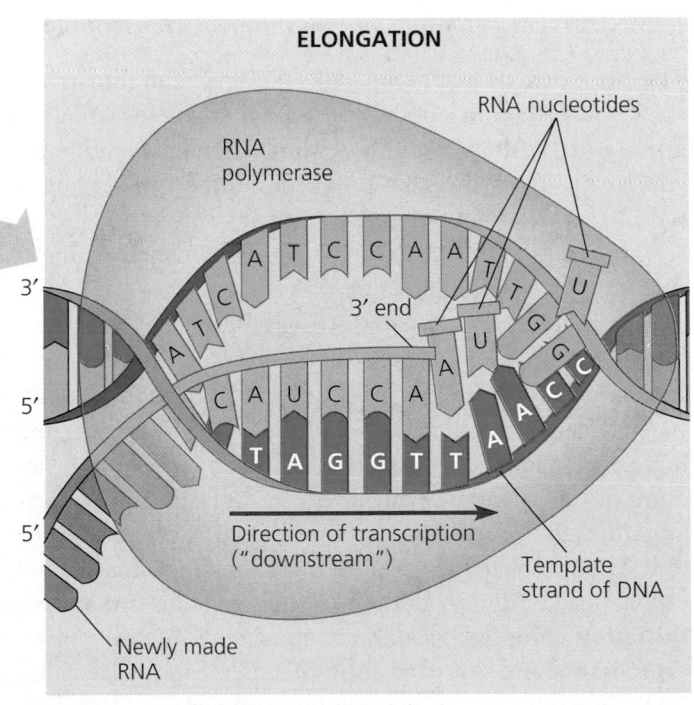

ELONGATION

RNA polymerase

RNA nucleotides

3' end

Direction of transcription ("downstream")

Template strand of DNA

Newly made RNA

FIGURE 17.6 The stages of transcription: initiation, elongation, and termination. RNA polymerase moves along a gene from the promoter (green) to just beyond the terminator (red), assembling an RNA molecule (transcript) complementary to the gene's template strand. In a prokaryote, the RNA transcript of a protein-coding gene is immediately usable as mRNA; in a eukaryote, it must first undergo processing, as described on pp. 311–312.

that function in DNA replication, RNA polymerases can add nucleotides only to the 3′ end of the growing polymer. Thus, an RNA molecule elongates in its 5′ → 3′ direction.

Specific sequences of nucleotides along the DNA mark where transcription of a gene begins and ends. The DNA sequence where RNA polymerase attaches and initiates transcription is known as the **promoter;** the sequence that signals the end of transcription is called the **terminator.** Molecular biologists refer to the direction of transcription as "downstream" and the other direction as "upstream." These terms are also used to describe the positions of nucleotide sequences within the DNA or RNA. Thus, the promoter sequence in DNA is said to be upstream from the terminator. The stretch of DNA that is transcribed into an RNA molecule is called a **transcription unit.**

Bacteria have a single type of RNA polymerase that synthesizes not only mRNA but also other types of RNA that function in protein synthesis. In contrast, eukaryotes have three types of RNA polymerase in their nuclei, numbered I, II, and III. The one used for mRNA synthesis is RNA polymerase II. In the discussion of transcription that follows, we start with the features of mRNA synthesis common to both prokaryotes and eukaryotes and then describe some key differences.

The three stages of transcription, as shown in FIGURE 17.6 and described next, are initiation, elongation, and termination of the RNA chain. Study FIGURE 17.6 to familiarize yourself with the stages of transcription and the terms used to describe them.

RNA Polymerase Binding and Initiation of Transcription

The promoter of a gene includes within it the transcription start point (the nucleotide where RNA synthesis actually begins) and typically extends several dozen nucleotide pairs "upstream" from the start point. In addition to serving as a binding site for RNA polymerase and determining where transcription starts, the promoter determines which of the two strands of the DNA helix is used as the template.

Certain sections of a promoter are especially important for binding RNA polymerase. In prokaryotes, the RNA polymerase itself specifically recognizes and binds to the promoter. In eukaryotes, a collection of proteins called **transcription factors** mediate the binding of RNA polymerase and the initiation of transcription. Only after certain transcription factors are attached to the promoter does the RNA polymerase bind to it. The completed assembly of transcription factors and RNA polymerase bound to the promoter is called a **transcription initiation complex.** FIGURE 17.7 shows the role of transcription factors and a crucial promoter DNA sequence called a **TATA box** in forming the initiation complex.

The interaction between eukaryotic RNA polymerase and transcription factors is an example of the special importance of protein-protein interactions in controlling eukaryotic transcription (as we will discuss further in Chapter 19). Once the polymerase is firmly attached to the promoter DNA, the two DNA strands unwind there, and the enzyme starts transcribing the template strand.

Elongation of the RNA Strand

As RNA polymerase moves along the DNA, it continues to untwist the double helix, exposing about 10 to 20 DNA bases at a time for pairing with RNA nucleotides (see FIGURE 17.6). The enzyme adds nucleotides to the 3′ end of the growing RNA molecule as it continues along the double helix. In the wake of this advancing wave of RNA synthesis, the DNA double helix

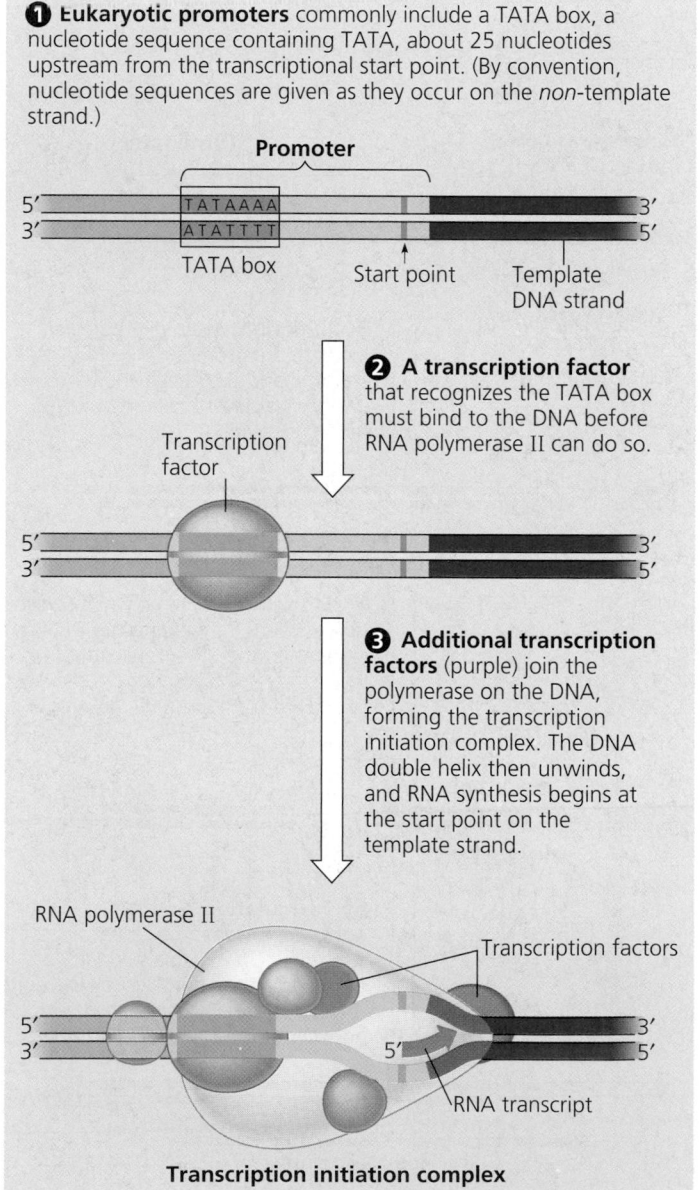

❶ **Eukaryotic promoters** commonly include a TATA box, a nucleotide sequence containing TATA, about 25 nucleotides upstream from the transcriptional start point. (By convention, nucleotide sequences are given as they occur on the *non*-template strand.)

Promoter

5′ — TATAAAA — 3′
3′ — ATATTTT — 5′

TATA box   Start point   Template DNA strand

❷ **A transcription factor** that recognizes the TATA box must bind to the DNA before RNA polymerase II can do so.

Transcription factor

5′ — 3′
3′ — 5′

❸ **Additional transcription factors** (purple) join the polymerase on the DNA, forming the transcription initiation complex. The DNA double helix then unwinds, and RNA synthesis begins at the start point on the template strand.

RNA polymerase II

Transcription factors

5′ — 3′
3′ — 5′
5′

RNA transcript

Transcription initiation complex

FIGURE 17.7 The initiation of transcription at a eukaryotic promoter. In eukaryotic cells, proteins called transcription factors mediate the initiation of transcription by RNA polymerase. The enzyme that transcribes protein-coding genes in eukaryotic cells is called RNA polymerase II.

re-forms and the new RNA molecule peels away from its DNA template. Transcription progresses at a rate of about 60 nucleotides per second in eukaryotes.

A single gene can be transcribed simultaneously by several molecules of RNA polymerase following each other like trucks in a convoy. A growing strand of RNA trails off from each polymerase, with the length of each new strand reflecting how far along the template the enzyme has traveled from the start point (see FIGURE 17.22). The congregation of many polymerase molecules simultaneously transcribing a single gene increases the amount of mRNA transcribed from it, which helps the cell make the encoded protein in large amounts.

Termination of Transcription

Transcription proceeds until after the RNA polymerase transcribes a terminator sequence in the DNA. The transcribed terminator—an *RNA* sequence—functions as the actual termination signal. There are several different mechanisms of transcription termination, the details of which are still somewhat murky. In the prokaryotic cell, transcription usually stops right at the end of the termination signal; when the polymerase reaches that point, it releases both the RNA and the DNA. By contrast, in the eukaryotic cell, the polymerase continues for hundreds of nucleotides past the termination signal, which is an AAUAAA sequence in the pre-mRNA (see FIGURE 17.8). But then, at a point about 10 to 35 nucleotides past the AAUAAA, the pre-mRNA is cut free from the enzyme. The cleavage site on the RNA is also the site for the addition of a poly(A) tail—one step of RNA processing, our next topic.

Eukaryotic cells modify RNA after transcription

Enzymes in the eukaryotic nucleus modify pre-mRNA in various ways before the genetic messages are dispatched to the cytoplasm. During this RNA processing, both ends of the primary transcript are usually altered. In most cases, certain interior sections of the molecule are then cut out and the remaining parts spliced together.

Alteration of mRNA Ends

Each end of a pre-mRNA molecule is modified in a particular way. The 5′ end, the end made first during transcription, is immediately capped off with a modified form of a guanine (G) nucleotide. This **5′ cap** has at least two important functions. First, it helps protect the mRNA from degradation by hydrolytic enzymes. Second, after the mRNA reaches the cytoplasm, the 5′ cap functions as part of an "attach here" sign for ribosomes. The other end of an mRNA molecule, the 3′ end, is also modified before the message exits the nucleus. At the 3′ end, an enzyme makes a **poly(A) tail** consisting of some 50 to 250 adenine nucleotides. Like the 5′ cap, the poly(A) tail inhibits degradation of the RNA and probably helps ribosomes attach to it. The poly(A) tail also seems to facilitate the export of mRNA from the nucleus. FIGURE 17.8 shows a eukaryotic mRNA molecule with cap and tail; it also shows the nontranslated *leader* and *trailer* segments of RNA to which they are attached.

Split Genes and RNA Splicing

The most remarkable stage of RNA processing in the eukaryotic nucleus is the removal of a large portion of the RNA molecule that is initially synthesized—a cut-and-paste job called **RNA splicing** (FIGURE 17.9, p. 312). The average length of a transcription unit along a eukaryotic DNA molecule is about 8,000 nucleotides, so the primary RNA transcript is also that long. But it takes only about 1,200 nucleotides to code for an average-sized protein of 400 amino acids. (Remember, each amino acid is encoded by a *triplet* of nucleotides.) This means that most eukaryotic genes and their RNA transcripts have long noncoding stretches of nucleotides, regions that are not translated. Even more surprising is that most of these noncoding sequences are interspersed between coding segments of

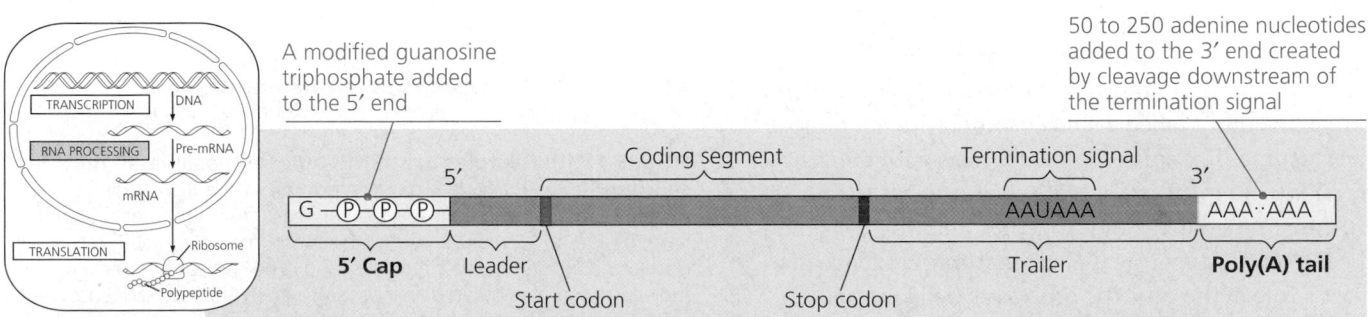

FIGURE 17.8 RNA processing: addition of the 5′ cap and poly(A) tail. Enzymes modify the two ends of a eukaryotic pre-mRNA molecule. The modified ends help protect the RNA from degradation, and the poly(A) tail may promote the export of mRNA from the nucleus. When the mRNA reaches the cytoplasm, the modified ends, in conjunction with certain cytoplasmic proteins, facilitate ribosome attachment. The leader and trailer are not translated, nor is the poly(A) tail.

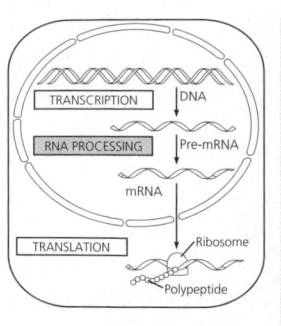

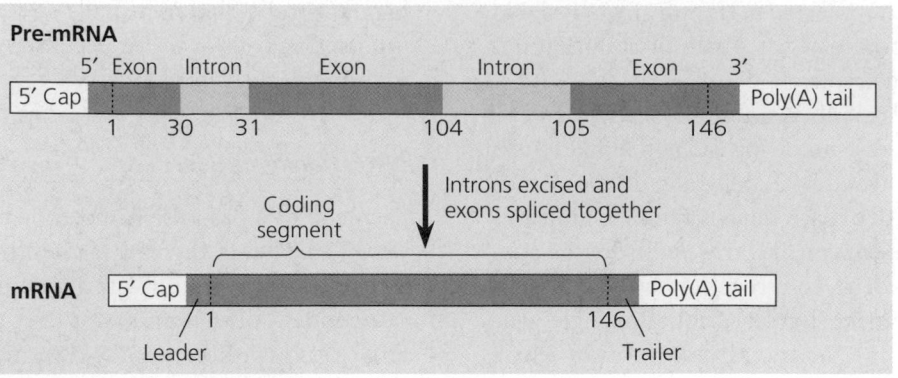

Pre-mRNA

5′ Exon Intron Exon Intron Exon 3′

5′ Cap   Poly(A) tail

1 30 31 104 105 146

Coding segment

Introns excised and exons spliced together

mRNA 5′ Cap   Poly(A) tail

1 146

Leader   Trailer

FIGURE 17.9 RNA processing: RNA splicing. The RNA molecule shown here codes for β-globin, one of the polypeptides of hemoglobin. The numbers under the RNA refer to codons; β-globin is 146 amino acids long. The β-globin gene and its pre-mRNA transcript have three regions, called exons, that consist mostly of coding sequences; exons are separated by noncoding regions, called introns. During RNA processing, the introns are excised and the exons are spliced together.

the gene and thus between coding segments of the pre-mRNA. In other words, the sequence of DNA nucleotides that codes for a eukaryotic polypeptide is not continuous. The noncoding segments of nucleic acid that lie between coding regions are called intervening sequences, or **introns** for short. The other regions are called **exons,** because they are eventually expressed, usually by being translated into amino acid sequences. (Exceptions include the leader and trailer portions of the exons at the ends of the RNA. Because of these exceptions, you may find it helpful to think of exons as DNA that *exits* the nucleus.) The terms *intron* and *exon* are used for both DNA and RNA. Richard Roberts and Phillip Sharp, who independently found evidence of "split genes" in 1977, shared a Nobel Prize in 1993 for this discovery.

In making a primary transcript from a gene, RNA polymerase transcribes both introns and exons from the DNA, but the mRNA molecule that enters the cytoplasm is an abridged version. The introns are cut out from the molecule and the exons joined together to form an mRNA molecule with a continuous coding sequence. This is the process of RNA splicing.

How is pre-mRNA splicing carried out? Researchers have learned that the signals for RNA splicing are short nucleotide sequences at the ends of introns. Particles called *small nuclear ribonucleoproteins,* or *snRNPs* (pronounced "snurps"), recognize these splice sites. As the name implies, snRNPs are located in the cell nucleus and are composed of RNA and protein molecules. The RNA in a snRNP particle is called a *small nuclear RNA (snRNA);* each molecule is about 150 nucleotides long. Several different snRNPs join with additional proteins to form an even larger assembly called a **spliceosome,** which is almost as big as a ribosome. The spliceosome interacts with the splice sites at the ends of an intron. It cuts at specific points to release the intron, then immediately joins together the two exons that flanked the intron (FIGURE 17.10). There is strong evidence that snRNA plays a role in the catalytic process as well as in spliceosome assembly and splice site recognition. The idea of a catalytic role for snRNA arose from the discovery of **ribozymes,** RNA molecules that function as enzymes.

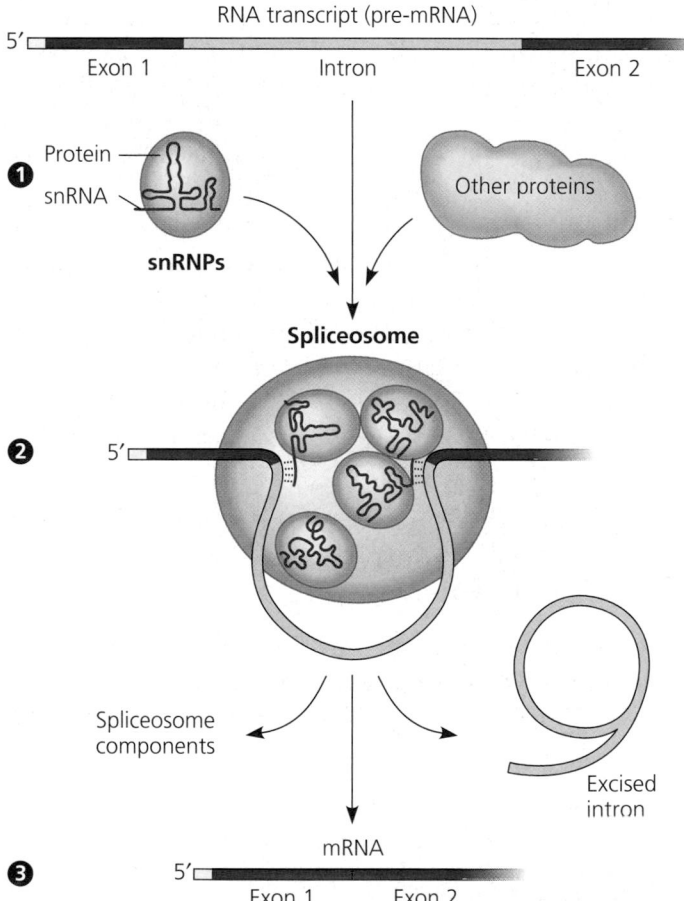

FIGURE 17.10 The roles of snRNPs and spliceosomes in mRNA splicing. The diagram shows only a portion of the RNA transcript; additional introns and exons lie downstream from the ones pictured here. ❶ Pre-mRNA containing exons and introns combines with small nuclear ribonucleoproteins (snRNPs) and other proteins to form a molecular complex called a spliceosome. ❷ Within the spliceosome, snRNA base-pairs with nucleotides at the ends of the intron. ❸ The RNA transcript is cut to release the intron, and the exons are spliced together. The spliceosome then comes apart, releasing mRNA, which now contains only exons.

Ribozymes

Like pre-mRNA, other kinds of primary transcripts may also be spliced, but by diverse mechanisms that do not involve spliceosomes. However, as in the mRNA case, RNA is often involved in catalyzing the reactions. In a few cases, the splicing occurs completely without proteins or even extra RNA molecules: The intron RNA catalyzes its own excision! For example, in the protozoan *Tetrahymena*, self-splicing occurs in the production of an RNA component of the organism's ribosomes (ribosomal RNA). Like enzymes, ribozymes function as catalysts. The discovery of ribozymes rendered obsolete the statement "All biological catalysts are proteins."

The Functional and Evolutionary Importance of Introns

What are the biological functions of introns and RNA splicing? One idea is that introns play regulatory roles in the cell. At least some introns contain sequences that control gene activity in some way, and the splicing process itself may help regulate the passage of mRNA from nucleus to cytoplasm.

One established benefit of split genes is to enable a single gene to encode more than one kind of polypeptide. A number of genes are known to give rise to two or more different polypeptides, depending on which segments are treated as exons during RNA processing; this is called **alternative RNA splicing** (see FIGURE 19.11). The fruit fly provides an interesting example: Sex differences in this animal are largely due to differences in how males and females splice the RNA transcribed from certain genes. Early results from the Human Genome Project (discussed in Chapter 20) suggest that alternative RNA splicing may be one reason humans can get along with a relatively small number of genes—only about twice as many as a fruit fly.

Split genes may also facilitate the evolution of new and potentially useful proteins. Proteins often have a modular architecture consisting of discrete structural and functional regions called **domains.** One domain of an enzymatic protein, for instance, might include the active site, while another might attach the protein to a cellular membrane. In many cases, different exons code for the different domains of a protein (FIGURE 17.11). Introns increase the probability of potentially beneficial crossing over between genes. Simply by providing more places where crossing over can occur (and without interfering with coding sequences), introns increase the opportunity for recombination between two alleles of a gene and raise the probability that a crossover will switch one version of an exon for another version found on the homologous chromosome. We can also imagine the occasional mixing and matching of exons between completely different (nonallelic) genes. Exon shuffling of either sort could lead to new proteins with novel combinations of functions.

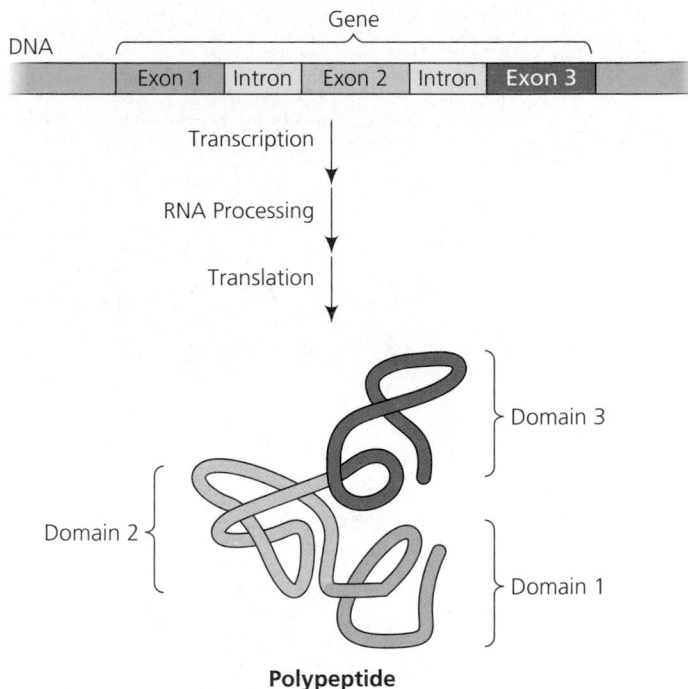

FIGURE 17.11 Correspondence between exons and protein domains. In a number of genes, different exons encode separate domains of the protein product. Each domain, an independently folding part of the protein, performs a different function. Such correspondence between exons and domains suggests that new proteins can evolve by exon shuffling among genes.

THE SYNTHESIS OF PROTEIN

We will now examine more closely how genetic information flows from mRNA to protein—the process of translation. As we did for transcription, we'll concentrate on the basic steps of translation that occur in both prokaryotes and eukaryotes while pointing out key differences.

Translation is the RNA-directed synthesis of a polypeptide: *a closer look*

In the process of translation, a cell interprets a genetic message and builds a protein accordingly. The message is a series of codons along an mRNA molecule, and the interpreter is called **transfer RNA (tRNA).** The function of tRNA is to transfer amino acids from the cytoplasm's amino acid pool to a ribosome. A cell keeps its cytoplasm stocked with all 20 amino acids, either by synthesizing them from other compounds or by taking them up from the surrounding solution. The ribosome adds each amino acid brought to it by tRNA to the growing end of a polypeptide chain (FIGURE 17.12, p. 314).

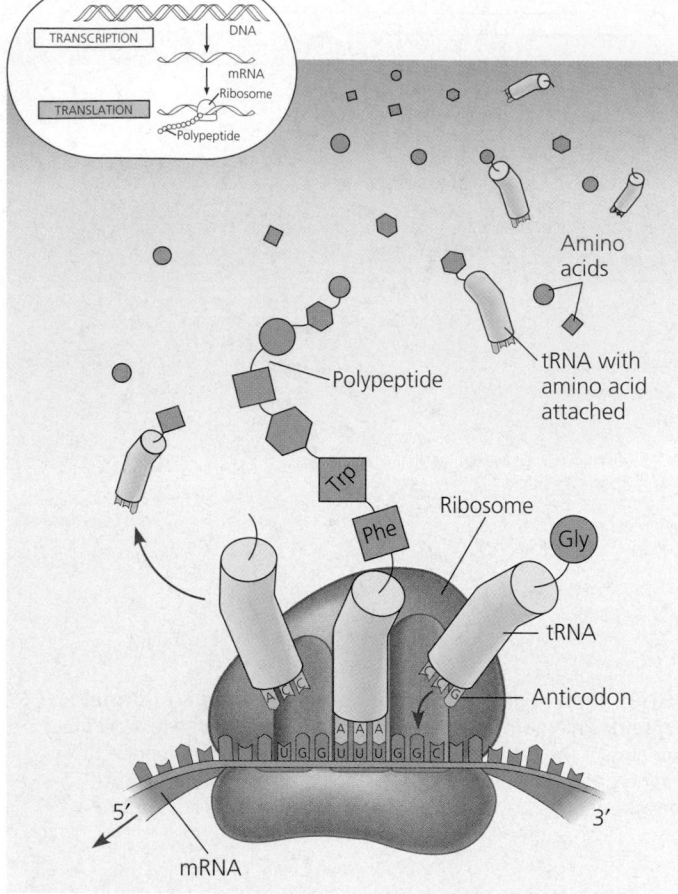

FIGURE 17.12 Translation: the basic concept. As a molecule of mRNA is moved through a ribosome, codons are translated into amino acids, one by one. The interpreters are tRNA molecules, each type with a specific anticodon at one end and a certain amino acid at the other end. A tRNA adds its amino acid cargo to a growing polypeptide chain when the anticodon bonds to a complementary codon on the mRNA. The figures that follow show some of the details of translation in the prokaryotic cell.

Molecules of tRNA are not all identical. The key to translating a genetic message into a specific amino acid sequence is that each type of tRNA molecule links a particular mRNA codon with a particular amino acid. As a tRNA molecule arrives at a ribosome, it bears a specific amino acid at one end. At the other end is a nucleotide triplet called an **anticodon,** which base-pairs with a complementary codon on mRNA. For example, consider the mRNA codon UUU, which is translated as the amino acid phenylalanine (see FIGURE 17.4). The tRNA that plugs into this codon by hydrogen bonding has AAA as its anticodon and carries phenylalanine at its other end. As an mRNA molecule is moved through a ribosome, phenylalanine will be added to the polypeptide chain whenever the codon UUU is presented for translation. Codon by codon, the genetic message is translated as tRNAs deposit amino acids in the order prescribed, and the ribosome joins the amino acids into a chain. The tRNA mole-

cule is like a flash card with a nucleic acid word (anticodon) on one side and a protein word (amino acid) on the other.

Translation is simple in principle but complex in its biochemistry and mechanics, especially in the eukaryotic cell. In dissecting translation, we'll concentrate on the slightly less complicated version of the process that occurs in prokaryotes. Let's first look at some of the major players in this cellular drama, then see how they act together to make a polypeptide.

The Structure and Function of Transfer RNA

Like mRNA and other types of cellular RNA, transfer RNA molecules are transcribed from DNA templates. In a eukaryotic cell, tRNA, like mRNA, is made in the nucleus and must travel from the nucleus to the cytoplasm, where translation occurs. In both prokaryotic and eukaryotic cells, each tRNA molecule is used repeatedly, picking up its designated amino acid in the cytosol, depositing this cargo at the ribosome, and then leaving the ribosome to pick up another load.

As illustrated in FIGURE 17.13, a tRNA molecule consists of a single RNA strand that is only about 80 nucleotides long (compared to hundreds of nucleotides for most mRNA molecules). This RNA strand folds back upon itself to form a molecule with a three-dimensional structure reinforced by interactions between different parts of the nucleotide chain. Nucleotide bases in certain regions of the tRNA strand form hydrogen bonds with complementary bases of other regions. Flattened into one plane to reveal this base pairing, a tRNA molecule looks like a cloverleaf. The tRNA actually twists and folds into a compact three-dimensional structure that is roughly L-shaped. The loop protruding from one end of the L includes the anticodon, the specialized base triplet that binds to a specific mRNA codon. From the other end of the L-shaped tRNA molecule protrudes its 3′ end, which is the attachment site for an amino acid. Thus, the structure of a tRNA molecule fits its function.

If one tRNA variety existed for each of the mRNA codons that specifies an amino acid, there would be 61 tRNAs (see FIGURE 17.4). The actual number is smaller: about 45. This number is sufficient because some tRNAs have anticodons that can recognize two or more different codons. Such versatility is possible because the rules for base pairing between the third base of a codon and the corresponding base of a tRNA anticodon are not as strict as those for DNA and mRNA codons. For example, the base U of a tRNA anticodon can pair with either A or G in the third position of an mRNA codon. This relaxation of the base-pairing rules is called **wobble.** The most versatile tRNAs are those with inosine (I), a modified base, in the wobble position of the anticodon. Inosine is formed by enzymatic alteration of adenine after tRNA is synthesized. When anticodons associate with codons, the base I can hydrogen-bond with any one of three bases: U, C, or A. Thus, the tRNA molecule that has CCI as its anticodon can bind to the codons GGU, GGC, and

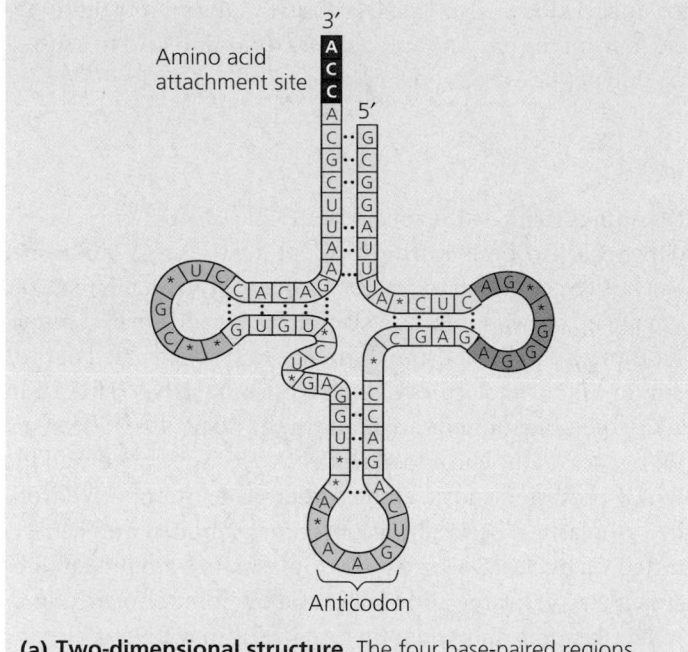

(a) Two-dimensional structure. The four base-paired regions and three loops are characteristic of all tRNAs, as is the base sequence of the amino acid attachment site, at the 3′ end. The anticodon triplet is unique to each tRNA type. (The asterisks mark bases that have been chemically modified, a characteristic of tRNA.)

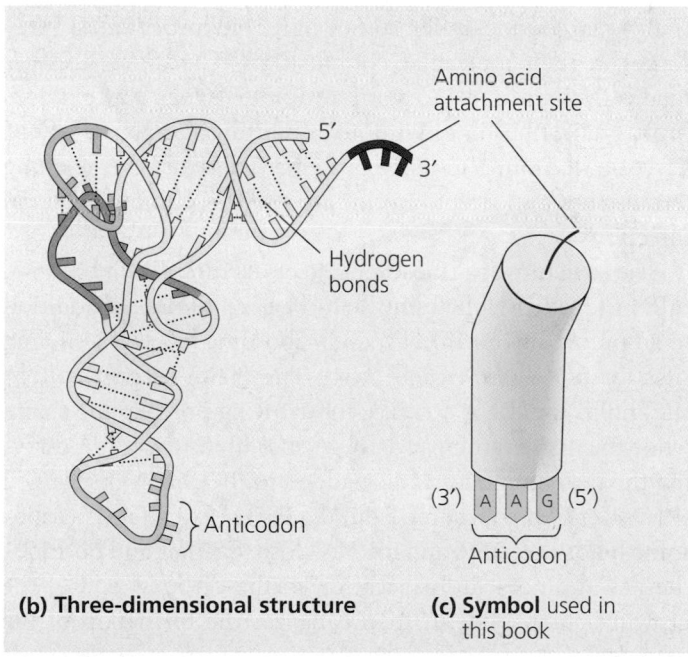

(b) Three-dimensional structure

(c) Symbol used in this book

FIGURE 17.13 The structure of transfer RNA (tRNA). Anticodons are conventionally written 3′ ⟶ 5′ to align properly with codons written 5′ ⟶ 3′ (see FIGURE 17.12). For base pairing, RNA strands must be antiparallel, like DNA (see FIGURE 16.12). For example, anticodon 3′-AAG-5′ pairs with mRNA codon 5′-UUC-3′.

GGA, all of which code for the amino acid glycine. Wobble explains why the synonymous codons for a given amino acid can differ in their third base, but usually not in their other bases.

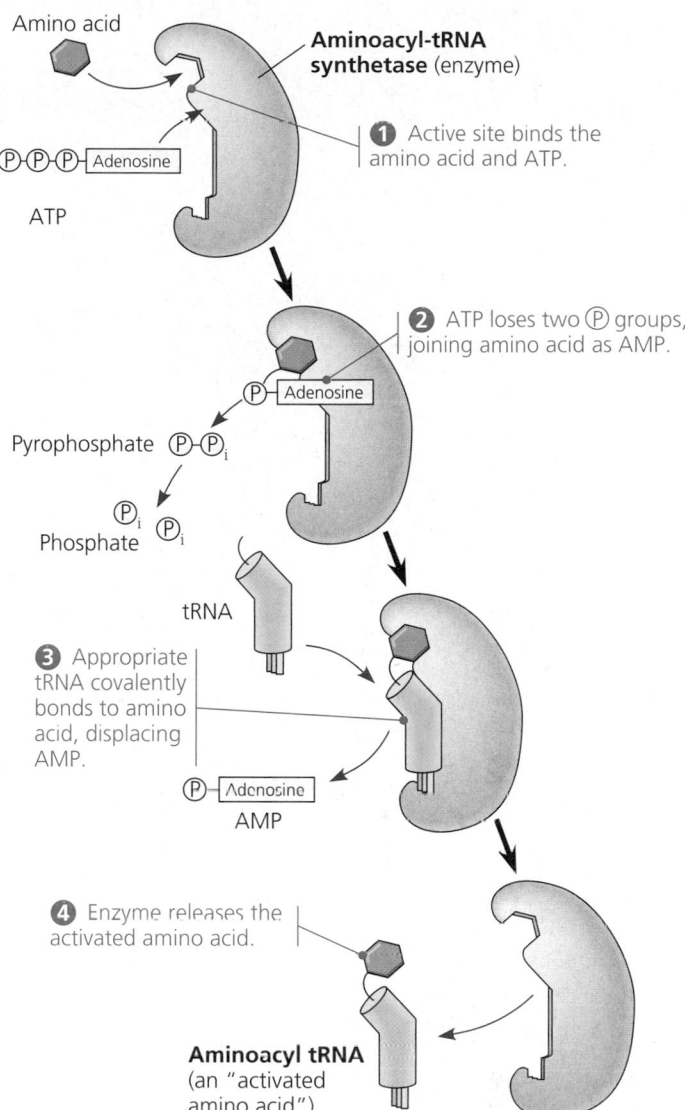

FIGURE 17.14 An aminoacyl-tRNA synthetase joins a specific amino acid to a tRNA. Linkage of the tRNA and amino acid is an endergonic process that occurs at the expense of ATP. The ATP loses two phosphate groups, becoming AMP (adenosine monophosphate).

Aminoacyl-tRNA Synthetases

Codon-anticodon bonding is actually the second of two recognition steps required for the accurate translation of a genetic message. It must be preceded by a correct match between tRNA and an amino acid. A tRNA that binds to an mRNA codon specifying a particular amino acid must carry *only* that amino acid to the ribosome. Each amino acid is joined to the correct tRNA by a specific enzyme called an **aminoacyl-tRNA synthetase.** There are 20 of these enzymes in the cell, one enzyme for each amino acid. The active site of each type of aminoacyl-tRNA synthetase fits only a specific combination of amino acid and tRNA. The synthetase catalyzes the covalent attachment of the amino acid to its tRNA in a process driven by the hydrolysis of ATP (FIGURE 17.14). The resulting

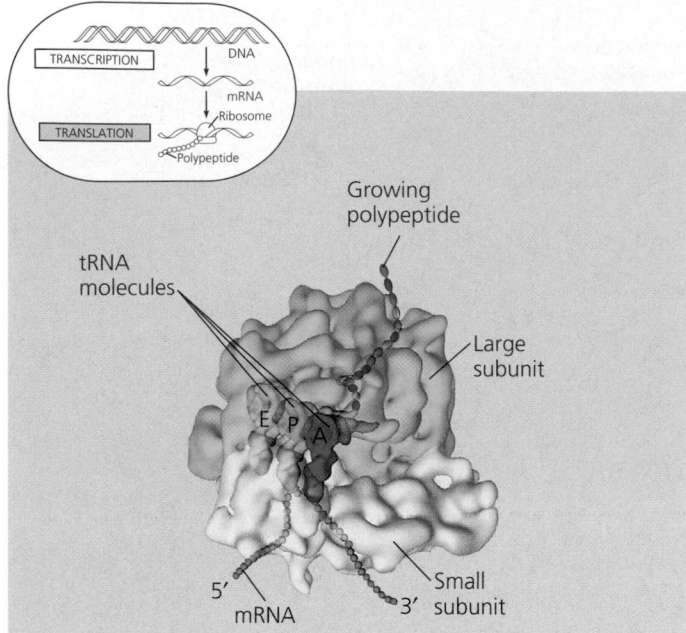

(a) Computer model of functioning ribosome. This is a model of a bacterial ribosome, showing its overall shape. The eukaryotic ribosome is roughly similar. A ribosomal subunit is an aggregate of ribosomal RNA molecules and proteins.

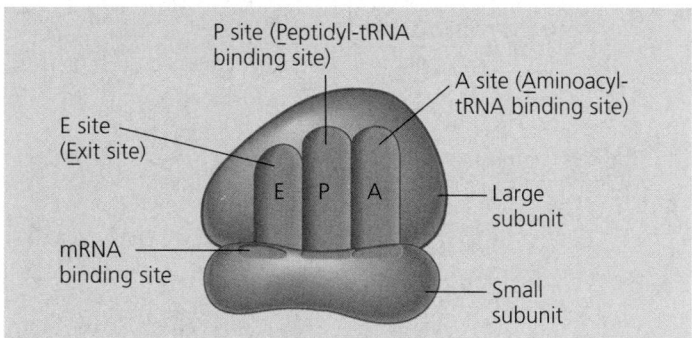

(b) Schematic model showing binding sites. A ribosome has an mRNA binding site and three tRNA binding sites, known as the P, A, and E sites. This schematic ribosome will appear in later diagrams.

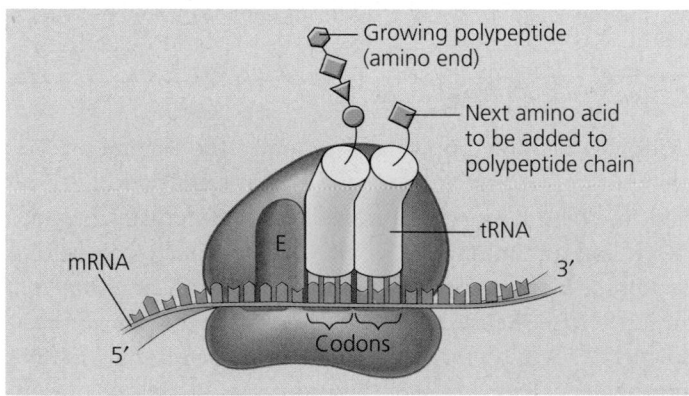

(c) Schematic model with mRNA and tRNA. A tRNA fits into a binding site when its anticodon base-pairs with an mRNA codon. The P site holds the tRNA attached to the growing polypeptide. The A site holds the tRNA carrying the next amino acid to be added to the polypeptide chain. Discharged tRNA leaves via the E site.

FIGURE 17.15 The anatomy of a functioning ribosome.

aminoacyl tRNA, also called an activated amino acid, is released from the enzyme and delivers its amino acid to a growing polypeptide chain on a ribosome.

Ribosomes

Ribosomes facilitate the specific coupling of tRNA anticodons with mRNA codons during protein synthesis. A ribosome, which is large enough to be seen with an electron microscope, is made up of two subunits, called the large and small subunits (FIGURE 17.15). The ribosomal subunits are constructed of proteins and RNA molecules named **ribosomal RNA (rRNA).** In eukaryotes, the subunits are made in the nucleolus. Ribosomal RNA genes on the chromosomal DNA are transcribed, and the RNA is processed and assembled with proteins imported from the cytoplasm. The resulting ribosomal subunits are then exported via nuclear pores to the cytoplasm. In both prokaryotes and eukaryotes, large and small subunits join to form a functional ribosome only when they attach to an mRNA molecule. About two-thirds of the mass of a ribosome is rRNA. Because most cells contain thousands of ribosomes, rRNA is the most abundant type of RNA.

Although the ribosomes of prokaryotes and eukaryotes are very similar in structure and function, those of eukaryotes are slightly larger and differ somewhat from prokaryotic ribosomes in their molecular composition. The differences are medically significant. Certain antibiotic drugs can paralyze prokaryotic ribosomes without inhibiting the ability of eukaryotic ribosomes to make proteins. These drugs, including tetracycline and streptomycin, are used to combat bacterial infections.

The structure of a ribosome reflects its function of bringing mRNA together with amino acid–bearing tRNAs. In addition to a binding site for mRNA, each ribosome has three binding sites for tRNA (see FIGURE 17.15). The **P site** (peptidyl-tRNA site) holds the tRNA carrying the growing polypeptide chain, while the **A site** (aminoacyl-tRNA site) holds the tRNA carrying the next amino acid to be added to the chain. Discharged tRNAs leave the ribosome from the **E site** (exit site). The ribosome holds the tRNA and mRNA close together and positions the new amino acid for addition to the carboxyl end of the growing polypeptide. It then catalyzes the formation of the peptide bond.

Four decades of genetic and biochemical research on ribosome structure recently culminated in the detailed structure of the bacterial ribosome. One view of the large subunit is shown in FIGURE 17.16, and a ribbon model of the entire ribosome appears at the beginning of this chapter. Ribosome structure strongly supports the hypothesis that rRNA, not protein, carries out the ribosome's functions. RNA is the main constituent of the interface between the two subunits and of the A and P sites, and it is the catalyst of peptide

THE PROCESS OF SCIENCE

FIGURE 17.16 Structure of the large ribosomal subunit at the atomic level. In 2000, researchers succeeded in determining the atomic structures of both subunits of the bacterial ribosome using X-ray crystallography (see FIGURE 5.27). This computer model shows the large subunit from a point of view different from that in FIGURE 17.15. Here we are looking at the "bottom" of the large subunit, as it would look from an attached small subunit. The rRNA molecules are colored orange or maroon (sugar-phosphate backbones) and gray (bases) and the proteins are purple. Notice that the proteins are mostly on the outside of the subunit, whereas the rRNA, responsible for the ribosome's functions, is mostly in the interior. The tRNA molecules are included for orientation.

bond formation. Thus, a ribosome can be regarded as one colossal ribozyme! The ribosome's proteins are largely on the exterior, apparently playing a mainly structural role.

Building a Polypeptide

We can divide translation, the synthesis of a polypeptide chain, into three stages (analogous to those of transcription): initiation, elongation, and termination. All three stages require protein "factors" that aid mRNA, tRNA, and ribosomes in the translation process. For chain initiation and elongation, energy is also required. It is provided by the hydrolysis of GTP (guanosine triphosphate), a molecule closely related to ATP.

Initiation. The initiation stage of translation brings together mRNA, a tRNA bearing the first amino acid of the polypeptide, and the two subunits of a ribosome (FIGURE 17.17). First, a small ribosomal subunit binds to both mRNA and a special initiator tRNA. The small ribosomal subunit attaches to the leader segment at the 5' (upstream) end of the mRNA. In bacteria, rRNA of the small subunit base-pairs with a specific sequence of nucleotides within the mRNA leader; in eukaryotes, the 5' cap first tells the small subunit to attach to the 5' end of the mRNA. Downstream on the mRNA is the initiation codon, AUG, which signals the start of translation. The initiator tRNA, which carries the amino acid methionine, attaches to the initiation codon.

The union of mRNA, initiator tRNA, and a small ribosomal subunit is followed by the attachment of a large ribosomal subunit, completing a translation initiation complex. Proteins called *initiation factors* are required to bring all these components together. The cell also spends energy in the form of a GTP molecule to form the initiation complex. At the completion of the initiation process, the initiator tRNA sits in the P site of the ribosome, and the vacant A site is ready for the next aminoacyl tRNA. The synthesis of a polypeptide is initiated at its amino end (see FIGURE 5.16b).

Elongation. In the elongation stage of translation, amino acids are added one by one to the preceding amino acid. Each addition involves the participation

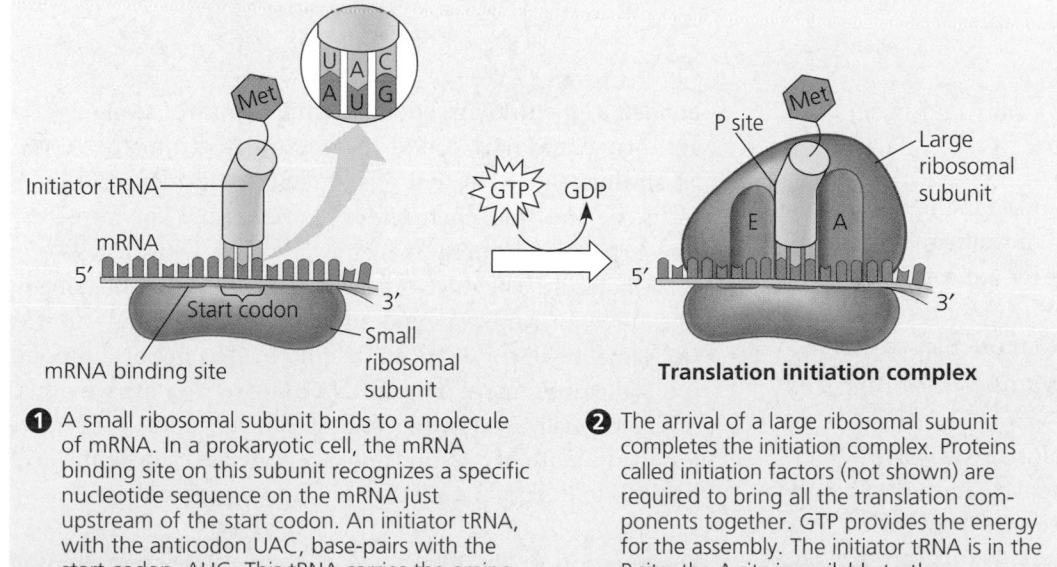

❶ A small ribosomal subunit binds to a molecule of mRNA. In a prokaryotic cell, the mRNA binding site on this subunit recognizes a specific nucleotide sequence on the mRNA just upstream of the start codon. An initiator tRNA, with the anticodon UAC, base-pairs with the start codon, AUG. This tRNA carries the amino acid methionine (Met).

❷ The arrival of a large ribosomal subunit completes the initiation complex. Proteins called initiation factors (not shown) are required to bring all the translation components together. GTP provides the energy for the assembly. The initiator tRNA is in the P site; the A site is available to the tRNA bearing the next amino acid.

FIGURE 17.17 The initiation of translation.

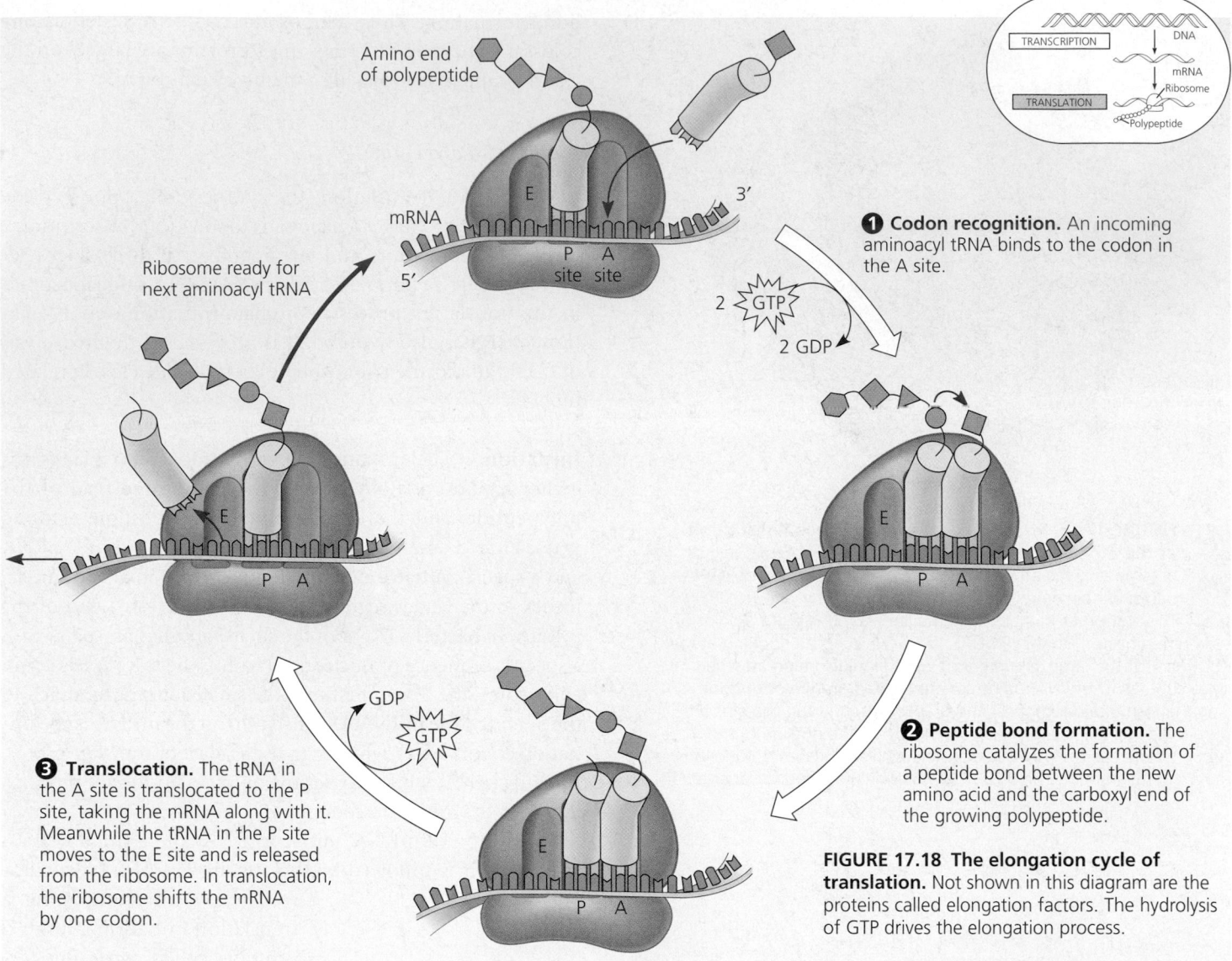

Amino end of polypeptide

Ribosome ready for next aminoacyl tRNA

mRNA

5′

3′

E

P site   A site

2 GTP → 2 GDP

1 Codon recognition. An incoming aminoacyl tRNA binds to the codon in the A site.

E

P   A

2 Peptide bond formation. The ribosome catalyzes the formation of a peptide bond between the new amino acid and the carboxyl end of the growing polypeptide.

E

P   A

GDP

GTP

3 Translocation. The tRNA in the A site is translocated to the P site, taking the mRNA along with it. Meanwhile the tRNA in the P site moves to the E site and is released from the ribosome. In translocation, the ribosome shifts the mRNA by one codon.

E

P   A

FIGURE 17.18 The elongation cycle of translation. Not shown in this diagram are the proteins called elongation factors. The hydrolysis of GTP drives the elongation process.

TRANSCRIPTION DNA

mRNA

TRANSLATION Ribosome

Polypeptide

of several proteins called *elongation factors* and occurs in a three-step cycle (FIGURE 17.18):

1 *Codon recognition.* The mRNA codon in the A site of the ribosome forms hydrogen bonds with the anticodon of an incoming molecule of tRNA carrying its appropriate amino acid. An elongation factor ushers the tRNA into the A site. This step requires the hydrolysis of two molecules of GTP.

2 *Peptide bond formation.* An rRNA molecule of the large ribosomal subunit, functioning as a ribozyme, catalyzes the formation of a peptide bond that joins the polypeptide extending from the P site to the newly arrived amino acid in the A site. In this step, the polypeptide separates from the tRNA to which it was attached, and the amino acid at its carboxyl end bonds to the amino acid carried by the tRNA in the A site.

3 *Translocation.* The ribosome now translocates (moves) the tRNA in the A site, with its attached polypeptide, to the P site. As the tRNA moves, its anticodon remains hydrogen-

bonded to the mRNA codon; the mRNA moves along with it and brings the next codon to be translated into the A site. Meanwhile, the tRNA that was in the P site is moved to the E (exit) site and from there leaves the ribosome. The translocation step requires energy, which is provided by hydrolysis of a GTP molecule. The mRNA is moved through the ribosome in one direction only, 5′ end first; this is equivalent to the ribosome moving 5′ ⟶ 3′ on the mRNA. The important point is that the ribosome and the mRNA move relative to each other, unidirectionally, codon by codon. The elongation cycle takes less than a tenth of a second and is repeated as each amino acid is added to the chain until the polypeptide is completed.

Termination. The final stage of translation is termination (FIGURE 17.19). Elongation continues until a stop codon in the mRNA reaches the A site of the ribosome. The special base triplets UAA, UAG, and UGA do not code for amino acids but instead act as signals to stop translation. A protein called a

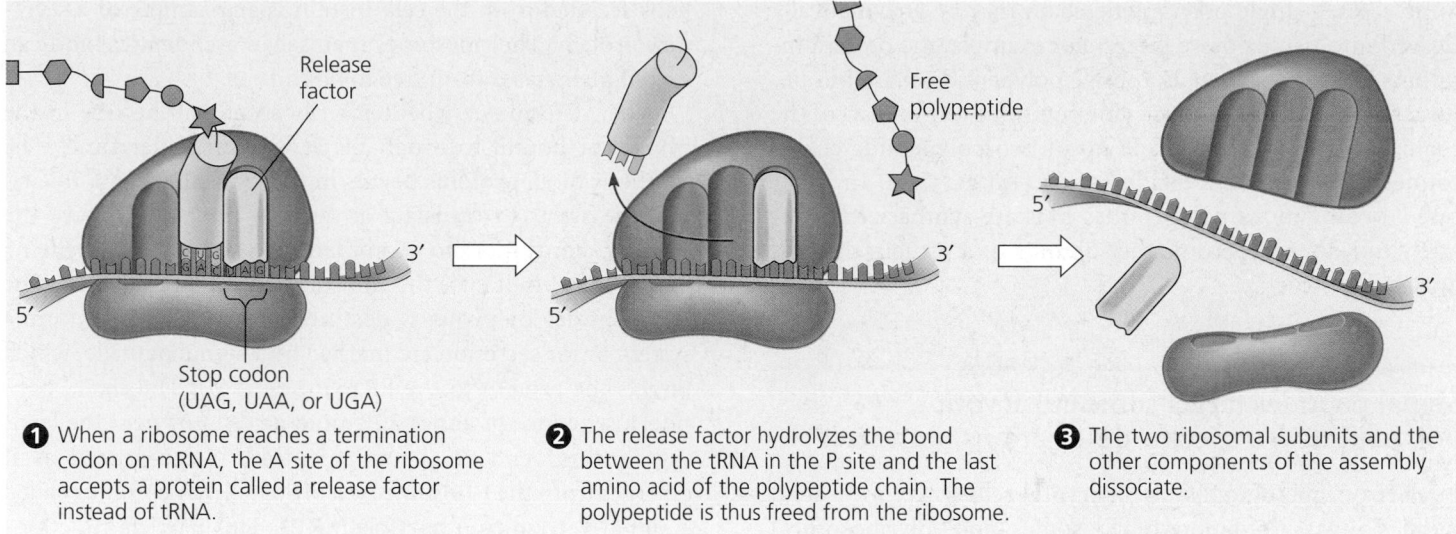

① When a ribosome reaches a termination codon on mRNA, the A site of the ribosome accepts a protein called a release factor instead of tRNA.

② The release factor hydrolyzes the bond between the tRNA in the P site and the last amino acid of the polypeptide chain. The polypeptide is thus freed from the ribosome.

③ The two ribosomal subunits and the other components of the assembly dissociate.

FIGURE 17.19 The termination of translation.

release factor binds directly to the stop codon in the A site. The release factor causes the addition of a water molecule instead of an amino acid to the polypeptide chain. This reaction hydrolyzes the completed polypeptide from the tRNA that is in the P site, freeing the polypeptide from the ribosome. The remainder of the translation assembly then comes apart.

Polyribosomes

A single ribosome can make an average-sized polypeptide in less than a minute. Typically, however, a single mRNA is used to make many copies of a polypeptide simultaneously, because a number of ribosomes work on translating the message at the same time. Once a ribosome moves past the initiation codon, a second ribosome can attach to the mRNA, and thus, multiple ribosomes may trail along the same mRNA. Such strings of ribosomes, called **polyribosomes,** can be seen with the electron microscope (FIGURE 17.20). Polyribosomes are found in

both prokaryotic and eukaryotic cells. They help a cell to make many copes of a polypeptide very quickly.

From Polypeptide to Functional Protein

During and after its synthesis, a polypeptide chain begins to coil and fold spontaneously, forming a functional protein of specific conformation: a three-dimensional molecule with secondary and tertiary structure (see FIGURE 5.24). A gene determines primary structure, and primary structure in turn determines conformation. In many cases, a chaperone protein helps the polypeptide fold correctly (see FIGURE 5.26).

Additional steps—*posttranslational modifications*—may be required before the protein can begin doing its particular job in the cell. Certain amino acids may be chemically modified by the attachment of sugars, lipids, phosphate groups, or other additions. Enzymes may remove one or more amino acids from the leading (amino) end of the polypeptide chain. In

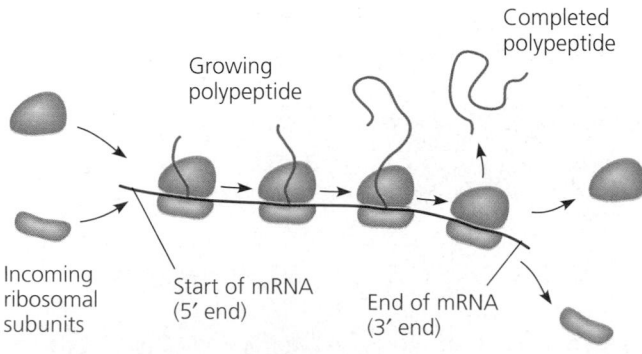

(a) An mRNA molecule is generally translated simultaneously by several ribosomes in clusters called polyribosomes.

FIGURE 17.20 Polyribosomes.

(b) This micrograph shows a large polyribosome in a prokaryotic cell (TEM).

some cases, a single polypeptide chain may be enzymatically cleaved into two or more pieces. For example, the protein insulin is first synthesized as a single polypeptide chain but becomes active only after an enzyme cuts out a central part of the chain, leaving a protein made up of two polypeptide chains connected by disulfide bridges (see FIGURE 5.22). In other cases, two or more polypeptides that are synthesized separately may join to become the subunits of a protein that has quaternary structure.

Signal peptides target some eukaryotic polypeptides to specific destinations in the cell

In electron micrographs of eukaryotic cells active in protein synthesis, two populations of ribosomes (and polyribosomes) are evident: free and bound (see FIGURE 7.10). Free ribosomes are suspended in the cytosol and mostly synthesize proteins that dissolve in the cytosol and function there. In contrast, bound ribosomes are attached to the cytosolic side of the endoplasmic reticulum (ER). They make proteins of the endomembrane system (the nuclear envelope, ER, Golgi apparatus, lysosomes, vacuoles, and plasma membrane) as well as pro-

teins secreted from the cell. Insulin is an example of a secretory protein. The ribosomes themselves are identical and can switch their status from free to bound.

What determines whether a ribosome will be free in the cytosol or bound to rough ER at any particular time? The synthesis of all proteins begins in the cytosol, when a free ribosome starts to translate an mRNA molecule. There the process continues to completion—*unless* the growing polypeptide itself cues the ribosome to attach to the ER. The polypeptides of proteins destined for the endomembrane system or for secretion are marked by a **signal peptide,** which targets the protein to the ER (FIGURE 17.21). The signal peptide, a sequence of about 20 amino acids at or near the leading (amino) end of the polypeptide, is recognized as it emerges from the ribosome by a protein-RNA complex called a **signal-recognition particle (SRP).** This particle functions as an adapter that brings the ribosome to a receptor protein built into the ER membrane. This receptor is part of a multiprotein complex. Polypeptide synthesis continues there, and the growing polypeptide snakes across the membrane into the cisternal space via a protein pore. The signal peptide is usually removed by an enzyme. The rest of the completed polypeptide, if it is to be a secretory protein, is released into

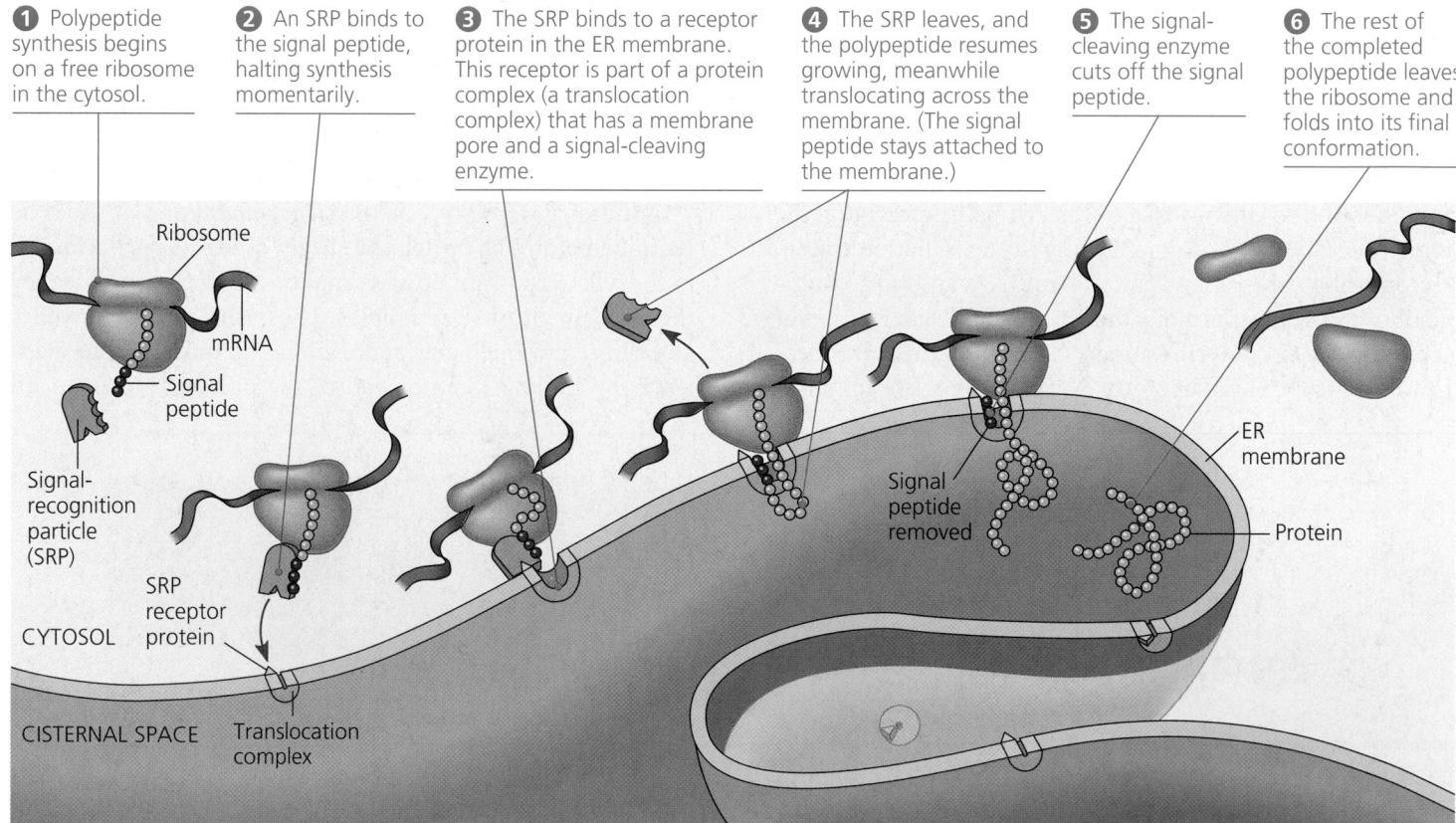

1 Polypeptide synthesis begins on a free ribosome in the cytosol.

2 An SRP binds to the signal peptide, halting synthesis momentarily.

3 The SRP binds to a receptor protein in the ER membrane. This receptor is part of a protein complex (a translocation complex) that has a membrane pore and a signal-cleaving enzyme.

4 The SRP leaves, and the polypeptide resumes growing, meanwhile translocating across the membrane. (The signal peptide stays attached to the membrane.)

5 The signal-cleaving enzyme cuts off the signal peptide.

6 The rest of the completed polypeptide leaves the ribosome and folds into its final conformation.

Ribosome

mRNA

Signal peptide

Signal-recognition particle (SRP)

SRP receptor protein

CYTOSOL

CISTERNAL SPACE

Translocation complex

Signal peptide removed

ER membrane

Protein

FIGURE 17.21 The signal mechanism for targeting proteins to the ER. A polypeptide destined for the endomembrane system or for secretion from the cell begins with a signal peptide, a series of amino acids that targets it for the ER. This figure shows the synthesis of a secretory protein and its simultaneous import into the ER. In the ER and then in the Golgi, the protein is further processed. Finally, a transport vesicle conveys it to the plasma membrane for release from the cell (see FIGURE 8.8).

solution within the cisternal space (as in FIGURE 17.21). Alternatively, if the polypeptide is to be a membrane protein, it remains partially embedded in the ER membrane.

Other kinds of signal peptides are used to target polypeptides to mitochondria, chloroplasts, the interior of the nucleus, and other organelles that are not part of the endomembrane system. The critical difference in these cases is that translation is completed in the cytosol before the polypeptide is imported into the organelle. The mechanisms of translocation also vary, but in all cases studied to date, the "postal" codes that address proteins to cellular locations are signal peptides of some sort.

RNA plays multiple roles in the cell: *a review*

As we have seen, the cellular machinery of protein synthesis (and ER targeting) is dominated by RNA of various kinds. In addition to mRNA, these include tRNA, rRNA, and, in eukaryotes, snRNA and SRP RNA (TABLE 17.1). The diverse functions of these molecules range from structural to informational to catalytic. The ability of RNA to perform so many different functions is based on two related characteristics of this kind of molecule: RNA can hydrogen-bond to other nucleic acid molecules (DNA or RNA), and it can assume a specific three-dimensional shape by forming hydrogen bonds between bases in different parts of its polynucleotide chain (you saw an example of this intramolecular bonding in tRNA, FIGURE 17.13). DNA may be the genetic material of all living cells today, but RNA is much more versatile. You will learn in Chapter 18 that many viruses even use RNA rather than DNA as their genetic material.

Comparing protein synthesis in prokaryotes and eukaryotes: *a review*

Although bacteria and eukaryotes carry out transcription and translation in very similar ways, we have noted certain differences in cellular machinery and in details of the processes. Prokaryotic and eukaryotic RNA polymerases are different, and those of eukaryotes depend on transcription factors. Transcription is terminated differently in the two kinds of cells. Also, prokaryotic and eukaryotic ribosomes are slightly different. The most important differences, however, arise from the eukaryotic cell's compartmental organization. Like a one-room workshop, a prokaryotic cell ensures a streamlined operation. In the absence of a nucleus, it can simultaneously transcribe and translate the same gene (FIGURE 17.22), and the newly made protein can quickly diffuse to its site of function.

Table 17.1 Types of RNA in a Eukaryotic Cell

Type of RNA	Functions
Messenger RNA (mRNA)	Carries information specifying amino acid sequences of proteins from DNA to ribosomes.
Transfer RNA (tRNA)	Serves as adapter molecule in protein synthesis; translates mRNA codons into amino acids.
Ribosomal RNA (rRNA)	Plays catalytic (ribozyme) roles and structural roles in ribosomes.
Primary transcript	Serves as a precursor to mRNA, rRNA, or tRNA and may be processed by splicing or cleavage. In eukaryotes, pre-mRNA commonly contains introns, noncoding segments that are spliced out as the primary transcript is processed. Some intron RNA acts as a ribozyme, catalyzing its own splicing.
Small nuclear RNA (snRNA)	Plays structural and catalytic roles in spliceosomes, the complexes of protein and RNA that splice pre-mRNA in the eukaryotic nucleus.
SRP RNA	Is a component of the signal-recognition particle (SRP), the protein-RNA complex that recognizes the signal peptides of polypeptides targeted to the ER.

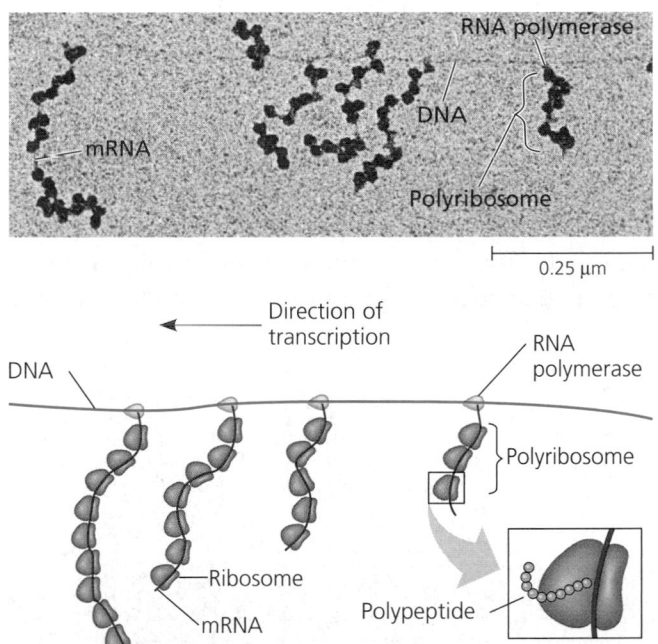

0.25 µm

FIGURE 17.22 Coupled transcription and translation in bacteria. In prokaryotic cells, the translation of mRNA can begin as soon as the leading (5') end of the mRNA molecule peels away from the DNA template. The micrograph shows a strand of *E. coli* DNA being transcribed by RNA polymerase molecules. Attached to each RNA polymerase molecule is a growing strand of mRNA, which is already being translated by ribosomes. The newly synthesized polypeptides are not visible in the micrograph (TEM).

Photo reprinted with permission from O. L. Miller, B. A. Hamkalo, and C. A. Thomas, Jr., *Science* 169 (1970): 392. Copyright © 1970 American Association for the Advancement of Science.

In contrast, the eukaryotic cell's nuclear envelope segregates transcription from translation and provides a compartment for extensive RNA processing. This processing stage provides additional steps whose regulation can help coordinate the eukaryotic cell's elaborate activities (see Chapter 19). Finally, as we have seen, eukaryotic cells have complicated mechanisms for targeting proteins to the appropriate cellular compartment (organelle).

Where did eukaryotes—and prokaryotes, for that matter—get the genes that encode the huge diversity of proteins they synthesize? For the past few billion years, the ultimate source of new genes has been the mutation of preexisting genes, the topic of the next section.

Point mutations can affect protein structure and function

Mutations are changes in the genetic material of a cell (or virus). In FIGURE 15.13, we considered large-scale mutations, chromosomal rearrangements that affect long segments of DNA. Now that you have learned about the genetic code and its translation, we can examine **point mutations,** which are chemical changes in just one base pair of a gene.

If a point mutation occurs in a gamete or in a cell that gives rise to gametes, it may be transmitted to offspring and to a succession of future generations. If the mutation has an adverse effect on the phenotype of a human or other animal, the mutant condition is referred to as a genetic disorder, or hereditary disease. For example, we can trace the genetic basis of sickle-cell disease to a mutation of a single base pair in the gene that codes for one of the polypeptides of hemoglobin. The change of a single nucleotide in the DNA's template strand leads to the production of an abnormal protein (FIGURE 17.23). In individuals who are homozygous for the mutant

allele, the sickling of red blood cells caused by the altered hemoglobin produces the multiple symptoms associated with sickle-cell disease (see FIGURE 14.15). Let's see how different types of point mutations translate into altered proteins.

Types of Point Mutations

Point mutations within a gene can be divided into two general categories: base-pair substitutions and base-pair insertions or deletions. While reading about how these mutations affect proteins, refer to the appropriate parts of FIGURE 17.24.

Substitutions. A **base-pair substitution** is the replacement of one nucleotide and its partner in the complementary DNA strand with another pair of nucleotides. Some substitutions are called *silent mutations* because, owing to the redundancy of the genetic code, they have no effect on the encoded protein. In other words, a change in a base pair may transform one codon into another that is translated into the same amino acid. For example, if CCG mutated to CCA, the mRNA codon that used to be GGC would become GGU, and a glycine would still be inserted at the proper location in the protein (see FIGURE 17.4). Other changes of a single nucleotide pair may switch an amino acid but have little effect on the protein. The new amino acid may have properties similar to those of the amino acid it replaces, or it may be in a region of the protein where the exact sequence of amino acids is not essential to the protein's function.

However, the base-pair substitutions of greatest interest are those that cause a readily detectable change in a protein. The alteration of a single amino acid in a crucial area of a protein—in the active site of an enzyme, for example—will significantly alter protein activity. Occasionally, such a mutation leads to an improved protein or one with novel capabilities that enhance the success of the mutant organism and its descendants. But much more often such mutations are detrimental, creating a useless or less active protein that impairs cellular function.

Substitution mutations are usually **missense mutations;** that is, the altered codon still codes for an amino acid and thus makes sense, although not necessarily the *right* sense. But if a point mutation changes a codon for an amino acid into a stop codon, translation will be terminated prematurely, and the resulting polypeptide will be shorter than the polypeptide encoded by the normal gene. Alterations that change an amino acid codon to a stop signal are called **nonsense mutations,** and nearly all nonsense mutations lead to nonfunctional proteins.

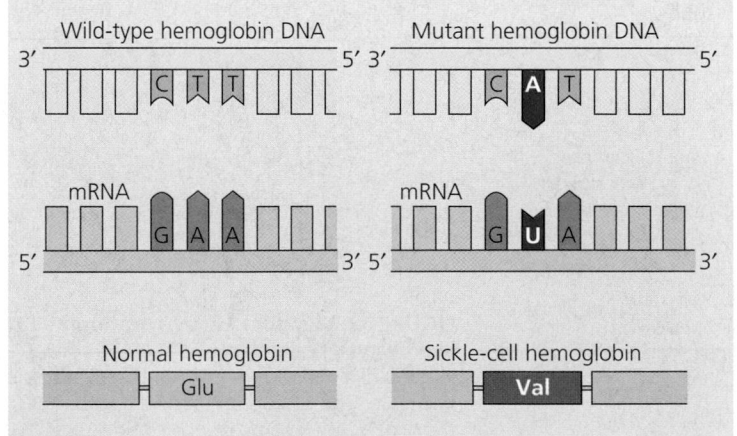

In the DNA, the mutant template strand has an A where the wild-type template has a T.

The mutant mRNA has a U instead of an A in one codon.

The mutant (sickle-cell) hemoglobin has a valine (Val) instead of a glutamic acid (Glu).

FIGURE 17.23 The molecular basis of sickle-cell disease: a point mutation. The allele that causes sickle-cell disease differs from the wild-type (normal) allele by a change in a single DNA base pair.

Insertions and Deletions. Insertions and **deletions** are additions or losses of nucleotide pairs in a gene. These mutations have a disastrous effect on the resulting protein more often than substitutions do. Because mRNA is read as a series of nucleotide triplets during translation, the insertion or deletion of nucleotides may alter the reading frame (triplet grouping) of the genetic message. Such a mutation, called a **frameshift mutation,** will occur whenever the number of nucleotides inserted or deleted is not a multiple of three. All the nucleotides that are downstream of the deletion or insertion will be improperly grouped into codons, and the result will be extensive missense probably ending sooner or later in nonsense—and premature termination. Unless the frameshift is very near the end of the gene, it will produce a protein that is almost certain to be nonfunctional.

Mutagens

The production of mutations can occur in a number of ways. Errors during DNA replication, repair, or recombination can lead to base-pair substitutions, insertions, or deletions, as well as to mutations affecting longer stretches of DNA. Mutations resulting from such errors are called *spontaneous mutations.*

A number of physical and chemical agents, called **mutagens,** interact with DNA to cause mutations. In the 1920s, Hermann Muller discovered that if he subjected fruit flies to X-rays, genetic changes increased in frequency. Using this method, Muller was able to obtain mutants of *Drosophila* that he could use in his genetic studies. But he also recognized an alarming implication of his discovery: Because they are mutagens, X-rays and other forms of high-energy radiation pose hazards to the genetic material of people as well as laboratory organisms.

Wild type

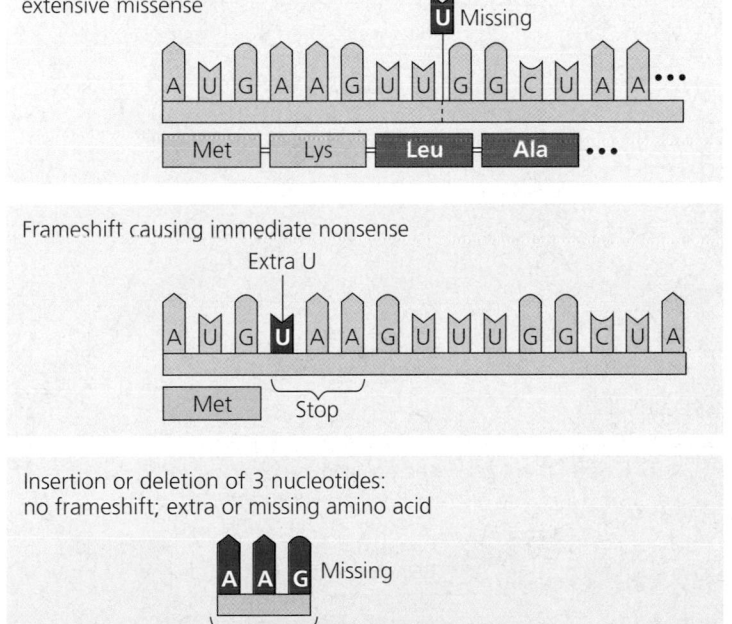

Base-pair insertion or deletion

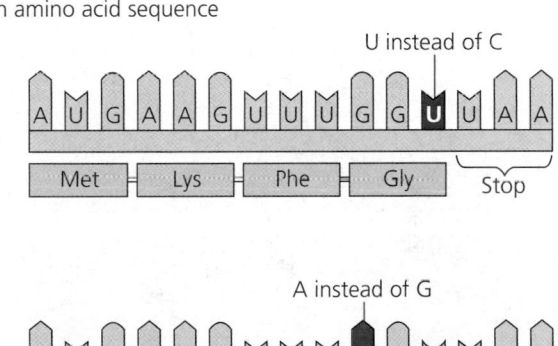

Base-pair substitution

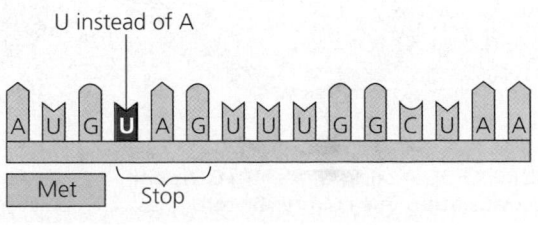

FIGURE 17.24 Categories and consequences of point mutations. Mutations are changes in DNA, but they are represented here as they are reflected in mRNA and its protein product. Strictly speaking, the example at the lower left is not a point mutation because it involves insertion or deletion of more than one nucleotide.

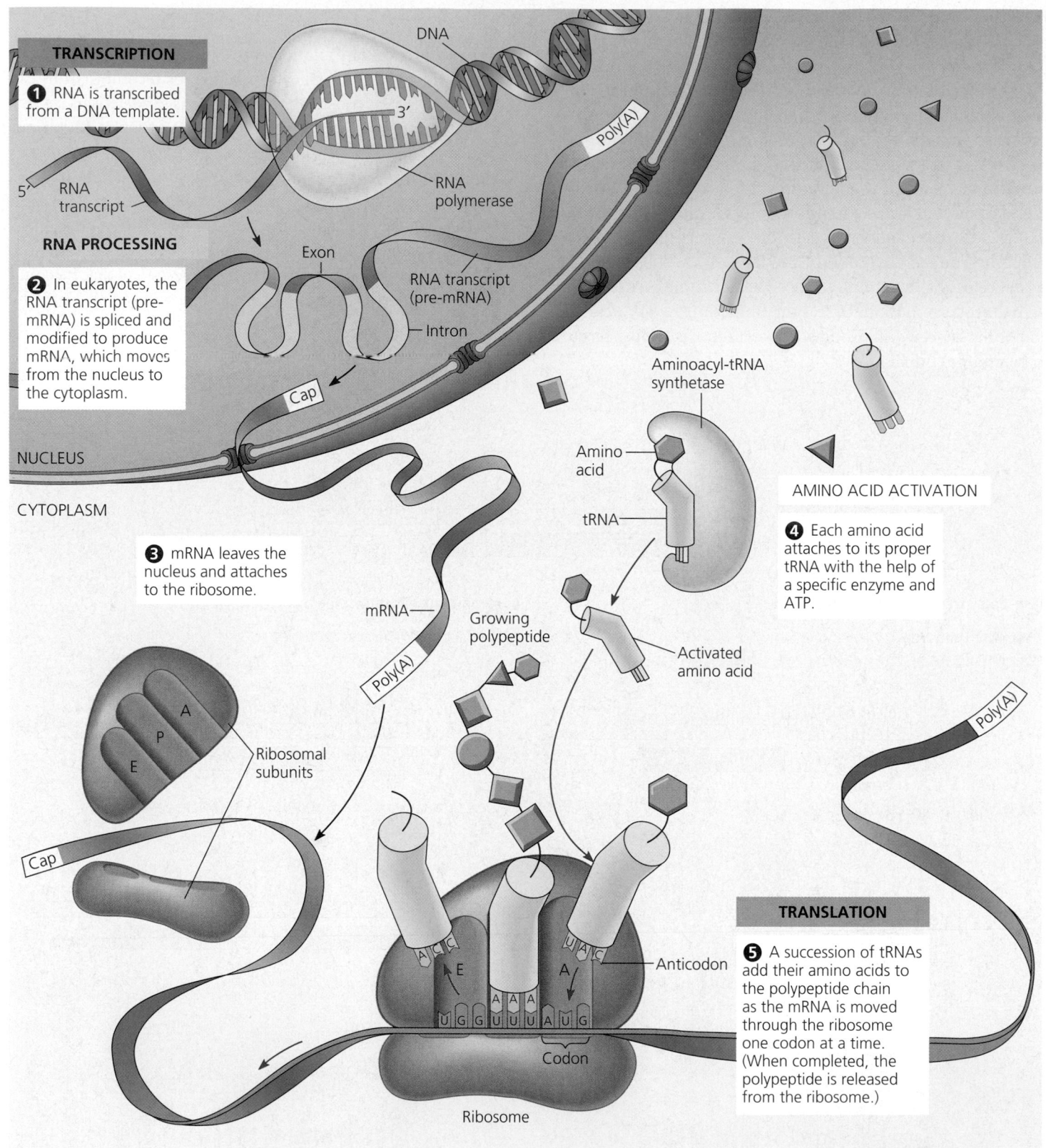

TRANSCRIPTION

DNA

❶ RNA is transcribed from a DNA template.

3′

RNA transcript

RNA polymerase

5′ RNA transcript

Poly(A)

RNA PROCESSING

❷ In eukaryotes, the RNA transcript (pre-mRNA) is spliced and modified to produce mRNA, which moves from the nucleus to the cytoplasm.

Exon

RNA transcript (pre-mRNA)

Intron

Cap

NUCLEUS

CYTOPLASM

❸ mRNA leaves the nucleus and attaches to the ribosome.

Aminoacyl-tRNA synthetase

Amino acid

tRNA

AMINO ACID ACTIVATION

❹ Each amino acid attaches to its proper tRNA with the help of a specific enzyme and ATP.

mRNA

Growing polypeptide

Activated amino acid

Poly(A)

Poly(A)

A
P
E

Ribosomal subunits

Cap

TRANSLATION

Anticodon

E

A

Codon

Ribosome

❺ A succession of tRNAs add their amino acids to the polypeptide chain as the mRNA is moved through the ribosome one codon at a time. (When completed, the polypeptide is released from the ribosome.)

FIGURE 17.25 A summary of transcription and translation in a eukaryotic cell. This diagram shows the path from one gene to one polypeptide. Keep in mind that each gene in the DNA can be transcribed repeatedly into many RNA molecules, and that each mRNA can be translated repeatedly to yield many polypeptides. (Also, remember that the final products of some genes are not polypeptides but RNA molecules, including tRNA and rRNA.) In general, the steps of transcription and translation are similar in prokaryotic and eukaryotic cells. The major difference is the occurrence of mRNA processing in the eukaryotic nucleus. Other significant differences are found in the initiation stages of both transcription and translation and in the termination of transcription.

Mutagenic radiation, a physical mutagen, includes ultraviolet (UV) light, which can produce disruptive thymine dimers in DNA (see FIGURE 16.17).

Chemical mutagens fall into several categories. Base analogues are chemicals that are similar to normal DNA bases but that pair incorrectly during DNA replication. Some other chemical mutagens interfere with correct DNA replication by inserting themselves into the DNA and distorting the double helix. Still other mutagens cause chemical changes in bases that change their pairing properties.

Researchers have developed various methods to test the mutagenic activity of different chemicals. A major application of these tests is the preliminary screening of chemicals to identify those that may cause cancer. This approach makes sense because most carcinogens (cancer-causing chemicals) are mutagenic and, conversely, most mutagens are carcinogenic.

■   ■   ■

What is a gene? *revisiting the question*

Our definition of a gene has evolved over the past few chapters, as it has through the history of genetics. We began with the Mendelian concept of a gene as a discrete unit of inheritance that affects a phenotypic character (Chapter 14). We saw that Morgan and his colleagues assigned such genes to specific loci on chromosomes and that geneticists sometimes use the term *locus* as a synonym for gene (Chapter 15). We went on to view a gene as a region of specific nucleotide sequence along the length of a DNA molecule (Chapter 16). Finally, in this chapter, we have moved toward a functional definition of a gene as a DNA sequence coding for a specific polypeptide chain. All these definitions are useful, depending on the con-

text in which genes are being studied. (FIGURE 17.25 summarizes the path from gene to polypeptide in a eukaryotic cell.)

Even the one gene–one polypeptide definition must be refined and applied selectively. Most eukaryotic genes contain noncoding segments (introns), so large portions of these genes have no corresponding segments in polypeptides. Molecular biologists also often include promoters and certain other regulatory regions of DNA within the boundaries of a gene. These DNA sequences are not transcribed, but they can be considered part of the functional gene because they must be present for transcription to occur. Our molecular definition of a gene must also be broad enough to include the DNA that is transcribed into rRNA, tRNA, and other RNAs that are not translated. These genes have no polypeptide products. Thus, we arrive at the following definition: *A gene is a region of DNA whose final product is either a polypeptide or an RNA molecule.*

For most genes, however, it is still useful to retain the one gene–one polypeptide idea. In this chapter, you have learned in molecular terms how a typical gene is expressed—by transcription into RNA and then translation into a polypeptide that forms a protein of specific structure and function. Proteins, in turn, bring about an organism's observable phenotype.

Genes are subject to regulation. The control of gene expression enables a bacterium, for example, to vary the amounts of particular enzymes as the metabolic needs of the cell change. In multicellular eukaryotes, the control of gene expression makes it possible for cells with the same DNA to diverge during their development into different cell types, such as muscle and nerve cells in animals. We will explore the regulation of gene expression in eukaryotes in Chapters 19 and 21. In the next chapter, we begin our discussion of gene regulation by focusing on the simpler molecular biology of bacteria and viruses.

CHAPTER 17 REVIEW

Go to the Campbell Biology website (www.campbellbiology.com) to explore an interactive version of the Chapter Review.

Summary of Key Concepts

THE CONNECTION BETWEEN GENES AND PROTEINS

■   The study of metabolic defects provided evidence that genes specify proteins (pp. 303–304, FIGURE 17.1) DNA controls metabolism by

directing cells to make specific enzymes and other proteins. Beadle and Tatum's experiments with mutant strains of *Neurospora* supported the one gene–one enzyme hypothesis, later modified to one gene–one polypeptide. In most cases, a gene determines the amino acid sequence of a polypeptide chain.
Web/CD Case Study in the Process of Science: *How Is a Metabolic Pathway Analyzed?*

■   Transcription and translation are the two main processes linking gene to protein: *an overview* (pp. 304–306, FIGURE 17.2) Both nucleic acids and proteins are informational polymers with linear sequences of monomers—nucleotides and amino acids, respectively. Transcription is the nucleotide-to-nucleotide transfer of information from DNA to RNA, while translation is the informational transfer from nucleotide sequence in RNA to amino acid sequence in a polypeptide.

Web/CD Activity 17A: *Overview of Protein Synthesis*

■ **In the genetic code, nucleotide triplets specify amino acids (pp. 306–308, FIGURES 17.3, 17.4)** A codon is a nucleotide triplet that in mRNA is either translated into an amino acid (61 codons) or serves as a translational stop signal (3 codons). The codon for methionine, AUG, also acts as a translational start signal.

■ **The genetic code must have evolved very early in the history of life (pp. 308–309)** The near universality of the genetic code suggests that it was present in ancestors common to all kingdoms of life.

THE SYNTHESIS AND PROCESSING OF RNA

■ **Transcription is the DNA-directed synthesis of RNA:** *a closer look* **(pp. 309–311, FIGURES 17.6, 17.7)** RNA synthesis on a DNA template is catalyzed by RNA polymerase. It follows the same base-pairing rules as DNA replication, except that in RNA, uracil substitutes for thymine. Promoters, specific nucleotide sequences at the start of a gene, signal the initiation of RNA synthesis. Transcription factors (proteins) help eukaryotic RNA polymerase recognize promoter sequences. Transcription continues until a particular RNA sequence signals termination.

Web/CD Activity 17B: *Transcription*

■ **Eukaryotic cells modify RNA after transcription (pp. 311–313, FIGURES 17.8–17.10)** Eukaryotic mRNA molecules are processed before leaving the nucleus by modification of their ends and by RNA splicing. The 5′ end receives a modified nucleotide cap, and the 3′ end a poly(A) tail. These seem to protect the molecule from degradation and enhance translation. Most eukaryotic genes have introns, noncoding regions interspersed among the coding regions, exons. In RNA splicing, introns are removed and exons joined. RNA splicing is catalyzed by small nuclear ribonucleoproteins (snRNPs), operating within larger assemblies called spliceosomes. In some cases, RNA alone catalyzes splicing. Catalytic RNA molecules are called ribozymes. The shuffling of exons by recombination may contribute to the evolution of protein diversity.

Web/CD Activity 17C: *RNA Processing*

THE SYNTHESIS OF PROTEIN

■ **Translation is the RNA-directed synthesis of a polypeptide:** *a closer look* **(pp. 313–320, FIGURES 17.17–17.19)** After picking up specific amino acids, transfer RNA (tRNA) molecules line up by means of their anticodon triplets at complementary codons on mRNA. The attachment of a specific amino acid to its particular tRNA is an ATP-driven process catalyzed by an aminoacyl-tRNA synthetase enzyme. Ribosomes coordinate the three stages of translation: initiation, elongation, and termination. Each ribosome is composed of two subunits made of protein and ribosomal RNA (rRNA). Ribosomes have a binding site for mRNA; P and A sites that hold adjacent tRNAs as amino acids are linked in the growing polypeptide chain; and an E site for release of tRNA. The formation of peptide bonds is catalyzed by one of the rRNA molecules. A number of ribosomes can work on a single mRNA molecule simultaneously, forming a polyribosome. After translation, the protein may be modified in ways that affect its three-dimensional shape.

Web/CD Activity 17D: *Translation*
Biology Labs On-Line: *TranslationLab*

■ **Signal peptides target some eukaryotic polypeptides to specific destinations in the cell (pp. 320–321, FIGURE 17.21)** Free ribosomes in the cytosol initiate the synthesis of all proteins, but proteins destined for membranes or for export from the cell complete their synthesis only after the ribosomes making them attach to the endoplasmic reticulum. In the latter case, a signal-recognition particle (SRP) binds to a signal sequence on the leading end of the growing polypeptide, enabling the ribosome to bind to the ER. Other signal sequences target proteins for mitochondria or chloroplasts.

■ **RNA plays multiple roles in the cell:** *a review* **(p. 321, TABLE 17.1)** More versatile than DNA, RNA performs structural, informational, and catalytic roles.

■ **Comparing protein synthesis in prokaryotes and eukaryotes:** *a review* **(pp. 321–322)** In a bacterial cell, which lacks a nuclear envelope, translation of an mRNA can begin while transcription is still in progress. In a eukaryotic cell, the nuclear envelope separates transcription from translation; extensive RNA processing occurs in the nucleus.

■ **Point mutations can affect protein structure and function (pp. 322–325, FIGURE 17.24)** Point mutations are changes in one base pair of DNA. Base-pair substitutions can cause missense or nonsense mutations, which are often detrimental to protein function. Base-pair insertions or deletions may produce frameshift mutations that disrupt the mRNA reading frame downstream of the mutation. Spontaneous mutations can occur during DNA replication or repair. Various chemical and physical mutagens can also alter genes.

■ **What is a gene?** *revisiting the question* **(p. 325, FIGURE 17.25)** A gene is usually a region of DNA encoding a polypeptide, but some genes have RNA molecules as their final products.

Self-Quiz

1. Base-pair substitutions involving the third base of a codon are unlikely to result in an error in the polypeptide. This is because
   a. base-pair substitutions are corrected before transcription begins.
   b. base-pair substitutions are restricted to introns, and these regions are later deleted from the mRNA.
   c. most tRNAs bind tightly to a codon with only the first two bases of the anticodon.
   d. a signal-recognition particle corrects coding errors before the mRNA reaches the ribosome.
   e. transcribed errors attract snRNPs, which then stimulate splicing and correction.

2. In eukaryotic cells, transcription cannot begin until
   a. the two DNA strands have completely separated and exposed the promoter.
   b. the appropriate transcription factors have bound to the promoter.
   c. the 5′ caps are removed from the mRNA.
   d. the DNA introns are removed from the template.
   e. DNA nucleases have isolated the transcription unit from the noncoding DNA.

3. Which of the following is *not* true of a codon?
   a. It consists of three nucleotides.
   b. It may code for the same amino acid as another codon does.
   c. It never codes for more than one amino acid.
   d. It extends from one end of a tRNA molecule.
   e. It is the basic unit of the genetic code.

4. Beadle and Tatum discovered several classes of *Neurospora* mutants that were able to grow on minimal medium with arginine added. Class I mutants were also able to grow on medium supplemented with either ornithine or citrulline, whereas class II mutants could grow on citrulline medium but not on ornithine medium. The metabolic pathway of arginine synthesis is is as follows:

$$Precursor \longrightarrow \underset{A}{Ornithine} \longrightarrow \underset{B}{Citrulline} \longrightarrow \underset{C}{Arginine}$$

From the behavior of their mutants, Beadle and Tatum could conclude that

a. one gene codes for the entire metabolic pathway.

b. the genetic code of DNA is a triplet code.

c. class I mutants have their mutations later in the nucleotide chain than do class II mutants and thus have more functional enzymes.

d. class I mutants have a nonfunctional enzyme at step A, and class II mutants have a nonfunctional enzyme at step B.

e. class I mutants have a nonfunctional enzyme at step B, and class II mutants have a nonfunctional enzyme at step C.

5. The anticodon of a particular tRNA molecule is

a. complementary to the corresponding mRNA codon.

b. complementary to the corresponding triplet in rRNA.

c. the part of tRNA that bonds to a specific amino acid.

d. changeable, depending on the amino acid that attaches to the tRNA.

c. catalytic, making the tRNA a ribozyme.

6. Which of the following is *not* true of RNA processing?

a. Exons are excised and hydrolyzed before mRNA moves out of the nucleus.

b. The presence of introns may facilitate crossing over between regions of a gene that code for polypeptide domains.

c. Ribozymes may function in RNA splicing.

d. RNA splicing may be catalyzed by spliceosomes.

e. A primary transcript is often much longer than the final RNA molecule that leaves the nucleus.

7. Which of the following is true of translation in both prokaryotes and eukaryotes?

a. Translation occurs simultaneously with transcription.

b. The product of transcription is directly translated.

c. The codon UUU codes for phenylalanine.

d. Ribosomes are affected by streptomycin.

e. The signal-recognition particle (SRP) binds to the first 20 amino acids of certain polypeptides.

8. Using the genetic code in FIGURE 17.4, identify a possible 5′ → 3′ sequence of nucleotides in the DNA template strand for an mRNA coding for the polypeptide sequence Phe-Pro-Lys.

a. UUU-GGG-AAA   d. CTT-CGG-GAA

b. GAA-CCC-CTT   e. AAA-CCC-UUU

c. AAA-ACC-TTT

9. Which of the following mutations would be *most* likely to have a harmful effect on an organism? Explain your answer.

a. a base-pair substitution

b. a deletion of three bases near the middle of a gene

c. a single base deletion near the middle of an intron

d. a single base deletion close to the end of the coding sequence

e. a single base insertion near the start of the coding sequence

10. Which component is not directly involved in the process known as translation?

a. mRNA   d. ribosomes

b. DNA   e. GTP

c. tRNA

11. An mRNA molecule contains the following nucleotide sequence:

5′-CCAUUUACG-3′

Using FIGURE 17.4, translate this sequence into the corresponding amino acid sequence.

12. How does RNA polymerase "know" where to start transcribing a gene?

13. What are the functions of the ribosome in polypeptide synthesis?

14. What would happen if a mutation changed a start codon to some other codon?

15. Once polypeptide synthesis starts, what are the three main steps by which it grows (elongates)?

16. What happens when one nucleotide pair is lost from the middle of the coding sequence of a gene?

Go to the website or CD-ROM for more quiz questions.

Evolution Connection

The genetic code (FIGURE 17.4) is rich with evolutionary implications. See what evolutionary deductions you can make from it by looking for patterns in the codons. For instance, notice that the 20 amino acids are not randomly scattered, but most amino acids are coded for by a similar set of codons. What evolutionary explanations can be given for this pattern? (*Hint:* there is one explanation relating to historical ancestry, and some less obvious ones of a "form-fits-function" type.)

The Process of Science

A biologist inserts a gene from a human liver cell into the chromosome of a bacterium. The bacterium then transcribes this gene into mRNA and translates the mRNA into protein. The protein produced is useless and is found to contain many more amino acids than does the protein made by the eukaryotic cell. Explain why.

Virtually reenact the Beadle-Tatum experiment in the Case Study in the Process of Science on the website and CD-ROM. You can also link to *TranslationLab* at the Biology Labs On-Line website.

Science, Technology, and Society

Our civilization generates many potentially mutagenic chemicals (pesticides, for example) and modifies the environment in ways that increase exposure to other mutagens, notably UV radiation. What role should government play in identifying mutagens and regulating their release to the environment?

Answers: 1. c; 2. b; 3. d; 4. d; 5. a; 6. a; 7. c; 8. d; 9. e; 10. b; 11. Pro-Phe-Thr. 12. It recognizes the gene's promoter, a specific nucleotide sequence. 13. The ribosome holds mRNA and tRNAs together, catalyzes the addition of amino acids from the tRNAs to the growing polypeptide chain, and, in translocation, moves the tRNAs and mRNA through. 14. Messenger RNA transcribed from the mutated gene, if any, would be nonfunctional because ribosomes would not initiate translation at the correct point. 15. Codon recognition by an incoming tRNA, peptide bond formation by the ribosome, and translocation of the tRNA and mRNA through the ribosome. 16. In the mRNA, the reading frame downstream from the deletion is shifted, leading to a long string of incorrect amino acids in the polypeptide and possibly premature termination. The polypeptide will be nonfunctional.

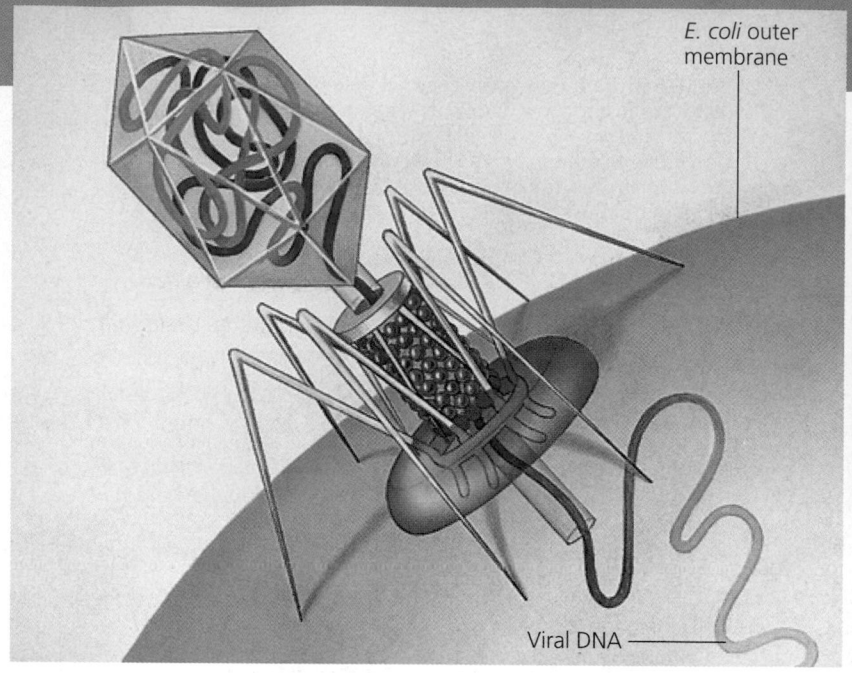

E. coli outer membrane

Viral DNA

CHAPTER 18

MICROBIAL MODELS: THE GENETICS OF VIRUSES AND BACTERIA

THE GENETICS OF VIRUSES

- Researchers discovered viruses by studying a plant disease
- A virus is a genome enclosed in a protective coat
- Viruses can reproduce only within a host cell: *an overview*
- Phages reproduce using lytic or lysogenic cycles
- Animal viruses are diverse in their modes of infection and replication
- Plant viruses are serious agricultural pests
- Viroids and prions are infectious agents even simpler than viruses
- Viruses may have evolved from other mobile genetic elements

THE GENETICS OF BACTERIA

- The short generation span of bacteria helps them adapt to changing environments
- Genetic recombination produces new bacterial strains
- The control of gene expression enables individual bacteria to adjust their metabolism to environmental change

The drawing *that opens this chapter dramatizes a remarkable event: the genetic takeover of a cell by a virus. In this case, the cell is the bacterium* E. coli, *and the virus, looking something like a miniature lunar landing craft, is the bacteriophage T4. The phage is infecting the cell by injecting its DNA. In* FIGURE 16.2, *you met a closely related virus, T2, which helped prove that DNA is the genetic material. Molecular biology was born in the laboratories of microbiologists studying such viruses and bacteria. Microbiologists provided most of the evidence that genes are made of DNA, and they worked out the major steps of DNA replication, transcription, and translation, the three main processes in the flow of genetic information. Viruses and bacteria are the simplest biological systems—microbial models where scientists find life's fundamental molecular mechanisms in their most basic, accessible forms.*

The value of viruses and bacteria as model systems in biological research is just one reason to learn about these microbes. While microbial models have helped biologists understand the molecular genetics of more complex organisms, viruses and bacteria also have unique features that make their genetics interesting in their own right. These specialized mechanisms have important applications for understanding how viruses and bacteria cause disease. In addition, techniques enabling scientists to manipulate genes and transfer them from one organism to another have emerged from the study of microbes. These techniques are having an important impact on both basic research and biotechnology (see Chapter 20).

*In this chapter, we explore the genetics of viruses and bacteria. Recall that bacteria are prokaryotic organisms, with cells much smaller and more simply organized than those of eukaryotes, such as plants and animals. Viruses are smaller and simpler still, lacking the structures and most of the metabolic machinery found in cells (*FIGURE 18.1). *In fact, most viruses are little more than aggregates of nucleic acid and protein—genes packaged in protein coats. It is with these simplest of all genetic systems that we begin.*

THE GENETICS OF VIRUSES

Researchers discovered viruses by studying a plant disease

Microbiologists were able to detect viruses indirectly long before they were actually able to see them. The story of how viruses were discovered begins in 1883 with Adolf Mayer, a German scientist seeking the cause of tobacco mosaic disease. This disease stunts the growth of tobacco plants and gives their leaves a mottled, or mosaic, coloration (see FIGURE 18.9a). Mayer discovered that the disease was contagious when he found he could transmit it from plant to

THE PROCESS OF SCIENCE

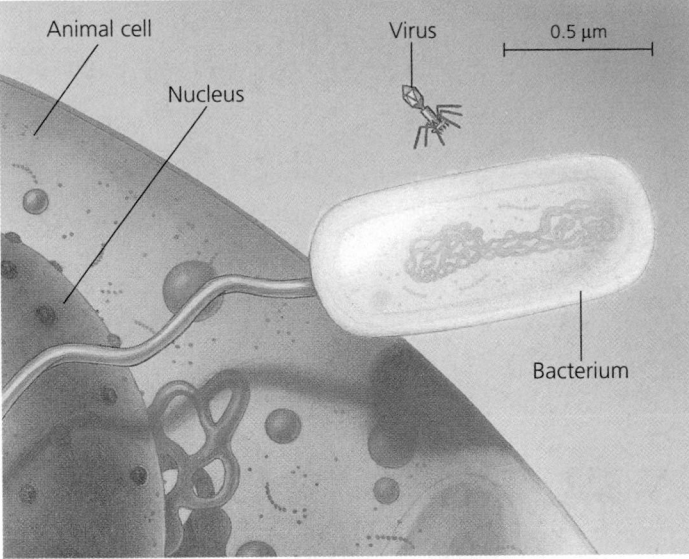

Animal cell

Nucleus

Virus

0.5 μm

Bacterium

FIGURE 18.1 Comparing the size of a virus, a bacterium, and a eukaryotic cell. By studying viruses and bacteria, the simplest biological systems, scientists caught their first glimpses of the elegant molecular mechanisms of heredity.

plant by spraying sap extracted from diseased leaves onto healthy plants. He searched for a microbe in the infectious sap but found none. Mayer concluded that the disease was caused by unusually small bacteria that could not be seen with the microscope. This hypothesis was tested a decade later by Dimitri Ivanowsky, a Russian who passed sap from infected tobacco leaves through a filter designed to remove bacteria. After filtering, the sap still produced mosaic disease.

Ivanowsky clung to the hypothesis that bacteria caused tobacco mosaic disease. Perhaps, he reasoned, the pathogenic bacteria were so small they could pass through the filter. Or perhaps the bacteria made a filterable toxin that caused the disease. This latter possibility was ruled out in 1897 when the Dutch botanist Martinus Beijerinck discovered that the infectious agent in the filtered sap could reproduce. Beijerinck sprayed plants with the filtered sap, and after these plants developed mosaic disease, he used their sap to infect more plants, continuing this process through a series of infections. The pathogen must have been reproducing, for its ability to cause disease was undiluted after several transfers from plant to plant.

In fact, the pathogen could reproduce only within the host it infected. Unlike bacteria, the mysterious agent of mosaic disease could not be cultivated on nutrient media in test tubes or petri dishes. Also, the pathogen was not inactivated by alcohol, which is generally lethal to bacteria. Beijerinck imagined a reproducing particle much smaller and simpler than bacteria. His suspicions were confirmed in 1935, when the American scientist Wendell Stanley crystallized the infectious particle, now known as tobacco mosaic virus (TMV). Subsequently, TMV and many other viruses were actually seen with the help of the electron microscope.

A virus is a genome enclosed in a protective coat

The tiniest viruses are only 20 nm in diameter—smaller than a ribosome. Millions could easily fit on a pinhead. Even the largest viruses can barely be resolved with the light microscope. Stanley's discovery that some viruses could be crystallized was exciting and puzzling news. Not even the simplest of cells can aggregate into regular crystals. But if viruses are not cells, then what are they? They are infectious particles consisting of nucleic acid enclosed in a protein coat and, in some cases, a membranous envelope. Let's examine the structure of viruses more closely; then we'll look at how viruses replicate.

Viral Genomes

We usually think of genes as being made of double-stranded DNA—the conventional double helix—but many viruses defy this convention. Their genomes may consist of double-stranded DNA, single-stranded DNA, double-stranded RNA, or single-stranded RNA, depending on the specific type of virus. A virus is called a DNA virus or an RNA virus, according to the kind of nucleic acid that makes up its genome. In either case, the genome is usually organized as a single linear or circular molecule of nucleic acid. The smallest viruses have only four genes, while the largest have several hundred.

Capsids and Envelopes

The protein shell that encloses the viral genome is called a **capsid.** Depending on the type of virus, the capsid may be rod-shaped (more precisely, helical), polyhedral, or more complex in shape. Capsids are built from a large number of protein subunits called *capsomeres,* but the number of different *kinds* of proteins is usually small. Tobacco mosaic virus, for example, has a rigid, rod-shaped capsid made from over a thousand molecules of a single type of protein (FIGURE 18.2a, p. 330). Adenoviruses, which infect the respiratory tracts of animals, have 252 identical protein molecules arranged into a polyhedral capsid with 20 triangular facets—an icosahedron (FIGURE 18.2b).

Some viruses have accessory structures that help them infect their hosts. Influenza viruses, as well as many other viruses found in animals, have **viral envelopes,** membranes cloaking their capsids (FIGURE 18.2c). These envelopes are derived from membrane of the host cell, but in addition to host cell phospholipids and proteins, they also contain proteins and glycoproteins of viral origin (glycoproteins are proteins with carbohydrate covalently attached). Some viruses carry a few viral enzyme molecules within their capsids.

The most complex capsids are found among viruses that infect bacteria. As you learned in Chapter 16, bacterial viruses are called **bacteriophages,** or simply **phages.** The first phages studied included seven that infect the bacterium *Escherichia coli.* These seven phages were named type 1 (T1), type 2 (T2),

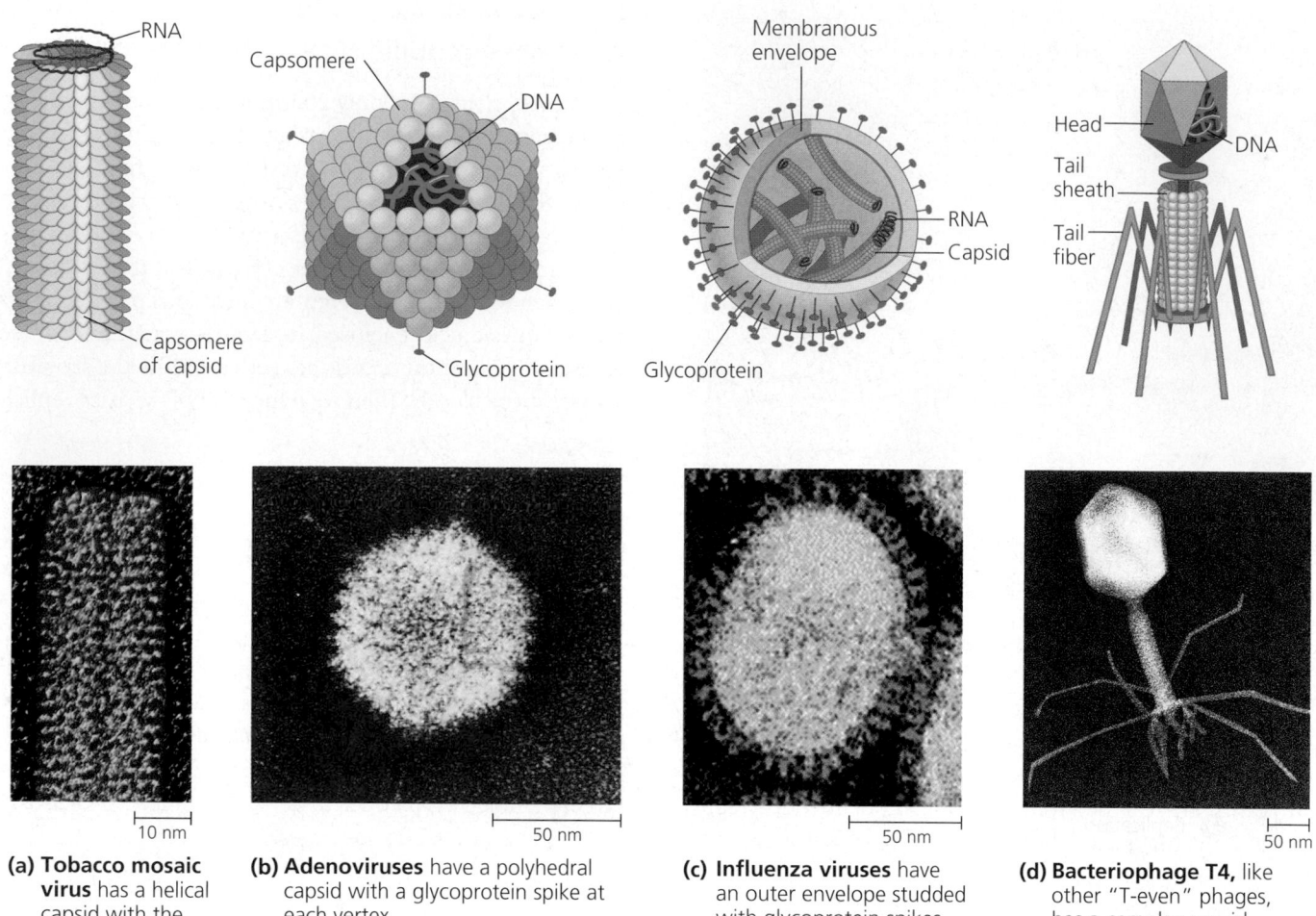

(a) Tobacco mosaic virus has a helical capsid with the overall shape of a rigid rod.

(b) Adenoviruses have a polyhedral capsid with a glycoprotein spike at each vertex.

(c) Influenza viruses have an outer envelope studded with glycoprotein spikes. The genome consists of eight different RNA molecules, each wrapped in a helical capsid.

(d) Bacteriophage T4, like other "T-even" phages, has a complex capsid consisting of a polyhedral head and a tail apparatus.

FIGURE 18.2 Viral structure. Viruses are made up of nucleic acid (DNA or RNA) enclosed in a protein coat (the capsid) and sometimes further wrapped in a membranous envelope. The individual protein subunits making up the capsid are called capsomeres. Although viruses are diverse in size and shape, there are common structural motifs, most of which appear in the four examples shown here. (All the micrographs are TEMs.)

and so forth, in the order of their discovery. By coincidence, the three T-even phages—T2, T4, and T6—turned out to be very similar in structure. Their capsids have elongated 20-sided heads that enclose their DNA. Attached to the head is a protein tail piece with tail fibers that the phages use to attach to a bacterium (FIGURE 18.2d).

Viruses can reproduce only within a host cell: *an overview*

Viruses are obligate intracellular parasites; that is, they can reproduce only within a host cell. An isolated virus is unable to reproduce—or do anything else, for that matter, except infect an appropriate host cell. Viruses lack the enzymes for metabolism and have no ribosomes or other equipment for making their own proteins. Thus, isolated viruses are merely packaged sets of genes in transit from one host cell to another.

Each type of virus can infect and parasitize only a limited range of host cells, called its **host range.** This host specificity depends on the evolution of recognition systems by the virus. Viruses identify their host cells by a "lock-and-key" fit between proteins on the outside of the virus and specific receptor molecules on the surface of the cell. (Presumably, the receptors first evolved because they carried out functions of benefit to the organism.) Some viruses have host ranges broad enough to include several species. Swine flu virus, for example, can infect both hogs and humans, and the rabies virus can infect a number of mammalian species, including raccoons, skunks, dogs, and humans. In other cases, viruses have host ranges so narrow that they infect only a single species. For instance, there are several phages that can parasitize only *E. coli.*

Viruses of eukaryotes are usually tissue specific. Human cold viruses infect only the cells lining the upper respiratory tract, ignoring other tissues. And the AIDS virus binds to a specific receptor on certain types of white blood cells.

A viral infection begins when the genome of a virus makes its way into a host cell (FIGURE 18.3). The mechanism by which this nucleic acid enters the cell varies, depending on the type of virus. For example, the T-even phages use their elaborate tail apparatus to inject DNA into a bacterium (see the chapter-opening drawing on p. 328). Once inside, the viral genome can commandeer its host, reprogramming the cell to copy the viral nucleic acid and manufacture viral proteins. Most DNA

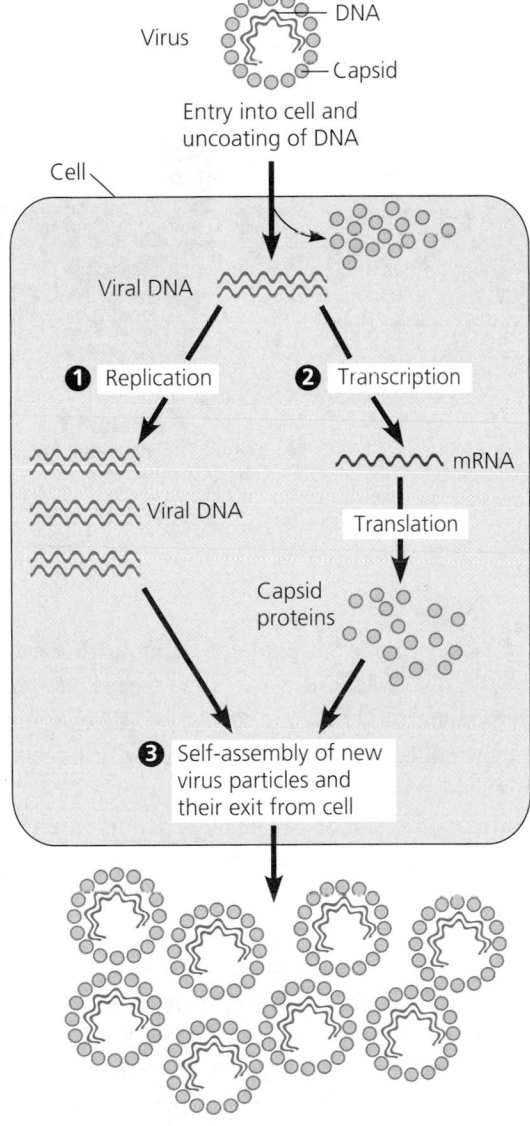

FIGURE 18.3 A simplified viral reproductive cycle. A virus is an obligate intracellular parasite that uses the equipment of its host cell to reproduce. In this simplest of viral cycles, the parasite is a DNA virus with a capsid consisting of a single type of protein. ❶ After entering the cell, the viral DNA uses host nucleotides and enzymes to replicate itself. ❷ The viral DNA uses other host resources to produce its capsid proteins by transcription and translation. ❸ The new viral DNA and capsid proteins assemble into new virus particles, which leave the cell.

viruses use the DNA polymerases of the host cell to synthesize new genomes along the templates provided by the viral DNA. In contrast, to replicate their genomes, RNA viruses must use special virus-encoded polymerases, ones that can use RNA as a template. (Cells generally have no native enzymes for carrying out such a process.) We will describe the replication of DNA and RNA viruses in more detail later in the chapter.

Regardless of the type of viral genome, the parasite diverts its host's resources for viral production. The host provides the nucleotides for nucleic acid synthesis. It also provides enzymes, ribosomes, tRNAs, amino acids, ATP, and other components needed for making the viral proteins dictated by viral genes.

After the viral nucleic acid molecules and capsomeres are produced, their assembly into new viruses is often a spontaneous process, a process of self-assembly. In fact, the RNA and capsomeres of TMV can be separated in the laboratory and then reassembled to form complete viruses simply by mixing the components together again. The simplest type of viral reproductive cycle is completed when hundreds or thousands of viruses emerge from the infected host cell. The cell is often destroyed in the process. In fact, some of the symptoms of human viral infections, such as colds and influenza, result from cellular damage and death and from the body's responses to this destruction. The viral progeny that exit a cell have the potential to infect additional cells, spreading the viral infection.

There are many variations on the simplified viral reproductive cycle we have traced in this overview. We will see several examples as we take a closer look at some bacterial viruses (phages), animal viruses, and plant viruses.

Phages reproduce using lytic or lysogenic cycles

The phages are the best understood of all viruses, although some of them are also among the most complex. Research on phages led to the discovery that some double-stranded DNA viruses can reproduce by two alternative mechanisms: the lytic cycle and the lysogenic cycle.

The Lytic Cycle

A phage reproductive cycle that culminates in death of the host cell is known as a **lytic cycle.** The term refers to the last stage of infection, during which the bacterium lyses (breaks open) and releases the phages that were produced within the cell. Each of these phages can then infect a healthy cell, and a few successive lytic cycles can destroy an entire bacterial colony in just hours. A phage that reproduces only by a lytic cycle is a **virulent phage.** FIGURE 18.4 (p. 332) uses the virulent phage T4 to illustrate the steps of a lytic cycle. The figure and legend describe the process, which you should study before proceeding.

After reading about the lytic cycle, you may wonder why phages haven't exterminated all bacteria. Actually, bacteria are

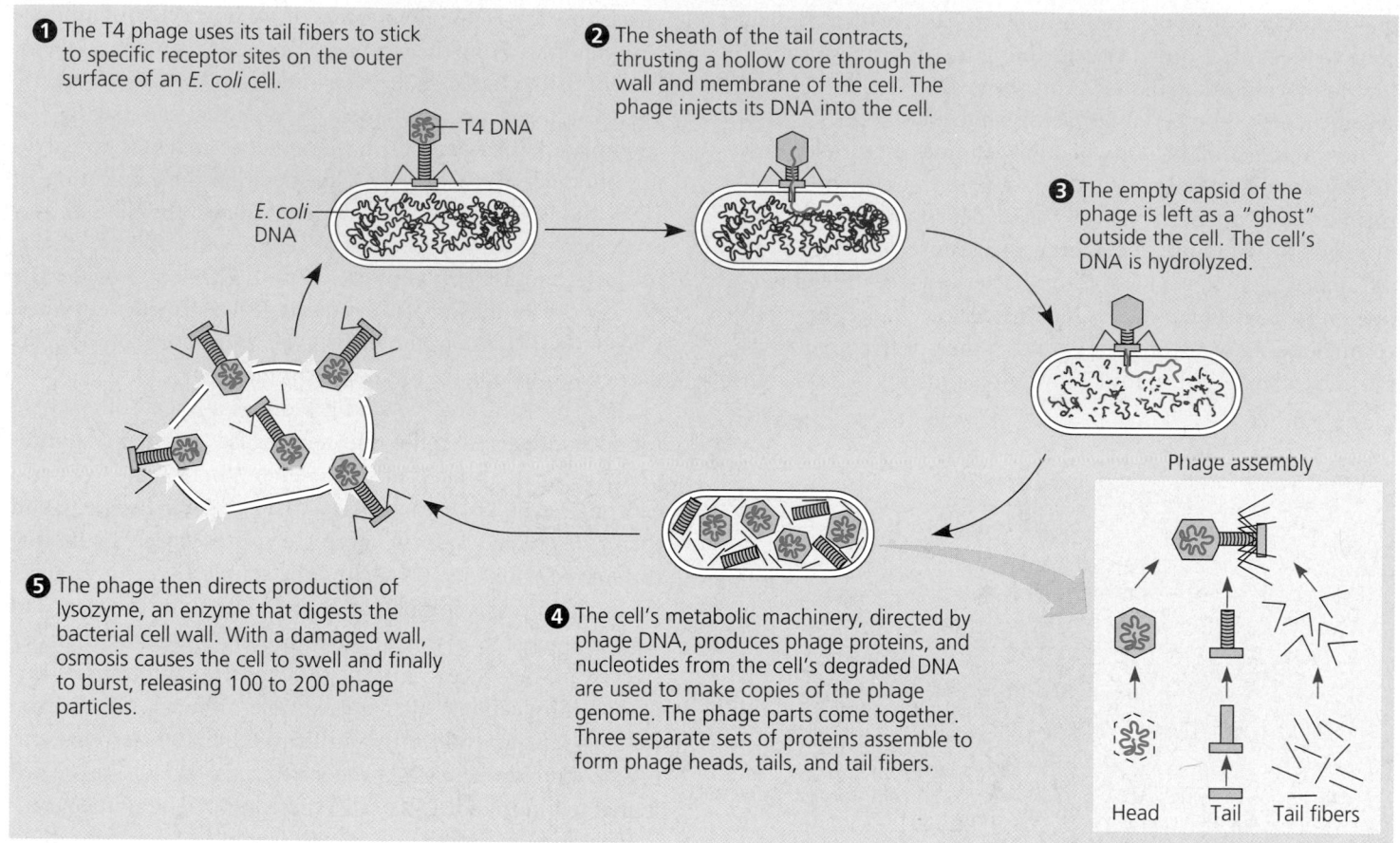

1 The T4 phage uses its tail fibers to stick to specific receptor sites on the outer surface of an *E. coli* cell.

T4 DNA

E. coli DNA

2 The sheath of the tail contracts, thrusting a hollow core through the wall and membrane of the cell. The phage injects its DNA into the cell.

3 The empty capsid of the phage is left as a "ghost" outside the cell. The cell's DNA is hydrolyzed.

Phage assembly

5 The phage then directs production of lysozyme, an enzyme that digests the bacterial cell wall. With a damaged wall, osmosis causes the cell to swell and finally to burst, releasing 100 to 200 phage particles.

4 The cell's metabolic machinery, directed by phage DNA, produces phage proteins, and nucleotides from the cell's degraded DNA are used to make copies of the phage genome. The phage parts come together. Three separate sets of proteins assemble to form phage heads, tails, and tail fibers.

Head   Tail   Tail fibers

FIGURE 18.4 The lytic cycle of phage T4. Phage T4 has about 100 genes, which are transcribed and translated using the host cell's machinery. One of the first phage genes translated after infection codes for an enzyme that chops up the host cell's DNA (step 3); the phage DNA is protected from breakdown because it contains a modified form of cytosine that is not recognized by the enzyme. The entire lytic cycle, from the phage's first contact with the cell surface to cell lysis, takes only 20–30 minutes at 37°C.

not defenseless. Natural selection favors bacterial mutants with receptor sites that are no longer recognized by a particular type of phage. And when phage DNA successfully enters a bacterium, various cellular enzymes may break it down. Enzymes called *restriction nucleases,* for example, recognize and cut up DNA that is foreign to the cell, including certain phage DNA. The bacterial cell's own DNA is chemically modified in a way that prevents attack by restriction enzymes. But just as natural selection favors bacteria with effective restriction enzymes, natural selection favors phage mutants that are resistant to these enzymes. Thus, the parasite-host relationship is in constant evolutionary flux.

There is still another important reason bacteria have been spared from extinction as a result of phage activity. Many phages can check their own destructive tendencies and, instead of lysing their host cells, coexist with them in what is called the lysogenic cycle.

The Lysogenic Cycle

In contrast to the lytic cycle, which kills the host cell, the **lysogenic cycle** replicates the phage genome without destroying the host. Phages that are capable of using both modes of reproducing within a bacterium are called **temperate phages.** To compare the lytic and lysogenic cycles, we will examine a temperate phage called lambda, written with the Greek letter λ. Phage λ resembles T4, but its tail has only one short tail fiber.

Infection of an *E. coli* cell by phage λ begins when the phage binds to the surface of the cell and injects its DNA (FIGURE 18.5). Within the host, the λ DNA molecule forms a circle. What happens next depends on the reproductive mode: lytic cycle or lysogenic cycle. During a lytic cycle, the viral genes immediately turn the host cell into a λ-producing factory, and the cell soon lyses and releases its viral products. The viral genome behaves differently during a lysogenic cycle. The λ DNA molecule is incorporated by genetic recombination (crossing over) into a specific site on the host cell's chromosome. It is then known as a **prophage.** One prophage gene codes for a protein that represses most of the other prophage genes. (This is the repressor protein Nancy Hopkins studied as a graduate student; see p. 232.) Thus, the phage genome is mostly silent within the bacterium. How, then, does the phage reproduce? Every time the *E. coli* cell prepares to divide, it

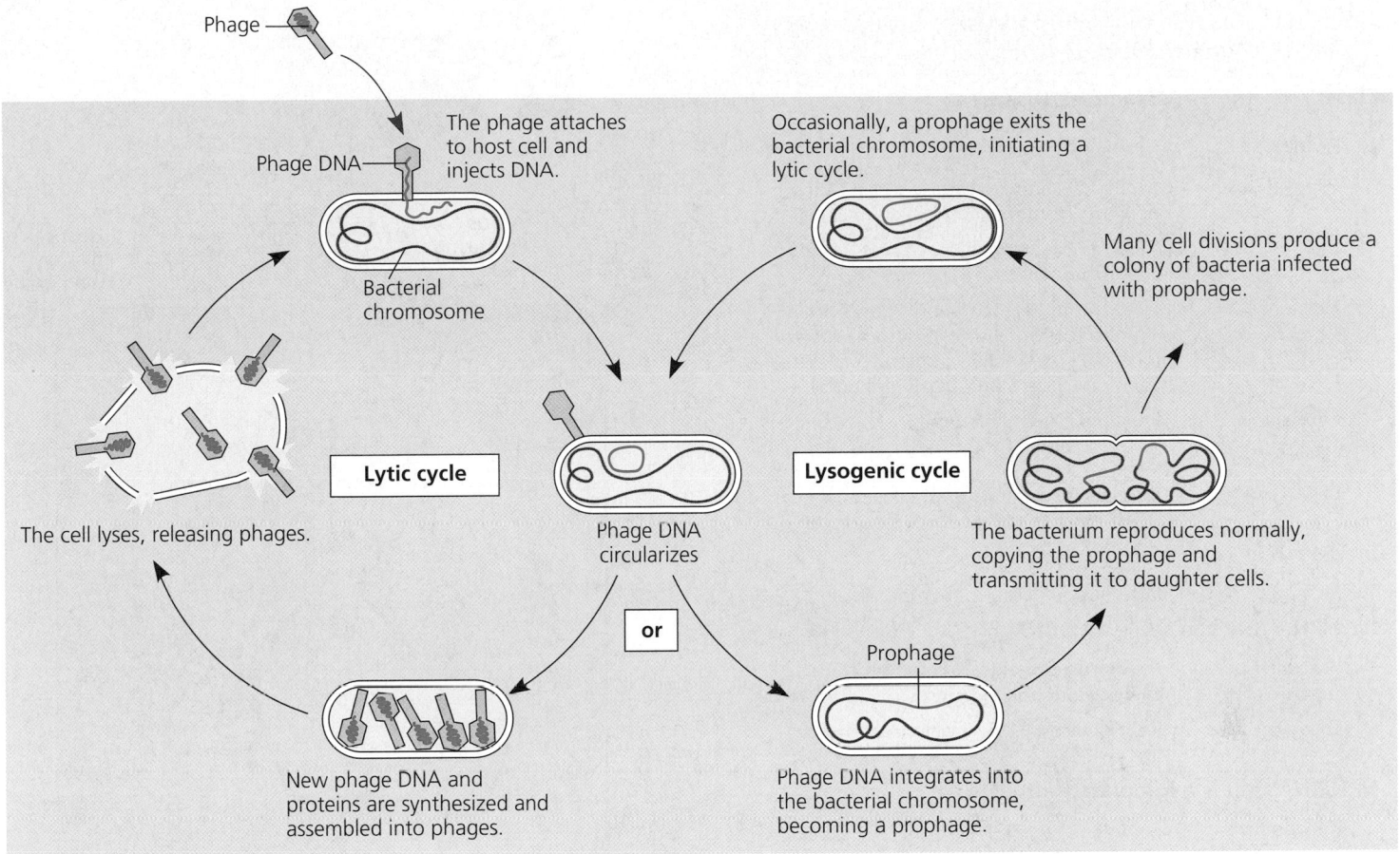

Phage

Phage DNA → | The phage attaches to host cell and injects DNA.

Bacterial chromosome

The cell lyses, releasing phages.

Lytic cycle

Phage DNA circularizes

or

New phage DNA and proteins are synthesized and assembled into phages.

Occasionally, a prophage exits the bacterial chromosome, initiating a lytic cycle.

Many cell divisions produce a colony of bacteria infected with prophage.

Lysogenic cycle

The bacterium reproduces normally, copying the prophage and transmitting it to daughter cells.

Prophage

Phage DNA integrates into the bacterial chromosome, becoming a prophage.

FIGURE 18.5 The lysogenic and lytic reproductive cycles of phage λ, a temperate phage. After entering the bacterial cell and circularizing, the λ DNA can either integrate into the bacterial chromosome (lysogenic cycle) or immediately initiate the production of a large number of progeny phages (lytic cycle). In most cases, the lytic pathway is followed, but once a lysogenic cycle begins, the prophage may be carried in the host cell's chromosome for many generations. Phage λ has a single, short tail fiber, not shown in this diagram.

replicates the phage DNA along with its own and passes the copies on to daughter cells. A single infected cell can quickly give rise to a large population of bacteria carrying the virus in prophage form. This mechanism enables viruses to propagate without killing the host cells on which they depend.

The term *lysogenic* implies that prophages are capable of giving rise to active phages that lyse their host cells. This occurs when, occasionally, the λ genome exits the bacterial chromosome. Once free in the cell, the λ genome initiates a lytic cycle. It is usually an environmental trigger, such as radiation or the presence of certain chemicals, that switches the virus from the lysogenic to the lytic mode.

In addition to the gene for the repressor protein, a few other prophage genes may also be expressed during lysogenic cycles, and the expression of these genes may alter the phenotype of the host bacteria. This phenomenon can have important medical significance. For example, the bacteria that cause the human diseases diphtheria, botulism, and scarlet fever would be harmless to humans if it were not for certain prophage genes that induce the host bacteria to make toxins.

Animal viruses are diverse in their modes of infection and replication

Everyone has suffered from viral infections, whether cold sores, influenza, or the common cold. TABLE 18.1 (p. 334) lists some important classes of animal viruses. Like all viruses, those that cause illness in humans and other animals can reproduce (replicate) only inside host cells.

Reproductive Cycles of Animal Viruses

Many variations on the basic scheme of viral infection and reproduction are represented among the animal viruses. One key variable is the type of nucleic acid that serves as a virus's genetic material (the basis for classification in TABLE 18.1). Another variable is the presence or absence of a membranous envelope. Rather than consider all the mechanisms of viral infection and reproduction, we will focus on the roles of viral envelopes and on the functioning of RNA as the genetic material of many viruses.

Table 18.1 Classes of Animal Viruses, Grouped by Type of Nucleic Acid

Class*	Examples/Diseases
I. dsDNA**	
Papovavirus	Papilloma (human warts, cervical cancer); polyoma (tumors in certain animals)
Adenovirus	Respiratory diseases; some cause tumors in certain animals
Herpesvirus	Herpes simplex I (cold sores); herpes simplex II (genital sores); varicella zoster (chicken pox, shingles); Epstein-Barr virus (mononucleosis, Burkitt's lymphoma)
Poxvirus	Smallpox; vaccinia; cowpox
II. ssDNA	
Parvovirus	Roseola; most parvoviruses depend on co-infection with adenoviruses for growth
III. dsRNA	
Reovirus	Diarrhea; mild respiratory diseases
IV. ssRNA that can serve as mRNA	
Picornavirus	Poliovirus; rhinovirus (common cold); enteric (intestinal) viruses
Togavirus	Rubella virus; yellow fever virus; encephalitis viruses
V. ssRNA that is a template for mRNA	
Rhabdovirus	Rabies
Paramyxovirus	Measles; mumps
Orthomyxovirus	Influenza viruses
VI. ssRNA that is a template for DNA synthesis	
Retrovirus	RNA tumor viruses (e.g., leukemia viruses); HIV (AIDS virus)

*The subclasses within each class differ mainly in capsid structure and in the presence or absence of a membranous envelope.

**ds = double-stranded; ss = single-stranded.

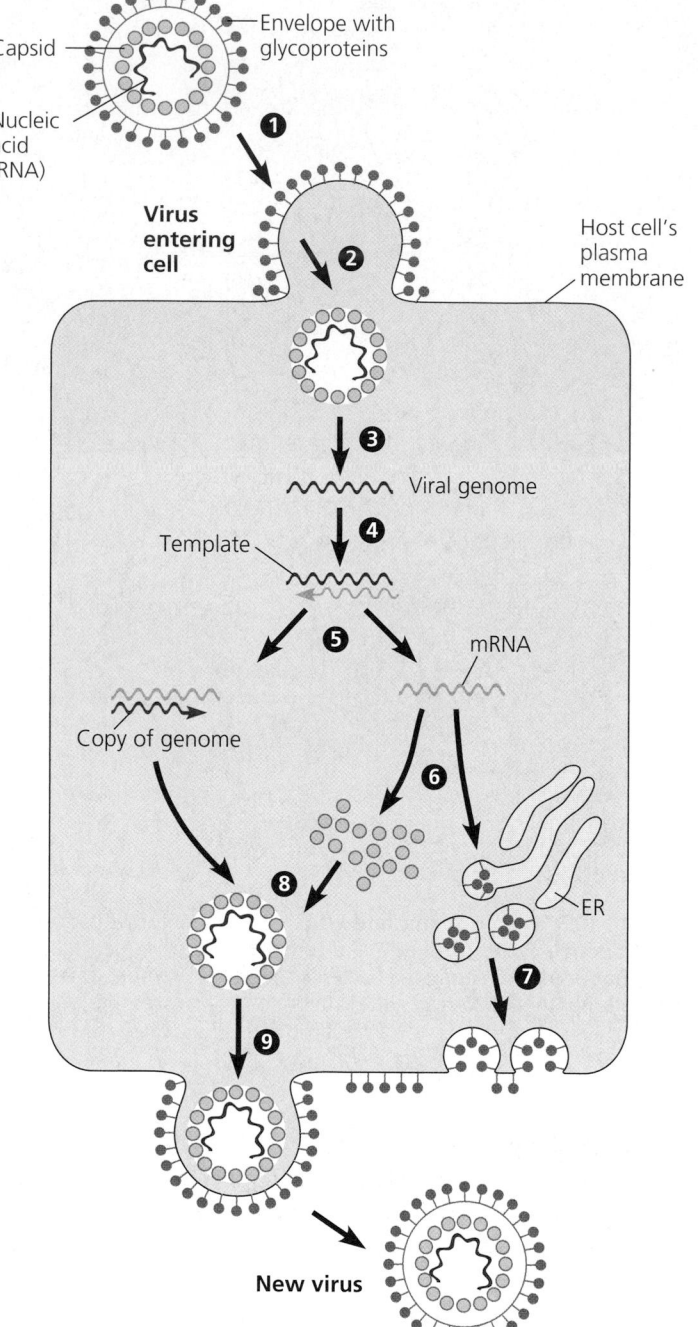

FIGURE 18.6 The reproductive cycle of an enveloped virus. The virus shown here has a genome of single-stranded RNA (class V in TABLE 18.1). ❶ Glycoproteins on the viral envelope recognize and bind to specific receptor molecules (not shown) on the surface of the host cell. ❷ The viral envelope fuses with the cell's plasma membrane, and the capsid and viral genome enter the cell. ❸ Cellular enzymes remove the capsid. ❹ The viral genome functions as a template for making complementary RNA strands (lighter color), which have two functions: ❺ They serve as templates for making new copies of genome RNA, and ❻ they serve as mRNA. The mRNA is translated into both capsid proteins and glycoproteins for the viral envelope. The cell's endoplasmic reticulum (ER) synthesizes the glycoproteins. ❼ Vesicles transport the glycoproteins to the cell's plasma membrane. ❽ A capsid assembles around each viral genome molecule. ❾ The virus buds from the cell. Its envelope, studded with viral glycoproteins, is derived from the cell's plasma membrane.

Viral Envelopes. An animal virus equipped with an outer membrane, or viral envelope, uses it to enter the host cell (FIGURE 18.6). This membrane is generally a lipid bilayer, like a cellular membrane, with glycoproteins protruding from the outer surface. The glycoprotein spikes bind to specific receptor molecules on the surface of a host cell. The viral envelope then fuses with the host's plasma membrane, transporting the capsid and viral genome into the cell. After cellular enzymes remove the capsid, the viral genome can replicate and direct the synthesis of viral proteins, including glycoproteins for new viral envelopes. The endoplasmic reticulum of the host cell makes these glycoproteins, which in most cases are transported to the plasma membrane. There they are clustered in patches that serve as exit points for the viral progeny. The new viruses bud from the cell surface at these points, in a process much like exocytosis, wrapping themselves in membrane as they go. In other words, the viral envelope is derived from the host cell's plasma membrane, although some of the molecules of this membrane are specified by viral genes. The enveloped viruses are now free to spread the infection to other cells. This reproductive cycle does not necessarily kill the host cell, in contrast to the lytic cycles of phages.

Some viruses have envelopes that are not derived from plasma membrane. Herpesviruses, for example, have envelopes derived from the nuclear membrane of the host. The genomes of herpesviruses are double-stranded DNA, and these viruses reproduce within the cell nucleus, using a combination of viral and cellular enzymes to replicate and transcribe their DNA. Copies of the herpesvirus DNA usually remain behind as mini-chromosomes in the nuclei of certain nerve cells. There they remain latent until some sort of physical or emotional stress triggers a new round of active virus production. The infection of other cells by these new viruses results in the blisters—cold sores or genital sores, for instance—characteristic of herpes. Once someone acquires a herpesvirus infection, flare-ups may recur throughout the person's life.

RNA as Viral Genetic Material.   Although some phages and most plant viruses are RNA viruses, the broadest variety of RNA genomes is found among the viruses that infect animals. As TABLE 18.1 indicates, RNA viruses are classified according to the strandedness of their RNA and how it functions in a host cell. Notice that there are three types of single-stranded RNA genomes (classes IV–VI). The genome of class IV viruses can directly serve as mRNA and thus can be translated into viral protein immediately after infection. (How do you suppose new RNA genomes are made?) FIGURE 18.6 shows a virus of class V, in which the RNA genome serves as a *template* for mRNA synthesis. The RNA genome is transcribed into a strand of complementary RNA, which serves both as mRNA and as template for the synthesis of additional copies of genome RNA. Like all viruses that require RNA → RNA synthesis to make mRNA, this one uses a viral enzyme that is packaged with the genome inside the capsid.

The RNA viruses with the most complicated reproductive cycles are the **retroviruses** (class VI). *Retro*, meaning "backward," refers to the reverse direction in which genetic information flows for these viruses. Retroviruses are equipped with an enzyme called **reverse transcriptase**, which transcribes DNA from an RNA template, providing an RNA → DNA information flow. The newly made DNA then integrates into a chromosome within the nucleus of the animal cell. The integrated viral DNA, called a **provirus,** remains a permanent resident of the host cell's genome. (Unlike a prophage, a provirus never leaves.) The host's RNA polymerase transcribes the viral DNA into RNA molecules, which can function both as mRNA for the synthesis of viral proteins and as genomes for new virus particles released from the cell. A retrovirus of particular importance is **HIV (human immunodeficiency virus)**, the virus that causes **AIDS (acquired immunodeficiency syndrome)**.

FIGURE 18.7a (p. 336) shows the structure of HIV, and FIGURE 18.7b traces its reproductive cycle, which is typical of a retrovirus. We will postpone a detailed discussion of AIDS until Chapter 43.

Causes and Prevention of Viral Diseases in Animals

The link between a viral infection and the symptoms it produces is often obscure. Some viruses damage or kill cells by causing the release of hydrolytic enzymes from lysosomes. Some viruses cause the infected cells to produce toxins that lead to disease symptoms, and some have molecular components that are toxic, such as envelope proteins. How much damage a virus causes depends partly on the ability of the infected tissue to regenerate by cell division. We usually recover completely from colds because the epithelium of the respiratory tract, which the viruses infect, can efficiently repair itself. In contrast, the poliovirus attacks nerve cells, which do not divide and cannot be replaced. Polio's damage to such cells, unfortunately, is permanent. Many of the temporary symptoms associated with viral infections, such as fever, aches, and inflammation, actually result from the body's own efforts at defending itself against the infection.

As we will see in Chapter 43, the immune system is a complex and critical part of the body's natural defense mechanisms. The immune system is also the basis for the major medical weapon for preventing viral infections—vaccines. **Vaccines** are harmless variants or derivatives of pathogenic microbes that stimulate the immune system to mount defenses against the actual pathogen.

The term *vaccine* is derived from *vacca*, the Latin word for cow; the first vaccine, against smallpox, consisted of cowpox virus. In the late 1700s, Edward Jenner, an English physician, learned from his patients in farm country that milkmaids who had contracted cowpox (a milder disease, which usually infects cows) were resistant to subsequent smallpox infections. In his famous experiment in 1796, Jenner scratched a farmboy with a needle bearing fluid from a sore of a milkmaid who had cowpox. When the boy was later exposed to smallpox, he resisted the disease.

The cowpox and smallpox viruses are so similar that the immune system cannot distinguish them. Vaccination with the cowpox virus sensitizes the immune system to react vigorously if it is ever exposed to actual smallpox virus. Vaccination has eradicated smallpox, which was once a devastating scourge in many parts of the world. Effective vaccines have also been developed against many other viral diseases, including polio, rubella, measles, mumps, and hepatitis B.

Although vaccines can prevent certain viral illnesses, medical technology can do little, at present, to cure most viral infections once they occur. The antibiotics that help us recover from bacterial infections are powerless against viruses. Antibiotics kill bacteria by inhibiting enzymes or processes specific to the pathogens, but viruses have few or no enzymes of their

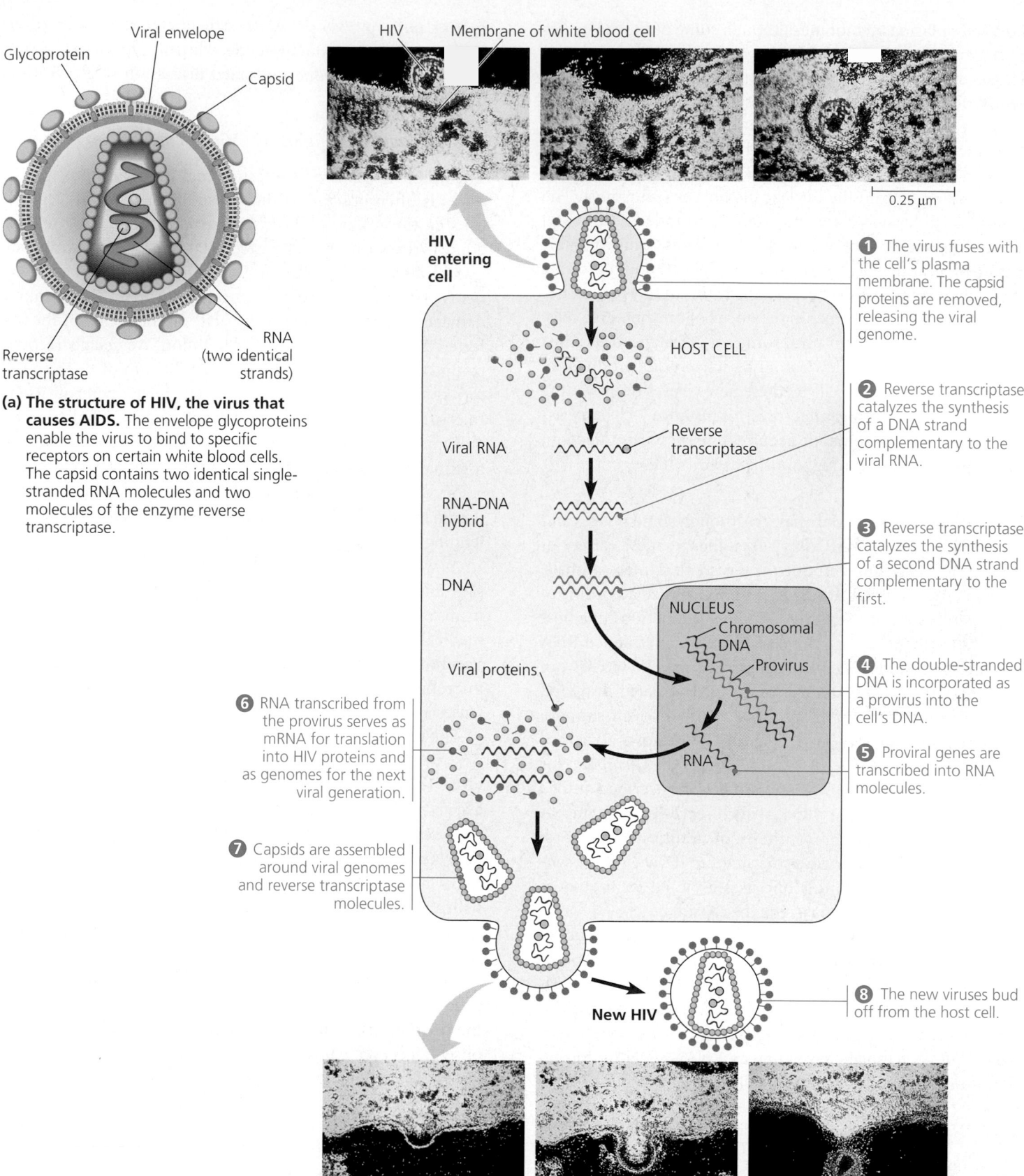

Glycoprotein | **Viral envelope** | **Capsid**

Reverse transcriptase | **RNA (two identical strands)**

(a) The structure of HIV, the virus that causes AIDS. The envelope glycoproteins enable the virus to bind to specific receptors on certain white blood cells. The capsid contains two identical single-stranded RNA molecules and two molecules of the enzyme reverse transcriptase.

HIV | Membrane of white blood cell

0.25 μm

HIV entering cell

1 The virus fuses with the cell's plasma membrane. The capsid proteins are removed, releasing the viral genome.

HOST CELL

Viral RNA — Reverse transcriptase

2 Reverse transcriptase catalyzes the synthesis of a DNA strand complementary to the viral RNA.

RNA-DNA hybrid

DNA

3 Reverse transcriptase catalyzes the synthesis of a second DNA strand complementary to the first.

NUCLEUS
— Chromosomal DNA
— Provirus

4 The double-stranded DNA is incorporated as a provirus into the cell's DNA.

Viral proteins

RNA

5 Proviral genes are transcribed into RNA molecules.

6 RNA transcribed from the provirus serves as mRNA for translation into HIV proteins and as genomes for the next viral generation.

7 Capsids are assembled around viral genomes and reverse transcriptase molecules.

New HIV

8 The new viruses bud off from the host cell.

(b) The reproductive cycle of HIV. Transmission electron micrographs (artificially colored) depict HIV entering (top) and leaving (bottom) a human white blood cell.

FIGURE 18.7 HIV, a retrovirus.

own. However, a few drugs have been found that do combat viruses, mostly by interfering with viral nucleic acid synthesis. One such drug is AZT, which inhibits HIV reproduction by interfering with the action of reverse transcriptase. Another is acyclovir, which inhibits herpesvirus DNA synthesis.

Emerging Viruses

HIV, the AIDS virus, seemed to make a sudden appearance in the early 1980s. In 1993, dozens of people in the southwestern United States died from hantavirus infection, which the news media first described as a "new" disease. The deadly Ebola virus (FIGURE 18.8a) has menaced the peoples of central Africa periodically since its initial recognition in 1976. The Ebola

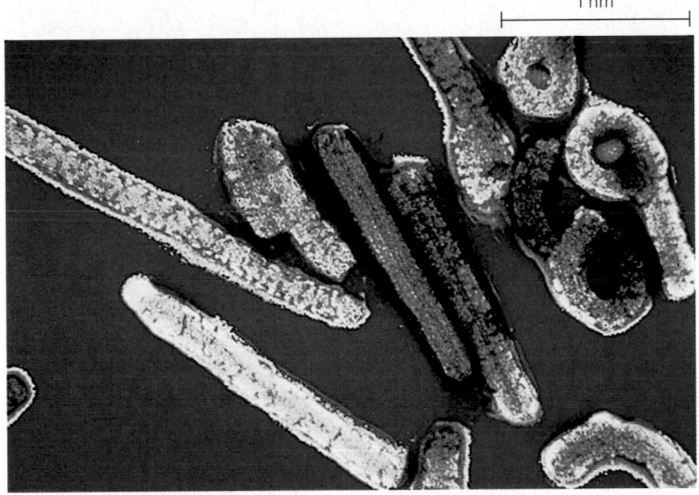

1 nm

(a) Ebola virus. Each filamentous virus particle is an enveloped thread of protein-coated RNA. The RNA is single-stranded (class V in TABLE 18.1).

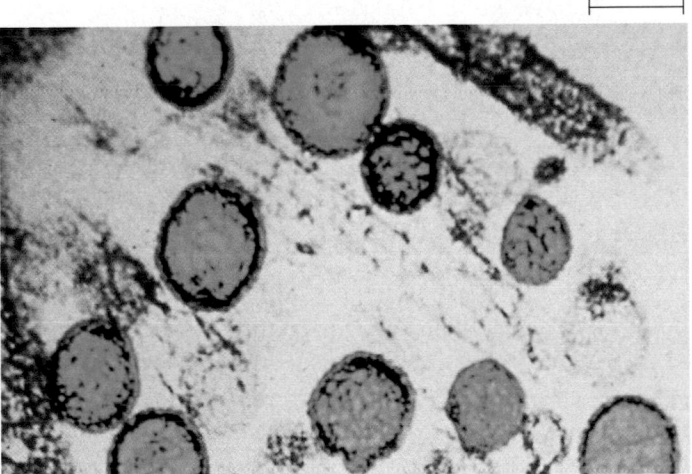

200 nm

(b) Hantavirus. Another enveloped RNA virus, hantavirus has a single-stranded RNA genome (class V in TABLE 18.1) that is in three pieces.

FIGURE 18.8 Emerging viruses. The term *emerging viruses* is used for viruses that have recently appeared or recently come to the attention of medical scientists. These colorized TEMs show two such viruses.

virus is only one of several recently recognized viruses that cause *hemorrhagic fever,* an often fatal syndrome characterized by fever, vomiting, massive bleeding, and circulatory system collapse. A number of other dangerous new viruses cause encephalitis, inflammation of the brain. One example is the Nipah virus, which in 1999 killed 105 people in Malaysia and destroyed the country's pig industry. And each year, new strains of influenza virus cause millions to miss work or classes, and deaths from flu are not uncommon. From where or what do these and other "emerging viruses" arise?

Three processes contribute to the emergence of viral diseases. First, the mutation of existing viruses is a major source of new viral diseases. RNA viruses tend to have an unusually high rate of mutation because the replication of their nucleic acid does not involve the proofreading steps of DNA replication. Some mutations enable existing viruses to evolve into new genetic varieties that can cause disease in individuals who had developed immunity to the ancestral virus. Flu epidemics are caused by viruses that are genetically different enough from earlier years' viruses that people have little immunity to them.

Another source of new viral diseases is the spread of existing viruses from one host species to another. Scientists estimate that about three-quarters of new human diseases have originated in other animals. For example, hantavirus (FIGURE 18.8b) is common in rodents, especially deer mice. The population of deer mice in the southwestern United States exploded in 1993 after unusually wet weather increased the rodents' food supply. Humans acquired hantavirus when they inhaled dust containing traces of urine and feces from infected mice.

Finally, the dissemination of a viral disease from a small, isolated population can lead to widespread epidemics. AIDS, for example, went unnamed and virtually unnoticed for decades before it began to spread around the world. In this case, technological and social factors, including affordable international travel, blood transfusion technology, sexual promiscuity, and the abuse of intravenous drugs, allowed a previously rare human disease to become a global scourge.

Thus, emerging viruses are generally not new, but are existing viruses that expand their host territory by evolving, by spreading to new host species, or by disseminating more widely in the current host species. Environmental change can increase the viral traffic responsible for emerging diseases. For example, new roads through remote areas can allow viruses to spread between previously isolated human populations. Another problem is the destruction of forests to expand crop land, an environmental disturbance that brings humans into contact with other animals that may host viruses capable of infecting humans.

Viruses and Cancer

Since 1911, when Peyton Rous discovered a virus that causes cancer in chickens, scientists have recognized that some viruses can cause cancer in animals. We know that these *tumor viruses* include members of the retrovirus, papovavirus, adenovirus, and herpesvirus groups (see TABLE 18.1).

There is strong evidence that viruses cause certain types of human cancer. The virus responsible for hepatitis B also seems to cause liver cancer in individuals with chronic hepatitis. And the Epstein-Barr virus, the herpesvirus that causes infectious mononucleosis, has been linked to several types of cancer prevalent in parts of Africa, notably Burkitt's lymphoma. Papilloma viruses (of the papovavirus group) have been associated with cancer of the cervix. Among the retroviruses, one called HTLV-1 causes a type of adult leukemia. All tumor viruses transform cells into cancer cells through the integration of viral nucleic acid into host cell DNA.

Scientists have identified a number of viral genes directly involved in triggering cancerous characteristics in cells. Many of these genes, called *oncogenes,* are not unique to tumor viruses or tumor cells; versions of these genes, called *proto-oncogenes,* are also found in normal cells. Proto-oncogenes generally code for proteins that affect the cell cycle, such as growth factors and proteins involved in growth factor action (for example, growth factor receptors). In some cases, the tumor virus lacks oncogenes and transforms the cell simply by turning on or increasing the expression of one or more of the cell's proto-oncogenes. Whatever the mechanism by which a particular virus causes cancer, there is evidence that more than one change must occur in a cell's genome to transform the cell into a fully cancerous state. It is likely that most tumor viruses cause cancer only in combination with other, mutagenic events, such as exposure to mutagens or mistakes in DNA replication or repair. (Cancer is covered further in Chapter 19.)

Plant viruses are serious agricultural pests

Plant viruses can stunt plant growth and diminish crop yields (FIGURE 18.9a). Most plant viruses discovered thus far are RNA viruses. Many of them, including the tobacco mosaic virus, have rod-shaped capsids with protein arranged in a spiral (see FIGURE 18.2a).

There are two major routes by which a plant viral disease can spread. By the first route, called *horizontal transmission,* a plant is infected from an external source of the virus. Since the invading virus must get past the plant's outer protective layer of cells (the epidermis), the plant becomes more susceptible to viral infections if it has been damaged by wind, chilling, injury, or insects. Insects are a double threat, because they often also act as carriers of viruses, transmitting disease from plant to plant. Farmers and gardeners may transmit plant viruses inadvertently on pruning shears and other tools. The other route of viral infection is *vertical transmission,* in which a plant inherits a viral infection from a parent. Vertical transmission can

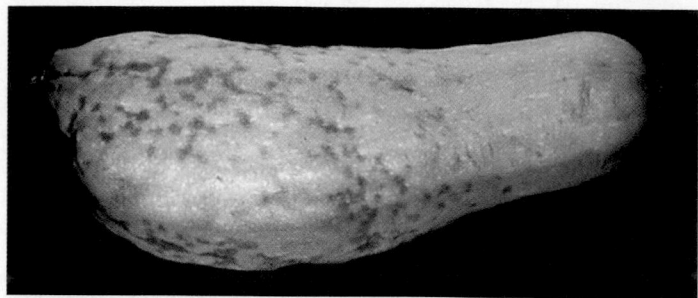

(a) Mosaic viruses caused the mottling of this summer squash (top) and tobacco leaf (bottom).

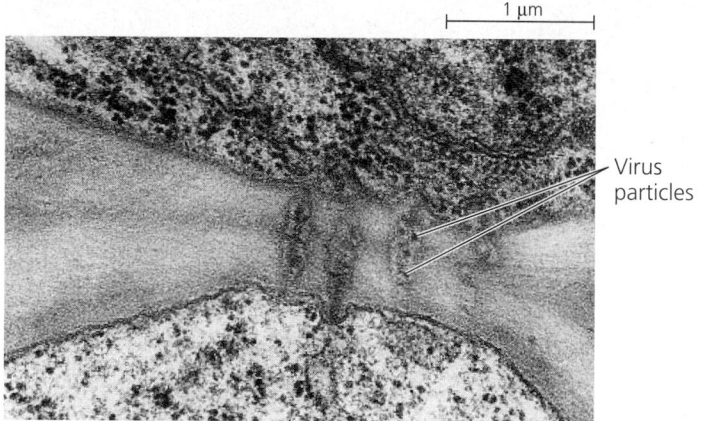

1 µm

Virus particles

(b) Viruses that infect a plant, such as these rice yellow mottle viruses, can spread throughout the plant body via the plasmodesmata that connect cells (TEM).

FIGURE 18.9 Viral infection of plants.

occur in asexual propagation (for example, by taking cuttings) or in sexual reproduction via infected seeds.

Once a virus enters a plant cell and begins reproducing, virus particles can spread throughout the plant by passing through plasmodesmata, the cytoplasmic connections that penetrate the walls between adjacent plant cells (FIGURE 18.9b). Agricultural scientists have not yet devised cures for most viral diseases of plants. Therefore, their efforts have focused largely on reducing the incidence and transmission of such diseases and on breeding genetic varieties of crop plants that are relatively resistant to certain viruses.

Viroids and prions are infectious agents even simpler than viruses

As small and simple as viruses are, they dwarf another class of pathogens, **viroids.** These are tiny molecules of naked circular RNA that infect plants. Only several hundred nucleotides long, viroids do not encode proteins but can replicate in host plant cells, apparently using cellular enzymes. Somehow, these RNA molecules can disrupt the metabolism of a plant cell and stunt the growth of the whole plant. One viroid disease has killed over 10 million coconut palms in the Philippines. Viroids seem to cause errors in the regulatory systems that control plant growth, and the symptoms that are typically associated with viroid diseases are abnormal development and stunted growth.

An important lesson from viroids is that a *molecule* can be an infectious agent that spreads a disease. But viroids are nucleic acid, whose ability to replicate is well known. More difficult to explain is the evidence for infectious *proteins,* called **prions.** Prions appear to cause a number of degenerative brain diseases, including scrapie in sheep, the "mad cow disease" that has plagued the European beef industry in recent years, and Creutzfeldt-Jakob disease in humans. How can a protein, which cannot replicate itself, be a transmissible pathogen? According to the leading hypothesis, a prion is a misfolded form of a protein normally present in brain cells. When the prion gets into a cell containing the normal form of the protein, the prion converts the normal protein to the prion version (FIGURE 18.10). In this way, prions may repeatedly trigger chain reactions that increase their numbers. American scientist Stanley Prusiner has long championed the prion hypothesis and in 1997 was awarded a Nobel Prize for his research in this area. Today, mounting evidence supports the pathogenic role of prions in animal diseases and the hypothesis depicted in FIGURE 18.10.

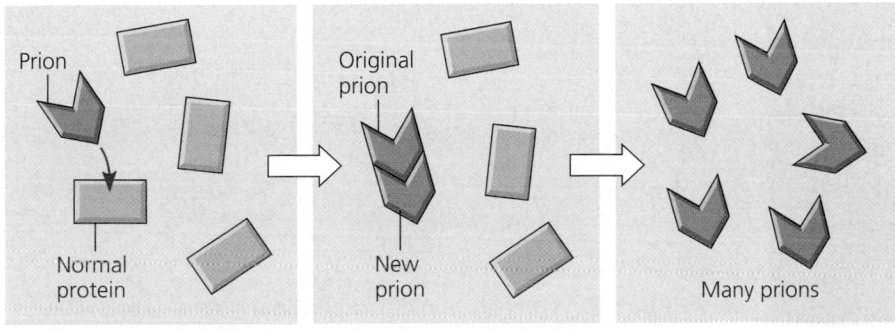

FIGURE 18.10 A hypothesis to explain how prions propagate. Prions are misfolded versions of normal brain proteins. When a prion contacts a normal "twin," it may induce the normal protein to assume the abnormal shape. The resulting chain reaction may continue until prions accumulate to dangerous levels, causing cellular malfunction and eventual degeneration of the brain.

Viruses may have evolved from other mobile genetic elements

Viruses are in the semantic fog between life and nonlife. Do we think of them as nature's most complex molecules or as the simplest forms of life? Either way, we must bend our usual definitions. An isolated virus is biologically inert, unable to replicate its genes or regenerate its own supply of ATP. Yet it has a genetic program written in the universal language of life. Although viruses are obligate intracellular parasites that cannot reproduce independently, it is hard to deny their evolutionary connection to the living world.

How did viruses originate? Because they depend on cells for their own propagation, it is reasonable to assume that viruses are not the descendants of precellular prototypes of life, but evolved *after* the first cells appeared. Most molecular biologists favor the hypothesis that viruses originated from fragments of cellular nucleic acids that could move from one cell to another. Consistent with this idea is the observation that a viral genome usually has more in common with the genome of its host than with the genomes of viruses infecting other hosts. Indeed, some viral genes are essentially identical to genes of the host, as in the case of oncogenes, for example. Perhaps the earliest viruses were naked bits of nucleic acid, similar to plant viroids, that made it from one cell to another via injured cell surfaces. The evolution of genes coding for capsid proteins may have facilitated the infection of undamaged cells.

Candidates for the original sources of viral genomes include two kinds of cellular genetic elements, plasmids and transposons. Plasmids are small, circular DNA molecules that are separate from chromosomes. They are found in bacteria and also in yeasts, which are unicellular eukaryotes. Plasmids, like most viruses, can replicate independently of the rest of the cell's genome and are occasionally transferred between cells. Transposons are DNA segments that can move from one location to another within a cell's genome. Thus, plasmids, transposons, and viruses all share an important feature: They are mobile genetic elements. (We will discuss plasmids and transposons in more detail later in the chapter.)

It is the evolutionary relationship between viruses and the genomes of their host cells that makes viruses such useful model systems in molecular biology. By studying how the replication of viruses is controlled, researchers are learning more about the mechanisms that regulate DNA replication and gene expression (transcription and translation) in cells. Bacteria are equally valuable as microbial models in genetics research, but for different reasons. Unlike viruses, bacteria are true cells. But as prokaryotic cells, bacteria provide researchers with the opportunity to investigate molecular genetics in the simplest organisms. In fact, *E. coli*, the bacterium sometimes called "the laboratory rat of molecular biology," is the most completely understood of all organisms at the molecular level. Let's now learn more about the genetics of bacteria.

THE GENETICS OF BACTERIA

The short generation span of bacteria helps them adapt to changing environments

Bacteria are very adaptable, in both the evolutionary sense of adaptation via natural selection and the physiological sense of adjustment to changes in the environment by individual bacteria. These sections on bacterial genetics will help clarify how these microbes can be so malleable.

The major component of the bacterial genome is one double-stranded, circular DNA molecule. Although we will refer to this structure as the bacterial chromosome, it is very different from eukaryotic chromosomes, which have linear DNA molecules associated with a large amount of protein. In the case of the well-studied intestinal bacterium *E. coli,* the chromosomal DNA consists of about 4.6 million nucleotide pairs, representing about 4,300 genes. This is 100 times more DNA than is found in a typical virus, but only about one-thousandth as much DNA as in an average eukaryotic cell. Still, this is a lot of DNA to be packaged in such a small container. Stretched out, the DNA of an *E. coli* cell would measure about a millimeter in length, 500 times longer than the cell. Within a bacterium, however, the chromosome is so tightly packed that it fills only part of the cell. This dense region of DNA, called the **nucleoid,** is not bounded by membrane like the nucleus of a eukaryotic cell. In addition to the chromosome, many bacteria also have plasmids, much smaller circles of DNA. Each plasmid has only a small number of genes, from just a few to several dozen. You will learn about the structure and function of plasmids in the next section.

Bacterial cells divide by binary fission, which is preceded by replication of the bacterial chromosome (see FIGURE 12.10). From a single origin of replication, DNA synthesis progresses in both directions around the circular chromosome (FIGURE 18.11). You can review the molecular details of the process in FIGURE 16.16.

Bacteria can proliferate very rapidly in a favorable environment, whether in a natural habitat or in a laboratory culture. For example, *E. coli* growing under optimal conditions can divide every 20 minutes. A laboratory culture started with a single cell can produce a colony of 10^7 to 10^8 bacteria overnight (12 hours). Reproductive rates in the organism's natural habitat, the large intestines (colons) of mammals, are just as impressive. For example, in the human colon, *E. coli* reproduces

rapidly enough to replace the 2×10^{10} bacteria lost each day in feces.

Because binary fission is an asexual process—the production of offspring from a single parent—most of the bacteria in a colony are genetically identical to the parent cell. As a result of mutation, however, some of the offspring *do* differ slightly in genetic makeup. For a given *E. coli* gene, the probability of a spontaneous mutation averages about 1×10^{-7} per cell division, only one in 10 million. But among the 2×10^{10} new *E. coli* cells that arise each day in a single human colon, there will be approximately $(2 \times 10^{10})(1 \times 10^{-7}) = 2,000$ bacteria that have a mutation in that gene. The total number of mutations when all 4,300 *E. coli* genes are considered is about $4,300 \times 2,000 = 9$ million per day per human host. The important

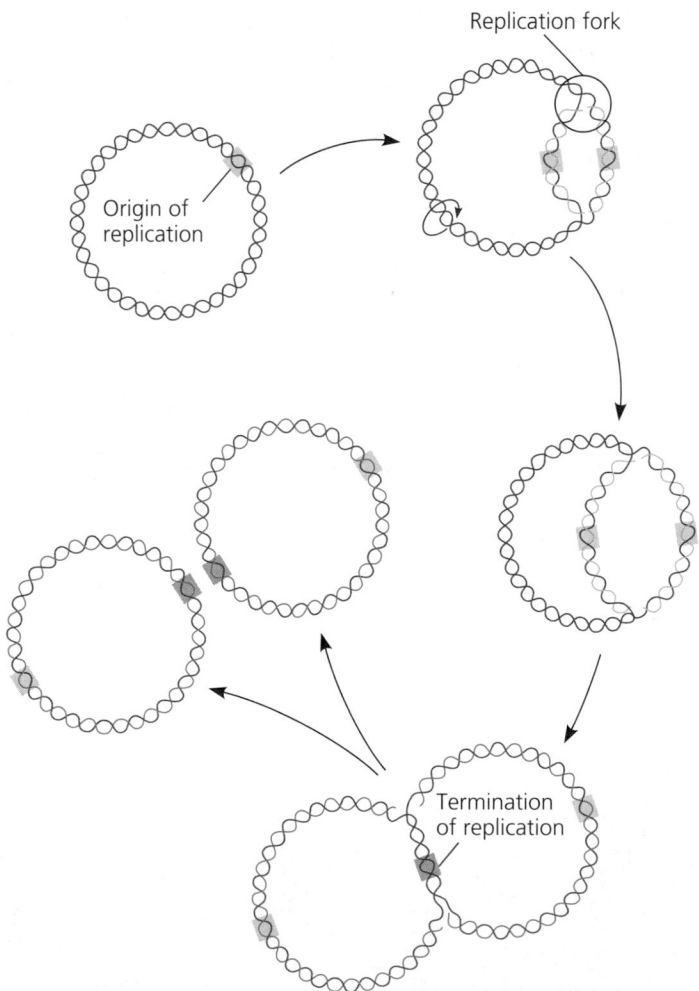

FIGURE 18.11 Replication of the bacterial chromosome. From one origin, DNA replication progresses in both directions around the circular chromosome until the entire chromosome has been reproduced. Enzymes that cut, twirl (magenta arrow), and reseal the double helix prevent the DNA from knotting. Keep in mind that although the *overall* direction of DNA replication is outward from the origin in both directions, one new strand at each replication fork is made discontinuously in the direction back toward the origin (see FIGURE 16.16).

point is that new mutations, though individually rare, can have a significant impact on genetic diversity when reproductive rates are very high because of short generation spans. This diversity, in turn, affects the evolution of bacterial populations: Individual bacteria that are genetically well equipped for the local environment clone themselves more prolifically than do less fit individuals.

In contrast, new mutations make a relatively small contribution to genetic variation in a population of slowly reproducing organisms, such as humans. Most of the heritable variation we observe in a human population is due not to the creation of novel alleles by *new* mutations, but to the sexual recombination of existing alleles (see Chapter 15). Even in bacteria, where new mutations *are* a major source of individual variation, genetic recombination adds still more diversity to a population, as we will now see.

Genetic recombination produces new bacterial strains

Natural selection depends on heritable variation among the individuals of a population (see Chapter 1). In addition to mutations, genetic recombination generates diversity within bacterial populations. We will define recombination here simply as the combining of DNA from two individuals into the genome of a single individual.

How can we detect genetic recombination in bacteria? Consider two mutant *E. coli* strains (genetic varieties), each unable to synthesize one of its required amino acids. Wild-type *E. coli* can grow on a minimal medium containing only glucose, as a source of organic carbon, and salts. The mutant strains cannot grow on this culture medium of minimal nutri-

ents because one of them cannot synthesize tryptophan and the other cannot synthesize arginine (FIGURE 18.12).

Let's suppose we mix bacteria from the two strains together in a liquid medium and allow them to incubate for an hour or so. We then spread a small sample of this culture on solid (agar-thickened) minimal medium in a petri dish and incubate the dish overnight. The next morning we observe numerous colonies of bacteria on the minimal medium. Each of these colonies must have started with a cell capable of making *both* tryptophan *and* arginine, but their number far exceeds what can be accounted for by mutation. Most of the cells that can synthesize both amino acids must have acquired one or more genes from the other strain. This is evidence that genetic recombination has occurred.

Bacteria are different from eukaryotes in the ways DNA from two individuals can come together in one cell. In eukaryotes, the sexual processes of meiosis and fertilization combine DNA from two individuals in a single zygote (see Chapter 13). But sex, as defined in eukaryotes, is absent in prokaryotes: Meiosis and fertilization do not occur. Instead, other processes bring together bacterial DNA from different individuals. These processes are transformation, transduction, and conjugation.

Transformation

In the context of bacterial genetics, **transformation** is the alteration of a bacterial cell's genotype by the uptake of naked, foreign DNA from the surrounding environment. For example, we saw in Chapter 16 that harmless *Streptococcus pneumoniae* bacteria could be transformed to pneumonia-causing cells by the uptake of naked DNA from a medium containing dead, broken-open cells of the pathogenic strain. This transformation occurs when a live nonpathogenic cell takes up a

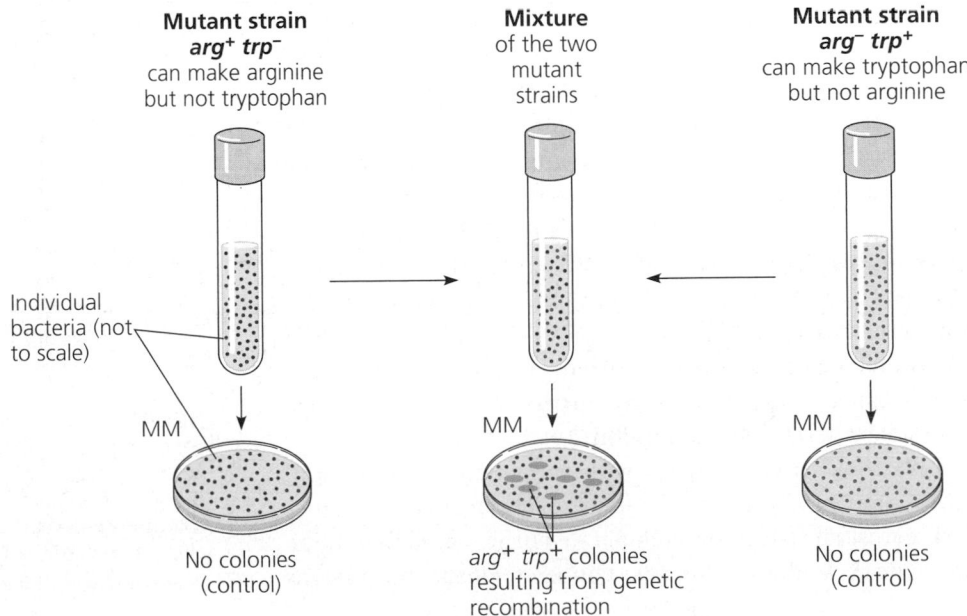

THE PROCESS OF SCIENCE

FIGURE 18.12 Detecting genetic recombination in bacteria. Each of the two mutant strains of *E. coli* in this experiment is unable to synthesize a particular amino acid, either arginine or tryptophan. These mutant bacteria cannot reproduce on minimal medium (MM), which is simply a solution of glucose and salts. But after bacteria of the two mutant strains are incubated together, some cells from the mixed culture are able to grow into colonies on a plate containing minimal medium solidified with agar. Those cells are recombinant bacteria: *arg⁺trp⁺* cells produced by gene transfer between cells of the two mutant types. (You will learn about the mechanism in FIGURE 18.15c and d.)

piece of DNA that happens to include the allele for pathogenicity (the gene for a cell coat that protects the bacterium from a host's immune system). The foreign allele is then incorporated into the bacterial chromosome, replacing the native allele (for the "coatless" condition, in this case). The process is genetic recombination—an exchange of DNA segments by crossing over. The cell is now a recombinant: Its chromosome contains DNA derived from two different cells.

For many years after transformation was discovered in laboratory cultures, most biologists believed the process to be too rare and haphazard to play an important role in natural bacterial populations. But researchers have since learned that many bacterial species possess on their surfaces proteins that are specialized for the uptake of naked DNA from the surrounding solution. These proteins specifically recognize and transport only DNA from closely related species of bacteria. Not all bacteria have such membrane proteins. For instance, *E. coli* does not seem to have any specialized mechanism for the uptake of foreign DNA. However, placing *E. coli* in a culture medium containing a relatively high concentration of calcium ions will artificially stimulate the cells to take up small pieces of DNA. In biotechnology, this technique is applied to introduce foreign genes into *E. coli*—genes coding for valuable proteins, such as human insulin and growth hormone.

Transduction

In the DNA transfer process known as **transduction,** phages (the viruses that infect bacteria) carry bacterial genes from one host cell to another. There are two forms of transduction: generalized transduction and specialized transduction. Both result from aberrations in phage reproductive cycles.

First let's consider generalized transduction, shown on the left side of FIGURE 18.13. Recall that near the end of a phage's lytic cycle, viral nucleic acid molecules are packaged within capsids, and the completed phages are released when the host cell lyses. Occasionally, a small piece of the host cell's degraded DNA is accidentally packaged within a phage capsid in place of the phage genome. Such a virus is defective because it lacks its own genetic material. However, after its release from the lysed host, the phage can attach to another bacterium and inject the piece of bacterial DNA acquired from the first cell. Some of this DNA can subsequently replace the homologous region of the second cell's chromosome. The cell's chromosome now has a combination of DNA derived from two cells; genetic recombination has occurred. This type of transduction is called **generalized transduction** because the phage transfers bacterial genes at random.

Let's contrast this process with specialized transduction, shown on the right side of FIGURE 18.13. This form of transduction requires infection by a temperate phage. Recall that in the lysogenic cycle, the genome of a temperate phage integrates as

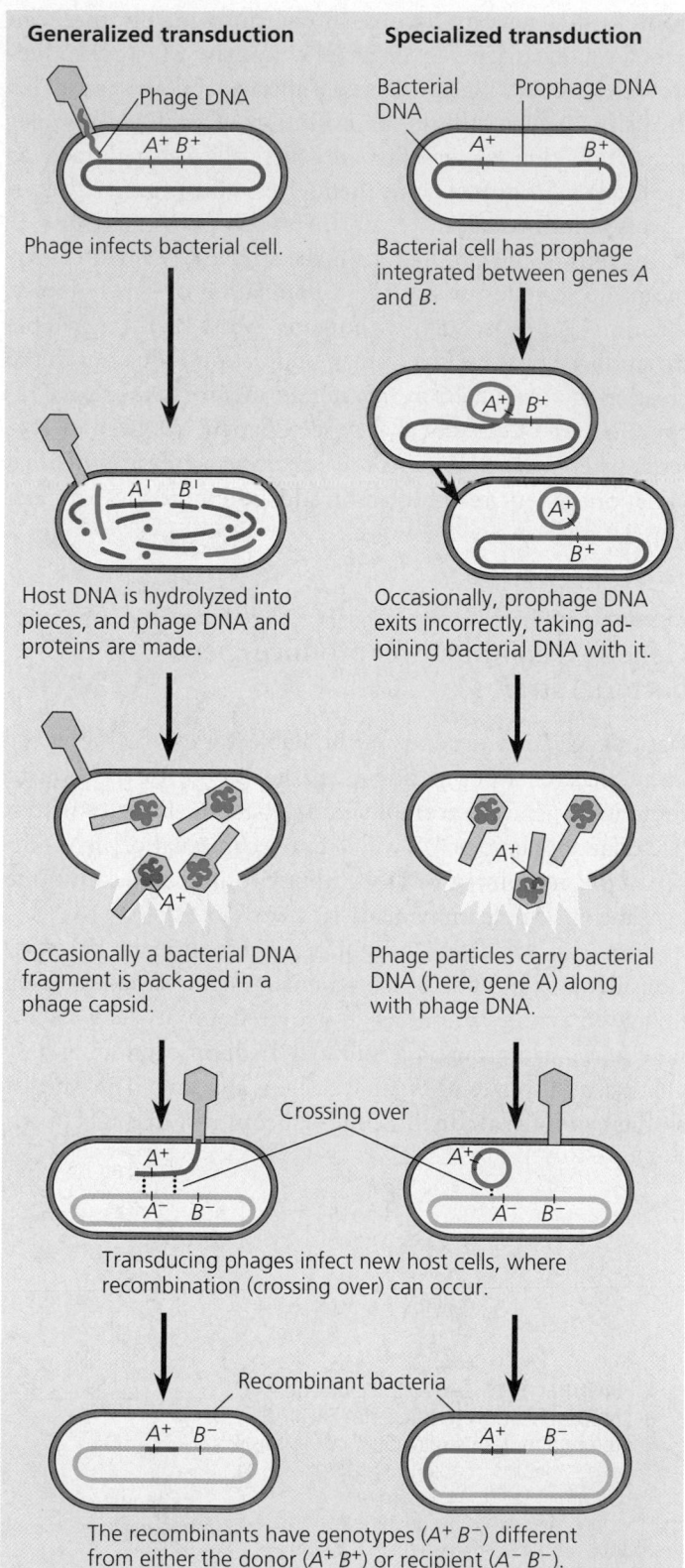

FIGURE 18.13 Transduction. Phages occasionally carry bacterial genes from one cell to another. In generalized transduction (left), random pieces of the host chromosome are packaged within a phage capsid. In specialized transduction (right), a prophage exits the chromosome in such a way that it carries adjacent bacterial genes along with it. In both types of transduction, the transferred DNA may recombine with the genome of the new host cell.

a prophage into the host bacterium's chromosome, usually at a specific site. Later, when the phage genome is excised from the chromosome, it sometimes takes with it a small region of the bacterial DNA that was adjacent to the prophage. When such a virus carrying bacterial DNA infects another host cell, the bacterial genes are injected along with the phage's genome. **Specialized transduction** transfers only certain genes, those near the prophage site on the bacterial chromosome.

Conjugation and Plasmids

Conjugation is the direct transfer of genetic material between two bacterial cells that are temporarily joined. This process, the bacterial version of sex, has been studied most extensively in *E. coli*. The DNA transfer is one-way: one cell donating DNA, and its "mate" receiving the genes. The DNA donor, referred to informally as the "male," uses appendages called sex pili to attach to the DNA recipient, the "female" (FIGURE 18.14). A sex pilus acts like a grappling hook: After contacting a female cell, it retracts, pulling the two cells together. A temporary cytoplasmic bridge then forms between the two cells, providing an avenue for DNA transfer. In most cases, "maleness," the ability to form sex pili and donate DNA during conjugation, results from the presence of a special piece of DNA called an **F factor** (F for fertility). An F factor can exist either as a segment of DNA within the bacterial chromosome or as a plasmid. Before discussing the role of the F factor in conjugation, we should examine plasmids more generally.

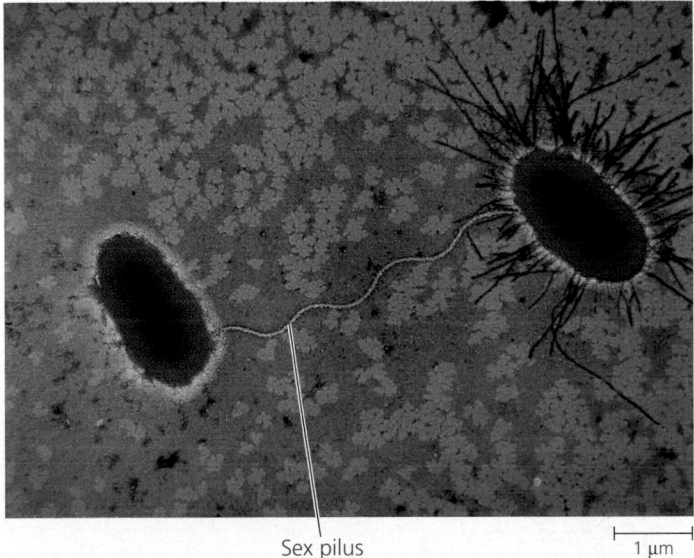

Sex pilus   1 μm

FIGURE 18.14 Bacterial mating. The *E. coli* "male" (right) extends sex pili, one of which is attached to a "female" cell. The two cells will be drawn close together, allowing a cytoplasmic bridge to form between them. Through this tube, the "male" will transfer DNA to the "female." This mechanism of DNA transfer is called conjugation (colorized TEM).

General Characteristics of Plasmids. A **plasmid** is a small, circular, self-replicating DNA molecule separate from the bacterial chromosome. Certain plasmids, such as F plasmids, can undergo reversible incorporation into the cell's chromosome. A genetic element that can exist either as a plasmid or as part of the bacterial chromosome is called an **episome.** In addition to some plasmids, temperate viruses, such as phage λ, also qualify as episomes. Recall that the genomes of these phages replicate independently during a lytic cycle and as an integral part of the bacterial chromosome during a lysogenic cycle. Of course, there are important differences between plasmids and viruses. Plasmids, unlike viruses, lack protein coats and do not normally exist outside the cell. And plasmids are generally beneficial to the bacterial cell, while viruses are parasites that usually harm their hosts.

A plasmid has only a small number of genes, and these genes are not required for the survival and reproduction of the bacterium under normal conditions. However, the genes of plasmids can confer advantages on bacteria living in stressful environments. For example, the F plasmid facilitates genetic recombination, which may be advantageous in a changing environment that no longer favors existing strains in a bacterial population.

The F Plasmid and Conjugation. The F factor and its plasmid form, the **F plasmid,** consist of about 25 genes, most required for the production of sex pili. Geneticists use the symbol F^+ to denote a cell that contains the F plasmid (a "male" cell). The F^+ condition is heritable: The F plasmid replicates in synchrony with the chromosomal DNA, and division of an F^+ cell usually gives rise to two offspring that are both F^+. Cells lacking the F factor in either form are designated F^-, and they function as DNA recipients ("females") during conjugation. The F^+ condition is "contagious" in the sense that an F^+ cell converts an F^- cell to F^+ when the two cells conjugate. The original F^+ cell remains F^+ because the F plasmid replicates within the "male" cell, and only one copy of the plasmid is transferred to the "female" through the conjugation tube joining the cells (FIGURE 18.15a, p. 344). In an $F^+ \times F^-$ mating, only an F plasmid is transferred.

Under what circumstances are genes of the bacterial chromosome transferred during conjugation? This can occur when the donor cell's F factor is integrated into the chromosome (FIGURE 18.15b). A cell with the F factor built into its chromosome is called an Hfr cell (for high frequency of recombination). Like an F^+ cell, an Hfr cell functions as a male during conjugation: It initiates DNA replication at a point on the F factor DNA and from that point starts to transfer the DNA copy to its F^- partner. But now, the leading end of the F factor is dragging a copy of chromosomal DNA along with it (FIGURE 18.15c). Random movements of the bacteria almost always disrupt conjugation long before an entire copy of the Hfr chromosome can be passed to the F⁻ cell. Temporarily, the recipient cell is a partial diploid, containing its own chromosome plus DNA copied from part of the donor's chromosome. Then recombination can occur: If part of the newly acquired DNA aligns with the homologous region of the F⁻ chromosome, segments of DNA can be exchanged (FIGURE 18.15d). Binary fission of this cell gives rise to a colony of recombinant bacteria with genes derived from two different cells. (This is what happened in the experiment in FIGURE 18.12, where one of the bacterial strains was an Hfr and the other an F⁻.)

R Plasmids and Antibiotic Resistance. In the 1950s, Japanese physicians began to notice that some hospital patients suffering from bacterial dysentery, which produces severe diarrhea, did not respond to antibiotics that had generally been effective in treating this type of infection. Apparently,

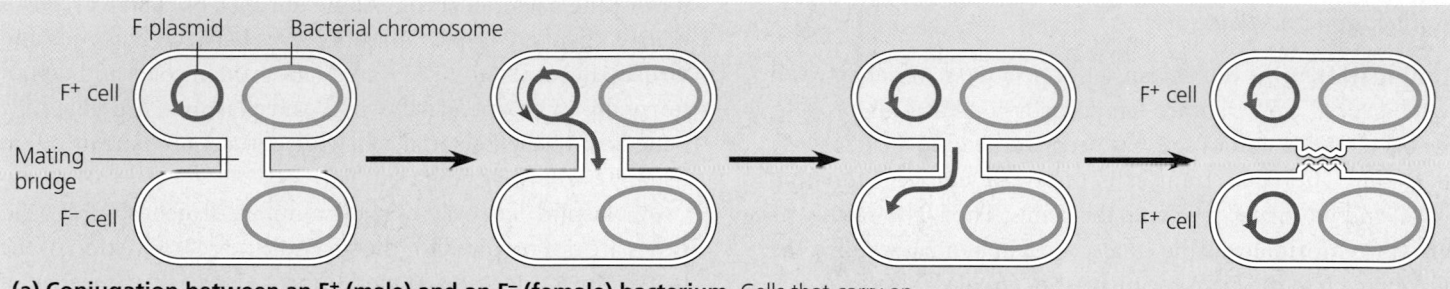

(a) Conjugation between an F⁺ (male) and an F⁻ (female) bacterium. Cells that carry an F plasmid are called F⁺ cells. They are "male" in that they can transfer an F plasmid to a "female" F⁻ cell. In this way, an F⁻ cell can become F⁺. The F plasmid replicates as it transfers, so that the donor cell remains F⁺. The black arrowhead marks the point where both replication and transfer begin. As replication proceeds and the copy peels off, the plasmid circle rotates.

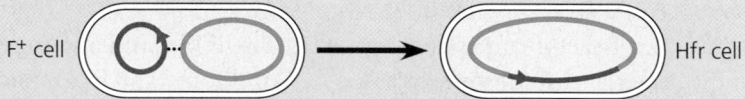

(b) Conversion of an F⁺ male into an Hfr male by integration of the F plasmid (an episome) into the chromosome. This process is similar to phage DNA joining the host chromosome as a prophage: Crossing over occurs between the two DNA circles at a specific site on each.

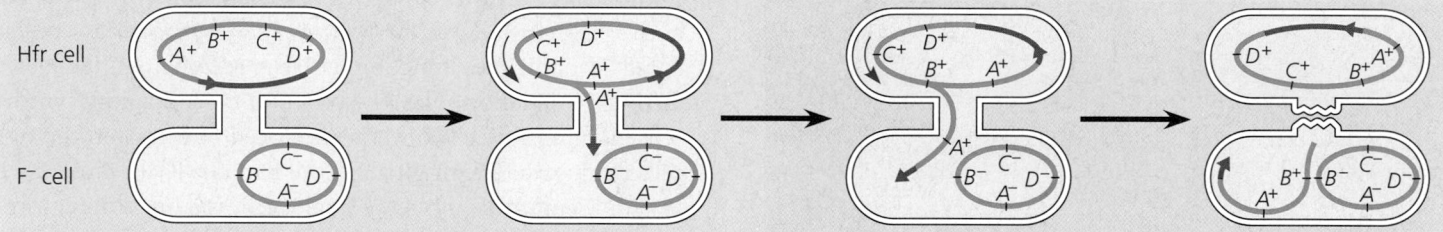

(c) Conjugation between an Hfr and an F⁻ bacterium. Replication and transfer of the Hfr chromosome begins at a fixed point (arrowhead) within the F factor. The location and orientation of the F factor in the chromosome determine the sequence in which genes are transferred during conjugation. In this *E. coli* strain, the transfer sequence for four genes is *A-B-C-D*. The conjugation bridge usually breaks well before the entire chromosome and most of the F factor are transferred.

(d) Recombination between the Hfr chromosome fragment and the F⁻ chromosome. Crossing over can occur between genes on the fragment of bacterial chromosome transferred from the Hfr cell and the same (homologous) genes on the recipient (F⁻) cell's chromosome. A recombinant F⁻ cell will result. Pieces of DNA ending up outside the bacterial chromosome will eventually be degraded by the cell's enzymes or lost in cell division.

FIGURE 18.15 Conjugation and recombination in *E. coli*. The DNA replication that accompanies transfer of an F plasmid or part of an Hfr bacterial chromosome is called *rolling-circle replication*.

resistance to these antibiotics had evolved in certain strains of *Shigella,* the pathogen. Eventually, researchers began to identify the specific genes that confer antibiotic resistance in *Shigella* and other pathogenic bacteria. Some of these genes, for example, code for enzymes that specifically destroy certain antibiotics, such as tetracycline or ampicillin. The genes conferring resistance, it turns out, are carried by plasmids, now known as **R plasmids** (R for resistance).

Exposure of a bacterial population to a specific antibiotic, whether in a laboratory culture or within a host organism, will kill antibiotic-sensitive bacteria but not those that happen to have R plasmids that counter that antibiotic. The theory of natural selection predicts that under these circumstances, the fraction of the bacterial population carrying genes for antibiotic resistance will increase, and that is exactly what happens. The medical consequences are also predictable: Resistant strains of pathogens are becoming more common, making the treatment of certain bacterial infections more difficult. The problem is compounded by the fact that R plasmids, like F plasmids, have genes that encode sex pili and enable plasmid transfer from one bacterial cell to another by conjugation. Making the problem still worse, some R plasmids carry as many as ten genes for resistance to that many antibiotics. How do so many antibiotic resistance genes become part of a single plasmid? The answer involves another type of mobile genetic element, called a transposon, which we investigate next.

Transposons

A **transposon,** also called a transposable genetic element, is a piece of DNA that can move from one location to another in a cell's genome. Unlike an episome or prophage, transposons never exist independently. Instead, transposon movement (transposition) occurs as a type of recombination between the transposon and another DNA site—a target site—that comes in contact with the transposon. In a bacterial cell, a transposon may move within the chromosome, from a plasmid to the chromosome (or vice versa), or from one plasmid to another. Transposons bring multiple genes for antibiotic resistance into a single R plasmid by moving the genes to that location from different plasmids.

Transposons are sometimes called "jumping genes," but the phrase is slightly misleading. Some transposons *do* jump from one genomic location to another, in what is called cut-and-paste transposition. However, in another type of transposition, called replicative transposition, the transposon replicates at its original site, and a *copy* inserts elsewhere; that is, the transposon is added at some new site without being lost from the old site.

Although transposons vary in their selectivity for target sites, most can move to many alternative locations in the DNA. This ability to scatter certain genes throughout the genome makes transposition fundamentally different from other mechanisms of genetic shuffling. The genetic recombi-

nation that occurs in bacterial transformation, generalized transduction, and conjugation (and during meiosis in eukaryotes as well) depends on base pairing between homologous regions of DNA, regions of identical or very similar base sequence. In contrast, the insertion of a transposon in a new site does not depend on complementary base sequences. A transposon can move genes to a site where genes of that sort have never before existed.

Insertion Sequences. The simplest bacterial transposons are **insertion sequences.** They consist of only the DNA necessary for the act of transposition. The one gene found in an insertion sequence codes for a transposase, an enzyme that catalyzes movement of the transposon from one location to another within the genome. The transposase gene is bracketed by a pair of DNA sequences called *inverted repeats,* noncoding sequences about 20 to 40 nucleotides long. Notice in FIGURE 18.16 that each base sequence is repeated in reverse along the opposite DNA strand of the inverted repeat at the other end of the transposon. Transposase recognizes these inverted repeats as the boundaries of the transposon. During transposition, molecules of the enzyme bind to the inverted repeats and to a target site elsewhere in the genome and catalyze the DNA cutting and resealing required for transposition. Other enzymes also participate in transposition. For example, DNA polymerase helps create identical regions of DNA, called *direct repeats,* that flank a transposon in its new site (FIGURE 18.17, p. 346).

Insertion sequences cause mutations when they happen to land within the coding sequence of a gene or within a DNA region that regulates gene expression. But notice that this mechanism of mutation is intrinsic to the cell, in contrast to mutagenesis by extrinsic factors such as environmental radiation and chemicals. Insertion sequences account for about 1.5% of the *E. coli* genome. However, mutation of a given gene by transposition occurs only rarely—about once in every 10 million generations. This is about the same as the spontaneous mutation rate due to other factors.

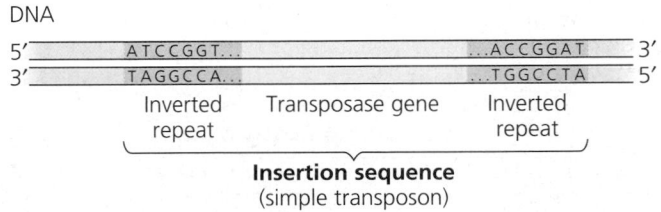

DNA

FIGURE 18.16 Insertion sequences, the simplest transposons. The one gene of an insertion sequence codes for transposase, which catalyzes the transposon's movement. The inverted repeats, about 20 to 40 nucleotide pairs long, are backward, upside-down versions of each other. In transposition, transposase molecules bind to the inverted repeats and catalyze the cutting and resealing of DNA required for insertion of the transposon at a target site. This diagram and the ones that follow are not to scale.

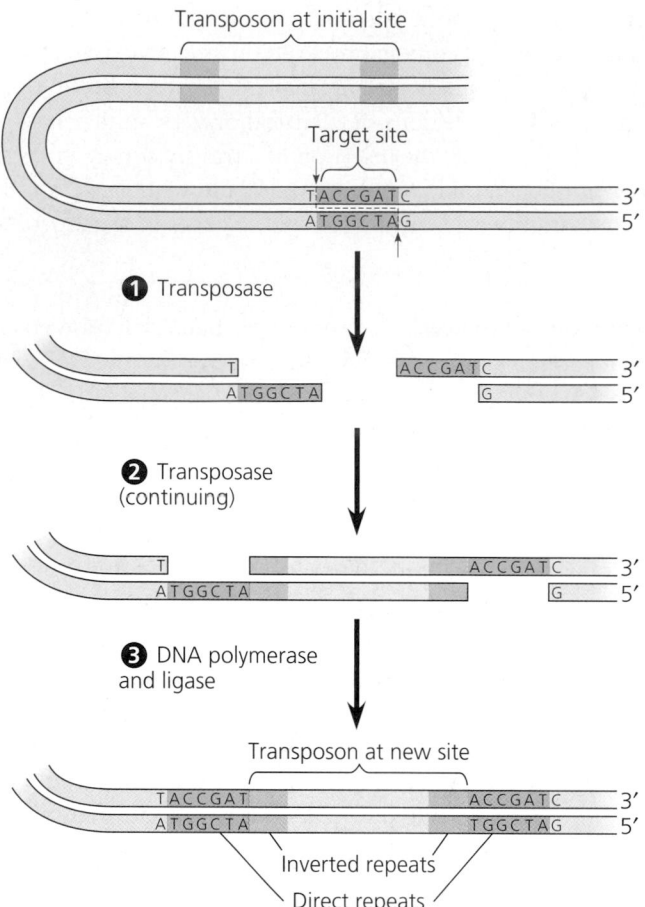

FIGURE 18.17 Insertion of a transposon and creation of direct repeats. ❶ First, the transposase enzyme makes staggered cuts (red arrows) in the two DNA strands at a target site, leaving short segments of unpaired DNA as shown. Meanwhile, the transposon is cut out or copied at its initial site. ❷ The transposon is then joined to the single-stranded ends at the target site. Presumably, the transposase holds all the components together during this process. ❸ Finally, the gaps in the DNA strands are filled in by DNA polymerase and sealed by ligase. This results in *direct repeats,* identical segments of DNA on either side of the transposon. (Distances along the DNA are not to scale).

FIGURE 18.18 Anatomy of a composite transposon. A composite transposon consists of one or more genes located between twin insertion sequences. The transposon here has a gene for resistance to an antibiotic, which is carried along as part of the transposon when the transposon is inserted at a new site in the genome.

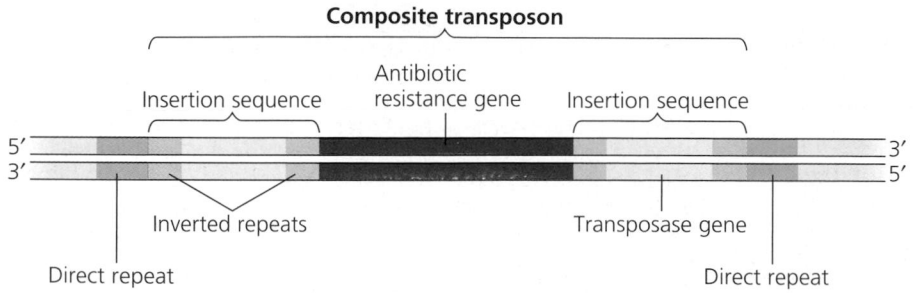

Composite Transposons. Transposons longer and more complex than insertion sequences also move about in the bacterial genome. In addition to the DNA required for transposition, these *composite transposons* (also called complex transposons) include extra genes that go along for the ride, such as genes for antibiotic resistance. The extra genes are sandwiched between two insertion sequences (FIGURE 18.18). It is as though two insertion sequences happened to land relatively close together in the genome and now travel together, along with all the DNA between them, as a single transposon. As is characteristic of transposons, there is an inverted repeat and a direct repeat at each end. (The direct repeat is not considered part of the transposon, however.)

In contrast to insertion sequences, which are not known to benefit bacteria in any specific way, composite transposons may help bacteria adapt to new environments. We mentioned earlier that a single R plasmid can carry several genes for resistance to different antibiotics. This is explained by composite transposons, which can add a gene for antibiotic resistance to a plasmid already carrying genes for resistance to other antibiotics. The transmission of this composite plasmid to other bacterial cells by cell division or conjugation can then spread resistance to a variety of antibiotics throughout a bacterial population. In an antibiotic-rich environment, natural selection favors bacterial clones that have built up composite R plasmids through a series of transpositions.

Transposable genetic elements are not unique to bacteria, but are important components of eukaryotic genomes as well. In fact, the first evidence for such wandering DNA segments came from the American geneticist Barbara McClintock's breeding experiments with Indian corn (maize) in the 1940s and 1950s. McClintock identified changes in the color of corn kernels that made sense only if she postulated the existence of mobile genetic elements capable of moving from other locations in the genome to the genes for kernel color. She called these mobile elements "controlling elements" because they seemed to insert next to the genes responsible for kernel color, either activating or inactivating those genes. McClintock's discovery received little attention until transposons were discovered in bacteria many years later and microbial geneticists learned more about the molecular basis of transposition. In 1983, more than 30 years after she discovered transposable genetic elements, Barbara McClintock was awarded a Nobel Prize, at age 81. McClintock continued her experiments at Cold Spring Harbor Laboratory in New York until her death in 1992.

You will learn more about transposons in eukaryotes in Chapter 19. We end this chapter by examining how bacterial genes are regulated in different environments.

THE PROCESS OF SCIENCE

The control of gene expression enables individual bacteria to adjust their metabolism to environmental change

Mutations and the various types of genetic transfer we have been studying generate the genetic variation that makes natural selection possible. And natural selection, acting over many generations, can increase the proportion of individuals in a bacterial population that are adapted to some new environmental condition. But how can an individual bacterium, locked into the genome it has inherited, cope with environmental fluctuation?

Think, for instance, of an *E. coli* cell living in the erratic environment of a human colon, dependent for its nutrients on the whimsical eating habits of its host. If the bacterium is deprived of the amino acid tryptophan, which it needs to survive, it responds by activating a metabolic pathway to make its own tryptophan from another compound. Later, if the human host eats a tryptophan-rich meal, the bacterial cell stops producing tryptophan for itself, thus saving the cell from squandering its resources to produce a substance that is available from the surrounding solution in prefabricated form. This is just one example of how bacteria tune their metabolism to changing environments.

Metabolic control occurs on two levels (FIGURE 18.19). First, cells can vary the numbers of specific enzyme molecules made; that is, they can regulate the expression of genes. Second, cells can adjust the activity of enzymes already present. The latter mode of control, which is more immediate, depends on the sensitivity of many enzymes to chemical cues that increase or decrease their catalytic activity (see Chapter 6). For example, activity of the first enzyme of the tryptophan synthesis pathway is inhibited by the pathway's end product. Thus, if tryptophan accumulates in a cell, it shuts down its own synthesis. Such feedback inhibition, typical of anabolic (biosynthetic) pathways, allows a cell to adapt to short-term fluctuations in levels of a substance it needs.

If, in our example, the environment continues to provide all the tryptophan the cell needs, the regulation of gene expression also comes into play: The cell stops making enzymes of the tryptophan pathway. This control of enzyme production occurs at the level of transcription, the synthesis of messenger RNA coding for these enzymes. More generally, many genes of the bacterial genome are switched on or off by changes in the metabolic status of the cell. The basic mechanism for this control of gene expression in bacteria, described as the operon model, was discovered in 1961 by François Jacob and Jacques Monod at the Pasteur Institute in Paris. Let's see what an operon is and how it works, using the control of tryptophan synthesis as our first example.

Operons: The Basic Concept

E. coli synthesizes tryptophan from a precursor molecule in a series of steps, each reaction catalyzed by a specific enzyme (see FIGURE 18.19). The five genes coding for the polypeptide chains that make up these enzymes are clustered together on the bacterial chromosome. A single promoter serves all five genes, which constitute a transcription unit. (Recall from Chapter 17 that a promoter is a site where RNA polymerase can bind to DNA and begin transcribing genes.) Thus, transcription gives rise to one long mRNA molecule representing all five genes for the tryptophan pathway. The cell can translate this transcript into separate polypeptides because the mRNA is punctuated with start and stop codons signaling where the coding sequence for each polypeptide begins and ends.

A key advantage of grouping genes of related function into one transcription unit is that a single "on-off switch" can control the whole cluster of functionally related genes. When an *E. coli* cell must make tryptophan for itself because the nutrient medium lacks this amino acid, all the enzymes for the metabolic pathway are synthesized at one time. The switch is a segment of DNA called an **operator.** Both its location and name suit its function: Positioned within the promoter or between the promoter and the enzyme-coding genes, the operator controls the access of RNA polymerase to the genes. All together, the operator, the promoter, and the genes they control—the entire stretch of DNA required for enzyme production for the

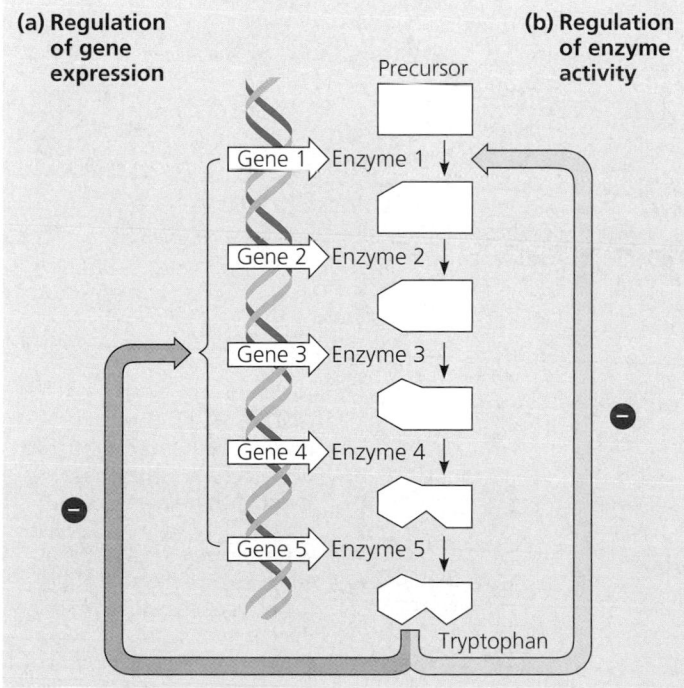

FIGURE 18.19 Regulation of a metabolic pathway. In the pathway for tryptophan synthesis, an abundance of tryptophan can both **(a)** repress expression of the genes for all the enzymes needed for the pathway, and **(b)** inhibit the activity of the first enzyme in the pathway (feedback inhibition). The symbol ⊖ stands for "inhibition."

tryptophan pathway—is called an **operon** (FIGURE 18.20). Here we are dissecting one of many operons that have been discovered in *E. coli*: the *trp* operon (*trp* for tryptophan).

If the operator is the control point for transcription, what determines whether the operator is in the on or off mode? By itself, the operator is on; RNA polymerase can bind to the promoter and transcribe the genes of the operon. The operon can be switched off by a protein called the **repressor.** The repressor binds to the operator and blocks attachment of RNA polymerase to the promoter, preventing transcription of the genes. Repressor proteins are specific; that is, they recognize and bind only to the operator of a certain operon. The repressor that switches off the *trp* operon has no effect on other operons in the *E. coli* genome.

The repressor is the product of a gene called a **regulatory gene.** The regulatory gene encoding the *trp* repressor, *trpR*, is located some distance away from the operon it controls and has its own promoter. Transcription of *trpR* produces an mRNA molecule that is translated into the repressor protein, which can then reach the operator of the *trp* operon by diffusion. Regulatory genes are transcribed continuously, although at a low rate, and a few *trp* repressor molecules are always present in the cell. Why, then, is the *trp* operon not switched off permanently? First, the binding of repressors to operators is reversible. An operator vacillates between the on and off modes, with the relative duration of each state depending on the number of active repressor molecules around. Secondly, the *trp* repressor, like most regulatory proteins, is an allosteric

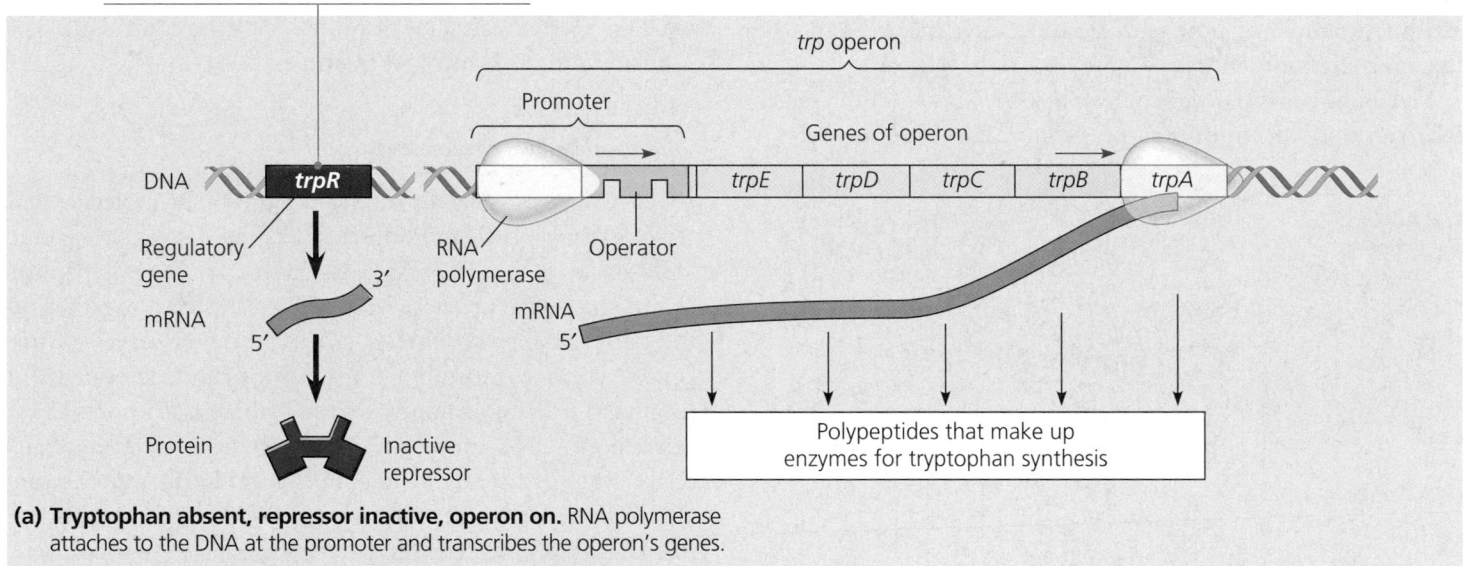

A regulatory gene outside the operon codes for an inactive repressor protein. This gene has its own promoter (not shown).

(a) Tryptophan absent, repressor inactive, operon on. RNA polymerase attaches to the DNA at the promoter and transcribes the operon's genes.

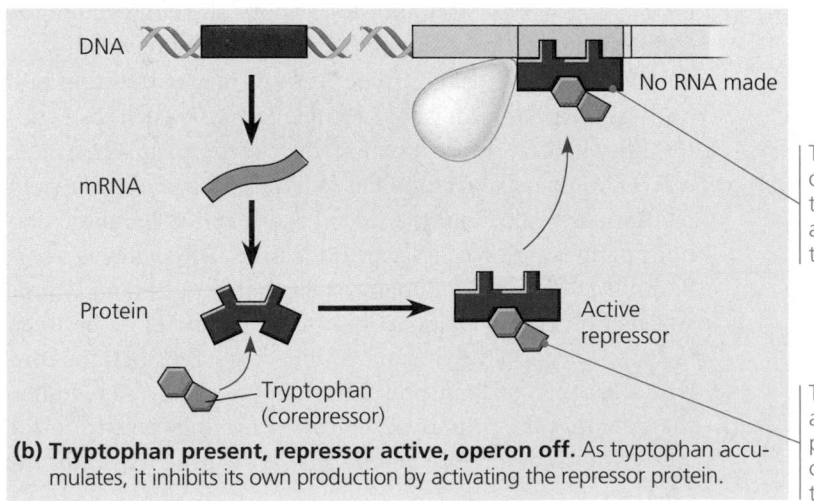

The repressor switches the operon off by binding to the operator and blocking access of RNA polymerase to the promoter.

Tryptophan binds to an allosteric site on the protein, causing its conformation to change to the active form.

(b) Tryptophan present, repressor active, operon off. As tryptophan accumulates, it inhibits its own production by activating the repressor protein.

FIGURE 18.20 The *trp* operon: regulated synthesis of repressible enzymes. Tryptophan is an amino acid produced by an anabolic pathway catalyzed by repressible enzymes. Five genes encoding the polypeptides that make up the enzymes of this pathway are grouped into an operon, along with a promoter and an operator. (The *trp* operator is located within the *trp* promoter.) Accumulation of tryptophan, the end product of the pathway, represses synthesis of the enzymes. The mechanism in *E. coli* is shown here.

protein, with two alternative shapes, active and inactive (see FIGURE 6.18). The *trp* repressor is synthesized in an inactive form with little affinity for the *trp* operator. Only if tryptophan binds to the repressor at an allosteric site does the repressor protein change to the active form that can attach to the operator, turning the operon off.

Tryptophan functions in this system as a **corepressor,** a small molecule that cooperates with a repressor protein to switch an operon off. As tryptophan accumulates, more tryptophan molecules associate with *trp* repressor molecules, which can then bind to the *trp* operator and shut down tryptophan production. If the cell's tryptophan level drops, transcription of the operon's genes resumes. This is one example of how gene expression responds rapidly to changes in the cell's internal and external environment.

Repressible Versus Inducible Operons: Two Types of Negative Gene Regulation

The *trp* operon is said to be a *repressible operon* because its transcription is *inhibited* when a specific small molecule (tryptophan) binds allosterically to a regulatory protein. In contrast, an *inducible operon* is *stimulated* (that is, induced) when a specific small molecule interacts with a regulatory protein. Let's investigate an example, which was actually the operon first worked out by Jacob and Monod (FIGURE 18.21).

The disaccharide lactose (milk sugar) is available to *E. coli* if the human host drinks milk. The bacteria can absorb the lactose and break it down for energy or use it as a source of organic carbon for synthesizing other compounds. Lactose metabolism begins with hydrolysis of the disaccharide into its

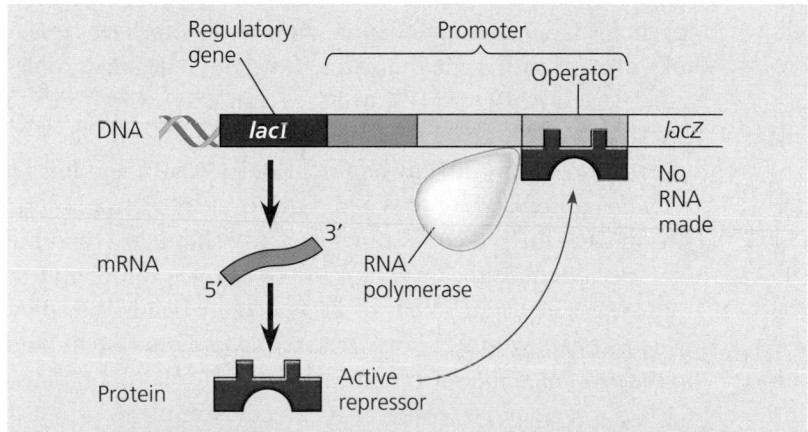

(a) Lactose absent, repressor active, operon off. The *lac* repressor is innately active, and in the absence of lactose it switches off the operon by binding to the operator.

FIGURE 18.21 The *lac* operon: regulated synthesis of inducible enzymes. *E. coli* uses three enzymes to take up and metabolize lactose. The genes for these three enzymes are clustered in the *lac* operon. One gene, *lacZ,* codes for β-galactosidase, which hydrolyzes lactose to glucose and galactose. The second gene, *lacY,* codes for a permease, the membrane protein that transports lactose into the cell. The third gene, *lacA,* codes for an enzyme called transacetylase, whose function in lactose metabolism is still unclear. The gene for the *lac* repressor, *lacI,* happens to be adjacent to the *lac* operon, an unusual situation. (The function of the upstream (left) end of the promoter is revealed in FIGURE 18.22.)

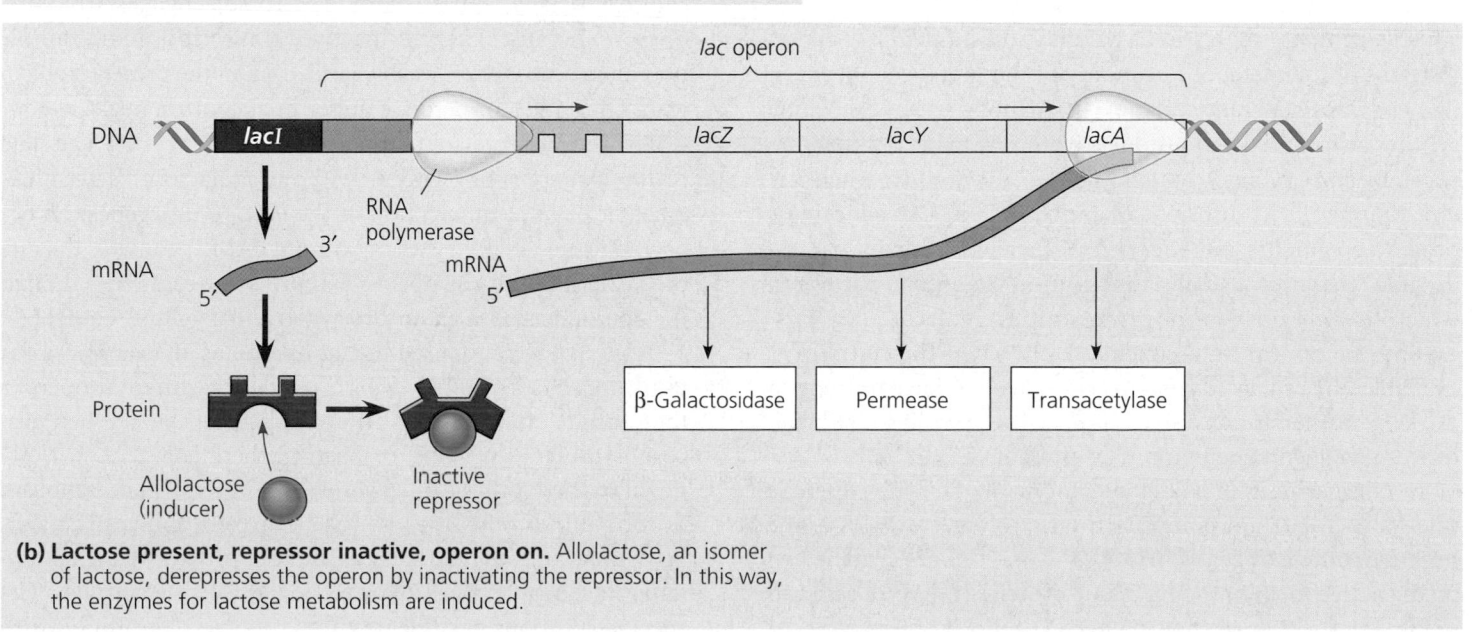

(b) Lactose present, repressor inactive, operon on. Allolactose, an isomer of lactose, derepresses the operon by inactivating the repressor. In this way, the enzymes for lactose metabolism are induced.

two component monosaccharides, glucose and galactose. The enzyme that catalyzes this reaction is called β-galactosidase. Only a few molecules of this enzyme are present in an *E. coli* cell that has been growing in the absence of lactose—in the intestines of a person who does not drink milk, for example. But if lactose is added to the bacterium's nutrient medium, it takes only about 15 minutes for the number of β-galactosidase molecules in the cell to increase a thousandfold.

The gene for β-galactosidase is part of an operon, the *lac* operon (*lac* for lactose metabolism), that includes two other genes coding for proteins that function in lactose metabolism (see FIGURE 18.21). This entire transcription unit is under the command of a single operator and promoter. The regulatory gene, *lacI*, located outside the operon, codes for an allosteric repressor protein that can switch off the *lac* operon by binding to the operator. So far, this sounds just like regulation of the *trp* operon, but there is one important difference. Recall that the *trp* repressor was innately inactive and required tryptophan as a corepressor in order to bind to the operator. The *lac* repressor, in contrast, is active all by itself, binding to the operator and switching the *lac* operon off. In this case, a specific small molecule, called an **inducer,** *inactivates* the repressor. For the *lac* operon, the inducer is allolactose, an isomer of lactose formed in small amounts from lactose that enters the cell. In the absence of lactose (and hence allolactose), the *lac* repressor is in its active configuration, and the genes of the *lac* operon are silenced. If lactose is added to the cell's nutrient medium, allolactose binds to the *lac* repressor and alters its conformation, nullifying the repressor's ability to attach to the operator. Now, on demand, the *lac* operon produces mRNA for the enzymes of the lactose pathway. In the context of gene regulation, these enzymes are referred to as inducible enzymes, because their synthesis is induced by a chemical signal (allolactose, in this case). Analogously, the enzymes for tryptophan synthesis are said to be repressible.

Let's compare repressible enzymes and inducible enzymes in terms of the metabolic economy of the *E. coli* cell. Repressible enzymes generally function in anabolic pathways, which synthesize essential end products from raw materials (precursors). By suspending production of an end product when it is already present in sufficient quantity, the cell can allocate its organic precursors and energy for other uses. In contrast, inducible enzymes usually function in catabolic pathways, which break a nutrient down to simpler molecules. By producing the appropriate enzymes only when the nutrient is available, the cell avoids making proteins that have nothing to do. Why bother, for example, to make the enzymes that break down milk sugar when no milk is present?

In comparing repressible and inducible enzymes, there is one more important point: Both systems are examples of the *negative* control of genes, because the operons are switched *off* by the active form of the repressor protein. It may be easier to see this in the case of the *trp* operon, but it is true for the *lac* operon as well. Allolactose induces enzyme synthesis not by acting directly on the genome, but by freeing the *lac* operon from the negative effect of the repressor. Technically, allolactose is more of a *derepressor* than an inducer of genes. Gene regulation is said to be positive only when an activator molecule interacts directly with the genome to switch transcription on. Let's look at an example, again involving the *lac* operon.

An Example of Positive Gene Regulation

For the enzymes that break down lactose to be synthesized in appreciable quantity, it is not enough that lactose be present in the bacterial cell. The other requirement is that the simple sugar glucose be in short supply. Given a choice of substrates for glycolysis and other catabolic pathways, *E. coli* preferentially uses glucose, the sugar most reliably present in its environment. The enzymes for glucose breakdown (glycolysis; see FIGURE 9.9) are continually present.

How does the *E. coli* cell sense the glucose concentration, and how is this information relayed to the genome? Again, the mechanism depends on the interaction of an allosteric regulatory protein with a small organic molecule. The small molecule is **cyclic AMP (cAMP),** which accumulates when glucose is scarce (see FIGURE 11.12 for the structure of cAMP). The regulatory protein is **cAMP receptor protein (CRP),** and it is an *activator* of transcription. When cAMP binds to the allosteric site on CRP, the protein assumes its active shape and can bind to a specific site at the upstream end of the *lac* promoter (FIGURE 18.22). The attachment of CRP actually bends the DNA, which somehow makes it easier for RNA polymerase to bind to the promoter and start transcription of the operon. Because CRP is a regulatory protein that directly stimulates gene expression, this mechanism qualifies as positive regulation.

If the amount of glucose in the cell increases, the cAMP concentration falls, and without it, CRP disengages from the operon. Because CRP is inactive, transcription of the *lac* operon proceeds at only a low level, even in the presence of lactose. Thus, the *lac* operon is under dual control: negative control by the *lac* repressor and positive control by CRP. The state of the *lac* repressor (with or without allolactose) determines whether or not transcription of the *lac* operon's genes can occur; the state of CRP (with or without cAMP) controls the rate of transcription if the operon is repressor-free. It is as though the operon has both an on-off switch and a volume control.

Although we have used the *lac* operon as an example, CRP, unlike repressor proteins, works on several different operons that encode enzymes used in catabolic pathways. When glucose is present and CRP is inactive, there is a general slowdown in the synthesis of enzymes required for the catabolism of compounds other than glucose. The cell's ability to catabolize other compounds, such as lactose, provides backup systems that enable a cell deprived of glucose to survive. The specific compounds present at the moment determine which

FIGURE 18.22 Positive control: cAMP receptor protein. RNA polymerase has a low affinity for the promoter of the *lac* operon unless helped by a regulatory protein called the cAMP receptor protein (CRP), which binds to a DNA site at the upstream end of the promoter. The CRP molecule can attach to the DNA only when associated with cyclic AMP (cAMP), whose concentration in the cell rises when the glucose concentration falls. Thus, even if lactose is available, the cell preferentially catabolizes glucose, using enzymes that are always present. This regulatory system ensures that *E. coli* will gear up for consumption of lactose and other secondary catabolites only when glucose is scarce.

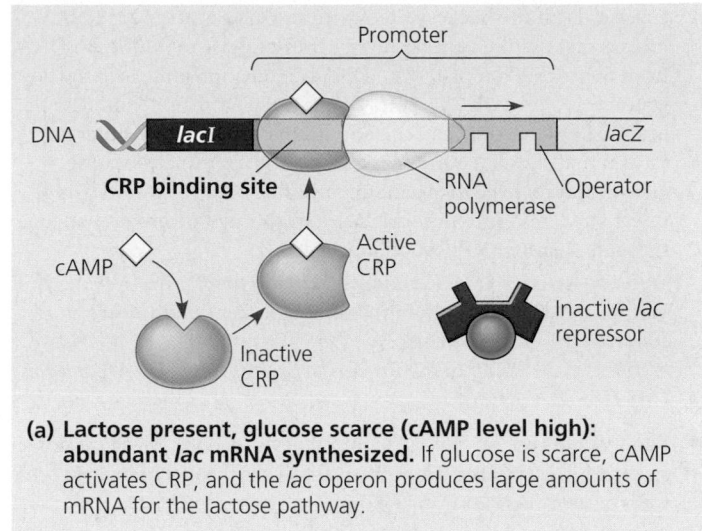

(a) Lactose present, glucose scarce (cAMP level high): abundant *lac* mRNA synthesized. If glucose is scarce, cAMP activates CRP, and the *lac* operon produces large amounts of mRNA for the lactose pathway.

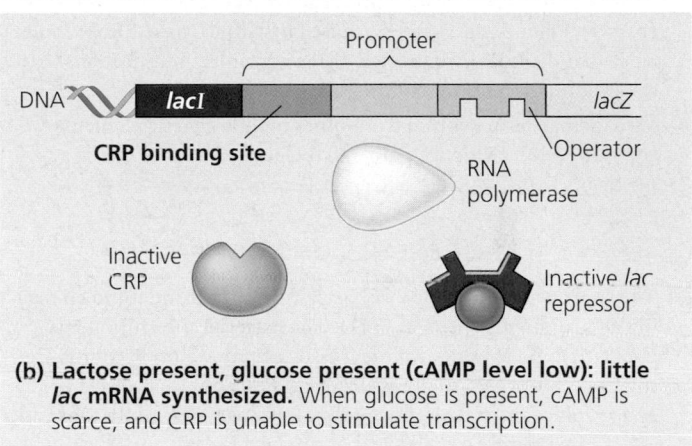

(b) Lactose present, glucose present (cAMP level low): little *lac* mRNA synthesized. When glucose is present, cAMP is scarce, and CRP is unable to stimulate transcription.

operons are switched on. These elaborate contingency mechanisms suit an organism that cannot control what its host eats. Bacteria are remarkable in their ability to adapt—over the long term by evolutionary changes in their genetic makeup and over the short term by the control of gene expression in individual cells. Of course, the various control mechanisms are also evolutionary products that exist because they have been favored by natural selection.

■   ■   ■

Molecular genetics was founded on the study of viruses and bacteria, the microbial models that have been the subjects of this chapter. Eukaryotic organisms and their genomes are much more complex, and only in the past decade or so have researchers begun to learn how the control of gene expression helps bring about this complexity. In the next chapter, we begin to explore this topic.

CHAPTER 18 REVIEW

Go to the Campbell Biology website (www.campbellbiology.com) to explore an interactive version of the Chapter Review.

Summary of Key Concepts

THE GENETICS OF VIRUSES

■ **Researchers discovered viruses by studying a plant disease (pp. 328–329)** In the late 1800s, they found an infectious agent much smaller than bacteria.

■ **A virus is a genome enclosed in a protective coat (pp. 329–330, FIGURE 18.2)** A virus is a small nucleic acid genome enclosed in a protein capsid and sometimes a membranous envelope. The genome may be single- or double-stranded DNA, or single- or double-stranded RNA.

■ **Viruses can reproduce only within a host cell:** *an overview* **(pp. 330–331, FIGURE 18.3)** Viruses use enzymes, ribosomes, and

small molecules of host cells to synthesize progeny viruses. Each type of virus has a characteristic host range, determined by specific receptors on host cells.

Web/CD Activity 18A: *Simplified Viral Reproductive Cycle*

■ **Phages reproduce using lytic or lysogenic cycles (pp. 331–333, FIGURE 18.5)** In the lytic cycle, injection of a phage genome into a bacterium programs destruction of host DNA, production of new phages, and digestion of the host's cell wall, releasing the progeny phages. In a lysogenic cycle, a temperate phage inserts its genome into the bacterial chromosome as a prophage, which is passed on to host daughter cells until it is stimulated to leave the chromosome and initiate a lytic cycle.

Web/CD Activity 18B: *Phage Lytic Cycle*
Web/CD Activity 18C: *Phage Lysogenic and Lytic Cycles*

■ **Animal viruses are diverse in their modes of infection and replication (pp. 333–338, FIGURES 18.6, 18.7)** Animal viruses often have an envelope acquired from host cell membrane, which allows them to enter and exit the host cell. Retroviruses (such as HIV) are RNA viruses that use the enzyme reverse transcriptase to synthesize DNA from their RNA template. The DNA can then integrate into the host genome as a provirus. Vaccines against specific viruses stimulate the immune system to defend the host against an infection. Emerging viruses that cause new outbreaks of disease are usually existing viruses that manage to expand their host territory. Tumor viruses insert viral DNA into host cell DNA, triggering cancerous changes through their own or host cell oncogenes.

Web/CD Activity 18D: *Retrovirus (HIV) Reproductive Cycle*
Web/CD Case Study in the Process of Science: *What Causes Infections in AIDS Patients?*
Web/CD Case Study in the Process of Science: *Why Do AIDS Rates Differ Across the U.S.?*

■ **Plant viruses are serious agricultural pests (p. 338)** Most are single-stranded RNA viruses. They enter plant cells through damaged cell walls or are inherited from a parent.

■ **Viroids and prions are infectious agents even simpler than viruses (p. 339)** Plant diseases can be caused by viroids, naked RNA molecules that disrupt plant growth. Prions are infectious proteins that cause brain diseases in mammals.

■ **Viruses may have evolved from other mobile genetic elements (pp. 339–340)** Evidence points to their origin as packaged fragments of cellular nucleic acid.

THE GENETICS OF BACTERIA

■ **The short generation span of bacteria helps them adapt to changing environments (pp. 340–341)** The bacterial chromosome is a circular DNA molecule with few associated proteins. Plasmids are smaller rings of DNA with accessory genes. Because bacteria proliferate rapidly and have a short generation span, new mutations can affect a population's genetic variation quickly.

■ **Genetic recombination produces new bacterial strains (pp. 341–346, FIGURES 18.12, 18.13, 18.15, 18.18)** The mechanisms of gene transfer between bacteria are transformation, transduction, and conjugation. In transformation, naked DNA enters the cell from the surroundings. In transduction, bacterial DNA is carried from one cell to another by phages. In conjugation, an F factor–containing "male" cell transfers DNA to an F$^-$ cell. F$^+$ cells transfer only the F plasmid. The F factor of an Hfr cell, which is integrated into the bacterial chromosome, brings some chromosomal DNA along with it when it transfers to an F$^-$ cell. R plasmids confer resistance to various antibiotics. Their transfer between bacterial cells poses medical problems. Transposons, DNA segments that can insert at multiple sites in a cell's DNA, also contribute to genetic shuffling in bacteria (and eukaryotes). Insertion sequences, the simplest bacterial transposons, can affect gene function as they move about. They consist of inverted repeats of DNA flanking a gene for transposase. Composite transposons have additional genes, such as those for antibiotic resistance.

Web/CD Case Study in the Process of Science: *What Are the Patterns of Antibiotic Resistance?*

■ **The control of gene expression enables individual bacteria to adjust their metabolism to environmental change (pp. 347–351, FIGURES 18.19 and 18.22)** Cells control metabolism by regulating enzyme activity or by regulating enzyme synthesis through activating or inactivating genes. In bacteria, coordinately regulated genes are often clustered into operons, with one promoter serving several adjacent genes. An operator site on the DNA switches the operon on or off. In a repressible operon, binding of a specific repressor protein to the operator shuts off transcription by blocking the attachment of RNA polymerase. The repressor is activated by binding a small corepressor molecule, usually the end product of an anabolic pathway. In an inducible operon, an innately active repressor is inactivated by binding an inducer, thereby turning on the genes of the operon only when necessary. Inducible enzymes usually function in catabolic pathways. Operons can also be subject to positive control via a stimulatory activator protein. For example, the active form of cAMP receptor protein (CRP) stimulates transcription by binding to a site next to the promoter and enhancing its ability to bind RNA polymerase.

Web/CD Activity 18E: *The* lac *Operon in* E. coli

Self-Quiz

1. Scientists have discovered how to put together a bacteriophage with the protein coat of phage T2 and the DNA of phage T4. If this composite phage were allowed to infect a bacterium, the phages produced in the host cell would have
   a. the protein of T2 and the DNA of T4.
   b. the protein of T4 and the DNA of T2.
   c. a mixture of the DNA and proteins of both phages.
   d. the protein and DNA of T2.
   e. the protein and DNA of T4.

2. Horizontal transmission of a plant viral disease could be caused by
   a. the movement of viral particles through plasmodesmata.
   b. the inheritance of an infection from a parent plant.
   c. the spread of an infection by vegetative (asexual) propagation.
   d. insects as vectors carrying virus particles between plants.
   e. the transmission of proviruses via cell division.

3. RNA viruses require their own supply of certain enzymes because
   a. the viruses are rapidly destroyed by host cell defenses.
   b. host cells do not have enzymes available that can replicate the viral genome.
   c. the enzymes translate viral mRNA into proteins.
   d. the viruses use these enzymes to penetrate host cell membranes.
   e. these enzymes cannot be made in host cells.

4. Which of the following is descriptive of an R plasmid?
   a. Its transfer converts an F$^-$ cell into an F$^+$ cell.
   b. It contains genes for antibiotic resistance and for sex pili.
   c. It is usually transferred between bacteria by transduction.
   d. It is a good example of a composite transposon.
   e. It makes bacteria resistant to phage.

5. Transposition differs from other mechanisms of genetic recombination because it
   a. occurs only in bacteria.
   b. moves genes between homologous regions of the DNA.
   c. plays little or no role in evolution.
   d. occurs only in eukaryotes.
   e. scatters genes to new loci in the genome.

6. A particular operon encodes enzymes that together manufacture an essential amino acid. If the regulation of this operon is like that of the *trp* operon,
   a. the amino acid inactivates the repressor.
   b. the enzymes produced are called inducible enzymes.
   c. the repressor binds to the operator in the absence of the amino acid.
   d. the amino acid acts as a corepressor.
   e. the amino acid turns on enzyme synthesis.

7. A mutation that makes the regulatory gene of an inducible operon nonfunctional would result in
   a. continuous transcription of the operon's genes.
   b. reduced transcription of the operon's genes.
   c. accumulation of large quantities of a substrate for the catabolic pathway controlled by the operon.
   d. irreversible binding of the repressor to the promoter.
   e. overproduction of cAMP receptor protein.

8. Which of the following information transfers is catalyzed by reverse transcriptase?
   a. RNA $\longrightarrow$ RNA
   b. DNA $\longrightarrow$ RNA
   c. RNA $\longrightarrow$ DNA
   d. protein $\longrightarrow$ DNA
   e. RNA $\longrightarrow$ protein

9. Which of the following characteristics or processes is common to *both* bacteria *and* viruses?
   a. nucleic acid as the genetic material
   b. binary fission
   c. mitosis
   d. ribosomes in the cytoplasm
   e. conjugation

10. Which of the following processes would never contribute to genetic variation within a bacterial population?
    a. transduction
    b. transformation
    c. conjugation
    d. mutation
    e. meiosis

11. During conjugation between an Hfr cell and an F⁻ cell,
    a. all the F⁻ cells become F⁺ cells.
    b. all the F⁻ cells become Hfr cells.
    c. the chromosome of the F⁻ cell is completely replaced by the chromosome of the Hfr cell.
    d. genes from the Hfr cell may replace genes of the F⁻ cell by recombination.
    e. DNA from the F⁻ cell transfers to the Hfr cell and DNA from the Hfr cell transfers to the F⁻ cell.

12. Emerging viruses arise by
    a. mutation of existing viruses.
    b. the spread of existing viruses to a new host species.
    c. broader dissemination of an existing virus within the current host population.
    d. all of the above
    e. none of the above

13. A certain mutation in *E. coli* makes the *lac* operator unable to bind the active repressor. How would this affect the cell?
14. Describe one way some viruses can perpetuate their genes without destroying the cells they infect.
15. How do some viruses reproduce without ever having DNA?
16. What are three ways that viruses get into a plant?
17. Why is HIV called a retrovirus?

Go to the website or CD-ROM for more quiz questions.

Evolution Connection

The success of viruses like HIV lies in their ability to evolve within the host's body. The HIV virus evades attack by the body's immune system by mutating at a rapid rate, producing generation after generation of progeny viruses that change before the body can mount a concerted attack, until eventually the viruses overcome the body's defenses. Thus, the viruses present at a late stage of infection are different from those that initially infected the body, though descended from them. Discuss this as an example of evolution in microcosm. Which viral lineages tend to survive?

The Process of Science

When bacteria infect an animal, the number of bacteria in the body increases gradually. A graph of the growth of the bacterial population is a smoothly increasing curve, as shown in graph (a). A viral infection shows a different pattern. For a while, there is no evidence of infection, and then there is a sudden rise in the number of viruses. The number stays constant for a time; then there is another sudden increase. The curve of viral population growth looks like a series of steps, as shown in graph (b). Explain the difference in the growth curves.

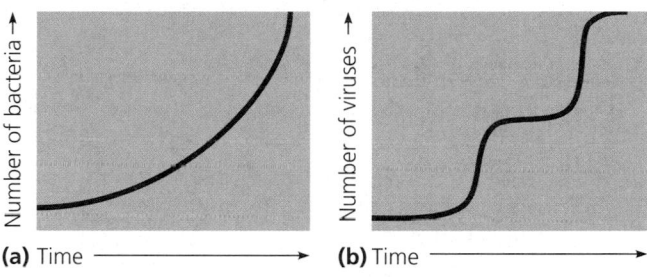

(a) Time   **(b)** Time

Investigate opportunistic infections, the epidemiology of AIDS, and antibiotic resistance in the Case Studies in the Process of Science, available on the website and CD-ROM.

Science, Technology, and Society

Explain how the excessive or inappropriate use of antibiotics poses a health hazard for a human population.

Answers: 1. e; 2. d; 3. b; 4. b; 5. e; 6. d; 7. a; 8. c; 9. a; 10. e; 11. d; 12. d; 13. The cell would wastefully produce the enzymes for lactose metabolism continuously, even in the absence of lactose. 14. Some viruses, such as phage λ and HIV, can insert their DNA into the DNA of the cell they infect. The viral DNA is replicated along with the cell's DNA every time the cell divides. 15. The genetic material of these viruses is RNA, which is replicated inside the infected cell by special enzymes encoded by the virus. The viral genome (or its complement) serves as mRNA for the synthesis of viral proteins. 16. Through lesions caused by injuries, through transfer by insects that feed on the plant, and through contaminated farming or gardening tools 17. Because it synthesizes DNA from its RNA genome. This is the reverse ("retro") of the usual DNA $\longrightarrow$ RNA information flow.

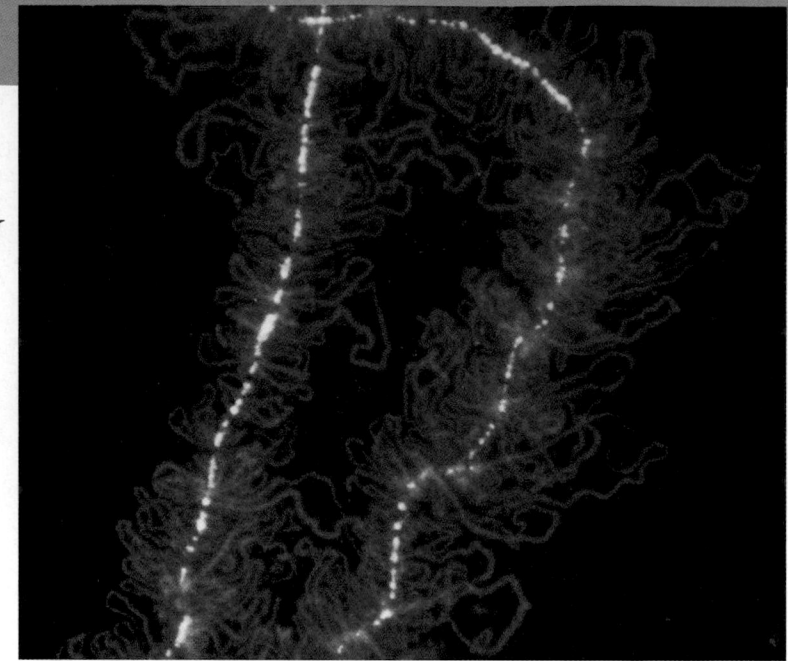

CHAPTER 19

THE ORGANIZATION AND CONTROL OF EUKARYOTIC GENOMES

Eukaryotic cells *face the same challenges as prokaryotic cells in expressing their genes, but with two main differences: the much greater size of the typical eukaryotic genome and the importance of cell specialization in multicellular eukaryotes. These two complications present a formidable information-processing task for the eukaryotic cell.*

Consider the genome of a human cell. It has an estimated 35,000 genes—more than ten times that of a typical bacterium. It also includes an enormous amount of DNA that does not program the synthesis of RNA or protein. The entire mass of DNA must be precisely replicated with each turn of the cell cycle. Managing so much DNA requires that it be elaborately organized. In both prokaryotes and eukaryotes, DNA associates with proteins to form chromatin, but in the eukaryotic cell, the chromatin is ordered into higher structural levels. The light micrograph above gives a sense of the complex organization of chromatin in a developing salamander ovum. Part of the chromatin (white) is packed into the main axis of each of the chromosome's two chromatids, while other parts (red loops) are being actively transcribed. As in prokaryotes, transcription is the stage at which eukaryotic gene expression is most often regulated. In this chapter, you will learn more about the structure of eukaryotic chromosomes, the organization of nucleotide sequences in eukaryotic genomes, and the mechanisms of eukaryotic gene control. You will also learn about DNA changes that disrupt gene regulation and lead to cancer.*

EUKARYOTIC CHROMATIN STRUCTURE

Let's start by looking at how eukaryotic cells pack their chromosomal DNA into chromatin.

Chromatin structure is based on successive levels of DNA packing

Even bacterial DNA has packing. Typically containing only several million nucleotide pairs, the bacterial genome was once thought to be a naked circle of DNA without any particular folding pattern. Now we know that the DNA of the bacterial chromosome is associated with specific proteins and that it is coiled and looped in a complex but orderly manner.

Nevertheless, eukaryotic chromatin (FIGURE 19.1) is considerably more complex than bacterial chromatin. Eukaryotic

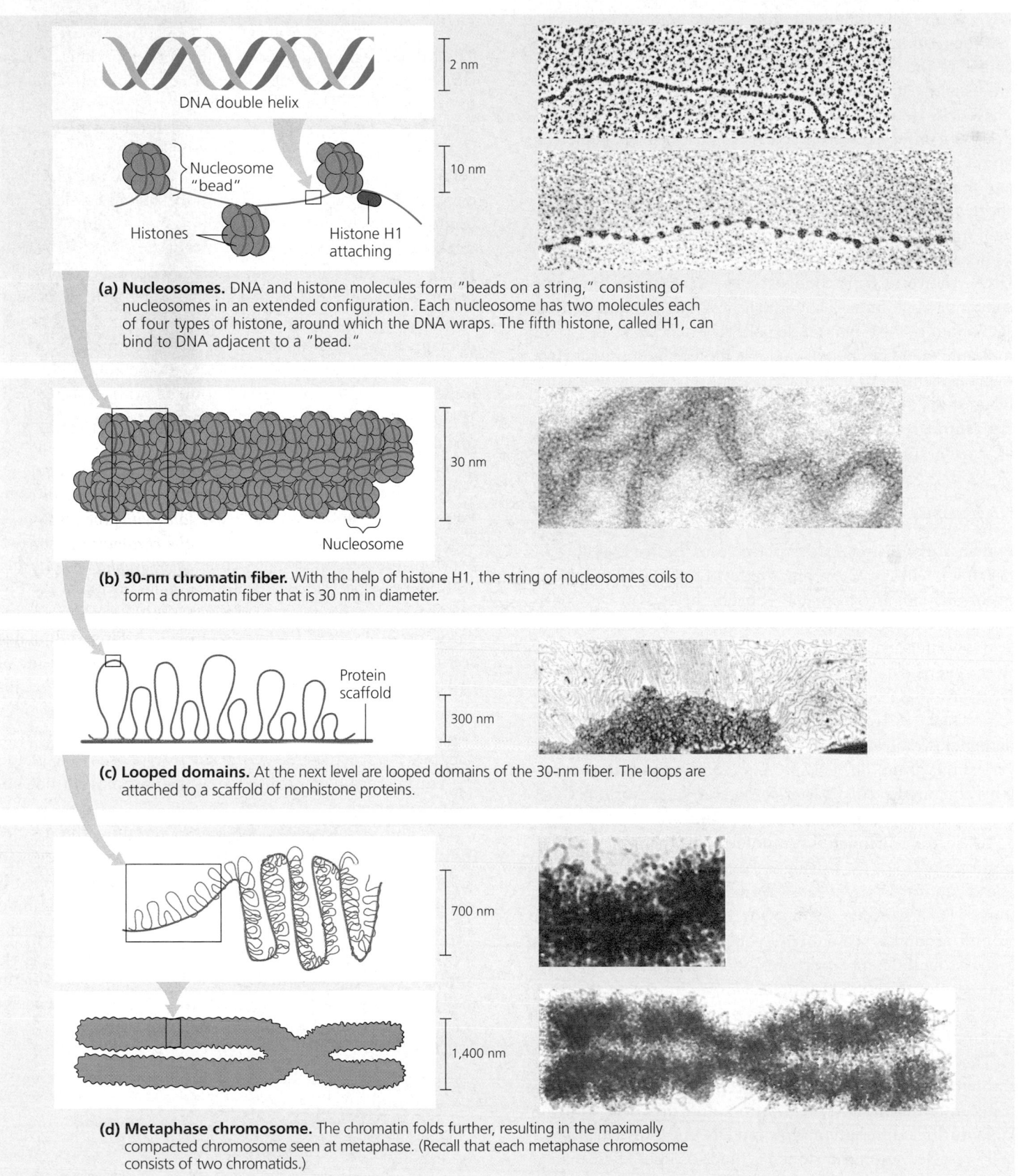

(a) Nucleosomes. DNA and histone molecules form "beads on a string," consisting of nucleosomes in an extended configuration. Each nucleosome has two molecules each of four types of histone, around which the DNA wraps. The fifth histone, called H1, can bind to DNA adjacent to a "bead."

(b) 30-nm chromatin fiber. With the help of histone H1, the string of nucleosomes coils to form a chromatin fiber that is 30 nm in diameter.

(c) Looped domains. At the next level are looped domains of the 30-nm fiber. The loops are attached to a scaffold of nonhistone proteins.

(d) Metaphase chromosome. The chromatin folds further, resulting in the maximally compacted chromosome seen at metaphase. (Recall that each metaphase chromosome consists of two chromatids.)

FIGURE 19.1 Levels of chromatin packing. This series of diagrams and transmission electron micrographs depicts a current model for the progressive stages of DNA coiling and folding that culminate in a highly condensed metaphase chromosome.

DNA is precisely combined with a large amount of protein, and the resulting chromatin undergoes striking changes in the course of the cell cycle. During interphase, chromatin fibers are usually highly extended within the nucleus. When interphase cells are stained for light microscopy, the chromatin appears as a diffuse, colored mass. As you learned in Chapter 12, however, when a cell prepares for mitosis, its chromatin coils and folds up (condenses) to form a characteristic number of short, thick chromosomes that are distinguishable with the light microscope.

Eukaryotic chromosomes contain an enormous amount of DNA relative to their condensed length. Each chromosome contains a single linear DNA double helix that, in humans, averages about 2×10^8 nucleotide pairs. If extended, such a DNA molecule would be about 6 cm long, thousands of times longer than the diameter of a cell nucleus. All this DNA—as well as the DNA of the other 45 human chromosomes—fits into the nucleus through an elaborate, multilevel system of DNA packing. FIGURE 19.1 (preceding page) shows a model for this packing.

Nucleosomes, or "Beads on a String"

Proteins called **histones** are responsible for the first level of DNA packing in eukaryotic chromatin. In fact, the mass of histone in chromatin is approximately equal to the mass of DNA. Histones have a high proportion of positively charged amino acids (lysine and arginine), and they bind tightly to the negatively charged DNA. (Recall that the phosphate groups of DNA give it a negative charge all along its length; see FIGURE 16.5.) The DNA-histone complex is chromatin in its most fundamental form. There are five types of histone. Histones are very similar from one eukaryote to another, and similar proteins are found even in bacteria. Apparently, the histone genes have been highly conserved during evolution.

In electron micrographs, unfolded chromatin has the appearance of beads on a string, as shown in FIGURE 19.1a. Each "bead" and its adjacent DNA form a **nucleosome,** the basic unit of DNA packing. The nucleosome bead consists of DNA wound around a protein core composed of two molecules each of four different types of histone: H2A, H2B, H3, and H4. A molecule of the fifth histone, called H1, attaches to the DNA near the bead when the chromatin undergoes the next level of packing.

The beaded string seems to remain essentially intact throughout the cell cycle. The histones leave the DNA only transiently during DNA replication, and they stay with the DNA during transcription. How can DNA be transcribed when it is wrapped around histones in nucleosomes? Researchers have learned that nucleosomes are dynamic structures. By changing shape and position, they can allow RNA-synthesizing polymerases to move along the DNA. Later in this chapter, we'll discuss some recent discoveries about the roles of nucleosomes in the regulation of gene expression.

Higher Levels of DNA Packing

The beaded string undergoes higher-order packing. This is clearest when the extended interphase chromatin condenses to produce the thickened, compact chromosomes we see during mitosis. In the laboratory, mitotic chromosomes can be isolated and unraveled to reveal several orders of chromatin coiling. FIGURE 19.1b–d illustrates the various structures in order of increasing compactness. With the aid of histone H1, the beaded string coils or folds to form a fiber roughly 30 nm in thickness, known as the *30-nm chromatin fiber* (see FIGURE 19.1b). The 30-nm fiber, in turn, forms loops called *looped domains,* which are attached to a chromosome scaffold made of nonhistone proteins. In a mitotic chromosome, the looped domains themselves coil and fold, further compacting all the chromatin to produce the characteristic metaphase chromosome you see in the micrograph at the bottom of FIGURE 19.1. We know that the packing steps are highly specific and precise, for particular genes always end up located at the same places in metaphase chromosomes.

Though interphase chromatin is generally much less condensed than the chromatin of mitotic chromosomes, it shows several of these same levels of higher-order packing. Much of the beaded string is compacted into a 30-nm fiber and then further folded into looped domains (as in the micrograph opening this chapter). Although an interphase chromosome lacks a discrete "scaffold," its looped domains seem to be attached to the nuclear lamina, on the inside of the nuclear envelope, and perhaps also to fibers of the nuclear matrix. These attachments may help organize regions of active transcription. The chromatin of each chromosome occupies a restricted area within the interphase nucleus, and the chromatin fibers of different chromosomes do not become entangled.

Even during interphase, portions of certain chromosomes in some cells exist in the highly condensed state represented in FIGURE 19.1d. This type of interphase chromatin, which is visible with a light microscope, is called **heterochromatin,** to distinguish it from the less compacted **euchromatin** ("true chromatin"). What is the function of this selective condensation in interphase cells? The formation of heterochromatin may be a sort of coarse adjustment in the control of gene expression, for heterochromatin DNA is not transcribed.

GENOME ORGANIZATION AT THE DNA LEVEL

Now we are ready to examine how genes and other DNA sequences are organized within the genome. Because most of this unit has focused on genes, you may be surprised to learn that genes make up only a tiny portion of the genomes of most multicellular eukaryotes.

Repetitive DNA and other noncoding sequences account for much of a eukaryotic genome

In prokaryotes, most of the DNA in a genome codes for protein (or tRNA and rRNA), with the small amount of noncoding DNA consisting mainly of regulatory sequences, such as promoters. The coding sequence of nucleotides along a prokaryotic gene proceeds from start to finish without interruption. In eukaryotic genomes, by contrast, most of the DNA—about 97% in humans—does *not* encode protein or RNA. What does this DNA consist of? Some of it is known to be regulatory sequences, but much of it consists of sequences whose functions, if any, are not yet understood. This DNA includes introns, the stretches of noncoding DNA that often interrupt the coding sequences of eukaryotic genes (see FIGURE 17.9). Even more of the noncoding DNA consists of **repetitive DNA,** nucleotide sequences that are present in many copies in a genome, usually not within genes.

Tandemly Repetitive DNA

TABLE 19.1 lists the main categories of repetitive DNA, which occur in different amounts in different species. In mammals, about 10–15% of the genome is *tandemly repetitive DNA,* short sequences repeated in series, as in the following example (showing one DNA strand only):

... GTTACGTTACGTTACGTTACGTTACGTTAC ...

The number of repetitions of the GTTAC unit at a site in the genome could be as high as several hundred thousand. Repeated units are up to ten base pairs long.

The nucleotide composition of tandemly repetitive DNA is often different enough from the rest of the cell's DNA to give the repetitive DNA an intrinsically different density, enabling researchers to isolate it by differential ultracentrifugation (as in the Meselson-Stahl experiment; see FIGURE 16.9). If the genomic DNA is cut into pieces before centrifugation, segments of different density migrate to different positions in the centrifuge tube. Repetitive DNA isolated in this way was originally called **satellite DNA,** because it appeared as a "satellite" band in the centrifuge tube, separate from the rest of the DNA. Now the term is often used for all tandemly repetitive DNA.

As TABLE 19.1 shows, satellite DNA is classified into three types, depending on the total length of the DNA at each site. Regular satellite DNA occurs in stretches over 100,000 base pairs long, while *minisatellite* and *microsatellite* subcategories have been created for sequences that appear in shorter stretches. Microsatellite DNA, with short units repeated only 10 to 100 times, has turned out to be extremely useful for DNA fingerprinting, as we'll discuss in Chapter 20.

A number of genetic disorders, including two you read about in Chapter 15, are caused by abnormally long stretches of tandemly repeated nucleotide triplets—a sort of microsatellite—within the affected gene. The first such "triplet repeat" sequence to be identified was the one responsible for fragile X syndrome, a major cause of mental retardation. In the normal allele for the fragile X gene, within the 5′ untranslated leader region of the first exon, the triplet CGG is repeated about 30 times; in a fragile X allele, the same triplet is repeated hundreds or even thousands of times, creating the fragile site. The lengthening of the repetitive DNA occurs in steps over a number of generations. Huntington's disease is also a triplet-repeat disorder, with a stretch of repeated CAG triplets that lengthens in a similar fashion. In this case, the triplet repeat is actually translated, and the resulting protein has a long string of glutamines, the amino acid encoded in DNA by CAG (recall that DNA sequences are given as they appear on the *non*template strand). The dozen triplet-repeat disorders identified to date all affect the nervous system, and in all cases the number of repeats seems to correlate with the severity of the disease and the age of onset. Scientists do not yet know exactly how the extra repeats cause disorders; the mechanism undoubtedly varies with the gene.

Much of a genome's regular satellite DNA is located at chromosomal telomeres and centromeres, suggesting that this DNA plays a structural role for chromosomes. The DNA at centromeres is essential for the separation of chromatids in cell division (see Chapter 12), and along with other tandemly repetitive sequences, it may help organize the chromatin within the interphase nucleus. We discussed the DNA of telomeres, the tips of chromosomes, in Chapter 16. Besides protecting genes from being lost as the DNA shortens with each round of replication, telomeric DNA protects the chromosome by binding proteins that protect the ends from degradation and from joining to other chromosomes.

The importance of telomeres and centromeres has been highlighted by the creation of *artificial chromosomes* in the

Table 19.1 Types of Repetitive DNA

Tandemly Repetitive DNA (Satellite DNA)

Repeated units at a site are usually identical

Proportion of mammalian DNA:	10–15%
Length of each repeated unit:	1–10 base pairs
Total length of repetitive DNA per site, in base pairs:	
Regular satellite DNA	100,000–10 million
Minisatellite DNA	100–100,000
Microsatellite DNA	10–100

Interspersed Repetitive DNA

"Copies" are very similar but not identical

Proportion of mammalian DNA:	25–40%
Length of each repeated unit:	100–10,000 base pairs
Number of repetitions per genome:	10–1 million

laboratory. The DNA used to make such chromosomes must, of course, include an origin of replication. But this is not sufficient; to function as a chromosome, it must also have a centromere and two telomeres. If an artificial chromosome has these features—no matter what the nature of its other DNA—a cell will be able to replicate it and distribute the copies properly to daughter cells.

Interspersed Repetitive DNA

In addition to tandemly repetitive DNA, eukaryotic genomes have huge amounts of *interspersed repetitive DNA* (see TABLE 19.1). The repeated units of this type of DNA are not next to each other; instead, they are scattered about the genome. A single unit is usually hundreds or even thousands of base pairs long, and the dispersed "copies" are similar but usually not identical to each other. Interspersed repetitive DNA makes up 25–40% of most mammalian genomes. In humans and other primates, a large portion of this DNA (at least 5% of the genome) consists of a family of similar sequences called *Alu* **elements,** for which each unit is about 300 nucleotide pairs long. Like the Huntington's disease triplet repeats, *Alu* elements are an exception to the notion that repetitive DNA is noncoding. Many *Alu* elements are transcribed into RNA molecules; their cellular function, if any, is unknown.

Although we know little about the functions of interspersed repetitive DNA, we do know how it comes to be both abundant and variable in location. Most such sequences, at least in mammals, seem to be transposons, the mobile genetic elements introduced in Chapter 18. We'll have more to say about eukaryotic transposons later in this chapter.

Gene families have evolved by duplication of ancestral genes

As in prokaryotes, most eukaryotic genes are present as unique sequences, with only one copy per haploid set of chromosomes. However, some genes are present in more than one copy, and others closely resemble each other in nucleotide sequence. A collection of identical or very similar genes is called a **multigene family.** It is likely that the members of each family evolved from a single ancestral gene. Multigene families can be thought of as repetitive DNA with very long (gene-length) repeating units. The members of a multigene family may be clustered or dispersed in the genome.

Some multigene families consist of *identical genes,* usually clustered tandemly. Although the genes for histone proteins are a notable exception, multigene families of identical genes usually consist of genes for RNA products. An example is the family of identical genes for the three largest ribosomal RNA (rRNA) molecules (FIGURE 19.2). These rRNA molecules are

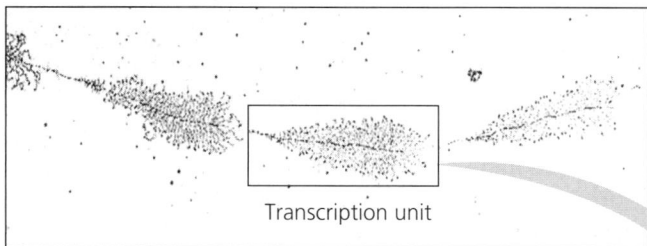

Transcription unit

Three of the hundreds of copies of rRNA transcription units in a salamander genome are shown above (TEM). Each transcription unit includes the genes for three types of rRNA. Each "feather" (right) corresponds to a single transcription unit being transcribed by about 100 molecules of RNA polymerase (the dark dots along the DNA), moving left to right. The growing RNA transcripts extend out from the DNA.

Transcription units are separated from one another by spacers of nontranscribed DNA.

The RNA transcripts are processed by cleavage to yield three kinds of rRNA molecules: 18S, 5.8S, and 28S. (The S designations refer to their sedimentation rates in the ultracentrifuge, which depend on molecular size.)

FIGURE 19.2 Part of a family of identical genes for ribosomal RNA.

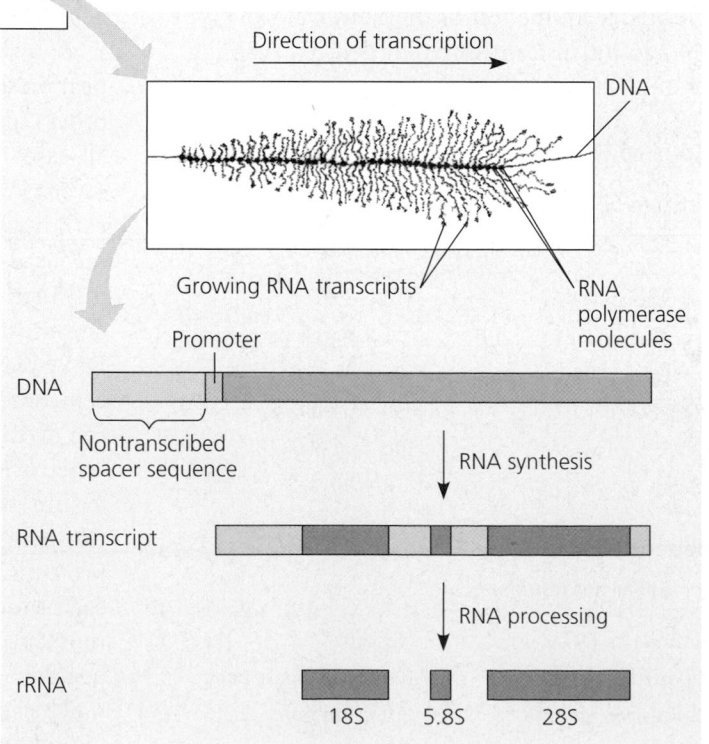

encoded in a single transcription unit that is repeated tandemly hundreds to thousands of times in the genome of a multicellular eukaryote. The many copies enable cells to make the millions of ribosomes needed for active protein synthesis. The primary transcript is cleaved to yield the three rRNA molecules. These are then combined with proteins and one other kind of rRNA to form ribosomal subunits.

The classic examples of multigene families of *nonidentical genes* are the two related families of genes that encode globins, the α and β polypeptide subunits of hemoglobin. One family, located on chromosome 16 in humans, encodes various versions of α-globin; the other, on chromosome 11, encodes versions of β-globin (FIGURE 19.3). Similarities in the sequences of the various globin genes indicate that the α-like globins and β-like globins all evolved from one common ancestral globin. The different versions of each globin subunit are expressed at different times in development, allowing hemoglobin to function effectively in the changing environment of the developing animal. In humans, for example, the embryonic and fetal forms of hemoglobin have a higher affinity for oxygen than the adult forms, ensuring the efficient transfer of oxygen from mother to developing fetus.

How do families of genes arise from a single gene? The most likely explanation is that they arise by repeated gene duplication, which can result from errors during DNA replication and recombination. The differences among the genes in families of nonidentical genes probably arise from mutations that accumulate in the gene copies over many generations. The existence of DNA segments called pseudogenes is evidence for this process of gene duplication and mutation. **Pseudogenes** have sequences very similar to real genes but do not yield functional products. Presumably, over evolutionary time, random mutations have destroyed their function. The globin gene families include several pseudogenes within the stretches of noncoding DNA between the functional genes.

The location of the α-globin and β-globin families on different human chromosomes, as well as the dispersion of the genes of certain other gene families, probably arose by transposition, one of the topics in the next section.

Gene amplification, loss, or rearrangement can alter a cell's genome during an organism's lifetime

We are used to the idea that, except for rare mutations, the nucleotide sequence of an organism's DNA is constant during its lifetime. However, there are a few important exceptions in which the DNA of somatic cells is altered in a systematic way. Because these changes do not affect gametes, they are not passed on to offspring, but they can and do have major effects on gene expression within particular cells and tissues.

Gene Amplification and Selective Gene Loss

Sometimes the number of copies of a gene or gene family temporarily increases in the cells of some tissues during a particular stage of development. For instance, consider the genes for

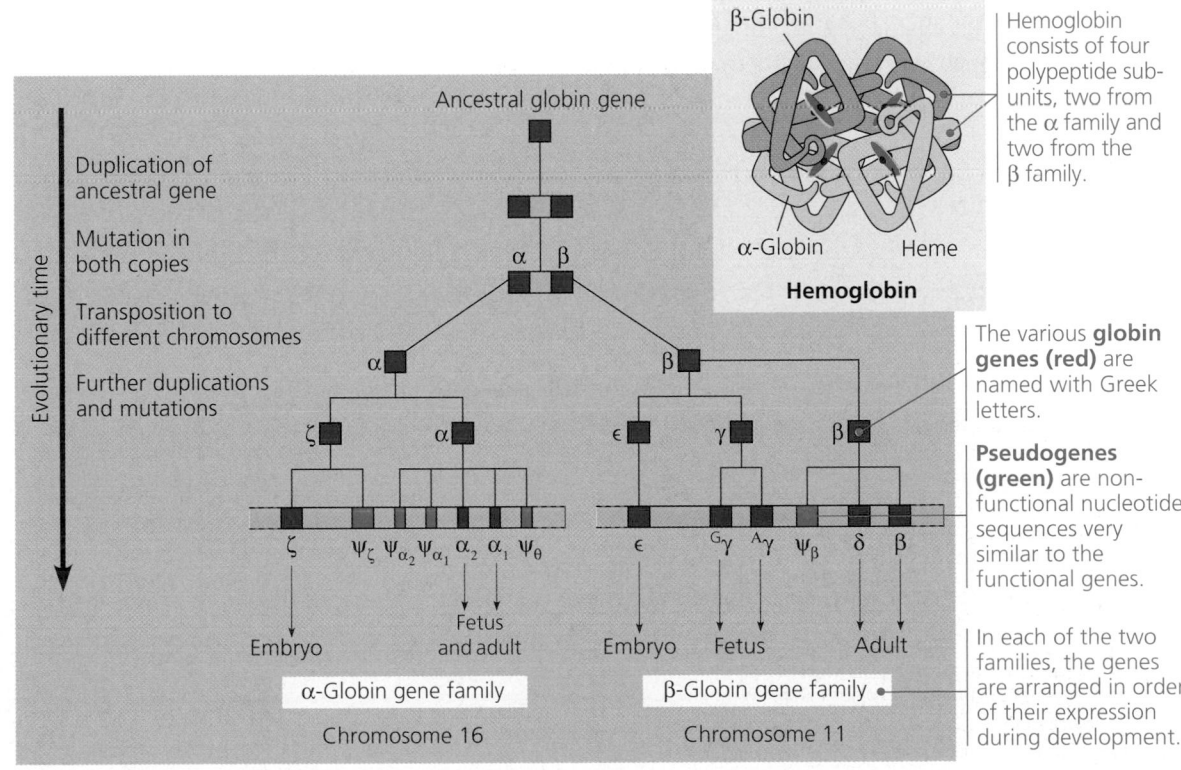

FIGURE 19.3 The evolution of human α-globin and β-globin gene families. The two kinds of polypeptides that make up hemoglobin are encoded by genes belonging to two gene families, which are found as clusters on two different chromosomes. Each family consists of a group of very similar, but not identical, genes. Noncoding DNA separates the functional genes within each family cluster, including pseudogenes (green). Shown here is a hypothesis for the evolution of the modern α- and β-globin gene families from a single ancestral globin gene.

ribosomal RNA in amphibians. As in most eukaryotes, multiple copies of these genes are built into the genome of every cell (the result of gene duplication in the evolutionary past). A developing ovum, however, synthesizes a million or more additional copies of the rRNA genes, which exist in nucleoli as tiny DNA circles separate from the chromosomes. This selective replication of certain genes, called **gene amplification,** is a potent way of increasing expression of the rRNA genes, enabling the developing egg cell to make enormous numbers of ribosomes. These ribosomes make possible a burst of protein synthesis once the egg is fertilized. The extra copies of the rRNA genes cannot replicate further and are broken down during early embryonic development.

Gene amplification has also been observed in cancer cells exposed to high concentrations of chemotherapeutic drugs. While such drugs may kill a great many cells in a tumor, invariably some cells are resistant. These cells often contain amplified segments of DNA carrying genes conferring drug resistance. In the laboratory, increasing the concentration of such drugs leads to increasing resistance in the cell population by selecting for cells that have amplified these genes.

In certain insects, genes are selectively *lost* in certain tissues (although not in the cells that give rise to gametes, of course). In fact, whole chromosomes or parts of chromosomes may be eliminated from certain cells early in insect development.

Rearrangements in the Genome

More common than gene amplification or loss is the shuffling of substantial stretches of DNA. Here we are talking not about the genetic recombination that goes on in meiosis but about rearrangements that change the loci of genes in somatic cells of an organism. Such rearrangements may have powerful effects on gene expression.

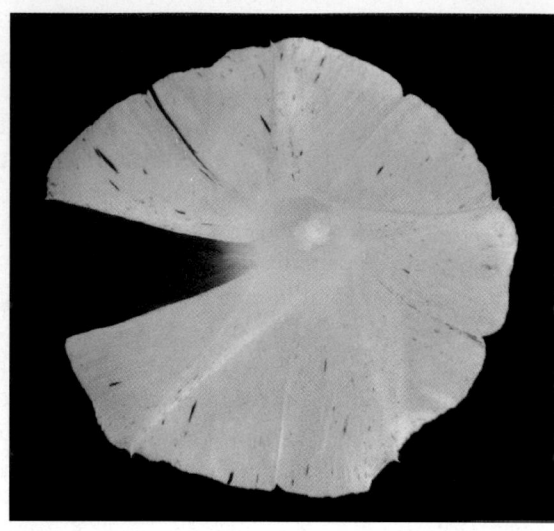

FIGURE 19.4 The effect of a transposon on flower color. Without the movement of a transposon, this morning glory flower would be entirely purple. The white portion of the flower (light blue-green in this photo) resulted from the movement of a transposon in the genome of one cell to a locus that determines purple flower color, destroying the function of the flower-color gene.

Courtesy of Evelyne Cudel-Epperson, MSU.

Transposons and Retrotransposons. All organisms seem to have transposons, stretches of DNA that can move from one location to another within the genome. Transposons were discussed in detail in Chapter 18 (see FIGURES 18.16–18.18), and transposition was mentioned as a source of scattered gene families earlier in this chapter. Recall that if a transposon "jumps" into the middle of a coding sequence of another gene, it prevents the normal functioning of the interrupted gene (FIGURE 19.4). If the transposon inserts within a sequence that is involved in regulating transcription, the transposition may increase or decrease the production of one or more proteins. In some cases, the transposon itself carries a gene that is activated

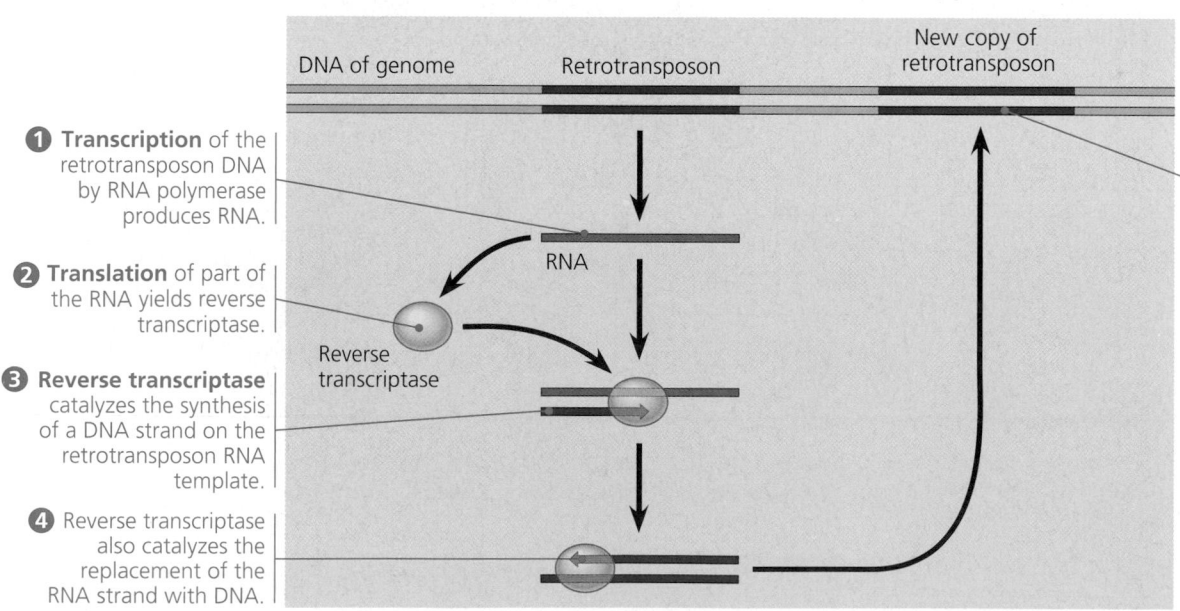

❶ **Transcription** of the retrotransposon DNA by RNA polymerase produces RNA.

❷ **Translation** of part of the RNA yields reverse transcriptase.

❸ **Reverse transcriptase** catalyzes the synthesis of a DNA strand on the retrotransposon RNA template.

❹ Reverse transcriptase also catalyzes the replacement of the RNA strand with DNA.

DNA of genome Retrotransposon New copy of retrotransposon

RNA

Reverse transcriptase

❺ **Insertion** of the double-stranded DNA version of the retrotransposon occurs at some other location.

FIGURE 19.5 Retrotransposon movement. Notice that the retrotransposition mechanism is essentially identical to part of the retrovirus reproductive cycle (see FIGURE 18.7). By replicative transposition, as shown here, a retrotransposon can populate the genome of a multicellular eukaryote in huge numbers.

when it is inserted just downstream from an active promoter. Barbara McClintock was the first to discover transposons when she found evidence for mobile genetic elements that affect the color of developing corn kernels.

Recent studies have revealed that transposons actually make up over 50% of the corn (maize) genome and 10% of the human genome. Most of these transposons are **retrotransposons,** transposable elements that move within a genome by means of an RNA intermediate, a transcript of the retrotransposon DNA (FIGURE 19.5). To insert at another site, the RNA retrotransposon must be converted back to DNA. This is accomplished by the enzyme reverse transcriptase, which is encoded in the retrotransposon itself, along with an enzyme that catalyzes the insertion at a new site. Thus, reverse transcriptase can be present in cells not infected with retroviruses. (In fact, retroviruses may have evolved from escaped and packaged retrotransposons.) The *Alu* elements, mentioned earlier, are retrotransposons that do not code for reverse transcriptase but can move using enzymes encoded by other retrotransposons in the genome.

Immunoglobulin Genes. In vertebrates, at least one set of genes normally undergoes permanent rearrangements of DNA segments. These rearrangements occur in the developing immune system during *cellular differentiation,* the specialization in cellular structure and function. Several kinds of genes become rearranged as cells of the immune system differentiate. Let's look at the genes encoding antibodies, or **immunoglobulins,** proteins that specifically recognize and help combat viruses, bacteria, and other invaders of the body.

Immunoglobulins are made by cells of the immune system called B lymphocytes, a type of white blood cell. B lymphocytes are highly specialized, with each differentiated cell and its descendants producing one specific type of antibody that attacks a specific invader. As an unspecialized cell differentiates into a B lymphocyte, functional antibody genes are pieced together using segments from DNA regions that are physically separated in the genome of an embryonic cell (FIGURE 19.6).

The basic immunoglobulin molecule is represented at the bottom right of FIGURE 19.6. It consists of four polypeptide

❶ The DNA of an immunoglobulin gene in an undifferentiated cell carries coding segments for hundreds of different antibody variable (*V*) regions, for several different junction (*J*) regions, and for one or more different constant (*C*) regions. This simplified diagram shows only three *V* segments, one *J* segment, and one *C* segment.

❷ During differentiation of a B lymphocyte, the deletion of a long stretch of DNA brings a *V* segment (in this case V_2) adjacent to a *J* segment and produces a gene that can be transcribed.

❸ The RNA transcript is processed in the usual way to remove introns, and the resulting mRNA is translated into one of the polypeptides for an immunoglobulin molecule. (The amino acids coded by the *J* segment are considered part of the variable region of the polypeptide.)

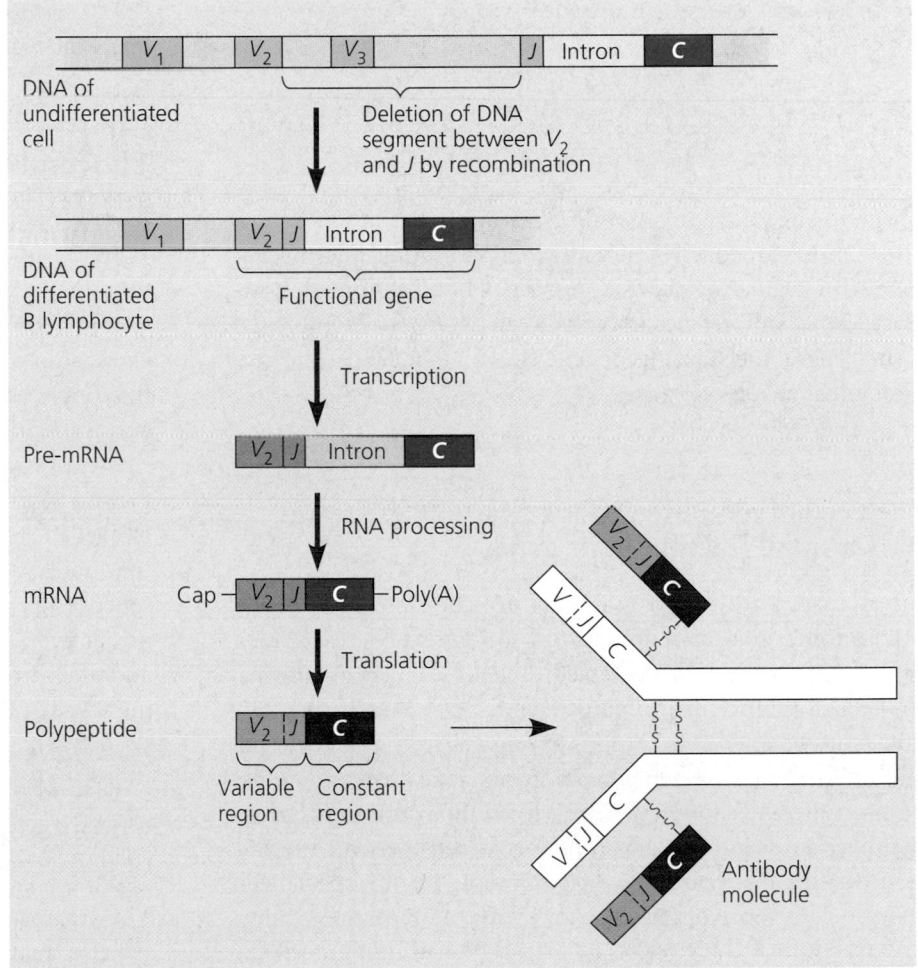

FIGURE 19.6 DNA rearrangement in the maturation of an immunoglobulin (antibody) gene. The joining of *V, J,* and *C* regions of DNA in random combinations plays a major role in arming the immune system with an enormous variety of antibody-producing lymphocytes.

chains held together by disulfide bridges. Each chain has two major parts: a constant (C) region, which is the same for all antibodies of a particular class, and a variable (V) region. The variable region gives a particular antibody its unique function—the ability to recognize and bind to a specific foreign molecule. In the genome of an embryonic cell, the DNA sequence coding for the constant part of an antibody polypeptide is separated by a long stretch of DNA from a location containing hundreds of segments coding for variable regions. As a B lymphocyte differentiates, a variable segment of the DNA is connected to a constant segment by the deletion of intervening DNA. This occurs by a type of genetic recombination within the DNA (remember that this is not RNA splicing!). The joined segments form the continuous sequence of nucleotides that functions as the gene for one of the immunoglobulin polypeptides. The combining of different variable and constant regions in this way creates an enormous variety of different polypeptides. And antibody diversity is increased still more by the combining of polypeptides to form complete antibody molecules. As a result, the mature immune system, with millions of subpopulations of B lymphocytes, can make millions of different kinds of antibody molecules.

THE CONTROL OF GENE EXPRESSION

The rearrangement of antibody genes and other processes that alter DNA sequences in somatic cells are bona fide mechanisms for regulating gene expression. They are limited, however, to particular genes in certain cells. We now examine the more general mechanisms of gene control that are used to regulate most eukaryotic genes.

Each cell of a multicellular eukaryote expresses only a small fraction of its genes

Interspersed within their masses of noncoding DNA, eukaryotic genomes may contain tens of thousands of genes. Which of these genes are to be expressed? Like unicellular organisms, the cells of multicellular organisms must continually turn certain genes on and off in response to signals from their external and internal environments. In addition, gene expression must be controlled on a long-term basis for **cellular differentiation,** the divergence in form and function as cells become specialized during an organism's development. Highly specialized cells, such as those of muscle or nervous tissue, express only a tiny fraction of their genes. In fact, a typical human cell expresses only 3–5% of its genes at any given time. The enzymes that transcribe DNA must locate the right genes at the right time, a task no easier than finding a needle in a haystack.

When gene expression goes awry, serious imbalances and diseases, including cancer, can arise. Thus, the question of how eukaryotic genes are regulated is paramount for medical research as well as for basic biology.

Only 35 years ago, an understanding of the mechanisms that control gene expression in eukaryotes seemed almost hopelessly out of reach. Since then, new research methods have empowered molecular biologists to begin to solve these mysteries. Equipped with DNA technology for isolating individual genes and sequencing DNA (discussed in Chapter 20), biologists are now uncovering many of the details of eukaryotic gene regulation.

Despite the extra "nuts and bolts" it requires, the control of gene activity in eukaryotes involves some of the same basic principles as prokaryotic gene regulation. In all organisms, the expression of specific genes is most commonly regulated at the level of transcription by DNA-binding proteins that also interact with other proteins and often with external signals. For that reason, the term *gene expression* is often equated with gene activity—that is, transcription—for both prokaryotes and eukaryotes. However, the greater complexity of eukaryotic cell structure and function provides opportunities for controlling gene expression at additional stages.

The control of gene expression can occur at any step in the pathway from gene to functional protein: *an overview*

FIGURE 19.7 is an overview of the entire process of gene expression in a eukaryotic cell. The figure highlights key stages in the expression of a protein-coding gene, from chromatin changes that unpack the DNA, through transcription, RNA processing, and translation, to various alterations of the protein product. The expression of a given gene will not necessarily involve every stage shown; for example, not every polypeptide is cleaved. The main lesson is that each stage is a potential control point where gene expression can be turned on or off, speeded up, or slowed down. The figure omits details and does not convey the web of control that connects *different* genes and their products. In the following three sections, we'll examine some of the important control points more closely.

Chromatin modifications affect the availability of genes for transcription

The organization of chromatin discussed earlier in the chapter serves a dual purpose. One function is to pack the DNA into a compact form that fits inside the nucleus of a cell. The other main function is regulatory: The physical state of DNA in or near a gene is important in helping control whether the gene is

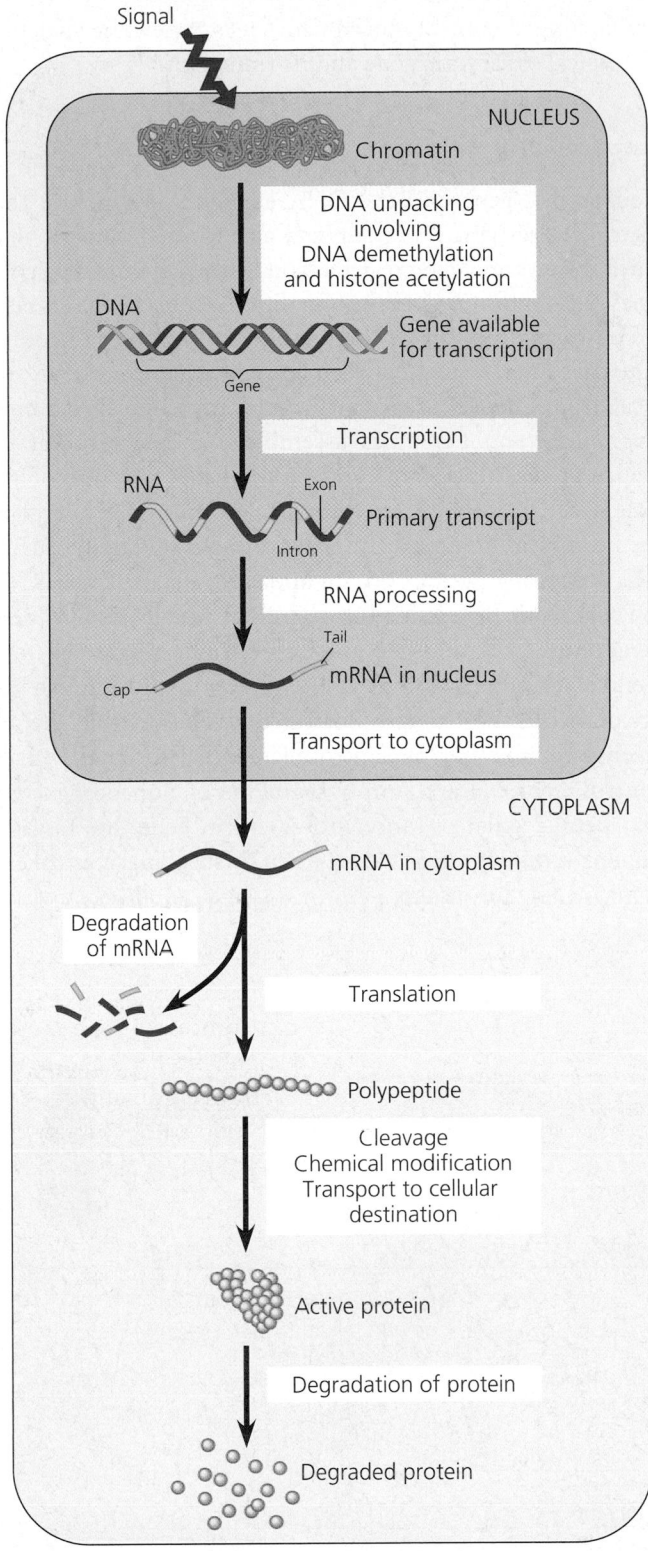

Signal

NUCLEUS

Chromatin

DNA unpacking
involving
DNA demethylation
and histone acetylation

DNA

Gene available
for transcription

Gene

Transcription

RNA

Exon

Primary transcript

Intron

RNA processing

Tail

Cap

mRNA in nucleus

Transport to cytoplasm

CYTOPLASM

mRNA in cytoplasm

Degradation
of mRNA

Translation

Polypeptide

Cleavage
Chemical modification
Transport to cellular
destination

Active protein

Degradation of protein

Degraded protein

FIGURE 19.7 Opportunities for the control of gene expression in eukaryotic cells. The nuclear envelope separating transcription from translation in eukaryotic cells offers opportunity for post-transcriptional control in the form of RNA processing that is absent in bacteria. In addition, eukaryotes have a greater variety of control mechanisms operating before transcription and after translation. In this diagram, the processes that offer opportunities for regulation are highlighted by white or yellow boxes. As in prokaryotes, transcription initiation is the most important control point.

available for transcription. Thus, the genes of heterochromatin, which is highly condensed, are usually not expressed, presumably because transcription proteins cannot reach the DNA. Also, a gene's location relative to nucleosomes and to the sites where the DNA attaches to the chromosome scaffold or nuclear lamina can affect whether it is transcribed. A flurry of recent research indicates that chemical modifications of chromatin play key roles in both chromatin structure and the regulation of transcription. Especially important are DNA methylation and histone acetylation. Both are catalyzed by specific enzymes.

DNA Methylation

DNA methylation is the attachment of methyl groups ($—CH_3$) to DNA bases after DNA is synthesized. The DNA of most plants and animals has methylated bases, usually cytosine. About 5% of the cytosine bases in methylated eukaryotic DNA have methyl groups. Inactive DNA, such as that of inactivated mammalian X chromosomes, is generally highly methylated compared to DNA that is actively transcribed, although there are exceptions. Comparison of the same genes in different types of tissues shows that the genes are usually more heavily methylated in cells where they are not expressed. In addition, demethylating certain inactive genes (removing their extra methyl groups) turns them on.

At least in some species, DNA methylation seems to be essential for the long-term inactivation of genes that occurs during cellular differentiation in the embryo. In organisms as different as mice and *Arabidopsis* (a plant), deficient DNA methylation—resulting from lack of a methylating enzyme, for example—causes abnormalities in embryonic development. Once methylated, genes usually stay that way through successive cell divisions. Methylation enzymes act at DNA sites where one strand is already methylated and thus correctly methylate the daughter strand after each round of DNA replication. In this way, methylation patterns are passed on, and cells forming specialized tissues keep a chemical record of what occurred during embryonic development. A methylation pattern maintained in this way also accounts for **genomic imprinting** in mammals, where methylation permanently turns off either the maternal or paternal allele of certain genes at the start of development (see FIGURE 15.15).

Histone Acetylation

There is mounting evidence that histone acetylation and deacetylation play a direct role in the regulation of gene transcription. **Histone acetylation** is the attachment of acetyl groups ($—COCH_3$) to certain amino acids of histone proteins; deacetylation is the removal of acetyl groups. When the histones of a nucleosome are acetylated, they change shape so that they grip the DNA less tightly. As a result, transcription proteins have easier access to genes in the acetylated region. Researchers have shown that some enzymes that acetylate or deacetylate histones are closely associated with or even components of the transcription factors that bind to promoters (see FIGURE 17.7). In other words, histone acetylation and the initiation of gene transcription seem to be coupled structurally as well as functionally. Moreover, researchers have discovered that certain proteins that bind to methylated DNA recruit histone deacetylation enzymes, thus providing a mechanism by which DNA methylation and histone deacetylation can cooperate to repress transcription. In summary, key enzymes that modify chromatin structure are apparently integral parts of the cell's machinery for regulating transcription.

Transcription initiation is controlled by proteins that interact with DNA and with each other

If chromatin-modifying enzymes provide a coarse adjustment of gene expression by making a region of DNA either more available or less available for transcription, then fine-tuning begins with the interaction of transcription factors with DNA sequences that control specific genes. Indeed, once a gene is "unpacked," the initiation of transcription is the most important and universally used control point in gene expression. Before looking at control mechanisms, let's review the structure of a typical eukaryotic gene and its transcript.

Organization of a Typical Eukaryotic Gene

A eukaryotic gene and the DNA elements (segments) that control it are typically organized as shown in FIGURE 19.8, which reviews and extends what you learned about eukaryotic genes in Chapter 17. One striking difference between this gene and a prokaryotic gene is the presence of introns, noncoding sequences interspersed between coding sequences, the exons. Recall from Chapter 17 that a cluster of proteins called a *transcription initiation complex* assembles on the promoter sequence at the "upstream" end of the gene, and one of the proteins, RNA polymerase, proceeds to transcribe the gene. The introns are removed from the primary transcript during RNA processing, so they do not appear in the mature mRNA. The eukaryotic processing of pre-mRNA usually also includes the addition of a modified guanosine triphosphate cap to its 5′ end and a poly(A) tail to its 3′ end. Another distinctive feature of the eukaryotic gene illustrated in FIGURE 19.8 is the relatively large number of control elements associated with it. **Control elements** are simply segments of noncoding DNA that help regulate transcription of a gene by binding proteins—*transcription factors*. (Strictly speaking, control elements include promoters.)

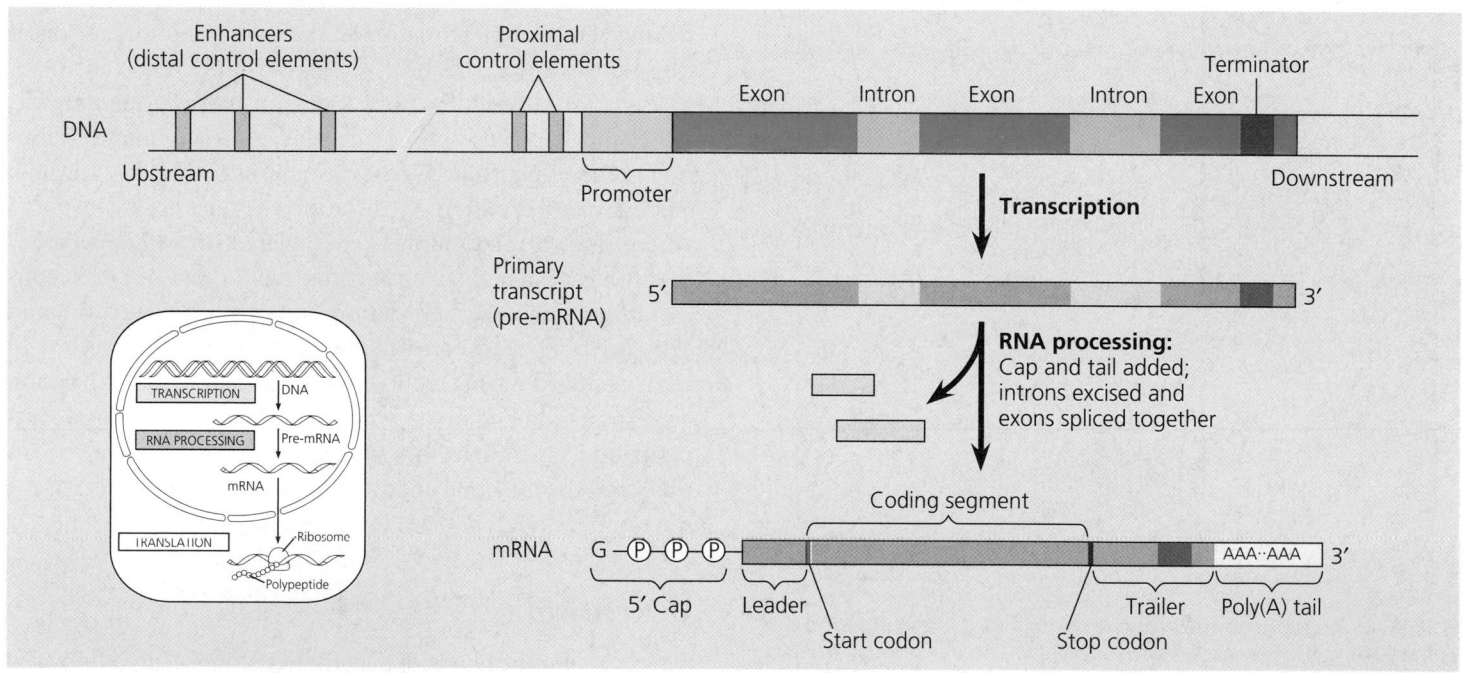

FIGURE 19.8 A eukaryotic gene and its transcript. Each functional eukaryotic gene has a promoter, a DNA sequence where RNA polymerase binds (with the help of transcription factors) and starts transcription, proceeding "downstream." The 3′ end of the transcript is determined by an RNA sequence encoded by a terminator located near the end of the last exon. RNA-processing enzymes immediately add a 5′ cap and then a poly(A) tail to the primary transcript, and a spliceosome excises the introns. The resulting mRNA, ready for export to the cytoplasm, includes leader and trailer regions that will not be translated. Involved in regulating the initiation of transcription are a number of control elements (orange), DNA sequences located near (*proximal* to) or far from (*distal* to) the promoter. Distal control elements called enhancers can lie either upstream or downstream of the gene.

The Role of Transcription Factors

As you learned in Chapter 17, eukaryotic RNA polymerase alone cannot initiate transcription of a gene; it is dependent on transcription factors. The transcription factors mentioned in Chapter 17 are essential for the transcription of *all* protein-coding genes (see FIGURE 17.7). Recall that only one of these transcription factors independently recognizes a DNA sequence, the TATA box within the promoter; the other transcription factors primarily recognize proteins, including each other and RNA polymerase. Protein-protein interactions are crucial to the initiation of eukaryotic transcription. Only when the complete initiation complex has assembled can the polymerase begin to move along the DNA template strand, producing a complementary strand of RNA.

However, the interaction of transcription factors and RNA polymerase with the promoter as summarized so far usually initiates transcription inefficiently, producing only a low rate of initiation and few RNA transcripts. The keys to high levels of eukaryotic transcription are the control elements. These DNA sequences greatly improve the efficiency of promoters by binding additional transcription factors.

As you can see in FIGURE 19.8, some of the control elements on the DNA are close to the promoter; in fact, some biologists include these *proximal control elements* in what they call the promoter. The more distant *distal control elements,* called **enhancers,** may be thousands of nucleotides away from the promoter and may even be downstream of the gene or within an intron. The associations between transcription factors and enhancers play an important role in the control of gene expression in eukaryotes. How do segments of DNA so far away from the promoter influence transcription? Apparently, bending of the DNA enables the transcription factors bound to enhancers to contact proteins of the transcription initiation complex at the promoter (FIGURE 19.9). A transcription factor that binds

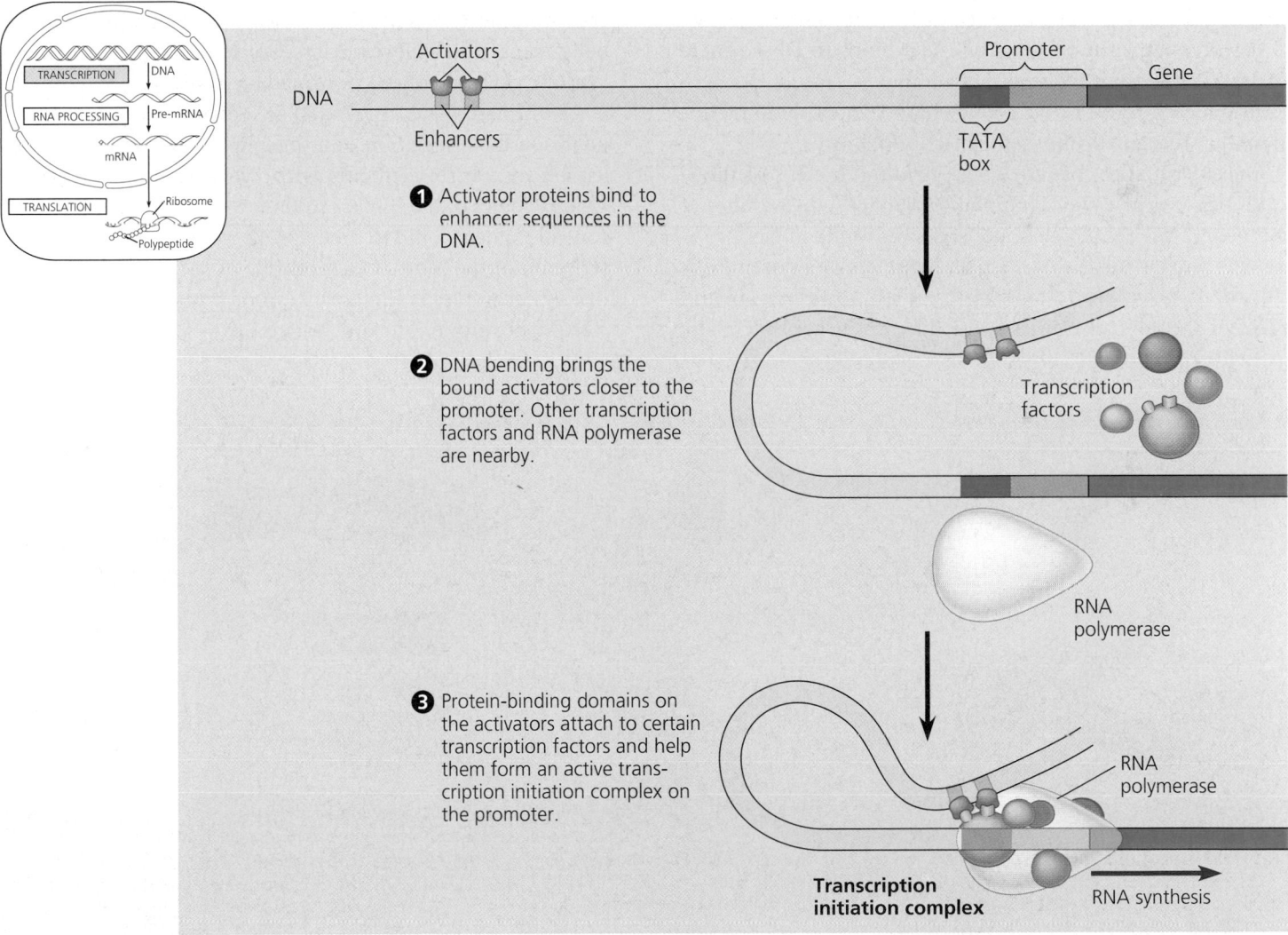

FIGURE 19.9 A model for enhancer action. Bending of the DNA enables enhancers to influence a promoter hundreds or even thousands of nucleotides away. Transcription factors called activators bind to the enhancer DNA sequences and then to proteins of the transcription initiation complex. In the process, they facilitate the correct positioning of the complex on the promoter and the initiation of RNA synthesis.

to an enhancer and stimulates transcription of a gene is termed an **activator**. Activators help position the initiation complex on the promoter.

Do eukaryotic cells also have transcription factors that function as *repressors,* as bacteria have? Yes, there is evidence for eukaryotic repressors. Some of them bind selectively to DNA control elements called *silencers,* which may be analogous to enhancers. Activators, though, are probably more important than such repressors, because the main regulatory mode in eukaryotic cells seems to be transcriptional activation of otherwise silent genes. The mechanisms responsible for eukaryotic gene silencing—that is, repression—may operate mostly at the level of chromatin modification (for example, DNA methylation).

In any case, the direct control of transcription largely depends on regulatory proteins that bind selectively to DNA and to other proteins. Hundreds of transcription factors have been discovered in eukaryotes. As numerous as these proteins are, however, they follow a few basic structural principles. A transcription factor generally has a **DNA-binding domain,** a part of its three-dimensional structure that binds to DNA. There are only a few major types of these domains (FIGURE 19.10). In addition, each transcription factor has a protein-binding domain that recognizes another transcription factor.

Considering the challenge of regulating the tens of thousands of genes in a typical animal or plant cell, the number of completely different nucleotide sequences found in DNA control elements is surprisingly small. Members of a dozen or so sequences about four to ten base pairs long appear again and again in the control elements for different genes. For many genes, the particular *combination* of control elements associated with the gene may be more important than the presence of a control element unique to the gene.

Coordinately Controlled Genes

How does the eukaryotic cell deal with genes of related function that need to be turned on or off at the same time? In Chapter 18, you learned that in prokaryotes, such coordinately controlled genes are often clustered into an operon; they lie adjacent to each other in the DNA molecule and share a promoter and other control elements located at the upstream end of the cluster. The genes of the operon are transcribed sequentially into a single mRNA molecule and are translated together. Such operons have not been found in eukaryotic cells, with rare exceptions. Genes coding for the enzymes of a metabolic pathway, for example, are often scattered over different chromosomes in a eukaryotic genome. Even when genes for related functions are located near one another on the same chromosome, each gene has its own promoter and is individually transcribed. Nevertheless, scattered collections of eukaryotic genes are often coordinately expressed.

Coordinate gene expression in eukaryotes probably depends on the association of a specific control element or collection of control elements with every gene of a dispersed group. Copies of the transcription factors that recognize these control elements bind to them, promoting simultaneous transcription of the genes. One example of such coordinate control in eukaryotes is the activation of a variety of genes by a steroid hormone. Steroid hormones—sex hormones, for

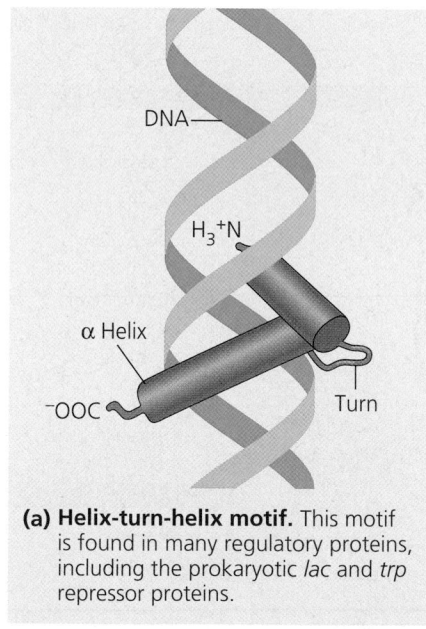

(a) Helix-turn-helix motif. This motif is found in many regulatory proteins, including the prokaryotic *lac* and *trp* repressor proteins.

(b) Zinc finger motif. Each "finger" consists of an α helix and a β sheet held together with the help of a zinc atom.

(c) Leucine zipper motif. Two α helices with regularly spaced leucines wrap around each other, joining two polypeptides together.

FIGURE 19.10 Three of the major types of DNA-binding domains in transcription factors. The parts of these structural motifs that interact with DNA are mostly α helices (shown as cylinders) that fit into the larger ("major") groove of the double helix. Variations in amino acid sequence within these helices are responsible for the recognition of specific DNA sequences by the protein.

example—have multiple effects on the body. As you may recall from FIGURE 11.10, a steroid hormone functions as a chemical signal by entering a cell and binding to a specific receptor protein in the cytoplasm or nucleus. When activated by steroid binding, a steroid receptor functions as a transcription activator. Every gene to be turned on by that hormone has a control element recognized by that activator.

Most other kinds of signal molecules, such as nonsteroid hormones or the extracellular signals exchanged by cells in a developing embryo, bind to receptors on the receiving cell's surface and never actually enter the cell. But they can control gene expression indirectly by triggering signal-transduction pathways that lead to the activation of particular transcription factors (see FIGURE 11.17). The principle of coordinate regulation remains the same: Genes with the same control elements are activated by the same chemical signals. Systems for coordinating gene regulation probably evolved by the duplication and distribution of control elements within the genome.

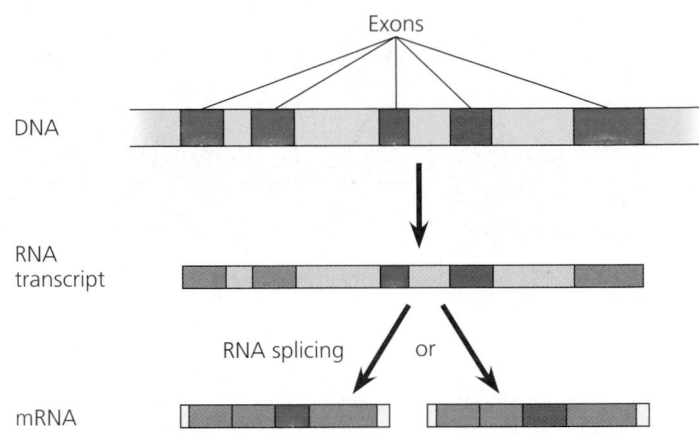

FIGURE 19.11 Alternative RNA splicing. The RNA transcripts of some genes can be spliced in more than one way, generating different mRNA molecules. Notice in this example that one mRNA molecule has ended up with the green exon and the other with the brown exon. With alternative splicing, an organism can get more than one type of polypeptide from a single gene.

Post-transcriptional mechanisms play supporting roles in the control of gene expression

Transcription alone does not constitute gene expression. The expression of a protein-coding gene is ultimately measured in terms of the amount of functional protein a cell makes, and much happens between the synthesis of the RNA transcript and the activity of the protein in the cell. In theory, gene expression may be blocked or stimulated at any post-transcriptional step (see FIGURE 19.7). By using regulatory mechanisms that operate after transcription, a cell can rapidly fine-tune gene expression in response to environmental changes without altering its transcription patterns.

RNA processing in the nucleus and the export of mature RNA to the cytoplasm provide several opportunities for controlling gene expression that are not available in bacteria. One example of regulation at the RNA-processing level is **alternative RNA splicing,** in which different mRNA molecules are produced from the same primary transcript, depending on which RNA segments are treated as exons and which as introns (FIGURE 19.11). Regulatory proteins specific to a cell type control intron-exon choices by binding to regulatory sequences within the primary transcript.

After RNA processing, other stages of gene expression that the cell may regulate are mRNA degradation, translation initiation, and protein processing and degradation.

Regulation of mRNA Degradation

The life span of mRNA molecules in the cytoplasm is an important factor in determining the pattern of protein synthesis in a cell. Prokaryotic mRNA molecules typically have very short lives; they are degraded by enzymes after only a few minutes. This is one reason bacteria can vary their patterns of protein synthesis so quickly in response to environmental changes. In contrast, mRNA lifetimes in multicellular eukaryotes are typically hours and can be days or even weeks. A striking example of long-lived mRNA is found in developing red blood cells, which are factories for the production of the protein hemoglobin. The mRNAs for the hemoglobin polypeptides (α-globin and β-globin) are unusually stable and are translated repeatedly in these cells.

Recent research on yeasts suggests that a common pathway of mRNA breakdown begins with the enzymatic shortening of the poly(A) tail. This helps trigger the action of enzymes that remove the 5′ cap (the two ends of the mRNA may be briefly held together by the proteins involved). The removal of the cap, a critical step, is also regulated by particular nucleotide sequences in the mRNA. Once the cap is removed, nuclease enzymes rapidly chew up the mRNA from its 5′ end.

Nucleotide sequences that affect mRNA stability are often found in the untranslated trailer region at the 3′ end of the molecule (see FIGURE 19.8). In one experiment, researchers transferred such a sequence from the short-lived mRNA for a growth factor to the 3′ end of a normally stable globin mRNA. The globin mRNA was quickly degraded.

Control of Translation

Most translational control mechanisms block the initiation stage of polypeptide synthesis, when ribosomal subunits and the initiator tRNA attach to an mRNA (see FIGURE 17.15). Translation of specific mRNAs can be blocked by regulatory proteins that bind to specific sequences or structures within the leader region at the 5′ end of the mRNA, preventing the

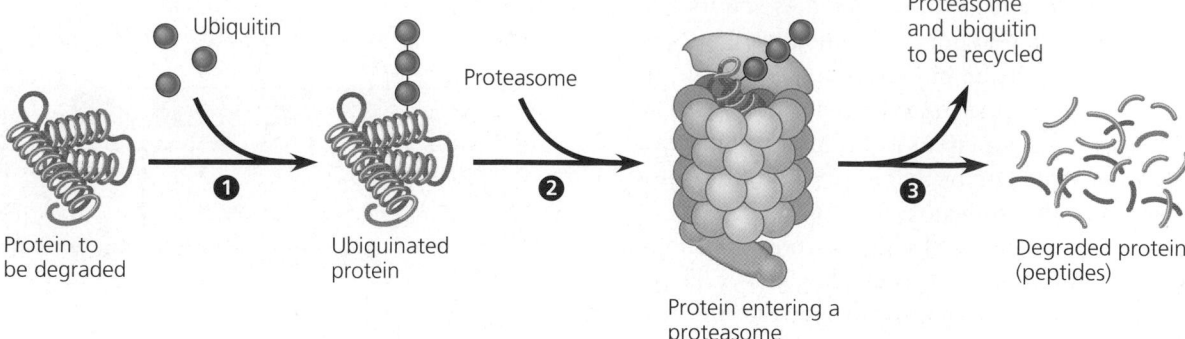

FIGURE 19.12 Degradation of a protein by a proteasome. A proteasome is an enormous protein complex, with a shape suggesting a trash can, that chops up unneeded proteins in the cell. In most cases, the proteins attacked by a proteasome have been tagged with short chains of ubiquitin, a small protein. ❶ Enzymes in the cytosol attach ubiquitin molecules to a protein. (It is not known how the protein is selected.) ❷ A proteasome recognizes the ubiquitinated protein, unfolds it, and sequesters it in its central cavity. ❸ Enzymatic components of the proteasome cut the protein into small peptides, which can be further degraded by cytosolic enzymes. Steps 1 and 3 require ATP. Eukaryotic proteasomes are as massive as ribosomal subunits and are distributed throughout the cell. Their barrel-like shape somewhat resembles that of chaperone proteins, which nurture protein structure rather than destroy it (see FIGURE 5.26).

attachment of ribosomes. This mechanism is important in embryonic development: The mother stores in the ovum a variety of mRNAs that are not translated until specific stages after fertilization.

Protein factors required to initiate translation in eukaryotes offer targets for simultaneously controlling the translation of *all* the mRNA in a cell. One example of such "global" control involves hemoglobin. A functional hemoglobin molecule has four heme groups, one attached to each polypeptide (see FIGURE 19.3). If hemes are in short supply in a developing red blood cell, a regulatory protein inactivates an essential translation initiation factor by phosphorylating it. This inhibits all translation, but mainly affects translation of hemoglobin mRNA, which constitutes most of the mRNA in the cell.

Global control of translation is important in embryonic development. For example, the egg cells of many organisms synthesize and store large numbers of mRNA molecules that are not translated until just after fertilization. At that point, translation is triggered by the sudden activation of translation initiation factors. The response is a burst of synthesis of particular proteins. Some plants and algae store mRNAs during periods of darkness; light then triggers the reactivation of the translational apparatus.

Protein Processing and Degradation

The final opportunities for controlling gene expression occur after translation. Often, eukaryotic polypeptides must be processed to yield functional protein molecules. The cleavage of the initial insulin polypeptide to form the active hormone is an example. In addition, many proteins require chemical modifications to function. For instance, proteins destined for the animal cell surface must acquire sugars, and regulatory proteins are commonly activated or inactivated by the reversible addition of phosphate groups. Moreover, polypeptides must often be transported to targeted destinations in the cell in order to function. Regulation might occur at any of the steps involved in modifying or transporting a protein.

Abnormal targeting of a protein can have serious consequences, as, for example, in cystic fibrosis. This disease results from mutations in the gene for a protein that functions as a chloride ion channel. The defective protein never reaches its final destination in the cell, the plasma membrane, and is rapidly degraded.

In addition to degrading defective and damaged proteins, the cell limits the lifetimes of normal proteins by selective degradation. Many proteins, such as the cyclins involved in regulating the cell cycle, must be relatively short-lived if the cell is to function appropriately (see FIGURE 12.14). To mark a particular protein for destruction, the cell commonly attaches molecules of a small protein called ubiquitin to the protein. Giant protein complexes called **proteasomes** then recognize the ubiquitin and degrade the tagged protein (FIGURE 19.12). The importance of proteasomes is underscored by the finding that mutations making cell cycle proteins impervious to proteasome degradation can lead to cancer.

THE MOLECULAR BIOLOGY OF CANCER

In Chapter 12, we considered cancer as a set of diseases in which cells escape from the control mechanisms normally limiting their growth. Now that we have discussed the molecular basis of gene expression and its regulation, we are ready to look at cancer more closely.

Cancer results from genetic changes that affect the cell cycle

Certain genes normally regulate cell growth and division—the cell cycle—and mutations that alter those genes in somatic cells can lead to cancer. The agent of such change can be random spontaneous mutation. However, it is likely that many cancer-causing mutations result from environmental influences such as chemical carcinogens, physical mutagens such as X-rays, or certain viruses. In fact, a breakthrough in understanding cancer came from the study of tumors induced by viruses. This research led to the discovery of cancer-causing genes called **oncogenes** in certain retroviruses (from the Greek *onco*, tumor). Subsequently, close counterparts of these oncogenes were found in the genomes of humans and other animals. The normal cellular genes, called **proto-oncogenes,** code for proteins that stimulate normal cell growth and division. (To review the cell cycle, see Chapter 12.)

How might a proto-oncogene—a gene that has an essential function in normal cells—become an oncogene, a cancer-causing gene? In general, an oncogene arises from a genetic change that leads to an increase in either the amount of the proto-oncogene's protein product or the intrinsic activity of each protein molecule. The genetic changes that convert proto-oncogenes to oncogenes fall into three main categories: movement of DNA within the genome, amplification of a proto-oncogene, and point mutation in a proto-oncogene (FIGURE 19.13). Malignant cells are frequently found to contain chromosomes that have broken and rejoined incorrectly, translocating fragments from one chromosome to another (see FIGURE 15.13). A proto-oncogene ending up in the joint region may now lie adjacent to an especially active promoter (or other control element) that increases transcription of the gene, making it an oncogene. An increase in gene expression can also arise when a proto-oncogene comes under the control of a more active promoter by transposition of either the gene or the promoter within a chromosome. The second main type of genetic change, amplification, increases the number of copies of the gene in the cell. The third possibility is a point mutation that changes the gene's protein product to one that is more active or more resistant to degradation than the normal protein. All these mechanisms can lead to abnormal stimulation of the cell cycle and put the cell on the path to malignancy.

In addition to mutations affecting growth-stimulating proteins, changes in genes whose normal products *inhibit* cell division also contribute to cancer. Such genes are called **tumor-suppressor genes** because the proteins they encode normally help prevent uncontrolled cell growth. Any mutation that decreases the normal activity of a tumor-suppressor protein may contribute to the onset of cancer, in effect stimulating growth through the absence of suppression. The protein products of tumor-suppressor genes have various functions. For instance, some tumor-suppressor proteins normally repair damaged DNA, a function that prevents the cell from accumulating cancer-causing mutations. Other tumor-suppressor proteins control the adhesion of cells to each other or to an extracellular matrix; proper cell anchorage is crucial in normal tissues—and often absent in cancers. Still other tumor-suppressor proteins are components of cell-signaling pathways that inhibit the cell cycle.

We'll encounter some important tumor-suppressor proteins and oncogene proteins in the next section, as we examine their roles in cell-signaling pathways.

Oncogene proteins and faulty tumor-suppressor proteins interfere with normal signaling pathways

Let's take a closer look at what the protein products of cancer genes do—or fail to do—in the cell. We will focus on the products of two key genes, the *ras* proto-oncogene and the *p53* tumor-suppressor gene. Mutations in these genes are very common in human cancers: *ras* is mutated in about 30% of human cancers; for *p53*, the frequency is close to 50%.

FIGURE 19.13 Genetic changes that can turn proto-oncogenes into oncogenes.

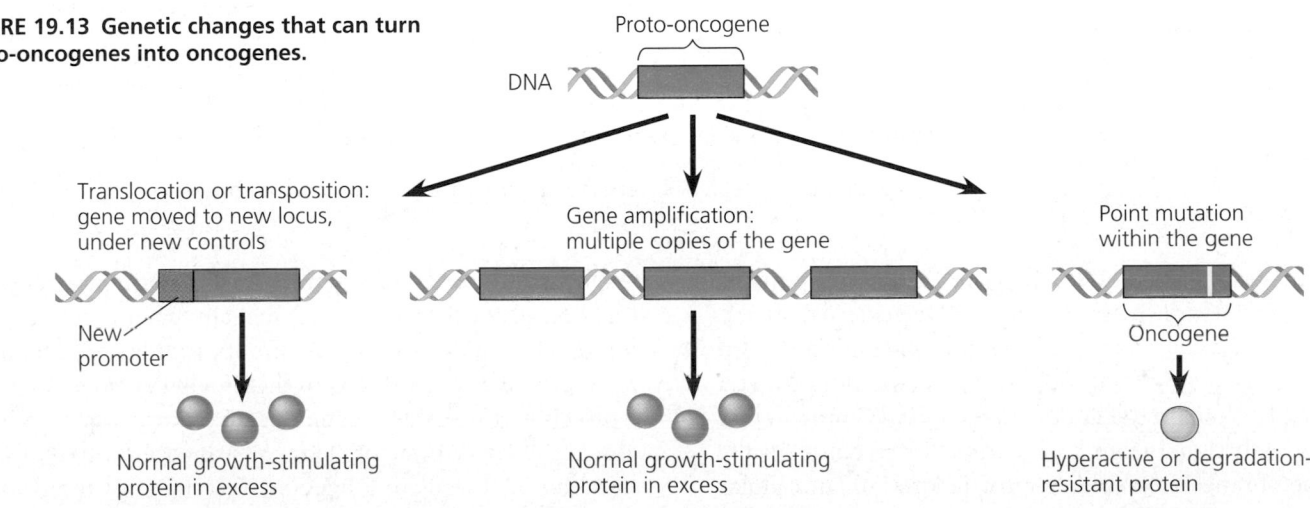

Proto-oncogene

DNA

Translocation or transposition: gene moved to new locus, under new controls

New promoter

Normal growth-stimulating protein in excess

Gene amplification: multiple copies of the gene

Normal growth-stimulating protein in excess

Point mutation within the gene

Oncogene

Hyperactive or degradation-resistant protein

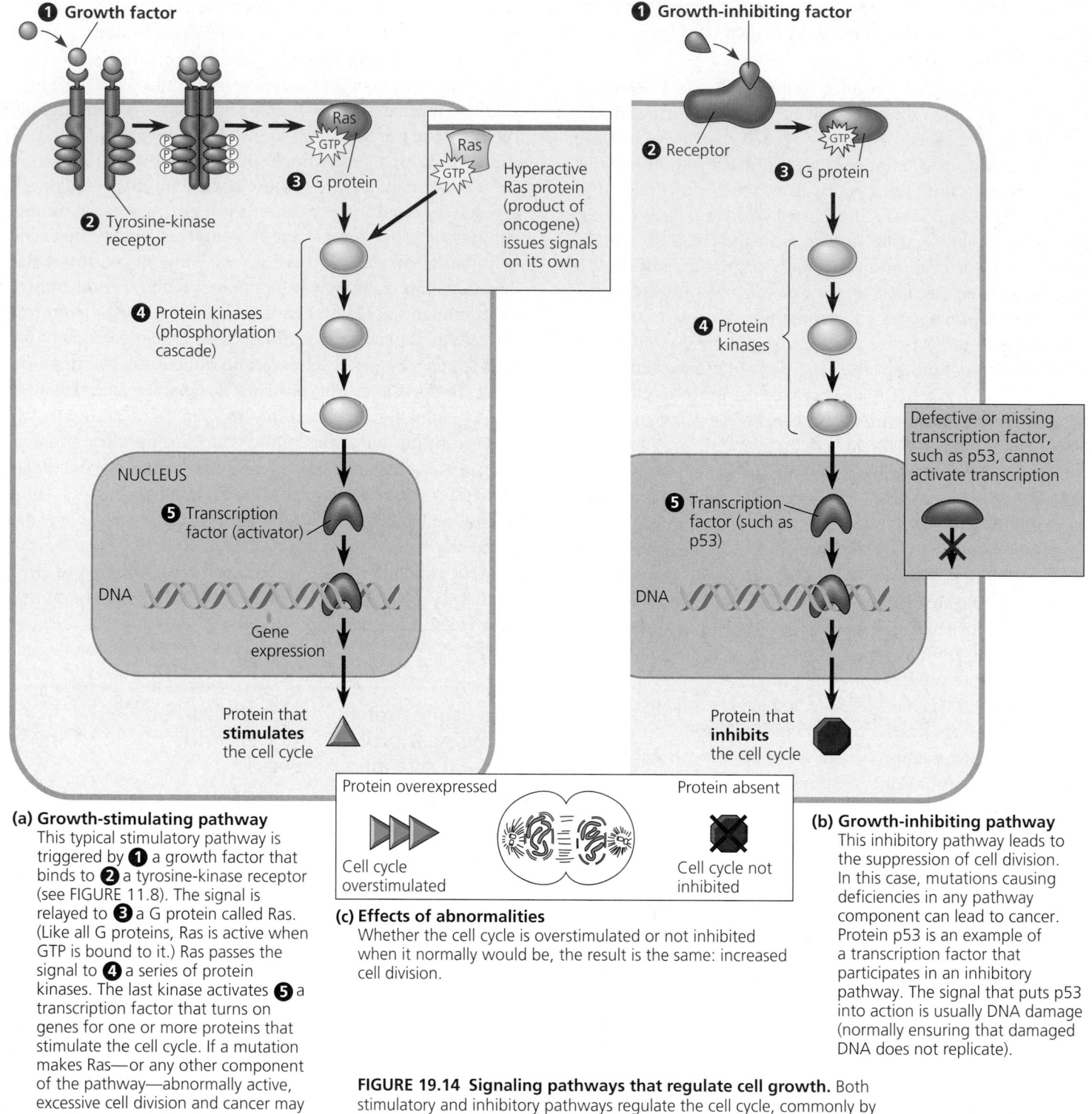

1 Growth factor

2 Tyrosine-kinase receptor

Ras GTP

3 G protein

Ras GTP Hyperactive Ras protein (product of oncogene) issues signals on its own

4 Protein kinases (phosphorylation cascade)

NUCLEUS

5 Transcription factor (activator)

DNA

Gene expression

Protein that **stimulates** the cell cycle

1 Growth-inhibiting factor

2 Receptor

GTP

3 G protein

4 Protein kinases

Defective or missing transcription factor, such as p53, cannot activate transcription

5 Transcription factor (such as p53)

DNA

Protein that **inhibits** the cell cycle

Protein overexpressed   Protein absent

Cell cycle overstimulated   Cell cycle not inhibited

(a) Growth-stimulating pathway
This typical stimulatory pathway is triggered by **1** a growth factor that binds to **2** a tyrosine-kinase receptor (see FIGURE 11.8). The signal is relayed to **3** a G protein called Ras. (Like all G proteins, Ras is active when GTP is bound to it.) Ras passes the signal to **4** a series of protein kinases. The last kinase activates **5** a transcription factor that turns on genes for one or more proteins that stimulate the cell cycle. If a mutation makes Ras—or any other component of the pathway—abnormally active, excessive cell division and cancer may result.

(c) Effects of abnormalities
Whether the cell cycle is overstimulated or not inhibited when it normally would be, the result is the same: increased cell division.

(b) Growth-inhibiting pathway
This inhibitory pathway leads to the suppression of cell division. In this case, mutations causing deficiencies in any pathway component can lead to cancer. Protein p53 is an example of a transcription factor that participates in an inhibitory pathway. The signal that puts p53 into action is usually DNA damage (normally ensuring that damaged DNA does not replicate).

FIGURE 19.14 Signaling pathways that regulate cell growth. Both stimulatory and inhibitory pathways regulate the cell cycle, commonly by influencing transcription. Cancer can result from aberrations in such pathways.

Both the Ras protein and the p53 protein are components of signal-transduction pathways that convey external signals to the DNA in the cell's nucleus. You saw a generalized version of such a pathway in FIGURE 11.17. Now, in FIGURE 19.14a you can see that Ras, the product of the *ras* gene, is a G protein that relays a growth signal from a growth factor receptor on the plasma membrane to a cascade of protein kinases. The cellular response at the end of the pathway is the synthesis of a protein that stimulates the cell cycle. Normally, such a pathway will not operate unless triggered by the appropriate growth factor. However, an oncogene protein that is a hyperactive version of a protein in the pathway can increase cell division even in the absence of growth factor. Many *ras* oncogenes have a point mutation that leads to a hyperactive version of the Ras protein, one that issues signals on its own. In fact, hyperactive versions or excess amounts of any of the pathway's components can have the same outcome: excessive cell division.

FIGURE 19.14b shows a comparable growth-*inhibiting* pathway, in which a growth-inhibiting signal leads to the synthesis of a protein that suppresses the cell cycle. Thus, the genes for the components of the pathway act as tumor-suppressor genes (as do genes whose products interfere with the growth-stimulating pathway). The tumor-suppressor protein encoded by the wild-type *p53* gene is a transcription factor that promotes the synthesis of growth-inhibiting proteins. That is why a mutation knocking out the *p53* gene can lead to excessive cell growth and cancer.

The modestly titled **p53 gene,** named for the 53,000-dalton molecular weight of its protein product, is often called the "guardian angel of the genome." Damage to the cell's DNA acts as a signal that leads to the expression of the *p53* gene. Once made, the p53 protein functions as a transcription factor for several genes. Often it activates a gene called *p21*, whose product halts the cell cycle by binding to cyclin-dependent kinases, allowing time for the cell to repair the DNA; the protein p53 can also turn on genes directly involved in DNA repair. When DNA damage is irreparable, p53 activates "suicide" genes, whose protein products cause cell death by a process called *apoptosis* (see FIGURE 21.18). Thus, in at least three ways, p53 prevents a cell from passing on mutations due to DNA damage. If mutations do accumulate and the cell survives through many divisions —as is more likely if the *p53* tumor-suppressor gene is defective or missing—cancer may ensue.

Multiple mutations underlie the development of cancer

More than one somatic mutation is generally needed to produce all the changes characteristic of a full-fledged cancer cell. This may help explain why the incidence of cancer increases greatly with age. If cancer results from an accumulation of mutations and if mutations occur throughout life, then the longer we live, the more likely we are to develop cancer.

The model of a multi-step path to cancer is well supported by studies of one of the best understood types of human cancer, colorectal cancer. About 135,000 new cases of colorectal cancer are diagnosed each year in the United States. Like most cancers, colorectal cancer develops gradually (FIGURE 19.15). The first sign is often a polyp, which is a small, benign growth in the colon lining. The cells of the polyp look normal, although they divide unusually frequently. The tumor grows and may eventually become malignant. The development of a malignant tumor is paralleled by a gradual accumulation of mutations that activate oncogenes and knock out tumor-suppressor genes. A *ras* oncogene and a mutated *p53* tumor-suppressor gene are often involved.

About a half dozen changes must occur at the DNA level for a cell to become fully cancerous. These usually include the appearance of at least one active oncogene and the mutation or loss of several tumor-suppressor genes. Furthermore, since mutant tumor-suppressor alleles are usually recessive, mutations must knock out *both* alleles in a cell's genome to block tumor suppression. (Most oncogenes, on the other hand, behave as dominant alleles.) Finally, in many malignant tumors, the gene for telomerase is activated. This enzyme prevents the erosion of the ends of the chromosomes, thus removing a natural limit on the number of times the cells can divide (see pp. 299–301).

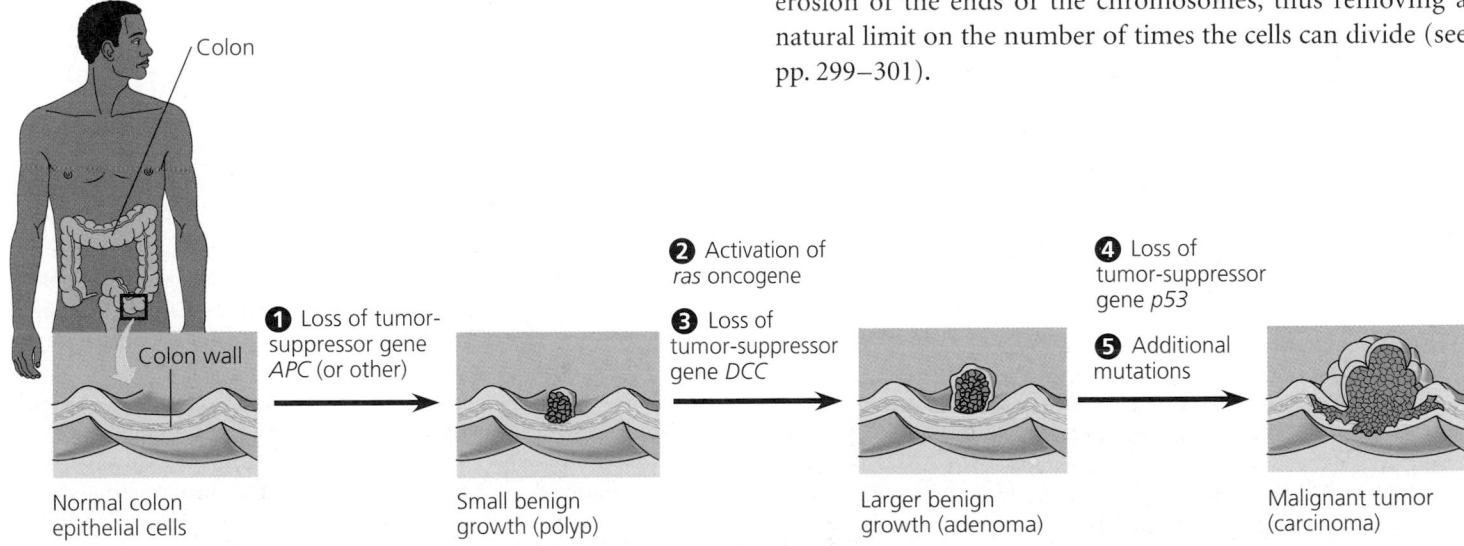

FIGURE 19.15 A multi-step model for the development of colorectal cancer. Affecting the colon and/or rectum, this type of cancer is one of the best understood. Changes in a tumor parallel a series of genetic changes, including mutations affecting several tumor-suppressor genes (such as *p53*) and the *ras* proto-oncogene. Mutations of tumor-suppressor genes often entail loss (deletion) of the gene. *APC* stands for "adenomatous polyposis coli"; mutant *APC* alleles are carried in many families predisposed to colorectal cancer. *DCC* stands for "deleted in colorectal cancer." Other mutation sequences can also lead to cancer.

Viruses seem to play a role in about 15% of human cancer cases worldwide. For example, retroviruses are involved in some types of leukemia, hepatitis viruses can cause liver cancer, and certain wart viruses promote the development of cancer of the cervix. Viruses contribute to cancer development by integrating their genetic material into the DNA of infected cells. By this process, a retrovirus may donate an oncogene to the cell. Alternatively, viral DNA may insert into a cellular genome in a way that disrupts a tumor-suppressor gene or converts a proto-oncogene to an oncogene.

The fact that multiple genetic changes are required to produce a cancer cell helps explain the predispositions to cancer that run in some families. An individual inheriting an oncogene or a mutant allele of a tumor-suppressor gene will be one step closer to accumulating the necessary mutations for cancer to develop.

Geneticists are now devoting much effort to finding inherited cancer alleles so that predisposition to certain cancers can be detected early in life. About 15% of colorectal cancers, for example, involve inherited mutations. Some of these affect DNA repair genes. Many others affect a tumor-suppressor gene called *APC* (see FIGURE 19.15). This gene seems to have multiple functions in the cell, including regulation of cell migration and adhesion. In 1997, researchers discovered a previ-

ously unrecognized mutation in *APC* carried by 6% of Ashkenazic Jews, making it the most common cancer-predisposing mutation yet found in a defined ethnic group.

Inherited alleles involved in breast cancer are also being scrutinized. Understanding the genetic basis of breast cancer is particularly important because it is the second most common type of cancer in the United States, striking over 180,000 women (and some men) each year. In 5–10% of breast cancer cases, there is evidence of a strong inherited predisposition. In 1994 and 1995, researchers identified two genes involved in these breast cancers, *BRCA1* and *BRCA2* (*BRCA* stands for BReast CAncer). Mutations in either gene increase the risk of developing breast cancer, and of ovarian cancer as well. Both genes are considered tumor-suppressor genes because their wild-type alleles protect against breast cancer and because their mutant alleles are recessive. However, determining what the normal products of *BRCA1* and *BRCA2* actually do in the cell has proved very difficult. One hypothesis is that one or both proteins are involved in DNA repair. Whatever the answer, the study of these and other genes associated with inherited cancer may lead to new methods for early diagnosis and treatment of all cancers. The DNA technologies you will learn about in the next chapter figure prominently in this research.

CHAPTER 19 REVIEW

Go to the Campbell Biology website (www.campbellbiology.com) to explore an interactive version of the Chapter Review.

Summary of Key Concepts

EUKARYOTIC CHROMATIN STRUCTURE

- **Chromatin structure is based on successive levels of DNA packing (pp. 354–356, FIGURE 19.1)** Eukaryotic chromatin is composed mostly of DNA and histone proteins that bind to the DNA to form nucleosomes, the most basic units of DNA packing. Additional folding leads ultimately to highly compacted heterochromatin, the form of chromatin in a metaphase chromosome. In interphase cells, most chromatin is in a highly extended form, euchromatin.

Web/CD Activity 19A: *DNA Packing*

GENOME ORGANIZATION AT THE DNA LEVEL

- **Repetitive DNA and other noncoding sequences account for much of a eukaryotic genome (pp. 357–358, TABLE 19.1)** Sequences of up to ten nucleotide pairs tandemly repeated thousands of times are especially prominent in the DNA of centromeres and telomeres, where

they probably play structural roles in the chromosome. The units of interspersed repetitive DNA are typically longer, and most may be transposons or retrotransposons.

- **Gene families have evolved by duplication of ancestral genes (pp. 358–359, FIGURE 19.3)** The ribosomal RNA genes, a tandemly arranged family of identical genes, enable the cell to make millions of ribosomes quickly. The nonidentical genes in the two globin gene families encode polypeptides used at different developmental stages of the animal.

- **Gene amplification, loss, or rearrangement can alter a cell's genome during an organism's lifetime (pp. 359–362, FIGURES 19.5, 19.6)** The selective replication of certain genes, such as rRNA genes, occurs during development in some species, further bolstering the cells' ability to make rRNA in large amounts. Some species selectively lose entire chromosomes or parts of chromosomes in certain cells. Rearrangements of DNA in a somatic cell by transposons or retrotransposons can alter the cell's genome in ways that affect gene expression. In vertebrates, a rearrangement and selective deletion of DNA segments in differentiating B lymphocytes accounts for antibody diversity.

Web/CD Activity 19B: *Gene Amplification, Loss, and Rearrangement*

THE CONTROL OF GENE EXPRESSION

■ **Each cell of a multicellular eukaryote expresses only a small fraction of its genes (p. 362)** In particular, selective control of genes is required for cellular differentiation.

■ **The control of gene expression can occur at any step in the pathway from gene to functional protein:** *an overview* (p. 362, FIGURE 19.7) Key stages include chromatin changes, initiation of transcription, RNA processing, translation, modification of the protein, and degradation of mRNA and protein.
Web/CD Activity 19C: *Overview: Control of Gene Expression*

■ **Chromatin modifications affect the availability of genes for transcription (pp. 362–364)** DNA methylation seems to diminish transcription of that DNA. Histone acetylation seems to loosen nucleosome structure and thereby enhance transcription.

■ **Transcription initiation is controlled by proteins that interact with DNA and with each other (pp. 364–367, FIGURES 19.8, 19.9)** Multiple DNA control elements that bind specific transcription factors regulate the formation of a transcription initiation complex at a eukaryotic promoter. But protein-protein interactions are as important as protein-DNA binding in regulating eukaryotic genes. Bending of DNA enables transcription factors bound to control elements (enhancers) distant from the promoter to contact proteins at the promoter. Unlike the genes of a prokaryotic operon, coordinately controlled eukaryotic genes each have a promoter and control elements; the same regulatory sequences are common to all the genes of a group, enabling recognition by the same transcription factors.
Web/CD Activity 19D: *Control of Transcription*
Web/CD Case Study in the Process of Science: *How Do You Design a Gene Expression System?*

■ **Post-transcriptional mechanisms play supporting roles in the control of gene expression (pp. 367–368, FIGURES 19.11, 19.12)** Regulation at the RNA-processing level is exemplified by alternative RNA splicing. Also, each mRNA has a characteristic life span, which sequences in the leader and trailer regions help determine. The initiation of translation can be controlled via regulation of initiation factors. After translation, various types of protein processing (such as cleavage and the addition of chemical groups) are subject to control, as is the degradation of the protein.
Web/CD Activity 19E: *Post-Transcriptional Control Mechanisms*
Web/CD Activity 19F: *Review: Control of Gene Expression*

THE MOLECULAR BIOLOGY OF CANCER

■ **Cancer results from genetic changes that affect the cell cycle (p. 369, FIGURE 19.13)** The products of proto-oncogenes and tumor-suppressor genes control cell division. A DNA change that makes a proto-oncogene excessively active converts it to an oncogene, which may promote excessive cell growth and cancer. A tumor-suppressor gene encodes a protein that inhibits abnormal cell division. The loss or mutation of such a gene has effects similar to the activation of an oncogene.

■ **Oncogene proteins and faulty tumor-suppressor proteins interfere with normal signaling pathways (pp. 369–371, FIGURE 19.14)** Typical components of both growth-stimulating and growth-inhibiting pathways include growth factors, membrane receptors, G proteins (such as Ras), protein kinases, and transcription activators. A hyperactive version of a protein in a stimulatory pathway, such as Ras (a G protein), functions as an oncogene protein. A defective version of a protein in an inhibitory pathway, such as p53 (a transcription activator), fails to function as a tumor suppressor.

■ **Multiple mutations underlie the development of cancer (pp. 371–372, FIGURE 19.15)** Tumor cells accumulate changes affecting proto-oncogenes and tumor-suppressor genes. Some of these mutations can be inherited, resulting in a predisposition to developing certain types of cancer.
Web/CD Activity 19G: *Causes of Cancer*

Self-Quiz

1. In a nucleosome, the DNA is wrapped around
   a. polymerase molecules.
   b. ribosomes.
   c. histones.
   d. the nucleolus.
   e. satellite DNA.

2. Apparently, our muscle cells are different from our nerve cells mainly because
   a. they express different genes.
   b. they contain different genes.
   c. they use different genetic codes.
   d. they have unique ribosomes.
   e. they have different chromosomes.

3. One of the unique characteristics of retrotransposons is that
   a. translation of their RNA transcript produces an enzyme that converts the RNA back to DNA.
   b. they are found only in animal cells.
   c. once removed from the DNA, the gene segments for an antibody variable region are rejoined to the constant region.
   d. they contribute a significant portion of the genetic variability seen within a population of gametes.
   e. their amplification is dependent on a concurrent retrovirus infection.

4. The functioning of enhancers is an example of
   a. transcriptional control of gene expression.
   b. a post-transcriptional mechanism for editing mRNA.
   c. the stimulation of translation by initiation factors.
   d. post-translational control that activates certain proteins.
   e. a eukaryotic equivalent of prokaryotic promoter functioning.

5. Multigene families are
   a. groups of enhancers that control transcription.
   b. usually clustered at the telomeres.
   c. equivalent to the operons of prokaryotes.
   d. collections of genes whose expression is controlled by the same regulatory proteins.
   e. identical or similar genes that have evolved by gene duplication.

6. Which of the following statements about the DNA in one of your brain cells is true?
   a. Some DNA sequences are present in multiple copies.
   b. Most of the DNA codes for protein.
   c. The majority of genes are likely to be transcribed.
   d. Each gene lies immediately adjacent to an enhancer that helps control transcription.
   e. Many genes are grouped into operon-like clusters.

7. Rearrangement of DNA segments is known to occur in genes coding for
   a. ribosomal RNA.
   b. most proteins in eukaryotes.
   c. hemoglobin.
   d. histone proteins.
   e. antibodies.

8. Which of the following is an example of a possible step in the post-transcriptional control of gene expression?
   a. the addition of methyl groups to cytosine bases of DNA
   b. the binding of transcription factors to a promoter
   c. the removal of introns and splicing together of exons
   d. gene amplification during a stage in development
   e. the folding of DNA to form heterochromatin

9. The amount of protein made from a given mRNA molecule depends partly on
   a. the degree of DNA methylation.
   b. the rate at which the mRNA is degraded.
   c. the presence of certain transcription factors.
   d. the number of introns present in the mRNA.
   e. the types of ribosomes present in the cytoplasm.

10. All of our cells contain proto-oncogenes, which can change into oncogenes that cause cancer. Which of the following is the best explanation for the presence of these potential time bombs in our cells?
    a. Proto-oncogenes first arose from viral infections.
    b. Proto-oncogenes normally help regulate cell division.
    c. Proto-oncogenes are genetic "junk."
    d. Proto-oncogenes are mutant versions of normal genes.
    e. Cells produce proto-oncogenes as a by-product of the aging process.

11. In general, how does dense packing of DNA in chromosomes prevent gene expression?

12. In stimulating transcription of a specific eukaryotic gene, an enhancer does not act directly on the gene's promoter, but has its effect via DNA-binding proteins called _____ _____.

13. Once mRNA encoding a particular protein reaches the cytoplasm, what are four mechanisms that can regulate the amount of the active protein in the cell?

14. In what sense is cancer always a genetic disease?

15. In what respect is the term *proto-oncogene* misleading?

16. How can a mutation in a tumor-suppressor gene contribute to the development of cancer?

17. Why is most breast cancer considered nonhereditary?

Go to the website or CD-ROM for more quiz questions.

Evolution Connection

One of the revelations of the human genome sequence draft announced in February 2001 was the presence of relict prokaryotic sequences—genes of prokaryotes incorporated into our genome but now defunct molecular fossils. What kinds of mutational changes may have occurred to maroon bacterial genes in our genome?

The Process of Science

Researchers have long puzzled over the observation that some prostate cancer cells thrive despite treatments that eliminate testosterone and other androgens usually necessary for prostate cell survival. It has recently been suggested that estrogen, often considered a female hormone, may be activating genes that are normally controlled by an androgen receptor. Describe one or more experiments to test this hypothesis. (The hormones mentioned are all steroids; to review their mode of action, see FIGURE 11.10.)

Design a eukaryotic gene expression system to investigate the role of the aldehyde reductase promoter in the Case Study in the Process of Science, available on the website and CD-ROM.

Science, Technology, and Society

A chemical called dioxin, or TCDD, is produced as a by-product of some chemical manufacturing processes. Trace amounts of this substance were present in Agent Orange, a defoliant sprayed on vegetation during the Vietnam War. There has been a continuing controversy over its effects on soldiers exposed to it during the war. Animal tests have suggested that dioxin can be lethal and can cause birth defects, cancer, liver and thymus damage, and immune system suppression. But its effects on humans are unclear, and even animal tests are equivocal; a hamster is not affected by a dose that can kill a guinea pig. Researchers have discovered that dioxin acts somewhat like a steroid hormone. It enters a cell and binds to a receptor protein, which in turn attaches to the cell's DNA. How might this mechanism help explain the variety of dioxin's effects on different body systems and in different animals? How might you determine whether a type of illness is related to dioxin exposure or whether a particular individual became ill as a result of exposure to dioxin? Which would be more difficult to demonstrate? Why?

Answers: 1. c; 2. a; 3. a; 4. a; 5. e; 6. a; 7. e; 8. c; 9. b; 10. b; 11. RNA polymerase and other proteins required for transcription do not have access to the DNA in tightly packed regions of a chromosome. 12. transcription factors 13. Breakdown of the mRNA; regulation of translation; activation of the protein (by polypeptide cutting, for example); and breakdown of the protein 14. It is caused by changes in genes. 15. The term does not describe the *normal* function of the gene, which is generally the regulation of cell division. The conversion of a proto-oncogene to an oncogene is an aberration. 16. A mutated tumor-suppressor gene may produce a defective protein unable to function in a pathway that normally inhibits cell division (that is, normally suppresses tumors). 17. Most breast cancers are associated with somatic mutations, not inherited mutations that are passed from parents to offspring via gametes.

DNA TECHNOLOGY AND GENOMICS

DNA CLONING

- DNA technology makes it possible to clone genes for basic research and commercial applications: *an overview*
- Restriction enzymes are used to make recombinant DNA
- Genes can be cloned in recombinant DNA vectors: *a closer look*
- Cloned genes are stored in DNA libraries
- The polymerase chain reaction (PCR) clones DNA entirely *in vitro*

DNA ANALYSIS AND GENOMICS

- Restriction fragment analysis detects DNA differences that affect restriction sites
- Entire genomes can be mapped at the DNA level
- Genome sequences provide clues to important biological questions

PRACTICAL APPLICATIONS OF DNA TECHNOLOGY

- DNA technology is reshaping medicine and the pharmaceutical industry
- DNA technology offers forensic, environmental, and agricultural applications
- DNA technology raises important safety and ethical questions

The photo on this page shows *a room of machines that are sequencing human DNA at the rate of 350,000 nucleotides per machine per day. The mapping and sequencing of the human genome, one of the great achievements of modern science, has been made possible by advances in DNA technology, starting with the invention of methods for making* **recombinant DNA**. *This is DNA in which genes from two different sources—often different species—are combined* in vitro *into the same molecule.*

The methods for making recombinant DNA are also central to **genetic engineering,** *the direct manipulation of genes for practical purposes. Applications include the manufacture of hundreds of products. Using DNA technology, scientists can make recombinant DNA and then introduce it into cultured cells that replicate the DNA and may express its genes, yielding a desired protein. The bacterium* E. coli *is often used as a "host"*

organism for recombinant DNA because it is easy to grow and its biochemistry is well understood.

DNA technology has launched a revolution in biotechnology. In broad terms, **biotechnology** *is the manipulation of organisms or their components to make useful products. Practices that go back centuries, such as the use of microbes to make wine and cheese and the selective breeding of livestock, are examples of biotechnology. Natural genetic processes, such as mutation and genetic recombination, have always been involved. But biotechnology based on the manipulation of DNA* in vitro *differs from earlier practices by enabling scientists to modify specific genes and move them between organisms as distinct as bacteria, plants, and animals.*

DNA technology is now applied in areas ranging from agriculture to criminal law, but its most important achievements have been in basic research. DNA technology has stimulated discoveries in virtually all fields of biology by giving researchers new tools for tackling age-old questions. And it is providing us with detailed knowledge of the genomes of humans and many other organisms, knowledge largely inaccessible only a few decades ago.

In this chapter, we describe the main techniques of DNA manipulation, discuss how genomes are analyzed at the DNA level, and survey the practical applications of DNA technology. In a concluding section, we consider some of the social and ethical issues that arise as DNA technology becomes more pervasive in our lives.

DNA CLONING

The molecular biologist studying a particular gene faces a challenge. Naturally occurring DNA molecules are very long, and a single molecule usually carries many genes. Moreover, genes may occupy only a small proportion of the chromosomal DNA, the rest being noncoding nucleotide sequences. A human gene, for example, might constitute only $\frac{1}{100,000}$ of a chromosomal DNA molecule. As a further complication, the distinctions between a gene and the surrounding DNA are subtle, consisting only of differences in nucleotide sequence. To work directly with specific genes, scientists needed to develop methods for preparing well-defined, gene-sized pieces of DNA in multiple identical copies. In other words, they needed techniques for **gene cloning.**

DNA technology makes it possible to clone genes for basic research and commercial applications: *an overview*

Most methods for cloning pieces of DNA share certain general features. In FIGURE 20.1, an overview of gene cloning and its applications, we consider an approach that uses bacteria and their plasmids. Recall from Chapter 18 that plasmids are small,

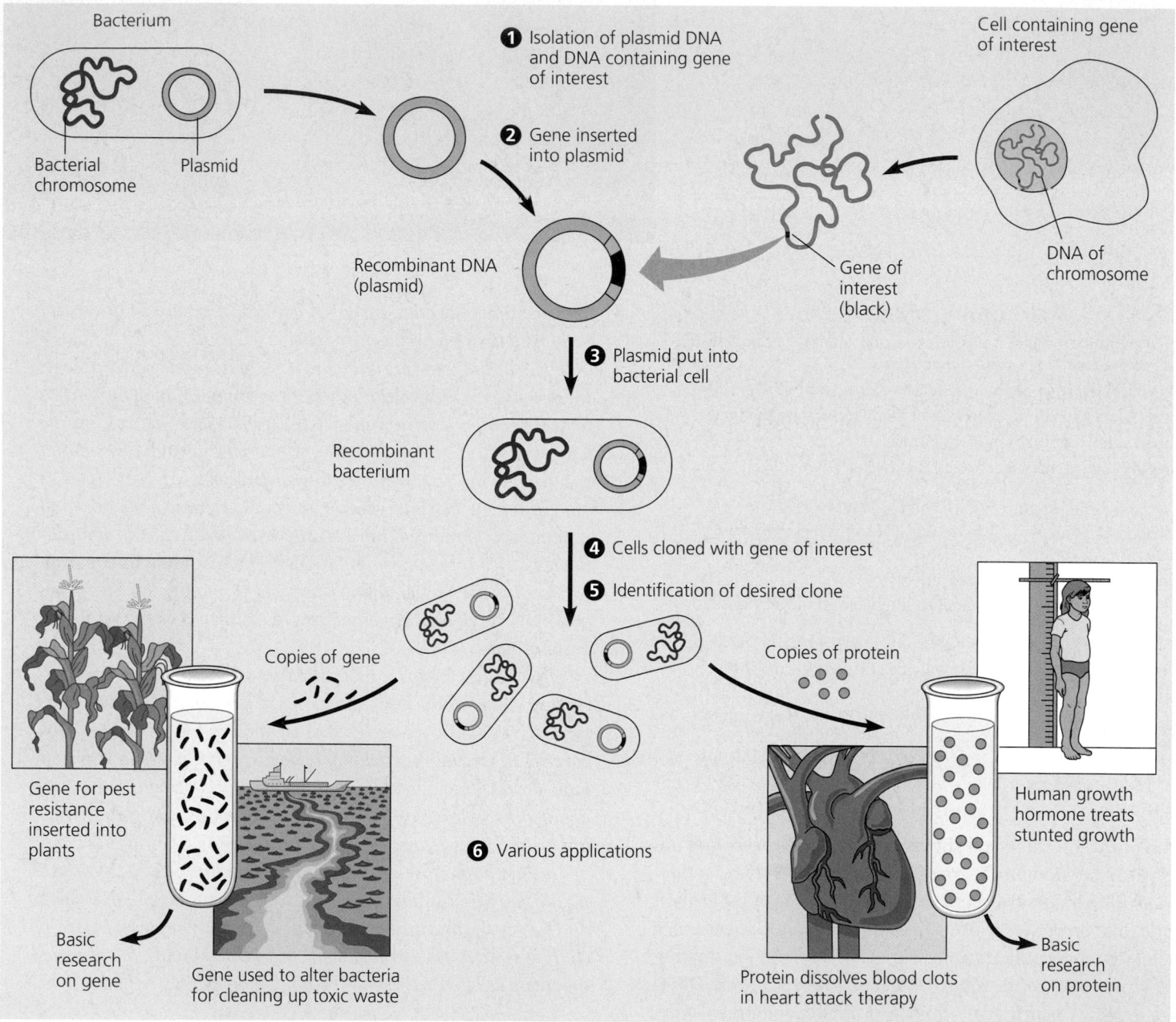

FIGURE 20.1 An overview of how bacterial plasmids are used to clone genes. ❶ A scientist isolates plasmid DNA from bacteria and DNA carrying a gene of interest from cells of another organism, such as an animal. ❷ A piece of DNA containing the gene is inserted into a plasmid, producing recombinant DNA, and ❸ the recombinant plasmid is returned to a bacterial cell. ❹ This cell is then grown in culture, forming a clone of cells. The foreign DNA spliced into the plasmid is replicated with the rest of the plasmid as the host cell multiplies. In this way, the gene of interest is "cloned." ❺ A critical step in gene cloning is the identification of the bacterial clone carrying the gene of interest. ❻ The drawings show some current applications of gene cloning in bacteria. In the examples on the left, a gene that originated in one organism is used to give another organism a new metabolic capability. In the cases on the right, useful protein products are harvested in large quantities from bacterial cultures. In this and later figures, the bacterial chromosomes and plasmids are not to scale; relative to the cells, the chromosomes are actually much larger and the plasmids much smaller.

circular DNA molecules that replicate within bacterial cells, separate from the bacterial chromosome. For cloning genes or other pieces of DNA, plasmids are first isolated from bacterial cells. FIGURE 20.1 follows one plasmid as a foreign gene is inserted into it. The plasmid is now a recombinant DNA molecule, combining DNA from two sources. The plasmid is returned to a bacterial cell, which then reproduces to form a cell clone. The foreign gene carried by the plasmid is "cloned" at the same time, for the dividing bacteria continue to replicate the recombinant plasmid and pass it on to their descendants. Under suitable conditions, the bacterial clone will make the protein encoded by the foreign gene.

The potential uses of cloned genes fall into two general categories. The goal may be to produce a protein product, either for study or for some practical use. For example, pharmaceutical companies use bacteria carrying the gene for human growth hormone to produce large quantities of the hormone for treating stunted growth. Alternatively, the goal may be to prepare many copies of the gene itself. A scientist may want to determine the gene's nucleotide sequence or use the gene to endow an organism with a new metabolic capability. For example, a cloned gene for pest resistance originating in one species of crop plant might be transferred into plants of another species. Most genes exist in only one copy per genome—something on the order of one part per million of DNA—so the ability to clone such rare DNA fragments is extremely valuable. In the remainder of this chapter, we take a closer look at the steps in FIGURE 20.1 and at related methods.

Restriction enzymes are used to make recombinant DNA

THE PROCESS OF SCIENCE Gene cloning and genetic engineering were made possible by the discovery of enzymes that cut DNA molecules at a limited number of specific locations. These enzymes, called **restriction enzymes,** were discovered in the late 1960s by researchers studying bacteria. In nature, these enzymes protect the bacteria against intruding DNA from other organisms, such as phages or other bacterial cells. They work by cutting up the foreign DNA, a process called *restriction.* Most restriction enzymes are very specific, recognizing short nucleotide sequences in DNA molecules and cutting at specific points within these sequences. The bacterial cell protects its own DNA from restriction by adding methyl groups ($—CH_3$) to adenines or cytosines within the sequences recognized by the restriction enzyme. Hundreds of different restriction enzymes have been identified and isolated.

The top of FIGURE 20.2 is a diagram of a DNA molecule containing a recognition sequence, or **restriction site,** for a particular restriction enzyme. As shown in this example, most restriction sites are symmetrical: The same $5' \rightarrow 3'$ sequence of four to eight nucleotides (here, six) is found on both strands, thus running in opposite directions (antiparallel). Restriction enzymes cut covalent phosphodiester bonds of both strands, often in a staggered way, as indicated in the diagram. Because the target sequence usually occurs (by chance) many times in a long DNA molecule, an enzyme will make many cuts. Copies of a DNA molecule always yield the same set of **restriction fragments** when exposed to that enzyme. In other words, a restriction enzyme cuts a DNA molecule in a reproducible way. (Later you will learn how the different fragments can be separated.)

Notice in FIGURE 20.2 that the restriction fragments are double-stranded DNA fragments with at least one single-stranded end, called a **sticky end.** These short extensions will form hydrogen-bonded base pairs with complementary single-stranded stretches on other DNA molecules cut with the same

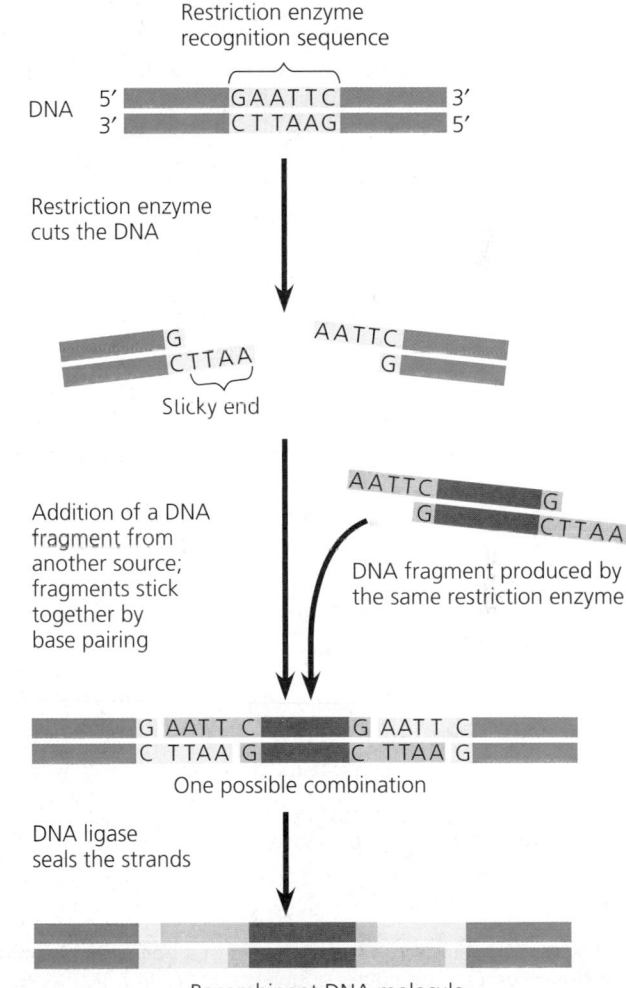

FIGURE 20.2 Using a restriction enzyme and DNA ligase to make recombinant DNA. The restriction enzyme in this example (called *Eco*RI) recognizes a specific six-base-pair sequence and makes staggered cuts in the sugar-phosphate backbone within this sequence. Complementary "sticky ends" will attach to each other by hydrogen bonding, transiently rejoining fragments in their original combinations or in new, recombinant arrangements. DNA ligase can then catalyze the formation of covalent bonds, joining their ends. If the fragments are from different sources, the result is recombinant DNA.

enzyme. The unions formed in this way are only temporary, because only a few hydrogen bonds hold the fragments together. The DNA fusions can be made permanent, however, by the enzyme **DNA ligase,** which seals the strands by catalyzing the formation of phosphodiester bonds. (Recall from Chapter 16 that DNA ligase is a key enzyme in DNA replication and repair.) We now have recombinant DNA, DNA that has been spliced together from two different sources.

Genes can be cloned in recombinant DNA vectors: *a closer look*

With a restriction enzyme and DNA ligase in our toolkit, we can make the recombinant plasmid depicted in step 2 of FIGURE 20.1. The original plasmid is called a **cloning vector,** defined as a DNA molecule that can carry foreign DNA into a cell and replicate there. Bacterial plasmids are widely used as cloning vectors. Recombinant plasmids produced by splicing restriction fragments from foreign DNA into plasmids isolated from bacteria can be returned relatively easily to bacteria. Then, as a bacterium carrying a recombinant plasmid reproduces, the plasmid replicates within it. The resulting cell clone, which appears as a colony on solid nutrient medium, contains multiple copies—a "clone"—of the foreign DNA.

Bacteria are the most commonly used host cells for gene cloning, primarily because of the ease with which DNA can be isolated from and reintroduced into such cells. Bacterial cultures also grow quickly, rapidly replicating any foreign genes they carry.

Procedure for Cloning a Eukaryotic Gene in a Bacterial Plasmid

Suppose we want to clone a particular eukaryotic gene using the method sketched out in steps 1–5 of FIGURE 20.1. The more detailed diagram in FIGURE 20.3 shows one way we might proceed. The step numbers in the text that follows correspond to the numbers in the figure.

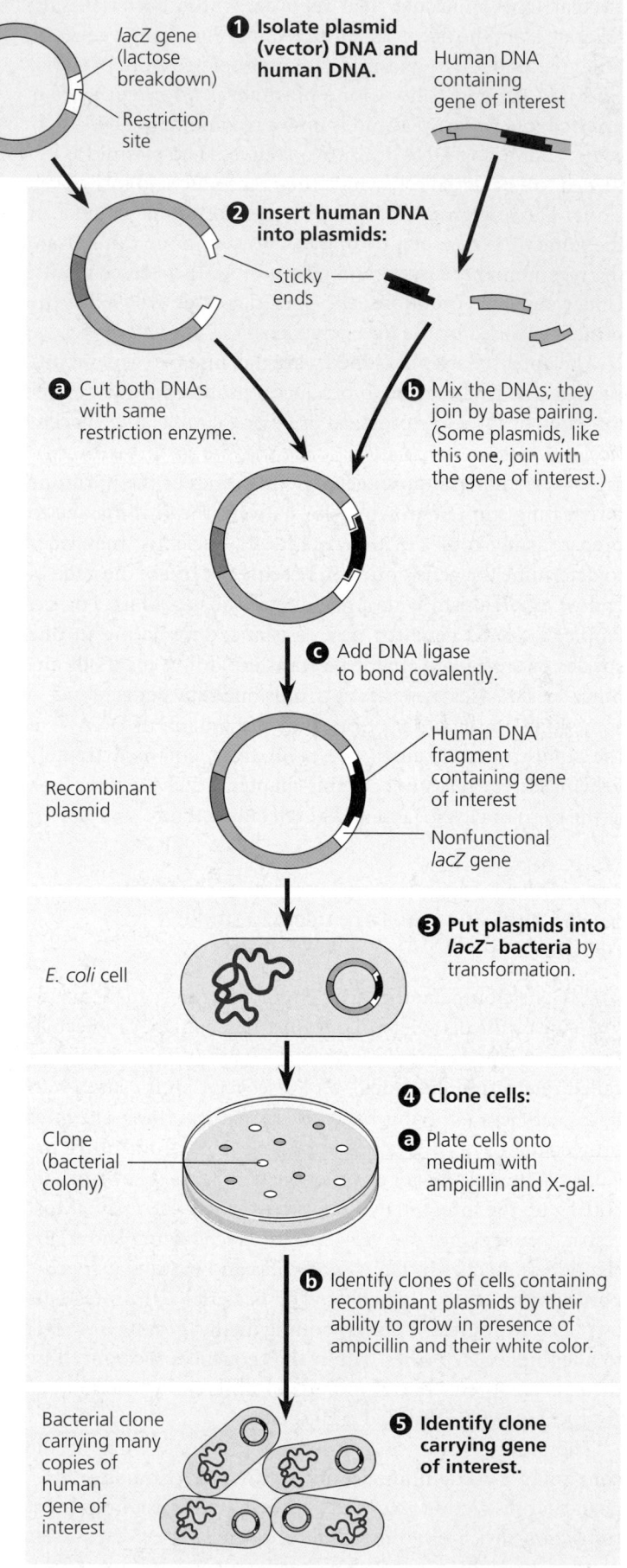

THE PROCESS OF SCIENCE

FIGURE 20.3 Cloning a human gene in a bacterial plasmid: a closer look. In this figure, double-stranded DNA is represented as a single band. (Refer to FIGURE 20.2 to see steps 2a–2c in more detail.) The circled numbers correspond to numbered steps in the text. The diagram focuses on a fragment of human DNA carrying a gene of interest (black), but steps 2a–2c actually produce an enormous variety of recombinant plasmids, each of which shows up as a separate colony in step 4. The method used to identify the clones of interest (step 4b) relies on the *amp*^R and *lacZ* genes of the vector used here; other vectors and identification methods can also be used. Identifying the desired colony, step 5, is often the greatest challenge in this "shotgun" procedure.

❶ Isolation of vector and gene-source DNA. We begin by preparing two kinds of DNA: the bacterial plasmid we'll use as a vector and DNA containing the gene of interest. This latter DNA comes from human tissue cells that we have grown in laboratory culture. The plasmid comes from the bacterium *E. coli* and carries two genes that will later prove useful: *amp^R*, conferring resistance to the antibiotic ampicillin on its *E. coli* host cell, and *lacZ,* encoding the enzyme β-galactosidase, which catalyzes the hydrolysis of the sugar lactose. The plasmid has a single recognition sequence for the restriction enzyme used, and the sequence lies within the *lacZ* gene.

❷ Insertion of DNA into the vector. In step 2a, we digest both the plasmid and the human DNA with the same restriction enzyme. The enzyme cuts the plasmid DNA at its single restriction site, disrupting the *lacZ* gene. It also cuts the human DNA, generating many thousands of fragments; one of these fragments carries the gene we want. In making the cuts, the restriction enzyme creates compatible sticky ends on both the human DNA fragments and the plasmid. For simplicity, FIGURE 20.3 shows the step-by-step processing of one human DNA fragment and one plasmid; but actually, millions of copies of the plasmid and a heterogeneous mixture of millions of human DNA fragments are treated simultaneously.

In step 2b, we mix the fragments of human DNA with the clipped plasmids. The sticky ends of a plasmid base-pair with the complementary sticky ends of the human DNA fragment we are following. (Many other combinations form as well, such as two plasmids together or a plasmid with several DNA fragments; and re-formed, nonrecombinant plasmids are also common.) In step 2c, we use the enzyme DNA ligase to join the DNA molecules by covalent bonds. The result is a mixture of recombinant DNA molecules, some of which are like the one shown.

❸ Introduction of the cloning vector into cells. In this step, bacterial cells take up the recombinant plasmids by transformation (the uptake of naked DNA from the surrounding solution; see pp. 341–342). The bacteria are *lacZ*⁻; a mutation in gene *lacZ* makes them unable to hydrolyze lactose. Some of the bacteria acquire the desired recombinant plasmid DNA; many other cells take up other DNA, both recombinant and nonrecombinant.

❹ Cloning of cells (and foreign genes). Finally, we arrive at the actual cloning step. We plate out the transformed bacteria on solid nutrient medium containing ampicillin and a sugar called X-gal (step 4a). Each reproducing bacterium forms a cell clone that appears as a colony on the nutrient medium. In the process, any human genes carried by recombinant plasmids are also cloned. We take advantage of the plasmid's own genes to select colonies of cells that carry recombinant plasmids. The ampicillin in the medium ensures that only plasmid-containing cells grow, for only they have the *amp^R* gene conferring ampi-

cillin resistance. The X-gal in the medium makes it easy to identify colonies of bacteria whose plasmids carry foreign DNA. X-gal is hydrolyzed by β-galactosidase to yield a blue product, so bacterial colonies containing plasmids with intact β-galactosidase genes will be blue. But if a plasmid has foreign DNA inserted into its *lacZ* gene, then a colony of cells containing it will be white, because the cells cannot produce β-galactosidase. In summary, with this method, bacteria with recombinant plasmids carrying foreign DNA will form white colonies on medium containing ampicillin and X-gal (step 4b).

Our procedure to this point will have cloned many different human DNA fragments, not just the one that interests us. The next step, screening colonies for the desired human gene, is the most difficult.

❺ Identification of cell clones carrying the gene of interest. How can we distinguish the colony containing our gene of interest from the many thousands of colonies carrying other pieces of human DNA? We can look either for the gene itself or for its protein product. All the methods for detecting the DNA of a gene directly depend on base pairing between the gene and a complementary sequence on another nucleic acid molecule, a process called **nucleic acid hybridization.** The complementary molecule, a short, single-stranded nucleic acid that can be either RNA or DNA, is called a **nucleic acid probe.** If we know at least part of the nucleotide sequence of our gene (perhaps from knowledge of the protein it encodes), we can synthesize a probe complementary to it.

We trace the probe, which will hydrogen-bond specifically to complementary single strands of the desired gene, by labeling it with a radioactive isotope or a fluorescent tag. FIGURE 20.4 (p. 380) shows how a number of bacterial clones, growing as colonies on a plate of solid medium (agar), can be simultaneously screened for the presence of DNA complementary to a DNA probe. An essential step is the **denaturation** of the cells' DNA—that is, the separation of its two strands. As with protein denaturation, DNA denaturation is routinely accomplished with chemicals or heat. The labeled probe tags the colonies carrying the targeted gene.

Once we've identified a cell clone carrying the desired gene, we can grow the cells in liquid culture in a large tank and then easily isolate large amounts of the gene. Also, we can use the cloned gene itself as a probe to identify similar or identical genes in DNA from other sources.

If the cells containing a desired gene translate the gene into protein, then it is sometimes possible to identify them by screening all clones for the protein. Detection of the protein can be based on either its activity (as with an enzyme) or its structure, using antibodies that bind specifically to it. However, for bacterial cells to make a eukaryotic protein, we need to take special measures, as discussed next.

Cloning and Expressing Eukaryotic Genes: Problems and Solutions

Getting a cloned eukaryotic gene to function in a prokaryotic setting can be difficult because certain details of gene expression are different in the two kinds of cells. To overcome differences in promoters and other DNA control sequences, scientists usually employ an **expression vector,** a cloning vector that contains the requisite prokaryotic promoter just upstream of a restriction site where the eukaryotic gene can be inserted. The bacterial host cell will then recognize the promoter and proceed to express the foreign gene that has been linked to it. Expression vectors allow the synthesis of many eukaryotic proteins in bacterial cells.

Another problem with cloning and expressing eukaryotic DNA in bacteria is the presence of long noncoding regions (introns) in most eukaryotic genes. Introns can make a eukaryotic gene very long and unwieldy, and they prevent correct expression of the gene by bacterial cells, which do not have RNA-splicing machinery. Fortunately, we can make artificial eukaryotic genes that lack introns (FIGURE 20.5). The starting material for this procedure is fully processed mRNA, which has been stripped of introns in the eukaryotic cell nucleus. We can extract this mRNA from cells and then use the enzyme reverse transcriptase (obtained from retroviruses) to make DNA transcripts of this RNA. Each DNA molecule produced carries the complete coding sequence for a gene but no

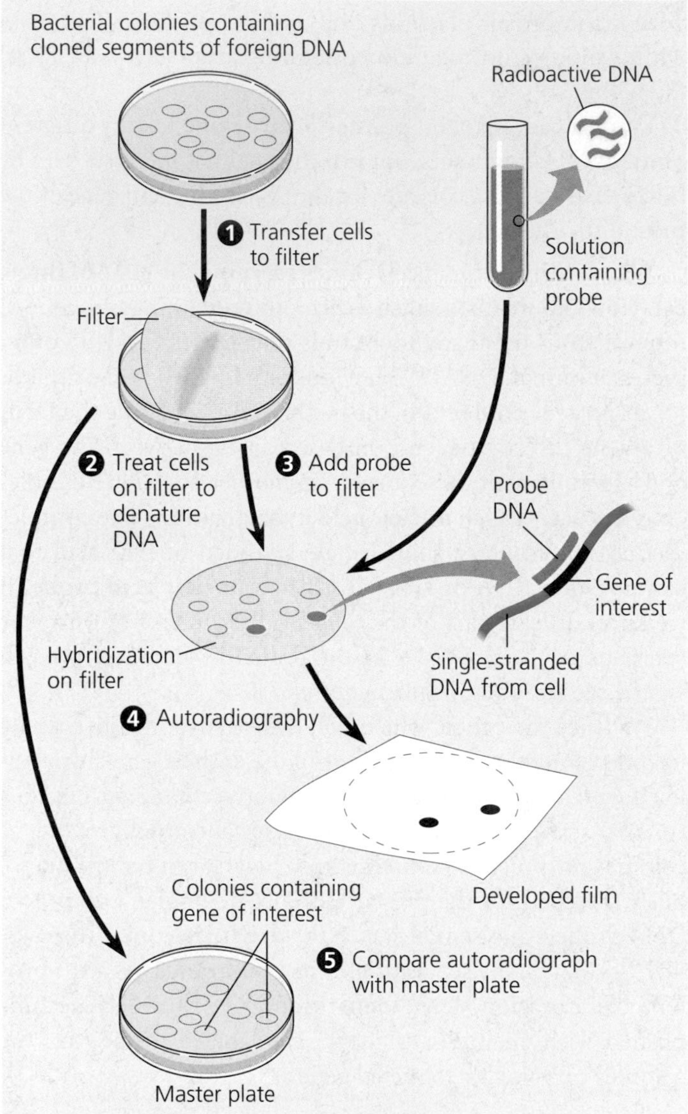

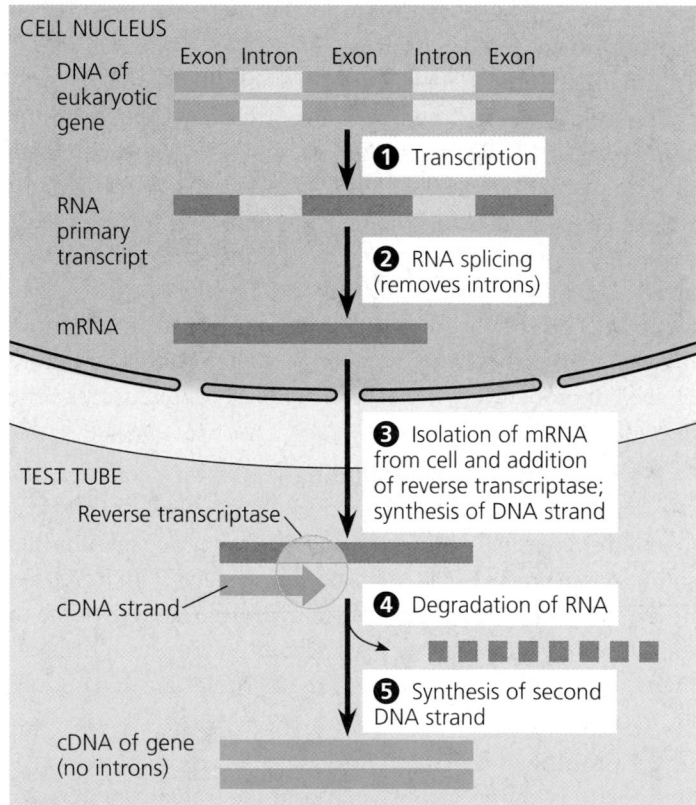

FIGURE 20.5 Making complementary DNA (cDNA) for a eukaryotic gene. Complementary DNA is DNA made *in vitro* using mRNA as a template and the enzyme reverse transcriptase. The synthesis and splicing of an mRNA molecule in a cell nucleus is shown in steps ❶ and ❷. In step ❸, we isolate the mRNA molecules from the cell and add reverse transcriptase, which makes a DNA strand using the RNA as a template. We provide ❹ another enzyme to degrade the RNA and ❺ DNA polymerase to synthesize a second DNA strand. The result is cDNA, which carries the complete coding sequence of the gene but no introns.

FIGURE 20.4 Using a nucleic acid probe to identify a cloned gene. This technique depends on the fact that complementary nucleotide sequences will base-pair, stuck together by hydrogen bonds. Here the cloned gene of interest is carried on a bacterial plasmid, and the probe is a short length of radioactive single-stranded DNA complementary to part of the gene. ❶ Bacterial colonies on agar are pressed against special filter paper, transferring cells to the filter. ❷ The filter is treated to break open the cells and denature their DNA; the resulting single-stranded DNA molecules stick to the filter. ❸ A solution of probe molecules is incubated with the filter. The probe DNA hybridizes (base-pairs) with any complementary DNA on the filter; excess DNA is rinsed off. ❹ The filter is laid on photographic film, allowing any radioactive areas to expose the film (autoradiography). ❺ The developed film, an autoradiograph, is compared with the master culture plate to determine which colonies carry the desired gene.

introns. This DNA is called **complementary DNA, or cDNA,** and can be attached to vector DNA for replication inside a cell. Bacteria can express a eukaryotic cDNA gene if the vector provides a bacterial promoter and any other control elements necessary for the gene's transcription and translation.

Molecular biologists can avoid eukaryotic-prokaryotic incompatibility by using eukaryotic cells rather than bacteria as hosts for cloning and/or expressing eukaryotic genes of interest. Yeast cells, single-celled fungi, offer two advantages: They are as easy to grow as bacteria, and they have plasmids, a rarity among eukaryotes. Scientists have even constructed recombinant plasmids that combine yeast and bacterial DNA and can replicate in either type of cell. Scientists have also constructed vectors called **yeast artificial chromosomes (YACs)** that combine the essentials of a eukaryotic chromosome—an origin for DNA replication, a centromere, and two telomeres—with foreign DNA. These chromosomes behave normally in mitosis, cloning the foreign DNA as the yeast cell divides. A YAC can carry much more DNA than can a plasmid vector, enabling very long pieces of DNA to be cloned.

Another reason to use eukaryotic host cells for expressing a cloned eukaryotic gene is that many eukaryotic proteins will not function unless they are modified after translation, for example by the addition of carbohydrate or lipid groups. Bacterial cells cannot carry out these modifications, and if the gene product requiring such processing is from a mammal, even yeast cells will not be able to modify the protein correctly. The use of host cells from an animal or plant cell culture may therefore be necessary.

Like bacteria, many kinds of eukaryotic cells can take up DNA from their surroundings but sometimes not very efficiently. To deal with this problem, scientists have developed a variety of more aggressive methods for introducing recombinant DNA into cells. In **electroporation,** they apply a brief electrical pulse to a solution containing cells. The electricity creates a temporary hole in the cell's plasma membrane, through which DNA can enter. (This technique is now commonly used for bacteria as well.) Alternatively, scientists can inject DNA directly into single eukaryotic cells using microscopically thin needles. And in a technique used primarily for plant cells, they can attach DNA to microscopic particles of metal and fire the particles into cells with a gun (see FIGURE 38.17). Once inside the cell, the DNA has a chance to be incorporated into the cell's DNA by natural genetic recombination.

Cloned genes are stored in DNA libraries

Because the gene-cloning procedure in FIGURE 20.3 starts with a mixture of fragments from the entire genome of an organism, it is called a "shotgun" approach—no single gene is targeted for cloning. Thousands of different recombinant plasmids are actually produced in step 2 of FIGURE 20.3, and a clone of

each ends up in a (white) colony in step 4. The complete set of thousands of recombinant plasmid clones, each carrying copies of a particular segment from the initial genome, is referred to as a **genomic library** (FIGURE 20.6a). A researcher can save such a library and use it as a source of other genes of interest or for genome mapping (as we'll discuss later).

In addition to plasmids, certain bacteriophages are also common cloning vectors for making genomic libraries. Fragments of foreign DNA can be spliced into a phage genome, as into a plasmid, by using a restriction enzyme and ligase. The recombinant phage DNA is then packaged into capsids *in vitro* and introduced into a bacterial cell through the normal infection process. Once inside the cell, the phage DNA replicates and produces new phage particles, each carrying the foreign DNA. A genomic library made using phage is stored as collections of phage clones (FIGURE 20.6b). Whatever the cloning vector, restriction enzymes do not respect gene boundaries in cutting up genomic DNA, so some genes in a genomic library may be divided up among two or more clones.

Researchers can make a more limited kind of gene library by using complementary DNA (cDNA). When they first isolate mRNA from cells (step 3 in FIGURE 20.5), they actually obtain a

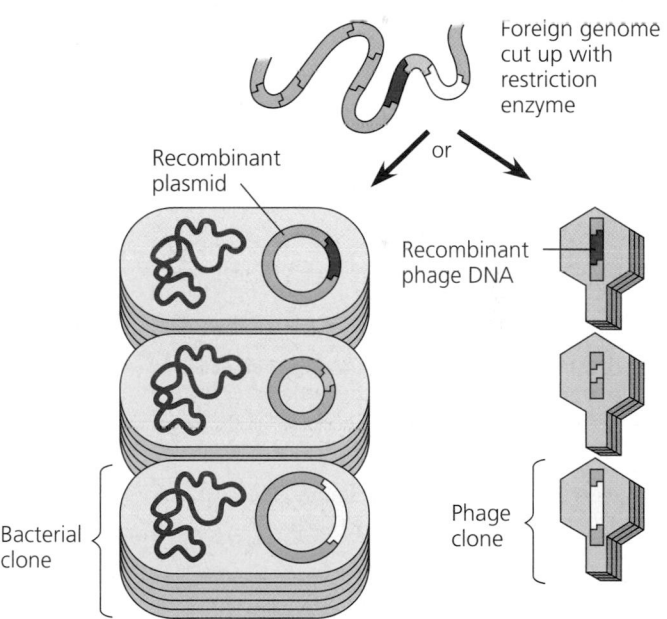

(a) Plasmid library. Shown are three of the thousands of "books" in the library. Each is a bacterial clone containing one particular variety of foreign genome fragment—red, orange, or yellow here—in its recombinant plasmid.

(b) Phage library. The same three foreign genome fragments are shown in three "books" of a phage library.

FIGURE 20.6 Genomic libraries. A genomic library is a collection of many bacterial or phage clones, each containing copies of a particular DNA segment from a foreign genome. In a complete genomic library, the foreign DNA segments represented cover the entire genome of an organism. This diagram shows parts of two genomic libraries.

mixture of all the mRNA molecules in the cell, transcribed from a number of different genes. Therefore, the cDNA that is made is a library containing a collection of genes. Such a **cDNA library** represents only part of a cell's genome—only the genes that were transcribed in the starting cells. This is an advantage if a researcher wants to study the genes responsible for specialized functions of a particular kind of cell, such as a brain or liver cell. Also, by making cDNA from cells of the same type at different times in the life of an organism, one can trace changes in patterns of gene expression.

The polymerase chain reaction (PCR) clones DNA entirely *in vitro*

DNA cloning in cells remains the best method for preparing large quantities of a particular gene or other DNA sequence. However, when the source of DNA is scanty or impure, a method called PCR is quicker and more selective. **PCR,** the **polymerase chain reaction,** is a technique by which any piece of DNA can be quickly amplified (copied many times) without using cells (FIGURE 20.7). The DNA is

The starting material for PCR is a solution of double-stranded DNA containing the nucleotide sequence that is "targeted" for copying. To this is added a heat-resistant type of DNA polymerase, a supply of all four nucleotides, and a supply of primers

The primers used to initiate DNA synthesis in PCR are short, synthetic molecules of single-stranded DNA complementary to the ends of the targeted DNA. The primers thus determine the particular segment of DNA to be amplified.

Each cycle of the PCR procedure takes only about 5 minutes. At the end of the cycle, the targeted DNA sequence—even one hundreds of base pairs long—has been doubled.

The solution is then heated again, starting the next cycle of strand separation, primer binding, and DNA synthesis.

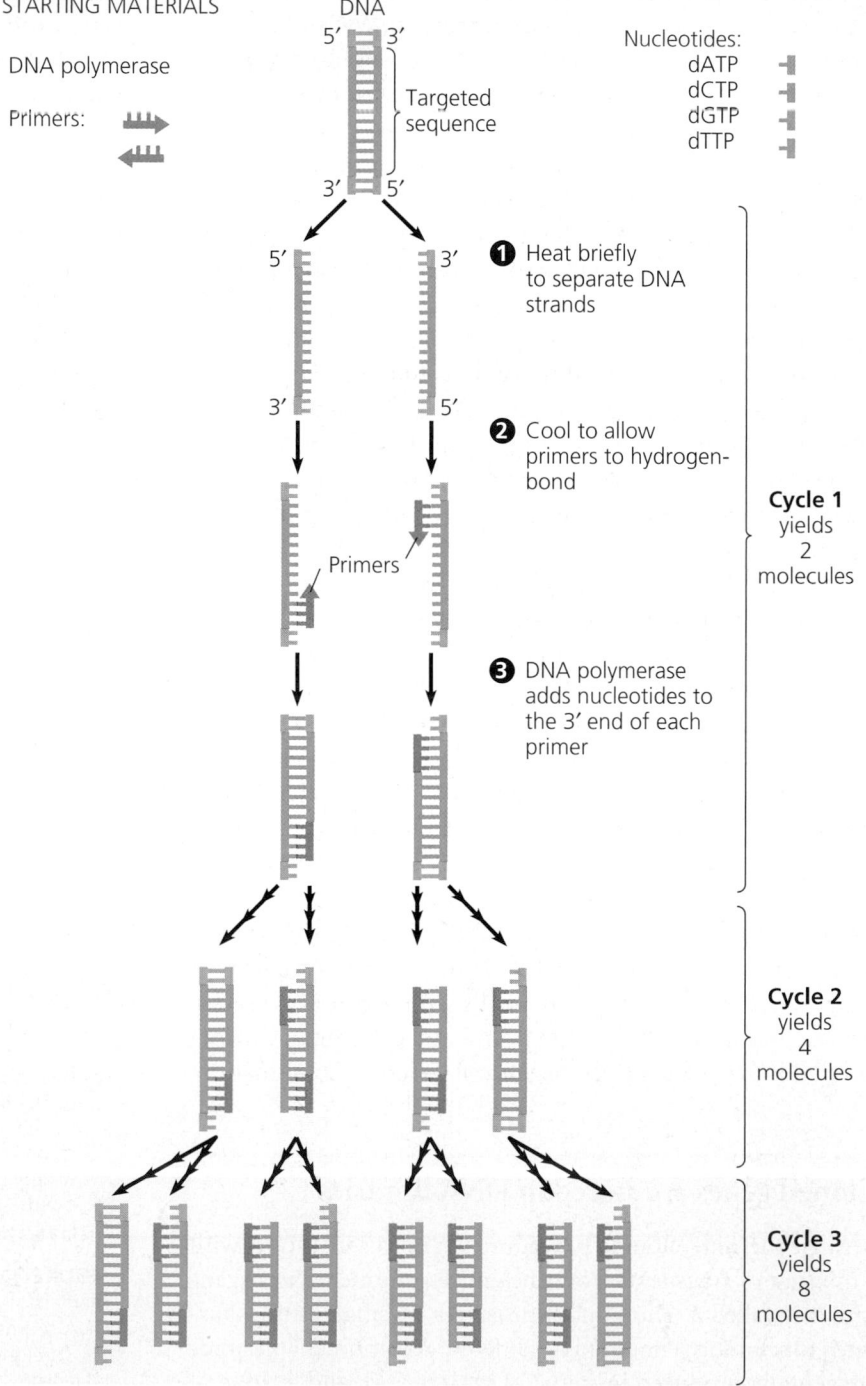

FIGURE 20.7 The polymerase chain reaction (PCR). PCR is a method for making many copies of a specific segment of DNA. It is much faster than gene cloning using a plasmid or phage vector and is performed completely *in vitro*. A PCR machine repeats the three-step cycle shown in this figure again and again, until the targeted sequence has been duplicated many times.

incubated in a test tube with a special kind of DNA polymerase, a supply of nucleotides, and short pieces of synthetic single-stranded DNA that serve as primers for DNA synthesis. (Recall from FIGURE 16.14 that DNA polymerases need primers because they can add nucleotides only to a preexisting nucleotide chain.) With automation, PCR can make billions of copies of a targeted segment of DNA in a few hours, significantly faster than the days it takes to clone a piece of DNA by making a recombinant plasmid and letting it replicate within bacteria. Amplifying a tiny DNA sample a billionfold does not produce much DNA, but it can be enough for some purposes, such as DNA fingerprinting for a murder trial.

In the PCR procedure (see FIGURE 20.7), a three-step cycle brings about a chain reaction that produces an exponentially growing population of DNA molecules. The key to easy PCR automation is an unusual DNA polymerase that was first isolated from bacteria living in hot springs. Unlike most proteins, this enzyme can withstand the heat needed to separate the DNA strands at the start of each cycle.

Just as impressive as the speed of PCR is its specificity. By their complementarity to sequences bracketing the targeted sequence, the primers determine the DNA sequence that is amplified. (In fact, PCR requires knowledge of such sequences.) PCR can be used, for example, to amplify a specific gene prior to further cloning in cells. The PCR makes the gene by far the most abundant DNA fragment, thus simplifying the later task of finding a clone carrying that gene. In fact, PCR is so specific and powerful that its starting material does not even have to be purified DNA. Only minute amounts of DNA need be present in the starting material, and this DNA can be in a partially degraded state. Note, however, that PCR cannot substitute for gene cloning in cells when large amounts of a gene are desired. Occasional errors during PCR replication impose limits on the number of good copies that can be made by this method.

Devised in 1985, PCR has had a major impact on biological research and biotechnology. PCR has been used to amplify DNA from a wide variety of sources: fragments of ancient DNA from a 40,000-year-old frozen woolly mammoth; DNA from tiny amounts of blood, tissue, or semen found at the scenes of violent crimes; DNA from single embryonic cells for rapid prenatal diagnosis of genetic disorders; and DNA of viral genes from cells infected with such difficult-to-detect viruses as HIV. We'll return to applications of PCR later in the chapter.

DNA ANALYSIS AND GENOMICS

Once we have techniques for preparing homogeneous samples of DNA, each containing large numbers of identical segments of a genome, we can begin to ask some far-ranging questions. Suppose we have cloned a DNA segment carrying a human gene of interest. Does the gene differ in different people, and are there certain alleles associated with a hereditary disorder? Where in the body and when is the gene expressed? What is the location of the gene within the genome? We can also address evolutionary questions by asking how the gene differs from species to species.

To answer such questions fully, we will eventually need to know the complete nucleotide sequence of the gene and its counterparts in other individuals and species. And ultimately, we want to have the sequences of entire genomes available so that we can use them as the basis for studying whole sets of genes and their interactions—an approach called **genomics.** But we can start by analyzing cloned DNA by means of more indirect methods that can rapidly provide useful comparative information. Most of these methods, and DNA sequencing as well, make use of a technique called **gel electrophoresis.** This technique, described in FIGURE 20.8 (p. 384), separates macromolecules—nucleic acids or proteins—on the basis of size, electrical charge, and other physical properties. For linear molecules of DNA, separation depends mainly on size. Gel electrophoresis sorts a mixture of DNA molecules into bands, each consisting of DNA molecules of the same length.

Restriction fragment analysis detects DNA differences that affect restriction sites

Restriction fragment analysis indirectly detects certain differences in the nucleotide sequences of DNA molecules. In this method, we use gel electrophoresis to sort by size the DNA fragments that result from treating a long DNA molecule with a restriction enzyme. Just viewing the stained pattern of bands on the gel can provide scientifically useful information. When the mixture of restriction fragments from a particular DNA molecule undergoes electrophoresis, it yields a band pattern characteristic of the starting molecule and the restriction enzyme used. In fact, the relatively small DNA molecules of viruses and plasmids can be identified simply by their restriction fragment patterns. (Larger DNA molecules, such as those of eukaryotic chromosomes, yield too many fragments to appear as distinct bands.) Because DNA can be recovered undamaged from gels, the procedure also provides a way to prepare pure samples of individual fragments.

Let's use restriction fragment analysis to compare two different DNA molecules representing, for example, different alleles of a gene. We start by digesting each DNA sample with the same restriction enzyme. Because the two alleles must differ slightly in DNA sequence, they may differ in one or more restriction sites. Thus, the collections of restriction fragments from the two alleles may produce different band patterns in gel electrophoresis. FIGURE 20.9 (p. 384) shows how restriction fragment analysis by electrophoresis can distinguish between two alleles of a gene that differ in sequence by only one base pair if a restriction site is affected.

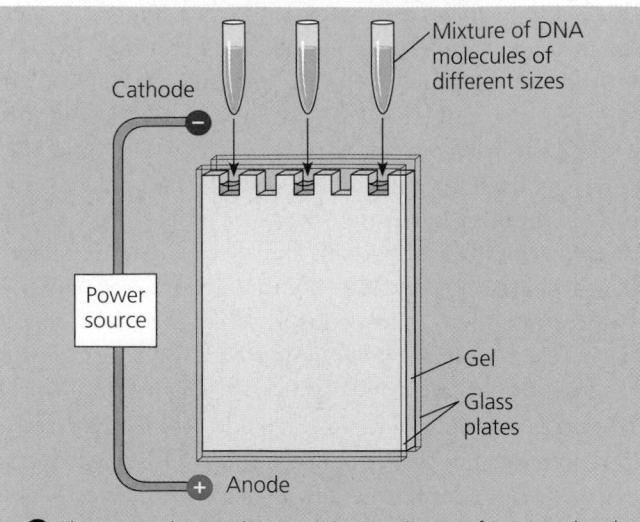

Cathode

Mixture of DNA molecules of different sizes

Power source

Gel

Glass plates

Anode

1 Three samples, each containing a mixture of DNA molecules, are placed in wells near one end of a thin slab of a polymeric gel. The gel is supported by glass plates and bathed in an aqueous solution. Electrodes are attached to both ends, and voltage is applied.

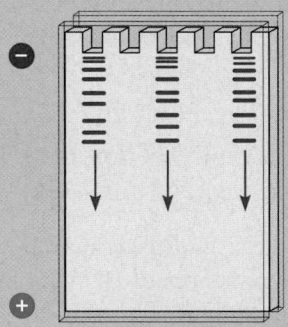

2 The DNA molecules, which are negatively charged, migrate toward the positive electrode, the anode. A molecule's rate of movement is determined mostly by its length; longer molecules travel more slowly through the gel.

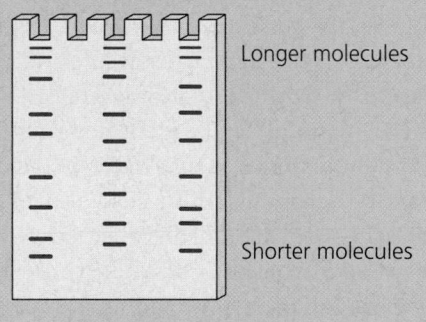

Longer molecules

Shorter molecules

Completed gel

3 When the current is turned off, the DNA molecules in each sample are arrayed in bands along a "lane," according to their size. The shortest molecules, having traveled the farthest, are in bands at the bottom of the gel.

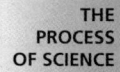
THE PROCESS OF SCIENCE

FIGURE 20.8 Gel electrophoresis of macromolecules. Gel electrophoresis separates macromolecules on the basis of their rate of movement through a gel in an electric field. For DNA, the migration rate—how far a molecule travels while the current is on—is inversely proportional to molecular size. Nucleic acids carry negative charges (on phosphate groups) proportional to their lengths, but the thicket of polymer fibers in the gel impedes longer fragments more than it does shorter ones.

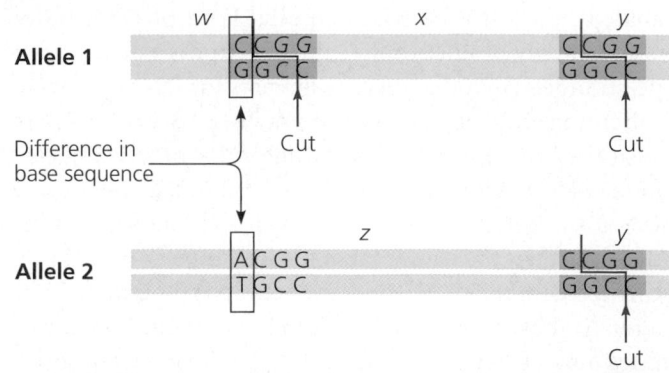

Allele 1

Difference in base sequence

Cut

Cut

Allele 2

Cut

(a) DNA from two alleles. Two homologous segments of DNA that carry different alleles of a gene are depicted; only the relevant bases are shown. A single base-pair difference results in allele 2 having one less recognition sequence (restriction site) for a particular restriction enzyme. This enzyme cuts the DNA from allele 1 into three pieces (w, x, and y) but cuts the DNA from allele 2 into only two pieces (z and y).

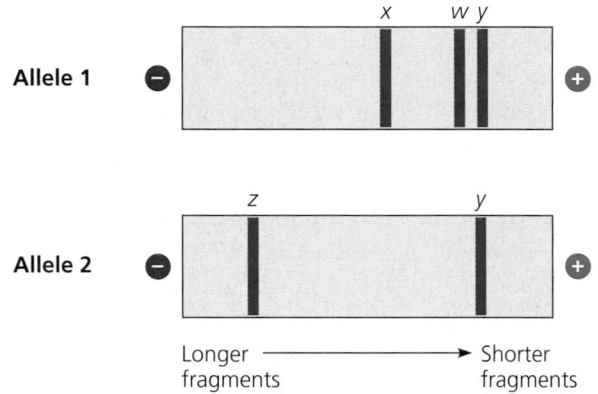

Allele 1

Allele 2

Longer fragments → Shorter fragments

(b) Electrophoresis of restriction fragments. Electrophoresis separates the restriction fragments formed from each allele. A clear difference between the two alleles is revealed by their band patterns on the gel. Allele 1 has three bands, corresponding to fragments w, x, and y; allele 2 has two bands, corresponding to z and y.

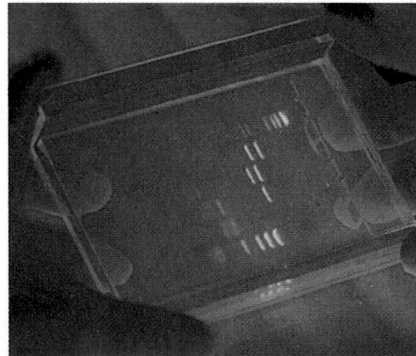

(c) Completed gel. After the addition of a DNA-binding dye, the bands fluoresce pink in ultraviolet light. In the gel shown here, six samples were run, each containing a mixture of fragments from a DNA preparation digested with a restriction enzyme. The pink bands correspond to DNA restriction fragments of different sizes.

FIGURE 20.9 Using restriction fragment patterns to distinguish DNA from different alleles.

THE PROCESS OF SCIENCE

The starting materials in FIGURE 20.9 are samples of cloned and purified genes. However, by combining gel electrophoresis with nucleic acid hybridization, we can make the same comparison without preliminary gene cloning. We can start with DNA of the whole genome. Although electrophoresis will yield too many bands to distinguish individually, we can use nucleic acid hybridization with a specific probe to label discrete bands that derive from our gene of interest.

The principle is the same as for the nucleic acid hybridization in FIGURE 20.4: A radioactive single-stranded nucleic acid probe selectively hydrogen-bonds with a targeted, complementary DNA sequence, which is then detected by autoradi-

ography. Now, however, we use a probe in combination with an additional technique called Southern hybridization or **Southern blotting.** FIGURE 20.10 outlines the entire procedure and demonstrates how it can be used to compare DNA samples from three individuals. This method goes beyond FIGURE 20.4 by revealing not only whether a particular sequence is present in a sample of DNA but also the restriction fragments that contain the sequence.

Southern blotting proved invaluable when biologists turned their attention to *noncoding* DNA, which comprises most of the DNA of animal and plant genomes. Are there differences in noncoding DNA sequences analogous to the

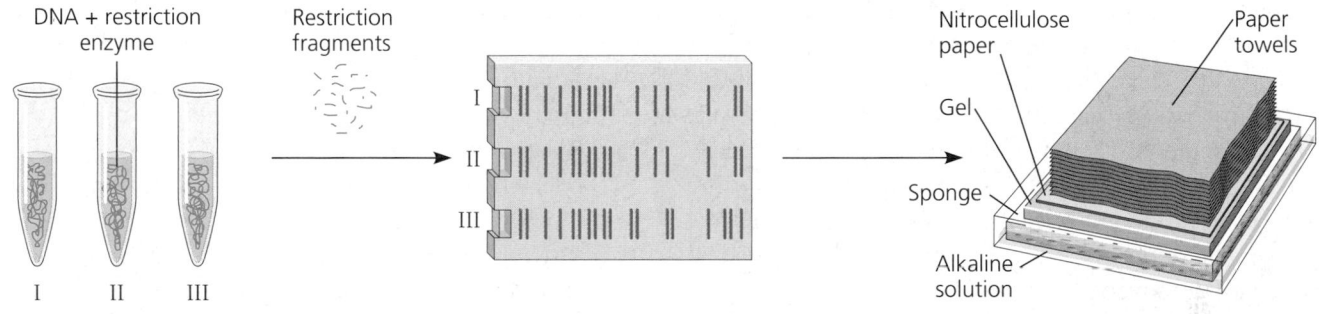

1 **Restriction fragment preparation.** DNA samples to be tested (in this case identified as samples I, II, and III) are prepared from the appropriate sources. A restriction enzyme is added to the three samples of DNA to produce restriction fragments.

2 **Electrophoresis.** The mixtures of restriction fragments from each sample are separated by electrophoresis. Each sample forms a characteristic pattern of bands. (There would actually be many more bands than shown here, and they would be invisible unless stained.)

3 **Blotting.** Capillary action pulls an alkaline solution upward through the gel and through a sheet of nitrocellulose paper laid on top of it, transferring the DNA to the paper and denaturing it in the process. The single strands of DNA stick to the paper, positioned in bands exactly as on the gel.

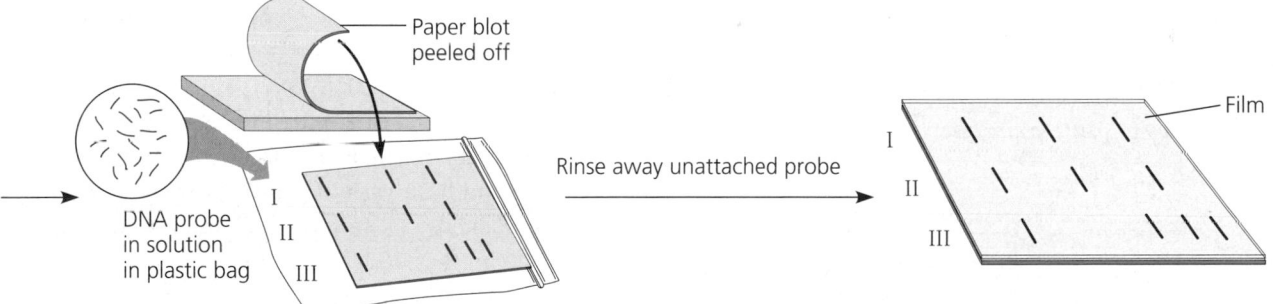

4 **Hybridization with radioactive probe.** The paper blot is exposed to a solution containing radioactively labeled probe. The probe is single-stranded DNA complementary to the DNA sequence of interest, and it attaches by base-pairing to restriction fragments of complementary sequence.

5 **Autoradiography.** A sheet of photographic film is laid over the paper. The radioactivity in the bound probe exposes the film to form an image corresponding to specific DNA bands—the bands containing DNA that base-pairs with the probe. The band patterns for samples I and II are identical, but III is different.

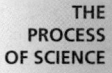
THE PROCESS OF SCIENCE

FIGURE 20.10 Restriction fragment analysis by Southern blotting. This procedure enables researchers to detect and analyze particular DNA sequences. The basis for detecting specific sequences is nucleic acid hybridization. The results can show not only whether certain sequences exist in different samples but also the number of such sequences within a genome and the size of the restriction fragments that contain them. In this way, it is possible to compare the DNA of different individuals or even species. Because of the selective power of nucleic acid hybridization, the starting material for analysis can be the entire genome of an organism. This huge length of DNA will produce so many restriction fragments that if all were made visible (with a dye, for example), they would appear as a smear in gel electrophoresis, rather than as discrete bands. Instead, only the DNA bands of interest are made visible by using a labeled probe. Before hybridization, the DNA being tested is transferred by capillary action (blotting) from the gel to a solid support, a sheet of nitrocellulose or nylon paper. The entire hybridization procedure is known as Southern blotting, after E.M. Southern, who developed it in 1975. The sensitivity of the hybridization can be adjusted either to detect only those sequences that are perfectly complementary to the probe or to detect similar (homologous) sequences as well.

differences in alleles of genes? Using noncoding DNA in procedures like that of FIGURE 20.9, researchers were excited to discover many differences in band patterns. Differences in DNA sequence on homologous chromosomes that can result in different restriction fragment patterns turn out to be scattered abundantly throughout genomes, including the human genome. Such differences have been named **restriction fragment length polymorphisms** (**RFLPs,** pronounced "Riflips"). This type of difference is conceptually the same as a difference in coding sequence; it, too, can serve as a genetic marker for a particular location (locus) in the genome. A given RFLP marker frequently occurs in numerous variants in a population. (The word *polymorphisms* comes from the Greek for "many forms.")

RFLPs are detected and analyzed by Southern blotting. The example shown in FIGURE 20.10 could as easily represent the detection of a RFLP in noncoding DNA as a difference in the coding sequences of two alleles. Because of the sensitivity of DNA hybridization, the entire genome can be used as the DNA starting material. (Samples of human DNA are typically obtained from white blood cells.) The result of the procedure shown in FIGURE 20.10 indicates that individuals I and II carry the same version of the marker (RFLP or gene), but individual III carries a different version.

Because RFLP markers are inherited in a Mendelian fashion, they can serve as genetic markers for making linkage maps. The geneticist uses the same reasoning you saw applied in FIGURE 15.6: The frequency with which two RFLP markers—or a RFLP marker and a certain allele for a gene—are inherited together is a measure of the closeness of the two loci on a chromosome. The discovery of RFLPs greatly increased the number of markers available for mapping the human genome. No longer were geneticists limited to genetic variations that lead to obvious phenotypic differences (such as genetic diseases) or even to differences in protein products.

Entire genomes can be mapped at the DNA level

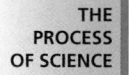

THE PROCESS OF SCIENCE

As early as 1980, molecular biologist David Botstein and colleagues proposed that the DNA variations reflected in RFLPs could serve as the basis for an extremely detailed map of the entire human genome. Since then, researchers have in fact used such markers for mapping the genomes of a number of organisms. For some organisms, they have already succeeded in bringing the map to the ultimate level of detail: the entire sequence of nucleotides in the DNA. Along the way, they have taken advantage of all the tools and techniques discussed in this chapter—restriction enzymes, DNA cloning, gel electrophoresis, labeled probes, and so forth.

The most ambitious research project made possible by DNA technology has been the **Human Genome Project,** offi-

cially begun in 1990. This is an effort to map the entire human genome, ultimately by determining the complete nucleotide sequence of the DNA of each human chromosome (the 22 autosomes and the X and Y sex chromosomes). As organized by an international, publicly funded consortium of 20 groups of researchers at universities and research institutes, the project has proceeded through three stages that focus on the DNA more and more closely: genetic (or linkage) mapping, physical mapping, and DNA sequencing. We'll describe these stages shortly.

In addition to mapping human DNA, the Human Genome Project includes mapping the genomes of other species important in biological research, including *E. coli* and other prokaryotes, *Saccharomyces cerevisiae* (yeast), *Caenorhabditis elegans* (nematode), *Drosophila melanogaster* (fruit fly), and *Mus musculus* (mouse). These genomes are of great interest in their own right and are also providing important insights of general biological significance, as we'll discuss later. In addition, the early work on these genomes was useful for developing the strategies, methods, and new technologies necessary for deciphering the human genome, which is much larger. The technological power for this daunting task has come in large part from advances in automation and from utilization of electronic technology, including computer software.

Genetic (Linkage) Mapping

In mapping a large genome, the first stage is to construct a linkage map of several thousand genetic markers spaced evenly throughout the chromosomes. As you learned in Chapter 15, the order of the markers on such a map and the relative distances between them are based on recombination frequencies. The markers can be genes (as in Chapter 15) or any other identifiable sequences in the DNA, such as RFLPs or the short repetitive sequences called microsatellites (see TABLE 19.1). Relying primarily on microsatellites, which are abundant in the human genome and have various "alleles" differing in length, researchers completed in a few years a human genetic map with some 5,000 markers. Such a map enables them to locate other markers, including genes, by testing for genetic linkage to the known markers. It is also valuable as a framework for organizing more detailed maps of particular regions.

Physical Mapping: Ordering DNA Fragments

In a physical map, the distances between markers are expressed in some physical measure, usually the number of nucleotides along the DNA. For whole-genome mapping, a physical map is made by cutting the DNA of each chromosome into a number of identifiable restriction fragments and then determining the original order of the fragments in the chromosomal DNA. The key is to make fragments that overlap and then use probes or automated nucleotide sequencing of the ends

FIGURE 20.11 Chromosome walking. The researcher starts with a known gene or other DNA segment—one that has already been cloned, mapped, and sequenced—and "walks" along the chromosomal DNA from that locus, producing a map of overlapping restriction fragments. To prepare the fragments, the scientist cuts two samples of chromosomal DNA with different restriction enzymes and then clones both sets of fragments to form two libraries. This and other methods of ordering DNA fragments involve determining the nucleotide sequences of short, unique segments within each fragment. The resulting physical map is thus a linear array of fragments, each identifiable by two or more sequenced regions. All the DNA in this diagram is shown as single strands.

❶ Prepare a probe to match the 3′ end of the known gene (probe 1).

❷ Cut the starting DNA with two restriction enzymes, and clone the fragments to make two libraries.

❸ Use probe 1 to screen library II for DNA fragments that overlap the known gene.

❹ Isolate DNA from the tagged clone, and prepare probe 2 to match the 3′ end of that segment.

❺ Use probe 2 to screen library I for an overlapping fragment farther along.

❻ Repeat steps 4 and 5, with new probes and alternating libraries, to "walk" down the DNA.

❼ The result is a DNA map with a series of known markers (sequences) in a known order and separated by known distances.

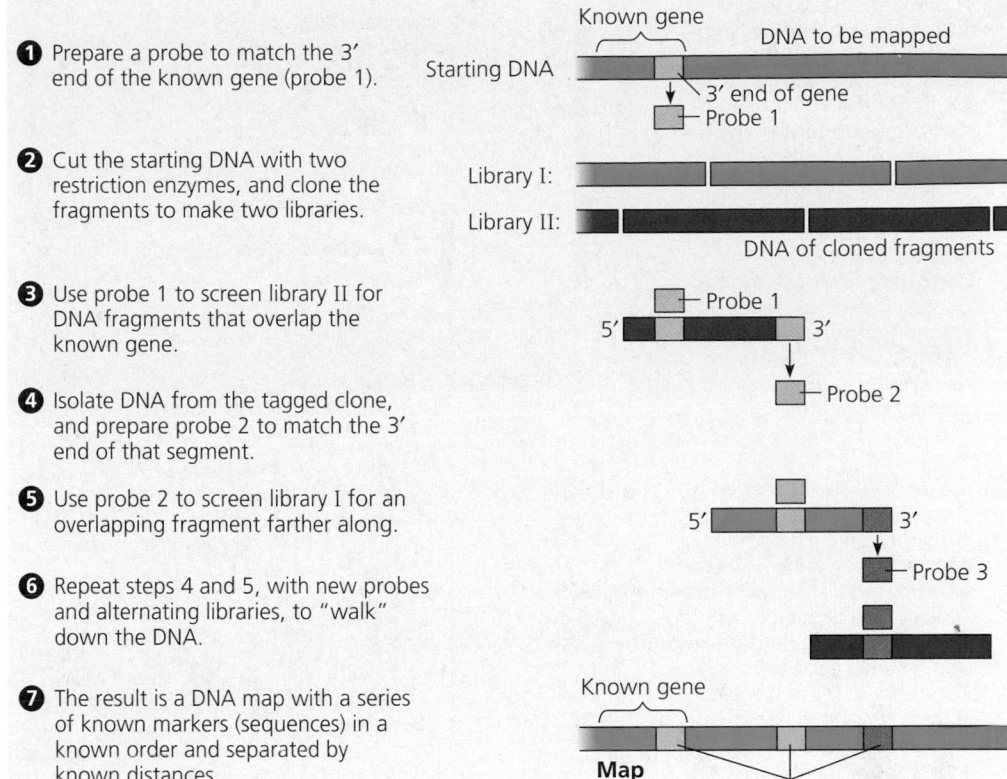

to find the overlaps. A method called **chromosome walking,** which uses probes, has been very useful (FIGURE 20.11).

Supplies of the DNA fragments used for physical mapping are prepared by cloning. In working with large genomes, researchers carry out several rounds of DNA cutting, cloning, and physical mapping. The first cloning vector is often a yeast artificial chromosome (YAC), which can carry inserted fragments a million base pairs long, or a **bacterial artificial chromosome (BAC),** an artificial version of a bacterial chromosome that can carry inserts of 100,000 to 500,000 base pairs. The scientists determine the order of these long fragments (by chromosome walking, for instance) and then cut each fragment into smaller pieces, which are cloned and ordered in turn. The final sets of fragments, cloned in plasmids or phage, are only about 1,000 base pairs long, short enough to be sequenced easily.

DNA Sequencing

The complete nucleotide sequence of a genome is the ultimate map. To repeat a key point, it is the technology for cloning fragments of DNA that has made possible the sequencing of the very long DNA molecules of genomes. Starting with a pure preparation of many copies of a relatively short DNA fragment, the nucleotide sequence of the fragment can be determined by a sequencing machine (see the photograph opening this chapter). The usual sequencing technique, described in FIGURE 20.12 (p. 388), combines DNA labeling, DNA synthesis

involving special chain-terminating nucleotides, and high-resolution gel electrophoresis. But even with automation, the sequencing of all 3.2 billion nucleotide pairs in a haploid set of human chromosomes is a formidable challenge. In fact, a major thrust of the Human Genome Project has been the development of technology for faster sequencing, along with more sophisticated computer software for analyzing and assembling the partial sequences.

FIGURE 20.13a (p. 389) reviews the three-stage, hierarchical approach to genome mapping that we've discussed so far. In practice, the three stages have overlapped in a way that our simplified version does not portray, but they have remained the overarching strategy of the public consortium of researchers.

Alternative Approaches to Whole-Genome Sequencing

In 1992, emboldened by advances in sequencing machines and computer technology, molecular biologist J. Craig Venter came up with an alternative approach to the sequencing of whole genomes. His idea was essentially to skip the genetic mapping and physical mapping stages and start directly with the sequencing of random DNA fragments. Powerful computer programs would then assemble the resulting very large number of overlapping short sequences into a single continuous one (FIGURE 20.13b). Despite the skepticism of many other scientists, Venter left the public consortium to follow

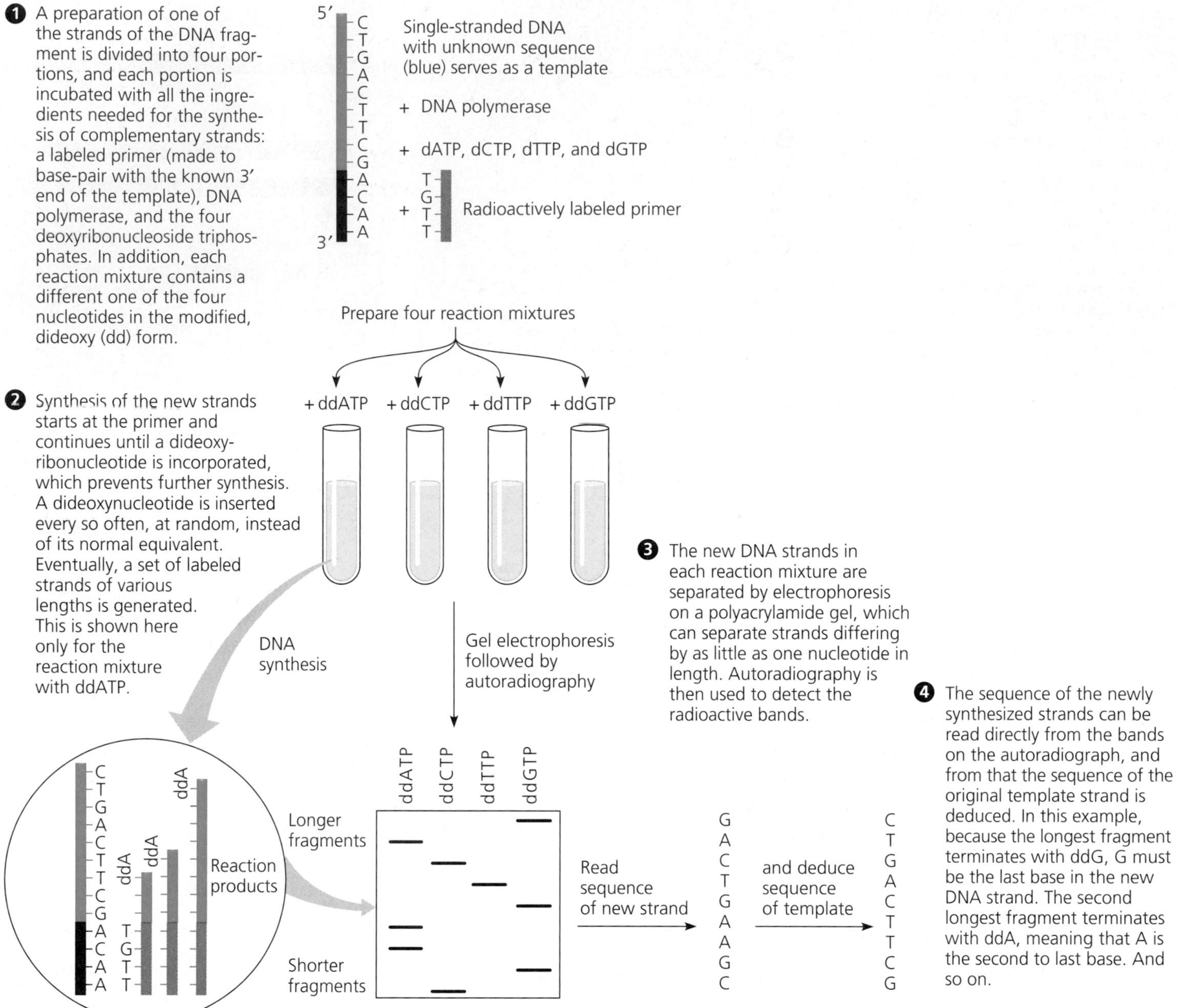

① A preparation of one of the strands of the DNA fragment is divided into four portions, and each portion is incubated with all the ingredients needed for the synthesis of complementary strands: a labeled primer (made to base-pair with the known 3' end of the template), DNA polymerase, and the four deoxyribonucleoside triphosphates. In addition, each reaction mixture contains a different one of the four nucleotides in the modified, dideoxy (dd) form.

5'─C
─T
─G
─A
─C
─T
─T
─C
─G
─A
─C
─A
3'─A

Single-stranded DNA with unknown sequence (blue) serves as a template

+ DNA polymerase

+ dATP, dCTP, dTTP, and dGTP

T─
G─
T─
T─

+ Radioactively labeled primer

Prepare four reaction mixtures

+ ddATP   + ddCTP   + ddTTP   + ddGTP

② Synthesis of the new strands starts at the primer and continues until a dideoxyribonucleotide is incorporated, which prevents further synthesis. A dideoxynucleotide is inserted every so often, at random, instead of its normal equivalent. Eventually, a set of labeled strands of various lengths is generated. This is shown here only for the reaction mixture with ddATP.

DNA synthesis

Gel electrophoresis followed by autoradiography

③ The new DNA strands in each reaction mixture are separated by electrophoresis on a polyacrylamide gel, which can separate strands differing by as little as one nucleotide in length. Autoradiography is then used to detect the radioactive bands.

─C
─T
─G
─A
─C
─T
─T
─C
─G
─A   T─   ddA
─C   G─
─A   T─
─A   T─   ddA   ddA

ddA

Reaction products

ddATP   ddCTP   ddTTP   ddGTP

Longer fragments

Shorter fragments

Read sequence of new strand

G
A
C
T
G
A
A
G
C

and deduce sequence of template

C
T
G
A
C
T
T
C
G

④ The sequence of the newly synthesized strands can be read directly from the bands on the autoradiograph, and from that the sequence of the original template strand is deduced. In this example, because the longest fragment terminates with ddG, G must be the last base in the new DNA strand. The second longest fragment terminates with ddA, meaning that A is the second to last base. And so on.

THE PROCESS OF SCIENCE

FIGURE 20.12 Sequencing of DNA by the Sanger method. Developed by British scientist Frederick Sanger, this method for determining the nucleotide sequence of DNA molecules involves synthesizing *in vitro* DNA strands complementary to one of the strands of the DNA being sequenced. The method is based on the random incorporation of a modified nucleotide, a dideoxyribonucleotide, which terminates a growing DNA chain because it lacks a 3'—OH for attachment of the next nucleotide (see FIGURE 16.11). DNA strands are synthesized that reflect all possible positions of the dideoxynucleotides and thus ultimately the entire DNA sequence. The molecules that are sequenced are cloned restriction fragments, typically several hundred nucleotides long.

Most of the work of DNA sequencing is now automated. Instead of labeling the fragments with radioactive primers, modern sequencing machines use fluorescent dyes to tag the dideoxyribonucleotides, one color for each of the four types of nucleotide. This approach allows the four reactions to be performed in a single tube; the dideoxy ends of the reaction product DNA strands can be distinguished by the color of their fluorescence (see chapter-opening photo).

his vision. The worth of his approach was demonstrated in 1995, at least for prokaryotic genomes, when he and colleagues reported the first complete sequence of an organism, the bacterium *Hemophilus influenzae*. In May 1998, he set up a company, Celera Genomics, and promised a completed human sequence in three years. His whole-genome shotgun approach to sequencing was further validated in March 2000 by the sequence of *Drosophila melanogaster*, carried out in collaboration with academic scientists. As promised, in February 2001, Celera announced the sequencing of over 90% of the human genome—simultaneously with a similar announcement by the public consortium.

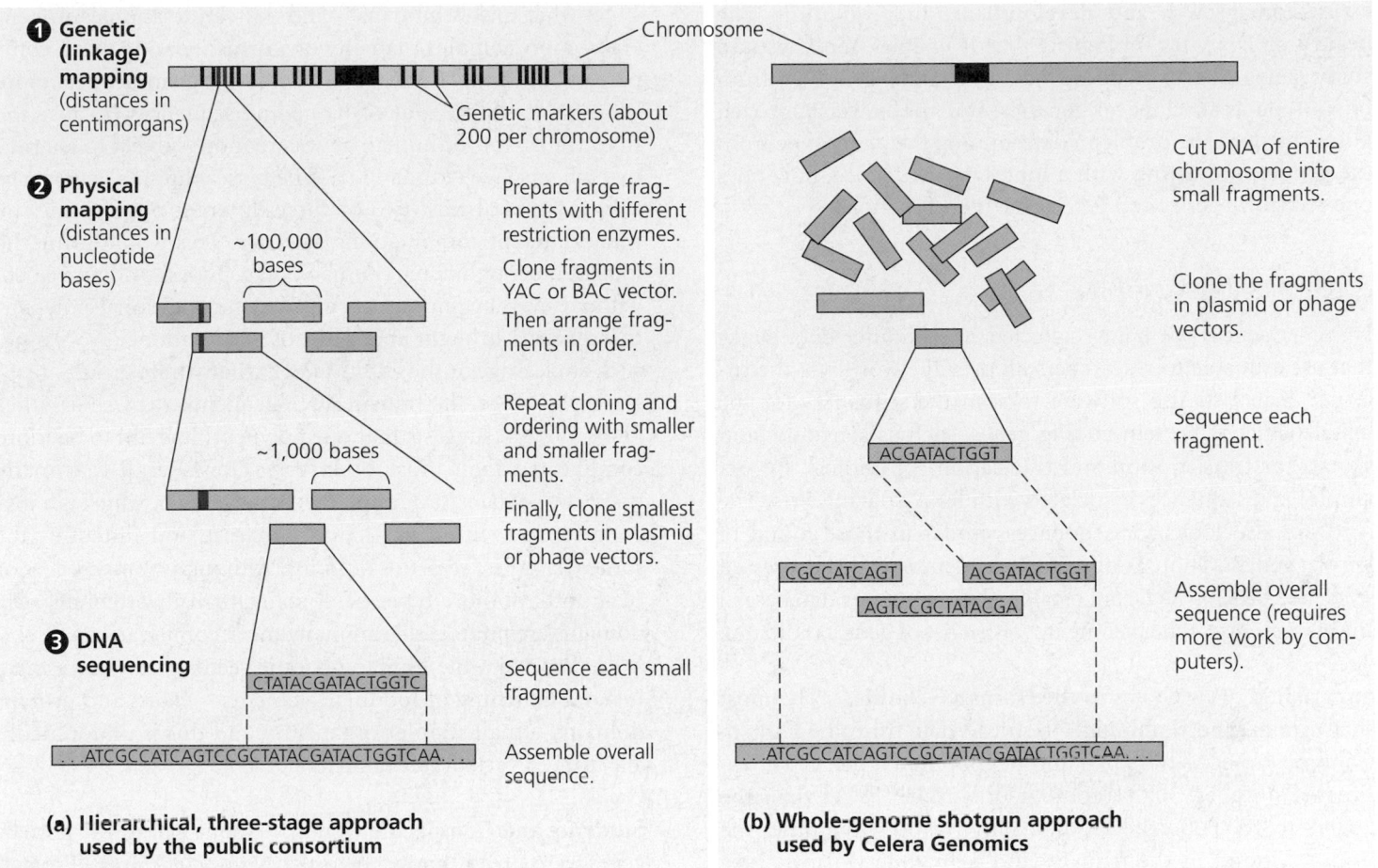

① Genetic (linkage) mapping (distances in centimorgans)

Chromosome

Genetic markers (about 200 per chromosome)

② Physical mapping (distances in nucleotide bases)

~100,000 bases

Prepare large fragments with different restriction enzymes.

Clone fragments in YAC or BAC vector.

Then arrange fragments in order.

Repeat cloning and ordering with smaller and smaller fragments.

~1,000 bases

Finally, clone smallest fragments in plasmid or phage vectors.

③ DNA sequencing

CTATACGATACTGGTC

Sequence each small fragment.

...ATCGCCATCAGTCCGCTATACGATACTGGTCAA...

Assemble overall sequence.

(a) Hierarchical, three-stage approach used by the public consortium

Cut DNA of entire chromosome into small fragments.

Clone the fragments in plasmid or phage vectors.

Sequence each fragment.

ACGATACTGGT

CGCCATCAGT   ACGATACTGGT
          AGTCCGCTATACGA

Assemble overall sequence (requires more work by computers).

...ATCGCCATCAGTCCGCTATACGATACTGGTCAA...

(b) Whole-genome shotgun approach used by Celera Genomics

FIGURE 20.13 Alternative strategies for sequencing an entire genome. In both cases, the arrangement of DNA fragments in order depends on their having overlapping regions.

Representatives of the public consortium have pointed out that Celera made heavy use of their maps and sequence data, which are all—unlike Celera's data—made immediately available for free use, and assert that the infrastructure established by their approach will greatly facilitate the finishing-up phase of the project. Venter, on the other hand, argues for the efficiency and economy of Celera's methods; and indeed, the public consortium has made some use of them. What is clear is that both approaches are valuable and that the competition between the two groups has hastened progress. (See FIGURE 20.13 to review the two strategies.)

As of mid-2001, the genomes of about 50 species have been completely (or almost completely) sequenced. Most are prokaryotes, including *E. coli*, a number of other bacteria (including ones of medical importance), and about 10 archaea. The yeast *Saccharomyces cerevisiae* was the first eukaryote to be completed, and the nematode *C. elegans*, a simple worm, the first multicellular organism. The plant *Arabidopsis thaliana*, another important research organism, has also been completed. Finally, the human genome is well on its way to completion, although there are still many gaps in the sequence.

Because of the presence of repetitive DNA and for other poorly understood reasons, certain parts of the chromosomes of multicellular organisms resist detailed mapping by the usual methods. However, as sequences for more organisms are assembled, the job gets easier. For example, the sequencing of the mouse genome, which is about 85% identical to the human genome, is being greatly aided by what is now known of the human sequence.

On one level these sequences are simply dry lists of nucleotide bases—millions of A's, T's, C's, and G's in mind-numbing succession. But on another level, reached with the help of computer analysis, these sequences are already the basis of some exciting discoveries, which we discuss next.

Genome sequences provide clues to important biological questions

Genomics, the study of genomes based on their DNA sequences, is yielding new insights into fundamental questions about genome organization, the control of gene expression, growth and development, and evolution. The beauty of DNA technology is that it enables geneticists to study genes directly, without having to infer genotype from phenotype as in classical genetics. But the newer approach poses the opposite problem, determining the phenotype from the genotype. Starting with a long DNA sequence, how does one recognize genes and determine their function?

Analyzing DNA Sequences

DNA sequences are being collected in computer data banks that are available to researchers all over the world via the Internet. Scientists use software to scan the sequences for the telltale signs of protein-coding genes, such as start and stop signals for transcription and translation (promoters, for example) and sequences associated with RNA-splicing sites. The software also looks for sequences similar to those found in known genes. (Thousands of such sequences, called expressed sequence tags, or *ESTs*, are cataloged in computer databases.) In this way, researchers come up with a list of gene candidates.

Surprisingly Few Genes in the Human Genome.   The most surprising—and humbling—result to date from the Human Genome Project is the small number of putative genes. The estimated number, 30,000 to 40,000, is much lower than the 50,000 to 100,000 expected and only two to three times the number found in the fruit fly and nematode worm (TABLE 20.1). Relative to the other organisms studied so far, a much smaller fraction of human DNA is genes. Much of the enormous amount of noncoding DNA in the human genome is repetitive DNA, but unusually long introns also contribute significantly. Human introns are typically ten times longer than those of the fly or worm.

Table 20.1 Genome Sizes and Numbers of Genes

Organism	Genome Size	Estimated Number of Genes	Genes per Mb*
H. influenzae (bacterium)	1.8 Mb*	1,700	940
S. cerevisiae (yeast)	12 Mb	6,000	500
A. thaliana (plant)	100 Mb	26,000	260
C. elegans (nematode)	97 Mb	19,000	200
D. melanogaster (fruit fly)	180 Mb	13,000	72
H. sapiens (human)	3,200 Mb	30,000–40,000	~10

*Mb = million base pairs

So what makes humans—and vertebrate animals in general—more complex than flies or worms? For one thing, compared with those of other organisms, our genome "gets more bang for the buck" out of its coding sequences because the RNA transcripts of human genes are more subject to alternative splicing (see FIGURE 19.11). The typical human gene probably specifies at least two or three different polypeptides by using different combinations of exons. This would bring the total number of different human polypeptides to about 90,000, without even beginning to consider the additional polypeptide diversity brought about by post-translational processing, such as cleavage or the addition of carbohydrate groups.

Furthermore, the human DNA sequence (as well as other kinds of data) suggests that our polypeptides tend to be more complicated than those of invertebrates. Recall that many polypeptides consist of more than one *domain,* which is a discrete section with a particular structure and function (for some examples, see FIGURE 19.10). Although humans do not seem to have more types of domains than invertebrates, the domains are put together in many more combinations.

In summary, the human genome seems to do more mixing and matching of modular elements—exons and protein domains—than simpler organisms, and this is undoubtedly a feature of vertebrates in general.

Studying and Comparing Genes.   About half of the human genes were already known before the Human Genome Project, but what of the others? Scientists compare the sequences of new gene candidates with those of known genes from various organisms. In some cases, a newly determined gene sequence will match, at least partially, a gene whose function is well known. For example, part of a new gene may match a known gene that encodes a protein kinase, suggesting that the new gene does, too. In other cases, the new gene sequence will be similar to a sequence encountered before but of unknown function. In still other cases, the sequence may be entirely unlike anything ever seen before. In the organisms that have been sequenced so far, many of the gene candidate sequences are entirely new to science. For example, about a third of the genes of *E. coli,* the best-studied of research organisms, are new to us!

Despite the high complexity of the human genome and the proteins it encodes, comparisons of genome sequences confirm very strongly the evolutionary connections between even distantly related organisms and the relevance of research on simpler organisms to our understanding of human biology. Similarities between genes of disparate organisms can be surprising, to the point that one researcher has said that he now views fruit flies as "little people with wings." Yeast, for example, has a number of genes close enough to the human versions that they can substitute for them in a human cell. In fact, researchers can sometimes work out what a human disease gene does by studying its normal counterpart in yeast. Bacterial sequences reveal unsuspected metabolic pathways that may have industrial or medical uses. Moreover, comparisons of the completed genome sequences of bacteria, archaea, and eukarya strongly support the theory that these are the three fundamental domains of life.

Studying Gene Expression

We study genomes not only to learn about individual genes and their evolution but to learn how genes act together to produce and maintain a functioning organism. A major part of the answer to how we get along with so few genes is likely to be unusually complex networks of interactions among genes and their products. While some scientists are laboring to sequence and analyze their structure, others are using the genome sequences already obtained to study patterns of gene expression from a new, global perspective. These researchers are asking which genes are transcribed in different situations. Their basic strategy is to isolate the mRNA made in particular cells, use these molecules as templates for making a cDNA library by reverse transcription, and then compare this cDNA with other collections of DNA by hybridization. Researchers can see what genes are active at different stages of development, in different tissues, or in tissues in different states of health.

DNA technology makes such gene expression studies possible; automation allows them to be easily performed on a large scale. Scientists can now detect and measure the expression of thousands of genes at one time. A revolutionary new method, described in FIGURE 20.14, uses **DNA microarray assays.** Tiny amounts of a large number of single-stranded DNA fragments representing different genes are fixed to a glass slide in a tightly spaced array (grid). (The array is also called a *DNA chip* by analogy to a computer chip.) Ideally, these fragments represent all the genes of an organism, as is already possible for organisms whose genomes have been completely sequenced. The fragments are tested for hybridization with various samples of cDNA molecules, which have been labeled with fluorescent dyes. An early test of this method compared the genes expressed in the roots and leaves of *Arabidopsis*. The researchers went on to sequence the genes that were more highly expressed in one tissue than in the other. The results were as predicted;

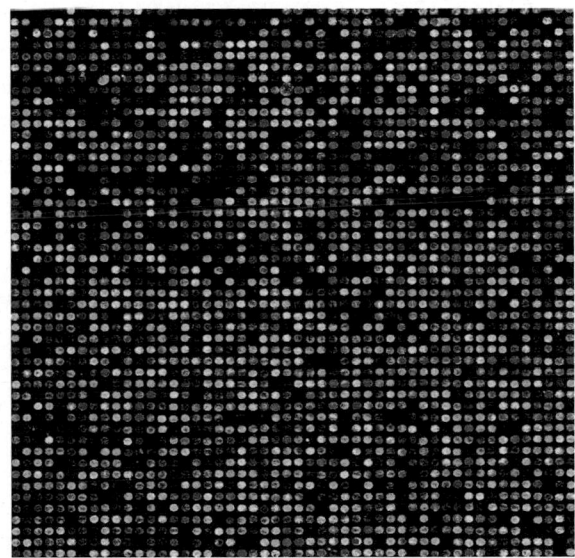

(b) Enlarged photo of a microarray representing 2,400 human genes

FIGURE 20.14 DNA microarray assay for gene expression. With this method, a researcher can simultaneously test all the genes expressed in a particular tissue for hybridization with an array of short DNA sequences representing thousands of genes, even all the genes of the organism. The cDNA molecules made in part (a) step 2 and added to every well of the microarray in step 3 represent the genes that were active in the tissue. In step 4, the spots where any of the cDNA hybridizes fluoresce with an intensity indicating the relative amount of the mRNA that was in the tissue. Molecules of cDNA representing more than one tissue or the same kind of tissue under different conditions can be tested together by using a different colored label for each.

THE PROCESS OF SCIENCE

Tissue sample

❶ Isolate mRNA.

mRNA molecules

❷ Make cDNA by reverse transcription, using fluorescently labeled nucleotides.

Labeled cDNA molecules (single strands)

❸ Hybridization: Apply the cDNA mixture to a DNA microarray.

Fixed to each spot on a microscope slide are copies of a short single-stranded DNA molecule representing one gene of the organism, a different gene in each spot.

❹ Rinse off excess cDNA; scan microarray for fluorescence.

Each fluorescent spot represents a gene expressed in the tissue sample.

Size of an actual DNA microarray with all the genes of yeast (6,400 spots)

(a) Procedure using labeled cDNA prepared from a tissue sample

genes encoding known photosynthetic enzymes, for instance, were turned on in the leaves but not in the roots. But more important than confirming predictions is the method's potential for uncovering new genes and gene interactions and providing clues to gene function. For example, DNA microarray assays are being used to compare cancerous with noncancerous tissues. Studying the differences in gene expression may lead researchers to new diagnostic techniques and biochemically targeted treatments, as well as a fuller understanding of cancer. Ultimately, information from microarray assays should provide us a grander view: how ensembles of genes interact to form a living organism.

Determining Gene Function

Perhaps the most interesting genes discovered in genome sequencing and gene expression studies are those whose functions are completely mysterious. How do researchers discover the functions of these genes? The main idea is to disable the gene in some way and hope that the consequences in the cell or organism will provide clues to the gene's normal function. One powerful approach is **in vitro mutagenesis,** a technique that can be used to introduce specific changes into the sequence of a cloned gene. Such mutations can alter or destroy the function of the protein product. Then, when the mutated gene is returned to a cell, it may be possible to determine the function of the missing normal protein by examining the phenotype of the mutant. Researchers can even put such a mutated gene into cells from the early embryo of a multicelled organism (such as a mouse) to study the role of the gene in the development and functioning of the whole organism.

Some researchers working with nonmammalian organisms have recently begun using a simpler and faster method to silence the expression of selected genes. Called **RNA interference (RNAi),** the method uses synthetic double-stranded RNA molecules matching the sequence of a particular gene to trigger breakdown of the gene's messenger RNA. How the double-stranded RNA brings about this effect in the cell is still largely a mystery, but the process is a natural one that probably evolved to protect cells from viruses and restrain the movement of retrotransposons (which can disrupt essential genes). Scientists have only recently achieved some success in using the method to silence genes in mammalian cells, but in other kinds of organisms, such as the nematode, RNAi has already proved valuable.

Future Directions in Genomics

The success in mapping and sequencing genomes is encouraging scientists to move ahead to **proteomics,** the systematic study of the full protein sets *(proteomes)* encoded by genomes. Proteomics is a new kind of challenge. Because of alternative splicing of RNA and modifications of proteins after transla-

tion, the number of proteins in humans and our close relatives is likely to far exceed the number of genes. And collecting all the proteins will be difficult because a cell's proteins differ with the cell type and its state. Moreover, unlike DNA, proteins are extremely varied in structure and chemical and physical properties. However, because proteins are the molecules that actually carry out the activities of the cell, we must study them to learn how cells and organisms function. Ongoing technical advances, such as the recent development of protein microarrays for studying protein interactions, will be necessary to meet the challenge.

Genomics and proteomics are giving biologists an increasingly global perspective on the study of life. In an essay in *Science,* noted MIT biologists Eric Lander and Robert Weinberg predict that the availability of "complete parts lists of organisms"—that is, catalogs of genes and proteins—will change the discipline of biology dramatically:

> The 21st century discipline will focus increasingly on the study of entire biological systems, by attempting to understand how component parts collaborate to create a whole. For the first time in a century, reductionists [are yielding] ground to those trying to gain a holistic view of cells and tissues.[*]

Advances in **bioinformatics,** the application of computer science and mathematics to genetic and other biological information, will play a crucial role in dealing with the enormous masses of data involved.

As Landers and Weinberg point out elsewhere in their essay, we are now poised to understand the spectrum of genetic variation in humans. Because the history of the human species is so short—we are probably all descended from a small population living in Africa 150,000 to 200,000 years ago—the amount of DNA variation in humans is small compared to that of other species. Most of our diversity seems to be in the form of **single nucleotide polymorphisms** (**SNPs,** pronounced "snips"), which are single base-pair variations in the genome.[†]

In the human genome, SNPs occur on average about once in 1,000 base pairs. In other words, if you could compare your personal DNA sequence with that of the person sitting next to you—or with that of someone on the other side of the world—you would find them to be 99.9% identical.

Scientists are already well on their way to identifying the locations of the 3 million or so SNP sites in the human genome. These will be useful genetic markers for studying human evolution, the differences between human popula-

---

[*] E. S. Lander and R. A. Weinberg, "Genomics: Journey to the Center of Biology," *Science* 287 (March 10, 2000): 1777.

[†] You may wonder about the relationship between SNPs and RFLPs. Some RFLPs are SNPs—for example, the single base-pair difference in FIGURE 20.9a. But some SNPs do not affect restriction sites; they must be identified by DNA sequencing. And some RFLPs involve variations in more than one base pair.

tions, and the migratory routes that led from Africa to the other continents. They will also be valuable markers for identifying disease genes and genes that affect our health in more subtle ways by influencing susceptibility to certain diseases or sensitivities to environmental toxins or drugs. This is likely to change the practice of medicine later in the 21st century. However, applications of DNA research and technology are already affecting our lives in many ways. We devote the rest of the chapter to a discussion of some of these applications and related issues.

PRACTICAL APPLICATIONS OF DNA TECHNOLOGY

Hardly a day goes by that DNA technology is not in the news. More often than not, the topic of the story is a new and promising application to a medical problem.

DNA technology is reshaping medicine and the pharmaceutical industry

Modern biotechnology is making enormous contributions to medicine, both in the diagnosis of diseases and in the development of pharmaceutical products. One obvious benefit of DNA technology and of the Human Genome Project is the identification of genes whose mutation is responsible for genetic diseases, for these discoveries could lead to ways of diagnosing, treating, and perhaps even preventing those conditions.

Equally important are the potential benefits of DNA technology for dealing with other kinds of diseases, from arthritis to AIDS. Susceptibility to many "nongenetic" diseases is influenced by a person's genes. Furthermore, diseases of all sorts involve changes in gene expression within the affected cells and often within the patient's immune system. By using DNA microarray assays or other techniques to compare gene expression in healthy and diseased tissues, researchers hope to find many of the genes that are turned on or off in particular diseases. These genes and their products are potential targets for prevention or therapy.

Diagnosis of Diseases

A new chapter in the diagnosis of infectious diseases has been opened by DNA technology, in particular the use of PCR and labeled nucleic acid probes to track down certain pathogens. For example, because the sequence of HIV DNA is known, PCR can be used to amplify, and thus detect, HIV DNA in blood or tissue samples. This is often the best way to detect an otherwise elusive infection.

Medical scientists can now diagnose hundreds of human genetic disorders using DNA technology. Increasingly, they

can identify individuals with genetic diseases before the onset of symptoms, even before birth. It is also possible to identify symptomless carriers of potentially harmful recessive alleles. Genes have been cloned for many human diseases, including hemophilia, cystic fibrosis, and Duchenne muscular dystrophy.

Hybridization analysis makes it possible to detect abnormal allelic forms of genes present in DNA samples. Even in cases where the gene has not yet been cloned, the presence of an abnormal allele can be diagnosed with reasonable accuracy if a closely linked RFLP marker has been found (FIGURE 20.15). Alleles for Huntington's disease and a number of other genetic diseases were first detected in this indirect way. The underlying principle is that the closeness of the marker to the gene itself makes it very unlikely that crossing over will occur between the marker and the gene during gamete formation; therefore, the marker and gene will almost always stay together in inheritance. (The same principle obviously applies to all kinds of markers, including the SNPs that may soon come into diagnostic use.) Once a gene is mapped precisely, it can be cloned for study and for use as a probe for finding identical or similar DNA, as is now the case for Huntington's disease, cystic fibrosis, and many other diseases.

Human Gene Therapy

Techniques for manipulating DNA hold great potential for treating a variety of diseases by **gene therapy**—the alteration of an afflicted individual's genes. In people with disorders traceable to a single defective gene, it should theoretically be possible to replace or supplement the defective gene with a normal allele. The new allele could be inserted into the somatic cells of the tissue affected by the disorder.

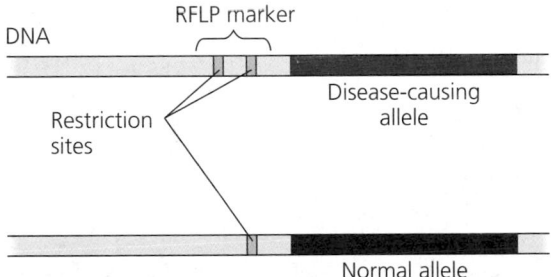

FIGURE 20.15 RFLP markers close to a gene. Even if a disease-causing allele has not been cloned and its precise locus is unknown, its presence can sometimes be detected with high (though not perfect) accuracy by testing for the presence of RFLP markers that are very close to the gene in question. This diagram depicts homologous segments of DNA from a family in which some members have a genetic disease. In this family, different versions of a RFLP marker are associated with the different alleles, allowing the test to be applied. If a family member has inherited the version of the RFLP marker with two restriction sites (rather than one), there is a high probability that the individual has also inherited the disease-causing allele.

For gene therapy of somatic cells to be permanent, the cells that receive the normal allele must be ones that multiply throughout the patient's life. Bone marrow cells, which include the *stem cells* that give rise to all the cells of the blood and immune system, are prime candidates. FIGURE 20.16 outlines one possible procedure for a situation in which bone marrow cells are failing to produce a vital enzyme because of a single defective gene. Some bone marrow cells are removed from the patient, the normal allele inserted by a viral vector, and the modified cells returned to the patient. If the procedure succeeds, the cells will multiply throughout the patient's life and express the normal gene. The engineered cells will supply the missing protein, and the patient will be cured.

But despite "hype" in the news media over the past decade, there has been very little scientifically strong evidence of effective gene therapy reported. Even when genes are successfully and safely transferred and are being expressed in their new host, their activity typically diminishes after a short period.

For this reason and because of concerns about safety, most of the current gene therapy trials now under way in humans are not aimed at correcting genetic defects. Instead, researchers are looking for ways to use gene therapy in the fight against major killers such as heart disease and cancer. The most promising trials are those in which a very limited period of activity by the transferred genes is not only sufficient but desirable. For example, one idea is to help treat coronary artery disease by introducing into the heart muscle a gene encoding a growth factor that stimulates new blood vessels to grow around blocked arteries. After success with pigs, two groups of researchers are now carrying out preliminary trials designed to assess safety in humans. One group is using the same vector used in the pig studies: an adenovirus (a major cause of the common cold) modified so it cannot cause disease; the other group is injecting the naked growth factor gene directly into the muscle. The viral vectors are rapidly destroyed by the immune system, and the direct injection method is very inefficient. But the goal is simply to get the heart cells to produce enough growth factor to trigger a brief period of blood vessel growth. Preliminary results are encouraging.

Many technical questions are posed by gene therapy. For example, how can the activity of the transferred gene be controlled so that cells make appropriate amounts of the gene product at the right time and in the right place? How can we be sure that the insertion of the therapeutic gene does not harm some other necessary cell function? Information about DNA control elements from genome sequencing and large-scale gene expression studies that teach us about gene interactions may help answer such questions.

Gene therapy raises some difficult ethical and social questions. Some critics suggest that tampering with human genes in any way, even in somatic cells and even to treat individuals who have life-threatening diseases, is wrong. They argue that it will inevitably lead to the practice of eugenics, a deliberate effort to control the genetic makeup of human populations. Other observers see no fundamental difference between genetic engineering of somatic cells and other conventional medical interventions to save lives. They compare the transplantion of genes to the transplantation of organs.

The most difficult ethical question is whether we should try to treat human germ-line cells in the hope of correcting the defect in future generations. In laboratory mice at least, transferring foreign genes into the germ line (egg cells) is now a routine procedure. Thus, despite the challenges, it is clear that the technical problems relating to similar genetic engineering in humans will eventually be solved. We will then have to face the question of whether it is advisable, under any circumstances, to alter the genomes of human germ lines or embryos. Should we interfere with evolution in this way?

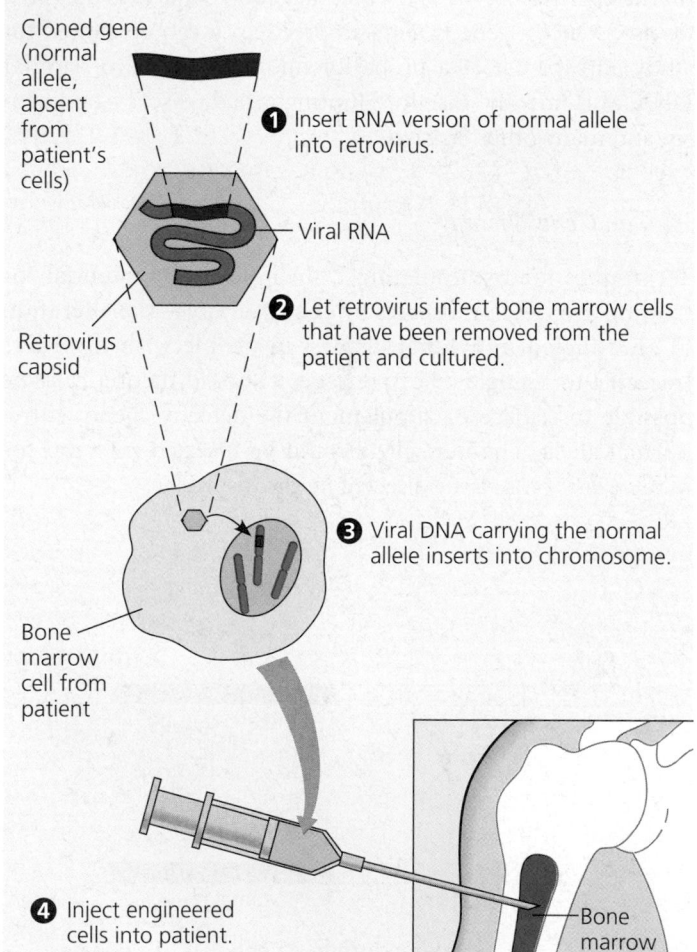

Cloned gene (normal allele, absent from patient's cells)

Viral RNA

Retrovirus capsid

Bone marrow cell from patient

❶ Insert RNA version of normal allele into retrovirus.

❷ Let retrovirus infect bone marrow cells that have been removed from the patient and cultured.

❸ Viral DNA carrying the normal allele inserts into chromosome.

❹ Inject engineered cells into patient.

Bone marrow

FIGURE 20.16 One type of gene therapy procedure. A retrovirus vector that has been rendered harmless is used to introduce a normal allele of a gene into the cells of a patient who lacks it. The method takes advantage of the fact that a retrovirus inserts a DNA transcript of its RNA genome into the chromosomal DNA of its host cell (see FIGURE 18.7). If the viral nucleic acid includes a foreign gene and that gene is expressed, the cell and its descendants will possess the gene product and may be cured. Cells that reproduce throughout life, such as bone marrow cells, are ideal candidates for gene therapy.

From a biological perspective, the elimination of unwanted alleles from the gene pool could backfire. Genetic variety is a necessary ingredient for the survival of a species as environmental conditions change with time. Genes that are damaging under some conditions may be advantageous under other conditions (one example is the sickle-cell allele, discussed in Chapter 14). Are we willing to risk making genetic changes that could be detrimental to our species in the future? We may have to face this question soon.

Pharmaceutical Products

DNA technology has been used to create many useful pharmaceutical products, mostly proteins. By transferring the gene for a desired protein product into a bacterium, yeast, or other kind of cell that is easy to grow in culture, one can produce large quantities of proteins that are present naturally in only minute amounts.

Using DNA technology to put highly active promoters (and other gene control elements) into vector DNA, scientists create expression vectors that enable the host cell to make large amounts of the product of a gene inserted into the vector. In addition, host cells can be engineered to secrete a protein as it is made, thereby simplifying the task of purifying it by traditional biochemical methods.

One of the first practical applications of gene splicing was the production of mammalian hormones and other mammalian regulatory proteins in bacteria. Human insulin and human growth hormone (HGH) were among the earliest examples. The insulin produced in this way has greatly benefited the 2 million diabetics in the United States who depend on insulin treatment to control their disease; previously they had to rely on insulin from pigs and cattle, which is not identical to human insulin. Human growth hormone has been a boon to children born with hypopituitarism, a form of dwarfism caused by inadequate amounts of HGH, and may prove to have other uses, such as the healing of injuries.

Another important pharmaceutical product produced by genetic engineering is tissue plasminogen activator (TPA). This protein helps dissolve blood clots and reduces the risk of subsequent heart attacks if administered very shortly after an initial attack. However, TPA illustrates a problem with many genetically engineered products: Because the development costs were high and the market is relatively limited, it is expensive.

The most recent developments in pharmaceutical products involve truly novel ways to fight certain diseases that do not respond to traditional drug treatments. One approach is the use of genetically engineered proteins that either block or mimic surface receptors on cell membranes. One such experimental drug mimics a receptor protein that HIV binds to in entering white blood cells. The HIV binds to the drug molecules instead and fails to enter the blood cells.

For many viral diseases for which there are no effective drug treatments, prevention by vaccination is virtually the only way to fight the disease. A **vaccine** is a harmless variant or derivative of a pathogen that stimulates the immune system to fight the pathogen. Traditional vaccines for viral diseases are of two types: particles of a virulent virus that have been inactivated by chemical or physical means, and active virus particles of an attenuated (nonpathogenic) viral strain. In both cases, the virus particles are similar enough to the active pathogen to trigger an immune response (see Chapter 43).

Recombinant DNA techniques can generate large amounts of a specific protein molecule normally found on the surface of a pathogen. If the protein, referred to as a subunit, is one that triggers an immune response against the intact pathogen, then it can be used as a vaccine. Alternatively, genetic engineering methods can be used to modify the genome of the pathogen to attenuate it. A vaccine consisting of an attenuated microbe is often more effective than a subunit vaccine because it usually triggers a greater response by the immune system. Pathogens attenuated by gene-splicing techniques may be safer than the natural mutants traditionally used.

A completely new development in the use of recombinant DNA technology by the pharmaceutical industry is the genetic modification of plants to produce vaccines and rare human proteins of medical importancee. We'll discuss this application, often called "pharming," in the next section.

DNA technology offers forensic, environmental, and agricultural applications

Forensic Uses of DNA Technology

In violent crimes, blood or small amounts of other tissue may be left at the scene or on the clothes or other possessions of the victim or assailant. If rape is involved, small amounts of semen may be recovered from the victim's body. If enough tissue or semen is available, forensic laboratories can determine the blood type or tissue type by using antibodies to test for specific cell surface proteins. However, such tests require fairly fresh tissue in relatively large amounts. Also, because there are many people in the population with the same blood or tissue type, this approach can only exclude a suspect; it cannot provide strong evidence of guilt.

DNA testing, on the other hand, can identify the guilty individual with a much higher degree of certainty, because the DNA sequence of every person is unique (except for identical twins). RFLP analysis by Southern blotting is a powerful method for the forensic detection of similarities and differences in DNA samples and requires only tiny amounts of blood or other tissue (about 1,000 cells). In a murder case, for example, this method can be used to compare DNA samples from the suspect, the victim, and a small amount of blood found at the crime scene. Radioactive probes mark the electrophoresis bands that contain certain RFLP markers. The forensic scientist usually tests for about five markers; in other words, only a few selected portions of the DNA are tested. However, even such a small set of markers from an individual can provide a **DNA fingerprint,** or specific pattern of bands, that is of forensic use, because the probability that two people (who are not identical twins) would have the exact same set of RFLP markers is very small. The autoradiograph in FIGURE 20.17 resembles the type of evidence presented (with explanation) to juries in murder trials.

The forensic use of DNA fingerprinting extends beyond violent crimes. For instance, comparing the DNA of a mother, her child, and the purported father can conclusively settle a question of paternity. Sometimes paternity is of historical interest: Recently, DNA fingerprinting provided strong evidence that Thomas Jefferson fathered at least one of the children of his slave Sally Hemings.

Today, instead of RFLPs, variations in the lengths of satellite DNA are increasingly used as markers for DNA fingerprinting. Recall from Chapter 19 that satellite DNA consists of tandemly repeated base sequences within the genome. The most useful satellite sequences for forensic purposes are microsatellites, which are roughly 10 to 100 base pairs long, have repeating units of only a few base pairs, and are highly variable from person to person. For example, one individual may have the unit ACA repeated 65 times at one genome locus, 118 times at a second locus, and so on, whereas another individual is likely to have different numbers of repeats at these loci. Such polymorphic genetic loci are usually called **simple tandem repeats (STRs).** Restriction fragments containing STRs vary in size among individuals because of differences in STR lengths, rather than because of different numbers of restriction sites within that region of the genome, as in RFLP analysis. The greater the number of markers examined in a DNA sample, the more likely it is that the DNA fingerprint is unique to one individual. PCR is often used to selectively amplify particular STRs or other markers before electrophoresis. Because of its selective power, PCR is especially valuable when the DNA is in poor condition or available only in minute quantities. A tissue sample as small as 20 cells can be sufficient for PCR.

Just how reliable is DNA fingerprinting? The DNA fingerprint of an individual would be truly unique if it were feasible to perform restriction fragment analysis on the person's entire genome. In practice, as already mentioned, forensic DNA tests focus on only about five tiny regions of the genome. However, the DNA regions chosen are ones known to be highly variable from one person to another. In most forensic cases, the probability of two people having identical DNA fingerprints is between one chance in 100,000 and one in a billion. The exact figure depends on the number of markers compared and on the frequency of those markers in the population. Information on how common various markers are in different ethnic

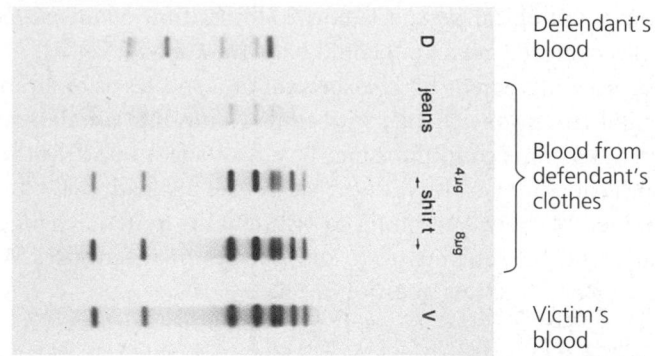

FIGURE 20.17 DNA fingerprints from a murder case. As revealed by RFLP analysis, DNA from bloodstains on the defendant's clothes matches the DNA fingerprint of the victim but differs from the DNA fingerprint of the defendant. This is evidence that the blood on the defendant's clothes came from the victim, not the defendant. Ten different RFLP markers appear at different positions on this one "autorad" (autoradiograph). It is now more common to make separate autorads for each marker tested. The DNA bands resulting from electrophoresis are exposed to various probes in succession, with the previous probe washed off before the next one is applied.
Cellmark Diagnostics, Germantown, MD

groups is key because these marker frequencies may be very different from frequencies in the population as a whole. Such data now enable forensic scientists to make extremely accurate statistical calculations. Thus, despite problems that can still arise from insufficient statistical data, human error, or flawed evidence, DNA fingerprints are now accepted as compelling evidence by legal experts and scientists alike. Many argue that DNA evidence is more reliable than eyewitnesses in placing a suspect at the scene of a crime.

Environmental Uses of DNA Technology

Increasingly, genetic engineering is being applied to environmental work. The ability of microorganisms to transform chemicals is remarkable, and scientists are now engineering these metabolic capabilities into organisms that will help cope with some environmental problems. For example, many bacteria can extract heavy metals, such as copper, lead, and nickel, from their environments and incorporate the metals into compounds such as copper sulfate or lead sulfate, which are readily recoverable. Genetically engineered microbes may become important in both mining minerals (especially as ore reserves are depleted) and cleaning up highly toxic mining wastes.

The metabolic diversity of microbes is also employed in dealing with wastes from other sources. Sewage treatment plants rely on the ability of microbes to degrade many organic compounds into nontoxic form. However, an increasing number of potentially harmful compounds being released into the environment are not readily degraded by naturally occurring microbes; chlorinated hydrocarbons are a prime example. Biotechnologists are trying to engineer microbes to degrade these compounds. These microbes could be used in wastewater treatment plants or by manufacturers before the compounds are ever released into the environment.

A related research area is the identification and engineering of microbes capable of detoxifying specific toxic wastes found in spills and waste dumps. For example, bacterial strains have been developed that can degrade some of the compounds released during oil spills. The ability to move the genes responsible for these transformations into different organisms allows the development of strains that can survive the harsh conditions of these environmental disasters and still help detoxify the wastes.

Agricultural Uses of DNA Technology

Scientists are working to learn more about the genomes of agriculturally important plants and animals, and for a number of years they have been using DNA technology to try to improve agricultural productivity.

Animal Husbandry and "Pharm" Animals. DNA technology is now routinely used to make vaccines and growth hormones for treating farm animals and, on a still largely experimental basis, to make **transgenic organisms,** whose genomes carry genes from another species. The goals of creating a transgenic animal are often the same as the goals of traditional breeding— for instance, to make a sheep with better quality wool, a pig with leaner meat, or a cow that will mature in a shorter time. Scientists might, for example, identify and clone a gene that causes the development of larger muscles (muscles make up most of the meat we eat) in one variety of cattle and transfer it to other cattle or even to sheep.

Another type of transgenic animal is one engineered to be a pharmaceutical "factory"—a producer of a large amount of an otherwise rare biological substance for medical use (FIGURE 20.18). In most cases to date, a gene for a desired human protein, such as a hormone or blood-clotting factor, has been added to the genome of a farm mammal in such a way that the gene's product is secreted in the animal's milk. It can then be purified, usually more easily than from a cell culture or a transgenic plant.

Human proteins produced by farm animals may or may not be structurally identical to the natural human protein, so they have to be tested very carefully to make sure they will not cause allergic reactions or other adverse effects in patients receiving them. Also, the health and welfare of farm animals carrying genes from humans and other foreign species are important issues; problems such as low fertility or increased susceptibilty to disease are not uncommon.

How is a transgenic animal created? Scientists first remove egg cells from a female and fertilize them *in vitro*. Meanwhile, they clone the desired gene from another organism. They then

FIGURE 20.18 "Pharm" animals. These transgenic sheep carry a gene for a human blood protein, which they secrete in their milk. This protein inhibits an enzyme that contributes to lung damage in patients with cystic fibrosis and some other chronic respiratory diseases. Easily purified from the sheep's milk, the protein is currently being tested as a treatment for cystic fibrosis.

inject the cloned DNA directly into the nuclei of the eggs. Some of the cells integrate the foreign DNA into their genomes and are able to express the foreign gene. The engineered eggs are then surgically implanted in a surrogate mother. If an embryo develops successfully, the result is a transgenic animal, containing a gene from a third "parent" that may even be of another species.

Genetic Engineering in Plants. Agricultural scientists have already provided a number of crop plants with genes for desirable traits, such as delayed ripening and resistance to spoilage and disease. In one striking way, plants are easier to engineer than most animals. For many plant species, a single tissue cell grown in culture can give rise to an adult plant (see FIGURE 21.5). Thus, genetic manipulations can be performed on a single cell and the cell then used to generate an organism with new traits.

The vector used to introduce new genes into plant cells is most often a plasmid from the soil bacterium *Agrobacterium tumefaciens*. This is the **Ti plasmid,** so called because in nature it induces tumors (called crown galls) in plants infected by the bacterium. The Ti plasmid integrates a segment of its DNA, known as T DNA, into the chromosomal DNA of its host plant cells. For vector purposes, researchers work with a version of the plasmid that does not cause disease.

Foreign genes can be inserted into the Ti plasmid using recombinant DNA techniques. The recombinant plasmid is either put back into *Agrobacterium*, which can then be used to infect plant cells growing in culture, or introduced directly into plant cells, where it inserts itself into the plant's chromosomes. Then, taking advantage of the capacity of those cells to regenerate whole plants, it is possible to produce plants that

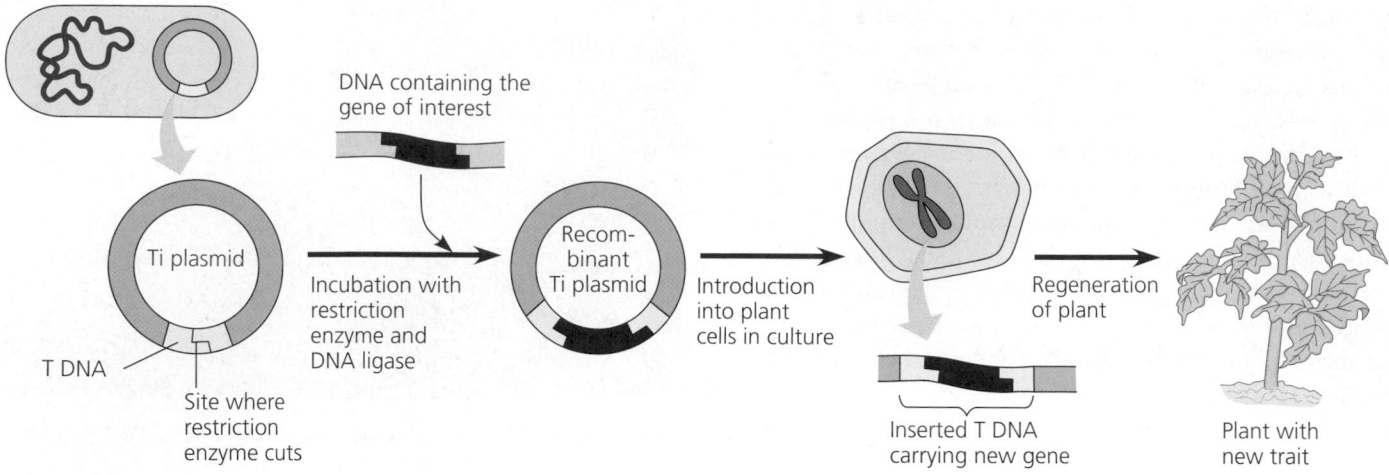

Agrobacterium tumefaciens

DNA containing the gene of interest

Ti plasmid

T DNA

Site where restriction enzyme cuts

Incubation with restriction enzyme and DNA ligase

Recombinant Ti plasmid

Introduction into plant cells in culture

Regeneration of plant

Inserted T DNA carrying new gene

Plant with new trait

❶ The Ti plasmid is isolated from the bacterium *Agrobacterium tumefaciens*, and a fragment of foreign DNA is inserted into its T region by standard recombinant DNA techniques.

❷ When the recombinant plasmid is introduced into cultured plant cells, the T DNA integrates into the plant's chromosomal DNA.

❸ As the plant cell divides, each of its descendants receives a copy of the T DNA and any foreign genes it carries. If an entire plant is regenerated, all its cells will carry—and may express—the new genes.

FIGURE 20.19 Using the Ti plasmid as a vector for genetic engineering in plants.

contain and express the foreign gene and pass it on to their offspring (FIGURE 20.19).

A major drawback to using the Ti plasmid as a vector is that only dicots (plants with two seed leaves) are susceptible to infection by *Agrobacterium.* Monocots, including agriculturally important grasses such as corn and wheat, cannot be infected by *Agrobacterium.* Fortunately, scientists can use newer techniques, such as electroporation and DNA guns (see FIGURE 38.17), to get DNA into the cells of these plants.

Genetic engineering is quickly replacing traditional plant-breeding programs, especially in cases where useful traits are determined by one or only a few genes. In the past few years, roughly half of the American crops of soybeans and corn have been grown from genetically modified seed. Many of the transgenic plants now being grown have received genes for herbicide resistance. For example, the development of cotton plants that carry a bacterial gene that makes the plants resistant to weed-killing herbicides has made it easier to grow crops while still ensuring that weeds are destroyed. In addition, a number of crop plants are being engineered to resist infectious microbes and pest insects (see p. 798). Growing insect-resistant plants reduces the use of chemical insecticides.

Scientists are also using gene transfer to improve the nutritional value of crop plants. One exciting recent advance has been the development of a transgenic rice plant that produces yellow rice grains containing beta-carotene, which our body uses to make vitamin A (FIGURE 20.20). This rice could help prevent vitamin A deficiency in the half of the world's population that depends on rice as a staple food. Vitamin A deficiency is a common problem. Currently, 70% of children under the age of 5 in Southeast Asia suffer from this condition, which leads to vision impairment and increases susceptibility to disease.

An important potential use of DNA technology in improving the nutrition of the world's population involves nitrogen fixation. Nitrogen fixation is the conversion of atmospheric nitrogen gas, which plants cannot use, to nitrogen compounds that plants can convert to essential nitrogen-containing molecules, such as amino acids. In nature, nitrogen fixation is performed by certain bacteria that live in the soil or in plant roots.

Genetically modified rice

Ordinary rice

FIGURE 20.20 "Golden" rice contrasted with ordinary rice. Beta-carotene in the golden rice kernels in this photograph is responsible for both their color and their increased nutritional value. The genes that give the plant its ability to make this vitamin inside its kernels come from daffodils and a bacterium. The vector was the Ti plasmid of *Agrobacterium fascians.*

Even so, the level of nitrogen compounds in the soil is often so low that fertilizers must be applied for crops to grow. Nitrogen-providing fertilizers are costly and contribute to water pollution. DNA technology offers ways to increase bacterial nitrogen fixation and eventually, perhaps, to engineer crop plants to fix nitrogen themselves.

Finally, we arrive at the surprising new alliance between the pharmaceutical industry and agriculture mentioned earlier. The pharmaceutical industry has long made use of plants as sources of drugs. Now, however, DNA technology has made it possible to create plants that produce human proteins for medical use and viral proteins for use as vaccines. Several such "pharm" products are now being tested in clinical trials, including vaccines for hepatitis B and an antibody produced in transgenic tobacco plants that blocks the bacteria that cause tooth decay. Large amounts of these proteins might be made more economically by plants than by cultured cells.

DNA technology raises important safety and ethical questions

As soon as scientists realized the power of DNA technology, they began to worry about potential dangers. Early concerns focused on the possibility that recombinant DNA technology might create hazardous new pathogens. What might happen, for instance, if cancer cell genes were transferred into bacteria or viruses? Scientists developed a set of guidelines that in the United States and some other countries have become formal government regulations.

One type of safety measure is a set of strict laboratory procedures designed to protect researchers from infection by engineered microbes and to prevent the microbes from accidentally leaving the laboratory. In addition, strains of microorganisms to be used in recombinant DNA experiments are genetically crippled to ensure that they cannot survive outside the laboratory. Finally, certain obviously dangerous experiments have been banned. Today, most public concern about possible hazards centers not on recombinant microbes but on **genetically modified (GM) organisms** used in agriculture. In common parlance, a "GM organism" is one that has acquired one or more genes by artificial means; the gene need not be from another species.

Animals that have been genetically modified by artificial means are still not part of our food supply, but GM crop plants are. In 1999, controversy about the safety of these foods exploded in the United Kingdom—where one of the more extreme headlines warned of "the mad forces of genetic darkness"—and soon spread through Europe. In response to these concerns, the European Union suspended the introduction of new GM crops pending new legislation and started considering the possibility of banning the import of all GM foodstuffs. In the United States and other countries where the GM revolution had been proceeding more quietly, the labeling of GM foods as such is now being debated.

Early in 2000, negotiators from 130 countries (including the United States) agreed on a Biosafety Protocol that requires exporters to identify GM organisms present in bulk food shipments and allows importing countries to decide whether they pose environmental or health risks. This agreement has been hailed as a breakthrough by environmentalists.

Advocates of a cautious approach fear that crops carrying genes from other species might somehow be hazardous to human health or cause ecological harm to the environment. A major concern is that transgenic plants might pass their new genes to close relatives in nearby wild areas. We know that lawn and crop grasses, for example, commonly exchange genes with wild relatives via pollen transfer. If domestic plants carrying genes for resistance to herbicides, diseases, or insect pests pollinated wild ones, the offspring might become "superweeds" very difficult to control. However, researchers may be able to prevent the escape of such plant genes by engineering plants so that they cannot hybridize. In April 2000, the U.S. National Academy of Sciences released a study finding no scientific evidence that crops genetically modified to resist pests pose any special health or environmental risks, but the authors of the study also recommended more stringent regulations than now exist. To date, there is little good data on either side; more study is needed.

Today, governments and regulatory agencies throughout the world are grappling with how to facilitate the use of biotechnology in argiculture, industry, and medicine while ensuring that new products and procedures are safe. In the United States, all projects are evaluated for potential risks by various regulatory agencies, including the Food and Drug Administration, the Environmental Protection Agency, the National Institutes of Health, and the Department of Agriculture. These agencies are under increasing pressure from some consumer groups. Meanwhile, these same agencies must consider some of the ethical questions raised by the new biotechnology.

As with all new technologies, developments in DNA technology have ethical overtones. Obtaining a complete map of the human genome, for example, opens the door to significant ethical questions. Who should have the right to examine someone else's genes? How should that information be used? Should a person's genome be a factor in suitability for a job or eligibility for insurance? Ethical considerations, as well as concerns about potential environmental and health hazards, will likely slow the application of the products of the new biotechnology. There is always a danger that too much regulation will stifle basic research and its potential benefits. However, the power of DNA technology and genetic engineering—our ability to profoundly and rapidly alter species that have been evolving for millennia—demands that we proceed with humility and caution.

CHAPTER 20 REVIEW

Go to the Campbell Biology website (www.campbelbiology.com) to explore an interactive version of the Chapter Review.

Summary of Key Concepts

DNA CLONING

- **DNA technology makes it possible to clone genes for basic research and commercial applications:** *an overview* (pp. 376–377, FIGURE 20.1) DNA technology is a powerful set of techniques that enables biologists to manipulate and analyze DNA. It can help make useful new products and organisms.
  Web/CD Activity 20A: *Applications of DNA Technology*

- **Restriction enzymes are used to make recombinant DNA** (pp. 377–378, FIGURE 20.2) A variety of bacterial restriction enzymes recognize short, specific nucleotide sequences in DNA and cut the sequences at specific points on both strands to yield a set of double-stranded DNA fragments with single-stranded sticky ends. The sticky ends readily form base pairs with complementary single-stranded segments on other DNA molecules. The enzyme DNA ligase can seal the strands to produce recombinant DNA molecules.
  Web/CD Activity 20B: *Restriction Enzymes*

- **Genes can be cloned in recombinant DNA vectors:** *a closer look* (pp. 378–381, FIGURES 20.3–20.5) Plasmids can serve as vectors (carriers) to introduce foreign genes into host bacteria. Recombinant DNA is made by inserting restriction fragments from DNA containing a gene of interest into the vector DNA, which has been cut open by the same enzyme. Gene cloning results when the foreign genes replicate inside the host bacterial cell as part of the recombinant vector. Eukaryotic cells can also serve as host cells for gene cloning. Cell clones carrying the gene of interest can be identified with a radioactively labeled nucleic acid probe, which has a sequence complementary to the gene.
  Web/CD Activity 20C: *Cloning a Gene in Bacteria*
  Web/CD Case Study in the Process of Science: *How Can Antibiotic-Resistant Plasmids Transform* E. coli?

- **Cloned genes are stored in DNA libraries** (pp. 381–382, FIGURE 20.6) When the starting material for DNA (gene) cloning is an entire genome, the resulting collection of recombinant vector clones is called a genomic library. Alternatively, a cDNA (complementary DNA) library can be made by cloning DNA made *in vitro* by reverse transcription of all the mRNA produced by a particular kind of cell. Libraries of cDNA are especially useful for working with eukaryotic genes (whose introns are not present in the cDNA versions) and for studying gene expression.

- **The polymerase chain reaction (PCR) clones DNA entirely** *in vitro* (pp. 382–383, FIGURE 20.7) For quickly making many copies of a particular segment of DNA, this method uses primers that bracket the desired sequence and a heat-resistant DNA polymerase.

DNA ANALYSIS AND GENOMICS

- **Restriction fragment analysis detects DNA differences that affect restriction sites** (pp. 383–386, FIGURES 20.8–20.10) Gel electrophoresis makes it possible to separate and isolate DNA restriction fragments of different lengths. Restriction fragment length polymor-

phisms (RFLPs) are differences in DNA sequence on homologous chromosomes that result in different patterns of restriction fragment lengths. These patterns are visualized as bands on gel electrophoresis. Specific fragments can be identified by Southern blotting, using labeled probes that hybridize to the DNA stuck to a "blot" of the gel. RFLPs are prevalent genetic markers, present throughout eukaryotic noncoding DNA. RFLP analysis has many applications, including genetic mapping and diagnosis of genetic disorders.
  Web/CD Activity 20D: *Gel Electrophoresis of DNA*
  Web/CD Activity 20E: *Analyzing DNA Fragments Using Gel Electrophoresis*
  Web/CD Case Study in the Process of Science: *How Can Gel Electrophoresis Be Used to Analyze DNA?*

- **Entire genomes can be mapped at the DNA level** (pp. 386–389, FIGURES 20.11–20.13) An international research effort, the Human Genome Project involves linkage mapping, physical mapping, and DNA sequencing of the human genome and the genomes of other organisms. An alternative approach starts with sequencing of random DNA fragments, relying especially heavily on computer power to assemble the sequences. The human genome is thought to have 30,000 to 40,000 genes, fewer than once thought.
  Web/CD Activity 20F: *The Human Genome Project: Genes on Human Chromosome 17*

- **Genome sequences provide clues to important biological questions** (pp. 389–393, FIGURE 20.14) Genome sequences are helping researchers find new genes, probe details of gene organization and control, and answer questions about evolution. DNA microarrays allow researchers to compare patterns of gene expression in different tissues and under different conditions. Genomics is the systematic study of entire genomes; proteomics is the systematic study of all the proteins encoded by a genome. Single nucleotide polymorphisms (SNPs) provide useful markers for studying human genetic variation.

PRACTICAL APPLICATIONS OF DNA TECHNOLOGY

- **DNA technology is reshaping medicine and the pharmaceutical industry** (pp. 393–395; FIGURES 20.15, 20.16) Medical applications of DNA technology include diagnostic tests for genetic and other diseases; safer, more effective vaccines; the large-scale production of many new, and some previously scarce, pharmaceutical products; and the prospect of treating or even curing certain genetic disorders.

- **DNA technology offers forensic, environmental, and agricultural applications** (pp. 395–399; FIGURES 20.17–20.20) DNA "fingerprints" obtained from RFLP or STR analysis of tissue found at the scenes of violent crimes provide evidence in trials; such fingerprints are also useful in parenthood disputes. Genetic engineering can modify the metabolism of microorganisms so that they can be used to extract minerals from the environment or degrade waste materials. In agriculture, transgenic plants and animals are being designed to improve food productivity and quality.
  Web/CD Activity 20G: *DNA Fingerprinting*

- **DNA technology raises important safety and ethical questions** (p. 399) Several U.S. government agencies are responsible for setting policies about and regulating recombinant DNA technology. The potential benefits of genetic engineering must be carefully weighed against the potential hazards of creating products or developing procedures that are harmful to humans or the environment.

Web/CD Activity 20H: *Making Decisions About DNA Technology: Golden Rice*

Self-Quiz

1. Which of the following tools of recombinant DNA technology is *incorrectly* paired with its use?
   a. restriction enzyme—production of RFLPs
   b. DNA ligase—enzyme that cuts DNA, creating the sticky ends of restriction fragments
   c. DNA polymerase—used in a polymerase chain reaction to amplify sections of DNA
   d. reverse transcriptase—production of cDNA from mRNA
   e. electrophoresis—DNA sequencing

2. Which of the following would *not* be true of cDNA produced using human brain tissue as the starting material?
   a. It could be amplified by the polymerase chain reaction.
   b. It could be used to create a complete genomic library.
   c. It is produced from mRNA using reverse transcriptase.
   d. It could be used as a probe to locate a gene of interest.
   e. It lacks the introns of the human genes and thus can probably be introduced into phage vectors.

3. Plants are more readily manipulated by genetic engineering than are animals because
   a. plant genes do not contain introns.
   b. more vectors are available for transferring recombinant DNA into plant cells.
   c. a somatic plant cell can often give rise to a complete plant.
   d. genes can be inserted into plant cells by microinjection.
   e. plant cells have larger nuclei.

4. A paleontologist has recovered a bit of tissue from the 400-year-old preserved skin of an extinct dodo (a bird). The researcher would like to compare DNA from the sample with DNA from living birds. Which of the following would be most useful for increasing the amount of dodo DNA available for testing?
   a. RFLP analysis
   b. polymerase chain reaction (PCR)
   c. electroporation
   d. gel electrophoresis
   e. Southern hybridization

5. Expression of a cloned eukaryotic gene in a prokaryotic cell involves many difficulties. The use of mRNA and reverse transcriptase is part of a strategy to solve the problem of
   a. post-transcriptional processing
   b. electroporation
   c. post-translational processing
   d. nucleic acid hybridization
   e. restriction fragment ligation

6. DNA technology has many medical applications. Which of the following is *not yet* done routinely?
   a. production of hormones for treating diabetes and dwarfism
   b. production of viral subunits for vaccines
   c. introduction of genetically engineered genes into human gametes
   d. prenatal identification of genetic disease genes
   e. genetic testing for carriers of harmful alleles

7. Which of the following has the largest genome size and the smallest number of genes per million base pairs?
   a. *H. influenzae* (bacterium)
   b. *S. cerevisiae* (yeast)
   c. *A. thaliana* (plant)
   d. *D. melanogaster* (fruit fly)
   e. *H. sapiens* (human)

8. Which of the following sequences in double-stranded DNA is most likely to be recognized as a cutting site for a restriction enzyme?
   a. AAGG / TTCC
   b. AGTC / TCAG
   c. GGCC / CCGG
   d. ACCA / TGGT
   e. AAAA / TTTT

9. In recombinant DNA methods, the term *vector* can refer to
   a. the enzyme that cuts DNA into restriction fragments.
   b. the sticky end of a DNA fragment.
   c. a RFLP marker.
   d. a plasmid used to transfer DNA into a living cell.
   e. a DNA probe used to identify a particular gene.

10. In its sequencing of the human genome, Celera carried out
    a. linkage mapping of each chromosome.
    b. extensive physical mapping of each chromosome, starting with large chromosomal fragments.
    c. DNA sequencing of small fragments and then assembly of the fragments to determine overall nucleotide sequence.
    d. a and b.
    e. a, b, and c.

11. The human genome seems to have only about 30,000 to 40,000 genes, but there is evidence for well over 100,000 different human polypeptides. How might we account for this discrepancy?

12. What are DNA microarrays mainly used for?

13. On average, how much difference is there between the DNA sequences of two people? Why isn't there more?

14. What would be the advantage of using stem cells for gene therapy?

15. List at least three different properties that have been acquired by crop plants via genetic engineering.

Go to the website or CD-ROM for more quiz questions.

Evolution Connection

If DNA-based technologies become widely used, how might they change the way evolution proceeds, as compared with the natural evolutionary mechanisms of the past 4 billion years?

The Process of Science

You hope to study a gene that codes for a neurotransmitter protein in human brain cells. You know the amino acid sequence of the protein. Explain how you might (a) identify the genes expressed in a specific type of brain cell, (b) identify the gene for the neurotransmitter, (c) produce multiple copies of the gene for study, and (d) produce a quantity of the neurotransmitter for evaluation as a potential medication.

Try your hand at analyzing gel electrophoresis results and transforming *E. coli* using antibiotic-resistant plasmids in the Case Studies in the Process of Science, available on the website and CD-ROM.

Science, Technology, and Society

Is there danger of discrimination based on testing for "harmful" genes? What policies can you suggest that would prevent such abuses?

Answers: 1. b; 2. b; 3. c; 4. b; 5. a; 6. c; 7. e; 8. c; 9. d; 10. c; 11. Alternative RNA splicing. 12. Determining what genes are expressed in a particular kind of cell. 13. About 1 difference per 1000 base pairs; the human species has not existed long enough for more differences to arise. 14. They continue to reproduce themselves. 15. Herbicide resistance, pest resistance, disease resistance, delayed ripening, improved nutritional value, and others.

CHAPTER 21

THE GENETIC BASIS OF DEVELOPMENT

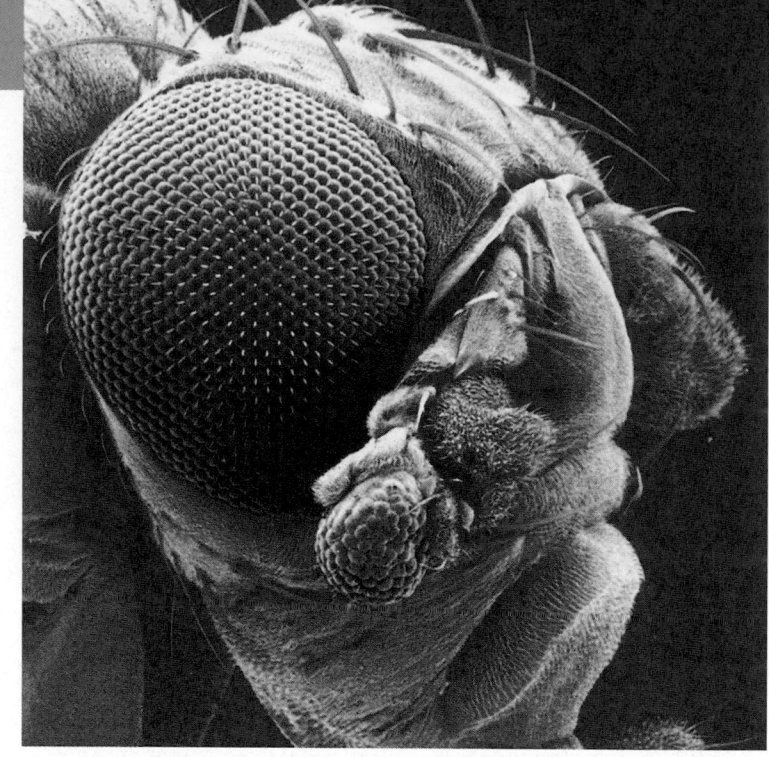

FROM SINGLE CELL TO MULTICELLULAR ORGANISM

- Embryonic development involves cell division, cell differentiation, and morphogenesis
- Researchers study development in model organisms to identify general principles

DIFFERENTIAL GENE EXPRESSION

- Different types of cells in an organism have the same DNA
- Different cell types make different proteins, usually as a result of transcriptional regulation
- Transcriptional regulation is directed by maternal molecules in the cytoplasm and signals from other cells

GENETIC AND CELLULAR MECHANISMS OF PATTERN FORMATION

- Genetic analysis of *Drosophila* reveals how genes control development: *an overview*
- Gradients of maternal molecules in the early embryo control axis formation
- A cascade of gene activations sets up the segmentation pattern in *Drosophila: a closer look*
- Homeotic genes direct the identity of body parts
- Homeobox genes have been highly conserved in evolution
- Neighboring cells instruct other cells to form particular structures: cell signaling and induction in the nematode
- Plant development depends on cell signaling and transcriptional regulation

This chapter applies *much of what you've learned about molecules, cells, and genes to one of biology's most important questions—how a complex multicellular organism develops from a single cell. The application of genetic analysis and DNA technology to the study of development has brought about a revolution in the field. In much the same way that researchers have used mutations to deduce pathways of cellular metabolism, they now use mutations to dissect developmental pathways. In one striking*

example, *Swiss researchers demonstrated in 1995 that a particular gene functions as a master switch that triggers the development of the eye in* Drosophila. *The scanning electron micrograph on this page shows the head of an abnormal fly with small extra eyes on its antennae. Expression of the master gene for eye development in an abnormal location in the fly caused the extra eyes. A similar gene triggers eye development in mice and other mammals. In fact, developmental biologists are discovering remarkable similarities in the mechanisms that shape diverse organisms.*

The scientific study of development got under way about a century ago, around the same time as genetics. But for decades the two disciplines proceeded along mostly separate paths. We have seen how geneticists advanced from Mendel's laws to an understanding of the molecular basis of inheritance. Meanwhile, developmental biologists focused on embryology, the study of the stages of development leading from fertilized egg to fully formed organism. Only in recent years have the concepts and tools of molecular genetics reached the point where a real synthesis has been possible. The synthesis is a challenge, for it means relating the linear information in genes to a process of development that takes place in four dimensions, three of space and one of time.

This chapter introduces some of the basic genetic and cellular mechanisms that control development. It focuses on principles that apply to both animals and plants, with emphasis on two invertebrate animals: the fruit fly Drosophila melanogaster *and the nematode* Caenorhabditis elegans. *In later chapters, you will learn much more about the development of plants (Chapter 35) and animals (Chapter 47).*

A capstone to the genetics unit, this chapter also serves as a bridge to the rest of the book, for understanding development is crucial to understanding the evolution, diversity, structure, function, and ecology of organisms.

FROM SINGLE CELL TO MULTICELLULAR ORGANISM

In the development of most multicellular organisms, a single-celled zygote (fertilized egg) gives rise to cells of many different types, each type with a different structure and corresponding function. For example, an animal will have muscle cells that enable it to move and nerve cells that transmit signals to the muscle cells; a plant will have mesophyll cells that carry out photosynthesis and stomatal cells that regulate the passage of gases into and out of leaves (see FIGURE 10.2). As you know, cells are only one level in the hierarchy of biological order within a multicellular organism (see FIGURE 2.1). Cells of similar types are organized into tissues, tissues into organs, organs into organ systems, and organ systems into the whole organism. Thus, the process of embryonic development must give rise not only to cells of different types but to higher-level structures arranged in a particular way in three dimensions.

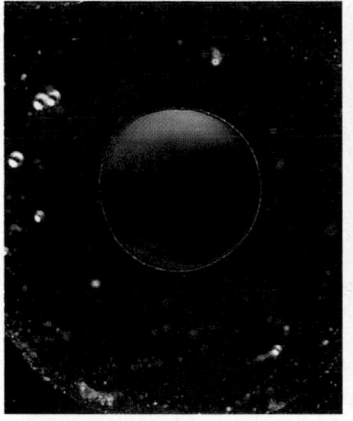 

(a) Fertilized egg of a frog   **(b)** Tadpole hatching from egg

FIGURE 21.1 From fertilized egg to animal: what a difference a week makes. It took just one week for cell division, differentiation, and morphogenesis to transform this fertilized frog egg **(a)** into a hatching tadpole **(b)**.

Embryonic development involves cell division, cell differentiation, and morphogenesis

An organism arises from a fertilized egg cell as the result of three interrelated processes: cell division, cell differentiation, and morphogenesis (FIGURE 21.1). Through a succession of mitotic divisions, the zygote gives rise to a large number of cells. Cell division alone, however, would produce only a great ball of identical cells, nothing like an animal or plant. During embryonic development, cells not only increase in number, but also undergo **differentiation,** becoming specialized in structure and function. Moreover, the different kinds of cells aren't just mixed up randomly but are organized into tissues and organs. The physical processes that give an organism its shape constitute **morphogenesis,** meaning "creation of form."

The processes of cell division, differentiation, and morphogenesis overlap in time. Early events of morphogenesis lay out the basic body plan very early in embryonic development, establishing, for example, which end of an animal embryo will be the head or which end of a plant embryo will become the roots. Cell division and differentiation play important roles in morphogenesis in all organisms, as does the programmed death of certain cells. However, the overall schemes of morphogenesis in animals and plants are very different, and the mechanisms differ in two major ways (FIGURE 21.2, p. 404). First, in animals, but not in plants, *movements* of cells and tissues are necessary to transform the early embryo into the characteristic three-dimensional form of the organism. We will discuss these morphogenetic movements in Chapter 47. The second major difference is that in plants, morphogenesis and growth in overall size are not limited to embryonic and juvenile periods but occur throughout the life of the plant. **Apical meristems,** which are perpetually embryonic regions in

the tips of shoots and roots, are responsible for the plant's continual growth and formation of new organs, such as leaves and roots. In animals, ongoing development in adults is restricted to the differentiation of cells, such as blood cells, that must be continually replenished throughout the animal's lifetime.

The importance of precise regulation of morphogenesis is evident in human disorders that result from morphogenesis gone awry. For example, cleft palate, in which the upper wall of the mouth cavity fails to close completely, is a defect of morphogenesis.

As humans, we may naturally be most interested in developmental processes in our own species. However, many aspects of development are much easier to study in other kinds of organisms.

Researchers study development in model organisms to identify general principles

THE PROCESS OF SCIENCE

Much of the early research on animal development focused on animals that lay their eggs in water, with amphibians, such as frogs, getting particular attention. Frogs have large eggs (2–3 mm in diameter) that are easy to observe and manipulate, and fertilization and development occur outside the mother's body. By studying these animals and others, biologists were able to work out a description of animal development at the macroscopic and microscopic levels, making a number of important discoveries in the process (see Chapter 47). Research on a variety of plants led to a basic understanding of plant development (see Chapter 35). When the primary research goal is to understand broad biological principles—of animal or plant development in this case—the organism chosen for study is called a **model organism.** Researchers select model organisms that lend themselves to the

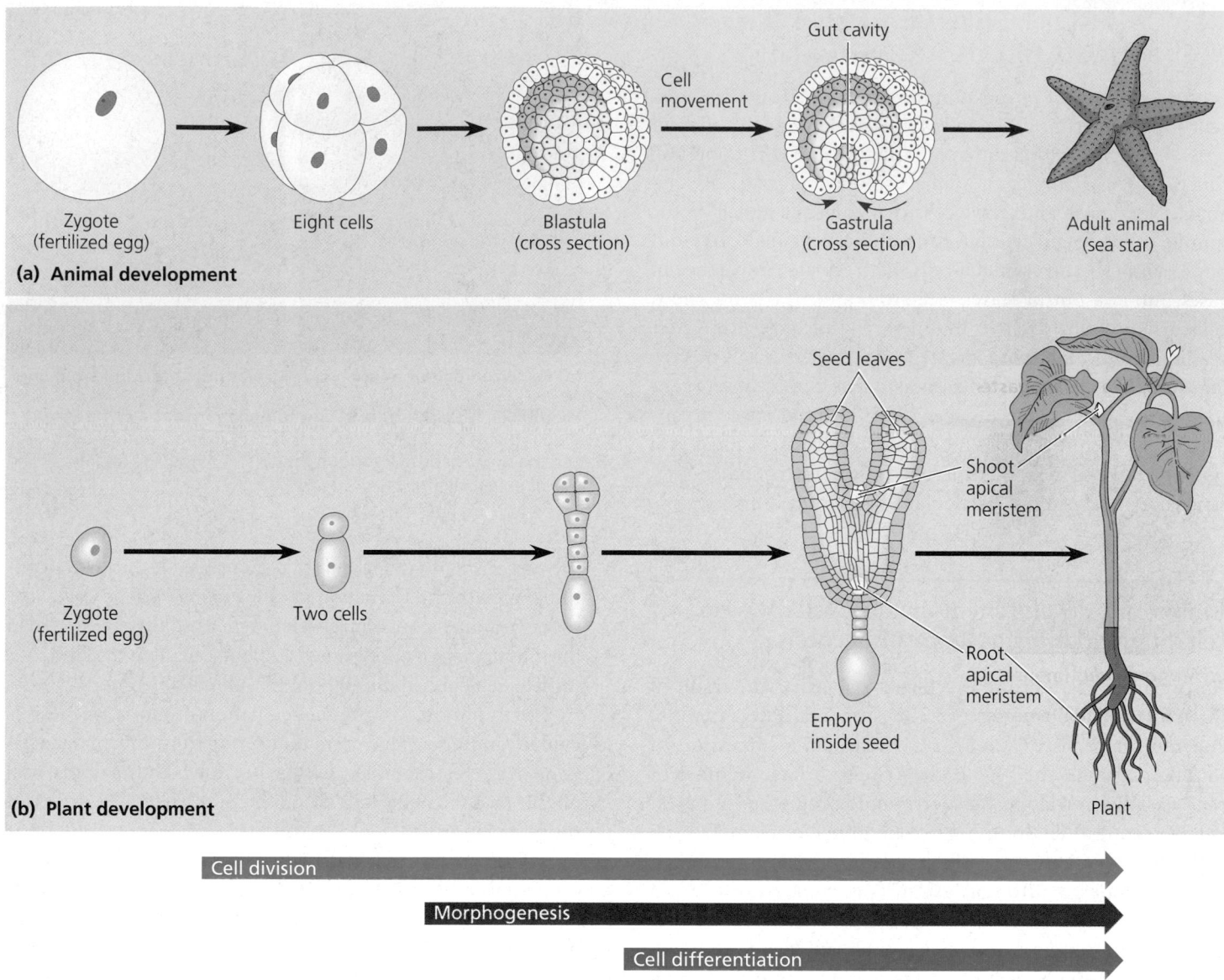

Cell division

Morphogenesis

Cell differentiation

FIGURE 21.2 Some key stages of development in animals and plants. Cell division, morphogenesis, and cell differentiation occur in both animal development and plant development. **(a)** Most animals go through some variation of the blastula and gastrula stages shown here in simplified form. The blastula is a sphere of cells surrounding a fluid-filled cavity. In forming the gastrula, a region of the blastula folds inward, forming a rudimentary gut cavity. The movement of cells and tissues plays an important role in animal morphogenesis. Biochemical events that will lead to cellular differentiation actually start before the gastrula forms. Once the animal is mature, differentiation occurs in only a limited way—for the replacement of damaged or lost cells. **(b)** In plants with seeds, a complete embryo develops within the seed. Morphogenesis, which does not involve cell or tissue movement, occurs throughout the plant's lifetime. Apical meristems (yellow) continuously arise and develop into the various plant organs as the plant grows to an indeterminate size.

study of a particular question and that are representative of a larger group. Frogs, for example, are useful model organisms for elucidating the role of cell movement in morphogenesis because frog development is easy to observe and fairly typical of vertebrate animals.

However, for the more recent research efforts aimed at uncovering the connections between genes and development, many developmental biologists have turned to organisms that are more convenient for genetic analysis. For developmental genetics, the criteria for choosing a model organism include, in addition to readily observable embryos, short generation times, relatively small genomes, and, ideally, preexisting knowledge about the organism and its genes. Several organisms have emerged as favorites, including *Drosophila*, the nematode *C. elegans*, the mouse, the zebrafish, and the plant *Arabidopsis* (FIGURE 21.3).

First chosen as a model organism by the pioneering geneticist T. H. Morgan and intensively studied by generations of geneticists after him, the fruit fly *Drosophila melanogaster* is small and easily grown in the laboratory. As you read in Chapter 15, *Drosophila* has a generation time of only two weeks and produces many offspring. Embryos develop outside

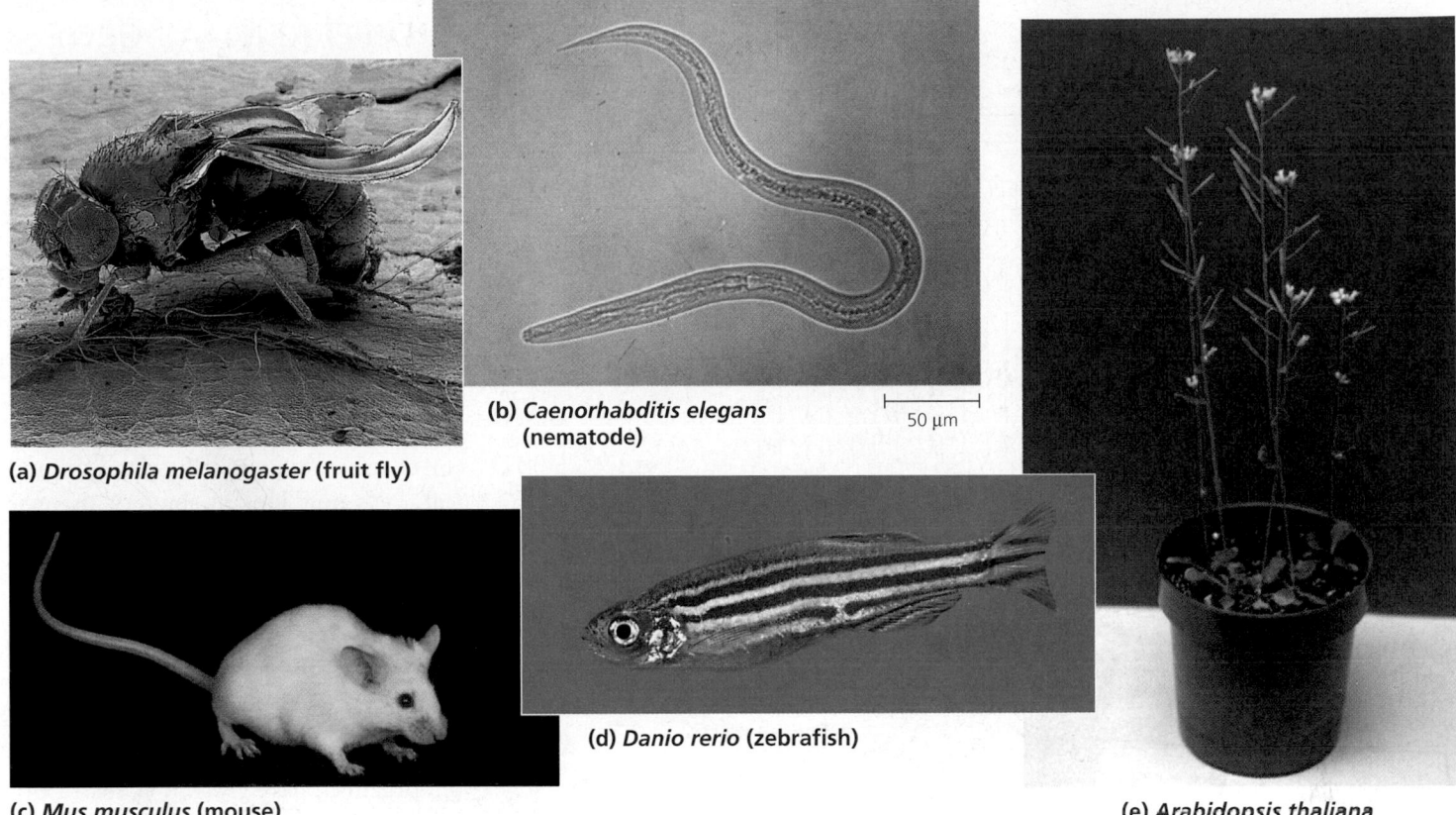

(a) *Drosophila melanogaster* (fruit fly)

(b) *Caenorhabditis elegans* (nematode)

50 μm

(c) *Mus musculus* (mouse)

(d) *Danio rerio* (zebrafish)

(e) *Arabidopsis thaliana* (common wall cress)

FIGURE 21.3 Model organisms. Each of these organisms offers particular advantages for research on the genetics of development.

the mother's body, an asset for developmental studies. And researchers can draw on a vast amount of information and experience relating to this animal's genes and other aspects of its biology. (The DNA sequence of its genome was completed in 2000.) One disadvantage of the fruit fly as a model organism for developmental research is that its early development is at least superficially quite different from the process depicted in FIGURE 21.2: The first rounds of mitosis occur without cytokinesis, leading to an early blastula containing a large number of nuclei within a single mass of cytoplasm. Nevertheless, research on *Drosophila* development has yielded deep insights into basic principles of animal development, as we will see.

The nematode *Caenorhabditis elegans* normally lives in soil but is easily grown in the laboratory in petri dishes. It is only about a millimeter long, has a simple, transparent body with only a few types of cells, and grows from zygote to mature adult in only three and a half days. One of its advantages for genetic studies is that its genome has been sequenced. Another is that most individuals are hermaphrodites, which produce both eggs and sperm. Hermaphrodites are convenient for genetic studies because recessive mutations are easy to detect. As with any diploid organism, an individual with a recessive mutation in only one copy of a gene will have the wild-type phenotype. But a researcher can quickly detect the mutation by allowing the worm to self-fertilize: One-fourth

of the offspring will be homozygous for the mutant allele and have a mutant phenotype.

For developmental biologists, a further important feature of *C. elegans* is that every adult hermaphrodite has exactly 959 somatic cells (some of which are fused together), and these cells arise from the zygote in virtually the same way for every individual. Using a microscope to follow all the cell divisions starting immediately after a zygote forms, biologists have been able to reconstruct the entire ancestry of every cell in the adult body, the organism's complete **cell lineage.** A cell lineage diagram like the one in FIGURE 21.4 (p. 406) is a type of *fate map,* a representation of the fates of various parts of a developing embryo. (You'll see more fate maps in Chapter 47.)

Among vertebrates, two in particular lend themselves to the genetic analysis of development, the mouse and the zebrafish. The mouse *Mus musculus* has a long history as a mammalian model, and much is known about its biology, including its genes. Moreover, researchers are now adept at manipulating mouse genes to make transgenic mice and mice in which particular genes are "knocked out" by mutation. But mice are complex animals with a genome as large as ours, and their embryos develop in the mother's uterus, hidden from view.

Many of the disadvantages of the mouse are absent in a newer vertebrate model, the zebrafish *Danio rerio.* These small fish (2–4 cm long) are easy to breed in the laboratory in

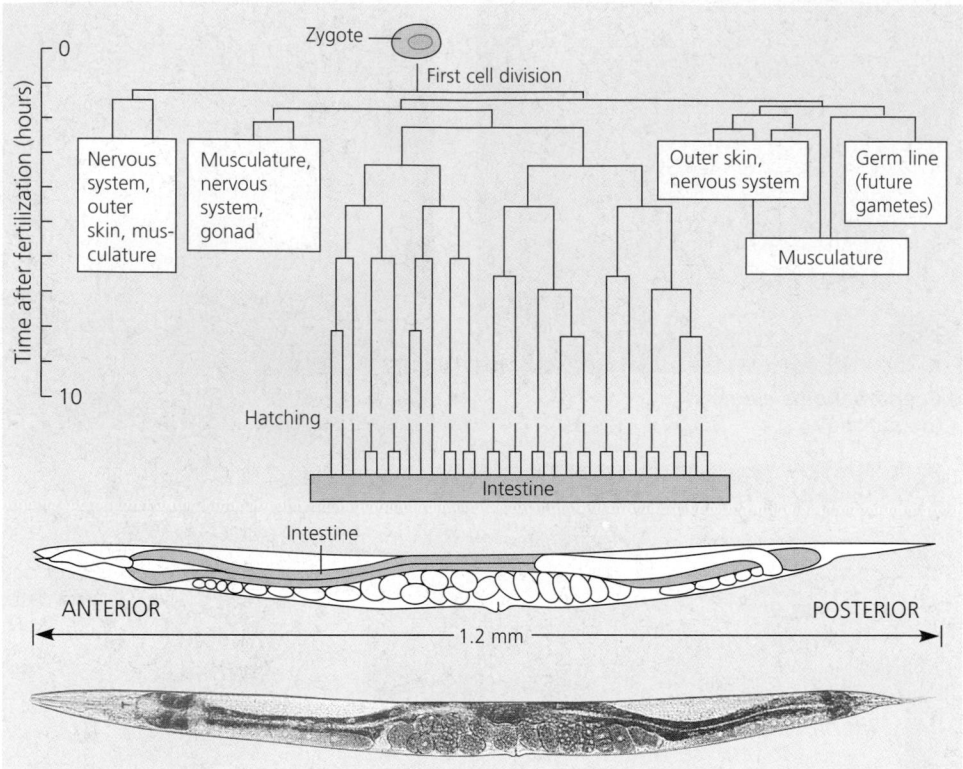

FIGURE 21.4 Cell lineage in *C. elegans*. The nematode *Caenorhabditis elegans* is transparent at all stages of its development, making it possible for researchers to trace the lineage of every cell, from the zygote to the adult worm (LM). The diagram shows a detailed lineage only for the intestine, which is derived exclusively from one of the first four cells formed from the zygote. The intestinal cell lineage does not happen to include any programmed cell death, an important aspect of the lineages for some other parts of the animal.

large numbers, and transparent embryos develop outside the mother's body. Although the generation time is relatively long (two to four months), the early stages of development proceed quickly: By 24 hours after fertilization, most of the tissues and early versions of the organs have formed, and by two days a tiny fish hatches out of the egg case. Although the zebrafish genome is still being mapped and sequenced, researchers such as Nancy Hopkins (see the interview at the opening of this unit) have already identified many genes involved in the animal's development.

For studying the molecular genetics of plant development, researchers are focusing on a small weed called *Arabidopsis thaliana* (the common wall cress, a member of the mustard family). One of these plants can grow in a test tube and produce thousands of progeny after eight to ten weeks; as in Mendel's pea plants, each flower makes both ova and sperm (in pollen). For gene manipulation research, scientists can grow *Arabidopsis* cells in culture and get the cells to take up foreign DNA (genetic transformation). Another advantage of *Arabidopsis* is that it has a relatively small genome, about 100 million nucleotide pairs, which has already been sequenced.

Later in this chapter, you will learn about important discoveries that researchers have made using some of these model organisms.

DIFFERENTIAL GENE EXPRESSION

We have stated on several previous occasions that differences between cells in a multicellular organism come almost entirely from differences in gene *expression,* not from differences in the cells' genomes. (There are a few exceptions, such as antibody-producing cells; see FIGURE 19.6.) Furthermore, we have mentioned that these differences arise during development, as regulatory mechanisms turn specific genes on and off. Let's now look at some of the evidence for this assertion.

Different types of cells in an organism have the same DNA

THE PROCESS OF SCIENCE

Much evidence supports the conclusion that nearly all the cells of an organism have *genomic equivalence*—that is, they all have the same genes. What happens to these genes as a cell begins to differentiate? We can shed some light on this question by asking whether genes are irreversibly inactivated during differentiation. For example, does an epidermal cell in your finger contain a viable gene specifying eye color, or has the eye-color gene been destroyed or permanently inactivated there?

Totipotency in Plants

One experimental approach to the question of genomic equivalence is to try to generate a whole organism from differentiated cells of a single type. In many plant species, whole new individuals *can* develop from differentiated somatic cells. This was first demonstrated during the 1950s by F. C. Steward and his students at Cornell University, working with carrot plants (FIGURE 21.5). They found that differentiated cells removed from the root (the carrot) and placed in culture medium could grow into normal adult plants, each genetically identical to the "parent" plant. Using a single somatic cell from a multicellular organism to make one or more genetically identical individuals is called **cloning,** and each new individual is popularly called a **clone.** (The meanings of these terms vary with context; for clarification, see the Glossary.) The fact that a mature plant cell can dedifferentiate (reverse its differentiation) and then give rise to all the different kinds of specialized cells of a new plant shows that differentiation does not necessarily involve irreversible changes in the DNA. In plants, at least, cells

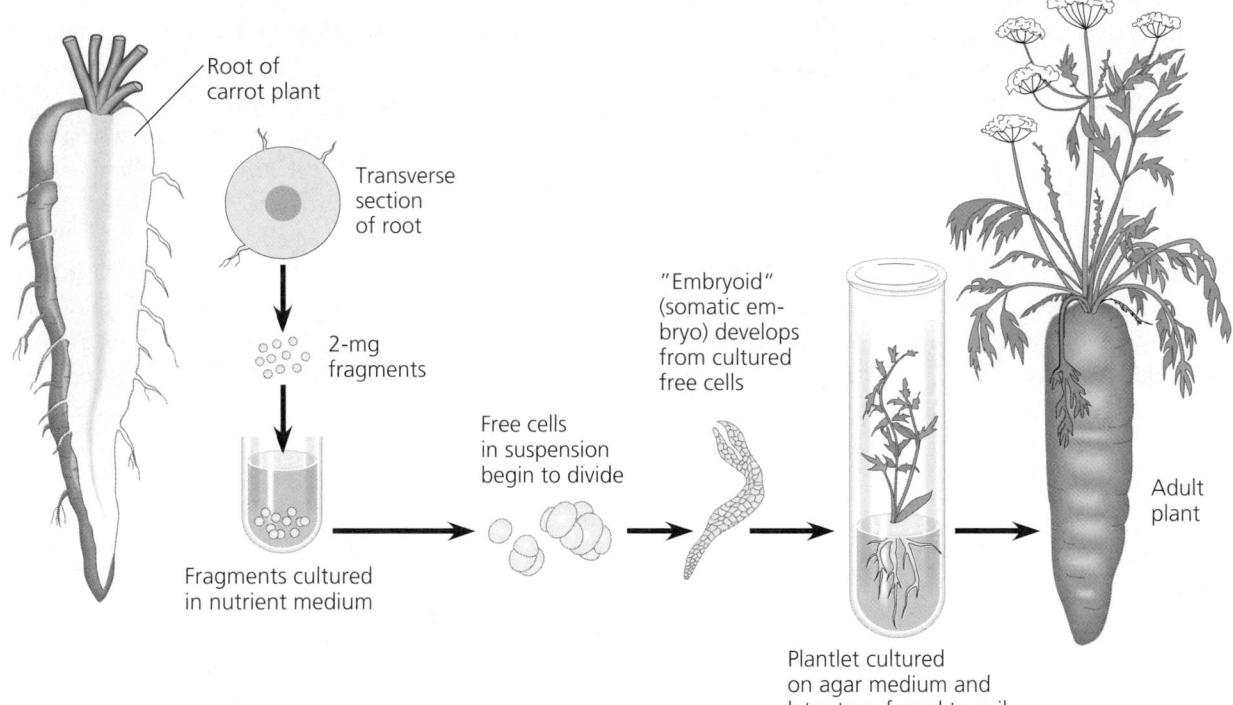

FIGURE 21.5 Test-tube cloning of carrots. In classic experiments conducted during the 1950s, F. C. Steward and his students at Cornell University demonstrated that whole plants could be regenerated from somatic (nonreproductive) cells dissected from a carrot. The new plants that result are genetic duplicates (clones) of the parent plant.

can remain **totipotent;** that is, they can retain the zygote's potential to form all parts of the mature organism. Plant cloning is now used extensively in agriculture.

Nuclear Transplantation in Animals

Differentiated cells from animals will often fail to divide in culture, much less develop into a new organism. Therefore, animal researchers have approached the genomic equivalence question by replacing the nucleus of an unfertilized egg cell or zygote with the nucleus of a differentiated cell. Can a nucleus derived from a differentiated cell direct development of an organism with all the proper tissues and organs? The pioneering experiments in nuclear transplantation were carried out by American embryologists Robert Briggs and Thomas King during the 1950s and were later extended by British embryologist John Gurdon. These investigators removed or destroyed the nucleus of a frog egg cell, then transplanted a nucleus from an embryonic or tadpole cell of the same species into the enucleated eggs (FIGURE 21.6). The ability of the transplanted nucleus to support normal development turned out to be inversely related to the age of the donor. In the case of nuclei from the relatively undifferentiated cells of an early embryo, most of the recipient eggs developed into tadpoles. But with nuclei from the differentiated intestinal cells of a tadpole, fewer than 2% of the eggs developed into

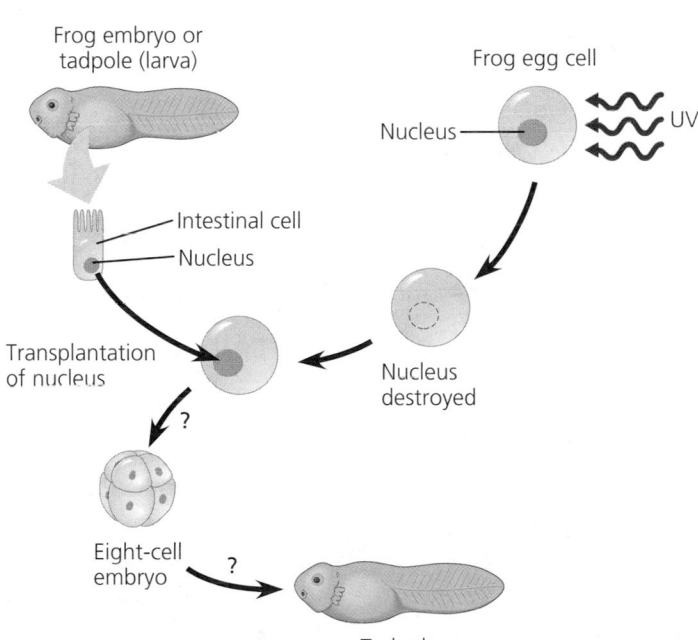

FIGURE 21.6 Nuclear transplantation. After the frog egg nucleus is destroyed by ultraviolet (UV) radiation, a nucleus from a more advanced developmental stage is inserted into the egg to test whether nuclei change irreversibly as cells begin to differentiate. The earlier the developmental stage from which the nucleus comes, the more likely it will support development. Nuclei from very early embryonic stages frequently prove to be totipotent, whereas nuclei from late developmental stages (such as a tadpole) rarely are.

normal tadpoles, and most of the embryos failed to make it through even the earliest stages of development.

Developmental biologists agree on several conclusions about these results. First, nuclei *do* change in some way as cells differentiate. Although the base sequence of the DNA usually does not change, chromatin structure alters in specific ways (for example, the methylation of the DNA may change; see p. 363). In frogs and most other animals, nuclear "potency" tends to be restricted more and more as embryonic development and cell differentiation progress. However, biologists also agree that these chromatin changes are sometimes reversible and that the nuclei of most differentiated animal cells probably have all the genes required for making the entire organism. In other words, biolo-

gists believe that the cells of the body differ in structure and function not because they contain different genes, but because they express different portions of a common genome.

Researchers working with mammals have long been able to clone animals using nuclei or cells from a variety of early embryos, but until recently, it was not known whether the restriction of genomic potential in differentiated cells of an adult mammal could be reversed. However, in 1997, Scottish researcher Ian Wilmut and his colleagues captured newspaper headlines with the announcement that they had cloned an adult sheep by transplanting the nucleus from an udder (mammary) cell into an unfertilized egg cell from another sheep (FIGURE 21.7). They achieved the necessary dedifferenti-

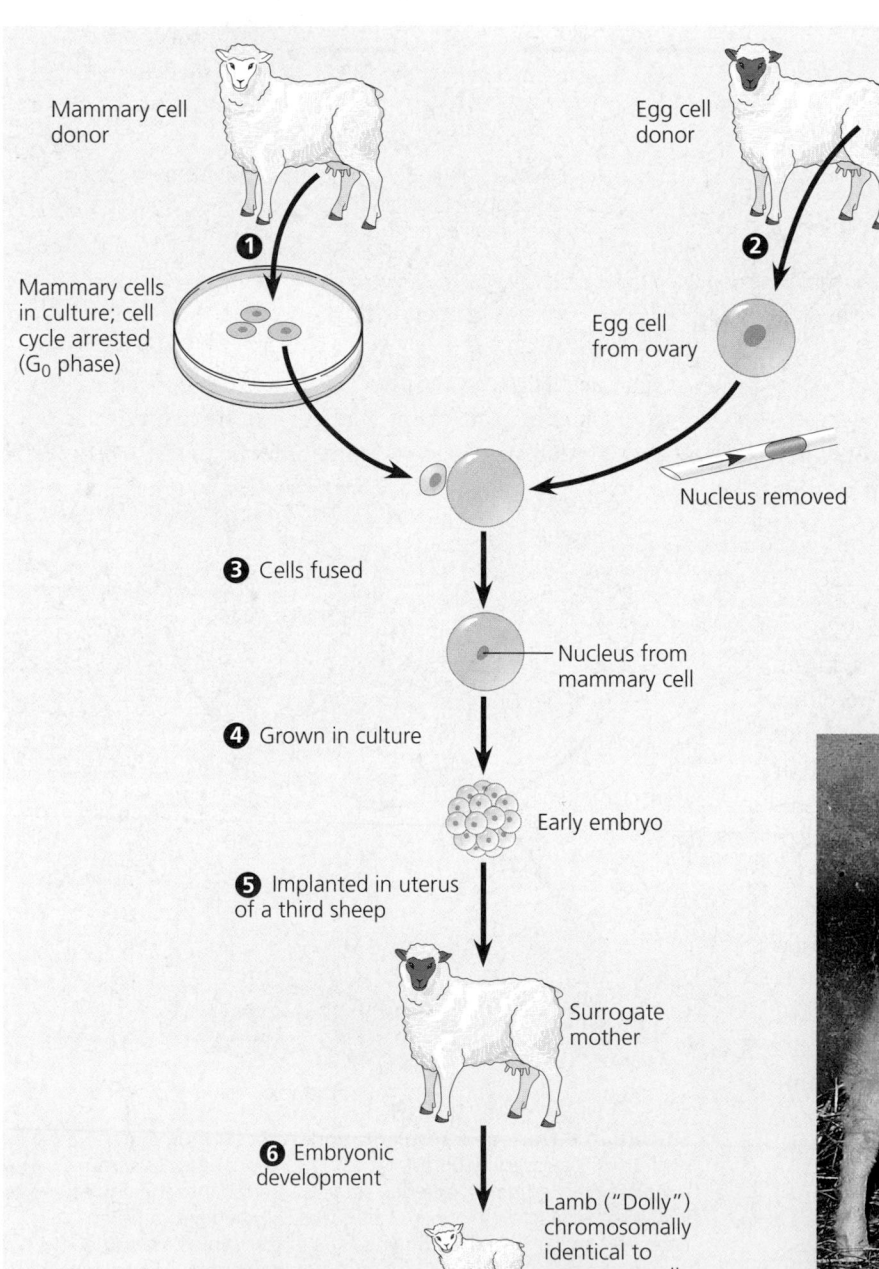

FIGURE 21.7 Cloning a mammal.

❶ Mammary cells were taken from the udder of a sheep and grown in culture with only low levels of nutrients. This semistarvation arrested the cell cycle in G_0 and apparently allowed the cells to dedifferentiate. ❷ Meanwhile, an egg cell was taken from another sheep and its nucleus removed. ❸ A mammary cell in G_0 was fused with the enucleated egg cell by subjecting the two cells to pulses of electrical current, which also stimulated the egg cell to start dividing. ❹ After growing in culture for six days, the embryo ❺ was implanted in the uterus of a third sheep, one similar to the egg cell donor. ❻ The result after gestation was a lamb, Dolly, identical in appearance and in chromosome makeup to the sheep that had donated the mammary cell. (However, Dolly's genes cannot be *completely* identical to those of the mammary cell donor because Dolly's mitochondrial DNA comes from the egg cell donor.) Dolly was the first reported case of a mammal "cloned" using the nucleus of a differentiated cell. The photograph shows Dolly with her surrogate mother.

Mammary cell donor

Egg cell donor

Mammary cells in culture; cell cycle arrested (G_0 phase)

Egg cell from ovary

Nucleus removed

❸ Cells fused

Nucleus from mammary cell

❹ Grown in culture

Early embryo

❺ Implanted in uterus of a third sheep

Surrogate mother

❻ Embryonic development

Lamb ("Dolly") chromosomally identical to mammary cell donor

ation of the nucleus by culturing mammary cells in nutrient-poor medium, which led to arrest of the cell cycle (see FIGURE 12.13). The cycle stopped at the G_1 checkpoint, and the cells entered a G_0 "resting" phase. The researchers then fused these cells with sheep egg cells whose nuclei had been removed. The resulting diploid cells divided to form early embryos, which the researchers implanted into surrogate mothers. One of several hundred of these embryos, they reported, successfully completed normal development. DNA analyses have since shown that the chromosomal DNA of this sheep, "Dolly," is indeed identical to that of the nucleus donor. (Later experiments showed that Dolly's mitochondrial DNA came from the egg cell donor, as we would expect.)

In July 1998, researchers in Hawaii reported cloning mice using nuclei from mouse ovary cells, and since then, cloning has been demonstrated in numerous mammals, many important in agriculture. Cloning of farm animals raises safety issues for human consumers, and the possibility of human cloning raises unprecedented ethical issues. However, problems with the cloning process have bought us a little more time for thought. In most cases, only a tiny percentage of cloned embryos develop normally. Why is this? Recently, scientists have found a clue: The DNA of many cloned embryos is improperly methylated, often having extra methyl groups. Because DNA methylation helps regulate gene expression, and appropriate gene expression is key to embryonic development, it makes sense that misplaced methyl groups could interfere with development. Cloning difficulties highlight the fact that we still have much to learn about the basic principles and processes of development.

The Stem Cells of Animals

Another hot research area that depends on the genetic potential retained by animal cells during development involves **stem cells.** These are cells with two important properties: As relatively unspecialized cells, they continually reproduce themselves; and under appropriate conditions, they differentiate into specialized cells of one or more types. The adult body has various kinds of stem cells, which serve to replace nonreproducing specialized cells as needed. For example, stem cells in the bone marrow give rise to all the different kinds of blood cells (see FIGURE 42.15). Another example, whose discovery surprised the scientific world very recently, is stem cells in the adult brain that continue to produce certain kinds of nerve cells there. Stem cells that can give rise to multiple cell types are said to be multipotent or, more often, *pluripotent.*

Although adults have only tiny numbers of stem cells, scientists are learning to identify and isolate these cells from various tissues and, in some cases, to grow them in culture. Much easier to culture are cells from early embryos, which can give rise to differentiated cells of any type. Cultures of these *embryonic stem cells* are "immortal"; the cells reproduce indefinitely (using telomerase to maintain their chromosomal telomeres). Taking this research further, scientists have recently demonstrated that with the right culture conditions (for instance, the addition of specific growth factors), they can make cultured stem cells derived from either source differentiate into specialized cells. Surprisingly, adult stem cells can sometimes be made to differentiate into a wider range of cell types than they normally do in the animal (FIGURE 21.8).

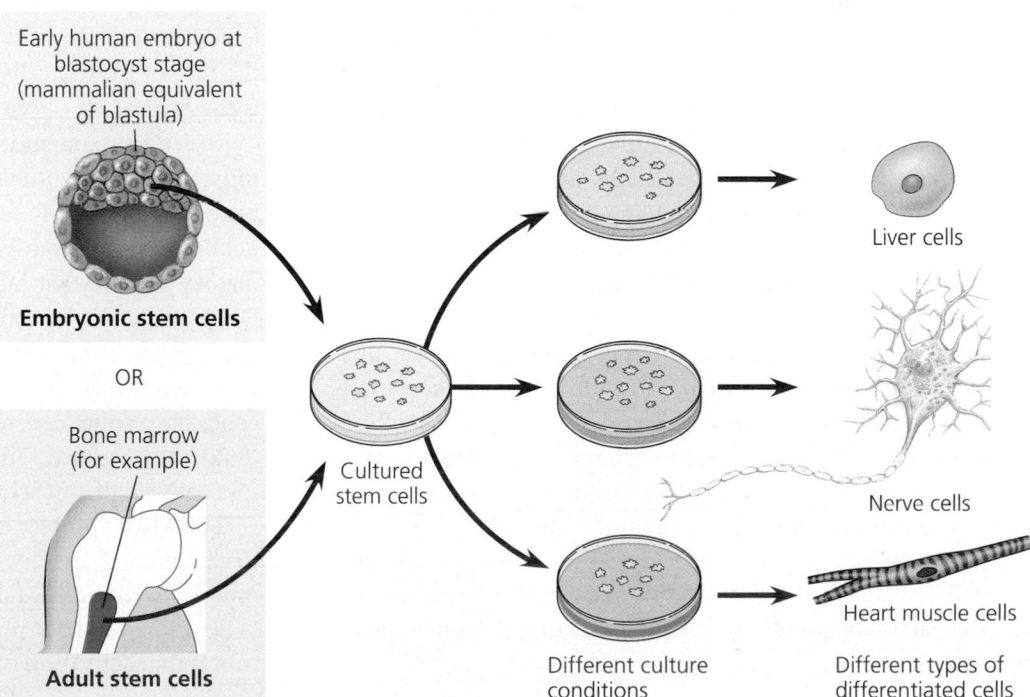

THE PROCESS OF SCIENCE

FIGURE 21.8 Working with stem cells. Animal stem cells—self-perpetuating, relatively undifferentiated cells that can give rise to specialized cells of various types—can be isolated from early embryos or adult tissues and grown in culture. Researchers are seeking to discover the growth conditions that direct stem cells to differentiate into particular cell types. Embryonic stem cells are easier to grow than adult stem cells and can theoretically give rise to *all* types of cells.

Early human embryo at blastocyst stage (mammalian equivalent of blastula)

Embryonic stem cells

OR

Bone marrow (for example)

Adult stem cells

Cultured stem cells

Different culture conditions

Liver cells

Nerve cells

Heart muscle cells

Different types of differentiated cells

In addition to providing a valuable means of studying differentiation, stem cell research has enormous potential for medical applications. The ultimate aim is to be able to supply cells for the repair of damaged or diseased organs. For example, providing insulin-producing pancreatic cells to diabetics or certain kinds of brain cells to people with Parkinson's disease or Huntington's disease could conceivably cure these diseases. At the present time, embryonic cells are more promising than adult stem cells for such applications, but because the cells are derived from human embryos (commonly, surplus embryos donated by patients undergoing infertility treatment), their use raises ethical and political issues.

In the next section, we look at the molecular basis of cell differentiation.

Different cell types make different proteins, usually as a result of transcriptional regulation

As the tissues and organs of an embryo take shape, differentiation of their cells becomes apparent; the cells become obviously different in structure and function. Cellular differentiation is actually the outcome of a cell's developmental history extending back to the first mitotic divisions of the zygote. However, the earliest changes that set a cell on a path to specialization are subtle ones, showing up only at the molecular level. Before biologists knew much about the molecular changes occurring in embryos, they coined the term **determination** to refer to the process that leads up to the observable differentiation of a cell. At the point in the process when the cell is irreversibly committed to its final fate, it is said to be "determined." Today we understand determination in terms of molecular changes. The outcome of determination—differentiation—is heralded by the expression of genes that encode *tissue-specific proteins,* which are found only in a certain type of cell and give the cell its characteristic structure and function. The first evidence of differentiation is the appearance of the mRNA for these proteins. Eventually, differentiation is observable with a microscope as changes in cellular structure. In most cases, the pattern of gene expression in a differentiated cell—what proteins the cell makes—is controlled at the level of transcription.

Differentiated cells are specialists at making tissue-specific proteins. These are the proteins that allow them to carry out their specialized roles in the organism. Developing lens cells in vertebrates, for example, synthesize large quantities of crystallins, proteins that aggregate to form transparent fibers that give the lens the ability to transmit and focus light. Because no other vertebrate cell type makes crystallins, these proteins are signposts of lens cell differentiation. Lens cells devote 80% of their capacity for protein synthesis to making this one type of protein.

The differentiation of skeletal muscle cells is another instructive example. The "cells" of skeletal muscle, which we use for walking and lifting and other voluntary movements, are long fibers containing many nuclei within a single plasma membrane. They contain very high concentrations of proteins specific to muscle tissue, such as muscle-specific versions of the contractile proteins myosin and actin and membrane receptor proteins that detect signals from nerve cells. Muscle cells develop from embryonic precursor cells that have the potential to develop into a number of alternative cell types, including cartilage cells or fat cells, but particular conditions commit them to becoming muscle cells. Although the committed cells appear unchanged under the microscope, determination has occurred, and they are now *myoblasts.* Eventually, myoblasts start to churn out large amounts of muscle-specific proteins and fuse to form mature, elongated, multinucleated skeletal muscle cells (FIGURE 21.9).

What actually happens in muscle cell determination? Researchers have been able to answer this question by growing myoblasts in culture and applying some of the techniques you learned about in Chapter 20. To test the hypothesis that certain muscle-specific regulatory genes are active in myoblasts, researchers isolated mRNA from cultured myoblasts and used reverse transcriptase to prepare a library of genes in cDNA form (see FIGURES 20.5 and 20.6)—intron-lacking versions of the genes expressed in myoblasts. In cloning the cDNA genes, the researchers positioned them next to a viral promoter that would turn on transcription in any kind of cell. The researchers then inserted each of the cloned genes into a separate embryonic precursor cell and looked for differentiation into myoblasts and muscle cells. In this way, they identified several crucial muscle determination genes, "master regulatory genes" that, when transcribed and translated, commit the cells to becoming skeletal muscle. Thus, in the case of muscle cells, the molecular basis of determination is the expression of one or more of these regulatory genes.

To understand more about how commitment occurs in muscle cell differentiation, let's focus on the master regulatory gene called *myoD.* Researchers learned that the protein product of *myoD,* called MyoD, is a transcription factor. It is a regulatory protein that binds to specific control elements in the DNA and stimulates the transcription of various genes, including some encoding still other muscle-specific transcription factors (see pp. 365–366). Presumably, all these target genes have enhancers recognized by MyoD and are thus coordinately controlled. Finally, the secondary transcription factors activate the muscle protein genes.

The MyoD protein is powerful. Researchers have been able to use it to change some kinds of fully differentiated non-muscle cells, such as fat cells and liver cells, into muscle cells. Why doesn't it work on *all* kinds of cells? One likely explanation is that activation of the muscle-specific genes is not solely

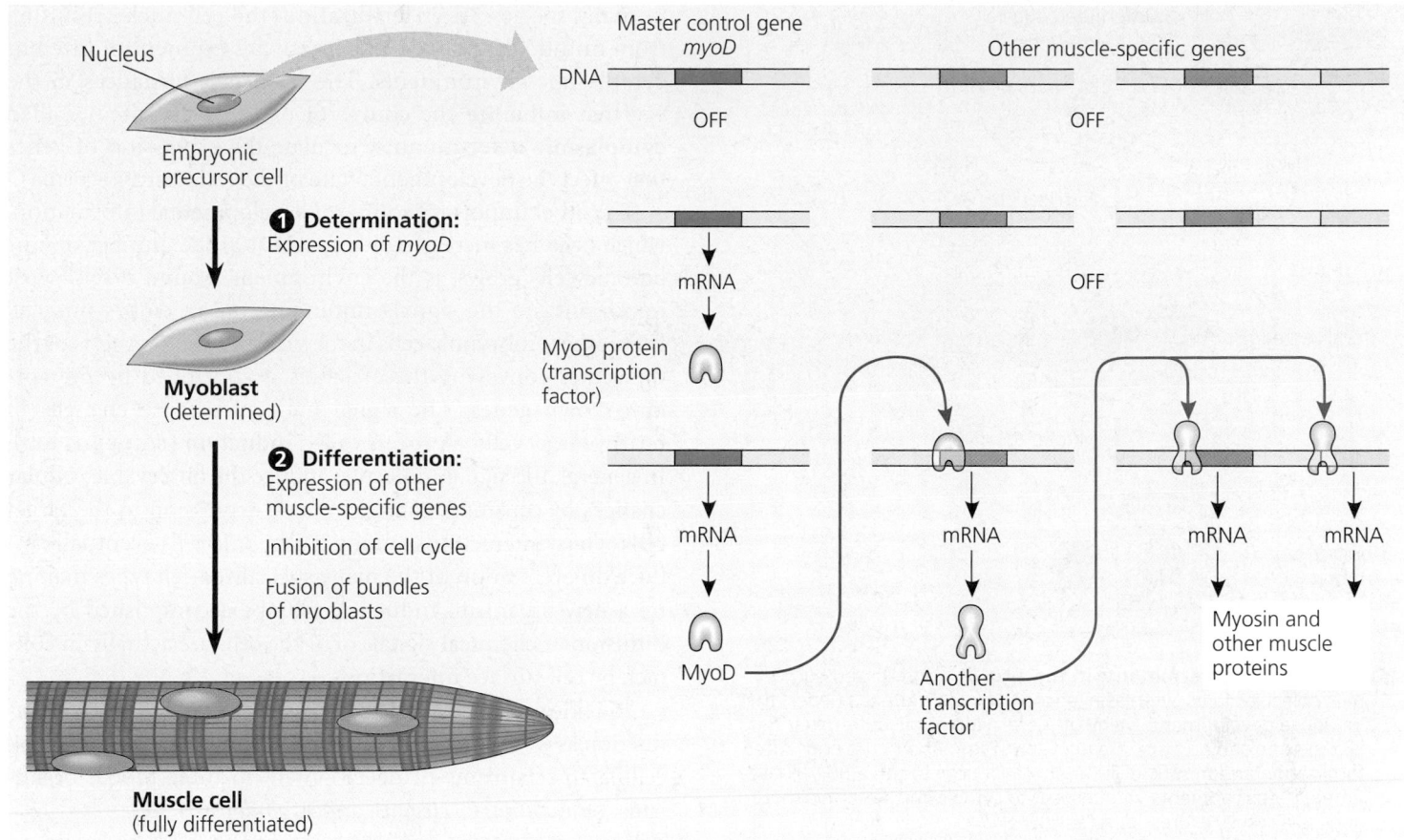

FIGURE 21.9 Determination and differentiation of muscle cells. This figure depicts a simplified version of how skeletal muscle cells arise from ordinary-looking embryonic cells that resemble fibroblasts (see the photo in FIGURE 12.15). ❶ When this kind of embryonic precursor cell receives certain signals from other cells, a master control gene called *myoD* is activated, and the cell makes the MyoD protein. Although the appearance of the cell in the microscope does not change, determination has occurred: The activation of *myoD* commits the cell, now called a myoblast, to becoming a skeletal muscle cell.
❷ The MyoD protein is a transcription factor that activates genes encoding other muscle-specific transcription factors. MyoD also turns on genes such as *p21* that block the cell cycle and thus stop cell division. The various muscle-specific transcription factors activate the genes for muscle proteins such as myosin and actin. Meanwhile, the myoblasts fuse to become multinucleated mature muscle cells, also called muscle fibers.

dependent on MyoD but requires a particular *combination* of regulatory proteins, some of which are lacking in cells that don't respond to MyoD. The determination and differentiation of other kinds of tissues may play out in a similar fashion.

Transcriptional regulation is directed by maternal molecules in the cytoplasm and signals from other cells

Explaining the role of *myoD* in muscle cell differentiation is a long way from explaining the development of an organism. The *myoD* story immediately raises the question of what triggers the expression of *that* gene and then raises a series of similar questions leading back to the zygote. What generates the *first* differences that arise among the cells in an early embryo? And what controls morphogenesis and the differentiation of all the different cell types as development proceeds? As we saw in the case of muscle cells, this question comes down to which genes are transcribed in the cells of a developing organism. Two sources of information "tell" a cell which genes to express at any given time.

One important source of information that operates early in development is the cytoplasm of the unfertilized egg cell, which contains both RNA and protein molecules encoded by the mother's DNA. The cytoplasm of an egg cell and even its cytosolic fluid are not homogeneous. Messenger RNA, proteins, other substances, and organelles are distributed unevenly in the unfertilized egg, and this heterogeneity has a profound impact on the development of the future embryo

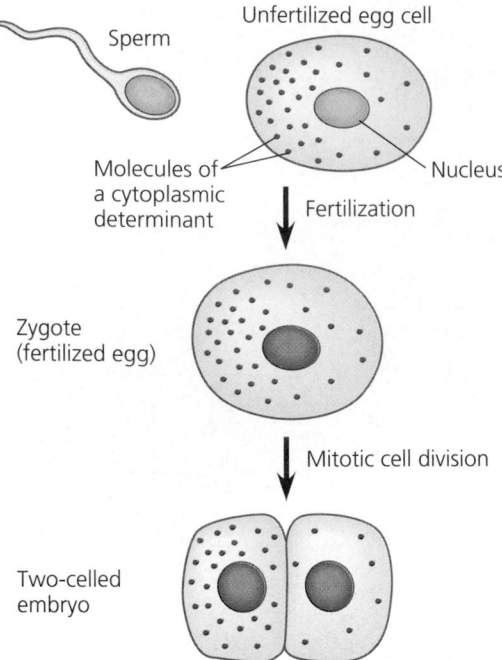

(a) Cytoplasmic determinants in the egg. The unfertilized egg cell has molecules in its cytoplasm, encoded by the mother's genes, that influence development. Many of these cytoplasmic determinants, like the one shown here, are unevenly distributed in the egg. After fertilization and mitotic division, the cell nuclei of the embryo have different environments with respect to cytoplasmic determinants and, as a result, express different genes.

(b) Induction by nearby cells. The teal cells at the top of the early embryo depicted here are releasing chemicals that signal nearby cells to change their gene expression.

FIGURE 21.10 Sources of developmental information for the early embryo.

in many species. After fertilization, the cell nuclei resulting from mitotic division of the zygote are exposed to different cytoplasmic environments. The maternal substances in the egg that influence the course of early development, called **cytoplasmic determinants,** regulate the expression of genes that affect the developmental fate of the cells (FIGURE 21.10a).

The other important source of developmental information, which becomes increasingly important as the number of embryonic cells grows, is the environment around a cell. Most important are the signals impinging on an embryonic cell from other embryonic cells in the vicinity. The synthesis of the molecules conveying these signals is controlled by the embryo's own genes. The signal molecules cause changes in nearby target cells, a process called **induction** (FIGURE 21.10b). In general, the signal molecules induce the observable cellular changes by causing a change in gene expression in the target cells. Thus, interactions among embryonic cells eventually induce differentiation of the many specialized cell types making up a new organism. Induction may be accomplished by the diffusion of chemical signals or, if the cells are actually in contact, by cell surface interactions.

You'll learn more about cytoplasmic determinants and induction as we take a closer look at some important genetic and cellular mechanisms of development in three model organisms: *Drosophila, C. elegans,* and *Arabidopsis.*

GENETIC AND CELLULAR MECHANISMS OF PATTERN FORMATION

How do cytoplasmic determinants, inductive signals, and their effects on embryonic cells contribute to morphogenesis, the shaping of the organism and its parts? We'll explore this question in the context of **pattern formation,** the development of a *spatial organization* in which the tissues and organs of an organism are all in their characteristic places. In the life of a plant, pattern formation occurs continually, in the apical meristems (see FIGURE 21.2b). In animals, pattern formation is mostly limited to embryos and juveniles.

Pattern formation in animals begins in the early embryo, when the animal's basic body plan—its overall three-dimensional arrangement—is established. Just as the outline of a building is laid out before construction begins, the major axes of an animal are established very early. Before specialized tissues or organs appear, the relative positions of the animal's head and tail, for example, are established. The molecular cues that control pattern formation, collectively called **positional information,** tell a cell its location relative to the body axes and to neighboring cells and determine how the cell and its progeny will respond to future molecular signals.

Genetic analysis of *Drosophila* reveals how genes control development: *an overview*

Pattern formation has been most extensively studied in *Drosophila melanogaster,* where genetic approaches have had spectacular success. These studies have established that genes control development and have led to an understanding of the key roles that specific molecules play in defining position and directing differentiation. Combining anatomical, genetic, and biochemical approaches to the study of *Drosophila* development, researchers have discovered developmental principles common to many other species, including humans.

The Life Cycle of *Drosophila*

Fruit flies and other arthropods have a modular construction, an ordered series of segments. These segments make up the body's three major parts: the head, the thorax (midbody, from which the wings and legs extend), and the abdomen (see FIGURE 21.11, bottom). Like other bilaterally symmetrical animals, *Drosophila* has an anterior-posterior (head-tail) axis and a dorsal-ventral (back-belly) axis. In *Drosophila,* cytoplasmic determinants that are present in the unfertilized egg provide positional information for the placement of the two axes even before fertilization. After fertilization, positional information operating on a finer and finer scale establishes a specific number of correctly oriented segments and finally triggers the formation of each segment's characteristic structures.

The developmental stages of *Drosophila* are illustrated in FIGURE 21.11. The egg cell develops in the mother's ovary, surrounded by ovarian cells called nurse cells and follicle cells. These supply the egg cell with nutrients and other substances needed for development and make the egg shell. ❶ Following fertilization and laying of the egg, mitosis begins. The early mitotic divisions have two notable features. First, the amount of cytoplasm does not change; the first ten divisions, which occur very quickly, consist of S and M phases only, with no growth. Second, cytokinesis does not occur; the early *Drosophila* embryo is one big multinucleated cell (in contrast to vertebrate embryos; see, for example, FIGURE 47.8). ❷ At the tenth nuclear division, the nuclei begin to migrate to the periphery of the embryo, and ❸ at division 13, plasma membranes finally partition the 6,000 or so nuclei into separate cells.

FIGURE 21.11 Key developmental events in the life cycle of *Drosophila.* In the top drawing, the yellow egg cell is surrounded by other cells, which form a structure called the follicle within one of the mother's ovaries. The egg cell grows as it matures and eventually fills the egg shell that is secreted by the follicle cells; the nurse cells shrink and disappear. The egg is fertilized within the mother and then laid. The embryo develops within the protective egg shell, as described in the text. The cell layer forming the equivalent of the blastula in *Drosophila* is called the blastoderm.

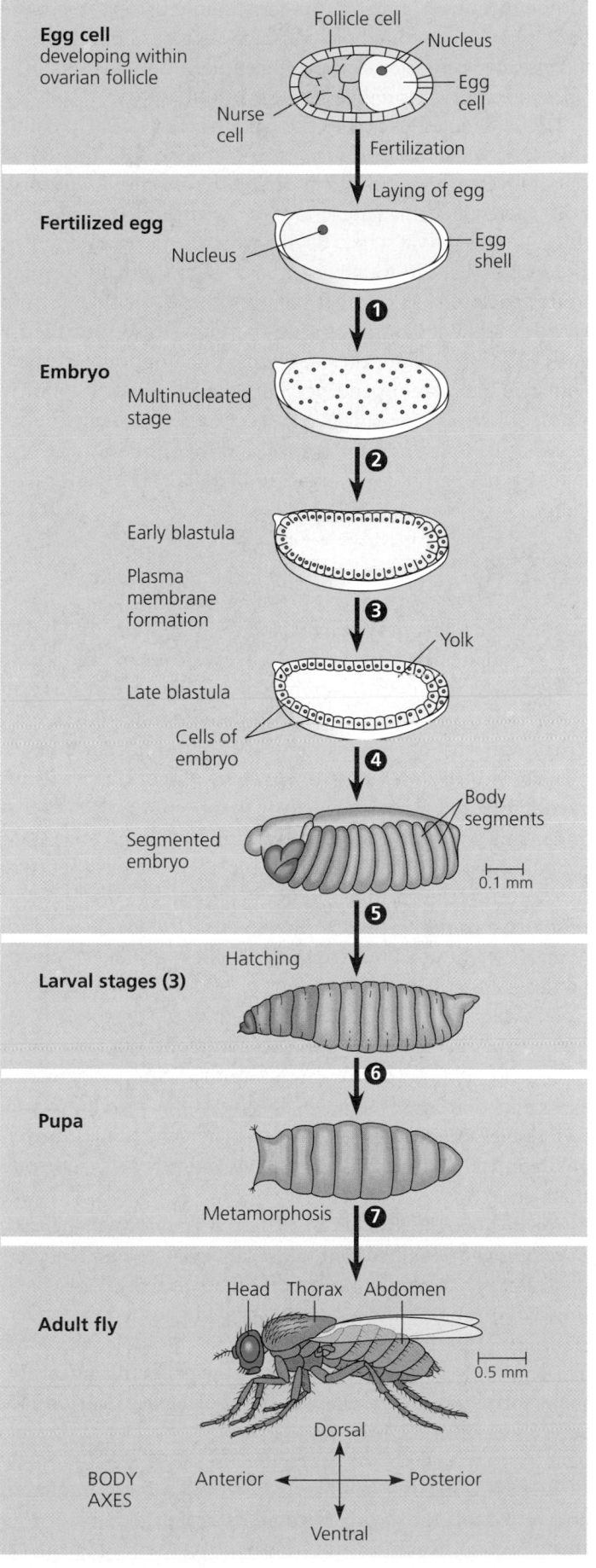

Although not yet apparent under the microscope, the basic body plan—including body axes and segment boundaries—has already been determined by this time. A central yolk nourishes the embryo, and the egg shell continues to protect it.

❹ Subsequent events in the embryo create clearly visible segments, which at first look very much alike. ❺ Then, some cells move to new positions, organs form, and a wormlike larva (juvenile form) hatches out of the shell. *Drosophila* goes through three larval stages, during which the larva eats, grows, and molts (sheds its tough outer layer), ❻ eventually forming a pupa enclosed in a case. ❼ Metamorphosis, the change from larva to adult fly, occurs inside the pupal case, and the fly emerges. In the adult, each segment is anatomically distinct, with characteristic appendages. For example, the first thoracic segment bears a pair of legs, the second thoracic segment has a pair of legs plus a pair of wings, and the third thoracic segment bears a pair of legs plus a pair of balancing organs called halteres.

Genetic Analysis of Early Development in Drosophila

THE PROCESS OF SCIENCE

During the first half of the 20th century, biologists made detailed anatomical observations of embryonic development in a number of species and performed experiments in which they manipulated embryonic tissues. This research laid the groundwork for understanding the mechanisms of development. However, these "classical" embryologists were unable to identify the specific molecules that guide development or determine how patterns are established. In the 1940s, one visionary biologist, Edward B. Lewis, of Caltech, showed that a genetic approach—the study of mutants—could be used to investigate *Drosophila* development. Lewis studied bizarre mutants, flies with developmental defects that led to extra wings or legs in the wrong places (see FIGURE 21.14). He located the mutations on the fly's genetic map, thus connecting the developmental abnormalities to specific genes. This research supplied the first concrete evidence that genes somehow direct the developmental processes studied by embryologists. The genes Lewis discovered control development in the late embryo; we'll return to them shortly.

Meanwhile, the mystery of pattern formation during early development remained. But in the late 1970s, two researchers in Germany, Christiane Nüsslein-Volhard and Eric Wieschaus, undertook an ambitious quest that finally pushed the understanding of early pattern formation to the molecular level. These scientists set out to identify *all* the genes that affect segment formation in *Drosophila*. The project was daunting for several reasons. The first was the sheer number of *Drosophila* genes, about 13,000. The genes affecting segmentation might be just a few needles in a haystack or might be so numerous and varied that the scientists would be unable to make sense of them. Second, mutations affecting a process as fundamental as segmentation would surely be **embryonic lethals,** mutations with phenotypes leading to death at the embryo or larval stage. Because such mutant organisms never reproduce, they cannot be bred for study. Third, because cytoplasmic determinants in the egg were known to play a role in axis formation, the researchers would have to study maternal genes as well as those of the embryo.

Nüsslein-Volhard and Wieschaus dealt with the problem of embryonic lethality by focusing their search on recessive mutations, which could be propagated in heterozygous flies. Their basic strategy was to expose flies to a mutagenic chemical to create mutations in the flies' gametes and then look among the flies' descendants for dead embryos or larvae with abnormal segmentation. By doing the appropriate crosses, they would be able to identify living heterozygotes carrying embryonic lethal mutations. To find as many segmentation genes as possible, they planned a *saturation screen:* They would make enough mutants to "saturate" the fly genome with mutations. The researchers hoped that the segmentation abnormalities visible in the dead embryos would suggest how the affected genes normally functioned.

After a year of laboriously performing thousands of crosses and examining thousands of dead embryos, Nüsslein-Volhard and Wieschaus succeeded in identifying about 1,200 genes essential for embryonic development, of which about 120 were essential for pattern formation leading to normal segmentation. Over several years, they were able to group the genes by general function, to map them, and to clone many of them. The result was a detailed molecular understanding of the early steps in pattern formation in *Drosophila*. When their results were combined with Lewis's earlier work, a coherent picture of *Drosophila* development emerged. In recognition of their discoveries, Nüsslein-Volhard, Wieschaus, and Lewis were awarded a Nobel Prize in 1995. Before we discuss how the segmentation genes function, we need to back up and look at the cytoplasmic determinants deposited in the egg by the mother, for these control the expression of the segmentation genes.

Gradients of maternal molecules in the early embryo control axis formation

As previously mentioned, cytoplasmic determinants are the substances that initially establish the axes of the *Drosophila* body. Already present in the unfertilized egg, they are encoded by genes of the mother called **maternal effect genes.** A maternal effect gene is a gene that, when mutant in the mother, results in a mutant phenotype in the offspring, regardless of their genotype. In fruit fly development, maternal effect genes encode proteins or mRNA that are placed in the egg while it is still in the mother's ovary. When the mother has a mutation in such a gene, she makes a defective gene product (or none at all), and her eggs are defective; when these eggs are fertilized, they fail to develop properly.

Because they control the orientation (polarity) of the egg and consequently of the fly, maternal effect genes are also called **egg-polarity genes.** One group of these genes sets up the anterior-posterior axis of the embryo, while a second group establishes the dorsal-ventral axis. Like mutations in segmentation genes, mutations in these genes are generally embryonic lethals.

How do products of maternal effect genes determine the body axes of the offspring? Let's focus on one important egg-polarity gene, *bicoid*, and see how it works. The term bicoid means "two-tailed," and an embryo whose mother was defective in this gene lacks the front half of its body. Instead, the embryo has duplicate posterior structures at both ends (FIGURE 21.12a). This phenotype suggested the hypothesis that the product of the mother's *bicoid* gene is essential for setting up the anterior end of the fly and that the gene's product is concentrated at the future anterior end. This hypothesis is a more specific version of the *gradient hypothesis* first proposed by embryologists a century ago. According to this idea, gradients of substances called **morphogens** establish an embryo's axes and other features of its form.

DNA technology and other modern biochemical methods enabled researchers to test the hypothesis that the *bicoid* product is in fact a morphogen that determines the anterior end of the fly. The researchers cloned the *bicoid* gene and used it as a DNA probe to learn the location of *bicoid* mRNA in the eggs produced by wild-type female flies. As predicted by the hypothesis, the *bicoid* mRNA is concentrated at the extreme anterior end of the egg cell (FIGURE 21.12b). After the egg is fertilized, the mRNA is translated into protein, which diffuses from the anterior end toward the posterior, resulting in a gradient of protein within the early embryo. These results were consistent with the hypothesis that bicoid protein was responsible for specifying the fly's anterior end. To test the hypothesis more specifically, the scientists injected pure *bicoid* mRNA into various regions of early embryos. The protein that resulted from its translation caused anterior structures to form at the injection sites.

The *bicoid* research is important for several reasons. First, it led to the identification of a specific protein required for some of the earliest steps in pattern formation. Second, it increased our understanding of the mother's role in the development of an embryo. (As one developmental biologist has put it, "Mom tells Junior which way is up.") Finally, the principle that a gradient of molecules can determine polarity and position has proved to be a key developmental concept, just as early embryologists had thought. In *Drosophila*, gradients of specific proteins determine the posterior end as well as the anterior and also are responsible for establishing the dorsal-ventral axis. Saturation screening has led to the identification of most of the genes and proteins involved.

FIGURE 21.12 The effect of the *bicoid* gene, a maternal effect (egg-polarity) gene in *Drosophila*.

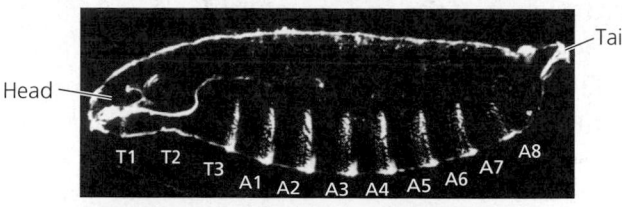

Wild-type larva

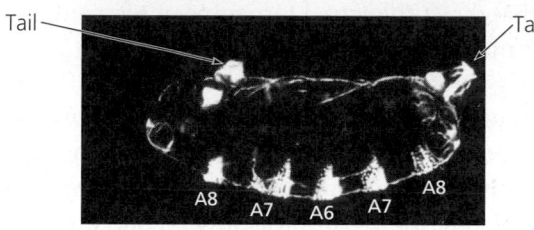

Mutant larva (*bicoid*)

(a) *Drosophila* larvae with wild-type and bicoid mutant phenotypes. Because of a mutation in the mother's *bicoid* gene, the bottom larva has tail structures at both ends. The numbers refer to the thoracic and abdominal segments that are present.

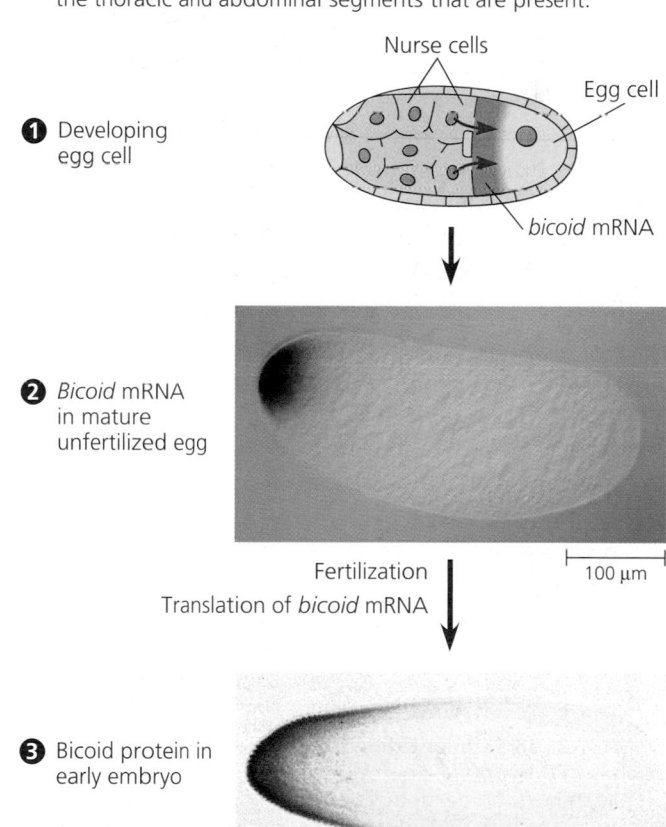

❶ Developing egg cell

Nurse cells

Egg cell

bicoid mRNA

❷ *Bicoid* mRNA in mature unfertilized egg

100 μm

Fertilization
Translation of *bicoid* mRNA

❸ Bicoid protein in early embryo

Anterior end

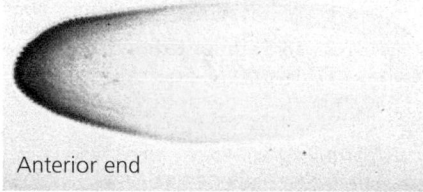

(b) Gradients of *bicoid* mRNA and protein in normal egg and early embryo. ❶ Transcribed from the maternal *bicoid* gene in nurse cells, *bicoid* mRNA reaches the egg cell via cytoplasmic bridges. It becomes anchored to the cytoskeleton at the anterior end of the egg as the egg grows and the nurse cells disappear. ❷ Labeled *bicoid* DNA (purple) was used as a probe to locate *bicoid* mRNA at the anterior end of the egg. ❸ After fertilization, the mRNA is translated. In this early embryo, the gradient of color reveals a gradient of bicoid protein. Bicoid protein is only one of several morphogens involved in axis specification.

A cascade of gene activations sets up the segmentation pattern in *Drosophila*: a closer look

The bicoid protein and other morphogens that are products of egg-polarity genes are transcription factors, proteins that regulate the activity (transcription) of some of the embryo's own genes. Gradients of these morphogens bring about regional differences in the expression of **segmentation genes,** the genes of the embryo that direct the actual formation of segments after the embryo's major axes are defined.

In a cascade of gene activations, sequential activation of three sets of segmentation genes provides the positional information for increasingly fine details of the animal's modular body plan. First, products of the **gap genes** map out the basic subdivisions along the anterior-posterior axis of the embryo (FIGURE 21.13a). Mutations in these genes cause "gaps" in the animal's segmentation. For example, one gap mutation results in an embryo lacking six abdominal segments. **Pair-rule genes** are the second set of segmentation genes to act. They define the modular pattern in terms of pairs of segments (FIGURE 21.13b). Mutations in pair-rule genes result in embryos having half the normal segment number because every other segment (odd or even, depending on the mutation) fails to develop. The third set of segmentation genes to act are the **segment polarity genes,** which set the anterior-posterior axis of each segment (FIGURE 21.13c). Embryos with mutations in segment polarity genes have the normal number of segments, but a part of each segment is replaced by a mirror-image repetition of some other part of the segment.

The products of many of the segmentation genes, like those of egg-polarity genes, are transcription factors that directly activate the next set of genes in the hierarchical scheme of pattern formation. Other segmentation genes operate more indirectly, supporting the functioning of the transcription factors in various ways. For example, some are components of cell-signaling pathways, including signal molecules used in cell-cell communication and the membrane receptors that recognize them (see Chapter 11). Cell-signaling molecules are critically important once plasma membranes have divided the embryo into separate cellular compartments.

Working together, the products of egg-polarity genes regulate the regional expression of gap genes, which control the localized expression of pair-rule genes, which in turn activate specific segment polarity genes in different parts of each segment. The boundaries and axes of the segments are now set. In the hierarchy of gene activations responsible for pattern formation, the next genes to be expressed determine the specific anatomy of each segment along the embryo.

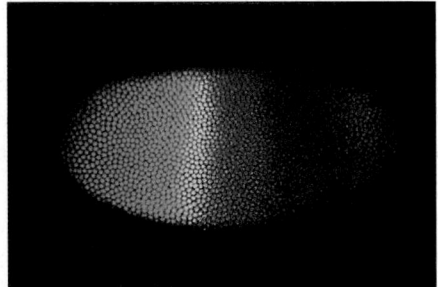

(a) Products of two gap genes. Gap genes, the first set of segmentation genes turned on, produce broad bands of gene-regulating proteins, prescribing a coarse subdivision of the embryo. The green and red colors of the bands in this micrograph represent products of two different gap genes; the yellow band is the region of overlap.

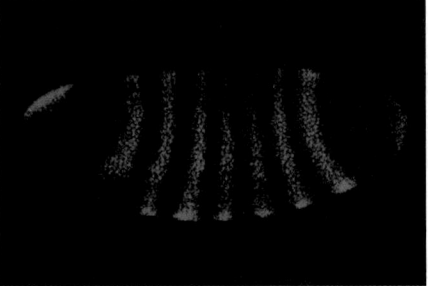

(b) Products of a pair-rule gene. Localized products of gap genes turn on a second set of segmentation genes, the pair-rule genes. The protein product of a pair-rule gene produced these green bands. The pair-rule genes prescribe further subdivision of the embryo.

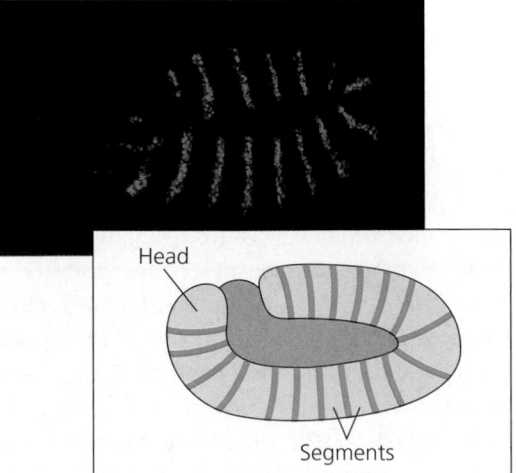

(c) Product of a segment polarity gene. The final directions for subdividing the embryo into segments are in place when the pair-rule proteins bring about localized expression of various segment polarity genes. The product of a segment polarity gene accounts for the green bands here. Each of the compartments between these protein bands represents a body segment of the embryo, which at this stage is folded over on itself.

THE PROCESS OF SCIENCE

FIGURE 21.13 Segmentation genes in *Drosophila*. Soon after fertilization, once the mother's egg-polarity genes have set up the body axes, bicoid protein and other maternal gene products initiate a cascade of activity by segmentation genes in the embryo's nuclei. These micrographs of developing fly embryos illustrate successive bands of regulatory proteins encoded by segmentation genes. The proteins are DNA-binding transcription factors that direct the division of the body into the segments characteristic of flies and other arthropods. The colors result from fluorescent dyes on antibodies bound to the segmentation gene proteins.

Homeotic genes direct the identity of body parts

In a normal fly, structures such as antennae, legs, and wings develop on the appropriate segments. The anatomical identity of the segments is set by master regulatory genes called **homeotic genes.** These are the genes discovered by Edward Lewis. Once the segmentation genes have staked out the fly's segments, homeotic genes specify the types of appendages and other structures that each segment will form. Mutations in homeotic genes produce flies with such strange traits as legs growing from the head in place of antennae (FIGURE 21.14). Thus, as Lewis found, homeotic mutations cause structures characteristic of a particular part of the animal to arise in the wrong place.

Like many of the egg-polarity and segmentation genes preceding them in the developmental cascade, the homeotic genes encode transcription factors. These regulatory proteins control the expression of the genes responsible for specific anatomical structures. For example, a homeotic protein made in the cells of a particular thoracic segment may selectively activate genes that bring about leg development. In contrast, a homeotic protein active in a certain head segment specifies "antennae go here." A mutant version of this protein may label the segment as "thoracic" instead of "head," causing legs to develop in place of antennae. Scientists are now busy identifying the genes activated by the homeotic proteins—the genes specifying the proteins that actually build the fly structures. The following flowchart summarizes the cascade of gene activity in the *Drosophila* embryo:

Hierarchy of Gene Activity in *Drosophila* Development

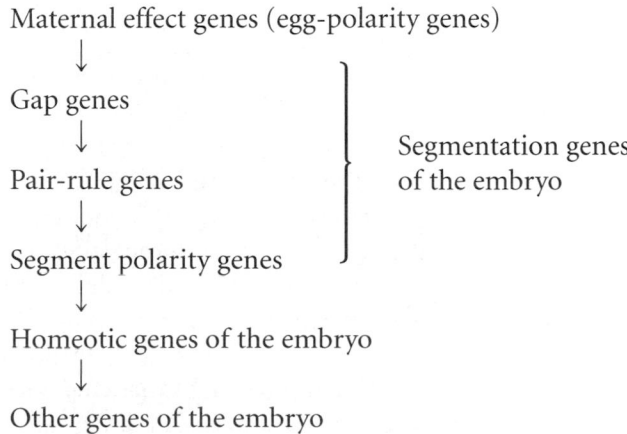

Amazingly, many of the molecules and mechanisms revealed by research on fly pattern formation have turned out to

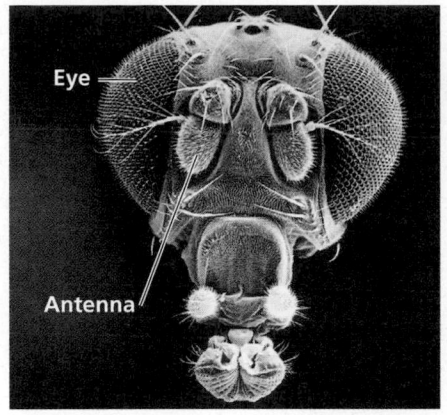
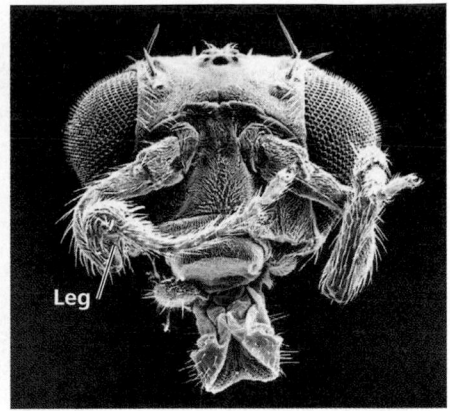

FIGURE 21.14 Homeotic mutations and abnormal pattern formation in *Drosophila*. Homeotic mutations cause a misplacement of structures in an animal. These micrographs contrast the heads of two fruit flies (SEMs). Where small antennae are located in the normal fruit fly (left photo), one homeotic mutant has legs (right photo).

have close counterparts throughout the animal kingdom. And nowhere are the similarities more striking than in the homeotic genes and their products.

Homeobox genes have been highly conserved in evolution

The homeotic genes of *Drosophila* all include a 180-nucleotide sequence called the **homeobox,** which specifies a 60-amino-acid *homeodomain.* An identical or very similar sequence of nucleotides has been discovered in genes of many other animals, including other insects, nematodes, mollusks, fish, frogs, birds, and mammals, including humans. (Homeobox-containing genes are often called *Hox* genes, especially in mammals.) Furthermore, related sequences have been found in regulatory genes of much more distantly related eukaryotes, such as yeast, and even in prokaryotes. From these similarities we can deduce that the homeobox DNA sequence evolved very early in the history of life and that it is sufficiently valuable to organisms to have been conserved in animals virtually unchanged for hundreds of millions of years. In fact, the vertebrate genes homologous to the homeotic genes of fruit flies have even kept their chromosomal arrangement (FIGURE 21.15, p. 418).

Not all homeobox-containing genes are homeotic genes; that is, some do not directly control the identity of body parts. However, most of these genes are associated with development, suggesting their ancient and fundamental importance in that process. For example, in *Drosophila*, homeoboxes are present not only in the homeotic genes but also in the egg-polarity gene *bicoid,* in several of the segmentation genes, and in the master regulatory gene for eye development (see the photograph at the beginning of the chapter).

What is the role of the homeodomain in a protein? This polypeptide segment is the part of the protein that binds to

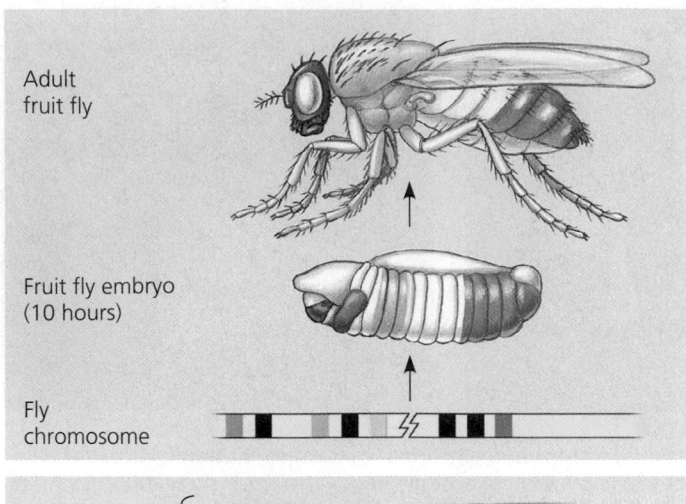

Adult
fruit fly

Fruit fly embryo
(10 hours)

Fly
chromosome

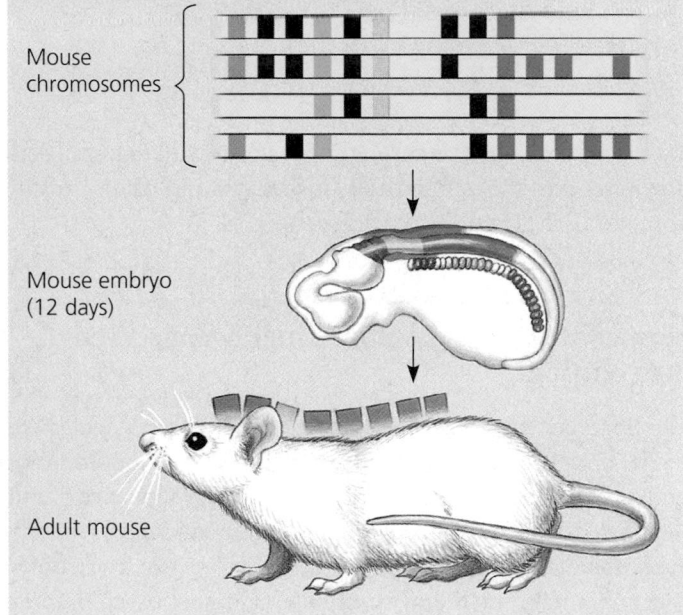

Mouse
chromosomes

Mouse embryo
(12 days)

Adult mouse

FIGURE 21.15 Homologous genes that affect pattern formation in a fruit fly and a mouse. Homeotic homeobox-containing genes that control the form of the anterior and posterior structures of the body occur in the same linear sequence on chromosomes in *Drosophila* and mice. Each small box on the chromosomes shown here represents a homeobox-containing gene. In fruit flies, all of these genes are found on one chromosome. The mouse and other mammals have the same or similar sets of genes on four chromosomes. The color code indicates the parts of the embryos in which these genes are expressed and the adult body regions that result. Purple, green, gray, and orange boxes represent genes with homeoboxes that are essentially identical in flies and mice. Black boxes represent genes with homeoboxes that are less similar in the two animals. In vertebrates, homeobox-containing genes are often called *Hox genes*.

DNA when the protein functions as a transcription factor. (The homeodomain forms three α helices, one of which fits neatly into the major groove of the DNA helix.) However, the shape of the homeodomain allows it to bind to any DNA segment; by itself it cannot select a specific sequence. It is other, more variable domains of such a protein that determine which genes the protein regulates. These latter domains interact with other transcription factors that help the protein recognize specific

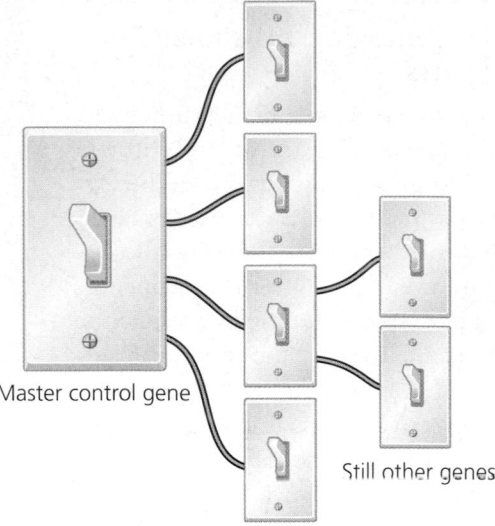

Master control gene

Still other genes

Battery of other genes

FIGURE 21.16 Homeobox-containing genes as switches. How a particular cell differentiates in the animal's body depends on how many of these switches are thrown—that is, which homeobox proteins are made within the cell. A single master switch may control a number of subordinate switches, which in turn control others. Here the wires represent the actions of the proteins that are the products of the genes.

enhancers or promoters in the DNA. Proteins with homeodomains probably regulate development by coordinating the transcription of batteries of developmental genes, switching them on or off (FIGURE 21.16). In *Drosophila* embryos, different combinations of homeobox genes are active in different parts of the embryo. This selective expression of regulatory genes, varying over time and space, is central to pattern formation.

Neighboring cells instruct other cells to form particular structures: cell signaling and induction in the nematode

The development of a multicellular organism requires close communication among cells. Even when the *Drosophila* embryo is still one multinucleated "cell," signaling between cells has already played a key role. For example, an exchange of signals between the unfertilized egg cell and neighboring follicle cells guided the establishment of the egg's anterior end; it was a follicle cell signal that triggered the localization of *bicoid* mRNA there. Once the embryo is truly multicellular, signaling among the embryo's own cells becomes increasingly important. In the process called induction, cells signal nearby cells to change in some specific way, often by expressing particular genes. As we've seen, the ultimate basis for the differences among cells is transcriptional regulation—the turning on and off of specific genes. It is induction, signaling from one group of cells to an adjacent group, that brings about differentiation.

Induction in Vulval Development

The nematode *C. elegans* has proved to be a very useful model organism for investigating the roles of cell signaling and induction in development. Particularly revealing has been research on the development of the nematode *vulva,* the tiny opening through which the worm lays its eggs. Researchers have combined genetic, biochemical, and embryological approaches to learn how this organ forms.

The pathway from fertilized egg to an adult nematode capable of reproduction involves four larval stages (the larvae look much like smaller versions of the adult). Already present on the ventral surface of the second-stage larva are six cells from which the vulva will arise (FIGURE 21.17a). A single cell of the embryonic gonad, the *anchor cell,* initiates a cascade of signals that establishes the fates of the vulval precursor cells. If an experimenter destroys the anchor cell with a laser beam, the vulva fails to form and the precursor cells simply become part of the worm's epidermis.

Researchers have figured out the molecular details of vulval induction by isolating mutants in which vulval development is abnormal. Such mutants can grow to adulthood because a normal egg-laying apparatus is not essential for viability. The phenotypes of the mutants range from multiple vulvae to none at all. (In the latter mutants, offspring develop internally within self-fertilizing hermaphrodites, eventually eating their way out of the parent's body!) From studying the mutants, researchers have identified a number of genes involved in vulval development and have worked out where and how their products function (FIGURE 21.17b).

The anchor cell secretes an inducer, a signal protein that binds to a receptor protein in the plasma membrane of vulval precursor cells. (An example of the "unity of life," this signal protein is very similar to the *epidermal growth factor (EGF)* of humans and other mammals.) At the start, all the precursor cells are equivalent: All make a basic set of proteins necessary for vulval formation, including receptors for the anchor cell's signal molecules. However, these cells do not all respond to the inducer. The cell that normally becomes the inner part of the vulva receives high levels of inducer, since it is closest to the anchor cell. The high levels of inducer probably produce two effects: (1) division and differentiation of the cell to form the inner part of the vulva and (2) activation of a gene for a cell surface signaling protein, which acts as a second inducer. Receptors on the two adjacent vulval precursor cells bind the second inducer, which stimulates these cells to divide and develop into the outer vulva. The three remaining vulval precursor cells are too far away to receive either signal; they give rise to epidermal cells.

FIGURE 21.17 Cell signaling and induction in the development of the nematode vulva. The vulva is the opening through which the nematode lays its eggs.

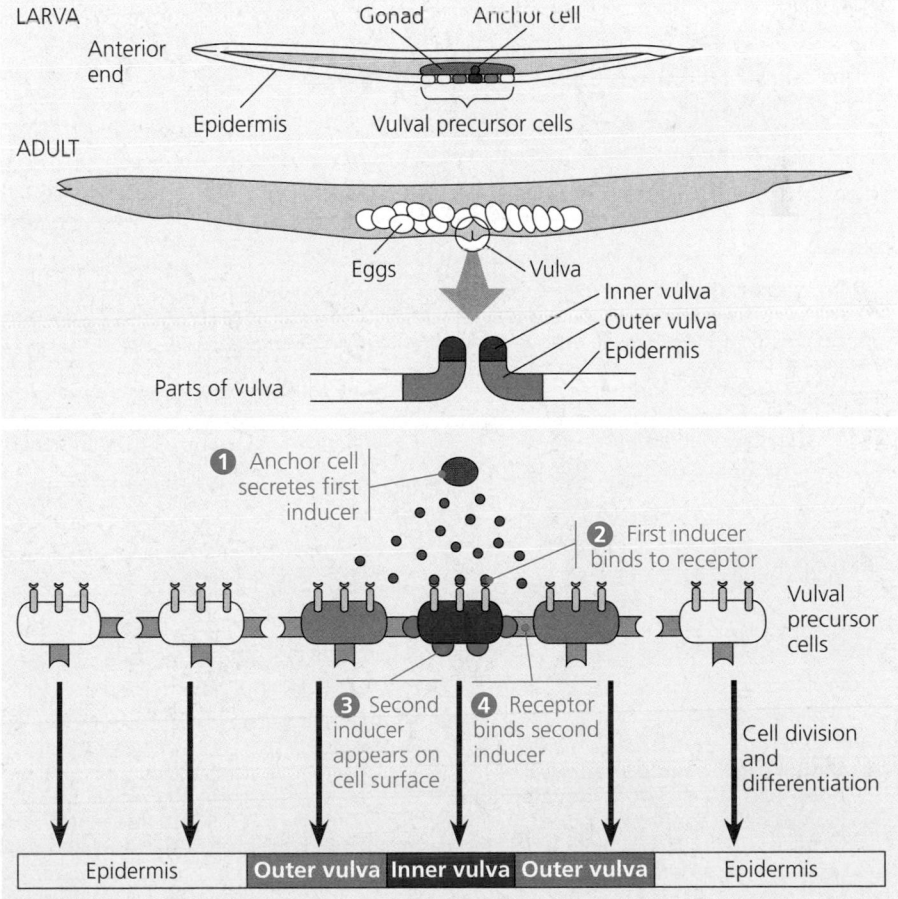

(a) Anatomy of vulval development. The vulval precursor cells of the larva have three possible fates: They can develop into the inner part of the vulva, the outer part, or the adjacent epidermis.

(b) Cell signaling and induction. The actual fates of these cells are determined by a series of inductions initiated by ❶ a chemical signal from a nearby cell called the anchor cell. Orange circles represent this first inducer. ❷ The binding of these molecules to receptors on the nearest vulval precursor cell triggers a signal-transduction pathway leading to ❸ the synthesis of a second inducer, a cell surface protein (green). ❹ This inducer, in turn, binds to specific receptors (purple) on the two adjacent vulval precursor cells.

The inducer released by the anchor cell and the signal-transduction pathway it triggers in the closest vulval precursor cell are similar to ones you have encountered in previous chapters. The signal is a growth-factor-like protein, and it is transduced within its target cell via a tyrosine-kinase receptor, a Ras protein, and a cascade of protein kinases—a common pathway leading to transcriptional regulation in many organisms (see FIGURE 19.14a). (The nematode research provided the first *in vivo* evidence linking the Ras pathway with a growth factor.)

In summary, vulval development in the nematode illustrates a number of important concepts that apply elsewhere in the development of *C. elegans* and many other animals:

- In the developing embryo, sequential inductions drive the formation of organs.
- The effect of an inducer can depend on its concentration (just as we saw with cytoplasmic determinants in *Drosophila*).
- Inducers produce their effects via signal-transduction pathways similar to those operating in adult cells.
- The induced cell's response is often the activation (or inactivation) of genes—transcriptional regulation—

which, in turn, establishes the pattern of gene activity characteristic of a particular kind of differentiated cell.
- Genetics is a powerful approach for elucidating the mechanisms of development.

Programmed Cell Death (Apoptosis)

Lineage analysis of *C. elegans* has highlighted another outcome of cell signaling that is crucial in animal development—programmed cell death, or **apoptosis.** The timely suicide of cells occurs exactly 131 times in the course of *C. elegans*'s normal development. At precisely the same points in each new worm's cell lineage, signals trigger the activation of a cascade of "suicide" proteins in the cells destined to die. The cells shrink, their nuclei condense and then break down, and neighboring cells quickly engulf and digest the remains, leaving no trace (FIGURE 21.18a).

Genetic screening of *C. elegans* has revealed two key apoptosis genes, *ced-3* and *ced-4* (*ced* stands for "cell death"), which encode proteins essential for apoptosis. The proteins are called Ced-3 and Ced-4, respectively. These and most other proteins

FIGURE 21.18 Apoptosis (programmed cell death).

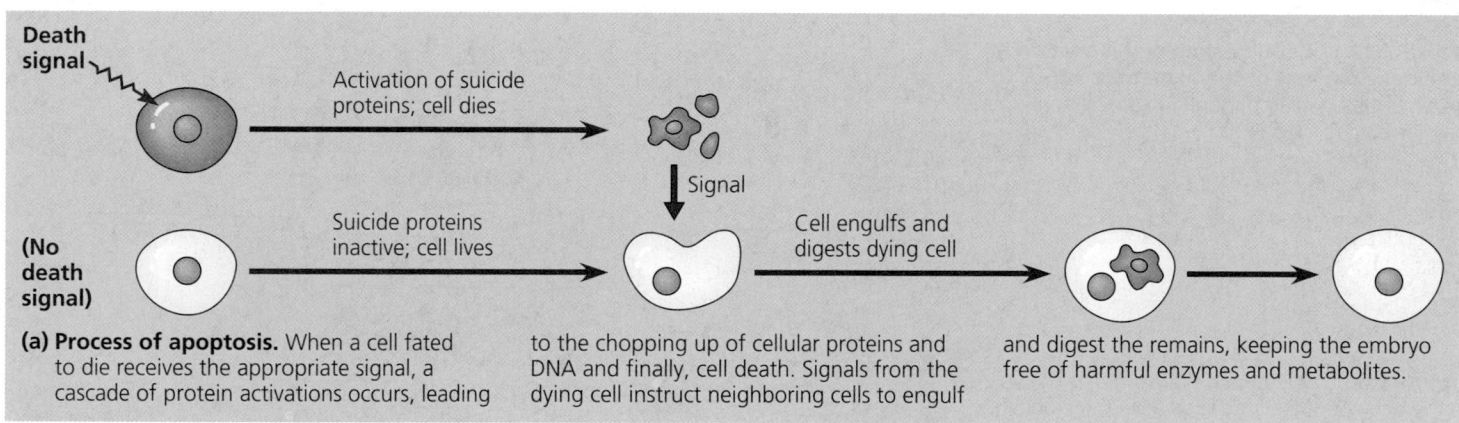

(a) Process of apoptosis. When a cell fated to die receives the appropriate signal, a cascade of protein activations occurs, leading to the chopping up of cellular proteins and DNA and finally, cell death. Signals from the dying cell instruct neighboring cells to engulf and digest the remains, keeping the embryo free of harmful enzymes and metabolites.

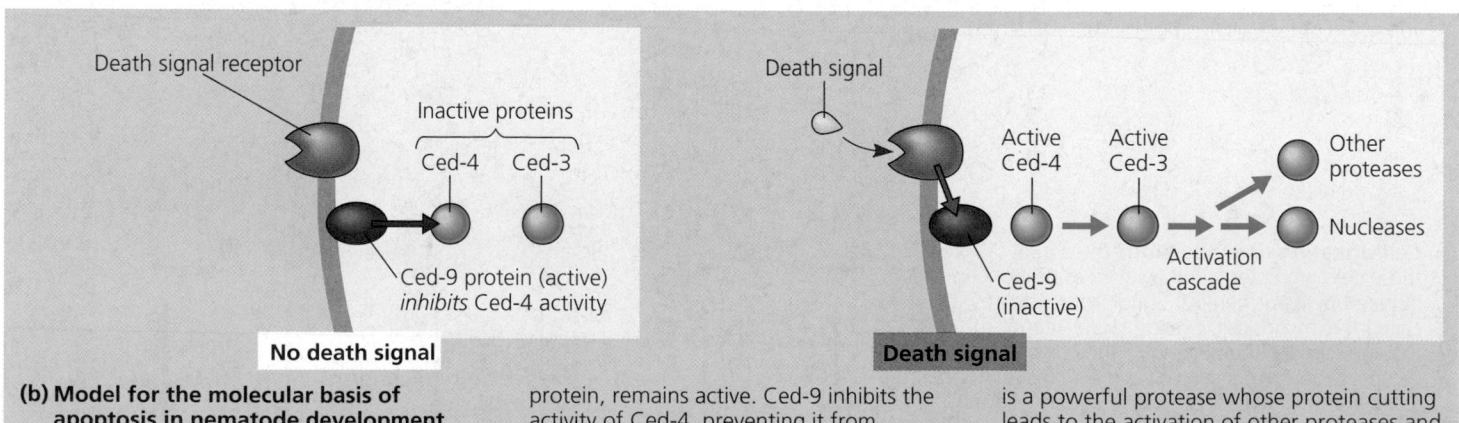

(b) Model for the molecular basis of apoptosis in nematode development. In *C. elegans,* apoptosis involves three main proteins: Ced-9, Ced-4, and Ced-3. A cell remains alive as long as Ced-9, a regulator protein, remains active. Ced-9 inhibits the activity of Ced-4, preventing it from activating Ced-3. "Death signals" received by the cell inactivate Ced-9, allowing both Ced-4 and Ced-3 to be active. Active Ced-3 is a powerful protease whose protein cutting leads to the activation of other proteases and nucleases. Together these enzymes kill the cell. The proteases of apoptosis are called caspases.

involved in apoptosis are continually present in cells, but in inactive form; thus, protein *activity* is regulated in this case, not transcription or translation. A protein called Ced-9 (the product of gene *ced-9*) is the master regulator of apoptosis (FIGURE 21.18b). The apoptosis pathway activates proteases and nucleases, enzymes that cut up the proteins and DNA of the cell. The main proteases of apoptosis are called *caspases;* in the nematode, the chief caspase is Ced-3.

Apoptosis pathways in humans and other mammals are more complicated. Notably, research on mammals has revealed a prominent role for mitochondria in apoptosis, at least in these more complex animals. Apoptosis pathway proteins or other signals somehow cause the mitochondrial outer membrane to leak, releasing proteins that promote apoptosis. (Surprisingly, these include cytochrome *c,* whose cellular function was thought to have been limited to electron transport; see FIGURE 9.15.) Still controversial is whether mitochondria play a central role in apoptosis or only a subsidiary role. The answer may vary with the circumstance. In any case, a cell must make life-or-death "decisions" by somehow integrating the signals it receives, both "death" signals and "life" signals such as growth factors.

A built-in cell suicide mechanism is essential to development in all animals, and similarities between apoptosis genes in nematodes and mammals indicate that the basic mechanism evolved early in animal evolution. The timely activation of apoptosis proteins in some cells functions in normal development and growth in both embryos and adults. In vertebrates, programmed cell death is essential for normal development of the nervous system, for normal operation of the immune system, and for the normal morphogenesis of human hands and feet. In the last case, the failure of normal cell death can result in webbed fingers and toes. Researchers are also investigating the possibility that certain degenerative diseases of the nervous system result from the inappropriate activation of apoptosis genes and that some cancers result from a failure of cell suicide. Cells that have suffered irreparable damage, including DNA damage that could lead to cancer, normally generate *internal* signals that trigger apoptosis.

Plant development depends on cell signaling and transcriptional regulation

THE PROCESS OF SCIENCE

The last common ancestor of plants and animals was probably a single-celled microbe living hundreds of millions of years ago, so the processes of development must have evolved independently in the two lineages of organisms. Plants evolved with rigid cell walls that make the movement of cells and tissue layers virtually impossible, ruling out a morphogenetic mechanism that is important in animals. Plant morphogenesis relies more heavily on differing planes of cell division and on selective cell enlargement. (You will learn about these processes in Chapter 35.)

But despite the differences between plants and animals, there are some basic similarities in their molecular, cellular, and genetic mechanisms of development—legacies of their shared cellular origins. In particular, plant development, like that of animals, depends on cell signaling (induction) and transcriptional regulation.

We are just beginning to understand the molecular basis of plant development in detail. Thanks to DNA technology, clues from animal research, and model organisms such as *Arabidopsis,* plant research is now progressing rapidly. The embryonic development of most plants occurs inside the seed and is relatively inaccessible to study (a mature seed already contains a fully formed embryo). However, other important aspects of plant development are observable throughout the plant's life in its meristems, and in particular in the apical meristems at the tips of shoots. It is there that cell division, morphogenesis, and differentiation give rise to new organs, such as leaves or the petals of flowers. We'll discuss two examples of research on floral meristems, the apical meristems that produce flowers.

Cell Signaling in Flower Development

Environmental signals, such as day length and temperature, trigger signal-transduction pathways that convert ordinary shoot meristems to floral meristems. Researchers have combined a genetic approach with tissue transplantation to study induction in the development of tomato flowers. As shown in FIGURE 21.19a (p. 422), a floral meristem is a "bump" consisting of three layers of cells (L1–L3). All three layers participate in the formation of a flower, a reproductive structure with four types of organs: carpels (containing egg cells), petals, stamens (containing sperm-bearing pollen), and sepals (leaflike structures outside the petals).

Researchers discovered that tomato plants homozygous for a mutant allele called *fasciated (f)* produce flowers with an abnormally large number of organs. Taking advantage of the totipotency of plant cells, they performed an experiment in which they grafted stems from this mutant onto wild-type plants and then grew new plants from the shoots that emerged at the graft sites. The new plants were **chimeras,** organisms with a mixture of genetically different cells. Some of the chimeras produced floral meristems in which the three cell layers did not all come from the same "parent" (FIGURE 21.19b). The researchers could identify the parental sources of the meristem layers by monitoring other genetic markers, such as an unrelated mutation that caused yellow leaves. They found that the number of organs per flower depends on the genes of the L3 (innermost) cell layer, which somehow induces the overlaying layers L2 and L1 to form that number of organs. The mechanism of cell-cell signaling leading to this induction is not yet known.

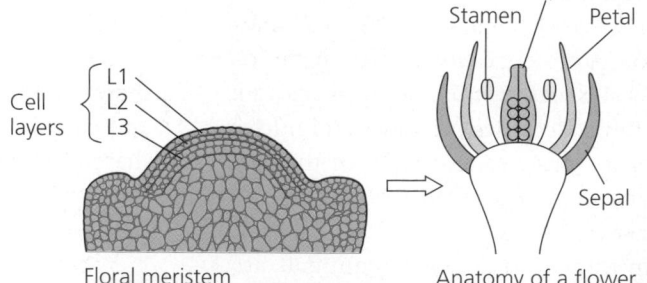

FIGURE 21.19 Induction in flower development.

(a) Flower development (side view). A flower develops from a meristem with three layers of cells, L1–L3. A specific pattern of cell division, differentiation, and enlargement produces the flower. A flower consists of four types of organs arranged in concentric circles: carpels (containing egg cells), stamens (containing sperm-bearing pollen), petals, and sepals.

Cell layers { L1, L2, L3 }

Floral meristem

Carpel
Stamen
Petal
Sepal

Anatomy of a flower

Tomato flower

(b) Evidence of induction in the floral meristem. Experiments with chimeras of the tomato plant (*Lycopersicum esculentum*) have provided evidence that induction between cell layers of the meristem helps determine the number of organs that develop in a flower. Made by grafting together plants of two different genotypes, some of the chimeras had floral meristems with layers originating from different "parents." The phenotypes of the resulting flowers showed that the L1 and L2 layers develop according to the genotype of the L3 layer. Thus, the L3 cells must somehow signal the L1 and L2 layers and induce them to form flowers with a particular number of organs.

PLANT	GENOTYPES OF FLORAL MERISTEM LAYERS			FLOWER PHENOTYPE
	L1	L2	L3	
Wild-type "parent"	++	++	++	Wild type
Fasciated "parent"	ff	ff	ff	Extra organs (fasciated)
Chimera 1	++	ff	ff	Extra organs
Chimera 2	++	++	ff	Extra organs
Chimera 3	++	ff	++	Wild type
Chimera 4	ff	++	++	Wild type

Organ Identity Genes in Plants

In contrast to genes controlling organ *number* in flowers are genes controlling organ *identity*. An **organ identity gene** determines the type of structure that will grow from a meristem—for instance, whether a particular outgrowth from a floral meristem becomes a petal or a stamen. Most of what we know about plant organ identity genes comes from research on flower development in *Arabidopsis*. Organ identity genes are analogous to homeotic genes in animals and are often referred to as plant homeotic genes. Just as a mutation in a fruit fly homeotic gene can cause legs to grow in place of antennae, a mutation in a plant organ identity gene can cause carpels to grow in place of sepals.

By collecting and studying mutants with abnormal flowers, researchers have been able to identify and clone a number of floral organ identity genes, and they are beginning to determine how they act. To follow their work, we need a top view of the floral meristem (FIGURE 21.20a). Viewed from above, the meristem can be divided into four concentric circles, or whorls, each of which develops into a circle of identical organs. The organ identity genes first identified fell into three classes (*A*, *B*, and *C*), each of which affects two adjacent whorls. A simple model, diagrammed in FIGURE 21.20a, explains how these three kinds of genes can direct the formation of the four types of organs.

This model (hypothesis) predicts that each organ identity gene will be transcribed in two specific whorls of the floral meristem, and once the genes were cloned, the predictions could be tested. Using nucleic acid from cloned genes as probes, researchers showed that the mRNA resulting from the transcription of each class of organ identity gene is present in the appropriate whorls of the developing floral meristem. For example, nucleic acid from a *C* gene hybridized appreciably only to cells in whorls 3 and 4 because only those cells were making RNA complementary to it (FIGURE 21.20b).

The model also can account for the phenotypes of mutants lacking *A*, *B*, or *C* gene activity with one addition: Where *A* gene activity is present, it inhibits *C*, and vice versa; if either *A* or *C* is missing, the other, not inhibited, takes its place. FIGURE 21.20c shows the floral patterns of mutants lacking each of the three classes of organ identity genes and explains how the model accounts for the floral phenotypes.

Further testing of this model and other hypotheses led in 2001 to the discovery of a fourth class of organ identity genes, termed class *E*. This discovery, combined with studies of the genes and gene products at the molecular level, is allowing researchers to extend and refine the older model.

Presumably the organ identity genes are acting as master regulatory genes, each controlling the activity of a battery of other genes that more directly bring about an organ's structure and function. Indeed, like the homeotic genes of animals, the organ identity genes of plants encode transcription factors that regulate other genes by binding to specific enhancers or promoters in the DNA. The plant genes, however, do not contain the homeobox DNA sequence that characterizes homeotic genes in animals. Instead, they have a sequence that encodes a different DNA-binding domain. Possibly as ancient as the homeobox, this sequence is also present in some transcription factor genes in yeast and animals.

THE PROCESS OF SCIENCE

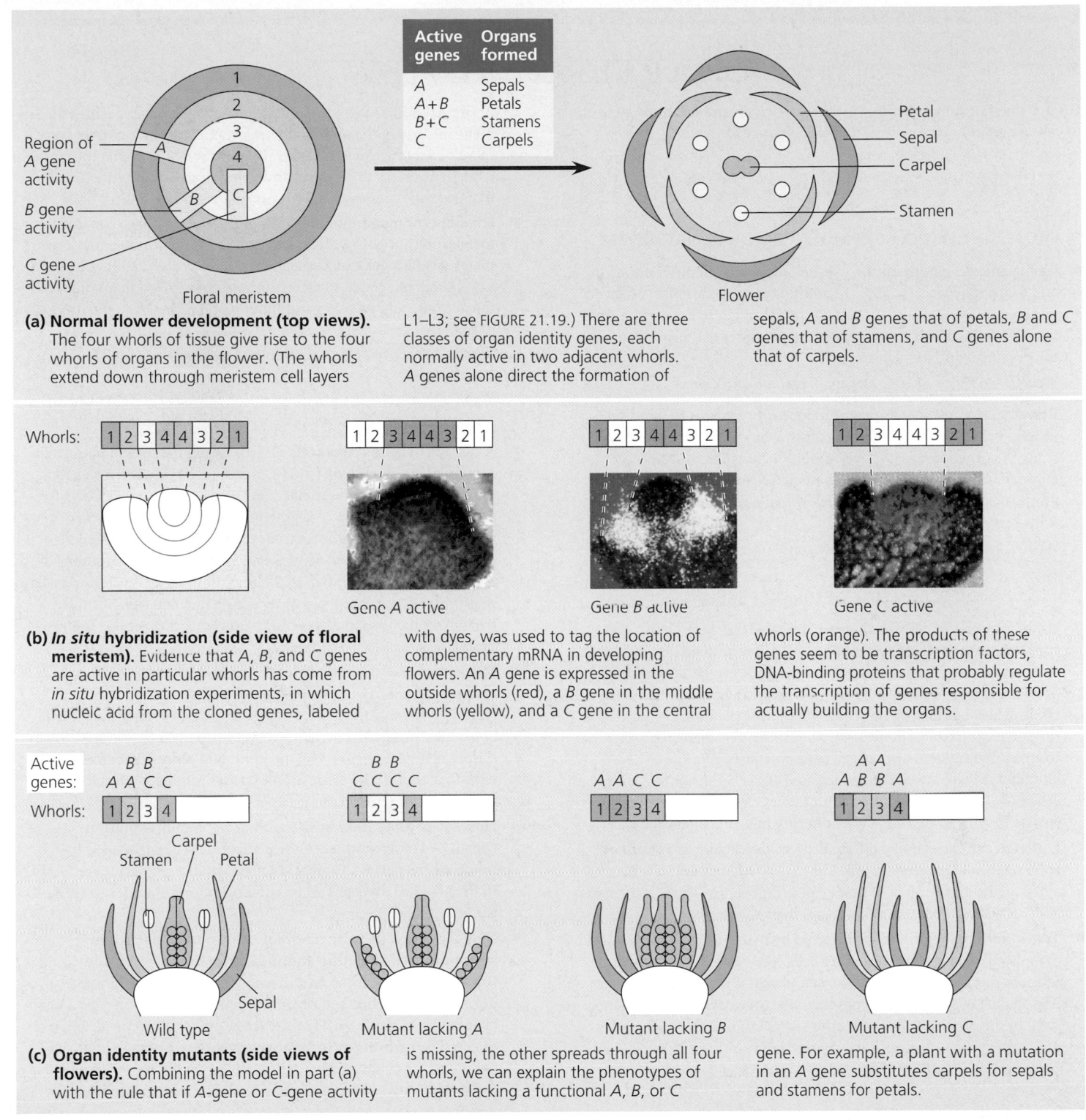

Active genes	Organs formed
A	Sepals
A+B	Petals
B+C	Stamens
C	Carpels

Region of A gene activity
B gene activity
C gene activity
Floral meristem

Petal
Sepal
Carpel
Stamen
Flower

(a) Normal flower development (top views). The four whorls of tissue give rise to the four whorls of organs in the flower. (The whorls extend down through meristem cell layers L1–L3; see FIGURE 21.19.) There are three classes of organ identity genes, each normally active in two adjacent whorls. A genes alone direct the formation of sepals, A and B genes that of petals, B and C genes that of stamens, and C genes alone that of carpels.

Whorls: 1 2 3 4 4 3 2 1

Gene A active
Gene B active
Gene C active

(b) In situ hybridization (side view of floral meristem). Evidence that A, B, and C genes are active in particular whorls has come from in situ hybridization experiments, in which nucleic acid from the cloned genes, labeled with dyes, was used to tag the location of complementary mRNA in developing flowers. An A gene is expressed in the outside whorls (red), a B gene in the middle whorls (yellow), and a C gene in the central whorls (orange). The products of these genes seem to be transcription factors, DNA-binding proteins that probably regulate the transcription of genes responsible for actually building the organs.

Active genes:
B B
A A C C
Whorls: 1 2 3 4

B B
C C C C
1 2 3 4

A A C C
1 2 3 4

A A
A B B A
1 2 3 4

Carpel
Stamen
Petal
Sepal
Wild type

Mutant lacking A
Mutant lacking B
Mutant lacking C

(c) Organ identity mutants (side views of flowers). Combining the model in part (a) with the rule that if A-gene or C-gene activity is missing, the other spreads through all four whorls, we can explain the phenotypes of mutants lacking a functional A, B, or C gene. For example, a plant with a mutation in an A gene substitutes carpels for sepals and stamens for petals.

THE PROCESS OF SCIENCE

FIGURE 21.20 Organ identity genes and pattern formation in flower development. This is a model for how genes specify the floral organs in *Arabidopsis*, with some supporting evidence.

The more scientists learn about the molecules and mechanisms of development in different organisms, the more striking are the underlying similarities that emerge. The similarities reflect the unity—the common ancestry—of life on Earth. But the differences are also crucial, for they have created the huge diversity of organisms that have evolved.

The remainder of the book goes beyond molecules, cells, and genes to explore life on the organismal level.

CHAPTER 21 REVIEW

Go to the Campbell Biology website (www.campbellbiology.com) to explore an interactive version of the Chapter Review.

Summary of Key Concepts

FROM SINGLE CELL TO MULTICELLULAR ORGANISM

- **Embryonic development involves cell division, cell differentiation, and morphogenesis (p. 403)** In addition to mitosis, embryonic cells undergo differentiation, becoming specialized in structure and function. Morphogenesis encompasses the processes that give shape to the organism and its various parts.

  Web/CD Activity 21A: C. elegans *Development Video*

- **Researchers study development in model organisms to identify general principles (pp. 403–406, FIGURE 21.3)** Organisms used to study the genetic basis of development include the fruit fly *Drosophila melanogaster,* the nematode *Caenorhabditis elegans,* the mouse *Mus musculus,* the zebrafish *Danio rerio,* and the common wall cress *Arabidopsis thaliana.*

  Web/CD Activity 21B: *Adult* C. elegans *Video*

DIFFERENTIAL GENE EXPRESSION

- **Different types of cells in an organism have the same DNA (pp. 406–410; FIGURES 21.5, 21.6)** Cells differ in structure and function not because they contain different genes, but because they express different portions of a common genome; they have genomic equivalence. Differentiated cells from mature plants are often totipotent, capable of generating a complete new plant. The nucleus from a differentiated animal cell can sometimes give rise to a new animal if transplanted to an enucleated egg cell. Pluripotent stem cells from animal embryos or adult tissues can reproduce and differentiate *in vitro* as well as *in vivo,* offering hope for medical use.

- **Different cell types make different proteins, usually as a result of transcriptional regulation (pp. 410–411)** Differentiation is heralded by the appearance of tissue-specific proteins. These proteins allow differentiated cells to carry out their specialized roles.

- **Transcriptional regulation is directed by maternal molecules in the cytoplasm and signals from other cells (pp. 411–412, FIGURE 21.10)** Cytoplasmic determinants in the cytoplasm of the unfertilized egg regulate the expression of genes that affect the developmental fate of cells. In the process called induction, signal molecules from embryonic cells cause transcriptional changes in nearby target cells.

  Web/CD Activity 21C: *Signal-Transduction Pathways*

GENETIC AND CELLULAR MECHANISMS OF PATTERN FORMATION

- Pattern formation, the development of a spatial organization of tissues and organs, occurs continually in plants but is mostly limited to embryos and juveniles in animals. Positional information, the molecular cues that control pattern formation, tell a cell its location relative to the body's axes and to other cells.

- **Genetic analysis of *Drosophila* reveals how genes control development: *an overview* (pp. 413–414, FIGURE 21.11)** After fertilization, positional information on an increasingly fine scale specifies the segments in *Drosophila* and finally triggers the formation of each segment's characteristic structures. By analyzing *Drosophila* mutants, Christiane Nüsslein-Volhard and Eric Wieschaus identified and mapped the genes essential for normal segmentation.

- **Gradients of maternal molecules in the early embryo control axis formation (pp. 414–415, FIGURE 21.12)** Studies of one important maternal effect gene in *Drosophila, bicoid,* established that its product is a morphogen, a substance whose concentration gradient influences an embryo's axes and other features. In effect, the products of maternal genes tell the embryo "which way is up."

  Web/CD Activity 21D: *Role of* bicoid *Gene in* Drosophila *Development*
  Web/CD Case Study in the Process of Science: *How Do* bicoid *Mutations Alter Development*?

- **A cascade of gene activations sets up the segmentation pattern in *Drosophila: a closer look* (p. 416, FIGURE 21.13)** Gradients of morphogens encoded by maternal effect genes produce regional differences in the expression of segmentation genes, the products of which direct the actual formation of segments. A cascade sequentially activates three sets of segmentation genes. First, products of gap genes map out subdivisions along the anterior-posterior axis of the embryo. Next, pair-rule genes define the pattern in terms of pairs of segments. Finally, segment polarity genes set the anterior-posterior axis of each segment.

- **Homeotic genes direct the identity of body parts (p. 417)** The anatomical identity of *Drosophila* segments is set by master regulatory genes called homeotic genes, which specify the type of appendages and other structures that form on each segment. Transcription factors encoded by the homeotic genes are regulatory proteins that control the expression of genes responsible for specific anatomical structures.

- **Homeobox genes have been highly conserved in evolution (pp. 417–418, FIGURE 21.15)** A nucleotide sequence called the homeobox is part of homeotic genes. Similar sequences are found in the genes of diverse animals and of more distantly related organisms as well.

- **Neighboring cells instruct other cells to form particular structures: cell signaling and induction in the nematode (pp. 418–421, FIGURES 21.17, 21.18)** In *C. elegans,* an inducer produced by the anchor cell of the embryonic gonad initiates a chain of inductions that results in the formation of the vulva. Thus, sequential inductions in the embryo can drive organ formation. In apoptosis, or programmed cell death, precisely timed signals trigger the activation of a cascade of "suicide" proteins in the cells destined to die.

- **Plant development depends on cell signaling and transcriptional regulation (pp. 421–423, FIGURES 21.19, 21.20)** Cell-cell signaling (induction) seems to be involved in determining the number of floral organs that develop from a floral meristem. Organ identity genes determine the type of structure (stamen, carpal, sepal, or petal) that grows from each whorl of a floral meristem. The organ identity genes apparently act as master regulatory genes, each controlling the activity of other genes that more directly bring about an organ's structure and function.

1. The establishment of the dorsal-ventral axis in a developing fruit fly embryo is a crucial aspect of
   a. pattern formation.
   b. transcriptional regulation.
   c. apoptosis.
   d. cell division.
   e. induction.

2. The criteria for a good model organism for studying development would probably include all of the following *except*
   a. observable embryonic development.
   b. short generation time.
   c. a relatively small genome.
   d. preexisting knowledge of the organism's life history.
   e. abundant local populations for specimen collection.

3. Totipotency is demonstrated when
   a. mutations in homeotic genes result in the development of misplaced appendages.
   b. a cell isolated from a plant leaf grows into a normal adult plant.
   c. an embryonic cell divides and differentiates.
   d. the replacement of the nucleus of an unfertilized egg with that of an intestinal cell converts the egg to an intestinal cell.
   e. segment-specific organs develop along the anterior-posterior axis of a *Drosophila* embryo.

4. Cell differentiation always involves
   a. the production of tissue-specific proteins, such as muscle actin.
   b. the formation of a gastrula.
   c. the transcription of the *myoD* gene.
   d. the selective loss of certain genes from the genome.
   e. the cell's sensitivity to environmental cues such as light or heat.

5. The development of *Drosophila* is somewhat unusual in that
   a. the early mitotic divisions proceed without cytokinesis.
   b. metamorphosis occurs during the larval stage rather than the pupal stage, as with other insects.
   c. homeotic genes are mutated.
   d. cell migration within the embryo does not occur.
   e. the initial cell divisions have lengthy G_1 phases.

6. In *Drosophila*, which genes initiate a cascade of gene activation that includes all other genes in the list?
   a. homeotic genes
   b. gap genes
   c. pair-rule genes
   d. egg-polarity genes
   e. segment polarity genes

7. Absence of *bicoid* mRNA from a *Drosophila* egg leads to the absence of anterior larval body parts and mirror-image duplication of posterior parts. This is evidence that the product of the *bicoid* gene
   a. is an inducer.
   b. contains a homeobox.
   c. is a morphogen.
   d. is a transcription factor.
   e. is a caspase.

8. Homeotic genes
   a. encode transcription factors that control the expression of genes responsible for specific anatomical structures.
   b. are found only in *Drosophila* and other arthropods.
   c. specify the anterior-posterior axis for each fruit fly segment.
   d. create the basic subdivisions of the anterior-posterior axis of the fly embryo.
   e. are responsible for the programmed cell death occurring during morphogenesis.

9. The embryonic development of *C. elegans* illustrates all of the following developmental concepts *except*:
   a. An inducer's effect can depend on its concentration gradient.
   b. The response of an induced cell involves the establishment of a unique pattern of gene activity.
   c. The signal-transduction pathways activated by inducers are unique to embryonic cells.
   d. Sequential inductions direct the formation of complex structures in the developing embryo.
   e. Inducers promote their effects via the activation or inactivation of genes that code for transcriptional regulators.

10. Although quite different in structure, plants and animals share some basic similarities in their development, such as
    a. the importance of cell and tissue movements.
    b. the importance of selective cell enlargement.
    c. the importance of signals from the environment.
    d. the retention of meristematic tissues in the adult.
    e. master regulatory genes that encode DNA-binding proteins.

11. Why can't a single embryonic stem cell develop into an embryo?

12. The signal molecules released by an embryonic cell can induce changes in a neighboring cell without entering the cell. How?

13. If the DNA sequences called homeoboxes, which help homeotic genes direct development, are common to flies and mice, then why aren't these animals more alike?

14. Why are fruit-fly maternal effect genes also called egg-polarity genes?

15. If you clone a carrot, will all the progeny plants ("clones") look identical? Why or why not?

Go to the website or CD-ROM for more quiz questions.

Evolution Connection

Genes important in the embryonic development of animals, such as homeobox-containing genes, have been relatively well conserved during evolution; that is, they are more similar between different species than are many other genes. Why is this?

The Process of Science

Stem cells in an adult organism send some daughter cells down a differentiation pathway, while maintaining their own relatively undifferentiated cell population. During a given mitotic division, one daughter cell remains a stem cell and the other initiates a differentiation pathway. Propose one or more hypotheses to explain how this can happen. (*Note:* There is no easy answer to this question, but it is well worth considering.)

Perform experiments on the cytoplasm of *Drosophila* embryos in the Case Study in the Process of Science, available on the CD-ROM and website.

Science, Technololgy, and Society

Government funding of embryonic stem cell research has been a contentious political issue. Why has this debate been so heated? Summarize the arguments for and against embryonic stem cell research, and explain your own position on the issue.

Answers: 1. a; 2. e; 3. b; 4. a; 5. a; 6. d; 7. c; 8. a; 9. c; 10. e; 11. Information deposited by the mother in the egg (cytoplasmic determinants) is required for embryonic development. 12. By binding to a receptor on the receiving cell's surface and triggering a signal-transduction pathway that affects gene expression. 13. Homeotic genes differ in their nonhomeobox sequences, which affect function. Also, the genes regulated by the homeotic genes differ in the two organisms. 14. Because their products, made by the mother, determine the head and tail ends of the egg (and eventually the adult fly). 15. No, primarily because of subtle (and perhaps not so subtle) differences in their environments.

MECHANISMS OF EVOLUTION

The hypothesis that the beaks of the diverse finch species on the Galápagos Islands had adapted to different food sources through natural selection went untested for almost 150 years. Then came the classic research of Peter and Rosemary Grant, who have been measuring ongoing beak evolution during their 30-year study of the Galápagos finches. The Grants shared some highlights when we interviewed them in their Princeton University offices.

What is the nature of your scientific partnership?

PG: I was trained mainly as an ecologist, but I was also interested in behavior, evolutionary biology, and genetics. Rosemary's training was rather the reverse. She was trained in genetics, but also had an interest in ecology and behavior. Over our years of working on the Galápagos, our interests have converged—or rather, we've expanded into each other's fields of expertise.

How did your shared interest in variation within populations become focused on the Galápagos finches?

RG: We wanted a place that was pristine—where there had been little human disturbance—and where the populations were quite isolated and small enough that we could actually keep track of individual birds—and where individuals in the population would vary enough that we would have a chance of measuring this variation. We knew from the work of earlier researchers that some of the most isolated islands of the Galápagos, such as Daphne Major and Genovesa, were promising for such studies. And Peter corresponded with Ian and Lynette Abbott, scientists from Australia who had been studying competition between finch populations in the Galápagos. So this convinced us that it was worth trying to get the money to go down to the Galápagos. Peter went first in March 1973 with the Abbotts and banded about 60 or 70 medium ground finches on Daphne. They also banded birds on other islands. Then we went down as a family later that year, in November.

Interview
PETER & ROSEMARY GRANT

How many trips have you made to the Galápagos?

RG: This year was our 29th year. We normally go for two or three months each year. The longest we've stayed is six months.

What are the camping conditions on Daphne?

PG: We've been kidded by our colleagues that we go down to the Galápagos and lie on the beach the whole time under swaying palm trees. But there is no beach on Daphne. There's just steep rocks. To land on the island, you have to find some little platform that the waves have cut out of the rock and then climb on from the boat when there are no waves. Then you climb up until you reach a slope where you can actually stand up and walk. And you have to get supplies up there too—something on the order of 30 5-gallon water jugs, cans of food, packets of rice, sugar, and other basics. Plus a stove and cylinder of gas for cooking as well as other camping supplies. Our tent fits on an area that's almost flat, about the size of a large table. We store our food and cook in a little cave, close to the landing point. Many people think we're tough. But it's not really too tough if you're accustomed to camping in wild places.

How old were your two daughters when your family started these annual expeditions to the Galápagos?

RG: They were 6 and 8. They loved it! It was just perfect. They each did little research projects themselves. One studied doves, and the other studied mockingbirds. Later, with Peter's help, they actually wrote two little articles on their research. And of course we took them out of school for as long as six months, and so they had to keep up with their schoolwork, too. They're grown up now—both mothers. And they still think their childhood experiences on the Galápagos were great. Nicola is a physician now. When she was in medical school doing her anatomy, she noticed how surprised other students were that the paths of nerves and other things in their cadavers didn't match up exactly to the drawings and photos in their textbooks. It didn't surprise Nicola at all that all the cadavers were different. By studying different families of mockingbirds on the Galápagos when she was a child, she gained an appreciation for variation. And of course, each human has his or her own idiosyncrasies, too. My father was a physician, and I can remember him telling me how every human responds differently to the same medicine.

PG: Okay, so you've told a story about Nicola. The story about Thalia, our other daughter, who is an artist, is that she was so impressed by her experience on the Galápagos that she has decided to live there! She's raising her two children on Santa Cruz, the main island.

When you initiated your research, did you realize that your family would make so many trips and spend so much time on the Galápagos?

PG: Our first study was designed for just one trip to the islands. But one thing led to another. Initially, we wanted to find out if beak sizes of the populations on different islands matched the differences in food supplies for the finches. In particular, we knew from earlier studies that there was a lot of variation in beak size within the population of medium ground finches, *Geospiza fortis*. We were looking for an ecological association between beak size and something in the environment. And it had to be food, because the main function of the beaks is to pick up food and crush it. For birds at opposite extremes in a distribution of morphology in the population, the small-beaked birds and large-beaked birds are likely to exploit the food in the environment in different ways.

What happened to change a one-time measurement of variation in beak size to a long-term study of evolutionary change in beak size?

PG: After our first trip, a colleague advised us that if we returned later in the year during the dry season, when seeds are in short supply, the variations in beak size would be more pronounced. And that is indeed what we found when the whole family made that second trip in November. We could find almost all the birds we marked when I visited Daphne earlier in the year. So we thought that if we wanted to follow the fates of individual birds with different beak sizes, this is the place. So we decided to make this a three-year study.

You decided on three years so that you could make measurements on the _off-spring_ of the birds you had banded earlier, and on their offspring?

PG: Yes. Peter Boag, a Ph.D. student, joined the study very early. He observed that the size of a ground finch's beak was correlated with the size and hardness of seeds it ate. And by measuring the beaks of parents and their offspring, Peter Boag showed that the variation in beak size was heritable. We wondered if we would be so lucky as to see natural selection acting on this variation. We _were_ lucky. It happened in 1977. It was an exceptionally dry year—a drought. There were far fewer seeds for the ground finches to eat. After consuming the smaller seeds, the finches had to eat larger, harder seeds. By the end of 1977, the population of _Geospiza fortis_ on Daphne had dropped by over 80%. And the average beak size of the survivors was larger than our earlier measurements of the population. And then the next question was, well, given the fact that beak size variation is heritable, shouldn't we observe an evolutionary response in the following generation? So we had to follow the fate of the survivors through the following breeding season and

observe whether their offspring had larger beaks than the preceding average, as expected. And indeed, we found that, too. The average beak size _had_ increased in the population. And that led to the next question: Was the beak measurement going to stay this high, or increase, or go back to the preceding state? For that, we needed long-term data. And, as I say, one thing led to another. And the questions started to accumulate faster than the answers.

Including your questions about hybridization?

PG: Somewhere along the line, it became obvious to us that _Geospiza fortis_ on Daphne was occasionally hybridizing with the cactus finch, _Geospiza scandens_. We found that the hybrids were surviving, and they were fertile. So we wondered if there were circumstances where they would _not_ survive so well.

RG: One big question about the hybrids is why the two species interbreed in the first place. It can happen when a male learns to sing the song of a different species. The birds learn their songs during a short sensitive period when they are in the nest and when they're being fed by their parents. So most sons sing their father's song—at least they do about 80% of the time. But if the father is not around or does not sing very much, their sons simply imprint on their neighbors' song, even if it's a different finch species. Or, perhaps the father dies, and his sons imprint on the song they hear. Or, a rare situation—but it does happen—is when a finch takes over the nest of a different species and kicks out all the eggs, except for one, and then lays its own eggs. So you have a _fortis_ growing up in a _scandens_ brood, learning to sing a _scandens_ song. And when these _fortis_ males are mature, they occasionally breed with a _scandens_ because they sing its song.

So what keeps the two finch species from amalgamating into one?

RG: The hybrids have bill sizes intermediate between the medium ground finch and the cactus finch. The hybrids can only survive during wet years, when there are a lot of small,

soft seeds. During dry years, the hybrids can't crack the larger, harder seeds that are on the ground, and they can't compete with the cactus finches for cactus seeds. Here's how I think of it: In these very new species that are coming together in secondary contact, there is this occasional hybridization through a breakdown of a learned cultural trait, the song. And so you get this balance between an input of genes and then selection, during drought years, keeping the populations on divergent trajectories in spite of the episodes of hybridization.

PG: A large part of the answer to the question of why the medium ground finches on Daphne are so variable is hybridization. The variation in characters such as beak size is broadened by the occasional input of genes from different species. And then the other part of the story is that the ecological environment favors particularly large beaks or particularly small beaks in the population at different times. Selection can act upon that increased variation and produce faster evolutionary responses as the environment changes. Also—this is conjectural, but certainly possible—perhaps hybrids occasionally disperse off one island to another island that has neither the hybrids nor the parent species. The hybrids could start a new population with a range of genetic variation different from the parent species on the island they just left.

Do you think this helps explain the adaptive radiation of the Galápagos finches?

PG: I see no reason why hybridization hasn't been important right from the beginning, from the first divergence of the ancestral finch stock that reached the islands. We don't have the early stages, but that's the big challenge of evolutionary biology—trying to infer from modern clues what happened in the past.

DESCENT WITH MODIFICATION: A DARWINIAN VIEW OF LIFE

THE HISTORICAL CONTEXT FOR EVOLUTIONARY THEORY

- Western culture resisted evolutionary views of life
- Theories of geologic gradualism helped clear the path for evolutionary biologists
- Lamarck placed fossils in an evolutionary context

THE DARWINIAN REVOLUTION

- Field research helped Darwin frame his view of life
- *The Origin of Species* developed two main points: the occurrence of evolution and natural selection as its mechanism
- Examples of natural selection provide evidence of evolution
- Other evidence of evolution pervades biology
- What is theoretical about the Darwinian view of life?

ON

THE ORIGIN OF SPECIES

BY MEANS OF NATURAL SELECTION,

OR THE

PRESERVATION OF FAVOURED RACES IN THE STRUGGLE FOR LIFE.

BY CHARLES DARWIN, M.A.,

FELLOW OF THE ROYAL, GEOLOGICAL, LINNÆAN, ETC., SOCIETIES;
AUTHOR OF 'JOURNAL OF RESEARCHES DURING H. M. S. BEAGLE'S VOYAGE ROUND THE WORLD.'

LONDON:
JOHN MURRAY, ALBEMARLE STREET.
1859.

The right of Translation is reserved.

Biology came of age *on November 24, 1859, the day Charles Darwin published* On the Origin of Species by Means of Natural Selection *(see upper right). Darwin's book drew a cohesive picture of life by connecting the dots of what had once seemed a bewildering array of unrelated facts. The Origin of Species focused biologists' attention on the great diversity of organisms—their origins and relationships, their similarities and differences, their geographic distribution, and their adaptations to surrounding environments. An understanding of evolution continues to inform every field of biology, from exploring life's molecules to analyzing ecosystems. And applications of evolutionary biology are transforming medicine, agriculture, biotechnology, and conservation biology. Because evolution integrates all of biology, it is the main thematic thread woven throughout this book. This unit of chapters explores the mechanisms by which life evolves.*

Darwin made two points in The Origin of Species. *First, he argued from evidence that the species of organisms inhabiting Earth today descended from ancestral species. Second, he pro-*

posed a mechanism for evolution, which he termed **natural selection.** *The basic idea of natural selection is that a population of organisms can change over the generations if individuals having certain heritable traits leave more offspring than other individuals. The result of natural selection is* **evolutionary adaptation,** *a prevalence of inherited characteristics that enhance organisms' survival and reproduction in specific environments. In modern terms, we would say that the genetic composition of the population had changed over time, and that is one way of defining* **evolution.** *But we can also use the term* evolution *on a much grander scale to mean all of biological history, from the earliest microbes to the enormous diversity of modern organisms.*

In this chapter, you will learn about the Darwinian view of life and its historical development.

THE HISTORICAL CONTEXT FOR EVOLUTIONARY THEORY

To put the Darwinian view in perspective, we must compare it with earlier ideas about Earth and its life. The impact of an intellectual revolution such as Darwinism depends on timing as well as logic. Let's explore the historical context of Darwin's life and ideas (FIGURE 22.1).

THE PROCESS OF SCIEN

Western culture resisted evolutionary views of life

The Origin of Species was truly radical for its time; not only did it challenge prevailing scientific views, but it also shook the deepest roots of Western culture. Darwin's view of life contrasted sharply with the conventional paradigm of an Earth only a few thousand years old, populated by unchanging forms of life that had been individually made during the single week in which the Creator formed the entire universe. Darwin's book challenged a worldview that had been taught for centuries.

The Scale of Nature and Natural Theology

A number of classical Greek philosophers had ideas about the gradual evolution of life. But the philosophers who influenced Western culture most, Plato (427–347 B.C.) and his student Aristotle (384–322 B.C.), held opinions that opposed any concept of evolution. Plato believed in two worlds: a real world that is ideal and eternal and an illusory world of imperfection that we perceive through our senses. Evolution would be counterproductive in a world where ideal organisms were already perfectly adapted to their environments.

Aristotle believed that all living forms could be arranged on a scale, or ladder, of increasing complexity, later called the *scala naturae* ("scale of nature"). Each form of life had its allotted rung on this ladder, and every rung was taken. In this view of life, which prevailed for over 2,000 years, species are permanent, are perfect, and do not evolve.

In Judeo-Christian culture, the Old Testament account of creation fortified the idea that species were individually designed and nonevolving. In the 1700s, biology in Europe and America was dominated by **natural theology,** a philosophy dedicated to discovering the Creator's plan by studying nature. Natural theologians saw the adaptations of organisms as evidence that the Creator had designed each and every species for a particular purpose. A major objective of natural theology was to classify species in order to reveal the steps of the scale of life that God had created.

Carolus Linnaeus (1707–1778), a Swedish physician and botanist, sought to discover order in the diversity of life "for the greater glory of God." Linnaeus specialized in **taxonomy,** the branch of biology concerned with naming and classifying the diverse forms of life. He developed the two-part, or binomial, system of naming organisms according to genus and species that is still used today. In addition, Linnaeus adopted a system for grouping similar species into a hierarchy of increasingly general categories. For example, similar species are grouped in the same genus, similar genera (plural of genus) are grouped in the same family, and so on (see FIGURE 1.10). To Linnaeus, clustering similar species together implied no evolutionary kinship, but a century later his taxonomic system would become a focal point in Darwin's arguments for evolution.

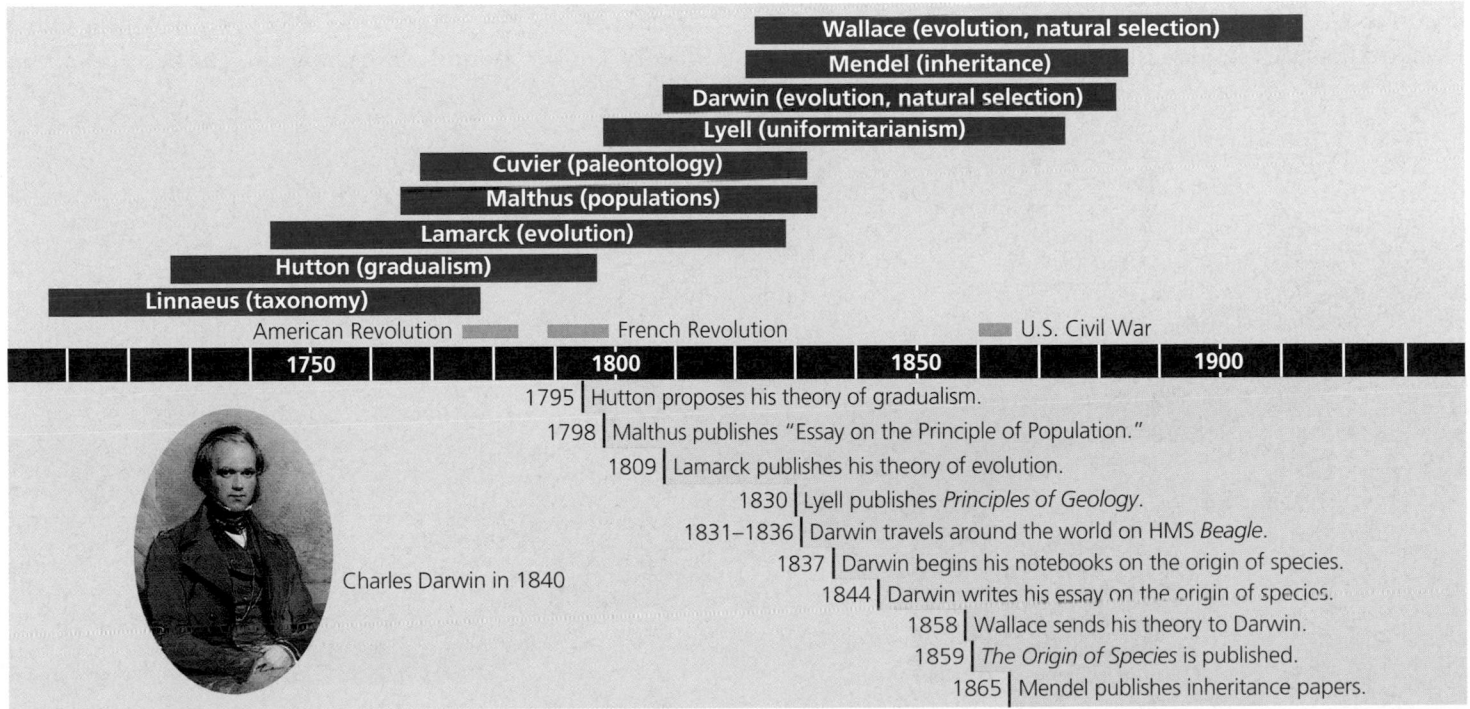

FIGURE 22.1 The historical context of Darwin's life and ideas.

FIGURE 22.2 Fossils of trilobites, animals that lived in the seas hundreds of millions of years ago.

of fossil species and that the deeper (older) the stratum, the more dissimilar the fossils are from modern life. Cuvier even recognized that extinction had been a common occurrence in the history of life. From stratum to stratum, new species appear and others disappear. Yet Cuvier was a staunch opponent of the evolutionists of his day. Instead, he advocated **catastrophism**, speculating that each boundary between strata corresponded in time to a catastrophe, such as a flood or drought, that had destroyed many of the species living there at that time. He proposed that these periodic catastrophes were usually confined to local geographic regions and that the ravaged region was repopulated by species immigrating from other areas.

Cuvier, Fossils, and Catastrophism

The study of fossils also helped lay the groundwork for Darwin's ideas. **Fossils** are relics or impressions of organisms from the past, preserved in rock (FIGURE 22.2). Most fossils are found in **sedimentary rocks** formed from the sand and mud that settle to the bottom of seas, lakes, and marshes. New layers of sediment cover older ones and compress them into superimposed layers of rock called strata. Later, erosion may scrape or carve through upper (younger) strata and reveal more ancient strata that had been buried. Fossils within the layers show that a succession of organisms has populated Earth throughout time (FIGURE 22.3).

Paleontology, the study of fossils, was largely developed by French anatomist Georges Cuvier (1769–1832). Realizing that the history of life is recorded in strata containing fossils, he documented the succession of fossil species in the Paris Basin. He noted that each stratum is characterized by a unique group

Theories of geologic gradualism helped clear the path for evolutionary biologists

Competing with Cuvier's theory of catastrophism was a very different idea of how geologic processes had shaped Earth's crust. In 1795, Scottish geologist James Hutton (1726–1797) proposed that it was possible to explain the various landforms by looking at mechanisms *currently* operating in the world. For example, he suggested that canyons were formed by rivers cutting down through rocks and that sedimentary rocks with marine fossils were built of particles that had eroded from the land and been carried by rivers to the sea (FIGURE 22.4). Hutton explained Earth's geologic features by the theory of **gradualism,** which holds that profound change is the cumulative product of slow but continuous processes.

The leading geologist of Darwin's era, a Scot named Charles Lyell (1797–1875), incorporated Hutton's gradualism into a theory known as **uniformitarianism.** The term refers to

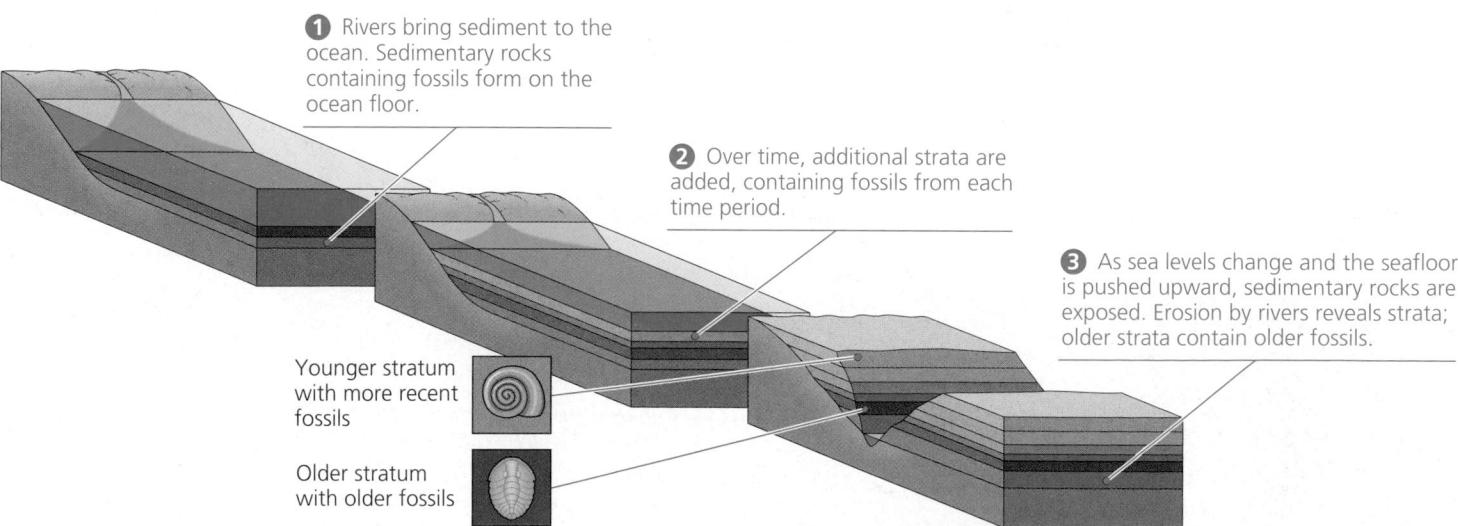

❶ Rivers bring sediment to the ocean. Sedimentary rocks containing fossils form on the ocean floor.

❷ Over time, additional strata are added, containing fossils from each time period.

❸ As sea levels change and the seafloor is pushed upward, sedimentary rocks are exposed. Erosion by rivers reveals strata; older strata contain older fossils.

Younger stratum with more recent fossils

Older stratum with older fossils

FIGURE 22.3 Formation of sedimentary rock and deposition of fossils from different time periods.

FIGURE 22.4 Strata of sedimentary rock at the Grand Canyon. The Colorado River has cut through over 2,000 meters of rock, exposing sedimentary layers that are like huge pages from the book of life. Scan the canyon wall from rim to floor, and you look back through hundreds of millions of years. Each layer entombs fossils that represent some of the organisms from that period of Earth's history.

Lyell's idea that geologic processes have not changed throughout Earth's history. Thus, for example, the forces that build mountains and erode mountains and the rates at which these forces operate are the same today as in the past.

Darwin was strongly influenced by two conclusions that followed directly from the observations of Hutton and Lyell. First, if geologic change results from slow, continuous actions rather than sudden events, then Earth must be very old, certainly much older than the 6,000 years assigned by many theologians on the basis of biblical inference. Second, very slow and subtle processes persisting over a long period of time can add up to substantial change. Darwin was not the first to apply the principle of gradualism to biological evolution, however.

Lamarck placed fossils in an evolutionary context

Toward the end of the 18th century, several naturalists, including Erasmus Darwin, Charles Darwin's grandfather, suggested that life had evolved as environments changed. But only one of Charles Darwin's predecessors developed a comprehensive model that attempted to explain how life evolves: Jean Baptiste Lamarck.

Lamarck published his theory of evolution in 1809, the year Darwin was born. Lamarck was in charge of the invertebrate collection at the Natural History Museum in Paris. By comparing current species with fossil forms, Lamarck could see what appeared to be several lines of descent, each a chronological series of older to younger fossils leading to a modern species.

Lamarck is remembered most for the mechanism he proposed to explain how specific adaptations evolve. It incorporates two ideas that were popular during Lamarck's era. The first was use and disuse, the idea that those parts of the body used extensively to cope with the environment become larger and stronger while those that are not used deteriorate. Among the examples Lamarck cited were a blacksmith developing a bigger bicep in the arm that wields the hammer and a giraffe stretching its neck to reach leaves on high branches. The second idea Lamarck adopted was called the inheritance of acquired characteristics. In this concept of heredity, the modifications an organism acquires during its lifetime can be passed along to its offspring. The long neck of the giraffe, Lamarck reasoned, evolved gradually as the cumulative product of a great many generations of ancestors stretching ever higher.

There is, however, no evidence that acquired characteristics can be inherited. Blacksmiths may increase strength and stamina by a lifetime of pounding with a heavy hammer, but these acquired traits do not change genes transmitted by gametes to offspring. Even though the Lamarckian theory of evolution is ridiculed often today because of its erroneous assumption that acquired characteristics are inherited, in Lamarck's time that concept of inheritance was generally accepted (and, indeed, Darwin could offer no acceptable alternative). To most of Lamarck's contemporaries, however, the mechanism of evolution was an irrelevant issue because they firmly believed that species were fixed and that no theory of evolution could be taken seriously. Lamarck was vilified, especially by Cuvier, who denied that species ever evolve. In retrospect, Lamarck deserves much credit for his theory, which was visionary in many respects: in its claim that evolution is the best explanation for both the fossil record and the current diversity of life; in its recognition of the great age of Earth; and especially in its emphasis on *adaptation to the environment* as a primary product of evolution.

THE DARWINIAN REVOLUTION

We have set the scene for the Darwinian revolution. Natural theology still dominated the intellectual climate as the 19th century dawned. A few clouds of doubt about the permanence of species were beginning to gather, but no one could have forecast the thundering storm just over the horizon.

Charles Darwin (1809–1882) was born in Shrewsbury in western England. Even as a boy he had a consuming interest in nature. When he was not reading nature books, he was fishing, hunting, and collecting insects. Darwin's father, an eminent physician, could see no future for a naturalist and sent Charles to the University of Edinburgh to study medicine. Only 16 years old at the time, Charles found medical school boring and distasteful. He left Edinburgh without a degree and shortly thereafter enrolled at Christ College at Cambridge University, with the intent of becoming a clergyman. At that time in Great Britain, most naturalists and other scientists belonged to the clergy, and nearly all saw the world in the context of natural theology. Darwin became the protégé of the Reverend John Henslow, professor of botany at Cambridge. Soon after Darwin received his B.A. degree in 1831, Professor Henslow recommended the young graduate to Captain Robert FitzRoy, who was preparing the survey ship *Beagle* for a voyage around the world. Darwin would pay his own way and serve as a conversation companion to the young captain. FitzRoy chose Darwin because of his education and because he was of the same social class and about the same age as the captain.

Field research helped Darwin frame his view of life

The Voyage of the Beagle

Darwin was 22 years old when he sailed from Great Britain aboard HMS *Beagle* in December 1831 (FIGURE 22.5). The primary mission of the voyage was to chart poorly known stretches of the South American coastline. While the ship's crew surveyed the coast, Darwin spent most of his time on shore, observing and collecting thousands of specimens of South American plants and animals. As the ship worked its way around the continent, Darwin observed the various adaptations of plants and animals that inhabited such diverse environments as the Brazilian jungles, the expansive grasslands of the Argentine pampas, the desolate lands of Tierra del Fuego near Antarctica, and the towering heights of the Andes Mountains.

Darwin noted that plants and animals he studied had definite South American characteristics, very distinct from those of Europe. That in itself may not have been surprising. But Darwin also noted that the plants and animals in temperate

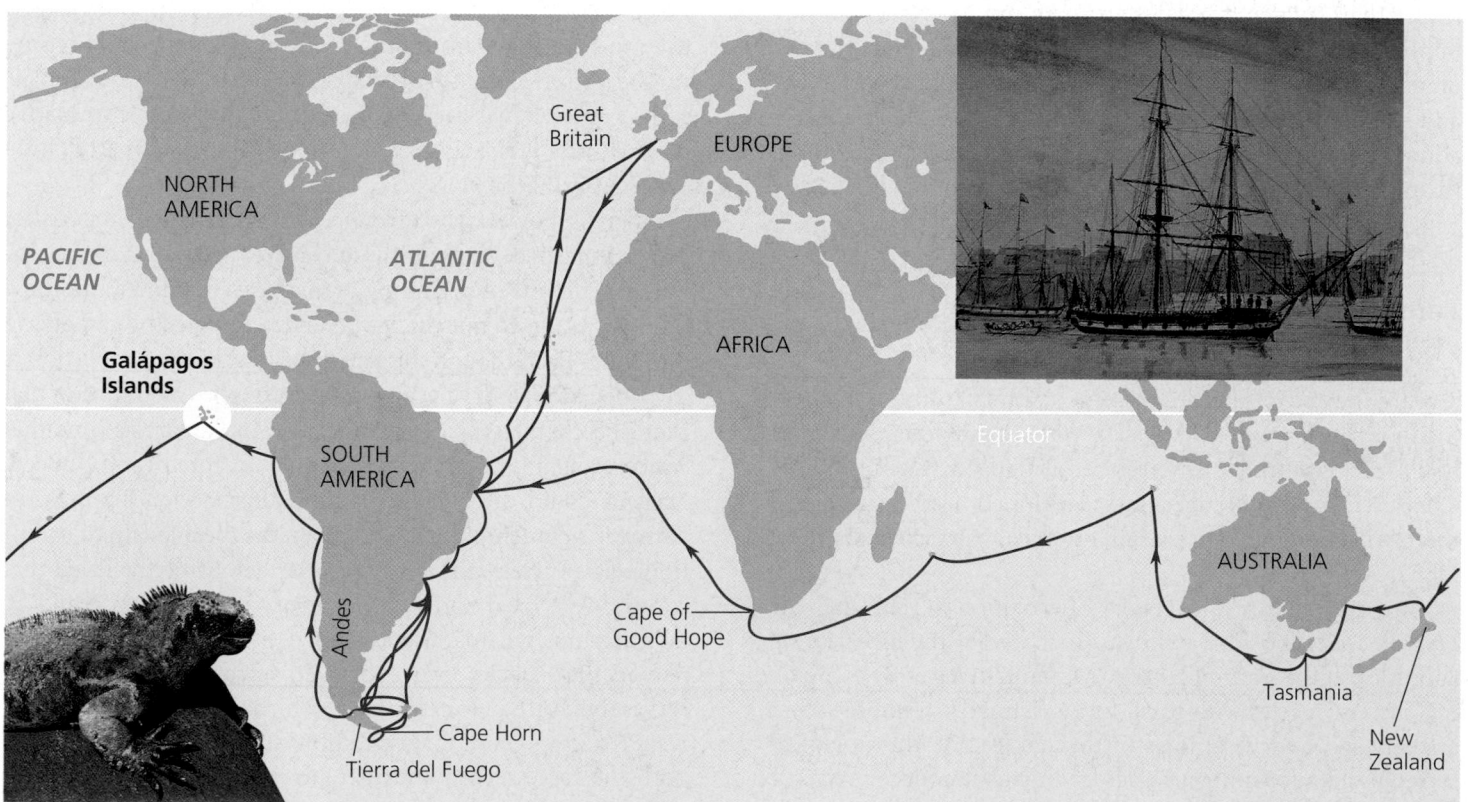

FIGURE 22.5 The Voyage of HMS *Beagle*. The two insets illustrate the ship and a marine iguana, one of the unique animals that evolved on the Galápagos Islands.

regions of South America were more closely related to species living in tropical regions of that continent than to species in temperate regions of Europe. Furthermore, the South American fossils that Darwin found, though clearly different from modern species, were distinctly South American in their resemblance to the living plants and animals of that continent.

The geographic distribution of species interested Darwin. For example, he was curious about the fauna of the Galápagos, islands of relatively recent volcanic origin that lie on the equator about 900 km west of the South American coast (see FIGURE 22.5). Darwin did not grasp the full significance of the Galápagos animals until he studied his collections after returning to England in 1836. He then learned that most of the animal species on the Galápagos live nowhere else in the world, although they resemble species living on the South American mainland. It was as though the islands had been colonized by plants and animals that strayed from the South American mainland and then diversified on the different islands. Among the birds Darwin collected on the Galápagos were several types of finches that, although quite similar, seemed to be different species. Some were unique to individual islands, while other species were distributed on two or more islands that were close together.

Darwin read Lyell's *Principles of Geology* while on board the *Beagle*. Lyell's ideas, together with his own experiences on the Galápagos, had Darwin doubting the traditional view that Earth was static and had been created only a few thousand years ago. By acknowledging that Earth was very old and constantly changing, Darwin took an important step toward recognizing that life on Earth had also evolved.

Darwin's Focus on Adaptation

Soon after returning to Great Britain in 1836, Darwin started reassessing all that he had observed during the voyage of the *Beagle*. He began to perceive the origin of new species and adaptation to the environment as closely related processes. Could a new species arise from an ancestral form by the gradual accumulation of adaptations to a different environment? From studies made years after Darwin's voyage, biologists have concluded that this is what happened to the Galápagos finches. Among the differences between the finches are their beaks, which are adapted to the specific foods available on their home islands (FIGURE 22.6). Darwin anticipated that explaining how such adaptations arise was essential to understanding evolution.

By the early 1840s, Darwin had worked out the major features of his theory of natural selection as the mechanism of evolution. However, he had not yet published his ideas. He was in poor health, and he rarely left home. Despite his reclusiveness, Darwin was not isolated from the scientific community. Already famous as a naturalist because of the letters and specimens he sent to Great Britain during the voyage of the *Beagle,* Darwin had frequent correspondence and visits from Lyell, Henslow, and other scientists.

In 1844, Darwin wrote a long essay on the origin of species and natural selection. However, Darwin was reluctant to introduce his theory publicly, apparently because he anticipated the uproar it would cause. Darwin asked his wife to publish his essay if he died before writing a more thorough dissertation on evolution. While he procrastinated, he continued to compile evidence in support of his theory. Lyell, not himself yet convinced of evolution, nevertheless advised Darwin to publish on the subject before someone else came to the same conclusions and published first.

In June 1858, Lyell's prediction came true. Darwin received a letter from Alfred Wallace (1823–1913), a young British naturalist working in the East Indies. The letter was accompanied by a manuscript in which Wallace developed a theory of natural selection essentially identical to Darwin's. Wallace asked Darwin to

(a) Seed eater. The large ground finch (*Geospiza magnirostris*) has a large beak adapted for cracking seeds that fall from plants to the ground.

(b) Insect eater. The small tree finch (*Camarhynchus parvulus*) uses its beak to grasp insects.

(c) Tool-using insect eater. The woodpecker finch (*Camarhynchus pallidus*) uses a cactus spine or small twig as a tool to probe for termites and other wood-boring insects.

FIGURE 22.6 Galápagos finches. The Galápagos Islands have a total of 14 species of closely related finches, some found only on a single island. The most striking difference among species is their beaks, which are adapted for specific diets. See also FIGURE 1.17b.

evaluate the paper and forward it to Lyell if it merited publication. Darwin complied, writing to Lyell: "Your words have come true with a vengeance. . . . I never saw a more striking coincidence . . . so all my originality, whatever it may amount to, will be smashed." Lyell and a colleague presented Wallace's paper, along with extracts from Darwin's unpublished 1844 essay, to the Linnaean Society of London on July 1, 1858. Darwin quickly finished *The Origin of Species* and published it the next year. Although Wallace wrote up his ideas for publication first, Darwin developed and supported the theory of natural selection so much more extensively that he is known as its main architect. And Darwin's notebooks prove that he formulated his theory of natural selection 15 years before reading Wallace's manuscript.

Within a decade, Darwin's book and its proponents had convinced the majority of biologists that biological diversity was the product of evolution. Darwin succeeded where previous evolutionists had failed, partly because science was beginning to shift away from natural theology, but mainly because he convinced his readers with immaculate logic and an avalanche of evidence in support of evolution.

The Origin of Species developed two main points: the occurrence of evolution and natural selection as its mechanism

Darwinism has a dual meaning. It refers to evolution as the explanation for life's unity and diversity, and it also refers to the Darwinian concept of natural selection as the cause of adaptive evolution.

Descent with Modification

In the first edition of *The Origin of Species,* Darwin did not use the word *evolution* until the last paragraph, referring instead to **descent with modification,** a phrase that condensed his view of life. Darwin perceived unity in life, with all organisms related through descent from some unknown ancestor that lived in the remote past. As the descendants of that ancestral organism spilled into various habitats over millions of years, they accumulated diverse modifications, or adaptations, that fit them to specific ways of life.

In the Darwinian view, the history of life is like a tree, with multiple branching and rebranching from a common trunk all the way to the tips of the youngest twigs, symbolic of the diversity of living organisms. At each fork of the evolutionary tree is an ancestor common to all lines of evolution branching from that fork. Closely related species, such as the Asian elephant and the African elephant, are very similar because they share the same line of descent until a relatively recent divergence from a common ancestor (FIGURE 22.7). Most branches of evolution, even some major ones, are dead ends; about 99% of all species that have ever lived are extinct.

Ironically, Linnaeus, who apparently believed that species are fixed, provided Darwin with a connection to evolution by recognizing that the great diversity of organisms could be ordered into "groups subordinate to groups" (Darwin's phrase). FIGURE 1.10 introduced the major taxonomic categories; let's review them here: kingdom > phylum > class > order > family > genus > species.

To Darwin, the natural hierarchy of the Linnaean scheme reflected the branching history of the tree of life, with organisms at the different taxonomic levels related through descent

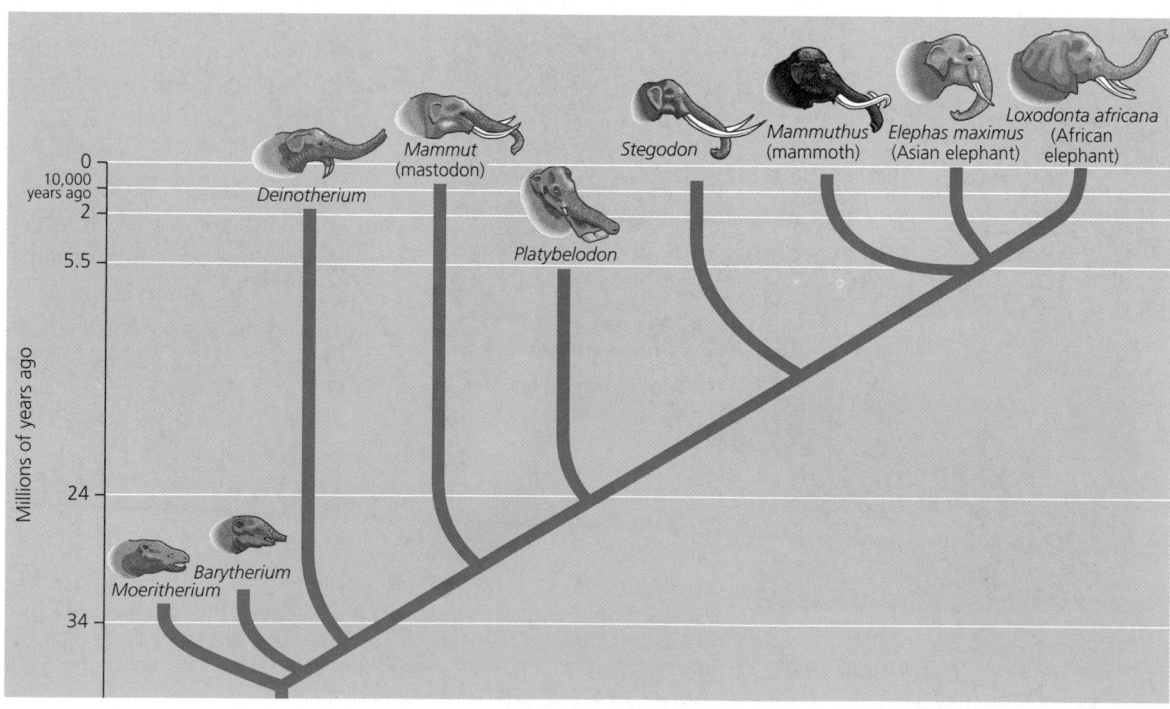

FIGURE 22.7 Descent with modification. This evolutionary tree of the elephant family is based mainly on evidence from fossils—their anatomy, their order of appearance in geologic time, and their geographic distribution. (Timeline not to scale.)

from common ancestors. Two species, such as lions and tigers, that are grouped in the same family (family Felidae) share a more recent common ancestor than two species, such as lions and elephants, that belong to different families within the same class (class Mammalia).

Natural Selection and Adaptation

How does natural selection work? And how does natural selection explain adaptation? Evolutionary biologist Ernst Mayr has dissected the logic of Darwin's theory of natural selection into three inferences based on five observations:*

OBSERVATION #1: All species have such great potential fertility that their population size would increase exponentially if all individuals that are born reproduced successfully (FIGURE 22.8).

OBSERVATION #2: Populations tend to remain stable in size, except for seasonal fluctuations.

OBSERVATION #3: Environmental resources are limited.

INFERENCE #1: Production of more individuals than the environment can support leads to a struggle for existence among individuals of a population, with only a fraction of offspring surviving each generation.

OBSERVATION #4: Individuals of a population vary extensively in their characteristics; no two individuals are exactly alike (FIGURE 22.9).

OBSERVATION #5: Much of this variation is heritable.

INFERENCE #2: Survival in the struggle for existence is not random, but depends in part on the hereditary constitution of the individuals. Those individuals whose inherited traits best fit them to their environment are likely to leave more offspring than less fit individuals.

INFERENCE #3: This unequal ability of individuals to survive and reproduce will lead to a gradual change in a population, with favorable characteristics accumulating over the generations.

We can summarize Darwin's main ideas as follows:

Natural selection is differential success in reproduction (unequal ability of individuals to survive and reproduce).

Natural selection occurs through an interaction between the environment and the variability inherent among the individual organisms making up a population.

The product of natural selection is the adaptation of populations of organisms to their environment (FIGURE 22.10, p. 436).

Let's elaborate on the important connections Darwin perceived between natural selection, the struggle for existence,

---

* Adapted from E. Mayr, *The Growth of Biological Thought: Diversity, Evolution and Inheritance* (Cambridge, MA: Harvard University Press, 1982).

FIGURE 22.8 Overproduction of offspring. A cloud of millions of spores is exploding from this puffball, a type of fungus. The wind will disperse the spores far and wide. Only a tiny fraction of the spores will actually give rise to offspring that survive and reproduce.

FIGURE 22.9 A few of the color variations in a population of Asian lady beetles.

and the capacity of organisms to "overreproduce." Darwin apparently began to recognize the struggle for existence after he read an essay on human population written in 1798 by Thomas Malthus (see FIGURE 22.1). Malthus contended that much of human suffering—disease, famine, homelessness, and war—was the inescapable consequence of the potential for the human population to increase faster than food supplies and other resources. The capacity to overproduce seems to be characteristic of all species. Of the many eggs laid, young born, and seeds spread, only a tiny fraction complete their development and leave offspring of their own (see FIGURE 22.8). The rest are eaten, frozen, starved, diseased, unmated, or unable to reproduce for some other reason.

In each generation, environmental factors filter heritable variations, favoring some over others. Differential reproduction—whereby organisms with traits favored by the environment produce more offspring than do organisms without those traits—results in the favored traits being disproportionately represented in the next generation. This increasing frequency of the favored traits in a population is evolution.

(a) A flower mantid in Malaysia

(b) A Trinidad tree mantid that mimics dead leaves

(c) A Central American mantid that resembles a green leaf

FIGURE 22.10 Camouflage as an example of evolutionary adaptation. Related species of insects called mantids have diverse shapes and colors that evolved in different environments.

Darwin illustrated the power of selection as a force in evolution with examples from **artificial selection,** the breeding of domesticated plants and animals. Humans have modified other species over many generations by selecting individuals with the desired traits as breeding stock. The plants and animals we grow for food often bear little resemblance to their wild ancestors (FIGURE 22.11). The power of selective breeding is especially apparent in our pets, which have been bred more for fancy than for utility.

If so much change can be achieved by artificial selection in a relatively short period of time, Darwin reasoned, then natural selection should be capable of considerable modification of species over hundreds or thousands of generations. Even if the

advantages of some heritable traits over others are slight, the advantageous variations will accumulate in the population after many generations of natural selection, eliminating less favorable variations.

Darwin incorporated gradualism, a concept so important in Lyell's geology, into evolutionary theory. He envisioned life as evolving by a gradual accumulation of minute changes, and he postulated that natural selection operating in varying contexts over vast spans of time could account for the entire diversity of life.

We can now summarize the two main features of the Darwinian view of life: (1) The diverse forms of life have arisen by descent with modification from ancestral species

(a) Cattle breeders of ancient Africa. About 5,000 years ago, the artist of this rock painting portrayed North African people with their various breeds of cattle. Thousands of such paintings document a keen awareness of variations in the physical characteristics of domesticated animals. Early farmers propagated certain variations by selectively breeding livestock and crops.

FIGURE 22.11 Artificial selection.

(b) Diverse vegetables derived from wild mustard. Cabbage, kale, kohlrabi, brussels sprouts, cauliflower, and broccoli have a common ancestor in one species of wild mustard (inset). By selecting different parts of the plant to accentuate, breeders have obtained these divergent results.

and (2) the mechanism of modification has been natural selection working over enormous tracts of time.

Some Subtleties of Natural Selection. At this point, we need to emphasize several subtleties of natural selection. One is the importance of populations in evolution. For now, we will define a population as a group of interbreeding individuals belonging to a particular species and sharing a common geographic area. A population is the smallest unit that can evolve. Natural selection occurs through interactions between individual organisms and their environment, but individuals do not evolve. Evolution can be measured only as changes in relative proportions of heritable variations in a population over a succession of generations.

Another key point about natural selection is that it can amplify or diminish *only* heritable variations. As we have seen, an organism may become modified through its own experiences during its lifetime, and such acquired characteristics may even adapt the organism to its environment, but there is no evidence that characteristics acquired during a lifetime can be inherited. We must distinguish between adaptations an organism acquires by its own actions and inherited adaptations that evolve in a population over many generations as a result of natural selection.

It must also be emphasized that the specifics of natural selection are situational; environmental factors vary from place to place and from time to time. An adaptation in one situation may be useless or even detrimental in different circumstances. Some examples will reinforce this situational quality of natural selection.

Examples of natural selection provide evidence of evolution

Natural selection and the adaptive evolution it causes are observable phenomena. As described in the interview at the beginning of this unit, Peter and Rosemary Grant of Princeton University are documenting natural selection and evolution in populations of finches in the Galápagos. We will now look at two additional examples of natural selection as a pervasive mechanism of evolution in populations.

Natural Selection in Action: The Evolution of Insecticide-Resistant Insects

A classic and unsettling example of natural selection is the evolution of insecticide resistance in hundreds of insect species. Insecticides are poisons used to kill insects that are pests in farmlands, swamps, backyards, and homes. Examples are DDT, now banned in many countries, and malathion. We have used insecticides to control insects that eat our crops, transmit diseases such as malaria, or just annoy us around the house or

campground, but these chemical weapons have proved to be double-edged swords. These poisons are not specific for the intended targets, and their widespread use has produced some colossal environmental problems, which we'll examine in Chapter 54. Our focus here is the evolutionary outcome of introducing these chemicals into the environments of insects.

Whenever a new type of insecticide is used to control agricultural pests, the story is usually the same. Early results are encouraging. A relatively small amount of the poison dusted onto a crop may kill 99% of the insects. But subsequent sprayings are less and less effective. One response is to increase the amount of the poison, but that brings high monetary (not to mention environmental) costs. Another strategy is for the farmer to switch to a different insecticide until it, too, becomes ineffective.

It is natural selection that causes the evolution of resistance to insecticides. The relatively few survivors of the first insecticide wave are insects with genes that somehow enable them to resist the chemical attack (FIGURE 22.12). In some cases, the

FIGURE 22.12 Evolution of insecticide resistance in insect populations.

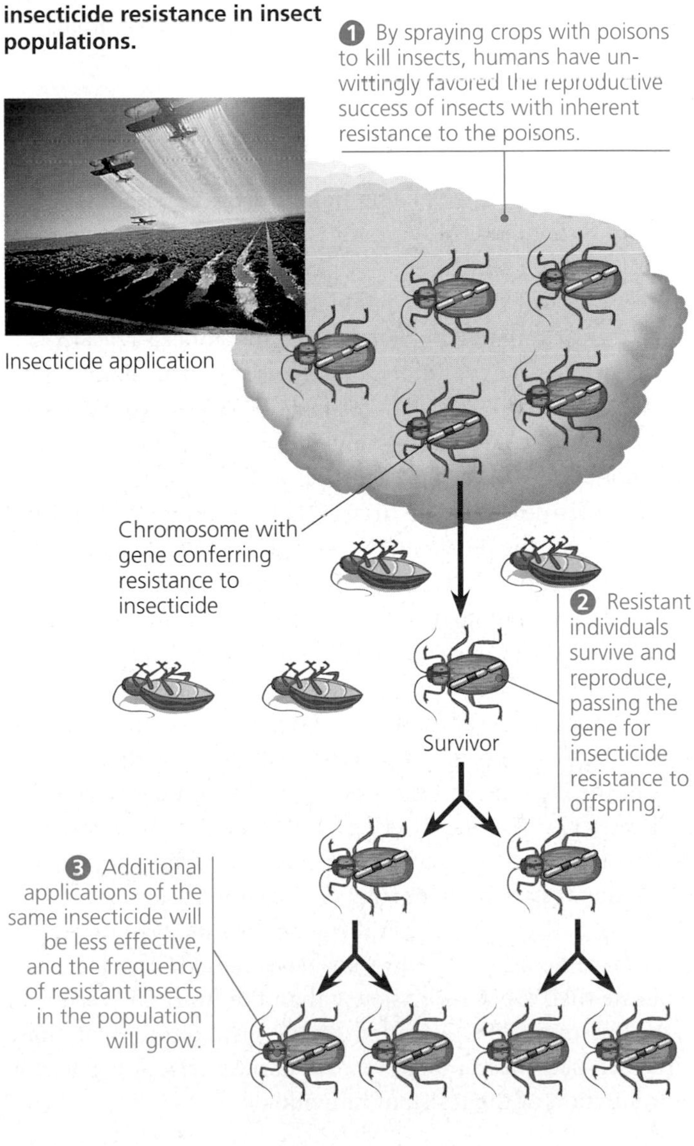

Insecticide application

1 By spraying crops with poisons to kill insects, humans have unwittingly favored the reproductive success of insects with inherent resistance to the poisons.

Chromosome with gene conferring resistance to insecticide

2 Resistant individuals survive and reproduce, passing the gene for insecticide resistance to offspring.

Survivor

3 Additional applications of the same insecticide will be less effective, and the frequency of resistant insects in the population will grow.

lucky few carry genes coding for enzymes that destroy the insecticide. The poison kills most members of the insect population, leaving the resistant individuals to reproduce. And their offspring inherit the genes for insecticide resistance. In each generation, the proportion of insecticide-resistant individuals in the insect population increases. The population has adapted to a change in its environment.

This example of insect adaptation to insecticides highlights two key points about natural selection. First, notice that natural selection is more a process of editing than it is a creative mechanism. An insecticide does not create resistant individuals, but selects for resistant insects that were already present in the population. Second, note again that natural selection is contingent on time and place. It favors those characteristics in a varying population that fit the current, local environment. What is adaptive in one situation may be useless or even detrimental in different circumstances. For example, some genetic mutations that endow houseflies with resistance to the insecticide DDT also reduce a fly's growth rate. Before DDT was introduced to environments, those particular genes were a handicap. But the appearance of DDT changed the environmental arena and favored insecticide-resistant individuals.

Natural Selection in Action: The Evolution of Drug-Resistant HIV

Researchers have developed numerous drugs to combat the human immunodeficiency virus (HIV), the pathogen that causes AIDS (see pp. 336 and 919–921). Often, resistance to a drug evolves rapidly in the HIV population of an individual patient soon after treatment with that drug begins. For example, FIGURE 22.13 illustrates the evolution of resistance to a drug named 3TC. Notice that the 3TC-resistant forms of HIV begin to increase in number almost immediately and make up 100% of the total HIV population in each patient after just a few weeks.

Scientists designed the drug 3TC to interfere with reverse transcriptase, the enzyme HIV uses to copy its RNA genome into the DNA of the human host cell (see FIGURE 18.7). DNA, remember, is a polymer of four kinds of nucleotides, abbreviated A, G, T, and C (see Chapter 16). The drug 3TC mimics the C (cytosine) nucleotide of DNA. The HIV's reverse transcriptase will pick up a 3TC molecule instead of C and insert it into a growing DNA chain. This error terminates further elongation of the DNA and thus blocks reproduction of the HIV.

The 3TC-resistant variety of HIV has a slightly different version of reverse transcriptase that is able to discriminate between the drug and the normal C nucleotide. Members of the HIV population that inherit the gene for this form of the enzyme have no advantage in the absence of 3TC; in fact, they replicate their DNA more slowly than the "normal" variety of HIV. But once 3TC is added to the environment of these viruses, it becomes a potent force in natural selection, favoring reproduction of the resistant individuals.

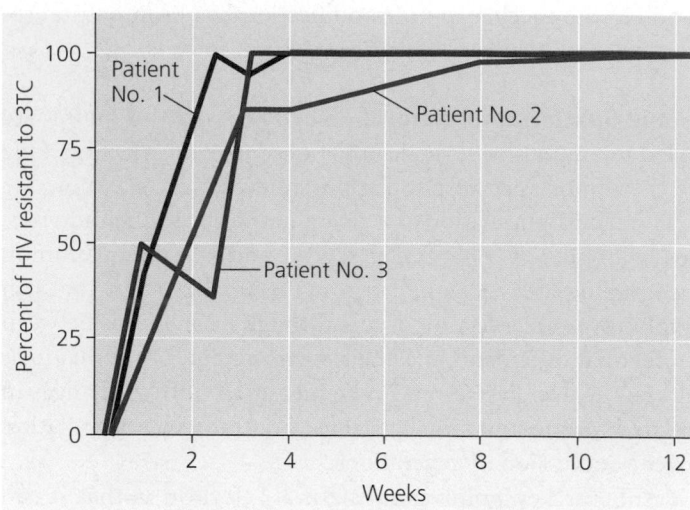

FIGURE 22.13 **Evolution of drug resistance in HIV.** In patients being treated with the anti-HIV drug 3TC, a very small number of drug-resistant viruses are present at the beginning of the treatment, but natural selection increases their frequency over time. Though the exact time course differs between patients, 3TC-resistant HIV makes up 100% of the virus populations within a few weeks in all cases.

Other evidence of evolution pervades biology

We have examined cases of evolution by natural selection that occur rapidly enough to be directly observed. However, the much grander changes of biological diversity documented by the fossil record occur on a time scale spanning hundreds of millions of years. Evidence that the diversity of life is a product of evolution prevades every research field of biology. And, as biology progresses, new discoveries, including the revelations of molecular biology, continue to validate the Darwinian view of life.

Homology

Descent with modification, Darwin's term for evolution, means that new species descend from ancestral species by the accumulation of modifications as populations adapt to new environments. The novel features that characterize a new species are not entirely new, but are altered versions of ancestral features. Species with common ancestry should display underlying similarities, even in features that no longer match in function. Similarity in characteristics resulting from common ancestry is known as **homology.**

Anatomical Homologies. Descent with modification is indeed evident in anatomical similarities between species grouped in the same taxonomic category. For example, many of the same skeletal elements make up the forelimbs of humans, cats, whales, bats, and all other mammals, although these appendages have very different functions (FIGURE 22.14). Surely, the best way to construct the infrastructure of a bat's

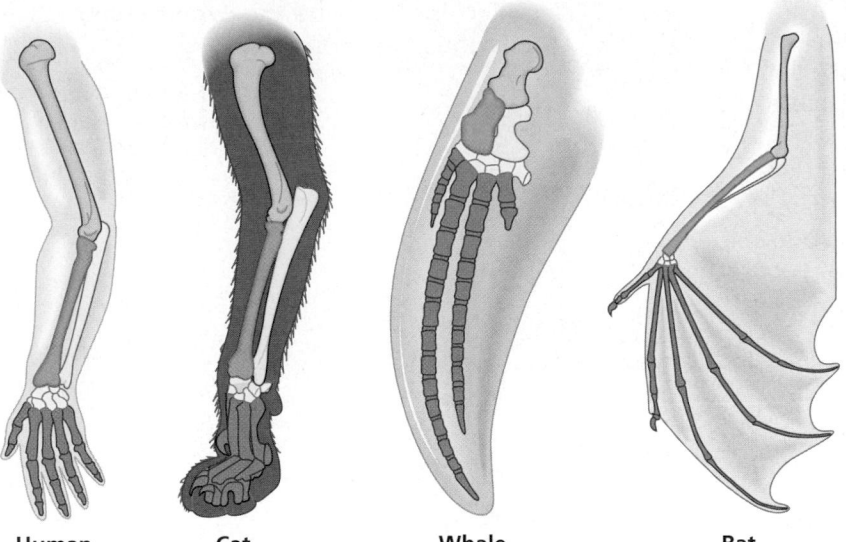

Human   **Cat**   **Whale**   **Bat**

FIGURE 22.14 Homologous structures: anatomical signs of descent with modification. The forelimbs of all mammals are constructed from the same skeletal elements (color-coded in these diagrams). The hypothesis that all mammals descend from a common ancestor predicts that their forelimbs, though diversely adapted, would be variations on a common anatomical theme.

wing is not also the best way to build a whale's flipper. Such anatomical peculiarities make no sense if the structures are uniquely engineered and unrelated. A more likely explanation is that the basic similarity of these forelimbs is the consequence of the descent of all mammals from a common ancestor. The forelegs, wings, flippers, and arms of different mammals are variations on a common structural theme. In taking on different functions in each species, the basic structures were modified. Such anatomical signs of evolution are called **homologous structures.**

Comparative anatomy, the comparison of body structures between species, confirms that evolution is a remodeling process. The historical constraints of this retrofitting are evident in anatomical imperfections. For example, the human knee joint and spine were derived from ancestral structures that supported four-legged mammals. Almost none of us will reach old age without experiencing knee or back problems. If these structures had first taken form specifically to support our bipedal posture, we would expect them to be less subject to injury. The anatomical remodeling that stood us up was apparently constrained by our evolutionary history.

Some of the most interesting homologous structures are **vestigial organs,** structures of marginal, if any, importance to the organism. Vestigial organs are historical remnants of structures that had important functions in ancestors. For instance, the skeletons of some snakes retain vestiges of the pelvis and leg bones of walking ancestors. We would not expect to see these structures if snakes had an origin separate from other vertebrate animals.

Embryological Homologies. Sometimes, homologies that are not obvious in adult organisms become evident when we look at embryonic development. For example, all vertebrate embryos have structures called pharyngeal pouches in their throat regions at some stage in their development. These embryonic structures develop into homologous structures with very different functions, such as the gills of fish or the Eustachian tubes that connect the middle ear with the throat in humans and other mammals.

Molecular Homologies. Anatomical homology cannot help us link such distantly related organisms as plants and animals, which have no anatomy in common. However, plants and animals, along with all other organisms, do share certain characteristics at the molecular level: For example, all species of life use the same basic genetic machinery of DNA and RNA, and the genetic code is essentially universal (see Chapter 17). Evidently, the language of the genetic code has been passed along through all branches of the tree of life ever since the code's inception in an early life-form. Molecular biology provides new tools for exploring evolutionary relationships in the diversity of life.

Homologies and the Tree of Life. Homologies mirror the taxonomic hierarchy of the tree of life. Some homologies, such as the genetic code, are shared by all life because they date back to the deep ancestral past. Homologies that evolved more recently are shared only by smaller branches of the tree of life. For example, all tetrapods (from the Greek *tetra,* "four," and *pod,* "foot"), the vertebrate branch consisting of amphibians, reptiles, birds, and mammals, share the same basic five-digit limb structure illustrated for mammals in FIGURE 22.14. Thus, homologies form a layered pattern, with all life sharing the deepest layer and each smaller group adding fresh homologies to those they share with larger groups. This hierarchical pattern is exactly what we would expect if life evolved and diversified from a common ancestor, but not what we would see if each species arose separately.

If homologies reflect evolutionary history, we should expect to find similar patterns whether we are comparing molecules or bones or any other characteristics. The new tools of molecular biology have generally corroborated rather than contradicted evolutionary trees based on comparative anatomy and other methods. Evolutionary relationships among species are documented in their DNA and proteins—in their genes and gene products. If two species have libraries of genes and proteins with sequences that match closely, the sequences have probably been copied from a common ancestor. (If two long paragraphs match except for the substitution of a letter here and there, we would surely attribute them both to a single source.) TABLE 22.1 compares the amino acid sequence of human hemoglobin, the oxygen-transporting protein of blood, with the hemoglobin of other vertebrates. The data show the same pattern of evolutionary relationships that researchers find when they compare other proteins or assess relationships based on nonmolecular methods, such as skeletal anatomy. The Darwinian view of life predicts that different kinds of homologies—anatomical, embryological, and molecular—will fall into the same hierarchical pattern because they have all evolved during the same branching pattern of evolutionary history.

Biogeography

The geographic distribution of species—**biogeography**—first suggested evolution to Darwin. Species tend to be more closely related to other species from the same area than to other species with the same way of life but living in different areas. For example, Australia is the home of a group of mammals—the marsupials—that are distinct from another group of mammals—the eutherians—that live elsewhere on Earth. (Eutherians are mammals that complete their embryonic development in the uterus, while marsupials are born as embryos and complete their development in an external pouch.) Some Australian marsupials have eutherian look-alikes with similar adaptations living on other continents. For example, a forest-dwelling marsupial called the sugar glider is superficially very similar to flying squirrels, eutherians that live in North American forests (FIGURE 22.15). These two mammals have adapted to the same way of life, but they evolved independently from different ancestors. The sugar glider is distinctly marsupial, much more closely related to kangaroos and other Australian marsupials than to flying squirrels or any other eutherian mammals. The sugar glider is a marsupial not because that is a requirement for its gliding lifestyle but simply because its ancestors were marsupials. The unique fauna of Australia diversified on that island continent after it became isolated from the landmasses on which placental mammals diversified. The resemblance between sugar gliders and flying squirrels is an example of what biologists call convergent evolution (we'll take a closer look at convergence in Chapter 25).

Islands are showcases of biogeographic evidence for evolution. They generally have many species of plants and animals that are **endemic,** which means they are found nowhere else in the world. And yet, as Darwin observed when he reassessed his collections from the voyage of the *Beagle,* most island species are closely related to species from the nearest mainland or neighboring island. This explains why two islands with

Table 22.1 Molecular Data and the Evolutionary Relationships of Vertebrates

Species	Number of Amino Acids That Differ from a Human Hemoglobin Polypeptide (Total Chain Length = 146 Amino Acids)
Human	0
Rhesus monkey	8
Mouse	27
Chicken	45
Frog	67
Lamprey	125

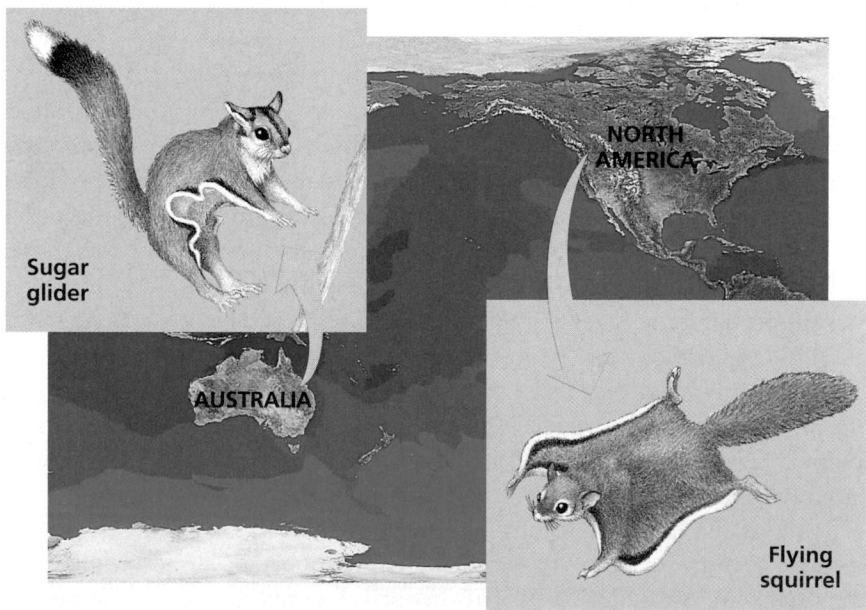

FIGURE 22.15 Different geographic regions, different mammalian "brands." The sugar glider is an example of the diverse marsupial life that evolved in isolation on the island continent of Australia. The resemblance of sugar gliders to the eutherian flying squirrels of North America is not due to a close evolutionary relationship but to convergent evolution in similar environmental contexts.

similar environments in different parts of the world are populated not by closely related species but by species taxonomically affiliated with the plants and animals of the nearest mainland, where the environment is often quite different. Island chains, or archipelagos, are especially interesting in their biogeography. If a species that disperses from a mainland to an island succeeds in its new environment, it may give rise to several new species as populations spread to other islands in the archipelago. The example of finches on the Galápagos archipelago came up earlier in the chapter. FIGURE 22.16 illustrates another example, the evolution of fruit flies *(Drosophila)* on the Hawaiian archipelago.

The Fossil Record

The succession of fossil forms is compatible with what is known from other types of evidence about the major branches of descent in the tree of life. For instance, evidence from biochemistry, molecular biology, and cell biology places prokaryotes as the ancestors of all life and predicts that prokaryotes should precede all eukaryotic life in the fossil record. Indeed, the oldest known fossils are prokaryotes. Another example is the chronological appearance of the different classes of vertebrate animals in the fossil record. Fossil fishes predate all other vertebrates, with amphibians next, followed by reptiles, then mammals and birds. This sequence is consistent with the history of vertebrate descent as revealed by many other types of evidence. In contrast, the idea that all species were individually created at about the same time predicts that all vertebrate classes would make their first appearance in the fossil record

in rocks of the same age, a prediction at odds with what paleontologists actually observe.

The Darwinian view of life also predicts that evolutionary transitions should leave signs in the fossil record. Paleontologists have discovered fossils of many transitional forms that link even older fossils to modern species. For example, a series of fossils documents the changes in skull shape and size that occurred as mammals evolved from reptiles. Every year, paleontologists turn up other important links between modern forms and their ancestors. In the past few years, for instance, researchers have found fossilized whales that link these aquatic mammals to their terrestrial predecessors (FIGURE 22.17, p. 442).

Thus, the Darwinian view of life endures in biology because it is supported by independent types of evidence: evolutionary patterns of homology that match patterns in space (biogeography) and time (the fossil record).

What is theoretical about the Darwinian view of life?

Some people dismiss Darwinism as "just a theory." This tactic for nullifying the evolutionary view of life has two flaws. First, it fails to separate Darwin's two claims: that modern species evolved from ancestral forms and that natural selection is the main mechanism for this evolution. The conclusion that life has evolved is based on historical evidence—the signs of evolution discussed in the previous section.

What, then, is theoretical about evolution? Theories are our attempts to explain facts and integrate them with overarching

FIGURE 22.16 The evolution of fruit fly *(Drosophila)* species on the Hawaiian archipelago. Geologists have determined the ages of these volcanic islands, which are progressively younger from Kauai (the oldest) to the big island of Hawaii (the youngest, still growing as active volcanoes add lava rock to the shoreline). The islands have about 500 endemic species of the fruit fly genus *Drosophila,* all descended from a common ancestor that managed to reach Kauai over 5 million years ago. The arrows trace the history of just a few of the species in one evolutionary branch. The vintage of each species closely matches the age of its island home.

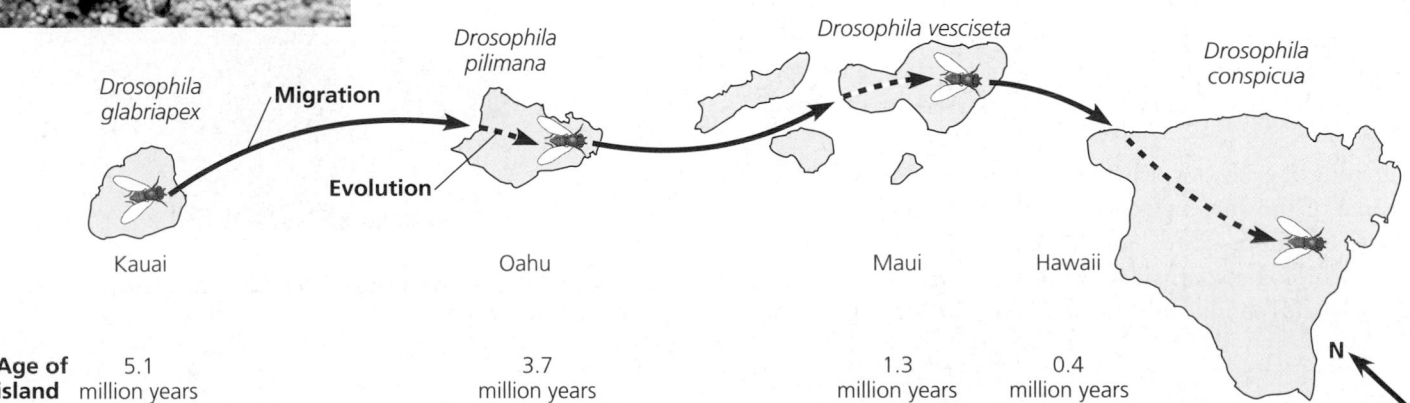

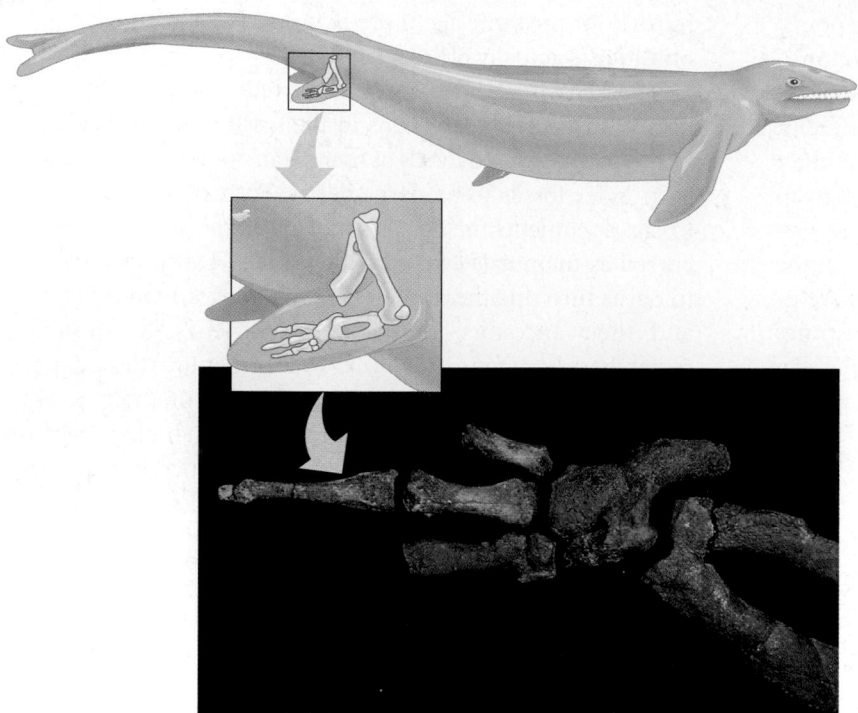

FIGURE 22.17 A transitional fossil linking past and present. The hypothesis that whales evolved from terrestrial (land-dwelling) ancestors predicts a four-limbed beginning for whales. Paleontologists digging in Egypt and Pakistan have identified extinct whales that had hind limbs. Shown here are the fossilized leg bones of *Basilosaurus,* one of those ancient whales. These whales were already aquatic animals that no longer used their legs to support their weight. The leg bones of an even older fossilized whale named *Ambulocetus* are heftier. *Ambulocetus* may have split its time between living on land and in water.

concepts. To biologists, Darwin's theory of evolution is natural selection—the mechanism Darwin proposed to explain the historical facts of evolution documented by fossils, biogeography, and other types of evidence.

So the "just a theory" argument concerns only Darwin's second point, his theory of natural selection. This brings us to the second flaw in the "just a theory" case. The term *theory* has a very different meaning in science than in everyday use. The colloquial use of the word *theory* comes close to what scientists mean by a hypothesis. In science, a theory is more comprehensive than a hypothesis. A theory, such as Newton's theory of gravitation or Darwin's theory of natural selection, accounts for many facts and attempts to explain a great variety of phenomena. Such a unifying theory does not become widely accepted in science unless its predictions stand up to thorough and continual testing by experiments and observations (see Chapter 1). Even then, good scientists do not allow theories to become dogma. For example, many evolutionary biologists now question whether natural selection alone accounts for the evolutionary history observed in the fossil record.

The study of evolution is livelier than ever, and we will evaluate some of the current debates in the next three chapters. But these questions about *how* life evolves in no way imply that most biologists consider evolution itself to be "just a theory." Debates about evolutionary theory are like arguments over competing theories about gravity; we know that objects keep right on falling while we debate the cause.

By attributing the diversity of life to natural causes rather than to supernatural creation, Darwin gave biology a sound, scientific basis (FIGURE 22.18). Nevertheless, the diverse products of evolution are elegant and inspiring. As Darwin said in the closing paragraph of *The Origin of Species,* "There is grandeur in this view of life."

FIGURE 22.18 Charles Darwin in 1859, the year *The Origin of Species* was published.

CHAPTER 22 REVIEW

Go to the Campbell Biology website (www.campbellbiology.com) to explore an interactive version of the Chapter Review.

Summary of Key Concepts

THE HISTORICAL CONTEXT FOR EVOLUTIONARY THEORY

■ **Western culture resisted evolutionary views of life (pp. 429–430, FIGURE 22.1)** Darwin's theory that natural selection is responsible for evolutionary change was a radical departure from the dominant religious and philosophical climate of Western culture.

■ **Theories of geologic gradualism helped clear the path for evolutionary biologists (pp. 430–431, FIGURE 22.4)** Geologists Hutton and Lyell perceived that changes in Earth's surface can result from slow, continuous actions.
Web/CD Activity 22A: *Grand Canyon Video*

■ **Lamarck placed fossils in an evolutionary context (p. 431)** Lamarck helped pave the way for Darwin by emphasizing the interactions between organisms and their environment.

THE DARWINIAN REVOLUTION

■ **Field research helped Darwin frame his view of life (pp. 432–434, FIGURES 22.5 and 22.6)** Darwin's experiences during the voyage of HMS *Beagle* provided much of the background for his idea that new species originate from ancestral forms by the gradual accumulation of adaptations.
Web/CD Activity 22B: *Darwin and the Galápagos Islands*
Web/CD Activity 22C: *Videos of the Galápagos Islands*
Web/CD Activity 22D: *The Voyage of the Beagle: Darwin's Trip Around the World*
Biology Labs On-Line: *EvolutionLab*

■ **The Origin of Species developed two main points: the occurrence of evolution and natural selection as its mechanism (pp. 434–437, FIGURES 22.7–22.11)** Natural selection is based on differential success in reproduction, made possible because of heritable variation among the individuals of any population and the tendency for a population to produce many more offspring than the environment can support. Natural selection results in adaptation, the presence in living things of heritable traits well suited to the local environment.
Web/CD Activity 22E: *Evolutionary Adaptation: Sea Horse Camouflage Video*

■ **Examples of natural selection provide evidence of evolution (pp. 437–438, FIGURES 22.12, 22.13)** Natural selection and the adaptive evolution it causes are observable phenomena.
Web/CD Case Study in the Process of Science: *How Do Environmental Changes Affect a Population?*
Web/CD Case Study in the Process of Science: *What Are the Patterns of Antibiotic Resistance?*

■ **Other evidence of evolution pervades biology (pp. 438–441, FIGURES 22.14–22.17)** Evolution is validated by evidence from homologies, similarities between species that are due to common ancestry. Biogeography and the fossil record generally support the evolutionary deductions based on homologies.
Web/CD Activity 22F: *Reconstructing Forelimbs*

■ **What is theoretical about the Darwinian view of life? (pp. 441–442, FIGURE 22.18)** Darwin's theory of natural selection unified biology and placed it in the realm of science.

Self-Quiz

1. The ideas of Hutton and Lyell that Darwin incorporated into his theory pertained to
   a. the age of Earth and gradual change.
   b. extinctions evident in the fossil record.
   c. adaptation of species to the environment.
   d. a hierarchical classification of organisms.
   e. the inheritance of acquired characteristics.

2. Which of the following is *not* an observation or inference on which natural selection is based?
   a. There is heritable variation among individuals.
   b. Poorly adapted individuals never leave offspring.
   c. There is a struggle for limited resources, and only a fraction of offspring survive.
   d. Individuals whose inherited characteristics best fit them to the environment will generally leave more offspring.
   e. Organisms interact with their environments.

3. Which of the following would provide the best information for distinguishing phylogenetic relationships between several species that are almost identical in anatomy?
   a. the fossil record
   b. homologous structures
   c. comparative anatomy
   d. comparative embryology
   e. molecular comparisons of DNA and amino acid sequences

4. Which of the following observations helped Darwin shape his concept of descent with modification?
   a. Species diversity declined as distance from the equator increased.
   b. Fewer species were found living on islands than on the nearest continents.
   c. Birds could be found on islands more distant from the mainland than their maximum nonstop flight distance.
   d. South American temperate plants were more similar to the tropical plants of South America than to the temperate plants of Europe.
   e. Finches on the Galápagos fed on seeds, whereas finches on the South American mainland were insectivorous.

5. Darwin synthesized information from several sources in developing his theory of evolution by natural selection. Which of the following did *not* influence his thinking?
   a. Linnaeus' hierarchical classification of species, which could imply evolutionary relationships
   b. Lyell's *Principles of Geology,* which described the gradualness and uniformity of geologic changes over long periods of time
   c. Mendel's paper describing the basic principles of inheritance
   d. examples of major changes in domesticated species produced by artificial selection
   e. the biogeographic distribution of species that he observed on the Galápagos Islands and during his journey around South America

6. In science, the term *theory* generally applies to an idea that
   a. is a speculation lacking supportive observations or experiments.
   b. attempts to explain many related phenomena.
   c. is synonymous with what biologists mean by a hypothesis.
   d. is so widely accepted that it is considered a law of nature.
   e. cannot be tested.

7. Within a few weeks of treatment with the drug 3TC, a patient's HIV population consists entirely of 3TC-resistant viruses. How can this result best be explained?
   a. HIV has the ability to change its surface proteins and resist vaccines.
   b. The patient must have become reinfected with 3TC-resistant viruses.
   c. HIV began making drug-resistant versions of reverse transcriptase in response to the drug.
   d. A few drug-resistant viruses were present at the start of treatment, and natural selection increased their frequency.
   e. Some viruses developed drug resistance and then passed their resistant genes to all the patient's viruses.

8. The smallest biological unit that can evolve over time is
   a. a cell.
   b. an individual organism.
   c. a population.
   d. a species.
   e. an ecosystem.

9. Which of the following ideas is common to both Darwin's and Lamarck's theories of evolution?
   a. Adaptation results from differential reproductive success.
   b. Evolution drives organisms to greater and greater complexity.
   c. Evolutionary adaptation results from interactions between organisms and their environments.
   d. Adaptation results from the use and disuse of anatomical structures.
   e. The fossil record supports the view that species are fixed.

10. Which of the following pairs of structures is least likely to represent homology?
    a. the wings of a bat and the forelimbs of a human
    b. the hemoglobin of a baboon and the hemoglobin of a gorilla
    c. the mitochondria of a plant and those of an animal
    d. the bark of a tree and the protective covering of a lobster
    e. the brain of a frog and the brain of a dog

11. What were the two main points in Darwin's *The Origin of Species*?

12. Darwin's phrase for evolution, _____ with _____, captured the idea that an ancestral species could diversify into many descendant species by the accumulation of different _____ to various environments.

13. Why are older fossils generally in deeper rock layers than younger fossils?

14. What is homology?

15. Define natural selection.

16. Explain why the following statement is incorrect: "Pesticides have created pesticide resistance in insects."

Go to the website or CD-ROM for more quiz questions.

Evolution Connection

An important maxim of historical geology states, "the present is the key to the past." What does this mean, and how can you apply this idea to organisms? Conversely, how does the past help us understand the present diversity of life?

The Process of Science

Darwin's argument for the idea that evolution has occurred is largely inductive, while his argument for the mechanism of natural selection is essentially deductive. Summarize in your own words the inductive and deductive components of Darwin's theory. (You can review induction and deduction in Chapter 1.)

In the Case Studies in the Process of Science section of the CD-ROM or website, explore the effects of various environmental changes on a population of leafhoppers and investigate case studies of antibiotic resistance. Also link to the EvolutionLab of Biology Labs On-Line to model evolutionary changes in finch populations.

Science, Technology, and Society

Is the concept of natural selection relevant in a political or economic context? In other words, if a particular nation or corporation achieves success or dominance, does this mean that it is fitter than its competitors and that unregulated dominance is justified? Why or why not?

Answers: 1. a; 2. b; 3. e; 4. d; 5. c; 6. b; 7. d; 8. c; 9. c; 10. d; 11. Descent of diverse species from common ancestors; natural selection as the mechanism of evolution. 12. descent . . . modification . . . adaptations. 13. Sedimentation adds younger rock layers on top of older ones. 14. Similarity between species that is due to shared ancestry. 15. Natural selection is the differential reproductive success among a population's varying individuals. 16. An environmental factor does not create new traits such as pesticide resistance, but rather selects among the traits that are already represented in the population.

CHAPTER 23

THE EVOLUTION OF POPULATIONS

POPULATION GENETICS

- The modern evolutionary synthesis integrated Darwinian selection and Mendelian inheritance
- A population's gene pool is defined by its allele frequencies
- The Hardy-Weinberg theorem describes a nonevolving population

CAUSES OF MICROEVOLUTION

- Microevolution is a generation-to-generation change in a population's allele frequencies
- The two main causes of microevolution are genetic drift and natural selection

GENETIC VARIATION, THE SUBSTRATE FOR NATURAL SELECTION

- Genetic variation occurs within and between populations
- Mutation and sexual recombination generate genetic variation
- Diploidy and balanced polymorphism preserve variation

A CLOSER LOOK AT NATURAL SELECTION AS THE MECHANISM OF ADAPTIVE EVOLUTION

- Evolutionary fitness is the relative contribution an individual makes to the gene pool of the next generation
- The effect of selection on a varying characteristic can be directional, diversifying, or stabilizing
- Natural selection maintains sexual reproduction
- Sexual selection may lead to pronounced secondary differences between the sexes
- Natural selection cannot fashion perfect organisms

One obstacle to understanding evolution *is the common misconception that individual organisms evolve, in the Darwinian sense, during their lifetimes. In fact, natural selection does act on individuals; their characteristics affect their chances of survival and their reproductive success. But the evolutionary impact of this natural selection is only apparent in tracking how a population of organisms changes over time. Consider, for example, representatives from a population of marine snails* (Liguus fascitus) *in the photograph above. Their different patterns of* coloration represent genetic variations within that population. If predators feed preferentially on snails having a particular coloration, then the proportion of individuals with that coloration probably will decline from one generation to the next because such snails will produce fewer offspring. Thus, it is the population, not its individuals, that evolves; some characteristics become more common within the overall population, while other characteristics decline.* FIGURE 23.1 *(p. 446) illustrates another example of this principle. Evolution on the smallest scale, or microevolution, can be defined as a change in the allele frequencies of a population (see p. 249 to review the definition of alleles). We begin our study of microevolution by tracing how biologists finally began to understand Darwin's theory of natural selection during the first half of the 20th century.*

POPULATION GENETICS

The Origin of Species convinced most biologists that species are products of evolution, but Darwin was not nearly so successful in gaining acceptance for his idea that natural selection is the main mechanism of evolution. Natural selection requires hereditary processes that Darwin could not explain. His theory was based on what seems like a paradox of inheritance: Like begets like—but not exactly. What was missing in Darwin's explanations was an understanding of inheritance that could explain how chance variations arise in a population while also accounting for the precise transmission of these variations from parents to offspring. Although Gregor Mendel and Charles Darwin were contemporaries, Mendel's discoveries were unappreciated at the time, and apparently no one noticed that he had elucidated the very principles of inheritance that could have resolved Darwin's paradox and given credibility to natural selection.

445

FIGURE 23.1 Individuals are selected, but populations evolve. The bent grass *(Agrostis tenuis)* in the foreground is growing on the tailings of an abandoned mine in Wales. These plants tolerate concentrations of heavy metals that are toxic to other plants of the same species growing just meters away, in the pasture on the other side of the fence. Each year, many seeds land on the mine tailings, but most are unable to grow successfully there. The only plants that germinate, grow, and reproduce are those that inherited genes enabling them to tolerate metallic soil. Thus, this adaptation does not evolve by individual plants becoming more metal-tolerant during their lifetimes. We can only see the evolution of this population by observing the proportions of metal-tolerant plants in successive generations.

The modern evolutionary synthesis integrated Darwinian selection and Mendelian inheritance

THE PROCESS OF SCIENCE Ironically, when Mendel's research article was rediscovered and reassessed at the beginning of the 20th century, many geneticists believed that the laws of inheritance were at odds with Darwin's theory of natural selection. Darwin considered the raw material for natural selection to be quantitative characters, those characteristics in a population that vary along a continuum, such as fur length in mammals or the speed with which an animal can flee from a predator. We know today that quantitative characters are influenced by multiple genetic loci. (To review polygenic inheritance and quantitative characters, see Chapter 14.) But Mendel (and later the geneticists of the early 20th century) recognized only discrete "either-or" traits, such as purple or white flowers in pea plants, as heritable. Thus, there seemed to be no genetic basis for natural selection to work on the more subtle variations within a population that were central to Darwin's theory.

An important turning point for evolutionary theory was the birth of **population genetics,** which emphasizes the extensive genetic variation within populations and recognizes the importance of quantitative characters. With progress in population genetics in the 1930s, Mendelism and Darwinism were reconciled, and the genetic basis of variation and natural selection was worked out.

A comprehensive theory of evolution that became known as the **modern synthesis** began to take form in the early 1940s. It is called a synthesis because it integrates discoveries and ideas from many different fields, including paleontology, taxonomy, biogeography, and, of course, population genetics. The architects of this modern synthesis included geneticists Theodosius Dobzhansky (1900–1975) and Sewall Wright (1889–1988), biogeographer and taxonomist Ernst Mayr (1904–), paleontologist George Gaylord Simpson (1902–1984), and botanist G. Ledyard Stebbins (1906–2000). The modern synthesis emphasizes the importance of populations as the units of evolution, the central role of natural selection as the most important mechanism of evolution, and the idea of gradualism to explain how large changes can evolve as an accumulation of small changes occurring over long periods of time. No scientific paradigm is likely to endure without modification for half a century. Many evolutionary biologists are now challenging some of the assumptions of the modern synthesis. Still, 20th-century biology has been profoundly affected by the modern synthesis, which has shaped most of our ideas about how populations evolve.

A population's gene pool is defined by its allele frequencies

A **population** is a localized group of individuals belonging to the same species. For now, we will define a **species** as a group of populations whose individuals have the potential to interbreed and produce fertile offspring in nature (this definition will be examined more critically in Chapter 24). Each species is distributed over a certain geographic range, but within this range individuals are usually concentrated in several localized populations. A population may be isolated from other populations of the same species, exchanging genetic material only rarely. Such isolation is particularly common for populations confined to widely separated islands, unconnected lakes, or mountain ranges separated by lowlands. However, populations are not always isolated, nor do they necessarily have sharp boundaries. One dense population center may blur into another in an intermediate region where members of the species occur but are less numerous. Although these populations are not isolated, individuals are still concentrated in centers and are more likely to breed with members of the same population than with members of other populations. Therefore, individuals near a population center are, on average, more closely related to one another than to members of other populations (FIGURE 23.2).

The total aggregate of genes in a population at any one time is called the population's **gene pool.** It consists of all alleles at all gene loci in all individuals of the population. For a diploid species, each locus is represented twice in the genome of an individual, who may be either homozygous or heterozygous for those homologous loci. (Recall that homozygous individ-

(a) Two dense populations of Douglas fir trees

(b) Human population centers in the eastern United States

FIGURE 23.2 Population distribution. (a) Separated by a river bottom, where firs are uncommon, the two Douglas fir populations are not totally isolated. Interbreeding occurs when wind blows pollen between the populations. Nonetheless, trees are more likely to breed with members of the same population than with trees on the other side of the river. **(b)** This nighttime satellite view of the eastern United States shows the lights of human population centers. People move around the country, of course, but people are more likely to choose mates locally.

uals have two identical alleles for a given character, whereas heterozygous individuals have two different alleles for that character.) If all members of a population are homozygous for the same allele, that allele is said to be *fixed* in the gene pool. Often, however, there are two or more alleles for a gene, each having a relative frequency (proportion) in the gene pool.

An example will make the concept of allele frequency in a gene pool less abstract. Imagine a wildflower population with two varieties contrasting in flower color. An allele for red flowers, which we will symbolize by *R*, is completely dominant to an allele for white flowers, symbolized by *r*. For our simplified situation, these are the only two alleles for this locus in the population. Suppose an imaginary population has 500 plants, and 20 of these plants have white flowers because they are homozygous for the recessive allele; their genotype is *rr*. The other 480 plants have red flowers; some of them will be homozygous (*RR*) and others will be heterozygous (*Rr*). Suppose that 320 plants are *RR* homozygotes and 160 are *Rr* heterozygotes. Because these are diploid organisms, there are a total of 1,000 copies of genes for flower color in the population of 500 individuals. The dominant allele (*R*) accounts for 800 of these genes (320 × 2 = 640 for *RR* plants, plus 160 × 1 = 160 for *Rr* individuals). Thus, the frequency of the *R* allele in the gene pool of this population is 800/1,000 = 0.8 = 80%. And because there are only two allelic forms of the gene, the *r* allele must have a frequency of 0.2, or 20%.

The Hardy-Weinberg theorem describes a nonevolving population

Before we consider the mechanisms that cause a population to evolve, it will be helpful to examine, for comparison, the gene pool of a *nonevolving* population. Such a gene pool is described by the **Hardy-Weinberg theorem,** named for the two scientists who derived the principle independently in 1908. The theorem states that the frequencies of alleles and genotypes in a population's gene pool remain constant over the generations unless acted upon by agents other than Mendelian segregation and recombination of alleles. Put another way, the shuffling of alleles due to meiosis and random fertilization has no effect on the overall gene pool of a population.

To apply the Hardy-Weinberg theorem, let's return to our imaginary wildflower population of 500 plants (FIGURE 23.3a, p. 448). Recall that 80% (0.8) of the flower-color loci in the gene pool have the *R* allele and 20% (0.2) have the *r* allele. How will meiosis during sexual reproduction affect the frequencies of the two alleles in the next generation of our wildflower population? We will assume that the union of sperm and ova in the population is completely random; that is, all male-female mating combinations are equally likely. The situation is analogous to mixing all gametes in a sack and then drawing them randomly, two at a time, to determine the genotype for each zygote (fertilized egg). Each gamete has one allele for flower color, and the allele frequencies of the gametes will be the same as the allele frequencies in the parent population. Every time a gamete is drawn from the pool at random, the chance that the gamete will bear an *R* allele is 0.8, and the chance that the gamete will have an *r* allele is 0.2.

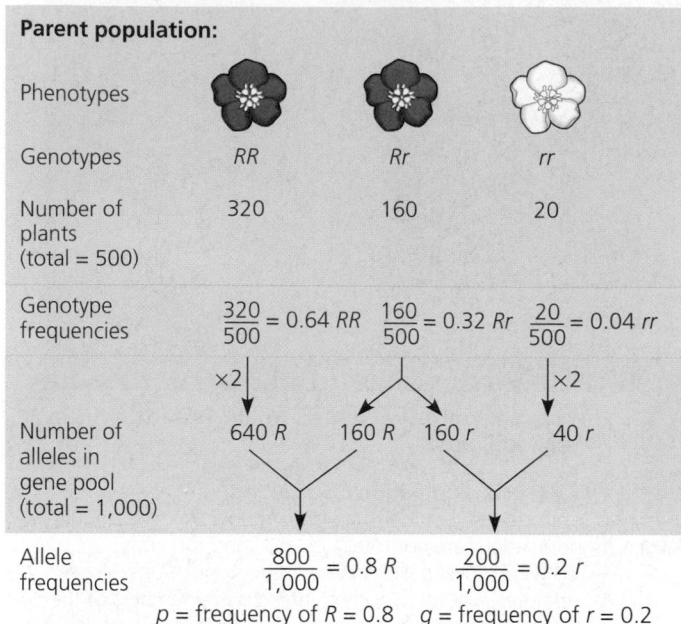

Parent population:

Phenotypes			
Genotypes	RR	Rr	rr
Number of plants (total = 500)	320	160	20
Genotype frequencies	$\frac{320}{500} = 0.64$ RR	$\frac{160}{500} = 0.32$ Rr	$\frac{20}{500} = 0.04$ rr

×2 →   ×2 →

Number of alleles in gene pool (total = 1,000)	640 R	160 R	160 r	40 r

Allele frequencies: $\frac{800}{1,000} = 0.8$ R $\frac{200}{1,000} = 0.2$ r

p = frequency of R = 0.8 q = frequency of r = 0.2

(a) Gene pool of parent population

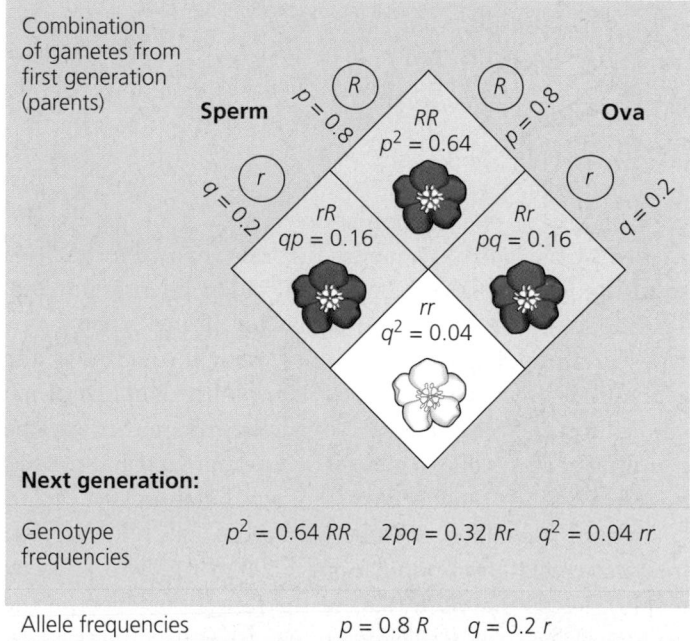

Combination of gametes from first generation (parents)

Sperm $p = 0.8$ $q = 0.2$

Ova $p = 0.8$ $q = 0.2$

RR $p^2 = 0.64$

rR $qp = 0.16$

Rr $pq = 0.16$

rr $q^2 = 0.04$

Next generation:

Genotype frequencies	$p^2 = 0.64$ RR $2pq = 0.32$ Rr $q^2 = 0.04$ rr
Allele frequencies	$p = 0.8$ R $q = 0.2$ r

(b) Gene pool of next generation

FIGURE 23.3 The Hardy-Weinberg theorem. The gene pool of a nonevolving population remains constant over the generations; Mendelian segregation alone will not alter the relative frequencies of alleles or genotypes. In this example (a hypothetical flower population), note that the frequencies of alleles and genotypes remain the same between one generation (a) and the next generation (b).

Using the rule of multiplication (see Chapter 14), we can calculate the frequencies of the three possible genotypes in the next generation of the population (FIGURE 23.3b). The probability of picking two R alleles from the pool of gametes is $0.8 \times 0.8 = 0.64$. Thus, about 64% of the plants in the next generation will have the genotype RR. The frequency of rr individuals will be about 0.04 ($0.2 \times 0.2 = 0.04$), or 4%. And 32%, or 0.32, of the plants will be heterozygous—that is, Rr or rR, depending on whether it is the sperm or ovum that supplies the dominant allele (frequency of Rr = $0.8 \times 0.2 = 0.16$; frequency of rR = $0.2 \times 0.8 = 0.16$; frequency of Rr + rR = 0.32).

Hardy-Weinberg Equilibrium

Notice in FIGURE 23.3 that the sexual processes of meiosis and random fertilization have maintained the same allele and genotype frequencies that existed in the previous generation of the wildflower population. For the flower-color locus, the population's gene pool is in a state of equilibrium—referred to as **Hardy-Weinberg equilibrium.** Theoretically, the allele frequencies could remain constant at 0.8 for R and 0.2 for r forever (though in reality, some other factor always intervenes). The Hardy-Weinberg theorem describes how the Mendelian system has no tendency to alter allele frequencies. For instance, the dominant allele (R) has no tendency to increase in frequency from one generation to the next relative to the recessive allele (r). The system operates somewhat like shuffling a deck of cards: No matter how many times the deck is reshuffled to deal out new hands, the deck itself remains the same. Aces do not grow more numerous than jacks. And the repeated shuffling of a population's gene pool over the generations cannot, in itself, increase the frequency of one allele relative to another.

The Hardy-Weinberg Equation

We can use our imaginary wildflower population to describe the Hardy-Weinberg theorem in more general terms. We will restrict our analysis to the simplest case of only two alleles, one dominant over the other. However, the Hardy-Weinberg theorem also applies to situations in which there are three or more alleles for a particular locus and no clear-cut dominance.

For a gene locus where only two alleles occur in a population, population geneticists use the letter p to represent the frequency of one allele and the letter q to represent the frequency of the other allele. In our imaginary wildflower population, $p = 0.8$ and $q = 0.2$ (see FIGURE 23.3). Note that $p + q = 1$; the combined frequencies of all possible alleles must add to 100% for that locus in the population. If there are only two alleles and we know the frequency of one, the frequency of the other can be calculated:

$$\text{If } p + q = 1 \quad \text{then} \quad p = 1 - q \quad \text{and} \quad q = 1 - p$$

When gametes combine their alleles to form zygotes, the probability of generating an RR genotype is p^2 (an application of the rule of multiplication). In our wildflower population, $p = 0.8$, and $p^2 = 0.64$, the probability of an R sperm fertilizing an R ovum to produce an RR zygote. The frequency of individuals homozygous for the other allele (rr) is q^2, or $0.2 \times 0.2 = 0.04$ for the wildflower population. There are two ways in which an Rr genotype can arise, depending on which parent contributes the dominant allele. Therefore, the frequency of heterozygous individuals in the population is $2pq$ ($2 \times 0.8 \times 0.2 = 0.32$ in our example). If we have included all possible genotypes, the genotype frequencies add up to 1:

p^2	$+$	$2pq$	$+$	q^2	$= 1$
Frequency of RR genotype		Frequency of Rr plus rR genotype		Frequency of rr genotype	

For our wildflowers, this is $0.64 + 0.32 + 0.04 = 1$.

Population geneticists refer to this general formula as the **Hardy-Weinberg equation.** The equation enables us to calculate frequencies of alleles in a gene pool if we know frequencies of genotypes, and vice versa.

Population Genetics and Health Science

We can use the Hardy-Weinberg equation to estimate the percentage of the human population that carries the allele for a particular inherited disease. For instance, one out of approximately 10,000 babies in the United States is born with phenylketonuria (PKU), a metabolic disorder that, left untreated, results in mental retardation and other problems. (Newborn babies are now routinely tested for PKU, and symptoms can be prevented by following a strict diet.) The disease is caused by a recessive allele; thus, the frequency of individuals in the U.S. population born with PKU corresponds to q^2 in the Hardy-Weinberg equation ($q^2 =$ frequency of the homozygous recessive genotype). Given one PKU occurrence per 10,000 births, $q^2 = 0.0001$. Therefore, assuming Hardy-Weinberg proportions, the frequency of the recessive allele for PKU in the population is

$$q = \sqrt{0.0001} = 0.01$$

and the frequency of the dominant allele is

$$p = 1 - q = 1 - 0.01 = 0.99$$

The frequency of carriers, heterozygous people who do not have PKU but may pass the PKU allele on to offspring, is

$$2pq = 2 \times 0.99 \times 0.01 = 0.0198 \text{ (approximately 2\%)}$$

Thus, about 2% of the U.S. population carries the PKU allele.

The Hardy-Weinberg Theorem and Genetic Variation

The Hardy-Weinberg theorem is important conceptually and historically because it shows how Mendel's theory of inheritance plugs a hole in Darwin's theory of natural selection. Natural selection requires genetic variation; it cannot act in a genetically uniform population. The Hardy-Weinberg theorem explains how Mendelian inheritance preserves genetic variation from one generation to the next. Pre-Mendelian theories of inheritance were mainly "blending" theories, in which the hereditary factors in the offspring were thought to be a blend of the hereditary factors inherited from the two parents. If a red flower mates with a white one, blending theory predicts that the offspring will be a paler red and will now have hereditary factors for this paler red color. Genetic variation has been eliminated, since the two kinds of factors in the parents have been reduced to only one kind in the offspring. Such a hereditary mechanism would soon produce a uniform population. In Mendelian inheritance, however, the hereditary mechanism has no tendency by itself to reduce genetic variation. The set of alleles inherited by each generation from its parents are in turn passed on when that generation breeds. This nonblending mechanism of inheritance preserves the genetic variation upon which natural selection acts.

The Assumptions of the Hardy-Weinberg Theorem

For a population to be in Hardy-Weinberg equilibrium, it must satisfy five main conditions:

1. *Very large population size.* In a population of finite size, especially if that size is small, genetic drift, which is chance fluctuation in the gene pool, can cause genotype frequencies to change over time.
2. *No migration.* Gene flow, the transfer of alleles between populations due to the movement of individuals or gametes, can increase the frequency of any genotype that is in high frequency among the immigrants.
3. *No net mutations.* By changing one allele into another, mutations alter the gene pool.
4. *Random mating.* If individuals pick mates with certain genotypes, then the random mixing of gametes required for Hardy-Weinberg equilibrium does not occur.
5. *No natural selection.* Differential survival and reproductive success of genotypes will alter their frequencies and may cause a detectable deviation from frequencies predicted by the Hardy-Weinberg equation.

Thus, we do not really expect a natural population to be in Hardy-Weinberg equilibrium. And a deviation from the stability of a gene pool—and from Hardy-Weinberg equilibrium—usually results in evolution.

CAUSES OF MICROEVOLUTION

Microevolution is a generation-to-generation change in a population's allele frequencies

The Hardy-Weinberg theorem is useful when we are studying evolution because it provides a baseline against which we can compare the allele and genotype frequencies of an evolving population. If the frequencies of alleles or genotypes deviate from values predicted by the Hardy-Weinberg equation, it is usually because the population is evolving.

We can now refine our definition of evolution at the population level: *Evolution is a generation-to-generation change in a population's frequencies of alleles.* Because such change in a gene pool is evolution on the smallest scale, it is referred to more specifically as **microevolution.**

Microevolution is occurring even if the frequencies of alleles are changing for only a single genetic locus. If we track allele and genotype frequencies in a population over a succession of generations, some loci may be at equilibrium while allele frequencies at other loci are changing. Such a population is evolving. For example, our imaginary wildflower population would be evolving if the frequencies of the red-flower and white-flower alleles were changing from generation to generation, even if Hardy-Weinberg equilibrium were maintained for all other genetic loci.

The two main causes of microevolution are genetic drift and natural selection

The main factors that can act to alter the allele frequencies in a population are genetic drift, natural selection, gene flow, and mutation. Of these four, genetic drift and natural selection are more important, but we will look at mutation and gene flow as well. We will see that all four of these causes of microevolution represent departures from the conditions required for Hardy-Weinberg equilibrium. (We will not consider departures from the Hardy-Weinberg assumption of random mating; non-random mating can affect the relative frequencies of homozygous versus heterozygous genotypes, but often has no effect on allele frequencies.)

Natural selection is the only cause of microevolution that generally adapts a population to its environment. The other agents of microevolution may affect populations in positive, negative, or neutral ways. Natural selection always has a positive effect, because selection favors the disproportionate propagation of favorable traits.

Genetic Drift

Flip a coin 1,000 times, and a result of 700 heads and 300 tails would make you very suspicious about that coin. But flip a coin ten times, and an outcome of seven heads and three tails would seem within reason. The smaller the sample, the greater the chance of deviation from an idealized result—an equal number of heads and tails, in the case of a sample of coin tosses. This deviation from the expected result, due to the sample size being finite in size rather than infinite, is called sampling error.

Let's apply coin toss logic to a population's gene pool. If a new generation draws its alleles at random from the previous generation, then the larger the population (the sample size), the more closely the new generation will represent the gene pool of the previous generation. Thus, one requirement for a gene pool to maintain the status quo—Hardy-Weinberg equilibrium—is a large population size (actually infinitely large, an ideal never met). The gene pool of a small population may not be accurately represented in the next generation because of sampling error. It is analogous to the erratic outcome from a small sample of coin tosses.

FIGURE 23.4 applies this concept of sampling error to a small population of wildflowers. Chance causes the frequencies of the alleles for red (*R*) and white (*r*) flowers to change over the generations. And that change fits our definition of microevolution. This evolutionary mechanism, a change in a population's allele frequencies due to chance, is called **genetic drift.** Two situations that can shrink populations down to a small size—small enough for genetic drift to have large effects—are known as the bottleneck effect and the founder effect.

The Bottleneck Effect. Disasters such as earthquakes, floods, droughts, and fires may reduce the size of a population drastically. The small surviving population may not be representative of the original population's gene pool. By chance, certain alleles will be overrepresented among the survivors. Other alleles will be underrepresented. And some alleles may be eliminated altogether. Genetic drift may continue to change the gene pool for many generations until the population is again large enough for sampling errors to be less significant. The analogy in FIGURE 23.5 illustrates why genetic drift due to a drastic reduction in population size is called the **bottleneck effect.**

Bottlenecking usually reduces the overall genetic variability in a population because at least some alleles are likely to be lost from the gene pool. An important example of this concept is the potential loss of individual variation, and hence adaptability, in bottlenecked populations of endangered species, such as the cheetah. The fastest of all running animals, cheetahs are magnificent cats that were once widespread in Africa and Asia. Like many African mammals, their numbers fell drastically during the last ice age, some 10,000 years ago. At that time, the species may have suffered a severe bottleneck, possibly as a result of disease, human hunting, and periodic droughts. Some researchers think that the South African cheetah population suffered a second bottleneck during the 19th century when

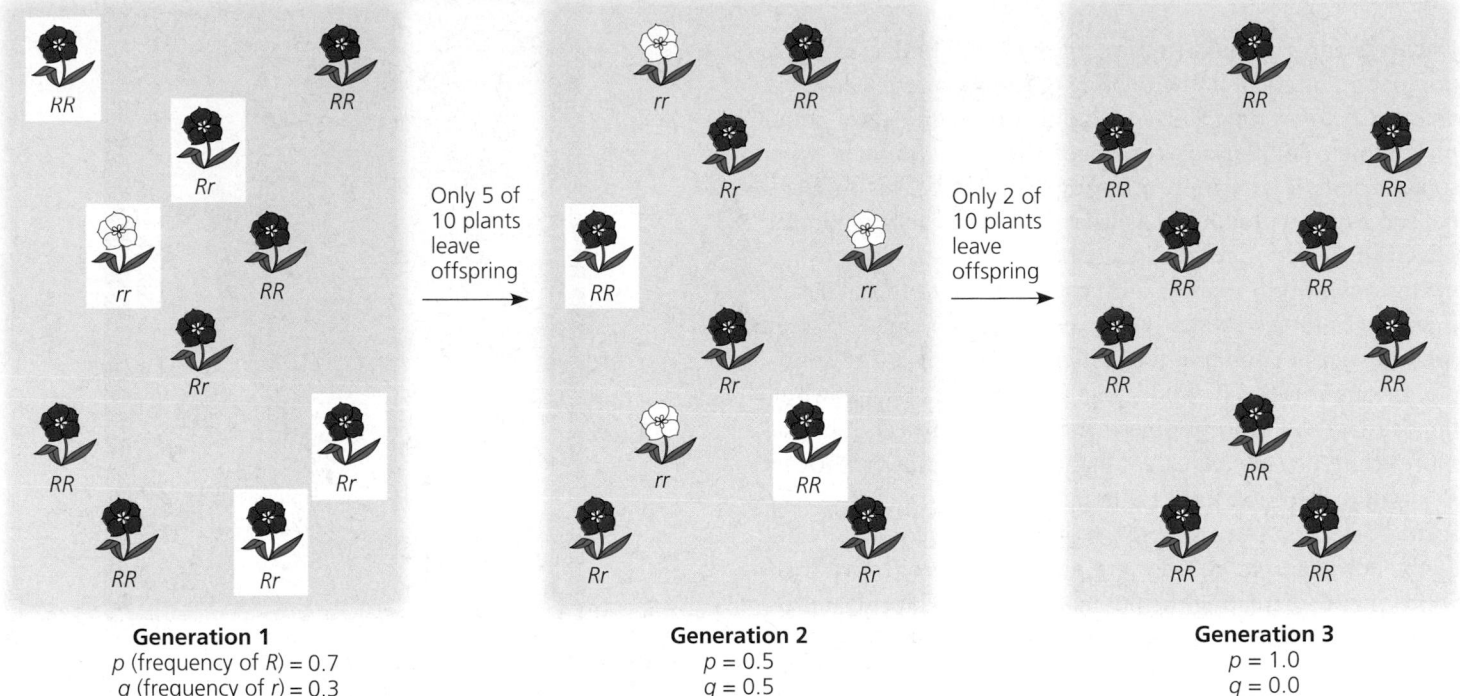

Generation 1
p (frequency of R) = 0.7
q (frequency of r) = 0.3

Only 5 of
10 plants
leave
offspring

Generation 2
p = 0.5
q = 0.5

Only 2 of
10 plants
leave
offspring

Generation 3
p = 1.0
q = 0.0

FIGURE 23.4 Genetic drift. This small wildflower population has a stable size of only ten plants. For generation 1, only the five boxed plants produce fertile offspring. Only two plants of generation 2 manage to leave fertile offspring. Over the generations, genetic drift can completely eliminate some alleles, as is the case for the r allele in generation 3 of this imaginary population.

South African farmers hunted the animals to near extinction. Today, only three small populations of cheetahs exist in the wild. Genetic variability in these populations is very low compared to populations of other mammals. In fact, genetic uniformity in cheetahs rivals that of highly inbred varieties of laboratory mice!

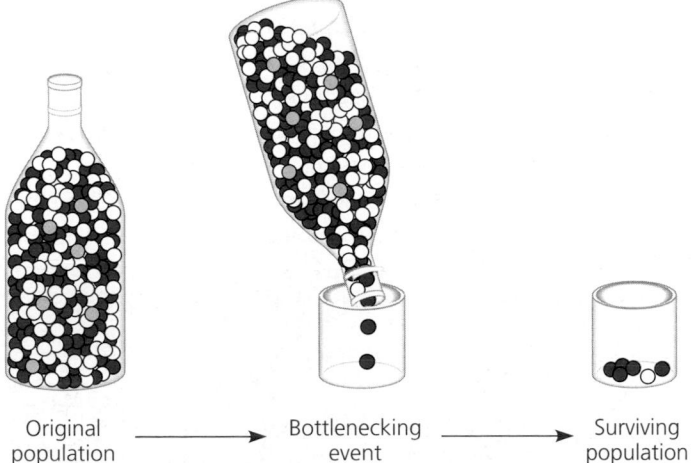

Original
population

Bottlenecking
event

Surviving
population

FIGURE 23.5 The bottleneck effect: an analogy. Shaking just a few of the marbles through the bottleneck is like drastically reducing the size of a population following some environmental disaster. By chance, blue marbles are overrepresented in the new population, and gold marbles are absent. Similarly, bottlenecking a population of organisms tends to reduce variability.

The Founder Effect. Genetic drift is also likely whenever a few individuals from a larger population colonize an isolated island, lake, or some other new habitat. The smaller the sample size, the less the genetic makeup of the colonists will represent the gene pool of the larger population they left. The most extreme case would be the founding of a new population by one pregnant animal or a single plant seed. If the colony is successful, genetic drift will continue to affect the frequency of alleles in the gene pool until the population is large enough for sampling errors from generation to generation to be minimal. Genetic drift in a new colony is known as the **founder effect.**

The founder effect probably accounts for the relatively high frequency of certain inherited disorders among human populations established by a small number of colonists. In 1814, 15 people founded a British colony on Tristan da Cunha, a group of small islands in the Atlantic Ocean midway between Africa and South America. Apparently, one of the colonists carried a recessive allele for retinitis pigmentosa, a progressive form of blindness that afflicts homozygous individuals. Of the 240 descendants who still lived on the island in the late 1960s, four had retinitis pigmentosa, and at least nine others were known to be carriers, based on pedigree analysis. The frequency of this allele is much higher on Tristan da Cunha than in the populations from which the founders came. In addition to inherited diseases, the founder effect also alters the frequencies of many alleles that affect more subtle characteristics.

Natural Selection

Hardy-Weinberg equilibrium requires that all individuals in a population be equal in their ability to survive and produce viable, fertile offspring. This condition is probably never completely met. Populations consist of varied individuals, with some variants leaving more offspring than others. This differential success in reproduction is what Darwin meant by **natural selection.** Selection results in alleles being passed along to the next generation in numbers disproportionate to their relative frequencies in the present generation. One possible cause of such a disproportion might be differential survival. For example, in our imaginary wildflower population, perhaps white flowers (*rr*) are more visible to herbivorous insects, so that more white flowers are eaten. Plants with red flowers (*RR or Rr*) would therefore have more opportunity to produce offspring. Selection can also affect reproductive success more directly. For example, red flowers may be more effective than white ones in attracting the pollinators required for seed production. This difference would disturb Hardy-Weinberg equilibrium; the frequency of the *R* allele would increase in the gene pool, and the frequency of the *r* allele would decline.

Of all the agents of microevolution that change a gene pool, only selection is likely to adapt a population to its environment. Natural selection accumulates and maintains favorable genotypes in a population.

Natural selection and genetic drift cause most of the changes in allele frequencies that we observe in evolving populations. However, allele frequencies can also be changed by migration between populations or by mutation, and in some cases these factors are important.

Gene Flow

A population may gain or lose alleles by **gene flow,** genetic exchange due to the migration of fertile individuals or gametes between populations. Perhaps, for example, a population near our hypothetical wildflower population consists entirely of white-flowered individuals (*rr*). A windstorm may blow pollen from the *rr* population to our wildflower population, and allele frequencies may change in the next generation.

Gene flow tends to reduce differences between populations. If it is extensive enough, gene flow can eventually amalgamate neighboring populations into a single population with a common gene pool. As humans began to move about the world more freely, gene flow undoubtedly became an important agent of microevolutionary change in populations that were previously quite isolated (FIGURE 23.6).

Mutation

A **mutation** is a change in an organism's DNA (see Chapter 17). A new mutation that is transmitted in gametes can im-

FIGURE 23.6 Gene flow and human evolution. The migration of people throughout the world is transferring alleles between populations that were once isolated. This magazine cover celebrates America's changing gene pools and culture with a computer-generated image blending facial features from several immigrant groups.

mediately change the gene pool of a population by substituting one allele for another. For any one gene locus, mutation alone does not have much quantitative effect on a large population in a single generation, because a mutation at any given gene locus is a very rare event. If some new allele produced by mutation increases its frequency by a significant amount in a population, it is not because mutation is generating the allele in abundance, but because individuals carrying the mutant allele are producing a disproportionate number of offspring as a result of natural selection or genetic drift.

Although mutations at a particular gene locus are rare, the cumulative impact of mutations at *all* loci can be significant. This is because each individual has thousands of genes, and many populations have thousands or millions of individuals. Certainly over the long term, mutation is, in itself, very important to evolution because it is the original source of the genetic variation that serves as raw material for natural selection.

GENETIC VARIATION, THE SUBSTRATE FOR NATURAL SELECTION

Heritable variation is at the heart of Darwin's theory of evolution, for variation provides the raw material—the substrate—on which natural selection works.

Genetic variation occurs within and between populations

You have no trouble recognizing your friends in a crowd. Each person has a unique genome, reflected in individual variations of appearance and temperament. Individual variation occurs in populations of all species of sexually reproducing organisms. We are very conscious of human diversity. We are generally less sensitive to individuality in populations of other animals and of plants, and the diversity may escape our notice because the variations are subtle. But these slight differences between individuals in a population are the variations Darwin wrote most about as the raw material for natural selection. And in addition to the differences we can see, populations have extensive variation that can only be observed at the molecular level. For example, you cannot tell a person's ABO blood group (A, B, AB, or O) just by looking at him or her.

Not all the variation we observe in a population is heritable. Phenotype is the cumulative product of an inherited genotype and a multitude of environmental influences. For example, bodybuilders alter their phenotypes dramatically. FIGURE 23.7 illustrates a striking example of environmentally induced variation in a butterfly population. It is important to remember that only the genetic component of variation can have evolutionary consequences as a result of natural selection, because it is the only component that transcends generations.

Variation Within Populations

Both quantitative and discrete characters contribute to variation *within* a population. Most heritable variation consists of *quantitative characters* that vary along a continuum within a population. For example, plant height may vary continuously in our hypothetical wildflower population, from very short individuals to very tall individuals and everything in between. Quantitative variation usually indicates polygenic inheritance, an additive effect of two or more genes on a single phenotypic character (see Chapter 14). *Discrete characters,* such as red versus white flowers, can be classified on an either-or basis, usually because they are determined by a single gene locus with different alleles that affect distinct phenotypes.

Polymorphism. When two or more forms of a discrete character are represented in a population, the different forms are called *morphs*—as in the red-flowered and white-flowered morphs of our wildflower population. A population is said to be **polymorphic** for a character if two or more distinct morphs are each represented in high enough frequencies to be readily noticeable. (Obviously, this definition is arbitrary, but a population is not considered polymorphic if it consists almost exclusively of a single morph, with other morphs extremely rare.) Look again at the photo on page 445 to see a striking example of polymorphism in a snail population. Polymorphism is extensive in human populations, both in physical characters, such as the presence or absence of freckles, and in biochemical characters, such as ABO blood groups (for which there are four morphs: type A, type B, type AB, and type O). Polymorphism applies only to discrete characters, not to characters such as human height, which varies among people in a continuum.

Measuring Genetic Variation. Population geneticists measure genetic variation at both the level of whole genes (gene diversity) and at the molecular level of DNA (nucleotide diversity). Let's consider a population of fruit flies (*Drosophila*). The genome of a fruit fly has about 13,000 loci. The **gene diversity** of *Drosophila* is the average percent of these loci that are heterozygous. On average, a fruit fly is homozygous (has two matching alleles) at about 86% of its loci. In other words, the flies, on average, are heterozygous (have two different alleles) at about 14% of their gene loci. We would therefore say that the fly population has a gene diversity of 14%, meaning that fruit flies are heterozygous at about 1,800 of their 13,000 gene loci and homozygous at all the rest.

Population geneticists measure **nucleotide diversity** by comparing the nucleotide sequences of DNA samples from two individuals and then pooling the data from many such comparisons of two individuals. The fruit fly genome has about 180 million nucleotides, and researchers have measured a nucleotide diversity in a population of about 1%. In other

(a) Map butterflies that emerge in spring: orange and brown

(b) Map butterflies that emerge in late summer: black and white

FIGURE 23.7 A nonheritable difference within a population. These two European insects, called map butterflies (*Araschnia levana*), are seasonal forms of the same species: **(a)** Individuals that emerge in the spring are orange and brown; **(b)** individuals that emerge in late summer are black and white. Seasonal differences in hormones are responsible for these different phenotypes; the two forms you see here are genetically identical at the loci for coloration. Therefore, if these two forms differed in survival and fertility, that in itself would not lead to any change in the frequencies of the two phenotypes in the next generation.

words, on average, two flies from the same population have different nucleotides at about 1.8 million nucleotide sites in their DNA.

Population geneticists have also measured gene diversity and nucleotide diversity in humans. It turns out that we have relatively little genetic variation compared to most other species. Gene diversity is 14% in humans, about the same as in fruit flies. But nucleotide diversity is only about 0.1% in humans, a tenth of the diversity found in fruit fly populations. You and your neighbor have the same nucleotide at 999 out of every 1,000 nucleotide sites in your DNA. We are clearly much more genetically alike than we are different. Still, this is enough genetic variation to account for the hereditary component of the enormous individuality we notice in how people look and act (not to mention the biochemical differences, such as blood group, that are not outwardly visible).

Variation Between Populations

Most species exhibit **geographic variation,** differences in gene pools between populations or subgroups of populations. Because at least some environmental factors are likely to be different from one place to another, natural selection can contribute to geographic variation. For example, perhaps one population of our now-familiar wildflower species has a higher frequency of recessive alleles at the flower-color locus than other populations because of a local prevalence of pollinators that prefer white flowers (recessive homozygotes). Genetic drift can also cause chance variations among different populations.

Geographic variation can also occur on a more local scale—that is, *within* a population—either because the environment has patchlike diversity or because the population is differentiated into subpopulations resulting from the limited dispersal of individuals.

One particular type of geographic variation, called a **cline,** is a graded change in some trait along a geographic axis. In some cases, a cline may represent a graded region of overlap where individuals of neighboring populations are interbreeding. In other cases, a gradation in some environmental variable may produce a cline. For example, the average body size of many North American species of birds and mammals increases gradually with increasing latitude. Presumably, the reduced ratio of surface area to volume that accompanies larger size is an adaptation that helps animals conserve body heat in cold environments. Experimental studies of some clines confirm the role of genetic variation—rather than strictly environmental effects—in the spatial differences of phenotype (FIGURE 23.8).

In contrast to a cline, with its gradation of some characteristic, geographic variation between isolated populations often consists of discrete differences. FIGURE 23.9 illustrates one example, a study of geographic variation in house mice reported by researchers in 2000.

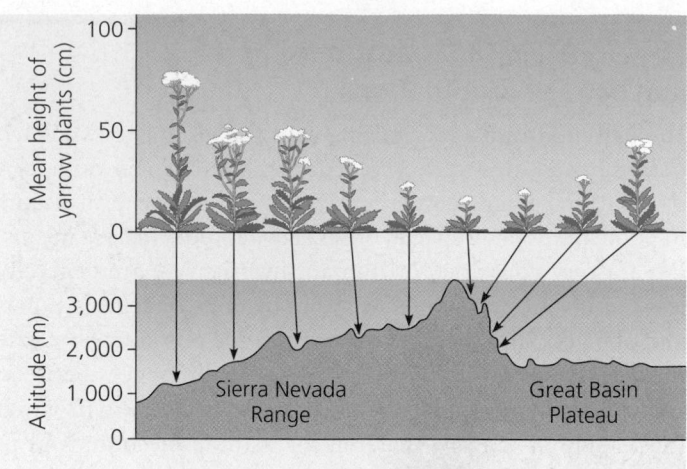

FIGURE 23.8 Clinal variation in a plant. Average size of yarrow plants *(Achillea)* growing on the slopes of the Sierra Nevada mountains gradually decreases with increasing elevation. Although the environment affects growth rates directly to some extent, some of the variation has a genetic basis. Researchers collected seeds at different elevations and grew plants in a common garden; the average sizes of the plants were correlated with the altitude at which the seeds were collected.

THE PROCESS OF SCIENCE

Mutation and sexual recombination generate genetic variation

Two random processes, mutation and sexual recombination (see Chapter 15), create variation in the gene pool of a population.

Mutation

New alleles originate only by mutation, or change in the nucleotide sequence of DNA. A mutation affecting any gene locus is an accident that is rare and random. Most mutations occur in somatic cells and are lost when the individual dies. Only mutations that occur in cell lines that produce gametes can be passed along to offspring.

A mutation is like a shot in the dark: Chance determines where it will strike and how it will alter a gene. Most point mutations, those affecting a single base in DNA, are probably relatively harmless. Much of the DNA in the eukaryotic genome does not code for protein products, and it is uncertain how a change of a single nucleotide base in this silent DNA will affect the well-being of the organism. Even mutations of structural genes, which do code for proteins, may occur with little or no effect on the organism, partly because of redundancy in the genetic code. Of course, a single point mutation *can* have a significant impact on phenotype, as in sickle-cell disease, for example (see FIGURE 5.19).

A mutation that alters a protein enough to affect its function is more often harmful than beneficial. Organisms are the refined products of thousands of generations of past selection, and a random change is not likely to improve the genome any

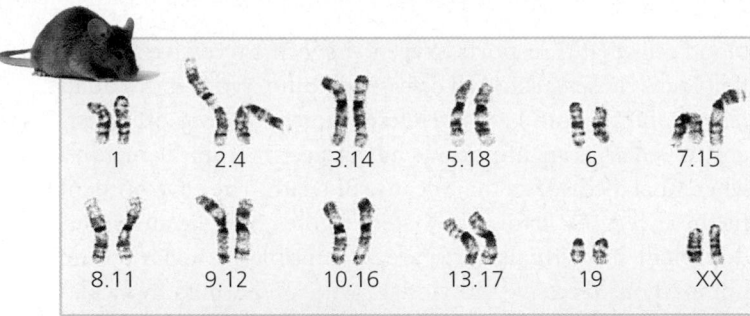

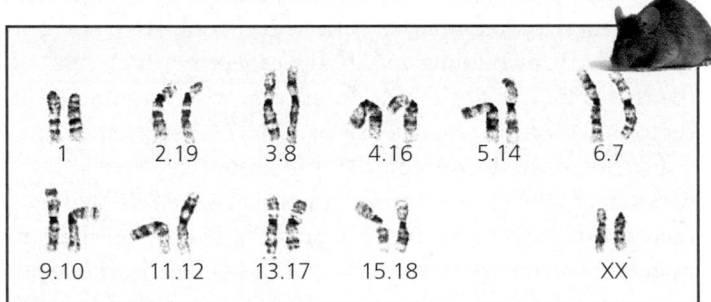

FIGURE 23.9 Geographic variation between isolated populations of house mice. When the Portuguese first settled the small Atlantic island of Madeira during the 15th century, they inadvertently brought house mice *(Mus musculus)* with them. Because of the mountains that separated many of the early Portuguese settlements, several populations of the mice have evolved in isolation from one another. Today, differences can be seen in the karyotypes (chromosome sets) of some of these isolated populations. For example, researchers have found that various chromosomes in the Madeira populations are fused, reducing the chromosome count from the species' standard diploid chromosome number of 40 ($2n = 40$). However, the pattern of fusion in one mouse population is different from the pattern of fusion in another. Mice that live in the areas indicated by the gold dots have the karyotype shown at the top of the figure; mice in the locales with the red dots have the lower karyotype. This geographic variation in karyotypes is probably due mainly to genetic drift.

more than blindly firing a gunshot through the hood of a car is likely to improve engine performance. On rare occasions, however, a mutant allele may actually fit its bearer to the environment better and enhance the reproductive success of the individual. This is not especially likely in a stable environment but becomes more probable when the environment is changing and mutations that were once selected against are now favorable under the new conditions. For example, mutations that happen to endow HIV with resistance to antiviral drugs also slow the reproductive rate of the virus (see Chapter 22). Once the drugs were in the environment, the mutant alleles were favored, and natural selection increased their frequency in HIV populations.

Chromosomal mutations that disrupt many gene loci are almost certain to have negative effects on the development of the organism. However, when chromosomal rearrangements leave genes intact, their effects on organisms may be neutral, as in FIGURE 23.9. Rearrangements of chromosomes may in rare instances bring benefits. For example, the translocation of a chromosomal piece could link alleles that affect the organism in some positive way when they are inherited together as a package.

Duplications of chromosome segments, like other chromosomal mutations, are nearly always harmful. But if the repeated segment does not disrupt genetic balance severely, it can persist over the generations and provide an expanded genome with superfluous loci that may eventually take on new functions by mutation. New genes may also arise from transpositions of existing DNA sequences or by the shuffling of exons within the genome, either within a single locus or between loci (see Chapter 19).

In microorganisms with very short generation spans, mutation generates genetic variation very rapidly. For example, HIV has a generation span of about two days. In an AIDS patient, the HIV infection produces 10^{10} or more new viruses per day. Each replication provides a chance for errors—mutations—to occur. In addition, HIV has an RNA genome, which has a much higher mutation rate than DNA genomes. The combined mutation and replication rates mean that in a single day, the HIV population in one human body will generate mutations in every site in the HIV genome. For this reason, single-drug treatments will probably never be effective for long against HIV. Even double-drug treatments are rarely effective for long, because individual viruses with double mutations conferring resistance to *both* drugs arise daily. This explains why the most effective AIDS treatments are drug "cocktails," combinations of several drugs. Compared to single and double mutations, it is far less probable that multiple mutations against *all* the drugs will turn up in individual viruses in a short time period.

Bacterial populations can also evolve rapidly by the explosive asexual expansion of mutant clones favored by the local environment. However, on a generation-to-generation time scale, animals and plants depend mainly on sexual recombination for the genetic variation that makes adaptation possible.

Sexual Recombination

Members of a sexually reproducing population owe nearly all their genetic differences to the unique recombinations of existing alleles each individual receives from the gene pool. (Of course, this allele variation has its ultimate basis in past mutations.)

Sex shuffles alleles and deals them at random to determine individual genotypes. During meiosis, homologous chromosomes, one inherited from each parent, trade some of their genes by crossing over, and then the homologous chromosomes and the alleles they carry segregate randomly into separate gametes. Gametes from one individual vary extensively in their genetic makeup, and each zygote made by a mating pair has a unique assortment of alleles resulting from the random union of a sperm and an ovum (see Chapter 13). A population, of course, contains a vast number of possible mating combinations, each bringing together the gametes of individuals that are likely to have different genetic backgrounds. Sexual reproduction recombines old alleles into fresh assortments every generation.

Diploidy and balanced polymorphism preserve variation

What prevents natural selection from extinguishing a population's variation by culling unfavorable genotypes? The tendency for natural selection to reduce variation is countered by mechanisms that preserve or restore variation.

Diploidy

The diploid nature of most eukaryotes hides a considerable amount of genetic variation from selection in the form of recessive alleles in heterozygotes. Recessive alleles that are less favorable than their dominant counterparts, or even harmful in the present environment, can persist in a population through their propagation by heterozygous individuals. This latent variation is exposed to selection only when both parents carry the same recessive allele and combine two copies in one zygote. This happens only rarely if the frequency of the recessive allele is very low. The rarer the recessive allele, the greater the degree of protection from natural selection. Heterozygote protection maintains a huge pool of alleles that may not be suitable for present conditions but that could bring new benefits when the environment changes.

Balanced Polymorphism

Selection itself may preserve variation at some gene loci. This ability of natural selection to maintain stable frequencies of two or more phenotypic forms in a population is called **balanced polymorphism.** Natural selection preserves variation by two mechanisms. One such mechanism is **heterozygote advantage.** If individuals who are heterozygous at a particular locus have greater survivorship and reproductive success than any type of homozygote, then two or more alleles will be maintained at that locus by natural selection.

An example of heterozygote advantage is seen at the locus in humans for one chain of hemoglobin, the protein of red blood cells that transports oxygen. A specific recessive allele at that locus causes sickle-cell disease in homozygous individuals (see FIGURES 5.19 and 14.15). Heterozygotes, however, are resistant to malaria, an important advantage in tropical regions, where that disease is a major cause of death. The environment in these regions favors the heterozygotes over homozygous dominant individuals, who are susceptible to malaria, and homozygous recessive individuals, who are harmed by sickle-cell disease. The frequency of the sickle-cell allele in Africa is generally highest in areas where the malaria parasite is most common (FIGURE 23.10). In some tribes, the recessive allele accounts for 20% of the hemoglobin loci in the gene pool, a very high frequency for an allele that is disastrous in homozygotes. But at this frequency ($q = 0.2$), 32% of the population consists of heterozygotes resistant to malaria ($2pq$), and only 4% of the population suffers from sickle-cell disease (q^2).

A second mechanism promoting balanced polymorphism is **frequency-dependent selection,** in which the survival and reproduction of any one morph declines if that phenotypic form becomes too common in the population. The relationships between parasites and their hosts may involve such frequency-dependent selection. Internal parasites generally recognize cells of their hosts by binding specifically to receptor molecules on the host cells. Imagine a host population with variation in receptor molecules for a particular parasite. Let's say that an individual host has either receptor A or receptor B on its cells. If the parasite recognizes receptor A as a signal to penetrate host cells, then natural selection will favor those individuals in the host population that have receptor B—the individuals less susceptible to infection by this parasite. The frequency of individuals with B receptors increases in the population. But the parasite population is also evolving by natural selection. Parasites that

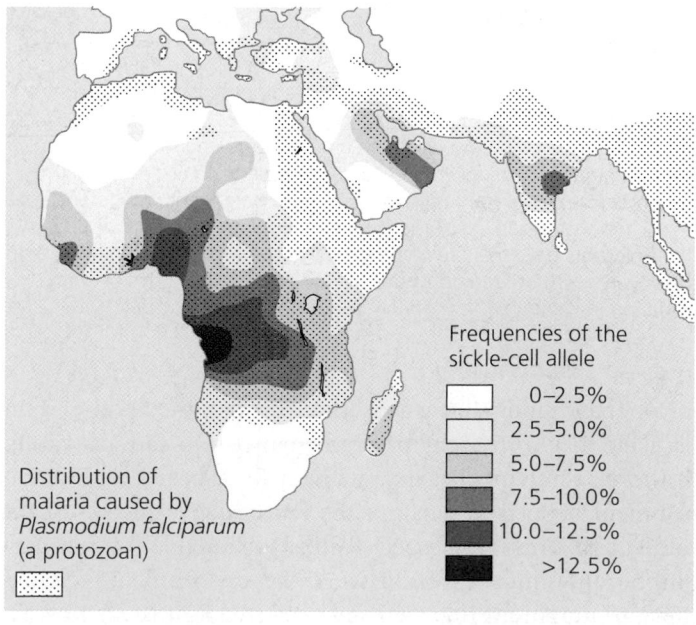

Distribution of malaria caused by *Plasmodium falciparum* (a protozoan)

Frequencies of the sickle-cell allele

- 0–2.5%
- 2.5–5.0%
- 5.0–7.5%
- 7.5–10.0%
- 10.0–12.5%
- >12.5%

FIGURE 23.10 Mapping malaria and the sickle-cell allele.

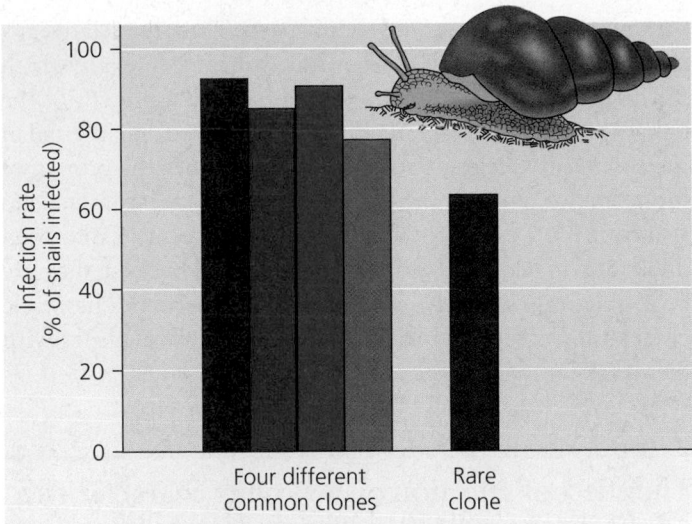

FIGURE 23.11 Frequency-dependent selection in a host-parasite relationship. The bar graph compares the rates at which a parasitic worm infects five different clones that researchers collected from Lake Poerua in New Zealand. (This snail population exists as diverse clones because the snails reproduce asexually.) Infection rates are higher for common clones than for a less common clone, suggesting that frequency-dependent selection helps maintain polymorphism in the snail population.

recognize what is now the most frequent variant of hosts, those with the B receptor, become favored and their frequency increases. This adaptation stimulates a new counteradaptation of the host population, with individuals having the A receptor again gaining advantage. The teeter-totter effect of frequency-dependent selection helps maintain a repertoire of variants in populations. In the year 2000, researchers reported an example of such frequency-dependent selection in populations of snails infected by a parasitic worm (FIGURE 23.11).

Neutral Variation

Some of the genetic variations observed in populations are probably trivial in their impact on reproductive success. The diversity of human fingerprints is an example of what is called **neutral variation,** which seems to confer no selective advantage for some individuals over others. Much of the protein and DNA variation detectable by methods such as electrophoresis may represent chemical "fingerprints" that are neutral in their adaptive qualities. The relative frequencies of neutral variations will not be affected by natural selection; some neutral alleles will increase in the gene pool and others will decrease by the chance effects of genetic drift.

There is no consensus among evolutionary biologists on how much genetic variation is neutral or even if any variation can be considered truly neutral. Variations that appear to be neutral may in fact influence survival and reproductive success in ways that are difficult to measure. It is possible to show that a particular allele is detrimental, but it is impossible to demonstrate

that an allele brings no benefits at all to an organism. Furthermore, a variation may be neutral in one environment but not in another. We can never know the degree to which genetic variation is neutral. But we can be certain that even if only a fraction of the extensive variation in a gene pool significantly affects the organisms, that is still an enormous reservoir of raw material for natural selection and the adaptive evolution it causes.

A CLOSER LOOK AT NATURAL SELECTION AS THE MECHANISM OF ADAPTIVE EVOLUTION

Adaptive evolution is a blend of chance and sorting—chance in the creation of new genetic variations by mutation and sexual recombination, and sorting in the workings of selection as it favors the propagation of some chance variations over others. From the range of variations available to it, natural selection increases the frequencies of certain genotypes and fits organisms to their environments. In this section, we take a closer look at natural selection as the mechanism of evolutionary adaptation.

Evolutionary fitness is the relative contribution an individual makes to the gene pool of the next generation

The phrases "struggle for existence" and "survival of the fittest" are misleading if we take them to mean direct competitive contests among individuals. There *are* animal species in which individuals, usually the males, lock horns or otherwise do combat to determine mating privilege. But reproductive success is generally more subtle and passive. A barnacle may produce more eggs than its neighbors because it is more efficient at collecting food from the water. In a population of moths, certain variants may average more offspring than others because their body and wing colors hide them from predators better. Plants in a wildflower population may differ in reproductive success because some are better able to attract pollinators, owing to slight variations in flower color, shape, or fragrance. These examples point to a biological definition of fitness: *Darwinian fitness is the contribution an individual makes to the gene pool of the next generation relative to the contributions of other individuals.*

In a more quantitative approach to natural selection, population geneticists define **relative fitness** as the contribution of a genotype to the next generation compared to the contributions of alternative genotypes for the same locus. For example, consider our wildflower population, in which *RR* and *Rr* plants have red flowers and *rr* plants have white flowers. Let's assume that, on average, individuals with red flowers produce more offspring than those with white flowers. The relative fitness of the most reproductively successful variants is set at 1 as a basis for comparison; so in this case, the relative fitness of an *RR* or *Rr* plant is 1. If plants with white flowers average only 80% as many offspring, their relative fitness is 0.8.

Survival alone does not guarantee reproductive success. Relative fitness is zero for a sterile plant or animal, even if it is robust and outlives other members of the population. But, of course, survival is a prerequisite for reproducing, and longevity increases fitness if it results in certain individuals leaving more descendants than other individuals leave. Then again, an individual that matures quickly and becomes fertile at an early age may have a greater reproductive potential than individuals that live longer but mature late. Thus, many factors that affect both survival and fertility determine an individual's evolutionary fitness.

An organism exposes its phenotype—its physical traits, metabolism, physiology, and behavior—not its genotype, to the environment. Acting on phenotypes, selection indirectly adapts a population to its environment by increasing or maintaining favorable genotypes in the gene pool.

The entity subjected to natural selection is the whole organism, which is an integrated composite of its many phenotypic features, not a collage of individual parts. Thus, the relative fitness of an allele depends on the entire genetic context in which it works. For example, alleles that enhance the growth of the trunk and limbs of a tree may be useless or even detrimental in the absence of alleles at other loci that enhance the growth of roots required to support the tree. On the other hand, alleles that contribute nothing to an organism's success, or may even be slightly maladaptive, may be perpetuated because they are present in individuals whose overall fitness is high. The whole baseball team wins the league pennant, even the player with the worst batting average and the most errors.

The effect of selection on a varying characteristic can be directional, diversifying, or stabilizing

Natural selection can affect the frequency of a heritable trait in a population in three different ways, depending on which phenotypes in a varying population are favored. These three selection trends are directional selection, diversifying selection, and stabilizing selection (FIGURE 23.12).

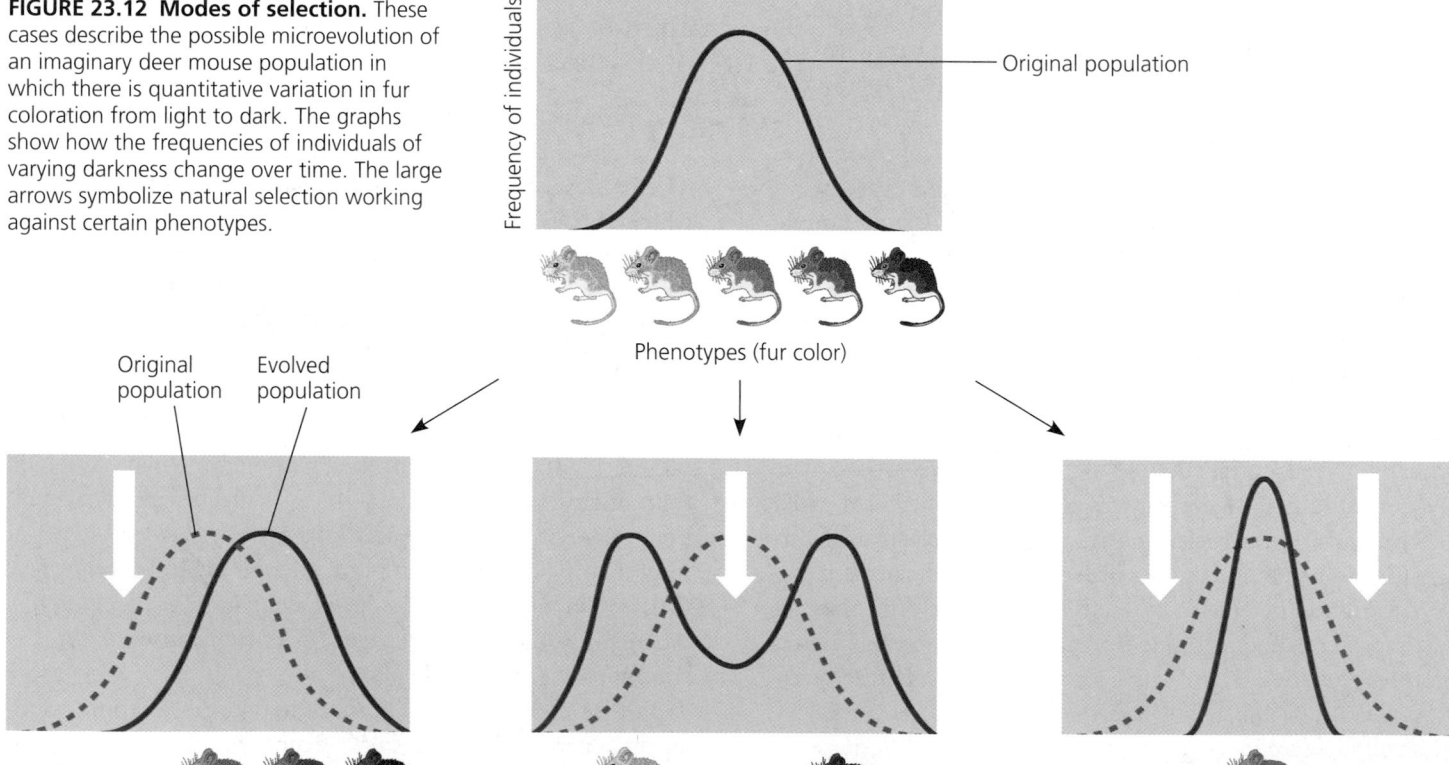

FIGURE 23.12 Modes of selection. These cases describe the possible microevolution of an imaginary deer mouse population in which there is quantitative variation in fur coloration from light to dark. The graphs show how the frequencies of individuals of varying darkness change over time. The large arrows symbolize natural selection working against certain phenotypes.

Frequency of individuals →

Original population

Phenotypes (fur color)

Original population — Evolved population

(a) Directional selection shifts the overall makeup of the population by favoring variants of one extreme. In this case, the trend is toward darker color, perhaps because the landscape has been shaded by the growth of trees.

(b) Diversifying selection favors variants of opposite extremes. Here, the relative frequencies of very light and dark mice have increased. Perhaps the mice colonized a patchy habitat where light soil is studded with dark rocks.

(c) Stabilizing selection culls extreme variants from the population, in this case eliminating individuals that are unusually light or dark. The trend is toward reduced phenotypic variation and maintenance of the status quo.

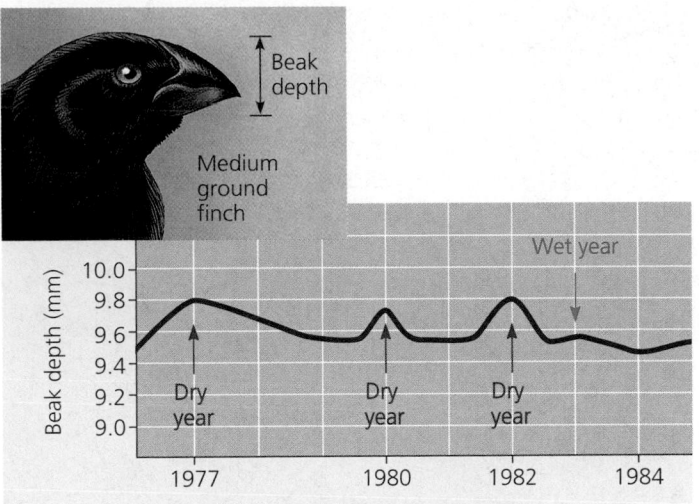

THE PROCESS OF SCIENCE

FIGURE 23.13 Directional selection for beak size in a Galápagos population of the medium ground finch.

FIGURE 23.14 Diversifying selection in a finch population. Two distinctly different beak sizes occur in a single population of black-bellied seedcrackers, a species of finch that lives in Cameroon, West Africa. Small-billed individuals (left) feed mainly on soft seeds, whereas large-billed birds specialize in cracking hard seeds. A reasonable hypothesis is that natural selection selects against intermediate-sized bills, which crack both classes of seeds relatively inefficiently.

Directional selection is most common during periods of environmental change or when members of a population migrate to some new habitat with different environmental conditions. Directional selection shifts the frequency curve for variations in some phenotypic character in one direction or the other by favoring what are initially relatively rare individuals that deviate from the average for that character. For instance, fossil evidence indicates that the average size of black bears in Europe increased with each glacial period of the ice ages, only to decrease again during the warmer interglacial periods. Over much shorter time intervals, Peter and Rosemary Grant documented directional evolution of beak size in a population of finches on Daphne Major in the Galápagos Islands (see interview on pages 426–427). The medium ground finch (*Geospiza fortis*) uses its strong beak to crush seeds. Given a choice of small seeds or large seeds, the birds eat mostly small ones, which are easier to crush. During wet years, small seeds are produced in such abundance that ground finches consume relatively few large seeds. However, during dry years all seeds are in short supply, and the birds eat proportionally more large seeds. This change in diet is correlated with a change in the average depth (top-to-bottom dimension) of the birds' beaks in the subsequent generation. This trait is inherited rather than acquired (by exercising the beak on large seeds, for example). The most likely explanation is that those birds that happen to have stronger beaks have a feeding advantage and so have greater reproductive success during droughts. Thus, they pass the genes for thicker beaks on to their offspring (FIGURE 23.13).

Diversifying selection occurs when environmental conditions are varied in a way that favors individuals on both extremes of a phenotypic range over intermediate phenotypes (FIGURE 23.14).

Stabilizing selection acts against extreme phenotypes and favors the more common intermediate variants. This mode of selection reduces variation and maintains the status quo for a particular phenotypic character. For example, stabilizing selection keeps the majority of human birth weights in the range of 3–4 kg. For babies much smaller or larger than this, infant mortality is greater.

Although we refer to these three selection trends as "modes of selection," the basic mechanism of natural selection is the same for each case. Selection favors certain heritable traits via differential reproductive success.

Natural selection maintains sexual reproduction

Sex is an evolutionary enigma. It is far inferior to asexual reproduction as measured by reproductive output. Consider, for example, a population of insects in which half the females reproduce only sexually and half reproduce only asexually. Even if both types of females produced the same number of offspring each generation, the asexual condition would increase in frequency because all of the females' offspring would be daughters that would themselves produce more reproductive daughters. In contrast, half of the offspring of the sexual females would be males, which would be required for reproduction but would themselves produce no offspring. FIGURE 23.15 (p. 460) diagrams this "twofold disadvantage" of sex.

Clearly, sex confers some benefit in addition to reproduction; otherwise, sexual individuals would soon be outcompeted in a population whenever asexual individuals appeared by mutation or migration. That is, were it not for some sexual bonus in Darwinian fitness, natural selection would act against alleles associated with sexual reproduction in favor of alleles that enable individuals to reproduce asexually. In fact,

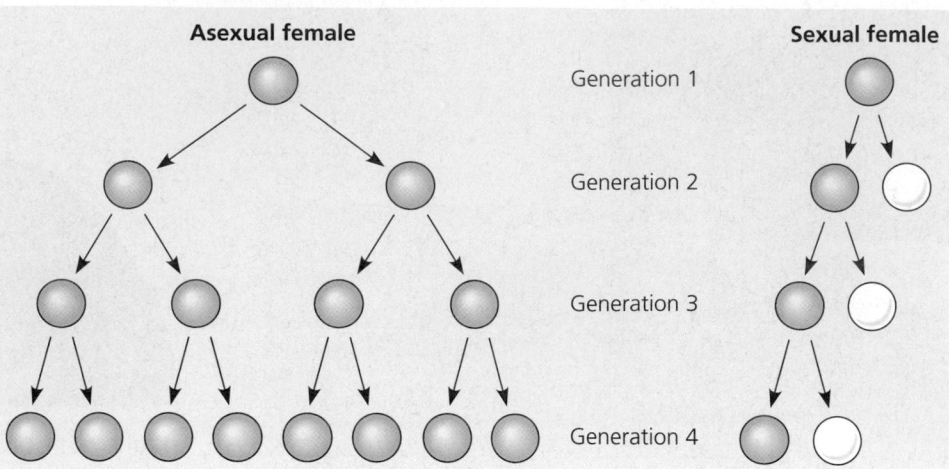

FIGURE 23.15 The twofold disadvantage of sex. These diagrams contrast the reproductive output of females (blue circles) over four generations for asexual versus sexual reproduction. In the sexual version, males, which do not produce offspring, are symbolized by the white circles. (The diagrams assume two offspring per female.)

Asexual female

Sexual female

Generation 1

Generation 2

Generation 3

Generation 4

sex is maintained in the vast majority of eukaryotic species, even in those that can also reproduce asexually.

The "textbook" explanation for the advantage of sex is that the processes of meiosis and fertilization generate the genetic variation upon which natural selection can act as the agent of adaptation. (For this textbook's explanation, see Chapter 13.) But the notion that natural selection sustains sex in spite of its reproductive disadvantage because genetic variation enables future adaptation in an unpredictable world is an idea that is difficult to defend. Natural selection acts in the here and now, favoring the reproductive success of those individuals of a varying population that work best in the current, local environment. The most attractive hypotheses for how natural selection maintains sex are those that place value on genetic variation over the short term—on a generation-to-generation time scale.

One such hypothesis emphasizes the importance of genetic variation in resistance to disease. As we discussed earlier, many pathogens such as viruses, bacteria, and parasites recognize and infect their specific hosts by attaching to receptor molecules on the host's cells. There should be an advantage in producing offspring that vary in their resistance to different diseases. For example, one offspring variant may have cellular markers that make it resistant to virus A, while another may be resistant to virus B. This hypothesis predicts that the diversity of alleles would be particularly extensive at those gene loci that code for the molecules on cell surfaces to which pathogens attach. And that seems to be the case. In humans, for example, there are hundreds of alleles for each of two gene loci that code for the proteins that give cell surfaces their molecular fingerprints (you'll learn more about these cellular markers in Chapter 43). Of course, because most pathogens evolve very rapidly in their ability to key on specific host receptors, resistance of a particular host genotype to a given disease is not permanent. But sex provides a mechanism for "changing the locks" and varying

them among offspring. Some biologists call this coevolution in a host-parasite relationship the "Red Queen effect," referring to the Red Queen in Lewis Carroll's *Through the Looking Glass*, who admonished Alice that she would have to run as fast as she could to keep from going backward.

Sexual selection may lead to pronounced secondary differences between the sexes

The males and females of an animal species obviously differ in their reproductive organs. But in many species, there are also marked differences, called secondary sexual characteristics, that are not directly associated with reproduction. This distinction in appearance is called **sexual dimorphism.** It is often manifested in a size difference, usually one in which males are larger than females. But sexual dimorphism can also be evident in such features as colorful plumage of male birds, manes on male lions, antlers on male deer, and other adornments. In fact, males are usually the showier sex in most cases of sexual dimorphism, at least among vertebrates.

Sexual dimorphism is a product of what Darwin called sexual selection. Today, biologists distinguish between *intra*sexual selection and *inter*sexual selection. Meaning selection "within the same sex," **intrasexual selection** is a direct competition among individuals of one sex (usually the males in vertebrates) for mates of the opposite sex. Males may use secondary sexual equipment such as antlers to battle competitors. This is especially common in species where a single male garners a harem of females. These males may gain their status by defeating smaller, weaker, or less fierce males in combat; more often, they are the victors in ritualized displays that discourage would-be competitors (see Chapter 51).

Intrasexual selection is a familiar sight in nature films, where rams butt heads, and in cartoons of lion kings claiming their thrones. However, intersexual selection is probably more common and, in a way, more intriguing. In **intersexual selection,** also called mate choice, individuals of one sex (usually females) are choosy in selecting their mates from individuals of the other sex. Apparently, males with the most impressive masculine features are the most attractive to females. A peacock strutting in front of hens with his tail feathers spread into a showy fan is an example of this "choose *me*" statement (FIGURE 23.16). What intrigued Darwin about such behavior is that some of the features that appear to help attract mates do not seem to be adaptive in any other way and may in fact pose some risk in natural environments. For example, showy plumage may make male birds more visible to predators. But if such secondary sexual characteristics help a male gain a mate, then they will be reinforced over the generations for the most Darwinian of reasons—because they enhance reproductive success. Every time a female chooses a mate based on a certain appearance or behavior, she perpetuates the alleles that caused her to make that choice and allows a male with a particular phenotype to perpetuate his alleles.

THE FAR SIDE® By GARY LARSON

"Don't encourage him, Sylvia."

FIGURE 23.16 The Far Side of sexual selection.

When females choose mates, what is the benefit of focusing on secondary sexual characteristics such as showy tail fans? It is unlikely that this is some kind of aesthetic preference. Several researchers are now testing the hypothesis that these sexual advertisements reflect the general health of a male. For example, male birds with serious infections may have relatively dull, disheveled plumage. They don't usually win many females. And for the female that chooses a healthy mate, even if the inclination is just a prewired response to visual signals, the benefit is a greater probability of having healthy offspring.

Natural selection cannot fashion perfect organisms

There are at least four reasons why natural selection cannot produce perfection.

1. *Evolution is limited by historical constraints.* Each species has a legacy of descent with modification from a long line of ancestral forms. Evolution does not scrap ancestral anatomy and build each new complex structure from scratch, but co-opts existing structures and adapts them to new situations. For example, the excruciating back problems some humans endure result in part because the skeleton and musculature modified from the anatomy of four-legged ancestors are not fully compatible with upright posture.

2. *Adaptations are often compromises.* Each organism must do many different things. A seal spends part of its time on rocks; it could probably walk better if it had legs instead of flippers, but it would not swim nearly as well. We humans owe much of our versatility and athleticism to our prehensile hands and flexible limbs, which also make us prone to sprains, torn ligaments, and dislocations; structural reinforcement has been compromised for agility.

3. *Not all evolution is adaptive.* Chance probably affects the genetic structure of populations to a greater extent than was once believed. For instance, when a storm blows insects hundreds of miles over an ocean to an island, the wind does not necessarily transport the specimens that are best suited to the new environment. Thus, not all alleles fixed by genetic drift in the gene pool of the small founding population are better suited to the environment than alleles that are lost.

4. *Selection can only edit existing variations.* Natural selection favors only the fittest variations from the phenotypes that are available, which may not be the ideal traits. New alleles do not arise on demand.

With all these constraints, we cannot expect evolution to craft perfect organisms. Natural selection operates on a "better than" basis. We can see evidence for evolution in the subtle imperfections of the organisms it produces.

CHAPTER 23 REVIEW

Go to the Campbell Biology website (www.campbellbiology.com) to explore an interactive version of the Chapter Review.

Summary of Key Concepts

POPULATION GENETICS

- The modern evolutionary synthesis integrated Darwinian selection and Mendelian inheritance (p. 446) The modern synthesis focuses on populations as units of evolution.

- A population's gene pool is defined by its allele frequencies (pp. 446–447) A population, a localized group of organisms belonging to the same species, is united by its gene pool, the aggregate of all alleles in the population.

- The Hardy-Weinberg theorem describes a nonevolving population (pp. 447–449, FIGURE 23.3) The frequencies of alleles in a population will remain constant if Mendelian segregation is the only process that affects the gene pool. If p and q represent the relative frequencies of the dominant and recessive alleles of a two-allele locus, respectively, then $p^2 + 2pq + q^2 = 1$, where p^2 and q^2 are the frequencies of the homozygous genotypes and $2pq$ is the frequency of the heterozygous genotype. For Hardy-Weinberg equilibrium to apply, the population must be very large, be totally isolated, have no net mutations, show random mating, and have equal reproductive success for all individuals.

Web/CD Case Study in the Process of Science: *How Can Frequency of Alleles Be Calculated?*
Biology Labs On-Line: *PopulationGeneticsLab*

CAUSES OF MICROEVOLUTION

- Microevolution is a generation-to-generation change in a population's allele frequencies (p. 450) Microevolution can occur when one or more of the conditions required for Hardy-Weinberg equilibrium are not met.

Web/CD Activity 23A: *Causes of Microevolution*

- The two main causes of microevolution are genetic drift and natural selection (pp. 450–452, FIGURES 23.4, 23.5) Natural selection and chance effects, called genetic drift, can change allele frequencies. Migration and mutation also influence allele frequencies in a population.

GENETIC VARIATION, THE SUBSTRATE FOR NATURAL SELECTION

- Genetic variation occurs within and between populations (pp. 453–454, FIGURES 23.7–23.9) Genetic variation includes individual variation in discrete and quantitative characters within a population, as well as geographic variation between populations.

Biology Labs On-Line: *EvolutionLab*

- Mutation and sexual recombination generate genetic variation (pp. 454–456) Most mutations have no effect or are harmful, but some are adaptive. Sexual recombination produces most of the genetic variation that makes adaptation possible in populations of sexually reproducing organisms.

Web/CD Activity 23B: *Genetic Variation from Sexual Recombination*

- Diploidy and balanced polymorphism preserve variation (pp. 456–457, FIGURES 23.10, 23.11) Diploidy maintains a reservoir of latent variation in heterozygotes. Balanced polymorphism may maintain variation at some gene loci as a result of heterozygote advantage or frequency-dependent selection.

A CLOSER LOOK AT NATURAL SELECTION AS THE MECHANISM OF ADAPTIVE EVOLUTION

- Evolutionary fitness is the relative contribution an individual makes to the gene pool of the next generation (pp. 457–458) One genotype has a greater relative fitness than another if it leaves more descendants. Selection favors certain genotypes in a population by acting on the phenotype of individual organisms.

- The effect of selection on a varying characteristic can be directional, diversifying, or stabilizing (pp. 458–459, FIGURE 23.12) Natural selection can favor relatively rare individuals on one end of the phenotypic range (directional selection), can favor individuals at both extremes of the range over intermediate phenotypes (diversifying selection), or can act against extreme phenotypes (stabilizing selection).

- Natural selection maintains sexual reproduction (pp. 459–460, FIGURE 23.15) Enhanced disease resistance based on genetic variation may help explain how sex can overcome its twofold disadvantage compared to asexual reproduction.

- Sexual selection may lead to pronounced secondary differences between the sexes (pp. 460–461) Sexual selection leads to the evolution of secondary sex characteristics, which can give individuals an advantage in mating.

- Natural selection cannot fashion perfect organisms (p. 461) Structures result from modified ancestral anatomy; adaptations are often compromises; the gene pool can be affected by genetic drift; and natural selection can act only on available variation.

Self-Quiz

1. A gene pool consists of
   a. all the alleles exposed to natural selection.
   b. the total of all alleles present in a population.
   c. the entire genome of a reproducing individual.
   d. the frequencies of the alleles for a gene locus within a population.
   e. all the gametes in a population.

2. In a population with two alleles for a particular locus, B and b, the allele frequency of B is 0.7. What would be the frequency of heterozygotes if the population were in Hardy-Weinberg equilibrium?
   a. 0.7
   b. 0.49
   c. 0.21
   d. 0.42
   e. 0.09

3. In a population in Hardy-Weinberg equilibrium, 16% of the individuals show the recessive trait. What is the frequency of the dominant allele in the population?
   a. 0.84
   b. 0.36
   c. 0.6
   d. 0.4
   e. 0.48

4. The average length of jackrabbit ears decreases the farther north the rabbits live. This variation is an example of
   a. a cline.
   b. discrete variation.
   c. polymorphism.
   d. genetic drift.
   e. diversifying selection.

5. Which of the following is an example of polymorphism in humans?
   a. variation in height
   b. variation in intelligence
   c. the presence or absence of a widow's peak (see FIGURE 14.14a)
   d. variation in the number of fingers
   e. variation in fingerprints

6. Selection acts *directly* on
   a. phenotype.
   b. genotype.
   c. the entire genome.
   d. each allele.
   e. the entire gene pool.

7. Male swallows with the longest tails were found to attract more mates than those with shorter tails. This observation is an example of
   a. genetic drift, in which tail length increases as a result of small population size.
   b. selection for sexual reproduction, through which high genetic variation is maintained within a population.
   c. intersexual selection in which females are more likely to choose mates with long tails.
   d. intrasexual selection in which males with the longest tails are most successful in fighting with other males over access to females.
   e. directional selection in which longer tail length increases flying ability and thus ability to forage for food over longer distances.

8. Most of the variation we see in coat coloration and pattern in a population of wild mustangs in any generation is probably due to
   a. new mutations that occurred in the preceding generation.
   b. sexual recombination of alleles.
   c. genetic drift due to the small size of the population.
   d. geographic variation within the population.
   e. environmental effects.

9. A founder event favors microevolution in the founding population mainly because
   a. mutations are more common in a new environment.
   b. a small founding population is subject to extensive sampling error in the composition of its gene pool.
   c. the new environment is likely to be patchy, favoring diversifying selection.
   d. gene flow increases.
   e. members of a small population tend to migrate.

10. In a particular bird species, individuals with average-sized wings survive severe storms more successfully than other birds in the same population with longer or shorter wings. This illustrates
    a. the founder effect.
    b. stabilizing selection.
    c. artificial selection.
    d. gene flow.
    e. diversifying selection.

11. Compare and contrast the bottleneck effect and the founder effect as causes of genetic drift.

12. Why might new diseases pose a greater threat to cheetah populations than to mammalian populations having more genetic variation?

13. Which mechanism of microevolution has been most affected by the ease of human travel resulting from new modes of transportation?

14. What is the best measure of Darwinian fitness?

15. Explain what is meant by the "twofold disadvantage of sex."

16. What is incorrect about describing evolution by natural selection as a random process?

Go to the website or CD-ROM for more quiz questions.

Evolution Connection

In what sense is evolutionary history inherent in the imperfections of living organisms?

The Process of Science

Let's return to the wildflowers we used to illustrate the Hardy-Weinberg theorem. The frequency of *R*, the dominant allele for red flowers, is 0.8, and the frequency of *r*, the recessive allele for white flowers, is 0.2. In one starting population, the frequencies of genotypes do not conform to Hardy-Weinberg equilibrium: 60% of the plants are *RR* and 40% of the plants are *Rr*. (At this point, the population has no plants with white flowers.) Assuming that all conditions for the Hardy-Weinberg theorem are met, prove that genotypes will reach equilibrium in the next generation.

Estimate frequencies of alleles in the Case Study in the Process of Science, available on the website and CD-ROM. Also link to the Population-GeneticsLab and EvolutionLab at the Biology Labs On-Line website.

Science, Technology, and Society

To what extent are humans in a technological society exempt from natural selection? Explain your answer.

Answers: 1. b; 2. d; 3. c; 4. a; 5. c; 6. a; 7. c; 8. b; 9. b; 10. b 11. Both processes result in populations being small enough for significant sampling error in the gene pool. A bottleneck reduces the size of an existing population. The founder effect is a new, small population consisting of individuals from a larger population. 12. Because cheetah populations have so little variation, there is the potential for some new disease against which no individuals are resistant. 13. Gene flow 14. The number of fertile offspring an individual leaves 15. Only *half* of the offspring (females) of a sexual population produce offspring, versus *all* offspring being reproductive in an asexual population. 16. While genetic variation depends on the chance processes of mutation and Mendelian segregation, the selection of which variations are most successful depends on specific environmental factors.

THE ORIGIN OF SPECIES

WHAT IS A SPECIES?

- The biological species concept emphasizes reproductive isolation
- Prezygotic and postzygotic barriers isolate the gene pools of biological species
- The biological species concept has some major limitations
- Biologists have proposed several alternative concepts of species

MODES OF SPECIATION

- Allopatric speciation: Geographic barriers can lead to the origin of species
- Sympatric speciation: A new species can originate in the geographic midst of the parent species
- The punctuated equilibrium model has stimulated research on the tempo of speciation

FROM SPECIATION TO MACROEVOLUTION

- Most evolutionary novelties are modified versions of older structures
- "Evo-devo": Genes that control development play a major role in evolution
- An evolutionary trend does not mean that evolution is goal oriented

When Darwin *saw that the geologically young Galápagos Islands had already become populated with many plants and animals known nowhere else in the world, he realized that he was visiting a place of genesis. The islands are named for the giant tortoises, such as the* Geochelone elephantopus *shown above, that are among the unique inhabitants. (Galápago is the Spanish word for tortoise.) After visiting the Galápagos, Darwin wrote in his diary: "Both in space and time, we seem to be brought somewhat near to that great fact—that mystery of mysteries—the first appearance of new beings on this Earth." The beginning of new forms of life—the origin of species—is at the focal point of evolutionary theory, for it is in new species that biological diversity arises. It is not enough to explain how adaptations evolve in populations, a topic covered in Chapter 23. Evolutionary theory*

must also explain **macroevolution,** *the origin of new taxonomic groups (new species, new genera, new families, even new kingdoms).* **Speciation** *(the origin of new species) is the key process because any genus, family, or higher taxon originates with a new species that is novel enough to be the inaugural member of the higher taxon.*

The fossil record chronicles two patterns of speciation: anagenesis and cladogenesis (FIGURE 24.1). Anagenesis (from the Greek ana, up, and genesis, origin), also known as phyletic evolution, is

(a) Anagenesis   **(b) Cladogenesis**

FIGURE 24.1 Two patterns of speciation. (a) Anagenesis (phyletic evolution) is the accumulation of heritable changes in a population, transforming that population into a new species. **(b)** Cladogenesis is branching evolution, in which a new species arises from a population that buds from a parent species. Cladogenesis is the basis for biological diversity.

the accumulation of changes associated with the transformation of one species into another. Cladogenesis (from the Greek clados, branch), also called branching evolution, is the budding of one or more new species from a parent species that continues to exist. Only cladogenesis can promote biological diversity by increasing the number of species.

Our objectives in this chapter are to evaluate definitions of species and mechanisms of speciation and to examine the possible origins of some novel features that define higher taxonomic groups. Our first step is to appraise the assumption that species actually exist in nature as discrete biological units distinct from all others.

WHAT IS A SPECIES?

Species is a Latin word meaning "kind" or "appearance." Indeed, we learn to distinguish between the kinds of plants or animals—between dogs and cats, for instance—from differences in their appearance. In addition to comparing morphology (body form), modern taxonomists also consider differences in body functions, biochemistry, behavior, and genetic makeup. However, dividing organisms into different species based on comparative data is only part of an extensive effort to better understand the nature of species and the factors that maintain their distinctiveness in nature.

The biological species concept emphasizes reproductive isolation

A classical species definition known as the biological species concept was first enunciated by evolutionary biologist Ernst Mayr in 1942. Mayr's definition addresses the question, What factors divide biological diversity into separate forms that we identify as species?

The **biological species concept** defines a species as a population or group of populations whose members have the potential to interbreed with one another in nature to produce viable, fertile offspring, but who cannot produce viable, fertile offspring with members of other species (FIGURE 24.2). In other words, a biological species is the largest set of populations in which genetic exchange is possible and that is genetically isolated from other such populations. Members of a biological species are united by being reproductively compatible, at least potentially. A businesswoman in Manhattan has little probability of producing offspring with a dairyman in Mongolia, but if the two should get together, they could have viable babies that develop into fertile adults. All humans belong to the same biological species. In contrast, humans and chimpanzees remain distinct biological species even where they share territory, because the two species do not interbreed. Thus, the biological species concept hinges on reproductive

(a) Similarity between different species. The eastern meadowlark (*Sturnella magna,* left) and the western meadowlark (*Sturnella neglecta*) have very similar body shapes and colorations, but they represent different species. Their songs are distinct, a behavioral difference that helps prevent interbreeding between the two species.

(b) Diversity within a species. As diverse as we may be in appearance, all humans belong to a single species (*Homo sapiens*), defined by our capacity to interbreed.

FIGURE 24.2 The biological species concept is based on interfertility rather than physical similarity.

isolation, with each species isolated by factors (barriers) that prevent interbreeding, thereby blocking genetic mixing with other species.

Prezygotic and postzygotic barriers isolate the gene pools of biological species

Any factor that impedes two species from producing viable, fertile hybrids contributes to reproductive isolation. No single barrier may be completely impenetrable to genetic exchange, but many species are genetically sequestered by more than one type of barrier. Here, we are considering only biological barriers to reproduction, which are intrinsic to the organisms. Of course, if two species are geographically segregated, they cannot possibly interbreed, but a geographic barrier is not what biologists mean by reproductive isolation because it is not intrinsic to the organisms themselves. Reproductive isolation prevents populations belonging to different species from interbreeding, even if their ranges overlap.

Clearly, a fly will not mate with a frog or a fern, but what prevents biological species that are very similar—that is, closely related—from interbreeding? The various reproductive barriers that isolate the gene pools of species can be categorized as prezygotic or postzygotic, depending on whether they function before or after the formation of zygotes, or fertilized eggs.

FIGURE 24.3 Courtship ritual as a behavioral barrier between species. These blue-footed boobies, inhabitants of the Galápagos, will mate only after a specific ritual of courtship displays. Part of the "script" calls for the male to high-step, a behavior that advertises the bright blue feet characteristic of the species.

Prezygotic Barriers

Prezygotic barriers impede mating between species or hinder the fertilization of ova if members of different species attempt to mate.

Habitat Isolation.   Two species that live in different habitats within the same area may encounter each other rarely, if at all, even though they are not technically geographically isolated. For example, two species of garter snakes in the genus *Thamnophis* occur in the same areas, but one lives mainly in water and the other is primarily terrestrial. Habitat isolation also affects parasites, which are generally confined to certain plant or animal host species. Two species of parasites living on different hosts will not have a chance to mate.

Behavioral Isolation.   Special signals that attract mates, as well as elaborate behavior unique to a species, are probably the most important reproductive barriers among closely related animals. Male fireflies of various species signal to females of their kind by blinking their lights in particular rhythms. The females respond only to signals characteristic of their own species, flashing back and attracting the males. The eastern and western meadowlarks shown in FIGURE 24.2a are almost identical in shape, coloration, and habitat, and their ranges overlap in the central United States. Yet they remain two separate biological species, partly because of the differences in their songs, which enable them to recognize individuals of their own kind. Behavioral isolation often depends on the elaborate courtship rituals of a particular species (FIGURE 24.3).

Temporal Isolation.   Two species that breed during different times of the day, different seasons, or different years cannot mix their gametes. The geographic ranges of the western spotted skunk (*Spilogale gracilis*) and the eastern spotted skunk (*Spilogale putorius*) overlap, but these two very similar species do not interbreed because *S. gracilis* mates in late summer and

S. putorius mates in late winter. Three species of the orchid genus *Dendrobium* living in the same rain forest do not hybridize because they flower on different days. Pollination of each species is limited to a single day because the flowers open in the morning and wither that evening.

Mechanical Isolation.   Closely related species may attempt to mate but fail to consummate the act because they are anatomically incompatible. For example, mechanical barriers contribute to reproductive isolation of flowering plants that are pollinated by insects or other animals. Floral anatomy is often adapted to certain pollinators that transfer pollen only among plants of the same species. In another example of mechanical isolation, if insects of closely related species attempt to mate, the male and female copulatory organs may not fit together and no sperm would be transferred.

Gametic Isolation.   Even if the gametes of different species meet, they rarely fuse to form a zygote. For animals whose eggs are fertilized within the female reproductive tract (internal fertilization), the sperm of one species may not be able to survive in the environment of the female reproductive tract of another species. Many aquatic animals release their gametes into the surrounding water, where the eggs are fertilized (external fertilization). Even when two closely related species release their gametes at the same time in the same place, cross-specific fertilization is uncommon. Gamete recognition may be based on the presence of specific molecules on the coats around the egg, which adhere only to complementary molecules on sperm cells of the same species. A similar mechanism of molecular recognition enables a flower to discriminate between pollen of the same species and pollen of different species.

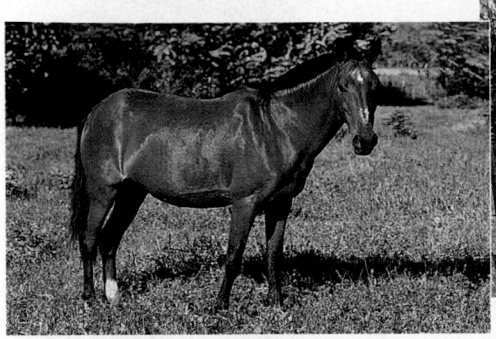

Mule (hybrid)

Horse

Donkey

FIGURE 24.4 Hybrid sterility, a postzygotic barrier. Horses and donkeys remain separate species because their hybrid offspring, mules, are sterile.

Postzygotic Barriers

If a sperm cell from one species does fertilize an ovum of another species, then **postzygotic barriers** usually prevent the hybrid zygote from developing into a viable, fertile adult.

Reduced Hybrid Viability. When prezygotic barriers are crossed and hybrid zygotes are formed, genetic incompatibility between the two species may abort development of the hybrid at some embryonic stage. Of the numerous species of frogs belonging to the genus *Rana,* some live in the same regions and habitats, where they may occasionally hybridize. But the hybrids generally do not complete development, and those that do are frail.

Reduced Hybrid Fertility. Even if two species mate and produce hybrid offspring that are vigorous, reproductive isolation is intact if the hybrids are completely or largely sterile. Since the infertile hybrid cannot backbreed with either parental species, genes cannot flow freely between the species. One cause of this barrier is a failure of meiosis to produce normal gametes in the hybrid if chromosomes of the two parent species differ in number or structure. A familiar case of a sterile hybrid is the mule, a robust cross between a horse and a donkey; horses and donkeys remain distinct species because, except very rarely, mules cannot backbreed with either species (FIGURE 24.4).

Hybrid Breakdown. In some cases when species crossmate, the first-generation hybrids are viable and fertile, but when these hybrids mate with one another or with either parent species, offspring of the next generation are feeble or sterile. For example, different cotton species can produce fertile hybrids, but breakdown occurs in the next generation when offspring of the hybrids die as seeds or grow into weak and defective plants.

FIGURE 24.5 summarizes the reproductive barriers between closely related species.

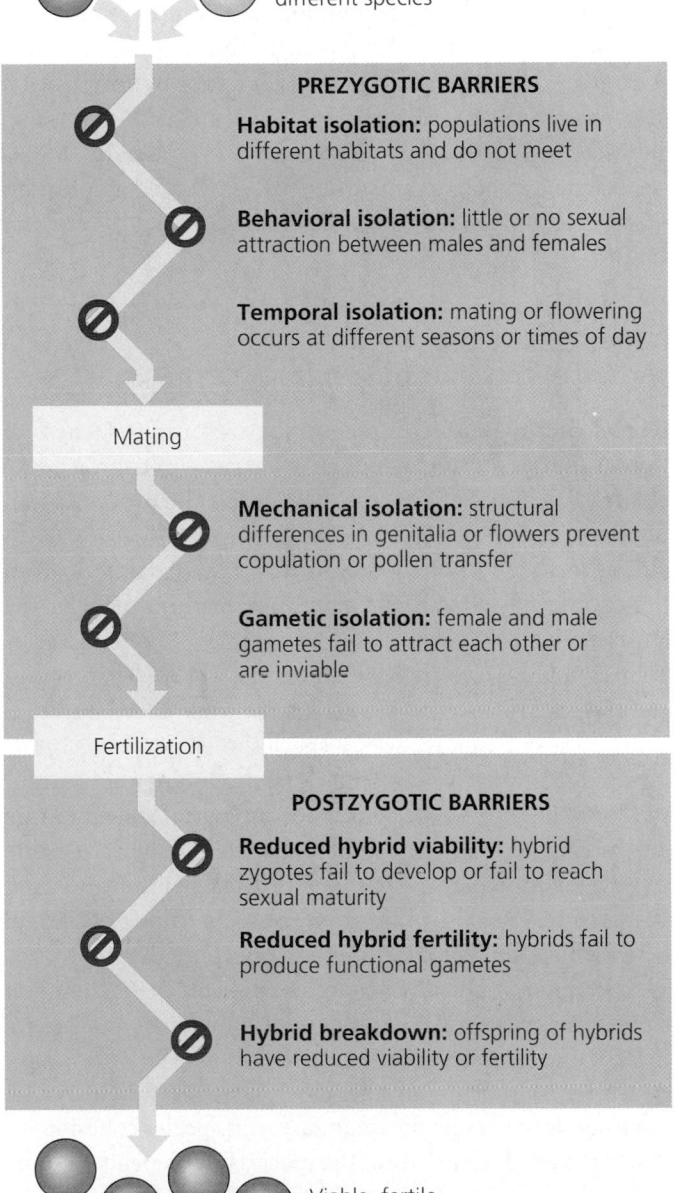

Individuals of different species

PREZYGOTIC BARRIERS

Habitat isolation: populations live in different habitats and do not meet

Behavioral isolation: little or no sexual attraction between males and females

Temporal isolation: mating or flowering occurs at different seasons or times of day

Mating

Mechanical isolation: structural differences in genitalia or flowers prevent copulation or pollen transfer

Gametic isolation: female and male gametes fail to attract each other or are inviable

Fertilization

POSTZYGOTIC BARRIERS

Reduced hybrid viability: hybrid zygotes fail to develop or fail to reach sexual maturity

Reduced hybrid fertility: hybrids fail to produce functional gametes

Hybrid breakdown: offspring of hybrids have reduced viability or fertility

Viable, fertile offspring

FIGURE 24.5 A summary of reproductive barriers between closely related species.

The biological species concept has some major limitations

In its emphasis on reproductive barriers as the boundaries between species, the biological species concept has had an important impact on evolutionary theory and how we think of species as discrete forms of life. However, as a criterion for actually distinguishing species in nature, the biological species concept does not work in all situations and is in fact impractical for demarcating species in most cases. For example, there is no way to check interbreeding in the extinct forms represented by fossils. Biologists classify fossils into species based on differences in morphology. Even most living species are distinguished by comparative morphology because we lack the information about interbreeding necessary to apply the biological species concept as the criterion. The biological species concept also has no utility at all for life-forms that are entirely asexual, such as bacteria. (Many bacteria do transfer genes by conjugation and other processes—see Chapter 18—but there is nothing akin to the sexual union of gametes from two parents.) Biologists assign asexual organisms to species based mainly on structural and biochemical characteristics.

Biologists have proposed several alternative concepts of species

The biological species concept emphasizes a separateness of different species due to reproductive barriers. In contrast, several alternative species concepts emphasize the processes that unite the members of a species. The distinction is analogous to defining a country not by the borders that separate it from other countries but by the set of cultural features that give the country an identity.

The **ecological species concept** defines a species in terms of its ecological niche, the set of environmental resources a species uses. Put another way, a species' niche depends on its unique adaptations to a particular role in a biological community. (We will examine the concept of ecological niche in more detail in Chapter 53.) For example, a species that is a parasite may be defined in part by its adaptations to a specific host organism. Note that this species concept accommodates asexual species, which the biological species concept does not.

According to a **pluralistic species concept,** the factors that are most important for the cohesion of individuals as a species vary. In some cases, reproductive isolation may be a key unifying factor for a species. In other cases, adaptation to a specific ecological niche may be the main factor in species cohesion. In still other cases, the integrity of the species may depend on some combination of reproductive isolation and a unique niche.

The biological, ecological, and pluralistic species concepts are all "explanatory" concepts—attempts to explain the very existence of species as discrete units in the diversity of life.

None, however, is especially useful in actually identifying various species in nature. For that, taxonomists still depend mainly on morphological characteristics. Thus, although it doesn't really explain why species exist, the **morphological species concept,** which characterizes each species in terms of a unique set of structural features, remains the way we distinguish most species.

Many evolutionary biologists are also evaluating a **genealogical species concept,** which defines a species as a set of organisms with a unique genetic history—that is, as one tip on the branching tree of life. The sequencing of nucleic acids and proteins provides data that researchers are now using to define each species in terms of unique genetic markers. Like the morphological species concept, the genealogical one is more concerned with identifying the various species than with a theoretical explanation for their existence as separate forms of life.

Each species concept has utility, depending on the situation and the types of questions we are asking. The biological species concept, in its emphasis on reproductive barriers to gene flow, is particularly useful in thinking about how species originate. Next, we examine the processes that can lead to such speciation.

MODES OF SPECIATION

There are two general modes of speciation based on how gene flow among populations is initially interrupted (FIGURE 24.6). In a speciation mode termed **allopatric speciation** (from the

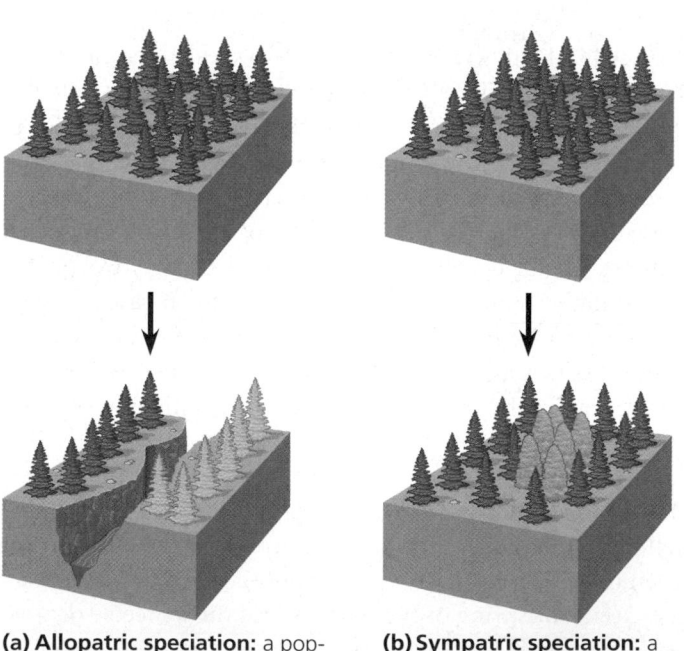

(a) Allopatric speciation: a population forms a new species while geographically isolated from its parent population

(b) Sympatric speciation: a small population becomes a new species without geographic separation from its parent population

FIGURE 24.6 Two modes of speciation. These sketches simplify the geographic relationships of new species to their parent species.

FIGURE 24.7 Allopatric speciation of squirrels in the Grand Canyon. Two species of antelope squirrels inhabit opposite rims of the Grand Canyon. On the south rim is Harris's antelope squirrel (*Ammospermophilus harrisi*). Just a few miles away on the north rim is the closely related white-tailed antelope squirrel (*Ammospermophilis leucurus*). Birds and other organisms that can disperse easily across the canyon have not diverged into different species on opposite rims.

Greek *allos*, other, and from the Latin *patria*, homeland), speciation takes place in populations with geographically separate ranges. Gene flow is initially interrupted or reduced between two populations because they are separated in space. In the second speciation mode, called **sympatric speciation** (from the Greek *sym*, together), speciation takes place in geographically overlapping populations. Here biological factors, such as chromosomal changes and nonrandom mating, reduce gene flow.

Allopatric speciation: Geographic barriers can lead to the origin of species

Conditions for Allopatric Speciation

Several geologic processes can fragment a population into two or more isolated populations. A mountain range may emerge and gradually split a population of organisms that can inhabit only lowlands. A land bridge, such as the Isthmus of Panama, may form and separate the marine life on either side. A large lake may subside until there are several smaller lakes with their populations now isolated. Even without such geologic remodeling, geographic isolation and allopatric speciation can occur if individuals colonize a new, geographically remote area and become isolated from the parent population. An example is the speciation that occurred on the Galápagos Islands following colonization by mainland organisms.

How formidable must a geographic barrier be to keep allopatric populations apart? The answer depends on the ability of the organisms to move about. Birds, mountain lions, and coyotes can cross mountain ranges, rivers, and canyons. Nor do such barriers hinder the windblown pollen of pine trees, and the seeds of many plants may be carried back and forth on animals. In contrast, small rodents may find a deep canyon or a wide river a formidable barrier (FIGURE 24.7).

The likelihood of allopatric speciation increases when a population is both small and isolated. A small, isolated population is more likely than a large population to have its gene pool changed substantially by genetic drift and natural selection. For example, in less than 2 million years, small populations of stray animals and plants from the South American mainland that managed to colonize the Galápagos Islands gave rise to all the new species that now inhabit the islands. But for each small, isolated population that becomes a new species, many more simply perish in their new environment. Life on the frontier is harsh, and most pioneer populations probably become extinct.

A key question about allopatric populations is whether they have become different enough that they can no longer interbreed and produce fertile offspring if they come back into contact (FIGURE 24.8). In some cases, researchers evaluate whether

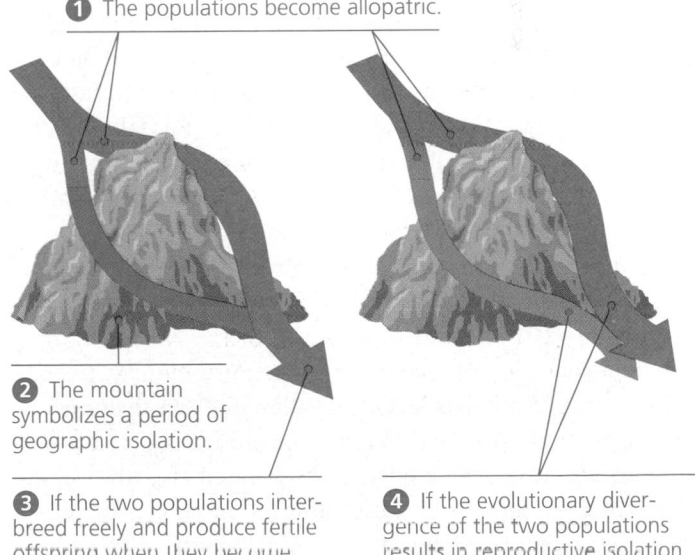

① The populations become allopatric.

② The mountain symbolizes a period of geographic isolation.

③ If the two populations interbreed freely and produce fertile offspring when they become sympatric again, their gene pools merge. Speciation has not occurred.

④ If the evolutionary divergence of the two populations results in reproductive isolation, then they will not interbreed, even if they come back into contact. Speciation has occurred.

FIGURE 24.8 Has speciation occurred during geographic isolation?

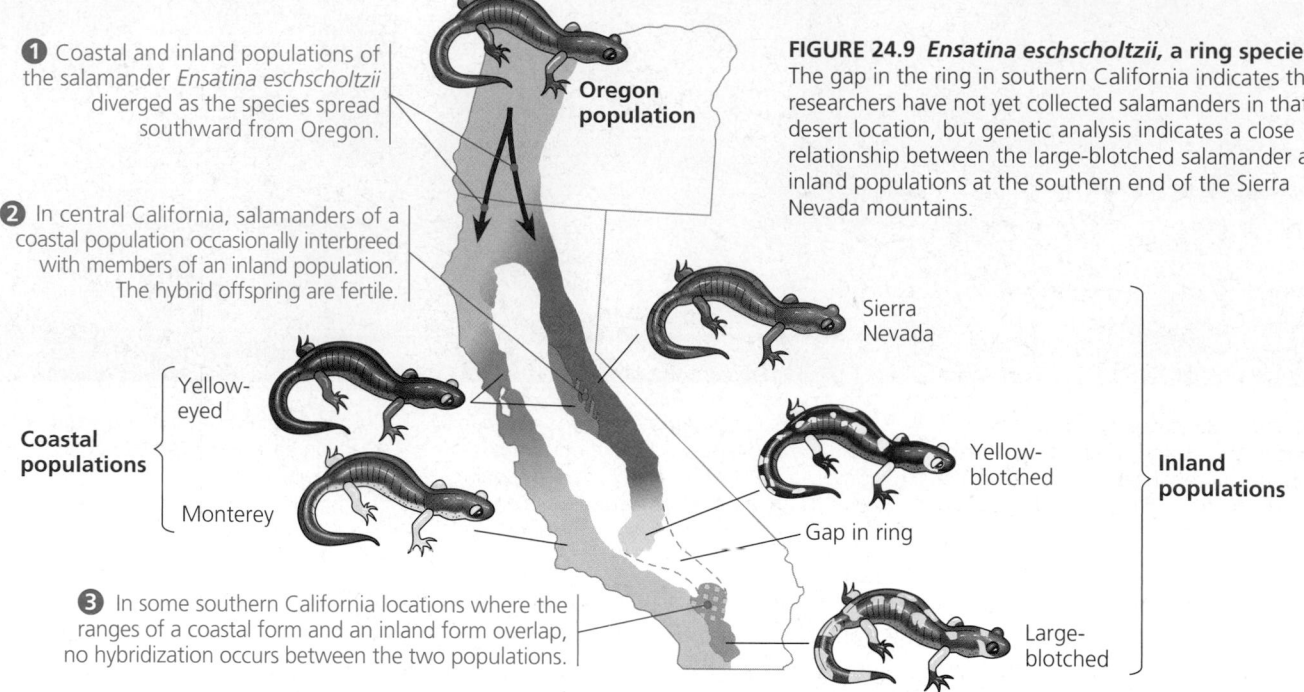

❶ Coastal and inland populations of the salamander *Ensatina eschscholtzii* diverged as the species spread southward from Oregon.

Oregon population

❷ In central California, salamanders of a coastal population occasionally interbreed with members of an inland population. The hybrid offspring are fertile.

Coastal populations { Yellow-eyed / Monterey

Sierra Nevada

Yellow-blotched

Gap in ring

Inland populations

❸ In some southern California locations where the ranges of a coastal form and an inland form overlap, no hybridization occurs between the two populations.

Large-blotched

FIGURE 24.9 *Ensatina eschscholtzii,* a ring species. The gap in the ring in southern California indicates that researchers have not yet collected salamanders in that desert location, but genetic analysis indicates a close relationship between the large-blotched salamander and inland populations at the southern end of the Sierra Nevada mountains.

speciation has occurred by artificially bringing together members of separated populations in a laboratory setting. In other cases, biologists can investigate allopatric speciation where separated populations have come back into contact naturally.

Ring Species: Allopatric Speciation in Progress?

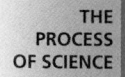

THE PROCESS OF SCIENCE

Some species have geographic distributions that provide evolutionary biologists with examples of what seem to be populations at various stages in their gradual divergence from common ancestors. Consider a so-called ring species, which is distributed around some geographic barrier, with the populations that have diverged the most in their evolution eventually meeting where the ring closes. FIGURE 24.9 illustrates populations of a North American salamander, *Ensatina eschscholtzii.* The species probably expanded southward into California from Oregon. The California population eventually split into a coastal population and an inland one, giving rise to two separate chains of interbreeding populations. One chain extends down the coastal mountains and one extends down the inland mountains (Sierra Nevada range). The dual chains of salamander populations gradually formed a ring around California's central valley (the San Joaquin Valley). Salamanders of the different populations contrast in coloration, and researchers have demonstrated that the coastal and inland populations exhibit more and more genetic differences the farther south the comparison is made.

In the northern and middle portions of the ring, the salamander populations interbreed as a single species (as defined by the biological species concept). About halfway down the ring, the coastal and inland gene pools occasionally "leak" across the central valley as members of the two populations interbreed. But near the ring's southern end in San Diego County, no hybridization occurs in some of the locales where the ranges of coastal and inland populations overlap. Based on the criterion of reproductive isolation, it is legitimate to designate the two southern populations as separate species.

Adaptive Radiation on Island Chains

Islands are living laboratories for the study of speciation. Flurries of allopatric speciation have occurred on island chains where organisms that have strayed or become passively dispersed from their parent populations have founded new populations that evolved in isolation (FIGURE 24.10). The

FIGURE 24.10 Long-distance dispersal. Seeds of a plant named *Pisonia* cling like Velcro® to this black noddy tern as the migratory seabird walks on an island off the coast of Australia. This is one mechanism that can transport terrestrial organisms to isolated islands.

many indigenous species of the Galápagos descended from stragglers that floated, flew, or were blown over the sea from the South American mainland. For example, consider the Galápagos finches. A single dispersal event may have seeded one island with a small population of the ancestral finch, which gave rise to a new species. Later, a few individuals of this island species may have reached neighboring islands, where geographic isolation permitted additional speciation episodes. After diverging on one of these other islands, a new species could recolonize the island from which its founding population emigrated and coexist there with its parent species or form still another species. Multiple colonizations by species from neighboring islands would eventually lead to the coexistence of several species on each island. The islands are far enough apart to permit populations to evolve in isolation, but close enough together for occasional dispersion events to occur. Such evolution of many diversely adapted species from a common ancestor is called **adaptive radiation** (FIGURE 24.11).

The Hawaiian Archipelago is one of the world's great showcases of evolution (see FIGURE 22.16). The volcanic islands are about 3,500 km from the nearest continent. They become progressively younger to the southeast, terminating with the youngest and largest island, Hawaii, which is less than a million years old and still has active volcanoes. Each island was born naked and was gradually populated by species derived from strays that rode the ocean currents and winds either from distant islands and continents or from older islands of the archipelago itself. The physical diversity of each island, including a range of altitudes and extensive differences in rainfall, provides many environmental opportunities for evolutionary divergence by natural selection. Multiple invasions and allopatric speciations have ignited an explosion of adaptive radiation; most of the thousands of species of plants and animals that now inhabit the islands are found nowhere else in the world. In contrast, there are no indigenous species on the Florida Keys. Apparently, those islands are so close to the mainland that founding populations are not sufficiently sequestered for their gene pools to become isolated from the steady stream of immigrants from the parent populations on the mainland.

How Do Reproductive Barriers Evolve?

This is a good time to address two potential misunderstandings about allopatric speciation. First, it is important to understand that geographic isolation, while obviously preventing interbreeding between allopatric populations, does not qualify as reproductive isolation in the biological sense. The reproductive barriers relevant to the biological species concept are intrinsic to the organisms themselves and prevent interbreeding even in the absence of geographic isolation (see FIGURE 24.5 to review reproductive barriers). Second, it is important to realize that speciation is not due to some drive to erect reproductive barriers around a population. In most cases, reproductive barriers are probably coincidental to changes in gene pools due to natural selection and genetic drift as allopatric populations evolve separately. Let's examine a couple of examples.

Example of Evolution of a Prezygotic Barrier. THE PROCESS OF SCIENCE Diane Dodd, of Yale University, designed laboratory experiments to test the hypothesis that reproductive barriers between allopatric populations can evolve as a by-product of the populations' adaptive divergence in different environments. She divided a sample of fruit flies (*Drosophila pseudoobscura*) into several laboratory populations that were then cultured for several generations on different nutrient sources. Some populations were

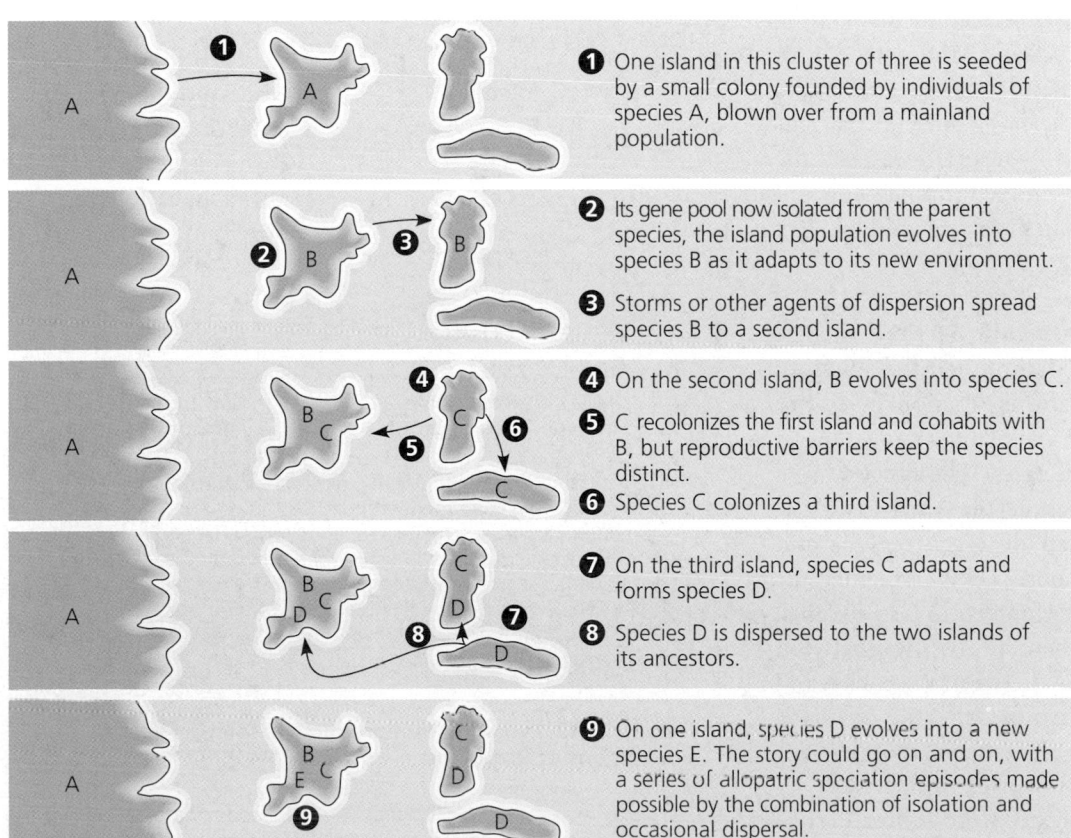

① One island in this cluster of three is seeded by a small colony founded by individuals of species A, blown over from a mainland population.

② Its gene pool now isolated from the parent species, the island population evolves into species B as it adapts to its new environment.

③ Storms or other agents of dispersion spread species B to a second island.

④ On the second island, B evolves into species C.

⑤ C recolonizes the first island and cohabits with B, but reproductive barriers keep the species distinct.

⑥ Species C colonizes a third island.

⑦ On the third island, species C adapts and forms species D.

⑧ Species D is dispersed to the two islands of its ancestors.

⑨ On one island, species D evolves into a new species E. The story could go on and on, with a series of allopatric speciation episodes made possible by the combination of isolation and occasional dispersal.

FIGURE 24.11 A model for adaptive radiation on island chains.

nourished with a starch medium, while others were nourished with a maltose (malt sugar) medium (FIGURE 24.12). Acting over several generations, natural selection favored those individuals that were best suited to the available nutrient; populations reared on starch improved in their efficiency at starch digestion, while the "maltose" populations improved in their efficiency at malt sugar digestion.

After the two types of fly populations diverged in their evolution for several generations, Dodd combined flies from various populations in mate-choice experiments. Female "maltose flies" were more likely to mate with male maltose flies than with male "starch flies," even if the maltose flies came from a different maltose fly population. The female starch flies also seemed to discriminate in favor of other starch flies as mates. This is an example of a prezygotic barrier—in this case, a behavioral obstacle to interbreeding. The reproductive barrier was not absolute—some mating between maltose flies and starch flies did occur—but reproductive isolation was apparently well under way after several generations of evolutionary divergence.

Why would a reproductive barrier between two populations be a consequence of their divergent adaptation to different environments? Specifically, how could adaptation to a certain diet affect mate choice? One hypothesis is based on pleiotropy, where one allele has multiple effects on phenotype—in this case affecting both nutrient digestion and mate choice (see Chapter 14). Mating in fruit flies follows an elaborate courtship ritual that includes a song produced by the buzzing of the wings, a dance with specific "steps," and detection of specific odors emitted from the potential mate's cuticle, the hard covering of insects. Perhaps the same allele(s) that enhances digestion of a certain carbohydrate also affects the chemical composition of the cuticle and determines which of these "bouquet" molecules function in mate recognition. How could you test this hypothesis?

THE PROCESS OF SCIENCE

Example of Evolution of a Postzygotic Barrier. Robert Vickery, of the University of Utah, tested the ability of plants from different populations of monkey flower (*Mimulus glabratus*) to interbreed. (The plant is named for its flower, which resembles a monkey's face.) The species has a very large range that extends throughout the Americas. Vickery imported *Mimulus* plants from different regions and cross-pollinated them in his greenhouse. He then planted the seeds of the hybrids to see if they developed into fertile plants. Most of the offspring were fertile when the cross was between plants from nearby populations—say, between Wisconsin and Michigan *Mimulus*. But the proportion of fertile offspring dropped off when Vickery crossed plants from more distant populations. In fact, the offspring of hybrids from crosses between plants from Wisconsin and Mexico were almost all sterile. Note that this is a postzygotic barrier (hybrid breakdown; see FIGURE 24.5).

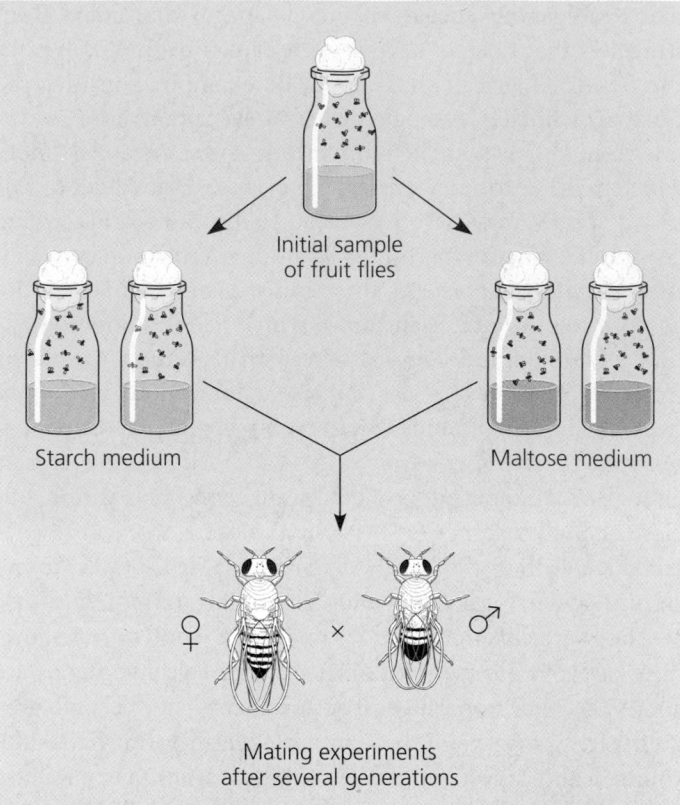

(a) Experimental design. After several generations of divergent adaptation to different diets, the researcher measured mating frequencies by combining flies from the same or different populations in mating cages.

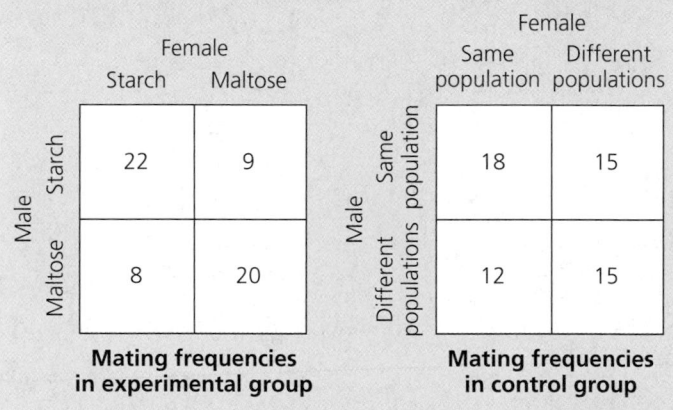

	Female	
	Starch	Maltose
Male Starch	22	9
Male Maltose	8	20

Mating frequencies in experimental group

	Female	
	Same population	Different populations
Male Same population	18	15
Male Different populations	12	15

Mating frequencies in control group

(b) Results. In the experimental cages, flies reared on the starch medium were combined with flies reared on the maltose medium. Note the greater tendency for the flies to mate with like partners. In each control cage, flies adapted to the same medium, but from different populations, were combined. Note that the flies do not show much mating preference for the flies from their own population compared to like-adapted flies from other populations.

FIGURE 24.12 Evolution of reproductive isolation in lab populations of *Drosophila*.

Again, it is important to dispel the notion that such a reproductive barrier evolves because of some kind of drive toward speciation. In fact, natural selection probably maintains reproductive compatibility between neighboring *Mimulus* populations because hybridization between these populations occurs often enough in nature that alleles hampering reproductive success of the hybrids would be eliminated from the gene pool. But a Wisconsin monkey flower would never hybridize naturally with one from Mexico, and thus there is no selection pressure keeping the genomes of such distant populations compatible enough for successful interbreeding. It's not that selection favors a reproductive barrier between allopatric populations, but that there is no longer any advantage to the absence of such barriers.

Summary of Allopatric Speciation

In allopatric speciation, a new species forms while geographically isolated from its ancestor. As the isolated population's gene pool evolves by genetic drift and natural selection, reproductive isolation from the ancestral species may evolve as a by-product of the genetic change. Such reproductive barriers prevent interbreeding with the ancestor, even if the populations come back into contact. We have examined four examples of how biologists are dissecting the process of allopatric speciation: the evolutionary divergence of populations of a ring species (the salamander *Ensatina eschscholtzii*); the adaptive radiation of island species; the origin of a prezygotic barrier to interbreeding in lab populations of the fruit fly *Drosophila pseudoobscura;* and the origin of a postzygotic barrier between distant populations of the monkey flower *Mimulus glabratus.* Let's turn next to mechanisms that can produce a new species *without* geographic isolation from its ancestor.

Sympatric speciation: A new species can originate in the geographic midst of the parent species

In sympatric speciation, new species arise within the range of parent populations rather than in geographically separate populations (see FIGURE 24.6b). How can reproductive barriers between sympatric populations evolve? Let's examine examples in both the plant and animal kingdoms.

Polyploid Speciation in Plants

Some plant species have their origins in accidents during cell division that result in extra sets of chromosomes, a mutant condition called **polyploidy.** An **autopolyploid** (from the Greek *autos,* self) is an individual that has more than two chromosome sets, all derived from a single species. For example, a failure of meiosis during gamete production can double chromosome number from the diploid count ($2n$) to a tetraploid number ($4n$) (FIGURE 24.13). The tetraploid can then fertilize itself (self-pollinate) or mate with other tetraploids. However, the mutants cannot interbreed successfully with diploid plants of the original population; the offspring, which would be triploid ($3n$), would be sterile because unpaired chromosomes result in abnormal meiosis. In just one generation, a postzygotic barrier has caused reproductive isolation and has interrupted gene flow between a tiny population of tetraploids (maybe only a single plant initially) and the parent diploid population that surrounds it. Sympatric speciation by autopolyploidy was first discovered in the early 1900s by geneticist Hugo de Vries when his experiments produced a new species of evening primrose (FIGURE 24.14, p. 474).

Another type of polyploid species, much more common than autopolyploids, is called an **allopolyploid,** referring to the contribution of two *different* species to a polyploid hybrid. The origin of an allopolyploid begins when two different species interbreed and combine their chromosomes. Interspecific hybrids are usually sterile because the haploid set of chromosomes from one species cannot pair during meiosis with the haploid set from the other species. Though infertile, a hybrid may actually be more vigorous than its parents and propagate itself asexually (which many plants can do). Various mechanisms can then change a sterile hybrid into a fertile polyploid (FIGURE 24.15 on p. 474 illustrates one such mechanism). The polyploid hybrids are fertile with each other but cannot interbreed with either parental species.

The origin of new polyploid plant species is common enough and rapid enough that scientists have documented several such speciations. For example, two new species of plants called goatsbeards originated in the Pacific Northwest in the mid-1900s. The goatsbeard genus, *Tragopogon,* is native to Europe, but three species were introduced by humans to the Americas in the early 1900s. These species, *T. dubius, T. pratensis,* and

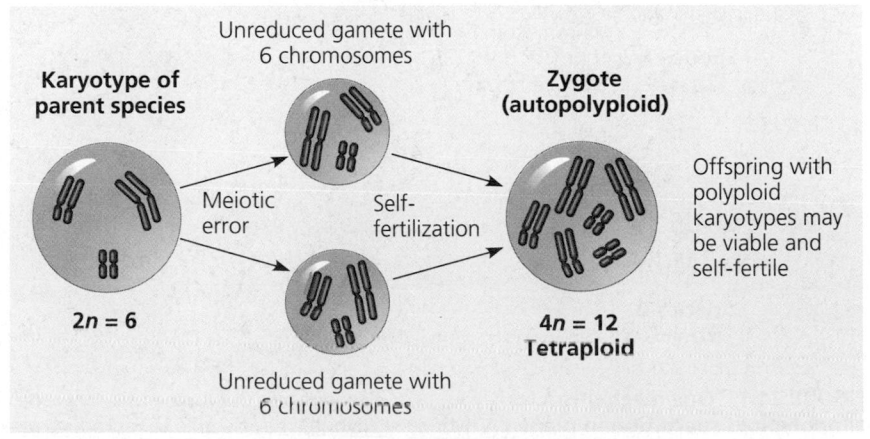

FIGURE 24.13 Sympatric speciation by autopolyploidy in plants.

O. gigas

O. lamarckiana

THE PROCESS OF SCIENCE

FIGURE 24.14 Botanist Hugo de Vries and his new primrose species. Working in the early 1900s, de Vries studied variation in evening primroses. In his breeding experiments with *Oenothera lamarckiana* (lower left), a diploid species of primrose with 14 chromosomes, he produced a new variety with 28 chromosomes. It was a tetraploid, which could not interbreed with its parent species. De Vries named the new primrose species *Oenothera gigas*, for its large size (upper right).

T. porrifolius, are now common weeds in urban wastelands such as abandoned parking lots. In the 1950s, botanists identified two new species of *Tragopogon* in regions of Idaho and eastern Washington where all three European species are also found. One new species, *T. miscellus*, is a tetraploid hybrid of *T. dubius* and *T. pratensis*; the other new species, *T. mirus*, is also an allopolyploid, but its ancestors are *T. dubius* and *T. porrifolius*. While the *T. mirus* population grows mainly by reproduction of its own members, additional episodes of hybridization between the ancestral species also add to the *T. mirus* population. This is an ongoing speciation process we can observe today.

Many of the plants we grow for food are polyploids. Oats, cotton, potatoes, tobacco, and wheat are among the polyploid species important to agriculture. The wheat used for bread, *Triticum aestivum*, is an allopolyploid that probably originated about 8,000 years ago as a spontaneous hybrid of a cultivated wheat and a wild grass. Plant geneticists now hybridize plants and use chemicals that induce meiotic and mitotic errors to help create new polyploids with special qualities. For example, artificial hybrids combine the high yield of wheat with the ability of rye to resist disease.

Sympatric Speciation in Animals

Although polyploid speciation is less common in animals than in plants, it does occur. However, there are other mechanisms that can lead to sympatric speciation in animals. For example, animals may become reproductively isolated within the geographic range of a parent population if genetic factors cause them to become dependent on resources not used by the parent population. Consider, for example, the wasps that pollinate figs. Each fig species is pollinated by a particular species

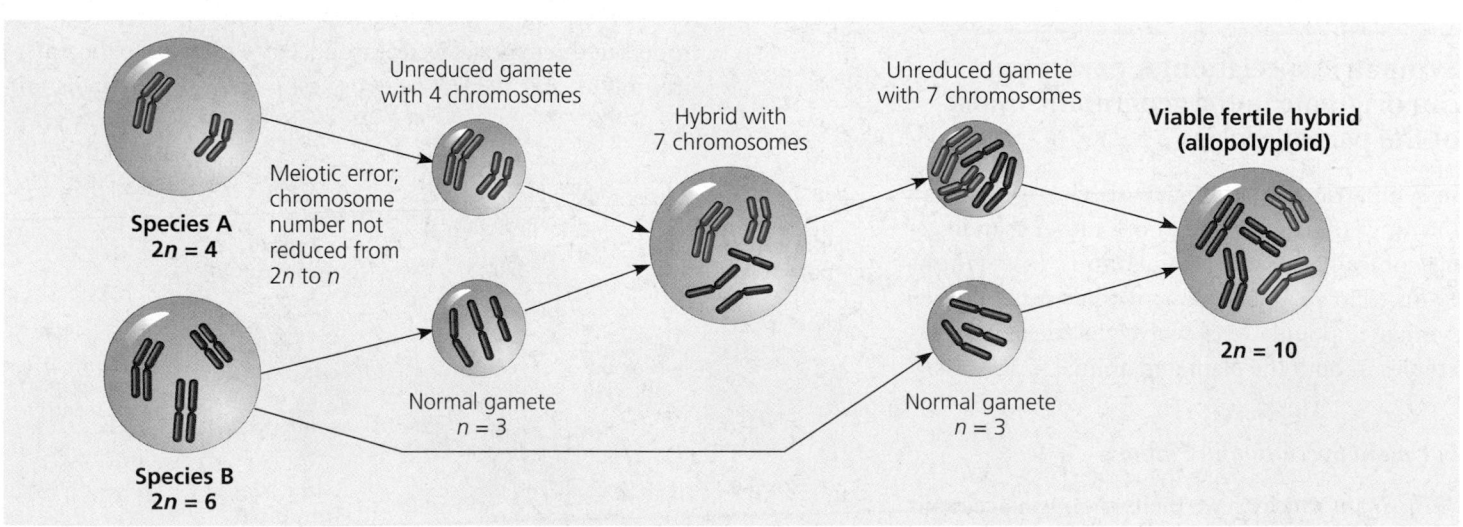

FIGURE 24.15 One mechanism for allopolyploid speciation in plants. A hybrid between two species is normally sterile because its chromosomes are not homologous and cannot pair during meiosis. However, the hybrids may be able to reproduce asexually. This diagram traces one mechanism that can produce fertile hybrids as new polyploid species. The new species has a chromosome number equal to the sum of the chromosomes in the two parent species.

of wasp, which mates and lays its eggs in the figs. A genetic change that caused wasps to select a different fig species would segregate mating individuals of this new phenotype from the parent population. This would set the stage for further evolutionary divergence.

One of Earth's hot spots of animal speciation is Lake Victoria, in East Africa. The lake, which is less than a million years old, is home to almost 200 species of closely related fishes belonging to the cichlid family. The subdivision of populations into groups specialized for exploiting different food sources and other resources in the lake is one process that probably contributed to the adaptive radiation of these fishes. But in at least some cases, nonrandom mating, in which females select males having a certain appearance, was probably a key factor in the sympatric speciation of cichlids. Consider two closely related sympatric species of cichlids that differ mainly in coloration: *Pundamilia pundamilia* has a blue back, and *Pundamilia nyererei* has a red back. It is a reasonable hypothesis that a preference for mates of like coloration functions as a behavioral barrier to interbreeding between the two species. In 1998, biologists from Holland's University of Leiden published experiments testing this hypothesis (FIGURE 24.16). In an aquarium with natural light, females of each species mated only with males of their own species. But in an aquarium illuminated with a monochromatic orange lamp, which makes the two cichlid species appear identical in color, females of each species mated indiscriminately with males of both species. The hybrids from the errant *P. pundamilia* × *P. nyererei* matings were viable and fertile. We can infer that mate choice based on coloration is the main reproductive barrier that normally keeps the gene pools of these two species separate. And we can also infer, since the genetic divergence is small enough that the species can still interbreed when the prezygotic barrier is

breached, that speciation occurred relatively recently. Perhaps the ancestral population was polymorphic for coloration and divergence began with mate choice by two subsets of females preferring one or the other color of male. Sexual selection would then reinforce the color difference as females mated preferentially with males having genes for the least mistakable coloration (see Chapter 23 to review sexual selection).

Summary of Sympatric Speciation

Sympatric speciation requires the emergence of some type of reproductive barrier that isolates the gene pool of a subset of a population without geographic separation from the parent population. In plants, the mechanism that is best understood is hybridization between closely related species coupled with errors during cell division that lead to polyploid individuals that are fertile. In animals, sympatric speciation can result from some subset of a population becoming reproductively isolated because of a switch to a habitat, food source, or other resource not used by the parent population (speciation in fig-eating wasps is an example); or it can result from the emergence of a rigid mating preference by females for specific male forms in a polymorphic population (as in our example of cichlid speciation).

The punctuated equilibrium model has stimulated research on the tempo of speciation

Traditional evolutionary trees that diagram the descent of species from ancestral forms sprout branches that diverge gradually, with species evolving continuously over long spans of time (FIGURE 24.17a, p. 476). Such trees are based on the idea that big changes occur by the accumulation of many small ones. This concept extends the processes of microevolution to the divergence of species. However, paleontologists rarely find gradual transitions of fossil forms. Instead, they often observe species appearing as new forms rather suddenly (in geologic terms) in a layer of rocks, persisting essentially unchanged for their tenure on Earth, and then disappearing from the fossil record as suddenly as they appeared. Darwin was bewildered by the dearth of connecting fossils and wrote: "Although each species must have passed through numerous transitional stages, it is probable that the periods during which each underwent modification, though many and long as measured by years, have been short in comparison with the periods during which each remained in an unchanged condition."

We can account for the "sudden" appearances of new forms in the fossil record partly by applying the allopatric model of speciation. In allopatric speciation, a new species arises as a splinter population in a separate place from where the ancestral species lives. If the new species later extends its range into that of the ancestral species, what we see in the fossil record is

P. pundamilia

P. nyererei

(a) Normal light　　　**(b) Monochromatic orange light**

FIGURE 24.16 Mate choice in two species of Lake Victoria cichlids. (a) In normal lighting, two sympatric species of the cichlid genus *Pundamilia* are noticeably different in coloration. Females of each species mate only with males of their own species. **(b)** With monochromatic lighting in laboratory experiments, females apparently cannot distinguish males of the two species and mate indiscriminately, producing fertile hybrids.

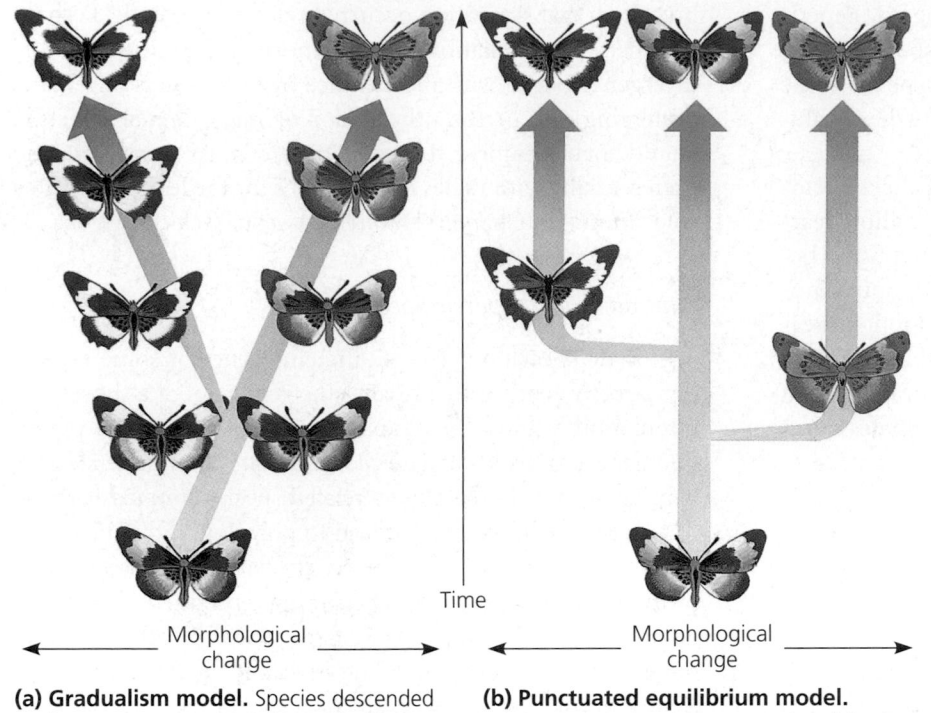

Time

← Morphological change →

(a) Gradualism model. Species descended from a common ancestor gradually diverge more and more in morphology as they acquire unique adaptations.

← Morphological change →

(b) Punctuated equilibrium model. A new species changes most as it buds from a parent species, and then changes little for the rest of its existence.

FIGURE 24.17 Two models for the tempo of speciation.

the geologically sudden appearance of a new species in a locale where there are also fossils of the ancestral species, though maybe in different strata. By then, the ancestor may have been extinct (or competition from the new species could contribute to extinction). Alternatively, the new species could coexist in the same range as the ancestor. Either way, the new species does not appear in the fossil record at the site of its ancestor until it has diverged in form during its period of geographic separation.

Advocates of a model called **punctuated equilibrium** incorporate ideas about the *tempo* of speciation in their explanations of what we see in the fossil record. According to this model, species diverge in spurts of relatively rapid change, instead of slowly and gradually (FIGURE 24.17b). In other words, species undergo most of their morphological modification as they first bud from parent species and then change little, even as they give rise to additional species. The term *punctuated equilibrium* is derived from the idea of long periods of stasis (equilibrium) punctuated by episodes of speciation.

How can speciation in a few thousand generations, which may require several thousand years, be called an abrupt episode? The fossil record indicates that successful species last for a few million years, on average. Suppose that a particular species survives for 5 million years, but most of its morphological changes occurred during the first 50,000 years of its existence. In this case, the evolution of the species-defining characteristics was compressed into just 1% of the lifetime of

the species. On the time scale that can generally be determined in fossil strata, the species will appear suddenly in rocks of a certain age and then linger with little or no change before becoming extinct. During its formative millennia, the species may have accumulated its modifications gradually, but relative to the overall history of the species, its inception was abrupt.

Once it is acknowledged that "sudden" may be many thousands of years on the vast scale of geologic time, the debate over the rate of speciation is muted somewhat. The degree to which a species changes after its origin is another issue. If the species is adapted to an environment that stays the same, then natural selection would counter changes in the gene pool. In this view, stabilizing selection tends to hold a population in a long period of stasis (see Chapter 23).

Some gradualists retort that stasis is an illusion. Many species may continue to change after they come into existence, but in ways that cannot be detected from fossils. By necessity, paleontologists base their theories of descent almost entirely on external anatomy and skeletons. Changes in internal anatomy and physiology may go unnoticed, as would modifications in behavior.

FROM SPECIATION TO MACROEVOLUTION

Speciation is at the boundary between microevolution and macroevolution. Microevolution is a change over the generations in a population's allele frequencies, mainly by genetic drift and natural selection. Speciation occurs when a population's genetic divergence from its ancestral population results in reproductive isolation. The change in morphology during that divergence may be very conspicuous when we compare two closely related species, but usually the differences are more subtle, such as the color difference between the two cichlid species in FIGURE 24.16. Yet the cumulative change during millions of speciation episodes over vast tracts of time must account for macroevolution, the level of change that is evident over the time scale of the fossil record. How, for example, did feathers and other flight equipment evolve during the descent of birds from reptiles? Put more generally, how do the evolutionary novelties that define taxonomic groups above the species level, such as classes and phyla, evolve? In this last section, we look at some of the mechanisms behind macroevolution.

Most evolutionary novelties are modified versions of older structures

We can extend the Darwinian concept of "descent with modification" to account for the major morphological transformations of macroevolution. In some cases, very complex structures evolved in increments from much simpler versions that had the same basic function. For example, the human eye is a refined optical organ constructed from multiple parts that work together in forming an image and sending that visual information to the brain. Many people find it hard to believe that such complex organs could be products of gradual evolution rather than finished designs created especially for humans. If the eye needs all its parts to work, the argument goes, how could a partial eye be of any use as an evolutionary stage? The fallacy is in the starting assumption that eyes have to be this complicated to be useful to an animal. The most basic versions of eyes are just clusters of photoreceptor cells, pigmented cells sensitive to light. Only slightly more refined are the eyes of the flatworms called planarians, which have their photoreceptor cells lining cup-shaped indentations on the head. These eyecups have no lenses or other equipment for focusing images, but they do enable the animal to distinguish light from dark; planarians move away from light, a behavioral adaptation that probably reduces the risk of being eaten.

Complex eyes of various types evolved independently from simpler ones many times in the animal kingdom. For example, some mollusks (members of the invertebrate phylum Mollusca), including squids and octopuses, have eyes every bit as complex as those of humans and other vertebrates. Among living mollusks, we can find eyes ranging in complexity from clusters of photoreceptors to camera-like eyes with lenses (FIGURE 24.18). Considering the long success of many so-called "primitive" animals with simple eyecups, such as flatworms and certain mollusks, it is clear that eyecups work fine for what these animals have to do to survive and reproduce. In those animals that do have complex eyes, the organs evolved from simpler ones *not* in one quantum evolutionary jump, but by incremental adaptation of organs that worked and benefited their owners at each stage of this macroevolution.

The evolution of the eye refined organs that retained the basic function of vision. But evolutionary novelty can also arise by the gradual refinement of existing structures for new functions. Such structures that evolve in one context but become co-opted for another function are called **exaptations.** This concept does not imply that a structure somehow evolves in anticipation of future use. Natural selection cannot predict the future and can only improve a structure in the context of its current utility. The lightweight, honeycombed bones of birds (see FIGURE 1.6b) are homologous to the bones of the earthbound ancestors of birds. However, honeycombed bones could not have evolved in the ancestors as an adaptation for upcoming flights. If light bones predated flight, as is clearly

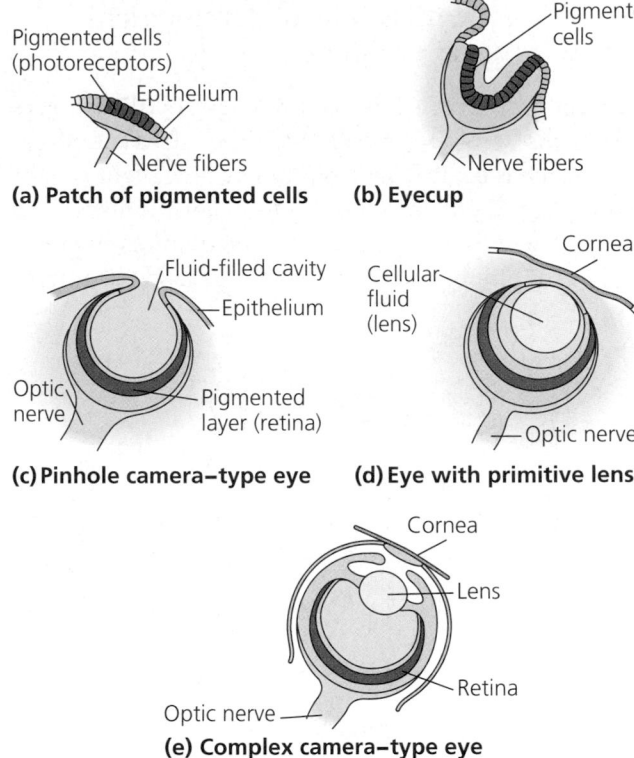

FIGURE 24.18 A range of eye complexity among mollusks.
(a) The limpet *Patella* has a simple patch of pigmented cells (photoreceptors). **(b)** The slit shell mollusk *Pleurotomaria* has an eyecup. **(c)** The *Nautilus* eye functions like a pinhole camera (an early type of camera lacking a lens). A thick fluid in a cavity helps focus light onto the retina, a layer of photoreceptors. **(d)** The eye of *Murex,* a marine snail, has a primitive lens consisting of a mass of crystal-like cells. The cornea is a transparent region of epithelium (outer skin) that protects the eye and helps focus light. **(e)** The squid *Loligo* has a complex camera–type eye with a cornea, lens, and retina.

indicated by the fossil record, then they must have had some function on the ground. The probable ancestors of birds were relatively small, agile, bipedal dinosaurs that also would have benefited from a light frame. It is possible that winglike forelimbs and feathers, which increased the surface area of these forelimbs, were also co-opted for flight after functioning in some other capacity, such as social displays—in courtship, for example. The first flights may have been only extended hops in pursuit of prey or escape from a predator. Once flight itself became an advantage, natural selection would have remodeled feathers and wings to better fit their additional function.

Exaptation offers one explanation for how novel features can arise gradually through a series of intermediate stages, each of which has some function in the organism's current context. Harvard zoologist Karel Liem puts it this way: "Evolution is like modifying a machine while it's running."

"Evo-devo": Genes that control development play a major role in evolution

The interface between evolutionary biology and the study of how organisms develop is called "evo-devo." This interdisciplinary research is beginning to illuminate how slight genetic divergences can become magnified into major morphological differences between species. Genes that program development control the rate, timing, and spatial pattern of changes in an organism's form as it is transfigured from a zygote into an adult.

The shape of an organism depends in part on the relative growth rates of its different parts during development. This proportioning that helps give a body its specific form is called **allometric growth** (from the Greek *allos,* other, and *metron,* measure). FIGURE 24.19a tracks how allometric growth alters human body proportions during development. Change these relative rates of growth even slightly, and you change the adult form substantially. For example, different allometric patterns contribute to the contrasting shapes of human and chimpanzee skulls (FIGURE 24.19b).

Evolution of morphology that arises by a modification in allometric growth is an example of **heterochrony,** which is evolutionary change in the rate or timing of developmental events. We can see another example of heterochrony in the evolution of salamander feet. Most salamanders are ground-dwelling, but some species live in trees. The feet of these tree-dwelling salamanders are adapted for climbing vertically rather than walking horizontally on the ground, with shorter digits and more webbing than the feet of ground-dwelling salamanders (FIGURE 24.20). The basis for this adaptation was

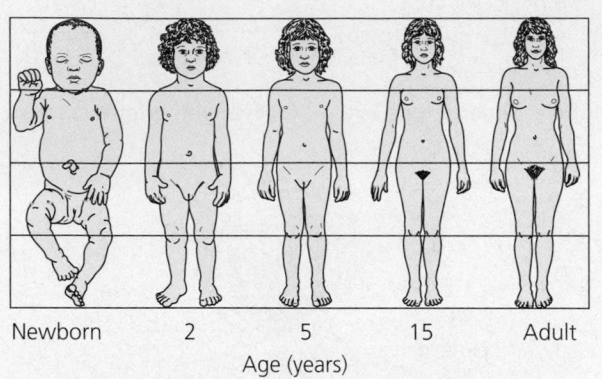

(a) Differential growth rates in a human. The arms and legs grow faster than the head and trunk, as can be seen in this conceptualization of different-aged individuals all rescaled to the same height.

Newborn   2   5   15   Adult
Age (years)

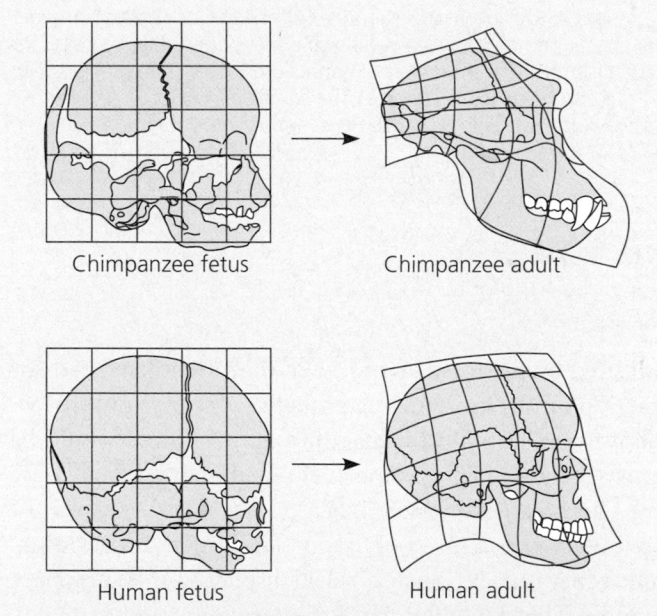

Chimpanzee fetus   Chimpanzee adult

Human fetus   Human adult

(b) Comparison of chimpanzee and human skull growth. The fetal skulls of humans and chimpanzees are similar in shape. Allometric growth transforms the rounded skull of a newborn chimpanzee into the sloping skull characteristic of adult apes as the jaw grows faster than other parts of the skull. The same basic allometric pattern occurs in humans, but with a less accelerated elongation of the jaw relative to the rest of the skull.

FIGURE 24.19 Allometric growth. Different growth rates for different parts of the body determine body proportions.

(a) Ground-dwelling salamander. A longer period of time for foot growth results in longer digits and less webbing.

(b) Tree-dwelling salamander. Foot growth ends sooner. This evolutionary change in the timing of foot development accounts for the shorter digits and more extensive webbing that help adapt tree-dwelling salamanders for climbing vertically.

FIGURE 24.20 Heterochrony and the evolution of salamander feet among closely related species.

FIGURE 24.21 Paedomorphosis. Some species retain as adults features that were juvenile in ancestors. This salamander is an axolotl, which grows to full size, becomes sexually mature, and reproduces while retaining certain larval (tadpole) characteristics, including gills.

probably selection for mutant alleles of genes controlling the timing of foot development. According to this hypothesis, the foot of the ancestral salamander grew until the products of certain regulatory genes switched off growth, resulting in a foot of a certain size. A mutation in one or more of these regulatory genes could have switched off foot growth sooner, resulting in the stunted feet of tree-dwelling salamanders. In this way, a relatively small amount of genetic change can be amplified into the substantial morphological change that marks macroevolution.

Heterochrony can also change the timing of reproductive development relative to development of somatic (nonreproductive) organs. If the rate of reproductive development accelerates compared to somatic development, the sexually mature stage of a species may retain body features that were juvenile structures in an ancestral species. This is called **paedomorphosis** (from the Greek *paedos,* child, and *morphosis,* formation). For example, most salamander species have a larval stage that undergoes metamorphosis to become an adult. But some species grow to adult size and become sexually mature while retaining gills and certain other larval features (FIGURE 24.21). Such an evolutionary alteration of developmental timing can produce animals that appear very different from their ancestors, even though the overall genetic change may be small.

In short, heterochrony affects the evolution of morphology by altering the rates of development of various body parts or by changing the timing of onset or completion of a particular part's development.

Macroevolution can also result from changes in genes that control the placement and spatial organization of body parts. For example, genes called **homeotic** genes determine such basic features as where a pair of wings and a pair of legs will develop on a bird or how a plant's flower parts are arranged (see Chapter 21).

The products of one class of homeotic genes called *Hox* genes provide positional information in an animal embryo. This information about position prompts cells to develop into structures appropriate for a particular location. Changes in *Hox* genes can have a profound impact on morphology. Consider, for example, the evolution of tetrapods (the terrestrial vertebrates: amphibians, reptiles, birds, and mammals) from fishes, which are aquatic vertebrates. One of the major transitions in this vertebrate history was the evolution of the walking legs of tetrapods from the fins of fishes. Unlike the fish fin, the tetrapod limb has digits (fingers and toes in humans) that extend skeletal support to the tip of the limb. During the development of a tetrapod, a *Hox* gene is expressed at the outer edge of the limb bud, the embryonic structure that develops into a leg. The product of this *Hox* gene apparently provides positional information about how far outward bones should grow in the limb (FIGURE 24.22). A related *Hox* gene is expressed during the development of a fish fin, but in a smaller region back from the tip of the fin bud. A mutation in this *Hox* gene that expanded its region of expression to the tip of the bud probably contributed to the evolution of skeletal extensions that made it possible for limbs to support vertebrates on land. The evolution of vertebrates from invertebrate animals was an even bigger episode in macroevolution, and it, too, was probably associated with changes in *Hox* genes (see FIGURE 24.23, p. 480).

In producing evolutionary novelties, changes in developmental dynamics, both temporal (heterochrony) and spatial, have undoubtedly played important roles in macroevolution.

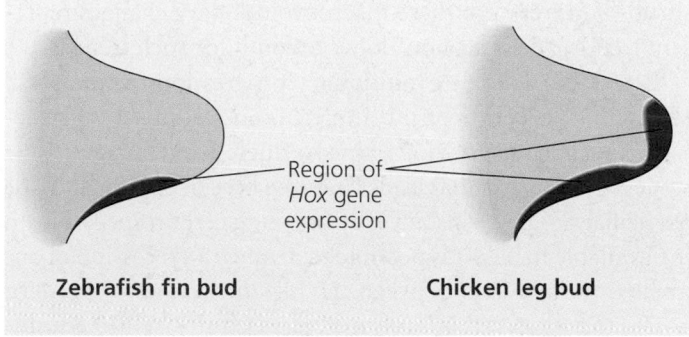

FIGURE 24.22 *Hox* genes and the evolution of tetrapod limbs. The red zones in these diagrams of embryonic appendages indicate regions where a *Hox* gene involved in skeleton development is expressed.

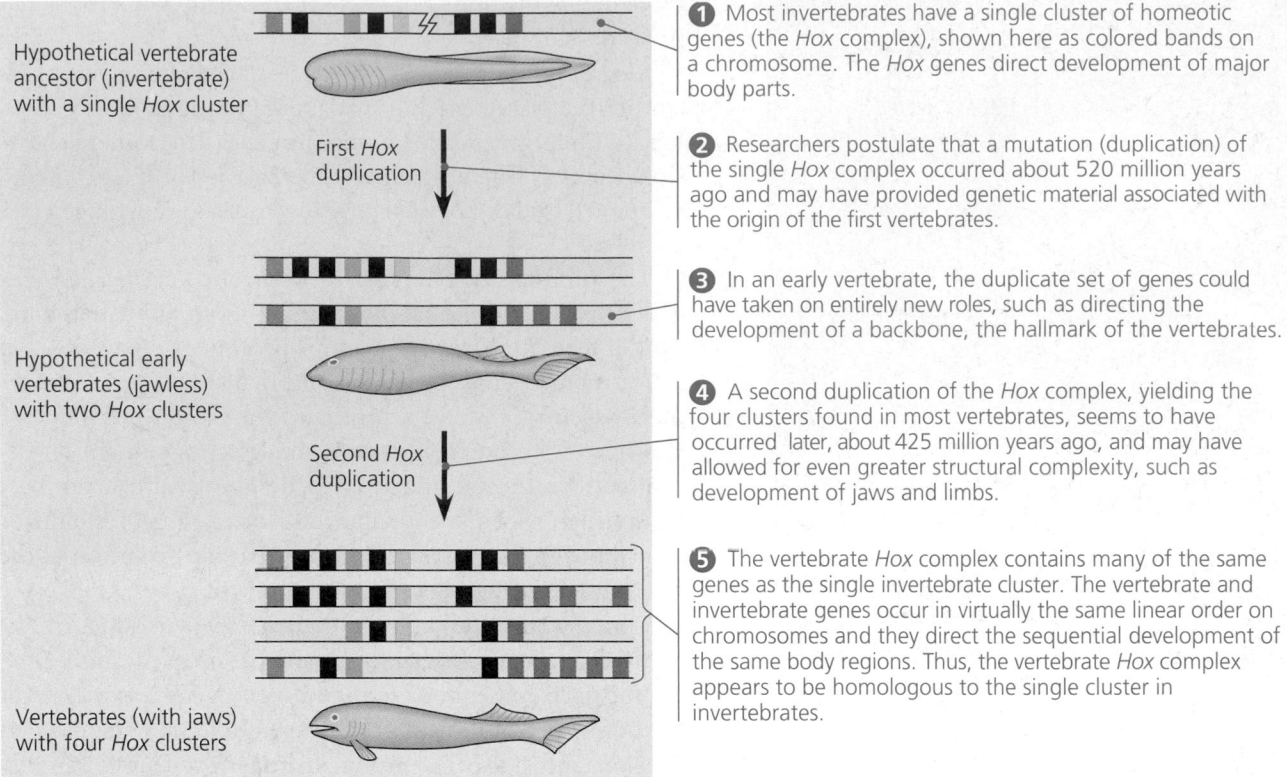

1 Most invertebrates have a single cluster of homeotic genes (the *Hox* complex), shown here as colored bands on a chromosome. The *Hox* genes direct development of major body parts.

Hypothetical vertebrate ancestor (invertebrate) with a single *Hox* cluster

First *Hox* duplication

2 Researchers postulate that a mutation (duplication) of the single *Hox* complex occurred about 520 million years ago and may have provided genetic material associated with the origin of the first vertebrates.

3 In an early vertebrate, the duplicate set of genes could have taken on entirely new roles, such as directing the development of a backbone, the hallmark of the vertebrates.

Hypothetical early vertebrates (jawless) with two *Hox* clusters

Second *Hox* duplication

4 A second duplication of the *Hox* complex, yielding the four clusters found in most vertebrates, seems to have occurred later, about 425 million years ago, and may have allowed for even greater structural complexity, such as development of jaws and limbs.

5 The vertebrate *Hox* complex contains many of the same genes as the single invertebrate cluster. The vertebrate and invertebrate genes occur in virtually the same linear order on chromosomes and they direct the sequential development of the same body regions. Thus, the vertebrate *Hox* complex appears to be homologous to the single cluster in invertebrates.

Vertebrates (with jaws) with four *Hox* clusters

FIGURE 24.23 *Hox* **mutations and the origin of vertebrates.**

An evolutionary trend does not mean that evolution is goal oriented

The fossil record seems to reveal trends in the evolution of many species and lineages. For instance, some lineages exhibit a trend toward larger or smaller body size. A case in point is the evolution of the modern horse, which is a descendant of a much smaller ancestor named *Hyracotherium*. About the size of a large dog, *Hyracotherium* browsed in woodlands some 40 million years ago. In comparison to this ancestor, modern horses (genus *Equus*) are larger, the number of toes has been reduced from four on each foot to one, and the teeth have become modified for grazing on grasses rather than browsing on shrubs and trees. Do these macroevolutionary changes represent trends, and if so, how do we account for such trends?

Extracting a single evolutionary progression from a fossil record that is likely to be incomplete is misleading; it is like describing a bush as growing toward a single point by tracing the system of branches that leads from the base of the bush to one particular twig. For instance, by selecting certain species from the available fossils, it is possible to arrange a succession of animals intermediate between *Hyracotherium* and modern horses that shows trends toward increased size, reduced number of toes, and modification of teeth for grazing (yellow line in FIGURE 24.24). We might interpret this series of fossils as an unbranched lineage leading directly from *Hyracotherium* to modern horses through a continuum of intermediate stages. If

we include all fossil horses known today, however, the illusion of coherent, progressive evolution leading directly to modern horses vanishes. The genus *Equus* is the only surviving twig of an evolutionary tree that is so branched that it is more like a bush. *Equus* descended through a series of speciation episodes that included several adaptive radiations, not all of which led to large, one-toed, grazing horses. For instance, notice in FIGURE 24.24 that only those lineages derived from *Parahippus* include grazers; other lineages derived from *Miohippus* produced browsers.

Branching evolution can produce a macroevolutionary trend even if some new species counter the trend. In a view of macroevolution enunciated by Steven Stanley, of Johns Hopkins University, species are analogous to individuals: Speciation is their birth, extinction is their death, and new species are their offspring. According to this model, an evolutionary trend is produced by **species selection,** which is analogous to the production of a trend within a population by natural selection. The species that endure the longest and generate the greatest number of new species determine the direction of major evolutionary trends. This concept suggests that differential speciation plays a role in macroevolution similar to the role of differential reproduction in microevolution.

To the extent that speciation rates and species longevity reflect success, the analogy to natural selection is even stronger. But qualities unrelated to the overall success of organisms in specific environments may be equally important in species

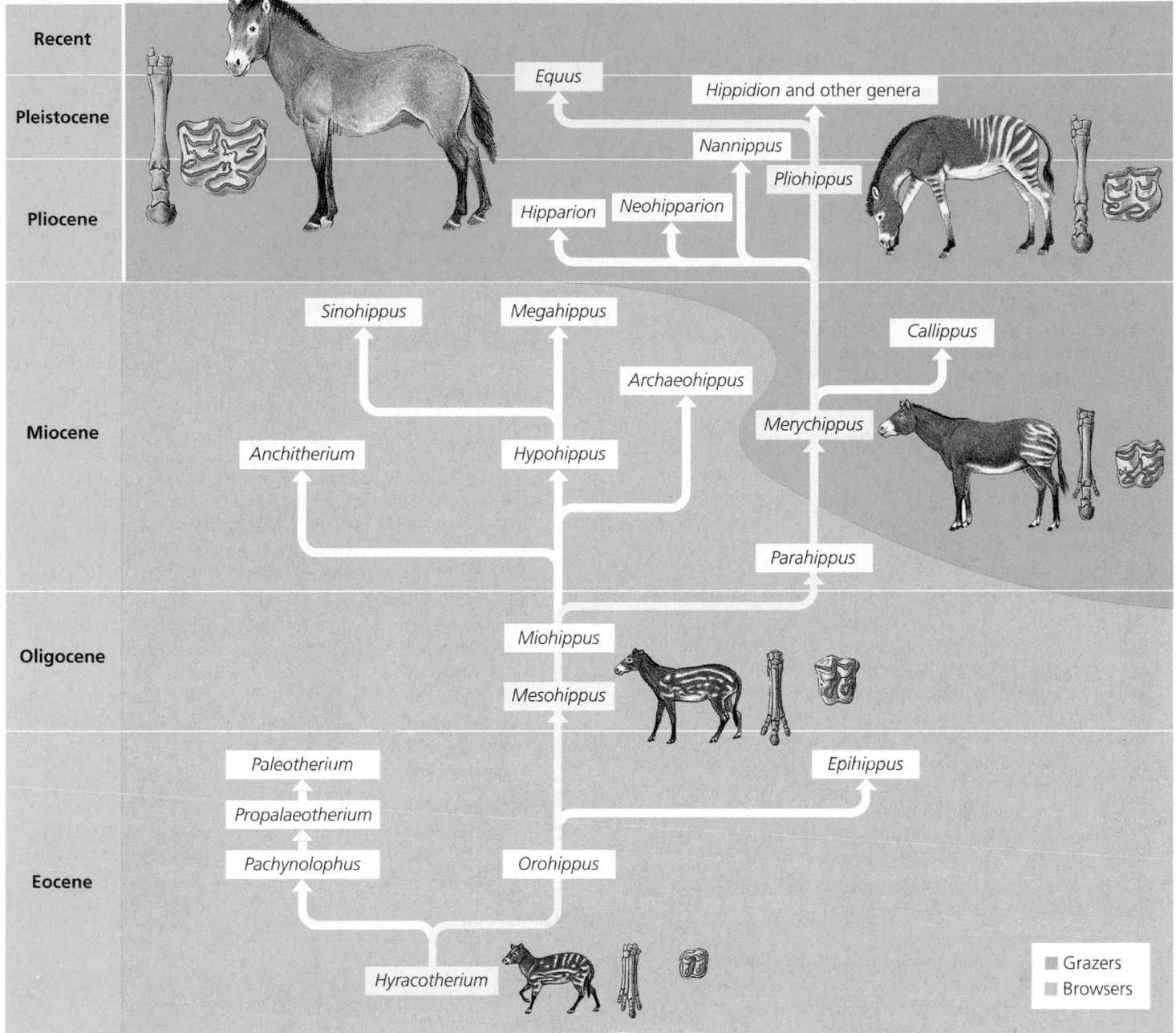

FIGURE 24.24 The branched evolution of horses. If we use a yellow highlighter to trace the sequence of fossil horses that are intermediate in form between the modern horse and its Eocene ancestor *Hyracotherium,* we create the illusion of a progressive trend toward larger size, reduced number of toes, and teeth modified for grazing. In fact, the modern horse (*Equus*) is the only surviving twig of an evolutionary bush with many divergent trends.

selection. For example, the ability of a species to disperse to new locations may contribute to its giving rise to a large number of "daughter species." The species selection model has many critics who argue that evolutionary trends more commonly result from the gradual modification of populations in response to environmental change. The value of such debates is that they stimulate research, and many paleontologists and other evolutionary biologists are now focusing on the question of what produces evolutionary trends.

Whatever its cause, the appearance of an evolutionary trend does not imply that there is some intrinsic drive toward a preordained state of being. Evolution is a response to interactions between organisms and their current environments. If conditions change, an evolutionary trend may cease or even reverse itself.

■   ■   ■

In the next chapter, we will continue our study of macroevolution with a more detailed look at the fossil record and some of the effects of significant environmental change on species and speciation.

CHAPTER 24 REVIEW

Go to the Campbell Biology website (www.campbellbiology.com) to explore an interactive version of the Chapter Review.

Summary of Key Concepts

■ Macroevolution is the origin of new species and other taxonomic groups. Two patterns of species change chronicled by the fossil record are anagenesis (phyletic evolution), the accumulation of changes associated with the transformation of one species into another, and cladogenesis, branching evolution (FIGURE 24.1).

Web/CD Activity 24A: *Overview of Macroevolution*

WHAT IS A SPECIES?

■ **The biological species concept emphasizes reproductive isolation (p. 465 and FIGURE 24.2)** The biological species concept defines a species as a group of populations whose individuals have the potential to interbreed and produce fertile offspring with each other but not with members of other species.

■ **Prezygotic and postzygotic barriers isolate the gene pools of biological species (pp. 465–467, FIGURE 24.5)** Prezygotic barriers prevent mating or fertilization between species. Species that occupy the same geographic area often live in separate habitats (habitat isolation); possess unique, exclusive mating signals and courtship behaviors (behavioral isolation); breed at different times (temporal isolation); and/or have anatomically incompatible reproductive organs (mechanical isolation) or incompatible sex cells (gametic isolation). Even if two different species manage to mate, postzygotic barriers usually prevent the interspecific hybrids from developing into adults, breeding with either parent species, or producing viable, fertile offspring.

■ **The biological species concept has some major limitations (p. 468)** For instance, it is not applicable to fossils or to organisms that reproduce only asexually.

■ **Biologists have proposed several alternative concepts of species (p. 468)** Alternative concepts include the ecological species concept, which explains the similarity among members of a species by their adaptation to exploit a particular set of ecological resources (or niche); the morphological species concept, which defines species by phenotypic characteristics; pluralistic species concept, suggesting that different factors operate in different creatures to explain why species exist; and a genealogical species concept, which defines a species by the close genetic relatedness among its members.

MODES OF SPECIATION

■ **Allopatric speciation: Geographic barriers can lead to the origin of species (pp. 469–473, FIGURES 24.6–24.12)** Allopatric speciation may occur when two populations of one species become geographically separated from each other. One or both populations may undergo evolutionary change and become reproductively isolated as a consequence of that change.

■ **Sympatric speciation: A new species can originate in the geographic midst of the parent species (pp. 473–475, FIGURES 24.13–24.16)** Many plant species have evolved by polyploidy (multiplications of the chromosome number). Autopolyploids are species derived this

way from one ancestral species. Allopolyploids are species with multiple sets of chromosomes derived from two different species. Examples of processes that can result in sympatric speciation in animals are host-switching by parasites and nonrandom mating in polymorphic populations.

Web/CD Case Study in the Process of Science: *How Do New Species Arise by Genetic Isolation?*

■ **The punctuated equilibrium model has stimulated research on the tempo of speciation (pp. 475–476, FIGURE 24.17)** The punctuated equilibrium model suggests that species change most as they bud from an ancestral species, after which they undergo relatively little change for the rest of their existence.

FROM SPECIATION TO MACROEVOLUTION

■ **Most evolutionary novelties are modified versions of older structures (p. 477, FIGURE 24.18)** Most novel biological structures evolve in many stages from previously existing structures. In some cases, such as the eye, the function of the organ has probably been constant during all stages of its evolution. In others, such as feathers, the function of the organ has changed.

■ **"Evo-devo": Genes that control development play a major role in evolution (pp. 477–479, FIGURES 24.19–24.23)** Many macroevolutionary changes may have been associated with mutations in genes that regulate development. Such changes can affect the timing of developmental events (heterochrony) or the spatial organization of body parts, as in mutations of homeotic genes.

Web/CD Activity 24B: *Allometric Growth*

■ **An evolutionary trend does not mean that evolution is goal oriented (pp. 480–481, FIGURE 24.24)** Long-term evolutionary trends may arise because of adaptation to a changing environment. Or, according to the species selection hypothesis, trends may result when species with certain characteristics endure longer and speciate more often than those with other characteristics.

Self-Quiz

1. Most of biological diversity has probably arisen by
   a. anagenesis.   d. hybridization.
   b. cladogenesis.   e. sympatric speciation.
   c. phyletic evolution.

2. The *largest* unit in which gene flow is possible is a
   a. population.   d. subspecies.
   b. species.   e. phylum.
   c. genus.

3. Bird guides once listed the myrtle warbler and Audubon's warbler as distinct species, but applying the biological species concept, recent books show them as eastern and western forms of a single species, the yellow-rumped warbler. Experts must have found that the two kinds of warblers
   a. live in the same areas.
   b. successfully interbreed in nature.
   c. look enough alike to be considered one species.
   d. are reproductively isolated from each other.
   e. are allopatric.

4. Among allopatric species of *Anopheles* mosquito, some live in brackish water, some in running fresh water, and others in stagnant water. What type of reproductive barrier is most obviously separating these different species?
   a. habitat isolation
   b. temporal isolation
   c. behavioral isolation
   d. gametic isolation
   e. postzygotic barriers

5. A genetic change that caused a certain *Hox* gene to be expressed along the tip of a vertebrate limb bud instead of further back made possible the evolution of the tetrapod limb. This type of change is illustrative of
   a. heterochrony, a change in the timing of developmental events.
   b. allopolyploidy, an increase in chromosome number.
   c. paedomorphosis, or retention of ancestral juvenile structures in an adult organism.
   d. a change in a homeotic developmental gene that altered the spatial organization of body parts.
   e. allopatric speciation.

6. According to advocates of the punctuated equilibrium model,
   a. natural selection is unimportant as a mechanism of evolution.
   b. given enough time, most existing species will branch gradually into new species.
   c. a new species accumulates most of its unique features as it comes into existence, then changes little for the rest of its duration as a species.
   d. most evolution is anagenic.
   e. speciation is usually due to a single mutation.

7. Which of the following species concepts identifies species based on their shared genetic histories and thus unique genetic markers?
   a. biological
   b. ecological
   c. genealogical
   d. morphological
   e. pluralistic

8. Which of the following factors would *not* contribute to allopatric speciation?
   a. A population becomes geographically isolated from the parent population.
   b. The separated population is small, and genetic drift occurs.
   c. The isolated population is exposed to different selection pressures than the ancestral population.
   d. Gene flow between the two populations is minimal or does not occur.
   e. The different environments of the two populations create different mutations.

9. Plant species A has a diploid number of 12. Plant species B has a diploid number of 16. A new species, C, arises as an allopolyploid from hybridization of A and B. The diploid number of C would probably be
   a. 12.   d. 28.
   b. 14.   e. 56.
   c. 16.

10. The speciation episode described in question 9 is most likely a case of
   a. allopatric speciation.
   b. sympatric speciation.
   c. speciation based on sexual selection.
   d. adaptive radiation.
   e. anagenic speciation.

11. Contrast microevolution with macroevolution.

12. Explain why nonbranching evolution (anagenesis) cannot increase the number of species.

13. Why would allopatric speciation be less common on an island close to a mainland than on a more isolated island of the same size?

14. How does the punctuated equilibrium model account for the relative rarity of transitional fossils linking newer species to older ones?

15. Explain why the concept of exaptation does not imply that a structure evolves in anticipation of some future environmental change.

16. How can heterochrony cause the evolution of body form by altering allometric growth?

Go to the website or CD-ROM for more quiz questions.

Evolution Connection

In the margin of one of his notebooks, Darwin scrawled a note to remind himself never to apply the terms "higher" or "lower" to species. It was, and is still, very common for people to think of some species or species groups as more or less evolved than others. This probably stems from a notion of evolutionary "progress." Is there such a thing as evolutionary progress? Why or why not? Defend your position as if you were debating someone holding the opposite view.

The Process of Science

Cultivated American cotton plants have a total of 52 chromosomes ($2n = 52$). In each cell, 13 pairs of chromosomes are smaller than the rest. Old World cotton plants have 26 chromosomes ($2n = 26$), all large. Wild American cotton plants have 26 chromosomes, all small. Propose a hypothesis to explain how cultivated American cotton may have originated. How could you test your hypothesis?

Learn about sympatric speciation through a series of plant experiments in the Case Study in the Process of Science, available on the website and CD-ROM.

Science, Technology, and Society

What is the biological basis for assigning all human populations to a single species? Explain why it is unlikely that a second human species could arise by cladogenesis in the future.

Answers: 1. b; 2. b; 3. b; 4. a; 5. d; 6. c; 7. c; 8. e; 9. d; 10. b; 11. Microevolution is a change in the gene pool of a population, often associated with adaptation; macroevolution is a change in life-form, such as the origin of a new species, that is noticeable enough to be evident in the fossil record. 12. In nonbranching evolution, an ancestral species changes gradually to form a new species; thus, there is no effect on species number. 13. Continued gene flow between mainland populations and those on nearby islands reduces the chance of enough genetic divergence for speciation. 14. According to this model, the time required for speciation in most cases is relatively short compared to the overall duration of the species' existence. Thus, on the vast geologic time scale of the fossil record, the transition of one species to another seems abrupt. 15. Although an exaptation is co-opted for new or additional functions in a new environment, it existed because it worked as an adaptation to the old environment. 16. If heterochrony alters allometric growth, then a change in the differential growth rates of different body parts alters morphology.

PHYLOGENY AND SYSTEMATICS

THE FOSSIL RECORD AND GEOLOGIC TIME

- Sedimentary rocks are the richest source of fossils
- Paleontologists use a variety of methods to date fossils
- The fossil record is a substantial, but incomplete, chronicle of evolutionary history
- Phylogeny has a biogeographic basis in continental drift
- The history of life is punctuated by mass extinctions

SYSTEMATICS: CONNECTING CLASSIFICATION TO PHYLOGENY

- Taxonomy employs a hierarchical system of classification
- Modern phylogenetic systematics is based on cladistic analysis
- Systematists can infer phylogeny from molecular data
- The principle of parsimony helps systematists reconstruct phylogeny
- Phylogenetic trees are hypotheses
- Molecular clocks may keep track of evolutionary time
- Modern systematics is flourishing with lively debate

Evolutionary biology *is about both process and history. We have already studied natural selection and other processes that change populations (Chapter 23), and we have examined processes that can lead to the origin of new species (Chapter 24). But evolutionary biology also seeks to reconstruct the history of life on Earth.*

*This chapter describes how biologists trace **phylogeny** (from the Greek, phylon, tribe, and genesis, origin), the evolutionary history of a species or group of related species. Reconstructing phylogeny is part of the scope of systematics, the study of biological diversity in an evolutionary context. Systematists also name and classify species. Our study of phylogeny and systematics begins with the fossils left by past life, such as the ancient fish in the photo above. We then examine the techniques through which systematists test and refine their hypotheses about phylogeny and classification. In the process, we see how molecular biology is changing systematics, as it is changing every field of biology.*

THE FOSSIL RECORD AND GEOLOGIC TIME

Fossils, the preserved remnants or impressions left by organisms that lived in the past, are the historical documents of biology. The **fossil record** is the ordered array in which fossils appear within layers, or strata, of sedimentary rocks that mark the passing of geologic time. Paleontologists collect and interpret fossils (FIGURE 25.1). We begin this section by studying how fossils form and how paleontologists determine their age. We then turn to the contributions and limitations of the fossil record in the study of phylogeny.

Sedimentary rocks are the richest source of fossils

Sedimentary rocks form from layers of minerals that settle out of water or as dust and sand on land (see FIGURE 22.3). Sand and silt that are weathered and eroded from the land are carried by rivers to seas and swamps, where the particles settle to the bottom. Deposits pile up and compress the older sediments below into rock—sand into sandstone and mud into shale. When aquatic life-forms and terrestrial organisms swept into the seas and swamps die, they settle along with the sediments. A tiny fraction of them are then preserved as fossils. At any particular location, sedimentation is not continuous but occurs in intervals when the sea level changes or lakes and swamps dry up and refill. Even when a region is submerged, the rate of sedimentation and the types of sedimentary particles vary over time. As a result of these different periods of sedimentation, the rock forms in strata (see FIGURE 22.4).

The organic substances of a dead organism buried in sediments usually decay rapidly. However, hard parts that are rich

FIGURE 25.1 A gallery of fossils.

(a) Dinosaur bones

(b) Skull

(c) Petrified trees

(d) Leaf

(e) Brachiopod casts

(f) Dinosaur tracks

(g) Scorpion in amber

(h) Mammoth tusks

(a) Sedimentary rocks are the richest hunting grounds for paleontologists. This researcher is excavating a fossilized dinosaur skeleton from sandstone in Dinosaur National Monument, located in Utah and Colorado. **(b)** The hard parts of organisms are the most common fossils. This is a skull of *Homo erectus,* a human ancestor that lived about 1.5 million years ago. **(c)** Fossils may become even harder if minerals replace their organic matter. These petrified (stone) trees in Arizona are about 190 million years old. **(d)** Some sedimentary fossils, such as this 40-million-year-old leaf, retain organic material, including DNA, which scientists can analyze. **(e)** Buried organisms that decay may leave molds that may be filled by minerals dissolved in water. The casts that form when the minerals harden are replicas of the organisms, as in the case of these casts of animals called brachiopods. **(f)** Trace fossils are footprints, burrows, and other remnants of an ancient organism's behavior. This boy is standing in a 150-million-year-old dinosaur track in Colorado. **(g)** This 30-million-year-old scorpion is embedded in amber (hardened resin from a tree). **(h)** These tusks belong to a whole 23,000-year-old mammoth, which scientists discovered in Siberian ice in 1999.

in minerals, such as the shells of many invertebrates and protists and the bones and teeth of vertebrates, may remain as fossils (FIGURE 25.1a and b). Paleontologists have unearthed nearly complete skeletons of dinosaurs and other forms, but more often the finds consist of parts of skulls, bone fragments, or teeth. Many of these relics are hardened even more and preserved by chemical changes; under the right conditions, minerals dissolved in groundwater seep into the tissues of a dead organism and replace its organic material. The plant or animal turns to stone (see FIGURE 25.1c).

Rarer than mineralized fossils are those that retain organic material. They are sometimes discovered as thin films pressed between layers of sandstone or shale. For example, paleontologists have discovered plant leaves millions of years old that are still green with chlorophyll and well enough preserved for their organic composition to be analyzed and the ultrastructure of their cells to be explored with the electron microscope (see FIGURE 25.1d). The most common fossilized plant material is pollen, which has a hard organic case that resists degradation.

The fossils that paleontologists find in many of their digs are not the actual remnants of organisms at all, but rocks that form as replicas of the organisms. These fossils result when a dead organism captured in sediment decays and leaves an empty mold that becomes filled with minerals dissolved in water. The minerals may subsequently crystallize, forming a cast in the shape of the organism (see FIGURE 25.1e).

Trace fossils consist of footprints, animal burrows, or other impressions left in sediments by the activities of animals. These rocks are in essence fossilized behavior; they tell paleontologists something about how the animals that left the trace fossils lived. For example, dinosaur tracks provide clues about the animal's locomotion—its gait (pattern of leg movements), stride length, and speed (see FIGURE 25.1f).

If an organism happens to die in a place where bacteria and fungi cannot decompose the corpse, the entire body, including soft parts, may be preserved as a fossil. For example, the scorpion in FIGURE 25.1g got stuck in a drop of resin from a tree about 30 million years ago. The resin eventually hardened into amber, entombing the animal. There are other mechanisms that preserve whole organisms. Explorers have discovered mammoths, bison, and even prehistoric humans frozen in ice or preserved in acid bogs, where conditions retard decomposition (see FIGURE 25.1h). Such rare discoveries make the news, but biologists rely mainly on more common sedimentary fossils to reconstruct the history of life.

Paleontologists use a variety of methods to date fossils

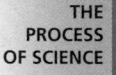

THE PROCESS OF SCIENCE

Fossils are reliable historical data only if we can determine their ages. Here we examine the methods for determining where fossils fit on the geologic time scale.

Relative Dating

The trapping of dead organisms in sediments freezes fossils in time. Thus, the fossils in each stratum of sedimentary rock are a local sample of the organisms that existed at the time that sediment was deposited. Because younger sediments are superimposed upon older ones, this book of sedimentary pages tells the relative ages of fossils.

The strata at one location can often be correlated with strata at another location by the presence of similar fossils, known as index fossils. The best index fossils for correlating strata that are far apart are the shells of sea animals that were widespread. At any one location where a roadcut or canyon wall reveals layered rocks, there are likely to be gaps in the sequence. That area may have been above sea level during different periods, and thus no sedimentation occurred; or some of the sedimentary layers that were deposited when the area

was submerged may have been scraped away by subsequent periods of erosion.

By studying many different sites, geologists have established a **geologic time scale** with a consistent sequence of historical periods (TABLE 25.1). These periods are grouped into four eras: the Precambrian, Paleozoic, Mesozoic, and Cenozoic eras. Each era represents a distinct age in the history of Earth and its life. For example, the Mesozoic era is sometimes called the "age of reptiles" because of its abundance of reptilian fossils, including dinosaurs. The boundaries between the eras correspond to times of mass extinctions, when many forms of life disappeared and were replaced by diversification of the survivors. Lesser extinctions also mark the boundaries of many of the periods that make up each era. The periods within each era are further subdivided into finer intervals called epochs (only the epochs of the current era, the Cenozoic, appear in TABLE 25.1). The timeline to the left of TABLE 25.1 shows that the geologic eras were not equal in duration. The periods also varied in length. For example, the Jurassic period lasted almost twice as long as the Triassic period. Scientists have not divided geologic time in an arbitrary manner, but have located boundaries in the record of the rocks that correspond to times of great change.

The record of the rocks is a serial that chronicles the *relative* ages of fossils; it tells us the order in which groups of species present in a sequence of strata evolved. However, the series of sedimentary rocks does not tell us the *absolute* ages of the embedded fossils. The difference is analogous to peeling the layers of wallpaper from the walls of a very old house that has been inhabited by many owners. You could determine the sequence in which the papers had been applied, but not the year that each layer was added.

Absolute Dating

"Absolute" dating does not mean errorless dating, but only that age is given in years instead of relative terms such as *before* and *after, early* and *late.* **Radiometric dating,** the measurement of certain radioactive isotopes in fossils or rocks, is the method most often used to determine the ages of rocks and fossils on a scale of absolute time. Fossils contain isotopes of elements that accumulated in the organisms when they were alive. For example, the carbon in a living organism includes both the most common carbon isotope, carbon-12, and a less common radioactive isotope, carbon-14, in the same ratio as is present in the atmosphere. Once an organism dies, it stops accumulating carbon. The radioactive carbon-14 that it contained at the time of death slowly decays and becomes another element, nitrogen-14, causing the proportion of carbon-14 to total carbon content to decline. Each radioactive isotope has a fixed rate of decay. An isotope's **half-life,** the number of years it takes for 50% of the original sample to decay, is unaffected

Table 25.1   The Geologic Time Scale

Relative Time Span of Eras	Era	Period	Epoch	Age (Millions of Years Ago)	Some Important Events in the History of Life
Cenozoic	Cenozoic	Quaternary	Recent		Historical time
Mesozoic				0.01	
			Pleistocene		Ice ages; humans appear
Paleozoic				1.8	
		Tertiary	Pliocene		Apelike ancestors of humans appear
				5	
			Miocene		Continued radiation of mammals and angiosperms
				23	
			Oligocene		Origins of many primate groups, including apes
				35	
			Eocene		Angiosperm dominance increases; continued radiation of most modern mammalian orders
				57	
			Paleocene		Major radiation of mammals, birds, and pollinating insects
				65	
Pre-cambrian	Mesozoic	Cretaceous			Flowering plants (angiosperms) appear; many groups of organisms, including dinosaurs, become extinct at end of period (Cretaceous extinctions)
				144	
		Jurassic			Gymnosperms continue as dominant plants; dinosaurs abundant and diverse
				206	
		Triassic			Cone-bearing plants (gymnosperms) dominate landscape; radiation of dinosaurs
				245	
	Paleozoic	Permian			Extinction of many marine and terrestrial organisms (Permian mass extinction); radiation of reptiles; origins of mammal-like reptiles and most modern orders of insects
				290	
		Carboniferous			Extensive forests of vascular plants; first seed plants; origin of reptiles; amphibians dominant
				363	
		Devonian			Diversification of bony fishes; first amphibians and insects
				409	
		Silurian			Diversity of jawless fishes; first jawed fishes; diversification of early vascular plants
				439	
		Ordovician			Marine algae abundant; colonization of land by plants and arthropods
				510	
		Cambrian			Radiation of most modern animal phyla (Cambrian explosion)
				543	
	Precambrian			600	Diverse soft-bodied invertebrate animals; diverse algae
				2,200	Oldest fossils of eukaryotic cells
				2,700	Atmospheric oxygen begins to accumulate
				3,500	Oldest fossils of cells (prokaryotes)
				3,800	Earliest traces of life
				4,600	Approximate time of origin of Earth

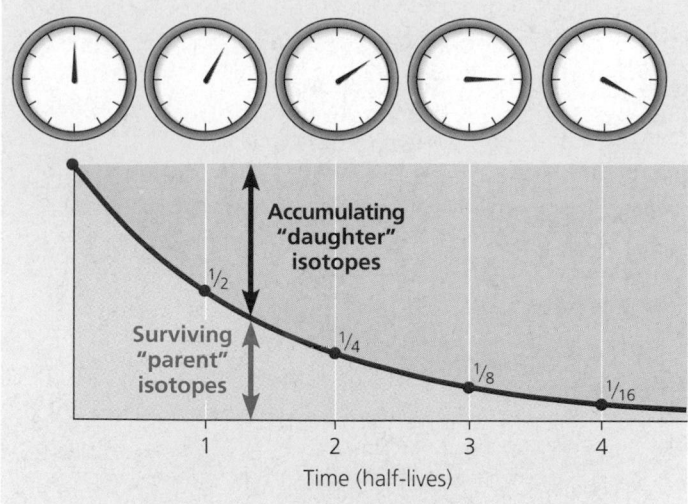

FIGURE 25.2 Radiometric dating. The graph plots the decay of "parent" isotopes, usually radioactive isotopes, to form "daughter" isotopes. An example is the clocklike decay of carbon-14 to form nitrogen-14. Each type of radioactive isotope has a set half-life, the time required for 50% of the isotope to decay. For example, carbon-14 has a half-life of 5,730 years; uranium-238 has a half-life of 4.5 billion years. Paleontologists use such radiometric clocks as one method to date fossils or to infer the ages of fossils by dating surrounding rock.

by temperature, pressure, and other environmental variables (FIGURE 25.2). Carbon-14 has a half-life of 5,730 years, a reliable rate of decay that can be used to date relatively young fossils. Thus, by measuring either the amount of nitrogen-14 in a fossil or the amount of carbon-14 remaining, we can determine the fossil's age.

Paleontologists use radioactive isotopes with longer half-lives to date older fossils. For example, uranium-238, which has a half-life of 4.5 billion years, has been used as a radiometric clock to infer the ages of fossils in rocks of the Cambrian period (see TABLE 25.1). Unlike carbon-14, uranium-238 is not present in living organisms. It occurs in molten lava and the volcanic rock that forms as the lava cools. After the volcanic rock forms, no more uranium-238 is incorporated, and the rock's original stock of the isotope decays slowly into lead-206 atoms. Thus, researchers can use the concentration of lead-206 to date volcanic rocks. And paleontologists can use these measurements to date very old fossils found near the volcanic rocks. For example, if fossils of animals are found in a rock layer sandwiched between two layers of volcanic rock that are 530 and 520 million years old, respectively, we can infer that the animals lived about 525 million years ago.

Methods other than radiometric dating can be used on some fossils. Amino acids exist in two isomers with either left-handed or right-handed symmetry, designated the L and D forms, respectively. Organisms synthesize only L-amino acids, which are incorporated into proteins. After an organism dies, however, its population of L-amino acids is slowly converted, resulting in a mixture of L- and D-amino acids. In a fossil, the ratio of L- and

D-amino acids can be measured. Knowing the rate at which this chemical conversion, called racemization, takes place, we can determine how long the organism has been dead. For instance, archaeologists have used this method to date ostrich eggshells found along with fossils of early humans. The humans probably ate the eggs and used the shells for water bowls. Racemization, unlike radioactive decay, is temperature sensitive, meaning that past changes in climate made the racemization clock run faster or slower. But for fossils found in locations where climate apparently has not changed significantly since the fossils formed, the two dating methods agree closely on the age of the fossils.

The fossil record is a substantial, but incomplete, chronicle of evolutionary history

The discovery of a fossil is the culmination of a sequence of improbable coincidences. First, the organism had to die in the right place at the right time for burial conditions to favor fossilization. Then the rock layer containing the fossil had to escape geologic processes that destroy or severely distort rocks, such as erosion, pressure from superimposed strata, or the melting of rocks that occurs at some locations. If the fossil *was* preserved, there is only a slight chance that a river carving a canyon or some other process will expose the rock containing the fossil. The chance that someone will find the fossil is even more remote, although, of course, discovery is more probable for people who are purposefully looking for fossils. No wonder the fossil record is incomplete: A substantial fraction of species that have lived probably left no fossils, most fossils that formed have been destroyed, and only a fraction of the existing fossils have been discovered.

The fossil record, far from being a complete sample of organisms of the past, is slanted in favor of species that existed for a long time, were abundant and widespread, and had shells or hard skeletons. Paleontologists, like all historians, must reconstruct the past from incomplete records. Even with its limitations, however, the fossil record is a remarkably detailed document of phylogeny over the vast scale of geologic time. The order of sedimentary strata records the sequence of biological change, and dating methods tell us approximately how long ago the changes occurred. Also recorded in rocks is the chronology of environmental changes associated with evolutionary changes in life. We now turn to some of the major changes in Earth's environments that have caused significant transformations of life.

Phylogeny has a biogeographic basis in continental drift

Evolution has dimension in space as well as in time. Indeed, it was biogeography, even more than fossils, that first nudged

Darwin and Wallace toward an evolutionary view of life. The history of Earth helps explain the current geographic distribution of species. For example, the emergence of volcanic islands such as the Galápagos opens new environments for founders that reach the outposts, and adaptive radiation fills many of the available niches with new species. On a global scale, the drifting of continents is the major geographic factor correlated with the spatial distribution of life and with such evolutionary episodes as mass extinctions and explosive increases in biological diversity.

The continents are not fixed, but drift about Earth's surface like passengers on great plates of crust floating on the hot, underlying mantle (FIGURE 25.3a). Unless two landmasses are embedded in the same plate, their positions relative to each other change. For example, North America and Europe are presently drifting apart at a rate of about 2 cm per year. Many important geologic processes, including mountain building, volcanism, and earthquakes, occur at plate boundaries (FIGURE 25.3b). California's infamous San Andreas Fault is part of a border where two plates slide past each other.

Plate movements rearrange geography incessantly, but two chapters in the continuing saga of continental drift had an especially strong influence on life. About 250 million years ago, near the end of the Paleozoic era, plate movements brought all the landmasses together into a supercontinent that has been named **Pangaea,** meaning "all land" (FIGURE 25.4, p. 490). Imagine some of the possible effects on life. Species that had been evolving in isolation came together and competed. When the landmasses coalesced, the total amount of shoreline was reduced, and there is evidence that the ocean basins increased in depth, which lowered sea level and drained the shallow coastal seas. Then, as now, most marine species inhabited shallow waters, and the formation of Pangaea destroyed a considerable amount of that habitat. It was probably a long, traumatic period for terrestrial life as well. The continental interior, which has a drier and more erratic climate than coastal regions, increased in area substantially when the land came together. Changing ocean currents also would have affected land life as well as sea life. The formation of Pangaea surely had a tremendous environmental impact that reshaped biological

FIGURE 25.3 Earth's crustal plates and plate tectonics (geologic processes resulting from plate movements).

(a) The modern continents are passengers on crustal plates that are swept across Earth's surface by convection currents of the hot mantle below. The red arrowheads indicate zones of violent tectonic events; many of these are subduction zones, where the edge of one plate is being pushed over the edge of an adjacent plate.

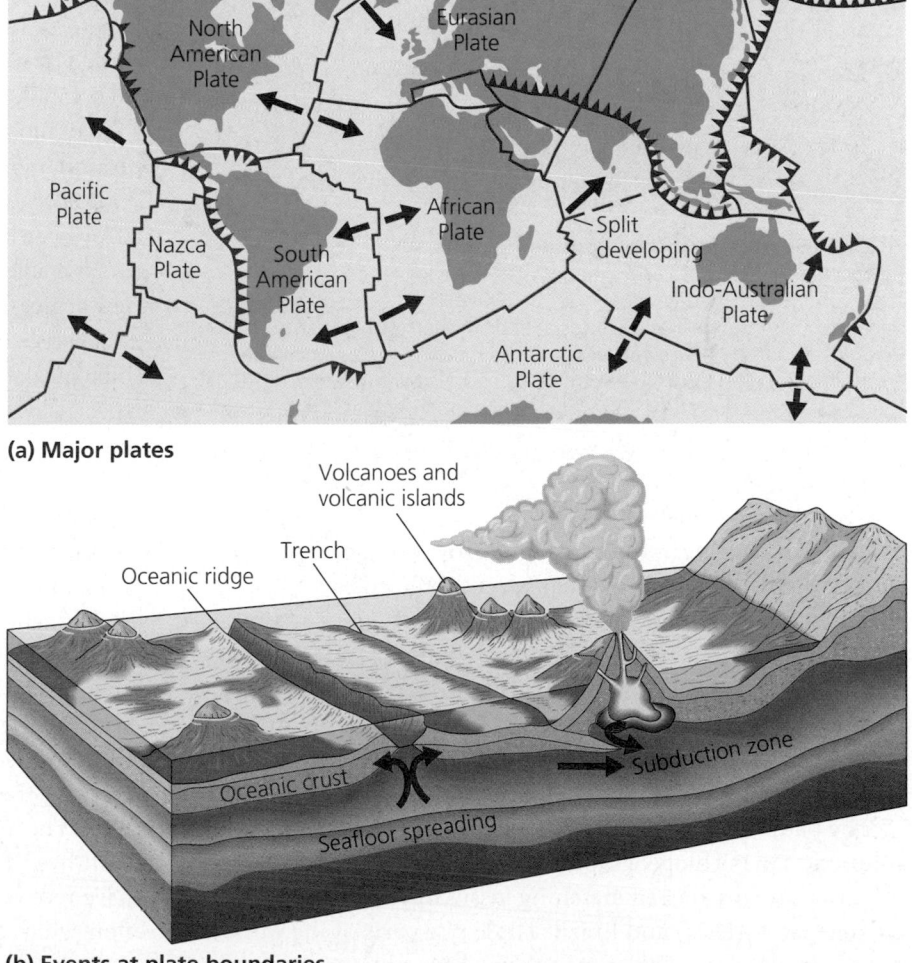

(a) Major plates

(b) At some plate boundaries, such as oceanic ridges (indicated by paired, opposing black arrows in part a), the plates separate, and molten rock wells up in the gap. The rock solidifies and adds crust symmetrically to both plates, a phenomenon called seafloor spreading. At subduction zones, where plates move toward each other, the denser plate dives below the less dense one, creating a trench. The abrasion at subduction zones causes earthquakes and volcanic eruptions. When continents riding on different plates collide, they pile up and build mountains.

(b) Events at plate boundaries

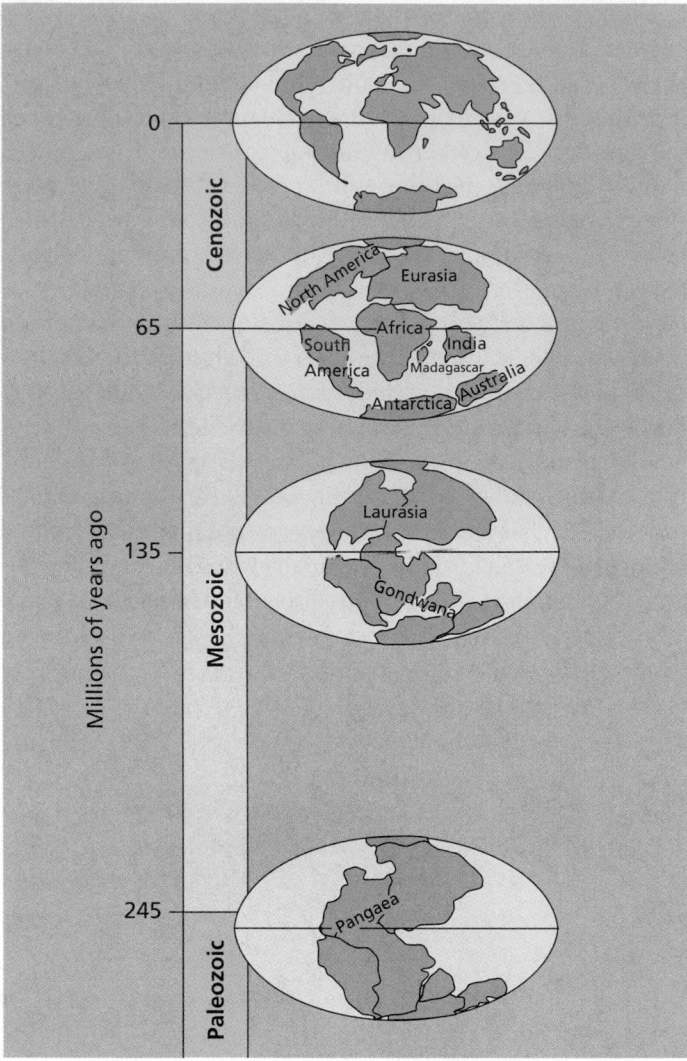

FIGURE 25.4 The history of continental drift. About 250 million years ago, all of Earth's landmasses were locked together in a supercontinent named Pangaea. About 180 million years ago, Pangaea began to split into northern (Laurasia) and southern (Gondwana) landmasses, which later separated into the modern continents. India collided with Eurasia just 10 million years ago, forming the Himalayas, the tallest and youngest of Earth's major mountain ranges. The continents continue to drift.

diversity by causing extinctions and providing new opportunities for taxonomic groups that survived the crisis.

The second dramatic chapter in the history of continental drift was written about 180 million years ago, during the Mesozoic era. Pangaea began to break up, causing geographic isolation of colossal proportions. As the continents drifted apart, each became a separate evolutionary arena, and the faunas and floras of the different biogeographic realms diverged.

The pattern of continental mergings and separations is the solution to many biogeographic puzzles. For example, paleontologists have discovered matching fossils of Triassic reptiles in Ghana (West Africa) and Brazil. These two parts of the world, now separated by 3,000 km of ocean, were contiguous during

the early Mesozoic era. Continental drift also explains much about the current distribution of organisms, such as why the Australian fauna and flora contrast so sharply with that of the rest of the world. The great diversity of marsupials (pouched mammals), which fill ecological roles in Australia analogous to those filled by eutherian (placental) mammals on other continents, is just one example of Australia's unique collection of species. Marsupials probably evolved first in what is now North America and reached Australia via South America and Antarctica while the continents were still joined. The subsequent breakup of the southern continents set Australia "afloat" like a great ark of marsupials. On Australia, marsupials evolved and diversified, while eutherian mammals evolved and diversified on other continents. Australia has been isolated for 50 million years. Bats, rats, mice, and humans (and their domesticated animals) are the only eutherian mammals that have managed to populate the island continent.

The history of life is punctuated by mass extinctions

The fossil record reveals an episodic history, with long, relatively quiescent periods punctuated by briefer intervals when the turnover in species composition was much more extensive. These biological upheavals were times of mass extinctions followed by extensive diversification of certain taxonomic groups that escaped extinction. We will emphasize episodes of radiation of biological diversity in the next unit of chapters. Here we examine some examples of mass extinctions.

A species may become extinct because its habitat has been destroyed or because the environment has changed in a direction unfavorable to the species. If ocean temperatures fall by even a few degrees, many species that are otherwise beautifully adapted will perish. Even if physical factors in the environment are stable, biological factors may change; the environment in which a species lives includes the other organisms that live there, and evolutionary change in one species is likely to have some impact on other species in the community. For example, the evolution by some Cambrian animals of hard body parts, such as jaws and shells, may have made some organisms lacking hard parts more vulnerable to predation and thereby more prone to extinction.

Extinction is in fact inevitable in a changing world, and there have been crises in the history of life when global environmental changes have been so rapid and disruptive that a majority of species were swept away. Mass extinctions are known primarily from the decimation of hard-bodied animals that lived in shallow seas, the organisms for which the fossil record is most complete. The fossil record chronicles a number of mass extinctions, with five to seven of them particularly extensive. The two that have received the most attention are the Permian mass extinction, which occurred about 250 million years ago, and the Cretaceous mass extinction of about 65 million years ago.

The Permian mass extinction, which defines the boundary between the Paleozoic and Mesozoic eras, claimed about 90% of the species of marine animals. Terrestrial life also crashed; for example, 8 out of 27 orders of Permian insects did not survive into the Triassic, the next geologic period. This mass extinction occurred in less than 5 million years—possibly much less—an instant in the context of geologic time.

Several factors may have combined to cause radical environmental change during the late Permian. That was about the time the continents merged to form Pangaea, which disturbed many marine and terrestrial habitats and altered climate. Massive volcanic eruptions also occurred in what is now Siberia, constituting the most extreme episode of volcanism of the past half-billion years. Besides spewing lava and ash into the atmosphere, the Siberian volcanoes may have produced enough carbon dioxide to warm the global climate. Reduced temperature differences between the equator and the poles would have slowed the mixing of ocean water, which in turn would have reduced the amounts of oxygen available to marine organisms. An oxygen deficit in the oceans may have played a large role in the Permian extinctions.

THE PROCESS OF SCIENCE

The Cretaceous mass extinction of 65 million years ago delineates the boundary between the Mesozoic and Cenozoic eras (FIGURE 25.5). That debacle doomed more than half the marine species and exterminated many families of terrestrial plants and animals, including the dinosaurs. The climate became cooler at that time, and shallow seas receded from continental lowlands. Large volcanic eruptions in what is now

India may have contributed to the cooling by releasing material into the atmosphere that blocked sunlight. However, many scientists now favor the so-called impact hypothesis, which postulates that the main cause of the Cretaceous extinctions was a collision of an asteroid or large comet with Earth. Separating Mesozoic from Cenozoic sediments is a thin layer of clay enriched in iridium, an element very rare on Earth but common in meteorites and other extraterrestrial debris that occasionally fall to Earth. Walter and Luis Alvarez and their colleagues at the University of California, Berkeley, studied the clay and suggested that it is fallout from a huge cloud of debris that billowed into the atmosphere when an asteroid hit Earth. The Alvarezes proposed that the great cloud would have blocked sunlight and severely disturbed climate for several months.

The impact hypothesis really has two parts: that such a collision occurred and that the event caused the Cretaceous extinctions. Much evidence in addition to the iridium layer supports the first part of the hypothesis—that a large comet or small asteroid crashed into Earth 65 million years ago. Earth is pocked with enough craters to tell us that many large objects have fallen on the planet in the past. Recent research has focused on the Chicxulub crater, a 65-million-year-old scar located beneath sediments on the Yucatán coast of Mexico. About 180 km in diameter, the Chicxulub crater is about the right size to have been caused by an asteroid with a diameter of about 10 km.

Critical evaluation of the impact hypothesis now focuses on the second claim—that the collision caused the Cretaceous extinctions. Advocates of the impact hypothesis argue that the

FIGURE 25.5 Diversity of life and periods of mass extinction. This graph, based on the fossil record of terrestrial and marine organisms, reveals a general increase in the diversity of organisms over time (red line and right vertical axis). Mass extinctions on land and in the oceans interrupted the buildup of diversity during many periods of geologic time (blue line and left vertical axis). In this figure, extinction events are estimated as percentages of taxonomic families that died out relative to the number extant in each period of geologic time. Of the two mass extinction periods labeled in the figure, the Permian mass extinction claimed more than 90% of species on land and in the seas, whereas the extinctions at the end of the Cretaceous probably wiped out more than half of all species, including the dinosaurs.

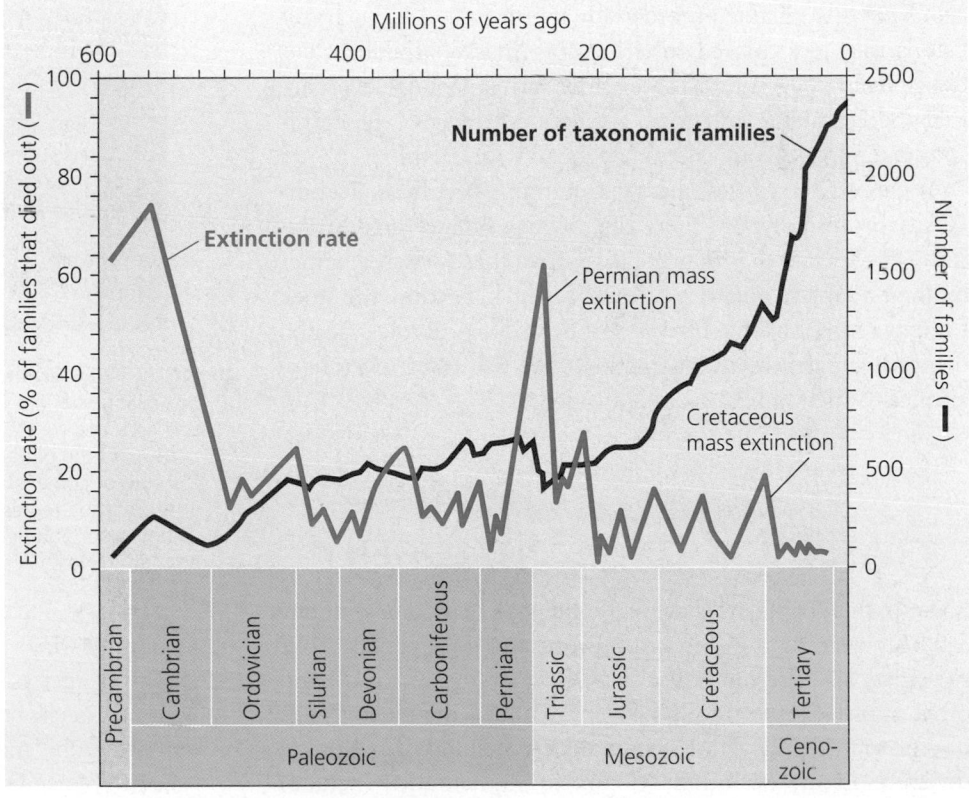

impact was large enough to darken the Earth for years, not months, and that the resulting reduction of photosynthesis would have lasted long enough for food chains to collapse. Some of the minerals in the dust cloud would also have caused severe acid precipitation (see Chapter 3). The crater's horseshoelike shape, along with evidence from sediments deposited during the late Cretaceous, indicates to some scientists that the Chicxulub asteroid struck Earth at a low angle from the southeast (FIGURE 25.6). Such an impact would have sent white-hot debris to the northwest, setting off a firestorm that within minutes would have killed most land plants and animals on the North American continent. This scenario could explain fossil evidence that extinction rates of species in North America were more severe and occurred in a briefer time than elsewhere. Less severe global effects would have developed more slowly after the initial catastrophe, which would explain why extinction rates during the late Cretaceous do not seem to have been uniform around the globe.

Although the debate over the impact hypothesis has muted somewhat, researchers maintain a healthy skepticism about the link between the Chicxulub impact event and the period of mass extinction at the end of the Cretaceous. The original Alvarez hypothesis triggered a strong research effort, fueled in part by its opponents, who contend that changes in climate due to continental drift, increased volcanism, and other processes on Earth, rather than extraterrestrial causes, could have caused mass extinction 65 million years ago. As some researchers point out, hypotheses about the Cretaceous extinctions may not be mutually exclusive. It is possible that an asteroid impact was the sudden final blow in an environmental assault on late Cretaceous life that included more gradual processes. And such collisions may have played roles in other mass extinctions. For example, in 2001, a research team reported geologic evidence that the Permian extinctions, the greatest dying of all, corresponded in time to an asteroid collision with Earth.

Whatever the causes, mass extinctions affect biological diversity profoundly. But there is a creative side to the destruction. The species that manage to survive these crises, whether by their adaptive qualities or by sheer luck, become the stock for new radiations that fill many of the biological roles vacated or newly created by the extinctions. We will trace several examples of such radiations in Unit Five.

SYSTEMATICS: CONNECTING CLASSIFICATION TO PHYLOGENY

So far in this chapter, we have explored phylogeny, or evolutionary history, in light of the fossil record and geologic time, but biologists also use molecular data, comparative anatomy, and other approaches to trace phylogeny. And tracing phylogeny is one of the main goals of **systematics,** the study of biological diversity in an evolutionary context. Systematics includes

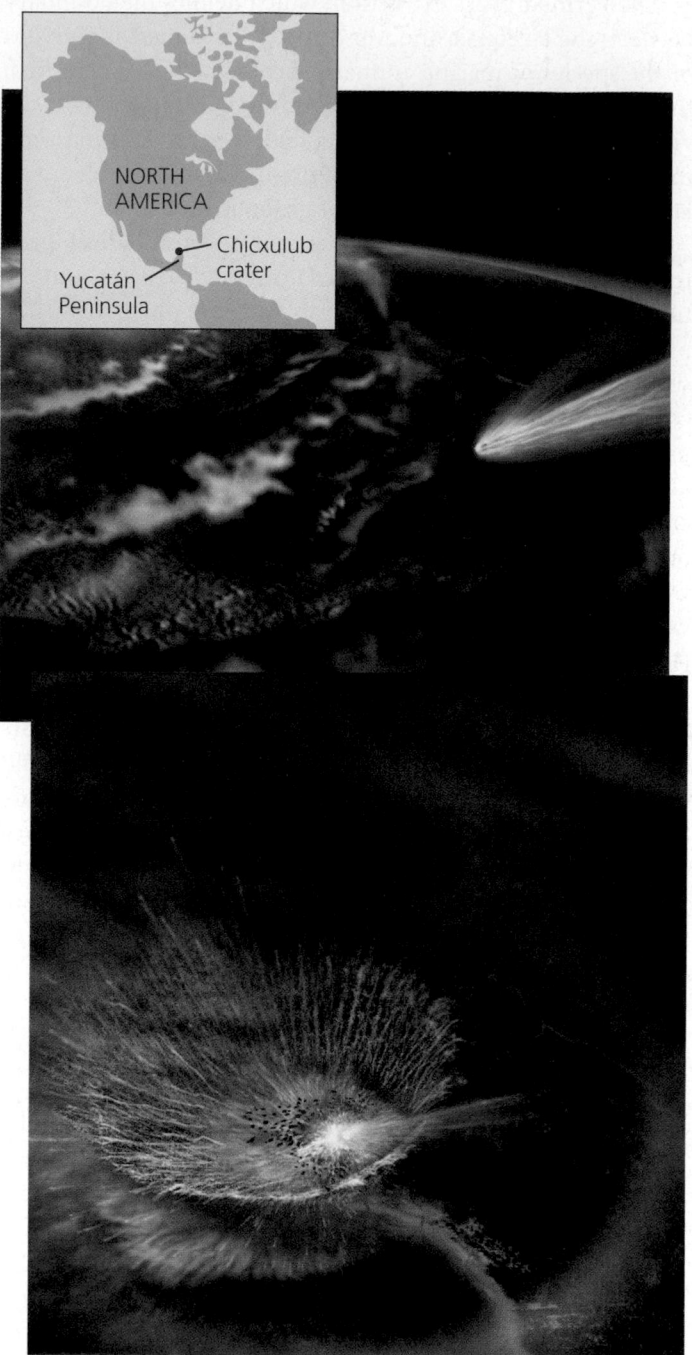

FIGURE 25.6 Trauma for planet Earth and its Cretaceous life. The 65-million-year-old Chicxulub impact crater is located in the Caribbean Sea near the Yucatán Peninsula of Mexico. The horseshoe shape of the crater and the pattern of debris in sedimentary rocks indicate that an asteroid or comet struck at a low angle from the southeast. This artist's interpretation represents the impact and its immediate effect—a cloud of hot vapor and debris that could have killed most of the plants and animals in North America within minutes.

taxonomy, which is the naming and classification of species and groups of species. Formal taxonomy began in the 18th century with Linnaeus's book *Systema naturae*, which translates as "System of Nature," meaning a classification of all known natural life. A century later, Darwin both anticipated the future direction of taxonomy and introduced the new science of systematics when he wrote in *The Origin of Species* that "our classifications will come to be, as far as they can be so made, genealogies." In this section, we examine the connection between phylogeny and classification that is at the center of modern systematics.

Taxonomy employs a hierarchical system of classification

The Linnaean system has two main characteristics: a two-part name for each species and a hierarchical classification of species into broader and broader groups of organisms.

The Binomial

Taxonomists assign to each species a two-part latinized name, or **binomial.** The first part of a binomial is the **genus** (plural, *genera*) to which the species belongs. The second part of a binomial, the **specific epithet,** refers to one **species** within the genus. An example of a binomial is *Panthera pardus,* the scientific name for the large cat we commonly call the leopard. Notice that the first letter of the genus is capitalized and that the whole binomial is italicized and latinized. (You can name a bug you discover after a friend, but you must add the appropriate Latin ending.)

Common names—such as cat, bear, finch, and lilac—often work well in informal communication, but they can be ambiguous because there are many species of each of these kinds of organisms. When biologists publish their research, they refer to the organisms they have studied with scientific names (binomials) to avoid ambiguity. Many of the scientific names in use today date back to Linnaeus, who assigned binomials to over 11,000 species of plants and animals. In fact, perhaps in a show of optimism, Linnaeus assigned to humans the scientific name *Homo sapiens,* which means "wise man."

Hierarchical Classification

In addition to identifying and naming species, a major objective of systematics is to group species into broader taxonomic categories. The first step of such a hierarchical classification is built into the binomial for a species. We group species that are closely related into the same genus. For example the leopard, *Panthera pardus,* belongs to a genus that also includes the African lion *(Panthera leo)* and the tiger *(Panthera tigris).* Grouping species is natural for us—a way for us to structure our view of the world. We lump together several trees we know as oaks and distinguish them from several other species of trees we call maples. Indeed, oaks and maples belong to separate genera. Biology's taxonomic scheme formalizes our tendency to group related objects.

Beyond the grouping of species within genera, taxonomy extends to progressively broader categories of classification. It

places related genera in the same **family,** puts families into **orders,** orders into **classes,** classes into **phyla** (singular, *phylum*), phyla into **kingdoms,** and kingdoms into **domains.** Each taxonomic level is more comprehensive than the previous one. All species of cats are mammals, but not all mammals are cats. The named taxonomic unit at any level is called a **taxon** (plural, taxa). For example, *Pinus* is a taxon at the genus level, the generic name for the various species of pine trees. Mammalia, a taxon at the class level, includes all the many orders of mammals. Only the genus name and specific epithet are italicized, and all taxa at the genus level and beyond are capitalized.

FIGURE 25.7 places the leopard in this taxonomic scheme of groups within groups. Classifying a species by kingdom,

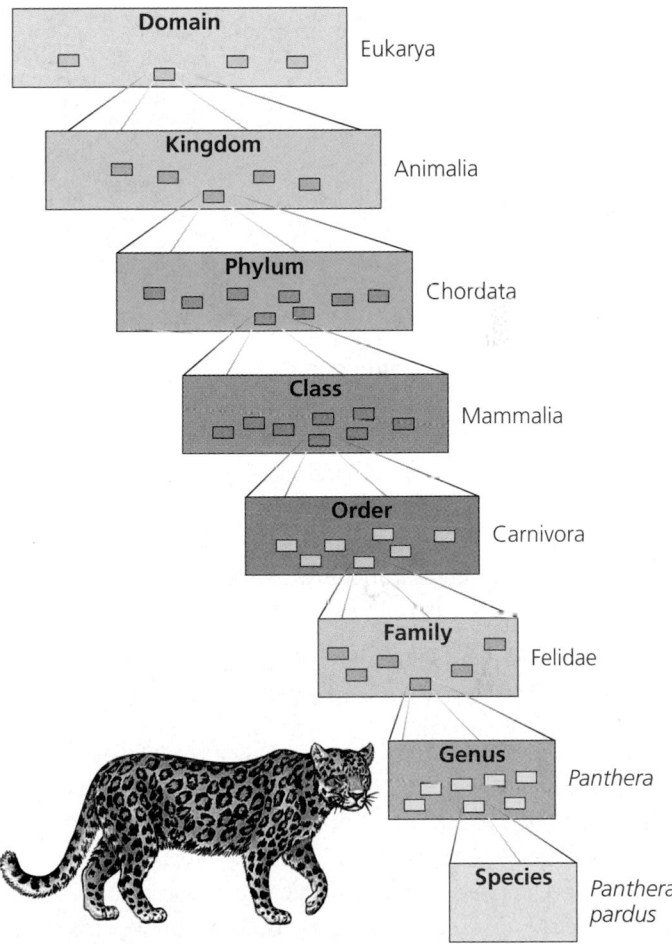

FIGURE 25.7 Hierarchical classification. The taxonomic scheme classifies species into groups belonging to more comprehensive groups. The leopard *(Panthera pardus)* belongs to the genus *Panthera,* which also includes the African lion and tiger. These wild felines belong to the cat family (Felidae), along with the genus *Felis,* which includes the domestic cat and several closely related wild cats, such as the lynx. Family Felidae belongs to the order Carnivora, which also includes the dog family, Canidae, and several other families. Order Carnivora, the carnivores, is grouped with many other orders in the class Mammalia, the mammals. Mammalia is one of several classes belonging to the phylum Chordata in the kingdom Animalia. The domain is an even broader taxonomic level. Domain Eukarya includes Kingdom Animalia and all other eukaryotic organisms.

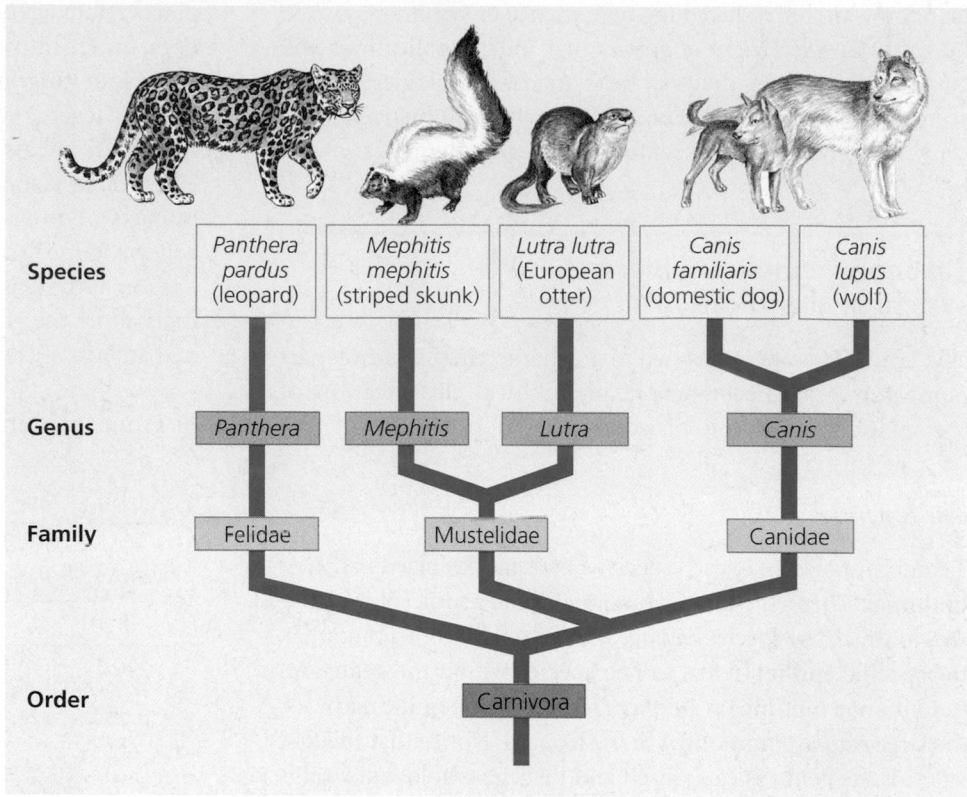

FIGURE 25.8 The connection between classification and phylogeny. Hierarchical classification is reflected in the finer and finer branching of phylogenetic trees. This tree suggests possible evolutionary relationships among some of the taxa within the order Carnivora, itself a branch of the class Mammalia.

phylum, and so on, is analogous to sorting mail, first by zip code and then by street, house number, and specific member of the household. The **phylogenetic trees** that systematists construct reflect the hierarchical classification of taxonomic groups nested within more inclusive groups. FIGURE 25.8 illustrates the connection between phylogeny and classification with a simplified tree that includes the leopard.

Modern phylogenetic systematics is based on cladistic analysis

THE PROCESS OF SCIENCE

Classification based on evolutionary history is called phylogenetic systematics. So far, we have assumed that the phylogeny of a group of species is already known, and then we have applied that knowledge to a hierarchical classification of the species (see FIGURE 25.8). But how do we reconstruct evolutionary history to draw phylogenetic trees in the first place? The fossil record helps, of course. Assessing relationships between living species by comparing their anatomy also provides data. And comparing two species' DNA (or proteins) gets to their hereditary relationships at the molecular level. But how do systematists evaluate all this information and apply it to tree building and classification? Today, most systematists practice what is called cladistic analysis, or simply cladistics. We can trace the seeds of this revolution in systematics to a book that German entomologist (insect specialist) Willi Hennig wrote 50 years ago. Let's examine some of the key elements of cladistic analysis.

Clades: Monophyletic Groups

A phylogenetic diagram based on cladistics is called a **cladogram.** It is a tree constructed from a series of dichotomies, or two-way branch points. Each branch point represents the divergence of two species from a common ancestor. For example, we could represent a branching within the cat family this way:

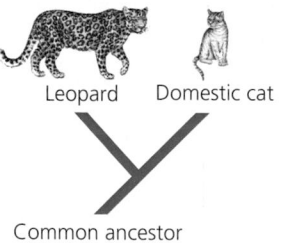

Leopard   Domestic cat

Common ancestor

We can also diagram dichotomous branching of taxa more inclusive than species, such as families or orders:

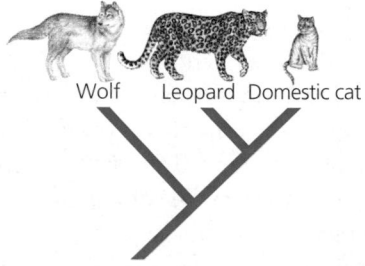

Wolf   Leopard   Domestic cat

The "deeper" branch point represents evolutionary divergence from an ancestor common to both the cat and dog families.

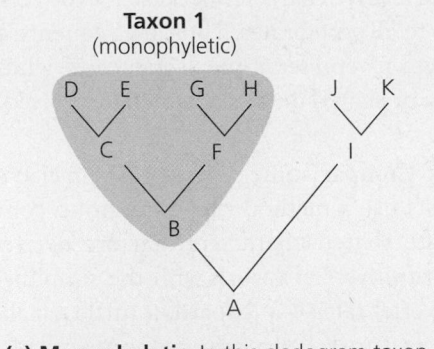

Taxon 1 (monophyletic)

(a) Monophyletic. In this cladogram taxon 1, consisting of the seven species B–H, is a monophyletic group, or clade. A mono- phyletic taxon is made up of an ancestral species (species B in this case) and *all* of its descendant species. Only monophyletic groups qualify as legitimate taxa derived from cladograms.

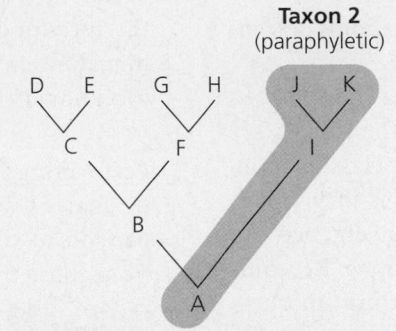

Taxon 2 (paraphyletic)

(b) Paraphyletic. Taxon 2 does not meet the cladistic criterion: it is paraphyletic, which means it consists of an ancestor (A in this case) and *some*, but not all, of that ancestor's descendants. (Taxon 2 includes the descendants I, J, and K, but excludes B–H, which also descended from A.)

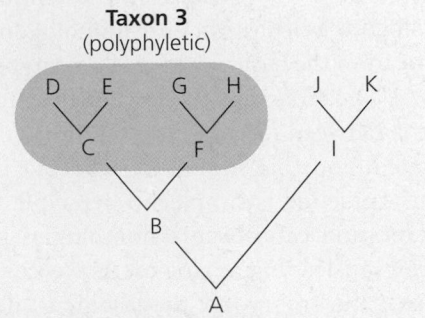

Taxon 3 (polyphyletic)

(c) Polyphyletic. Taxon 3 also fails the cladistic test. It is polyphyletic, which means that it lacks the common ancestor (B) that would unite the species as a monophyletic group.

FIGURE 25.9 Monophyletic versus paraphyletic and polyphyletic groups.

The sequence of branching symbolizes historical chronology; the last ancestor common to both the cat and dog families lived longer ago than the last ancestor shared by leopards and domestic cats.

Each evolutionary branch in a cladogram is called a **clade** (from the Greek *clados*, branch). Notice that clades, like taxo- nomic levels, can be nested within larger clades. For example, the cat family represents a clade within a larger clade that also includes the dog family. But not all groupings of organisms qualify as clades. A clade consists of an ancestral species and all of its descendants. Such a group of species, be it a genus, fam- ily, or some higher taxa, is said to be **monophyletic,** which means "single tribe." FIGURE 25.9 contrasts a monophyletic group, or clade, with groupings of species that are unaccept- able in the practice of cladistics.

Constructing a Cladogram

How do systematists decide on the sequence of branching in a cladogram? Put another way, how can we determine the simi- larities between species that are relevant for grouping the species as a clade? This can be quite a challenge. The first com- plication is that not all similarity represents common ancestry.

Sorting Homology from Analogy. Recall from Chapter 22 that likeness attributed to shared ancestry is called **homology.** The forelimbs of mammals are homologous; that is, the simi- larity in the intricate skeleton that supports the limbs has a ge- nealogical basis (see FIGURE 22.14).

There is a wild card in this game of making evolutionary connections by evaluating similarity: Not all likeness qualifies as homology. Species from different evolutionary branches may

come to resemble one another if they have similar ecological roles and natural selection has shaped analogous adaptations. This is called **convergent evolution,** and similarity due to con- vergence is called **analogy,** not homology (FIGURE 25.10). The wings of bats and those of birds, for example, are analogous flight equipment. The fossil record documents that bat wings and bird wings evolved independently from walking forelimbs of different ancestors. (Note that homology, like taxonomy,

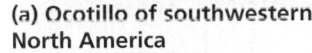

(a) Ocotillo of southwestern North America

(b) *Alluaudia* of Madagascar

FIGURE 25.10 Convergent evolution and analogous structures. **(a)** The ocotillo of southwestern North America looks remarkably similar to **(b)** the *Alluaudia* found in Madagascar. The plants are not closely related and owe their resemblance to analogous adaptations that evolved independently in response to similar environmental pressures.

can be hierarchical; though the forelimbs of bats and birds are analogous as *wings,* they are homologous as forelimbs descendant from the basic vertebrate prototype limb.)

As a general rule, the greater the number of homologous parts between two species, the more closely the species are related, and this should be reflected in their classification. This guideline is simpler in principle than it is in practice. Adaptation can obscure homologies, and convergence can create misleading analogies. As we saw in Chapter 22, comparing the embryonic development of the features in question can often expose homology that is not apparent in the mature structures.

There is another clue to identifying homology and sorting it from analogy: The more complex two similar structures are, the less likely it is they evolved independently. Consider the skulls of a human and a chimpanzee, for example. The skulls are not single bones, but a fusion of many, and the two skulls match almost perfectly, bone for bone. It is highly improbable that such complex structures matching in so many details could have separate origins. Most likely, the genes required to build these skulls were inherited from a common ancestor.

Identifying Shared Derived Characters. Beyond separating homologous from analogous similarity, systematists must sort through the homologies to identify what are called shared derived characters. "Character" here refers to any feature that a particular taxon possesses. The characters that are relevant to phylogeny, of course, are the homologous ones. For example, hair is a character shared by all mammals. But so is a backbone. The presence of a backbone cannot help us distinguish mammals from other vertebrates because non-mammalian vertebrates such as fishes and reptiles also have backbones. In other words, the backbone is a homology that predates the branching of the mammalian clade from the vertebrate tree; it is a **shared primitive character,** a homology common to a taxon more inclusive than the one we're trying to define. In contrast, hair is a homology found only in the vertebrates called mammals; in comparing mammals to other vertebrates, hair is a **shared derived character,** meaning an evolutionary novelty unique to a particular clade. The following flowchart summarizes the types of similarities between species:

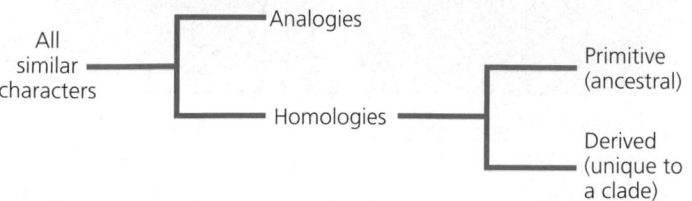

Note that the backbone can also qualify as a shared derived character, but at a deeper branch point that distinguishes *all* vertebrates from other animals. *Among* vertebrates, the

backbone is a shared primitive character because it evolved in the ancestor common to all vertebrates. Thus, the sequence of branching in a cladogram represents the sequence in which evolutionary novelties, or shared derived characters, evolved.

Performing Outgroup Comparison. As a key step in cladistic analysis, systematists use a method called outgroup comparison to differentiate shared characters that are derived from those that are primitive. Let's work with the simplified case of ordering five vertebrates—a leopard, a turtle, a salamander, a tuna, and a creature called a lamprey—into a cladogram. As a basis of comparison, we need to designate an **outgroup,** which is a species or group of species that is closely related to the species we are studying, but known to be less closely related than any study-group members are to each other (based on evidence from paleontology, analysis of embryonic development, and molecular biology, for example). A good choice of an outgroup in our example is an animal called a lancelet. The five vertebrates we are studying make up the **ingroup,** and we can begin building our cladogram by comparing this ingroup with the outgroup.

Outgroup comparison is based on the assumption that homologies present in both the outgroup and the ingroup must be primitive characters that were already present in the ancestor common to both groups. In our study case, an example of such a character is a structure called a notochord, a flexible rod running the length of the animal. (In vertebrates, the notochord is present in embryos, but the only vestiges of the structure in adults of many vertebrates are the rubbery disks between the vertebrae of the backbone.) The species making up the ingroup display a mixture of shared primitive and shared derived characters. The outgroup comparison enables us to focus on just those characters that were derived at the various branch points of vertebrate evolution. FIGURE 25.11a tabulates examples of these characters. Note that *all* the vertebrates in the ingroup have backbones; thus, this is a shared primitive character that was present in the ancestral vertebrate, though not in the outgroup. Now note that the presence of jaws is a character absent in lampreys but present in other members of the ingroup. Thus, this character helps us identify the earliest branch point in the vertebrate clade. FIGURE 25.11b illustrates how the data in our table of homologies translate into a cladogram.

Summary of Cladistic Analysis

A cladogram represents the chronological sequence of branching during the evolutionary history of a set of organisms. Systematists infer this branching sequence by analyzing homologies, identifying the shared derived characters unique to each clade. In FIGURE 25.11b, for example, we can infer from cladistic analysis that the turtle-leopard clade includes a common ancestor that is more recent than the ancestor of the salamander-turtle-leopard clade.

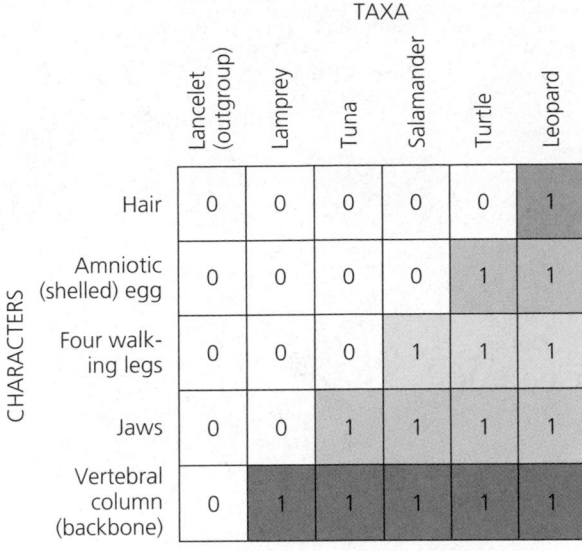

TAXA

CHARACTERS	Lancelet (outgroup)	Lamprey	Tuna	Salamander	Turtle	Leopard
Hair	0	0	0	0	0	1
Amniotic (shelled) egg	0	0	0	0	1	1
Four walking legs	0	0	0	1	1	1
Jaws	0	0	1	1	1	1
Vertebral column (backbone)	0	1	1	1	1	1

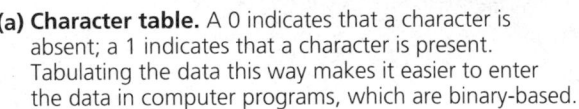

(a) Character table. A 0 indicates that a character is absent; a 1 indicates that a character is present. Tabulating the data this way makes it easier to enter the data in computer programs, which are binary-based.

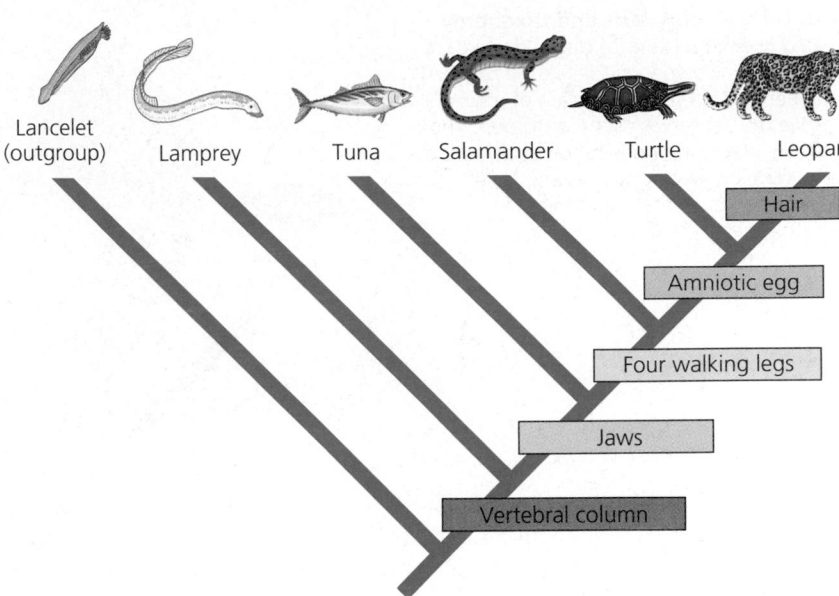

(b) Cladogram. Analyzing the taxonomic distribution of these homologies enables us to identify the sequence in which derived characters evolved during vertebrate phylogeny.

FIGURE 25.11
Constructing a cladogram.

THE PROCESS OF SCIENCE

Do not confuse this chronology of phylogenetic branching with the vintages of the organisms we are comparing. For example, the cladogram in FIGURE 25.11b does *not* indicate that all turtles evolved more recently than all salamanders, but only that their common ancestor preceded the ancestor of the turtle-leopard clade. A particular species of turtle may be a very old member of that clade, while the salamander species used in the cladistic analysis may be a relatively recent member of an older clade. This is analogous to an early building in a relatively new city being older than a new building in a more ancient city. Note also that the chronology represented in a cladogram is relative (earlier versus later) rather than absolute (so many millions of years ago). Many of the evolutionary trees you will see in the next unit of chapters and elsewhere in this book are based on a combination of cladistic analysis and the use of fossil evidence and other data to place the phylogenetic branch points in the context of geologic time. (See FIGURE 29.1 for an example.)

Systematists can use the cladograms they construct to place species in the taxonomic hierarchy of groups within groups, reflecting the nesting of clades within more comprehensive clades. FIGURE 25.12, page 498, applies this principle to a group of mammals.

Some systematists now argue that the hierarchical system of classification is antiquated because it is so disruptive to rearrange taxa when a cladogram is revised based on new evidence. These systematists propose replacing the Linnaean system with a strictly cladistic classification called **phylocode**.

Advocates of this alternative taxonomy simply name clades without the hierarchical tags, such as class, order, and family. For example, two clades we can extract from FIGURE 25.11b might be named clade Vertebrata and clade Amniota (the vertebrates with amniotic, or shelled, eggs). So far, most biologists still prefer a hierarchical system of taxonomic levels as a more useful way of organizing the diversity of life.

Each diagram of phylogeny, be it a cladogram or an evolutionary tree that incorporates a time scale, represents a hypothesis or a set of hypotheses about how the organisms in the tree are related. New evidence can compel systematists to revise their trees. And such reassessment has accelerated with the application of molecular methods for comparing species and tracing phylogeny.

Systematists can infer phylogeny from molecular data

When tracing phylogeny and classifying organisms, it is useful to compare macromolecules as well as anatomical characters. If homology reflects common ancestry, then comparing the genes and gene products (proteins) of organisms gets right to the heart of their evolutionary relationships. Sequences of nucleotides in DNA are inherited, and they program corresponding sequences of amino acids in proteins. At the molecular level, the evolutionary divergence of species parallels the accumulation of differences in their genomes.

THE PROCESS OF SCIENCE

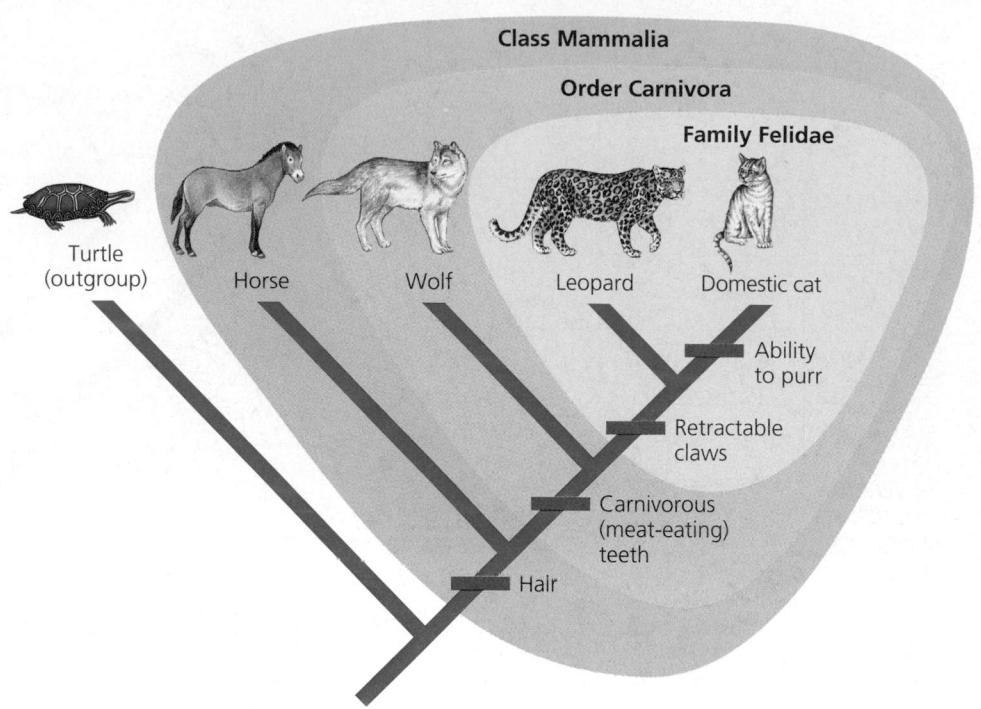

FIGURE 25.12 Cladistics and taxonomy. We can use the pattern of clades nested within clades to order species into a hierarchy of higher taxa. In this cladogram of mammals, a reptile (turtle) serves as the outgroup. The crossbars preceding branch points along the cladogram are labeled with examples of derived characters that evolved in the ancestor of each clade.

The more recently two species have branched from a common ancestor, the more similar their DNA and amino acid sequences should be. Today, both the amino acid sequences for many proteins and the nucleotide sequences for a rapidly expanding archive of DNA from diverse species are in databases available via the Internet. This has catalyzed a boom in systematics, as researchers use the databases to trace phylogeny by comparing the hereditary information of different species.

Molecular systematics makes it possible to assess phylogenetic relationships that cannot be measured by comparative anatomy and other nonmolecular methods. One application is comparison of species too closely related to display much divergence in morphology. At the other extreme, molecular systematics makes it possible to trace evolutionary relationships of species that are so different that there is little morphological homology; for example, it would be impossible to tell from morphology that fungi are more closely related to animals than they are to plants. Thus, molecular biology helps extend phylogenetic systematics to the extremes of evolutionary relationships, to the major branches and the finest twigs of the tree of life.

Phylogenetic Data from DNA Sequences

Systematists use a variety of methods to compare the proteins and nucleic acids of species, but most research is now based on comparison of nucleotide sequences in DNA or RNA. Each nucleotide position along a stretch of DNA represents an inherited character in the form of one of the four DNA bases: A (adenine), G (guanine), C (cytosine), or T (thymine). Thus, homologous regions of DNA from two species that are 1,000 nucleotides long provide 1,000 points of comparison. And a systematist may compare several DNA regions to assess the relationship between two species. Automated DNA sequencers enable researchers to collect an enormous amount of data, and computers make it possible to analyze the extensive quantity of genetic information. This DNA sequence analysis provides a quantitative tool for constructing cladograms with branch points defined by mutations in DNA sequence that mark each clade—the shared derived characters of molecular systematics.

The rates of change in DNA sequences over evolutionary time vary from one part of the genome to another. The DNA coding for ribosomal RNA (rRNA) changes relatively slowly, so comparisons of DNA sequences in these genes (or their products, rRNA) is useful for investigating relationships between taxa that diverged hundreds of millions of years ago. For example, systematists have focused on rRNA genes to help sort out the phylogeny of the animal phyla. In contrast, the DNA in mitochondria (mtDNA) evolves relatively rapidly. Thus, some systematists compare mtDNA sequences to assess the phylogeny of species that are relatively closely related or even populations of the same species. For example, one research team used mtDNA sequences to trace the relationships between different groups of Native Americans. Their results corroborate evidence that the Pima of Arizona, the Maya of Mexico, and the Yanomami of Venezuela are closely related, probably descending from the first of three waves of immigrants to cross the Bering Land Bridge from Asia to the Americas during the glaciation of the late Pleistocene epoch.

Aligning the DNA Sequences

DNA comparisons pose technical challenges. The first step in analyzing the genetic data is to align homologous DNA sequences from the two species we are comparing. If the two species have diverged from a common ancestor relatively recently, the sequences of homologous regions of DNA will probably be identical in length. Of course, along this homologous sequence, the two species may differ in which base is present at one or a few sites. In contrast, less closely related species may have homologous DNA sequences that differ not only in the bases at certain sites but also in their total length. This is because of an accumulation over evolutionary time of mutations such as insertions and deletions that alter the lengths of genes (see Chapter 17). Suppose, for example, that two DNA sequences were very similar, but a deletion mutation eliminated the first base of the sequence in one of the species. This would shift the remaining sequence forward a notch, and a point-by-point comparison of the two sequences would then distort what in fact is a very good match. FIGURE 25.13 illustrates a simplified example of how systematists use computer programs to align homologous DNA segments that differ in length.

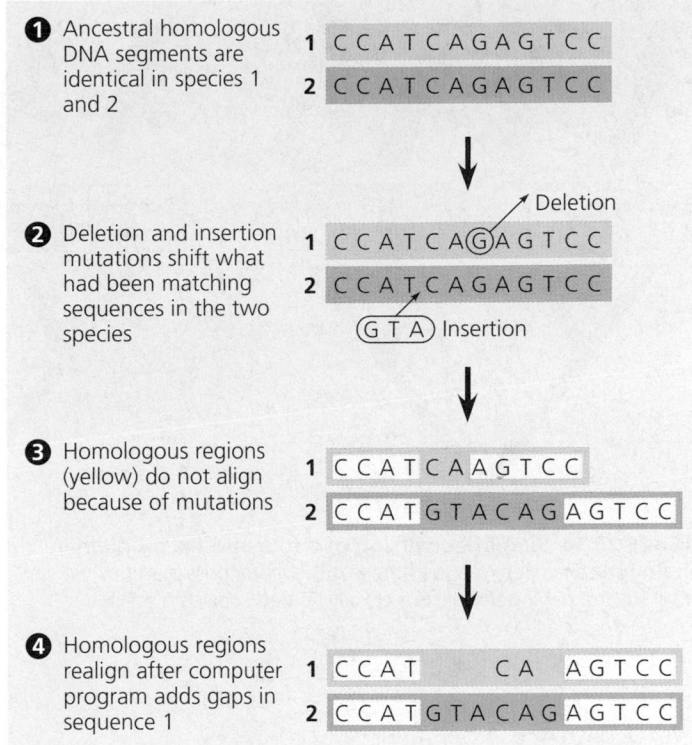

FIGURE 25.13 Aligning segments of DNA. This simplified example illustrates how insertion and deletion mutations can alter the length of DNA segments and make point-by-point comparison between two species challenging. Systematists use computer programs that search for similar sequences along the DNA segments from the two species. The programs then reestablish alignment by inserting gaps in one DNA segment to compensate for mutations that altered sequence length. This now makes it possible to compare sequences in homologous regions along the DNA segments. (In this example, the homologous sequences are identical.)

The principle of parsimony helps systematists reconstruct phylogeny

THE PROCESS OF SCIENCE

We can use DNA sequence data to learn more about general systematic theory. After sequence alignment and the point-by-point comparisons of base sequences in homologous DNA from a set of species, systematists may convert the data to phylogenetic trees. In making the sequence comparisons, each base change—say, a point mutation that changes a G to an A—counts as one evolutionary event. Challenged with many such base changes throughout a set of species, a systematist has quite a phylogenetic puzzle to solve. If there are only a few species in the set, then the number of ways to construct a phylogenetic tree based on the DNA sequence data is not too large (FIGURE 25.14, p. 500). But let's say a systematist is trying to construct a phylogeny for 50 species. About 3×10^{76} trees are possible for a 50-species problem! Even a supercomputer would take too long to search for the tree that best fits the DNA data. But for more manageable data sets, systematists can screen the tree variations by using a principle called parsimony.

Applied to science, the principle of **parsimony** states that a theory about nature should be the simplest explanation that is consistent with the facts. In other words, "keep it simple." The parsimony principle is sometimes called "Occam's Razor," in honor of William of Occam, a 14th-century English philosopher who is held to have advocated this minimalist approach to problem solving. (The "razor" refers to the shaving off of unneeded complications.) Six centuries later, the philosopher Bertrand Russell reinforced the principle of parsimony with these words: "It is vain to do with more what can be done with fewer."

Systematists apply the principle of parsimony to construct phylogenetic trees that represent the smallest number of evolutionary changes. In the case of cladograms based on morphological characters, the most parsimonious tree is the one that requires the fewest evolutionary events in the form of shared derived characters. FIGURE 25.15 (pp. 500–501) walks you through the construction of the most parsimonious tree based on DNA sequence data for the relatively simple case of the four-species problem introduced in FIGURE 25.14. For species sets much larger, we would need a computer to serve as an electronic razor to shave away all the possible trees that are unnecessarily complicated in our quest for parsimony.

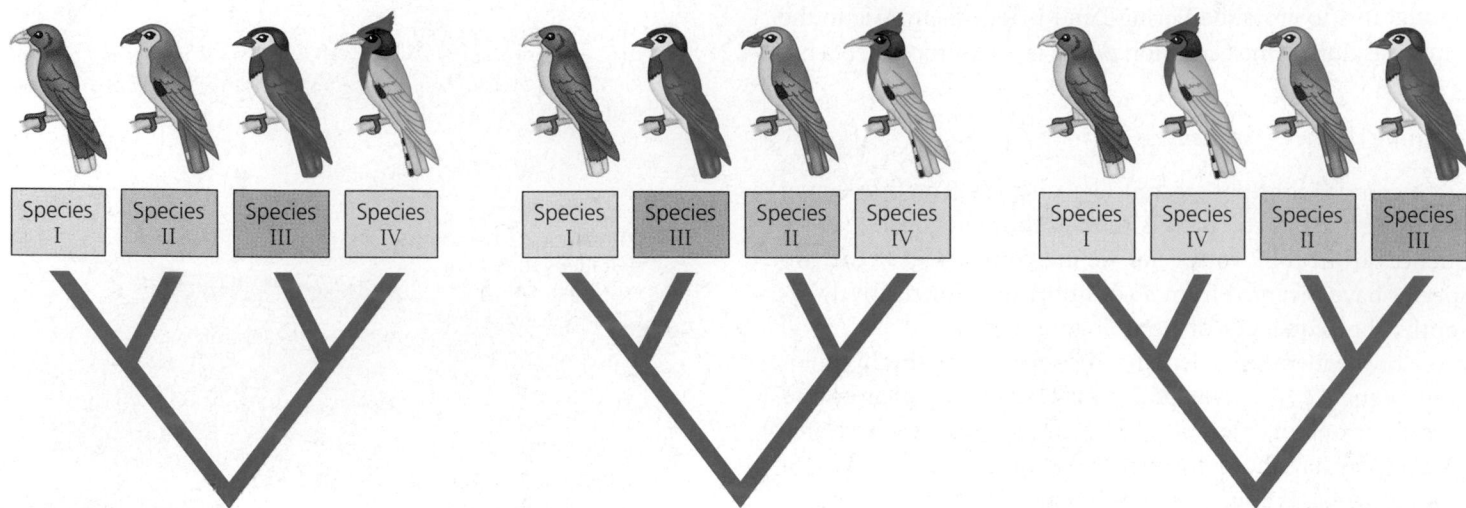

FIGURE 25.14 Simplified version of a four-species problem in phylogenetics. Here are just three of the many possible trees we can draw for the phylogeny of four closely related species of birds.

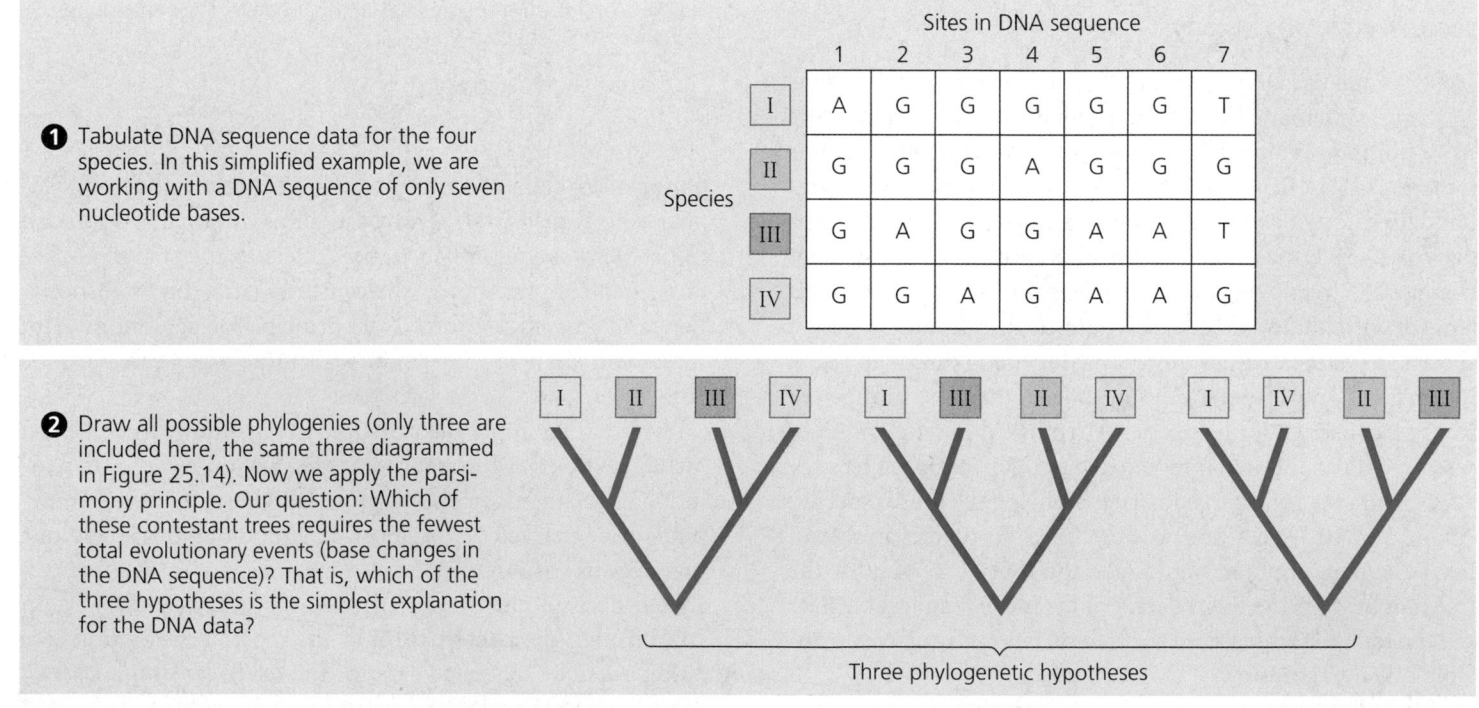

❶ Tabulate DNA sequence data for the four species. In this simplified example, we are working with a DNA sequence of only seven nucleotide bases.

	Sites in DNA sequence						
Species	1	2	3	4	5	6	7
I	A	G	G	G	G	G	T
II	G	G	G	A	G	G	G
III	G	A	G	G	A	A	T
IV	G	G	A	G	A	A	G

❷ Draw all possible phylogenies (only three are included here, the same three diagrammed in Figure 25.14). Now we apply the parsimony principle. Our question: Which of these contestant trees requires the fewest total evolutionary events (base changes in the DNA sequence)? That is, which of the three hypotheses is the simplest explanation for the DNA data?

Three phylogenetic hypotheses

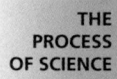

THE PROCESS OF SCIENCE

FIGURE 25.15 Parsimony and molecular systematics. Follow the steps to search for the most parsimonious tree representing the phylogeny of the four species introduced in FIGURE 25.14. (We are evaluating only three of the possible trees.)

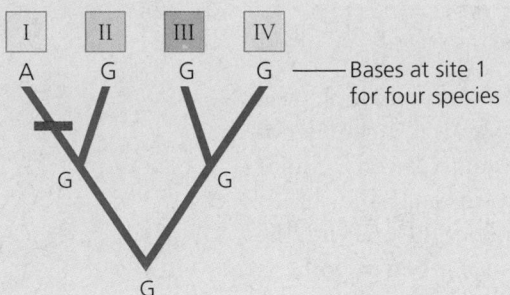

3 We focus first on site 1 in the sequence. A single base-change event, marked by the crossbar in the branch leading to species I, can account for the site 1 data.

Bases at site 1 for four species

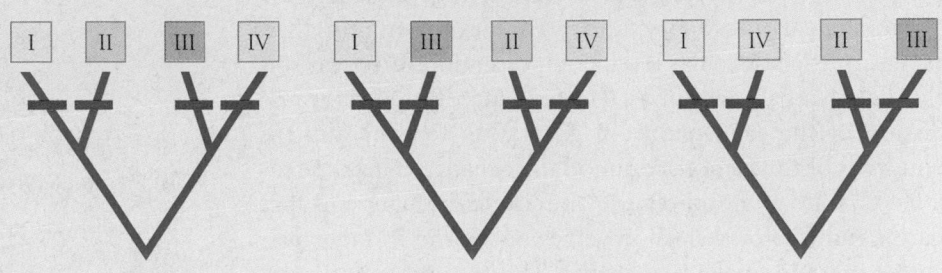

4 Continuing this site-by-site analysis for sites 2–4, we find that each contestant tree requires a total of four evolutionary events (again indicated by the crossbars on the trees) to account for the data over the first four base sites. Thus, the first four sites in the DNA sequence won't help us pick the most parsimonious tree.

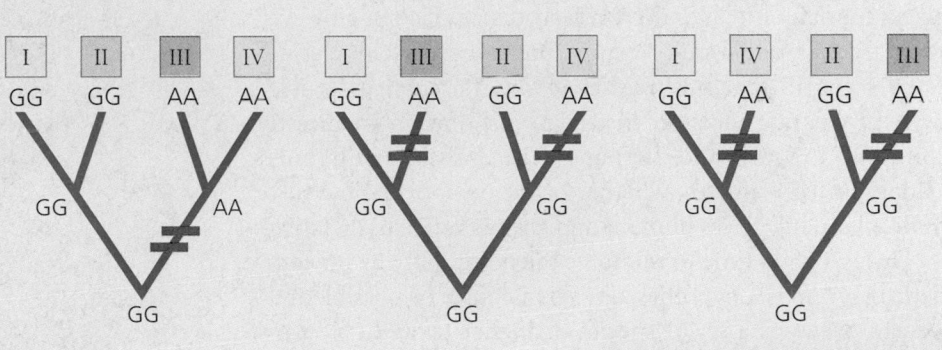

5 Moving on to sites 5 and 6, we can see that the first tree (on the left) requires fewer evolutionary events than the other two trees (two base changes versus four changes). In these diagrams, we're beginning with a deep ancestor having GG at sites 5 and 6. But if we started instead with an AA deep ancestor, we would still find that the first phylogeny requires only two changes, while four changes are required to make the other two hypothetical phylogenies work. Parsimony depends only on the total number of events, not on the exact nature of the events (the types of base changes).

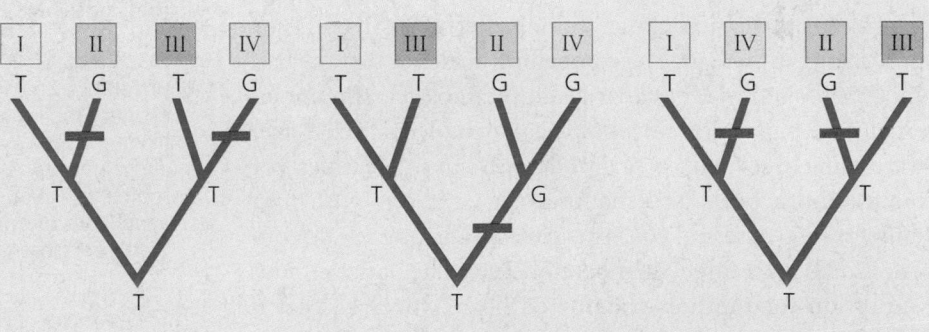

6 For site 7, the three trees also differ in the number of evolutionary events required to explain the DNA data.

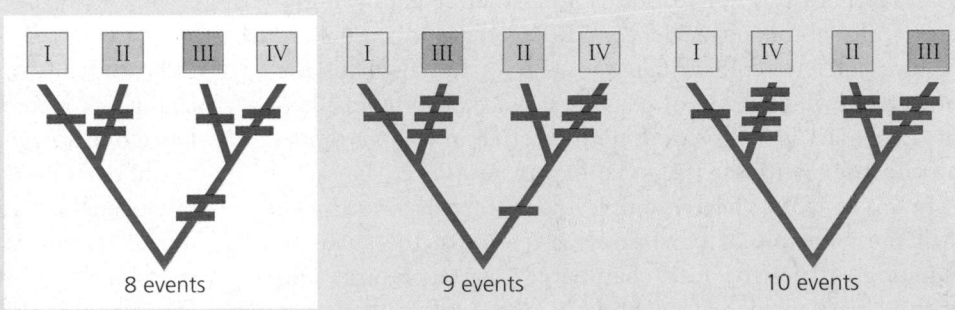

7 Now we add up the total events from steps 3–6. Don't forget to include the changes required for the first four sites in the DNA sequence (steps 3 and 4). The tree on the left is the most parsimonious of these three phylogenies.

8 events   9 events   10 events

FIGURE 25.15, *continued*

Phylogenetic trees are hypotheses

Considering how most of us struggle to understand biological complexity, especially in molecular biology, it is certainly not intuitive that parsimony should be our guide. Given a choice of many possible phylogenetic trees that we could draw based on DNA sequence data, why should we prefer the one with the fewest number of evolutionary changes as our best hypothesis for the relationships among a set of species? The rationale is that for any of a species' characters, whether molecular or morphological, hereditary fidelity is more common than change. At the molecular level, point mutations do occasionally change a base within a DNA sequence. But exact transmission of the sequence from generation to generation is thousands of times more common than change. The same rationale applies to morphology. For example, humans and apes share a number of skeletal novelties not found in other primates (the mammalian order that includes monkeys, apes, and humans). We *could* construct a primate phylogeny that places humans and apes on relatively distant clades. However, such a cladogram would assume the unnecessarily complicated scenario that a whole set of skeletal changes evolved twice in separate ancestors instead of just once in an ancestor common to apes and humans. The most parsimonious cladogram of primates, whether based on morphological or molecular data, places humans and apes as very close relatives.

This is a good time to reinforce the point that any phylogenetic diagram is a hypothesis. Given a choice of possible trees we can draw for a set of species or higher taxa, the best hypothesis is the one that is the best fit for all the available data. In the absence of conflicting information, the most parsimonious tree is the logical choice among alternative hypotheses. But sometimes there may be compelling evidence that the best hypothesis is a phylogeny that is *not* the most parsimonious. Perhaps the particular morphological or molecular character we are using to sort taxa actually *did* evolve multiple times. For example, both birds and mammals have hearts with four chambers (two atria and two ventricles; see Chapter 42). In contrast, lizards have three-chambered hearts. The parsimonious assumption for the four-chambered heart would be that it evolved once and was present in an ancestor common to birds and mammals but not to lizards and other reptiles (FIGURE 25.16a). But abundant evidence indicates that birds are more closely related to reptiles than they are to mammals. Thus, the four-chambered heart evolved at least twice in vertebrate history. This fact is inconsistent with the tree in FIGURE 25.16a, but consistent with the tree in FIGURE 25.16b.

In this vertebrate heart example, the problem is not so much with the principle of parsimony as it is with the analogy-homology issue; the four-chambered hearts of birds and mammals are analogous, not homologous. A matching change of bases in DNA sequences in two species can also occur

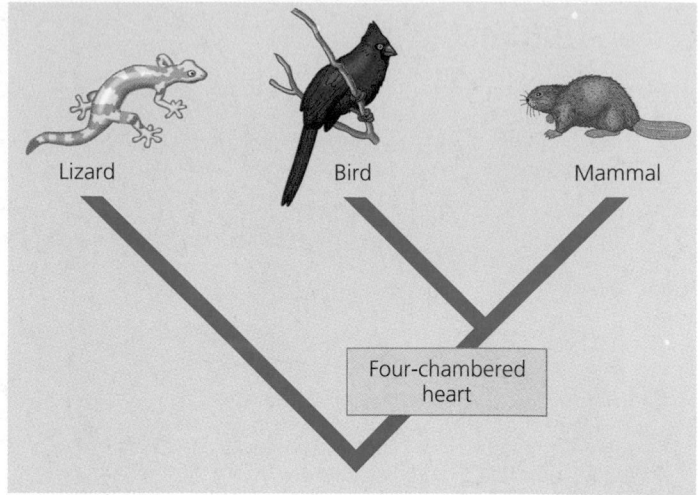

(a) Mammal–bird clade

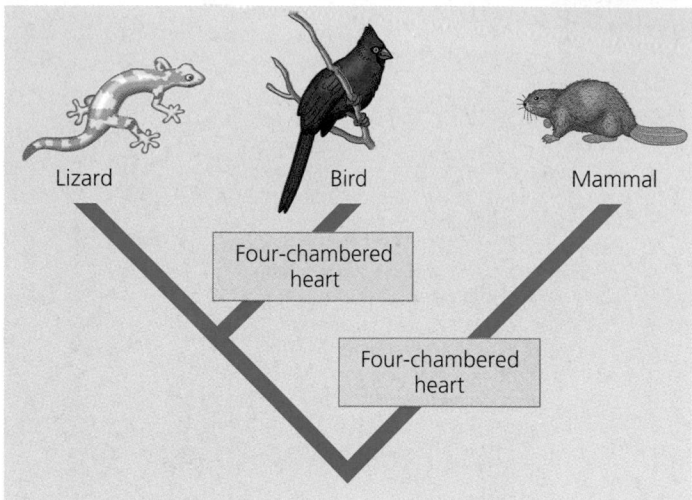

(b) Lizard–bird clade

FIGURE 25.16 Parsimony and the analogy-versus-homology pitfall. The tree in (a) places a bird and a mammal on a clade that excludes the lizard. The tree in (b) recognizes a lizard-bird clade. If we made the mistake of interpreting the four-chambered hearts of birds and mammals as homologous instead of analogous, we might opt for the tree in (a) as our most parsimonious hypothesis. In fact, an abundance of evidence supports the hypothesis that birds and lizards are closer relatives than birds and mammals.

independently; parsimony only suggests that such double changes in molecular evolution are less likely than inheritance of a single change from a common ancestor. The practice of parsimony in molecular systematics is more reliable if a phylogenetic tree is based on a large database of DNA sequence comparisons for the set of species in the tree. Similarly, occasionally misjudging an analogous similarity in morphology as a shared derived homology is less likely to distort a phylogenetic tree if each clade in the tree is defined by several derived characters. And the strongest phylogenetic hypotheses of all are those supported by both the molecular and morphological evidence, and, if it exists, by the fossil evidence.

Molecular clocks may keep track of evolutionary time

THE PROCESS OF SCIENCE

The timing of evolutionary history in TABLE 25.1 is based mainly on the fossil record. In the past few decades, however, molecular biology has provided an independent method for placing the origin of taxonomic groups in time. Called **molecular clocks,** these new timing methods are based on the observation that at least some regions of genomes evolve at constant rates. If certain homologous DNA sequences or their protein products are compared for taxa that are known to have diverged from common ancestors during certain periods in the past, the number of nucleotide and amino acid substitutions is proportional to the time that has elapsed since the lineages branched. For example, the homologous proteins of bats and dolphins are much more alike than are those of sharks and tuna. This is consistent with the fossil evidence that sharks and tuna have been on separate evolutionary paths much longer than bats and dolphins. In this case, molecular divergence has kept better track of the time than have changes in morphology.

Using Molecular Clocks to Measure Absolute Time

It's one thing to use molecular clocks to help assess the *relative* chronology of branching in phylogeny—that sharks and tuna diverged earlier than bats and dolphins did, for example. But how accurate are molecular clocks for keeping *absolute* time? No genes mark time with precise tick-tock accuracy in their rates of base sequence changes. In fact, some regions of genomes evolve in fits and starts that are not at all clocklike. And even those genes that make good molecular clocks are accurate only in the statistical sense of a fairly smooth *average* rate of change; over time, there may still be chance deviations above and below that average rate.

For a gene having a reliable average rate of evolution, the molecular clock can be calibrated in actual time. This is done by graphing the number of amino acid or nucleotide differences against the times for a series of evolutionary branch points known from the fossil record. The graph line representing the evolution rate of the molecular clock can then be used to estimate how long ago certain evolutionary episodes that cannot be discerned from the fossil record occurred—the origin of a species or some higher taxa, for example.

Some biologists remain skeptical about the accuracy of molecular clocks. The regularity of some genes as molecular clocks implies that much of the change in DNA sequences is due to genetic drift and that the changes are mostly neutral— neither adaptive nor detrimental. Molecular evolution due to natural selection favoring certain DNA changes over others would probably be too irregular to mark time accurately. Thus, the skepticism of some scientists about molecular clocks partly reflects a more general debate about the extent to which neutral genetic variation can account for DNA diversity. Biologists who doubt some of the evolutionary conclusions based on molecular clocks are also critical of extrapolation to time spans beyond what has been calibrated to the fossil record. For example, molecular clocks that have been calibrated to fossil history extending back a few hundred million years have on occasion been used to attach time to evolutionary divergence that occurred a billion or more years ago. If molecular clocks are used judiciously, however, they will continue to help evolutionary biologists reconstruct the past.

Using a Molecular Clock to Date the Origin of HIV

In the year 2000, a research team at Los Alamos National Laboratory in New Mexico used a molecular clock to date the origin of HIV infections in humans. The pathogen HIV is the virus that causes AIDS. The virus descended from related viruses that infect chimpanzees and another primate called the sooty mangabey (the viruses do not cause any AIDS-like diseases in these nonhuman hosts). When did the virus jump from its simian hosts to first infect humans? The question does not have a simple answer because the virus has spread to humans more than once. These multiple origins of HIV are reflected in the various major strains (genetic types) of the virus that cause AIDS in humans. The most widespread strain in the global AIDS epidemic is HIV-1 M. Because HIV evolves very rapidly, samples of the virus differ even when taken from patients who were infected with HIV-1 M just a few years apart. To pinpoint the time of the earliest HIV-1 M infection, the Los Alamos researchers calibrated a molecular clock by comparing DNA sequences in a specific HIV gene obtained from patients sampled at different times in the history of the epidemic. The data enabled the scientists to project backward to the 1930s as the probable time of the first HIV-1 M invasion of humans (FIGURE 25.17, p. 504).

Modern systematics is flourishing with lively debate

THE PROCESS OF SCIENCE

Systematics is thriving today at the interface of modern evolutionary biology and current taxonomic theory. Biologists have been accumulating phenotypic information, mainly morphological, about diverse species, living and extinct, for centuries. The development of cladistics provided

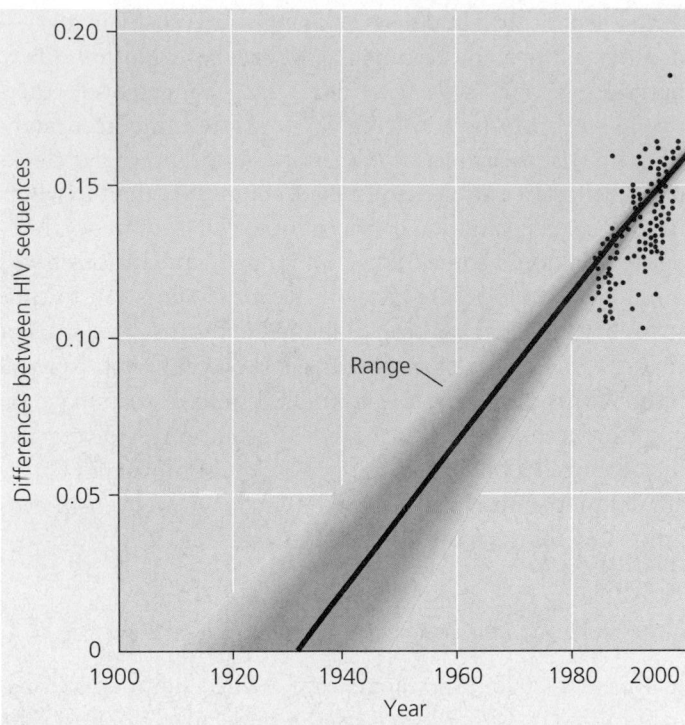

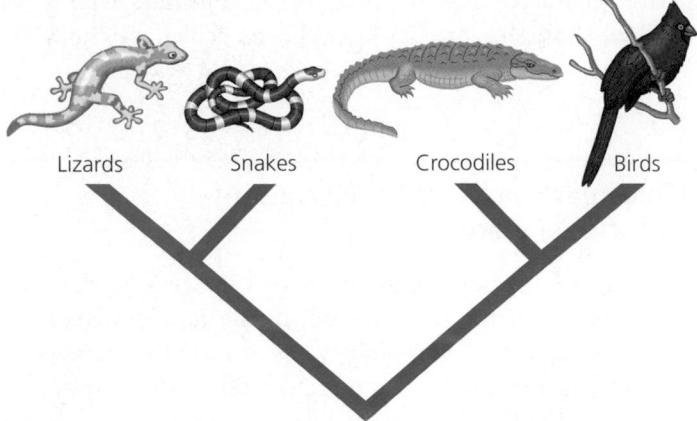

FIGURE 25.18 Modern systematics is shaking some phylogenetic trees. In our traditional vertebrate taxonomy, crocodiles, snakes, lizards, and other reptiles are classified in the class Reptilia, while birds are placed in a separate class (Aves). But most vertebrate systematists now agree that crocodiles are actually more closely related to birds than they are to lizards and snakes. Thus, the class Reptilia, in its traditional form, is paraphyletic, not monophyletic (see FIGURE 25.9).

FIGURE 25.17 Dating the origin of HIV-1 M with a molecular clock. The numerous dots in the upper right-hand corner of this graph are based on DNA sequences for a specific HIV gene in blood samples collected from patients at different known times. The scale on the *y* axis is a measure of the number of base changes in the gene serving as the molecular clock. Note that during the span of time that the HIV samples were collected from the patients in this data set, between the early 1980s and the late 1990s, the gene evolved at a relatively consistent rate. If we project that rate backward in time, we intersect the time axis (*x* axis) of the graph during the 1930s. The team that published this research concluded that the HIV-1 M strain first infected humans in the 1930s.

more objective methods for comparing morphology and incorporating the data into phylogenetic hypotheses, or cladograms. Molecular systematics added a powerful new tool in comparative biology, extending the analysis of phylogenetic relationships down to the level of DNA.

Cladistic analysis and molecular systematics, complemented by a revival of interest in paleontology and comparative biology in the past few decades, are stimulating a reassessment of phylogeny that is bringing us closer to understanding the history of life on Earth. In many cases, independent approaches, such as paleontology and DNA sequencing, converge in supporting a particular phylogenetic hypothesis. For example, the fossil record, comparative anatomy, and molecular comparisons all concur that crocodiles are more

closely related to birds than to lizards and snakes, a conclusion that probably would have surprised Linnaeus and Darwin (FIGURE 25.18).

In other cases, molecular data are at odds with other evidence, such as the fossil record. One such debate centers on the origin of the major groups (orders) of mammals. The oldest fossils of mammals date back 220 million years, into the Triassic period (see TABLE 25.1). However, fossils documenting the origin of most modern mammalian orders are much younger, dating to the early Tertiary period, about 60 million years ago, after the extinction of the dinosaurs. In apparent contradiction of the fossil evidence, molecular clocks push the origin of the major mammalian orders back closer to 100 million years ago. Many researchers place more trust in the fossil evidence and express doubts about whether the molecular clocks are reliable. The counterargument is that the paleontologists have not yet documented an earlier origin for most mammalian orders because the fossil record is incomplete. Between these two extremes is a **phylogenetic fuse** hypothesis: Perhaps the modern mammalian orders *originated* about 100 million years ago, but did not proliferate extensively enough to be noticeable in the fossil record until

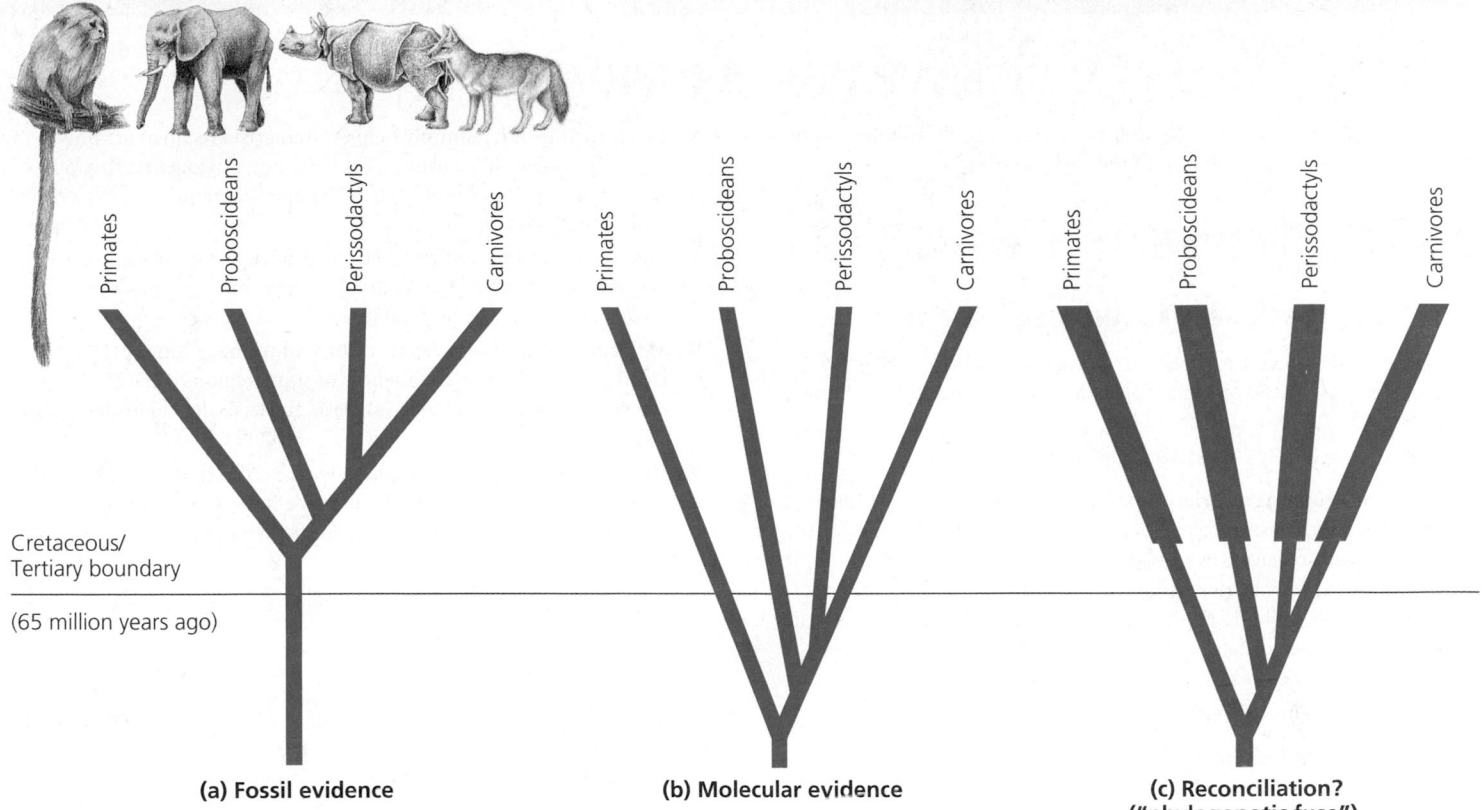

Primates **Proboscideans** **Perissodactyls** **Carnivores** **Primates** **Proboscideans** **Perissodactyls** **Carnivores** **Primates** **Proboscideans** **Perissodactyls** **Carnivores**

Cretaceous/
Tertiary boundary

(65 million years ago)

(a) Fossil evidence   **(b) Molecular evidence**   **(c) Reconciliation?**
("phylogenetic fuse")

FIGURE 25.19 When did most major mammalian orders originate? These simplified trees include just four orders of mammals. **(a)** According to the fossil record, the orders are tens of millions of years younger than the age inferred from **(b)** molecular clocks. **(c)** A possible reconciliation is the "phylogenetic fuse" hypothesis of an earlier origin followed by a later diversification within the orders. According to this hypothesis, the "explosion," symbolized in the diagram by the thicker lines, made the animals abundant and widespread enough to show up in the fossil record.

after the extinction of the dinosaurs almost 40 million years later (FIGURE 25.19). Additional research may help to resolve the debate about mammalian history.

Note again that the cladistic approach focuses only on the sequence of phylogenetic branching in evolution, not on the degree of evolutionary divergence between branches. To appreciate this distinction, return to FIGURE 25.18. A strict practitioner of cladistics would argue that a taxonomic separation of crocodiles and birds in the vertebrate classes Reptilia and Aves, respectively, is obsolete. Other evolutionary biologists counter that a useful taxonomy should reflect *both* features of phylogenetic trees: the sequence of branching and the extent of divergence between branches. In the case of birds, many biologists still prefer to assign them to a

class separate from all reptiles because the evolutionary adaptations of birds associated with flight have been so extensive. We will examine many such debates about classification in the next unit of chapters, which surveys the diversity of life in an evolutionary context.

■   ■   ■

In this unit, we have seen how the Darwinian theme of descent with modification has shaped all of biology. We have also seen that evolutionary theory itself has evolved as new methods, new data, and new ideas have continued to refine our view of life. Vigorous debate is vital to that progress and is a healthy sign that evolutionary biology is a robust science.

CHAPTER 25 REVIEW

Go to the Campbell Biology website (www.campbellbiology.com) to explore an interactive version of the Chapter Review.

Summary of Key Concepts

THE FOSSIL RECORD AND GEOLOGIC TIME

■ **Sedimentary rocks are the richest source of fossils (pp. 484–486, FIGURE 25.1)** The fossil record provides the historical archives biologists use to study the history of life.

Web/CD Activity 25A: *Grand Canyon Video*

■ **Paleontologists use a variety of methods to date fossils (pp. 486–488, TABLE 25.1, FIGURE 25.2)** Sedimentary strata reveal the relative ages of fossils in successive geologic periods. The absolute ages of fossils in years can be determined by radiometric dating and other methods. The geologic eras and periods correspond to major transitions in the composition of fossil species. The chronology of geologic periods and eras makes up the geologic time scale.

Web/CD Activity 25B: *A Scrolling Geologic Time Scale*

■ **The fossil record is a substantial, but incomplete, chronicle of evolutionary history (p. 488)** The fossil record favors species that existed for a long time, were abundant and widespread, and had shells or hard skeletons.

■ **Phylogeny has a biogeographic basis in continental drift (pp. 488–490, FIGURES 25.3, 25.4)** Continental drift has had a significant impact on the history of life by causing major geographic rearrangements affecting biogeography and evolution. The formation of the supercontinent Pangaea during the late Paleozoic era and its subsequent breakup during the early Mesozoic era explain many biogeographic puzzles.

■ **The history of life is punctuated by mass extinctions (pp. 490–492, FIGURES 25.5, 25.6)** Evolutionary history has been characterized by long, relatively stable periods interrupted by intervals of extensive species turnover—mass extinctions followed by grand episodes of adaptive radiation.

SYSTEMATICS: CONNECTING CLASSIFICATION TO PHYLOGENY

■ **Taxonomy employs a hierarchical system of classification (pp. 493–494, FIGURE 25.7)** Systematics, the study of biological diversity in an evolutionary context, includes taxonomy, the identification and classification of species. The hierarchy of taxa should reflect the branching nature of phylogeny.

■ **Modern phylogenetic systematics is based on cladistic analysis (pp. 494–497, FIGURES 25.11, 25.12)** A clade is a monophyletic taxon, an evolutionary ancestor and all its descendants. In cladistic analysis, clades are defined by their evolutionary novelties, or shared derived characters.

■ **Systematists can infer phylogeny from molecular data (pp. 497–499, FIGURE 25.13)** Species-to-species comparisons of the amino acid sequences in proteins and the base sequences in nucleic acids reveal phylogenetic relationships.

Web/CD Case Study in the Process of Science: *How Is Phylogeny Determined by Comparing Proteins?*

■ **The principle of parsimony helps systematists reconstruct phylogeny (pp. 499–501, FIGURE 25.15)** Among phylogenetic hypotheses, the most parsimonious tree is the one that requires the fewest evolutionary changes.

■ **Phylogenetic trees are hypotheses (pp. 502–503, FIGURE 25.16)** The best phylogenetic hypotheses are those that incorporate extensive molecular and morphological data.

■ **Molecular clocks may keep track of evolutionary time (p. 503, FIGURE 25.17)** The base sequences of some regions of DNA change at a rate consistent enough to serve as clocks to date episodes in past evolution.

■ **Modern systematics is flourishing with lively debate (pp. 503–505, FIGURE 25.19)** The progress in our understanding of phylogeny is built on research stimulated by scientific debates.

Self-Quiz

1. A paleontologist estimates that when a particular rock formed, it contained 12 mg of the radioactive isotope potassium-40. The rock now contains 3 mg of potassium-40. The half-life of potassium-40 is 1.3 billion years. About how old is the rock?
   a. 0.4 billion years   d. 2.6 billion years
   b. 0.3 billion years   e. 5.2 billion years
   c. 1.3 billion years

2. If humans and pandas belong to the same class, then they must also belong to the same
   a. order.   d. genus.
   b. phylum.   e. species.
   c. family.

3. In the case of comparing birds to other vertebrates, having four appendages is
   a. a shared primitive character.
   b. a shared derived character.
   c. a character useful for distinguishing the birds from other vertebrates.
   d. an example of analogy rather than homology.
   e. a character useful for sorting the avian (bird) class into orders.

4. The animals and plants of India are almost completely different from the species in nearby Southeast Asia. Why might this be true?
   a. They have become separated by convergent evolution.
   b. The climates of the two regions are completely different.
   c. India is in the process of separating from the rest of Asia.
   d. Life in India was wiped out by ancient volcanic eruptions.
   e. India was a separate continent until relatively recently.

5. How would one apply the principle of parsimony to the construction of a phylogenetic tree?
   a. Choose the tree in which the branch points are based on as few shared derived characters as possible.
   b. Choose the tree in which the branch points are based on as many shared derived characters as possible.
   c. Base phylogenetic trees only on the fossil record, as this provides the simplest explanation for evolution.
   d. Choose the tree that represents the fewest evolutionary changes, either in DNA sequence comparisons or morphological characters.
   e. Choose the tree with the fewest branch points and thus the fewest taxa.

6. What would be the best source of data for determining phylogenetic relationships of lineages of protists that diverged hundreds
of millions of years ago?
   a. fossils from the Precambrian era
   b. morphological characters that are shared and derived
   c. amino acid sequences of their various chlorophyll molecules
   d. DNA sequences for mitochondrial genes
   e. DNA sequences for ribosomal RNA

7. If you were using cladistic analysis to build a phylogenetic tree of
cats, which of the following would be the best choice for an outgroup?
   a. lion   d. leopard
   b. domestic cat   e. tiger
   c. wolf

8. Which of the following would be most useful for constructing a
phylogenetic tree emphasizing evolutionary branchings among
several fish species?
   a. several analogous characteristics shared by all the fishes
   b. a single homologous characteristic shared by all the fishes
   c. the total degree of morphological similarity among various fish
      species
   d. several characteristics thought to have evolved after different
      fishes diverged from one another
   e. a single characteristic that is different in all the fishes

9. Molecular clocks indicate that most modern mammalian orders
originated about 100 million years ago, whereas fossil evidence
dates the origin of these orders about 60 million years ago. What
explanation does the phylogenetic fuse hypothesis offer for this
discrepancy?
   a. Molecular clocks have not been shown to be reliable for any-
      thing more than giving the sequence of divergence.
   b. The radiometric method of dating fossils has a large margin of
      error, so these time estimates are actually quite close.
   c. Mammals did not become abundant and widespread enough to
      appear in the fossil record until after the extinction of the
      dinosaurs.
   d. The genes that were used in the molecular clock comparison did
      not have a consistent rate of evolution, but must have evolved
      more rapidly during the early adaptive radiation of mammals.
   e. When in doubt, always trust the fossil record over molecular
      comparisons, because DNA sequence comparisons cannot
      account for homologous segments that do not align properly.

10. The recent estimate that HIV-1 M first jumped from chimpanzees
to humans in the 1930s is based on
   a. the first clinical evidence of AIDS recorded in local village
      records in Africa.
   b. a molecular clock that plotted changes in sequences of an HIV
      gene sampled from patients over the past 20 years and then
      projected backward to an estimated origin.
   c. a comparison of homologous genes in HIV found in
      chimpanzees and in humans.
   d. a parsimonious explanation of the evolutionary relationships
      among the various strains of the virus found in humans.
   e. the recent discovery of HIV in a blood sample saved from that
      period.

11. Use TABLE 25.1 to estimate how many hundreds of millions of years
prokaryotes inhabited Earth before eukaryotes evolved.

12. Your measurements indicate that a fossilized skull you unearthed
has a ^{14}C-to-^{12}C ratio about one-sixteenth that of the atmosphere.
What is the approximate age of the skull?

13. The Andes Mountains are associated with tectonic activity near
which plate boundary? (Refer to FIGURE 25.3.)

14. How much of the classification in FIGURE 25.7 do we share with the
leopard?

15. Our forearms and the wings of a bat are derived from the same
ancestral prototype; thus, they are _____. In contrast, the
wings of a bat and the wings of a bee evolved from totally unrelated
structures; thus, they are _____.

16. If proteins are not genes, then how can comparisons between the
proteins of two species yield data about their evolutionary
relationship?

17. To distinguish a particular clade of mammals within the larger
clade that corresponds to class Mammalia, why is hair not a useful
characteristic?

Go to the website or CD-ROM for more quiz questions.

Evolution Connection

With characteristic insight, Darwin suggested looking at close relatives
of a species of interest to gain insight into what its ancestors may have
been like. How does his suggestion anticipate use of outgroups in modern cladistic analysis?

The Process of Science

Some nucleotide changes during molecular evolution cause amino acid
substitutions in the encoded protein (nonsynonymous changes), and
others do not (synonymous changes). In a comparison of rodent and
human genes, rodents were found to accumulate synonymous changes
2.0 times faster than humans, and nonsynonymous substitutions 1.3
times as fast. What factors could explain this difference? How do such
data complicate the use of molecular clocks in absolute dating?

**Deduce evolutionary and taxonomic relationships between species by
comparing amino acid sequences of proteins in the Case Study in the
Process of Science, available on the website and CD-ROM.**

Science, Technology, and Society

Experts estimate that human activities cause the extinction of hundreds
of species every year. The natural rate of extinction is thought to be a
few species per year. As we continue to alter the global environment,
especially by cutting down tropical rain forests, the resulting extinction
will probably rival that at the end of the Cretaceous period. Most biologists are alarmed at this prospect. What are some reasons for their concern? Consider that life has endured numerous mass extinctions and
has always bounced back. How is the present mass extinction different
from previous extinctions? Why? What might be some of the consequences for the surviving species?

Answers: 1. d; 2. b; 3. a; 4. e; 5. d; 6. e; 7. c; 8. d; 9. c; 10. b; 11. About 2,000 million
years, or 2 billion years. 12. 22,920 years (four half-life reductions). 13. Boundary
between the Nazca and South American Plates. 14. We are classified the same
down to the class level: Both the leopard and human are mammals. We do not belong to the same order. 15. homologous; analogous. 16. Although proteins are not
genes, they are gene products, their amino acid sequences determined by nucleotide sequences in the DNA. 17. Hair is a shared primitive character common
to all mammals and cannot be helpful in distinguishing different mammalian
subgroups.

THE EVOLUTIONARY HISTORY OF BIOLOGICAL DIVERSITY

Paleontologist Paul Sereno, professor of organismal biology and anatomy at the University of Chicago, focuses on dinosaur and early avian evolution. During expeditions to remote regions of Africa and South America, his teams have unearthed numerous fossils that have helped scientists understand the evolution of dinosaurs after the supercontinent Pangaea broke apart. We were scheduled to meet with Dr. Sereno in Chicago on September 14, 2001, but had to conduct the interview by phone instead because planes were still grounded after the tragic events of September 11. During a week that was otherwise very dark, our spirits were brightened by Professor Sereno's stories about his expeditions and his science outreach projects with school kids.

Interview
PAUL SERENO

Which of your expeditions so far was the most challenging and memorable?

The expedition that most defined my career was the first one to Niger in 1993. I took a very young crew on a wild trail that had us packing six Land Rovers in London, ferrying across to Algeria—which was a chancy place to travel at the time—and then threading with our overpacked Land Rovers through more than a thousand miles of desert to Niger, which had just elected its first president. Many of our people were seeing desert for the first time, and we encountered all sorts of delays and problems due to political situations. Half of my team departed because they feared for their lives. I then continued with the remainder—a wily group of very young students—and the ten of us discovered and excavated 6 tons of dinosaur bones. We returned across that very desert and managed to get all 6 tons of dinosaur bones onto a ferry, making all our future work in Africa possible. The story of that expedition is truly a great one, about which I'm writing a book. When you come out of something like that, you realize the great thrill of coming to the edge. And the edge was when I actually canceled the expedition halfway through because I thought it was all over. But it came back to life, and we went into the desert and made it work. I also met my future wife on that expedition! So it had all the elements.

When you're recruiting your expedition teams, what qualities do you look for in students and other crew members?

Well, it's one thing if you're going to a place like Wyoming. But if you're going to places like the center of the Sahara, then you need to find a team that can focus their physical and mental energies—about equal amounts of each—on a project that, I think it's fair to say, will be the most daunting and challenging in their lives up to that point. And to join these kinds of expeditions, you need to be a certain kind of person—a person who is adventuresome, preferably with some experience in fossil recovery. But most important, you need to be a person who thinks about other people—the kind of person who lets someone else go first in line—because what you find in an expedition, when the going gets tough, is that the most critical factor is a person's ability to work in a group, as a team member. And, of course, on these kinds of expeditions, you need to be a person who can withstand physical and mental stress at a level you have probably not experienced before. You'll find yourself in a very foreign environment and set upon by very trying weather conditions, such as extreme heat and sandstorms, eating foods you're not used to over long periods of time, and exerting yourself in these circumstances. But as a leader, I have to fix in my team members' minds what I believe is the truth, which is that this is one of the most exciting things you can do in a lifetime—you are really making history.

How do you know where to look for dinosaurs?

Your first clues about location are based on what past explorers have found. And the geography and geology of most areas have been mapped well enough so that you know roughly where there is rock of dinosaur age exposed. But there's usually a huge area to cover, so the team has to be very organized. It may mean five or six prospecting episodes over huge areas of desert in 120° heat. Sometimes you just have to blanket cover as much area as you can. Other times you begin to see a pattern of rich sites that you can focus on. Handheld satellite navigation devices have made the whole process more efficient. It allows us to return precisely to a site we found a month earlier. This means we can charge out into the field and locate many things we might want to collect and then organize our collecting after the initial exploration instead of trying to find something we walked away from on a rather featureless landscape.

How do you extract a dinosaur skeleton when you find one?

We mostly use very simple tools, from the small to the large: dental picks, awls, hammers, sledgehammers and chisels, and picks. Occasionally, we've had to use jackhammers, but they're usually not very useful. Then we have to have block and tackle to lift big fossils and rocks.

If it's a newly described dinosaur, how do you choose a name?

It's up to the whole group to come up with a name. It has to be a name that captures something about the animal—about its history or form. But the best name is also an aesthetic thing, sonorous and short and catchy enough for a kid to say. And of course, it has to make sense when translated from its classical word roots. Combining all those criteria, we've had the most interesting names. For example, we got to choose the name for the first reconstructed predatory dinosaur

from the Cretaceous in Africa. We called it *Afrovenator*, which means "the African hunter." We also found a big, long-necked, plant-eating dinosaur—23 m long—that was a contemporary of *Afrovenator*. Even before we left the field, a graduate student came up with *Jobaria* for this giant. It was based on "Jobar," the name the indigenous Tuareg people in Niger use for a mythical giant creature.

When you find a dinosaur, who has rights to the skeleton?

That situation has changed, both for domestic paleontology and for expeditions to other countries. Before, say, 50 years ago, paleontologists might take fossils out of Montana or some other state out West without regard to museums in those states. Today, you can't find a *T. rex* on public land in Montana and just drag it back to your favorite institution on the East Coast. And you certainly can't go to another country—not legally, at least—and just dig up a dinosaur and take it home. Even to dig in foreign countries, and certainly to take national treasures out of the country even temporarily, requires diplomacy with local scientists, politicians, nomads, villagers, and ministers. All sorts of situations arise. In Niger, for example, scientists are trying to establish a new museum for a dinosaur collection. Most specimens are on temporary loan. The bulk of the material will go back to the country of origin. To have all of these fossils lodged on a shelf somewhere here, thousands of miles away from where they were found, with no opportunity for people in the country of origin to study or even look at the specimens—that's not a situation we want to promote.

How did you get started in paleontology?

It was certainly not a talent that was apparent early on. Although I enjoyed nature and liked the outdoors as a kid, I was hardly a stellar student. In fact, when I was in the sixth grade, I couldn't even imagine myself graduating from high school. I was a very unsettled, unkempt student even into my first couple of years of high school. Then I found myself studying art. Through art, I gained confidence that I could actually apply my talents and energy to something. And I was able to improve my studies enough to go to a state university, thinking I would become a studio artist, but also taking an interest in various sciences—mainly in the field of anatomy, which was related to what I was doing in art. We were drawing human bodies, and I often found myself drawing skeletons. The one very dramatic event in my pathway was tagging along with my older brother, who was applying to graduate school and had an interview about studying paleontology at the American Museum of Natural History in New York. I was just totally blown away by all the things that were happening there—the lore of the historical work in the field of paleontology and the scientists who had come before, the great expeditions, all the active work that was ongoing—I just found the whole field intriguing. So it's a wild and woolly tale of just discovering what talents reside in you and what you're really interested in and then combining your interest and energy. And the discovery of those talents led me to paleontology and the things I do today.

As a paleontologist teaching skeletal anatomy to medical students, do you connect evolution to lessons about the human body?

Oh, yes! Many of the clinical problems, such as deteriorating joints, occur because we are not perfect designs, but animals with evolutionary backgrounds. Evolutionary solutions to things like walking and bending over are imperfect. And the changes that evolved most recently in humans are perhaps the ones most poorly integrated into our anatomy. Those are often the points where we have problems.

Speaking of teaching, did your personal educational experience inspire your interest in helping children discover their own talents and interests through Project Exploration, the science outreach group initiated by you and your wife, Gabrielle Lyon?

Yes, you come away realizing that there are all these undiscovered talents in any human being. You will only be able to discover a very few of them. There are things you thought you couldn't do, only to discover the talent at a different point in life. And there are things you'll never even be able to test, be it musical skills, acting, or some other talents that are inside you; you may never be able to experience those because of the particular trajectory of your mind. So we hope we get to discover at least a few of our talents. And that's what I try to impress on kids through Project Exploration. I feel I have sort of a calling or mission that way because of where I've come from and from not following a completely conventional pathway. I can relate to kids who are struggling in the classroom format to find something that interests them, because I was very intimidated in that setting and, for whatever reason, it wasn't working in my case. So in my spare time, I enjoy including kids in my research life. We just returned from a Wyoming trip with kids from the inner city who got to experience camping on a ranch. And we actually made a major discovery in the field. I collected a 5,000-pound block of tyrannosaur with a bunch of junior high and high school kids. Those kids found out that the natural world is exciting and that science is adventuresome, and that it involves a variety of talents. And once you discover those talents, it's a matter of organizing and focusing that energy. That's all it was in my life. And I get a huge amount of enjoyment in seeing kids inspired to go on toward careers in whatever they discover they want to do.

CHAPTER 26

EARLY EARTH AND THE ORIGIN OF LIFE

INTRODUCTION TO THE HISTORY OF LIFE

- Life on Earth originated between 3.5 and 4.0 billion years ago
- Prokaryotes dominated evolutionary history from 3.5 to 2.0 billion years ago
- Oxygen began accumulating in the atmosphere about 2.7 billion years ago
- Eukaryotic life began by 2.1 billion years ago
- Multicellular eukaryotes evolved by 1.2 billion years ago
- Animal diversity exploded during the early Cambrian period
- Plants, fungi, and animals colonized the land about 500 million years ago

THE ORIGIN OF LIFE

- The first cells may have originated by chemical evolution on a young Earth: *an overview*
- Abiotic synthesis of organic monomers is a testable hypothesis
- Laboratory simulations of early-Earth conditions have produced organic polymers
- RNA may have been the first genetic material
- Protobionts can form by self-assembly
- Natural selection could refine protobionts containing hereditary information
- Debate about the origin of life abounds

THE MAJOR LINEAGES OF LIFE

- The five-kingdom system reflected increased knowledge of life's diversity
- Arranging the diversity of life into the highest taxa is a work in progress

Life is a continuum *extending from the earliest organisms through various phylogenetic branches to the great variety of forms alive today. In Unit Five, we will survey the diversity of life today and trace the evolution of this diversity over 3.8 billion years of history.*

One of this book's ten themes is the interaction between organisms and their environments (see Chapter 1). We will see examples throughout this unit in the connections between biological history and geologic history. Geologic events that alter environments change the course of biological evolution. The formation and subsequent breakup of the supercontinent Pangaea, for instance, had a tremendous impact on the diversity of life (see Chapter 25). Conversely, life has changed the planet it inhabits. For example, the evolution of photosynthetic organisms that release oxygen into the air had a dramatic impact on Earth's atmosphere. (This early photosynthetic life included prokaryotes similar to those in the dense mats that resemble stepping stones in the painting of early Earth on this page.) Much more recently, the emergence of Homo sapiens *has changed the land, water, and air on a scale and at a rate unprecedented for a single species. The histories of Earth and its life are inseparable.*

These chapters also emphasize key junctures in evolution that have punctuated the history of biological diversity. Geologic history and biological history have been episodic, marked by what were in essence revolutions that opened many new ways of life.

Historical study of any sort is an inexact discipline that depends on the preservation, reliability, and interpretation of past records. The fossil record of past life is generally less and less complete the farther into the past we delve. Fortunately, each organism alive today carries traces of its evolutionary history in its molecules, metabolism, and anatomy. As we saw in Unit Four, such traces are clues to the past that augment the fossil record. Still, the evolutionary episodes of greatest antiquity are generally the most obscure.

This chapter begins with an overview of life's history. We then take a closer look at the origin of life. That discussion is the most speculative in the entire unit, for no fossil record of that seminal episode exists. The last section of the chapter introduces the main branches of life as a prelude to the survey of biological diversity in Chapters 27–34.

INTRODUCTION TO THE HISTORY OF LIFE

This unit of chapters compresses the history of life into a few weeks' worth of reading—like playing the "tape" of life in *very* fast forward! This overview is an even briefer look at life's major episodes. FIGURE 26.1 diagrams the chronology of these episodes in the form of a phylogenetic tree, and FIGURE 26.2, on page 512, presents the same chronology with a clock analogy.

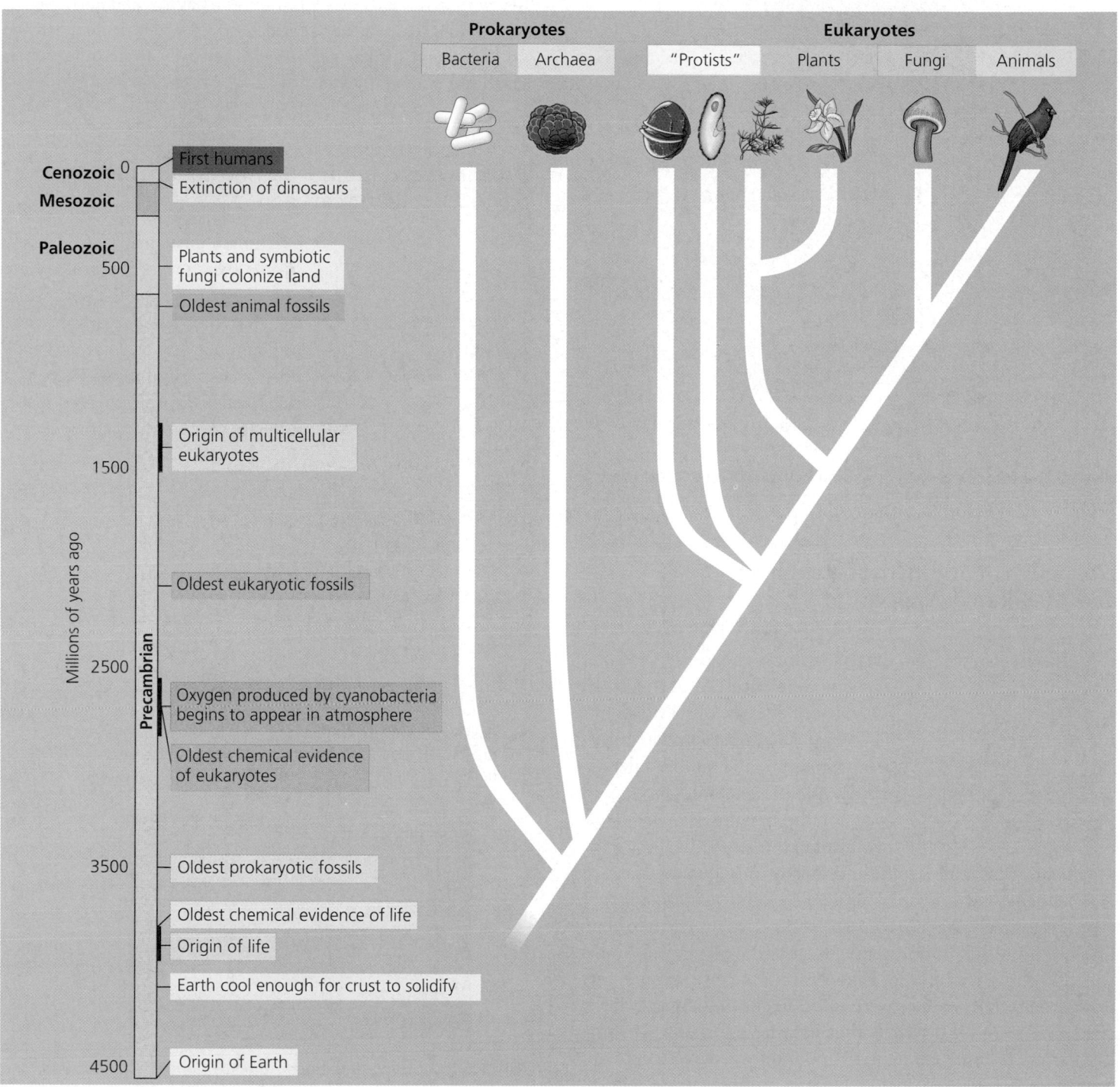

FIGURE 26.1 Some major episodes in the history of life. The timing of some events is based on fossil evidence. Other "dates" are based on chemical evidence or molecular clocks (see Chapter 25). Systematists continue to evaluate the evolutionary history of life's major branches, including the phylogeny of the diverse eukaryotes informally called "protists."

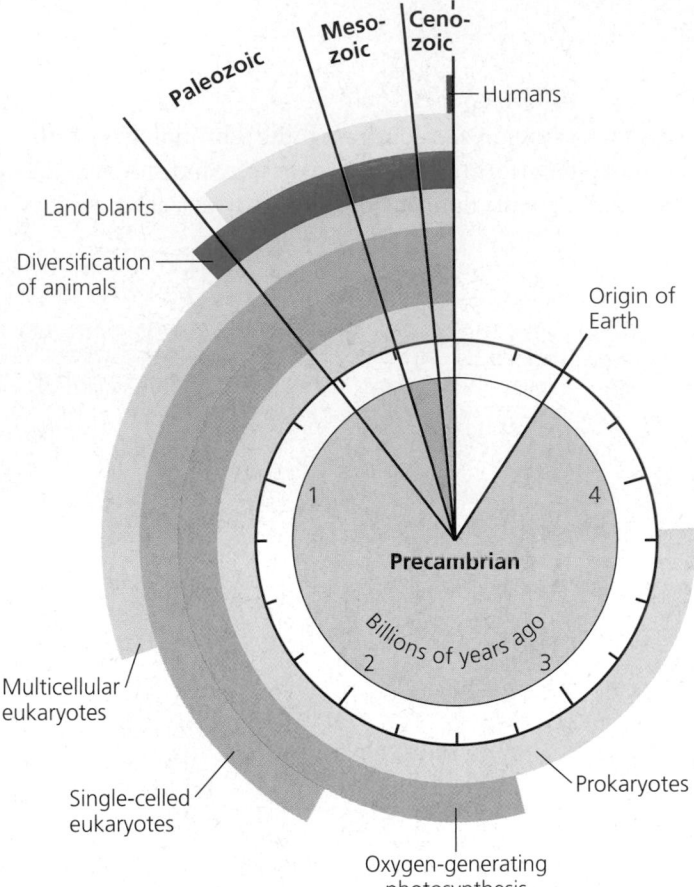

FIGURE 26.2 Clock analogy for some key events in evolutionary history. The clock ticks down from the origin of Earth to the present.

Life on Earth originated between 3.5 and 4.0 billion years ago

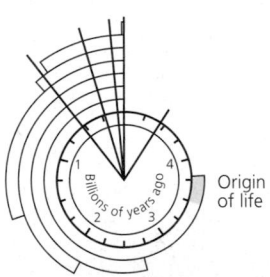

Life began remarkably early in Earth's history, and those first organisms were ancestral to the great diversity of life we observe today. The organisms most familiar to us are macroscopic and multicellular—mainly plants and animals. However, for the first three-quarters of evolutionary history, Earth's only organisms were microscopic and mostly unicellular.

The planet Earth formed about 4.5 billion years ago. However, it is unlikely that life could have survived on Earth for the first few hundred million years, because the planet was bombarded by huge rock bodies left over from the formation of the solar system. The impacts were colossal; one of them may have dislodged a chunk of Earth that became the moon. During this period of bombardment, the pounding generated enough heat to vaporize all the available water and prevent seas from forming. Most geologists now agree that this bombardment phase ended about 3.9 billion years ago.

The oldest known rocks on Earth's surface, located at a site called Isua, in Greenland, are 3.8 billion years old. Although there are some chemical clues in these rocks that life may have existed at the time, no one has yet found fossils of microorganisms in rocks so old. The oldest fossils of organisms that biologists have found so far are embedded in rocks from western Australia that are 3.5 billion years old. These microfossils resemble certain bacteria that still exist today (FIGURE 26.3). For bacteria so complex to have evolved by 3.5 billion years ago, it is a reasonable hypothesis that life originated much earlier, perhaps as early as 3.9 billion years ago, when Earth began to cool to a temperature at which liquid water could exist. We do know that prokaryotic life was already flourishing when Earth was still relatively young.

Prokaryotes dominated evolutionary history from 3.5 to 2.0 billion years ago

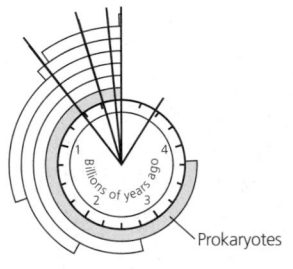

One might guess from the relatively simple structure of the prokaryotic cell, compared with the eukaryotic cell, that the earliest organisms were prokaryotes. The fossil record supports that presumption. There is a rich fossil history of prokaryotic life, without solid evidence of eukary-

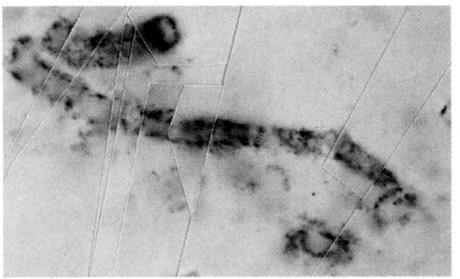

(a) Fossilized ancient bacterium. This fossilized filamentous prokaryote, about 3.5 billion years old, is an example of ancient bacteria collected in western Australia (LM).

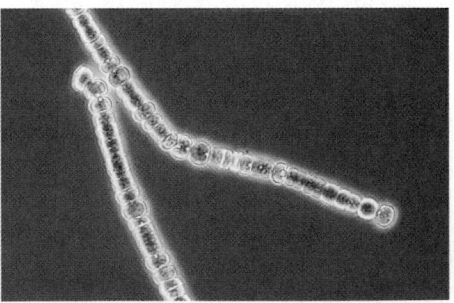

(b) Living *Leptolyngbya* bacterium. *Leptolyngbya*, a filamentous bacterium alive today, is remarkably similar to the fossilized prokaryote in (a) (LM).

FIGURE 26.3 Early and modern prokaryotes. Fossils in rocks older than about 500 million years escaped notice until about 40 years ago, partly because they are microscopic.

(a) Lynn Margulis and Kenneth Nealson, who study the history of life, are collecting bacterial mats in a Baja California lagoon. The mats are sedimentary structures produced by colonies of bacteria and cyanobacteria that live, uncropped by predators, in environments inhospitable to most other life.

(b) The bands seen in this section of a mat are layers of sediment that adhere to the sticky prokaryotes, which produce the succession of layers by migrating upward.

FIGURE 26.4 Bacterial mats and stromatolites.

(c) Fossilized mats known as stromatolites resemble the layered structures formed by modern-day bacterial colonies (shown in b). This stromatolite is a western Australian specimen that is about 3.5 billion years old. The prokaryote in FIGURE 26.3a was found in such a stromatolite.

otes, spanning 1.5 billion years, from about 3.5 to 2.0 billion years ago. Relatively early in that prokaryotic world, two main evolutionary branches, the bacteria and the archaea, diverged. Diverse species of these two main prokaryotic groups continue to thrive in various environments today.

Many of the oldest fossils of prokaryotes are found in **stromatolites,** fossilized mats similar to layered microbial mats that certain groups of prokaryotes still form today in salt marshes and warm lagoons (FIGURE 26.4). Researchers have also discovered fossils of prokaryotes in Australian sediments that formed around hydrothermal vents about 3.2 billion years ago. Hydrothermal vents are hot volcanic outlets in the deep-sea floor. Prokaryotes that inhabit such vents today are very different in their metabolism from prokaryotes making up algal mats such as those in FIGURE 26.4b. We'll explore the metabolically diverse prokaryotes in detail in Chapter 27. For now, the important point is that considerable metabolic diversity among prokaryotes living in various environments had already evolved over 3 billion years ago.

photosynthesis first evolved, the free O_2 from the cyanobacteria probably dissolved in the surrounding water until the seas and lakes became saturated with the oxygen. Additional O_2 would then react with dissolved iron and precipitate as iron oxide. These marine sediments were the source of banded iron formations, red layers of rock rich in the iron oxide that is a valuable source of iron ore today (FIGURE 26.5). Once all the

Oxygen began accumulating in the atmosphere about 2.7 billion years ago

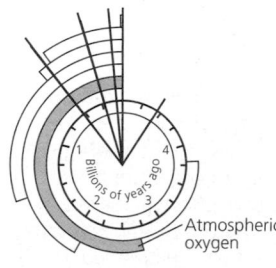

Atmospheric oxygen

Photosynthesis probably evolved very early in prokaryotic history, but in metabolic versions that did not split water and liberate oxygen (see Chapter 10). We'll see examples of such nonoxygenic photosynthesis among modern prokaryotes in Chapter 27. The only photosynthetic prokaryotes that generate O_2 are called cyanobacteria. Plentiful and diverse today, they probably evolved over 2.7 billion years ago.

Most atmospheric oxygen is of biological origin, from the water-splitting step of photosynthesis. When this oxygenic

FIGURE 26.5 Banded iron formations are evidence of the vintage of oxygenic photosynthesis. These bands of iron oxide at Jasper Knob in Michigan are about 2 billion years old.

dissolved iron had precipitated, additional O_2 finally began to "gas out" of the seas and lakes to accumulate in the atmosphere. This change left its mark in the rusting of terrestrial rocks rich in iron that began oxidizing about 2.7 billion years ago. This chronology implies that cyanobacteria may have originated as early as 3.5 billion years ago, when the microbial mats that left stromatolites began forming.

The accumulation of atmospheric O_2 was gradual from about 2.7 to 2.2 billion years ago, but then shot up relatively rapidly to more than 10% of its present level. This oxygen revolution had an enormous impact on life. The "corrosive" O_2, which attacks chemical bonds, doomed many prokaryotic groups. Some species survived in habitats that remained anaerobic, where we find their descendants still living today as obligate anaerobes (see p. 533). Among other survivors, a diversity of adaptations to the changing atmosphere evolved, including cellular respiration, which uses oxygen to help harvest energy stored in organic molecules.

The early rise in atmospheric O_2 was associated with the photosynthesis of early cyanobacteria. But what caused the accelerated rise in O_2 a few hundred million years later? One hypothesis is that it followed the evolution of eukaryotic algae containing chloroplasts.

Eukaryotic life began by 2.1 billion years ago

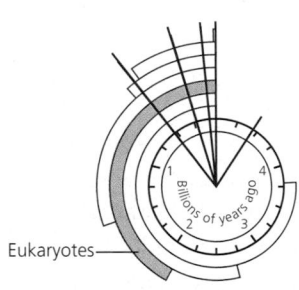

Eukaryotes

Eukaryotic cells are generally larger and much more complex than prokaryotic cells (see Chapter 7). In Chapter 28, we'll examine evidence that eukaryotic cells evolved from a symbiotic community of prokaryotes living within larger prokaryotic cells. The mitochondria of our cells and those of all other eukaryotes are descendants of some of those "endosymbionts" (symbiotic cells living within larger host cells). And so are the chloroplasts of plants and algae.

The oldest putative fossils of eukaryotes are 2.2-billion-year-old corkscrew-shaped organisms that look like relatively simple single-celled algae. The oldest fossils that are large enough to be eukaryotic to the satisfaction of most researchers are only about 2.1 billion years old. However, some researchers postulate a much earlier eukaryotic origin based on certain chemical traces dating back 2.7 billion years. This range of vintages places the earliest eukaryotes during a time when the oxygen revolution was changing Earth's environments dramatically. Chloroplasts may be part of the explanation for this temporal correlation. And another eukaryotic organelle, the mitochondrion, turned the accumulating O_2 to metabolic advantage through the process of cellular respiration.

Multicellular eukaryotes evolved by 1.2 billion years ago

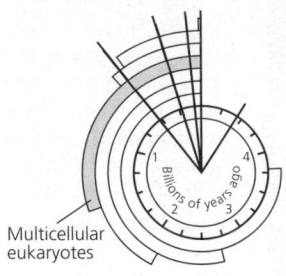

Multicellular eukaryotes

The origin of the more complex cells set the stage for a diversification of eukaryotic life. A great range of unicellular forms evolved, giving rise to a diversity of protists, and their single-celled descendants, that continue to flourish today. But multicellular forms also evolved. Their descendants include a variety of multicellular algae, plants, fungi, and animals. A multicellular organism, such as an animal, generally develops from a single cell—the fertilized egg, or zygote, in the case of sexual reproduction (see FIGURE 13.4). Cell division and cell differentiation help transform the single cell to a multicellular adult with many types of specialized cells.

Molecular clocks date the common ancestor of multicellular eukaryotes back to 1.5 billion years ago. However, the oldest known fossils of multicellular eukaryotes are of relatively small algae that lived about 1.2 billion years ago (FIGURE 26.6). Larger organisms, including animals such as jellies ("jellyfishes") and worms, do not appear in the fossil record until several hundred million years later, in the late Precambrian era about 600 million years ago (see FIGURE 26.1). Chinese paleontologists recently described a particularly rich fossil site 570 million years old containing a diversity of algae and animals, including some beautifully preserved embryos (FIGURE 26.7).

Geologists have recently reported evidence of a severe ice age enduring from 750 to 570 million years ago that may help to explain why multicellular eukaryotes were relatively limited in diversity and distribution until the very late Precambrian. This **snowball Earth** hypothesis is based partly on geologic clues that glaciers covered the planet's landmasses from pole to pole. The seas also would have been iced over. Thus, most life

50 µm

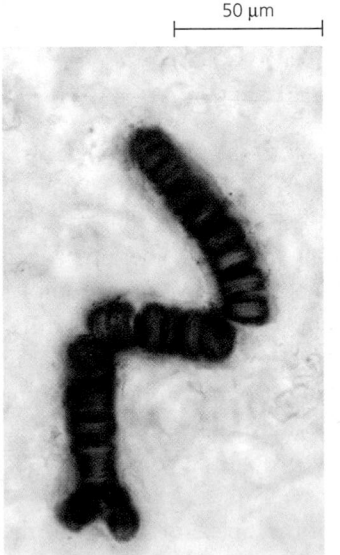

FIGURE 26.6 Fossilized alga about 1.2 billion years old. Such fossils are the earliest evidence of organisms with more than one type of cell. At one end of this filamentous eukaryote is the bi-lobed "holdfast," which probably anchored the alga to its substratum. This fossil is among numerous such specimens collected on Somerset Island in arctic Canada.

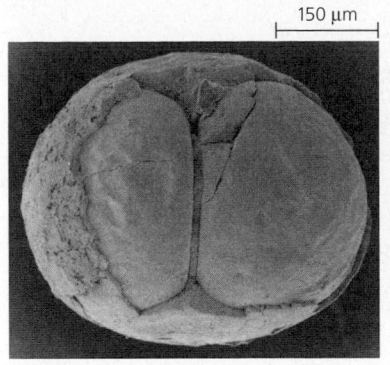

(a) Two-celled stage of embryonic development (SEM)

(b) Later embryonic stage (SEM)

FIGURE 26.7 Fossilized animal embryos from Chinese sediments 570 million years old.

would have been confined to areas near deep-sea vents and hot springs or to those sparse locales where enough ice melted for sunlight to penetrate the surface waters of the seas. According to this hypothesis, the fossil record for the first major diversification of multicellular eukaryotes corresponds in time to the thawing of snowball Earth. A second radiation of eukaryotic forms produced most of the major groups of animals during the early Cambrian period, the first period of the Paleozoic era.

Animal diversity exploded during the early Cambrian period

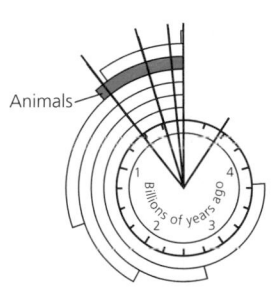

Although the history of animal life dates back to the late Precambrian era, a much greater diversity of animals evolved just after the dawn of the Paleozoic era 543 million years ago. Cnidarians (the phylum that includes jellies) and poriferans (sponges) were already present in the late Precambrian. However, most of the major groups (phyla) of animals make their first fossil appearances during the relatively short span of the Cambrian period's first 20 million years (FIGURE 26.8). In Chapter 32, we'll evaluate a few hypothetical explanations for the so-called Cambrian explosion.

Plants, fungi, and animals colonized the land about 500 million years ago

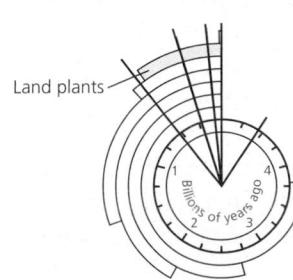

The colonization of land was one of the pivotal milestones in the history of life. There is fossil evidence that cyanobacteria and other photosynthetic prokaryotes coated damp terrestrial surfaces well over a billion years ago. However, macroscopic life in the form of plants, fungi, and

animals did not colonize land until about 500 million years ago, during the early Paleozoic era. This gradual evolutionary venture out of ancestral aquatic environments was associated with adaptations that helped prevent dehydration and that made it possible to reproduce on land. For example, plants, which evolved from green algae, have a waterproof coating of wax on their leaves that slows the loss of water.

Plants colonized land in the company of fungi. Even today, the roots of most plants are associated with fungi that aid in the absorption of water and minerals from the soil. The fungi, in turn, obtain their organic nutrients from the plant. Such symbiotic associations of plants and fungi are evident in some of the oldest fossilized roots, dating this relationship back to the early spread of life onto land. You will learn more about the evolution of plants and fungi in Chapters 29–31.

Plants transformed the landscape, creating new opportunities for all life, including herbivorous (plant-eating) animals and their predators. Although many animal groups are represented in terrestrial environments, the most widespread and diverse land animals are certain arthropods (in the form of insects and spiders) and certain vertebrates (in the form of amphibians such as frogs and salamanders, reptiles such as lizards and snakes, birds, and mammals). The terrestrial vertebrates are called tetrapods, a reference to the four walking limbs that distinguish them from the various aquatic vertebrates we call fishes. An excellent fossil record documents the

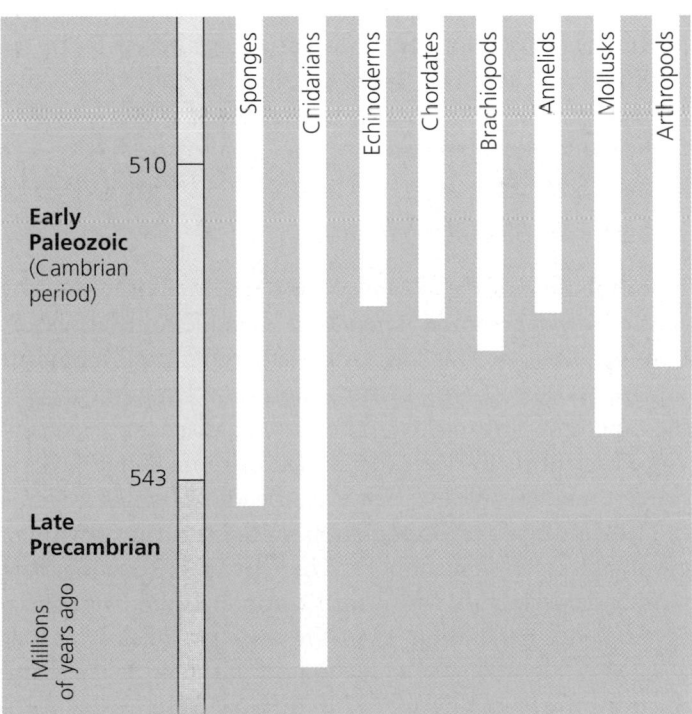

FIGURE 26.8 The Cambrian radiation of animals. The bars in this diagram extend back to the earliest appearances of a small sample of animal groups in the fossil record. You'll learn more about the Cambrian explosion and the major animal phyla in Chapters 32–34.

descent of amphibians from fishes. Reptiles evolved from amphibians, and both birds and mammals evolved from reptiles. Most of the groups (orders) of modern mammals were established by about 50–60 million years ago, including the primates, the order that includes apes and humans. The human lineage diverged from other primates only about 5 million years ago—the most recent second in a clock of life's history rescaled to represent an hour. But our history, like that of every other species in the rich diversity of life, actually begins with the first organisms to live on the primordial Earth over 3.5 billion years ago.

THE ORIGIN OF LIFE

The question of how life began is more specifically about the genesis of prokaryotes. Sometime between about 4.0 billion years ago, when Earth's crust began to solidify, and 3.5 billion years ago, when the planet was inhabited by bacteria advanced enough to build stromatolites, the first organisms came into being. What was their origin? We will never know for sure, of course, how life on Earth began. But science seeks natural causes for natural phenomena, and that is the approach that must guide scientific inquiry about the origin of life.

The first cells may have originated by chemical evolution on a young Earth: *an overview*

Most biologists favor the hypothesis that life on Earth developed from nonliving materials that became ordered into molecular aggregates that were eventually capable of self-replication and metabolism.

Resolving the Biogenesis Paradox

From the time of the ancient Greeks until well into the 19th century, it was common "knowledge" that life could arise from nonliving matter. This idea of life emerging from inanimate material is called **spontaneous generation.** Experiments with flies and other organisms in the late Renaissance period convinced scientists to reject the notion of spontaneous generation for macroscopic life. However, the idea persisted well into the 19th century as an explanation for the rapid growth of microorganisms in spoiled foods. Then, in 1862, Louis Pasteur's famous experiments with broth completed the overturn of spontaneous generation, even for microorganisms (FIGURE 26.9). As far as we know, all life today arises only by the reproduction of preexisting life. This "life-from-life" principle is called **biogenesis.**

But what about the *first* organisms? If *they* arose by biogenesis, then they couldn't have been the first organisms.

Although there is no evidence that spontaneous generation occurs today, conditions on the early Earth were very different. For instance, there was relatively little atmospheric oxygen to tear apart complex molecules. And such energy sources as lightning, volcanic activity, and ultraviolet sunlight were all more intense than what we experience today. The resolution to the biogenesis paradox is that life did not begin on a planet anything like the modern Earth, but on a young Earth that was a very different world.

A Four-Stage Hypothesis for the Origin of Life

Most biologists now think that it is at least a credible hypothesis that chemical and physical processes in Earth's primordial environment eventually produced very simple cells through a sequence of stages. There is much debate about the nature of those stages.

According to one hypothetical scenario, the first organisms were products of chemical evolution in four stages: (1) the abiotic (nonliving) synthesis of small organic molecules, such as amino acids and nucleotides; (2) the joining of these small molecules (monomers) into polymers, including proteins and nucleic acids; (3) the origin of self-replicating molecules that eventually made inheritance possible; and (4) the packaging of all these molecules into "protobionts," droplets with membranes that maintained an internal chemistry different from the surroundings. This is all speculative, of course, but what makes it science is that the hypothesis leads to predictions that can be tested in the laboratory. Let's take a closer look at some of the evidence for each of these four stages.

Abiotic synthesis of organic monomers is a testable hypothesis

In the 1920s, A. I. Oparin, of Russia, and J. B. S. Haldane, of Great Britain, independently postulated that conditions on the primitive Earth favored chemical reactions that synthesized organic compounds from inorganic precursors present in the early atmosphere and seas. This cannot happen in the modern world, Oparin and Haldane reasoned, because the present atmosphere is rich in oxygen produced by photosynthetic life. The oxidizing atmosphere of today is not conducive to the spontaneous synthesis of complex molecules because the oxygen attacks chemical bonds, extracting electrons. Before oxygen-producing photosynthesis, Earth had a much less oxidizing atmosphere, derived mainly from volcanic vapors. Such a reducing (electron-adding) atmosphere would have enhanced the joining of simple molecules to form more complex ones. Even with a reducing atmosphere, making organic molecules would require considerable energy, which was probably provided by lightning and the intense UV radiation

① Pasteur began each experiment by heating a nutrient medium (beef broth) to kill any microorganisms that were already present. If flasks of this sterilized broth were then left open, it took just a few days for them to become contaminated with dense growths of microorganisms. Were the microbes spontaneously generated from the broth, or were they generated by the reproduction of microorganisms that rained into the broth from above?

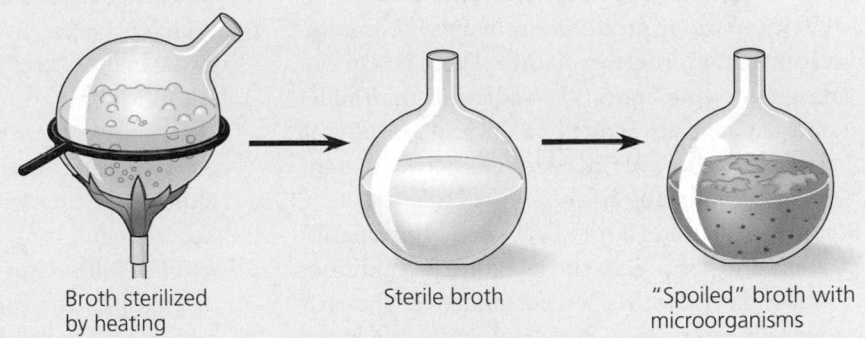

Broth sterilized by heating

Sterile broth

"Spoiled" broth with microorganisms

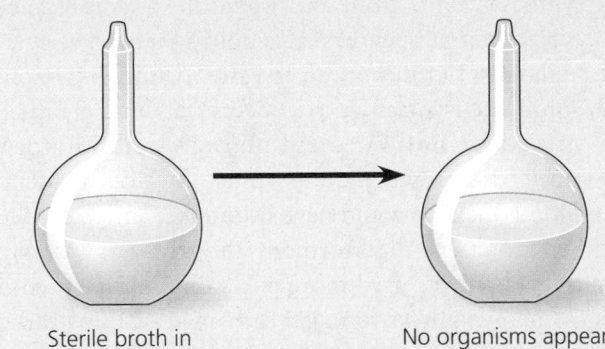

Sterile broth in sealed flask

No organisms appear

② The broth remained sterile for months if it was kept in sealed flasks after heating. However, some of Pasteur's critics argued that sealing the flasks isolated the broth from an airborne "life force" that was supposedly required for spontaneous generation to occur.

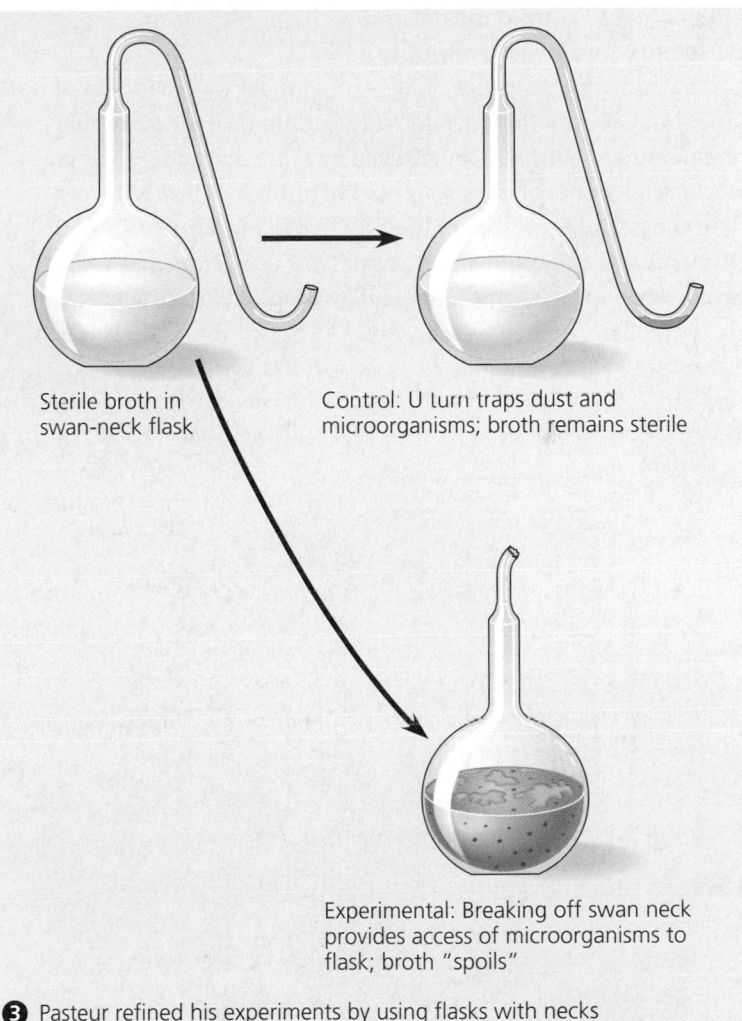

Sterile broth in swan-neck flask

Control: U turn traps dust and microorganisms; broth remains sterile

Experimental: Breaking off swan neck provides access of microorganisms to flask; broth "spoils"

③ Pasteur refined his experiments by using flasks with necks shaped like swans' necks. The bend in the swan neck traps particles of dust and microbes but allows outside air access to the broth.

FIGURE 26.9 Pasteur and biogenesis of microorganisms. In the early 1860s, Louis Pasteur (inset) conducted a series of experiments to test whether microorganisms emerge by spontaneous generation or by reproduction of existing microorganisms (biogenesis). His research contributed to the germ theory of disease, which connected infections to the spread of microorganisms and led to improvements in hospital hygiene and public sanitation. The legacy of these famous experiments is also manifest in the term *pasteurization*. Pasteurized milk, for example, has been heated to destroy potentially harmful microorganisms and then sealed to maintain the sterility.

that penetrated the primitive atmosphere. The modern atmosphere has a layer of ozone produced from oxygen, and this ozone shield screens out most UV radiation. There is also evidence that young suns emit more UV radiation than older suns. Oparin and Haldane envisioned an ancient world with the chemical conditions and energy resources needed for the abiotic synthesis of organic molecules.

In 1953, Stanley Miller and Harold Urey tested the Oparin-Haldane hypothesis by creating, in the laboratory, conditions comparable to those that scientists had postulated for the early Earth. Their apparatus produced a variety of amino acids and other organic compounds found in living organisms today (FIGURE 26.10; also see FIGURE 4.1).

The atmosphere in the Miller-Urey model was made up of H_2O, H_2, CH_4 (methane), and NH_3 (ammonia), the gases that researchers in the 1950s believed prevailed in the ancient world. This atmosphere was probably more strongly reducing than the actual atmosphere of early Earth. Modern volcanoes emit CO, CO_2, N_2, and water vapor, and it is likely that these gases were abundant in the ancient atmosphere. Hydrogen gas was probably not a major component, and traces of O_2 may even have been present, formed from reactions among other gases as they baked under the powerful UV radiation. Many laboratories have repeated the Miller experiment using a variety of recipes for the atmosphere. Abiotic synthesis of organic compounds occurred in these modified models, although yields were generally smaller than in the original experiment.

The Miller-Urey experiments still stimulate debate on the origin of Earth's early stockpile of organic ingredients. Today, one line of research focuses on where chemicals needed for organic syntheses came from and where the reactions most likely occurred. Some scientists now doubt that the early atmosphere played a significant role in early chemical reactions. Instead, submerged volcanoes and deep-sea vents—gaps in Earth's crust where hot water and minerals gush into deep oceans—may have provided the essential resources. Evidence is also building that life could have begun in a much simpler chemical environment than formerly thought. For instance, the first cells may have used inorganic sulfur and iron compounds as energy sources to make their own ATP instead of taking it up from their surroundings.

It is also plausible that some organic compounds reached Earth from space. In 2000, Indian scientists reported computer models showing how molecules such as adenine, an ingredient of DNA, could form by reactions of cyanide in the clouds of gas between stars. These simulations would explain why some meteorites that have crashed to Earth contain organic molecules. But whether the primordial Earth was stocked with organic monomers made here or elsewhere, the key point is that the molecular ingredients of life were probably present very early.

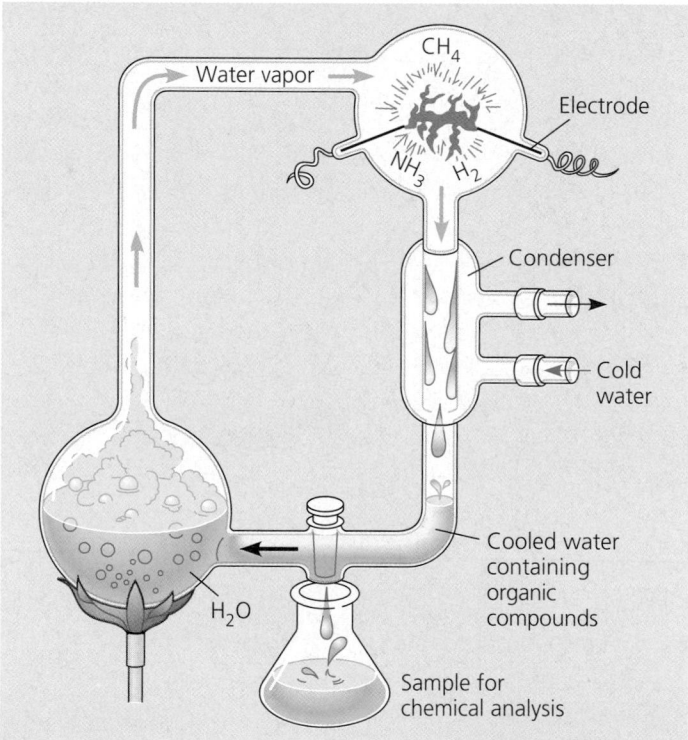

FIGURE 26.10 The Miller-Urey experiment. A warmed flask of water simulated the primeval sea. The "atmosphere" consisted of H_2O, H_2, CH_4, and NH_3. Sparks were discharged in the synthetic atmosphere to mimic lightning. A condenser cooled the atmosphere, raining water and any dissolved compounds back to the miniature sea. As material circulated through the apparatus, the solution in the flask changed from clear to murky brown. After one week, Miller and Urey analyzed the contents of the solution and found a variety of organic compounds, including some of the amino acids that make up the proteins of organisms.

Laboratory simulations of early-Earth conditions have produced organic polymers

The hypothesis of an abiotic origin of life makes another prediction that can be tested in the laboratory. If the hypothesis is correct, then it should be possible to link organic monomers to form polymers such as proteins and nucleic acids without the help of enzymes and other cellular equipment. Researchers have observed such polymerization after dripping solutions of organic monomers onto hot sand, clay, or rock. The heat vaporizes the water in the solutions and concentrates the monomers on the underlying substance. Some of the monomers then spontaneously bond together to form polymers, including polypeptides, the chains of amino acids that make up proteins. On the early Earth, raindrops or waves may have splashed dilute solutions of organic monomers onto fresh lava or other hot rocks and then rinsed polypeptides and other polymers back into the sea. Alternatively, deep-sea vents, where gases and superheated water with dissolved minerals escape from Earth's interior, may have been locales for the abiotic synthesis of both organic monomers and polymers.

RNA may have been the first genetic material

Life is defined partly by the process of inheritance, which is based on self-replicating molecules. Today's cells store their genetic information as DNA. They transcribe the information into RNA and then translate RNA messages into specific enzymes and other proteins (see Chapter 17). This mechanism of information flow probably emerged gradually through a series of refinements to much simpler processes. In fact, many researchers now favor the hypothesis that the first hereditary material was not DNA, but RNA, which may also have functioned as the first enzymes. (This helps resolve the "chicken and egg" paradox of which came first, genes or enzymes.) According to this hypothesis, the molecular biology of today was preceded by an "RNA world."

Molecular Replication in an RNA World

Several scientists have tested the hypothesis of RNA self-replication. Short polymers of ribonucleotides have been produced abiotically in laboratory experiments. If such RNA is added to a solution containing monomers for making more RNA, sequences about five to ten nucleotides long are copied from the template according to the base-pairing rules (FIGURE 26.11). If zinc is added as a catalyst, sequences up to 40 nucleotides long are copied with less than 1% error.

In the 1980s, Thomas Cech revolutionized thinking about the evolution of life when he discovered that RNA molecules are important catalysts in modern cells. This finding disproved the long-held view that only proteins (enzymes) serve as biological catalysts. Cech and other researchers found that modern cells use RNA catalysts, called **ribozymes,** to remove introns from RNA (see Chapter 17). Ribozymes also help catalyze the synthesis of new RNA, notably rRNA, tRNA, and mRNA. Thus, RNA is autocatalytic, and in the prebiotic world, before there were enzymes (proteins) or DNA, RNA molecules may have been fully capable of ribozyme-catalyzed replication.

Natural Selection in an RNA World

Natural selection on the molecular level has been observed operating on RNA populations in the laboratory. Unlike double-stranded DNA, which takes the form of a uniform helix, single-stranded RNA molecules assume a variety of specific three-dimensional shapes mandated by their nucleotide sequences. The molecule thus has both a genotype (its nucleotide sequence) and a phenotype (its conformation, which interacts with surrounding molecules in specific ways). In a particular environment, RNA molecules of certain base sequences are more stable and replicate faster and with fewer errors than other sequences. Beginning with a diversity of RNA molecules that must compete for monomers to replicate, the sequence best suited to the temperature, salt concentration, and other features of the surrounding solution and having the greatest autocatalytic activity will prevail. Its descendants will not be a single RNA species but will be a family of closely related sequences because of copying errors. Selection screens mutations in the original sequence, and occasionally a copying error results in a molecule that folds into a shape that is even more stable or more adept at self-replication than the ancestral sequence. Similar selection events may have occurred in the prebiotic RNA world.

The rudiments of RNA-directed protein synthesis may have been the weak binding of specific amino acids to bases along RNA molecules, which functioned as simple templates holding a few amino acids together long enough for them to be linked. (Indeed, this is one function of rRNA in modern ribosomes.) If RNA happened to synthesize a short polypeptide that in turn behaved as an enzyme helping the RNA molecule to replicate, then the early chemical dynamics included molecular cooperation as well as competition. These first steps toward the replication and translation of genetic information

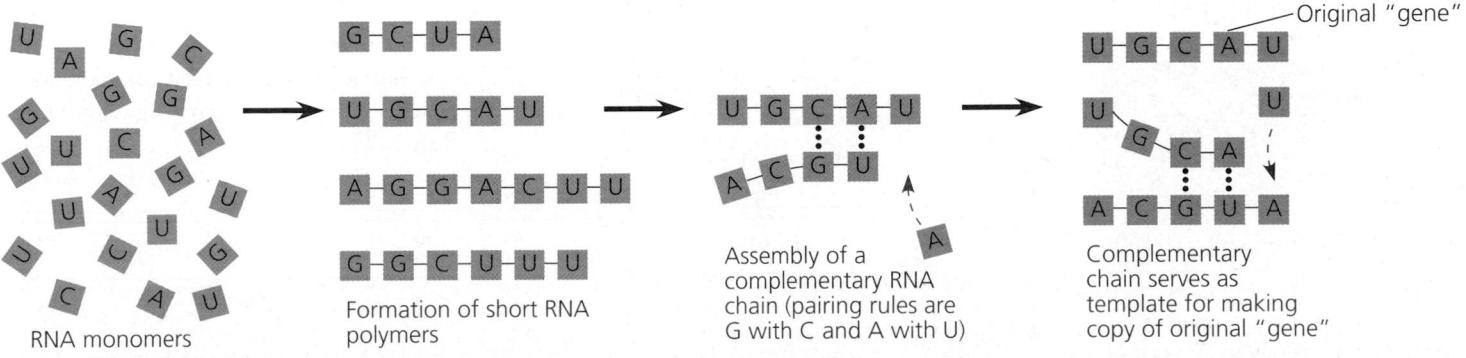

RNA monomers

Formation of short RNA polymers

Assembly of a complementary RNA chain (pairing rules are G with C and A with U)

Complementary chain serves as template for making copy of original "gene"

Original "gene"

FIGURE 26.11 Abiotic replication of RNA.

may have been taken by molecular evolution even before RNA and polypeptides became packaged within membranes.

Protobionts can form by self-assembly

The properties of life emerge from an interaction of molecules organized into higher levels of order. Living cells may have been preceded by **protobionts,** aggregates of abiotically produced molecules. Protobionts are not capable of precise reproduction, but they maintain an internal chemical environment different from their surroundings and exhibit some of the properties associated with life, including metabolism and excitability.

Laboratory experiments demonstrate that protobionts could have formed spontaneously from abiotically produced organic compounds. For example, droplets called liposomes form when the organic ingredients include certain lipids. These lipids organize into a molecular bilayer at the surface of the droplet, much like the lipid bilayer of cell membranes. Because the membrane is selectively permeable, the liposomes undergo osmotic swelling or shrinking when placed in solutions of different salt concentrations. Some of these protobionts also store energy in the form of a membrane potential, a voltage across the surface. The protobionts can discharge the voltage in nervelike fashion; such excitability is characteristic of all life (which is not to say that liposomes are alive, but only that they display *some* of the properties of life). Liposomes behave dynamically, sometimes growing by engulfing smaller liposomes and then splitting, other times "giving birth" to smaller liposomes (FIGURE 26.12a). If enzymes are included among the ingredients, they are incorporated into the droplets. The protobionts are then able to absorb substrates from their surroundings and release the products of the reactions catalyzed by the enzymes (FIGURE 26.12b).

Unlike some laboratory models, protobionts that formed in the ancient seas would not have possessed refined enzymes, which are made in cells according to inherited instructions. Some molecules produced abiotically, however, do have weak catalytic capacities, and there could well have been protobionts that had a rudimentary metabolism that allowed them to modify substances they took in across their membranes.

Natural selection could refine protobionts containing hereditary information

The packaging of primitive RNA genes and their polypeptide products within a membrane would have been a significant milestone in the early history of life. Once this occurred, protobionts could have evolved as units, and molecular cooperation could be refined because components that interacted in ways favorable to the success of the protobiont as a whole were concentrated together in a microscopic volume rather than being spread throughout the surroundings (FIGURE 26.13). Suppose, for example, that an RNA molecule ordered amino acids into a primitive enzyme that extracted energy from inorganic sulfur compounds taken up from the surroundings. This energy could be used for other reactions within the protobiont, including replication of RNA. Natural selection could favor such a gene only if its product were kept close by, rather than being shared with competing RNA sequences in its environment. The most successful protobionts would grow and split, distributing copies of their genes to offspring. Even if only one such protobiont arose initially by the abiotic processes that have been described, its descendants would vary because of mutations, errors in the copying of RNA.

Evolution in the true Darwinian sense—differential reproductive success of varied individuals—presumably accumulated many refinements to primitive metabolism and inheritance. One trend apparently led to DNA becoming the hereditary material. Initially, RNA could have provided the template on which DNA nucleotides were assembled. But DNA is a much more stable repository for genetic information than RNA, and once DNA appeared, RNA molecules would have begun to take on their modern roles as intermediates in the translation of genetic programs. The "RNA world" gave way to a "DNA world."

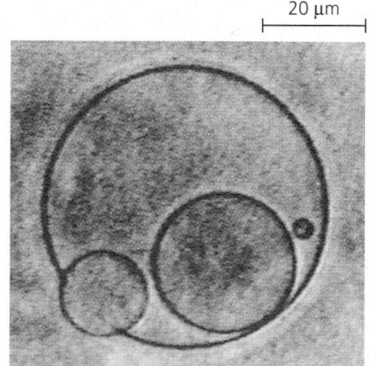

(a) Simple reproduction. This liposome is "giving birth" to smaller liposomes. (LM)

FIGURE 26.12 Laboratory versions of protobionts.

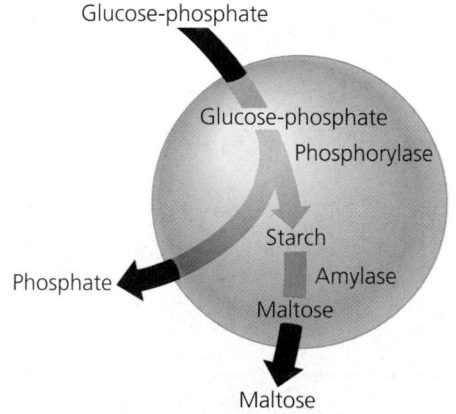

20 µm

Glucose-phosphate

Glucose-phosphate
Phosphorylase

Starch

Phosphate

Amylase

Maltose

Maltose

(b) Simple metabolism. If enzymes—in this case, phosphorylase and amylase—are included in the solution from which the droplets self-assemble, some protobionts can carry out simple metabolic pathways.

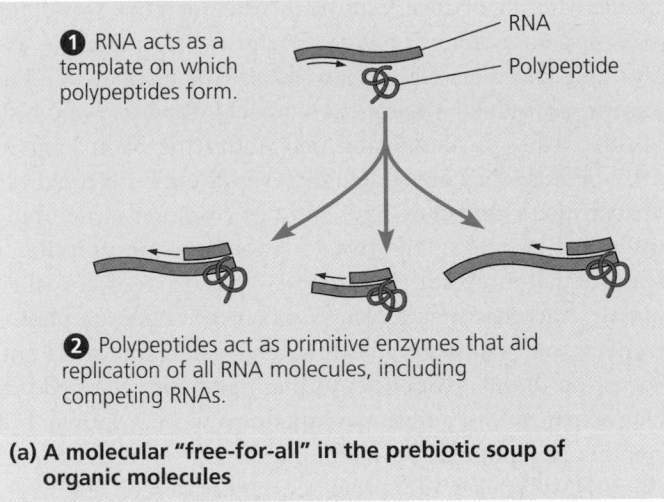

1 RNA acts as a template on which polypeptides form.

RNA

Polypeptide

2 Polypeptides act as primitive enzymes that aid replication of all RNA molecules, including competing RNAs.

(a) A molecular "free-for-all" in the prebiotic soup of organic molecules

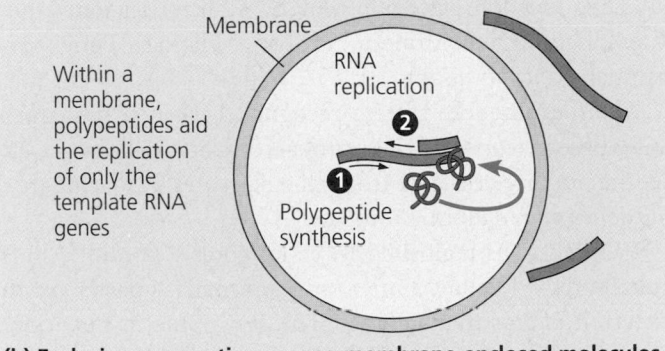

Within a membrane, polypeptides aid the replication of only the template RNA genes

Membrane

RNA replication

Polypeptide synthesis

(b) Exclusive cooperation among membrane-enclosed molecules

FIGURE 26.13 Hypotheses for the beginnings of molecular cooperation.

as a first step in the origin of life. As we discussed earlier, it is possible that at least some organic compounds reached the early Earth from space on meteorites and comets.

Where life began is another issue. Until recently, most researchers favored shallow water or moist sediments as the most likely sites for life's origin. Some scientists now question this view, arguing that Earth's surface was very inhospitable during the period when life probably began. Perhaps life began on the less exposed seafloor. The discovery of deep-sea vents in the late 1970s raised the possibility that earlier vents supplied the energy and chemical precursors for the origin of protobionts (FIGURE 26.14). Molecular phylogenetic analyses indicate that the ancestors of modern prokaryotes thrived in very hot conditions and may have lived on inorganic sulfur compounds that are common in deep-sea vent environments. Laboratory studies by Günter Wachtershäuser and colleagues also point to the environments of deep-sea vents and volcanoes as abiotic sources of some of the organic compounds, such as acetyl coenzyme A, that cells use in energy metabolism (see FIGURE 9.11). Sulfides of iron and nickel, common in volcanic and deep-sea vent areas, catalyze the formation of acetic acid and precursors of acetyl coenzyme from CO and H_2S. Wachtershäuser considers it likely that life arose in an environment where these reactions were common and then adopted them as the earliest form of metabolism.

Debate about the origin of life abounds

Laboratory simulations cannot prove that the kind of chemical evolution that has been described here actually created life on the primitive Earth, but only that some of the key steps *could* have happened. The origin of life remains a matter of speculation, and there are alternative views of how several key processes occurred.

Some researchers question whether abiotic synthesis of organic monomers on Earth was necessary

THE PROCESS OF SCIENCE

FIGURE 26.14 A window to early life? In this porthole view from the deep-sea research submarine *Alvin,* you can see a water-testing wand mounted to a robotic arm. This 2000 expedition to the Sea of Cortés sampled the environments around deep-sea hydrothermal vents at depths of greater than a mile. The column of hot effluent seen rising from this vent included hydrogen sulfide and iron sulfide, which react to produce pyrite (fool's gold) and H_2 gas. Prokaryotes that live around the vents use the H_2 as an energy source. Although we may consider such environments extreme by modern-Earth criteria, many researchers favor the hypothesis that life began in such places.

On a broader scale, the hypothesis that life is not restricted to Earth is becoming more accessible to scientific testing. Photographs taken by the Galileo spacecraft of the ice-covered surface of Europa, one of Jupiter's moons, have led to hypotheses that liquid water lies beneath the surface and may support prokaryotic life. The surface of Mars is a cold, dry, lifeless desert, but billions of years ago it was probably relatively warm and may have held liquid water and had a CO_2-rich atmosphere. It is possible that prebiotic chemistry, similar to that on early Earth, also occurred on Mars before the planet became less hospitable. Scientists anticipate that exploration in the next decade will provide clues about whether microbial life on Mars once existed. Did life evolve there and then die out, or was prebiotic chemistry snuffed out by planetary changes before any life forms developed? Many scientists also see Mars as an ideal place to test hypotheses about Earth's prebiotic chemistry.

Debate about the origin of terrestrial and extraterrestrial life abounds, and we have sampled only a few of the issues. Whatever took place in the prebiotic world, the leap from an aggregate of molecules that reproduces to even the simplest prokaryotic cell is immense, and change must have occurred in many smaller evolutionary steps. The point at which we stop calling membrane-enclosed compartments that metabolize and replicate their genetic programs protobionts and begin calling them living cells is as fuzzy as our definitions of life. We do know that prokaryotes were already flourishing at least 3.5 billion years ago and that all the lineages of life arose from those ancient prokaryotes.

THE MAJOR LINEAGES OF LIFE

THE PROCESS OF SCIENCE In Chapter 25, we looked at systematics as the study of biological diversity in an evolutionary context. Now that we have gone back in time to the very origin of life on Earth, systematics is once again relevant as we attempt to reconstruct evolutionary relationships among the immense diversity of forms that arose from those early organisms.

The five-kingdom system reflected increased knowledge of life's diversity

Systematists have traditionally considered the kingdom to be the highest—the most inclusive—taxonomic category. Many of us grew up with the notion that there are only two kingdoms of life—plants and animals—because we live in a macroscopic, terrestrial realm where we rarely see organisms that do not fit neatly into a plant-animal dichotomy. The two-kingdom scheme also had a long tradition in formal taxonomy; Linnaeus divided all known forms of life between the plant and animal kingdoms.

Even with the discovery of the diverse microbial world, the two-kingdom system persisted. Bacteria were placed in the plant kingdom, their rigid cell walls cited as justification. Eukaryotic unicellular organisms with chloroplasts were also considered plants. Fungi, too, fell under the plant banner, partly because they are sedentary, even though no fungi are photosynthetic and they have little in common structurally with green plants. In the two-kingdom system, unicellular creatures that move and ingest food—protozoa—were called animals. Microbes such as *Euglena* that move but are photosynthetic were claimed by both botanists and zoologists and showed up in the taxonomies of plant and animal kingdoms alike. Schemes with additional kingdoms were proposed, but none became popular with the majority of biologists until Robert H. Whittaker, of Cornell University, argued effectively for a five-kingdom system in 1969. Whittaker designated these five kingdoms as Monera, Protista, Plantae, Fungi, and Animalia (FIGURE 26.15).

The five-kingdom system recognized the two fundamentally different types of cells, prokaryotic and eukaryotic, and set the prokaryotes apart from all eukaryotes by placing them in their own kingdom, Monera.

Whittaker distinguished three kingdoms of multicellular eukaryotes—Plantae, Fungi, and Animalia—partly on the criterion of nutrition. Plants are autotrophic in nutritional mode, making their food by photosynthesis. Fungi are heterotrophic organisms that are absorptive in nutritional mode. Most fungi are decomposers that live embedded in their food source, secreting digestive enzymes and absorbing the small organic molecules that are the products of digestion. Most animals live by ingesting food and digesting it within specialized cavities.

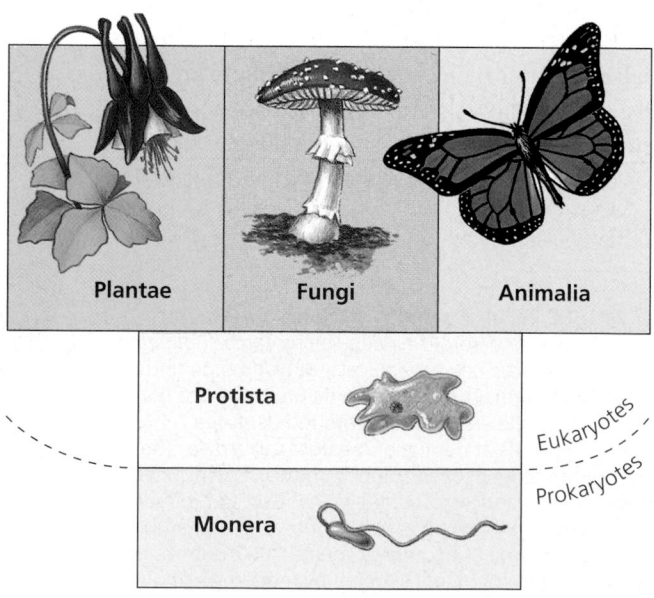

FIGURE 26.15 Whittaker's five-kingdom system.

We are left with the kingdom Protista. In Whittaker's five-kingdom system, Protista consisted of all eukaryotes that did not fit the definition of plants, fungi, or animals. Most protists are unicellular forms. However, the boundaries of Whittaker's kingdom Protista were expanded to include some multicellular organisms, such as seaweeds, because of their relationships to certain unicellular protists. With such refinements, the five-kingdom system prevailed in biology for over 20 years.

Arranging the diversity of life into the highest taxa is a work in progress

Like any classification scheme, the five-kingdom system is not a natural fact but a human construct. It is one attempt to order the diversity of life into a scheme that is useful and, hopefully, phylogenetically reasonable. During the past three decades, systematists applying cladistic analysis, including the construction of cladograms based on molecular data (see Chapter 25), have been pointing out problems with the traditional five-kingdom scheme.

One challenge to the five-kingdom system is the evidence that there are two distinct lineages of prokaryotes. These new data led to a **three-domain system** (FIGURE 26.16a and b). The three domains, Bacteria, Archaea, and Eukarya, are essentially superkingdoms, a taxonomic level even higher than the kingdom level. Note that the three-domain system makes the kingdom Monera obsolete, since it would have members in two different domains. In fact, many microbiologists now divide each of the two prokaryotic domains into multiple kingdoms based on cladistic analysis of molecular data (see FIGURE 26.16c).

The second major challenge to the five-kingdom system is being mounted by systematists who are sorting out the phylogeny of the diverse eukaryotes formerly collected in the kingdom Protista. Molecular systematics and cladistics have revealed that Protista is not a monophyletic grouping (see Chapter 25 to review cladistics). The specialists who study these organisms now split most of them into five or more newly designated kingdoms, but have also assigned certain groups that used to be included in Protista to Plantae, Fungi, and Animalia (see FIGURE 26.16c).

Clearly, taxonomy at the highest levels is a work in progress. It may seem ironic that systematists are generally more confident in their groupings of species into lower taxa such as genera and families than they are about the evolutionary relationships among the major groups of organisms. But this makes sense when you consider our own family genealogies. It's easier to sort out our relationships to siblings and cousins than it is to identify more distant relatives, such as third or fourth cousins, by tracing our genealogies back to ancestors who lived long ago. Similarly, tracing phylogeny at the kingdom level takes us back to the evolutionary branching that occurred in Precambrian seas a billion or more years ago.

There will be much more research before there is anything close to a new consensus for how the three domains of life are related and how many kingdoms in each domain of life are required to have a classification that reflects evolutionary history. And even then, new data will undoubtedly lead to further taxonomic remodeling. As we survey the diversity of life in the chapters that follow, keep in mind that phylogenetic trees and taxonomic groupings are hypotheses that fit the best available data. It is this continuing scrutiny of testable hypotheses that validates evolutionary biology as a natural science.

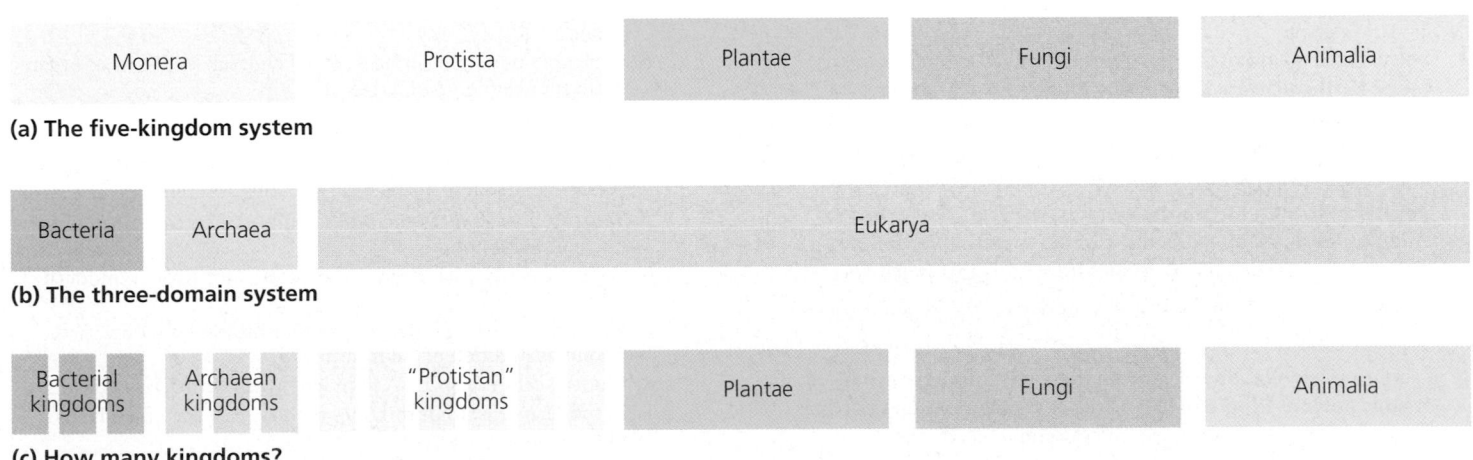

(a) The five-kingdom system

| Monera | Protista | Plantae | Fungi | Animalia |

(b) The three-domain system

| Bacteria | Archaea | Eukarya |

(c) How many kingdoms?

| Bacterial kingdoms | Archaean kingdoms | "Protistan" kingdoms | Plantae | Fungi | Animalia |

FIGURE 26.16 Our changing view of biological diversity.

CHAPTER 26 REVIEW

Go to the Campbell Biology website (www.campbellbiology.com) to explore an interactive version of the Chapter Review.

Summary of Key Concepts

INTRODUCTION TO THE HISTORY OF LIFE

- **Life on Earth originated between 3.5 and 4.0 billion years ago** (pp. 511–512, FIGURE 26.2) Earth formed 4.5 billion years ago. The oldest fossils of prokaryotes are 3.5 billion years old.

- **Prokaryotes dominated evolutionary history from 3.5 to 2.0 billion years ago** (pp. 512–513, FIGURE 26.4) The two domains of prokaryotes, Bacteria and Archaea, diversified as a variety of metabolic types living near hydrothermal vents and in shallow water communities that left fossils called stromatolites.

- **Oxygen began accumulating in the atmosphere about 2.7 billion years ago** (pp. 513–514, FIGURE 26.5) Oxygenic photosynthesis evolved in cyanobacteria. As O_2 accumulated in the atmosphere, the reactive molecule posed an environmental challenge for life.

- **Eukaryotic life began by 2.1 billion years ago** (p. 514) The oldest fossils of eukaryotic cells date back 2.1 billion years. The eukaryotic cell evolved from a prokaryotic ancestor that hosted smaller internal prokaryotes.

- **Multicellular eukaryotes evolved by 1.2 billion years ago** (pp. 514–515, FIGURES 26.6, 26.7) There are fossils of multicellular algae dating back 1.2 billion years. The oldest fossils of animals are about 600 million years old.

- **Animal diversity exploded during the early Cambrian period** (p. 515, FIGURE 26.8) Most phyla of animals make their first fossil appearance during a relatively brief span from about 540 to 520 million years ago.

- **Plants, fungi, and animals colonized the land about 500 million years ago** (pp. 515–516) A symbiotic relationship of plants with fungi contributed to the move onto land. Herbivorous animals and their predators followed.

Web/CD Activity 26A: *The History of Life*

THE ORIGIN OF LIFE

- **The first cells may have originated by chemical evolution on a young Earth:** *an overview* (p. 516, FIGURE 26.9) Though life today arises by biogenesis, the first cells may have been products of prebiotic chemistry.

- **Abiotic synthesis of organic monomers is a testable hypothesis** (pp. 516–518, FIGURE 26.10) Laboratory experiments performed under conditions simulating those of the primitive Earth have produced diverse organic molecules from inorganic precursors.

Web/CD Case Study in the Process of Science: *How Did Life Begin on Early Earth?*

- **Laboratory simulations of early-Earth conditions have produced organic polymers** (pp. 518–519) Small organic molecules polymerize when they are concentrated on hot sand, rock, or clay.

- **RNA may have been the first genetic material** (pp. 519–520, FIGURE 26.11) The first genes may have been abiotically produced RNA, whose base sequences served as templates for both alignment of amino acids in polypeptide synthesis and alignment of complementary nucleotide bases in a primitive form of self-replication.

- **Protobionts can form by self-assembly** (p. 520, FIGURE 26.12) Organic molecules synthesized in the laboratory have spontaneously assembled into a variety of droplets with some of the properties associated with life.

- **Natural selection could refine protobionts containing hereditary information** (p. 520, FIGURE 26.13) The molecular aggregates most effective at using resources from the environment and at reproducing would increase their proportions among a population of varying protobionts.

- **Debate about the origin of life abounds** (pp. 521–522) Researchers continue to debate how and where life originated.

Web/CD Activity 26B: *Tubeworm Video*

THE MAJOR LINEAGES OF LIFE

- **The five-kingdom system reflected increased knowledge of life's diversity** (pp. 522–523, FIGURE 26.15) The traditional five-kingdom system classifies organisms as Monera (prokaryotes), Protista (relatively simple eukaryotes), Plantae, Fungi, and Animalia.

- **Arranging the diversity of life into the highest taxa is a work in progress** (p. 523, FIGURE 26.16) A three-domain system (Bacteria, Archaea, and Eukarya) and the splitting of the prokaryotes and protists into many kingdoms are two departures from the five-kingdom system.

Web/CD Activity 26C: *Classification Schemes*

Self-Quiz

1. The *main* explanation for the lack of a continuing abiotic origin of life on Earth today is that
   a. there is not sufficient lightning to provide an energy source.
   b. our oxidizing atmosphere is not conducive to the spontaneous formation of complex molecules.
   c. much less visible light is reaching Earth to serve as an energy source.
   d. there are no molten surfaces on which weak solutions of organic molecules would polymerize.
   e. all habitable places are already filled.

2. Which statement does *not* support the hypothesis that RNA functioned as the first genetic material of early protobionts?
   a. Short RNA sequences can self-assemble when combined with nucleotide monomers.
   b. Catalytic activity has been demonstrated for RNA in modern cells.
   c. Variations in base sequences produce molecules with variable stabilities in different environments.
   d. Modern cells use an RNA template when synthesizing proteins.
   e. In modern cells, RNA provides the template on which DNA nucleotides are assembled.

3. Fossilized mats called stromatolites
   a. date from 3.5 billion years ago and contain fossils that resemble modern filamentous prokaryotes.
   b. formed around deep-sea vents and provide the first evidence of life on Earth.
   c. contain layers of iron oxide that provide evidence for the oxygenic photosynthesis of cyanobacteria around 2.7 billion years ago.
   d. provide evidence that plants moved onto land in the company of fungi around 500 million years ago.
   e. contain the first undisputed fossils of eukaryotes and date from 2.1 billion years ago.

4. The oxygen revolution changed Earth's environment dramatically. Which of the following adaptations took advantage of this change?
   a. the evolution of chloroplasts when early protists engulfed photosynthetic cyanobacteria
   b. the persistence of some animal groups in anaerobic habitats
   c. the evolution of photosynthetic pigments that protected early algae from the corrosive effects of oxygen
   d. the evolution of cellular respiration, which used oxygen to help harvest energy from fuel molecules
   e. the evolution of multicellular eukaryotic colonies from symbiotic communities of prokaryotes

5. The oldest known fossils of multicellular eukaryotes
   a. are dated by molecular clocks to be 1.5 billion years old.
   b. are corkscrew-shaped algae and date from 2.2 billion years ago.
   c. are filamentous algae that date from 1.2 billion years ago.
   d. are fossilized embryos that have been found in Chinese sediments 570 million years old.
   e. first appear in the fossil record in the late Precambrian after the thawing of snowball Earth.

6. Competition among various protobionts may have led to evolutionary improvement only when
   a. they were first able to catalyze chemical reactions.
   b. some kind of heredity mechanism developed.
   c. they were able to grow and split in two.
   d. photosynthesis evolved.
   e. DNA first appeared.

7. Which of the following represents a probable order in the biological history of Earth?
   a. metabolism before mitosis
   b. an oxidizing atmosphere followed by a reducing atmosphere
   c. eukaryotes before prokaryotes
   d. DNA genes before RNA genes
   e. animals before algae

8. One current debate raises the issue that, rather than beginning in shallow pools, life could have begun
   a. on dry land.
   b. near deep-sea vents.
   c. from viruses.
   d. in northern Africa.
   e. when chunks that broke off from the moon bombarded Earth.

9. Which of the following steps has *not* yet been accomplished by scientists studying the origin of life?
   a. abiotic synthesis of small RNA polymers
   b. abiotic synthesis of polypeptides
   c. formation of molecular aggregates with selectively permeable membranes
   d. formation of protobionts that use DNA to direct the polymerization of amino acids
   e. abiotic synthesis of organic monomers

10. Current debates about the number and boundaries of the kingdoms of life center *mainly* on which groups of organisms?
    a. plants and animals
    b. plants and fungi
    c. prokaryotes and single-celled eukaryotes
    d. fungi and animals
    e. amphibians and reptiles

11. What was the hypothesis that Stanley Miller and Harold Urey were testing with their experiments?

12. What is a ribozyme?

13. Why was the origin of membranes enclosing protein–nucleic acid cooperatives a key step in the onset of Darwinian evolution (natural selection)?

14. Put the following events in order, from the earliest to the most recent: diversification of animals (Cambrian explosion), evolution of eukaryotic cells, first humans, colonization of land by plants and fungi, origin of prokaryotes, evolution of land animals, evolution of multicellular eukaryotes.

15. What is the relationship between stromatolites and microbial mats?

16. What are the two taxonomic domains of prokaryotes?

Go to the website or CD-ROM for more quiz questions.

Evolution Connection

Describe the minimum structural, metabolic, and genetic equipment of a postprotobiont that you would consider to be a true primitive cell.

The Process of Science

Discovery of life elsewhere in the solar system would be momentous at many levels, but one of the biological questions that would immediately arise would be whether terrestrial and extraterrestrial life had independent origins. If the physical and chemical attributes of such new life could be studied, what kinds of evidence would support a single origin? What questions would be raised by evidence of a single origin?

Virtually perform the Miller-Urey experiment and model the chemistry of early Earth in the Case Study in the Process of Science, available on the website and CD-ROM.

Science, Technology, and Society

During the next decade, several nations (including the United States) plan to explore the Martian environment with orbiting laboratories and landers. One of the main goals is to test the hypothesis that chemical conditions on ancient Mars were similar to those on ancient Earth and may have led to prokaryotic life. In what ways do you think our perspective would be changed by convincing evidence of extraterrestrial life?

Answers: 1. b; 2. e; 3. a; 4. d; 5. c; 6. b; 7. a; 8. b; 9. d; 10. c; 11. That conditions on the early Earth favored synthesis of organic molecules from inorganic ingredients. 12. An RNA molecule that functions as a catalyst. 13. In contrast to random mingling of molecules in an open solution, segregation of molecular systems by membranes resulted in selection for the most successful self-replicating aggregates. 14. Origin of prokaryotes, evolution of eukaryotic cells, evolution of multicellular eukaryotes, diversification of animals (Cambrian explosion), colonization of land by plants and fungi, evolution of land animals, first humans. 15. Stromatolites are fossils of microbial mats. 16. Archaea and Bacteria.

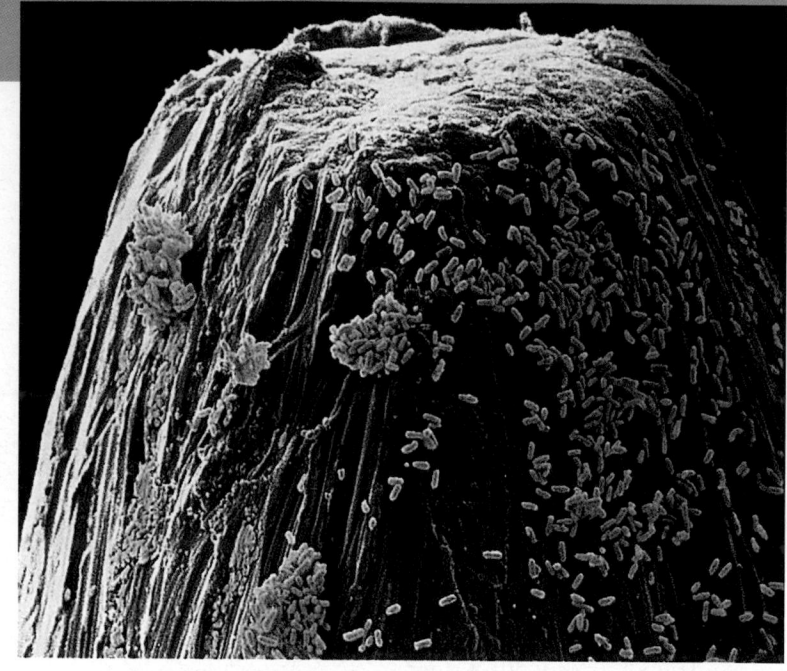

CHAPTER 27

PROKARYOTES AND THE ORIGINS OF METABOLIC DIVERSITY

THE WORLD OF PROKARYOTES

- They're (almost) everywhere! An overview of prokaryotic life
- Bacteria and archaea are the two main branches of prokaryote evolution

THE STRUCTURE, FUNCTION, AND REPRODUCTION OF PROKARYOTES

- Nearly all prokaryotes have a cell wall external to the plasma membrane
- Many prokaryotes are motile
- The cellular and genomic organization of prokaryotes is fundamentally different from that of eukaryotes
- Populations of prokaryotes grow and adapt rapidly

NUTRITIONAL AND METABOLIC DIVERSITY

- Prokaryotes can be grouped into four categories according to how they obtain energy and carbon
- Photosynthesis evolved early in prokaryotic life

A SURVEY OF PROKARYOTIC DIVERSITY

- Molecular systematics is leading to a phylogenetic classification of prokaryotes
- Researchers are identifying a great diversity of archaea in extreme environments and in the oceans
- Most known prokaryotes are bacteria

THE ECOLOGICAL IMPACT OF PROKARYOTES

- Prokaryotes are indispensable links in the recycling of chemical elements in ecosystems
- Many prokaryotes are symbiotic
- Pathogenic prokaryotes cause many human diseases
- Humans use prokaryotes in research and technology

The history of prokaryotic life is a success story spanning more than 3.5 billion years. Prokaryotes were the earliest organisms, and they lived and evolved all alone on Earth for 1.5 billion years. They have continued to adapt and flourish on an evolving Earth, and in turn they have helped to change the Earth.

In this chapter, you will become more familiar with prokaryotes by studying their structure and function, their origins and evolution, their diversity, and their ecological significance.

THE WORLD OF PROKARYOTES

They're (almost) everywhere! An overview of prokaryotic life

Today, prokaryotes still dominate the biosphere. Their collective biological mass (biomass) outweighs all eukaryotes combined by at least tenfold. More prokaryotes inhabit a handful of fertile soil or the mouth or skin of a human than the total number of people who have ever lived. In the colorized scanning electron micrograph on this page, the tiny orange rods are bacteria on the point of a pin. (You can see from this micrograph why a pin prick can cause an infection, and why it's important to flame the tip of a needle to kill bacteria before using the needle to remove a splinter.)

Prokaryotes are found wherever there is life. They also thrive in habitats too cold, too hot, too salty, too acidic, or too alkaline for any eukaryote (FIGURE 27.1). In 1999, biologists even discovered prokaryotes growing on the walls of a gold mine 2 miles below Earth's surface.

Though individual prokaryotes are relatively small organisms, they are giants in their collective impact on Earth and its life. We hear most about a minority of species that cause serious illness. During the 14th century, Black Death—bubonic plague, a bacterial disease—spread across Europe, killing an estimated 25% of the human population. Tuberculosis, cholera, many sexually transmissible diseases, and certain types of food poisoning are other human diseases caused by bacteria.

However, prokaryotic life is no rogues' gallery. Far more common than harmful bacteria are those that are benign or beneficial. Bacteria in our intestines provide us with important vitamins, and others living in our mouth prevent harmful fungi from growing there. Prokaryotes also cycle carbon and other vital chemical elements back and forth between organic matter and the soil and atmosphere. For example, there are prokaryotes that decompose dead organisms. Found in soil and at the bottom of lakes, rivers, and oceans, these decomposers return chemical elements to the environment in the form of inorganic compounds that can be used by plants, which in turn feed animals. If prokaryotic decomposers were to disappear, the chemical cycles that sustain life would break down. All forms of eukaryotic life would also be doomed. In contrast, prokaryotic life would undoubtedly persist in the absence of eukaryotes, as it once did for 1.5 billion years.

Prokaryotes often live in close associations with other prokaryotes and with eukaryotes in what are called symbiotic relationships. In the most historically important case of such symbiosis, mitochondria and chloroplasts evolved from prokaryotes that became residents within larger host cells. Thus, animals, plants, fungi, and protists evolved from symbiotic associations of ancestral cells. (We will examine this theory of eukaryotic origins further in Chapter 28.)

Modern prokaryotes are diverse in structure and metabolism. About 5,000 species of prokaryotes are known, and estimates of actual prokaryotic diversity range from about 400,000 to 4 million species. As Harvard University biologist E. O. Wilson puts it, a true sense of biodiversity requires a "downward adjustment of scale."

Bacteria and archaea are the two main branches of prokaryote evolution

In the traditional five-kingdom system of classification, prokaryotes make up the kingdom Monera, and the four eukaryotic kingdoms are Protista, Plantae, Fungi, and Animalia (see FIGURE 26.15). This scheme emphasizes the structural differences between the cells of prokaryotes and eukaryotes. In the past two decades, however, systematists have determined that a single kingdom incorporating all prokaryotes is not consistent with evolutionary history. By comparing ribosomal RNA and the completely sequenced genomes of several extant (living today) species, researchers have identified two major branches of prokaryote evolution. The names for these two groups are bacteria and archaea. The term *archaea* refers to the antiquity of the group's origin from the earliest cells (from the Greek *archaio,* ancient). Many species of archaea inhabit extreme environments, such as hot springs and salt ponds. Few, if any, other modern organisms can survive in these environments, which may resemble habitats on the early Earth. Archaea differ from bacteria in many key structural,

FIGURE 27.1 "Heat-loving" prokaryotes. The vivid reds, oranges, and yellows that paint these rocks are colonies of prokaryotes that thrive in the extremely hot water (up to 104°C) that flows from this Nevada geyser. Ranchers accidentally tapped the geyser when they were drilling for water in 1916.

biochemical, and physiological characteristics, differences that will be highlighted later in this chapter.

When researchers led by Carl Woese, of the University of Illinois, first recognized the distinction between bacteria and archaea, they proposed a six-kingdom system: two prokaryotic kingdoms along with the four eukaryotic kingdoms. Because bacteria and archaea diverged so early in the history of life and are so fundamentally different, Woese and many other systematists now favor organizing the diversity of life into three **domains,** a taxonomic level above kingdom (FIGURE 27.2). In this view, prokaryotes account for two of the domains: domain Bacteria and domain Archaea.

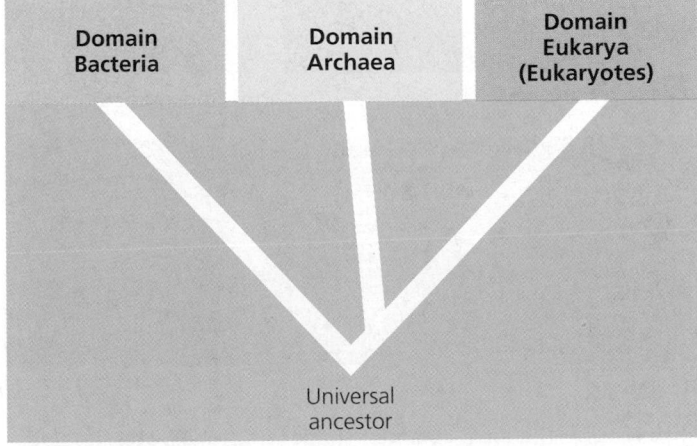

FIGURE 27.2 The three domains of life. This phylogenetic tree represents the hypothesis that the two major prokaryotic groups diverged very early and that Archaea is more closely related to Eukarya than to Bacteria. In Chapter 28, we'll see how new data are stimulating some remodeling of this hypothesis.

Taxonomic issues aside, bacteria and archaea are both structurally organized at the prokaryotic level, which is the rationale for combining them in this chapter. The distinction between bacteria and archaea will become more apparent after we examine some of the structural, genetic, and metabolic adaptations that contribute to the pervasiveness of prokaryotes on Earth.

THE STRUCTURE, FUNCTION, AND REPRODUCTION OF PROKARYOTES

Most prokaryotes are unicellular. However, some species tend to aggregate transiently in groups of two or more cells. Others have the form of true colonies, which are permanent aggregates of identical cells. And in some species of prokaryotes, there is even a division of labor between two or more specialized types of cells.

There is a diversity of cell shapes among prokaryotes, the three most common being spheres (cocci), rods (bacilli), and helices (including the bacteria known as spirilla and spirochetes). An important step in identifying prokaryotes is determining their shape by microscopic examination (FIGURE 27.3).

Most prokaryotes have diameters in the range of 1–5 μm, compared to 10–100 μm for the majority of eukaryotic cells. There are, however notable exceptions; the largest prokaryote discovered so far measures about 0.75 mm in diameter, dwarfing most eukaryotic cells (FIGURE 27.4).

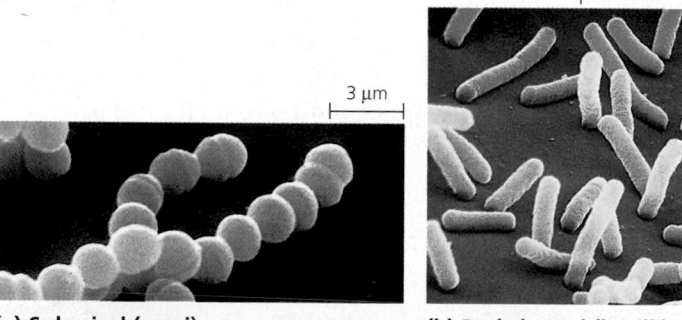

(a) Spherical (cocci)   **(b) Rod-shaped (bacilli)**

(c) Helical

FIGURE 27.3 The most common shapes of prokaryotes. (a) Cocci (singular, coccus), or spherical prokaryotes, occur singly, in pairs (diplococci), in chains of many cells (streptococci, shown here), and in clusters resembling bunches of grapes (staphylococci). **(b)** Rod-shaped prokaryotes, or bacilli (singular, bacillus), are usually solitary, but in some forms the rods are arranged in chains. **(c)** Helical prokaryotes include the spirilla and the corkscrew-shaped spirochetes. (All colorized SEMs.)

FIGURE 27.4 The largest known prokaryote. The bright ball in the upper left is a marine bacterium with cells almost as large as a fruit fly's head (included in the photo for scale). The two smaller cells above the bright one are dead. Researchers who discovered the giant prokaryote in 1997 named it *Thiomargarita namibiensis,* which means "sulfur pearl of Namibia." It uses sulfur compounds in its metabolism and inhabits coastal sediments in the African country of Namibia.

Nearly all prokaryotes have a cell wall external to the plasma membrane

Present in nearly all prokaryotes, the cell wall maintains the shape of the cell, affords physical protection, and prevents the cell from bursting in a hypotonic environment (see Chapter 8). Like other walled cells, however, prokaryotes plasmolyze (shrink away from their wall) and may die in a hypertonic medium, which is why heavily salted meat can be kept so long without being spoiled by prokaryotes. The cell walls of prokaryotes differ in molecular composition and construction from those of plants, fungi, and protists.

The presence of a cell wall is one reason prokaryotes were grouped with plants in the old two-kingdom system. Instead of cellulose, the staple of plant walls, most bacterial walls contain a unique material called **peptidoglycan,** which consists of polymers of modified sugars cross-linked by short polypeptides that vary from species to species (the walls of archaea lack peptidoglycan). The effect is a single molecular network enclosing and protecting the entire cell. External to this fabric are other substances that also differ from species to species.

One of the most valuable tools for identifying specific bacteria is the **Gram stain,** which can be used to separate many species into two groups based on differences in their cell walls. **Gram-positive** bacteria have simpler walls, with a relatively large amount of peptidoglycan. The walls of **gram-negative** bacteria have less peptidoglycan and are structurally more complex. An outer membrane on the gram-negative cell wall contains lipopolysaccharides, carbohydrates bonded to lipids (FIGURE 27.5).

Among pathogenic, or disease-causing, bacteria, gram-negative species are generally more threatening than gram-positive species. The lipopolysaccharides on the walls of gram-negative bacteria are often toxic, and the outer membrane helps protect the pathogens against the defenses of their hosts. Furthermore, gram-negative bacteria are commonly more resistant than gram-positive species to antibiotics because the outer membrane impedes entry of the drugs.

Many antibiotics, including penicillins, inhibit the synthesis of cross-links in peptidoglycan and prevent the formation of a functional wall, particularly in gram-positive species. These drugs are like selective bullets that cripple many species of infectious bacteria without adversely affecting humans and other eukaryotes, which do not make peptidoglycan.

Many prokaryotes secrete sticky substances that form still another protective layer called a **capsule** outside the cell wall. Capsules enable the organisms to adhere to their substratum and provide additional protection, including increased resistance of pathogenic prokaryotes to host defenses. Gelatinous capsules glue together the cells of many prokaryotes that live as colonies.

Another way some prokaryotes adhere to one another or to some substratum is by means of surface appendages called **pili** (singular, *pilus;* FIGURE 27.6, p. 530). For example, *Neisseria gonorrhoeae,* the pathogen that causes gonorrhea, uses pili to fasten itself to mucous membranes of its host. Some pili are specialized for holding prokaryotes together long enough for the cells to transfer DNA during conjugation (see Chapter 18).

Many prokaryotes are motile

About half of all prokaryotes are capable of directional movement. Some species exceed speeds of 50 μm/sec, or about 100 times their body length per second.

Flagellar action is the most common mechanism of movement among prokaryotes. Flagella may be scattered over the entire cell surface or concentrated at one or both ends of the cell. The flagella of prokaryotes and eukaryotes differ in both

(a) Gram-positive.
Gram-positive bacteria have cell walls with a large amount of peptidoglycan that traps the violet dye (LM).

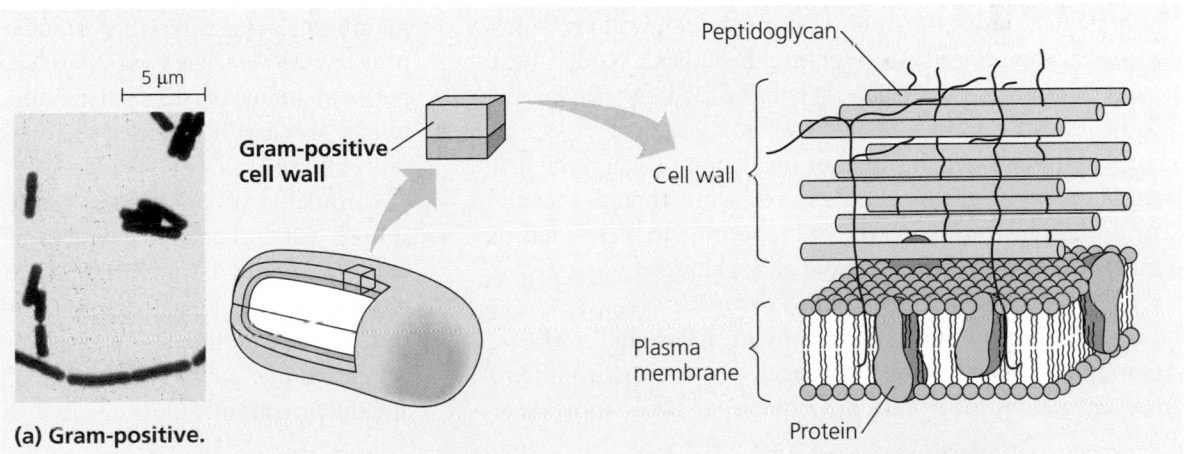

(b) Gram-negative.
Gram-negative bacteria have less peptidoglycan, which is located in a periplasmic gel between the plasma membrane and an outer membrane. The violet dye is easily rinsed from gram-negative bacteria, but the cells retain the red dye (LM).

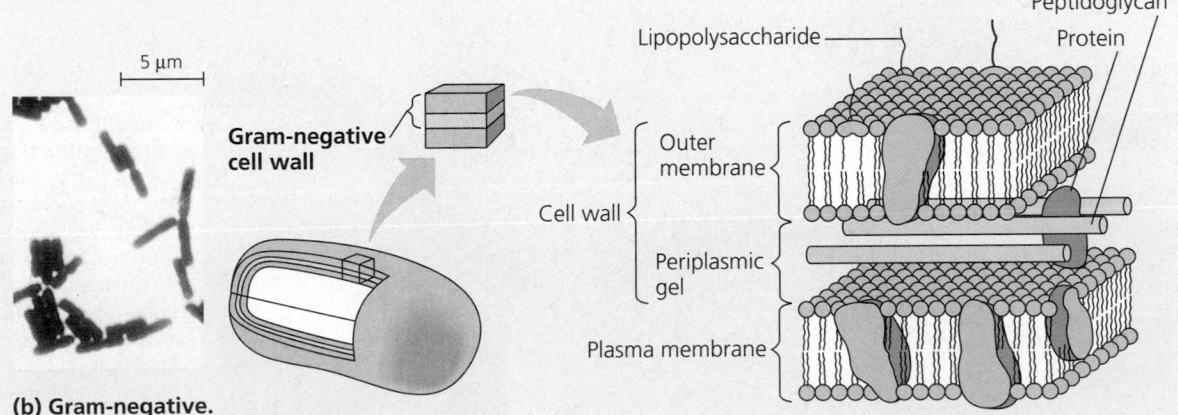

FIGURE 27.5 Gram-positive and gram-negative bacteria. Gram-staining, named for Hans Christian Gram, a Danish physician who developed the technique in the late 1800s, distinguishes between two different kinds of bacterial cell walls. Bacteria are stained with a violet dye and iodine, rinsed in alcohol, and then stained again with a red dye. The structure of the cell wall determines the staining response. (The colors in the membrane diagrams do not represent the stains.)

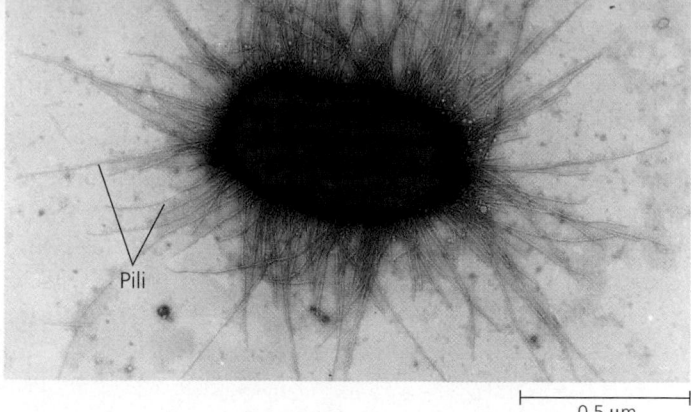

FIGURE 27.6 Pili. These appendages enable some prokaryotes to attach to surfaces or to other prokaryotes (colorized TEM). Some pili function during conjugation, holding partners together while DNA is transferred (see FIGURE 18.14).

toward or away from a stimulus (from the Greek *taxis*, to arrange). With chemotaxis, for example, prokaryotes respond to chemical stimuli, perhaps moving toward food or oxygen (a positive chemotaxis) or away from some toxic substance (a negative chemotaxis). Several kinds of receptor molecules that detect specific substances are located on the surfaces of chemotactic prokaryotes. Motile prokaryotes that are photosynthetic generally display a positive phototaxis, a behavior that keeps them in the light. Some prokaryotes even contain a row of tiny magnetic particles that allow the cells to orient in Earth's magnetic field. These particles may help cells distinguish up from down and cause the prokaryotes to migrate toward the nutrient-rich sediments at the bottoms of ponds and shallow seas.

structure and function. (To review eukaryotic flagella, see Chapter 7.) Prokaryotic flagella are one-tenth the width of eukaryotic flagella and are not covered by an extension of the plasma membrane (FIGURE 27.7).

A second motility mechanism characterizes a group of helix-shaped bacteria called spirochetes. Two or more helical filaments just under the outer layer of the cell wall are much like prokaryotic flagella in structure. Each has a basal motor attached at one end of the cell. When the filaments rotate, the flexible cell moves like a corkscrew.

In a third mechanism of motility, some prokaryotes that form filamentous chains of cells secrete slimy threads that anchor to the substratum. As the cells continue to secrete jets of slime, the filamentous prokaryote glides along at the growing end of the threads.

In a relatively uniform environment, flagellated prokaryotes may wander randomly. In a heterogeneous environment, however, many prokaryotes are capable of **taxis,** movement

The cellular and genomic organization of prokaryotes is fundamentally different from that of eukaryotes

Recall that prokaryotes are named for their lack of true nuclei enclosed by membranes (see FIGURE 7.4). The cells of prokaryotes lack the extensive compartmentalization by internal membranes characteristic of eukaryotes. However, various prokaryotes do have a variety of specialized membranes that perform many of their metabolic functions. These membranes are usually infolded regions of the plasma membrane (FIGURE 27.8).

Compared to eukaryotes, prokaryotes also have smaller, simpler genomes. On average, prokaryotes have only about one-thousandth as much DNA as a eukaryotic cell. In most prokaryotes, the DNA is concentrated as a snarled fiber in a **nucleoid region** that stains less densely than the surrounding cytoplasm in electron micrographs. The mass of fibers is actually the prokaryotic chromosome, one double-stranded DNA

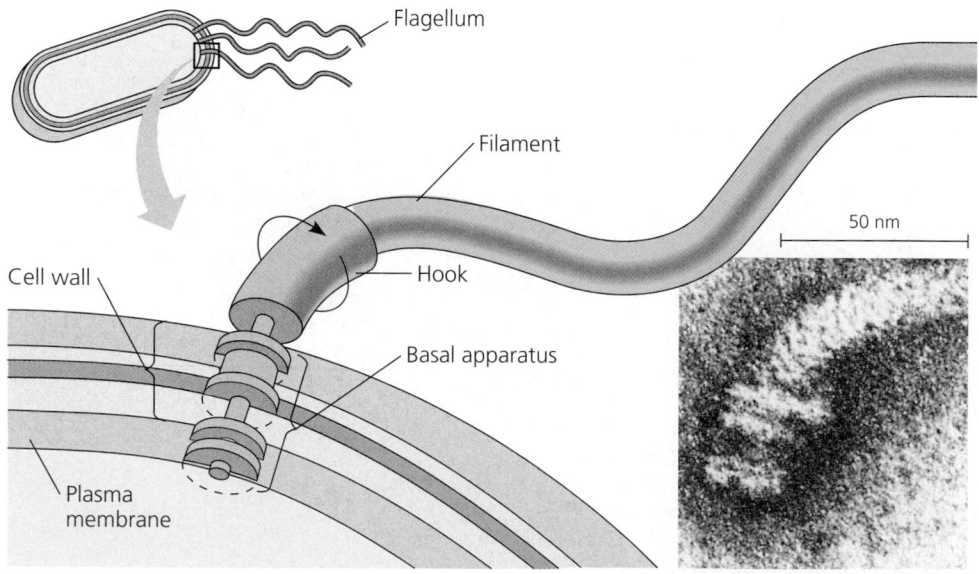

FIGURE 27.7 Form and function of prokaryotic flagella. Chains of a globular protein, flagellin, are wound in a tight spiral to form a semirigid, helical filament. The filament is attached to another protein that forms a curved hook, which is inserted into a basal apparatus. Composed of about 35 different proteins, the basal apparatus includes a system of rings in the layers of the cell wall (TEM). (The structures shown in this figure are characteristic of gram-negative bacteria.) The basal apparatus—the flagellar motor—rotates the filament, which propels the cell. The motor is powered by the diffusion of protons (H^+) into the cell after they have been pumped outward by ATP-powered proton pumps embedded in the plasma membrane.

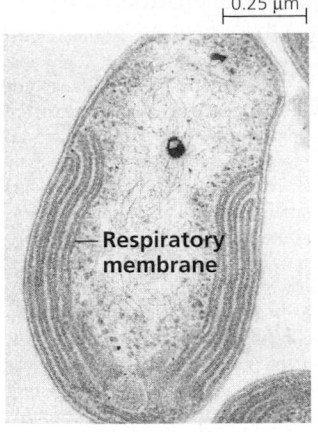

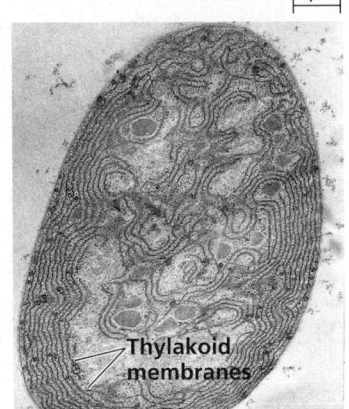

(a) Aerobic prokaryote   **(b) Photosynthetic prokaryote**

FIGURE 27.8 Specialized membranes of prokaryotes. (a) These infoldings of the plasma membrane, reminiscent of the cristae of mitochondria, function in cellular respiration of some aerobic prokaryotes (TEM). **(b)** Photosynthetic prokaryotes called cyanobacteria have thylakoid membranes, much like those in chloroplasts (TEM).

molecule in the form of a ring. The DNA has relatively little protein associated with it. The term *genophore* is sometimes used for the bacterial chromosome to distinguish it from eukaryotic chromosomes, which have a very different structure. The genome of eukaryotes consists of linear DNA molecules packaged along with proteins into multiple chromosomes.

In addition to its one major chromosome, the prokaryotic cell may also have much smaller rings of DNA called **plasmids,** most consisting of only a few genes. In most environments, prokaryotes can survive without their plasmids because all essential functions are programmed by the chromosome. However, plasmids endow the cell with genes for resistance to antibiotics, for the metabolism of unusual nutrients not present in the normal environment, and for other special contingencies. Plasmids replicate independently of the main chromosome, and many can be readily transferred between partners when prokaryotes conjugate (see FIGURE 18.15).

Although the general processes for DNA replication and the translation of genetic messages into proteins are alike for eukaryotes and prokaryotes, some of the details differ. For example, the prokaryotic ribosome is slightly smaller than the eukaryotic version and differs in its protein and RNA content. The disparity is great enough that selective antibiotics, including tetracycline and chloramphenicol, bind to the ribosomes and block protein synthesis in many prokaryotes but not in eukaryotes.

Populations of prokaryotes grow and adapt rapidly

Prokaryotes reproduce only asexually by the mode of cell division called **binary fission,** synthesizing DNA almost continuously (see FIGURE 12.10). A single prokaryotic cell in a favorable environment will give rise by repeated divisions to a colony of offspring (FIGURE 27.9). Note that neither mitosis nor meiosis occurs among prokaryotes; this is another fundamental difference between prokaryotes and eukaryotes.

Although prokaryotes are never sexual in the eukaryotic manner of the meiosis-fertilization cycle, they do have three mechanisms that transfer genes between individuals (see Chapter 18). In **transformation,** a prokaryotic cell takes up genes from the surrounding environment, allowing for considerable genetic transfer between prokaryotes, even across species lines. **Conjugation** is the direct transfer of genes from one prokaryote to another. And in **transduction,** viruses transfer genes between prokaryotes. These processes, however, involve the unilateral passage of a variable amount of DNA—nothing like the meiotic sex of eukaryotes, in which two parents each contribute homologous genomes to a zygote. Mutation is the major source of genetic variation in prokaryotes. Because generation times are measured in minutes or hours, a favorable mutation can be rapidly propagated to a large number of offspring. This enables prokaryotic populations to adapt very rapidly to environmental change, as natural selection screens new mutations and novel genomes resulting from gene transfer.

The word *growth* as applied to prokaryotes refers more to the multiplication of cells and an increase in population size than to the enlargement of individual cells. The conditions for optimal growth—temperature, pH, salt concentrations, nutrient sources, and so on—vary according to species. Refrigeration retards food spoilage because most microorganisms grow only very slowly at such low temperatures.

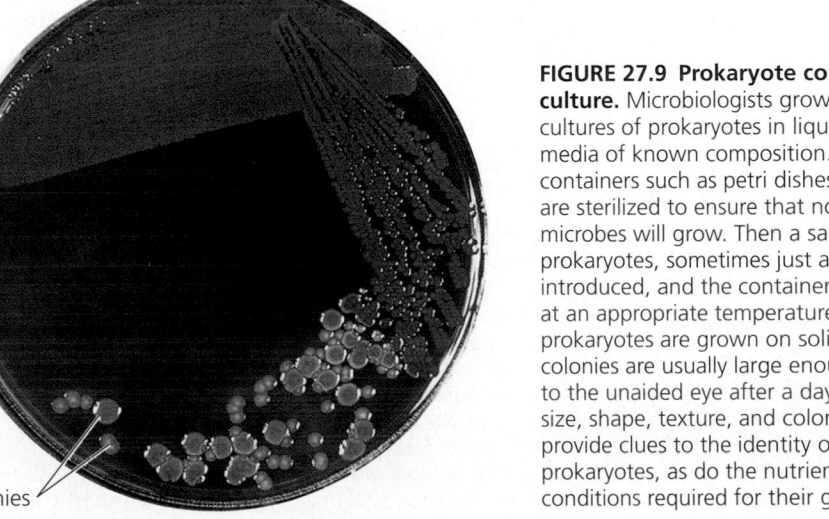

Colonies

FIGURE 27.9 Prokaryote colonies in culture. Microbiologists grow laboratory cultures of prokaryotes in liquid or solid media of known composition. The media, in containers such as petri dishes or test tubes, are sterilized to ensure that no unwanted microbes will grow. Then a sample of prokaryotes, sometimes just a single cell, is introduced, and the containers are incubated at an appropriate temperature. When the prokaryotes are grown on solid media, colonies are usually large enough to be visible to the unaided eye after a day or two. The size, shape, texture, and color of a colony provide clues to the identity of the prokaryotes, as do the nutrients and physical conditions required for their growth.

In an environment with unlimited resources, the growth of prokaryotes is effectively geometric: One cell divides to form 2, which divide again to produce a total of 4 cells, then 8, 16, and so on, the number of cells in a colony doubling with each generation. Many prokaryotes have generation times in the range of 1–3 hours, but some species can double every 20 minutes in an optimal environment. If the latter growth rate were sustained, a single cell would give rise to a colony outweighing Earth in just three days! Obviously, prokaryotic growth both in the laboratory and in nature is usually checked at some point, when the cells exhaust some nutrient or the colony poisons itself with an accumulation of metabolic wastes. Still, prokaryotes manage to reach astounding population densities—a hundred billion prokaryotes per milliliter of fluid in the human colon, for example.

The ability of some prokaryotes to withstand harsh conditions is also impressive. Some bacteria form resistant cells called **endospores** (FIGURE 27.10). The original cell replicates its chromosome, and one copy becomes surrounded by a durable wall. The outer cell disintegrates, leaving the highly resistant endospore. Boiling water is not hot enough to kill most endospores in a relatively short length of time. Home canners and the food-canning industry, therefore, must take extra precautions to kill endospores of dangerous bacteria. To sterilize media, glassware, and utensils in the laboratory, microbiologists use an appliance called an autoclave, a pressure cooker that kills even endospores by heating to a temperature of 120°C. In less hostile environments, endospores may remain dormant for centuries or more. If placed in a hospitable environment, they will hydrate and revive to the vegetative (colony-producing) state. For example, in 2000, researchers revived a bacterial spore that had apparently been encased for 250 million years in a salt formation within New Mexico's Carlsbad Caverns.

In most natural environments, prokaryotes must compete for space and nutrients. Many microorganisms (including certain species of prokaryotes, protists, and fungi) release **antibiotics,** chemicals that inhibit the growth of other microorganisms. Humans have discovered some of these compounds and use them to combat pathogenic bacteria.

NUTRITIONAL AND METABOLIC DIVERSITY

Metabolic diversity is greater among prokaryotes than among all eukaryotes combined. Every type of nutrition observed in eukaryotes is represented among prokaryotes, along with some nutritional modes unique to prokaryotes.

Prokaryotes can be grouped into four categories according to how they obtain energy and carbon

Nutrition refers here to how an organism obtains two resources from the environment: energy and a carbon source to build the organic molecules of cells. Species that use light energy are called *phototrophs. Chemotrophs* obtain their energy from chemicals taken from the environment. If an organism needs only the inorganic compound CO_2 as a carbon source, it is called an *autotroph. Heterotrophs* require at least one organic nutrient—glucose, for instance—as a source of carbon for making other organic compounds. We can combine the phototroph-versus-chemotroph (energy source) and autotroph-versus-heterotroph (carbon source) criteria to group prokaryotes according to four major modes of nutrition:

1. **Photoautotrophs** are photosynthetic organisms that harness light energy to drive the synthesis of organic compounds from carbon dioxide. Among the diverse groups of photoautotrophic prokaryotes are the cyanobacteria. All photosynthetic eukaryotes—plants and algae—also fit into this nutritional category.
2. **Chemoautotrophs** need only CO_2 as a carbon source, but instead of using light for energy, these prokaryotes obtain energy by oxidizing inorganic substances. Chemical energy is extracted from hydrogen sulfide (H_2S), ammonia (NH_3), ferrous ions (Fe^{2+}), or some other chemical, depending on the species. This mode of nutrition is unique to certain prokaryotes. Some of the species that extract energy by oxidizing minerals in stone are "eating away" at some of the world's greatest statues.

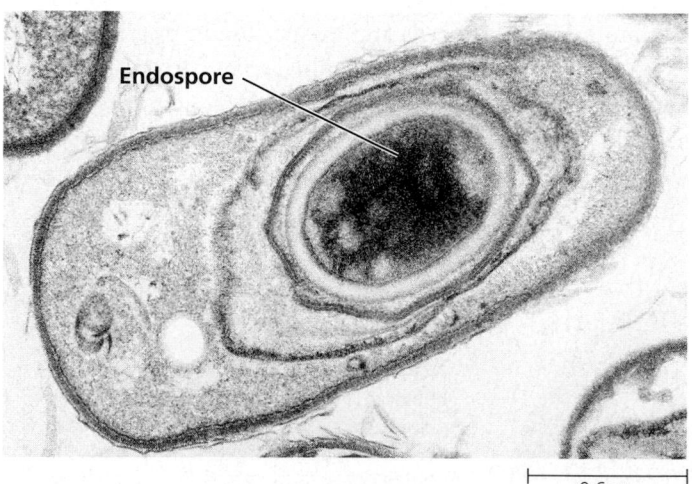

Endospore

0.6 μm

FIGURE 27.10 An anthrax endospore. This prokaryote is *Bacillus anthracis,* the notorious bacterium that produces the deadly disease called anthrax in cattle, sheep, and humans (TEM). The outer cell produced the specialized inner cell, called an endospore. The endospore has a thick, protective coat. Its cytoplasm is dehydrated, and the cell does not metabolize. Under harsh conditions, the outer cell may disintegrate, but the endospore survives all sorts of trauma, including lack of water and nutrients, extreme heat or cold, and most poisons. When the environment becomes more hospitable, the endospore absorbs water and resumes growth. In late 2001, one or more bioterrorists disseminated anthrax spores through the U.S. postal system.

Table 27.1 Major Nutritional Modes

Mode of Nutrition	Energy Source	Carbon Source	Types of Organisms
Autotroph			
Photo-autotroph	Light	CO_2	Photosynthetic prokaryotes, including cyanobacteria; plants; certain protists (algae)
Chemo-autotroph	Inorganic chemicals	CO_2	Certain prokaryotes (for example, *Sulfolobus*)
Heterotroph			
Photo-heterotroph	Light	Organic compounds	Certain prokaryotes
Chemo-heterotroph	Organic compounds	Organic compounds	Many prokaryotes and protists; fungi; animals; some parasitic plants

3. **Photoheterotrophs** can use light to generate ATP but must obtain their carbon in organic form. This mode of nutrition is restricted to certain prokaryotes.
4. **Chemoheterotrophs** must consume organic molecules for both energy and carbon. This nutritional mode is found widely among prokaryotes, protists, fungi, animals, and even some parasitic plants.

TABLE 27.1 reviews the four major modes of nutrition.

Nutritional Diversity Among Chemoheterotrophs

The majority of known prokaryotes are chemoheterotrophs. This category includes **saprobes,** decomposers that absorb their nutrients from dead organic matter, and **parasites,** which absorb their nutrients from the body fluids of living hosts.

The specific organic nutrients needed for growth vary extensively among chemoheterotrophic prokaryotes. Some species are very exacting in their requirements; for example, bacteria of the genus *Lactobacillus* will grow well only in a medium containing all 20 amino acids, several vitamins, and other organic compounds. Among species less particular in their nutritional needs, *E. coli* can grow on a medium containing glucose as the only organic ingredient, and the organism's metabolism is so versatile that many other compounds can substitute for glucose as the sole organic nutrient.

There is such a diversity of chemoheterotrophs that almost any organic molecule can serve as food for at least some species. For example, some bacteria are capable of metabolizing petroleum; they are used to clean up oil spills. Those few classes of synthetic organic compounds (including some kinds of plastics) that cannot be broken down by any chemoheterotrophs are said to be nonbiodegradable.

Nitrogen Metabolism

Nitrogen metabolism is another area of nutritional diversity among prokaryotes. Nitrogen is an essential component of proteins and nucleic acids. While animals, plants, and other eukaryotes are limited in the forms of nitrogen they can use, diverse prokaryotes are able to metabolize most nitrogenous compounds.

Key steps in the cycling of nitrogen through ecosystems are performed only by prokaryotes. (For an overview of the nitrogen cycle, see FIGURE 54.18.) Some chemoautotrophic bacteria in the soil, such as *Nitrosomonas,* convert ammonium (NH_4^+) to nitrite (NO_2^-). Other bacteria, such as a few species of *Pseudomonas,* "denitrify" soil nitrite or nitrate (NO_3^-), a metabolism that returns nitrogen gas (N_2) to the atmosphere. And diverse species of prokaryotes, including some cyanobacteria, are able to use atmospheric nitrogen directly as a source of nitrogen. In this process, called **nitrogen fixation,** prokaryotes convert atmospheric nitrogen (N_2) to ammonium (NH_4^+). Nitrogen fixation, unique to certain prokaryotes, is the only biological mechanism that makes atmospheric nitrogen available to organisms for incorporation into organic compounds. In terms of nutrition, nitrogen-fixing cyanobacteria are the most self-sufficient of all organisms. These photoautotrophs require only light energy, CO_2, N_2, water, and some minerals in order to grow (FIGURE 27.11).

Metabolic Relationships to Oxygen

Another metabolic variation among prokaryotes is in the effect that oxygen has on growth (see Chapter 9). **Obligate aerobes** use O_2 for cellular respiration and cannot grow without it. **Facultative anaerobes** will use O_2 if it is present but can also grow by fermentation in an anaerobic environment. **Obligate anaerobes** are poisoned by O_2. Some obligate anaerobes live exclusively by fermentation; other species extract chemical energy by **anaerobic respiration,** in which inorganic

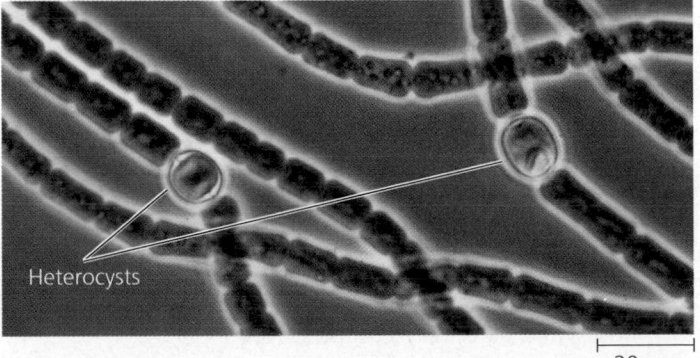

Heterocysts

20 µm

FIGURE 27.11 One of the most independent organisms on Earth. A photoautotroph that can use atmospheric nitrogen (N_2) as its nitrogen source, *Anabaena* is an example of the metabolic virtuosity of cyanobacteria. The specialized cells with the enzymes for nitrogen fixation are called heterocysts (LM).

molecules other than O_2 accept electrons at the "downhill" end of electron transport chains.

Now that we have surveyed variation in nutrition and metabolism among prokaryotes, let's trace the evolutionary roots of this metabolic diversity.

Photosynthesis evolved early in prokaryotic life

All forms of nutrition and nearly all metabolic pathways evolved among prokaryotes before there were any eukaryotes. As early prokaryotes evolved, they were met with constantly changing physical and biological environments. In response to these changes, new metabolic capabilities evolved that in turn changed the environment faced by the next community of prokaryotes. All the major metabolic capabilities seen among prokaryotes (and hence among eukaryotes) probably evolved in the first billion years of life. And during this expansion of metabolic diversity, photosynthesis evolved relatively early—a hypothesis supported by molecular systematics, comparisons of energy metabolism among extant prokaryotes, and geologic evidence.

The energy metabolism of even the simplest photoautotrophs is relatively complex. Thus, it seems reasonable to postulate that the *very* first prokaryotes were heterotrophs that obtained their energy and carbon skeletons from the pool of organic molecules available in the "primordial soup" of early Earth (see Chapter 26). Glycolysis, which can extract energy from organic fuels to generate ATP in anaerobic environments, was probably one of the first metabolic pathways. That would account for the existence of glycolysis in almost every group of modern organisms. As heterotrophs depleted the supply of organic nutrients in the environment, natural selection would have favored any prokaryotes that could harness the energy of sunlight to drive the synthesis of ATP and generate reducing power to synthesize organic compounds from CO_2.

THE PROCESS OF SCIENCE What evidence is there for an early evolution of photosynthesis? First, photosynthetic groups are scattered among diverse branches of prokaryote phylogeny. It is possible that photosynthesis evolved independently many times in various prokaryotic lineages (FIGURE 27.12a). This seems unlikely, however, since the molecular machinery required for photosynthesis is very complex. Applying the principle of parsimony we dis-

cussed in Chapter 25, the most reasonable hypothesis is that photosynthesis originated once in an ancestor common to the diverse prokaryotic groups where we find photosynthesis today. The heterotrophic groups that are closely related to photosynthetic ones could represent a loss of photosynthetic ability during evolution (FIGURE 27.12b). This would explain molecular data suggesting that in nutritionally diverse taxa of prokaryotes, the autotrophic species are generally older lineages than the heterotrophic ones. Although the very first organisms may have been heterotrophs from which autotrophs evolved, the diversity of heterotrophs we observe today probably descended secondarily from photosynthetic ancestors.

Further evidence for an early origin of photosynthesis is the antiquity of cyanobacteria. These are the only autotrophic prokaryotes that release O_2 by splitting water during their light reactions. The geologic evidence for accumulation of atmospheric oxygen beginning at least 2.7 billion years ago suggests that cyanobacteria were already a major part of the biosphere by then. And fossils of prokaryotes that look very much like modern cyanobacteria have been found in stromatolites as old as 3.5 billion years, pushing the possible origin of oxygenic photosynthesis back considerably further.

Oxygenic photosynthesis is especially complex because it requires two cooperative photosystems (see FIGURE 10.12). Some groups of modern prokaryotes perform a simpler mode of photosynthesis, using a single photosystem to extract electrons from compounds such as H_2S instead of splitting water. But certain components of the photosynthetic machinery are common to such nonoxygenic photosynthesis and the oxygenic version in cyanobacteria. A logical inference is that cyanobacteria evolved from ancestors with simpler, nonoxygenic photosynthesis. This would push the origin of photosynthesis back very close to our earliest fossil evidence of life.

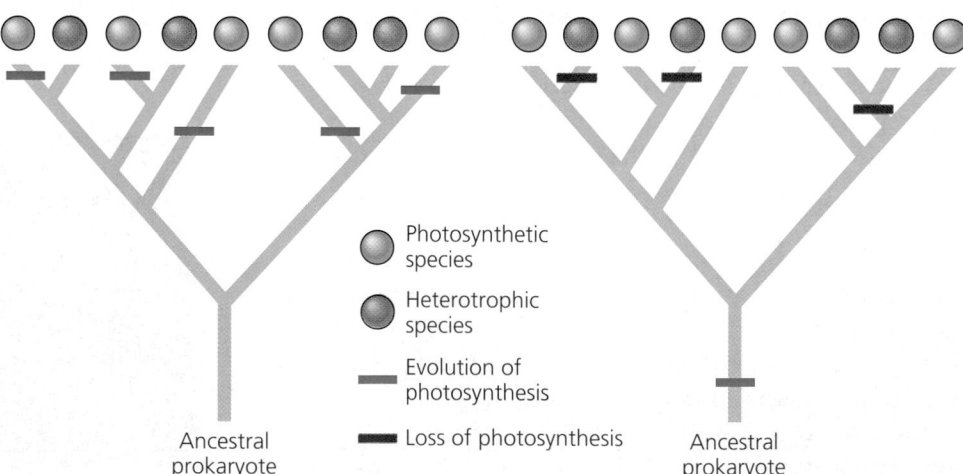

FIGURE 27.12 Contrasting hypotheses for the taxonomic distribution of photosynthesis among prokaryotes.

Hypothesis (a): Photosynthesis evolved many times. This hypothesis assumes that heterotrophic nutrition is the primitive condition and thus requires that the complex process of photosynthesis evolved independently several separate times.

Hypothesis (b): Photosynthesis evolved once. The more parsimonious hypothesis assumes that photosynthesis evolved once very early in prokaryotic life and that the loss of photosynthesis in certain lineages explains the scattering of nutritional types throughout the phylogenetic tree.

The evolution of cyanobacteria changed the Earth in a radical way, gradually transforming its atmosphere from a reducing one to an oxidizing one. In Chapter 25, we discussed how the oxygen revolution posed an environmental crisis for life, which originated in a mostly anaerobic world. The most elegant adaptation to the changing atmosphere was the evolution of cellular respiration, which uses the oxidizing power of O_2 to increase the efficiency of fuel consumption. Photosynthesis and cellular respiration are actually closely related, both using electron transport chains to generate proton gradients that power ATP synthase machines (see FIGURE 10.15). Given the early origin of photosynthesis, it is likely that cellular respiration evolved by modification of the photosynthetic equipment for a new function.

Now that we have applied this book's theme of evolution to the origin of metabolic diversity in prokaryotes, let's survey the diverse groups of archaea and bacteria that continue to have enormous impact on Earth and its life.

A SURVEY OF PROKARYOTIC DIVERSITY

In this section, we'll take a closer look at the distinction between archaea and bacteria and then examine a few of the major groups within these two domains of prokaryotic organisms.

Molecular systematics is leading to a phylogenetic classification of prokaryotes

THE PROCESS OF SCIENCE A classification of prokaryotic life that reflects evolutionary history was impossible until the development of molecular systematics. Neither the fossil record nor comparative morphology are of much help in working out details of prokaryotic phylogeny. The major lineages diverged during the very early history of life and left a limited fossil record. And because prokaryotes are so structurally simple compared to macroscopic, eukaryotic organisms, comparing the structure of modern prokaryotic species does not turn up much in the way of derived characters that are useful in cladistic analysis (see Chapter 25).

The breakthrough came when Carl Woese and his colleagues, mentioned earlier, began to cluster contemporary prokaryotes into taxonomic groups based on comparisons of nucleic acid sequences. The molecular material that turned out to be particularly useful is the RNA found in the smaller of the two subunits of ribosomes. Because all cells have ribosomes, Woese reasoned that the gene for the small-subunit ribosomal RNA (SSU-rRNA) has a history extending from the very first prokaryotes to the entire diversity of modern life. Each group of organisms has **signature sequences,** regions of its SSU-rRNA that have unique nucleotide sequences acquired by an accumulation of mutations in the ancestor of that taxonomic group. Thus, Woese found a way to define the very

deepest branchings in the tree of life. FIGURE 27.13 (p. 536) diagrams the phylogeny of some of the major groups of prokaryotes as sorted by molecular systematics.

Before molecular systematics, prokaryotic taxonomy was based on such phenotypic characters as nutritional mode and gram-positive versus gram-negative staining behavior (see FIGURE 27.5). Such criteria remain useful for the identification of pathogenic bacteria in the clinical labs of hospitals. The telltale signs of a particular bacterial species cultured from a patient's blood or other body fluid may be its shape as viewed with a microscope, its response to the Gram stain, whether or not the pathogen is motile, and what nutrients are required to grow the bacteria on a culture medium. For the most part, however, these clinically useful phenotypes are a poor guide to phylogeny. We have seen, for example, how the various nutritional modes are scattered around the phylogenetic tree of prokaryotes. However, a few of the traditional phenotype-based groups *do* persist in a phylogenetic classification based on molecular methods. For instance, note in FIGURE 27.13 that the cyanobacteria and the spirochetes cluster into two distinct lineages in the prokaryotic phylogeny. And although the gram-positive bacteria are very diverse, they all cluster into a single major prokaryotic branch. However, the gram-negative bacteria are scattered throughout several lineages; there are even some species that are gram-negative in staining behavior but are now included in the taxon called "gram-*positive,*" so named because *most* of its members *are* gram-positive.

Researchers have now sequenced the complete genomes of several prokaryotes, producing an enormous database that has supported most of the taxonomic conclusions based on SSU-rRNA comparison but has also produced some surprises. For example, in Chapter 28 we'll discuss how rampant gene swapping within early communities of prokaryotes left a record in the genomes of all modern organisms, including humans.

Researchers are identifying a great diversity of archaea in extreme environments and in the oceans

Note again in FIGURE 27.13 the very early divergence into two main prokaryotic lineages that are now designated the domains Archaea and Bacteria. TABLE 27.2 (p. 537) summarizes some differences between these two main prokaryotic groups. The table includes the third domain of life, Eukarya, to reinforce the point made earlier that the archaea have at least as much in common with eukaryotes as they do with bacteria. However, the archaea also have many unique characteristics, which we would expect of a taxon that has followed a separate evolutionary path for so long.

Systematists have sorted most of the known species of archaea into two major taxa, the Euryarchaeota and Crenarchaeota named in FIGURE 27.13. However, much of the research

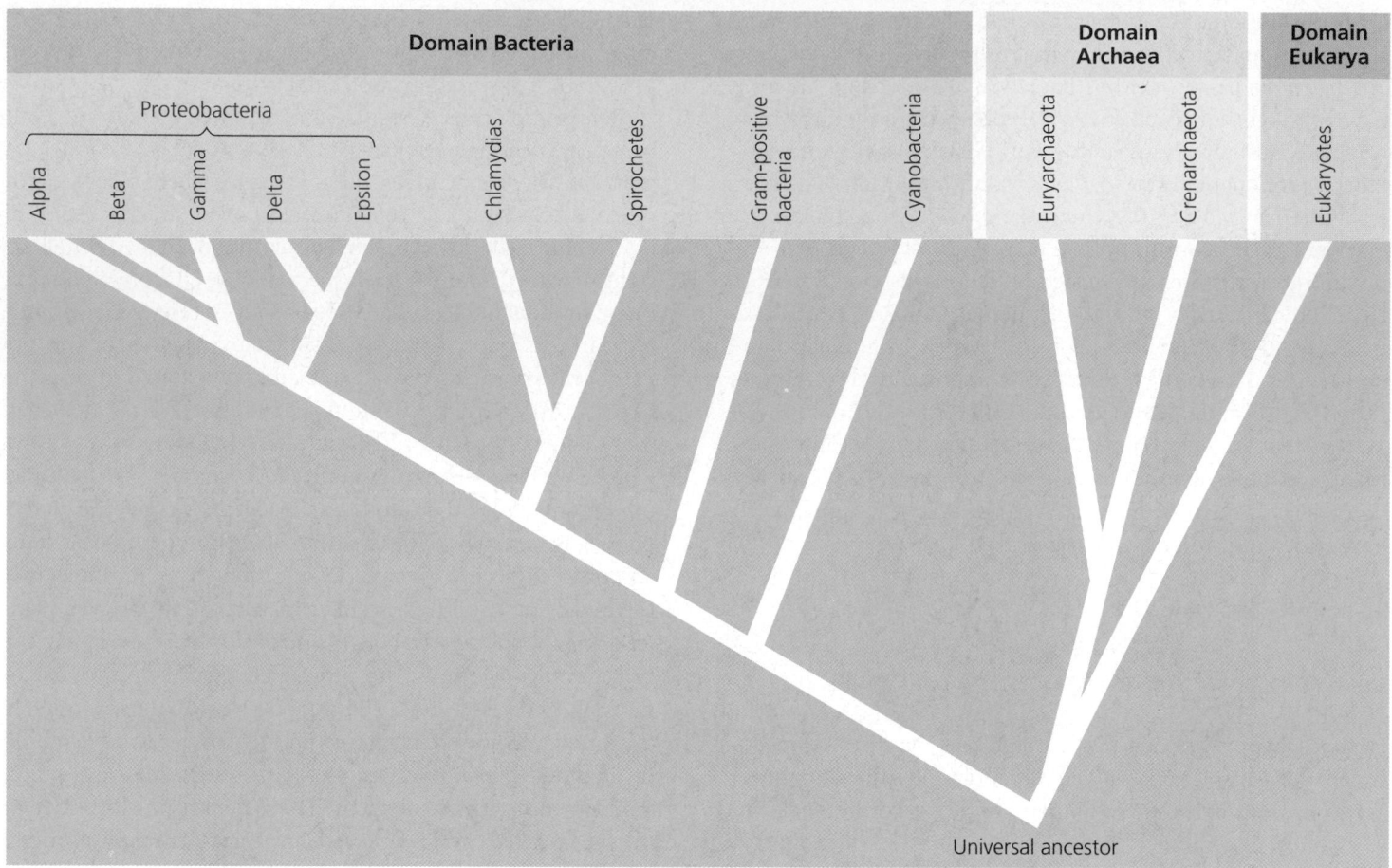

FIGURE 27.13 Some major groups of prokaryotes. This tentative phylogeny, like all such trees, is a hypothesis about evolutionary history. This simplified tree—consisting only of those prokaryotic groups that are discussed in this chapter—is based on molecular systematics, particularly comparisons of SSU-rRNA signature sequences. The groups shown here might be considered kingdoms within domains Bacteria and Archaea. However, many systematists prefer to wait for more data before assigning the groups to ranked taxonomic categories, such as kingdoms and phyla. This tree includes a "universal ancestor" common to all life, but we'll see evidence in Chapter 28 that the deepest branchings of life may have emerged *not* from a single ancestor but from an ancestral community of gene-swapping prokaryotes.

so far on the archaea has focused not on their phylogeny, but on their ecology—on the ability of many archaea to live where no other life can. We can think of these creatures as **extremophiles,** "lovers" of extreme environments, such as the hot-water geyser shown in FIGURE 27.1. In addition to phylogenetic classification, it is useful to classify the extremophiles into three groups based on environmental criteria: methanogens, extreme halophiles, and extreme thermophiles.

The **methanogens** are named for the unique way they obtain energy by using CO_2 to oxidize H_2, producing methane (CH_4) as a waste product. Methanogens, among the strictest of anaerobes, are poisoned by oxygen. They live in swamps and marshes where other microbes have consumed all the oxygen; the methane that bubbles out at these sites is known as marsh gas. Methanogens are also important decomposers used in sewage treatment. Some farmers have experimented with the use of these microbes to convert garbage and dung to methane, a valuable fuel. Other species of methanogens inhabit the anaerobic environment within the guts of animals, playing an important role in the nutrition of cattle, termites,

and other herbivores that subsist mainly on a diet of cellulose. This gas production does not normally bloat cows because the animals belch out large volumes of the methane. On a global scale, methanogens produce enough methane to play a major role in the greenhouse effect, an atmospheric phenomenon you will learn about in Chapter 54.

The **extreme halophiles** (from the Greek *halo,* salt, and *philos,* lover) live in such saline places as the Great Salt Lake and the Dead Sea. Some species merely tolerate salinity, whereas others actually require an environment ten times saltier than seawater to grow (FIGURE 27.14). Colonies of halophiles form a purple-red scum that owes its color to **bacteriorhodopsin,** a photosynthetic pigment very similar to the visual pigments in the retinas of our eyes.

As their name implies, the **extreme thermophiles** thrive in hot environments. The optimal conditions for most of these archaea are temperatures of 60–80°C. *Sulfolobus* inhabits hot sulfur springs in Yellowstone National Park, obtaining its energy by oxidizing sulfur. Another sulfur-metabolizing thermophile lives in the 105°C water near deep-sea hydrothermal vents.

Some researchers postulate that the very earliest prokaryotes to evolve were extreme thermophiles that inhabited hot environments similar to deep-sea vents. This would make it more historically accurate to consider most forms of life, including humans, as "cold adapted," rather than viewing the thermophilic archaea, which have continued to inhabit hot environments, as "extreme."

Until recently, researchers thought that archaea were prevalent organisms mainly in extreme environments where other life-forms were essentially absent. However, in the past decade, scientists have discovered that an abundance of marine archaea live among other forms of life in the more moderate habitats of the oceans.

To return to the phylogenetic classification in FIGURE 27.13, all the methanogens and halophiles fit into **Euryarchaeota.** (The word root "eury" means "broad," a reference here to the variety and habitat range of these prokaryotes.) This archaean group also includes some thermophiles, though most of the thermophilic species belong to **Crenarchaeota.** (The word root "cren" means "spring," as in hydrothermal vents.) Each of these two taxa also includes some of the newly discovered marine archaea.

FIGURE 27.14 Extreme halophiles. The colors of these seawater evaporating ponds at the edge of San Francisco Bay result from a dense growth of extreme halophiles that thrive in the ponds when the water reaches a salinity of 15–20%. (Before evaporation, the salinity of seawater is about 3%.) The ponds are used for commercial salt production; the salt-loving archaea are harmless.

Table 27.2 A Comparison of the Three Domains of Life

CHARACTERISTIC	DOMAIN		
	Bacteria	Archaea	Eukarya
Nuclear envelope	Absent	Absent	Present
Membrane-enclosed organelles	Absent	Absent	Present
Peptidoglycan in cell wall	Present	Absent	Absent
Membrane lipids	Unbranched hydrocarbons	Some branched hydrocarbons	Unbranched hydrocarbons
RNA polymerase	One kind	Several kinds	Several kinds
Initiator amino acid for start of protein synthesis	Formyl-methionine	Methionine	Methionine
Introns (noncoding parts of genes)	Rare	Present in some genes	Present
Response to the antibiotics streptomycin and chloramphenicol	Growth inhibited	Growth not inhibited	Growth not inhibited
Histones associated with DNA	Absent	Present	Present
Circular chromosome	Present	Present	Absent
Ability to grow at temperatures >100°C	No	Some species	No

Because bacteria have been the prokaryotic subjects of most research throughout the history of microbiology, much more is known about them than about the archaea. Now that the evolutionary and ecological importance of the archaea have come into focus, we can expect research on this domain to turn up many more surprises about the history of life and the roles of microbes in ecosystems.

Most known prokaryotes are bacteria

The term *bacteria* was once synonymous with "prokaryotes," but molecular systematics changed that when it became clear that prokaryotic life is represented by two distinct domains. It's a bit confusing that the name Bacteria was reassigned as the title of just one prokaryotic domain, Archaea being the other. True, most of the *known* prokaryotes *are* bacteria, but the archaea may catch up now that there is so much interest in describing their diversity.

Every major mode of nutrition and metabolism is represented among the thousands of known species of bacteria. And as mentioned earlier, multiple nutritional types may be represented within a particular taxonomic group of bacteria. If you return to FIGURE 27.13, you'll see five major bacterial groups identified (we've omitted several other groups). As with the two main groups of archaea, the major bacterial taxa are now accorded kingdom status by some prokaryotic systematists. We've organized our description of these monophyletic groups (clades) into TABLE 27.3 (pp. 538–539).

Table 27.3 Five of the Major Clades of Bacteria

Group/Description   **Example**

Proteobacteria

This large and diverse clade of gram-negative bacteria includes photoautotrophs, chemoautotrophs, and heterotrophs. Proteobacteria include both anaerobic and aerobic species. Molecular systematists recognize five subgroups of proteobacterial species.

Alpha Proteobacteria

Many of the species in this group are closely associated with eukaryotic hosts, either as mutual symbionts or as parasites. For example, *Rhizobium* species live in nodules within the roots of legumes (plants of the pea/bean family), where the bacteria convert atmospheric N_2 to compounds the host plant can use to make proteins. Species of the genus *Agrobacterium* are pathogens that cause tumors to form in plants; genetic engineers use these bacteria to carry foreign DNA into the genomes of crop plants (see FIGURE 20.19). The rickettsias, tiny even by bacterial standards, are pathogens that live within the cells of animals, causing such diseases as Rocky Mountain spotted fever in humans. Mitochondria, the sites of cellular respiration in eukaryotic cells, evolved from an aerobic alpha that inhabited a larger host cell, a phenomenon called endosymbiosis, which we will examine in detail in Chapter 28.

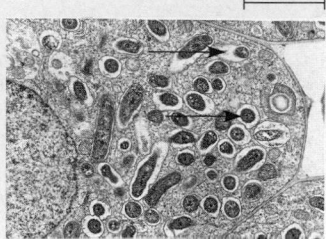

Rhizobium; the arrows point to the bacteria inside root cells of a legume (TEM)

Beta Proteobacteria

This nutritionally diverse group includes *Nitrosomonas,* soil bacteria that play an important role in nitrogen recycling within ecosystems by oxidizing ammonium (NH_4^+), producing nitrite (NO_2^-) as a waste product.

Gamma Proteobacteria

Among the photosynthetic members of this group are sulfur bacteria such as *Chromatium,* which split H_2S as a source of electrons to make organic molecules. The yellow globules in the *Chromatium* cells in the photo are sulfur wastes from photosynthesis. Among the heterotrophic gammas are some pathogens, including *Legionella,* named for the discovery that it causes Legionnaires' disease. Other gammas are enterics (bacteria that inhabit animal intestines), including *Salmonella,* one of the microbes that causes food poisoning; *Vibrio cholerae,* the pathogen that causes cholera; and the famous *Escherichia coli,* a species from the human intestine. The presence of *E. coli* in the water supply is a sign of contamination with feces. You may remember *E. coli* as one of the most important research organisms in molecular biology and DNA technology (see Chapters 16 and 20).

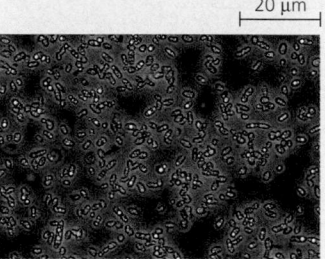

Chromatium; the yellow globules are sulfur wastes from H_2S-splitting photosynthesis (LM)

Delta Proteobacteria

One group of deltas, the myxobacteria, form the most elaborate colonies of all prokaryotes. The cells secrete a slimy substratum on which they glide through soil (*myxa* is Greek for "mucus"). When the soil dries out or food becomes scarce, the cells congregate and erect a bulbous "fruiting" body, which may be brightly colored and as large as a millimeter in diameter. The fruiting body releases resistant spores that become active and found new colonies in favorable environments. Another delta group, the bdellovibrios, are predators that attack other bacteria. A bdellovibrio charges its prey at a speed of 100 μm per second, which, relatively speaking, is like a human running about 600 km per hour, half the speed of sound! Then the predator turns into a bacterial drill, boring into its prey by spinning at 100 revolutions per second.

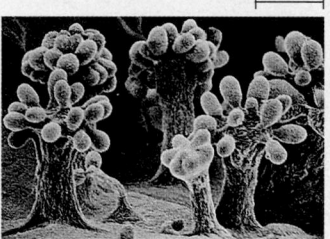

Myxobacteria: Fruiting bodies of *Chondromyces crocatus* (SEM)

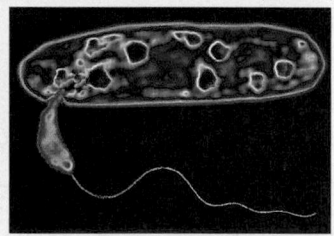

Bdellovibrio bacteriophorus attacking a larger bacterium (colorized TEM)

Epsilon Proteobacteria

A group closely related to the deltas, the epsilons include *Helicobacter pylori,* the bacterium that causes stomach ulcers.

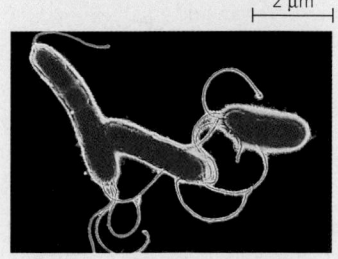

Helicobacter pylori, which causes stomach ulcers (colorized TEM)

Table 27.3 Five of the Major Clades of Bacteria (*continued*)

Chlamydias

These parasites can survive only within the cells of animals, depending on their hosts for resources as basic as ATP. The gram-negative walls of chlamydias are unusual among bacteria in that they lack peptidoglycan. One species, *Chlamydia trachomatis,* is the most common cause of blindness in the world and also causes nongonococcal urethritis, the most common sexually transmitted disease (STD) in the United States.

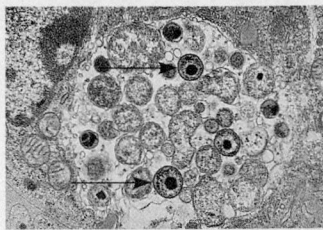

Chlamydias (arrows) living inside an animal cell (colorized TEM)

Spirochetes

Some of these helical heterotrophs are 0.25 mm long, though too thin to be visible without a microscope. Rotation of internal flagellum-like filaments produces a corkscrewlike movement. Many spirochetes are free-living, but the group also includes some notorious pathogens: *Treponema pallidum* causes the STD syphilis, and *Borrelia burgdorferi* is the pathogen of Lyme disease.

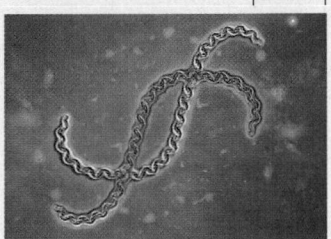

Leptospira, a spirochete (colorized TEM)

Gram-Positive Bacteria

All gram-positive bacteria cluster in this clade, but the group also includes some closely related gram-negative species. This branch of bacteria rivals the proteobacteria in diversity.

One gram-positive subgroup, the actinomycetes, forms colonies with branched chains of cells (the *mycetes* part of the name means "fungus," for which these bacteria were once mistaken). Two actinomycetes of medical importance are the species that cause tuberculosis and leprosy. But most actinomycetes are free-living species that help decompose the organic litter in soil; their secretions are partly responsible for the "earthy" odor of rich soil. Soil-dwelling species of the genus *Streptomyces* are cultured by pharmaceutical companies as a source of many antibiotics, including streptomycin.

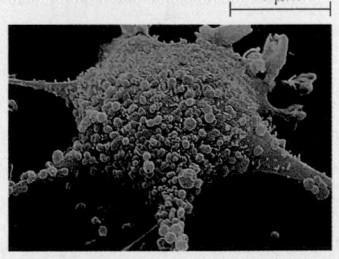

Streptomyces, source of many antibiotics (colorized SEM)

In addition to the colonial actinomycetes, the gram-positive group also includes a diversity of solitary species. There are spore formers such as *Bacillus* and *Clostridium*. Anthrax is caused by *Bacillus anthracis* (see FIGURE 27.10). *Clostridium botulinum* produces the toxin that causes the potentially fatal disease botulism. The various species of *Staphylococcus* and *Streptococcus* are also gram-positive bacteria.

The most structurally unusual members of the gram-positive clade are the mycoplasmas, which are the only bacteria lacking cell walls and which are the smallest of all known cells, with diameters as tiny as 0.1 μm, only about five times the diameter of a ribosome. Many mycoplasmas are soil bacteria, but there are also pathogens, including a species that causes "walking pneumonia" in humans.

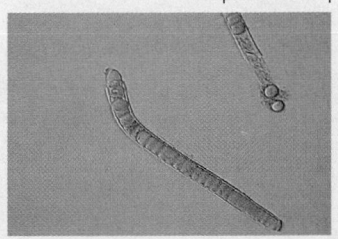

Hundreds of mycoplasmas cover this human fibroblast cell (colorized SEM)

Cyanobacteria

These photoautotrophs are the only prokaryotes with plantlike, oxygenic photosynthesis. (In fact, chloroplasts evolved from a cyanobacterium that lived as an endosymbiont within a larger host cell; see Chapter 28). Both solitary and colonial cyanobacteria are abundant wherever there is water, providing an enormous amount of food for freshwater and marine ecosystems. Some filamentous colonies have cells specialized for nitrogen fixation, the metabolic process that converts atmospheric N_2 to compounds that can be incorporated into proteins and other organic molecules (see FIGURE 27.11).

Cyanobacteria: *Scytonema* (LM)

THE ECOLOGICAL IMPACT OF PROKARYOTES

One of the themes of this unit is changing life on a changing planet—the relationship between geologic history and evolving biological forms. Organisms as pervasive, abundant, and diverse as the prokaryotes have had a tremendous impact on Earth and all its inhabitants.

Prokaryotes are indispensable links in the recycling of chemical elements in ecosystems

Not too long ago, in geologic terms, the atoms of the organic molecules in our bodies were parts of the inorganic compounds of soil, air, and water, as they will be again. Ongoing life depends on the recycling of chemical elements between the biological and physical components of ecosystems. Prokaryotes play essential roles in these chemical cycles (to be discussed in detail in Chapter 54). If it were not for such **decomposers,** carbon, nitrogen, and other elements essential to life would become locked in the organic molecules of corpses and waste products.

Prokaryotes also mediate the return of elements from the nonliving components of the environment (air, inorganic soil, and water) to the pool of organic compounds. Autotrophic prokaryotes make organic compounds from CO_2, supporting food chains through which organic nutrients pass from the prokaryotes to prokaryote eaters, then on to secondary consumers.

Because of their many unique metabolic capabilities, prokaryotes are the only organisms able to metabolize inorganic molecules containing elements such as iron, sulfur, nitrogen, and hydrogen. Cyanobacteria not only synthesize food and restore oxygen to the atmosphere, but also fix nitrogen, stocking the soil and water with nitrogenous compounds that other organisms can use to make proteins. And when plants and the animals that eat them die, soil prokaryotes return the nitrogen to the atmosphere. All life on Earth depends on prokaryotes and their unparalleled metabolic diversity.

Many prokaryotes are symbiotic

Prokaryotes rarely function singly in the environment. More often they interact in groups, which often include other species of prokaryotes or eukaryotes with complementary metabolisms.

Symbiosis (from the Greek word for "living together") is the term used to describe ecological relationships between organisms of different species that are in direct contact. If one of the symbiotic organisms is much larger than the other, the larger is called the **host.** There are three categories of symbiotic relationships: mutualism, commensalism, and parasitism. In **mutualism,** both symbiotic organisms benefit (FIGURE 27.15). In **commensalism,** one organism receives benefits while nei-

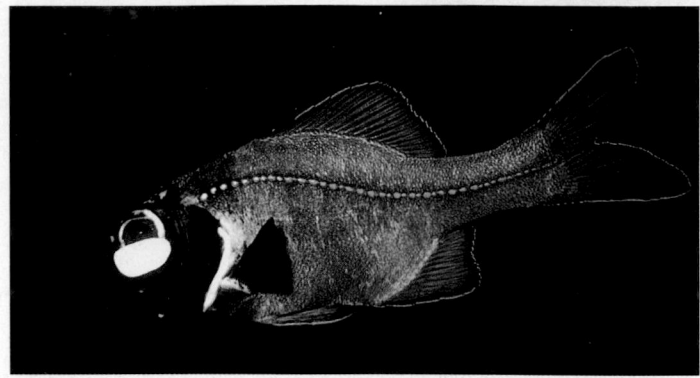

FIGURE 27.15 Mutualism: bacterial "headlights." The lighted oval under the eye of this deep-sea fish is an organ harboring symbiotic bacteria that are bioluminescent. A chemical reaction powered by ATP emits the light. The fish uses its headlamps to lure light-seeking prey and to signal potential mates.

ther harming nor helping the other in any significant way. In **parasitism,** one symbiotic organism, called a **parasite** in this case, benefits at the expense of the host.

Prokaryotes are represented in all three categories of symbiosis with eukaryotes. For example, plants of the legume family (peas, beans, alfalfa, and others) have lumps on their roots called nodules, which are home to mutualistic prokaryotes that fix nitrogen used by the host (see *Rhizobium* in TABLE 27.3). The plant reciprocates by providing the prokaryotes with a steady supply of sugar and other organic nutrients. Bacteria inhabiting the inner and outer surfaces of the human body consist mostly of commensal species, but some species are mutualistic symbionts. For instance, fermenting bacteria living in the vagina produce acids that maintain a pH between 4.0 and 4.5, suppressing the growth of yeast and other potentially harmful microorganisms. Among parasitic bacteria are those classified as pathogens because they cause disease in their hosts.

Pathogenic prokaryotes cause many human diseases

Most of us are well most of the time because our defenses check the growth of pathogens to which we are exposed. Occasionally, the balance shifts in favor of a pathogen, and we become ill. To be pathogenic, a parasite must invade the host, resist internal defenses long enough to begin growing, then harm the host in some way. Pathogenic prokaryotes cause about half of all human diseases (FIGURE 27.16).

Some pathogens are **opportunistic,** meaning they are normal residents of a host but can cause illness when the host's defenses are weakened by such factors as poor nutrition or a recent bout with the flu. For example, *Streptococcus pneumoniae* (a gram-positive bacterium) lives in the throats of most healthy people, but this opportunist can multiply and cause pneumonia when the host's defenses are down.

Louis Pasteur, Joseph Lister, and other scientists began linking disease to pathogenic microbes in the late 1800s. The first to ac-

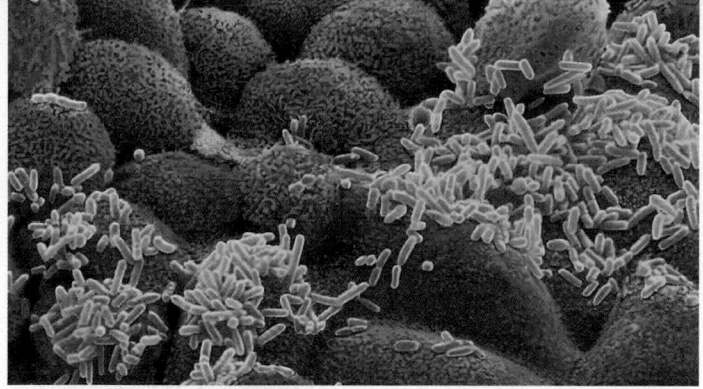

FIGURE 27.16 A very harmful "germ." The yellow rods are *Haemophilus influenzae* bacteria on skin cells lining the interior of a human nose (colorized SEM). These pathogens are transmitted through the air. *H. influenzae,* not to be confused with influenza (flu) viruses, causes pneumonia and other lung infections that kill about 4 million people worldwide per year. Most victims are children in developing countries, where malnutrition lowers resistance to all pathogens.

tually connect certain diseases to specific bacteria was Robert Koch, a German physician who identified the bacteria responsible for anthrax and tuberculosis. His methods established four criteria, now called **Koch's postulates,** that are still the guidelines for medical microbiology. To establish that a specific pathogen is the cause of a disease, the researcher must (1) find the same pathogen in each diseased individual investigated, (2) isolate the pathogen from a diseased subject and grow the microbe in a pure culture, (3) induce the disease in experimental animals by transferring the pathogen from the culture, and (4) isolate the same pathogen from the experimental animals after the disease develops. The postulates work for most pathogens, but exceptions do occur. For example, no one has yet cultured the spirochete *Treponema pallidum* on artificial media, but circumstantial evidence leaves no doubt that this organism causes syphilis.

How do pathogenic prokaryotes produce symptoms of disease? Some species, such as the gram-positive actinomycete that causes tuberculosis, disrupt the health of the host by invading tissues. More commonly, pathogens cause illness by producing poisons called exotoxins and endotoxins.

Exotoxins are proteins secreted by prokaryotes. Exotoxins can produce disease symptoms even without the prokaryotes being present. For example, when the gram-positive bacterium *Clostridium botulinum* (see TABLE 27.3) grows anaerobically in improperly canned foods, one of the by-products of its fermentation is an exotoxin that causes the potentially fatal disease botulism. Some exotoxins are among the most potent poisons known; just 1 g of botulism toxin would be sufficient to kill a million people. Another exotoxin-producing species is the enteric proteobacterium *Vibrio cholerae,* which causes cholera, a dangerous disease characterized by severe diarrhea. Resulting from the consumption of water contaminated with human feces, cholera is often epidemic during wars and famines. Even *E. coli* can be an exotoxin-releasing culprit. Some cases of traveler's diarrhea result from toxins released by strains of *E. coli* obtained from another person via contaminated food or water.

In contrast to exotoxins, **endotoxins** are components of the outer membranes of certain gram-negative bacteria. Examples of endotoxin-producing bacteria include nearly all members of the genus *Salmonella,* which are not normally present in healthy animals. *Salmonella typhi* causes typhoid fever, and several other species of *Salmonella,* some of which are commonly found in poultry, cause food poisoning.

Since the discovery in the 19th century that "germs" cause disease, improved sanitation has significantly reduced infant mortality and extended life expectancy in developed countries. Antibiotics have also saved many lives and reduced disease incidence. More than half of our antibiotics (including streptomycin, neomycin, erythromycin, aureomycin, and tetracycline) come from soil bacteria of the genus *Streptomyces* (an actinomycete; see TABLE 27.3). In nature, these compounds prevent encroachment by competing microbes.

Bacterial diseases are a continuing threat. Their general decline over the past century is probably due more to public health policies and education than to "wonder drugs." A case in point is Lyme disease, currently the most widespread pest-carried disease in the United States. The disease is caused by a spirochete bacterium carried by ticks that live on deer and field mice (FIGURE 27.17). Lyme disease usually starts as a red rash shaped like a bull's-eye around a tick bite. Antibiotics can cure the disease if administered within about a month after exposure. If untreated, Lyme disease can cause debilitating arthritis, heart disease, and nervous disorders. A vaccine is now available, but it does not give full protection. The best defense is public education about avoiding tick bites and seeking treatment if a rash develops. When walking through brush, it is advisable to use insect repellent and wear light-colored clothing to reduce contact with ticks.

Today, the rapid evolution of antibiotic-resistant strains of pathogenic bacteria is a serious health threat aggravated by imprudent and excessive antibiotic use.

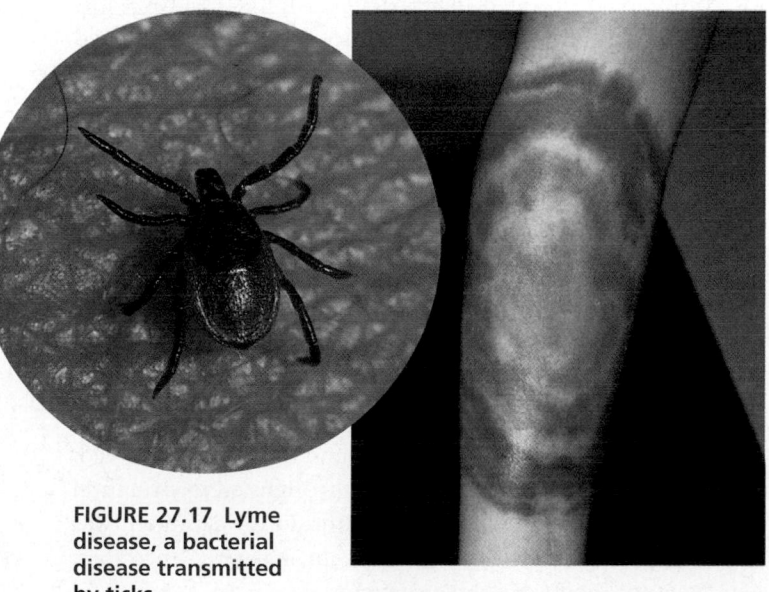

FIGURE 27.17 Lyme disease, a bacterial disease transmitted by ticks.

Although declared illegal by the United Nations, the selective culturing and stockpiling of deadly bacterial disease agents for use as biological weapons remains a threat to world peace.

Humans use prokaryotes in research and technology

Humans have learned many ways of exploiting the diverse metabolic capabilities of prokaryotes, both for scientific research and for practical purposes. Much of what we know about metabolism and molecular biology has been learned in laboratories using prokaryotes as relatively simple model systems. In fact, *E. coli,* the prokaryotic "white rat" of so many research labs, is the best understood of all organisms. And we are just beginning to explore the great potential prokaryotes have for helping us solve some of our environmental problems.

The use of organisms to remove pollutants from water, air, and soil is called **bioremediation.** The most familiar example of bioremediation is the use of prokaryotic decomposers to treat our sewage. Raw sewage is first passed through a series of screens and shredders, and solid matter is allowed to settle out from the liquid waste. This solid matter, called sludge, is then gradually added to a culture of anaerobic prokaryotes, including both bacteria and archaea. The microbes decompose the organic matter in the sludge, converting it to material that can be used as landfill or fertilizer after chemical sterilization. Liquid wastes are treated separately from the sludge (FIGURE 27.18). In other examples of bioremediation, soil bacteria called pseudomonads (proteobacteria) are used to decompose petroleum compounds on beaches and other sites of oil spills and to decompose a variety of synthetic compounds, such as pesticides (FIGURE 27.19). Genetic engineers are improving the bioremediation powers of certain prokaryotes.

Humans also put bacteria to work as metabolic "factories" for commercial products. The chemical industry grows immense cultures of bacteria that produce acetone, butanol, and several other products. Pharmaceutical companies culture bacteria that make vitamins and antibiotics. Bacteria containing magnetic particles may soon become a commercial source for manufacturing magnetic tapes and other recording devices. The food industry uses bacteria to convert milk to yogurt and various kinds of cheese. And DNA technology has opened a new era in the commercial use of prokaryotes (see Chapter 20).

■   ■   ■

In this chapter, we have surveyed the prokaryotes and traced their history. On an ancient Earth inhabited only by prokaryotes, all the diverse forms of nutrition and metabolism evolved. Most subsequent evolutionary breakthroughs were structural rather than metabolic. The most significant development was the origin of eukaryotic cells from prokaryotic ancestors, which we will explore in the next chapter.

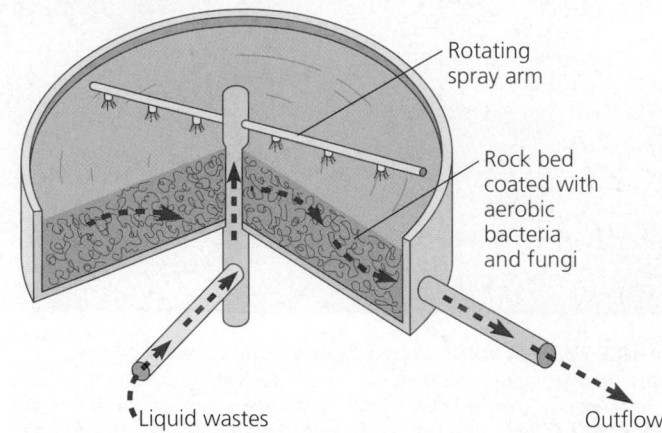

FIGURE 27.18 Putting prokaryotes to work in sewage treatment facilities. This is a trickling filter system, one type of mechanism for treating liquid wastes after sludge is removed. The long horizontal pipes rotate slowly, spraying liquid wastes through the air onto a thick bed of rocks. Bacteria and fungi growing on the rocks remove much of the organic material dissolved in the waste. Outflow from the rock bed is sterilized and then released, usually into a river or ocean.

FIGURE 27.19 Bioremediation for an oil spill. A worker is spraying fertilizers on an oil-soaked beach in Alaska. The fertilizers stimulate growth of indigenous bacteria that can initiate the breakdown of the oil—in some cases, speeding the natural process of breakdown some fivefold. This technique is the fastest and least expensive way yet devised to clean up oil spills on beaches.

CHAPTER 27 REVIEW

Go to the Campbell Biology website (www.campbellbiology.com) to explore an interactive version of the Chapter Review.

Summary of Key Concepts

THE WORLD OF PROKARYOTES

- **They're (almost) everywhere! An overview of prokaryotic life (pp. 526–527)** Prokaryotes were the first organisms, and they persist today as the most numerous and pervasive of all living things.
- **Bacteria and archaea are the two main branches of prokaryote evolution (pp. 527–528, FIGURE 27.2)** The two prokaryotic domains diverged early in the history of life.

THE STRUCTURE, FUNCTION, AND REPRODUCTION OF PROKARYOTES

- Prokaryotes are generally single-celled organisms, although some occur as aggregates, colonies, or simple multicellular forms. Most prokaryotes are spherical (cocci), rod-shaped (bacilli), or helical in shape (p. 528, FIGURE 27.3).
- **Nearly all prokaryotes have a cell wall external to the plasma membrane (pp. 528–529, FIGURE 27.5)** Gram-positive and gram-negative bacteria differ in the structure of their walls and other surface layers. Many species have capsules and pili outside the cell wall, which help the cells adhere to one another or to a substratum. Some pili are specialized for conjugation.

  Web/CD Activity 27A: *Prokaryotic Cell Structure and Function*

- **Many prokaryotes are motile (pp. 529–530, FIGURE 27.7)** Motile bacteria propel themselves by flagella, use flagellum-like filaments positioned inside the cell wall (spirochetes), or glide ahead of their slime secretions.
- **The cellular and genomic organization of prokaryotes is fundamentally different from that of eukaryotes (pp. 530–531, FIGURE 27.8)** The cells of prokaryotes are not compartmentalized by internal membranes, but infoldings of the plasma membrane in some species provide internal membrane surfaces. The genome of prokaryotes is a circular DNA molecule unbounded by a membrane. Smaller rings of DNA called plasmids code for special metabolic pathways and resistance to antibiotics in some species.
- **Populations of prokaryotes grow and adapt rapidly (pp. 531–532, FIGURES 27.9, 27.10)** Prokaryotes grow in number by binary fission. Genetic variation occurs through mutation and through gene transfer by transformation, conjugation, or viral transduction.

NUTRITIONAL AND METABOLIC DIVERSITY

- **Prokaryotes can be grouped into four categories according to how they obtain energy and carbon (pp. 532–534, TABLE 27.1)** Photoautotrophs use light energy, and chemoautotrophs use inorganic substances to synthesize their organic compounds from carbon dioxide. Photoheterotrophs use light energy and require organic molecules. Most known prokaryotes are chemoheterotrophs, which require organic molecules as a source of both energy and carbon. Obligate aerobes require O_2, obligate anaerobes are poisoned by it, and facultative anaerobes can survive with or without O_2.

  Web/CD Case Study in the Process of Science: *What Are the Modes of Nutrition in Prokaryotes?*

- **Photosynthesis evolved early in prokaryotic life (pp. 534–535, FIGURE 27.12)** An early evolution of photosynthesis explains the diffuse occurrence of this nutritional mode in prokaryotic phylogeny. Heterotrophic lineages lost the ability to photosynthesize. The first cyanobacteria, which released O_2 as a by-product of photosynthesis, changed Earth's ancient atmosphere drastically.

A SURVEY OF PROKARYOTIC DIVERSITY

- **Molecular systematics is leading to a phylogenetic classification of prokaryotes (p. 535, FIGURE 27.13)** Signature sequences in SSU-rRNA have enabled systematists to identify major clades in prokaryotic diversity.
- **Researchers are identifying a great diversity of archaea in extreme environments and in the oceans (pp. 535–537, FIGURE 27.14)** The extremophiles include halophiles, extreme thermophiles, and methanogens.
- **Most known prokaryotes are bacteria (p. 537, TABLE 27.3)** Diverse nutritional types are scattered among the major clades of bacteria. The two largest clades are the proteobacteria and the gram-positive bacteria.

  Web/CD Activity 27B: *Classification of Prokaryotes*

THE ECOLOGICAL IMPACT OF PROKARYOTES

- **Prokaryotes are indispensable links in the recycling of chemical elements in ecosystems (p. 540)** Nitrogen fixation is just one example of this ecosystem service.
- **Many prokaryotes are symbiotic (p. 540, FIGURE 27.15)** Many prokaryotes live with other species in symbiotic relationships: mutualism, commensalism, or parasitism.
- **Pathogenic prokaryotes cause many human diseases (pp. 540–542, FIGURES 27.16, 27.17)** Pathogenic bacteria cause human disease by invading tissues or poisoning with endotoxins or exotoxins.
- **Humans use prokaryotes in research and technology (p. 542, FIGURES 27.18, 27.19)** Bioremediation and antibiotic production are just two examples of applications of microbiology.

Self-Quiz

1. Home canners pressure-cook vegetables as a precaution primarily against
   a. mycoplasmas.
   b. endospore-forming bacteria.
   c. enteric bacteria.
   d. cyanobacteria.
   e. actinomycetes.

2. Photoautotrophs use
   a. light as an energy source and can use water or hydrogen sulfide as a source of electrons for producing organic compounds.
   b. light as an energy source and oxygen as an electron source.
   c. inorganic substances for energy and CO_2 as a carbon source.
   d. light to generate ATP but need organic molecules for a carbon source.
   e. light as an energy source and CO_2 to reduce organic nutrients.

3. Which of the following statements about the domains of prokaryotes is *not* true?
   a. The lipid composition of the plasma membrane found in archaea is different from that of bacteria.
   b. Archaea and bacteria probably diverged very early in evolutionary history.
   c. Both archaea and bacteria have cell walls, but the walls of archaea lack peptidoglycan.
   d. Of the two groups, bacteria are more closely related to eukaryotes.
   e. Bacteria include the cyanobacteria.

4. Several diverse branches of prokaryotes contain both photosynthetic and heterotrophic species. The most parsimonious interpretation of this observation is that
   a. photosynthesis evolved several times during prokaryotic history.
   b. all these lineages evolved from a primitive heterotrophic ancestor.
   c. all these lineages evolved from a photosynthetic ancestor, and some groups lost the ability to photosynthesize.
   d. photosynthesis evolved at the same time as aerobic respiration.
   e. glycolysis is the oldest metabolic pathway because it is present in all prokaryotic groups.

5. Which of the following is *not* descriptive of the domain Archaea?
   a. It includes the taxa (kingdoms) Euryarchaeota and Crenarchaeota.
   b. It includes the methanogens, extreme halophiles, extreme thermophiles, and some marine groups.
   c. This domain is believed to include the universal ancestor of all life.
   d. It has characteristics in common with domain Bacteria (such as no nuclear envelope) and domain Eukarya (such as histone proteins associated with DNA) and characteristics it shares with neither (such as the ability to grow in harsh habitats).
   e. It is separated from domain Bacteria based on signature sequences and other molecular evidence of an early divergence between the two groups.

6. Which of the following groups is mismatched with its members?
   a. Proteobacteria—diverse gram-negative bacteria
   b. Chlamydias—intracellular parasites
   c. Spirochetes—helical heterotrophs
   d. Gram-positive bacteria—diverse pathogens whose endotoxins are components of their outer membrane
   e. Cyanobacteria—solitary and filamentous colonies exhibiting oxygenic photosynthesis

7. Which of the following statements about demonstrating the pathogenicity of a particular bacterial species is *not* true?
   a. The bacteria must be capable of inducing the disease when transferred to an experimental host.
   b. The bacteria isolated from a diseased host must be grown in pure culture.
   c. The same bacteria must be present in each diseased host investigated.
   d. The bacteria isolated from the experimental host must be capable of reinducing the disease when returned to the original host.
   e. The bacteria must be identified in the artificially infected experimental host after the disease develops.

8. Penicillins function as antibiotics mainly by inhibiting the ability of some bacteria to
   a. form spores.
   b. replicate DNA.
   c. synthesize normal cell walls.
   d. produce functional ribosomes.
   e. synthesize ATP.

9. Plantlike photosynthesis that releases O_2 occurs in the
   a. cyanobacteria.   d. actinomycetes.
   b. chlamydias.   e. chemoautotrophic bacteria.
   c. archaea.

10. An example of bioremediation is
    a. the use of prokaryotes to treat sewage or clean up oil spills.
    b. the use of antibiotics produced by cultured prokaryotes.
    c. the genetic engineering of bacteria to produce human proteins and useful chemical products.
    d. the introduction of parasitic bacteria to kill other bacteria.
    e. all of the above.

11. Upon microscopic examination, how do you think you could distinguish the cocci that cause "staph" infections from those that cause "strep" throat?

12. Why are some archaea referred to as "extremophiles"?

13. Why do microbiologists autoclave their laboratory instruments and glassware?

14. Contrast exotoxins with endotoxins.

15. How do bacteria help restore the atmospheric CO_2 required by plants for photosynthesis?

Go to the website or CD-ROM for more quiz questions.

Evolution Connection

Health officials worldwide are concerned by a resurgence of diseases caused by bacteria that are resistant to standard antibiotics. For instance, antibiotic-resistant bacteria are now causing an epidemic of tuberculosis (TB), a lung disease spread by airborne droplets. Drugs can relieve TB symptoms in a few weeks, but it takes much longer to halt the infection, and patients are likely to discontinue treatment while bacteria are still present. Why can prokaryotes quickly reinfect a patient if they are not wiped out? How might this result in the evolution of drug-resistant pathogens?

The Process of Science

You learned in this chapter that the former taxon "Monera" is undergoing modification, with at least 21 kingdoms erected in some schemes in addition to the super-kingdom level of the domain. Proponents of this new classification argue it better reflects evolutionary history, but why do you think this is desirable? Can you suggest some questions or hypotheses that can be framed with the new classification but could or would not even be asked with the former classification?

Experiment with the nutritional needs of various prokaryotes in the Case Study in the Process of Science on the CD-ROM and website.

Science, Technology, and Society

Many local newspapers publish a weekly list of restaurants that have been cited by inspectors for poor sanitation. Locate such a report and highlight the cases that are likely associated with potential food contamination by pathogenic prokaryotes.

Answers: 1. b; 2. a; 3. d; 4. c; 5. c; 6. d; 7. d; 8. c; 9. a; 10. a; 11. By the arrangement of the cell aggregates: grapelike clusters for staphylococcus and chains of cells for streptococcus. 12. Because they can thrive in extreme environments too hot, too salty, or too acidic for other organisms. 13. To kill bacterial endospores, which can survive boiling water. 14. Exotoxins are poisons secreted by pathogenic bacteria; endotoxins are components of the cell walls of pathogenic bacteria. 15. By decomposing the organic molecules of dead organisms and organic refuse such as leaf litter, the metabolism of bacteria releases carbon from the organic matter in the form of CO_2.

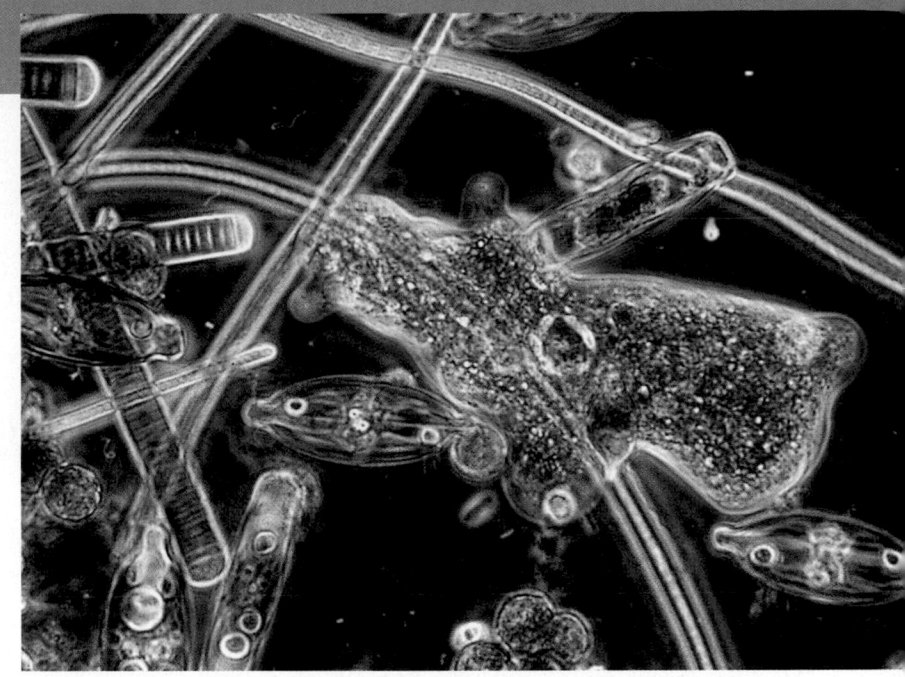

CHAPTER 28

THE ORIGINS OF EUKARYOTIC DIVERSITY

INTRODUCTION TO THE PROTISTS

- Systematists have split protists into many kingdoms
- Protists are the most diverse of all eukaryotes

THE ORIGIN AND EARLY DIVERSIFICATION OF EUKARYOTES

- Endomembranes contributed to larger, more complex cells
- Mitochondria and plastids evolved from endosymbiotic bacteria
- The eukaryotic cell is a chimera of prokaryotic ancestors
- Secondary endosymbiosis increased the diversity of algae
- Research on the relationships between the three domains is changing ideas about the deepest branching in the tree of life
- The origin of eukaryotes catalyzed a second great wave of diversification

A SAMPLE OF PROTISTAN DIVERSITY

- Diplomonadida and Parabasala: Diplomonads and parabasalids lack mitochondria
- Euglenozoa: The euglenozoa includes both photosynthetic and heterotrophic flagellates
- Alveolata: The alveolates are unicellular protists with subsurface cavities (alveoli)
- Stramenopila: The stramenopile clade includes the water molds and the heterokont algae
- Structural and biochemical adaptations help seaweeds survive and reproduce at the ocean's margins
- Some algae have life cycles with alternating multicellular haploid and diploid generations
- Rhodophyta: Red algae lack flagella
- Chlorophyta: Green algae and plants evolved from a common photoautotrophic ancestor
- A diversity of protists use pseudopodia for movement and feeding
- Mycetozoa: Slime molds have structural adaptations and life cycles that enhance their ecological roles as decomposers
- Multicellularity originated independently many times

"**N**o more pleasant sight *has met my eye than this of so many thousands of living creatures in one small drop of water*," wrote Anton van Leeuwenhoek *after his discovery of the microbial world more than three centuries ago. It is a world every biology student rediscovers by peering through a microscope into a droplet of pond water filled with diverse creatures we call protists (see the photograph on this page).*

Protists are eukaryotic, and thus even the simplest protists are much more complex than the prokaryotes. The first eukaryotes to evolve from prokaryotic ancestors were unicellular. The primal eukaryotes were not only the predecessors of the great variety of modern protists but were also ancestral to all other eukaryotes—plants, fungi, and animals. Two of the most significant chapters in the history of life—the origin of the eukaryotic cell and the subsequent emergence of multicellular eukaryotes—unfolded during the evolution of protists.

This chapter traces the origins of eukaryotic cells, examines the diversity, evolution, and ecology of protists, and considers the evolutionary origins of multicellular organization.

INTRODUCTION TO THE PROTISTS

In Chapter 26, we discussed eukaryotic fossils that are 2.1 billion years old and "chemical signatures" characteristic of eukaryotes dating back 2.7 billion years. For about 2 billion years before there were animals, fungi, and plants, the early eukaryotes gave rise to a great diversity of the mostly microscopic organisms we know by the informal name "protists."

Systematists have split protists into many kingdoms

Molecular systematics and cladistics are combining to chip away at the traditional five-kingdom system of classification (see FIGURE 26.15). A single prokaryotic kingdom, Monera, is incompatible with the new three-domain system that divides the prokaryotes into two superkingdoms, or domains: the domain Bacteria and the domain Archaea (see Chapter 27). Within the third domain, Eukarya, we find all of the eukaryotes, which were divided in the five-kingdom system into four kingdoms: Protista, Plantae, Fungi, and Animalia. The plant, fungus, and animal kingdoms are surviving the taxonomic remodeling so far, though their boundaries have expanded somewhat to include certain groups formerly classified as protists. However, the kingdom Protista has been crumbled beyond repair by the methods of modern systematics.

A protist kingdom was always a bit of a problem, even when the five-kingdom system was in its prime. Protista was defined partly by its members' structure—containing mostly unicellular eukaryotes—and partly by its role as a receptacle for all those eukaryotes that do not fit our definitions of plants, fungi, or animals. While most organisms classified as protists are single-celled and microscopic, the kingdom also included some relatively simple multicellular forms and even some relatively complex giants, such as seaweeds. Thus, organisms as diverse as amoebas and kelp (brown seaweeds) were classified in the same kingdom (FIGURE 28.1). The main thing they had in common was that they were not animals, fungi, or true plants (as we'll define them in the next chapter). The kingdom Protista was convenient, but it has not stood up to phylogenetic scrutiny. Certain protists are far less closely related to each other than plants are to animals. And certain protists are more closely related to plants, fungi, or animals than they are to other protists. To use the terminology of cladistics, the kingdom Protista was paraphyletic (FIGURE 28.2; also see FIGURE 25.9).

As biologists learned more about various groups of protists, the inclusion of such diverse lineages within a single kingdom became phylogenetically untenable. Systematists have split what was once the kingdom Protista into as many as 20 kingdoms. However, most biologists use "protist" as an informal term for this great diversity of eukaryotic kingdoms.

Protists are the most diverse of all eukaryotes

Given the taxonomic fragmentation of the protists into multiple kingdoms, you shouldn't be surprised that protists are so diverse that few general characteristics can be cited without exceptions. In fact, protists vary in structure and function more than any other group of organisms. Most of the approximately 60,000 known species of extant protists are unicellular, but

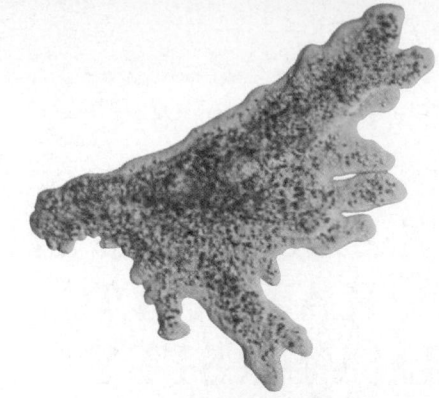

(a) *Amoeba proteus,* a unicellular "protozoan" (LM)　　25 μm

(b) A diatom, a unicellular "alga" (SEM)　　10 μm

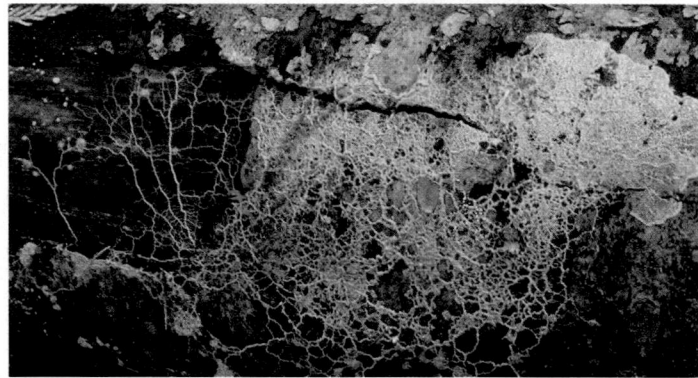

(c) A slime mold (*Physarum polychalum*)　　4 cm

(d) Australian bull kelp (*Durvillea potatorum*)

FIGURE 28.1 Too diverse for one kingdom: a small sample of protists. Though systematists have split the five-kingdom system's Protista into many kingdoms, "protist" is still a convenient informal term for the great diversity of eukaryotes that are not plants, fungi, or animals.

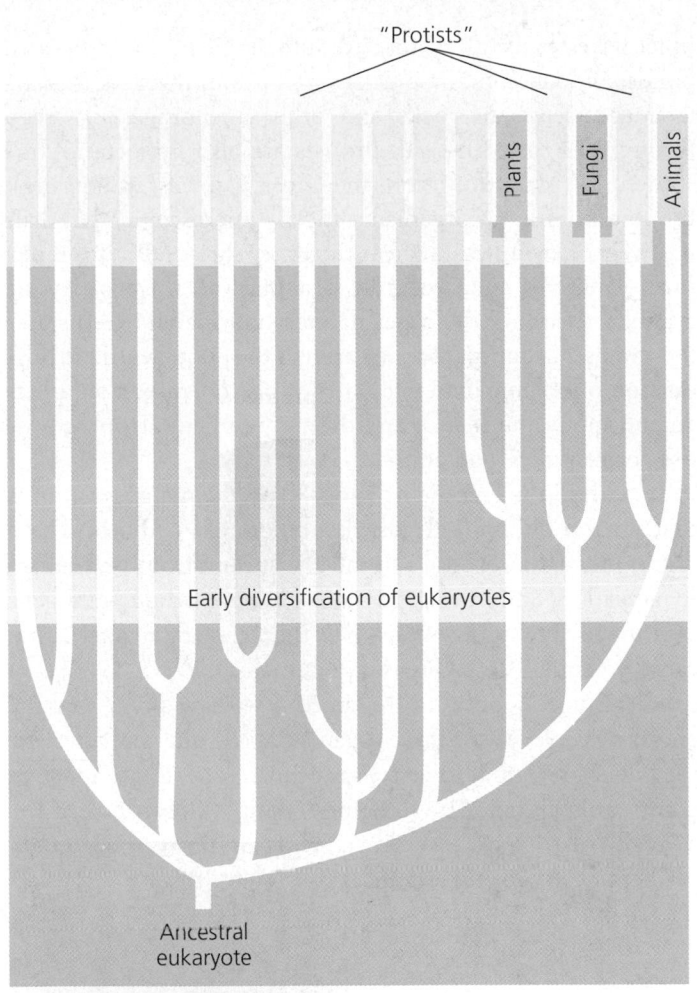

"Protists"

Plants
Fungi
Animals

Early diversification of eukaryotes

Ancestral
eukaryote

FIGURE 28.2 The kingdom Protista problem. In this simplified tree of eukaryotes, the diverse lineages that were once combined in the kingdom Protista are highlighted by the yellow background. It was a paraphyletic taxon (see FIGURE 25.9).

there are also some colonial and multicellular species (see FIG- URE 28.1). Because most protists are unicellular, they are justifiably considered the simplest eukaryotic organisms. But at the cellular level, many protists are exceedingly complex—the most elaborate of all cells. We should expect this of organisms that must carry out within the boundaries of a single cell all the basic functions performed by the specialized cells that make up the bodies of plants and animals. Each unicellular protist is not at all analogous to a single cell from a human or other multicellular organism, but is itself an organism as complete as any whole plant or animal.

Nutrition

Protists are the most nutritionally diverse of all eukaryotes. Most protists are aerobic in their metabolism, using mitochondria for cellular respiration (a few lack mitochondria and either live in anaerobic environments or contain mutualistic respiring bacteria). Some protists are photoautotrophs, containing chloroplasts; some are heterotrophs, absorbing organic molecules or ingesting larger food particles; and still others, called **mixotrophs**, combine photosynthesis *and* heterotrophic nutrition (an example is *Euglena,* shown in FIGURE 28.3).

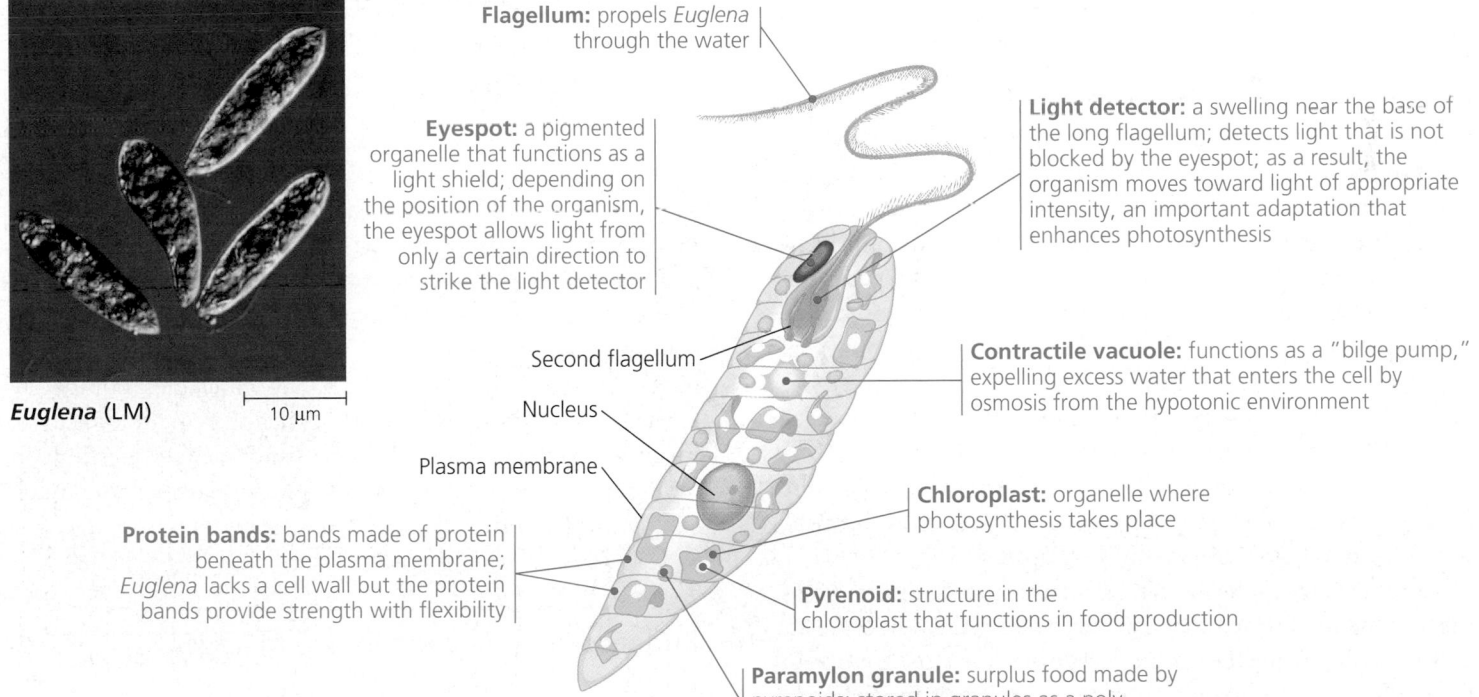

Euglena (LM)   10 μm

Flagellum: propels *Euglena* through the water

Eyespot: a pigmented organelle that functions as a light shield; depending on the position of the organism, the eyespot allows light from only a certain direction to strike the light detector

Light detector: a swelling near the base of the long flagellum; detects light that is not blocked by the eyespot; as a result, the organism moves toward light of appropriate intensity, an important adaptation that enhances photosynthesis

Second flagellum

Nucleus

Plasma membrane

Contractile vacuole: functions as a "bilge pump," expelling excess water that enters the cell by osmosis from the hypotonic environment

Protein bands: bands made of protein beneath the plasma membrane; *Euglena* lacks a cell wall but the protein bands provide strength with flexibility

Chloroplast: organelle where photosynthesis takes place

Pyrenoid: structure in the chloroplast that functions in food production

Paramylon granule: surplus food made by pyrenoids; stored in granules as a poly- saccharide called paramylon

FIGURE 28.3 *Euglena:* an example of a single-celled protist. This protist, one of the most common inhabitants of murky pond water, is mixotrophic: In light, its chloroplasts are photosynthetic; if placed in the dark, the organism can live as a heterotroph by absorbing organic nutrients from the environment. Some related species lack chloroplasts and ingest food by phagocytosis.

The different modes of nutrition are scattered throughout the various phylogenetic lineages of protists. For example, there are photosynthetic species in many different protistan groups. And the same group may include photosynthetic species, heterotrophic species, and mixotrophs. Thus, nutrition is not a reliable taxonomic characteristic. However, nutrition *is* useful in an ecological context to help us understand the adaptations of protists and the roles they play in biological communities. In this ecological context, we can divide protists into three categories: ingestive (animal-like) protists, or **protozoa** (singular, *protozoan*); absorptive (funguslike) protists (these have no other general name); and photosynthetic (plantlike) protists, or **algae** (singular, *alga*). The term *alga* refers to relatively simple aquatic photoautotrophs, including some organisms that some biologists include in the plant kingdom. Though commonly used, the terms *protozoa* and *algae* have no basis in phylogeny and no significance in taxonomy.

Motility

Most protists are motile, usually having flagella or cilia at some time in their life cycles. It is important to understand that prokaryotic and eukaryotic versions of flagella are not homologous structures. Prokaryotic flagella are attached to the cell surface (see FIGURE 27.7). In contrast, eukaryotic flagella and cilia are extensions of the cytoplasm, with bundles of microtubules covered by the plasma membrane (see FIGURE 7.24). Eukaryotic cilia and flagella have the same basic ultrastructure (the 9 + 2 arrangement of microtubules), but cilia are shorter and more numerous.

Life Cycles

Reproduction and life cycles are highly varied among protists. Mitosis occurs in almost all protists, but there are many variations in the process unknown in other eukaryotes. Some protists are exclusively asexual; others can also reproduce sexually or at least employ the sexual processes of meiosis and **syngamy** (the union of two gametes), thereby shuffling genes between two individuals that then go on to reproduce asexually. In Chapter 13, you learned about three basic types of sexual life cycles that differ in the timing of meiosis and syngamy (see FIGURE 13.5). All three types are represented among protists, along with some variations that do not quite fit any of the three basic life cycle patterns. However, the haploid stage is the main vegetative (feeding and growing) stage of most protists, with the only diploid cell being the zygote.

At some point in the life cycle of many protists, resistant cells called **cysts** are formed that can survive harsh conditions. Microfossils resembling the ruptured cysts of some algae are among the oldest remnants of eukaryotic life in the Precambrian era.

Protistan Habitats

Most protists are aquatic organisms. They are found almost anywhere there is water, including damp soil, leaf litter, and other terrestrial habitats that are sufficiently moist. In oceans, ponds, and lakes, many protists are bottom dwellers that attach themselves to rocks and other anchorages or creep through the sand and silt. Protists are also important constituents of **plankton** (from the Greek *planktos*, wandering), the communities of organisms, mostly microscopic, that drift passively or swim weakly near the water surface. **Phytoplankton** (planktonic eukaryotic algae, along with the prokaryotic cyanobacteria) are the bases of most marine and freshwater food webs. Accounting for at least half the photosynthetic production of organic material globally, phytoplankton supports an enormous abundance and diversity of heterotrophic protists, prokaryotes, and animals.

In addition to free-living protists are the many symbionts that inhabit the body fluids, tissues, or cells of hosts. These symbiotic relationships span the continuum from mutualism to parasitism. Some parasitic protists are important pathogens of animals, including many, like the agent of malaria, that cause potentially fatal diseases in humans.

In this introduction to the protists, we have previewed the enormous diversity of these organisms. The one characteristic they do share is that their cells are eukaryotic, as are those of plants, animals, and fungi. Before taking a closer look at the diversity of protists, let's consider the origin of the domain Eukarya from prokaryotic ancestors.

THE ORIGIN AND EARLY DIVERSIFICATION OF EUKARYOTES

Even a very small protist such as *Euglena*, shown in FIGURE 28.3, is far more complex in structure than any prokaryote. During the genesis of protists, the cellular structures and processes unique to eukaryotic cells arose: a membrane-enclosed nucleus, the endomembrane system, mitochondria, chloroplasts, the cytoskeleton, 9 + 2 flagella, multiple chromosomes consisting of linear DNA molecules compactly arranged with proteins, and life cycles that include mitosis, meiosis, and sex. Among the most fundamental questions in biology is how the complex eukaryotic cell evolved from much simpler prokaryotic cells.

Endomembranes contributed to larger, more complex cells

The small size and relatively simple construction of a prokaryote confer many advantages but also impose limits on the number of different metabolic activities that can be handled at one time. The relatively small size of the prokaryotic genome limits the number of genes coding for the enzymes that control these activities. This is not to say that prokaryotes are less successful than eukaryotes. Prokaryotes have been evolving and adapting since the dawn of life, and they are the most widespread organisms even today.

In at least some prokaryotic groups, increasing complexity —higher levels of organization with emergent properties— evolved. One trend was the evolution of multicellular prokaryotes, such as some filamentous cyanobacteria, where different cell types are specialized for different functions (see FIGURE 27.8). A second trend was the evolution of complex communities of prokaryotes, where each species benefited from the metabolic specialties of other species. A third trend was the compartmentalization of different functions within single cells, an evolutionary solution that contributed to the origin of eukaryotes.

How did compartmental organization of the eukaryotic cell evolve from the simpler prokaryotic condition? In one process, the endomembrane system of eukaryotic cells—the nuclear envelope, endoplasmic reticulum, Golgi apparatus, and related structures—may have evolved from specialized infoldings of the prokaryotic plasma membrane (FIGURE 28.4). Another process, called endosymbiosis, probably led to mitochondria, plastids, and perhaps some of the other features of eukaryotic cells.

Mitochondria and plastids evolved from endosymbiotic bacteria

THE PROCESS OF SCIENCE

The hypothesis we just referred to for the origin of the endomembrane system of eukaryotic cells fits the traditional Darwinian concept of gradual refinement of existing structures through natural selection. It seems like a reasonable guess that eukaryotes evolved from a single prokaryotic ancestor that gradually accumulated greater structural complexity. But the evidence is now overwhelming that the eukaryotic cell originated from a symbiotic coalition of multiple prokaryotic ancestors, not just one. We'll examine how this merger of organisms gave rise to two key eukaryotic organelles: mitochondria and **plastids** (the general term for the class of eukaryotic organelles that includes chloroplasts, as well as other types of plastids, both photosynthetic and nonphotosynthetic).

An idea originated by the early 20th century Russian biologist C. Mereschkovsky and developed extensively by Lynn Margulis, of the University of Massachusetts, the theory of **serial endosymbiosis** proposes that mitochondria and chloroplasts were formerly small prokaryotes living within larger cells. The term *endosymbiont* is used for the cell that lives within another cell, called the host cell. The proposed ancestors of mitochondria were aerobic heterotrophic prokaryotes that became endosymbionts. The proposed ancestors of chloroplasts were photosynthetic prokaryotes that became endosymbionts.

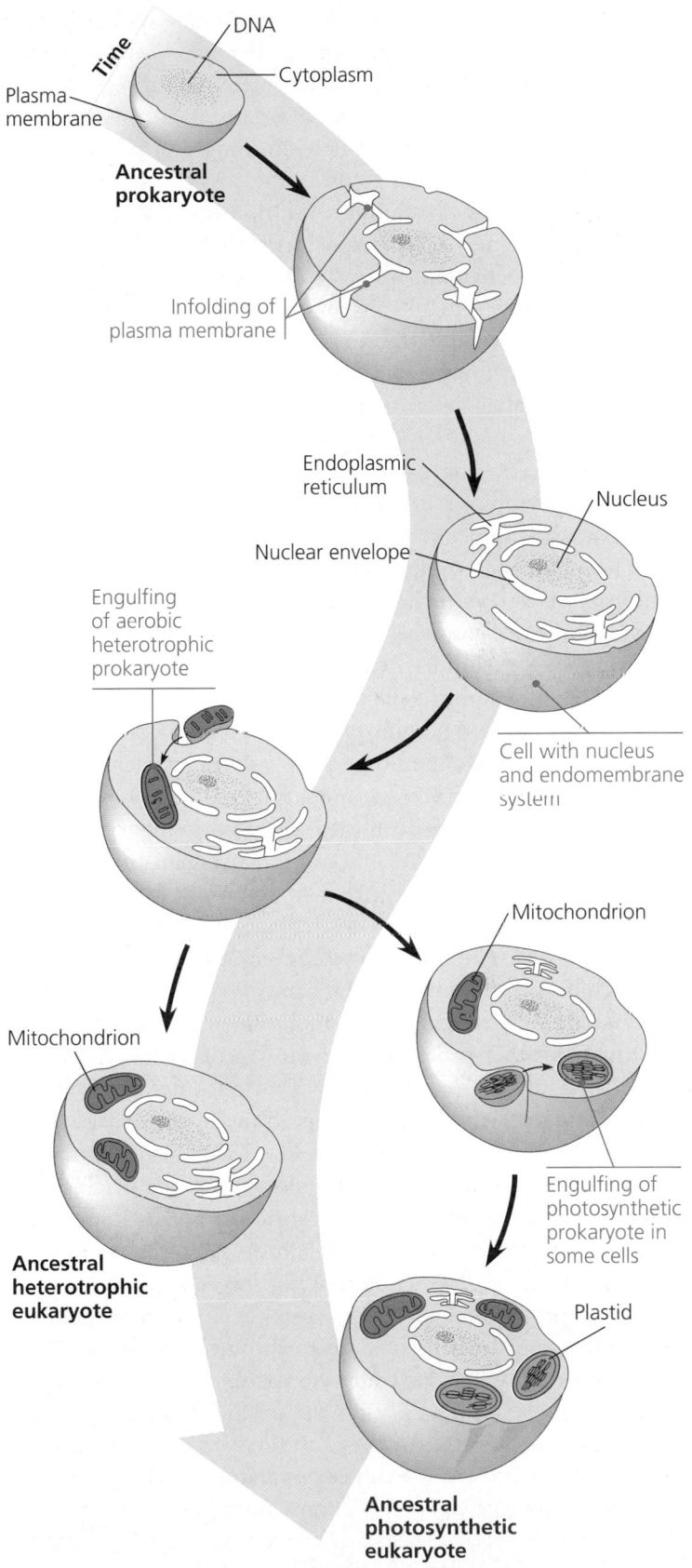

FIGURE 28.4 A model of the origin of eukaryotes.

The prokaryotic ancestors of mitochondria and chloroplasts probably gained entry to the host cell as undigested prey or internal parasites (see FIGURE 28.4). Such a mechanism suggests an earlier evolution of an endomembrane system and a cytoskeleton that made it possible for the larger cell to engulf the smaller prokaryotes and package them within vesicles. By whatever means the relationships began, it is not hard to imagine the symbiosis eventually becoming mutually beneficial. A heterotrophic host could use nutrients released from photosynthetic endosymbionts. And in a world that was becoming increasingly aerobic, a cell that was itself an anaerobe would have benefited from aerobic endosymbionts that turned the oxygen to advantage. In the process of becoming more interdependent, the host and endosymbionts would have become a single organism, its parts inseparable. All eukaryotes, whether heterotrophic or autotrophic, have mitochondria or genetic remnants of these organelles. However, not all eukaryotes have chloroplasts. Thus, the hypothesis of serial endosymbiosis (a sequence of endosymbiotic events) supposes that mitochondria evolved before chloroplasts.

The evidence supporting an endosymbiotic origin of chloroplasts and mitochondria includes the existence of endosymbiotic relationships in the modern world. Another line of evidence is the similarity between bacteria and the chloroplasts and mitochondria of eukaryotes. Both chloroplasts and mitochondria are the appropriate size to be descendants of bacteria. The inner membranes of chloroplasts and mitochondria, derived from the membranes of endosymbiotic prokaryotes, according to the endosymbiotic theory, do indeed have several enzymes and transport systems that resemble those found on the plasma membranes of modern prokaryotes. Mitochondria and chloroplasts replicate by a splitting process reminiscent of binary fission in bacteria. Each chloroplast and mitochondrian contains a genome consisting of a single circular DNA molecule not associated with histones or other proteins, as in most prokaryotes. The organelles contain the transfer RNAs, ribosomes, and other equipment needed to transcribe and translate their DNA into proteins. In terms of size, biochemical characteristics, and sensitivity to certain antibiotics, the ribosomes of chloroplasts are more similar to prokaryotic ribosomes than they are to the ribosomes outside the chloroplast in the cytosol of the eukaryotic cell. Mitochondrial ribosomes vary extensively from one group of eukaryotes to another, but they are generally more similar to prokaryotic ribosomes than to their counterparts in the eukaryotic cytoplasm.

A comprehensive theory for the origin of the eukaryotic cell must also account for the evolution of the cytoskeleton, including the 9 + 2 microtubule apparatus of eukaryotic flagella and cilia (see Chapter 7). Some researchers have speculated that eukaryotic flagella and cilia evolved from symbiotic bacteria (see discussion of spirochetes in TABLE 27.3), but the evidence for this is not strong.

Related to the evolution of the eukaryotic flagellum is the origin of mitosis and meiosis, processes unique to eukaryotes that also employ microtubules. Mitosis made it possible to reproduce the large genomes of the eukaryotic nucleus, and the closely related mechanics of meiosis became an essential process in eukaryotic sex. The question of cytoskeleton origin clearly demonstrates how much more researchers have to learn about the genesis of eukaryotic life.

The eukaryotic cell is a chimera of prokaryotic ancestors

The chimera of Greek mythology was a monster that was part goat, part lion, and part serpent. The eukaryotic cell is a chimera of prokaryotic parts, its mitochondria derived from one type of bacteria, its plastids from another, and its nuclear genome from at least one other prokaryote, the cell that hosted the endosymbiotic ancestors of the organelles. Once a consensus for the endosymbiotic theory developed from an accumulation of evidence, many researchers turned their attention to a search for the closest living prokaryotic relatives of the chimeric eukaryotic cell.

The Ancestors of Mitochondria and Chloroplasts

Systematists assessing relationships between organisms as different as modern prokaryotes and eukaryotes must compare molecules, for there are no morphological homologies connecting species so diverse (see Chapter 25). Molecular systematists working on the origin of eukaryotes first focused on the nucleotide sequence of the RNA in the smaller of the two ribosomal subunits. The gene for this small ribosomal subunit RNA (SSU-rRNA, for short) is present in all organisms, making it a good choice for studying the deepest branching in evolutionary history (see Chapter 27). By comparing sequences of SSU-rRNA from various prokaryotes and mitochondria, researchers have determined that the closest prokaryote relatives of mitochondria are the alpha proteobacteria (see TABLE 27.3). The sequences of SSU-rRNA from the plastids of various photosynthetic eukaryotes cluster with the prokaryotes called cyanobacteria. This corroborates other evidence of a relationship between cyanobacteria and plastids, namely that cyanobacteria are the only autotrophic prokaryotes with the water-splitting, oxygenic version of photosynthesis that characterizes chloroplasts and other plastids (see Chapter 27).

Of course, cyanobacteria and proteobacteria have continued to evolve as diverse prokaryotic groups since eukaryotic life began over 2 billion years ago. But somewhere, sometime, in that murky past, cyanobacteria share an ancestor common with the plastids of plants and eukaryotic algae. And proteobacteria share a common ancestor with the mitochondria of all eukaryotes; our own cells depend on the descendants of these bacteria for their ATP supply.

THE PROCESS OF SCIENCE

Gene Transfer to the Nucleus

While it is true that plastids and mitochondria contain DNA and the equipment for building proteins, they are not genetically self-sufficient. Some of the proteins within mitochondria and plastids are indeed encoded by the organelles' own DNA. However, the genes for other proteins in the organelles reside in the eukaryotic cell's nucleus. And still other proteins in the organelles are molecular chimeras of polypeptides made in the organelles and polypeptides imported from the cytoplasm, where they are translated from messenger RNA transcribed in the nucleus. An example is the mitochondrial ATP synthase, the protein complex that generates ATP during cellular respiration (see Chapter 9).

If mitochondria and plastids are descendants of bacteria that had their own genomes, then how do we explain the modern collaboration between the genomes of the organelles and nucleus? A reasonable hypothesis is that the endosymbionts transferred some of their DNA to the host cell's genome during the evolutionary transition from a symbiotic community of prokaryotes to an integrated eukaryotic organism. This hypothesis is consistent with the observation that the transfer of DNA between modern prokaryotic species is common—by transformation, for example, when a prokaryote takes up DNA from its surroundings and incorporates it into its own genome (see Chapter 18). It is fair to say that the eukaryotic cell, though chimeric in origin, now has only *one* genome, mostly nuclear, but complemented by DNA that has remained in the mitochondria and plastids.

Secondary endosymbiosis increased the diversity of algae

We mentioned in our overview of protists that taxonomic groups with plastids—plants and different kinds of algae—are scattered throughout the phylogenetic tree of eukaryotes. The diverse algae display various types of plastids that differ in their ultrastructure, as revealed by electron microscopes. For example, the chloroplasts of plants and a protistan group called green algae have envelopes consisting of two membranes (see FIGURE 7.18). In contrast, some algal groups have plastid envelopes with three or four membranes. For example, the plastids of *Euglena* (see FIGURE 28.3) have a three-membrane envelope. We also mentioned that some algae are closely related to heterotrophic species. For instance, *Euglena* belongs to a group of protists that also includes flagellated, heterotrophic forms lacking plastids.

How can we explain the diversity of plastids and the phylogenetic discontinuity of photosynthesis among protists? Support is building for the hypothesis that plastids were acquired independently several times during the early evolution of eukaryotes. The plastids of some algal groups, those with envelopes having more than two membranes, were acquired by

secondary endosymbiosis. It was by *primary* endosymbiosis that certain eukaryotes first acquired the ancestors of plastids by engulfing cyanobacteria. Secondary endosymbiosis occurred after a heterotrophic protist engulfed an alga containing plastids—one eukaryote taking in another eukaryote. In FIGURE 28.5 you see these two modes of plastid acquisition, primary and secondary. How does secondary endosymbiosis

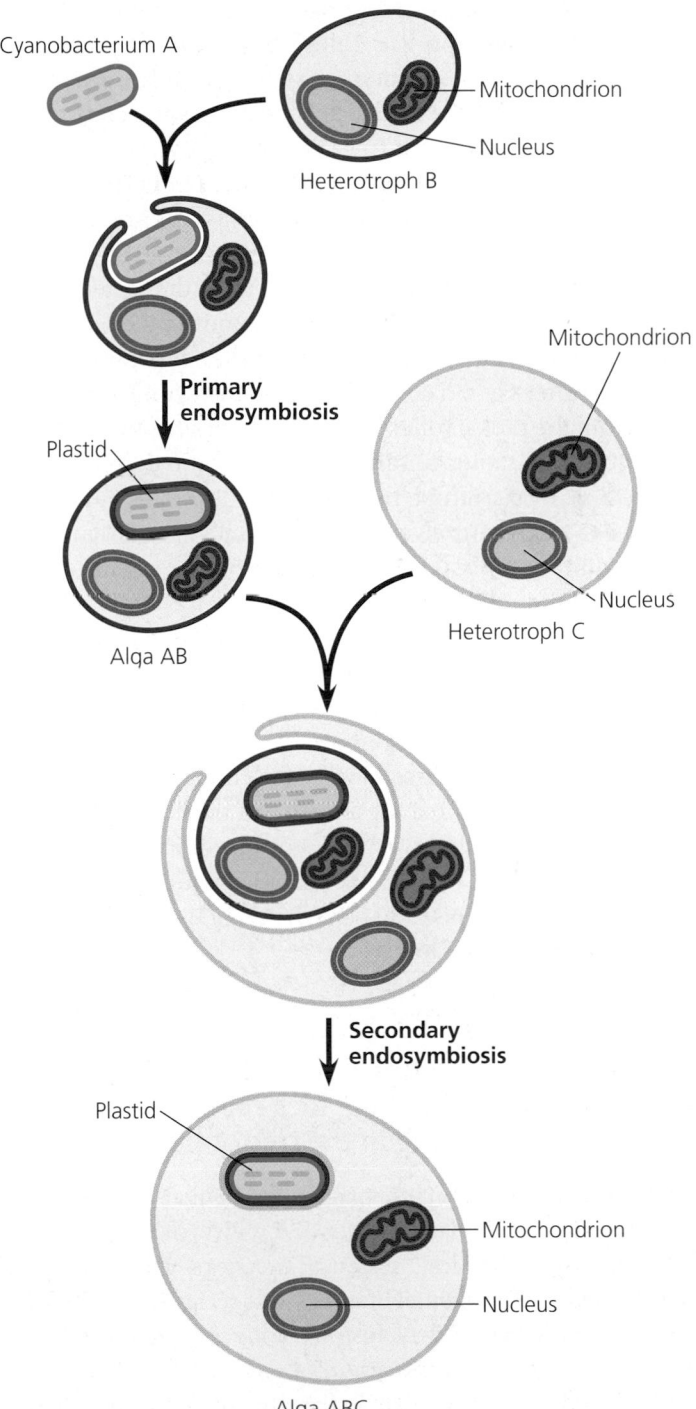

FIGURE 28.5 Secondary endosymbiosis and the origin of algal diversity. Algal species ABC in this diagram contains plastids derived from endosymbionts that were eukaryotic algae that in turn contained plastids derived from cyanobacteria.

explain the observation that some algae have plastids with more than two outer membranes? Each endosymbiotic event adds a membrane derived from the vacuole membrane of the host cell that engulfed the endosymbiont.

In most cases of secondary plastid acquisition, the endosymbiont gradually lost most of its parts, except, of course, its plastid, which became a plastid within the host cell. In some algae, however, the secondary endosymbionts retained other remnants of their independent past. Consider, for example, a group of flagellated protists called cryptomonad algae. The plastids of these algae contain a structure called a nucleomorph, a vestige of the nucleus of the eukaryotic ancestor of the plastid. There is also a trace of the endosymbiont's cytoplasm, complete with ribosomes. In fact, the cryptomonad plastid contains two distinct populations of ribosomes: ribosomes of the eukaryotic type derived from those of the algal endosymbiont's cytoplasm, and ribosomes of the bacterial type derived from a cyanobacterial endosymbiont of the secondary endosymbiont. Thus, a cryptomonad is like a box containing a box containing a box. The main nucleus and cytoplasm are derived from the prokaryotic host cell that engulfed the prokaryotic ancestor of mitochondria. The eukaryotic descendant of that cell was a heterotroph that later became photosynthetic by taking in a eukaryotic alga. But that secondary endosymbiont itself acquired its plastid earlier in its evolutionary history by taking in a cyanobacterium. Thus, we can trace the *ultimate* origin of all plastids to cyanobacteria. But some algae acquired that photosynthetic equipment secondhand (or even thirdhand, in the case of algae containing endosymbionts of endosymbionts of endosymbionts).

Through all of this discussion of the chimeric origin of eukaryotes, we have ignored the ancestry of the host cell that got it all started by engulfing the ancestor of mitochondria. We'll see in the next subsection that the search for that ancestor turned up some surprises that are changing the way biologists think about the tree of life.

Research on the relationships between the three domains is changing ideas about the deepest branching in the tree of life

THE PROCESS OF SCIENCE Note how the chimeric origin of the eukaryotic cell contrasts with the classical Darwinian view of lineal descent through a "vertical" series of ancestors. The eukaryotic cell evolved by "horizontal" fusions of species from different phylogenetic lineages. If the history of life is represented by an evolutionary tree, then the origin of eukaryotes would be depicted by cross-branches connecting the main branches of the tree. Trees, of course, aren't like that. Our metaphor of an evolutionary tree, still useful for thinking about diversification within certain lineages such as the plant and animal kingdoms, starts to break down when it comes to the origin of eukaryotes

and other very early evolutionary episodes. And now, researchers who are dissecting the chimeric makeup of the nuclear genome of eukaryotes may actually uproot the tree of life even more by changing our view of how eukaryotes are related to the two prokaryotic domains, Bacteria and Archaea. The key question is, What was the identity of the host cell that engulfed the ancestors of mitochondria during the origin of eukaryotes?

FIGURE 28.6 reviews the conventional model of how the three domains are related (also see FIGURE 27.2). This tree reflects molecular evidence that archaea are more closely related to eukaryotes than they are to prokaryotes. For example, archaea and eukaryotes have quite similar proteins functioning in the information-processing steps of transcription and translation (see Chapter 27). As such phylogenetic evidence accumulated, the case strengthened for the hypothesis that the host cell in the endosymbiotic origin of eukaryotes was derived from an early archaean.

The hypothesis diagrammed in FIGURE 28.6 predicts that the only DNA of bacterial origin we should find in the nucleus of eukaryotes are genes that were transferred there from the endosymbionts that gave rise to mitochondria and plastids. The rest of the genome should be uniquely eukaryotic or derived from archaeal DNA via the host cell of the endosymbiotic scenario. Thus, systematists were surprised to find that the

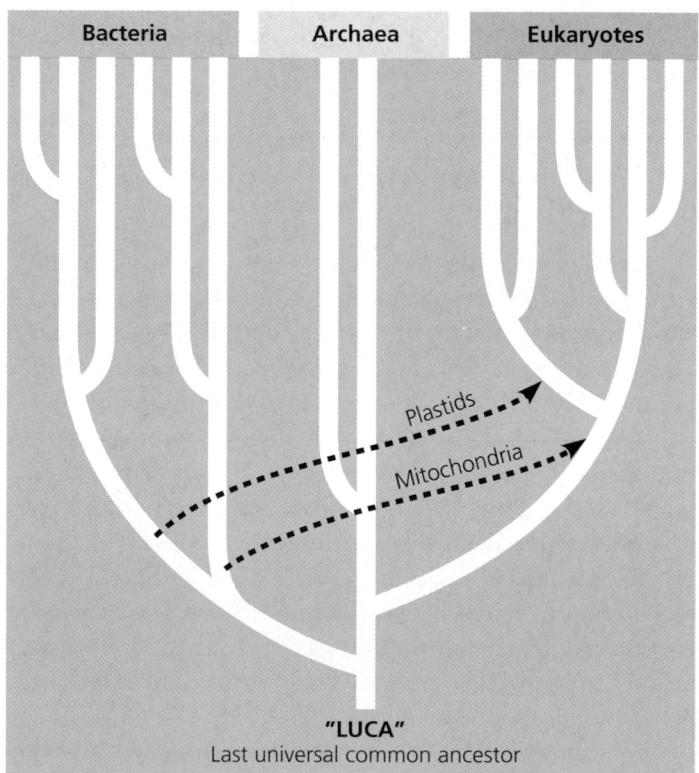

FIGURE 28.6 Traditional hypothesis for how the three domains of life are related. Tracing the domains back in time, the two prokaryotic branches converge on "LUCA," the last universal common ancestor. The hypothetical tree also depicts the origin of eukaryotes from an archaean-derived host cell with bacterial endosymbionts.

nuclear genome of eukaryotes includes many DNA sequences of bacterial origin that have nothing to do with the functions of mitochondria or plastids. The traditional diagram for the deepest branching of the tree of life also predicts that bacterial genes migrated only to the eukaryotic lineage, not to the branch giving rise to modern archaea. Another surprise: Modern archaea have many genes of bacterial origin. All three domains—Bacteria, Archaea, and Eukarya—seem to have genomes that are chimeric mixes of DNA that was transferred across the boundaries of the domains.

Many researchers who are turning up these surprising relationships among the domains' genomes suggest replacing the classical tree shown in FIGURE 28.6 with a weblike phylogeny like the one in FIGURE 28.7. The key departure from the traditional tree is the absence of a single ancestor common to the three domains (the "root" or "trunk" in the metaphor of an evolutionary tree). Instead, this new model has the three domains arising from an ancestral *community* of primitive cells

that swapped DNA promiscuously. This would explain the chimeric nature of genomes in modern representatives of all three domains. Such gene transfer across species lines still occurs among prokaryotes (see Chapter 18). This doesn't seem to be the case for modern eukaryotes. The chimeric makeup of the eukaryotic cell is a vestige of gene transfers and endosymbiotic events that occurred in prokaryotic communities over 2 billion years ago.

The origin of eukaryotes catalyzed a second great wave of diversification

An orchestra can play a greater variety of musical compositions than a violin soloist can. Put simply, increased complexity makes more variations possible. The origin of the eukaryotic cell catalyzed the evolution of much more structural diversity than was possible for the simpler prokaryotic cells. This built on the first great adaptive radiation, the metabolic diversification of the prokaryotes. A third wave of diversification followed the origin of multicellular bodies in several eukaryotic lineages (a topic we will consider near the end of this chapter).

FIGURE 28.8 (p. 554) is a hypothetical phylogeny for some of the major groups of eukaryotes. The diversity ranges from a great variety of unicellular forms to such macroscopic, multicellular groups as brown algae (a seaweed group), plants, fungi, and animals.

The clustering of eukaryotic forms into the clades of FIGURE 28.8 is based on comparisons of cell structure, life cycles, and molecules (nucleic acid and protein sequences). The molecular data include SSU-rRNA sequences and amino acid sequences for a few of the proteins of the cytoskeleton—proteins unique to eukaryotes. The protein comparisons have enabled systematists to resolve some classification issues that were not clear from SSU-rRNA data and cell structure.

FIGURE 28.8 brings up the "kingdoms problem" again. If plants, fungi, and animals are designated as kingdoms in this phylogenetic classification of eukaryotes, then each of the other major clades of eukaryotic life probably deserves kingdom status as well. This taxonomic trend was mentioned earlier when we discussed the splitting of the old kingdom Protista into many kingdoms. However, as protistan systematics is still so unsettled, FIGURE 28.8 simply names major clades without committing to a ranking of groups into kingdoms and phyla; any attempt here to specify the number of kingdoms and their names would just make the tree obsolete before this book's ink had time to dry! In fact, some of the best-known protists, such as the single-celled amoebas, are not even included in this tentative phylogeny because it is still so uncertain where they fit into the overall eukaryotic tree.

As tentative as our eukaryotic tree is, we'll use it to organize the survey of protistan diversity that follows.

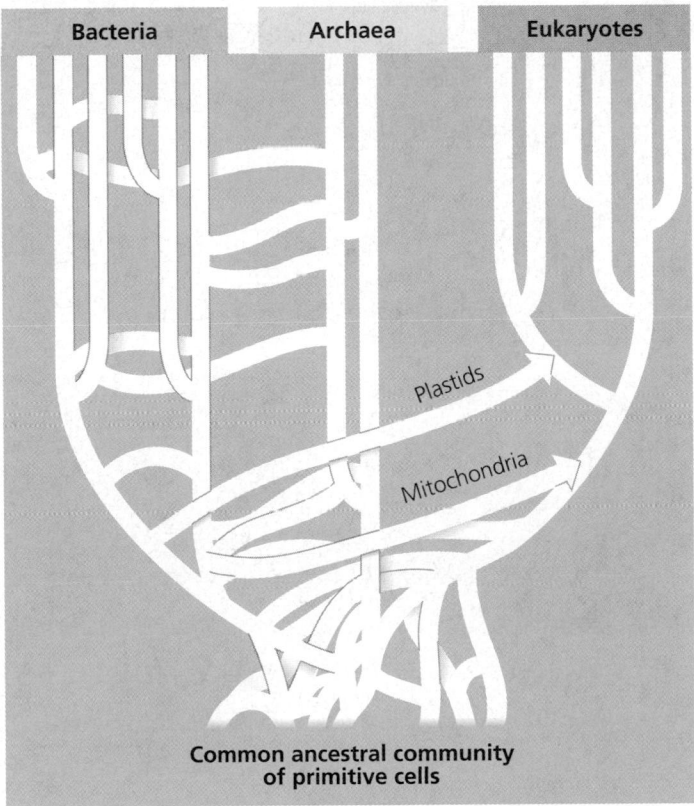

Bacteria   Archaea   Eukaryotes

Plastids

Mitochondria

**Common ancestral community
of primitive cells**

FIGURE 28.7 An alternative hypothesis for how the three domains of life are related. Based on evidence of extensive gene transfer between prokaryotic lineages today and throughout the past, this hypothesis replaces "LUCA" with an ancestral community of primitive cells that swapped genes. The diversity of organisms that evolved from that community consolidated into three main phylogenetic branches, the domains Bacteria, Archaea, and Eukarya. In the case of the endosymbiosis that gave rise to the mitochondria and chloroplasts, whole organisms of different evolutionary branches fused. However, most of the horizontal strands fusing with the vertical branches in this phylogeny symbolize the transfer of genes, not whole organisms, between different lineages.

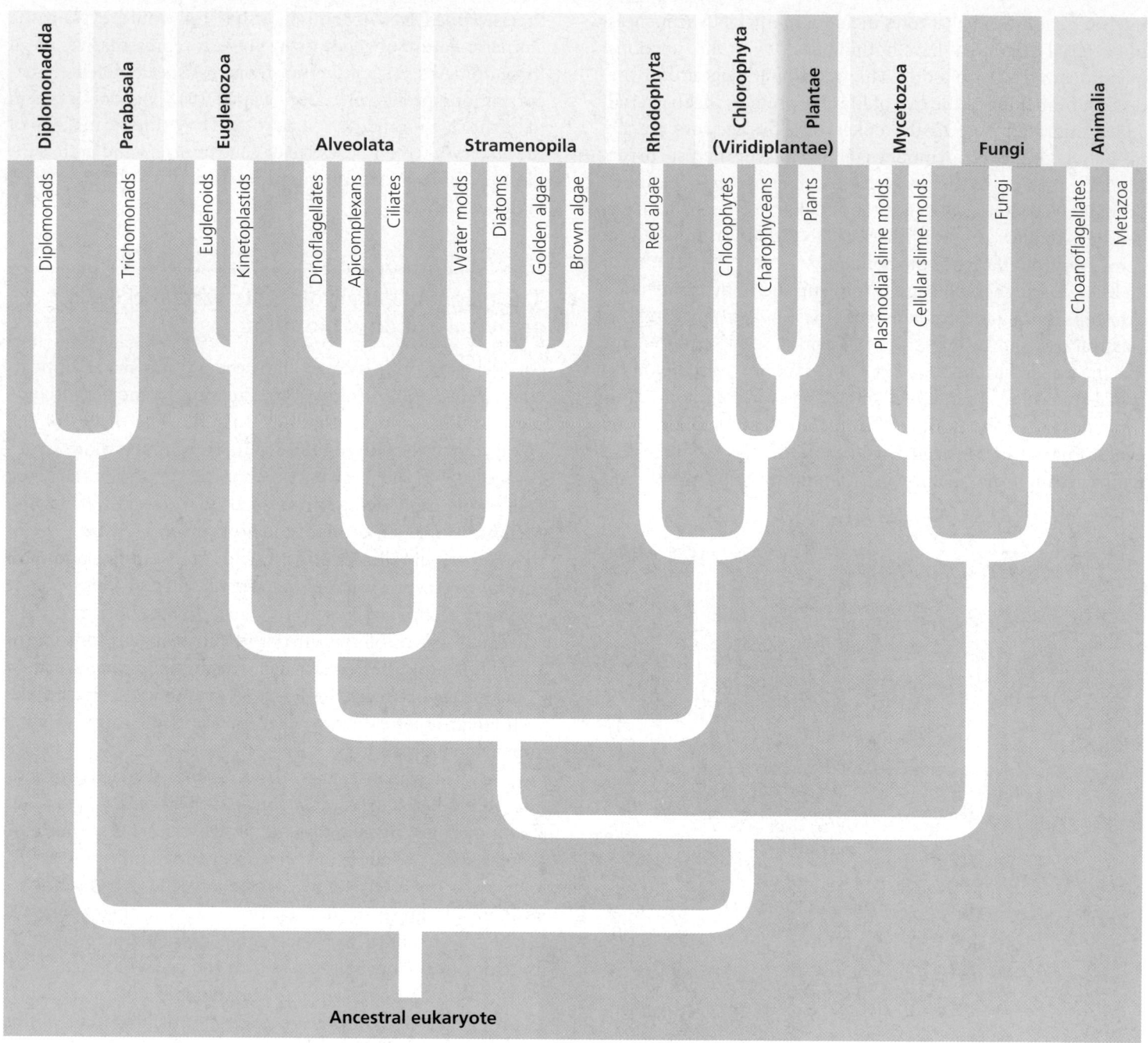

FIGURE 28.8 A tentative phylogeny of eukaryotes. The eukaryotes that are labeled on the branches are grouped into larger clades that are named in the boxes at the top of the tree. The kingdoms Fungi, Animalia, and Plantae have survived from the five-kingdom system of classification, although their boundaries have changed somewhat. Most of the major groups that used to be included in the kingdom Protista are color-coded yellow. Most protistan systematists now assign kingdom status to each of these groups. In fact, the trend is to split some of the more diverse groups, such as Alveolata and Stramenopila, into multiple kingdoms. Thus, we will make no attempt to predict the outcome of future research by ranking the protistan clades into kingdoms and phyla. As we survey the diversity of protists, each group will be introduced with a miniature version of this diagram so that you can place the group into the overall phylogeny of eukaryotes. However, our diagram omits several protistan groups that are not discussed in this chapter. And the tree even omits some protists that *are* discussed, such as the forams, because their phylogeny is still uncertain.

A SAMPLE OF PROTISTAN DIVERSITY

Diplomonadida and Parabasala: Diplomonads and parabasalids lack mitochondria

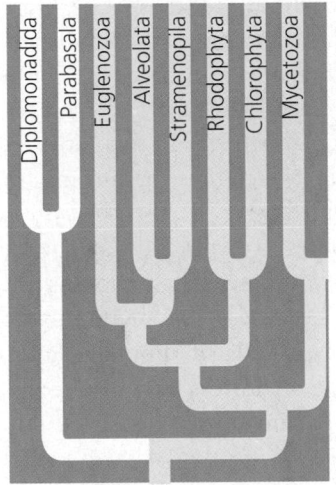

A few protistan groups lack mitochondria. Two of these groups are the diplomonads and the parabasalids. According to what is called the "archaezoa hypothesis," these protists are the modern forms of ancient eukaryotic lineages that evolved before the acquisition of the endosymbiotic bacteria that gave rise to mitochondria. Most protistan systematists have recently rejected that hypothesis based on the discovery of mitochondrial genes in the nuclear genomes of the diplomonads and parabasalids. In other words, the evidence now supports the hypothesis that these protists *lost* their mitochondria during their evolution from ancestors in which mitochondria were present. However, other details of cell structure combined with data from molecular systematics still place the diplomonads and parabasalids on the phylogenetic branch that diverged earliest in eukaryotic history (see FIGURE 28.8).

The **diplomonads** have multiple flagella, two separate nuclei, a relatively simple cytoskeleton (compared with other eukaryotes), no plastids, and, as already pointed out, no mitochondria. An infamous example is *Giardia lamblia,* a parasite that infects the human intestine, causing abdominal cramps and severe diarrhea (FIGURE 28.9). People most often pick up *Giardia* by drinking water contaminated with human feces

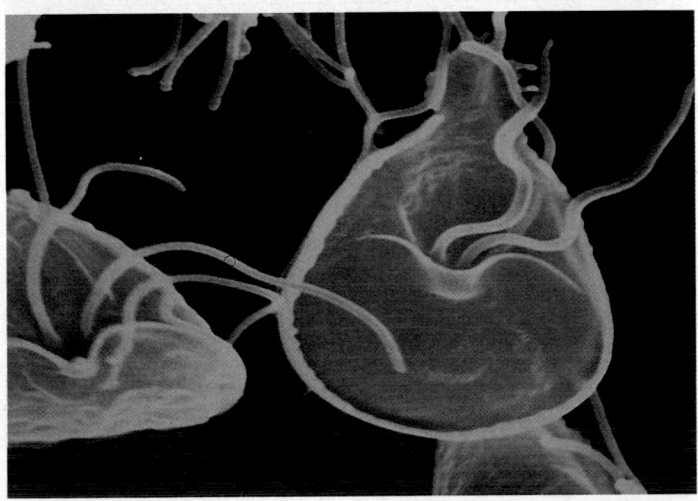

FIGURE 28.9 *Giardia lamblia,* a diplomonad (colorized SEM).

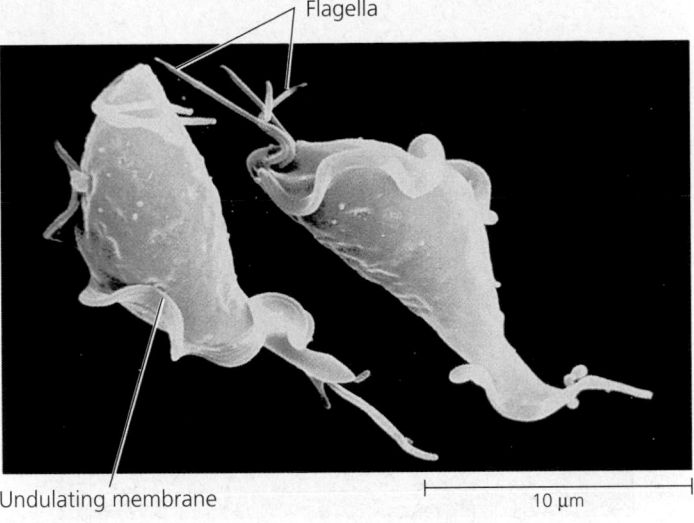

FIGURE 28.10 *Trichomonas vaginalis,* a parabasalid (colorized SEM).

containing the parasite in a dormant cyst stage. Drinking such water from what seems like a pristine stream or river can ruin a camping trip. Boiling drinking water kills the cysts.

The **parabasalids,** another group lacking mitochondria, include the protists called trichomonads. The best known species is *Trichomonas vaginalis,* a common inhabitant of the vagina of human females. It can proliferate and infect the vaginal lining by overcoming beneficial microbial populations if the normal acidity of the vagina is disturbed. Such infections also occur in the urethras of males, though often without symptoms. Sexual transmission can spread the infection. Note in FIGURE 28.10 that *T. vaginalis* has both flagella and an undulating membrane, structures that enable these protists to move along the mucus-coated skin within the reproductive and urinary tracts of their human hosts.

Euglenozoa: The euglenozoa includes both photosynthetic and heterotrophic flagellates

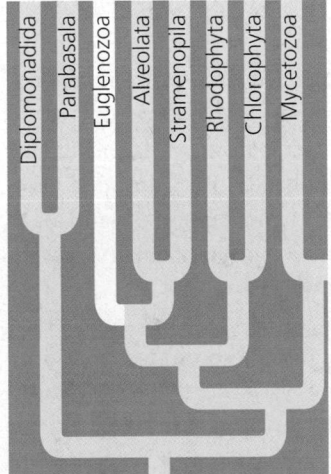

Several protistan groups are equipped with flagella that function in locomotion. Two of these groups of flagellates, the euglenoids and the kinetoplastids, make up the clade labeled Euglenozoa in FIGURE 28.8.

The **euglenoids** (Euglenophyta, such as *Euglena* and its close relatives) are characterized by an anterior pocket, or chamber, from which one or two flagella emerge. Paramylon, a glucose polymer that functions as a storage molecule, is also characteristic of euglenoids. *Euglena,* the organism in FIGURE 28.3,

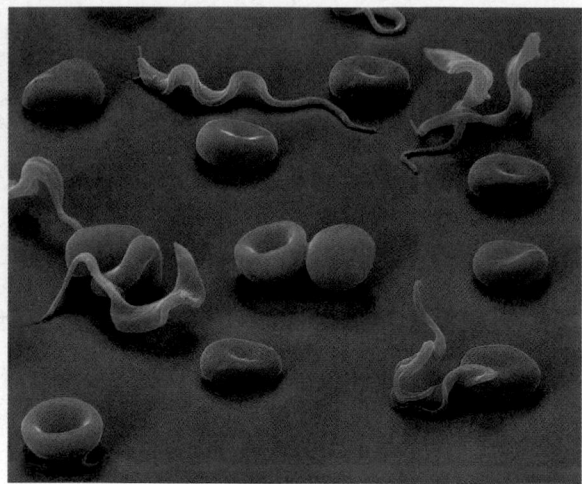

FIGURE 28.11 *Trypanosoma,* the kinetoplastid that causes sleeping sickness. The squiggles among these red blood cells are the trypanosomes (colorized SEM).

9 µm

is chiefly autotrophic. However, many euglenoids are mixotrophic or heterotrophic, absorbing organic molecules from their surroundings or engulfing prey by phagocytosis.

The **kinetoplastids** (Kinetoplastida) have a single large mitochondrion associated with a unique organelle, the kinetoplast, that houses extranuclear DNA. The kinetoplastids are symbiotic, and some are pathogenic to their hosts. For example, species of *Trypanosoma* cause African sleeping sickness, a human disease that is spread by the bite of the tsetse fly (FIGURE 28.11). Sleeping sickness is a debilitating disease common in parts of Africa. Trypanosomes escape being killed by the host's defenses by being quick-change artists. They alter the molecular structure of their coats frequently, thus preventing immunity from developing in the host.

Alveolata: The alveolates are unicellular protists with subsurface cavities (alveoli)

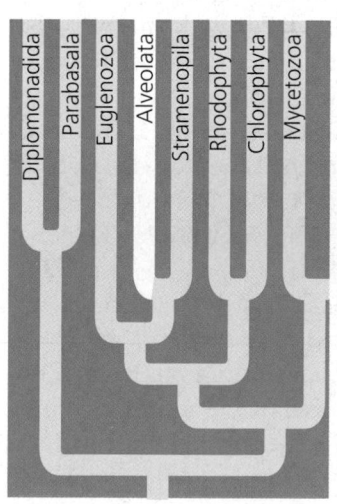

Another protistan clade that is emerging from molecular systematics, **Alveolata,** draws together a group of flagellates (the dinoflagellates), a group of parasites (apicomplexans), and a distinctive group of eukaryotes that move by means of cilia (the ciliates). Alveolates have small membrane-bounded cavities (alveoli) under their cell surfaces. The function of the alveoli is unknown; they may help stabilize the cell surface and regulate the cell's water and ion content.

Dinoflagellata (Dinoflagellates)

Dinoflagellates are abundant components of the vast aquatic pastures of phytoplankton that are suspended near the water surface and provide the foundation of most marine and many freshwater food webs. There are also heterotrophic species of dinoflagellates.

Of the several thousand known dinoflagellate species, most are unicellular, but some are colonial forms. Each dinoflagellate species has a characteristic shape reinforced in some species by internal plates of cellulose (FIGURE 28.12). The action of two flagella in perpendicular grooves in this "armor" produces a spinning movement for which these organisms are named (from the Greek *dinos,* whirling).

Dinoflagellate blooms—episodes of explosive population growth—cause red tides in coastal waters. These blooms are brownish red or pinkish orange because of the predominant pigments (xanthophylls) in the plastids of dinoflagellates. Toxins produced by some red-tide organisms have resulted in massive invertebrate and fish kills and can be deadly to humans as well.

One especially dangerous dinoflagellate called *Pfiesteria piscicida* is actually carnivorous. During blooms, its toxin stuns fish, and the dinoflagellates feed on their prey's body fluids. In the past decade, the frequency of *Pfiesteria* blooms and fish kills have increased in coastal waters of the U.S. mid-Atlantic states, possibly a result of pollution of the water with fertilizers (nitrates and phosphates).

Some dinoflagellates are spectacularly bioluminescent. An ATP-driven chemical reaction gives off light, creating an eerie ocean glow at night when waves, boats, or human swimmers agitate water containing dense populations of the dinoflagellates. A possible function of the bioluminescence: When the surrounding water is agitated by small predators that feed on phytoplankton, the light emitted by the dinoflagellates attracts

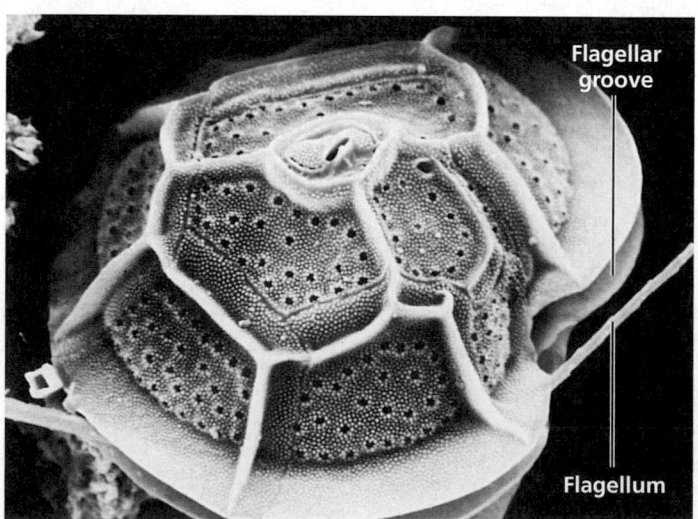

Flagellar groove

Flagellum

FIGURE 28.12 A dinoflagellate. Flagellar action spins this toplike form (SEM).

10 µm

fish that eat those smaller predators. How would you test this hypothesis?

Some dinoflagellates live as mutualistic symbionts of animals called cnidarians that build coral reefs; the photosynthetic output of these dinoflagellates is the main food source for reef communities.

Apicomplexa (Apicomplexans)

All **apicomplexans** are parasites of animals. Some cause serious human diseases. The parasites disseminate as tiny infectious cells called **sporozoites.** As seen with the electron microscope, one end (the *apex*) of the sporozoite cell contains a *complex* of organelles specialized for penetrating host cells and tissues; thus the name Apicomplexa. Most apicomplexans have intricate life cycles with both sexual and asexual stages, and these cycles often require two or more different host species for completion. An example is *Plasmodium*, the parasite that causes malaria (FIGURE 28.13). The incidence of malaria was greatly diminished in the 1960s by the use of insecticides that reduced populations of *Anopheles* mosquitoes, which spread the disease, and by drugs that killed the parasites

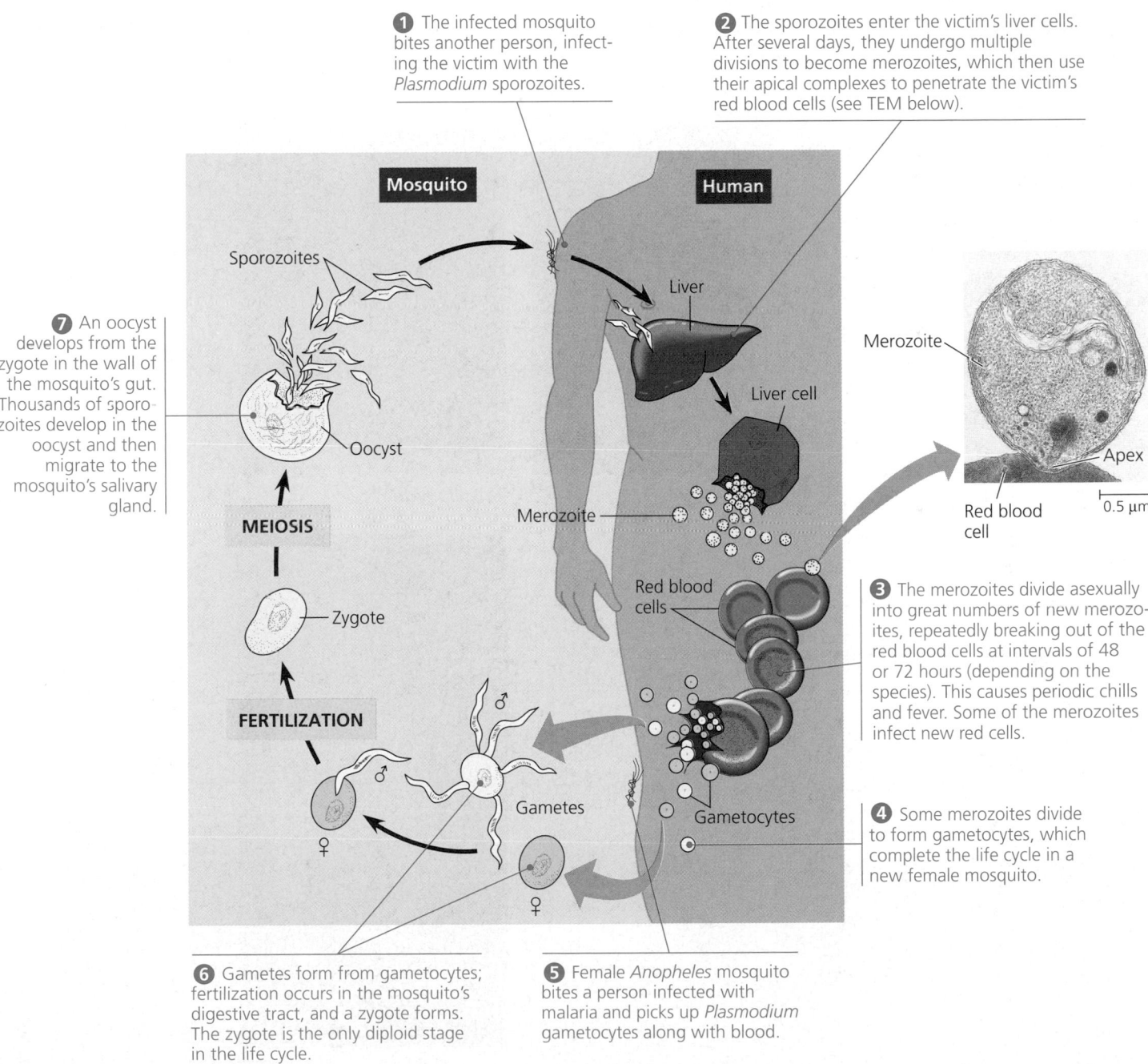

❶ The infected mosquito bites another person, infecting the victim with the *Plasmodium* sporozoites.

❷ The sporozoites enter the victim's liver cells. After several days, they undergo multiple divisions to become merozoites, which then use their apical complexes to penetrate the victim's red blood cells (see TEM below).

❼ An oocyst develops from the zygote in the wall of the mosquito's gut. Thousands of sporozoites develop in the oocyst and then migrate to the mosquito's salivary gland.

❸ The merozoites divide asexually into great numbers of new merozoites, repeatedly breaking out of the red blood cells at intervals of 48 or 72 hours (depending on the species). This causes periodic chills and fever. Some of the merozoites infect new red cells.

❹ Some merozoites divide to form gametocytes, which complete the life cycle in a new female mosquito.

❻ Gametes form from gametocytes; fertilization occurs in the mosquito's digestive tract, and a zygote forms. The zygote is the only diploid stage in the life cycle.

❺ Female *Anopheles* mosquito bites a person infected with malaria and picks up *Plasmodium* gametocytes along with blood.

FIGURE 28.13 The two-host life history of *Plasmodium*, the apicomplexan that causes malaria. (Colors are not true to life.)

in humans. However, the multiplication of resistant varieties of both the mosquitoes and the *Plasmodium* species have caused a resurgence of the disease. About 300 million people are now infected in the tropics, and up to 2 million die each year from the disease.

Considerable research on possible malarial vaccines has been carried out, but with little success. *Plasmodium* is an extremely evasive parasite. It spends most of its time inside human liver and blood cells, hiding from the host's immune system. The problem is compounded by changes in the surface proteins of *Plasmodium;* the parasite continually changes the "face" that it shows to the infected person's immune system. Recent identification of a gene that seems to confer resistance to chloroquine, an important antimalarial drug, may lead to ways of blocking drug resistance in *Plasmodium.* Another re-

cent discovery has exciting evolutionary and medical implications. A team of molecular biologists led by David Roos, at the University of Pennsylvania, found that *Plasmodium* and several other apicomplexans contain a plastid, now nonphotosynthetic, that an apicomplexan ancestor probably acquired from a green or red alga by secondary endosymbiosis. Once researchers discover the critical functions the plastids perform for their host cells, it may be possible to develop new antimalarial drugs that target those functions.

Ciliophora (Ciliates)

This diverse group of protists is named for their use of cilia to move and feed (FIGURE 28.14). Most **ciliates** live as solitary

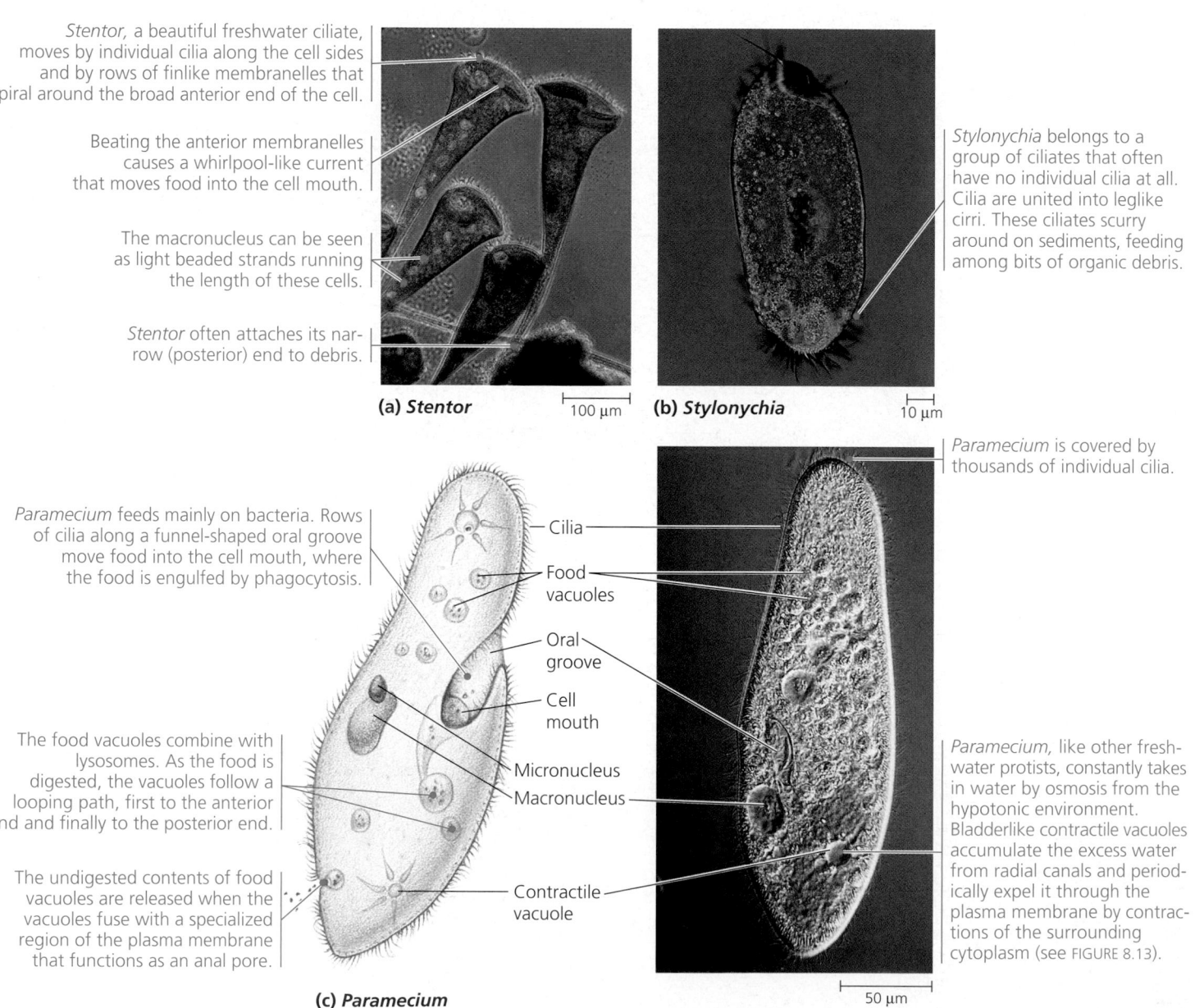

Stentor, a beautiful freshwater ciliate, moves by individual cilia along the cell sides and by rows of finlike membranelles that spiral around the broad anterior end of the cell.

Beating the anterior membranelles causes a whirlpool-like current that moves food into the cell mouth.

The macronucleus can be seen as light beaded strands running the length of these cells.

Stentor often attaches its narrow (posterior) end to debris.

(a) *Stentor* | 100 μm

Stylonychia belongs to a group of ciliates that often have no individual cilia at all. Cilia are united into leglike cirri. These ciliates scurry around on sediments, feeding among bits of organic debris.

(b) *Stylonychia* | 10 μm

Paramecium feeds mainly on bacteria. Rows of cilia along a funnel-shaped oral groove move food into the cell mouth, where the food is engulfed by phagocytosis.

The food vacuoles combine with lysosomes. As the food is digested, the vacuoles follow a looping path, first to the anterior end and finally to the posterior end.

The undigested contents of food vacuoles are released when the vacuoles fuse with a specialized region of the plasma membrane that functions as an anal pore.

Cilia
Food vacuoles
Oral groove
Cell mouth
Micronucleus
Macronucleus
Contractile vacuole

(c) *Paramecium*

Paramecium is covered by thousands of individual cilia.

Paramecium, like other freshwater protists, constantly takes in water by osmosis from the hypotonic environment. Bladderlike contractile vacuoles accumulate the excess water from radial canals and periodically expel it through the plasma membrane by contractions of the surrounding cytoplasm (see FIGURE 8.13).

50 μm

FIGURE 28.14 Ciliates (all LMs).

cells in fresh water. In contrast to most flagella, cilia are relatively short. They are associated with a submembrane system of microtubules that may coordinate the movement of the thousands of cilia.

Some ciliates are completely covered by rows of cilia, whereas others have their cilia clustered into fewer rows or tufts. The specific arrangements adapt the ciliates for their diverse lifestyles. Some species, for instance, scurry about on leglike structures constructed from many cilia bonded together. Other forms, such as *Stentor,* have rows of tightly packed cilia that function collectively as locomotor membranelles. Ciliates are among the most complex of all cells.

A unique feature of ciliate genetics is the presence of two types of nuclei, a large macronucleus and usually several tiny micronuclei. The macronucleus has 50 or more copies of the genome. The genes are not distributed in typical chromosomes but are instead packaged into a much larger number of small units, each with hundreds of copies of just a few genes. The macronucleus controls the everyday functions of the cell by synthesizing RNA and is also necessary for asexual reproduction. Ciliates generally reproduce by binary fission, during which the macronucleus elongates and splits, rather than undergoing mitotic division. Some species of *Paramecium* have from 1 to as many as 80 micronuclei, which do not function in growth, maintenance, and asexual reproduction of the cell but are required for sexual processes that generate genetic variation. The sexual shuffling of genes occurs during the process known as **conjugation** (FIGURE 28.15). Notice in the diagram that in ciliates, sexual mechanisms of meiosis and syngamy are processes separate from reproduction.

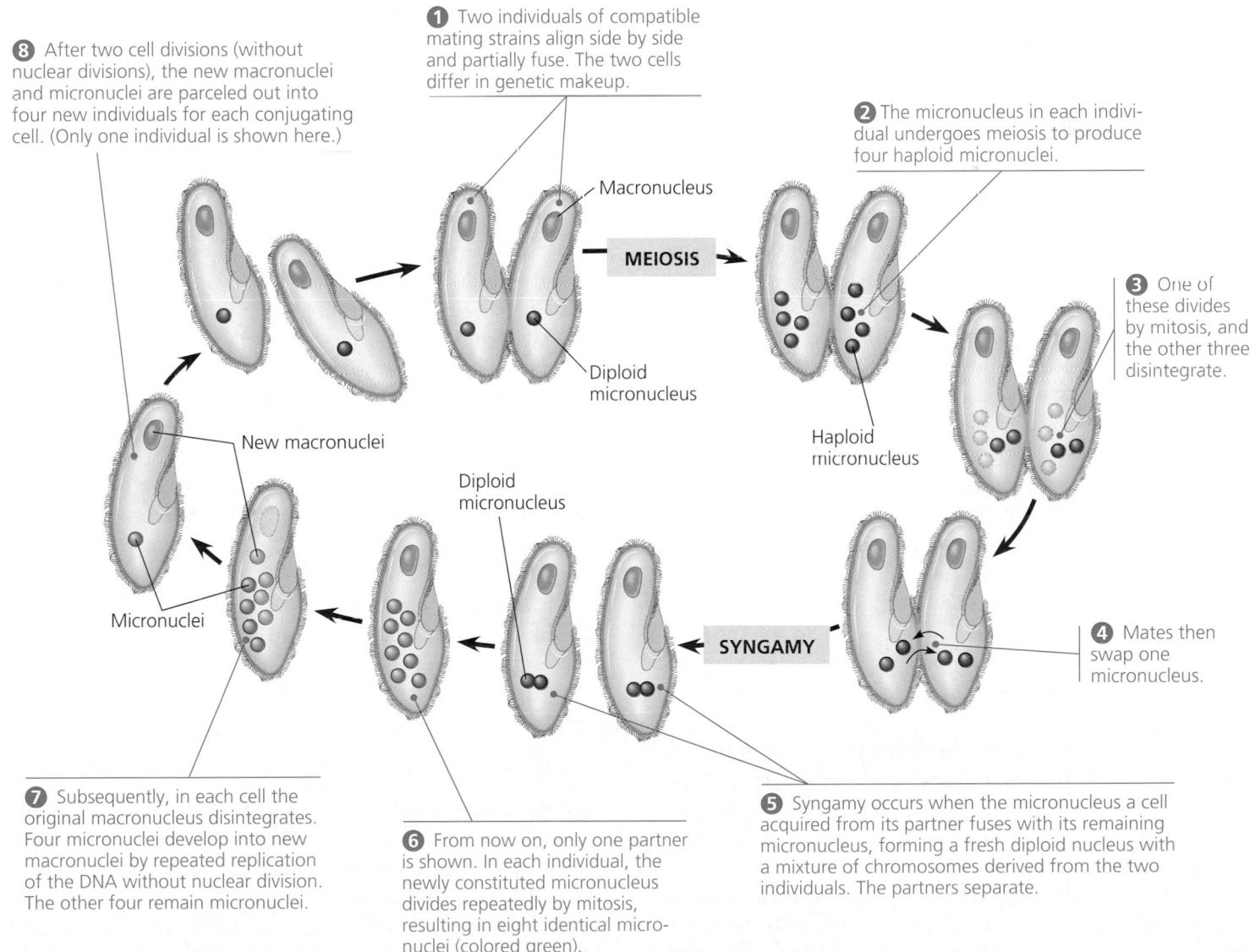

8 After two cell divisions (without nuclear divisions), the new macronuclei and micronuclei are parceled out into four new individuals for each conjugating cell. (Only one individual is shown here.)

1 Two individuals of compatible mating strains align side by side and partially fuse. The two cells differ in genetic makeup.

Macronucleus

2 The micronucleus in each individual undergoes meiosis to produce four haploid micronuclei.

MEIOSIS

3 One of these divides by mitosis, and the other three disintegrate.

Diploid micronucleus

Haploid micronucleus

New macronuclei

Diploid micronucleus

Micronuclei

SYNGAMY

4 Mates then swap one micronucleus.

7 Subsequently, in each cell the original macronucleus disintegrates. Four micronuclei develop into new macronuclei by repeated replication of the DNA without nuclear division. The other four remain micronuclei.

6 From now on, only one partner is shown. In each individual, the newly constituted micronucleus divides repeatedly by mitosis, resulting in eight identical micronuclei (colored green).

5 Syngamy occurs when the micronucleus a cell acquired from its partner fuses with its remaining micronucleus, forming a fresh diploid nucleus with a mixture of chromosomes derived from the two individuals. The partners separate.

FIGURE 28.15 Conjugation and genetic recombination in *Paramecium caudatum.*

Stramenopila: The stramenopile clade includes the water molds and the heterokont algae

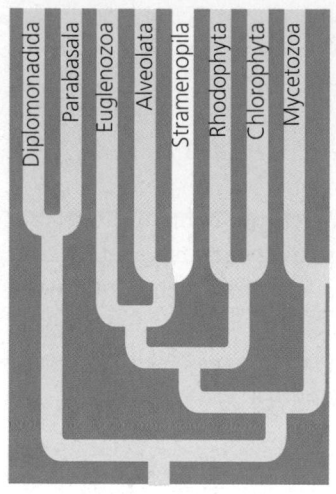

A diverse protistan clade, **Stramenopila** includes several heterotrophic groups as well as a variety of photosynthetic protists (algae). The term *stramenopila* refers to the numerous fine, hairlike projections on the flagella that are characteristic of these organisms (from the Latin *stramen,* straw, flagellum, and *pilos,* hair). In most cases, the "hairy" flagellum is paired with a smooth (nonhairy) flagellum. In the life cycles of most stramenopile groups, the only flagellated stages are motile reproductive cells.

Oomycota (Water Molds and Their Relatives)

Oomycotes include the **water molds, white rusts,** and **downy mildews,** all examples of heterotrophic stramenopiles that lack chloroplasts. Some of these organisms are unicellular; others consist of hyphae (fine, branching filaments) that are multi-nucleated (coencytic). Hyphae in this group are analogous to the branching filaments (also called hyphae) of true fungi. However, water molds and their relatives typically have cell walls made of cellulose, while the walls of true fungi are made of another polysaccharide, chitin. The diploid condition, which is reduced in true fungi, prevails in the life cycles of most members of Oomycota. Biflagellated cells occur in the life cycles of oomycotes, while almost all true fungi lack flagella. Molecular systematics has confirmed that water molds and other oomycotes are not closely related to fungi. The similar "body plan"—a mass of thin filaments (hyphae)—is an example of convergent evolution. In both the oomycotes and fungi, the filamentous bodies have an extensive surface area, which enhances the absorption of nutrients.

Oomycota means "egg fungi," a reference to the mode of sexual reproduction in water molds. A relatively large egg cell is fertilized by a smaller "sperm nucleus," forming a resistant zygote (FIGURE 28.16).

Most water molds are decomposers that grow as cottony masses on dead algae and animals, mainly in fresh water. There are also parasitic water molds, such as those that grow on the skin and gills of fish in ponds or aquariums, but they usually attack only injured tissue.

White rusts and downy mildews generally live on land as parasites of plants. They are dispersed primarily by wind-blown spores, but they also form flagellated zoospores at some point during their life cycles. One devastating plant pathogen, a downy mildew, threatened the French vineyards in the 1870s. Another oomycote species, *Phytophthora infestans,* causes late potato blight, which contributed to the Irish famine in the 19th century.

Overview of the Heterokont Algae

The stramenopile taxa with mostly photosynthetic members are collectively called the heterokont algae, with the "hetero" referring to the two different types of flagella ("hairy" and "nonhairy"), which are characteristic of the whole stramenopile group.

The plastids of heterokont algae evolved by secondary endosymbiosis. This explains the three-membrane envelope and the presence of a small amount of eukaryotic cytoplasm within the plastid (see FIGURE 28.5). The probable ancestor of the plastids in heterokont algae was a red alga, its own plastid derived from a cyanobacterium by primary endosymbiosis. The heterokont algae include the diatoms, golden algae, and brown algae.

Bacillariophyta (Diatoms)

Yellow or brown in color, **diatoms** have unique glasslike walls consisting of hydrated silica embedded in an organic matrix. Each wall is in two parts that overlap like a shoe box and lid (FIGURE 28.17).

Most of the year, diatoms reproduce asexually by mitotic cell divisions, with each daughter cell receiving half of the cell wall of its parent and regenerating a new second half. Cysts are formed by some species as resistant stages. Sexual stages are not common and involve the formation of eggs and sperm; sperm cells are amoeboid or flagellated, depending on the species.

Both freshwater and marine plankton are rich in diatoms; a bucket of water scooped from the surface of the sea may have millions of these microscopic algae. In common with golden algae and brown algae, diatoms store food reserves in the form of a glucose polymer called laminarin. Some diatoms also store food in the form of oil.

Massive accumulations of fossilized diatom walls are major constituents of the sediments known as diatomaceous earth, which is mined for its quality as a filtering medium and for many other uses.

Chrysophyta (Golden Algae)

Golden algae, or chrysophytes (from the Greek *chrysos,* golden), are named for their color, which results from yellow and brown carotene and xanthophyll accessory pigments. Their cells are typically biflagellated, with both flagella attached near one end of the cell. Many golden algae live among freshwater and marine plankton. Some species are mixotrophic, absorbing dissolved organic compounds or ingesting food particles and

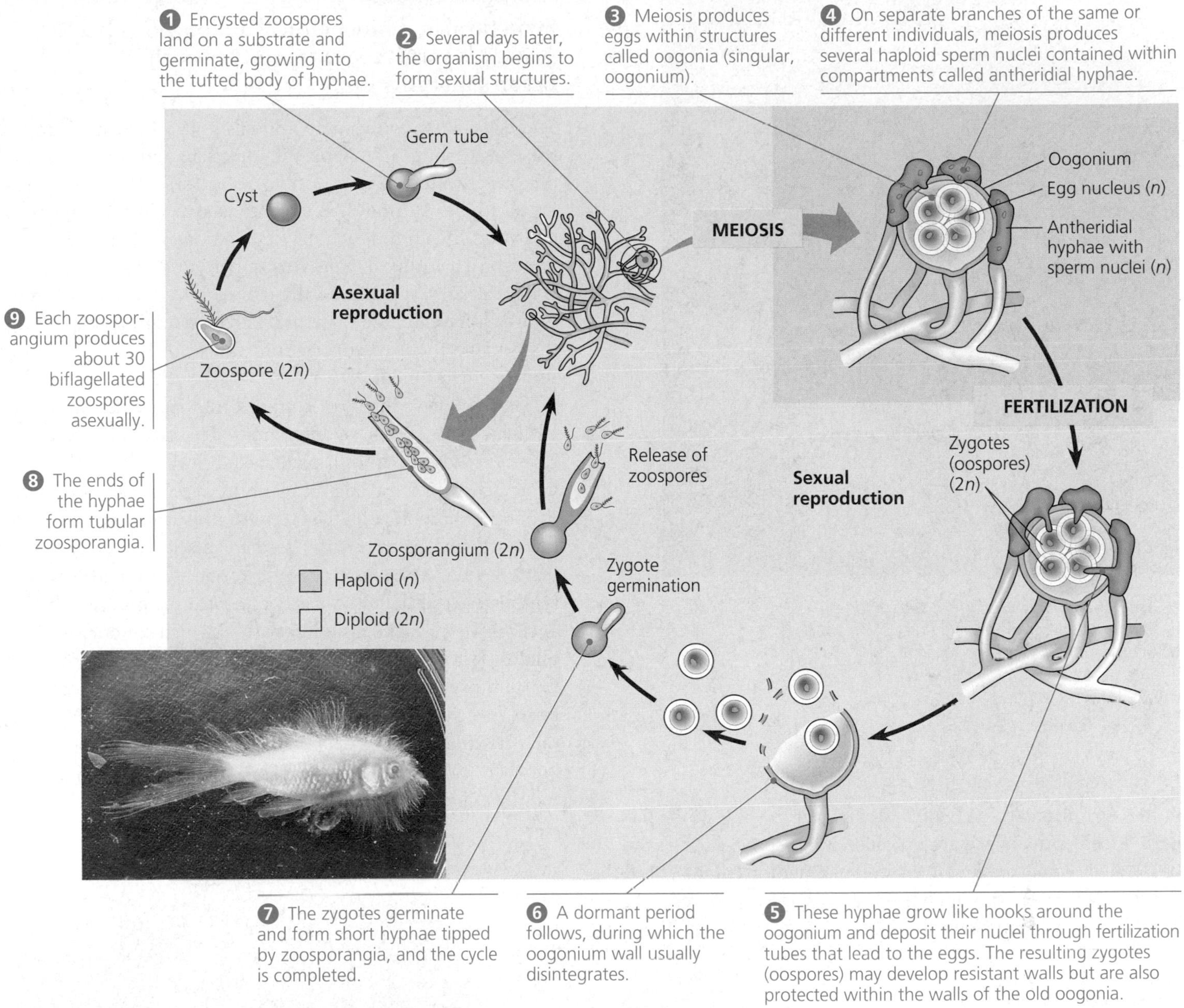

1 Encysted zoospores land on a substrate and germinate, growing into the tufted body of hyphae.

2 Several days later, the organism begins to form sexual structures.

3 Meiosis produces eggs within structures called oogonia (singular, oogonium).

4 On separate branches of the same or different individuals, meiosis produces several haploid sperm nuclei contained within compartments called antheridial hyphae.

Germ tube

Cyst

Oogonium
Egg nucleus (n)
Antheridial hyphae with sperm nuclei (n)

MEIOSIS

Asexual reproduction

9 Each zoosporangium produces about 30 biflagellated zoospores asexually.

Zoospore ($2n$)

FERTILIZATION

Zygotes (oospores) ($2n$)

8 The ends of the hyphae form tubular zoosporangia.

Release of zoospores

Sexual reproduction

Zoosporangium ($2n$)

Zygote germination

☐ Haploid (n)

☐ Diploid ($2n$)

7 The zygotes germinate and form short hyphae tipped by zoosporangia, and the cycle is completed.

6 A dormant period follows, during which the oogonium wall usually disintegrates.

5 These hyphae grow like hooks around the oogonium and deposit their nuclei through fertilization tubes that lead to the eggs. The resulting zygotes (oospores) may develop resistant walls but are also protected within the walls of the old oogonia.

FIGURE 28.16 The life cycle of a water mold. Water molds help decompose dead insects, fish, and other animals submerged in fresh water (see the hyphal mass on the goldfish in the inset).

FIGURE 28.17 Diatoms. The glasslike shells consist of two halves that fit like the bottom and lid of a shoe box. Tiny pores in the ornate shells allow for the exchange of gases and other substances between the cell and its environment. During mitosis, each daughter cell keeps half of the parent cell's shell and builds a new complementary half.

(a) Diatom diversity (LM)

50 μm

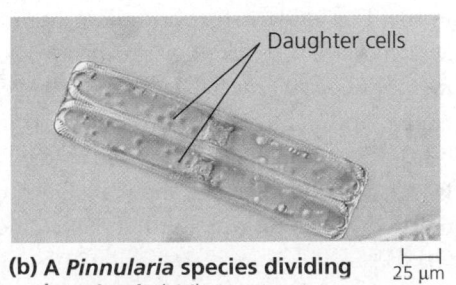

Daughter cells

(b) A *Pinnularia* species dividing by mitosis (LM)

25 μm

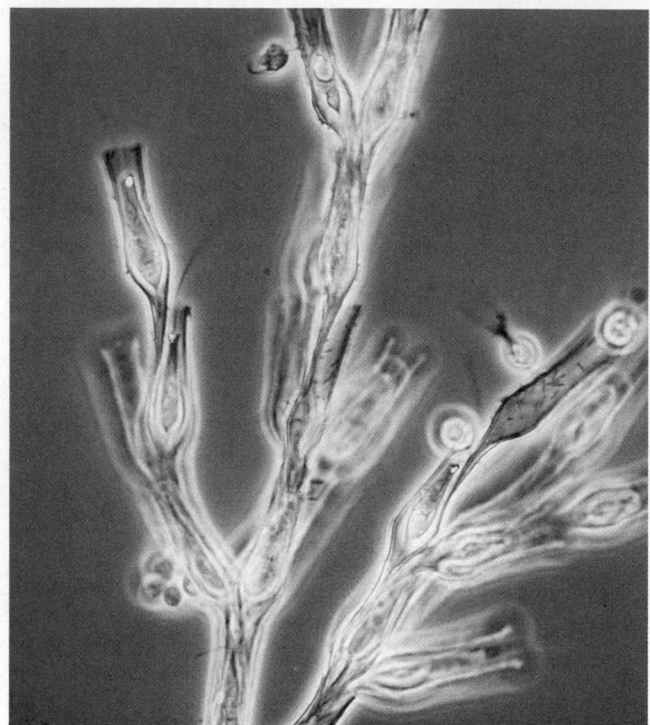

FIGURE 28.18 A golden alga. *Dinobryon* is a freshwater member of this algal group. This species is colonial, but most golden algae are unicellular (LM).

`25 μm`

bacteria by phagocytosis, which occurs near the base of the flagella. Most golden algae are unicellular, but some, such as the freshwater genus *Dinobryon,* are colonial (FIGURE 28.18). If cell density reaches a certain high level, many species form resistant cysts that can remain viable for decades.

Phaeophyta (Brown Algae)

The largest and most complex algae are brown algae, or phaeophytes (from the Greek *phaios,* dusky, brown). All are multicellular, and most are marine. Brown algae are especially common along temperate coasts, where the water is cool. They owe their characteristic brown or olive color to accessory pigments in the plastids. The plastid structure and pigment composition of brown algae are homologous to the photosynthetic equipment of golden algae and diatoms.

Many of the eukaryotes commonly called seaweeds are brown algae. *Seaweeds* are large marine algae, and the largest members of this important ecological group are brown algae. Two other groups, the red algae and green algae, also include seaweeds. We will examine these algal groups after we look at some of the adaptations, human uses, and life cycles of the seaweeds.

Structural and biochemical adaptations help seaweeds survive and reproduce at the ocean's margins

Seaweeds, along with many animals and other heterotrophs these algae support, inhabit the intertidal and subtidal zones of coastal waters. The intertidal zone presents unique challenges to life. At times it is violently active, churned by waves and wind. Two times each day, at low tide, intertidal seaweeds are exposed to the drying atmosphere and rays of the sun. Twice each day, at high tide, the same seaweeds are submerged. Seaweeds have unique structural and biochemical adaptations that enable them to survive and thrive in this rough-and-tumble environment.

Seaweeds have the most complex multicellular anatomy of all algae. Some even have differentiated tissues and organs that resemble those we find in plants. The similarities evolved independently in the algal and plant lineages and are thus analogous, not homologous. The term **thallus** (plural, *thalli;* from the Greek *thallos,* sprout) is used for a seaweed body that is plantlike, but a thallus lacks true roots, stems, and leaves. A typical seaweed thallus consists of a rootlike **holdfast,** which anchors the alga, and a stemlike **stipe,** which supports leaflike **blades** (FIGURE 28.19). The blades provide most of the surface for photosynthesis. Some brown algae are equipped with floats, which keep the blades near the water surface. Beyond the intertidal zone in deeper waters live the giant seaweeds known as kelps (FIGURE 28.20). The stipes of these brown algae may be as long as 60 m.

FIGURE 28.19 Seaweeds: adapted to life at the ocean's margins. The sea palm, *Postelsia,* lives on rocks in the crashing surf along the coast of the northwestern United States and Canada. Its thallus is well adapted to maintaining a firm foothold in this extreme environment. *Postelsia* is a brown alga (Phaeophyta).

FIGURE 28.20 A kelp forest. The great kelp beds of temperate coastal waters provide habitat and food for a variety of organisms, including many species of fish caught by humans. The kelps, brown algae (Phaeophyta), are prodigiously productive. This brown alga, *Macrocystis,* common along the U.S. Pacific coast, grows to a length of more than 60 m in a single season, the fastest linear growth of any organism. Kelp is a renewable resource reaped by special boats that cut and collect the tops of the algae.

In addition to these structural adaptations of thalli, some seaweeds are also endowed with biochemical adaptations to intertidal and subtidal conditions. For example, their cell walls are composed of cellulose and gel-forming polysaccharides, accounting for the slimy and rubbery feel of many seaweeds. These substances help cushion the thalli against the agitation of the waves.

Seaweeds are an important source of food and other commodities for humans. Coastal people, particularly in Asia, harvest seaweeds for food. For example, in Japan and Korea, the brown alga *Laminaria* is used in soups (Japanese "kombu"), and the red alga *Porphyra* is used to wrap sushi (Japanese "nori"). Marine algae are rich in iodine and other essential minerals, but much of their organic material consists of unusual polysaccharides that humans cannot digest, which prevents seaweeds from becoming staple foods. They are used mostly for their rich tastes and unusual textures. The gel-forming substances in their cell walls (algin in brown algae, agar and carageenan in red algae) are extracted in commercial operations. These substances are widely used in the manufacture of thickeners for such processed foods as puddings and salad dressing and as lubricants in oil drilling, and agar is used as the gel-forming base for microbiological culture media.

Some algae have life cycles with alternating multicellular haploid and diploid generations

A variety of life cycles have evolved among the multicellular brown, red, and green algae. The most complex cycles include an **alternation of generations,** the alternation of multicellular haploid forms and multicellular diploid forms. (Notice that haploid and diploid conditions alternate in *all* sexual life cycles—human gametes, for example, are a haploid stage— but the term *alternation of generations* is reserved for life cy-

cles that include haploid and diploid stages that are both multicellular organisms.) As we will see in Chapter 29, alternation of generations also evolved in the life cycles of all plants.

We can examine the brown alga *Laminaria* as an example of a complex life cycle with an alternation of generations. The diploid individual is called the **sporophyte** because it produces reproductive cells called spores (zoospores). The haploid individual is called the **gametophyte,** which is named for its production of gametes. Notice in FIGURE 28.21 (p. 564) that these two generations alternate—that is, they take turns producing one another. Spores released from the sporophyte develop into gametophytes, which in turn produce gametes. The union of two gametes (fertilization, or syngamy) results in a diploid zygote, which gives rise to a new sporophyte. In the case of *Laminaria,* the two generations are **heteromorphic,** meaning that the sporophytes and gametophytes are structurally different. Other algal life cycles have an alternation of **isomorphic** generations, meaning that the sporophytes and gametophytes look alike, although they differ in chromosome number.

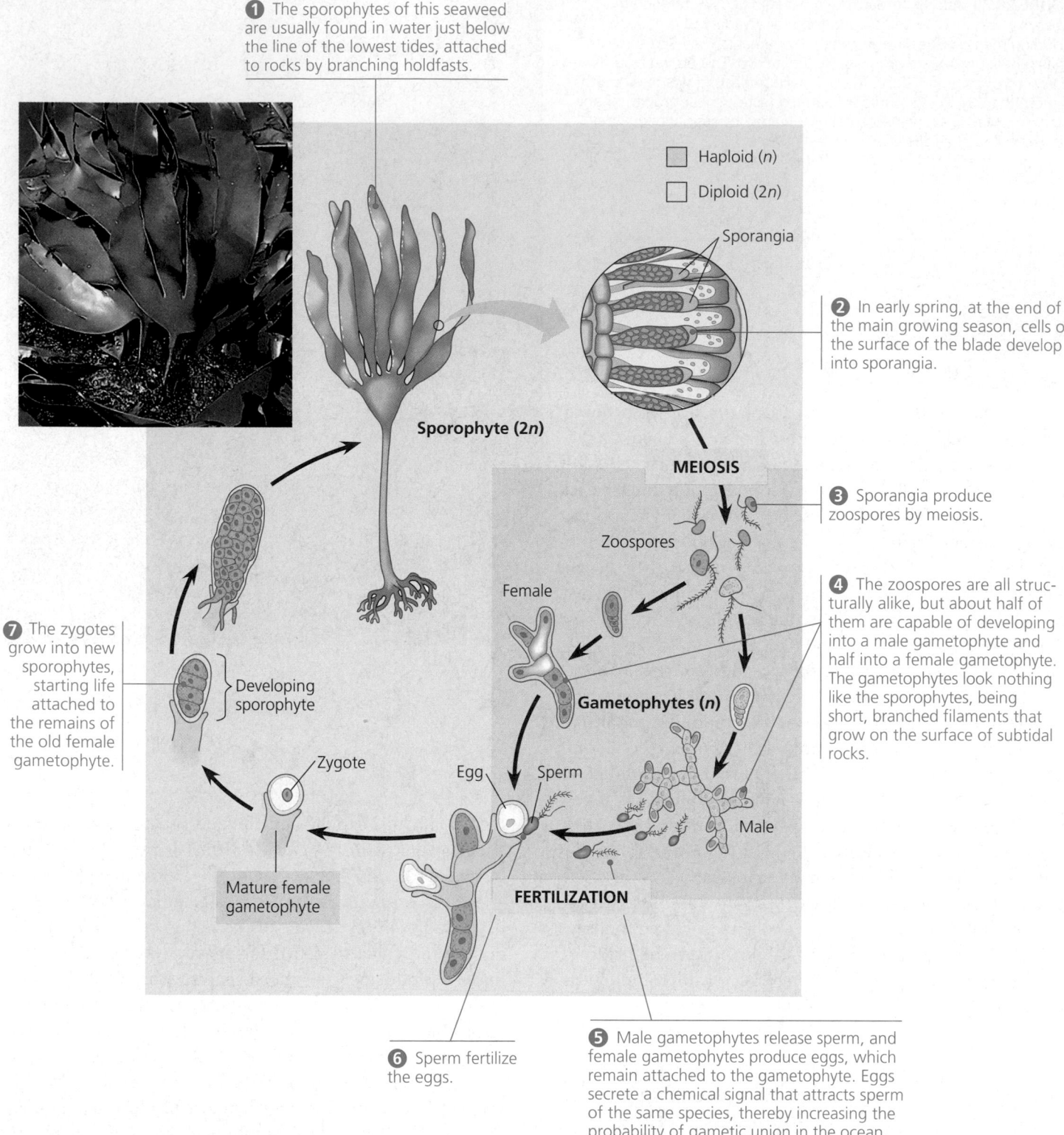

① The sporophytes of this seaweed are usually found in water just below the line of the lowest tides, attached to rocks by branching holdfasts.

Haploid (*n*)

Diploid (2*n*)

Sporangia

Sporophyte (2*n*)

② In early spring, at the end of the main growing season, cells on the surface of the blade develop into sporangia.

MEIOSIS

③ Sporangia produce zoospores by meiosis.

Zoospores

Female

④ The zoospores are all structurally alike, but about half of them are capable of developing into a male gametophyte and half into a female gametophyte. The gametophytes look nothing like the sporophytes, being short, branched filaments that grow on the surface of subtidal rocks.

Gametophytes (*n*)

⑦ The zygotes grow into new sporophytes, starting life attached to the remains of the old female gametophyte.

Developing sporophyte

Zygote

Egg   Sperm

Male

Mature female gametophyte

FERTILIZATION

⑥ Sperm fertilize the eggs.

⑤ Male gametophytes release sperm, and female gametophytes produce eggs, which remain attached to the gametophyte. Eggs secrete a chemical signal that attracts sperm of the same species, thereby increasing the probability of gametic union in the ocean.

FIGURE 28.21 The life cycle of *Laminaria*: an example of alternation of generations.

Rhodophyta: Red algae lack flagella

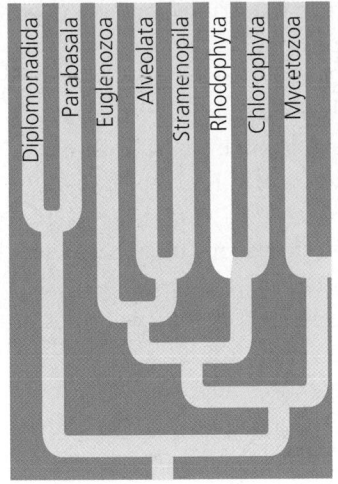

Unlike other eukaryotic algae, **red algae** (Rhodophyta; from the Greek *rhodos,* red) have no flagellated stages in their life cycle. They are commonly reddish because of an accessory pigment called phycoerythrin. It belongs to a family of pigments known as phycobilins, also found in cyanobacteria. In fact, the plastids of red algae evolved from cyanobacteria by primary endosymbiosis.

Despite their name, not all rhodophytes are red. Species adapted to different water depths differ in their proportions of accessory pigments. Rhodophytes may be almost black in deep water, bright red at more moderate depths, and greenish in very shallow water because less phycoerythrin masks the green of chlorophyll. Some species lack pigmentation altogether and function heterotrophically as parasites on other red algae.

Red algae are the most abundant large algae in the warm coastal waters of tropical oceans, but there are also some freshwater and soil species. The phycobilins and other accessory pigments allow some species to absorb the light wavelengths (blues and greens) that penetrate down to deep water. A species of red alga has recently been discovered living near the Bahamas at a depth of more than 260 m.

Most red algae are multicellular, and the largest share the designation "seaweeds" with the brown algae, although none of the reds are as big as the giant browns (kelps). The thalli of many red algae are filamentous, often branched and interwoven in delicate lacy patterns (FIGURE 28.22). The base of the thallus is usually differentiated as a simple holdfast.

Life cycles are especially diverse among the red algae. Lacking flagella, these algae depend on water currents to bring gametes together for fertilization. Alternation of generations is common in red algae.

Chlorophyta: Green algae and plants evolved from a common photoautotrophic ancestor

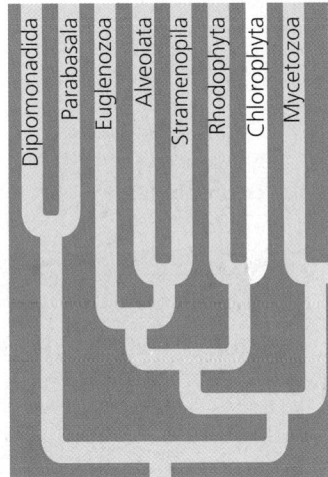

Green algae (chlorophytes and charophyceans in FIGURE 28.8) are named for their grass-green chloroplasts, which are much like those of organisms we traditionally call plants in ultrastructure and pigment composition. Molecular systematics and cellular morphology leave little doubt that green algae and land plants are closely related. In fact, some systematists now advocate inclusion of green algae in an expanded kingdom, the *Viridiplantae* of

(a) Dulse (*Palmaria*), an edible species with a "leafy" form.

(b) *Bonnemaisonia hamifera,* a filamentous red alga.

(c) A coralline alga. The cell walls of coralline algae are hardened by calcium carbonate. Some coralline algae are members of the biological communities called coral reefs.

FIGURE 28.22 Red algae.

FIGURE 28.8 (Latin *viridis*, green). The common ancestor of green algae and plants probably had chloroplasts derived from cyanobacteria by primary endosymbiosis. Among green algae, those called charophyceans are especially closely related to land plants, and we will discuss them along with the plants in the next chapter. Here our survey of green algae features the *chlorophytes* (from the Greek *chloros*, green).

More than 7,000 species of chlorophytes have been identified. Most live in fresh water, but there are also many marine species. Various species of unicellular chlorophytes live as plankton or inhabit damp soil or snow; some species live symbiotically within other eukaryotes, contributing portions of their photosynthetic products to the food supply of the hosts. Chlorophytes are among the algae that live symbiotically with fungi in the mutualistic collectives known as **lichens** (see FIGURE 31.16).

The simplest chlorophytes are biflagellated unicells such as *Chlamydomonas*, which resemble the gametes and zoospores of more complex chlorophytes. In addition to unicellular chlorophytes, there are colonial species, as well as many multicellular filamentous forms that contribute to the stringy masses known as pond scum. There are also multicellular chlorophytes, with bodies large and complex enough that marine species qualify as seaweeds along with the large brown algae and red algae.

Larger size and greater complexity have evolved by three different mechanisms: (1) the formation of colonies of individual cells, as seen in species of *Volvox* (FIGURE 28.23a); (2) the repeated division of nuclei with no cytoplasmic division, as seen in multinucleate filaments of *Caulerpa* (FIGURE 28.23b); and (3) the formation of true multicellular forms by cell division and cell differentiation, as in *Ulva* (FIGURE 28.23c).

Most green algae have complex life histories, with both sexual and asexual reproductive stages. Nearly all reproduce sexually by way of biflagellated gametes having cup-shaped chloroplasts (FIGURE 28.24). The exceptions are the conjugating algae, such as *Spirogyra*, which produce amoeboid

FIGURE 28.23 Colonial and multicellular chlorophytes.

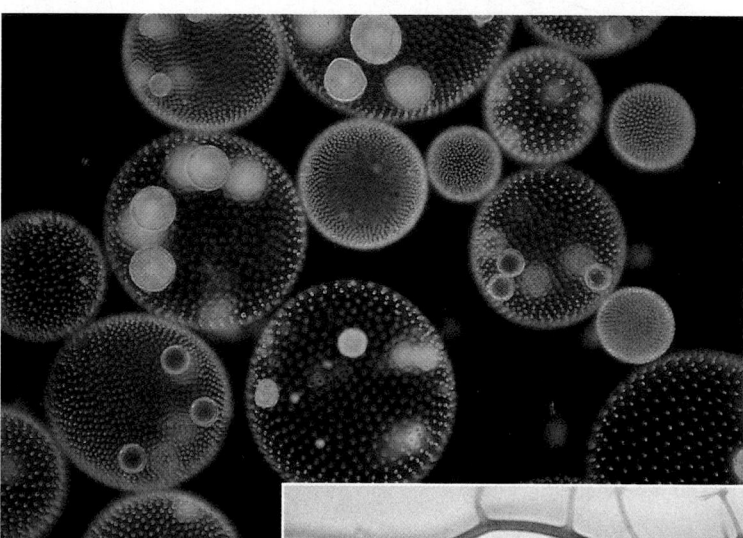

50 μm

(a) *Volvox.* This colonial chlorophyte inhabits fresh water. The colony is a hollow ball, with its wall composed of hundreds or thousands of biflagellated cells embedded in a gelatinous matrix. The cells are usually connected by strands of cytoplasm; if isolated, these cells cannot reproduce. The large colonies seen here will eventually release the small "daughter" colonies within them (LM).

(b) *Caulerpa.* This chlorophyte is found in the marine intertidal zone. The branched filaments lack cross-walls and thus are multinucleate. In effect, the thallus is one huge "supercell."

(c) *Ulva*, or sea lettuce. This edible seaweed has a multicellular thallus differentiated into leaflike blades and a rootlike holdfast that anchors the alga against turbulent waves and tides.

gametes. Alternation of generations evolved in the life cycles of some green algae, including the *Ulva* of FIGURE 28.23c.

Throughout our survey of protists so far, we have encountered algae, or photosynthetic protists, as members of several different clades, often alongside heterotrophic members of the same clades. Recall from FIGURE 28.5 that different episodes of secondary endosymbiosis account for the diversity of protists with plastids. As a follow-up to that illustration, FIGURE 28.25 (p. 568) places the algal groups we've discussed in the context of their plastids' origins.

A diversity of protists use pseudopodia for movement and feeding

The three groups we discuss in this section represent some of the immense diversity of eukaryotes that move and often feed by means of cellular extensions called **pseudopodia.** Most of these organisms are heterotrophs that actively seek and consume bacteria, other protists, and detritus (dead organic matter). There are also symbiotic species, including some parasites

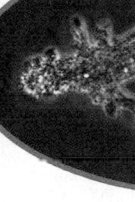

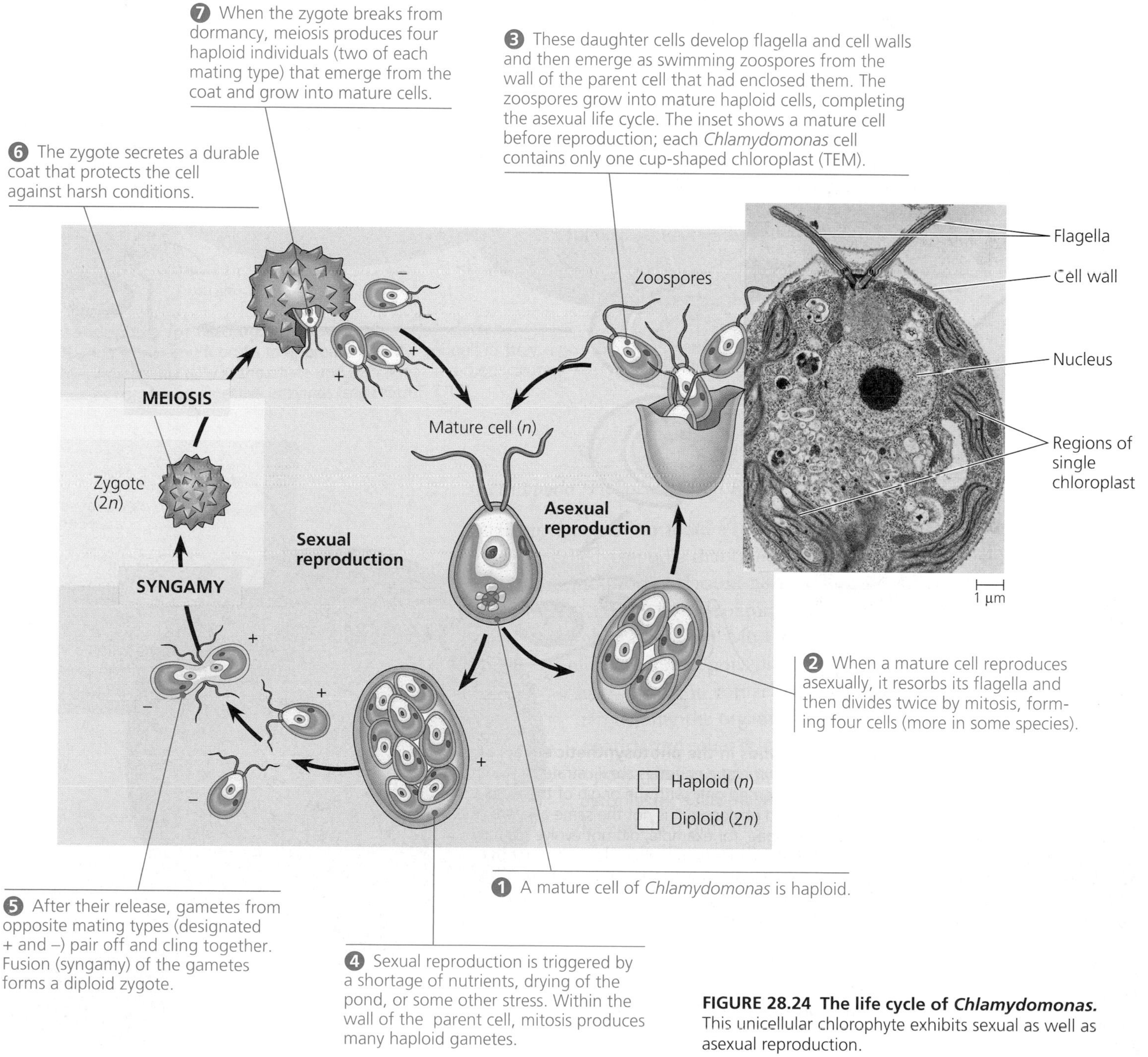

7 When the zygote breaks from dormancy, meiosis produces four haploid individuals (two of each mating type) that emerge from the coat and grow into mature cells.

6 The zygote secretes a durable coat that protects the cell against harsh conditions.

3 These daughter cells develop flagella and cell walls and then emerge as swimming zoospores from the wall of the parent cell that had enclosed them. The zoospores grow into mature haploid cells, completing the asexual life cycle. The inset shows a mature cell before reproduction; each *Chlamydomonas* cell contains only one cup-shaped chloroplast (TEM).

Zoospores

Flagella

Cell wall

Nucleus

Regions of single chloroplast

MEIOSIS

Mature cell (*n*)

Zygote (2*n*)

SYNGAMY

Sexual reproduction

Asexual reproduction

1 μm

2 When a mature cell reproduces asexually, it resorbs its flagella and then divides twice by mitosis, forming four cells (more in some species).

Haploid (*n*)

Diploid (2*n*)

1 A mature cell of *Chlamydomonas* is haploid.

5 After their release, gametes from opposite mating types (designated + and −) pair off and cling together. Fusion (syngamy) of the gametes forms a diploid zygote.

4 Sexual reproduction is triggered by a shortage of nutrients, drying of the pond, or some other stress. Within the wall of the parent cell, mitosis produces many haploid gametes.

FIGURE 28.24 The life cycle of *Chlamydomonas.* This unicellular chlorophyte exhibits sexual as well as asexual reproduction.

Foraminifera (Forams)

Foraminiferans, or **forams,** are almost all marine. Most species live in the sand or attach themselves to rocks and algae, but some are also abundant in plankton. Foraminiferans are named for their porous shells (from the Latin *foramen,* little hole, and *ferre,* to bear). The shells are generally multichambered and consist of organic material hardened with calcium carbonate. Strands of cytoplasm (pseudopodia) extend through the pores, functioning in swimming, shell formation, and feeding (FIGURE 28.28). Many forams also derive nourishment from the photosynthesis of symbiotic algae that live within the shells.

Ninety percent of all identified species of forams are fossils. Along with the calcareous remains of other protists, the fossilized shells of forams are components of marine sediments, including sedimentary rocks that are now land formations. Foram fossils are excellent markers for correlating the ages of sedimentary rocks in different parts of the world.

FIGURE 28.28 Foraminiferan. *Globigerina* has a snail-like shell. The largest forams, though single-celled, grow to diameters of several centimeters. The calcium carbonate shells of these protists have left an excellent fossil record in limestone sediments (LM).

20 µm

Mycetozoa: Slime molds have structural adaptations and life cycles that enhance their ecological roles as decomposers

Mycetozoa translates as "fungus animals," though these protists are neither fungi nor animals. The common name, slime molds, reinforces the misconception that these organisms are true fungi. But any resemblance to fungi is analogous, not homologous—an evolutionary convergence of adaptations for decomposing leaf litter and other organic refuse. In fact, if we were classifying protists strictly on the basis of their modes of motility, it would make sense to group the slime molds with the amoebas. Slime molds use pseudopodia for movement and feeding. And slime molds may in fact be phylogenetically close to some of the other amoeboid protists, especially forms such as *Amoeba.* But systematists are now nearing a consensus that slime molds represent a distinct eukaryotic kingdom, Mycetozoa. And a comparison of protein sequences places the slime molds relatively close to the fungi and animals in the eukaryotic tree (see FIGURE 28.8).

We will see that slime molds have structures that maximize exposure to their food sources, as well as complex life cycles that contribute to survival in changing habitats and facilitate dispersal to new food sources. Mycetozoa consists of two main groups: the plasmodial slime molds and the cellular slime molds.

Myxogastrida (Plasmodial Slime Molds)

The **plasmodial slime molds,** or myxogastrids, are more attractive than their common name implies. Many species are brightly pigmented, usually yellow or orange, but all are heterotrophic. The feeding stage of the life cycle is an amoeboid mass called a **plasmodium,** which may grow to a diameter of several centimeters (FIGURE 28.29). Large as it is, the plasmodium is not multicellular; it is a single mass of cytoplasm that is undivided by membranes and that contains many nuclei. This "supercell" results from mitotic divisions of the nuclei which are not followed by cytokinesis, the division of cytoplasm. In most species, the nuclei of the plasmodium are diploid and divisions are synchronous, with each of thousands of nuclei going through each phase of mitosis at the same time. Because of this characteristic, plasmodial slime molds have been used to study the molecular details of the cell cycle.

Within the fine channels of the plasmodium, cytoplasm streams first one way, then the other, in pulsing flows that are beautiful to watch through a microscope. The cytoplasmic streaming apparently helps distribute nutrients and oxygen. The plasmodium engulfs food particles by phagocytosis as it grows by extending pseudopodia through moist soil, leaf mulch, or rotting logs. If the habitat of a slime mold begins to dry up or there is no food left, the plasmodium ceases growth and differentiates into a stage of the life cycle that functions in sexual reproduction.

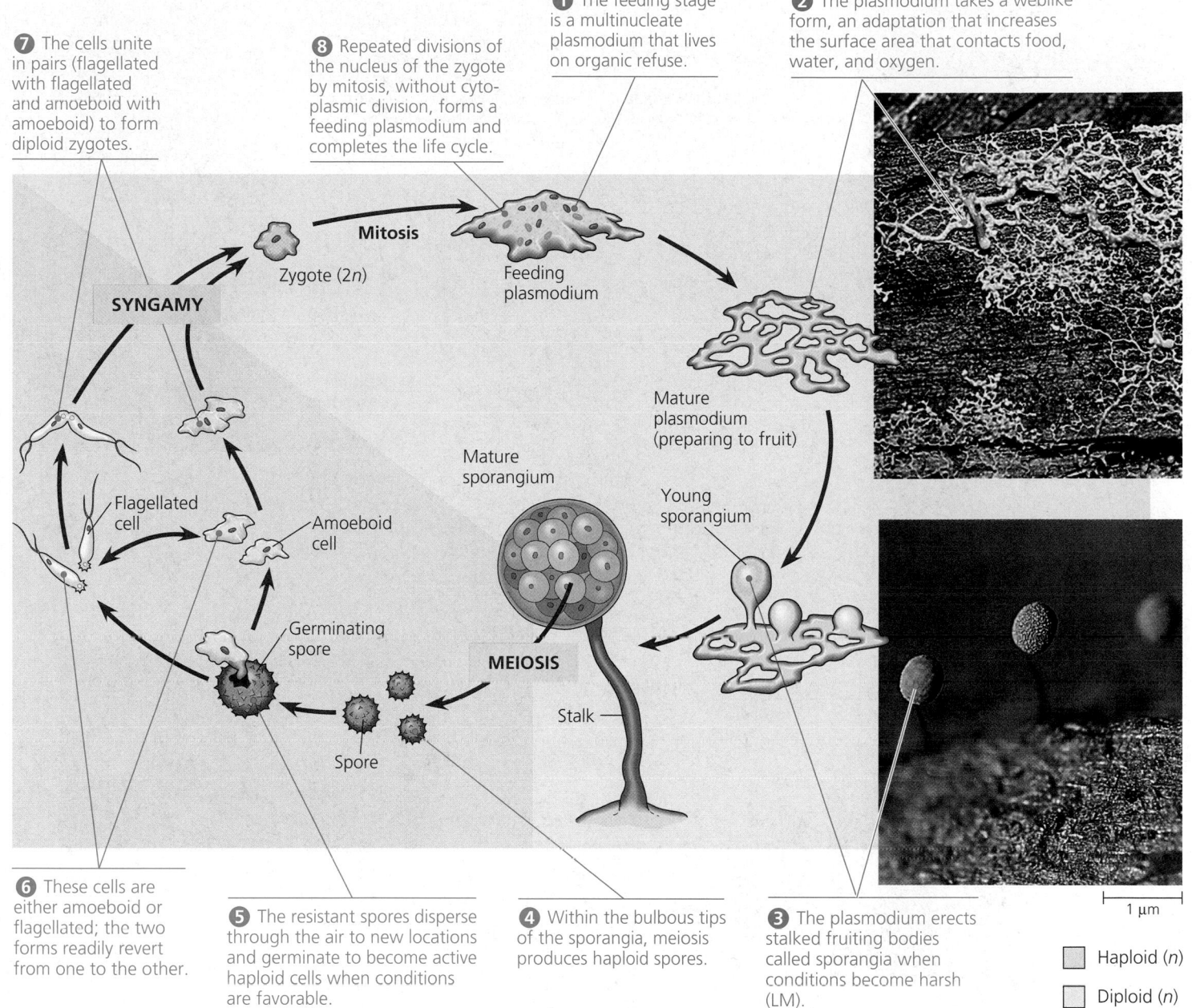

1 The feeding stage is a multinucleate plasmodium that lives on organic refuse.

2 The plasmodium takes a weblike form, an adaptation that increases the surface area that contacts food, water, and oxygen.

7 The cells unite in pairs (flagellated with flagellated and amoeboid with amoeboid) to form diploid zygotes.

8 Repeated divisions of the nucleus of the zygote by mitosis, without cytoplasmic division, forms a feeding plasmodium and completes the life cycle.

Mitosis

Zygote (2n)

Feeding plasmodium

SYNGAMY

Mature plasmodium (preparing to fruit)

Young sporangium

Flagellated cell

Amoeboid cell

Mature sporangium

Germinating spore

MEIOSIS

Stalk

Spore

6 These cells are either amoeboid or flagellated; the two forms readily revert from one to the other.

5 The resistant spores disperse through the air to new locations and germinate to become active haploid cells when conditions are favorable.

4 Within the bulbous tips of the sporangia, meiosis produces haploid spores.

3 The plasmodium erects stalked fruiting bodies called sporangia when conditions become harsh (LM).

1 μm

☐ Haploid (n)

☐ Diploid (n)

FIGURE 28.29 The life cycle of a plasmodial slime mold, such as *Physarum*.

Dictyostelida (Cellular Slime Molds)

Cellular slime molds, or dictyostelids, pose a semantic question about what it means to be an individual organism. Although the feeding stage of the life cycle consists of solitary cells that function individually, when food is depleted the cells form an aggregate that functions as a unit (FIGURE 28.30, p. 572). Though the mass of cells resembles a plasmodial slime mold, the important distinction is that the cells of a cellular slime mold maintain their identity and remain separated by their membranes.

Cellular slime molds also differ from plasmodial slime molds in being haploid organisms (only the zygote is diploid), whereas the diploid condition predominates in the life cycles of most plasmodial slime molds (compare FIGURES 28.29 and 28.30). Cellular slime molds have fruiting bodies that function in asexual reproduction. Also, most cellular slime molds have no flagellated stages.

This completes our survey of diversity and phylogenetic relationships among the eukaryotes traditionally classified as protists. We complete this chapter with a brief discussion of the origins of multicellularity.

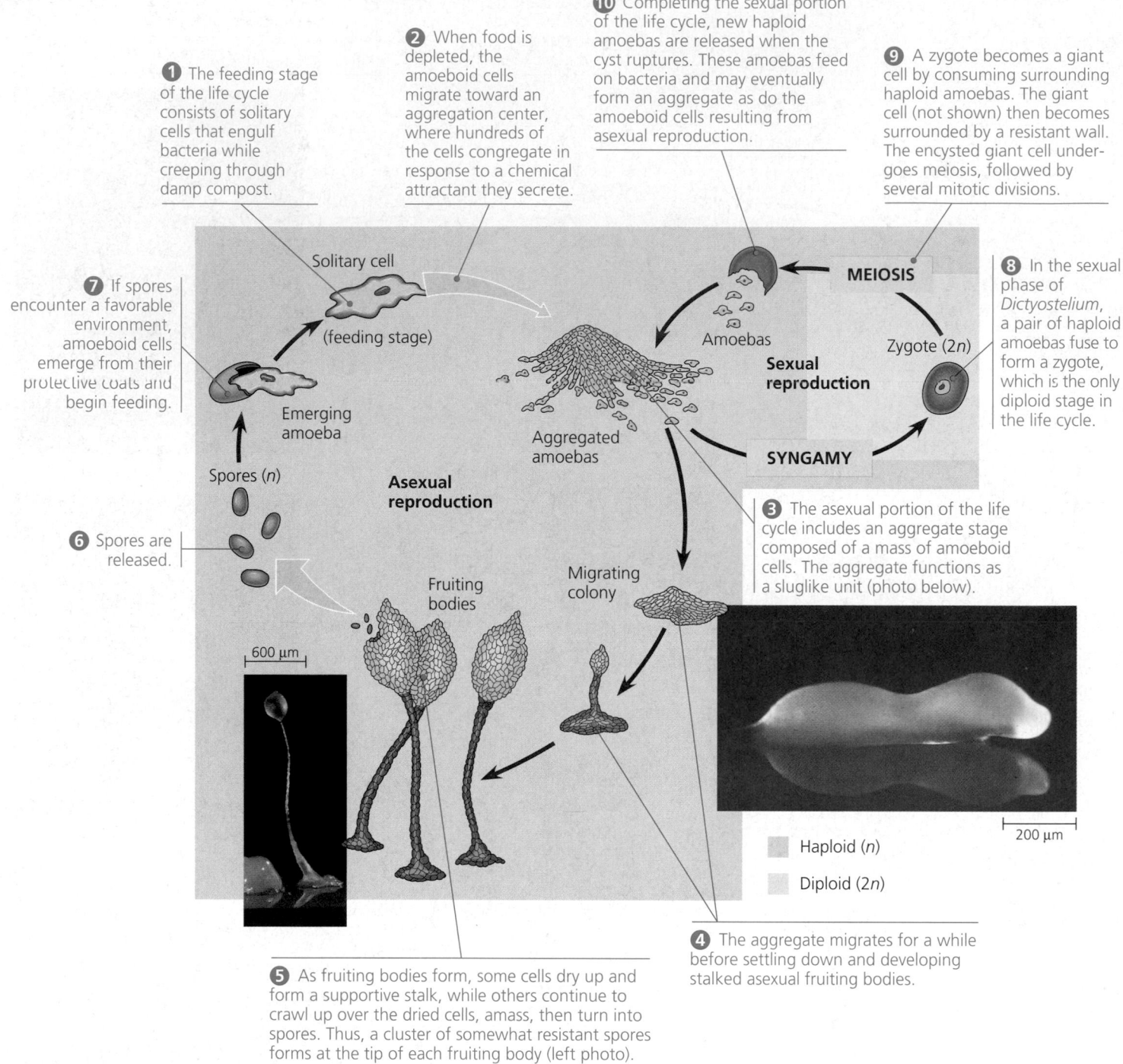

1 The feeding stage of the life cycle consists of solitary cells that engulf bacteria while creeping through damp compost.

2 When food is depleted, the amoeboid cells migrate toward an aggregation center, where hundreds of the cells congregate in response to a chemical attractant they secrete.

10 Completing the sexual portion of the life cycle, new haploid amoebas are released when the cyst ruptures. These amoebas feed on bacteria and may eventually form an aggregate as do the amoeboid cells resulting from asexual reproduction.

9 A zygote becomes a giant cell by consuming surrounding haploid amoebas. The giant cell (not shown) then becomes surrounded by a resistant wall. The encysted giant cell undergoes meiosis, followed by several mitotic divisions.

7 If spores encounter a favorable environment, amoeboid cells emerge from their protective coats and begin feeding.

8 In the sexual phase of *Dictyostelium*, a pair of haploid amoebas fuse to form a zygote, which is the only diploid stage in the life cycle.

Solitary cell

(feeding stage)

MEIOSIS

Amoebas

Zygote (2*n*)

Sexual reproduction

Emerging amoeba

Spores (*n*)

Asexual reproduction

Aggregated amoebas

SYNGAMY

6 Spores are released.

Fruiting bodies

Migrating colony

3 The asexual portion of the life cycle includes an aggregate stage composed of a mass of amoeboid cells. The aggregate functions as a sluglike unit (photo below).

600 μm

200 μm

Haploid (*n*)

Diploid (2*n*)

5 As fruiting bodies form, some cells dry up and form a supportive stalk, while others continue to crawl up over the dried cells, amass, then turn into spores. Thus, a cluster of somewhat resistant spores forms at the tip of each fruiting body (left photo).

4 The aggregate migrates for a while before settling down and developing stalked asexual fruiting bodies.

FIGURE 28.30 The life cycle of a cellular slime mold (*Dictyostelium*).

Multicellularity originated independently many times

We have seen that the origin of eukaryotes ignited an explosion of biological diversification; unicellular eukaryotes are much more diverse in structure than the simpler prokaryotes. The evolution of multicellular bodies broke through another threshold in structural organization and provided the stock for new waves of evolutionary diversification. Multicellularity evolved several times among early eukaryotes, with diverse multicellular algae such as seaweeds as examples of the results. The evolution of multicellularity also gave rise to plants, fungi, and animals, and we will discuss the origins of those kingdoms in the following chapters.

CHAPTER 28 REVIEW

Go to the Campbell Biology website (www.campbellbiology.com) to explore an interactive version of the Chapter Review.

Summary of Key Concepts

INTRODUCTION TO THE PROTISTS

- **Systematists have split protists into many kingdoms (p. 546, FIGURES 28.1, 28.2)** Protistan systematists are beginning to sort out several monophyletic clades.

- **Protists are the most diverse of all eukaryotes (pp. 546–548)** All protists are eukaryotes, and the diversity of protists represents different "experiments" in the evolution of eukaryotic organization. Most protists are unicellular, but colonial and simple multicellular forms also exist. Protists are found wherever there is water, living as plankton, submerged bottom dwellers, or inhabitants of moist soil or the body fluids of other organisms. Of all eukaryotes, protists are the most nutritionally diverse; photoautotrophs, heterotrophs, and mixotrophs are all represented.

THE ORIGIN AND EARLY DIVERSIFICATION OF EUKARYOTES

- **Endomembranes contributed to larger, more complex cells (pp. 548–549, FIGURE 28.4)** The endomembrane system of eukaryotes may have evolved from specialized infoldings of the plasma membrane of ancestral prokaryotes.

- **Mitochondria and plastids evolved from endosymbiotic bacteria (pp. 549–550, FIGURE 28.4)** Chloroplasts and mitochondria are descendants of cyanobacteria and aerobic, heterotrophic prokaryotes, respectively, that took up residence within evolving eukaryotic cells.

- **The eukaryotic cell is a chimera of prokaryotic ancestors (pp. 550–551)** In the eukaryotic genome, some of the genes of the ancestral endosymbionts have been transferred from the organelles to the nucleus.

- **Secondary endosymbiosis increased the diversity of algae (pp. 551–552, FIGURE 28.5)** Some algal groups have plastids derived from endosymbionts that were themselves eukaryotic algae.

- **Research on the relationships between the three domains is changing ideas about the deepest branching in the tree of life (pp. 552–553, FIGURES 28.6, 28.7)** The base of the tree of life may not be represented by a "trunk" (a single common ancestor) but by a prokaryotic community in which genes were transferred extensively.

- **The origin of eukaryotes catalyzed a second great wave of diversification (pp. 553–554, FIGURE 28.8)** A diversity of new kingdoms evolved during the early history of eukaryotes, though the number and names of the kingdoms are still matters of debate.

Web/CD Activity 28A: *Tentative Phylogeny of Eukaryotes*

Use TABLE 28.1 to review the protistan groups surveyed in this chapter. Each major clade represents a "candidate kingdom" in the remodeling of protist taxonomy that is currently occurring. Many of the subgroups of each clade represent traditional protistan phyla and may retain phylum status as the single kingdom Protista is split into multiple kingdoms based on cladistic analysis.

A SAMPLE OF PROTISTAN DIVERSITY

- **Structural and biochemical adaptations help seaweeds survive and reproduce at the ocean's margins (pp. 562–563, FIGURES 28.19, 28.20)** Seaweeds include thallus-forming marine species among the brown, red, and green algae.

- **Some algae have life cycles with alternating multicellular haploid and diploid generations (pp. 563–564, FIGURE 28.21)** Haploid gametophytes and diploid sporophytes take turns reproducing one another.

- **Multicellularity originated independently many times (p. 572)** In addition to seaweeds and other multicellular protists, multicellularity evolved in the ancestors of plants, fungi, and animals.

Web/CD Case Study in the Process of Science: *What Kinds of Protists Do Various Habitats Support?*

Table 28.1 A Sample of Protistan Diversity

Major Clade	A Key Characteristic	Example
1 Diplomonadida-Parabasala	Secondary loss of mitochondria	
A. Diplomonadida (Diplomonads)	Two separate nuclei	*Giardia*
B. Parabasala (Trichomonads and other parabasalids)	Undulating membrane	*Trichomonas*
2 Euglenozoa	Photosynthetic, heterotrophic, and mixotrophic flagellates	
A. Euglenophyta (euglenoids)	Paramylon as storage polysaccharide	*Euglena*
B. Kinetoplastida (kinetoplastids)	Kinetoplast, a unique organelle	*Trypanosoma*
3 Alveolata	Subsurface alveoli (membrane-bound cavities)	
A. Dinoflagellata (dinoflagellates)	Armor of cellulose plates	*Pfiesteria*
B. Apicomplexa (apicomplexans)	Apical complex functioning in penetration of host cells	*Plasmodium*
C. Ciliophora (ciliates)	Cilia functioning in movement and feeding	*Paramecium*
4 Stramenopila	"Hairy" flagella	
A. Oomycota (oomycotes)	Hyphae that absorb nutrients	Water molds, rusts, downy mildews
B. Bacillariophyta (diatoms)	Glassy, two-part walls	*Pinnularia*
C. Chrysophyta (golden algae)	Biflagellate cells; xanthophyll pigments	*Dinobryon*
D. Phaeophyta (brown algae)	Brown color from accessory pigments	*Laminaria*
5 Rhodophyta (red algae)	No flagellated stages; phycoerythrin pigment	*Porphyra* ("Nori")
6 Viridiplantae (includes the green algal group Chlorophyta)	Plant-type chloroplasts	*Chlamydomonas*
7 Mycetozoa	Decomposers having complex life cycles with amoeboid stages	
A. Myxogastrida (plasmodial slime molds)	Netlike plasmodium as feeding stage	*Physarum*
B. Dictyostelida (cellular slime molds)	Amoeboid feeding cells that aggregate to form reproductive colonies	*Dictyostelium*
Pseudopod-equipped protists of uncertain phylogeny	Pseudopodia that function in movement and feeding	
A. Rhizopoda (rhizopods)	Lobe-like pseudopodia	*Amoeba*
B. Actinopoda (actinopods)	Ray-like pseudopodia (axopodia)	Heliozoans and radiolarians
C. Foraminifera (forams)	Porous shells	*Globigerina*

Self-Quiz

1. The hypothesis that life's three taxonomic domains originated from a community of gene-swapping prokaryotes is based mainly on
   a. comparisons of SSU-rRNA nucleotide sequences.
   b. evidence of gene transfer from plastids to mitochondria.
   c. the appearance of photosynthetic organisms in diverse prokaryotic and protist lineages.
   d. the presence of bacterial genes in the genomes of eukaryotes and modern archaeans.
   e. new fossil evidence.

2. Which of the following organisms are *incorrectly* paired with their description?
   a. rhizopods—naked and shelled amoebas
   b. actinopods—planktonic with slender, raylike axopodia
   c. forams—flagellated algae, free-living or symbiotic
   d. apicomplexans—parasites with complex life cycles
   e. diplomonads—protists lacking mitochondria

3. Which of the following algal groups is mismatched with its description?
   a. Dinoflagellates—glassy, two-part shells
   b. Green algae—closest relative of green plants
   c. Red algae—has no flagellated stages in life cycle
   d. Brown algae—includes the largest seaweeds
   e. Diatoms—examples of stramenopiles

4. Plastids that are surrounded by more than two membranes are evidence of
   a. evolution from mitochondria.
   b. phagocytosis of algal cells by another autotrophic algae, after which their plastids fused.
   c. origin of the plastids from archaea.
   d. secondary endosymbiosis of an algal protist by a heterotrophic protist, which left the new endosymbiont wrapped in a vacuole membrane.
   e. budding of the plastids from the nuclear envelope.

5. Biologists suspect that endosymbiosis gave rise to mitochondria before plastids because
   a. the products of photosynthesis could not be metabolized without mitochondrial enzymes.
   b. almost all eukaryotes have mitochondria, while only autotrophic eukaryotes have plastids.
   c. mitochondrial DNA is less similar to prokaryotic DNA than is the DNA from plastids.
   d. without mitochondrial CO_2 production, photosynthesis could not occur.
   e. plastids utilize their own ribosomes, while mitochondrial proteins are synthesized in the cytosol.

6. Which of the following characteristics supports molecular evidence for combining the dinoflagellates, apicomplexans, and ciliates in the monophyletic clade Alveolata?
   a. Their flagella or cilia are organized with the 9 + 2 microtubular ultrastructure.
   b. All are pathogenic.
   c. All are found exclusively in freshwater or marine habitats.
   d. All possess mitochondria.
   e. The three groups have small membrane-bound alveoli under their cell surfaces.

7. Which of the following is an *incorrect* statement about the possible endosymbiotic origins of plastids and mitochondria?
   a. They are the appropriate size to be descendants of bacteria.
   b. They contain their own genome and produce all their own proteins.
   c. They contain circular DNA molecules.
   d. Their membranes have enzymes and transport systems that resemble those found in the plasma membranes of prokaryotes.
   e. Their ribosomes are more similar to those of bacteria than to those of eukaryotes.

8. The organism that contributed to the Irish potato famine is
   a. an actinopod.
   b. a ciliate.
   c. an oomycote.
   d. a plasmodial slime mold.
   e. a cellular slime mold.

9. You find a colorful, weblike, amoeboid mass on a rotting log. After a dry period, you note stalked fruiting bodies growing up from this multinucleate mass. What is the probable identity of this organism?
   a. euglenoid
   b. plasmodial slime mold
   c. cellular slime mold
   d. foram
   e. water mold

10. Which of the following groups probably represents the earliest branch in the evolution of eukaryotes?
   a. archaea
   b. diplomonads
   c. fungi
   d. amoebas
   e. diatoms

11. Why are protists especially important to biologists investigating the evolution of eukaryotic life?

12. What three main modes of locomotion occur among protists?

13. What is kelp?

14. Why doesn't the sexual life cycle of humans, which has haploid and diploid stages, qualify as an example of alternation of generations?

Go to the website or CD-ROM for more quiz questions.

Evolution Connection

Based on arguments from molecular sytematics and cladistics, explain why kingdom Protista is probably an obsolete taxon.

The Process of Science

Applying the "If . . . then" logic of science (see Chapter 1), what are a few of the predictions that arise from the hypothesis that plants evolved from green algae? Put another way, how could you test this hypothesis?

Examine, identify, and classify protists collected from various habitats in the Case Study in the Process of Science, available on the website and CD-ROM.

Science, Technology, and Society

The ability of the pathogen *Plasmodium* to evade the immune system is one reason it is difficult to develop a malaria vaccine. An additional problem is that fewer scientists are engaged in malaria research, and less money is spent on it, than on diseases (such as cystic fibrosis) that affect far fewer people than does malaria. What are the possible reasons for this imbalance in research effort?

Answers: 1. d; 2. c; 3. a; 4. d; 5. b; 6. e; 7. b; 8. c; 9. b; 10. b; 11. Because the first eukaryotes were protists, and protists were ancestral to all other eukaryotes, including plants, fungi, animals, and modern protists 12. Movement using flagella, cilia, and pseudopodia 13. Giant brown seaweed 14. Because the haploid stage (gametes) is unicellular; in alternation of generations, there are multicellular haploid *and* diploid forms during the life cycle.

PLANT DIVERSITY I: HOW PLANTS COLONIZED LAND

AN OVERVIEW OF LAND PLANT EVOLUTION

- Evolutionary adaptations to terrestrial living characterize the four main groups of land plants
- Charophyceans are the green algae most closely related to land plants
- Several terrestrial adaptations distinguish land plants from charophycean algae

THE ORIGIN OF LAND PLANTS

- Land plants evolved from charophycean algae over 500 million years ago
- Alternation of generations in plants may have originated by delayed meiosis
- Adaptations to shallow water preadapted plants for living on land
- Plant taxonomists are reevaluating the boundaries of the plant kingdom
- The plant kingdom is monophyletic

BRYOPHYTES

- The three phyla of bryophytes are mosses, liverworts, and hornworts
- The gametophyte is the dominant generation in the life cycles of bryophytes
- Bryophyte sporophytes disperse enormous numbers of spores
- Bryophytes provide many ecological and economic benefits

THE ORIGIN OF VASCULAR PLANTS

- Additional terrestrial adaptations evolved as vascular plants descended from mosslike ancestors
- A diversity of vascular plants evolved over 400 million years ago

PTERIDOPHYTES: SEEDLESS VASCULAR PLANTS

- Pteridophytes provide clues to the evolution of roots and leaves
- A sporophyte-dominant life cycle evolved in seedless vascular plants
- Lycophyta and Pterophyta are the two phyla of modern seedless vascular plants
- Seedless vascular plants formed vast "coal forests" during the Carboniferous period

When you view *a lush landscape, such as the Tasmanian forest scene in the photo above, it is difficult to picture the land without any macroscopic organisms. But that is how we must imagine Earth for almost the first 90% of the time life has existed. Life was cradled in the seas and ponds, and there it evolved in confinement for 3 billion years. Paleobiologists have discovered fossils of cyanobacteria that coated moist soil about 1.2 billion years ago, but the long evolutionary pilgrimage of more complex organisms onto land did not begin until about 500 million years ago. The terrestrial communities founded by plants transformed the biosphere. Consider, for example, that humans would not exist had it not been for the chain of evolutionary events that began when certain descendants of green algae first colonized land.*

The evolutionary history of the plant kingdom is a story of adaptation to changing terrestrial conditions. That is the historical context in which this chapter and the next survey the current diversity of plants and trace their origins.

AN OVERVIEW OF LAND PLANT EVOLUTION

More than 280,000 species of plants inhabit Earth today. Though some species, such as sea grasses, have returned to aquatic habitats during their evolution, most plants live in terrestrial environments, such as deserts, grasslands, and forests. For now, we will refer to all of these organisms as *land plants,* even those that are now aquatic, to distinguish them from the algae we discussed as photosynthetic protists in Chapter 28. But land plants did indeed evolve from certain green algae called charophyceans, and we will explore this phylogenetic connection later in the chapter. This section introduces the diversity of modern land plants in the context of their terrestrial adaptations.

Evolutionary adaptations to terrestrial living characterize the four main groups of land plants

There are four main groups of land plants: bryophytes, pteridophytes, gymnosperms, and angiosperms. The most common bryophytes are the mosses. The pteridophytes include the ferns. Pines and other conifers (cone-bearing plants) are the most familiar gymnosperms. The angiosperms are the flowering plants. Studying these four plant groups provides perspective for understanding the history of life in general, since each plant group originated at a distinct time in Earth's past.

Mosses and other **bryophytes** are distinguished from algae by several features derived during evolutionary adaptation to living on land. Many of these adaptations are reproductive. For example, when bryophytes and all other land plants reproduce, the offspring develop from multicellular embryos that remain attached to the "mother" plant, which protects and nourishes the embryos.

Additional adaptations distinguish the other major groups of land plants from the bryophytes. An example is the vascular tissue that evolved in an ancestor common to pteridophytes, gymnosperms, and angiosperms. This derived character and others unite these three plant groups in a clade known as the **vascular plants** (see Chapter 25 to review derived homologies and cladistics). In **vascular tissue,** cells are joined into tubes that transport water and nutrients throughout the plant body. Most bryophytes lack vascular tissue and hence are sometimes referred to as "nonvascular plants," in contrast to the vascular plants. This distinction is not entirely accurate, however, for water-conducting tubes *are* present in some bryophytes. (Later, we will examine some derived features that more accurately mark the "vascular plant" clade.)

Among vascular plants, ferns and other **pteridophytes** are sometimes called *seedless plants* because there is no seed stage in their life cycles. In contrast, gymnosperms and angiosperms are *seed plants*. The evolution of the seed in an ancestor common to gymnosperms and angiosperms added an adaptation that facilitated reproduction on land.

A **seed** consists of a plant embryo packaged along with a food supply within a protective coat. The first vascular plants with seeds evolved about 360 million years ago, near the end of the Devonian period. Their seeds were not enclosed in any specialized chambers. These early seed plants gave rise to the diversity of present-day **gymnosperms** (from the Greek *gymnos,* naked, and *sperma,* seed), including the conifers.

The evolution of flowers led to further diversification of plants beginning in the early Cretaceous period about 130 million years ago. The flower is a complex reproductive structure that bears seeds within protective chambers called ovaries. This contrasts with the bearing of naked seeds by gymnosperms. The great majority of modern-day plant species are flowering plants, or **angiosperms** (from the Greek *angion,* container, and *sperma,* seed).

Thus, we can trace the bryophytes, pteridophytes, gymnosperms, and angiosperms of today to four great evolutionary episodes in the history of land plants: the origin of bryophytes from algal ancestors; the origin and diversification of vascular plants; the origin of seeds; and the evolution of flowers. The hypothetical phylogeny in FIGURE 29.1 places the phyla of land plants in this evolutionary context. The taxonomic scheme we'll use in this chapter and the next recognizes a total of ten plant phyla (TABLE 29.1). Before we take a closer look at these diverse plant phyla, let's distinguish land plants from other photosynthetic organisms and then continue our exploration of the terrestrial adaptations that made it possible for plants to colonize land.

Charophyceans are the green algae most closely related to land plants

How can we define land plants? What features do bryophytes, pteridophytes, gymnosperms, and angiosperms have in common that distinguish them from other organisms? Plants are multicellular, eukaryotic, photosynthetic autotrophs. However, red seaweeds and brown seaweeds also fit this description (see Chapter 28). Cell walls made of cellulose and the presence of chlorophylls *a* and *b* within chloroplasts are often cited as defining characteristics of plants. But cellulose walls are also present in several algal groups, including dinoflagellates and

Table 29.1 Ten Phyla of Extant Plants

	Common Name	Approximate Number of Extant Species
Bryophytes		
❶ Phylum Hepatophyta	Liverworts	6,500
❷ Phylum Anthocerophyta	Hornworts	100
❸ Phylum Bryophyta	Mosses	12,000
Vascular Plants		
Seedless Vascular Plants (Pteridophytes)		
❹ Phylum Lycophyta	Lycophytes	1,000
❺ Phylum Pterophyta	Ferns, horsetails, and whisk ferns	12,000
Seed Plants		
Gymnosperms		
❻ Phylum Ginkgophyta	Ginkgo	1
❼ Phylum Cycadophyta	Cycads	100
❽ Phylum Gnetophyta	Gnetae	70
❾ Phylum Coniferophyta	Conifers	550
Angiosperms		
❿ Phylum Anthophyta	Flowering plants	250,000

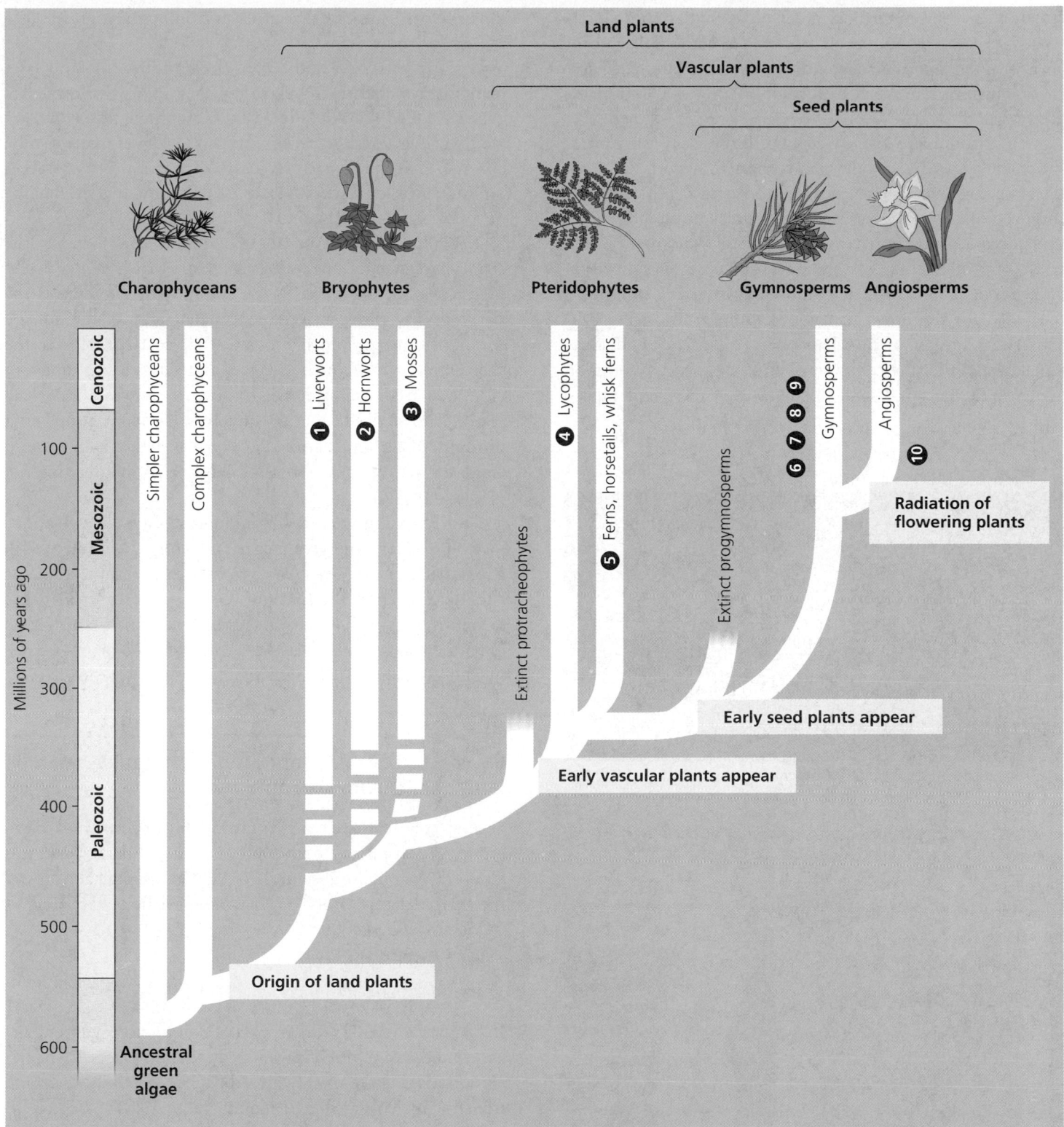

FIGURE 29.1 Some highlights of plant evolution. In this hypothetical phylogeny, the labeling at the top identifies four main groups of plant phyla. The gymnosperm lineage actually consists of a few phyla, but they are collected here in a single branch because the gymnosperms are monophyletic. (TABLE 29.1 identifies the various gymnosperm phyla.) In contrast, neither the bryophyte nor pteridophyte groups are monophyletic, which is why they are represented here as multiple phyla branching from the plant tree. We made the bases of the bryophyte branches broken lines because the chronology of their origin is uncertain. The branches and groups of branches here are labeled with informal names. TABLE 29.1 provides the formal names of the ten plant phyla we will examine in this chapter and the next. The circled numbers will help you match the table to this tree.

brown algae. And even those green algae that are not closely related to land plants have chloroplasts with both chlorophylls *a* and *b*, as do the euglenoids and some dinoflagellates. So we must look to other features to distinguish land plants from algae.

An electron microscopic view of cells reveals two key ultrastructural features that land plants share only with the green algae that are their closest relatives, the algal group called the **charophyceans** (FIGURE 29.2). First, the plasma membranes of land plants and charophyceans are equipped with rose-shaped arrays of proteins that synthesize the cellulose microfibrils of the cell walls. These **rosette cellulose-synthesizing complexes** contrast with the linear arrays of cellulose-producing proteins characteristic of noncharophycean algae. This distinction suggests that the cellulose walls of an ancestor common to charophyceans and land plants evolved independently of the cellulose walls of other algae, including most green algae (except charophyceans).

A second derived homology in charophyceans and land plants is a match in the enzymes contained within the organelles called **peroxisomes,** which are often closely associated with chloroplasts (see FIGURE 7.19). These peroxisome enzymes help minimize the loss of organic product due to photorespiration (see Chapter 10). Except for the charophyceans, algal peroxisomes lack these enzymes.

Two additional derived homologies link land plants to the charophycean algae. In those land plants that have flagellated sperm cells, the structure of the sperm closely resembles the sperm of charophyceans. And certain details of cell division are common only to land plants and the most complex of the charophycean algae, such as *Chara* and *Coleochaete,* shown in FIGURE 29.2. For example, the synthesis of new cross-walls (cell plates) during cell division involves the formation of a **phragmoplast,** an alignment of cytoskeletal elements and Golgi-derived vesicles across the midline of the dividing cell (see FIGURE 12.9).

Clearly, we are starting to build a case for a close relationship between charophycean algae and land plants, and we will strengthen our case when we explore the origin of land plants later in the chapter. But for now we search for features unique to land plants—characteristics that distinguish land plants from their closest relatives—for those are the plant features most likely to be associated with terrestrial adaptation.

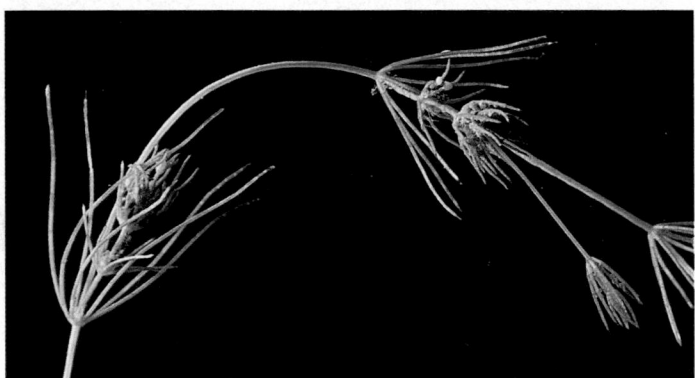

(a) *Chara,* a pond organism (LM). This charophycean is popular with researchers because of its giant cells (the cylinder between consecutive whorls of "branches" is a single cell). For example, physiologists studying the role of electricity in membrane transport impale these giant cells with electrodes to measure voltage changes.

| 10 mm

(b) *Coleochaete orbicularis,* a disk-shaped charophycean (LM)

| 40 μm

FIGURE 29.2 Charophyceans, closest algal relatives of the plant kingdom.

Several terrestrial adaptations distinguish land plants from charophycean algae

Several characteristics are common to all four land plant groups—bryophytes, pteridophytes, gymnosperms, and angiosperms—but are absent in even their closest algal relatives, the charophyceans. Here we'll examine five derived characters unique to land plants.

Apical Meristems, Producers of a Plant's Tissues

In terrestrial habitats, the resources that a photosynthetic organism needs are found in two very different places. Light and carbon dioxide are mainly available aboveground; water and mineral nutrients are found mainly in the soil. Thus, the complex bodies of plants show varying degrees of structural specialization for subterranean and aerial organs—roots and leaf-bearing shoots, respectively, in most plants.

Though plants cannot move from place to place, the elongation and branching of their shoots and roots maximize their exposure to environmental resources. This growth in length is sustained throughout the life of a plant by the activity of **apical meristems,** localized regions of cell division at the tips of shoots and roots (FIGURE 29.3). Cells produced by these meristems

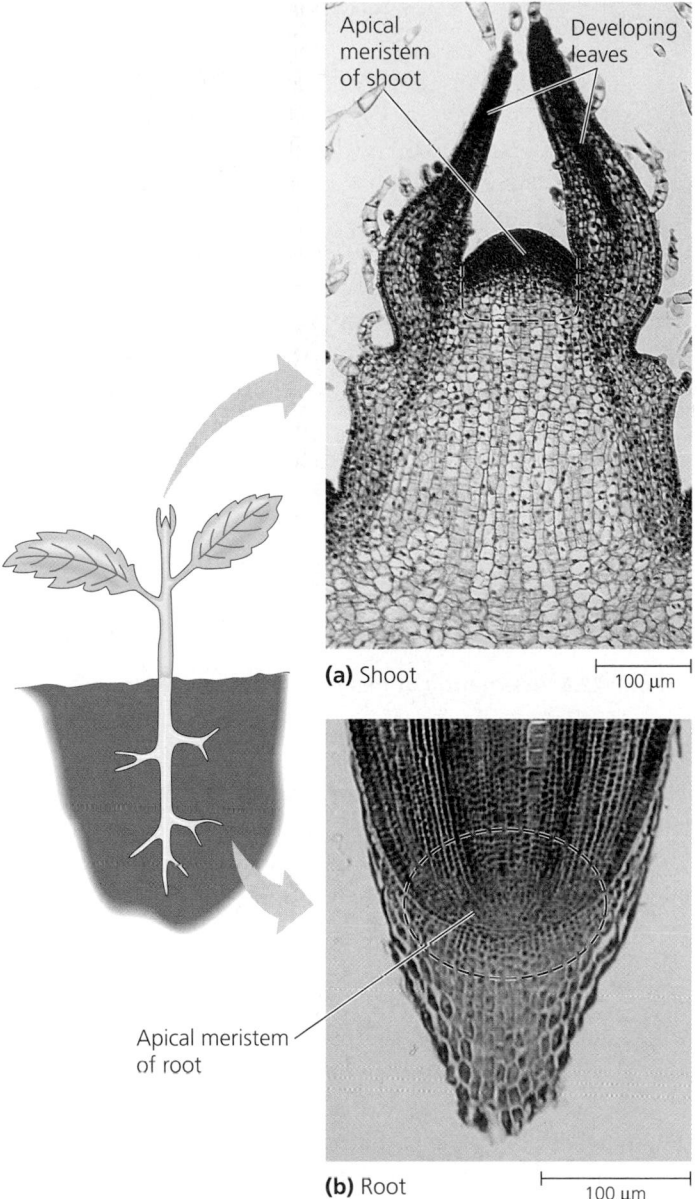

(a) Shoot

100 µm

(b) Root

100 µm

FIGURE 29.3 Apical meristems of plant shoots and roots. The light micrographs are longitudinal sections at the tips of a shoot and root.

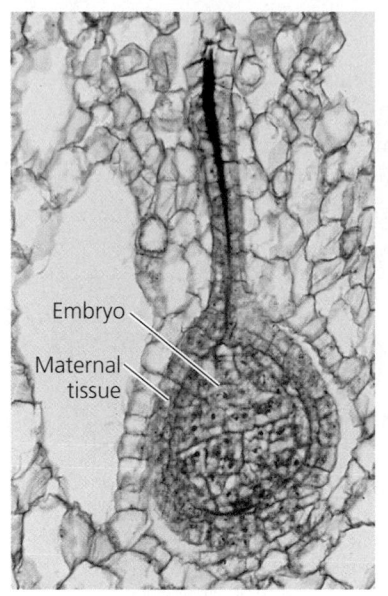
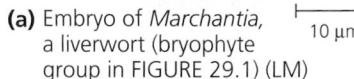
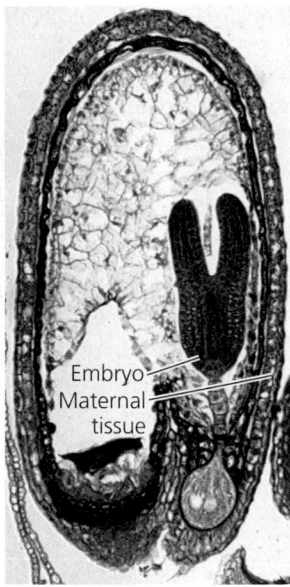

(a) Embryo of *Marchantia,* a liverwort (bryophyte group in FIGURE 29.1) (LM) 10 µm

(b) Embryo of shepherd's purse, an angiosperm (LM) 150 µm

FIGURE 29.4 Embryos of land plants.

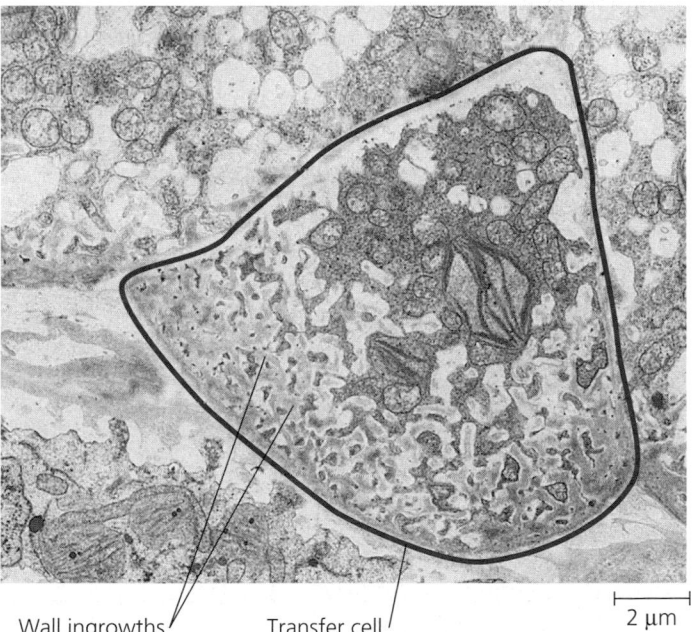

Wall ingrowths Transfer cell 2 µm

FIGURE 29.5 Placental transfer cell in a liverwort (a bryophyte). In these specialized cells at the embryo-maternal tissue interface, elaborate ingrowths of the cell surface (plasma membrane and cell wall) increase the surface area for the transfer of nutrients (TEM).

differentiate into the plant's various tissues, including a surface epidermis that protects the body and several types of internal tissues. Shoot meristems also generate leaves in most plants.

Multicellular, Dependent Embryos

Multicellular plant embryos develop from zygotes that are retained within tissues of the female parent (FIGURE 29.4). The parental tissues provide the developing embryo with nutrients, such as sugars and amino acids. The embryo has specialized **placental transfer cells,** sometimes present in the adjacent maternal tissue as well, which enhance the transfer of nutrients from parent to embryo (FIGURE 29.5). This interface

is analogous to the nutrient-transferring embryo-mother interface of eutherian (placental) mammals. Land plants are also known as **embryophytes,** a distinction that recognizes multicellular, dependent embryos as a derived characteristic common to the land plant clade.

Alternation of Generations

During the life cycles of all land plants, two multicellular body forms alternate, each form producing the other. This type of reproductive cycle, called **alternation of generations,** also evolved in various groups of algae (see FIGURE 28.24). However, alternation of generations does *not* occur in the charophyceans, the algae most closely related to land plants. We can infer from this observation that alternation of generations evolved independently as a derived characteristic of land plants that was not present in the ancestor common to land plants and charophyceans.

The two multicellular body forms that alternate in the life cycles of land plants are the gametophyte and sporophyte generations. The cells of the **gametophyte** are haploid, meaning that they have a single set of chromosomes (see Chapter 13). The gametophyte is named for its production of gametes—eggs and sperm. Fusion of eggs and sperm during the process of fertilization forms diploid zygotes. Mitotic division of the zygote produces the multicellular sporophyte. Thus, the cells of a **sporophyte** are diploid, having two sets of chromosomes, one set contributed by each gamete. Meiosis in a mature sporophyte produces haploid reproductive cells called spores (hence the name sporophyte). A **spore** is a reproductive cell that can develop into a new organism without fusing with another cell (in contrast to gametes, which cannot develop directly into a multicellular organism; they must first fuse to form zygotes). Mitotic division of a plant spore produces a new multicellular gametophyte. And so the alternation of generations continues, sporophytes producing spores that develop into gametophytes, and gametophytes producing gametes that unite to form the zygotes that develop into sporophytes (FIGURE 29.6).

Don't confuse alternation of generations with the more general occurrence of haploid and diploid stages in the life cycles of *all* sexually reproducing organisms, including animals (see FIGURE 13.4). In humans, for example, meiosis in the gonads (ovaries and testes) produces haploid gametes, which unite to form diploid zygotes. In this case, the only haploid stage in the life cycle is the gamete, which is single-celled. What distinguishes alternation of generations as a special type of haploid ↔ diploid sexual cycle is that *both* stages are represented by multicellular bodies.

In some algae with alternation of generations, the multicellular sporophyte and gametophyte look alike to the unaided eye, though microscopic examination would reveal the difference in chromosome count. However, in other algal groups and *all* land plants, the sporophyte and gametophyte generations are very different in their morphology.

The relative size and complexity of the sporophyte and gametophyte depend on the plant group. In bryophytes, the gametophyte is the "dominant" generation—that is, the organism that is larger and more conspicuous. In contrast, the sporophyte form is the dominant generation in the life cycles

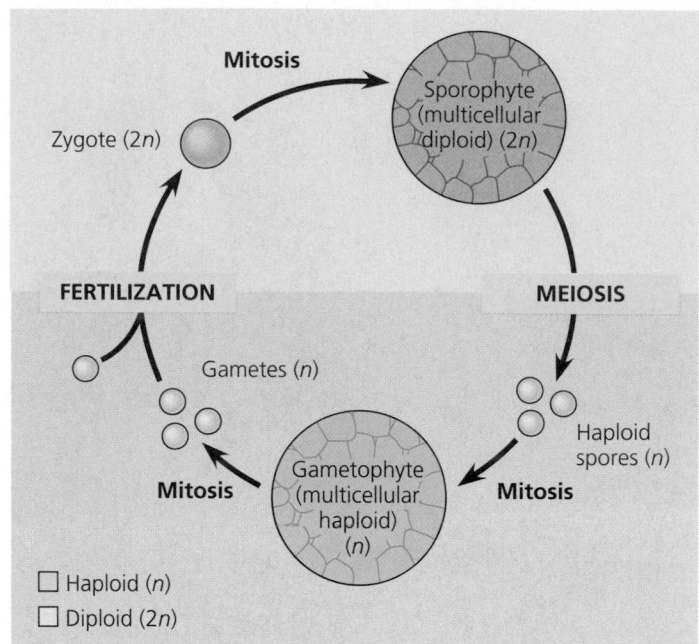

FIGURE 29.6 Alternation of generations: a generalized scheme.

of pteridophytes, gymnosperms, and angiosperms. For example, the fern plant most of us are familiar with—the plant with the fronds as leaves—is the diploid sporophyte. The gametophyte is a tiny plant on the forest floor.

Walled Spores Produced in Sporangia

Plant spores are haploid reproductive cells that have the potential to grow into multicellular, haploid gametophytes by mitosis (see FIGURE 29.6). A polymer called **sporopollenin,** the most durable organic material known, makes the walls of plant spores very tough and resistant to harsh environments (FIGURE 29.7). This chemical adaptation makes it possible for wind-carried spores to disperse through dry air without harm. All four major plant groups produce spores.

Multicellular organs called **sporangia** (singular, *sporangium*), found on the sporophyte generation of a plant, produce the spores (FIGURE 29.8). Within a sporangium, **spore mother cells** undergo meiosis and generate the haploid spores. The outer tissues of the sporangium protect developing spores until they are ready to be released into the air.

Sporangia and resistant spores with sporopollenin-enriched walls are key terrestrial adaptations of

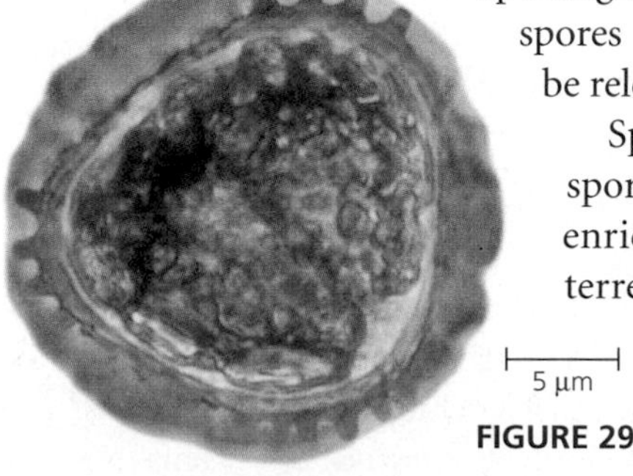

5 μm

FIGURE 29.7 A fern spore (LM).

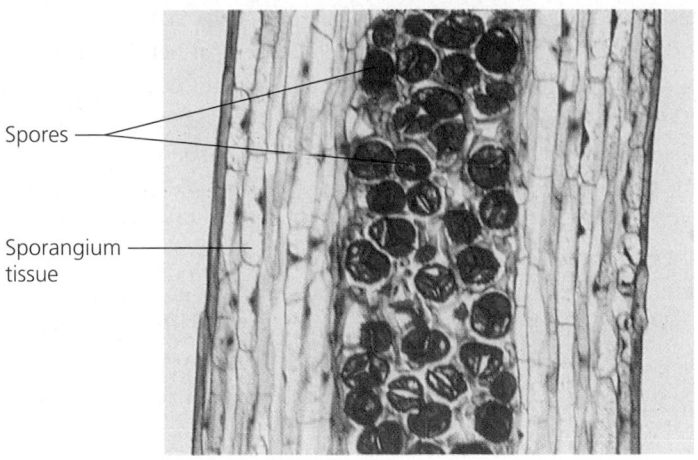

Spores

Sporangium
tissue

FIGURE 29.8 Sporangium of a hornwort (a bryophyte) sporophyte (LM).

50 μm

land plants. Although the charophyceans produce spores, these algae lack multicellular sporangia, and the flagellated, water-dispersed spores lack sporopollenin.

Multicellular Gametangia

The gametophyte forms of bryophytes, pteridophytes, and gymnosperms all produce their gametes within multicellular organs called **gametangia** (FIGURE 29.9). The female gametangia are called **archegonia** (singular, *archegonium*). Each archegonium is a vase-shaped organ that produces a single egg cell and retains the egg within the base of the organ. Male gametangia, called **antheridia,** produce many sperm cells that are released to the environment when mature. The sperm cells of bryophytes, pteridophytes, and some gymnosperms bear flagella and swim through water droplets or water films to eggs. Eggs are fertilized within archegonia, where the zygote begins to develop into an embryo.

Other Terrestrial Adaptations Common to Many Land Plants

In addition to the characteristics we have discussed so far that distinguish the land plant clade from charophycean algae, most plants have additional terrestrial adaptations. These include adaptations for acquiring, transporting, and conserving water; adaptations for reducing the harmful effects of UV radiation, which is much more intense on land than in aquatic habitats; and adaptations for repelling terrestrial herbivores and resisting pathogens.

Adaptations for Water Conservation.   The epidermis (outer cell layers) of leaves and other aerial parts of most land plants is coated with a **cuticle,** a layer consisting of polymers called polyesters and waxes (FIGURE 29.10). The cuticle helps protect

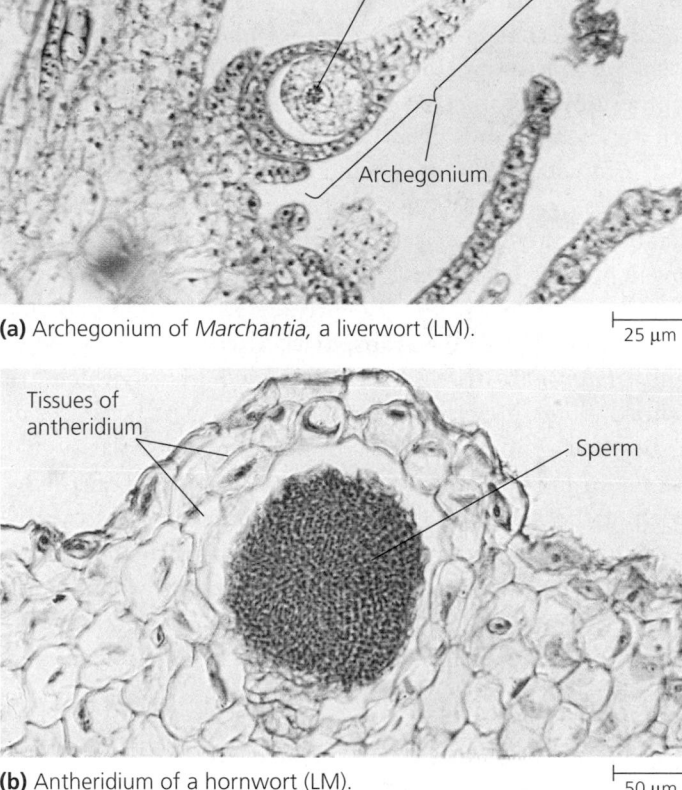

Egg

Archegonium

(a) Archegonium of *Marchantia,* a liverwort (LM).

25 μm

Tissues of antheridium

Sperm

(b) Antheridium of a hornwort (LM).

50 μm

FIGURE 29.9 Gametangia.

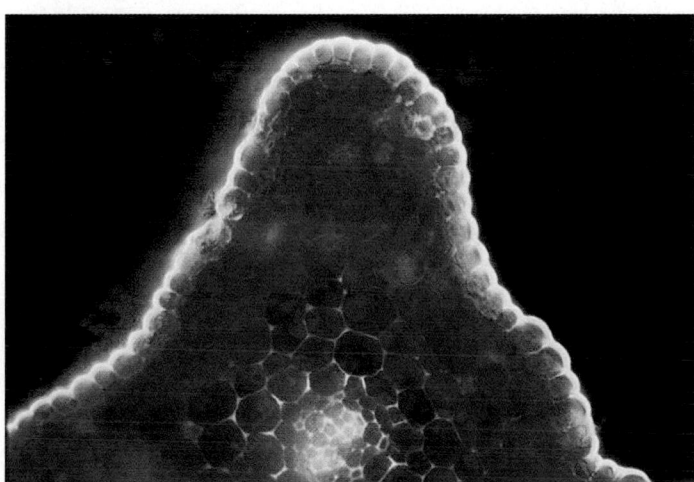

FIGURE 29.10 Cuticle of a stem from *Psilotum* (a pteridophyte). The yellow "glow" on the surface is caused by fluorescence of the cuticle's waxy chemicals when they are illuminated with ultraviolet light in a special type of light microscope.

100 μm

the plant from microbial attack, and its waxy nature acts as waterproofing that helps prevent excessive water loss from the aboveground plant organs.

The epidermis of leaves and other photosynthetic organs has pores called **stomata** (singular, *stoma*) that support photosynthesis by allowing the exchange of carbon dioxide and oxygen between the outside air and the leaf interior (see FIGURE 10.2). Stomata are also the main avenues by which water exits the leaves by evaporation. Changes in the shapes of the cells that border the stomata can close the pores to minimize water loss in hot, dry conditions.

Adaptations for Water Transport. Except for bryophytes, land plants have true roots, stems, and leaves, which are defined by the presence of vascular tissues. The two types of tissues that conduct materials in the plant's vascular system are **xylem** and **phloem** (FIGURE 29.11). Tube-shaped cells in the xylem carry water and minerals up from roots. These water-conducting cells are actually dead, with only their walls remaining to provide a system of microscopic water pipes. Phloem is a living tissue with nutrient-conducting cells arranged into tubes that distribute sugars, amino acids, and other organic products throughout the plant.

Secondary Compounds as Terrestrial Adaptations. Land plants produce many unique molecules called secondary compounds. They are so named because they are products of "secondary" metabolic pathways, side branches off the primary metabolic pathways that produce lipids, carbohydrates, and the other compounds common to all organisms.

Examples of secondary compounds in plants are alkaloids, terpenes, tannins, and phenolics such as flavonoids. Various al-kaloids, terpenes, and tannins have bitter tastes, strong odors, or toxic effects that help defend land plants against herbivorous animals. Flavonoids absorb harmful UV radiation, and some of these compounds also act as signals that function in symbiotic relationships with beneficial soil microbes (see FIGURE 37.13). Some phenolics deter attack by pathogenic microbes. Other phenolics play key structural roles in land plants. For example, the phenolic polymer called lignin hardens the cell walls of "woody" tissues in vascular plants, supporting even the tallest of trees.

Humans have found many uses, including medicinal applications, for secondary compounds extracted from plants. Just one example is the alkaloid quinine, used to help prevent malaria.

Now that we have examined many terrestrial adaptations that define land plants, let's take a closer look at how these descendants of aquatic algae colonized land.

THE ORIGIN OF LAND PLANTS

Land plants evolved from charophycean algae over 500 million years ago

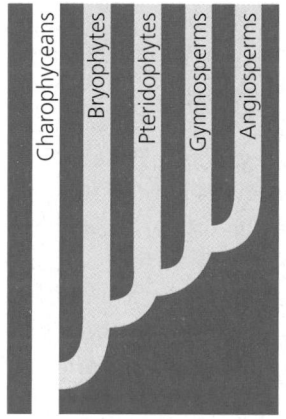

Let's review the evidence for a phylogenetic connection between land plants and green algae, especially the charophyceans:

- *Homologous chloroplasts.* The algal plastids most like plant chloroplasts are those of green algae and algal groups such as euglenoids, which acquired green algae as secondary endosymbionts (see FIGURE 28.25). The similarities include chlorophyll *b* and beta-carotene as accessory pigments and thylakoids stacked as grana. Comparisons of chloroplast DNA place the charophyceans as the green algae most closely related to land plants.

- *Homologous cellulose walls.* Though cellulose walls are found in a diversity of algae, the walls of charophyceans are the most plantlike, with cellulose making up 20–26% of the wall material in both plants and charophyceans. The cellulose-manufacturing rosettes found only in the plasma membranes of charophyceans and land plants are further evidence of cell wall homology.

- *Homologous Peroxisomes.* Charophyceans are the only algae with their anti-photorespiration enzymes packaged in peroxisomes, as they are in plants.

- *Phragmoplasts.* These assemblies occur during cell division only in plants and charophyceans.

- *Homologous sperm.* Many plants have flagellated sperm, which match charophycean sperm closely in ultrastructure.

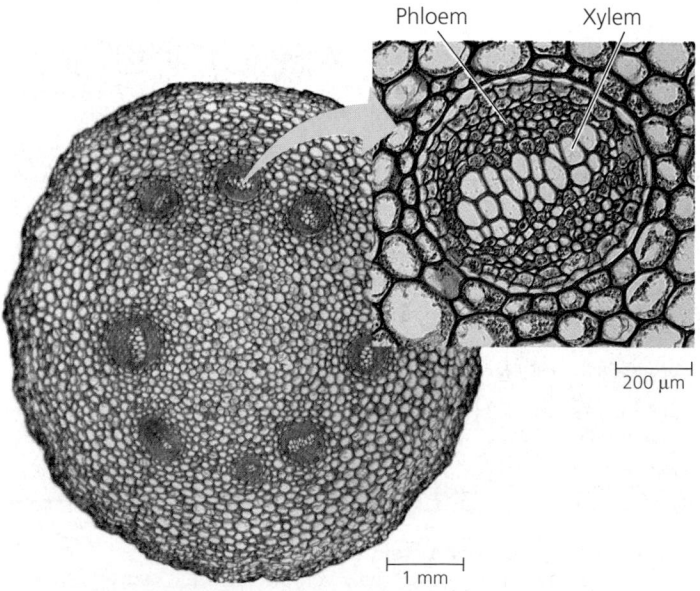

FIGURE 29.11 Xylem and phloem in the stem of *Polypodium*, a fern (a pteridophyte) (LM).

Phloem   Xylem

200 µm

1 mm

- *Molecular systematics.* In addition to comparing the chloroplast DNA of charophyceans and plants, molecular systematists have also compared key nuclear genes, such as those for ribosomal RNA and cytoskeletal proteins. The results support all the other evidence connecting charophyceans and plants to a common ancestor and point especially to the most complex charophyceans, such as *Chara* and *Coleochaete* (see FIGURE 29.2) as the algae most closely related to plants. However, it is important to understand that *modern Chara and Coleochaete* are not the ancestors of plants.

Fossils place plants on land over 500 million years ago. The oldest known traces of land plants are encased in mid-Cambrian rocks about 550 million years old. They are microscopic fossils of what paleobotanists interpret to be plant spores. Fossilized plant spores are plentiful in mid-Ordovician (460-million-year-old) deposits from around the world, suggesting that by then plants were abundant and widespread. Some of these fossilized spores occur in aggregates of four, which is the way some modern bryophytes release their spores. The tough walls of sporopollenin explain why spores are the plant structures most likely to leave a fossil record. However, fossilized spores are sometimes associated with scraps of tissue that may represent the remains of the sporophytes that produced the spores (FIGURE 29.12).

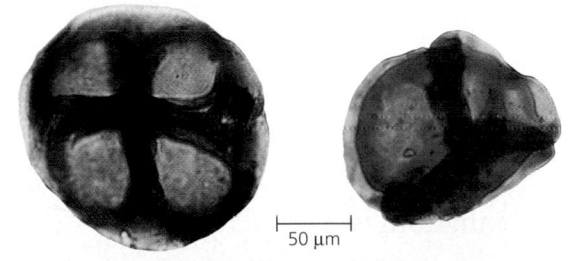

(a) Fossilized spores in tetrads (four-spore aggregates) (LM)

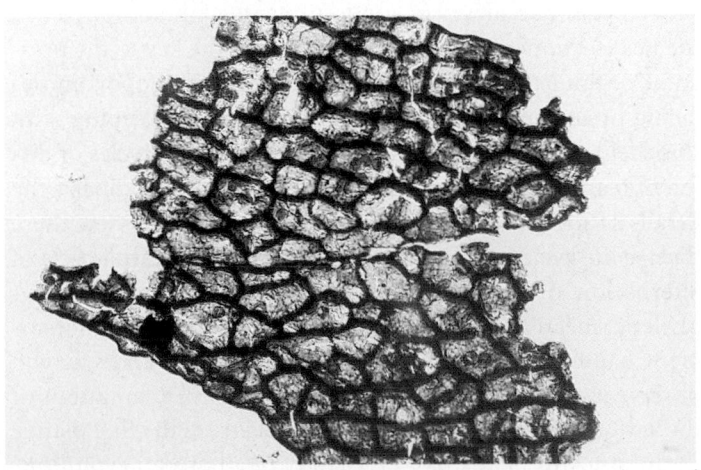

(b) Putative fossilized tissue of a sporophyte (LM)   50 μm

FIGURE 29.12 Fossilized spores and sporophyte tissue.

Alternation of generations in plants may have originated by delayed meiosis

The charophyceans *Chara* and *Coleochaete* are haploid organisms. There is no sporophyte generation, though these algae do retain and seem to nourish zygotes on the parental bodies, as do the gametophytes of land plants. The main difference is that the zygote of a charophycean undergoes *meiosis* to produce haploid spores, while the zygote of a plant undergoes *mitosis* to produce a multicellular sporophyte, which in turn produces haploid spores by meiosis. Thus, a reasonable hypothesis for the origin of plant sporophytes is that genetic change (mutations) in an ancestral charophycean caused a delay in meiosis until one or more mitotic divisions of the zygote had occurred (FIGURE 29.13). The result would be a multicellular, diploid sporophyte. This would increase the number of cells that undergo meiosis and thus increase the number of spores produced per zygote. Some botanists speculate that such an evolutionary breakthrough was an important adaptation for maximizing the output of sexual reproduction from each zygote on land, where a shortage of water decreased the probability of swimming sperm fertilizing eggs.

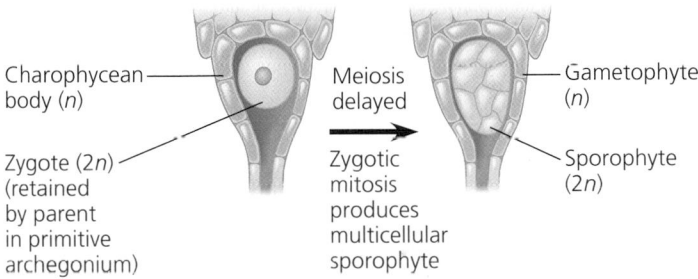

FIGURE 29.13 A hypothetical mechanism for the origin of alternation of generations in the ancestor of plants. Meiosis in a multicellular sporophyte, rather than direct meiosis by the zygote, would increase the number of haploid offspring (gametophytes) possible from each sexual union of sperm and ovum.

Adaptations to shallow water preadapted plants for living on land

Many species of charophycean algae inhabit shallow waters around the edges of ponds and lakes, where they are subject to occasional drying. In such environments, natural selection favors individual algae that can survive through periods when they are not completely submerged in water. A layer of sporopollenin prevents exposed charophycean zygotes from drying out until they are in water again. This chemical adaptation may have been the precursor to the tough spore walls so important to the terrestrial survival of plants. Such adaptations accumulated by at least one population of charophyceans enabled their descendants—the first plants—to live permanently above the waterline. The evolutionary novelties of these first land plants opened an expanse of terrestrial habitat previously occupied only by films of bacteria. The new frontier was spacious, the bright sunlight was unfiltered by water and algae, the atmosphere had an abundance of carbon dioxide, the soil was rich in mineral nutrients, and, at least at first, there were relatively few herbivores and pathogens. But these environmental benefits only became available with the evolution of the adaptations that enable plants to survive and reproduce on land, a very different place than the aquatic habitats of the ancestral algae.

Plant taxonomists are reevaluating the boundaries of the plant kingdom

THE PROCESS OF SCIENCE From our chapters on prokaryotes and protists (see Chapters 27 and 28), you are already aware that there is a revolution under way in how biologists view taxonomy as an outcome of phylogenetic analysis. That taxonomic turmoil also applies to the plant kingdom. Plant taxonomists are combining molecular systematics with cladistic analysis of the molecular data, as well as data from comparing the morphology, life cycles, and cell ultrastructure of organisms, to reevaluate the classification of plants. One international initiative, called **"deep green,"** is focusing on the deepest phylogenetic branching within the plant kingdom to identify and name the major plant clades (monophyletic groups). For the most part, these clades correspond to the ten phyla we will survey in these chapters on plant diversity.

Even "deeper" down the phylogenetic tree of plants is the branching of the whole land plant clade from its algal relatives. Because a phylogenetic tree consists of clades nested within clades, a debate about where to draw boundaries in a hierarchical taxonomy is inevitable. In particular, which clades do we want to include in a plant kingdom?

FIGURE 29.14 contrasts three versions of the plant kingdom. The traditional one equates the plant kingdom with the clade of embryophytes, or plants with embryos—what we have been calling land plants: bryophytes, pteridophytes, gymnosperms, and angiosperms. But some plant biologists now argue that the boundaries of the plant kingdom should be expanded to include the green algae most closely related to plants—the charophyceans and a few related groups. **Kingdom Streptophyta** is the name that has been proposed for this version of a plant kingdom. And an even broader definition of plants includes chlorophytes (the noncharophycean green algae) in a **kingdom Viridiplantae** (see Chapter 28). Exercising caution while the experts continue to debate, this text adopts the traditional embryophyte definition of the plant kingdom, and we'll use **kingdom Plantae** as the formal name for the taxon.

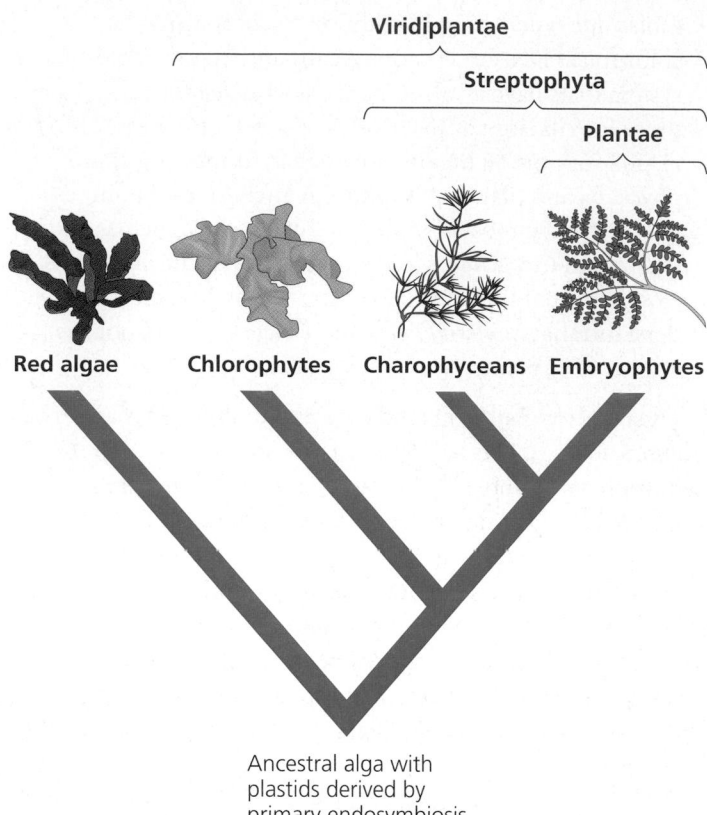

FIGURE 29.14 Three clades that are candidates for designation as the plant kingdom. This textbook adopts the embryophyte definition of plants and uses the name Plantae for the kingdom.

The plant kingdom is monophyletic

As we begin our survey of modern plants with bryophytes in the next section, remember that the past is the key to the present. Continue to think about the problems and opportunities facing organisms that began living on land. In keeping with this theme of evolution, we will compare the life cycles of different plant phyla. It is *not* important for you to memorize the details of these life cycles. What *is* important is to view these diverse life cycles as variations of a common ancestral cycle of alternation of generations. The plant kingdom is monophyletic, meaning that it is derived from a common ancestor (see Chapter 25). Thus, we can interpret the differences we will observe in the life cycles as special reproductive adaptations of the various plant phyla as they diversified from the first plants. In the study of plants, as in the study of all life, an evolutionary perspective connects what would otherwise be an overwhelming collection of seemingly unrelated facts.

BRYOPHYTES

The three phyla of bryophytes are mosses, liverworts, and hornworts

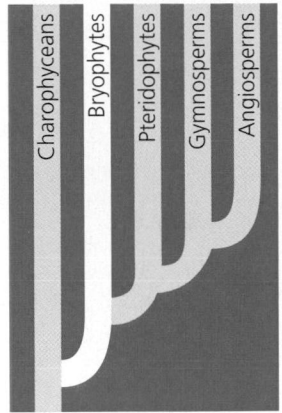

Bryophytes are represented today by three separate phyla: **phylum Hepatophyta,** commonly called **liverworts**; **phylum Anthocerophyta (hornworts)**; and **phylum Bryophyta (mosses)** (FIGURE 29.15). The suffix *wort* is an old Anglo-Saxon word meaning "herb," and liverworts and hornworts are indeed small herbaceous (nonwoody) plants named for their shapes. Mosses are the most familiar bryophytes. But some organisms commonly known as "mosses" are not really mosses or even bryophytes; these include Irish moss (a red seaweed), reindeer moss (a lichen), club mosses (pteridophytes), and Spanish moss (a flowering plant).

Note that the terms *Bryophyta* and *bryophyte* are not synonymous. Bryophyta is the formal taxonomic name for a certain phylum, the mosses. The term *bryophyte* refers informally to *all* the nonvascular plants—the phyla Hepatophyta, Anthocerophyta, and Bryophyta. These diverse bryophytes are not a monophyletic grouping. Gene sequence data, comparative cell structure, and other evidence indicate that liverworts, hornworts, and mosses diverged independently early in plant evolution, before the origin of vascular plants (see FIGURE 29.1). And as noted earlier, fossil evidence supports the hypothesis that bryophytes were the earliest plants. Though the identity of the modern bryophyte group most similar to the earliest plants is still uncertain, liverworts and hornworts may be the most reasonable models of what early plants were like. Most plant systematists now regard mosses as the bryophyte group most closely related to vascular plants (pteridophytes, gymnosperms, and angiosperms).

The gametophyte is the dominant generation in the life cycles of bryophytes

In all three bryophyte phyla, gametophytes are the most conspicuous, dominant phase of life history; sporophytes are typically smaller and present only part of the time.

If bryophyte spores are dispersed to a favorable habitat, such as moist soil or tree bark, they may germinate and grow into gametophytes by mitosis (FIGURE 29.16, p. 586). Germinating moss spores typically produce a mass of green, branched, one-cell-thick filaments, known as a **protonema,** that is sometimes mistaken for algae. Protonemata (plural, meaning "first threads")

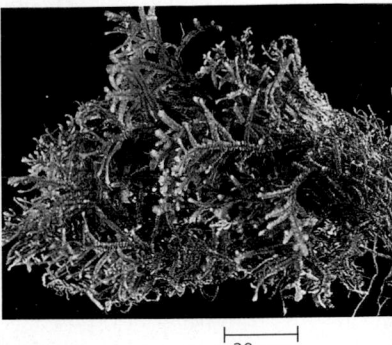

(a) Liverworts: Phylum Hepatophyta. The gametophytes of these two liverwort genera are very different in appearance. *Pallavicinia* (left) is a "thalloid" liverwort and *Porella* (right) is a "leafy" liverwort.

(b) Hornworts: Phylum Anthocerophyta. Hornworts are named for their hornlike sporophytes, which you can see here growing from their parental gametophytes.

(c) Moss: Phylum Bryophyta. The gametophytes of this star moss, *Polytrichum commune,* have produced the stalked sporophytes, which have caplike sporangia at their tips.

FIGURE 29.15 Bryophytes.

have a large surface area, enhancing absorption of water and minerals. When sufficient resources are available, a protonema produces buds having tissue-producing meristems. These meristems generate the mature, gamete-producing structure, which is known as a **gametophore** ("gamete bearer"). Together, the protonema and gametophores make up the gametophyte body of a moss.

Bryophyte gametophytes are generally only one or a few cells thick, which places all cells close to water and dissolved minerals. Most bryophytes lack conducting tissues that can distribute water and organic compounds within thick tissues. Though some bryophytes do have specialized tissues that function in water and solute conduction, cell walls in these tissues lack the lignin coating that is characteristic of vascular plant xylem. The absence of lignified vascular tissues also

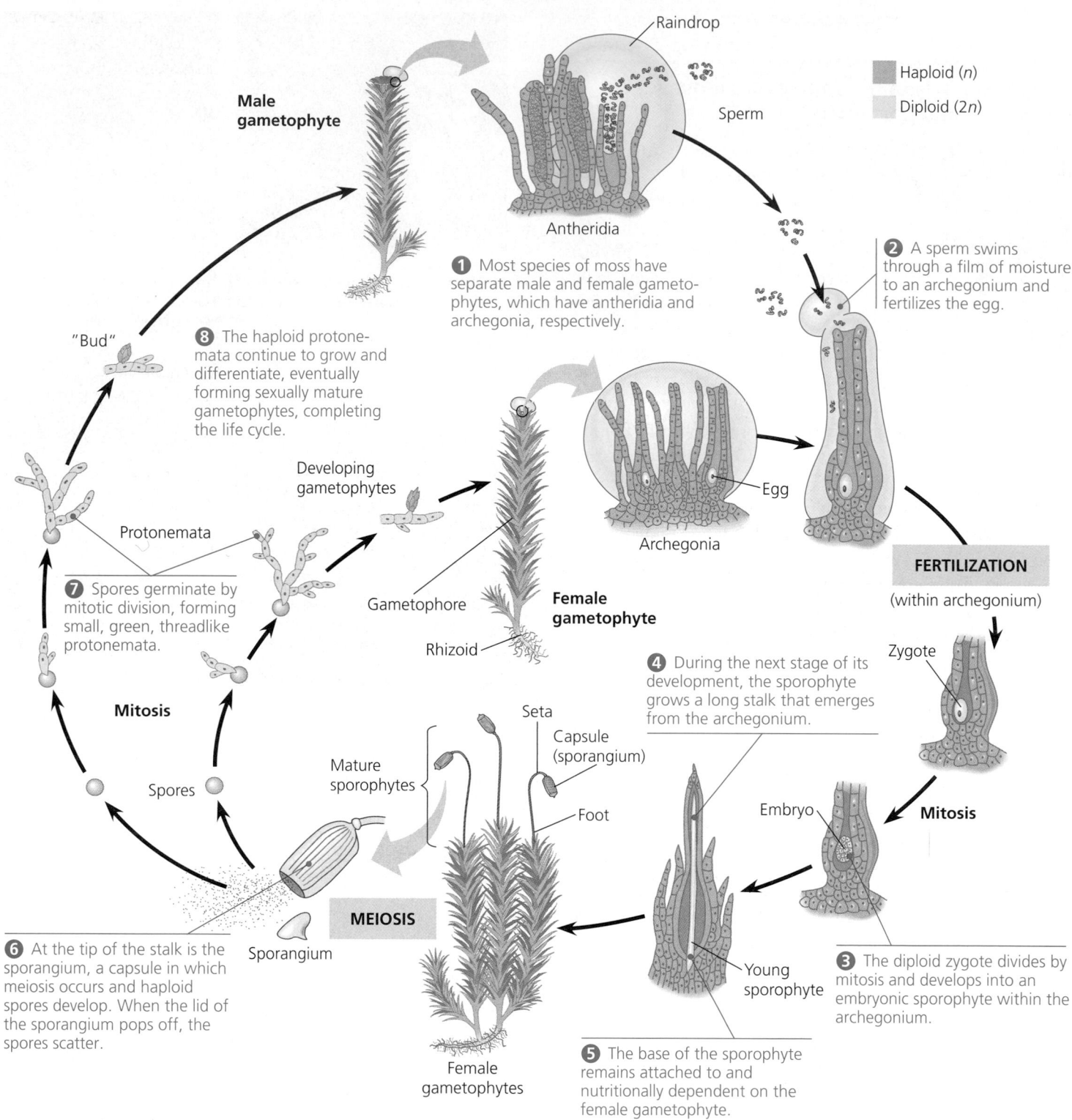

Raindrop

Sperm

Haploid (n)
Diploid (2n)

Male gametophyte

Antheridia

1 Most species of moss have separate male and female gametophytes, which have antheridia and archegonia, respectively.

2 A sperm swims through a film of moisture to an archegonium and fertilizes the egg.

"Bud"

8 The haploid protonemata continue to grow and differentiate, eventually forming sexually mature gametophytes, completing the life cycle.

Developing gametophytes

Egg

Protonemata

Archegonia

FERTILIZATION
(within archegonium)

7 Spores germinate by mitotic division, forming small, green, threadlike protonemata.

Gametophore

Female gametophyte

Zygote

4 During the next stage of its development, the sporophyte grows a long stalk that emerges from the archegonium.

Rhizoid

Mitosis

Seta

Embryo

Capsule (sporangium)

Mitosis

Mature sporophytes

Foot

MEIOSIS

Spores

6 At the tip of the stalk is the sporangium, a capsule in which meiosis occurs and haploid spores develop. When the lid of the sporangium pops off, the spores scatter.

Sporangium

Young sporophyte

3 The diploid zygote divides by mitosis and develops into an embryonic sporophyte within the archegonium.

5 The base of the sporophyte remains attached to and nutritionally dependent on the female gametophyte.

Female gametophytes

FIGURE 29.16 The life cycle of *Polytrichum,* a moss. The gametophyte is the prevalent generation in all bryophyte life cycles, a characteristic that contrasts with other plants.

limits the height of bryophytes, most of which are only a few centimeters tall. Thus, most bryophytes grow close to the ground, anchored by delicate, colorless **rhizoids.** Rhizoids are long, tubular single cells (in liverworts and hornworts) or filaments of cells (in mosses). Rhizoids are not composed of tissues, they lack specialized conducting cells, and they do not play a primary role in water and mineral absorption. In these ways, rhizoids differ from the roots of vascular plants.

The gametophytes of hornworts and some liverworts are flattened (see FIGURE 29.15a and b) and grow close to the ground. Moss gametophytes and those of some liverworts are described as "leafy" because they have stemlike structures that bear many leaflike appendages (see FIGURE 29.15a and c). These are not true stems and leaves because they lack lignin-coated vascular cells. Moss "leaves" are usually only one cell thick and often lack a cuticle, features that enhance water and mineral absorption from the moist environment. However, the common "hairy cap" moss *Polytrichum* (FIGURE 29.15c) and its close relatives have more complex "leaves" with ridges that enhance absorption of sunlight. These leaf ridges are coated with cuticle. *Polytrichum* and some other mosses also possess conducting tissues in the center of their "stems," and a few mosses can grow as tall as 2 m as a result. Plant biologists are still trying to determine whether the conducting tissues of mosses are homologous with vascular plant xylem and phloem or are instead the analogous result of evolutionary convergence.

When the gametophores of bryophytes are mature, they produce gametes in gametangia. Eggs are produced singly in vase-shaped archegonia, and large numbers of sperm are produced in elongate antheridia (see FIGURE 29.16). Jackets of protective tissue enclose both types of gametangia. Bryophyte gametophytes can produce many archegonia and antheridia; in mosses these are typically borne on separate male and female plants. Flagellated sperm are released into water films in which they swim toward eggs, passing down the openings of archegonia in response to chemical attractants. Eggs are not released, but remain within the bases of archegonia. Gamete fusion results in the formation of zygotes, which are also retained by the archegonia during their development into young sporophytes. Layers of placental nutritive cells help transport materials from parent gametophyte to embryos. These nutrients help support the development of embryos into mature sporophytes.

Bryophyte sporophytes disperse enormous numbers of spores

The cells of bryophyte sporophytes contain plastids that are usually green and photosynthetic when the sporophytes are young. Even so, bryophyte sporophytes are not capable of living apart from their maternal gametophytes. A bryophyte sporophyte remains attached to its parental gametophyte throughout the sporophyte's lifetime, dependent on the gametophyte for supplies of sugars, amino acids, minerals, and water.

Bryophytes have the smallest and simplest sporophytes of all modern plant groups, consistent with the hypothesis that sporophytes started small and simple, with larger and more complex sporophytes evolving only later in vascular plants. Among bryophytes, liverworts have the simplest sporophytes. You would need to use a magnifying glass to see their tiny bodies, which consist of a short stalk bearing a round sporangium with a protective epidermis enclosing developing spores. At the other end of the stalk, a nutritive foot is embedded in gametophyte tissues (FIGURE 29.17). Hornwort and moss sporophytes are larger and more complex. For example, hornwort sporophytes, which superficially resemble grass blades (see FIGURE 29.15b), possess a cuticle. The sporophytes of hornworts and mosses possess epidermal stomata similar to those of vascular plants.

Moss sporophytes are fairly conspicuous and familiar to most people. Though green and photosynthetic when young, moss sporophytes turn tan or brownish red when they are ready to release their spores, often remaining visible for months after spore discharge has occurred (see FIGURE 29.15c). Moss sporophytes consist of a **foot,** an elongated stalk known as a **seta,** and a spore-producing organ, the **sporangium,** or **capsule** (see FIGURE 29.16). The foot gathers sugars, amino acids, water, and minerals from the parent gametophyte via transfer cells. The seta conducts these materials to the capsule, which uses the resources to produce spores. In most mosses, the seta becomes elongated, which elevates the capsule and enhances spore dispersal.

The moss capsule (sporangium) is the site of meiosis and spore production; up to 50 million spores can be generated by one capsule. When the capsule is immature, it is covered with

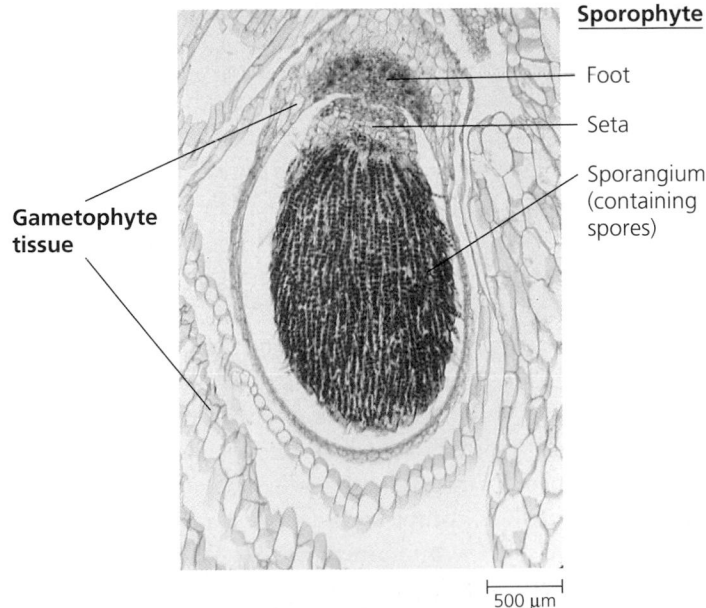

FIGURE 29.17 Sporophyte of *Marchantia*, a liverwort. The tiny spore-producing sporophyte remains attached to the tissues of the parent gametophyte, which continues to nourish the sporophyte (LM).

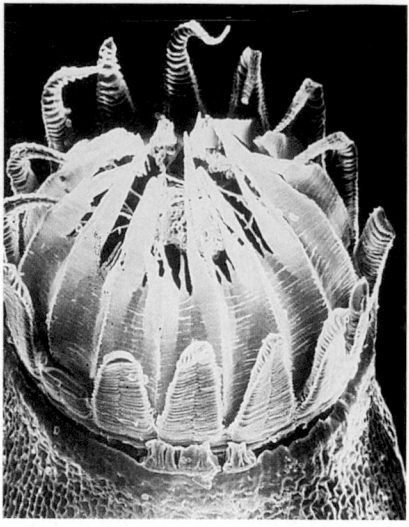

FIGURE 29.18 A moss sporangium. At the tip of the sporangium of the moss *Brachythecium,* two rings of "teeth" regulate the gradual release of spores from the capsule. When conditions are moist, the two rings of teeth interlock and keep the spores inside the capsule. Drying causes the teeth to separate, allowing spores to disperse (SEM).

a protective cap of gametophyte tissue, the **calyptra,** but this cap is lost when the capsule is ready to release spores. The upper part of the capsule, known as a **peristome,** is often specialized for gradual spore discharge (FIGURE 29.18). As a result, the capsule disperses spores gradually rather than all at once. This gradual release allows mosses to take advantage of periodic wind gusts that can carry spores long distances.

Bryophytes provide many ecological and economic benefits

Wind dispersal of lightweight spores has distributed bryophytes around the world. These plants are particularly common and diverse in moist alpine, boreal, temperate, and tropical forests, as well as wetlands, where they form essential habitats for a variety of tiny animals. Some mosses even inhabit such extreme environments as mountaintops, arctic and antarctic tundra, and deserts; similarly harsh habitats likely challenged the earliest plants. Mosses are able to exist in very cold or dry habitats because many are able to lose most of their body water without dying, then rehydrate and reactivate their cells when moisture again becomes available. In contrast, few vascular plants can survive the same degree of desiccation. Moreover, phenolic compounds in moss cell walls absorb damaging levels of UV and other short-wavelength radiation present in deserts or at high altitudes and latitudes.

One moss genus is especially abundant and widespread. *Sphagnum* is a wetland moss that forms extensive deposits of undecayed organic material known as **peat.** Thus, *Sphagnum*

(a) A peat bog in Oneida County, Wisconsin. *Sphagnum* is the low-profile plant growing near the water.

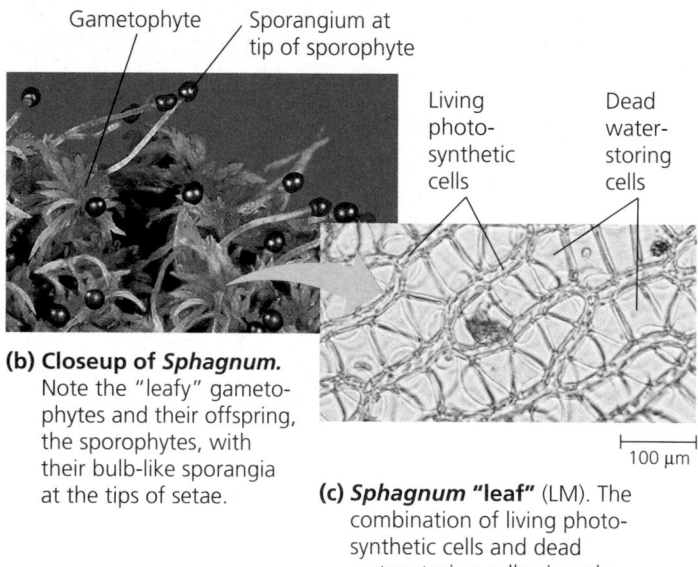

(b) Closeup of *Sphagnum.* Note the "leafy" gametophytes and their offspring, the sporophytes, with their bulb-like sporangia at the tips of setae.

(c) *Sphagnum* "leaf" (LM). The combination of living photosynthetic cells and dead water-storing cells gives the moss its spongy quality.

FIGURE 29.19 *Sphagnum,* or peat moss.

is also called peat moss. Boggy regions dominated by this moss are known as peat bogs (FIGURE 29.19), and extensive high-latitude boreal regions occupied by *Sphagnum* are among the wetlands called peatlands. *Sphagnum* and the peat formed from it do not decay readily because of the resistant phenolic compounds embedded in the moss's cell walls. *Sphagnum* moss also secretes acidic and phenolic compounds that may reduce bacterial activity. Low temperatures and nutrient levels typical of peat bogs also inhibit microbial decay.

The world's peatlands store an estimated 400 billion tons of organic carbon. As carbon reservoirs, peatlands play an important role in stabilizing Earth's atmospheric carbon dioxide concentrations, and hence climate (the role of atmospheric CO_2 in climate is discussed in Chapter 54).

In the past, *Sphagnum* moss was widely used by aboriginal people for diapers and during wartime as a naturally antiseptic packing material for wounds. Today, *Sphagnum* is widely harvested for use as a soil conditioner and for packing plant roots during shipment. The usefulness of this moss is based on the presence of large, dead, water-absorbing cells, in addition to green, photosynthetic cells (see FIGURE 29.19c). This enables dry moss to absorb 20 times its weight in water. Some ecologists are concerned that overharvesting of *Sphagnum* will reduce its beneficial ecological effects.

Bryophytes were probably Earth's only plants for the first 100 million years that terrestrial communities existed. Then vegetation began to take on a taller profile with the evolution of vascular plants.

THE ORIGIN OF VASCULAR PLANTS

You learned earlier in the chapter that modern vascular plants include the ferns and related plants (pteridophytes), gymnosperms, and flowering plants (angiosperms). And recall that vascular plants are equipped with food-transporting phloem as well as water-conducting xylem tissue with lignified cells. Vascular plants also differ from bryophytes in having a dominant sporophyte generation and **branched sporophytes** that become independent of parental gametophytes. Pteridophytes are described as **seedless vascular plants** because, unlike gymnosperms and angiosperms, they do not produce seeds. The first vascular plants were seedless (see FIGURE 29.1).

Additional terrestrial adaptations evolved as vascular plants descended from mosslike ancestors

Vascular plants likely inherited tissue-producing meristems, gametangia, embryos and the sporophytes that develop from them, stomata, cuticles, and sporopollenin-walled spores from mosslike ancestors. Branched, independent sporophytes probably evolved next. The evidence for this is provided by a group of Silurian fossils with the long descriptive name **protracheophyte** ("before vascular plants") **polysporangiophytes** ("plants producing many sporangia"). We'll abbreviate by just using the first part of the name.

Protracheophytes were like bryophytes in lacking lignified vascular tissue, but were different in having branched sporophytes that were not dependent on gametophytes for their growth. Branched sporophytes make more complex bodies possible and also enable individual plants to produce many more spores. And more complex forms with multiple sporangia would stand a better chance of surviving and reproducing in spite of losing some of their vegetation and sporangia to herbivores. In contrast to modern vascular plants, which have much reduced gametophytes, fossils of early species indicate life cycles with gametophytes and sporophytes that were about equal in size.

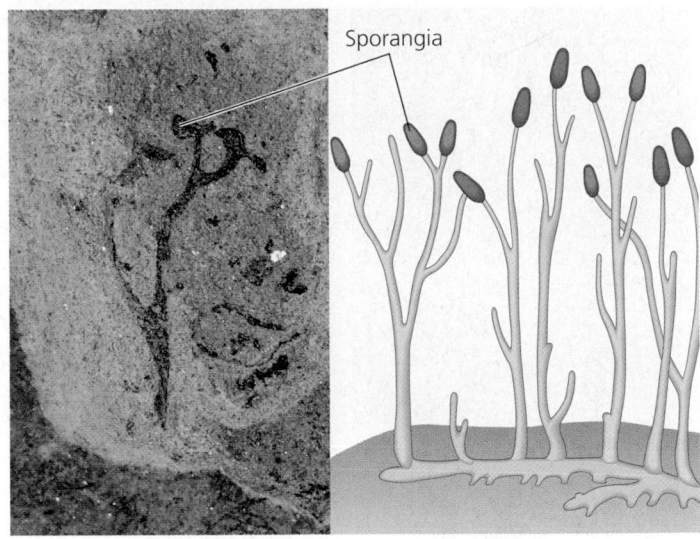

FIGURE 29.20 *Cooksonia*, a vascular plant of the Silurian. The dichotomous (forked like a Y) branching and terminal sporangia characteristic of *Cooksonia* are evident in the photograph of the fossil on the left. Water-conducting cells were present in true stems, but true roots and leaves were absent. The plant was anchored by a rhizome, a horizontal stem. *Cooksonia* grew in dense stands around marshes. The largest species was about 50 cm tall.

A diversity of vascular plants evolved over 400 million years ago

Cooksonia, named for the paleobotanist Isabel Cookson, is an extinct plant that left fossils over 408 million years old in the Silurian rocks of Europe and North America. It is the earliest-known vascular plant (FIGURE 29.20). Though its branched sporophytes stood only a few centimeters to 50 or so centimeters tall, *Cooksonia* had small lignified cells similar to the water-conducting cells in the xylem of modern pteridophytes (seedless vascular plants). Some of *Cooksonia*'s branches terminated in bulbous sporangia, the sporophyte organs that produce spores.

By Devonian times, 362–408 million years ago, a diversity of vascular plants had evolved, as revealed in the fossil record. The seven extant vascular plant phyla listed in TABLE 29.1 are their descendants, including the two modern phyla of seedless vascular plants.

PTERIDOPHYTES: SEEDLESS VASCULAR PLANTS

The two phyla of seedless vascular plants represented in the modern flora are the phylum Lycophyta (lycophytes) and phylum Pterophyta (ferns, whisk ferns, and horsetails). These two phyla probably had separate origins from different ancestors among the early vascular plants. Thus, the term "pteridophyte" is an informal one for all seedless vascular plants, while "Pterophyta" is the formal name for one seedless plant

(a) Phylum Lycophyta: _Lycopodium_, a club "moss." Club mosses (not really mosses) are common inhabitants of forest floors in the northeastern United States. The small plant has a horizontal rhizome that gives rise to roots and vertical branches and has true leaves containing strands of vascular tissue. The sporangia of lycophytes are borne by specialized leaves called sporophylls. In this lycophyte, the sporophylls are clustered at the tips of branches into club-shaped structures called strobili (hence the common name club mosses).

Strobili (sporophyll clusters)

(b) Phylum Pterophyta: _Psilotum_, a whisk fern. The scales of these dichotomously branching (Y-like branching) stems are not true leaves; they lack vascular tissue. The knobs along the stems are sporangia.

(c) Phylum Pterophyta: _Equisetum_, a horsetail. _Equisetum_ has an underground rhizome from which vertical stems arise. The straight, hollow stems are jointed, and whorls of small leaves or branches emerge at the joints. At the tips of some stems of _Equisetum_ are conelike structures bearing sporangia. The epidermis, the outer layer of cells, is embedded with silica, which gives the plants a gritty texture. Before scouring pads, people used the abrasive stems of horsetails to scrub pots and pans, which is why these plants are also known as scouring rushes.

(d) Phylum Pterophyta: _Polypodium vulgare_, a fern. Among modern flora, ferns are by far the most diverse and widespread of the seedless vascular plants.

FIGURE 29.21 Examples of pteridophytes (seedless vascular plants). The phyla Lycophyta and Pterophyta probably evolved from different ancient vascular plants. Based on recent evidence that whisk ferns, horsetails, and ferns make up a monophyletic group of seedless vascular plants, many plant systematists now place them all in the single phylum Pterophyta. All of the plants illustrated here are sporophytes, the dominant forms in the life cycles of all vascular plants.

phylum, the ferns and their relatives. FIGURE 29.21 illustrates examples of pteridophytes.

Pteridophytes provide clues to the evolution of roots and leaves

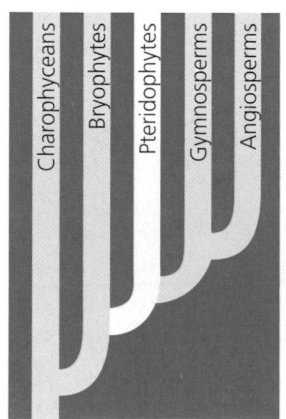

One reason plant biologists study the morphology of ferns and other seedless plants is to explore some of the terrestrial adaptations that evolved in vascular plants, including true roots and leaves.

Most pteridophytes have true roots with lignified vascular tissue. The vascular system in these roots resembles that of the stems of early vascular plants, some of which left fossils so well preserved that researchers can study their internal tissues. This suggests that pteridophyte roots probably evolved from the lowermost, subterranean portions of stems of ancient vascular plants. It is uncertain whether roots of seed plants (gymnosperms and angiosperms) arose independently, rather than being homologous to pteridophyte roots.

Lycophytes, the modern vascular plants that diverged earliest in plant phylogeny, have small leaves with only a single unbranched vein. These leaves probably evolved from tissue flaps on the surface of stems into which a strand of vascular tissue grew. Such leaves are known as **microphylls** (meaning "small leaves") (FIGURE 29.22a). In contrast, leaves of other modern vascular plants are known as **megaphylls** because they are typically much larger (FIGURE 29.22b). Their increased size is possible because such leaves are served by a highly branched vascular system. A branched vascular system can supply an expanded leaf with water and minerals and can export sugar from the photosynthetic tissue of the larger leaves. Thus, the branched vascular systems of megaphylls support a greater photosynthetic productivity compared to microphylls.

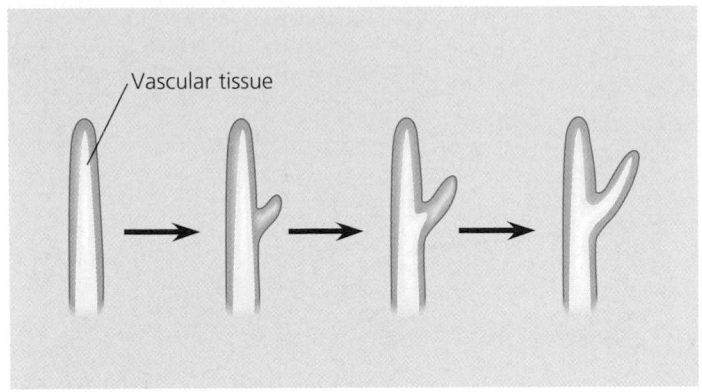

(a) Microphylls, such as those of lycophytes, probably originated as small stem outgrowths supported by single, unbranched strands of vascular tissue.

FIGURE 29.22 Hypotheses for the evolution of leaves.

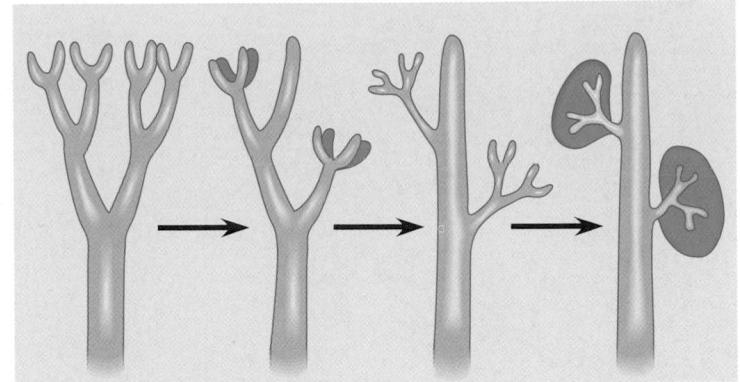

(b) Megaphylls, which have branched vascular systems, evolved by the fusion of branched stems.

The study of fossils suggests that megaphylls evolved from a series of branches lying close together on a stem. Hypothetical stages in such a megaphyll origin were a flattening of the branch system and the development of tissue webbing that joined branches (see FIGURE 29.22b). According to this hypothesis, megaphylls could not have evolved in plants that lacked branched sporophytes and lignified vascular tissue. Thus, the evolution of true, branched stems likely preceded the origin of large leaves and roots (which probably evolved from the bases of stems).

A sporophyte-dominant life cycle evolved in seedless vascular plants

Though pteridophytes lack seeds, they provide clues to the evolution of life cycle adaptations that enhanced the reproductive success of vascular plants on land.

From *Cooksonia* and the other early vascular plants to all of the vascular plants that live today, the sporophyte (diploid) generation is the larger and more complex plant in the alternation of generations. For example, the leafy fern plants familiar to us are the sporophytes. You would have to get down on your hands and knees and explore with careful hands and sharp eyes to find fern gametophytes, tiny plants growing on or just below the soil surface. Until you have a chance to do that, you can study the sporophyte-dominant life cycle of seedless vascular plants in FIGURE 29.23, page 592, which uses a fern as an example. Then, for review, compare this life cycle with FIGURE 29.16, which illustrates the gametophyte-dominated life cycles of bryophytes. In Chapter 30, you will see that the gametophyte generation became even more reduced during the evolution of seed plants, and it is there that we will consider this trend as an adaptation to living on land.

We can also use ferns to illustrate a key variation among vascular plants: the distinction between homosporous and heterosporous plants. The sporophyte of a **homosporous** plant produces a single type of spore. The fern depicted in FIGURE 29.23 is an example. Notice that each spore develops into a bisexual gametophyte having both female and male sex organs, the gametangia called archegonia and antheridia, respectively. In contrast, the sporophyte of a **heterosporous** plant produces two kinds of spores: **Megaspores** develop into female gametophytes bearing archegonia; **microspores** develop into male gametophytes with antheridia. Among ferns, those that returned to aquatic habitats during their evolution—the water ferns—are the only heterosporous species. However, we will see in Chapter 30 that the heterosporous condition was very important in the evolution of seeds.

The following diagrams will help you compare the homosporous and heterosporous conditions:

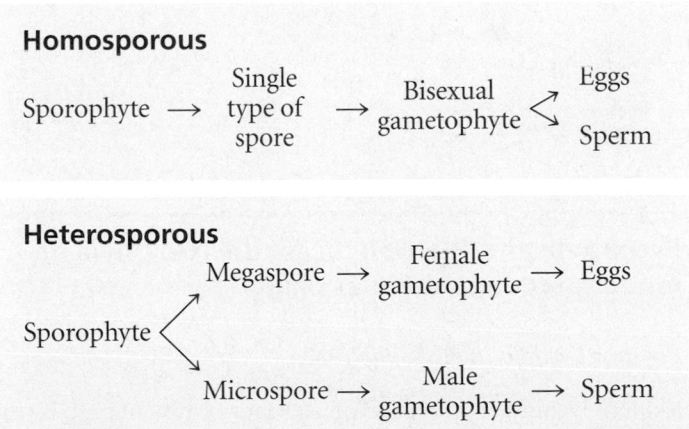

The life cycle in FIGURE 29.23 illustrates one more point: The sperm cells of ferns and all other seedless vascular plants (and even some seed plants) are flagellated and must swim through a film of water to reach eggs, a characteristic shared with bryophytes. With their swimming sperm and fragile gametophytes, seedless vascular plants are most common in relatively damp habitats.

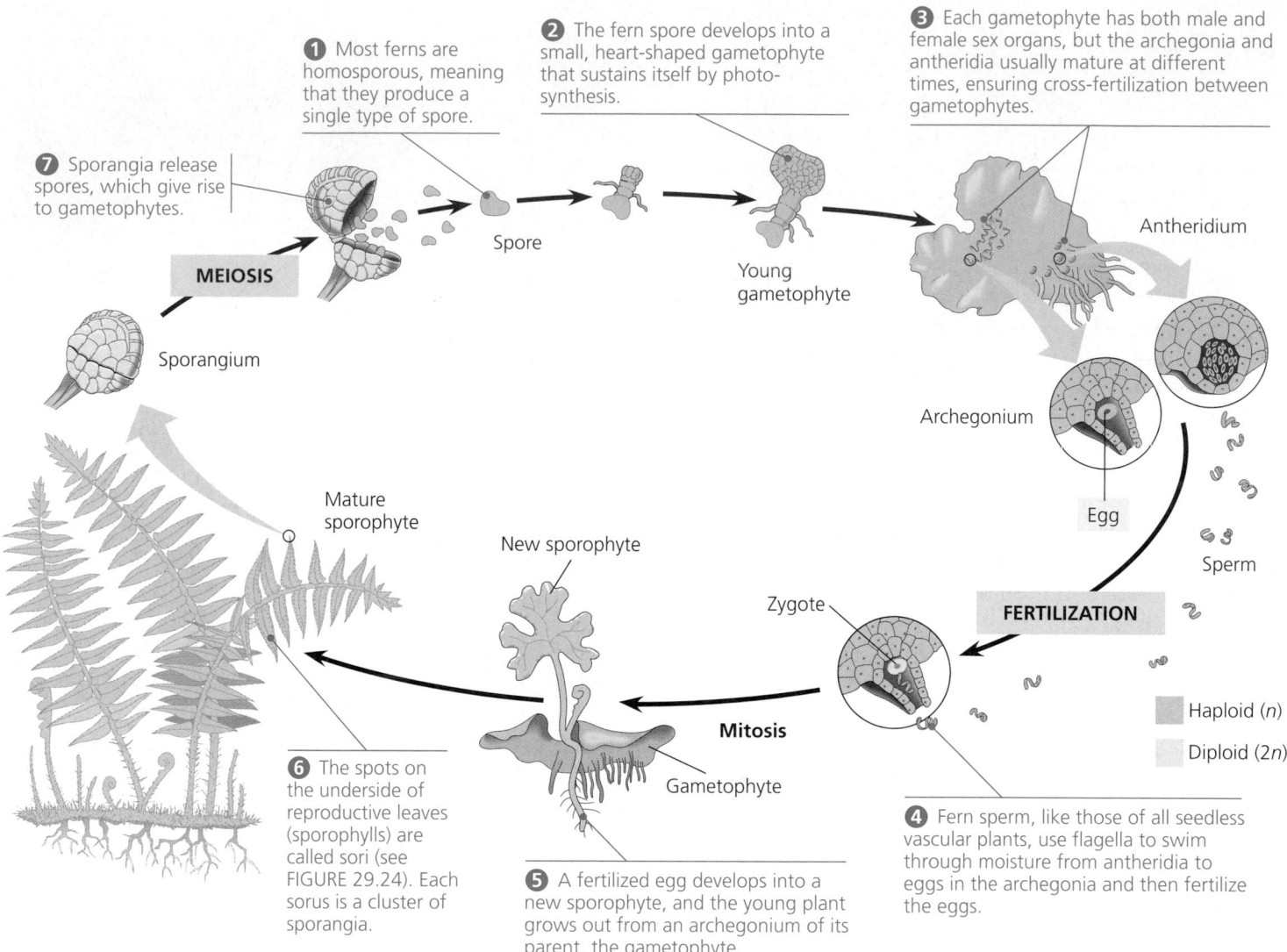

1 Most ferns are homosporous, meaning that they produce a single type of spore.

2 The fern spore develops into a small, heart-shaped gametophyte that sustains itself by photo-synthesis.

3 Each gametophyte has both male and female sex organs, but the archegonia and antheridia usually mature at different times, ensuring cross-fertilization between gametophytes.

7 Sporangia release spores, which give rise to gametophytes.

Antheridium

MEIOSIS

Spore

Young gametophyte

Sporangium

Archegonium

Egg

Sperm

Mature sporophyte

New sporophyte

Zygote

FERTILIZATION

Mitosis

Haploid (*n*)

Diploid (2*n*)

6 The spots on the underside of reproductive leaves (sporophylls) are called sori (see FIGURE 29.24). Each sorus is a cluster of sporangia.

Gametophyte

4 Fern sperm, like those of all seedless vascular plants, use flagella to swim through moisture from antheridia to eggs in the archegonia and then fertilize the eggs.

5 A fertilized egg develops into a new sporophyte, and the young plant grows out from an archegonium of its parent, the gametophyte.

FIGURE 29.23 The life cycle of a fern.

Lycophyta and Pterophyta are the two phyla of modern seedless vascular plants

Phylum Lycophyta (Lycophytes)

Modern lycophytes are relics of a far more eminent past. By the Carboniferous period, there were two evolutionary lines of lycophytes—one composed of relatively small, herbaceous plants and another line of giant woody trees having diameters of more than 2 m and heights of more than 40 m. The giant lycophytes thrived for millions of years in swamps common in the warm, moist Carboniferous climate, but became extinct when the climate became cooler and drier at the end of that geologic period. The small lycophytes survived, and they are represented today by about a thousand species. Common names for some lycophytes include "club mosses" or "ground pine," though they are neither mosses nor pines (see FIGURE 29.21a).

Many species of lycophytes are tropical plants that grow on trees as epiphytes—plants that use other organisms as a substratum but are not parasites. Other lycophyte species grow close to the ground on forest floors in temperate regions, including the northern United States. Lycophyte sporophytes have upright stems bearing many small green leaves (microphylls) and horizontal stems that grow along the ground surface. Roots extend downward from the horizontal stems. Groups of specialized leaves called **sporophylls,** which bear sporangia, are clustered to form club-shaped cones (also known as strobili). When they are mature, the oil-rich flammable spores are released in clouds. (Early photographers ignited lycophyte spores to provide a flash of light.) After the spores disperse, they develop into inconspicuous haploid gametophytes. Depending on the species, these tiny haploid plants may be either photosynthetic plants above ground or nonphotosynthetic plants underground. The underground species may live for as long as ten years, nurtured by symbiotic fungi.

Phylum Pterophyta (Ferns and Their Relatives)

Psilophytes. Plant biologists once considered *Psilotum*, the whisk fern, to be a "living fossil" (see FIGURE 29.21b). The dichotomous (Y-like) branching is reminiscent of ancient vascular plants, such as the *Cooksonia* in FIGURE 29.20. Whisk ferns also lack true leaves and roots, which reinforced the idea that they were surviving remnants of an early lineage of vascular plants. Thus, until recently, plant biologists placed *Psilotum* and a closely related genus in their own phylum. However, comparisons of DNA sequences and analysis of certain ultrastructural features, such as sperm structure, has convinced most plant systematists that the psilophytes are close relatives of the plants we commonly call ferns. According to this view, the lack of true roots and leaves evolved secondarily during the divergence of the psilophytes from the fern lineage.

Sphenophytes (Horsetails). Sphenophytes are commonly called horsetails because of their often brushy appearance (see FIGURE 29.21c). As with psilophytes, the sphenophytes were once granted their own phylum; but recent molecular data suggest that they are closely related to ferns and should be classified with them. During the Carboniferous period, sphenophytes were very diverse and grew as tall as 15 m, but today there exist only about 15 species of a single, widely distributed genus, *Equisetum.*

Horsetails are often found in marshy places or alongside streams and sandy roadsides. They possess both upright green stems and root-bearing horizontal stems (rhizomes) that extend along the ground. The erect stems are green and jointed, with a ring of tiny leaves or branches emerging from the joints. The upright stem actually serves as the major photosynthetic organ. Internally, horsetail stems are penetrated by large air canals that allow movement of oxygen down into rhizomes and roots, which often grow in waterlogged, low-oxygen soils. This adaptation probably first arose long ago in Carboniferous swamps, but it is still useful. Reproductive horsetail stems produce cones at their tips. *Equisetum*'s cones, like those of certain lycophytes, are composed of clusters of sporophylls—leaves bearing sporangia.

Ferns. From their Devonian beginning, ferns radiated extensively and today include more than 12,000 species. Ferns grew alongside tree lycophytes and horsetails in the great swamp forests of the Carboniferous period, but are by far the most widespread and diverse of the pteridophytes today. Ferns are most diverse in the tropics, but there are many representatives in temperate forests, and a few are able to grow in arid habitats. Ferns often have horizontal rhizomes from which grow large leaves with extensively branched vascular systems. Fern leaves are commonly known as fronds and are often compound—that is, divided into many leaflets (see FIGURE 29.21d). The fern frond grows as its coiled tip, the fiddlehead, unfurls. Tree ferns have tall erect stems bearing a crown of fronds at the top.

Ferns produce clusters of sporangia known as **sori** (singular, *sorus*) on the undersides of the green leaves or on special, nongreen leaves (sporophylls). Sori can be arranged in various patterns, such as parallel lines or dots, that are useful in fern identification (FIGURE 29.24). Most fern sporangia are equipped

(a) The fern *Polypodium.*

(b) On the underside of this sporophyll, you can see the sori, clusters of sporangia.

(c) Light micrograph of a sorus (singular)

200 μm

FIGURE 29.24 Fern sporophyll, a leaf specialized for spore production.

FIGURE 29.25 Artist's conception of a Carboniferous forest based on fossil evidence.
Most of the large trees with straight trunks are lycophytes. On the left, the tree with
numerous feathery branches is a horsetail. Tree ferns, though not featured in this painting,
were also abundant in the "coal forests" of the Carboniferous. Animals, including giant
dragonflies like the one near the horsetail, also thrived.

with springlike devices that catapult spores several meters. Once
airborne, spores can be blown by the wind far from their origin.
Spores, protected by sporopollenin, are the main means of dispersal for all seedless plants.

Seedless vascular plants formed vast "coal forests" during the Carboniferous period

The phyla Lycophyta and Pterophyta represent the modern
lineages of seedless vascular plants that formed forests during
the Carboniferous period about 290–360 million years ago
(FIGURE 29.25). Seedless vascular plants of the Carboniferous
forests left not only living relicts and fossils, but also fossil fuel
in the form of coal.

Coal formed during several geologic periods, but the most
extensive beds of coal are found in strata deposited during the
Carboniferous, a time when most of the continents were
flooded by shallow swamps. Europe and North America, located near the equator at that time, were covered by tropical
swamp forests. Dead plants did not completely decay in the
stagnant waters, and great depths of organic remains—
peat—accumulated by·a process similar to that occurring today in peat bogs. The swamps were later covered by the sea,
and marine sediments piled up on top of the peat. Heat and
pressure gradually converted the peat to coal, a "fossil fuel."
Coal powered the Industrial Revolution, and some politicians
advocate a resurgence in coal use as we continue to deplete oil
and gas reserves. However, burning more coal will increase the
industrial output of CO_2 and other "greenhouse gases" that
contribute to global warming (see Chapter 55). Energy conservation and the development of alternative (nonfossil fuel)
energy sources seem like prudent options instead.

■   ■   ■

Growing along with the seedless plants in Carboniferous
swamps were primitive seed plants. Though seed plants were
not the dominant plants at that time, they rose to prominence
after the swamps began to dry up at the end of the Carboniferous period. The next chapter traces the origin and diversification of seed plants, continuing our theme of terrestrial
adaptation.

CHAPTER 29 REVIEW

Go to the Campbell Biology website (www.campbellbiology.com) to explore an interactive version of the Chapter Review.

Summary of Key Concepts

AN OVERVIEW OF LAND PLANT EVOLUTION

■ **Evolutionary adaptations to terrestrial living characterize the four main groups of land plants (pp. 575–576, FIGURE 29.1, TABLE 29.1)** The four main plant groups are bryophytes (including mosses), pteridophytes (seedless vascular plants, including ferns), gymnosperms (including conifers), and angiosperms (flowering plants).
Web/CD Activity 29A: *Highlights of Plant Phylogeny*

■ **Charophyceans are the green algae most closely related to land plants (pp. 576–578, FIGURE 29.2)** The derived homologies that plants and charophyceans share include rosette cellulose-synthesizing complexes, peroxisomes, and phragmoplasts during cell division.

■ **Several terrestrial adaptations distinguish land plants from charophycean algae (pp. 578–582, FIGURES 29.3–29.11)** Five characteristics unique to plants are tissue-producing apical meristems; multicellular embryos dependent on the parent plant; alternation of generations that evolved independently of similar life cycles in some algae; sporangia that produce walled spores; and gametangia. In many plants, additional terrestrial adaptations, such as vascular tissues and secondary compounds, also evolved.
Web/CD Activity 29B: *Terrestrial Adaptations of Plants*

THE ORIGIN OF LAND PLANTS

■ **Land plants evolved from charophycean algae over 500 million years ago (pp. 582–583, FIGURE 29.12)** The oldest putative fossilized spores date back 550 million years.

■ **Alternation of generations in plants may have originated by delayed meiosis (p. 583, FIGURE 29.13)** The sporophyte may have had its origins in mitotic division of a zygote before the meiotic production of spores, which would have amplified the output of offspring from each fertilization event.

■ **Adaptations to shallow water preadapted plants for living on land (pp. 583–584)** Periodic drops in water level would have selected for characteristics, such as sporopollenin-protected spores, that contributed to eventual success on land.

■ **Plant taxonomists are reevaluating the boundaries of the plant kingdom (p. 584, FIGURE 29.14)** This text equates the plant kingdom with the embryophytes, but alternative boundaries extend to the charophyceans or even to all green algae.

■ **The plant kingdom is monophyletic (p. 584)** Many adaptations of the ancestral life cycle have evolved.

BRYOPHYTES

■ **The three phyla of bryophytes are mosses, liverworts, and hornworts (p. 585, FIGURE 29.15)** Mosses (Bryophyta) are the most diverse and widespread bryophytes. Liverworts (Hepatophyta) are named for their shape. Hornworts (Anthocerophyta) are named for their hornlike sporophytes.

■ **The gametophyte is the dominant generation in the life cycles of bryophytes (pp. 585–587, FIGURE 29.16)** A mat of moss consists of haploid gametophytes, with the diploid sporophytes growing out of archegonia, dependent on the gametophyte for nourishment. Flagellated sperm require a film of water to reach eggs.
Web/CD Activity 29C: *Moss Life Cycle*

■ **Bryophyte sporophytes disperse enormous numbers of spores (pp. 587–588, FIGURES 29.17, 29.18)** Spores are the main dispersal stage in the life histories of bryophytes.

■ **Bryophytes provide many ecological and economic benefits (pp. 588–589, FIGURE 29.19)** *Sphagnum* covers great expanses of land as peat bogs and peatlands, playing an important role in Earth's carbon cycle.

THE ORIGIN OF VASCULAR PLANTS

■ **Additional terrestrial adaptations evolved as vascular plants descended from mosslike ancestors (p. 589)** The study of fossils called protracheophytes suggests that branched sporophytes preceded vascular tissue as terrestrial adaptations.

■ **A diversity of vascular plants evolved over 400 million years ago (p. 589, FIGURE 29.20)** *Cooksonia*, the oldest known fossil of a vascular plant, dates back to the Silurian period.

PTERIDOPHYTES: SEEDLESS VASCULAR PLANTS

■ **Pteridophytes provide clues to the evolution of roots and leaves (pp. 590–591, FIGURE 29.22)** True roots possibly evolved from the bases of stems. Microphylls are small leaves with unbranched veins. Megaphylls, which have branched vascular tissue, probably evolved from clusters of nearby branches.

■ **A sporophyte-dominant life cycle evolved in seedless vascular plants (p. 591, FIGURE 29.23)** One variation in this life cycle is the contrast between homosporous and heterosporous plants. The ancestral condition of flagellated sperm is retained by all seedless vascular plants.
Web/CD Activity 29D: *Fern Life Cycle*
Web/CD Case Study in the Process of Science: *What Are the Different Stages of a Fern Life Cycle?*

■ **Lycophyta and Pterophyta are the two phyla of modern seedless vascular plants (pp. 592–594, FIGURE 29.24)** Lycophytes include the club mosses, with some shoots ending in clusters of sporophylls. Psilophytes (whisk ferns) and sphenophytes (horsetails) are now classified with the ferns in phylum Pterophyta.

■ **Seedless vascular plants formed vast "coal forests" during the Carboniferous period (p. 594, FIGURE 29.25)** Coal formed from peat, the partially decomposed bodies of swamp plants.

1. Which of the following characteristics of plants is absent in their closest relatives, the charophycean algae?
   a. chlorophyll *b*
   b. cellulose in cell walls
   c. alternation of multicellular generations
   d. sexual reproduction
   e. formation of a cell plate during cytokinesis

2. All bryophytes (mosses, liverworts, and hornworts) share certain characteristics. These are
   a. reproductive cells in gametangia; embryos
   b. branched sporophytes
   c. vascular tissues, true leaves, and a waxy cuticle
   d. seeds
   e. lignified walls

3. Which of the following is *not* common to all phyla of vascular plants?
   a. the development of seeds
   b. alternation of generations
   c. dominance of the diploid generation
   d. xylem and phloem
   e. the addition of lignin to cell walls

4. A heterosporous plant is one that
   a. produces a gametophyte that bears both antheridia and archegonia.
   b. produces microspores and megaspores, which give rise to male and female gametophytes.
   c. produces spores all year long instead of during just one season.
   d. produces two kinds of spores, one asexually by mitosis and the other sexually by meiosis.
   e. reproduces only sexually.

5. During the Carboniferous period, the dominant plants, which later formed the great coal beds, were mainly
   a. giant lycophytes, horsetails, and ferns.
   b. conifers.
   c. angiosperms.
   d. charophyceans.
   e. early seed plants.

6. A land plant that produces flagellated sperm and has a diploid-dominant generation is most likely a
   a. fern.
   b. moss.
   c. liverwort.
   d. charophycean.
   e. hornwort.

Questions 7–10.
For each of the following structures or life cycle stages, indicate whether the cells are haploid or diploid in chromosome number.

7. The body of a charophycean
8. The nonreproductive cells that line the gametangia of a moss
9. The cells that make up the stalk of a moss sporophyte
10. The spores produced by the sporophyte of a fern

11. What are three adaptations of plants for living on land?
12. Why are coal, oil, and natural gas called "fossil" fuels?
13. Why does the sexual life cycle of humans not qualify as "alternation of generations"?
14. In terms of the alternation of generations, what is the biggest difference between the life cycle of a moss and the life cycle of a fern?
15. What is the evidence that the cellulose walls of plants and charophycean algae had a common evolutionary origin?

Go to the website or CD-ROM for more quiz questions.

Draw a cladogram that includes a moss, a fern, and a gymnosperm. Use a charophycean alga as the out-group. (See Chapter 25 if you need to review cladistics.) Label each branch of the cladogram with at least one derived characteristic unique to that clade.

In April 1986, an accident at a nuclear power plant in Chernobyl, Ukraine, scattered radioactive fallout for hundreds of miles. In assessing the biological effects of the radiation, researchers found mosses to be especially valuable as organisms for monitoring the damage. Radiation damages organisms by causing mutations. Explain why it is faster to observe the genetic effects of radiation on bryophytes than on plants from other groups. Imagine that you are conducting tests shortly after a nuclear accident. Using potted moss plants as your experimental organisms, design an experiment to test the hypothesis that the frequency of mutations decreases with the organism's distance from the source of radiation.

Explore different stages in the life cycle of the fern in the Case Study in the Process of Science, available on the website and CD-ROM.

Nonvascular plants and seedless vascular plants are extremely common, and several of them have important economic and ecological uses. And yet, essentially no member of these ancient plant groups is of agricultural importance. Why? Can you think of attributes that these groups lack that may be precisely why so many seed plants are of agricultural significance to society? And attributes they possess that limit their agricultural utility?

Answers: 1. c; 2. a; 3. a; 4. b; 5. a; 6. a; 7. Haploid; 8. Haploid; 9. Diploid; 10. Haploid; 11. Any three of the following: cuticle; stomata; vascular tissue, gametangia that protect gametes; protected embryos; sporangia that produce sporopollenin-protected spores; meristems; differentiation of the body into a subterranean root system and stems and leaves above ground. 12. Because they are derived from ancient organisms that did not decay completely after dying. 13. Because the human life cycle does not include a haploid stage that is multicellular. 14. Gametophyte-dominant life cycle in a moss; sporophyte-dominant life cycle in a fern. 15. The presence of rosette cellulose-producing complexes in the plasma membranes of plants and charophyceans, but not in other algae with cellulose walls.

PLANT DIVERSITY II: THE EVOLUTION OF SEED PLANTS

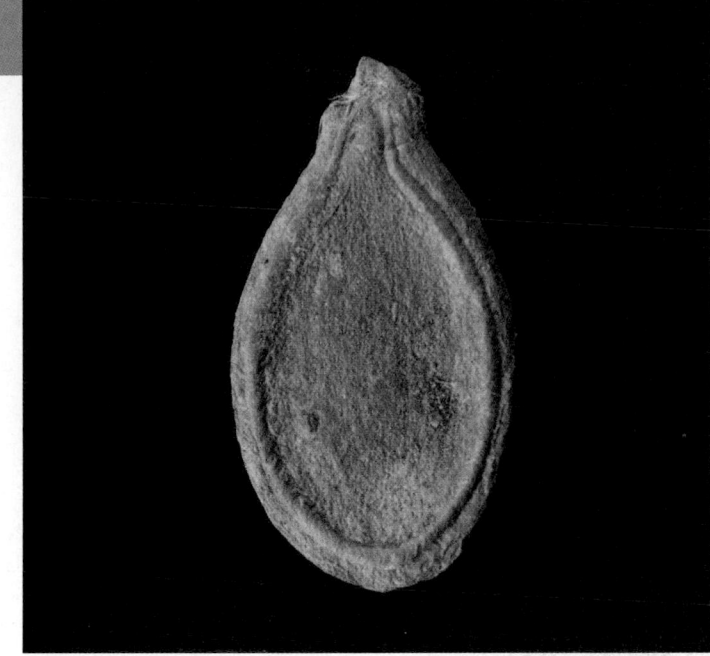

OVERVIEW OF SEED PLANT EVOLUTION

- Reduction of the gametophyte continued with the evolution of seed plants
- Seeds became an important means of dispersing offspring
- Pollen eliminated the liquid-water requirement for fertilization
- The two clades of seed plants are gymnosperms and angiosperms

GYMNOSPERMS

- The Mesozoic era was the age of gymnosperms
- The four phyla of extant gymnosperms are ginkgo, cycads, gnetophytes, and conifers
- The life cycle of a pine demonstrates the key reproductive adaptations of seed plants

ANGIOSPERMS (FLOWERING PLANTS)

- Systematists are identifying the angiosperm clades
- The flower is the defining reproductive adaptation of angiosperms
- Fruits help disperse the seeds of angiosperms
- The life cycle of an angiosperm is a highly refined version of the alternation of generations common to all plants
- The radiation of angiosperms marks the transition from the Mesozoic era to the Cenozoic era
- Angiosperms and animals have shaped one another's evolution

PLANTS AND HUMAN WELFARE

- Agriculture is based almost entirely on angiosperms
- Plant diversity is a nonrenewable resource

Why was this discovery front-page news for many newspapers and scientific journals? The cave had been occupied by humans some 8,000–10,000 years ago, and the seeds and pieces of squash fruit found there differ enough from wild varieties of the same species to suggest that humans had already begun cultivating the plants, perhaps more for gourds than for food. If this interpretation is correct, then agriculture started at about the same time— approximately 10,000 years ago—in Asia, Europe, and the Americas. (Before this discovery, scientists had been able to trace American agriculture back only about 5,000 years.) The invention of agriculture, which depends almost entirely on the cultivation and harvest of seed plants, was the single most important cultural change in the history of humanity, for it made possible the transition from hunter-gatherer societies to permanent settlements.

More generally, the seeds and other adaptations of gymnosperms and angiosperms enhanced the ability of plants to survive and reproduce in diverse terrestrial environments, and these plants became the main producers supporting the food webs of most ecosystems on land. Our study of gymnosperms and angiosperms begins with an overview of some of the key terrestrial adaptations that seed plants added to those already in place in the bryophytes and seedless vascular plants you learned about in Chapter 29.

This chapter continues *the saga of how plants adapted to land and transformed Earth. The beautifully preserved squash seed in the photo, discovered in 1997 in a cave in Oaxaca, Mexico, symbolizes two important landmarks in the evolution of plants: (1) the evolution of seed plants, which led to the gymnosperms and angiosperms, the plants that dominate most modern landscapes, and (2) the emergence of the importance of seed plants to animals, specifically to humans in this case.*

OVERVIEW OF SEED PLANT EVOLUTION

Seed plants are vascular plants that produce seeds. Modifications in life cycles contributed to the success of seed plants as terrestrial organisms. Here we examine the three most important reproductive adaptations: continued reduction of the gametophyte, the advent of the seed, and the evolution of pollen.

Reduction of the gametophyte continued with the evolution of seed plants

In Chapter 29, we discussed an important distinction between the life cycles of mosses and other bryophytes and the life cycles of ferns and other seedless vascular plants: a gametophyte-dominant life cycle for bryophytes versus a sporophyte-dominant life cycle for seedless vascular plants. That trend continued with the evolution of seed plants—the gymnosperms and angiosperms. The gametophytes of seed plants are even more reduced than the gametophytes of seedless vascular plants such as ferns. And in seed plants, the miniature female gametophytes develop from spores that are retained within the sporangia of the parental sporophyte. One advantage of this arrangement is that the delicate female gametophytes do not have to cope with many environmental stresses. These female gametophytes and the young embryos they produce after fertilization are sheltered from drought and harmful UV radiation by their enclosure within the moist reproductive tissues of the parental sporophyte generation. This relationship also makes it possible for the gametophytes to obtain nutrients from their parents. In contrast, the free-living gametophytes of seedless vascular plants must fend for themselves (see FIGURE 29.23).

This protection of female gametophytes within sporophyte tissues required extreme miniaturization of the gametophytes. The free-living gametophytes of seedless vascular plants (pteridophytes), though small compared to the sporophytes, are visible to the unaided eye. But the gametophytes of seed plants are microscopic. FIGURE 30.1 reviews this terrestrial adaptation of seed plants by contrasting the sporophyte-gametophyte relationships for different plant groups.

Why has the gametophyte generation not been completely eliminated from the plant life cycle? One hypothesis is that haploid gametophytes provide a mechanism for "screening" alleles, including new mutations. Gametophytes with deleterious mutations that affect basic metabolism and cell division will not survive to produce gametes that could combine to start new sporophytes.

Another possible reason that the gametophyte has not gone completely "out of style" in seed plants is that all sporophyte embryos are dependent, at least to some extent, on tissues of the maternal gametophyte. You learned, for example, that in bryophytes the embryonic sporophyte is nourished by a gametophyte as it grows from the archegonium. You'll soon learn that even in seed plants, the gametophyte continues to play a role in nourishing the sporophyte embryo, at least during its early development.

FIGURE 30.1 Three variations on gametophyte/sporophyte relationships.

Gametophyte (*n*)

Sporophyte (2*n*)

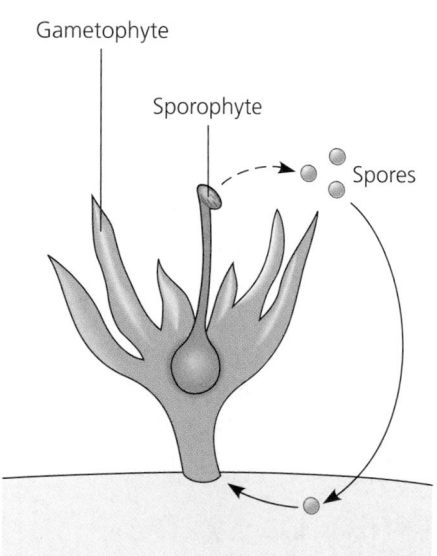

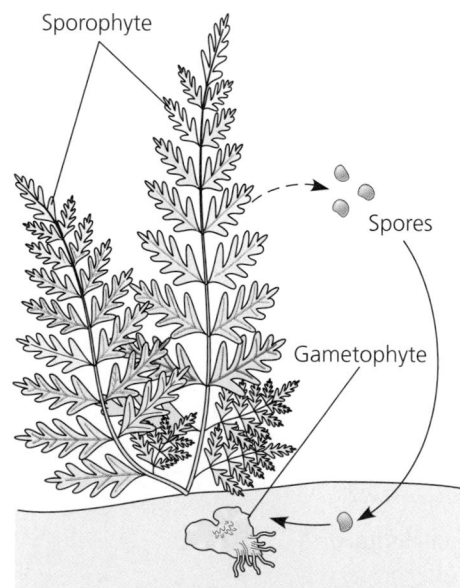

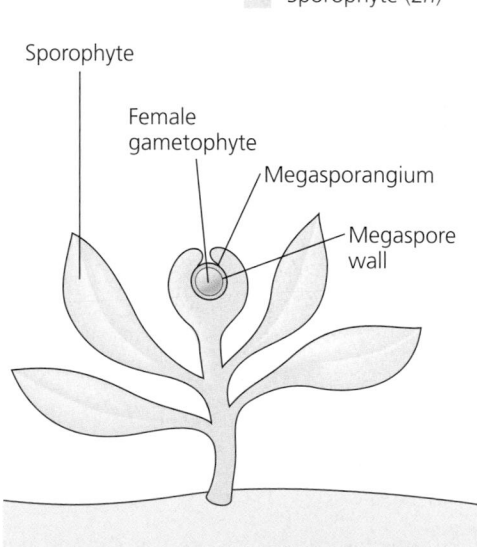

(a) Sporophyte dependent on gametophyte (e.g., bryophytes). In the gametophyte-dominant life cycle of mosses and other bryophytes, the dependent sporophyte is nourished by the gametophyte as it grows out of the archegonium.

(b) Large sporophyte and small, independent gametophyte (e.g., ferns). The sporophyte is the dominant generation in the life cycles of all vascular plants. The gametophytes of most ferns, though small, are photosynthetic and free-living (not dependent on the sporophyte for their nutrition).

(c) Reduced gametophyte dependent on sporophyte (seed plants). The gametophytes of seed plants are surrounded by tissues of the sporophyte, from which the gametophyte derives its nutrition. Note that the gametophyte develops within the confines of a spore.

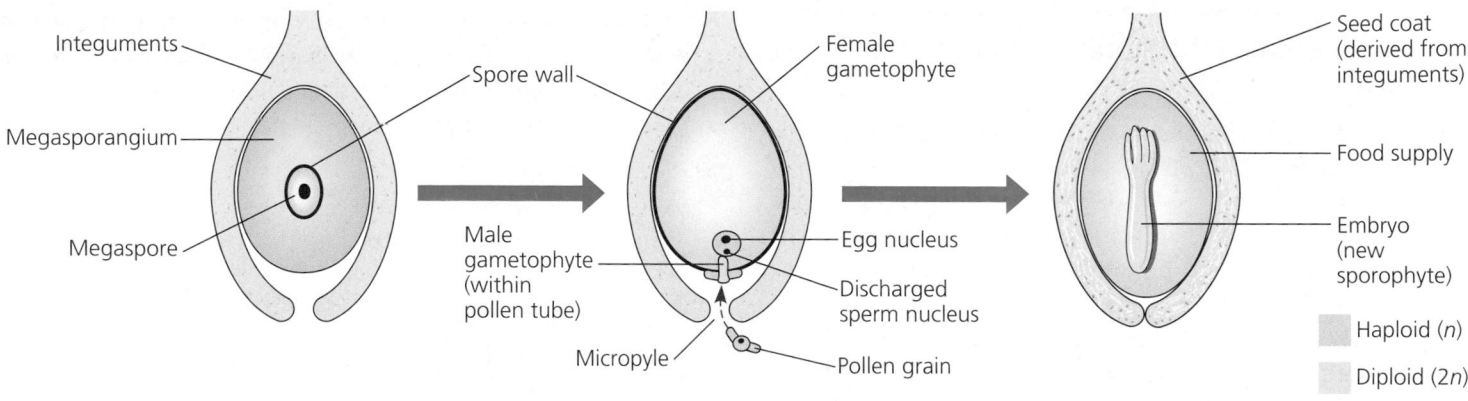

(a) Unfertilized ovule. In this sectional view through a generalized ovule of a gymnosperm, a fleshy megasporangium is surrounded by protective layers of tissue called the integuments.

(b) Fertilized ovule. A megaspore develops into a multicellular female gametophyte. The micropyle, the only opening through the integuments, allows entry of a pollen grain. The pollen grain contains a male gametophyte, which develops a pollen tube that discharges sperm.

(c) Gymnosperm seed. Fertilization initiates the transformation of the ovule into a seed, which consists of a sporophyte embryo, a food supply, and a protective seed coat derived from the integuments.

FIGURE 30.2 From ovule to seed.

Seeds became an important means of dispersing offspring

In bryophytes and seedless vascular plants such as ferns, spores produced by sporophytes are the resistant stage in the life cycle—that is, the stage that can withstand harsh environments. (A spore, recall from Chapter 29, is a resistant cell that can develop into a new organism.) For example, the spores of a moss might be able to survive even if the local environment becomes too cold, too hot, or too dry for the moss plants themselves to live. And because of their tiny size, the spores might also be dispersed in a dormant state to a new area, where they will germinate to give rise to new moss gametophytes if and when the environment is favorable enough for them to break dormancy. Spores were the main way that plants spread over the Earth for the first 200 million years of plant life on land.

The seed represents a different solution to resisting harsh environments and dispersing offspring. In contrast to a spore, which is single-celled, a seed is a resistant structure that is multicellular and much more complex. A **seed** consists of a sporophyte embryo packaged along with a food supply within a protective coat. There are evolutionary and developmental relationships between spores and seeds. Recall from the previous section that the reduced gametophytes of seed plants develop within tissues of the parental sporophyte. This occurs because the parent sporophyte does not release its spores, but instead retains them within its sporangia. Not only are the spores retained within the sporangia instead of being released, but the gametophyte develops within the confines of the spore from which it is derived (see FIGURE 30.1c).

All seed plants are heterosporous, meaning they have two different types of sporangia that produce two types of spores: Megasporangia produce megaspores, which give rise to female (egg-containing) gametophytes; microsporangia produce microspores, which give rise to male (sperm-containing) gametophytes. And recall from Chapter 29 that some seedless vascular plants, including certain ferns, are also heterosporous. The gametophytes of those plants develop within the spore walls, as they do in seed plants. What distinguishes seed plants is that the megaspores, and hence the female gametophytes, are retained on the parent sporophyte.

In seed plants, layers of sporophyte tissues called **integuments** envelop the megasporangium. Thus a megaspore formed within the megasporangium is very well protected. The whole structure—integuments, megasporangium, and megaspore—is called an **ovule** (FIGURE 30.2a). A female gametophyte develops inside a megaspore and produces one or more egg cells. And if an egg cell is fertilized by a sperm (FIGURE 30.2b), the zygote develops into a sporophyte embryo. The whole ovule develops into a seed (FIGURE 30.2c).

A seed's protective seed coat is derived from the integuments of the ovule. Once released from the parent plant, a seed may remain dormant for days, months, or even years. Under favorable conditions, the seed can then germinate, its sporophyte embryo emerging through the seed coat as a seedling. Some seeds drop close to their parents; others are carried far by wind or animals (FIGURE 30.3).

FIGURE 30.3 Winged seed of a White Pine (*Pinus strobus*).

Pollen eliminated the liquid-water requirement for fertilization

We have seen the relationship of the megasporangium to the ovule and the seed, but what goes on in a microsporangium of a seed plant? Microspores develop into pollen grains, which mature to become the male gametophytes of seed plants. The pollen grains, protected by tough sporopollenin-containing coats (see p. 580), can be carried away by wind or animals after their release from the microsporangium.

The transfer of pollen to ovules is called **pollination.** If a pollen grain lands in the vicinity of an ovule, it will elongate a tube that discharges one or more sperm into the female gametophyte within the ovule (see FIGURE 30.2b). In some gymnosperms, the sperm cells retain the ancestral flagellated condition. But in the most common gymnosperms (the conifers) and in all angiosperms (flowering plants), the sperm lack flagella.

This mechanism for transfer of sperm contrasts sharply with what we observed in seedless plants. Recall that in bryophytes and pteridophytes such as ferns, flagellated sperm released from antheridia must swim through a film of water to reach egg cells in archegonia. The distance for this sperm transport rarely exceeds a few centimeters. In seed plants, the use of resistant, far-traveling, airborne pollen to bring gametes together is a terrestrial adaptation that led to even greater success and diversity of plants on land.

The two clades of seed plants are gymnosperms and angiosperms

Progress in biology due to new methods, new data, and new ideas is what makes this subject so much fun. Each new edition of this textbook requires revision of most of the phylogenetic trees to reflect the ongoing revolution in systematics. Phylogenetic trees and the classifications they inform, remember, are hypotheses about the history of life. And in the case of seed plant phylogeny, the data now support a tree with two main monophyletic branches—the gymnosperms and the angiosperms—as depicted in FIGURE 30.4. Gymnosperms and angiosperms probably evolved from different ancestors in an extinct group of plants called **progymnosperms,** some of which had seeds.

The two key reproductive adaptations of seed plants—seeds and pollen—will seem less abstract once we have looked more closely at the diversity of gymnosperms and angiosperms.

GYMNOSPERMS

The most familiar gymnosperms are the conifers, the cone-bearing plants such as pines. Gymnosperms (the term means "naked seeds") lack the enclosed chambers (ovaries) in which

angiosperm ovules and seeds develop. Rather, gymnosperm ovules and seeds develop on the surfaces of specialized leaves called **sporophylls.** Gymnosperms appear in the fossil record much earlier than angiosperms.

The Mesozoic era was the age of gymnosperms

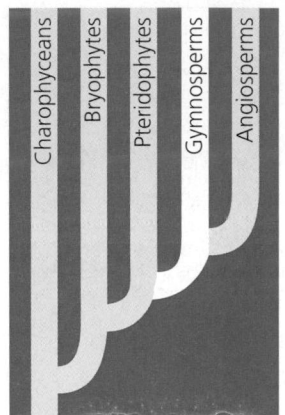

The gymnosperms probably descended from progymnosperms, a group of Devonian plants (see FIGURE 30.4). The earliest progymnosperms were seedless plants, but by the end of the Devonian period, seeds had evolved in some progymnosperm species. Adaptive radiation during the Carboniferous and early Permian produced the various phyla of gymnosperms.

In the history of life, the Permian period was one of great crises. Formation of the supercontinent Pangaea (see Chapter 25) may have been one reason that continental interiors became warmer and drier as the Permian progressed. The flora and fauna of Earth changed dramatically, as many groups of organisms disappeared and others emerged as their successors. The changeover was most pronounced in the seas, but terrestrial life was affected as well. In the animal kingdom, amphibians decreased in diversity and were replaced by reptiles, which were especially well adapted to the arid conditions. Similarly, the lycophytes, horsetails, and ferns that dominated the Carboniferous swamps were largely replaced by gymnosperms, which were more suited to the drier climate. The world and its life had changed so markedly that geologists use the end of the Permian period, about 245 million years ago, as the boundary between the Paleozoic and Mesozoic eras. (This boundary was originally defined by the changeover in marine fossils.) The terrestrial animals of the Mesozoic, including the dinosaurs, were supported by a vegetation consisting mostly of conifers and great palmlike cycads, two phyla of gymnosperms. At the end of the Mesozoic, a powerful meteorite or comet impact and intensive volcanic activity contributed to a cooler climate, and the dinosaurs became extinct. Many of the gymnosperms persisted, however, and are still an important part of Earth's flora.

The four phyla of extant gymnosperms are ginkgo, cycads, gnetophytes, and conifers

Of the ten plant phyla in the taxonomic scheme adopted by this textbook (see TABLE 29.1), four are grouped as gymnosperms (see FIGURE 30.4). Three are relatively small phyla: Ginkgophyta, Cycadophyta, and Gnetophyta. *Ginkgo biloba* is the only

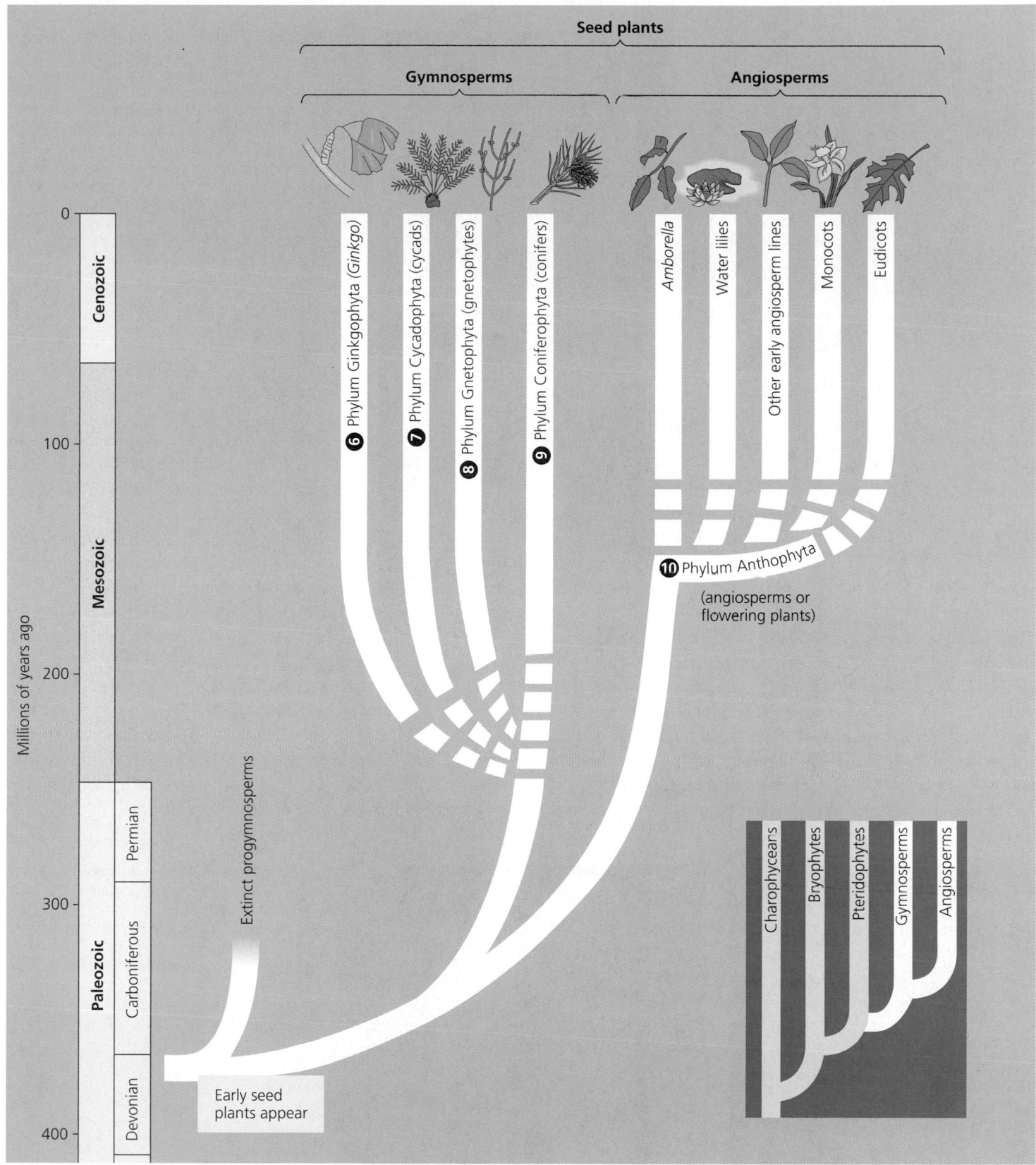

FIGURE 30.4 Hypothetical phylogeny of the seed plants. Recent research supports the division of the seed plants into two distinct clades (monophyletic branches), the gymnosperms and angiosperms. However, evolutionary relationships among the major groups within each of the two seed-plant clades is less certain, which is why the bases of those subbranches are broken lines in this diagram. The numbering of the seed-plant phyla in this tree corresponds to our introduction to plant diversity in Chapter 29, reviewed here in the inset. Note that the taxonomy adopted by this text recognizes four phyla of gymnosperms, but groups all angiosperms into a single phylum, Anthophyta.

like leaves. Plants of the second genus, *Gnetum,* grow in the tropics as trees or vines. *Ephedra* (Mormon tea), the third genus of Gnetophyta, is a shrub of the American deserts.

By far the largest of the four gymnosperm phyla is **phylum Coniferophyta,** the conifers. The term **conifer** (from the Latin *conus,* cone, and *ferre,* to carry) comes from the reproductive structure of these plants, the cone, which is a cluster of scalelike sporophylls. Pines, firs, spruces, larches, yews, junipers, cedars, cypresses, and redwoods all belong to this phylum of gymnosperms (FIGURE 30.8, p. 604). Many are large trees. Although there are only about 550 species of conifers, a few of those species dominate vast forested regions of the Northern Hemisphere, where the growing season is relatively short because of latitude or altitude.

extant species of **phylum Ginkgophyta.** It has fanlike leaves that turn gold and are deciduous in autumn (FIGURE 30.5). *Gingko* is the source of a popular over-the-counter herbal preparation that some people believe improves memory. Cycads (**phylum Cycadophyta**) superficially resemble palms, which are actually flowering plants (FIGURE 30.6). **Phylum Gnetophyta** consists of three genera that are very different in appearance (FIGURE 30.7). One, *Welwitschia,* has giant strap-

(a) Cycads

(b) Cycad cone

FIGURE 30.6 Phylum Cycadophyta: cycads. (a) The palm-like plants in this garden are actually cycads, which are gymnosperms (true palms are angiosperms). **(b)** Cycad seeds develop on the surfaces of sporophylls, specialized reproductive leaves. The sporophylls are packed closely together to form cones, as shown in this interior view of a female cone.

— Seed

— Sporophyll

(a) Welwitschia. *Welwitschia* lives only in the deserts of southwestern Africa. Its straplike leaves are among the largest leaves known.

(b) Gnetum. *Gnetum* species are tropical trees or vines.

(c) Ephedra. Species of *Ephedra* inhabit arid regions throughout the world.

FIGURE 30.7 Phylum Gnetophyta.

Most conifers are evergreens, meaning they retain leaves throughout the year. Even during winter, a limited amount of photosynthesis occurs on sunny days. And when spring comes, conifers already have fully developed leaves that can take advantage of the sunnier days. However, some conifers have deciduous leaves that drop in the autumn. The dawn redwood and tamarack are two examples.

The needle-shaped leaves of some conifers, such as pines and firs, are adapted to dry conditions. A thick cuticle covers the leaf, and the stomata are located in pits, further reducing water loss.

We get much of our lumber and paper pulp from the wood of conifers. What we call wood is actually xylem tissue, which gives the tree structural support.

Coniferous trees are among the largest and oldest organisms on Earth. Redwoods, found only in a narrow coastal strip of northern California, grow to heights of more than 110 m; only certain eucalyptus trees in Australia are taller. Bristlecone pines, another species of California conifer, are among the oldest organisms alive. One bristlecone, named Methuselah, is more than 4,600 years old; it was a young tree when humans invented writing.

The life cycle of a pine demonstrates the key reproductive adaptations of seed plants

You learned earlier in the chapter that the evolution of seed plants added three key terrestrial adaptations in reproduction: the increasing dominance of the sporophyte generation; the advent of the seed as a resistant, dispersible stage in the life cycle; and the evolution of pollen as an airborne agent that brings gametes together. Examining the life cycle of a pine, one of the most familiar of conifers and gymnosperms, will reinforce your understanding of these adaptations (FIGURE 30.9, p. 605).

The pine tree is a sporophyte. Its sporangia are located on scalelike sporophylls that are packed densely in structures called cones. The female gametophyte generation develops from haploid spores that are retained within the sporangia. Conifers, like all seed plants, are heterosporous; male and female gametophytes develop from different types of spores produced by separate cones (FIGURE 30.10, p. 606). Each tree usually has both types of cones. Small pollen cones produce microspores that develop into the male gametophytes, or pollen grains. Larger ovulate cones make megaspores that develop into female gametophytes. From the time young cones appear on the tree, it takes nearly three years to produce the male and female gametophytes, bring them together via pollination, and form mature seeds from the fertilized ovules. The scales of the ovulate cone then separate, and the seeds travel on the wind. A seed that lands in a habitable place germinates, its embryo emerging as a pine seedling.

(a) Douglas fir. Expansive stands of Douglas fir (*Pseudotsuga menziesii*) dominate this Oregon forest. "Doug fir" provides more timber than any other North American tree species. Uses for the wood include house framing, plywood production, pulpwood for paper, railroad ties, and boxes and crates.

(b) Sequoia. This giant sequoia (*Sequoiadendron giganteum*) in California's Sequoia National Park dwarfs the tourists at the bottom of this scene. It weighs about 2,500 metric tons, equivalent to about 14 blue whales (the largest animals), or 40,000 people.

(c) Cypress. The "Lone Cypress" (*Cupressus macrocarpa*) in Monterey, California, is one of the world's most photographed trees.

(e) Common juniper. The "berries" of the common juniper (*Juniperus communis*) are actually ovule-producing cones consisting of fleshy sporophylls. An extract from juniper "berries" gives gin its distinctive flavor.

(d) Pacific yew. The bark of Pacific yew (*Taxa brevifolia*) is a source of taxol, a compound used to treat women with ovarian cancer. The leaves of a European yew species produce a similar compound, which can be harvested without destroying the plants. And pharmaceutical companies are now refining techniques for synthesizing drugs with taxol-like properties.

(g) Wollemia pine. A survivor of a conifer group once known only from fossils, the Wollemia pine (*Wollemia nobilis*) was discovered alive in 1994 in a national park only 150 kilometers from Sydney, New South Wales, Australia. The species has only 40 known individuals in two small groves. The inset compares the leaves of this "living fossil" with actual fossils.

(f) A pine farm. This loblolly pine (*Pinus taeda*) farm in South Carolina is a clone of fast-growing trees cultivated from cell cultures that produce seedlings (inset).

FIGURE 30.8 Phylum Coniferophyta: A sampling of conifer diversity.

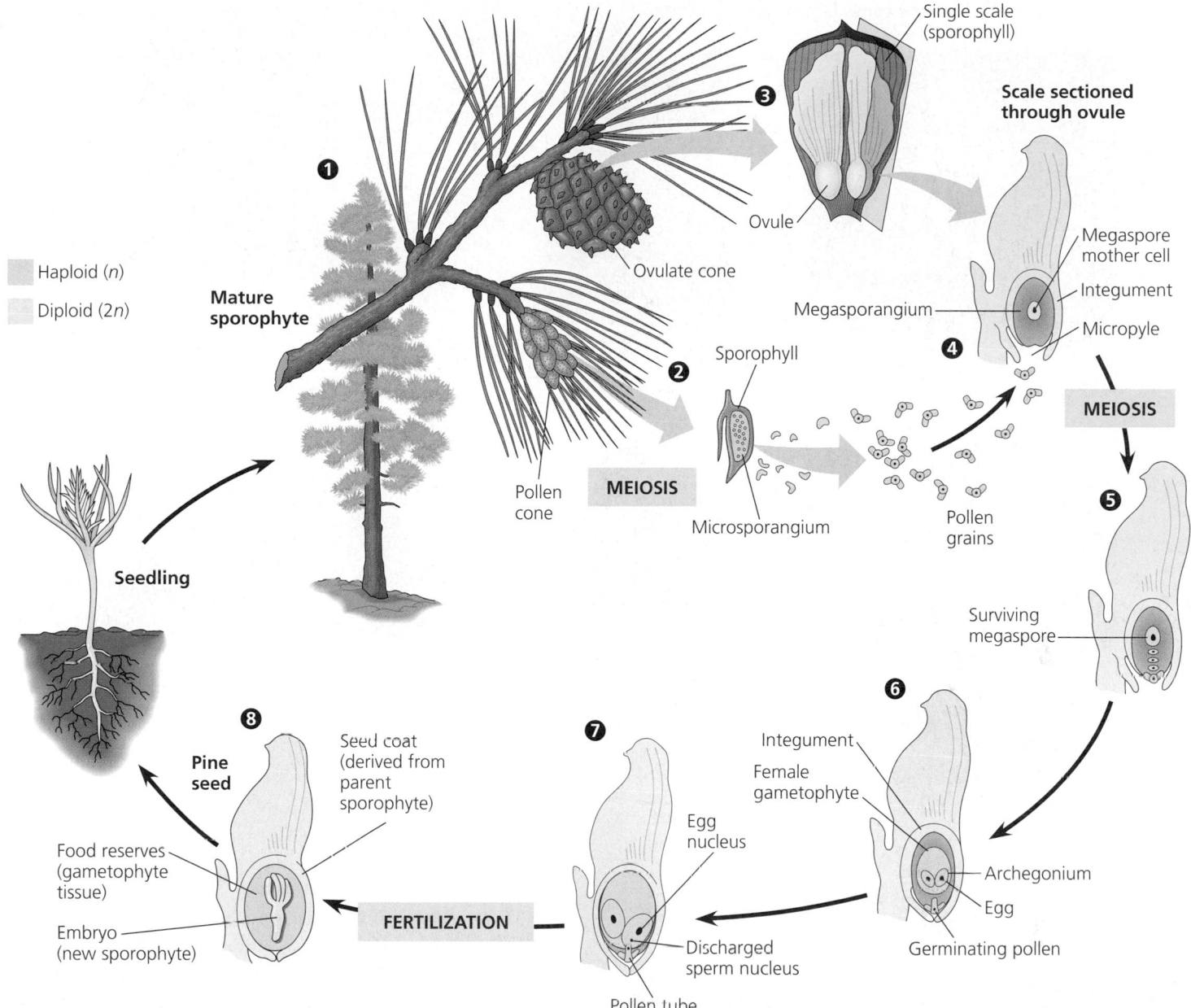

Haploid (*n*)

Diploid (2*n*)

Single scale (sporophyll)

❸

Scale sectioned through ovule

Ovule

Ovulate cone

Mature sporophyte

❶

Megaspore mother cell

Integument

Megasporangium

Micropyle

❹

Sporophyll

❷

MEIOSIS

Pollen cone

Microsporangium

Pollen grains

MEIOSIS

❺

Seedling

Surviving megaspore

❻

Integument

Female gametophyte

Archegonium

Egg

Germinating pollen

Seed coat (derived from parent sporophyte)

Pine seed

❽

❼

Egg nucleus

Food reserves (gametophyte tissue)

FERTILIZATION

Embryo (new sporophyte)

Discharged sperm nucleus

Pollen tube

Figure 30.9 The life cycle of a pine.
❶ Trees (sporophytes) of most species bear both pollen cones and ovulate cones.
❷ A pollen cone contains hundreds of microsporangia held in small reproductive leaves, or sporophylls. Cells in the microsporangia undergo meiosis, giving rise to haploid microspores that develop into pollen grains (immature male gametophytes).
❸ An ovulate cone consists of many scales, each with two ovules. Each ovule includes a megasporangium. ❹ During pollination, windblown pollen falls on the ovulate cone and is drawn into the ovule through the micropyle. The pollen grain germinates in the ovule, forming a pollen tube that begins to digest its

way through the megasporangium. Fertilization usually occurs more than a year after pollination. During that year, ❺ a megaspore mother cell undergoes meiosis to produce four haploid cells. One of these cells survives as a megaspore, which grows and divides repeatedly, giving rise to the immature female gametophyte. Notice that the gametophyte develops within the spore. ❻ Two or three archegonia, each with an egg, then develop within the gametophyte. ❼ By the time eggs are ready to be fertilized, two sperm cells have developed in the male gametophyte (pollen grain), and the pollen tube has extended to the female gametophyte. Fertilization occurs when one of the sperm nuclei, injected into an egg cell by the pollen

tube, unites with the egg nucleus. All the eggs in an ovule may be fertilized, but usually only one zygote develops into an embryo. ❽ The pine embryo, the new sporophyte, has a rudimentary root and several embryonic leaves. A food supply, consisting of the female gametophyte, surrounds and nourishes the embryo. The ovule has developed into a pine seed, which consists of an embryo (new sporophyte), its food supply (derived from gametophyte tissue), and a surrounding seed coat derived from the integuments of the parent tree (parent sporophyte). Notice that three plant generations—one gametophyte and two sporophyte generations—are represented in a gymnosperm seed.

FIGURE 30.10 A closer look at pine cones (*Pinus* sp.).

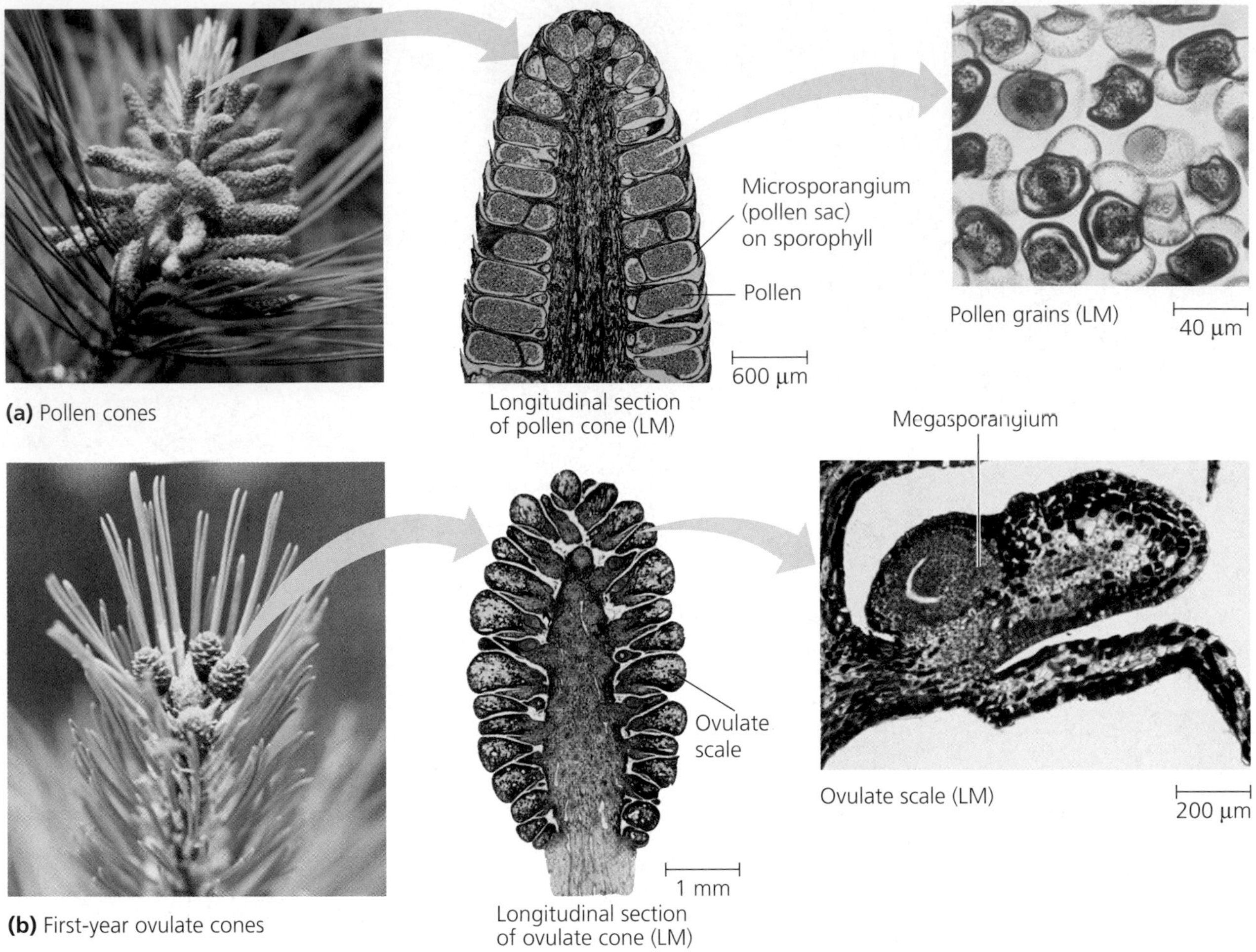

Microsporangium
(pollen sac)
on sporophyll
Pollen
600 μm
Pollen grains (LM)
40 μm
(a) Pollen cones
Longitudinal section
of pollen cone (LM)
Megasporangium
Ovulate
scale
Ovulate scale (LM)
200 μm
1 mm
(b) First-year ovulate cones
Longitudinal section
of ovulate cone (LM)

ANGIOSPERMS (FLOWERING PLANTS)

Charophyceans
Bryophytes
Pteridophytes
Gymnosperms
Angiosperms

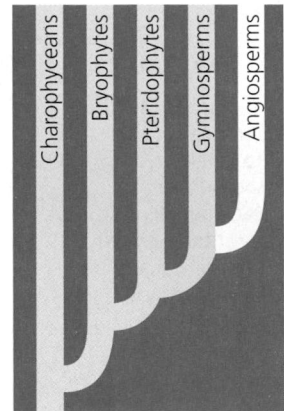

Angiosperms, better known as flowering plants, are vascular seed plants that produce the reproductive structures called flowers and fruits. Today, angiosperms are by far the most diverse and geographically widespread of all plants. There are about 250,000 known species of angiosperms, compared with about 720 gymnosperm species.

Systematists are identifying the angiosperm clades

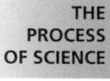

THE PROCESS OF SCIENCE

All angiosperms are placed in a single phylum, the **phylum Anthophyta** (from the Greek *antho,* flower). If you return to FIGURE 30.4, you'll find a division into different angiosperm lineages that contrasts sharply with what we would have diagrammed three years ago. Until the late 1990s, most plant taxonomists divided the angiosperms into two main classes, the **monocots** and the **dicots,** which differ in several anatomical and morphological details. For example, most monocots have leaves with veins running parallel (think of a grass blade), while most dicots have netlike venation in their leaves (think of an oak leaf). A combination of molecular systematics and cladistic analysis has upheld the monocots as a monophyletic group—a clade. Examples of monocots are lilies, orchids, yuccas, palms, and grasses, including lawn grasses, sugar cane, and grain crops (corn, wheat, rice, and others). However, comparisons of DNA revealed that not all plants having the dicot type of anatomy fall into a single monophyletic group. One clade, the **eudicots,** *does* include the majority of dicots. Examples of eudicots are roses, peas, buttercups, sunflowers, oaks, and maples. But some of the other dicots actually belong to angiosperm lineages that diverged earlier than the origin of either monocots or eudicots. For example, the branch in FIGURE 30.4 labeled "other early angiosperm lines" includes a dicot called the star anise. And the water lilies belong to an even older angiosperm lineage. But the

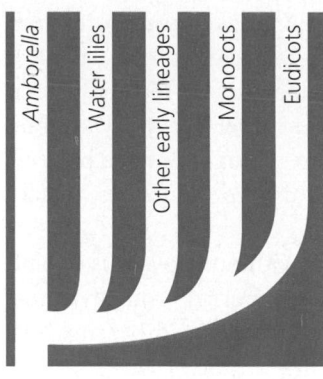

(a) Flowering plants

(b) *Amborella*

(c) Water lily

(d) Star anise

FIGURE 30.11 Representatives of major angiosperm clades.
(a) Hypothetical tree of flowering plants. **(b)** *Amborella* is the only survivor of a branch at the very base of the angiosperm tree. **(c)** Water lilies, such as this *Nymphaea carulea*, are modern members of a clade predated only by the *Amborella* lineage. **(d)** Other early angiosperm lineages include species such as this *Illicium floridanum*, or star anise. **(e)** This orchid (*Paphiopedilum furheyo*) is an example of a monocot. There are about 65,000 species of monocots. **(f)** The California poppy (*Eschscholzia californica*) is just one of 165,000 species of eudicots.

(e) Orchid (monocot)　　　　**(f) California poppy (eudicot)**

oldest angiosperm branch of all is represented by a single species, *Amborella trichopoda* (FIGURE 30.11b).

Amborella was an unlikely candidate to become the center of attention among biologists interested in plant phylogeny. A small shrub with tiny flowers and tiny fruit, *Amborella* is found only on New Caledonia, a South Pacific island. In 1999, four independent research teams published evidence that *Amborella* is the most "primitive" known angiosperm—primitive in the sense that the molecular evidence places it on the oldest branch of angiosperm evolution. A plant most botanists didn't know existed a few years ago has become a focal point in research efforts to trace the evolution of flowering plants.

What evolutionary "innovations" contributed most to the enormous success of angiosperms? Refinements in vascular tissue, especially xylem, probably played a role in the spread of angiosperms into diverse terrestrial habitats. Like gymnosperms, angiosperms are equipped with xylem cells called tracheids. These are long, tapered cells that function in both mechanical support and water transport (FIGURE 30.12). In addition to these tracheids, the xylem of angiosperms also has fiber cells, which are specialized for support. A third type of xylem cell, called a vessel element, is also found in most angiosperms. Compared to tracheids, vessel elements are shorter and wider, and they are arranged end-to-end into continuous tubes—xylem vessels—that are more efficient than tracheids in transporting water. *Amborella* has tracheids, but not vessels, suggesting that true vessel elements evolved in angiosperms after *Amborella* branched off as a distinct lineage.

Whatever roles evolutionary refinements in vascular tissue and other structural characteristics played in plant history, it

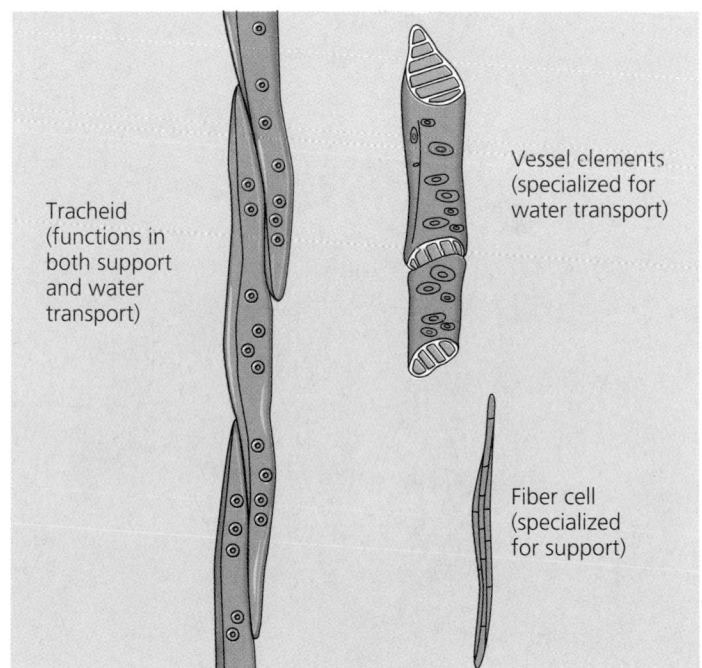

FIGURE 30.12 Xylem cells in angiosperms.

was probably the reproductive adaptations associated with flowers and fruits that contributed most to making angiosperms so successful.

The flower is the defining reproductive adaptation of angiosperms

The **flower** is an angiosperm structure specialized for reproduction. In many angiosperms, insects and other animals transfer pollen from one flower to female sex organs on another flower, which makes pollination less random than the wind-dependent pollination of most gymnosperms. However, some flowering plants *are* wind-pollinated, a characteristic especially common for plants that occur in dense populations, such as grasses and tree species in temperate forests.

A flower is a specialized shoot with four circles of modified leaves: sepals, petals, stamens, and carpels (FIGURE 30.13). Starting at the bottom of the flower are the **sepals,** which are usually green. They are modified leaves that enclose the flower before it opens (think of a rosebud). Above the sepals are the **petals,** brightly colored in most flowers. They aid in attracting insects and other pollinators. Flowers that are wind-pollinated generally lack bright-colored parts. The sepals and petals are sterile floral parts, meaning that they are not directly involved in reproduction. Within the ring of petals are the fertile sporophylls, the leaf-derived parts that produce spores. The two rings of sporophylls are the stamens and carpels. **Stamens** are the male reproductive organs, the sporophylls that produce microspores that give rise to male gametophytes. **Carpels** are

the female sporophylls, the organs that make megaspores and their products, female gametophytes. A stamen consists of a stalk called the **filament** and a terminal sac, the **anther,** where pollen is produced. At the tip of the carpel is a sticky **stigma** that receives pollen. A **style** leads to the **ovary** at the base of the carpel. Protected within the ovary are the ovules, which develop into seeds after fertilization.

Recall that the enclosure of seeds within the ovary is one of the features that distinguishes angiosperms from gymnosperms. The carpel probably evolved from a seed-bearing leaf (sporophyll) that became rolled into a tube (FIGURE 30.14). Some angiosperms, such as garden peas, have flowers with single carpels. Others, such as magnolias, have several separate carpels. Still other species, such as lilies, have two or more fused carpels, usually forming an ovary with multiple ovule-containing chambers.

Fruits help disperse the seeds of angiosperms

A **fruit** is a mature ovary. As seeds develop from ovules after fertilization, the wall of the ovary thickens. A pea pod is an example of a fruit, with seeds (mature ovules, the peas) encased in the ripened ovary (the pod) (FIGURE 30.15). Fruits protect dormant seeds and aid in their dispersal.

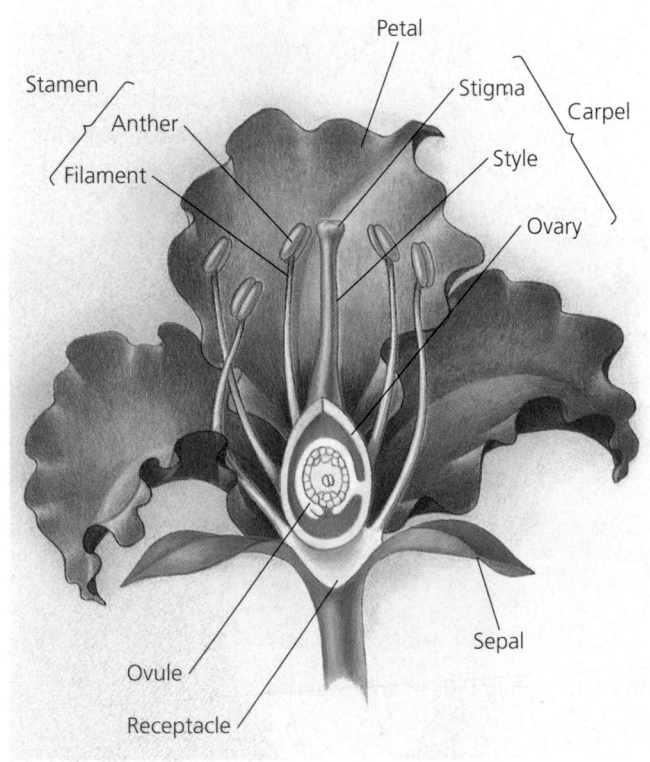

(a)

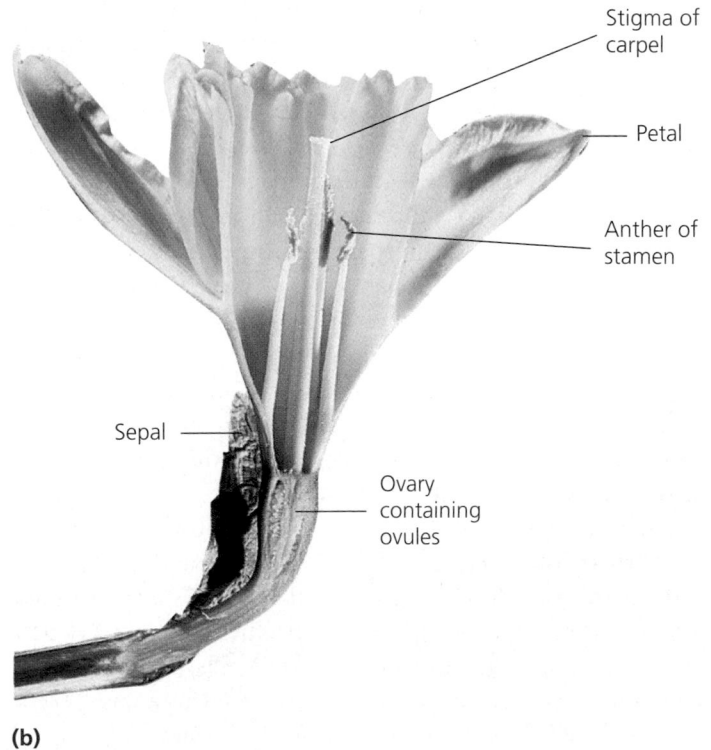

(b)

FIGURE 30.13 The structure of a flower. (a) The parts of an idealized flower. **(b)** The photo shows a cutaway view of a daffodil flower.

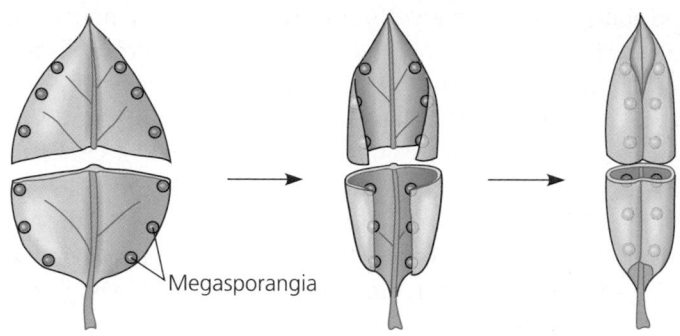

FIGURE 30.14 Hypothesis for the origin of the carpel from a reproductive leaf (sporophyll). The enclosure of ovules (and seeds) within an ovary may have resulted from the rolling of leaves bearing megasporangia on their margins. A similar process may have transformed sporophylls with microsporangia into the anthers of stamens.

Megasporangia

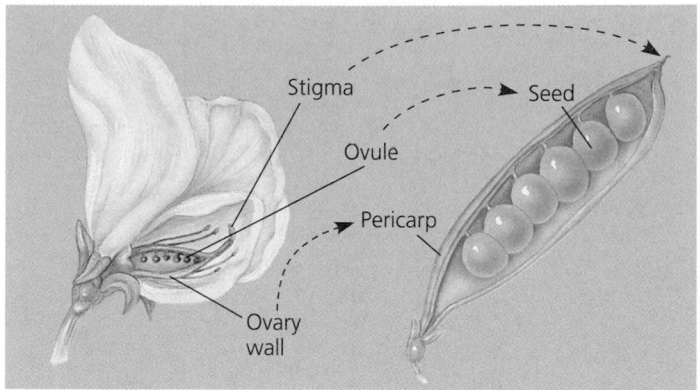

Stigma

Seed

Ovule

Pericarp

Ovary wall

FIGURE 30.15 Relationship between a pea flower and a fruit (pea pod). At the same time that the ovules are transformed into seeds, the ovary and other flower parts are transformed into a fruit we call a pod.

Various modifications in fruits help disperse seeds (FIGURE 30.16). Some flowering plants, such as dandelions and maples, have seeds within fruits that function like kites or propellers, adaptations that enhance dispersal by wind. However, many angiosperms use animals to carry seeds. Some of these plants have fruits modified as burrs that cling to animal fur (or to the clothes of humans). Other angiosperms produce edible fruits. Such fruits are usually nutritious, often sweet-tasting, and commonly vividly colored, which advertises their ripeness to certain animals. When it eats the fruit, the animal digests the fleshy part, but the tough seeds usually pass unharmed through the digestive tract. Mammals and birds may deposit seeds, along with a supply of fertilizer, miles from where the fruit was eaten.

The fruit begins to develop after pollination triggers hormonal changes that cause the ovary to grow (see FIGURE 30.15). The wall of the ovary becomes the **pericarp,** the thickened wall of the fruit. As the ovary grows, the other parts of the flower wither away in many plants. If a flower has not been pollinated, fruit usually does not develop, and the entire flower withers and falls away.

Fruits are classified into several types, depending on their developmental origin (TABLE 30.1, p. 610). A fruit derived from a single ovary is called a **simple fruit.** A simple fruit may be fleshy, such as a cherry, or dry, such as a soybean pod. An **aggregate fruit,** such as a blackberry, results from a single flower that has several carpels. A **multiple fruit,** such as a pineapple, develops from an inflorescence, a group of flowers tightly clustered together. When the walls of the many ovaries start to thicken, they fuse together and become incorporated into one fruit.

By selectively breeding plants, humans have capitalized on the production of edible fruits. The apples, oranges, and other fruits in grocery stores are exaggerated versions of much smaller natural varieties of fleshy fruits. However, the staple foods for humans are the dry, wind-dispersed fruits of grasses, which are harvested while still on the parent plant. The cereal grains of wheat, rice, corn, and other grasses are easily mistaken for seeds, but each is actually a fruit with a dry pericarp that adheres tightly to the seed coat of the single seed within.

Interactions with animals that transport pollen and seeds have helped angiosperms become the most successful plants on Earth. However, as we will see, the angiosperm life cycle was not an evolutionary "invention," but was built upon adaptive themes we have tracked in our study of plant diversity.

(a) Dandelion fruit dispersed by the wind

(b) Cockleburs (fruits) carried by animal fur

(c) A mouse eating a berry containing seeds that will be dispersed later with the animal's feces

FIGURE 30.16 Fruit adaptations that enhance seed dispersal.

Table 30.1 Classification of Fleshy Fruits

Type of Fruit	Floral Origin	Example
Simple	Single ovary of one flower	Cherry
Aggregate	Many ovaries of one flower	Raspberry
Multiple	Many ovaries of many clustered flowers	Pineapple

The life cycle of an angiosperm is a highly refined version of the alternation of generations common to all plants

Angiosperms are heterosporous, a characteristic they share with all seed plants. The flower of the sporophyte produces microspores that form male gametophytes and megaspores that form female gametophytes (FIGURE 30.17). The immature male gametophytes are contained within **pollen grains,** which develop within the anthers of stamens. Each pollen grain has two haploid cells. **Ovules,** which develop in the ovary, contain the female gametophyte, also known as the **embryo sac.** It consists of only a few cells, one of which is the egg. (We'll describe the development of pollen and the embryo sac in more detail in Chapter 38.) Notice that the evolutionary trend toward reduction of the gametophyte generation in vascular plants continued with the angiosperms.

After its release from the anther, the pollen is carried to the sticky stigma at the tip of a carpel. Although some flowers self-pollinate, most have mechanisms that ensure **cross-pollination,** the transfer of pollen from flowers of one plant to flowers of another plant of the same species. In some cases, stamens and carpels of a single flower may mature at different times, or the organs may be so arranged within the flower that self-pollination is unlikely.

The pollen grain germinates (begins growing) after it adheres to the stigma of a carpel. The pollen grain, now containing a mature male gametophyte, extends a tube that grows down within the style of the carpel. After it reaches the ovary, the pollen tube penetrates through the micropyle, a pore in the integuments of the ovule, and discharges two sperm cells into the female gametophyte (embryo sac). One sperm nucleus unites with the egg, forming a diploid zygote. The other sperm nucleus fuses with the two nuclei in the large center cell of the female gametophyte. This central cell now has a triploid ($3n$) nucleus. This phenomenon, known as **double fertilization,** is characteristic of angiosperms. (Double fertilization also evolved independently in some gymnosperms of phylum Gnetophyta.)

After double fertilization, the ovule matures into a seed. The zygote develops into a sporophyte embryo with a rudimentary root and either one or two seed leaves, the **cotyledons** (monocots have one seed leaf and dicots have two, hence their names). The triploid nucleus in the center of the embryo sac divides repeatedly, giving rise to a triploid tissue called **endosperm,** rich in starch and other food reserves. Monocot seeds such as corn store most of their food in the endosperm. Beans and many other dicots transfer most of the nutrients from the endosperm to the developing cotyledons.

What is the function of double fertilization? According to one hypothesis, double fertilization synchronizes the development of food storage in the seed with development of the embryo. If a particular flower is not pollinated or sperm cells are not discharged into the embryo sacs, fertilization does not occur and neither embryo nor endosperm forms. Perhaps double fertilization is an adaptation that prevents flowering plants from squandering nutrients on infertile ovules.

The seed consists of the embryo, endosperm, sporangium, and a seed coat derived from the integuments (outer layers of the ovule). An ovary develops into a fruit as its ovules develop into seeds. After being dispersed by wind or animals, a seed germinates if environmental conditions are favorable. The coat ruptures and the embryo emerges as a seedling, using the food stored in the endosperm and cotyledons.

The radiation of angiosperms marks the transition from the Mesozoic era to the Cenozoic era

Earth's landscapes changed dramatically with the origin and radiation of flowering plants. The ancestry of angiosperms is still uncertain, but, as discussed earlier, cladistic analysis of gene-sequence data points to *Amborella* and water lilies such as *Nymphaea* as the closest living relatives of early angiosperms (see FIGURE 30.11). The oldest fossils that are widely accepted as angiosperms are found in rocks of the early Cretaceous period that are about 130 million years old. In such rock strata, angiosperms are sparsely represented among a much greater abundance of ferns and gymnosperms. By the end of the Cretaceous, 65 million years ago, following a period of environmental disturbance, the angiosperms had radiated

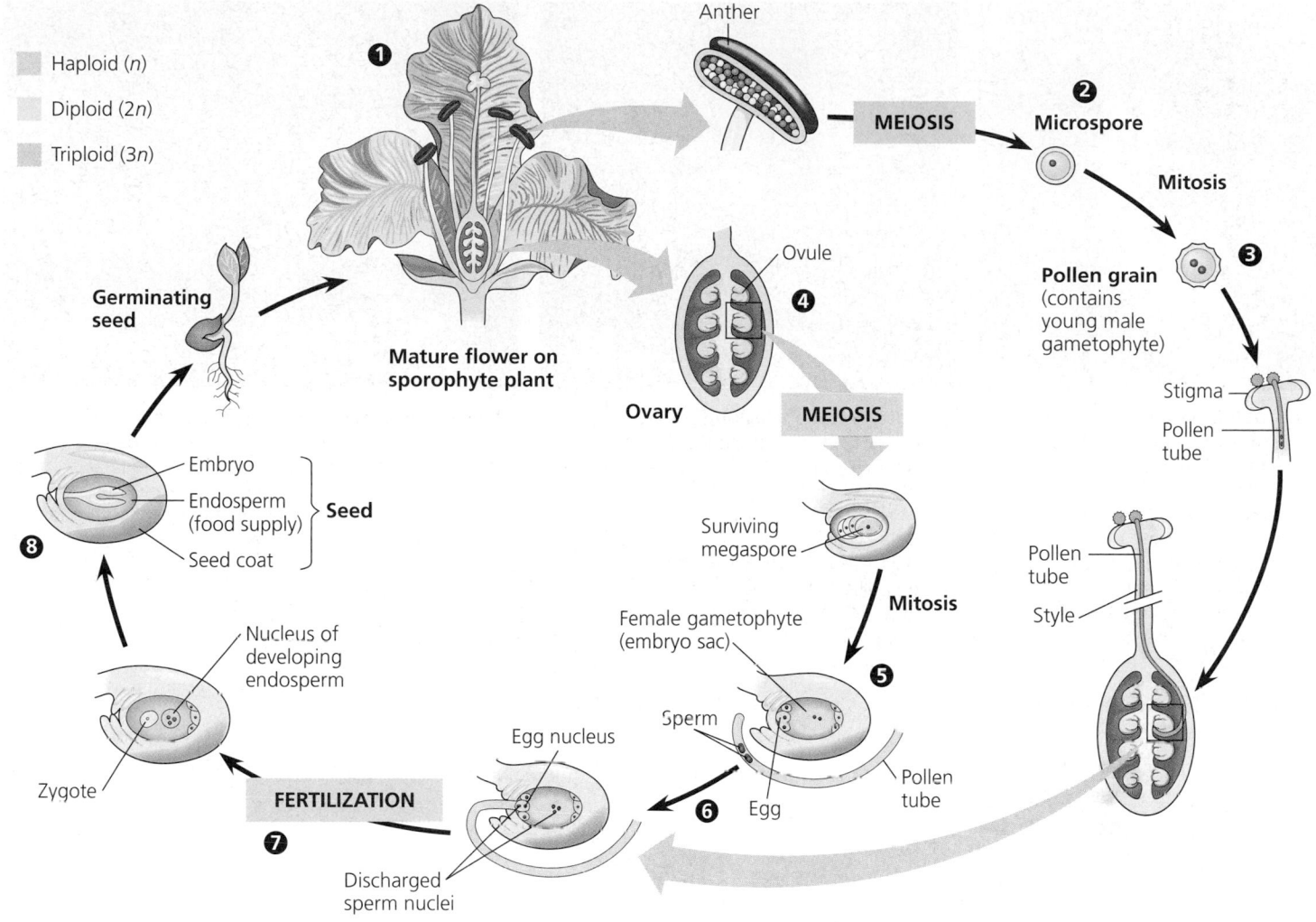

Haploid (n)
Diploid (2n)
Triploid (3n)

1 Mature flower on sporophyte plant
Germinating seed
Seed
Embryo
Endosperm (food supply)
Seed coat
8
Nucleus of developing endosperm
Zygote
FERTILIZATION
7
Discharged sperm nuclei
Egg nucleus
Sperm
Egg
Pollen tube
Female gametophyte (embryo sac)
5
Mitosis
Surviving megaspore
MEIOSIS
Ovule
4
Ovary
Anther
MEIOSIS
2 Microspore
Mitosis
Pollen grain (contains young male gametophyte)
3
Stigma
Pollen tube
Pollen tube
Style
6

FIGURE 30.17 The life cycle of an angiosperm. 1 The anthers of the flower produce **2** microspores that form **3** male gametophytes (pollen). **4** The ovules produce megaspores that form **5** female gametophytes (embryo sacs). **6** Pollination brings the gametophytes together in the ovary. **7** Double fertilization occurs, and **8** zygotes develop into sporophyte embryos that are packaged along with food into seeds. (The fruit tissues surrounding the seed are omitted in this diagram.) When a seed germinates, the embryo grows and develops into a sporophyte.

and become the dominant plants on Earth, as they are today. The change in fossils during the late Cretaceous is so extreme that geologists use the end of that period as the boundary between the Mesozoic and Cenozoic eras.

Angiosperms and animals have shaped one another's evolution

Ever since they colonized the land, animals have influenced the evolution of terrestrial plants, and vice versa. The fact that animals must eat affects the natural selection of both animals and plants. For instance, with animals crawling and foraging for food on the forest floor, natural selection must have favored plants that kept their spores and gametophytes far above ground, rather than dropping these crucial structures to within reach of hungry animals on the ground. This, in turn, may have been a selection factor in the evolution of flying in-

sects. On the other hand, as plants with flowers and fruits evolved, some herbivores became beneficial to the plants by carrying the pollen and seeds of plants they used as food. Certain animals became specialists at these tasks, feeding on specific plants. Natural selection reinforced these interactions, for they improved the reproductive success of both partners: The plant got pollinated and the animal got fed. The mutual evolutionary influence between two species is termed **coevolution.** (This definition is refined in Chapter 53.)

Pollinator-plant relationships are partly responsible for the diversity of flowers (FIGURE 30.18, p. 612). In most cases, relationships between plants and their pollinators are less specific than in the extreme coevolution between one plant species and one animal species. For example, the flowers of a particular plant species may be adapted for attracting insects rather than birds, but many different insect species may serve as pollinators. Conversely, a single animal species—a honeybee species, for example—may pollinate many different plant species. But

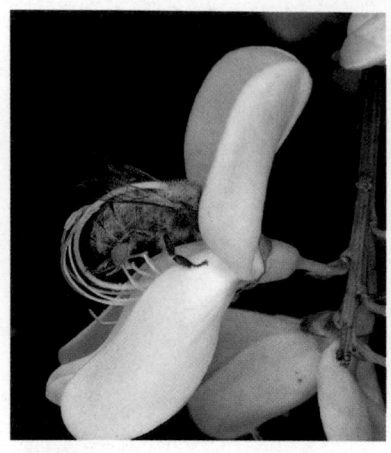

(a) Scottish broom flower and honeybee. This honeybee is harvesting pollen and nectar (a sugary solution secreted by flower glands) from a Scottish broom flower. The flower has a tripping mechanism that arches the stamens over the bee and dusts it with pollen, some of which will rub off onto the stigma of the next flower the bee visits.

(b) Floral tubes and hummingbird. The long, thin beak and tongue of this rufous hummingbird enable the animal to probe flowers that secrete nectar deep within floral tubes. Before the hummer leaves, anthers will dust its beak and head feathers with pollen. Many flowers that are pollinated by birds are red or pink, colors to which bird eyes are especially sensitive.

(c) Fragrance and nighttime pollinators. Some angiosperms, such as this baobab tree, depend mainly on nocturnal pollinators, including bats. Common adaptations of such plants include large, light-colored, highly fragrant flowers that nighttime pollinators can locate.

FIGURE 30.18 Flower-pollinator relationships.

even in these less specific relationships, flower color, fragrance, and structure often reflect specialization for a particular taxonomic *group* of pollinators, such as diverse species of bees or hummingbirds.

Relationships between angiosperms and animals are also evident in the edible fruits of angiosperms, as discussed earlier in the chapter. Again, we see that one of the keys to angiosperm success has been interaction with animals.

PLANTS AND HUMAN WELFARE

The absolute dependence of humans on Earth's flora is a specific and highly refined case of the more general connection between animals and plants. Of course, we humans, like most of Earth's species, depend on photosynthetic organisms for their ecological "services" of food production and oxygen release. What is special about our relationship with plants is our development of technology to maximize the harvest of plant products for human use. The lumber industry is an example. But the most important example is agriculture.

Agriculture is based almost entirely on angiosperms

A visit to the produce section of a market is a colorful reminder that even city folk depend on farms. Flowering plants provide nearly all our food. All of our fruit and vegetable crops

are angiosperms. Corn, rice, wheat, and the other grains are grass fruits. The endosperm of the grain seeds is the main food source for most of the people of the world and their domesticated animals. We also grow angiosperms for fiber, medications, perfumes, and decoration.

Like other animals, early humans probably collected wild seeds and fruits. Agriculture was gradually developed as humans began sowing seeds and cultivating plants to have a more dependable food source. As they domesticated certain plants, such as the squash represented by the seed in the photo that introduced this chapter, humans began to intervene in plant evolution by selective breeding designed to improve the quantity and quality of the foods the crops produced. Agriculture is a unique kind of evolutionary relationship between plants and animals.

Plant diversity is a nonrenewable resource

The exploding human population and its demand for space and natural resources are extinguishing plant species at an unprecedented rate. The problem is especially critical in the tropics, where more than half the human population lives and population growth is fastest. Tropical rain forests are being destroyed at a frightening pace. The most common cause of this destruction is slash-and-burn clearing of the forest for agricultural use. Fifty million acres, an area about the size of the state of Washington, are cleared each year, a rate that would completely eliminate Earth's tropical forests within 25 years. As the forest disappears, so do thousands of plant species. Extinction is irrevocable; plant diversity is a nonrenewable resource. Insects and other rain forest animals that depend on these plants are also vanishing. In all, researchers estimate that the destruction of habitat in the rain forest and other ecosystems is claiming hundreds of species each year. The toll is greatest in the tropics, because that is where most species live; but environmental assault is a generically human tendency. Europeans eliminated most of their forests centuries ago, and habitat destruction is endangering many species in North America (FIGURE 30.19).

Many people have ethical concerns about contributing to the extinction of living forms. But there are also practical reasons to be concerned about the loss of plant diversity. We depend on plants for thousands of products, including food, building materials, and medicines. TABLE 30.2 lists only a few examples of how we use the unique secondary compounds of plants (see p. 582 to review secondary compounds). So far, we have explored the potential uses of only a tiny fraction of the 250,000 known plant species. For example, almost all our food is based on the cultivation of only about two dozen species. More than 25% of prescription drugs are extracted from plants, and many more medicinal compounds were first discovered in plants and then synthesized artificially. However, researchers have investigated fewer than 5,000 plant species as potential sources of medicine. Pharmaceutical companies were led to most of these species by local people who use the plants in preparing their traditional medicines.

The tropical rain forest may be a medicine chest of healing plants that could be extinct before we even know they exist.

Table 30.2 A Sampling of Medicines Derived from Plants

Compound	Example of Source		Example of Use
Atropine	Belladonna plant		Pupil dilator in eye exams
Digitalin	Foxglove		Heart medication
Menthol	Eucalyptus tree		Ingredient in cough medicines
Morphine	Opium poppy		Pain reliever
Quinine	Quinine tree		Malaria preventive
Taxol	Pacific yew		Ovarian cancer drug
Tubocurarine	Curare tree		Muscle relaxant during surgery
Vinblastine	Periwinkle		Leukemia drug

Source: Adapted from Randy Moore et al., *Botany,* 2nd ed. Dubuque, IA: Brown, 1998. Table 2.2, p. 37.

This is only one reason to value what is left of plant diversity and to search for ways to slow the loss. The solutions we propose must be economically realistic. If the goal is only profit for the short term, then we will continue to slash and burn until the forests are gone. If, however, we begin to see rain forests and other ecosystems as living treasures that can regenerate only slowly, we may learn to harvest their products at sustainable rates. What else can we do to preserve plant diversity? Few questions are as important.

(a) Clear-cutting of an old-growth forest in Oregon to provide lumber for housing.

FIGURE 30.19 Deforestation is an international practice.

(b) Slash-and-burn clearing of a tropical rain forest in the Amazon Basin to provide temporary farmland that will become unproductive in just a few years because of depletion of soil nutrients.

CHAPTER 30 REVIEW

Go to the Campbell Biology website (www.campbellbiology.com) to explore an interactive version of the Chapter Review.

Summary of Key Concepts

OVERVIEW OF SEED PLANT EVOLUTION

- **Reduction of the gametophyte continued with the evolution of seed plants (p. 598, FIGURE 30.1)** The gametophytes of seed plants develop within the walls of spores retained within tissues of the parent sporophyte.

- **Seeds became an important means of dispersing offspring (p. 599, FIGURES 30.2 and 30.3)** A seed, which is derived from a fertilized ovule, consists of a sporophyte embryo packaged along with a food supply within a seed coat.

- **Pollen eliminated the liquid-water requirement for fertilization (p. 600)** A pollen grain, containing a male gametophyte, can be dispersed through the air by wind or transported by animals.

- **The two clades of seed plants are gymnosperms and angiosperms (p. 600, FIGURE 30.4)** Gymnosperms and angiosperms are monophyletic groups that evolved from progymnosperm ancestors.

GYMNOSPERMS

- **The Mesozoic era was the age of gymnosperms (p. 600)** Gymnosperms bear their seeds "naked" on the surfaces of sporophylls.

- **The four phyla of extant gymnosperms are ginkgo, cycads, gnetophytes, and conifers (pp. 600–603, FIGURES 30.5–30.8)** The cone-bearing conifers, including pines, firs, and spruces, are by far the most diverse gymnosperms today.

- **The life cycle of a pine demonstrates the key reproductive adaptations of seed plants (pp. 603–606, FIGURES 30.9 and 30.10)** Dominance of the sporophyte generation, the development of seeds from fertilized ovules, and the role of pollen in transferring sperm to ovules are key features of the life cycle of a conifer.
  Web/CD Activity 30A: *Pine Life Cycle*

ANGIOSPERMS (FLOWERING PLANTS)

- **Systematists are identifying the angiosperm clades (pp. 606–607, FIGURE 30.11)** Two clades, the monocots and the eudicots, include the vast majority of plants.
  Web/CD Case Study in the Process of Science: *How Are Trees Identified by Their Leaves?*

- **The flower is the defining reproductive adaptation of angiosperms (p. 608, FIGURES 30.13 and 30.14)** Sepals, petals, stamens (which produce pollen), and carpels (which produce ovules) are the whorls of modified leaves that make up flowers.

- **Fruits help disperse the seeds of angiosperms (pp. 608–609, FIGURES 30.15 and 30.16, TABLE 30.1)** Ovaries ripen into fruits, which are often carried by wind or animals to new locations.

- **The life cycle of an angiosperm is a highly refined version of the alternation of generations common to all plants (p. 610, FIGURE 30.17)** Double fertilization occurs when a pollen tube discharges two sperm into the female gametophyte (embryo sac) within an ovule.

One sperm fertilizes the egg, while the other combines with two nuclei in the center cell of the female gametophyte to initiate development of food-storing endosperm. Endosperm supports the development of the embryo.
Web/CD Activity 30B: *Angiosperm Life Cycle*

- **The radiation of angiosperms marks the transition from the Mesozoic era to the Cenozoic era (pp. 610–611)** An adaptive radiation of angiosperms occurred during the Cretaceous period.

- **Angiosperms and animals have shaped one another's evolution (pp. 611–612, FIGURE 30.18)** Pollination of flowers by animals and transport of seeds by animals are two important relationships in terrestrial ecosystems.

PLANTS AND HUMAN WELFARE

- **Agriculture is based almost entirely on angiosperms (p. 612)** Human cultures depend on the cultivation and harvest of angiosperms, especially the fruits of grains.

- **Plant diversity is a nonrenewable resource (pp. 612–613, FIGURE 30.19, TABLE 30.2)** Destruction of habitat is causing extinction of many plant species and the animal species they support.

Self-Quiz

1. Where would you find a megasporangium in an angiosperm?
   a. at the base of a sporophyll in an ovulate cone
   b. producing a megaspore within the archegonium of the female gametophyte
   c. enclosed in the stigma of a flower
   d. within an ovule contained within an ovary of a flower
   e. packed into pollen sacs within the anthers found on a stamen

2. A fruit is most commonly
   a. a mature ovary.
   b. a thickened style.
   c. an enlarged ovule.
   d. a modified root.
   e. a mature female gametophyte.

3. Which angiosperm cell is *incorrectly* paired with its chromosome count (n or $2n$)?
   a. egg cell—n d. zygote—$2n$
   b. megaspore—$2n$ e. sperm—n
   c. microspore—n

4. Plant diversity is greatest in
   a. tropical forests.   d. temperate forests.
   b. deserts.   e. farmlands.
   c. salt marshes.

5. Evidence that *Amborella* may represent the only survivor of the oldest branch of the angiosperm lineage comes from
   a. the fossil record.
   b. the lack of flowers in this primitive plant.
   c. the lack of vessels in its xylem tissue.
   d. molecular systematics.
   e. Both c and d provide evidence for the early divergence of its ancestors.

6. Gymnosperms and angiosperms have the following in common *except*

  a. seeds.
  b. pollen.
  c. vascular tissue.
  d. ovaries.
  e. ovules.

Questions 7–10
In the following cladogram, match the derived characters (see Chapter 25) with the correct branch points:

  a. flowers
  b. embryos
  c. seeds
  d. vascular tissue

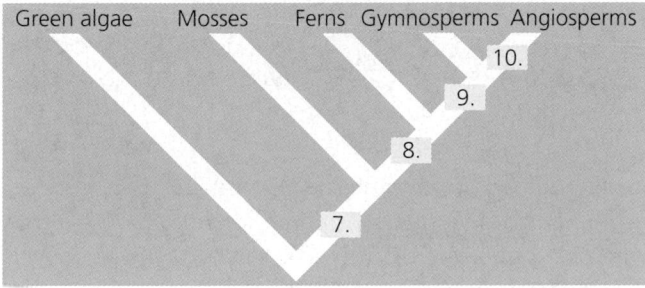

11. How does the evergreen nature of pines and other conifers adapt the plants for living where the growing season is very short?
12. Contrast the mode of sperm delivery in ferns (see Chapter 29) with sperm delivery in conifers.
13. What is a fruit?
14. In terms of evolutionary origin, what do a pine cone scale and an angiosperm carpel have in common?
15. Why is it incorrect to describe *Amborella* as the ancestor of flowering plants?
16. What is double fertilization?

Go to the website or CD-ROM for more quiz questions.

Evolution Connection

The history of life has been punctuated by several mass extinctions. The impact of a meteorite may have wiped out the dinosaurs and many forms of marine life at the end of the Cretaceous period. Fossils indicate that plants were much less severely affected by this and other mass extinctions. What adaptations may have enabled plants to withstand these disasters better than animals?

The Process of Science

Suggest some ways to test the hypothesis that a particular angiosperm species is pollinated exclusively by beetles.

Learn how to identify trees using leaf characteristics in the Case Study in the Process of Science, available on the CD-ROM and website.

Science, Technology, and Society

Why are tropical rain forests being destroyed at such an alarming rate? What kinds of social, technological, and economic factors are responsible? Most forests in developed Northern Hemisphere countries have already been cut. Do the developed nations have a right to ask the developing nations in the Southern Hemisphere to slow or stop the destruction of their forests? Defend your answer. What kinds of benefits, incentives, or programs might slow the assault on rain forests?

Answers: 1. d; 2. a; 3. b; 4. a; 5. e; 6. d, 7. b; 8. d; 9 c; 10, a; 11. Because the plants do not lose their leaves during autumn and winter, the leaves are already fully developed for photosynthesis when the short growing season begins in spring. 12. The flagellated sperm of ferns must swim through a film of water to reach eggs. In contrast, the airborne pollen of conifers brings gametes together without need of an aqueous route, and the pollen releases sperm near the egg after pollination occurs. 13. A fruit is the ripened ovary of a flower. The fruit protects and aids in the dispersal of seeds contained in the fruit. 14. Both are derived from sporophylls, leaves specialized for spore production. 15. *Amborella* is contemporary with other modern flowering plants, though it belongs to the earliest lineage of flowering plants. 16. Double fertilization occurs in angiosperms when the pollen tube releases two sperm into the female gametophyte. One sperm fertilizes the egg; the other fertilizes a large cell that develops into nutritive endosperm.

CHAPTER 31

FUNGI

The words *fungus* and *mold* *may evoke some unpleasant images. Fungi rot timbers, attack living plants, spoil food, and afflict humans with athlete's foot and worse maladies. However, ecosystems would be in trouble without fungi to decompose dead organisms, fallen leaves, feces, and other organic materials, thus recycling vital chemical elements back to the environment in forms other organisms can assimilate. Most plants depend on mutualistic fungi that help their roots absorb minerals and water from the soil. In addition to these ecological roles, fungi have been* used by humans in various ways for centuries. We eat some fungi (mushrooms, for instance), culture fungi to produce antibiotics and other drugs, add them to dough to make bread rise, and use them to ferment beer and wine. Many fungi are also quite colorful and beautiful, as you can see from Mary Elizabeth Banning's 19th-century watercolor of indigo milk cap (Lactarius indigo). Whatever our initial subjective perceptions may be, fungi are fascinating as objects of study. They are a form of life so distinctive that they have been accorded their own taxonomic kingdom.

In this chapter we will characterize the members of the kingdom Fungi, survey their diversity, discuss their ecological and commercial impact, and consider hypotheses about their evolutionary origin. As we did with the plant kingdom, we will look in some detail at life cycles, mainly for what they tell us about the phylogeny and evolutionary adaptations of fungi.

INTRODUCTION TO THE FUNGI

Fungi are eukaryotes, and most are multicellular. Although fungi were once grouped with plants, they are unique organisms that generally differ from other eukaryotes in nutritional mode, structural organization, growth, and reproduction. In fact, molecular studies indicate that animals, not plants, are the closest relatives of fungi (see FIGURE 28.8).

Absorptive nutrition enables fungi to live as decomposers and symbionts

Fungi are heterotrophs that acquire their nutrients by **absorption.** In this mode of nutrition, small organic molecules are absorbed from the surrounding medium. A fungus digests food outside its body by secreting powerful hydrolytic enzymes into the food. These **exoenzymes,** as they are called, decompose complex molecules to the simpler compounds that the fungus can absorb and use.

The absorptive mode of nutrition is associated with the ecological roles of fungi as decomposers (saprobes), parasites, or mutualistic symbionts. Saprobic fungi absorb nutrients from nonliving organic material, such as fallen logs, animal corpses, or the wastes of live organisms. In the process of this saprobic nutrition, fungi decompose the organic material. Parasitic fungi absorb nutrients from the cells of living hosts. Some of these fungi, such as certain species infecting the lungs of humans, are pathogenic. Pathogenic fungi cause about 80% of plant diseases. Mutualistic fungi also absorb nutrients from a host organism, but they reciprocate with functions beneficial to their partners in some way, such as aiding a plant in the uptake of minerals from the soil.

Extensive surface area and rapid growth adapt fungi for absorptive nutrition

The vegetative (nutritionally active) bodies of most fungi are usually hidden, being diffusely organized around and within the tissues of their food sources. Except for yeasts, which are unicellular, the bodies of fungi are constructed of tiny filaments called **hyphae** (singular, **hypha**). Hyphae are composed of tubular walls surrounding plasma membranes and cytoplasm. The hyphae form an interwoven mat called a **mycelium** (plural, **mycelia**), the "feeding" network of a fungus (FIGURE 31.1).

Fungal mycelia can be huge, although they usually escape our notice—because they are subterranean, for instance. In 2000, scientists discovered the mycelium of one giant individual of the fungus *Armillaria ostoyae* in Oregon that is 3.4 miles in diameter and spreads through 2,200 acres of forest, equivalent to over 1,600 football fields. This fungus is at least 2,400 years old and hundreds of tons in weight, qualifying it as one of Earth's oldest and largest organisms.

Most fungi are multicellular with hyphae divided into cells by cross-walls, or **septa** (singular, **septum**). The septa generally have pores large enough to allow ribosomes, mitochondria, and even nuclei to flow from cell to cell (FIGURE 31.2a, p. 618). The cell walls of fungi differ from the cellulose walls of plants. Most fungi build their cell walls mainly of **chitin,** a strong but flexible nitrogen-containing polysaccharide identical to the chitin found in the external skeletons of insects and other arthropods. Some fungi are aseptate; that is, their hyphae are not divided into cells by cross-walls. Known as **coenocytic** fungi, they consist of a continuous cytoplasmic mass with hundreds or thousands of nuclei (FIGURE 31.2b). The coenocytic condition results from the repeated division of nuclei without cytoplasmic division. Parasitic fungi usually have some of their hyphae modified as **haustoria,** nutrient-absorbing hyphal tips that penetrate the tissues of the host (FIGURE 31.2c). There are even fungi with hyphae adapted for preying on animals (FIGURE 31.2d).

The correlation between structure and function is a fundamental theme of biology. The filamentous structure of the mycelium provides an extensive surface area that suits the absorptive nutrition of fungi. Ten cubic centimeters of rich organic soil may contain as much as 1 km of hyphae having a fungal surface area of over 300 cm^2 interfacing with the soil.

FIGURE 31.1 Fungal mycelia.

Reproductive structure. The mushroom functions in reproduction by producing tiny cells called spores.

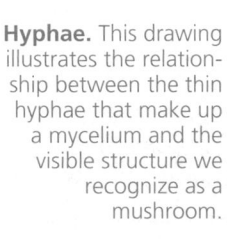

Hyphae. This drawing illustrates the relationship between the thin hyphae that make up a mycelium and the visible structure we recognize as a mushroom.

Spore-producing structures

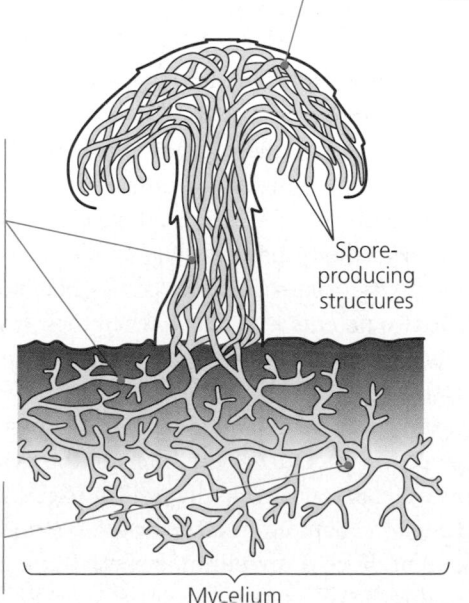

A fungal mycelium begins with the germination of a fungal spore in a suitable habitat.

Mycelium

This photo shows the mushrooms (*Mycena*) that grow up from the mycelium each fall.

This photo shows the vegetative part of a mycelium (white, cottony threads) decomposing brown conifer needles.

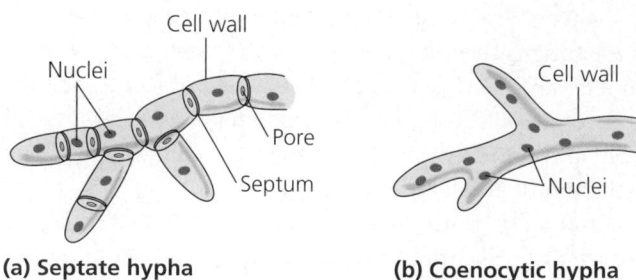

(a) Septate hypha

(b) Coenocytic hypha

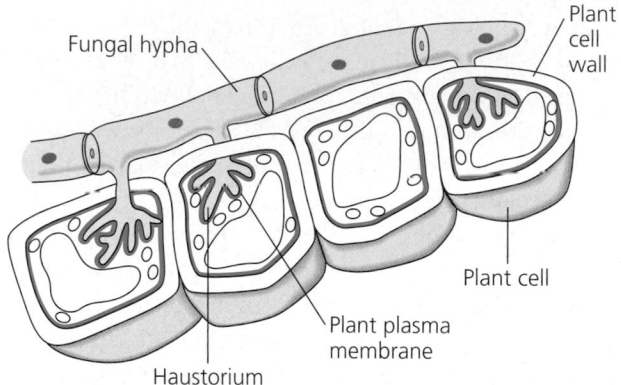

(c) Haustoria. Specialized hyphae called haustoria parasitize plant cells from *outside*, separated from the host cell's cytoplasm by the plasma membrane of the plant cell (blue). (The effect is like sinking your fingers into an underinflated balloon.)

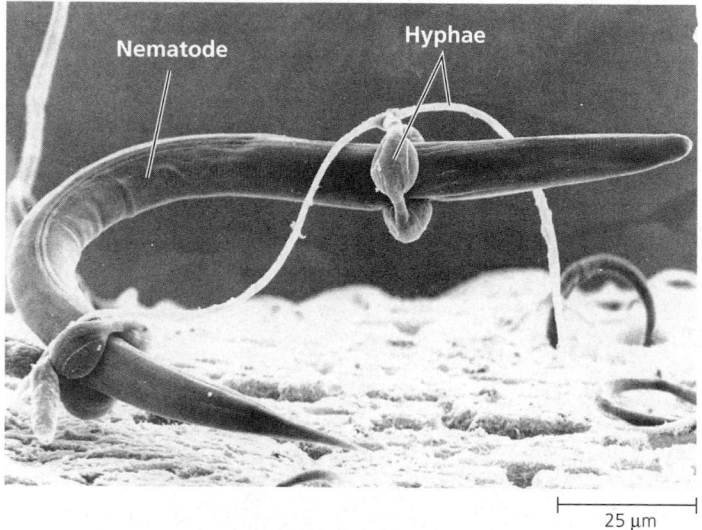

25 μm

(d) Hyphae adapted for trapping and killing prey. In *Arthrobotrys,* a soil fungus, portions of the hyphae are modified as hoops that constrict around nematodes (roundworms) in less than a second, when a worm rubs the inside of the hoop. The fungus then penetrates its prey with hyphae and digests its inner tissues (SEM).

FIGURE 31.2 Examples of fungal hyphae.

A fungal mycelium grows rapidly, adding as much as a kilometer of hyphae each day as it branches throughout a food source. Such fast growth is possible because proteins and other materials synthesized by the entire mycelium are channeled by cytoplasmic streaming to the tips of the extending hyphae.

The fungus concentrates its energy and resources on adding hyphal length and thus overall absorptive surface area, rather than girth. Fungal mycelia are nonmotile; they cannot run, swim, or fly in search of food or mates. But the mycelium makes up for the lack of mobility by swiftly extending the tips of its hyphae into new territory.

Fungi disperse and reproduce by releasing spores that are produced sexually or asexually

Fungi reproduce by releasing spores that are produced either sexually or asexually. The output of spores is enormous. For example, puffballs, which are reproductive structures of certain fungi, can puff out clouds containing trillions of spores (see FIGURE 22.8). Carried by wind or water, spores germinate to produce mycelia if they land in a moist place where there is food. Spores thus function in dispersal and account for the wide geographic distribution of many species of fungi. The airborne spores of fungi have even been found more than 160 km (100 miles) above Earth. Closer to home, try leaving a slice of bread out for a week or two and you will observe the furry mycelia that grow from the invisible spores raining down from the surrounding air.

Many fungi have a heterokaryotic stage

The nuclei of fungal hyphae and spores of most species are haploid, except for transient diploid stages that form during sexual life cycles. However, some mycelia may become genetically heterogeneous through the fusion of two hyphae that have genetically different nuclei. Such a mycelium is said to be a **heterokaryon,** meaning "different nuclei." In some cases, the different nuclei stay in separate parts of the same mycelium, which is then a mosaic in terms of genotype and phenotype. In other cases, the different nuclei mingle and may even exchange chromosomes and genes in a process similar to crossing over. This heterokaryon condition has some of the advantages of diploidy; one haploid genome may be able to compensate for harmful mutations in the other nucleus, and vice versa.

In many fungi with sexual life cycles, the union of partners occurs in two distinct stages called plasmogamy and karyogamy (FIGURE 31.3). First comes **plasmogamy,** the fusion of the two parents' cytoplasm when their mycelia come together. The second stage, **karyogamy,** is the fusion of the haploid nuclei contributed by the two parents. Plasmogamy and karyogamy may be separated in time by hours, days, or even years or centuries. During the interim, the hybrid mycelium exists as a heterokaryon, its haploid nuclei still separate. In some fungi, the haploid nuclei pair off, two to a cell, one from each parent. Such a mycelium, a special case of a heterokaryon, is said to be **dikaryotic,** meaning "two nuclei." Without fusing,

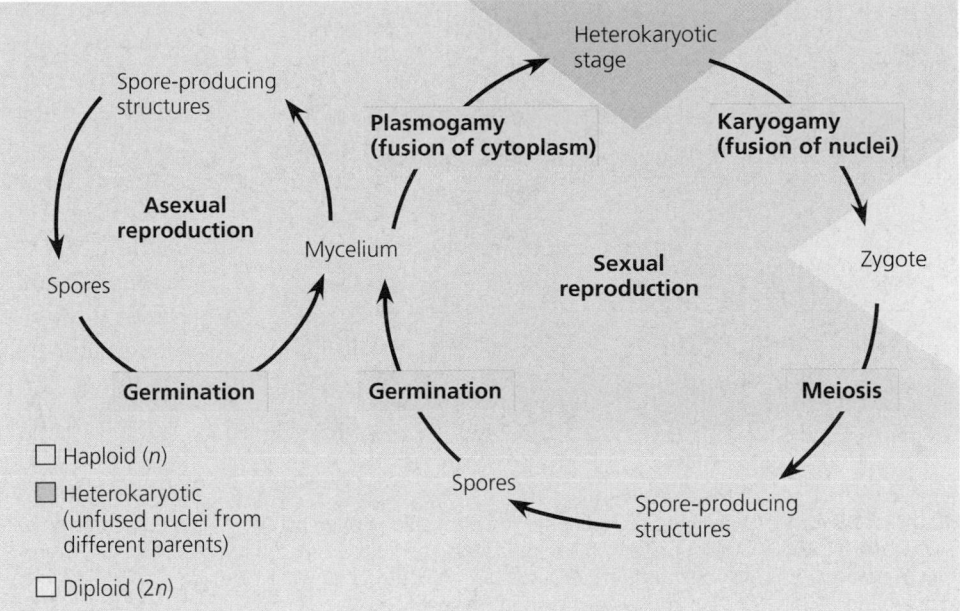

FIGURE 31.3 Generalized life cycle of fungi. Not all fungi use both sexual and asexual modes of reproduction; some fungi reproduce only asexually, while some are entirely sexual.

the two nuclei in each cell divide in tandem as the mycelium grows, until karyogamy finally occurs. In most fungi, the zygotes or transient structures formed by karyogamy are the only diploid stages in the life cycle. Meiosis restores the haploid condition before specialized reproductive structures of the mycelium produce and disperse spores. Of course, the sexual processes of karyogamy and meiosis have generated genetic variation, upon which adaptive evolution depends (see Chapters 13 and 23 to review sex as a mechanism that increases genetic diversity within a population).

DIVERSITY OF FUNGI

More than 100,000 species of fungi are known, and mycologists (biologists who study fungi) estimate that there are actually about 1.5 million species worldwide. The taxonomic scheme used in this chapter classifies fungi into four phyla (FIGURE 31.4).

Phylum Chytridiomycota: Chytrids may provide clues about fungal origins

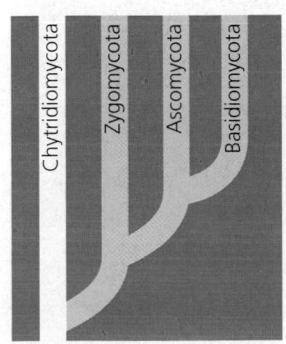

The **chytrids** are mainly aquatic. Some are saprobes; others parasitize protists, plants, and animals. Parasitic chytrids may be contributing to a worldwide decline in the number of amphibians.

Until recently, some systematists emphasized the absence of flagellated cells as a membership requirement for the kingdom Fungi. By that criterion, chytrids were excluded and placed instead in the kingdom Protista (of the five-kingdom system) because they form uniflagellated spores called zoospores (FIGURE 31.5, p. 620). However, in the past decade, molecular systematists comparing the sequences of proteins and nucleic acids uncovered strong support for including the chytrids with the fungi

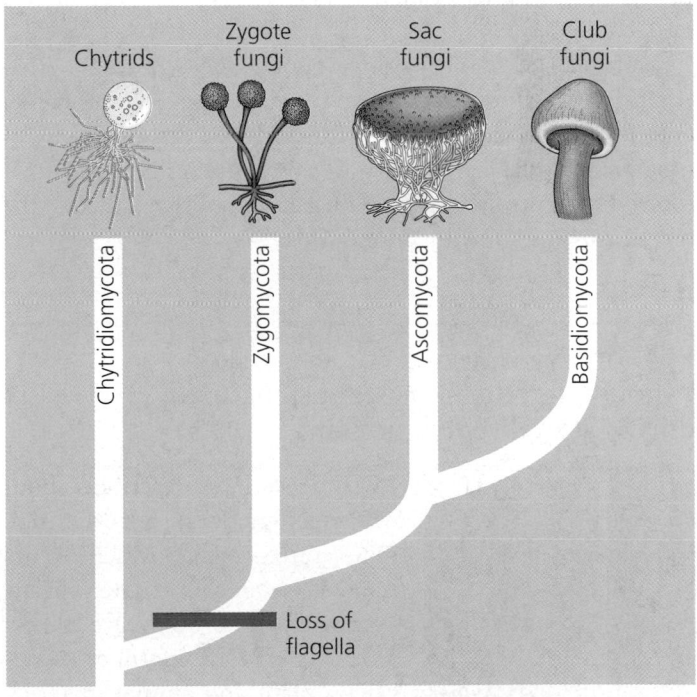

FIGURE 31.4 Phylogeny of fungi. This phylogenetic tree, supported primarily by molecular evidence, indicates probable evolutionary relationships among the four phyla of the kingdom Fungi. Representing the oldest lineage of fungi, the chytrids are mainly aquatic and have flagellated cells. Fungi in the other three phyla lack flagellated stages.

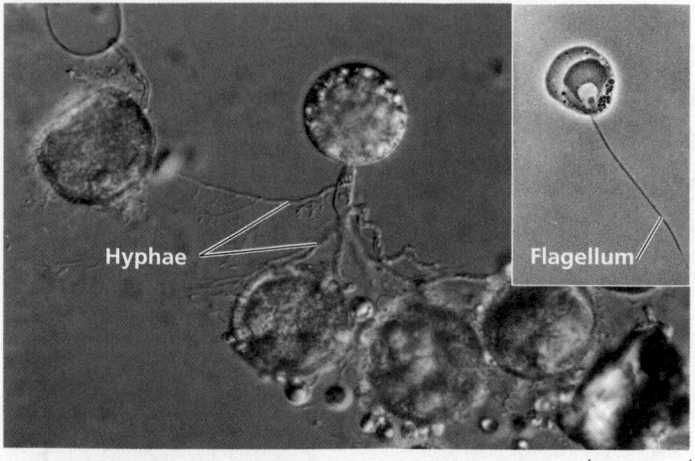

FIGURE 31.5 Chytridiomycota (chytrids). The branched hyphae of *Chytridium* expose a large surface to the surrounding medium, from which the organism absorbs nutrients. Chytrids are the only fungi with a flagellated stage, the zoospore shown in the inset (TEM).

as a monophyletic branch of the eukaryotic tree (see FIGURE 31.4). Other key fungal characteristics of chytrids are an absorptive mode of nutrition and cell walls made of chitin. Most chytrids form coenocytic hyphae, although some are unicellular. Chytrids also have some key enzymes and metabolic pathways that are common among fungi but are not found in the so-called funguslike protists (slime molds and water molds; see Chapter 28).

Molecular evidence also supports the hypothesis that chytrids are the most primitive fungi, meaning that they belong to the lineage that diverged earliest in the phylogeny of fungi. A reasonable extension of this hypothesis is that fungi evolved from protists that had flagella, a feature retained in the fungal kingdom only by the chytrids.

Phylum Zygomycota:
Zygote fungi form resistant structures during sexual reproduction

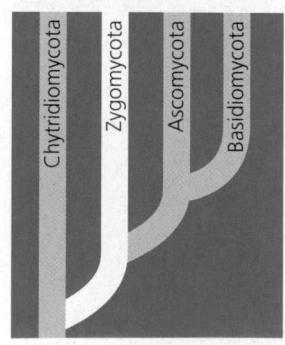

Mycologists have described about 600 zygomycetes, or **zygote fungi.** (The suffix -*mycete*, which occurs many times in this chapter, means "fungus.") These fungi are mostly terrestrial and live in soil or on decaying plant and animal material. One group of major importance forms **mycorrhizae,** mutualistic associations with the roots of

plants (see FIGURE 31.18). Zygomycete hyphae are coenocytic, with septa found only where reproductive cells are formed.

A common zygomycete is black bread mold, *Rhizopus stolonifer,* still an occasional household pest despite the addition of preservatives to most processed foods (FIGURE 31.6). Horizontal hyphae spread out over the food, penetrate it, and absorb nutrients. In the asexual phase, bulbous black sporangia develop at the tips of upright hyphae. Within each sporangium, hundreds of haploid spores develop and are dispersed through the air. Spores that happen to land on moist food germinate, growing into new mycelia. If environmental conditions deteriorate—for instance, if all the food is used up—this species of *Rhizopus* reproduces sexually. The parents in a sexual union are mycelia of opposite mating types, identical in appearance but different in the chemical markers that mates recognize. Plasmogamy produces a resistant structure called a **zygosporangium,** in which karyogamy and then meiosis occur (FIGURE 31.7). Note that a zygosporangium, while representing the zygote ($2n$) stage in the life cycle, is not a zygote in the usual sense of a cell with one diploid nucleus. Rather, the zygosporangium is a multinucleate structure, first heterokaryotic with many nuclei from the two parents, then with many diploid nuclei after karyogamy.

The zygosporangia, for which zygomycetes are named, are resistant to freezing and drying and are metabolically inactive. When conditions improve, the zygosporangia release genetically diverse haploid spores that colonize the new substrate. Some zygomycetes, such as *Pilobolus,* can actually aim their spores (FIGURE 31.8).

FIGURE 31.6 The common mold *Rhizopus* decomposing strawberries.

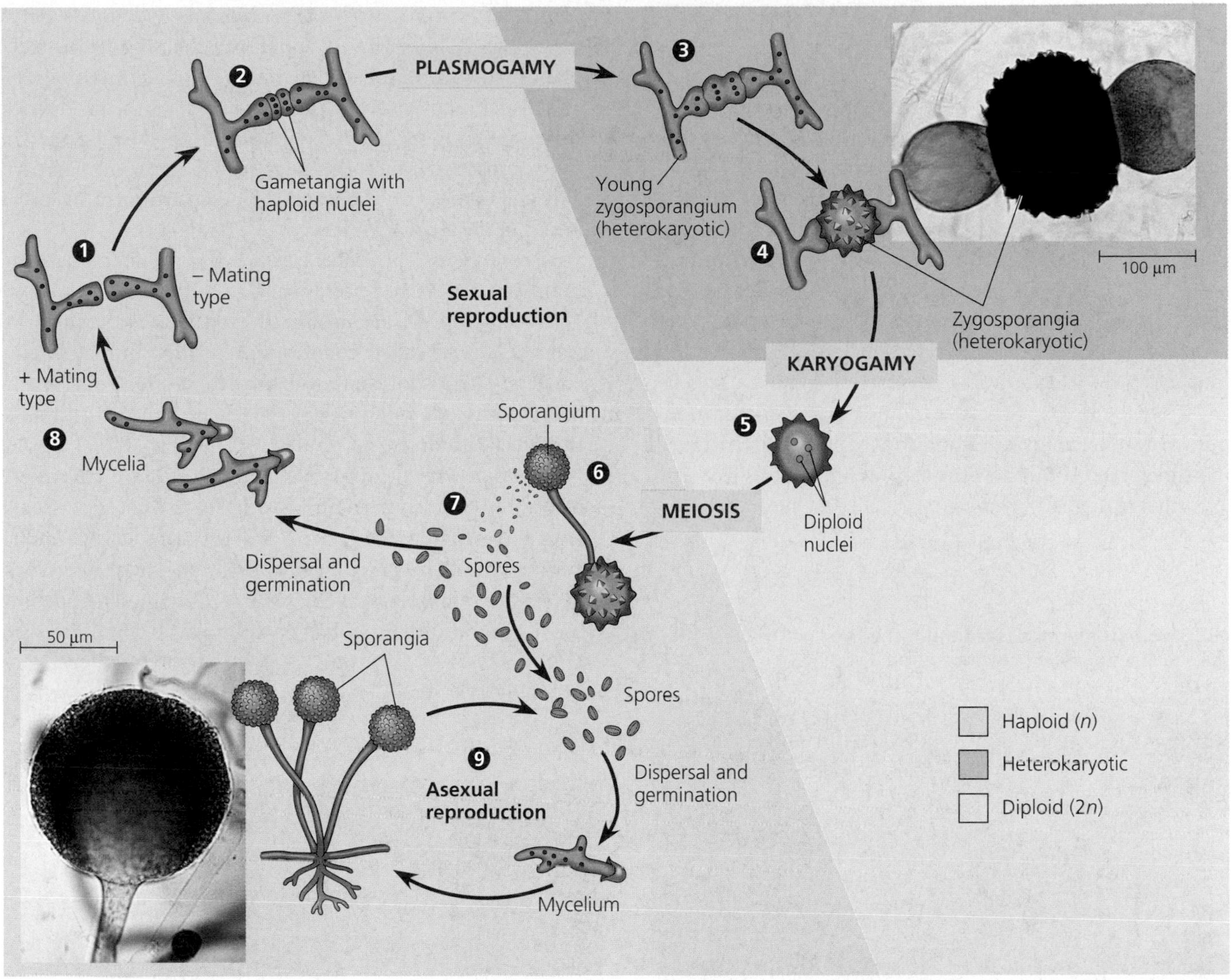

The diagram labels (Figure 31.7):

PLASMOGAMY

Gametangia with haploid nuclei

2 3

Young zygosporangium (heterokaryotic)

100 μm

Zygosporangia (heterokaryotic)

– Mating type

1

Sexual reproduction

KARYOGAMY

+ Mating type

8

Mycelia

Sporangium

6

5

7

MEIOSIS

Diploid nuclei

Dispersal and germination

Spores

50 μm

Sporangia

Spores

9

Asexual reproduction

Dispersal and germination

Mycelium

Mycelium

Haploid (*n*)

Heterokaryotic

Diploid (2*n*)

FIGURE 31.7 The life cycle of the zygomycete *Rhizopus* (black bread mold). ❶ Neighboring mycelia of opposite mating types (designated + and -) ❷ form hyphal extensions called gametangia, each walled off around several haploid nuclei by a septum. ❸ The gametangia undergo plasmogamy (fusion of cytoplasm), forming a heterokaryotic zygosporangium containing multiple haploid nuclei from the two parents. ❹ This cell develops a rough, thick-walled coating (upper right LM) that can resist dry conditions and other harsh environments for months. ❺ When conditions are favorable again, karyogamy occurs. Paired nuclei fuse, followed by meiosis. ❻ The zygosporangium then breaks dormancy, germinating into a short sporangium that ❼ disperses genetically diverse, haploid spores. ❽ These spores germinate and grow into new mycelia. ❾ Mycelia of *Rhizopus* can also reproduce asexually by forming sporangia (lower left LM) that produce genetically identical haploid spores.

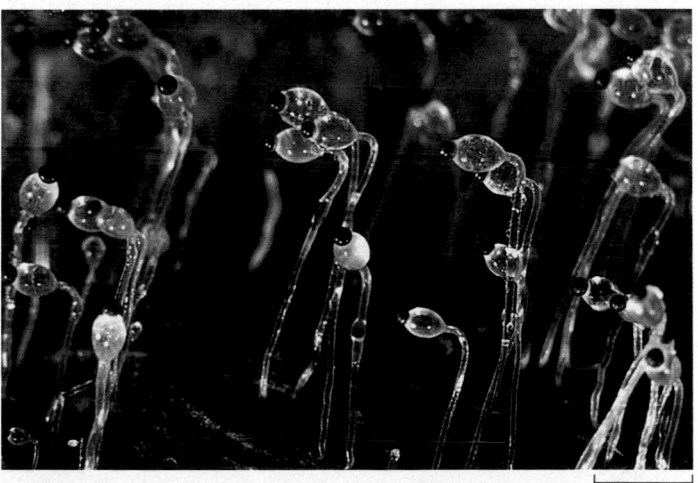

0.5 mm

FIGURE 31.8 *Pilobolus* aiming its sporangia. This zygomycete decomposes animal dung. The mycelium bends its sporangium-bearing hyphae toward bright light, where grass is likely to be growing. The fungus then shoots its sporangia like cannonballs; they can land and stick to grass as far as 2 m away. Grazing animals such as cows scatter the spores in feces.

Phylum Ascomycota: Sac fungi produce sexual spores in saclike asci

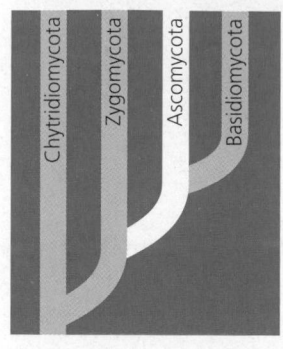

Mycologists have described over 60,000 species of ascomycetes, or **sac fungi,** from a variety of marine, freshwater, and terrestrial habitats. They range in size and complexity from unicellular yeasts to minute leaf-spot fungi to elaborate cup fungi and morels (FIGURE 31.9). Ascomycetes include some of the most devastating plant pathogens, which will be discussed later in the chapter. However, many are important saprobes, particularly of plant material. About half the ascomycete species live with algae in the mutualistic associations called lichens. Some as-comycetes form mycorrhizae with plants. Others live between mesophyll cells in leaves, apparently helping to protect the plant tissues from insects by releasing toxic compounds.

The defining feature of the Ascomycota is the production of sexual spores in saclike **asci** (singular, **ascus**). Unlike the zygote fungi, most sac fungi bear their sexual stages in macroscopic fruiting bodies, or **ascocarps.** The spore-forming asci are found in the ascocarps.

Ascomycetes reproduce asexually by producing enormous numbers of asexual spores, which are often dispersed by wind. The asexual spores are produced externally at the tips of specialized hyphae called conidiophores, often in long chains or clusters. These spores are not formed inside sporangia, as in the Zygomycota. Such naked spores are called **conidia,** from the Greek for "dust."

Compared to zygomycetes, ascomycetes are characterized by a more extensive heterokaryotic stage, which is associated with the formation of ascocarps (FIGURE 31.10). Plasmogamy between specialized regions of two parental hyphae produces a heterokaryotic bulge called the ascogonium. The coenocytic ascogonium extends hyphae partitioned by septa into dikary-

FIGURE 31.9 Ascomycetes (sac fungi). The "fruiting" (sexual) structures of ascomycetes are called ascocarps.

(a) The ascocarps of scarlet cup *(Hygrophorus coccineal)*

(b) Truffles are ascocarps that "fruit" underground and emit strong odors, which attract animals that eat the fungi and disperse the ascospores. *Tuber melanosporum* is highly prized for its flavor by gourmet cooks, who pay over $600 a pound for this truffle.

(c) The edible ascocarp of *Morchella esculenta,* the succulent morel, is often found under trees in orchards.

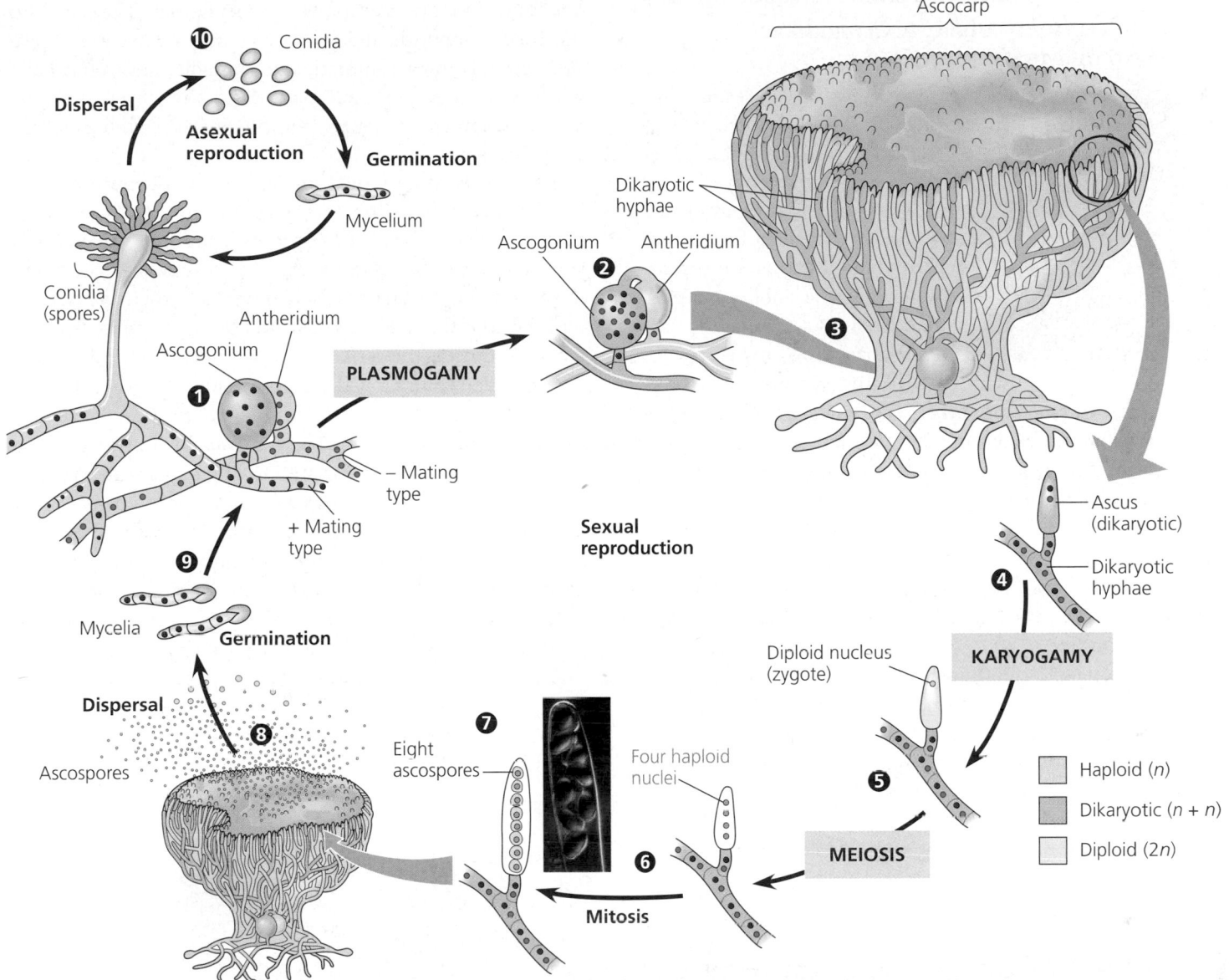

FIGURE 31.10 The life cycle of an ascomycete. (In this case, the key to color coding relates to structures rather than to portions of the life cycle.) ❶ Haploid mycelia of opposite mating types become intertwined and form an ascogonium and an antheridium. ❷ A cytoplasmic bridge forms, allowing plasmogamy (cytoplasmic fusion) to occur. The ascogonium acts as a "female," receiving haploid nuclei from the "male" antheridium. The ascogonium then has a pool of nuclei from both parents, but karyogamy (fusion of nuclei)

does not occur at this time. ❸ The ascogonium gives rise to dikaryotic hyphae that are incorporated into an ascocarp, the cup of a cup fungus (see FIGURE 31.9a). ❹ The tips of the ascocarp's dikaryotic hyphae are partitioned into asci (singular, ascus). ❺ Karyogamy occurs within these asci, and the diploid nucleus divides by meiosis, ❻ yielding four haploid nuclei. ❼ Each of these haploid nuclei divides once by mitosis, and the ascus now contains eight nuclei. Cell walls develop around these nuclei to form ascospores. (The LM shows a

mature ascus with eight ascospores.) ❽ When mature, all ascospores in an ascus are dispersed at once out the end of the ascus. A collapsing ascus jars neighboring asci and causes them to release their spores. The chain reaction releases a visible cloud of spores with an audible hiss. ❾ Germinating ascospores give rise to new haploid mycelia. ❿ Ascomycetes can also reproduce asexually by producing airborne spores called conidia.

otic cells, each with two haploid nuclei representing the two parental mycelia. It is the cells at the tips of these dikaryotic hyphae that develop into asci. Within an ascus, karyogamy combines the two parental genomes, and then meiosis forms four genetically varied ascospores. One mitotic division doubles the ascospore number to eight. In many asci, the eight as-

cospores are lined up in a row in the order in which they formed from a single zygotic nucleus. This arrangement provides geneticists with a unique opportunity to study genetic recombination. Genetic differences between mycelia grown from ascospores taken from one ascus reflect crossing over and independent assortment of chromosomes during meiosis.

Phylum Basidiomycota: Club fungi have long-lived dikaryotic mycelia

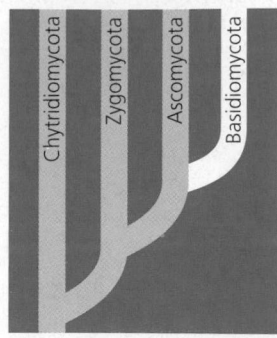

Approximately 25,000 fungi, including mushrooms, shelf fungi, puffballs, and rusts, are classified in the phylum Basidiomycota (FIGURE 31.11). The name derives from the **basidium** (L., "little pedestal"), a transient diploid stage in the organism's life cycle. The clublike shape of the basidium also gives rise to the common name **club fungus.**

Basidiomycetes are important decomposers of wood and other plant material. The phylum also includes mycorrhiza-forming mutualists and plant parasites. Of all fungi, the saprobic basidiomycetes are best at decomposing the complex polymer lignin, an abundant component of wood. Many shelf fungi (FIGURE 31.11b) break down the wood of weak or damaged trees and continue to decompose the wood after the trees die. Two groups of basidiomycetes, the rusts and smuts, include particularly destructive plant parasites.

The life cycle of a club fungus usually includes a long-lived dikaryotic mycelium (FIGURE 31.12). Periodically, in response to environmental stimuli, this mycelium reproduces sexually by producing elaborate fruiting bodies called **basidiocarps** (see FIGURE 31.11). The numerous basidia of a basidiocarp are the sources of sexual spores. Asexual reproduction in basidiomycetes is much less common than in ascomycetes.

A mushroom is an example of a basidiocarp. The cap of the mushroom supports and protects a large surface area of basidia on gills; each common, store-bought mushroom has a gill surface area of about 200 cm². Such a mushroom may release a billion basidiospores, which drop beneath the cap and are blown away.

By concentrating growth in the hyphae of mushrooms, a basidiomycete mycelium can erect the fruiting structures in just a few hours. A ring of mushrooms, popularly called a fairy ring, may appear on a lawn overnight (FIGURE 31.13). Although the grass in the center of the ring is normal, you may notice after a few days that the grass beneath the ring is stunted and the grass just outside the garland of mushrooms is especially lush. As the underground mycelium grows outward, its center portion and the mushrooms above the center portion die because the mycelium has consumed all the available nutrients. Thus, the living mycelium is an expanding ring that produces mushrooms above it. The grass beneath the mushrooms is stunted because it cannot compete for minerals with the active mycelium. But the advancing mycelium secretes digestive agents ahead of it that decompose the organic matter in the soil, producing a lush growth of grass that absorbs the minerals that have become available. The fairy ring slowly increases in diameter as the mycelium advances at a rate of about 30 cm per year. Some giant fairy rings may be centuries old.

TABLE 31.1 on page 626 reviews the four phyla of fungi. Let's now turn to a *different* way of classifying certain fungi.

(a) Miniature waxy cap (*Hygrophorus*). This species is mycorrhizal with oaks.

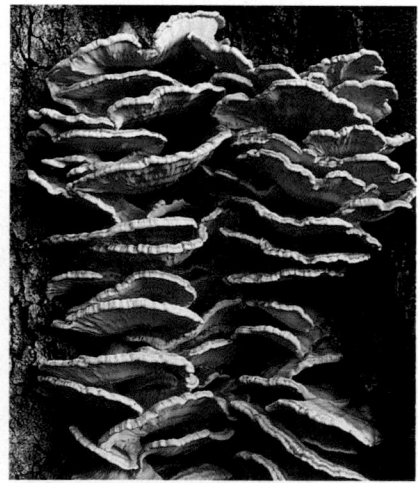

(b) Shelf fungus growing on the trunk of a tree. Shelf fungi are important decomposers of wood.

(c) Stinkhorn fungus (*Dictyophora duplicatal*). This fungus emits a rotten-meat odor that attracts such carrion-eating insects as flies, which you can see on the slimy, stinky fruiting body of the fungus. Spores stick to the bristles on the legs of the flies, which then disperse the spores. This adaptation is analogous to flowering plants using insects to spread pollen.

FIGURE 31.11 Basidiomycetes (club fungi). These photographs showcase diverse basidiocarps, the fruiting bodies that produce sexual spores.

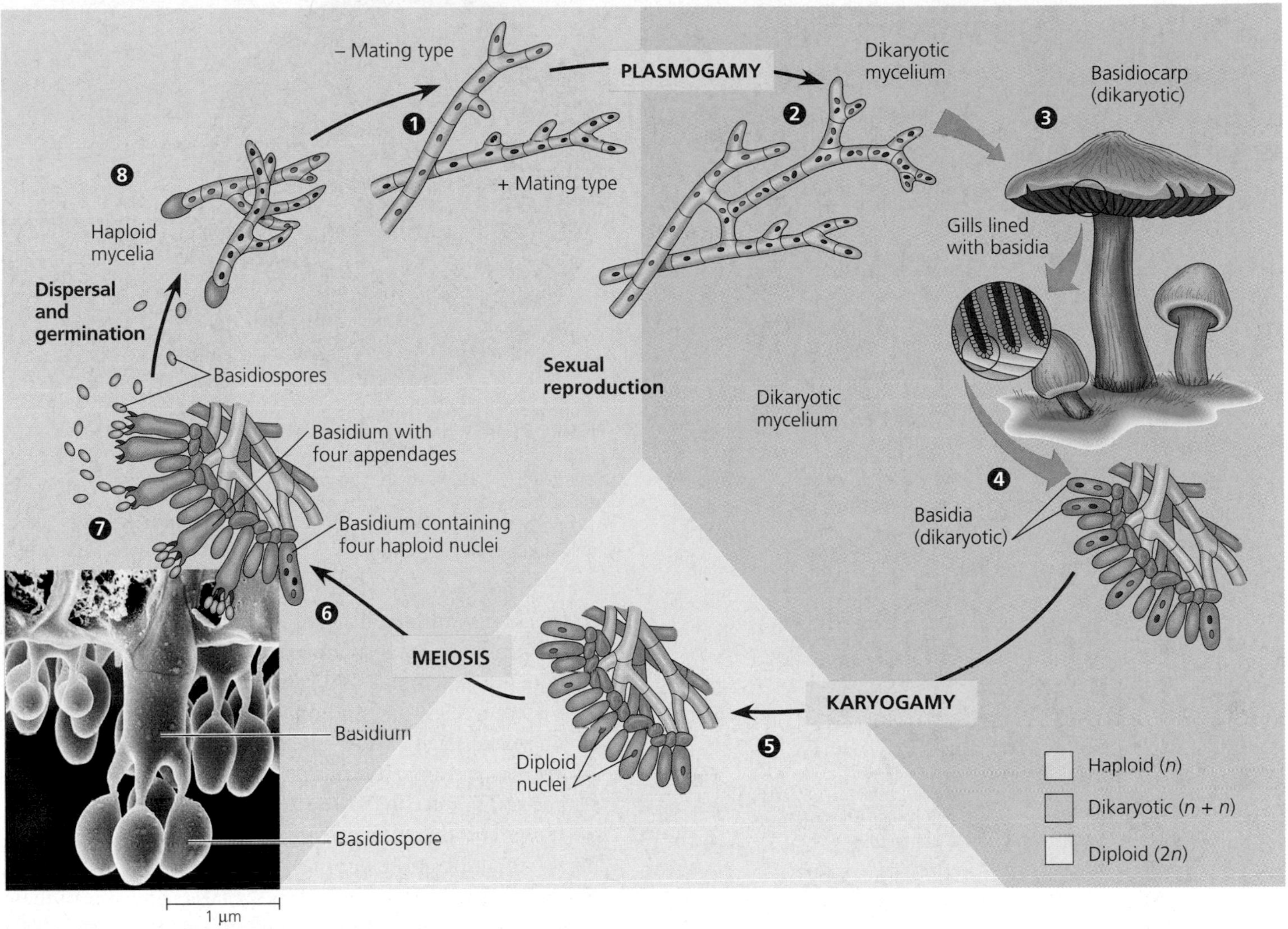

– Mating type

PLASMOGAMY

Dikaryotic
mycelium

Basidiocarp
(dikaryotic)

❶

❷

❸

+ Mating type

❽

Haploid
mycelia

Gills lined
with basidia

**Dispersal
and
germination**

Basidiospores

**Sexual
reproduction**

Dikaryotic
mycelium

Basidium with
four appendages

Basidium containing
four haploid nuclei

❼

❻

Basidia
(dikaryotic)

❹

MEIOSIS

Basidium

Basidiospore

Diploid
nuclei

KARYOGAMY

❺

1 μm

☐ Haploid (*n*)

☐ Dikaryotic (*n* + *n*)

☐ Diploid (2*n*)

FIGURE 31.12 The life cycle of a mushroom-forming basidiomycete.

❶ Two haploid mycelia of opposite mating type undergo plasmogamy, ❷ creating a dikaryotic mycelium that grows faster than, and ultimately crowds out, the parent haploid mycelia. ❸ Environmental cues such as rain or temperature changes induce the dikaryotic mycelium to form compact masses that develop into basidiocarps (mushrooms, in this case). Cytoplasm streaming in from the mycelium swells the hyphae of mushrooms, causing them to "pop up" overnight. The dikaryotic mycelia of basidiomycetes are long-lived, generally producing a new crop of basidiocarps each year. ❹ The surfaces of the basidiocarp's gills are lined with terminal dikaryotic cells called basidia. ❺ Karyogamy produces diploid nuclei, which then undergo meiosis, ❻ each yielding four haploid nuclei. Each basidium grows four appendages, and one haploid nucleus enters each appendage and develops into a basidiospore (SEM inset). ❼ When mature, the basidiospores are propelled slightly (by electrostatic forces) into the spaces between the gills. After the spores drop below the cap, they are dispersed by the wind. ❽ The haploid basidiospores germinate in a suitable environment and grow into short-lived haploid mycelia.

FIGURE 31.13 A fairy ring. The legendary explanation of these circles of fungi is that mushrooms spring up where fairies have danced in a ring on moonlit nights. Afterward, the tired fairies sit down on the mushrooms, but some of the mushrooms are sat on by toads. The mushrooms the fairies choose are edible by humans, but the "toadstools" are poisonous. See the text on page 624 to read an alternate explanation of fairy rings.

Table 31.1 Review of Fungal Phyla

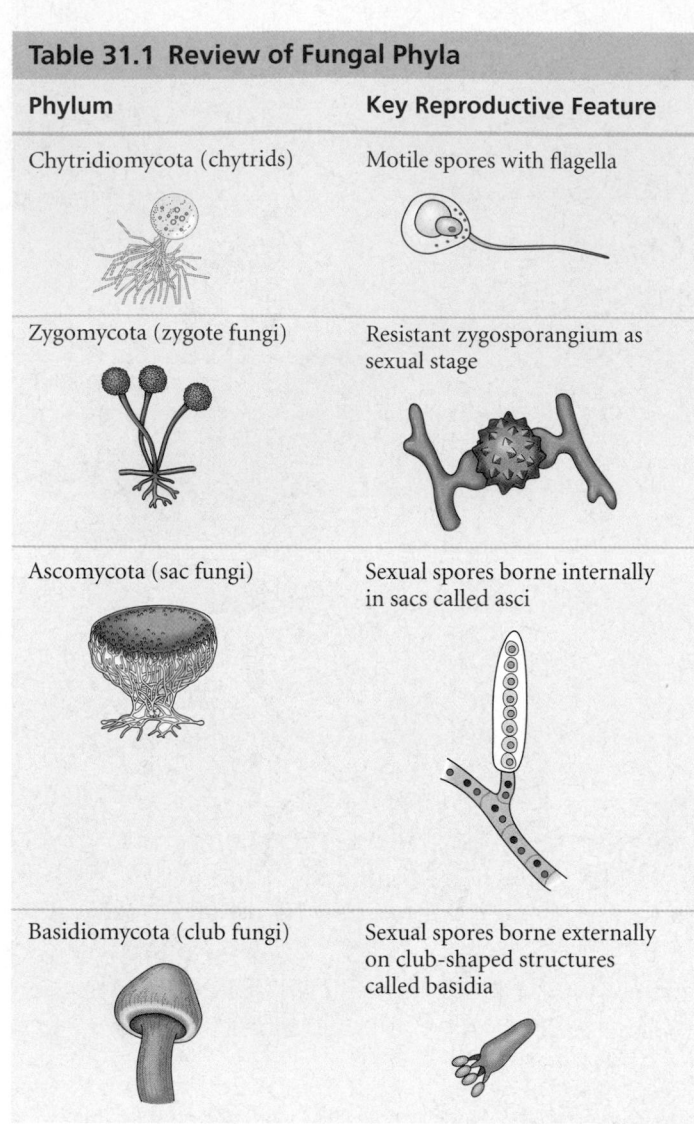

Phylum	Key Reproductive Feature
Chytridiomycota (chytrids)	Motile spores with flagella
Zygomycota (zygote fungi)	Resistant zygosporangium as sexual stage
Ascomycota (sac fungi)	Sexual spores borne internally in sacs called asci
Basidiomycota (club fungi)	Sexual spores borne externally on club-shaped structures called basidia

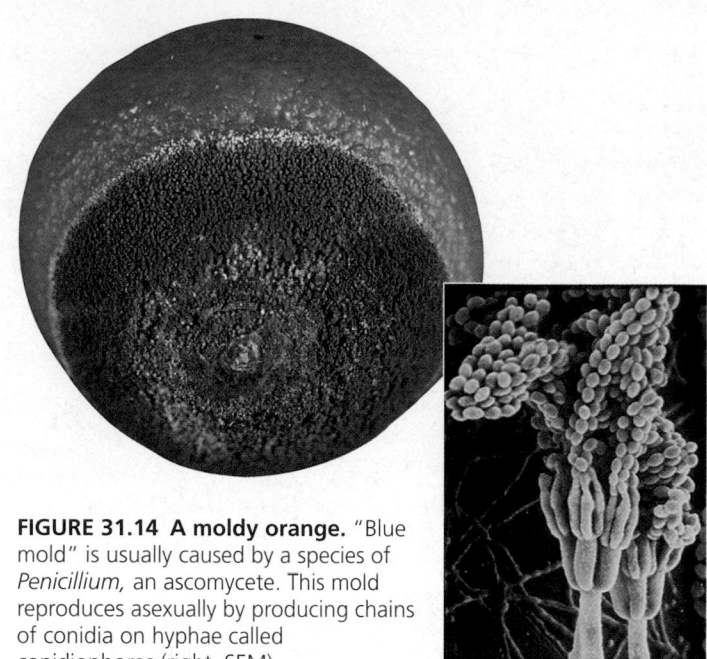

FIGURE 31.14 A moldy orange. "Blue mold" is usually caused by a species of *Penicillium*, an ascomycete. This mold reproduces asexually by producing chains of conidia on hyphae called conidiophores (right, SEM).

2.5 µm

ual stages (FIGURE 31.14). Later, the same fungus *may* reproduce sexually, producing zygosporangia, ascocarps, or basidiocarps.

There are also molds that cannot be classified as zygomycetes, ascomycetes, or basidiomycetes because they have no known sexual stages. Those molds are collectively called deuteromycetes, or **imperfect fungi** (from the botanical use of the term *perfect* to refer to the sexual stages of life cycles). Imperfect fungi reproduce asexually by producing spores. Note that this is an informal grouping without phylogenetic basis. Whenever a mycologist discovers a sexual stage in one of these fungi, the species is moved from the imperfect category to a particular phylum, depending on the type of sexual structures.

Molds, yeasts, lichens, and mycorrhizae are specialized lifestyles that evolved independently in diverse fungal phyla

Certain ways of living that require both morphological and ecological adaptations have evolved independently among the zygote fungi, sac fungi, and club fungi. This section explores four fungal forms with highly specialized ways of life: molds, yeasts, lichens, and mycorrhizae.

Molds

Mention of fungi may bring the ubiquitous molds to mind. A **mold** is a rapidly growing, asexually reproducing fungus. The mycelia of these fungi grow as saprobes or parasites on a great variety of substrates. You are already familiar with one example, bread mold (*Rhizopus*; see FIGURE 31.6). Molds may go through a series of different reproductive stages. Early in life, a mold produces asexual spores. The term *mold* applies only to these asex-

Yeasts

Yeasts are unicellular fungi that inhabit liquid or moist habitats, including plant sap and animal tissues. Yeasts reproduce asexually, by simple cell division or by the pinching of small "bud cells" off a parent cell (FIGURE 31.15). Some yeasts repro-

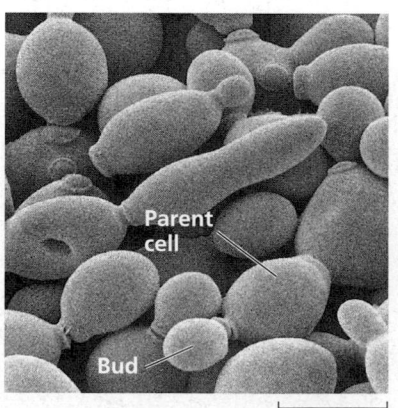

Parent cell

Bud

10 µm

FIGURE 31.15 Budding yeast. This micrograph shows *Saccharomyces cerevisiae* in various stages of budding (SEM).

duce sexually, by forming asci or basidia, and are classified as Ascomycota or Basidiomycota. Others are placed in the imperfect fungi because no sexual stages are known. Some fungi can grow as either single cells (yeasts) or as a filamentous mycelium, depending on the availability of nutrients.

Humans have used yeasts to raise bread and ferment alcoholic beverages for thousands of years. Only relatively recently have the yeasts involved been separated into pure culture for more controlled human use. The yeast *Saccharomyces cerevisiae,* an ascomycete, is the most important of all domesticated fungi (see FIGURE 31.15). The tiny yeast cells, available as many strains of baker's yeast and brewer's yeast, are very active metabolically. The cells release small bubbles of CO_2 that leaven dough. Cultured anaerobically in breweries and wineries, *Saccharomyces* ferments sugars to alcohol. Researchers use *Saccharomyces* to study the molecular genetics of eukaryotes because these microbes are easy to culture and manipulate (see Chapter 19).

Some yeasts cause problems for humans. A pink yeast, *Rhodotorula,* grows on shower curtains and other moist surfaces in our homes. Another yeast is *Candida,* one of the normal inhabitants of moist human epithelial tissue, such as the vaginal lining. Certain circumstances can cause *Candida* to become pathogenic by growing too rapidly and releasing harmful substances ("yeast infections"). This can occur, for example, with an environmental change such as a pH shift, or when the immune system of the human host is compromised—by AIDS, for instance.

Lichens

At a distance, lichens are often mistaken for mosses or other simple plants growing on rocks, rotting logs, trees, and roofs (FIGURE 31.16). In fact, lichens are *not* mosses or any other kind of plant, nor are they even individual organisms. A **lichen** is a symbiotic association of millions of photosynthetic microorganisms held in a mesh of fungal hyphae. The fungal component is most commonly an ascomycete, but several

FIGURE 31.16 Lichens. Lichens representing three growth forms inhabit the bark of this maple branch. *Parmelia* (lower left) is foliose (having a flattened, leaflike appearance). *Ramalina* (upper right) is fruticose (shrublike). Elsewhere, several crustose (paint smearlike) species in the genera *Lecanora* and *Bacidia* are forming disk-shaped ascocarps. The minute orange lichen is *Xanthoria,* a foliose form.

basidiomycete lichens are known. The photosynthetic partners are usually unicellular or filamentous green algae or cyanobacteria. The merger of fungus and alga is so complete that lichens are actually given genus and species names, as though they were single organisms. Over 25,000 species have been described.

The fungus usually gives the lichen its overall shape and structure, and tissues formed by hyphae account for most of the lichen's mass. The algal component usually occupies an inner layer below the lichen surface (FIGURE 31.17). In most cases that have been examined, each partner provides things the other could not obtain on its own. The alga provides the fungus with food. Cyanobacteria in lichens fix nitrogen (see Chapter 27) and provide organic nitrogen. The fungus provides the alga with a suitable physical environment for growth. Lichens absorb most of the minerals they need either from dust in the air or from rain. The physical arrangement of hyphae retains water and minerals, allows for gas exchange, and protects the algae. Fungal pigments help shade the algae from intense sunlight. Some fungal compounds are toxic and prevent lichens

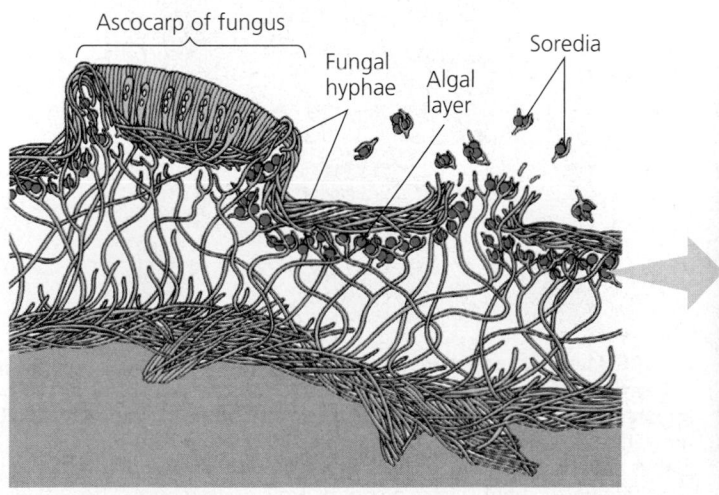

Ascocarp of fungus
Fungal hyphae
Algal layer
Soredia

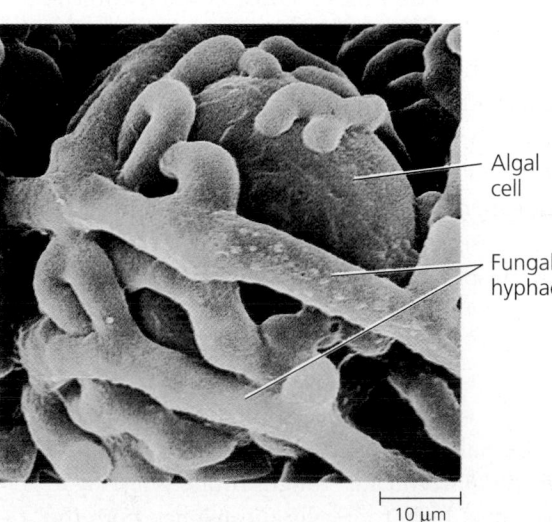

FIGURE 31.17 Anatomy of a lichen.

Algal cell

Fungal hyphae

10 μm

from being eaten by consumers. The fungi also secrete acids, which aid in the uptake of minerals.

The fungi of many lichens reproduce sexually by forming ascocarps or basidiocarps. Lichen algae reproduce independently of the fungus by asexual cell division. As might be expected of "dual organisms," asexual reproduction as symbiotic units also occurs commonly, either by fragmentation of the parental lichen or by the formation of specialized structures called **soredia** (see FIGURE 31.17). Soredia are small clusters of hyphae with embedded algae.

The nature of the lichen symbiosis is complex and is probably best described as mutual exploitation instead of mutual benefit. Lichens are able to live in environments where neither fungi nor algae could live alone. While the fungal components do not grow alone in the wild, some lichen algae also occur as free-living organisms. The fungi and algae of some lichens have been experimentally separated and cultured. Such cultures look like free-living molds and algae; the cultured fungi do not produce lichen compounds, and the cultured lichen algae do not "leak" carbohydrates from their cells as they do in lichens. In some lichens, the fungus invades the algal cells with haustoria and may kill some of them, though not as fast as the alga replenishes its numbers by reproduction.

Lichens are important pioneers on newly cleared rock and soil surfaces, such as burned forests and volcanic flows. Physical penetration of the outer crystals of rocks and chemical attack of rock by lichen acids help break down the rock and establish soil-trapping lichens. This process makes it possible for a succession of plants to grow. Nitrogen-fixing lichens also add organic nitrogen to some ecosystems.

Some lichens tolerate severe cold. In the arctic tundra, herds of caribou and reindeer graze on carpets of reindeer lichen (reindeer "moss") at times of the year when other foods are unavailable. Lichens can also survive desiccation. When it is foggy or rainy, lichens may absorb more than ten times their weight in water. Photosynthesis begins almost immediately when the water content reaches 65% to 75%. In dry air, lichens rapidly dehydrate, and photosynthesis stops. Thus, in arid climates lichens grow very slowly, often less than a millimeter per year. Some lichens are thousands of years old, rivaling the oldest plants as the elder organisms on Earth.

As tough as lichens are, many do not stand up very well to air pollution. Their passive mode of mineral uptake from rain and moist air makes them particularly sensitive to sulfur dioxide and other aerial poisons. The death of sensitive lichens and an increase in hardier species in an area can be an early warning that air quality is deteriorating.

Mycorrhizae

Mycorrhizae are mutualistic associations of plant roots and fungi. The word *mycorrhizae* means "fungus roots," referring to the structures formed by both root cells and hyphae from the

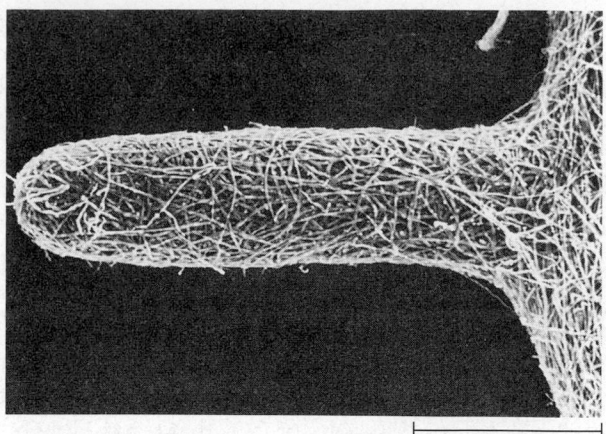

FIGURE 31.18 Mycorrhizae. Mutualistic associations between fungi and roots, mycorrhizae enhance the absorption of minerals. This scanning electron micrograph of fungal hyphae on a small root of a eucalyptus reveals the intimate association between the symbiotic partners of mycorrhizae . Some mycorrhizal fungi surround the living cells of the root, while others actually penetrate the root cells.

associated fungus (FIGURE 31.18). The anatomy of this symbiosis varies, depending on the type of fungus. The extensions of the fungal mycelium from the hyphae forming the mycorrhizae greatly increase the absorptive surface of the plant roots. The partners exchange minerals accumulated from the soil by the fungus for organic nutrients synthesized by the plant.

Mycorrhizae are enormously important in natural ecosystems and agriculture (FIGURE 31.19). Almost all vascular plants have mycorrhizae. The fungi involved are permanent associates with their hosts and periodically form fruiting bodies (structures for sexual reproduction). Basidiomycota, Ascomycota, and Zygomycota all have members that form mycorrhizae. In

FIGURE 31.19 An experimental test of the benefits of mycorrhizae. The soybean plant on the left lacks mycorrhizae. It represents an experimental group of plants that researchers grew in soil treated with fungicide (poison that kills fungi). The plant's growth is stunted, probably due to a phosphorus deficiency. The plant on the right is from a control group with mycorrhizae, which enhance the uptake of phosphate and other minerals.

THE
PROCESS
OF SCIENCE

fact, almost half of all species of mushroom-forming basidiomycetes live as mycorrhizae with oak, birch, and pine trees. The mushrooms that sprout around the bases of these trees are the surface evidence of the underground symbiotic relationship between plants and fungi. (The structure and physiology of mycorrhizae are described in more detail in Chapter 36.)

ECOLOGICAL IMPACTS OF FUNGI

Though we rarely notice, fungi are all around us as major forces in the dynamics of all ecosystems.

Ecosystems depend on fungi as decomposers and symbionts

Fungi and bacteria are the principal decomposers that keep ecosystems stocked with the inorganic nutrients essential for plant growth. Without decomposers, carbon, nitrogen, and other elements would become tied up in organic matter. Plants and the animals they feed would starve because elements taken from the soil would not be returned (see Chapter 54).

Fungi are well adapted as decomposers of plant material. Their invasive hyphae enter the tissues and cells of dead organic matter, secreting exoenzymes that hydrolyze polymers, including the cellulose and lignin of plant cell walls. A succession of fungi, in concert with bacteria and, in some environments, invertebrate animals, are responsible for the complete breakdown of plant litter. The air is so loaded with fungal spores that as soon as a leaf falls or an insect dies, it is covered with spores and is soon infiltrated by saprobic hyphae.

We may applaud fungi that decompose forest litter or dung, but it is a different story when molds attack our fruit or our shower curtains. Between 10% and 50% of the world's fruit harvest is lost each year to fungal attack. Ethylene, a plant hormone that causes fruit to ripen, also stimulates fungal spores on the fruit surface to germinate. This timing mechanism allows fungi to invade when fruit is most vulnerable and nutritious. A wood-digesting saprobe does not distinguish between a fallen oak limb and the oak planks of a boat. During the Revolutionary War, the British lost more ships to fungal rot than to enemy attack. Soldiers stationed in the tropics during World War II watched their tents, clothing, boots, and binoculars be destroyed by molds. Some fungi can even decompose certain plastics.

Some fungi are pathogens

Of the 100,000 known species of fungi, about 30% make their living as parasites, mostly on or in plants. For example, *Ophiostoma ulmi,* the ascomycete that causes Dutch elm disease, has drastically changed the landscape of the northeastern United States (FIGURE 31.20a). Accidentally introduced to the United States on logs that were sent from Europe to help pay World War I debts, the fungus is carried from tree to tree by bark beetles. Another ascomycete, *Cryphonectria parasitica,* has almost eliminated the native American chestnut. Fungi are also serious agricultural pests. Some species infect grain crops and cause tremendous economic losses each year. For example, the basidiomycete *Puccinia graminis* causes wheat rust (FIGURE 31.20b).

(a) American elm trees killed by a fungus

(b) Black stem rust on wheat

(c) Ergots on rye

FIGURE 31.20 Examples of fungal diseases of plants.

Some of the fungi that attack food crops are toxic to humans. For example, some species of the mold *Aspergillus* contaminate improperly stored grain and peanuts by secreting compounds called aflatoxins, which are carcinogenic. In another example, one type of ascomycete, *Claviceps purpurea*, forms purple structures called ergots on rye (FIGURE 31.20c). If diseased rye is inadvertently milled into flour and consumed, poisons from the ergots cause gangrene, nervous spasms, burning sensations, hallucinations, and temporary insanity. One epidemic in about 944 A.D. killed more than 40,000 people in France. One of the hallucinogens that has been isolated from ergots is lysergic acid, the raw material from which LSD is made. On the other hand, toxins extracted from fungi often have medicinal uses when administered in weak doses. For example, an ergot compound is helpful in treating high blood pressure and stopping maternal bleeding after childbirth.

Animals are much less susceptible to parasitic fungi than are plants. Only about 50 species of fungi are known to be parasitic in humans and other animals. Their damage to hosts, however, is disproportionately large compared to their taxonomic diversity. In humans, fungi cause infections ranging from annoyances such as athlete's foot to deadly lung diseases.

The general term for a fungal infection is **mycosis.** Skin mycoses include the disease ringworm, so named because it appears as circular red areas on the skin. The ascomycetes that cause ringworm can infect almost any skin surface. Most commonly, they grow on the feet, causing the intense itching and blisters known as athlete's foot. Though highly contagious, athlete's foot and other ringworm infections can be treated with various fungicidal lotions and powders. Systemic mycoses, usually very serious, are fungal infections that spread throughout the body, usually from spores that are inhaled. Two examples are histoplasmosis and coccidioidomycosis, which are both very serious diseases with tuberculosis-like symptoms. The yeast *Candida albicans* is an example of an opportunistic pathogen. That means it is a normal inhabitant of the human body that only causes problems such as vaginal yeast infections when some other change in the body's microbiology, chemistry, or immunology allows the yeast to grow unchecked. Opportunistic infections, including mycoses, have increased in the past few decades, partly because of AIDS, which compromises the immune system.

Fungi are commercially important

It would not be fair to fungi to end our discussion with an account of diseases. Far more important are the benefits we derive from these interesting eukaryotes. We depend on them as decomposers and recyclers of organic matter.

Fungi also have a number of practical uses for humans. Most of us have eaten mushrooms, although we may not have realized that we were ingesting the fruiting bodies (basidio-

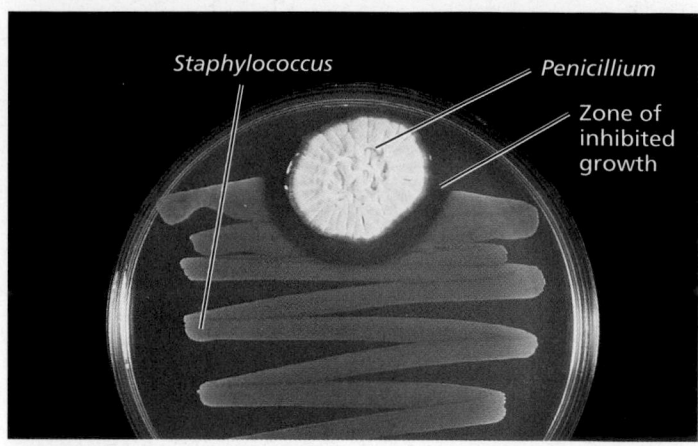

FIGURE 31.21 Fungal production of an antibiotic. In this petri dish, the clear area between the mold and the bacterial colony is where the antibiotic produced by *Penicillium* inhibits the growth of the bacteria, a species of *Staphylococcus*.

carps) of subterranean fungi. And mushrooms are not the only fungi we eat. The distinctive flavors of certain kinds of cheeses, including Roquefort and blue cheese, come from the fungi used to ripen them. The soft drink industry uses a species of the ascomycete mold *Aspergillus* to produce citric acid for colas. Highly prized by gourmets are truffles, the fruiting bodies of certain mycorrhizal ascomycetes associated with tree roots (see FIGURE 31.9b). Their complex flavor is variously described as nutty, musky, cheesy, or all three. The fruiting bodies (ascocarps) release strong odors that attract mammal and insect consumers that excavate the truffles and disperse their spores. In some cases, the odors mimic sex attractants of certain mammals. Truffle hunters traditionally used pigs to locate their prizes, although dogs are now more commonly used.

More important in food production are unicellular fungi, the yeasts. As discussed on page 627, yeasts are used in baking, brewing, and winemaking. Fungi are medically valuable as well. Some fungi produce antibiotics used to treat bacterial diseases. In fact, the first antibiotic discovered was penicillin, made by the common mold *Penicillium* (FIGURE 31.21).

EVOLUTION OF FUNGI

Fungi colonized land with plants

The fossil record indicates that from their inception, terrestrial communities have been dependent on fungi as decomposers and symbionts. Much of the diversity of fungi we observe today may have had its phylogenetic origin in adaptive radiation when life began to colonize land. The oldest undisputed fossils of fungi date back 460 million years, about the time plants began to colonize land. Fossils of the first vascular plants from the late Silurian period have petrified mycorrhizae. Plants probably moved onto land in the company of fungi.

Molecular evidence supports the widely held view that the four fungal divisions are monophyletic (see FIGURE 31.4). The occurrence of flagella in the chytrids, representing the oldest lineage of fungi, indicates that fungal ancestors were aquatic flagellated organisms. Flagellated cells were probably lost in the chytrid lineage that led to the Zygomycota, as ancestral fungi became increasingly adapted for life on land. Many of the differences among the phyla Zygomycota, Ascomycota, and Basidiomycota represent different solutions to the problem of reproducing and dispersing on land. These fungi may have diverged during the transition from aquatic to terrestrial habitats.

Fungi and animals evolved from a common protistan ancestor

Animals probably evolved from aquatic flagellated organisms too, and there is compelling evidence that animals and fungi diverged from a common protistan ancestor. Comparisons of several proteins and ribosomal RNA indicate that fungi are more closely related to animals than to plants (see FIGURE 28.8). We discuss animal phylogeny in more detail in Chapter 32.

CHAPTER 31 REVIEW

Go to the Campbell Biology website (www.campbellbiology.com) to explore an interactive version of the Chapter Review.

Summary of Key Concepts

INTRODUCTION TO THE FUNGI

- **Absorptive nutrition enables fungi to live as decomposers and symbionts (pp. 616–617)** All fungi are heterotrophs (decomposers and symbionts), acquiring their nutrients by absorption.

- **Extensive surface area and rapid growth adapt fungi for absorptive nutrition (pp. 617–618, FIGURES 31.1, 31.2)** The vegetative bodies of fungi consist of mycelia, netlike collections of branched hyphae adapted for absorption. Most fungi have cell walls made of chitin. Although aseptate (coenocytic) forms occur, most fungi have their hyphae partitioned into cells by septa, with pores allowing cell-to-cell continuity.

- **Fungi disperse and reproduce by releasing spores that are produced either sexually or asexually (p. 618)** Fungi produce reproductive structures (spores) by sexual and asexual means.
  Web/CD Activity 31A: *Fungal Reproduction and Nutrition*

- **Many fungi have a heterokaryotic stage (pp. 618–619, FIGURE 31.3)** The sexual cycle involves cell fusion (plasmogamy) and nuclear fusion (karyogamy), with an intervening heterokaryotic stage (cells with haploid nuclei from two parents). The diploid phase is short-lived and rapidly undergoes meiosis to produce haploid spores.

DIVERSITY OF FUNGI

- **Phylum Chytridiomycota: Chytrids may provide clues about fungal origins (pp. 619–620, FIGURE 31.5)** Chytrids, which produce flagellated spores, may represent a link between the fungal kingdom and the protists.

- **Phylum Zygomycota: Zygote fungi form resistant structures during sexual reproduction (pp. 620–621, FIGURES 31.6–31.8)** Zygote fungi, such as black bread mold, are named for their sexually produced zygosporangia, which are heterokaryotic structures capable of persisting through unfavorable conditions.

- **Phylum Ascomycota: Sac fungi produce sexual spores in saclike asci (pp. 622–623, FIGURES 31.9, 31.10)** Sexual reproduction in the sac fungi involves the formation of spores in sacs, or asci, at the ends of heterokaryotic hyphae, usually in ascocarps.

- **Phylum Basidiomycota: Club fungi have long-lived dikaryotic mycelia (pp. 624–625, FIGURES 31.11–31.13)** Mycelia of club fungi can grow for years in the heterokaryotic stage. Sexual reproduction involves the formation of spores on club-shaped basidia at the ends of dikaryotic hyphae in fruiting bodies, such as mushrooms.
  Web/CD Activity 31B: *Fungal Life Cycles*

- **Molds, yeasts, lichens, and mycorrhizae are specialized lifestyles that evolved independently in diverse fungal phyla (pp. 626–629, FIGURES 31.14–31.19)** Molds are rapidly growing, asexually reproducing fungi. Yeasts are unicellular fungi adapted to life in liquids such as plant saps. Lichens are such highly integrated mutualistic associations of algae and fungi that they are classified as single organisms. Mycorrhizae are mutualistic associations of fungi with the roots of vascular plants.

ECOLOGICAL IMPACTS OF FUNGI

- **Ecosystems depend on fungi as decomposers and symbionts (p. 629)** Without fungi and bacteria as decomposers, biological communities would be deprived of the essential recycling of chemical elements between the biological and nonbiological world.
  Web/CD Case Study in the Process of Science: *How Does the Fungus* Pilobolus *Succeed as a Decomposer?*

- **Some fungi are pathogens (pp. 629–630, FIGURE 31.20)** Some fungi cause disease, harming humans with a variety of ills. Plants are especially vulnerable to fungal infections.

- **Fungi are commercially important (p. 630, FIGURE 31.21)** Fungi are important in the food and drug industries.

EVOLUTION OF FUNGI

- **Fungi colonized land with plants (pp. 630–631)** Kingdom Fungi is a monophyletic taxon. Early fossils of land plants have mycorrhizae.

- **Fungi and animals evolved from a common protistan ancestor (p. 631)** Molecular evidence supports the hypothesis that fungi and animals diverged from a common ancestor, probably a flagellated aquatic protist.

Self-Quiz

1. *All* fungi share which one of the following characteristics?
   a. symbiotic
   b. heterotrophic
   c. flagellated
   d. pathogenic
   e. saprobic

2. Which feature seen in chytrids supports the hypothesis that they represent the most primitive fungi?
   a. the absence of chitin within the cell wall
   b. coenocytic hyphae
   c. flagellated spores
   d. formation of resistant zygosporangia
   e. all representatives are parasitic

3. Which of the following cells or structures are associated with *asexual* reproduction in fungi?
   a. ascospores
   b. basidiospores
   c. conidia
   d. zygosporangia
   e. ascogonia

4. Which of the following is an example of an opportunistic pathogen that can cause a mycosis?
   a. *Claviceps pururea*, which produces ergots on rye that can cause serious symptoms in humans if milled into flour
   b. *Ophiostoma ulmi*, which causes Dutch elm disease
   c. the ascomycetes that cause ringworm
   d. *Candida albicans*, which causes vaginal yeast infections
   e. the mold *Penicillium*, an ascomycete that is now grown in liquid culture to produce antibiotics

5. The adaptive advantage associated with the filamentous nature of the mycelium is primarily related to
   a. the ability to form haustoria and parasitize other organisms.
   b. avoiding sexual reproduction until environmental change occurs.
   c. the potential to inhabit almost all terrestrial habitats.
   d. the increased probability of contact between different mating types.
   e. an extensive surface area well suited for absorptive nutrition.

6. Sporangia on erect hyphae that produce asexual spores are characteristic of
   a. Ascomycota.
   b. Basidiomycota.
   c. Ascocarps.
   d. Zygomycota.
   e. lichens.

7. The basidiomycetes differ from other fungal phyla in that they
   a. have no known sexual reproduction stage.
   b. have long-lived dikaryotic mycelia.
   c. produce resistant sporangia that are initially heterokaryotic before karyogamy and meiosis occur.
   d. have members that are mutualistic partners with algae in lichens.
   e. form eight spores that line up in a sac in the order they were formed in meiosis, allowing geneticists to study genetic recombination.

8. Which of the following is the best description of a mold?
   a. a deuteromycete, or imperfect fungus, which has no known sexual stage
   b. a coenocytic, rapidly growing mycelium
   c. mycorrhizae that envelope plant roots and reproduce without forming spores
   d. unicellular fungi that grow rapidly in liquid or moist habitats
   e. the fast-growing mycelia of any asexually reproducing fungus

9. The photosynthetic symbiont of a lichen is most commonly a(n)
   a. moss
   b. green alga
   c. red alga
   d. ascomycete
   e. small vascular plant

10. The closest relatives of fungi are probably
    a. animals
    b. vascular plants
    c. mosses
    d. brown algae
    e. slime molds

11. What are mycorrhizae?

12. Contrast the heterotrophic nutrition of a fungus with your own heterotrophic nutrition.

13. Use the terms *plasmogamy, karyogamy,* and *heterokaryotic* in one sentence that correctly applies to fungal life cycles.

14. Why is the health of lichens an indicator of air quality?

15. What is athlete's foot?

16. What do you think is the function of the antibiotics that fungi produce in their natural environments?

Go to the website or CD-ROM for more quiz questions.

Evolution Connection

The fungal-algal symbiosis producing lichens is thought to have evolved several times independently in different fungal groups. However, lichens fall into three well-defined growth forms. What research could you perform to distinguish these two hypotheses: Hypothesis #1—Crustose, foliose, and fruiticose lichens each represent a monophyletic group; Hypothesis #2—Each lichen growth form represents convergent evolution by taxonomically diverse fungi.

The Process of Science

 Lichens colonize gravestones, such as the one in this photo, almost as soon as the stones are placed and then continue to grow for decades or even centuries. Explain how you could collect data for a particular lichen species at an old cemetery and use the data to calculate the growth rate of that species.

Explore how the fungus *Pilobolus* disperses its spores in the Case Study in the Process of Science, available on the website and CD-ROM.

Science, Technology, and Society

American chestnut trees once made up more than 25% of the hardwood forests of the eastern United States. These trees were wiped out by a fungus accidentally introduced on imported Asian chestnuts, which are not affected. More recently, a fungus has killed large numbers of dogwood trees from New York to Georgia; some experts suspect the parasite was accidentally introduced from elsewhere. Why are plants particularly vulnerable to fungi imported from other regions? What kinds of human activities might contribute to the spread of plant diseases? Do you think the introduction of plant pathogens like chestnut blight is more likely to occur in the future, or less likely to occur? Why?

Answers 1. b; 2. c; 3. c; 4. d; 5. e; 6. d; 7. b; 8. e; 9. b; 10. a; 11. Mycorrhizae are root-fungus associations that enhance the uptake of water and minerals by the plant and provide organic nutrients to the fungus. 12. A fungus digests its food externally by secreting digestive juices into the food and then absorbing the small nutrients that result from digestion. In contrast, humans and most other animals "eat" relatively large pieces of food and digest the food within their bodies. 13. Just one example: "In the sexual cycles of many fungi, there is a heterokaryotic stage between plasmogamy and karyogamy." 14. Lichens are very sensitive to air pollution because they absorb mineral compounds, both nutrients and pollutants, from the air. 15. Athlete's foot is an infection of the foot's skin with ringworm fungus. 16. The antibiotics probably block the growth of microorganisms, especially bacteria, that compete with the fungi for nutrients and other resources.

INTRODUCTION TO ANIMAL EVOLUTION

WHAT IS AN ANIMAL?
- Structure, nutrition, and life history define animals
- The animal kingdom probably evolved from a colonial, flagellated protist

TWO VIEWS OF ANIMAL DIVERSITY
- The remodeling of phylogenetic trees illustrates the process of scientific inquiry
- The traditional phylogenetic tree of animals is based mainly on grades in body "plans"
- Molecular systematists are moving some branches around on the phylogenetic tree of animals

THE ORIGINS OF ANIMAL DIVERSITY
- Most animal phyla originated in a relatively brief span of geologic time
- "Evo-devo" may clarify our understanding of the Cambrian diversification

Animal life began *in Precambrian seas with the evolution of multicellular forms that lived by eating other organisms. This new way of life allowed the exploitation of previously untapped resources and led to an evolutionary radiation of diverse forms. Early animals populated the seas, fresh water, and eventually the land. In the above photo of a coral reef, the diver, the fish, and the various invertebrates (animals without backbones) are just a few examples of the diverse forms derived during the past half-billion years of animal evolution.*

This chapter begins with the general characteristics of animals. We then discuss possible relationships among the animal phyla and examine hypotheses about the origin and early diversification of animals. This overview provides an orientation for our closer look at animal phyla in the next two chapters.

WHAT IS AN ANIMAL?

Structure, nutrition, and life history define animals

Constructing a good definition of an animal is not as easy as it might first appear. There are exceptions to nearly every criterion for distinguishing an animal from other life forms. However, when taken together, the following characteristics of animals will serve our purposes.

1. Animals are multicellular, heterotrophic eukaryotes. In contrast to the autotrophic nutrition of plants and algae, animals must take into their bodies preformed organic molecules; they cannot construct them from inorganic chemicals. Most animals do this by **ingestion**—eating other organisms or organic material that is decomposing.

2. Animal cells lack the cell walls that provide strong support in the bodies of plants and fungi. The multicellular bodies of animals are held together by structural proteins, the most abundant being collagen (see FIGURES 7.29 and 40.2). In addition to collagen, which is found mainly in extracellular matrices, animal tissues have unique types of intercellular junctions—tight junctions, desmosomes, and gap junctions—that are composed of other structural proteins (see FIGURE 7.30).

3. Also unique among animals are two types of tissues responsible for impulse conduction and movement: nervous tissue and muscle tissue.

4. A few key features of life history also distinguish animals. Most animals reproduce sexually, with the diploid stage usually

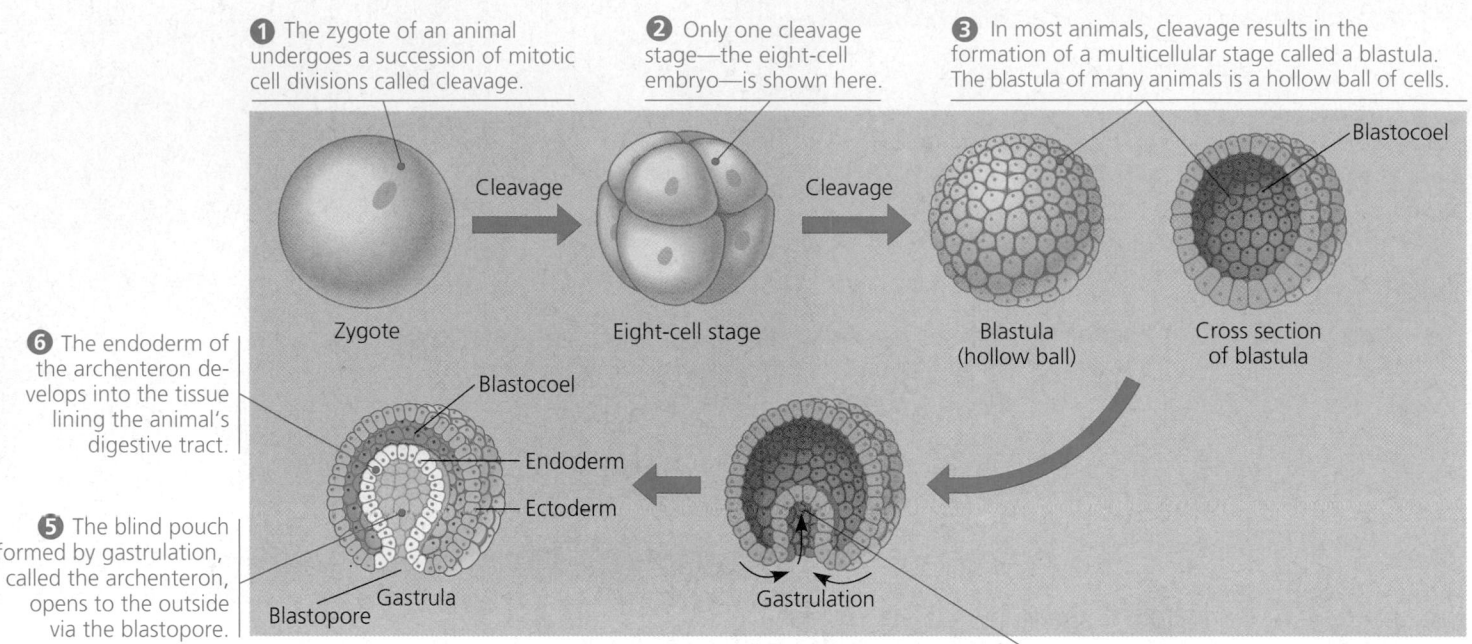

① The zygote of an animal undergoes a succession of mitotic cell divisions called cleavage.

② Only one cleavage stage—the eight-cell embryo—is shown here.

③ In most animals, cleavage results in the formation of a multicellular stage called a blastula. The blastula of many animals is a hollow ball of cells.

Blastocoel

Zygote → Cleavage → Eight-cell stage → Cleavage → Blastula (hollow ball) — Cross section of blastula

⑥ The endoderm of the archenteron develops into the tissue lining the animal's digestive tract.

Blastocoel
Endoderm
Ectoderm

⑤ The blind pouch formed by gastrulation, called the archenteron, opens to the outside via the blastopore.

Gastrula
Blastopore

Gastrulation

④ Most animals also undergo gastrulation, a rearrangement of the embryo where one end of the embryo folds inward, expands, and eventually fills the blastocoel, producing layers of embryonic tissues: the ectoderm (outer layer) and the endoderm (inner layer).

FIGURE 32.1 Early embryonic development.

dominating the life cycle. In most species, a small flagellated sperm fertilizes a larger, nonmotile egg to form a diploid zygote. The zygote then undergoes **cleavage,** a succession of mitotic cell divisions. During the development of most animals, cleavage leads to the formation of a multicellular stage called a **blastula,** which in many animals takes the form of a hollow ball. Following the blastula stage is the process of **gastrulation,** during which layers of embryonic tissues that will develop into adult body parts are produced. The resulting developmental stage is called a **gastrula** (FIGURE 32.1). Some animals develop directly through transient stages of maturation into adults, but the life cycles of many animals include larval stages. The **larva** is a sexually immature form. It is morphologically distinct from the adult stage, usually eats different food, and may even have a different habitat than the adult, as in the case of a frog tadpole. Animal larvae eventually undergo **metamorphosis,** a resurgence of development that transforms the animal into an adult.

5. The transformation of a zygote to an animal of specific form depends on the controlled expression in the developing embryo of special regulatory genes called *Hox* genes. All eukaryotes have genes that regulate the expression of other genes. And many of these regulatory genes contain common "modules" of DNA sequences called homeoboxes. But among eukaryotes outside the animal kingdom, homeoboxes are not found in homeotic genes, those regulatory genes that function in the development of body form. (You can review homeoboxes, homeotic genes, and pattern formation in Chapter 21.) So far, genes that are *both* homeobox-containing in structure and homeotic in function—*Hox* genes—have been discovered only in animals. And all animals, from the simplest

sponges to the most complex insects and vertebrates, have *Hox* genes containing homeoboxes of DNA sequences that are clearly related (see FIGURE 21.15). In general, the number of *Hox* genes is correlated with the complexity of the animal's anatomy. More specifically, variation in when and where the various *Hox* genes are expressed in a developing embryo provides the genetic basis for the great diversity of animal forms that have evolved from a common ancestor.

The animal kingdom probably evolved from a colonial, flagellated protist

Most systematists now agree that the animal kingdom is monophyletic; that is, if we could trace all animal lineages back to their origin, they would converge on a common ancestor. That ancestor was most likely a colonial flagellated protist that lived over 700 million years ago in the Precambrian era. This protist was probably related to choanoflagellates, a group that arose about a billion years ago. Modern choanoflagellates are tiny, stalked organisms inhabiting shallow ponds, lakes, and marine environments (FIGURE 32.2). FIGURE 32.3 diagrams one hypothesis for how such an ancestor may have evolved into simple animals with specialized cells arranged in two or more layers.

TWO VIEWS OF ANIMAL DIVERSITY

Zoologists recognize about 35 phyla of animals, though our survey of animal diversity will include only 15 of these major animal groups. For the past century, there

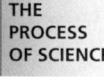

THE PROCESS OF SCIENCE

was broad consensus among systematists for at least the major branching in a hypothetical tree of the animal kingdom. This traditional tree was based mainly on the anatomical features of the animals and on certain details of their embryonic development. However, the molecular systematics of the past decade is challenging some of these long-held ideas about the phylogenetic relationships among the animal phyla.

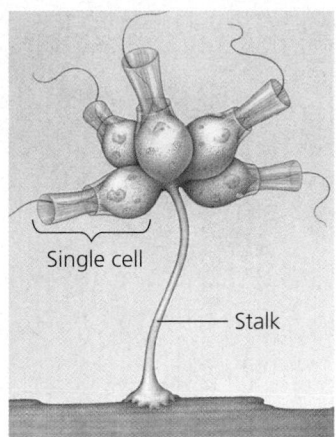

FIGURE 32.2
A choanoflagellate colony
(about 0.02 mm high).

Throughout this unit of chapters on the diversity of life, we have seen many cases where molecular data are enabling systematists to refine hypotheses about the history of life. Just a few examples are the splitting of prokaryotes into the archaea and the bacteria (Chapter 27), the reassessment of phylogenetic relationships among the protists (Chapter 28), reinforcement of the hypothesis that plants descended from the green algae called charophyceans (Chapter 29), progress in identifying the major clades (evolutionary branches) of flowering plants (Chapter 30), and the revelation that Fungi is the kingdom most closely related to the animal kingdom (Chapter 31). Thus, it probably doesn't surprise you that molecular systematics is beginning to reshape our view of animal diversity.

The remodeling of phylogenetic trees illustrates the process of scientific inquiry

On one hand, many biology students must find it frustrating that the phylogenetic trees pressed onto textbook pages cannot be memorized as fossilized truths. On the other hand, the current revolution in systematics is a healthy reminder that science is a process of inquiry that makes even the so-called "classical" fields, such as systematics and taxonomy, dynamic. As emerging technologies such as molecular biology and fresh approaches such as cladistics produce new data or stimulate scientists to reconsider old data, hypotheses sometimes bend or even break under the pressure of the closer scrutiny. New hypotheses or refinements of the old ones represent the latest versions of what we understand about nature based on the best available evidence. And *evidence* is the key word in this disclaimer that even our most cherished ideas in science are probationary. Science is partly distinguished from other ways of knowing because its ideas *can* be falsified through testing with experiments or observation. The more testing a hypothesis withstands, the more credible it becomes.

In comparing the traditional phylogenetic tree of animals based on anatomy and embryology with the remodeled tree based on molecular biology, we'll see agreement on some issues and disagreement on others. And though the new data from molecular systematics are compelling, the traditional view of animal phylogeny still offers some important advantages for helping us understand the diversity of animal body plans. That is the rationale for looking at both the "old" tree and the "new" tree in some detail as an overview of animal diversity.

The traditional phylogenetic tree of animals is based mainly on grades in body "plans"

FIGURE 32.4 (p. 636) illustrates a phylogenetic tree in which relationships among the animal phyla are based mainly on key characteristics of body plan and embryonic development. Each major branch represents what systematists call a **grade,** which is defined by certain body-plan features shared by the animals belonging to that branch. The circled numbers on the tree mark four deep branch points distinguishing the major grades. For example, the first branch point splits the grade of animals

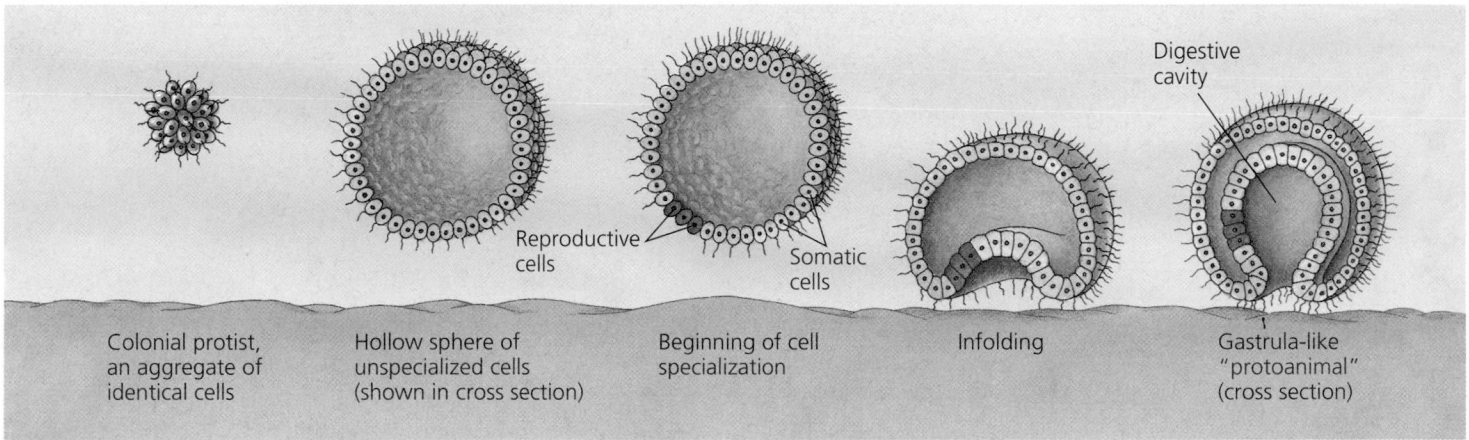

Colonial protist, an aggregate of identical cells

Hollow sphere of unspecialized cells (shown in cross section)

Beginning of cell specialization

Infolding

Gastrula-like "protoanimal" (cross section)

Reproductive cells

Somatic cells

Digestive cavity

FIGURE 32.3 One hypothesis for the origin of animals from a flagellated protist.

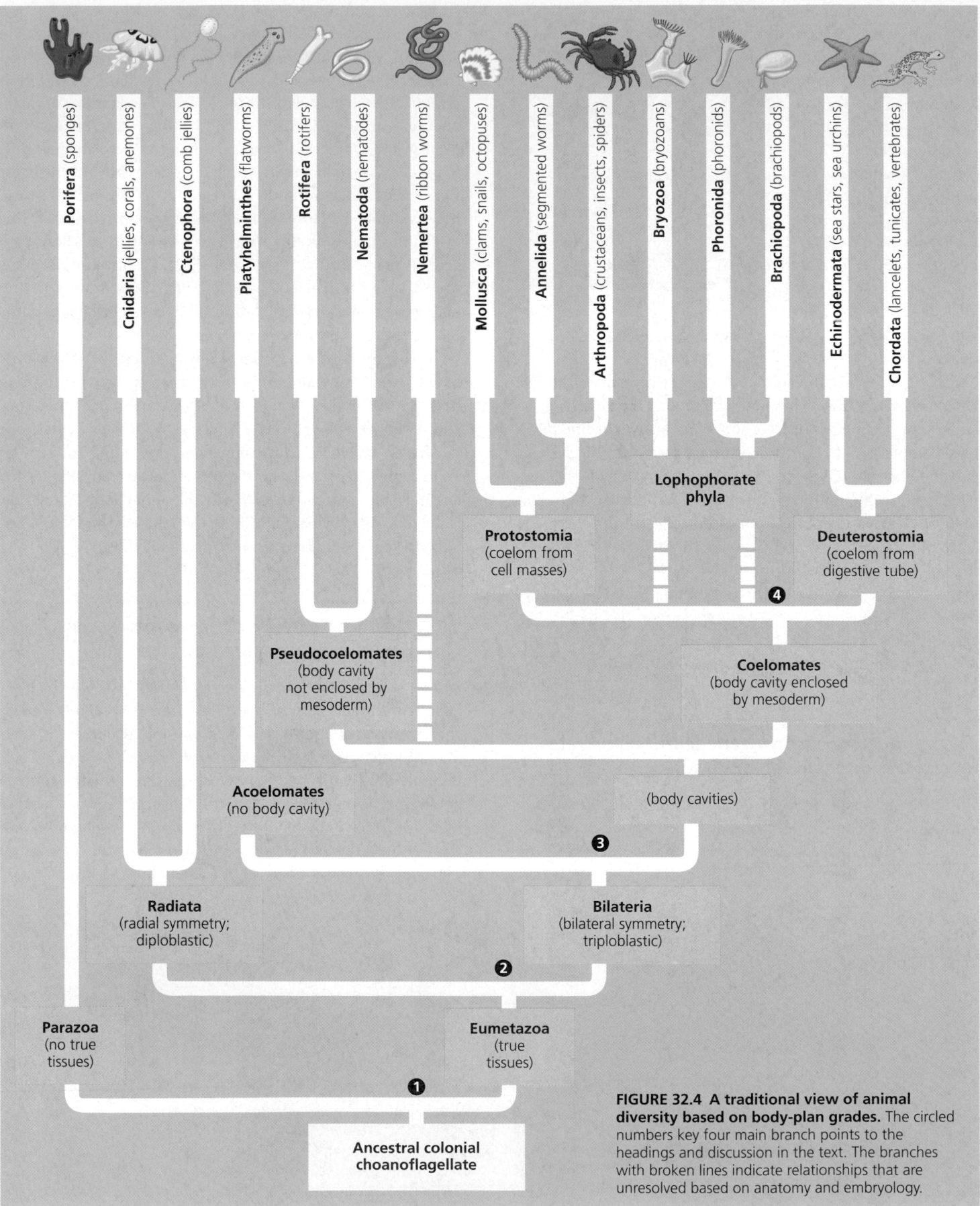

FIGURE 32.4 A traditional view of animal diversity based on body-plan grades. The circled numbers key four main branch points to the headings and discussion in the text. The branches with broken lines indicate relationships that are unresolved based on anatomy and embryology.

Labels in figure:

Porifera (sponges)
Cnidaria (jellies, corals, anemones)
Ctenophora (comb jellies)
Platyhelminthes (flatworms)
Rotifera (rotifers)
Nematoda (nematodes)
Nemertea (ribbon worms)
Mollusca (clams, snails, octopuses)
Annelida (segmented worms)
Arthropoda (crustaceans, insects, spiders)
Bryozoa (bryozoans)
Phoronida (phoronids)
Brachiopoda (brachiopods)
Echinodermata (sea stars, sea urchins)
Chordata (lancelets, tunicates, vertebrates)

Lophophorate phyla

Protostomia (coelom from cell masses)

Deuterostomia (coelom from digestive tube)

Pseudocoelomates (body cavity not enclosed by mesoderm)

Coelomates (body cavity enclosed by mesoderm)

Acoelomates (no body cavity)

(body cavities)

Radiata (radial symmetry; diploblastic)

Bilateria (bilateral symmetry; triploblastic)

Parazoa (no true tissues)

Eumetazoa (true tissues)

Ancestral colonial choanoflagellate

with no true tissues (the parazoa) from the grade of animals with no true tissues (the parazoa) from the grade of animals with true tissues (the eumetazoa). Let's take a closer look at how the diverse body plans of animals can be organized into grades by focusing on each of these four main dichotomies (branchings).

❶ The Parazoa-Eumetazoa Dichotomy

Among the extant phyla, sponges (phylum Porifera) represent an early branch of the animal kingdom. Sponges have unique development and a structural simplicity that separates them from all other animal phyla. They lack true tissues and are called the **parazoans** (meaning "beside the animals"). Tissues are a basic feature of nearly all other animal phyla, collectively called the **eumetazoans.**

❷ The Radiata-Bilateria Dichotomy

The eumetazoans are divided into two major branches, partly on the basis of body symmetry. Members of phylum Cnidaria (hydras; jellies, also called "jellyfishes"; sea anemones; and their relatives) and phylum Ctenophora (comb jellies) have **radial symmetry** (FIGURE 32.5a) and are collectively called **radiata.** A radial animal has a top and bottom, or an oral (mouth) and an aboral side, but no head end and rear end and no left and right. The other major branch of eumetazoan evolution led to animals with **bilateral** (two sided) **symmetry** (FIGURE 32.5b). A bilateral animal has not only a **dorsal** (top) side and a **ventral** (bottom) side, but also an **anterior** (head) end and a **posterior** (tail) end and a left and right side. Animals of this body-plan grade are collectively called **bilateria.**

Associated with bilateral symmetry is **cephalization,** an evolutionary trend toward the concentration of sensory equipment on the anterior end, the end of a traveling animal that is usually first to encounter food, danger, and other stimuli. In most bilateral animals, cephalization also includes the development of a central nervous system concentrated in the head and extending toward the tail as a longitudinal nerve cord. A head end is an adaptation for movement, such as crawling, burrowing, or swimming. The symmetry of an animal generally fits its lifestyle. Many radial animals are sessile forms (attached to a substratum) or plankton (drifting or weakly swimming aquatic forms). Their symmetry equips them to meet the environment equally well from all sides. Most animals that move actively from place to place are bilateral. These two fundamentally different kinds of symmetry probably arose very early in the history of animal life.

Another difference in body plan helps define the radiata-bilateria split: In all animals except sponges, the embryo becomes layered through the process of gastrulation (see FIGURE 32.1). As development progresses, these concentric layers, called **germ layers,** form the various tissues and organs of the body. **Ectoderm,** covering the surface of the embryo, gives rise to the outer covering of the animal and, in some phyla, to the central nervous system. **Endoderm,** the innermost germ layer, lines the developing digestive tube, or **archenteron,** and gives rise to the lining of the digestive tract and organs derived from it, such as the liver and lungs of vertebrates. All eumetazoans except cnidarians and ctenophores (the radiata) have a third germ layer, the **mesoderm,** between the ectoderm and endoderm. Mesoderm forms the muscles and most other organs between the digestive tube and the outer covering of the animal. Cnidarians and ctenophores have only two germ layers (ectoderm and endoderm) or have a third layer that is not homologous with the mesoderm of bilateral animals. As a group, the radiata are said to be **diploblastic** (having two germ layers). All other eumetazoans, the bilateria, are **triploblastic** (having three germ layers).

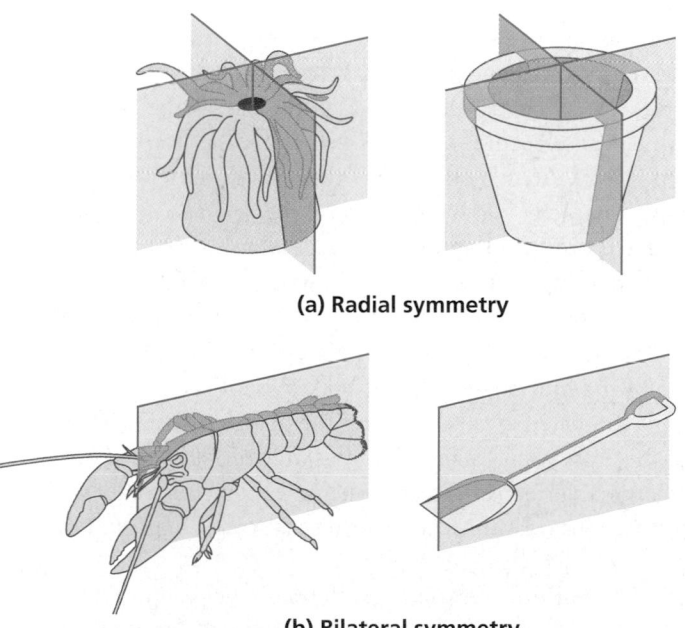

(a) Radial symmetry

(b) Bilateral symmetry

FIGURE 32.5 Body symmetry. The pail and shovel are included to help you remember the radial-bilateral distinction. **(a)** The parts of a radial animal, such as this sea anemone (phylum Cnidaria), radiate from the center. Any imaginary slice through the central axis would divide the animal into mirror images. **(b)** A bilateral animal, such as a lobster (phylum Arthropoda), has a left and right side. Only one imaginary cut would divide the animal into mirror-image halves.

❸ The Acoelomate, Pseudocoelomate, and Coelomate Grades

Triploblastic animals with solid bodies—that is, without a cavity between the digestive tract and outer body wall—are collectively called **acoelomates** (from the Greek *a,* without, and *koilos,* a hollow). This group includes phylum Platyhelminthes, the flatworms (FIGURE 32.6a, p. 638). In contrast to the acoelomates, most phyla of bilateral, triploblastic animals have tube-within-a-tube body plans, with a **body cavity,** a fluid-filled space separating the digestive tract from the outer body wall.

Among animals with a body cavity, there are differences in how the cavity develops. If the cavity is not completely lined by tissue derived from mesoderm, it is termed a **pseudocoelom.** Animals with this body plan, such as rotifers (phylum Rotifera)

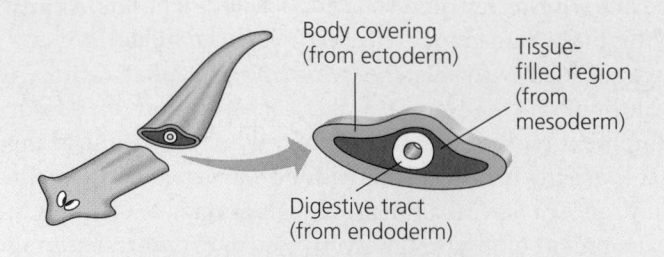

(a) Acoelomate (e.g., a flatworm). Acoelomates lack a body cavity between the digestive tract and outer body wall.

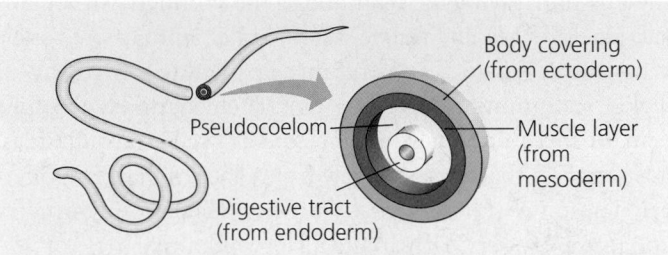

(b) Pseudocoelomate (e.g., a nematode). Pseudocoelomates have a body cavity only partially lined by mesodermally derived tissue.

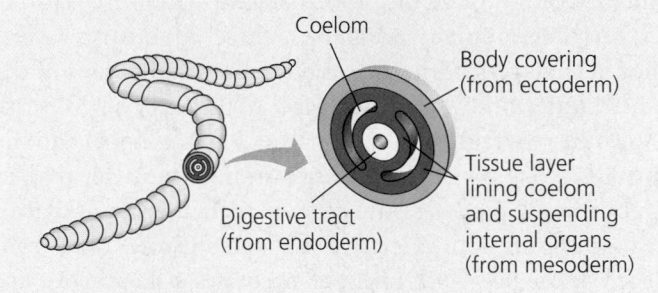

(c) Coelomate (e.g., an annelid). Coelomates have a true coelom, a body cavity completely lined by mesodermally derived tissue.

FIGURE 32.6 Body plans of the bilateria. The various organ systems of the animal develop from the three germ layers that form in the embryo. Colors traditionally used to represent the germ layers and their derivatives are blue for ectoderm, red for mesoderm, and yellow for endoderm.

and roundworms (phylum Nematoda), are called **pseudo-coelomates** (FIGURE 32.6b). **Coelomates** are animals with a true **coelom,** a fluid-filled body cavity completely lined by tissue derived from mesoderm. The inner and outer layers of tissue that surround the cavity connect dorsally and ventrally to form mesenteries, which suspend the internal organs (see FIGURE 32.6c).

A body cavity has many functions. Its fluid cushions the suspended organs, helping to prevent internal injury. In soft-bodied coelomates such as earthworms, the noncompressible fluid of the body cavity is under pressure and functions as a hydrostatic skeleton against which muscles can work. The cavity also enables the internal organs to grow and move independently of the outer body wall. If it were not for your coelom, every beat of your heart or ripple of your intestine could deform your body surface, and exercise would distort the shapes of the internal organs.

❹ *The Protostome-Deuterostome Dichotomy Among Coelomates*

The coelomate phyla are divided into two distinct grades: **Protostomia** and **Deuterostomia.** Mollusks, annelids, arthropods, and several other phyla represent one of these grades and are collectively called **protostomes.** Echinoderms, chordates, and some other phyla, collectively called **deuterostomes,** represent the other grade. (Some zoologists include the acoelomates and pseudocoelomates with the protostomes in the traditional phylogenetic tree, but this text uses the term protostome *only* as a subgroup of coelomate animals.) Protostomes and deuterostomes are distinguished by several fundamental differences in their development.

Cleavage. The pattern of cleavage divisions during early development generally differs between the two coelomate branches, though there are many exceptions to this distinction. Many protostomes undergo **spiral cleavage,** in which planes of cell division are diagonal to the vertical axis of the embryo. As seen in the eight-cell stage resulting from spiral cleavage, small cells lie in the grooves between larger, underlying cells (FIGURE 32.7a). Furthermore, the so-called **determinate cleavage** of some protostomes rigidly casts the developmental fate of each embryonic cell very early. A cell isolated at the four-cell stage from a snail, for example, forms an inviable embryo that lacks parts.

In contrast to the spiral cleavage pattern, the zygotes of many deuterostomes undergo **radial cleavage.** Here, the cleavage planes are either parallel or perpendicular to the vertical axis of the egg; as seen in the eight-cell stage, the tiers of cells are aligned, one directly above the other. Most deuterostomes are further characterized by **indeterminate cleavage,** meaning that each cell produced by early cleavage divisions retains the capacity to develop into a complete embryo. For example, if the cells of a sea star embryo are separated at the four-cell stage, each will go on to form a normal larva. It is the indeterminate cleavage of the human zygote that makes identical twins possible. This characteristic also explains the developmental versatility of the embryonic "stem cells" that may provide new ways to overcome a variety of diseases, including juvenile diabetes, Parkinson's disease, and Alzheimer's disease (see Chapter 21).

Coelom Formation. Another difference between protostomes and deuterostomes is apparent later in development. In gastrulation, the developing digestive tube of an embryo initially forms as a blind pouch, the archenteron. As the archenteron forms in a protostome, initially solid masses of mesoderm split to form the coelomic cavities; this is called **schizocoelous** development (from the Greek *schizo,* split). Development of the body cavities of deuterostomes is termed **enterocoelous:** The

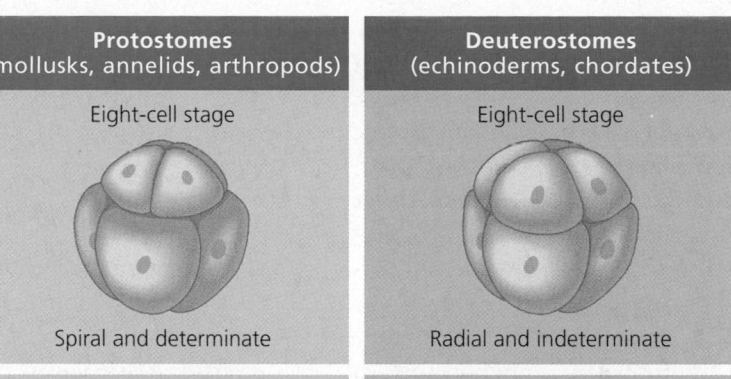

Protostomes
(mollusks, annelids, arthropods)

Deuterostomes
(echinoderms, chordates)

(a) Cleavage. Most protostomes have spiral, determinate cleavage; most deuterostomes have radial, indeterminate cleavage.

Eight-cell stage

Eight-cell stage

Spiral and determinate

Radial and indeterminate

FIGURE 32.7 A comparison of early development in protostomes and deuterostomes. These are useful general distinctions, though there are many variations and exceptions to these patterns of development.

(b) Coelom formation. Coelom formation begins in the gastrula stage. In the schizocoelous development of protostomes, the coelom forms from splits in the mesoderm. In the enterocoelous development of deuterostomes, the coelom forms from mesodermal outpocketings of the archenteron (blue = ectoderm, red = mesoderm, yellow = endoderm).

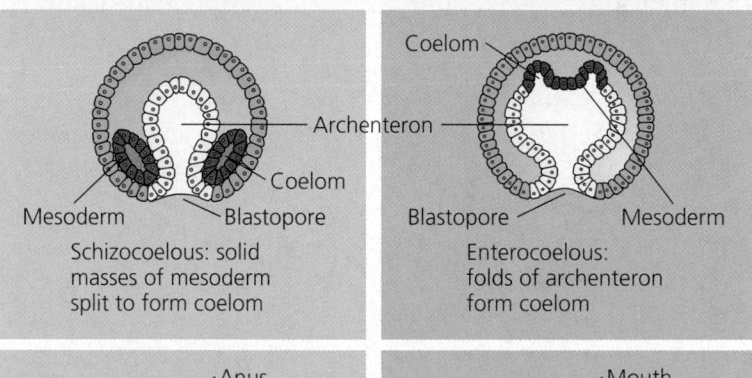

Coelom

Archenteron

Mesoderm — Coelom — Blastopore

Coelom

Blastopore — Mesoderm

Schizocoelous: solid masses of mesoderm split to form coelom

Enterocoelous: folds of archenteron form coelom

(c) Fate of blastopore. The blastopore forms the mouth in protostomes; the mouth forms from a secondary opening in deuterostomes.

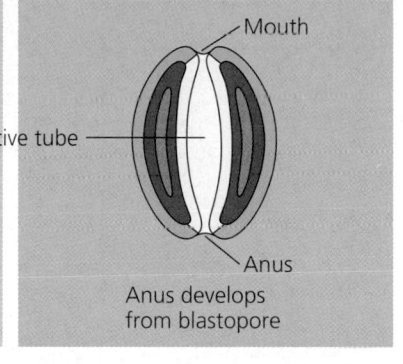

Anus

Mouth

Digestive tube

Mouth

Anus

Mouth develops from blastopore

Anus develops from blastopore

mesoderm buds from the wall of the archenteron and hollows to become the coelomic cavities (FIGURE 32.7b).

Blastopore Fate. A third fundamental difference between protostomes and deuterostomes is in the fate of the **blastopore,** the opening of the archenteron. After the archenteron develops, a second opening forms at the opposite end of the gastrula. Ultimately, the blastopore and this second opening become the two ends of the digestive tube (the mouth and the anus). The mouth of many protostomes develops from the first opening, the blastopore, and it is for this characteristic that the protostome grade is named (from the Greek *protos,* first, and *stoma,* mouth). In contrast, the mouth of a deuterostome (from the Greek *deuteros,* second) is derived from the secondary opening, and the blastopore usually forms the anus, not the mouth (FIGURE 32.7c).

Let us now compare and contrast the traditional phylogenetic tree of animals based on grades in body plan (see FIGURE 32.4) with an emerging view of animal phylogeny based mainly on molecular systematics.

Molecular systematists are moving some branches around on the phylogenetic tree of animals

Modern phylogenetic systematics is based on the identification of clades, which are monophyletic sets of taxa as defined by shared-derived characters unique to those taxa and their common ancestor (see Chapter 25). Based on cladistic methods, a phylogenetic tree takes shape as a hierarchy of clades nested within larger clades—the finer branches and major branches of the tree, respectively. The traditional phylogenetic tree of animals in FIGURE 32.4 is based on the assumption that grades in body plan are good indicators of clades, as long as the key anatomical and embryological homologies that define a grade are unique to the phyla placed on that evolutionary branch.

Molecular systematics has added a new set of shared-derived characters in the form of unique monomer sequences within certain genes and their products. And these molecular data can be used to identify the clusters of monophyletic taxa that make up clades. There wouldn't be much of a story here if molecular systematics simply reinforced the traditional animal tree based on comparative anatomy and embryology. But that is not the case.

The phylogenetic tree in FIGURE 32.8 is based on nucleotide sequences in the small subunit ribosomal RNA (SSU-rRNA), the gene product that is so commonly analyzed by molecular systematists (see, for example, Chapters 27 and 28). Researchers have also sequenced some of the *Hox* genes in various animals, and so far those sequences support the phylogenetic tree based on SSU-rRNA analysis. Let's examine how this tree agrees with the traditional one, and how the two trees differ.

How Are the Two Views of Animal Phylogeny Alike?

Agreement on the Deepest Branches. First, there is no dispute about the very deepest branches of animal phylogeny. Molecular systematics supports the traditional hypotheses of the Parazoa-Eumetazoa and Radiata-Bilateria dichotomies (see branch points ❶ and ❷ in FIGURE 32.4).

Agreement on the Deuterostome Clade. Second, the molecular evidence reinforces the hypothesis that the deuterostomes, which include echinoderms such as sea stars and chordates such as vertebrates, make up a clade—a monophyletic sub-branch of the coelomate bilateria (see branch point ❹ in FIGURE 32.4).

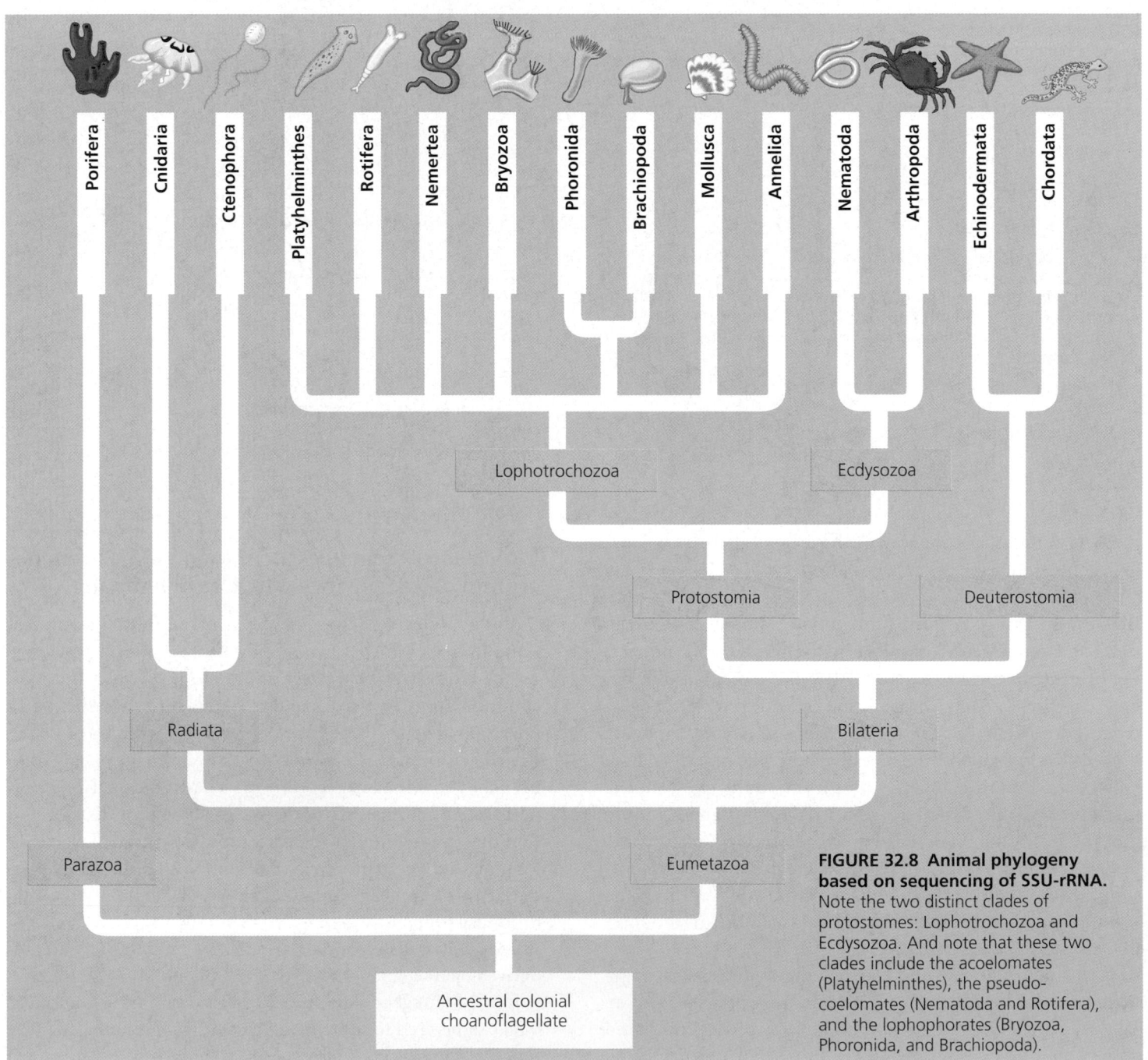

FIGURE 32.8 Animal phylogeny based on sequencing of SSU-rRNA. Note the two distinct clades of protostomes: Lophotrochozoa and Ecdysozoa. And note that these two clades include the acoelomates (Platyhelminthes), the pseudocoelomates (Nematoda and Rotifera), and the lophophorates (Bryozoa, Phoronida, and Brachiopoda).

How Are the Two Views of Animal Phylogeny Different?

You can see where the traditional and molecular-based phylogenetic trees clash by focusing on the protostome branch of the bilateria as represented in FIGURE 32.8.

Two Main Protostome Clades.   First, note that the molecular evidence resolves two distinct clades within the protostomes: **Lophotrochozoa,** which includes the annelids (segmented worms) and the mollusks (such as clams and snails); and **Ecdysozoa,** which includes the arthropods. (We'll derive these very intimidating terms shortly.)

Based on comparative anatomy and embryology alone, the relationships among the annelids, mollusks, and arthropods were uncertain. Some zoologists favored an annelid-arthropod lineage, partly because both annelids and arthropods have segmented bodies (think of an earthworm, which is an annelid, and the underside of the tail of a lobster, which is an arthropod). But other zoologists argued that certain features linked the annelids closer to the mollusks than to the arthropods. This hypothesis was based in part on the observation that many annelids and mollusks go through a similar larval stage called the **trochophore larva** (FIGURE 32.9). The molecular data add weight to the hypothesis of an annelid-mollusk clade.

Relocation of the Acoelomates and Pseudocoelomates.   In the grade-based phylogeny of FIGURE 32.4, the acoelomate phylum Platyhelminthes (flatworms) branches from the tree before the origin of body cavities. But the molecular data place the flatworms among the protostomes, specifically within the lophotrochozoan clade. If this turns out to be correct, the implication is that flatworms are not the primitive "pre-coelomate" animals of the traditional phylogeny, but are protostomes in which the body plan became simplified by loss of the coelom later in evolution. Similarly, the molecular-based phylogeny also places the pseudocoelomate phyla Rotifera (rotifers) and Nematoda (nematodes, or round worms) within the proto-

FIGURE 32.10 Ecdysis. This molting cricket is in the process of escaping from its old exoskeleton. The animal will secrete a new, larger exoskeleton.

stome clade. Rotifers cluster with the lophotrochozoan phyla, while nematodes fit among the ecdysozoans.

The name Ecdysozoa refers to a characteristic shared by nematodes, arthropods, and some of the other ecdysozoan phyla (which are not included in our survey). These animals secrete external skeletons (exoskeletons)—the armor of a lobster is an example. As the animal grows, it molts, squirming out of its old exoskeleton and secreting a new, larger one. The shedding of the old exoskeleton is called ecdysis, the process for which the ecdysozoans are named (FIGURE 32.10). Though named for this characteristic, the clade is actually defined mainly by the molecular evidence for common ancestry.

Assignment of the Lophophorate Phyla.   Take one more look at the traditional tree in FIGURE 32.4. Among the coelomates are three phyla called the lophophorate phyla. The animals of these three phyla all have a structure called a **lophophore,** a horseshoe-shaped crown of ciliated tentacles used for feeding (FIGURE 32.11). The lophophorate phyla share certain characteristics with protostomes and other features with deuterostomes—hence the dashed line indicating uncertain phylogeny in the traditional tree. If the molecular data stand up, they settle debate about the affinities of the lophophorate phyla by placing

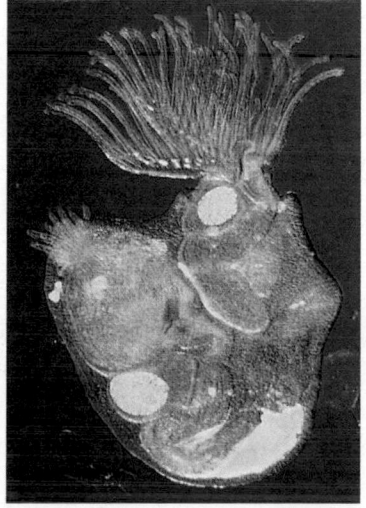

FIGURE 32.11 A lophophorate. This bryozoan uses its lophophore, the crown of ciliated tentacles, for feeding.

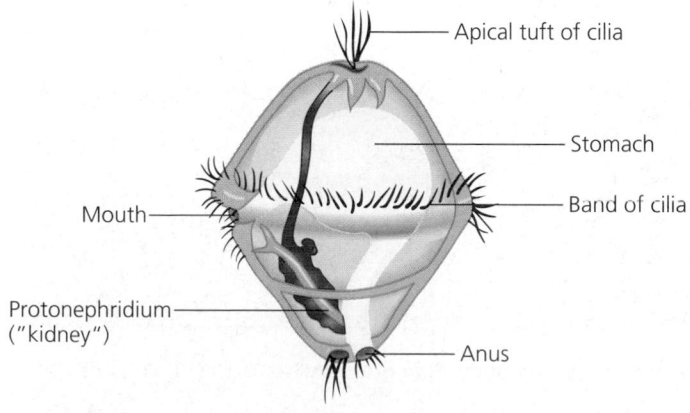

Apical tuft of cilia

Stomach

Band of cilia

Mouth

Protonephridium ("kidney")

Anus

FIGURE 32.9 A trochophore larva.

them as protostomes among the lophotrochozoans. Thus, we can now dissect this long word as the name for a group that unites the lophophore-bearing phyla with phyla having a trochophore larva. Note, however, that while the name of this clade may be based on anatomical and developmental characteristics, the animals are grouped mainly because they share certain DNA sequences.

Summary of the Two Views of Animal Diversity

We can summarize the differences between the traditional and molecular-based animal phylogenies this way: the molecular data recognize two distinct clades within the protostomes—Lophotrochozoa and Ecdysozoa—and distribute the acoelomates, pseudocoelomates, and lophophorate phyla among those two protostome clades. FIGURE 32.12 will help you review the two views of animal phylogeny.

Though we will base our survey of animal phyla in the next two chapters on the newer, molecular-based phylogeny, we do so with two caveats. First, we will not discard the concept of body-plan grades because it is still a very useful way to think about the diversity of animal forms that have evolved. Second, the phylogenetic tree of animals built from molecular data, like all such trees, represents a set of hypotheses about the history of life, and is thus tentative. The molecular phylogeny is based on just a very few genes—and mainly on one, the gene for SSU-rRNA. Many zoologists will probably stick with the grade-based phylogeny unless much more molecular evidence builds a convincing case for a new tree. In the meantime, the hypothetical tree based on molecules will inform continuing research, including additional studies of anatomy, embryology, cell structure, and other non-molecular analyses. Ideally, such research will eventually square the molecular data with the data from other approaches. Continued exploration of the fossil record, of course, will also help to reconstruct the history of the animal kingdom. But as we'll see in this chapter's last major section, there is a reason that it has not been possible so far for paleontologists to sort out the evolutionary relationships among the animal phyla.

THE ORIGINS OF ANIMAL DIVERSITY

Most animal phyla originated in a relatively brief span of geologic time

The fossil record and molecular studies concur that the diversification that produced the many animal phyla occurred rapidly on the vast scale of geologic time. This relatively brief evolutionary episode probably lasted about 40 million years (about 565 to 525 million years ago) during the late Precambrian and early Cambrian (which began about 543 million years ago).

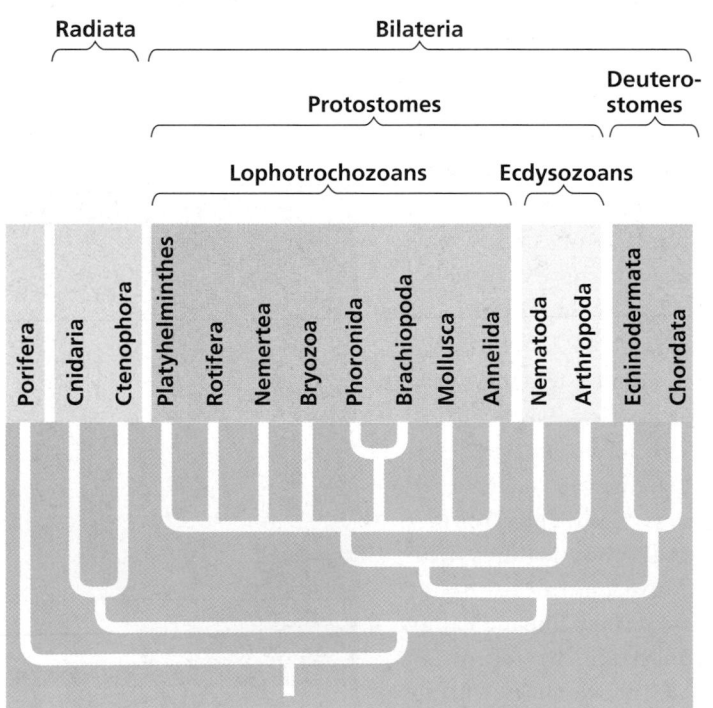

(a) Tree based on molecular comparisons. (See FIGURE 32.8 for details.)

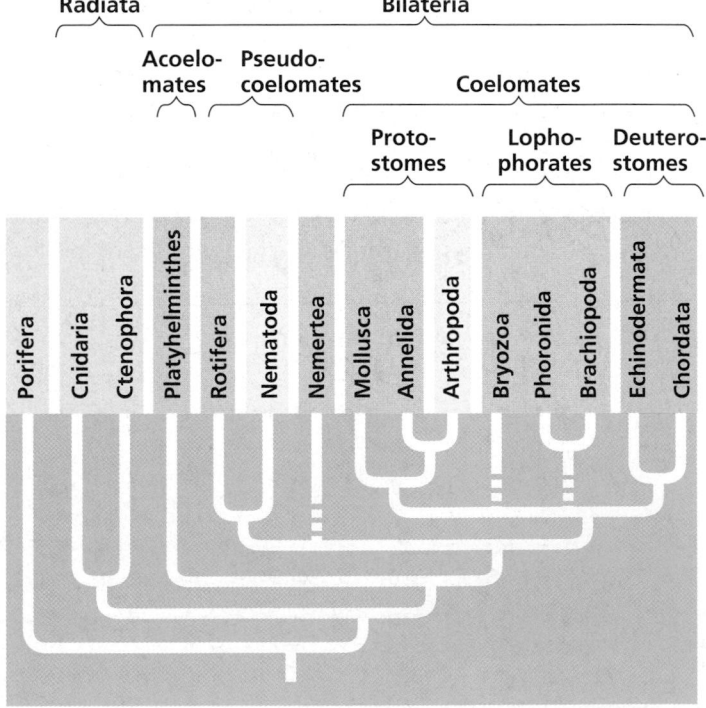

(b) Tree based on body-plan grades. (See FIGURE 32.4 for details.)

FIGURE 32.12 Comparing the molecular-based and grade-based trees of animal phylogeny. The color-coded blocks will help you compare the locations of the phyla in the two trees.

Paleontologists have named the last period of the Precambrian era the **Ediacaran period,** for the Ediacara Hills of Australia, where fossils of Precambrian animals were first discovered. The fossils range in age from about 565 to 543 million years old. Similar animals of the same vintage have since been found on other continents. There is some evidence that animal life began earlier, and maybe much earlier. In 2000, researchers reported the discovery of fossilized animal embryos in Chinese strata that are 570 million years old (see FIGURE 26.7). And in 1998, a team of paleontologists discovered what could be fossilized burrows of animals in rocks that are 1.1 billion years old. By themselves, these putative trace fossils of animals would not convince many biologists to push back the origin of animals so far. However, the data of molecular systematics also suggest an animal origin about a billion years ago. That would mean that the genesis of animals was part of an early diversification of multicellular eukaryotes, known mainly from fossils of algae (see FIGURE 26.6). Until there is more solid evidence for such ancient animals, however, all we know for sure is that a diversity of animals had evolved by the time of the Ediacaran period. Most of the Ediacaran fossils appear to represent cnidarians (animals similar to hydras), but soft-bodied mollusks (similar to a modern group called the chitons) were also present, and numerous fossilized burrows and tracks indicate the activities of several forms of worms.

In contrast to the relatively limited variety of Ediacaran animals, nearly all the major animal body plans appear in Cambrian rocks dating from 543 to 525 million years ago. During this relatively short span, a burst of animal origins called the **Cambrian explosion** left a rich fossil assemblage that includes the first animals with hard, mineralized skeletons (see FIGURE 26.8). The Burgess Shale in British Columbia, Canada, is the most famous fossil bed documenting the diversity of Cambrian animals. Two other fossil sites, one in Greenland and the other in the Yunnan region of China, predate the Burgess Shale by more than 10 million years. Burgess Shale fossils are rather bizarre-looking in the context of the marine animals we know today (FIGURE 32.13). Some of these Cambrian forms may represent extinct "experiments" in animal diversity. However, most of the Cambrian fossils, as strange as they may appear to us, are simply ancient variations within the taxonomic boundaries of phyla still represented in the modern fauna. Indeed, the number of exclusively Cambrian phyla seems to be dropping as the fossils are studied more closely and are classified in extant phyla.

On the scale of geologic time, animals diversified so rapidly that it is difficult from the fossil record to sort out the sequence of branching in animal phylogeny. And that is why, when reconstructing the evolutionary history of animal phyla, systematists depend largely on clues from the comparative anatomy, embryology, developmental genetics, and molecular systematics of extant species.

FIGURE 32.13 A sample of some of the animals that evolved during the Cambrian explosion. This drawing is based on fossils collected from the Burgess Shale in British Columbia, Canada.

"Evo-devo" may clarify our understanding of the Cambrian diversification

What sparked the Cambrian explosion? There are three main hypotheses for what caused the diversification of animals:

1. *Ecological Causes:* The main variation on this hypothesis emphasizes the emergence during the Cambrian of predator-prey relationships. Such a change in the dynamics of biological communities would lead to a diversity of evolutionary adaptations such as various kinds of protective shells and diverse modes of locomotion.
2. *Geologic Causes:* Perhaps, for example, atmospheric oxygen finally reached a high enough concentration during the Cambrian to support the more active metabolism required for the feeding and other activities of mobile animals.
3. *Genetic Causes:* Much of the diversity in body form we observe among the 35 or so animal phyla is associated with variations in the spatial and temporal expression of *Hox* genes within developing embryos (see FIGURE 21.15). Thus, it is a reasonable hypothesis that the diversification of animals was associated with the evolution of the *Hox* complex of regulatory genes, which led to variation in morphology during embryonic development. In fact, the Cambrian explosion is a major interest of many of the biologists working in the field of "evo-devo," the new synthesis of evolutionary biology and developmental biology.

These three major hypothetical causes of the Cambrian explosion are not mutually exclusive. The relatively rapid radiation of animal phyla over a half-billion years ago may have been a product of multiple causes, both external (geologic change and ecological change) and internal (genetic change).

Some of the systematists studying animal phylogeny interpret the molecular data to mean that the Cambrian radiation was actually *three* explosions, not just one. Within each of the three main branches of bilateral animals—Lophotrochozoa, Ecdysozoa, and Deuterostomia—the relationships among the phyla are difficult to resolve based on differences in SSU-rRNA. For example, one cannot tell from the molecular data whether, among the lophotrochozoans, flatworms are more closely related to mollusks or to annelids. This tight clustering of the nucleic acid sequences is consistent with a rapid diversification of phyla *within* each of the three major clades of bilateria. However, the molecular differences *between* those three great clades are relatively extensive and suggest that they branched apart very early, probably in the Precambrian. This deep branching of the bilateria may have been associated with the evolution of the *Hox* complex, but then geologic and/or ecological changes during the early Cambrian may have driven the diversification of phyla within each major clade (FIGURE 32.14).

Apparently, by the end of the Cambrian radiation, the animal phyla were locked into developmental patterns that constrained evolution enough that no additional phyla evolved after that period. Of course, this does not imply that animal evolution came to a halt; variations in developmental patterns continue to allow subtle changes in body structures and func-

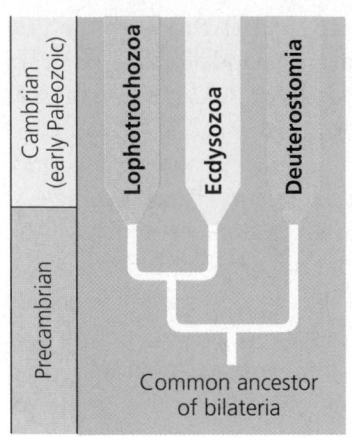

FIGURE 32.14 One Cambrian explosion, or three? The Cambrian explosion may actually have been a secondary diversification of each of three major body plans that evolved during the Precambrian.

tions, leading to speciation and the origin of taxa below the phylum level.

Continuing research will help test these hypotheses. But even as the Cambrian explosion becomes less mysterious, it will seem no less wonderful. In the last half-billion years, animal evolution has mainly generated new variations on old "designs." In Chapters 33 and 34 we take a closer look at the extant animal phyla and their evolutionary history.

CHAPTER 32 REVIEW

Go to the Campbell Biology website (www.campbellbiology.com) to explore an interactive version of the Chapter Review.

Summary of Key Concepts

WHAT IS AN ANIMAL?

- **Structure, nutrition, and life history define animals (pp. 633–634, FIGURE 32.1)** Animals are multicellular, heterotrophic eukaryotes that ingest their food. Animal cells lack walls. Nervous tissue and muscle tissue are unique to animals. Embryonic development usually includes a blastula stage. Gastrulation follows the blastula, resulting in formation of embryonic tissue layers. All animals, and only animals, have *Hox* genes that regulate the development of body form.

- **The animal kingdom probably evolved from a colonial, flagellated protist (pp. 634–635, FIGURES 32.2, 32.3)** The probable ancestor of animals was a colonial choanoflagellate.

TWO VIEWS OF ANIMAL DIVERSITY

- **The remodeling of phylogenetic trees illustrates the process of scientific inquiry (p. 635)** Phylogenetic trees are sets of hypotheses that are refined to accommodate new data.

- **The traditional phylogenetic tree of animals is based mainly on grades in body "plans" (pp. 635–639, FIGURES 32.4–32.7)** Based on body-plan grades, there are four main branchings in the animal

kingdom: (1) the Parazoa-Eumetazoa dichotomy; (2) the Radiata-Bilateria dichotomy; (3) the presence vs. absence of body cavities; and (4) the Protostome-Deuterostome dichotomy among animals with true coeloms.

Web/CD Activity 32A: *Traditional Animal Phylogenetic Tree*

- **Molecular systematists are moving some branches around on the phylogenetic tree of animals (pp. 639–642, FIGURES 32.8–32.12)** Based on molecular data, many animal systematists now split the protostomes into two main clades—Lophotrochozoa and Ecdysozoa.

Web/CD Case Study in the Process of Science: *How Do Molecular Data Fit Traditional Phylogenies?*

THE ORIGINS OF ANIMAL DIVERSITY

- **Most animal phyla originated in a relatively brief span of geologic time (pp. 642–643, FIGURE 32.13)** The causes of the Cambrian explosion may have been one or some combination of ecological changes, geologic changes, or genetic changes associated with the evolution of *Hox* genes.

- **"Evo-devo" may clarify our understanding of the Cambrian diversification (pp. 643–644, FIGURE 32.14)** Diversification of body plans based on the evolution of *Hox* genes probably accounts for the early branching of bilateria into the deuterostome and two protostome clades.

Self-Quiz

1. The distinction between the parazoans and eumetazoans is based mainly on the absence versus the presence of
   a. body cavities.
   b. a complete digestive tract.
   c. true tissues.
   d. a circulatory system.
   e. mesoderm.

2. As a group, acoelomates are characterized by
   a. the absence of a brain.
   b. the absence of mesoderm.
   c. deuterostome development.
   d. a coelom that is not completely lined with mesoderm.
   e. a solid body without a cavity surrounding internal organs.

3. What is the main basis for placing the arthropods and nematodes in the Ecdysozoa?
   a. Animals in both groups are segmented.
   b. Animals in both groups undergo ecdysis.
   c. They both have radial, determinate cleavage, and their embryonic development is similar.
   d. The fossil record has revealed a common ancestor to these two phyla.
   e. Their SSU-rRNA sequences are quite similar, and these sequences differ from those of the lophotrochozoans and deuterostomes.

4. How does the molecular-based phylogenetic tree differ from the grade-based tree?
   a. placement of the acoelomates and pseudocoelomates within the Protostomia
   b. division of the protostomes into clades Lophotrochozoa and Ecdysozoa
   c. grouping of arthropods and annelids (both segmented animals) in Ecdysozoa and assignment of mollusks to Lophotrochozoa
   d. both a and b are correct
   e. a, b, and c are correct

5. Bilateral symmetry in the animal kingdom is best correlated with
   a. an ability to sense equally in all directions.
   b. the presence of a skeleton.
   c. motility and active predation and escape.
   d. development of a true coelom.
   e. adaptation to terrestrial environments.

6. A direct consequence of indeterminate cleavage is
   a. formation of the archenteron.
   b. the ability of cells isolated from the early embryo to develop into viable individuals.
   c. the arrangement of cleavage planes perpendicular to the egg's vertical axis.
   d. the unpredictable formation of either a schizocoelous or enterocoelous body cavity.
   e. a mouth that forms in association with the blastopore.

7. Which of the following was the *least* likely factor in the Cambrian explosion?
   a. the emergence of predator–prey relationships between animals
   b. the accumulation of diverse adaptations such as shells and different modes of locomotion
   c. the movement of animals onto land
   d. the evolution of *Hox* genes that controlled development
   e. the accumulation of sufficient atmospheric oxygen to support the more active metabolism of mobile animals

8. Among the characteristics unique to animals is
   a. gastrulation.
   b. multicellularity.
   c. sexual reproduction.
   d. flagellated sperm.
   e. heterotrophic nutrition.

9. Which of the following combinations of phylum and description is incorrect?
   a. Echinodermata—branch bilateria, coelom from archenteron
   b. Nematoda—roundworms, pseudocoelomate
   c. Cnidaria—radial symmetry, diploblastic
   d. Platyhelminthes—flatworms, acoelomate
   e. Porifera—coelomate, mouth from blastopore

10. Which of the following subdivisions of the animal kingdom encompasses all the others in the list?
    a. protostomes
    b. bilateria
    c. pseudocoelomates
    d. coelomates
    e. deuterostomes

11. What is the function of *Hox* genes in animals?

12. Contrast the grade-based and the molecular-based phylogenies in their placement of the acoelomates (flatworms).

13. Why have paleontologists not been able to deduce the sequence of phylogenetic branching among animal phyla from the fossil record?

Go to the website or CD-ROM for more quiz questions.

Evolution Connection

Lynn Margulis of the University of Massachusetts has suggested that observing an explosion of animal diversity in Cambrian strata is like viewing Earth from a satellite over a long period of time and noticing the emergence of cities only after they are large enough to be evident at that distance. What do you think Dr. Margulis was saying about the Cambrian "explosion"? How does this idea relate to the main point of FIGURE 32.14?

The Process of Science

In applying cladistic analysis to construct a phylogeny of the animal kingdom, why would the presence of flagella be a poor choice of a characteristic for grouping the phyla into clades?

Examine how recent data from ribosomal DNA may change traditional invertebrate phylogenies in the Case Study in the Process of Science, available on the website and CD-ROM.

Science, Technology, and Society

The study of animal phylogeny is sometimes viewed as "science for science's sake," and some organizations that fund scientific research tend to favor projects that have more obvious applications to human needs. On the other hand, general interest in the history of life seems to remain high, with articles on new discoveries frequently appearing in popular magazines. Suppose you have the opportunity to join a research team studying various aspects of the Cambrian explosion. Write a few paragraphs that could convince nonbiologists that this research is worth funding.

Answers: 1. c; 2. e; 3. e; 4. d; 5. c; 6. b; 7. c; 8. a; 9. e; 10. b; 11. They regulate the development of an embryo's body plan. 12. In the grade-based phylogeny, the acoelomate condition is considered to be primitive, and the flatworms thus branch from the tree before the origin of the coelom. The molecular-based tree moves the acoelomates up to the protostome branch, suggesting that the acoelomate condition evolved secondarily from an ancestor with a coelom. 13. On the vast scale of geologic time, the origins of all the animal phyla are compressed into a very short span of time—the Cambrian explosion—thus the sequence of their appearance cannot be resolved.

INVERTEBRATES

PARAZOA

- Phylum Porifera: Sponges are sessile with porous bodies and choanocytes

RADIATA

- Phylum Cnidaria: Cnidarians have radial symmetry, a gastrovascular cavity, and cnidocytes
- Phylum Ctenophora: Comb jellies possess rows of ciliary plates and adhesive colloblasts

PROTOSTOMIA: LOPHOTROCHOZOA

- Phylum Platyhelminthes: Flatworms are acoelomates with gastrovascular cavities
- Phylum Rotifera: Rotifers are pseudocoelomates with jaws, crowns of cilia, and complete digestive tracts
- The lophophorate phyla: Bryozoans, phoronids, and brachiopods are coelomates with ciliated tentacles around their mouths
- Phylum Nemertea: Proboscis worms are named for their prey-capturing apparatus
- Phylum Mollusca: Mollusks have a muscular foot, a visceral mass, and a mantle
- Phylum Annelida: Annelids are segmented worms

PROTOSTOMIA: ECDYSOZOA

- Phylum Nematoda: Roundworms are nonsegmented pseudocoelomates covered by tough cuticles
- Arthropods are segmented coelomates with exoskeletons and jointed appendages

DEUTEROSTOMIA

- Phylum Echinodermata: Echinoderms have a water vascular system and secondary radial anatomy
- Phylum Chordata: The chordates include two invertebrate subphyla and all vertebrates

More than a million extant species *of animals are known, and at least as many more will probably be identified by future generations of biologists. Animals are grouped into about 35 phyla, the exact number depending on the views of different systematists. Our survey of animal diversity focuses on just 15 of these phyla.*

Animals inhabit nearly all environments on Earth, but most phyla consist mainly of aquatic species. The seas, where the first animals probably arose, are still home to the greatest number of animal phyla. The freshwater fauna is extensive, but not nearly as rich in diversity as the marine fauna.

Terrestrial habitats pose special problems for animals, as they do for plants (see Chapter 29), and few animal phyla have made successful evolutionary treks onto land. Earthworms (phylum Annelida) and land snails (phylum Mollusca) are generally confined to moist soil and vegetation. Only the vertebrates and arthropods, including insects and spiders, are represented by a great diversity of animal species adapted to various terrestrial environments.

*Living as we do on land, our sense of animal diversity is biased in favor of vertebrates, the animals with backbones, which are well represented in terrestrial environments. But vertebrates make up one subphylum within phylum Chordata, less than 5% of all animal species. If we were to sample the animals inhabiting a tidepool (as in the photograph on this page), a coral reef, or the rocks on a stream bottom, we would find ourselves in the realm of **invertebrates,** the animals without backbones. The diversity of invertebrates is the main subject of this chapter.*

Our survey of invertebrate phyla will follow the branching pattern of the phylogenetic tree in FIGURE 33.1. *This tree is based mainly on molecular data, as discussed in Chapter 32 (see* FIGURE 32.8, *p. 640). However, much of our discussion of the diverse phyla will center on key differences in body "plan," the grades in* FIGURE 32.4 *on page 636.*

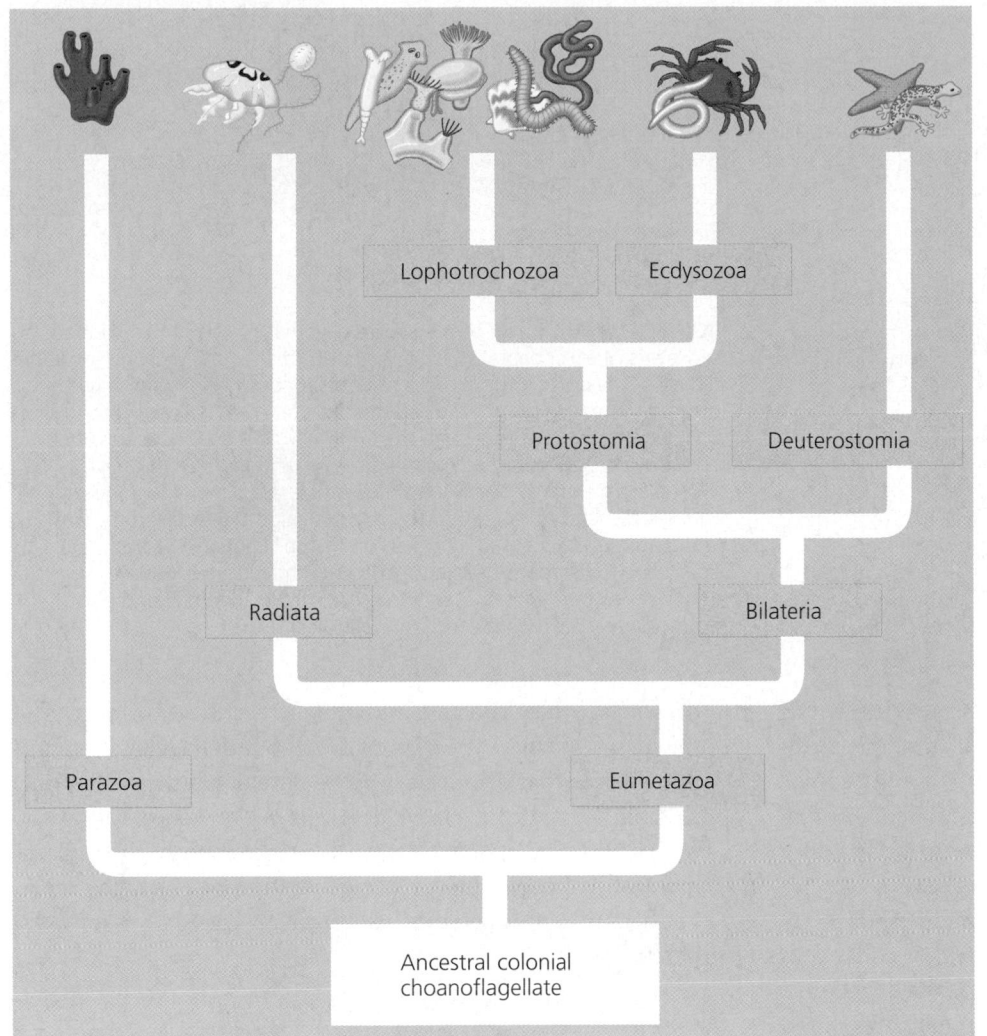

FIGURE 33.1 Review of animal phylogeny.
This tree is a streamlined version of FIGURE 32.8 (p. 640). We'll use an even simpler miniature of the tree throughout this chapter, with individual branches sequentially highlighted to help place each clade in the overall context of animal phylogeny. Your instructor may prefer to connect his or her survey of animal phyla to FIGURE 32.4 (p. 636), which is based on evolutionary grades in the body plans of animals.

have no nerves or muscles, but the individual cells can sense and react to changes in the environment.

Sponges range in height from about 1 cm to 2 m. Of the 9,000 or so species of sponges, only about 100 live in fresh water; the rest are marine. The body of a simple sponge resembles a sac perforated with holes (*Porifera* means pore bearer; FIGURE 33.2). Water is drawn through the pores into a central cavity, the **spongocoel,** then flows out of the sponge through a larger opening called the **osculum.** More complex sponges have folded body walls, and many contain branched water canals and several oscula. Under certain conditions, the cells around the pores and osculum contract, closing the openings.

PARAZOA

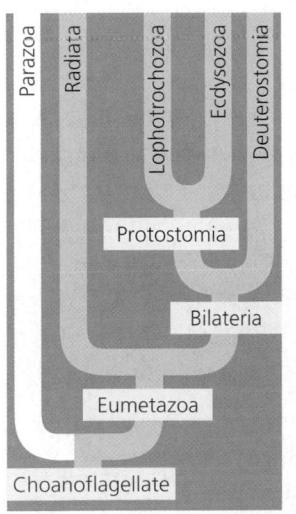

Of all the animals we will discuss, sponges (on branch Parazoa of the phylogenetic tree) represent the lineage closest to the colonial choanoflagellates that gave rise to the animal kingdom (see FIGURE 32.4). The cell layers of sponges are loose federations of cells, which are not really tissues because the cells are relatively unspecialized.

Phylum Porifera: Sponges are sessile with porous bodies and choanocytes

Sponges are sessile animals that appear so sedate to the human eye that the ancient Greeks believed them to be plants. Sponges

FIGURE 33.2 A sponge. Sessile animals without specialized organs and tissues, sponges filter food from water pumped through their porous bodies. The diverse species of sponges vary in shape, color, and structural complexity. Some species are brightly pigmented by symbiotic algae. This is an azure vase sponge, *Callyspongia plicifera*.

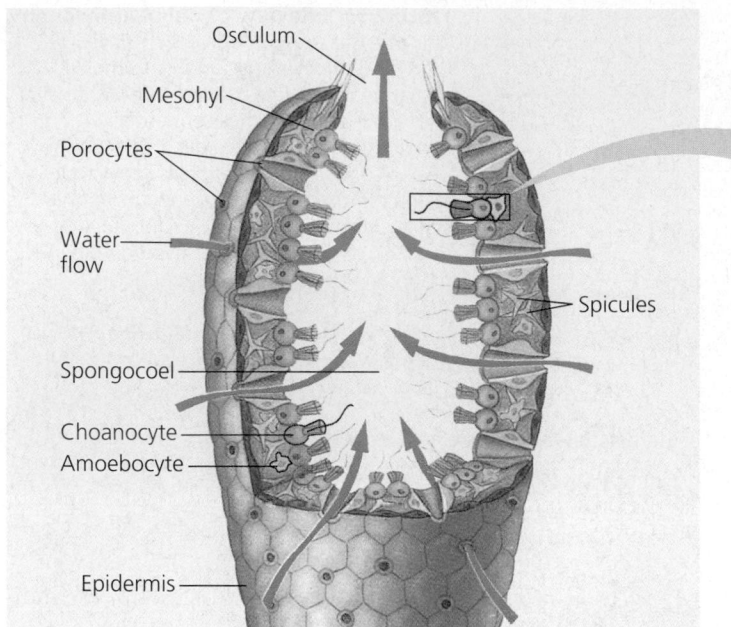

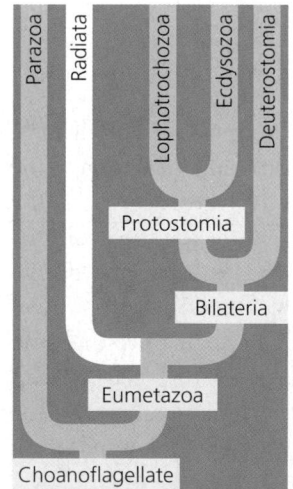

FIGURE 33.3 Anatomy of a sponge. The wall of this simple sponge has two layers of cells separated by a gelatinous matrix, the mesohyl ("middle matter"). The outer layer consists of tightly packed epidermal cells. The incurrent pores are channels through porocytes, cells shaped like elongated doughnuts that span the body wall. The spongocoel is lined mainly by choanocytes, each with a flagellum ringed by a collar of fingerlike projections coated with mucus. Beating flagella sweep large volumes of water (arrows) into the body through the incurrent pores. Food particles are trapped in mucus on the collar, phagocytosed, and digested within choanocytes and adjacent amoebocytes. Mobile amoebocytes transport nutrients to other cells of the body and also produce materials for skeletal fibers (spicules).

Nearly all sponges are suspension-feeders (also known as filter-feeders), which are animals that collect food particles from water passed through some type of food-trapping equipment. Sponges trap food from the water circulated through the porous body. To obtain enough food to grow by 100 g (about 3 ounces), a sponge must filter 1,000 kg (about 275 gallons) of seawater. Lining the inside of the spongocoel or internal water chambers are flagellated **choanocytes,** or collar cells (for the membranous collar around the base of the flagellum). The flagella generate a water current, and the collars trap food particles that the choanocytes then ingest by phagocytosis (FIGURE 33.3). The similarity between choanocytes and the cells of choanoflagellates reinforces the molecular evidence in supporting the hypothesis of an animal origin from a choanoflagellate ancestor.

The body of a sponge consists of two layers of cells separated by a gelatinous region called the **mesohyl.** Wandering through the mesohyl are cells called **amoebocytes,** named for their use of pseudopodia. Amoebocytes have many functions. They take up food from the water and from choanocytes, digest it, and carry nutrients to other cells. Amoebocytes also form tough skeletal fibers within the mesohyl. In some groups of sponges, these fibers are sharp spicules made from calcium carbonate or silica; other sponges produce more flexible fibers composed of a collagen protein called spongin. We use these pliant, honeycombed skeletons as bath sponges.

Most sponges are **hermaphrodites** (from the Greek *Hermes,* the god, and *Aphrodite,* the goddess), meaning that each individual functions as both male and female in sexual reproduction by producing sperm *and* eggs. Gametes arise from choanocytes or amoebocytes. Eggs reside in the mesohyl, but sperm cells are carried out of the sponge by the water current. Cross-fertilization results from some of the sperm being drawn into neighboring individuals. Fertilization occurs in the mesohyl, where the zygotes develop into flagellated, swimming larvae that disperse from the parent. Upon settling on a suitable substratum, a larva develops into the sessile adult.

Sponges are capable of extensive regeneration, the replacement of lost parts. They use regeneration not only for repair but also to reproduce asexually from fragments broken off a parent sponge.

RADIATA

All animals except sponges belong to a clade called Eumetazoa, the animals with true tissues (see Chapter 32). Among the eumetazoans, the oldest clade is Radiata, consisting of animals with radial symmetry and diploblastic embryos (with ectoderm and endoderm, but no mesoderm). The two phyla of Radiata, Cnidaria and Ctenophora, may have had separate origins from different parazoan ancestors.

Phylum Cnidaria: Cnidarians have radial symmetry, a gastrovascular cavity, and cnidocytes

Lacking mesoderm, the cnidarians (hydras, jellies, sea anemones, and coral animals) have a relatively simple body construction. Nonetheless, they are a diverse group with over 10,000 living species, most of which are marine.

The basic body plan of a cnidarian is a sac with a central digestive compartment, the **gastrovascular cavity.** A single opening to this cavity functions as both mouth and anus. This basic body plan has two variations: the sessile polyp and the floating medusa (FIGURE 33.4). **Polyps** are cylindrical forms that adhere to the substratum by the aboral end of the body and extend their tentacles, waiting for prey. Examples of the polyp form are hydras and sea anemones. A **medusa** is a flattened, mouth-down version of the polyp. It moves freely in the water by a combination of passive drifting and contractions of its bell-shaped body. The animals we generally call jellies are medusas. The tentacles of a jelly dangle from the oral surface, which points downward. Some cnidarians exist only as polyps, others only as medusas, and still others pass sequentially through both a medusa stage and a polyp stage in their life cycle.

Cnidarians are carnivores that use tentacles arranged in a ring around the mouth to capture prey and push the food into the gastrovascular cavity, where digestion begins. The undigested remains are egested through the mouth/anus. The tentacles are armed with batteries of **cnidocytes,** unique cells that function in defense and the capture of prey (FIGURE 33.5). Cnidocytes contain cnidae, capsule-like organelles capable of

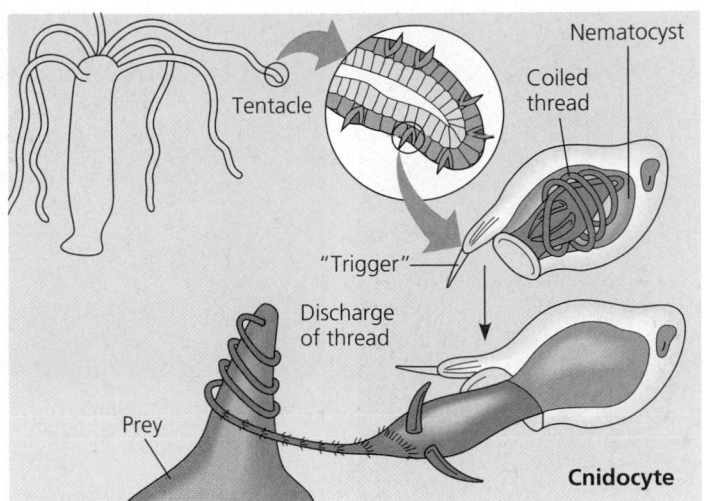

FIGURE 33.5 A cnidocyte of a hydra. This type of cnidocyte contains a stinging capsule, the nematocyst, which itself contains an inverted thread. When a "trigger" is stimulated by touch or by certain chemicals, the thread shoots out from the nematocyst, puncturing and injecting poison into prey. Other kinds of cnidocytes have longer threads that stick to prey or entangle small animals that bump into the tentacles.

everting, giving phylum Cnidaria its name (from the Greek *cnide,* nettle). Cnidae called **nematocysts** are stinging capsules.

Muscles and nerves occur in their simplest forms in cnidarians. Cells of the epidermis (outer layer) and gastrodermis (inner layer) have bundles of microfilaments arranged into contractile fibers (see Chapter 7). True muscle tissue develops from mesoderm and does not appear in diploblastic animals. The gastrovascular cavity acts as a hydrostatic skeleton against which the contractile cells can work. When the animal closes its mouth, the volume of the cavity is fixed, and contraction of selected cells causes the animal to change shape. Movements are coordinated by a nerve net. Cnidarians have no brain, and the noncentralized nerve net is associated with simple sensory receptors that are distributed radially around the body. Thus, the animal can detect and respond to stimuli equally from all directions.

The phylum Cnidaria is divided into three major classes: Hydrozoa, Scyphozoa, and Anthozoa (TABLE 33.1; FIGURE 33.6, p. 650).

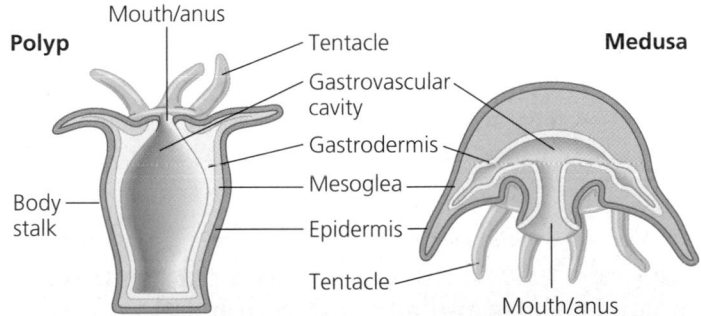

(a) Sea anemone: a polyp   **(b) Jelly: a medusa**

FIGURE 33.4 Polyp and medusa forms of cnidarians. The body wall of **(a)** a polyp or **(b)** a medusa has two layers of cells, an outer layer of epidermis (from ectoderm) specialized for protection and an inner layer of gastrodermis (from endoderm) for digestion. Digestion begins in the gastrovascular cavity and is completed within the gastrodermal cells in food vacuoles. Flagella on the gastrodermal cells keep the contents of the gastrovascular cavity agitated and help distribute nutrients. Sandwiched between the epidermis and gastrodermis is a gelatinous layer of mesoglea. In many medusas, the mesoglea is thick and jellylike—thus the name jellies.

Table 33.1 Classes of Phylum Cnidaria

Class and Examples	Main Characteristics
Hydrozoa (Portuguese man-of-war, hydras, *Obelia,* some corals) (see FIGURES 33.6a and 33.7)	Most marine, a few freshwater; both polyp and medusa stages in most species; polyp stage often colonial
Scyphozoa (jellies, sea wasp, sea nettle) (see FIGURE 33.6b)	All marine; polyp stage reduced; free-swimming; medusas up to 2 m in diameter
Anthozoa (sea anemones, most corals, sea fans) (see FIGURE 33.6c and d)	All marine; medusa stage completely absent; sessile, many colonial

(b) Scyphozoans (jellies)

(c) Anthozoan (sea anemone)

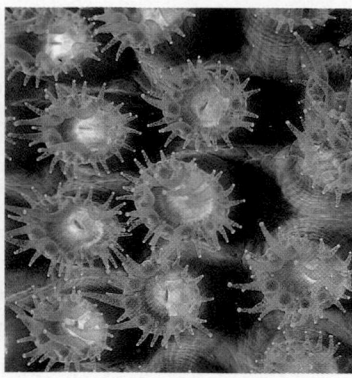

(d) Anthozoans (coral animals)

FIGURE 33.6 Representatives of the cnidarian classes. (a) Polyps of a colonial species belonging to class Hydrozoa. **(b)** Jellies of class Scyphozoa. The medusa is the conspicuous stage of the scyphozoan life cycle. Many species of jellies, including this one, glow in the dark (bioluminescence). The largest scyphozoan species have tentacles over 100 m long dangling from umbrellas up to 2 m in diameter. **(c)** Sea anemones and other members of class Anthozoa exist only as polyps. **(d)** A colony of coral polyps (class Anthozoa). Many coral animals harbor symbiotic algae that contribute to the food supply of the polyps. Coral reefs, which provide habitats for an enormous variety of invertebrates and fishes, are restricted to warm, shallow seas. This is a star coral.

(a) Hydrozoans

Class Hydrozoa

Most hydrozoans alternate polyp and medusa forms, as in the life cycle of *Obelia* (FIGURE 33.7). The polyp stage, a colony of interconnected polyps in the case of *Obelia,* is more conspicuous than the medusas. Hydras, among the few cnidarians found in fresh water, are unusual members of class Hydrozoa in that they exist only in the polyp form. When environmental conditions are favorable, a hydra reproduces asexually by budding, the formation of outgrowths that pinch off from the parent to live independently (see FIGURE 13.1). When environmental conditions deteriorate, hydras reproduce sexually, forming resistant zygotes that remain dormant until conditions improve.

Class Scyphozoa

The medusa generally prevails in the life cycle of class Scyphozoa. The medusas of most species live among the plankton as jellies. Most coastal scyphozoans go through a small polyp stage during their life cycle, but jellies that live in the open ocean generally lack the sessile polyp.

Class Anthozoa

Sea anemones and corals belong to class Anthozoa ("flower animals"). They occur only as polyps. Coral animals live as solitary or colonial forms and secrete hard external skeletons of calcium carbonate. Each polyp generation builds on the skeletal remains of earlier generations to construct "rocks"

with shapes characteristic of the species. It is these skeletons that we call coral.

In tropical seas, coral reefs provide habitat for a great diversity of invertebrates and fishes. Coral reefs in many parts of the world are currently being damaged by environmental changes—global warming is one suspect. We'll examine this problem in Chapter 54.

Phylum Ctenophora: Comb jellies possess rows of ciliary plates and adhesive colloblasts

Comb jellies, or ctenophores, superficially resemble cnidarian medusas. However, the relationship between ctenophores and cnidarians is uncertain. There are only about 100 species of comb jellies, all of which are marine. Ctenophores range in diameter from about 1 to 10 cm. Most are spherical or ovoid, but there are elongate and ribbonlike forms up to 1 m long. *Ctenophora* means "comb-bearer," and these animals are named for their eight rows of comblike plates composed of fused cilia. They are the largest animals to use cilia for locomotion. An aboral sensory organ functions in orientation, and nerves running from the sensory organ to the combs of cilia coordinate movement. Most comb jellies have a pair of long, retractable tentacles (FIGURE 33.8). The tentacles bear adhesive structures called **colloblasts** (also called lasso cells). When prey (mostly small plankton) contact a tentacle, colloblasts burst open. A sticky thread released by each colloblast captures the food, which is then wiped off the tentacle into the mouth.

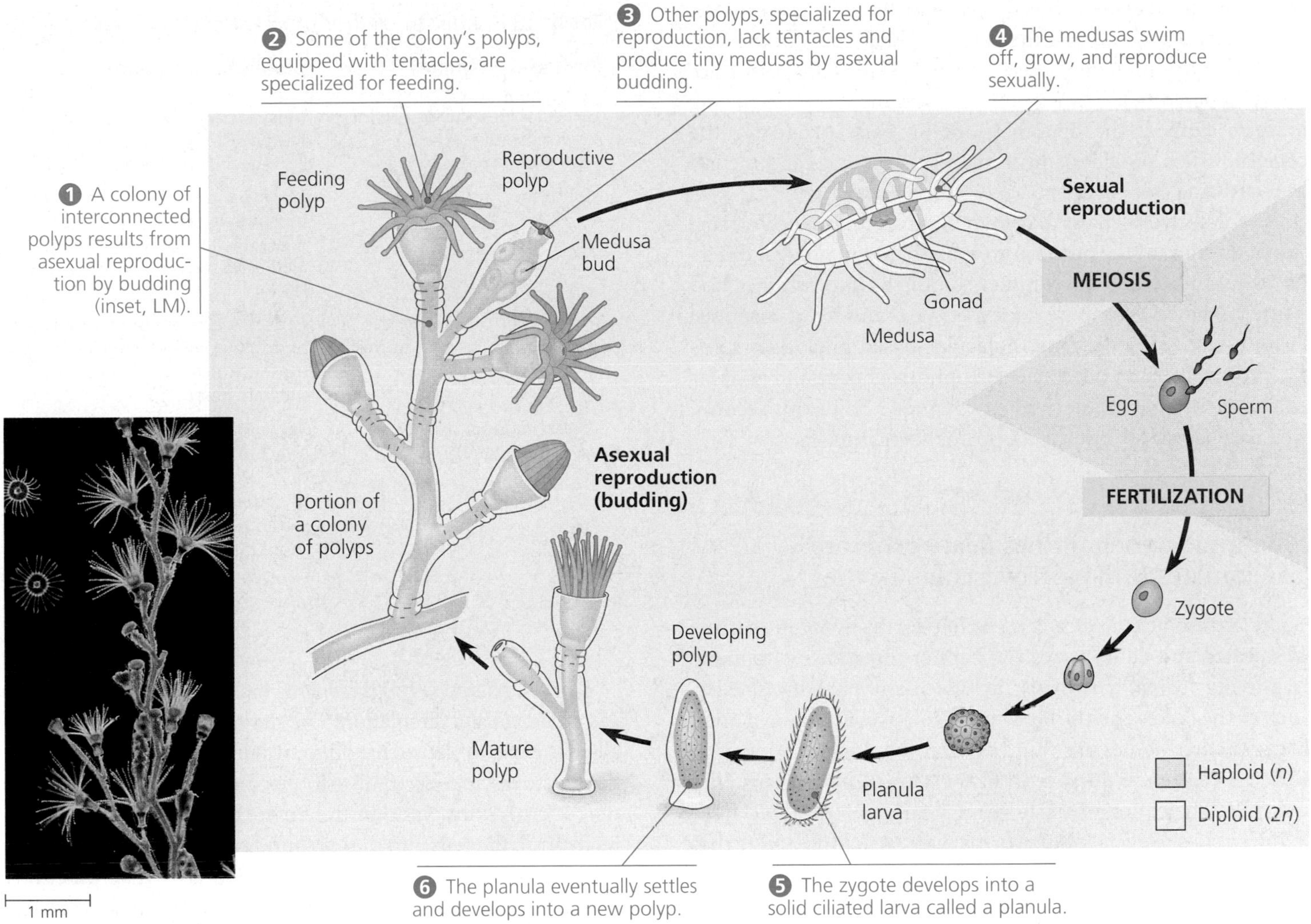

❷ Some of the colony's polyps, equipped with tentacles, are specialized for feeding.

❸ Other polyps, specialized for reproduction, lack tentacles and produce tiny medusas by asexual budding.

❹ The medusas swim off, grow, and reproduce sexually.

❶ A colony of interconnected polyps results from asexual reproduction by budding (inset, LM).

Feeding polyp

Reproductive polyp

Medusa bud

Portion of a colony of polyps

Sexual reproduction

MEIOSIS

Gonad

Medusa

Egg   Sperm

Asexual reproduction (budding)

FERTILIZATION

Zygote

Developing polyp

Mature polyp

Planula larva

Haploid (*n*)

Diploid (2*n*)

1 mm

❻ The planula eventually settles and develops into a new polyp.

❺ The zygote develops into a solid ciliated larva called a planula.

FIGURE 33.7 The life cycle of the hydrozoan *Obelia.* The polyp stage is asexual, the medusa stage is sexual, and these two stages alternate, one producing the other. But do not confuse this with the alternation of generations that occurs in the plant kingdom. Both polyp and medusa are diploid organisms. (Typical of animals, only the gametes of *Obelia* are haploid.) By contrast, one plant generation is haploid, and the other is diploid.

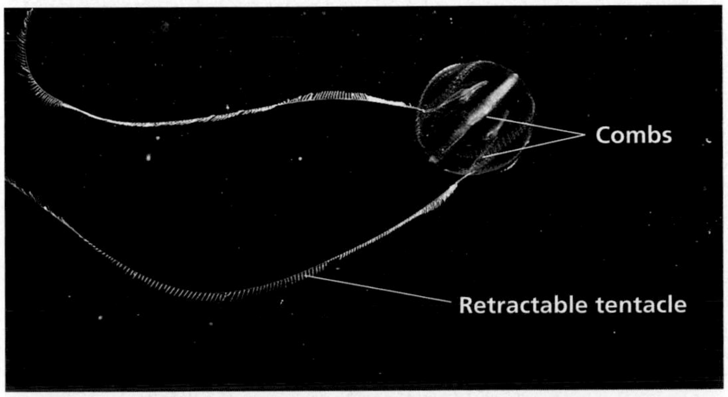

Combs

Retractable tentacle

FIGURE 33.8 A ctenophore, or comb jelly. This planktonic marine animal is named for its eight combs of cilia, used for locomotion. The retractable tentacles capture food. Ctenophores and cnidarians are radiate, diploblastic animals, but these two phyla may have originated independently from different parazoan ancestors.

PROTOSTOMIA: LOPHOTROCHOZOA

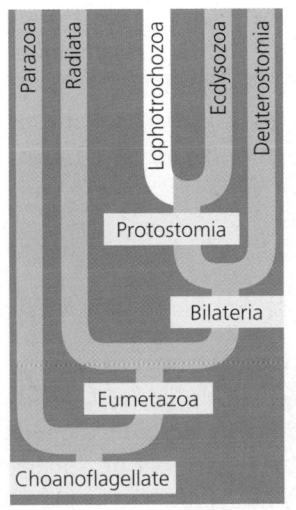

Parazoa

Radiata

Lophotrochozoa

Ecdysozoa

Deuterostomia

Protostomia

Bilateria

Eumetazoa

Choanoflagellate

The molecular-based phylogeny in FIGURE 33.1 includes the hypothesis that the bilateral animals make up a monophyletic group (clade Bilateria) derived from a common ancestor. And this hypothesis implies that the original bilateral animals—called **urbilateria** ("ur" is a prefix meaning earliest)—were relatively complex animals with true body cavities (coeloms). In this view of animal phylogeny, simpler bilaterians lacking coeloms (acoelomates) and those having pseudocoeloms (not completely lined by mesoderm) evolved secondarily from coelomates. The molecular data place the origin of bilaterians in the Precambrian period, before the Cambrian explosion. And trace fossils apparently left in Precambrian sediments by burrowing animals support the hypothesis of an early origin for the coelom, which provides a hydraulic skeleton that can function in burrowing (see Chapter 32).

The data from molecular systematics reinforce the traditional division of the bilateral animals, based mainly on differences in embryonic development, into the protostomes and deuterostomes (compare FIGURES 32.4 and 32.8; also, see FIGURE 32.7). However, the molecular-based phylogeny splits the protostomes into two clades: Lophotrochozoa and Ecdysozoa (these names are derived in Chapter 32). In this section, we take a closer look at the lophotrochozoan phyla.

Phylum Platyhelminthes: Flatworms are acoelomates with gastrovascular cavities

There are about 20,000 species of flatworms living in marine, freshwater, and damp terrestrial habitats. In addition to many free-living forms, flatworms include many parasitic species, such as the flukes and the tapeworms. Flatworms are so named because their bodies are thin between the dorsal and ventral surfaces (flattened dorsoventrally; *platyhelminth* means "flat worm"). They range in size from nearly microscopic free-living species to certain tapeworms over 20 m long. (Note that "worm" is not a formal taxonomic name, but a general term for animals with long, thin bodies.)

In contrast to the radiate animals (cnidarians and ctenophores), flatworms and all other bilateratians are triploblastic. The middle embryonic tissue layer, mesoderm, contributes to the development of more complex organs and organ systems and to true muscle tissue. Thus the flatworms are structurally more complex than cnidarians or ctenophores. However, in common with the radiate animals, flatworms have a gastrovascular cavity with only one opening. (Tapeworms lack a digestive tract altogether and absorb nutrients across their body surface.) Flatworms are also simpler than other bilateratians in lacking a body cavity; flatworms are acoelomates.

Flatworms are divided into four classes: Turbellaria (mostly free-living flatworms), Monogenea (monogeneans), Trematoda (trematodes, or flukes), and Cestoidea (tapeworms) (TABLE 33.2).

Class Turbellaria

Turbellarians are nearly all free-living (nonparasitic) and mostly marine (FIGURE 33.9). Members of the genus *Dugesia*, commonly known as planarians, abound in unpolluted ponds and streams. **Planarians** are carnivores that prey on smaller animals or feed on dead animals (FIGURE 33.10).

Table 33.2 Classes of Phylum Platyhelminthes

Class and Examples	Main Characteristics
Turbellaria (mostly free-living flatworms; e.g., *Dugesia*) (see FIGURE 33.9 and 33.10)	Most marine, some freshwater, a few terrestrial, predators and scavengers; body surface ciliated
Monogenea (monogeneans)	Marine and freshwater parasites; most infect external surfaces of fishes; life history simple; a ciliated larva starts an infection on a host
Trematoda (trematodes, also called flukes) (see FIGURE 33.11)	Parasites, almost always of vertebrates; two suckers attach to host; most life histories include intermediate hosts
Cestoidea (tapeworms) (see FIGURE 33.12)	Parasites of vertebrates; scolex attaches to host; proglottids produce eggs and break off after fertilization; no head or digestive system; life history with one or more intermediate hosts

Planarians and other flatworms lack organs specialized for gas exchange and circulation. The flat shape of the body places all cells close to the surrounding water, and fine branching of the gastrovascular cavity distributes food throughout the animal. Nitrogenous waste in the form of ammonia diffuses directly from the cells into the surrounding water. Flatworms also have a relatively simple excretory apparatus that functions mainly to maintain osmotic balance between the animal and its surroundings. This system consists of ciliated cells called flame cells that waft fluid through branched ducts opening to the outside (see FIGURE 44.18). The evolution of osmoregulatory structures was a major factor in allowing some turbellarians to invade freshwater and even moist terrestrial environments.

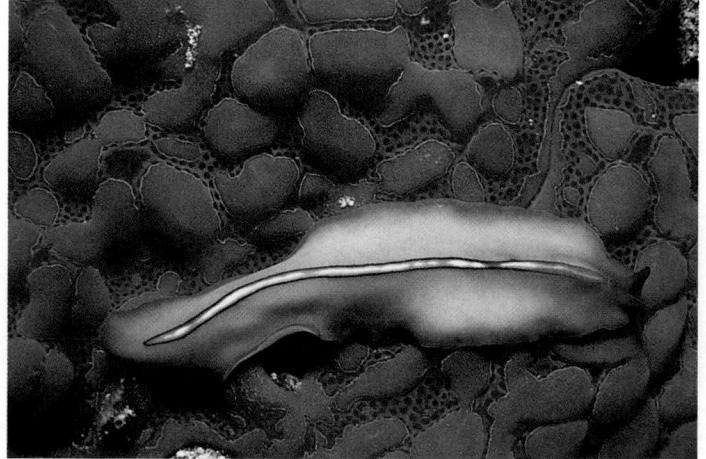

FIGURE 33.9 A flatworm. Class Turbellaria consists mainly of free-living marine flatworms, such as this colorful species.

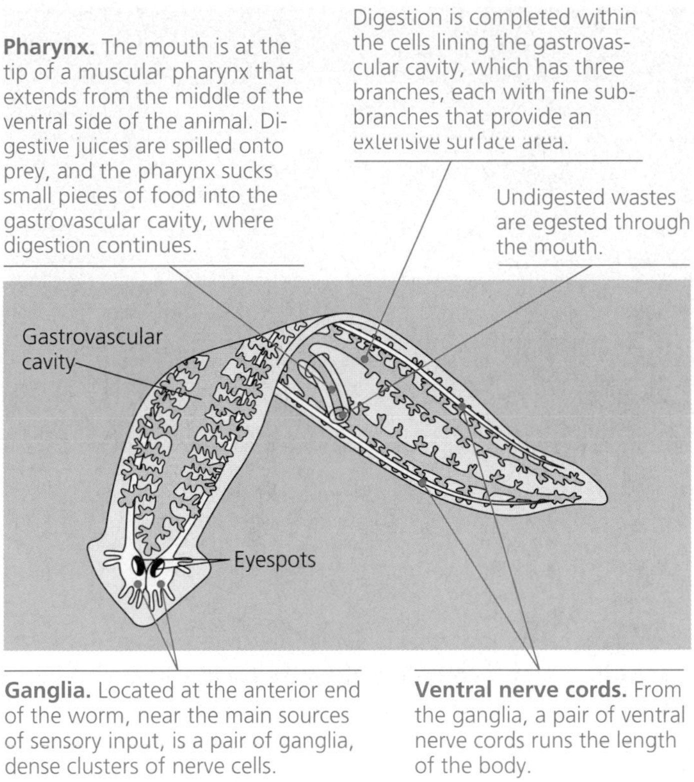

FIGURE 33.10 Anatomy of a planarian.

Planarians move by using cilia on the ventral epidermis, gliding along a film of mucus they secrete. Some turbellarians also use their muscles to swim through water with an undulating motion.

A planarian has a head (cephalized) with a pair of eyespots that detect light and lateral flaps that function mainly for smell. The planarian nervous system is more complex and centralized than the nerve nets of cnidarians. Planarians can learn to modify their responses to stimuli.

Planarians can reproduce asexually through regeneration. The parent constricts in the middle, and each half regenerates the missing end. Sexual reproduction also occurs. Although planarians are hermaphrodites, copulating mates cross-fertilize.

Classes Monogenea and Trematoda

The monogeneans and the trematodes (sometimes called flukes) live as parasites in or on other animals. Many have suckers for attaching to internal organs or to the outer surfaces of the host. A tough covering helps protect the parasites. Reproductive organs nearly fill the interior of these worms.

As a group, trematodes parasitize a wide range of hosts, and most species have complex life cycles with an alternation of sexual and asexual stages. Many trematodes require an intermediate host in which larvae develop before infecting the final host (usually a vertebrate), where the adult worm lives. For ex-

ample, trematodes that parasitize humans spend parts of their life histories in snails (FIGURE 33.11). The 200 million people around the world who are infected with blood flukes (*Schistosoma*) suffer body pains, anemia, and dysentery.

Most monogeneans are external parasites of fishes. Their life cycle is relatively simple, with a ciliated, free-swimming larva starting an infection on a host. Although monogeneans have been traditionally aligned with the trematodes, some structural and chemical evidence suggests they are more closely related to tapeworms.

Class Cestoidea

Tapeworms (class Cestoidea) are also parasitic. The adults live mostly in vertebrates, including humans. The tapeworm head, or scolex, is armed with suckers and often menacing hooks

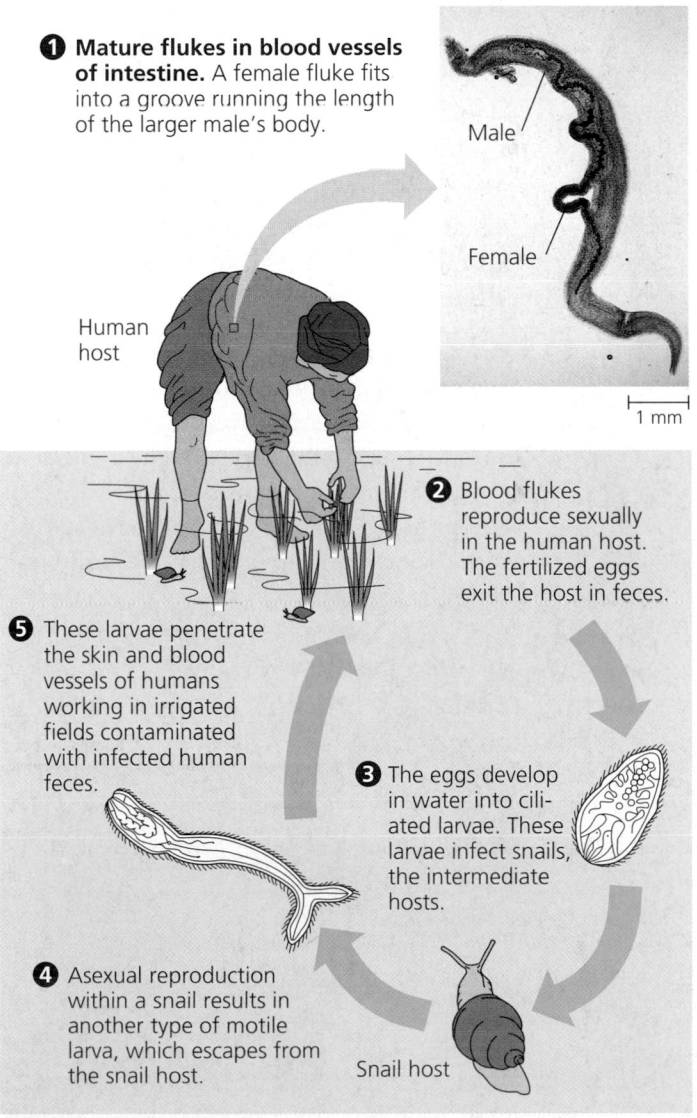

❶ **Mature flukes in blood vessels of intestine.** A female fluke fits into a groove running the length of the larger male's body.

Male

Female

1 mm

Human host

❷ Blood flukes reproduce sexually in the human host. The fertilized eggs exit the host in feces.

❸ The eggs develop in water into ciliated larvae. These larvae infect snails, the intermediate hosts.

❹ Asexual reproduction within a snail results in another type of motile larva, which escapes from the snail host.

Snail host

❺ These larvae penetrate the skin and blood vessels of humans working in irrigated fields contaminated with infected human feces.

FIGURE 33.11 The life history of a blood fluke (*Schistosoma mansoni*).

that lock the worm to the intestinal lining of the host (FIGURE 33.12). Posterior to the scolex is a long ribbon of units called proglottids, which are little more than sacs of sex organs. Lacking a digestive tract, the tapeworm absorbs food predigested by the host.

Mature proglottids, loaded with thousands of eggs, are released from the posterior end of a mature tapeworm and leave the host's body with feces. In one type of life cycle, human feces contaminate the food or water of intermediate hosts, such as pigs or cattle, and the tapeworm eggs develop into larvae that encyst in muscles of these animals. Humans acquire the larvae by eating undercooked meat contaminated with cysts, and the worms develop into mature adults within the human. Large tapeworms, which may be 20 m or more in length, can cause intestinal blockage and can rob enough nutrients from the human host to cause nutritional deficiencies. An orally administered drug named niclosamide kills the adult worms.

Phylum Rotifera: Rotifers are pseudocoelomates with jaws, crowns of cilia, and complete digestive tracts

Rotifers (about 1,800 species) are tiny animals that mainly inhabit fresh water, although some live in the sea or in damp soil. Ranging in size from only about 0.05 to 2.0 mm, smaller than many protists, rotifers are nevertheless truly multicellular and have specialized organ systems (FIGURE 33.13). In contrast

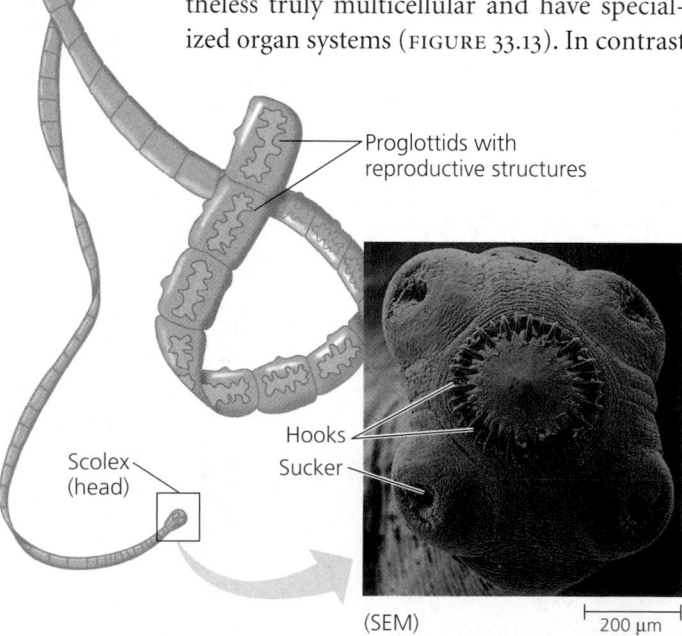

FIGURE 33.12 **Anatomy of a tapeworm.**

Proglottids with reproductive structures

Hooks
Sucker

Scolex (head)

(SEM) 200 μm

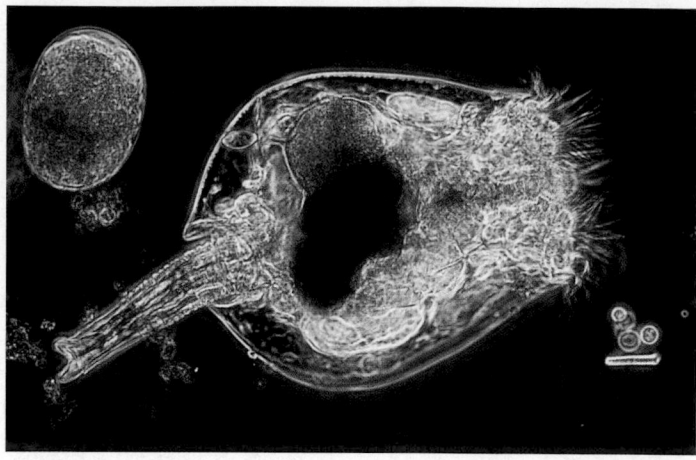

FIGURE 33.13 A rotifer. These pseudocoelomates, smaller than many protists, are much more anatomically complex than flatworms (LM). 0.1 mm

to cnidarians and flatworms, which have gastrovascular cavities, rotifers have a **complete digestive tract,** which is a digestive tube with a separate mouth and anus. Internal organs lie within the pseudocoelom, a body cavity that is not completely lined by mesoderm (see FIGURE 32.6b). Fluid in the pseudocoelom serves as a hydrostatic skeleton and as a medium for the internal transport of nutrients and wastes in these tiny animals. Movement of a rotifer's body distributes the fluid within the pseudocoelom, and thus this body cavity and its fluid function as a circulatory system.

The word *rotifer*, derived from Latin, means "wheel-bearer," a reference to the crown of cilia that draws a vortex of water into the mouth. Posterior to the mouth, a region of the digestive tract called the pharynx bears jaws (trophi) that grind food, mostly microorganisms suspended in the water.

Rotifer reproduction is unusual. Some species consist only of females that produce more females from unfertilized eggs, a type of reproduction called **parthenogenesis.** Other species produce two types of eggs that develop by parthenogenesis, one type forming females and the other type developing into degenerate males that cannot even feed themselves. The males survive long enough to produce sperm that fertilize eggs, forming resistant zygotes that can survive when a pond dries up. When conditions are favorable again, the zygotes break dormancy and develop into a new female generation that then reproduces by parthenogenesis until conditions become unfavorable again.

The lophophorate phyla: Bryozoans, phoronids, and brachiopods are coelomates with ciliated tentacles around their mouths

In subdividing bilaterians into protostomes and deuterostomes based on embryological criteria, the lophophorate phyla, including Bryozoa, Phoronida, and Brachiopoda, were never a

(a) A bryozoan   **(b) A brachiopod**

FIGURE 33.14 The lophophorate animals. The most distinctive characteristic of these phyla is the lophophore, an organ that functions in suspension-feeding. **(a)** Bryozoans, such as this common sea mat *(Membranipora membranacea)*, are colonial lophophorates, many with hard exoskeletons. **(b)** Brachiopods are lophophorates with a hinged shell. The two parts of the shell are dorsal and ventral, in contrast to the lateral shells of bivalve mollusks (such as clams).

good fit in either coelomate branch. However, molecular data place the lophophorates squarely in the protostome branch. These phyla are collectively called the **lophophorate animals** after the most distinctive structure they share, the lophophore (FIGURE 33.14). The **lophophore** is a horseshoe-shaped or circular fold of the body wall bearing ciliated tentacles that surround the mouth (see FIGURE 32.11). The anus lies outside the whorl of tentacles. The cilia draw water toward the mouth between the tentacles, which help trap food particles for these suspension-feeders. The common occurrence of this complex apparatus in the lophophorate animals suggests that the three phyla are related. Other similarities, such as a U-shaped digestive tract and the absence of a distinct head, are adaptations to a sessile existence. In contrast to flatworms, which lack body cavities, and rotifers, which have pseudocoeloms, lophophorates have true coeloms completely lined by mesoderm (see FIGURE 32.6c).

Bryozoans are colonial animals that superficially resemble mosses. (*Bryozoa* means "moss animals.") In most species, the colony is encased in a hard exoskeleton with pores through which the lophophores of the animals extend (FIGURE 33.14a). Of the 5,000 species of bryozoans, most live in the sea, where they are among the most widespread and numerous sessile animals. Several species are important reef builders.

Phoronids are tube-dwelling marine worms ranging from 1 mm to 50 cm in length. Some live buried in the sand within tubes made of chitin, extending their lophophore from the opening of the tube and withdrawing it into the tube when threatened. There are only about 15 species of phoronid worms in two genera.

Brachiopods, or lamp shells, superficially resemble clams and other bivalve mollusks, but the two halves of the brachiopod shell are dorsal and ventral to the animal rather than lateral, as in clams (FIGURE 33.14b). A brachiopod lives attached to its substratum by a stalk, opening its shell slightly to allow water to flow between the shells and the lophophore. All brachiopods are marine. The living brachiopods are remnants of a much richer past; only about 330 extant species are known, but there are 30,000 species of Paleozoic and Mesozoic fossils. A tie to the past is *Lingula,* a living brachiopod genus that has changed little in 400 million years.

Phylum Nemertea: Proboscis worms are named for their prey-capturing apparatus

Members of the phylum Nemertea are called proboscis worms or ribbon worms (FIGURE 33.15). A proboscis worm's body is structurally acoelomate, like that of a flatworm, but it contains a small fluid-filled sac that may be a reduced version of a true coelom. The sac and fluid hydraulically operate an extensible proboscis by which the worm captures prey.

Proboscis worms range in length from less than 1 mm to more than 30 m. Nearly all of the 900 or so members of this phylum are marine, but a few species inhabit fresh water or damp soil. Some are active swimmers, and others burrow in the sand.

Proboscis worms and flatworms have similar excretory, sensory, and nervous systems. But, in addition to the unique proboscis apparatus, two anatomical features not found in flatworms have evolved in the phylum Nemertea: a complete digestive tract (with mouth and anus) and a **closed circulatory system**—the blood is contained in vessels and is therefore distinct from fluid in the body cavity. Proboscis worms have no heart, but the blood is propelled by muscles squeezing the vessels.

FIGURE 33.15 A proboscis worm, phylum Nemertea.

Phylum Mollusca: Mollusks have a muscular foot, a visceral mass, and a mantle

Snails and slugs, oysters and clams, and octopuses and squids are all mollusks. In all, the phylum Mollusca has more than 150,000 known species. Most mollusks are marine, though some inhabit fresh water, and there are snails and slugs that live on land. Mollusks are soft-bodied animals (from the Latin *molluscus,* soft), but most are protected by a hard shell made of calcium carbonate. Slugs, squids, and octopuses have reduced shells, most of which are internal, or they have lost their shells completely during their evolution.

Despite their apparent differences, all mollusks have a similar body plan (FIGURE 33.16). The body has three main parts: a muscular **foot,** usually used for movement; a **visceral mass** containing most of the internal organs; and a **mantle,** a fold of tissue that drapes over the visceral mass and secretes a shell (if one is present). In many mollusks, the mantle extends beyond the visceral mass, producing a water-filled chamber, the **mantle cavity,** which houses the gills, anus, and excretory pores. Many mollusks feed by using a straplike rasping organ called a **radula** to scrape up food.

Most mollusks have separate sexes, with gonads (ovaries or testes) located in the visceral mass. Many snails, however, are hermaphrodites. The life cycle of many marine mollusks includes a ciliated larva called the **trochophore,** also characteristic of marine annelids (segmented worms) and some other lophotrochozoans (see FIGURE 32.9).

The basic body plan of mollusks has evolved in various ways in the different classes of the phylum. Of the eight classes, we examine four here: Polyplacophora (chitons), Gastropoda (snails and slugs), Bivalvia (clams, oysters, and other bivalves), and Cephalopoda (squids, octopuses, and nautiluses) (TABLE 33.3).

Table 33.3 Major Classes of Phylum Mollusca

Class and Examples	Main Characteristics
Polyplacophora (chitons) (see FIGURE 33.17)	Marine; shell with eight plates; foot used for locomotion; head reduced
Gastropoda (snails, slugs) (see FIGURES 33.18 and 33.19)	Marine, freshwater, or terrestrial; asymmetric body, usually with a coiled shell; shell reduced or absent in some; foot for locomotion; radula present
Bivalvia (clams, mussels, scallops, oysters) (see FIGURES 33.20 and 33.21)	Marine and freshwater, flattened shell with two valves; head reduced; paired gills; most are filter-feeders; mantle forms siphons
Cephalopoda (squids, octopuses, chambered nautiluses) (see FIGURE 33.22)	Marine; head surrounded by grasping tentacles, usually with suckers; shell external, internal, or absent; mouth with or without radula; locomotion by jet propulsion using siphon made from mantle

Class Polyplacophora

Chitons are marine animals with oval shapes and shells divided into eight dorsal plates (the body itself, however, is unsegmented). You can find chitons clinging to rocks along the shore during low tide (FIGURE 33.17). Try to dislodge a chiton by hand, and you will be surprised at how well its foot, acting as a suction cup, grips the rock. Using this muscular foot, a chiton can creep slowly over the rock surface. Chitons are grazers that use their radulas to cut and ingest algae.

FIGURE 33.16 The basic body plan of mollusks. Three hallmarks of the phylum are the mantle, visceral mass, and foot. A mantle cavity houses gills in many species. The long digestive tract is coiled in the visceral mass. Most mollusks have an open circulatory system with a dorsal heart that pumps circulatory fluid (hemolymph) through arteries into sinuses (body spaces) bathing the organs. Excretory organs called nephridia remove metabolic wastes from the hemolymph. The nervous system consists of a nerve ring around the esophagus, from which nerve cords extend. The enlargement shows the mouth region with the radula, a rasplike feeding organ present in many mollusks. The radula is a belt of backward-curved teeth that extends from the mouth and slides back and forth, scraping and scooping like a backhoe.

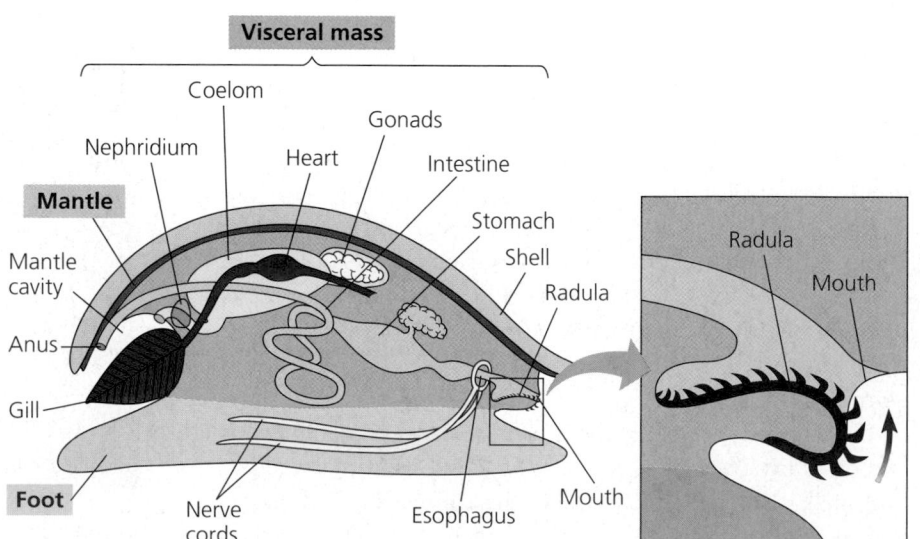

FIGURE 33.17 A chiton. Clinging tenaciously to rocks in the intertidal zone, this chiton (class Polyplacophora) displays the eight-plate shell characteristic of this class of mollusks.

(a) Gastropod shells. Shell collectors find delight in the variety of Gastropoda, one of the most diverse animal classes.

FIGURE 33.19 Gastropods.

(b) Sea slug. Nudibranchs, or sea slugs, have lost the shell during their evolution.

Class Gastropoda

The largest of the molluscan classes, Gastropoda has more than 40,000 living species. Most gastropods are marine, but there are also many freshwater species. Garden snails and slugs have adapted to land.

The most distinctive characteristic of the class Gastropoda is a process known as **torsion**. During embryonic development, an asymmetrical muscle forms, and contraction of the muscle and uneven growth causes the visceral mass to rotate up to 180 degrees, so that the anus and mantle cavity are placed above the head in the adult (FIGURE 33.18). Some zoologists speculate that the advantage of torsion is to place the visceral mass and heavy shell more centrally over the snail's body.

Most gastropods are protected by single, spiraled shells into which the animals can retreat when threatened (FIGURE 33.19). The shell is often conical, but abalones and limpets have somewhat flattened shells. Many gastropods have distinct heads with eyes at the tips of tentacles. Gastropods inch along literally at a snail's pace by a rippling motion of the elongated foot. Most gastropods use their radula to graze on algae or plant material. Several groups, however, are predators, and the radula is modified for boring holes in the shells of other mollusks or for tearing apart tough animal tissue. In one group, the cone snails, the teeth of the radula form separate poison darts, which penetrate prey, including fishes.

Gastropods are among the few invertebrate groups to have successfully populated the land. Terrestrial snails lack the gills typical of most aquatic gastropods, and instead the lining of the mantle cavity functions as a lung, exchanging respiratory gases with the air.

Class Bivalvia

The mollusks of class Bivalvia include many species of clams, oysters, mussels, and scallops. Bivalves have shells divided into two halves (FIGURE 33.20). The two parts of the shell are hinged at the mid-dorsal line, and powerful adductor muscles draw the two halves tightly together to protect the soft-bodied animal. When the shell is open, the bivalve may extend its hatchet-shaped foot for digging or anchoring.

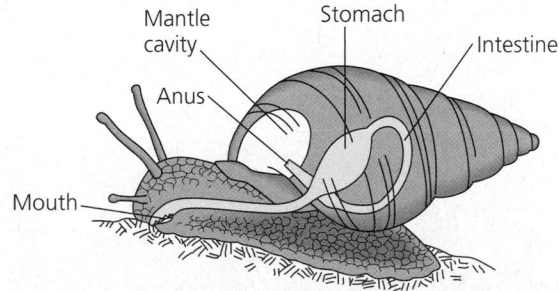

FIGURE 33.18 The results of torsion in a gastropod. Because of torsion (twisting of the visceral mass) during embryonic development, the digestive tract is coiled and the anus is near the mouth at the head end of the animal. After torsion, some of the organs that were bilateral are reduced in size or are lost on one side of the body. Torsion should not be confused with the formation of a coiled shell, which is an independent developmental process.

FIGURE 33.20 A bivalve. This scallop has many eyes peering out between the two halves of the hinged shell.

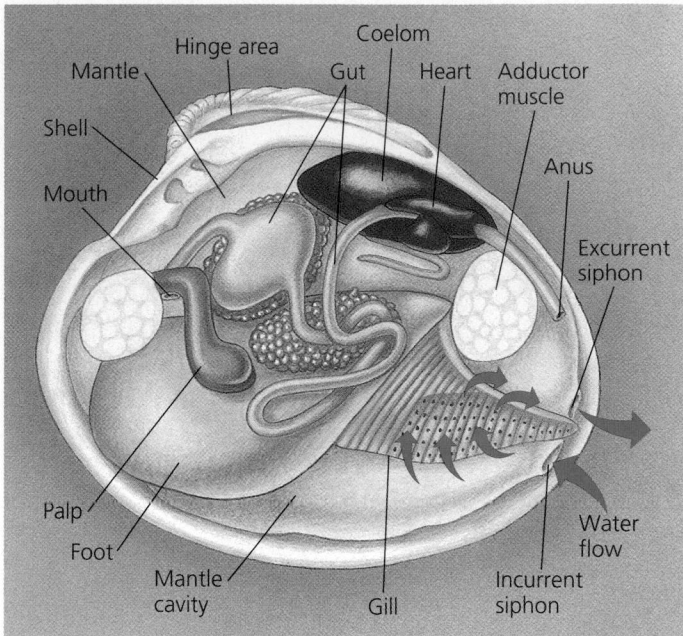

FIGURE 33.21 Anatomy of a clam. The left half of the bivalve shell has been removed. Food particles suspended in water that enters through the incurrent siphon are collected by the gills and passed via cilia and elongated flaps called palps to the mouth.

The mantle cavity of a bivalve contains gills that are used for feeding as well as gas exchange (FIGURE 33.21). Most bivalves are suspension-feeders. They trap fine food particles in mucus that coats the gills, and cilia then convey the particles to the mouth. Water flows into the mantle cavity through an incurrent siphon, passes over the gills, and then exits the mantle cavity through an excurrent siphon. Bivalves have no distinct head, and the radula has been lost.

Being suspension-feeders, most bivalves lead rather sedentary lives. Sessile mussels secrete strong threads that tether them to rocks, docks, boats, and the shells of other animals. Clams can pull themselves into the sand or mud, using the muscular foot for an anchor. In addition to digging, scallops can also skitter along the seafloor by flapping their shells, rather like the mechanical false teeth sold in novelty shops.

Class Cephalopoda

Cephalopods are built for speed, an adaptation that fits their carnivorous diet. Squids and octopuses use beaklike jaws to bite their prey; they then inject poison to immobilize the victim. The mouth is at the center of several long tentacles. A mantle covers the visceral mass, but the shell is reduced and internal (squids) or missing altogether (many octopuses) (FIGURES 33.22a and b). One small group of shelled cephalopods, the chambered nautiluses, survives today (FIGURE 33.22c).

A squid darts about, usually backward, by drawing water into its mantle cavity and then firing a jet stream of water through the excurrent siphon that points anteriorly. The animal steers by pointing the siphon in different directions. The foot of a cephalopod has become modified into this muscular siphon and parts of the tentacles and head. (*Cephalopod* means "head foot.") Most species of squid are less than 75 cm long, but there are also giant squids, the largest of all invertebrates. The biggest specimen on record was 17 m long (including the tentacles) and weighed about 2 tons.

Rather than swimming as squids do in the open seas, most octopuses live on the seafloor, where they creep and scurry about in search of crabs and other food.

Cephalopods are the only mollusks with a closed circulatory system. They also have a well-developed nervous system with a complex brain. The ability to learn and behave in a complex manner is probably more critical to fast-moving

FIGURE 33.22 Cephalopods.

(a) Squids, such as this California market squid *(Loligo opalescens),* are speedy carnivores with beaklike jaws and well-developed eyes.

(b) Octopuses are believed to be among the most intelligent invertebrates.

(c) Chambered nautiluses are the only extant cephalopods with external shells.

predators than to sedentary animals such as clams. Squids and octopuses also have well-developed sense organs.

The ancestors of octopuses and squids were probably shelled mollusks that took up a predaceous lifestyle, the loss of the shell occurring in later evolution. Shelled cephalopods called **ammonites,** some of them as large as truck tires, were the dominant invertebrate predators of the seas for hundreds of millions of years until their disappearance during the mass extinctions at the end of the Cretaceous period (see Chapter 25).

Phylum Annelida: Annelids are segmented worms

Annelida means "little rings," and a segmented body resembling a series of fused rings is a hallmark of the annelid worms. There are about 15,000 annelid species, ranging in length from less than 1 mm to the 3-m length of a giant Australian earthworm. Annelids live in the sea, most freshwater habitats, and damp soil. We can describe the anatomy of annelids in terms of a well-known member of the phylum, the earthworm (FIGURE 33.23).

The coelom of the earthworm is partitioned by septa, but the digestive tract, longitudinal blood vessels, and nerve cords penetrate the septa and run the length of the animal (the major vessels have segmental branches). The digestive system has several specialized regions: the pharynx, the esophagus, the crop, the gizzard, and the intestine. The closed circulatory system consists of a network of vessels containing blood with oxygen-carrying hemoglobin. Dorsal and ventral vessels are connected by segmental pairs of vessels. The dorsal vessel and five pairs of vessels that circle the esophagus of an earthworm are muscular and pump blood through the circulatory system. Tiny blood vessels are abundant in the earthworm's skin, which functions as its respiratory organ.

In each segment of the worm is a pair of excretory tubes called **metanephridia** with ciliated funnels, called nephrostomes, that remove wastes from the blood and coelomic fluid. The metanephridia lead to exterior pores, through which the metabolic wastes are discharged.

A brainlike pair of cerebral ganglia lies above and in front of the pharynx. A ring of nerves around the pharynx connects to a sub-pharyngeal ganglion, from which a fused pair of nerve cords runs posteriorly. All along these ventral nerve cords are segmental ganglia, also fused.

Earthworms are hermaphrodites, but they cross-fertilize. Two earthworms mate by aligning themselves in such a way that they exchange sperm, and then they separate. The received sperm cells are stored temporarily while a special organ, the clitellum, secretes a mucous cocoon. The cocoon slides along the worm, picking up the eggs and then the stored sperm. The cocoon then slips off the worm's head and resides in the soil while the embryos develop. Some earthworms can also reproduce asexually by fragmentation followed by regeneration.

Some aquatic annelids swim in pursuit of food, but most are bottom-dwellers that burrow in the sand and silt; earthworms, of course, are burrowers.

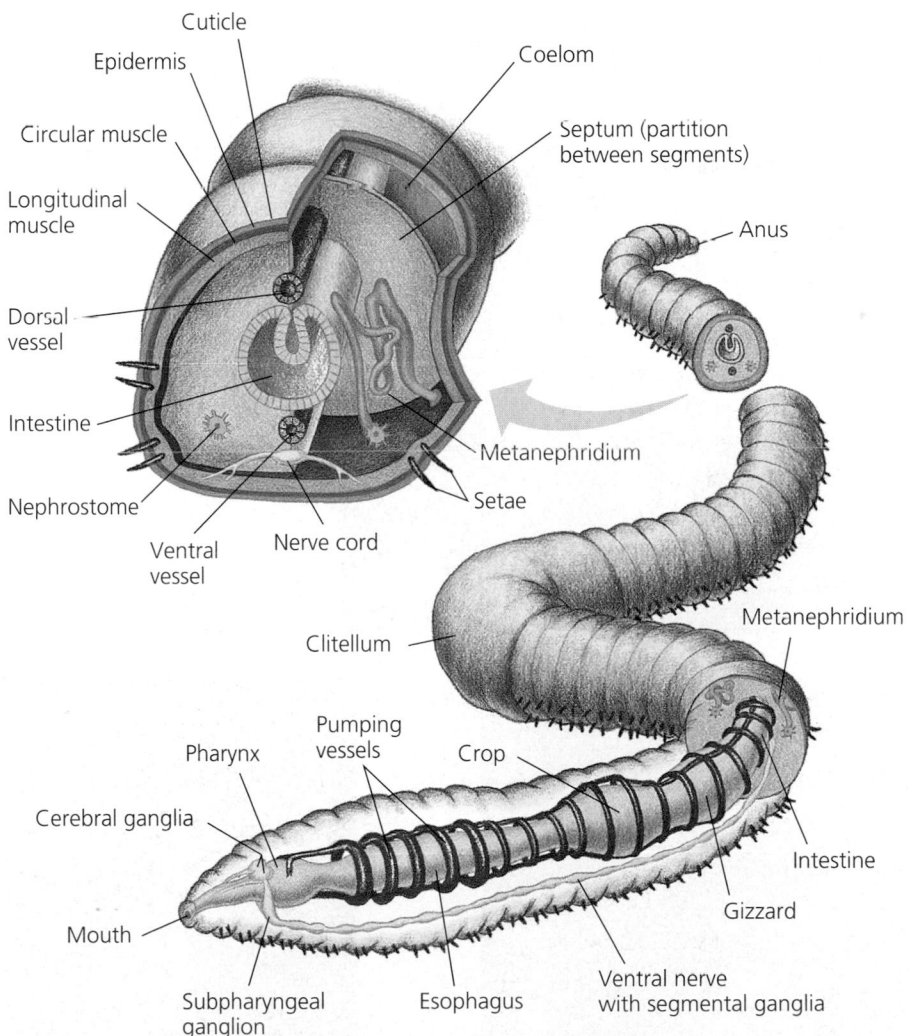

FIGURE 33.23 Anatomy of an earthworm. Annelids are segmented both externally and internally. Many of the internal structures are repeated, segment by segment. Externally, each segment has four pairs of setae, bristles that provide traction for burrowing. Earthworms and many other annelids creep along or burrow by coordinating two sets of muscles, one longitudinal and the other circular (see FIGURE 49.27). These muscles work against the noncompressible coelomic fluid, a hydrostatic skeleton.

The phylum Annelida is divided into three classes: Oligochaeta (earthworms and their relatives), Polychaeta (polychaetes), and Hirudinea (leeches) (TABLE 33.4).

Class Oligochaeta

This class of segmented worms includes the earthworms and a variety of aquatic species (FIGURE 33.24a). An earthworm eats its way through the soil, extracting nutrients as the soil passes through the digestive tube. Undigested material, mixed with mucus secreted into the digestive tract, is egested as castings through the anus. Farmers value earthworms because the animals till the earth, and the castings improve the texture of the soil. Darwin estimated that 1 acre of British farmland had about 50,000 earthworms that produced 18 tons of castings per year.

Table 33.4 Classes of Phylum Annelida

Class and Examples	Main Characteristics
Oligochaeta (terrestrial and freshwater segmented worms; e.g., earthworms) (see FIGURES 33.23 and 33.24a)	Reduced head; no parapodia, but setae present
Polychaeta (mostly marine segmented worms) (see FIGURE 33.24b and c)	Well-developed head; each segment usually has parapodia with setae; tube-dwelling and free-living
Hirudinea (leeches) (see FIGURE 33.24d)	Body usually flattened, with reduced coelom and segmentation; setae absent; suckers at anterior and posterior ends; parasites, predators, and scavengers

FIGURE 33.24 Annelids, the segmented worms. (a) Class Oligochaeta: Some giant Australian earthworms are bigger than snakes. **(b)** Class Polychaeta: Most polychaetes are marine worms. Each segment has a pair of lateral flaps that function in movement and as gills for the exchange of respiratory gases with the surrounding water. **(c)** Fanworms are tube-dwellers that use their feathery headdresses for gas exchange and to extract suspended food particles from the seawater. This species is known as a Christmas-tree worm. **(d)** Class Hirudinea: A nurse applied this medicinal leech *(Hirudo medicinalis)* to a patient's sore thumb to drain blood from a hematoma (abnormal accumulation of blood around an internal injury).

(a) Australian earthworm

(b) Polychaete

(c) Christmas-tree worm

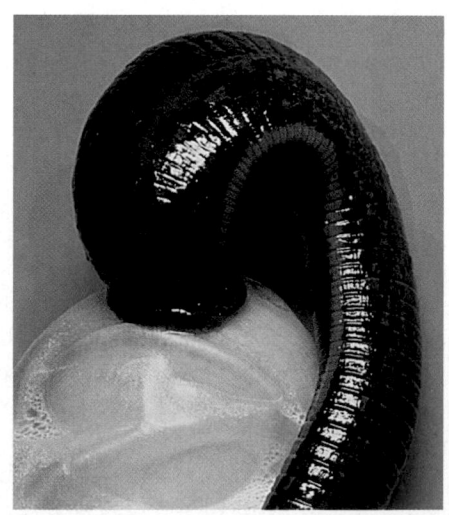

(d) Leech

Class Polychaeta

Each segment of a polychaete (meaning "many setae," for the bristles on each segment) has a pair of paddlelike or ridgelike structures called parapodia ("almost feet") that function in locomotion. Each parapodium has several setae made of the polysaccharide chitin. In many polychaetes the parapodia are richly supplied with blood vessels and function as gills (FIGURE 33.24b).

Most polychaetes are marine. A few drift and swim among the plankton, many crawl on or burrow in the seafloor, and many others live in tubes, which the worms make by mixing mucus with bits of sand and broken shells. The tube-dwellers include the brightly colored fanworms, which trap microscopic food particles in feathery tentacles that extend from the opening of the tube (FIGURE 33.24c).

Class Hirudinea

The majority of leeches inhabit fresh water, but there are also land leeches that move through moist vegetation. Many leeches feed on other invertebrates, but some are blood-sucking parasites that feed by attaching temporarily to other animals, including humans. Leeches range in length from about 1 to 30 cm. Some parasitic species use bladelike jaws to slit the skin of the host, whereas others secrete enzymes that digest a hole through the skin. The host is usually oblivious to this attack because the leech secretes an anesthetic. After making the incision, the leech secretes another chemical, hirudin, which keeps the blood of the host from coagulating. The parasite then sucks as much blood as it can hold, often more than ten times its own weight. After this gorging, a leech can last for months without another meal.

Until this century, leeches were frequently used by physicians for bloodletting. Leeches are still used for treating bruised tissues and for stimulating the circulation of blood to fingers or toes that have been sewn back to hands or feet after accidents (FIGURE 33.24d).

■   ■   ■

Before leaving annelids, let's highlight two evolutionary adaptations that are well developed in this phylum: the coelom and segmentation. The evolutionary significance of the coelom cannot be overemphasized. In addition to providing a hydrostatic skeleton and allowing new and diverse methods of locomotion, it also provides body space for storage and for complex organ development. The coelom also serves as a cushion that protects internal structures, and it allows a functional separation of the action of body wall muscles from those of internal organs, such as the muscles of the digestive tract.

Segmentation allows a high degree of specialization of body regions. This regional specialization—groups of segments modified for different functions—is seen to some degree in annelids but is a true hallmark of the body plan of arthropods, as we will see in the next section.

PROTOSTOMIA: ECDYSOZOA

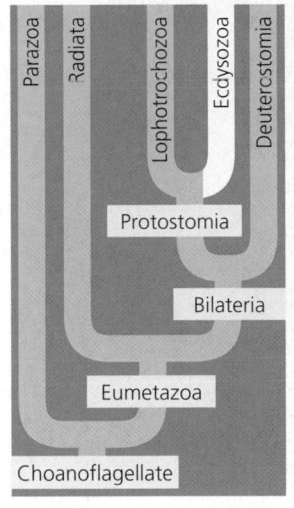

Though the ecdysozoans are considered a clade based mainly on the data of molecular systematics, this branch of protostomes is named for a characteristic of life history: ecdysis, the shedding of an exoskeleton outgrown by the animal. Of the several ecdysozoan phyla, we will examine only the nematodes and the arthropods.

Phylum Nematoda: Roundworms are nonsegmented pseudocoelomates covered by tough cuticles

Among the most widespread of all animals, roundworms (nematodes) are found in most aquatic habitats, in wet soil, in the moist tissues of plants, and in the body fluids and tissues of animals. About 90,000 species are known, and perhaps ten times that number actually exist. In contrast to annelids, the bodies of nematodes are not segmented. The cylindrical bodies of roundworms range from less than 1 mm to more than a meter in length, tapering to a fine tip at the posterior end and to a more blunt tip at the anterior (head) end (FIGURE 33.25a, p. 662). A tough exoskeleton called a cuticle covers the body; as the worm grows, it periodically sheds its old cuticle (molting, or ecdysis) and secretes a new, larger one. Nematodes have a complete digestive tract. They lack a circulatory system, but nutrients are transported throughout the body via fluid in the pseudocoelom (a body cavity only partially lined by mesoderm—see FIGURE 32.6b). The muscles of nematodes are all longitudinal, and their contraction produces a thrashing motion.

Nematode reproduction is usually sexual. The sexes are separate in most species, females generally being larger than males. Fertilization is internal, and a female may deposit 100,000 or more fertilized eggs per day. The zygotes of most species are resistant cells capable of surviving harsh conditions.

Great numbers of nematodes live in moist soil and in decomposing organic matter on the bottoms of lakes and oceans. These extremely numerous free-living worms play an important role in decomposition and nutrient cycling, but little is known about most species. One species of soil nematode, *Caenorhabditis elegans*, is widely cultured and has become a model research organism in developmental biology (see Chapter 21).

Phylum Nematoda also includes many important agricultural pests that attack the roots of plants. Other species of

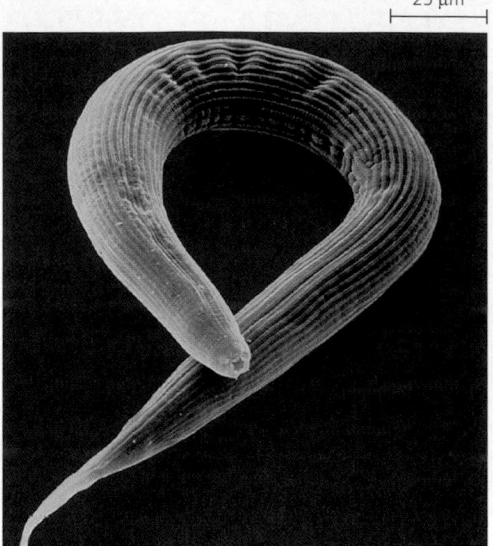

25 μm

(a) **Free-living nematode.** This free-living nematode has the classic roundworm shape: cylindrical with tapered ends. You can see the mouth at the end that is more blunt. Not visible is the anus at the other end of a complete digestive tract. What looks like a corduroy coat is the worm's ridged cuticle (colorized SEM).

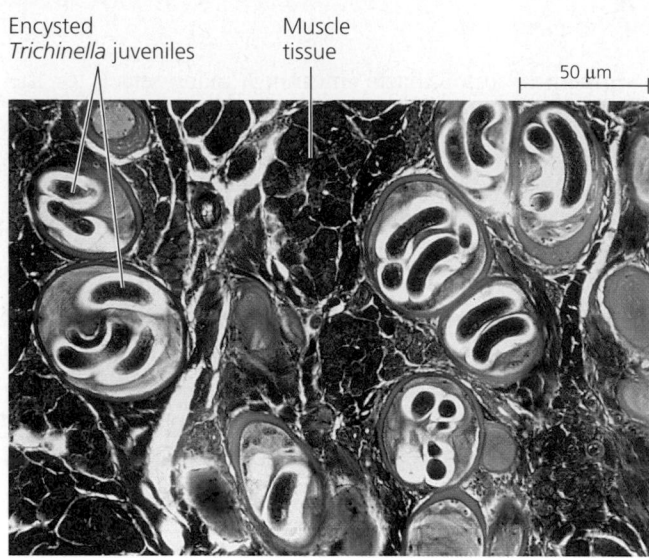

Encysted *Trichinella* juveniles   Muscle tissue

50 μm

(b) **Parasitic nematode.** These juveniles of *Trichinella spiralis* are encysted in human muscle tissue. This parasitic nematode causes trichinosis, characterized by severe nausea and sometimes death when large numbers of the juveniles penetrate heart muscle (LM).

FIGURE 33.25 Nematodes.

roundworms parasitize animals. Humans host at least 50 nematode species, including various pinworms and hookworms. One notorious nematode is *Trichinella spiralis,* the worm that causes trichinosis (FIGURE 33.25b). Humans acquire this nematode by eating undercooked infected pork or other meat with juvenile worms encysted in the muscle tissue. Within the human intestine, the juveniles develop into sexually mature adults. Females burrow into the intestinal muscles and produce more juveniles, which bore through the body or travel in lymph vessels to encyst in other organs, including skeletal muscles.

Arthropods are segmented coelomates with exoskeletons and jointed appendages

Zoologists estimate that the arthropod population of the world, including crustaceans, spiders, and insects, numbers about a billion billion (10^{18}) individuals. Nearly 1 million arthropod species have been described, mostly insects. In fact, two out of every three organisms known are arthropods, and the phylum is represented in nearly all habitats of the biosphere. On the criteria of species diversity, distribution, and sheer numbers, arthropods must be regarded as the most successful of all animal phyla.

General Characteristics of Arthropods

The diversity and success of **arthropods** are largely related to their segmentation, hard exoskeleton, and jointed appendages. (*arthropod* means "jointed feet.") Groups of body segments

and their appendages have become specialized for a great variety of functions. This evolutionary flexibility resulted not only in great diversification but also in an efficient body plan by the division of labor among regions. For example, the appendages are variously modified for walking, feeding, sensory reception, copulation, and defense. FIGURE 33.26 illustrates the diverse appendages and other arthropod characteristics of a lobster.

The body of an arthropod is completely covered by the **cuticle,** an **exoskeleton** (external skeleton) constructed from layers of protein and chitin. The cuticle can be a thick, hard armor over some parts of the body and paper-thin and flexible in other locations, such as the joints. The exoskeleton protects the animal and provides points of attachment for the muscles that move the appendages. The skeleton of arthropods is both strong and relatively impermeable to water. As we will see, both of these qualities were largely responsible for the move onto land by various arthropod groups. The rigid exoskeleton also posed some problems. For example, in order to grow, an arthropod must occasionally shed its old exoskeleton and secrete a larger one. This **molting** (ecdysis) is energetically expensive and leaves the animal temporarily vulnerable to predators and other dangers.

Arthropods tune in to their environment with well-developed sensory organs, including eyes, olfactory receptors for smell, and antennae for touch and smell. Cephalization is extensive, with most sensory organs concentrated at the anterior end of the animal.

Arthropods have **open circulatory systems** in which fluid called hemolymph is propelled by a heart through short arteries and then into spaces called sinuses surrounding the tissues and

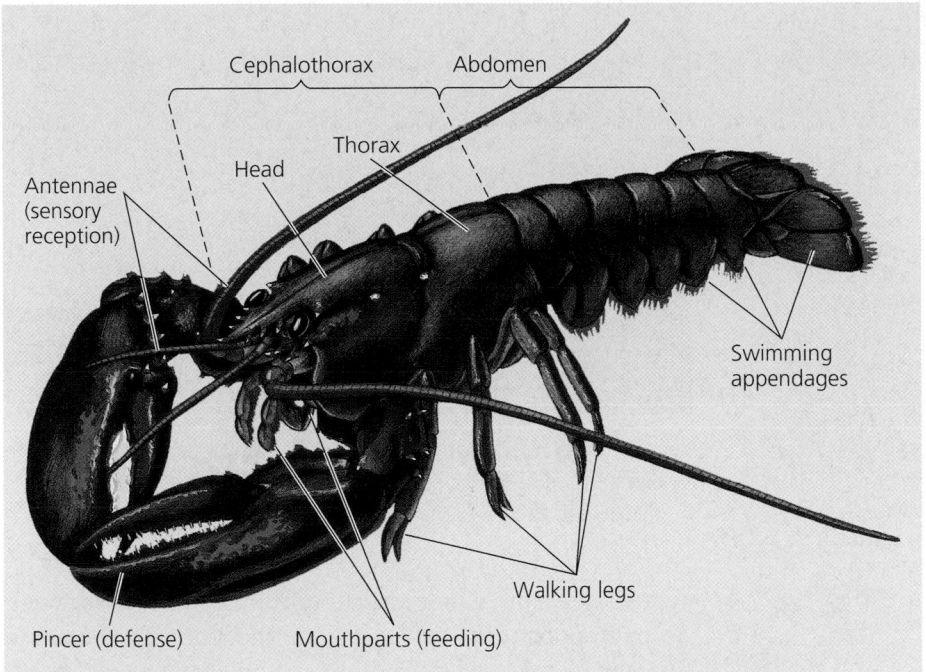

FIGURE 33.26 External anatomy of an arthropod. Many of the distinctive features of arthropods are apparent in this dorsal view of a lobster, along with some uniquely crustacean characteristics. The whole body, including appendages, is covered by an exoskeleton. Distinct regions of the body are the head and thorax (together forming the cephalothorax) and the abdomen. The body is segmented, but this characteristic is obvious only in the abdomen. All body regions bear jointed appendages (including pincers, mouthparts, walking legs, and swimming appendages). The two pairs of sensory antennae are also jointed appendages. The head bears a pair of eyes, each situated on a movable stalk.

organs. (The term *blood* is best reserved for fluid in a closed circulatory system.) Hemolymph reenters the arthropod heart through pores that are usually equipped with valves. The body sinuses are collectively called the hemocoel, which is not part of the coelom. In most arthropods, the coelom that forms in the embryo becomes much reduced as development progresses, and the hemocoel becomes the main body cavity in adults. Although this condition resembles the open circulatory system of mollusks, the two systems probably arose independently.

A variety of organs specialized for gas exchange have evolved in arthropods. These organs must allow the diffusion of respiratory gases in spite of the exoskeleton. Most aquatic species have gills with thin feathery extensions that place an extensive surface area in contact with the surrounding water. Terrestrial arthropods generally have internal surfaces specialized for gas exchange. Most insects, for instance, have tracheal systems, branched air ducts leading into the interior from pores in the cuticle.

Arthropod Phylogeny and Classification

Molecular systematics generally supports evidence from fossils and comparative anatomy that the arthropods diverged early in their history into four main evolutionary lineages: **trilobites** (all extinct); **chelicerates** (horseshoe crabs, scorpions, ticks, spiders,

and an extinct group called the eurypterids); **uniramians** (centipedes, millipedes, and insects); and **crustaceans** (crabs, lobsters, shrimps, barnacles, and many others).

Crustaceans are primarily aquatic and are believed to have evolved in the ocean. Insects, millipedes, centipedes, and most extant chelicerates diversified on land.

Chelicerates (from the Greek *cheilos*, lips, and *cheir*, arm) are named for clawlike feeding appendages called **chelicerae.** In contrast, uniramians and crustaceans have jawlike **mandibles.** They are also distinguished from chelicerates in having one or two pairs of sensory **antennae** and usually a pair of **compound eyes** (multifaceted eyes with many separate focusing elements). Uniramians have one pair of antennae and uniramous (unbranched) appendages; crustaceans have two pairs of antennae and typically biramous (branched) appendages (note the branched walking legs of the lobster in FIGURE 33.26). Chelicerates lack antennae, and most have simple eyes (eyes with a single lens).

The move onto land by arthropods was made possible in part by the exoskeleton. When the exoskeleton first evolved in the seas, its main functions were probably protection and anchorage for muscles, but it eventually helped certain arthropods live on land by solving the problems of water loss and structural support. The arthropod cuticle is relatively impermeable to water, helping prevent desiccation. The firm exoskeleton also solved the problem of support when arthropods left the buoyancy of water. Chelicerates, insects, millipedes, and centipedes diversified on land during the late Silurian and early Devonian periods, following the colonization by plants. The oldest fossil signs of terrestrial animals are tracks of extinct chelicerates (apparently eurypterids that spent at least some of their time on land) about 450 million years old. Fossilized arachnids almost as old have also been found.

Taxonomists traditionally grouped all the arthropods in one phylum, the phylum **Arthropoda.** TABLE 33.5, page 664, lists a few of the major arthropod classes in the context of this one-phylum tradition. However, many zoologists now prefer to split the arthropods into multiple phyla corresponding to the four great lineages: **phylum Trilobita, phylum Chelicerata, phylum Uniramia,** and **phylum Crustacea.** This trend toward splitting former taxa into a larger number of new ones, based mainly on cladistic analysis, extends to lower taxa. For example, many systematists now split what was the uniramian class Insecta into several classes. And there are other modern taxonomic issues. For instance, some systematists now question

Table 33.5 Some Major Arthropod Classes (based on a traditional taxonomy that places all arthropods in a single phylum, Arthropoda)

Class and Examples	Main Characteristics
Arachnida (spiders, scorpions, ticks, mites)	Body having one or two main parts; six pairs of appendages (chelicerae, pedipalps, and four pairs of walking legs); mostly terrestrial
Diplopoda (millipedes)	Body with distinct head bearing antennae and chewing mouthparts, segmented body with two pairs of walking legs per segment; terrestrial; herbivorous
Chilopoda (centipedes)	Body with distinct head bearing large antennae and three pairs of mouthparts; appendages of first body segment modified as poison claws; trunk segments bear one pair of walking legs each; terrestrial; carnivorous
Insecta (insects)	Body divided into head, thorax, and abdomen; antennae present; mouthparts modified for chewing, sucking, or lapping; usually with two pairs of wings and three pairs of legs; mostly terrestrial
Crustacea (crabs, lobsters, crayfish, shrimp)	Body of two or three parts; antennae present; chewing mouthparts; three or more pairs of legs; mostly marine

FIGURE 33.27 A trilobite fossil. Trilobites were prevalent arthropods throughout the Paleozoic era. Paleontologists have described about 4,000 trilobite species.

appendages showed little variation from segment to segment. As arthropods continued to evolve, the segments tended to fuse and become fewer in number, and the appendages became specialized for a variety of functions. (Compare the trilobite in FIGURE 33.27 with the lobster in FIGURE 33.26.)

Spiders and Other Chelicerates

The trilobites were outlasted by the **eurypterids,** or water scorpions. These mainly marine and freshwater predators, up to 3 m long, were chelicerates. A chelicerate has an anterior cephalothorax and a posterior abdomen. The appendages are more specialized than those of trilobites, and the most anterior appendages are modified as chelicerae (either pincers or fangs). Most of the marine chelicerates, including all of the eurypterids, are extinct; only four marine species survive today, one being the horseshoe crab (FIGURE 33.28).

whether Uniramia is a monophyletic clade. These scientists interpret the molecular and anatomical data used to construct cladograms to mean that insects are actually more closely related to crustaceans than they are to centipedes and millipedes.

As we have seen throughout our survey of the diversity of life, systematics, and the taxonomy it informs, is currently one of the most vibrant fields in biology. We *do* want you to be aware that phylogenetic trees and classifications represent hypotheses about the history of life that are presently being reconstructed in light of molecular data and other new approaches. However we *don't* want the current taxonomic turmoil to detract from your appreciation of life's diverse forms. The following survey of arthropods showcases a few major groups, which can be variously classified as phyla, subphyla, or classes, depending on the taxonomic scheme.

Trilobites

Among the early arthropods were the **trilobites** (FIGURE 33.27). They were common denizens of the shallow seas throughout the Paleozoic era but disappeared with the great Permian extinctions that ended that era, about 250 million years ago. Trilobites had pronounced segmentation, but their

FIGURE 33.28 Horseshoe crabs (Limulus polyphemus). This "living fossil," which has changed little in hundreds of millions of years, has survived from a rich diversity of chelicerates that once filled the seas. The horseshoe crab is common on the Atlantic and Gulf coasts of the United States.

(a) Scorpions, which hunt by night, were among the first terrestrial carnivores, preying on other arthropods that fed on the early land plants. The pedipalps of scorpions are pincers specialized for defense and the capture of food. The tip of the tail bears a poisonous stinger.

100 µm

(b) This magnified house-dust mite is a ubiquitous scavenger in human dwellings (colorized SEM). Unlike some mites that carry disease-causing bacteria, dust mites are harmless except to people who are allergic to them.

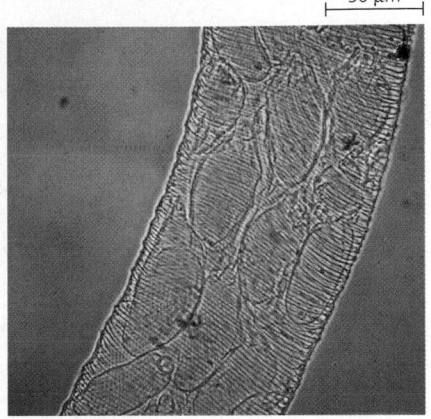

50 µm

(c) Many arthropods parasitize other arthropods. In this light micrograph, the trachea (air tube) of a honeybee (an insect) is filled with parasitic mites.

FIGURE 33.29 Arachnids.

The bulk of modern chelicerates are found on land in the form of **class Arachnida,** which includes scorpions, spiders, ticks, and mites (FIGURE 33.29). Ticks and many mites are among a large group of parasitic arthropods. Nearly all ticks are blood-sucking parasites on the body surfaces of reptiles, birds, or mammals. Parasitic mites live on or in a wide variety of vertebrates and invertebrates, including other arthropods (FIGURE 33.29c).

Arachnids have a cephalothorax with six pairs of appendages: the chelicerae, a pair of appendages called pedipalps that usually function in sensing or feeding, and four pairs of walking legs (FIGURE 33.30). Spiders use their fanglike chelicerae, equipped with poison glands, to attack prey. As the chelicerae masticate (chew) the prey, the spider spills digestive juices onto the torn tissues. The food softens, and the spider sucks up the liquid meal.

In most spiders, gas exchange is carried out by **book lungs,** stacked plates contained in an internal chamber (see FIGURE 33.30b). The extensive surface area of these respiratory organs is a structural adaptation that enhances the exchange of O_2 and CO_2 between the hemolymph and air.

A unique adaptation of many spiders is the ability to catch flying insects by stringing webs of silk, a protein produced as a liquid by special abdominal glands. The silk is spun by organs called spinnerets into fibers that solidify. Each spider engineers a style of web characteristic of its species and constructs the

(a) Spiders are generally most active during the daytime, when they hunt for prey or trap insects in webs.

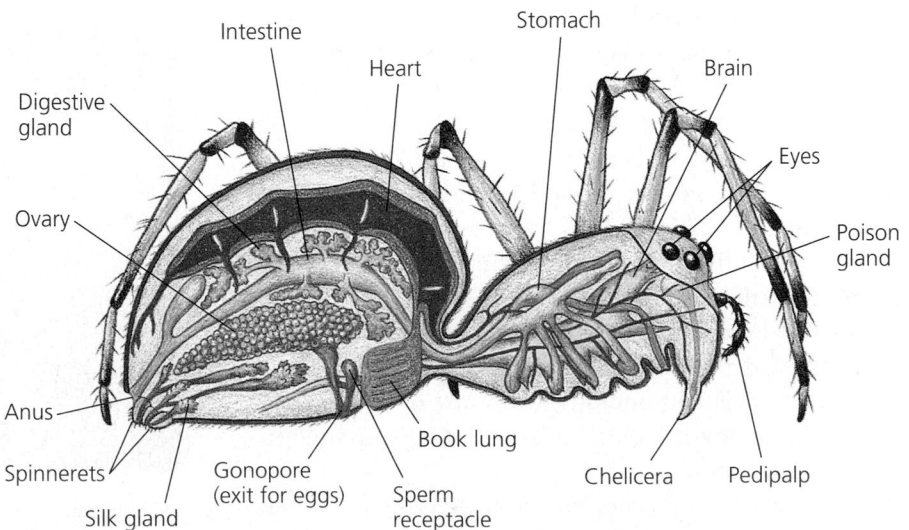

(b) Anatomy of a spider

FIGURE 33.30 Spiders (class Arachnida).

FIGURE 33.31 Class Diplopoda (millipedes) and class Chilopoda (centipedes).

(a) Millipedes feed on decaying plant matter.

(b) The house centipede (*Scutigera coleoptrata*), a fast-moving carnivore, feeds on insects, including cockroaches, and other small invertebrates.

web perfectly on the first try. This complex behavior is apparently inherited. Besides building their webs from silk, various spiders use these fibers in other ways: as droplines for rapid escape, as cloth that covers eggs, and even as "gift wrapping" for food that certain male spiders offer females during courtship.

Millipedes and Centipedes

Millipedes (**class Diplopoda**) are wormlike, with a large number of walking legs (two pairs per segment), though fewer than the thousand their name implies (FIGURE 33.31a). They eat decaying leaves and other plant matter. Millipedes may have been among the earliest animals on land, living on mosses and primitive vascular plants.

Centipedes (**class Chilopoda**) are terrestrial carnivores. The head has a pair of antennae and three pairs of appendages modified as mouthparts, including the jawlike mandibles. Each segment of the trunk region has one pair of walking legs (FIGURE 33.31b). Centipedes have poison claws on the anterior-most trunk segment that paralyze prey and aid in defense.

Insects

In species diversity, insects (**class Insecta**) outnumber all other forms of life combined. They live in almost every terrestrial habitat and in fresh water, and flying insects fill the air. Insects are rare, though not absent, in the seas, where crustaceans are the dominant arthropods. Class Insecta is divided into about 26 orders, some of which are described in TABLE 33.6, pages 668–669. **Entomology,** the study of insects, is a vast field with many subspecialties, including physiology, ecology, and taxonomy.

The oldest insect fossils date back to the Devonian period, which began about 400 million years ago. However, when flight evolved during the Carboniferous and Permian periods, it spurred an explosion in insect variety. A fossil record of diverse insect mouthparts indicates that specialized feeding

on gymnosperms and other Carboniferous plants also contributed to the adaptive radiation of insects. A widely held hypothesis is that the greatest diversification of insects paralleled the evolutionary radiation of flowering plants during the Cretaceous and early Tertiary periods about 60 to 65 million years ago. This view is challenged by new research suggesting that insects diversified extensively before the angiosperm radiation. Thus, during the coevolution of flowering plants and the herbivorous insects that pollinated them, insect diversity may have been more a cause of angiosperm radiation than an effect.

Flight is obviously one key to the great success of insects. An animal that can fly can escape many predators, find food and mates, and disperse to new habitats much faster than an animal that must crawl about on the ground. Many insects have one or two pairs of wings that emerge from the dorsal side of the thorax (FIGURE 33.32). Because the wings are exten-

FIGURE 33.32 Insect flight. Insect wings, such as this dragonfly's, are not modified appendages but extensions of the cuticle. Some insects beat their wings at speeds of several hundred cycles per second by using muscles to warp the shape of the entire cuticle covering the thorax. As the wings flap, they change angles, producing lift on both the up and down strokes.

sions of the cuticle and not true appendages, insects can fly without sacrificing any walking legs. By contrast, the flying vertebrates—birds and bats—have one of their two pairs of walking legs modified for wings and are generally quite clumsy on the ground.

Insect wings may have first evolved as extensions of the cuticle that helped the insect body absorb heat, only later becoming organs for flight. Other views suggest that wings allowed the animals to glide from vegetation to the ground, or even served as gills in aquatic insects. Still another hypothesis is that insect wings functioned for swimming before they functioned for flight.

Dragonflies, with two similar pairs of wings, were among the first insects to fly (see FIGURE 33.32). Several insect orders that evolved later than dragonflies have modified flight equipment. The wings of bees and wasps, for instance, are hooked together and move as a single pair. Butterfly wings operate in a similar fashion because the anterior pair overlaps the posterior wings. In beetles, the posterior wings function in flight, while the anterior ones are modified as covers that protect the flight wings when the beetle is on the ground or when the beetle is burrowing.

The internal anatomy of an insect includes several complex organ systems (FIGURE 33.33). The complete digestive system is regionally specialized, with discrete organs functioning in the breakdown of food and the absorption of nutrients. Like other arthropods, an insect has an open circulatory system, with a heart pumping hemolymph through the sinuses of the hemocoel. Metabolic wastes are removed from the hemolymph by unique excretory organs called **Malpighian** tubules, which are outpocketings of the digestive tract. Gas exchange in insects is accomplished by a **tracheal system** of branched, chitin-lined tubes that infiltrate the body and carry oxygen directly to cells. The tracheal system opens to the outside of the body through spiracles, pores that can open or close to regulate air flow and limit water loss.

The insect nervous system consists of a pair of ventral nerve cords with several segmental ganglia. The two cords meet in the head, where the ganglia of several anterior segments are fused into a cerebral ganglion (brain) close to the antennae, eyes, and other sense organs concentrated on the head.

Many insects undergo metamorphosis in their development. In the **incomplete metamorphosis** of grasshoppers and some other orders, the young resemble adults but are smaller

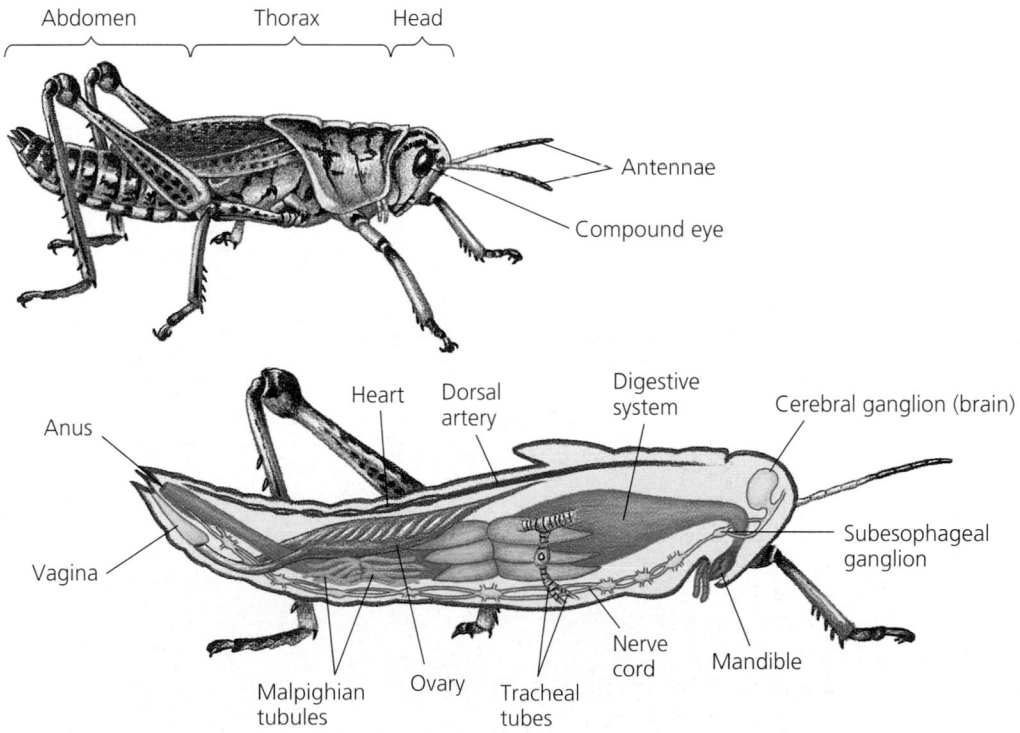

FIGURE 33.33 Anatomy of a grasshopper, an insect. The insect body has three regions: a head, a thorax, and an abdomen (top drawing). Segmentation is apparent along the thorax and abdomen, but the head segments are fused. On the insect head are one pair of antennae and a pair of compound eyes. Several pairs of appendages modified for chewing (as in grasshoppers) or for lapping, piercing, or sucking (in certain other insects) form the mouthparts. The thorax of the insect bears three pairs of walking legs. The bottom drawing identifies major internal organs.

Table 33.6 Some Major Orders of Insects

Order	Approximate Number of Species	Main Characteristics	Examples	
Anoplura	2,400	Wingless ectoparasites; sucking mouthparts; small with flattened body, reduced eyes; legs with clawlike tarsi for clinging to skin; incomplete metamorphosis; very host-specific	Sucking lice	Human body louse
Coleoptera	500,000	Two pairs of wings (one pair thick and leathery, one pair membranous), armored exoskeleton; biting and chewing mouthparts; complete metamorphosis	Beetles	Japanese beetle
Dermaptera	1,000	Two pairs of wings (one pair leathery, one pair membranous) or wingless; biting mouthparts; large posterior pincers; incomplete metamorphosis	Earwigs	Earwig
Diptera	120,000	One pair of wings and halteres (balancing organs); sucking, piercing, or lapping mouthparts; complete metamorphosis	Flies, mosquitoes	Horsefly
Hemiptera	55,000	Two pairs of wings (one pair partly leathery, one pair membranous); piercing or sucking mouthparts; incomplete metamorphosis	True bugs; assassin bug, bedbug, chinch bug	Leaf-footed bug
Hymenoptera	100,000	Two pairs of membranous wings; head mobile; chewing or sucking mouthparts; posterior stinging organ on females; complete metamorphosis; many species social	Ants, bees, wasps	Cicada-killer wasp

Table 33.6 Some Major Orders of Insects

Order	Approximate Number of Species	Main Characteristics	Examples	
Isoptera	2,000	Two pairs of membranous wings (some stages wingless); chewing mouthparts; highly social; incomplete metamorphosis	Termites	Termite
Lepidoptera	140,000	Two pairs of wings covered with tiny scales; long coiled tongue for sucking; complete metamorphosis	Butterflies, moths	Swallowtail butterfly
Odonata	5,000	Two pairs of membranous wings; biting mouthparts; incomplete metamorphosis	Damselflies, dragonflies	Dragonfly
Orthoptera	30,000	Two pairs of wings (one pair leathery, one pair membranous); biting and chewing mouthparts; incomplete metamorphosis	Crickets, roaches, grasshoppers, mantids	Katydid
Siphonaptera	2,000	Wingless, laterally compressed; adults are bloodsuckers on birds and mammals; piercing and sucking mouthparts; jumping legs; complete metamorphosis	Fleas	Flea
Trichoptera	7,000	Two pairs of hairy wings; chewing or lapping mouthparts; complete metamorphosis; aquatic larvae build silken nets or cases (of sand, gravel, and wood) bound together by silk	Caddisflies	Caddisfly

and have different body proportions. The animal goes through a series of molts, each time looking more like an adult, until it reaches full size. Insects with **complete metamorphosis** have larval stages specialized for eating and growing that are known by such names as maggot, grub, or caterpillar. The larval stage looks entirely different from the adult stage, which is specialized for dispersal and reproduction. Metamorphosis from the larval stage to the adult occurs during a pupal stage (FIGURE 33.34, p. 670).

Reproduction in insects is usually sexual, with separate male and female individuals. Adults come together and recognize each other as members of the same species by advertising with bright colors (butterflies), sound (crickets), or odors (moths). Fertilization is generally internal. In most species, sperm cells are deposited directly into the female's vagina at the time of copulation, though in some species the male deposits a sperm packet outside the female, and the female picks it up. An internal structure in the female called the spermatheca stores the sperm, usually enough to fertilize more than one batch of eggs. Many insects mate only once in a lifetime. After mating, a female lays her eggs on an appropriate food source where the next generation can begin eating as soon as it hatches.

(a) Larva (caterpillar)

(b) Pupa

(c) Pupa

(d) Emerging adult

(e) Adult

FIGURE 33.34 Metamorphosis of a butterfly. (a) The larva (caterpillar) spends its time eating and growing, molting as it grows. **(b)** After several molts, the larva encases itself in a cocoon and becomes a pupa. **(c)** Within the pupa, the larval tissues are broken down, and the adult is built by the division and differentiation of cells that were quiescent in the larva. **(d)** Finally, the adult begins to emerge from the cocoon. **(e)** Fluid is pumped into veins of the wings and then withdrawn, leaving the hardened veins as struts supporting the wings. Now the insect flies off and reproduces, deriving much of its nourishment from the calories stored by the feeding larva.

Animals so numerous, diverse, and widespread as insects are bound to affect the lives of all other terrestrial organisms, including humans. On one hand, we depend on bees, flies, and many other insects to pollinate our crops and orchards. On the other hand, insects are carriers for many diseases, including malaria (spread by mosquitos that carry *Plasmodium*; see FIGURE 28.13) and African sleeping sickness (spread by tsetse flies that carry *Trypanosoma*; see FIGURE 28.11). Furthermore, insects compete with humans for food. In parts of Africa, for instance, insects claim about 75% of the crops. Trying to minimize their losses, farmers in the United States spend billions of dollars each year on pesticides, spraying crops with massive doses of some of the deadliest poisons ever invented. Try as they may, not even humans have challenged the preeminence of insects and their arthropod kin. Or, as Cornell University's Thomas Eisner, interviewed on page 24, puts it: "Bugs are not going to inherit the Earth. They own it now. So we might as well make peace with the landlord."

Crustaceans

While arachnids and insects thrived on land, crustaceans, for the most part, remained in marine and freshwater environments, where they are now represented by about 40,000 species. Crabs, lobsters, crayfish, and shrimp are the most familiar crustaceans (FIGURE 33.35).

The multiple appendages of crustaceans are extensively specialized. Lobsters and crayfish, for instance, have a toolkit of 19 pairs of appendages (see FIGURE 33.26). Crustaceans are the only arthropods with two pairs of antennae. Three or more pairs of appendages are modified as mouthparts, including the hard mandibles. Walking legs are present on the thorax, and, unlike insects, crustaceans have appendages on the abdomen. A lost appendage can be regenerated.

Small crustaceans exchange gases across thin areas of the cuticle, but larger species have gills. The circulatory system is open, with a heart pumping hemolymph through arteries into sinuses that bathe the organs. Crustaceans excrete nitrogenous wastes by diffusion through thin areas of the cuticle, but a pair of glands regulates the salt balance of the hemolymph.

Sexes are separate in most crustaceans. In the case of lobsters and crayfish, the male uses a specialized pair of appendages to transfer sperm to the reproductive pore of the female during copulation. Most aquatic crustaceans go through one or more swimming larval stages.

One of the largest groups of crustaceans (about 10,000 species), the **isopods** are mostly small marine species. Some are numerous at the bottom of deep oceans. Isopods also include the land-dwelling pill bugs, or wood lice, common on the undersides of moist logs and leaves.

Another group of small crustaceans, the **copepods,** are among the most numerous of all animals. They are important

(a)

(b)

(c)

FIGURE 33.35 Crustaceans. (a) This is a Sally lightfoot crab from the Galápagos Islands. **(b)** Planktonic crustaceans known as krill are consumed by the ton by whales and other large suspension-feeders. **(c)** Barnacles are sessile crustaceans with a shell (exoskeleton) hardened by calcium carbonate. The jointed appendages projecting from the shell capture small plankton and organic particles suspended in the water.

members of marine and freshwater plankton communities, eating protists and bacteria, and being eaten by many fishes.

Lobsters, crayfish, crabs, and shrimp are all relatively large crustaceans called **decapods** (FIGURE 33.35a). The exoskeleton, or cuticle, is hardened by calcium carbonate; the portion that covers the dorsal side of the cephalothorax forms a shield called the carapace. Most decapods are marine. Crayfish, however, live in fresh water, and some tropical crabs live on land.

The larvae of many larger-bodied crustaceans are also planktonic. A marine group known as krill are shrimplike planktonic organisms that grow to about 3 cm long (FIGURE 33.35b). A major food source for many species of whales, krill are now being harvested in great numbers by humans for food and agricultural fertilizer.

Barnacles are sessile crustaceans with parts of their cuticles hardened into shells by calcium carbonate. They feed by using their appendages to strain food from the water (FIGURE 33.35c).

How Many Times Did Segmentation Evolve in the Animal Kingdom?

Before we leave the arthropods, let's consider the evolution of one of their key anatomical characteristics, the segmentation of the body. (You can see many examples of this segmentation if you reexamine some of the arthropods in the preceding illustrations—for example, note the segmentation of the scorpion in FIGURE 33.29a and of the millipede in FIGURE 33.31a.) Until recently, based partly on this segmented body plan, the majority of biologists favored the hypothesis that arthropods evolved from an annelid ancestor, or from a segmented ancestor common to both annelids and arthropods. In placing annelids and arthropods on divergent clades of Protostomia—Lophotrochozoa for annelids, Ecdysozoa for arthropods—molecular systematists have challenged the long-held view of a close kinship between the segmented worms and the arthropods. However, many biologists do not find the existing molecular data convincing enough to reject the hypothesis of a closer annelid-arthropod relationship. The debate has focused interest on the evolutionary origin of segmentation.

The segmented bodies of arthropods, annelids, and certain other animals represent a special case of a more general phenomenon: the blocking-out of an embryo into regions where certain body parts will develop. Each bilaterally symmetrical animal has a particular linear arrangement of anatomical features along its anterior (head) to posterior axis. Eyes, for example, are located at the anterior end. Differential expression of various regulatory genes that code for transcription factors plays a key role in this blocking-out of anterior $\longrightarrow$ posterior anatomy in the developing embryo. In those phyla with segmented bodies, certain genes first determine the segmentation, and then genes of the *Hox* complex determine what organs will develop at each segment (see FIGURE 21.15). For example, differential expression of various *Hox* genes along the length of a lobster embryo cause antennae to develop on a certain segment and walking legs to develop on other segments. But even in nonsegmented animals, such as flatworms, *Hox* genes determine where certain organs, such as eyes, develop along the animal's length. In fact, sponges have at least one *Hox* gene and cnidarians such as jellies have several. In a cnidarian, for example, expression of a *Hox* gene determines where tentacles will develop in the embryo.

Thus, as developmental determinants of morphology, *Hox* genes predate the origin of the bilaterians. The mechanism for the development of segmented bodies in certain animal phyla is a variation on a basic regulatory scheme that dates back to the first animals. But an increase in *Hox* gene number through gene duplications and mutations, along with adaptation of *Hox* gene function for development of segmented bodies, made it possible for a great diversity of complex animals to evolve.

Body segmentation evolved in several of the 35 animal phyla, including the annelids, arthropods, and chordates—the phylum that includes humans and other vertebrates. (The chain of vertebrae that make up your backbone is an example of this chordate segmentation.) This means that segmented animals occur in all *three* major clades of bilaterians: annelids are lophotrochozoans; arthropods are ecdysozoans; and chordates are deuterostomes. Each clade also includes phyla of nonsegmented animals.

Three different hypotheses can account for the scattered distribution of segmentation in the phylogenetic tree of animals. In the first hypothesis, segmentation had separate evolutionary origins in each bilaterian clade (FIGURE 33.36a). In the second hypothesis, there were two separate origins of segmentation, one for the protostomes (lophotrochozoans + ecdysozoans) and one for the deuterostomes (FIGURE 33.36b). And in the third hypothesis, segmentation evolved just once, in an ancestor common to all three bilaterian lineages (FIGURE 33.36c). Note that hypotheses 2 and 3 require evolutionary loss of segmentation in the majority of animal phyla. The principle of parsimony in cladistics would seem to favor hypothesis 1 because it requires the fewest evolutionary changes (see Chapter 25 to review cladistics and parsimony). However, application of parsimony is merely an analytical aid in cladistics, not some law of evolution that life always follows. Thus, we cannot rule out hypotheses 2 and 3 as plausible explanations for the scattered distribution of segmentation among animal phyla.

"Evo-devo," the research field at the interface of evolutionary biology and developmental biology, may illuminate the origin(s) of segmentation. Several lab groups are investigating how various *Hox* genes function during the development of segmented bodies in diverse phyla. By comparing and contrasting details in how these regulatory genes block-out the segments of annelid, arthropod, and chordate embryos, this research may bring us closer to knowing whether segmentation evolved once, twice, or three times in the animal kingdom. And these studies of how segmentation develops will help systematists test various hypotheses about animal phylogeny.

DEUTEROSTOMIA

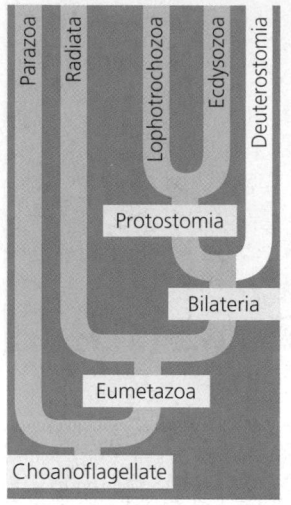

At first glance, sea stars and other echinoderms may seem to have little in common with the phylum Chordata, which includes the vertebrates: fishes, amphibians, reptiles, birds, and mammals. These animals, however, share features characteristic of deuterostomes: radial cleavage, development of the coelom from the archenteron, and formation of the mouth at the end of the embryo opposite the blastopore (see FIGURE 32.7). Molecular systematics has reinforced Deuterostomia as a clade of bilaterian animals.

Phylum Echinodermata: Echinoderms have a water vascular system and secondary radial anatomy

Sea stars and most other **echinoderms** (from the Greek *echin*, spiny, and *derma*, skin) are sessile or slow-moving animals. The internal and external parts of the animal radiate from the center, often as five spokes. A thin skin covers an endoskeleton of hard calcareous plates. Most echinoderms are prickly from skeletal bumps and spines that have various functions. Unique to echinoderms is the **water vascular system,** a network of hydraulic canals branching into extensions called **tube feet** that function in locomotion, feeding, and gas exchange.

Sexual reproduction of echinoderms usually involves separate male and female individuals that release their gametes

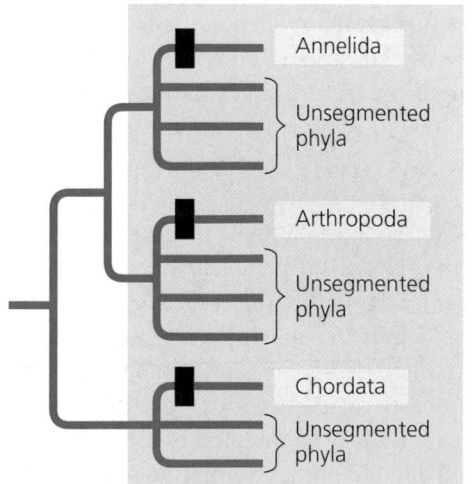

Hypothesis 1: Three origins of segmentation

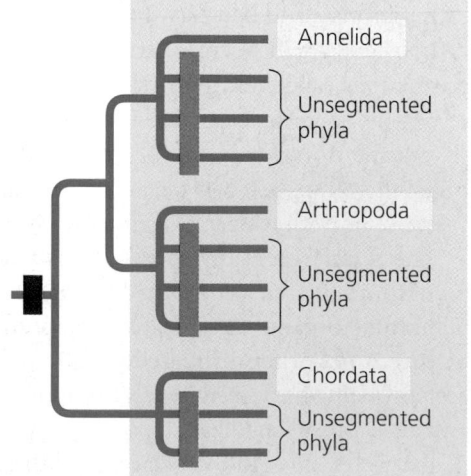

Hypothesis 2: Two origins of segmentation

Hypothesis 3: One origin of segmentation

FIGURE 33.36 Three hypotheses for the origin of segmentation. The purple bars represent origins of segmentation, while the orange bars represent losses of segmentation.

into the seawater. The radial adults develop by metamorphosis from bilateral larvae.

Given our division of eumetazoans into Radiata and Bilateria, we should explain the radial appearance of most echinoderms. The echinoderms are definitely bilaterians, not radiate animals related to cnidarians. Echinoderm larvae have bilateral symmetry. The radial anatomy of adult echinoderms is a secondary adaptation to a sessile lifestyle. And even echinoderm adults are not truly radial in their anatomy.

For example, the opening (madreporite) of a sea star's water vascular system is not central, but to one side of the "star" (see FIGURE 33.38).

The 7,000 or so echinoderms, all marine, are divided into six classes (FIGURE 33.37): Asteroidea (sea stars), Ophiuroidea (brittle stars), Echinoidea (sea urchins and sand dollars), Crinoidea (sea lilies and feather stars), Holothuroidea (sea cucumbers), and Concentricycloidea (sea daisies). The sea daisies, discovered only recently, live on waterlogged wood in the deep sea.

FIGURE 33.37 Echinoderms

(a) A sea star (class Asteroidea) on coral

(b) A sea star eating a clam

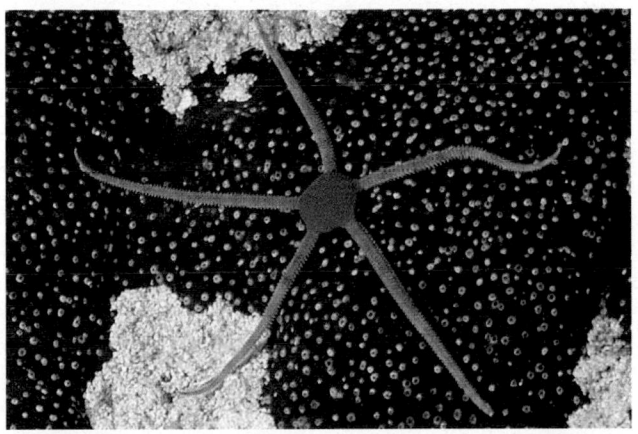

(c) A brittle star (class Ophiuroidea)

(d) A sea urchin (class Echinoidea)

(e) A sea lily (class Crinoidea)

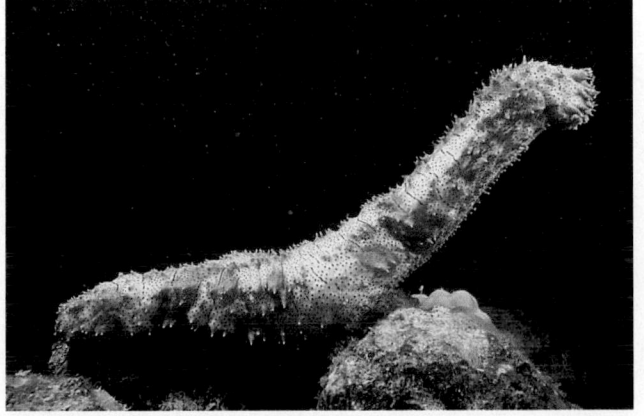

(f) A sea cucumber (class Holothuroidea)

Class Asteroidea

Sea stars have five arms (sometimes more) radiating from a central disk (FIGURE 33.37a). The undersurfaces of the arms bear tube feet, each of which can act like a suction disk. By a complex set of hydraulic and muscle actions, the suction can be created or released (FIGURE 33.38). The sea star coordinates its tube feet to adhere firmly to rocks or to creep along slowly as the tube feet extend, grip, contract, release, extend, and grip again. Sea stars also use their tube feet to grasp prey, such as clams and oysters (FIGURE 33.37b). The arms of the sea star embrace the closed bivalve, hanging on tightly by the tube feet. The sea star then turns its stomach inside-out, everting it through its mouth and into the narrow opening between the shells of the bivalve. The digestive system of the sea star secretes juices that begin digesting the soft body of the mollusk within its own shell.

Sea stars and some other echinoderms are capable of regeneration. Sea stars can regrow lost arms, and members of one genus can even regrow an entire body from a single arm.

Class Ophiuroidea

Brittle stars have distinct central disks, and the arms are long and flexible (FIGURE 33.37c). Their tube feet lack suckers, and they move by serpentine lashing of the arms. Some species are suspension-feeders; others are predators or scavengers.

Class Echinoidea

Sea urchins and sand dollars have no arms, but they do have five rows of tube feet that function in slow movement (FIGURE 33.37d). Sea urchins also have muscles that pivot their long spines, which aids in moving. The mouth of an urchin is ringed by complex jawlike structures adapted for eating sea-weeds and other food. Sea urchins are roughly spherical in shape, whereas sand dollars are flattened and disk-shaped.

Class Crinoidea

Sea lilies live attached to the substratum by stalks; feather stars crawl about by using their long, flexible arms. As a group, crinoids use their arms in suspension-feeding (FIGURE 33.37e). The arms circle the mouth, which is directed upward, away from the substratum. Crinoidea is an ancient class whose evolution has been very conservative; fossilized sea lilies some 500 million years old could pass for modern members of the class.

Class Holothuroidea

On casual inspection, sea cucumbers do not look much like other echinoderms (FIGURE 33.37f). They lack spines, and the hard endoskeleton is much reduced. Sea cucumbers are elongated in the oral-aboral axis, giving them the shape for which they are named and further disguising their relationship to sea stars and sea urchins. Closer examination, however, reveals five rows of tube feet, part of the water vascular system found only in echinoderms. Some of the tube feet around the mouth are developed as feeding tentacles.

Phylum Chordata: The chordates include two invertebrate subphyla and all vertebrates

This phylum, to which we belong, consists of two subphyla of invertebrate animals plus the subphylum Vertebrata, the animals with backbones. Grouping the chordates with echinoderms as deuterostomes on the basis of similarities in early embryonic development is not meant to imply that one phylum evolved from the other. Chordates and echinoderms have

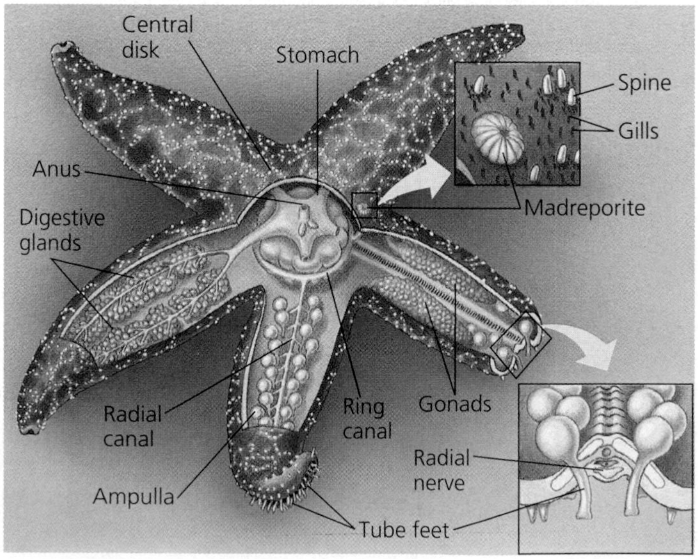

FIGURE 33.38 Anatomy of a sea star. The surface of a sea star is covered by spines that help defend against predators and by small gills for gas exchange. Internal organs are suspended by mesenteries in a well-developed coelom. A short digestive tract runs from the mouth on the bottom of the central disk to the anus on the top of the disk. Digestive glands secrete digestive juices and aid in the absorption and storage of nutrients. The central disk has a nerve ring and nerve cords radiating from the ring into the arms. The water vascular system consists of a ring canal in the central disk and five radial canals, each running the length of an arm in a groove. The system connects to the outside by way of the sievelike madreporite. Branching from each radial canal are hundreds of hollow, muscular tube feet filled with fluid continuous with the rest of the water vascular system. Each tube foot consists of a bulblike ampulla and suckered podium (foot portion). The podium expands and extends to contact the substratum when the ampulla squeezes water into it. The podium shortens and bends when muscles in its wall contract, forcing water back into the ampulla.

existed as distinct phyla for at least half a billion years; if the developmental similarities stem from shared ancestry, then the evolutionary paths of the two phyla must have diverged very early. We will trace the phylogeny of chordates in Chapter 34, focusing on the history of vertebrates.

TABLE 33.7 summarizes the animal phyla we have discussed in this chapter.

Table 33.7 Animal Phyla

Category	Phyla		Description of Phyla
Kingdom Animalia Parazoa	Porifera (sponges)		Choanocytes (collar cells—unique flagellated cells that ingest bacteria and tiny food particles); cells tend to be totipotent (retain zygote's potential to form the whole animal)
Eumetazoa Radiata	Cnidaria (hydras, jellies, sea anemones, corals)		Unique stinging structures (cnidae), each housed in a specialized cell (cnidocyte); gastrovascular cavity (incomplete digestive tract with a mouth but no anus)
	Ctenophora (comb jellies)		Colloblasts (adhesive structures) for prey capture; eight rows of comblike ciliary plates; gastrovascular cavity
Bilateria Protostomia: Lophotrochozoa	Platyhelminthes (flatworms)		Dorsoventrally flattened, unsegmented acoelomates; gastrovascular cavity or no digestive tract
	Rotifera (rotifers)		Pseudocoelomates with complete digestive tracts; jaws in pharynx structures (trophi); head with a ciliated crown (corona); no circulatory system
	Lophophorates: Bryozoa, Brachiopoda, Phoronida		Coelomates with lophophore (feeding structure bearing ciliated tentacles)
	Nemertea (proboscis worms)		Unique anterior proboscis surrounded by fluid-filled cavity (rhynchocoel); complete digestive tract (mouth and anus); circulatory system with closed vessels
	Mollusca (clams, snails, squids)		Coelomates with three main body parts (muscular foot, visceral mass, mantle); coelom reduced; main body cavity is a hemocoel
	Annelida (segmented worms)		Coelomates with body wall and internal organs (except digestive tract) segmented
Protostomia: Ecdysozoa	Nematoda (roundworms)		Cylindrical, unsegmented pseudocoelomates with tapered ends; no circulatory system
	Arthropoda (crustaceans, insects, spiders)		Coelomates with segmented body, jointed appendages, exoskeleton from ectoderm
Deuterostomia	Echinodermata (sea stars, sea urchins)		Coelomates with secondary radial anatomy (larvae bilateral; adults radial); unique water vascular system; endoskeleton
	Chordata (lancelets, tunicates, vertebrates)		Coelomates with notochord; dorsal hollow nerve cord; pharyngeal slits; muscular postanal tail

CHAPTER 33 REVIEW

Go to the Campbell Biology website (www.campbellbiology.com) to explore an interactive version of the Chapter Review.

Summary of Key Concepts

PARAZOA

- **Phylum Porifera: Sponges are sessile with porous bodies and choanocytes (pp. 647–648, FIGURES 33.2, 33.3)** Sponges lack tissues and organs. They filter-feed by drawing water through pores; choanocytes (flagellated collar cells) ingest bacteria and particulate food suspended in the water.

RADIATA

- **Phylum Cnidaria: Cnidarians have radial symmetry, a gastrovascular cavity, and cnidocytes (pp. 648–651, FIGURES 33.4–33.7, TABLE 33.1)** Cnidarians are mainly marine carnivores possessing tentacles armed with cnidocytes that aid in defense and the capture of prey. Two body forms are sessile polyps and floating medusas. The digestive tract (gastrovascular cavity) is incomplete (has a single opening, the mouth, that also functions as an anus). Class Hydrozoa usually alternates polyp and medusa forms, although the polyp is more conspicuous. In class Scyphozoa, jellies (medusas) are the prevalent forms of the life cycle. Class Anthozoa contains the sea anemones and corals, which occur only as polyps.

- **Phylum Ctenophora: Comb jellies possess rows of ciliary plates and adhesive colloblasts (pp. 650–651, FIGURE 33.8)** Comb jellies use retractable tentacles to capture food.

PROTOSTOMIA: LOPHOTROCHOZOA

- **Phylum Platyhelminthes: Flatworms are acoelomates with gastrovascular cavities (pp. 652–654, FIGURES 33.9–33.12, TABLE 33.2)** Most flatworms are ribbonlike animals with a gastrovascular cavity. Class Turbellaria is made up of mostly free-living, primarily marine species. Members of the classes Trematoda and Monogenea live as parasites in or on animals. Class Cestoidea (tapeworms), all parasites, lack a digestive tract.

- **Phylum Rotifera: Rotifers are pseudocoelomates with jaws, crowns of cilia, and complete digestive tracts (p. 654, FIGURE 33.13)** Found mainly in fresh water, many rotifer species are parthenogenetic.

- **The lophophorate phyla: Bryozoans, phoronids, and brachiopods are coelomates with ciliated tentacles around their mouths (pp. 654–655, FIGURE 33.14)** The lophophore is a horseshoe-shaped, suspension-feeding organ bearing ciliated tentacles.

- **Phylum Nemertea: Proboscis worms are named for their prey-capturing apparatus (p. 655, FIGURE 33.15)** The proboscis worms have a unique retractable tube (proboscis) used for defense and prey capture. A fluid-filled cavity surrounds the proboscis.

- **Phylum Mollusca: Mollusks have a muscular foot, a visceral mass, and a mantle (pp. 656–659, FIGURES 33.16–33.22, TABLE 33.3)** Class Polyplacophora is composed of the chitons, oval-shaped marine animals encased in an armor of dorsal plates. Most members of class Gastropoda, the snails and their relatives, possess a single, spiraled shell; embryonic torsion of the body is a distinctive characteristic; sea slugs lack a shell. Class Bivalvia (clams and their relatives) have hinged shells divided into two halves. Class Cephalopoda includes squids and octopuses, carnivores with beaklike jaws surrounded by tentacles of the modified foot.

- **Phylum Annelida: Annelids are segmented worms (pp. 659–660, FIGURES 33.23, 33.24, TABLE 33.4)** Class Oligochaeta includes earthworms and various aquatic species. Members of class Polychaeta possess paddlelike parapodia that function as gills and aid in locomotion. Class Hirudinea consists of the leeches.

PROTOSTOMIA: ECDYSOZOA

- **Phylum Nematoda: Roundworms are nonsegmented pseudocoelomates covered by tough cuticles (pp. 661–662, FIGURE 33.25)** Among the most widespread and numerous animals, nematodes inhabit most aquatic habitats. Some species are important parasites of animals and plants.

- **Arthropods are segmented coelomates with exoskeletons and jointed appendages (pp. 662–672, FIGURES 33.26–33.36, TABLES 33.5, 33.6)** There are more known species of arthropods than all other phyla combined. Chelicerates, arthropods with pincer or fanglike feeding appendages, include class Arachnida (spiders, ticks, scorpions, and mites). A traditional classification scheme groups the insects (class Insecta), centipedes (class Chilopoda), and millipedes (class Diplopoda) as uniramians—all with one pair of antennae and unbranched (uniramous) appendages. Crustaceans (lobsters, crayfish, crabs, shrimps, and barnacles) are primarily aquatic arthropods with two pairs of antennae and branched appendages. Research on the origin of segmentation in arthropods and other phyla of segmented animals will help systematists test hypotheses about phylogeny.

  Web/CD Case Study in the Process of Science: *How Are Insect Species Identified?*

DEUTEROSTOMIA

- **Phylum Echinodermata: Echinoderms have a water vascular system and secondary radial anatomy (pp. 672–674, FIGURES 33.37, 33.38)** Sea stars and their relatives make up six classes of the marine phylum Echinodermata. The radial anatomy of many species evolved secondarily from bilateral ancestors. The vascular system ending in tube feet is used for locomotion and feeding. A thin, bumpy, or spiny skin covers a calcareous endoskeleton.

  Web/CD Activity 33A: *Characteristics of Invertebrates*

- **Phylum Chordata: The chordates include two invertebrate subphyla and all vertebrates (pp. 674–675)** Chordates share many features of embryonic development with the echinoderms.

Self-Quiz

1. Which two clades branch from the earliest Eumetazoan ancestor?
   a. Parazoa and Bilateria
   b. Parazoa and Radiata
   c. Radiata and Bilateria
   d. Protostomia and Deuterostomia
   e. Lophotrochozoa and Ecdysozoa

2. Choose the phylum characterized by animals that have segmented bodies:
   a. Cnidaria
   b. Platyhelminthes
   c. Porifera
   d. Arthropoda
   e. Mollusca

3. The water vascular system of echinoderms
   a. functions as a circulatory system that distributes nutrients to body cells.
   b. functions in locomotion, feeding, and gas exchange.
   c. is bilateral in organization, even though the adult animal has radial anatomy.
   d. moves water through the animal's body for suspension feeding.
   e. is analogous to the hydrostatic skeleton of annelids.

4. Water movement through a sponge would follow what path?
   a. porocyte $\longrightarrow$ spongocoel $\longrightarrow$ osculum
   b. blastopore $\longrightarrow$ gastrovascular cavity $\longrightarrow$ protostome
   c. choanocyte $\longrightarrow$ mesohyl $\longrightarrow$ spongocoel
   d. porocyte $\longrightarrow$ choanocyte $\longrightarrow$ mesohyl
   e. colloblast $\longrightarrow$ coelom $\longrightarrow$ porocyte

5. Although a diverse group, all cnidarians are characterized by
   a. a gastrovascular cavity.
   b. an alteration between a medusa and a polyp stage.
   c. some degree of cephalization.
   d. muscle tissue of mesodermal origin.
   e. the complete absence of asexual reproduction.

6. A land snail, a clam, and an octopus all share
   a. a mantle.
   b. a radula.
   c. gills.
   d. embryonic torsion.
   e. distinct cephalization.

7. Which of the following is not a characteristic of most members of the phylum Annelida?
   a. hydrostatic skeleton
   b. segmentation
   c. metanephridia
   d. pseudocoelom
   e. closed circulatory system

8. Which of the following is *not* true of the chelicerates?
   a. They have antennae.
   b. Their body is divided into a cephalothorax and an abdomen.
   c. The horseshoe crab is one surviving marine member.
   d. They include ticks, scorpions, and spiders.
   e. Their anterior appendages are modified as pincers or fangs.

9. Which of the following combinations of phylum and description is *incorrect*?
   a. Echinodermata—bilateral and radial symmetry, coelom from archenteron
   b. Nematoda—roundworms, pseudocoelomate
   c. Cnidaria—radial symmetry, polyp and medusa body forms
   d. Platyhelminthes—flatworms, gastrovascular cavity, acoelomate
   e. Porifera—gastrovascular cavity, coelom present

10. Which of the following characteristics is probably *most* responsible for the incredible diversification of insects on land?
    a. segmentation   d. metamorphosis
    b. antennae   e. flight
    c. tracheal system

11. Your circulatory system distributes nutrients, wastes, and oxygen between different parts of your body. How do sponges accomplish these transport functions?

12. Flatworms and cnidarians differ in symmetry, with flatworms being _____ and cnidarians being _____, but the animals of both phyla have _____ cavities.

13. Why is it risky for someone who eats pork chops to order them "rare" in a restaurant?

14. As representatives of classes of mollusks, a garden snail is an example of a _____; a clam is an example of a _____; and a squid is an example of a _____.

15. What is the main difference between the digestive tract of an earthworm and the gastrovascular cavity of a sea anemone?

16. Contrast the skeleton of an echinoderm with that of an arthropod.

Go to the website or CD-ROM for more quiz questions.

Evolution Connection

Horseshoe crabs are called "living fossils" because the fossil record shows they have remained essentially unchanged in morphology for many millions of years. Why might this organism have retained the same morphology for this vast time period? What other aspects of its biology, less obvious than structure, do you think may have evolved?

The Process of Science

A marine biologist has dredged up an unknown animal from the seafloor. Describe some of the characteristics she should look at to determine the animal phylum to which the creature should be assigned.

Use a dichotomous key to identify insects in the Case Study in the Process of Science, available on the CD-ROM and website.

Science, Technology, and Society

Construction of a dam and irrigation canals in an African country has enabled farmers to increase the amount of food they can grow. In the past, crops were planted only after spring floods; the fields were too dry the rest of the year. Now fields can be watered year-round. Improvement in crop yield has had an unexpected cost—a tremendous increase in the incidence of schistosomiasis, or blood fluke disease. Look at the blood fluke life cycle in FIGURE 33.11 and imagine that your Peace Corps assignment is to help local health officials control the disease. Why do you think the irrigation project increased the incidence of schistosomiasis? It is difficult and expensive to control the disease with drugs. Suggest three other methods that could be tried to prevent people from becoming infected.

Answers: 1. c; 2. d; 3. b; 4. a; 5. a; 6. a; 7. d; 8. a; 9. e; 10. e; 11. Transport is carried out mainly by the mobile cells called amoebocytes. 12. Bilateral; radial; gastrovascular. 13. Incomplete cooking doesn't kill nematodes and other parasites that might be present in the meat. 14. Gastropod; bivalve; cephalopod. 15. A complete digestive tract with separate mouth and anus for the earthworm; a digestive sac with a single opening for the sea anemone. 16. An echinoderm has an endoskeleton; an arthropod has an exoskeleton.

VERTEBRATE EVOLUTION AND DIVERSITY

Most of us are curious *about our genealogies. On the personal level, we wonder about our family ancestry. As biology students, we are interested in retracing human ancestry within the broader scope of the evolutionary history of the entire animal kingdom. The questions we must ask are the following: What were our ancestors like? How are we related to other animals? What are our closest relatives? In this chapter, we trace the evolution of the vertebrates, the group that includes humans and their closest relatives. Mammals, birds, lizards, snakes, turtles, amphibians, and the various classes of fishes are all examples of* **vertebrates.** *They share many features unique to the vertebrates, including the backbone, a series of vertebrae for which the group is named. You can see this vertebrate hallmark in the above photograph of a snake skeleton. Our first step in tracking the vertebrate genealogy is to determine where vertebrates fit in the animal kingdom.*

INVERTEBRATE CHORDATES AND THE ORIGIN OF VERTEBRATES

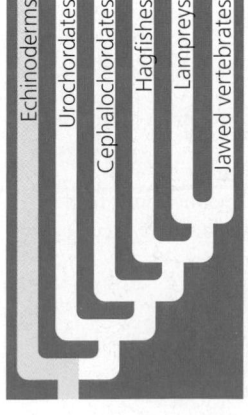

The vertebrates are part of a phylogenetic branch of the animal kingdom consisting of deuterostomes (see FIGURE 32.4). The deuterostome branch has two main modern phyla: the **chordates** (which include the vertebrates) and the echinoderms. An echinoderm such as the sea star is the closest living relative of human beings among the *well-known* groups of invertebrates. However, the phylum Chordata itself also includes two groups of less well-known invertebrate animals: the urochordates and the cephalochordates. The urochordates and cephalochordates are therefore more closely related to us than is a sea star. FIGURE 34.1 is a cladogram of some of the animal groups we will examine during the first part of this chapter.

Four anatomical features characterize the phylum Chordata

Although chordates vary widely in appearance, they are distinguished as a phylum by the presence of four anatomical structures that appear at some point during the animal's life-time, often only during embryonic development. These four chordate characteristics are a notochord; a dorsal, hollow nerve cord; pharyngeal slits; and a muscular, postanal tail (FIGURE 34.2, p. 680).

1. Notochord

Chordates are named for a skeletal structure, the notochord, present in all chordate embryos. The **notochord** is a longitudinal, flexible rod located between the digestive tube and the nerve cord. Composed of large, fluid-filled cells encased in fairly stiff, fibrous tissue, it provides skeletal support throughout most of the length of the animal. The notochord persists

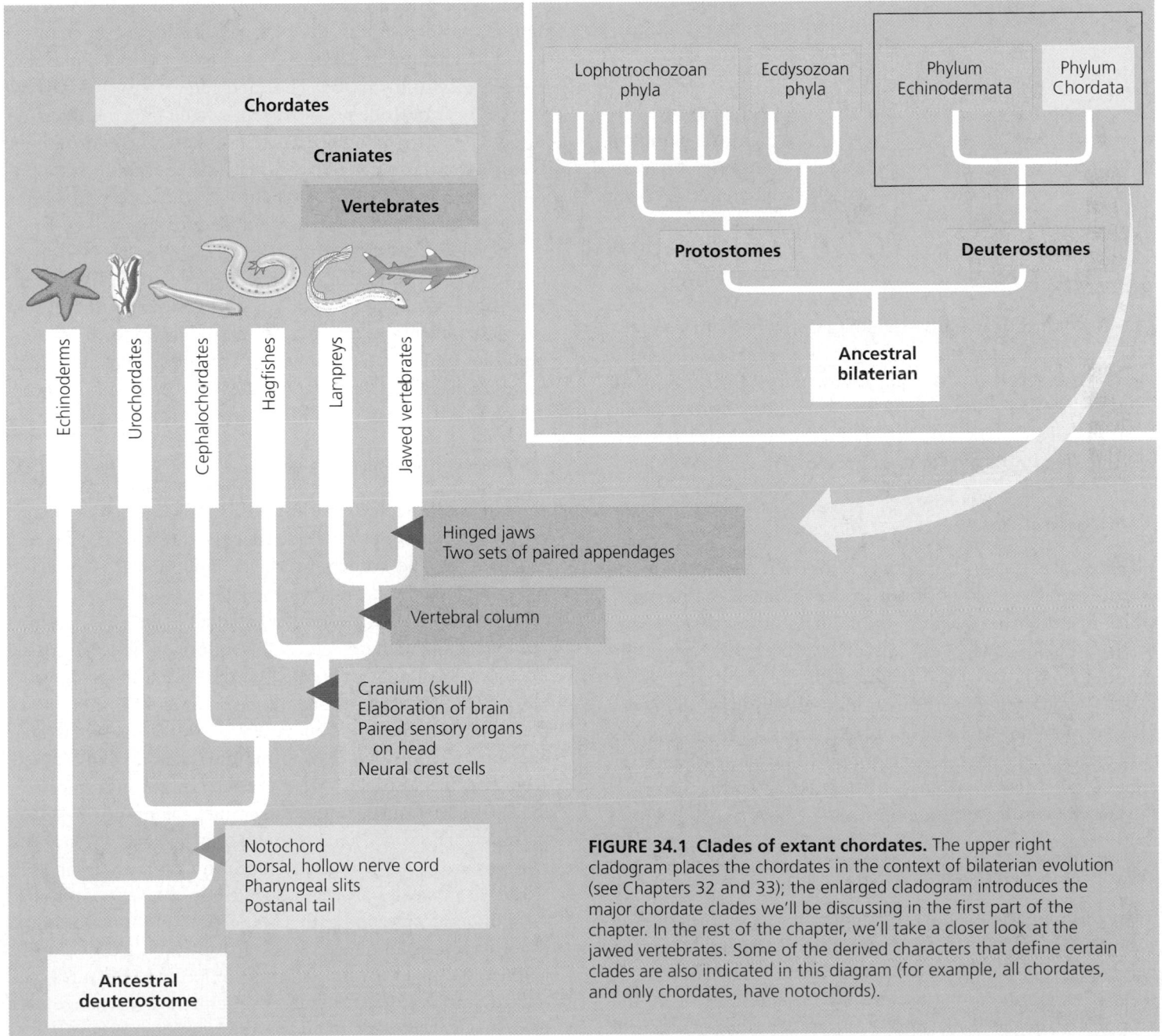

FIGURE 34.1 Clades of extant chordates. The upper right cladogram places the chordates in the context of bilaterian evolution (see Chapters 32 and 33); the enlarged cladogram introduces the major chordate clades we'll be discussing in the first part of the chapter. In the rest of the chapter, we'll take a closer look at the jawed vertebrates. Some of the derived characters that define certain clades are also indicated in this diagram (for example, all chordates, and only chordates, have notochords).

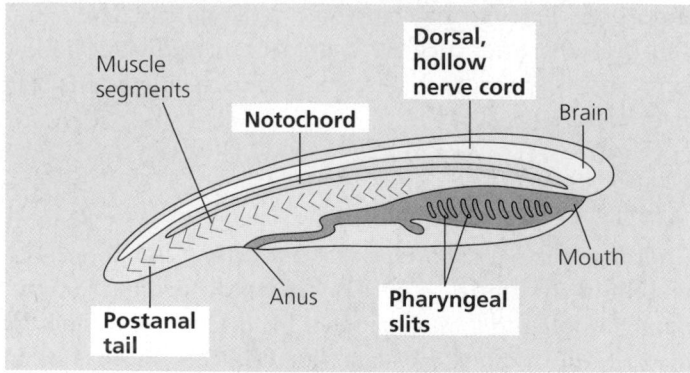

FIGURE 34.2 Chordate characteristics. All chordates possess the four trademarks of the phylum: a notochord; a dorsal, hollow nerve cord; pharyngeal slits; and a muscular, postanal tail.

in the adult of some invertebrate chordates and primitive vertebrates. However, in most vertebrates a more complex, jointed skeleton develops, and the adult retains only remnants of the embryonic notochord—as gelatinous material of the disks between the vertebrae of humans, for example.

2. Dorsal, Hollow Nerve Cord

The nerve cord of a chordate embryo develops from a plate of ectoderm that rolls into a tube located dorsal to the notochord. The result is a dorsal, hollow nerve cord unique to chordates. Other animal phyla have solid nerve cords, usually ventrally located. The nerve cord of a chordate embryo develops into the central nervous system: the brain and spinal cord.

3. Pharyngeal Slits

The digestive tube of chordates extends from the mouth to the anus. The region just posterior to the mouth is the pharynx, which opens to the outside of the animal through several pairs of slits. These pharyngeal slits allow water that enters the mouth to exit without continuing through the entire digestive tract. The pharyngeal slits function as suspension-feeding devices in many invertebrate chordates. The slits and structures supporting them have become modified for gas exchange (in aquatic vertebrates), jaw support, hearing, and other functions during vertebrate evolution.

4. Muscular, Postanal Tail

Most chordates have a tail extending posterior to the anus. By contrast, most nonchordates have a digestive tract that extends nearly the whole length of the body. The chordate tail contains skeletal elements and muscles and provides much of the propulsive force in many aquatic species.

Invertebrate chordates provide clues to the origin of vertebrates

Members of the two subphyla of invertebrate chordates, Urochordata and Cephalochordata, illustrate the chordate body plan in its most "stripped-down" versions—without the additional features that evolved in vertebrates. The study of urochordates and cephalochordates also provides clues to the origin of vertebrates.

Subphylum Urochordata

Urochordates are commonly called **tunicates.** Most tunicates are sessile marine animals that adhere to rocks, docks, and boats (FIGURE 34.3a). Others are planktonic. Some species are colonial. Seawater enters the animal through an incurrent siphon, passes through the pharyngeal slits into a chamber called the atrium, and exits through an excurrent siphon, or atriopore (FIGURE 34.3b). The food filtered from this water current by a mucous net is passed by cilia into the intestine. The anus empties into the excurrent siphon. The entire animal is cloaked in a tunic made of a celluloselike carbohydrate. Because they shoot a jet of water through the excurrent siphon when molested, tunicates are also called sea squirts.

The adult tunicate scarcely resembles a chordate. It displays no trace of a notochord, nor is there a nerve cord or tail. Only the pharyngeal slits suggest a link to other chordates. But all four chordate trademarks are manifest in the larval form of some groups of tunicates (FIGURE 34.3c). The larva swims until it attaches by its head to a surface and undergoes metamorphosis, during which most of its chordate characteristics disappear.

Subphylum Cephalochordata

Known as **lancelets** because of their bladelike shape, **cephalochordates** closely resemble the idealized chordate in FIGURE 34.2. The notochord; dorsal, hollow nerve cord; numerous gill slits; and postanal tail all persist into the adult stage (FIGURE 34.4). Lancelets are small animals, only a few centimeters long, that live in the sand at the bottom of the sea in coastal regions. Globally they are rare, but they live at huge densities (over 5,000 lancelets per square meter) in a small number of places, including one at Tampa Bay, Florida. Lancelets wriggle backward into the sand, leaving only their anterior end exposed. A mucous net secreted across the pharyngeal slits traps tiny food particles from seawater drawn into the mouth by ciliary pumping. The water exits through the slits, and the trapped food passes down the digestive tube. Animals such as lancelets that feed on small particles suspended in the water are called suspension feeders, and early chordates, until the evolution of fishes with jaws and teeth, were all suspension feeders. In the lancelet, the pharynx and gill slits are feeding structures, and only to a minor extent respiratory structures; gas exchange takes place mainly across regions of the external body surface.

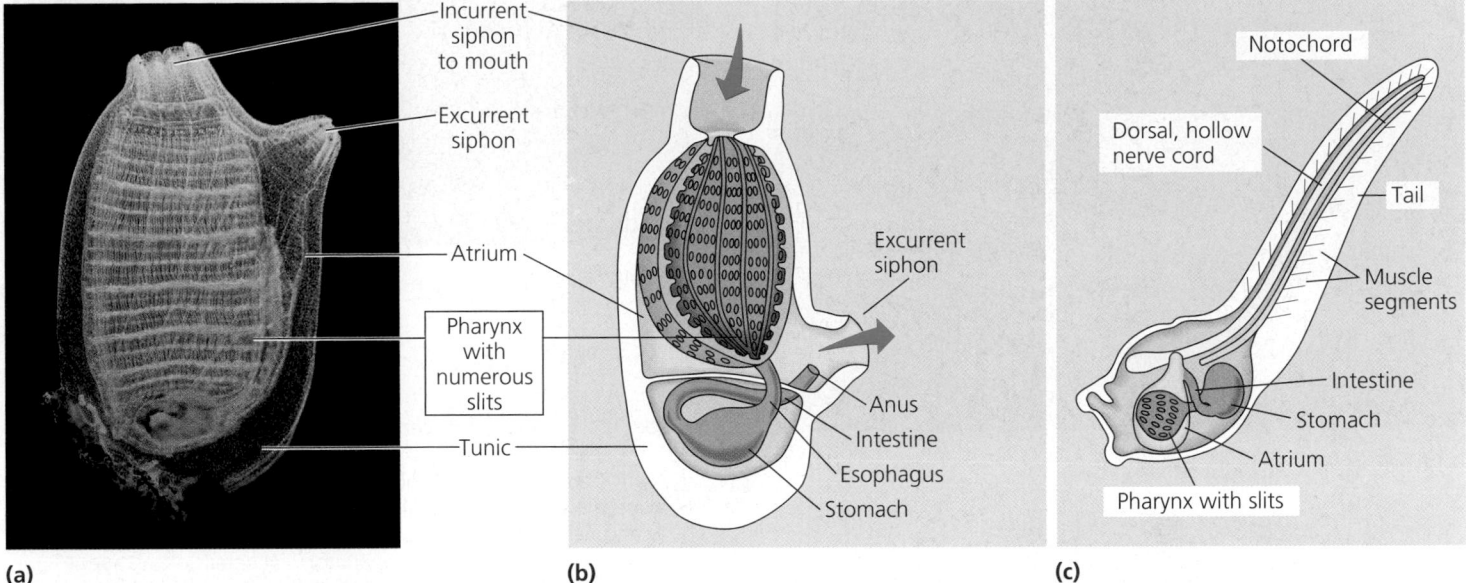

(a)

(b)

(c)

FIGURE 34.3 Subphylum Urochordata: a tunicate. (a) An adult tunicate, or sea squirt, is a sessile animal commonly arranged in a U shape (photo is approximately life-sized).

(b) In the adult, prominent pharyngeal slits function in suspension feeding, but the other chordate characteristics are not obvious. **(c)** In the tunicate larva, which is a free-

swimming but nonfeeding "tadpole," the chordate characteristics are evident: It has a notochord, dorsal nerve tube, and a tail with muscle segments.

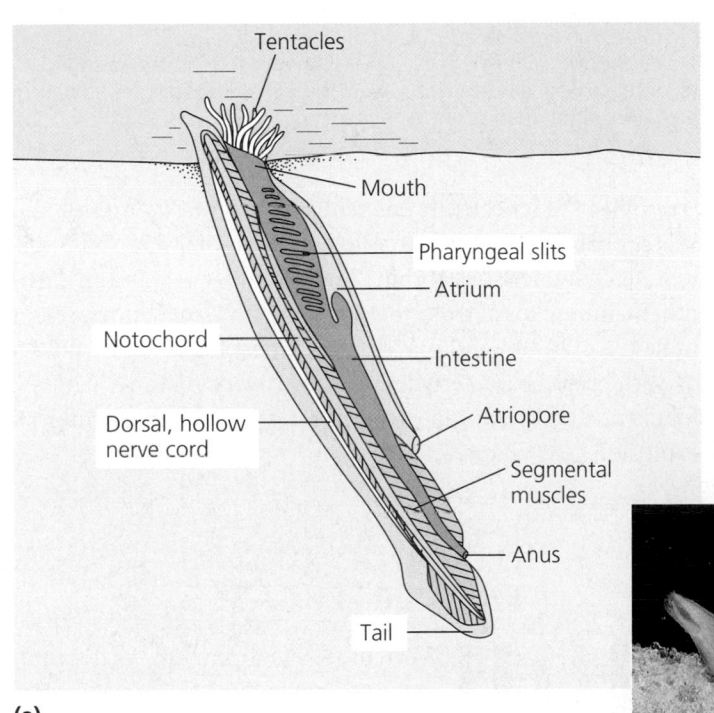

(a)

(b)

FIGURE 34.4 Subphylum Cephalochordata: the lancelet Branchiostoma. (a) This small invertebrate displays all four chordate characteristics. The pharyngeal slits function in suspension feeding. Water passes into the pharynx and through slits into the atrium, a chamber that vents to the outside via the atriopore. Food particles trapped by mucus are swept by cilia into the digestive tract. **(b)** The muscle segments you can see in this photo of a lancelet produce the sinusoidal swimming of these animals.

A lancelet frequently leaves its burrow and swims to a new location. Though feeble swimmers, these invertebrate chordates display, in simple form, the swimming mechanism of fishes. Coordinated contraction of muscles serially arranged like rows of chevrons (<<<<) along the sides of the notochord flexes the notochord, producing side-to-side undulations that thrust the body forward. This serial musculature is evidence of the lancelet's segmentation. The muscle segments develop from blocks of mesoderm called **somites** arranged along each side of the notochord of a chordate embryo. Chordates are segmented animals.

The Relationship Between Invertebrate Chordates and Vertebrates

Molecular evidence suggests that the cephalochordates are the vertebrates' closest relatives, and urochordates are their next closest relatives (see FIGURE 34.1). According to one hypothesis, there were two stages in the evolution of vertebrates from invertebrates: In the first stage, an ancestral cephalochordate evolved from a form resembling a modern urochordate larva; in the second stage, a vertebrate evolved from a cephalochordate. The first stage may have been preceded by **paedogenesis,** the precocious development of sexual maturity in a larva (see Chapter 24). Notice that a cephalochordate appears more akin

(a) *Haikouella*

(b) *Haikouichthys*

5 mm

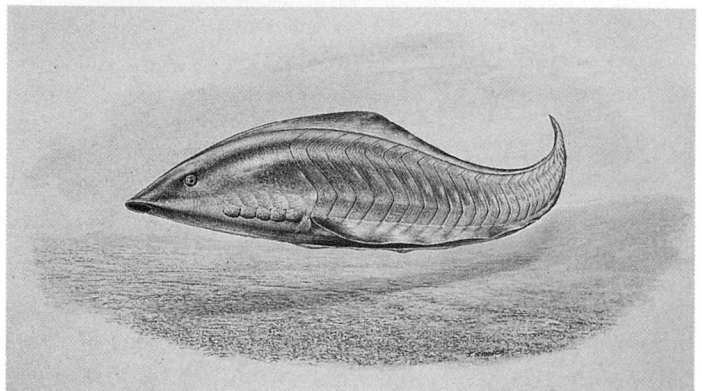

(c) *Myllokunmingia*

FIGURE 34.5 Early fossil vertebrates. These fossils were discovered in 1999 near the town of Haikou in southern China. They date back to the early Cambrian period, about 530 million years ago. **(a)** Fossils of *Haikouella* are about 2–3 cm long. Based on detailed study of over 300 specimens fossilized in a variety of body positions, researchers have been able to describe many anatomical details of this species. *Haikouella* resembled a modern cephalochordate, but with such vertebrate characters as eyes, mineralized toothlike structures, and what seems to be an enlarged brain. However, there is no evidence of a cranium (braincase), a character found in all true modern vertebrates. Thus, *Haikouella* may represent animals that were transitional in the evolution of vertebrates from invertebrate chordates. **(b)** *Haikouichthys*, about the same size as *Haikouella*, probably had a cranium, making it a full-fledged vertebrate. **(c)** Based on fossils from the same Chinese sediments, an artist reconstructed *Myllokunmingia*, which is very similar to *Haikouichthys*. These earliest known vertebrates, in contrast to lancelets (cephalochordates), were more mobile and were probably predators rather than suspension feeders.

to a urochordate larva than to an adult urochordate (compare FIGURES 34.3 and 34.4). Changes in genes controlling development can alter the timing of developmental events, such as the maturation of gonads. Perhaps such a change occurred in the ancestor of cephalochordates and vertebrates, causing the gonads to mature in swimming larvae before the onset of metamorphosis to the sessile adult form. If reproducing larvae were very successful, natural selection may have reinforced paedogenesis and eliminated metamorphosis.

There is no fossil evidence to support or contradict the paedogenetic hypothesis; it is deduced from comparing modern forms. But fossil evidence does tell us something about the second stage, from cephalochordate to vertebrate. In 1999, paleontologists reported that fossils from a site in China support the idea that cephalochordates are the closest relatives of vertebrates. The 530-million-year-old fossils appear to blur the distinction between cephalochordates and vertebrates by showing a series of "missing links" (FIGURE 34.5). Vertebrates, as we consider further below, differ from lancelets in having a much more elaborate brain—an enlargement of the nerve cord at the anterior end—and a braincase, or cranium. One Chinese fossil, *Haikouella*, seems to have a brain but may lack a cranium (skull). It also (unlike the lancelet) has eyes, and hardened structures called "denticles" in its pharynx, which may have functioned somewhat like teeth. Otherwise, *Haikouella*

resembles the lancelet. It has tentacles around its mouth and was probably a suspension feeder. This animal had some vertebrate characteristics (brain) but not others (cranium). Another Chinese fossil from the same area, *Haikouichthys*, seems to have a cranium and may be the earliest fossil with a full set of vertebrate characteristics. The discovery of these Chinese fossils pushes the origin of vertebrates back to the Cambrian explosion (see Chapter 32).

INTRODUCTION TO THE VERTEBRATES

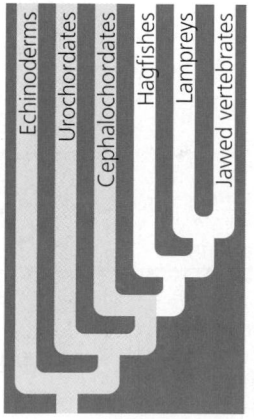

Vertebrates (subphylum Vertebrata) retain the primitive chordate characteristics but have additional specializations, shared derived features that distinguish the subphylum from invertebrate chordates. Many of the features that distinguish vertebrates are associated with larger size and an active lifestyle.

Neural crest, pronounced cephalization, a vertebral column, and a closed circulatory system characterize the subphylum Vertebrata

The dorsal, hollow nerve cord found in all chordates develops when the edges of an ectodermal plate on the surface of the embryo roll together to form the neural tube. In vertebrates, a group of embryonic cells called the **neural crest** forms near the dorsal margins of the closing neural tube (FIGURE 34.6). A distinctive feature of vertebrates, the neural crest contributes to the formation of certain skeletal elements, such as some of the bones and cartilage of the cranium (braincase), and many other structures that distinguish vertebrates from other chordates.

The vertebrate cranium and brain (which is the enlarged anterior end of the dorsal, hollow nerve cord), along with the eyes, ears, and nose, are evidence of an important evolutionary feature of vertebrates—a high degree of cephalization, the concentration of sensory and neural equipment in the head (see Chapter 32).

Note in FIGURE 34.1 that these vertebrate characteristics are present in the animals called hagfishes, but the hagfishes lack a backbone of vertebrae. Therefore, the cladogram in FIGURE 34.1 nests the vertebrate clade within a larger clade, **Craniata,** which is named for the cranium and includes the hagfishes. However, we will include hagfishes in our survey of vertebrates, with the caveat that they represent an ancient branch of Craniata that predates the origin of the vertebral column.

The cranium and vertebral column, surrounding and protecting the nerve cord, are parts of the vertebrate axial skeleton, the main support structure for the axis, or central trunk, of the body. The axial skeleton helps make large body size and strong, fast movement possible. The axial skeleton of most vertebrates also includes ribs, which anchor muscles and protect internal organs. Most vertebrates also have an appendicular skeleton, supporting two pairs of appendages (fins, legs, or arms).

The vertebrate endoskeleton is made of bone or more flexible cartilage, or most commonly some combination of these two materials. Although the skeleton consists mostly of a nonliving matrix, living cells within the skeleton secrete and maintain the materials of the matrix. The endoskeleton of a vertebrate can grow with the animal, unlike the nonliving exoskeleton of arthropods, which must be periodically shed and rebuilt as the animal grows.

When vertebrates move in pursuit of food or when escaping predators, they regenerate their ATP supply mainly by cellular respiration, which consumes oxygen. Adaptations of the vertebrate respiratory and circulatory systems support busy mitochondria in muscle cells and other active tissues. Vertebrates have a closed circulatory system, with a ventral, chambered heart that pumps blood through arteries to microscopic vessels called capillaries that branch throughout every tissue in the body. The blood is oxygenated as it passes through capillaries in gills or lungs.

A vigorous lifestyle also requires a relatively large supply of organic fuel. Vertebrate adaptations for feeding, digestion, and nutrient absorption help support active behavior. For example, muscles in the walls of the digestive tract propel food from organ to organ along the tract. These are all examples of vertebrate form and function having their historical basis in the transition from a relatively sedentary lifestyle to a more active one.

An overview of vertebrate diversity

FIGURE 34.7 (p. 684) is a hypothetical cladogram of extant (living) vertebrates based on a combination of anatomical, molecular, and fossil evidence. Two groups, the hagfishes and the lampreys, lack hinged jaws. All the other vertebrates are grouped in FIGURE 34.7 within the clade identified as the **gnathostomes** (which means "jawed mouth"). In addition to jaws, gnathostomes also have two sets of paired appendages. Among these jawed vertebrates, there are various classes of aquatic animals that we generally call "fishes": the cartilaginous fishes (sharks and rays) and three classes of bony fishes (the ray-fins, lobe-fins, and lungfishes). In the case of fishes, the two sets of paired appendages are fins that function in swimming. All the other gnathostomes are **tetrapods** ("four footed"), in which the two sets of paired appendages are modified as legs that can support the animals on land. The tetrapods include the amphibians (frogs and salamanders) along with the clade identified as **amniotes** in FIGURE 34.7. The amniotes are named for the amniotic egg, which is a shelled, water-retaining egg (think of grocery store eggs, which are laid by birds). The amniotic egg functions as a "self-contained pond" that enables amniotes to complete their life

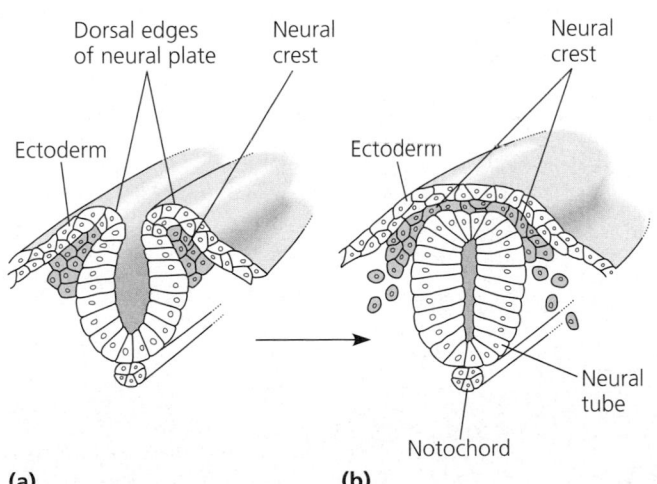

(a)   **(b)**

FIGURE 34.6 The neural crest, embryonic source of many unique vertebrate characters. (a) The neural crest consists of bilateral bands of cells near the margins of the embryonic folds that meet to form the dorsal, hollow nerve cord (neural tube). **(b)** Cells from the neural crest migrate to distant sites in the embryo, where they give rise to some of the anatomical structures unique to vertebrates, including some of the bones and cartilage of the skull.

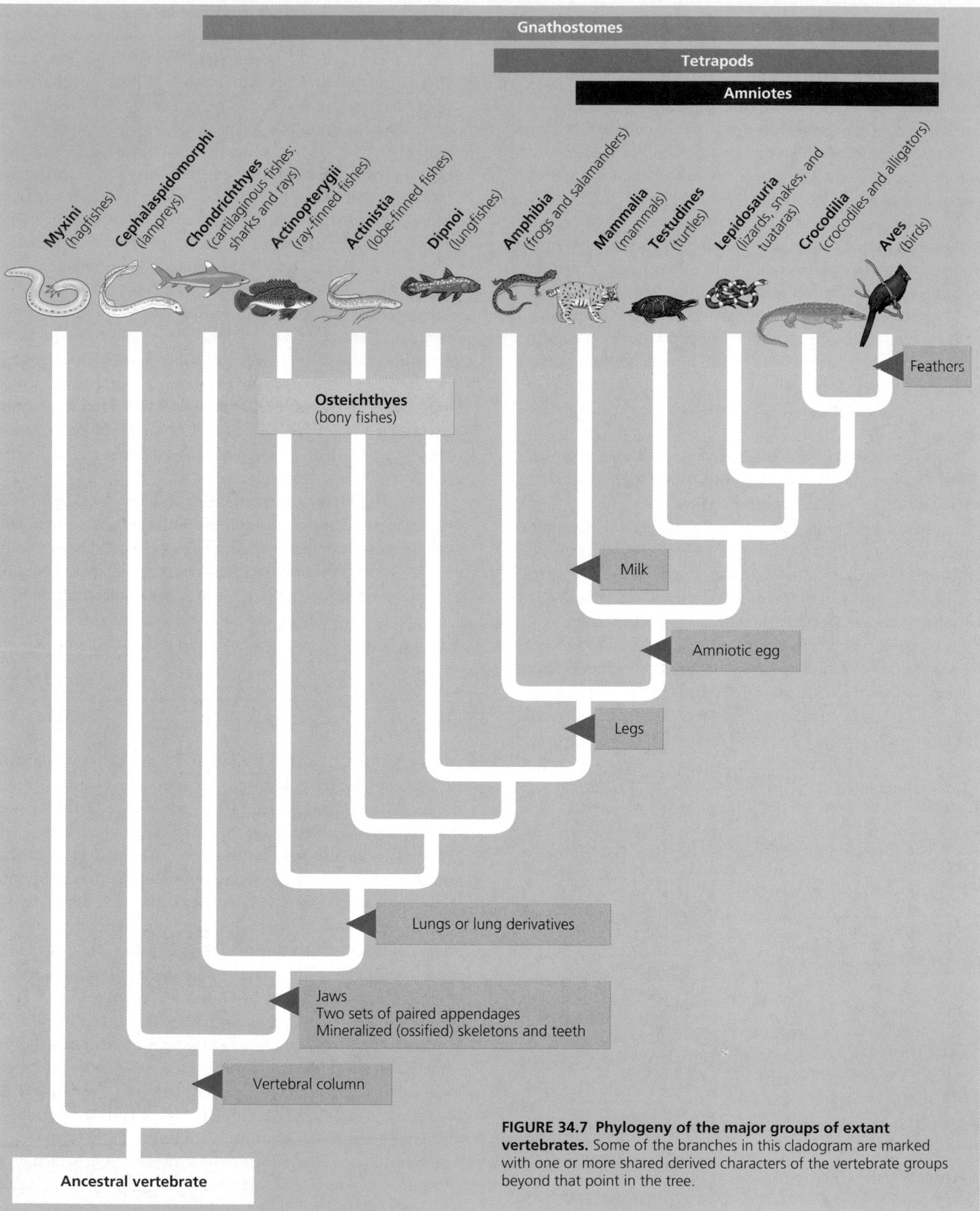

FIGURE 34.7 Phylogeny of the major groups of extant vertebrates. Some of the branches in this cladogram are marked with one or more shared derived characters of the vertebrate groups beyond that point in the tree.

cycles on land. Although most modern mammals don't lay eggs, they retain many other key features of the amniotic mode of reproduction. Note that the amniotes we customarily lump together as "reptiles" (turtles, snakes, lizards, crocodiles, and alligators) do not represent a clade (monophyletic group)—*unless* we also include the birds.

With the cladogram in FIGURE 34.7 as our guide, let's now take a closer look at the major vertebrate groups.

JAWLESS VERTEBRATES

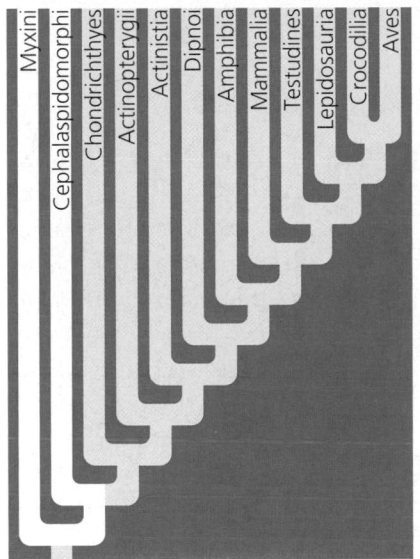

The two extant classes of jawless vertebrates, or **agnathans** (which means "without jaws"), are the hagfishes and the lampreys. These animals are eel-like in shape, but the true eels are actually bony fishes. In fact, the agnathans represent ancient vertebrate lineages that predate the origin of paired fins, teeth, and bones hardened by mineralization (ossification). Thus, agnathans provide clues about early stages in the evolution of vertebrates.

Class Myxini: Hagfishes are the most primitive living "vertebrates"

There are about 30 species of hagfishes, all of them marine. They are mainly bottom-dwelling scavengers, ingesting such food as worms and sick or dead fish. Rows of slime glands on a hagfish's flanks secrete a gooey substance (FIGURE 34.8). The glands produce small amounts of the slime when a hagfish is feeding, a behavior that may repulse other scavengers. But when a hagfish is attacked by a potential predator, it can secrete enough water-absorbing goo to form several liters of slime in less than a minute. The slime can coat the gills of a predatory fish, sending it into retreat or even suffocating it.

The skeleton of a hagfish is made entirely of cartilage, a rubbery connective tissue (see FIGURE 40.2). In addition to the cartilaginous cranium (skull), a strong but flexible rod of cartilage (which is actually the notochord) extends the length of the body, much like the notochords of lancelets (cephalochordates). The hagfish notochord provides support and a skeleton against which muscles can exert force to drive the animal's serpentine (snakelike) swimming.

Slime glands

FIGURE 34.8 A hagfish.

Hagfishes do not have vertebrae, and so, as we pointed out earlier, these chordates are only included with vertebrates if we equate the subphylum Vertebrata with the clade marked Craniata in FIGURE 34.1. Until recently, a popular hypothesis was that hagfishes lost their vertebrae secondarily during their descent from an ancestor with a true vertebral column. However, more detailed study has convinced most zoologists that the hagfish skeleton is truly primitive. In fact, hagfishes are now considered to be the most primitive living vertebrates (or craniates), diverging from the vertebrate lineage about 530 million years ago, during the early Cambrian.

Class Cephalaspidomorphi: Lampreys provide clues to the evolution of the vertebral column

There are about 35 species of lampreys inhabiting both marine and freshwater environments. The sea lamprey (FIGURE 34.9) feeds by clamping its round mouth onto the flank of a live fish, using a rasping tongue to penetrate the skin of its prey, and ingesting the prey's blood. Sea lampreys live as larvae for years in freshwater streams and then migrate to the sea or lakes as they mature into adults. The larva is a suspension feeder that resembles the lancelets (cephalochordates). Some species of lampreys

FIGURE 34.9 A sea lamprey. Acting as both a predator and a parasite, a lamprey uses this rasping mouth (inset) to bore a hole in the side of a fish, living on the blood and other tissues of its host.

feed only as larvae. Following several years in streams, they attain sexual maturity, reproduce, and die within a few days.

The notochord of lampreys persists as the main axial skeleton in the adult animal, as it does in hagfishes. However, a lamprey also has a cartilaginous pipe surrounding the rodlike notochord. Along the length of this pipe, pairs of cartilaginous projections extend upward (dorsally), partially enclosing the nerve cord with what may be a vestige of an early stage in the evolution of the vertebral column.

Note again that both hagfishes and lampreys not only lack skeleton-supported jaws but also lack paired appendages. And the skeletons of hagfishes and lampreys are made entirely of cartilage. In contrast, the skeletons of most jawed vertebrates are ossified (hardened with minerals) as bone. In the gnathostomes (jawed vertebrates), the notochord is a larval structure, largely replaced during development by the segmental vertebrae that make up the backbone.

Thus, anatomical comparison of agnathans with gnathostomes indicates that the brain and braincase (cranium) evolved first in the vertebrate lineage, followed by the vertebral column, with the jaws, ossified skeleton, and paired appendages evolving later. Analysis of the early Cambrian fossils in Chinese strata is consistent with this interpretation (see FIGURE 34.5).

Some extinct jawless vertebrates had ossified teeth and bony armor

Jawless vertebrates are much more diverse and common in the fossil record than they are among today's fauna. A diversity of taxa informally called **ostracoderms** thrived from about 450 to 375 million years ago, from the Ordovician period to the late Devonian. Most species were small, less than 50 cm in length. Most lacked paired fins and apparently were bottom dwellers that wiggled along streambeds or the seafloor, but there were also some more active species with paired fins. Their mouths were circular or slitlike openings that lacked jaws. The majority of ostracoderms were probably mud-suckers or suspension feeders that took in sediments or suspended organic debris through their mouths and then passed it through the gill slits, where food was trapped. Thus, the pharyngeal apparatus retained the primitive feeding function, although gills in agnathans had probably also evolved into the major sites of gas exchange. The ostracoderms and most other agnathan groups declined and finally disappeared during the Devonian period.

Fossils of the extinct agnathans provide evidence that the process of mineralization of certain body structures evolved early in vertebrate history. Ostracoderm means "shelled skin," a reference to the armor of bony plates that encased these animals. The plates may represent an early evolutionary stage of ossification, the hardening of connective tissue that occurs when specialized cells secrete calcium and phosphate, which precipitate as calcium phosphate, a hard mineral salt. Even earlier evidence of ossification is found in fossils of ancient vertebrates called **conodonts,** which date back as far as 510 million years. These vertebrates are named for their cone-shaped toothlike structures, which were ossified. In contrast, the mouths of hagfishes contain toothlike structures made mainly of keratin, a structural protein.

Based on cladistic analysis, most vertebrate systematists now agree that hagfishes and lampreys, though still represented in the modern fauna, are actually more primitive as vertebrates than were the conodonts and ostracoderms. The jawed vertebrates probably evolved from one of the many lineages of ostracoderms.

FISHES AND AMPHIBIANS

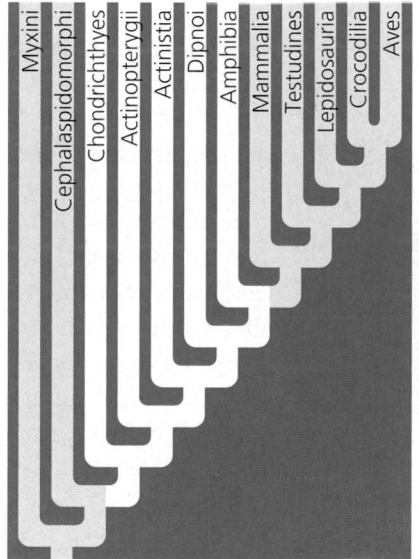

During the late Silurian period and early Devonian period, gnathostomes largely replaced the agnathans. The two extant groups of fishes—Chondrichthyes (the cartilaginous fishes: sharks and rays) and Osteichthyes (the bony fishes: ray-finned fishes, lobe-finned fishes, and lungfishes)—first appeared during that time. There is also a substantial fossil record of the armored, jawed fishes called **placoderms** ("plate skinned"), which are completely extinct. Most placoderms were less than a meter long, though some giants measured more than 10 m.

In addition to jaws, fishes also have two pairs of fins, whereas agnathans either lacked paired appendages or had a single pair. By studying the developmental genetics of extant vertebrates, biologists in the "evo-devo" field have found that differential expression of some of the *Hox* genes may determine whether one or two sets of paired appendages develop in a vertebrate embryo (see Chapters 21 and 25).

Jaws and paired fins were major evolutionary breakthroughs in vertebrate history. Jaws are hinged structures that, especially with the help of teeth, enable the animal to grip food items firmly and slice them up. A jawed fish can exploit food supplies that were unavailable to earlier agnathans, which fed mainly on food particles suspended in the water or located on the sea bottom. Paired fins, along with the tail, enable fishes to maneuver accurately while swimming, analogous to an airplane's movements of wing flaps and tail rudder for ascending, descending, and turning. With their jaws and paired fins, many species of fishes were active predators, capable of chasing prey and biting off chunks of flesh. Thus, these modifications of the early vertebrate body plan allowed the diversification of both lifestyles and nutrient sources, which may explain why jawed fishes replaced most agnathans during the Devonian period.

Vertebrate jaws evolved from skeletal supports of the pharyngeal slits

Vertebrate jaws evolved by modification of the skeletal rods that had previously supported the anterior pharyngeal (gill) slits (FIGURE 34.10). The remaining gill slits, no longer required for suspension feeding, remained as the major sites of respiratory gas exchange with the external environment. The origin of vertebrate jaws from these skeletal parts illustrates a general feature of evolutionary change: New adaptations usually evolve by the modification of existing structures. As a mechanism of adaptation, evolution is limited by the raw material with which it must work. Evolution is generally more of a remodeling process than a creative one.

The Devonian period (about 360 to 400 million years ago) has been called the "age of fishes," when the placoderms and another group of jawed fishes called **acanthodians** radiated and many new forms evolved in both fresh and salt water. Placoderms and acanthodians dwindled and disappeared almost completely by the beginning of the Carboniferous period, about 360 million years ago. An ancestor common to the placoderms and acanthodians may also have given rise to sharks and bony fishes some 425 to 450 million years ago. Sharks and bony fishes are currently the main vertebrates in the watery two-thirds of the surface of Earth.

Class Chondrichthyes: Sharks and rays have cartilaginous skeletons

The vertebrates of the class **Chondrichthyes,** sharks and their relatives, are called cartilaginous fishes because they have relatively flexible endoskeletons made of cartilage rather than bone. However, in most species, parts of the skeleton are strengthened by mineralized granules, and the teeth are bony. There are about 750 extant species in this class. Jaws and paired fins are well developed in the cartilaginous fishes. The largest and most diverse subclass consists of the sharks and rays. A second subclass is composed of a few dozen species of unusual fishes called chimaeras, or ratfishes.

The cartilaginous skeleton of these fishes is a derived characteristic, not a primitive one; that is, the ancestors of Chondrichthyes had bony skeletons, and the cartilaginous skeleton characteristic of the class evolved secondarily. During the development of most vertebrates, the skeleton is first cartilaginous and then becomes bony (ossified) as hard calcium phosphate matrix replaces the rubbery matrix of cartilage. Apparently, some modification in the developmental process of cartilaginous fishes prevents the replacement process.

Most sharks have streamlined bodies (FIGURE 34.11a, p. 688). They are swift swimmers, but they do not maneuver very well. Powerful swimming muscles, especially in the caudal (tail) fin, propel them forward. The dorsal fins function mainly as stabilizers, and the paired pectoral (fore) and pelvic (hind) fins provide lift in the water. Although a shark gains additional buoyancy by storing a large amount of oil in its huge liver, the animal is still denser than water, and it sinks if it stops swimming. Continual swimming also ensures that water will flow into the mouth and out through the gills, where gas exchange occurs. However, some sharks and many skates and rays spend a good deal of time resting on the seafloor. When doing so, these fishes use muscles of the jaws and pharynx to pump water over the gills.

The largest sharks and rays are suspension feeders that feed on plankton. Most sharks, however, are carnivores that swallow their prey whole or use their powerful jaws and sharp teeth to tear flesh from animals too large to swallow in one piece (FIGURE 34.11b). Shark teeth probably evolved from the jagged scales that cover the abrasive skin. The digestive tract of many sharks is proportionately shorter than the digestive tube of many other vertebrates. Within the shark intestine is a **spiral valve,** a corkscrew-shaped ridge that increases surface area and prolongs the passage of food along the short digestive tract.

Acute senses are adaptations that go along with the active, carnivorous lifestyle of sharks. Sharks have sharp vision but cannot distinguish colors. The nostrils of sharks, like those of most fishes, open into dead-end cups. They function only for olfaction (smelling), not for breathing. Along with eyes and nostrils, the shark head also has a pair of regions in the skin that

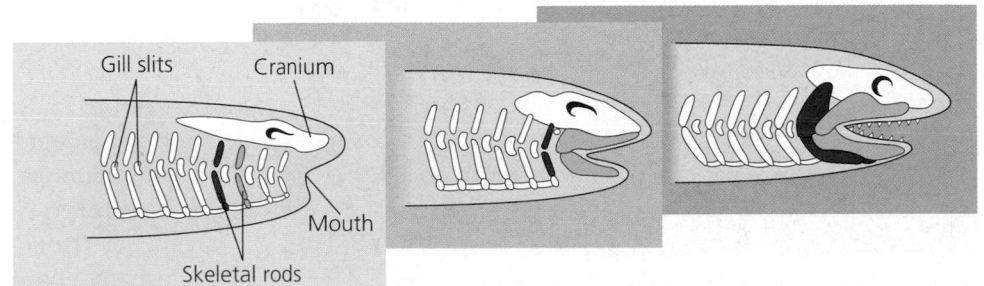

FIGURE 34.10 Hypothesis for the evolution of vertebrate jaws. The skeleton of the jaws and their supports evolved from two pairs of skeletal rods located between gill slits that were near the mouth. Pairs of rods anterior to those that formed the jaws were either lost or incorporated into the jaws.

(a) Blacktip reef shark. Fast swimmers with acute senses, sharks, such as this blacktip reef shark, have paired pectoral and pelvic fins.

(b) Great white shark. A great white shark displays its hinged jaws and toothy mouth, which complement the animal's mobility and sensory equipment as adaptations for a predatory way of life.

(c) Blue-spotted stingray. Most rays, such as the blue-spotted stingray, are flattened bottom dwellers that crush mollusks and crustaceans for food. Some other species cruise in open water, scooping food into their gaping mouths.

FIGURE 34.11 Cartilaginous fishes (class Chondrichthyes).

can detect electrical fields generated by the muscle contractions of nearby animals. Running the length of each flank of the shark is the **lateral line system,** a row of microscopic organs sensitive to changes in the surrounding water pressure. A primitive feature and present in most species of aquatic vertebrates, the lateral line system enables the animal to detect minor vibrations. Sharks and other fishes have no eardrums, structures that terrestrial vertebrates use to transmit sound waves traveling through air into the ear toward the auditory organs. Sound reaches the shark through water, and the animal's entire body transmits the sound to the hearing organs of the inner ear.

Shark eggs are fertilized internally. The male has a pair of claspers on its pelvic fins that transfer sperm into the reproductive tract of the female. Some species of sharks are **oviparous;** they lay eggs that hatch outside the mother's body. These sharks release their eggs after encasing them in protective coats. Other species are **ovoviviparous;** they retain the fertilized eggs in the oviduct. Nourished by the egg yolk, the embryos develop into young that are born after hatching within the uterus. A few species are **viviparous;** the young develop within the uterus, nourished prior to birth by nutrients received from the mother's blood through a placenta. The reproductive tract of the shark empties along with the excretory system and digestive tract into the **cloaca,** a common chamber that expels through a single vent.

Although rays are closely related to sharks, they have adopted a very different lifestyle. Most rays are flattened bottom dwellers that feed by using their jaws to crush mollusks and crustaceans (FIGURE 34.11c). A ray's pectoral fins are greatly enlarged and used like water wings to propel the animal through the water. The tail of many rays is whiplike and, in some species, bears venomous barbs that function in defense.

Osteichthyes: The extant classes of bony fishes are the ray-finned fishes, the lobe-finned fishes, and the lungfishes

Of all vertebrate groups, bony fishes are the most numerous, both in individuals and in species (about 30,000). Ranging in size from about 1 cm to more than 6 m long, bony fishes are abundant in the seas and in nearly every freshwater habitat (FIGURE 34.12).

Until recently, zoologists combined all bony fishes into a single vertebrate class (**Osteichthyes**), and we'll continue to use that name as a general, informal one for the great diversity of bony fishes. However, based on cladistic analysis, most vertebrate systematists now recognize *three* extant classes of bony fishes: the ray-finned fishes, the lobe-finned fishes, and the lungfishes (see FIGURE 34.7).

Nearly all bony fishes have an ossified endoskeleton with a hard matrix of calcium phosphate. The skin is often covered by flattened, bony scales that differ in structure from the tooth-

Rays in dorsal fin

Operculum

Lateral line

Pectoral fins   Pelvic fins

(a) Yellow perch

FIGURE 34.12 Ray-finned fishes (class Actinopterygii).

(b) Long-snouted sea horse

density of the fish. Thus, many bony fishes, in contrast to most sharks, can conserve energy by remaining almost motionless. Swim bladders evolved from balloonlike lungs, which the ancestral bony fishes may have used to supplement gas exchange by gills in shallow water.

Bony fishes are generally maneuverable swimmers, their flexible fins better for steering and propulsion than the stiffer fins of sharks. The fastest bony fishes can swim in short bursts of up to 80 km/hr.

Details about the reproduction of bony fishes vary extensively. Most species are oviparous, reproducing by external fertilization after the female sheds large numbers of small eggs. However, internal fertilization and birthing characterize other species.

like scales of sharks. Glands in the skin of a bony fish secrete a mucus that gives the animal its characteristic sliminess, an adaptation that reduces drag during swimming. In common with sharks, bony fishes have a lateral line system clearly evident as a row of tiny pits in the skin on either side of the body (FIGURE 34.13).

Bony fishes breathe by drawing water over four or five pairs of gills located in chambers covered by a protective flap called the **operculum.** Water is drawn into the mouth, through the pharynx, and out between the gills by movement of the operculum and contraction of muscles surrounding the gill chambers. This process enables a bony fish to breathe while stationary.

Another adaptation of most bony fishes not found in sharks is the **swim bladder,** an air sac that helps control the buoyancy of the fish. The transfer of gases between the swim bladder and the blood varies the inflation of the bladder and adjusts the

Both cartilaginous and bony fishes diversified extensively during the Devonian and Carboniferous periods, but whereas sharks arose in the sea, bony fishes probably originated in fresh water. Swim bladders were modified from simple lungs that had augmented the gills in gas exchange, perhaps in stagnant swamps with low oxygen content. The three classes of bony fishes that exist today had diverged by the end of the Devonian.

Nearly all the families of fishes familiar to us are **ray-finned fishes (class Actinopterygii;** Gr. *aktin,* "ray," and *pteryg,* "wing" or "fin"). The various species of bass, trout, perch, tuna, and herring are examples. The fins, supported mainly by long flexible rays, are modified for maneuvering, defense, and other functions (see FIGURE 34.12a). Ray-finned fishes spread from fresh water to the seas during their long history. (Adaptations that solve the osmotic problems of this move to salt water are

FIGURE 34.13 Anatomy of a trout, a representative ray-finned fish.

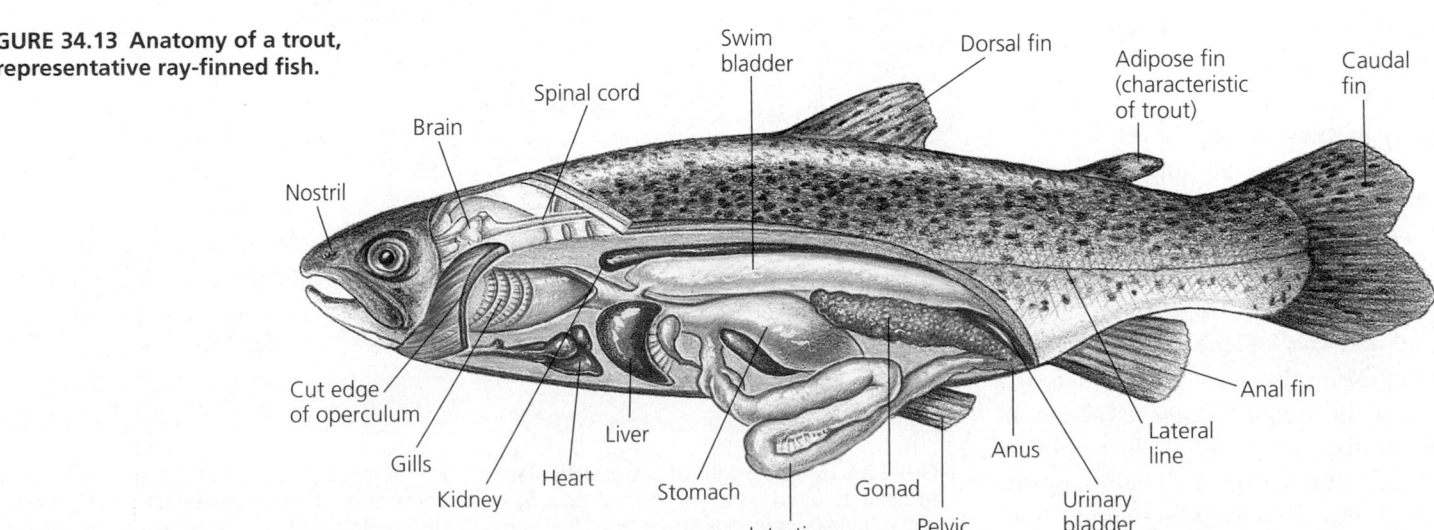

Brain   Spinal cord   Swim bladder   Dorsal fin   Adipose fin (characteristic of trout)   Caudal fin

Nostril

Cut edge of operculum   Gills   Heart   Kidney   Liver   Stomach   Intestine   Gonad   Pelvic fin   Anus   Urinary bladder   Lateral line   Anal fin

FIGURE 34.14 A coelocanth (_Latimeria_), the only extant lobe-finned genus.

discussed in Chapter 44.) Numerous species of ray-finned fishes returned to fresh water at some point in their evolution. Some of these, including salmon and sea-run trout, replay their evolutionary round-trip from fresh water to seawater back to fresh water during their life cycle.

In contrast to the ray-fins, lobe-finned fishes (**class Actinistia**) have muscular pectoral and pelvic fins supported by extensions of the bony skeleton. Known mainly from the fossil record, many lobe-fins were large, apparently bottom dwellers that may have used their paired, muscular fins as aids to "walking" on the substrate under water. Lobe-finned fishes are represented today by only one known genus, the coelacanth (_Latimeria_) (FIGURE 34.14). A population of coelacanths inhabiting waters near Indonesia may represent a different species than the only other known populations, which live in deep waters off the coasts of Madagascar and South Africa. Although most Devonian lobe-fins were probably freshwater animals with lungs, _Latimeria_ belongs to a lineage that entered the seas at some point in its evolution.

Three genera of lungfishes (**class Dipnoi**) live today in the Southern Hemisphere. They generally inhabit stagnant ponds and swamps, surfacing to gulp air into lungs connected to the pharynx of the digestive tract. Lungfishes also have gills, which are the main organs for gas exchange in Australian lungfishes. When ponds shrink during the dry season, some lungfishes can burrow in the mud and aestivate (wait in a state of torpor).

Although most lineages of lungfishes died out, these groups were dominant predators in shallow freshwater habitats during the Devonian. They were also of great importance in vertebrate evolution; the ancestor of amphibians and all other tetrapods was probably one of these Devonian bony fishes.

Tetrapods evolved from specialized fishes that inhabited shallow water

Amphibians were the first tetrapods to spend a substantial portion of their time on land. But if we define tetrapods as vertebrates that have relatively sturdy, skeleton-supported legs instead of paired fins, then the first animals to qualify were highly specialized fishes that lived in shallow aquatic habitats (FIGURE 34.15). And for the name _tetrapod_ to identify a

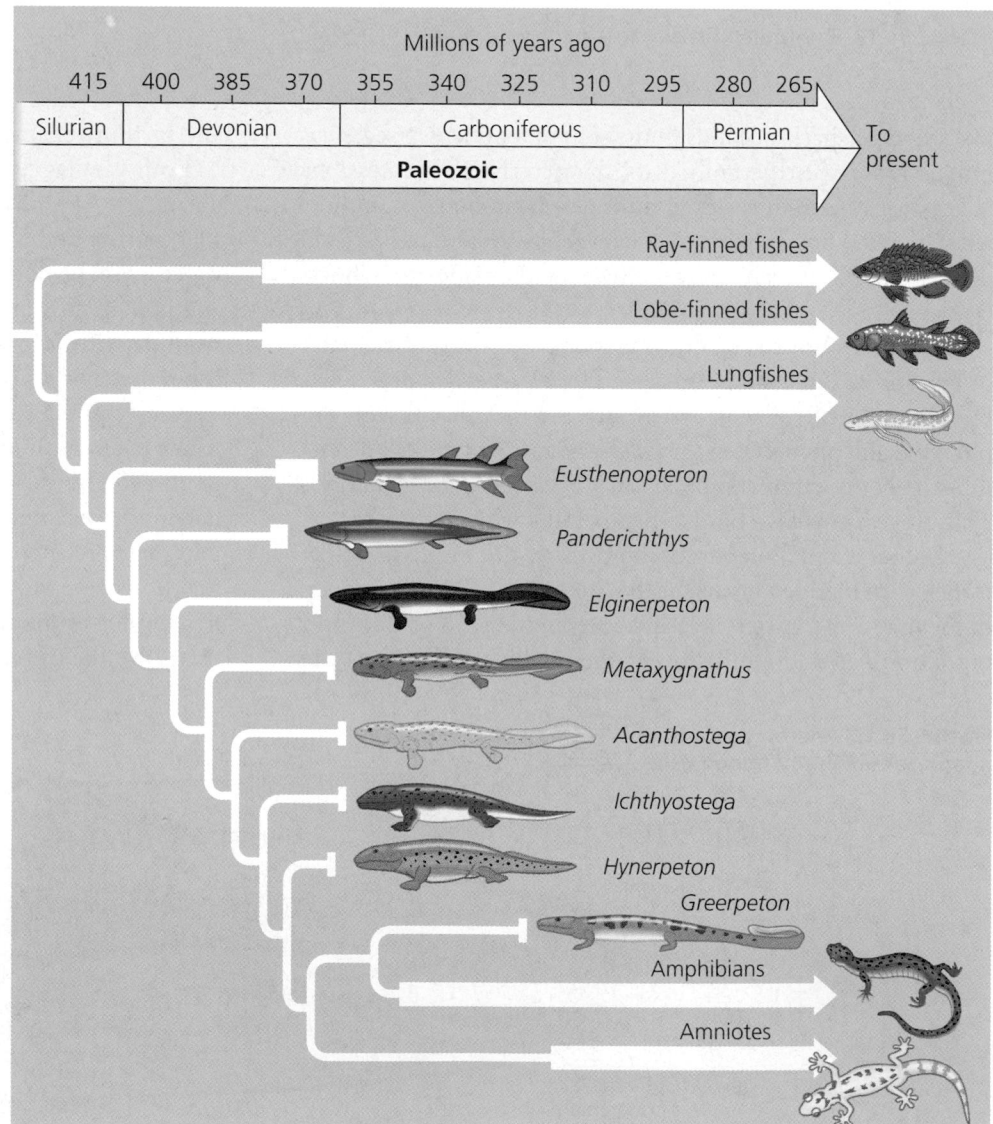

FIGURE 34.15 The origin of tetrapods. Many of the intermediate fossils on which this cladogram is based have been discovered only recently. The bars on the branches place the known fossils in time (arrows indicate lineages that extend to today). In this artist's recreation of the extinct animals, the body forms are based on fossilized skeletons, but the colors are fanciful artistic touches.

monophyletic lineage (clade), with legs as the key derived homology, then we must include these ancestral fishes.

During the Devonian period, a diversity of plants and arthropods already inhabited the land, and the evolution of trees and other large vegetation was transforming these terrestrial ecosystems. Plants rooted at the margins of ponds and swamps and organic material that dropped into the water from the terrestrial ecosystems also created new living conditions and new food supplies for fishes living near water's edge. A diversity of fishes resembling modern lobe-fins and lungfishes had already evolved. Both modern lungfishes and frogs breathe air by a mechanism called buccal pumping rather than using the rib cage, as we do. In buccal breathing, the animal drops the floor of the mouth, which sucks air in, then closes the mouth and raises the floor, forcing air into the lungs. In addition to lungs that supplemented gills for gas exchange at water's edge, leglike appendages were probably better equipment than fins for paddling and crawling through the dense vegetation in shallow water. Thus, the lungs and appendages of tetrapods evolved in certain specialized fishes tens of millions of years before this equipment was used by the earliest amphibians to live on land.

The fossil record chronicles this transition, over a 50-million-year period from 400 to 350 million years ago, from sturdy-finned fishes such as *Eusthenopteron* to tetrapod fishes to amphibians (see FIGURE 34.15). For example, fossils of *Acanthostega* have the bony supports of gills but also have four appendages with the same basic skeletal elements as the walking limbs of amphibians, reptiles, and mammals, including humans (FIGURE 34.16). Though *Acanthostega* was aquatic—it was a fish—it represents a period of vertebrate evolution when adaptations that equipped certain fishes for shallow water preadapted one lineage of those fishes for a gradual transition to spending more and more time walking and breathing on the terrestrial side of water's edge.

As the earliest terrestrial tetrapods, amphibians benefited from an abundance of food and relatively little competition. An adaptive radiation gave rise to a great diversity of amphib-

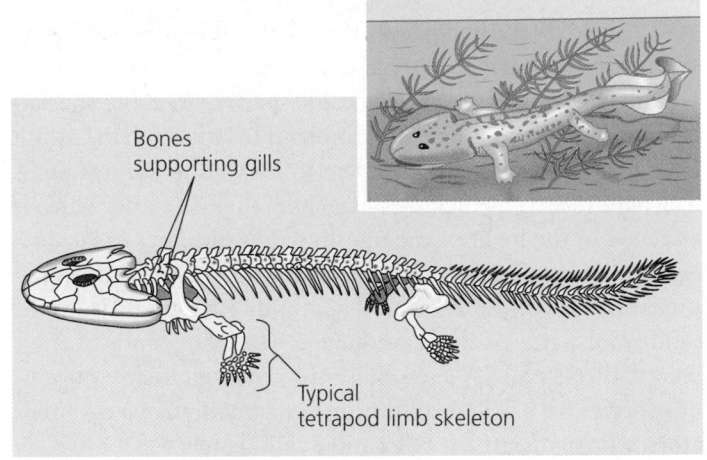

FIGURE 34.16 Skeleton of *Acanthostega*, a Devonian tetrapod fish. This vertebrate was a fish—an aquatic vertebrate with gills as an adult—but its appendages had the tetrapod skeletal structure. Early amphibians had very similar skeletons.

ians during the early Carboniferous period, and in fact, the Carboniferous is sometimes called the "age of amphibians." Amphibians began to decline in numbers and diversity during the late Carboniferous. As the Mesozoic era dawned with the Triassic period, about 245 million years ago, most surviving lineages of amphibians resembled modern species.

Class Amphibia: Salamanders, frogs, and caecilians are the three extant amphibian orders

Today the amphibians (**class Amphibia**) are represented by a total of about 4,800 species of salamanders (**order Urodela,** which means "tailed ones"), frogs (**order Anura,** "tail-less ones"), and caecilians (**order Apoda,** meaning "legless ones," describing the wormlike bodies of these amphibians).

There are only about 500 species of urodeles. Some are entirely aquatic, but others live on land as adults or throughout life (FIGURE 34.17a). Most salamanders that live on land walk

(a) Order Urodela. Urodeles (salamanders) retain their tails as adults. Some are entirely aquatic, but others live on land. This is a spotted salamander.

FIGURE 34.17 Amphibian orders.

(b) Order Anura. Anurans, such as this poison arrow frog, lack tails as adults. Poison arrow frogs inhabit tropical forests; their skin glands secrete deadly nerve toxins used by Central and South American natives to coat arrow tips.

(c) Order Apoda. Apodans, also called caecilians, are legless, mainly burrowing amphibians.

with a side-to-side bending of the body that may resemble the swagger of the early terrestrial tetrapods.

Anurans, numbering nearly 4,200 species, are more specialized than urodeles for moving on land (FIGURE 34.17b). Adult frogs use their powerful hind legs to hop along the terrain. A frog nabs insects by flicking out its long sticky tongue, which is attached to the front of the mouth. Frogs display a great variety of adaptations that help them avoid being eaten by larger predators. In common with other amphibians, many frogs exhibit color patterns that camouflage. The skin glands of frogs secrete distasteful, or even poisonous, mucus. Many poisonous species have bright coloration that apparently warns predators, who associate the coloration with danger.

Apodans, the caecilians (about 150 species), are legless and nearly blind, and superficially resemble earthworms (FIGURE 34.17c). The reduction of legs evolved secondarily from a legged ancestor. Caecilians inhabit tropical areas where most species burrow in moist forest soil. A few South American apodans live in freshwater ponds and streams.

Amphibian means "two lives," a reference to the metamorphosis of many frogs (FIGURE 34.18). The tadpole, the larval stage of a frog, is usually an aquatic herbivore with gills, a lateral line system resembling that of fishes, and a long finned tail. The tadpole lacks legs and swims by undulating like its fishlike ancestors. During the metamorphosis that leads to the "second life," legs develop, and the gills and lateral line system disappear. The young frog with air-breathing lungs, a pair of external eardrums, and a digestive system adapted to a carnivorous diet crawls onto shore and begins life as a terrestrial hunter. In spite of the name *amphibian*, however, many frogs do not go through the aquatic tadpole stage, and many amphibians do not live a dualistic—aquatic and terrestrial—life. There are some strictly aquatic and strictly terrestrial frogs, salamanders, and caecilians. Moreover, salamander and caecilian larvae look much like adults, and typically both the larvae and the adults are carnivorous. Paedomorphosis is common among some groups of salamanders; the mudpuppy *(Necturus),* for instance, retains gills and other larval features when sexually mature (see FIGURE 24.21).

Most amphibians maintain close ties with water and are most abundant in damp habitats such as swamps and rain forests. Even those frogs that are adapted to drier habitats spend much of their time in burrows or under moist leaves, where the humidity is high. Most amphibians rely heavily on their moist skin to carry out gas exchange with the environment. Some terrestrial species lack lungs and breathe exclusively through their skin and oral cavity.

Amphibian eggs lack a shell and dehydrate quickly in dry air. Fertilization is external in most species, with the male grasping the female and spilling his sperm over the eggs as the female sheds them (see FIGURE 34.18a). Amphibians generally lay their eggs in ponds or swamps or at least in moist environments. Some species lay vast numbers of eggs in temporary pools, and mortality is high. In contrast, some species display various types of parental care and lay relatively few eggs. Depending on the species, either males or females may house eggs on their back, in the mouth, or even in the stomach. Certain tropical tree frogs stir their egg masses into moist foamy nests that resist drying. There are also some ovoviviparous and viviparous species that retain the eggs in the female reproductive tract, where embryos can develop without drying out.

Many amphibians exhibit complex and diverse social behavior, especially during their breeding seasons. Frogs are usually quiet creatures, but many species fill the air with their

(a)

FIGURE 34.18 The "dual life" of a frog *(Rana temporaria).* (a) The male grasps the female, stimulating her to release eggs. The eggs are laid and fertilized in water. They have a jelly coat but lack shells and would desiccate in air. **(b)** The tadpole is an aquatic herbivore with a fishlike tail and internal gills. **(c)** During metamorphosis, the gills and tail are resorbed, and walking legs develop.

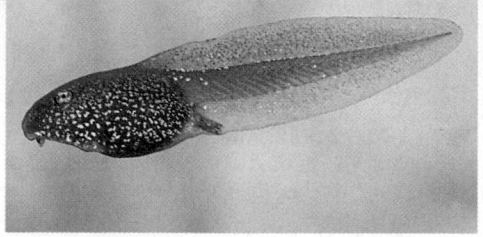

(b)

(c)

mating calls during the breeding season. Males may vocalize to defend breeding territory or to attract females. In some terrestrial species, migrations to specific breeding sites may involve vocal communication, celestial navigation, or chemical signaling.

For the past 25 years, zoologists have been documenting a rapid and alarming decline in amphibian populations throughout the world. The causes may be multiple, including environmental degradation and spread of a pathogen, a chytrid fungus. The environmental assaults include acid precipitation, which is especially damaging to amphibians because of their dependence on wet places for completing their life cycles.

AMNIOTES

Evolution of the amniotic egg expanded the success of vertebrates on land

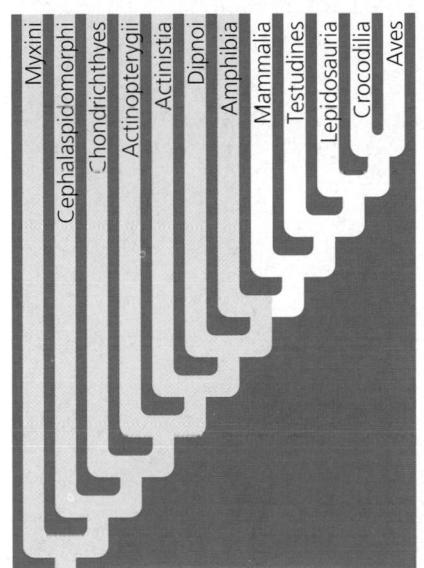

The amniote clade consists of the mammals, the birds, and the vertebrates we commonly call reptiles, including turtles, lizards, snakes, and crocodiles.

The evolution of amniotes from an amphibian ancestor involved many adaptations for terrestrial living. The amniotic egg, a reproductive adaptation that enabled terrestrial vertebrates to complete their life cycles on land and sever their last ties with their aquatic origins, was a particularly important breakthrough. (Seeds played a similar role in the evolution of land plants, as you learned in Chapter 29.) In contrast to the shell-less eggs of amphibians, amniotic eggs (of birds, most reptiles, and some mammals) have a shell that retains water and can therefore be laid in a dry place. The shells of bird eggs are calcareous (made of calcium carbonate) and inflexible, while the shells of many reptile eggs are leathery and flexible. Most mammals have dispensed with the shell; instead, the embryo implants in the wall of the uterus and obtains its nutrients from the mother.

Reptiles, birds, and mammals all have specialized membranes within the amniotic egg (FIGURE 34.19). Called **extraembryonic membranes** because they are not part of the body of the developing animal, these structures function in gas exchange, waste storage, and the transfer of stored nutrients to the embryo. They develop from tissue layers that grow out from the embryo. The amniotic egg is named for one of these membranes, the amnion, which encloses a compartment of amniotic fluid that bathes the embryo and acts as a hydraulic shock absorber. The amniotes also show other adaptations to terrestrial life, including waterproof skin and increasing use of the rib cage to ventilate the lungs.

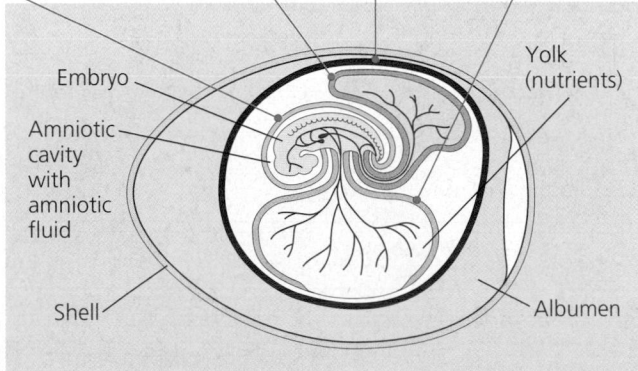

Extraembryonic membranes

Allantois. The allantois functions as a disposal sac for certain metabolic wastes produced by the embryo. The membrane of the allantois also functions with the chorion as a respiratory organ.

Chorion. The chorion and the membrane of the allantois exchange gases between the embryo and the surrounding air. O_2 and CO_2 diffuse freely across the egg's shell.

Amnion. The amnion protects the embryo in a fluid-filled cavity that prevents dehydration and cushions mechanical shock.

Yolk sac. The yolk sac expands over the yolk, a stockpile of nutrients stored in the egg. Blood vessels in the yolk sac membrane transport nutrients from the yolk into the embryo. Other nutrients are stored in the albumen (the "egg white").

Embryo

Amniotic cavity with amniotic fluid

Shell

Yolk (nutrients)

Albumen

FIGURE 34.19 The amniotic egg. The embryos of reptiles, birds, and mammals form four extraembryonic membranes: the amnion, yolk sac, allantois, and chorion. The shell is waterproof, an adaptation that prevents the embryo and the fluid compartments around it from drying out on land. This drawing represents the extraembryonic membranes in the shelled egg of a reptile or bird. Mammals have the four extraembryonic membranes, but in most species the embryos develop without shells in the reproductive tract of the mother.

Vertebrate systematists are reevaluating the classification of amniotes

FIGURE 34.20 diagrams the phylogeny of amniotes, with some of the clades identified as boxed names on the branches. Note that the amniotes are a monophyletic group (clade), with all modern reptiles, birds, and mammals sharing a common ancestor.

An evolutionary radiation of amniotes during the early Mesozoic era gave rise to three main groups, called **synapsids, anapsids,** and **diapsids**—names based on key differences between the three groups in their skull anatomy. The synapsids included mammal-like reptiles called therapsids, from which

mammals evolved. The anapsid lineage is probably totally extinct. Until recently, systematists considered turtles, based on skull anatomy, to be the only surviving anapsids. However, molecular comparisons place the turtles with the diapsids (the dotted branch in FIGURE 34.20 represents the still-uncertain ancestry of turtles). The diapsids include most or all (depending on the position of turtles) groups of modern reptiles, as well as a diversity of extinct swimming, flying, and land-based reptiles.

During the early Mesozoic radiation of amniotes, the diapsid lineage split into two evolutionary branches, the **lepidosaurs** (which include lizards, snakes, and two species of New Zealand animals called tuataras) and the **archosaurs** (which include

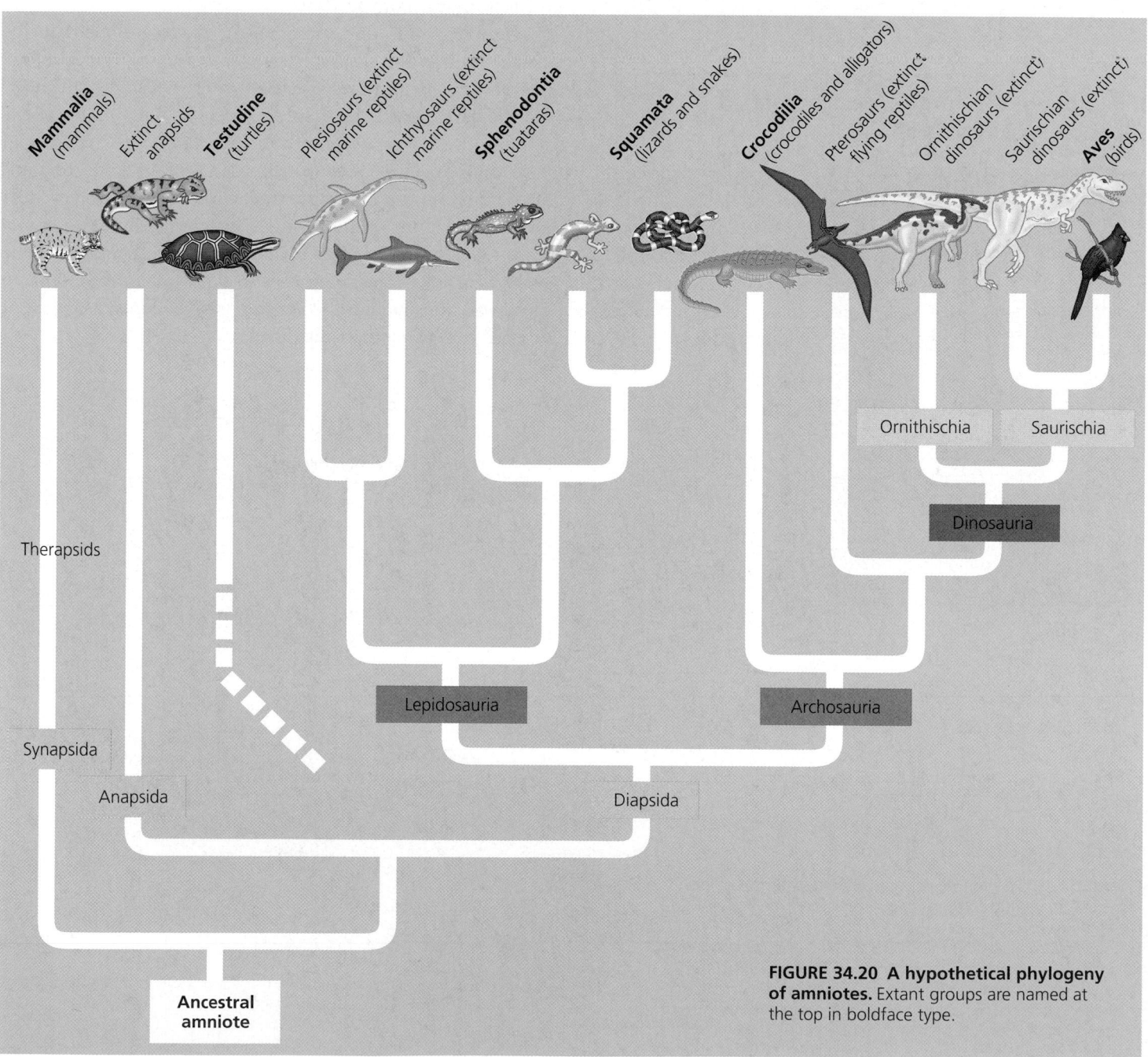

FIGURE 34.20 A hypothetical phylogeny of amniotes. Extant groups are named at the top in boldface type.

crocodiles and alligators, dinosaurs, and birds). It's important to note the position of birds in the amniote phylogeny. Birds are diapsids, specifically members of the archosaur clade. Their closest extant relatives are the crocodiles and alligators, but they are even more closely related to a branch of the extinct reptiles known as dinosaurs. In fact, neither dinosaurs nor reptiles represent monophyletic taxa unless we also include birds.

FIGURE 34.21 compares the classical classification of the amniotes with three of several alternatives that are based on strict application of cladistic conventions, in which each taxon must represent a monophyletic group. The classical taxonomy (FIGURE 34.21a) divides the amniotes into three classes, Reptilia, Aves (birds), and Mammalia. It is a classification that highlights the distinctive adaptations that evolved in the avian (bird) lineage. However, another look at the cladogram in FIGURE 34.20 will reinforce the point made earlier that the traditional class Reptilia is paraphyletic because it does not include the birds. Researchers have proposed various cladistics-based alternatives for placing the birds in the taxonomy of amniotes. These options illustrate an enduring difference of taxonomic preference, between "lumpers," who prefer to minimize the number of taxa at each taxonomic level, and "splitters," who prefer a finer-grained taxonomy having more taxa at each level. In the most extreme lumpers' version of amniote classification, the bird problem is solved by simply including birds in the class Reptilia (FIGURE 34.21b). In one splitters' version, class Aves is preserved

by also granting class status to other clades of reptiles (FIGURE 34.21c, which includes only extant groups). Another alternative identifies the two major diapsid branches as taxonomic classes, Lepidosauria and Archosauria (FIGURE 34.21d).

Whichever classification your instructor prefers, "reptile" is still a useful *informal* category for discussing all the amniotes except the birds and mammals. We will also take the liberty of extending the "reptile" appellation to include the fossilized animals of the earliest amniote radiation, which branched from an amphibian ancestor. In other words, *all* modern amniotes, including mammals and birds, evolved from forms that most of us would probably call "reptiles" if we saw them walking around today.

A reptilian heritage is evident in all amniotes

Reptilian Characteristics

Reptiles have several adaptations for terrestrial living not generally found in amphibians. Scales containing the protein keratin waterproof the skin of a reptile, helping prevent dehydration in dry air. Keratinized skin is the vertebrate analogue of the chitinized cuticle of insects and the waxy cuticle of land plants. Reptiles cannot breathe through their keratinized dry skin, and they obtain all their oxygen with lungs. Many turtles also use the moist surfaces of their cloaca for gas exchange.

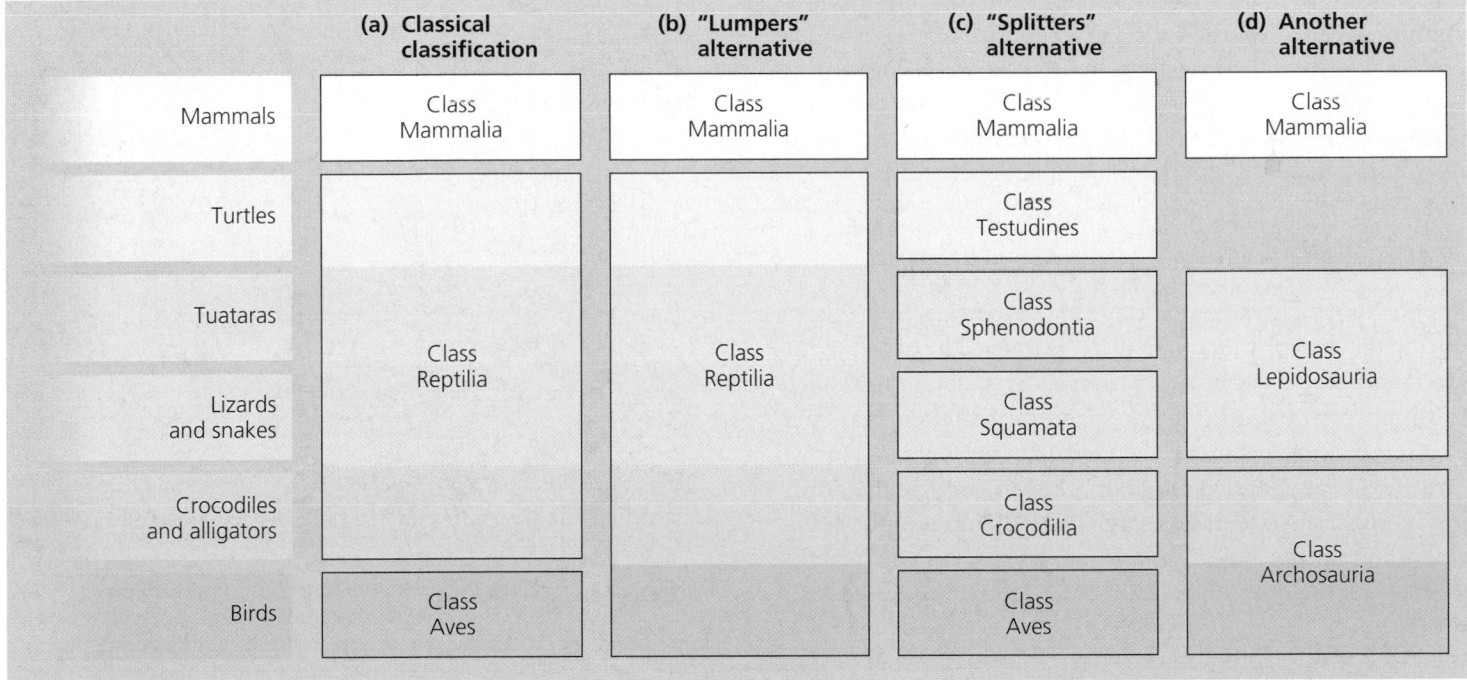

FIGURE 34.21 Taxonomic classes of amniotes. Many zoologists prefer the classical taxonomy, which is pragmatic, though it breaks from the strict cladistic requirement that named taxa must be monophyletic groups. The turtles are omitted from alternative (d) because their phylogeny is still uncertain.

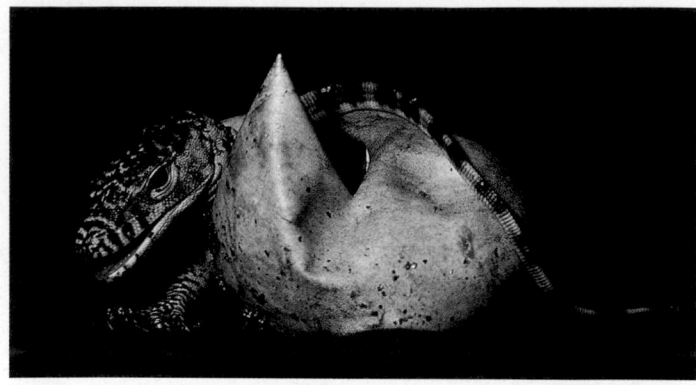

FIGURE 34.22 A hatching reptile. This Komodo dragon is breaking out of a parchmentlike shell, a common type of shell among reptiles. The shells of some reptiles are harder because of an abundance of calcium carbonate, as in bird eggs.

Most reptiles lay shelled amniotic eggs on land (FIGURE 34.22). Fertilization in reptiles must occur internally, before the shell is secreted as the egg passes through the reproductive tract of the female. Some species of snakes and lizards are viviparous, their extraembryonic membranes forming a placenta that enables the embryo to obtain nutrients from its mother.

Reptiles are sometimes labeled "cold-blooded" animals because they do not use their metabolism extensively to control body temperature. But reptiles do regulate body temperature by using behavioral adaptations. For example, many lizards regulate their internal temperature by basking in the sun when the air is cool and seeking shade when the air is too warm. Because they absorb external heat rather than generating much of their own, reptiles are said to be **ectotherms,** a term more appropriate than cold-blooded. (The control of body temperature is discussed in more detail in Chapter 44.) By heating directly with solar energy rather than through the metabolic breakdown of food, a reptile can survive on less than 10% of the calories required by a mammal of equivalent size.

The Origin and Evolutionary Radiation of Reptiles

Reptiles were far more widespread, numerous, and diverse during the Mesozoic era than they are today. The oldest reptilian fossils are found in rocks from Kansas, dating to the late Carboniferous period; they are about 300 million years old. Their ancestor was among the Devonian amphibians. In two great waves of adaptive radiation, reptiles became the dominant terrestrial vertebrates in a dynasty that lasted more than 200 million years.

The first major reptilian radiation occurred by the dawn of the Permian, the last period of the Paleozoic era, giving rise to the three main evolutionary branches introduced in FIGURE 34.20: Synapsida, Anapsida, and Diapsida.

The second great reptilian radiation was under way by late Triassic times (a little more than 200 million years ago) and was marked mainly by the origin and diversification of two groups of reptiles: the **dinosaurs,** which lived on land, and the **pterosaurs,** or flying reptiles. These groups were the dominant vertebrates on Earth for millions of years. Pterosaurs had wings formed from a membrane of skin stretched from the body wall, along the forelimb, to the tip of an elongated finger. Stiff fibers provided support for the skin of the wing. Dinosaurs, an extremely diverse group varying in body shape, size, and habitat, included the largest animals ever to inhabit land. Fossils of gigantic dinosaurs that were 45 m long have recently been unearthed in New Mexico and Utah. And Paul Sereno, the paleontologist you met on pages 508–509, has led teams that have unearthed many newly described dinosaurs that lived on land masses now located in the Southern Hemisphere. There were two main dinosaur lineages: the ornithischians, mostly herbivores; and the saurischians, with both herbivorous and carnivorous dinosaurs, including the ancestors of birds.

Contradicting the long-standing view that dinosaurs were slow, sluggish creatures, there is increasing evidence that many dinosaurs were agile, fast-moving, and, in some species, social. Paleontologists have also discovered signs of parental care among dinosaurs (FIGURE 34.23). There is continuing debate about whether dinosaurs were **endothermic,** capable of keeping the body warm through metabolism. Some anatomical evidence supports this hypothesis, but many experts are

FIGURE 34.23 Dinosaur social behavior and parental care. This artist's re-creation of duck-billed dinosaurs is based on 80-million-year-old fossils discovered in Montana. Dinosaurs of this species were relatively large, about 7 m in length. They used their bills to browse on plants. Fossilized footprints indicate that duck-billed dinosaurs traveled in large social groups, or herds. Fossilized nests contain both eggs and offspring up to a few months old, evidence that these dinosaurs took care of their young. In fact, the genus name for this dinosaur, *Maiasaura,* means "good mother lizard."

skeptical. The Mesozoic climate was relatively warm and consistent, and behavioral adaptations such as basking may have been sufficient for maintaining a suitable body temperature, especially for the land-dwelling dinosaurs. Also, large dinosaurs had low surface-to-volume ratios that reduced the effects of daily fluctuations in air temperature on the internal temperature of the animal. However, given the enormous diversity of dinosaurs, the clade could have included some endothermic species, especially among smaller forms inhabiting relatively cool climates. The dinosaur that gave rise to birds was *certainly* endothermic, as are all birds.

At the end of the Cretaceous, the last period of the Mesozoic era, the dinosaurs became extinct. Some paleontologists argue that a few species survived into the early Cenozoic. The fossil record of terrestrial vertebrates in the Cretaceous is relatively poor, and it is uncertain whether the dinosaurs were undergoing a decline before they were finished off by the asteroid impact we discussed in Chapter 25. Either way, the greatest age of the reptiles had come to an end.

Modern Reptiles

There are about 6,500 species of extant reptiles. Taxonomists classify them into four groups: **Testudines** (turtles); **Sphenodontia** (tuataras); **Squamata** (lizards and snakes), and **Crocodilia** (alligators and crocodiles). In the traditional classification scheme of FIGURE 34.21a, these four taxa are orders within the class Reptilia; in the alternative classification of FIGURE 34.21c, the four taxa are elevated to the class level.

Turtles evolved during the Mesozoic era and have scarcely changed since (FIGURE 34.24a). The usually hard shell, an adaptation that protects against predators, has certainly contributed to this long success. Those turtles that returned to water during their evolution crawl ashore to lay their eggs.

Lizards are the most numerous and diverse reptiles alive today (FIGURE 34.24b). Most are relatively small; perhaps they were able to survive the Cretaceous "crunch" by nesting in crevices and decreasing their activity during cold periods, a practice many modern lizards use.

(a) Eastern box turtle

(b) Australian frillneck lizard

(c) Western diamondback rattlesnake

(d) American alligator

FIGURE 34.24 Extant reptiles. (a) Turtles have changed little since their origin early in the Mesozoic era. This species is an eastern box turtle. **(b)** Lizards, such as this Australian frillneck, are the most numerous and diverse of the extant reptiles. A frillneck only spreads its frill when threatened, a response that probably discourages many predators. **(c)** Snakes may have evolved from lizards that adapted to a burrowing lifestyle. Note the loosely-hinged jaw of this western diamondback rattlesnake, photographed during an attempted strike. **(d)** Crocodiles and alligators are the reptiles most closely related to dinosaurs and birds. An American alligator is shown here.

Snakes are probably descendants of lizards that adapted to a burrowing lifestyle. Today, most snakes live above the ground, but they have retained the limbless condition. Vestigial pelvic and limb bones in primitive snakes such as boas, however, are evidence that snakes evolved from reptiles with legs.

Snakes are carnivorous, and a number of adaptations aid them in hunting and eating prey. They have acute chemical sensors, and though they lack eardrums, snakes are sensitive to ground vibrations, which helps them detect the movements of prey. Heat-detecting organs between the eyes and nostrils of pit vipers, including rattlesnakes, are sensitive to minute temperature changes, enabling these night hunters to locate warm animals. Poisonous snakes inject their toxin through a pair of sharp hollow or grooved teeth. The flicking tongue is not poisonous but helps fan odors toward olfactory organs on the roof of the mouth. Loosely articulated jaws enable most snakes to swallow prey larger than the diameter of the snake itself (FIGURE 34.24c).

Crocodiles and alligators (crocodilians) are among the largest living reptiles (some turtles are heavier) (FIGURE 34.24d). They spend most of their time in water, breathing air through their upturned nostrils. Crocodilians are confined to the warm regions of Africa, China, Indonesia, India, Australia, South America, and the southeastern United States, where alligators have made a strong comeback after spending years on the endangered species list. Among the modern animals traditionally classified as reptiles, crocodilians are the most closely related to the dinosaurs (see FIGURE 34.20). However, as we pointed out earlier, the modern animals that seem to share the most recent ancestor with the dinosaurs are the birds.

Birds began as feathered reptiles

Birds evolved during the great reptilian radiation of the Mesozoic era (see FIGURE 34.20). Amniotic eggs and scales on the legs are just two of the reptilian features we see in birds. But modern birds look quite different from modern reptiles because of their feathers and other distinctive flight equipment.

Characteristics of Birds

Almost every part of a typical bird's anatomy is modified in some way that enhances flight. The bones have an internal structure that is honeycombed, making them strong but light (FIGURE 34.25). The skeleton of a frigate bird, for instance, has a wingspan of more than 2 m but weighs only about 113 g (4 oz). Another adaptation reducing the weight of birds is the absence of some organs. Females, for instance, have only one ovary. Also, modern birds are toothless, an adaptation that trims the weight of the head. Food is not chewed in the mouth but ground in the gizzard, a digestive organ near the stomach. (Crocodiles also have gizzards, as did some dinosaurs.) The bird's beak, made of keratin, has proven to be very adaptable during avian evolution, taking on a great variety of shapes suitable for different diets.

Flying requires a great expenditure of energy from an active metabolism. Birds are endothermic; they use their own metabolic heat to maintain a warm, constant body temperature. Feathers and in some species a layer of fat provide insulation that enables birds to retain their metabolically generated heat. An efficient respiratory system and a circulatory system with a

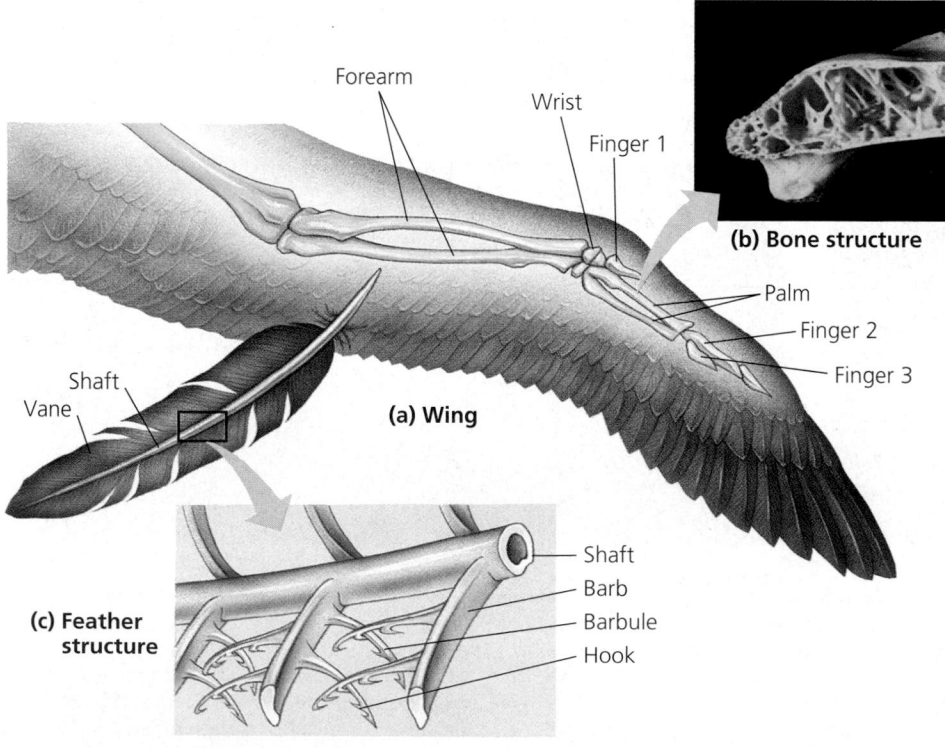

(b) Bone structure

(a) Wing

(c) Feather structure

FIGURE 34.25 Form fits function: the avian wing and feather. (a) A wing is supported by a remodeled version of the tetrapod forelimb. (b) The honeycombed anatomy of the bones is a flight adaptation that reduces weight. (c) A feather consists of a central hollow shaft, from which radiate the vanes. The vanes are made up of barbs, in turn bearing even smaller branches called barbules. Birds have contour feathers and downy feathers. Contour feathers are the stiff ones that contribute to the aerodynamic shapes of the wings and body. Their barbules have hooks that cling to barbules on neighboring barbs. When a bird preens, it runs the length of a feather through its bill, engaging the hooks and uniting the barbs into a precisely shaped vane. Downy feathers lack hooks, and the free-form arrangement of barbs produces a fluffiness that provides excellent insulation because of the trapped air.

FIGURE 34.26 A bald eagle in flight. Bird wings are airfoils, structures whose shape creates lift by altering air currents. The leading edge of an airfoil is thicker than the trailing edge, its upper surface is somewhat convex, and its under surface is flattened or concave. This shape makes the air passing over the wing travel farther than the air passing under the wing. As a result, air molecules are spaced farther apart (expanded) above the wing than below it. Thus, the air pressure pushing downward on the wing is less than the pressure pushing upward on the wing's bottom surface. This pressure differential provides the "lift" for flight.

four-chambered heart keep tissues well supplied with oxygen and nutrients, supporting a high rate of metabolism. The lungs have tiny tubes leading to and from elastic air sacs that help dissipate heat and reduce the density of the body.

For safe flight, senses, especially vision, must be acute. Birds have excellent eyes, perhaps the best of all the vertebrates. The visual areas of the brain are well developed, as are the motor areas; flight also requires excellent coordination.

With brains proportionately larger than those of reptiles and amphibians, birds generally display very complex behavior. Avian behavior is particularly intricate during breeding season, when birds engage in elaborate rituals of courtship. Because eggs are shelled when laid, fertilization must be internal. Copulation involves contact between the mates' vents, the openings to their cloacas. After eggs are laid, the avian embryo must be kept warm through brooding by the mother, father, or both, depending on the species.

A bird's most obvious adaptation for flight is its wings. Bird wings are airfoils that illustrate the same principles of aerodynamics as the wings of an airplane (FIGURE 34.26). Providing power for flight, birds flap their wings by contractions of large pectoral (breast) muscles anchored to a keel on the sternum (breastbone). Some birds, such as eagles and hawks, have wings adapted for soaring on air currents and flap their wings only occasionally; other birds, including hummingbirds, must flap continuously to stay aloft. In either case, it is the shape and arrangement of the feathers that form the wing into an airfoil. The fastest birds are the appropriately named swifts, which can fly 170 km/hr.

In being both extremely light and strong, feathers are among the most remarkable of vertebrate adaptations. Feath-

ers are made of keratin, the same protein that forms our hair and fingernails and the scales of reptiles. Feathers may have functioned first as insulation during the evolution of endothermy, only later being co-opted as flight equipment.

The evolution of flight required radical alteration in body form, but flight provides many benefits. It enhances hunting and scavenging; many birds exploit flying insects, an abundant, highly nutritious food resource. Flight also provides ready escape from earthbound predators and enables some birds to migrate great distances to utilize different food resources and seasonal breeding areas. The bird that travels farthest in its annual migration is the arctic tern, which flies round-trip between the North Pole and South Pole each year.

The Origin of Birds

Cladistic analyses of fossilized skeletons support the hypothesis that the closest reptilian relatives of birds were the **theropods,** a group of relatively small, bipedal, carnivorous dinosaurs. (The velociraptors of *Jurassic Park* fame were theropods.) Most researchers agree that the ancestor of birds was a feathered theropod. However, some scientists place the origin of birds much earlier, from an ancestor common to both birds and dinosaurs.

THE PROCESS OF SCIENCE

The most famous Mesozoic bird is *Archaeopteryx,* known from fossils discovered in Bavarian limestone in Germany. This ancient bird lived about 150 million years ago, during the late Jurassic period. Unlike modern birds, but similar to reptiles, *Archaeopteryx* had clawed forelimbs, teeth, and a long tail containing vertebrae (FIGURE 34.27). Thus, *Archaeopteryx* was

FIGURE 34.27 *Archaeopteryx,* **a Jurassic bird-reptile.**

even more anatomically similar to dinosaurs than are modern birds. In fact, if it were not for the preservation of its feathers, *Archaeopteryx* would probably be classified as a theropod dinosaur. The skeletal anatomy of *Archaeopteryx* indicates that it was a weak flyer, perhaps mainly a tree-dwelling glider. A combination of gliding downward and jumping into the air from the ground may have been the precursor of actual flight in the evolution of birds. However, *Archaeopteryx* is not considered to be the ancestor of modern birds but probably represents an extinct side branch of avian evolution. Nonetheless, the evolutionary branch that included *Archaeopteryx* probably stemmed from an ancestor that also gave rise to the lineage from which modern birds evolved.

In 1998, paleontologists described a diversity of fossils in Chinese sediments that may help fill in the gap between dinosaurs and early birds such as *Archaeopteryx*. The Chinese fossils are feathered but flightless dinosaurs (FIGURE 34.28). Such anatomy is consistent with a hypothesis that feathers evolved as adaptations with functions other than flight, such as thermoregulation or courtship displays for mating. However, the Chinese fossils are about 20 million years *younger*

than *Archaeopteryx*. Do the Chinese fossils represent descendants of feathered, flightless dinosaurs that survived for tens of millions of years, even as *Archaeopteryx* and other birds also evolved from those flightless ancestors? This hypothesis makes the prediction that paleontologists may one day unearth flightless dinosaurs with feathers that predate *Archaeopteryx*. Or did these flightless, feathered animals evolve secondarily from early flying birds, analogous to the abandonment of flight in such modern birds as ostriches? Certain skeletal features support this second hypothesis.

The Chinese sediments also include younger fossils with wings, such as an abundant Cretaceous bird named *Confuciusornis*. It had many features suggesting a closer kinship to modern birds than can be claimed for *Archaeopteryx*. For example, the weight-trimming adaptations of *Confuciusornis* included the absence of teeth, the replacement of certain skull elements by a horny bill, and a tail reduced to a short stub.

The intense current interest in the origin of birds will undoubtedly bring us closer to understanding how these masters of the sky evolved from nonflying reptiles.

(a) *Sinoauropteryx*

(b) *Caudipteryx*

FIGURE 34.28 Cretaceous theropod dinosaurs with putative feathers from Chinese sediments. (a) *Sinoauropteryx* was a flightless dinosaur with a fringe of filaments along its back that may have been "protofeathers." **(b)** *Caudipteryx,* also flightless, bore true feathers (inset).

Modern Birds

There are about 8,600 extant species of birds classified in about 28 orders. Flight is typical of birds, but there are several flightless species. Among flightless birds, the ostrich, kiwi, and emu are called **ratites** (Latin, "flat-bottomed") because their breastbone lacks a keel and the large breast muscles that attach to it in flying birds (FIGURE 34.29a). In contrast to ratites, other birds have a sternal keel supporting their large breast muscles. These muscles provide flight power in flying birds.

The demands of flight have rendered the general body form of many birds similar to one another, yet experienced bird watchers can distinguish diverse species by their body profile. Birds also exhibit great variety in feather colors, beak and foot shape, behavior, and flying style (FIGURE 34.29b–d). Among the most unusual birds are the penguins, which do not fly, but use their powerful breast muscles in swimming.

Nearly 60% of the living bird species belong in one order called the passeriformes, or perching birds. These are the familiar jays, swallows, sparrows, warblers, and many others (FIGURE 34.29d).

Mammals diversified extensively in the wake of the Cretaceous extinctions

With the extinction of the dinosaurs and the fragmentation of continents that occurred at the close of the Mesozoic era, mammals underwent an extensive adaptive radiation. There are about 4,500 species of mammals on Earth today. As mammals ourselves, we naturally have a special interest in this class of vertebrates. Let's examine some of the features we share with all other mammals.

(a) Emus. These ratite birds live in Australia.

(b) Harlequin duck. In common with many species, the harlequin duck, which inhabits mountain streams and coastal areas of the Pacific Northwest, exhibits pronounced color differences between the sexes.

(c) Laysan albatrosses. Most birds, as exhibited here by a pair of Laysan albatrosses, have specific courtship and mating behavior.

FIGURE 34.29 A small sample of birds.

(d) Barn swallow. The barn swallow is a member of the order Passeriformes. Passeriformes are called perching birds because the toes of their feet can lock around a tree branch, enabling the bird to rest in place for long periods of time.

Mammalian Characteristics

Vertebrates of the **class Mammalia** were first defined by Linnaeus by their possession of mammary glands. Mammary glands, which produce milk, are a distinctively mammalian characteristic: All mammalian mothers nourish their babies with milk, a balanced diet rich in fats, sugars, proteins, minerals, and vitamins. Hair is another mammalian characteristic. Like the feathers of birds, hair is made of keratin, although zoologists are uncertain about its evolutionary origin. Mammals are endothermic and most mammals have an active metabolism. Efficient respiratory and circulatory systems (including a four-chambered heart) support a high metabolic rate. A sheet of muscle called the diaphragm helps ventilate the lungs. Hair and a layer of fat under the skin help the body retain metabolic heat.

Most mammals are born rather than hatched. Fertilization is internal, and the embryo develops inside the uterus of the female's reproductive tract. In eutherian (placental) mammals and marsupials, the lining of the mother's uterus and extraembryonic membranes arising from the embryo collectively form a **placenta,** where nutrients diffuse into the embryo's blood. Eutherians are commonly called placental mammals because, compared to marsupials, their placentas are more complex and provide a more intimate and long-lasting association between the mother and her developing young.

Mammals generally have larger brains than other vertebrates of equivalent size, and many species are capable learners. The relatively long duration of parental care extends the time for offspring to learn important survival skills by observing their parents.

Differentiation of teeth is another important mammalian trait. Whereas the teeth of reptiles are generally conical and uniform in size, the teeth of mammals come in a variety of sizes and shapes adapted for chewing many kinds of foods. Our own dentition, for example, includes teeth modified for shearing (incisors and canine teeth) and for crushing and grinding (premolars and molars). Another part of the feeding apparatus, the jaw, was also remodeled during the evolution of mammals from reptiles, and two of the bones formerly making up the jaw joint were incorporated into the mammalian inner ear (FIGURE 34.30).

The Evolution of Mammals

Mammals evolved from reptilian stock even earlier than birds. The oldest fossils believed to be mammalian date back 220 million years, into the Triassic period. The ancestor of mammals was among the **therapsids,** part of the synapsid branch of reptilian phylogeny (see FIGURE 34.20). An extensive record of fossils from the Permian and Triassic connects mammals with their reptilian ancestors. These fossils show that the distinctively mammalian structure of the legs, skull, jaws, and teeth evolved in many small stages. The therapsids disappeared during the reign of the dinosaurs, but their mammalian descendants coexisted with the dinosaurs throughout the Mesozoic era. Most Mesozoic mammals were very small—about the size of shrews—and most probably ate insects. A variety of evidence, such as the size of the eye sockets, suggests that these tiny mammals were nocturnal.

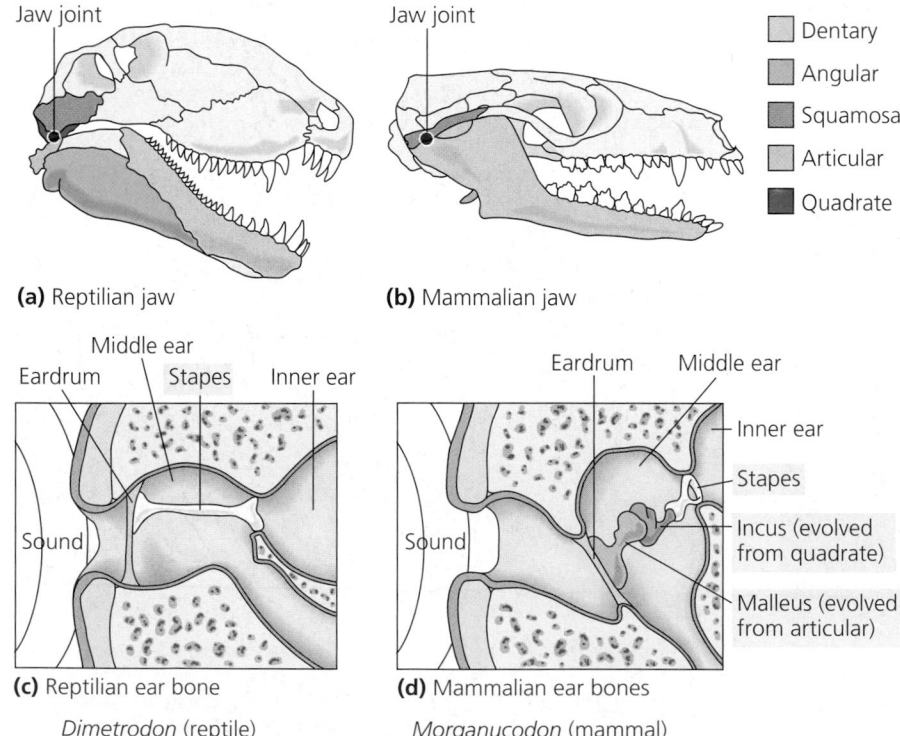

(a) Reptilian jaw

(b) Mammalian jaw

Dentary
Angular
Squamosal
Articular
Quadrate

(c) Reptilian ear bone

Dimetrodon (reptile)

(d) Mammalian ear bones

Morganucodon (mammal)

FIGURE 34.30 The evolution of the mammalian jaw and ear bones. The reptile *Dimetrodon* (left) was an early synapsid, the lineage that eventually gave rise to the mammals. *Morganucodon* (right) was one of the first mammals. **(a)** Notice that the lower jaw of the reptile is composed of several fused bones. In the reptile, two small bones, the quadrate and articular, form part of the jaw joint. **(b)** The mammalian lower jaw is reduced to a single bone, the dentary, and the location of the jaw joint has shifted. The single-bone lower jaw of mammals is stronger than the multibone jaw of reptiles. **(c)–(d)** During the evolutionary remodeling of the mammalian skull, the quadrate and articular bones became incorporated into the middle ear as two of the three bones that transmit sound from the eardrum to the inner ear. The steps in this evolutionary remodeling are evident in a succession of fossils of mammal-like reptiles.

As the Cenozoic era dawned in the wake of the Cretaceous extinctions, mammals were in the midst of a great adaptive radiation (though the major lineages may have originated earlier—see FIGURE 25.19). That diversity is represented today in the three major groups: monotremes (egg-laying mammals), marsupials (mammals with pouches), and eutherian (placental) mammals.

Monotremes

Monotremes—the platypuses and the echidnas (spiny anteaters)—are the only living mammals that lay eggs (FIGURE 34.31a). The egg, which is reptilian in structure and development, contains enough yolk to nourish the developing embryo. Monotremes have hair and produce milk, two of the most important trademarks of Mammalia. On the belly of a monotreme mother are specialized glands that secrete milk. After hatching, the baby sucks the milk from the fur of the mother, who has no nipples. The mixture of ancestral reptilian characters and derived characters of mammals suggests that monotremes descended from a very early branch in the mammalian genealogy. Today, monotremes are found only in Australia and New Guinea.

Marsupials

Opossums, kangaroos, bandicoots, and koalas are examples of marsupial mammals. A **marsupial** is born very early in its development and completes its embryonic development while nursing (FIGURE 34.31b). In most species, the nursing young are held within a maternal pouch called a marsupium. A red kangaroo, for instance, is about the size of a honeybee at its birth, just 33 days after fertilization. Its hind legs are merely buds, but the forelimbs are strong enough for the offspring to crawl from the exit of the reproductive tract to the mother's pouch, a journey lasting a few minutes.

In Australia, marsupials have radiated and filled niches occupied by eutherian (placental) mammals in other parts of the world. Convergent evolution has resulted in a diversity of marsupials that resemble their eutherian counterparts occupying similar ecological roles (FIGURE 34.32, p. 704).

The opossums of North and South America are the most diverse of only three families of extant marsupials outside the Australian region, though South America had a diverse marsupial fauna throughout the Tertiary period. After the breakup of Pangaea, South America and Australia became island continents, and their marsupials diversified in isolation from the placental mammals that began an adaptive radiation on the northern continents. Australia has not been in contact with

FIGURE 34.31 Australian monotremes and marsupials.

(a) Short-beaked echidna, a monotreme. Monotremes are the only mammals that lay eggs (inset). Monotremes have hair and milk, but no nipples.

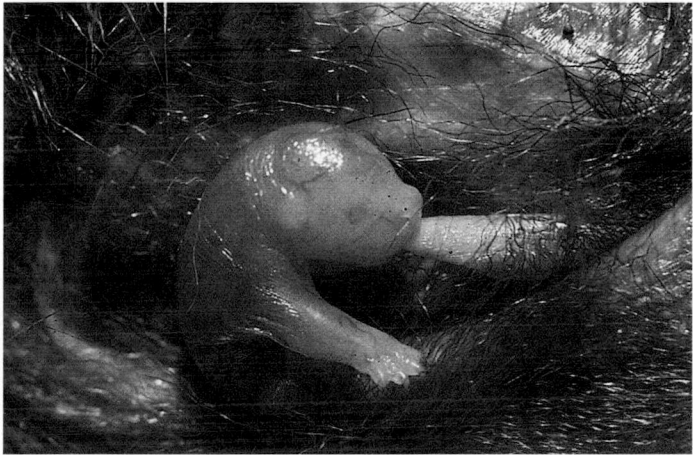

(b) A young marsupial. The young of marsupials, such as this Australian brushtail possum, are born very early in their development. They finish their growth while nursing from a nipple (in their mother's pouch in most species).

(c) Long-nosed bandicoot, a marsupial. Bandicoots have a pouch that opens to the rear, whereas in other marsupials, such as the kangaroos, it opens to the front. Most bandicoots are diggers and burrowers that eat mainly insects but also some small vertebrates and plant material. Their rear-opening pouch helps protect the young from dirt as the mother digs.

Marsupial mammals	Eutherian mammals
Plantigale	Deer mouse
Marsupial mole	Mole
Sugar glider	Flying squirrel
Wombat	Woodchuck
Tasmanian devil	Wolverine
Kangaroo	Patagonian cavy

FIGURE 34.32 Evolutionary convergence of marsupial and eutherian (placental) mammals. Adaptive radiation in Australia has fit marsupials to many of the ecological roles filled by eutherian mammals on other continents. Convergent evolution has produced a number of remarkable look-alikes. (Drawings are not to scale.)

another continent since early in the Cenozoic era, about 65 million years ago. The South American fauna, meanwhile, has not remained cloistered. Placental mammals reached South America throughout the Cenozoic. The most important migrations occurred about 12 million years ago and then again about 3 million years ago, when North and South America joined at the Panamanian isthmus. Extensive two-way traffic of animals took place over the land bridge. The biogeography of mammals is another example of the interplay between biological and geologic evolution.

Eutherian (Placental) Mammals

Compared to marsupials, **eutherian mammals** (placentals) have a longer period of pregnancy. Young eutherians complete their embryonic development within the uterus, joined to the mother by the placenta.

Adaptive radiation during the Cretaceous and early Tertiary periods produced the orders of eutherian mammals we recognize today (TABLE 34.1). Marsupials and eutherians are more closely related to each other than either is to monotremes.

Table 34.1 Major Orders of Mammals (Orders are color-coded according to the clades in FIGURE 34.33)

Order Examples	Main Characteristics		Order Examples	Main Characteristics	
Monotremata Platypuses, echidnas	Lay eggs; have no nipples; suck milk from fur of mother	Echidna	**Carnivora** Dogs, wolves, bears, cats, weasels, otters, seals, walruses	Carnivorous; possess sharp, pointed canine teeth and molars for shearing	Coyote
Marsupialia Kangaroos, opossums, koalas	Embryonic development completed in marsupial pouch	Koala	**Cetartiodactyla**		
			Artiodactyls Sheep, pigs, cattle, deer, giraffes	Possess hooves with an even number of toes on each foot; herbivorous	Bighorn sheep
Proboscidea Elephants	Have a long, muscular trunk; thick, loose skin; upper incisors elongated as tusks	African elephant	**Cetaceans** Whales, dolphins, porpoises	Marine forms with fish-shaped bodies, paddlelike forelimbs and no hind limbs; thick layer of insulating blubber	Pacific white-sided porpoise
Sirenia Sea cows (manatees)	Aquatic herbivores; possess finlike forelimbs and no hindlimbs	Manatee	**Perissodactyla** Horses, zebras, tapirs, rhinoceroses	Possess hooves with an odd number of toes on each foot; herbivorous	Indian rhinoceros
Edentata Sloths, anteaters, armadillos	Have reduced or no teeth	Tamandua	**Chiroptera** Bats	Adapted for flying; possess a broad skinfold that extends from elongated fingers to body and legs	Frog-eating bat
Rodentia Squirrels, beavers, rats, porcupines, mice	Possess chisel-like, continuously growing incisor teeth	Red squirrel	**Insectivora** "Core insectivores": some moles, some shrews	Insect-eating mammals	Star-nosed mole
Lagomorpha Rabbits, hares, pikas	Possess chisel-like incisors, hind legs longer than forelegs and adapted for running and jumping	Jackrabbit			
Primates Lemurs, monkeys, apes, humans	Opposable thumb; forward-facing eyes; well-developed cerebral cortex; omnivorous	Golden lion tamarin			

Fossil evidence indicates that marsupials and eutherians may have diverged from a common ancestor about 80 to 100 million years ago, but the molecular clock suggests an older date, at least 125 million years ago. The common ancestor of monotremes and the other two groups may have lived more like 200 million years ago.

Molecular systematics has helped to clarify the evolutionary relationships among the orders of eutherian mammals, though there is still no broad consensus on a phylogenetic tree. The main current hypothesis clusters the eutherian orders into four main clades (FIGURE 34.33).

One branch includes the elephants, together with several less well-known African mammals, such as the aardvarks and hyraxes, as well as manatees. This eutherian clade is referred to as the **Afrotheria.**

A second branch underwent a radiation in South America, and consists of the sloths, anteaters, and armadillos—the order Edentata in TABLE 34.1.

A third clade includes the bats (Chiroptera), certain insectivorous mammals (the "core insectivores," such as some of the animals that have the common names shrews and moles), carnivores, two ungulate (hoofed) orders—artiodactyls (pigs, cows, camels, and hippos) and perissodactyls (horses and rhinoceroses)—together with the cetaceans (dolphins and whales). One of the most striking results of recent phylogenetic research on mammals is that the cetaceans probably form a branch within the artiodactyls. Thus, hippos are more

closely related to whales and dolphins than they are to other artiodactyls, such as sheep and camels. In late 2001, in fact, researchers reported an Eocene whale fossil *(Rodhocetus)* in Pakistan that has a foot skeleton very similar to the foot anatomy of hippos and pigs. The hypothetical mammalian phylogeny based on molecular systematics predicts that whales with hoof-like swimming appendages once lived, and now the fossil record has corroborated the molecular evidence. The traditional order Artiodactyla is therefore not monophyletic, and researchers suggest that the cetaceans and artiodactyls should be combined in the order Cetartiodactyla (see TABLE 34.1). The order Carnivora includes cats, dogs, raccoons, skunks, and the pinnipeds (seals, sea lions, and walruses). Seals and their relatives apparently evolved from middle Cenozoic carnivorans that became adapted for swimming.

The fourth eutherian branch is by far the largest in terms of numbers of modern species. It contains the lagomorphs (rabbits and relatives) and rodents, together with the primates. Order Rodentia includes rats, mice, squirrels, and beavers. With about 1,770 species, this is by far the largest order of mammals. The name *Rodentia,* from the Latin word for "gnawing," refers to one of the distinctive features of the group: Both the upper jaw and the lower jaw have a pair of large front teeth (incisors) that resist heavy wear by growing continually. Order Primates, which we discuss next, includes monkeys, apes, and humans.

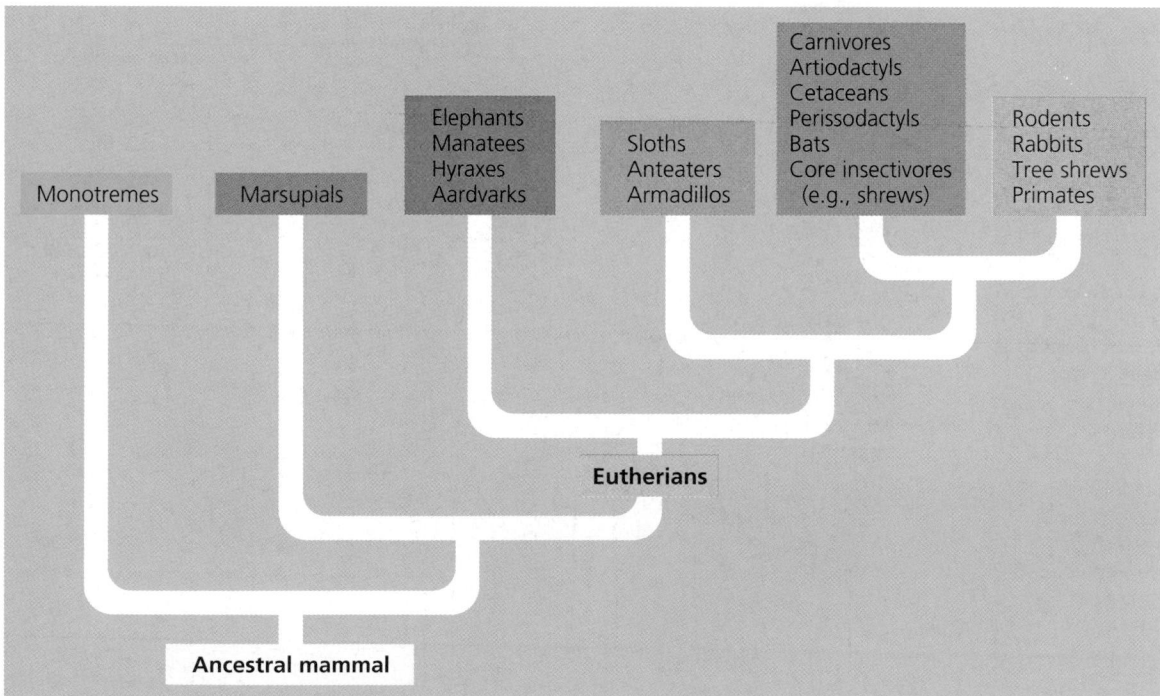

FIGURE 34.33 Hypothetical cladogram of mammals.

PRIMATES AND THE EVOLUTION OF *Homo sapiens*

Primate evolution provides a context for understanding human origins

We have tracked vertebrate evolution through the mammalian orders and can now begin to trace our own specific ancestry. Order Primates includes *Homo sapiens* and its closest kin. We are primates.

Some General Primate Characteristics

Primates are difficult to define unambiguously in terms of morphological attributes, but most primates have hands and feet adapted for grasping, and, relative to other mammals, they have large brains and short jaws, giving them a flat face. Primates have forward-looking eyes, close together on the front of the face. They also have flat nails on their digits rather than the narrow claws of other mammals. There are other changes in the hands and feet, such as the evolution of the skin ridges on fingers (these account for our fingerprints). Primates, perhaps related to their large brains, also have relatively well-developed parental care and relatively complex social behavior.

The earliest primates were probably tree dwellers, shaped through natural selection to the demands of living in the trees. For example, the grasping hands and feet of primates are adaptations for hanging on to tree branches. All modern primates, except *Homo,* have a big toe that is widely separated from the other toes, enabling them to grasp onto branches with their feet. A similar separation of the thumb from the other fingers evolved in the hands of primates. Humans have a fully **opposable thumb;** that is, we can touch the ventral surface (fingerprint side) of the tip of all four fingers with the ventral surface of the thumb of the same hand. The thumb is relatively mobile and separate from the fingers in *all* primates, but a *fully* opposable thumb is found only in anthropoid primates (monkeys, apes, and humans). The opposable thumb, like the grasping foot, evolved first for gripping branches. In monkeys and apes, the opposable thumb functions in a grasping "power grip," but in humans, the opposable thumb has adaptations that enable more precise manipulation; we have a distinctive bone structure at the base of our thumbs associated with our "precision grip." The unique dexterity of humans represents descent with modification from ancestral hands adapted for life in the trees. Some of the other primate features also originated as adaptations for tree dwelling. For instance, the overlapping fields of vision of the two eyes enhance depth perception, an obvious advantage when brachiating (swinging). Excellent eye-hand coordination is also important for ar-

boreal maneuvering. Humans, of course, do not live in trees, but we retain in modified form many of the traits that originally evolved there.

Modern Primates

The two subgroups of the Primates are the Prosimii and Anthropoidea. The **prosimians** ("premonkeys") probably resemble early arboreal primates (FIGURE 34.34). The lemurs of Madagascar and the lorises, pottos, and tarsiers that live in tropical Africa and southern Asia are examples of prosimians. The **anthropoids** include monkeys, apes, and humans. The oldest known anthropoid fossils, discovered in China in mid-Eocene strata about 45 million years old, support the hypothesis that tarsiers are the prosimians most closely related to anthropoids (FIGURE 34.35, p. 708).

The fossil record indicates that by 40 million years ago, monkeys were established both in the Old World (in Africa and Asia) and in the New World (in South America). By that time, South America and Africa had drifted apart. The first monkeys probably evolved in the Old World, and may have reached South America by rafting on logs or other debris from Africa. What is certain is that New World monkeys and Old

FIGURE 34.34 Prosimians: crowned lemurs.

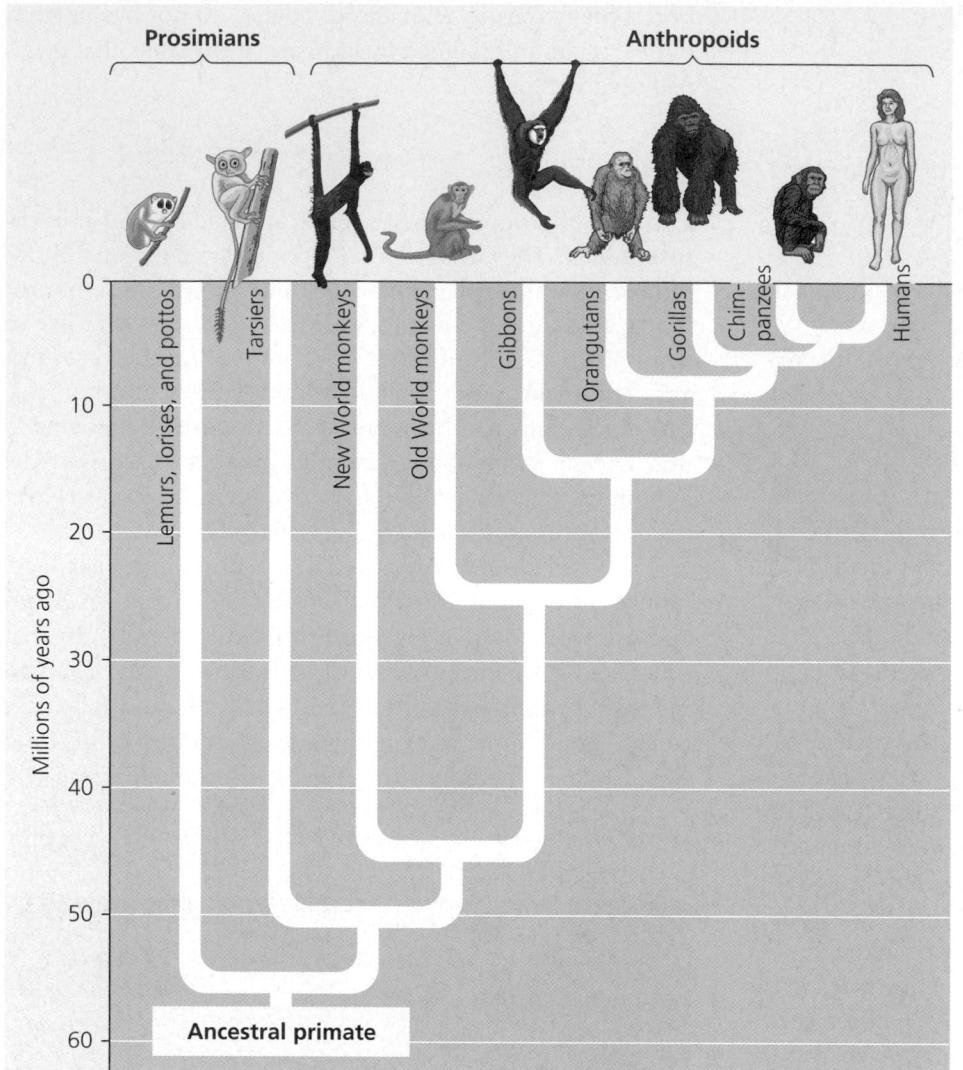

FIGURE 34.35 A phylogenetic tree of primates. The fossil record indicates that the prosimian and anthropoid lineages were diverging by about 50 million years ago. Among the anthropoids, New World monkeys, Old World monkeys, and the apes (represented today by several species of gibbons, the orangutan, the gorilla, and chimpanzees) have been evolving on separate lineages for over 40 million years. Molecular evidence shows that chimpanzees are more closely related to humans than are any other apes. The time of divergence of the hominid lineage (human branch of the evolutionary tree) is somewhere between 5 and 7 million years ago.

World monkeys underwent separate adaptive radiations during their many millions of years of separation (FIGURE 34.36). All New World monkeys are arboreal, whereas Old World monkeys include ground-dwelling as well as arboreal species. Most monkeys of both groups are diurnal (active during the day) and usually live in bands held together by social behavior.

In addition to monkeys, the anthropoid suborder also includes the four genera of apes, shown in FIGURE 34.37: *Hylobates* (gibbons), *Pongo* (orangutans), *Gorilla* (gorillas), and *Pan* (chimpanzees and bonobos). Modern apes are confined exclusively to tropical regions of the Old World; the ancestors of the apes evolved from Old World monkeys about 25–30 million

(a) New World monkey (white-faced capuchin)

(b) Old World monkeys (pig-tailed macaques)

FIGURE 34.36 New World monkeys and Old World monkeys compared. **(a)** New World monkeys, such as spider monkeys, squirrel monkeys, and capuchins, have prehensile tails and nostrils that open to the sides. **(b)** Old World monkeys lack prehensile tails, and their nostrils open downward. The tough seat pad is unique to the Old World group, which includes macaques, mandrills, baboons, and rhesus monkeys.

(a) Gibbon

(b) Orangutan

(c) Gorilla with infant

(d) Chimpanzee

(e) Bonobo with infant

FIGURE 34.37 Apes. (a) Gibbons, such as this white-handed gibbon, which is brachiating, have long arms and are among the most acrobatic of all primates. These Asian primates are also the only monogamous apes. **(b)** The orangutan is a shy and solitary ape that lives in the rain forests of Sumatra and Borneo. Orangutans spend most of their time in trees, but they do venture onto the forest floor occasionally. Note the foot adapted for grasping and the opposable thumb **(c)** Gorillas are the largest apes, with some males almost 2 m tall and weighing about 200 kg. These herbivores are confined to Africa, where they usually live in small groups of about 10 to 20 individuals. **(d)** Chimpanzees live in tropical Africa. They feed and sleep in trees but also spend a great deal of time on the ground. Chimpanzees are intelligent, communicative, and social. **(e)** Bonobos, smaller than chimpanzees, survive today only in the African nation of Congo.

years ago. With the exception of gibbons, modern apes are larger than monkeys, with relatively long arms and short legs and no tails. Although all the apes are capable of brachiation (traveling by swinging from branch to branch in trees), only gibbons and orangutans are primarily arboreal. Social organization varies among the genera of apes; gorillas and chimpanzees are highly social. Apes have larger brains proportionate to body size than monkeys, and their behavior is more flexible.

Humanity is one very young twig on the vertebrate tree

In the continuum of life spanning over 3.5 billion years, humans and apes have shared ancestry for all but the last few million years. **Paleoanthropology**, the study of human origins and evolution, focuses on this tiny fraction of geologic time during which humans and chimpanzees diverged from a common ancestor.

Paleoanthropologists use two words that are easy to confuse. **Hominoid** is a broader term, referring to great apes and humans. (Note that "anthropoid" is an even broader term, since it includes monkeys.) If a fossil is called "hominoid," it is related to us but may be more closely related to chimpanzees, gorillas, or orangutans. **Hominid** is a word with narrower meaning, referring to the twigs of the evolutionary tree that are more closely related to us than to any other living species. If a fossil is called a "hominid," this implies the hypothesis that the species is closer to us than it is to a chimpanzee or a gorilla. There are two main groups of hominids: the australopithecines, which came first and are all extinct, and members of the genus *Homo*, with all species extinct except one: *Homo sapiens.*

Some Common Misconceptions

Paleoanthropology has a checkered history. Until about 25 years ago, researchers often gave new names to fossil forms that were undoubtedly the same species as fossils found by

skull bones indicate that *A. afarensis* walked on two legs (FIGURE 34.39a). Fossils discovered in the early 1990s extend the longevity of *A. afarensis* as a species to a span of at least one million years.

At the risk of oversimplifying, one could say that *A. afarensis* was more apelike than human *above* the neck, but more humanlike *below*. Fossilized footprints in Laetoli, Tanzania, corroborate the skeletal evidence that homonids living at the time of *A. afarensis* were bipedal (FIGURE 34.39b). However, the skeletons also suggest a capacity for arboreal locomotion, with arms relatively long in proportion to body size compared to the proportions for modern humans. *A. afarensis* may have been able to move well both in the trees and on the ground, suggesting a mixed forest-savanna habitat.

In the past few years, paleoanthropologists have found hominid species that predate *A. afarensis*. The oldest fossil that is unambiguously human—that is, definitely more closely related to humans than to apes based on skeletal analysis—is *Australopithecus anamensis*, which lived just over 4 million years ago. And note in FIGURE 34.38 that there are even older fossils of putative hominids going back 6 million years. Thus, the australopithecine fossils indicate that hominids were bipedal at least 4 million years ago, and perhaps hundreds of thousands of years earlier. The fossil record of humanity is creeping closer to the ape-human split that molecular systematists estimate occurred about 5–7 million years ago.

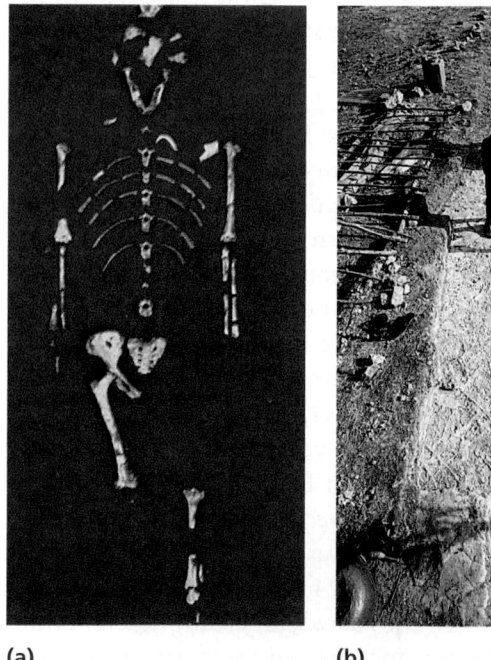
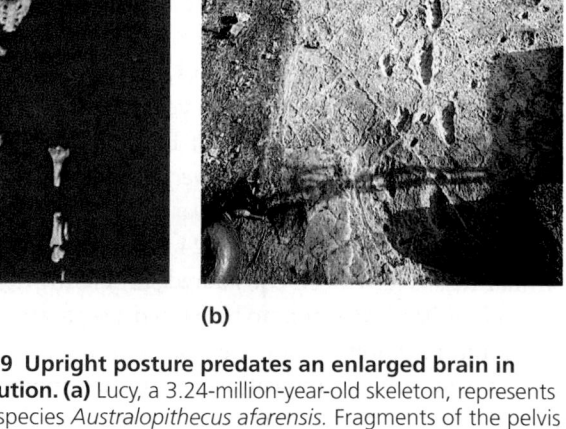

(a)   (b)

FIGURE 34.39 Upright posture predates an enlarged brain in human evolution. (a) Lucy, a 3.24-million-year-old skeleton, represents the hominid species *Australopithecus afarensis*. Fragments of the pelvis and skull show that *A. afarensis* was bipedal, though the arms indicate that Lucy was also adapted for arboreal locomotion. **(b)** The Laetoli footprints, over 3.5 million years old, confirm that upright posture evolved quite early in hominid history.

One of the key questions in paleoanthropology is which of the australopithecines were evolutionary dead ends and which were either on, or close to, the phylogenetic lineage that eventually sprouted the *Homo* branch. For the million years that *A. afarensis* is represented in the fossil record, it changed little. Then, beginning about 3 million years ago, what was apparently an adaptive radiation produced several new hominid species, including *A. africanus*, the human first discovered by Dart 80 years ago. Either *A. africanus* or some closely related species was probably ancestral to two hominid branches known from later fossils. One lineage consisted of the "robust" australopithecines. They had sturdy skulls with powerful jaws and large teeth, adapted for grinding and chewing hard, tough foods. These robust forms contrast with other australopithecines, including the earlier *A. afarensis* and *A. africanus*, which were what anthropologists call "gracile" (slender) australopithecines. The gracile forms had lighter feeding equipment adapted for softer foods. Most researchers agree that the robust australopithecines were an evolutionary dead end, and that the ancestors of *Homo* were among the gracile australopithecines.

Homo: *The Evolution of Larger Brains and the Global Dispersion of Humans*

The earliest fossils that anthropologists place in our genus, *Homo*, are classified as *Homo habilis*. These fossils, ranging in age from about 2.5 to 1.6 million years old, show clear signs of some modern hominid characters above the neck. Compared to the australopithecines, *H. habilis* had less prognathic jaws and larger brains, about 600–750 cm^3 compared to the average of 500 cm^3 for *Australopithecus africanus*. In some cases, anthropologists have found sharp stone tools with fossils of *Homo habilis*, which means "handy man." After walking upright for at least 2 million years, some hominids had started to use their brains and hands to fashion tools.

A remarkably complete fossil of a young hominid known as "Turkana Boy" indicates that even larger brains had evolved by 1.6 million years ago (FIGURE 34.40). The boy had a brain that would probably be over 900 cm^3 in an adult of his species. This brain size is between that of *H. habilis* and another species, *Homo erectus*. It is a reasonable hypothesis that Turkana Boy represents hominid forms linking *H. habilis* to *H. erectus* in our phylogeny.

Homo erectus was the first hominid species to migrate out of Africa. These humans colonized Asia, including the Indonesian archipelago, leaving fossils known by such names as "Beijing Man" and "Java Man." *H. erectus* lived from about 1.8 million years ago to 500,000 years ago, and had already populated Asia by at least 1.5 million years ago. Compared to *H. habilis*, *H. erectus* was taller and had a larger brain, averaging about 1,100 cm^3. And *H. erectus* males were only about 1.2 times the size of females in the same population, a sexual dimorphism in size matching that of modern humans. Some anthropologists

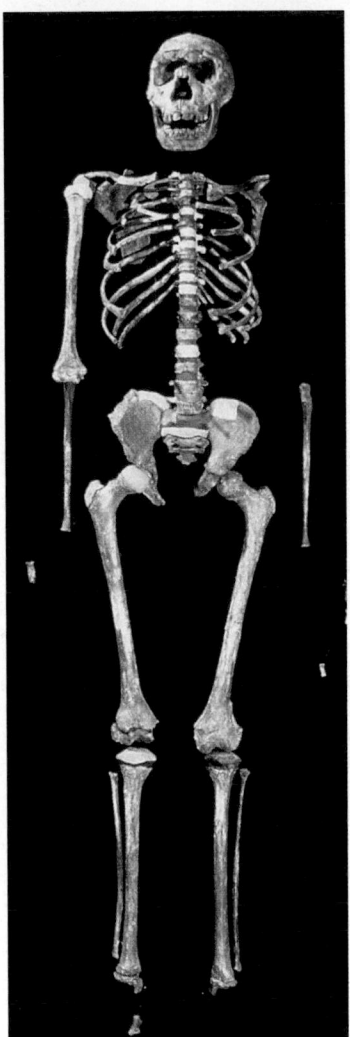

FIGURE 34.40 The Turkana boy. This 1.7 million-year-old fossil is anatomically intermediate between *H. habilis* and *H. erectus*. Some anthropologists classify the specimen as *H. erectus*, while others assign it to a separate species *(H. ergaster)*, hypothetically ancestral to both *H. erectus* and *H. sapiens*.

to 100,000 years ago. These fossils represent the various regional descendants of *H. erectus*, including Neanderthals. One school of researchers refers to all of these regional forms as "archaic *Homo sapiens*," classifying them in the same species, but with subspecies names for the regional variants. In the view of these scientists, for example, the scientific name for Neanderthals should be *Homo sapiens neanderthalensis*. The other school restricts the name *Homo sapiens* to later fossils and gives separate species names to the regional fossils of the earlier period—for example, *Homo neanderthalensis* for the European fossils. Though this controversy may seem like trivial semantic quibbling, it is really a debate between advocates of alternative hypotheses for the origin of modern humans.

The Origin of Anatomically Modern Humans

When and where did fully modern humans—what paleoanthropologists call **anatomically modern humans**—originate?

THE PROCESS OF SCIENCE

One hypothesis for the origin of fully modern humans is that *Homo sapiens* evolved in each region from the local populations of *H. erectus*. This model of parallel evolution of modern humans is called the **multiregional hypothesis** (FIGURE 34.41a, p. 714). It is mainly advocates of this hypothesis who refer to the regional derivatives of *H. erectus* as "archaic *Homo sapiens*," giving the geographic variants of fossils subspecies names, such as *H. sapiens neanderthalensis* for European forms. In this view, the great genetic similarity of all modern people is the product of occasional interbreeding between neighboring populations that has provided corridors for gene flow throughout the geographic range of humans.

On the other side of a very lively debate about human origins are the proponents of the **"Out of Africa" hypothesis**, also called the **replacement hypothesis**. According to this hypothesis, all *Homo sapiens* throughout the world evolved from a second major migration out of Africa that occurred about 100,000 years ago—a migration of anatomically modern humans that completely replaced all the regional populations of *Homo* derived from the first hominid migrations of *H. erectus* out of Africa about 1.5 million years ago (FIGURE 34.41b). Most advocates of the replacement hypothesis prefer to give separate species names to the regional hominids that were not anatomically modern, such as *Homo neanderthalensis* for the Neanderthals in Europe.

Note that both hypotheses recognize the fossil evidence for humanity's African origin. The debate centers on the vintage of the most recent hominid ancestor in Africa common to all the world's modern populations. The multiregional hypothesis places that last common ancestor in Africa over 1.5 million years ago, when *H. erectus* began migrating to other parts of the world. But according to the replacement hypothesis, the diverse human populations of the world are much more closely related. In this view, all of the world's populations diverged from

interpret this to mean that monogamy based on pair-bonding had evolved in *H. erectus* societies, replacing a more polygamous system where the largest, strongest males generally outcompeted smaller males and left the most offspring.

In addition to dispersing from Africa to Asia, *H. erectus* also spread into Europe, though the timing of that migration is less certain than for Asia. In Europe, *H. erectus* gave rise to the humans known as Neanderthals.

Neanderthals are named for the locale where their fossils were first discovered, the Neander Valley of Germany. Anthropologists now use the name for humans who lived throughout Europe from about 200,000 to 30,000 years ago. Fossilized skulls indicate that Neanderthals had brains as large as ours, though somewhat different in shape. Neanderthals were also generally more heavily built than modern humans. Despite these differences, it is unlikely that an appropriately dressed Neanderthal misplaced in time would stand out from the crowd on a city street today.

Controversy surrounds the classification of fossils of the humans that lived in Europe, Asia, and Africa from about 500,000

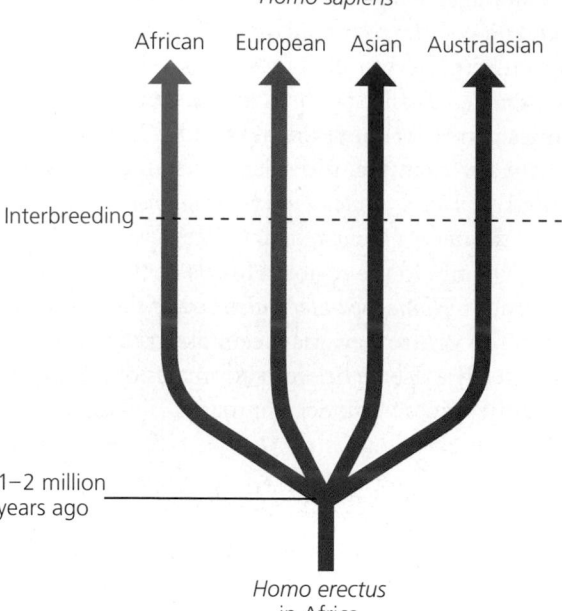

(a) Multiregional hypothesis. According to the multiregional hypothesis, modern humans evolved in many parts of the world from regional descendants of *Homo erectus*, who dispersed from Africa between 1 and 2 million years ago. The dashed line symbolizes interbreeding and gene flow between regional populations.

FIGURE 34.41 Two hypotheses for the origin of anatomically modern humans.

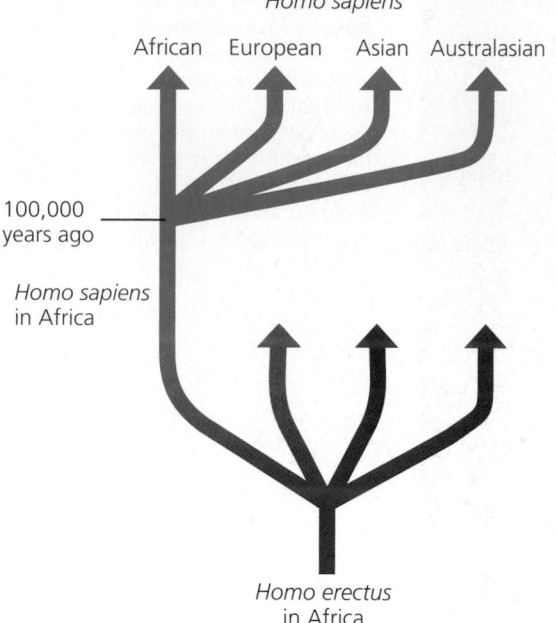

(b) "Out of Africa" hypothesis (replacement hypothesis). According to the "Out of Africa" hypothesis, only the African descendants of *Homo erectus* gave rise to modern humans. All other regional descendants of *H. erectus*, including Neanderthals, became extinct without contributing significantly to the gene pool of modern humanity. Advocates of this hypothesis argue that anatomically modern humans began spreading from Africa just 100,000 years ago, giving rise to all the diverse human populations on Earth today.

anatomically modern *Homo sapiens* that evolved from an African *H. erectus* population and then migrated throughout the world beginning only 100,000 years ago. These anatomically modern humans—*Homo sapiens*—completely replaced all the other regional descendants of *H. erectus,* including Neanderthals, which are therefore evolutionary dead ends in the view of the replacement hypothesis.

A third alternative to the multiregional and replacement hypotheses suggests that *Homo sapiens* dispersing from Africa 100,000 years ago interbred with the regional descendants of the earlier *H. erectus* migration. This hypothesis predicts that the genomes of the indigenous people around the globe today should reflect a complex mix of ancestries. More generally, the key to our past, and a way to test the predictions made by the competing hypotheses, is to assess our current genetic diversity.

So far, the genetic data have mostly supported the replacement hypothesis. One approach has been to compare the mitochondrial DNA (mtDNA) between population samples from various parts of the world. Using changes in the mtDNA as a molecular clock (see Chapter 25), researchers have reported a time of genetic divergence of about 100,000 years ago, consistent with the replacement hypothesis. Some other researchers question the reliability of this approach. However,

studies of genetic markers from the nucleus have corroborated the mtDNA data in tracing the geographic branching of *Homo sapiens* to about 100,000 years ago. And in the late 1990s, the replacement hypothesis accrued further support from researchers who sequenced mtDNA extracted from Neanderthal bones. The multiregional hypothesis predicts that the mtDNA of modern Europeans should be more similar to this Neanderthal mtDNA than to mtDNA from modern humans indigenous to other geographic regions. But the mtDNA from four Neanderthals all fall completely outside the range of mtDNA for modern Europeans. These data suggest that Neanderthals contributed nothing to the ancestry of anatomically modern humans in Europe. So far, however, there have been no such comparisons of DNA between local fossils and local indigenous people in regions outside Europe, such as Asia.

In testing the predictions of contesting hypotheses on the origin of modern humans, perhaps the most important genetic data so far are comparisons of Y chromosomes that a large research team published in 2001. In contrast to other human chromosomes, copies of a Y chromosome transmitted from male to male through the generations of a family retain their genetic identity because there is no crossing over for the Y chromosome during meiosis (except for a very small portion, which may cross over with the X chromosome). The diversity among Y chromosomes is thus limited to mutations, which serve as markers for tracing the ancestry and relationships among males alive today. By comparing the Y chromosomes of males from various geographic regions, researchers were able to infer divergence from a common African ancestor less than 100,000 years ago. As one former advocate of the multiregional hypothesis put it when he learned of these Y-chromosome data: "I have undergone a conversion—a sort of epiphany. There are no old Y chromosome lineages [in modern humans]. There are no old mtDNA lineages. Period. It was total replacement." However, not all multiregionalists find the genetic data so conclusive.

So far, the fossil evidence has been less one-sided than the genetic data in testing the alternative explanations for the origin of modern humans. The fossil evidence in western Europe is consistent with a total replacement of Neanderthals about 40,000 years ago by anatomically modern humans, known as Cro-Magnons in the case of the European fossils. There is no evidence for a transition in skeletal morphology that would suggest interbreeding between Neanderthals and the later arrivals. Thus, for western Europe, at least, the fossil evidence corroborates the genetic evidence in support of the replacement hypothesis. However, the fossil data outside of Europe are more ambiguous. For example, some paleoanthropologists interpret certain Asian fossils to be intermediate between older fossils of *H. erectus* in Asia and skeletal features of modern Asians, a result predicted by the multiregional hypothesis (or implying interbreeding with *H. erectus* after modern humans arrived in Asia).

Scientific debates about our evolutionary history, including the origin of anatomically modern humans, will certainly continue to make paleoanthropology one of the most exciting research fields in science.

■   ■   ■

Though we are ending this unit of chapters on the evolution of biological diversity with the origin of modern humans, we do not mean to reinforce the common misconception of a "ladder of life leading from lowly microbes to lofty humanity." Biological diversity is the product of branching phylogeny, not ladderlike "progress," however we choose to measure it. The fact that there are more species of bony fishes alive today than all other vertebrates combined is a clear indication that our finned relatives are not outmoded underachievers that failed to get out of the water. The tetrapods—amphibians, reptiles, birds, and mammals—are derived from one fish population. As tetrapods diversified on land, fishes continued their branching evolution in the greatest portion of the biosphere's volume. Similarly, the ubiquitous presence of diverse prokaryotes throughout the biosphere today is a reminder of the enduring ability of these relatively simple organisms to keep up with the times through adaptive evolution. Biology exalts life's diversity, past and present.

CHAPTER 34 REVIEW

Go to the Campbell Biology website (www.campbellbiology.com) to explore an interactive version of the Chapter Review.

Summary of Key Concepts

INVERTEBRATE CHORDATES AND THE ORIGIN OF VERTEBRATES

■ **Four anatomical features characterize the phylum Chordata** (pp. 679–680, FIGURE 34.2) These chordate characteristics are a notochord; a dorsal, hollow nerve cord; pharyngeal slits; and a postanal tail.

■ **Invertebrate chordates provide clues to the origin of vertebrates** (pp. 680–682, FIGURES 34.3–34.5) Subphylum Urochordata includes the marine, suspension-feeding tunicates, or sea squirts. Members of the subphylum Cephalochordata (the lancelets) are exemplary chordates. Fossil evidence shows some transitional stages in the evolution of vertebrates from invertebrate chordates.

INTRODUCTION TO THE VERTEBRATES

■ **Neural crest, pronounced cephalization, a vertebral column, and a closed circulatory system characterize the subphylum Vertebrata** (p. 683, FIGURE 34.6) A mass of embryonic cells called the neural crest gives rise to many of the uniquely vertebrate characteristics. Vertebrates have a well-developed head with a brain and braincase (skull). Their segmentally arranged vertebrae enclose the dorsal hollow nerve cord (spinal cord). Vertebrate organ systems, such as their closed circulatory system, support an active metabolism.

■ **An overview of vertebrate diversity** (pp. 683–685, FIGURE 34.7) Except for a few species that lack jaws, all extant vertebrates have hinged jaws. Jawed vertebrates living today are classified in the various extant classes of fishes and tetrapods.

JAWLESS VERTEBRATES

■ **Class Myxini: Hagfishes are the most primitive living "vertebrates"** (p. 685, FIGURE 34.8) Hagfishes are jawless marine "vertebrates" that have cartilaginous skulls and an axial skeleton derived from the notochord, but hagfishes have no column of vertebrae.

- **Class Cephalaspidomorphi: Lampreys provide clues to the evolution of the vertebral column (pp. 685–686, FIGURE 34.9)** Lampreys are jawless vertebrates that have cartilage segments surrounding the notochord and arching partly over the nerve cord.

- **Some extinct jawless vertebrates had ossified teeth and bony armor (p. 686)** A diversity of fossilized fishes called ostracoderms suggest that ossified skeletons began to evolve in some lineages of early Paleozoic jawless vertebrates.

FISHES AND AMPHIBIANS

- **Vertebrate jaws evolved from skeletal supports of the pharyngeal slits (p. 687, FIGURE 34.10)** Jaws and teeth opened up new food resources for gnathostomes.

- **Class Chondrichthyes: Sharks and rays have cartilaginous skeletons (pp. 687–688, FIGURE 34.11)** This skeleton is not primitive, but evolved secondarily from ancestors with bone.

- **Osteichthyes: The extant classes of bony fishes are the ray-finned fishes, the lobe-finned fishes, and the lungfishes (pp. 688–690, FIGURES 34.12–34.14)** The bony fishes have skeletons reinforced by calcium phosphate. In contrast to the cartilaginous fishes, bony fishes have opercula (bony gill covers) and can adjust their density and thus control their buoyancy by means of a swim bladder.

- **Tetrapods evolved from specialized fishes that inhabited shallow water (pp. 690–691, FIGURES 34.15, 34.16)** Fossil evidence suggests that the tetrapod limb, now mainly used for walking on land, first evolved for paddling in water.

- **Class Amphibia: Salamanders, frogs, and caecilians are the three extant amphibian orders (pp. 691–693, FIGURES 34.17, 34.18)** Attesting to their aquatic heritage, most modern amphibians have moist skin that complements the lungs in gas exchange. Most species lay their unshelled eggs in wet environments. Most frogs and their relatives undergo metamorphosis of an aquatic larval stage into a terrestrial adult. Extant amphibians are the urodeles (salamanders), anurans (frogs), and apodans (legless, burrowing amphibians).

AMNIOTES

- **Evolution of the amniotic egg expanded the success of vertebrates on land (p. 693, FIGURE 34.19)** The amniotic egg has extraembryonic membranes and fluids that protect and hydrate the embryo.

- **Vertebrate systematists are reevaluating the classification of amniotes (pp. 694–695, FIGURES 34.20, 34.21)** The traditional separation of birds and reptiles into separate classes is a paraphyletic classification. Various alternatives have been proposed for classification of amniotes based on clades (monophyletic groups).

- **A reptilian heritage is evident in all amniotes (pp. 695–698, FIGURES 34.22–34.24)** Reptiles, a diverse group represented today by lizards, snakes, turtles, and crocodilians, have numerous terrestrial adaptations, including lungs, waterproof scales, and a shelled amniotic egg.

- **Birds began as feathered reptiles (pp. 698–701, FIGURES 34.25–34.29)** Flight feathers are diagnostic of birds, whereas their amniotic eggs and scaled legs give testimony to their reptilian heritage. Birds probably descended from a group of small, carnivorous dinosaurs.
  Web/CD Case Study in The Process of Science: *How Does Bone Structure Shed Light on the Origin of Birds?*

- **Mammals diversified extensively in the wake of the Cretaceous extinctions (pp. 701–706, FIGURES 34.30–34.33; TABLE 34.1)** Hair and mammary glands are two derived characteristics of mammals. Monotremes are a small group of egg-laying mammals. Marsupials include opossums, kangaroos, and koalas, animals whose young complete their embryonic development inside a maternal pouch, the marsupium. The most widespread and diverse modern mammals are the eutherians (placentals), a group whose young complete their embryonic development attached to a placenta inside the mother's uterus.
  Web/CD Activity 34A: *Characteristics of Chordates*

PRIMATES AND THE EVOLUTION OF *Homo sapiens*

- **Primate evolution provides a context for understanding human origins (pp. 707–708, FIGURES 34.34–34.37)** The first primates were probably small arboreal animals. All primates have hands and (with the exception of humans) feet adapted for gripping. Two subgroups of modern primates are the prosimians—lemurs and their relatives—and the anthropoids. Anthropoids diverged early into New World and Old World monkeys. Modern apes—gibbons, orangutans, gorillas, chimpanzees, and bonobos—evolved from Old World monkeys.
  Web/CD Activity 34B: *Primate Diversity*

- **Humanity is one very young twig on the vertebrate tree (pp. 709–715, FIGURES 34.38–34.41)** Humans evolved in Africa. The three main features of human anatomical evolution are bipedal locomotion, reduction of the jaw, and expansion of the brain. Bipedal locomotion evolved before changes in the jaw and brain, as evident in fossils of *Australopithecus afarensis*. Around 2 to 2.5 million years ago, the genus *Homo* originated as a hominid lineage with larger brains and smaller jaws. Stone tools were associated with fossils of the earliest *Homo* species, *H. habilis*. A population of *H. habilis* probably gave rise to *H. erectus* about 1.8 million years ago, and *H. erectus* colonized Asia and Europe. There are two hypotheses for the evolution and geographic spread of anatomically modern humans: the "multiregional" and "Out of Africa" (replacement) hypotheses. Genetic evidence, including DNA sequences from Neanderthal fossils and analysis of Y chromosomes, supports the replacement hypothesis, but the multiregional hypothesis has some supporters, who mainly cite fossil evidence.
  Web/CD Activity 34C: *Human Evolution*

Self-Quiz

1. Vertebrates and tunicates may seem as different as two animal groups can be, yet they share
   a. jaws adapted for feeding.
   b. a high degree of cephalization.
   c. the formation of structures from the neural crest.
   d. an endoskeleton that includes a cranium.
   e. the presence of a notochord; a dorsal, hollow nerve cord; and pharyngeal slits.

2. Some 530-million-year-old Chinese fossils resemble lancelets but have a more elaborate brain and a brain case (cranium). These fossils may represent
   a. the first chordate.
   b. a "missing link" between the urochordates and cephalochordates.
   c. an early vertebrate.
   d. a primitive bony fish.
   e. a non-tetrapod gnathostome.

3. In addition to skeletal differences, cartilaginous fishes can be distinguished from bony fishes
   a. by the presence in bony fishes of a cranium.
   b. by the presence in bony fishes of a lateral line.
   c. by the presence in cartilaginous fishes of unpaired fins.
   d. by the absence in cartilaginous fishes of a swim bladder.
   e. by the absence in cartilaginous fishes of paired sensory organs.

4. Mammals and extant birds share all of the following characteristics *except*
   a. endothermy.
   b. descent from reptiles.
   c. a dorsal, hollow nerve cord.
   d. teeth specialized for diverse diets.
   e. the ability of some species to fly.

5. If you were to observe a monkey in a zoo, which characteristic would indicate a New World origin for that monkey species?
   a. distinct "seat pads"
   b. eyes close together on the front of the skull
   c. use of the tail to hang from a tree limb
   d. occasional bipedal walking
   e. downward orientation of the nostrils

6. Which of the following could be considered to be the first tetrapod?
   a. sturdy-finned, shallow-water lungfishes whose appendages had skeletal supports similar to those of terrestrial vertebrates
   b. armored, jawed placoderms that had two sets of paired appendages
   c. early ray-finned fishes that developed bony skeletal supports in their paired fins
   d. salamanders of the order Urodela that had legs supported by a bony skeleton but moved with the same side-to-side bending typical of fishes
   e. an early terrestrial caecilian line whose legless condition had evolved secondarily

7. Unlike eutherian (placental) mammals, both monotremes and marsupials
   a. lack nipples.
   b. have some embryonic development outside the mother's uterus.
   c. lay eggs.
   d. are found in Australia and Africa.
   e. include only insectivores and herbivores.

8. Which of the following is *not* thought to be ancestral to humans?
   a. a reptile   d. an amphibian
   b. a bony fish   e. a bird
   c. a primate

9. As humans diverged from other primates, which of the following most likely appeared first?
   a. the development of technology
   b. language
   c. an erect stance
   d. toolmaking
   e. an enlarged brain

10. The multiregional and replacement hypotheses for the origin of modern humans agree that
   a. *Homo erectus* had an African origin.
   b. modern *Homo sapiens* originated only in Africa.
   c. Neanderthals are the ancestors of modern humans in Europe.
   d. Australopithecines migrated out of Africa.
   e. North America had the first population of modern humans.

11. What four features do we share with invertebrate chordates such as lancelets?
12. What is an amniotic egg?
13. Birds and reptiles differ in their main source of body heat, with birds being _____ and reptiles being _____.
14. What are two hallmarks of mammals?
15. To which mammalian order do we belong? What are the two main subgroups of this order?
16. Arrange these clades from the most comprehensive to the least: primates, hominoids, chordates, mammals, hominids, vertebrates, deuterostomes, anthropoids, amniotes, gnathostomes.

Go to the website or CD-ROM for more quiz questions.

Evolution Connection

Provide one characteristic that qualifies humans for membership in each of the following taxa: domain Eukarya, kingdom Animalia, clade Deuterostomia, phylum Chordata, subphylum Vertebrata, clade Gnathostomata, clade Amniota, class Mammalia, order Primates.

The Process of Science

Scientific inquiry often proceeds by trying to make sense of an interesting observation. One such observation concerns patterns of genetic vs. morphological divergence in some vertebrate groups. Amphibian species, for example, show great similarity in morphology, yet are far more genetically divergent than the morphologically diverse birds. A related pattern is evident in chimpanzee-human comparisons: These two species are quite divergent morphologically, yet are nearly identical genetically. Propose one or more hypotheses that might explain these puzzling patterns.

Analyze bone cross-sections to determine whether their patterns support alternative hypotheses for the origin of birds in the Case Study in the Process of Science, available on the website and CD-ROM.

Science, Technology, and Society

While our biological evolution is Darwinian, our cultural evolution could be described as Lamarckian. Explain this distinction after reviewing Darwin and Lamarck in Chapter 22.

Answers: 1. e; 2. c; 3. d; 4. d; 5. c; 6. a; 7. b; 8. e; 9. c; 10. a; 11. Dorsal, hollow nerve chord; notochord; gill structures at some time during development; postanal tail at some time during development. 12. A shelled egg with the embryo contained in a fluid-filled sac, the amnion. 13. endothermic; ectothermic 14. Hair and mammary glands. 15. Primates; prosimians and anthropoids 16. deuterostomes > chordates > vertebrates > gnathostomes > amniotes > mammals > primates > anthropoids > hominoids > hominids

PLANT FORM AND FUNCTION

Joanne Chory is a professor of biology and Howard Hughes Medical Institute Investigator at The Salk Institute for Biological Studies in La Jolla, California. Her research has revealed key steps in the signal-transduction pathways by which light regulates the development of plants. Dr. Chory's dissection of such regulatory mechanisms also led to her discovery of the role of steroid hormones in this process and the identification of a plant steroid hormone receptor. In 1999, the central importance of Dr. Chory's research was recognized by her election to the National Academy of Sciences. As is often the case in biology, Dr. Chory's success has depended in part on selecting an appropriate organism as a research model—in this case, a tiny member of the mustard family named Arabidopsis, *the "laboratory mouse" of modern plant biology. Plant science reached a milestone in 2000, when an international team announced the complete sequence of the* Arabidopsis *genome. The significance of that achievement seemed like a good place to start our interview with Dr. Chory.*

**Interview
JOANNE CHORY**

Now that the *Arabidopsis* genome has been sequenced, how will this information affect plant biology?

First, it tells us how many genes it takes to be a plant. *Arabidopsis* has about 26,000 genes. We still don't know what most of those genes do, but now progress will be faster. It used to take two or three people several years to find and clone a gene. Now one graduate student can do this in just a few months.

How long do you think it will take to determine the functions of all the *Arabidopsis* genes?

There's a ten-year plan. About three years ago, the National Science Foundation could see that the sequencing of the *Arabidopsis* genome was almost completed, so the next phase would be to figure out what all these genes do. So a committee of plant biologists developed a plan called The 2010 Project, which maps goals and strategies to determine the functions of all 26,000 *Arabidopsis* genes by the year 2010. This includes understanding at what time during the life cycle of the plant each gene is expressed and in which types of cells. Eventually, we'd like to have a "virtual plant." A researcher should be able to go

to the computer and access the whole database of gene functions, with all the proteins produced in a particular part of the plant and at different times as the plant develops—a sort of four-dimensional map of the plant's gene expression during its life cycle.

How do you think researchers will use such a virtual plant?

Some people interpret the idea of a virtual plant to mean that we don't want to do experiments anymore. But that isn't true. What we want is to be able to design smarter experiments. The database of gene functions will help us formulate hypotheses for development of the plant—for example, how signaling networks interact in a specific cell at a specific time in development.

Why has so much effort focused on *Arabidopsis*, of all plants?

First, *Arabidopsis* has a very rapid life cycle. It takes only about seven weeks to go from seed to seed. Also, *Arabidopsis* is self-fertile, and each plant can produce 10,000 to 50,000 seeds, which means we can propagate a lot of genetically identical plants. *Arabidopsis* is also a good research plant because it has the smallest known plant genome. It doesn't have a lot of junk DNA. It would have taken much more time to sequence the genome of a crop plant, which may have 30 to 100 times more DNA than *Arabidopsis*. Although it's important

to understand the biology of various crop plants, plant molecular biologists decided in the mid-1980s that it would be a lot more useful to pick a good research organism and try to understand it fully. And it has proven to be a wise decision to have a model or reference plant. Understanding *Arabidopsis* will help us understand flowering plants in general.

Are there any examples of how an understanding of *Arabidopsis* has had agricultural applications?

Yes. One example is what we've learned from *Arabidopsis* mutants about how the hormone ethylene functions in fruit ripening. The same genes responsible for the ethylene pathway in *Arabidopsis* are found in such fruits as tomatoes, and understanding how these genes work enables us to control the ripening process. Another application is that identifying genes in *Arabidopsis* can help breeders of crop plants understand how to use certain mutations in the process of selective breeding for useful varieties. For instance, sorghum would not normally grow in Texas, but breeders have selected for a mutation affecting a photoreceptor in the plant that we know, based on *Arabidopsis* research, would allow sorghum to complete its life cycle in Texas fields. So, this kind of connecting back and forth between a reference plant and crop plants has been really useful.

With the goal of improving human nutrition?

Yes, I think we need another green revolution because of the pace at which world population is growing. The predictions are for a population of 10 billion people in the year 2050. Even now, with 6 billion people, 800 million people suffer from chronic malnutrition. We have to figure out how to distribute food better, but that is more a question for governments than for plant biologists. For our part, I think the only way to improve nutrition in the world is to increase crop yields. This includes such crop traits as better resistance to pests. And I think crop improvement will depend on applications of molecular genetics, either to make breeding more effective or by producing genetically modified plants.

What about the concerns that many people have about genetically modified organisms in food?

I don't think there's a single response, because the objections to genetically modified crops are so diverse. For example, one whole group is critical because it just doesn't like the whole economic trend of globalization and worries that agricultural technology will make it possible for big multinational companies to control the food supply. And then there are the concerns about how safe it is for people to eat foods made from genetically modified organisms. But we've had genetically modified organisms in food for about 18 years already, and there are no data that indicate any danger. However, I think it's a good thing if these concerns result in government guidelines for assessing the safety of these products. Genetically modified crops will play a big role in the future, as long as people are confident that these products are safe and good for society. The farmers are completely on board about it, but the consumers need to see the benefits.

Much of your own research centers on how light regulates plants. Other than photosynthesis, how is light important in the life of a plant?

It's fascinating what plants can actually read from their light environment. Plants get such clues from light as what season it is and what time of day it is. For example, there's more red light compared to far-red light in the middle of the day, and a plant can tell this because it has a photoreceptor called phytochrome that can measure the ratio of red to far-red light. Light controls the development of plants in ways that optimize their performance as photosynthetic machines. So, if one plant is shaded by another, for example, it will elongate stems at the expense of expanding its leaves and grow toward light.

What mechanism do plants use to detect and respond to light?

There's an array of photoreceptors that function in different cell types throughout a plant's development. Once light is detected, these photoreceptors change their shapes and trigger signal pathways that affect transcription of specific genes in the nucleus.

How do you study these signal pathways?

We dissect the steps of a pathway by genetic analysis—by producing and identifying mutants that affect responses to light. So far, we've identified almost 50 mutant varieties of *Arabidopsis* that are affected abnormally by light. We know now that these signaling pathways are not linear sequences of events, but networks of interactions between proteins. It's going to be a big challenge to sort out these complicated webs of interactions, but I think it's worth it because, after all, responding to light is one of the most important things plants do.

You're also studying steroid hormones in plants. What are you learning?

Plant steroids, which are called brassinosteroids, do a lot of the same kinds of things as sex steroids do in humans. The more steroid a plant has, the bigger and tougher and more robust it is. When plants don't make a steroid—because of a mutation, for example—they are dwarfy things. Steroids also regulate sexual reproduction in plants. I think it's interesting how a certain group of molecules began functioning in diverse organisms as signaling molecules. Many of the enzymes a plant uses to make its steroids are also found in animals that make their own types of steroids. So some of the genes for these enzymes have probably been conserved since plants and animals diverged from a common ancestor over a billion years ago. However, the molecules of the signaling pathway for responses to steroids are very different in plants and animals.

How does plant research fit into the two institutions with which you're associated, the Salk Institute and the Howard Hughes Medical Institute?

Here at the Salk, there are only three plant scientists among the 57 faculty. And with Hughes, I think there are two plant scientists among 350 investigators. So in these organizations, plant scientists are still a little bit on the fringe. But I think there's an appreciation that it's important to study organisms that have different life strategies, which are of interest to all of us who study biology.

How did your interest in science and plants develop?

I was a late bloomer in science. I went to college at Oberlin not knowing what I wanted to major in, but I was good at science and math, so I kept these subjects in my curriculum. I took a biology course, and then genetics and microbiology courses. I did my Ph.D. in microbiology, but then became interested in plants for post-doctoral research. I was fascinated by how much we don't know about plants. In a way this makes it difficult to articulate a specific, sophisticated question or hypothesis. But that's also the fun of it. You get to work on really general questions that can lead you to discoveries.

Based on your personal experience, what advice do you have for undergraduates who are developing an interest in science?

I encourage an undergraduate who's interested in biology to seek an active researcher and volunteer to work on a project in the lab. There's nothing like the feeling of discovery, even if it's something really small. Although I liked science classes and lab courses, it wasn't until I did an honors thesis that I felt like I had my own little project. We have lots of undergraduates from UC San Diego doing honors projects in my lab here at Salk. The graduate students mentor the undergraduate students, and this is also a good experience for the grad students. The Salk Institute also has an outreach program for high school students to work in the lab during the summer. So my advice to students who are thinking of becoming scientists is to personalize science by actually doing it.

PLANT STRUCTURE AND GROWTH

THE PLANT BODY

- Both genes and environment affect plant structure
- Plants have three basic organs: roots, stems, and leaves
- Plant organs are composed of three tissue systems: dermal, vascular, and ground
- Plant tissues are composed of three basic cell types: parenchyma, collenchyma, and sclerenchyma

THE PROCESS OF PLANT GROWTH AND DEVELOPMENT

- Meristems generate cells for new organs throughout the lifetime of a plant: *an overview of plant growth*
- Primary growth: Apical meristems extend roots and shoots by giving rise to the primary plant body
- Secondary growth: Lateral meristems add girth by producing secondary vascular tissue and periderm

MECHANISMS OF PLANT GROWTH AND DEVELOPMENT

- Molecular biology is revolutionizing the study of plants
- Growth, morphogenesis, and differentiation produce the plant body
- Growth involves both cell division and cell expansion
- Morphogenesis depends on pattern formation
- Cellular differentiation depends on the control of gene expression
- Clonal analysis of the shoot apex emphasizes the importance of a cell's location in its developmental fate
- Phase changes mark major shifts in development
- Genes controlling transcription play key roles in a meristem's change from a vegetative to a floral phase

his unit examines *the biology of the flowering plants, or angiosperms. (The structure of algae, mosses, ferns, and gymnosperms was covered in Unit Five, along with the evolutionary relationships of these plant groups to the angiosperms.) With about 280,000 known species, the angiosperms are by far the most diverse and widespread group of land plants. As primary producers, flowering plants are at the base of the food web of* nearly *every terrestrial ecosystem. Most land animals, including humans, depend on plants directly or indirectly for sustenance. Science has increased agricultural productivity to such an extent that most of us are no longer directly involved in the growing of foods, but this is a recent development; most of us do not have to trace our ancestry too far back to find a farmer. Setting the stage for our study of plant biology, this chapter introduces the structural organization of flowering plants and how this organization develops from a single cell, the zygote.*

THE PLANT BODY

Both genes and environment affect plant structure

In their long evolutionary journey from water onto land, plants became adapted by natural selection to the specific problems posed by terrestrial environments (see Chapters 29 and 30). A plant's structure reflects interactions with the environment on two time scales. Over the long term, entire plant species have, by natural selection, accumulated morphological adaptations that enhance survival and reproductive success in the environments in which they grow. For example, some desert plants, such as cacti, have leaves that are so highly reduced that the stem is actually the primary photosynthetic organ. This reduction in leaf size, and thus in surface area, is a morphological adaptation that reduces water loss. Over the short term, individual plants, far more than individual animals, exhibit structural responses to their specific environment. For example, look at how submersion in water affects leaf development in *Cabomba*, the aquatic plant illustrated in the opening illustration of this chapter. Those leaves that developed while the plant was submerged have a feathery appearance—a morphological adaptation that enhances photosynthesis by increasing the surface area of the leaf for the uptake of bicarbonate (HCO_3^-), the form in which CO_2 occurs in water. In contrast, those leaves that develop when the growing plant reaches the surface are arrowhead-shaped pads that aid in flotation (leaves to right of flower). The architecture of a plant is a dynamic process, continuously shaped by the plant's genetically directed growth pattern along with fine-tuning to the environment. In contrast, animal form is much less adjustable.

Even faster than a plant's structural responses to environmental changes are its physiological (functional) adjustments. Unlike cacti, most plants are rarely exposed to severe drought conditions and rely mainly on physiological adaptations to cope with drought stress. In the most common response, the plant produces a hormone that causes the stomata, the pores in the leaves through which most water is lost, to close. In our study of such responses, remember that structure and function are closely related. For example, we cannot understand the function of stomata unless we also examine the structure of the guard cells that border these pores.

Plants have three basic organs: roots, stems, and leaves

The plant body is a hierarchy of structural levels, with emergent properties arising from the ordered arrangement and interactions of component parts (see Chapter 1). As in multicellular animals, the plant body consists of organs that are composed of different tissues, and these tissues are teams of different types of cells. Although our exploration of the plant body will emphasize features common to all angiosperms, we will also note some important variations among plants. In particular, the two plant groups called the monocots and the dicots differ in many anatomical details (FIGURE 35.1). The dicot group includes the largest class of angiosperms, the eudicots, along with some smaller classes in which the dicot-type anatomy evolved independently (see FIGURE 30.4).

The basic morphology of plants reflects their evolutionary history as terrestrial organisms that must simultaneously inhabit and draw resources from two very different environments—soil and air. Soil provides water and minerals, but air is the main source of CO_2, and light does not penetrate far into the soil. The evolutionary solution to this separation of resources was differentiation of the plant body into two main systems:

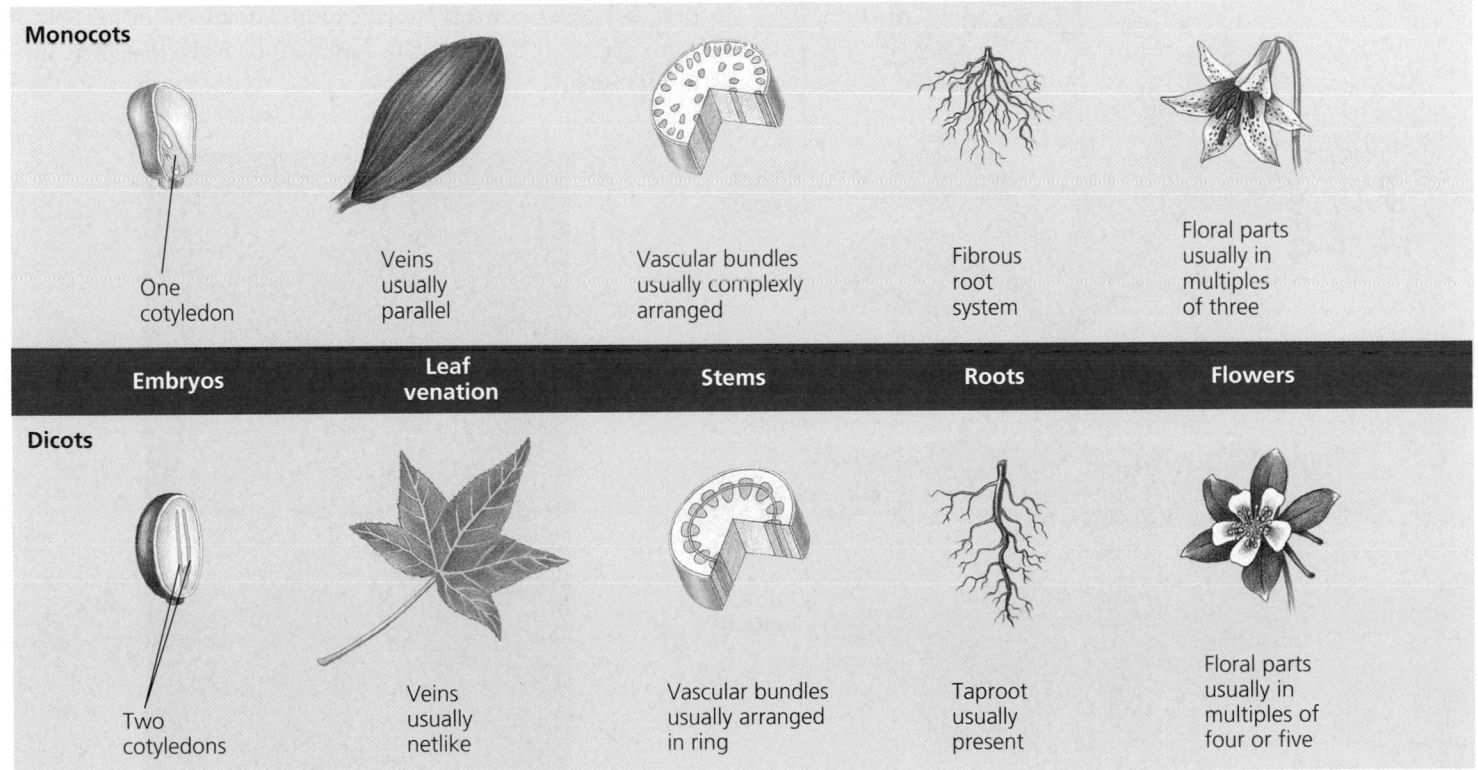

FIGURE 35.1 A comparison of monocots and dicots. These two groups of angiosperms are named for the number of cotyledons, or seed leaves, present on the embryo of the plant. Monocots include orchids, bamboos, palms, lilies, and yuccas, as well as the grasses, such as wheat, corn, and rice. A few examples of dicots are roses, beans, sunflowers, and oaks (which are all eudicots, the largest class of angiosperms with the dicot-type anatomy).

a subterranean **root system** and an aerial **shoot system** consisting of stems and leaves (FIGURE 35.2). (Flowers are shoots consisting of leaves and stems highly modified for sexual reproduction). Neither system can live without the other. Lacking chloroplasts and living in the dark, roots would starve without sugar and other organic nutrients imported from the photosynthetic tissues of the shoot system. Conversely, the shoot system depends on water and minerals absorbed from the soil by the roots. As we take a closer look at the morphology of roots and shoots, try to view these systems from the evolutionary perspective of adaptations to living on land.

The Root System

Roots anchor the plant in the soil, absorb minerals and water, and store food. Monocots, including grasses, generally have **fibrous root** systems consisting of a mat of thin roots that spread out below the soil surface. (Large monocots, such as palms and bamboo, are exceptional in having thicker roots.) The fibrous root system extends the plant's exposure to soil water and minerals and anchors it tenaciously to the ground (see FIGURE 35.1). Because their root systems are concentrated in the

upper few centimeters of the soil, grasses hold the topsoil in place and make excellent ground cover for preventing erosion.

Many dicots have a **taproot** system, consisting of one large, vertical root (the taproot) that produces many smaller lateral, or branch, roots (see FIGURES 35.1 and 35.2). If you have ever tried to pull up a dandelion, then you probably appreciate that one of the primary functions of taproots is to firmly anchor the plant in the soil. In addition, taproots often store food. The plant consumes the food reserves during flowering and fruit production. For this reason, root crops, such as carrots, turnips, and sugar beet, are harvested before they flower. Taproots are particularly long in certain desert plants, which "tap" water sources located far belowground.

Although the entire root system helps anchor a plant, most absorption of water and minerals in both monocots and dicots occurs near the root tips, where vast numbers of tiny **root hairs** increase the surface area of the root enormously (FIGURE 35.3). Root hairs are extensions of individual epidermal cells on the root surface, not to be confused with lateral (branch) roots, which are multicellular organs. (You will learn about symbiotic relationships between plant roots and fungi and bacteria in Chapters 36 and 37.)

In addition to roots that extend from the base of the shoot, some plants have roots arising aboveground from stems or even from leaves. Such roots are said to be **adventitious** (from the Latin *adventicius*, extraneous), a term that describes any plant part that grows in an atypical location. The adventitious roots of some plants, including corn, function as props that help support tall stems.

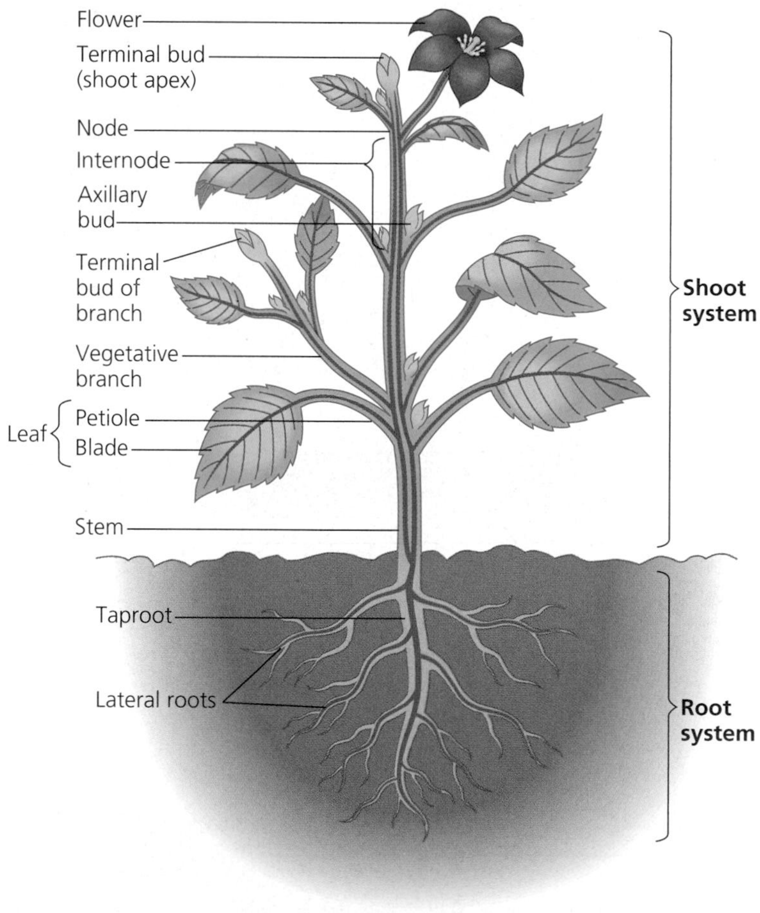

FIGURE 35.2 Morphology of a flowering plant: an overview. The plant body is divided into a root system and a shoot system, connected by vascular tissue (purple strands in this diagram) that is continuous throughout the plant. The plant shown in this diagram is an idealized dicot.

Labels in figure: Flower; Terminal bud (shoot apex); Node; Internode; Axillary bud; Terminal bud of branch; Vegetative branch; Leaf — Petiole, Blade; Stem; Taproot; Lateral roots; Shoot system; Root system

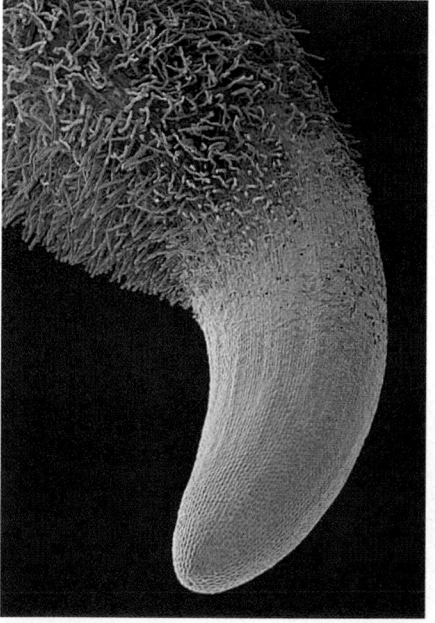

FIGURE 35.3 Radish root hairs. Growing by the thousands just behind the tip of each root, the hairs increase the surface area for the absorption of water and minerals by the roots (colorized SEM).

500 μm

The Shoot System: Stems and Leaves

Shoots consist of stems and leaves. Shoot systems may be vegetative (leaf bearing) or reproductive (flower bearing). Here, we focus on the morphology of vegetative shoots. We'll discuss the transition of vegetative shoots into reproductive shoots later in the chapter.

Stems. A stem is an alternating system of **nodes,** the points at which leaves are attached, and **internodes,** the stem segments between nodes (see FIGURE 35.2). In the angle (axil) formed by each leaf and the stem is an **axillary bud,** a structure that has the potential to form a vegetative branch. Most axillary buds of a young shoot are dormant (not growing). Thus, growth of a young shoot is usually concentrated at its apex (tip), where there is a **terminal bud** with developing leaves and a compact series of nodes and internodes.

The presence of the terminal bud is partly responsible for inhibiting the growth of axillary buds, a phenomenon called **apical dominance.** By concentrating resources on growing taller, apical dominance is an evolutionary adaptation that increases the plant's exposure to light. But what if an animal eats the top of the plant? Or what if, because of obstructions, light is more intense to the side of a plant than directly above it? Under such conditions, axillary buds break dormancy—that is, they start growing. A growing axillary bud gives rise to a vegetative branch complete with its own terminal bud, leaves, and axillary buds. Removing the terminal bud usually stimulates the growth of axillary buds. This is the rationale for pruning trees and shrubs and "pinching back" houseplants to make them bushy.

Modified shoots with diverse functions have evolved in many plants. These modified shoots, which include stolons, rhizomes, tubers, and bulbs, are often mistaken for roots (FIGURE 35.4). Stolons, such as the "runners" of strawberry plants, grow on the surface of the ground and enable a plant to colonize large areas asexually when the single parent plant fragments into many smaller offspring. Rhizomes, such as those of ginger plants, are horizontal stems similar to stolons except that they grow underground. Tubers, including potatoes, are the swollen ends of rhizomes specialized for storing food. Bulbs, such as onions, are vertical, underground shoots consisting mostly of the swollen bases of leaves that store food.

Leaves. Leaves are the main photosynthetic organs of most plants, although green stems also perform photosynthesis. Leaves vary extensively in form, but they generally consist of a flattened **blade** and a stalk, the **petiole,** which joins the leaf to a node of the stem (see FIGURE 35.2). Grasses and many other monocots lack petioles; instead, the base of the leaf forms a sheath that envelops the stem. Some monocots, including palm trees, do have petioles.

The leaves of monocots and dicots differ in the arrangement of their major veins (see FIGURE 35.1). Most monocots have parallel major veins that run the length of the leaf blade.

**FIGURE 35.4
Modified shoots.**

(a) Stolons, shown here on a strawberry plant, grow on the surface of the ground. These "runners" enable a plant to colonize a large area and to reproduce asexually if the single parent plant fragments into many smaller offspring.

(b) Rhizomes, like the edible base of this ginger plant, are horizontal stems that grow underground.

(c) Tubers, such as these white potatoes, are swollen ends of rhizomes specialized for storing food. The "eyes" arranged in a spiral pattern around a potato are clusters of axillary buds that mark the nodes.

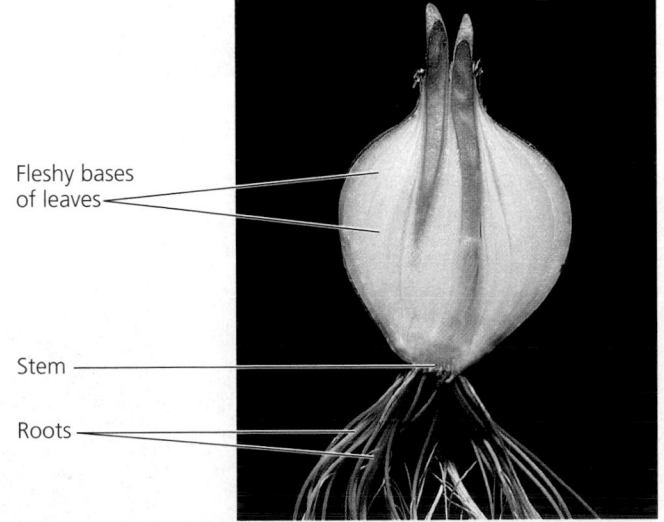

Fleshy bases of leaves

Stem

Roots

(d) Bulbs are vertical, underground shoots consisting mostly of the swollen bases of leaves that store food. You can see the many layers of modified leaves attached to the short stem by slicing an onion bulb lengthwise.

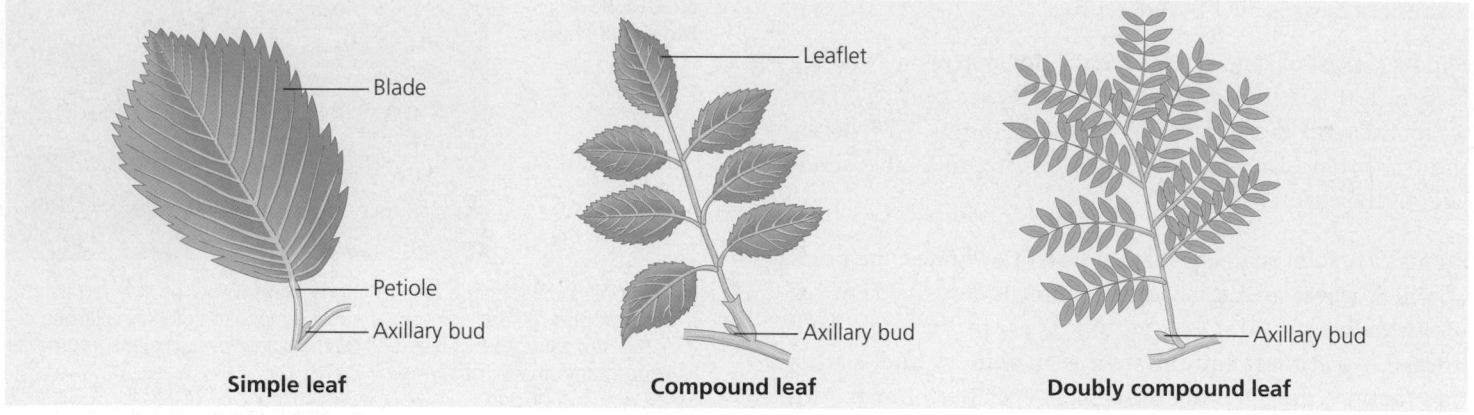

Blade	Leaflet	
Petiole		
Axillary bud	Axillary bud	Axillary bud
Simple leaf	**Compound leaf**	**Doubly compound leaf**

FIGURE 35.5 Simple versus compound leaves. A simple leaf has a single, undivided blade. The blade of a compound leaf is divided into several leaflets, which are themselves divided in a doubly compound leaf. You can distinguish a compound leaf from a stem with several closely spaced simple leaves by examining the locations of axillary buds. There is only one axillary bud per leaf. Thus, a compound leaf has a bud where its petiole attaches to the stem, but not at the bases of the individual leaflets.

In contrast, dicot leaves generally have a multibranched network of major veins.

Because leaf morphology varies extensively among plant species, plant taxonomists use characteristics such as leaf shape, spatial arrangement of leaves on a stem, and the pattern of a leaf's veins to help identify and classify plants. FIGURE 35.5 illustrates one variation: simple versus compound leaves. Most very large leaves are compound or doubly compound. This structural adaptation enables large leaves to withstand strong wind with less tearing and also confines some pathogens that invade the leaf to a single leaflet, rather than allowing the pathogens to spread to the entire leaf.

Although most leaves are specialized for photosynthesis, some plants have leaves that have become adapted by evolution for other functions (FIGURE 35.6).

FIGURE 35.6 Modified leaves.

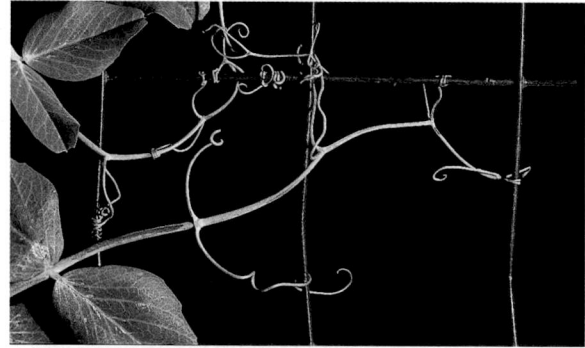

(a) The tendrils used by this pea plant to cling to supports are modified leaflets.

Plant organs are composed of three tissue systems: dermal, vascular, and ground

Each organ of a plant—leaf, stem, or root—has three tissue systems: the dermal, vascular, and ground tissue systems. Each tissue system is continuous throughout the plant body, although the specific characteristics of the tissues and their spatial relationships to one another vary in different organs of the plant (FIGURE 35.7). Here we survey the three tissue systems as they occur in a young, nonwoody plant.

(b) The spines of cacti, such as this prickly pear, are actually leaves, and photosynthesis is carried out mainly by the fleshy green stems.

(c) Most succulents, such as this ice plant, have leaves modified for storing water.

(d) In many plants, brightly colored leaves help attract pollinators to the flower. The red "petals" of the poinsettia are actually leaves that surround a group of flowers.

The **dermal tissue,** or **epidermis,** is generally a single layer of tightly packed cells that covers and protects all young parts of the plant—the "skin" of the plant. In addition to the general function of protection, the epidermis has more specialized characteristics consistent with the function of the particular organ it covers. For example, the root hairs so important in the absorption of water and minerals are extensions of epidermal cells near the tips of roots. The epidermis of leaves and most stems secretes a waxy coating called the **cuticle** that helps the aerial parts of the plant retain water—an important adaptation to living on land (see p. 581).

Vascular tissue, continuous throughout the plant, is involved in the transport of materials between roots and shoots. The two types of vascular tissue are **xylem,** which conveys water and dissolved minerals upward from roots into the shoots, and **phloem,** which transports food made in mature leaves to the roots and to nonphotosynthetic parts of the shoot system, such as developing leaves and fruits. Both xylem and phloem are composed of a variety of cell types, but those directly in-

volved in long-distance transport are so highly modified as to warrant special discussion.

The water-conducting elements of xylem, the **tracheids** and **vessel elements,** are elongated cells that are dead at functional maturity. The term *functional maturity* refers to the stage in a cell's development when it is fully specialized for its function. When the living interior of a tracheid or vessel element disintegrates, the cell's thickened cell walls remain behind, forming a nonliving conduit through which water can flow (FIGURE 35.8).

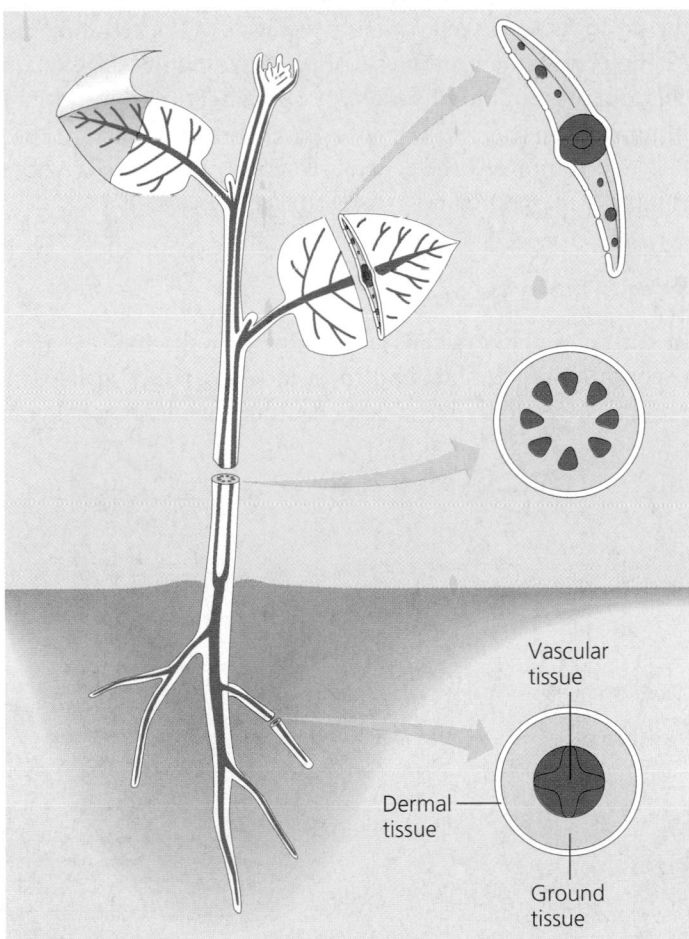

FIGURE 35.7 The three tissue systems. The dermal tissue system—epidermis (white) in the case of a young, nonwoody organ—is a single layer of cells that covers the entire body of a young plant. The vascular tissue system (purple) is also continuous throughout the plant, but is arranged differently in each organ. The ground tissue system (yellow), responsible for most of the plant's metabolic functions, is located between the dermal tissue and the vascular tissue in each organ.

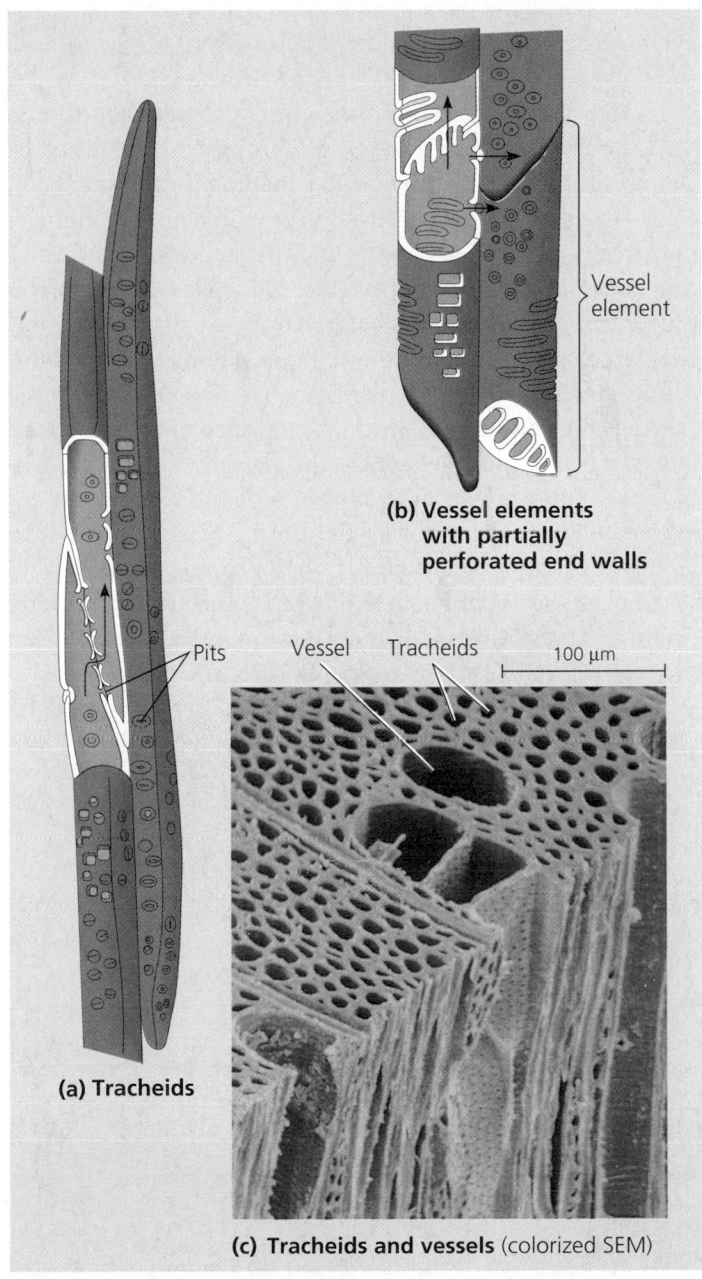

FIGURE 35.8 Water-conducting cells of xylem. Arrows indicate the flow of water. **(a)** Tracheids are spindle-shaped cells with pits through which water flows from cell to cell. **(b)** Vessel elements are linked together end to end, forming long tubes, or xylem vessels. Water streams from cell to cell through perforated end walls. Water can also migrate laterally between neighboring vessels through pits. **(c)** Wood is composed mostly of tracheids and vessels (SEM).

Both tracheids and vessels form in parts of the plant that are no longer elongating. Their secondary walls are interrupted only by **pits,** thinner regions where only primary walls are present (see FIGURE 7.28 to review primary and secondary walls). Tracheids are long, thin cells with tapered ends. Water moves from cell to cell mainly through the pits, where the water does not have to cross thick secondary walls. Because their secondary walls are hardened with lignin, tracheids function in support as well as water transport. Vessel elements are generally wider, shorter, thinner walled, and less tapered than tracheids. Vessel elements are aligned end to end, forming long micropipes, the **xylem vessels.** The end walls of vessel elements are perforated, enabling water to flow freely through xylem vessels.

In the phloem, sucrose, other organic compounds, and some mineral ions are transported through tubes formed by chains of cells called **sieve-tube members** (FIGURE 35.9). Sieve-tube members are alive at functional maturity, although they lack such organelles as the nucleus, ribosomes, and a distinct vacuole. In angiosperms, the end walls between sieve-tube members, called **sieve plates,** have pores that presumably facilitate the flow of fluid from cell to cell along the sieve tube. Alongside each sieve-tube member is a nonconducting cell called a **companion cell,** which is connected to the sieve-tube member by numerous channels, the plasmodesmata (see FIGURE 7.8). The nucleus and ribosomes of the companion cell may serve not only that cell but also the adjacent sieve-tube member, which has no nucleus or ribosomes of its own. In some plants, companion cells in leaves also help load sugar produced in the leaf into the sieve-tube members. The phloem then transports the sugar to other parts of the plant.

Ground tissue is tissue that is neither dermal nor vascular (see FIGURE 35.7). In dicot stems, ground tissue is divided into **pith,** internal to the vascular tissue, and **cortex,** external to the vascular tissue. Ground tissue is more than just filler; among its diverse functions are photosynthesis, storage, and support. In fact, ground tissue illustrates the point that each tissue system of a plant consists of a variety of cell types. The cortex of a dicot stem, for example, typically consists of both fleshy storage cells and thick-walled support cells. All plant tissues consist of a variety of the three basic plant cell types that we discuss in the next section.

Plant tissues are composed of three basic cell types: parenchyma, collenchyma, and sclerenchyma

A multicellular organism is characterized by a division of labor among cell types specialized for different functions. As you consider each major type of plant cell, notice the structural adaptations that make specific functions possible. In some cases, we will find distinguishing characteristics within the **protoplast,** the cell contents exclusive of the cell wall. For example, only the protoplasts of photosynthetic cells contain chloroplasts. But modifications of cell walls are also important in how the specialized cells of a plant function. FIGURE 35.10 will help you review the general structure of plant cells before you proceed to our survey of specific cell types.

Parenchyma Cells

Mature **parenchyma cells** have primary walls that are relatively thin and flexible, and most lack secondary walls. The protoplast generally has a large central vacuole. Parenchyma

FIGURE 35.9 Food-conducting cells of the phloem. Sieve-tube members transport a sugar-rich sap from areas of sugar production (such as leaves) to areas of sugar consumption (such as growing root and shoot tips). **(a)** This diagram of a longitudinal view shows how the sieve-tube members are arranged end to end with porous walls (sieve plates) between them. Alongside each sieve-tube member is a nucleated companion cell. **(b)** A transverse section showing two sieve-tube members sectioned through sieve plates.

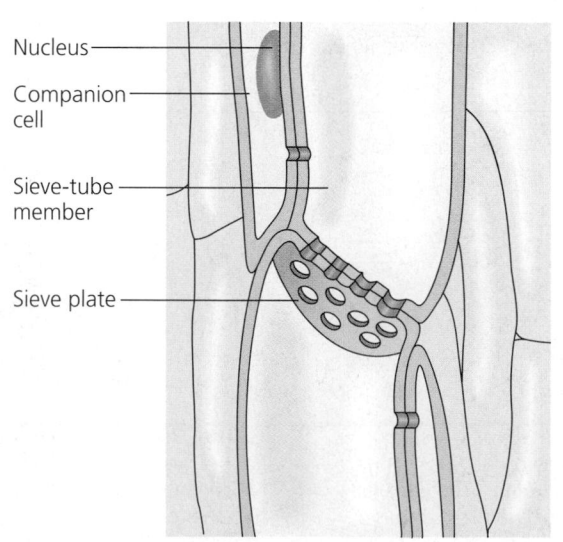

Nucleus
Companion cell
Sieve-tube member
Sieve plate

(a) Longitudinal view

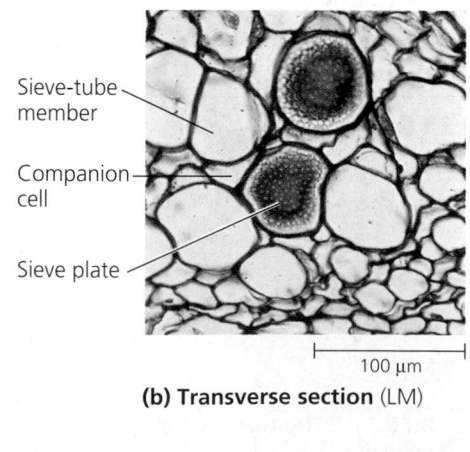

Sieve-tube member
Companion cell
Sieve plate

100 μm

(b) Transverse section (LM)

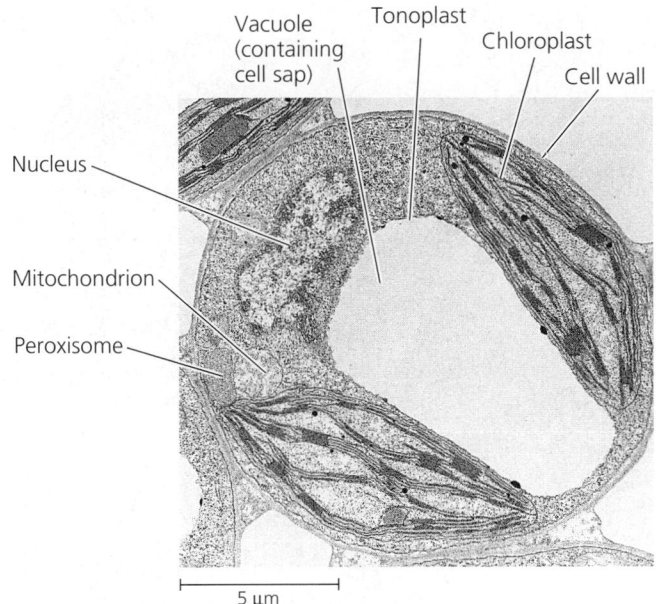

(a) **Plant cell.** In addition to general features of eukaryotic cells, you can see three structures of plant cells *not* found in animal cells: chloroplasts, the sites of photosynthesis; a central vacuole containing a fluid called cell sap and bounded by the tonoplast, a specialized membrane that regulates the traffic of molecules between the cell sap and the cytosol; and a cell wall external to the plasma membrane. The part of the cell interior to the cell wall—the plasma membrane, the cytoplasm, and the nucleus—is called the protoplast. Thus, a plant cell consists of a protoplast enclosed by a cell wall (TEM).

FIGURE 35.10 Review of general plant cell structure.

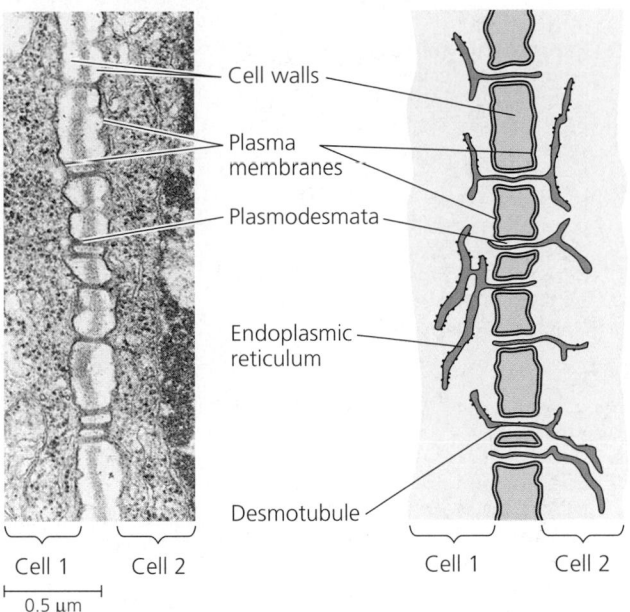

(b) **Plasmodesmata.** The protoplasts of neighboring cells are generally connected by plasmodesmata, cytoplasmic channels that pass through pores in the walls. The endoplasmic reticulum is also continuous through the plasmodesmata in the form of structures called desmotubules (TEM).

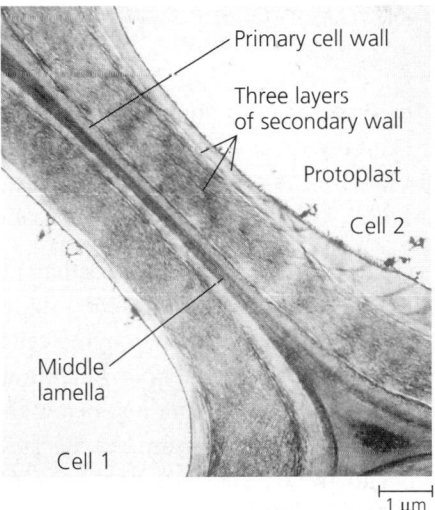

(c) **Cell walls.** An adhesive layer called the middle lamella cements together the cell walls of adjacent cells. All plant cells have a primary cell wall secreted as the cell grows and develops. In addition, many specialized plant cells produce a secondary wall. Notice that the secondary wall is closer to the protoplast than the primary wall, because it forms later, usually after the cell has stopped growing. Most cell walls are relatively porous at the molecular level, with water and many small solutes moving freely through the spaces between cellulose fibrils (TEM).

cells are often depicted as "typical" plant cells because they generally are the least specialized (FIGURE 35.11a, p. 728), but there are exceptions. For example, the highly specialized cells that function in the transport of sugar sap in the phloem—the sieve-tube members—are thin walled and living and are examples of parenchyma cells.

Parenchyma cells perform most of the metabolic functions of the plant, synthesizing and storing various organic products. For example, photosynthesis occurs within the chloroplasts of parenchyma cells in the leaf. Some parenchyma cells in stems and roots have colorless plastids that store starch. And the fleshy tissue of most fruit is composed mostly of parenchyma cells.

Developing plant cells of all types are parenchyma cells before specializing further in structure and function. Those cells that retain the less specialized condition and become mature parenchyma cells do not generally undergo cell division. Most of them, however, retain the ability to divide and differentiate into other types of plant cells under special conditions—during the repair and replacement of organs after injury to the plant, for example. It is even possible in the laboratory to regenerate an entire plant from a single parenchyma cell.

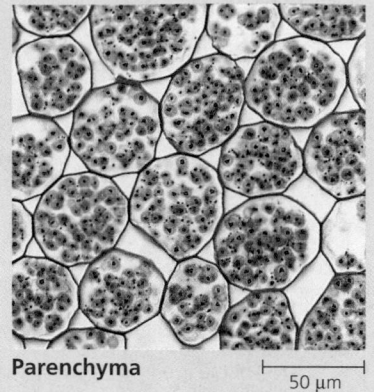

(a) Parenchyma cells are relatively unspecialized, with thin, flexible primary walls. These cells carry on most of the plant's metabolic functions. These parenchyma cells are from the root of a buttercup, and the purple granules are starch-storing plastids.

Parenchyma

50 μm

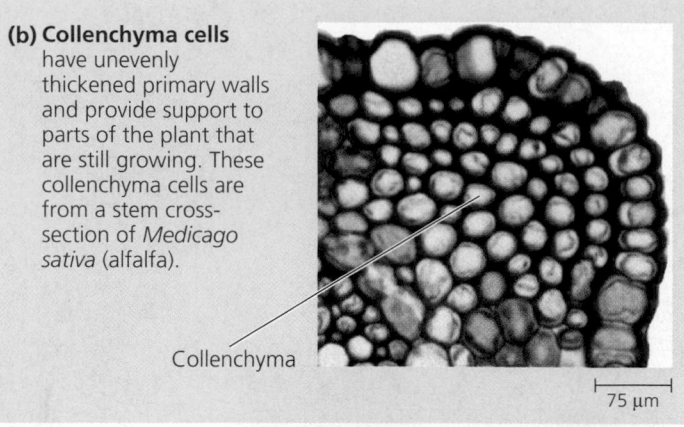

(b) Collenchyma cells have unevenly thickened primary walls and provide support to parts of the plant that are still growing. These collenchyma cells are from a stem cross-section of *Medicago sativa* (alfalfa).

Collenchyma

75 μm

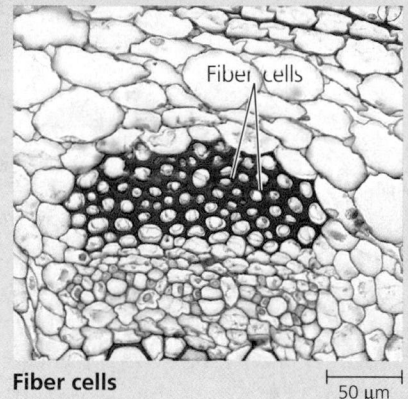

(c) Sclerenchyma cells, specialized for support, have secondary walls hardened with lignin and may be dead (lacking protoplasts) at functional maturity. The **fiber cells** in the left micrograph are elongated sclerenchyma cells, though this is not apparent in this cross section. **Sclereids** (right) are irregularly shaped sclerenchyma cells with very thick, lignified secondary walls. These sclereids, or stone cells, are from a pear fruit. The tracheids and vessel elements of xylem are also sclerenchyma cells.

Fiber cells

Fiber cells

50 μm

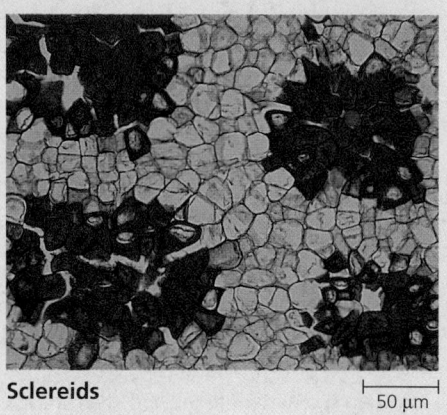

Sclereids

50 μm

FIGURE 35.11 The three major categories of plant cells.

Collenchyma Cells

Collenchyma cells have thicker primary walls than parenchyma cells, though the walls are unevenly thickened (FIGURE 35.11b). Grouped in strands or cylinders, **collenchyma cells** help support young parts of the plant shoot. Young stems and petioles often have a cylinder of collenchyma just below their surface (the "strings" of a celery stalk, for example). Because they lack secondary walls and the hardening agent lignin is absent in their primary walls, collenchyma cells provide support without restraining growth. Unlike sclerenchyma cells, which we discuss next, mature, functioning collenchyma cells are living and flexible and elongate with the stems and leaves they support.

Sclerenchyma Cells

Also functioning as supporting elements in the plant, but with thick secondary walls usually strengthened by lignin, **sclerenchyma cells** are much more rigid than collenchyma cells. Mature sclerenchyma cells cannot elongate, and they occur in regions of the plant that have stopped growing in length. So specialized are sclerenchyma cells for support that many are dead at functional maturity, but they produce secondary walls before the protoplast dies. The rigid walls remain as a "skeleton" that supports the plant. In parts of the plant that are still elongating, the secondary walls of sclerenchyma are deposited unevenly in spiral or ring patterns like the wire helix that reinforces the wall of a vacuum hose. These forms of cell wall thickenings enable the cell wall to stretch like a spring as the cell grows.

The water-conducting vessel elements and tracheids in the xylem are sclerenchyma cells with dual functions: support and transport. But there are also two types of sclerenchyma cells called **fibers** and **sclereids** that specialize entirely in support (see FIGURE 35.11c). Long, slender, and tapered, fibers usually occur in groups. Some plant fibers are used commercially, such as hemp fibers for making rope and flax fibers for weaving into linen. Sclereids, which are shorter than fibers and irregular in shape, impart the hardness to nutshells and seed coats and the gritty texture to pear fruits.

So far, our description of the types of plant tissues and cells has focused on their structure and arrangement in mature organs of the plant. But how does this organization arise? A major difference between plants and most animals is that the growth and development of plants is not just limited to an embryonic or juvenile period, but occurs throughout the life of the plant. At any given instance, a typical plant consists of embryonic organs, developing organs, and mature organs.

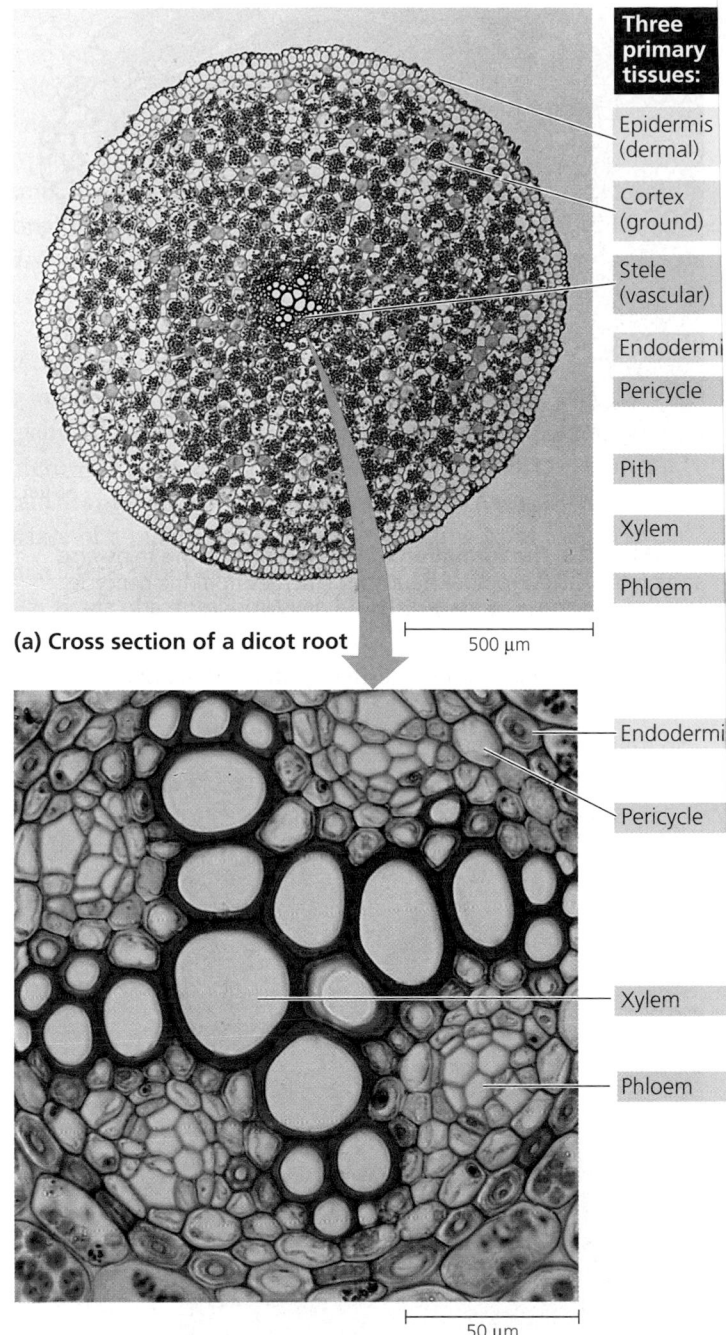

Three primary tissues:

- Epidermis (dermal)
- Cortex (ground)
- Stele (vascular)
- Endodermi
- Pericycle
- Pith
- Xylem
- Phloem

(a) Cross section of a dicot root

500 μm

- Endodermi
- Pericycle
- Xylem
- Phloem

50 μm

The **zone of cell division** includes the apical meristem and its derivatives, called primary meristems. The apical meristem, at the heart of the zone of cell division, produces the cells of the primary meristems and also replaces cells of the root cap that are sloughed off. Near the center of the apical meristem is the **quiescent center,** a population of cells that divide much more slowly than the other meristematic cells. Cells of the quiescent center are relatively resistant to damage from radiation and toxic chemicals, and they may function as reserves that can be recruited to restore the meristem if it is somehow damaged. In experiments where part of the apical meristem is removed, cells of the quiescent center become more mitotically active and produce a new meristem.

THE PROCESS OF PLANT GROWTH AND DEVELOPMENT

A plant's continuous growth and development depend on processes that shape organs and generate specific patterns of specialized cells and tissues within those organs. **Growth** is the irreversible increase in mass that results from cell division and cell expansion. **Development** is the sum of all of the changes that progressively elaborate an organism's body. The early stages of growth and development—germination of the seed and emergence of the seedling—are among the topics of Chapter 38. Here we will study how plants continue to grow after their shoot and root systems are established.

Meristems generate cells for new organs throughout the lifetime of a plant: an overview of plant growth

Most plants continue to grow as long as the plant lives, a condition known as indeterminate growth. In contrast, most animals and certain plant organs, such as flowers and leaves, undergo determinate growth; that is, they cease growing after reaching a certain size. Indeterminate growth does not imply immortality. Although they continue to grow throughout their lives, plants, of course, do die. Plants known as **annuals** complete their life cycle—from germination through flowering and seed production to death—in a single year or less. Many wildflowers are annuals, as are the most important food crops, including the cereal grains and legumes. A plant is called a **biennial** if its life generally spans two years. In many cases, plants with this biennial life cycle are those that live through an intervening cold period (winter) between vegetative growth (first spring/summer) and flowering (second spring/summer). Beets and carrots are biennials, but we rarely leave them in the ground long enough to see them flower. Plants that live many years, including trees, shrubs, and some grasses, are known as **perennials.** Some of the buffalo grass of the North American plains is believed to have been growing for 10,000 years from seeds that sprouted at the close of the last ice age. When a perennial plant finally dies, it is not usually from old age, but from an infection or some environmental trauma, such as fire or severe drought.

For as long as it survives, a plant is capable of indeterminate growth because it has perpetually embryonic tissues called **meristems** in its regions of growth. Meristematic cells divide to generate additional cells. Some of the products of this division remain in the meristematic region to produce still more cells, while others become specialized and are incorporated into the tissues and organs of the growing plant. Cells that remain as wellsprings of new cells in the meristem are called initials. The new cells that are displaced from the meristem, called derivatives, continue to divide for some time, until the cells they produce begin to specialize within developing tissues.

The pattern of plant growth depends on the locations of the meristems (FIGURE 35.12). **Apical meristems,** located at the tips of roots and in the buds of shoots, supply cells for the plant to grow in length. This elongation, called **primary growth,** enables roots to ramify throughout the soil and shoots to increase their exposure to light and carbon dioxide. In herbaceous (nonwoody) plants, only primary growth occurs. In woody plants, however, there is also **secondary growth,** a progressive thickening of the roots and shoots formed earlier by primary growth. Secondary growth is the product of **lateral meristems,** cylinders of dividing cells extending along the length of roots and shoots. One lateral meristem replaces the epidermis with a secondary dermal tissue, such as bark, that is thicker and tougher. A second lateral meristem adds layers of vascular tissue. Wood is the secondary xylem that accumulates over the years.

In woody plants, primary and secondary growth occur simultaneously but in different locations. Primary growth is restricted to the youngest parts of the plant—the tips of roots and shoots, where the apical meristems are located. The lateral meristems develop in slightly older regions of the roots and shoots, some distance away from the tips. There, secondary growth adds girth to the organs. The oldest region of a root or shoot—the base of a tree branch, for example—has the greatest accumulation of secondary tissues formed by the lateral meristems. Each growing season, primary growth produces

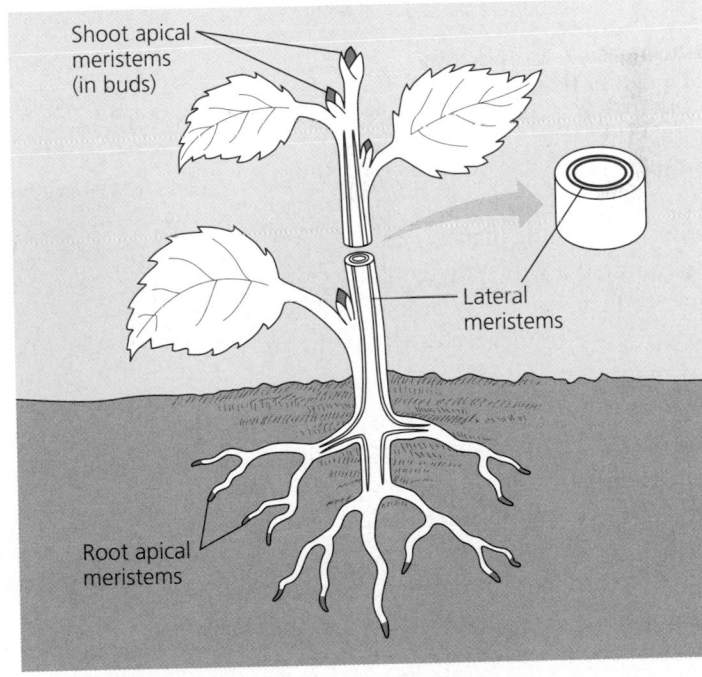

FIGURE 35.12 Locations of major meristems: an overview of plant growth. Meristems are self-renewing populations of cells that divide and provide cells for plant growth. Apical meristems (blue in this diagram) are located near the tips of roots and shoots and are responsible for primary growth, or growth in length. Woody plants also have lateral meristems (red here) that function in secondary growth, which adds girth to roots and shoots.

Labels in figure:
- Shoot apical meristems (in buds)
- Lateral meristems
- Root apical meristems

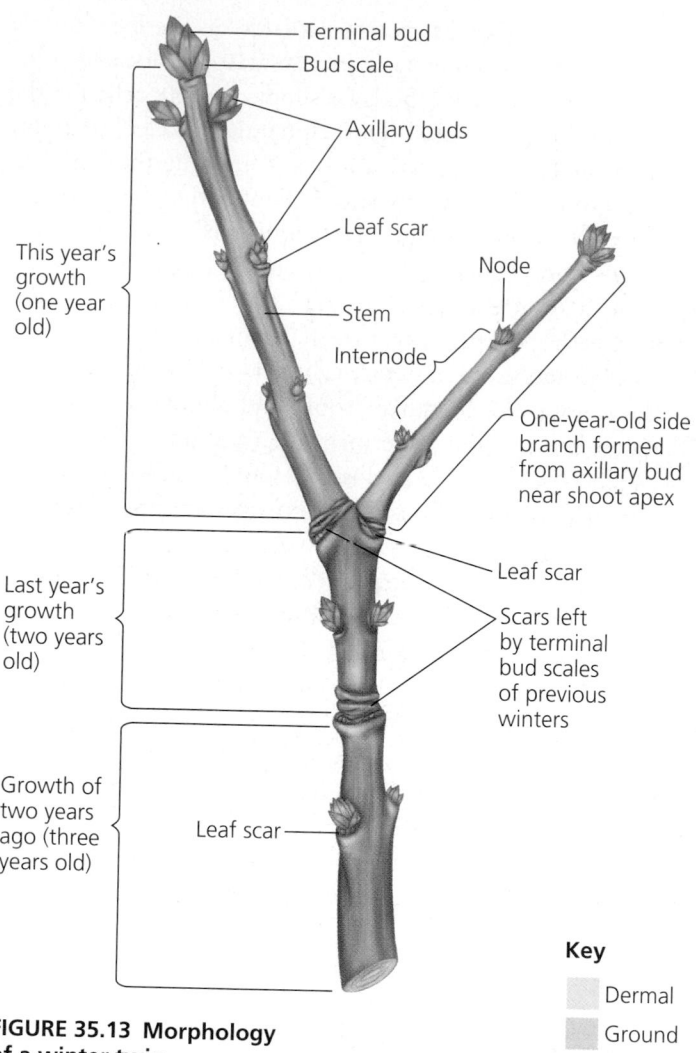

FIGURE 35.13 **Morphology of a winter twig.**

Terminal bud
Bud scale
Axillary buds
Leaf scar
Node
Stem
Internode
One-year-old side branch formed from axillary bud near shoot apex
Leaf scar
Scars left by terminal bud scales of previous winters
Leaf scar

This year's growth (one year old)

Last year's growth (two years old)

Growth of two years ago (three years old)

Key
Dermal
Ground
Vascular

young extensions of roots and shoots, while secondary growth thickens and strengthens the older parts of the plant.

Examine a winter twig of a deciduous tree, and you can see the relationship between primary and secondary growth. At the tip is the dormant terminal bud, enclosed by scales that protect its apical meristem (FIGURE 35.13). In spring, the bud will shed its scales and begin a new spurt of primary growth, producing a series of nodes and internodes. Along each growth segment, nodes are marked by scars left when leaves fell during autumn. Above each leaf scar is either an axillary bud or a branch twig formed by an axillary bud. Farther down the twig are whorls of scars left by the scales that enclosed the terminal bud during the previous winter. Each spring and summer, as primary growth extends the shoot, secondary growth thickens the parts of the shoot that formed in previous years.

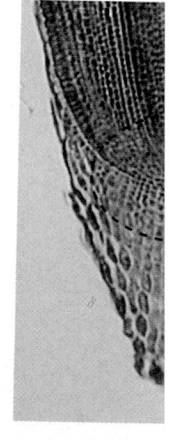

FIGURE 35.14 |
into the tip of a
the apical merist
apical meristem
that are slough
and restores the
concentrated in
of maturation. T

FIGURE 35.19 **Leaf anatomy. (a)** This cutaway drawing of a leaf illustrates the organization of the three tissue systems: dermal tissue (epidermis), vascular tissue, and ground tissue (the mesophyll region consisting of palisade parenchyma and spongy parenchyma). **(b)** This surface view of a *Tradescantia* leaf shows the cells of the epidermis and stomata, with their guard cells (LM). **(c)** Palisade and spongy regions of mesophyll are present within the leaf of a lilac, a dicot (LM).

Key
Dermal
Ground
Vascular

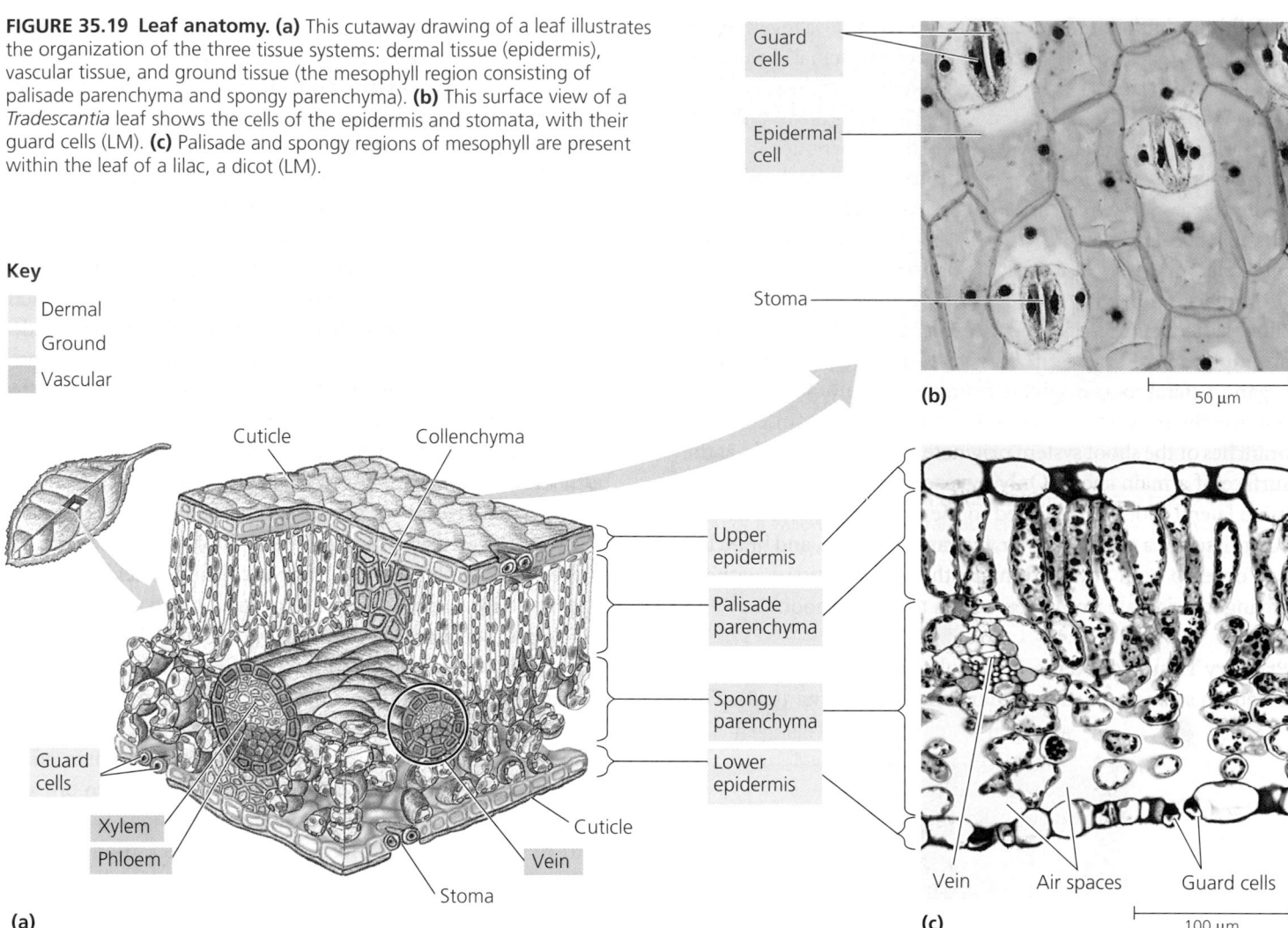

Cuticle
Collenchyma
Upper epidermis
Palisade parenchyma
Spongy parenchyma
Lower epidermis
Guard cells
Xylem
Phloem
Cuticle
Vein
Stoma
(a)

Guard cells
Epidermal cell
Stoma
(b)
50 µm

Vein
Air spaces
Guard cells
(c)
100 µm

Tissue Organization of Leaves. The leaf epidermis is composed of cells tightly interlocked like pieces of a puzzle (FIGURE 35.19). Like our own skin, the leaf epidermis is a first line of defense against physical damage and pathogenic organisms. Also, the waxy cuticle of the epidermis is a barrier to the loss of water from the plant. The epidermal barrier is interrupted only by the **stomata,** tiny pores flanked by specialized epidermal cells called **guard cells.** Each stoma is actually a gap between a pair of guard cells. The stomata allow gas exchange between the surrounding air and the photosynthetic cells inside the leaf. Stomata are also major avenues for the evaporative loss of water from the plant—a process called transpiration.

The ground tissue of a leaf is sandwiched between the upper and lower epidermis in the region called **mesophyll** (from the Greek *mesos,* middle, and *phyll,* leaf). It consists mainly of parenchyma cells equipped with chloroplasts and specialized for photosynthesis. The leaves of many dicots have two distinct regions of mesophyll. On the upper part of the leaf are one or more layers of palisade parenchyma, made up of cells that are columnar in shape. Below the palisade region is the spongy parenchyma, which gets its name from the labyrinth of

air spaces through which carbon dioxide and oxygen circulate around the irregularly shaped cells and up to the palisade region. The air spaces are particularly large in the vicinity of stomata, where gas exchange with the outside air occurs.

The vascular tissue of a leaf is continuous with the xylem and phloem of the stem. Leaf traces, which are branches from vascular bundles in the stem, pass through petioles and into leaves. Within a leaf, veins subdivide repeatedly and branch throughout the mesophyll. This brings xylem and phloem into close contact with the photosynthetic tissue, which obtains water and minerals from the xylem and loads its sugars and other organic products into the phloem for shipment to other parts of the plant. The vascular infrastructure also functions as a skeleton that reinforces the shape of the leaf.

Secondary growth: Lateral meristems add girth by producing secondary vascular tissue and periderm

The stems and roots, but not the leaves, of most dicots increase in girth by secondary growth. The **secondary plant body** consists of the tissues produced during this secondary growth in diameter. Two lateral meristems function in secondary growth: the **vascular cambium,** which produces secondary xylem (wood) and secondary phloem, and the **cork cambium,** which produces a tough, thick covering for stems and roots that replaces the epidermis. Secondary growth occurs in all gymnosperms. Among angiosperms, secondary growth takes place in most dicot species but is rare in monocots.

Secondary Growth of Stems

Vascular Cambium and the Production of Secondary Vascular Tissue.

The vascular cambium is a cylinder of meristematic cells that forms secondary vascular tissue (see FIGURE 35.12). The accumulation of this secondary vascular tissue over the years accounts for most of the increase in the diameter of a woody plant. The vascular cambium produces secondary xylem to its interior and secondary phloem to its exterior (FIGURE 35.20). A tree grows in girth as the diameter of the cylinder of vascular cambium increases over time, laying down successive layers of secondary tissues, each with a larger diameter than the last.

It is important to understand that primary and secondary growth occur simultaneously but in different regions of a stem. While the apical meristem serves to elongate the stem, secondary growth commences farther down the shoot. How does the primary plant body of a young shoot make the transition from primary to secondary growth? The vascular cambium forms from parenchyma cells that regain the capacity to divide; that is, the cells become meristematic. This meristem forms in a layer between the primary xylem and primary phloem of each vascular bundle and in the rays of ground tissue between the bundles. The meristematic bands within the vascular bundles and rays unite to form the vascular cambium as a continuous cylinder of dividing cells surrounding the primary xylem and pith of the stem (FIGURE 35.21, p. 736).

Viewed in cross section, the cylinder of vascular cambium appears as a ring. If we trace around the ring, there are alternating regions of cambium cells called ray initials and fusiform initials. The **ray initials** are cambium cells that produce radial files of parenchyma cells known as xylem rays and phloem rays. These rays separate wedge-shaped sections of secondary vascular tissue. Xylem and phloem rays, consisting mainly of parenchyma, function as living avenues for the radial transport of water and nutrients within a woody stem and in the storage of starch and other reserves. The cambium cells within the vascular bundles are the **fusiform initials,** a name that refers to the shape of these cells, which have tapered (fusiform) ends and are elongated along the axis of the stem. Fusiform initials produce new vascular tissue, forming secondary xylem to the inside of the vascular cambium and forming secondary phloem to the outside (see FIGURE 35.20).

As secondary growth continues over the years, layer upon layer of secondary xylem accumulates, producing the tissue we call wood. Wood consists mainly of tracheids, vessel elements (in angiosperms), and fibers. These cells, dead at functional maturity, have thick, lignified walls that give wood its hardness and strength. In temperate regions of the world, secondary growth in perennial plants is interrupted each year when the vascular cambium becomes dormant during winter. When secondary growth resumes in the spring, the first tracheids and vessel cells to develop usually have relatively large diameters and thin walls compared to the secondary xylem produced later in the summer. Thus, it is usually possible to distinguish early wood (usually produced during spring) from late wood (usually produced during summer) (see FIGURE 35.21, p. 736).

The structure of the early wood maximizes delivery of water to new, expanding leaves during the start of the growth season. Though the thick-walled cells of late wood do not transport as much water as the cells of early wood, they add more physical support to the tree than do the thinner-walled cells. The annual growth rings that are evident in the cross sections of most tree trunks in temperate regions result from the yearly activity of the vascular cambium: cambium dormancy, early wood production, and late wood production. The boundary between one year's growth and the next is usually quite conspicuous, allowing us to estimate the age of a tree by counting its annual rings.

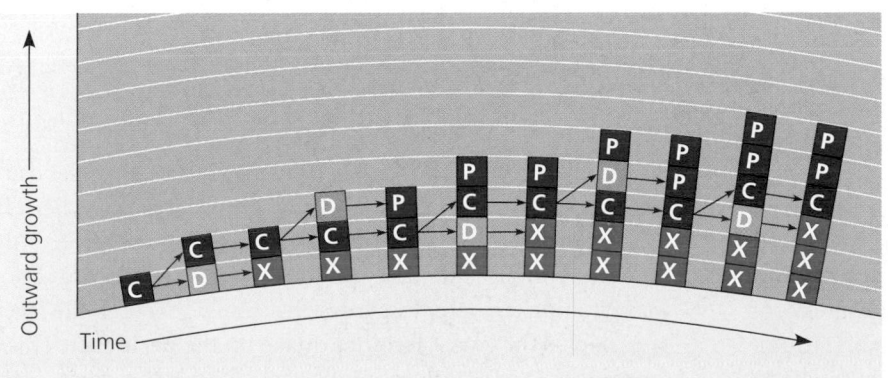

FIGURE 35.20 Production of secondary xylem and phloem by the vascular cambium. This diagram traces the radial file of cells that develops from the meristematic activity of a single cell of the vascular cambium. The cambium cell (C) gives rise to secondary xylem (X) on the inside and secondary phloem (P) on the outside. Each time a cambium cell divides, one daughter cell retains its status as an initial, and the other, the derivative (D), differentiates into a xylem or phloem cell. As layers of xylem are added, vascular cambium itself increases in diameter.

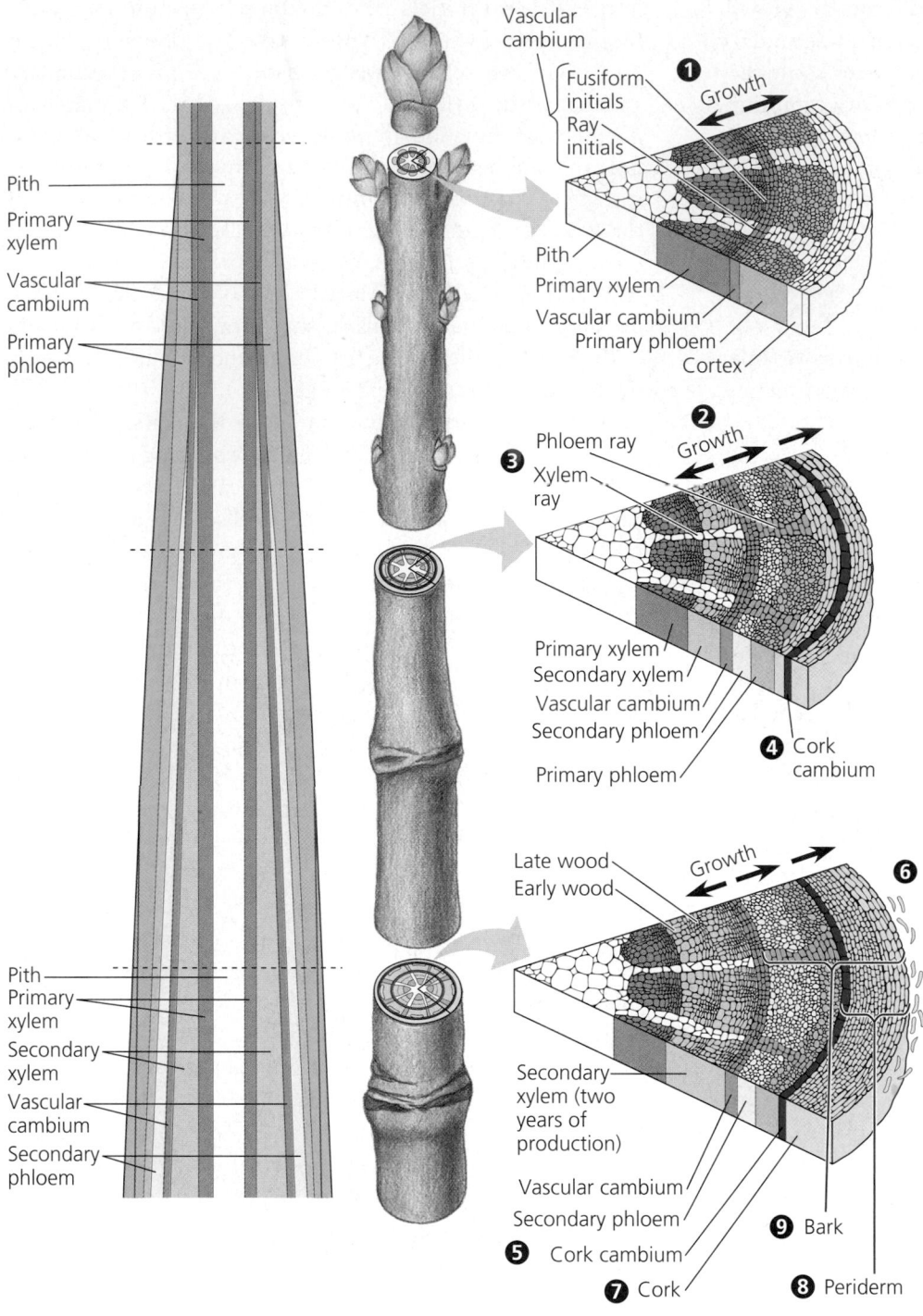

Pith
Primary xylem
Vascular cambium
Primary phloem

Vascular cambium
Fusiform initials
Ray initials
❶ Growth

Pith
Primary xylem
Vascular cambium
Primary phloem
Cortex

Phloem ray
Growth
❸ Xylem ray

Primary xylem
Secondary xylem
Vascular cambium
Secondary phloem
❹ Cork cambium
Primary phloem

Late wood
Early wood
Growth
❻

Secondary xylem (two years of production)

Vascular cambium
Secondary phloem
❾ Bark
❺ Cork cambium
❼ Cork
❽ Periderm

Pith
Primary xylem
Secondary xylem
Vascular cambium
Secondary phloem

❶ In the youngest part of the stem, you can see the primary plant body, as formed by the apical meristem during primary growth. The vascular cambium is beginning to develop.

❷ As primary growth continues to elongate the stem, the portion of the stem formed earlier the same year has already started its secondary growth. This portion of the stem increases in girth as fusiform initials of the vascular cambium form secondary xylem to the inside and secondary phloem to the outside.

❸ The ray initials of the cambium give rise to the xylem and phloem rays.

❹ As the diameter of the vascular cambium increases, the secondary phloem and other tissues external to the cambium cannot keep pace with the expansion because the cells no longer divide. A second lateral meristem, the cork cambium, develops from parenchyma cells in the stem's cortex. The cork cambium produces cork cells, which replace the epidermis as the protective covering at the stem's surface.

❺ In year 2 of secondary growth, the vascular cambium adds to the secondary xylem and phloem, and the cork cambium produces cork.

❻ As the diameter of the stem continues to increase, the outermost tissues exterior to the cork cambium rupture and slough off from the stem.

❼ Cork cambium re-forms in progressively deeper layers of the cortex. When none of the original cortex is left, the cork cambium develops from parenchyma cells in the secondary phloem.

❽ The epidermis of the primary plant body has been replaced by the protective tissue of the secondary plant body, which is called the periderm. Periderm consists of the cork cambium and the cork.

❾ Bark consists of all tissues exterior to the vascular cambium.

FIGURE 35.21 Secondary growth of a stem. You can track the progress of secondary growth by examining the sections through sequentially older parts of the stem. (You would observe the same changes if you could follow the youngest region, near the apex, for the next three years.)

Cork Cambium and the Production of Periderm.   During the early stages of secondary growth, the epidermis produced by primary growth splits, dries, and falls off the stem. It is replaced by new protective tissues produced by the cork cambium, a cylinder of meristematic tissue that first forms in the outer cortex of the stem and later in the secondary phloem (FIGURE 35.22). Cork cambium produces cork cells, which accumulate to

the cambium's exterior. As the cork cells mature, they deposit a waxy material called suberin in their walls and then die. The cork tissue then functions as a barrier that helps protect the stem from physical damage and pathogens. And because cork is waxy, it impedes water loss from the stems. Together, the layers of cork plus the cork cambium make up the **periderm.** This is the protective coat of the secondary plant body that replaces the

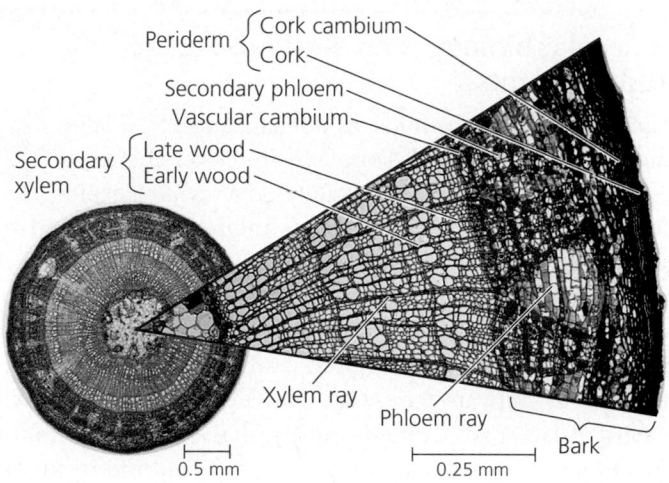

FIGURE 35.22 **Anatomy of a three-year-old stem** (LM).

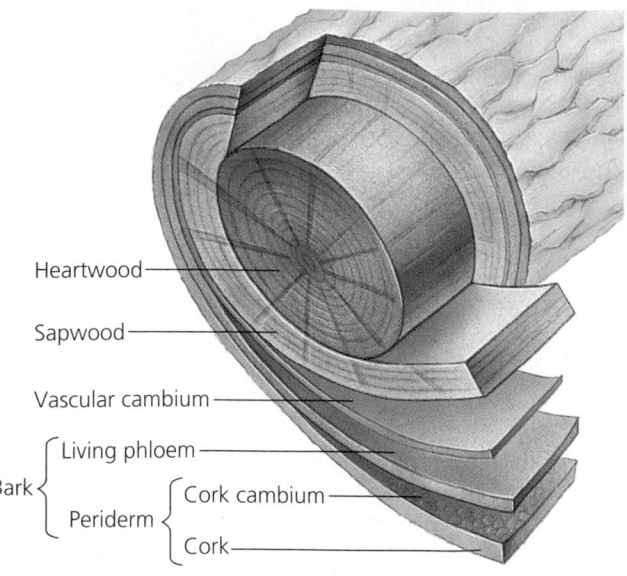

FIGURE 35.23 **Anatomy of a tree trunk.**

epidermis of the primary body. In localized areas, the periderm may split open because the cork cambium is more active than elsewhere. These regions, called **lenticels,** make it possible for living cells within the trunk to exchange gases with the outside air for cellular respiration.

The term **bark,** more inclusive than periderm, refers to all tissues external to the vascular cambium. Thus, in an outward direction, bark consists of the secondary phloem, the cork cambium, and cork. Put another way, bark is phloem plus periderm (see FIGURES 35.21 lower right and 35.22).

Unlike the vascular cambium, which grows in diameter, the original cork cambium is a cylinder of fixed size. After a few weeks of cork production, the cork cambium loses its meristematic activity, and its remaining cells differentiate into cork. Expansion of the stem splits the original periderm. How is it renewed to keep pace with continued secondary growth? New cork cambium forms deeper and deeper in the cortex. Eventually, no cortex is left, and the cork cambium then develops from parenchyma cells in the secondary phloem.

Only the youngest secondary phloem, which is internal to the cork cambium, functions in sugar transport. The older secondary phloem, outside the cork cambium, dies and helps protect the stem until it is sloughed off as part of the bark

during later seasons of secondary growth. This is why secondary phloem does not accumulate as extensively over the years as the secondary xylem does.

Examining an old tree trunk in cross section enables us to see the result of many years of secondary stem growth. Beginning at the center of the tree and tracing outward, we can distinguish several zones (FIGURE 35.23). Heartwood and sapwood both consist of secondary xylem. Heartwood is older and no longer functions in water transport; the lignified walls of its dead cells form a central column that supports the tree. This wood owes its rich color to resins and other compounds that clog the cell cavities and help protect the core of the tree from fungi and wood-boring insects. Sapwood is so named because its secondary xylem still functions in the upward transport of water and minerals (xylem sap). Because each new layer of secondary xylem has a larger circumference, secondary growth enables the xylem to transport more sap each year, providing water and minerals to an increasing number of leaves.

FIGURE 35.24 summarizes the relationships among the primary and secondary tissues of a woody shoot.

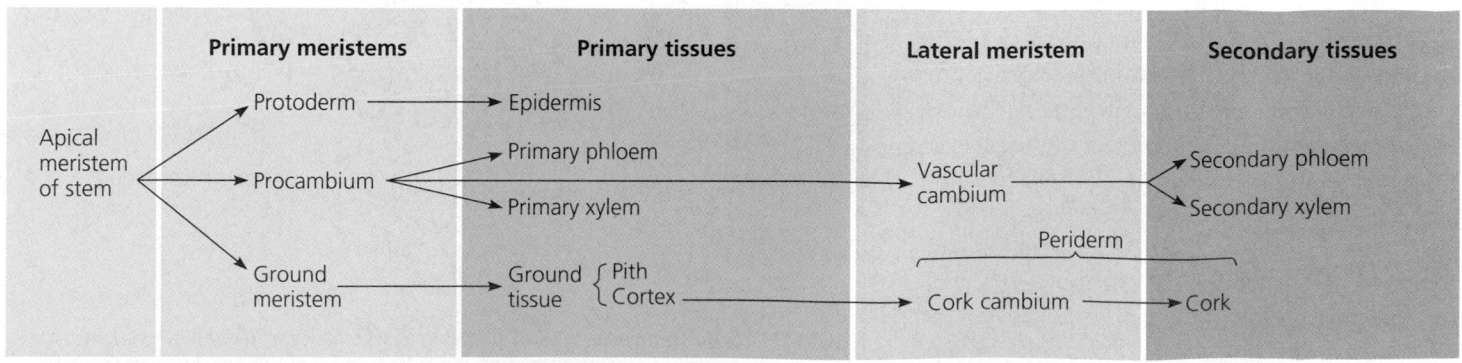

FIGURE 35.24 **A summary of primary and secondary growth in a woody stem.**

Secondary Growth of Roots

The two lateral meristems, vascular cambium and cork cambium, also develop and produce secondary growth in roots. The vascular cambium forms within the stele and produces secondary xylem to its inside and secondary phloem to its outside. As the stele grows in diameter, the cortex and epidermis are split and shed. A cork cambium forms from the pericycle of the stele and produces the periderm, which becomes the secondary dermal tissue. Unlike the primary epidermis of a younger root, periderm is impermeable to water. Therefore, it is only the youngest parts of roots, those representing the primary plant body, that absorb water and minerals from the soil. Older parts of roots, with secondary growth, function mainly to anchor the plant and to transport water and solutes between the younger roots and the shoot system.

Over the years, the root becomes woodier, and annual rings are evident in the secondary xylem. The tissues external to the vascular cambium form a thick, tough bark. After extensive secondary growth, old stems and old roots are quite similar.

So far we have described the development of the plant body from meristems. In the next section, we'll move from a *description* of plant growth and development to the *mechanisms* that underlie these processes.

MECHANISMS OF PLANT GROWTH AND DEVELOPMENT

A plant begins life as a single cell, the zygote formed by the union of egg and sperm. (Fertilization and embryonic development in plants are described in Chapter 38.) The main objective of plant developmental biology is to understand how this single cell gives rise to a multicellular plant of particular form with functionally integrated cells, tissues, and organs. With each meristematic cell possessing the same genetic information, why should one cell become a sieve-tube member in the phloem while another becomes a vessel element in the xylem? Or why should a particular region of the shoot apex give rise to leaves at one time and later to flowers? In considering these questions, we must bear in mind that plants have tremendous developmental plasticity. A plant's form, including height, branching patterns, and reproductive output, is greatly influenced by environmental factors, and a broad range of morphologies can result from the same genotype. (Chapter 39 considers how plants respond to environmental factors.)

Three developmental processes acting in concert are responsible for transforming the fertilized egg into a plant: growth, morphogenesis, and differentiation. The modern techniques of molecular biology are helping plant biologists explore how these processes give rise to plant form.

Molecular biology is revolutionizing the study of plants

Plant biology is in the midst of a renaissance. New laboratory and field methods coupled with clever choices of experimental organisms have catalyzed a research explosion. For example, many scientists interested in the genetic control of plant development, including Joanne Chory, whom you met in the interview on pages 718–719, are focusing their research on *Arabidopsis thaliana*, a little weed of the mustard family (FIGURE 35.25). *Arabidopsis* is small enough to allow researchers to cultivate thousands of the plants in a few square meters of lab space. *Arabidopsis* also has a short generation span of about six weeks, making it an excellent model for genetic studies. Plant biologists are also attracted to *Arabidopsis* by its tiny genome; the amount of DNA per cell ranks among the least of all known plants. Because of these attributes, *Arabidopsis* was the first plant to have its entire genome sequenced, a multinational effort that took six years to complete.

Arabidopsis possesses about 26,000 genes, but many of these are duplicates—there are probably fewer than 15,000 different types of genes. This level of complexity is similar to that found in the fly *Drosophila*. Knowing what some of the *Arabidopsis* genes do has already expanded our understanding of plant development. However, the functions of about 45% of *Arabidopsis*'s genes are unknown (FIGURE 35.25).

Now that the DNA sequence of the *Arabidopsis* genome is completely known, plant biologists have launched an ambitious quest to determine the function of every one of the plant's genes by the year 2010. By identifying each gene's function and tracking every chemical pathway, researchers aim to establish a blueprint for how plants are built. One of the key tasks will be

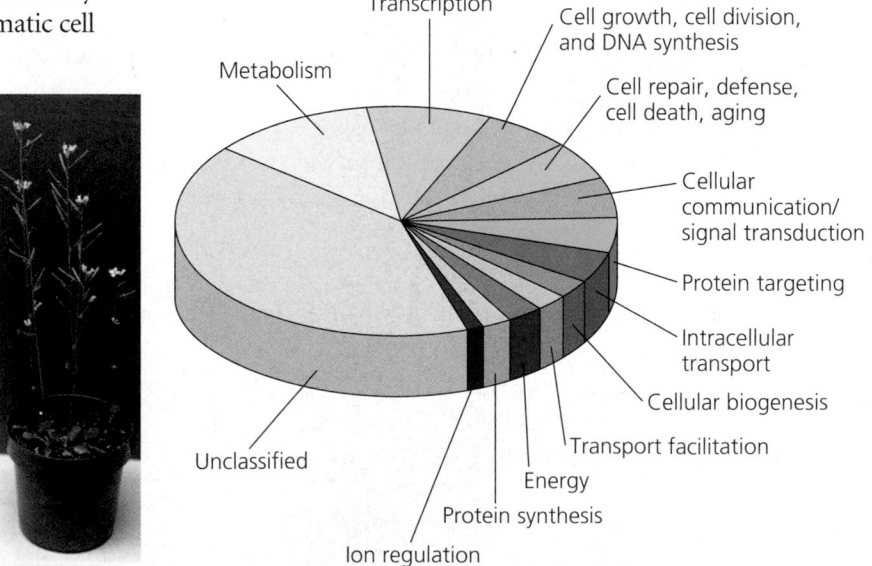

FIGURE 35.25 *Arabidopsis thaliana.* Owing to its small size, rapid life cycle, and small genome, *Arabidopsis* (inset) was the first plant to have its entire genome sequenced (some 26,000 genes). The pie chart represents the proportion of *Arabidopsis* genes in different functional categories.

to identify exactly which cells are manufacturing which gene products and at what stages in the plant's life. Researchers also plan to create mutants for every gene in the genome. As Joanne Chory predicted, it may one day be possible to create a computer-generated "virtual plant" that will enable researchers to visualize which plant genes are activated in different parts of the plant during the entire course of development.

Growth, morphogenesis, and differentiation produce the plant body

Consider a typical annual weed. It may consist of billions of cells—some large, some small, some highly specialized and others not, but all derived from a single fertilized egg. The increase in mass, or growth, that occurs during the life of the plant results from both cell division and cell expansion, but what controls these processes? Why do leaves stop growing upon reaching a certain size, whereas apical meristems divide perpetually? Also, note that the billions of cells in our hypothetical weed are not an undifferentiated clump of cells. They are organized into recognizable tissues and organs. Leaves arise from nodes; roots (unless adventitious) do not. Epidermis forms on the exterior of the leaf, and vascular tissue in the interior— never vice versa. This development of body form and organization is called **morphogenesis.** Each cell in the plant body

contains the same set of genes, exact copies of the genome present in the fertilized egg. But how can this same set of genetic instructions produce such a diversity of cell types—guard cells, mesophyll, sieve-tube members, root hairs, and so on? This generation of cellular diversity is called **differentiation** (see p. 403).

Next, let's take a closer look at the molecular mechanisms that underlie growth, morphogenesis, and cellular differentiation in plants.

Growth involves both cell division and cell expansion

Cell division in meristems, by increasing cell number, increases the potential for growth, but it is cell expansion that accounts for the actual increase in plant mass. The process of plant cell division is described more fully in Chapter 12 (see FIGURE 12.9) and the process of cell elongation in Chapter 39 (see FIGURE 39.7). Here we are more concerned with how these processes contribute to plant form.

The Plane and Symmetry of Cell Division

The plane (direction) of cell division is an important determinant of plant form (FIGURE 35.26a). Imagine a single cell poised to undergo mitosis. If the planes of division of its descendants

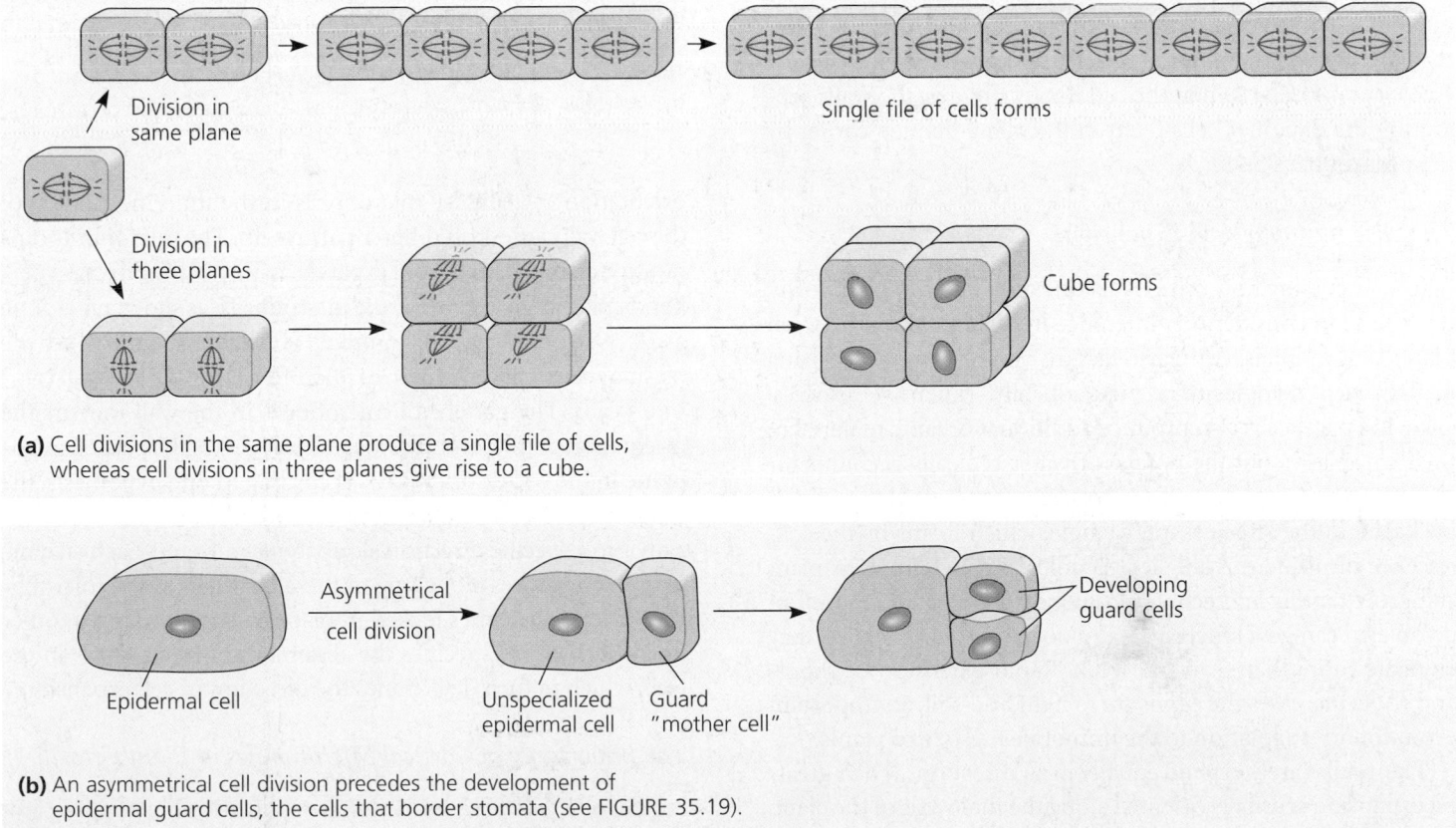

(a) Cell divisions in the same plane produce a single file of cells, whereas cell divisions in three planes give rise to a cube.

(b) An asymmetrical cell division precedes the development of epidermal guard cells, the cells that border stomata (see FIGURE 35.19).

FIGURE 35.26 The plane and symmetry of cell division influence development of form.

are parallel to the plane of the first cell division, a single file of cells will be produced (FIGURE 35.26). At the other extreme, if the planes of cell division of the descendant cells are randomly aligned, an unorganized clump of cells will result. The plane of cell division is immensely important in determining form, but so is the symmetry of cell division. Although in earlier chapters we have emphasized the symmetry of chromosomal redistribution during mitosis, the same does not necessarily hold for the cytoplasm. **Asymmetrical cell division,** in which one daughter cell receives more cytoplasm than the other during mitosis, is fairly common in plant cells and usually signals a key event in development. For example, the formation of guard cells typically involves both an asymmetrical cell division and a change in the plane of the cell division. An epidermal cell divides asymmetrically to form a large cell that will remain an unspecialized epidermal cell and a small cell that will become the guard cell "mother cell." This small mother cell then undergoes another cell division, in a plane perpendicular to the first, to form the guard cells (FIGURE 35.26b).

The plane in which a cell will divide is determined during late interphase. The first sign of this spatial orientation is a rearrangement of the cytoskeleton. Microtubules in the cortex (outer cytoplasm) of the cell become concentrated into a ring called the **preprophase band** (FIGURE 35.27). The band disappears before metaphase, but it has already set the future plane of cell division. The "imprint" consists of an ordered array of actin microfilaments that remain after the microtubules of the preprophase band disperse. These microfilaments hold the nucleus in a fixed orientation until the spindle forms, and later they direct movement of the vesicles that produce the cell plate (see FIGURE 12.8b). When the cell finally divides, the walls separating the daughter cells form in the plane defined earlier by the preprophase band.

The Orientation of Cell Expansion

Before discussing how cell expansion contributes to plant form, it is useful to consider the difference in cell expansion between plants and animals. Animal cells grow mainly by synthesizing protein-rich cytoplasm, a metabolically expensive process. Growing plant cells also produce additional organic material in their cytoplasm, but the uptake of water typically accounts for about 90% of a plant cell's expansion. Most of this water is packaged in the large central vacuole, which forms by the coalescence of numerous smaller vacuoles as a cell grows. A plant can grow rapidly and economically because a small amount of cytoplasm can go a long way. Bamboo shoots, for instance, may elongate more than 2 m per week. Rapid extension of shoots and roots increases the exposure to light and soil, an important evolutionary adaptation to the immobile lifestyle of plants.

Plant cells rarely expand equally in all directions. Their greatest expansion is usually oriented along the main axis of the plant. For example, cells near the root tip may elongate to 20 times their original length, with relatively little increase in width. The

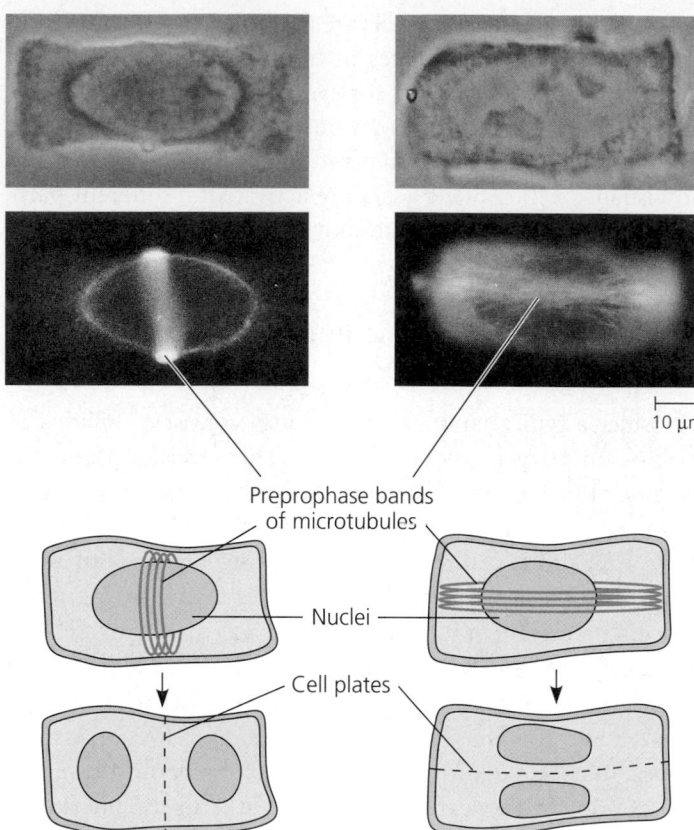

10 µm

Preprophase bands of microtubules

Nuclei

Cell plates

FIGURE 35.27 The preprophase band and the plane of cell division. The location of the preprophase band predicts the plane of cell division. Although the cells shown on the left and right are similar in shape, they will divide in different planes. Each cell is represented by two light micrographs, one (top) unstained and the other (bottom) stained with a fluorescent dye that binds specifically to microtubules. The stained microtubules form a "halo" (preprophase band) around the nucleus in the outer cytoplasm.

orientation of cellulose microfibrils in the innermost layers of the cell wall causes this differential growth. The microfibrils cannot stretch much, so the cell expands mainly in the direction perpendicular to the "grain" of the microfibrils, as shown in FIGURE 35.28. A rosette-shaped complex of enzymes built into the plasma membrane synthesizes microfibrils for the cell wall (FIGURE 35.29). The pattern of microfibrils in the wall mirrors the orientation of microtubules located just across the plasma membrane in the cortex of the cell. According to one hypothesis, the microtubules confine the flow of the cellulose-producing enzymes to a specific direction along the membrane. Each enzyme complex advances along one of these channels as the microfibril it extends becomes locked in place by cross-linking to other microfibrils. This specifies the alignment of microfibrils in the wall, which in turn determines the direction of cell expansion.

The Importance of Cortical Microtubules in Plant Growth

Studies of *Arabidopsis* mutants have confirmed the importance of cortical microtubules in both cell division and expansion. As an example, let's consider what are called *fass*

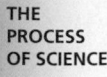

THE PROCESS OF SCIENCE

FIGURE 35.28 The orientation of plant cell expansion. Growing plant cells expand mainly through water uptake. In a growing cell, enzymes weaken cross-links in the cell wall, allowing it to expand as water flows in by osmosis. The orientation of cell growth is mainly in the plane perpendicular to the orientation of cellulose microfibrils in the wall. The microfibrils are embedded in a matrix of other (noncellulose) polysaccharides, some of which form the cross-links visible in the micrograph (TEM). Loosening of the wall occurs when hydrogen ions secreted by the cell activate cell wall enzymes that break the cross-links between polymers in the wall. This reduces restraint on the turgid cell, which can take up more water and expand. Small vacuoles, which accumulate most of this water, coalesce and form the cell's central vacuole.

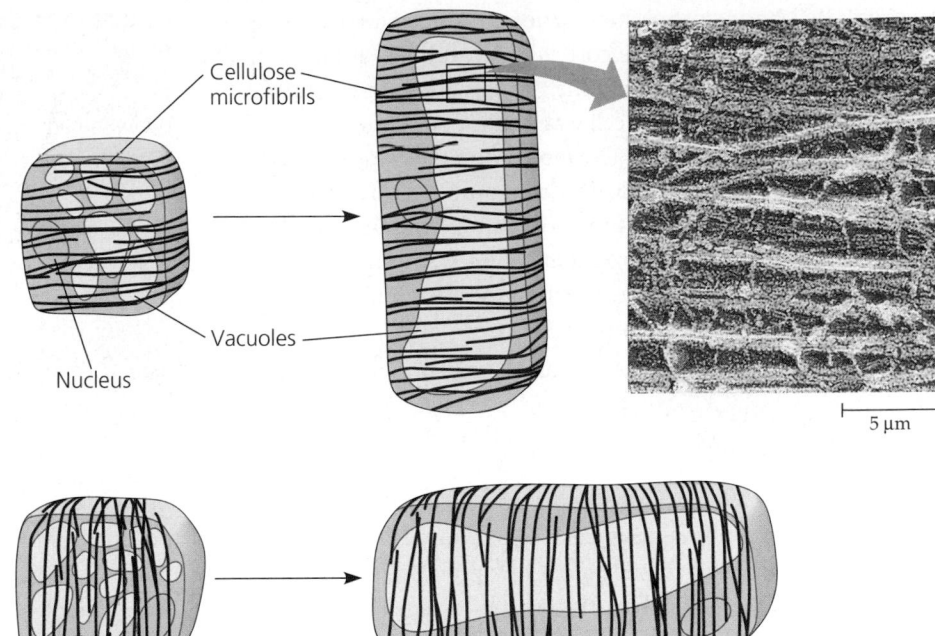

mutants of *Arabidopsis. Fass* mutants have cells that are unusually squat and that seemingly align their planes of division in a random fashion. In the roots and stems of these mutants, the ordered cell files and layers normally present are completely absent. Despite these abnormalities, *fass* mutants do develop into tiny adult plants with all their parts, including flowers, but these organs are compressed longitudinally (FIGURE 35.30).

The cortical microtubular organization of *fass* mutants is abnormal. Although the microtubules involved in chromosome movements and in cell plate deposition are normal, the

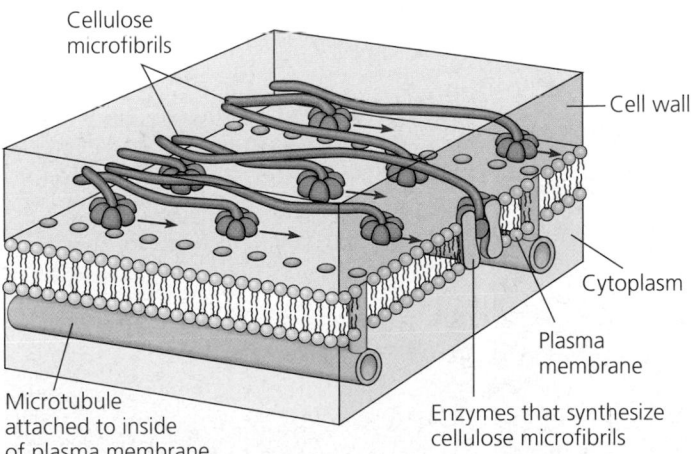

FIGURE 35.29 A hypothetical mechanism for how microtubules orient cellulose microfibrils. Cellulose microfibrils are synthesized at the cell surface by complexes of enzymes that can move in the plane of the plasma membrane. According to one hypothesis, microtubules form "banks" that confine the movement of the enzymes to channels of specified direction. Each enzyme complex advances along one of these channels as the microfibril it extends becomes locked in place by cross-linking to other microfibrils.

(b) *fass* seedling

(c) Mature *fass* mutant

(a) Wild-type seedling

FIGURE 35.30 The *fass* mutant of *Arabidopsis* confirms the importance of cortical microtubules to plant growth. The squat body of the *fass* mutant results from cell division and cell elongation being randomly oriented instead of orienting in the direction of the normal plant axis.

preprophase bands do not form prior to mitosis (see FIGURE 35.27). In interphase cells, the cortical microtubules are not organized in arrays but are randomly positioned, so the cellulose microfibrils deposited in the cell wall cannot be arranged to determine the direction of the cell's elongation (see FIGURES 35.28 and 35.29). This defect in cortical microtubular organization gives rise to cells that expand in all directions equally and that divide in a haphazard arrangement—hence the stout stature and disorganized tissue arrangement of these mutants.

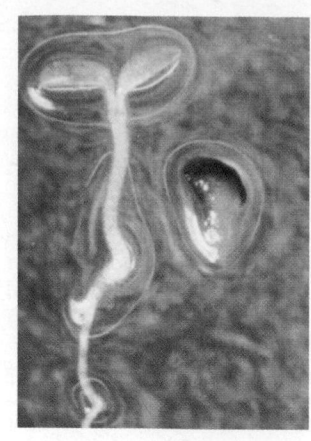

FIGURE 35.31 Establishment of axial polarity. The normal *Arabidopsis* seedling (left) has a shoot end and a root end. In the *gnom* mutant (right), the first division of the zygote was not asymmetrical; as a result, the plant is ball-shaped and lacks cotyledons (seed leaves) and roots.

Morphogenesis depends on pattern formation

A plant's body is more than a collection of dividing and expanding cells. Morphogenesis must occur for development to proceed properly; that is, cells must be organized into multicellular arrangements such as tissues and organs. The development of specific structures in specific locations is called **pattern formation.**

Pattern formation depends to a large extent on **positional information,** signals of some kind that indicate each cell's location within an embryonic structure, such as a shoot tip. Within a developing organ, each cell continues to detect positional information and responds by differentiating into a particular type of cell. Developmental biologists are accumulating evidence that gradients of specific molecules, generally proteins, provide positional information. For example, a substance diffusing from a shoot's apical meristem may "inform" the cells below of their distance from the shoot tip. We can imagine that the cells gauge their radial positions within the developing organ by detecting a second chemical signal that emanates from the outermost cells. The gradients of these two substances would be sufficient for each cell to "get a fix" on its position relative to the longitudinal and radial axes of the rudimentary organ. This idea of diffusible chemical signals is just one of several alternative hypotheses that developmental biologists are testing to learn how an embryonic cell detects its location.

One type of positional information is associated with **polarity.** In plants, there is typically a well-developed axis, but the two ends of this axis are not the same; there is a root end and a shoot end. Such polarity is most obvious in morphological differences, but it is also manifest in several physiological properties, including the unidirectional movement of certain hormones (see Chapter 39) and the emergence of adventitious roots and shoots from the appropriate ends of "cuttings." Adventitious roots form from the root end of a stem or a root cutting, and shoots from the shoot end.

The first division of a plant zygote is normally asymmetrical, initiating polarization of the plant body into shoot and root. Once this polarity has been induced, it becomes exceedingly difficult to reverse. Thus, the proper establishment of axial polarity is a critical step in a plant's morphogenesis. In the *gnom* mutant of *Arabidopsis,* the establishment of polarity is defective. The first cell division of the zygote is abnormal in being symmetrical, and the resulting ball-shaped plant has neither roots nor cotyledons (seed leaves) (FIGURE 35.31).

Let's examine another example of how genes regulate pattern formation and morphogenesis in plants. Plants, like other multicellular organisms, have master regulatory genes called homeotic genes (see Chapter 21) that mediate many of the major events in an individual's development, such as the initiation of an organ. For example, the protein product of the *KNOTTED-1* homeotic gene, found in many plant species, is important in the development of leaf morphology, including the production of compound leaves. If the *KNOTTED-1* gene is overexpressed in tomato plants, the normally compound leaves become "supercompound" (FIGURE 35.32).

FIGURE 35.32 Too much "volume" from a homeotic gene. *KNOTTED-1* is a homeotic gene involved in leaf and leaflet formation. Its overexpression in tomato plants results in leaves that are "supercompound" (right) compared to normal leaves (left).

Cellular differentiation depends on the control of gene expression

It is remarkable that cells as diverse as guard cells, sieve-tube members (phloem), and xylem vessel elements all share the same DNA and all descend from a common cell, the zygote. This cellular differentiation occurs continuously throughout a plant's life, as meristems sustain indeterminate growth. Differentiation reflects the synthesis of different proteins in different types of cells (see Chapter 19). For example, two distinct cell types are formed in the root epidermis of *Arabidopsis*: root hair cells and hairless epidermal cells. Cell fate is associated with the position of the epidermal cells. Immature epidermal cells that are in contact with two underlying cortical cells differentiate into root hair cells, whereas those in contact with only one cortical cell differentiate into mature, hairless cells. A homeotic gene called *GLABRA-2* is required for appropriate root hair distribution. *GLABRA-2* is normally expressed only in epidermal cells that will not develop root hairs. If *GLABRA-2* is rendered dysfunctional by mutation, *every* root epidermal cell develops a root hair. Researchers demonstrated this by coupling the *GLABRA-2* gene to a "reporter gene" that causes every cell expressing *GLABRA-2* in the root to turn blue following a certain treatment (FIGURE 35.33).

When epidermal cells border a single cortical cell, the homeotic gene *GLABRA-2* is selectively expressed, and these cells will remain hairless. (The blue color in this light micrograph indicates cells in which *GLABRA-2* is expressed.)

Here an epidermal cell borders two cortical cells, *GLABRA-2* is not expressed, and the cell will develop a root hair.

Cortical cells

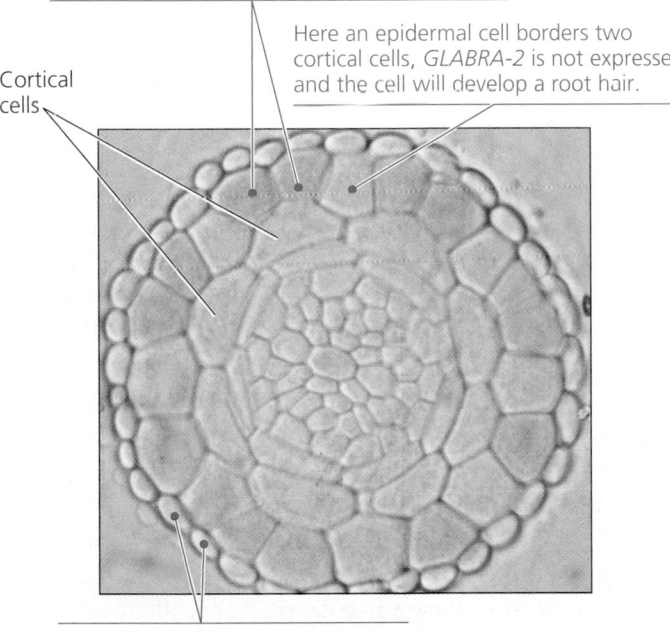

The ring of cells external to the epidermal layer is composed of root cap cells that will be sloughed off as the root hairs start to differentiate.

FIGURE 35.33 Example of cellular differentiation. Two distinct cell types are formed in the root epidermis of *Arabidopsis*: root hair cells and hairless epidermal cells.

What makes differentiation so fascinating is that the cells of a developing organism synthesize different proteins and diverge in structure and function even though they share a common genome. The cloning of whole plants from somatic cells supports the conclusion that the genome of a differentiated cell remains intact (see FIGURE 21.5). If a mature cell removed from a root or leaf can "dedifferentiate" in tissue culture and give rise to the diverse cell types of a plant, then it must possess all the genes necessary to make these many kinds of cells. This means that cellular differentiation depends, to a large extent, on the control of gene expression—the regulation of transcription and translation leading to specific proteins. Cells with the same genomes follow different developmental pathways because they selectively express certain genes at specific times during their differentiation. A guard cell has the genes that program the self-destruction of a xylem protoplast, but it does not express those genes. A xylem vessel cell *does* express them, but only at a specific time in its differentiation, after the cell has elongated and produced its secondary wall. Researchers are beginning to unravel the molecular mechanisms that switch specific genes on and off at critical times in a cell's development (see Chapters 19 and 21).

Clonal analysis of the shoot apex emphasizes the importance of a cell's location in its developmental fate

In the process of shaping a rudimentary organ, patterns of cell division and cell expansion affect the differentiation of cells by placing them in specific locations relative to other cells. Thus, positional information underlies all the processes of development: growth, morphogenesis, and differentiation. One approach to studying the relationships among these processes is clonal analysis, in which the cell lineages (clones) derived from each cell in an apical meristem are mapped as organs develop. Researchers can do this by using radiation or chemicals to induce somatic mutations that alter chromosome number or otherwise tag a cell in some way that distinguishes it from its neighbors in the shoot tip. The lineage of cells derived by mitosis from the mutant meristematic cell will also be "marked." For example, a single cell in the shoot apical meristem may undergo a somatic mutation that prevents chlorophyll from being produced. This cell and all of its descendants will be "albino," and they will appear as a linear file of colorless cells running down the long axis of the otherwise green shoot. One of the important questions that clonal analysis can address is: How early is the developmental fate of a cell determined by its position in an embryonic structure? To some extent, the developmental fates of cells in the shoot apex are predictable. For example, almost all the cells derived from division of the outermost meristematic cells end up as part of the dermal tissue of leaves and stems. But it is not possible to pinpoint precisely which cells of the meristem will give rise to specific tissues and organs. Apparently random changes in rates and planes of cell division can reorganize the meristem. For example, the outermost cells *usually* divide in a plane perpendicular to the surface of the shoot tip, resulting in the addition of cells to the surface layer. But occasionally, a cell at the surface divides in a plane parallel to this meristematic layer, placing one daughter cell beneath the surface among cells derived from different lineages. Thus, the cells of the meristem are not dedicated early to forming specific tissues and organs. Put another way, a cell's developmental fate is not determined by its membership in a lineage derived from a particular meristematic cell. Rather, it is the cell's *final* position in an emerging organ that determines what kind of cell it will become, presumably as a result of positional information.

Phase changes mark major shifts in development

From what you have learned so far about the shoot apex and primary growth, it would seem as if the meristem lays down an identical repeating pattern of stems and leaves. In fact, the apical meristem can change from one developmental phase to another during its history—a process called a **phase change**. One of these phase changes is a gradual transition in vegetative (leaf-producing) growth from a juvenile state to a mature state in some species. Usually, the most obvious sign of this phase change is a change in the morphology of the leaves produced. The leaves of juvenile versus mature shoot regions differ in shape and other features (FIGURE 35.34). Once the meristem has laid down juvenile nodes and internodes, they retain that status even as the shoot continues to elongate and the meristem eventually changes to the mature phase. If axillary buds give rise to branches, those shoots reflect the developmental phase of the main shoot region from which they arise. Though the main shoot apex may have made the transition from the juvenile to the mature phase, an older region of the shoot below the apex will continue to give rise to branches bearing juvenile leaves if that shoot region was laid down when the main apex was still in the juvenile phase. Ironically, this means that a branch with juvenile leaves may actually be *older* than a branch with mature leaves. This is analogous to repairing a vintage automobile with newly manufactured body parts; the parts are new, but they must be of the old style to go with the rest of the car.

The juvenile-to-mature phase transition is another case where it is misleading to compare plant and animal development. In an animal, this transition occurs at the level of the entire organism—as when a larva develops into an adult animal, for example. In plants, phase changes during the history of apical meristems can result in juvenile and mature regions coexisting along the axis of each shoot.

(a) **(b)**

FIGURE 35.34 Phase change in the shoot system of *Eucalyptus*. A silver-dollar eucalyptus (*Eucalyptus polyanthemos*) has both **(a)** juvenile leaves (the round "silver dollars") and **(b)** mature leaves (the lance-shaped leaves). This dual foliage reflects a phase change in the development of the apical meristem of each shoot. In its juvenile vegetative phase, a meristem lays down modules on which round leaves develop. As the meristem changes gradually to the mature vegetative phase, the leaves become more and more lance-shaped. Once a module forms, its developmental phase—juvenile or mature— is fixed; that is, round leaves do not mature into lanceolate leaves.

Genes controlling transcription play key roles in a meristem's change from a vegetative to a floral phase

Another particularly striking phase change in plant development is the transition of a vegetative shoot tip into a floral meristem. A combination of environmental cues, such as day length, and internal signals, such as hormones, trigger this transition. (You will learn more about the control of flowering in Chapter 39.) Unlike vegetative growth, which is self-renewing, the production of a flower by an apical meristem terminates primary growth of that shoot tip; the apical meristem is consumed in the production of the flower's organs. The transition from vegetative growth to flowering is associated with the switching on of floral **meristem identity genes.** The protein products of these genes are transcription factors that in turn help activate the genes required for development of the floral meristem.

Once a shoot meristem is induced to flower, positional information commits each primordium that arises on the flanks of the shoot tip to develop into an organ of specific structure and function—for example, an anther-bearing stamen (see FIGURE 30.13 to review flower structure). Plant biologists have identified some of the genes that are regulated by positional information and that function in this development of floral pattern. Mutations in these **organ identity genes** substitute one type of floral organ where another would normally form (FIGURE 35.35). For example, a particular mutation may cause an extra whorl of sepals to develop in a flower where

THE PROCESS OF SCIENCE

there ought to be petals. The inference is that the wild-type alleles for these organ identity genes are responsible for the development of normal floral pattern.

Organ identity genes code for transcription factors (see p. 422). Positional information determines which organ identity genes are expressed in a particular floral-organ primordium. The resulting transcription factors probably induce the expression of those genes responsible for building an organ of specific structure and function. FIGURE 35.36 illustrates a hypothesis for how three regulatory products of organ identity genes are responsible for normal flower development. (Chapter 21 explores this hypothesis for flower development in much greater detail—see FIGURE 21.20.) By constructing such hypotheses and designing experiments to test them, researchers are tracing the genetic basis of plant development.

■   ■   ■

In dissecting the plant to examine its parts, as we have done in this chapter, we must remember that the whole plant functions as an integrated organism. In the following chapters, you will learn more about how materials are transported within the plant (Chapter 36), how plants obtain nutrients (Chapter 37), how plants reproduce (Chapter 38), and how the various functions of the plant are coordinated (Chapter 39). Remembering that structure fits function and that plant anatomy and physiology reflect evolutionary adaptation to the problems of living on land will enhance your understanding of plants.

(a)

(b)

FIGURE 35.35 Organ identity genes and pattern formation in flower development. **(a)** *Arabidopsis* normally has four whorls of flower parts: sepals (Se), petals (Pe), stamens (St), and carpels (Ca). **(b)** Researchers have identified several mutations of organ identity genes that cause abnormal flowers to develop. This flower has an extra set of petals in place of stamens and an internal flower where normal plants have carpels.

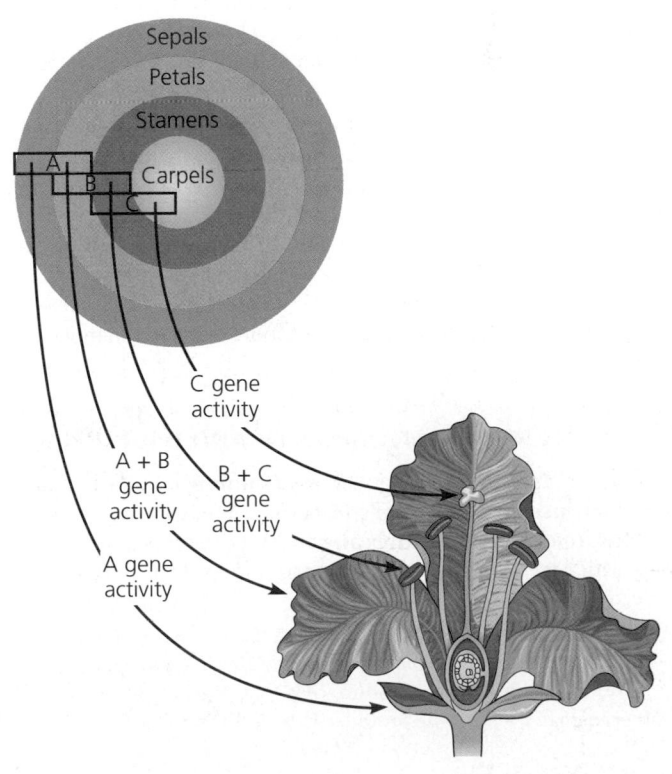

FIGURE 35.36 The ABC hypothesis for the functioning of organ identity genes in flower development. Based on their studies of mutations that affect floral morphology in *Arabidopsis* and other plants (a generalized flower is illustrated here), researchers have determined that three classes of organ identity genes are responsible for the spatial pattern of floral organs. (Chapter 21 presents the experimental support for this hypothesis.) The three classes of organ identity genes are designated A, B, and C in the upper drawing, a schematic diagram of a floral meristem in transverse view. The products of these genes are transcription factors that regulate expression of other genes responsible for the development of the specific floral organs: sepals, petals, stamens, and carpels. Sepals develop from the region of the meristem where only the A genes are active. Petals develop from the region where both A and B genes are expressed. Stamens arise from the meristem where both the B and C genes are active. And carpels are derived from the region where only the C genes are expressed.

CHAPTER 35 REVIEW

Go to the Campbell Biology website (www.campbellbiology.com) to explore an interactive version of the Chapter Review.

Summary of Key Concepts

THE PLANT BODY

- **Both genes and environment affect plant structure (pp. 720–721)** The correlation between structure and function and the evolutionary context of adaptation to the environment illuminate our study of plants.

- **Plants have three basic organs: roots, stems, and leaves (pp. 721–724, FIGURES 35.1–35.6)** Roots anchor the plant, absorb and conduct water and minerals, and store food. The shoot system consists of stems, leaves, and flowers. Leaves are attached by their petioles to the nodes of the stem, with internodes of the stem separating the nodes. Axillary buds, located in the axils of petioles and stems, have the potential to extend as vegetative or floral branches. The two classes of angiosperms, monocots and eudicots, differ in anatomical details.

  Web/CD Activity 35A: *Root, Stem, and Leaf Sections*

- **Plant organs are composed of three tissue systems: dermal, vascular, and ground (pp. 724–726, FIGURES 35.7–35.9)** Dermal tissue (epidermis), vascular tissue (xylem and phloem), and ground tissue are continuous throughout the plant, although in the various plant organs the three tissues differ in arrangement and in some specialized functions. Vascular tissues integrate the parts of the plant. Water and minerals move up from roots in the xylem. Sugar is exported from leaves or storage organs in the phloem. Tracheids and vessel elements, the water-conducting cells of xylem, have thick walls and are dead at functional maturity. Sieve-tube members are the sugar-transporting cells of phloem. Though alive at functional maturity, sieve-tube cells depend on the services of neighboring companion cells.

- **Plant tissues are composed of three basic cell types: parenchyma, collenchyma, and sclerenchyma (pp. 726–728, FIGURES 35.10, 35.11)** Parenchyma cells, relatively unspecialized cells that retain the ability to divide, perform most of the plant's metabolic functions of synthesis and storage. Collenchyma cells, which have unevenly thickened walls, support young, growing parts of the plant. Sclerenchyma cells—fibers, sclereids, and the water-conducting cells of xylem—have thick, lignified walls that help support mature, nongrowing parts of the plant.

THE PROCESS OF PLANT GROWTH AND DEVELOPMENT

- **Meristems generate cells for new organs throughout the lifetime of a plant:** *an overview of plant growth* **(pp. 729–730, FIGURES 35.12, 35.13)** Apical meristems elongate shoots and roots through primary growth. Lateral meristems add girth to woody plants through secondary growth.

- **Primary growth: Apical meristems extend roots and shoots by giving rise to the primary plant body (pp. 730–734, FIGURES 35.14–35.19)** Apical meristems produce cells that continue to divide as meristematic cells of the protoderm, procambium, and ground meristem. These primary meristems give rise to the dermal, vascular, and ground tissues of the primary plant body. In roots, the apical meristem is located near the tip, where it regenerates the root cap as well as producing the primary meristems. The apical meristem of a shoot is located in the terminal bud, where it gives rise to a repetition of internodes and leaf-bearing nodes.

  Web/CD Case Study in the Process of Science: *What Are Functions of Monocot Tissues?*

- **Secondary growth: Lateral meristems add girth by producing secondary vascular tissue and periderm (pp. 734–738, FIGURES 35.20–35.24)** The vascular cambium develops from parenchyma cells into a meristematic cylinder that produces secondary xylem and secondary phloem. The cork cambium gives rise to the secondary plant body's protective covering, or periderm, which consists of the cork cambium plus the layers of cork cells it produces. Bark is periderm plus secondary phloem—all the tissues external to the vascular cambium.

  Web/CD Activity 35B: *Primary and Secondary Growth*

MECHANISMS OF PLANT GROWTH AND DEVELOPMENT

- **Molecular biology is revolutionizing the study of plants (pp. 738–739, FIGURE 35.25)** New techniques and new model systems, including *Arabidopsis*, are catalyzing explosive progress in our understanding of plants. *Arabidopsis* is the first plant to have had its entire genome sequenced.

- **Growth, morphogenesis, and differentiation produce the plant body (p. 739)** Morphogenesis occurs throughout the plant's life as apical meristems continuously divide to add new organs.

- **Growth involves both cell division and cell expansion (pp. 739–742, FIGURES 35.26–35.30)** Cell division and cell expansion are primary determinants of growth and form. A preprophase band determines where a cell plate will form in a dividing cell. The orientation of the cytoskeleton also affects the direction of cell elongation by controlling the orientation of cellulose microfibrils in the wall.

- **Morphogenesis depends on pattern formation (p. 742, FIGURES 35.31, 35.32)** Development of tissues and organs in specific locations depends on the ability of cells to detect and respond to positional information.

- **Cellular differentiation depends on the control of gene expression (p. 743, FIGURE 35.33)** The challenge of understanding cellular differentiation is explaining how cells with matching genomes diverge into various cell types.

- **Clonal analysis of the shoot apex emphasizes the importance of a cell's location in its developmental fate (pp. 743–744)** A cell's position in a developing organ determines its pathway of differentiation.

- **Phase changes mark major shifts in development (p. 744, FIGURE 35.34)** Internal or environmental cues may cause a plant to switch from one developmental phase to another—for example, from development of juvenile leaves to development of mature leaves.

- **Genes controlling transcription play key roles in a meristem's change from a vegetative to a floral phase (pp. 744–745, FIGURES 35.5, 35.6)** Research on organ identity genes in developing flowers provides an important model system for the study of pattern formation.

1. Which structure is *incorrectly* paired with its tissue system?

   a. root hair—dermal tissue

   b. palisade parenchyma—ground tissue

   c. guard cell—dermal tissue

   d. companion cell—ground tissue

   e. tracheid—vascular tissue

2. A vessel cell would likely lose its protoplast in which zone of growth in a root?

   a. zone of cell division   d. root cap

   b. zone of elongation   e. quiescent center

   c. zone of maturation

3. Wood consists of

   a. bark.   d. secondary phloem.

   b. periderm.   e. cork.

   c. secondary xylem.

4. Which of the following is *not* part of an older tree's bark?

   a. cork   d. secondary xylem

   b. cork cambium   e. secondary phloem

   c. lenticels

5. The phase change of an apical meristem from the juvenile to mature vegetative phase is often signaled by

   a. a change in the morphology of the leaves that are produced.

   b. the initiation of secondary growth.

   c. the transcription of different organ-identity genes within primordia on the flanks of the shoot tip.

   d. a change in the orientation of both the preprophase bands and cortical microtubules within dividing cells of the lateral meristems.

   e. the activation of floral meristem identity genes.

6. A tree will eventually die if it is girdled, meaning that a ringlike cut has been made all the way around the trunk to a depth just below the bark. The cause of death is mainly

   a. destruction of the procambium.

   b. destruction of axillary buds.

   c. the killing of bark cells.

   d. destruction of the cork cambium, phloem, and vascular cambium.

   e. destruction of the plant's ability to continue primary growth.

7. Which of the following cell types is least likely to have a secondary wall?

   a. sclerenchyma cell   d. tracheid

   b. parenchyma cell   e. sclereid

   c. fiber cell

8. _____ is to primary xylem as vascular cambium is to _____.

   a. Primary phloem; secondary xylem

   b. Tracheid; vessel cell

   c. Procambium; secondary xylem

   d. Apical meristem; lateral meristem

   e. Stele; primary phloem

9. The type of mature cell that a particular embryonic plant cell will become appears to be determined mainly by

   a. the selective loss of genes.

   b. the cell's final position in a developing organ.

   c. the cell's pattern of migration.

   d. the cell's age.

   e. the cell's particular meristematic lineage.

10. Based on the hypothesis presented in FIGURES 21.20 and 35.36, predict the floral morphology of a mutant lacking activity of the B genes.

    a. carpel-petal-petal-carpel

    b. petal-petal-petal-petal

    c. sepal-sepal-carpel-carpel

    d. sepal-carpel-carpel-sepal

    e. carpel-carpel-carpel-carpel

11. Explain why pruning certain types of fruit trees increases future fruit harvest.

12. The "eyes" of a white potato mark nodes with buds. If those buds break dormancy and the potato "sprouts," are the resulting appendages root branches or shoot branches? Explain your answer.

13. Biologists generally define an "animal tissue" as a "group of cells with common structure and function." How does this definition of a tissue contrast with what biologists call a "tissue system" in plants?

14. You have cells in the lower layers of your skin that continue dividing, replacing dead cells that are sloughed from your surface. Why is it inaccurate to compare such regions of active cell division in your body to a plant meristem?

15. Explain why a mutation such as *fass* in *Arabidopsis,* which prevents orderly alignment of cortical microtubules, results in a stubby plant rather than a normal elongated one.

Go to the website or CD-ROM for more quiz questions.

Evolution Connection

The evolution of plant structure can be explored by looking at alternative growth strategies of related plants that occur in different environments. In this respect, Darwin was one of the earliest observers to note that many plant species with herbaceous growth forms on continental mainlands have woody tree-like relatives on remote oceanic islands. In the Hawaiian Islands, for example, one can find tree lobelias and tall, woody violets, groups that occur as small herbs in North America. Suggest an evolutionary hypothesis for this trend: Why is it so common for woody tree-like forms to evolve from herbaceous ancestors that colonize isolated islands?

The Process of Science

Write a paragraph explaining why certain mutants are so useful to researchers who investigate the regulation of plant development. Your paragraph should include at least one specific example.

Identify functions of monocot tissues in the Case Study in the Process of Science, available on the website and CD-ROM.

Science, Technology, and Society

Make a list of the plants and plant products you use in a typical day. How do you use these various plant products? Do you think the number of plants and plant products used in everyday life has increased or decreased in the last century? Do you think the number is likely to increase or decrease in the future? Why?

Answers: 1. d; 2. c; 3. c; 4. d; 5. a; 6. d; 7. b; 8. c; 9. b; 10. c; 11. Removal of terminal buds from major branches results in more branching because of less inhibition of axillary buds. More branches results in more flowers and hence more fruit. 12. Shoot branches; the potato tuber is a modified stem, part of the shoot system. 13. A plant tissue system may consist of several types of specialized cells, such as the different cell types of the vascular tissue system. 14. Your dividing cells are normally limited in the types of cells they can form. In contrast, the products of cell division in plant meristem differentiate into all the diverse cell types of a plant organ. 15. An orderly arrangement of cortical microtubules determines both the plane of cell division (by setting the preprophase band) and the main direction of cell elongation (by controlling the orientation of cellulose microfibrils).

TRANSPORT IN PLANTS

AN OVERVIEW OF TRANSPORT MECHANISMS IN PLANTS

- Transport at the cellular level depends on the selective permeability of membranes
- Proton pumps play a central role in transport across plant membranes
- Differences in water potential drive water transport in plant cells
- Aquaporins affect the rate of water transport across membranes
- Vacuolated plant cells have three major compartments
- Both the symplast and the apoplast function in transport within tissues and organs
- Bulk flow functions in long-distance transport

ABSORPTION OF WATER AND MINERALS BY ROOTS

- Root hairs, mycorrhizae, and a large surface area of cortical cells enhance water and mineral absorption
- The endodermis functions as a selective sentry between the root cortex and vascular tissue

TRANSPORT OF XYLEM SAP

- The ascent of xylem sap depends mainly on transpiration and the physical properties of water
- Xylem sap ascends by solar-powered bulk flow: *a review*

THE CONTROL OF TRANSPIRATION

- Guard cells mediate the photosynthesis-transpiration compromise
- Xerophytes have evolutionary adaptations that reduce transpiration

TRANSLOCATION OF PHLOEM SAP

- Phloem translocates its sap from sugar sources to sugar sinks
- Pressure flow is the mechanism of translocation in angiosperms

The algal ancestors *of plants were completely immersed in water and dissolved minerals, and none of their cells were far from these ingredients. The evolutionary journey onto land involved the differentiation of the plant body into roots, which absorb water and minerals from soil, and shoots, which are exposed to light and atmospheric CO_2. This body plan enabled plants to survive in an environment where chemical resources are divided between two media, soil and air. But the morphological solution to a dual environment posed a new problem: the need to transport materials between roots and shoots, sometimes over long distances. For example, the leaves of the eucalyptus trees in the photograph above are more than 100 m from the roots. These remote organs are bridged by vascular tissues that transport sap throughout the plant body. The mechanisms responsible for this internal transport are the subject of this chapter.*

AN OVERVIEW OF TRANSPORT MECHANISMS IN PLANTS

Transport in plants occurs on three levels: (1) the uptake and loss of water and solutes by individual cells, such as the absorption of water and minerals from the soil by cells of a root; (2) short-distance transport of substances from cell to cell at the level of tissues and organs, such as the loading of sugar from photosynthetic cells of a mature leaf into the sieve tubes of phloem; and (3) long-distance transport of sap within xylem and phloem at the level of the whole plant. FIGURE 36.1 provides an overview of the major transport processes in whole plants.

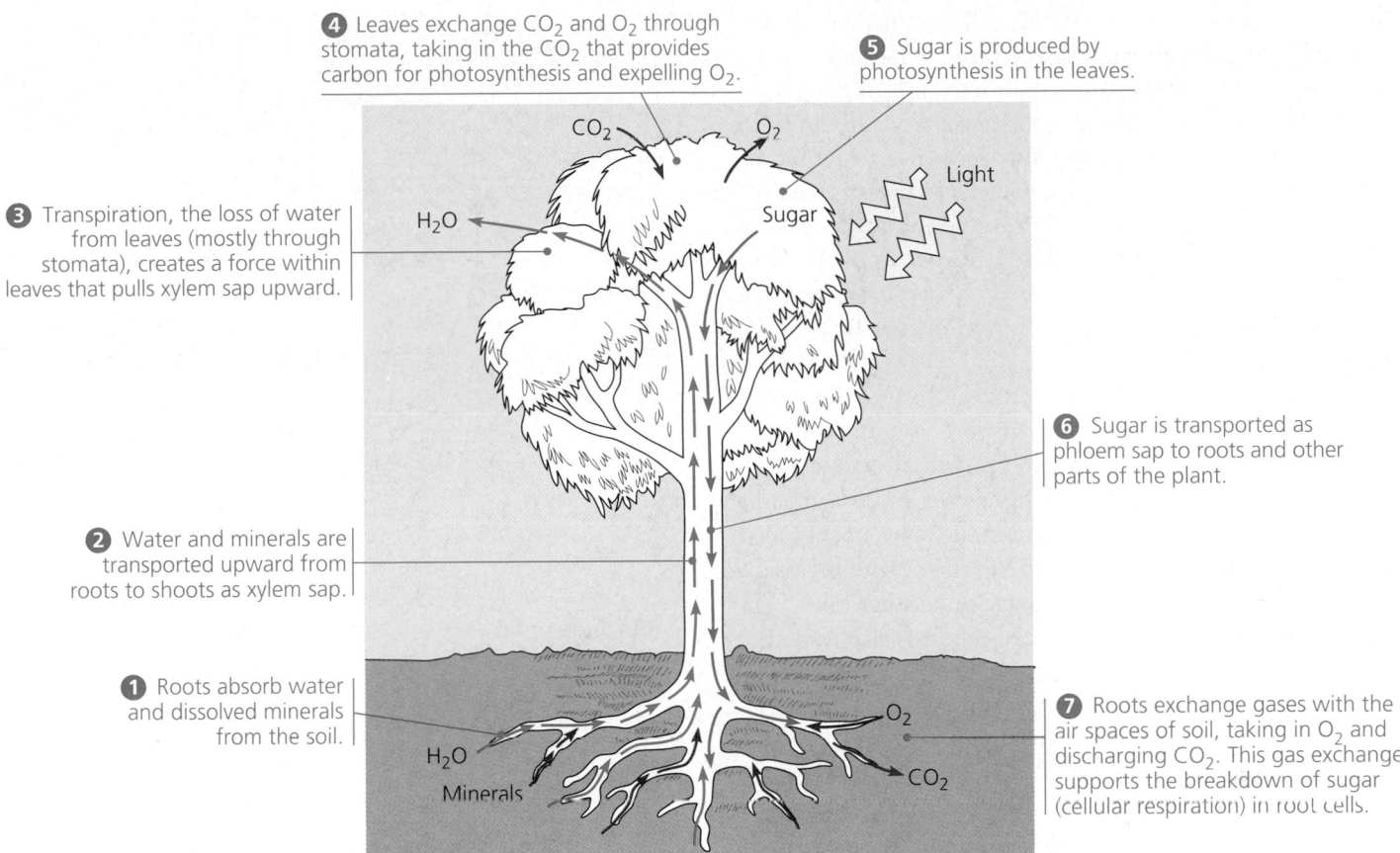

FIGURE 36.1 **An overview of transport in whole plants.**

④ Leaves exchange CO_2 and O_2 through stomata, taking in the CO_2 that provides carbon for photosynthesis and expelling O_2.

⑤ Sugar is produced by photosynthesis in the leaves.

③ Transpiration, the loss of water from leaves (mostly through stomata), creates a force within leaves that pulls xylem sap upward.

⑥ Sugar is transported as phloem sap to roots and other parts of the plant.

② Water and minerals are transported upward from roots to shoots as xylem sap.

① Roots absorb water and dissolved minerals from the soil.

⑦ Roots exchange gases with the air spaces of soil, taking in O_2 and discharging CO_2. This gas exchange supports the breakdown of sugar (cellular respiration) in root cells.

Transport at the cellular level depends on the selective permeability of membranes

The transport of solutes and water across biological membranes was covered in detail in Chapter 8. Here we reexamine a few of these transport processes in the specific context of plant cells.

The selective permeability of a plant cell's plasma membrane controls the movement of solutes between the cell and the extracellular solution. Recall from Chapter 8 that solutes tend to diffuse down their gradients, and when this occurs across a membrane, the process is termed passive transport—*passive* because it happens without the direct expenditure of metabolic energy by the cell. Most solutes, however, cross a membrane very slowly unless they can pass through **transport proteins** embedded in the membrane. Some of these facilitate diffusion by binding selectively to a solute on one side of the membrane and releasing the solute on the opposite side. Transfer of the solute across the membrane involves a conformational (shape) change by the transport protein. Other transport proteins function as **selective channels,** which are simply selective passageways across the membrane. For example, the membranes of most plant cells have potassium channels that allow potassium ions (K^+) to pass, but not similar ions, such as sodium (Na^+). Some channels are gated; that is,

certain environmental or biochemical stimuli can cause the channels to open or close. Later in this chapter, we will see how gated K^+ channels in the membranes of guard cells function in the opening and closing of stomata.

Recall also from Chapter 8 that active transport is the pumping of solutes across membranes against their electrochemical gradients, the combined effects of the solute's concentration gradient and the voltage (charge difference) across the membrane. It is termed *active* because the cell must expend metabolic energy, usually in the form of ATP, to transport a solute "uphill"—that is, counter to the direction in which the solute diffuses. Transport proteins that simply facilitate diffusion, such as selective channels, cannot perform active transport. The active transporters are a special class of membrane proteins, each responsible for pumping specific solutes.

Proton pumps play a central role in transport across plant membranes

The most important active transporter in the plasma membranes of plant cells is the **proton pump,** which hydrolyzes ATP and uses the released energy to pump hydrogen ions (H^+) out of the cell. This results in a proton gradient with a higher H$^+$ concentration outside the cell than inside (FIGURE 36.2a). The gradient is a form of stored energy because the hydrogen ions tend to diffuse "downhill" back into the cell. And because the proton pump moves positive charge, in the form of H$^+$, out of the cell, the pump also generates a membrane potential. Membrane potential is a voltage, a separation of opposite charges across a membrane. Proton pumping makes the inside of a plant cell negative in charge relative to the outside. This voltage is called a membrane *potential* because the charge separation is a form of potential (stored) energy that can be harnessed to perform cellular work.

Plant cells use energy stored in the proton gradient and membrane potential to drive the transport of many different solutes. For example, the membrane potential generated by proton pumps contributes to the uptake of potassium ions (K$^+$) by root cells (FIGURE 36.2b). And in the mechanism called **cotransport,** a transport protein couples the downhill passage of one solute (H$^+$) to the uphill passage of another (NO$_3$$^-$, in the case of FIGURE 36.2c). This "coattail" effect is also responsible for the uptake of the sugar sucrose by plant cells (FIGURE 36.2d). A membrane protein cotransports sucrose with the H$^+$ that is moving down its gradient through the protein.

The role of proton pumps in the transport processes of plant cells is a specific application of the general mechanism called **chemiosmosis** (see FIGURE 9.15). The key feature of chemiosmosis is a transmembrane proton gradient, which links energy-releasing processes to energy-consuming processes in cells. For example, you learned in Chapters 9 and 10 that mitochondria and chloroplasts use proton gradients generated by electron transport chains (which release energy) to drive ATP synthesis (which consumes energy). The ATP synthases that couple H$^+$ diffusion to ATP synthesis during cellular respiration and photosynthesis function somewhat like the proton pumps in the plasma membranes of plant cells. But in contrast to ATP synthases, proton pumps normally run in reverse, using ATP energy to pump H$^+$ against its gradient. In both cases, proton gradients are the metabolic gears that enable one process to drive another. Chemiosmosis is a unifying principle of cellular energetics.

Differences in water potential drive water transport in plant cells

The survival of plant cells depends on their ability to balance water uptake and loss. The net uptake or loss of water by a cell occurs by **osmosis,** the passive transport of water across a membrane (see FIGURE 8.11). How can we predict the direction of osmosis when a cell is surrounded by a particular solution? In the case of an animal cell, it is enough to know whether the extracellular solution is hypotonic (has a lower solute concentration) or hypertonic (has a higher solute concentration) relative to the cell; water will move by osmosis in the hypotonic → hypertonic direction. But in the case of a plant

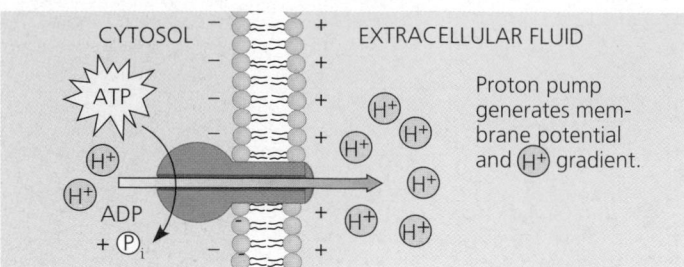

(a) Proton pump

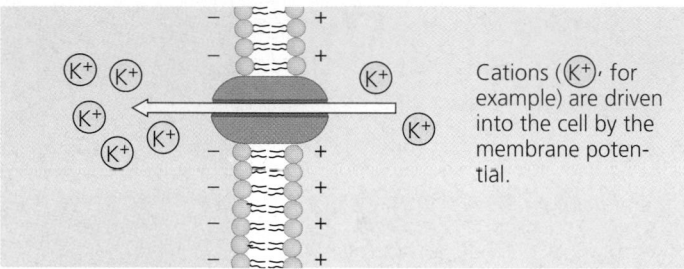

(b) Cation uptake

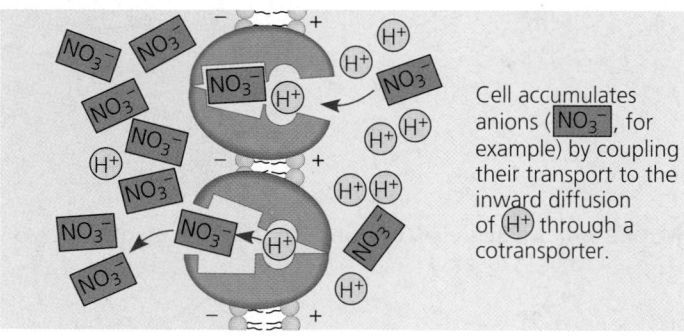

(c) Anion uptake

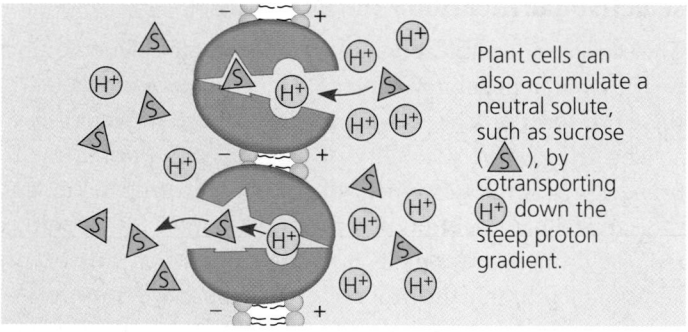

(d) Transport of a neutral solute

FIGURE 36.2 A chemiosmotic model of solute transport in plant cells.

cell, the presence of a cell wall adds a second factor affecting osmosis: physical pressure. The combined effects of these two factors—solute concentration and pressure—are incorporated into a single measurement called **water potential,** abbreviated by the Greek letter psi (ψ).

The most important thing for you to learn about water potential is that water will move across a membrane from the solution with the higher water potential to the solution with the lower water potential. For example, if a plant cell is immersed in a solution having a higher water potential than the cell, osmotic uptake of the water will cause the cell to swell. By moving, water can perform work (expanding a cell, for instance). The *potential* in water potential refers to this potential energy, the capacity to perform work when water moves from a region of higher ψ to a region of lower ψ. It is a special case of the general tendency of systems to change spontaneously to a state of lowest free energy (see FIGURE 6.5).

Plant biologists measure ψ in units of pressure called **megapascals** (abbreviated **MPa**), where 1 MPa is equal to about 10 atmospheres of pressure. (An atmosphere is the pressure exerted at sea level by an imaginary column of air—about 1 kg of pressure per square centimeter.) A couple of nonbiological examples will give you some idea of the magnitude of a megapascal: A car tire is usually inflated to a pressure of about 0.2 MPa; the water pressure in home plumbing is about 0.25 MPa.

How Solutes and Pressure Affect Water Potential

Let's see how solute concentration and pressure affect water potential. For purposes of comparison, the water potential of pure water in a container open to the atmosphere is zero ($\psi = 0$ MPa). The addition of solutes lowers the water potential. This is because the water molecules that form shells around a solute have less freedom to move than they do in pure water. Because ψ is standardized as 0 MPa for pure water, any solution at atmospheric pressure has a negative water potential owing to the presence of solutes. For instance, a 0.1-molar (M) solution of any solute has a water potential of −0.23 MPa. If this solution is separated from pure water by a selectively permeable membrane, water will move by osmosis into the solution, from the region of higher ψ (0 MPa) to the region of lower ψ (−0.23 MPa). But to predict the direction of water movement, we also have to consider the influence of physical pressure on ψ.

In contrast to the *inverse* relationship of ψ to solute concentration, water potential is *directly* proportional to pressure; increasing pressure raises ψ. Remember that ψ measures the relative tendency for water to leave one location in favor of another. Physical pressure—pressing the plunger of a syringe filled with water, for example—causes water to escape via any available exit. If a solution is separated from pure water by a selectively permeable membrane, external pressure on the solution can counter its tendency to take up water due to the presence of solutes. In fact, even greater pressure will force water across the membrane from the solution to the compartment containing pure water. It is also possible to create a negative pressure, or **tension,** on water or solutions. For example, if you pull up on the plunger of a syringe, the negative pressure within the syringe draws a solution through the needle.

Quantitative Analysis of Water Potential

The combined effects of pressure and solute concentration on water potential are incorporated into the following equation:

$$\psi = \psi_P + \psi_S$$

where ψ_P is the pressure potential (physical pressure on a solution) and ψ_S is the solute potential, which is proportional to the solute concentration of a solution. (ψ_S is also called osmotic potential.)

Let's see how this equation works with the help of FIGURE 36.3. A 0.1 M solution has a ψ_S of −0.23 MPa. Thus, in the absence of a physical pressure ($\psi_P = 0$), water potential, as stated

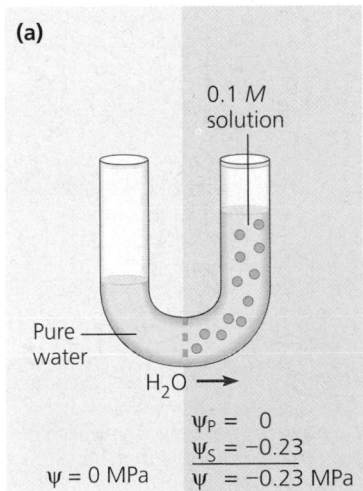
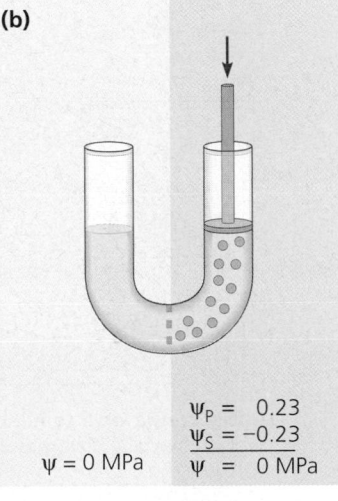
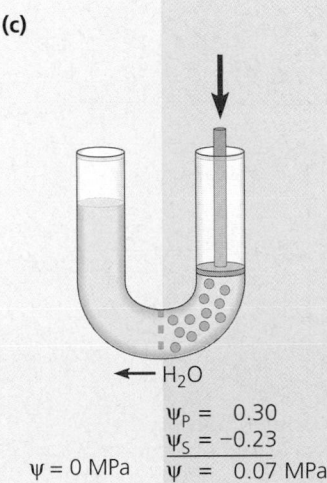
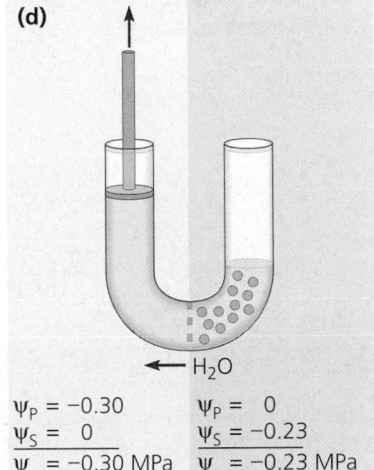

FIGURE 36.3 Water potential and water movement: a mechanical model. In this U-shaped apparatus, a selectively permeable membrane separates pure water from a 0.1 molar (M) solution containing a particular solute that cannot pass freely across the membrane. Water moves across a selectively permeable membrane from where water potential is higher to where it is lower. The water potential (ψ) of pure water at atmospheric pressure is 0 MPa. If we know the values of pressure potential (ψ_P) and solute potential (ψ_S), we can calculate water potential: ($\psi = \psi_P + \psi_S$). The values for ψ and ψ_S in the left and right arms of the U-tube are given for *initial* conditions, *before* any net movement of water. **(a)** The addition of solutes reduces water potential (to a negative value). **(b, c)** Application of physical pressure increases water potential. **(d)** A negative pressure (tension) decreases water potential.

earlier, is -0.23 MPa for a 0.1 M solution:

$$\psi = \psi_P + \psi_S = 0 + (-0.23) = -0.23 \text{ MPa}$$

Water would move into this solution from a compartment of pure water separated from the solution by a selectively permeable membrane (FIGURE 36.3a). But if we now apply a physical pressure of $+0.23$ MPa to this solution, we raise its water potential from a negative value to 0 MPa ($\psi = 0.23 - 0.23$). There will now be no net flow of water between this pressurized solution and the compartment of water (FIGURE 36.3b). In fact, if we increase ψ_P to $+0.3$ MPa, then the solution has a water potential of $+0.07$ MPa ($\psi = 0.3 - 0.23$), and this solution will actually *lose* water to a compartment containing pure water (FIGURE 36.3c). Finally, imagine using a plunger to pull upward on the water instead of pushing downward on the solution. A negative pressure (tension) of -0.3 MPa on the water compartment would be sufficient to draw water from the solution, which has a water potential of -0.23 MPa (FIGURE 36.3d). In dissecting water potential to see these opposing effects of pressure and solutes, it is important to remember the key point: Water will move across a membrane in the direction of lower (more negative) water potential.

Let's now apply what we have learned about water potential to the uptake and loss of water by plant cells. First, imagine a **flaccid** cell, one that has a ψ_P of 0 and is therefore not firm. Let's bathe this flaccid cell in a solution of higher solute concentration than the cell itself (FIGURE 36.4a). Since the external solution has the lower (more negative) water potential, water will leave the cell by osmosis, and the cell will **plasmolyze,** or shrink and pull away from its wall. Now let's place the same

flaccid cell in pure water ($\psi = 0$; FIGURE 36.4b). The cell has a lower water potential because of the presence of solutes, and water enters the cell by osmosis. The cell begins to swell and push against the wall, producing a **turgor pressure.** The partially elastic wall pushes back against the pressurized cell. When this wall pressure is great enough to offset the tendency for water to enter because of the solutes in the cell, then ψ_P and ψ_S are equal in magnitude and thus $\psi = 0$. This matches the water potential of the extracellular environment—in this example, 0 MPa. A dynamic equilibrium has been reached, and there is no further net movement of water, although an equal exchange of water across the membrane continues.

In contrast to a flaccid cell, a walled cell that has a greater solute concentration than its surroundings will be **turgid,** referring to the turgor pressure that keeps it firm. Healthy plant cells are turgid most of the time. Their turgor contributes to support in nonwoody parts of the plant. You can see the consequences of turgor loss in a wilted tomato plant, which has flaccid cells (FIGURE 36.5).

Aquaporins affect the rate of water transport across membranes

THE PROCESS OF SCIENCE

We have been examining water potential as the force that moves water across the membranes of plant cells, but how do the water molecules actually cross the membranes? Because water molecules are so small, they move relatively freely across the lipid bilayer of membranes, even though the middle zone of that bilayer is hydrophobic (see FIGURE 8.1). Until

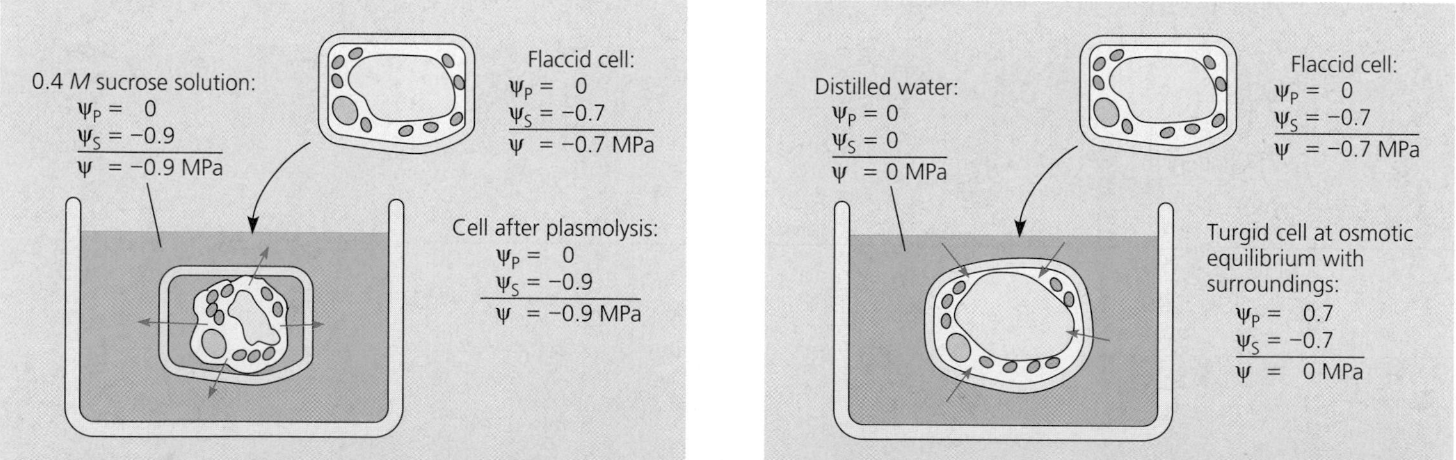

0.4 M sucrose solution:
$$\psi_P = 0$$
$$\psi_S = -0.9$$
$$\psi = -0.9 \text{ MPa}$$

Flaccid cell:
$$\psi_P = 0$$
$$\psi_S = -0.7$$
$$\psi = -0.7 \text{ MPa}$$

Cell after plasmolysis:
$$\psi_P = 0$$
$$\psi_S = -0.9$$
$$\psi = -0.9 \text{ MPa}$$

(a) Initial conditions: cellular ψ > environmental ψ. In the concentrated sucrose solution, the cell initially has a greater water potential than its surroundings. The cell loses water and plasmolyzes. After plasmolysis is complete, the water potentials of the cell and its surroundings are the same. Plasmolysis kills most plant cells.

Distilled water:
$$\psi_P = 0$$
$$\psi_S = 0$$
$$\psi = 0 \text{ MPa}$$

Flaccid cell:
$$\psi_P = 0$$
$$\psi_S = -0.7$$
$$\psi = -0.7 \text{ MPa}$$

Turgid cell at osmotic equilibrium with surroundings:
$$\psi_P = 0.7$$
$$\psi_S = -0.7$$
$$\psi = 0 \text{ MPa}$$

(b) Initial conditions: cellular ψ < environmental ψ. In distilled water, the cell initially has a lower water potential than its surroundings. There is a net uptake of water by osmosis, causing the cell to become turgid. When this tendency for water to enter is offset by the back pressure of the elastic wall, water potentials are equal for the cell and its surroundings. (The volume change of the cell is exaggerated in this diagram. The osmotic uptake of a relatively small amount of water does not actually increase the volume of the cell much. This explains why the solute potential, ψ_S, does not change by a significant amount when a cell becomes more turgid.)

FIGURE 36.4 Water relations of plant cells. In these experiments, identical cells, initially flaccid, are placed in two different environments. (The protoplasts of flaccid cells are in contact with their walls but lack turgor pressure.) The blue arrows indicate the initial direction of water movement.

FIGURE 36.5 A watered tomato plant regains its turgor.

Vacuolated plant cells have three major compartments

Outside the protoplast of a plant cell is a thick cell wall that helps maintain the cell's shape (see FIGURES 7.8 and 35.10). The cell wall itself, however, does not play a direct role in regulating the traffic of molecules into and out of the protoplast; that is the role of the selectively permeable plasma membrane. The plasma membrane serves as a barrier between two major compartments: the cell wall and the cytosol (the part of the cytoplasm contained within the plasma membrane but outside the intracellular organelles). Most mature plant cells have a third major compartment, the vacuole, a large organelle that can occupy as much as 90% of the protoplast's volume (FIGURE 36.6a). The membrane that bounds the vacuole, the **tonoplast**, regulates molecular traffic between the cytosol and the vacuolar contents, called cell sap. For example, the tonoplast has proton pumps

recently, most biologists accepted the hypothesis that leakage of water across the lipid bilayer was enough to account for water fluxes across membranes. That hypothesis was challenged in the early 1990s, when careful measurements indicated that water transport across biological membranes is too specific and too rapid to be explained entirely by diffusion through the lipid bilayer. This observation suggested the possibility of selective channels for water, and such channels have since been discovered in the membranes of both plant and animal cells.

These specific channels for passive traffic of water are transport proteins called **aquaporins**. Aquaporins do not affect the water potential gradient or the direction of water flow, but rather the *rate* at which water diffuses down its water potential gradient. The existence of aquaporins raises the possibility that a cell can regulate the rate of water uptake or loss when it has a water potential different from that of its surroundings. Aquaporins may form *gated* channels that open and close in response to variables such as turgor pressure of the cell. Many laboratories are now investigating how aquaporins work, and this research is likely to increase our understanding of how the cells of plants and other organisms regulate water balance.

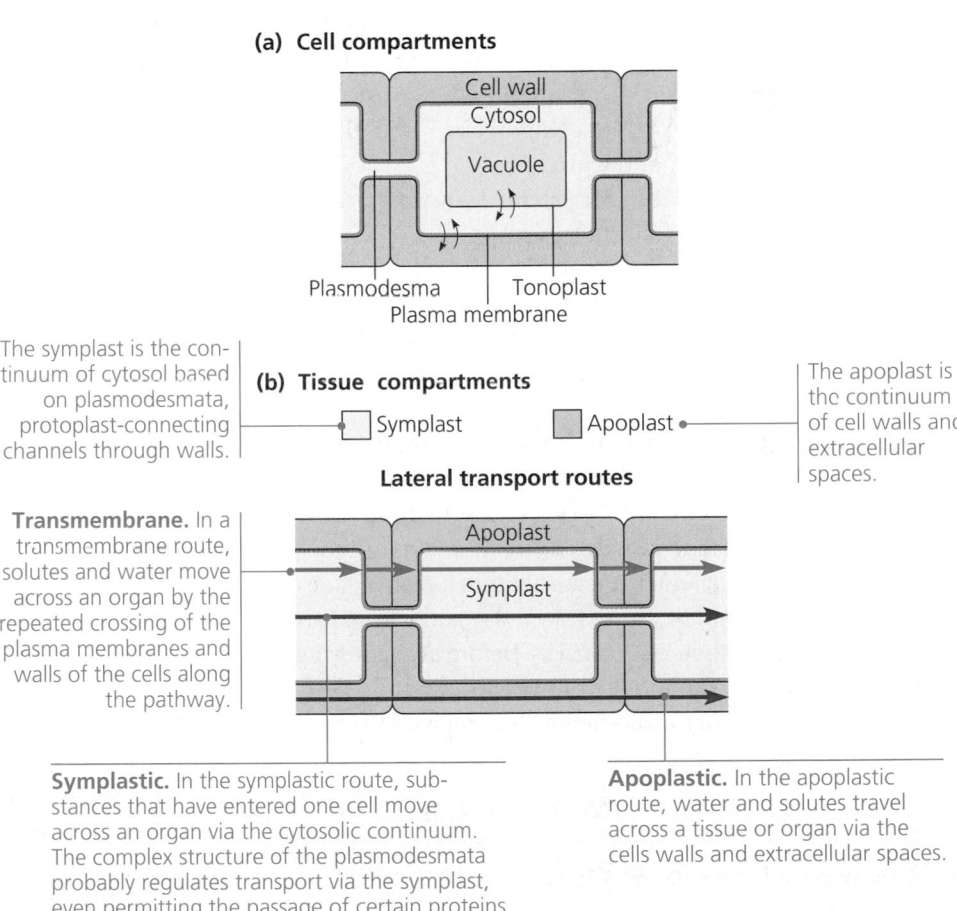

(a) Cell compartments

Cell wall
Cytosol
Vacuole

Plasmodesma — Tonoplast
Plasma membrane

The symplast is the continuum of cytosol based on plasmodesmata, protoplast-connecting channels through walls.

(b) Tissue compartments

☐ Symplast   ☐ Apoplast

The apoplast is the continuum of cell walls and extracellular spaces.

Lateral transport routes

Apoplast
Symplast

Transmembrane. In a transmembrane route, solutes and water move across an organ by the repeated crossing of the plasma membranes and walls of the cells along the pathway.

Symplastic. In the symplastic route, substances that have entered one cell move across an organ via the cytosolic continuum. The complex structure of the plasmodesmata probably regulates transport via the symplast, even permitting the passage of certain proteins and other large molecules between cells.

Apoplastic. In the apoplastic route, water and solutes travel across a tissue or organ via the cells walls and extracellular spaces.

FIGURE 36.6 Compartments of plant cells and tissues and routes for lateral transport. **(a)** The cell wall, cytosol, and vacuole are the three main compartments of most mature plant cells. Specific transport proteins embedded in the plasma membrane and tonoplast regulate traffic of molecules between the three compartments. **(b)** At the tissue level, there are two compartments, the symplast and the apoplast. This anatomy provides three routes for lateral transport in a plant tissue or organ. In this diagram, substances seem confined to one of the three routes; in fact, substances may transfer from one route to another during their commute across an organ.

that expel H^+ from the cytosol into the vacuole. This augments the ability of the proton pumps of the plasma membrane to maintain a low cytosolic concentration of H^+.

In most plant tissues, two of the three cellular compartments are continuous from cell to cell. Plasmodesmata connect the cytosolic compartments of neighboring cells, forming a continuous pathway for transport of certain molecules between cells. This cytoplasmic continuum is called the **symplast** (see FIGURE 36.6). The walls of adjacent plant cells are also in contact, forming a second compartment at the tissue level. This compartment, the continuum of cell walls, is called the **apoplast.** The third cellular compartment, the vacuole, is not shared with neighboring cells.

Both the symplast and the apoplast function in transport within tissues and organs

How do water and solutes move from one location to another within plant tissues and organs? For example, what mechanisms transport water and minerals absorbed by a root from the outer cells to the inner cells of the root? Such short-distance transport is sometimes called lateral transport because its usual direction is along the radial axis of plant organs, rather than up and down along the length of the plant.

Three routes are available for lateral transport (FIGURE 36.6b). By the first route, substances move out of one cell, across the cell wall, and into the neighboring cell, which may then pass the substances along to the next cell in the pathway by the same mechanism. This transmembrane route requires repeated crossings of plasma membranes, as the solutes exit one cell and enter the next.

The second route, via the symplast, the continuum of cytosol within a plant tissue, requires only one crossing of a plasma membrane. After entering one cell, solutes and water can then move from cell to cell via plasmodesmata.

The third route for lateral transport within a plant tissue or organ is along the apoplast, the extracellular pathway consisting of cell walls and extracellular spaces. Before ever entering a cell, water and solutes can move from one location to another within a root or other organ along the byways provided by the continuum of cell walls.

Bulk flow functions in long-distance transport

Diffusion in a solution is fairly efficient for transport over distances of cellular dimensions (less than 100 μm), but it is much too slow to function in long-distance transport within a plant— for example, in the transport of water and minerals from roots to leaves. Water and solutes move through xylem vessels and sieve tubes by **bulk flow,** the movement of a fluid driven by pressure. In phloem, for example, hydrostatic pressure is generated

at one end of a sieve tube, forcing sap to the opposite end of the tube. In xylem, it is actually tension (negative pressure) that drives long-distance transport. Transpiration, the evaporation of water from a leaf, reduces pressure in the leaf xylem. This creates a tension that pulls xylem sap upward from the roots.

If you have ever dealt with a partially clogged drain, you know that the rate of flow through a pipe depends on the pipe's internal diameter. Clogs reduce flow because they reduce the effective diameter of the drainpipe. Such household experiences help us understand how the unusual structures of plant cells specialized for bulk flow—sieve-tube members of the phloem and vessel elements and tracheids of the xylem—fit their function. Recall from Chapter 35 that sieve-tube members are almost entirely devoid of internal organelles and that vessel elements and tracheids are dead at maturity. Like unplugging a kitchen drain, the developmental loss of cytoplasmic content in a plant's "plumbing" allows for the efficient bulk flow of materials through the conduits formed by the specialized cells. The porous plates that connect contiguous sieve-tube members (see FIGURE 35.9) and the perforated end walls of xylem vessel elements (see FIGURE 35.8) also enhance bulk flow.

Now that we have an overview of the basic mechanisms of transport at the cellular, tissue, and whole-plant levels, we are ready to take a closer look at how these mechanisms work together in the overall transport functions that enable a plant to survive on land. For example, bulk flow due to a pressure difference is the mechanism of long-distance transport of phloem sap, but it is active transport of sugar at the cellular level that maintains this pressure difference. The four transport functions we will examine in more detail are the absorption of water and minerals by roots, the ascent of xylem sap, the control of transpiration, and the transport of organic nutrients within phloem.

ABSORPTION OF WATER AND MINERALS BY ROOTS

Water and mineral salts from soil enter the plant through the epidermis of roots, cross the root cortex, pass into the stele, and then flow up xylem vessels to the shoot system. In this section, we focus on the soil → epidermis → root cortex → xylem segments of this transport pathway. Use FIGURE 36.7 to reinforce the key concepts as you read this section.

Root hairs, mycorrhizae, and a large surface area of cortical cells enhance water and mineral absorption

Much of the absorption of water and minerals occurs near root tips, where the epidermis is permeable to water and where root hairs are located. Root hairs, which are extensions

FIGURE 36.7 Lateral transport of minerals and water in roots. Minerals are absorbed from the soil solution by the root surface, especially by root hairs and mycorrhizae (symbiotic associations of roots and fungi, not shown in this diagram). The water and minerals then move across the root cortex to the stele by a combination of the apoplastic and symplastic routes (see FIGURE 36.6).

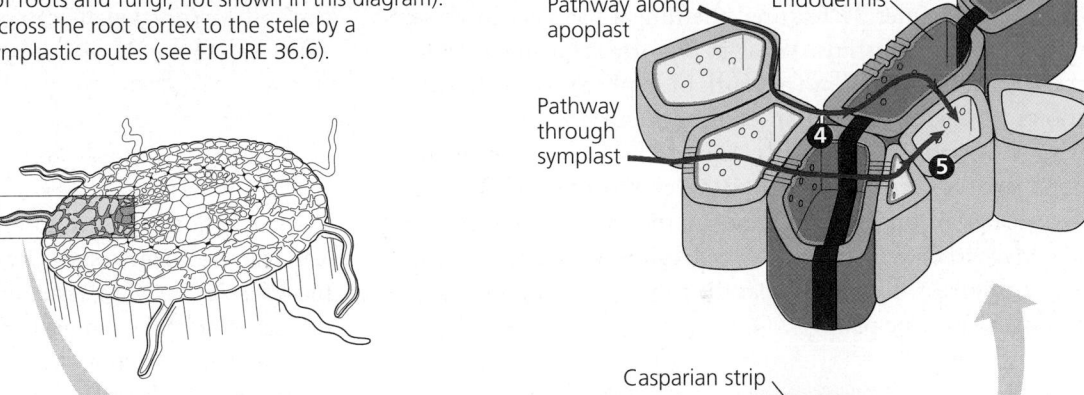

❶ The uptake of soil solution by the hydrophilic walls of the epidermis provides access to the apoplast, and water and minerals can soak into the cortex along this matrix of walls.

❷ Minerals and water that cross the plasma membranes of root hairs enter the symplast.

❸ As soil solution moves along the apoplast, some water and minerals are transported into cells of the epidermis and cortex and then move inward via the symplast.

❹ Water and minerals that move all the way to the endodermis along cell walls cannot continue into the stele via the apoplastic route. Within the wall of each endodermal cell is the Casparian strip, a belt of waxy material (purple band) that blocks the passage of water and dissolved minerals. Only minerals that are already in the symplast or enter that pathway by crossing the plasma membrane of an endodermal cell can detour around the Casparian strip and pass into the stele.

❺ Endodermal cells and parenchyma cells within the stele discharge water and minerals into their walls (apoplast). Since the vessel elements of a xylem vessel are dead cells, their walls and internal cavities are both part of the apoplast. The xylem vessels transport the water and minerals upward into the shoot system.

---

of epidermal cells, account for much of the surface area of roots (see FIGURE 35.14). Soil particles, which are usually coated with water and dissolved minerals, adhere tightly to the hairs. The soil solution flows into the hydrophilic walls of epidermal cells and passes freely along the apoplast into the root cortex. This exposes all the parenchyma cells of the cortex to soil solution, providing a much greater surface area of membrane than the surface area of the epidermis alone.

As the soil solution moves along the apoplast into the roots, cells of the epidermis and cortex take up water and certain solutes into the symplast. The soil solution is usually very dilute, and roots can accumulate essential minerals to concentrations that are hundreds of times higher than the concentrations of these minerals in soil. For example, selective transport proteins of the plasma membrane and tonoplast enable root cells to extract K^+, an essential mineral nutrient. In contrast, the cells exclude most Na^+, which may be much more concentrated than K^+ in the soil solution.

In their essential business of absorbing water and minerals from the soil, most plants are not on their own but have partners in the form of symbiotic fungi. "Infected" roots form **mycorrhizae,** symbiotic structures consisting of the plant's roots united with the hyphae (filaments) of the fungi (FIGURE 36.8). The hyphae absorb water and selected minerals, transferring

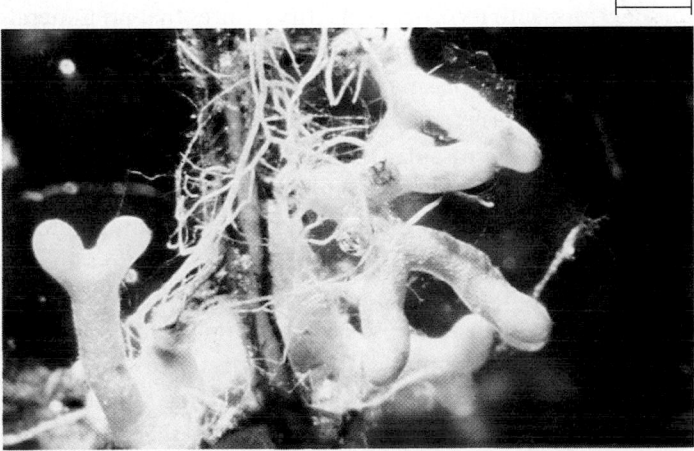

FIGURE 36.8 Mycorrhizae, symbiotic associations of fungi and roots. The white mycelium of the fungus ensheathes these roots of a red pine tree. The fungal hyphae provide an extensive surface area for the absorption of water and minerals.

much of these resources to the host plant. Chapter 37 highlights the role of mycorrhizae in the mineral nutrition of plants, and Chapter 31 featured the fungal partners in these mutualistic relationships. What is important to understand here is that the mycelium (network of hyphae) of the fungus endows mycorrhizae with an enormous surface area for absorption of water and minerals. As much as 3 m of hyphae can extend from each centimeter along the length of a root, reaching a far greater volume of soil than the root alone could penetrate. Mycorrhizae enable even older regions of roots, far from the abundance of root hairs near the root tips, to supply water and minerals to the plant.

The endodermis functions as a selective sentry between the root cortex and vascular tissue

Water and minerals that pass from the soil into the root cortex cannot be transported to the rest of the plant until they enter the xylem of the stele. The **endodermis,** the innermost layer of cells in the root cortex, surrounds the stele and functions as a last checkpoint for the selective passage of minerals from the cortex into the vascular tissue (see FIGURE 36.7). Minerals already in the symplast when they reach the endodermis continue through the plasmodesmata of the endodermal cells and pass into the stele. These minerals were already screened by the selective membrane they had to cross to enter the symplast in the epidermis or cortex. Those minerals that reach the endodermis via the apoplast encounter a dead end that blocks their passage into the stele: In the wall of each endodermal cell is the **Casparian strip,** a belt made of suberin, a waxy material that is impervious to water and dissolved minerals. Thus, water and minerals cannot cross the endodermis and enter vascular tissue via the apoplast. The only way past this barrier is for the water and minerals to cross the plasma membrane of an endodermal cell and enter the stele via the symplast. The endodermis, with its Casparian strip, ensures that no minerals can reach the vascular tissue of the root without crossing a selectively permeable plasma membrane. If minerals do not enter cells in the cortex, they must enter endodermal cells or be excluded from the vascular tissue. The structure of the endodermis and its strategic location in the root fit its function as sentry of the cortex-stele border, a function that contributes to the ability of roots to preferentially transport certain minerals from the soil into the xylem.

The last segment in the soil $\rightarrow$ xylem pathway is the passage of water and minerals into the tracheids and vessel elements of the xylem. These water-conducting cells lack protoplasts, and thus the lumen of these cells and the cell walls are part of the apoplast. Endodermal cells as well as parenchyma cells within the stele discharge minerals into their walls. Both diffusion and active transport are probably involved in this transfer of solutes from symplast to apoplast, and the water

and minerals are now free to enter the tracheids and xylem vessels. The water and mineral nutrients we have tracked from the soil to root xylem can now be transported upward as xylem sap to the shoot system.

TRANSPORT OF XYLEM SAP

Xylem sap flows upward to veins that branch throughout each leaf, placing xylem vessels close to every cell. Leaves depend on this efficient delivery system for their supply of water. Plants lose an astonishing amount of water by **transpiration,** the loss of water vapor from leaves and other aerial parts of the plant. An average-sized maple tree, for instance, loses more than 200 L of water per hour during the summer. Unless the transpired water is replaced by water transported up from the roots in xylem, leaves wilt and eventually die. The flow of xylem sap upward in the plant also brings mineral nutrients to the shoot system.

The ascent of xylem sap depends mainly on transpiration and the physical properties of water

Xylem sap rises against gravity, without the help of any mechanical pump, to reach heights of more than 100 m in the tallest trees. Is the sap *pushed* upward from the roots, or is it *pulled* upward by the leaves? Let's evaluate the relative contributions of these two possible mechanisms.

Pushing Xylem Sap: Root Pressure

At night, when transpiration is very low or zero, the root cells are still expending energy to pump mineral ions into the xylem. The endodermis surrounding the stele of the root helps prevent the leakage of these ions back out of the stele. The accumulation of minerals in the stele lowers water potential there. Water flows in from the root cortex, generating a positive pressure that forces fluid up the xylem. This upward push of xylem sap is called **root pressure.**

Root pressure causes **guttation,** the exudation of water droplets that can be seen in the morning on tips of grass blades or the leaf margins of some small, herbaceous (nonwoody) dicots. During the night, when the rate of transpiration is low, the roots of some plants keep accumulating minerals, and root pressure pushes xylem sap into the shoot system. More water enters leaves than is transpired, and the excess is forced out as guttation fluid (FIGURE 36.9).

In most plants, root pressure is not the major mechanism driving the ascent of xylem sap. At most, root pressure can force water upward only a few meters, and many plants generate no root pressure at all. Even in plants that display guttation, root pressure cannot keep pace with transpiration after sunrise.

FIGURE 36.9 Guttation. Root pressure is forcing excess water from this strawberry leaf.

For the most part, xylem sap is not pushed from below by root pressure but pulled upward by the leaves themselves.

Pulling Xylem Sap: The Transpiration-Cohesion-Tension Mechanism

If we want to move solid material upward, we can either push it from below or pull it from above. It is less obvious that something as seemingly fluid as water could be pulled up a pipe. Nevertheless, that is what happens in the xylem vessels of plants. As we investigate this mechanism of transport, we will see that transpiration provides the pull, and the cohesion of water due to hydrogen bonding transmits the upward pull along the entire length of the xylem to the roots.

Transpirational Pull. Stomata, the microscopic pores on the surface of a leaf, lead to a maze of internal air spaces that expose the mesophyll cells to the carbon dioxide they need for photosynthesis. The air in these spaces is saturated with water vapor because it is in contact with the moist walls of the cells. On most days, the air outside the leaf is drier; that is, it has a lower water concentration than the air inside the leaf. Therefore, gaseous water, diffusing down its concentration gradient, exits the leaf via the stomata. It is this loss of water vapor from the leaf that we call transpiration.

How is transpiration translated into a pulling force for the movement of water in a plant? The mechanism depends on the generation of a negative pressure (tension) in the leaf due to the unique physical properties of water. Evaporation from the thin film of water that coats the mesophyll cells replaces the water vapor that is lost from the leaf's air spaces by transpiration. As water evaporates, the remaining film of liquid water retreats into the pores of the cell walls, attracted by adhesion to the hydrophilic walls (FIGURE 36.10). At the same time, cohesive forces in the water resist an increase in the surface area of the film (a surface-tension effect; see Chapter 3). The combination of the two forces acting on the water—adhesion to the wall and surface tension—causes the surface of the water film to form a meniscus, or concave shape. In a sense, the water is being "pulled on" by the adhesive and cohesive forces. Thus, the water film at the surface of leaf cells has a negative pressure, a pressure less than atmospheric pressure. And the more concave the meniscus, the more negative the pressure of the water film. This negative pressure, or tension, is the pulling force that draws water out of the leaf xylem, through the mesophyll, and toward the cells and surface film bordering the air spaces near stomata.

This mechanism of water flow fits with what you learned earlier about water potential. In the equation for water potential, a tension (negative pressure) *lowers* the potential. And

FIGURE 36.10 The generation of transpirational pull in a leaf. Water vapor diffuses from the moist air spaces of the leaf to the drier air outside via stomata. Evaporation from the water film coating the mesophyll cells maintains the high humidity of the air spaces. This loss of water causes the water film to form menisci that become more and more concave as the rate of transpiration increases. A meniscus has a tension that is inversely proportional to the radius of the curved water surface. Thus, as the water film recedes and its menisci become more concave, the tension of the water film increases. Tension is a negative pressure—a force that pulls water from locations where hydrostatic pressure is greater. The tension of water lining the air spaces of the leaf is the physical basis of transpirational pull, which draws water out of xylem.

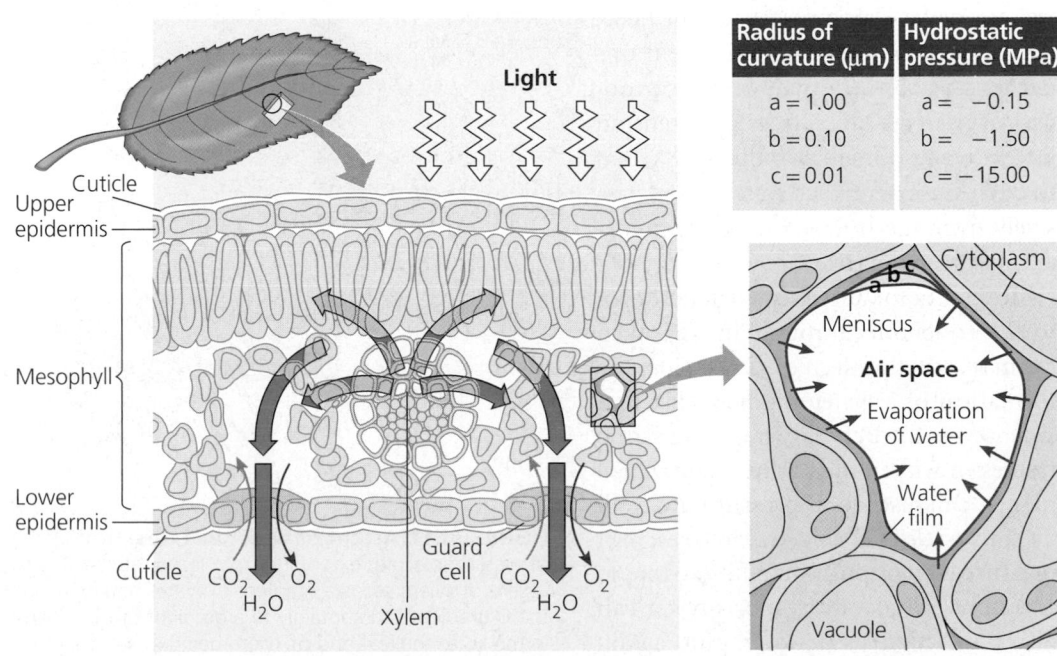

Radius of curvature (μm)	Hydrostatic pressure (MPa)
a = 1.00	a = −0.15
b = 0.10	b = −1.50
c = 0.01	c = −15.00

since water moves from where its potential is higher to where it is lower, mesophyll cells will lose water to the surface film lining air spaces, which in turn loses water by transpiration. The water lost via the stomata is replaced by water that is pulled out of the leaf xylem. The negative water potential of leaves provides the "pull" in transpirational pull.

Cohesion and Adhesion in the Ascent of Xylem Sap. The transpirational pull on xylem sap is transmitted all the way from the leaves to the root tips and even into the soil solution (FIGURE 36.11). The cohesion of water due to hydrogen bonding makes it possible to pull a column of sap from above without the water separating. Water molecules exiting the xylem in the leaf tug on adjacent water molecules, and this pull is relayed, molecule by molecule, down the entire column of water in the xylem. Also helping to fight gravity is the strong adhesion of water molecules (again by hydrogen bonds) to the hydrophilic walls of the xylem cells. The very small diameter of tracheids and vessel elements exposes a large proportion of the water to the hydrophilic walls, thus enhancing adhesion in overcoming the downward force of gravity.

The upward pull on the cohesive sap creates tension within the xylem. Pressure will cause an elastic pipe to swell, but tension will pull the walls of the pipe inward. You can actually measure a decrease in the diameter of a tree trunk on a warm day, when transpirational pull puts the xylem under tension. The rings of secondary walls prevent xylem vessels from collapsing, much as wire rings maintain the shape of a vacuum hose. Transpirational pull puts the xylem under tension all the way down to the root tips, even in the tallest trees. This tension lowers water potential in the root xylem to such an extent that water flows passively from the soil, across the root cortex, and into the stele.

Transpirational pull can extend down to the roots only through an unbroken chain of water molecules. Cavitation, the formation of a water vapor pocket in a xylem vessel, such as when xylem sap freezes in winter, breaks the chain. Small plants can use root pressure to refill xylem vessels in spring, but in trees, root pressure cannot push water to the top, so a vessel with a water vapor pocket can never function as a water pipe again.

However, the transpiration stream can detour around the water vapor pocket through pits between adjacent xylem vessels, and secondary growth adds a layer of new xylem vessels each year. In some angiosperm trees, including oaks and elms, the youngest, outermost growth ring of xylem transports most of the water. The older xylem functions to support the tree (see Chapter 35).

Xylem sap ascends by solar-powered bulk flow: *a review*

The transpiration-cohesion-tension mechanism that transports xylem sap against gravity is an excellent example of how physical principles apply to biological problems. Long-distance

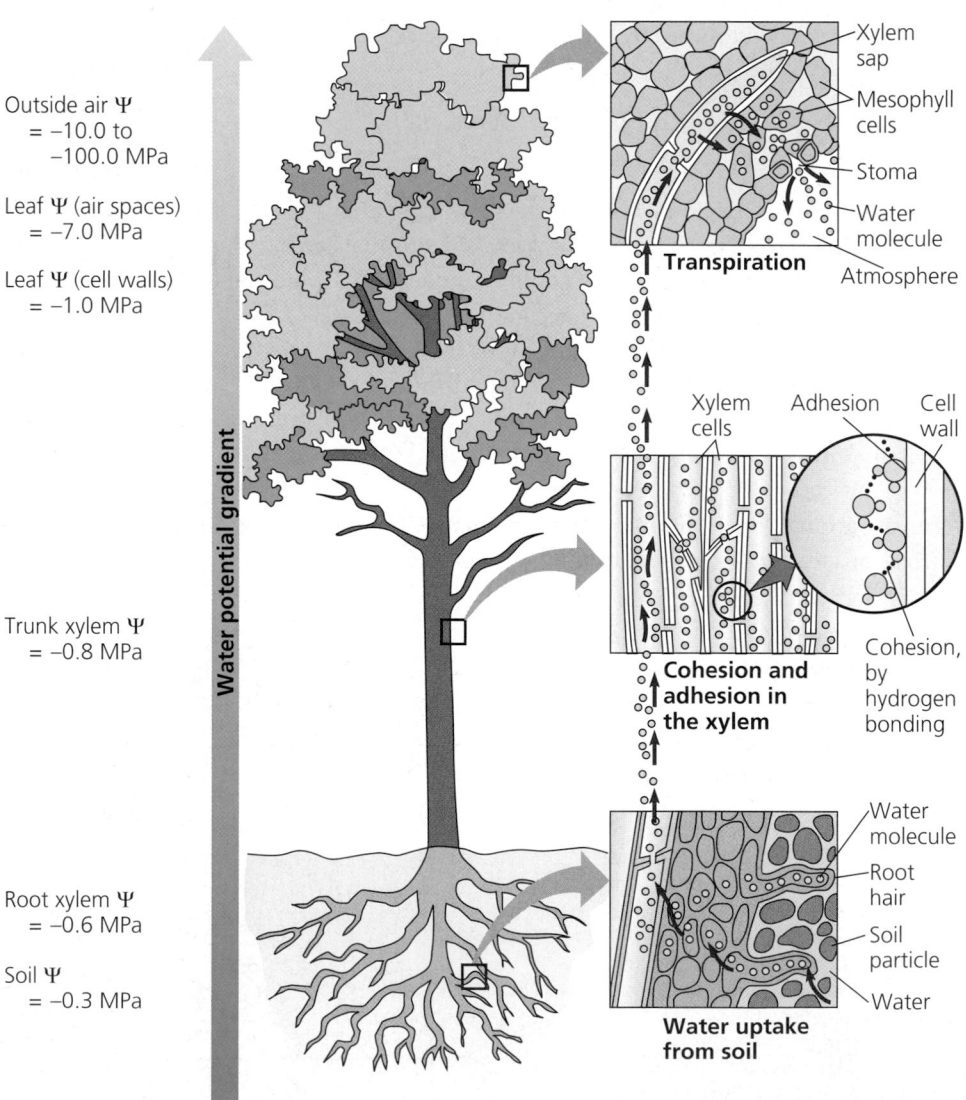

Outside air Ψ
= −10.0 to
−100.0 MPa

Leaf Ψ (air spaces)
= −7.0 MPa

Leaf Ψ (cell walls)
= −1.0 MPa

Trunk xylem Ψ
= −0.8 MPa

Root xylem Ψ
= −0.6 MPa

Soil Ψ
= −0.3 MPa

Water potential gradient

Xylem sap

Mesophyll cells

Stoma

Water molecule

Transpiration

Atmosphere

Xylem cells

Adhesion

Cell wall

Cohesion and adhesion in the xylem

Cohesion, by hydrogen bonding

Water molecule

Root hair

Soil particle

Water

Water uptake from soil

FIGURE 36.11 Ascent of water in a tree. Hydrogen bonding forms an unbroken chain of water molecules extending from leaves all the way to the soil. The force that drives the ascent of xylem sap is a gradient of water potential (ψ). For the bulk flow over long distance, the ψ gradient is due mainly to a gradient of the pressure potential ($ψ_P$). Transpiration results in the $ψ_P$ at the leaf end of xylem being lower than the $ψ_P$ at the root end.

transport of water from roots to leaves occurs by bulk flow, the movement of fluid driven by a pressure difference at opposite ends of a conduit. In a plant, the conduits are xylem vessels or chains of tracheids. The pressure difference is generated at the leaf end by transpirational pull, which lowers pressure (increases tension) at the "upstream" end of the xylem.

On a smaller scale, gradients of water potential drive the osmotic movement of water from cell to cell within root and leaf tissue (see FIGURE 36.11). Differences in both solute concentration and pressure contribute to this microscopic transport. In contrast, bulk flow, the mechanism for long-distance transport up xylem vessels, depends only on pressure. Another contrast with osmosis, which moves only water, is that bulk flow moves the whole solution, water plus minerals and any other solutes dissolved in the water.

The plant expends none of its own metabolic energy to lift xylem sap up to the leaves by bulk flow. The absorption of sunlight drives transpiration by causing water to evaporate from the moist walls of mesophyll cells and by maintaining a high humidity in the air spaces within a leaf. Thus, the ascent of xylem sap is ultimately solar powered.

THE CONTROL OF TRANSPIRATION

Guard cells mediate the photosynthesis-transpiration compromise

A leaf may transpire more than its weight in water each day. Leaves are kept from wilting by a transpiration stream in xylem vessels that flows as fast as 75 cm/min, about the speed of the tip of a second hand sweeping around a wall clock. A plant's tremendous requirement for water is part of the cost of making food by photosynthesis. Guard cells, by controlling the size of stomata, help balance the plant's need to conserve water with its requirement for photosynthesis (FIGURE 36.12).

The Photosynthesis-Transpiration Compromise

To make food, a plant must spread its leaves to the sun and obtain CO_2 from the air. Carbon dioxide diffuses into the leaf through the stomata, and oxygen produced as a by-product of photosynthesis diffuses out of the leaf via the stomata (see FIGURE 36.10). Upon diffusing through the stomata, CO_2 enters a honeycomb of air spaces formed by the irregularly shaped spongy parenchyma cells (see FIGURE 35.11). Because of the irregular shape of these cells,

the internal surface area of the leaf may be 10 to 30 times greater than the external surface area we see when we look at the leaf. This structural feature of leaves supports photosynthesis by increasing exposure to CO_2, but it also increases the surface area for the evaporation of water, which exits the plant freely through open stomata. About 90% of the water a plant loses escapes through stomata, though these pores account for only 1–2% of the external leaf surface. The waxy cuticle limits water loss through the remaining surface of the leaf.

One way to evaluate how efficiently a plant uses water is to determine its **transpiration-to-photosynthesis ratio,** the amount of water lost per gram of CO_2 assimilated into organic material by photosynthesis. For many plant species, this ratio is about 600:1, meaning the plant transpires 600 g of water for each gram of CO_2 that becomes incorporated into carbohydrate. However, corn and other plants that assimilate atmospheric CO_2 by the C_4 pathway have transpiration-to-photosynthesis ratios of 300:1 or less. With the same concentration of CO_2 within the air spaces of the leaf, C_4 plants can assimilate that CO_2 at a greater rate than C_3 plants can (see Chapter 10). Because water loss is the trade-off for allowing CO_2 to diffuse into the leaf, the photosynthetic return for each gram of water sacrificed is greater for C_4 plants, which can assimilate CO_2 faster than C_3 plants when stomata are partially closed.

In addition to supplying water to leaves, the transpiration stream assists in the delivery of minerals and other substances from the roots to the shoots and leaves. Transpiration also results in evaporative cooling, which can lower the temperature of a leaf by as much as 10–15°C compared with the surrounding air. This prevents the leaf from reaching temperatures that could denature various enzymes involved in photosynthesis and other metabolic processes. Cacti and other desert succulents, which have low rates of transpiration, can tolerate high leaf temperatures; in this case, the loss of water due to transpiration is a greater threat than overheating.

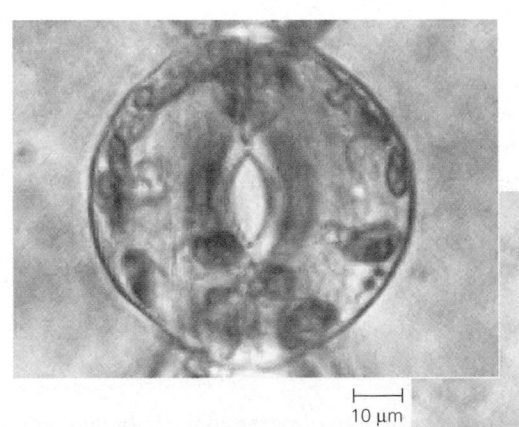
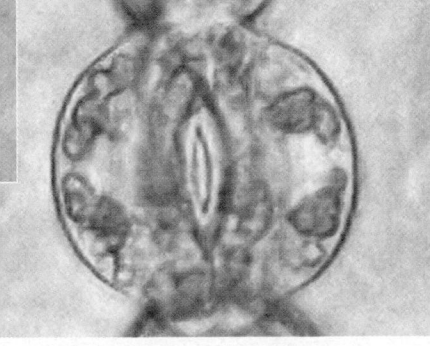

10 μm

FIGURE 36.12 An open (left) and closed (right) stoma of a spider plant (*Chlorophytum colosum*) leaf.

When transpiration exceeds the delivery of water by xylem, as when the soil begins to dry out, the leaves begin to wilt as their cells lose turgor pressure (see FIGURE 36.5). The potential rate of transpiration will be greatest on a day that is sunny, warm, dry, and windy, because these are the environmental factors that increase the evaporation of water. Plants are not helpless against the elements, however, for they are capable of adjusting to their environment. In the photosynthesis-transpiration compromise, mechanisms that regulate the size of the stomatal openings strike a balance.

How Stomata Open and Close

Each stoma is flanked by a pair of guard cells, which are kidney-shaped in dicots and dumbbell-shaped in monocots. The guard cells are suspended by their epidermal neighbors over an air chamber, which leads to a honeycomb of air spaces.

Guard cells control the diameter of the stoma by changing shape, thereby widening or narrowing the gap between the two cells (FIGURE 36.13a). When guard cells take in water by osmosis, they become more turgid and swell. In most dicots, the cell walls of guard cells are not uniformly thick, and the cellulose microfibrils are oriented in a direction that causes the guard cells to buckle outward when they are turgid. This increases the size of the gap between the cells. When the cells lose water and become flaccid, they become less bowed and close the space between them. This basic mechanism also applies to the stomata of monocots.

The changes in turgor pressure that open and close stomata result primarily from the reversible uptake and loss of potassium ions (K^+) by the guard cells. Stomata open when guard cells actively accumulate K^+ from neighboring epidermal cells (FIGURE 36.13b). This uptake of solute causes the water potential to become more negative within the guard cells, and the cells become more turgid as water enters by osmosis. Most of the K^+ and water are stored in the vacuole, and thus the tonoplast also plays a role. Stomatal closing results from an exodus of K^+ from guard cells, which leads to an osmotic loss of water. Regulation of aquaporins may also be involved in the swelling and shrinking of guard cells by varying the permeability of the membranes to water.

The K^+ fluxes across the guard cell membrane are probably passive, being coupled to the generation of membrane potentials by proton pumps. Stomatal opening correlates with active transport of H^+ out of the guard cell. The resulting voltage (membrane potential) drives K^+ into the cell through specific membrane channels (see FIGURE 36.2). Plant physiologists are using a method called patch clamping to study regulation of the guard cell's proton pumps and K^+ channels (FIGURE 36.14).

In general, stomata are open during the day and closed at night. This prevents the plant from needlessly losing water when it is too dark for photosynthesis. At least three cues contribute to stomatal opening at dawn. First, light itself stimulates guard cells to accumulate potassium and become turgid. This response is triggered by the illumination of a blue-light receptor in a guard cell, perhaps built into the plasma membrane. Activation of these blue light receptors stimulates the activity of ATP-powered proton pumps in the plasma

FIGURE 36.13 The mechanism of stomatal opening and closing.

Cells turgid/Stoma open   **Cells flaccid/Stoma closed**

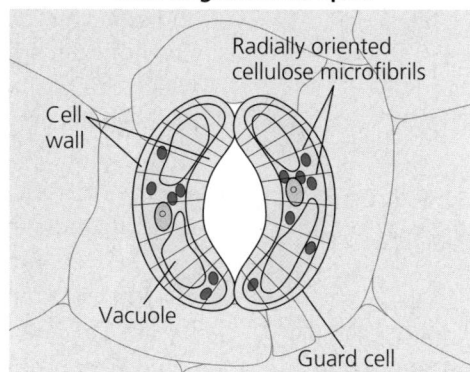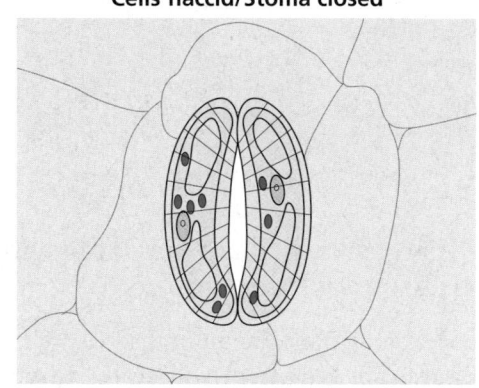

(a) Changes in guard cell shape and stomatal opening and closing (surface view). Guard cells of a dicot are illustrated in their turgid (stoma open) and flaccid (stoma closed) states. The pair of guard cells buckle outward when turgid. Cellulose microfibrils in the walls resist stretching and compression in the direction parallel to the microfibrils. Thus, the radial orientation of the microfibrils causes the cells to increase in length more than width when turgor increases. The two guard cells are attached at their tips, so the increase in length causes buckling.

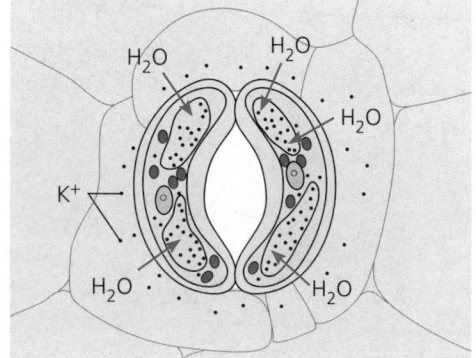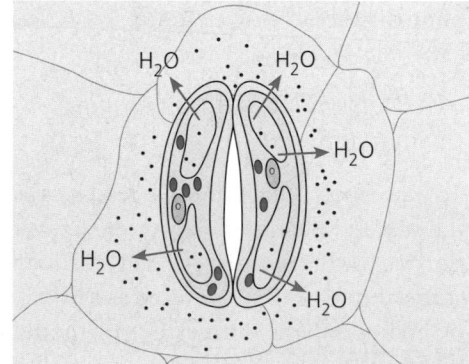

(b) Role of potassium in stomatal opening and closing. The transport of K^+ (potassium ions) across the plasma membrane and tonoplast causes the turgor changes of guard cells. Stomata open when guard cells accumulate potassium (red dots), which lowers the cells' water potential and causes them to take up water by osmosis. The cells become turgid. An exodus of K^+ from guard cells causes stomatal closure.

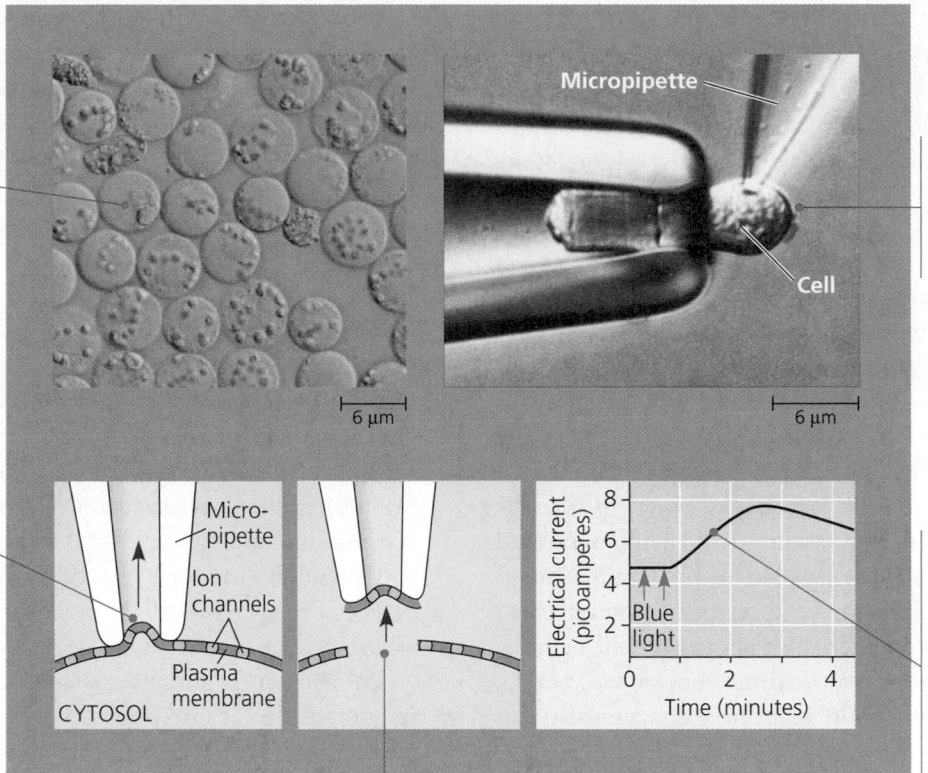

❶ This is a light micrograph of guard cell protoplasts (walls removed) from tobacco (*Nicotiana*) leaves.

❷ The tool for making a patch is a micropipette with a tip only about 1 μm in diameter. A larger suction pipette is used to hold the cell in place.

❸ Slight suction draws a membrane "blister" into the opening of the micropipette, and the rim of the micropipette seals to the membrane.

❺ This graph shows a recording for H⁺ transport across a membrane patch of a guard cell. Passage of H⁺ across the membrane represents an electrical current that can be measured. The graph indicates current due to H⁺ flux during a recording that lasts several minutes.

❹ This suction partitions a small patch from the rest of the plasma membrane. The micropipette, hooked up to appropriate equipment, now functions as an electrode to record ion fluxes across the tiny patch of membrane. Because the patch is so small, and because the ions in the bathing solutions are known, it is possible to record the transport of a single kind of ion through selective channels or pumps in the membrane.

THE PROCESS OF SCIENCE

FIGURE 36.14 A patch-clamp study of guard cell membranes. Patch clamping makes it possible to isolate a very tiny "patch" of membrane and study ion movement through a single type of channel— K⁺ or H⁺ channels, for instance. This particular experiment measures the effect of blue light, which increases H⁺ current across the membrane. Along with other evidence, patch-clamp studies support the hypothesis that blue light and other stimuli regulate guard cells by affecting the proton (H⁺) pumps of the membrane.

membrane of the guard cells, which in turn promotes the uptake of K⁺ (see FIGURE 36.14). Light may also stimulate stomatal opening by driving photosynthesis in guard cell chloroplasts, making ATP available for the active transport of hydrogen ions. (Guard cells are the only epidermal cells equipped with chloroplasts.)

A second stimulus causing stomata to open is depletion of CO_2 within air spaces of the leaf, which occurs when photosynthesis begins in the mesophyll. We can actually trick a plant into opening its stomata at night by placing it in a chamber devoid of CO_2.

A third cue in stomatal opening is an internal clock located in the guard cells. Even if you keep a plant in a dark closet, stomata will continue their daily rhythm of opening and closing. All eukaryotic organisms have internal clocks that somehow keep track of time and regulate cyclic processes. Cycles that have intervals of approximately 24 hours are called **circadian**

rhythms. You will learn more about circadian rhythms and the biological clocks that control them in Chapter 39.

Environmental stresses of various kinds can cause stomata to close during the daytime. When the plant is suffering a water deficiency, guard cells may lose turgor. In addition, a hormone called abscisic acid, which is produced in the mesophyll cells in response to water deficiency, signals guard cells to close stomata. This response reduces further wilting, but it also slows down photosynthesis; this is one reason droughts reduce crop yields. High temperatures also induce stomatal closure, probably by stimulating cellular respiration and increasing CO_2 concentration within the air spaces of the leaf. High temperature and excessive transpiration may combine to cause stomata to close briefly during mid-day. Thus, guard cells arbitrate the photosynthesis-transpiration compromise on a moment-to-moment basis by integrating a variety of internal and external stimuli.

Xerophytes have evolutionary adaptations that reduce transpiration

Plants adapted to arid climates, called **xerophytes,** have various leaf modifications that reduce the rate of transpiration. Many xerophytes have small, thick leaves, an adaptation that limits water loss by reducing surface area relative to leaf volume. A thick cuticle gives some of these leaves a leathery consistency. The stomata are concentrated on the lower (shady) leaf surface, and they are often located in depressions that shelter the pores from the dry wind (FIGURE 36.15). During the driest months, some desert plants shed their leaves. Others, such as cacti, subsist on water the plant stores in its fleshy stems during the rainy season. (These modified stems are the photosynthetic organs of cacti; the spines are modified leaves.)

One of the most elegant adaptations to arid habitats is found in ice plants and other succulent plants of the family Crassulaceae and in representatives of many other plant families. These plants assimilate CO_2 by an alternative photosynthetic pathway known as CAM, for crassulacean acid metabolism (see FIGURE 10.19). Mesophyll cells in a CAM plant have enzymes that can incorporate CO_2 into organic acids during the night. During the daytime, the organic acids are broken down to release CO_2 in the same cells, and sugars are synthesized by the conventional (C_3) photosynthetic pathway. Because the leaf takes in its CO_2 at night, the stomata can close during the day, when transpiration is most severe.

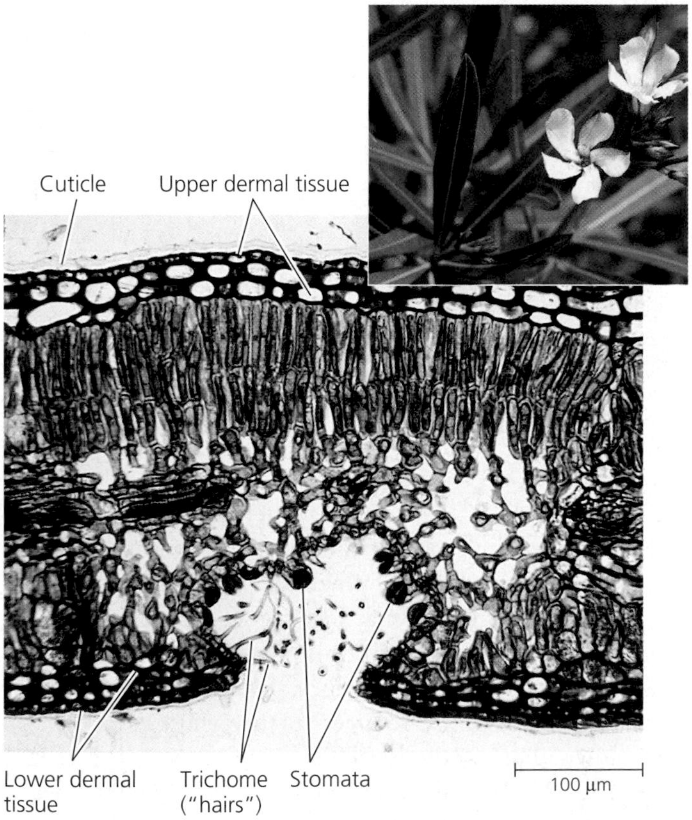

TRANSLOCATION OF PHLOEM SAP

Xylem sap flows in a direction that generally does not allow it to function in exporting sugar from leaves to other parts of the plant. A second vascular tissue, the phloem, transports the organic products of photosynthesis throughout the plant. This transport of food in the plant is called **translocation.** In angiosperms, the specialized cells of phloem that function in translocation are the sieve-tube members, which are arranged end to end to form long sieve tubes. Between the cells are sieve plates, porous cross-walls that allow the flow of sap along the sieve tube (see FIGURE 35.9).

Phloem sap is an aqueous solution that differs markedly in composition from xylem sap. By far the prevalent solute in phloem sap is sugar, primarily the disaccharide sucrose in most plant species. The sucrose concentration may be as high as 30% by weight, giving the sap a syrupy thickness. Phloem sap may also contain minerals, amino acids, and hormones in transit from one part of the plant to another.

Phloem translocates its sap from sugar sources to sugar sinks

In contrast to the unidirectional transport of xylem sap from roots to leaves, the direction that phloem sap travels is variable. The one generalization that holds is that sieve tubes carry food from a sugar source to a sugar sink. A **sugar source** is a plant organ in which sugar is being produced by either photosynthesis or the breakdown of starch. Mature leaves are the primary sugar sources. A **sugar sink** is an organ that is a net consumer or storer of sugar. Growing roots, shoot tips, stems, and fruit are sugar sinks supplied by phloem. A storage organ, such as a tuber or a bulb, may be either a source or a sink, depending on the season. When the storage organ is stockpiling carbohydrates during the summer, it is a sugar sink. After breaking dormancy in the early spring, however, the storage organ becomes a source as its starch is broken down to sugar, which is carried away in the phloem to the growing buds of the shoot system.

Other solutes may be transported to sinks along with sugar. For example, minerals that reach leaves in xylem may later be transferred in the phloem to developing fruit.

A sugar sink usually receives its sugar from the sources nearest to it. The upper leaves on a branch may send sugar to the growing shoot tip, whereas the lower leaves on the same branch export sugar to roots. A growing fruit requires so much food

Cuticle   Upper dermal tissue

Lower dermal tissue   Trichome ("hairs")   Stomata   100 μm

FIGURE 36.15 Structural adaptations of a xerophyte leaf. Oleander (inset) is commonly planted along desert highways. The leaves have a thick cuticle and multiple-layered dermal tissue that reduce water loss. Stomata are recessed in "crypts," an adaptation that reduces the rate of transpiration by protecting the stomata from hot, dry wind. The trichomes ("hairs") also help minimize transpiration by breaking up the flow of air, allowing the chamber of the crypt to have a higher humidity than the surrounding atmosphere (LM).

that it may monopolize the sugar sources all around it. One sieve tube in a vascular bundle may carry phloem sap in one direction while sap in a different tube in the same bundle flows in the opposite direction. For each sieve tube, the direction of transport depends only on the locations of the source and sink connected by that tube. This direction may change with the season or developmental stage of the plant.

Phloem Loading and Unloading

Sugar from the mesophyll cells of a leaf and other sources must be loaded into sieve-tube members before it can be exported to sugar sinks. In some species, sucrose moves all the way from mesophyll cells to sieve-tube members via the symplast, passing from cell to cell through plasmodesmata. In other species, sucrose reaches sieve-tube members by a combination of symplastic and apoplastic pathways (FIGURE 36.16a). For example, in corn leaves, sucrose diffuses through the symplast from mesophyll cells into small veins. Much of the sugar then moves out of the cells into the apoplast (walls) in the vicinity of sieve-tube members and companion cells. This sucrose is accumulated from the apoplast (walls) directly by the sieve-tube members or by their companion cells. Companion cells pass the sugar they accumulate into the sieve tube members through plasmodesmata linking the cells. In some plants, companion cells have numerous ingrowths of their walls, an adaptation that increases the cells' surface area and enhances the transfer of solutes between apoplast and symplast. Such modified cells are called **transfer cells** (see FIGURE 29.5).

In corn and many other plants, sieve-tube members accumulate sucrose to concentrations two to three times higher than concentrations in mesophyll, and thus phloem loading requires active transport. Proton pumps do the work that enables the cells to accumulate sucrose (FIGURE 36.16b).

Downstream, at the sink end of a sieve tube, phloem unloads its sucrose. Phloem unloading is a highly variable process; its mechanism depends on the plant species and the type of organ. Regardless of its exact mechanism, the concentration of free sugar in the sink is lower than that in the sieve tube because the unloaded sugar is either consumed during growth and metabolism of the sink cells or is converted into insoluble polymers such as starch. As a result of this sugar concentration gradient, sugar molecules diffuse from the phloem into the sink tissues, and water follows by osmosis.

Pressure flow is the mechanism of translocation in angiosperms

Phloem sap flows from source to sink at rates as great as 1 m/hr, which is much too fast to be accounted for by either diffusion or cytoplasmic streaming. Phloem sap moves by bulk flow, which is driven by pressure (thus the synonym *pressure flow*). Phloem loading results in a high solute concentration at the source end of a sieve tube, which lowers the water potential and causes water to flow into the tube (FIGURE 36.17, p. 764). Hydrostatic pressure develops within the sieve tube, and the pressure is greatest at the source end of the tube. At the sink end, the pressure is relieved by the loss of water, owing to water potential being lowered outside the sieve tube by the exodus of sucrose. The building of pressure at one end of the tube (source) and reduction of that pressure at the opposite end

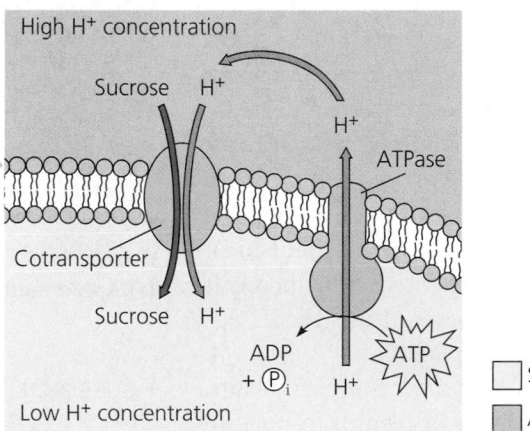

(a) Sucrose manufactured in mesophyll cells can travel via the symplast (blue arrows) to sieve-tube members. In some species, sucrose exits (magenta arrow) the symplast near sieve tubes and is actively accumulated from the apoplast by sieve-tube members and their companion cells.

(b) A chemiosmotic mechanism is responsible for the active transport of sucrose into companion cells and sieve-tube members. Proton pumps (ATPase) generate an H+ gradient, which drives sucrose accumulation with the help of a cotransport protein that couples sucrose transport to the diffusion of H+ back into the cell.

FIGURE 36.16 Loading of sucrose into phloem.

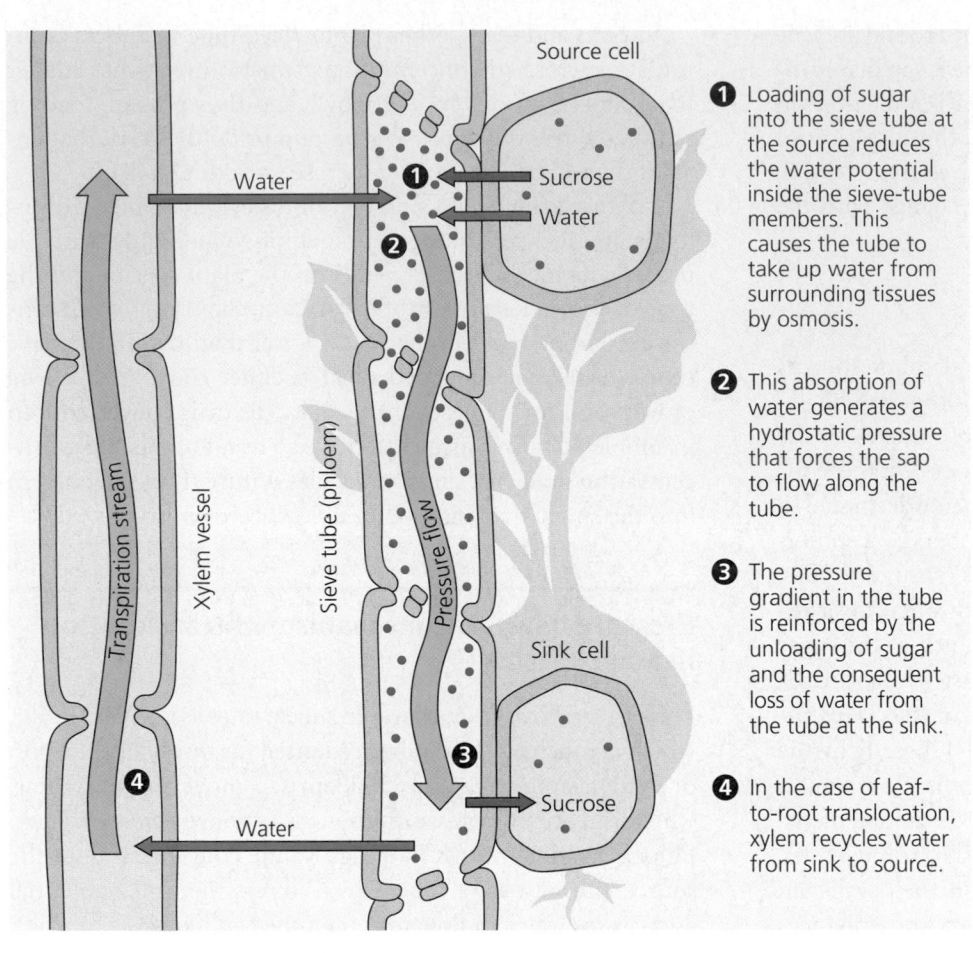

FIGURE 36.17 Pressure flow in a sieve tube. In this example, the source is a leaf and the sink is a root.

1 Loading of sugar into the sieve tube at the source reduces the water potential inside the sieve-tube members. This causes the tube to take up water from surrounding tissues by osmosis.

2 This absorption of water generates a hydrostatic pressure that forces the sap to flow along the tube.

3 The pressure gradient in the tube is reinforced by the unloading of sugar and the consequent loss of water from the tube at the sink.

4 In the case of leaf-to-root translocation, xylem recycles water from sink to source.

the long-distance level of transport between organs (bulk flow within sieve tubes). An increased understanding of these transport processes will be central to agricultural advancement in the future. Consider the fact that it is not photosynthesis that limits yield in a plant growing under ideal conditions, but the ability of the plant to transport sugar away from the leaf. Thus, the genetic engineering of higher-yielding crop plants may depend on gaining a better understanding of the factors that limit the bulk flow of sugars through the phloem.

■   ■   ■

Plant physiologists still have much to learn about the mechanisms of transport in the vascular systems of xylem and phloem. William Harvey, the great 17th-century physiologist, speculated that plants and animals have similar circulatory systems. The idea was abandoned after careful dissection failed to turn up a heart in plants. We are only now beginning to understand how a plant keeps sap flowing through its veins without the help of moving parts.

(sink) cause water to flow from source to sink, carrying the sugar along. Water is recycled back from sink to source by xylem vessels.

The pressure flow model explains why phloem sap always flows from a sugar source to a sugar sink, regardless of their locations in the plant. Researchers have devised several kinds of experiments to test the model. FIGURE 36.18 illustrates an innovative experiment that takes advantage of natural phloem probes: aphids that feed on phloem sap. The case for pressure flow as the mechanism of translocation in angiosperms is convincing. It is not yet known, however, if this model applies to other vascular plants.

In our study of how sugar moves in plants, we have seen examples of plant transport on three levels: at the cellular level of transport across plasma membranes (sucrose accumulation by active transport in phloem cells); at the short-distance level of lateral transport within organs (sucrose migration from mesophyll to phloem via the symplast and apoplast); and at

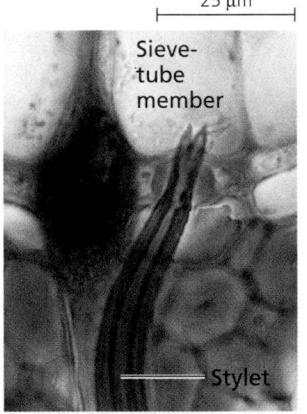

(a) Aphid feeding

(c) Severed stylet exuding phloem sap

(b) Stylet of aphid in sieve-tube member

FIGURE 36.18 Tapping phloem sap with the help of an aphid. (a) The "honeydew" droplet exuded from the anus of this aphid consists of phloem sap minus some nutrients absorbed by the insect. **(b)** The aphid inserts a modified mouthpart called a stylet into the plant and probes until the tip of this hypodermic-like organ penetrates a single sieve-tube member (LM). The pressure within the sieve tube force-feeds the aphid, swelling it to several times its original size. **(c)** While the aphid is feeding, it can be anesthetized and severed from its stylet, which then serves the researcher as a miniature tap that exudes phloem sap for hours. The closer the stylet is to a sugar source, the faster the sap will flow out and the greater its sugar concentration. These results are predicted by the pressure flow hypothesis.

THE PROCESS OF SCIENCE

CHAPTER 36 REVIEW

Go to the Campbell Biology website (www.campbellbiology.com) to explore an interactive version of the Chapter Review.

Summary of Key Concepts

AN OVERVIEW OF TRANSPORT MECHANISMS IN PLANTS

- **Transport at the cellular level depends on the selective permeability of membranes (p. 749)** Specific transport proteins enable plant cells to maintain an internal environment different from their surroundings.

- **Proton pumps play a central role in transport across plant membranes (pp. 749–750, FIGURE 36.2)** The membrane potential and H^+ gradient generated by proton pumps are harnessed to drive the transport of a variety of solutes.

- **Differences in water potential drive water transport in plant cells (pp. 750–752, FIGURES 36.3–36.5)** Solutes decrease water potential, while pressure increases water potential. Water flows by osmosis from a compartment with a higher water potential to one with a lower potential.
  Web/CD Case Study in the Process of Science: How Are Water and Solute Potentials Calculated?

- **Aquaporins affect the rate of water transport across membranes (pp. 752–753)** Aquaporins are water-specific channels in membranes that might help regulate the rate of osmosis.

- **Vacuolated plant cells have three major compartments (pp. 753–754, FIGURE 36.6)** The plasma membrane regulates transport between the cytosol and the wall solution, while the tonoplast regulates transport between the cytosol and the vacuole.

- **Both the symplast and the apoplast function in transport within tissues and organs (p. 754, FIGURE 36.6)** The symplast is the continuum of cytosol linked by plasmodesmata. The apoplast is the continuum of cell walls.

- **Bulk flow functions in long-distance transport (p. 754)** Transport of xylem sap and phloem sap is due to pressure differences at opposite ends of conduits—the xylem vessels and sieve tubes.

ABSORPTION OF WATER AND MINERALS BY ROOTS

- **Root hairs, mycorrhizae, and a large surface area of cortical cells enhance water and mineral absorption (pp. 754–756, FIGURE 36.8)** Root hairs are the most important avenues of absorption near root tips, but mycorrhizae, symbiotic associations of fungi and roots, are responsible for most absorption by the whole root system. Once soil solution enters the root, the extensive surface area of cortical cell membranes enhances uptake of water and selected minerals.

- **The endodermis functions as a selective sentry between the root cortex and vascular tissue (p. 756, FIGURE 36.7)** Water can cross the cortex via the symplast or apoplast, but minerals that reach the endodermis via the apoplast must finally cross the selective membranes of endodermal cells. The waxy Casparian strip of the endodermal wall blocks apoplastic transfer of minerals from the cortex to the stele.

TRANSPORT OF XYLEM SAP

- **The ascent of xylem sap depends mainly on transpiration and the physical properties of water (pp. 756–758, FIGURE 36.10)** Loss of water vapor (transpiration) lowers water potential in the leaf by producing a negative pressure (tension). This low water potential draws water from the xylem. Cohesion and adhesion of the water transmits the pulling force all the way down to the roots.

- **Xylem sap ascends by solar-powered bulk flow: *a review* (pp. 758–759, FIGURE 36.11)** The movement of xylem sap against gravity is maintained by transpiration.
  Web/CD Activity 36A: Transport of Xylem Sap

THE CONTROL OF TRANSPIRATION

- **Guard cells mediate the photosynthesis-transpiration compromise (pp. 759–761, FIGURES 36.12–36.14)** Stomata support photosynthesis by allowing CO_2 and O_2 exchange between the leaf and atmosphere, but these pores are also the main avenues for transpirational loss of water from the plant. Turgor changes in guard cells, which depend on K^+ and water transport into and out of the cells, regulate the size of the stomatal openings.
  Web/CD Case Study in the Process of Science: How Is the Rate of Transpiration Calculated?

- **Xerophytes have evolutionary adaptations that reduce transpiration (p. 762, FIGURE 36.15)** Protection of stomata within leaf indentations and other structural adaptations enable certain plants to survive in arid environments.

TRANSLOCATION OF PHLOEM SAP

- **Phloem translocates its sap from sugar sources to sugar sinks (pp. 762–763, FIGURE 36.16)** Mature leaves are the main sources, though storage organs such as bulbs can be sources during certain seasons. Developing roots and shoot tips are examples of sugar sinks. Phloem loading and unloading depend on active transport of sucrose. The sucrose is cotransported along with H^+, which is diffusing down a gradient generated by proton pumps.
  Web/CD Activity 36B: Translocation of Phloem Sap

- **Pressure flow is the mechanism of translocation in angiosperms (pp. 763–764, FIGURES 36.17, 36.18)** Loading of sugar at the source end of a sieve tube and unloading at the sink end maintain a pressure difference that keeps the sap flowing through the tube.

Self-Quiz

1. Which of the following would *not* contribute to water uptake by a plant cell?
   a. an increase in the water potential (ψ) of the surrounding solution
   b. a decrease in pressure on the cell exerted by the wall
   c. the uptake of solutes by the cell
   d. a decrease in ψ of the cytoplasm
   e. an increase in tension on the surrounding solution

2. Stomata open when guard cells
   a. sense an increase in CO_2 in the air spaces of the leaf.
   b. flop open because of a decrease in turgor pressure.
   c. become more turgid because of an influx of K^+, followed by the osmotic entry of water.
   d. close aquaporins, preventing uptake of water.
   e. accumulate water by active transport.

3. Which of the following is *not* part of the transpiration-cohesion-tension mechanism for the ascent of xylem sap?

   a. the loss of water from the mesophyll cells, which initiates a pull of water molecules from neighboring cells

   b. the transfer of transpirational pull from one water molecule to the next owing to the cohesion caused by hydrogen bonds

   c. the hydrophilic walls of the narrow tracheids and xylem vessels that help maintain the column of water against the force of gravity

   d. the active pumping of water into the xylem of roots

   e. the reduction of water potential in the surface film of mesophyll cells due to transpiration

4. Which of the following does *not* appear to involve active transport across membranes?

   a. the movement of mineral nutrients from the apoplast to the symplast

   b. the movement of sugar from mesophyll cells into sieve-tube members in corn

   c. the movement of sugar from one sieve-tube member to the next

   d. K^+ uptake by guard cells during stomatal opening

   e. the movement of mineral nutrients into cells of the root cortex

5. The movement of sap from a sugar source to a sugar sink

   a. occurs through the apoplast of sieve-tube members.

   b. may translocate sugars from the breakdown of stored starch in a root up to developing shoots.

   c. is similar to the flow of xylem sap in depending on tension, or negative pressure.

   d. depends on the active pumping of water into sieve tubes at the source end.

   e. results mainly from diffusion.

6. The productivity of a crop declines when leaves begin to wilt mainly because

   a. the chlorophyll of wilting leaves decomposes.

   b. flaccid mesophyll cells are incapable of photosynthesis.

   c. stomata close, preventing CO_2 from entering the leaf.

   d. photolysis, the water-splitting step of photosynthesis, cannot occur when there is a water deficiency.

   e. an accumulation of CO_2 in the leaf inhibits the enzymes required for photosynthesis.

7. Imagine cutting a live twig from a tree and examining the cut surface of the twig with a magnifying glass. You locate the vascular tissue and observe a growing droplet of fluid exuding from the cut surface. This fluid is probably

   a. phloem sap.

   b. xylem sap.

   c. guttation fluid.

   d. fluid of the transpiration stream.

   e. cell sap from the broken vacuoles of cells.

8. Which structure or compartment is *not* part of the plant's apoplast?

   a. the lumen of a xylem vessel   d. the cell wall of a transfer cell

   b. the lumen of a sieve tube   e. the cell wall of a root hair

   c. the cell wall of a mesophyll cell

9. Which of the following is *not* an adaptation that enhances the uptake of water and minerals by roots?

   a. mycorrhizae, the symbiotic associations of roots and fungi

   b. root hairs, which increase surface area near root tips

   c. selective uptake of minerals by xylem vessels

   d. selective uptake of minerals by cortical cells

   e. plasmodesmata, which facilitate symplastic transport from the cortex into the stele

10. A plant cell with a solute potential of -0.65 MPa maintains a constant volume when bathed in a solution that has a solute potential of -0.3 MPa and is in an open container. What do we know about the cell?

    a. The cell has a pressure potential of $+0.65$ MPa.

    b. The cell has a water potential of -0.65 MPa.

    c. The cell has a pressure potential of $+0.35$ MPa.

    d. The cell has a pressure potential of $+0.3$ MPa.

    e. The cell has a water potential of 0 MPa.

11. What is the main function of the Casparian strip?

12. Some leaf molds, fungi that parasitize plants, secrete a chemical that causes guard cells to accumulate potassium ions. How does this adaptation enable the mold to infect the plant?

13. Compare and contrast the forces that move phloem sap versus the forces that move xylem sap over long distance.

14. Describe the environmental conditions that would minimize the transpiration-to-photosynthesis ratio for a C_3 plant, like an oak.

15. A tip for helping cut flowers last longer without wilting is to cut off the ends of the stems underwater and then transfer the flowers to a vase while drops of water are still present on the cut ends of the stems. Explain why this works.

Go to the website or CD-ROM for more quiz questions.

Evolution Connection

Analysis of preserved leaves shows that the density of stomata per unit area of leaf has decreased over the past 200 years. Suggest a hypothesis that connects this evolutionary trend to environmental change.

The Process of Science

Barrel cacti (*Ferocactus*) of the Sonoran Desert do not grow straight up, but tilt southward at about a 45° angle. Suggest a hypothesis for this evolutionary adaptation. How could you test your hypothesis?

Calculate rates of transpiration and water and solute potentials in the Case Studies in the Process of Science, available on the website and CD-ROM.

Science, Technology and Society

Water use is a serious social and environmental issue in the arid southwestern United States. In recent years, there has been growing criticism of water-intensive ornamental landscapes like lawns and golf greens. These areas are maintained artificially by diverting water from rivers and streams or pumping from ancient subterranean aquifers. Is this form of water use something society should limit or even eliminate in such areas? Or, should property owners be free to landscape as they choose? Defend your side in this debate.

Answers: 1. e; 2. c; 3. d; 4. c; 5. b; 6. c; 7. a; 8. b; 9. c; 10. c; 11. It regulates the passage of minerals into the xylem by blocking access via cell walls and requiring all minerals to cross a selectively permeable membrane. 12. Accumulation of potassium by guard cells results in osmotic water uptake, and the turgid condition of the cells keeps the stomata open. This enables the mold to grow into the leaf interior via the stomata. 13. In both cases, the long-distance transport is a bulk flow driven by a pressure difference at opposite ends of tubes. Pressure is generated at the source end of a sieve tube by the loading of sugar and resulting osmotic flow of water into the phloem, and this pressure *pushes* sap from the source end to the sink end of the tube. In contrast, transpiration generates a negative pressure (tension) as a force that *pulls* the ascent of xylem sap. 14. A sunny, warm, but not hot, day; high humidity; low wind speed. 15. After the flowers are cut, transpiration from any leaves and from the petals (which are modified leaves) will continue to draw water up the xylem. If the cut flowers are transferred directly to a vase, the air pockets in the xylem vessels will prevent delivery of water from the vase to the flowers. Cutting the stems again underwater, a few centimeters from the original cut, will sever the xylem above the air pocket. The water droplets will prevent another air pocket from forming while placing the flowers in a vase.

CHAPTER 37

PLANT NUTRITION

Every organism *is an open system connected to its environment by a continuous exchange of energy and materials. In the energy flow and chemical cycling that keep an ecosystem alive, plants and other photosynthetic autotrophs perform the key step of transforming inorganic compounds into organic ones. Autotrophic does not mean autonomous, however. Plants need sunlight as the energy source for photosynthesis. And to synthesize organic matter,* plants also require raw materials in the form of inorganic substances: carbon dioxide, water, and a variety of minerals present as inorganic ions in the soil. With its ramifying root system and shoot system (see the photograph of a hyacinth above), a plant is extensively networked with its environment—the soil and air, which are the reservoirs of the plant's inorganic nutrients. In this chapter, you will learn more about the nutritional requirements of plants and examine some of the structural and physiological adaptations for plant nutrition that have evolved.*

NUTRITIONAL REQUIREMENTS OF PLANTS

The chemical composition of plants provides clues to their nutritional requirements

Watch a large plant grow from a tiny seed, and you cannot help wondering where all the mass comes from. Aristotle thought that soil provided the substance for plant growth, because plants seemed to spring from the ground. Leaves, he believed, functioned only to shade the developing fruit. In the 17th century, a Belgian physician named Jean-Baptiste van Helmont performed an experiment to test the hypothesis that plants grow by absorbing soil. He planted a willow seedling in a pot that contained 90.9 kg of soil. After five years, the willow had grown into a tree weighing 76.8 kg, but only 0.06 kg of soil had disappeared from the pot. Van Helmont concluded that the willow had grown mainly from the water he had added regularly. A century later, Stephen Hales, an English physiologist, postulated that plants are nourished mostly by air.

As it turns out, none of the early ideas about plant nutrition is entirely incorrect. Plants *do* extract minerals from the soil.

THE PROCESS OF SCIENCE

Mineral nutrients are essential chemical elements absorbed from the soil in the form of inorganic ions. For example, plants need nitrogen, which they acquire from the soil mainly in the form of nitrate ions (NO_3^-). However, as we can conclude from van Helmont's data, mineral nutrients from the soil make only a small contribution to the overall mass of the plant. About 80–85% of a herbaceous (nonwoody) plant is water, and plants grow mainly by accumulating water in the central vacuoles of their cells. Furthermore, water can truly be considered a nutrient because it supplies most of the hydrogen atoms and some of the oxygen atoms that are incorporated into organic compounds by photosynthesis (see FIGURE 10.3). However, only a small fraction of the water that enters a plant contributes atoms to organic molecules. Generally, more than 90% of the water absorbed by plants is lost by transpiration, and most of the water retained by the plant functions as a solvent, provides most of the mass for cell elongation, and helps maintain the form of soft tissue by keeping cells turgid. By weight, the bulk of the organic material of a plant is derived not from water or soil minerals, but from the CO_2 assimilated from the atmosphere (FIGURE 37.1).

We can measure water content by comparing the weight of plant material before and after it is dried. We can then analyze the chemical composition of the dry residue. Organic substances account for about 95% of the dry weight, with inorganic substances making up the remaining 5%. Most of the organic material is carbohydrate, including the cellulose of cell walls. Thus, carbon, oxygen, and hydrogen, the ingredients of carbohydrates, are the most abundant elements in the dry weight of a plant. Because some organic molecules contain nitrogen, sulfur, or phosphorus, these elements are also relatively abundant in plants.

More than 50 chemical elements have been identified among the inorganic substances present in plants, but it is unlikely that all these elements are essential. Roots are able to absorb minerals somewhat selectively, enabling the plant to accumulate essential elements that may be present in the soil in very minute quantities. To a certain extent, however, the minerals in a plant reflect the composition of the soil in which the plant is growing. Plants growing on mine tailings, for instance, may contain gold or silver. Studying the chemical composition of plants provides clues about their nutritional requirements, but we must distinguish elements that are essential from those that are merely present in the plant.

Plants require nine macronutrients and at least eight micronutrients

A particular chemical element is considered an **essential nutrient** if it is required for a plant to grow from a seed and complete the life cycle, producing another generation of seeds. Researchers can use a method known as hydroponic culture to determine which of the mineral elements are actually essential nutrients (FIGURE 37.2). Such studies have helped identify 17

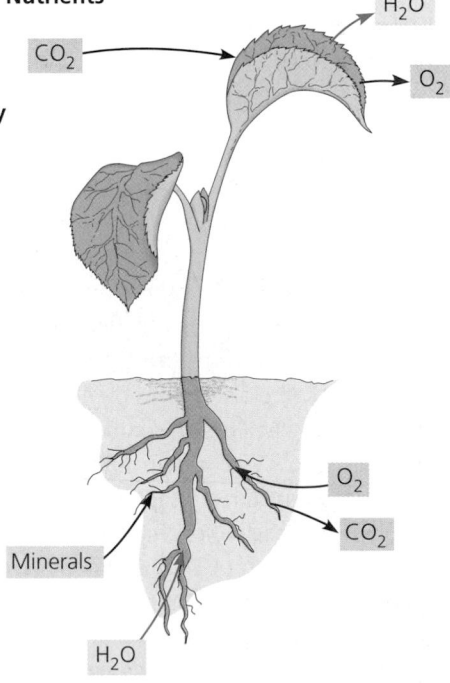

FIGURE 37.1 The uptake of nutrients by a plant: an overview. Roots absorb water and minerals from the soil, with mycorrhizae and root hairs greatly increasing surface area for absorption. Carbon dioxide, the source of carbon for photosynthesis, diffuses into leaves from the surrounding air through stomata. (Plants also need O_2 for cellular respiration, although the plant is a net producer of O_2.) From these inorganic nutrients the plant can produce all of its own organic material.

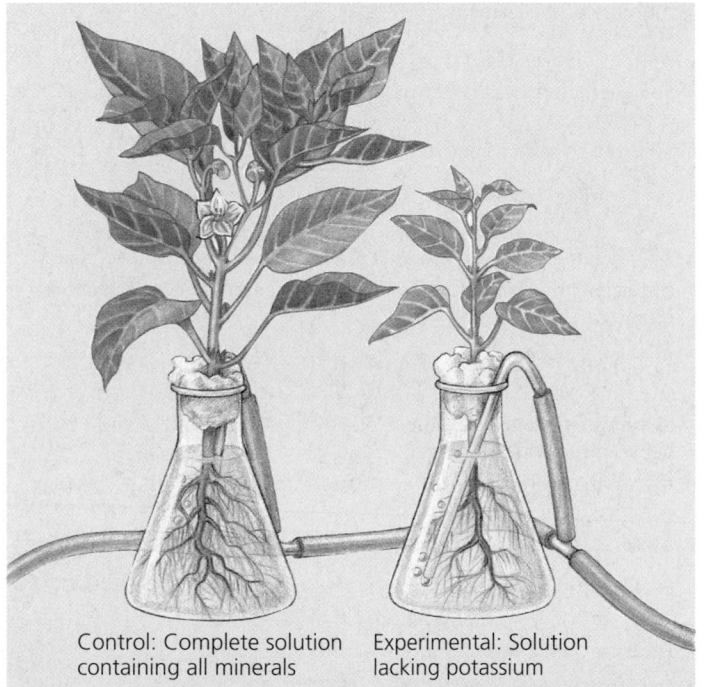

Control: Complete solution containing all minerals

Experimental: Solution lacking potassium

FIGURE 37.2 Using hydroponic culture to identify essential nutrients. A researcher bathes the roots of plants in solutions of various minerals dissolved in known concentrations. Aerating the water provides the roots with oxygen for cellular respiration. A particular mineral, such as potassium, can be omitted from the culture medium to test whether it is essential to the plants. If the element deleted from the mineral solution is an essential nutrient, then the incomplete medium will cause plants to become abnormal in appearance compared with controls grown on a complete mineral medium. The most common symptoms of a mineral deficiency are stunted growth and discolored leaves.

THE PROCESS OF SCIENCE

Table 37.1 Essential Nutrients in Plants

Element	Form Available to Plants	Major Functions
Macronutrients		
Carbon	CO_2	Major component of plant's organic compounds
Oxygen	CO_2	Major component of plant's organic compounds
Hydrogen	H_2O	Major component of plant's organic compounds
Nitrogen	NO_3^-, NH_4^+	Component of nucleic acids, proteins, hormones, and coenzymes
Sulfur	SO_4^{2-}	Component of proteins, coenzymes
Phosphorus	$H_2PO_4^-$, HPO_4^{2-}	Component of nucleic acids, phospholipids, ATP, several coenzymes
Potassium	K^+	Cofactor that functions in protein synthesis; major solute functioning in water balance; operation of stomata
Calcium	Ca^{2+}	Important in formation and stability of cell walls and in maintenance of membrane structure and permeability; activates some enzymes; regulates many responses of cells to stimuli
Magnesium	Mg^{2+}	Component of chlorophyll; activates many enzymes
Micronutrients		
Chlorine	Cl^-	Required for water-splitting step of photosynthesis; functions in water balance
Iron	Fe^{3+}, Fe^{2+}	Component of cytochromes; activates some enzymes
Boron	$H_2BO_3^-$	Cofactor in chlorophyll synthesis; may be involved in carbohydrate transport and nucleic acid synthesis
Manganese	Mn^{2+}	Active in formation of amino acids; activates some enzymes; required for water-splitting step of photosynthesis
Zinc	Zn^{2+}	Active in formation of chlorophyll; activates some enzymes
Copper	Cu^+, Cu^{2+}	Component of many redox and lignin-biosynthetic enzymes
Molybdenum	MoO_4^{2-}	Essential for nitrogen fixation; cofactor that functions in nitrate reduction
Nickel	Ni^{2+}	Cofactor for an enzyme functioning in nitrogen metabolism

elements that are essential nutrients in all plants and a few other elements that are essential to certain groups of plants. Most research has involved crop plants; little is known about the specific nutritional needs of uncultivated plants, even some of the most commercially important conifers that provide lumber.

Elements required by plants in relatively large amounts are called **macronutrients.** There are nine macronutrients in all, including the six major ingredients of organic compounds: carbon, oxygen, hydrogen, nitrogen, sulfur, and phosphorus. The other three macronutrients are potassium, calcium, and magnesium. (TABLE 37.1 lists some of their functions.)

Elements that plants need in very small amounts are called **micronutrients.** The eight micronutrients are iron, chlorine, copper, manganese, zinc, molybdenum, boron, and nickel. These elements function in plants mainly as cofactors of enzymatic reactions (see Chapter 6). Iron, for example, is a metallic component of cytochromes, the proteins that function in the electron transport chains of chloroplasts and mitochondria. It is because micronutrients generally play catalytic roles that plants need only minute quantities of these elements. The requirement for molybdenum, for example, is so modest that there is only one atom of this rare element for every 16 million atoms of hydrogen in dried plant material. Yet a deficiency of molybdenum or any other micronutrient can weaken or kill a plant.

The symptoms of a mineral deficiency depend on the function and mobility of the element

The symptoms of a mineral deficiency depend partly on the function of that nutrient in the plant. For example, a deficiency of magnesium, an ingredient of chlorophyll, causes yellowing of the leaves, or chlorosis (FIGURE 37.3). In some cases,

FIGURE 37.3 Magnesium deficiency in a tomato plant. Yellowing of the leaves (chlorosis) is the result of an inability to synthesize chlorophyll, which contains magnesium.

FIGURE 37.4 Hydroponic farming. In this apparatus, a nutrient solution flows over the roots of lettuce growing on a slat. Perhaps astronauts living in a space station will one day grow their vegetables hydroponically, but because of the expense, it is unlikely that this type of farming will relieve hunger here on Earth.

the relationship between a mineral deficiency and its symptoms is less direct. For instance, iron deficiency can cause chlorosis even though chlorophyll contains no iron, because this metal is required as a cofactor in one of the steps of chlorophyll synthesis.

Mineral deficiency symptoms depend not only on the role of the nutrient in the plant but also on its mobility within the plant. If a nutrient moves about freely from one part of the plant to another, symptoms of a deficiency will show up first in older organs. This is because young, growing tissues have more "drawing power" than old tissues for nutrients that are in short supply. (The mechanism for this preferential routing is the source-to-sink translocation in phloem that you learned about in Chapter 36.) A plant starved for magnesium, for example, will show signs of chlorosis first in its older leaves. Magnesium, which is relatively mobile in the plant, is shunted preferentially to young leaves. In contrast, a deficiency of a nutrient that is relatively immobile within a plant will affect young parts of the plant first. Older tissues may have adequate amounts of the mineral, which they are able to retain during periods of short supply. For example, iron does not move freely within a plant, and a deficiency of iron will cause yellowing of young leaves before any effect on older leaves is visible.

The symptoms of a mineral deficiency are often distinctive enough for a plant physiologist or farmer to diagnose its cause. One way to confirm the diagnosis of a specific deficiency is to analyze the mineral content of the plant and soil. Deficiencies of nitrogen, potassium, and phosphorus are the most common problems. Shortages of micronutrients are less common and tend to be geographically localized because of differences in soil composition. The amount of a micronutri-

ent needed to correct a deficiency is usually quite small. For example, a zinc deficiency in fruit trees can usually be cured by hammering a few zinc nails into each tree trunk. Moderation is important because overdoses of some micronutrients can be toxic to plants.

One way to ensure optimal mineral nutrition is to grow plants hydroponically on nutrient solutions that can be precisely regulated (FIGURE 37.4). Hydroponics is currently practiced commercially, but only on a limited scale, because the requirements for equipment and labor make hydroponic farming relatively expensive compared with growing crops in soil.

Mineral deficiencies are not limited to terrestrial ecosystems, nor are they unique to plants among photosynthetic organisms. Vast "pastures" of algae in the world's southern oceans are capable of explosive blooms, restrained only by deficiencies of iron in the seawater. In a limited trial in the relatively unproductive seas between Tasmania and Antarctica, researchers demonstrated that dispersing small amounts of iron produces large algal blooms that pull carbon dioxide out of the air. Seeding the ocean with iron might help slow the increase in global carbon dioxide levels that has occurred because of our burning of fossil fuels (and which contributes to global warming by the greenhouse effect; see Chapter 54). Despite the possible benefits, widespread adoption of iron fertilization must be approached warily, given the many uncertain ecological and climatological effects it could have.

THE ROLE OF SOIL IN PLANT NUTRITION

Soil characteristics are key environmental factors in terrestrial ecosystems

The texture and chemical composition of soil are major factors determining what kinds of plants can grow well in a particular location, be it a natural ecosystem or an agricultural region. (Climate, of course, is another important factor.) Plants that grow naturally in a certain type of soil are adapted to its mineral content and texture and are able to absorb water and extract essential nutrients from that soil. In interacting with the soil that supports their growth, plants, in turn, affect the soil, as we will soon see. The soil-plant interface is a critical component of the chemical cycles that sustain terrestrial ecosystems.

Texture and Composition of Soils

Soil has its origin in the weathering of solid rock. Water that seeps into crevices and freezes in winter fractures the rock, and acids dissolved in the water also help to break down the rock. Once organisms are able to invade the rock, they accelerate the breakdown. Lichens, fungi, bacteria, mosses, and the roots of

The A horizon is the topsoil, a mixture of broken-down rock of various textures, living organisms, and decaying organic matter.

The B horizon contains much less organic matter than the A horizon and is less weathered.

The C horizon, composed mainly of partially broken-down rock, serves as the "parent" material for the upper layers of soil.

FIGURE 37.5 Soil horizons. This researcher is photographing a vertical profile of three soil layers, or horizons, in a Tennessee cotton field.

vascular plants all secrete acids, and the expansion of roots growing in fissures cracks rocks and pebbles. The eventual result of all this activity is **topsoil,** a mixture of particles derived from rock, living organisms, and **humus,** a residue of partially decayed organic material. The topsoil and other distinct soil layers, or **horizons,** are often visible in vertical profile where there is a roadcut or deep hole (FIGURE 37.5).

The texture of topsoil depends on the size of its particles, which are classified in a range from coarse sand to microscopic clay particles. The most fertile soils are usually **loams,** made up of roughly equal amounts of sand, silt (particles of intermediate size), and clay. Loamy soils have enough fine particles to provide a large surface area for retaining minerals and water, which adhere to the particles. But loams also have enough coarse particles to provide air spaces containing oxygen that can be used by roots for cellular respiration. If soil does not drain adequately, roots suffocate because the air spaces are replaced by water; the roots may also be attacked by molds favored by the soaked soil. These are common hazards for houseplants that are overwatered in pots with poor drainage. Some plants, however, are adapted to waterlogged soil. For example, mangroves that inhabit swamps and marshes have some of their roots modified as hollow tubes that grow upward and function as snorkels, bringing down oxygen from the air.

Topsoil is home to an astonishing number and variety of organisms. A teaspoon of soil has about 5 billion bacteria that cohabit with various fungi, algae and other protists, insects, earthworms, nematodes, and the roots of plants. The activities of all these organisms affect the physical and chemical properties of the soil. Earthworms, for instance, aerate the soil by their burrowing and add mucus that holds fine soil particles together. The metabolism of bacteria alters the mineral composition of the soil. Plant roots extract water and minerals but also affect soil pH and reinforce the soil against erosion.

Humus, an important component of topsoil, is the decomposing organic material formed by the action of bacteria and fungi on dead organisms, feces, fallen leaves, and other organic refuse. Humus prevents clay from packing together and builds a crumbly soil that retains water but is still porous enough for the adequate aeration of roots. Humus is also a reservoir of mineral nutrients that are returned gradually to the soil as microorganisms decompose the organic matter.

The Availability of Soil Water and Minerals

After a heavy rainfall, water drains away from the larger spaces of the soil, but smaller spaces retain water because of its attraction for the soil particles, which have electrically charged surfaces. Some of this water adheres so tightly to the hydrophilic soil particles that it cannot be extracted by plants. The film of water bound less tightly to the particles is the water generally available to plants (FIGURE 37.6a, p. 772). It is not pure water, but a soil solution containing dissolved minerals. Roots absorb this soil solution.

Many minerals in soil—especially those that are positively charged, such as potassium (K^+), calcium (Ca^{2+}), and magnesium (Mg^{2+})—adhere by electrical attraction to the negatively charged surfaces of clay particles. The presence of clay in a soil helps prevent the leaching (draining away) of mineral nutrients during heavy rain or irrigation because the finely divided particles provide so much surface area for binding minerals. However, clay particles must release their bound minerals to the soil solution in order for roots to absorb the nutrients. Minerals that are negatively charged, such as nitrate (NO_3^-), phosphate ($H_2PO_4^-$), and sulfate (SO_4^{2-}), are usually not bound tightly to

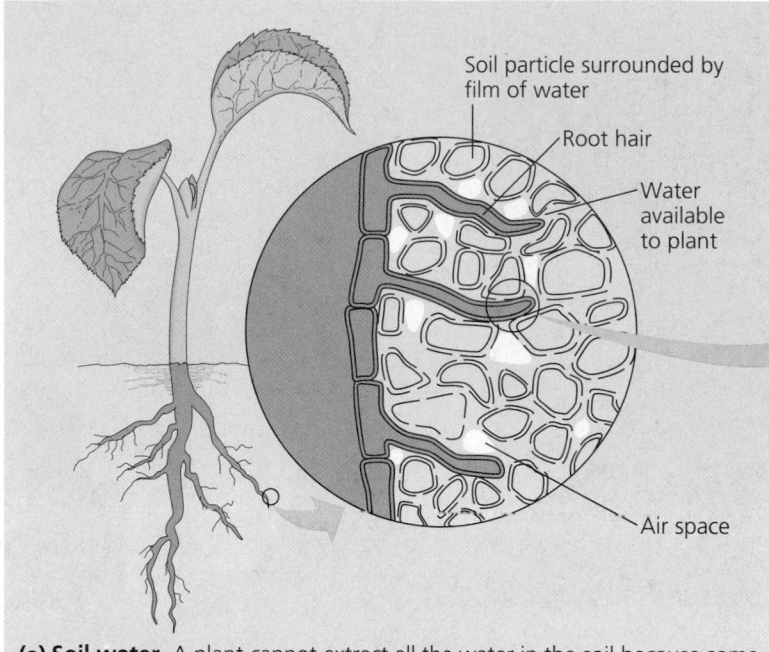

Soil particle surrounded by film of water

Root hair

Water available to plant

Air space

(a) Soil water. A plant cannot extract all the water in the soil because some of it is tightly held by hydrophilic soil particles. Water bound less tightly to soil particles can be imbibed by the root.

FIGURE 37.6 The availability of soil water and minerals.

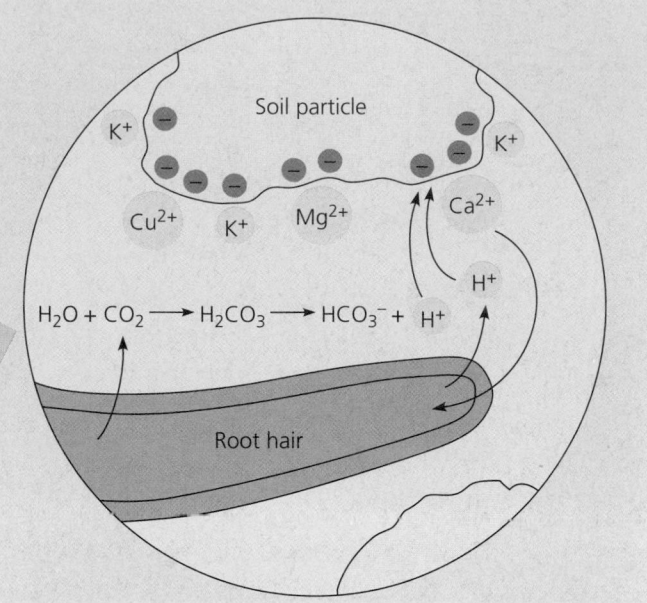

Soil particle

K^+ Cu^{2+} K^+ Mg^{2+} Ca^{2+} K^+ H^+

$H_2O + CO_2 \longrightarrow H_2CO_3 \longrightarrow HCO_3^- + H^+$

Root hair

(b) Cation exchange in soil. Hydrogen ions in the soil solution help make certain nutrients available to plants by displacing positively charged minerals (cations such as Ca^{2+}) that were bound tightly to the surface of fine, negatively charged soil particles. Plants contribute to the pool of H^+ in the soil in two ways: First, by secreting H^+ and second, by the process of cellular respiration in roots. This respiration releases CO_2 to the soil solution, where the CO_2 reacts with water to form carbonic acid (H_2CO_3). Dissociation of this acid adds hydrogen ions to the soil.

soil particles and thus tend to leach away more quickly. Positively charged minerals are made available to the plant when hydrogen ions in the soil displace the mineral ions from the clay particles. This process, called **cation exchange,** is stimulated by the roots themselves, which secrete H^+ and compounds that form acids in the soil solution (FIGURE 37.6b).

Soil conservation is one step toward sustainable agriculture

It may take centuries for a soil to become fertile through the breakdown of rock and the accumulation of organic material, but human mismanagement can destroy that fertility within a few years. Soil mismanagement has been a recurring problem in human history. For example, the Dust Bowl was an ecological and human disaster that took place in the southwestern Great Plains region of the United States in the 1930s. Before the arrival of farmers, the region was covered by hardy grasses that held the soil in place in spite of the long recurrent droughts and torrential rains characteristic of the region. However, in the 30 years before World War I, a large number of homesteaders settled in the region, planting wheat and raising cattle. Both of these land uses left the soil exposed to the danger of erosion by the winds that constantly sweep over the area (FIGURE 37.7). Bad luck in the form of a few years of drought made the problem worse. In many places, 8–10 cm of topsoil were blown away. Millions of hectares of farmland became useless, and hundreds of thou-

sands of people were forced to abandon their homes and land. Better soil conservation practices could have preserved the soil's fertility and sustained agricultural productivity.

To understand soil conservation, we must begin with the premise that agriculture is unnatural. In forests, grasslands, and other natural ecosystems, mineral nutrients are usually recycled by the decomposition of dead organic material in the soil. In contrast, when we harvest a crop, essential elements are diverted from the chemical cycles going on in that location. In general, agriculture depletes the mineral content of the soil. To grow a ton of wheat grain, the soil gives up 18.2 kg of nitrogen, 3.6 kg of phosphorus, and 4.1 kg of potassium. Each year, the fertility of the soil diminishes unless fertilizers are applied to replace the lost minerals. Many crops also use far more water than the natural vegetation that once grew on that land, forcing farmers to irrigate. Prudent fertilization, thoughtful irrigation, and the prevention of erosion are three of the most important goals of soil conservation.

Fertilizers

Prehistoric farmers may have started fertilizing their fields after noticing that grass grew faster and greener where animals had defecated. The Romans used manure to fertilize their crops, and Native Americans buried fish along with seeds when they planted

FIGURE 37.7 Poor soil conservation has contributed to ecological disasters such as the Dust Bowl. The widespread planting of wheat and the raising of cattle by homesteaders in the southwestern plains states in the early part of the 20th century, in combination with recurrent droughts, left the land susceptible to wind erosion. The organic matter, clay, and silt in the soil were carried great distances by the winds, in some cases darkening the sky as far as the Atlantic coast. Sand and heavier materials drifted against houses, fences, and barns. In the 1930s, hundreds of thousands of people abandoned their homesteads. Many became destitute migrant laborers in California. Their plight was immortalized as a symbol of the Great Depression in John Steinbeck's *The Grapes of Wrath*.

an essential element may be abundant in the soil, plants may be starving for that element because it is bound too tightly to clay or is in a chemical form the plant cannot absorb. Managing the pH of soil is tricky; a change in hydrogen ion concentration may make one mineral more available to the plant while causing another mineral to become less available. At pH 8, for instance, the plant can absorb calcium, but iron is almost completely unavailable. The pH of the soil should be matched to the specific mineral needs of the crop. If the soil is too alkaline, sulfate can be added to lower the pH. Soil that is too acidic can be adjusted by liming (adding calcium carbonate or calcium hydroxide).

A major problem with acid soils, particularly in tropical areas, is that aluminum dissolves in the soil at low pH and becomes toxic to roots. Some plants are able to cope with high aluminum levels in the soil by secreting certain organic anions that bind the aluminum and render it harmless.

corn. In developed nations today, most farmers use commercially produced fertilizers containing minerals that are either mined or prepared by industrial processes. These fertilizers are usually enriched in nitrogen, phosphorus, and potassium, the three mineral elements most commonly deficient in farm and garden soils. Commercial fertilizers, such as those you can buy in a garden shop, are labeled with a three-number code that indicates the mineral content. A fertilizer marked "10-12-8," for instance, is 10% nitrogen (as ammonium or nitrate), 12% phosphorus (as phosphoric acid), and 8% potassium (as the mineral potash).

Manure, fishmeal, and compost are referred to as "organic" fertilizers because they are of biological origin and contain organic material that is in the process of decomposing. However, before the elements in compost can be of any use to plants, the organic material must be decomposed to the inorganic nutrients that roots can absorb. In the end, the minerals a plant extracts from the soil are in the same form whether they came from organic fertilizer or from a chemical factory. Compost releases minerals gradually, however, whereas the minerals in commercial fertilizers are available immediately but may not be retained by the soil for long. Excess minerals not taken up by the plants are usually wasted because they are often leached from the soil by rainwater or irrigation. To make matters worse, this mineral runoff may pollute the groundwater and eventually reach streams and lakes. Agricultural researchers are attempting to develop ways to reduce the use of fertilizers while maintaining crop yields.

To fertilize judiciously, a farmer must pay close attention to the pH of the soil. Soil pH not only affects cation exchange but also influences the chemical form of all minerals. Even though

Irrigation

Even more than mineral deficiencies, the unavailability of water most often limits the growth of plants. Irrigation can transform a desert into a garden, but farming in arid regions is a huge drain on water resources. Many of the rivers in the southwestern United States have been reduced to trickles by the diversion of water for irrigation. (Quenching the thirst of growing cities adds to the problem.) Another problem is that irrigation in an arid region can gradually make the soil so salty that it becomes completely infertile. Salts dissolved in the irrigation water accumulate in the soil as the water evaporates. Eventually, the salt makes the water potential of the soil solution lower than that of root cells, which then lose water instead of absorbing it (see Chapter 36).

As the world population continues to grow, more and more acres of arid land will have to be cultivated. New methods of irrigation may reduce the risks of running out of water or losing farmland to salinization (salt accumulation). For instance, drip irrigation is now used as an alternative to flooding fields for many of the crops and orchards in Israel and the western United States. In another approach to solving some of the problems of dryland farming, plant breeders are working to develop varieties of plants that require less water.

Erosion

Topsoil from thousands of acres of farmland is lost to water and wind erosion each year in the United States alone. Certain precautions can help reduce these losses. Rows of trees dividing

FIGURE 37.8 Contour tillage. These crops in Wisconsin are planted in rows that go around, rather than up and down, the hills. Contour tillage helps slow the runoff of water and erosion of topsoil after heavy rains.

fields make effective windbreaks, and terracing a hillside can prevent the topsoil from washing away in a heavy rain (FIGURE 37.8). Such crops as alfalfa and wheat provide good ground cover and protect the soil better than corn and other crops that are usually planted in rows.

If managed properly, soil is a renewable resource in which farmers can grow food for generations to come. The goal is **sustainable agriculture,** a commitment embracing a variety of farming methods that are conservation-minded, environmentally safe, and profitable.

Phytoremediation

Some areas have become unfit for agriculture or wildlife as a result of human activities that contaminate the soil or groundwater with toxic heavy metals or organic pollutants. Nonbiological technologies to detoxify soils, such as the removal and storage of contaminated soil in landfills, are costly and disruptive to the landscape. **Phytoremediation** is an emerging, nondestructive technology that seeks to reclaim such contaminated areas cheaply by taking advantage of the remarkable ability of some plant species to extract heavy metals and other pollutants from the soil and to concentrate them in easily harvested portions of the plant. For example, alpine pennycress (*Thlaspi caerulescens*) can accumulate zinc in its shoots at concentrations that are 300 times the level that most plants can tolerate. The use of such plants for phytoremediation shows promise for cleaning up areas contaminated by smelters, mining operations, or nuclear testing. Phytoremediation is just part of the more general technology of bioremediation, which includes the use of prokaryotes for detoxifying polluted sites (see Chapters 27 and 55).

THE SPECIAL CASE OF NITROGEN AS A PLANT NUTRIENT

Of all mineral elements, nitrogen is the one that most often limits the growth of plants and the yields of crops. Plants require nitrogen as an ingredient of proteins, nucleic acids, and other important organic molecules.

The metabolism of soil bacteria makes nitrogen available to plants

It is ironic that plants sometimes suffer nitrogen deficiencies, for the atmosphere is nearly 80% nitrogen. This atmospheric nitrogen, however, is gaseous N_2, and plants cannot use nitrogen in that form. For plants to absorb nitrogen, it must first be converted to ammonium (NH_4^+) or nitrate (NO_3^-). In contrast to other minerals, the NH_4^+ and NO_3^- in soil are not derived from the breakdown of parent rock. Over the short term, the main source of nitrogenous minerals is the decomposition of humus by microbes, including ammonifying bacteria (FIGURE 37.9). In this way, nitrogen present in organic compounds, such as proteins, is repackaged in inorganic compounds that can be recycled when they are absorbed as minerals by roots. However, nitrogen is lost from this local cycle when soil microbes called denitrifying bacteria convert NO_3^- to N_2, which diffuses from the soil to the atmosphere. Still other bacteria, called **nitrogen-fixing bacteria,** restock nitrogenous minerals in the soil by converting N_2 to NH_3 (ammonia), a metabolic process called **nitrogen fixation.** The complex cycling of nitrogen in ecosystems is traced in detail in Chapter 54. Here we focus on nitrogen fixation and the other steps that lead directly to nitrogen assimilation by plants.

All life on Earth depends on nitrogen fixation, a process performed only by certain prokaryotes. Soil is populated by several species of free-living bacteria that are among the nitrogen-fixing prokaryotes. (Other species of nitrogen-fixing bacteria live in plant roots in symbiotic relationships you will learn more about in the next section.) The conversion of atmospheric nitrogen (N_2) to ammonia (NH_3) is a complicated, multi-step process, but we can simplify nitrogen fixation by just indicating the reactants and products:

$$N_2 + 8\,e^- + 8\,H^+ + 16\,ATP \longrightarrow 2\,NH_3 + H_2 + 16\,ADP + 16\,\circledP_i$$

One enzyme complex, called **nitrogenase,** catalyzes the entire reaction sequence, which reduces N_2 to NH_3 by adding electrons along with hydrogen ions. Notice that nitrogen fixation is very expensive in terms of metabolic energy, costing the bacteria eight ATP molecules for each ammonia molecule synthesized. Nitrogen-fixing bacteria are most abundant in soils rich in organic material, which provides fuel for cellular respiration.

In the soil solution, ammonia picks up another hydrogen ion to form ammonium (NH_4^+), which plants can absorb.

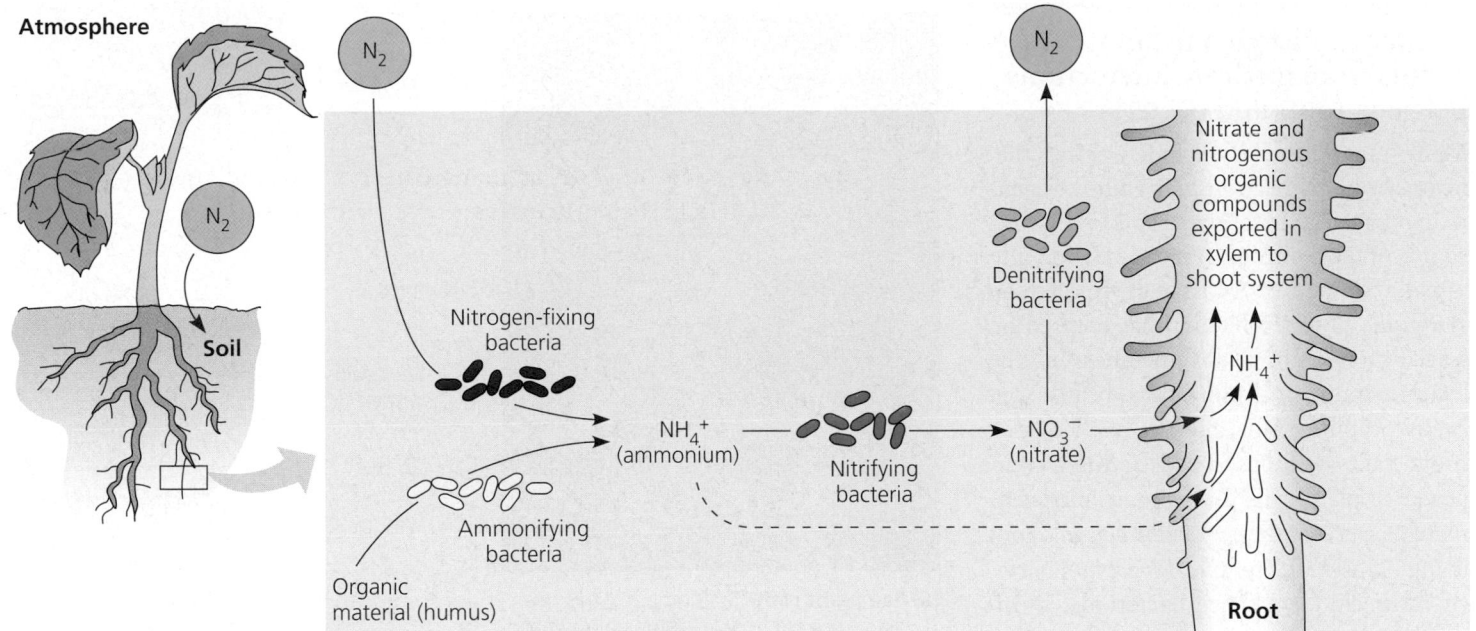

FIGURE 37.9 The role of soil bacteria in the nitrogen nutrition of plants. Ammonium is made available to plants by two types of soil bacteria: those that fix atmospheric N_2 (nitrogen-fixing bacteria) and those that decompose organic material (ammonifying bacteria). Although plants absorb some ammonium from the soil, they absorb mainly nitrate, which is produced from ammonium by nitrifying bacteria. Plants reduce nitrate back to ammonium before incorporating the nitrogen into organic compounds. Xylem transports nitrogen from roots to shoots in the form of nitrate, amino acids, and various other organic compounds, depending on the species.

However, plants acquire their nitrogen mainly in the form of nitrate (NO_3^-), which is produced in the soil by nitrifying bacteria that oxidize ammonium (see FIGURE 37.9). After nitrate is absorbed by roots, a plant enzyme can reduce the nitrate back to ammonium, which other enzymes can then incorporate into amino acids and other organic compounds. Most plant species export nitrogen from roots to shoots, via the xylem, in the form of nitrate or organic compounds that have been synthesized in the roots.

Improving the protein yield of crops is a major goal of agricultural research

The ability of plants to incorporate fixed nitrogen into proteins and other organic substances has a major impact on human welfare; the most common form of malnutrition in humans is protein deficiency. Either by choice or by economic necessity, the majority of people in the world—particularly in developing countries—have a predominantly vegetarian diet and thus depend mainly on plants for protein. Unfortunately, many plants have a low protein content, and the proteins that are present may be deficient in one or more of the amino acids that humans need from their diet. Improving the quality and quantity of proteins in crops is a major goal of agricultural research.

Plant breeding has resulted in new varieties of corn, wheat, and rice that are enriched in protein. However, many of these "super" varieties have an extraordinary demand for nitrogen, which is usually supplied in the form of commercial fertilizer. The industrial production of ammonia and nitrate from atmospheric nitrogen is, like biological nitrogen fixation, very expensive in energy costs. A chemical factory making fertilizer consumes large quantities of fossil fuels. Generally, the countries that most need high-protein crops are the ones least able to afford to pay the fuel bill. The use of new catalysts based on the mechanism by which nitrogenase fixes nitrogen may make commercial fertilizer production less costly in the future. Biochemists determined the structure of *Rhizobium* nitrogenase several years ago, providing a model for chemical engineers to design catalysts by imitating nature. Another strategy that could potentially increase protein yields of crops is to improve the productivity of symbiotic nitrogen fixation, a process we examine in the next section.

NUTRITIONAL ADAPTATIONS: SYMBIOSIS OF PLANTS AND SOIL MICROBES

The roots of plants belong to subterranean communities that include a diversity of other organisms. Among those organisms are certain species of bacteria and fungi that have coevolved with specific plants, forming symbiotic relationships with roots that enhance the nutrition of both partners. The two most important examples are symbiotic nitrogen fixation (roots and bacteria) and the formation of mycorrhizae (roots and fungi).

Symbiotic nitrogen fixation results from intricate interactions between roots and bacteria

Many plant families include species that form symbiotic relationships with nitrogen-fixing bacteria that give roots a built-in source of fixed nitrogen for assimilation into organic compounds. Most of the research on symbiotic nitrogen fixation has focused on agriculturally important members of the legume family, including peas, beans, soybeans, peanuts, alfalfa, and clover. A legume's roots have swellings called **nodules** composed of plant cells that contain nitrogen-fixing bacteria of the genus *Rhizobium* ("root living"). Inside the nodule, *Rhizobium* bacteria assume a form called **bacteroids,** which are contained within vesicles formed by the root cell (FIGURE 37.10). Each legume is associated with a particular species of *Rhizobium.* FIGURE 37.11 describes the steps in the development of root nodules after bacteria enter through what is called an infection thread.

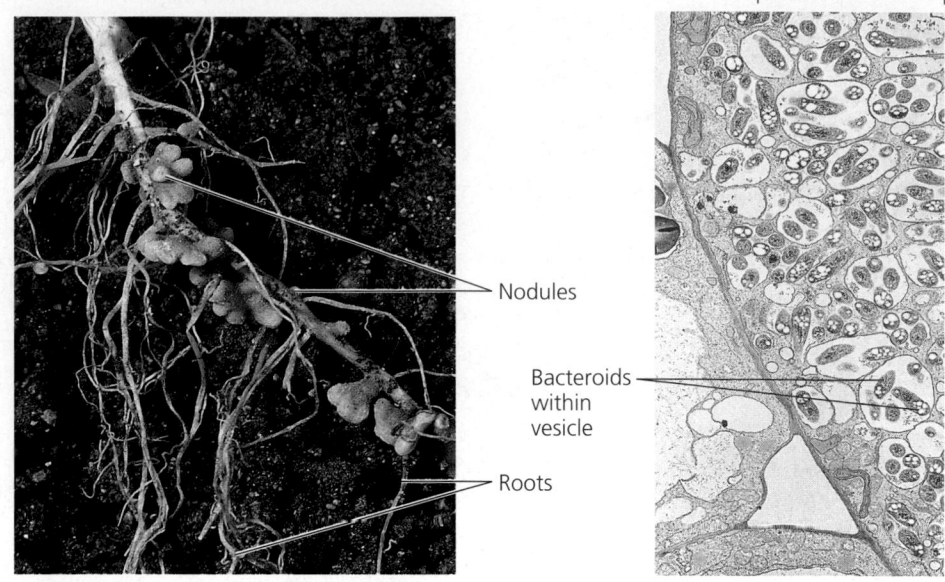

5 μm

Nodules

Roots

Bacteroids within vesicle

(a) Pea plant root. The bumps on this pea plant root are nodules containing symbiotic bacteria. The bacteria fix nitrogen and obtain photosynthetic products supplied by the plant.

(b) Bacteroids in a soybean root nodule. In this TEM, a cell from a root nodule of soybean is filled with bacteroids in vesicles. The cell on the left is uninfected.

FIGURE 37.10 Root nodules on legumes.

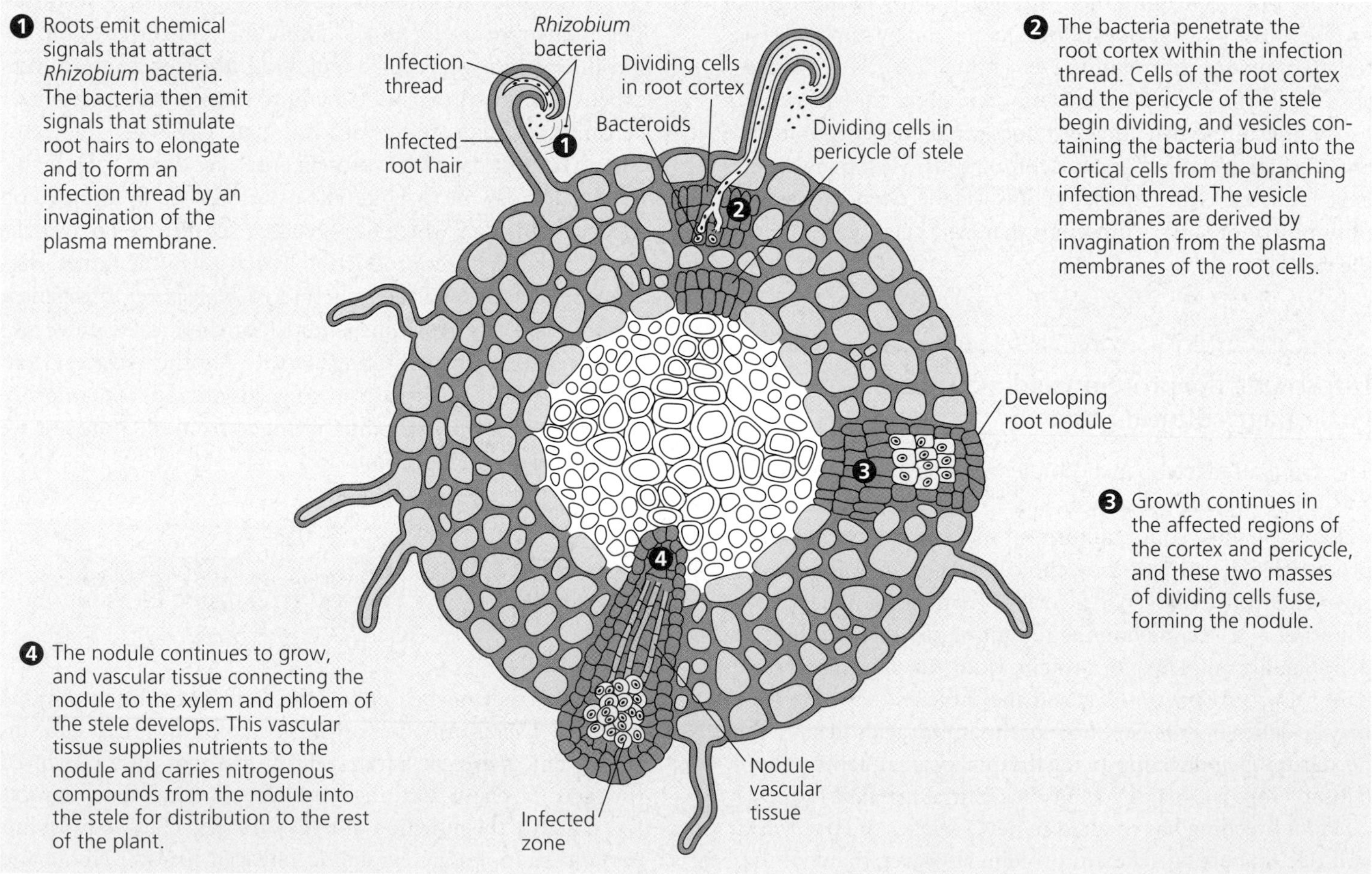

❶ Roots emit chemical signals that attract *Rhizobium* bacteria. The bacteria then emit signals that stimulate root hairs to elongate and to form an infection thread by an invagination of the plasma membrane.

❷ The bacteria penetrate the root cortex within the infection thread. Cells of the root cortex and the pericycle of the stele begin dividing, and vesicles containing the bacteria bud into the cortical cells from the branching infection thread. The vesicle membranes are derived by invagination from the plasma membranes of the root cells.

❸ Growth continues in the affected regions of the cortex and pericycle, and these two masses of dividing cells fuse, forming the nodule.

❹ The nodule continues to grow, and vascular tissue connecting the nodule to the xylem and phloem of the stele develops. This vascular tissue supplies nutrients to the nodule and carries nitrogenous compounds from the nodule into the stele for distribution to the rest of the plant.

Rhizobium bacteria

Infection thread

Dividing cells in root cortex

Bacteroids

Dividing cells in pericycle of stele

Infected root hair

Developing root nodule

Infected zone

Nodule vascular tissue

FIGURE 37.11 Development of a soybean root nodule.

The symbiotic relationship between a legume and nitrogen-fixing bacteria is mutualistic, with both partners benefiting. The bacteria supply the legume with fixed nitrogen, and the plant provides the bacteria with carbohydrates and other organic compounds. Most of the ammonium produced by symbiotic nitrogen fixation is used by the nodules to make amino acids, which are then transported to the shoot and leaves via the xylem.

Some root nodules have a reddish color owing to a molecule called leghemoglobin. Leghemoglobin is an iron-containing protein that, like the hemoglobin of human red blood cells, binds reversibly to oxygen (*leg*- is for legume). The root nodule's leghemoglobin acts as an oxygen "buffer," regulating the supply of oxygen for the intense respiration required by the bacteria to produce ATP for nitrogen fixation.

Symbiotic Nitrogen Fixation and Agriculture

Now that you have learned about symbiotic nitrogen fixation, you can understand the agricultural practice of crop rotation. One year a nonlegume such as corn is planted, and the following year alfalfa or some other legume is planted to restore the concentration of fixed nitrogen in the soil. Instead of being harvested, the legume crop is often plowed under so that it will decompose as "green manure" (FIGURE 37.12). To ensure that the legume encounters its specific *Rhizobium*, the seeds are soaked in a culture of the bacteria or dusted with bacterial spores before sowing.

Many plant families besides legumes include species that benefit from symbiotic nitrogen fixation. For example, alders and certain tropical grasses host nitrogen-fixing bacteria of the actinomycete group (see Chapter 27). Rice, a crop of great commercial importance, benefits indirectly from symbiotic nitrogen fixation. Rice farmers culture a water fern called *Azolla* in their

FIGURE 37.12 Crop rotation and "green manure." The "green manure" being mulched into the soil of this Washington State farm is sweet clover, a legume with root nodules containing nitrogen-fixing bacteria. Every third year, clover is planted and the crop is plowed under. This improves the physical structure and nitrogen content of the soil for growing wheat and other crops during the other two years of the crop rotation cycle. Crop rotation, especially when the legume is mulched instead of harvested, reduces the need for manufactured fertilizers.

paddies. The fern has symbiotic cyanobacteria that fix nitrogen and increase the fertility of the rice paddy. The growing rice eventually shades and kills the *Azolla*, and decomposition of this organic material adds more nitrogenous minerals to the paddy.

The Molecular Biology of Root Nodule Formation in Legumes

How does a legume species recognize a certain species of *Rhizobium* among the many bacterial species inhabiting a root's soil environment? And how does an encounter with that specific *Rhizobium* lead to the development of a nodule? These two questions have led researchers to a chemical dialogue between the bacteria and the root, with each partner responding to the chemical signals from the other by expressing certain genes whose products contribute to nodule formation (FIGURE 37.13). The plant initiates the communication when its

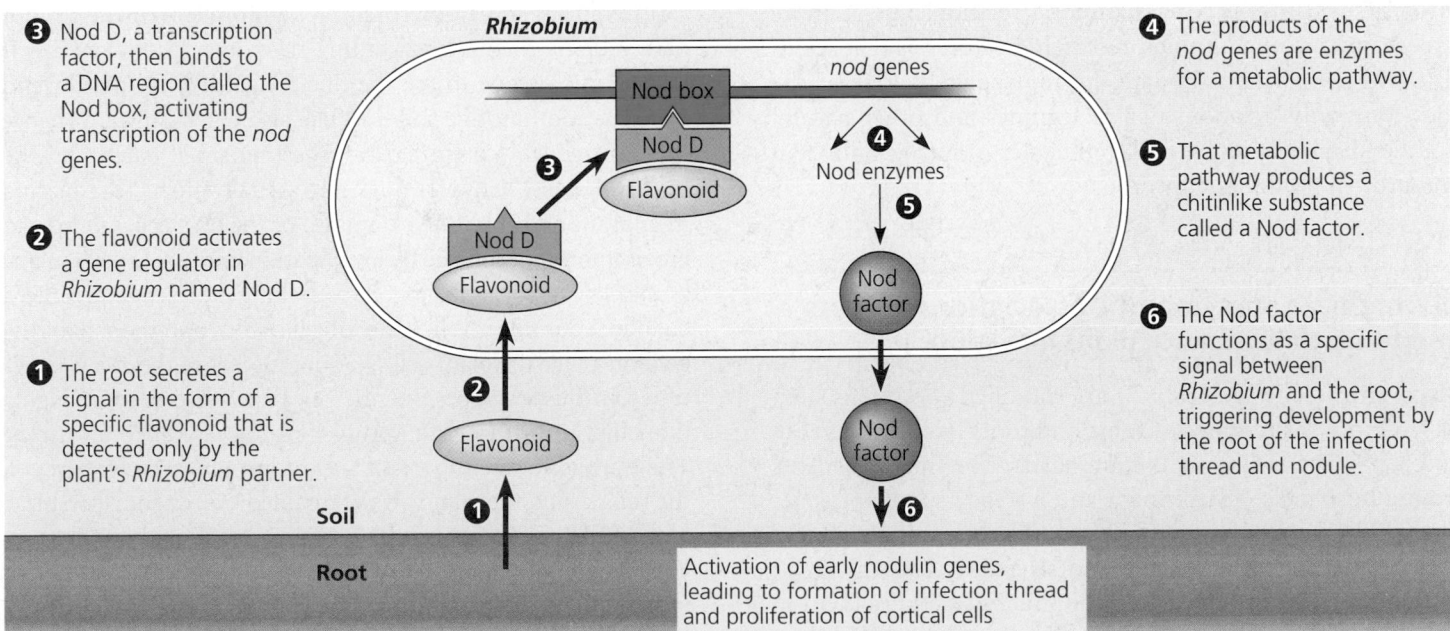

❸ Nod D, a transcription factor, then binds to a DNA region called the Nod box, activating transcription of the *nod* genes.

❷ The flavonoid activates a gene regulator in *Rhizobium* named Nod D.

❶ The root secretes a signal in the form of a specific flavonoid that is detected only by the plant's *Rhizobium* partner.

❹ The products of the *nod* genes are enzymes for a metabolic pathway.

❺ That metabolic pathway produces a chitinlike substance called a Nod factor.

❻ The Nod factor functions as a specific signal between *Rhizobium* and the root, triggering development by the root of the infection thread and nodule.

Rhizobium

Nod box   *nod* genes

Nod D

Flavonoid

Nod enzymes

Nod D

Flavonoid

Nod factor

Soil

Flavonoid

Root

Nod factor

Activation of early nodulin genes, leading to formation of infection thread and proliferation of cortical cells

FIGURE 37.13 Molecular biology of root nodule formation.

roots secrete molecules called flavonoids, which enter *Rhizobium* cells living in the vicinity of the roots. The specificity of this signal arises from variations in flavonoid structure, with a particular legume species secreting a type of flavonoid that only a certain *Rhizobium* species will detect and absorb. The plant's signal triggers the production of an answering molecule by the bacterium. Specifically, the plant's signal molecule activates a gene-regulating protein, which switches on a cluster of bacterial genes called *nod,* for "nodulation" genes. The products of these genes are enzymes that catalyze production of species-specific molecules called Nod factors. Secreted by the bacterial cells, the Nod factors return the "hello" and signal the root to form the infection thread the *Rhizobium* will enter and to begin forming the root nodule (see FIGURE 37.11). The plant's responses require activation of genes called early nodulin genes, probably by gene-regulating signal-transduction pathways, such as those you learned about in Chapter 11.

THE PROCESS OF SCIENCE Researchers have analyzed the molecular structure of the Nod factors for clues to these bacterial molecules' ability to influence genes within the cells of a plant. An early discovery, puzzling at the time, was that Nod factors are very similar to chitins, the main substances in the cell walls of fungi and the exoskeletons of arthropods (see Chapter 5). But now there is evidence that plants themselves produce chitin-like substances that probably function as growth regulators. This discovery led to the hypothesis that Nod factors mimic certain plant growth regulators in stimulating the roots to grow new organs—nodules, in this case. Researchers have also learned that the plant genes that must be expressed for nodules to form are genes that also function in many other developmental processes in plants. This suggests that researchers might someday learn to induce *Rhizobium* uptake and nodule formation in crop plants that do not normally form such nitrogen-fixing symbiotic relationships. A more likely outcome of continued research, however, is that scientists will learn to manipulate the molecular biology of the root-*Rhizobium* relationships of legumes and other natural nodule formers to improve the efficiency of nitrogen fixation and protein production by crops.

Mycorrhizae are symbiotic associations of roots and fungi that enhance plant nutrition

Mycorrhizae ("fungus roots") are modified roots consisting of symbiotic associations of fungi and roots (see FIGURES 31.18 and 36.8). The symbiosis is mutualistic. The fungus benefits from a hospitable environment and a steady supply of sugar donated by the host plant. In return, the fungus increases the surface area for water uptake and selectively absorbs phosphate and other minerals from the soil and supplies them to the plant. The fungi of mycorrhizae also secrete growth factors that stimulate roots to grow and branch. And the fungi produce antibiotics that may help protect the plant from pathogenic bacteria and pathogenic fungi in the soil.

Mycorrhizae are not oddities; they are formed by almost all plant species. In fact, this plant-fungus symbiosis might have been one of the evolutionary adaptations that made it possible for plants to colonize land in the first place; fossilized roots from some of the earliest plants include mycorrhizae (see Chapter 31). When terrestrial ecosystems were young, the soil was probably not very rich in nutrients. The fungi of mycorrhizae, which are more efficient at absorbing minerals than the roots themselves, would have helped nourish the pioneering plants. Even today, the plants that first become established on nutrient-poor soils, such as abandoned farmland or eroded hillsides, are usually well endowed with mycorrhizae.

The Two Main Types of Mycorrhizae

The modified roots formed from the symbiosis of fungi and plants take two major forms: ectomycorrhizae and endomycorrhizae. In **ectomycorrhizae,** the mycelium (mass of branching hyphae; see Chapter 31) forms a dense sheath, or mantle, over the surface of the root (FIGURE 37.14a). Hyphae extend from the mantle into the soil, greatly increasing the surface area for water and mineral absorption. Fungal hyphae also grow into the cortex of the root. These hyphae do not penetrate the root cells but form a network in the extracellular spaces that facilitates nutrient exchange between the fungus and the plant. Compared with "uninfected" roots, ectomycorrhizae are generally thicker, shorter, and more branched. Ectomycorrhizae do not form root hairs, which would be superfluous given the extensive surface area of the fungal mycelium. Ectomycorrhizae are especially common in woody plants, including trees of the pine, spruce, oak, walnut, birch, willow, and eucalyptus families.

In contrast to ectomycorrhizae, **endomycorrhizae** do not have a dense mantle ensheathing the root (FIGURE 37.14b). It takes a microscope to see the fine fungal hyphae that extend from the root into the soil. Hyphae also extend inward (hence the term *endo*mycorrhizae) by digesting small patches of the root cell walls. A hypha does not actually pierce the plasma membrane and enter the cytoplasm of the host cell, but instead grows into a tube formed by invagination of the root cell's membrane. The action is analogous to poking a finger gently into a balloon; your finger is like the fungal hypha, and the skin of the balloon is like the plant cell's membrane. Once they have penetrated in this way, some of the fungal hyphae become highly branched to form dense knotlike structures called arbuscles. These arbuscles are important sites of nutrient transfer between the fungus and the plant. To the unaided eye, endomycorrhizae look like "normal" roots with root hairs, but a microscope reveals a symbiotic relationship of enormous importance to plant nutrition. Endomycorrhizae, much more common than ectomycorrhizae, are found in over 90% of plant species, including important crop plants such as corn, wheat, and legumes.

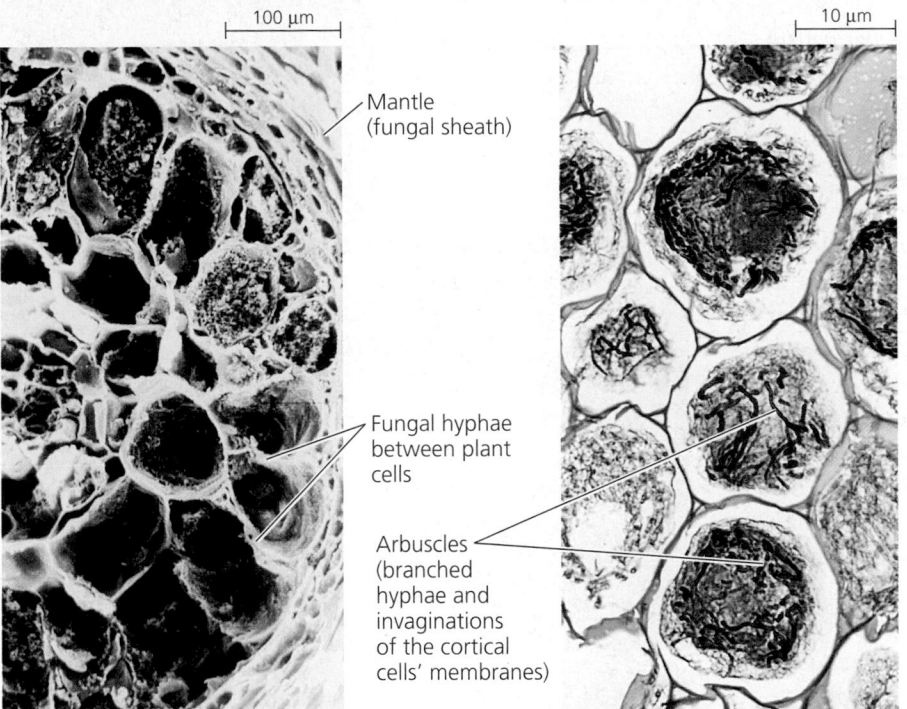

(a) Ectomycorrhizae. The mantle of the fungal mycelium ensheathes this aspen root. Fungal hyphae extend from the mantle into the soil, absorbing water and minerals, especially phosphate. Hyphae also extend into the extracellular spaces of the root cortex, providing extensive surface area for nutrient exchange between the fungus and its host plant (colorized SEM).

(b) Endomycorrhizae. No mantle forms around the root, although microscopic hyphae of the fungus do extend into the soil. Within the root cortex the fungus makes extensive contact with the plant through branching of hyphae to form the knotlike arbuscles. The arbuscles, visible in these sectioned cortical cells, provide an enormous surface area for nutrient swapping between the symbionts (LM).

FIGURE 37.14 Mycorrhizae.

Agricultural Importance of Mycorrhizae

Roots can be transformed into mycorrhizae only if they are exposed to the appropriate species of fungus. In most natural ecosystems, these fungi are present in soil, and seedlings develop mycorrhizae. But if seeds are collected in one environment and planted in foreign soil, the plants may show signs of malnutrition resulting from the absence of the plants' mycorrhizal partners. Researchers observe similar results in experiments in which soil fungi are poisoned. Farmers and foresters are already applying the lessons of such research. For example, inoculating pine seeds with spores of mycorrhizal fungi promotes the formation of mycorrhizae by the seedlings. Pine seedlings so infected grow more vigorously than trees without the fungal association.

Mycorrhizae and root nodules may have an evolutionary relationship

There is growing evidence that the molecular biology of root nodule formation is closely related to mechanisms that evolved first in mycorrhizae. In 1997, researchers reported that the nodulin genes activated in the plant during the early stages of root nodule formation are the very same genes activated during the early development of endomycorrhizae. In fact, mutations in these early nodulin genes block development of both root nodules and mycorrhizae in legumes that form both structures. In addition, the signal-transduction pathways that relay messages from the microorganisms to the plant's gene-regulating equipment involved in the development of both root nodules and mycorrhizae may share at least some components. For example, experimental application of plant hormones called cytokinins to root cells of legumes activates expression of the early nodulin genes even in the absence of the bacterial or fungal symbionts. (You will learn more about cytokinins and other plant hormones in Chapter 39.) "Infection" by either the bacteria or fungi causes the concentration of cytokinins in root tissue to increase naturally. These experiments suggest that the hormone is one of the links between the "I'm here" announcement of the microbes and the changes in gene expression in the plant that lead to structural modification of the roots. Even the chemical cues from the two kinds of microbes may be similar. Recall that the Nod factors secreted by *Rhizobium* bacteria are related to chitins, the same compounds that make up the cell walls of fungi. A reasonable hypothesis is that root cells have a family of closely related receptors that detect their particular bacterial and fungal symbionts.

Mycorrhizae, as you have learned, evolved very early, probably over 400 million years ago in the earliest vascular plants. In contrast, the root nodules of legumes originated only 65–150 million years ago, during the early evolution of angiosperms. The recent experiments revealing common molecular mechanisms in roots' two major symbiotic relationships suggest that root nodule development is at least partly adapted from a signaling pathway that was already in place in mycorrhizae. This is one more example of how evolution can co-opt existing equipment for new functions.

NUTRITIONAL ADAPTATIONS: PARASITISM AND PREDATION BY PLANTS

Symbiotic nitrogen fixation and mycorrhizae underscore the relationship between plants and their environment, which includes the other organisms that interact with plants. We conclude this chapter by exploring predation and parasitism as two other types of plant adaptations that enhance nutrition through interactions with other organisms.

Parasitic plants extract nutrients from other plants

The mistletoe we find tacked above doorways during the holiday season lives in nature as a parasite on oaks and other trees. Mistletoe is photosynthetic, but it supplements its nutrition by using projections called haustoria to siphon xylem sap from the vascular tissue of the host tree. Some parasitic plants, such as dodder (FIGURE 37.15a), do not perform photosynthesis at all, drawing all their nutrients from other plants by tapping into the host's vascular tissue. In another version of parasitism, Indian pipe obtains its nutrients from trees indirectly via its association with fungal hyphae of the host tree's mycorrhizae (FIGURE 37.15b).

Plants called epiphytes (from the Greek *epi*, upon, and *phyton*, plant) are sometimes mistaken for parasites. An epiphyte is an autotrophic plant that nourishes itself but grows on the surface of another plant, usually on the branches or trunks of trees. An epiphyte is anchored to its living substratum, but it absorbs water and minerals mostly from rain that falls on its

FIGURE 37.16 Carnivorous plants. (a) The Venus flytrap is a modified leaf with two lobes that close together rapidly enough to capture an insect. Prey that enters the trap touches sensory hairs, initiating an electrical impulse that triggers closure of the trap. Glands in the trap then secrete digestive enzymes, and nutrients are later absorbed by the modified leaf. In spite of its name, the flytrap catches more ants and grasshoppers than it does flies. **(b)** Pitcher plants use a pitfall to capture insects. Insects slip into a long water-filled funnel. After the insect drowns, it is digested by enzymes secreted into the water.

leaves. Examples of epiphytes are staghorn ferns, some mosses, Spanish moss (actually an angiosperm), and many species of bromeliads and orchids.

Dodder

Haustorium

Host's phloem

0.5 mm

(a) Dodder growing on a California pickleweed. The inset micrograph, a transverse section of a host stem supporting dodder (the orange "strings"), shows a haustorium (modified root) of the parasite tapping the host plant's vascular tissue for water and nutrients (LM).

FIGURE 37.15 Parasitic plants.

(b) Indian pipe. Indian pipe is nutritionally bridged to its host tree by fungal hyphae extending from the host's mycorrhizae.

Carnivorous plants supplement their mineral nutrition by digesting animals

Living in acid bogs and other habitats where soil conditions are poor (especially in nitrogen) are plants that fortify themselves by occasionally feeding on animals. These carnivorous plants make their own carbohydrates by photosynthesis, but they obtain some of their nitrogen and minerals by killing and digesting insects and other small animals. Various kinds of insect traps have evolved by the modification of leaves (FIGURE 37.16). The traps are usually equipped with glands that secrete digestive juices. Fortunately for the animal kingdom, this ironic turnabout is a relatively rare exception to the standard ecosystem dynamics of animals eating plants!

CHAPTER 37 REVIEW

Go to the Campbell Biology website (www.campbellbiology.com) to explore an interactive version of the Chapter Review.

Summary of Key Concepts

NUTRITIONAL REQUIREMENTS OF PLANTS

- The chemical composition of plants provides clues to their nutritional requirements (pp. 767–768, FIGURE 37.1) Plants derive most of their organic mass from the CO_2 of air, but they also depend on soil nutrients in the form of water and minerals.

- Plants require nine macronutrients and at least eight micronutrients (pp. 768–769, TABLE 37.1, FIGURE 37.2) Macronutrients include carbon, hydrogen, oxygen, nitrogen, and other major ingredients of organic compounds. Many micronutrients have catalytic functions as cofactors of enzymes.

- The symptoms of a mineral deficiency depend on the function and mobility of the element (pp. 769–770, FIGURE 37.3) Deficiency of a mobile nutrient usually affects older organs more than younger ones; the reverse is true for nutrients that are less mobile within a plant.

THE ROLE OF SOIL IN PLANT NUTRITION

- Soil characteristics are key environmental factors in terrestrial ecosystems (pp. 770–772, FIGURES 37.5, 37.6) Various sizes of particles derived from the breakdown of rock are found in soil, along with organic material (humus) in various stages of decomposition. Acids derived from roots contribute to a plant's uptake of minerals when H^+ displaces mineral cations from clay particles.
  Web/CD Activity 37A: *How Plants Obtain Minerals from Soil*
  Web/CD Case Study in the Process of Science: *How Does Acid Precipitation Affect Mineral Deficiency?*

- Soil conservation is one step toward sustainable agriculture (pp. 772–774, FIGURES 37.7, 37.8) In contrast to natural ecosystems, agriculture depletes the mineral content of soil, taxes water reserves, and encourages erosion. The goal of soil conservation strategies is to minimize this damage.

THE SPECIAL CASE OF NITROGEN AS A PLANT NUTRIENT

- The metabolism of soil bacteria makes nitrogen available to plants (pp. 774–775, FIGURE 37.9) Nitrogen-fixing bacteria convert atmospheric N_2 to nitrogenous minerals that plants can absorb as a nitrogen source for organic synthesis.
  Web/CD Activity 37B: *The Nitrogen Cycle*

- Improving the protein yield of crops is a major goal of agricultural research (p. 775) Such research addresses the most widespread form of human malnutrition: protein deficiency.

NUTRITIONAL ADAPTATIONS: SYMBIOSIS OF PLANTS AND SOIL MICROBES

- Symbiotic nitrogen fixation results from intricate interactions between roots and bacteria (pp. 776–778, FIGURES 37.10–37.13) The development of nitrogen-fixing root nodules depends on chemical cross-talk between *Rhizobium* bacteria and root cells of their specific plant hosts. The bacteria of a nodule obtain sugar from the plant and supply the plant with fixed nitrogen.

- Mycorrhizae are symbiotic associations of roots and fungi that enhance plant nutrition (pp. 778–779, FIGURE 37.14) The fungal hyphae of both ectomycorrhizae and endomycorrhizae absorb water and minerals, which they supply to their plant hosts.

- Mycorrhizae and root nodules may have an evolutionary relationship (p. 779) There is evidence that root nodule development depends on molecular mechanisms of signaling and root cell responses that evolved first in mycorrhizae.

NUTRITIONAL ADAPTATIONS: PARASITISM AND PREDATION BY PLANTS

- Parasitic plants extract nutrients from other plants (p. 780, FIGURE 37.15) They do so either directly by tapping into the host's vascular tissue or indirectly via mycorrhizae.

- Carnivorous plants supplement their mineral nutrition by digesting animals (p. 780, FIGURE 37.16) This predation is most common in ecosystems with nutrient-poor soil.

Self-Quiz

1. Most of the mass of organic material of a plant comes from
   a. water.
   b. carbon dioxide.
   c. soil minerals.
   d. atmospheric oxygen.
   e. nitrogen.

2. Micronutrients are needed in very small amounts because
   a. most of them are mobile in the plant.
   b. most function as cofactors of enzymes.
   c. most are supplied in large enough quantities in seeds.
   d. they play only a minor role in the health of the plant.
   e. only the growing regions of the plants require them.

3. Two groups of tomatoes were grown under laboratory conditions, one with humus added to the soil and one a control without the humus. The leaves of the plants grown without humus were yellowish (less green) than those of the plants growing in humus-enriched soil. The best explanation for this difference is that
   a. the healthy plants used the food in the decomposing leaves of the humus for energy to make chlorophyll.
   b. the humus made the soil more loosely packed, so the plants' roots grew with less resistance.
   c. the humus contained minerals such as magnesium and iron, needed for the synthesis of chlorophyll.
   d. the heat released by the decomposing leaves of the humus caused more rapid growth and chlorophyll synthesis.
   e. the plants absorbed chlorophyll from the humus.

4. We would expect the greatest difference in size and general appearance between two groups of plants of the same species, one group with mycorrhizae and one without, in an environment
   a. where nitrogen-fixing bacteria are abundant.
   b. that has soil with poor drainage.
   c. that has hot summers and cold winters.
   d. in which the soil is relatively deficient in mineral nutrients.
   e. that is near a body of water, such as a pond or river.

5. A mineral deficiency is likely to affect older leaves more than younger leaves if
   a. the mineral is a micronutrient.
   b. the mineral is very mobile within the plant.
   c. the mineral is required for chlorophyll synthesis.
   d. the deficiency persists for a long time.
   e. the older leaves are in direct sunlight.

6. Carnivorous adaptations of plants mainly compensate for soil that has a relatively low content of
   a. potassium.
   b. nitrogen.
   c. calcium.
   d. water.
   e. phosphate.

7. Based on our retrospective view, the most reasonable conclusion to draw from van Helmont's famous experiment on the growth of a willow tree is that
   a. the tree increased in mass mainly by producing its own matter.
   b. the increase in the mass of the tree could not be accounted for by the consumption of soil.
   c. most of the increase in the mass of the tree was due to the uptake of O_2.
   d. soil simply provides physical support for the tree without providing nutrients.
   e. trees do not require water to grow.

8. It is valid to consider water a plant nutrient because
   a. plants die without a water source.
   b. cell elongation depends mainly on the osmotic absorption of water by cells.
   c. hydrogen atoms from water molecules are incorporated into organic molecules.
   d. transpiration depends on a continuous supply of water to leaves.
   e. most of a plant's mass of organic compounds is derived from water.

9. The specific relationship between a legume and its symbiotic *Rhizobium* species probably depends on
   a. each legume having a specific set of early nodulin genes.
   b. each *Rhizobium* species having a form of nitrogenase that only works in the appropriate legume host.
   c. each legume being found where the soil has only the *Rhizobium* specific to that legume.
   d. specific recognition between the chemical signals and signal receptors of the *Rhizobium* and legume species.
   e. destruction of all incompatible *Rhizobium* species by enzymes secreted from the legume's roots.

10. Mycorrhizae enhance plant nutrition mainly by
    a. absorbing water and minerals through the fungal hyphae.
    b. providing sugar to the root cells, which have no chloroplasts of their own.
    c. converting atmospheric nitrogen to ammonia.
    d. enabling the roots to parasitize neighboring plants.
    e. stimulating the development of root hairs.

11. Imagine conducting experiments like the one diagrammed in FIGURE 37.2 to test whether a particular chemical element is required as a micronutrient by a certain plant species. Why might even the slightest contamination of glassware by dirt lead you to the erroneous conclusion that the element is not required by the plant?

12. How do roots actively increase the availability of mineral nutrients that are cations?

13. Why does irrigating by repeatedly flooding fields often lead to the soil becoming too salty?

14. Why do organic fertilizers generally contaminate water resources less than industrially produced inorganic fertilizers?

15. How do the nitrogen-fixing bacteria of root nodules benefit from their symbiotic relationship with plants?

16. Why are the missions to understand nitrogen metabolism of crop plants and to apply those lessons in agriculture so relevant to human health on a global scale?

Go to the website or CD-ROM for more quiz questions.

Evolution Connection

Imagine taking the plant out of the picture in FIGURE 37.9. Write a paragraph explaining how the soil bacteria could sustain the recycling of nitrogen *before* land plants evolved.

Test the effects of acid precipitation on mineral deficiency in the Case Study in the Process of Science, available on the website and CD-ROM.

The Process of Science

Acid rain contains an abnormally high concentration of hydrogen ions (H^+). One effect of acid rain is to deplete the soil of plant nutrients such as calcium (Ca^{2+}), potassium (K^+), and magnesium (Mg^{2+}). Suggest a hypothesis to explain why acid rain washes these nutrients from the soil. How might you test your hypothesis?

Science, Technology, and Society

About 10% of U.S. cropland is irrigated. Agriculture is by far the biggest user of water in arid western states, including Colorado, Arizona, and California. The populations of these states are growing, and there is an ongoing conflict between cities and farm regions over water. To ensure water supplies for urban growth, cities are purchasing water rights from farmers. This is often the least expensive way for a city to obtain more water, and it is possible for some farmers to make more money selling water rights than growing crops. Discuss the possible consequences of this trend. Is this the best way to allocate water for all concerned? Why or why not?

PLANT REPRODUCTION AND BIOTECHNOLOGY

SEXUAL REPRODUCTION

- Sporophyte and gametophyte generations alternate in the life cycles of plants: *a review*
- Flowers are specialized shoots bearing the reproductive organs of the angiosperm sporophyte
- Male and female gametophytes develop within anthers and ovaries, respectively: Pollination brings them together
- Plants have various mechanisms that prevent self-fertilization
- Double fertilization gives rise to the zygote and endosperm
- The ovule develops into a seed containing an embryo and a supply of nutrients
- The ovary develops into a fruit adapted for seed dispersal
- Evolutionary adaptations of seed germination contribute to seedling survival

ASEXUAL REPRODUCTION

- Many plants clone themselves by asexual reproduction
- Sexual and asexual reproduction are complementary in the life histories of many plants
- Vegetative propagation of plants is common in agriculture

PLANT BIOTECHNOLOGY

- Neolithic humans created new plant varieties by artificial selection
- Biotechnology is transforming agriculture
- Plant biotechnology has incited much public debate

I t has been said *that an oak is an acorn's way of making more acorns. Indeed, in a Darwinian view of life, the fitness of an organism is measured only by its ability to replace itself with healthy, fertile offspring. Consider the century plant (Agave) in the photograph on this page. It lives for decades without flowering, and then one spring it grows a floral stalk as tall as a telephone pole. That season the plant produces seeds and then withers and dies, its food reserves, minerals, and water spent in the formation of its massive bloom. Although not all flowering plants are as completely consumed as the century plant in leaving offspring, most of their other* functions—*for example, photosynthesis and growth—can be interpreted, in the broadest Darwinian sense, as mechanisms contributing to propagation. Sexual reproduction, as in the case of oak and Agave, however, is not the sole means by which flowering plants reproduce: Many species can also reproduce asexually, creating offspring that are genetically identical to themselves.*

In Chapters 29 and 30, we approached plant reproduction from an evolutionary perspective, tracing the descent of angiosperms and other land plants from their aquatic ancestors, charophycean algae. In this chapter, we explore the reproductive biology of flowering plants (angiosperms) in much greater detail because they are the most important group of plants in most terrestrial ecosystems and in agriculture. In addition to examining both sexual and asexual reproduction of angiosperms, we will place plant biotechnology in the broader context of human intervention in the propagation and genetics of the angiosperm varieties we have been cultivating for millennia.

SEXUAL REPRODUCTION

Sporophyte and gametophyte generations alternate in the life cycles of plants: *a review*

In Chapter 30 you learned about the life cycle of a flowering plant from an evolutionary perspective. The life cycles of angiosperms and other plants are characterized by an **alternation of generations,** in which haploid (*n*) and diploid (2*n*) generations take turns producing each other (see FIGURE 29.6). The diploid plant, which is called the **sporophyte,** produces haploid spores by meiosis. These spores divide by mitosis, giving rise to multicellular male and female haploid plants—the

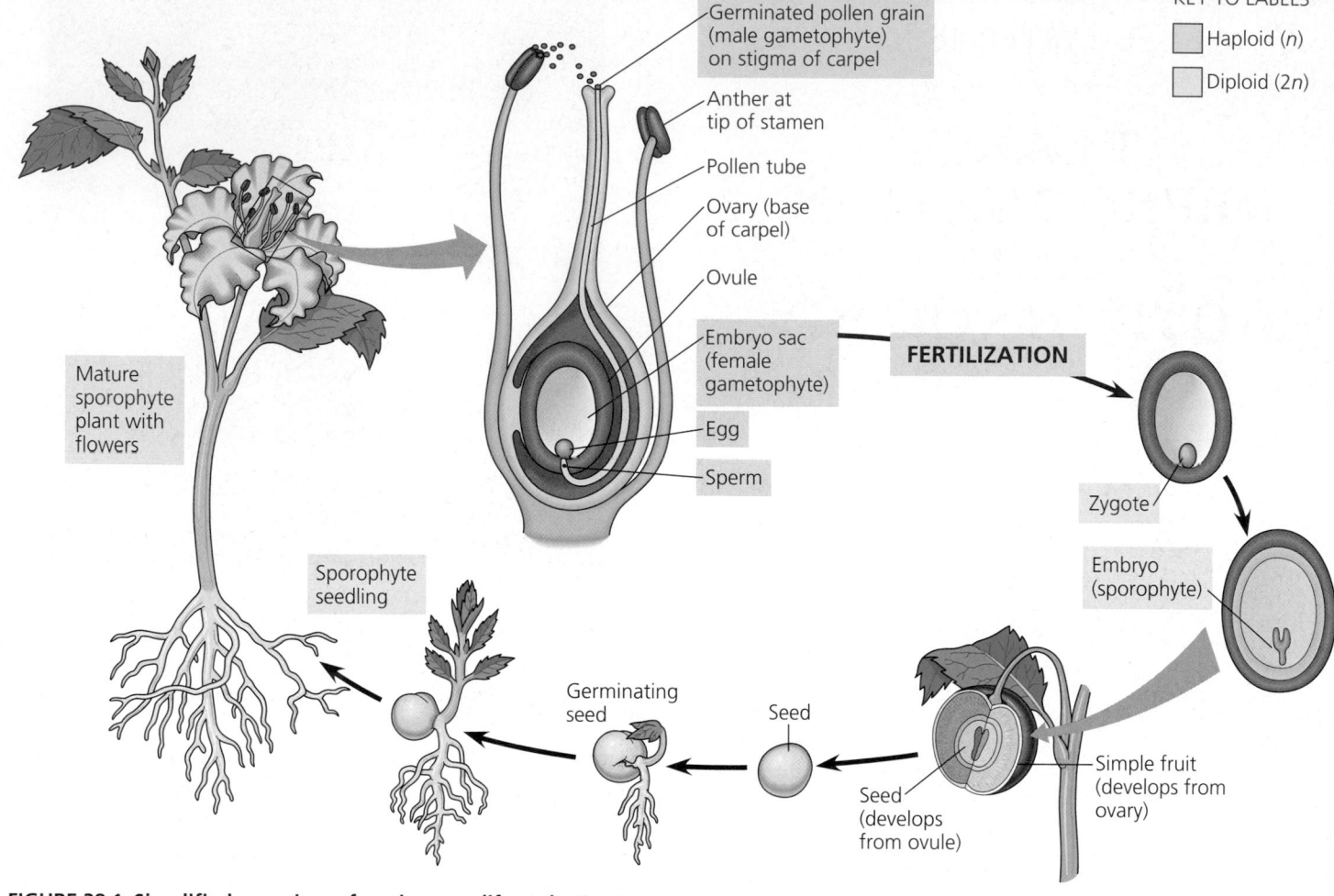

KEY TO LABELS

☐ Haploid (*n*)

☐ Diploid (2*n*)

Germinated pollen grain (male gametophyte) on stigma of carpel

Anther at tip of stamen

Pollen tube

Ovary (base of carpel)

Ovule

Embryo sac (female gametophyte)

Egg

Sperm

FERTILIZATION

Zygote

Embryo (sporophyte)

Mature sporophyte plant with flowers

Sporophyte seedling

Germinating seed

Seed

Seed (develops from ovule)

Simple fruit (develops from ovary)

FIGURE 38.1 Simplified overview of angiosperm life cycle. See FIGURE 30.17 for a more detailed version of the life cycle.

gametophytes. By mitosis and cellular differentiation, the gametophytes develop and produce gametes—sperm and eggs. Fertilization results in diploid zygotes, which divide by mitosis and form new sporophytes. FIGURE 38.1 follows the main stages of the angiosperm life cycle.

In angiosperms, the sporophyte is the dominant generation in the sense that it is the conspicuous plant we see. Over the course of seed plant evolution, gametophytes became reduced in size and dependent upon their sporophyte parents: Angiosperm gametophytes are the most reduced of all, consisting of only a few cells. The sporophyte also develops a reproductive structure unique to angiosperms—the flower. Male and female gametophytes develop within the anthers and ovaries, respectively, of a sporophyte flower (see FIGURE 38.1). Pollination by wind or animals brings a male gametophyte (pollen grain) to a female gametophyte; union of their gametes (fertilization) takes place within the ovary, as does the development of the seeds containing sporophyte embryos. The ovary itself becomes a fruit (another unique structure of the angiosperms).

Flowers are specialized shoots bearing the reproductive organs of the angiosperm sporophyte

Flowers, the reproductive shoots of the angiosperm sporophyte, are typically composed of four whorls of highly modified leaves called floral organs, which are separated by very short internodes. Unlike vegetative shoots, which grow indeterminately, flowers are determinate shoots, meaning that they cease growing once the flower and fruit are formed.

The four kinds of floral organs, in sequence from the outside to the inside of the flower, are the **sepals, petals, stamens,** and **carpels.** Their site of attachment to the stem is the **receptacle** (FIGURE 38.2). Stamens and carpels are the male and female reproductive organs, respectively, whereas sepals and petals are nonreproductive organs. Sepals, which enclose and protect the floral bud before it opens, are usually green and more leaflike in appearance than the other floral organs. In many angiosperms, the petals are more brightly colored than sepals and advertise the flower to insects and other pollinators.

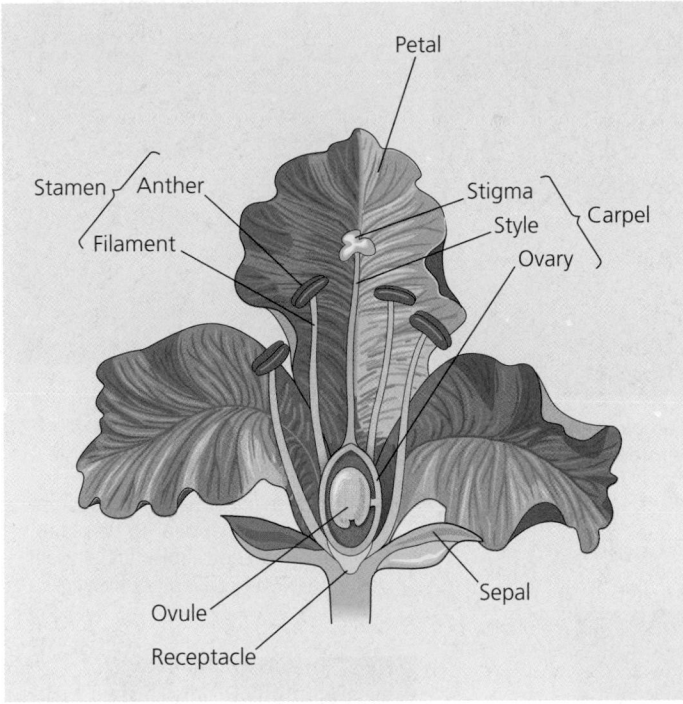

FIGURE 38.2 Review of an idealized flower.

A stamen consists of a stalk called the filament and a terminal structure called the **anther;** within the anther are chambers called pollen sacs, in which pollen is produced. A carpel has an **ovary** at its base and a slender neck called the style. At the top of the style is a sticky structure called the stigma that serves as a landing platform for pollen. Within the ovary are one or more **ovules** (the number depends on the species). The flower shown in FIGURE 38.2 has a single carpel, but the flowers of many species have multiple carpels. In some species, several carpels are fused into a single structure; the result is an ovary with two or more chambers, each containing one or more ovules.

The stamen and carpels of flowers contain the sporangia, the structures where first the spores and then the gametophytes develop. The male gametophytes are sperm-producing structures called **pollen grains,** which form within the pollen sacs of anthers. The female gametophytes are egg-producing structures called **embryo sacs,** which form within the ovules in ovaries.

For the egg within the embryo sac to be fertilized, the male and female gametophytes must meet and their gametes must unite. Pollination occurs when pollen released from anthers and carried by wind or animals lands on a stigma (not necessarily on the same flower or plant). Each pollen grain produces a structure called a pollen tube, which grows down into the ovary via the style and discharges sperm into the embryo sac,

resulting in fertilization of the egg (see FIGURE 38.1). The zygote gives rise to an embryo, and as the embryo grows, the ovule that contains it develops into a seed. The entire ovary, meanwhile, develops into a fruit containing one or more seeds, depending on the species. Fruits, carried by wind or by animals, help disperse seeds some distance from their source plants. If deposited in sufficiently moist soil, seeds germinate; that is, their embryos start growing into seedlings, a new generation of flowering sporophytes.

Numerous floral variations have evolved during the 130 million years of angiosperm history (FIGURE 38.3, p. 786). In certain flowers, one or more of the four basic floral organs—sepals, petals, stamens, and carpels—have been eliminated. Plant biologists distinguish between **complete flowers,** those having all four organs, and **incomplete flowers,** those lacking one or more of the four floral parts. For example, most grasses have incomplete flowers lacking petals.

The presence or absence of reproductive organs further distinguishes flowers. A **bisexual flower** (in older terminology termed a "perfect" flower) is equipped with both stamens and carpels. Complete flowers such as the *Trillium* shown in FIGURE 38.3a are always bisexual, because they have all four types of floral organs, but an incomplete flower that lacks sepals or petals may also be bisexual.

A **unisexual flower** (or "imperfect" flower in older terminology) is missing either stamens or carpels. Unisexual flowers are called staminate or carpellate, depending on which set of reproductive organs is present. If staminate and carpellate flowers are located on the same individual plant, then that plant species is said to be **monoecious** (from a Greek word meaning "one house"). Maize and other cereal varieties are examples. The "ears" of a maize plant are derived from clusters of carpellate flowers, while the tassels consist of staminate flowers (FIGURE 38.3e). In contrast, a **dioecious** ("two houses") species, such as *Sagittaria*, has staminate flowers and carpellate flowers on separate plants (FIGURE 38.3f). Date palms are dioecious. Because dates develop only on the carpellate (female) palms, commercial date growers plant mostly carpellate individuals. A few staminate male plants provide enough pollen for hundreds of females.

In addition to these differences based on the presence of floral organs, flowers have many variations in size, shape, and color (see FIGURE 38.3). Much of this diversity represents adaptations of flowers to different animal pollinators. Indeed, the presence of animals in the environment was a key factor in angiosperm evolution (see Chapter 30).

The following concepts describe the development of angiosperm gametophytes, then trace how pollination leads to fertilization and the development of embryos, seeds, and fruits. Keep in mind, however, that there are many variations in the details of these processes, depending on the species of flowering plant.

(a) Trillium. This *Trillium* flower is complete, meaning that sepals, petals, stamens, and carpels are all present.

(b) Lupines. Lupines are examples of plants with inflorescences, clusters of flowers.

(c) Sunflowers. What appears in this sunflower to be a single flower is actually a collection of hundreds. The central disk of this composite inflorescence consists of tiny complete flowers. What appear to be petals ringing the central disk are actually sterile flowers called ray flowers.

(d) Hibiscus. The diverse shapes, colors, and odors of flowers also reflect adaptations to different modes of pollination. For example, *Hibiscus* pollinators include hummingbirds, which are attracted to the red color.

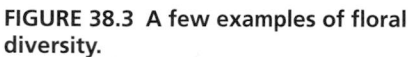

FIGURE 38.3 A few examples of floral diversity.

(e) Maize. Maize is a monoecious species with inflorescences of carpellate (female) and staminate (male) flowers on the same individual plant. An "ear" of maize is a collection of kernels (one-seeded fruits) that develops from an inflorescence of fertilized carpellate flowers. The maize "silk" is composed of numerous long styles (left). The staminate inflorescences are the tassels (right).

(f) Sagittaria. *Sagittaria* is dioecious, its staminate (left) and carpellate (right) flowers on separate plants.

Male and female gametophytes develop within anthers and ovaries, respectively: Pollination brings them together

Development of the Male Gametophyte (Pollen Grain)

Within the sporangia (pollen sacs) of an anther are numerous diploid cells called microsporocytes. Each microsporocyte undergoes meiosis, forming four haploid **microspores,** each of which can eventually give rise to a haploid male gametophyte (FIGURE 38.4a).

A microspore divides once by mitosis and produces two cells, a generative cell and a tube cell. The generative cell will eventually produce sperm. The tube cell, which encloses the generative cell, will produce the pollen tube, a structure essential for sperm delivery to the egg. The two-celled structure is encased in a thick, resistant wall that becomes sculptured into an elaborate pattern unique to the particular plant species (FIGURE 38.5). Together, the two cells and their wall constitute a pollen grain, which at this stage of its development is an immature male gametophyte.

A pollen grain becomes a mature male gametophyte when the generative cell divides by mitosis to form two sperm cells. In most species, this process occurs after the pollen grain lands on the stigma of a carpel and the pollen tube begins to form (see FIGURE 38.1).

Development of the Female Gametophyte (Embryo Sac)

Ovules, each containing a single sporangium, form within the chambers of the ovary. One cell in the sporangium of each ovule, the megasporocyte, grows and then goes through meiosis, producing four haploid **megaspores** (FIGURE 38.4b).

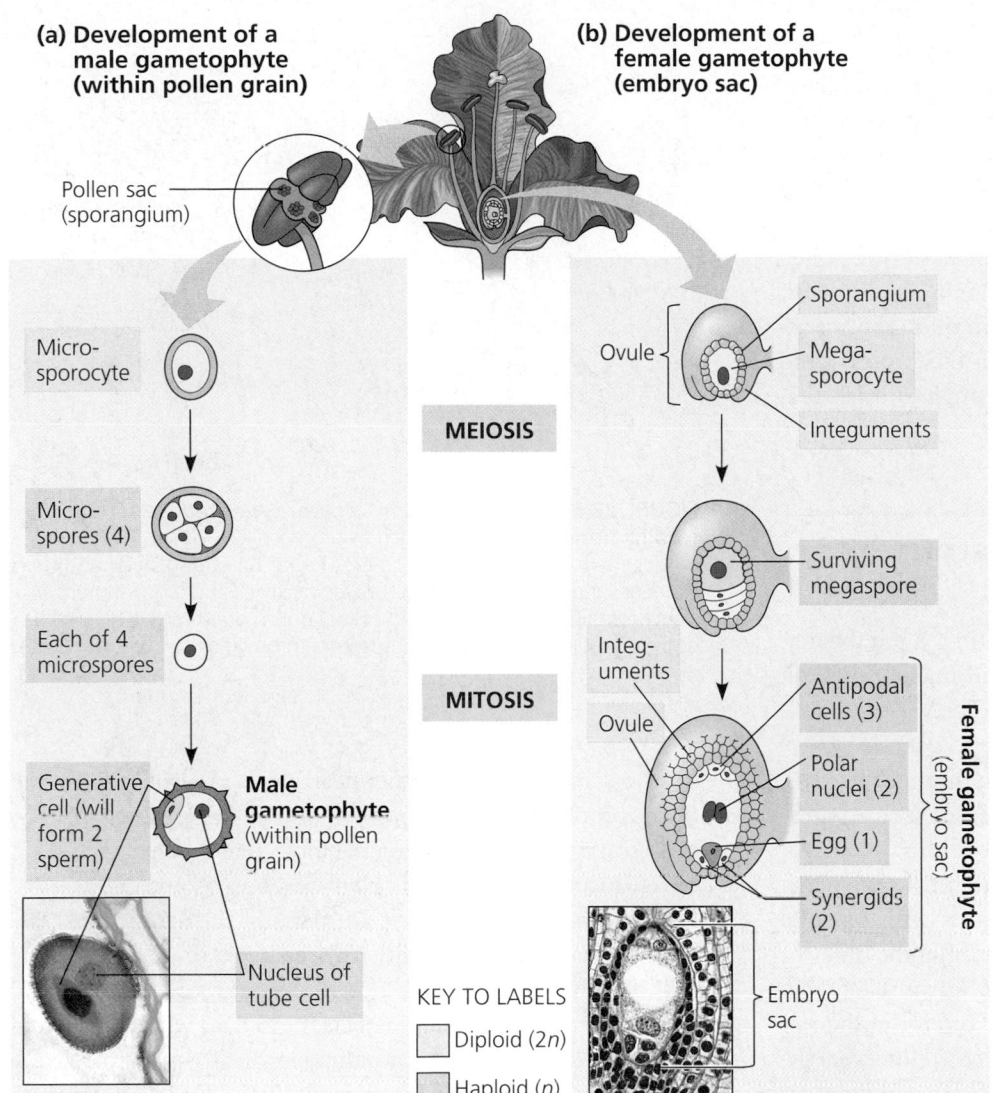

(a) Development of a male gametophyte (within pollen grain)

Pollen sac (sporangium)

Micro-sporocyte

MEIOSIS

Micro-spores (4)

Each of 4 microspores

MITOSIS

Generative cell (will form 2 sperm)

Male gametophyte (within pollen grain)

Nucleus of tube cell

KEY TO LABELS

Diploid (2n)

Haploid (n)

(b) Development of a female gametophyte (embryo sac)

Ovule

Sporangium

Mega-sporocyte

Integuments

Surviving megaspore

Integ-uments

Ovule

Antipodal cells (3)

Polar nuclei (2)

Egg (1)

Synergids (2)

Female gametophyte (embryo sac)

Embryo sac

FIGURE 38.4 The development of angiosperm gametophytes (pollen and embryo sacs). (a) Pollen grains (immature male gametophytes) develop within the sporangia (pollen sacs) of anthers at the tips of stamens. The pollen grain becomes a *mature* male gametophyte when its generative cell divides to form two sperm. This usually occurs after a pollen grain lands on the stigma of a carpel and the pollen tube begins to grow (see FIGURE 38.1). **(b)** The embryo sac (female gametophyte) develops within an ovule, itself enclosed by the ovary at the base of a carpel. Within the ovule's sporangium is a large, diploid cell called the megasporocyte. The megasporocyte in this species divides by meiosis and gives rise to four haploid cells, but only one of these survives as the megaspore. (This contrasts with pollen formation, in which all four products of meiosis go on to form gametophytes.) Three mitotic divisions of the megaspore form the embryo sac, a multicellular female gametophyte. The ovule now consists of the embryo sac along with the surrounding integuments (protective tissues).

The details of the next steps vary extensively, depending on the species. In many angiosperms (such as the one illustrated in FIGURE 38.4), only one of the megaspores survives. This megaspore continues to grow, and its nucleus divides by mitosis three times, resulting in one large cell with eight haploid nuclei. Membranes then partition this mass into a multicellular female gametophyte—the embryo sac. At one end of the embryo sac are three cells: the egg cell, or female gamete, and two cells called synergids. The synergids flank the egg cell and function in the attraction and guidance of the pollen tube. At the opposite end of the embryo sac are three antipodal cells of unknown function. The remaining two nuclei, called polar nuclei, are not partitioned into separate cells but share the cytoplasm of the large central cell of the embryo sac. The ovule, which will eventually become a seed, now consists of the embryo sac and the surrounding integuments (protective layers of sporophytic tissue).

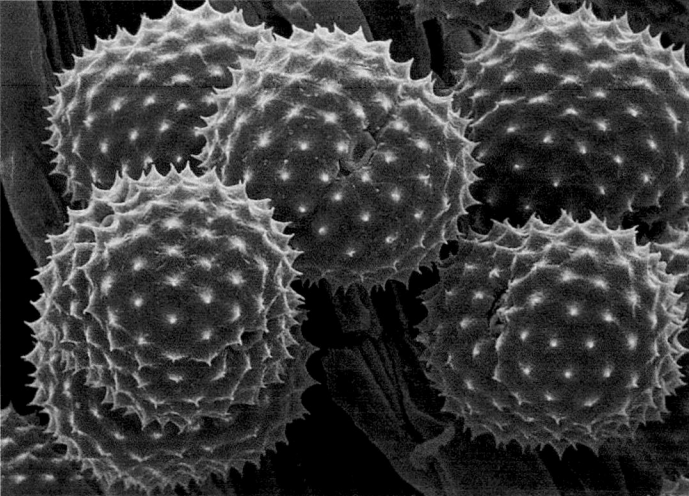

30 μm

FIGURE 38.5 Pollen grains have tough, ornate, and distinctive walls. The resilient wall of a pollen grain protects the male gameto-phyte. This is ragweed pollen (colorized SEM). Because of their tough walls, pollen grains are preserved well in the fossil record. The species of fossilized pollen found in an area can provide insights into the past climates of the area.

Pollination

Pollination, which brings the male and female gametophytes together, is the first step in the chain of events that leads to fertilization. This step is accomplished in various ways. Some angiosperms, including grasses and many trees, use wind as a pollinating agent. They compensate for the randomness of this dispersal by releasing enormous quantities of tiny pollen grains. At certain times of the year, the air is loaded with pollen, as anyone plagued with pollen allergies can attest. Many angiosperms, however, do not rely on the aimless wind to carry pollen but interact with insects or other animals that transfer pollen directly between flowers.

Plants have various mechanisms that prevent self-fertilization

Some flowers self-fertilize (this is called "selfing"), but the majority of angiosperms have mechanisms that make it difficult or impossible for a flower to fertilize itself. The various barriers that prevent self-fertilization contribute to genetic variety by ensuring that sperm and eggs come from different parents. (The evolutionary and ecological importance of sexual reproduction is discussed in Chapter 23.) Dioecious plants, of course, cannot self-fertilize because they are unisexual, being either staminate or carpellate. In some plants with bisexual flowers, the stamens and carpels mature at different times or are structurally arranged in such a way that it is unlikely that an animal pollinator could transfer pollen from the anthers to the stigma of the same flower (FIGURE 38.6). However, the most common anti-selfing mechanism in flowering plants is **self-incompatibility,** the ability of a plant to reject its own pollen and the pollen of closely related individuals. If a pollen grain from an anther happens to land on a stigma of a flower

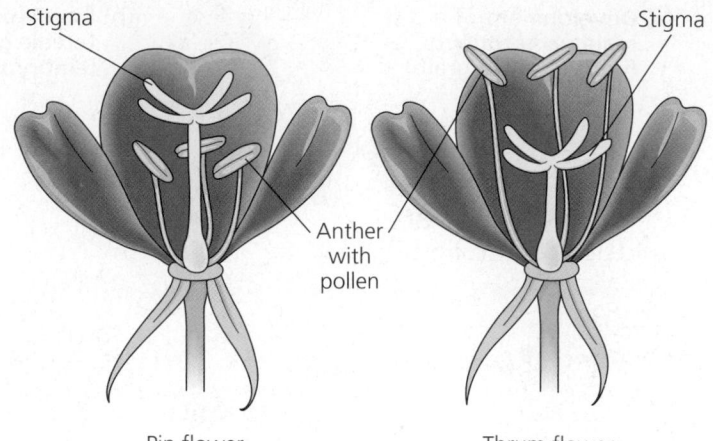

FIGURE 38.6 "Pin" and "thrum" flower types reduce self-fertilization. Some species, such as primrose (*Oenothera biennis*), produce two types of flowers: "pins," which have long styles and short stamens, and "thrums," which have short styles and long stamens. An insect foraging for nectar would collect pollen on different parts of its body; pin pollen would be deposited on thrum stigmas, and vice versa.

on the same plant, a biochemical block prevents the pollen from completing its development and fertilizing an egg.

Researchers are unraveling the molecular mechanisms of self-incompatibility. This plant response is analogous to the immune response of animals, in the sense that both are based on the ability of organisms to distinguish the cells of "self" from those of "nonself." The key difference is that the animal immune system rejects nonself, as when the system mounts a defense against a pathogen or attempts to reject a transplanted organ. Self-incompatibility in plants, in contrast, is a rejection of self.

Recognition of "self" pollen is based on genes for self-incompatibility, called S-genes (FIGURE 38.7). In a particular

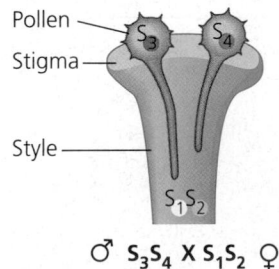

(a) In this cross, the pollen comes from a plant with an S_3S_4 genotype. (Segregation during meiosis results in half of the haploid pollen having the S_3 allele and the other half having the S_4 allele.) The pollen alleles do not match alleles of the stigma, and the pollen "germinates" (grows pollen tubes that can deliver sperm to ovules in the ovary at the base of the carpel).

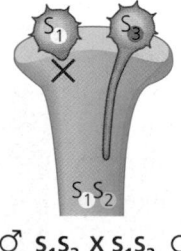

(b) In this cross, half of the pollen grains have an S_1 allele that matches an allele in the stigma, and those pollen grains fail to germinate.

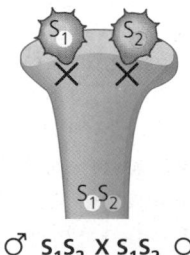

(c) In a cross between plants having the same S genotypes, none of the pollen can germinate.

FIGURE 38.7 Genetic basis of self-incompatibility. In the gene pool of a plant population, there can be dozens of alleles of the S-gene. If a pollen grain has an allele that matches an allele of the stigma upon which it lands, then a pollen tube fails to grow.

plant population, as many as 50 different alleles can occur at the S-locus. If a pollen grain and the carpel's stigma upon which it lands have matching alleles at the S-locus, then the pollen grain fails to initiate or complete formation of a pollen tube, and thus no fertilization occurs. The pollen grain is haploid, and it will be recognized as "self" if its one S-allele matches either of the two S-alleles of the diploid stigma.

Although self-incompatibility genes are all referred to as S-loci, such genes have actually evolved independently in various plant families. As a consequence, self-recognition blocks pollen tube growth by different molecular mechanisms. In some cases, the block occurs in the pollen grain itself; this is called gametophytic self-incompatibility. For example, in some members of the tobacco, rose, and bean (legume) families, self-recognition leads to enzymatic destruction of RNA within a rudimentary pollen tube. The RNA-hydrolyzing enzymes, or RNases, are present in the style of the carpel, but apparently they can enter a pollen tube and attack its RNA only if the pollen is of a "self" type.

In other cases, the block is a response by the cells of the carpel's stigma; this is called sporophytic self-incompatibility (because the carpel is part of the sporophyte). In members of the mustard family, for example, self-recognition activates a signal-transduction pathway in epidermal cells of the stigma that prevents germination of the pollen grain (FIGURE 38.8). Recent research suggests that one of the responses of the stigma to self-pollen is the opening of aquaporins, membrane proteins that function in water transport (see Chapter 36). Uptake of additional water by the stigma cell might block pollen germination by preventing the stigma from hydrating the relatively dry pollen, a step required for growth of a pollen tube. This model of self-incompatibility will certainly evolve as researchers learn more about how certain plants reject their own pollen.

Basic research on mechanisms of self-incompatibility may lead to agricultural applications. Plant breeders sometimes hybridize different varieties of a crop plant to combine the best traits of the varieties and counter the loss of vigor that can result from excessive inbreeding (see Chapter 14). Many agriculturally important plants are self-*compatible.* To maximize the number of hybrid seeds, breeders currently must prevent self-fertilization by laboriously removing the anthers from the parent plants that provide the seeds. Eventually, it may be possible instead to impose self-incompatibility on crop species that are normally self-compatible.

Double fertilization gives rise to the zygote and endosperm

After landing on a receptive stigma, a pollen grain absorbs moisture and germinates—that is, it produces a pollen tube that extends down between the cells of the style toward the ovary (FIGURE 38.9, p. 790). The generative cell divides by mitosis and forms two sperm, the male gametes. The germinated pollen grain is the mature male gametophyte. Directed by a chemical attractant, possibly calcium, the tip of the pollen tube enters the ovary, probes through the micropyle (a gap in the integuments of the ovule), and discharges its two sperm within the embryo sac.

The events that follow are a distinctive feature of the angiosperm life cycle (shared only with a few gymnosperms, in which the process probably evolved independently). One sperm fertilizes the egg to form the zygote. The other sperm combines with the two polar nuclei to form a triploid ($3n$) nucleus in the center of the large central cell of the embryo sac. This large cell will give rise to the **endosperm,** a food-storing tissue of the seed. The union of two sperm cells with different nuclei of the

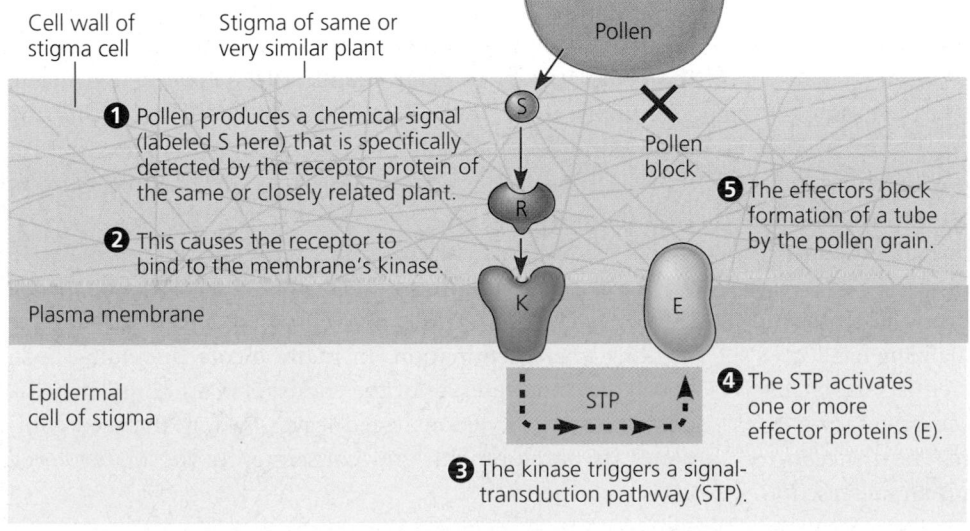

Cell wall of stigma cell

Stigma of same or very similar plant

Pollen

❶ Pollen produces a chemical signal (labeled S here) that is specifically detected by the receptor protein of the same or closely related plant.

Pollen block

❷ This causes the receptor to bind to the membrane's kinase.

❺ The effectors block formation of a tube by the pollen grain.

Plasma membrane

Epidermal cell of stigma

STP

❹ The STP activates one or more effector proteins (E).

❸ The kinase triggers a signal-transduction pathway (STP).

FIGURE 38.8 A possible mechanism of sporophytic self-incompatibility. In plants of the mustard family, there are at least two protein products of the S-locus. (The genes coding for these proteins are so tightly linked that they are inherited as though a single gene.) One of the protein products, labeled R in this diagram, is a receptor protein located in the extracellular matrix (wall) of the stigma epidermal cell. Another product of the S-locus is a protein kinase (K) embedded in the plasma membrane of the stigma cell. (A protein kinase, recall from Chapter 11, is an enzyme that activates other proteins by phosphorylating them.) These proteins interact according to the steps in this diagram of a hypothetical mechanism for self-incompatibility.

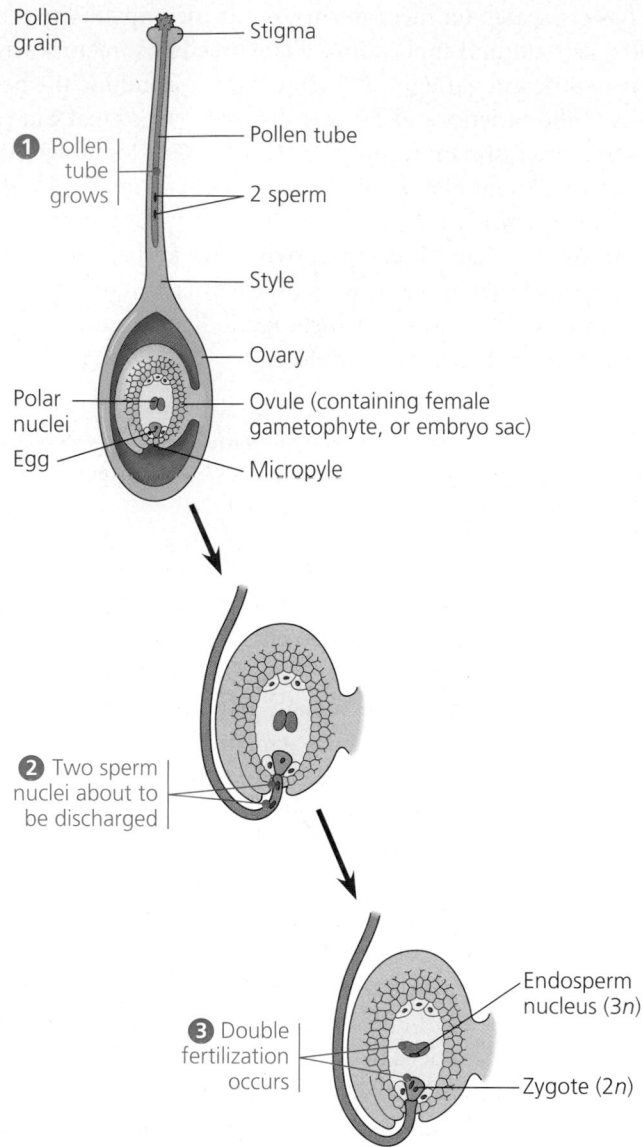

Pollen grain — Stigma

1 Pollen tube grows

Pollen tube

2 sperm

Style

Ovary

Polar nuclei

Ovule (containing female gametophyte, or embryo sac)

Egg

Micropyle

2 Two sperm nuclei about to be discharged

3 Double fertilization occurs

Endosperm nucleus (3*n*)

Zygote (2*n*)

FIGURE 38.9 Growth of the pollen tube and double fertilization. After a pollen grain is carried by wind or an animal to the stigma, a long pollen tube begins growing down the style toward the ovary. The tube discharges two sperm into the embryo sac of an ovule. One sperm fertilizes the egg, forming the zygote. The other combines with the two polar nuclei of the embryo sac's large central cell and forms a triploid cell that will develop into a nutritive tissue called endosperm.

embryo sac is termed **double fertilization.** Double fertilization ensures that the endosperm will develop only in ovules where the egg has been fertilized, thereby preventing angiosperms from squandering nutrients.

The nonreproductive tissues surrounding the embryo sac have prevented researchers from visualizing fertilization in plants. Recently, however, scientists isolated sperm cells from pollen grains and eggs from embryo sacs and observed the merging of plant gametes *in vitro.* The first cellular event that takes place after gamete fusion is an increase in

the cytoplasmic Ca^{2+} levels of the egg, as also occurs during animal gamete fusion (see Chapter 47). Another similarity to animals is the establishment of a block to polyspermy, the fertilization of an egg by more than one sperm cell. Maize (*Zea mays*) sperm cannot fuse with zygotes *in vitro.* This barrier is established as early as 45 seconds after the initial sperm fusion with the egg in maize. The block to polyspermy in animals occurs in two stages; a fast block within milliseconds reinforced by a slow block to polyspermy that requires many seconds. In the case of plants, one hypothesis for the slow block to polyspermy is a deposition of cell wall material that mechanically impedes sperm. Such a mechanism would be analogous to the fertilization envelope that leads to a slow block to polyspermy in several animal species. The opening of ion channels in the plasma membrane of the egg might also lead to a fast block to polyspermy, as occurs in several animal species (see Chapter 47).

The ovule develops into a seed containing an embryo and a supply of nutrients

After double fertilization, the ovule develops into a seed, and the ovary develops into a fruit enclosing the seed(s). As the embryo develops from the zygote, the seed stockpiles proteins, oils, and starch. This is why seeds are such major nutrient sinks (see Chapter 36). Initially, these nutrients are stored in the endosperm, but later in seed development in many species, the storage function of the endosperm is more or less taken over by the swelling storage leaves (cotyledons) of the embryo itself.

Endosperm Development

Endosperm development usually precedes embryo development. After double fertilization, the triploid nucleus of the ovule's central cell divides, forming a multinucleate "supercell" having a milky consistency. This liquid mass, the endosperm, becomes multicellular when cytokinesis partitions the cytoplasm by forming membranes between the nuclei. Eventually, these "naked" cells produce cell walls, and the endosperm becomes solid. Coconut "milk" is an example of liquid endosperm; coconut "meat" is an example of solid endosperm.

The endosperm is rich in nutrients, which it provides to the developing embryo. In most monocots and some dicots, the endosperm also stores nutrients that can be used by the seedling after germination. In many dicots (including bean seeds), the food reserves of the endosperm are completely exported to the cotyledons (seed leaves) before the seed completes its development, and consequently the mature seed lacks endosperm.

Embryo Development

The first mitotic division of the zygote is transverse, splitting the fertilized egg into a basal cell and a terminal cell (FIGURE 38.10). The terminal cell eventually gives rise to most of the embryo. The basal cell continues to divide transversely, producing a thread of cells called the suspensor, which anchors the embryo to its parent. The suspensor functions in the transfer of nutrients to the embryo from the parent plant and, in some plants, from the endosperm. Meanwhile, the terminal cell divides several times and forms a spherical proembryo attached to the suspensor. The cotyledons (seed leaves) begin to form as bumps on the proembryo. A dicot, with its two cotyledons, is heart-shaped at this stage. Only one cotyledon develops in monocots.

Soon after the rudimentary cotyledons appear, the embryo elongates. Cradled between the cotyledons is the apical meristem of the embryonic shoot. At the opposite end of the embryo's axis, where the suspensor attaches, is the apex of the embryonic root, also with a meristem. After the seed germinates, the apical meristems at the tips of shoot and root will sustain primary growth as long as the plant lives (see Chapter 35). The three primary meristems—protoderm, ground meristem, and procambium—are also present in the embryo. Thus, development of the embryo establishes two features of plant form: the root-shoot axis, with meristems at opposite ends; and a radial pattern of protoderm, ground meristem, and procambium, set to give rise to the three tissue systems (dermal, ground, and vascular tissues).

Structure of the Mature Seed

During the last stages of its maturation, the seed dehydrates until its water content is only about 5–15% of its weight. The embryo, surrounded by its enlarged cotyledons or endosperm, or both, stops growing until the seed germinates. The embryo and its food supply are enclosed by a protective **seed coat** formed from the integuments of the ovule.

We can take a closer look at one type of dicot seed by splitting open the seed of a common bean. At this stage, the embryo consists of an elongate structure, the embryonic axis, attached to fleshy cotyledons (FIGURE 38.11a, p. 792). Below the point at which the cotyledons are attached, the embryonic axis is called the **hypocotyl** (from the Greek *hypo,* "under"). The hypocotyl terminates in the **radicle,** or embryonic root. The portion of the embryonic axis above the cotyledons is the **epicotyl** (from the Greek *epi,* "on" or "over"). At its tip is the plumule, consisting of the shoot tip with a pair of miniature leaves.

The cotyledons of the common bean are fleshy before the seed germinates because they absorbed food from the endosperm when the seed developed. However, the seeds of some dicots, such as castor beans *(Ricinus communis),* retain their food supply in the endosperm and have cotyledons that

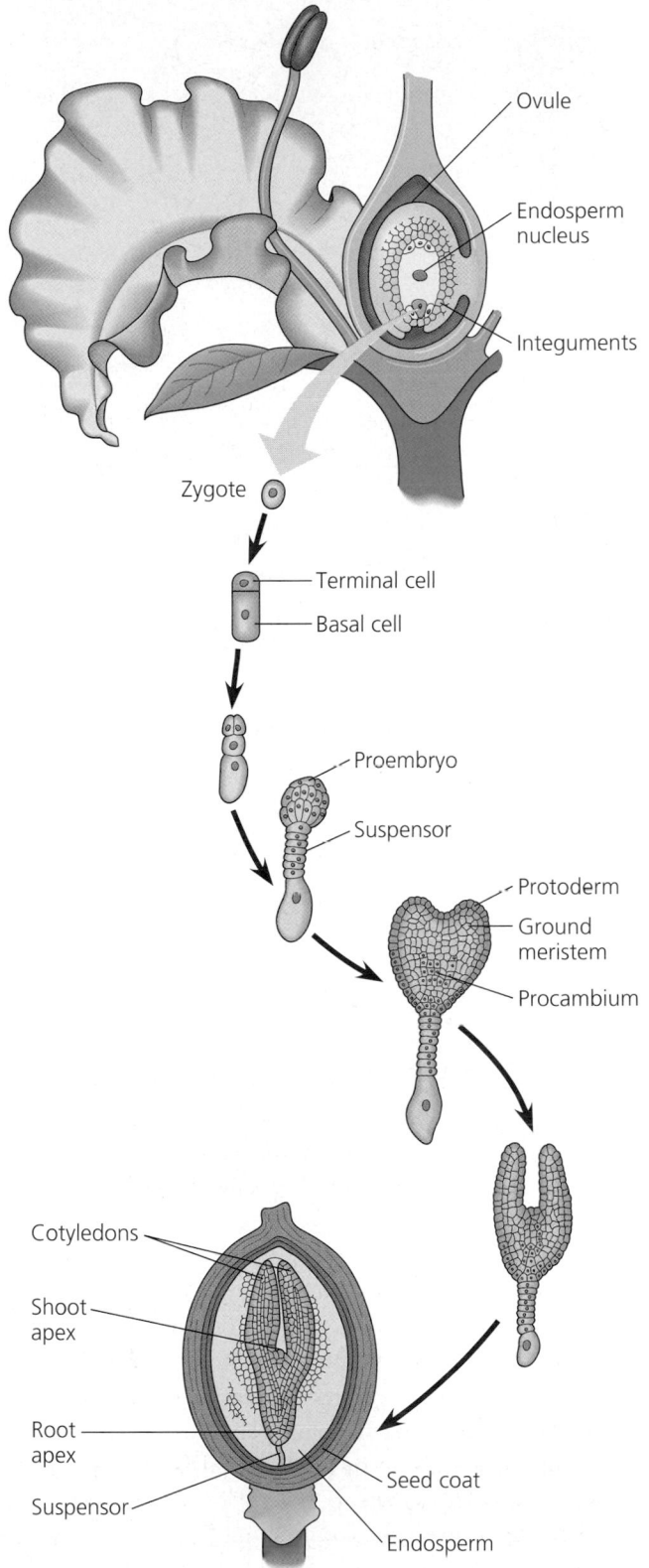

FIGURE 38.10 The development of a dicot plant embryo.
By the time the ovule becomes a mature seed and the integuments harden and thicken to form the seed coat, the zygote has given rise to an embryonic plant with rudimentary organs.

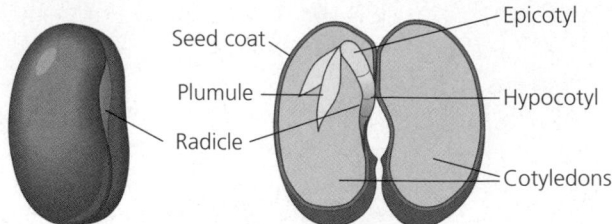

(a) Common bean. The fleshy cotyledons of the common garden bean, a dicot, store food that was absorbed from the endosperm when the seed developed.

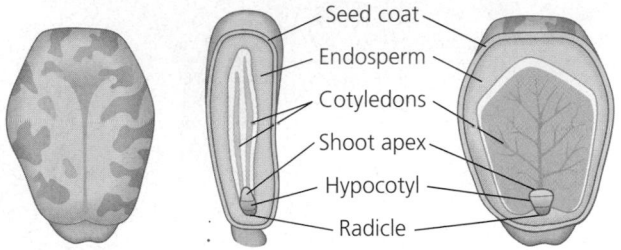

(b) Castor bean. The castor bean has membranous cotyledons that will absorb food from the endosperm when the seed germinates.

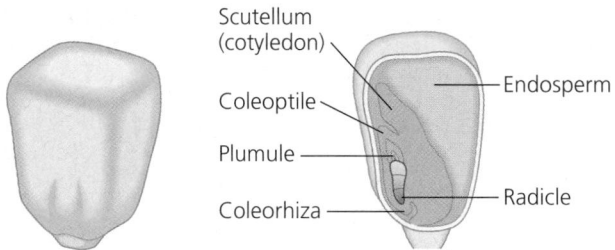

(c) Corn. Like other monocots, corn has only one cotyledon (the scutellum). The rudimentary shoot is sheathed in a structure called the coleoptile.

FIGURE 38.11 Seed structure.

are very thin (FIGURE 38.11b). The cotyledons will absorb nutrients from the endosperm and transfer them to the embryo when the seed germinates.

The seed of a monocot has a single cotyledon (FIGURE 38.11c). Members of the grass family, including corn and wheat, have a specialized type of cotyledon called a **scutellum** (L. *scutella,* "small shield," a reference to the scutellum's shape). The scutellum is very thin, with a large surface area pressed against the endosperm, from which the scutellum absorbs nutrients during germination. The embryo of a grass seed is enclosed by two sheathes, a **coleorhiza,** which covers the young root, and a **coleoptile,** which covers the young shoot.

The ovary develops into a fruit adapted for seed dispersal

As the seeds are developing from ovules, the ovary of the flower is developing into a **fruit,** which protects the enclosed seeds and aids in their dispersal by wind or animals. Pollina-

tion triggers hormonal changes that cause the ovary to begin its transformation into a fruit (FIGURE 38.12). If a flower has not been pollinated, fruit usually does not develop, and the entire flower withers and falls away.

During fruit development, the wall of the ovary becomes the **pericarp,** the thickened wall of the fruit. As the ovary grows, the other parts of the flower wither and are shed. (The pointed tip of the pea pod in FIGURE 38.12 is the withered remains of the pea flower's stigma.) However, in some angiosperms, other floral parts also contribute to what we call a fruit in grocery store vernacular. In apple flowers, for example, the ovary is embedded in the receptacle (see FIGURE 38.2), and the fleshy part of the fruit is derived mainly from the swollen receptacle; only the core of the apple fruit develops from the ovary. Fruits are classified into several types, depending on their developmental origin (see TABLE 30.1).

The fruit usually ripens about the same time that its seeds are completing their development. For a dry fruit such as a soybean pod, ripening is little more than senescence (aging) of the fruit tissues, which allows the fruit to open and release the seeds. The ripening of a fleshy fruit is more elaborate, its steps controlled by the complex interactions of hormones. In this case, ripening results in an edible fruit that serves as an enticement to the animals that help spread the seeds. The "pulp" of the fruit becomes softer as a result of enzymes digesting

FIGURE 38.12 Development of a pea fruit (pod). (Also, see FIGURE 30.15.) These photographs illustrate the changes in a pea plant leading to pod formation. **(a)** Soon after pollination, **(b)** the flower drops its petals, and hormonal changes make the ovary start growing. The ovary expands, and its wall thickens, **(c)** forming the pod, or fruit.

components of the cell walls. There is usually a color change from green to some other color such as red, orange, or yellow. The fruit becomes sweeter as organic acids or starch molecules are converted to sugar, which may reach a concentration of as much as 20% in a ripe fruit.

By selectively breeding plants, humans have capitalized on the production of edible fruits. The apples, oranges, and other fruits in grocery stores are exaggerated versions of much smaller natural varieties of fleshy fruits. However, the staple foods for humans are the dry, wind-dispersed fruits of grasses, which are harvested while still on the parent plant. The cereal grains of wheat, rice, maize, and other grasses are easily mistaken for seeds, but each is actually a fruit with a dry pericarp that adheres tightly to the seed coats of the single seed within.

Evolutionary adaptations of seed germination contribute to seedling survival

As a seed matures, it dehydrates and enters a phase referred to as **dormancy** (from the Latin word meaning "to sleep"), a condition of extremely low metabolic rate and a suspension of growth and development. A plant resumes the growth and development that were temporarily suspended when the seed and its embryo became dormant. Conditions required to break dormancy vary between plant species. Some seeds germinate as soon as they are in a suitable environment. Other seeds remain dormant and will not germinate, even if sown in a favorable place, until a specific environmental cue causes them to break dormancy.

Seed Dormancy

Seed dormancy increases the chances that germination will occur at a time and place most advantageous to the seedling. Breaking dormancy generally requires certain environmental conditions. Seeds of many desert plants, for instance, germinate only after a substantial rainfall. If they were to germinate after mild drizzle, the soil might soon be too dry to support the seedlings. Where natural fires are common, many seeds require intense heat to break dormancy; seedlings are therefore most abundant after fire has cleared away competing vegetation. Where winters are harsh, seeds may require extended exposure to cold; seeds sown during summer or fall do not germinate until the following spring. This assures a long growth season before the next winter. Very small seeds, such as those of some lettuce varieties, require light

for germination and will break dormancy only if they are buried shallow enough for the seedlings to poke through the soil surface. Some seeds have coats that must be weakened by chemical attack as they pass through an animal's digestive tract, and thus are likely to be carried some distance before germinating.

The length of time a dormant seed remains viable and capable of germinating varies from a few days to decades or even longer, depending on the species and environmental conditions. Most seeds are durable enough to last a year or two until conditions are favorable for germinating. Thus, the soil has a pool of nongerminated seeds that may have accumulated for several years. This is one reason vegetation reappears so rapidly after a fire, drought, flood, or some other environmental disruption.

From Seed to Seedling

Germination of seeds depends on the physical process of imbibition, the uptake of water due to the low water potential of the dry seed. Imbibing water causes the seed to expand and rupture its coat and also triggers metabolic changes in the embryo that enable it to resume growth (FIGURE 38.13). Following hydration, enzymes begin digesting the storage materials of the endosperm or cotyledons, and the nutrients are transferred to the growing regions of the embryo (FIGURE 38.13).

The first organ to emerge from the germinating seed is the radicle, the embryonic root. Next, the shoot tip must break through the soil surface. In garden beans and many other dicots,

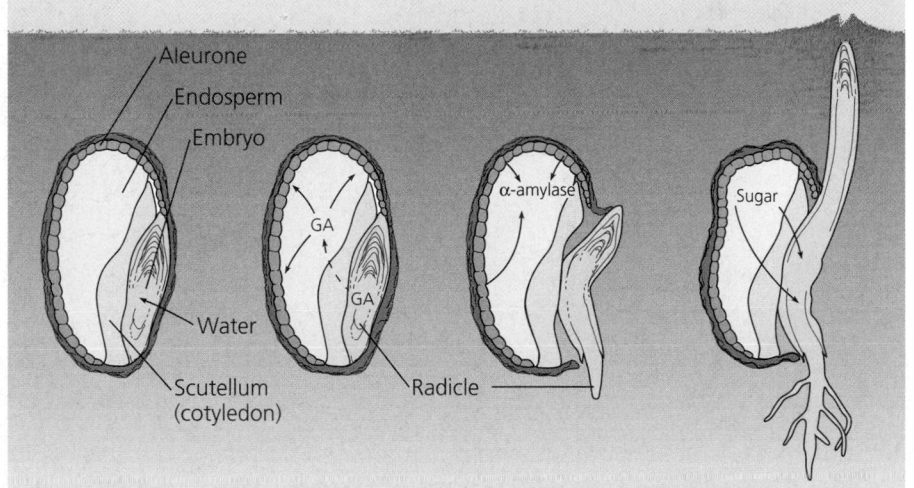

FIGURE 38.13 Mobilization of nutrients during the germination of a barley seed. After the seed imbibes water, the embryo releases hormones called gibberellins (GA) as signals to the aleurone, the thin outer layer of the endosperm. The aleurone responds by synthesizing and secreting digestive enzymes that hydrolyze stored foods in the endosperm, producing small, soluble molecules. One example is α-amylase, an enzyme that hydrolyzes starch. (A similar enzyme in our saliva helps us digest bread and other foods made from the starchy endosperm of ungerminated seeds.) Sugars and other nutrients absorbed from the endosperm by the scutellum are consumed during growth of the embryo into a seedling.

a hook forms in the hypocotyl, and growth pushes the hook aboveground (FIGURE 38.14a). Stimulated by light, the hypocotyl straightens, raising the cotyledons and epicotyl. Thus, the delicate shoot apex and bulky cotyledons are pulled aboveground, rather than being pushed tip-first through the abra-

sive soil. The epicotyl now spreads its first foliage leaves (true leaves, so called to distinguish them from cotyledons, or "seed leaves"). The foliage leaves expand, become green, and begin making food by photosynthesis. The cotyledons shrivel and fall away from the seedling, their food reserves having been consumed by the germinating embryo.

Light seems to be the main cue that tells the seedling it has broken ground. We can trick a bean seedling into behaving as though it is still buried by germinating the seed in darkness. The seedling extends an exaggerated hypocotyl with a hook at its tip, and the foliage leaves fail to green. After it exhausts its food reserves, the spindly seedling stops growing and dies.

Peas, although in the same family as beans, have a different style of germinating (FIGURE 38.14b). A hook forms in the epicotyl rather than the hypocotyl, and the shoot tip is lifted gently out of the soil by elongation of the epicotyl and straightening of the hook. Pea cotyledons, unlike those of beans, remain behind underground.

Corn and other grasses, which are monocots, use yet a different method for breaking ground when they germinate (FIGURE 38.14c). The coleoptile, the sheath enclosing and protecting the embryonic shoot, pushes upward through the soil and into the air. The shoot tip then grows straight up through the tunnel provided by the tubular coleoptile.

Seed germination is a precarious stage in a plant's life. The tough seed gives rise to a fragile seedling that will be exposed to predators, parasites, wind, and other hazards. In the wild, only a small fraction of seedlings endure long enough to become parents themselves. Production of enormous numbers of seeds compensates for the odds against individual survival and gives natural selection ample genetic variations to screen. However, this is a very expensive means of reproduction in terms of the resources consumed in flowering and fruiting. Asexual reproduction, generally simpler and less hazardous for offspring than sexual reproduction, is an alternative means of plant propagation.

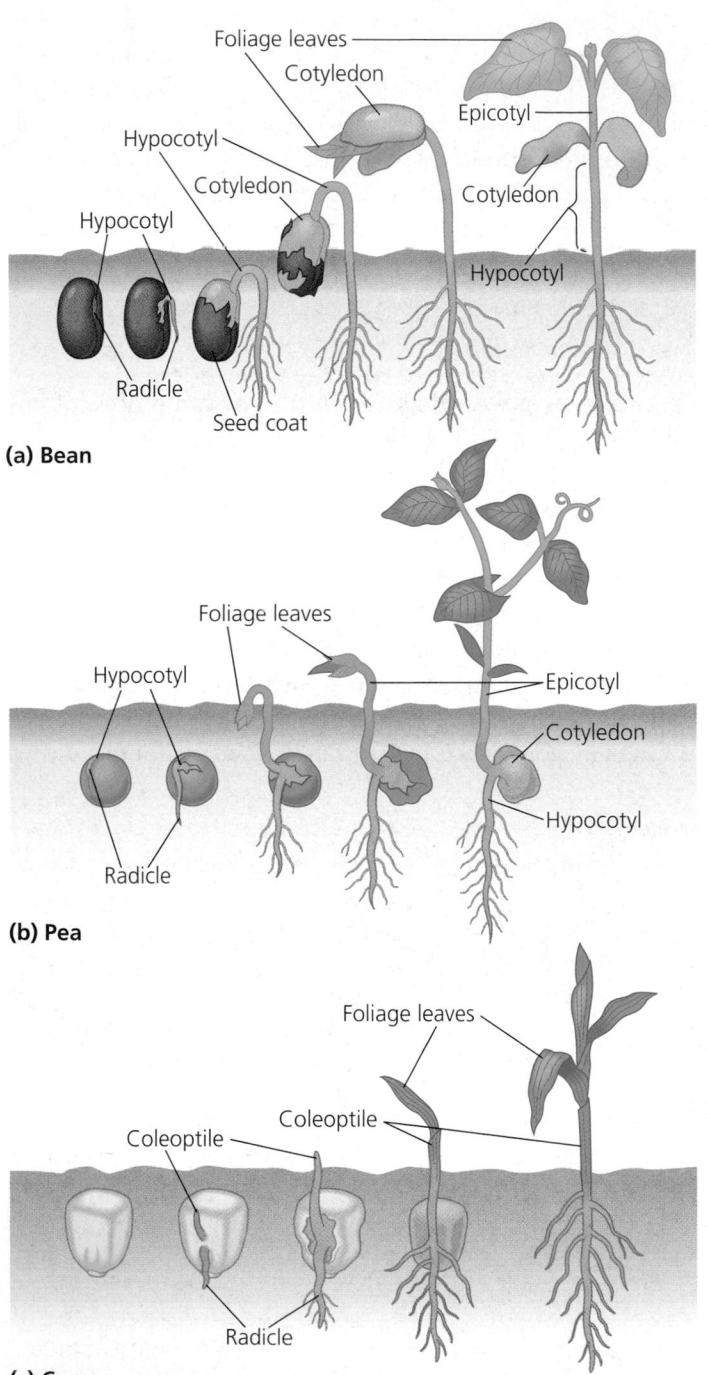

(a) Bean

(b) Pea

(c) Corn

FIGURE 38.14 Seed germination. The radicle, the root of the embryo, emerges from the seed first. Then the shoot breaks the soil surface by one of the following mechanisms. **(a)** In beans, straightening of a hook in the hypocotyl pulls the cotyledons from the soil. **(b)** In peas, the hook is above the cotyledons on the epicotyl, and the cotyledons remain underground. **(c)** In corn and other grasses, the shoot grows straight up through the tube of the coleoptile.

ASEXUAL REPRODUCTION

Many plants clone themselves by asexual reproduction

Imagine some of your fingers separating from your body, taking up life on their own, and eventually developing into entire copies of yourself. This would be an example of asexual reproduction: offspring derived from a single parent without genetic recombination (which, of course, does not actually occur in humans). The result would be a clone, a population of asexually produced, genetically identical organisms. Many plant species clone themselves by asexual reproduction, also called **vegetative reproduction.**

Asexual reproduction is an extension of the capacity of plants for indeterminate growth. Plants, remember, have meristematic tissues of dividing, undifferentiated cells that can sustain or renew growth indefinitely. In addition, parenchyma cells throughout the plant can divide and differentiate into the various types of specialized cells, enabling plants to regenerate lost parts. Detached fragments of some plants can develop into whole offspring; a severed stem, for instance, may develop adventitious roots and become a whole plant. **Fragmentation,** the separation of a parent plant into parts that re-form whole plants, is one of the most common modes of vegetative reproduction (FIGURE 38.15a). A variation of this process occurs in some species of dicots, in which the root system of a single parent gives rise to many adventitious shoots that become separate shoot systems. The result is a clone formed by asexual reproduction from one parent (FIGURE 38.15b). Such asexual propagation has produced the oldest of all known plant clones, a ring of creosote bushes in the Mojave Desert of California, believed to be at least 12,000 years old.

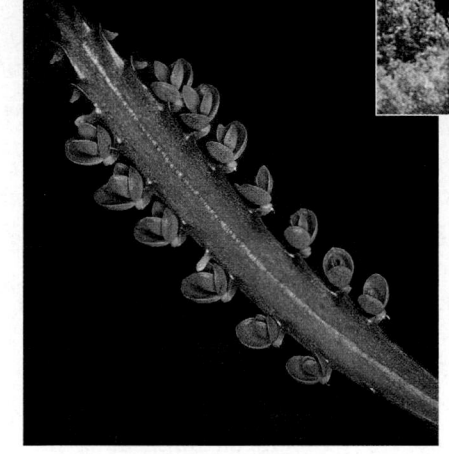

(a)

(b)

FIGURE 38.15 **Natural mechanisms of vegetative reproduction. (a)** *Kalanchoe* is known as the maternity plant because of the numerous plantlets it produces along its leaf margins. The asexually produced plantlets fragment from their parent and become independent plants. **(b)** Some aspen groves, such as those shown here, are actually clones of thousands of trees descended by asexual reproduction from the root system of one parent. Notice that genetic differences among the clones result in different timing for the development of fall color and the loss of leaves.

An entirely different mechanism of asexual reproduction has evolved in dandelions and some other plants, which produce seeds without their flowers being fertilized. This asexual production of seeds is called **apomixis.** A diploid cell in the ovule gives rise to the embryo, and the ovules mature into seeds, which in the dandelion are dispersed by windblown fruit. Thus, though these plants clone themselves by an asexual process, they also have the advantage of seed dispersal, an adaptation usually associated with sexual reproduction.

self rapidly. Moreover, the offspring of vegetative reproduction, usually mature fragments of the parent plant, are not as frail as the seedlings produced by sexual reproduction. A sprawling clone of prairie grass may cover an area so thoroughly that seedlings of the same or other species have little chance of competing. But in the soil is a pool of seeds, waiting in the wings for some cue to germinate. After a fire, drought, or some other disturbance clears patches of the turf, seedlings can finally get a foothold when conditions improve. The seedlings are unequal in their traits, for their genotypes are products of the sexual recombination of genes. A new competition ensues, in which certain plants excel and clone themselves. Both modes of reproduction, sexual and asexual, have contributed to the success of plants in diverse terrestrial environments.

Sexual and asexual reproduction are complementary in the life histories of many plants

Many plants are capable of both sexual and asexual reproduction, and each offers advantages in certain situations. Sex generates variation in a population, an asset in an environment where evolving pathogens and other variables affect survival and reproductive success. An additional benefit of sexual reproduction in plants is the seed, which can disperse to new locations and can also wait to grow until hostile environmental conditions have improved.

An advantage of asexual reproduction is that a plant well suited to a particular environment can clone many copies of it-

Vegetative propagation of plants is common in agriculture

With the objective of improving crops, orchards, and ornamental plants, various methods have been devised for propagating plants asexually. Most are based on the ability of plants to form adventitious roots or shoots.

Clones from Cuttings

Most houseplants, woody ornamentals, and orchard trees are asexually reproduced from plant fragments called cuttings. In some cases, shoot or stem cuttings are used. At the cut end of the shoot a mass of dividing, undifferentiated cells called a **callus** forms, and then adventitious roots develop from the callus. If the shoot fragment includes a node, then adventitious roots form without a callus stage. Some plants, including African violets, can be propagated from single leaves rather than stems. For still other plants, cuttings are taken from specialized storage stems. For example, a potato can be cut up into several pieces, each with a vegetative bud, or "eye," that regenerates a whole plant.

In a modification of vegetative reproduction from cuttings, a twig or bud from one plant can be grafted onto a plant of a closely related species or a different variety of the same species. Grafting makes it possible to combine the best qualities of different species or varieties into a single plant. The graft is usually done when the plant is young. The plant that provides the root system is called the **stock;** the twig grafted onto the stock is referred to as the **scion.** For example, scions from French varieties of vines that produce superior wine grapes are grafted onto root stock of American varieties, which are more resistant to certain soil pathogens. The quality of the fruit, determined by the genes of the scion, is not diminished by the genetic makeup of the stock. In some cases of grafting, however, the stock can alter the characteristics of the shoot system that develops from the scion. For example, dwarf fruit trees are made by grafting normal twigs onto dwarf stock varieties that retard the vegetative growth of the shoot system. Because seeds are produced by the part of the plant derived from the scion, they would give rise to plants of the scion species if planted.

Test-Tube Cloning and Related Techniques

THE PROCESS OF SCIENCE

Plant biotechnologists have adopted *in vitro* methods to create and clone novel plant varieties. It is possible to grow whole plants by culturing small explants (pieces of tissue cut from the parent), or even single parenchyma cells, on an artificial medium containing nutrients and hormones (FIGURE 38.16). The cultured cells divide and form an undifferentiated callus. When the hormonal balance is manipulated in the culture medium, the callus can sprout shoots and roots with fully differentiated cells. The test-tube plantlets can then be transferred to soil, where they continue their growth. A single plant can be cloned into thousands of copies by subdividing calluses as they grow. This method is used for propagating orchids and also for cloning pine trees that deposit wood at unusually fast rates (see FIGURE 30.8f).

Plant tissue culture also facilitates genetic engineering in plants. Most techniques for the introduction of foreign genes into plants require the use of small

(a)   **(b)**

FIGURE 38.16 Test-tube cloning of carrots. (a) Just a few parenchyma cells from a carrot gave rise to this callus, a mass of undifferentiated cells. **(b)** The callus differentiates into an entire plant, with leaves, stems, and roots. (See FIGURE 21.5)

pieces of plant tissue or single plant cells as the starting material. Test-tube culture makes it possible to regenerate genetically modified (transgenic) plants from a single plant cell into which the foreign DNA has been incorporated. For example, researchers have used recombinant DNA technology to transfer a gene for bean protein into cultured cells from a sunflower plant. The experiment improved the protein quality of sunflower seeds harvested from the transgenic plants. Firing DNA-coated pellets from a gun is one method researchers use to insert foreign DNA into plant cells (FIGURE 38.17). The techniques of genetic engineering are discussed in more detail in Chapter 20.

Some researchers are coupling a technique known as **protoplast fusion** with tissue culture methods to actually invent new

(a)

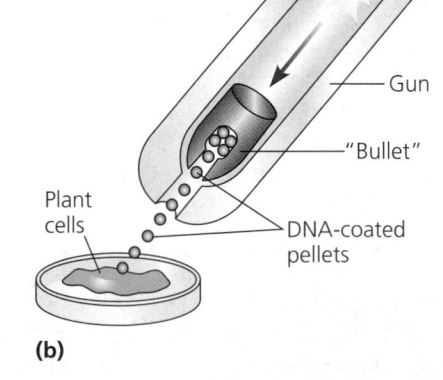

(b)

FIGURE 38.17 A DNA gun. (a) This researcher is preparing to use a modified .22-caliber gun to shoot foreign DNA into cultured plant cells. **(b)** The gun fires a plastic bullet loaded with tiny metallic pellets coated with DNA. (A different type of DNA gun uses a burst of gas rather than an explosion to propel pellets.) When a plate stops the bullet shell at the end of the gun, the pellets continue toward the cellular targets. (This drawing represents the pellets as much larger than they actually are, compared to the size of the gun and petri dish.) The projectiles penetrate cell walls and membranes, introducing foreign DNA to the nuclei of some cells. A cell that integrates this DNA into its genome can be cultured to produce a plantlet, and it is then possible to clone the transgenic plant.

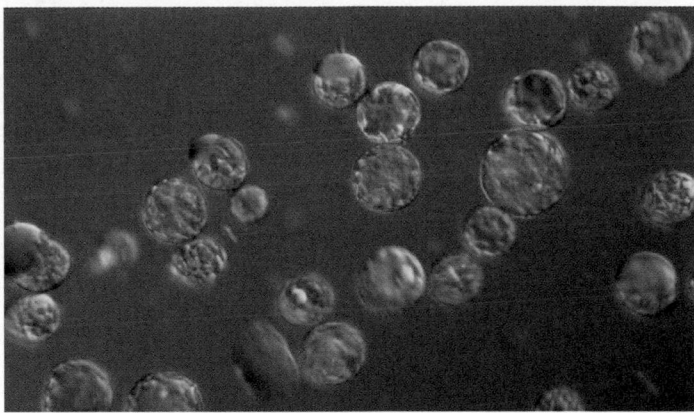

FIGURE 38.18 Protoplasts. These wall-less plant cells are prepared by treating plant cells or tissues with purified wall-degrading enzymes isolated from certain types of fungi. Researchers can fuse protoplasts from different species to make hybrids and can also culture the hybrid cells to produce a new plant.

50 μm

plant varieties that can be cloned. Protoplasts are plant cells that have had their cell walls enzymatically removed by treatment with cell-wall-degrading enzymes (cellulases and pectinases) isolated from fungi (FIGURE 38.18). Before they are cultured, the protoplasts can be screened for mutations that may improve the agricultural value of the plant. It is also possible in some cases to fuse two protoplasts from different plant species that would otherwise be reproductively incompatible, and then culture the hybrid protoplasts. Each of the many protoplasts can regenerate a wall and eventually form a hybrid plantlet. One success of this method has been a hybrid between a potato and a wild relative called black nightshade. The nightshade is resistant to an herbicide that is commonly used to kill weeds. The hybrids are also resistant, and this makes it possible to "weed" a potato field with the herbicide without killing the potato plants.

The *in vitro* culturing of plant cells and tissues is fundamental to most types of plant biotechnology. The other basic process behind modern plant biotechnology is the ability to produce transgenic plants through various methods of genetic engineering. In this chapter's last section, we take a closer look at plant biotechnology.

PLANT BIOTECHNOLOGY

THE PROCESS OF SCIENCE

Plant biotechnology has two meanings. In the general sense, it refers to innovations in the use of plants, or of substances obtained from plants, to make products of use to humans—an endeavor that began in prehistory. Used in a more specific sense, biotechnology refers to the use of genetically modified (GM) organisms in agriculture and industry. Indeed, in the last two decades, genetic engineering has become such a powerful—some would say too powerful—force in biotechnology that the terms *genetic engineering* and *biotechnology*

have become synonymous in the media. In this last section, we examine the question of how humans have altered plants to suit their purposes from the birth of agriculture to the present. (This section expands upon the discussion on pp. 397–399.)

Neolithic humans created new plant varieties by artificial selection

Humans have intervened in the reproduction and genetic makeup of plants for thousands of years. It is no exaggeration to say that maize is an unnatural monster created by humans. Left on its own in nature, maize would soon become extinct for the simple reason that it cannot spread its seeds. Maize kernels not only are permanently attached to the central axis (the "cob") but also are permanently protected by tough, overlapping leaf sheaths (the "husk"). Obviously, these attributes, so useful to people, arose not by natural selection but by artificial selection by humans (FIGURE 38.19). (See Chapter 22 to review the basic concept of artificial selection.) Indeed, Neolithic (late Stone Age) humans domesticated virtually all of our crop species over a relatively short period about 10,000 years ago. But genetic modification began long before humans started altering crops by artificial selection. For example, the wheat groups we rely on for much of our food supply are the result of natural hybridization between different species of grasses. Such hybridization is common in plants, and has long been exploited by plant breeders to introduce additional genetic variation for artificial selection and crop improvement. Let's consider one modern example, the selective breeding of maize.

Maize is a major staple in many developing countries. Nevertheless, the most common varieties are relatively poor sources of protein, thus requiring that diets be supplemented with other protein sources such as beans. The proteins that are stored in these popular maize varieties are very low in lysine

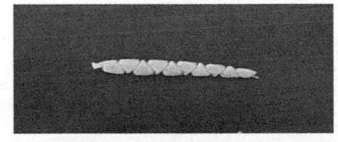

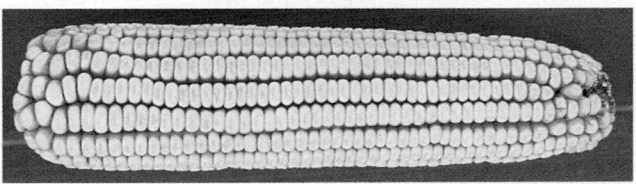

FIGURE 38.19 Maize: a product of artificial selection. Modern maize (bottom) was derived from teosinte (top). Teosinte kernels are tiny, and each one has a husk that must be removed to get at the kernel. The cob shatters during seed dispersal, which probably made harvesting difficult for early farmers. Neolithic farmers selected for larger cob and kernel size as well as the permanent attachment of seeds to the cob and the encasing of the entire cob by a tough husk.

and tryptophan, two of the eight essential amino acids that humans cannot synthesize and must obtain from their diet (see FIGURE 41.4). Forty years ago, researchers discovered a mutant maize known as *opaque-2* that had much higher levels of lysine and tryptophan. But, as is often the case in plant breeding, a highly desirable trait turned out to be closely associated with several undesirable ones. The *opaque-2* maize kernels have a soft endosperm that makes them difficult to harvest and more vulnerable to attack by pests. But, of course, this maize variety is more nutritious. Swine fed with *opaque-2* maize gain weight three times faster than those fed with normal maize. Using the conventional breeding methods of hybridization and artificial selection, plant breeders were able to convert the original soft *opaque-2* endosperm into a more desirable hard endosperm type. This transition took nearly 20 years. Had modern methods of genetic engineering been available then, the genes responsible for the high lysine and tryptophan condition could have been inserted directly into hard-endosperm maize varieties.

Unlike traditional plant breeders, modern plant biotechnologists, using the techniques of genetic engineering, are not limited solely to the transfer of genes between closely related species or varieties of the same species. For example, traditional plant breeding techniques could not be used to insert a desired gene from daffodil into rice, but genetic engineering makes this possible. The term **transgenic** is used to describe organisms that have been genetically engineered to express a foreign gene from another species. The reason that traditional plant breeders cannot introduce daffodil genes into rice is that the many intermediate species between rice and daffodil and their common ancestor are extinct. In theory, if plant breeders did have the intermediate species at their disposal, they could, probably over the course of several centuries, introduce a daffodil gene into rice by traditional hybridization and breeding methods. With genetic engineering technology, this gene transfer can be accomplished without the need for intermediate species.

Biotechnology is transforming agriculture

Eight hundred million people on earth suffer from nutritional deficiencies. Forty thousand people die each day of malnutrition, half of them children. Disagreements exist over the causes of such hunger. Some argue that food shortages arise not from a lack of food production but from inequities in distribution: The dire poor simply cannot afford food. Others regard food shortages as evidence that the world is overpopulated—that the human species has exceeded the carrying capacity of the planet (see Chapter 52). Whatever the social and demographic causes of starvation, increasing food production seems like a humane objective. Because land and water are the most limiting resources for food production, the best option will be to increase yields on the available land. Indeed, there is very little "extra" land that can be put to the plow, especially if the few remaining pockets of wilderness are to be conserved. Based on conservative estimates of population growth, the world's farmers will have to produce 40% more grain per hectare to feed the human population in 2020. Plant biotechnology could help make these crop yields possible.

Already, the commercial adoption by farmers of transgenic crops has been one of the most rapid cases of technology transfer in the history of agriculture. Between 1996 and 1999, the area planted commercially with transgenic crops increased from 1.7 to 39.9 million hectares. These crops include transgenic varieties and hybrids of cotton, maize, and potatoes that contain genes from the bacterium *Bacillus thuringiensis*. These "transgenes" encode for a protein (*Bt* toxin) that effectively controls a number of serious insect pests. The use of such plant varieties greatly reduces the need for spraying crops with chemical insecticides. Considerable progress also has been made in the development of transgenic plants of cotton, maize, soybeans, sugar beet, and wheat that are tolerant to a number of herbicides. The cultivation of these plants could lower production costs and enable farmers to "weed" crops with herbicides (which won't damage the transgenic crop plants) instead of by heavy tillage, which can cause erosion of soil. Researchers are also engineering transgenic plants with enhanced resistance to disease. In one case, a transgenic papaya resistant to a ringspot virus was introduced into Hawaii, thereby saving the papaya industry (FIGURE 38.20). The nutritional quality of plants is also being improved. "Golden Rice," a transgenic variety with a few daffodil genes that increase quantities of vitamin A, was specifically designed to prevent the blindness that develops in many of the world's poor whose diet is chronically deficient in vitamin A (see FIGURE 20.20).

FIGURE 38.20 Genetically engineered papaya. A ringspot virus devastated papaya cultivation worldwide. A transgenic papaya variety rescued the industry. The genetically engineered papaya on the right is more resistant to ringspot virus than the nontransgenic papaya on the left.

Plant biotechnology has incited much public debate

Many people, including some scientists, are concerned about the unknown risks associated with the release of GM organisms into the environment. Much of the animosity regarding GM organisms in agriculture is political, economic, or ethical in nature, and outside the scope of this book. But we *should* consider the biological concerns about GM crops. The most fundamental debate centers on the extent to which GM organisms are an unknown risk that could potentially cause harm to human health or to the environment. Those who want to proceed more slowly with agricultural biotechnology, or stop it entirely, are concerned about the unstoppable nature of the "experiment." If a drug trial produces unanticipated harmful results, the trial is stopped: We cannot stop the "trial" of novel organisms that have been introduced into the biosphere. Chapter 20 introduced the key concerns. Here, we take a closer look at a few of the controversies surrounding plant biotechnology in agriculture.

One of the specific concerns is that genetic engineering could potentially transfer allergens, which are molecules to which some humans are allergic, from a gene source to a plant used for food. So far, there is no credible evidence that any GM plants specifically designed for human consumption have had any adverse effect on human health. Nevertheless, because of these health concerns, activists continue to lobby for the clear labeling of all foods made wholly or in part from products of GM organisms. Some activists also argue for strict regulations against the mixing of GM foods with non-GM foods during food transport, storage, and processing. Some biotechnology advocates, however, point out that similar demands were not raised when "transgenic" crops produced by traditional plant breeding techniques were put on the market. For example, triticale, a completely new crop, was artificially synthesized a few decades ago by combining the genomes of wheat and rye—two distinct genera that do not interbreed in nature. Triticale is now grown on over three million acres worldwide.

Many environmentalists are concerned that the growing of GM crops might have unforeseen effects on nontarget organisms. One recent study indicated that the larvae (caterpillars) of monarch butterflies responded adversely and even died following their consumption in the laboratory of milkweed leaves (their preferred food) heavily dusted with pollen from transgenic maize that produces the *Bt* toxin. This effect is not too surprising if one considers that *Bt* toxin was engineered into crop plants specifically because it is toxic to pests closely related to monarchs. Debate about this specific issue now centers on just how big an effect *Bt* maize pollen has on monarchs in nature, in contrast to the extreme exposure to the *Bt* toxin in the laboratory experiments. Maize pollen is produced for only a short time during the growing season, and most of it is not wind-dispersed too far from the field. A recent study provided evidence that only those milkweed plants bordering a maize field would potentially be dusted heavily enough with *Bt* pollen to adversely affect nontarget larvae. Moreover, in considering the negative effects of *Bt* pollen on monarch butterflies, one must also weigh the effects of the most likely alternative to the cultivation of *Bt* maize—the spraying of non-*Bt* maize with chemical pesticides. Such spraying may be even more harmful to the nearby monarch population.

Perhaps the most serious concern that some scientists raise about GM crops is the possibility of the introduced genes escaping from a transgenic crop into related weeds through crop-to-weed hybridization. Perhaps, for example, spontaneous hybridization between a crop engineered for herbicide resistance and a wild relative might give rise to a "superweed" that would be much more difficult to control. Crop-to-weed transgene escape does occur. Its likelihood depends on the ability of the crop and weed to hybridize and on how the transgenes affect the overall fitness of the hybrids. In some cases, a desirable crop trait—a dwarf phenotype, for example—would be disadvantageous to a weed growing in the wild. In other instances, there are no weedy relatives nearby with which to hybridize; soybean, for example, has no wild relatives in the United States. However, many crops, such as cabbage and broccoli, are members of the mustard family and hybridize readily with wild mustard (generally considered to be a weed). In such cases, the planting of nontransgenic plant borders around fields of transgenic crops can help to "brush" the pollen off insect pollinators, thereby reducing crop-to-weed gene transfer. Efforts are also under way to breed male sterility into transgenic crops. These plants will still produce seeds and fruit if they are pollinated by nearby nontransgenic bisexual plants, but they will produce no viable pollen of their own. Another approach is to engineer the transgene into the chloroplast DNA of the crop; chloroplast DNA is inherited strictly from the maternal plant, so transgenes in the chloroplast cannot be transferred by pollen.

The continuing debate about GM organisms in agriculture exemplifies one of this textbook's themes: the relationships of science and technology to society. Technological advances almost always involve some risk that unintended outcomes could occur. In the case of plant biotechnology, zero risk is unrealistic and probably unattainable. Scientists and the public need to assess the possible benefits of transgenic products versus the risks society is willing to take on a case-by-case basis. The best scenario would be for these discussions and decisions to be based on sound scientific information and testing rather than on reflexive fear or blind optimism.

CHAPTER 38 REVIEW

Go to the Campbell Biology website (www.campbellbiology.com) to explore an interactive version of the Chapter Review.

Summary of Key Concepts

SEXUAL REPRODUCTION

■ **Sporophyte and gametophyte generations alternate in the life cycles of plants: *a review*** (pp. 783–784, FIGURE 38.1) The sporophyte, the dominant generation, produces spores that develop within flowers into male gametophytes (pollen grains) and female gametophytes (embryo sacs).
   Web/CD Activity 38A: *Angiosperm Life Cycle*

■ **Flowers are specialized shoots bearing the reproductive organs of the angiosperm sporophyte** (pp. 784–785, FIGURES 38.2, 38.3) The four floral organs are sepals, petals, stamens, and carpels.

■ **Male and female gametophytes develop within anthers and ovaries, respectively: Pollination brings them together** (pp. 786–788, FIGURES 38.4, 38.5) Pollen develops from microspores within the sporangia of anthers; embryo sacs develop from megaspores within ovules. Pollination, which precedes fertilization, is the placing of pollen on the stigma of a carpel.

■ **Plants have various mechanisms that prevent self-fertilization** (pp. 788–789, FIGURES 38.6, 38.8) Some plants reject pollen that has an S-allele matching an allele in the stigma cells. Recognition of "self" pollen triggers a signal-transduction pathway leading to a block in growth of a pollen tube.

■ **Double fertilization gives rise to the zygote and endosperm** (pp. 789–790, FIGURE 38.9) The pollen tube discharges two sperm into the embryo sac; one fertilizes the egg, while the other fertilizes the polar cell, which gives rise to the food-storing endosperm.

■ **The ovule develops into a seed containing an embryo and a supply of nutrients** (pp. 790–792, FIGURES 38.10, 38.11) The seed coat encloses the embryo, along with a food supply stocked in either the cotyledons or the endosperm.
   Web/CD Activity 38B: *Seed and Fruit Development*

■ **The ovary develops into a fruit adapted for seed dispersal** (pp. 792–793, FIGURE 38.12) The fruit protects the enclosed seeds and aids in wind dispersal or in the attraction of seed-dispersing animals.

■ **Evolutionary adaptations of seed germination contribute to seedling survival** (pp. 793–794, FIGURES 38.13, 38.14) Seed dormancy assures that seeds germinate only when conditions for seedling survival are optimal. The breaking of dormancy often requires environmental cues, such as temperature or lighting changes.
   Web/CD Case Study in the Process of Science: *What Tells Desert Seeds When to Germinate?*

ASEXUAL REPRODUCTION

■ **Many plants clone themselves by asexual reproduction** (pp. 794–795, FIGURE 38.15) One important mode of asexual reproduction is the fragmentation of a parent plant into parts that re-form whole plants.

■ **Sexual and asexual reproduction are complementary in the life histories of many plants** (p. 795) Asexual reproduction enables successful clones to spread; sexual reproduction generates the genetic variation that makes evolutionary adaptation possible.

■ **Vegetative propagation of plants is common in agriculture** (pp. 795–797, FIGURES 38.16, 38.18) Cloning plants from cuttings is an ancient practice. We can now clone plants from single cells, which can be genetically manipulated before cloning.

PLANT BIOTECHNOLOGY

■ **Neolithic humans created new plant varieties by artificial selection** (pp. 797–798, FIGURE 38.19) Interspecific hybridization of plants is common in nature, and has been used by breeders, ancient and modern, to introduce new genes into crops.

■ **Biotechnology is transforming agriculture** (p. 798, FIGURE 38.20) Genetically modified plants have the potential of increasing the quality and quantity of food worldwide.
   Web/CD Activity 38C: *Making Decisions About DNA Technology: Golden Rice*

■ **Plant biotechnology has incited much public debate** (p. 799) There are many people concerned about the unknown risks of releasing genetically modified organisms into the environment, but the potential benefits of transgenic crops should also be considered.

Self-Quiz

1. Which of the following would *definitely* be a unisexual flower? A flower that
   a. is also incomplete.
   b. lacks sepals.
   c. is self-compatible.
   d. is staminate.
   e. cannot self-pollinate.

2. Germinated pollen grain is to _____ as _____ is to female gametophyte.
   a. male gametophyte; embryo sac
   b. embryo sac; ovule
   c. ovule; sporophyte
   d. anther; seed
   e. petal; sepal

3. A seed develops from
   a. an ovum.
   b. a pollen grain.
   c. an ovule.
   d. an ovary.
   e. an embryo.

4. A fruit is a(an)
   a. mature ovary.
   b. mature ovule.
   c. seed plus its integuments.
   d. fused carpel.
   e. enlarged embryo sac.

5. Which of the following conditions is needed by almost all seeds to break dormancy?
    a. exposure to light
    b. imbibition
    c. abrasion of the seed coat
    d. exposure to cold temperatures
    e. covering of fertile soil

6. A plant that is self-incompatible has a genotype of S_5S_9 for the S-locus. It receives pollen from a plant that is S_3S_9. Which of the following is most likely to occur?
    a. All of the pollen will germinate, forming pollen tubes.
    b. None of the pollen will germinate.
    c. About half of the pollen will germinate.
    d. Fertilization will occur in about half of the flowers of the pollinated plant.
    e. Pollen from the S_3S_9 plant will secrete ribonuclease that destroys epidermal cells of the S_5S_9 stigma.

7. Plant biotechnologists use protoplast fusion mainly to
    a. culture plant cells *in vitro*.
    b. asexually propagate desirable plant varieties.
    c. introduce bacterial genes into a plant genome.
    d. study the early events following the fertilization of egg and sperm.
    e. produce new hybrid species.

8. The basal cell formed from the first division of a plant zygote will eventually develop into
    a. the suspensor that anchors the embryo and transfers nutrients.
    b. the proembryo, in which develop the procambium, protoderm, and ground meristem.
    c. the endosperm that nourishes the developing embryo.
    d. the root apex of the embryo.
    e. two cotyledons in dicots, but only one cotyledon in monocots.

9. The introduction of genes that code for *Bt* toxin from *Bacillus thuringiensis* into cotton, maize, and potatoes has raised concerns because
    a. these crops have been shown to be toxic to humans.
    b. pollen from these crops has been shown to harm monarch butterfly larvae in laboratory experiments.
    c. if these genes "escape" to related weed species, herbicides will not be able to control their growth.
    d. this bacterium is a pathogen of humans.
    e. the toxin reduces the nutrional quality of crops.

10. "Golden Rice" is a transgenic variety that
    a. is resistant to various herbicides and thus rice fields can be weeded with those herbicides.
    b. is resistant to a virus that commonly attacks rice fields.
    c. includes bacterial genes that produce a toxin that reduces damage from insect pests.
    d. produces much larger, golden grains that increase crop yields.
    e. contains daffodil genes that increase the vitamin A content of the rice.

11. What is the function of a developing seed's endosperm?

12. Explain why self-fertilization cannot occur in a dioecious plant species.

13. Explain why it is not quite accurate to describe grains such as corn and oats as "seed crops."

14. A germinating pea plant kept in a dark closet will actually grow several centimeters high before dying. What nutrient supply sustains this growth?

15. What are two advantages, compared with sexual reproduction, of asexual reproduction in plants?

16. What is *Bt* maize?

Go to the website or CD-ROM for more quiz questions.

With the self-incompatibility gene system in mind, review the ways that evolutionary forces can preserve variation in a population. Which evolutionary force is acting in the self-incompatibility system? The S-alleles are one of the most polymorphic and rapidly evolving gene systems in plants. Think of an evolutionary hypothesis to explain this fact.

With respect to sexual reproduction, some plant species are fully self-fertile, others are fully self-incompatible, and some have adopted a mixed strategy with partial self-incompatibility. These reproductive strategies differ in their implications for evolutionary potential. How, for example, might a self-incompatible species fare as small founder populations or remnant populations in a severe population bottleneck, as compared with a self-fertile species?

In agriculture, many varieties of fruits, vegetables, and ornamental plants are derived from "sports," spontaneously arising variants that often appear as side shoots on "normal" plant varieties. What cellular and genetic processes underlie this phenomenon in plants? What role might "sports" play in plant evolution?

The Process of Science

One concern about GM crops is the possibility of introduced genes moving from a transgenic crop into natural populations by gene flow and hybridization. If you wanted to develop measures to ensure that engineered crops could not cross with wild relatives, what approaches might you employ? Why would your approach depend on whether a crop is harvested for a vegetative structure versus a fruit?

Conduct experiments on the seeds of the desert plant *Pectis papposa* to find out what causes germination in the Case Study in the Process of Science, available on the website and CD-ROM.

Science, Technology, and Society

Humans have engaged in genetic manipulation for millennia, producing many plant and animal varieties through selective breeding and hybridization processes that can modify the genomes of organisms on a significant level. Why do you think modern genetic engineering, which often entails introducing or modifying only one or a few genes, has met with so much public opposition? Should some forms of genetic engineering be of greater concern than others? If so, what forms, and why?

Answers: 1. d; 2. a; 3. c; 4. a; 5. b; 6. c; 7. e; 8. a; 9. b; 10. e; 11. The endosperm stockpiles nutrients and provides them to the developing embryo. In some plant species, the endosperm also provides nutrients to the seedling when a seed germinates; in other plant species, the endosperm transfers nutrients to the embryo's cotyledons during seed development. 12. A dioecious plant has staminate (male) and carpellate (female) flowers on separate individuals, and thus no single plant can fertilize its own flowers. 13. Because a corn kernel or an oat are not naked seeds, but dry fruits containing seeds. 14. Nutrients stored in the pea's two cotyledons. 15. Any two of the following: sexual reproduction can clone genotypes that are successful in a particular environment; asexual reproduction is generally faster than sexual reproduction; the progeny of asexual reproduction are generally less frail than the seedlings resulting from sexual reproduction. 16. Transgenic maize containing a bacterial gene that codes for a protein that is toxic to many insects that damage crops.

PLANT RESPONSES TO INTERNAL AND EXTERNAL SIGNALS

At every stage *in the life of a plant, sensitivity to the environment and coordination of responses are evident. One part of a plant can send signals to other parts. For example, the terminal bud at the apex of a shoot is able to suppress the growth of axillary buds that may be many meters away. Plants keep track of the time of day and the time of year. They can sense gravity and the direction of light. For example, the grass seedling in the photograph on this page is growing toward light. A plant's morphology and physiology are constantly tuned to its variable surroundings by complex interactions between environmental stimuli and internal signals.*

This chapter focuses on how plants respond to these external and internal cues. At the organismal *level, plants and animals respond to environmental stimuli by very different means. Animals, being mobile, respond mainly by behavioral mechanisms, moving toward positive stimuli and away from negative stimuli. Rooted to one location for life, a plant generally responds to environmental cues by adjusting its pattern of growth and development. For this reason, plants of the same species vary in body form much more than do animals of the same species. All lions have four limbs and approximately the same body proportions, but oak trees vary considerably in their number of limbs and their shapes. But at the* cellular *level, plants, animals, and all other eukaryotes are surprisingly similar in their signaling mechanisms. We begin our study of plant response with the role of signal transduction in plant cells.*

SIGNAL TRANSDUCTION AND PLANT RESPONSES

All organisms, including plants, have the ability to receive specific environmental and internal signals and respond to them in ways that enhance survival and reproductive success. Bees, which possess UV-sensitive photoreceptors in their eyes, can discern nectar-guiding patterns on flower petals—patterns that are completely invisible to humans. A dog sniffing a fire hydrant receives information about its olfactory world, the nature of which we, with our inferior sense of smell, can only surmise. Plants, too, have cellular receptors that they use to detect important changes in their environment, whether the change is an increase in the concentration of a growth hormone, an injury from a caterpillar munching on leaves, or a decrease in day length as winter approaches.

In order for an internal or external stimulus to elicit a physiological response, certain cells within the organism must possess an appropriate receptor, a molecule that is sensitive to and affected by the specific stimulus. For example, our blindness to UV-reflecting floral patterns is due to our eyes lacking the necessary UV photoreceptors. Upon receiving a stimulus, a receptor initiates a specific series of biochemical steps, a signal transduction pathway, which couples reception of the stimulus to the response of the organism. As you will learn, plants are sensitive to a wide range of internal and external stimuli, and each of these stimuli initiates a specific signal transduction pathway. We introduced general concepts of signal transduction in cells in Chapter 11. Here, we apply those concepts to specific examples in plants.

Signal-transduction pathways link internal and environmental signals to cellular responses

In the back corner of a food cabinet, a long forgotten potato (a modified underground stem, or tuber) has sprouted shoots from its "eyes" (axillary buds). These shoots, however, scarcely resemble those of a typical plant. Instead of broad green leaves and sturdy stems, these dark-sprouted shoots are ghostly pale, have stems that are long and thin, bear leaves that are unexpanded, and give rise to roots that are reduced (FIGURE 39.1a). Seedlings germinated

in the dark have a similar appearance. These morphological adaptations for growing in darkness make sense if we consider that a potato or a seedling, in nature, usually encounters continuous darkness when sprouting underground. In such circumstances, the pre-emergent plant is supported by the surrounding soil and would not benefit from a thick stem. Expanded leaves would be a hindrance to soil penetration and would be damaged as the shoots pushed upward. Because the leaves are unexpanded, there is little evaporative loss of water, and little requirement for an extensive root system to replace the water lost by transpiration. The energy expended in producing green chlorophyll would be wasted since there is no light for photosynthesis. Rather, a plant growing in the dark allocates as much energy as possible to the elongation of stems. This "strategy" enables the shoots to break ground before the stored food reserves in the tuber or seed are exhausted.

Once a shoot reaches the sunlight, its morphology and biochemistry undergo profound changes collectively called **greening:** the elongation rate of the stems slows, the leaves expand, the roots start to elongate, and the entire shoot begins to produce chlorophyll; in short, it begins to resemble a typical plant (FIGURE 39.1b). In the following discussion, we use this greening response as an example of how a plant cell receives a signal—in this case, light—and how this reception is transduced into a response (greening). Along the way we will explore how studies of mutants have provided valuable insights into the roles played by various molecules in the three stages of cell-signal processing: reception, transduction, and response (FIGURE 39.2).

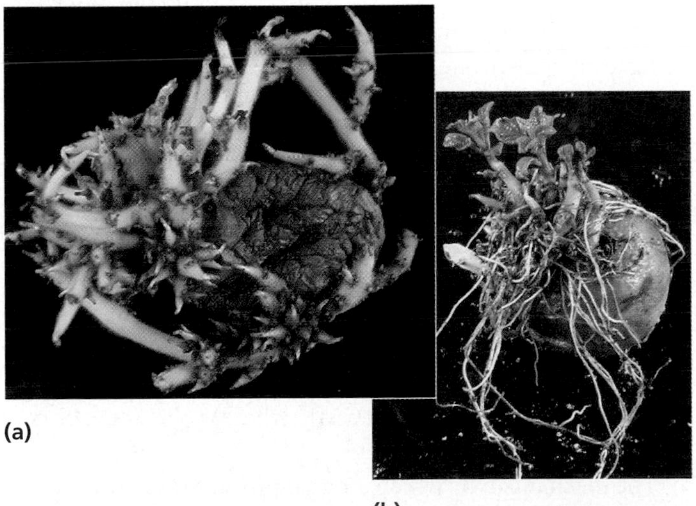

FIGURE 39.1 Light-induced greening of dark-sprouted potatoes. (a) A dark-grown potato has tall, spindly stems and nonexpanded leaves—a morphological adaptation that enables the shoots to penetrate the soil. The roots are short, but little water is lost by the shoots. **(b)** After a week's exposure to natural daylight, the potato plant begins to resemble a typical plant with broad green leaves, short sturdy stems, and long roots. This transformation begins with the reception of light by a specific pigment.

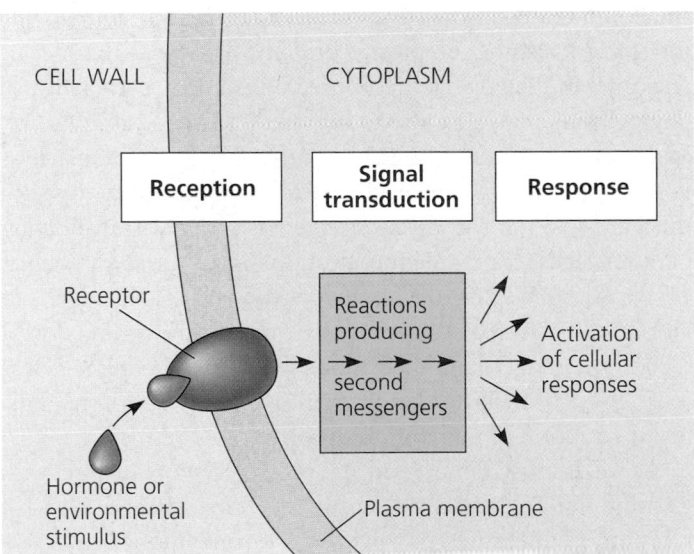

FIGURE 39.2 Review of a general model for signal-transduction pathways. A hormone or other signal binding to a specific receptor stimulates the cell to produce second messengers. Second messengers trigger the cell's various responses to the original signal. In this diagram, the receptor is on the surface of the target cell. In other cases, hormones enter cells and bind to specific receptors inside.

Reception

Signals, whether internal or external, are first detected by receptors, proteins that undergo conformational changes in response to a specific stimulus. The receptor involved in greening in plants is called phytochrome, which consists of a light-absorbing pigment attached to a specific protein. Unlike many receptors, which are built into the plasma membrane, the phytochrome that functions in greening occurs in the cytoplasm. Researchers have demonstrated the requirement for phytochrome in greening through studies of a tomato mutant called *aurea*. The mutant, which has lower-than-normal levels of phytochrome, greens less than wild-type tomatoes when exposed to light. (The name *aurea* comes from the Latin for "gold-colored." In the absence of chlorophyll, the yellow plant pigments called carotenoids are more obvious.) Researchers were able to produce a normal greening response by using a microneedle to inject phytochrome extracted from other plants into *aurea* leaf cells and then exposing the *aurea* to light. Such experiments support the hypothesis that phytochrome functions in light detection in the greening process.

Transduction

The greening response is triggered by extremely low levels of light. For example, light levels equivalent to a few seconds of moonlight are sufficient to cause a slowing of stem elongation in dark-grown oat seedlings. Receptors such as phytochrome are sensitive to very weak environmental and chemical signals. How is the information from these extremely weak signals amplified, and how is their reception transduced into a specific response by the plant? The answer is **second messengers**— small, internally produced chemicals that transfer and amplify the signal from the receptor to proteins that cause the specific response. In the greening response, for example, each activated phytochrome may give rise to hundreds of molecules of a second messenger, each of which, in turn, may lead to the activation of hundreds of molecules of a specific enzyme. By such mechanisms, the second messenger of a signal transduction pathway leads to a rapid amplification of the signal. In Chapter 11, we examined this role of second messengers in general (see FIGURE 11.12). Here, let's specifically consider the production of second messengers and their function in the greening response (FIGURE 39.3). Refer to the diagram frequently as you read the text description of this complex process.

As we discussed in Chapter 11, many receptors interact with guanine-binding proteins (G-proteins). Phytochrome is such a receptor. Light causes phytochrome to undergo a conformational change, and the phytochrome then interacts with a specific G-protein. During activation, guanosine diphosphate (GDP) that is bound to the inactive G-protein is displaced by guanosine triphosphate (GTP). The G-protein, now in its active form, in turn activates other enzymes in the signal-transduction pathway that leads to greening. For example, phytochrome-

activated G-proteins activate guanyl cyclase, the enzyme that produces cyclic GMP, a second messenger. In *aurea* tomato cells, G-protein inhibitors such as cholera toxin block greening after microinjection of phytochrome, whereas G-protein activators such as pertussis toxin stimulate the response.

The cyclic nucleotides are second messengers that include cyclic adenosine monophosphate (cyclic AMP) and cyclic guanosine monophosphate (cyclic GMP). In some cases, cyclic nucleotides activate specific protein kinases (enzymes that phosphorylate and activate other proteins—see FIGURE 11.11). Experiments indicate that cyclic GMP (cGMP) is involved in the greening process. The microinjection of cGMP into *aurea* tomato cells induces a partial greening response, even without addition of phytochrome.

Changes in cytosolic Ca^{2+} also play an important role in phytochrome signal transduction. The levels of Ca^{2+} are generally very low in the cytosol (about 10^{-7} M). However, a wide range of hormonal and environmental stimuli can cause brief increases in cytosolic Ca^{2+}. In many cases, Ca^{2+} then binds directly to small proteins called calmodulins. The Ca^{2+}-calmodulin complex then binds to and activates several enzymes, most notably certain types of protein kinases. Note in FIGURE 39.3 that phytochrome activation during the greening mechanism results in both cGMP and Ca^{2+}-calmodulin as second messengers.

Response

Ultimately, a signal-transduction pathway leads to the regulation of one or more cellular activities. In most cases, especially when changes in development are involved, these responses to stimulation involve the increased activity of certain enzymes. The two main mechanisms by which a signaling pathway can activate an enzyme are by stimulating transcription of mRNA for the enzyme or by activating existing enzyme molecules (post-translational modification).

Transcriptional Regulation. Transcription factors bind directly to specific regions of DNA and control the transcription of specific genes (see FIGURE 19.8). In the case of phytochrome-induced greening, several transcription factors are activated by phosphorylation in response to the appropriate light conditions. The activation of some of these transcription factors depends upon cyclic GMP, whereas the activation of others requires Ca^{2+}-calmodulin.

The mechanism by which a signal promotes a new developmental course may depend on the activation of positive transcriptional factors (proteins that *increase* transcription of specific genes) or the deactivation of negative transcriptional factors (proteins that decrease transcription), or both. For example, there are *Arabidopsis* mutants that, except for their pale color, have a light-grown morphology (expanded leaves and short, sturdy stems) when grown in the dark (they are not green because the final step in chlorophyll production requires

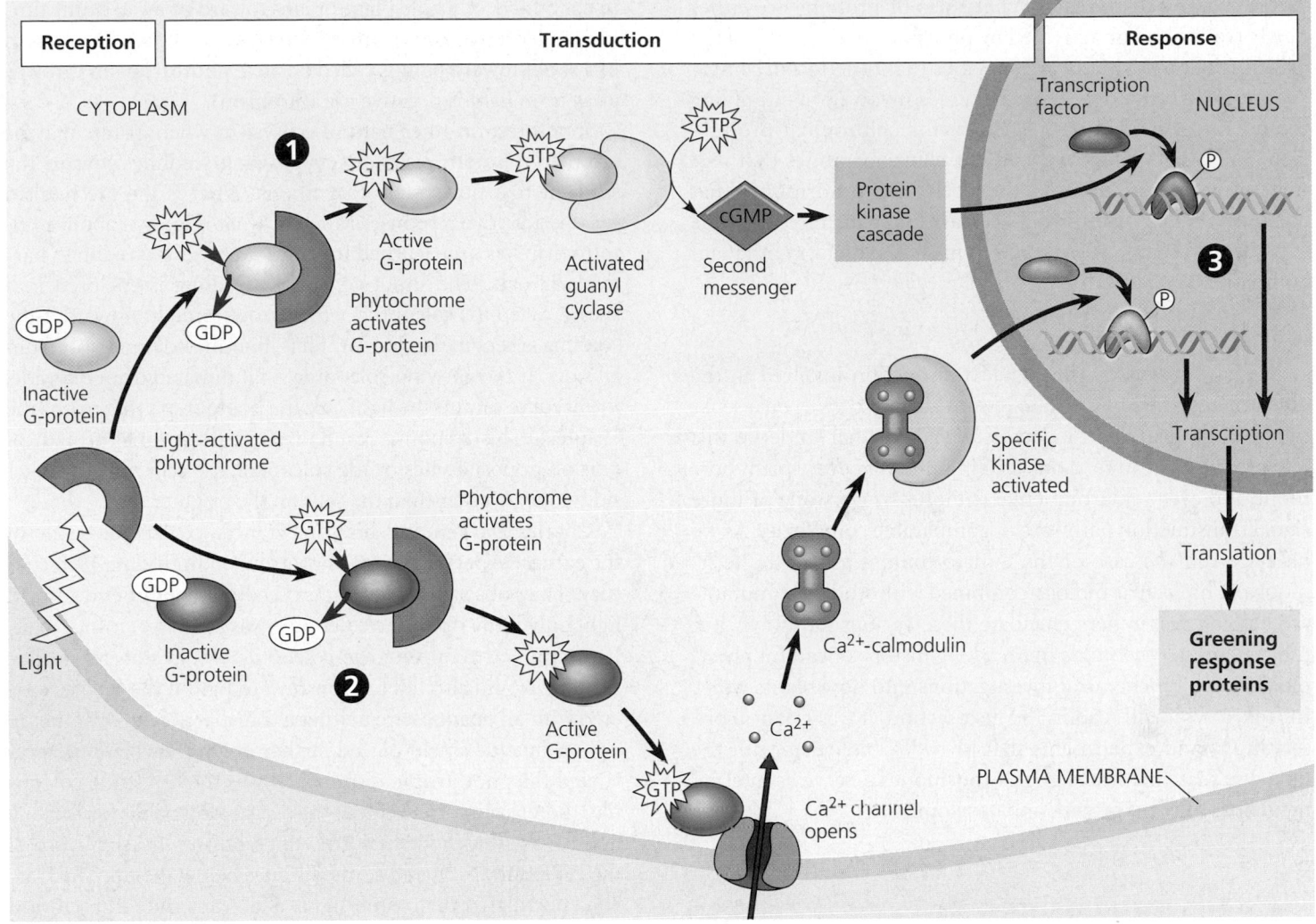

Reception	Transduction	Response

CYTOPLASM

NUCLEUS

Transcription factor

1

GTP

GTP

GTP

Active G-protein

Phytochrome activates G-protein

Activated guanyl cyclase

cGMP

Second messenger

Protein kinase cascade

GDP

Inactive G-protein

GDP

Light-activated phytochrome

3

P

P

Transcription

GTP

GDP

GDP

Inactive G-protein

Phytochrome activates G-protein

2

GTP

Active G-protein

GTP

Specific kinase activated

Translation

Light

Ca²⁺-calmodulin

Ca²⁺

Ca²⁺ channel opens

Greening response proteins

PLASMA MEMBRANE

FIGURE 39.3 An example of signal transduction in plants: the role of phytochrome in the greening response. The light signal is detected by the phytochrome receptor, which then activates at least two signal-transduction pathways involving G-proteins. **1** One pathway leads to cGMP as a second messenger that activates a protein kinase cascade (see FIGURE 11.11). **2** The other pathway leads to formation of a Ca²⁺-calmodulin complex that activates a specific protein kinase. **3** Both pathways lead to expression of genes for the proteins that function in the greening response. **We do not present such a complex diagram because its details are important for you to remember. The main point of this diagram: Signal transduction pathways in plants are variations of the general schemes you learned about in Chapter 11, involving such components as G-proteins, second messengers, and protein kinases.**

direct light). These mutants have defects in a negative transcriptional factor that inhibits the expression of other genes normally activated by light. When the negative factor is eliminated by mutation, the pathway that it normally blocks becomes activated: hence, these mutants, except for their pale color, appear to have been grown in the light.

Post-Translational Modification of Proteins. Although the synthesis of new proteins by transcription and translation are important molecular events associated with greening, so are post-translational modifications of existing proteins. Most often these existing proteins are modified by phosphorylation, the addition of a phosphate group onto the protein. The diverse proteins called protein kinases catalyze this phosphorylation of target proteins (see FIGURE 11.11). Many second messengers, such as cGMP, and some receptors themselves, including some forms

of phytochrome, activate protein kinases directly. About 2–3% of all plant genes may encode for protein kinases. Often one protein kinase will phosphorylate another protein kinase, which then phosphorylates another, and so on. Such kinase cascades may eventually link initial stimuli to responses at the level of gene expression, usually via the phosphorylation of transcription factors. By such mechanisms, many signal pathways ultimately regulate the synthesis of new proteins, usually by turning specific genes on or off (FIGURE 39.3).

Signal pathways must also have a means for turning off once the initial signal is no longer present—for example, what if we put the potato back into the cupboard? Protein phosphatases, enzymes that dephosphorylate specific proteins, are involved in these "switch-off" processes. At any given moment, the activities of a cell depend upon the balance of activity of many types of protein kinases and protein phosphatases.

The Greening Proteins. What sorts of proteins are either newly transcibed or activated by phosphorylation during the greening process? Many are enzymes that function in photosynthesis directly; others are enzymes involved in supplying the chemical precursors necessary for chlorophyll production; still others affect the levels of plant hormones that regulate growth. For example, the levels of two hormones that enhance stem elongation will decrease following phytochrome activation—hence, the reduction in stem elongation that accompanies greening.

■ ■ ■

We have discussed the signal transduction involved in the greening response of a potato in some detail to give you a sense of the complexity of biochemical changes that underlie this one process. As you read on, keep in mind that every plant hormone and every environmental stimulus triggers one or more signal transduction pathways of comparable complexity. As we have seen in the case of the tomato mutant *aurea,* the techniques of molecular biology combined with studies of mutants are helping researchers elucidate these various pathways. But molecular biology builds upon a long history of careful physiological and biochemical investigations into how plants work. Indeed, as you will read in the next section, it was classical observations and experiments that provided biologists with the first clue that chemical signals—hormones—serve as internal regulators of plant growth and development.

PLANT RESPONSES TO HORMONES

The word *hormone* is derived from a Greek verb meaning "to excite." Found in all multicellular organisms, **hormones** are chemical signals that coordinate the parts of the organism. By definition, a hormone is a compound that is produced by one part of the body and then transported to other parts of the body, where it binds to a specific receptor and triggers responses in target cells and tissues. Another important characteristic of hormones is that only minute amounts are required to induce substantial change in an organism. Hormone concentrations or the rates of their transport can change in response to environmental stimuli. Often the response of a plant is governed by the interaction of two or more hormones.

Research on how plants grow toward light led to the discovery of plant hormones

THE PROCESS OF SCIENCE

The concept of chemical messengers in plants emerged from a series of classic experiments on how stems respond to light. A houseplant on a windowsill grows toward light. If you rotate the plant, it will soon reorient its growth until its leaves again face the window. Any growth response that results in curvatures of whole plant organs toward or away from stimuli is called a **tropism** (from the Greek *tropos,* turn). The growth of a shoot toward light is called positive **phototropism** (growth away from light is negative phototropism).

In a forest or other natural ecosystem where plants may be crowded, phototropism directs growing seedlings toward the sunlight that powers photosynthesis. What is the mechanism for this adaptive response? Much of what is known about phototropism has been learned from studies of grass seedlings, particularly oats. The shoot of a grass seedling is enclosed in a sheath called the coleoptile, which grows straight upward if the seedling is kept in the dark or if it is illuminated uniformly from all sides. If the growing coleoptile is illuminated from one side, it will curve toward the light (see the photograph that opens the chapter). This response results from a differential growth of cells on opposite sides of the coleoptile; the cells on the darker side elongate faster than the cells on the brighter side.

Charles Darwin and his son, Francis, conducted some of the earliest experiments on phototropism in the late 19th century. They observed that a grass seedling could bend toward light only if the tip of the coleoptile was present (FIGURE 39.4). If the tip was removed, the coleoptile would not curve. The seedling would also fail to grow toward light if the tip was covered with an opaque cap; neither a transparent cap over the tip nor an opaque shield placed farther down the coleoptile prevented the phototropic response. It was the tip of the coleoptile, the Darwins concluded, that was responsible for sensing light. However, the actual growth response, the curvature of the coleoptile, occurred some distance below the tip. The Darwins postulated that some signal was transmitted downward from the tip to the elongating region of the coleoptile. A few decades later, Peter Boysen-Jensen of Denmark tested this hypothesis and demonstrated that the signal was a mobile chemical substance. He separated the tip from the remainder of the coleoptile by a block of gelatin, which would prevent cellular contact but allow chemicals to pass. These seedlings behaved normally, bending toward light. However, if the tip was experimentally segregated from the lower coleoptile by an impermeable barrier, no phototropic response occurred.

In 1926, F. W. Went, a Dutch graduate student, extracted the chemical messenger for phototropism by modifying the experiments of Boysen-Jensen (FIGURE 39.5). Went removed the coleoptile tip and placed it on a block of agar, a gelatinous material. The chemical messenger from the tip, Went reasoned, should diffuse into the agar, and the agar block should then be able to substitute for the coleoptile tip. Went placed the agar blocks on decapitated coleoptiles that were kept in the dark. A block that was centered on top of the coleoptile caused the stem to grow straight upward. However, if the block was placed off center, then the coleoptile began to bend away from the side with the agar block, as though growing toward light. Went concluded that the agar block contained a chemical produced in the coleoptile tip, that this chemical stimulated growth as it

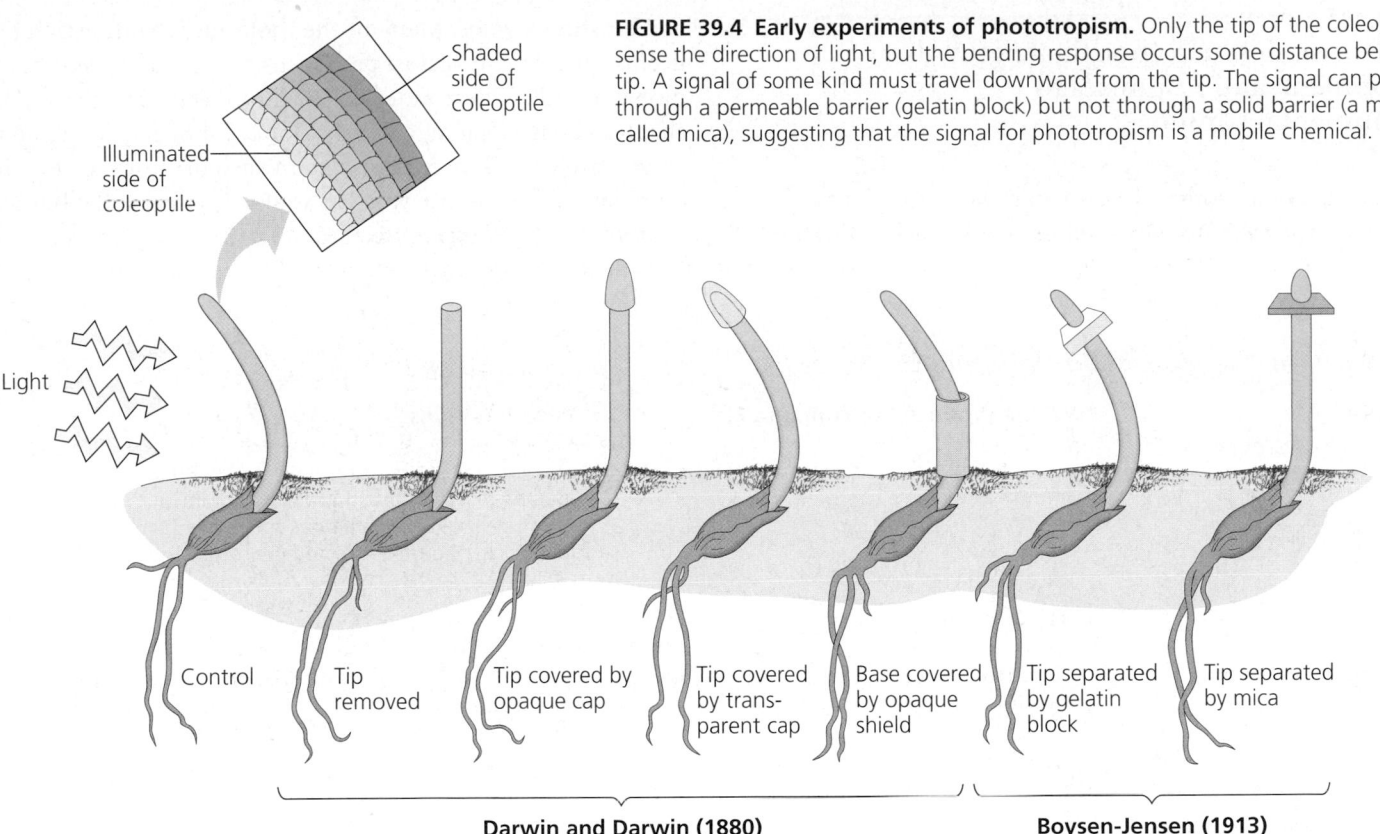

FIGURE 39.4 Early experiments of phototropism. Only the tip of the coleoptile can sense the direction of light, but the bending response occurs some distance below the tip. A signal of some kind must travel downward from the tip. The signal can pass through a permeable barrier (gelatin block) but not through a solid barrier (a mineral called mica), suggesting that the signal for phototropism is a mobile chemical.

Shaded side of coleoptile

Illuminated side of coleoptile

Light

Control

Tip removed

Tip covered by opaque cap

Tip covered by transparent cap

Base covered by opaque shield

Tip separated by gelatin block

Tip separated by mica

Darwin and Darwin (1880)

Boysen-Jensen (1913)

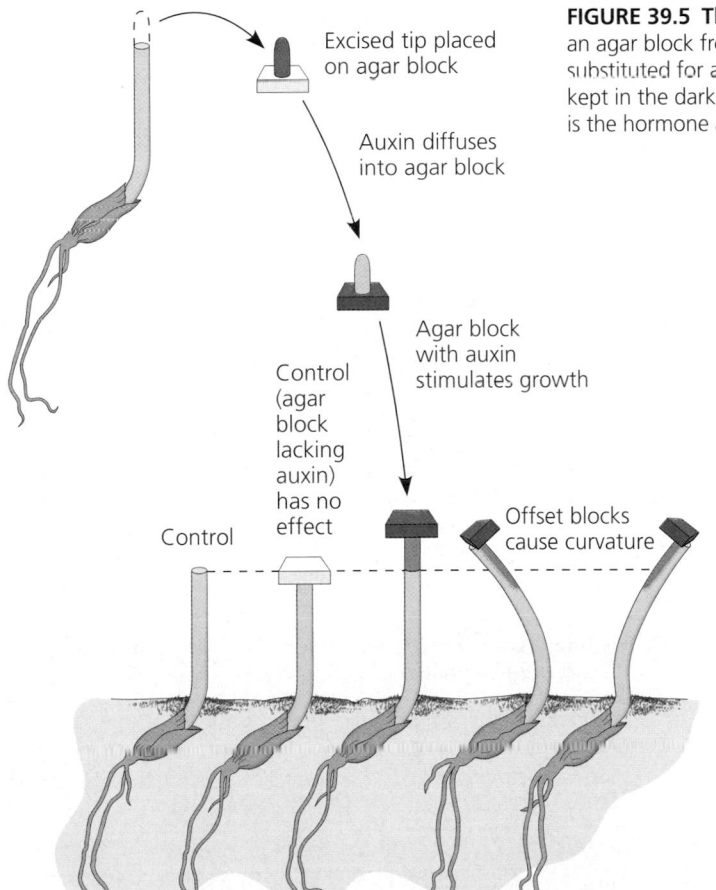

Excised tip placed on agar block

Auxin diffuses into agar block

Agar block with auxin stimulates growth

Control (agar block lacking auxin) has no effect

Control

Offset blocks cause curvature

FIGURE 39.5 The Went experiments. Some chemical (indicated by pink) that can pass into an agar block from a coleoptile tip stimulates elongation of the coleoptile when the block is substituted for a tip. If the block is placed off-center on the top of a decapitated coleoptile kept in the dark, the organ bends as if responding to illumination from one side. The chemical is the hormone auxin, which stimulates elongation of cells in the shoot.

passed down the coleoptile, and that a coleoptile curved toward light because of a higher concentration of the growth-promoting chemical on the darker side of the coleoptile. For this chemical messenger, or hormone, Went chose the name auxin (from the Greek *auxein*, to increase). Auxin was later purified and its structure determined by Kenneth Thimann and his colleagues at the California Institute of Technology.

The classical hypothesis for what causes grass coleoptiles to grow toward light, based on the work of the Darwins and Went, is that an asymmetrical distribution of auxin moving down from the coleoptile tip causes cells on the darker side to elongate faster than cells on the brighter side. However, studies of phototropism by organs other than grass coleoptiles provide less support for this idea. For example, there is no evidence that unilateral light causes an asymmetrical distribution of auxin in the stems of sunflowers, radishes, and other dicots. There *is*, however, an asymmetrical distribution of certain substances that may act as growth *inhibitors*, with these substances more concentrated on the lighted side of a stem. Still, the specific case of auxin's role in the phototropism of grass coleoptiles opened up the whole field of research on plant hormones.

Plant hormones help coordinate growth, development, and responses to environmental stimuli

TABLE 39.1 previews some of the major classes of plant hormones: auxin, cytokinins, gibberellins, abscisic acid, ethylene, and brassinosteroids. Many of the molecules that function in defense of the plant against pathogens are probably plant hormones as well. (These molecules will be discussed later in the chapter.) Notice that all of the hormones shown in TABLE 39.1 are relatively small molecules. Their transport from cell to cell often involves passage across cell walls, a pathway that blocks the movement of large molecules.

Table 39.1 An Overview of Plant Hormones

Hormone	Where Produced or Found in Plant	Major Functions
Auxin (IAA)	Embryo of seed, meristems of apical buds, young leaves	Stimulates stem elongation (low concentration only) root growth, cell differentiation, and branching; regulates development of fruit; enhances apical dominance; functions in phototropism and gravitropism.
Cytokinins (such as zeatin)	Synthesized in roots and transported to other organs	Affect root growth and differentiation; stimulate cell division and growth; stimulate germination; delay senescence.
Gibberellins (such as GA_3)	Meristems of apical buds and roots, young leaves, embryo	Promote seed and bud germination, stem elongation, and leaf growth; stimulate flowering and development of fruit; affect root growth and differentiation
Abscisic acid	Leaves, stems, roots, green fruit	Inhibits growth; closes stomata during water stress; counteracts breaking of dormancy
Ethylene	Tissues of ripening fruits, nodes of stems, aging leaves and flowers	Promotes fruit ripening, opposes some auxin effects; promotes or inhibits growth and development of roots, leaves, and flowers, depending on species
Brassinosteroids (such as brassinolide)	Seeds, fruits, shoots, leaves, and floral buds	Inhibits root growth; retards leaf abscission; promotes xylem differentiation

In general, hormones control plant growth and development by affecting the division, elongation, and differentiation of cells. Some hormones also mediate shorter-term physiological responses of plants to environmental stimuli. Each hormone has multiple effects, depending on its site of action, its concentration, and the developmental stage of the plant.

Plant hormones are produced in very low concentrations, but a minute amount of hormone can have a profound effect on the growth and development of a plant organ. This implies that the hormonal signal must be amplified in some way. A hormone may act by altering the expression of genes, by affecting the activity of existing enzymes, or by changing properties of membranes. Any of these actions could redirect the metabolism and development of a cell responding to a small number of hormone molecules. Signal transduction pathways amplify the hormonal signal and connect it to a cell's specific responses.

Response to a hormone usually depends not so much on the absolute amount of that hormone as on its relative concentration compared with other hormones. It is hormonal balance, rather than hormones acting in isolation, that may control the growth and development of the plant. These interactions will become apparent in the following survey of hormone function.

Auxin

The term **auxin** is used for any chemical substance that promotes the elongation of coleoptiles, although auxins actually have multiple functions in both monocots and dicots. The natural auxin occurring in plants is indoleacetic acid, or IAA, but several other compounds, including some synthetic ones, have auxin activity. Throughout this chapter, however, the name *auxin* is used specifically to refer to IAA. Although auxin was the first plant hormone to be discovered, much remains to be learned about auxin signal transduction and the regulation of auxin biosynthesis. Current evidence suggests that auxin is produced from the amino acid tryptophan in the shoot tips of plants.

The speed at which auxin is transported down the stem from the shoot apex is about 10 mm per hour—much too fast for diffusion, although slower than translocation in phloem. Auxin seems to be transported directly through parenchyma tissue, from one cell to the next. It moves only from shoot tip to base, not in the reverse direction. This unidirectional transport of auxin is called polar transport. Polar transport has nothing to do with gravity, for auxin travels upward in experiments where a stem or coleoptile segment is placed upside down. Polar auxin transport requires energy. FIGURE 39.6 illustrates how proton pumps in the plasma membrane, driven by ATP, couple metabolic energy to auxin transport. The mechanism of polar auxin transport is another example of cellular work driven by chemiosmosis, the harnessing of H^+ gradients generated by proton pumps (see Chapter 9).

Although auxin affects several aspects of plant development, one of its chief functions is to stimulate the elongation of cells in young developing shoots.

The Role of Auxin in Cell Elongation. The apical meristem of a shoot is a major site of auxin synthesis. As auxin from the shoot apex moves down to the region of cell elongation (see Chapter 35), the hormone stimulates growth of the cells, probably by binding to a receptor built into the plasma membrane. Auxin stimulates growth only over a certain concentration range, from about 10^{-8} to 10^{-4} M. At higher concentrations,

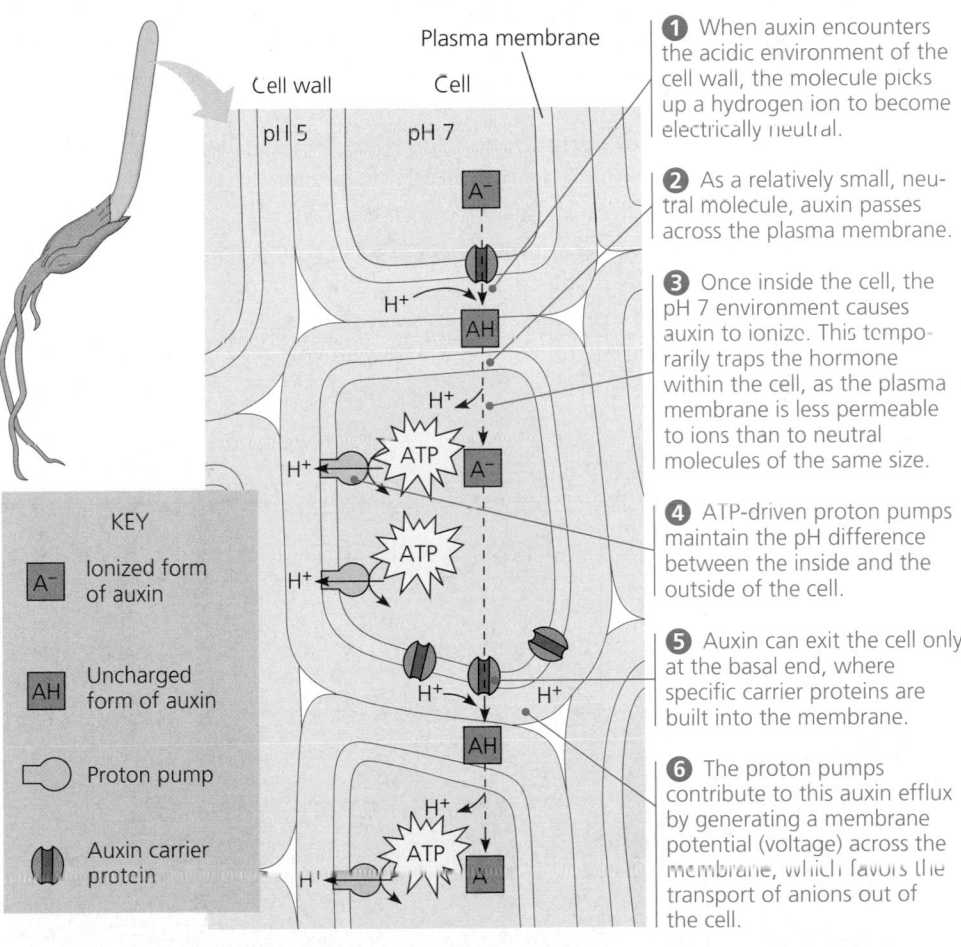

Plasma membrane

Cell wall   Cell

pH 5   pH 7

1 When auxin encounters the acidic environment of the cell wall, the molecule picks up a hydrogen ion to become electrically neutral.

2 As a relatively small, neutral molecule, auxin passes across the plasma membrane.

3 Once inside the cell, the pH 7 environment causes auxin to ionize. This temporarily traps the hormone within the cell, as the plasma membrane is less permeable to ions than to neutral molecules of the same size.

4 ATP-driven proton pumps maintain the pH difference between the inside and the outside of the cell.

5 Auxin can exit the cell only at the basal end, where specific carrier proteins are built into the membrane.

6 The proton pumps contribute to this auxin efflux by generating a membrane potential (voltage) across the membrane, which favors the transport of anions out of the cell.

KEY

A^- Ionized form of auxin

AH Uncharged form of auxin

Proton pump

Auxin carrier protein

FIGURE 39.6 Polar auxin transport: a chemiosmotic model. In growing shoots, auxin is transported unidirectionally, from the apex down the shoot. Along this pathway, the hormone enters a cell at the apical end, exits at the basal end, diffuses across the wall, and enters the apical end of the next cell.

auxin may inhibit cell elongation, probably by inducing the production of ethylene, a hormone that generally acts as an inhibitor of elongation. We will return to this hormonal interaction in the discussion of ethylene.

According to a proposal called the acid growth hypothesis, proton pumps play a major role in the growth response of cells to auxin. In a shoot's region of elongation, auxin stimulates plasma membrane proton pumps, an action that, within minutes, both increases the voltage across the membrane (membrane potential) and lowers the pH in the cell wall (FIGURE 39.7). The acidification of the wall activates enzymes called **expansins** that break the cross-links (hydrogen bonds) between cellulose microfibrils, loosening the fabric of the wall. (Expansins can even weaken the integrity of filter paper made of pure cellulose.) Increasing the membrane potential enhances ion uptake into the cell, which causes the osmotic uptake of water. The uptake of water, along with the increased plasticity of the walls, enables the cell to elongate.

Auxin also alters gene expression rapidly, causing cells in the region of elongation to produce new proteins within minutes. Some of these proteins are short-lived transcription factors that repress or activate the expression of other genes. For sustained growth after this initial spurt, cells must make more cytoplasm and wall material. Auxin also stimulates this sustained growth response.

Lateral and Adventitious Root Formation. Auxins are used commercially in the vegetative propagation of plants by cuttings. Treating a detached leaf or stem with rooting powder containing auxin often causes adventitious roots to form near the cut surface. Auxin is also involved in the branching of roots. Researchers found that an *Arabidopsis* mutant that exhibits extreme proliferation of lateral roots has an auxin concentration 17-fold higher than normal.

Auxins as Herbicides. Synthetic auxins, such as 2,4-dinitrophenol (2,4-D), are widely used as herbicides. Monocots, such as maize and turfgrass, can rapidly inactivate these synthetic auxins, but dicots cannot, and die from a hormonal overdose. Spraying cereal fields or turf with 2,4-D eliminates dicot (broadleaf) weeds such as dandelions.

Other Effects of Auxin. In addition to stimulating cell elongation for primary growth, auxin affects secondary growth by inducing cell division in the vascular cambium and by influencing the differentiation of secondary xylem (see Chapter 35). Developing seeds synthesize auxin, which promotes the growth of fruits in plants. Synthetic auxins sprayed on tomato vines induce fruit development without a need for pollination. This makes it possible to grow seedless tomatoes by substituting synthetic auxin for the auxin normally synthesized by the developing seeds.

Cytokinins

Trial-and-error attempts to find chemical additives that would enhance the growth and development of plant cells in tissue culture led to the discovery of **cytokinins.** In the 1940s, Johannes van Overbeek, working at the Cold Spring Harbor Laboratory in New York, found he could stimulate the growth of plant embryos by adding coconut milk, the liquid endosperm of a coconut's giant seed, to his culture medium. A decade later, Folke Skoog and Carlos O. Miller, at the University of Wisconsin, induced cultured tobacco cells to divide by adding degraded samples of DNA. The active ingredients of both experimental additives turned out to be modified forms of adenine, one of the components of nucleic acids. These growth regulators were named cytokinins because they stimulate cytokinesis, or cell division. Of the variety of cytokinins that occur

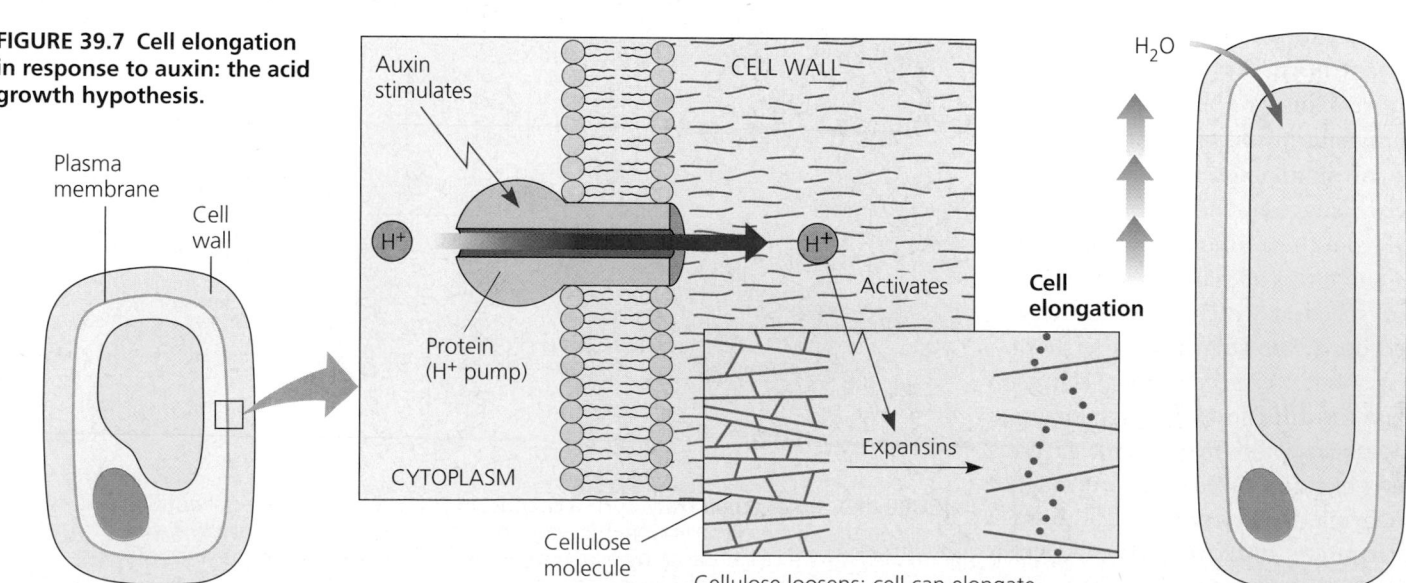

FIGURE 39.7 Cell elongation in response to auxin: the acid growth hypothesis.

Plasma membrane

Cell wall

Auxin stimulates

CELL WALL

Protein (H⁺ pump)

CYTOPLASM

Cellulose molecule

Activates

Expansins

Cellulose loosens; cell can elongate

H_2O

Cell elongation

naturally in plants, the most common is zeatin, so named because it was first discovered in maize (*Zea mays*).

Despite much effort, the enzyme that produces cytokinins has neither been purified from plants nor has the gene that encodes for it been identified. Mark Holland of Salisbury State University has proposed that plants may not even produce their own cytokinins: His hypothesis is that cytokinins are actually produced by prokaryotes called methylobacteria that live symbiotically inside actively growing plant tissues, even in *in vitro* cultures. Indeed, normal developmental processes are impaired when methylobacteria are eliminated, and these processes can be restored either by the reapplication of methylobacteria or the addition of cytokinins. Regardless of whether this provocative hypothesis gains support, the issue underscores the importance of genomic sequencing. Now that the sequencing of *Arabidopsis* is complete, the gene for a cytokinin-producing enzyme, if present, should be more easily identified.

Regardless of the source of cytokinins, plant cells do have cytokinin receptors. Some evidence points to the existence of two entirely different classes of cytokinin receptors, one intracellular and the other on the cell surface. The cytoplasmic receptor binds cytokinins directly and can stimulate transcription in isolated nuclei. In some plant cells, cytokinins open Ca^{2+} channels in the plasma membrane, causing an increase in cytosolic Ca^{2+}. Although much remains to be learned about cytokinin synthesis and signal transduction, some of the major effects that cytokinins have on the physiology and development of plants are well documented.

Control of Cell Division and Differentiation.
Cytokinins are produced in actively growing tissues, particularly in roots, embryos, and fruits. Cytokinins produced in the root reach their target tissues by moving up the plant in the xylem sap. Acting in concert with auxin, cytokinins stimulate cell division and influence the pathway of differentiation. The effects of cytokinins on cells growing in tissue culture provide clues about how this class of hormones may function in an intact plant. When a piece of parenchyma tissue from a stem is cultured in the absence of cytokinins, the cells grow very large but do not divide. Cytokinins alone have no effect. If, however, cytokinins are added along with auxin, the cells divide. The ratio of cytokinin to auxin controls the differentiation of the cells. When the concentrations of the two hormones are balanced, the mass of cells continues to grow, but it remains a cluster of undifferentiated cells called a callus. If cytokinin levels are raised, shoot buds develop from the callus. If auxin levels are raised, roots form.

Control of Apical Dominance.
Cytokinins, auxin, and other factors interact in the control of apical dominance, the ability of the terminal bud to suppress the development of axillary buds. Until recently, the leading hypothesis to explain the hormonal regulation of apical dominance—the direct inhibition hypothesis—proposed that auxin and cytokinin act antagonistically in regulating axillary bud growth. According to this view, auxin transported down the shoot from the terminal bud directly inhibited axillary buds from growing, causing a shoot to lengthen at the expense of lateral branching. Meanwhile, cytokinins entering the shoot system from roots counter the action of auxin by signaling axillary buds to begin growing. Thus, the ratio of auxin and cytokinin was viewed as the critical factor in controlling axillary bud inhibition. Many observations are consistent with the direct inhibition hypothesis. If the terminal bud, the primary source of auxin, is removed, the inhibition of axillary buds is removed and the plant becomes bushier (FIGURE 39.8). Application of auxin to the cut surface of the decapitated seedling resuppresses the growth of the lateral buds. Mutants that overproduce cytokinins or plants treated with cytokinins also tend to be bushier than normal. One prediction of the direct inhibition hypothesis not borne out by experiment is that decapitation, by removing the primary source of auxin, should lead to a decrease in the auxin levels of axillary buds. Biochemical studies, however, have revealed the opposite: auxin levels actually *increase* in the axillary buds of decapitated plants. The direct

Axillary buds

"Stump" after removal of apical bud

Lateral branches

(a) (b)

FIGURE 39.8 Apical dominance. (a) Auxin from the apical bud inhibits the growth of axillary buds. This favors elongation of the shoot's main axis. Cytokinins, which are transported upward from roots, counter auxin, stimulating the growth of axillary buds. This explains why, in many plants, axillary buds near the shoot tip are less likely to grow than those closer to the roots. **(b)** Removal of the apical bud from the same plant enabled lateral branches to grow.

inhibition hypothesis, therefore, does not account for all experimental findings: It is likely that plant biologists have not uncovered all the pieces of this puzzle.

Anti-Aging Effects.   Cytokinins retard the aging of some plant organs by inhibiting protein breakdown, by stimulating RNA and protein synthesis, and by mobilizing nutrients from surrounding tissues. If leaves removed from a plant are dipped in a cytokinin solution, they stay green much longer than otherwise. Cytokinins also slow the deterioration of leaves on intact plants. Because of this anti-aging effect, florists use cytokinin sprays to keep cut flowers fresh.

Gibberellins

THE PROCESS OF SCIENCE A century ago, farmers in Asia noticed that some rice seedlings in their paddies grew so tall and spindly that they toppled over before they could mature and flower. In 1926, the Japanese plant pathologist E. Kurosawa discovered that a fungus of the genus *Gibberella* caused this "foolish seedling disease" (FIGURE 39.9). By the 1930s, Japanese scientists had determined that the fungus produced hyperelongation of rice stems by secreting a chemical, which was given the name **gibberellin**. In the 1950s, researchers discovered that plants also make gibberellins. In the past 40 years, scientists have identified more than 100 different gibberellins that occur naturally in plants, although a much smaller number occur in each plant species. Foolish rice seedlings, it seems, suffer from an overdose of growth regulators normally found in plants in lower concentrations. Gibberellins have a variety of effects in plants.

Stem Elongation.   Roots and young leaves are major sites of gibberellin production. Gibberellins stimulate growth in both the leaves and the stem, but they have little effect on root growth. In stems, gibberellins stimulate cell elongation *and* cell division. Like auxin, gibberellins cause cell wall loosening, but not by acidifying the cell wall. One hypothesis proposes that gibberellins stimulate cell wall loosening enzymes that

FIGURE 39.9 "Foolish seedling disease" in rice. The spindly rice plants on the right are infected with the fungus *Gibberella*. The pathogen secretes gibberellins, a growth stimulant that uninfected plants (left) also produce, but in smaller quantity.

FIGURE 39.10 Treating pea dwarfism with a growth hormone. Contrast the untreated dwarf seedlings on the left (control) with the dwarf seedlings on the right, which were treated five days earlier with an application of 5μg of gibberellin.

facilitate the penetration of expansin proteins into the cell wall. Thus, in a growing stem, auxin, by acidifying the cell wall and activating expansins, and gibberellins, by facilitating the penetration of expansins, act in concert to promote elongation.

The effects of gibberellins in enhancing stem elongation are evident when certain dwarf varieties (mutants) of plants are treated with gibberellins. For instance, some dwarf pea plants (including the variety Mendel studied; see Chapter 14) grow to normal height if treated with gibberellins (FIGURE 39.10). If gibberellins are applied to plants of normal size, there is often no response. Apparently, these plants are already producing optimal dose of the hormone.

The most dramatic example of gibberellin-induced stem elongation is bolting, the rapid growth of the floral stalk. In their vegetative state, some plants, such as cabbage, develop in a rosette form; that is, they are low to the ground with very short internodes. As the plant switches to reproductive growth, a surge of gibberellins induces internodes to elongate rapidly, which elevates the floral buds that develop at the tips of the stems.

Fruit Growth.   In many plants, both auxin and gibberellins must be present for fruit to set. The most important commercial application of gibberellins is in the spraying of Thompson seedless grapes (FIGURE 39.11). The hormone makes the individual grapes grow larger, a trait prized by the consumer, and it makes the internodes of the grape bunch elongate, allowing more space for the individual grapes. This increase in space, by

FIGURE 39.11 The effect of gibberellin treatment on seedless grapes. The grape bunch on the left is an untreated control. The bunch on the right is growing from a vine that was sprayed with a gibberellin during fruit development.

enhancing air circulation between the grapes, also makes it harder for yeast and other microorganisms to infect the fruits.

Germination. The embryo of seeds is a rich source of gibberellins. After water is imbibed, the release of gibberellins from the embryo signals the seeds to break dormancy and germinate. Some seeds that require special environmental conditions to germinate, such as exposure to light or cold temperatures, will break dormancy if they are treated with gibberellins. Gibberellins support the growth of cereal seedlings by stimulating the synthesis of digestive enzymes such as α-amylase that mobilize stored nutrients (see FIGURE 38.13).

Abscisic Acid

THE PROCESS OF SCIENCE In the 1960s, one research group studying the chemical changes that precede bud dormancy and another team investigating chemical changes preceding leaf abscission (the dropping of autumn leaves) isolated the same compound, **abscisic acid** (ABA). Ironically, ABA is no longer thought to play a primary role in either bud dormancy or leaf abscission, but it is a plant hormone of great importance in other functions. Unlike the growth-stimulating hormones we have studied so far—auxin, cytokinins, and gibberellins—ABA generally slows down growth. Often ABA antagonizes the actions of the growth hormones, and it is the ratio of ABA to one or more growth hormones that determines the final physiological outcome. ABA has many effects on plants, but we will consider just two.

Seed Dormancy. Seed dormancy has great survival value because it ensures that the seed will germinate only when there are optimal conditions of light, temperature, and moisture (see Chapter 38). What prevents a seed dispersed in autumn

from germinating immediately only to be killed by winter? What mechanisms ensure that such seeds germinate in the spring? For that matter, what prevents seeds from germinating in the dark, moist interior of the fruit? The answer to these questions is ABA. The levels of ABA may increase 100-fold during seed maturation. The high levels of ABA in maturing seeds inhibit germination and induce the production of special proteins that help the seeds withstand the extreme dehydration that accompanies maturation.

Many types of dormant seeds will germinate when ABA is removed or inactivated in some way. The seeds of some desert plants break dormancy only when heavy rains wash ABA out of the seed. Other seeds require light or prolonged exposure to cold to trigger the inactivation of ABA. Often the ratio of ABA to gibberellins determines whether the seed will remain dormant or germinate, and the addition of ABA to seeds that are primed to germinate returns them to the dormant condition. A maize mutant that has seeds that germinate while still on the cob lacks a functional transcription factor required for ABA to induce expression of certain genes (FIGURE 39.12).

Drought Stress. ABA is the primary internal signal that enables plants to withstand drought. When a plant begins to wilt, ABA accumulates in leaves and causes stomata to close rapidly, reducing transpiration and preventing further water loss. ABA, through its effects on second messengers such as calcium, causes an increase in the opening of outwardly directed potassium channels in the plasma membrane of guard cells, leading to a massive loss of potassium from them. The accompanying osmotic loss of water leads to a reduction in guard cell

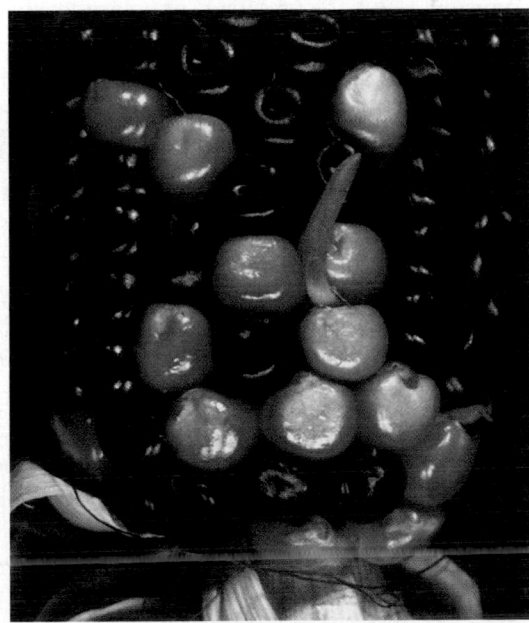

FIGURE 39.12 Precocious germination of mutant maize seeds. Abscisic acid induces dormancy in seeds. When its action is blocked—in this case, by a mutation affecting an abscisic acid–regulated transcription factor—precocious germination results.

turgor and a decrease in stomatal aperture (see FIGURE 36.13). In some cases, water shortage can stress the root system before the shoot system, and ABA transported from roots to leaves may function as an "early warning system." "Wilty" mutants, which are especially prone to wilting, are in many cases deficient in the production of ABA.

Ethylene

During the 19th century when coal gas was used for street illumination, the leakage of "illuminating gas" from gas mains caused nearby trees to drop their leaves prematurely. In 1901, a Russian scientist named Dimitry Neljubow demonstrated that the gas **ethylene** was the active factor in "illuminating gas." The idea that ethylene is a hormone that plants themselves produce only became widely accepted, however, when the advent of a technique called gas chromatography simplified the quantitative measurement of ethylene. Plants produce ethylene in response to stresses such as drought, flooding, mechanical pressure, injury, and infection. Ethylene production also occurs during fruit ripening and during programmed cell death. And ethylene is also produced in response to high concentrations of externally applied auxin. Indeed, many physiological effects previously ascribed to auxin, such as inhibition of root elongation, are now thought to be due to auxin-induced ethylene production. Ethylene has numerous effects on plants, but we shall focus on just four.

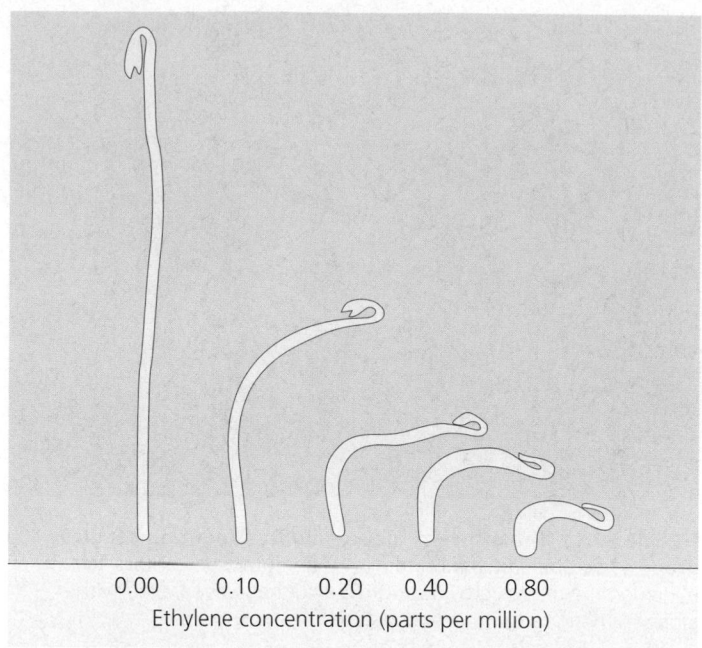

0.00 0.10 0.20 0.40 0.80
Ethylene concentration (parts per million)

FIGURE 39.13 Ethylene induces the triple response in pea seedlings. Either artificial treatment with the gaseous hormone ethylene or natural production of the hormone triggered by mechanical stress causes the stems of germinating pea seedlings to elongate less rapidly, to become thicker, and to grow horizontally—the triple response. This growth response is an adaptation that helps the seedling circumvent obstacles encountered during soil penetration. This experiment shows the extent of the triple response in pea seedlings treated with different concentrations of ethylene.

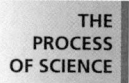

THE PROCESS OF SCIENCE

The Triple Response to Mechanical Stress: Using Mutants to Dissect a Signal-Transduction Pathway. Imagine a pea seedling pushing upward through the soil only to come up against the underside of an immovable object such as a stone. As the stem pushes against the obstacle, the mechanical stress in its delicate tip induces the seedling to start producing ethylene. Ethylene, in turn, instigates the seedling to perform a growth maneuver called the **triple response** that enables it to circumvent the obstacle. The three parts of this response, which you can see in FIGURE 39.13, are a slowing of stem elongation, a thickening of the stem (which makes it stronger), and a curvature that causes the stem to start growing horizontally. As the stem continues to grow, its tip touches upward intermittently. If these probings continue to detect a solid object above, then another pulse of ethylene is generated and the stem continues its horizontal progress. If, however, the upward touch detects no solid object, then ethylene production decreases, and the stem, now clear of the obstacle, resumes its normal upward growth.

It is ethylene that induces the stem to grow horizontally rather than the physical obstruction *per se*; when ethylene is applied to normal seedlings growing free of all physical impediments, they undergo the triple response (see FIGURE 39.13).

Researchers have studied *Arabidopsis* mutants with abnormal triple responses to investigate the signal transduction pathways for the response. Ethylene-insensitive (*ein*) mutants fail to undergo the triple response after exposure to ethylene (FIGURE 39.14a). Some types of *ein* mutants are insensitive to the presence of ethylene because they lack a functional ethylene receptor. Other mutants undergo the triple response even out of soil, in the air, where there are no physical obstacles. Some mutants of this type have a regulatory defect that causes them to produce ethylene at rates 20 times normal. The phenotype of such ethylene-overproducing (*eto*) mutants can be restored to wild-type by treating the seedlings with inhibitors of ethylene synthesis. Still other mutants, called constitutive triple-response (*ctr*) mutants, undergo the triple response in air but do not respond to inhibitors of ethylene synthesis (FIGURE 39.14b). In this case, ethylene signal transduction is permanently turned on even though there is no ethylene present. FIGURE 39.15 summarizes the responses of *ein, eto,* and *ctr* mutants to ethylene and ethylene synthesis inhibitors.

The affected gene in *ctr* mutants turns out to code for a protein kinase. The fact that this mutation *activates* the ethylene response suggests that the normal kinase product of the wild-type allele is a *negative* regulator of ethylene signal transduction. Here is one hypothesis for how the pathway works in wild-type plants: Binding of the hormone ethylene

FIGURE 39.14 Ethylene triple-response mutants.

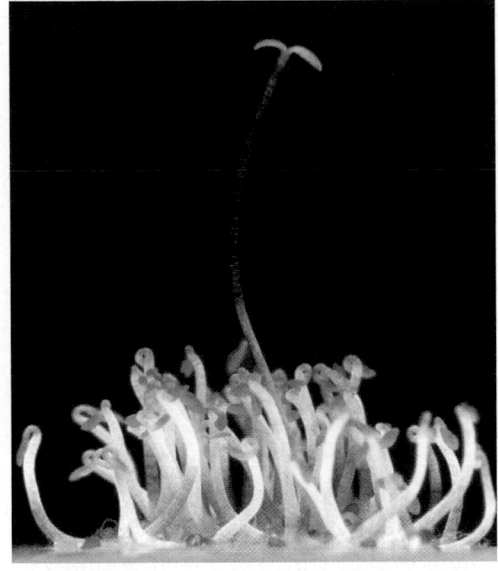

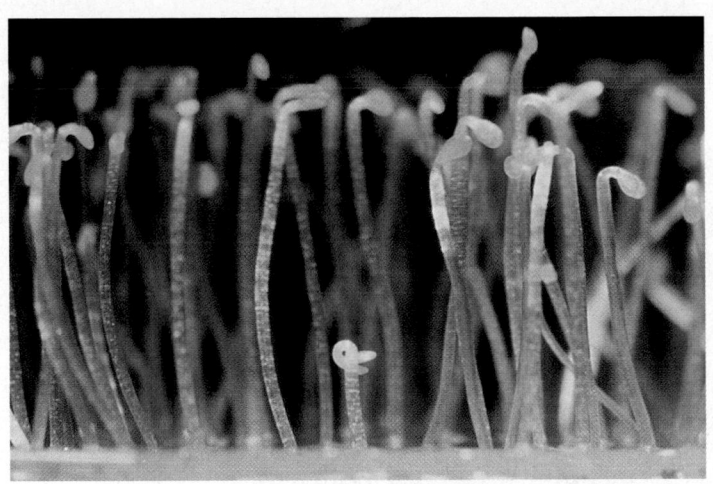

(a) *ein* **mutant.** An ethylene-insensitive (*ein*) mutant fails to undergo the triple response in the presence of ethylene.

(b) *ctr* **mutant.** A constitutive triple-response (*ctr*) mutant undergoes the triple response even in the absence of ethylene.

to the ethylene receptor leads to inactivation of the kinase; and inactivation of this negative regulator allows synthesis of the proteins required for the triple response.

Apoptosis: Programmed Cell Death. Consider the shedding of a leaf in autumn or the death of an annual after flowering. Or

FIGURE 39.15 Ethylene signal-transduction mutants can be distinguished by their different responses to experimental treatments.

think about the final step in the differentiation of a xylem vessel element, when its living contents are destroyed, leaving a hollow tube behind. All of these events involve the programmed death of certain cells or organs, or of the entire plant. The cells, organs and plants that are genetically programmed to die on a particular schedule do not simply shut down their cellular machinery and await death. Rather, the onset of programmed cell death, called **apoptosis**, is one of the busiest times in a cell's life, requiring new gene expression. During apoptosis, newly formed enzymes break down many chemical components, including chlorophyll, DNA, RNA, proteins, and membrane lipids. The plant may salvage the breakdown products. A burst of ethylene is almost always associated with this programmed destruction of cells, organs, or the whole plant.

Leaf Abscission. The loss of leaves each autumn is an adaptation that keeps deciduous trees from desiccating during winter when the roots cannot absorb water from the frozen ground. Before leaves abscise, many essential elements are salvaged from the dying leaves and are stored in stem parenchyma cells. These nutrients are recycled back to developing leaves the following spring. Fall color is a combination of new red pigments made during autumn and yellow and orange carotenoids (see Chapter 10) that were already present in the leaf but are rendered visible by the breakdown of the dark green chlorophyll in autumn.

When an autumn leaf falls, the breaking point is an abscission layer located near the base of the petiole. The small parenchyma cells of this layer have very thin walls, and there are no fiber cells around the vascular tissue. The abscission layer is further weakened when enzymes hydrolyze polysaccharides in the cell walls. Finally, the weight of the leaf, with the help of wind,

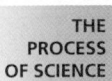

THE PROCESS OF SCIENCE

Twig   Protective   Abscission   Petiole   [0.5 mm]
        layer   layer

FIGURE 39.16 Abscission of a maple leaf. Abscission is controlled by a change in the balance of ethylene and auxin. The abscission layer can be seen here as a vertical band at the base of the petiole. After the leaf falls, a protective layer of cork becomes the leaf scar that helps prevent pathogens from invading the plant (LM).

causes a separation within the abscission layer. Even before the leaf falls, a layer of cork forms a protective scar on the twig's side of the abscission layer, preventing pathogens from invading the plant (FIGURE 39.16).

A change in the balance of ethylene and auxin controls abscission. An aging leaf produces less and less auxin and this makes the cells of the abscission layer more sensitive to ethylene. As the influence of ethylene on the abscission layer prevails, the cells produce enzymes that digest the cellulose and other components of cell walls.

Fruit Ripening.   Fruits help disperse the seeds of flowering plants (see Chapter 30). Immature (unripe) fruits that are tart, hard, and green become edible at the time of seed maturation. A burst of ethylene production in the fruit triggers this ripening. Enzymatic breakdown of cell wall components softens the fruit, and the conversion of starches and acids to sugars makes the fruit sweet. The production of new scents and colors helps

advertise fruits' ripeness to animals, who eat the fruits and disperse the seeds.

A chain reaction occurs during ripening: ethylene triggers ripening, and ripening, in turn, triggers even more ethylene production—one of the rare examples of positive feedback in physiology (see FIGURE 1.8). The result is a huge burst in ethylene production. Because ethylene is a gas, the signal to ripen even spreads from fruit to fruit: One bad apple, in fact, *does* spoil the lot. If you pick or buy green fruit, you may be able to speed ripening by storing the fruit in a plastic bag, so that ethylene gas will accumulate. On a commercial scale, many kinds of fruit are ripened in huge storage containers in which ethylene levels are enhanced. In other cases, measures are taken to retard ripening caused by natural ethylene. Apples, for instance, are stored in bins flushed with carbon dioxide. Circulating the air prevents ethylene from accumulating, and carbon dioxide inhibits synthesis of new ethylene. Stored in this way, apples picked in autumn can still be shipped to grocery stores the following summer.

Given the importance of ethylene in the post-harvest physiology of fruits, the genetic engineering of ethylene signal transduction pathways has potentially important commercial applications. For example, molecular biologists, by adding an antisense RNA that blocks the transcription of one of the genes required for ethylene synthesis, have created tomato fruits that ripen on demand. These fruits are picked while green and will not ripen unless ethylene gas is added. As such methods are refined, they will reduce spoilage of fruits and vegetables, a problem that currently ruins almost half the produce harvested in the United States.

Brassinosteroids

First isolated from *Brassica* pollen in 1979, **brassinosteroids** are steroids chemically similar to cholesterol and the sex hormones of animals. Brassinosteroids induce cell elongation and division in stem segments and seedlings at concentrations as low as 10^{-12} M. They also retard leaf abscission and promote xylem differentiation. These effects are so qualitatively similar to those of auxin that it took several years for plant physiologists to accept brassinosteroids as nonauxin hormones.

It was evidence from molecular biology that established brassinosteroids as plant hormones. Joanne Chory, our interviewee on pages 718–719, and her colleagues at the Salk Institute in San Diego were interested in *Arabidopsis* mutants that had morphological features similar to light-grown plants even though they were grown in the dark. The researchers discovered that the mutation affects a gene that normally codes for an enzyme similar to one involved in steroid synthesis in mammalian cells. The researchers also showed that the mutant plants could be restored to normal phenotype by the experimental application of brassinosteroids. The mutant studied by Chory was brassinosteroid-deficient.

As we conclude our survey of plant hormones, it is important to remember that these chemical signals are components of control systems that tune a plant's growth, development, reproduction, and physiology to the environment. For example, you learned that auxin functions in the phototropic bending of shoots toward light; that abscisic acid "holds" certain seeds dormant until the environment is suitable for germination; and that ethylene functions in leaf abscission as shorter days and cooler temperatures announce autumn. Such relationships between outside stimuli and hormonal signals are important to keep in mind as we examine how plants actually detect and respond to changes in their surroundings.

PLANT RESPONSES TO LIGHT

Light is an especially important environmental factor in the lives of plants. In addition to being required for photosynthesis, light also cues many key events in plant growth and development. These effects of light on plant morphology are what plant biologists call **photomorphogenesis.** Light reception is also important in allowing plants to measure the passage of days and seasons.

Plants detect not just the presence of light, but also its direction, intensity, and wavelengths (colors). Recall from Chapter 10 that an **action spectrum** is a graph that relates physiological response to wavelength of light. The action spectrum for photosynthesis, for example, has two peaks, one in the red and one in the blue (see FIGURE 10.8). This is because chlorophyll absorbs light primarily in the red and blue portions of the visible spectrum. Action spectra, however, are useful in the study of *any* process that depends on light. By comparing action spectra of different plant responses researchers can determine which responses are mediated by the same photoreceptor (pigment). Scientists can also compare action spectra to the absorption spectra of putative photoreceptor molecules. An absorption spectrum is a graph profiling the wavelengths of light a particular pigment absorbs. A close correspondence between an action spectrum of a plant response and the absorption spectrum of a purified pigment suggests that the pigment may be the photoreceptor involved in mediating the response.

Action spectra reveal that red and blue light are the most important colors in regulating a plant's photomorphogenesis. These observations led researchers to two major classes of light receptors: a heterogeneous group of blue-light photoreceptors and a family of photoreceptors called phytochromes that absorb mostly red light.

Blue-light photoreceptors are a heterogeneous group of pigments

The action spectra of many plant processes, including phototropism (the bending toward or away from light), the light-induced slowing of hypocotyl elongation when a seedling breaks ground during germination (see FIGURE 38.14), and the light-induced opening of stomata (see FIGURE 36.13), show that blue light is most effective in initiating these diverse responses. (FIGURE 39.17 illustrates phototropism as an example.) The biochemical identity of the blue-light photoreceptor was so elusive that in the 1970s plant physiologists began to

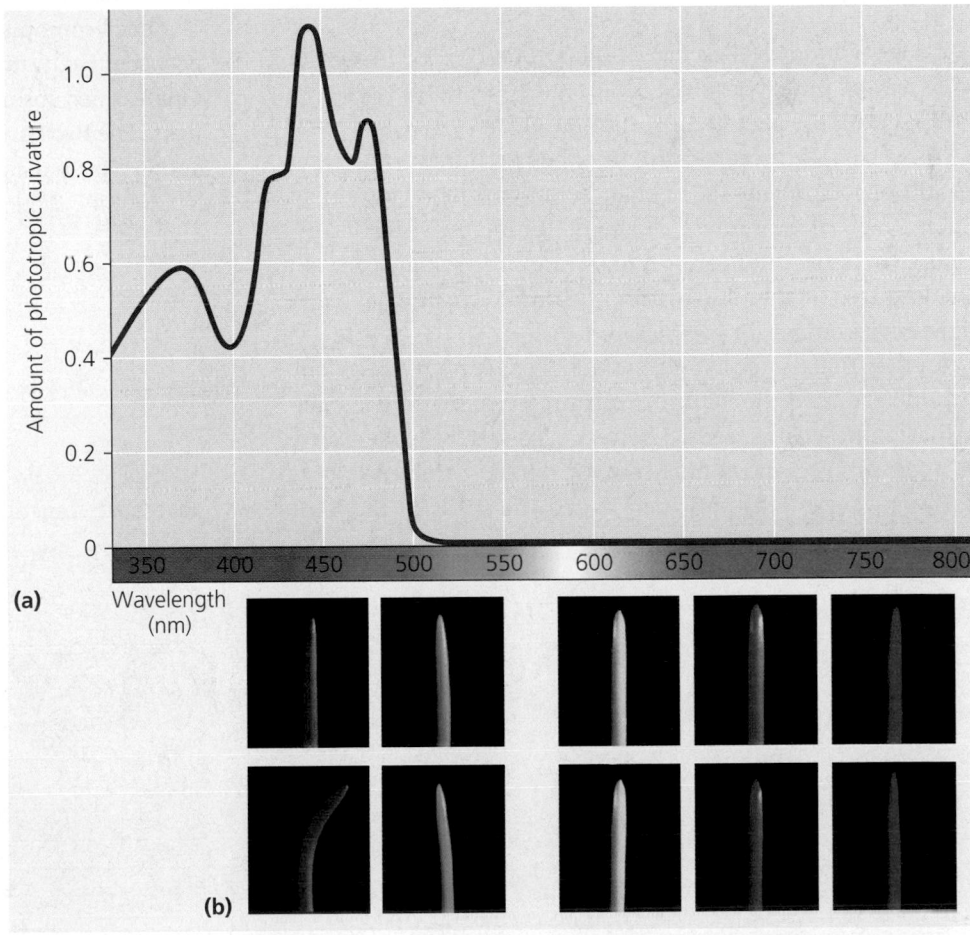

(a) Wavelength (nm)

(b)

FIGURE 39.17 Action spectrum for blue-light-stimulated phototropism. (a) Phototropic bending toward light is controlled by the blue-light photoreceptor called phototropin. In this action spectrum for phototropism in maize (*Zea mays*) coleoptiles, you can see that only light of wavelengths below 500 nm (blue and ultraviolet) are effective in inducing curvature. **(b)** The upper photograph in each pair was taken at the beginning of the experiment; the lower photos were taken after a 90-minute exposure to side lighting of the indicated colors.

refer to this putative receptor as cryptochrome (from the Greek *kryptos,* hidden, and *chrom,* pigment). In the 1990s, molecular biologists analyzing *Arabidopsis* mutants found that plants actually use at least three completely different types of pigments to detect blue light: **cryptochromes** (for the inhibition of hypocotyl elongation), **phototropin** (for phototropism), and a carotenoid-based photoreceptor called **zeaxanthin** (for stomatal opening).

Phytochromes function as photoreceptors in many plant responses to light

When we introduced signal transduction in plants earlier in the chapter, we discussed the role of a family of plant pigments called phytochromes in the greening process. Phytochromes regulate many of a plant's responses to light throughout its life from seed to flower. Let's look at a couple of examples.

The Phytochrome Switch and Seed Germination

It was studies of seed germination that led to the discovery of phytochromes. Because of their limited food reserves, the successful sprouting of many types of small seeds, such as lettuce, requires that they germinate only when conditions, especially the light environment, are near optimal. Such seeds often remain dormant for many years until a change in light conditions occurs. For example, the death of a shading tree or the plowing of a field may create a favorable light environment for germination.

In the 1930s, scientists at the U.S. Department of Agriculture determined the action spectrum for light-induced germination of lettuce seeds. They exposed water-swollen seeds to a few minutes of monochromatic (single-colored) light of various wavelengths and then stored the seeds in the dark.

After two days, the researchers scored the number of seeds that had germinated under each light regimen. The action spectrum revealed that red light of 660 nm wavelength increases the germination percentage of lettuce seeds the most compared to control seeds that had not been exposed to any light. In marked contrast, far-red light, that is light of wavelengths just near the edge of human visibility (730 nm), *inhibits* lettuce seed germination compared to the dark controls. What happens when the lettuce seeds are subjected to a flash of red (R) light followed by a flash of far-red (FR) light or, conversely, to FR light followed by R light? The answer is that the *last* flash of light determines the seeds' response (FIGURE 39.18). In other words, the effects of red and far-red light are reversible.

The photoreceptor responsible for these opposing effects of red and far-red light is a phytochrome. It consists of a protein covalently bonded to a nonprotein part that functions as a chromophore, the light-absorbing part of the molecule (FIGURE 39.19). So far, researchers have identified five different phytochromes in *Arabidopsis,* each with a slightly different protein component.

The chromophore of a phytochrome reverts back and forth between two isomeric forms (see Chapter 4 to review isomers). One isomer absorbs red light, and the other absorbs far-red light. The two variations of phytochrome—P_r (red absorbing) and P_{fr} (far-red absorbing)—are said to be photoreversible:

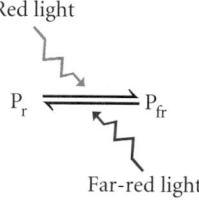

This $P_r \rightleftharpoons P_{fr}$ interconversion acts as a switching mechanism that controls various light-induced events in the life of

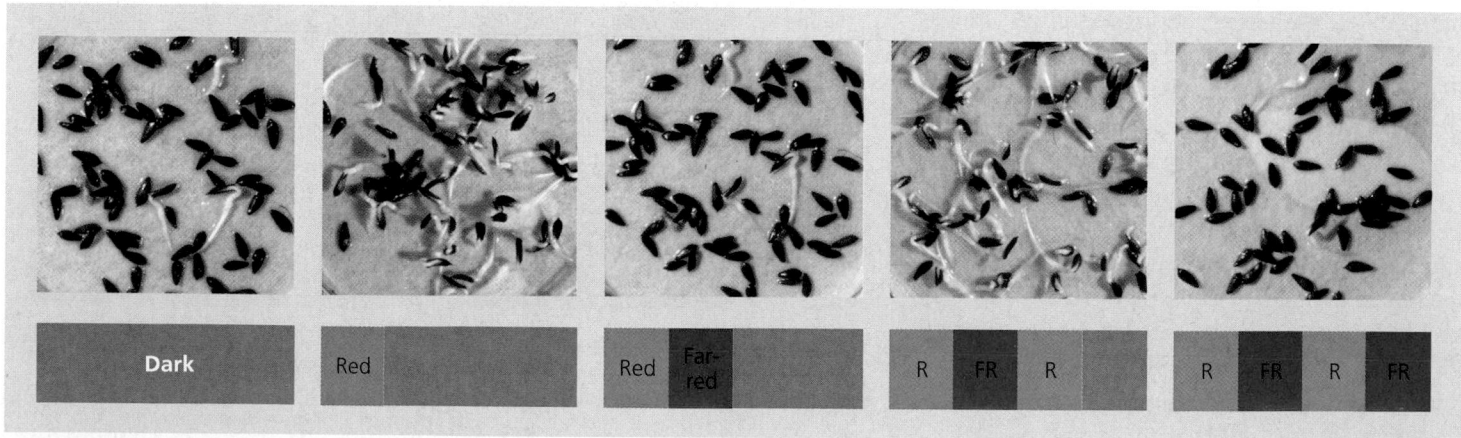

FIGURE 39.18 Phytochrome regulation of lettuce seed germination. The control seeds, on the left, were kept in the dark. The various batches of experimental seeds received flashes of light. The bars below the photos indicate the sequences of red and far-red light flashes.

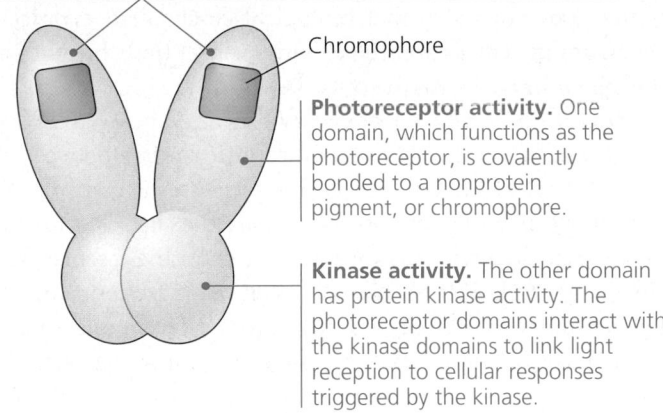

A phytochrome consists of two identical proteins joined to form one functional molecule. Each of these proteins has two domains.

Chromophore

Photoreceptor activity. One domain, which functions as the photoreceptor, is covalently bonded to a nonprotein pigment, or chromophore.

Kinase activity. The other domain has protein kinase activity. The photoreceptor domains interact with the kinase domains to link light reception to cellular responses triggered by the kinase.

FIGURE 39.19 Structure of a phytochrome.

the plant (FIGURE 39.20). P_{fr} is the form of phytochrome that triggers many of a plant's developmental responses to light. For example, in the seed germination experiments we just described, P_r in the lettuce seeds that were exposed to red light was converted to P_{fr}, stimulating the cellular responses that lead to germination. When the red-illuminated seeds were then exposed to far-red light, the P_{fr} was converted back to P_r, inhibiting the germination response.

How does phytochrome switching explain light-induced germination in nature? Plants synthesize phytochrome as P_r, and if seeds are kept in the dark the pigment remains almost entirely in the P_r form (FIGURE 39.20). But if the seeds are illuminated with sunlight, the phytochrome is exposed to red light (along with all the other wavelengths in sunlight), and much of the P_r is converted to P_{fr}. This appearance of P_{fr} is one

of the ways that plants detect sunlight. When seeds are exposed to adequate sunlight for the first time, it is the appearance of P_{fr} that triggers their germination.

The Phytochrome Switch and Shade Avoidance

The phytochrome system also provides the plant with information about the *quality* of light. Sunlight includes both red and far-red radiation. Thus, during the day the $P_r \rightleftharpoons P_{fr}$ photoreversion reaches a dynamic equilibrium, with the ratio of the two phytochrome forms indicating the relative amounts of red and far-red light. This sensing mechanism enables plants to adapt to changes in light conditions. Consider, for example, the "shade-avoidance" response of a tree that requires relatively high light intensity. If other trees in a forest shade this tree, the phytochrome ratio shifts in favor of P_r because the forest canopy screens out more red light than far-red light. (This is because the chlorophyll pigments in the leaves of the canopy absorb red light and allow far-red light to pass.) This shift in the ratio of red to far-red light induces the tree to allocate more of its resources to growing taller. In contrast, direct sunlight increases the proportion of P_{fr}, which stimulates branching and inhibits vertical growth.

In addition to enabling plants to detect light and distinguish its quality, phytochromes also function in a plant's ability to keep track of the passage of days and seasons. To understand phytochrome's role in these timekeeping processes, we must first examine the clock itself.

Biological clocks control circadian rhythms in plants and other eukaryotes

Many plant processes, such as transpiration and synthesis of certain enzymes, oscillate during the course of a day. Some of these cyclic variations are responses to the changes in light levels, temperature, and relative humidity that accompany the 24-hour cycle of day and night. However, one can eliminate these exogenous (external) factors by growing plants in growth chambers under rigidly maintained conditions of light, temperature, and humidity. Even under these artificially constant conditions, many physiological processes in plants, such as the opening and closing of stomata and the production of photosynthetic enzymes, continue to oscillate with a period of about 24 hours. For example, many legumes lower their leaves in the evening and raise them in the morning (FIGURE 39.21). A bean plant will

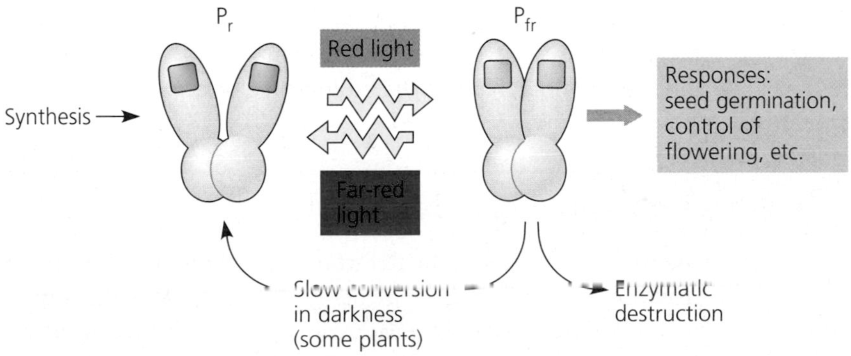

P_r Red light P_{fr}

Synthesis →

Responses: seed germination, control of flowering, etc.

Far-red light

Slow conversion in darkness (some plants)

Enzymatic destruction

FIGURE 39.20 Phytochrome: a molecular switching mechanism. Absorption of red light causes the bluish P_r to change to the blue-greenish P_{fr}. Far-red light reverses this conversion. In most cases, it is the P_{fr} form of the pigment that switches on physiological and developmental responses in the plant.

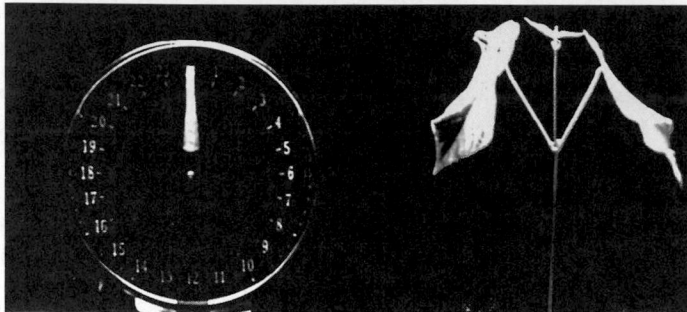

FIGURE 39.21 Sleep movements of a bean plant. Leaf position at noon (top) and at midnight. The movements are caused by reversible changes in the turgor pressure of cells on opposing sides of the pulvini, swollen regions of the petiole.

continue these "sleep movements" even if kept in constant light or constant darkness; the leaves are not simply responding to sunrise and sunset. Such physiological cycles with a frequency of about 24 hours and that are not directly paced by any known environmental variable are called **circadian rhythms** (from the Latin *circa*, approximately, and *dies*, day). Circadian rhythms are a ubiquitous feature of eukaryotic life. For example, your pulse, blood pressure, temperature, rate of cell division, blood cell count, alertness, urine composition, metabolic rate, sex drive, and responsiveness to medications all fluctuate in a circadian manner.

Are circadian rhythms truly prompted by an internal clock, or are they merely daily responses to some subtle but pervasive environmental cycle, such as geomagnetism or cosmic radiation? Organisms, including plants and humans, continue their rhythms when placed in the deepest mine shafts or when orbited in satellites, conditions that alter these subtle geophysical periodicities. All research thus far indicates that the oscillator for circadian rhythms is endogenous (internal). This internal clock, however, is entrained (set) to a period of precisely 24 hours by daily signals from the environment.

If an organism is kept in a constant environment, its circadian rhythms deviate from a 24-hour period (a "period" is the duration of one cycle). These free-running periods, as they are called, vary from about 21 to 27 hours, depending on the particular rhythmic response. The sleep movements of bean plants, for instance, have a period of 26 hours when the plants are kept under the free-running conditions of constant darkness. Deviation of the free-running period from exactly 24 hours does not mean that biological clocks drift erratically. Free-running clocks are still keeping perfect time, but they are not synchronized with the outside world.

How do biological clocks work? In attempting to answer this question, we must be careful to differentiate between the oscillator (clock) and the rhythmic processes it controls. The sweeping leaves of sleep movements are the "hands" of the biological clock, but these movements are not the essence of the clockwork itself. If the leaves of a bean plant are restrained for several hours so they cannot move, they will, on release, rush to the position appropriate for the time of day. We can interfere with a biological rhythm, but the clockwork goes right on ticking off the time.

Researchers are tracing the clock to a molecular mechanism that may be common to all eukaryotes. A leading hypothesis is that biological timekeeping may depend on synthesis of a protein that regulates its own production through feedback control. This protein may be a transcription factor that inhibits transcription of the gene that encodes for the transcription factor itself. The concentration of this transcription factor may accumulate during the first half of the circadian cycle, and then decline during the second half, due to self-inhibition of its own production.

Researchers have recently used a novel technique to identify clock mutants of *Arabidopsis*. One prominent circadian rhythm in plants is the daily production of certain photosynthesis-related proteins. Molecular biologists traced this rhythm to the promoter that regulates the transcription of the genes for these photosynthesis proteins. To identify clock mutants, scientists spliced the gene for an enzyme called luciferase to the promoter. Luciferase is the enzyme responsible for the bioluminescence of fireflies. When the biological clock turned on the promoter in the *Arabidopsis* genome, it also turned on the production of luciferase. The plants began to glow with a circadian periodicity. Clock mutants were then isolated by selecting specimens that glowed for a longer or shorter time than normal. The defective genes in some of these mutants affect proteins that normally bind photoreceptors. Perhaps these particular mutations disrupt a light-dependent mechanism that sets the biological clock.

THE PROCESS OF SCIENCE

Light entrains the biological clock

As we have discussed, the free running period of the circadian rhythm of bean leaf movements is 26 hours. Consider a bean plant placed at dawn in a dark cabinet for 72 hours: Its leaves would not rise again until 2 hours after natural dawn on the second day, and 4 hours after natural dawn on the third day, and so on. Shut off from environmental cues, the plant then becomes desynchronized with its natural environment. Desynchronization also happens when we cross several time zones in an airplane; when we reach our destination, the clocks on the wall are not synchronized with our internal clocks. All eukaryotes are probably prone to jetlag.

What entrains the biological clock to precisely 24 hours every day? The answer is light. Both phytochrome and blue-light photoreceptors can entrain circadian rhythms in plants, but our understanding of how phytochrome does this is more complete. As you might suspect, the mechanism involves turning cellular responses on and off by means of the $P_r \rightleftharpoons P_{fr}$ switch.

Consider once again the photoreversible system diagrammed in FIGURE 39.20. In darkness, the phytochrome ratio shifts gradually in favor of the P_r form. This is partly due to turnover in the overall phytochrome pool. The pigment is synthesized in the P_r form, and degradative enzymes destroy more P_{fr} than P_r. In addition, in some plant species, P_{fr} present at sundown slowly converts to P_r by a biochemical mechanism. In darkness, there is no means for the accumulating P_r to be reconverted to P_{fr}, but when the sun rises, the P_{fr} level suddenly increases again by rapid photochemical conversion from P_r. It is this sudden increase in P_{fr} each day at dawn that resets the biological clock: bean leaves always reach their most extreme night position 16 hours after dawn.

Interactions between phytochrome and the biological clock enable plants to measure the passage of night and day in nature. The relative lengths of night and day, however, change over the course of the year (except at the Equator). Plants use this change to mark the passage of seasons.

Photoperiodism synchronizes many plant responses to changes of season

Imagine the consequences if a plant produced flowers when its insect pollinators were not present or if a deciduous tree produced leaves in the middle of winter. Seasonal events are of critical importance in the life cycles of most plants. Seed germination, flowering, and the onset and breaking of bud dormancy are examples of stages in plant development that usually occur at specific times of the year. The environmental stimulus plants use most often to detect the time of year is the photoperiod, the relative lengths of night and day. A physiological response to photoperiod, such as flowering, is called **photoperiodism.**

Photoperiodism and the Control of Flowering

One of the earliest clues to how plants detect the progression of seasons came from a mutant variety of tobacco studied by W. W. Garner and H. A. Allard in 1920. This variety, named Maryland Mammoth, grew exceptionally tall but failed to flower during summer, when normal tobacco plants flowered. Maryland Mammoth finally bloomed in a greenhouse

in December. After trying to induce earlier flowering by varying temperature, moisture, and mineral nutrition, Garner and Allard learned that it was the shortening days of winter that stimulated Maryland Mammoth to flower. If the plants were kept in light-tight boxes so that lamps could be used to manipulate durations of "day" and "night," flowering occurred only if the day length was 14 hours or shorter. Maryland Mammoth did not flower during summer because, at Maryland's latitude, the days were too long during that season.

Garner and Allard termed Maryland Mammoth a **short-day plant,** because it apparently required a light period *shorter* than a critical length to flower. Chrysanthemums, poinsettias, and some soybean varieties are a few of the other short-day plants, which generally flower in late summer, fall, or winter. Another group of plants dependent on photoperiod will flower only when the light period is *longer* than a certain number of hours. These **long-day plants** generally flower in late spring or early summer. Spinach, for example, flowers when days are 14 hours or longer. Radish, lettuce, iris, and many cereal varieties are also long-day plants. Flowering in a third group, day-neutral plants, is unaffected by photoperiod. Tomatoes, rice, and dandelions are examples of **day-neutral plants,** which flower when they reach a certain stage of maturity, regardless of day length at that time.

Critical Night Length.   In the 1940s, researchers discovered that it is actually night length, not day length, that controls flowering and other responses to photoperiod. Many of these scientists worked with cocklebur, a short-day plant that flowers only when days are 16 hours or shorter (and nights are at least 8 hours long). These researchers found that if the daytime portion of the photoperiod is broken by a brief exposure to darkness, there is no effect on flowering. However, if the nighttime part of the photoperiod is interrupted by even a few minutes of dim light, cocklebur will not flower, and this turned out to be true for other short-day plants as well (FIGURE 39.22a, p. 822). Cocklebur is actually unresponsive to *day* length, but it requires at least 8 hours of *continuous darkness* to flower. Short-day plants are really long-night plants, but the older term is embedded firmly in the jargon of plant physiology. Similarly, long-day plants are actually short-night plants. A long-day plant grown on photoperiods of long nights that would not normally induce flowering will flower if the period of continuous darkness is interrupted by a few minutes of light (FIGURE 39.22b, p. 822). Notice that we distinguish long-day from short-day plants *not* by an absolute night length but by whether the critical night length sets a maximum (long-day plants) or minimum (short-day plants) number of hours of darkness required for flowering. In both cases, the actual number of hours in the critical night length is specific to each species of plant.

Red light is the most effective color in interrupting the nighttime portion of the photoperiod. Action spectra and photoreversibility experiments show that phytochrome is the pigment that receives the red light (FIGURE 39.23). For example, if a

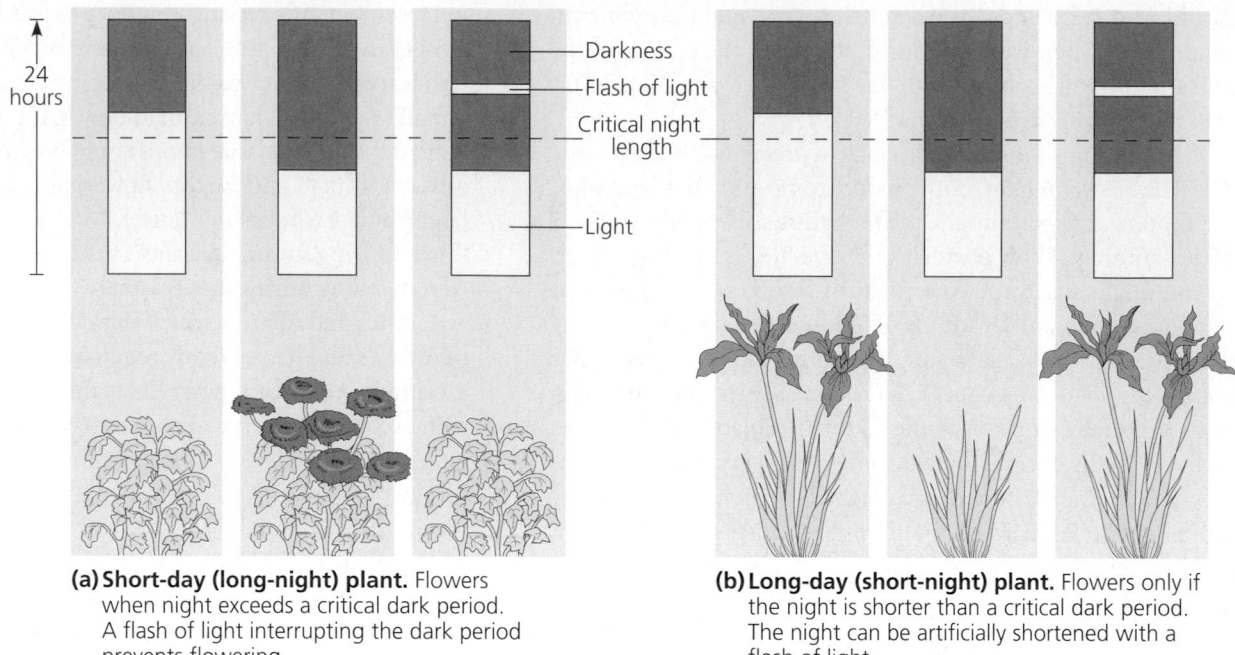

FIGURE 39.22 Photoperiodic control of flowering.

24 hours

Darkness
Flash of light
Critical night length
Light

(a) Short-day (long-night) plant. Flowers when night exceeds a critical dark period. A flash of light interrupting the dark period prevents flowering.

(b) Long-day (short-night) plant. Flowers only if the night is shorter than a critical dark period. The night can be artificially shortened with a flash of light.

flash of red light (R) during the dark period is followed by a flash of far-red (FR) light, then the plant detects no interruption of night length. As in the case of phytochrome-mediated seed germination, red/far-red photoreversibility can be demonstrated.

Plants measure night length very accurately; some short-day plants will not flower if night is even one minute shorter than the critical length. Some plant species always flower on the same day each year. It appears that plants use their biological clock, apparently entrained with the help of phytochrome, to tell the season of the year by measuring night length. The floriculture (flower-growing) industry has applied this knowledge to produce flowers out of season. Chrysanthemums, for instance, are short-day plants that normally bloom in fall, but their blooming can be stalled until Mother's Day in May by punctuating each long night with a flash of light, thus turning one long night into two short nights.

Some plants bloom after a single exposure to the photoperiod required for flowering. Other species need several successive days of the appropriate photoperiod. Still other plants will respond to a photoperiod only if they have been previously exposed to some other environmental stimulus, such as a period of cold temperatures. Winter wheat, for example, will not flower unless it has been exposed to several weeks of temperatures below 10°C.

This requirement for pretreatment with cold before flowering is called vernalization. Several weeks after winter wheat is vernalized, a photoperiod with long days (short nights) induces flowering.

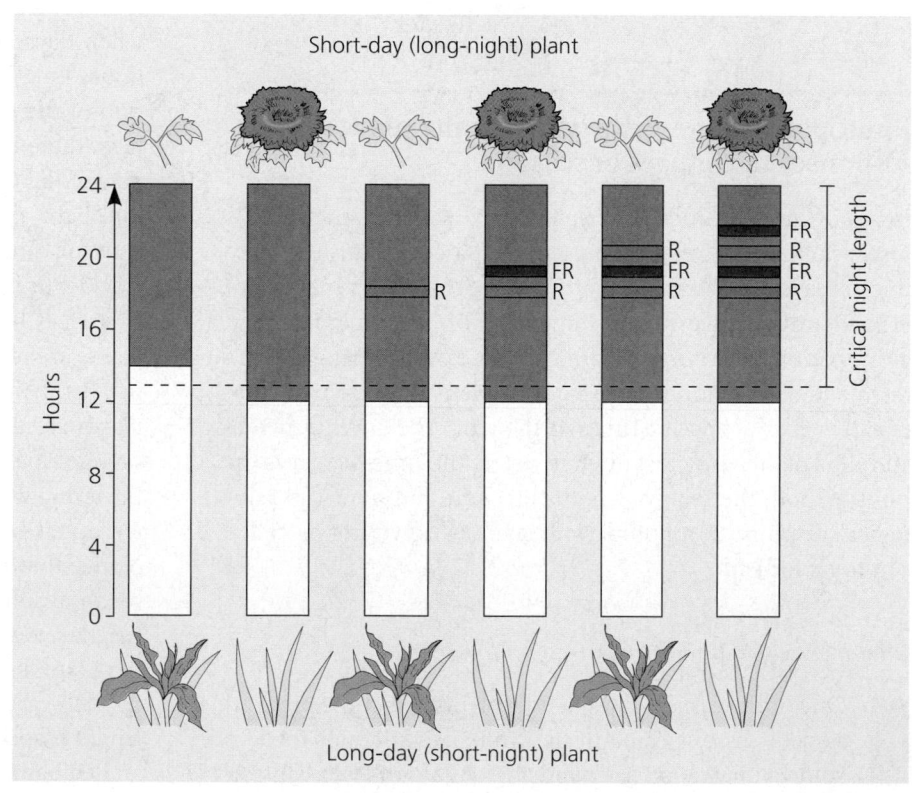

Short-day (long-night) plant

24
20
16
Hours
12
8
4
0

R
FR
R
R
FR
R
FR
R
FR
R
FR
R

Critical night length

Long-day (short-night) plant

FIGURE 39.23 Reversible effects of red and far-red light on photoperiodic response. A flash of red light shortens the dark period. A subsequent flash of far-red light cancels the effect of the red flash.

Is There a Flowering Hormone?

Buds produce flowers, but it is leaves that detect photoperiod. When the photoperiodic requirement for flowering is met, leaves must send a signal to the buds cuing them to develop as flowers. To induce a short-day plant or long-day plant to flower, it is enough in many species to expose a single leaf to the appropriate photoperiod. Indeed, if only one leaf is left attached to the plant, photoperiod is detected and floral buds are induced. If all leaves are removed, however, the plant is blind to photoperiod. Most plant physiologists believe the flowering signal is a hormone or some change in the relative concentrations of two or more hormones (FIGURE 39.24). The flowering stimulus appears to be the same for short-day and long-day plants, despite the difference in the photoperiodic conditions required for their leaves to send this signal. The evidence for hormonal regulation of flowering is compelling, but researchers have not yet identified the hormone(s).

Meristem Transition from Vegetative Growth to Flowering

Whatever combination of environmental cues (such as photoperiod or vernalization) and internal signals (such as hor-

mones) is necessary for flowering to occur, the outcome is the transition of a bud's meristem from a vegetative state to a flowering state. This transition requires changes in the expression of genes that regulate pattern formation. Meristem-identity genes that specify that the bud will now form a flower instead of a vegetative shoot must first be switched on. Then the organ-identity genes that specify the spatial organization of the floral organs—sepals, petals, stamens, and carpels—are activated in the appropriate regions of the meristem (see FIGURES 21.19 and 35.12). Research on flower development is progressing rapidly, and one goal is to identify the signal-transduction pathways that link such cues as photoperiod and hormonal changes to the gene expression required for flowering.

PLANT RESPONSES TO ENVIRONMENTAL STIMULI OTHER THAN LIGHT

Plants can neither migrate to a watering hole when water is scarce nor seek shelter when the weather is too windy. A seed, landing upside down in the soil, cannot maneuver itself into an upright position. Because of their immobility, plants must adjust to a wide range of environmental circumstances through developmental and physiological mechanisms. Natural selection has refined these responses. Light is so important in the life of a plant that we devoted the entire previous section to a plant's reception of and response to this one environmental factor. In this section, we examine responses to some of the other environmental stimuli that a plant commonly faces in its "struggle for existence."

Plants respond to environmental stimuli through a combination of developmental and physiological mechanisms

Responses to Gravity

Since plants are solar-powered organisms, it is not surprising that mechanisms for growing toward sunlight have evolved. But what environmental cue does the shoot of a young seedling use to grow upward when it is completely underground and there is no light for it to detect? Similarly, what environmental factor prompts the young root to grow downward? The answer to both questions is gravity.

Place a seedling on its side, and it will adjust its growth so that the shoot bends upward and the root curves downward. In their responses to gravity, or **gravitropism**, roots display positive gravitropism and shoots exhibit negative gravitropism. Gravitropism functions as soon as a seed germinates, ensuring that the root grows into the soil and the shoot reaches sunlight regardless of how the seed happens to be

Plant subjected to photoperiod that induces flowering

FIGURE 39.24 Experimental evidence for a flowering hormone(s). If a plant that has been induced to flower by photoperiod is grafted to a plant that has not been induced, both plants flower, indicating the transmission of a flower-inducing substance. This works in some cases even if one is a short-day plant and the other is a long-day plant.

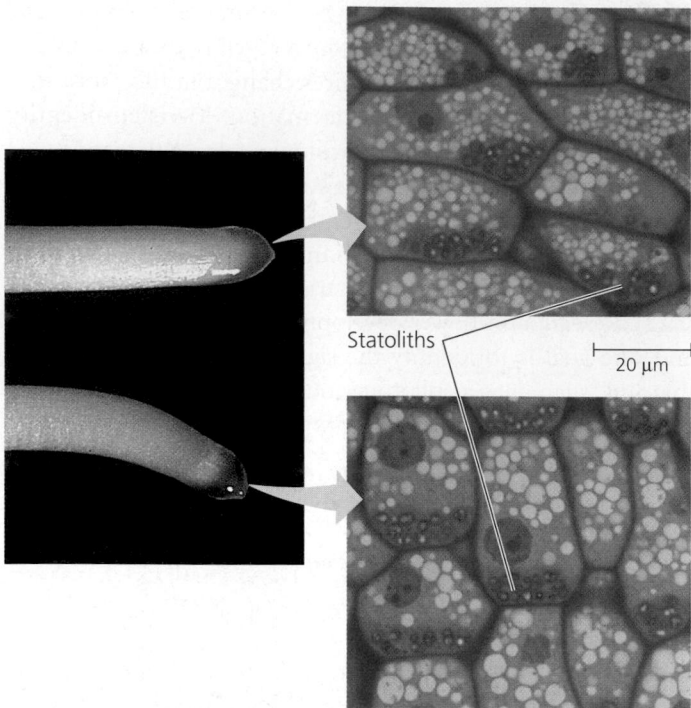

Statoliths

20 μm

FIGURE 39.25 The statolith hypothesis for root gravitropism.
These corn roots were placed on their sides and photographed before (top) and 1.5 hours after initiation of the gravitropic response. The light micrographs show the locations of statoliths, modified plastids, within cells of the root cap. The settling of statoliths to the low points of these cells may be a gravity-sensing mechanism that leads to the redistribution of auxin and the differential rates of elongation by cells on opposite sides of the root.

oriented when it lands. Auxin plays a major role in gravitropic responses.

Plants may tell up from down by the settling of **statoliths**, specialized plastids containing dense starch grains, to the lower portions of cells (FIGURE 39.25). In roots, statoliths are located in certain cells of the root cap. According to one hypothesis, the aggregation of statoliths at the low points of these cells triggers the redistribution of calcium, which in turn causes lateral transport of auxin within the root. The calcium and auxin accumulate on the lower side of the root's zone of elongation. (Because these chemicals are dissolved, they do not respond to gravity but must be actively transported to one side of the root.) At high concentration, auxin inhibits cell elongation, an effect that slows growth on the lower side of the root. The more rapid elongation of cells on the upper side causes the root to curve as it grows. This tropism continues until the root is growing straight down.

Plant physiologists are refining the "falling statolith" hypothesis of root gravitropism as they try new experiments. For example, mutants of *Arabidopsis* and tobacco that lack statoliths are still capable of gravitropism, though the response is slower than in wild-type plants. It could be that the entire cell helps the root sense gravity by mechanically pulling on proteins that tether the protoplast to the cell wall, stretching the

proteins on the "up" side and compressing the proteins on the "down" side of the root cells. Dense organelles (in addition to starch granules) may also contribute by distorting the cytoskeleton as they are pulled by gravity. Statoliths, because of their density, may enhance gravitational sensing by a mechanism that works more slowly in their absence.

Responses to Mechanical Stimuli

A tree growing on a windy mountain ridge will usually have a shorter, stockier trunk than a tree of the same species growing in a more sheltered location. The advantage of this stunting of growth is that it enables the plant to hold its ground against strong gusts of wind. The term **thigmomorphogenesis** (from the Greek *thigma*, touch) refers to the changes in form that result from mechanical perturbation. Plants are very sensitive to mechanical stress: Some researchers have found that even the act of measuring the length of a leaf with a ruler alters its subsequent growth.

Rubbing the stems of a young plant a few times results in plants that are shorter than controls (FIGURE 39.26). Mechanical stimulation activates a signal-transduction pathway involving an increase in cytoplasmic calcium that, in turn, mediates the activation of specific genes, some of which encode for proteins that affect cell wall properties.

Some plant species have become, over the course of their evolution, "touch specialists." Acute responsiveness to mechanical stimuli is an integral part of these plants' life "strategies." Most vines and other climbing plants have tendrils that coil rapidly around supports (see FIGURE 35.6a). These grasping organs usually grow straight until they touch something; the contact stimulates a coiling response caused by differential growth of

FIGURE 39.26 Altering gene expression by touch in *Arabidopsis*.
The shorter plant on the left was touched twice a day. The unmolested plant (right) grew much taller.

cells on opposite sides of the tendril. This directional growth in response to touch is called **thigmotropism,** and it allows the vines to take advantage of whatever mechanical supports it comes across as it climbs upward toward a forest canopy.

Other examples of touch specialists are plants that undergo rapid leaf movements in response to mechanical stimulation. For example, when the compound leaf of the sensitive plant *Mimosa* is touched, it collapses and its leaflets fold together (FIGURE 39.27). This response, which takes only a second or two, results from a rapid loss of turgor by cells within pulvini, specialized motor organs located at the joints of the leaf. The motor cells suddenly become flaccid after stimulation because they lose potassium, which causes water to leave the cells by osmosis. It takes about ten minutes for the cells to regain their turgor and restore the "unstimulated" form of the leaf. The function of the sensitive plant's behavior invites speculation. Perhaps by folding its leaves and reducing its surface area when jostled by strong winds, the plant conserves water. Or perhaps because the collapse of the leaves exposes thorns on the stem, the rapid response of the sensitive plant discourages herbivores.

A remarkable feature of rapid leaf movements is the mode of transmission of the stimulus through the plant. If one leaflet on a sensitive plant is touched, first that leaflet responds, then the adjacent leaflet responds, and so on until all the leaflet-pairs have folded together. From the point of stimulation, the signal that produces this response travels at a speed of about a centimeter per second. Traveling at the the same rate, an electrical impulse can be detected by attaching electrodes to the leaf. These impulses, called **action potentials,** resemble nervous-system messages in animals, though the action potentials of plants are thousands of times slower. Action potentials, which have been discovered in many species of algae and plants, may be widely used as a form of internal communication. Another example is the Venus flytrap, in which action potentials are transmitted from sensory hairs in the trap to the cells that respond by closing the trap (see FIGURE 37.16a). In the case of *Mimosa,* more violent stimuli, such as touching a leaf with a hot needle, causes *all* the leaves and leaflets on a plant to droop, but this systemic response involves the spread of chemical signals released from the injured area to other parts of the shoot.

Responses to Stress

Occasionally, factors in the environment change severely enough to have a potentially adverse effect on a plant's survival, growth, and repro- duction. Environmental stresses, such as flooding, drought, or extreme temperatures can have a devastating impact on crop yields in agriculture. In natural ecosystems, plants that cannot tolerate an environmental stress will either succumb or be outcompeted by other plants, and they will become locally extinct. Thus, environmental stresses are also important in de- termining the geographic ranges of plants. Here, we consider some of the more common abiotic (nonliving) stresses that plants encounter. In the next section, we will examine the de- fensive responses of plants to common biotic (biological) stresses such as pathogens and herbivores.

Drought.   On a bright, warm, dry day, a plant may be stressed by a water deficiency because it is losing water by transpiration faster than the water can be restored by uptake from the soil. Pro- longed drought can stress crops and the plants of natural ecosys- tems for weeks or months. Severe water deficit, of course, will kill a plant, as you may know from experience with neglected

(a) Unstimulated

(b) Stimulated

Leaflets after stimulation

Pulvinus (motor organ)

Side of pulvinus with flaccid cells

Side of pulvinus with turgid cells

Vein

(c) Motor organs

0.5 μm

FIGURE 39.27 Rapid turgor movements by the sensitive plant (*Mimosa pudica*). **(a)** In the unstimulated plant, leaflets are spread apart. **(b)** Within a second or two of being touched, the leaflets have folded together. **(c)** In these light micrographs of a leaflet pair in the closed (stimulated) state, you can see the motor cells in the sectioned pulvini (motor organs). The curvature of a pulvinus is caused by motor cells on one side losing water and becoming flaccid while cells on the opposite side retain their turgor.

houseplants. But plants have control systems that enable them to cope with less extreme water deficits.

Many of a plant's responses to water deficit help the plant conserve water by reducing the rate of transpiration. Water deficit in a leaf causes guard cells to lose turgor, a simple control mechanism that slows transpiration by closing stomata (see Chapter 36). Water deficit also stimulates increased synthesis and release of abscisic acid in the leaf, and this hormone helps keep stomata closed by acting on guard cell membranes. Leaves respond to water deficit in several other ways. Because cell expansion is a turgor-dependent process, a water deficit will inhibit the growth (expansion) of young leaves. This response minimizes the transpirational loss of water by slowing the increase in leaf surface. When the leaves of many grasses and other plants wilt from a water deficit, they roll into a shape that reduces transpiration by exposing less leaf surface to dry air and wind. While all of these responses of leaves help the plant conserve water, they also reduce photosynthesis. This is one reason a drought diminishes crop yield.

Root growth also responds to water deficit. During a drought, the soil usually dries from the surface down. This inhibits the growth of shallow roots, partly because cells cannot maintain the turgor required for elongation. Deeper roots surrounded by soil that is still moist continue to grow. Thus, the root system proliferates in a way that maximizes exposure to soil water.

Flooding. An overwatered houseplant may suffocate because the soil lacks the air spaces that provide oxygen for cellular respiration in the roots. Some plants are structurally adapted to very wet habitats. For example, the submerged roots of trees called mangroves, which inhabit coastal marshes, are continuous with aerial roots that provide access to oxygen. But how do plants less specialized for aquatic environments cope with oxygen deprivation in waterlogged soils? Oxygen deprivation stimulates the production of the hormone ethylene, which causes some of the cells in the root cortex to undergo apoptosis (programmed cell death). Enzymatic destruction of cells creates air tubes that function as "snorkels," providing oxygen to the submerged roots (FIGURE 39.28).

Salt Stress. An excess of sodium chloride or other salts in the soil threatens plants for two reasons. First, by lowering the water potential of the soil solution, salt can cause a water deficit in plants even though the soil has plenty of water. This is because in an environment with a water potential more negative than that of the root tissue, roots will lose water rather than absorb it (see Chapter 36). The second problem with saline soil is that sodium and certain other

ions are toxic to plants when their concentrations are relatively high. The selectively permeable membranes of root cells impede the uptake of most harmful ions, but this only aggravates the problem of acquiring water from soil that is rich in solutes. Many plants can respond to moderate soil salinity by producing compatible solutes, organic compounds that keep the water potential of cells more negative than that of the soil solution without admitting toxic quantities of salt. However, most plants cannot survive salt stress for long. The exceptions are halophytes, salt-tolerant plants with adaptations such as salt glands. These glands pump salts out across the leaf epidermis.

Heat Stress. Excessive heat can harm and eventually kill a plant by denaturing its enzymes and damaging its metabolism in other ways. One function of transpiration is evaporative cooling. On a warm day, for example, the temperature of a leaf may be 3–10°C below ambient air temperature. Of course, hot, dry weather also tends to cause water deficiency in many plants; the closing of stomata in response to this stress conserves water but sacrifices evaporative cooling. This dilemma is one of the reasons that very hot, dry days take such a toll on most plants.

Most plants have a backup response that enables them to survive heat stress. Above a certain temperature—about 40°C for most plants that inhabit temperate regions—plant cells begin synthesizing relatively large quantities of special proteins called **heat-shock proteins.** Researchers have also discovered this response in heat-stressed animals and microorganisms. Some heat-shock proteins are identical to chaperone proteins, which function in unstressed cells as temporary scaffolds that help other proteins fold into their functional shapes (see Chapter 5). In their roles as heat-shock proteins, perhaps these molecules embrace enzymes and other proteins and help prevent denaturation.

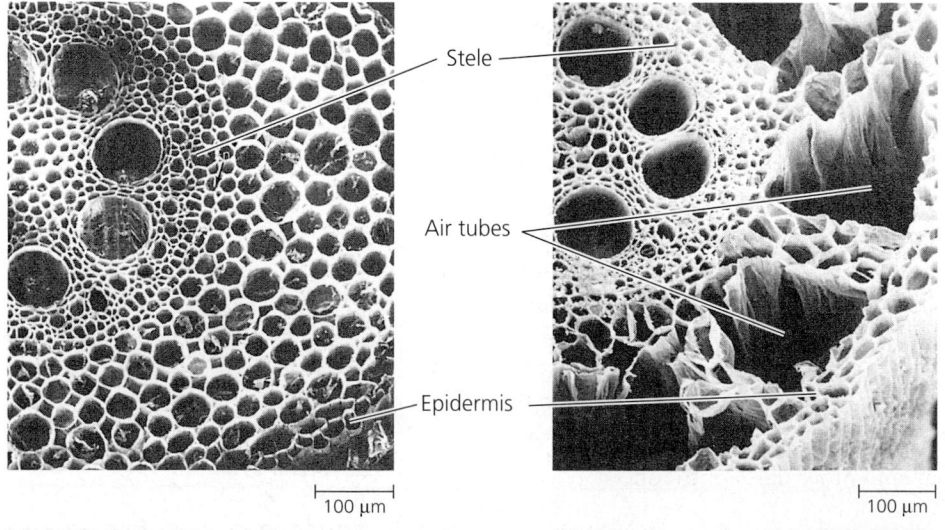

Stele

Air tubes

Epidermis

100 μm   100 μm

(a) Control root (aerated)   **(b) Experimental root (nonaerated)**

FIGURE 39.28 A developmental response of corn roots to flooding and oxygen deprivation. (a) A transverse section of a control root grown in an aerated hydroponic medium. **(b)** An experimental root grown in a nonaerated hydroponic medium. Ethylene-stimulated apoptosis (programmed cell death) creates the air tubes.

Cold Stress.   One problem plants face when the temperature of the environment falls is a change in the fluidity of cell membranes. Recall from Chapter 8 that a biological membrane is a fluid mosaic, with proteins and lipids moving laterally in the plane of the membrane. When a membrane cools below a critical point, it loses its fluidity as the lipids become locked into crystalline structures. This alters solute transport across the membrane and also adversely affects the functions of membrane proteins. Plants respond to cold stress by altering the lipid composition of their membranes. For example, membrane lipids increase in their proportion of unsaturated fatty acids, which have shapes that help keep membranes fluid at lower temperatures by impeding crystal formation (see FIGURE 8.4b). Such molecular modification of the membrane requires from several hours to days, which is one reason rapid chilling is generally more stressful to plants than the more gradual drop in air temperature that occurs seasonally.

Freezing is a more severe version of cold stress. At subfreezing temperatures, ice forms in the cell walls and intercellular spaces of most plants. (The cytosol generally does not freeze at the cooling rates encountered in nature because it contains more solutes than the very dilute solution found in the cell wall—solutes depress the freezing point of a solution.) The reduction in liquid water in the cell wall caused by ice formation lowers the extracellular water potential, causing water to leave the cytoplasm. The resulting increase in the concentration of ionic salts in the cytoplasm is harmful and can lead to cell death. Whether the cell survives depends to a large extent on how resistant it is to dehydration. Plants native to regions where winters are cold have special adaptations that enable them to cope with freezing stress. For example, before the onset of winter, the cells of many frost-tolerant species increase their cytoplasmic levels of specific solutes, such as sugars, that are better tolerated at high concentrations, and which help reduce the loss of water from the cell during extracellular freezing.

PLANT DEFENSE: RESPONSES TO HERBIVORES AND PATHOGENS

Plants do not exist in isolation, but interact with many other species in their communities. Some of these interspecific interactions—for example, the associations of plants with fungi in mycorrhizae (see Chapter 37) or with insect pollinators (see Chapter 38)—are mutually beneficial. Most of the interactions that plants have with other organisms, however, are not beneficial to the plant. As primary producers, plants are at the base of most food webs and are subject to attack by a wide range of plant-eating (herbivorous) animals. A plant is also subject to infection by a diversity of pathogenic viruses, bacteria, and fungi that have the potential to damage tissues or even kill the plant. Plants counter these threats with defense systems that deter herbivory and prevent infection or combat pathogens that do manage to infect the plant.

Plants deter herbivores with both physical and chemical defenses

Herbivory—animals eating plants—is a stress that plants face in any ecosystem. Plants counter excessive herbivory with both physical defenses, such as thorns, and chemical defenses, such as the production of distasteful or toxic compounds. For example, some plants produce an unusual amino acid called **canavanine**, named for one of its sources, the jackbean (*Canavalia ensiformis*). Canavanine resembles arginine, one of the 20 amino acids organisms incorporate into their proteins. If an insect eats a plant containing canavanine, the molecule is incorporated into the insect's proteins in place of arginine. Because canavanine is different enough from arginine to adversely affect the conformation and hence the function of the proteins, the insect dies.

At least some plants even recruit predatory animals that help defend the plant against specific herbivores. For example, insects called parasitoid wasps inject their eggs into their prey, including caterpillars feeding on plants. The eggs hatch within the caterpillars, and the larvae eat through their organic containers from the inside out. The plant, which benefits from the destruction of the herbivorous caterpillars, has an active role in this ecological drama. A leaf damaged by caterpillars releases volatile compounds that attract parasitoid wasps. The stimulus for this response is a combination of the physical damage to the leaf caused by the munching caterpillar and a specific compound present in the caterpillar's saliva (FIGURE 39.29).

The volatile molecules a plant releases in response to herbivore damage can also function as an "early warning system" for nearby plants of the same species. Lima bean plants infested with spider mites release volatile chemicals that

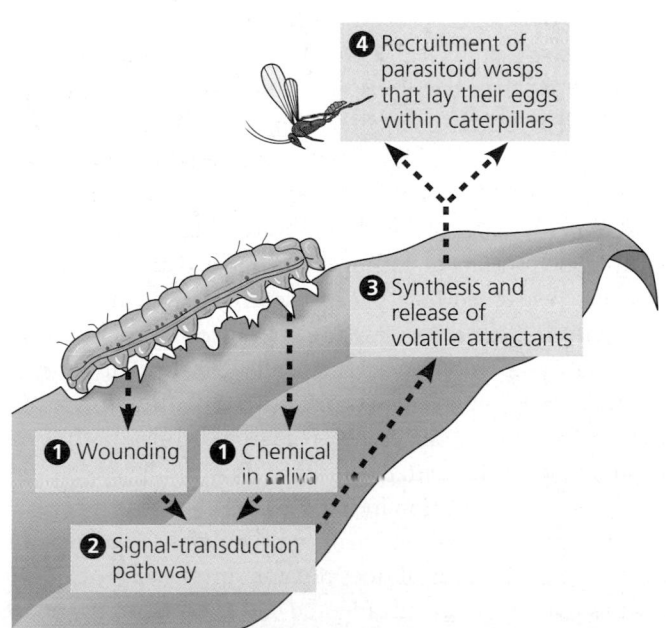

FIGURE 39.29 A corn leaf recruits a parasitoid wasp as a defensive response to an herbivore, an army-worm caterpillar.

signal "news" of the attack to neighboring, noninfested lima bean plants. In response to these volatiles, noninfested lima bean leaves activate defense genes. Volatiles released from artificially wounded leaves do not have the same effect. The expression pattern of the genes activated by the infestation-released volatiles is similar to that produced by exposure to **jasmonic acid,** an important molecule in plant defense. As a result of this gene activation, noninfested neighbors become less susceptible to spider mites and more attractive to another species of mite that preys upon spider mites.

Plants use multiple lines of defense against pathogens

A plant's first line of defense against infection is the physical barrier of the plant's "skin," the epidermis of the primary plant body and the periderm of the secondary plant body (see Chapter 35). This first defense system, however, is not impenetrable. Viruses, bacteria, and the spores and hyphae of fungi can enter the plant through injuries or through natural openings in the epidermis, such as stomata. Once a pathogen invades, the plant mounts a chemical attack as a second line of defense that kills the pathogens and prevents their spread from the site of infection. This second defense system is enhanced by the plant's inherited ability to recognize certain pathogens.

Gene-for-Gene Recognition

Plants are generally resistant to most pathogens. This is because plants have an innate ability to recognize invading pathogens and to mount successful defenses. In a converse manner, successful pathogens cause disease because they are able to evade recognition or suppress host defense mechanisms. Pathogens against which a plant has little specific defense are said to be virulent. They are the exceptions, for if they were not, then hosts and pathogens would soon perish together. A kind of "compromise" has coevolved between plants and most of their pathogens. In such cases, the pathogen gains enough access to its host to perpetuate itself without severely damaging or killing the plant. Such strains of pathogens are termed **avirulent.**

Specific resistance to a plant disease is based on what is called **gene-for-gene recognition,** because it depends on a precise match-up between a genetic allele in the plant and an allele in the pathogen. This occurs when a plant with specific dominant resistance alleles (*R*) recognizes those pathogens that possess complementary avirulence (*Avr*) alleles (FIGURE 39.30). Specific recognition induces expression of certain plant genes, the products of which mount a defense against the pathogen. If the plant host does not contain the appropriate *R* gene, the pathogen can invade and kill the plant. There are many pathogens, and plants have many *R* genes—*Arabidopsis* has at least several hundred.

It is not the *R* and *Avr* genes themselves that interact, of course, but their products. The product of an *R* gene is probably a specific receptor protein inside a plant cell or at its surface. The *Avr* gene probably leads to production of some "signal" molecule from the pathogen, a ligand capable of binding specifically to the plant cell's receptor. The *Avr* product undoubtedly has some function upon which the pathogen itself depends, but the plant is able to "key" on this molecule as an announcement of the pathogen's presence. Binding of the ligand to the receptor triggers a signal-transduction pathway leading to a defense response in the infected plant tissue. This defense includes both an enhancement of the localized response at the site of infection and a more general systemic response of the whole plant.

Hypersensitive Response

Even if a plant is infected by a virulent strain of a pathogen—one for which that particular plant has no genetic resistance—the plant is able to mount a localized chemical attack in re-

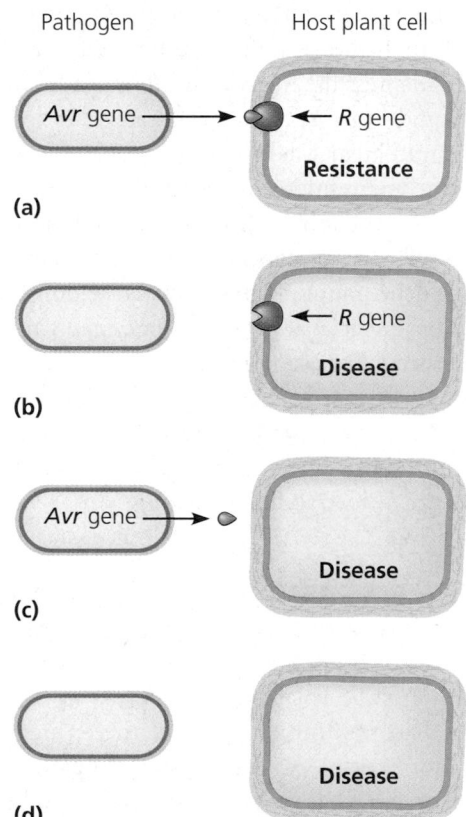

FIGURE 39.30 Gene-for-gene resistance of plants to pathogens.
(a) Resistance occurs when the plant has a particular dominant *R* allele that corresponds to a specific dominant *Avr* allele in the pathogen. *R* genes probably code for specific receptors. *Avr* genes produce compounds that function in the pathogen but also act as ligands that bind specifically to the host-plant cell's receptors. Disease occurs if there is no gene-for-gene recognition because **(b)** the pathogen has no dominant *Avr* allele matching an *R* allele in the plant, **(c)** the plant has no dominant *R* allele matching an *Avr* allele in the pathogen, or **(d)** both pathogen and plant lack alleles upon which recognition could be based.

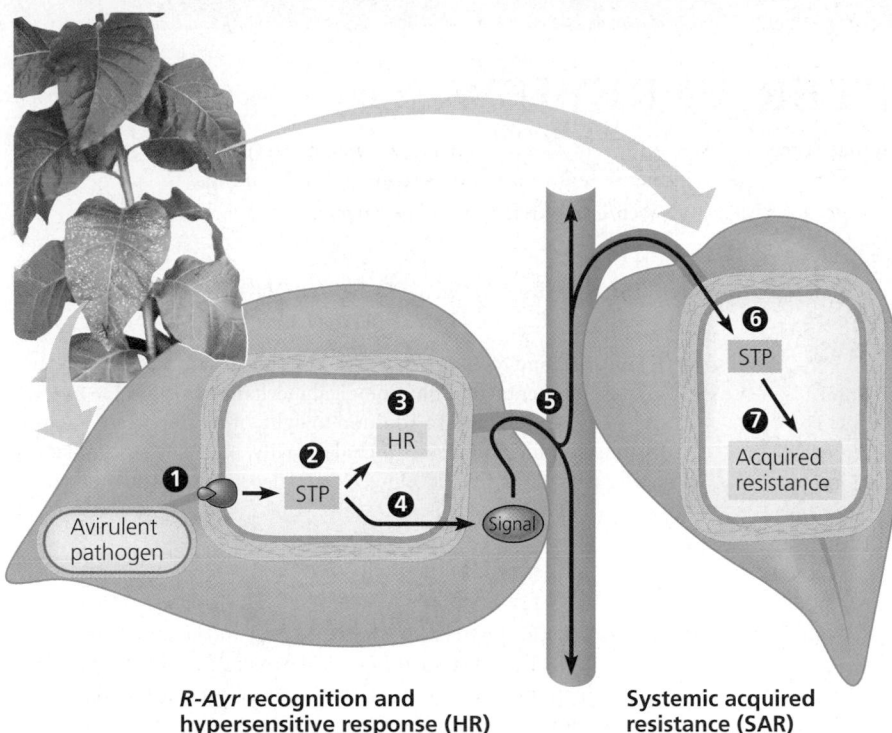

R-Avr recognition and hypersensitive response (HR)

Systemic acquired resistance (SAR)

FIGURE 39.31 Defense responses against an avirulent pathogen.

sponse to molecular signals released from cells damaged by the infection. Molecules called **elicitors**, often cellulose fragments called **oligosaccharins** released by cell-wall damage, induce the production of antimicrobial compounds called **phytoalexins.** Infection also activates genes that produce **PR proteins** (for pathogenesis-related). Some of these proteins are antimicrobial, attacking molecules in the cell wall of a bacterium, for example. Others may function as signals that spread "news" of the infection to nearby cells. Infection also stimulates cross-linking of molecules in the cell wall and deposition of lignin, responses that set up a local barricade that slows spread of the pathogen to other parts of the plant.

If the pathogen is avirulent based on an *R-Avr* match, the localized defense response is more vigorous and is called a **hypersensitive response** (abbreviated **HR**). There is an enhanced production of phytoalexins and PR proteins, and the "sealing" response that contains the infection is more effective. After cells at the site of infection mount their chemical defense and seal off the area, they destroy themselves. We can see the result of an HR as lesions on a leaf or other infected organ. As "sick" as such a leaf appears, it will survive, and its defense response will help protect the rest of the plant (FIGURE 39.31).

Systemic Acquired Resistance

The hypersensitive response, as you have learned, is localized and specific, a containment response based on gene-for-gene (*R-Avr*) recognition between host and pathogen. However, this defense response also includes production of chemical signals that "sound the alarm" of infection to the whole plant. Released from the site of infection, the alarm hormones are transported throughout the plant, stimulating production of phytoalexins and PR proteins. This response, called **systemic acquired resistance (SAR),** is nonspecific, providing protection against a diversity of pathogens for days (FIGURE 39.31).

A good candidate for one of the hormones responsible for activating SAR is **salicylic acid.** A modified form of this compound, acetylsalicylic acid, is the active ingredient in aspirin. Centuries before aspirin was sold as a pain reliever, some cultures had learned that chewing the bark of a willow tree *(Salix)* would lessen the pain of a toothache or headache. With the discovery of systemic acquired resistance, biologists have finally learned one function of salicylic acid in plants. Aspirin turns out to be a natural medicine in the plants that produce it, but with effects entirely different from the medicinal action in humans who consume the drug.

■   ■   ■

Plant biologists investigating disease resistance and other evolutionary adaptations of plants are getting to the heart of how a plant responds to internal and external signals. These scientists, along with thousands of other plant biologists working on other questions and millions of students experimenting with plants in biology courses, are all extending a centuries-old tradition of curiosity about the green organisms that feed the biosphere.

CHAPTER 39 REVIEW

Go to the Campbell Biology website (www.campbellbiology.com) to explore an interactive version of the Chapter Review.

Web/CD Activity 39B: *Flowering Lab*

SIGNAL TRANSDUCTION AND PLANT RESPONSES

- **Signal-transduction pathways link internal and environmental signals to cellular responses** (pp. 803–806, FIGURES 39.1–39.3) Hormones and environmental stimuli interact with specific receptors, thereby activating specific signal transduction pathways and inducing cellular responses.

PLANT RESPONSES TO HORMONES

- **Research on how plants grow toward light led to the discovery of plant hormones** (pp. 806–807, FIGURES 39.4, 39.5) Researchers discovered auxin by identifying the compound responsible for transmitting a signal downward through coleoptiles, from the tips to the elongating regions during phototropism.

- **Plant hormones help coordinate growth, development, and responses to environmental stimuli** (pp. 808–817, TABLE 39.1, FIGURES 39.6–39.16) This review cites one major function of each hormone. Produced primarily in the apical meristem of the shoot, auxin simulates cell elongation in different target tissues. Cytokinins, produced in actively growing tissues such as roots, embryos, and fruits, stimulate cell division. Gibberellins produced in roots and young leaves stimulate growth in leaves and stems. Abscisic acid maintains dormancy in seeds. Ethylene helps control fruit ripening. Brassinosteroids, chemically similar to the sex hormones of animals, induce cell elongation and division.

  Web/CD Activity 39A: *Leaf Abscission*
  Web/CD Case Study in the Process of Science: *What Plant Hormones Affect Organ Formation?*

PLANT RESPONSES TO LIGHT

- **Blue-light photoreceptors are a heterogeneous group of pigments** (pp. 817–818, FIGURE 39.17) Various blue-light photoreceptors control hypocotyl elongation, stomatal opening, and phototropism.

- **Phytochromes function as photoreceptors in many plant responses to light** (pp. 818–819, FIGURES 39.18–39.20) Phytochromes exist in two photoreversible states, with conversion of P_r to P_{fr} triggering many developmental responses.

- **Biological clocks control circadian rhythms in plants and other eukaryotes** (pp. 819–820, FIGURE 39.21) Free-running circadian cycles are approximately 24 hours long but are entrained to exactly 24 hours by the day/night cycle.

- **Light entrains the biological clock** (pp. 820–821) Phytochrome conversion marks sunrise and sunset, providing the clock with environmental cues.

- **Photoperiodism synchronizes many plant responses to changes of season** (pp. 821–823, FIGURES 39.22–39.24) Some developmental processes, including flowering in many plant species, require a certain photoperiod. For example, a critical night length sets a mini-

mum (in short-day plants) or maximum (in long-day plants) number of hours of darkness required for flowering.

Web/CD Activity 39B: *Flowering Lab*

PLANT RESPONSES TO ENVIRONMENTAL STIMULI OTHER THAN LIGHT

- **Plants respond to environmental stimuli through a combination of developmental and physiological mechanisms** (pp. 823–827, FIGURES 39.25–39.28) In addition to light, other important environmental stimuli and stresses include gravity, mechanical stimulation, water deficit, salinity, flooding, oxygen deprivation, heat, and cold.

PLANT DEFENSE: RESPONSES TO HERBIVORES AND PATHOGENS

- **Plants deter herbivores with both physical and chemical defenses** (pp. 827–828, FIGURE 39.29) Physical defenses include morphological adaptations such as thorns, chemical defenses such as distasteful or toxic compounds, and airborne attractants that bring animals that destroy herbivores.

- **Plants use multiple lines of defense against pathogens** (pp. 828–829, FIGURES 39.30, 39.31) A pathogen is avirulent if it has a specific dominant *Avr* gene corresponding to a particular *R* allele in the host plant. A hypersensitive response against an avirulent pathogen seals off the infection and kills both pathogen and host cells in the region of the infection. Salicylic acid is a signal molecule that triggers generalized defense responses in organs distant from the original site of infection (systemic acquired resistance, or SAR).

Self-Quiz

1. Which of the following plant hormones is incorrectly paired with its function?
   a. auxin—promotes stem growth through cell elongation
   b. cytokinins—initiate programmed cell death
   c. gibberellins—stimulate seed germination
   d. abscisic acid—promotes seed dormancy
   e. ethylene—inhibits cell elongation

2. Which of the following is *not* a typical component of a signal-transduction pathway such as the one involved in producing the greening response?
   a. G-proteins acting as transcription factors that activate specific genes
   b. activation of enzymes that produce second messengers such as cGMP
   c. activation of a G-protein by an activated receptor protein
   d. protein kinase cascades
   e. phosphorylation of transcription factors

3. Buds and sprouts often form on tree stumps. Which of the following hormones would you expect to stimulate their formation?
   a. auxin
   b. cytokinins
   c. abscisic acid
   d. ethylene
   e. gibberellins

4. Which of the following is *not* part of the acid-growth hypothesis?
   a. Auxin stimulates proton pumps in cell membranes.
   b. Lowered pH results in the breakage of cross-links between cellulose microfibrils.
   c. The wall fabric becomes looser (more plastic).
   d. Auxin-activated proton pumps stimulate cell division in meristems.
   e. The turgor pressure of the cell exceeds the restraining pressure of the loosened cell wall, and the cell takes up water and elongates.

5. The signal for flowering could be released earlier than normal in a long-day plant experimentally exposed to flashes of
   a. far-red light during the night.
   b. red light during the night.
   c. red light followed by far-red light during the night.
   d. far-red light during the day.
   e. red light during the day.

6. How might a plant respond to *severe* heat stress?
   a. orient leaves toward the sun to increase evaporative cooling
   b. produce ethylene that kills some cortex cells and creates air tubes for ventilation
   c. produce salicylic acid that initiates a systemic acquired resistance response
   d. increase the proportion of unsaturated fatty acids in cell membranes to reduce their fluidity
   e. produce heat-shock proteins that may protect the plant's proteins from denaturing

7. If a long-day plant has a critical night length of 9 hours, which of the following 24-hour cycles would prevent flowering?
   a. 16 hours light/8 hours dark
   b. 14 hours light/10 hours dark
   c. 15.5 hours light/8.5 hours dark
   d. 4 hours light/8 hours dark/4 hours light/8 hours dark
   e. 8 hours light/8 hours dark/light flash/8 hours dark

8. The probable role of salicylic acid in systemic acquired resistance of plants is to
   a. destroy pathogens directly.
   b. activate plant defenses throughout the plant before infection spreads.
   c. close stomata, thus preventing the entry of pathogens.
   d. activate heat-shock proteins.
   e. sacrifice infected tissues by hydrolyzing cells.

9. Auxin triggers the acidification of cell walls that results in rapid growth, but also stimulates sustained, long-term cell elongation. What best explains how auxin brings about this dual growth response?
   a. Auxin binds to different receptors in different cells.
   b. Different concentrations of auxin have different effects.
   c. Auxin causes second messengers to activate both proton pumps and certain genes.
   d. The dual effects are due to two different auxins.
   e. Other antagonistic hormones modify auxin's effects.

10. The subscripts in the following choices indicate specific *Avr* and *R* genes in pathogens and plant cells. Uppercase letters indicate dominant alleles, while lowercase symbolizes recessive alleles. In which of the situations would the pathogen be avirulent?
    a. $Avr_D–R_d$
    b. $Avr_E–R_G$
    c. $Avr_M–R_M$
    d. $Avr_g–R_g$
    e. $Avr_e–R_E$

11. How do the experiments illustrated in FIGURES 39.4 and 39.5 provide evidence that phototropism depends on a chemical signal, or hormone?

12. Imagine that your lab is equipped with a tiny pH electrode that enables you to measure the pH of a plant cell's wall. How could you use this device to test the hypothesis presented in FIGURE 39.7 for how auxins stimulate cell elongation?

13. Why are tropisms, such as phototropism and gravitropism, sometimes called "growth responses"?

14. The free-running period of the sleep movements of bean leaves is 26 hours. If a bean plant is kept in constant darkness, how many days will it take for its leaves to be in the "noon position" when the actual time is midnight?

15. A particular short-day plant won't flower in the spring. A flower grower tries to induce flowering by using a few minutes of darkness to split the long light period of spring into two short light periods. Based on the mechanism of photoperiodic control of flowering, what outcome do you predict for this experiment? How would you explain this result to the grower?

16. How does salicylic acid function in a plant's systemic acquired resistance (SAR) fit our definition of a hormone?

Go to the website or CD-ROM for more quiz questions.

Evolution Connection

Coevolution is defined as reciprocal adaptations between two species, each species adapting in how it interacts with the other. In this context of coevolution, write a paragraph explaining the relationship between a plant and an avirulent pathogen.

The Process of Science

A plant biologist observed a peculiar pattern when a tropical shrub was attacked by caterpillars. The scientist noticed that after a caterpillar ate a leaf, it would skip over nearby leaves and attack a leaf some distance away. The researcher found that when a leaf was eaten, nearby leaves started making a chemical that deterred the caterpillars. Simply removing a leaf did not deter the caterpillars from eating nearby leaves. The biologist suspected that a damaged leaf sent out a chemical that signaled other leaves. How could the researcher test this hypothesis?

Test how hormones affect tissue growth in plants in the Case Study in the Process of Science, available on the website and CD-ROM.

Science, Technology, and Society

Based on your study of this chapter, write a short essay explaining at least three examples of how knowledge about the control systems of plants is applied in agriculture or horticulture.

Answers: 1. b; 2. a; 3. b; 4. d; 5. b; 6. e; 7. b; 8. b; 9. c; 10. c; 11. The two key points: Light is detected by the shoot tip, but the bending response occurs some difference down from the tip; the signal is transmitted from tip to responding region through a barrier that prevents cell contact but allows chemicals to pass. 12. The hypothesis predicts that addition of auxins to the cell should stimulate proton pumps and lower (make more acidic) the pH of the wall. You could test this prediction with your pH electrode by measuring the wall pH for cells in the presence of (experimental group) or absence of (control group) auxins. 13. Because the bending of a plant organ toward or away from some environmental factor depends on cells on one side of the organ growing faster than cells on the opposite side. 14. About six days—the cycle in a constant environment is about 26 hours, two hours longer than a real day. Thus, it would take six days for the plant's leaf position to be 12 hours out of step with the actual time. 15. The plants still won't flower, because it is actually night length, not day length, that counts in the photoperiodic control of flowering. 16. The salicylic acid is a chemical signal produced by one part of the plant and transported to other parts of the plant, where it triggers responses in target cells.

ANIMAL FORM AND FUNCTION

Flossie Wong-Staal is a pioneer in research on AIDS, a topic discussed in this unit (see Chapter 43). She is credited with being the first to clone the retrovirus HIV and map its genes. This work, carried out in the laboratory headed by Robert Gallo at the National Cancer Institute (NCI), paved the way for the development of more sensitive and reliable tests for the presence of HIV in blood. After growing up in Hong Kong, Dr. Wong-Staal moved to Los Angeles, where she received her B.A. in bacteriology and Ph.D. in molecular biology from UCLA. She went to the NCI after postdoctoral work at the University of California, San Diego. Back at UCSD since 1990, she is the codirector of the AIDS Research Institute and a professor of biology and medicine, holding the Florence Riford Chair in AIDS Research.

Interview
FLOSSIE WONG-STAAL

What led you to retroviruses?

My interest in retroviruses first came about because of their value as tools in molecular biology. When I was in graduate school, RNA tumor viruses (retroviruses that cause cancer in certain animals) were generating a lot of excitement, particularly the discovery of their reverse transcriptase, which offered a potential tool for gene cloning and analysis. The most fascinating thing about these RNA tumor viruses was that many of them had essentially cloned a cellular gene, which acted as an oncogene—cancer-causing gene—in the cells it infected. Later we learned that some RNA tumor viruses do not carry oncogenes but produce cancer in other ways. Anyway, I became very interested in RNA tumor viruses.

After postdoctoral work at UCSD, I joined Robert Gallo's group at the NCI. Not only were they studying retroviruses but they were interested in determining if retroviruses were involved in human disease.

Was this unusual? Weren't other labs also looking for human retroviruses?

In those days, many people didn't think retroviruses had any role in human disease. Yes, the Rous sarcoma virus caused cancer in chickens, and other RNA tumor viruses caused cancer in inbred lab mice, but these viruses were regarded as flukes (accidents of nature). Most scientists working on retroviruses thought of them mainly as conven-

ient tools for doing molecular biology, not as important disease agents in the real world. Some people even referred to the viruses we were seeking as "human RNA *rumor* viruses" because they didn't believe such viruses existed. But our lab kept at it. We were encouraged by the appearance of evidence that retroviruses could cause leukemia in cats and cattle.

What were the results of this research?

The persistence of people in our group paid off with the discovery of the first human retrovirus, HTLV (human T cell leukemia virus), which causes a type of leukemia affecting the white blood cells called T cells. I was heading the molecular biology section of Gallo's group, and we immediately began to study HTLV. Soon it became clear that HTLV was unlike other known retroviruses in that it could turn on its own transcription in the cell—it was not solely dependent on cellular proteins.

It was now the late 1970s, and AIDS was becoming known as a specific disease that attacks the immune system. But what caused AIDS? Every causative agent imaginable was suggested, from chemicals to fungi. Bob Gallo and a few of his colleagues thought that a retrovirus might be involved. There are actually some parallels between AIDS and the leukemia caused by HTLV. Both diseases affect T cells, although leukemia causes T cell proliferation and

AIDS causes T cell depletion; and both diseases are transmitted by blood or other body fluids and can be passed from mother to fetus. In addition, the feline leukemia virus was known occasionally to cause depletion of white blood cells.

People who thought that the cause of AIDS was likely to be a virus formed an international task force. There was a spirit of cooperation and frequent collaboration. We exchanged reagents, as well as tissue and blood samples from patients. This work finally led to the isolation of HIV in our lab and also in the lab of Luc Montagnier at the Pasteur Institute in France.

In light of the fact that the relationship between the Gallo group and the Montagnier group became contentious, what are your thoughts about collaboration and competition in science?

A certain level of competition is healthy, I think. It certainly spurs you on and hastens discoveries. A good example in the area of art occurred during the Renaissance. Why were so many masterpieces of art produced within such a short period? Part of the answer, I think, was competition. Artists saw what others were doing and wanted to outdo them. Competition contributes to higher aspirations and thus to greater achievements.

But competition can be overdone. Unlike art, science requires collaboration, especially these days. Research questions are often very complicated and also multidisciplinary. No single scientist can do it all. A virologist, for example, might need to collaborate with a structural biologist, a clinician, a statistician, a protein chemist, or a geneticist.

What motivates you as a researcher?

For me, as for most researchers, the main motivation is simply the satisfaction of making discoveries, finding things out that no one knew before. However, we are now at an interesting time, when there is no longer a clear dividing line between basic, "pure," research and applied research. The case of AIDS and HIV is an obvious example of how basic research can quickly lead to important practical developments, such as diagnostic tests and treatments. It adds to

the joy of discovery to know that your work may make a difference in people's lives.

On the negative side, the potential for commercial application of a discovery can cause problems. The controversy surrounding the discovery of HIV was, I think, blown out of proportion because of the patent issue and the commercial implications of the discovery.

Why don't we have an AIDS vaccine yet?

Before AIDS emerged on the scene, there was a lot of optimism about conquering infectious diseases, to the point that some medical schools actually did away with infectious disease departments. It was thought that antibiotics and vaccines would quickly conquer all such diseases. And it was known that the immune system alone can overcome many infections.

A viral vaccine consists of inactivated ("dead") virus, viral proteins, or a mutant virus that isn't pathogenic. The vaccine primes the immune system without causing disease. Then, if the pathogenic virus comes along, the immune system destroys it.

Unfortunately, HIV differs in some major ways from most other viruses. First of all, as a retrovirus, it integrates its genetic information into the host's DNA and remains latent even when progeny viruses are not being made. This is a big problem for HIV patients undergoing drug treatment. You think the virus has completely gone away because there's no sign of any viral proteins. But, once you take away the drug, the virus can become rampant again.

The second problem is HIV's diversity. We're talking about millions of variations, which result from mutations. We would of course expect strains of the virus from different places to be different. But with HIV, even when we examine the viruses isolated from a single person at a single point in time, we see a very diverse population. The population continually changes in response to external factors. For instance, in the presence of a drug or antibodies produced by a vaccine, any viral variants that happen to be resistant will outreplicate the others and come to predominate. So the virus is a moving target. Any weapon used against it has to be so broadly effective that the virus cannot mutate to escape its effects.

The third problem is that HIV attacks the immune system, the very system that's supposed to combat it. And not only does it demolish immune cells, but many of the chemical signals involved in the operation of the immune system actually help the virus replicate! So we have a serious problem: If we try to counter HIV by stimulating the immune system, we also stimulate the virus!

Finally, there is the practical problem of not having a good animal model. HIV doesn't infect rodents, and even with monkeys we have to use a related virus rather than HIV itself. Chimpanzees are infected by HIV, but they don't get sick.

Right now, many people think that a vaccine to block infection completely may not be feasible, but that it may be possible to develop a vaccine that lowers the amount of virus to a point that is too low to cause disease. That would also decrease the transmission of the virus.

For the time being, I think AIDS prevention by education is probably the most powerful way to stem the epidemic. There has been progress along this line in some developing countries. Thailand, for example, has had considerable success with education campaigns and other preventive measures, such as condom distribution. Ultimately, however, a vaccine would certainly be the cheapest and most effective way to combat AIDS around the world.

Besides a vaccine or drugs, are there other approaches to prevention or treatment?

One approach we're working on is gene therapy. It's not likely to be the answer in the near term because gene therapy is not yet a mature field; there are technical problems we just can't overcome right now. However, eventually gene therapy

may be useful. The idea is simple: You introduce genes into the target cells of HIV to block infection. What might these introduced genes be? We are using synthetic genes whose transcripts are ribozymes that recognize and cut up HIV RNA, inactivating it. We have a lot of data showing that this method works at the cellular level; and, in collaboration with clinical researchers, we have even shown that it can work in T cells inside the body. But for the ribozyme genes to work in a patient for the long term, they would have to be put into stem cells and remain permanently active.

Another approach we are taking is to look for cellular genes that are essential for HIV replication. A treatment doesn't necessarily have to target the virus; instead, it could target cellular genes that are critically important for the virus but not for the cell. This approach won't immediately lead to anti-HIV drugs, but rather to the identification of molecular targets against which drugs might be developed.

What is your advice to undergraduates considering a career in medical research?

You need to have a passion for making discoveries because this is the most rewarding aspect of a scientific career. But you also have to realize that if you choose this career, you're in for a long haul. "Eureka" moments are few and far between. So before deciding, it's important to go into a research lab, both to get some experience with the scientific process and to see the kind of hard work that's required. It's also helpful to expose yourself to the passion and enthusiasm of outstanding scientists, by attending seminars, for instance. In this way, you can get a glimpse of the dedication that science requires and also the satisfaction it can provide.

AN INTRODUCTION TO ANIMAL STRUCTURE AND FUNCTION

It is natural *for us to be curious about how our bodies work. In these chapters on animal form and function, you will indeed learn much about human anatomy and physiology. But the scope of this unit of chapters is not limited to humans or even to vertebrates. The study of animal form and function is integrated by the common set of problems that all animals must solve. How, for example, do animals as different as hydras, halibut, and humans obtain oxygen from their environments? How do they nourish themselves? excrete waste products? move? Our main goal in this unit of chapters is to see how animals of diverse evolutionary history and varying complexity solve these general challenges of life. This chapter introduces the unit with some unifying concepts that apply across the animal kingdom.*

FUNCTIONAL ANATOMY: AN OVERVIEW

Animal form and function reflect biology's major themes

Animals provide vivid examples of biology's overarching theme of evolution. The adaptations we will observe in our comparative study of animals evolved by natural selection. The long, tonguelike proboscis of the hawk moth in the photo you see on this page is a structural adaptation for feeding. Recoiled when not in use, the proboscis extends as a straw through which the moth can suck nectar from deep within tube-shaped flowers.

A foraging hawk moth illustrates another major theme, one we introduced in Chapter 1: regulation. While natural selection provides a mechanism for long-term adaptation, organisms also have the capacity to adjust to environmental change over the short term by physiological responses (these short-term responses are themselves evolutionary adaptations). Many insects are inactive when it is cold. But *Manduca sexta*, the hawk moth species in our opening photo, can forage for nectar when air temperatures are as low as 5°C. The moth uses a shivering-like mechanism for preflight warm-up of its muscles. Once the moth takes off, metabolic activity of its flight muscles generates heat, and a variety of regulatory adaptations maintain a temperature of about 30°C in the muscles, even though the external temperature may be close to freezing. (You'll learn more about this regulation in Chapter 44.)

Searching for food, generating body heat and regulating internal temperature, sensing and responding to environmental stimuli, and all other animal activities require fuel in the form of chemical energy. Thus, we will apply concepts of bioenergetics—how organisms obtain, process, and use their energy resources—as another theme throughout our comparative study of animals.

And there's one more theme to guide our study of animals: the correlation of structure and function. Which is the better utensil: a spoon or a fork? The answer, of course, depends on what you are trying to eat. Given a choice of tools, you would not eat broth with a fork or a salad with a spoon. Form fits function. The same principle applies to life at its many levels, from molecules to organisms. Analyzing a biological structure such as a hawk moth's proboscis gives us clues about what it does and how it works. Conversely, knowing the function of a structure provides insight about its construction, such as the strawlike anatomy of *Manduca*'s proboscis. **Anatomy** is the study of the *structure* of an organism; **physiology** is the study of the *functions* an organism performs. The distinction blurs when we apply the structure-function theme, and "anatomy-and-physiology" then rolls off the tongue as though it were one big compound noun. We could use the term *functional anatomy* instead. But here, our analogy to human inventions such as eating utensils breaks down. Without any sort of goal, natural selection can fit structure to function by selecting, over many generations, for what works best among the available options presented by a varying population. Thus, the form-function principle is just another extension of biology's central theme of evolution.

Function correlates with structure in the tissues of animals

Life is characterized by hierarchical levels of organization, each with emergent properties (see Chapter 1). Animals are multicellular organisms with their specialized cells grouped into tissues. In most animals, combinations of various tissues make up functional units called organs, and groups of organs that work together form organ systems. For example, the human digestive system consists of a stomach, small intestine, large intestine, and several other organs, each a composite of different kinds of tissues.

Tissues are groups of cells with a common structure and function. Different types of tissues have different structures that are especially suited to their functions. A tissue may be held together by a sticky extracellular matrix that coats the cells (see FIGURE 7.29) or weaves them together in a fabric of fibers. Indeed, the term *tissue* is from a Latin word meaning "weave."

Tissues are classified into four main categories: epithelial tissue, connective tissue, nervous tissue, and muscle tissue. These are present to some extent in all but the simplest animals; the following survey emphasizes the tissues of vertebrates.

Epithelial Tissue

Occurring in sheets of tightly packed cells, **epithelial tissue** covers the outside of the body and lines organs and cavities within the body (FIGURE 40.1, p. 836). The cells of an epithelium are closely joined, with little material between them. In many epithelia, the cells are riveted together by tight junctions (see FIGURE 7.30). This tight packing enables the epithelium to function as a barrier protecting against mechanical injury, invasive microorganisms, and fluid loss. The free surface of the epithelium is exposed to air or fluid, whereas the cells at the base of the barrier are attached to a **basement membrane,** a dense mat of extracellular matrix.

Two criteria for classifying epithelia (plural) are the number of cell layers and the shape of the cells on the free surface. A **simple epithelium** has a single layer of cells, whereas a **stratified epithelium** has multiple tiers of cells. A pseudostratified epithelium is single-layered, but it appears to be stratified because the cells vary in length. The shape of the cells that are at the free surface of an epithelium may be **cuboidal** (like dice), **columnar** (like bricks on end), or **squamous** (flat like floor tiles). Combining the features of cell shape and number of layers, we get such terms as *simple cuboidal epithelium* and *stratified squamous epithelium* (see FIGURE 40.1).

As well as protecting the organs they line, some epithelia, called **glandular epithelia,** absorb or secrete chemical solutions. For example, glandular epithelia lining tubules in the thyroid gland secrete a hormone that regulates the body's rate of fuel consumption. The glandular epthelia that line the lumen (cavity) of the digestive and respiratory tracts form a **mucous membrane;** they secrete a slimy solution called mucus that lubricates the surface and keeps it moist. The free epithelial surfaces of some mucous membranes have beating cilia that move the film of mucus along the surface. For example, the ciliated epithelium of our respiratory tubes helps keep our lungs clean by trapping dust and other particles and sweeping them back up the trachea (windpipe).

Connective Tissue

Connective tissue functions mainly to bind and support other tissues. In contrast to epithelia, with their tightly packed cells, connective tissues have a sparse population of cells scattered through an extracellular matrix. The matrix generally consists of a web of fibers embedded in a uniform foundation that may be liquid, jellylike, or solid. In most cases, the substances of the matrix are secreted by the cells of the connective tissue.

Connective tissue fibers, which are made of protein, are of three kinds: collagenous fibers, elastic fibers, and reticular fibers. **Collagenous fibers** are made of collagen, probably the most abundant protein in the animal kingdom. Collagenous fibers are nonelastic and do not tear easily when pulled lengthwise. If you pinch and pull some skin on the back of your hand,

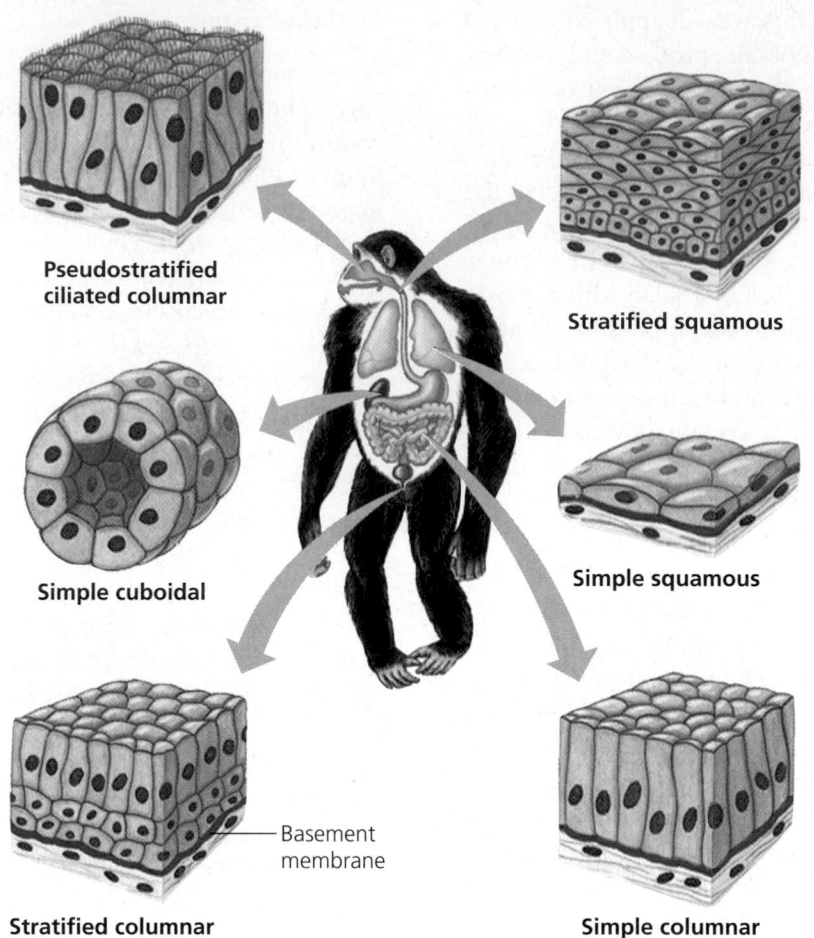

Columnar epithelia, having cells with relatively large cytoplasmic volumes, are often located where secretion or the active absorption of substances is an important function. For example, pseudostratified ciliated columnar epithelia line the nasal passages of many vertebrates.

Pseudostratified ciliated columnar

Cuboidal cells specialized for secretion make up the epithelia of kidney tubules and many glands, including the thyroid gland and salivary glands.

Simple cuboidal

Stratified columnar epithelia line the inner surface of the urethra, the tube through which urine exits the body.

Basement membrane

Stratified columnar

Stratified squamous epithelia regenerate rapidly by cell division near the basement membrane. The new cells are pushed to the free surface as replacements for cells that are continually sloughed off. This type of epithelium is commonly found on surfaces subject to abrasion, such as the outer skin and linings of the esophagus, anus, and vagina.

Stratified squamous

Simple squamous epithelia, which are thin and leaky, function in the exchange of material by diffusion. These epithelia line blood vessels and the air sacs of the lungs.

Simple squamous

The intestines are lined with **simple columnar epithelia** that secrete digestive juices and absorb nutrients.

Simple columnar

FIGURE 40.1 The structure and function of epithelial tissues. The structure of an epithelium fits its function.

it is mainly collagen that keeps the flesh from tearing away from the bone. **Elastic fibers** are long threads made of a protein called elastin. Elastic fibers provide a rubbery quality that complements the nonelastic strength of collagenous fibers. When you pinch the back of your hand and then let go, elastic fibers quickly restore your skin to its original shape. **Reticular fibers** are very thin and branched. Composed of collagen and continuous with collagenous fibers, they form a tightly woven fabric that joins connective tissue to adjacent tissues.

The major types of connective tissue in vertebrates are loose connective tissue, adipose tissue, fibrous connective tissue, cartilage, bone, and blood (FIGURE 40.2). Each has a structure correlated with its specialized functions.

The most widespread connective tissue in the vertebrate body is **loose connective tissue.** It binds epithelia to underlying tissues and functions as packing material, holding organs in place. This type of connective tissue gets its name from the loose weave of its fibers. Loose connective tissue has all three fiber types: collagenous, elastic, and reticular.

Among the cells scattered in the fibrous mesh of loose connective tissue, two types predominate: fibroblasts and

macrophages. **Fibroblasts** secrete the protein ingredients of the extracellular fibers. **Macrophages** are amoeboid cells that roam the maze of fibers, engulfing bacteria and the debris of dead cells by phagocytosis (see Chapter 8). They are weapons in an elaborate arsenal of defense you will learn more about in Chapter 43.

Adipose tissue is a specialized form of loose connective tissue that stores fat in adipose cells distributed throughout its matrix. Adipose tissue pads and insulates the body and stores fuel as fat molecules (see FIGURE 4.5). Each adipose cell contains a large fat droplet that swells when fat is stored and shrinks when the body uses fat as fuel.

Fibrous connective tissue is dense, due to its large numbers of collagenous fibers. The fibers are organized into parallel bundles, an arrangement that maximizes nonelastic strength. We find this type of connective tissue in **tendons,** which attach muscles to bones, and in **ligaments,** which join bones together at joints.

Cartilage has an abundance of collagenous fibers embedded in a rubbery matrix made of a substance called chondroitin sulfate, a protein-carbohydrate complex. Chondroitin sulfate

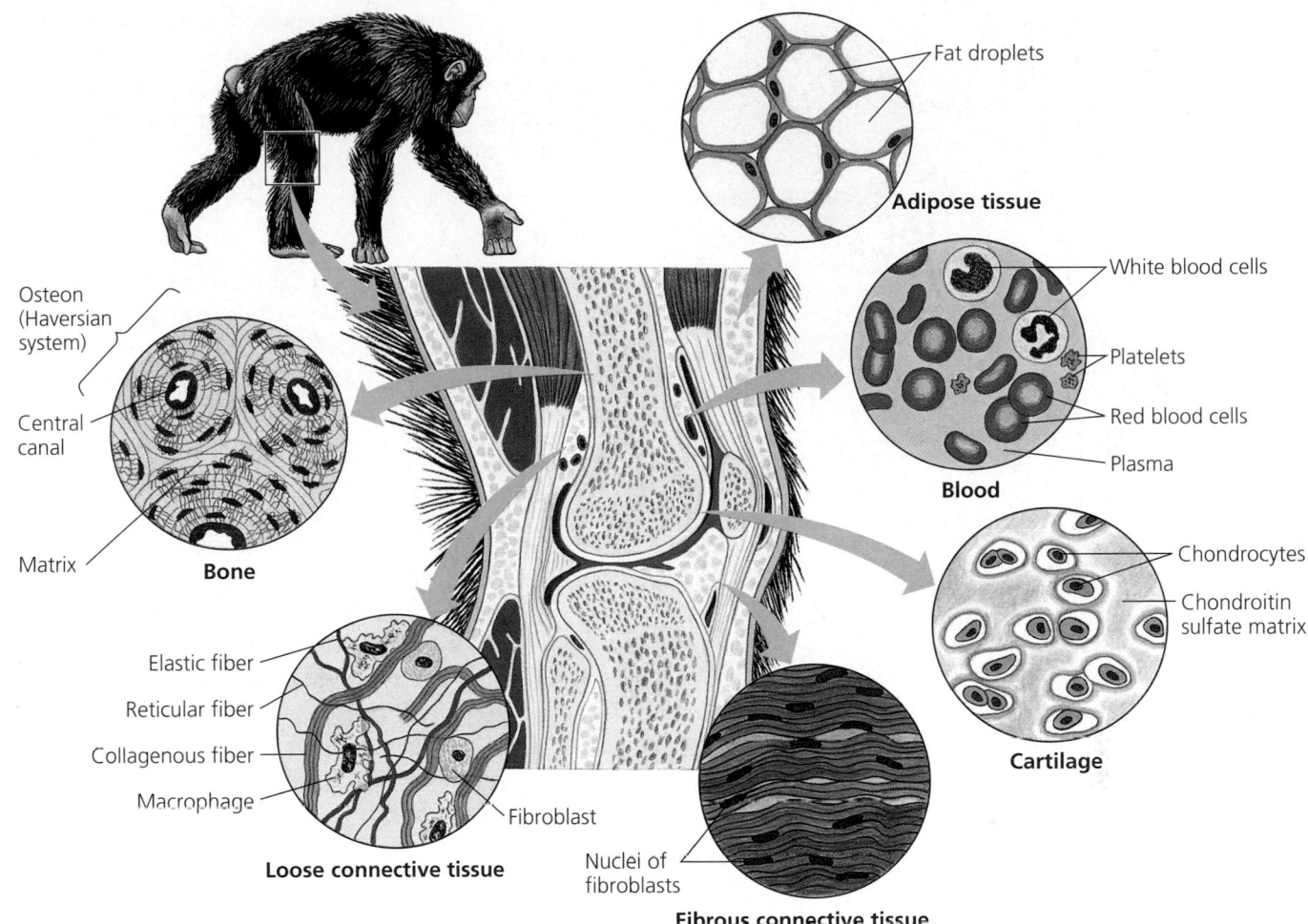

FIGURE 40.2 Some representative types of connective tissue. The area shown is the region around the knee joint.

Labels in figure:

- Fat droplets — **Adipose tissue**
- Osteon (Haversian system)
- Central canal
- Matrix — **Bone**
- White blood cells
- Platelets
- Red blood cells
- Plasma — **Blood**
- Chondrocytes
- Chondroitin sulfate matrix — **Cartilage**
- Elastic fiber
- Reticular fiber
- Collagenous fiber
- Macrophage
- Fibroblast — **Loose connective tissue**
- Nuclei of fibroblasts — **Fibrous connective tissue**

and collagen are secreted by cells called **chondrocytes** (see FIG-URE 40.2). The composite of collagenous fibers and chondroitin sulfate makes cartilage a strong yet somewhat flexible support material. The skeleton of a shark is made of cartilage. Other vertebrates, including humans, have cartilaginous skeletons during the embryo stage, but most of the cartilage is replaced by bone as the embryo matures. We nevertheless retain cartilage as flexible support in certain locations, such as the nose, the ears, the rings that reinforce the windpipe, the discs that act as cushions between our vertebrae, and the caps on the ends of some bones.

The skeleton supporting the body of most vertebrates is made of **bone,** a mineralized connective tissue. Bone-forming cells called **osteoblasts** deposit a matrix of collagen. Calcium, magnesium, and phosphate ions combine and harden within the matrix into the mineral hydroxyapatite. The combination of hard mineral and flexible collagen makes bone harder than cartilage without being brittle. The microscopic structure of hard mammalian bone consists of repeating units called **osteons** (or Haversian systems) (see FIGURE 40.2). Each osteon has concentric layers of the mineralized matrix, which are deposited

around a central canal containing blood vessels and nerves that service the bone. Once osteoblasts become trapped in their own secretions, they are called osteocytes. (We will examine bones and skeletons in more detail in Chapter 49.)

Although **blood** functions differently from other connective tissues, it does meet the criterion of having an extensive extracellular matrix. In this case, the matrix is a liquid called plasma, consisting of water, salts, and a variety of dissolved proteins. Suspended in the plasma are two classes of blood cells, erythrocytes (red blood cells) and leukocytes (white blood cells), and cell fragments called platelets. Red cells carry oxygen; white cells function in defense against viruses, bacteria, and other invaders; and platelets aid in blood clotting. Blood will be discussed in detail in Chapters 42 and 43.

Nervous Tissue

Nervous tissue senses stimuli and transmits signals from one part of the animal to another. The functional unit of nervous tissue is the **neuron,** or nerve cell, which is uniquely

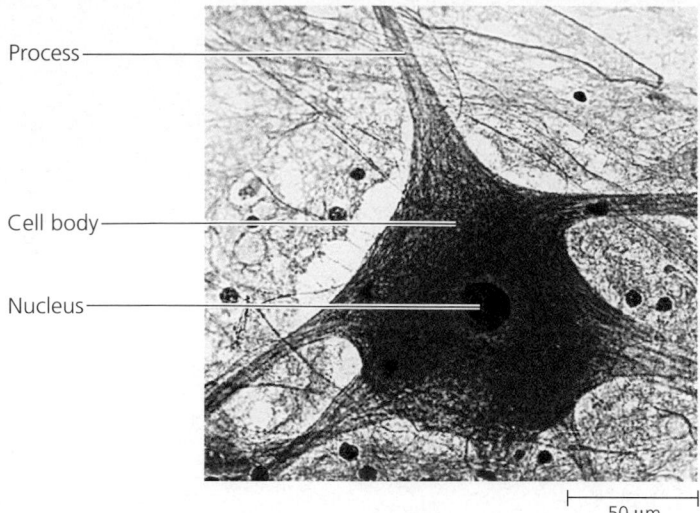

Process

Cell body

Nucleus

50 μm

FIGURE 40.3 The basic structure of a neuron. This nerve cell from the spinal cord has a large cell body with multiple processes that transmit electrical signals called impulses (LM).

specialized to transmit signals called nerve impulses (FIGURE 40.3). It consists of a cell body and two or more extensions, or processes, called dendrites and axons, which may be as long as a meter in humans. Dendrites transmit impulses from their

tips toward the rest of the neuron. Axons transmit impulses toward another neuron or toward an effector, a structure such as a muscle cell that carries out a body response. We will postpone a detailed discussion of the structure and function of neurons until Chapter 48.

Muscle Tissue

Muscle tissue is composed of long cells called muscle fibers that are capable of contracting when stimulated by nerve impulses. Arranged in parallel within the cytoplasm of muscle fibers are large numbers of myofibrils made of the contractile proteins actin and myosin. Muscle is the most abundant tissue in most animals, and muscle contraction accounts for much of the energy-consuming cellular work in an active animal.

In the vertebrate body, there are three types of muscle tissue: skeletal muscle, cardiac muscle, and smooth muscle (FIGURE 40.4). Attached to bones by tendons, **skeletal muscle** is responsible for voluntary movements of the body. Adults have a fixed number of muscle cells; weight lifting and other methods of building muscle do not increase the number of cells but simply enlarge those already present. Skeletal muscle is also called **striated muscle** because the arrangement of overlapping filaments gives the cells a striped (striated) appearance under the microscope.

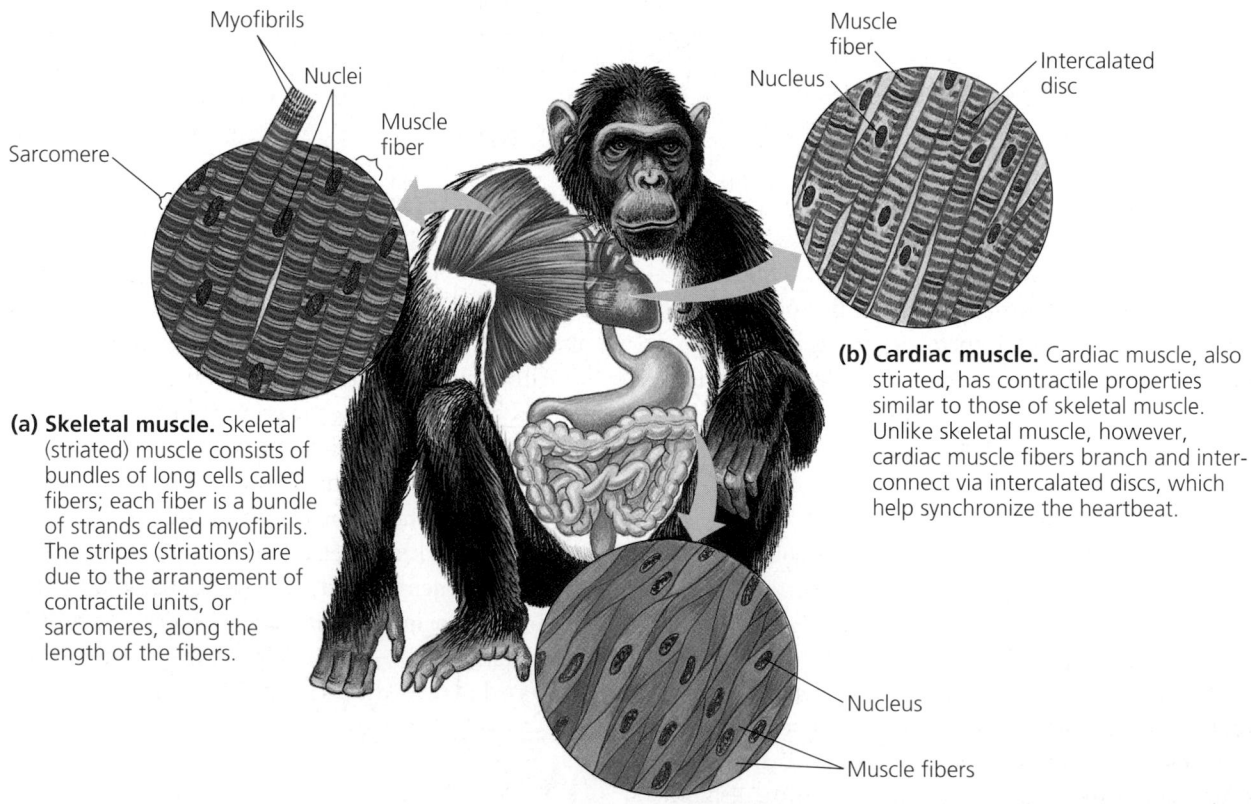

(a) Skeletal muscle. Skeletal (striated) muscle consists of bundles of long cells called fibers; each fiber is a bundle of strands called myofibrils. The stripes (striations) are due to the arrangement of contractile units, or sarcomeres, along the length of the fibers.

(b) Cardiac muscle. Cardiac muscle, also striated, has contractile properties similar to those of skeletal muscle. Unlike skeletal muscle, however, cardiac muscle fibers branch and interconnect via intercalated discs, which help synchronize the heartbeat.

(c) Smooth muscle. Smooth muscle consists of spindle-shaped cells lacking cross-striations.

Myofibrils

Nuclei

Sarcomere

Muscle fiber

Muscle fiber

Nucleus

Intercalated disc

Nucleus

Muscle fibers

FIGURE 40.4 Three kinds of vertebrate muscle.

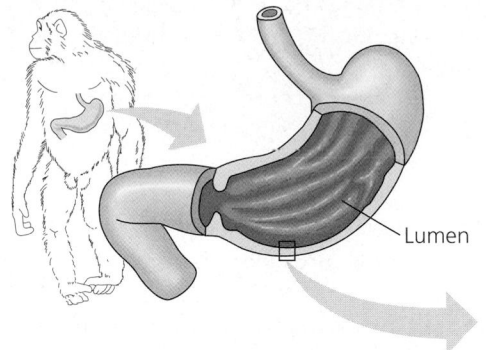

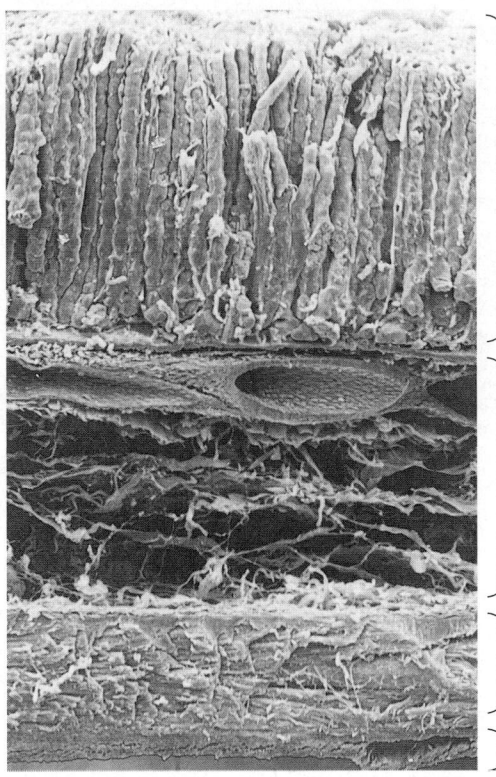

Lumen

Mucosa. The mucosa is an epithelial layer that lines the lumen.

Submucosa. The submucosa is a matrix of connective tissue that contains blood vessels and nerves.

Muscularis. The muscularis consists mainly of smooth muscle tissue.

Serosa. External to the muscularis is the serosa, a thin layer of connective tissue and epithelial tissue.

0.1 mm

FIGURE 40.5 Tissue layers of the stomach, a digestive organ. The wall of the stomach and other tubular organs of the digestive system has four main tissue layers.

SEM copyright by Richard G. Kessel and Randy H. Kardon, *Tissues and Organs: A Text-Atlas of Scanning Electron Microscopy.* W. H. Freeman, 1979, all rights reserved.

Cardiac muscle forms the contractile wall of the heart. It is striated like skeletal muscle, but cardiac cells are branched. The ends of the cells are joined by structures called intercalated discs, which relay signals from cell to cell during a heartbeat.

Smooth muscle, so named because it lacks striations, is found in the walls of the digestive tract, urinary bladder, arteries, and other internal organs. The cells are spindle-shaped. They contract more slowly than skeletal muscles but can remain contracted longer. Controlled by different kinds of nerves than those controlling skeletal muscles, smooth muscles are responsible for involuntary body activities, such as churning of the stomach or constriction of arteries. You will learn more about the control and contraction of muscles in Chapter 49.

The organ systems of an animal are interdependent

In all but the simplest animals (sponges and some cnidarians), different tissues are organized into **organs.** In some organs the tissues are arranged in layers. For example, the vertebrate stomach has four major tissue layers (FIGURE 40.5). A thick epithelium lines the lumen and secretes mucus and digestive juices into it. Outside this layer is a zone of connective tissue, surrounded by a thick layer of smooth muscle. Yet another layer of connective tissue encapsulates the entire stomach.

Many of the organs of vertebrates are suspended by sheets of connective tissue called **mesenteries** in body cavities moistened or filled with fluid. Mammals have a **thoracic cavity** housing the lungs and heart that is separated from the lower **abdominal cavity** by a sheet of muscle called the diaphragm.

A level of organization higher than organs, **organ systems** carry out the major body functions of most animals (TABLE 40.1, p. 840). Each organ system consists of several organs and has specific functions, but the efforts of all systems must be coordinated for the animal to survive. For instance, nutrients absorbed from the digestive tract are distributed throughout the body by the circulatory system. But the heart that pumps blood through the circulatory system depends on nutrients absorbed by the digestive tract and also on oxygen (O_2) obtained from the air or water by the respiratory system. Any organism, whether single-celled or an assembly of organ systems, is a coordinated living whole greater than the sum of its parts.

BODY PLANS AND THE EXTERNAL ENVIRONMENT

An animal's size and shape, features that biologists often call body plans or designs, are fundamental aspects of form and function that significantly affect the way an animal interacts with its environment. By using the terms *plan* and *design* here,

Table 40.1 Organ Systems: Their Main Components and Functions in Mammals

Organ System	Main Components	Main Functions
Digestive	Mouth, pharynx, esophagus, stomach, intestines, liver, pancreas, anus	Food processing (ingestion, digestion, absorption, elimination)
Circulatory	Heart, blood vessels, blood	Internal distribution of materials
Respiratory	Lungs, trachea, other breathing tubes	Gas exchange (uptake of oxygen; disposal of carbon dioxide)
Immune and Lymphatic	Bone marrow, lymph nodes, thymus, spleen, lymph vessels, white blood cells	Body defense (fighting infections and cancer)
Excretory	Kidneys, ureters, urinary bladder, urethra	Disposal of metabolic wastes; regulation of osmotic balance of blood
Endocrine	Pituitary, thyroid, pancreas, other hormone-secreting glands	Coordination of body activities (e.g., digestion, metabolism)
Reproductive	Ovaries, testes, and associated organs	Reproduction
Nervous	Brain, spinal cord, nerves, sensory organs	Coordination of body activities; detection of stimuli and formulation of responses to them
Integumentary	Skin and its derivatives (e.g., hair, claws, skin glands)	Protection against mechanical injury, infection, drying out
Skeletal	Skeleton (bones, tendons, ligaments, cartilage)	Body support, protection of internal organs
Muscular	Skeletal muscles	Movement, locomotion

we do not mean to imply that animal body forms are products of a conscious invention. The body plan or design of an animal results from a pattern of development programmed by the genome, itself the product of millions of years of evolution due to natural selection.

Physical laws constrain animal form

Imagine the horror of wading into a murky lake and feeling your legs engulfed by a squishy amoeba the size of a pro wrestler. Fortunately, you don't have to add this to your worry list. It will never happen. Physical requirements constrain what natural selection can "invent," including the size of single cells. An amoeba the size of a human could never move materials across its membrane fast enough to satisfy such a large blob of cytoplasm. This is just one example of how physical law—in this case, the math of surface-to-volume relations—affects the evolution of an organism's form.

Consider another example: how the laws of hydrodynamics constrain the shapes that are possible for aquatic animals that swim very fast. Tuna and other fast bony fishes can swim at speeds up to 80 kilometers per hour. Sharks, penguins (birds), and aquatic mammals such as dolphins, seals, and whales are also fast swimmers. And they all have the same basic body shape. It's called a fusiform shape, which means tapered on both ends (FIGURE 40.6). Water is about a thousand times denser than air, and thus the slightest bump that causes drag impedes a swimmer even more than it does a runner or a flyer. We should expect speedy fishes and marine mammals to have similar shapes, because the laws of hydrodynamics are

universal. This is an example of convergent evolution (see Chapter 25). Convergence occurs because natural selection shapes similar adaptations when diverse organisms face the same environmental challenge, such as the resistance of water to fast travel.

Body size and shape affect interactions with the environment

An animal's size and shape have a direct effect on how the animal exchanges energy and materials with its surroundings. As a requirement for maintaining the fluid integrity of the plasma membranes of its cells, an animal body must be arranged so that all of its living cells are bathed in an aqueous medium. Exchange with the environment occurs as dissolved substances diffuse and are transported across the plasma membranes between the cells and their aqueous surroundings. As shown in FIGURE 40.7a, a single-celled protist living in water has a sufficient surface area of plasma membrane to service its entire volume of cytoplasm because it is so small. A large cell has less surface area relative to its volume than a smaller cell of the same shape (see FIGURE 7.5). As you learned in the preceding concept, this is one of the physical constraints on the size of protists such as amoebas.

Multicellular animals are composed of microscopic cells, each with its own plasma membrane that functions as a loading and unloading platform for a modest volume of cytoplasm. But this only works if all the cells of the animal have access to a suitable aqueous environment. A hydra, built on the sac plan, has a body wall only two cell layers thick (FIGURE 40.7b).

(a) Tuna

(b) Shark

(c) Penguins

(d) Dolphins

(e) Seal

FIGURE 40.6 Evolutionary convergence on fusiform shapes in fast swimmers.

(f) Submarine

Because its gastrovascular cavity opens to the exterior, both outer and inner layers of cells are bathed in water. A flat body shape is another way to maximize exposure to the surrounding medium. For instance, a tapeworm may be several meters long, but because it is very thin, most of its cells are bathed in the intestinal fluid of the worm's vertebrate host, from which it obtains nutrients.

Two-layered sacs and flat shapes are designs that put a large surface area in contact with the environment, but these simple forms do not allow much complexity in internal organization. Most animals are more complex and made up of compact masses of cells; their outer surfaces are relatively small compared with their volume. As an extreme comparison, the surface-to-volume ratio of a whale is millions of times smaller than that of a protozoan, yet every cell in the whale must be bathed in fluid and have access to oxygen, nutrients, and other resources. Whales and most other animals have extensively folded or branched internal surfaces specialized for exchange with the environment (FIGURE 40.8, p. 842). The circulatory system shuttles materials among all the exchange surfaces within the animal.

Although exchange with the environment is a problem for animals whose cells are mostly internal, complex body forms have distinct benefits. Because the animal's external surface need not be bathed in water, it is possible for the animal to live on land. Also, because the immediate environment for the cells is the internal body fluid, the animal's organ systems can control the composition of the solution bathing its cells.

Diffusion

Mouth

Diffusion

Diffusion

Gastrovascular cavity

(a) Single cell

(b) Two cell layers

FIGURE 40.7 Contact with the environment. (a) In a unicellular organism, such as this amoeba, the entire surface area contacts the environment. Because of its small size, the cell has a large surface area relative to its volume through which to exchange materials with the external world. **(b)** A hydra is bilayered. Because the aqueous environment can circulate in and out of its mouth, virtually every one of its cells directly contacts the environment and exchanges materials with it.

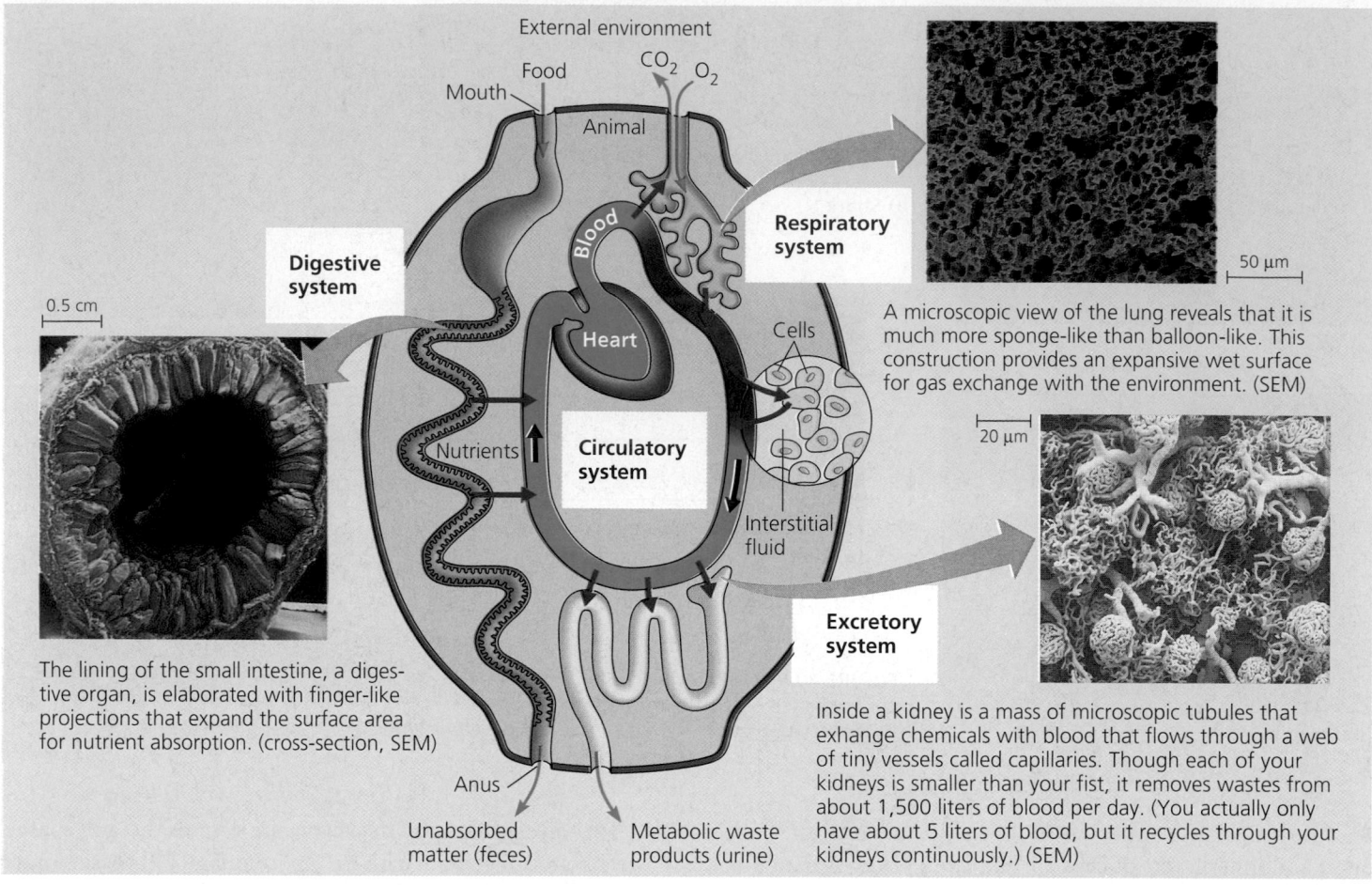

FIGURE 40.8 Internal exchange surfaces of complex animals. This diagrammatic animal illustrates logistics of chemical exchange with the environment by a mammal. Most animals have expansive surfaces that are specialized for exchanging certain chemicals with the surroundings. These exchange surfaces are usually internal, but are connected to the environment via openings on the body surface (the mouth, for example). The exchange surfaces are finely branched or folded, giving them a very large area. The digestive, respiratory, and excretory systems all have such specialized surfaces. Chemicals that are transported across these surfaces are carried throughout the body by the circulatory system.

Within the figure:

External environment

Food

Mouth

CO_2 O_2

Animal

Digestive system

0.5 cm

Blood

Respiratory system

50 μm

A microscopic view of the lung reveals that it is much more sponge-like than balloon-like. This construction provides an expansive wet surface for gas exchange with the environment. (SEM)

Heart

Cells

Nutrients

Circulatory system

20 μm

Interstitial fluid

The lining of the small intestine, a digestive organ, is elaborated with finger-like projections that expand the surface area for nutrient absorption. (cross-section, SEM)

Anus

Unabsorbed matter (feces)

Metabolic waste products (urine)

Excretory system

Inside a kidney is a mass of microscopic tubules that exhange chemicals with blood that flows through a web of tiny vessels called capillaries. Though each of your kidneys is smaller than your fist, it removes wastes from about 1,500 liters of blood per day. (You actually only have about 5 liters of blood, but it recycles through your kidneys continuously.) (SEM)

REGULATING THE INTERNAL ENVIRONMENT

Mechanisms of homeostasis moderate changes in the internal environment

More than a century ago, French physiologist Claude Bernard made the distinction between the external environment surrounding an animal and the internal environment in which the cells of the animal actually live. The internal environment of vertebrates is called the **interstitial fluid** (see FIGURE 40.8). This fluid, which fills the spaces between our cells, exchanges nutrients and wastes with blood contained in microscopic vessels called capillaries. Bernard also recognized that many animals tend to maintain relatively constant conditions in their internal environment, even when the external environment changes. A pond-dwelling hydra is powerless to affect the temperature of the fluid that bathes its cells, but the human body can maintain its "internal pond" at a more-or-less constant temperature of about 37°C. Our bodies also can control the pH of our blood and interstitial fluid to within a tenth of a pH unit of 7.4, and regulate the amount of sugar in our blood so that it does not fluctuate for long from a concentration of 0.1%. There are times, of course, during the development of an animal when major changes in the internal environment are programmed to occur. For example, the balance of hormones in human blood is altered radically during puberty and pregnancy. Still, the stability of the internal environment is remarkable.

Today, Bernard's "constant internal milieu" is incorporated into the concept of **homeostasis,** which means "steady state," or internal balance. One of the main objectives of modern physiology, and a theme of this unit, is to learn how animals maintain homeostasis. Actually, the internal environment of an animal always fluctuates slightly. Homeostasis is a dynamic state, an interplay between outside forces that tend to change the internal environment and internal control mechanisms that oppose such changes.

Homeostasis depends on feedback circuits

Any homeostatic control system has three functional components: a receptor, a control center, and an effector. The *receptor* detects a change in some variable of the animal's internal environment, such as change in body temperature. The *control center* processes information it receives from the receptor and directs an appropriate response by the *effector*. As a nonliving example of how these components interact, consider how the temperature of a room is controlled (FIGURE 40.9a). In this case, the control center, called a thermostat, also contains the receptor (a thermometer). When room temperature falls below a set point, say 20°C, the thermostat switches on the heater (the effector). When the thermometer detects a temperature above the set point, the thermostat switches the heater off. This type of control circuit is called **negative feedback,** because a change in the variable being monitored triggers the control mechanism to counteract further change in the same direction. Owing to a lag time between reception and response, the variable drifts slightly above and below the set point, but the fluctuations are moderate. Negative-feedback mechanisms prevent small changes from becoming too large. Most homeostatic mechanisms in animals operate on this principle of negative feedback.

Our own body temperature is kept close to a set point of 37°C by the cooperation of several negative-feedback circuits that regulate energy exchange with the environment (FIGURE 40.9b). One of these involves sweating as a means to dispose of metabolic heat and cool the body. A thermostat in the brain monitors the temperature of the blood. If the thermostat detects a rise in body temperature above the set point, it sends nerve impulses directing sweat glands to increase their production of sweat, thereby lowering body temperature by evaporative cooling (see FIGURE 3.4). When body temperature drops below the set point, the thermostat in the brain stops sending the signals to the glands, and the body retains more of the heat produced by metabolism. We will see several examples of negative feedback in the chapters that follow.

In contrast to negative feedback, **positive feedback** involves a change in some variable that triggers mechanisms that amplify rather than reverse the change. During childbirth, for instance, the pressure of the baby's head against sensors near the opening of the uterus stimulates uterine contractions, which cause greater pressure against the uterine opening, heightening the contractions, which causes still greater pressure. Positive feedback brings childbirth to completion, a very different sort of process from maintaining a steady state.

It is important not to overstate the concept of a constant internal environment. In fact, *regulated change* is essential to normal body functions. In some cases the changes are cyclical, such as the changes in hormone levels responsible for the menstrual cycle in women (see Chapter 46). In other cases a

FIGURE 40.9 An example of negative feedback: control of temperature. Regulating either **(a)** room or **(b)** body temperature depends on a control center that detects temperature change and activates mechanisms that reverse that change.

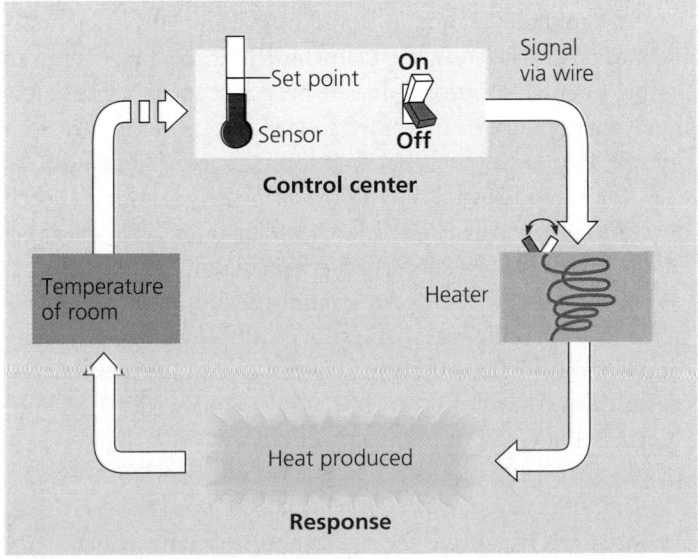

(a) Control of room temperature

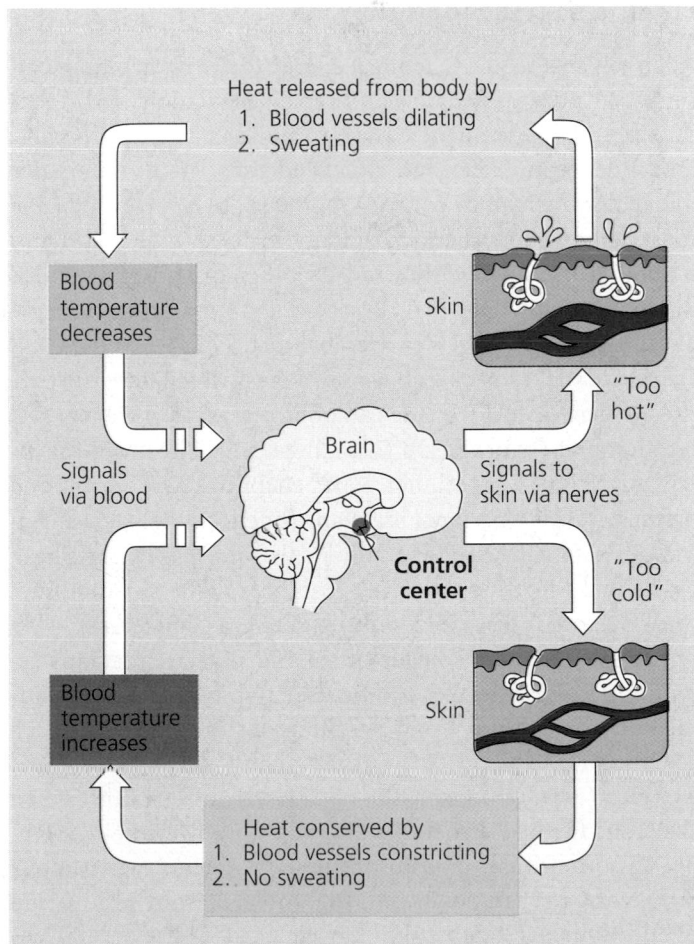

(b) Control of body temperature

regulated change is a reaction to a challenge to the body. For example, the human body reacts to certain infections by raising the set point for temperature to a slightly higher level, and the resulting fever helps fight the infection. Over the short term, homeostatic mechanisms keep body temperature close to a set point, whatever it is at that particular time. But over the longer term, homeostasis allows regulated change in the body's internal environment.

Internal regulation is expensive. Anyone who pays utility bills is aware of the energy costs for heating or cooling a home to maintain a comfortable interior temperature. Similarly, animals use a considerable portion of the energy from the food they eat to maintain favorable internal conditions. Animals must manage their energy resources not only for homeostasis, but for everything else they do, including movement, defense against disease, and reproduction. Let's take a closer look at some basic concepts of animal energetics.

INTRODUCTION TO THE BIOENERGETICS OF ANIMALS

Animals are heterotrophs that harvest chemical energy from the food they eat

All organisms require chemical energy for growth, physiological processes, maintenance and repair, regulation, and reproduction. Plants use light energy to build energy-rich organic molecules from water and CO_2, and then use those organic molecules for fuel. In contrast, animals are heterotrophs and must obtain their chemical energy in food, which contains organic molecules synthesized by other organisms. Food is digested by enzymatic hydrolysis (see FIGURE 5.2), and energy-containing fuel molecules are absorbed by body cells. Once absorbed, fuel molecules have several possible fates. Most are used to generate ATP by the catabolic processes of cellular respiration and fermentation (see Chapter 9). The chemical energy of ATP powers cellular work, enabling cells, organs, and organ systems to perform the many functions that keep an animal alive. Since the production and use of ATP generates heat, an animal must continuously lose heat to its surroundings (heat balance is discussed in more detail in Chapter 44).

After the energetic needs of staying alive are met, any remaining food molecules can be used in biosynthesis, including body growth and repair, synthesis of storage material such as fat, and production of reproductive structures, including gametes (FIGURE 40.10). Biosynthesis requires both carbon skeletons for new structures and ATP to power their assembly. In some situations, biosynthetic products (such as body fat) can be broken down into fuel molecules for production of additional ATP, depending on the needs of the animal (see FIGURE 9.19).

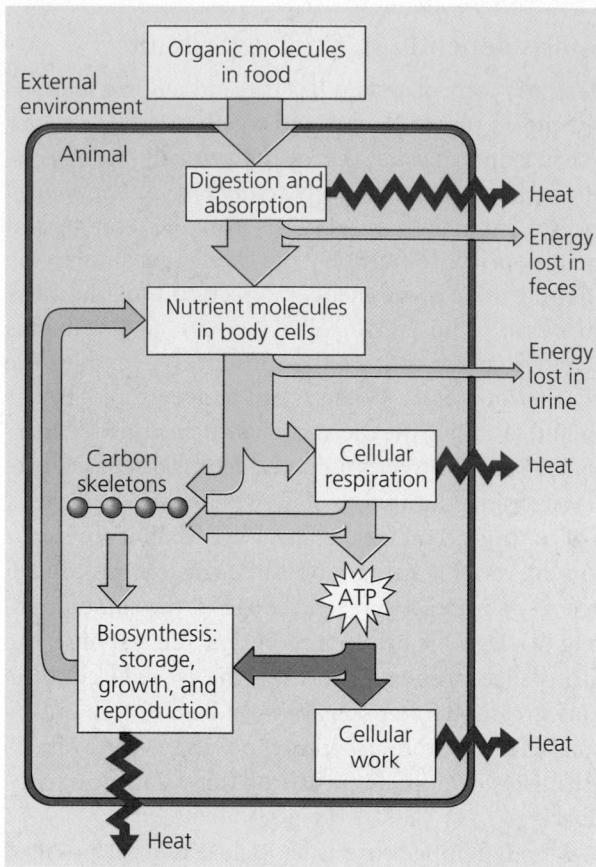

FIGURE 40.10 Bioenergetics of an animal: an overview.

Metabolic rate provides clues to an animal's bioenergetic "strategy"

The flow of energy through an animal—the animal's bioenergetics—ultimately sets the limits to the animal's behavior, growth, and reproduction and determines how much food it needs. An understanding of bioenergetics tells us a great deal about an animal's adaptations and how it fits into its environment as an energy consumer. How much of the total energy an animal obtains from food does it need just to stay alive? How much energy must be expended to walk, run, swim, or fly from one place to another? What fraction of energy intake can be used for reproduction? Physiologists obtain answers to such questions by measuring the rates at which animals use chemical energy and how these rates change in different circumstances.

The amount of energy an animal uses in a unit of time is called its **metabolic rate**—the sum of all the energy-requiring biochemical reactions occurring over a given time interval. Energy is measured in calories (cal) or kilocalories (kcal). (A kilocalorie is 1,000 calories. The term *Calorie*, with a capital C, as used by many nutritionists, is actually a kilocalorie.)

Metabolic rate can be determined in several ways. Because nearly all of the chemical energy used in cellular respiration eventually appears as heat, metabolic rate

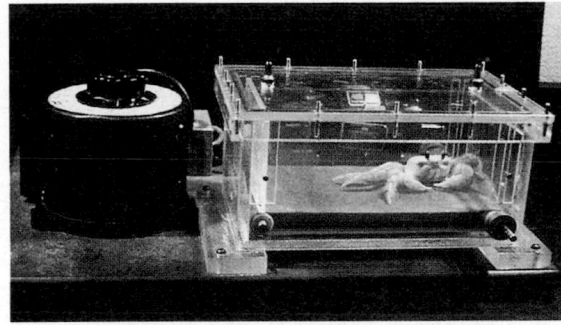

(a) A ghost crab running on a treadmill in a respirometer

FIGURE 40.11 Measuring metabolic rate.

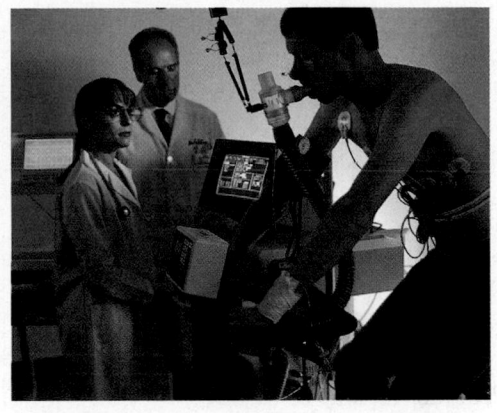

(b) A man riding a stationary bike while attached to a plastic breathing apparatus

can be measured by monitoring an animal's rate of heat loss. An animal is placed in a calorimeter, which is a closed, insulated chamber equipped with a device that records the animal's heat loss. Calorimeters are used most often with small animals, since they are cumbersome with large animals. A more indirect way to measure metabolic rate is to determine the amount of oxygen consumed or carbon dioxide produced by an animal's cellular respiration (FIGURE 40.11). Over long periods, the rate of food consumption and the energy content of the food (about 4.5–5 kcal per gram of protein or carbohydrate and about 9 kcal per gram of fat) can be used to estimate metabolic rate, but this method must account for the energy in food that cannot be used by the animal (the energy lost in feces and urine).

There are two basic bioenergetic "strategies" used by animals. Birds and mammals are mainly **endothermic,** meaning their bodies are warmed by heat generated by metabolism, and their body temperature must be maintained at a certain level to sustain life (see Chapter 34). Endothermy is a high-energy strategy that permits intense, long-duration activity over a wide range of environmental temperatures. In contrast, most fishes, amphibians, reptiles, and invertebrates are **ectothermic,** meaning they do not produce enough metabolic heat to have much effect on body temperature. The ectothermic strategy requires much less energy than is needed by endotherms, because of the energy cost of heating (or cooling) an endothermic body. However, ectotherms are generally incapable of intense activity over long periods.

Metabolic rate per gram is inversely related to body size among similar animals

One of animal biology's most intriguing but largely unanswered questions has to do with the relationship between body size and metabolic rate. By measuring the metabolic rates of many species of vertebrates and invertebrates, physiologists have shown that the amount of energy it takes to maintain each gram of body weight is inversely related to body size. Each gram of a mouse, for instance, consumes about 20 times more calories than a gram of an elephant (even though the whole elephant, of course, uses far more calories than the whole mouse). The higher metabolic rate of a smaller animal's tissues demands a proportionately greater rate of delivery of oxygen. And correlated with its higher metabolic rate, the smaller animal also has a higher breathing rate, blood volume (relative to its size), and heart rate (pulse) and must eat much more food per unit of body mass.

What explains the inverse relationship between metabolic rate and size? One hypothesis is that for endotherms, the smaller the animal, the greater the energy cost of maintaining a stable body temperature. This idea stems from the surface area to volume relationship: The smaller an animal, the greater its surface area to volume ratio, and thus the greater the loss of heat to (or gain of heat from) the surroundings. Logical as this hypothesis appears to be, it fails to explain the inverse relationship between metabolism and size in ectotherms (which do not use metabolic heat to maintain body temperature) and is not supported by various experimental tests. Researchers continue to search for causes underlying the inverse relationship between body size and metabolic rate.

Animals adjust their metabolic rates as conditions change

Every animal has a range of metabolic rates. Minimal rates power the basic functions that support life, such as cell maintenance, breathing, and heartbeat. The metabolic rate of a nongrowing endotherm at rest, with an empty stomach, and experiencing no stress, is called the **basal metabolic rate (BMR).** The BMR for humans averages about 1,600 to 1,800 kcal per day for adult males and about 1,300 to 1,500 kcal per day for adult females. These BMRs are about equivalent to the energy consumption of a 75-watt light bulb.

In ectotherms, body temperature changes with the temperature of the surroundings, and so does metabolic rate. Unlike BMRs, which can be determined within a range of environmental temperatures, the minimal metabolic rate of an ectotherm must be determined at a specific temperature. The metabolic rate of a resting, fasting, nonstressed ectotherm is called its **standard metabolic rate (SMR).**

For both ectotherms and endotherms, activity has a large effect on metabolic rate. Any behavior, even a person working quietly at a desk or an insect extending its wings, consumes energy beyond the BMR or SMR. Maximal metabolic rates (the highest rates of ATP utilization) occur during peak activity, such as lifting heavy weights, all-out running, or high-speed swimming.

In general, an animal's maximum possible metabolic rate is inversely related to the duration of activity. FIGURE 40.12 contrasts the ectothermic and endothermic "strategies" for sustaining activity over different time intervals. Both an alligator (ectotherm) and a human (endotherm) are capable of very intense exercise in short spurts of a minute or less. During these "sprints," the ATP present in muscle cells and ATP generated anaerobically by glycolysis can power the activity. Neither the ectotherm nor the endotherm can sustain their maximum metabolic rates and peak levels of activity over longer periods of exercise, though the endotherm has an advantage in such endurance tests. Sustained activity depends on the aerobic process of cellular respiration for ATP supply, and an endotherm's respiration rate is about 10 times greater than an ectotherm's. Only endotherms are capable of long-duration activities such as distance running.

Between the extremes of BMR or SMR and maximal metabolic rate, many factors influence energy requirements, including age, sex, size, body and environmental temperatures, the quality and quantity of food, activity level, oxygen availability, hormonal balance, and time of day. Birds, humans, and many insects are usually active (and therefore have their highest metabolic rates) during daylight hours. In contrast, bats, mice, and many other mammals generally are active and have their highest metabolic rates at night or during the hours of dawn and dusk. Metabolic rates measured when animals are performing a variety of activities give a better idea of the energy costs of everyday life. For most terrestrial animals (both ectotherms and endotherms), the average daily rate of energy consumption is 2–4 times BMR or SMR. Humans in most developed countries have an unusually low average daily metabolic rate of about 1.5 times BMR—an indication of relatively sedentary lifestyles.

Energy budgets reveal how animals use energy and materials

Different species of animals use the energy and materials in food in different ways, depending on their environment, behavior, size, and basic energy "strategy" of endothermy or ectothermy. For most animals, the majority of food is devoted to the production of ATP, and relatively little goes to growth or reproduction. However, the amount of energy used for BMR (or SMR), activity, and temperature control varies considerably between species. For example, consider the typical annual

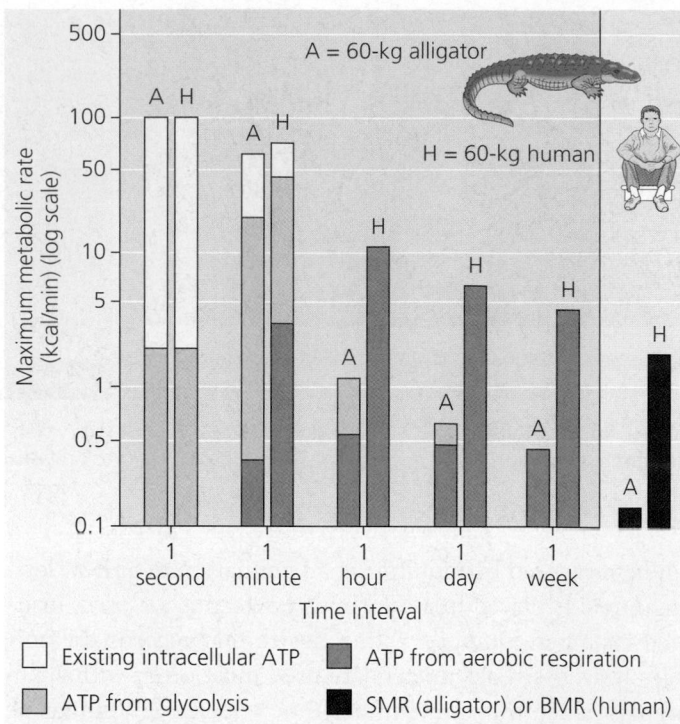

FIGURE 40.12 Maximum metabolic rates over different time spans. The two bars at the far right compare the resting metabolic rates for an ectotherm (alligator) versus an endotherm (human) of equivalent size. Note the much higher energy cost required for the endothermic "strategy." The other bars compare the two animals for their *maximum* potential metabolic rates and ATP sources over different durations of time.

energy "budgets" of four terrestrial vertebrates: a 25-gram female deer mouse, a 4-kg female python, a 4-kg male Adélie penguin, and a 60-kg female human (FIGURE 40.13). We will assume that all of these animals reproduce during the year in question.

The female human spends a large fraction of her energy budget for BMR and relatively little for activity and temperature regulation. The small amount of growth, about 1% of her annual energy budget, is equivalent to adding about 1 kg of body fat or 5–6 kg of other tissues. The cost of nine months of pregnancy and several months of breast-feeding amounts to only 5–8% of the mother's annual energy requirements.

A male penguin spends a much larger fraction of his energy expenditures for activity because he must swim to catch his food. Since the penguin is well insulated and fairly large, he has relatively low costs of temperature regulation in spite of living in the cold Antarctic environment. His reproductive costs, about 6% of annual energy expenditures, mainly come from incubating eggs and bringing food to his chicks. Penguins, like most birds, do not grow once they are adults.

The deer mouse spends a large fraction of her energy budget for temperature regulation. Because of the high surface-to-volume ratio that goes with small size, mice lose body heat rapidly to the environment and must constantly generate meta-

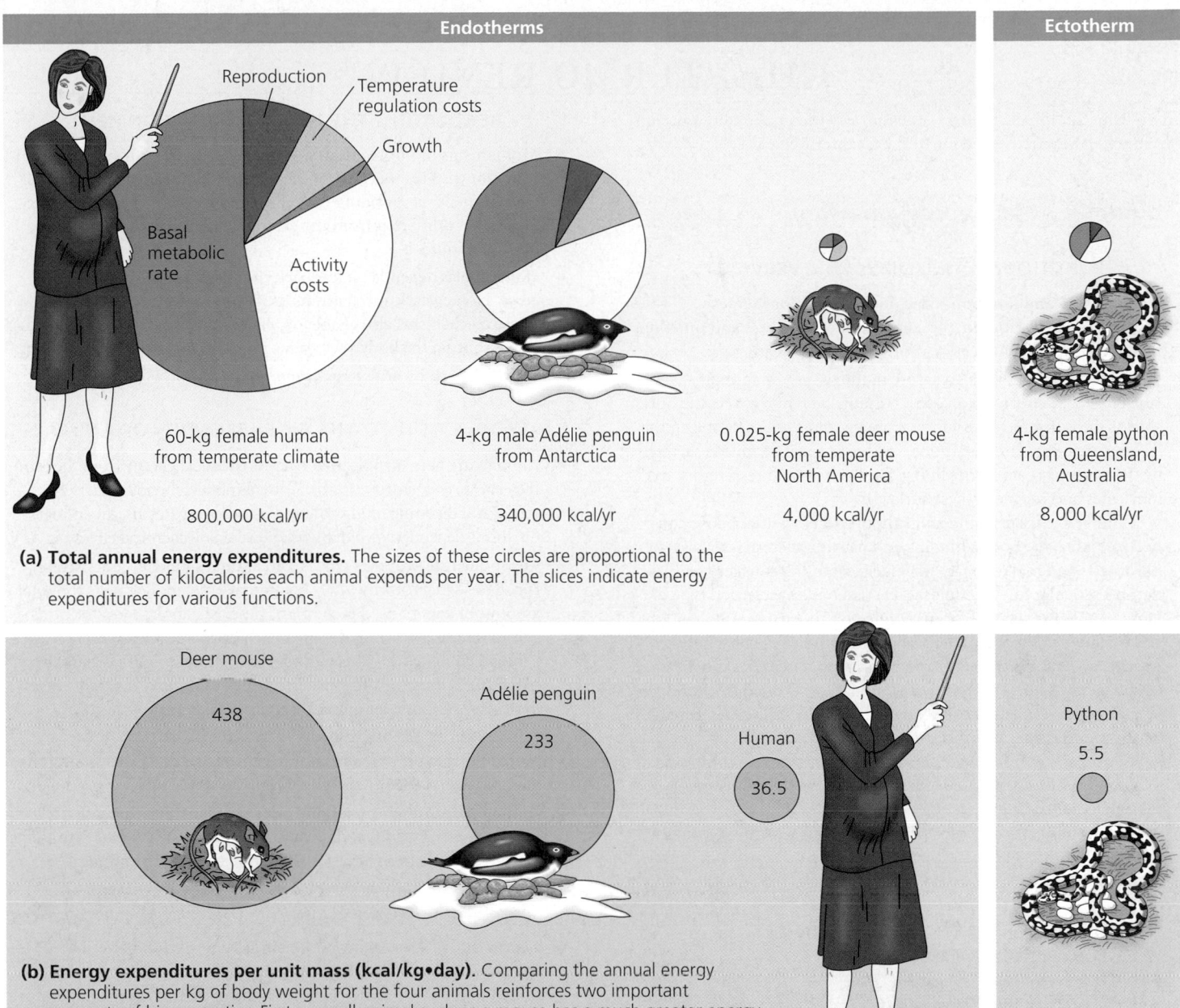

(a) Total annual energy expenditures. The sizes of these circles are proportional to the total number of kilocalories each animal expends per year. The slices indicate energy expenditures for various functions.

(b) Energy expenditures per unit mass (kcal/kg•day). Comparing the annual energy expenditures per kg of body weight for the four animals reinforces two important concepts of bioenergetics. First, a small animal such as a mouse has a much greater energy demand per kg than does a large animal of the same taxonomic class, such as a human (both mammals). Second, note again that an ectotherm, such as a python, requires much less energy per kg than does an endotherm of equivalent size, such as a penguin.

FIGURE 40.13 Annual energy budgets for four animals.

bolic heat to maintain body temperature. Female deer mice spend about 12% of their energy budgets on reproduction.

In contrast to these endothermic animals, the ectothermic python has no temperature regulation costs. Like most reptiles, she grows continuously throughout life. In the FIGURE 40.13 example, the snake added about 750 g of new body tissue. She also produced about 650 g of eggs. The python's economical ectothermic strategy is revealed by her very low annual energy expenditure, which is only ¹⁄₄₀ of the energy expended by the same-sized endothermic penguin (see FIGURE 40.13b).

Throughout our study of animal biology, we will encounter many other examples of how bioenergetics relates to the form and function of diverse animals.

▪   ▪   ▪

Having examined some general principles of animal biology, we are now ready to compare how diverse animals perform such activities as digestion, circulation, gas exchange, excretion of wastes, reproduction, and coordination—the topics of the next several chapters.

CHAPTER 40 REVIEW

Go to the Campbell Biology website (www.campbellbiology.com) to explore an interactive version of the Chapter Review.

Summary of Key Concepts

FUNCTIONAL ANATOMY: AN OVERVIEW

■ **Animal form and function reflect biology's major themes (pp. 834–835)** Evolution, the correlation of structure and function, regulation, and bioenergetics inform our study of animals.

■ **Function correlates with structure in the tissues of animals (pp. 835–839, FIGURES 40.1–40.4)** Epithelial tissue covers the outside of the body and lines internal organs and cavities. Some epithelia are specialized for absorption and secretion. The mucus secreted by the mucous membranes lining the digestive and respiratory tracts lubricates and moistens these surfaces.

Connective tissues bind and support other tissues. Loose connective tissue, the body's binding and packing material, consists of fibroblasts and macrophages interspersed among collagenous, elastic, and reticular fibers. Adipose (fat) tissue is a specialized type of loose connective tissue. Fibrous connective tissue, found in tendons and ligaments, is made of dense, parallel bundles of collagenous fibers. Cartilage is a strong yet flexible support material consisting of collagenous fibers in a rubbery foundation secreted by chondrocytes. The hard substance of bone is secreted by cells called osteoblasts. Blood is also a connective tissue.

Neurons, the functional units of nervous tissue, are composed of a cell body with extending dendrites and axons that transmit electrical signals called impulses.

Muscle tissue is composed of long cells (muscle fibers) containing parallel fibrils of contractile proteins. The three types of muscle tissue in vertebrates are skeletal, cardiac, and smooth muscles, which differ in shape, striation, and nervous control.

Web/CD Activity 40A: *Overview of Animal Tissues*
Web/CD Activity 40B: *Epithelial Tissue*
Web/CD Activity 40C: *Connective Tissue*
Web/CD Activity 40D: *Nervous Tissue*
Web/CD Activity 40E: *Muscle Tissue*

■ **The organ systems of an animal are interdependent (p. 839, TABLE 40.1, FIGURE 40.5)** The body functions as a whole, greater than the sum of its parts, because the activities of all tissues, organs, and organ systems are coordinated.

BODY PLANS AND THE EXTERNAL ENVIRONMENT

■ **Physical laws constrain animal form (p. 840, FIGURE 40.6)** Evolutionary convergence reflects independent adaptation to a similar environmental challenge.

■ **Body size and shape affect interactions with the environment (pp. 840–841, FIGURES 40.7 AND 40.8)** Each cell of a multicellular animal must have access to an aqueous environment. Simple two-layered sacs and flat shapes maximize exposure to the surrounding medium. More complex body plans have highly folded internal surfaces specialized for exchanging materials with the environment.

REGULATING THE INTERNAL ENVIRONMENT

■ **Mechanisms of homeostasis moderate changes in the internal environment (p. 842)** Because of regulatory mechanisms, the internal environment surrounding the cells making up an animal's body is usually very different from the external environment surrounding the entire animal.

■ **Homeostasis depends on feedback circuits (pp. 843–844, FIGURE 40.9)** Homeostatic mechanisms usually involve negative feedback. These mechanisms also enable regulated change as they react to occasional shifts in the body's set points for variables such as temperature.

Web/CD Activity 40F: *Regulation: Negative and Positive Feedback*

INTRODUCTION TO THE BIOENERGETICS OF ANIMALS

■ **Animals are heterotrophs that harvest chemical energy from the food they eat (p. 844, FIGURE 40.10)** Animals obtain chemical energy by eating and digesting food produced by other organisms. An animal's functions depend on cellular work powered by chemical energy in ATP.

■ **Metabolic rate provides clues to an animal's bioenergetic "strategy" (pp. 844–845, FIGURE 40.11)** An animal's metabolic rate is the total amount of energy used in a unit of time. Metabolic rates for birds and mammals, which maintain a fairly constant body temperature using metabolic heat (the endothermic "strategy"), are generally higher than those of most fishes, reptiles, amphibians, and invertebrates, whose body temperature changes with that of their surroundings (the ectothermic "strategy").

■ **Metabolic rate per gram is inversely related to body size among similar animals (p. 845)** This relationship holds for each animal taxon.

■ **Animals adjust their metabolic rates as conditions change (pp. 845–846, FIGURE 40.12)** Activity increases metabolic rate above the BMR (endotherms) or SMR (ectotherms) of an animal.

Web/CD Case Study in the Process of Science: *How Does Temperature Affect Metabolic Rate in* Daphnia?

■ **Energy budgets reveal how animals use energy and materials (pp. 846–847, FIGURE 40.13)** An animal's use of energy is partitioned to BMR (or SMR), activity, homeostasis (such as temperature regulation), growth, and reproduction.

Self-Quiz

1. Consider the energy budgets for a human, an elephant, a penguin, a mouse, and a python. The _____ would have the highest total annual energy expenditure, and the _____ would have the highest energy expenditure per unit mass.
   a. elephant; mouse
   b. elephant; human
   c. human; penguin
   d. mouse; python
   e. penguin; mouse

2. Which of the following structures or substances is *incorrectly* paired with a tissue?
   a. osteon—bone
   b. platelets—blood
   c. fibroblasts—skeletal muscle
   d. chondroitin sulfate—cartilage
   e. basement membrane—epithelium

3. For which of the following animals would the percent of its energy budget spent for homeostatic control be the largest?
   a. an amoeba in a freshwater pond
   b. a marine jellyfish
   c. a snake in a temperate forest
   d. a desert insect
   e. an arctic bird

4. The involuntary muscles that cause the wavelike contractions pushing food along our intestine are
   a. striated muscles.
   b. cardiac muscles.
   c. skeletal muscles.
   d. smooth muscles.
   e. intercalated muscles.

5. Which of the following is *not* considered to be a tissue?
   a. cartilage
   b. the mucous membrane lining the stomach
   c. blood
   d. the brain
   e. cardiac muscle

6. Which of the following statements about bioenergetics is true?
   a. Every animal has a specific metabolic rate that does not change.
   b. A BMR can be determined only at a specific temperature.
   c. Endotherms are warmed by metabolic heat.
   d. An SMR is best measured just after an ectotherm has eaten.
   e. Ectotherms and endotherms use the same basic energy "strategy."

7. Compared to a smaller cell, a larger cell of the same shape has
   a. less surface area.
   b. less surface area per unit of volume.
   c. the same surface-to-volume ratio.
   d. a smaller average distance between its mitochondria and the external source of oxygen.
   e. a smaller cytoplasm-to-nucleus ratio.

8. Which of the following vertebrate organ systems does *not* open directly to the external environment?
   a. digestive system
   b. circulatory system
   c. excretory system
   d. respiratory system
   e. reproductive system

9. Most of our cells are surrounded by
   a. blood.
   b. basement membranes.
   c. interstitial fluid.
   d. pure water.
   e. air.

10. Which of the following physiological responses is an example of *positive* feedback?
    a. An increase in the concentration of glucose in the blood stimulates the pancreas to secrete insulin, a hormone that lowers blood glucose concentration.
    b. A high concentration of carbon dioxide in the blood causes deeper, more rapid breathing, which expels carbon dioxide.
    c. Stimulation of a nerve cell causes sodium ions to leak into the cell, and the sodium influx triggers the inward leaking of even more sodium.

d. The body's production of red blood cells, which transport oxygen from the lungs to other organs, is stimulated by a low concentration of oxygen.
e. The pituitary gland secretes a hormone called TSH, which stimulates the thyroid gland to secrete another hormone called thyroxine; a high concentration of thyroxine suppresses the pituitary's secretion of TSH.

11. Why does blood qualify as a type of connective tissue?

12. Explain why a disease that damages connective tissue can impair most of the body's organs.

13. When the level of a sex hormone in your body reaches a set point, it turns off its own production by acting on a control center called the hypothalamus, which in turn regulates sex hormone production by your testes or ovaries. This control circuit is an example of _____.

14. How do anatomy and physiology complement one another?

15. Explain why a hummingbird at rest must use a greater proportion of its energy budget for temperature regulation on a cool day than a resting crow does.

Go to the website or CD-ROM for more quiz questions.

Evolution Connection

The biologist C. Bergmann noted that mammals and birds occurring at higher latitudes are on average larger and bulkier than related species found at lower latitudes. This observation, sometimes called Bergmann's Rule, has exceptions, but appears to hold true in most cases. Suggest an evolutionary hypothesis for this "rule" based on physical principles.

The Process of Science

Suggest your own hypothesis to explain the inverse relationship between body size and metabolic rate per gram of tissue. How could you test your hypothesis?

Explore how changes in the environmental temperature affect the metabolic rate of *Daphnia*, the water flea, in the Case Study in the Process of Science, available on the CD-ROM and website.

Science, Technology, and Society

Medical researchers are investigating the possibilities of artificial substitutes for various human tissues. Examples are a liquid that could serve as "artificial blood" and a fabric that could temporarily serve as artificial skin for victims of serious burns. In what other situations might artificial blood or skin be useful? What characteristics would these substitutes need in order to function effectively in the body? Why do real tissues work better? Why not use the real things if they work better? Can you think of other artificial tissues that might be useful? What problems do you anticipate in developing and applying them?

Answers: 1. a; 2. c; 3. e; 4. d; 5. d; 6. c; 7. b; 8. b; 9. c; 10. c; 11. Because it consists of a relatively sparse population of cells surrounded by a noncellular matrix, which is plasma in the case of blood. 12. Connective tissue is a component of most organs. 13. Feedback control—or better, negative feedback. 14. Since form fits function, the study of anatomy (structure of an organism) and physiology (functions of an organism) go hand-in-hand. 15. Birds are endotherms. A small bird such as a hummingbird has a higher surface-to-volume ratio than does a larger bird such as a crow. Thus, the small bird loses body heat faster and must use its metabolism to offset this heat loss.

ANIMAL NUTRITION

Every mealtime *is a reminder that we are heterotrophs dependent on a regular supply of food derived from other organisms. As a group, animals exhibit a great variety of nutritional adaptations. The snowshoe hare shown in the photograph on this page is adapted for life in northern forests. Able to obtain all their nutritional needs from plants alone, hares and rabbits have a large intestinal pouch housing prokaryotes and protists that digest cellulose. When deep snow covers the ground, a hare* can live on pine, fir, or spruce branches—often the only plant materials available.

For any animal, a nutritionally adequate diet is essential for homeostasis, a steady-state balance in body function. A balanced diet provides fuel for cellular work, as well as all the materials the body needs to construct its own organic molecules. In this chapter we will examine the nutritional requirements of animals and look at some of the diverse adaptations for obtaining and processing food.

NUTRITIONAL REQUIREMENTS

Animals are heterotrophs that require food for fuel, carbon skeletons, and essential nutrients: *an overview*

A nutritionally adequate diet satisfies three needs: fuel (chemical energy) for all the cellular work of the body; the organic raw materials animals use in biosynthesis (carbon skeletons to make many of their own molecules); and essential nutrients, substances that the animal cannot make for itself from *any* raw material and therefore must obtain in food in prefabricated form.

Homeostatic mechanisms manage an animal's fuel

The theme of bioenergetics is integral to our study of nutrition. As we saw in Chapter 40, the flow of food energy into and out of an animal can be viewed as a "budget," with the production of ATP accounting for the largest fraction by far of the energy budgets of most animals. The ATP powers basal or resting metabolism as well as activity and, in endotherms, temperature regulation. Nearly all of this ATP is derived from oxidation of organic fuel molecules—carbohydrates, proteins, and fats—in cellular respiration. The monomers of any of these substances can be used as fuel, though priority is usually given to carbohydrates and fats. Fats are especially rich in energy; the oxidation of fat liberates about twice the energy liberated from an equal amount of carbohydrate or protein.

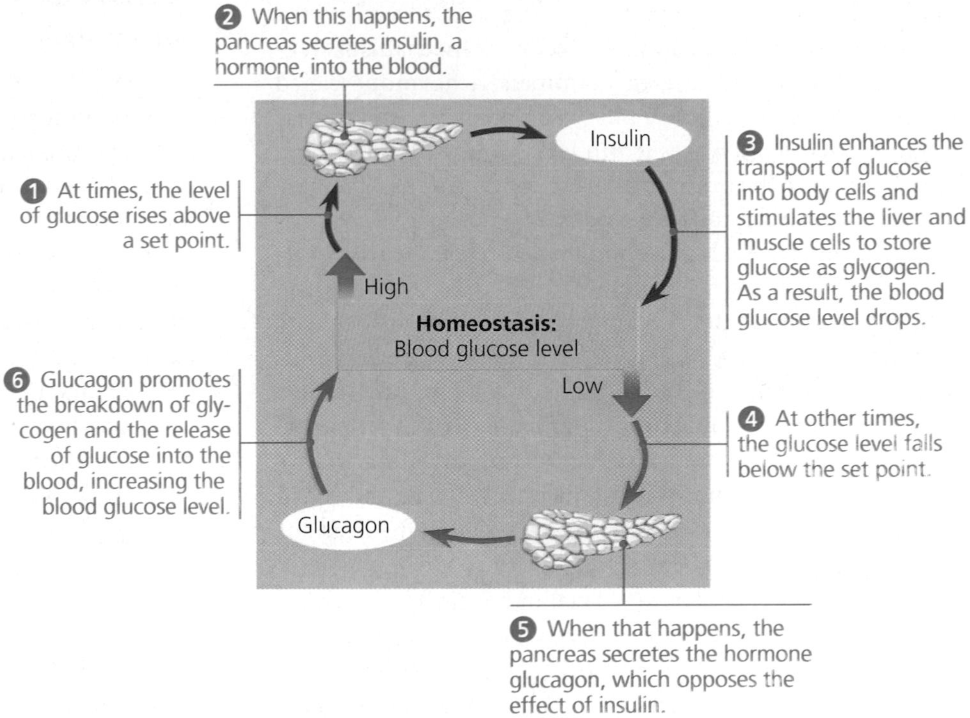

Glucose Regulation as an Example of Homeostasis in Nutrition

When an animal takes in more calories than it needs to produce ATP, the excess can be used for biosynthesis. If the animal isn't growing in size or reproducing, the body tends to store the surplus in energy depots. In humans, the liver and muscle cells store energy in the form of glycogen, a polymer made up of many glucose units (see FIGURE 5.6b). Glucose is a major fuel molecule for cells, and its metabolism, regulated by hormone action, is an important aspect of homeostasis (FIGURE 41.1). If glycogen stores are full and caloric intake still exceeds caloric expenditure, the excess is usually stored as fat.

When fewer calories are taken in than are expended—perhaps because of sustained heavy exercise or lack of food—fuel is taken out of storage depots and oxidized. This may cause an animal to lose weight. The human body generally expends liver glycogen first, and then draws on muscle glycogen and fat. Most healthy people—even if they are not obese—have enough stored fat to sustain them through several weeks of starvation (an average human's energy needs can be fueled by the oxidation of only 0.3 kg of fat per day).

Caloric Imbalance

Severe problems occur if the energy budget remains out of balance for long periods. If the diet of a person or other animal is chronically deficient in calories, **undernourishment** results. In this condition, the stores of glycogen and fat are used up, the body begins breaking down its own proteins for fuel, muscles begin to decrease in size, and the brain can become protein-deficient. If energy intake remains less than energy expenditures, death will eventually result. Even if a seriously undernourished person survives, some of the damage may be irreversible. Because a diet of a single staple such as rice or

① At times, the level of glucose rises above a set point.

② When this happens, the pancreas secretes insulin, a hormone, into the blood.

③ Insulin enhances the transport of glucose into body cells and stimulates the liver and muscle cells to store glucose as glycogen. As a result, the blood glucose level drops.

High

Homeostasis: Blood glucose level

Low

④ At other times, the glucose level falls below the set point.

⑤ When that happens, the pancreas secretes the hormone glucagon, which opposes the effect of insulin.

⑥ Glucagon promotes the breakdown of glycogen and the release of glucose into the blood, increasing the blood glucose level.

FIGURE 41.1 Homeostatic regulation of cellular fuel. The human body regulates the use and storage of glucose, a major cellular fuel. After a meal is digested, glucose and other monomers are absorbed into the blood from the digestive tract.

corn can often provide sufficient calories, undernourishment is generally common only where drought, war, or some other crisis has severely disrupted the food supply. Another cause of undernourishment is anorexia nervosa, an eating disorder associated with a compulsive aversion to body fat.

Detrimental effects also result from excessive food intake. In the United States and other affluent nations, **overnourishment,** or obesity, is a common problem. The human body hoards fat. It tends to store any excess fat molecules obtained from food instead of using them for fuel. In contrast, when we eat an excess of carbohydrates, the body tends to increase its rate of carbohydrate oxidation. Thus, the amount of fat in the diet can have a more direct effect on weight gain than the amount of dietary carbohydrates. Fat hoarding can be a liability today, but it probably provided a fitness advantage for our hunting/gathering ancestors. Individuals with genes promoting the storage of high-energy molecules during feasts may have been those that survived famines.

Obesity

Despite its propensity to store fat, the human body seems to impose limits on weight gain (and loss). Some people remain lean and hold a more-or-less constant weight no matter how much they eat. Even obese people usually attain a relatively stable weight, generally unaffected by how much they eat. Most dieters return to their former weight soon after they stop dieting. These observations, along with several recent discoveries, suggest that complex feedback mechanisms regulate fat storage and use. In mammals, a hormone called leptin, produced by adipose cells, is a key player. An increase in adipose tissue increases leptin levels in the blood. A high leptin level cues the brain to depress appetite and to increase energy-consuming muscular activity and body-heat production. Conversely, loss of body fat decreases leptin levels in the blood, signaling the brain to increase appetite and weight gain. Apparently these feedback mechanisms regulate body weight around a fairly rigid set point in some individuals and over a relatively wide range in others. Researchers have also identified some of the genes involved in fat homeostasis and several chemical signals that underlie the brain's regulatory roles (FIGURE 41.2). Some of the signals and signal antagonists are under development as potential medications for obesity.

Obesity may actually be beneficial in certain species. Small seabirds called petrels must fly long distances to find food. Most of the food petrel parents bring to their chicks is very rich in lipids. This minimizes the weight of the food the birds must carry during their long foraging trips (recall that fat has twice as many calories per gram as other fuels). However, in addition to energy, growing baby petrels need lots of protein for building new tissues. There is relatively little protein in their oily diet, so in order to get all the protein they need, young petrels have to consume many more calories than they burn in metabolism—and consequently they become very obese. In some petrel species, chicks at the end of the growth period weigh much more than their parents and are far too heavy to fly. These fat youngsters need to starve for several days to lose enough weight to be capable of flight. However, the fat depots in young petrels do serve an important function; the energy reserves help growing chicks survive periods when parents are unable to find enough food.

An animal's diet must supply essential nutrients and carbon skeletons for biosynthesis

In addition to providing fuel for ATP production, an animal's diet must also supply all the raw materials needed for biosynthesis. To build the complex molecules it needs to grow and maintain itself, an animal must obtain organic precursors (carbon skeletons) from its food. Given a source of organic carbon (such as sugar) and a source of organic nitrogen (usually in amino acids from the digestion of protein), animals can fabricate a great variety of organic molecules—carbohydrates, proteins, and lipids.

Besides fuel and carbon skeletons, an animal's diet must also supply **essential nutrients.** These are materials that must be obtained in preassembled form because the animal's cells cannot make them from *any* raw material (FIGURE 41.3). Some of these materials are essential for all animals, but others are needed only by certain species. For instance, ascorbic acid (vitamin C) is an essential nutrient for humans and other primates, guinea pigs, and some birds and snakes, but not for most other animals.

An animal whose diet is missing one or more essential nutrients is said to be **malnourished** (recall that *undernourished* refers to caloric deficiency). For example, cattle and other herbivorous animals may suffer mineral deficiencies if they graze on plants growing in soil lacking key minerals (see FIGURE 41.3).

FIGURE 41.3 Obtaining essential nutrients. A giraffe, an herbivore of East Africa, chews on old bones from a dead mammal. Bones contain calcium phosphate, and osteophagia ("bone-eating") is common among herbivores living where soils and plants are deficient in phosphorus. Animals require phosphorus as a mineral nutrient to make ATP, nucleic acids, phospholipids, and bones.

FIGURE 41.2 A ravenous rodent. This obese mouse on the left has a defect in a gene that normally produces an appetite-regulating protein. (The mouse on the right lacks the genetic defect.) Several other genes function in weight management in mammals, including humans.

Malnutrition is much more common than undernutrition in human populations, and it is even possible for an overnourished individual to be malnourished.

There are four classes of essential nutrients: essential amino acids, essential fatty acids, vitamins, and minerals.

Essential Amino Acids

Animals require 20 amino acids to make proteins, and most animal species can synthesize about half of these, as long as their diet includes organic nitrogen. The remaining ones, the **essential amino acids,** must be obtained from food in prefabricated form. Eight amino acids are essential in the adult human diet (a ninth, histidine, is essential for infants); the same amino acids are essential for most animals.

A diet that provides insufficient amounts of one or more essential amino acids causes a form of malnutrition known as protein deficiency. This is the most common type of malnutrition among humans. The victims are usually children, who, if they survive infancy, are likely to be retarded in physical and perhaps mental development.

The most reliable sources of essential amino acids are meat, eggs, cheese, and other animal products. The proteins in animal products are "complete," which means that they provide all the essential amino acids in their proper proportions. Most plant proteins are "incomplete," being deficient in one or more essential amino acids. Corn, for example, is deficient in the amino acid lysine. People forced by economic necessity or other circumstances to obtain nearly all their calories from corn would show symptoms of protein deficiency, as would those who eat only rice, wheat, or potatoes. This problem can be avoided by eating a combination of plant foods that complement one another to supply all essential amino acids. For example, beans supply the lysine that is missing in corn; and while beans are deficient in methionine, this essential amino acid is present in corn. Thus, a diet of beans and corn provides all the essential amino acids—as long as these vegetables are both consumed during the same day (FIGURE 41.4). Because the body cannot easily store amino acids, a deficiency of a single essential amino acid, even for a relatively short period, retards protein synthesis and limits the use of other amino acids. Most cultures have, by trial and error, developed balanced diets that prevent protein deficiency.

Some animals have special adaptations that get them through periods when their bodies demand extraordinary amounts of protein. For example, penguins can use their muscle protein as a source of amino acids to make new proteins when they replace their feathers after molting (FIGURE 41.5).

Essential Fatty Acids

Animals can synthesize most of the fatty acids they need. The **essential fatty acids,** the ones they cannot make, are certain

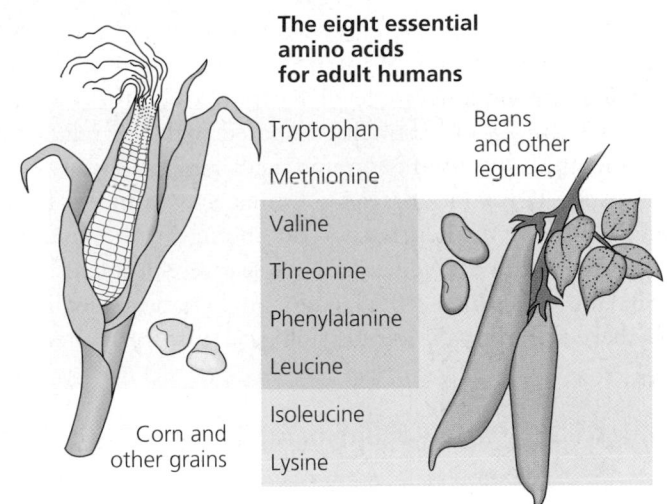

FIGURE 41.4 Essential amino acids from a vegetarian diet. An adult human can obtain all eight essential amino acids by eating a meal of corn and beans.

The eight essential amino acids for adult humans

Tryptophan
Methionine
Valine
Threonine
Phenylalanine
Leucine
Isoleucine
Lysine

Corn and other grains

Beans and other legumes

unsaturated fatty acids (fatty acids having double bonds; see FIGURE 5.11). In humans, for example, linoleic acid must be present in the diet. This essential fatty acid is required to make some of the phospholipids found in membranes. Most diets furnish ample quantities of essential fatty acids, and thus deficiencies are rare.

FIGURE 41.5 Storing protein for growth. Penguins, such as this Adélie from Antarctica, must make an abundance of new protein when they molt (grow new feathers). Penguins replace all their feathers at once during a two- to three-week period. Because of the temporary loss of their insulating coat of feathers, penguins cannot swim—or feed—when molting. What is the source of amino acids for production of feather protein? There is no special storage tissue for free amino acids, but it is possible to store amino acids in certain proteins. Before molting, a penguin greatly increases its muscle mass. The penguin then breaks down the extra muscle protein, which supplies the amino acids for growing new feathers.

Vitamins

Vitamins are organic molecules required in the diet in amounts that are quite small compared with the relatively large quantities of essential amino acids and fatty acids animals need. Tiny amounts of vitamins may suffice—from about 0.01 to 100 mg per day—depending on the vitamin. However, vitamin deficiencies can cause severe problems.

So far, 13 vitamins essential to humans have been identified. They have extremely diverse physiological functions. Vitamins are grouped into two categories: water-soluble vitamins and fat-soluble vitamins (TABLE 41.1). The water-soluble vitamins include the B complex, which consists of several compounds that generally function as coenzymes in key metabolic processes. Vitamin C, also water-soluble, is required for the production of connective tissue. Excesses of water-soluble vitamins are excreted in urine, and moderate overdoses of these vitamins are probably harmless.

The fat-soluble vitamins are A, D, E, and K. They have a wide variety of functions. Vitamin A is incorporated into vi-

Table 41.1 Vitamin Requirements of Humans

Vitamin	Major Dietary Sources	Some Major Functions in the Body	Possible Symptoms of Deficiency or Extreme Excess
Water-Soluble Vitamins			
Vitamin B_1 (thiamine)	Pork, legumes, peanuts, whole grains	Coenzyme used in removing CO_2 from organic compounds	Beriberi (nerve disorders, emaciation, anemia)
Vitamin B_2 (riboflavin)	Dairy products, meats, enriched grains, vegetables	Component of coenzymes FAD and FMN	Skin lesions such as cracks at corners of mouth
Niacin	Nuts, meats, grains	Component of coenzymes NAD^+ and $NADP^+$	Skin and gastrointestinal lesions, nervous disorders Flushing of face and hands, liver damage
Vitamin B_6 (pyridoxine)	Meats, vegetables, whole grains	Coenzyme used in amino acid metabolism	Irritability, convulsions, muscular twitching, anemia Unstable gait, numb feet, poor coordination
Pantothenic acid	Most foods: meats, dairy products, whole grains, etc.	Component of coenzyme A	Fatigue, numbness, tingling of hands and feet
Folic acid (folacin)	Green vegetables, oranges, nuts, legumes, whole grains (also made by colon bacteria)	Coenzyme in nucleic acid and amino acid metabolism	Anemia, gastrointestinal problems May mask deficiency of vitamin B_{12}
Vitamin B_{12}	Meats, eggs, dairy products	Coenzyme in nucleic acid metabolism; needed for maturation of red blood cells	Anemia, nervous system disorders
Biotin	Legumes, other vegetables, meats	Coenzyme in synthesis of fat, glycogen, and amino acids	Scaly skin inflammation, neuro-muscular disorders
Vitamin C (ascorbic acid)	Fruits and vegetables, especially citrus fruits, broccoli, cabbage, tomatoes, green peppers	Used in collagen synthesis (e.g., for bone, cartilage, gums); antioxidant; aids in detoxification; improves iron absorption	Scurvy (degeneration of skin, teeth, blood vessels), weakness, delayed wound healing, impaired immunity Gastrointestinal upset
Fat-Soluble Vitamins			
Vitamin A (retinol)	Provitamin A (beta-carotene) in deep green and orange vegetables and fruits; retinol in dairy products	Component of visual pigments; needed for maintenance of epithelial tissues; antioxidant; helps prevent damage to lipids of cell membranes	Vision problems; dry, scaling skin Headache, irritability, vomiting, hair loss, blurred vision, liver and bone damage
Vitamin D	Dairy products, egg yolk (also made in human skin in presence of sunlight)	Aids in absorption and use of calcium and phosphorus; promotes bone growth	Rickets (bone deformities) in children, bone softening in adults Brain, cardiovascular, and kidney damage
Vitamin E (tocopherol)	Vegetable oils, nuts, seeds	Antioxidant; helps prevent damage to lipids of cell membranes	None well documented in humans; possibly anemia
Vitamin K (phylloquinone)	Green vegetables, tea (also made by colon bacteria)	Important in blood clotting	Defective blood clotting Liver damage and anemia

sual pigments of the eye. Vitamin D aids in calcium absorption and bone formation. The function of vitamin E is not yet fully understood, but along with vitamin C it seems to protect the phospholipids in membranes from oxidation (you have probably encountered advertisements for dietary supplements containing vitamin E as an "antioxidant.") Vitamin K is required for blood clotting. Excesses of fat-soluble vitamins are not excreted but are deposited in body fat, so overconsumption may result in an accumulation of these compounds to toxic levels.

The subject of vitamin dosage has aroused heated scientific and popular debate. Some believe it is sufficient to meet recommended daily allowances (RDAs), the nutrient intakes proposed by nutritionists to maintain health. Others argue that RDAs are set too low for some vitamins, and a fraction of those people believe, probably mistakenly, that *massive* doses of vitamins confer health benefits. Research is far from complete, and debate continues, especially over optimal doses of vitamins C and E. At this time, all that can be said with any certainty is that people who eat a balanced diet are not likely to develop symptoms of vitamin deficiency.

Minerals

Minerals are simple inorganic nutrients, usually required in small amounts—from less than 1 mg to about 2,500 mg per day (TABLE 41.2). As with vitamins, mineral requirements vary with animal species. Humans and other vertebrates require

Table 41.2 Mineral Requirements of Humans

Mineral	Major Dietary Sources	Some Major Functions in the Body	Possible Symptoms of Deficiency*
Calcium (Ca)	Dairy products, dark green vegetables, legumes	Bone and tooth formation, blood clotting, nerve and muscle function	Retarded growth, possibly loss of bone mass
Phosphorus (P)	Dairy products, meats, grains	Bone and tooth formation, acid-base balance, nucleotide synthesis	Weakness, loss of minerals from bone, calcium loss
Sulfur (S)	Proteins from many sources	Component of certain amino acids	Symptoms of protein deficiency
Potassium (K)	Meats, dairy products, many fruits and vegetables, grains	Acid-base balance, water balance, nerve function	Muscular weakness, paralysis, nausea, heart failure
Chlorine (Cl)	Table salt	Acid-base balance, formation of gastric juice, nerve function, osmotic balance	Muscle cramps, reduced appetite
Sodium (Na)	Table salt	Acid-base balance, water balance, nerve function	Muscle cramps, reduced appetite
Magnesium (Mg)	Whole grains, green leafy vegetables	Cofactor; ATP bioenergetics	Nervous system disturbances
Iron (Fe)	Meats, eggs, legumes, whole grains, green leafy vegetables	Component of hemoglobin and of electron-carriers in energy metabolism; enzyme cofactor	Iron-deficiency anemia, weakness, impaired immunity
Fluorine (F)	Drinking water, tea, seafood	Maintenance of tooth (and probably bone) structure	Higher frequency of tooth decay
Zinc (Zn)	Meats, seafood, grains	Component of certain digestive enzymes and other proteins	Growth failure, scaly skin inflammation, reproductive failure, impaired immunity
Copper (Cu)	Seafood, nuts, legumes, organ meats	Enzyme cofactor in iron metabolism, melanin synthesis, electron transport	Anemia, bone and cardiovascular changes
Manganese (Mn)	Nuts, grains, vegetables, fruits, tea	Enzyme cofactor	Abnormal bone and cartilage
Iodine (I)	Seafood, dairy products, iodized salt	Component of thyroid hormones	Goiter (enlarged thyroid)
Cobalt (Co)	Meats and dairy products	Component of vitamin B_{12}	None, except as B_{12} deficiency
Selenium (Se)	Seafood, meats, whole grains	Enzyme cofactor; antioxidant functioning in close association with vitamin E	Muscle pain, possibly heart muscle deterioration
Chromium (Cr)	Brewer's yeast, liver, seafood, meats, some vegetables	Involved in glucose and energy metabolism	Impaired glucose metabolism
Molybdenum (Mo)	Legumes, grains, some vegetables	Enzyme cofactor	Disorder in excretion of nitrogen-containing compounds

*All of these minerals are also harmful when consumed in excess.

relatively large quantities of calcium and phosphorus for the construction and maintenance of bone. Calcium is also necessary for the normal functioning of nerves and muscles, and phosphorus is also an ingredient of ATP and nucleic acids. Iron is a component of the cytochromes that function in cellular respiration (see FIGURE 9.13) and of hemoglobin, the oxygen-binding protein of red blood cells. Magnesium, iron, zinc, copper, manganese, selenium, and molybdenum are cofactors built into the structure of certain enzymes; magnesium, for example, is present in enzymes that split ATP. Vertebrates need iodine to make thyroid hormones, which regulate metabolic rate. Sodium, potassium, and chlorine are important in nerve function and also have a major influence on the osmotic balance between cells and the interstitial fluid.

Most people ingest far more salt (sodium chloride) than they need. The average U.S. citizen eats enough salt to provide about 20 times the required amount of sodium. Ingesting an excess of salt or several other minerals can upset homeostatic balance and cause toxic side effects. For example, too much sodium is associated with high blood pressure, and excess iron can cause liver damage.

We now shift our attention to the kinds of foods that animals eat and how they process these foods to meet their nutritional needs.

FOOD TYPES AND FEEDING MECHANISMS

Most animals are opportunistic feeders

All animals eat other organisms—dead or alive, whole or by the piece. (We'll stretch "by the piece" to include parasites such as tapeworms that absorb organic molecules from host organisms across their body surface.) In general, animals fit into one of three dietary categories. **Herbivores,** such as gorillas, cows, hares, and many snails, eat mainly autotrophs (plants, algae). **Carnivores,** such as sharks, hawks, spiders, and snakes, eat other animals. **Omnivores** regularly consume animals as well as plant or algal matter. Omnivorous animals include cockroaches, crows, bears, raccoons, and humans, who evolved as hunters, scavengers, and gatherers.

The terms *herbivore, carnivore,* and *omnivore* represent the kinds of food an animal usually eats and the adaptations enabling it to obtain and process that food. However, most animals are opportunistic, eating foods that are outside their main dietary category when these foods are available. For example, cattle and deer, which are herbivores, may occasionally eat small animals or bird eggs, along with grass and other plants. Most carnivores obtain some nutrients from plant materials that remain in the digestive tract of the prey they eat. And all animals consume bacteria with other types of food.

FIGURE 41.6 Suspension-feeding: a baleen whale. The humpback whale and other baleen whales use comblike plates suspended from the upper jaw to sift small invertebrates from enormous volumes of water. The whale opens its mouth and fills an expandable oral pouch with water, then closes its mouth and contracts the pouch. This forces water out of the mouth through the baleen, leaving a mouthful of trapped food.

Diverse feeding adaptations have evolved among animals

The mechanisms by which animals ingest food are highly varied but fall into four main groups. Many aquatic animals are **suspension-feeders** that sift small food particles from the water. Clams and oysters, for example, use their gills to trap tiny morsels, which are then swept along with a film of mucus to the mouth by beating cilia. Baleen whales, the largest animals ever to live, are also suspension-feeders. They swim with their mouths agape, straining millions of small animals from huge volumes of water forced through screenlike plates attached to their jaws (FIGURE 41.6).

Substrate-feeders live in or on their food source, eating their way through the food. Examples are maggots that burrow into animal carcasses, and leaf miners, which are larvae of various insects that tunnel through the interior of leaves (FIGURE 41.7).

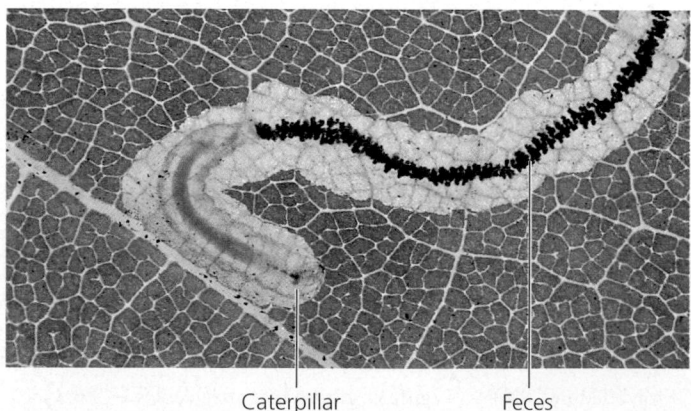

Caterpillar   Feces

FIGURE 41.7 Substrate-feeding: a leaf miner. Eating its way through the soft mesophyll of an oak leaf, this caterpillar, the larva of a moth, leaves a trail of dark feces in its wake.

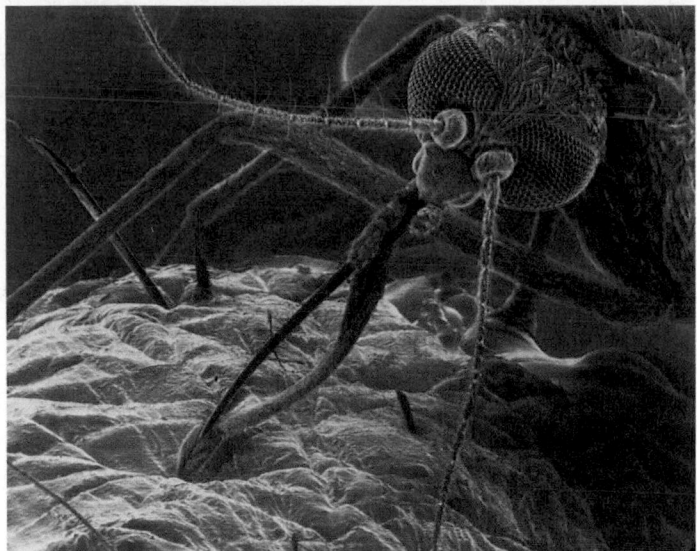

FIGURE 41.8 Fluid-feeding: a mosquito. This parasite has pierced the skin of its human host with hollow needlelike mouthparts and is filling its digestive tract with a blood meal (colorized SEM).

Earthworms are also substrate-feeders or, more specifically, **deposit-feeders.** Eating their way through the dirt, earthworms salvage partially decayed organic material consumed along with soil.

Fluid-feeders make their living by sucking nutrient-rich fluids from a living host (FIGURE 41.8). Mosquitoes and leeches suck blood from animals. Aphids tap the phloem sap of plants (see FIGURE 36.18). Because these fluid-feeding species harm their hosts, they are considered parasites. Hummingbirds and bees, however, are fluid-feeders that benefit their host plants, transferring pollen as they move from flower to flower to obtain nectar.

Most animals are **bulk-feeders** that eat relatively large pieces of food (FIGURE 41.9). Their adaptations include such diverse utensils as tentacles, pincers, claws, poisonous fangs, and jaws and teeth that kill their prey or tear off pieces of meat or vegetation.

The four main stages of food processing are ingestion, digestion, absorption, and elimination

Ingestion, the act of eating, is only the first stage of food processing. Nearly all animals, even many fluid-feeders, must deal with food that is "packaged" in bulk form and contains very complex arrays of molecules, including large polymers and various substances that may be difficult to process or may even be toxic. Organic material in food consists largely of proteins, fats, and carbohydrates in the form of starch and other polysaccharides. Animals cannot use these macromolecules directly, for two reasons. First, polymers are too large to pass through membranes and enter the cells of the animal. Second, the macromolecules that make up an animal are not identical to those of its food. In building their macromolecules, however, all organisms use common monomers—for example, soybeans, fruit flies, and humans all assemble their proteins from the same 20 amino acids.

Digestion, the second stage of food processing, is the process of breaking food down into molecules small enough for the body to absorb. Digestion cleaves macromolecules into their component monomers, which the animal then uses to make its own molecules or as fuel for ATP production. Polysaccharides and disaccharides are split into simple sugars, fats are digested to glycerol and fatty acids, proteins are broken down into amino acids, and nucleic acids are cleaved into nucleotides.

Recall from Chapter 5 that when a cell makes a macromolecule by linking together monomers, it does so by removing a molecule of water for each new covalent bond formed. Digestion reverses this process by breaking bonds with the enzymatic addition of water (see FIGURE 5.2). This splitting process is called **enzymatic hydrolysis.** A variety of hydrolytic enzymes

FIGURE 41.9 Bulk-feeding: a python. Most animals ingest relatively large pieces of food. This is especially true of snakes, which cannot chew their food into pieces and must swallow it whole—even if the prey is much bigger than the diameter of the snake. In this amazing scene, a rock python is beginning to ingest a gazelle it has captured and killed. After swallowing its prey, which may take more than an hour, the snake will spend two weeks or more in a quiet place digesting its meal.

catalyze the digestion of each of the classes of macromolecules found in food. This chemical digestion is usually preceded by mechanical fragmentation of the food—by chewing, for instance. Breaking food into smaller pieces increases the surface area exposed to digestive juices containing hydrolytic enzymes.

The last two stages of food processing occur after the food is digested. In the third stage, **absorption,** the animal's cells take up (absorb) small molecules such as amino acids and simple sugars from the digestive compartment. Finally, **elimination** occurs, as undigested material passes out of the digestive compartment.

Digestion occurs in specialized compartments

How do animals apply their digestive processes to food without digesting their own cells and tissues? After all, digestive enzymes hydrolyze the same biological materials (such as proteins, lipids, and carbohydrates) that animals are made of, and it is obviously important to avoid digesting oneself! Most animals reduce the risk of self-digestion by processing food in specialized compartments.

Intracellular Digestion

Food vacuoles—organelles in which hydrolytic enzymes break down food without digesting the cell's own cytoplasm—are the simplest digestive compartments. Heterotrophic protists digest their meals in food vacuoles, usually after engulfing food by phagocytosis or pinocytosis (see FIGURE 8.19a). Newly formed food vacuoles fuse with lysosomes, which are organelles containing hydrolytic enzymes. This mixes the food with the enzymes, allowing digestion to occur safely within a compartment that is enclosed by a protective membrane. This is termed **intracellular digestion** (FIGURE 41.10). Sponges are unusual among animals in that they digest their food entirely by the intracellular mechanism, as do heterotrophic protists (which are not animals) (see FIGURE 33.3).

Extracellular Digestion

In most animals, at least some hydrolysis occurs by **extracellular digestion,** the breakdown of food outside cells. Extracellular digestion occurs within compartments that are continuous with the outside of the animal's body. Having an extracellular cavity for digestion enables an animal to devour much larger prey than can be ingested by phagocytosis and digested intracellularly.

Many animals with simple body plans have digestive sacs with single openings. These pouches, called **gastrovascular cavities,** function in both digestion and distribution of nutrients throughout the body (hence, the *vascular* part of the term). The cnidarians called hydras provide a good example of how a gastrovascular cavity works. Hydras are carnivores that sting prey with specialized organelles called nematocysts and then use tentacles to stuff the food through the mouth into the gastrovascular cavity (FIGURE 41.11). With food in the cavity, specialized cells of the gastrodermis, the tissue layer that lines the cavity, secrete digestive enzymes that break the soft tissues of the prey into tiny pieces. Gastrodermal cells then engulf the food particles, and most of the actual hydrolysis of macromolecules occurs intracellularly, as in *Paramecium* and sponges. After a hydra has digested its meal, undigested materials remaining in the gastrovascular cavity, such as the exoskeletons of small crustaceans, are eliminated through the single opening, which functions in the dual role of mouth and anus. Many flatworms also have a gastrovascular cavity with a single opening (see FIGURE 33.10).

In contrast to cnidarians and flatworms, most animals—including nematodes, annelids, mollusks, arthropods, echinoderms, and chordates—have digestive tubes extending between two openings, a mouth and an anus. These tubes are called **complete digestive tracts** or **alimentary canals.** Because

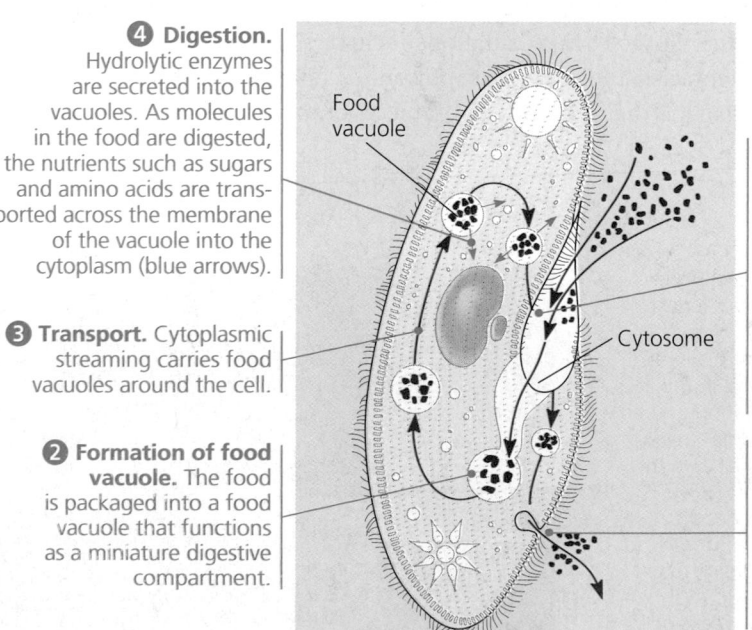

❹ **Digestion.** Hydrolytic enzymes are secreted into the vacuoles. As molecules in the food are digested, the nutrients such as sugars and amino acids are transported across the membrane of the vacuole into the cytoplasm (blue arrows).

❸ **Transport.** Cytoplasmic streaming carries food vacuoles around the cell.

❷ **Formation of food vacuole.** The food is packaged into a food vacuole that functions as a miniature digestive compartment.

Food vacuole

Cytosome

❶ **Food intake.** *Paramecium* has a specialized feeding structure called the oral groove. Cilia that line the groove draw water and suspended food particles, mostly bacteria, toward the "mouth" (cytosome).

❺ **Elimination.** Later, the vacuole fuses with an anal pore, a specialized region of the plasma membrane where undigested material can be eliminated.

FIGURE 41.10 Intracellular digestion in *Paramecium*.

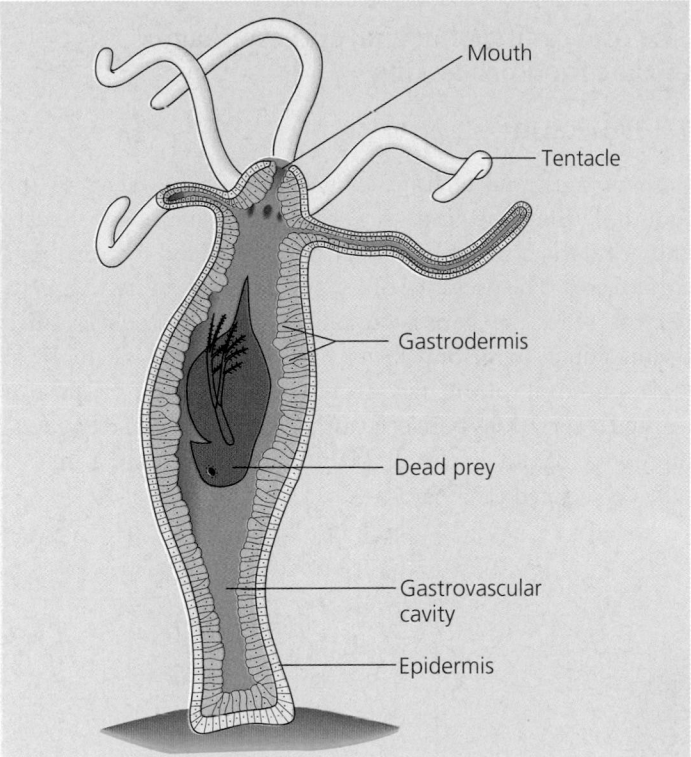

FIGURE 41.11 Extracellular digestion in a gastrovascular cavity. The outer epidermis of the cnidarian hydra has protective and sensory functions, whereas the inner gastrodermis is specialized for digestion. Digestion begins in the gastrovascular cavity and is completed intracellularly after small food particles are engulfed by the gastrodermal cells.

food moves along the canal in one direction, the tube can be organized into specialized regions that carry out digestion and nutrient absorption in a stepwise fashion (FIGURE 41.12). Food ingested through the mouth and pharynx passes through an esophagus that leads to a crop, gizzard, or stomach, depending on the species. Crops and stomachs usually serve as food storage organs (although some digestion may occur there), whereas gizzards grind and fragment food. The food next enters the intestine, where digestive enzymes hydrolyze the food molecules, and nutrients are absorbed across the lining of the tube into the blood. Undigested wastes are eliminated through the anus. Another advantage of a complete digestive tract is the ability to ingest additional food before earlier meals are completely digested—which may be difficult or inefficient for animals with gastrovascular cavities.

THE MAMMALIAN DIGESTIVE SYSTEM

The general principles of food processing are similar for a diversity of animals, so we can use the digestive system of mammals as a representative example. The mammalian digestive

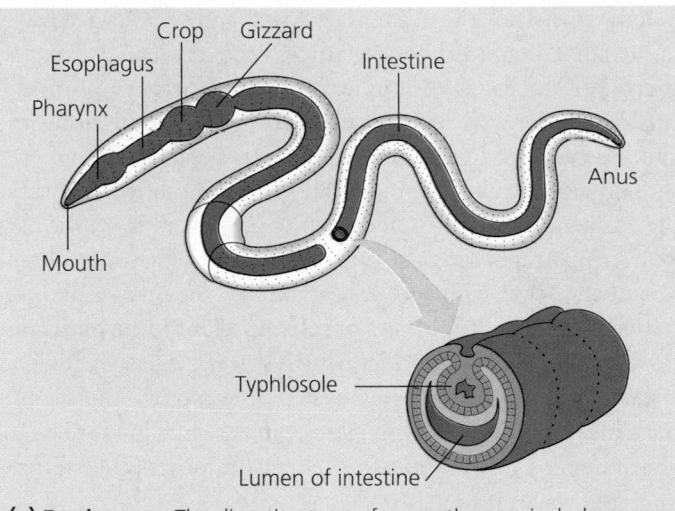

(a) Earthworm. The digestive tract of an earthworm includes a muscular pharynx that sucks food in through the mouth. Food passes through the esophagus and is stored and moistened in the crop. The muscular gizzard, which contains small bits of sand and gravel, pulverizes the food. Digestion and absorption occur in the intestine, which has a dorsal fold, the typhlosole, that increases the surface area for nutrient absorption.

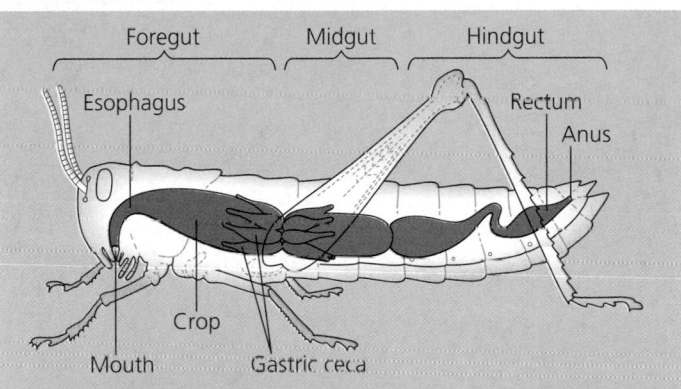

(b) Grasshopper. A grasshopper has several digestive chambers grouped into three main regions: a foregut, with an esophagus and crop; a midgut; and a hindgut. Food is moistened and stored in the crop, but most digestion occurs in the midgut. Gastric ceca, pouches extending from the midgut, absorb nutrients.

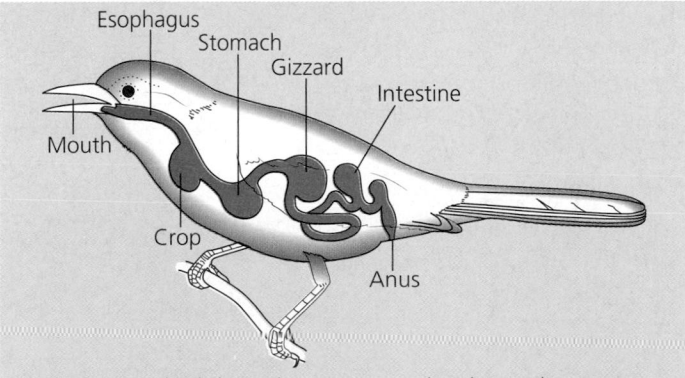

(c) Bird. Many birds have three separate chambers—the crop, stomach, and gizzard—where food is pulverized and churned before passing into the intestine. Some bird species lack the crop, the gizzard, or both. In most birds, chemical digestion and absorption of nutrients occur in the intestine.

FIGURE 41.12 Alimentary canals.

system consists of the alimentary canal and various accessory glands that secrete digestive juices into the canal through ducts. **Peristalsis,** rhythmic waves of contraction by smooth muscles in the wall of the canal, pushes the food along the tract. At some of the junctions between specialized segments of the digestive tube, the muscular layer is modified into ring-like valves called **sphincters,** which close off the tube like drawstrings, regulating the passage of material between chambers of the canal. The accessory glands of the mammalian digestive system are three pairs of **salivary glands,** the **pancreas,** the **liver,** and the **gallbladder,** which stores a digestive juice.

Using the human digestive system as a model, we now follow a meal through the alimentary canal, seeing in more detail what happens to the food in each of the processing stations along the way (FIGURE 41.13).

The oral cavity, pharynx, and esophagus initiate food processing

The Oral Cavity

Both physical and chemical digestion of food begin in the mouth. During chewing, teeth of various shapes cut, smash, and grind food, making it easier to swallow and increasing its surface area. The presence of food in the **oral cavity** triggers a nervous reflex that causes the salivary glands to deliver saliva through ducts to the oral cavity. Even before food is actually in the mouth, salivation may occur in anticipation because of learned associations between eating and the time of day, cooking odors, or other stimuli. In humans, more than a liter of saliva is secreted each day.

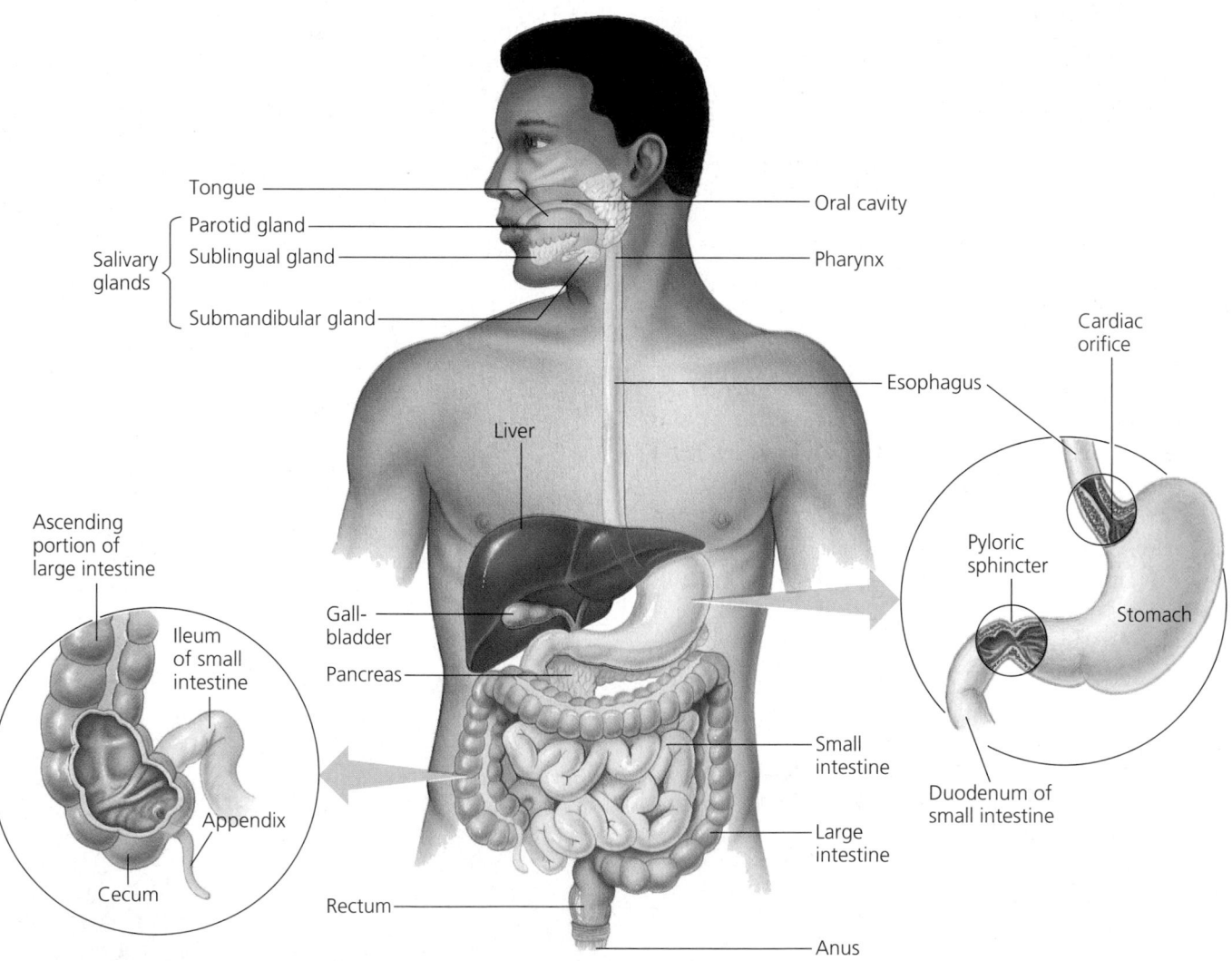

FIGURE 41.13 The human digestive system. After chewing and swallowing, it takes only 5 to 10 seconds for food to pass down the esophagus and into the stomach, where it spends 2 to 6 hours being partially digested. Final digestion and nutrient absorption occur in the small intestine over a period of 5 to 6 hours. In 12 to 24 hours, any undigested material passes through the large intestine, and feces are expelled through the anus.

Saliva contains a slippery glycoprotein (carbohydrate-protein complex) called mucin, which protects the soft lining of the mouth from abrasion and lubricates the food for easier swallowing. Saliva also contains buffers that help prevent tooth decay by neutralizing acid in the mouth. Antibacterial agents in saliva kill many of the bacteria that enter the mouth with food.

Chemical digestion of carbohydrates, a main source of chemical energy, begins in the oral cavity. Saliva contains **salivary amylase,** an enzyme that hydrolyzes starch (a glucose polymer from plants) and glycogen (a glucose polymer from animals). The main products of this enzyme's action are smaller polysaccharides and the disaccharide maltose.

The tongue tastes food, manipulates it during chewing, and helps shape the food into a ball called a **bolus.** During swallowing, the tongue pushes a bolus to the back of the oral cavity and into the pharynx.

The Pharynx

The region we call our throat is the **pharynx,** a junction that opens to both the esophagus and the windpipe (trachea). When we swallow, the top of the windpipe moves up so that its opening, the glottis, is blocked by a cartilaginous flap, the **epiglottis.** You can see this motion in the bobbing of the "Adam's apple"

during swallowing. This carefully controlled mechanism normally ensures that a bolus will be guided into the entrance of the esophagus (FIGURE 41.14, steps one and two). On rare occasions when food or liquids go "down the wrong pipe" because the swallowing reflex didn't close the opening of the windpipe in time, the blockage of airflow to the lungs can be fatal.

The Esophagus

The **esophagus** conducts food from the pharynx down to the stomach by peristalsis (FIGURE 41.14, step six). The muscles at the very top of the esophagus are striated (voluntary). Thus, the act of swallowing begins voluntarily, but then the involuntary waves of contraction by smooth muscles in the rest of the esophagus take over.

The stomach stores food and performs preliminary digestion

The **stomach** is located in the upper abdominal cavity, just below the diaphragm. With accordionlike folds and a very elastic wall, the stomach can stretch to accommodate about 2 L of food and fluid. Because this large organ can store an entire meal, we do not need to eat constantly. Besides storing food,

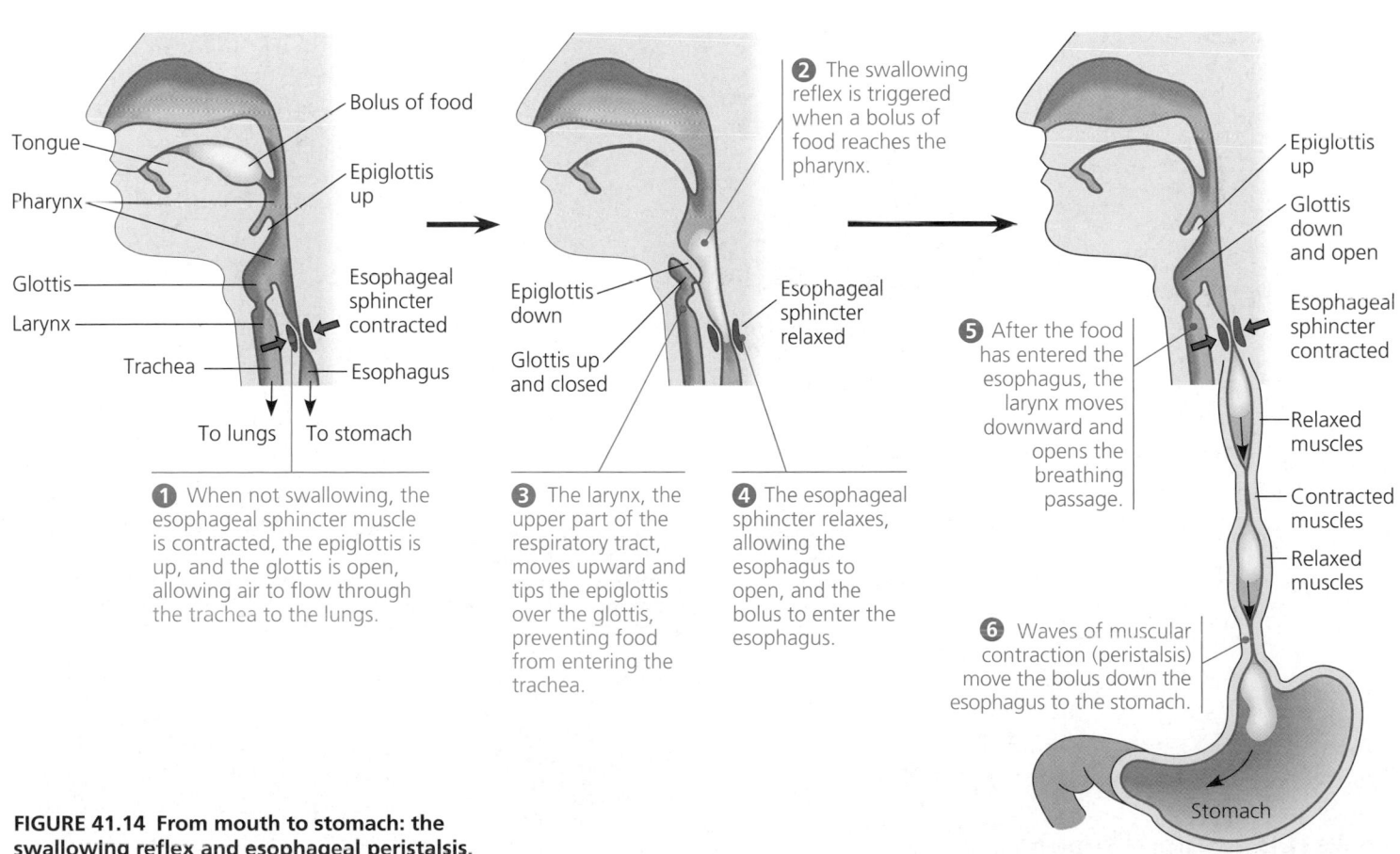

FIGURE 41.14 From mouth to stomach: the swallowing reflex and esophageal peristalsis.

the stomach performs important digestive functions: it secretes a digestive fluid called **gastric juice** and mixes this secretion with the food by the churning action of the smooth muscles in the stomach wall.

Gastric juice is secreted by the epithelium lining numerous deep pits in the stomach wall. With a high concentration of hydrochloric acid, gastric juice has a pH of about 2—acidic enough to dissolve iron nails. One function of the acid is to disrupt the extracellular matrix that binds cells together in meat and plant material. The acid also kills most bacteria that are swallowed with food. Also present in gastric juice is **pepsin,** an enzyme that begins the hydrolysis of proteins. Pepsin breaks peptide bonds adjacent to specific amino acids, cleaving proteins into smaller polypeptides. Pepsin is one of the few enzymes that works best in a strongly acidic environment. The low pH of gastric juice denatures (unfolds) the proteins in food, increasing exposure of their peptide bonds to pepsin.

What prevents pepsin from destroying the cells of the stomach wall? First, pepsin is secreted in an *inactive* form

called **pepsinogen** by specialized cells called chief cells located in gastric pits (FIGURE 41.15). Other cells, called parietal cells, also in the pits, secrete hydrochloric acid. The acid converts pepsinogen to active pepsin by removing a small portion of the molecule and exposing its active site. Because different cells secrete the acid and pepsinogen, the two ingredients do not mix—and pepsinogen is not activated—until they enter the lumen of the stomach. Activation of pepsinogen is an example of positive feedback. Once some pepsinogen is activated by acid, activation occurs at an increasingly rapid rate because pepsin itself can activate additional molecules of pepsinogen. Many other digestive enzymes are also secreted in inactive forms that become active within the lumen of the digestive tract.

The stomach's second defense against self-digestion is a coating of mucus, secreted by the epithelial cells, that helps protect the stomach lining. Still, the epithelium is constantly eroded, and mitosis generates enough cells to completely replace the stomach lining every three days. Gastric ulcers, lesions in the stomach lining, are caused mainly by the acid-tolerant

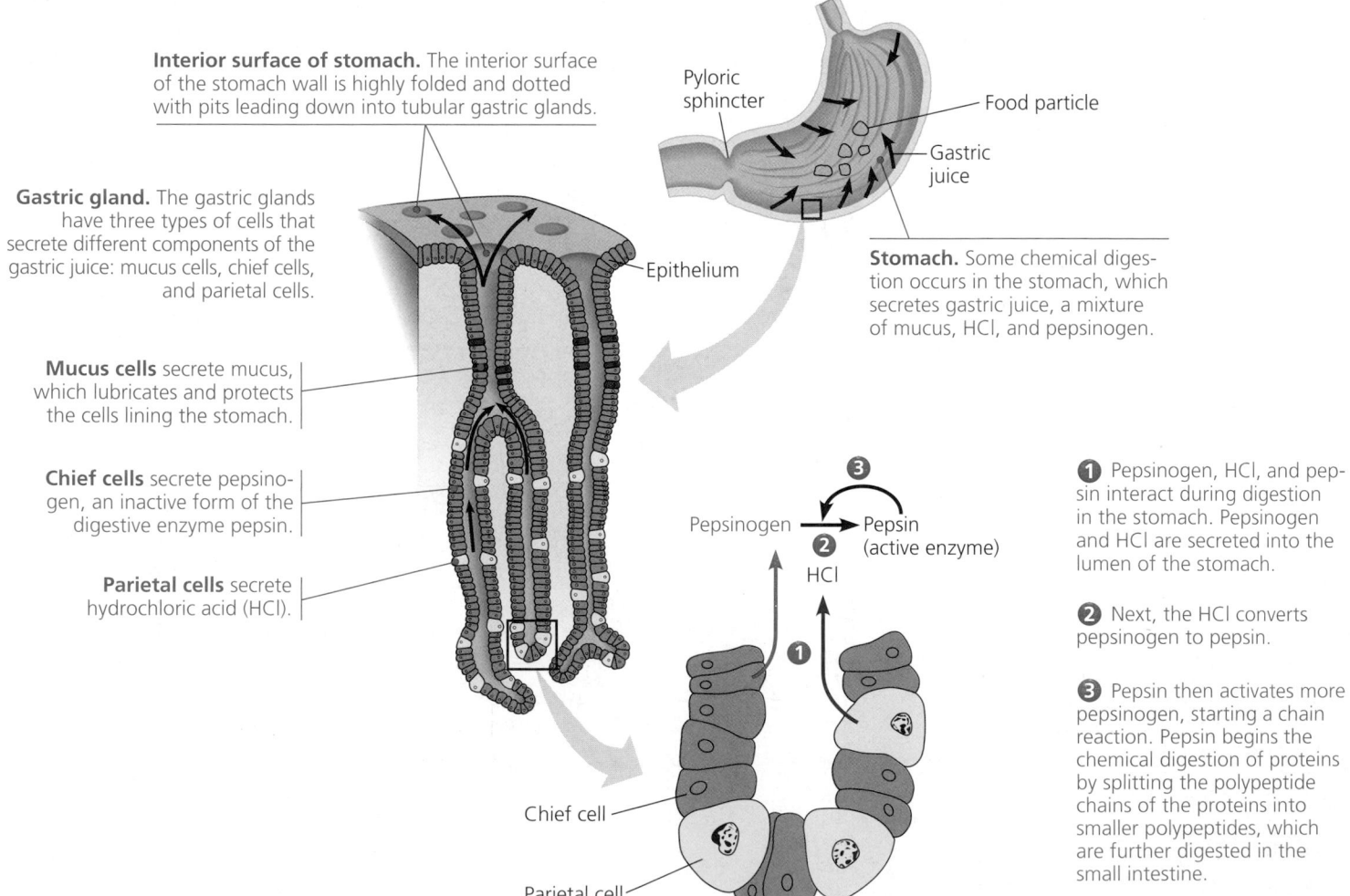

Interior surface of stomach. The interior surface of the stomach wall is highly folded and dotted with pits leading down into tubular gastric glands.

Gastric gland. The gastric glands have three types of cells that secrete different components of the gastric juice: mucus cells, chief cells, and parietal cells.

Mucus cells secrete mucus, which lubricates and protects the cells lining the stomach.

Chief cells secrete pepsinogen, an inactive form of the digestive enzyme pepsin.

Parietal cells secrete hydrochloric acid (HCl).

Pyloric sphincter

Food particle

Gastric juice

Epithelium

Stomach. Some chemical digestion occurs in the stomach, which secretes gastric juice, a mixture of mucus, HCl, and pepsinogen.

Pepsinogen → Pepsin (active enzyme)

HCl

❸

❷

❶

Chief cell

Parietal cell

❶ Pepsinogen, HCl, and pepsin interact during digestion in the stomach. Pepsinogen and HCl are secreted into the lumen of the stomach.

❷ Next, the HCl converts pepsinogen to pepsin.

❸ Pepsin then activates more pepsinogen, starting a chain reaction. Pepsin begins the chemical digestion of proteins by splitting the polypeptide chains of the proteins into smaller polypeptides, which are further digested in the small intestine.

FIGURE 41.15 Secretion of gastric juice.

bacterium *Helicobacter pylori* and are treated with antibiotics. However, gastric ulcers may worsen if pepsin and acid destroy the lining faster than it can regenerate.

About every 20 seconds, the stomach contents are mixed by the churning action of smooth muscles. You may feel hunger pangs when your empty stomach churns. (Sensations of hunger are also associated with brain centers that monitor the blood's nutritional status.) As a result of mixing and enzyme action, what begins in the stomach as a recently swallowed meal becomes a nutrient-rich broth known as **acid chyme.**

Most of the time, the stomach is closed off at either end (see FIGURE 41.13). The opening from the esophagus to the stomach, the cardiac orifice, normally dilates only when a bolus driven by peristalsis arrives. The occasional backflow of acid chyme from the stomach into the lower end of the esophagus causes heartburn. (If backflow is a persistent problem, an ulcer may develop in the esophagus.) At the opening from the stomach to the small intestine is the **pyloric sphincter,** which helps regulate the passage of chyme into the intestine. A squirt at a time, it takes about 2 to 6 hours after a meal for the stomach to empty.

The small intestine is the major organ of digestion and absorption

With a length of more than 6 m in humans, the **small intestine** is the longest section of the alimentary canal (its name is based on its small diameter, compared with the large intestine). Most of the enzymatic hydrolysis of food macromolecules and most of the absorption of nutrients into the blood occurs in the small intestine.

The first 25 cm or so of the small intestine is called the **duodenum.** It is here that acid chyme from the stomach mixes with digestive juices from the pancreas, liver, gallbladder, and gland cells of the intestinal wall itself (FIGURE 41.16).

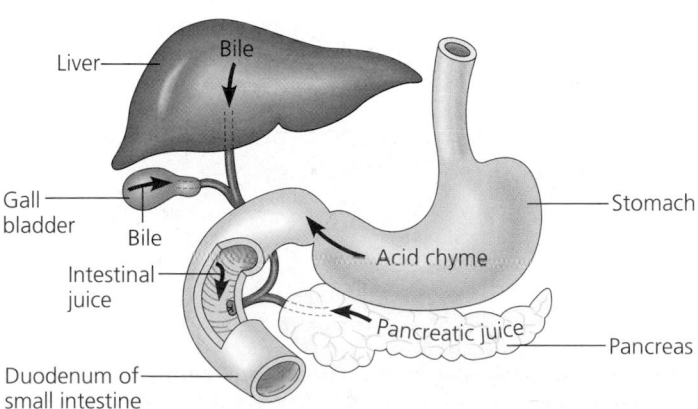

FIGURE 41.16 The duodenum. As peristalsis squeezes the mixture of digestive juices and partially-digested food along the small intestine, hydrolytic enzymes break the food molecules down to monomers.

The pancreas produces several hydrolytic enzymes and an alkaline solution rich in bicarbonate. The bicarbonate acts as a buffer, offsetting the acidity of chyme from the stomach.

The liver performs a wide variety of important functions in the body, including the production of **bile,** a mixture of substances that is stored in the gallbladder until needed. Bile contains no digestive enzymes, but it does contain bile salts, which act as detergents that aid in the digestion and absorption of fats. Bile also contains pigments that are by-products of red blood cell destruction in the liver; these bile pigments are eliminated from the body with the feces.

Enzymatic Action in the Small Intestine

Let's now follow the action of enzymes from the pancreas and intestinal (duodenal) wall in digesting macromolecules.

Carbohydrate Digestion. The digestion of starch and glycogen begun by salivary amylase in the oral cavity continues in the small intestine (FIGURE 41.17a, p. 864). Pancreatic amylases hydrolyze starch, glycogen, and smaller polysaccharides into disaccharides. The enzyme maltase completes the digestion of maltose, splitting it into two molecules of the simple sugar glucose. Maltase is one of a family of disaccharidases, each one specific for hydrolysis of a different disaccharide. Sucrase, for instance, hydrolyzes table sugar (sucrose), and lactase digests lactose, a sugar found in milk (adults usually have much less lactase and therefore less ability to digest milk sugar than children have). The disaccharidases are built into the membranes and extracellular matrix covering the intestinal epithelium, which is also the site of sugar absorption. Thus, the terminal steps in carbohydrate digestion—the steps that yield energy-rich monomers—occur where these monomers are absorbed into the blood.

Protein Digestion. Digestion of proteins in the small intestine completes the process begun by pepsin in the stomach (FIGURE 41.17b). Several enzymes in the duodenum dismantle polypeptides into their component amino acids or into small peptides that in turn are attacked by other enzymes. **Trypsin** and **chymotrypsin** are specific for peptide bonds adjacent to certain amino acids, and thus, like pepsin, break large polypeptides into shorter chains. Enzymes called **dipeptidases,** attached to the intestinal lining, split small peptides. **Carboxypeptidase** splits off one amino acid at a time, beginning at the end of the polypeptide that has a free carboxyl group. **Aminopeptidase** works in the opposite direction. Either aminopeptidase or carboxypeptidase alone could eventually completely digest a protein, but the teamwork among the various enzymes speeds up hydrolysis tremendously.

Many of the protein-digesting enzymes, such as aminopeptidase, are secreted by the intestinal epithelium, but trypsin, chymotrypsin, and carboxypeptidase are secreted in inactive

	(a) Carbohydrate digestion	(b) Protein digestion	(c) Nucleic acid digestion	(d) Fat digestion
Oral cavity, pharynx, esophagus	Polysaccharides (starch, glycogen) ↓ **Salivary amylase** Smaller polysaccharides, maltose			
Stomach		Proteins ↓ **Pepsin** Small polypeptides		
Lumen of small intestine	Polysaccharides ↓ **Pancreatic amylases** Maltose and other disaccharides	Polypeptides ↓ **Trypsin, Chymotrypsin** Smaller polypeptides ↓ **Aminopeptidase, Carboxypeptidase** Amino acids	DNA, RNA ↓ **Nucleases** Nucleotides	Fat globules ↓ **Bile salts** Fat droplets (emulsified) ↓ **Lipase** Glycerol, fatty acids, glycerides
Epithelium of small intestine (brush border)	↓ **Disaccharidases** Monosaccharides	Small peptides ↓ **Dipeptidases** Amino acids	**Nucleotidases** Nucleosides ↓ **Nucleosidases** Nitrogenous bases, sugars, phosphates	

FIGURE 41.17 Enzymatic digestion in the human digestive system.

forms by the pancreas. Another intestinal enzyme, called **enteropeptidase,** directly or indirectly triggers activation of these enzymes within the intestinal lumen (FIGURE 41.18).

Nucleic Acid Digestion.   The digestion of nucleic acids involves a hydrolytic assault similar to that mounted on proteins. A team of enzymes called **nucleases** hydrolyze DNA and RNA into their component nucleotides (FIGURE 41.17c). Other hydrolytic enzymes then break nucleotides down further into nucleosides, nitrogenous bases, sugars, and phosphates.

Fat Digestion.   Nearly all the fat in a meal reaches the small intestine completely undigested. Hydrolysis of fats is a special problem, because fat molecules are insoluble in water. However, bile salts from the gallbladder secreted into the duodenum coat tiny fat droplets and keep them from coalescing, a process called **emulsification.** Because the droplets are small, a large surface area of fat is exposed to **lipase,** an enzyme that hydrolyzes the fat molecules (FIGURE 41.17d).

Thus, the macromolecules from food are completely hydrolyzed to their component monomers as peristalsis moves the mixture of chyme and digestive juices along the small intestine. Most digestion is completed early in this journey, while the chyme is still in the duodenum. The remaining regions of

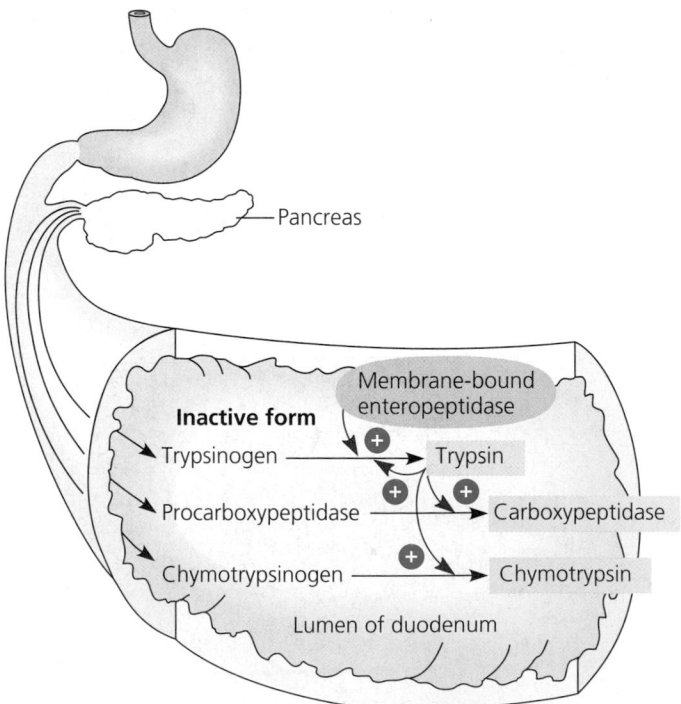

FIGURE 41.18 Activation of protein-digesting enzymes in the small intestine. The pancreas secretes protein-digesting enzymes in an inactive form into the lumen of the duodenum. An enzyme called enteropeptidase, which is bound to the intestinal epithelium, converts trypsinogen to trypsin. Trypsin then activates procarboxypeptidase and chymotrypsinogen. (⊕ indicates activation.)

the small intestine, the **jejunum** and **ileum,** function mainly in the absorption of nutrients and water.

Absorption of Nutrients

To enter the body, nutrients in the lumen must cross the lining of the digestive tract. A few nutrients are absorbed in the stomach and large intestine, but most absorption occurs in the small intestine. This organ has a huge surface area—300 m^2, roughly the size of a tennis court. Large circular folds in the lining bear fingerlike projections called **villi,** and each epithelial cell of a villus has many microscopic appendages called **microvilli** that are exposed to the intestinal lumen. This enormous microvillar surface is an adaptation that greatly increases the rate of nutrient absorption (FIGURE 41.19).

Penetrating the core of each villus is a net of microscopic blood vessels (capillaries) and a small vessel of the lymphatic system called a **lacteal.** (In addition to their circulatory system, vertebrates have an associated network of vessels—the lymphatic system—that carries a clear fluid called lymph, discussed in Chapter 43.) Nutrients are absorbed across the intestinal epithelium and then across the unicellular epithelium of the capillaries or lacteals. Only these two single layers of epithelial cells separate nutrients in the lumen of the intestine from the bloodstream.

In some cases, transport of nutrients across the epithelial cells is passive. The simple sugar fructose, for example, apparently moves by diffusion down its concentration gradient from the lumen of the intestine into the epithelial cells, and then into capillaries. Other nutrients, including amino acids, small peptides, vitamins, and glucose and several other simple sugars, are pumped against concentration gradients by the epithelial membranes. This active transport allows the intestine to absorb a much higher proportion of the nutrients in the intestine than would be possible with passive diffusion.

Amino acids and sugars pass through the epithelium, enter capillaries, and are carried away from the intestine by the bloodstream. After glycerol and fatty acids are absorbed by epithelial cells, they are recombined into fats within those cells. The fats are then mixed with cholesterol and coated with special proteins, forming small globules called **chylomicrons,** most of which are transported by exocytosis out of the epithelial cells and into lacteals. The lacteals converge into the larger vessels of the lymphatic system. Lymph, containing chylomicrons, eventually drains from the lymphatic system into large veins that return blood to the heart.

In contrast to the lacteals, the capillaries and veins that drain the nutrients away from the villi all converge into the **hepatic portal vessel,** which leads directly to the liver. This ensures that the liver—which has the metabolic versatility to interconvert various organic molecules—has first access to amino acids and sugars absorbed after a meal is digested. Therefore, blood that leaves the liver may have a very different balance of these nutrients than the blood that entered via the hepatic portal vessel. For example, the liver helps regulate the level of glucose molecules in the blood, and blood exiting the liver usually has a glucose concentration very close to 0.1%, regardless

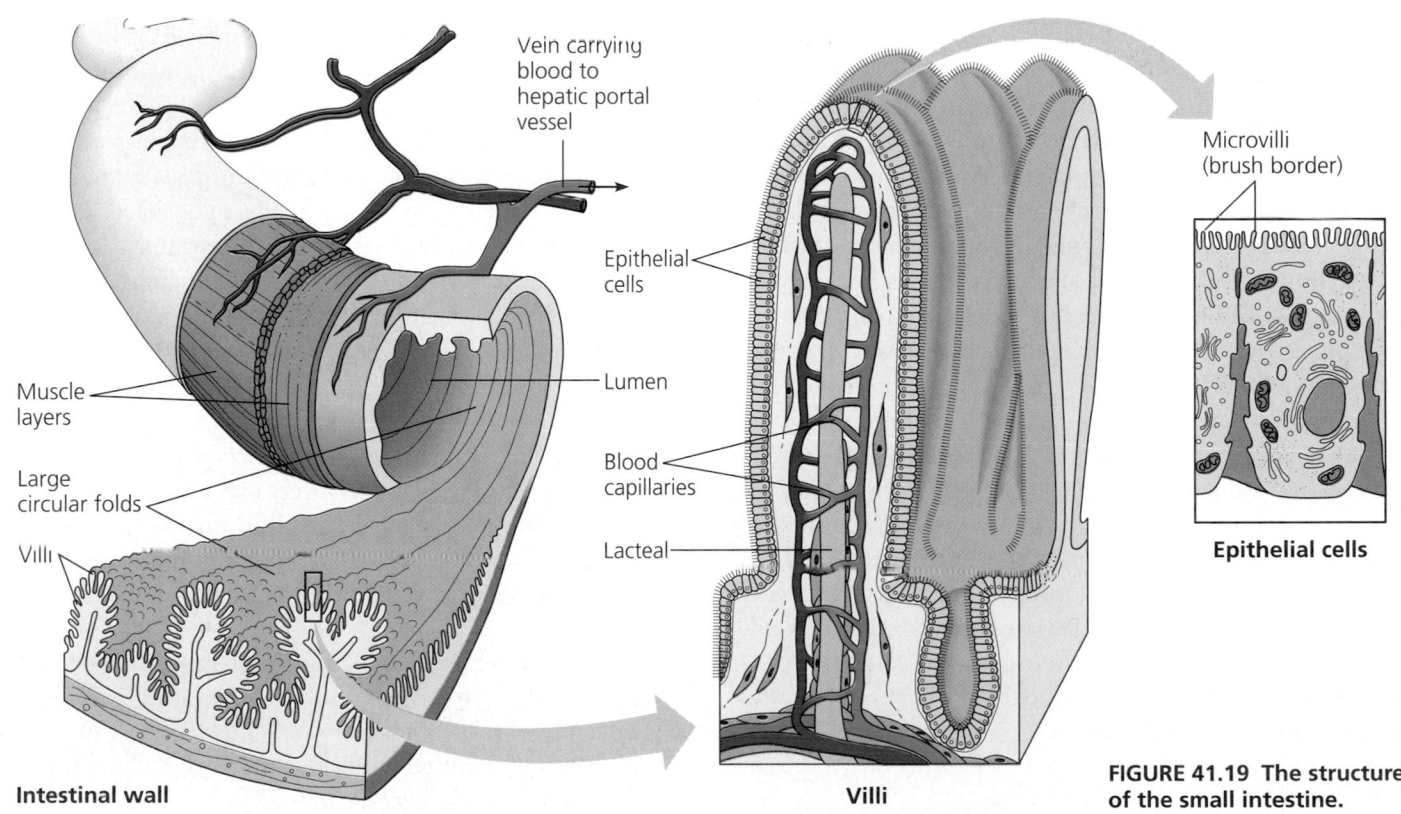

Vein carrying blood to hepatic portal vessel

Muscle layers

Large circular folds

Villi

Intestinal wall

Epithelial cells

Lumen

Blood capillaries

Lacteal

Villi

Microvilli (brush border)

Epithelial cells

FIGURE 41.19 The structure of the small intestine.

of the carbohydrate content of a meal (see FIGURE 41.1). From the liver, blood travels to the heart, which pumps the blood and the nutrients it contains to all parts of the body.

Digestive Efficiency and Cost

The digestive and absorptive processes are very effective in obtaining energy and nutrients. People eating the typical diets consumed in developed countries usually absorb 80 to 90 percent of the organic matter in their food. Much of the remaining undigestible material is cellulose from plant cell walls, and thus diets containing lots of fresh vegetables are not absorbed as completely as diets that are rich in meats, fats, or simple carbohydrates.

Because digestion and absorption depend on such active mechanisms as peristalsis, enzyme secretion, and active transport, there is a substantial energy cost to processing food. Depending on the animal species and the diet, digesting and absorbing a meal may require the animal to expend an amount of energy equal to between 3% and 30% of the chemical energy contained in the meal.

Hormones help regulate digestion

Many animals go for long intervals between meals and do not need to run their digestive systems continuously. Hormones released by the wall of the stomach and duodenum help ensure that digestive secretions are present only when needed. When we see, smell, or taste food, impulses from the brain to the stomach initiate the secretion of gastric juice. Then certain substances in the food stimulate the stomach wall to release the hormone **gastrin** into the circulatory system. As gastrin gradually recirculates in the bloodstream back to the stomach wall, the hormone stimulates further secretion of gastric juice. Thus, an initial burst of gastric secretion at mealtime is followed by a sustained secretion that continues to add gastric juice to the food for some time. If the pH of the stomach contents becomes too low, the acid will inhibit the release of gastrin, decreasing the secretion of gastric juice; this is an example of negative feedback.

Other hormones, collectively called **enterogastrones,** are secreted by the wall of the duodenum. The acidic pH of the chyme that enters the duodenum stimulates cells in the wall to release the hormone **secretin.** This enterogastrone signals the pancreas to release bicarbonate, which neutralizes the acid chyme. A second enterogastrone, **cholecystokinin (CCK),** is secreted in response to the presence of amino acids or fatty acids. CCK causes the gallbladder to contract and release bile into the small intestine. CCK also triggers the release of pancreatic enzymes. The chyme, particularly if rich in fats, also causes the duodenum to release other enterogastrones that inhibit peristalsis in the stomach, thereby slowing down the entry of food into the small intestine.

Reclaiming water is a major function of the large intestine

The **large intestine,** or **colon,** is connected to the small intestine at a T-shaped junction, where a sphincter (a muscular valve) controls the movement of material. One arm of the T is a pouch called the **cecum** (see FIGURE 41.13). Compared to many other mammals, humans have a relatively small cecum with a fingerlike extension, the **appendix,** which is dispensable. (Lymphoid tissue in the appendix makes a minor contribution to body defense.) The main branch of the human colon is shaped like an upside-down U about 1.5 m long.

A major function of the colon is to recover water that has entered the alimentary canal as the solvent of the various digestive juices. About 7 L of fluid are secreted into the lumen of the digestive tract each day, which is much more liquid than most people drink. Most of this water is reabsorbed along with nutrient absorption in the small intestine. The colon reclaims much of the remaining water in the lumen. Together, the small intestine and colon reabsorb about 90% of the water that entered the alimentary canal. The wastes of the digestive tract, the **feces,** become more solid as they are moved along the colon by peristalsis. The movement is sluggish, and it generally takes about 12 to 24 hours for material to travel the length of the organ. If the lining of the colon is irritated—by a viral or bacterial infection, for instance—less water than normal may be reabsorbed, resulting in diarrhea. The opposite problem, constipation, occurs when peristalsis moves the feces along too slowly. An excess of water is reabsorbed, and the feces become compacted.

Living in the large intestine is a rich flora of mostly harmless bacteria. One of the common inhabitants of the human colon is *Escherichia coli,* a favorite research organism of molecular biologists (see Chapter 18). The presence of *E. coli* in lakes and streams is an indication of contamination by untreated sewage. Intestinal bacteria live on unabsorbed organic material. As byproducts of their metabolism, many colon bacteria generate gases, including methane and hydrogen sulfide. Some of the bacteria produce vitamins, including biotin, folic acid, vitamin K, and several B vitamins. These vitamins, absorbed into the blood, supplement our dietary intake of vitamins.

Feces contain masses of bacteria, as well as cellulose and other undigested materials. Although cellulose fibers have no caloric value to humans, their presence in the diet helps move food along the digestive tract. Feces may also contain an abundance of salts. For instance, when iron and calcium concentrations in the blood get too high, the colon lining excretes salts of these elements into the lumen, and they are eliminated in the feces.

The terminal portion of the colon is called the **rectum,** where feces are stored until they can be eliminated. Between the rectum and the anus are two sphincters, one involuntary and the other voluntary. Once or more each day, strong contractions of the colon create an urge to defecate.

EVOLUTIONARY ADAPTATIONS OF VERTEBRATE DIGESTIVE SYSTEMS

Structural adaptations of digestive systems are often associated with diet

The digestive systems of mammals and other vertebrates are variations on a common plan, but there are many intriguing adaptations, often associated with the animal's diet. We will examine just a few.

Dentition, an animal's assortment of teeth, is one example of structural variation reflecting diet. Particularly in mammals, evolutionary adaptation of teeth for processing different kinds of food is one of the major reasons this vertebrate class has been so successful. Compare the dentition of carnivorous, herbivorous, and omnivorous mammals in FIGURE 41.20.

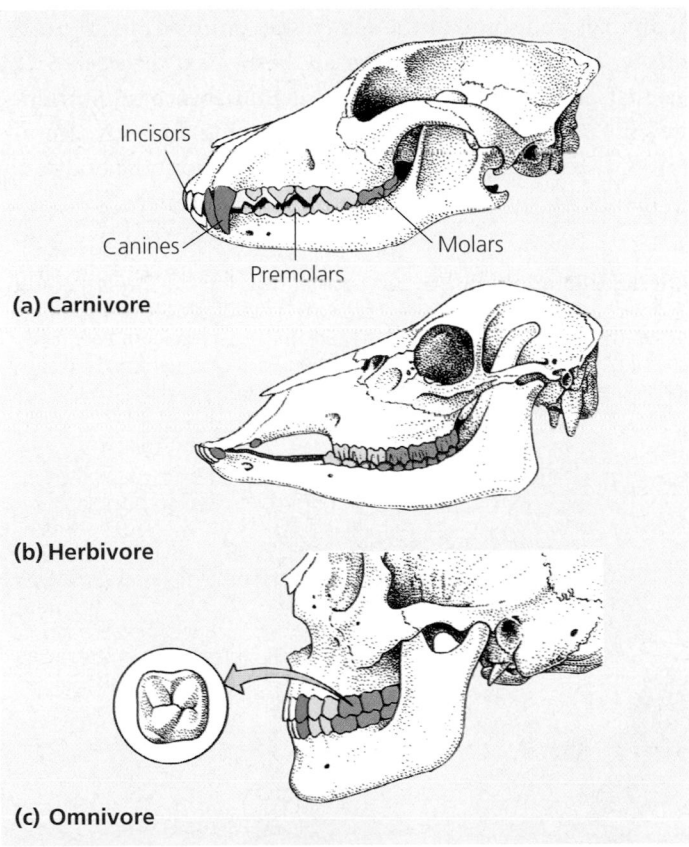

(a) Carnivore

(b) Herbivore

(c) Omnivore

FIGURE 41.20 Dentition and diet. (a) Carnivores, such as members of the dog and cat families, generally have pointed incisors and canines that can be used to kill prey and rip or cut away pieces of flesh. The jagged premolars and molars crush and shred food. **(b)** In contrast, herbivorous mammals, such as cows, usually have teeth with broad, ridged surfaces that grind tough plant material. The incisors and canines are generally modified for biting off pieces of vegetation. **(c)** Humans, being omnivores adapted for eating both vegetation and meat, have a relatively unspecialized dentition. The permanent (adult) set of teeth is 32 in number. Beginning at the midline of the upper and lower jaw are two blade-like incisors for biting, a pointed canine for tearing, two premolars for grinding, and three molars for crushing.

Nonmammalian vertebrates generally have less specialized dentition, but there are interesting exceptions. For example, poisonous snakes, such as rattlesnakes, have fangs, modified teeth that inject venom into prey. Some fangs are hollow, like syringes, while others drip the poison along grooves on the surfaces of the teeth. All snakes have another important anatomical adaptation associated with feeding: They swallow their prey whole, with no chewing, and the lower jaw is loosely hinged to the skull by an elastic ligament that permits the mouth and throat to open very wide for swallowing impressively large prey (once again, witness the astonishing episode recorded in FIGURE 41.9). Large, expandable stomachs are common in carnivores, which may go for a long time between meals and therefore must eat as much as they can when they do catch prey. For example, a 200-kg African lion can consume 40 kg of meat in one meal.

The length of the vertebrate digestive system is also correlated with diet. In general, herbivores and omnivores have longer alimentary canals relative to their body size than carnivores (FIGURE 41.21). Vegetation is more difficult to digest than meat because it contains cell walls. A longer tract furnishes more time for digestion and more surface area for the absorption of nutrients.

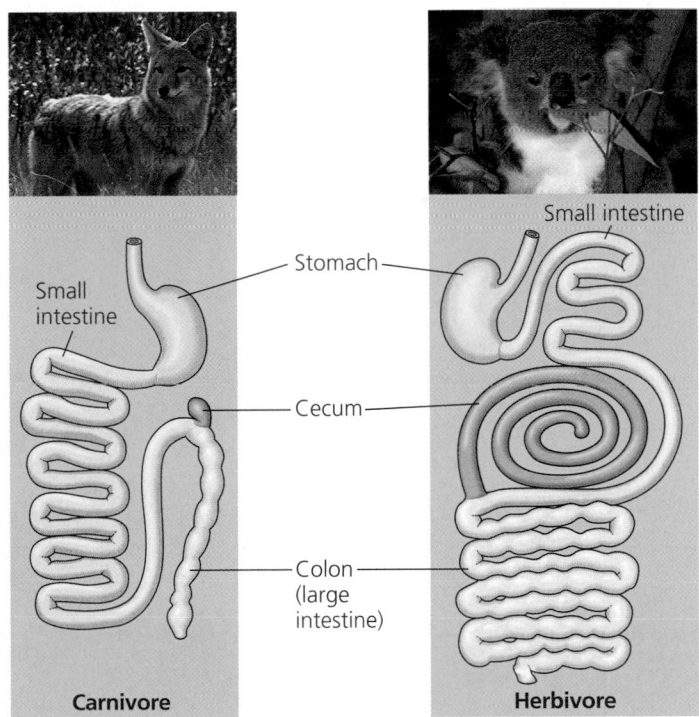

Carnivore

Herbivore

FIGURE 41.21 The digestive tracts of a carnivore (coyote) and a herbivore (koala) compared. Although these two mammals are about the same size, the koala's intestines are much longer, an adaptation that enhances processing of fibrous, protein-poor eucalyptus leaves from which it obtains virtually all its food and water. Extensive chewing chops the leaves into very small pieces, increasing exposure of the food to digestive juices. The koala's cecum—at 2 m, the longest of any animal of equivalent size—functions as a fermentation chamber where symbiotic bacteria convert the shredded leaves into a more nutritious diet.

Symbiotic microorganisms help nourish many vertebrates

Herbivorous animals face a special challenge: much of the chemical energy in their diets is contained in the cellulose of plant cell walls, but animals do not produce enzymes that hydrolyze cellulose. Many vertebrates (as well as termites, whose wood diets are largely cellulose) solve this problem by housing large populations of symbiotic bacteria and protists in special fermetation chambers in their alimentary canals. These microorganisms *do* have enzymes that can digest cellulose to simple sugars and other compounds that the animal can absorb. In many cases the microorganisms can also use the sugars from digested cellulose along with minerals to make a variety of nutrients essential to the animal, such as vitamins and amino acids.

The location of symbiotic microbes in herbivores' digestive tracts varies depending on the type of animal. The hoatzin, an herbivorous bird that lives in South American rain forests, has a large, muscular crop (an esophageal pouch) that houses symbiotic microorganisms. Hard ridges in the wall of the crop grind plant leaves into small fragments, and the microorganisms break down cellulose. Many herbivorous mammals, including horses, house symbiotic microorganisms in a large cecum, the pouch where the small and large intestines connect. The symbiotic bacteria of rabbits and some rodents live in the large intestine as well as in the cecum. Since most nutrients are absorbed in the small intestine, nourishing by-products of fermentation by bacteria in the large intestine are initially lost with the feces. Rabbits and rodents recover these nutrients by eating some of their feces and passing the food through the alimentary canal a second time. (The familiar rabbit "pellets," which are not reingested, are the feces eliminated after food has passed through the digestive tract twice.) The koala, an Australian marsupial, also has an enlarged cecum, where symbiotic bacteria ferment finely shredded eucalyptus leaves (see FIGURE 41.21). The most elaborate adaptations for a herbivorous diet have evolved in the **ruminants,** which include deer, cattle, and sheep (FIGURE 41.22).

■   ■   ■

Our focus in this chapter has been on provisioning the animal body to support both metabolic production of ATP and biosynthesis. We have examined the nutritional requirements of animals and some of the many ways animals obtain nutrients for fuel and building materials. In the next chapter we will see that obtaining food, digesting it, and absorbing nutrients are only parts of a larger story. Provisioning the body also involves distributing nutrients to cells throughout the body and exchanging respiratory gases with the environment.

FIGURE 41.22 Ruminant digestion. The stomach of a ruminant has four chambers. Because of the microbial action in the chambers, the diet from which a ruminant actually absorbs its nutrients is much richer than the grass the animal originally ate. In fact, a ruminant eating grass or hay obtains many of its nutrients by digesting the symbiotic microorganisms, which reproduce rapidly enough in the rumen to maintain a stable population.

❶ **Rumen.** When the cow first chews and swallows a mouthful of grass, boluses (green arrows) enter the rumen.

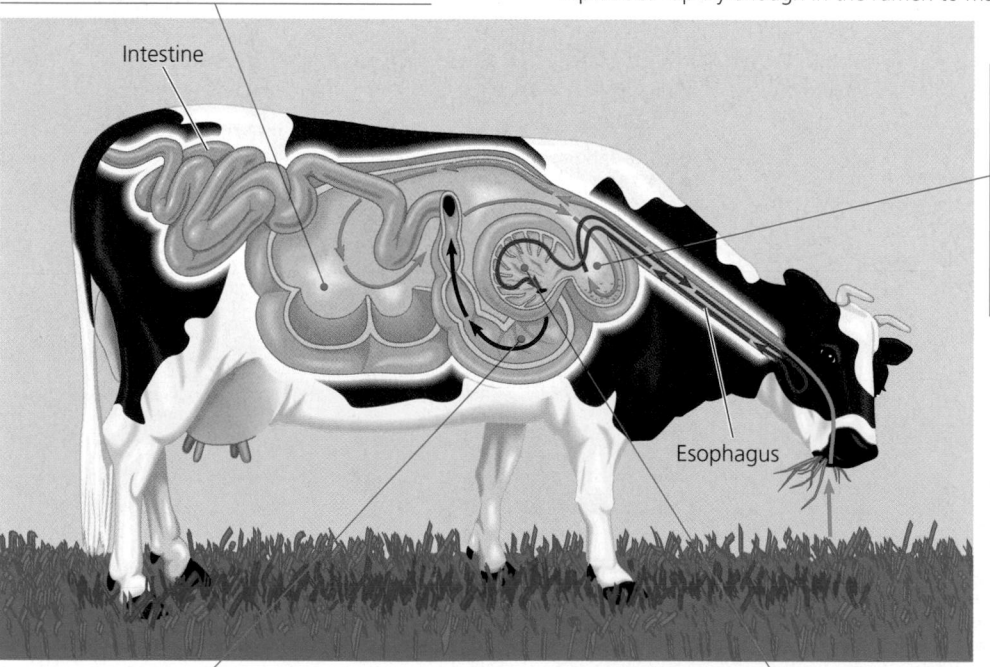

❷ **Reticulum.** Some boluses also enter the reticulum. In both the rumen and the reticulum, symbiotic prokaryotes and protists (mainly ciliates) go to work on the cellulose-rich meal. As by-products of their metabolism, the microorganisms secrete fatty acids. The cow periodically regurgitates and rechews the cud (red arrows), which further breaks down the fibers, making them more accessible to further microbial action.

❹ **Abomasum.** The cud, containing great numbers of microorganisms, finally passes to the abomasum for digestion by the cow's own enzymes (black arrows).

❸ **Omasum.** The cow then reswallows the cud (blue arrows), which moves to the omasum, where water is removed.

CHAPTER 41 REVIEW

Go to the Campbell Biology website (www.campbellbiology.com) to explore an interactive version of the Chapter Review.

Summary of Key Concepts

NUTRITIONAL REQUIREMENTS

- Animals are heterotrophs that require food for fuel, carbon skeletons, and essential nutrients: *an overview* (p. 850)

- Homeostatic mechanisms manage an animal's fuel (pp. 850–852, FIGURES 41.1 AND 41.2) Animals store excess calories as glycogen in the liver and muscles and as fat. Undernourished animals have diets deficient in calories.

- An animal's diet must supply essential nutrients and carbon skeletons for biosynthesis (pp. 852–856, FIGURES 41.3 AND 41.5, TABLES 41.1, 41.2) Carbon skeletons are required in biosynthesis. Essential nutrients must be supplied in preassembled form. Essential amino acids are those an animal cannot make. Essential fatty acids are unsaturated. Vitamins are organic molecules required in small amounts. Minerals are inorganic nutrients.

FOOD TYPES AND FEEDING MECHANISMS

- Most animals are opportunistic feeders (p. 856) Herbivores eat mainly plants, and carnivores mainly eat other animals. Omnivores regularly eat both plant and animal matter.

- Diverse feeding adaptations have evolved among animals (pp. 856–857, FIGURES 41.6–41.9) Many aquatic animals are suspension-feeders, sifting small particles from the water. Substrate-feeders tunnel through their food, eating as they go. Fluid-feeders suck nutrient-rich fluids from a living host. Most animals are bulk-feeders, eating relatively large pieces of food.

OVERVIEW OF FOOD PROCESSING

- The four main stages of food processing are ingestion, digestion, absorption, and elimination (pp. 857–858) Food processing in animals involves ingestion (the act of eating), digestion (enzymatic breakdown of the macromolecules of food into their monomers), absorption (the uptake of nutrients by body cells), and elimination (the passage of undigested materials out of the body in feces).

- Digestion occurs in specialized compartments (pp. 858–859, FIGURES 41.10–41.12) In intracellular digestion, food particles are engulfed by endocytosis and digested within food vacuoles. Most animals use extracellular digestion, with enzymatic hydrolysis occurring outside cells in a gastrovascular cavity or alimentary canal.

THE MAMMALIAN DIGESTIVE SYSTEM

- The oral cavity, pharynx, and esophagus initiate food processing (pp. 860–861, FIGURES 41.13 AND 41.14) Food is lubricated and digestion begins in the oral cavity, where teeth chew food into smaller particles that are exposed to salivary amylase, initiating the breakdown of glucose polymers. The pharynx is the intersec-

tion leading to the trachea and the esophagus. The esophagus conducts food from the pharynx to the stomach by involuntary peristaltic waves.

- The stomach stores food and performs preliminary digestion (pp. 861–863, FIGURE 41.15) The stomach stores food and secretes gastric juice, which converts a meal to acid chyme. Gastric juice includes hydrochloric acid and the enzyme pepsin.

- The small intestine is the major organ of digestion and absorption (pp. 863–866, FIGURES 41.16–41.19) Acid chyme from the stomach mixes in the duodenum with intestinal juice, bile, and pancreatic juice. Diverse enzymes complete the hydrolysis of food molecules to monomers, which are absorbed into the blood across the lining of the small intestine.
Web/CD Activity 41A: *Digestive System Function*
Web/CD Case Study in the Process of Science: *What Role Does Amylase Play in Digestion?*

- Hormones help regulate digestion (p. 866) Nerve impulses and the hormone gastrin stimulate gastric motility and secretion of gastric juice. A class of intestinal hormones called enterogastrones regulates the activities of the pancreas and gallbladder.
Web/CD Activity 41B: *Hormonal Control of Digestion*

- Reclaiming water is a major function of the large intestine (p. 866) The large intestine (colon) aids the small intestine in reabsorbing water and houses bacteria, some of which synthesize vitamins. Feces pass through the rectum and out the anus.

EVOLUTIONARY ADAPTATIONS OF VERTEBRATE DIGESTIVE SYSTEMS

- Structural adaptations of digestive systems are often associated with diet (p. 867, FIGURES 41.20, 41.21) A mammal's dentition is generally correlated with its diet. Herbivores generally have longer alimentary canals, reflecting the longer time needed to digest vegetation.

- Symbiotic microorganisms help nourish many vertebrates (p. 868, FIGURE 41.22) Many herbivorous animals have special fermentation chambers where symbiotic microorganisms digest cellulose.

Self-Quiz

1. Which of the following animals is *incorrectly* paired with its feeding mechanism?
   a. lion—substrate-feeder
   b. baleen whale—suspension-feeder
   c. aphid—fluid-feeder
   d. earthworm—deposit-feeder
   e. snake—bulk-feeder

2. The mammalian trachea and esophagus both open into the
   a. large intestine.
   b. stomach.
   c. pharynx.
   d. rectum.
   e. epiglottis.

3. Our oral cavity, with its dentition, is most functionally analogous to an earthworm's
   a. intestine.
   b. pharynx.
   c. gizzard.
   d. stomach.
   e. anus.

4. Which of the following enzymes has the lowest pH optimum?
   a. salivary amylase
   b. trypsin
   c. pepsin
   d. pancreatic amylase
   e. pancreatic lipase

5. After surgical removal of an infected gallbladder, a person must be especially careful to restrict his or her dietary intake of
   a. starch.
   b. protein.
   c. sugar.
   d. fat.
   e. water.

6. Enteropeptidase, an enzyme bound to the intestinal epithelium, has which of the following actions?
   a. inhibits bile secretion
   b. inhibits duodenal secretion
   c. activates pancreatic enzymes
   d. inhibits peristalsis in the stomach
   e. increases the pH of chyme

7. Individuals whose diet consists primarily of corn would likely become
   a. obese.
   b. anorexic.
   c. overnourished.
   d. undernourished.
   e. malnourished.

8. Which of the following organs is *incorrectly* paired with its function?
   a. stomach—protein digestion
   b. oral cavity—starch digestion
   c. large intestine—bile production
   d. small intestine—nutrient absorption
   e. pancreas—enzyme production

9. If you were to jog a mile a few hours after lunch, which stored fuel would you probably tap?
   a. muscle proteins
   b. muscle and liver glycogen
   c. fat stored in the liver
   d. fats stored in adipose tissue
   e. blood proteins

10. The symbiotic microbes that help nourish a ruminant live mainly in specialized regions of the
    a. large intestine.
    b. liver.
    c. small intestine.
    d. pharynx.
    e. stomach.

11. What is peristalsis and what is its function in our digestive system?

12. When we start coughing because food or drink "went down the wrong pipe," the material has entered the _____ instead of the _____.

13. If you added pepsinogen to a test tube containing protein dissolved in distilled water, not much protein would be digested. What inorganic substance could you add to the tube to accelerate protein digestion?

14. Explain why treatment of chronic infection with antibiotics for an extended period of time can cause a vitamin K deficiency.

15. What is an "essential nutrient"?

16. Why are vitamins required in such small doses compared to other essential organic nutrients, such as essential amino acids?

Go to the website or CD-ROM for more quiz questions.

Evolution Connection

The human esophagus and trachea share a common passage leading from the mouth and nasal passages, a "design" that occasionally contributes to death by choking. Explain the historical (evolutionary) basis for this "imperfect" anatomy.

The Process of Science

Design a controlled experiment to test the hypothesis that human salivary amylase digests starch faster at 37°C (body temperature) than it does at either 0°C (freezing point of water) or 100°C (boiling point of water). Your only materials and equipment are a source of human saliva, distilled water, starch, an iodine reagent that stains starch dark purple, ice, beakers, and a Bunsen burner (source of heat).

Conduct enzyme assays to determine the role of amylase in digestion in the Case Study in the Process of Science, available on the CD-ROM and website.

Science, Technology, and Society

The media report numerous claims and counterclaims about the benefits and dangers of certain foods. Just a few examples are debates about vitamin doses, advocacy of diets enriched in certain food molecules such as carbohydrates or proteins, and publicity about new products such as cholesterol-lowering margarine. Have you modified your eating habits on the basis of nutritional information disseminated by the media? Why or why not? How should a person evaluate whether such nutritional claims are valid?

Answers 1. a; 2. c; 3. c; 4. c; 5. d; 6. c; 7. e; 8. c; 9. b; 10. e; 11. Peristalsis is the wavelike contraction of smooth muscles that moves food along our alimentary canal. 12. Trachea (windpipe); esophagus. 13. Hydrochloric acid or some other acid that will convert inactive pepsinogen to active pepsin. 14. The treatment can kill symbiotic bacteria that synthesize vitamin K in the colon. 15. A substance an organism requires but cannot make by its own metabolism. 16. Because vitamins generally have catalytic functions as coenzymes, and thus each vitamin molecule can repeat its function many times.

CHAPTER 42

CIRCULATION AND GAS EXCHANGE

CIRCULATION IN ANIMALS

- Transport systems functionally connect the organs of exchange with the body cells: *an overview*
- Most invertebrates have a gastrovascular cavity or a circulatory system for internal transport
- Vertebrate phylogeny is reflected in adaptations of the cardiovascular system
- Double circulation in mammals depends on the anatomy and pumping cycle of the heart
- Structural differences of arteries, veins, and capillaries correlate with their different functions
- Physical laws governing the movement of fluids through pipes affect blood flow and blood pressure
- Transfer of substances between the blood and the interstitial fluid occurs across the thin walls of capillaries
- The lymphatic system returns fluid to the blood and aids in body defense
- Blood is a connective tissue with cells suspended in plasma
- Cardiovascular diseases are the leading cause of death in the United States and most other developed nations

GAS EXCHANGE IN ANIMALS

- Gas exchange supplies oxygen for cellular respiration and disposes of carbon dioxide: *an overview*
- Gills are respiratory adaptations of most aquatic animals
- Tracheal systems and lungs are respiratory adaptations of terrestrial animals
- Control centers in the brain regulate the rate and depth of breathing
- Gases diffuse down pressure gradients in the lungs and other organs
- Respiratory pigments transport gases and help buffer the blood
- Deep-diving air-breathers stockpile oxygen and deplete it slowly

Every organism *must exchange materials and energy with its environment, and this exchange ultimately occurs at the cellular level. Cells live in aqueous surroundings; the resources they need, such as nutrients and oxygen, move across the plasma* membrane into the cytoplasm, and metabolic wastes, such as carbon dioxide, move out of the cell.

The feathery external gills projecting from the salmon shown above present an expansive surface area to the outside environment. A network of tiny blood vessels (capillaries) lies close to the outside surface of the gills. Oxygen dissolved in the surrounding water diffuses across the thin epithelium covering the gills and into the blood, while carbon dioxide diffuses out into the water.

Salmon and most other animals have organ systems specialized for exchanging materials with the environment, and many have an internal transport system that conveys fluid (blood or interstitial fluid) throughout the body.

In this chapter you will learn about mechanisms of internal transport in animals. You will also learn about one of the most important cases of chemical transfer between animals and their environment: the exchange of the gases oxygen (O_2) and carbon dioxide (CO_2), which is essential to cellular respiration and bioenergetics.

CIRCULATION IN ANIMALS

Transport systems functionally connect the organs of exchange with the body cells: *an overview*

Diffusion alone is not adequate for transporting substances over long distances in animals—for example, for moving glucose from the digestive tract and oxygen from the lungs to the brain of a mammal. Diffusion is insufficient over distances of more than a few millimeters, because the time it takes for a substance to diffuse from one place to another is proportional to the *square* of the distance. For example, if it takes 1 second for a given quantity of glucose to diffuse 100 μm, it will take 100 seconds for the same quantity to diffuse 1 mm and almost three hours to diffuse 1 cm. The circulatory system solves this problem by ensuring that no substance must diffuse very far to enter or leave a cell. By rapidly transporting fluid in bulk throughout the body, it functionally connects the aqueous environment of the body cells to the organs that exchange gases, absorb nutrients, and dispose of wastes. In the lungs of a mammal, for example, oxygen from inhaled air diffuses across a thin epithelium and into the blood, while carbon dioxide diffuses in the opposite direction. Bulk fluid movement in the circulatory system, powered by the heart, then quickly carries the oxygen-rich blood to all parts of the body. As the blood streams through the tissues within microscopic vessels called capillaries, chemicals are transported between the blood and the interstitial fluid that directly bathes the cells.

Internal transport and gas exchange are functionally related in most animals; thus, we focus on both the circulatory and respiratory systems in this chapter. We also highlight the role of these two organ systems in homeostasis (see Chapter 40)— for example, in regulating the interstitial fluid's content of nutrients and wastes. Let's turn first to some of the ways fluids circulate within animals.

Most invertebrates have a gastrovascular cavity or a circulatory system for internal transport

Gastrovascular Cavities

The body plan of a hydra and other cnidarians makes a circulatory system unnecessary. A body wall only two cells thick encloses a central gastrovascular cavity, which serves for both digestion and for distribution of substances throughout the body (see FIGURE 41.11). The fluid inside the cavity is continuous with the water outside through a single opening; thus, both inner and outer tissue layers are bathed by fluid. Thin branches of the gastrovascular cavity extend into the tentacles of a hydra, and some cnidarians have even more elaborate gastrovascular cavities (FIGURE 42.1). Since digestion begins in the cavity, only the cells of the inner layer have direct access to nutrients, but the nutrients have only a short distance to diffuse to the cells of the outer layer.

Planarians and most other flatworms also have gastrovascular cavities that exchange materials with the environment through a single opening (see FIGURE 33.10). The flat shape of

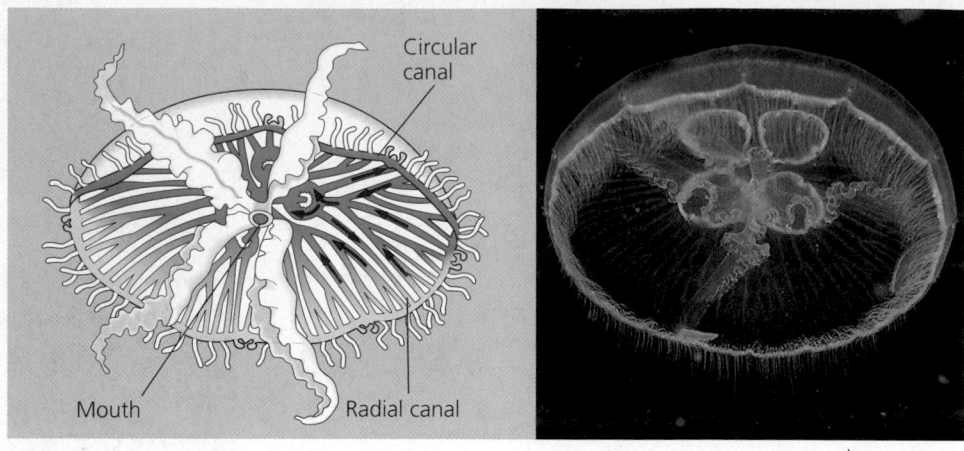

FIGURE 42.1 Internal transport in the cnidarian *Aurelia*. The mouth leads to an elaborate gastrovascular cavity (shown in blue) that has branches radiating to and from a circular canal. Ciliated cells lining the canals circulate fluid in the directions indicated by the arrows. The animal is viewed here from its underside (oral surface).

the body and the branching of the gastrovascular cavity throughout the animal ensure that all cells are bathed by a suitable medium and diffusion distances are short.

Open and Closed Circulatory Systems

For animals with many cell layers, gastrovascular cavities are insufficient for internal transport because diffusion distances are too great for adequate exchange of nutrients and wastes. In these more complex animals, two types of circulatory systems that overcome the limitations of diffusion have evolved: open circulatory systems and closed circulatory systems. Both have three basic components: a circulatory fluid (**blood**), a set of tubes (**blood vessels**) through which the blood moves through the body, and a muscular pump (the **heart**). The heart powers circulation by using metabolic energy to elevate the hydrostatic pressure of the blood, which then flows down a pressure gradient through its circuit and then back to the heart. This **blood pressure** is the motive force for fluid movement in the circulatory system.

In insects, other arthropods, and most mollusks, blood bathes the organs directly in an **open circulatory system** (FIGURE 42.2a). There is no distinction between blood and interstitial fluid, and the general body fluid is more correctly termed **hemolymph.** One or more hearts pump the hemolymph into an interconnected system of **sinuses,** which are spaces surrounding the organs. Here, chemical exchange occurs between the hemolymph and body cells. In insects and other arthropods, the heart is an elongated tube located dorsally. When the heart contracts, it pumps hemolymph through vessels out into sinuses. When the heart relaxes, it draws hemolymph into the circulatory system through pores called ostia. Body movements that squeeze the sinuses help circulate the hemolymph.

In a **closed circulatory system,** blood is confined to vessels and is distinct from the interstitial fluid (FIGURE 42.2b). One

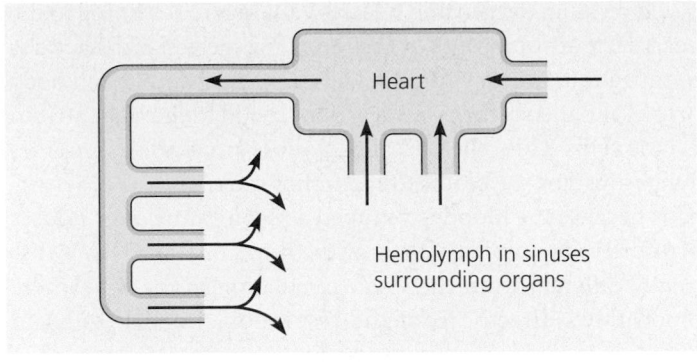

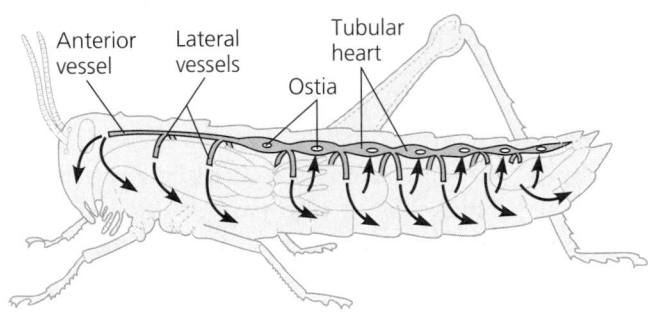

(a) Open circulatory system. In an open circulatory system, such as that of a grasshopper, blood and interstitial fluid are the same, and this fluid is called hemolymph. The heart pumps hemolymph through vessels into sinuses, where materials are exchanged between the hemolymph and cells. Hemolymph returns to the heart through ostia, which are equipped with valves that close when the heart contracts.

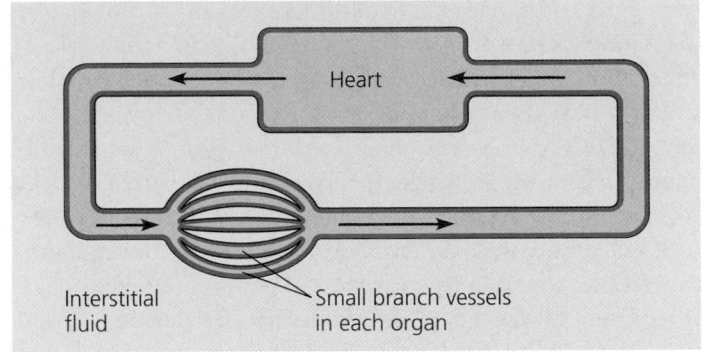

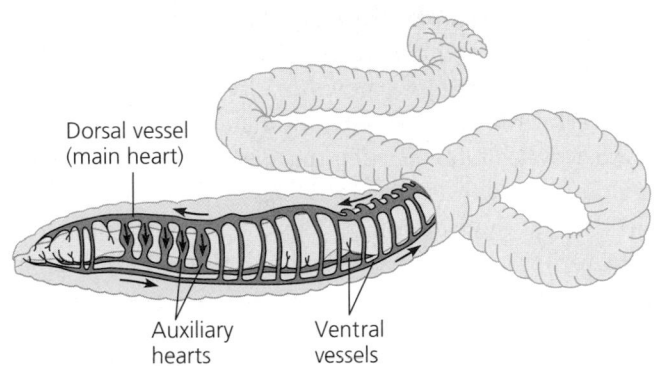

(b) Closed circulatory system. Closed circulatory systems circulate blood entirely within vessels, distinct from the interstitial fluid. As blood passes through small vessels in organs, chemical exchange occurs between the blood and the interstitial fluid, and between the interstitial fluid and body cells. In an earthworm, three major vessels, one dorsal and two ventral, branch into smaller vessels that supply blood to the various organs. The dorsal vessel functions as the main heart, pumping blood forward by peristalsis. Near the worm's anterior end, five pairs of vessels loop around the digestive tract, connecting the dorsal and a ventral vessel. The paired vessels function as auxiliary hearts, propelling blood ventrally.

FIGURE 42.2 Open and closed circulatory systems.

or more hearts pump blood into large vessels that branch into smaller ones coursing through the organs. Here, materials are exchanged by diffusion between the blood and the interstitial fluid bathing the cells. Earthworms, squids, octopuses, and vertebrates have closed circulatory systems.

Vertebrate phylogeny is reflected in adaptations of the cardiovascular system

Humans and other vertebrates have a closed circulatory system, often called the **cardiovascular system.** The heart has one **atrium** or two **atria** (plural), the chambers that receive blood returning to the heart, and one or two **ventricles,** the chambers that pump blood out of the heart. **Arteries, veins,** and **capillaries** are the three main kinds of blood vessels, which in the human body extend a total distance of about 100,000 km. Arteries carry blood away from the heart to organs throughout the body. Within organs, arteries branch into **arterioles,** small vessels that convey blood to the capillaries.

Capillaries are microscopic vessels with very thin, porous walls. Networks of these vessels, called **capillary beds,** infiltrate each tissue. It is across the thin walls of capillaries that chemicals, including dissolved gases, are exchanged by diffusion between the blood and the interstitial fluid surrounding the cells. At their "downstream" end, capillaries converge into **venules,** and venules converge into veins, which return blood to the heart. Notice that arteries and veins are distinguished by the *direction* in which they carry blood, not by the characteristics of the blood they contain. All arteries carry blood from the heart *toward* capillaries, and veins return blood to the heart *from* capillaries. The cardiovascular systems of different vertebrate taxa are variations of this general scheme, but they have been modified by natural selection.

Metabolic rate is an important factor in the evolution of cardiovascular systems. In general, animals with high metabolic rates have more complex circulatory systems and more powerful hearts than animals with low metabolic rates. Similarly, the complexity and number of blood vessels in a particular organ are correlated with that organ's metabolic requirements.

Perhaps the most fundamental differences in cardiovascular adaptations are associated with gill breathing in aquatic vertebrates compared with lung breathing in terrestrial vertebrates.

A fish heart has two main chambers, one atrium and one ventricle (FIGURE 42.3a). Blood pumped from the ventricle travels first to the gills (the **gill circulation**), where it picks up oxygen and disposes of carbon dioxide across capillary walls (see photo on p. 871). The gill capillaries converge into a vessel that carries oxygen-rich blood to capillary beds in all other parts of the body (the **systemic circulation**). Blood then returns in veins to the atrium of the heart. Notice that in a fish, blood must pass through *two* capillary beds during each circuit. When blood flows through a capillary bed, blood pressure—the motive force for circulation—drops substantially (for reasons that will be explained shortly). Therefore, oxygen-rich blood leaving the gills flows to the systemic circulation quite slowly (although the process is aided by body movements during swimming). This constrains the delivery of oxygen to body tissues, and hence the maximum aerobic metabolic rate of fishes.

Frogs and other amphibians have a three-chambered heart, with two atria and one ventricle (FIGURE 42.3b). The ventricle pumps blood into a forked artery that splits the ventricle's output into the **pulmocutaneous** and **systemic circulations.** The pulmocutaneous circulation leads to capillaries in the gas-exchange organs (the lungs and skin in a frog), where the blood picks up oxygen and releases CO_2 before returning to the heart's left atrium. Most of the returning oxygen-rich blood is pumped into the systemic circulation, which supplies all body organs and then returns oxygen-poor blood to the right atrium via the veins. This scheme, called **double circulation,** provides a vigorous flow of blood to the brain, muscles, and other organs because the blood is pumped a second time after it loses pressure in the capillary beds of the lungs or skin. This is distinctly different from the single circulation in the fish, where blood flows directly from the respiratory organs (gills) to other organs under reduced pressure.

In the ventricle of the frog, there is some mixing of oxygen-rich blood that has returned from the lungs with oxygen-poor blood that has returned from the rest of the body. However, a ridge within the ventricle diverts most of the oxygen-rich blood from the left atrium into the systemic circuit and most of the oxygen-poor blood from the right atrium into the pulmocutaneous circuit. Reptiles also have double circulation with **pulmonary** (lung) and systemic circuits, but there is even less mixing of oxygen-rich and oxygen-poor blood than in amphibians. Although the reptilian heart is three-chambered, the ventricle is partially divided.

In one order of reptiles, the crocodilians (crocodiles and alligators), and in all birds and mammals, the ventricle is completely divided into separate right and left chambers (FIGURE 42.3c). In this arrangement the left side of the heart receives

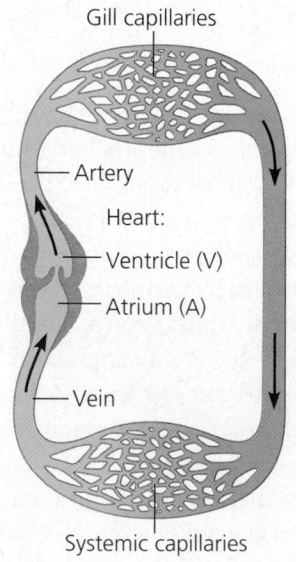

(a) Fish

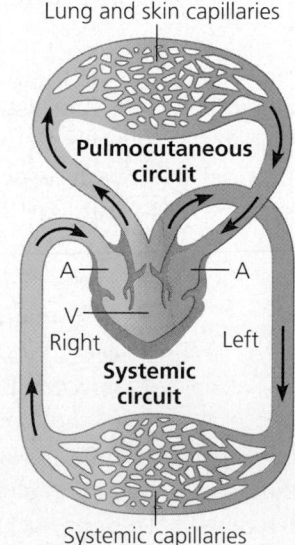

(b) Amphibian

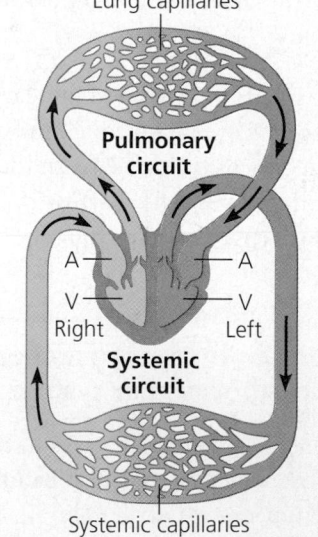

(c) Mammal

FIGURE 42.3 Generalized circulatory schemes of vertebrates. Red symbolizes oxygen-rich blood, and blue represents oxygen-poor blood. **(a)** Fishes have a two-chambered heart and a single circuit of blood flow. **(b)** Amphibians have a three-chambered heart and two circuits of blood flow: pulmocutaneous and systemic. This double circulation delivers blood to systemic organs under high pressure. In the single ventricle, there is some mixing of oxygen-rich blood with oxygen-poor blood. (It is customary to illustrate cardiovascular systems with the right side of the heart on the left and the left side on the right, as if the body is facing you from the page.) **(c)** Mammals have a four-chambered heart and double circulation. Within the heart, oxygen-rich blood is kept completely segregated from oxygen-poor blood.

and pumps only oxygen-rich blood, while the right side handles only oxygen-poor blood. Oxygen delivery is enhanced because there is no mixing of oxygen-rich and oxygen-poor blood, and double circulation restores pressure to the systemic circuit after blood has passed through the lung capillaries.

The evolution of a powerful four-chambered heart was an essential adaptation in support of the endothermic way of life characteristic of birds and mammals. Endotherms use about ten times as much energy as equal-sized ectotherms, and therefore their circulatory systems need to deliver about ten times as much fuel and oxygen to their tissues (and remove ten times as much CO_2 and other wastes). This large traffic of substances is made possible by their separate and independent systemic and pulmonary circulations, and by having large and powerful hearts to pump the necessary volume of blood. Birds and mammals descended from different reptilian ancestors, and their powerful four-chambered hearts evolved independently—an example of convergent evolution.

Double circulation in mammals depends on the anatomy and pumping cycle of the heart

A more detailed diagram of blood flow through the mammalian cardiovascular system is shown in FIGURE 42.4, which has numbers keyed to the circled numbers in this text discussion. Beginning our tour with the pulmonary (lung) circuit, ❶ the right ventricle pumps blood to the lungs via ❷ the pulmonary arteries. As the blood flows through ❸ capillary beds in the right and left lungs, it loads oxygen and unloads carbon dioxide. Oxygen-rich blood returns from the lungs via the pulmonary veins to ❹ the left atrium of the heart. Next, the oxygen-rich blood flows into ❺ the left ventricle, as the ventricle opens and the atrium contracts. The left ventricle pumps the oxygen-rich blood out to body tissues through the systemic circuit. Blood leaves the left ventricle via ❻ the aorta, which conveys blood to arteries leading throughout the body. The first branches from the aorta are the coronary arteries (not shown), which supply blood to the heart muscle itself. Then come branches leading to capillary beds ❼ in the head and arms (or forelimbs). The aorta continues in a posterior direction, supplying oxygen-rich blood to arteries leading to ❽ arterioles and capillary beds in the abdominal organs and legs. Within the capillaries, blood gives up much of its oxygen and picks up the carbon dioxide produced by cellular respiration. Capillaries rejoin to form venules, which convey blood to veins. Oxygen-poor blood from the head, neck, and forelimbs is channeled into a large vein called ❾ the anterior (or superior) vena cava. Another large vein called ❿ the posterior (or inferior) vena cava drains blood from the trunk and hind limbs. The two venae cavae empty their blood into ⓫ the right atrium, from which the oxygen-poor blood flows into the right ventricle.

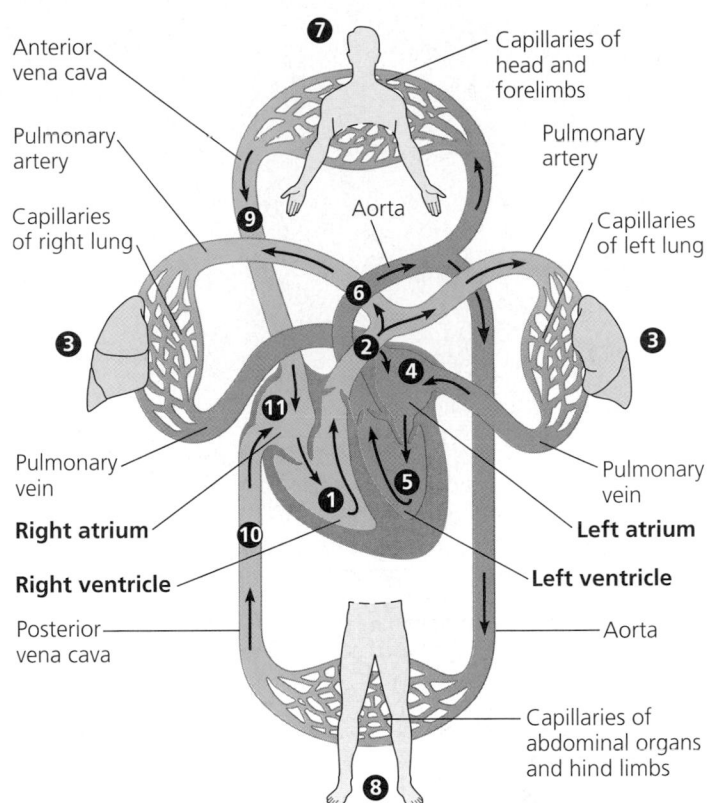

FIGURE 42.4 The mammalian cardiovascular system: an overview. The circled numbers are keyed to the discussion in the text. It is as though we are tracking a single blood cell as it flows first through the pulmonary circuit and then through the systemic circuit. However, if we consider the circulatory system as a whole instead of focusing on the path of an individual cell, it is important to understand that the dual circuits operate simultaneously, not in the serial fashion that the numbering in the diagram suggests. The two ventricles pump almost in unison; while some blood is traveling in the pulmonary circuit, the rest of the blood is flowing in the systemic circuit.

The Mammalian Heart: A Closer Look

A closer look at the mammalian heart (using the human heart as an example) provides a better understanding of how double circulation works (FIGURE 42.5, p. 876). Located beneath the breastbone (sternum), the human heart is about the size of a clenched fist and consists mostly of cardiac muscle (see FIGURE 40.4). The two atria have relatively thin walls and function as collection chambers for blood returning to the heart, pumping blood only the short distance to the ventricles. The ventricles have thicker walls and contract much more strongly than the atria—especially the left ventricle, which must pump blood to all body organs through the systemic circuit.

The heart contracts and relaxes in a rhythmic cycle. When it contracts, it pumps blood; when it relaxes, its chambers fill with blood. One complete sequence of pumping and filling is called the **cardiac cycle.** The contraction phase of the cycle is

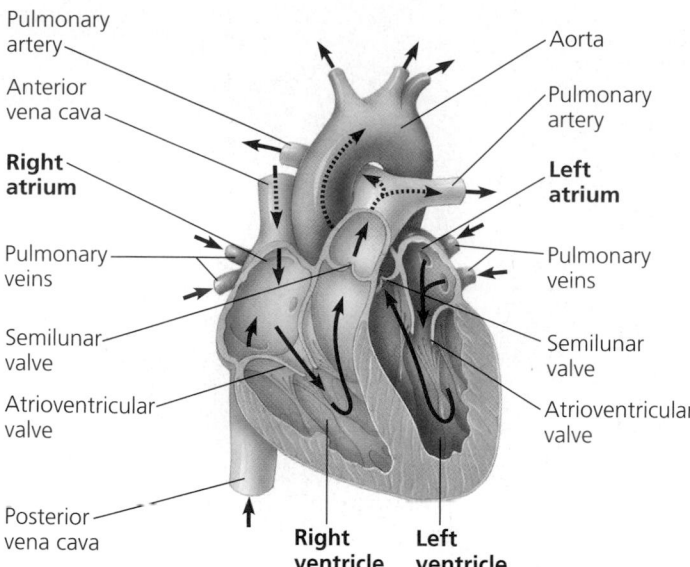

FIGURE 42.5 The mammalian heart: a closer look. In this detailed view of the structure of the heart, notice the valves, which prevent backflow of blood within the heart, and the relative thickness of the walls of the heart chambers. The atria, which pump blood only into the ventricles, have thinner walls than the ventricles, which pump blood to the pulmonary and systemic circuits.

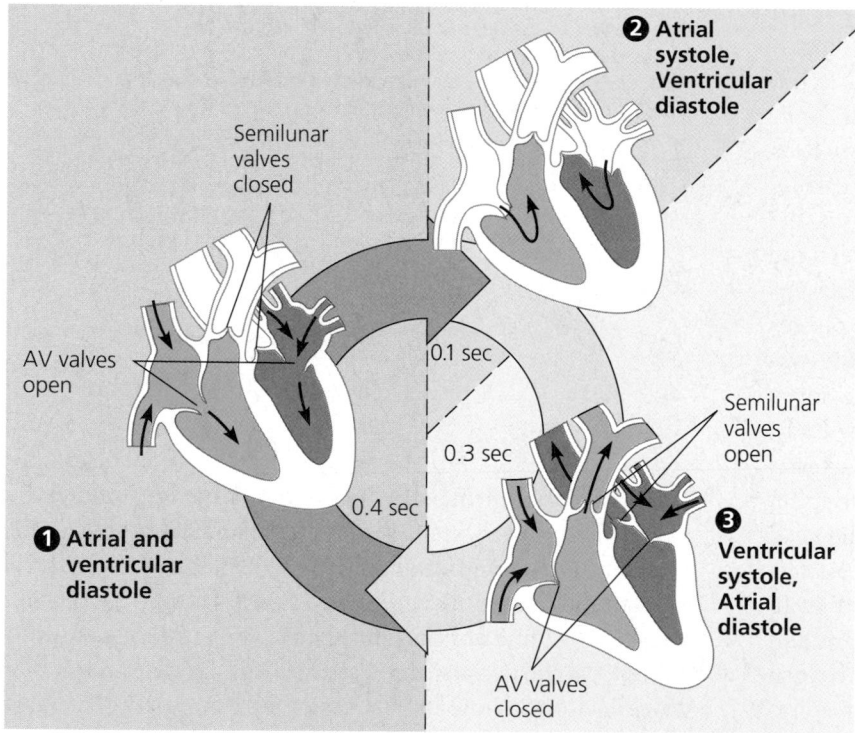

FIGURE 42.6 The cardiac cycle. The heart contracts (systole) and relaxes (diastole) in a rhythmic cycle. For an adult human at rest with a pulse of about 75 beats per minute, one complete cardiac cycle takes about 0.8 sec. ❶ During a relaxation phase (atria and ventricles in diastole) lasting about 0.4 sec, blood returning from the large veins flows into the atria and ventricles. ❷ A brief period (about 0.1 sec) of atrial systole then forces all remaining blood out of the atria into the ventricles. ❸ During the remaining 0.3 sec of the cycle, ventricular systole pumps blood into the large arteries. Note that seven-eighths of the time—all but 0.1 sec of the cardiac cycle—the atria are relaxed and filling with blood returning in the veins.

called **systole,** and the relaxation phase is called **diastole** (FIG-URE 42.6). The volume of blood per minute that the left ventricle pumps into the systemic circuit is called **cardiac output.** Cardiac output depends on two factors: the rate of contraction, or **heart rate** (number of beats per minute) and **stroke volume,** the amount of blood pumped by the left ventricle in each contraction. The average stroke volume for a human is about 75 mL. A person with this stroke volume and a heart rate during rest of 70 beats per minute has a cardiac output of 5.25 L/min. This is about equivalent to the total volume of blood in the human body. Cardiac output can increase about fivefold during heavy exercise. This is equivalent to pumping an amount of blood matching an average person's body weight every 2–3 minutes.

Four valves in the heart, each consisting of flaps of connective tissue, prevent backflow and keep blood moving in the correct direction (see FIGURE 42.5). Between each atrium and ventricle is an **atrioventricular (AV) valve.** The AV valves are anchored by strong fibers that prevent them from turning inside out. Pressure generated by the powerful contraction of the ventricles closes the AV valves, keeping blood from flowing back into the atria. **Semilunar valves** are located at the two exits of the heart, where the aorta leaves the left ventricle and the pulmonary artery leaves the right ventricle. These valves are forced open by pressure created by ventricular contraction. When the ventricles relax, blood starts to flow back toward the heart, closing the semilunar valves, which prevents blood from flowing back into the ventricles. The elastic walls of the arteries expand when they receive the blood expelled from the ventricles. By taking your **pulse**—the rhythmic stretching of arteries caused by the pressure of blood driven by the powerful contractions of the ventricles—you can measure your heart rate.

The heart sounds we can hear with a stethoscope are caused by the closing of the valves. (Even without a stethoscope, you can hear these sounds by pressing your ear tightly against the chest of a friend—a *close* friend.) The sound pattern is "lub-dup, lub-dup, lub-dup." The first heart sound ("lub") is created by the recoil of blood against the closed AV valves. The second sound ("dup") is the recoil of blood against the semilunar valves.

A defect in one or more of the valves causes a condition known as a **heart murmur,** which may be detectable as a hissing sound when a stream of blood squirts backward through a valve. Some people are born with heart murmurs, while others have their valves damaged by infection (from rheumatic fever, for instance). Most heart murmurs do not reduce the efficiency of blood flow enough to warrant surgery.

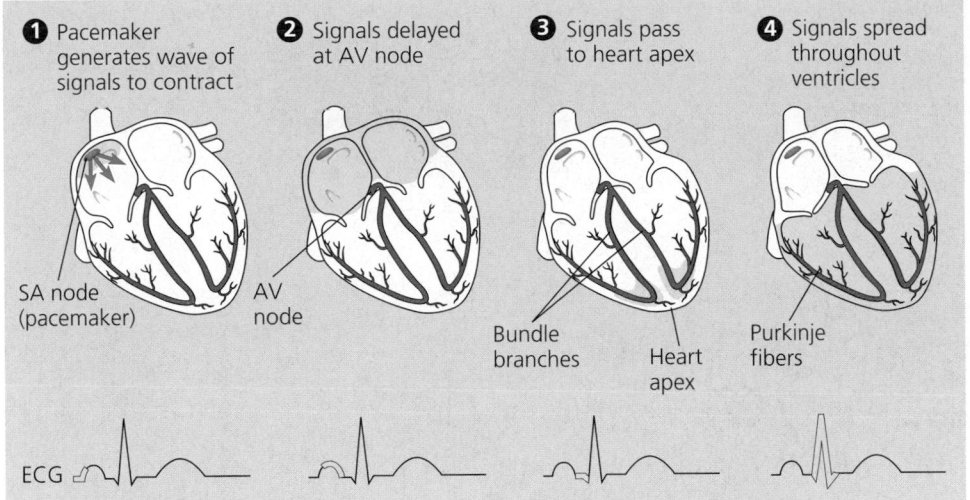

① Pacemaker
generates wave of
signals to contract

② Signals delayed
at AV node

③ Signals pass
to heart apex

④ Signals spread
throughout
ventricles

SA node
(pacemaker)

AV
node

Bundle
branches

Heart
apex

Purkinje
fibers

ECG

FIGURE 42.7 The control of heart rhythm. ① The SA node, or pacemaker, sets the tempo of the heartbeat by generating electrical signals (gold) that ② spread through both atria, making them contract simultaneously. The signals to contract are delayed at the AV node for about 0.1 sec, during which blood in the atria empties into the ventricles. ③ Specialized muscle fibers called bundle branches and Purkinje fibers then conduct the signals to the apex of the heart and ④ throughout the ventricular walls. The signals trigger powerful contractions of both ventricles from the apex toward the atria, driving blood into the large arteries. Gold color in the graphs at the bottom indicates the components of an electrocardiogram (ECG) corresponding to the sequence of electrical events in the heart. In step 4, the black portion of the ECG to the right of the gold "spike" represents electrical activity after the ventricles contract; during this phase of the ECG, the ventricles become re-primed electrically and thus are able to conduct the next round of contraction signals.

Maintaining the Heart's Rhythmic Beat

The timely delivery of oxygen to the body's organs is critical. For example, brain cells die within a few minutes if oxygen supply is interrupted. Thus, maintaining heart function is crucial for survival. Several mechanisms that assure the continuity and control of heartbeat have evolved.

Certain cells of vertebrate cardiac muscle are self-excitable, meaning they contract without any signal from the nervous system, even if removed from the heart and placed in tissue culture. Each of these cells has its own intrinsic contraction rhythm. How are their contractions coordinated in the intact heart? A region of the heart called the **sinoatrial (SA) node,** or **pacemaker,** sets the rate and timing at which all cardiac muscle cells contract. Composed of specialized muscle tissue, the SA node is located in the wall of the right atrium, near the point where the superior vena cava enters the heart (FIGURE 42.7).

The SA node generates electrical impulses much like those produced by nerve cells. Because cardiac muscle cells are electrically coupled (by the intercalated discs between adjacent cells; see FIGURE 40.4), impulses from the SA node spread rapidly through the walls of the atria, making them contract in unison. The impulses also pass to another region of specialized muscle tissue, a relay point called the **atrioventricular (AV) node,** located in the wall between the right atrium and right ventricle. Here the impulses are delayed for about 0.1 sec before spreading to the walls of the ventricles, which ensures

that the atria empty completely before the ventricles contract.

The impulses that travel through cardiac muscle during the heart cycle produce electrical currents that are conducted through body fluids to the skin, where the currents can be detected by electrodes and recorded as an **electrocardiogram (ECG or EKG).**

The SA node sets the tempo for the entire heart, but is influenced by a variety of physiological cues. Two sets of nerves affect heart rate; one set speeds up the pacemaker, and the other set slows it down. Heart rate is a compromise regulated by the opposing actions of these two sets of nerves. The pacemaker is also influenced by hormones secreted into the blood by glands. For example, epinephrine, the "fight-or-flight" hormone from the adrenal glands, increases heart rate (see Chapter 45). Body temperature is another factor that affects the pacemaker. An increase of only 1°C raises the heart rate by about 10 beats/min. This is the reason your pulse increases substantially when you have a fever. Heart rate also increases with exercise, an adaptation that enables the circulatory system to provide the additional oxygen needed by muscles hard at work.

Structural differences of arteries, veins, and capillaries correlate with their different functions

All blood vessels are built of similar tissues. The walls of both arteries and veins, for instance, have three similar layers (FIGURE 42.8, p. 878). On the outside, a layer of connective tissue with elastic fibers allows the vessel to stretch and recoil. A middle layer contains smooth muscle and more elastic fibers. Lining the lumen of all blood vessels, including capillaries, is an **endothelium,** a single layer of flattened cells that provides a smooth surface that minimizes resistance to the flow of blood.

Structural differences correlate with the different functions of arteries, veins, and capillaries. Capillaries lack the two outer layers, and their very thin walls consist only of endothelium and its basement membrane. This facilitates the exchange of substances between the blood and the interstitial fluid that bathes the cells. Arteries have thicker middle and outer layers than veins. Blood flows through the vessels of the circulatory system at uneven speeds and pressures. The thicker walls of arteries provide strength to accommodate blood pumped rapidly and at high pressure by the heart, and their elasticity helps

FIGURE 42.8 The structure of blood vessels. In the micrograph (SEM), an artery can be seen next to a thinner-walled vein.

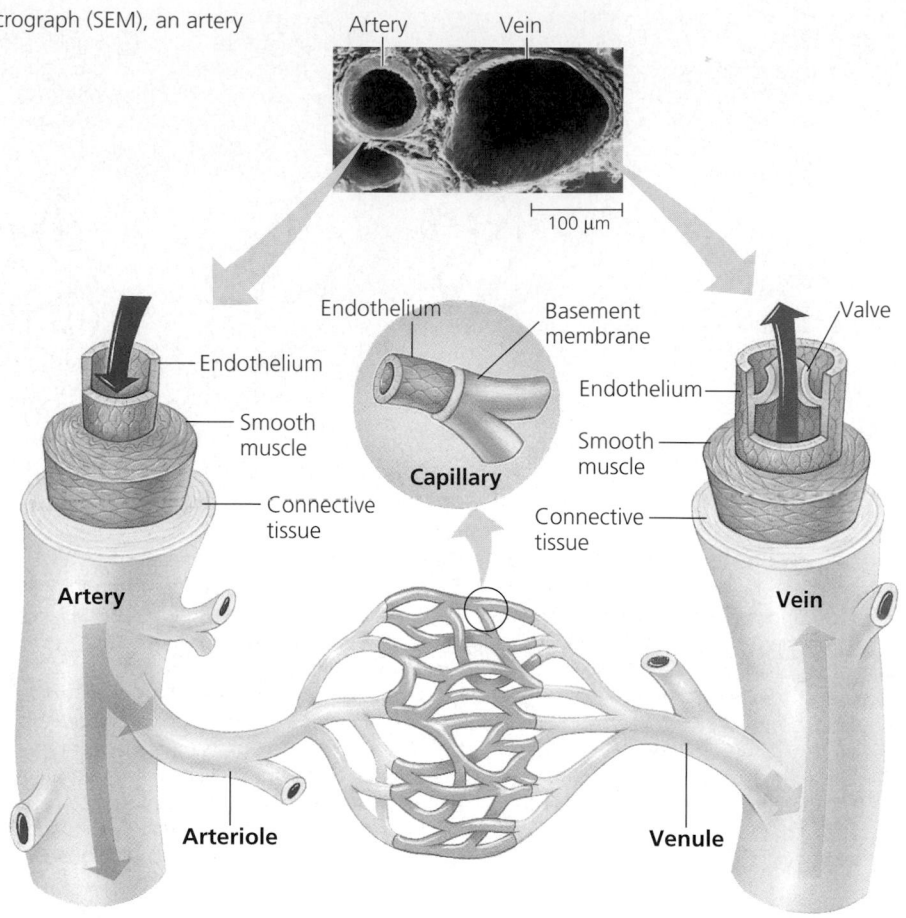

maintain blood pressure even when the heart relaxes between contractions. The thinner-walled veins convey blood back to the heart at low velocity and pressure. Blood flows through the veins mainly as a result of muscle action; whenever we move, our skeletal muscles pinch our veins and squeeze blood through them. Within large veins, flaps of tissue act as one-way valves that allow blood to flow only toward the heart (FIGURE 42.9).

Physical laws governing the movement of fluids through pipes affect blood flow and blood pressure

Blood Flow Velocity

Blood travels over a thousand times faster in the aorta (about 30 cm/sec on average) than in capillaries (about 0.026 cm/sec). This velocity change follows from the *law of continuity,* which describes fluid movement through pipes. If a pipe's diameter changes over its length, a fluid will flow through narrower segments of the pipe faster than it flows through wider segments. The *volume* of flow per second must be constant through the entire pipe, so the fluid must flow faster as the cross-sectional area of the pipe narrows. For instance, compare the velocity of water squirted by a hose with and without a nozzle.

Based on the law of continuity, it may seem that blood should travel faster through capillaries than through arteries,

because the diameter of capillaries is very small. However, it is the *total* cross-sectional area of capillaries that determines flow rate. Each artery conveys blood to such an enormous number of capillaries that the total cross-sectional area is much greater in capillary beds than in any other part of the circulatory system. For this reason, the blood slows substantially as it enters the arterioles from arteries, and slows further in the capillary beds. Capillaries are the only vessels with walls thin enough to permit the transfer of substances between the blood and interstitial fluid, and the slower flow of blood through these tiny vessels enhances this exchange. As blood leaves the capillary beds and passes to the venules and veins, it speeds up again as a result of the reduction in total cross-sectional area (FIGURE 42.10).

Blood Pressure

Fluids exert a force called hydrostatic pressure against surfaces they contact, and it is that pressure that drives fluids through pipes. Fluids always flow from areas of high pressure to areas of lower pressure. The hydrostatic force that blood exerts against the wall of a vessel and that propels blood is blood pressure. Blood pressure is much greater in arteries than in veins and is highest in arteries when the heart contracts during ventricular systole (**systolic pressure;** see FIGURE 42.10).

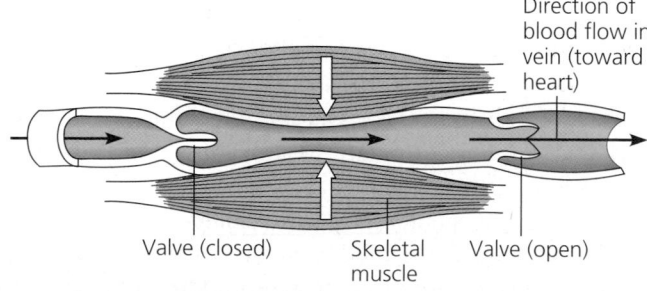

FIGURE 42.9 Blood flow in veins. Contracting skeletal muscles squeeze the veins. Flaps of tissue within the veins act as one-way valves that keep blood moving only toward the heart. If we sit or stand too long, the lack of muscular activity causes our feet to swell with stranded blood unable to return to the heart.

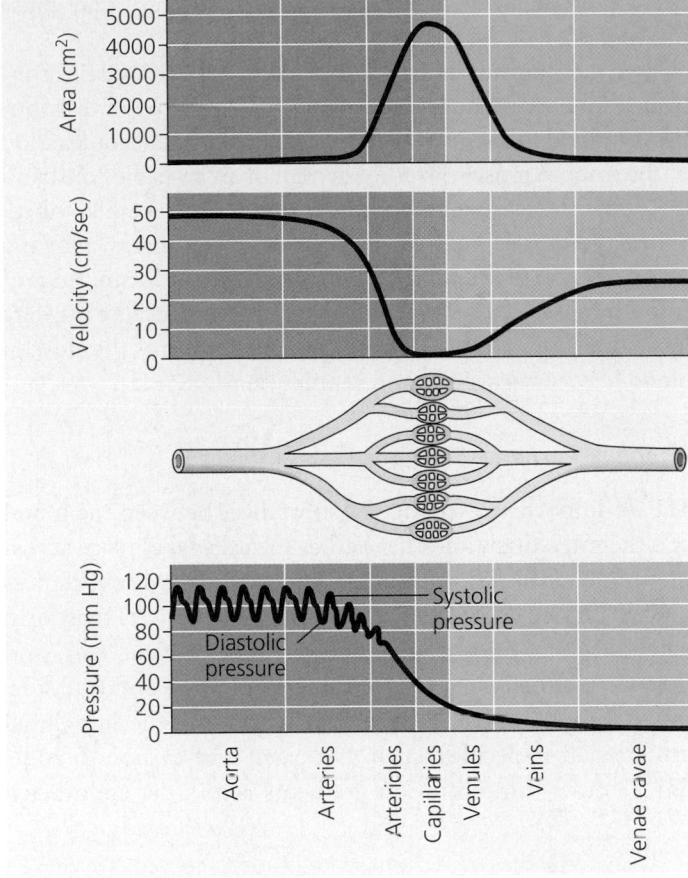

FIGURE 42.10 The interrelationship of blood flow velocity, cross-sectional area of blood vessels, and blood pressure. Blood flow velocity decreases markedly in the arterioles and is slowest in the capillaries, owing to an increase in total cross-sectional area. Blood pressure, the main force driving blood from the heart to the capillaries, is highest in the arteries. Peaks in blood pressure corresponding to ventricular systole alternate with lower blood pressures corresponding to diastole (elasticity in arterial walls keeps diastolic pressure from falling to zero). Resistance to flow through the arterioles and capillaries, due to contact of the blood with a greater surface area of endothelium, reduces blood pressure and eliminates pressure peaks.

When you take your pulse by placing your fingers on your wrist, you can feel an artery bulge with each heartbeat. The surge of pressure is partly due to the narrow openings of arterioles impeding the exit of blood from the arteries. Thus, when the heart contracts, blood enters the arteries faster than it can leave, and the vessels stretch from the pressure. The elastic walls of the arteries snap back during diastole, but the heart contracts again before enough blood has flowed into the arterioles to completely relieve pressure in the arteries. This impedance by the arterioles is called **peripheral resistance**. As a consequence of the elastic arteries working against peripheral resistance, there is a substantial blood pressure even during diastole (**diastolic pressure**), and blood flows into arterioles and capillaries continuously (see FIGURE 42.10). As shown in FIGURE 42.11 on page 880, the arterial blood pressure of a healthy resting human oscillates between about 120 mm Hg at systole and 70 mm Hg at diastole.

Blood pressure is determined partly by cardiac output and partly by peripheral resistance. Contraction of smooth muscles in the walls of the arterioles constricts the tiny vessels, increases peripheral resistance, and therefore increases blood pressure upstream in the arteries. When the smooth muscles relax, the arterioles dilate, blood flow through the arterioles increases, and pressure in the arteries falls. Nerve impulses, hormones, and other signals control these arteriole wall muscles. Stress, both physical and emotional, can raise blood pressure by triggering nervous and hormonal responses that will constrict blood vessels.

Cardiac output is adjusted in concert with changes in peripheral resistance. This coordination of regulatory mechanisms maintains adequate blood flow as the demands on the circulatory system change. For example, during heavy exercise the arterioles in working muscles dilate. This response admits a greater flow of oxygen-rich blood to the muscles and also decreases peripheral resistance. By itself, this would cause a fall in blood pressure (and therefore blood flow) in the body as a whole. However, cardiac output increases, maintaining blood pressure and supporting the necessary increase in blood flow.

In large land animals, another factor that affects blood pressure is gravity. Besides the force needed to overcome peripheral resistance, additional pressure is necessary if blood must be pushed above the level of the heart. In a standing human, blood must rise about 0.35 m to get from the heart to the brain. This demands an extra 27 mm Hg of pressure, which requires the heart to expend more energy in its contraction cycle. The pumping challenge is much greater for animals with long necks. Standing giraffes need to pump blood as much as 2.5 m above the heart. That requires about 190 mm Hg of additional blood pressure in the left ventricle, and a giraffe's normal systolic pressure near the heart is over 250 mm Hg. (Systolic pressure that high would be extremely dangerous in a human.) Special check valves and sinuses, along with feedback mechanisms that reduce cardiac output, prevent this high pressure from damaging the giraffe's brain when it puts its head down to drink—a body position that causes blood to flow downhill almost 2 m from the heart, adding an extra 150 mm Hg of blood pressure in the arteries leading to the brain. Some physiologists speculate about blood pressure and cardiovascular adaptations in dinosaurs—some of which had necks almost 10 m long, requiring a systolic pressure of 760 mm Hg just to pump blood to the brain when the head was fully raised.

By the time blood reaches the veins, its pressure is not affected much by the action of the heart. This is because the blood encounters so much resistance as it passes through the millions of tiny arterioles and capillaries that the pressure generated by the pumping heart has been dissipated and can no longer propel the blood in the veins. How does blood return to the heart, especially when it must travel from the lower extremities against gravity? Rhythmic contractions of smooth muscles in the walls of venules and veins account for some movement of the blood. More importantly, the activity of skeletal muscles during exercise squeezes blood through the veins (see FIGURE 42.9). Also, when we inhale, the change in pressure within the thoracic (chest) cavity causes the venae cavae and other large veins near the heart to expand and fill with blood.

Transfer of substances between the blood and the interstitial fluid occurs across the thin walls of capillaries

Blood Flow Through Capillary Beds

At any given time, only about 5–10% of the body's capillaries have blood flowing through them. However, each tissue has many capillaries, so every part of the body is supplied with blood at all times. Capillaries in the brain, heart, kidneys, and liver are usually filled to capacity, but in many other sites, the blood supply varies over time as blood is diverted from one destination to another. After a meal, for instance, blood supply to the digestive tract increases. During strenuous exercise, blood is diverted from the digestive tract and supplied more generously to skeletal muscles and skin. This is one reason

heavy exercise immediately after eating a big meal may cause indigestion.

Two mechanisms, both dependent on smooth muscles controlled by nerve signals and hormones, regulate the distribution of blood in capillary beds. In one mechanism, contraction of the smooth muscle layer in the wall of an arteriole constricts the vessel, decreasing blood flow through it to a capillary bed. When the muscle layer relaxes, the arteriole dilates, allowing blood to enter the capillaries. In the other mechanism, rings of smooth muscle, called precapillary sphincters because they are located at the entrance to capillary beds, control the flow of blood between arterioles and venules (FIGURE 42.12).

Capillary Exchange

The all-important exchange of substances between the blood and the interstitial fluid that bathes the cells takes place across the thin endothelial walls of the capillaries. Some substances may be carried across an endothelial cell in vesicles that form by endocytosis on one side of the cell (see FIGURE 8.19) and then release their contents by exocytosis on the opposite side; others simply diffuse between the blood and the interstitial fluid. Small molecules, such as oxygen and carbon dioxide, diffuse down concentration gradients across the endothelial

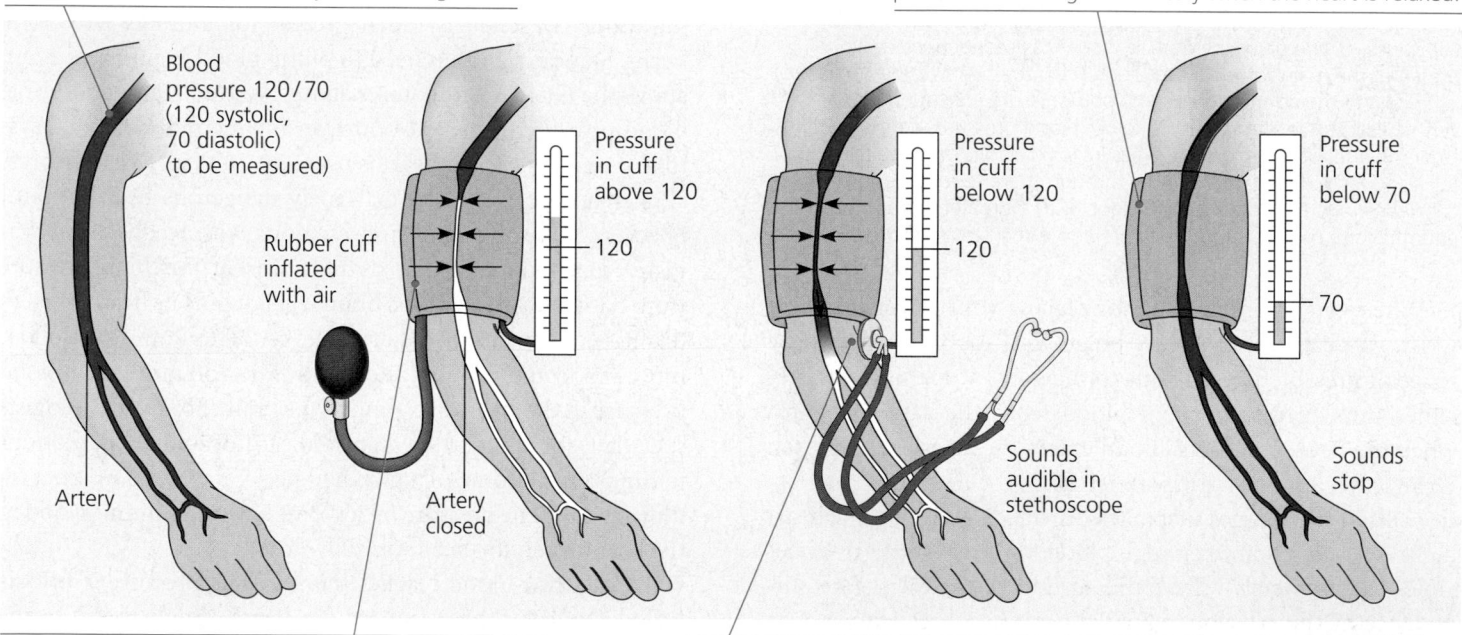

① A typical blood pressure reading for a 20-year-old is 120/70. The units for these numbers are mm of mercury (Hg); a blood pressure of 120 is a force that can support a column of mercury 120 mm high.

④ The cuff is loosened further until the blood flows freely through the artery and the sounds below the cuff disappear. The pressure at this point is the diastolic pressure remaining in the artery when the heart is relaxed.

② A sphygmomanometer, an inflatable cuff attached to a pressure gauge, measures blood pressure in an artery. The cuff is wrapped around the upper arm and inflated until the pressure closes the artery, so that no blood flows past the cuff. When this occurs, the pressure exerted by the cuff exceeds the pressure in the artery.

③ A stethoscope is used to listen for sounds of blood flow below the cuff. If the artery is closed, there is no pulse below the cuff. The cuff is gradually deflated until blood begins to flow into the forearm, and sounds from blood pulsing into the artery below the cuff can be heard with the stethoscope. This occurs when the blood pressure is greater than the pressure exerted by the cuff. The pressure at this point is the systolic pressure.

FIGURE 42.11 Measurement of blood pressure. Blood pressure is recorded as two numbers separated by a slash; the first number is the systolic pressure; the second is the diastolic pressure.

cells. Diffusion can also occur through the clefts between adjoining cells. However, transport through these clefts occurs mainly by bulk flow due to fluid pressure. Blood pressure within the capillary pushes fluid (water and small solutes such as sugars, salts, oxygen, and urea) through the capillary clefts. This causes a net loss of fluid from the upstream end of the capillary (near an arteriole). Blood cells suspended in blood and most proteins dissolved in the blood are too large to pass readily through the endothelium and remain in the capillaries, causing an increase in blood osmolarity (solute concentration) as blood loses fluid during its passage through a capillary. The resulting osmotic gradient pulls water into the capillary by osmosis near the downstream (venule) end (FIGURE 42.13). About 85% of the fluid that leaves the blood at the arterial end of a capillary bed reenters from the interstitial fluid at the venous end, and the remaining 15% is eventually returned to the blood by the vessels of the lymphatic system.

The lymphatic system returns fluid to the blood and aids in body defense

So much blood passes through the capillaries that the cumulative loss of fluid adds up to about 4 L per day. There is also some leakage of blood proteins, even though the capillary wall is not very permeable to large molecules. The lost fluid and proteins return to the blood via the **lymphatic system.** Fluid enters this system by diffusing into tiny lymph capillaries intermingled among capillaries of the cardiovascular system. Once inside the lymphatic system, the fluid is called **lymph;** its composition is about the same as that of interstitial fluid. The lymphatic system drains into the circulatory system near the junction of the venae cavae with the right atrium (see FIGURE 43.4).

Lymph vessels, like veins, have valves that prevent the backflow of fluid toward the capillaries. Rhythmic contractions of the vessel walls help draw fluid into lymphatic capillaries. Also like veins, lymph vessels depend mainly on the movement of skeletal muscles to squeeze fluid toward the heart.

Along a lymph vessel are organs called **lymph nodes.** By filtering the lymph and attacking viruses and bacteria, lymph nodes play an important role in the body's defense. Inside a lymph node is a honeycomb of connective tissue with spaces filled by white blood cells specialized for defense. When the body is fighting an infection,

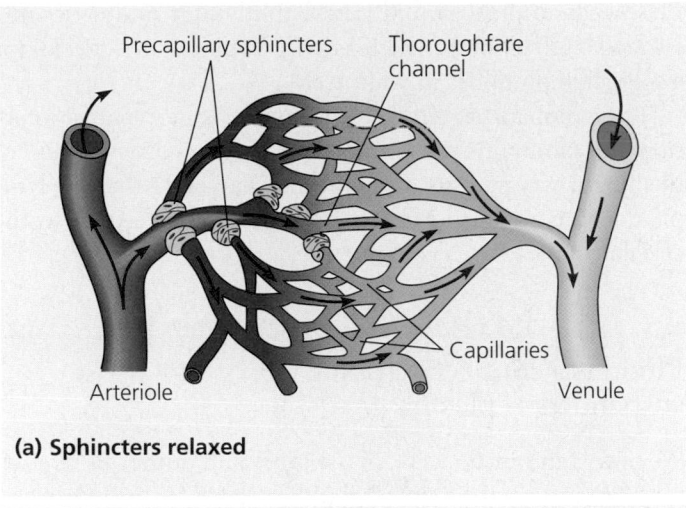

(a) Sphincters relaxed

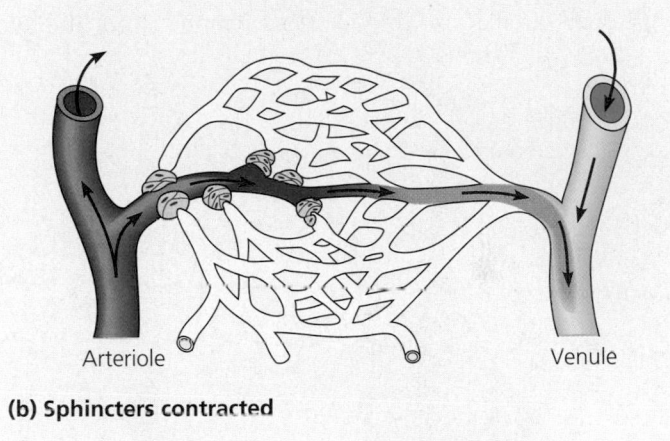

(b) Sphincters contracted

FIGURE 42.12 Blood flow in capillary beds. Precapillary sphincters regulate the passage of blood into capillary beds. Some blood flows directly from arterioles to venules through capillaries called thoroughfare channels, which are always open.

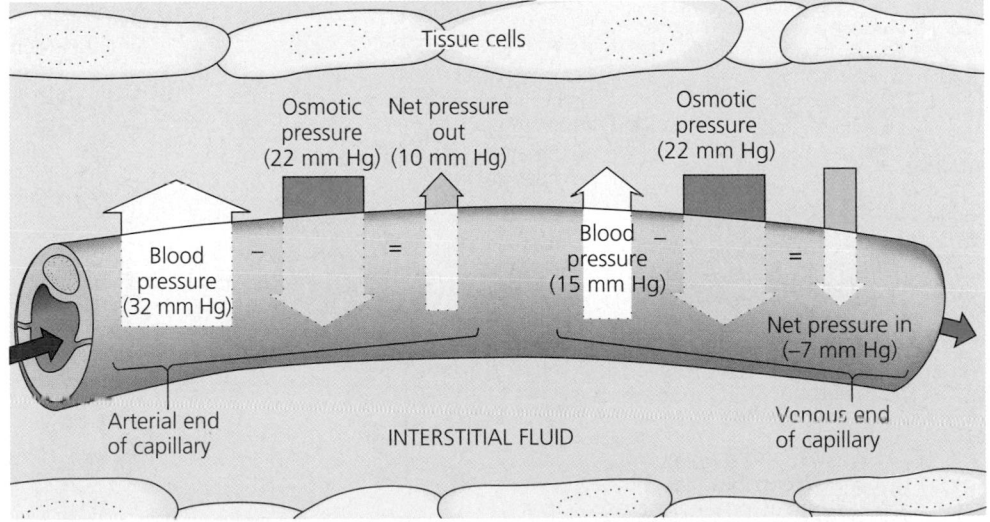

FIGURE 42.13 The movement of fluid between capillaries and the interstitial fluid. Fluid flows out of a capillary at the upstream end near an arteriole and reenters a capillary downstream near a venule. The direction of fluid movement across the capillary wall at any point depends on the difference between two opposing forces: blood pressure and osmotic pressure.

these cells multiply rapidly, and the lymph nodes become swollen and tender (which is why your doctor checks for swollen lymph nodes in your neck).

The lymphatic system helps defend the body against infection and maintains the volume and protein concentration of the blood. You may also recall from Chapter 41 that the lymphatic system transports fats from the digestive tract to the circulatory system.

Blood is a connective tissue with cells suspended in plasma

We now shift our focus from the tubes and pumps of circulatory systems to the fluids being circulated. In invertebrates with open circulations, blood (hemolymph) is not different from interstitial fluid. However, blood in the closed circulatory systems of vertebrates is a specialized connective tissue consisting of several kinds of cells suspended in a liquid matrix called **plasma.** If a blood sample is taken, the cells can be separated from plasma by spinning the whole blood in a centrifuge (an anticoagulant must be added to prevent the blood from clotting). The cellular elements (cells and cell fragments), which occupy about 45% of the volume of blood, settle to the bottom of the centrifuge tube, forming a dense red pellet. Above this cellular pellet is the transparent, straw-colored plasma (FIGURE 42.14).

Plasma

Blood plasma is about 90% water. Among its many solutes are inorganic salts in the form of dissolved ions, sometimes re-

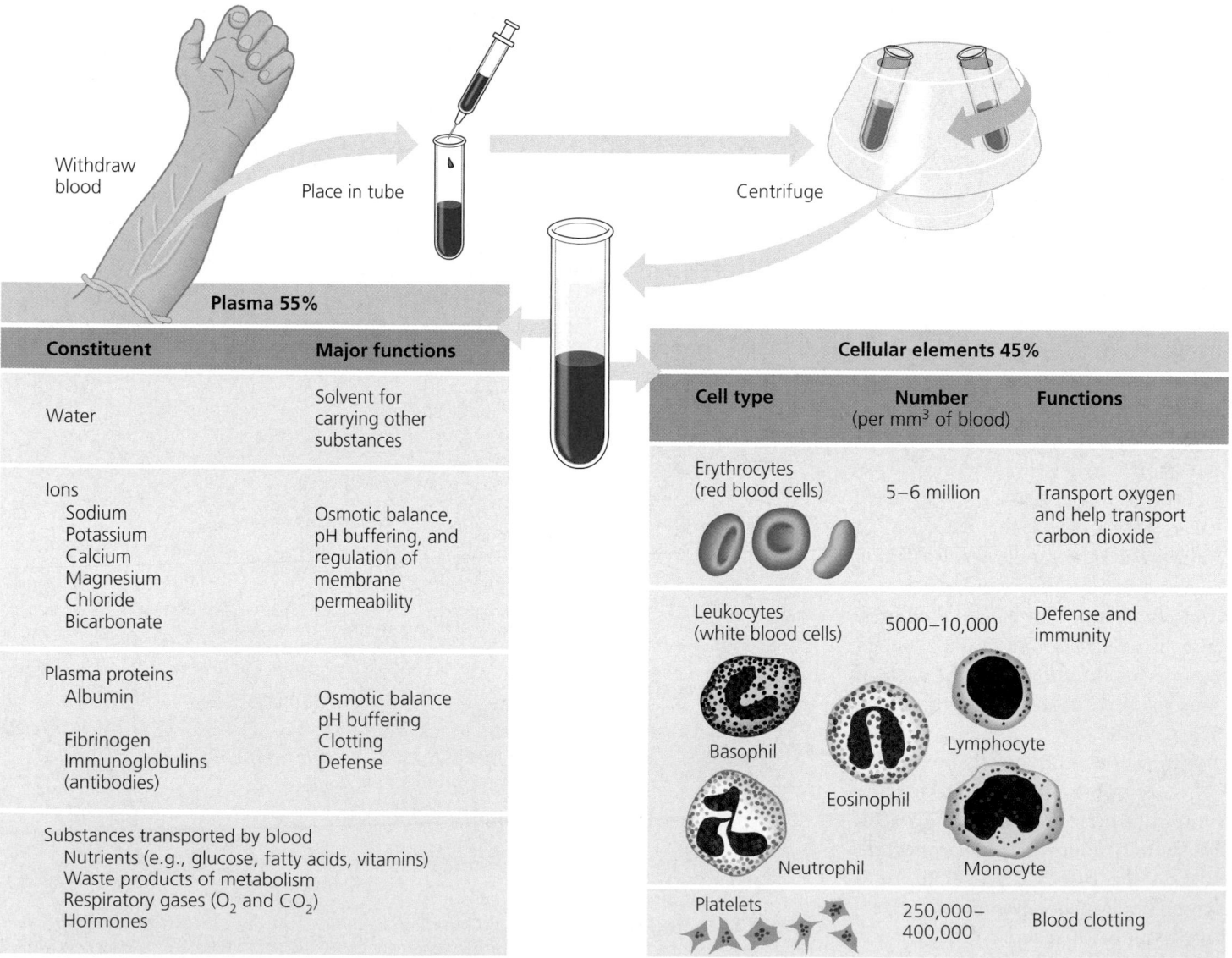

Plasma 55%	
Constituent	**Major functions**
Water	Solvent for carrying other substances

Ions
  Sodium
  Potassium
  Calcium
  Magnesium
  Chloride
  Bicarbonate
	Osmotic balance, pH buffering, and regulation of membrane permeability

Plasma proteins
  Albumin
 	
  Fibrinogen
  Immunoglobulins
  (antibodies)
	Osmotic balance pH buffering
	Clotting
	Defense

Substances transported by blood
  Nutrients (e.g., glucose, fatty acids, vitamins)
  Waste products of metabolism
  Respiratory gases (O_2 and CO_2)
  Hormones	

Cellular elements 45%		
Cell type	**Number** (per mm^3 of blood)	**Functions**
Erythrocytes (red blood cells)	5–6 million	Transport oxygen and help transport carbon dioxide
Leukocytes (white blood cells)	5000–10,000	Defense and immunity
Platelets	250,000–400,000	Blood clotting

Withdraw blood

Place in tube

Centrifuge

Basophil

Lymphocyte

Eosinophil

Neutrophil

Monocyte

FIGURE 42.14 The composition of mammalian blood.

ferred to as blood electrolytes. The combined concentration of these ions is important in maintaining osmotic balance of the blood. Some ions also help buffer the blood, which normally has a pH of 7.4 in humans. And the normal functioning of muscles and nerves depends on the concentration of key ions in the interstitial fluid, which reflects their concentration in plasma. The kidney maintains plasma electrolytes at precise concentrations, an example of homeostasis.

Another important class of solutes is the plasma proteins, which have many functions. Collectively, they act as buffers against pH changes, help maintain the osmotic balance between blood and interstitial fluid, and contribute to the blood's viscosity (thickness). The various types of plasma proteins also have specific functions. Some are escorts for lipids, which are insoluble in water and can travel in blood only when bound to proteins. Another class of proteins, the immunoglobulins or antibodies, help combat viruses and other foreign agents that invade the body (see Chapter 43). And the plasma proteins called fibrinogens are clotting factors that help plug leaks when blood vessels are injured. Blood plasma that has had these clotting factors removed is called serum.

Plasma also contains a wide variety of substances in transit from one part of the body to another, including nutrients, metabolic wastes, respiratory gases, and hormones. Blood plasma and interstitial fluid are similar in composition, except that plasma has a much higher protein concentration (capillary walls, remember, are not very permeable to proteins).

Cellular Elements

Suspended in blood plasma are two classes of cells: **red blood cells,** which transport oxygen, and **white blood cells,** which function in defense. A third cellular element, **platelets,** are pieces of cells that are involved in clotting.

Red blood cells, or **erythrocytes,** are by far the most numerous blood cells. Each cubic millimeter of human blood contains 5 to 6 million red cells, and there are about 25 trillion of these cells in the body's 5 L of blood.

Red blood cells are an excellent example of the close relationship between structure and function. Their main function of oxygen transport depends on rapid diffusion of oxygen across the red cells' plasma membranes. Human erythrocytes are small biconcave disks (about 7–8.5 μm in diameter), thinner in the center than at the edges. Their small size and biconcave shape create a large surface area for the total population of red cells, and the greater the total area of red cell membrane in a given volume of blood, the more rapidly oxygen diffusion can occur. Mammalian erythrocytes lack nuclei, an unusual characteristic that leaves more space in the tiny cells for **hemoglobin,** the iron-containing protein that transports oxygen (see FIGURE 5.23b). Red blood cells also lack mitochondria and generate their ATP exclusively by anaerobic metabolism. Oxygen transport by erythrocytes would be less efficient

if their own metabolism were aerobic and consumed some of the oxygen they carry.

Despite its small size, an erythrocyte contains about 250 million molecules of **hemoglobin.** Since each hemoglobin binds up to four molecules of O_2, an erythrocyte can transport about a billion oxygen molecules. Researchers have recently found that hemoglobin binds the gaseous molecule nitric oxide (NO) as well as O_2. As red blood cells pass through the capillary beds of lungs, gills, or other respiratory organs, oxygen diffuses into the erythrocytes and hemoglobin binds O_2 and NO. In the systemic capillaries, hemoglobin unloads oxygen, and it then diffuses into body cells. The NO relaxes the capillary walls, allowing them to expand, which probably helps deliver O_2 to the cells.

There are five major types of white blood cells, or **leukocytes:** monocytes, neutrophils, basophils, eosinophils, and lymphocytes (see FIGURE 42.14). Their collective function is to fight infections. For example, monocytes and neutrophils are phagocytes, which engulf and digest bacteria and debris from our own dead cells. As we will see in Chapter 43, lymphocytes develop into specialized B cells and T cells, which produce the immune response against foreign substances. White blood cells spend most of their time outside the circulatory system, patrolling through interstitial fluid and the lymphatic system, where most of the battles against pathogens are waged. Normally, a cubic millimeter of human blood has about 5,000 to 10,000 leukocytes, but their numbers increase temporarily whenever the body is fighting an infection.

The third cellular element of blood, **platelets,** are fragments of cells about 2 to 3 μm in diameter. They have no nuclei and originate as pinched-off cytoplasmic fragments of large cells in the bone marrow. Platelets then enter the blood and function in the important process of blood clotting.

Stem Cells and the Replacement of Cellular Elements

The cellular elements of blood (erythrocytes, leukocytes, and platelets) wear out and are replaced constantly throughout a person's life. Erythrocytes, for example, usually circulate for only about 3 to 4 months and then are destroyed by phagocytic cells in the liver and spleen. Enzymes digest the old cell's macromolecules, and biosynthetic processes construct new macromolecules using many of the monomers, such as amino acids, obtained from old blood cells, as well as new materials and energy from food. Many of the iron atoms derived from the hemoglobin in old red blood cells are built into new hemoglobin molecules.

Erythrocytes, leukocytes, and platelets all develop from a common source, a single population of cells called **pluripotent stem cells** in the red marrow of bones, particularly the ribs, vertebrae, breastbone, and pelvis (FIGURE 42.15, p. 884). "Pluripotent" means these cells have the potential to differentiate into any type of blood cell or into cells that produce

and large total cross-sectional area of the arterioles and capillaries. This slower flow enhances the exchange of substances between the blood and interstitial fluid. Blood pressure, the hydrostatic force blood exerts against the wall of a vessel, is determined by the cardiac output and peripheral resistance due to variable constriction of the arterioles.

Biology Labs On-Line: *CardioLab*

■ **Transfer of substances between the blood and the interstitial fluid occurs across the thin walls of capillaries (pp. 880–881, FIGURES 42.12, 42.13)** The supply of blood to different organs is determined by variable constriction of arterioles and precapillary sphincters. Substances traverse the endothelium of capillaries in endocytotic/exocytotic vesicles, by diffusion, or dissolved in fluids forced out by blood pressure at the arterial end of the capillary.

Web/CD Activity 42C: *Mammalian Cardiovascular System Function*

■ **The lymphatic system returns fluid to the blood and aids in body defense (pp. 881–882)** Fluid reenters the circulation directly at the venous end of the capillary and indirectly through the lymphatic system. White blood cells concentrated in lymph nodes help fight infections.

■ **Blood is a connective tissue with cells suspended in plasma (pp. 882–885, FIGURES 42.14–42.16)** Whole blood consists of cellular elements (cells and pieces of cells called platelets) suspended in a liquid matrix called plasma. Plasma is a complex aqueous solution of inorganic electrolytes, proteins, nutrients, metabolic waste products, respiratory gases, and hormones. Plasma proteins influence blood pH, osmotic pressure, and viscosity, and function in lipid transport, immunity (antibodies), and blood clotting (fibrinogens). Red blood cells, or erythrocytes, transport oxygen. Five types of white blood cells, or leukocytes, function in defense by phagocytosing bacteria and debris or by producing antibodies. Pluripotent stem cells in red bone marrow give rise to all types of blood cells. Platelets function in blood clotting, a cascade of complex reactions that converts plasma fibrinogen to fibrin.

■ **Cardiovascular diseases are the leading cause of death in the United States and most other developed nations (pp. 884–886, FIGURE 42.17)** Cardiovascular disease is a deterioration of the heart and blood vessels. Gradual plaque buildup during atherosclerosis or arteriosclerosis narrows the diameter of blood vessels and may be associated with vessel blockage and consequent heart attack or stroke.

Web/CD Case Study in the Process of Science: *How Is Cardiovascular Fitness Measured?*

GAS EXCHANGE IN ANIMALS

■ **Gas exchange supplies oxygen for cellular respiration and disposes of carbon dioxide: an overview (pp. 886–887, FIGURE 42.18)** Animals require large, moist respiratory surfaces for the adequate diffusion of respiratory gases (O_2 and CO_2) between their cells and the respiratory medium, either air or water.

■ **Gills are respiratory adaptations of most aquatic animals (pp. 887–888, FIGURES 42.19–42.21)** Gills are outfoldings of the body surface specialized for gas exchange. The effectiveness of gas exchange in some gills, including those of fishes, is increased by ventilation and countercurrent flow of blood and water.

■ **Tracheal systems and lungs are respiratory adaptations of terrestrial animals (pp. 889–892, FIGURES 42.22–42.25)** The tracheae of insects are tiny branching tubes that penetrate the body, bringing O_2 directly to cells. Most terrestrial vertebrates, land snails, and spiders have internal lungs. In mammals, air inhaled through the nostrils passes through the pharynx into the trachea, bronchi, bronchioles, and dead-end alveoli, where gas exchange occurs. Lungs must be ventilated by breathing. Frogs ventilate by positive pressure, pumping air into their lungs. Mammals ventilate with negative pressure by contracting and relaxing rib muscles and the diaphragm, which changes the volume and hence the pressure of the chest cavity and lungs relative to the atmosphere. Birds have one-way ventilation of the lungs, made possible by a system of air sacs and tubular parabronchi in the lungs.

Web/CD Activity 42D: *The Human Respiratory System*

■ **Control centers in the brain regulate the rate and depth of breathing (pp. 892–893, FIGURE 42.26)** A control center in the medulla oblongata of the brain sets the basic breathing rhythm. Sensors detect blood pH (reflecting CO_2 concentration) and O_2 levels in the blood, and the medulla adjusts the rate and depth of breathing to match the metabolic demands of the body.

■ **Gases diffuse down pressure gradients in the lungs and other organs (pp. 893–894, FIGURE 42.27)** O_2 and CO_2 diffuse from where their partial pressures are higher to where they are lower.

■ **Respiratory pigments transport gases and help buffer the blood (pp. 894–896, FIGURES 42.28, 42.29)** Respiratory pigments greatly increase the amount of oxygen that blood can carry. Arthropods and many mollusks have copper-containing hemocyanin; vertebrates have hemoglobin. A hemoglobin molecule has four iron-containing subunits, each capable of binding a molecule of oxygen. Most CO_2 generated during metabolism is transported in the form of bicarbonate ions.

Web/CD Activity 42E: *Transport of Respiratory Gases*

Biology Labs On-Line: *HemoglobinLab*

■ **Deep-diving air-breathers stockpile oxygen and deplete it slowly (pp. 896–897, FIGURE 42.30)** Diving mammals have an unusually large volume of blood and store additional oxygen in their muscles.

Self-Quiz

1. Which of the following respiratory systems is not closely associated with a blood supply?
   a. vertebrate lungs
   b. fish gills
   c. tracheal systems of insects
   d. the outer skin of an earthworm
   e. the parapodia of a polychaete worm

2. Blood returning to the mammalian heart in a pulmonary vein will drain first into the
   a. vena cava.
   b. left atrium.
   c. right atrium.
   d. left ventricle.
   e. right ventricle.

3. Pulse is a direct measure of
   a. blood pressure.
   b. stroke volume.
   c. cardiac output.
   d. heart rate.
   e. breathing rate.

4. When you hold your breath, which of the following blood gas changes first leads to the urge to breathe?
   a. rising O_2
   b. falling O_2
   c. rising CO_2
   d. falling CO_2
   e. rising CO_2 and falling O_2

5. In negative-pressure breathing, inhalation results from
   a. forcing air from the throat down into the lungs.
   b. contracting the diaphragm.
   c. relaxing the muscles of the rib cage.
   d. using muscles of the lungs to expand the alveoli.
   e. contracting the abdominal muscles.

6. The conversion of fibrinogen to fibrin
   a. occurs when fibrinogen is released from broken platelets.
   b. occurs within red blood cells.
   c. is linked to hypertension and may damage artery walls.
   d. is likely to occur too often in an individual with hemophilia.
   e. is the final step of a clotting process that involves multiple clotting factors.

7. A decrease in the pH of human blood caused by exercise would
   a. decrease breathing rate.
   b. increase heart rate.
   c. decrease the amount of O_2 unloaded from hemoglobin.
   d. decrease cardiac output.
   e. decrease CO_2 binding to hemoglobin.

8. Compared to the interstitial fluid that bathes active muscle cells, blood reaching these cells in arteries has a
   a. higher P_{O_2}.
   b. higher P_{CO_2}.
   c. greater bicarbonate concentration.
   d. lower pH.
   e. lower osmotic pressure.

9. Which of the following reactions prevails in red blood cells traveling through pulmonary capillaries? (Hb = hemoglobin)
   a. $Hb + 4\,O_2 \longrightarrow Hb(O_2)_4$
   b. $Hb(O_2)_4 \longrightarrow Hb + 4\,O_2$
   c. $CO_2 + H_2O \longrightarrow H_2CO_3$
   d. $H_2CO_3 \longrightarrow H^+ + HCO_3^-$
   e. $Hb + 4\,CO_2 \longrightarrow Hb(CO_2)_4$

10. The relationship between blood pressure (bp), cardiac output (co), and peripheral resistance (pr) can be expressed as $bp = co \times pr$. All of the following changes would result in an increase in blood pressure *except*
   a. increase in the stroke volume.
   b. increase in the heart rate.
   c. increase in the duration of ventricular diastole.
   d. contraction of the arteriolar smooth muscle.
   e. reduction in arteriolar diameter.

11. Some babies are born with a small hole in the septum between the left and right ventricles. Explain how this would affect the oxygen content of the blood pumped out of the heart into the systemic circuit.

12. A slight decrease in blood pH causes the pacemaker to speed up. What is the function of this control mechanism?

13. Explain why going for a vigorous swim right after eating a heavy meal is probably more likely to cause indigestion than cramping of the muscles.

14. Explain how edema, the accumulation of fluid in body tissues, can result from a decrease in blood plasma protein concentration due to severe protein deficiency in the diet.

15. About how many red blood cells does the bone marrow of a human produce per day, assuming a total red cell count of 25 trillion (2.5×10^{13}) and an average longevity of 4 months for the cells?

16. How is it advantageous to have a stock of fibrinogen in the blood compared to what would happen if the body had to synthesize fibrin protein when there is an injury?

17. In contrasting gills to lungs as respiratory organs, what is the main difference in their spatial relationships to the rest of the animals' bodies?

18. Explain how hyperventilation subverts the control of breathing.

Go to the website or CD-ROM for more quiz questions.

Evolution Connection

One of the many mutant opponents that the movie monster Godzilla contends with is Mothra, a giant moth-like creature with a wingspan of some dozens of feet. Science fiction creatures like these can be critiqued on the grounds of biomechanical and physiological principles. Focusing on respiration and the principles of gas exchange you learned about in this chapter, what physiological problems does Mothra face? The largest insects that have ever lived are Paleozoic dragonflies with half-meter wingspans. Why do you think truly giant insects are improbable?

The Process of Science

The hemoglobin of a human fetus differs from adult hemoglobin. Compare the dissociation curves of the two hemoglobins in the graph below. Propose a hypothesis for the *function* of this difference between these two versions of hemoglobin.

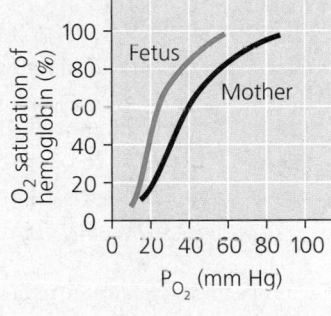

Learn to measure cardiovascular fitness in the Case Study in the Process of Science, available on the CD-ROM and website. Explore CardioLab and HemoglobinLab at Biology Labs On-Line.

Science, Technology, and Society

Hundreds of studies have linked smoking with cardiovascular and lung disease. According to most health authorities, smoking is the leading cause of preventable, premature death in the United States. Antismoking and health groups have proposed that cigarette advertising in *all* media be banned *entirely*. What are some arguments in favor of a total ban on cigarette advertising? What are arguments in opposition? Do you favor or oppose such a ban? Defend your position.

Answers: 1. c; 2. b; 3. d; 4. c; 5. b; 6. e; 7. b; 8. a; 9. a; 10. c; 11. This reduces the oxygen content by mixing O_2-depleted blood returned to the right ventricle from the systemic circuit with the O_2-rich blood of the left ventricle. 12. Rising CO_2 concentration lowers pH by the formation of carbonic acid. The increased heart rate in response to lower pH enhances delivery of the CO_2-rich blood to the lungs for removal. 13. The vigorous exercise routes more blood to the capillary beds of muscles, resulting in less blood flow to the capillaries of the digestive tract. 14. Decreased protein concentration reduces capillary uptake of fluid from interstitial fluid by reducing the osmotic gradient across the capillary wall. 15. About 200 billion, or 2.08×10^{11}, calculated by dividing the total number of cells, 2.5×10^{13}, by 120 days. 16. Fibrinogen can be activated to fibrin much faster than fibrin can be synthesized. 17. The extensive respiratory surface of gills extends outward from the body into the surrounding environment (water); in contrast, lungs extend from the body surface into the interior of the animal. 18. A slight drop in pH, caused by rising carbonic acid (and hence CO_2) stimulates inhalation by acting on the brain's breathing center. By purging the blood of CO_2, hyperventilation temporarily suspends breathing.

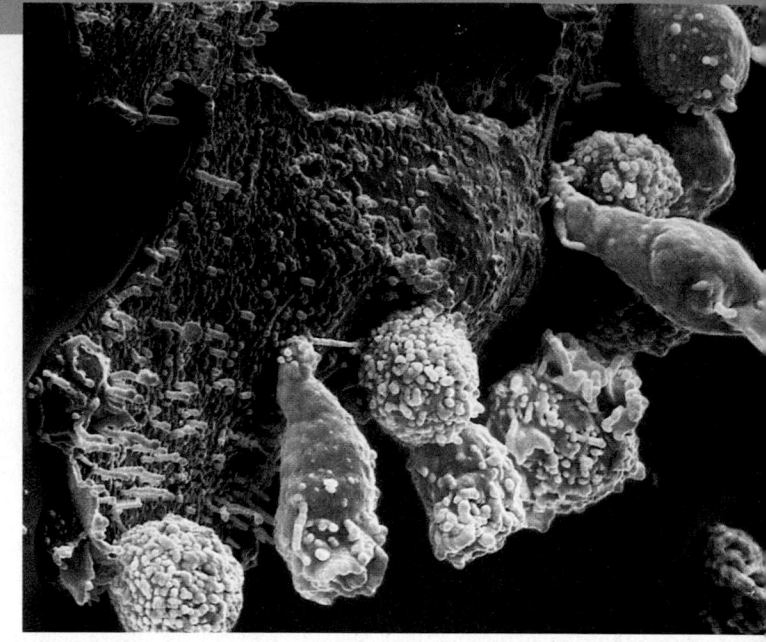

CHAPTER 43

THE BODY'S DEFENSES

NONSPECIFIC DEFENSES AGAINST INFECTION

■ The skin and mucous membranes provide first-line barriers to infection
■ Phagocytic cells, inflammation, and antimicrobial proteins function early in infection

HOW SPECIFIC IMMUNITY ARISES

■ Lymphocytes provide the specificity and diversity of the immune system
■ Antigens interact with specific lymphocytes, inducing immune responses and immunological memory
■ Lymphocyte development gives rise to an immune system that distinguishes self from nonself

IMMUNE RESPONSES

■ Helper T lymphocytes function in both humoral and cell-mediated immunity: *an overview*
■ In the cell-mediated response, cytotoxic T cells counter intracellular pathogens: *a closer look*
■ In the humoral response, B cells make antibodies against extracellular pathogens: *a closer look*
■ Invertebrates have a rudimentary immune system

IMMUNITY IN HEALTH AND DISEASE

■ Immunity can be achieved naturally or artificially
■ The immune system's capacity to distinguish self from nonself limits blood transfusion and tissue transplantation
■ Abnormal immune function can lead to disease
■ AIDS is an immunodeficiency disease caused by a virus

An animal must defend itself *against unwelcome intruders, the many potentially dangerous viruses, bacteria, and other pathogens it encounters in the air, in food, and in water. It must also deal with abnormal body cells, which, in some cases, may develop into cancer. Three cooperative lines of defense that counter these threats have evolved (*FIGURE 43.1*). Two of these are nonspecific—that is, they do not distinguish one infectious agent from another. The first line of nonspecific defense is external, consisting of epithelial tissues that cover and line our bodies (skin and mucous membranes) and the secretions they produce. The second line of nonspecific defense is internal: It is triggered by chemical signals and involves phagocytic cells and antimicrobial proteins that indiscriminately attack invaders that penetrate the body's outer barriers. The appearance of inflammation is a sign that this second line of defense has been deployed.*

*The third line of defense is the immune system. The immune system comes into play simultaneously with the second line of defense, but it responds in a specific way to particular microorganisms, aberrant body cells, toxins, and other substances marked by foreign molecules. The body's three lines of defense are somewhat analogous to the defenses of a besieged city: first the city walls, then ordinary soldiers, and finally intelligence officers who identify and track down specific dangerous infiltrators. The immune response, which includes the production of specific defensive proteins called antibodies, involves a diverse group of white blood cells called lymphocytes (see *FIGURES 42.14* and 42.15). The photograph on this page (a colorized SEM) shows specialized lymphocytes (light green) attacking a cancer cell (brown). This chapter examines how an animal's nonspecific and specific defenses work together to protect the body. Our main focus is on the defense mechanisms of vertebrates, which have a highly developed immune system.*

NONSPECIFIC DEFENSES AGAINST INFECTION

An invading microbe must penetrate the external barrier formed by the skin and mucous membranes, which cover the surface and line the openings of an animal's body. If it succeeds in doing so, the pathogen encounters the second line of nonspecific defense, interacting mechanisms that include phagocytosis, the inflammatory response, and antimicrobial proteins.

The skin and mucous membranes provide first-line barriers to infection

Intact skin is a barrier that cannot normally be penetrated by bacteria or viruses, although even minute abrasions may allow their passage. Likewise, the mucous membranes that line the digestive, respiratory, and genitourinary tracts bar the entry of potentially harmful microbes (see FIGURE 40.1). Beyond their role as a physical barrier, the skin and mucous membranes counter pathogens with chemical defenses. In humans, for example, secretions from sebaceous and sweat glands give the skin a pH ranging from 3 to 5, which is acidic enough to prevent colonization by many microbes. (Bacteria that make up the skin's normal flora are adapted to its acidic, relatively dry environment.) Microbial colonization is also inhibited by the washing action of saliva, tears, and mucous secretions that continually bathe the surfaces of exposed epithelia. In addition, all these secretions contain antimicrobial proteins. One of these protective proteins is the enzyme **lysozyme** (see FIGURE 5.17), which digests the cell walls of many bacteria and thus destroys many bacteria entering the upper respiratory tract and the openings around the eyes.

Mucus, the viscous fluid secreted by cells of mucous membranes, also traps microbes and other particles that contact it. In the trachea, ciliated epithelial cells sweep out mucus with its trapped microbes, preventing them from entering the lungs (FIGURE 43.2). Microbes present in food or water, or those in swallowed mucus, must contend with the highly acidic environment of the stomach. The acid destroys many microbes before they can enter the intestinal tract. There are exceptions,

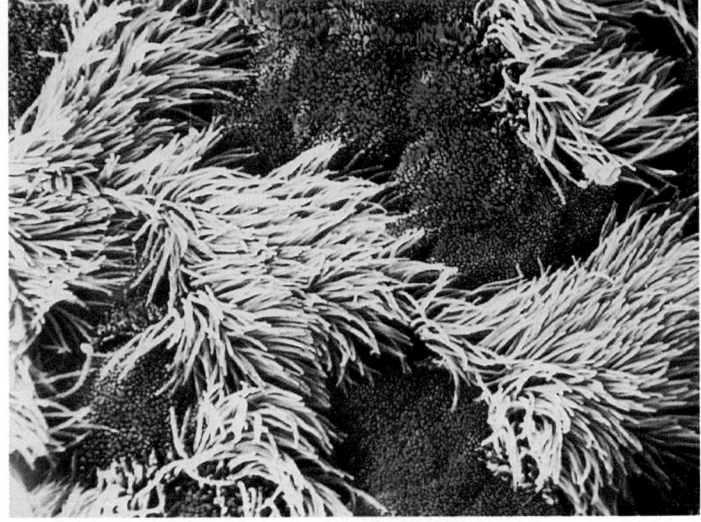

FIGURE 43.2 First-line respiratory defenses. In the lining of the trachea, shown here, specialized cells (orange) produce mucus that traps microbes before the invaders can enter the lungs. The mucous membrane is also equipped with ciliated cells (yellow). Synchronized beating of the cilia expels mucus and the trapped microbes upward into the pharynx (colorized SEM).

however: Hepatitis A virus, for example, is one of many pathogens that survives gastric acidity and gains access to the body via the digestive tract.

Phagocytic cells, inflammation, and antimicrobial proteins function early in infection

Microbes that penetrate the first line of defense, such as those that enter through a break in the skin, face the second line of defense. The body's internal mechanisms of nonspecific defense depend mainly on **phagocytosis,** the ingestion of invading organisms by certain types of white cells (see FIGURES 8.19a and 42.15). As you will see, phagocyte function is intimately associated with an effective inflammatory response and also with certain antimicrobial proteins. These nonspecific mechanisms help limit the spread of microbes in advance of specific immune responses.

Phagocytic and Natural Killer Cells

The phagocytic cells called **neutrophils** constitute about 60%–70% of all white blood cells (leukocytes). Cells damaged by invading microbes release chemical signals that attract neutrophils from the blood. The neutrophils enter the infected tissue, engulfing and destroying microbes there. (This migration toward the source of a chemical attractant is called *chemotaxis.*) However, neutrophils tend

Nonspecific defense mechanisms		Specific defense mechanisms (immune system)
First line of defense	Second line of defense	Third line of defense
• Skin • Mucous membranes • Secretions of skin and mucous membranes	• Phagocytic white blood cells • Antimicrobial proteins • The inflammatory response	• Lymphocytes • Antibodies

FIGURE 43.1 An overview of the body's defenses.

to self-destruct as they destroy foreign invaders, and their average life span is only a few days.

Monocytes, although they constitute only about 5% of leukocytes, provide an even more effective phagocytic defense. New monocytes circulate in the blood for only a few hours, then migrate into tissues, developing into large **macrophages** ("big eaters"). Macrophages, the largest phagocytic cells, are especially effective, long-lived phagocytes. These cells extend long pseudopodia that can attach to polysaccharides on a microbe's surface (FIGURE 43.3). A macrophage engulfs a microbe in a vacuole that fuses with a lysosome (like the "food vacuole" in FIGURE 7.14). The lysosome has two ways of killing the trapped microbe. First, it can generate toxic forms of oxygen. Two of these toxic molecules, superoxide anion and nitric oxide, are now thought to be the main antimicrobial agents within phagocytes. Second, lysosomal enzymes, including lysozyme, digest microbial components. Interestingly, some microbes have evolved mechanisms for evading phagocytic destruction. Some bacteria have outer capsules to which a macrophage cannot attach.

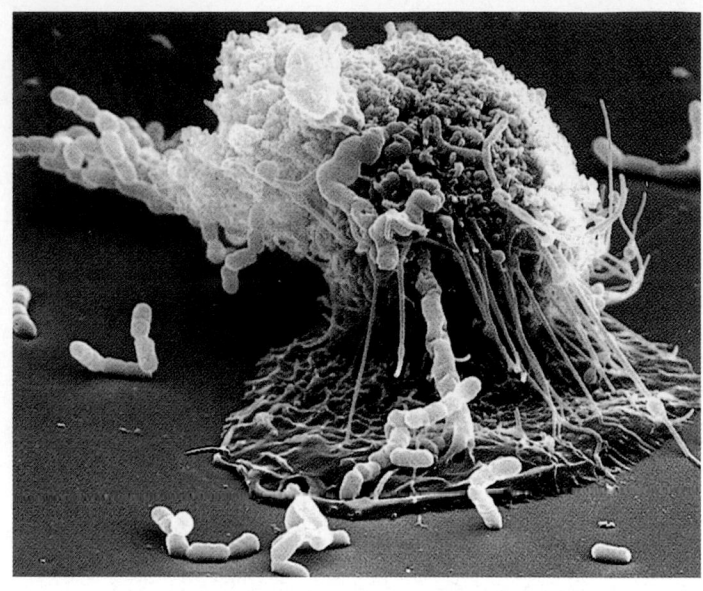

FIGURE 43.3 Phagocytosis by a macrophage. This micrograph shows fibril-like pseudopodia of a macrophage attaching to rod-shaped bacteria, which will be ingested and destroyed (colorized SEM).

FIGURE 43.4 The human lymphatic system.

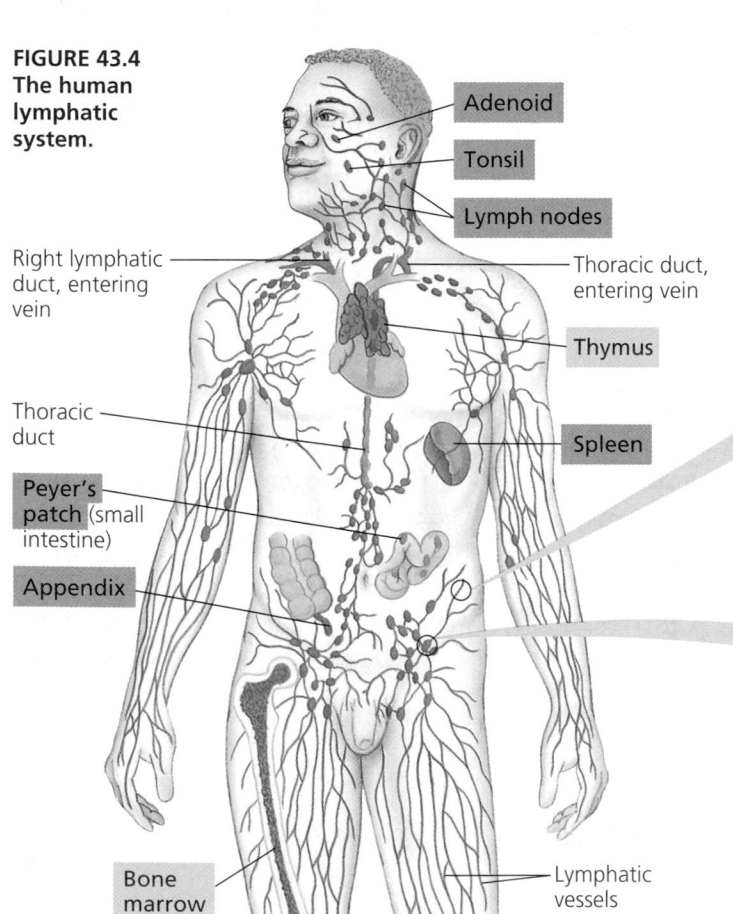

Adenoid

Tonsil

Lymph nodes

Right lymphatic duct, entering vein

Thoracic duct, entering vein

Thymus

Thoracic duct

Spleen

Peyer's patch (small intestine)

Appendix

Bone marrow

Lymphatic vessels

(a) The lymphatic vessels return fluid and leukocytes from the interstitial spaces throughout the body to the blood. The other organs of the lymphatic system are of two types: (1) structures that trap "foreign" molecules and particles, including the spleen, lymph nodes, and other structures labeled here in pink; and (2) structures in which leukocytes develop, labeled in green.

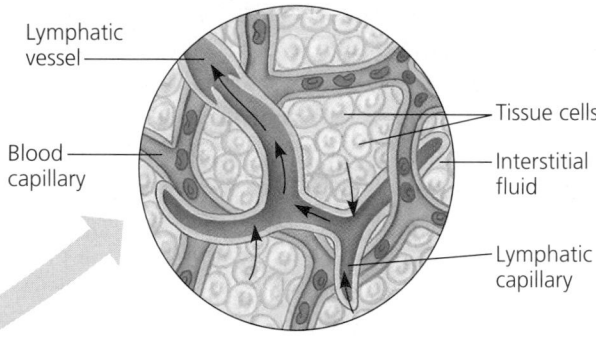

Lymphatic vessel

Blood capillary

Tissue cells

Interstitial fluid

Lymphatic capillary

(b) The interstitial fluid that bathes tissues, along with leukocytes in it, is continually taken up by lymphatic capillaries. The fluid, now called lymph, flows through the system of vessels, eventually returning to the blood circulatory system near the shoulders, where the right lymphatic duct and the thoracic duct drain into veins.

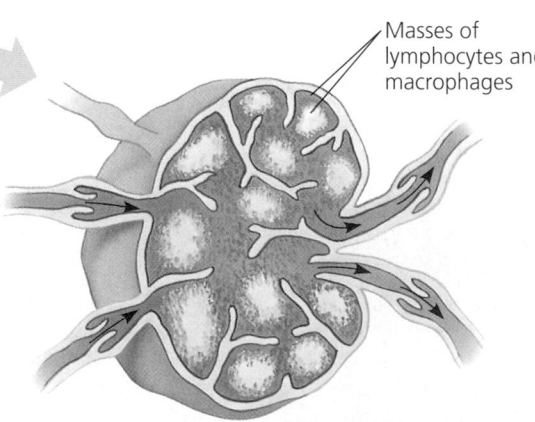

Masses of lymphocytes and macrophages

(c) Along the way, lymph must pass through numerous lymph nodes, where any pathogens present in lymph encounter macrophages and lymphocytes, another class of white blood cells with defensive functions.

Others, like *Mycobacterium tuberculosis,* are readily engulfed but are resistant to lysosomal destruction and can even reproduce inside a macrophage. These microorganisms are a particular problem for both nonspecific and specific defenses of the body.

Some macrophages migrate throughout the body, while others reside permanently in certain tissues: in lung (alveolar macrophages), liver (Kupffer's cells), kidney (mesangial cells), brain (microglial cells), connective tissues (histiocytes), and especially in lymph nodes and the spleen, key organs of the lymphatic system (FIGURE 43.4). The fixed macrophages in the spleen, lymph nodes, and other lymphatic tissues are particularly well located to contact infectious agents. Microorganisms, microbial fragments, and foreign molecules that enter the blood encounter macrophages as they become trapped in the netlike architecture of the spleen, while those in tissue fluid flow into lymph and are filtered through the lymph nodes.

About 1.5% of all leukocytes are **eosinophils.** Their main contribution to defense is against larger parasitic invaders, such as the blood fluke *Schistosoma mansoni* (see FIGURE 33.10). Eosinophils position themselves against the external wall of a parasite and discharge destructive enzymes from cytoplasmic granules. These cells have only limited phagocytic activity.

Nonspecific defense also includes **natural killer (NK) cells.** NK cells do not attack microorganisms directly; instead, they destroy virus-infected body cells (as well as abnormal body cells that could become cancerous). NK cells are not phagocytic; rather, they mount an attack on the cell's membrane, causing the cell to lyse (burst open).

The Inflammatory Response

Damage to tissue by a physical injury (such as a cut) or by the entry of microorganisms triggers a localized **inflammatory response** (FIGURE 43.5). In the injured area, precapillary arterioles dilate and postcapillary venules constrict, increasing the local blood supply (see FIGURE 42.7). These events are responsible for the characteristic redness and heat of inflammation (L. *inflammo,* "to set on fire"). The blood-engorged capillaries leak fluid into neighboring tissues, causing the edema (swelling) also associated with inflammation.

The inflammatory response is initiated by chemical signals. Some of these signals arise from the invading organism itself. Others, such as **histamine,** are released by cells of the body in response to tissue injury. Histamine is produced by circulating leukocytes called **basophils** and by **mast cells** found in connective tissue. When injured, these cells release histamine, triggering both dilation and increased permeability of nearby capillaries. Leukocytes and damaged tissue cells also discharge *prostaglandins* (see discussion of local regulators in Chapter 45) and other substances that further promote blood flow to the site of injury. Enhanced blood flow and vessel permeability aid in delivering clotting elements to the injured area. Blood clotting marks the beginning of the repair process and helps block the spread of microbes to other parts of the body (see FIGURE 42.16).

Increased local blood flow and capillary permeability also enhance the migration of phagocytic cells from the blood into the injured tissues. Probably the most important element of inflammation—indeed, of nonspecific defense—is phagocytosis. Phagocyte migration to the damage site usually begins within an hour after injury. Chemotactic factors released from invading bacteria and injured tissues attract phagocytes. In addition, molecules called **chemokines,** secreted by blood vessel endothelial cells and monocytes, also attract phagocytes to the area. Chemokines constitute a group of about 50 different proteins that bind to receptors on many types of leukocytes and induce numerous other changes central to inflammation. For example,

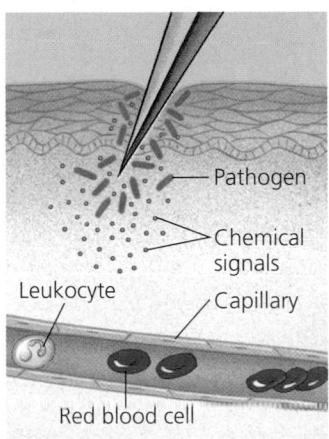

❶ Cells of a tissue injured by physical damage or bacteria release chemical signals such as histamine and prostaglandins.

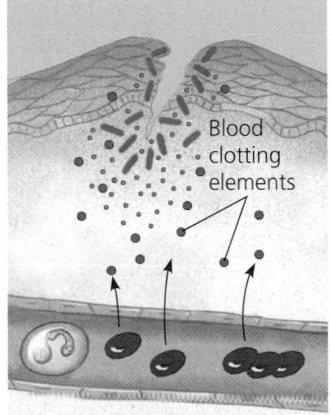

❷ In response to the signals, nearby capillaries dilate and become more permeable. Fluid and clotting elements move from the blood to the site, and clotting begins.

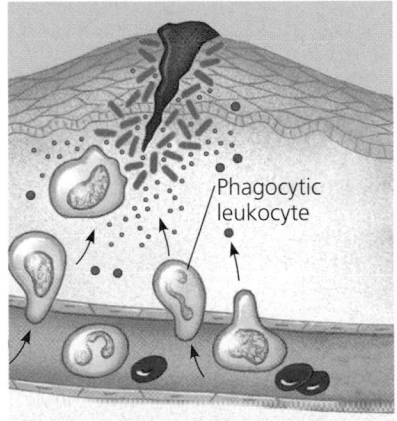

❸ Chemokines and other chemotactic factors released by various kinds of cells attract phagocytic cells from the blood.

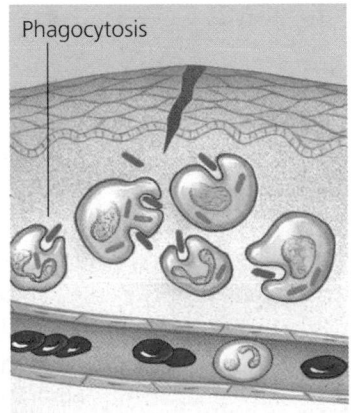

❹ When the phagocytic cells arrive at the site of injury, they consume pathogens and cell debris, and the tissue heals.

FIGURE 43.5 A simplified view of the inflammatory response.

they induce the production of toxic forms of oxygen in phagocyte lysosomes and the release of histamine from basophils.

Neutrophils are the first phagocytes to arrive at the point of assault, followed by macrophages that have developed from migrating monocytes. Macrophages not only phagocytose pathogens and their products, but also clean up damaged tissue cells and the remains of neutrophils destroyed in the phagocytic process. The pus that accumulates at the site of some infections consists mostly of dead phagocytic cells and the fluid and proteins that leaked from the capillaries during the inflammatory response. Usually, the pus is absorbed by the body within a few days.

A sliver or other minor injury causes localized inflammation, as you have just seen, but the body may also mount a systemic (widespread) nonspecific response to severe tissue damage or infection. Injured cells often put out a call for reinforcements, emitting chemicals that stimulate the release of more neutrophils from the bone marrow. In a severe infection such as meningitis or appendicitis, the number of leukocytes in the blood may increase severalfold within a few hours of the initial inflammatory events. Another systemic response to infection is fever. Toxins produced by pathogens may trigger the fever, but certain leukocytes also release molecules called **pyrogens,** which set the body's thermostat at a higher temperature. A very high fever may be dangerous, but a moderate fever contributes to defense by inhibiting the growth of some microorganisms. Fever may also facilitate phagocytosis and, by speeding up body reactions, may speed the repair of tissues.

Certain bacterial infections can induce an overwhelming systemic inflammatory response leading to a condition known as *septic shock.* Characterized by high fever and low blood pressure, septic shock is the most common cause of death in U.S. critical care units. Clearly, while local inflammation is an essential step toward healing, widespread inflammation can be devastating.

Antimicrobial Proteins

A variety of proteins function in nonspecific defense either by attacking microbes directly or by impeding their reproduction. You have already learned about lysozyme, an antimicrobial enzyme present in tears, saliva, and mucous secretions. Other antimicrobial agents include a set of about 20 serum proteins, known collectively as the **complement system,** that carry out a cascade of steps leading to the lysis of microbes. Some complement components also function along with chemokines in attracting phagocytic cells to sites of infection. Complement proteins are an essential part of both nonspecific and specific defense. We will learn about them in detail later in this chapter.

Another set of proteins that provide nonspecific defense are the **interferons,** which are secreted by virus-infected cells. While they do not seem to benefit the infected cell, these antiviral proteins diffuse to neighboring cells and induce them to produce other chemicals that inhibit viral reproduction. In this

way, interferons limit the cell-to-cell spread of viruses in the body, helping to control viral infections such as colds and influenza. The defense is not virus-specific; interferons produced in response to one virus may confer short-term resistance to unrelated viruses. In addition to its role as an antiviral agent, one type of interferon activates phagocytes, enhancing their ability to ingest and kill microorganisms. Interferons can now be mass-produced by recombinant DNA technology and are being tested clinically for the treatment of viral infections and cancer.

Let's review the body's nonspecific forms of defense: The first line of defense, the skin and mucous membranes, prevents most microbes from entering the body; the second line of defense uses phagocytes, natural killer cells, inflammation, and antimicrobial proteins to defend against microbes that have managed to enter the body. These two lines of defense are nonspecific in that they do not distinguish among pathogens.

HOW SPECIFIC IMMUNITY ARISES

While microorganisms are under assault by phagocytic cells, the inflammatory response, and antimicrobial proteins, they inevitably encounter lymphocytes, the key cells of the immune system—the body's third line of defense. Lymphocytes respond to such contacts by generating efficient and selective immune responses that work throughout the body to eliminate the particular invaders. Keep in mind that the cells of the immune system respond similarly to transplanted cells and even cancer cells, which they detect as foreign.

Lymphocytes provide the specificity and diversity of the immune system

The vertebrate body is populated by two main types of lymphocytes: **B lymphocytes (B cells)** and **T lymphocytes (T cells).** Like macrophages, both types of lymphocytes circulate throughout the blood and lymph and are concentrated in the spleen, lymph nodes, and other lymphatic tissues (see FIGURE 43.4). Because lymphocytes recognize and respond to particular microbes and foreign molecules, they are said to display *specificity.* A foreign molecule that elicits a specific response by lymphocytes is called an **antigen.** Antigens include molecules belonging to viruses, bacteria, fungi, protozoa, and parasitic worms. Antigenic molecules are also found on the surfaces of foreign materials such as pollen and transplanted tissue. B cells and T cells specialize in different types of antigens, and they carry out different, but complementary, defensive actions, as we will see later. One way that an antigen elicits an immune response is by activating B cells to secrete proteins called **antibodies.** The term *antigen* is a contraction of *anti*body-*gen*erator: Each antigen has a particular molecular shape and stimulates certain B cells to secrete antibodies that interact specifically with it. In fact, B and T cells can

FIGURE 43.6 Clonal selection. The B cells and T cells in lymph nodes and other lymphatic organs collectively recognize an enormous number of antigens, but each individual cell recognizes only one type of antigen. In this diagram, a B cell is "selected" by an antigen to proliferate and differentiate into memory cells and antibody-secreting cells called plasma cells. For simplicity, we show the antigen as free molecules rather than as part of a microbe.

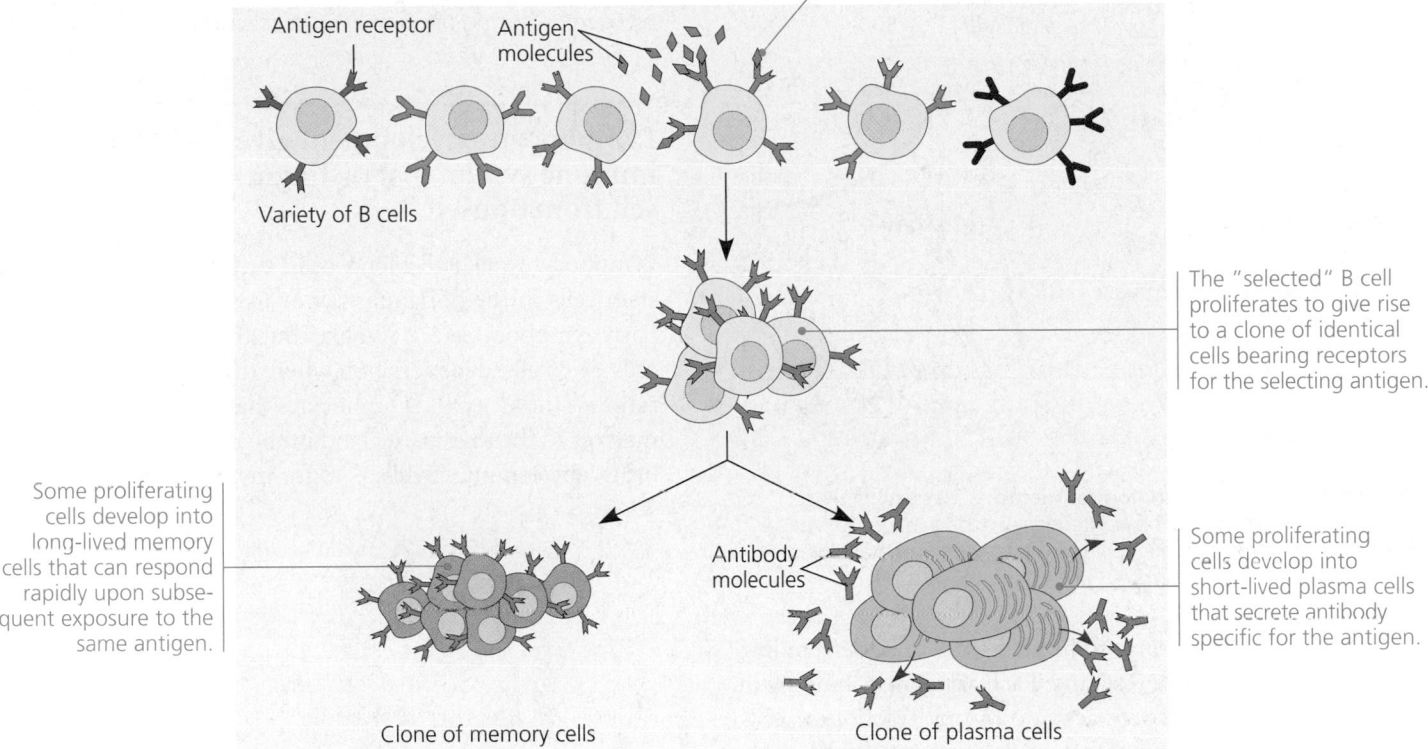

Antigen molecules bind to the antigen receptors of only one of the six B cells shown (note the variation in the shapes of the antigen receptors on the six different B cells).

Antigen receptor

Antigen molecules

Variety of B cells

The "selected" B cell proliferates to give rise to a clone of identical cells bearing receptors for the selecting antigen.

Some proliferating cells develop into long-lived memory cells that can respond rapidly upon subsequent exposure to the same antigen.

Antibody molecules

Some proliferating cells develop into short-lived plasma cells that secrete antibody specific for the antigen.

Clone of memory cells

Clone of plasma cells

distinguish among antigens with molecular shapes that are only slightly different. So, in contrast to the nonspecific defenses, the immune system targets specific invaders.

The means by which B cells and T cells recognize specific antigens are their plasma membrane–bound **antigen receptors.** Antigen receptors on a B cell are actually transmembrane versions of antibody molecules and are often referred to as *membrane antibodies* (or membrane immunoglobulins). The antigen receptors on a T cell, called **T cell receptors,** are structurally related to membrane antibodies, and they recognize antigens just as specifically. But, unlike antibodies, T cell receptors are never produced in a secreted form. A single T or B lymphocyte bears about 100,000 receptors for antigen, all with exactly the same specificity. The particular structure of a lymphocyte's receptors is determined by genetic events that occur in the lymphocyte during its early development (see FIGURE 19.6). As an unspecialized cell differentiates into a B or T lymphocyte, segments of antibody genes or receptor genes are linked together by a type of genetic recombination, generating a single functional gene for each polypeptide of an antibody or receptor protein. This process, which occurs before any contact with foreign antigens, creates an enormous variety of B and T

cells in the body, each bearing antigen receptors of particular specificity. With this diversity of lymphocytes, the immune system has the capacity to respond to millions of different antigenic molecules (even ones that do not yet exist)—and thus to millions of different potential pathogens.

Antigens interact with specific lymphocytes, inducing immune responses and immunological memory

Although it encounters a large repertoire of B cells and T cells in the body, a microorganism interacts only with lymphocytes bearing receptors specific for its various antigenic molecules. The "selection" of a lymphocyte by one of the microbe's antigens activates the lymphocyte, stimulating it to divide and to differentiate. Eventually the lymphocyte forms two clones of cells. One clone consists of a large number of **effector cells,** short-lived cells that combat the same antigen. The other clone consists of **memory cells,** long-lived cells bearing receptors specific for the same antigen. This antigen-driven cloning of lymphocytes is called **clonal selection** (FIGURE 43.6). The

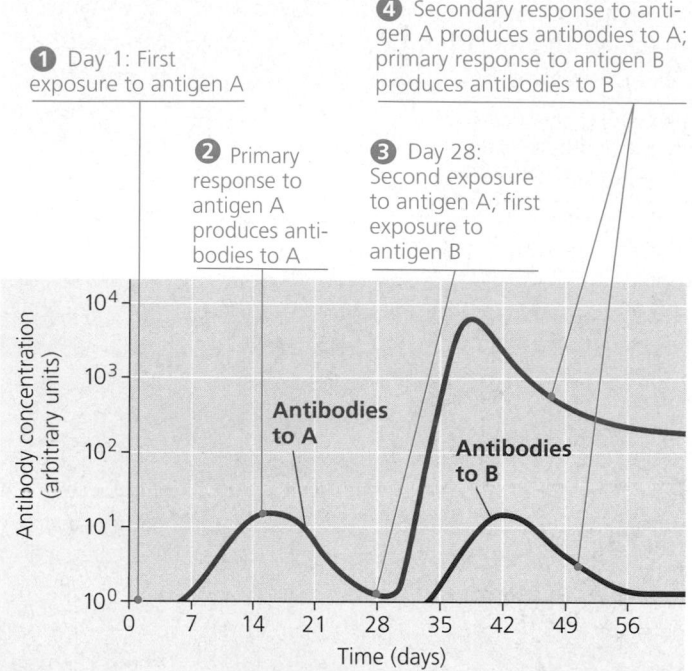

① Day 1: First exposure to antigen A

② Primary response to antigen A produces antibodies to A

③ Day 28: Second exposure to antigen A; first exposure to antigen B

④ Secondary response to antigen A produces antibodies to A; primary response to antigen B produces antibodies to B

Antibodies to A

Antibodies to B

FIGURE 43.7 Immunological memory. This experiment demonstrates that the heightened secondary response, which is due to long-lived memory cells, is specific for a particular antigen—here, antigen A.

concept of clonal selection is so fundamental to understanding immunity that it is worth restating: Each antigen, by binding to specific receptors, selectively activates a tiny fraction of cells from the body's diverse pool of lymphocytes; this relatively small number of selected cells gives rise to clones of thousands of cells, all specific for and dedicated to eliminating that antigen.

The selective proliferation and differentiation of lymphocytes that occurs the first time the body is exposed to an antigen is the **primary immune response.** In the primary response, about 10 to 17 days are required from the initial exposure to antigen for selected lymphocytes to generate the maximum effector cell response. During this period, selected B cells and T cells generate antibody-producing effector B cells, called **plasma cells,** and effector T cells, respectively. While these effector cells are developing, a stricken individual may become ill. Eventually, symptoms of illness diminish and disappear as antibodies and effector T cells clear the antigen from the body. If that individual is exposed to the same antigen at some later time, the response is faster (only 2 to 7 days), of greater magnitude, and more prolonged. This is the **secondary immune response.** Measures of antibody concentrations in the blood serum over time show clearly the difference between primary and secondary immune responses (FIGURE 43.7). In addition to being more numerous, antibodies produced in the secondary response tend to have greater affinity for the antigen than those secreted in the primary response. The immune system's capacity to generate secondary immune responses is called *immunological memory.* As you have seen, an initial exposure to

an antigen gives rise not only to effector cells but also to clones of long-lived T and B memory cells (see FIGURE 43.6). The memory cells are poised to proliferate and differentiate rapidly when they later contact the same antigen. The long-term protection developed after exposure to a pathogen was recognized 2,400 years ago by Thucydides of Athens, who described how those sick and dying of plague were cared for by others who had recovered, "for no one was ever attacked a second time."

Lymphocyte development gives rise to an immune system that distinguishes self from nonself

Lymphocytes, like all blood cells, originate from pluripotent stem cells in the bone marrow or liver of a developing fetus. Early lymphocytes are all alike, but they later develop into T cells or B cells, depending on where they continue their maturation (FIGURE 43.8). Lymphocytes that migrate from the bone marrow to the thymus, a gland in the thoracic cavity above the heart, develop into T cells ("T" for thymus). Lymphocytes that

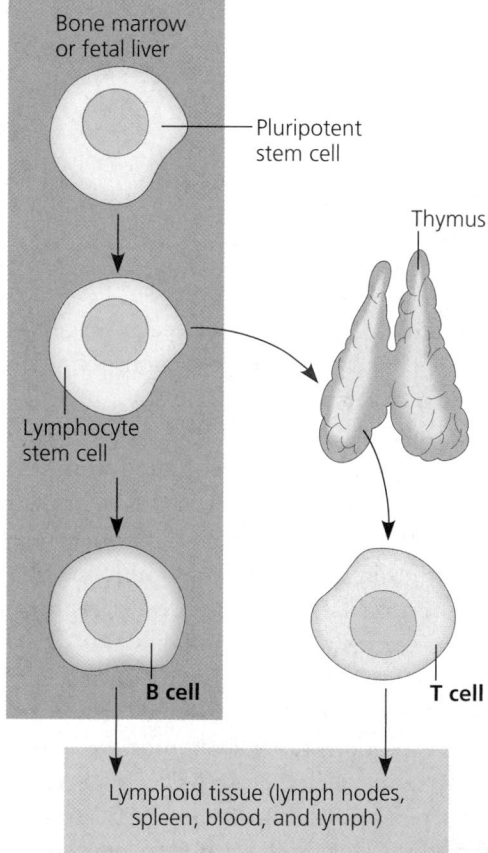

Bone marrow or fetal liver

Pluripotent stem cell

Thymus

Lymphocyte stem cell

B cell

T cell

Lymphoid tissue (lymph nodes, spleen, blood, and lymph)

FIGURE 43.8 The development of lymphocytes. Like other blood cells, lymphocytes differentiate from pluripotent stem cells in bone marrow (see FIGURE 42.14). Lymphocytes that continue their maturation in bone marrow develop into B cells, while lymphocytes that move to the thymus and complete their maturation there differentiate into T cells. When mature, lymphocytes of both classes migrate to the other lymphatic organs, where they encounter antigens (see FIGURE 43.4).

remain in the bone marrow and continue their maturation there become B cells. The "B" actually stands for the bursa of Fabricius, an organ unique to birds where avian B cells mature, and the place where B lymphocytes were first discovered. But since the B cells of all other vertebrates develop in bone marrow, you can equate "B" with "bone" as well as with "bursa."

Immune Tolerance for Self

While B cells and T cells are maturing in the bone marrow and thymus, their antigen receptors are tested for potential self-reactivity. For the most part, lymphocytes bearing receptors specific for molecules already present in the body are either rendered nonfunctional or destroyed by apoptosis (programmed cell death; see FIGURE 21.18), leaving only lymphocytes that react to foreign molecules. This *capacity to distinguish self from non-self* continues to develop even as the cells migrate to lymphatic organs. Thus, the body normally has no mature lymphocytes that react against self components: The immune system exhibits the critical feature of *self-tolerance.* Failure of self-tolerance can lead to autoimmune diseases such as multiple sclerosis. As you will see in the next section, certain body cell surface molecules are essential in the development of T cell self-tolerance as well as in T cell activity.

The Role of Cell Surface Markers in T Cell Function and Development

Lymphocytes do not react to most self antigens, but T cells do have a crucial interaction with one important group of native molecules. These are a collection of cell surface glycoproteins (proteins with attached sugar chains) encoded by a family of genes called the **major histocompatibility complex (MHC).** In humans, the MHC glycoproteins are also known as the *HLA* (for *human leukocyte antigens*). Two main classes of MHC

molecules mark body cells as "self." **Class I MHC molecules** are found on almost all nucleated cells—that is, on almost every cell of the body. **Class II MHC molecules** are restricted to a few specialized cell types, including macrophages, B cells, activated T cells, and the cells that make up the interior of the thymus.

For a vertebrate species, there are numerous different possible alleles for each class I and class II MHC gene; the MHC proteins are the most polymorphic known. As a result of the large number of different alleles in the human population, for example, most of us are heterozygous for every one of our MHC genes. Moreover, it is extremely unlikely that any two people, except identical twins, will have exactly the same set of MHC molecules. Thus, the major histocompatibility complex provides a biochemical fingerprint virtually unique to each individual. In fact, the discovery of the MHC occurred in the process of studying the phenomena of skin graft rejection and acceptance; the *histo* in histocompatibility refers to tissues.

The MHC and its role in the body's rejection of tissue grafts initially puzzled scientists: Why would the vertebrate body have evolved markers that prevent members of the same species from sharing tissues? We now know that MHC molecules are varied from person to person because of their central role in the immune response. Through a process known as **antigen presentation,** an MHC molecule cradles a fragment of an intracellular protein antigen in its hammocklike groove, carries it to the cell surface, and "presents" it to an antigen receptor on a nearby T cell. Thus T cells are alerted to an infectious agent after it has been internalized by a cell (through phagocytosis or receptor-mediated endocytosis), or after it has entered and replicated within a cell (through virus infection). There are two main types of T cells, and each responds to one of the two classes of MHC molecule. **Cytotoxic T cells (T_C)** have antigen receptors that bind to protein fragments (peptides) displayed by the body's class I MHC molecules (FIGURE 43.9a). **Helper T cells (T_H)** have receptors that bind to peptides displayed by the body's class II MHC molecules (FIGURE 43.9b).

Whether or not T cells respond to a pathogen, then, depends on the ability of MHC molecules to present a fragment of it. Fortunately, any one MHC molecule can present a variety of peptides that are structurally similar, and, because of the heterozygosity of our MHC genes, we

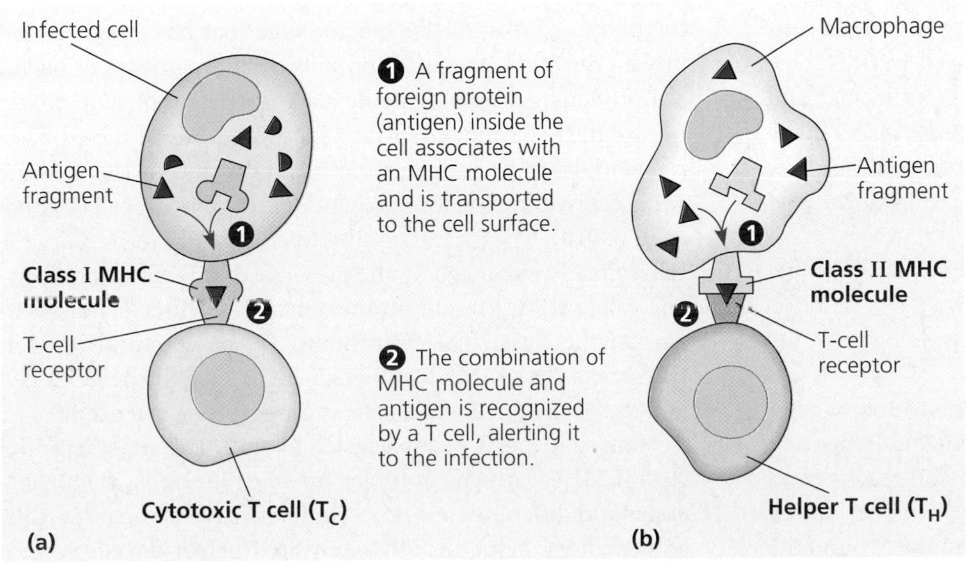

(a) Infected cell · Antigen fragment · **Class I MHC molecule** · T-cell receptor · **Cytotoxic T cell (T_C)**

❶ A fragment of foreign protein (antigen) inside the cell associates with an MHC molecule and is transported to the cell surface.

❷ The combination of MHC molecule and antigen is recognized by a T cell, alerting it to the infection.

(b) Macrophage · Antigen fragment · **Class II MHC molecule** · T-cell receptor · **Helper T cell (T_H)**

FIGURE 43.9 The interaction of T cells with MHC molecules. Molecules of the major histocompatibility complex (MHC) on the surfaces of body cells "present" antigen fragments to T cells. **(a)** Cytotoxic T cells have receptors for antigen-bearing class I MHC molecules. **(b)** Helper T cells have receptors for antigen-bearing class II MHC molecules.

each make two different MHC polypeptides per gene. So chances are good that at least one of our MHC molecules will be able to present at least one fragment of a particular pathogen to our T cells—and generate an immune response against it. In addition, human populations benefit from having hundreds of different MHC alleles in the gene pool, so that individual collections of MHC molecules vary from person to person. This polymorphism is adaptive because it increases the likelihood that at least some individuals of a population will survive an epidemic.

Let's look at the distribution and roles of class I and class II MHC molecules in the body, in terms of how they defend us against infection. Class I MHC molecules, found in almost all cells, are poised to present fragments of proteins made by infecting microbes, usually viruses, to cytotoxic T cells. As we will see in detail later, cytotoxic T cells respond by killing the infected cells. Because all of our cells are vulnerable to infection by one or another virus, the wide distribution of class I MHC molecules is critical to our health. Class II MHC molecules, on the other hand, are made by only a few cell types, chiefly macrophages and B cells. These cells, called **antigen-presenting cells (APCs)** in this context, ingest bacteria (and viruses) and then destroy them. Class II MHC molecules in these cells collect peptide remnants of this degradation and present them to helper T cells. In response, the helper T cells send out chemical signals that incite other cell types to fight the pathogen.

MHC proteins also play a key role in the development of T-cell self-tolerance. During the development of T cells in the thymus, developing T cells interact with other thymic cells, which have high levels of both class I and class II MHC molecules. Only T cells bearing receptors with affinity for self MHC proteins reach maturity. Developing T cells having receptors with affinity for class I MHC become cytotoxic T cells. Those having receptors with affinity for class II MHC become helper T cells.

Now let's review what you've learned so far about the immune system: The immune responses of B and T lymphocytes exhibit four attributes that characterize the immune system as a whole: specificity, diversity, memory, and the capacity to distinguish self from nonself. A critical component of the immune response is the MHC: Proteins encoded by this gene complex display a combination of self (MHC molecule) and nonself (antigen fragment) that is recognized by specific T cells. In the following section you will learn more about how lymphocytes recognize and respond to foreign substances, and how they generate immunity.

IMMUNE RESPONSES

The immune system can mount two types of responses to antigens: a humoral response and a cell-mediated response. **Humoral immunity** involves B cell activation and results from the production of antibodies that circulate in the blood plasma and lymph, fluids that were long ago called humors. Around the end of the nineteenth century, researchers performed an experiment in which they transferred such fluids from animals that had recovered from an infection to others that had not been exposed to it. For a short time the latter animals were protected from the infection. The investigators had transferred humoral immunity (antibodies) from animal to animal. They also found that immunity to some infections could be passed along only if cells, later identified as T lymphocytes, were transferred. This second type of immunity, which depends on the action of T cells, became known as **cell-mediated immunity.**

The circulating antibodies of the humoral response defend mainly against free bacteria, toxins, and viruses present in body fluids. In contrast, the T cells of the cell-mediated response are active against viruses and bacteria within infected body cells and against fungi, protozoa, and parasitic worms. Cell-mediated immunity is crucial in the body's response against transplanted tissue and cancer cells, both of which are perceived as "nonself." FIGURE 43.10 provides an overview of the humoral and cell-mediated responses, the two branches of the immune system, which we will explore shortly. In addition to introducing these two branches, the figure shows the connections linking them—cell-signaling interactions among the lymphocytes. Central to this network of cell-signaling is the helper T cell, which responds to antigen presented by a macrophage and stimulates both B cells and other T cells.

Helper T lymphocytes function in both humoral and cell-mediated immunity: *an overview*

Before we look at the function of helper T lymphocytes, we must return to the MHC and its role in antigen presentation. Recall that class II MHC molecules, the ones recognized by helper T cells, are found only on certain cell types, mainly those that engulf foreign antigens. These antigen-presenting cells (APCs), including macrophages and some B cells, tell the immune system, via helper T cells, that foreign antigen is in the body. For example, a macrophage that has engulfed and broken down a bacterium contains small fragments of bacterial proteins. As a newly made class II MHC molecule moves toward the macrophage surface, it captures one of these bacterial peptides in its antigen-binding groove and carries it to the surface, revealing the foreign peptide to a helper T cell (FIGURE 43.11, p. 910). The interaction between an APC and a helper T cell is greatly enhanced by the presence of a T cell surface protein called **CD4.** Present on most helper T cells, CD4 binds to part of the class II MHC protein. The interaction between CD4 and a class II MHC molecule helps keep the helper T cell and the APC joined while activation of the T_H cell occurs.

When a helper T cell is selected by specific contact with the class II MHC–antigen complex on an APC, the T_H cell proliferates and differentiates into a clone of activated helper T cells and memory helper T cells. Activated helper T cells secrete

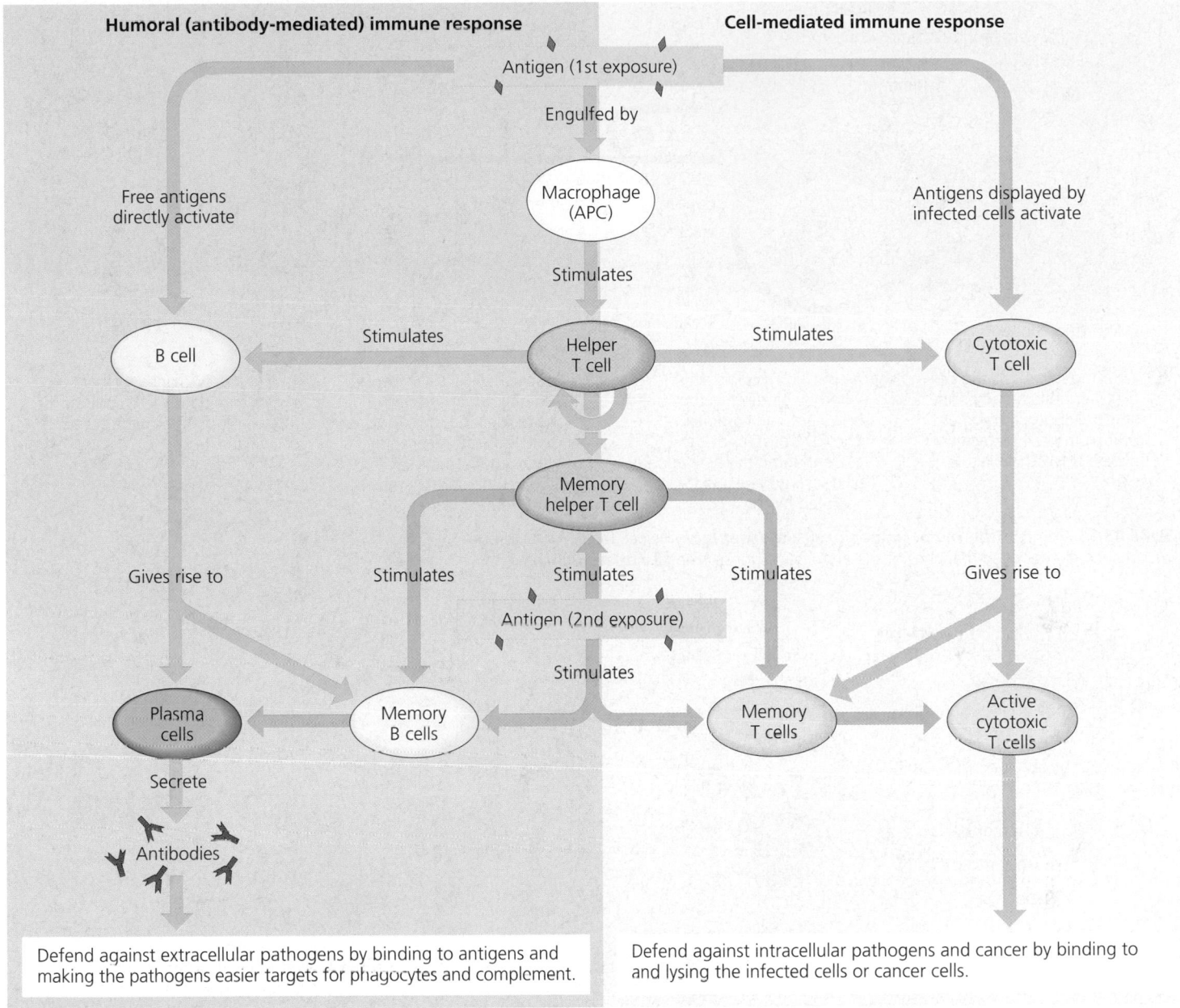

Humoral (antibody-mediated) immune response

Cell-mediated immune response

Antigen (1st exposure)

Engulfed by

Macrophage (APC)

Free antigens directly activate

Antigens displayed by infected cells activate

Stimulates

Stimulates

Stimulates

Stimulates

B cell

Helper T cell

Cytotoxic T cell

Memory helper T cell

Gives rise to

Stimulates

Stimulates

Stimulates

Gives rise to

Antigen (2nd exposure)

Stimulates

Plasma cells

Memory B cells

Memory T cells

Active cytotoxic T cells

Secrete

Antibodies

Defend against extracellular pathogens by binding to antigens and making the pathogens easier targets for phagocytes and complement.

Defend against intracellular pathogens and cancer by binding to and lysing the infected cells or cancer cells.

FIGURE 43.10 An overview of the immune responses. In this simplified flowchart, green arrows track the primary response, and blue arrows track the secondary response. Notice the connections between the humoral response and the cell-mediated response and the central role of the helper T cell. APC stands for antigen-presenting cell.

several different **cytokines,** proteins or peptides that stimulate other lymphocytes. For example, the cytokine **interleukin-2 (IL-2)** helps B cells that have contacted antigen differentiate into antibody-secreting plasma cells. IL-2 also helps cytotoxic T cells become active killers.

The helper T cell itself is also subject to regulation by cytokines. As a macrophage phagocytoses and presents antigen, the macrophage is stimulated to secrete a cytokine called **interleukin-1 (IL-1).** IL-1, in combination with the presented antigen, is what activates the helper T cell to produce IL-2 and other cytokines. Also, in an example of positive feedback, IL-2 secreted by the helper T cell stimulates that same cell to prolifer-

ate more rapidly and to become an even more active cytokine producer. In these ways, helper T cells modulate both humoral (B cell) and cell-mediated (cytotoxic T cell) immune responses.

In the cell-mediated response, cytotoxic T cells counter intracellular pathogens: *a closer look*

Antigen-activated cytotoxic T lymphocytes kill cancer cells and cells infected by viruses or other intracellular pathogens. Before we look at these events, we must return to class I MHC proteins and their role in presenting antigen to cytotoxic T cells. Recall

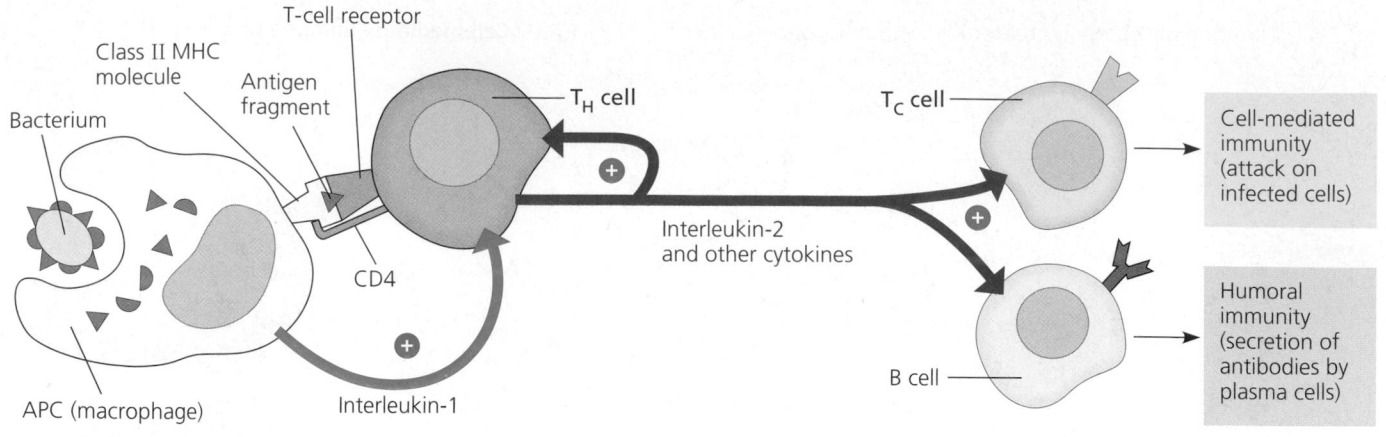

- **Bacterium**
- **Class II MHC molecule**
- **Antigen fragment**
- **T-cell receptor**
- **T_H cell**
- **CD4**
- **APC (macrophage)**
- **Interleukin-1**
- **Interleukin-2 and other cytokines**
- **T_C cell**
- **B cell**
- **Cell-mediated immunity (attack on infected cells)**
- **Humoral immunity (secretion of antibodies by plasma cells)**

❶ An antigen-presenting cell (APC) engulfs a bacterium and transports a fragment of it to the cell surface via a class II MHC molecule.

❷ A specific T_H cell is activated by binding to the MHC-antigen complex. The CD4 protein of the T_H cell enhances the activation, as does interleukin-1 secreted by the APC.

❸ The activated T_H cell proliferates, giving rise to a clone of identical cells (not shown), all with receptors keyed to the same MHC-antigen combination. These cells secrete cytokines.

❹ The cytokines further stimulate the T_H cells and also help activate B cells and T_C cells.

FIGURE 43.11 The central role of helper T cells: a *closer look.* Helper T cells mobilize both humoral and cell-mediated branches of the immune response. The ⊕ indicates stimulation.

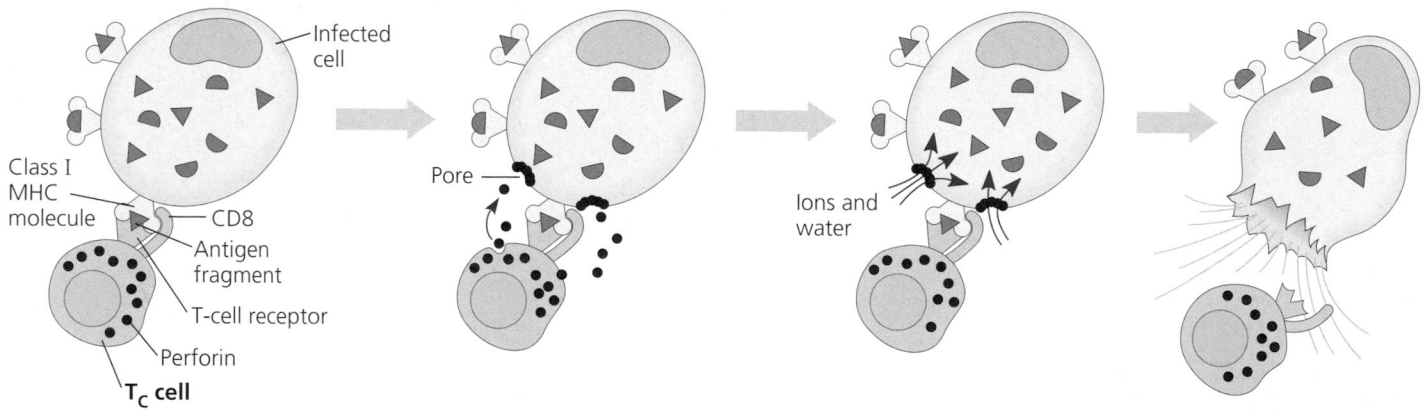

- **Infected cell**
- **Class I MHC molecule**
- **CD8**
- **Antigen fragment**
- **T-cell receptor**
- **Perforin**
- **T_C cell**
- **Pore**
- **Ions and water**

❶ An infected cell (or cancer cell) displays an antigen fragment on its surface using a class I MHC molecule. A specific T_C cell is activated by binding to the MHC-antigen complex. The CD8 protein of the T_C cell enhances the activation, along with interleukin-2 from helper T cells (not shown).

(a)

❷ The activated T_C cell discharges perforin molecules, which create pores in the membrane of the infected cell.

❸ Water and ions flow into the infected cell, and the cell lyses.

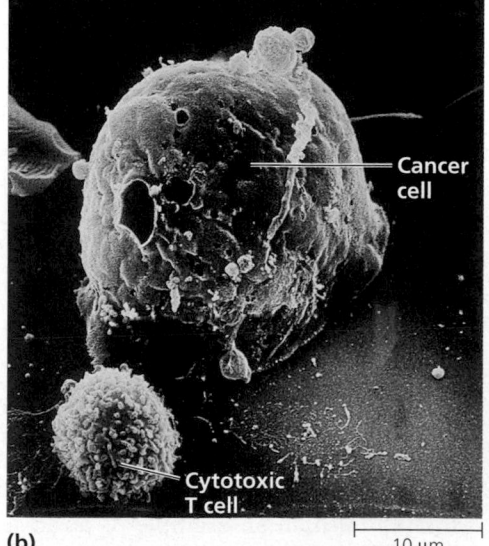

- **Cancer cell**
- **Cytotoxic T cell**
- 10 μm

(b)

FIGURE 43.12 The functioning of cytotoxic T cells. (a) A cytotoxic T cell (T_C cell) responds to an infected cell (or cancer cell) that displays an antigen fragment recognized by its T-cell receptor. A class I MHC molecule presents the antigen fragment. **(b)** In this SEM, the T_C cell has lysed the cancer cell.

that all nucleated cells of the body continuously produce class I MHC molecules. As a newly synthesized class I molecule moves toward the cell surface, it captures a small fragment of one of the other proteins synthesized by that cell. If that cell happens to contain a replicating virus, peptide fragments of viral proteins are captured and transported to the cell surface. In this way, class I MHC molecules expose foreign proteins that are synthesized in infected or abnormal cells to cytotoxic T cells. The interaction between the antigen-presenting infected cell and a cytotoxic T cell is greatly enhanced by the presence of a T cell surface protein called **CD8.** Present on most cytotoxic T cells, CD8 binds to a portion of the class I MHC molecule. The CD8–class I MHC interaction helps keep the two cells in contact while activation of the T_C cell is occurring (FIGURE 43.12). Thus, the roles of class I MHC molecules and CD8 are similar to those of class II MHC molecules and CD4, except that different cells are involved.

A cytotoxic T cell, when activated by specific contacts with class I MHC–antigen complexes on an infected cell and further stimulated by IL-2 from a helper T cell, differentiates into an active killer. It kills its target cell—the antigen-presenting cell—primarily by releasing **perforin,** a protein that forms pores in the target cell's membrane. As ions and water flow into the target cell, it swells and eventually lyses (see FIGURE 43.12a). The death of the infected cell not only deprives the pathogen of a place to reproduce but also exposes it to circulating antibodies, which mark it for disposal. After destroying an infected cell, the T_C cell moves on to kill other cells infected with the same pathogen.

In the same way, T_C cells defend against malignant tumors. Because tumor cells carry distinctive molecules not found on normal cells, they are identified as foreign by the immune system. Class I MHC molecules on a tumor cell present fragments of **tumor antigen** to T_C cells. Interestingly, certain cancers and viruses (such as Epstein-Barr virus) actively reduce the amount of class I MHC protein on affected cells so that they escape detection by T_C cells. The body has a backup defense: Natural killer (NK) cells, part of the body's nonspecific defenses, also lyse virus-infected and cancer cells.

In the humoral response, B cells make antibodies against extracellular pathogens: *a closer look*

You have learned that the humoral immune response is initiated when B cells bearing antigen receptors are selected by binding with specific antigens. You also know that B cell activation is aided by IL-2 and other cytokines secreted from helper T cells activated by the same antigen. Stimulated by both antigen and cytokines, the B cell proliferates and differentiates into a clone of antibody-secreting plasma cells and a clone of memory B cells. Antigens that evoke this type of B cell response are known as **T-dependent antigens** because they can stimulate antibody production only with help from T_H cells (FIGURE 43.13). Most protein antigens are T-dependent.

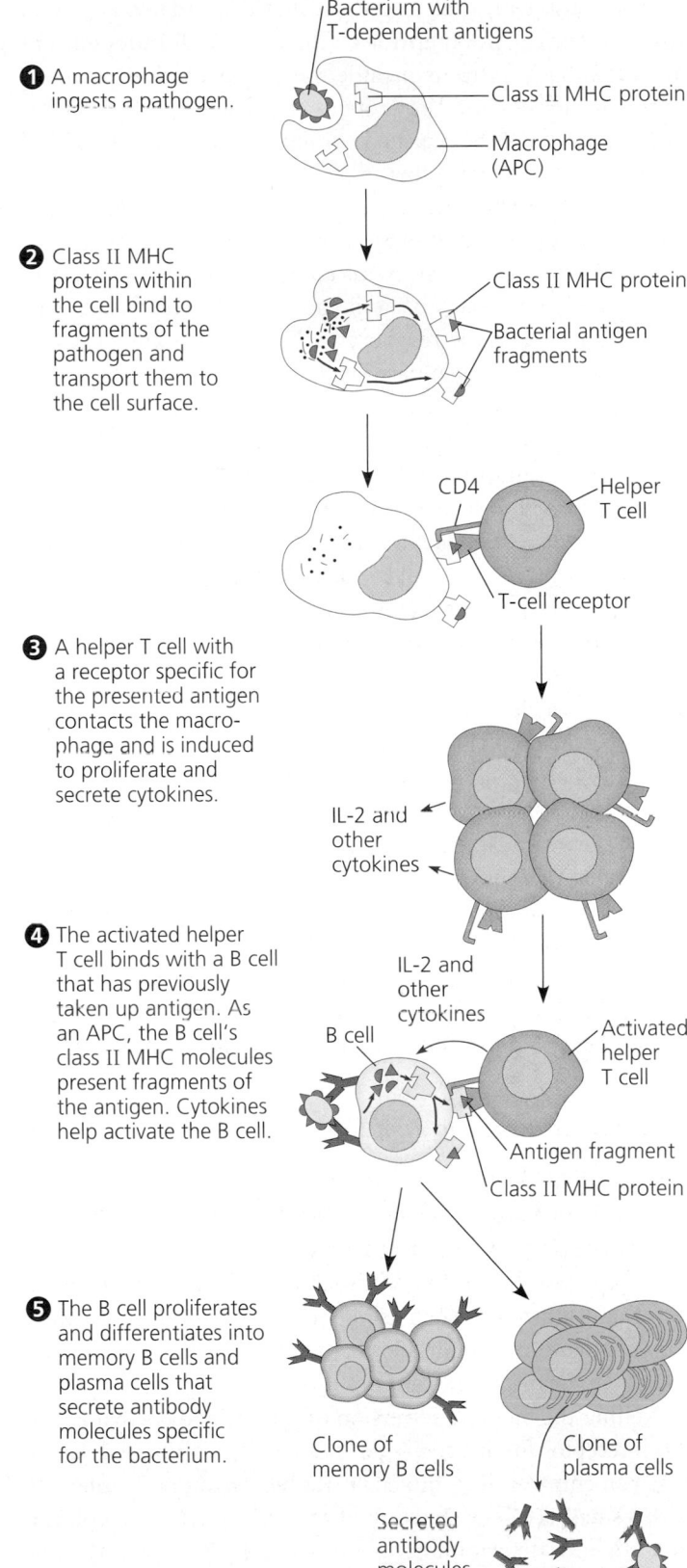

① A macrophage ingests a pathogen.

② Class II MHC proteins within the cell bind to fragments of the pathogen and transport them to the cell surface.

③ A helper T cell with a receptor specific for the presented antigen contacts the macrophage and is induced to proliferate and secrete cytokines.

④ The activated helper T cell binds with a B cell that has previously taken up antigen. As an APC, the B cell's class II MHC molecules present fragments of the antigen. Cytokines help activate the B cell.

⑤ The B cell proliferates and differentiates into memory B cells and plasma cells that secrete antibody molecules specific for the bacterium.

Bacterium with T-dependent antigens
Class II MHC protein
Macrophage (APC)
Class II MHC protein
Bacterial antigen fragments
CD4
Helper T cell
T-cell receptor
IL-2 and other cytokines
IL-2 and other cytokines
B cell
Activated helper T cell
Antigen fragment
Class II MHC protein
Clone of memory B cells
Clone of plasma cells
Secreted antibody molecules

FIGURE 43.13 Humoral response to a T-dependent antigen. Many antigens can trigger a humoral (antibody-mediated) immune response by B cells only with the participation of helper T cells. Such antigens are called T-dependent antigens, and most protein antigens are of this type.

Other antigens, such as polysaccharides and proteins with many identical polypeptides, function as **T-independent antigens.** Such antigens include the polysaccharides of many bacterial capsules and the proteins that make up bacterial flagella. Apparently, the repeated subunits of these antigens bind simultaneously to a number of membrane antibodies on the B cell surface. This provides enough stimulus to the B cell to generate antibody-secreting plasma cells without the help of IL-2. The response to T-independent antigens is very important in defending against many bacteria; however, the response is generally weaker than the response to T-dependent antigens, and no memory cells are generated in T-independent responses.

Before we leave our discussion of B cell function, it is important to recall that B cells bear class II MHC molecules: They are antigen presenting cells. When antigen first binds to membrane antibodies, the B cell takes in a few of the foreign molecules by receptor-mediated endocytosis (see FIGURE 8.19c). In a process very similar to presentation by macrophages, the B cell presents antigen to a helper T cell. However, although a macrophage can engulf and present peptide fragments from a wide variety of antigens, a B cell internalizes and presents peptides of only the antigen to which it specifically binds. Therefore, immunologists think that macrophages are the main APCs in the primary response (when B cells specific for a particular antigen are rare), whereas B cells, specifically memory B cells, are more important as APCs in secondary responses.

In any given humoral response, the processes just discussed stimulate a variety of different B cells, each giving rise to a clone of thousands of plasma cells. Each plasma cell is estimated to secrete about 2,000 antibody molecules per second over the cell's 4- to 5-day life span. Next we look more closely at antibodies and how they bind to and mediate the disposal of antigens.

Antibody Structure and Function

Antigens that elicit a humoral immune response are typically the protein and polysaccharide surface components of microbes, incompatible transplanted tissue, or incompatible transfused blood cells. In addition, for some of us, the proteins of foreign substances such as pollen or bee venom act as antigens that induce an allergic, or hypersensitive, humoral response (to be discussed later).

Neither the membrane version of the antibody—that is, the B cell receptor for antigen—nor the secreted antibody actually binds an entire antigen molecule. Rather, an antibody interacts with a small, accessible portion of the antigen called an **epitope** or antigenic determinant (FIGURE 43.14). A single antigen such as a bacterial surface protein usually has several effective epitopes, each capable of inducing the production of specific antibody. It is therefore easy to imagine the entire surface of a bacterium coated with many different kinds of antibodies, each specific for a particular epitope. It is estimated that a bacterium can be bound by 4 million antibody molecules!

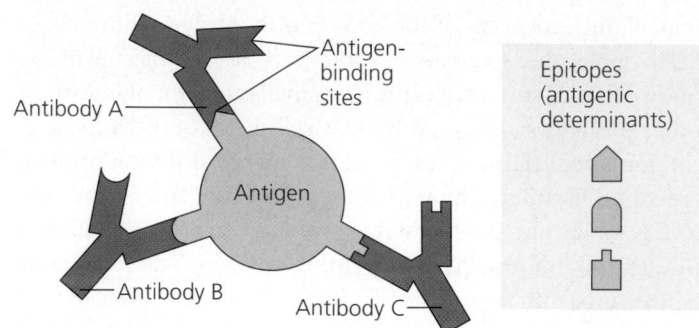

FIGURE 43.14 Epitopes (antigenic determinants). Antibodies bind to epitopes on the surface of an antigen. In this example, three different antibody molecules react with different epitopes on the same large antigen molecule.

Antibodies constitute a group of globular serum proteins called **immunoglobulins (Igs).** A typical antibody molecule has two identical antigen-binding sites specific for the epitope that provoked its production. Each molecule consists of four polypeptide chains, two identical **heavy chains** and two identical **light chains,** joined by disulfide bridges to form a Y-shaped molecule (FIGURE 43.15a). At the two tips of the Y-shaped molecule are the variable regions (V) of the heavy and light chains, so named because the amino acid sequences in these regions vary extensively from antibody to antibody (see FIGURE 19.6). As shown in FIGURE 43.15b, a heavy-chain V region and a light-chain V region together form the unique contours of an antibody's antigen-binding site. The interaction between an antigen-binding site and its epitope resembles an enzyme-substrate interaction: Multiple noncovalent bonds form between chemical groups on the respective molecules, resulting in a highly stable complex.

The power of antibody specificity and antigen–antibody binding has been harnessed for use in laboratory research, clinical diagnosis, and the treatment of diseases. Some of these antibody tools are *polyclonal:* They are the products of many different clones of B cells, each specific for a different epitope. Others are *monoclonal:* They are prepared from a single clone of B cells grown in culture. Because they are all identical, the **monoclonal antibodies** produced by such a culture are specific for the same epitope on an antigen. In both basic research and medicine, antibodies are useful for tagging specific molecules. For example, certain types of cancer have been treated with tumor-specific antibodies covalently coupled to toxin molecules. The toxin-linked antibodies carry out a precise search-and-destroy mission, selectively attaching to and killing tumor cells.

THE PROCESS OF SCIENCE

While the antigen-binding sites on an antibody are responsible for its ability to identify a specific antigen, the tail of the Y-shaped antibody, formed by the constant regions (C) of the heavy chains (see FIGURE 43.15a), is responsible for the antibody's distribution in the body and for the mechanisms by which it mediates antigen disposal. There are five major types

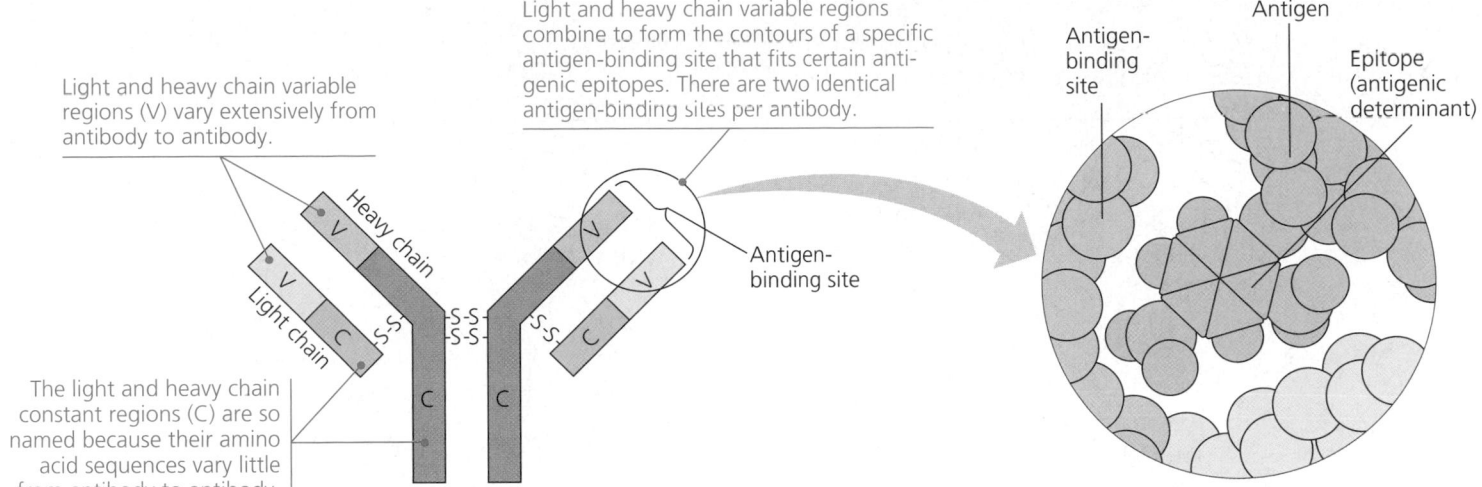

Light and heavy chain variable regions (V) vary extensively from antibody to antibody.

Light and heavy chain variable regions combine to form the contours of a specific antigen-binding site that fits certain antigenic epitopes. There are two identical antigen-binding sites per antibody.

The light and heavy chain constant regions (C) are so named because their amino acid sequences vary little from antibody to antibody.

Antigen-binding site

(a) Basic structure of an antibody molecule. The Y-shaped molecule is composed of two light and two heavy chains linked together by disulfide bridges (S—S).

Antigen-binding site

Antigen

Epitope (antigenic determinant)

(b) Close-up view of an antigen-binding site with bound antigen

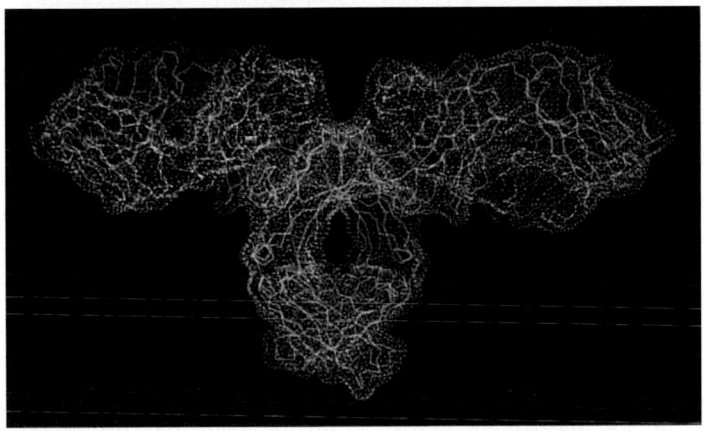

(c) A computer graphic image of an antibody molecule

FIGURE 43.15 The structure of a typical antibody molecule.

of heavy-chain constant regions, and these determine the five major classes of antibodies: IgM, IgG, IgA, IgD, and IgE. The structures and functions of each of the five immunoglobulin classes are summarized in TABLE 43.1. Note that each light chain also has a constant region. It is not a part of the antibody tail and does not contribute to Ig class functions.

Antibody-Mediated Disposal of Antigen

The binding of antibodies to antigens to form antigen-antibody complexes is the basis of several antigen disposal mechanisms (FIGURE 43.16, p. 914). The simplest of these is **neutralization,** in which the antibody binds to and blocks the activity of the antigen. For example, antibodies neutralize a virus by attaching to the molecules that the virus must use to infect its host cell. Similarly, antibodies may bind to the surface of a pathogenic bacterium. These microbes, now coated by antibodies, are readily

Table 43.1 The Five Classes of Immunoglobulins	
IgM (pentamer) 	IgMs are the first circulating antibodies to appear in response to an initial exposure to an antigen; their concentration in the blood then declines rapidly. Thus the presence of IgM usually indicates a current infection. IgM consists of five Y-shaped monomers arranged in a pentagonal structure. The numerous antigen-binding sites make it very effective in agglutinating antigens and in reactions involving complement. IgM is too large to cross the placenta and does not confer maternal immunity.
IgG (monomer)	IgG is the most abundant of the circulating antibodies. It readily crosses the walls of blood vessels and enters tissue fluids. IgG also crosses the placenta and confers passive immunity on the fetus. IgG protects against bacteria, viruses, and toxins in the blood and lymph, and triggers action of the complement system.
IgA (dimer) 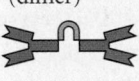	IgA is produced by cells in mucous membranes. The main function of IgA is to prevent the attachment of viruses and bacteria to epithelial surfaces. IgA is also found in many body secretions, such as saliva, perspiration, and tears. Its presence in the first milk produced helps protect the infant from gastrointestinal infections.
IgD (monomer)	IgD antibodies do not activate the complement system and cannot cross the placenta. They are mostly found on the surfaces of B cells, probably functioning as antigen receptors that help initiate the differentiation of B cells into plasma cells and memory B cells.
IgE (monomer)	IgE molecules are slightly larger than IgG and represent only a small fraction of the antibodies in the blood. The tails attach to mast cells and basophils and, when triggered by an antigen, cause the cells to release histamine and other chemicals that cause an allergic reaction.

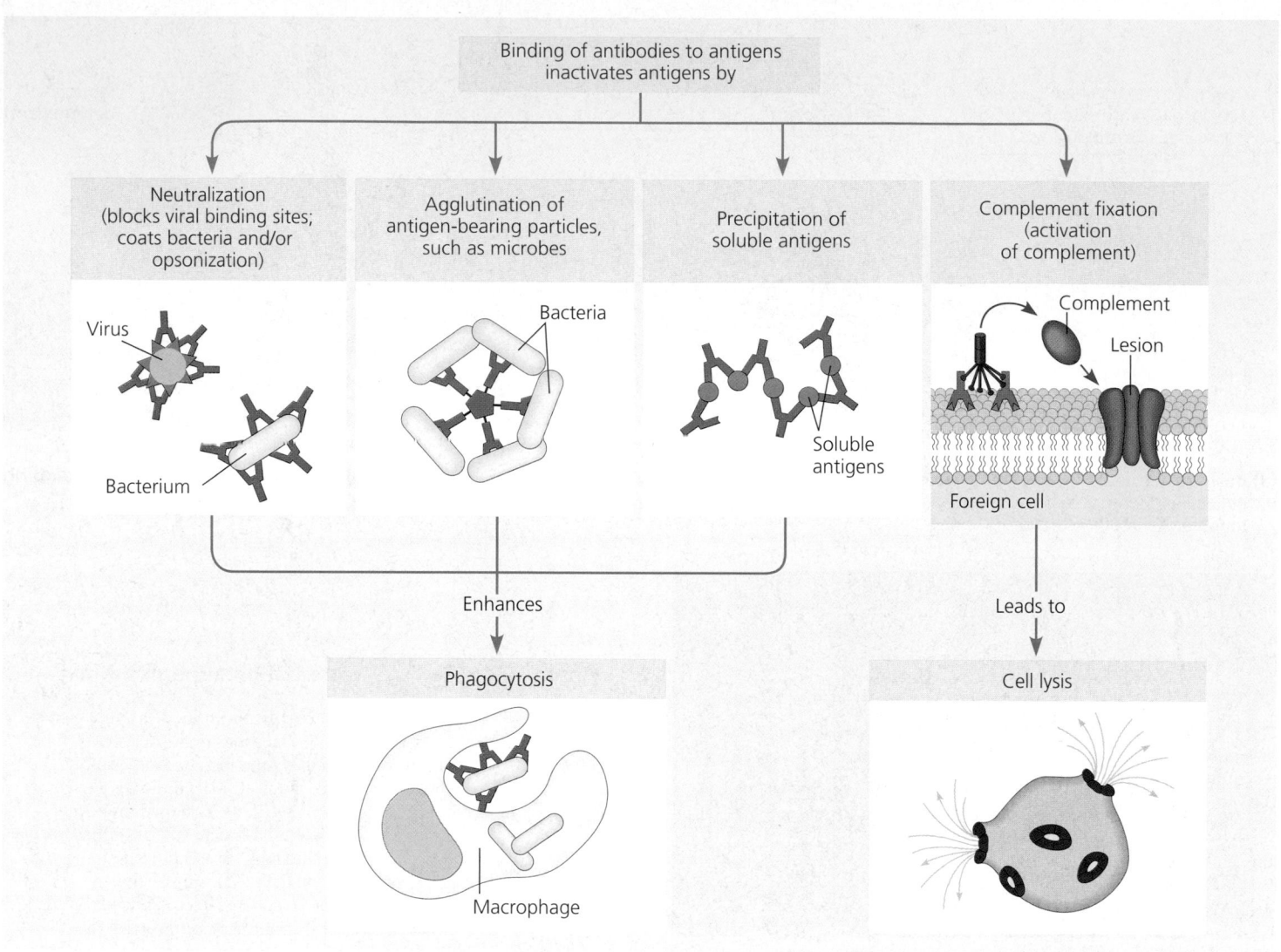

FIGURE 43.16 Effector mechanisms of humoral immunity. The binding of antibodies to antigens tags foreign cells and molecules for destruction by phagocytes or the complement system of proteins.

eliminated by phagocytosis. In a process called **opsonization,** the bound antibodies enhance macrophage attachment to, and thus phagocytosis of, the microbes.

Antibody-mediated **agglutination** (clumping) of bacteria or viruses effectively neutralizes and opsonizes the microbes. Agglutination is possible because each antibody molecule has at least two antigen-binding sites. IgG, for example, can bind to identical epitopes on two bacterial cells or viral particles, linking them together. IgM can link together five or more viruses or bacteria (as shown in FIGURE 43.16). These large complexes are readily phagocytosed by macrophages. A similar mechanism is precipitation, the cross-linking of soluble antigen molecules—molecules dissolved in body fluids—to form immobile precipitates that are disposed of by phagocytes.

One of the most important antibody-mediated disposal mechanisms is **complement fixation,** the activation of the complement system by antigen-antibody complexes. Recall that complement consists of about 20 different serum proteins that,

in the absence of infection, are inactive. In an infection, however, the first in the series of complement proteins is activated, triggering a cascade of activation steps, each component activating the next in the series. Completion of the complement cascade results in the lysis of many types of viruses and pathogenic cells. Lysis by complement can be achieved in two ways. The *classical pathway* (so called because it was discovered first) is triggered by antibodies bound to antigen and is therefore important in the humoral immune response. The *alternative pathway* is triggered by substances that are naturally present on many bacteria, yeast, viruses, and protozoan parasites; it does not involve antibodies and thus is an important nonspecific defense.

The classical pathway can begin when IgM or IgG antibodies bind to a pathogen, such as a bacterial cell (FIGURE 43.17). The first complement component links two bound antibodies and is thus activated, initiating the cascade. Ultimately, complement proteins generate a **membrane attack complex (MAC),** which forms a pore in the bacterial membrane 7–10 nm in diameter.

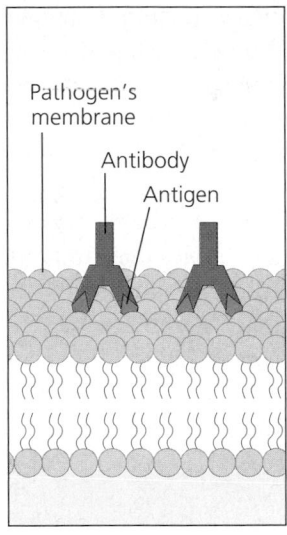  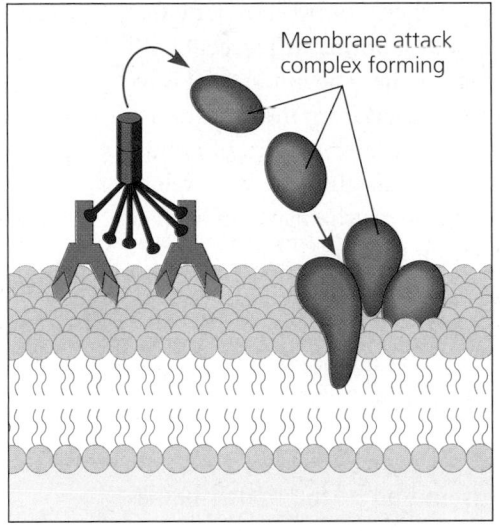 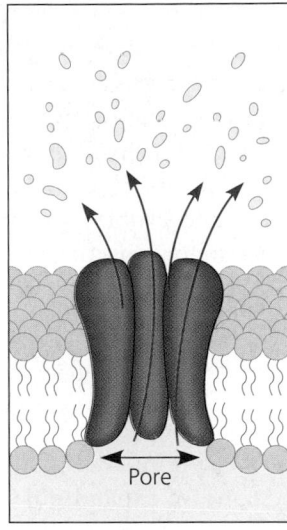

1 Antibody molecules attach to antigens on pathogen's plasma membrane.

2 Complement proteins link two antibody molecules.

3 Activated complement proteins attach to pathogen's membrane in step-by-step sequence, forming a membrane attack complex (MAC).

4 MAC pores in the membrane causes cell lysis.

FIGURE 43.17 The classical complement pathway, resulting in lysis of a target cell.

Ions and water rush into the cell, causing it to swell and lyse. The MAC pore is similar to the perforin pore generated by cytotoxic T cells.

In both the classical and alternative pathways, many activated complement proteins contribute to inflammation. By binding to basophils and mast cells, some trigger the release of histamine, the injury-signaling molecule that triggers dilation and increased permeability of blood vessels. Several active complement proteins also attract phagocytes to the site. In addition, one of the activated complement proteins can cause opsonization: Copies of this protein coat bacterial surfaces and, like antibodies, stimulate phagocytosis. In a final example of teamwork in the body's defense systems, antibodies, complement, and phagocytes function together in a phenomenon called **immune adherence.** Microbes coated with antibodies and complement proteins adhere to blood vessel walls, making the pathogens easier prey for phagocytic cells circulating in the blood.

You have seen that antibodies contribute in several ways to phagocytosis by macrophages. Keep in mind that phagocytosis enables macrophages to act as APCs, stimulating helper T cells, which, in turn, stimulate the very B cells whose antibodies contribute to phagocytosis. Thus, positive feedback links together the nonspecific and specific immune systems, resulting in a coordinated and effective response to an active infection.

Invertebrates have a rudimentary immune system

Although this chapter focuses on the immune mechanisms of vertebrates, invertebrate animals also exhibit highly effective mechanisms of host defense, which undoubtedly have contributed to their evolutionary success. The ability to make the distinction between self and nonself is seen in animals as ancient as the sponges. If cells from two sponges of the same species are mixed, the cells from each sponge sort themselves and reaggregate, each excluding cells from the other individual. Invertebrates also have the ability to dispose of what is not self, which they accomplish primarily by phagocytosis. In sea stars, for example, amoeboid cells called *coelomocytes* phagocytose foreign matter. Furthermore, immunologists have begun to find cytokines in invertebrates. For example, like mammalian macrophages, sea star coelomocytes produce interleukin-1 as they engulf foreign material. The IL-1 enhances the animal's defensive response by stimulating coelomocyte proliferation and attracting more coelomocytes to the area.

Invertebrates depend on innate, nonspecific mechanisms of defense rather than acquired, antigen-specific mechanisms like those that vertebrates accomplish with lymphocytes. However, some invertebrates possess lymphocyte-like cells that produce antibody-like molecules. Insects, for example, have a hemolymph protein, called *hemolin*, that binds to microbes and assists in their disposal. Hemolin is a member of the immunoglobulin superfamily, a large group of proteins that are structurally related to antibodies. Hemolin molecules do not exhibit diversity, but they are likely evolutionary precursors of vertebrate antibodies.

By and large, invertebrates do not exhibit the hallmark of acquired immunity—immunological memory. Sea star coelomocytes, for example, respond to a particular microbe with the same speed no matter how many times they have encountered that invader before. However, earthworms do appear to have a kind of immunological memory. When a portion of body wall from one worm is grafted onto another, the recipient's phagocytic cells attack the foreign tissue. The initial graft is rejected in about two weeks; a second graft from the same donor is rejected in just a few days. Comparative studies of animal immune systems continue to provide insight into the development and evolution of host defense mechanisms, as well as clues about evolutionary patterns in general.

IMMUNITY IN HEALTH AND DISEASE

As we continue to learn more about the vertebrate immune system, we expand our understanding of the body's battles against infections and cancer, its establishment of long-term immunity, and its reaction to blood transfusions and tissue transplants. We also gain insight into the pathogenesis, prevention, and treatment of numerous diseases, including disorders of the immune system itself.

Immunity can be achieved naturally or artificially

Immunity conferred by recovering from an infectious disease such as chicken pox is called **active immunity** because it depends on the response of the infected person's own immune system. As is the case with all such infections, this immunity is naturally acquired. Active immunity can also be acquired artificially, by **immunization**, also known as **vaccination.** Vaccines include inactivated bacterial toxins, killed microbes, parts of microbes, and viable but weakened microbes. These agents can no longer cause disease, but they retain the ability to act as antigens, stimulating an immune response, and more important, immunological memory. A vaccinated person who encounters the actual pathogen will have the same quick secondary response based on memory cells as a person who has had the disease. Routine immunization of infants and children has dramatically reduced the incidence of infectious diseases such as measles and whooping cough, and has led to the eradication of smallpox, a disfiguring and often fatal viral disease. Unfortunately, not all infectious agents are easily managed by vaccination. For example, although researchers are working intensively to develop a vaccine for HIV, they face many problems, such as the antigenic variability of the virus.

Antibodies can be transferred from one individual to another, providing **passive immunity.** This occurs naturally when IgG antibodies of a pregnant woman cross the placenta to her fetus. In addition, IgA antibodies are passed from mother to nursing infant in breast milk, especially in the early secretions called colostrum. Passive immunity persists only as long as these antibodies last (a few weeks to a few months), but it provides

protection from infections until the baby's own immune system has matured. Passive immunity can also be transferred artificially by injecting antibodies from an animal that is already immune to a disease into another animal, conferring short-term but immediate protection against that disease. Thus, for example, a person bitten by a rabid animal may be injected with antibodies from other people who have been vaccinated against rabies. This measure is important because rabies may progress rapidly, and the response to an active immunization could take too long to save the life of the victim. Actually, most people infected with rabies virus are given both passive and active immunizations: The injected antibodies fight the virus for a few weeks, and then the person's own immune response, induced by the immunization and the infection itself, takes over.

The immune system's capacity to distinguish self from nonself limits blood transfusion and tissue transplantation

In addition to distinguishing between the body's own cells and pathogens such as bacteria, the immune system, given the opportunity, wages war against cells from other individuals. For example, skin transplanted from one person to a non-identical person will look healthy for a day or two, but it will then be destroyed by immune responses. Interestingly, however, a pregnant woman does not reject the fetus as a foreign body. Apparently, the structure of the placenta is the key to this acceptance (see FIGURE 46.17).

In this section you will learn about the potential problems associated with blood transfusions and tissue transplants. Keep in mind that the body's hostile reaction to an incompatible transfusion or transplant is not a disorder of the immune system, but a normal reaction of a healthy immune system exposed to foreign antigens.

Blood Groups and Blood Transfusion

The genetic basis for the **ABO blood groups** was discussed in Chapter 14. An individual with type A blood has A antigens on the surface of his or her red blood cells. The A molecule is referred to as an antigen because it may be identified as foreign if placed in the body of another person; the A antigen is not antigenic to its "owner." Similarly, B antigens are found on type B red blood cells, A and B antigens are found on type AB red blood cells, and neither antigen is found on type O red blood cells.

An individual with type A blood will not, of course, produce antibodies to A antigens. However, a person with type A blood has antibodies to the B antigen, even if the person has never been exposed to type B blood. You may find it odd that antibodies to foreign blood group antigens exist in the body even in the absence of an incompatible blood transfusion.

These antibodies arise in response to bacteria (normal flora) that have epitopes very similar to blood group antigens. Thus an individual with type A blood does not make antibodies to A-like bacterial epitopes—the immune system considers these as self—but that person does make antibodies to B-like bacterial epitopes. Therefore, if a person with type A blood receives a transfusion of type B blood, the preexisting anti-B antibodies will induce an immediate and devastating transfusion reaction. The ABO blood types and corresponding antibodies are summarized in FIGURE 14.10.

Because blood group antigens are polysaccharides, they induce T-independent responses, which elicit no memory cells. As a result, each response is like a primary response, and it generates IgM anti-blood-group antibodies, not IgG. This is fortunate; because IgM does not cross the placenta, no harm comes to a developing fetus with a blood type different from its mother's. However, another red blood cell antigen, the **Rh factor,** is able to cause trouble because antibodies produced to it are of the IgG class. A potentially dangerous situation can arise when a mother is Rh-negative (lacks the Rh factor) but has a fetus that is Rh-positive, having inherited the factor from the father. If small amounts of fetal blood cross the placenta, as may happen late in pregnancy or during delivery of the baby, the mother mounts a T-dependent humoral response against the Rh factor. The danger occurs in subsequent Rh-positive pregnancies, when the mother's Rh-specific memory B cells are exposed to the Rh factor. These B cells produce IgG antibodies, which can cross the placenta and destroy the red cells of the fetus. To prevent this, the mother is injected with anti-Rh antibodies after delivering her first Rh-positive baby. She is, in effect, passively immunized (artificially) to eliminate the Rh antigen before her own immune system responds and generates immunological memory against the Rh factor, endangering her future Rh-positive babies.

Tissue Grafts and Organ Transplantation

The major histocompatibility complex (MHC), which encodes the protein fingerprint unique to each individual, is responsible for stimulating the rejection of tissue grafts and organ transplants. Foreign MHC molecules are antigenic, inducing immune responses against the donated tissue or organ. To minimize rejection, attempts are made to match the MHC of the tissue donor and recipient as closely as possible. In the absence of an identical twin, siblings usually provide the closest tissue-type match. In addition to a close match, various medicines are necessary to suppress the immune response to the transplant. However, this strategy leaves the recipient more susceptible to infection and cancer during the course of treatment. More selective drugs such as cyclosporin A and FK506, which suppress helper T cell activation without crippling non-specific defense or T-independent humoral responses, have greatly improved the success of organ transplants.

In one type of life-giving transplant, that of bone marrow, the graft itself, rather than the host, is the source of potential immune rejection. Bone marrow transplants are used to treat leukemia and other cancers, as well as various hematological (blood cell) diseases. As in any transplant, the MHC of donor and recipient are matched as closely as possible. Prior to transplant, the recipient is typically treated with irradiation to eliminate his or her own bone marrow cells, including any abnormal cells. This treatment effectively obliterates the recipient's immune system, leaving little chance of graft rejection. However, the great danger in bone marrow transplants is that the donated marrow, containing lymphocytes, will react against the recipient. This **graft versus host reaction** is limited if the MHC molecules of the donor and recipient are well matched. Bone marrow donor programs around the world continually seek volunteers; because of the great variability of the MHC, a diverse pool of potential donors is essential.

Abnormal immune function can lead to disease

The highly regulated interplay of lymphocytes with foreign substances, with each other, and with other body cells provides us with extraordinary protection from many diseases. However, if this delicate balance is disrupted by an immune system malfunction, the effects on the individual can range from the minor inconvenience of some allergies to the serious and often fatal consequences of certain autoimmune and immunodeficiency diseases.

Allergies

Allergies are hypersensitive (exaggerated) responses to certain environmental antigens, called allergens. One hypothesis to explain the origin of allergies is that they are evolutionary remnants of the immune system's response to parasitic worms. The humoral mechanism that combats worms is similar to the allergic response that causes such disorders as hay fever and allergic asthma.

The most common allergies involve antibodies of the IgE class (see TABLE 43.1). Hay fever, for example, occurs when plasma cells secrete IgE specific for pollen allergens. Some of the IgE antibodies attach by their tails to mast cells present in connective tissues, without binding to the pollen. Later, when pollen grains enter the body, they attach to the antigen-binding sites of mast cell–associated IgE, cross-linking adjacent antibody molecules. This event induces the mast cell to *degranulate*—that is, to release histamine and other inflammatory agents from vesicles called granules (FIGURE 43.18, p. 918). Recall that histamine causes dilation and increased permeability of small blood vessels. These inflammatory events lead to typical allergy symptoms: sneezing, runny nose, tearing eyes, and smooth muscle contractions that can result in breathing

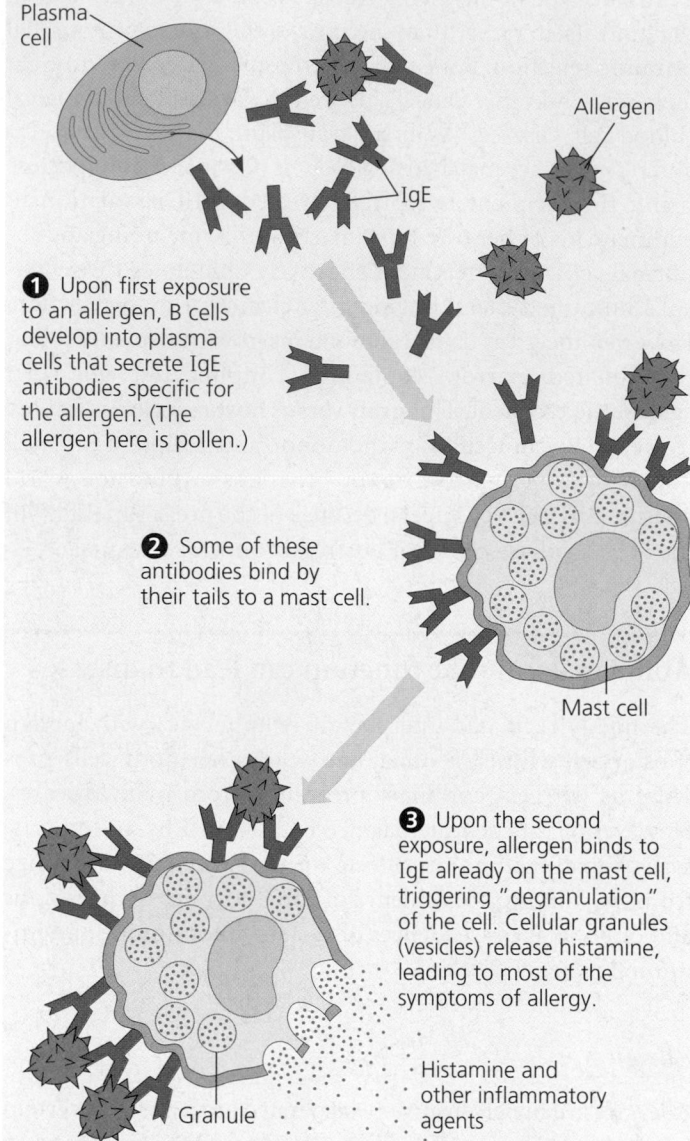

1 Upon first exposure to an allergen, B cells develop into plasma cells that secrete IgE antibodies specific for the allergen. (The allergen here is pollen.)

2 Some of these antibodies bind by their tails to a mast cell.

3 Upon the second exposure, allergen binds to IgE already on the mast cell, triggering "degranulation" of the cell. Cellular granules (vesicles) release histamine, leading to most of the symptoms of allergy.

Plasma cell

Allergen

IgE

Mast cell

Granule

Histamine and other inflammatory agents

FIGURE 43.18 Mast cells, IgE, and the allergic response.

difficulty. Antihistamines diminish allergy symptoms by blocking receptors for histamine.

Sometimes, an acute allergic response can result in **anaphylactic shock,** a life-threatening reaction to injected or ingested allergens. Anaphylactic shock occurs when widespread mast cell degranulation triggers abrupt dilation of peripheral blood vessels, causing a precipitous drop in blood pressure. Death may occur within a few minutes. Allergic responses to bee venom or penicillin can lead to anaphylactic shock in people who are extremely allergic to these substances. Likewise, people very allergic to peanuts, fish, or other foods have died from eating only tiny amounts of these allergens. Some individuals with severe hypersensitivities carry syringes containing the hormone epinephrine, which counteracts this allergic response.

Autoimmune Diseases

Sometimes the immune system loses tolerance for self and turns against certain molecules of the body, causing one of the many autoimmune diseases. In *systemic lupus erythematosus (lupus),* for example, the immune system generates antibodies (known as autoantibodies) against all sorts of self molecules, even histones and DNA released by the normal breakdown of body cells. Lupus is characterized by skin rashes, fever, arthritis, and kidney dysfunction. Another antibody-mediated autoimmune disease, *rheumatoid arthritis,* leads to damage and painful inflammation of the cartilage and bone of joints. In *insulin-dependent diabetes mellitus,* the insulin-producing beta cells of the pancreas are the targets of autoimmune cell-mediated responses. A final example is *multiple sclerosis (MS),* the most common chronic neurological disease in developed countries. In MS, T cells reactive against myelin infiltrate the central nervous system and destroy the myelin of neurons (see FIGURE 48.2). People with MS experience a number of serious neurological abnormalities.

The mechanisms that lead to autoimmunity are not fully understood. For a long time, it was thought that people with autoimmune diseases differed from healthy people in having self-reactive lymphocytes that happened to escape elimination during their development. We now know that healthy people also have lymphocytes with the capacity to react against self. However, these cells are prevented from inducing autoimmune reactions by a number of regulatory mechanisms. So, autoimmune disease likely arises from some failure in immune regulation. One intriguing finding is that the inheritance of particular MHC alleles is associated with susceptibility to certain autoimmune diseases, such as insulin-dependent diabetes mellitus.

Immunodeficiency Diseases

There are almost as many immunodeficiency diseases as there are components of the immune system. Many inborn deficiencies affect the function of either humoral or cell-mediated immune defenses. In *severe combined immunodeficiency (SCID),* both branches of the immune system fail to function. For people with this genetic disease, long-term survival usually requires a bone marrow transplant that will continue to supply functional lymphocytes. For one type of SCID, caused by deficiency of the enzyme adenosine deaminase (ADA), medical scientists have been working to develop a gene therapy in which the individual's own cells are removed, provided with a functional ADA gene, and returned to the body. This treatment would eliminate the danger of graft versus host disease. However, the results to date are equivocal because the patients are also being given additional supplies of the enzyme. Gene therapy for several patients suffering from another type of SCID has also had equivocal results.

Immunodeficiency is not always an inborn condition; an individual may develop immune dysfunction later in life. For example, certain cancers suppress the immune system, especially Hodgkin's disease, which damages the lymphatic system. Another well-known and devastating acquired immune deficiency is AIDS, described in the next section.

Healthy immune function appears to depend on both the endocrine system and the nervous systems. Nearly 2,000 years ago the Greek physician Galen recorded that people suffering from depression were more likely than others to develop cancer. In fact, there is growing evidence that physical and emotional stress can harm immunity. Hormones secreted by the adrenal glands during stress affect the numbers of white blood cells and may suppress the immune system in other ways.

The association between emotional stress and immune function also involves the nervous system. Some neurotransmitters secreted when we are relaxed and happy may enhance immunity. In one study, college students were examined just after a vacation and again during final exams. Their immune systems were impaired in various ways during exam week; for example, interferon levels were lower. These and other observations indicate that general health and state of mind affect immunity. Physiological evidence also points to an immune system–nervous system link: Receptors for neurotransmitters have been discovered on the surfaces of lymphocytes, and a network of nerve fibers penetrates deep into the thymus.

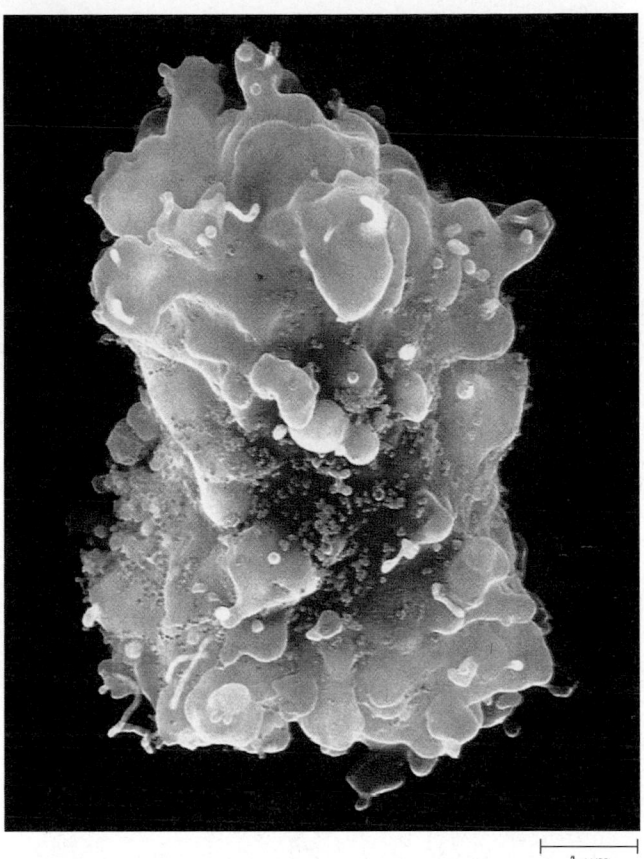

1 μm

FIGURE 43.19 A T cell infected with HIV. The viruses (blue) bud continuously from the surface of the T cell (orange; colorized SEM). The cell will die, but only after it produces many copies of its viral killer.

AIDS is an immunodeficiency disease caused by a virus

In 1981, health care workers in the United States noticed an increasing number of cases of Kaposi's sarcoma, a cancer of the skin and blood vessels, and of pneumonia caused by *Pneumocystis carinii,* a protozoan. The increased rates were notable because of the rarity of these diseases among the general population; they were known to occur mainly in severely immunosuppressed individuals. These observations led to the recognition of what came to be known as **acquired immunodeficiency syndrome,** or **AIDS.** People with AIDS are highly susceptible to *opportunistic diseases,* infections and cancers that take advantage of an immune system in collapse. The *Pneumocystis* protozoan is a ubiquitous organism, yet it does not cause pneumonia in a person with a healthy immune system. In people with AIDS, opportunistic diseases, neurologic damage, and physiological wasting lead to death.

By 1983, a retrovirus, now called **human immunodeficiency virus (HIV),** had been identified as the causative agent of AIDS (FIGURE 43.19; also see FIGURE 18.7 and the interview that opens this unit). With the AIDS mortality rate close to 100%, HIV is the most lethal pathogen ever encountered. Molecular studies of HIV reveal that the virus probably evolved from another HIV-like virus in chimpanzees in central Africa and appeared in hu-

mans sometime between 1915 and 1940, causing rare cases of infection and AIDS that went unrecognized.

There are two major strains of the virus, HIV-1 and HIV-2. HIV-1 is the more widely distributed and more virulent strain. Both strains infect cells that bear surface CD4 molecules. As you know, CD4 molecules are located on helper T cells and enhance the binding between those cells and class II MHC–bearing antigen-presenting cells. Because CD4 also functions as the major receptor for the virus, helper T cells are highly susceptible to infection. Other cell types that bear fewer CD4 molecules, such as macrophages, some B lymphocytes, and brain cells, are also among the cells infected by HIV.

The entry of the virus requires not only CD4 on the surface of the susceptible cell but also a second protein molecule, a *coreceptor.* The coreceptors identified so far include fusin (also called CXCR4), found on helper T cells, and CCR5, found on macrophages. Fusin and CCR5 normally function as receptors for chemokines. In fact, these molecules were first recognized as HIV coreceptors after it was discovered that chemokines can suppress HIV-1 infection. Apparently, chemokines bind to these receptors and block HIV-1 entry. Some people who are innately resistant to HIV-1 owe their resistance to defective

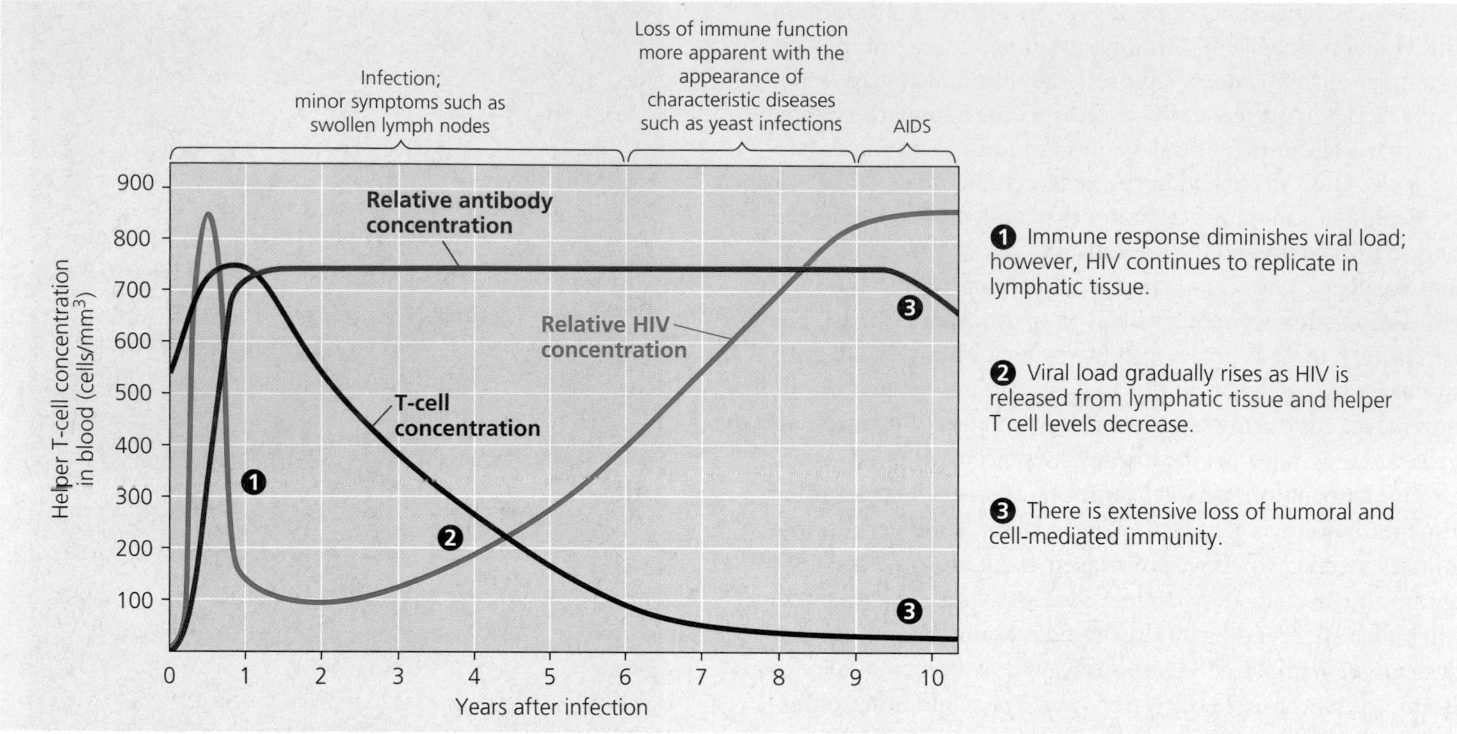
Loss of immune function more apparent with the appearance of characteristic diseases such as yeast infections
Infection; minor symptoms such as swollen lymph nodes
AIDS
Relative antibody concentration
Relative HIV concentration
T-cell concentration
Helper T-cell concentration in blood (cells/mm³)
900
800
700
600
500
400
300
200
100
0
1
2
3
4
5
6
7
8
9
10
Years after infection
1
2
3
3
1 Immune response diminishes viral load; however, HIV continues to replicate in lymphatic tissue.
2 Viral load gradually rises as HIV is released from lymphatic tissue and helper T cell levels decrease.
3 There is extensive loss of humoral and cell-mediated immunity.

FIGURE 43.20 The stages of HIV infection.

chemokine receptors. They are not infected because their coreceptors for HIV are faulty.

Once inside a cell, HIV RNA is reverse-transcribed, and the product DNA is integrated into the host cell genome. In this provirus form, the viral genome directs the production of new virus particles (see FIGURE 18.7). Because a retrovirus exists as a provirus for the life of the infected cell, immune responses fail to eradicate it from the body. Even more challenging to both humoral and cell-mediated responses, however, are the frequent mutational changes that occur in each round of virus replication. Indeed, most HIV particles produced in an infected person differ at least slightly from the original infecting virus.

In spite of these challenges, the immune system engages in a prolonged battle against HIV. After an initial peak, the number of viruses in the blood falls while anti-HIV antibody rises (FIGURE 43.20). The decrease in the number of viruses in the blood results from the early immune response to HIV. Detection of the anti-HIV antibodies, which appear in the blood about 1 to 12 months after infection, is the most common method for identifying infected individuals. A person who is *HIV-positive* is infected, having tested positive for the presence of antibodies to the virus. The HIV antibody test has also been used to screen all blood supplies in the United States since 1985. Because of the chronic presence of virus, a person continues to have anti-HIV antibodies, perhaps until late in AIDS, when both branches of immunity collapse with the loss of helper T cells.

The early fall in the level of HIV in the blood is misleading, however. While the number of circulating viruses may be low, viruses continue to be produced in cells of the lymph nodes, causing structural and functional damage there. In time, the concentration of HIV in the blood (the viral load) increases. Causes of this increase include the breakdown of lymphatic tissue function, the release of virus from these tissues, and diminishing responses to the infection because of the depletion of helper T cells (see FIGURE 43.20).

THE PROCESS OF SCIENCE

Recent studies have revealed that, in HIV infection, helper T cells die because they become infected with the virus. This may sound obvious; cells often die as a result of virus infection. However, other models for T cell depletion have been proposed to explain the apparent loss of helper T cells not infected by HIV. One model, for example, suggests that HIV-mediated interactions induce a T cell to undergo inappropriately timed apoptosis, a process that is normally highly regulated (see FIGURE 21.18). But while such a mechanism may contribute to helper T cell depletion, it is now thought that direct infection is the primary culprit. In fact, the half-life of an actively infected helper T cell (one producing new copies of HIV) is less than 1.5 days.

The time required for an HIV infection to progress to severe helper T cell depletion and AIDS varies greatly, but it currently averages about ten years. During most of this time the individual exhibits only moderate hints of illness, such as swollen lymph nodes (indicating ongoing virus activity) and occasional fever. People with HIV and their physicians follow changes in the level of T cells as an indication of disease progression. However, it has been shown that measures of viral load are a better indicator of disease prognosis and of the effectiveness of anti-HIV treatment.

At this time, HIV infection cannot be cured, and the progression to AIDS cannot be prevented. New drugs show promise in slowing this progression but are very expensive and not available to all those with HIV infection. The evolution of drug-resistant strains of HIV is also a problem (see p. 438); it is addressed by using combinations of drugs. Drugs that seem to slow viral replication when used in various combinations include DNA-synthesis inhibitors, reverse transcriptase inhibitors (such as AZT and ddI), and protease inhibitors. Protease inhibitors prevent a key step in the synthesis of HIV proteins. Used simultaneously, these kinds of drugs decrease viral load and therefore allow the number of helper T cells to rise. Also important to people with AIDS are the numerous medicines used to treat the opportunistic diseases they develop.

Transmission of HIV requires the transfer of body fluids containing infected cells, such as semen or blood, from person to person. Unprotected sex (that is, without a condom) among male homosexuals and transmission via nonsterile needles (typically among intravenous drug users) account for most of the AIDS cases reported thus far in the United States and Europe. However, transmission of HIV among heterosexuals is rapidly increasing as a result of unprotected sex with infected partners. In Africa and Asia, transmission has been primarily by heterosexual sex, especially where there is a high incidence of other sexually transmitted diseases that result in genital lesions. These sores facilitate the transmission of HIV, as the skin barrier (first line of defense) is breached and HIV-susceptible cells, mainly macrophages and helper T cells, are attracted to the area by the inflammatory response.

HIV is not transmitted by casual contact. So far, only one case of HIV transmission by kissing has been reported, and both the person who transmitted the virus and the recipient had bleeding gums. Although this case is isolated, it is important to remember that the virus can be transmitted whenever blood or body secretions are passed from one person to another. Transmission of HIV from mother to child can happen in two ways: during fetal development (as occurs in about 25% of HIV-infected mothers) and during nursing. HIV antibody screening has virtually eliminated blood transfusions as a route of transmission in developed countries. This test, however, will never completely guarantee a safe blood supply because a person may be infected by the virus for several weeks or months before anti-HIV antibodies become detectable.

As of 2000 the Joint United Nations Program on AIDS estimates that 30 to 40 million people worldwide are living with HIV or HIV/AIDS. Of these, approximately 70% reside in sub-Saharan Africa. The number of people with AIDS is expected to grow by nearly 20% per year. The best approach for slowing the spread of HIV is to educate people about the practices that transmit the virus, such as using nonsterile needles and having sex without a condom. Although condoms do not completely eliminate the risk of transmitting HIV (or other similarly transmitted viruses, such as the hepatitis B virus), they do reduce it. Anyone who has sex—vaginal, oral, or anal—with a partner who had unprotected sex with another person during the past two decades risks exposure to HIV.

■   ■   ■

The immune response is one of many adaptations that enable animals to adjust to the adversities of the environment. The next chapter describes several other processes that help maintain favorable conditions within animals as they cope with varying external environments.

CHAPTER 43 REVIEW

Go to the Campbell Biology website (www.campbellbiology.com) to explore an interactive version of the Chapter Review.

Summary of Key Concepts

NONSPECIFIC DEFENSES AGAINST INFECTION

■ **The skin and mucous membranes provide first-line barriers to infection (p. 901, FIGURE 43.1)** The first line of nonspecific defense consists of the intact skin and mucous membranes, mucus, ciliated cells lining the upper respiratory system, lysozyme, and gastric juices.

■ **Phagocytic cells, inflammation, and antimicrobial proteins function early in infection (pp. 901–904, FIGURE 43.5)** The second line of nonspecific defense depends primarily upon neutrophils and macrophages, phagocytic white cells in the blood and tissues. Natural killer cells mediate lysis of virus-infected cells and tumor cells. Tissue damage triggers a local inflammatory response. Injured cells release histamine, a chemical signal that causes dilation and increased permeability of blood vessels, allowing fluid and large numbers of phagocytic white blood cells to enter the tissues. The most important antimicrobial proteins in the blood and tissues are the proteins of the complement system, involved in both nonspecific and specific defense, and interferons. Secreted by virus-infected cells, interferons inhibit virus production in neighboring cells.

HOW SPECIFIC IMMUNITY ARISES

- **Lymphocytes provide the specificity and diversity of the immune system (pp. 904–905)** A substance that elicits an immune response is called an antigen. The immune system recognizes specific antigens (molecules belonging to microbes, toxins, transplanted tissue, or cancer cells) and develops an immune response that inactivates or destroys that substance. B lymphocytes and T lymphocytes recognize antigens via surface antigen receptors: membrane antibodies for B cells, and T-cell receptors for T cells. Lymphocytes circulate throughout the blood and lymph and are found in high numbers in lymphatic tissues. The great diversity of lymphocytes, each with receptors of one particular specificity, gives the immune system the capacity to respond to virtually any antigen.

- **Antigens interact with specific lymphocytes, inducing immune responses and immunological memory (pp. 905–906, FIGURE 43.6)** Clonal selection occurs when an antigen activates a lymphocyte by binding to a specific receptor. In the primary immune response (to the body's first exposure to an antigen), the lymphocyte proliferates and differentiates, forming a clone of short-lived, infection-fighting effector cells and a clone of long-lived memory cells, all specific for the antigen. Secondary immune responses to that same antigen, which involve memory cells, are faster and often protective.

- **Lymphocyte development gives rise to an immune system that distinguishes self from nonself (pp. 906–908, FIGURES 43.8 and 43.9)** Lymphocytes develop from pluripotent stem cells in the bone marrow. B cells mature in the marrow, while T cells mature in the thymus. Self-tolerance develops as lymphocytes bearing receptors specific for native molecules are destroyed or rendered nonresponsive. Major histocompatibility complex (MHC) molecules are crucial to T cell function. Class I MHC molecules, located on all nucleated cells of the body, present antigen fragments to cytotoxic T cells. Class II MHC molecules, found mainly on macrophages and B cells, present antigen fragments to helper T cells. Developing T cells are exposed to class I and II MHC molecules on cells of the thymus. Only T cells bearing receptors with affinity for self-MHC molecules reach maturity.

IMMUNE RESPONSES

- **Helper T lymphocytes function in both humoral and cell-mediated immunity:** *an overview* **(pp. 908–909, FIGURES 43.10 and 43.11)** Humoral, or B cell, immunity, based on circulation of antibodies in the blood and lymph, defends against free viruses, bacteria, and other extracellular threats. Cell-mediated, or T cell, immunity defends against intracellular pathogens by destroying infected cells; it also defends against transplanted tissue and cancer cells. A CD4-bearing helper T cell is activated when its receptor binds specifically to a class II MHC–antigen complex on the surface of an antigen-presenting cell (APC). The T cell then secretes interleukin-2 and other cytokines, which help activate B cells and cytotoxic T cells.

- **In the cell-mediated response, cytotoxic T cells counter intracellular pathogens:** *a closer look* **(pp. 909–911, FIGURE 43.12)** Most cytotoxic T cells are activated by cytokines and specific binding to class I MHC–antigen complexes on a target (infected, transplanted, or cancerous) cell. The T cell then secretes perforins, which form pores in the target cell membrane, causing the cell to lyse.

- **In the humoral response, B cells make antibodies against extracellular pathogens:** *a closer look* **(pp. 911–915, FIGURES 43.13–43.16)** B cells are activated by cytokines and specific binding of their membrane antibodies to extracellular antigens. Most of these antigens are proteins or large polysaccharides, each with multiple epitopes. Antibodies, also called immunoglobulin (Ig) molecules, are serum proteins. The variable regions of an Ig molecule bind to a specific epitope; the constant regions determine the antibody's class. The five major immunoglobulin classes are IgG, IgM, IgA, IgD, and IgE. An antibody does not destroy an antigen directly but neutralizes it or targets it for elimination by opsonization, agglutination, precipitation, or complement fixation. Opsonization, agglutination, and precipitation enhance phagocytosis of the antigen-antibody complex; complement fixation leads to lysis of a complement protein–bound bacterium or virus.

- **Invertebrates have a rudimentary immune system (pp. 915–916)** Invertebrates have the ability to distinguish between self and nonself. In many invertebrates, amoeboid cells called coelomocytes can identify and destroy foreign substances. Experiments with earthworms show that their defense systems form memory against tissue grafts.

Web/CD Activity 43A: *Immune Responses*

IMMUNITY IN HEALTH AND DISEASE

- **Immunity can be achieved naturally or artificially (p. 916)** Active immunity occurs when the immune system responds to a foreign antigen acquired either by natural infection or artificially, as by immunization. In immunization, a nonpathogenic form of a microbe or part of a microbe generates an immune response to and immunological memory for that microbe. Passive immunity occurs when antibodies are transferred from one individual to another. It occurs naturally, when IgG passes from mother to fetus or when IgA passes from mother to infant in breast milk, or artificially, when antibodies from an animal immune to a disease are injected into another animal, conferring short-term protection.

- **The immune system's capacity to distinguish self from nonself limits blood transfusion and tissue transplantation (pp. 916–917)** Certain antigens on red blood cells determine whether a person has type A, B, AB, or O blood. Antibodies to nonself blood types (generally IgM) already exist in the body. If incompatible blood is transfused, the transfused cells are killed by antibody- and complement-mediated lysis. The Rh factor, another red blood cell antigen, creates difficulties when an Rh-negative mother carries successive Rh-positive fetuses. During delivery of the first Rh-positive infant, the mother's immune system develops anti-Rh IgG, which can cross the placenta and may attack the red blood cells of a subsequent Rh-positive fetus. The chances of success in organ or tissue transplantation are improved if the donor and recipient MHC tissue types are well matched. In addition, immunosuppressive drugs help prevent rejection. In bone marrow transplantation, there is danger of a graft versus host reaction.

- **Abnormal immune function can lead to disease (pp. 917–919, FIGURE 43.18)** In allergies such as hay fever, an allergen, such as pollen, triggers histamine release from mast cells, inducing vascular changes and typical symptoms. Sometimes the immune system loses tolerance for self, leading to autoimmune diseases such as rheumatoid arthritis and insulin-dependent diabetes. Some people are naturally deficient in humoral or cell-mediated immune defenses, or both.

- **AIDS is an immunodeficiency disease caused by a virus (pp. 919–921, FIGURE 43.20)** Acquired immunodeficiency syndrome (AIDS) is caused by the direct and indirect destruction of CD4-bearing T cells by HIV, the human immunodeficiency virus, over a period of years. AIDS, the final stage of this process, is marked by low helper T cell levels and opportunistic diseases characteristic of a deficient cell-mediated immune response.

Web/CD Activity 43B: *HIV Reproductive Cycle*

Web/CD Case Study in the Process of Science: *What Causes Infections in AIDS Patients?*
Web/CD Case Study in the Process of Science: *Why Do AIDS Rates Differ Across the U.S.?*

Self-Quiz

1. An Rh-positive baby is born to an Rh-negative mother. The mother is treated with antibodies specific for the Rh factor in order to
   a. protect her from an inappropriate immune response.
   b. prevent her from generating memory B cells specific for the Rh factor.
   c. protect her future Rh-positive babies.
   d. induce an immune response to Rh antibodies.
   e. both b and c.

2. Which of the following results in long-term immunity?
   a. the passage of maternal antibodies to her developing fetus
   b. the inflammatory response to a splinter
   c. the administration of serum obtained from people immune to rabies
   d. the administration of the chicken pox vaccine
   e. the passage of maternal antibodies to her nursing infant

3. Which of the following is not part of the body's nonspecific defense system?
   a. natural killer (NK) cells
   b. inflammation
   c. phagocytosis by neutrophils
   d. phagocytosis by macrophages
   e. antibodies

4. Which of the following molecules is *incorrectly* paired with a source?
   a. lysozyme—tears
   b. interferons—virus-infected cells
   c. interleukin-1—macrophages
   d. perforins—cytotoxic T cells
   e. immunoglobulins—helper T cells

5. HIV targets include all of the following *except*
   a. macrophages.
   b. cytotoxic T cells.
   c. helper T cells.
   d. cells bearing CD4 and fusin.
   e. cells bearing CD4 and CCR5.

6. Which of the following best describes the difference in the way B cells and cytotoxic T cells respond to invaders?
   a. B cells confer active immunity; cytotoxic T cells confer passive immunity.
   b. B cells kill viruses directly; cytotoxic T cells kill virus-infected cells.
   c. B cells secrete antibodies against a virus; cytotoxic T cells kill virus-infected cells.
   d. B cells accomplish cell-mediated immunity; cytotoxic T cells accomplish humoral immunity.
   e. B cells respond the first time the invader is present; cytotoxic T cells respond subsequent times.

7. Which of the following is a characteristic of the early stages of local inflammation?
   a. precapillary arteriole constriction
   b. fever
   c. attack by cytotoxic T cells
   d. release of histamine
   e. antibody-complement–mediated lysis of microbes

8. An epitope associates with which part of an antibody?
   a. the antibody-binding site
   b. the heavy-chain constant regions only
   c. the variable regions of a heavy chain and light chain combined
   d. the light-chain constant regions only
   e. the antibody tail

9. Which of the following is *not* true about helper T cells?
   a. They function in both cell-mediated and humoral immune responses.
   b. They recognize polysaccharide fragments presented by class II MHC molecules.
   c. They bear surface CD4 molecules.
   d. They are subject to infection by HIV.
   e. When activated, they secrete IL-2 and other cytokines.

10. Indicate whether each of the following choices is descriptive of a B cell (B), cytotoxic T cell (T_C), helper T cell (T_H), or macrophage (M). A single feature may be descriptive of more than one type of cell.
    a. develops into an antibody-secreting plasma cell
    b. is phagocytic
    c. bears antigen receptors called immunoglobulins
    d. bears the surface molecule CD4
    e. bears the surface molecule CD8
    f. is an important component of nonspecific responses
    g. produces cytokines such as interleukin-2 that boost both humoral and cell-mediated responses
    h. mediates specific recognition of and response to a particular antigen
    i. bears surface TCR and CD3
    j. kills virus-infected cells

11. Indicate whether each of the following statements is consistent (C) or inconsistent (I) with your understanding of immune reactions.
    a. When antibodies bind to a bacterium, they directly kill it within seconds.
    b. Autoantibodies are a normal response to one's own biological molecules.
    c. Complement activation can result in bacterial cell lysis.
    d. Invertebrate immune systems can distinguish self from nonself.
    e. The secondary immune response is slower and weaker than the primary immune response.
    f. One way that antibodies mediate the death of bacteria is by activating the complement system.

12. Complete each of the following statements with the appropriate term or terms from this list: helper T cells, cytotoxic T cells, antibodies, B cells, class I MHC, class II MHC, cytokines, complement. Some terms may be used more than once; others may not be used at all.
    a. Gamma interferon induces virus-infected cells to put more than the usual number of class I MHC molecules on their surfaces. This improves the ability of _____ to detect the infected cells.
    b. The activation of _____ in culture could be measured by testing how much interleukin-2 they produce.
    c. The activation of B cells in response to an immunization can be measured by testing blood for increased levels of _____.
    d. When macrophages phagocytose viruses, they present viral antigen associated with _____ molecules to helper T cells.
    e. A type of severe combined immune deficiency disease results from the lack of functional interleukin-2 receptors. The severity of this disease is not surprising given that _____ require both binding to foreign antigen and interleukin-2 in order to be fully activated.

13. Matching:
    a. Carries out humoral immunity
    b. Contains histamines that trigger allergy symptoms
    c. Cell type most commonly infected by HIV
    d. Kills virus-infected body cells
    e. A phagocytic white blood cell
    f. A long-lived cell that provides for faster secondary immune responses
    g. A general term for leukocytes capable of specific antigen recognition

    Neutrophil
    Cytotoxic T cell
    B cell
    Lymphocyte
    Memory B cell
    Helper T cell
    Mast cell

14. Indicate the kind of defense to which each of the following terms or phrases apply, using N for nonspecific defense, H for humoral defense, and C for cell-mediated defense.
    a. the production of antibodies
    b. the specific recognition and direct killing of virus-infected cells
    c. phagocytosis by neutrophils
    d. T cells are mainly responsible
    e. B cells are mainly responsible
    f. fever
    g. inflammation
    h. phagocytosis by macrophages

15. For each statement about HIV and AIDS, indicate whether it is true (T) or false (F).
    a. Entry of HIV into T cells requires two coreceptors, CD4 and fusin.
    b. Only cells with CD4 are infected by HIV.
    c. HIV is a retrovirus.
    d. CCR4 and fusin normally act as cytokine receptors.
    e. A person with HIV might transmit it to another person through a handshake.
    f. Protease inhibitors block a step in the production of HIV.
    g. Opportunistic infections occur when immune responses are diminished, as in AIDS.
    h. Donated blood is routinely screened for the presence of antibodies to HIV.

Go to the website or CD-ROM for more quiz questions.

Evolution Connection

1. Invertebrates make up more than 90% of the Earth's species today. The reasons for their success undoubtedly include their systems of defending themselves against microbial invasion. Describe one mechanism by which invertebrates combat such invaders, and discuss how this mechanism comprises an evolutionary adaptation that is retained in the vertebrate immune system.

2. Cheetahs were an endangered species until programs of breeding in captivity increased their numbers. The current populations of cheetahs, however, appear to be more susceptible to infectious diseases than their large-cat relatives. Discuss MHC polymorphism as an evolutionary adaptation and how the cheetahs' loss of polymorphism at the MHC may impact the health and survival of this species.

The Process of Science

You know that gamma interferon has many effects on cells, one of which is to increase the number of class I MHC molecules on the cell surface. You have prepared a supply of gamma interferon using recombinant DNA technology, and you want to test it on laboratory animals with (a) virus infections and (b) cancers. Considering the multiple activities of gamma interferon, what predictions would you make about its effects on the animals' immune response against (a) virus-infected cells and (b) cancer cells?

Learn about opportunistic infections and the epidemiology of AIDS in the Case Studies in the Process of Science, available on the CD-ROM and website.

Science, Technology, and Society

Until a vaccine became available, poliomyelitis (polio) was one of the most feared diseases, resulting in paralysis when the virus destroyed motor neurons in the brain and spinal cord. Currently, both a live, attenuated oral and an injectable, inactivated (killed) poliovirus vaccine are available. The ease of administration has made the oral vaccine more popular, and it is generally credited with controlling the disease worldwide. However, the live virus may mutate to the more virulent form. In any given year approximately ten people in the United States contract vaccine-associated paralytic polio. Of those cases, the majority are in individuals who were healthy when immunized or who came into contact with someone recently vaccinated. Statistically, the probability of contracting vaccine-associated polio remains low, 1 in 12 million. Do you feel the risk is acceptable when compared to the benefits of oral vaccination? How might public health decisions of this type be made?

Answers: 1. e; 2. d; 3. e; 4. e; 5. b; 6. c; 7. d; 8. c; 9. b; 10. a. B; b. M; c. B; d. TH; e. TC; f. M; g. TH; h. B, TC, TH; i. TC, TH; j. TC; 11. a. I; b. I; c. C; d. C; e. I; f. C; 12. a. cytotoxic T cells; b. helper T cells; c. antibodies; d. class II MHC; e. B cells, cytotoxic T cells and helper T cells; 13. a. B cell; b. Mast cell; c. Helper T cell; d. Cytotoxic T cell; e. Neutrophil; f. Memory B cell; g. Lymphocyte; 14. a. H; b. C; c. N; d. C; e. H; f. N; g. N; h. N; 15. a. T; b. T; c. T; d. F; e. F; f. T; g. T; h. T

O ne of the most remarkable *characteristics of animals is that they can maintain physiologically favorable internal environments even as external conditions undergo dramatic shifts that would be lethal to individual cells. This ability of animals to regulate their internal environment is called* **homeostasis** *(see Chapters 1 and 40). An example is the regulation of body temperature. Humans may be exposed to substantial changes in outside temperatures but will die if their internal temperatures drift more than a few degrees above or below 37°C. Another mammal, the arctic wolf you see in the photo above, is so effective at regulating body temperature that it survives winters when temperatures drop as low as −50°C.*

This chapter focuses on: **thermoregulation,** *how animals maintain internal temperature within a tolerable range;* **osmoregulation,** *how they regulate solute balance and the gain and loss of water; and* **excretion,** *how they get rid of the nitrogen-containing waste products of metabolism such as urea. But first, we'll develop a couple of general concepts that provide additional context for our study of homeostasis.*

AN OVERVIEW OF HOMEOSTASIS

Regulating and conforming are the two extremes in how animals cope with environmental fluctuations

An animal is said to be a **regulator** for a particular environmental variable if it uses mechanisms of homeostasis to moderate internal change in the face of external fluctuation. For example, endothermic animals such as mammals and birds are thermoregulators, meaning they keep body temperature within narrow limits in spite of changes in environmental temperature. Let's consider another example. Pacific salmon spend part of their lives in salt water and part in fresh water. Throughout this change in "osmotic environment," a salmon uses mechanisms of osmoregulation to maintain a constant concentration of solutes in its blood and interstitial fluid.

In contrast to such regulators, many other animals, especially those that live in relatively stable environments, are **conformers** in their relationship to certain environmental changes. Such conformers allow some conditions within their bodies to vary with certain external changes (FIGURE 44.1a). Many marine invertebrates, such as spider crabs of the genus *Libinia,* live in environments where the salinity is relatively stable. These organisms do not osmoregulate, and if placed in water of varying salinity, they will lose or gain water to conform to the external environment even when this internal adjustment is extreme enough to cause death (FIGURE 44.1b).

Conforming and regulating represent extremes on a continuum, and no organisms are perfect regulators or conformers. For example, the salmon previously described can osmoregulate, but they conform to external temperatures.

Even for a particular environmental variable, a species may conform in one situation and regulate in another. Regulation requires the expenditure of energy, and in some environments the cost of regulation may outweigh the benefits of homeostasis. For example, temperature regulation would require the forest-dwelling lizard *Anolis cristatellus* to travel long distances (and risk capture by a predator) to find an exposed sunny perch. (Reptiles are ectotherms, but can regulate body temperature by moving to a location of favorable temperature.) The lizard is therefore likely to survive longer and produce more offspring by allowing its body temperature to conform to that of the forest environment. However, this same species uses a behavioral adaptation to thermoregulate in open habitats; the lizard seeks patches of sun where it can bask.

Homeostasis balances an animal's gains versus losses for energy and materials

Like all organisms, animals are open systems that must exchange energy and materials with their environments: Food must be eaten to provide nutrients and chemical energy, oxygen is needed for cellular respiration, CO_2 and other metabolic wastes must be removed, heat and water are exchanged, and so forth. These inward and outward flows of energy and materials are frequently rapid and often variable, but as they occur animals also need to maintain reasonably constant internal conditions. This means that rates of gain and loss must be closely matched over time; if they are not, life-threatening imbalances will eventually result. Normally, an animal's inputs of energy and materials only exceed its outputs when there is a net increase in organic matter due to growth or reproduction.

Consider some of the exchanges that occur during ten years in the life of a typical woman weighing 60 kg (shown symbolically in FIGURE 44.2). During that decade she will eat about 2 tons of food, drink 6 to 10 tons of water, use almost 2 tons of oxygen, and metabolically generate more than 7 million kilocalories of heat (enough to warm 90 tons of water from room temperature to the boiling point). The same quantity of materials and heat must be lost from the woman's body in order to maintain its size, temperature, and chemical composition.

If the woman produces two children (and breast-feeds each for two years) during this ten-year span, she will need to increase the total flow of energy and materials by only 4–5% compared to her basic maintenance needs. Reproduction is a larger part of the energy and materials flow in many other species—for example, 10–15% of the annual energy requirement in female mice rearing two litters per year—but regardless of reproductive costs, every animal's survival depends on accurate control of materials and energy exchange.

Because homeostasis requires such a careful balance of materials and energy, it can be viewed as a set of **budgets** of gains and losses (a heat budget, an energy budget, a water budget, and so on). Most energy and materials budgets are interconnected, with changes in the flux of one component affecting the exchanges of other components. For example, when terrestrial animals exchange gases with the air by breathing, they

FIGURE 44.1 Regulators and conformers.

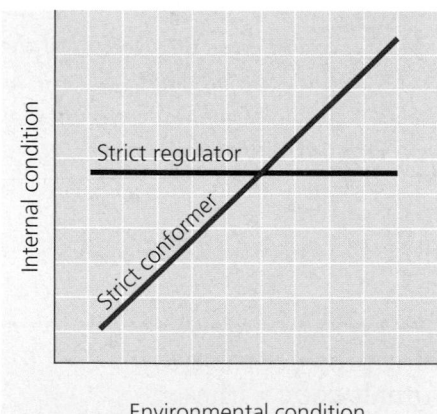

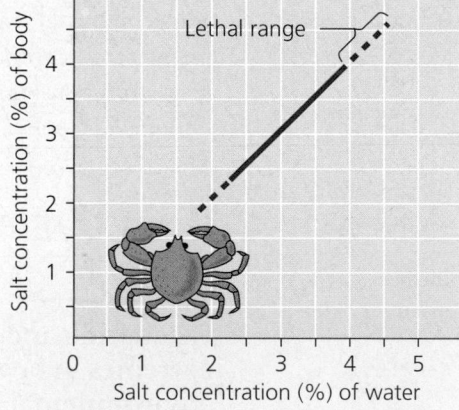

(a) With respect to a given environmental variable, organisms can be described as either regulators, which maintain a nearly constant internal environment over a range of external conditions, or conformers, which allow their internal environment to vary. Although the two "strategies" are idealized in the illustration, most organisms are neither strict regulators nor strict conformers.

(b) Spider crabs (*Libinia*) are osmoconformers with little ability to regulate internal salt concentration in the narrow range of salinity where they live. If exposed to slightly higher or lower salinity in the laboratory, the crabs continue to conform and soon die.

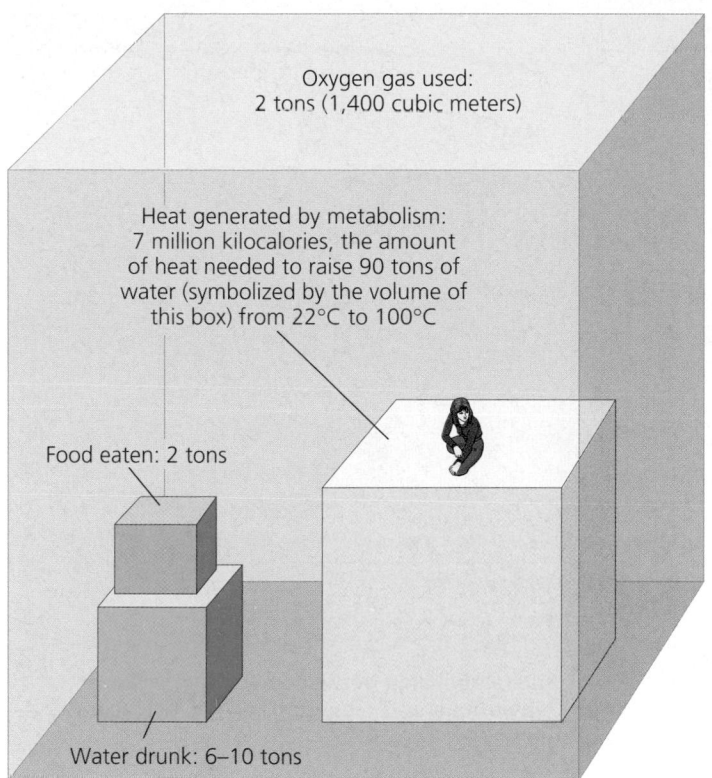

Oxygen gas used:
2 tons (1,400 cubic meters)

Heat generated by metabolism:
7 million kilocalories, the amount
of heat needed to raise 90 tons of
water (symbolized by the volume of
this box) from 22°C to 100°C

Food eaten: 2 tons

Water drunk: 6–10 tons

FIGURE 44.2 A partial energy and material bookkeeping for ten years in the life of a young woman. The accounting here is incomplete. For example, uptake of material in the form of food and O_2 must be balanced by the loss of material as wastes such as CO_2 and organic wastes in feces and urine.

also lose water by evaporation from the moist lung surfaces. That loss must be compensated by intake (in food or drink) of an equal amount of water. Also, evaporation removes heat from the body (see FIGURE 3.4), and that heat loss must be balanced by the gain of an equivalent amount of heat from some other source.

Let us now see how mechanisms of homeostasis balance an animal's energy and materials budgets, beginning with thermoregulation.

REGULATION OF BODY TEMPERATURE

Most biochemical and physiological processes are very sensitive to changes in body temperature. The rates of most enzyme-mediated reactions increase by a factor of 2–3 for every 10°C temperature increase, until temperature is high enough to begin to denature proteins. This is known as the **Q_{10} effect,** with Q_{10} being the multiple by which a particular enzymatic reaction or overall metabolic process increases with a 10°C increase in body temperature. For example, if the rate of glycogen hydrolysis in a frog is 2.5 times greater at 30°C than at 20°C, then the Q_{10} for that reaction is 2.5. The properties of membranes also change with temperature. These

thermal effects dramatically influence animal function and performance. For example, because the power and speed of muscle contraction is strongly temperature dependent, a body temperature change of only a few degrees may have a very large impact on an animal's ability to run, jump, or fly.

Although different species of animals are adapted to different environmental temperatures, each animal has an optimal temperature range. Within that range, many animals maintain a nearly constant internal temperature as the external temperature fluctuates. This thermoregulation helps keep body temperature within a range that enables cells to function most effectively. An animal that thermoregulates balances its heat budget over time in such a way that the rate of heat gain exactly matches the rate of heat loss. Therefore, to understand thermoregulation, we first need to consider the ways heat is exchanged between animals and their environments.

Four physical processes account for heat gain or loss

An organism, like any object, exchanges heat by four physical processes called conduction, convection, radiation, and evaporation (FIGURE 44.3). These processes account for the flow of heat within an organism and between an organism and its external environment.

Conduction is the direct transfer of thermal motion (heat) between molecules of objects in direct contact with each other, as when an animal sits in a pool of cold water or on a

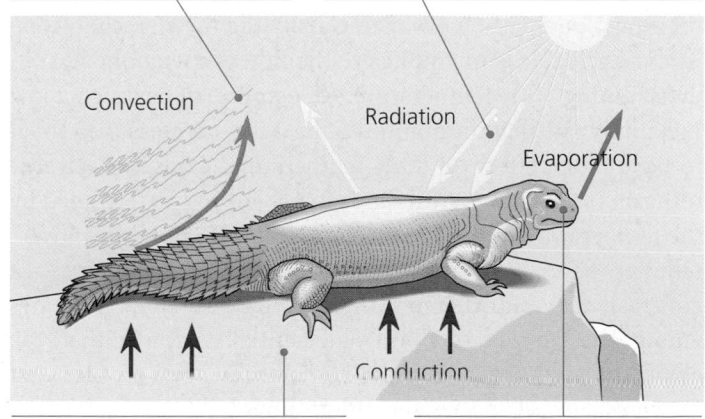

Convection may contribute to heat exchange if a current of air or water passes over an animal.

All objects warmer than absolute zero emit electromagnetic energy. Here, the lizard gains more heat by absorbing radiation from the sun than its body radiates to the environment.

Convection

Radiation

Evaporation

Conduction

An animal can elevate a low body temperature using heat conducted from a warm rock, a common practice among reptiles.

The animal loses some heat by evaporative cooling from moist surfaces exposed to the environment.

FIGURE 44.3 Heat exchange between an organism and its environment. Heat is conducted from an object of higher temperature to one of lower temperature.

hot rock. Heat is always conducted from an object of higher temperature to one of lower temperature, but the rate and amount of heat transfer varies with different materials. Water is 50 to 100 times more effective than air in conducting heat. This is one reason you can rapidly cool your body on a hot day just by standing in cold water.

Convection is the transfer of heat by the movement of air or liquid past a surface, as when a breeze contributes to heat loss from the surface of an animal with dry skin, or circulating blood moves heat from an animal's warm body core to the cooler extremities such as legs. The familiar "wind-chill factor" is an example of how convection compounds the harshness of low temperatures by increasing the rate of heat transfer.

Radiation is the emission of electromagnetic waves by all objects warmer than absolute zero, including an animal's body, the environment, and the sun. Radiation can transfer heat between objects that are not in direct contact, as when an animal absorbs heat radiating from the sun.

Evaporation is the removal of heat from the surface of a liquid that is losing some of its molecules as gas. Evaporation of water from an animal has a strong cooling effect, but it can only occur if the surrounding air is not saturated with water molecules (that is, if the relative humidity is less than 100%). This is the basis for the common complaint, "The heat is not as bad as the humidity."

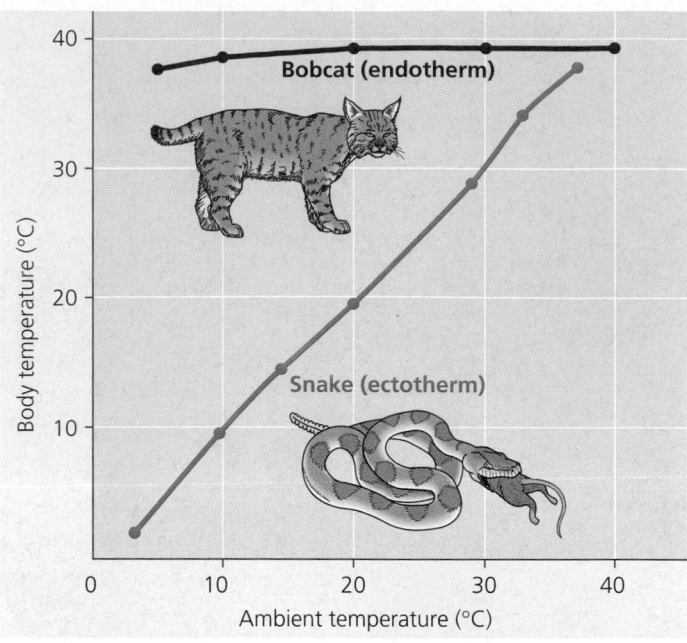

FIGURE 44.4 The relationship between body temperature and ambient (environmental) temperature in an ectotherm and an endotherm.

Modified from P. T. Marshall and G. M. Hughes, *Physiology of Mammals and Other Vertebrates*, 2nd ed. (Cambridge: Cambridge University Press, 1980). Reprinted with the permission of Cambridge University Press.

Ectotherms have body temperatures close to environmental temperature; endotherms can use metabolic heat to keep body temperature warmer than their surroundings

Although all animals exchange heat through some combination of conduction, convection, radiation, and evaporation, there are important differences in how various species manage their heat budgets. One way to classify the thermal characteristics of animals is to emphasize the role of metabolic heat in determining body temperature. An **ectotherm** has such a low metabolic rate that the amount of heat it generates is too small to have much effect on body temperature. Consequently, ectotherm body temperatures are almost entirely determined by the temperature of the surrounding environment. Most invertebrates, fishes, amphibians, and reptiles are ectotherms. In contrast, an **endotherm's** high metabolic rate generates enough heat to keep its body substantially warmer than the environment. Mammals, birds, some fishes, a few reptiles, and numerous insect species are endotherms.

Many endotherms, including humans, maintain high and very stable internal temperature even as the temperature of their surroundings fluctuates (FIGURE 44.4). However, it is not constant body temperature that distinguishes endotherms from ectotherms. For example, many ectothermic marine fishes and invertebrates inhabit water with such stable temperatures

that their body temperature varies less than that of humans and other endotherms. Also, many endotherms sustain high body temperatures only part of the time (for example, while they are active). And not all ectotherms have low body temperatures: When sitting in the sun, many ectothermic lizards have higher body temperatures than mammals. Therefore, most biologists do not use the familiar terms *cold-blooded* and *warm-blooded* because they are frequently misleading.

Endothermy has several important advantages. High and stable body temperature, along with other biochemical and physiological adaptations associated with endothermy (such as elaborate circulatory and respiratory systems), give these animals very high levels of aerobic metabolism (cellular respiration). This allows endotherms to perform vigorous activity for much longer than is possible for ectotherms (see FIGURE 40.13). Sustained intense activity, such as long-distance running or powered (flapping) flight, is usually only feasible for animals with an endothermic way of life. Endothermy also solves certain thermal problems of living on land, enabling terrestrial animals to maintain stable body temperature in the face of environmental temperature fluctuations that are generally more severe than in aquatic habitats. For example, no ectotherm can be active in the below-freezing weather that prevails during winter over much of Earth's surface, but many endotherms function very well in these conditions. Most of the time, endothermic vertebrates—birds and mammals—are warmer than their surroundings, but these animals also have mechanisms for cooling the body in a hot environment, and this enables them to withstand heat loads that are intolerable for most ectotherms.

Being endothermic is liberating, but it is also energetically expensive, especially in a cold environment. For example, at 20°C, a human at rest has a metabolic rate of 1,300 to 1,800 kcal per day (see Chapter 40). In contrast, a resting ectotherm of similar weight, such as an American alligator, has a metabolic rate of only about 60 kcal per day at 20°C. Thus, endotherms generally need to consume much more food than ectotherms of equivalent size—a serious disadvantage for endotherms if food supplies are limited. For this and other reasons, ectothermy is an extremely effective and successful "strategy" in many terrestrial environments, as shown by the abundance and diversity of ectothermic insects, spiders, amphibians, and reptiles.

Thermoregulation involves physiological and behavioral adjustments that balance heat gain and loss

For endotherms and for those ectotherms that thermoregulate, the essence of thermoregulation is management of the heat budget so that rates of heat gain are equal to rates of heat loss. If the heat budget gets out of balance, the animal will either become warmer or colder. Four general categories of adaptations help animals thermoregulate:

1. *Adjusting the rate of heat exchange between the animal and its surroundings.* Insulation, such as hair, feathers, and fat located just beneath the skin, reduces the flow of heat between an animal and its environment. Other mechanisms that regulate rates of heat exchange usually involve adaptations of the circulatory system. Many endotherms and some ectotherms can alter the amount of blood (and hence heat) flowing between the body core and the skin. Elevated blood flow in the skin normally results from **vasodilation,** an increase in the diameter of superficial blood vessels (those near the body surface) triggered by nerve signals that relax the muscles of the vessel walls. In endotherms, vasodilation usually warms the skin, increasing the transfer of body heat to a cool environment by radiation, conduction, and convection (see FIGURE 44.3). The reverse process, **vasoconstriction,** reduces blood flow and heat transfer by decreasing the diameter of superficial vessels.

Another circulatory adaptation is a special arrangement of blood vessels called a **countercurrent heat exchanger** that helps trap heat in the body core and is important in reducing heat loss in many endotherms (FIGURE 44.5). For example, marine mammals and many birds face the problem of losing large amounts of heat from their extremities, which are often immersed in cold water or in contact with ice or snow. Arteries carrying warm blood down the legs of a bird or the flippers of a dolphin are in close contact with veins conveying cool blood in the opposite direction, back toward the trunk of the

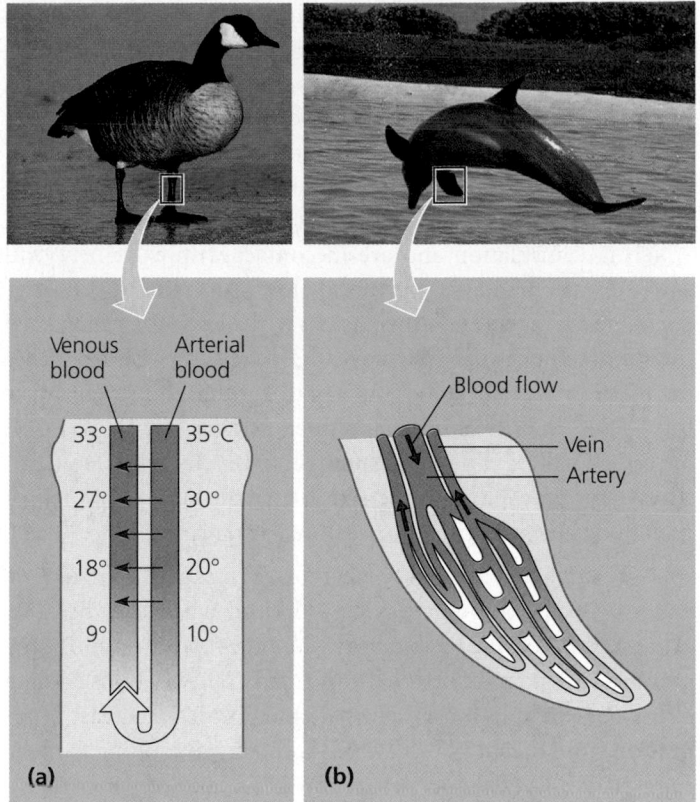

FIGURE 44.5 Countercurrent heat exchangers. (a) Many birds, such as this Canada goose, have countercurrent systems in their legs that reduce heat loss. The arteries carrying blood down the legs contact the veins that return blood to the body core. In a cold environment, the heat in warm arterial blood is transferred (black arrows) to venous blood returning to the body, instead of being lost to the environment. **(b)** In the flippers of marine mammals, such as this Pacific bottlenose dolphin, each artery is surrounded by several veins in a countercurrent arrangement that allows efficient heat exchange between arterial and venous blood.

body. This countercurrent arrangement facilitates heat transfer from arteries to veins along the entire length of the blood vessels. Near the end of the leg or flipper, where arterial blood has been cooled to far below the animal's core temperature, the artery can still transfer heat to the even colder blood of an adjacent vein (recall that heat moves from a warmer object to a cooler one). The venous blood can continue to absorb heat because it is passing warmer and warmer arterial blood traveling in the opposite direction. As the venous blood approaches the center of the body, it is almost as warm as the body core, minimizing the heat lost as a result of supplying blood to body parts immersed in cold water. In essence, heat in the arterial blood emerging from the body core is transferred directly to the returning venous blood, instead of being lost to the environment. In some species, blood can either go through the heat exchanger or bypass it in other blood vessels. The relative amount of blood that flows through the two different paths varies, adjusting the rate of heat loss as an animal's physiological state or environment changes.

Circulatory adaptations that reduce heat loss enable some endotherms to survive the most extreme winter conditions. For example, take another look at the arctic wolf in the photo on page 925. Wolves remain active in very cold weather. Thick fur coats help keep their bodies warm. Countercurrent heat exchangers keep a wolf's legs from freezing. But what mechanism restricts heat loss from the hairless, fleshy foot and toe pads, which lack insulation and are in contact with extremely cold snow or ice? By adjusting blood flow through the countercurrent exchangers and other vessels in the legs, wolves can keep their foot temperature just above 0°C in freezing weather—cool enough to reduce heat loss but warm enough to prevent frostbite—even in environmental temperatures as low as −50°C. If necessary, wolves can warm their feet to much higher temperatures—for example, if they need to lose the large quantities of metabolic heat produced during long-distance running.

2. *Cooling by evaporative heat loss.* Terrestrial animals lose water by evaporation across the skin and when they breathe. Water absorbs considerable heat when it evaporates, and some animals have adaptations that can greatly augment this cooling effect. For example, most mammals and birds can increase evaporation from the lungs by panting. Sweating or bathing to make the skin wet also enhances evaporative cooling (see FIGURE 3.4).

3. *Behavioral responses.* Both endotherms and ectotherms use behavioral responses, such as changes in posture or moving about in their environment, to control body temperature. Many terrestrial animals will bask in the sun or on warm rocks when cold or find cool, shaded, or damp areas when hot. Many ectotherms can maintain a very constant body temperature by these simple behaviors. More extreme behavioral adaptations in some animals include hibernation or migration to a more suitable climate.

4. *Changing the rate of metabolic heat production.* This fourth category of thermoregulatory adaptation applies only to endotherms, particularly mammals and birds. By means we will discuss in the next section, many species of mammals and birds can greatly increase their metabolic heat production when exposed to cold.

Most animals are ectothermic, but endothermy is widespread

In this section we survey several large taxa of animals, concentrating on the mechanisms by which they thermoregulate as ectotherms or endotherms.

Mammals and Birds

Mammals and birds generally maintain body temperatures within a narrow range (36–38°C for most mammals and 39–42°C for most birds) that is usually considerably warmer than the environment. Because heat always flows from a warm object to cooler surroundings, birds and mammals must counteract the constant heat loss. This maintenance of warm body temperature depends on several key adaptations. The most basic mechanism is the high metabolic rate of endothermy itself. Endotherms can produce large amounts of metabolic heat that replace the flow of heat to the environment, and they can vary heat production to match changing rates of heat loss. Heat production is increased by such muscle activity as moving or shivering. In some mammals, certain hormones can cause mitochondria to increase their metabolic activity and produce heat instead of ATP. This **nonshivering thermogenesis (NST)** takes place throughout the body, but some mammals also have a tissue called **brown fat** in the neck and between the shoulders that is specialized for rapid heat production. Through shivering and NST, mammals and birds in cold environments can increase their metabolic heat production by as much as 5 to 10 times above the minimal levels that occur in warm conditions.

Another major thermoregulatory adaptation that evolved in mammals and birds is insulation (hair, feathers, and fat layers), which reduces the flow of heat and lowers the energy cost of keeping warm. The insulating power of a layer of fur or feathers mainly depends on how much still air the layer traps. Most land mammals and birds react to cold by raising their fur or feathers, thereby trapping a thicker layer of air. Humans rely more on a layer of fat just beneath the skin as insulation (FIGURE 44.6); goose bumps are a vestige of hair-raising left over from our furry ancestors.

Vasodilation and vasoconstriction also regulate heat exchange and may contribute to regional temperature differences within the animal. For example, heat loss from a human is reduced when arms and legs cool to several degrees below the temperature of the body core, where most vital organs are located.

Hair loses most of its insulating power when wet. Marine mammals such as whales and seals have a very thick layer of insulating fat called blubber, just under the skin. Marine mammals swim in water colder than their body core temperature, and many species spend at least part of the year in nearly freezing polar seas. The loss of heat to water occurs 50 to 100 times more rapidly than heat loss to air, and the skin temperature of a marine mammal is close to water temperature. Even so, the blubber insulation is so effective that marine mammals maintain body core temperatures of about 36–38°C with metabolic rates about the same as those of land mammals of similar size. The flippers or tail of a whale or seal lack insulating blubber, but as we discussed, countercurrent heat exchangers greatly reduce heat loss in these extremities, as they do in the legs of many birds (see FIGURE 44.5).

Through metabolic heat production, insulation, and vascular adjustments, birds and mammals are capable of astonishing feats of thermoregulation. For example, small birds called chickadees, which weigh only 20 grams, can remain active and hold body temperature nearly constant at 40°C in environmental

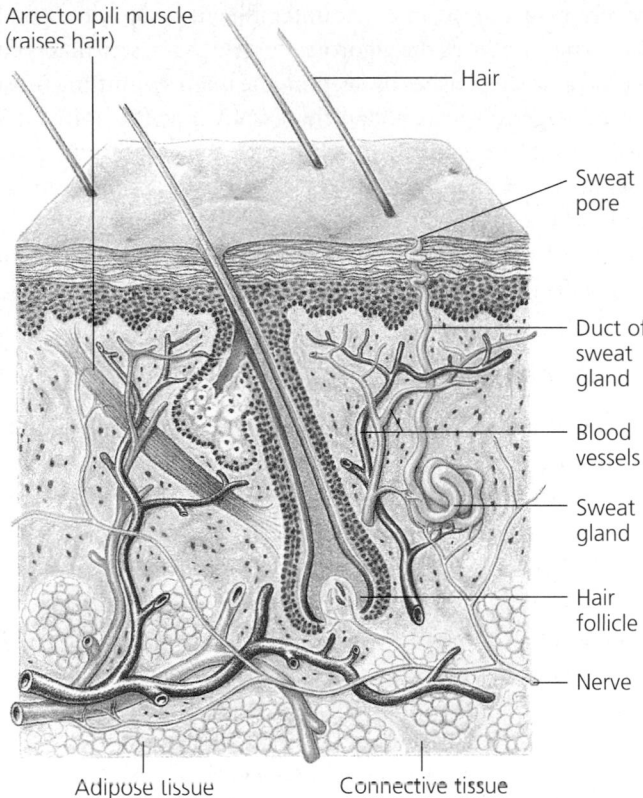

FIGURE 44.6 **Skin as an organ of thermoregulation.** Fat (adipose tissue) and hair help insulate mammals. Heat loss to the environment can be regulated by the constriction and dilation of superficial blood vessels and by the erection and compaction of fur. Sweat glands, under nervous control, function in evaporative cooling.

Labels on figure:
Arrector pili muscle (raises hair)
Hair
Sweat pore
Duct of sweat gland
Blood vessels
Sweat gland
Hair follicle
Nerve
Adipose tissue
Connective tissue

FIGURE 44.7 **A terrestrial mammal bathing, an adaptation that enhances evaporative cooling.**

temperatures as low as −40°C—as long as they have enough food to supply the large amount of energy necessary for heat production.

Many mammals and birds live in places where thermoregulation requires cooling as well as warming. For example, when a marine mammal moves into warm seas, as many whales do when they reproduce, excess metabolic heat is removed by vasodilation of numerous blood vessels in the outer layer of the skin. In hot climates or when vigorous exercise adds large amounts of metabolic heat to the body, many terrestrial mammals and birds may allow body temperature to rise by several degrees, which enhances heat loss by increasing the temperature gradient between the body and a warm environment. Evaporative cooling often plays a key role in dissipating the body heat (FIGURE 44.7). If environmental temperature is above body temperature, animals gain heat from the environment as well as from metabolism, and evaporation is the only way to keep body temperature from rising rapidly. Panting is important in birds and many mammals. Some birds have a pouch richly supplied with blood vessels in the floor of the mouth; fluttering the pouch increases evaporation. Pigeons can use evaporative cooling to keep body temperature close to 40°C in air temperatures as high as 60°C, as long as they have sufficient water. Many terrestrial mammals have sweat glands

controlled by the nervous system (see FIGURE 44.6). Other mechanisms that promote evaporative cooling include spreading saliva on body surfaces, an adaptation of some kangaroos and rodents for combating severe heat stress. Some bats use both saliva and urine to enhance evaporative cooling.

Amphibians and Reptiles

All amphibians and most reptiles are ectothermic, and their low metabolic rates have little influence on normal body temperature. The optimal temperature range for amphibians varies substantially with the species. For example, closely related species of salamanders have average body temperatures ranging from 7° to 25°C. When exposed to air, most amphibians lose heat rapidly by evaporation from their moist body surfaces, making it difficult to keep body temperature sufficiently warm. However, behavioral adaptations help these animals maintain a satisfactory temperature most of the time—by moving to a location where solar heat is available, for instance. When the surroundings are too warm, amphibians seek cooler microenvironments, such as shaded areas. Some species, such as bullfrogs, can vary the amount of mucus they secrete from their surface, a response that regulates evaporative cooling.

Like amphibians, reptiles control body temperature mainly by behavior. When cold, they seek warm places, orienting themselves toward heat sources to increase heat uptake and expanding the body surface exposed to a heat source. When hot, they move to cool areas or turn in another direction, thereby reducing the surface area exposed to the sun. Many reptiles keep their body temperatures very stable over the course of a day by shuttling back and forth between warm and cool spots.

Some reptiles also have physiological adaptations that regulate heat loss. For example, in the marine iguana, which inhabits the Galápagos Islands, body heat is conserved by vasoconstriction of superficial blood vessels, routing more blood to the central core of the body when the animal is swimming in the cold ocean. A few large reptiles become endothermic in special circumstances. For example, female pythons that are incubating eggs increase their metabolic rate by shivering, generating enough heat to keep their body (and egg) temperatures 5–7°C warmer than the surrounding air for weeks at a time. This temporary endothermy consumes considerable energy. Researchers continue to debate whether certain groups of dinosaurs were endothermic (see Chapter 34).

Fishes

In terms of body temperature, most fishes are conformers, with internal temperature usually within 1–2°C of the surrounding water temperature. The swimming muscles may be large and active, but nearly all the metabolic heat most fishes generate is lost to the surrounding water when blood passes through the gills. However, some specialized endothermic fishes, mainly large, powerful swimmers such as the bluefin tuna, swordfish, and the great white shark, have circulatory adaptations that retain metabolic heat in the body. Large arteries convey most of the cold blood from the gills to tissues just under the skin. Branches deliver blood to the deep muscles, where the small

vessels are arranged into a countercurrent heat exchanger. Endothermy enhances the vigorous, sustained activity that is characteristic of these fishes by keeping the main swimming muscles several degrees warmer than the tissues near the animal's surface, about the same temperature as the surrounding water (FIGURE 44.8). In a few fish species, special heat-generating organs, which may be evolutionarily derived from eye muscles, warm just the eyes or brain. These adaptations probably allow more effective functioning of the warmed organs.

Invertebrates

Aquatic invertebrates are mainly thermoconformers with little control over their body temperature. However, many terrestrial invertebrates can adjust internal temperature by the same behavioral mechanisms used by vertebrate ectotherms such as lizards. The desert locust, for example, must reach a certain temperature to become active, and on cold days it orients in a direction that maximizes the absorption of sunlight.

Many species of flying insects, such as bees and moths, are actually endothermic—the smallest of all endotherms. The

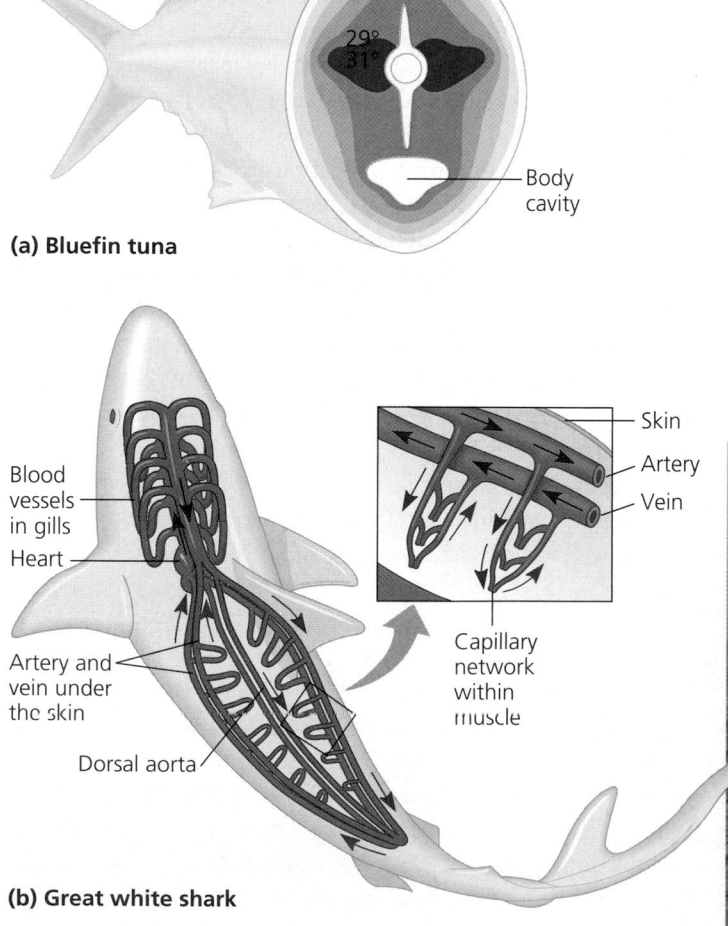

(a) Bluefin tuna

(b) Great white shark

FIGURE 44.8 Thermoregulation in large, active fishes.
(a) Unlike most fishes, the bluefin tuna maintains temperatures in its main swimming muscles that are much higher than the surrounding water (colors indicate swimming muscles cut in transverse section). These temperatures were recorded for a tuna in 19°C water.
(b) Like the bluefin tuna, the great white shark has a countercurrent heat exchanger in its swimming muscles that reduces the loss of metabolic heat. (The photo shows a great white leaping out of the water in pursuit of a decoy model seal.) All fishes lose heat to the surrounding water when their blood passes through the gills. However, endothermic fishes have a small dorsal aorta, and as a result relatively little cold blood from the gills goes directly to the core of the body. Instead, most of the blood leaving the gills is conveyed via large arteries just under the skin, keeping cool blood away from the body core. As shown in the enlargement, small arteries carrying cool blood inward from the large arteries under the skin are paralleled by small veins carrying warm blood outward from the inner body. This countercurrent flow retains heat in the muscles.

hawk moth is an example. The capacity of such endothermic insects to elevate body temperature depends on powerful flight muscles, which generate large amounts of heat when operating. Many endothermic insects use shivering—by contracting the flight muscles in synchrony, so that only slight wing movements occur but considerable heat is produced—to warm up before taking off. Chemical reactions, and hence cellular respiration, speed up in the warmed-up flight "motors" (the Q_{10} effect), enabling these insects to fly even on cold days or at night (FIGURE 44.9a). Many endothermic insects (bumblebees, honeybees, and some moths) have a countercurrent heat exchanger that helps maintain a high temperature in the thorax, where the flight muscles are located. For example, the heat exchanger keeps the thorax of certain moths at about 30°C during flight, even on cold, snowy nights (FIGURE 44.9b). In contrast, insects flying in hot weather run the risk of overheating because of the large amount of heat produced by working flight muscles. In some species the countercurrent mechanism can be shut down to allow muscle-produced heat to be lost from the thorax to the abdomen, and from there to the environment. Bumblebee queens use this means to incubate their eggs: They generate heat by shivering their flight muscles and then transfer the heat to the abdomen, which the bee presses against her eggs.

Honeybees use an additional thermoregulatory mechanism that depends on social behavior. In cold weather they increase heat production and huddle together, thereby retaining heat.

They maintain a relatively constant temperature by changing the density of the huddling. Individuals move between the cooler outer edges of the cluster and the warmer center, thus circulating and distributing the heat. Even when huddling, honeybees must expend considerable energy to keep warm during long periods of cold weather, and this is the main function of storing large quantities of fuel in the hive in the form of honey. Honeybees also control the temperature of their hive by transporting water to it in hot weather and fanning with their wings, which promotes evaporation and convection. Thus, a honeybee colony uses many of the mechanisms of thermoregulation seen in single organisms.

Feedback Mechanisms in Thermoregulation

The regulation of body temperature in humans and other mammals is a complex system facilitated by feedback mechanisms (see FIGURE 40.9). Nerve cells that control thermoregulation, as well as those that control many other aspects of homeostasis, are concentrated in a region of the brain called the hypothalamus (discussed in detail in Chapter 48). The hypothalamus contains a group of neurons that functions as a thermostat, responding to changes in body temperature above and below a set point (actually above or below a normal range) by activating mechanisms that promote heat loss or gain (FIGURE 44.10, p. 934). Nerve cells that sense temperatures are in the skin, the hypothalamus itself, and several other body regions. Warm receptors signal the hypothalamic thermostat when temperatures increase; cold receptors indicate temperature decrease. At

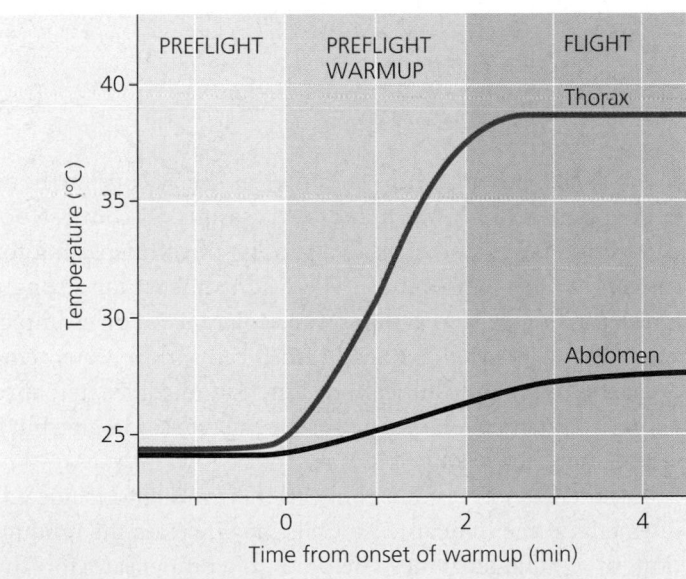

(a) Preflight warmup in the hawk moth. The hawk moth (*Manduca sexta*) is one of many insect species that use a shivering-like mechanism for preflight warmup of thoracic flight muscles. Warming up helps these muscles produce enough power to let the animal take off. Once airborne, flight muscle activity maintains a high thoracic temperature.

FIGURE 44.9a is reprinted with permission from B. Heinrich, *Science* 185 (1974): 747–756. Copyright © 1974 American Association for the Advancement of Science.

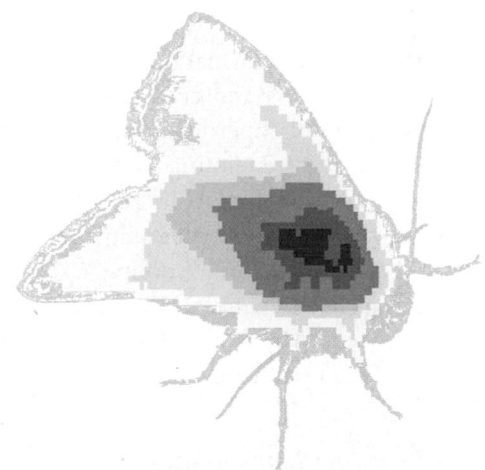

(b) Internal temperature in the winter moth. Various endothermic adaptations, including a countercurrent heat exchanger in the thorax, help keep the flight muscles of winter-active moths warmed to a temperature of 30°C, even though the external temperature may be subfreezing. This infrared map of a winter moth shows the heat distribution immediately after a flight. Red, located here in the thorax region, indicates the highest temperature. Moving outward from the thorax, the variously colored zones correspond to regions of increasingly cooler body temperatures.

FIGURE 44.9 Thermoregulation in moths.

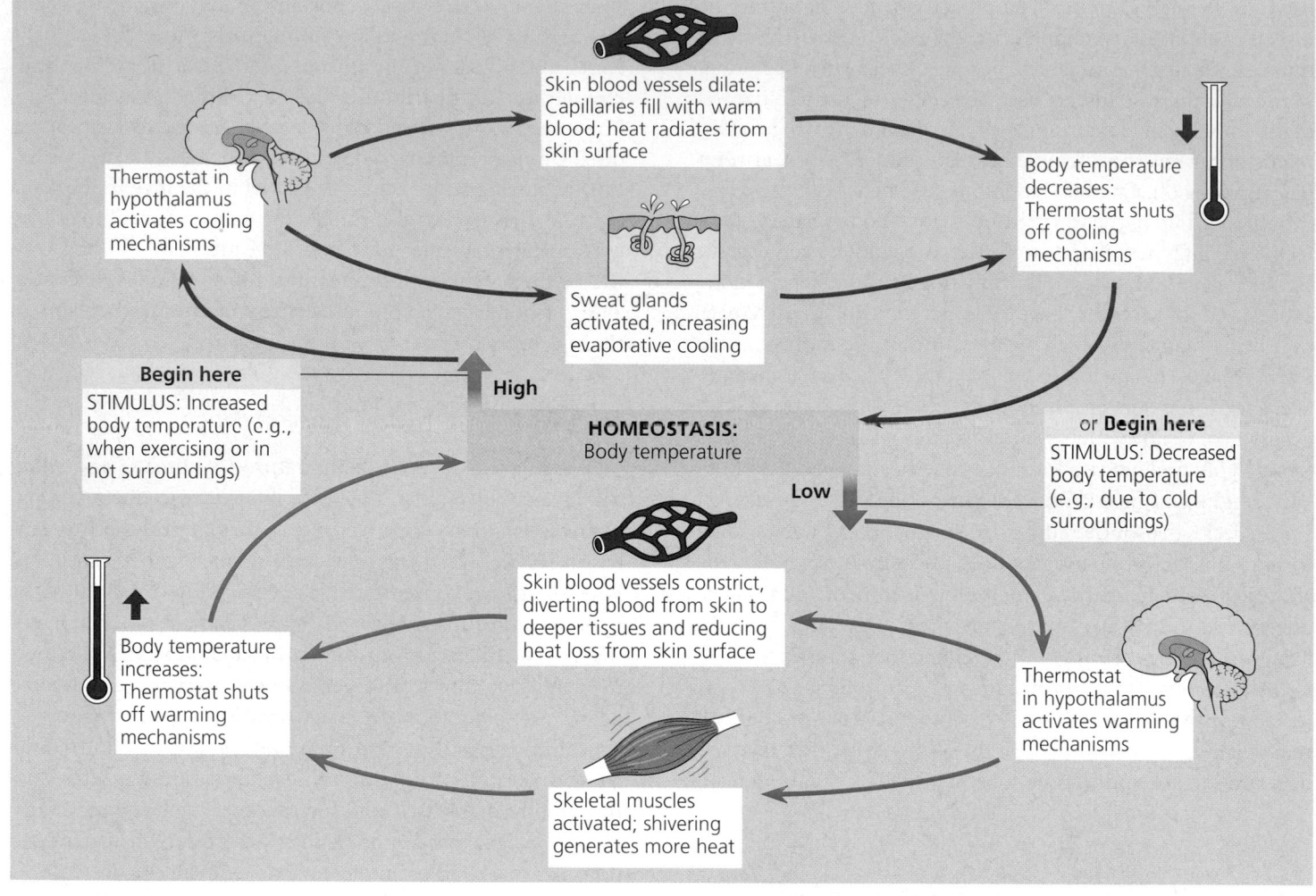

FIGURE 44.10 The thermostat function of the hypothalamus and feedback mechanisms in human thermoregulation.

Labels in figure:

Skin blood vessels dilate: Capillaries fill with warm blood; heat radiates from skin surface

Thermostat in hypothalamus activates cooling mechanisms

Body temperature decreases: Thermostat shuts off cooling mechanisms

Sweat glands activated, increasing evaporative cooling

Begin here
STIMULUS: Increased body temperature (e.g., when exercising or in hot surroundings)

High

HOMEOSTASIS: Body temperature

Low

or **Begin here**
STIMULUS: Decreased body temperature (e.g., due to cold surroundings)

Body temperature increases: Thermostat shuts off warming mechanisms

Skin blood vessels constrict, diverting blood from skin to deeper tissues and reducing heat loss from skin surface

Thermostat in hypothalamus activates warming mechanisms

Skeletal muscles activated; shivering generates more heat

body temperature below the normal range, the thermostat inhibits heat-loss mechanisms and activates heat-saving ones such as vasoconstriction of superficial vessels and erection of fur, while stimulating heat-generating mechanisms (shivering and nonshivering thermogenesis). In response to elevated body temperature, the thermostat shuts down heat-retention mechanisms and promotes body cooling by vasodilation, sweating, or panting. The thermostat can also respond to external temperature (sensed as skin temperature) even without changes in body core temperature.

Adjustment to Changing Temperatures

Many animals can adjust to a new range of environmental temperatures over a period of days or weeks, a physiological response called **acclimatization.** Both ectotherms and endotherms acclimatize, but in different ways. In birds and mammals, acclimatization often includes adjusting the amount of insulation—by growing a thicker coat of fur in the winter and shedding it in the summer, for example—and sometimes varying the capacity for metabolic heat production in different

seasons. These changes help endotherms keep a constant body temperature in both warm and cold seasons. In contrast, acclimatization in ectotherms is a process of compensating for *changes* in body temperature. These adjustments can strongly affect physiology and temperature tolerance. For example, summer-acclimatized bullhead catfish can survive water temperatures up to 36°C but cannot function in cold water; after winter acclimatization they can easily tolerate cold water, but a temperature above 28°C is lethal.

Acclimatization responses in ectotherms often include adjustments at the cellular level. Cells may increase the production of certain enzymes, helping to compensate for the lowered activity of each enzyme molecule at temperatures that are not optimal. In other cases, cells produce variants of enzymes that have the same function but different temperature optima. Membranes may also change the proportions of saturated and unsaturated lipids they contain, which helps keep membranes fluid at different temperatures (see FIGURE 8.4).

Some ectotherms that experience subzero body temperatures protect themselves by producing "antifreeze" compounds (cryoprotectants) that prevent ice formation in the cells. In arctic regions or cold mountain peaks, cryoprotectants in the body fluids let overwintering ectotherms, such as some frogs and many arthropods and their eggs, withstand body temperatures considerably below zero. Cryoprotectants are also found in certain species of fishes from arctic and antarctic seas, where water temperatures can be as cold as −1.8°C, well below the freezing point of unprotected body fluids (about −0.7°C).

Cells can often make rapid adjustments to temperature changes. For example, mammalian cells grown in laboratory cultures respond to a marked increase in temperature and to other forms of severe stress, such as toxins, rapid pH changes, and viral infections, by accumulating special molecules called **stress-induced proteins,** including **heat-shock proteins.** When "shocked" by a rapid change in temperature from 37°C to about 43°C, cultured mammalian cells begin synthesizing heat-shock proteins within minutes. These molecules help maintain the integrity of other proteins that would be denatured by severe heat. Found in bacteria, yeasts, and plant cells as well as in animals, stress-induced proteins help prevent cell death when an organism is challenged by severe changes in the cellular environment.

Torpor conserves energy during environmental extremes

Despite their many adaptations for homeostasis, animals may periodically encounter conditions that severely challenge their abilities to balance heat, energy, and materials budgets. For example, at certain seasons of the year (or certain times of day) temperatures may be extremely hot or cold, or food may be unavailable. One way animals can save energy while avoiding

difficult and dangerous conditions is to use **torpor,** a physiological state in which activity is low and metabolism decreases.

Hibernation is long-term torpor that evolved as an adaptation to winter cold and food scarcity. When vertebrate endotherms (birds and mammals) enter torpor or hibernation, their body temperatures decline—in effect, their body's thermostat is turned down. The temperature reduction may be dramatic: Some hibernating mammals cool to as low as 1–2°C, and a few even drop slightly below 0°C in a supercooled (unfrozen) state. The resulting energy savings due to lower metabolic rate and less heat production are huge; metabolic rates during hibernation can be several hundred times lower than if the animal attempted to maintain normal body temperatures of 36–38°C. This allows hibernators to survive for very long periods on limited supplies of energy, stored in the body tissues or as food cached in a burrow. For example, hibernation can take up as much as nine months of the year in some small mammals, leaving only three months for normal activity.

Certain ground squirrels are favorite research models for biologists interested in the physiology of hibernation. A Belding's ground squirrel living in the high mountains of California is active only during spring and summer, when it maintains a body temperature of about 37°C and a metabolic rate of about 85 kcal per day (FIGURE 44.11). In September the squirrel retreats to a safe burrow where it spends the next eight months hibernating. For most of the hibernation season, the squirrel's body temperature is only slightly above burrow temperature (which may be close to freezing), and its metabolic rate is extremely

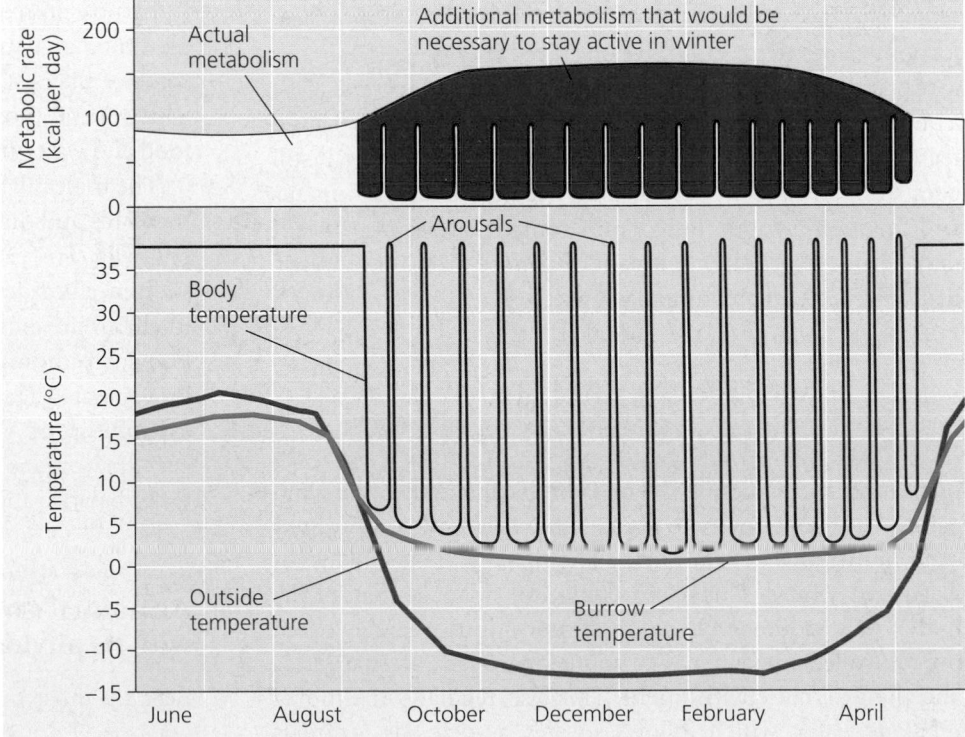

FIGURE 44.11 Body temperature and metabolism during hibernation of Belding's ground squirrel.

low. Every week or two it arouses for a few hours, using metabolic heat to warm up to about 37°C (these periodic arousals may be needed for maintenance functions that require high body temperature). In late spring, when outside temperature is climbing, the squirrel resumes normal endothermy. By hibernating, Belding's ground squirrels avoid severe cold and greatly reduce the amount of energy they need to survive the winter, when their normal food of grasses and seeds is not available. Instead of having to spend 150 kcal per day to maintain normal body temperatures in winter weather, a squirrel in its burrow spends an average of only 5–8 kcal per day (only about 1 kcal per day when hibernating) and can live on stored fat—without eating—for the entire hibernation season.

Estivation, or summer torpor, also characterized by slow metabolism and inactivity, enables animals to survive long periods of high temperatures and scarce water supplies. Hibernation and estivation are often triggered by seasonal changes in the length of daylight. As the days shorten, some hibernators prepare for winter by storing food in their burrows; other species eat huge quantities of food and fatten dramatically. For example, ground squirrels double their weight in a month of gorging.

Many small mammals and birds exhibit a **daily torpor** that seems to be adapted to their feeding patterns. For instance, most bats and shrews feed at night and go into torpor when they are inactive during daylight hours. Chickadees and hummingbirds feed during the day and often undergo torpor on cold nights; the body temperature of chickadees drops as much as 10°C at night, and that of hummingbirds can fall 25–30°C. All endotherms that use daily torpor are relatively small; when active they have high metabolic rates and thus very high rates of energy consumption. During hours when they cannot feed, torpor enables them to survive on stored energy.

An animal's daily cycle of activity and torpor appears to be a built-in rhythm controlled by the biological clock (see Chapter 48). Even if food is made available to a shrew all day, it still goes through its daily torpor. The need for sleep in humans and the slight drop in body temperature that accompanies it may be an evolutionary remnant of a more pronounced daily torpor in our early mammalian ancestors.

WATER BALANCE AND WASTE DISPOSAL

Just as thermoregulation depends on balancing heat loss and gain, an animal's ability to regulate the chemical composition of its body fluids depends on balancing the uptake and loss of water and solutes. This **osmoregulation** (management of the body's water content and solute composition) is largely based on controlled movements of solutes between internal fluids and the external environment. This also regulates the movement of water, which follows solutes by osmosis. Animals must also remove various metabolic waste products before they accumulate to harmful levels.

The ultimate function of osmoregulation is to maintain the composition of the cytoplasm of the body's cells, but most animals do this indirectly by managing the composition of an internal body fluid that bathes the cells. In insects and other animals with an open circulatory system, this fluid is the hemolymph (see Chapter 42). In vertebrates and other animals with a closed circulatory system, the cells are bathed in an interstitial fluid. The composition of interstitial fluid is controlled indirectly by managing the composition of blood. Animals often have complex organs, such as the kidneys of vertebrates, that are specialized for the maintenance of fluid composition.

Water balance and waste disposal depend on transport epithelia

In most animals, one or more kinds of **transport epithelium** —a layer or layers of specialized epithelial cells that regulate solute movements—are essential components of osmotic regulation and metabolic waste disposal. The most important feature of all transport epithelia is their ability to move specific solutes in controlled amounts in particular directions. Some transport epithelia directly face the outside environment, while others line channels connected to the outside by an opening on the body surface. Joined by impermeable tight junctions (see FIGURE 7.30), the cells of the epithelium form a barrier at the tissue-environment boundary. Like the Casparian strip in plant roots (see FIGURE 36.7), this arrangement ensures that any solutes moving between animal and environment must pass through selectively permeable membranes.

In most animals, transport epithelia are arranged into complex tubular networks with extensive surface areas. We find some of the best examples in the salt glands of marine birds, which spend months or years at sea and need to obtain both food and water from the ocean (FIGURE 44.12).

The molecular structure of plasma membranes determines the kinds and directions of solutes that move across transport epithelia. For example, salt-excreting glands remove excess sodium chloride from the blood. By contrast, transport epithelia in the gills of freshwater fishes move salts from the dilute surrounding water into the blood (this kind of solute movement—against a concentration gradient—requires the expenditure of ATP in active transport). Transport epithelia in excretory organs often have the dual functions of maintaining water balance and disposing of metabolic wastes.

An animal's nitrogenous wastes are correlated with its phylogeny and habitat

Because most metabolic wastes must be dissolved in water when they are removed from the body (a major exception being loss of carbon dioxide in air-breathing animals), the type and quantity of waste products may have a large impact

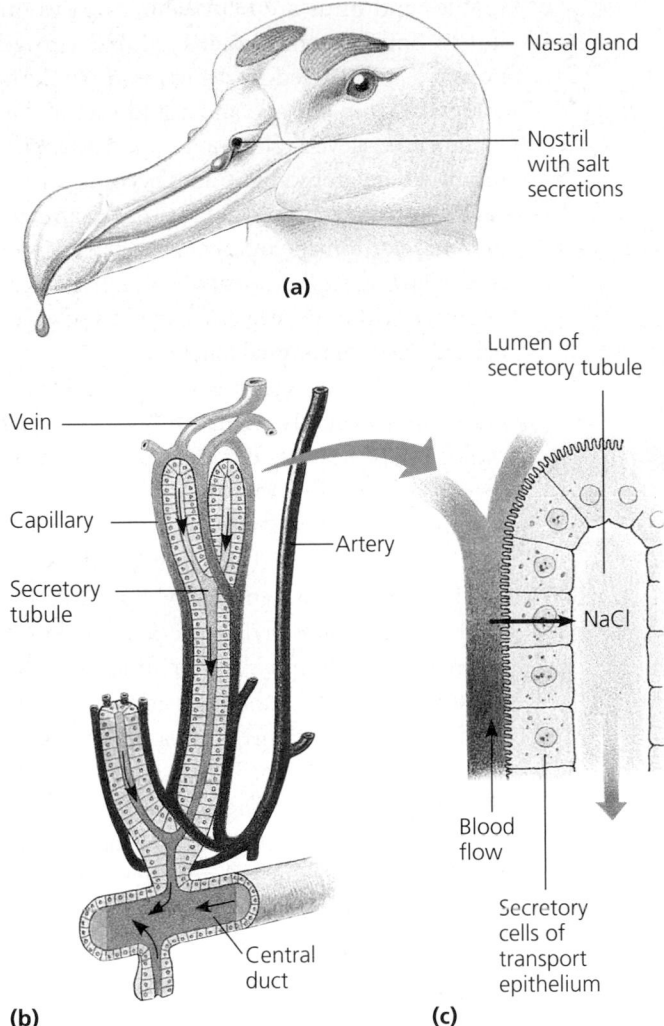

- Nasal gland
- Nostril with salt secretions

(a)

- Lumen of secretory tubule
- Vein
- Capillary
- Artery
- Secretory tubule
- NaCl
- Blood flow
- Secretory cells of transport epithelium
- Central duct

(b)   **(c)**

FIGURE 44.12 Salt-excreting glands in birds. (a) Many marine birds, such as this albatross, can survive long periods at sea by drinking seawater. They can do this because they have a pair of nasal glands that secrete an excretory fluid much more salty than the ocean. Thus, even though drinking seawater brings in a lot of salt, the bird achieves a net gain of water (in contrast, humans who drink seawater must use more water to excrete the salt load than was gained by drinking). A seabird's salt glands empty via a duct into the nostrils, and the salty solution either drips off the tip of the beak or is exhaled in a fine mist. **(b)** This diagram shows one of several thousand secretory tubules in a salt-excreting gland. Each tubule is lined by a transport epithelium surrounded by capillaries, and drains into a central duct. **(c)** The secretory cells of the transport epithelium actively transport salt from the blood into the tubules. Notice that blood flows counter to the flow of salt secretion. By maintaining a concentration gradient of salt in the tubule (graded blue shading), this countercurrent system enhances salt transfer from the blood to the lumen of the tubule (see Chapter 42).

on water balance. In terms of their effect on osmoregulation, among the most important wastes are the nitrogenous (nitrogen-containing) breakdown products of proteins and nucleic acids. When these macromolecules are broken apart for energy or converted to carbohydrates or fats, enzymes remove nitrogen in the form of **ammonia,** a small and very toxic molecule. Some animals excrete ammonia directly, but many species first convert it to compounds such as urea or uric acid, which are less toxic but require energy in the form of ATP to produce.

In general, the *kinds* of nitrogenous wastes excreted depend on an animal's evolutionary history and habitat—especially the availability of water (FIGURE 44.13, p. 938). The *amount* of nitrogenous waste produced is coupled to the energy budget, as it strongly depends on how much and what kind of food an animal eats. Because they use energy at high rates, endotherms eat more food—and therefore produce more nitrogenous wastes—than ectotherms. Predators, which derive much of their energy from dietary proteins, need to excrete more nitrogen than animals that rely mainly on lipids or carbohydrates as energy sources.

Ammonia

Because ammonia is very soluble but can only be tolerated at very low concentrations, animals that excrete nitrogenous wastes as ammonia need access to lots of water. Therefore, ammonia excretion is most common in aquatic species. Ammonia molecules easily pass through membranes and are readily lost by diffusion to the surrounding water. In many invertebrates, ammonia release occurs across the whole body surface. In fishes, most of the ammonia is lost as ammonium ions (NH_4^+) across the epithelium of the gills, with kidneys excreting only minor amounts of nitrogenous wastes. In freshwater fishes, the gill epithelium takes up Na^+ from the water in exchange for NH_4^+, which helps to maintain a much higher Na^+ concentration in body fluids than in the surrounding water.

Urea

Although it works well in many aquatic species, ammonia excretion is much less suitable for land animals. Because it is so toxic, ammonia can only be transported and excreted in large volumes of very dilute solutions, and most terrestrial animals and many marine species (which tend to lose water to their environment by osmosis) simply do not have access to sufficient water. Instead, mammals, most adult amphibians, and many marine fishes and turtles excrete mainly **urea,** a substance produced in the vertebrate liver by a metabolic cycle that combines ammonia with carbon dioxide. The circulatory system carries urea to the excretory organs, the kidneys.

The main advantage of urea is its low toxicity, about 100,000 times less than that of ammonia. This permits animals to transport and store urea safely at high concentrations. And this greatly reduces the amount of water required for nitrogen excretion, because much less water is lost when a given quantity of nitrogen is excreted in a concentrated solution of urea rather than a dilute solution of ammonia.

The main disadvantage of urea is that animals must expend energy to produce it from ammonia. From a bioenergetic standpoint, we would predict that animals that spend part of their lives in water and part on land would switch between

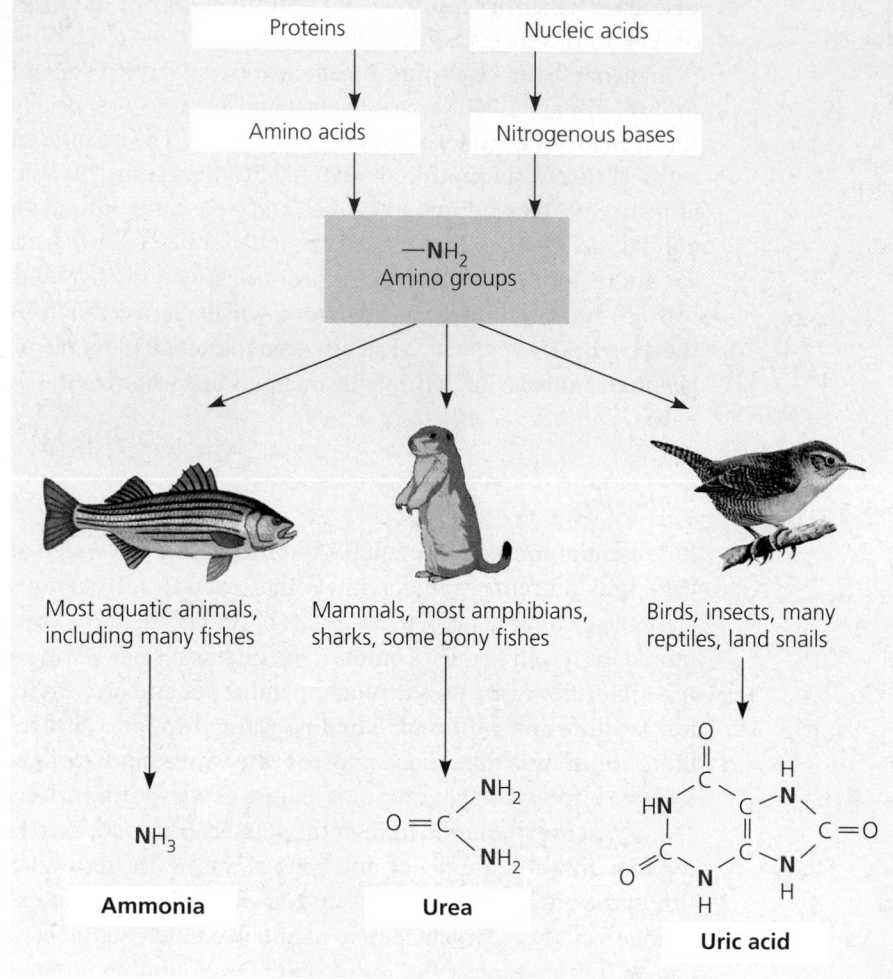

FIGURE 44.13 Nitrogenous wastes.

reptiles and birds are permeable to gases but not to liquids, which means soluble nitrogenous wastes released by an embryo would be trapped within the egg and could accumulate to dangerous levels (although urea is much less harmful than ammonia, it does become toxic at very high concentrations). In these animals natural selection apparently favored use of uric acid, which precipitates out of solution and can be stored within the egg as a harmless solid left behind when the animal hatches.

The type of nitrogenous waste produced by vertebrates depends on habitat as well as on evolutionary lineage. For example, terrestrial turtles (which often live in dry areas) excrete mainly uric acid, whereas aquatic turtles excrete both urea and ammonia. In some species, individuals can change their nitrogenous wastes when environmental conditions change. For example, certain tortoises that usually produce urea shift to uric acid production when temperature increases and water becomes less available. This is another example of how response to the environment occurs on two levels: Over generations, evolution determines the limits of physiological responses for a species, but during their lives individual organisms make adjustments within these evolutionary constraints.

excreting ammonia (thereby saving energy) and urea (reducing excretory water loss). In fact, many amphibians excrete mainly ammonia when they are aquatic tadpoles and switch largely to urea when they are land-dwelling adults.

Uric Acid

Land snails, insects, birds, and many reptiles excrete **uric acid** as the major nitrogenous waste. Like urea, uric acid is relatively nontoxic. But unlike either ammonia or urea, uric acid is largely insoluble in water and can be excreted as a semisolid paste with very little water loss. This is a great advantage for animals with little access to water, but there is a cost: Uric acid is even more energetically expensive to produce than urea, requiring considerable ATP for synthesis from ammonia.

Uric acid and urea represent different adaptations for excreting nitrogenous wastes with minimal water loss. One factor that seems to have been important in determining which of these alternatives evolved in a particular group of animals is the mode of reproduction. Soluble wastes can diffuse out of a shell-less amphibian egg or be carried away by the mother's blood in a mammalian embryo. However, the shelled eggs produced by

Cells require a balance between osmotic gain and loss of water

All animals—regardless of phylogeny, habitat, or type of nitrogenous waste—face the same central problem of osmoregulation: Over time, the rates of water uptake and loss must balance. Animal cells—which lack cell walls—swell and burst if there is a continuous net uptake of water or shrivel and die if there is a substantial net loss of water.

Water enters and leaves cells by osmosis. Recall from Chapter 8 that osmosis, a special case of diffusion, is the movement of water across a selectively permeable membrane. It occurs whenever two solutions separated by the membrane differ in osmotic pressure, or **osmolarity** (total solute concentration expressed as molarity, or moles of solute per liter of solution; see Chapter 3). The unit of measurement for osmolarity used in this chapter is milliosmoles per liter (mosm/L); 1 mosm/L is equivalent to a total solute concentration of 10^{-3} M. The osmolarity of human blood is about 300 mosm/L, while seawater has an osmolarity of about 1,000 mosm/L. Two solutions separated by a selectively permeable membrane are said to be isoosmotic if they have the same osmolarity.

There is no *net* movement of water by osmosis between isoosmotic solutions (although water molecules are continually crossing the membrane, they do so at equal rates in both directions). When two solutions differ in osmolarity, the one with the greater concentration of solutes is referred to as hyperosmotic and the more dilute solution as hypoosmotic. Water flows by osmosis from a hypoosmotic solution to a hyperosmotic one.*

Osmoregulators expend energy to control their internal osmolarity; osmoconformers are isoosmotic with their surroundings

There are two basic solutions to the problem of balancing water gain with water loss. One—available only to marine animals—is to be isoosmotic to the surroundings. Such an animal, which does not actively adjust its internal osmolarity, is known as an **osmoconformer.** Because an osmoconformer's internal osmolarity is the same as that of its environment, there is no tendency to gain or lose water. Although they do not compensate for changes in external osmolarity, osmoconformers often live in water that has a very stable composition and hence have a very constant internal osmolarity. In contrast, an **osmoregulator** is an animal that must control its internal osmolarity, because its body fluids are not isoosmotic with the outside environment. An osmoregulator must discharge excess water if it lives in a hypoosmotic environment or take in water to offset osmotic loss if it inhabits a hyperosmotic environment. Osmoregulation enables animals to live in environments that are uninhabitable for osmoconformers, such as freshwater and terrestrial habitats, and allows many marine animals to maintain internal osmolarities different from that of seawater.

Whenever animals maintain an osmolarity difference between the body and the external environment, osmoregulation has an energy cost. Because diffusion tends to equalize concentrations in a system, osmoregulators must expend energy to maintain the osmotic gradients that allow water to move in or out. They do so by using active transport to manipulate solute concentrations in their body fluids.

The energy cost of osmoregulation depends mainly on how different an animal's osmolarity is from its surroundings, how easily water and solutes can move across the animal's surface, and on how much membrane-transport work is required to pump solutes. Because of the difference in concentration between cytoplasm, fresh water (1–50 mosm/L), and seawater, osmoregulation accounts for nearly 5% of the resting metabolic rate of many marine and freshwater bony fishes. For brine shrimp, small crustaceans that live in the Great Salt Lake

in Utah and in other extremely salty lakes, the gradient between internal and external osmolarities is very large and the cost of osmoregulation is correspondingly high—as much as 30% of the resting metabolic rate.

Most animals, whether osmoconformers or osmoregulators, cannot tolerate substantial changes in external osmolarity and are said to be **stenohaline** (Gr. *stenos*, "narrow"; haline refers to salt). In contrast, **euryhaline** animals (Gr. *eurys*, "broad")—which include both certain osmoconformers and certain osmoregulators—can survive large fluctuations of external osmolarity. Familiar examples of euryhaline osmoregulators are the various species of salmon. A more extreme example is a bony fish called tilapia (a native of Africa grown widely for human food), which can adjust to any salt concentration between fresh water and 2,000 mosm/L (twice that of seawater).

Next, we'll take a closer look at some of the adaptations for osmoregulation that have evolved in marine, freshwater, and terrestrial animals.

Maintaining Water Balance in the Sea

Animals first evolved in the sea, and more animal phyla are found there than in any other environment. Most marine invertebrates are osmoconformers, as are the hagfishes (jawless vertebrates). Their total osmolarity (the sum of the concentrations of all dissolved substances) is the same as seawater. However, they differ considerably from seawater in their concentrations of most specific solutes. Thus, even an animal that conforms to the osmolarity of its surroundings does regulate its internal composition.

Except for the hagfishes, marine vertebrates are osmoregulators. For most of these animals, the ocean is a strongly dehydrating environment because it is much saltier than internal fluids and water tends to be lost from the body by osmosis. Marine fishes (class Osteichthyes) constantly lose water through their skin and especially their gills (FIGURE 44.14a, p. 940). To balance these water losses, these fishes obtain water in food and by drinking large amounts of seawater. The accompanying salt intake (along with salt that enters the body by diffusion) is disposed of by active transport out of the gills. Very little urine is produced, an adaptation that conserves water.

Marine sharks and most other cartilaginous fishes (class Chondrichthyes) use a different osmoregulatory "strategy." Like bony fishes, their internal salt concentration is much less than that of seawater, so in addition to salt intake in food, salt tends to diffuse into the body (especially across the gills). The kidneys of sharks remove some of this salt load, and the rest is excreted by an organ called the rectal gland or is lost in feces. Unlike bony fishes, and despite relatively low internal salt concentration, marine sharks do not experience a large and continuous osmotic water loss. The explanation is that sharks maintain high concentrations of urea in their body fluids, along with another organic solute, trimethylamine oxide (TMAO), which protects proteins from damage by urea. (If

---

* In this chapter, we use the terms *isoosmotic*, *hypoosmotic*, and *hyperosmotic*, which refer specifically to osmolarity, instead of the more familiar terms *isotonic*, *hypotonic* and *hypertonic*. The latter set of terms is more limited because it applies only to the response of animal cells—whether they swell or shrink—in solutions of known solute concentrations.

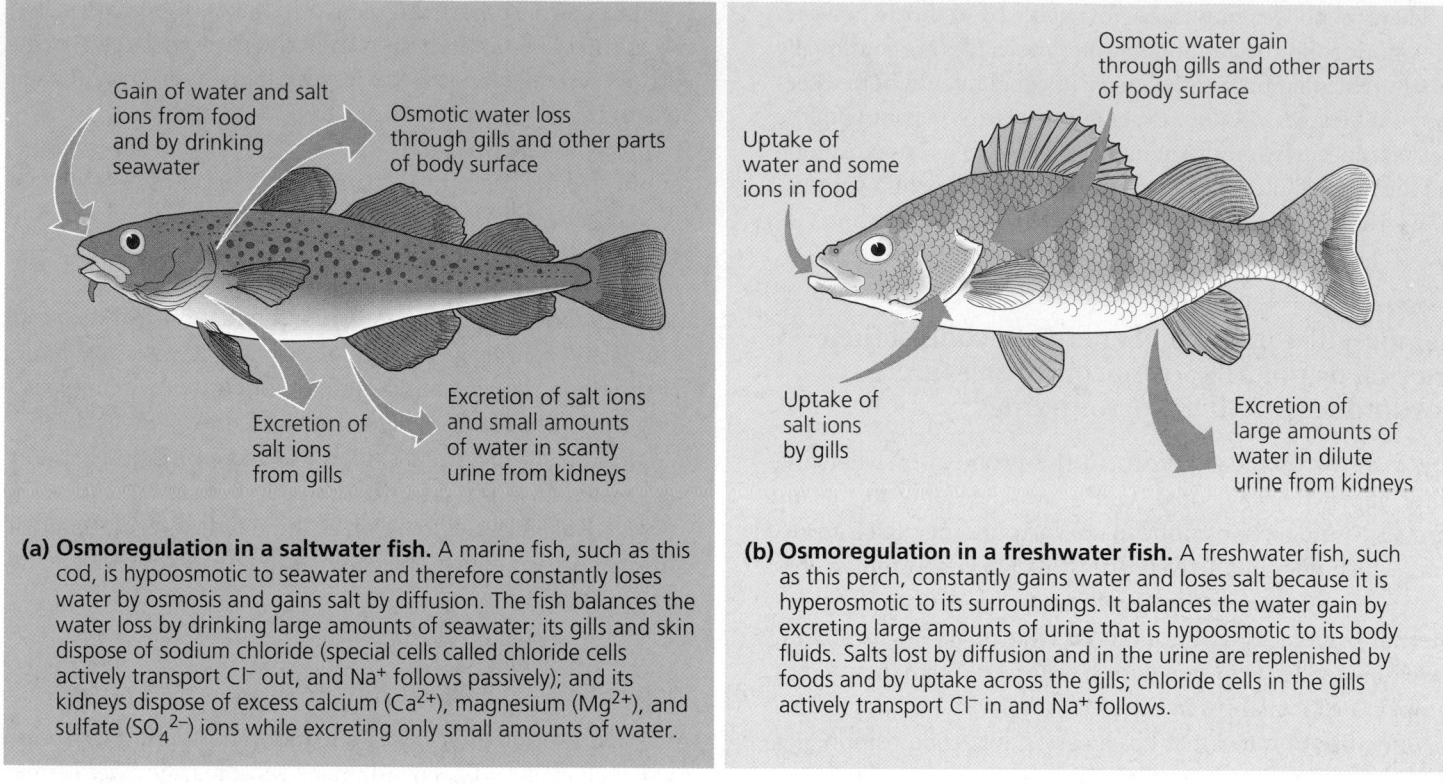

Gain of water and salt ions from food and by drinking seawater

Osmotic water loss through gills and other parts of body surface

Excretion of salt ions from gills

Excretion of salt ions and small amounts of water in scanty urine from kidneys

(a) Osmoregulation in a saltwater fish. A marine fish, such as this cod, is hypoosmotic to seawater and therefore constantly loses water by osmosis and gains salt by diffusion. The fish balances the water loss by drinking large amounts of seawater; its gills and skin dispose of sodium chloride (special cells called chloride cells actively transport Cl^- out, and Na^+ follows passively); and its kidneys dispose of excess calcium (Ca^{2+}), magnesium (Mg^{2+}), and sulfate (SO_4^{2-}) ions while excreting only small amounts of water.

Osmotic water gain through gills and other parts of body surface

Uptake of water and some ions in food

Uptake of salt ions by gills

Excretion of large amounts of water in dilute urine from kidneys

(b) Osmoregulation in a freshwater fish. A freshwater fish, such as this perch, constantly gains water and loses salt because it is hyperosmotic to its surroundings. It balances the water gain by excreting large amounts of urine that is hypoosmotic to its body fluids. Salts lost by diffusion and in the urine are replenished by foods and by uptake across the gills; chloride cells in the gills actively transport Cl^- in and Na^+ follows.

FIGURE 44.14 Osmoregulation in marine and freshwater bony fishes: a comparison.

you have ever prepared shark meat, you know it should be soaked in fresh water to remove urea before cooking.) The total solute concentration of body fluids (salts, urea, TMAO, and other compounds) is somewhat greater than 1,000 mosm/L and therefore slightly hyperosmotic to seawater. Consequently, water slowly *enters* the shark's body by osmosis and in food (sharks do not drink), and this small influx of water is disposed of in urine produced by the kidneys.

Maintaining Osmotic Balance in Fresh Water

The osmoregulatory problems of freshwater animals are opposite those of marine animals. Freshwater animals are constantly gaining water by osmosis and losing salts by diffusion because the osmolarity of their internal fluids is much higher than that of their surroundings. Freshwater protists such as *Amoeba* and *Paramecium* have contractile vacuoles that pump out excess water (see FIGURE 28.14). Many freshwater animals, including fishes, maintain water balance by excreting large amounts of very dilute urine and manage their salt budget by regaining lost salts in food and by active uptake from their surroundings (FIGURE 44.14b).

Salmon and other euryhaline fishes that migrate between seawater and fresh water undergo dramatic and rapid changes in osmoregulatory status. While in the ocean, salmon osmoregulate like other marine fishes by drinking seawater and excreting excess salt from the gills. When they migrate to fresh water, salmon cease drinking, begin to produce lots of dilute urine,

and their gills start taking up salt from the dilute environment— just like fishes that spend their entire lives in fresh water.

Special Problems of Living in Temporary Waters

Dehydration dooms most animals, but some aquatic invertebrates living in temporary ponds and films of water around soil particles can lose almost all their body water and survive in a dormant state when their habitats dry up. This remarkable adaptation is called **anhydrobiosis** ("life without water"). Among the most striking examples are the tardigrades, or water bears, tiny invertebrates less than 1 mm long (FIGURE 44.15). In their active, hydrated state, these animals contain about 85%

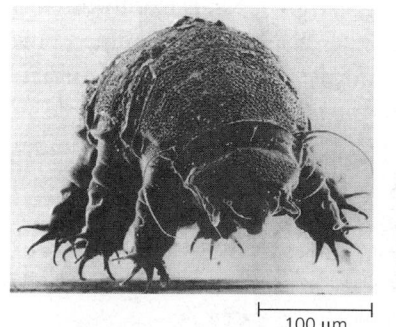

| 100 μm | 50 μm |

(a) Hydrated tardigrade   **(b) Dehydrated tardigrade**

FIGURE 44.15 Anhydrobiosis. Tardigrades (water bears) inhabit temporary ponds and droplets of water in soil and on moist plants (SEMs).

water by weight but can dehydrate to less than 2% water and survive in an inactive state, dry as dust, for a decade or more. Just add water, and within minutes the rehydrated tardigrades are moving about and feeding.

Anhydrobiotic animals must have adaptations that keep cell membranes intact. Researchers are just beginning to learn how tardigrades survive drying out, but studies of anhydrobiotic roundworms (class Nematoda) show that dehydrated individuals contain large amounts of sugars, especially a disaccharide called trehalose. Consisting of two glucose units, trehalose seems to protect the cells by replacing the water associated with membranes and proteins. Many insects that survive freezing in the winter also utilize trehalose as a membrane protectant.

Maintaining Osmotic Balance on Land

The threat of desiccation is perhaps the largest regulatory problem confronting terrestrial plants and animals. Humans die if they lose about 12% of their body water (mammals that evolved in dry enviroments, such as camels, can withstand about twice that level of dehydration). The severity of this challenge may be one reason only arthropods and vertebrates have colonized the land with great success (other phyla have terrestrial representatives, but most of their species are aquatic).

Adaptations that reduce water loss are key to survival on land. Much as a waxy cuticle contributes to the success of land plants, most terrestrial animals have body coverings that help prevent dehydration. Examples are the waxy layers of insect exoskeletons, the shells of land snails, and the multiple layers of dead, keratinized skin cells covering most terrestrial vertebrates. Many terrestrial animals, especially desert dwellers, are nocturnal; this reduces evaporative water loss by taking advantage of the lower temperatures and higher relative humidity of night air.

Despite these adaptations, most terrestrial animals lose considerable water from moist surfaces in their gas exchange organs, in urine and feces, and across the skin. Land animals balance their water budgets by drinking and eating moist foods and by using metabolic water (the water produced in mitochondria during cellular respiration). Some animals are so well adapted for minimizing water loss that they can survive in deserts without drinking. For example, many insect-eating desert birds and reptiles do not drink. Kangaroo rats lose so little water that they can recover 90% of the loss by using metabolic water (FIGURE 44.16), gaining the remaining 10% from the small amount of water in their diet of seeds.

EXCRETORY SYSTEMS

Although the problems of water balance on land or in salt water or fresh water are very different, their solutions all depend on the regulation of solute movement between internal fluids and the external environment. Much of this is handled by excretory systems, which are central to homeostasis because they dispose of metabolic wastes and control body fluid composition by adjusting the rates of loss of particular solutes.

Most excretory systems produce urine by refining a filtrate derived from body fluids: *an overview*

As we will see in the next section, excretory systems are diverse, but nearly all produce urine by a two-step process. First, body fluid (blood, coelomic fluid, or hemolymph) is collected, and then the composition of the collected fluid is adjusted by **selective reabsorption** or **secretion** of solutes (FIGURE 44.17, p. 942). The initial fluid collection usually involves **filtration** through the selectively permeable membranes of transport epithelia. These membranes retain cells as well as proteins and other large molecules in the body fluid; hydrostatic pressure (blood pressure in many animals) forces water and small solutes, such as salts, sugars, amino acids, and nitrogenous wastes, into the excretory system. This fluid is called the **filtrate.**

Fluid collection, even when filtration occurs, is largely nonselective, and thus it is important that essential small molecules are recovered from the filtrate and returned to the body fluids. Excretory systems use active transport to selectively reabsorb valuable solutes such as glucose, certain salts, and

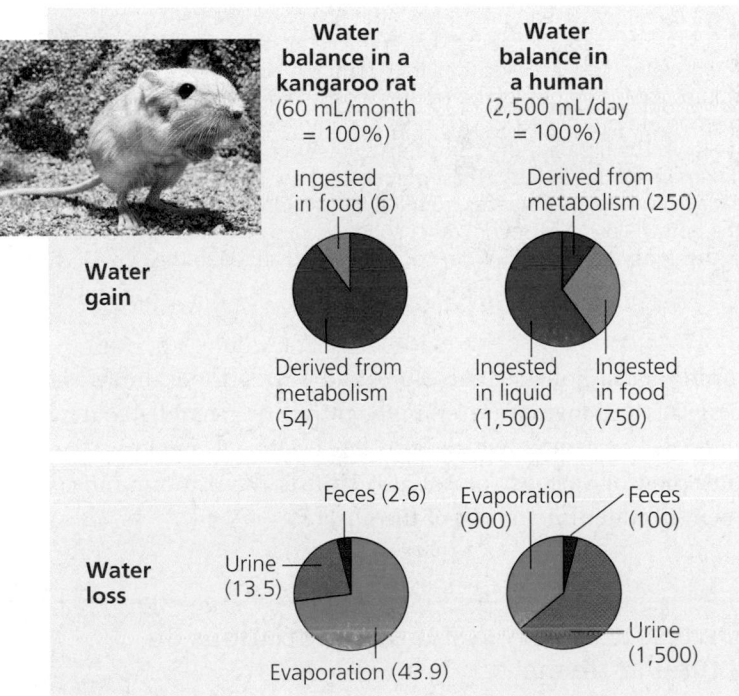

FIGURE 44.16 Water balance in two terrestrial mammals. Kangaroo rats, which live in the American southwest, eat mostly dry seeds and do not drink water. A kangaroo rat loses water mainly by evaporation during gas exchange and gains water mainly from cellular metabolism. In contrast, a human loses a large amount of water in urine, and regains it mostly in food and drink.

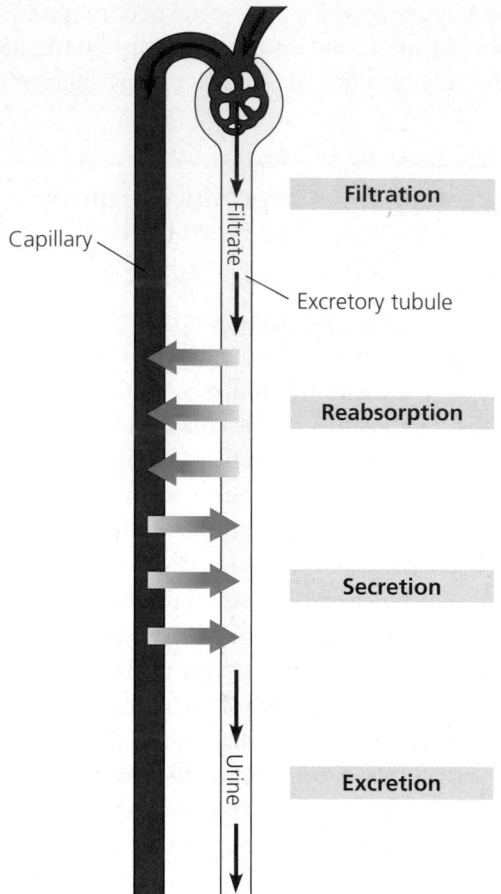

FIGURE 44.17 Key functions of excretory systems: an overview.
Most excretory systems produce a filtrate by pressure-filtering body fluids. In this diagram, modeled after the vertebrate excretory system, the excretory tubule collects a filtrate from the blood. Water and solutes are forced by blood pressure across the selectively permeable membranes of a cluster of capillaries and into the excretory tubule (yellow). The filtrate is then modified by the transport epithelium lining the tubule. In the process of reabsorption, the epithelium reclaims valuable substances from the filtrate and returns them to the body fluids. In secretion, other substances, such as toxins and excess ions, are extracted from body fluids and added to the contents of the excretory tubule.

amino acids. Nonessential solutes and wastes (for example, excess salts and toxins) are left in the filtrate or are added to it by selective secretion, which also uses active transport. The pumping of various solutes also adjusts the osmotic movement of water into or out of the filtrate.

Diverse excretory systems are variations on a tubular theme

Protonephridia: Flame-Bulb Systems

Flatworms (phylum Platyhelminthes) have excretory systems called protonephridia. A **protonephridium** is a network of dead-end tubules lacking internal openings (FIGURE 44.18).

The tubules branch throughout the body, and the smallest branches are capped by a cellular unit called a flame bulb. The flame bulb has a tuft of cilia projecting into the tubule. The beating of the cilia draws water and solutes from the interstitial fluid through the flame bulb (filtration) into the tubule system, and then moves the urine outward through the tubules until they empty to the external environment through openings called nephridiopores. Excreted urine is very dilute in freshwater flatworms, helping balance the osmotic uptake of water from the environment. Apparently, the tubules reabsorb most solutes before the urine exits the body.

The flame-bulb systems of freshwater flatworms seem to function mainly in osmoregulation; most metabolic wastes diffuse out of the animal across the body surface or are excreted into the gastrovascular cavity and eliminated through the mouth (see Chapter 33). However, in some parasitic flatworms, which are isoosmotic to the surrounding fluids of their host organisms, protonephridia function mainly to dispose of

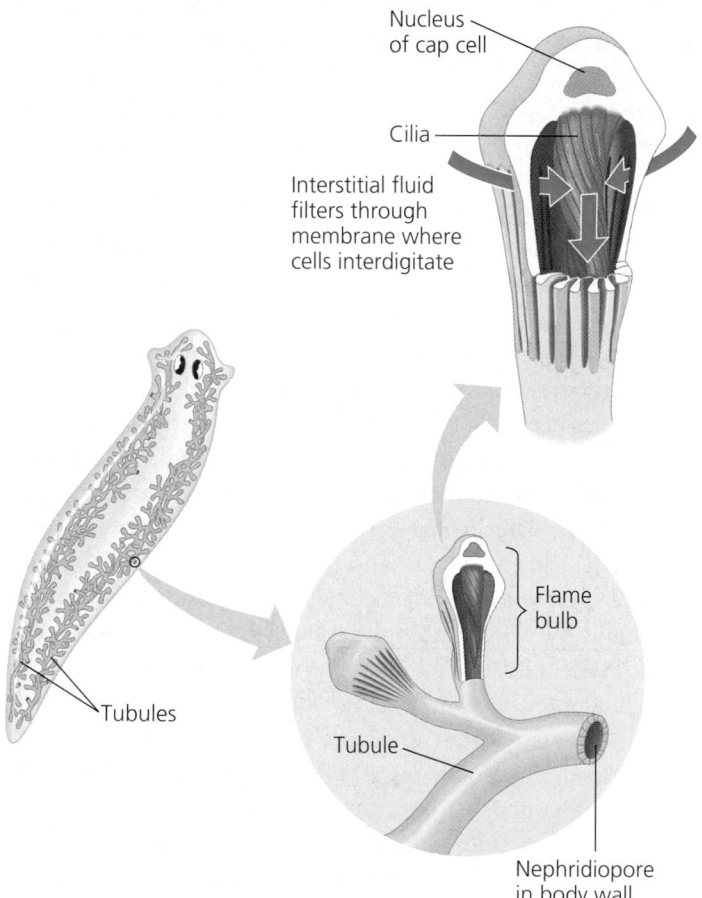

FIGURE 44.18 Protonephridia: the flame-bulb system of a planarian. Protonephridia are branching internal tubules that function mainly in osmoregulation. A single cell caps the internal end of each tubule and interlocks with a tubule cell, forming a flame bulb. Water and solutes from the interstitial fluid enter the lumen of the tubule across the interdigitating membranes of the cap cells and tubule cells. The beating of cilia on the cap cell keeps the fluid moving into and through the tubules to the nephridiopores. (The beating cilia resemble a flickering flame, thus the name flame bulb.)

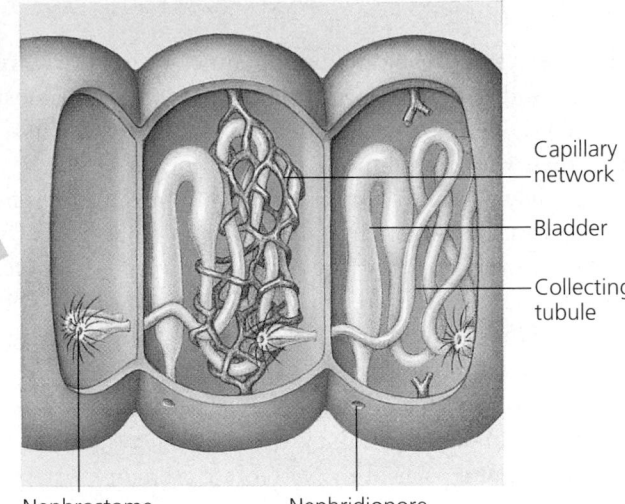

FIGURE 44.19 Metanephridia of an earthworm. Each segment of the worm contains a pair of metanephridia (shown in green and blue), which collect coelomic fluid from the adjacent anterior segment. Fluid enters the nephrostome and passes through the coiled collecting tubule, which includes a storage bladder that opens to the outside through the nephridiopore. Nitrogenous wastes remain in the fluid and are excreted, but certain salts are pumped back into the blood. An earthworm's urine is very dilute, with this loss of water balancing osmotic uptake of water through the skin.

Capillary network

Bladder

Collecting tubule

Nephrostome

Nephridiopore

nitrogenous wastes. This difference in function shows how structures common to a group of organisms can be adapted in diverse ways by evolution in different environments. Protonephridia are also found in rotifers, some annelids, the larvae of mollusks, and lancelets, which are invertebrate chordates. (See Chapters 33 and 34 to review these animal phyla.)

Metanephridia

Another type of tubular excretory system, the **metanephridium,** has internal openings that collect body fluids (FIGURE 44.19). Metanephridia are found in most annelids, including earthworms. Each segment of a worm has a pair of metanephridia, which are immersed in coelomic fluid and enveloped by a capillary network. The internal opening of a metanephridium is surrounded by a ciliated funnel, the nephrostome, which collects fluid from the coelom.

An earthworm's metanephridia have both excretory and osmoregulatory functions. As urine moves along the tubule, the transport epithelium bordering the lumen reabsorbs most solutes and returns them to the blood in the capillaries. Nitrogenous wastes remain in the tubule and are dumped to the outside. Earthworms inhabit damp soil and usually experience a net uptake of water by osmosis. Their metanephridia balance the water influx by producing dilute urine (hypoosmotic to the body fluids).

Malpighian Tubules

Insects and other terrestrial arthropods have organs called **Malpighian tubules** that remove nitrogenous wastes and also function in osmoregulation (FIGURE 44.20). Malpighian tubules open into the digestive tract and dead-end at tips that are immersed in hemolymph (circulatory fluid). The transport epithelium lining the tubules secretes certain solutes, including nitrogenous wastes, from the hemolymph

into the lumen of the tubule. Water follows the solutes into the tubule by osmosis, and the fluid then passes into the rectum, where most solutes are pumped back into the hemolymph. Water again follows the solutes, and the nitrogenous wastes (mainly insoluble uric acid) are eliminated as nearly dry matter along with the feces. The insect excretory system—highly effective in conserving water—is one of several key adaptations that contributed to the tremendous success of these animals on land.

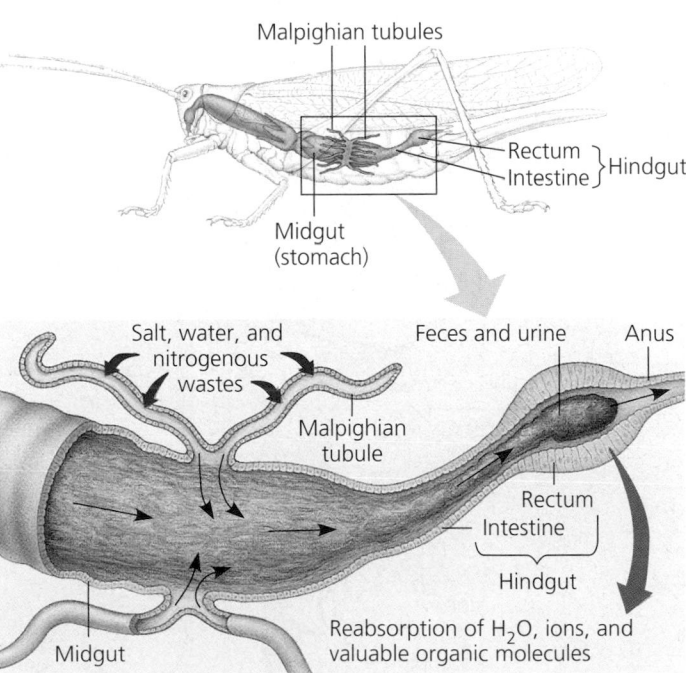

Malpighian tubules

Rectum
Intestine } Hindgut

Midgut (stomach)

Salt, water, and nitrogenous wastes

Feces and urine

Anus

Malpighian tubule

Rectum

Intestine

Hindgut

Midgut

Reabsorption of H_2O, ions, and valuable organic molecules

FIGURE 44.20 Malpighian tubules of insects. Malpighian tubules are outfoldings of the digestive tract. The tubules secrete nitrogenous wastes and salts from the hemolymph into the tubule lumen, and water follows these solutes by osmosis. Most of the salts and water are reabsorbed across the epithelium of the rectum, and the nearly dry nitrogenous wastes are eliminated with the feces.

Vertebrate Kidneys

The kidneys of vertebrates usually function in both osmoregulation and excretion. Like the excretory organs of most animal phyla, kidneys are built of tubules. The osmoconforming hagfishes, which are among the most primitive living vertebrates, have kidneys with segmentally arranged excretory tubules, and it is likely that the excretory structures of vertebrate ancestors were also segmented. However, the kidneys of most vertebrates are compact, nonsegmented organs containing numerous tubules arranged in a highly organized manner. A dense network of capillaries intimately associated with the tubules, along with ducts and other structures that carry urine out of the tubules and kidney and eventually out of the body, are also integral parts of the vertebrate excretory system.

We will focus first on the mammalian version of the vertebrate excretory system, using humans as our example. We will then compare the excretory organs of the various vertebrate classes to see evolutionary modifications that function in different environments.

Nephrons and associated blood vessels are the functional units of the mammalian kidney

Mammals have a pair of kidneys. Each kidney, bean-shaped and about 10 cm long in humans, is supplied with blood by a **renal artery** and a **renal vein** (FIGURE 44.21). Blood flow through the kidneys is voluminous. In humans the kidneys account for

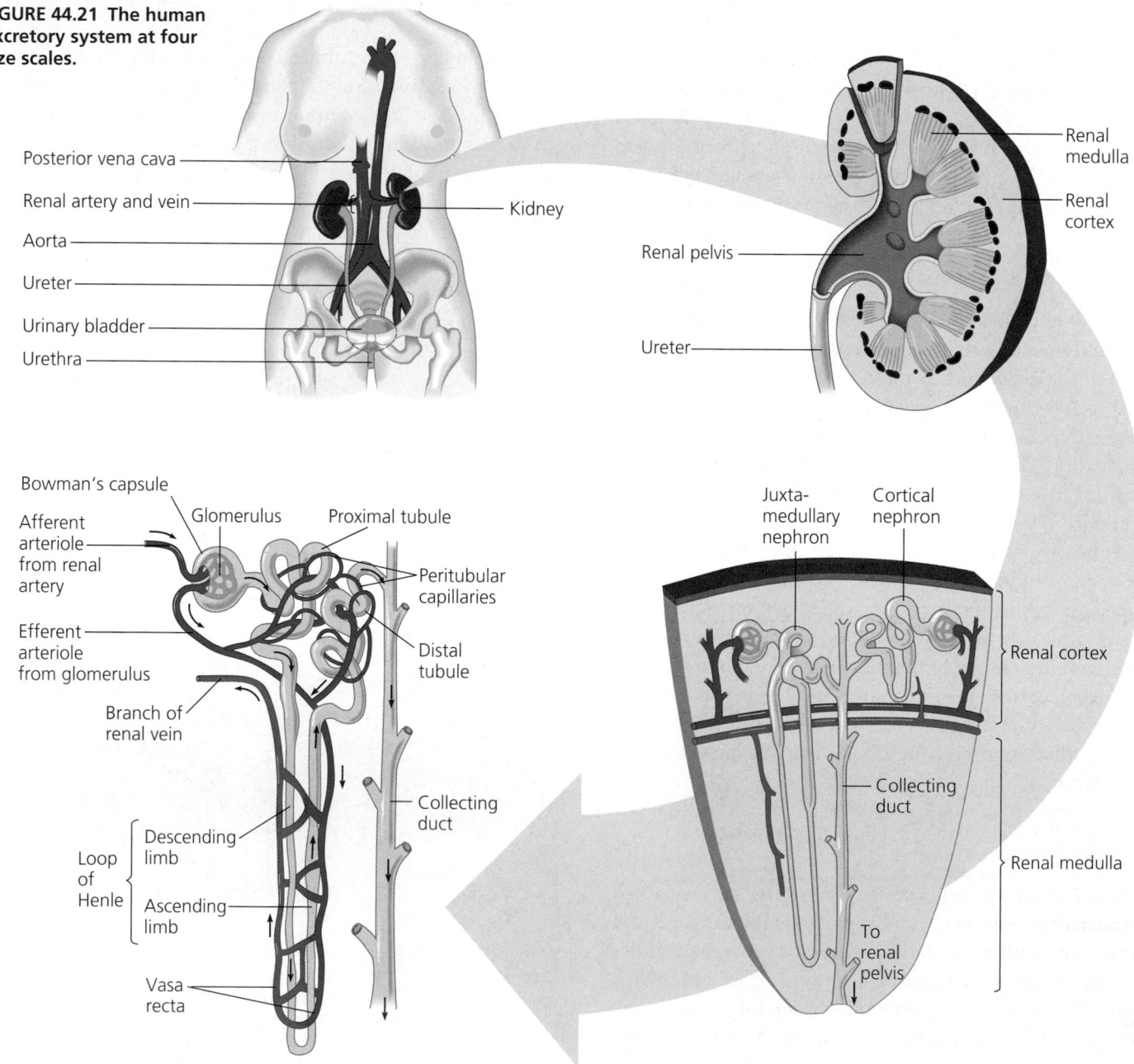

FIGURE 44.21 The human excretory system at four size scales.

Posterior vena cava
Renal artery and vein
Aorta
Ureter
Urinary bladder
Urethra
Kidney

Renal medulla
Renal cortex
Renal pelvis
Ureter

Bowman's capsule
Afferent arteriole from renal artery
Glomerulus
Proximal tubule
Peritubular capillaries
Efferent arteriole from glomerulus
Distal tubule
Branch of renal vein
Collecting duct
Loop of Henle
Descending limb
Ascending limb
Vasa recta

Juxta-medullary nephron
Cortical nephron
Renal cortex
Collecting duct
Renal medulla
To renal pelvis

less than 1% of body weight, but they receive about 20% of resting cardiac output. Urine exits each kidney through a duct called the **ureter,** and both ureters drain into a common **urinary bladder.** During urination, urine is expelled from the urinary bladder through a tube called the **urethra,** which empties to the outside near the vagina in females or through the penis in males. Sphincter muscles near the junction of the urethra and the bladder control urination.

Structure and Function of the Nephron and Associated Structures

The mammalian kidney has two distinct regions, an outer **renal cortex** and an inner **renal medulla** (see FIGURE 44.21). Packing both regions are microscopic excretory tubules and their associated blood vessels. The **nephron**—the functional unit of the vertebrate kidney—consists of a single long tubule and a ball of capillaries called the **glomerulus.** The blind end of the tubule forms a cup-shaped swelling, called **Bowman's capsule,** which surrounds the glomerulus. Each human kidney has about a million nephrons, with a total tubule length of 80 km.

Filtration of the Blood. Filtration occurs as blood pressure forces fluid from the blood in the glomerulus into the lumen of Bowman's capsule. The porous capillaries, along with specialized cells of the capsule called podocytes, are permeable to water and small solutes but not to blood cells or large molecules such as plasma proteins. Filtration of small molecules is nonselective and the filtrate in the Bowman's capsule contains salts, glucose, and vitamins; nitrogenous wastes such as urea; and other small molecules—a mixture that mirrors the concentrations of these substances in blood plasma.

Pathway of the Filtrate. From Bowman's capsule, the filtrate passes through three regions of the nephron: the **proximal tubule;** the **loop of Henle,** a hairpin turn with a descending limb and an ascending limb; and the **distal tubule.** The distal tubule empties into a **collecting duct,** which receives processed filtrate from many nephrons. The many collecting ducts of the kidney empty into the renal pelvis, which is drained by the ureter.

In the human kidney, about 80% of the nephrons, the **cortical nephrons,** have reduced loops of Henle and are almost entirely confined to the renal cortex. The other 20%, the **juxtamedullary nephrons,** have well-developed loops that extend deeply into the renal medulla (see FIGURE 44.21). Only mammals and birds have juxtamedullary nephrons; the nephrons of other vertebrates lack loops of Henle. It is the juxtamedullary nephrons that enable mammals to produce urine that is hyperosmotic to body fluids, an adaptation that is extremely important for water conservation.

The nephron and the collecting duct are lined by a transport epithelium that processes the filtrate to form the urine. One of this epithelium's most important tasks is reabsorption of solutes and water. From 1,100 to 2,000 L of blood flows through a pair of human kidneys each day, a volume about 275 times the total volume of blood in the body. From this enormous traffic of blood, the nephrons and collecting ducts process about 180 L of initial filtrate, equivalent to two or three times the body weight of an average person. Of this, nearly all of the sugar, vitamins, and other organic nutrients, and about 99% of the water, are reabsorbed into the blood, leaving only about 1.5 L of urine to be voided.

Blood Vessels Associated with the Nephrons. Each nephron is supplied with blood by an **afferent arteriole,** a branch of the renal artery that subdivides into the capillaries of the glomerulus. The capillaries converge as they leave the glomerulus, forming an **efferent arteriole.** This vessel subdivides again into a second network of capillaries, the **peritubular capillaries,** which surround the proximal and distal tubules. Additional capillaries extend downward to form the **vasa recta,** the capillary system that serves the loop of Henle. The vasa recta is also a loop, with a descending vessel and an ascending vessel conveying blood in opposite directions.

Although the excretory tubules and their surrounding capillaries are closely associated, they do not exchange materials directly. The tubules and capillaries are immersed in interstitial fluid, through which various substances diffuse between the plasma within capillaries and the filtrate within the nephron tubule.

From Blood Filtrate to Urine: A Closer Look

In this section we concentrate on how the filtrate becomes urine as it flows through the mammalian nephron and collecting duct. The circled numbers correspond to the numbers in FIGURE 44.22, page 946.

❶ **Proximal tubule.** Secretion and reabsorption in the proximal tubule substantially alter the volume and composition of filtrate. For example, the cells of the transport epithelium help maintain a constant pH in body fluids by the controlled secretion of hydrogen ions. The cells also synthesize and secrete ammonia, which neutralizes the acid and keeps the filtrate from becoming too acidic. The more acidic the filtrate, the more ammonia the cells produce and secrete, and the urine of a mammal usually contains some ammonia from this source (even though most nitrogenous waste is excreted as urea). The proximal tubules also reabsorb about 90% of the important buffer bicarbonate (HCO_3^-). Drugs and other poisons that have been processed in the liver pass from the peritubular capillaries into the interstitial fluid, and then are secreted across the epithelium of the proximal tubule into the nephron's

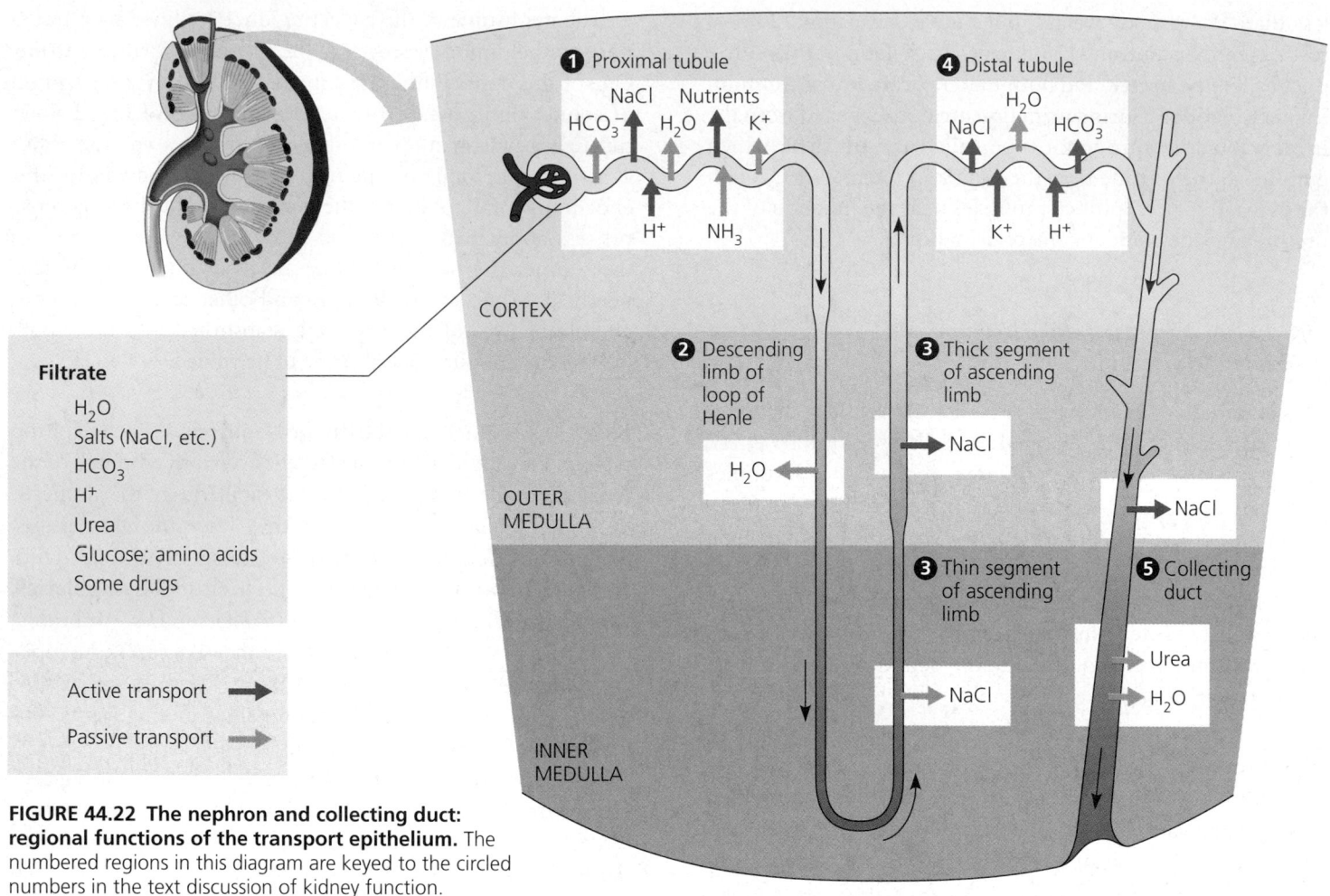

Filtrate

H₂O
Salts (NaCl, etc.)
HCO₃⁻
H⁺
Urea
Glucose; amino acids
Some drugs

Active transport ➡
Passive transport ➡

FIGURE 44.22 The nephron and collecting duct: regional functions of the transport epithelium. The numbered regions in this diagram are keyed to the circled numbers in the text discussion of kidney function.

lumen. Conversely, valuable nutrients, including glucose, amino acids, and potassium (K^+), are actively or passively transported from the filtrate to the interstitial fluid, and then move into the peritubular capillaries.

One of the most important functions of the proximal tubule is reabsorption of most of the NaCl (salt) and water from the huge initial filtrate volume. Salt in the filtrate diffuses into the cells of the transport epithelium, and the membranes of the cells actively transport Na^+ into the interstitial fluid. This transfer of positive charge is balanced by the passive transport of Cl^- out of the tubule. As salt moves from the filtrate to the interstitial fluid, water follows by osmosis. The exterior side of the epithelium has a much smaller surface area than the side facing the lumen, which minimizes leakage of salt and water back into the tubule. Instead, the salt and water now diffuse from the interstitial fluid into the peritubular capillaries.

❷ Descending limb of the loop of Henle. Reabsorption of water continues as the filtrate moves into the descending limb of the loop of Henle. Here the transport epithelium is freely permeable to water but not very permeable to salt and other small solutes. For water to move out of the tubule by osmosis,

the interstitial fluid bathing the tubule must be hyperosmotic to the filtrate. The osmolarity of the interstitial fluid does in fact become progressively greater from the outer cortex to the inner medulla of the kidney (the mechanism that maintains this gradient will be discussed shortly). Thus, filtrate moving downward from the cortex to the medulla within the descending limb of the loop of Henle continues to lose water to interstitial fluid of greater and greater osmolarity. As water departs by osmosis, the solute concentration of the filtrate increases.

❸ Ascending limb of the loop of Henle. The filtrate reaches the tip of the loop, deep in the renal medulla in the case of juxtamedullary nephrons, then moves back to the cortex within the ascending limb. In contrast to the descending limb, the transport epithelium of the ascending limb is permeable to salt but not to water. The ascending limb actually has two specialized regions: a thin segment near the loop tip and a thick segment adjacent to the distal tubule. As filtrate ascends in the thin segment, NaCl, which became concentrated in the descending limb, diffuses out of the permeable tubule into the interstitial fluid. This movement of salt adds to the high osmolarity of the interstitial fluid in the medulla. The exodus of salt from the filtrate continues in the thick segment of the ascending limb, but here the epithelium actively transports NaCl into the interstitial fluid. By losing salt without giving up water, the filtrate becomes progressively more dilute as it moves up to the cortex in the ascending limb of the loop.

❹ **Distal tubule.** The distal tubule is another important site of secretion and reabsorption. It plays a key role in regulating the K^+ and NaCl concentration of body fluids by varying the amount of the K^+ that is secreted into the filtrate and the amount of NaCl reabsorbed from the filtrate. Like the proximal tubule, the distal tubule also contributes to pH regulation, by the controlled secretion of H^+ and reabsorption of bicarbonate (HCO_3^-).

❺ **Collecting duct.** The collecting duct carries the filtrate through the medulla to the renal pelvis. By actively reabsorbing NaCl, the transport epithelium of the collecting duct plays a large role in determining how much salt is actually excreted in the urine. The epithelium is permeable to water but not to salt or (in the renal cortex) to urea. Thus, as the collecting duct traverses the gradient of osmolarity in the kidney, the filtrate becomes increasingly concentrated as it loses more and more water by osmosis to the hyperosmotic interstitial fluid. In the inner medulla, the duct becomes permeable to urea. Because of the high urea concentration in the filtrate at this point, some urea diffuses out of the duct and into the interstitial fluid. Along with NaCl, this interstitial urea is a major solute contributing to the high osmolarity of the interstitial fluid in the medulla. It is this high osmolarity that enables the mammalian kidney to conserve water by excreting urine that is hyperosmotic to the general body fluids.

The mammalian kidney's ability to conserve water is a key terrestrial adaptation

The osmolarity of human blood is about 300 mosm/L, but the kidney can excrete urine up to four times as concentrated—about 1,200 mosm/L. Some mammals can do even better. For example, Australian hopping mice, which live in dry desert regions, can produce urine concentrated to 9,300 mosm/L—9 times as concentrated as seawater and 25 times as concentrated as their body fluid.

In a mammalian kidney, the cooperative action and precise arrangement of the loops of Henle and the collecting ducts are largely responsible for the osmotic gradient that concentrates the urine. But even with this highly organized structure, the maintenance of osmotic differences and the production of hyperosmotic urine are only possible because considerable energy is expended for the active transport of solutes against concentration gradients. In essence, the nephrons—especially the loops of Henle—can be thought of as tiny energy-consuming machines whose function is to produce a region of high osmolarity in the kidney, which can then be used to extract water from the urine in the collecting duct. The two primary solutes in this osmolarity gradient are NaCl, which is deposited in the renal medulla by the loop of Henle, and urea, which leaks across the epithelium of the collecting duct in the inner medulla (see FIGURE 44.22).

Conservation of Water by Two Solute Gradients

To better understand the physiology of the mammalian kidney as a water-conserving organ, let's retrace the flow of filtrate through the excretory tubule, this time focusing on how the juxtamedullary nephrons maintain an osmolarity gradient in the kidney and use that gradient to excrete a hyperosmotic urine (FIGURE 44.23, p. 948). Filtrate passing from Bowman's capsule to the proximal tubule has an osmolarity of about 300 mosm/L, the same as blood. As the filtrate flows through the proximal tubule in the renal cortex, a large amount of water *and* salt is reabsorbed; thus, the volume of filtrate decreases substantially but its osmolarity remains about the same.

As the filtrate flows from cortex to medulla in the descending limb of the loop of Henle, water leaves the tubule by osmosis. The osmolarity of the filtrate increases as solutes, including NaCl, become more concentrated. The highest osmolarity occurs at the elbow of the loop of Henle. This maximizes the diffusion of salt out of the tubule as the filtrate rounds the curve and enters the ascending limb, which, remember, is permeable to salt but not to water. Thus, the two limbs of the loop of Henle cooperate in maintaining the gradient of osmolarity in the interstitial fluid of the kidney. The descending limb produces a progressively saltier filtrate, and the ascending limb exploits this concentration of NaCl to help maintain a high osmolarity in the interstitial fluid of the renal medulla.

Notice that the loop of Henle has several qualities of a countercurrent system, similar in principle to the countercurrent mechanisms that maximize oxygen absorption by fish gills (see FIGURES 42.19 and 42.20) or reduce heat loss from endotherm legs (see FIGURE 44.5). Although the two limbs of the loop are not in direct contact, they are close enough together to exchange substances through the interstitial fluid. The nephron can concentrate salt in the inner medulla largely because exchange between opposing flows in the descending and ascending limbs overcomes the tendency for diffusion to even out salt concentration throughout the kidney's interstitial fluid, thus confining the high salt concentration in the interior of the kidney.

What prevents the capillaries of the vasa recta from dissipating the gradient by carrying away the high concentration of NaCl in the medulla's interstitial fluid? Notice in FIGURE 44.21 that the vasa recta is also a countercurrent system, with descending and ascending vessels carrying blood in opposite directions through the kidney's osmolarity gradient. As the descending vessel conveys blood toward the inner medulla, water is lost from the blood and NaCl diffuses into it. These

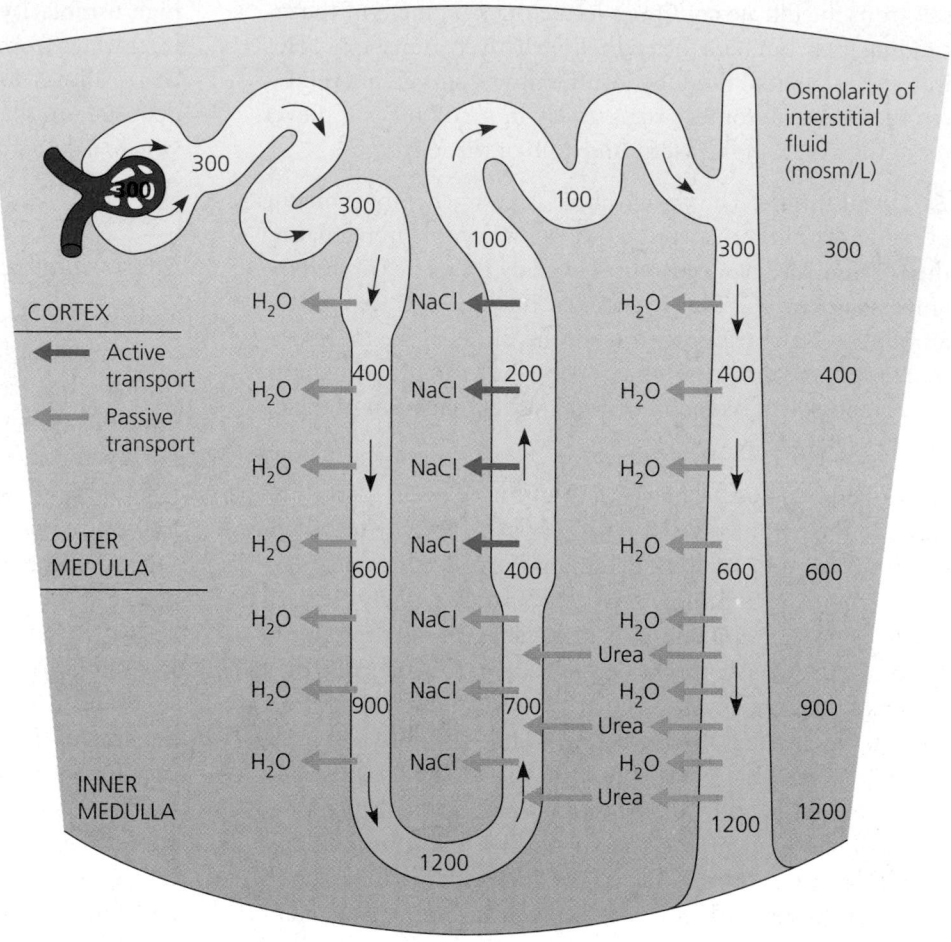

FIGURE 44.23 How the human kidney concentrates urine: the two-solute model. From the cortex to the inner medulla, the interstitial fluid increases in osmolarity from about 300 to 1,200 mosm/L. Two solutes contribute to this gradient: NaCl and urea. The loop of Henle maintains the interstitial gradient of NaCl. The filtrate concentration of this salt increases because of the loss of water from the descending limb, then the ascending limb leaks the salt into the interstitial fluid. Additional salt is actively transported out of the thick segment of the ascending limb. The countercurrent-like arrangement of descending and ascending limbs (and the associated capillaries; FIGURE 44.21) helps maintain the high salt concentration in the medulla. The second solute, urea, is added to the interstitial fluid of the medulla by diffusion out of the collecting duct (most of the urea in the filtrate remains in the collecting duct and is excreted). The filtrate makes a total of three trips between the cortex and medulla: first down, then up, and then down one more time in the collecting duct. As the filtrate flows in the collecting duct past interstitial fluid of increasing osmolarity, more and more water moves out of the duct by osmosis, thereby concentrating the solutes, including urea, that are left behind in the filtrate. Under conditions in which the kidney conserves as much water as possible, urine can reach an osmolarity of about 1,200 mosm/L, considerably hyperosmotic to blood (about 300 mosm/L). This ability to excrete nitrogenous wastes with a minimal loss of water is a key terrestrial adaptation of mammals.

fluxes are reversed as blood flows back toward the cortex in the ascending vessel, with water reentering the blood and salt diffusing out. Thus, the vasa recta can supply the kidney with nutrients and other important substances carried by blood without interfering with the osmolarity gradient that makes it possible for the kidney to excrete a hyperosmotic urine.

The countercurrent-like characteristics of the loop of Henle and the vasa recta make it easier to maintain the steep osmotic gradient between the medulla and cortex. However, any osmotic gradient in an animal will eventually be eliminated by diffusion unless energy is expended to preserve it. In the kidney this expenditure largely occurs in the thick segment of the ascending limb of the loop of Henle, where NaCl is actively transported out of the tubule. Even with the benefits of countercurrent exchange, this process—along with other renal active transport systems—consumes considerable ATP, and for its size the kidney has one of the highest metabolic rates of any organ.

By the time the filtrate reaches the distal tubule, it is actually hypoosmotic to body fluids because of active transport of NaCl out of the thick segment of the ascending limb. Now the filtrate descends again toward the medulla, this time in the collecting duct, which is permeable to water but not to salt. Therefore, osmosis extracts water from the filtrate as it passes from cortex to medulla and encounters interstitial fluid of increasing osmolarity. This concentrates salt, urea, and other solutes in the filtrate. Some urea leaks out of the lower portion of the collecting duct and contributes to the high interstitial osmolarity of the inner medulla. (This urea is recycled by diffusion into the loop of Henle, but continual leakage from the collecting duct maintains a high interstitial urea concentration.) Before leaving the kidney, the urine may attain the osmolarity of the interstitial fluid in the inner medulla, which can be as high as 1,200 mosm/L. Although *isoosmotic* to the inner medulla's interstitial fluid, the urine is *hyperosmotic* to blood and interstitial fluid elsewhere in the body. This high osmolarity allows the solutes remaining in the urine to be excreted from the body with minimal water loss.

The juxtamedullary nephron, with its urine-concentrating features, is a key adaptation to terrestrial life, enabling mammals to get rid of salts and nitrogenous wastes without squandering water. As we have seen, the remarkable ability of the mammalian kidney to produce hyperosmotic urine is completely dependent on the precise arrangement of the tubules and collecting ducts in the renal cortex and medulla. In this respect, the kidney is one of the clearest examples of how the function of an organ is inseparably linked to its structure.

How the Nervous System and Hormones Regulate Kidney Functions

One of the most important aspects of the mammalian kidney is its ability to adjust both the volume and osmolarity of urine, depending on the animal's water and salt balance and the rate of urea production. In situations of high salt intake and low water availability, a mammal can excrete urea and salt with minimal water loss in small volumes of hyperosmotic urine. But if salt is scarce and fluid intake is high, the kidney can get rid of the excess water with little salt loss by producing large volumes of hypoosmotic urine (as dilute as 70 mosm/L, compared to about 300 mosm/L for human blood). This versatility in osmoregulatory function is managed with a combination of nervous and hormonal controls.

One hormone important in regulating water balance is **antidiuretic hormone (ADH)** (FIGURE 44.24a, p. 950). It is produced in a part of the brain called the hypothalamus and is stored in and released from the pituitary gland, which is positioned just below the hypothalamus. Osmoreceptor cells in the hypothalamus monitor the osmolarity of blood; when it rises above a set point of 300 mosm/L (perhaps due to water loss from sweating in a hot environment), more ADH is released into the bloodstream and reaches the kidney. The main targets of ADH are the distal tubules and collecting ducts of the kidney, where the hormone increases the permeability of the epithelium to water. This amplifies water reabsorption, which reduces urine volume and helps prevent further increase of blood osmolarity above the set point. By negative feedback, the subsiding osmolarity of the blood reduces the activity of osmoreceptor cells in the hypothalamus, and less ADH is then secreted. But only the gain of additional water in food and drink can bring osmolarity all the way back down to 300 mosm/L. Conversely, if a large intake of water has reduced blood osmolarity below the set point, very little ADH is released. This decreases the permeability of the distal tubules and collecting ducts, so water reabsorption is reduced, resulting in an increased discharge of dilute urine. (Increased urination is called diuresis, and it is because ADH opposes this state that it is called *anti*diuretic hormone.) Alcohol can disturb water balance by inhibiting the release of ADH, causing excessive urinary water loss and dehydration (which may cause some of the symptoms of a hangover). Normally, blood osmolarity, ADH release, and water reabsorption in the kidney are all linked in a feedback loop that contributes to homeostasis.

A second regulatory mechanism involves a specialized tissue called the **juxtaglomerular apparatus (JGA),** located near the afferent arteriole that supplies blood to the glomerulus (FIGURE 44.24b). When blood pressure or blood volume in the afferent arteriole drops (sometimes as a result of reduced salt intake), the enzyme renin initiates chemical reactions that convert a plasma protein called angiotensinogen to a peptide called **angiotensin II.** Functioning as a hormone, angiotensin II increases blood pressure and blood volume in several ways. It raises blood pressure by constricting arterioles, decreasing blood flow to many capillaries, including those of the kidney. Angiotensin II also stimulates the proximal tubules of the nephrons to reabsorb more NaCl and water. This reduces the amount of salt and water excreted in the urine and consequently raises blood volume and pressure. Another effect of angiotensin II is stimulation of the adrenal glands, organs located atop the kidneys, to release a hormone called **aldosterone.** This hormone acts on the nephrons' distal tubules, making them reabsorb more sodium (Na^+) and water and increasing blood volume and pressure. In summary, the **renin-angiotensin-aldosterone system (RAAS)** is part of a complex feedback circuit that functions in homeostasis. A drop in blood pressure and blood volume triggers renin release from the JGA. In turn, the rise in blood pressure and volume resulting from the various actions of angiotensin II and aldosterone reduce the release of renin (see FIGURE 44.24b).

The functions of ADH and the RAAS may seem to be redundant, but this is not the case. Both increase water reabsorption, but they counter different osmoregulatory problems. The release of ADH is a response to an increase in the osmolarity of the blood, as when the body is dehydrated from excessive loss or inadequate intake of water. However, a situation that causes an excessive loss of salt and body fluids—an injury, for example, or severe diarrhea—will reduce blood volume *without* increasing osmolarity. This will not induce a change in ADH release, but the RAAS will detect the fall in blood volume and pressure and respond by increasing water and Na^+ reabsorption. Normally, ADH and the RAAS are partners in homeostasis; ADH alone would lower blood Na^+ concentration by stimulating water reabsorption in the kidney, but the RAAS helps maintain balance by stimulating Na^+ reabsorption.

Still another hormone, a peptide called **atrial natriuretic factor (ANF),** opposes the RAAS. The walls of the atria of the heart release ANF in response to an increase in blood volume and pressure. ANF inhibits the release of renin from the JGA, inhibits NaCl reabsorption by the collecting ducts, and reduces aldosterone release from the adrenal glands. These actions lower blood volume and pressure. Thus, ADH, the RAAS, and ANF provide an elaborate system of checks and balances that regulate the kidney's ability to control the osmolarity, salt concentration, volume, and pressure of blood.

The flexibility of the mammalian kidney enables it to adjust rapidly to contrasting osmoregulatory and excretory problems. The South American vampire bat (FIGURE 44.25, p. 951) illustrates this versatility. Bats of this species feed on the blood of large birds and mammals. With a feeding mechanism that is surprisingly stealthy, the bats use sharp teeth to make a small incision in the victim's skin and then lap up blood from the wound. Anticoagulents in the bat's saliva prevent the blood from clotting, but the prey animal is usually not seriously harmed. Because vampire bats often search for many hours

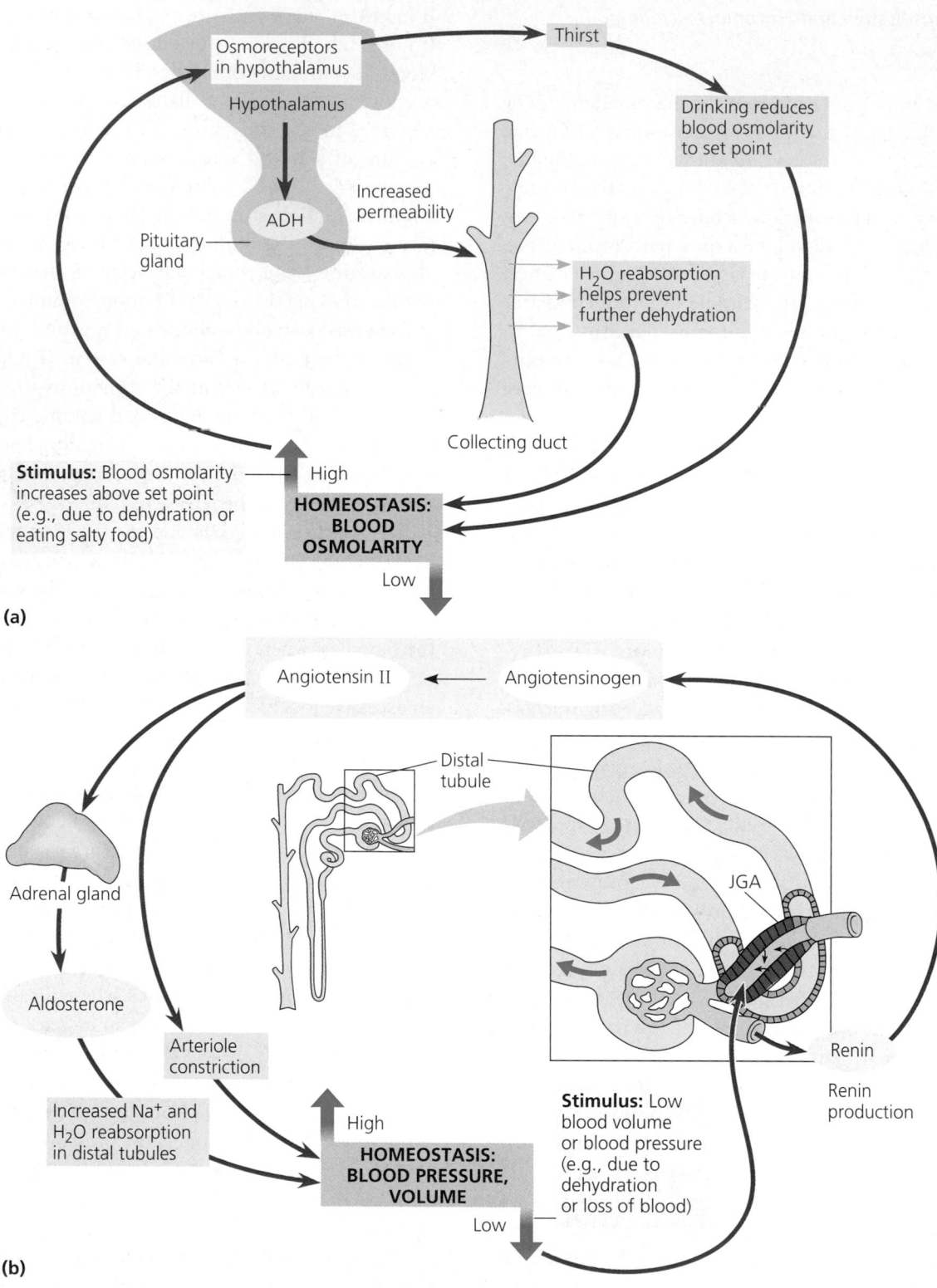

FIGURE 44.24 Hormonal control of the kidney by negative feedback circuits.
(a) Antidiuretic hormone (ADH) enhances fluid retention by making the kidneys reclaim more water. The release of ADH is triggered when osmoreceptor cells in the hypothalamus detect an increase in the osmolarity of the blood. In this situation, the osmoreceptor cells also promote thirst. Drinking reduces the osmolarity of the blood, which inhibits the secretion of ADH, completing the feedback circuit. **(b)** The renin-angiotensin-aldosterone system (RAAS) centers on the juxtaglomerular apparatus (JGA). The JGA responds to a decrease in blood pressure or blood volume by releasing the enzyme renin into the bloodstream (small black arrows). In the blood, renin initiates the conversion of angiotensinogen to angiotensin II. Angiotensin II increases blood pressure by causing arterioles to constrict. It also increases blood volume in two ways: by signaling the proximal tubules of the nephrons to reabsorb more NaCl and water (not illustrated) and by stimulating the adrenal glands to release aldosterone, a hormone that makes the distal tubules reabsorb more Na^+ and water. This leads to an increase in blood volume and pressure, completing the feedback circuit by suppressing the release of renin.

FIGURE 44.25 A vampire bat (*Desmodus rotundas*), a mammal with a unique excretory situation. While the bat feeds on a blood meal, its kidneys excrete dilute urine copiously, thus shedding weight for the flight home. While the bat roosts during the day, the kidneys conserve water by excreting small quantities of urine that is very concentrated in the urea that is produced during the metabolism of proteins from the blood meal.

and fly for long distances to locate a suitable victim, they benefit from consuming as much blood as possible when they do find prey—so much that after feeding, a bat could be too heavy to fly. This problem is solved by the bat using its kidneys to offload much of the water absorbed from a blood meal by excreting large volumes of dilute urine as it feeds, at rates of up to 24% of body mass per hour. Having lost enough weight to take off, the bat can fly back to its roost in a cave or hollow tree, where it spends the day. In the roost, the bat faces a very different regulatory problem. Its food is mostly protein, which generates large quantities of urea—but roosting bats don't have access to drinking water. Their kidneys shift to producing small quantities of highly concentrated urine (up to 4,600 mosm/L), an adjustment that disposes of the urea load while conserving as much water as possible. The vampire bat's ability to alternate rapidly between producing large amounts of dilute urine and small amounts of very hyperosmotic urine is an essential part of its adaption to an unusual food source.

Having considered the mammalian kidney and its regulation in detail, we can now compare the structures and functions of kidneys in other vertebrate classes.

Diverse adaptations of the vertebrate kidney have evolved in different habitats

Variations in nephron structure and function equip the kidneys of different vertebrates for osmoregulation in their various habitats. We have seen, for instance, that nephrons of the mammalian kidney can conserve water by making highly concentrated urine. Mammals that excrete the most hyperosmotic urine, such as hopping mice and other desert mammals, have exceptionally long loops of Henle. Long loops maintain steep osmotic gradients in the kidney, resulting in urine becoming very concentrated as it passes from cortex to medulla in the collecting ducts. In contrast, beavers, which spend much of

their time in fresh water and rarely face problems of dehydration, have nephrons with very short loops, resulting in a much lower ability to concentrate urine.

Birds, like mammals, have kidneys with juxtamedullary nephrons that specialize in conserving water. However, the nephrons of birds have much shorter loops of Henle than do mammalian nephrons. Bird kidneys, therefore, cannot concentrate urine to the osmolarities achieved by mammalian kidneys. Although they can produce a hyperosmotic urine, the main water conservation adaptation of birds is use of uric acid as the nitrogen excretion molecule.

The kidneys of reptiles, having only cortical nephrons, produce urine that is, at most, isoosmotic to body fluids. However, the epithelium of the cloaca (see Chapter 34) helps conserve fluid by reabsorbing some of the water present in urine and feces. Also, like birds, most terrestrial reptiles excrete nitrogenous wastes as uric acid.

In contrast to mammals and birds, a freshwater fish must excrete excess water because the animal is hyperosmotic to its surroundings. Instead of conserving water, the nephrons produce a large volume of very dilute urine. Freshwater fishes conserve salts by reabsorption of ions from the filtrate in the nephrons.

Amphibian kidneys function much like those of freshwater fishes. When in fresh water, the skin of the frog accumulates certain salts from the water by active transport, and the kidneys excrete dilute urine. On land, where dehydration is the most pressing problem of osmoregulation, frogs conserve body fluid by reabsorbing water across the epithelium of the urinary bladder.

Bony fishes that live in seawater, being hypoosmotic to their surroundings, have the opposite problem of their freshwater relatives. In many species, nephrons lack glomeruli and Bowman's capsules, and concentrated urine is formed by secreting ions into the excretory tubules. Thus, as mentioned previously, the kidneys of marine fishes excrete very little urine and function mainly to get rid of divalent ions such as Ca^{2+}, Mg^{2+}, and SO_4^{2-}, which the fish takes in by its incessant drinking of seawater. Its gills excrete mainly monovalent ions such as Na^+ and Cl^- and the bulk of its nitrogenous wastes in the form of NH_4^+ (ammonium ion).

Interacting regulatory systems maintain homeostasis

Numerous regulatory systems are involved in maintaining homeostasis in an animal's internal environment. As we have seen, the mechanisms that rid the body of nitrogenous wastes operate hand in hand with those involved in osmoregulation and are often closely linked with energy budgets and temperature regulation. Similarly, the regulation of body temperature directly affects metabolic rate and exercise capacity and is closely associated with mechanisms controlling blood pressure, gas exchange, and energy balance. Under some conditions, usually at the physical extremes compatible with life, the demands of one system might come into conflict with those of other systems. For instance, in hot, dry environments, water conservation often takes precedence over evaporative heat loss. Many desert animals tolerate occasional abnormally high body temperature, which helps them save water by reducing evaporation. However, if body temperature exceeds a critical upper limit, the animal will start vigorous evaporative cooling and risk dangerous dehydration. Normally, however, the various regulatory systems act together to maintain homeostasis in the internal environment.

Our discussion of homeostasis would be incomplete without mention of the liver, the vertebrate body's most functionally diverse organ. Liver functions are pivotal to homeostasis and involve interaction with most of the body's organ systems.

For example, liver cells interact with the circulatory system in taking up glucose from the blood. Liver cells store excess glucose as glycogen and, in response to the body's demand for fuel, convert glycogen back to glucose, releasing glucose to the blood. The liver also synthesizes plasma proteins important in blood clotting and in maintaining the osmotic balance of the blood. Assisting the excretory system, liver cells detoxify many chemical poisons and prepare metabolic wastes for disposal. The liver's diverse functions and interactions with other organs accentuate the point that homeostasis requires the coordinated action of several body systems.

■   ■   ■

In this chapter we have touched on some of the ways the nervous system and hormones regulate organs involved in homeostasis. We concentrate on the hormonal control of homeostasis in the next chapter.

CHAPTER 44 REVIEW

Go to the Campbell Biology website (www.campbellbiology.com) to explore an interactive version of the Chapter Review.

Summary of Key Concepts

AN OVERVIEW OF HOMEOSTASIS

■ Regulating and conforming are the two extremes in how animals cope with environmental fluctuations (pp. 925–926, FIGURE 44.1) Most animals use some combination of these two "strategies," depending on the environmental situation.

■ Homeostasis balances an animal's gains versus losses for energy and materials (pp. 926–927, FIGURE 44.2) Homeostasis can be viewed in terms of energy and chemical budgets.

REGULATION OF BODY TEMPERATURE

■ Four physical processes account for heat gain or loss (pp. 927–928, FIGURE 44.3) They are conduction, convection, radiation, and evaporation.

■ Ectotherms have body temperatures close to environmental temperature; endotherms can use metabolic heat to keep body temperature warmer than their surroundings (pp. 928–929, FIGURE 44.4) Most invertebrates, fishes, amphibians, and reptiles are ectotherms. Endothermy enables animals such as birds and mammals to maintain a relatively uniform body temperature and a high level of aerobic metabolism.

■ Thermoregulation involves physiological and behavioral adjustments that balance heat gain and loss (pp. 929–930, FIGURE 44.5) Ectotherms and endotherms adjust the rate of heat exchange with their surroundings by evaporative cooling and by behavioral responses. Birds and mammals can also change the rate of metabolic

heat production. Insulation, vasodilation, vasoconstriction, and countercurrent heat exchangers alter the rate of heat exchange. Panting, sweating, and bathing increase evaporation.

■ Most animals are ectothermic, but endothermy is widespread (pp. 930–935, FIGURES 44.6–44.10) Some large active insects and fishes generate metabolic heat by muscle contractions, and many retain it by countercurrent heat exchanges. Some invertebrates, amphibians, and reptiles maintain tolerable internal temperatures by behavioral means. Thermoregulatory mechanisms in mammals and birds include shivering and nonshivering thermogenesis; insulation by fat, hair, or feathers; panting; and countercurrent heat exchange.

■ Torpor conserves energy during environmental extremes (pp. 935–936, FIGURE 44.11) Torpor involves a decrease in metabolic, heart, and respiratory rates and enables the animal to temporarily withstand unfavorable temperatures or lack of food and water.

WATER BALANCE AND WASTE DISPOSAL

■ Water balance and waste disposal depend on transport epithelia (p. 936, FIGURE 44.12) Specialized cell layers regulate the solute movements required for waste disposal and for tempering changes in body fluids.

■ An animal's nitrogenous wastes are correlated with its phylogeny and habitat (pp. 936–938, FIGURE 44.13) Protein and nucleic acid metabolism generates ammonia, a toxic waste product excreted in three forms. Most aquatic animals excrete ammonia across the body surface or gill epithelia into the surrounding water. The liver of mammals and most adult amphibians converts ammonia to the less-toxic urea, which is carried to the kidneys, concentrated, and excreted with a minimal loss of water. Uric acid is an insoluble precipitate excreted in the pastelike urine of land snails, insects, birds, and many reptiles.

- **Cells require a balance between osmotic gain and loss of water (pp. 938–939)** Water uptake and loss are balanced by various mechanisms of osmoregulation in different environments.

- **Osmoregulators expend energy to control their internal osmolarity; osmoconformers are isoosmotic with their surroundings (pp. 939–941, FIGURES 44.14–44.16)** Osmoconformers, which do not regulate their osmolarity, include most marine invertebrates. Osmoregulators control water uptake and loss in a hyperosmotic or hypoosmotic environment. Sharks have an osmolarity slightly higher than seawater because they retain urea. Marine bony fishes lose water to their hyperosmotic environment and drink seawater. Marine vertebrates excrete excess salt through rectal glands, gills, salt-excreting glands, or kidneys. Freshwater animals, which constantly take in water from their hypoosmotic environment, excrete dilute urine. Salt loss is replaced by eating or by ion uptake by gills. Terrestrial animals combat desiccation by behavioral adaptations, water-conserving excretory organs, and by drinking and eating food with high water content.

EXCRETORY SYSTEMS

- **Most excretory systems produce urine by refining a filtrate derived from body fluids: an overview (pp. 941–942, FIGURE 44.17)** Key functions of most excretory systems are filtration (pressure-filtering of body fluids, producing a filtrate) and the production of urine from the filtrate by reabsorption, (reclaiming valuable solutes from the filtrate) and secretion (addition of toxins and other solutes from the body fluids to the filtrate).

- **Diverse excretory systems are variations on a tubular theme (pp. 942–944, FIGURES 44.18–44.20)** Extracellular fluid is filtered into the protonephridia of the flame-bulb system in flatworms; these tubules excrete a dilute fluid and also function in osmoregulation. Each segment of an earthworm has a pair of open-ended metanephridia, tubules that collect coelomic fluid and produce dilute urine for excretion. In insects, Malpighian tubules function in osmoregulation and removal of nitrogenous wastes from the hemolymph. Insects produce a relatively dry waste matter, an important adaptation to terrestrial life. Kidneys, the excretory organs of vertebrates, function in both excretion and osmoregulation.

- **Nephrons and associated blood vessels are the functional units of the mammalian kidney (pp. 944–947, FIGURES 44.21 and 44.22)** Excretory tubules (consisting of nephrons and collecting ducts) and associated blood vessels pack the kidney. Fluid from several nephrons flows into a collecting duct. A ureter conveys urine from the renal pelvis to the urinary bladder. Nephrons control the composition of the blood by filtration, secretion, and reabsorption. The descending limb of the loop of Henle is permeable to water but not to salt; water moves by osmosis into the hyperosmotic interstitial fluid. Salt diffuses out of the concentrated filtrate as it moves through the salt-permeable ascending limb of the loop of Henle.

Web/CD Activity 44A: *Structure of the Human Excretory System*
Web/CD Activity 44B: *Nephron Function*

- **The mammalian kidney's ability to conserve water is a key terrestrial adaptation (pp. 947–951, FIGURES 44.23 and 44.24)** The collecting duct, permeable to water but not to salt, carries the filtrate through the kidney's osmolarity gradient, and more water exits by osmosis. Urea also diffuses out of the tubule and, with salt, forms the osmotic gradient that enables the kidney to produce urine that is hyperosmotic to the blood. The osmolarity of the urine is regulated by nervous system and hormonal control of water and salt reabsorption in the kidneys. This regulation involves the actions of antidiuretic hormone (ADH), the renin-angiotensin-aldosterone system (RAAS), and atrial natriuretic factor (ANF).

Web/CD Activity 44C: *Control of Water Reabsorption*
Web/CD Case Study in the Process of Science: *What Affects Urine Production?*

- **Diverse adaptations of the vertebrate kidney have evolved in different habitats (p. 951)** The form and function of nephrons in the various vertebrate classes are related primarily to the requirements for osmoregulation in the animal's habitat.

- **Interacting regulatory systems maintain homeostasis (pp. 951–952)** The vertebrate liver performs diverse functions vital to homeostasis. Feedback circuits involving nervous system communication and hormones integrate homeostatic mechanisms.

Self-Quiz

1. *Unlike* an earthworm's metanephridia, a mammalian nephron
   a. is intimately associated with a capillary network.
   b. forms urine by changing the composition of fluid inside the tubule.
   c. functions in both osmoregulation and the excretion of nitrogenous wastes.
   d. processes blood instead of coelomic fluid.
   e. has a transport epithelium.

2. The majority of water and salt filtered into Bowman's capsule is reabsorbed by
   a. the transport epithelia of the proximal tubule.
   b. diffusion from the descending limb of the loop of Henle into the hyperosmotic interstitial fluid of the medulla.
   c. active transport across the transport epithelium of the thick upper segment of the ascending limb of the loop of Henle.
   d. selective secretion and diffusion across the distal tubule.
   e. diffusion from the collecting duct into the increasing osmotic gradient of the renal medulla.

3. The high osmolarity of the renal medulla is maintained by all of the following *except*
   a. diffusion of salt from the ascending limb of the loop of Henle.
   b. active transport of salt from the upper region of the ascending limb.
   c. the spatial arrangement of juxtamedullary nephrons.
   d. diffusion of urea from the collecting duct.
   e. diffusion of salt from the descending limb of the loop of Henle.

4. Select the pair in which the nitrogenous waste is incorrectly matched with the benefit of its excretion.
   a. urea—low toxicity relative to ammonia
   b. uric acid—can be stored as a precipitate
   c. ammonia—very soluble in water
   d. uric acid—minimal loss of water when excreted
   e. urea—very insoluble in water

5. Which of the following is *not* an adaptation for reducing the rate of heat exchange between an animal and its environment?
   a. feathers or fur
   b. vasoconstriction
   c. nonshivering thermogenesis
   d. countercurrent heat exchanger
   e. blubber or fat layer

6. An animal's inputs of energy and materials would exceed its outputs
   a. if the animal is an endotherm, which must always take in more energy because of its high metabolic rate.
   b. if it is actively foraging for food.
   c. if it is hibernating.
   d. if it is growing and increasing its biomass.
   e. never—homeostasis makes these energy and material budgets always balance.

7. Which of the following correctly describes a case of osmoregulation?
   a. body fluids that are isoosmotic with the external environment
   b. discharge of excess water in a hypoosmotic environment
   c. expenditure of energy to convert ammonia to less toxic wastes
   d. excretion of salt in a hypoosmotic environment
   e. secretion of drugs and reabsorption of nutrients by the proximal tubule

8. Which process in the nephron is *least* selective?
   a. secretion
   b. reabsorption
   c. active transport
   d. filtration
   e. salt pumping by the loop of Henle

9. You are studying a large tropical reptile that has a high and quite constant body temperature. How would you determine whether this animal is an endotherm or an ectotherm?
   a. You know from its high and constant body temperature that it must be an endotherm.
   b. You know that it is an ectotherm because it is not a bird or mammal.
   c. You subject this reptile to various temperatures in the lab and find that its body temperature and metabolic rate changes with the ambient temperature. You conclude that it is an ectotherm.
   d. You note that its environment has a high and constant temperature. Because its body temperature matches the environmental temperature, you conclude that it is an ectotherm.
   e. You measure the metabolic rate of the reptile, and, because it is higher than that of a related species that lives in temperate forests, you conclude that this reptile is an endotherm and its relative is an ectotherm.

10. The vertebrate liver functions in all of the following regulatory processes *except*
    a. osmoregulation by variable excretion of salts.
    b. maintenance of blood sugar concentration.
    c. detoxification of harmful substances.
    d. production of nitrogenous wastes.
    e. caloric storage in the form of glycogen.

11. Why doesn't your sleep qualify as a short hibernation?

12. Contrast osmoconforming and osmoregulating as mechanisms of homeostasis.

13. Why is ammonia safe as the main nitrogenous waste for most aquatic animals?

14. Some of the drugs classified as diuretics make the epithelium of the collecting duct less permeable to water. How would this affect kidney function?

15. What role does the liver play in the body's processing of nitrogenous waste?

Go to the website or CD-ROM for more quiz questions.

Evolution Connection

A large part of the evolutionary success of arthropods and vertebrates on land is attributable to their osmoregulatory capabilities. Compare and contrast the Malpighian tubule with the nephron in regard to anatomy, relationship to circulation, and physiological mechanisms for conserving body water.

The Process of Science

Eastern tent caterpillars (*Malacosoma americanum*) live in sizable groups in silk nests, or tents, which they construct in cherry trees. They are among the first insects to become active in the spring, emerging very early in the season—a time when the caterpillars regularly contend with large daily temperature fluctuations, from freezing to very hot conditions. Observing a colony over the course of a day, you notice striking differences in group behavior: Early in the morning, the black caterpillars rest in a tightly packed group on the east-facing surface of the tent. In midafternoon the group is found on the tent undersurface, each caterpillar individually hanging from the tent by just a few of its legs. Propose a hypothesis to explain this behavior. How could you test your hypothesis?

Explore how interstitial concentrations and hormones affect urine production in the Case Study in the Process of Science, available on the CD-ROM and website.

Science, Technology, and Society

Kidneys were the first organs to be successfully transplanted. A donor can live a normal life with a single kidney, making it possible for individuals to donate a kidney to an ailing relative or even an unrelated individual with a similar tissue type. In some countries, poor people *sell* kidneys to transplant recipients through organ brokers. What are some of the ethical issues associated with this organ commerce?

Answers : 1. d; 2. a; 3. e; 4. e; 5. c; 6. d; 7. b; 8. d; 9. c; 10. a 11. Because there is no large decrease in body temperature or basal metabolic rate. 12. Osmoconforming prevents large gains or losses of water due to the animal being isoosmotic to the surrounding water; osmoregulating expends energy to balance water uptake and loss for an animal that is not isoosmotic to its surroundings. 13. Because they can dispose of the very toxic ammonia continuously by secreting it across epithelia into the surrounding water. 14. The kidney medulla would reabsorb less water, and thus the diuretic would increase water loss in the urine. 15. The liver is the site of urea synthesis.

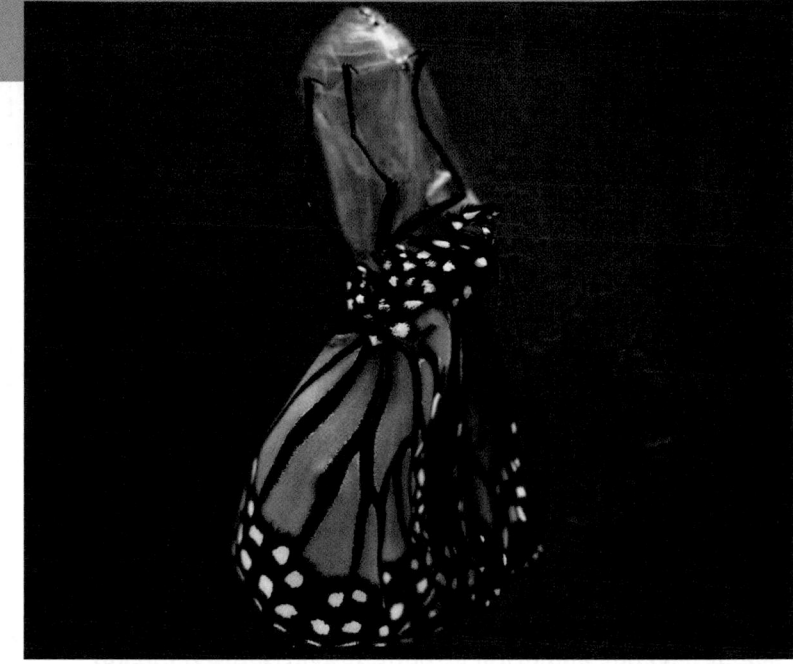

CHAPTER 45

CHEMICAL SIGNALS IN ANIMALS

People offer hormones *as an explanation for the howling of alley cats and the moodiness of teenagers. Over a million diabetics in the United States take the hormone insulin, and other hormones are used in cosmetics intended to keep the skin smooth or are added to livestock feed to fatten cattle. The monarch butterfly in the photograph that opens this chapter has just emerged from the silvery cocoon attached to the twig above it. In becoming an adult, the butterfly has undergone a complete change of body form, a metamorphosis regulated by hormones. Internal communication involving those hormones makes it possible for different parts of the insect's body to develop in concert.*

An animal **hormone** *(from the Greek* hormon, *excite) is a chemical signal that is secreted into body fluids, most often into the blood, and communicates regulatory messages within the body. Hormones may reach all parts of the body, but only certain types of cells, the* **target cells,** *are equipped to respond. Thus, a given hormone traveling in the bloodstream elicits specific responses—a change in metabolism, for example—from selected target cells, while other cell types ignore that particular hormone.*

This chapter's focus is internal chemical signals—what they are and how they function within an animal's body. Our overarching theme is the role of chemical signals in maintaining homeostasis, a dynamic, steady state in body functions.

AN INTRODUCTION TO REGULATORY SYSTEMS

Animals have two systems of internal communication and regulation, the nervous system and the endocrine system. The nervous system, which we will study in Chapters 48 and 49, conveys high-speed signals along specialized cells called neurons. These rapid messages function in such activities as the

A variety of local regulators affect neighboring target cells

At several points earlier in the book, we have mentioned chemical messengers that affect target cells adjacent to or near their point of secretion. For example, in Chapter 43 we described the roles of interleukins in immunity. In Chapters 11, 12, and 19 we introduced the local regulators called growth factors and their involvement in cancer.

Growth factors are peptides and proteins that stimulate cell proliferation. They must be present in the extracellular environment for many types of cells to grow, divide, and develop normally. Because growth factors are generally named for the first function discovered for them, their names can be misleading, for a given growth factor can have several kinds of target cells and a variety of functions. For instance, nerve growth factor (NGF), a protein that speeds the development of certain embryonic nerve cells, also affects developing white blood cells and several other kinds of cells.

Growth factors have been studied mainly in cell cultures, but many experiments show that growth factors work within the animal body as well as in culture. For instance, injecting epidermal growth factor (EGF) into fetal mice accelerates epidermal development. And a group of peptides known as insulinlike growth factors (IGFs), produced by the liver, is essential to skeletal development. It is likely that the interaction of numerous growth factors regulates cell behavior in developing tissues and organs of animals. Ongoing research also suggests that growth factors (specifically, transforming growth factor, TGF) can enhance the strength of synapses between neurons in the brain of a mature animal.

Another important kind of local regulator is a gas, **nitric oxide (NO).** Many types of cells produce NO, and it has multiple functions. Highly reactive and potentially toxic, NO usually triggers changes in a target cell within a few seconds of contact and then breaks down. Secreted by neurons, NO functions as a neurotransmitter (see Chapter 48); secreted by white blood cells, it kills certain bacteria and cancer cells in body fluids. And NO released by endothelial cells in blood vessels makes the adjacent smooth muscles relax, dilating the vessel walls. American pharmacologists Robert Furchgott, Louis Ignarro, and Ferid Murad shared the 1998 Nobel Prize in medicine for their pioneering work on the physiological effects of NO.

The local regulators called **prostaglandins (PGs)** are modified fatty acids, often derived from lipids to the plasma membrane. They are so named because they were first discovered in the components of semen produced by the human prostate gland. Prostaglandins in semen stimulate contraction of smooth muscles in the wall of the uterus, helping to convey sperm to the egg. Released from most types of cells into the interstitial fluid, prostaglandins function as local regulators affecting nearby cells in various ways. Some of the best-known actions of prostaglandins are on the female reproductive system.

For example, prostaglandins secreted by cells of the placenta cause chemical changes in the nearby muscles of the uterus, making them more excitable and thereby helping to induce labor during childbirth. This is an example of a positive feedback mechanism coordinating a body function (see FIGURE 46.19).

Prostaglandins also function as local regulators in vertebrate defense systems. Various prostaglandins help induce fever and inflammation and also intensify the sensation of pain (which can be thought of as contributing to the body's defense by sounding an alarm that something harmful is going on). The anti-inflammatory effects of aspirin and ibuprofen result from the inhibitory effects of these drugs on prostaglandin synthesis.

Two prostaglandins with very similar molecular structures, prostaglandin E (PGE) and prostaglandin F (PGF), have opposite effects on the smooth muscle cells in the walls of blood vessels serving the lungs. PGE causes the muscles to relax, which dilates the blood vessels and promotes oxygenation of the blood. PGF signals the muscles to contract, which constricts the vessels and reduces blood flow through the lungs. In other words, these two chemical signals are antagonistic. Shifts in their relative concentrations help maintain homeostasis in changing circumstances, making these local regulators another example of the use of antagonistic signals to counterbalance each other.

Most chemical signals bind to plasma-membrane proteins, initiating signal-transduction pathways

As we discussed in Chapter 11, a signal molecule has a specific shape that can be recognized by that signal's target cells. *Reception* of the signal occurs when the signal molecule binds to a specific receptor protein, which is either built into the plasma membrane of the target cell or located inside the target cell (FIGURE 45.3). The binding of a signal molecule to a receptor protein triggers events within the target cell—*signal transduction*—that result in a *response*, a change in the cell's behavior. Cells are unresponsive to a signal if they lack the appropriate receptors. How a chemical signal brings about changes in target cells depends on whether it binds to a receptor on the target cell's surface or inside the cell.

Almost all local regulators and a majority of hormones have plasma-membrane receptors (see FIGURE 45.3a). Chapter 11 discussed the discovery that the hormone epinephrine exerts its effects via membrane receptors. Another dramatic example of the role of membrane receptors involves changes in frog skin color, an adaptation that helps camouflage the frog as lighting changes. The darkness of the skin varies with the organelle arrangement within melanocytes, skin cells that contain the dark brown pigment melanin in cytoplasmic organelles called melanosomes. The frog's skin appears light when the melanosomes are clustered tightly around the cell nuclei and darker when the melanosomes spread throughout the cytoplasm. The arrangement of melanosomes is controlled by a

THE PROCESS OF SCIENCE

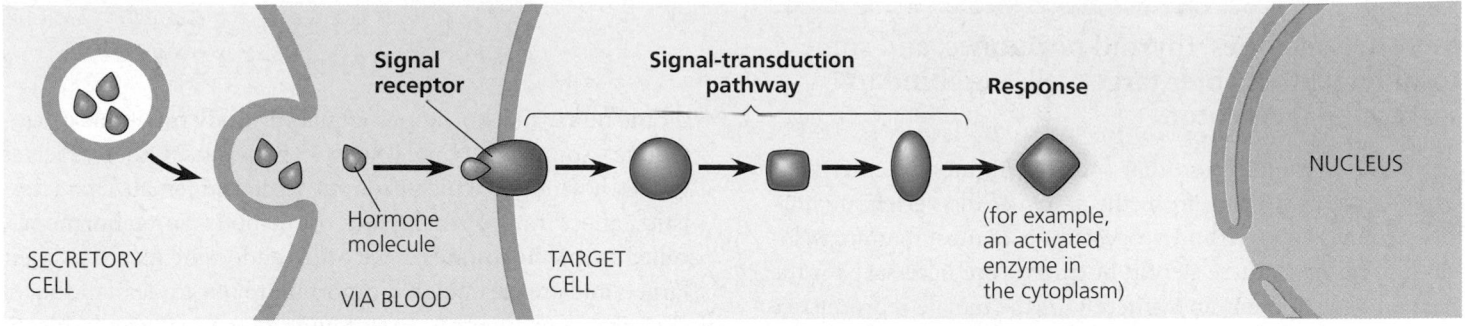

(a) Receptor in plasma membrane

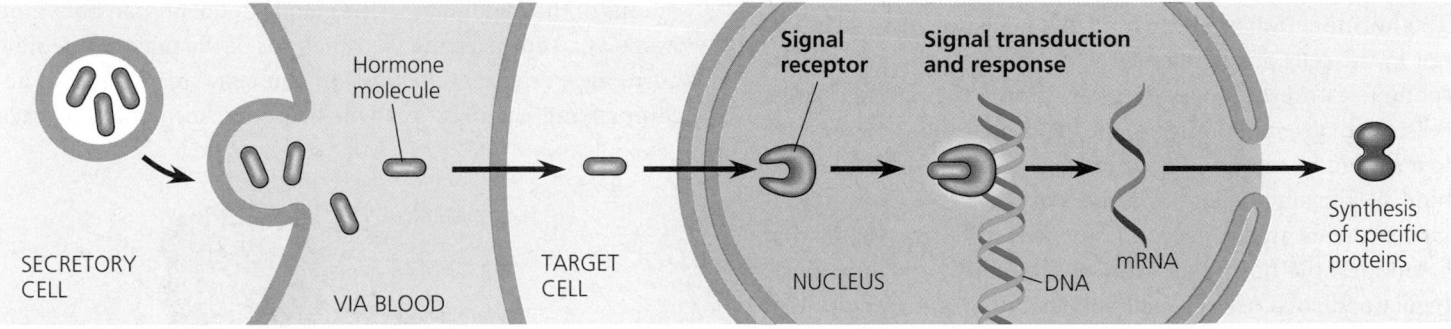

(b) Receptor in cell nucleus

FIGURE 45.3 Mechanisms of chemical signaling: a review. A chemical signal secreted by a cell either **(a)** binds to a receptor protein on the surface of a target cell, triggering a signal-transduction pathway, or **(b)** penetrates the target cell's plasma membrane and binds to a receptor inside the cell. When signal is bound to an intracellular receptor, the receptor acts as a transcription factor, causing a change in gene expression. The binding of signal to a surface receptor can lead to either a change in gene expression or a change in a cytoplasmic activity.

peptide hormone called melanocyte-stimulating hormone (MSH), secreted by the pituitary gland. When MSH is added to the interstitial fluid surrounding the pigment-containing cells, the melanosomes disperse. However, direct microinjection of MSH into individual melanocytes does not induce melanosome dispersion—evidence that interaction between the hormone and a *surface* receptor is required for hormone action.

A signal receptor located in the plasma membrane is generally the first component of a **signal-transduction pathway,** a series of molecular changes that converts an extracellular chemical signal to a specific intracellular response. You may wish to review the molecules and mechanisms of cell signaling in Chapter 11. Depending on the signal molecule and the molecules present in a target cell, a signal-transduction pathway may lead to responses in either the cytoplasm (the activation of an enzyme, for instance) or the nucleus (usually involving the regulation of specific genes). And because different types of cells have different collections of molecules (especially proteins), the same signal can bring about different responses in different target cells (FIGURE 45.4). Hormones are especially potent regulators, effective in minute amounts, and signal-transduction pathways can trigger elaborate enzyme cascades that amplify the hormonal signal (see FIGURE 11.16).

FIGURE 45.4 One chemical signal, different effects. A single type of signal molecule, in this case a neurotransmitter called acetylcholine, can produce different responses in different target cells. Different responses may result because the receptors are different [compare (a) with (b) and (c)] or because signal-transduction pathways within the target cells are different (see also FIGURE 11.18).

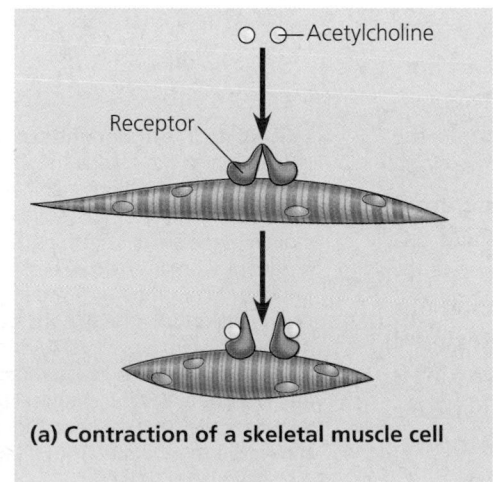

(a) Contraction of a skeletal muscle cell

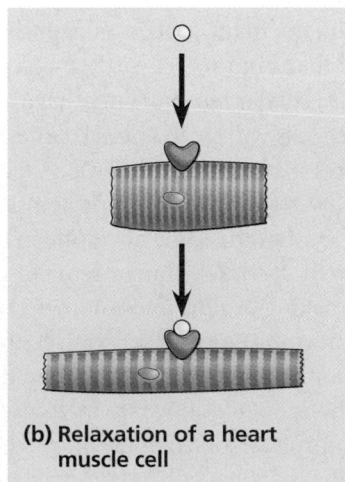

(b) Relaxation of a heart muscle cell

(c) Secretion by an endocrine cell

Steroid hormones, thyroid hormones, and some local regulators enter target cells and bind to intracellular receptors

The first indication that some hormones do enter their target cells came from the study of the vertebrate hormones estrogen and progesterone. In most mammals, including humans, these steroid hormones are necessary for the normal development and function of the female reproductive system. In the early 1960s, researchers demonstrated that cells in the reproductive tract of female rats accumulate estrogen. The hormone was found within the nuclei of these cells, but not in the cells of such tissues as the spleen, which do not respond to estrogen. Progesterone also enters the nuclei of target cells. Such observations led to the hypothesis that cells sensitive to a steroid hormone contain internal receptor molecules that bind specifically to that hormone. We now know that intracellular proteins are the receptors for steroid hormones, thyroid hormones, the hormonal form of vitamin D, and some local regulators, such as NO—all small nonpolar molecules that pass easily through cell membranes. Most intracellular receptors are located in the nucleus (see FIGURE 45.3b); others are in the cytoplasm, at least initially (see FIGURE 11.10).

Intracellular receptor proteins usually perform the entire task of transducing the signal within the *cell*. The chemical signal activates the receptor, which then directly triggers the cell's response. In almost all cases, the intracellular receptor activated by a hormone is a transcription factor, and the response is a change in gene expression. The active form of the factor—the hormone-receptor complex—binds to specific sites in the cell's DNA and either stimulates or represses the transcription of specific genes (see FIGURES 17.7 and 19.9). Newly made mRNA is then translated into new protein in the cytoplasm.

The arthropod molting hormone ecdysone (see FIGURE 45.2) is an example of a steroid that brings about the synthesis of new proteins. Ecdysone causes certain cells of an arthropod to synthesize the enzymes that catalyze the production of a new exoskeleton. In another example, estrogen induces cells in the reproductive system of a female bird to synthesize large amounts of ovalbumin, the main protein of egg white.

As with hormones that bind to cell-surface receptors, hormones that bind to intracellular receptors may have different effects on different target cells within an animal. For example, the estrogen that stimulates a bird's reproductive system to make ovalbumin causes its liver to make other proteins. And from that example you can see that the same hormone may have different effects in different *species*—human females do not respond to estrogen by making ovalbumin! Another example of a species difference in hormone response involves thyroxine, a thyroid hormone. In humans and other vertebrates, thyroxine is responsible for metabolic regulation. But in frogs, thyroxine also triggers the metamorphosis of a tadpole into an adult, stimulating resorption of the tadpole's tail and other changes.

THE VERTEBRATE ENDOCRINE SYSTEM

Of the numerous hormones regulating body functions of vertebrates, some affect only one or a few tissues. Others, such as the sex hormones, which promote male and female characteristics, affect most of the tissues of the body. Some hormones, called **tropic hormones,** have other endocrine glands as their targets and are particularly important to our understanding of chemical coordination. In studying hormonal regulation of the vertebrate body, you may find FIGURE 45.5, showing the locations of the major endocrine glands in the human body, and TABLE 45.1, summarizing the functions of the major vertebrate hormones, especially helpful. In the text, small sketches accompanying each section will help you keep track of each

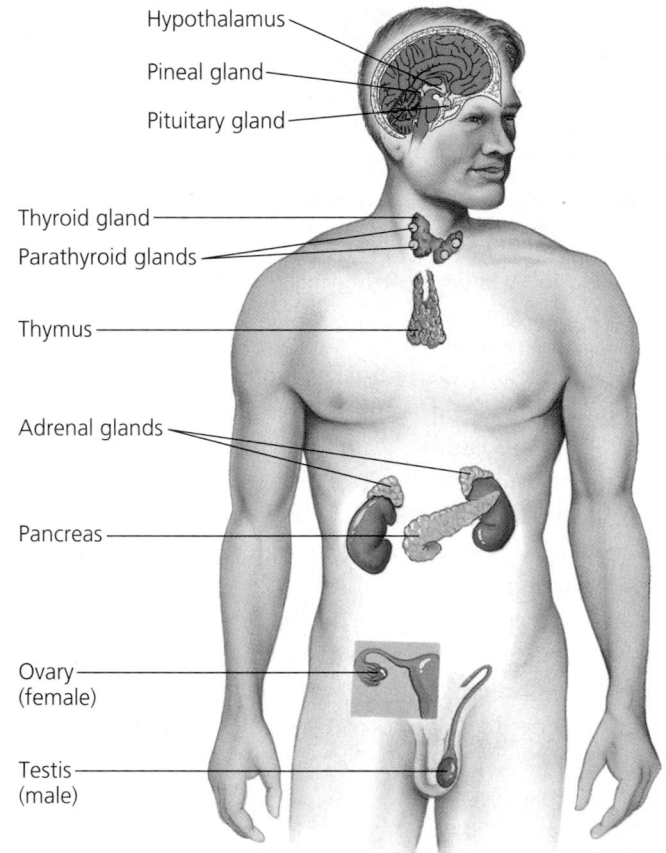

FIGURE 45.5 Human endocrine glands surveyed in this chapter. In addition to the glands shown here, many organs with primarily nonendocrine functions also have cells that secrete hormones. These organs, which are discussed in other chapters, include the heart, the stomach, the small intestines, the kidney, and the placenta of a pregnant woman. Also discussed elsewhere is the thymus gland, an important component of the immune system (Chapter 43). Quite large during childhood, the thymus begins to decline at puberty, when the immune system is well established. By adulthood it is largely replaced by adipose and fibrous tissue, but it continues to function throughout life. It secretes several chemical signals, including thymosin, that stimulate the development and differentiation of T lymphocytes after they leave the thymus. The other endocrine organs in this drawing are discussed in this chapter.

Table 45.1 Major Vertebrate Endocrine Glands and Some of Their Hormones

Gland	Hormone	Chemical Class	Representative Actions	Regulated By
Hypothalamus	Hormones released by the posterior pituitary and hormones that regulate the anterior pituitary (see below)			
Pituitary gland Posterior pituitary (releases hormones made by hypo-thalamus)	Oxytocin	Peptide	Stimulates contraction of uterus and mammary gland cells	Nervous system
	Antidiuretic hormone (ADH)	Peptide	Promotes retention of water by kidneys	Water/salt balance
Anterior pituitary	Growth hormone (GH)	Protein	Stimulates growth (especially bones) and metabolic functions	Hypothalamic hormones
	Prolactin (PRL)	Protein	Stimulates milk production and secretion	Hypothalamic hormones
	Follicle-stimulating hormone (FSH)	Glycoprotein	Stimulates production of ova and sperm	Hypothalamic hormones
	Luteinizing hormone (LH)	Glycoprotein	Stimulates ovaries and testes	Hypothalamic hormones
	Thyroid-stimulating hormone (TSH)	Glycoprotein	Stimulates thyroid gland	Thyroxine in blood; hypothalamic hormones
	Adrenocorticotropic hormone (ACTH)	Peptide	Stimulates adrenal cortex to secrete glucocorticoids	Glucocorticoids; hypothalamic hormones
Thyroid gland	Triiodothyronine (T_3) and thyroxine (T_4)	Amine	Stimulate and maintain metabolic processes	TSH
	Calcitonin	Peptide	Lowers blood calcium level	Calcium in blood
Parathyroid glands	Parathyroid hormone (PTH)	Peptide	Raises blood calcium level	Calcium in blood
Pancreas	Insulin	Protein	Lowers blood glucose level	Glucose in blood
	Glucagon	Protein	Raises blood glucose level	Glucose in blood
Adrenal glands Adrenal medulla	Epinephrine and norepinephrine	Amine	Raise blood glucose level; increase metabolic activities; constrict certain blood vessels	Nervous system
Adrenal cortex	Glucocorticoids	Steroid	Raise blood glucose level	ACTH
	Mineralocorticoids	Steroid	Promote reabsorption of Na^+ and excretion of K^+ in kidneys	K^+ in blood
Gonads Testes	Androgens	Steroid	Support sperm formation; promote development and maintenance of male secondary sex characteristics	FSH and LH
Ovaries	Estrogens	Steroid	Stimulate uterine lining growth; promote development and maintenance of female secondary sex characteristics	FSH and LH
	Progesterone	Steroid	Promotes uterine lining growth	FSH and LH
Pineal gland	Melatonin	Amine	Involved in biological rhythms	Light/dark cycles
Thymus	Thymosin	Peptide	Stimulates T lymphocytes	Not known

gland's location. We begin our discussion of the vertebrate endocrine system by looking at the hypothalamus and pituitary gland, which control much of the endocrine system.

The hypothalamus and pituitary integrate many functions of the vertebrate endocrine system

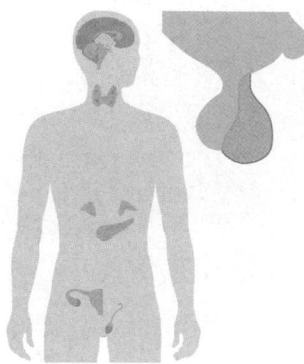

The **hypothalamus** plays an important role in integrating the vertebrate endocrine and nervous systems. This region of the lower brain receives information from nerves throughout the body and from other parts of the brain, then initiates endocrine signals appropriate to environmental conditions. In many vertebrates, for example, the brain passes sensory information about seasonal changes and the availability of a mate to the hypothalamus by means of nerve signals; the hypothalamus then triggers the release of reproductive hormones required for breeding.

The hormone-releasing cells in the hypothalamus are two sets of neurosecretory cells whose secretions are stored in or regulate the activity of the **pituitary gland,** a small organ with multiple endocrine functions. The pituitary gland was formerly called the "master gland" because so many of its hormones regulate other endocrine functions. However, the pituitary itself obeys hormonal orders from the hypothalamus.

No organ illustrates the close structural, functional, and developmental relationships of the endocrine and nervous systems better than the pituitary gland. Located at the base of the hypothalamus, it has two discrete parts that develop from two separate regions of the embryo and have very different functions (FIGURE 45.6). The **anterior pituitary,** or **adenohypophysis,** develops from a fold of tissue at the roof of the embryonic mouth that grows upward toward the brain and eventually loses its connection to the mouth. The anterior pituitary consists of endocrine cells that synthesize and secrete several hormones directly into the blood. A set of neurosecretory cells in the hypothalamus exerts control over the anterior pituitary by secreting two kinds of hormones into the blood. **Releasing hormones** make the anterior pituitary secrete its hormones. **Inhibiting hormones** from the hypothalamus make the anterior pituitary stop secreting hormone.

The hypothalamic releasing and inhibiting hormones are released into capillaries in a region at the base of the hypothalamus (FIGURE 45.6b). The capillaries drain into portal vessels, which are short blood vessels that subdivide into a second capillary bed within the anterior pituitary. In this way, the hypothalamic hormones have direct access to the gland they control. Every anterior pituitary hormone is controlled by at least one releasing hormone, and some have both a releasing hormone and an inhibiting hormone.

Unlike the anterior pituitary, the **posterior pituitary,** or **neurohypophysis,** is an extension of the brain. It develops from a small bulge of the hypothalamus that grows downward toward the mouth fold that forms the anterior pituitary. The posterior pituitary remains an extension of the hypothalamus. It stores and secretes two hormones that are made by a set of neurosecretory cells in the hypothalamus.

Posterior Pituitary Hormones

The two hormones released by the posterior pituitary, oxytocin and antidiuretic hormone (ADH), are made by the hypothalamus. **Oxytocin** acts on muscles of the uterus and ADH acts on the kidneys, rather than affecting other endocrine glands. Oxytocin induces contraction of the uterine muscles during childbirth and causes the mammary glands to eject milk during nursing. ADH acts on the kidneys, increasing water retention and thus decreasing urine volume.

Antidiuretic hormone (ADH) is part of an elaborate feedback scheme that helps regulate the osmolarity of the blood. This mechanism was introduced in Chapter 44, but it is worth reviewing here to illustrate how hormones contribute to homeostasis and how negative feedback controls hormone levels. Blood osmolarity is monitored by a group of nerve cells that function as osmoreceptors in the hypothalamus. When the osmolarity of the plasma increases, these cells shrink slightly (due to osmosis) and transmit nerve impulses to certain neurosecretory cells in the hypothalamus. These cells respond by releasing ADH from their tips, which are located in the posterior pituitary. The posterior pituitary stores ADH and releases it into the general circulation. When ADH reaches the kidneys, it binds to receptors on the surface of the cells lining the collecting ducts. This binding activates a signal-transduction pathway that increases the water permeability of the collecting duct. Water exits the collecting ducts and enters nearby capillaries, helping prevent a further increase in blood osmolarity above the set point. The brain's osmoreceptors also stimulate a thirst drive, and drinking water brings blood osmolarity back down to the set point. Thus, the hormonal reaction to high blood osmolarity is augmented by a behavioral response controlled by the nervous system. As the more dilute blood arrives at the brain, the hypothalamus responds to the reduction in osmolarity by slowing the release of ADH and diminishing the sensation of thirst. The effects of these hormonal and behavioral responses—increased water reabsorption in the kidneys and drinking, respectively—prevent overcompensation by shutting off further secretion of the hormone and quenching thirst. We see once again how negative feedback helps maintain homeostasis. This example also highlights the central role of the hypothalamus as a member of both the endocrine system and the nervous system.

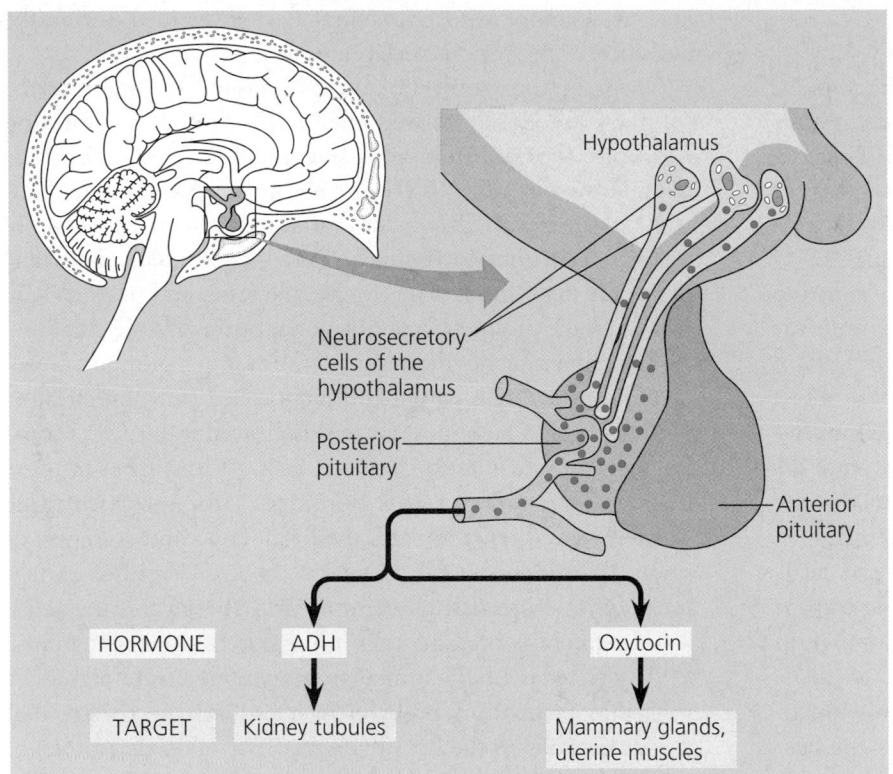

FIGURE 45.6 Hormones of the hypothalamus and pituitary glands. The pituitary gland, located at the base of the brain and surrounded by bone, consists of the posterior pituitary (neurohypophysis) and the anterior pituitary (adenohypophysis). Note that the posterior pituitary is actually an extension of the hypothalamus.

Hypothalamus

Neurosecretory cells of the hypothalamus

Posterior pituitary

Anterior pituitary

| HORMONE | ADH | Oxytocin |
| TARGET | Kidney tubules | Mammary glands, uterine muscles |

(a) The posterior pituitary. Neurosecretory cells in the hypothalamus synthesize antidiuretic hormone (ADH) and oxytocin, which are transported down the axons to the posterior pituitary, where they are stored. The posterior pituitary releases them into the blood circulation. ADH binds to target cells in the kidneys, oxytocin to target cells in the mammary glands and uterus.

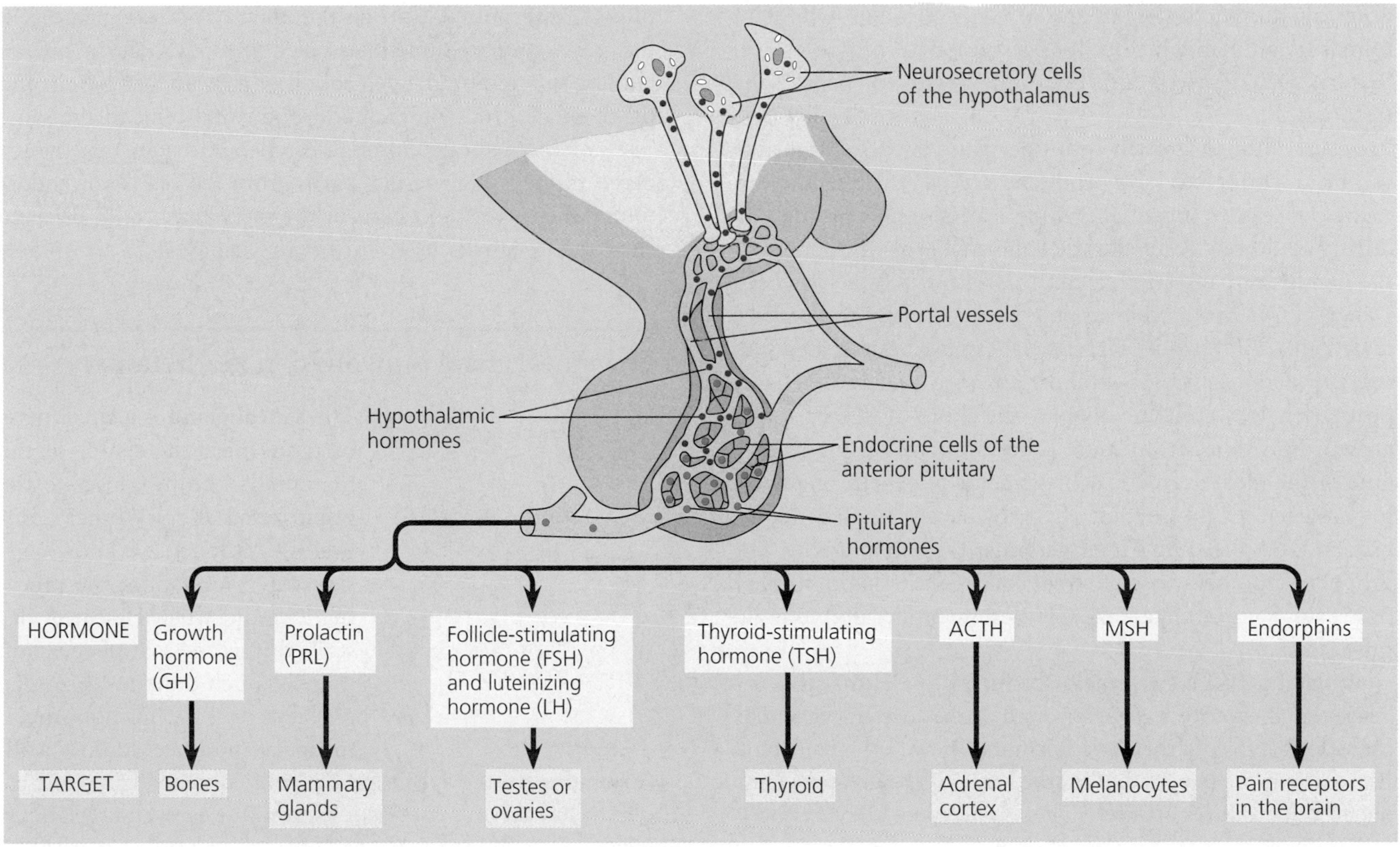

Neurosecretory cells of the hypothalamus

Portal vessels

Hypothalamic hormones

Endocrine cells of the anterior pituitary

Pituitary hormones

| HORMONE | Growth hormone (GH) | Prolactin (PRL) | Follicle-stimulating hormone (FSH) and luteinizing hormone (LH) | Thyroid-stimulating hormone (TSH) | ACTH | MSH | Endorphins |
| TARGET | Bones | Mammary glands | Testes or ovaries | Thyroid | Adrenal cortex | Melanocytes | Pain receptors in the brain |

(b) The anterior pituitary. The release of anterior pituitary hormones is controlled by the hypothalamus. Neurosecretory cells in the hypothalamus secrete releasing hormones and inhibiting hormones into a capillary network located above the stalk of the pituitary. The hormones travel through short portal vessels into a second capillary network within the anterior pituitary. In response to specific releasing hormones, endocrine cells in the anterior pituitary secrete certain hormones into the circulation.

The anterior pituitary produces many different hormones. Four of these are tropic hormones that stimulate the synthesis and release of hormones from other endocrine glands. Thyroid-stimulating hormone (TSH) regulates the release of thyroid hormones; adrenocorticotropic hormone (ACTH) controls the adrenal cortex; and follicle-stimulating hormone (FSH) and luteinizing hormone (LH) govern reproduction by acting on the gonads. Other hormones produced by the anterior pituitary are growth hormone (GH), prolactin (PRL), melanocyte-stimulating hormone (MSH), and the endorphins.

Growth hormone (GH), a protein of about 200 amino acids, affects a wide variety of target tissues and has both direct effects and tropic effects. GH promotes growth directly and stimulates the production of growth factors. For example, the ability of GH to stimulate the growth of bones and cartilage is partly due to the hormone's signaling the liver to produce **insulinlike growth factors (IGFs),** which circulate in blood plasma and directly stimulate bone and cartilage growth. (This endocrine response to growth hormone qualifies GH as a tropic hormone. And IGF secretion by the liver qualifies that organ as an endocrine gland, among its many other functions.) In the absence of GH, skeletal growth of an immature animal will stop. If GH is injected into an animal that has been deprived of its own GH, growth will be partially restored.

Several human growth disorders are related to abnormal GH production. Excessive production of GH during development can lead to gigantism, while excessive GH production during adulthood results in the abnormal growth of bones in the hands, feet, and head, a condition known as acromegaly. Deficient GH production in childhood can lead to pituitary dwarfism. Children with GH deficiency have been treated successfully with human growth hormones isolated from cadaver pituitaries. However, the supply falls short of demand, and growth hormones from most other animals are ineffective. One of the most dramatic achievements of genetic engineering has been the production of GH by bacteria with genes for human GH spliced into their genomes (see Chapter 20). The product is now being used to treat children with hypopituitary dwarfism. Some athletes take GH (legally or illegally) to build muscles.

Prolactin (PRL) is a protein so similar to GH that it is believed that their genes evolved from the same ancestral gene. The physiological roles of these two hormones, however, are different. Prolactin's most remarkable characteristic is the great diversity of effects it produces in different vertebrate species. For example, PRL stimulates mammary gland growth and milk synthesis in mammals; regulates fat metabolism and reproduction in birds; delays metamorphosis in amphibians, where it may also function as a larval growth hormone; and regulates salt and water balance in freshwater fishes. This list suggests that prolactin is an ancient hormone whose functions have diversified during the evolution of the various vertebrate classes.

Three of the tropic hormones secreted by the anterior pituitary are closely related chemically. **Follicle-stimulating hormone (FSH), luteinizing hormone (LH),** and **thyroid-stimulating hormone (TSH)** are all similar glycoproteins, protein molecules with carbohydrate attached to them. FSH and LH are also called **gonadotropins** because they stimulate the activities of the male and female gonads, the testes and ovaries. TSH stimulates the production of hormones by the thyroid gland.

The remaining hormones from the anterior pituitary are all made from a single parent molecule called pro-opiomelanocortin. This is a large protein that is cleaved into several short fragments inside anterior pituitary cells. At least three of these fragments are active peptide hormones. **Adrenocorticotropic hormone (ACTH)** stimulates the production and secretion of steroid hormones by the adrenal cortex. As described earlier, **melanocyte-stimulating hormone (MSH)** regulates the activity of pigment-containing cells in the skin of some vertebrates. MSH secreted by the mammalian pituitary seems to play a key role in fat metabolism, probably by a feedback mechanism that targets neurons in the hypothalamus. The other derivatives of pro-opiomelanocortin are a class of hormones called **endorphins.** These molecules are also produced by certain neurons in the brain. They are sometimes called the body's natural opiates because they inhibit the perception of pain. In fact, heroin and other opiate drugs mimic endorphins and bind to the same receptors in the brain. Some researchers speculate that the so-called runner's high results partly from the release of endorphins when stress and pain in the body reach critical levels. (Endorphins are discussed further in Chapter 48.)

The pineal gland is involved in biorhythms

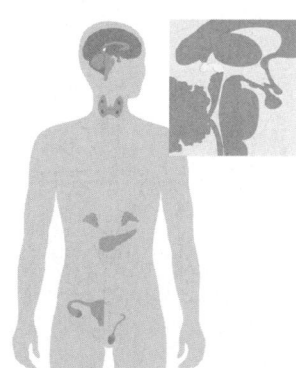

The **pineal gland** is a small mass of tissue near the center of the mammalian brain (closer to the brain surface in some other vertebrates). Although we know considerably more about the pineal than we did when Descartes described it as the seat of the soul, there is much more to learn. The pineal secretes the hormone **melatonin,** a modified amino acid. Depending on the species, the pineal contains light-sensitive cells or has nervous connections from the eyes, and melatonin regulates functions related to light and to seasons marked by changes in day length. For example, melatonin, like melanocyte-stimulating hormone (MSH), affects skin pigmentation in many vertebrates. Most of the pineal's functions, though, are related to biological rhythms associated with reproduction.

Melatonin is secreted at night, and the amount secreted depends on the length of the night. In winter, for example, the days are short and the nights are long, so more melatonin is secreted. Thus, melatonin production is a link between a biological clock and daily or seasonal activities, such as reproduction. Recent evidence suggests that the main target cells of melatonin are in the part of the brain called the suprachiasmatic nucleus (SCN), which functions as a biological clock. Melatonin seems to decrease the activity of neurons in the SCN, and this may be related to its role in mediating rhythms. However, much remains to be learned about the precise role of melatonin and about biological clocks in general.

Thyroid hormones function in development, bioenergetics, and homeostasis

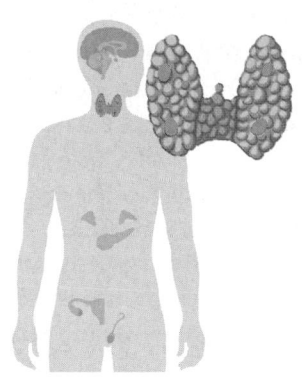

In humans and other mammals, the **thyroid gland** consists of two lobes located on the ventral surface of the trachea (see FIGURE 42.23a). In many other vertebrates the two halves of the gland are separated on the two sides of the pharynx. The thyroid gland produces two very similar hormones derived from the amino acid tyrosine: **triiodothyronine (T$_3$),** which contains three iodine atoms, and tetraiodothyronine, or **thyroxine (T$_4$),** which contains four iodine atoms (FIGURE 45.7). In mammals, the thyroid secretes mainly T$_4$, but target cells convert most of it to T$_3$. T$_3$ has greater affinity for the hormone receptor, which is located in the cell nucleus. Thus it is mostly T$_3$ that brings about responses in target cells. The secretion of thyroid hormones is controlled by the hypothalamus and pituitary in a complex negative feedback system (FIGURE 45.8).

The thyroid gland plays a crucial role in vertebrate development and maturation. A striking example is the thyroid's control of metamorphosis of a tadpole into a frog, which involves massive reorganization of many different tissues. The thyroid is equally important in human development. An inherited condition of thyroid deficiency known as cretinism results in markedly retarded skeletal growth and poor mental development. These defects can often be overcome, at least partially, if treatment with thyroid hormones is begun early in life. Studies with nonhuman animals have shown that thyroid hormones are required for the normal functioning of bone-forming cells and for the branching of nerve cells during embryonic development of the brain.

The thyroid gland also plays a vital role in homeostasis. In adult mammals, for instance, thyroid hormones help maintain normal blood pressure, heart rate, muscle tone, digestion, and reproductive functions. Throughout the body, T$_3$ and T$_4$ are

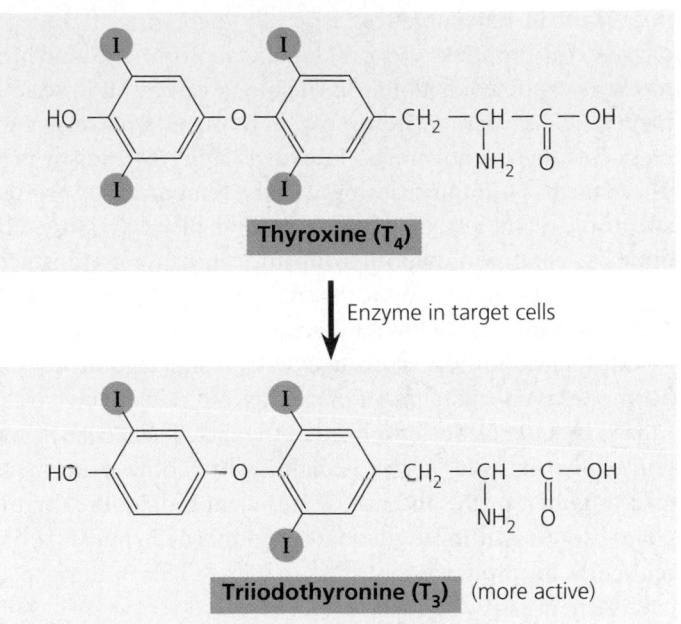

FIGURE 45.7 Two thyroid hormones. Structurally identical, except that T$_3$ has three iodine atoms and T$_4$ has four, these hormones regulate metabolism in most body cells. The thyroid secretes mainly T$_4$, most of which is converted to T$_3$ by an enzyme in target cells. T$_3$ binds more avidly than T$_4$ to receptors in the target cells.

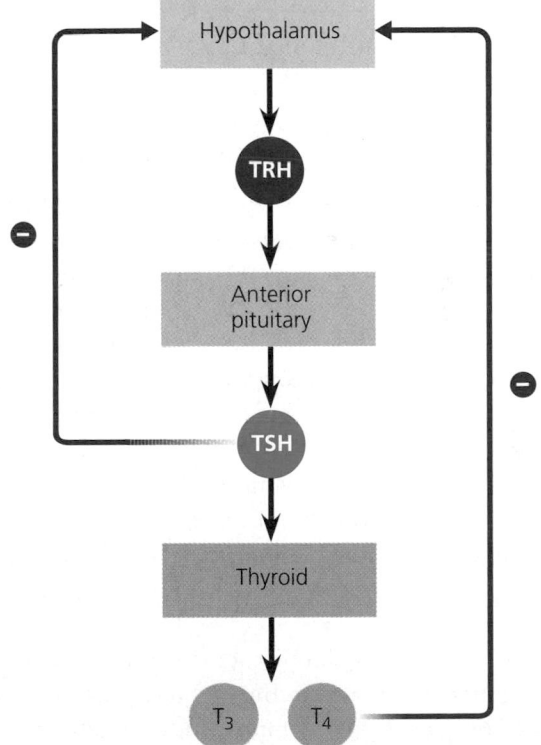

FIGURE 45.8 Feedback control loops regulating the secretion of thyroid hormones T$_3$ and T$_4$. The hypothalamus secretes TRH (TSH-releasing hormone), which stimulates the anterior pituitary to secrete TSH (thyroid-stimulating hormone). When TSH binds to specific receptors in the thyroid gland, a signal-transduction pathway involving cAMP as a second messenger triggers the synthesis and release of the thyroid hormones T$_3$ and T$_4$. The system is balanced by negative feedback loops. In the two main ones (red arrows), high levels of T$_3$, T$_4$, and TSH in the blood inhibit TRH secretion by the hypothalamus.

important in bioenergetics, generally increasing the rate of oxygen consumption and cellular metabolism. Too much or too little of these hormones in the blood can result in serious metabolic disorders. For example, in humans, an excessive secretion of thyroid hormones, known as hyperthyroidism, produces such symptoms as high body temperature, profuse sweating, weight loss, irritability, and high blood pressure. The opposite condition, hypothyroidism, can cause cretinism in infants and produce symptoms such as weight gain, lethargy, and intolerance to cold in adults.

Another condition associated with a shortage of thyroid hormones is an enlargement of the thyroid called goiter, often caused by a deficiency of iodine in the diet. The feedback system shown in FIGURE 45.8 explains why iodine deficiencies lead to goiter. In the absence of sufficient iodine, the thyroid gland cannot synthesize adequate amounts of T_3 and T_4. Consequently, the pituitary continues to secrete TSH, leading to an enlargement of the thyroid.

In addition to having cells that secrete T_3 and T_4, the mammalian thyroid gland contains endocrine cells that secrete **calcitonin.** This hormone lowers calcium (Ca^{2+}) levels in the blood as part of calcium homeostasis, described in the next section.

Parathyroid hormone and calcitonin balance blood calcium

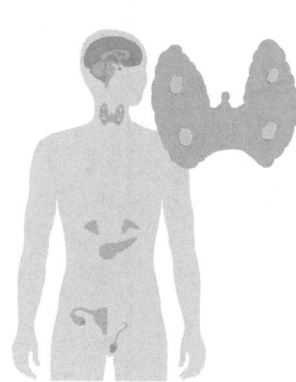

The four **parathyroid glands,** embedded in the surface of the thyroid, function in the homeostasis of calcium ions. They secrete **parathyroid hormone (PTH),** which raises blood levels of Ca^{2+} and thus has an effect opposite to that of the thyroid hormone calcitonin. A lack of PTH causes blood levels of calcium to drop dramatically, leading to convulsive contractions of the skeletal muscles. If unchecked, this condition, known as tetany, is fatal. Calcium ions, in appropriate concentration, are essential to the normal functioning of all cells.

PTH raises the level of blood Ca^{2+} by direct and indirect effects on bone and kidney. In bone, PTH induces specialized cells called osteoclasts to decompose the mineralized matrix of bone and release Ca^{2+} to the blood. In the kidneys, PTH has two effects: It stimulates reabsorption of Ca^{2+} through the renal tubules, and it activates the conversion of vitamin D to its active form. An inactive form of **vitamin D,** a steroid-derived molecule, is obtained from food or synthesized in the skin and then activated sequentially in the liver and the kidneys. The active form of vitamin D functions as a hormone. It acts in concert with PTH in bone, and it also affects the intestines, where it stimulates the uptake of Ca^{2+} from food. For hormonal action, vitamin D binds to receptors in the nuclei of target cells and regulates gene transcription.

When blood calcium rises above a certain level, calcitonin comes into play. It has effects on bone and kidneys opposite to those of PTH and thus decreases blood Ca^{2+}. The control of blood calcium level is an example of how homeostasis is often maintained by the balancing of two antagonistic hormones—in this case, PTH and calcitonin. In classic feedback fashion, these two hormones balance each other's effects, thereby minimizing fluctuations in the concentration of calcium in the blood (FIGURE 45.9).

Endocrine tissues of the pancreas secrete insulin and glucagon, antagonistic hormones that regulate blood glucose

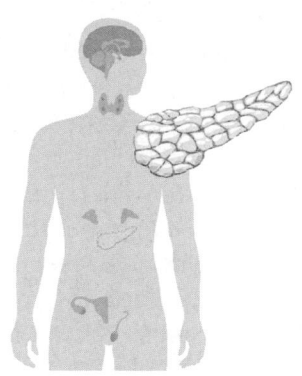

The **pancreas** is one of many organs that perform both endocrine and exocrine functions. Endocrine cells make up only 1–2% of the weight of the pancreas. The rest of the organ is exocrine tissue that produces bicarbonate ions and digestive enzymes that are carried to the small intestine via the pancreatic duct (see FIGURE 41.16). Scattered throughout this exocrine tissue are the **islets of Langerhans,** clusters of endocrine cells that secrete two major hormones directly into the circulatory system. Each islet has a population of **alpha cells,** which secrete the peptide hormone **glucagon,** and a population of **beta cells,** which secrete the hormone **insulin.**

Insulin and glucagon are antagonistic hormones that regulate the concentration of glucose in the blood. This is a critical bioenergetic and homeostatic function, because glucose is a major fuel for cellular respiration and a key source of carbon skeletons for the synthesis of other organic compounds. Metabolic balance depends on the maintenance of blood glucose at a concentration near a set point, which is about 90 mg/100 mL in humans. When blood glucose exceeds this level, insulin is released and acts to lower the glucose concentration. When blood glucose drops below the set point, glucagon increases glucose concentration. By negative feedback, blood glucose concentration determines the relative amounts of insulin and glucagon secreted by the islet cells (FIGURE 45.10, p. 968).

Insulin and glucagon both influence blood glucose concentration by multiple mechanisms. Insulin lowers blood glucose levels by stimulating virtually all body cells except those of the brain to take up glucose from the blood. (Brain cells are unusual in being able to take up glucose without insulin; as a result, the brain has access to circulating fuel molecules almost

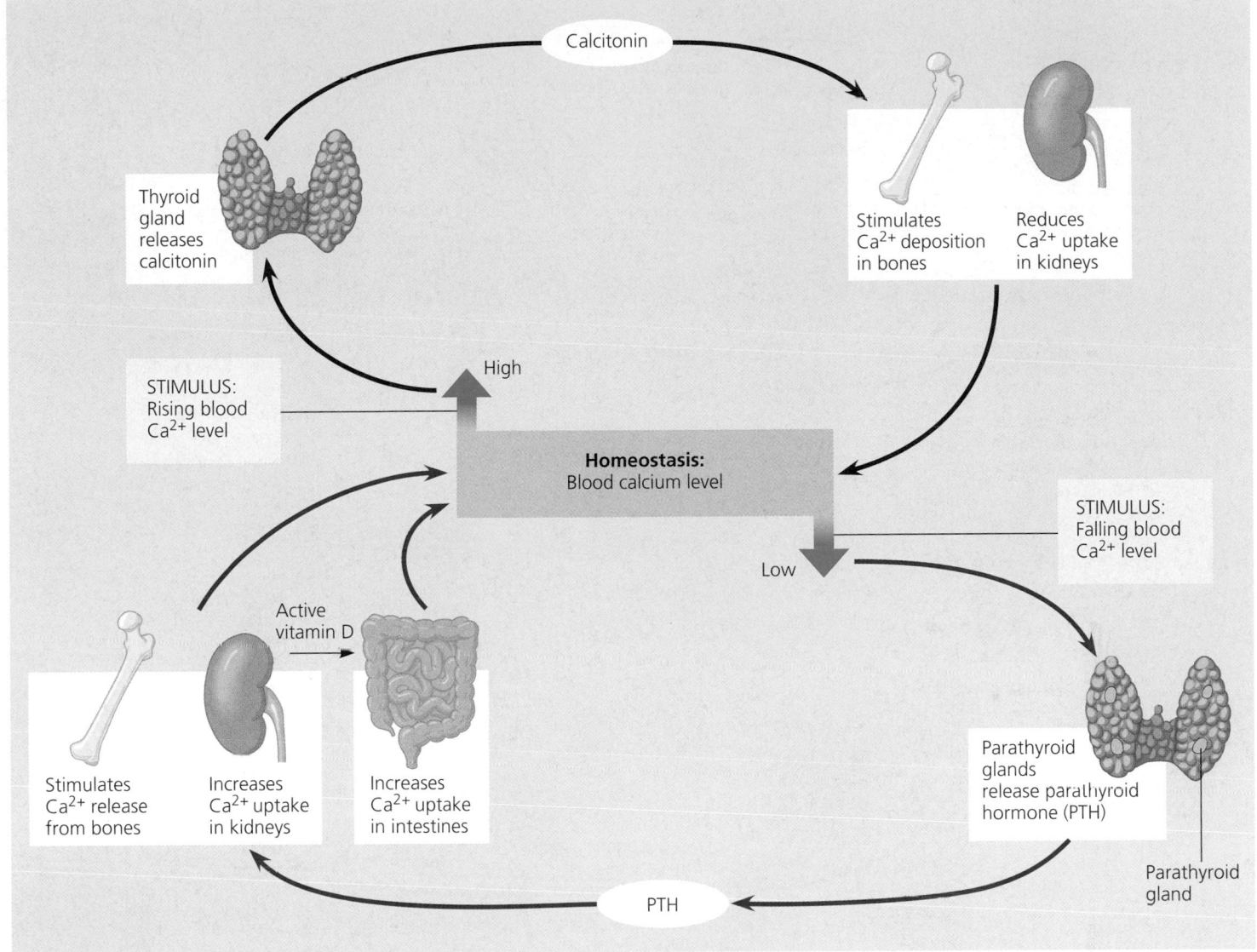

Calcitonin

Thyroid gland releases calcitonin

Stimulates Ca²⁺ deposition in bones

Reduces Ca²⁺ uptake in kidneys

STIMULUS: Rising blood Ca²⁺ level

High

Homeostasis: Blood calcium level

STIMULUS: Falling blood Ca²⁺ level

Low

Active vitamin D

Stimulates Ca²⁺ release from bones

Increases Ca²⁺ uptake in kidneys

Increases Ca²⁺ uptake in intestines

Parathyroid glands release parathyroid hormone (PTH)

Parathyroid gland

PTH

FIGURE 45.9 Hormonal control of calcium homeostasis in mammals. A negative feedback system involving two antagonistic hormones, calcitonin and parathyroid hormone (PTH), maintains the concentration of calcium in blood very close to 10 mg/100mL. A rise in blood Ca²⁺ induces the thyroid gland to secrete calcitonin, which lowers the Ca²⁺ concentration by increasing bone deposition and reducing reabsorption in the kidneys. (By interfering with PTH, the calcitonin also helps reduce Ca²⁺ uptake by the intestines, not shown.) These effects are reversed by PTH, which is secreted from the parathyroid glands when the concentration of blood Ca²⁺ falls below the set point. Blood calcium levels begin to increase as target cells in the bones and kidneys respond to PTH. In addition to stimulating Ca²⁺ uptake in these organs directly, PTH also acts indirectly by helping activate vitamin D in the kidneys. The active form of vitamin D then stimulates the intestines to increase Ca²⁺ uptake from food. But blood Ca²⁺ will rise only so far before the thyroid counters by secreting more calcitonin.

all the time.) Insulin also decreases blood glucose by slowing glycogen breakdown in the liver and inhibiting the conversion of amino acids and glycerol (from fats) to sugar.

The liver, skeletal muscles, and adipose tissues store large amounts of fuel molecules and are especially important in bioenergetics. The liver and muscles store sugar as glycogen, whereas adipose tissue cells convert sugars to fats. The liver is a key fuel-processing center because only liver cells are sensitive to glucagon. Normally, glucagon starts having an effect before blood glucose levels even drop below the set point. In fact, as soon as excess glucose is cleared from the blood, glucagon signals the liver cells to increase glycogen hydrolysis,

convert amino acids and glycerol to glucose, and start slowly releasing glucose back into the circulation.

The antagonistic effects of glucagon and insulin are vital to glucose homeostasis, a mechanism that precisely manages both fuel storage and fuel use by body cells. We will revisit the topic of fuel management and see that it involves additional hormones when we study the adrenal glands in the next section. The liver's ability to perform its vital roles in glucose homeostasis results from the metabolic versatility of its cells and its access to absorbed nutrients via the hepatic portal vessel, which carries blood directly from the small intestine to the liver.

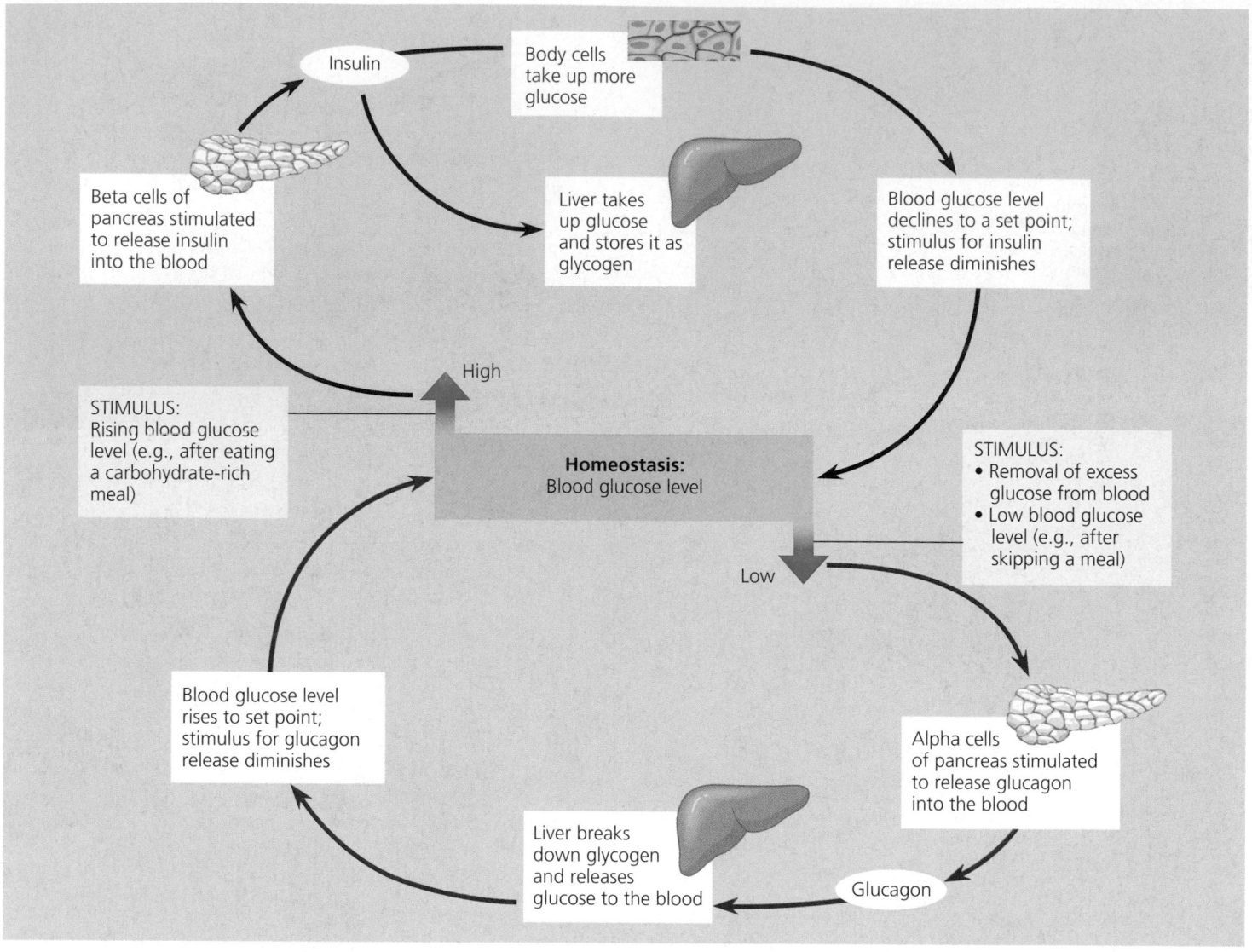

FIGURE 45.10 Glucose homeostasis maintained by insulin and glucagon. A rise in blood glucose above the set point (about 90 mg/100 mL in humans) stimulates the pancreas to secrete insulin, which triggers its target cells to take up the excess glucose from the blood. Once the excess is removed and blood glucose concentration dips below the set point, the pancreas responds by secreting glucagon, which acts on the liver to raise the blood glucose level.

When the mechanisms of glucose homeostasis go awry, there are serious consequences. Diabetes mellitus, perhaps the best-known endocrine disorder, is caused by a deficiency of insulin or a loss of response to insulin in target tissues. The result is high blood glucose—so high, in fact, that the diabetic's kidneys excrete glucose, which explains why the presence of sugar in urine is one test for diabetes. As more glucose concentrates in the urine, more water is excreted with it, resulting in excessive volumes of urine and persistent thirst. (*Diabetes*, from the Greek, refers to this copious urination, and *mellitus* is from the Greek word for "honey," referring to the presence of sugar in the urine.) Because glucose is unavailable to most body cells as a major fuel source for diabetics, fat must serve as the main substrate for cellular respiration. In severe cases of diabetes, acidic metabolites formed during fat breakdown accumulate in the blood, threatening life by lowering blood pH.

There are actually two major forms of diabetes with very different causes. **Type I diabetes mellitus** (insulin-dependent diabetes) is an autoimmune disorder, in which the immune system mounts an attack on the cells of the pancreas. (Chapter 43 discusses possible causes of autoimmune reactions.) This disorder usually occurs rather suddenly during childhood, destroying the person's ability to produce insulin. Treatment consists of insulin injections, which are usually taken several times daily. Until recently, insulin for injections was extracted from animal pancreases, but genetic engineering has provided a relatively inexpensive source of human insulin by inserting DNA encoding the hormone into bacteria (see FIGURE 20.1). **Type II diabetes mellitus** (non-insulin-dependent diabetes) is characterized either by a deficiency of insulin or, more commonly, by reduced responsiveness in target cells due to some change in insulin receptors. Type II diabetes usually occurs after about age 40, becoming more likely with increasing age. More than 90% of diabetics are type II, and many can manage their blood glucose solely by exercise and dietary control, although helpful drugs are now available. Both heredity and obesity are major factors in type II diabetes.

The adrenal medulla and adrenal cortex help the body manage stress

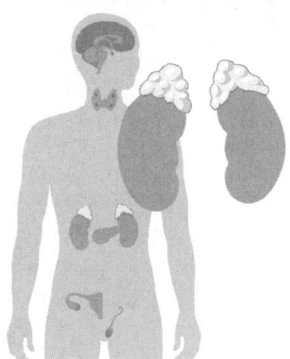

The **adrenal glands** are adjacent to the kidneys. In mammals, each adrenal gland is actually made up of two glands with different cell types, functions, and embryonic origins: the **adrenal cortex,** or outer portion, and the **adrenal medulla,** or central part. Nonmammalian vertebrates have quite different arrangements of the same tissues.

The adrenal medulla has close developmental and functional ties with the nervous system. Secretory cells of the adrenal medulla are derived from cells of the neural crest (see FIGURE 34.6). Some of the neural crest cells in the abdominal region of a vertebrate embryo may differentiate into either endocrine cells of the adrenal medulla or neurons, depending on chemical signals in the vicinity (FIGURE 45.11).

What makes your heart beat faster and your skin develop goose bumps when you sense danger or approach a stressful situation, like speaking in public? These reactions are part of the fight-or-flight response stimulated by two hormones of the adrenal medulla, **epinephrine** (also known as adrenaline) and **norepinephrine** (noradrenaline). These hormones are members of a class of compounds, the **catecholamines,** that are synthesized from the amino acid tyrosine (FIGURE 45.12).

Epinephrine, norepinephrine, and other catecholamines are secreted in response to positive or negative stress—everything from extreme pleasure to increased cold to life-threatening danger. Their release into the blood gives the body a rapid bioenergetic boost, increasing the basal metabolic rate and having dramatic effects on several targets. Epinephrine and norepinephrine increase the rate of glycogen breakdown in the liver and skeletal muscles and glucose release into the blood by liver cells. They also stimulate the release of fatty acids from fat cells. The fatty acids may be used by cells for energy. In addition to increasing the availability of energy sources, epinephrine and

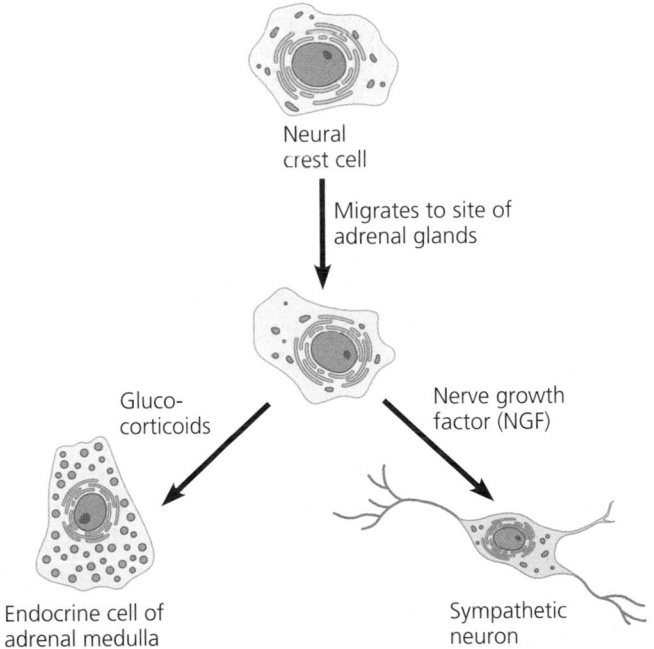

FIGURE 45.11 Derivation of endocrine cells of the adrenal medulla and neurons from neural crest cells. Studies of neural crest cells in laboratory cultures indicate that the fate of these cells in embryonic vertebrates depends on chemical signals. Embryonic chicken neural crest cells that have migrated to the region of the body where the adrenal glands develop may differentiate into hormone-secreting cells of the adrenal medulla if they are given glucocorticoids (hormones secreted by the adrenal cortex). If the same neural crest cells are exposed to nerve growth factor (NGF), they develop into nerve cells called sympathetic neurons (you'll learn about the function of these neurons in Chapter 48).

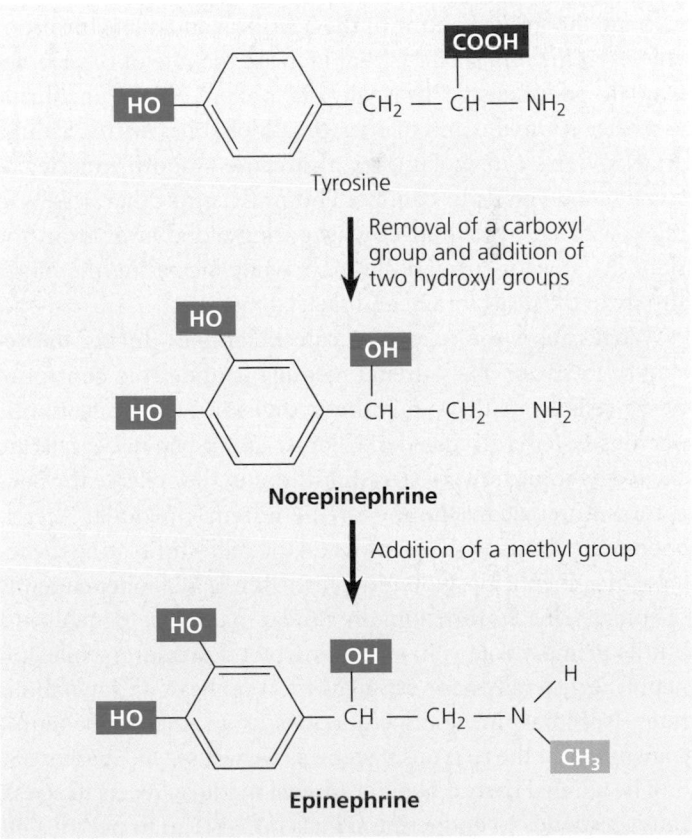

FIGURE 45.12 The synthesis of catecholamine hormones. Cells in the adrenal medulla synthesize the catecholamines norepinephrine and epinephrine from the amino acid tyrosine.

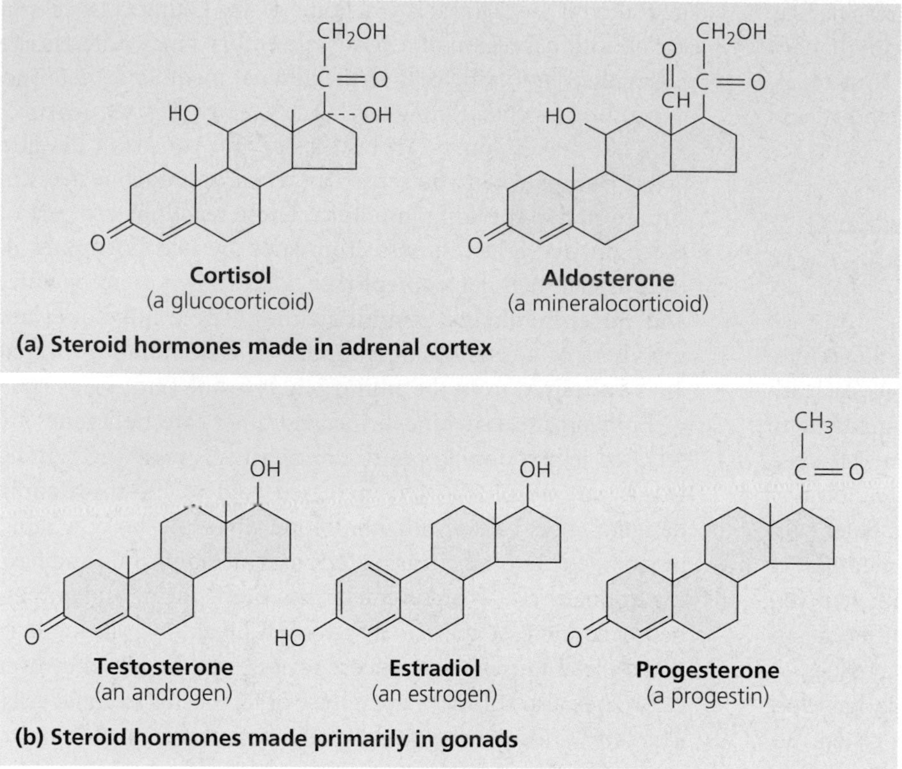

FIGURE 45.13 Steroid hormones from the adrenal cortex and gonads. (a) Cortisol (a glucocorticoid) and aldosterone (a mineralocorticoid), both made in the adrenal cortex, are structurally similar to **(b)** the sex hormones testosterone (an androgen), estradiol (an estrogen), and progesterone (a progestin). The precursor for the synthesis of all steroid hormones is cholesterol (see FIGURE 5.14). Most of the androgens (male hormones) that circulate in the blood are made by the testes, and most of the estrogens and progestins (female hormones) are produced by the ovaries; however, small amounts of these sex hormones are also made by the adrenal cortex.

norepinephrine have profound effects on the cardiovascular and respiratory systems. For example, they increase both the rate and the stroke volume of the heartbeat and dilate the bronchioles in the lungs, effects that increase the rate of oxygen delivery to body cells. (This is why doctors prescribe epinephrine as a heart stimulant and to open breathing tubes during asthma attacks.) The catecholamines also cause smooth muscles of some blood vessels to contract and muscles of other vessels to relax, with an overall effect of shunting blood away from the skin, digestive organs, and kidneys while increasing the blood supply to the heart, brain, and skeletal muscles.

What causes the release of catecholamines during the response to stress? The adrenal medulla is under the control of nerve cells from the sympathetic division of the autonomic nervous system (discussed in Chapter 48). When nerve cells are excited by some form of stressful stimulus, they release the neurotransmitter acetylcholine in the adrenal medulla. Acetylcholine combines with receptors on the cells, stimulating the release of epinephrine. Norepinephrine is released independently of epinephrine. Its functions are similar to those of epinephrine, but its primary role is in sustaining blood pressure, while epinephrine generally has a stronger effect on heart and metabolic rates. Norepinephrine also functions as an important neurotransmitter in the nervous system, as we will see in Chapter 48.

The adrenal cortex, like the adrenal medulla, reacts to stress. But it responds to endocrine signals rather than to nervous input. Stressful stimuli cause the hypothalamus to secrete a releasing hormone that stimulates the anterior pituitary to release the tropic hormone ACTH. When it reaches its target via the blood-stream, ACTH stimulates cells of the adrenal cortex to synthesize and secrete a family of steroids called **corticosteroids.** In another case of negative feedback, elevated levels of corticosteroids in the blood suppress the secretion of ACTH.

Many corticosteroids have been isolated from the adrenal cortex. The two main types in humans are the **glucocorticoids,** such as cortisol, and the **mineralocorticoids,** such as aldosterone (FIGURE 45.13).

The primary effect of glucocorticoids is on bioenergetics, specifically on glucose metabolism. Augmenting the fuel-mobilizing effects of glucagon from the pancreas, glucocorticoids promote the synthesis of glucose from noncarbohydrate sources, such as proteins, making more glucose available as fuel. Glucocorticoids act on skeletal muscle, causing a breakdown of muscle proteins. The resulting carbon skeletons are transported to the liver and kidney, where they are converted to glucose and released into the blood. The synthesis of glucose from muscle proteins is a homeostatic mechanism providing circulating fuel when body activities require more than the liver can mobilize from its glycogen stores. It can also be part of a broader role of the glucocorticoids, that of helping the body withstand long-term environmental challenge.

Abnormally high doses of glucocorticoids administered as medication suppress certain components of the body's immune system—for example, the inflammatory reaction that occurs at the site of an infection. Glucocorticoids are used to treat diseases in which excessive inflammation is a problem. Cortisone, for instance, was once thought to be a miracle drug that could cure serious inflammatory conditions such as

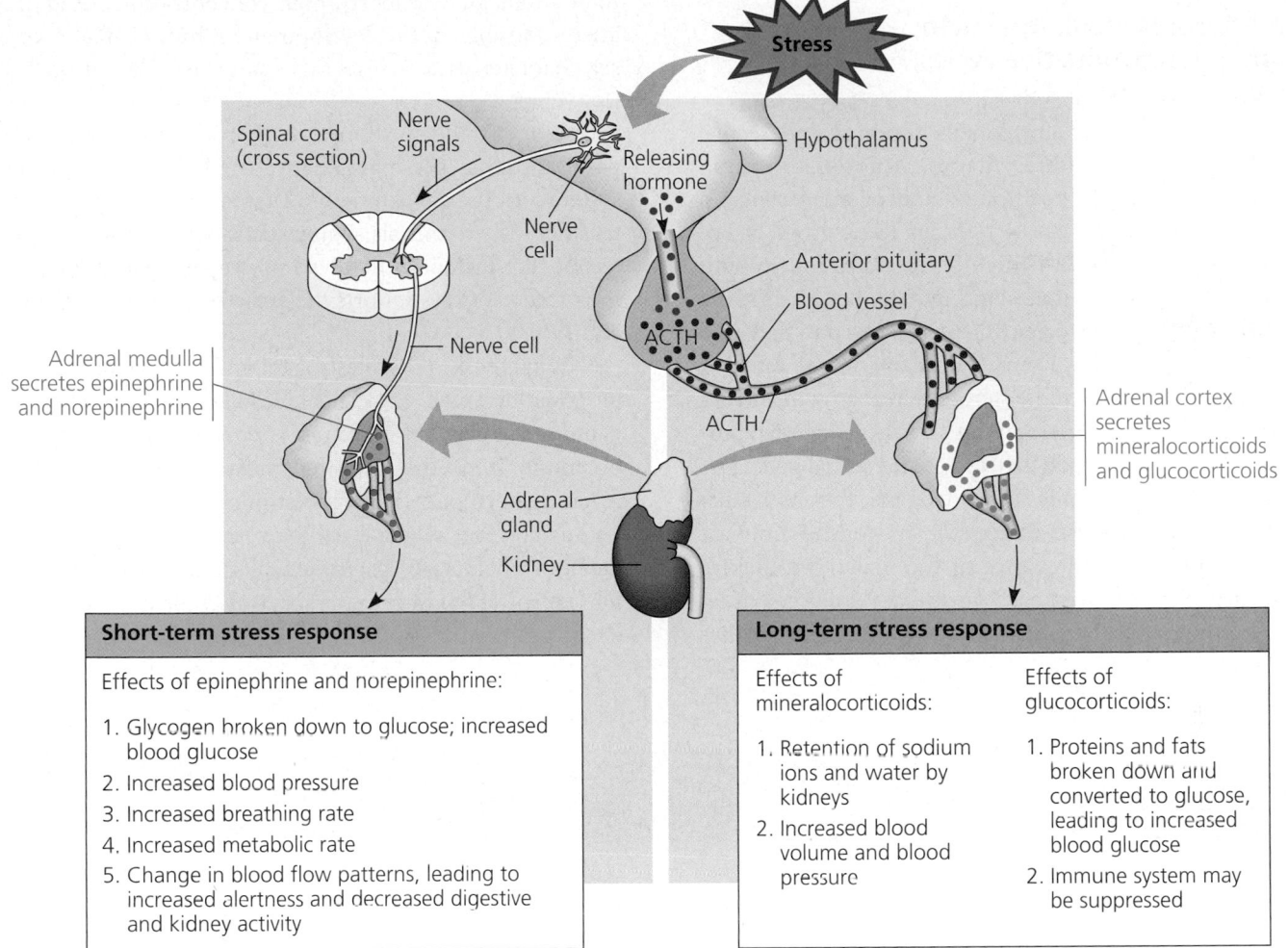

FIGURE 45.14 Stress and the adrenal gland. Stressful stimuli cause the hypothalamus to activate the adrenal medulla via nerve impulses and the adrenal cortex via hormonal signals. The adrenal medulla mediates short-term responses to stress by secreting the catecholamine hormones epinephrine and norepinephrine. The adrenal cortex controls more prolonged responses by secreting steroid hormones.

Labels in figure:
Stress
Spinal cord (cross section)
Nerve signals
Releasing hormone
Hypothalamus
Nerve cell
Anterior pituitary
Blood vessel
ACTH
Nerve cell
ACTH
Adrenal medulla secretes epinephrine and norepinephrine
Adrenal cortex secretes mineralocorticoids and glucocorticoids
Adrenal gland
Kidney

Short-term stress response

Effects of epinephrine and norepinephrine:

1. Glycogen broken down to glucose; increased blood glucose
2. Increased blood pressure
3. Increased breathing rate
4. Increased metabolic rate
5. Change in blood flow patterns, leading to increased alertness and decreased digestive and kidney activity

Long-term stress response

Effects of mineralocorticoids:

1. Retention of sodium ions and water by kidneys
2. Increased blood volume and blood pressure

Effects of glucocorticoids:

1. Proteins and fats broken down and converted to glucose, leading to increased blood glucose
2. Immune system may be suppressed

arthritis. It has become clear, however, that long-term use of corticosteroids can result in increased susceptibility to infection and disease because of the immunosuppressive effects of these powerful drugs.

Mineralocorticoids have their major effects on salt and water balance. Aldosterone, for example, stimulates cells in the kidney to reabsorb sodium ions and water from the filtrate, raising blood pressure and volume. Aldosterone (of the renin-angiotensin-aldosterone system, or RAAS), ADH from the posterior pituitary, and atrial natriuretic factor (ANF) from the heart form a regulatory complex with multiple feedback loops that underlie the kidney's ability to maintain the ion and water homeostasis of the blood (see FIGURE 44.21). Aldosterone secretion is largely regulated by the RAAS itself in response to changes in plasma ion concentration. However, when an individual is under severe stress, the hypothalamus tends to secrete more releasing hormones that increase the

rate of secretion of ACTH by the anterior pituitary. The rise in ACTH in the blood then increases the rate of secretion of aldosterone by the adrenal cortex.

Evidence is mounting that both glucocorticoids and mineralocorticoids help maintain homeostasis when the body experiences stress over an extended period of time. FIGURE 45.14 compares the long-term stress-induced actions of corticosteroids with the short-term ones of epinephrine and norepinephrine.

A third group of corticosteroids are sex hormones, mainly androgens (male hormones) similar to testosterone produced by the testes and small amounts of estrogens and progesterone (female hormones). There is evidence that adrenal androgens account for the sex drive in adult females, but otherwise the physiological roles of the adrenal sex hormones are not well understood. We will discuss the sex hormones from the gonads in the next section.

Gonadal steroids regulate growth, development, reproductive cycles, and sexual behavior

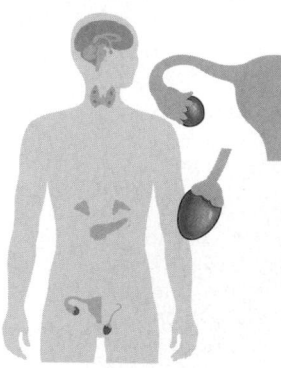

The gonads produce and secrete three major categories of steroid hormones: androgens, estrogens, and progestins (see FIGURE 45.13). All three types are found in both males and females but in different proportions. Produced in the testes of males and ovaries of females, these steroids affect growth and development and also regulate reproductive cycles and sexual behavior.

The testes primarily synthesize **androgens,** the main such hormone being **testosterone.** In general, androgens stimulate the development and maintenance of the male reproductive system. Androgens produced early in the development of an embryo determine that the fetus will develop as a male rather than a female. At puberty, high concentrations of androgens are responsible for the development of human male secondary sex characteristics, such as male patterns of hair growth and a low voice.

Estrogens, the most important of which is estradiol, have a parallel role in the maintenance of the female reproductive system and the development of female secondary sex characteristics. In mammals, **progestins,** which include progesterone, are primarily involved in preparing and maintaining the uterus, which supports the growth and development of an embryo.

The synthesis of both estrogens and androgens is controlled by gonadotropins, FSH and LH, from the anterior pituitary gland. FSH and LH secretion is controlled by one releasing hormone from the hypothalamus, GnRH (gonadotropin-releasing hormone). We will examine the complex feedback relationships that regulate the secretion of gonadal steroids in detail in Chapter 46. There we will see that the endocrine system is central not only to the survival of the individual but also to the propagation of the species.

CHAPTER 45 REVIEW

Go to the Campbell Biology website (www.campbellbiology.com) to explore an interactive version of the Chapter Review.

Summary of Key Concepts

AN INTRODUCTION TO REGULATORY SYSTEMS

- Animals have two regulatory systems that coordinate internal body functions: the nervous system and the endocrine system. Chemical messengers called hormones regulate the activity of target organs at distant sites.

- **The endocrine system and the nervous system are structurally, chemically, and functionally related (p. 956, FIGURE 45.1)** The endocrine and nervous systems often function together in maintaining homeostasis, development, and reproduction. Neurosecretory cells of the nervous system secrete many hormones; several hormones, such as norepinephrine, are used as signals by both the endocrine and the nervous system; and many body functions are regulated by both systems, often by feedback mechanisms.

- **Invertebrate regulatory systems clearly illustrate endocrine and nervous system interactions (pp. 956–957, FIGURE 45.2)** Diverse hormones regulate different aspects of homeostasis in invertebrates. In insects, molting and development are controlled by an interplay between ecdysone, juvenile hormone, and brain hormone secreted by neurosecretory cells.

CHEMICAL SIGNALS AND THEIR MODES OF ACTION

- Hormones convey information to target cells via the bloodstream. Local regulators, which also function in homeostasis, reach their target cells, in the immediate vicinity, by diffusion through interstiital fluid. Pheromones act as communication signals between different individuals of the same species.

- **A variety of local regulators affect neighboring target cells (p. 958)** Neurotransmitters are local regulators in the synapses of the nervous system. Growth factors and prostaglandins, released by a large variety of cells into their surroundings, are other important local regulators. The local regulator nitric oxide (NO) is a gas.

- **Most chemical signals bind to plasma-membrane proteins, initiating signal-transduction pathways (pp. 958–959, FIGURES 45.3a, 45.4)** Most chemical signals bind to specific receptors on the plasma membrane of target cells. These signals trigger changes in target cells through signal-transduction pathways.

 Web/CD Activity 45A: *Overview of Cell Signaling*
 Web/CD Activity 45B: *Peptide Hormone Action*

- **Steroid hormones, thyroid hormones, and some local regulators enter target cells and bind to intracellular receptors (p. 960, FIGURE 45.3b)** Chemical signals that actually enter target cells bind to specific protein receptors in the cytosol or nucleus. The hormone-receptor complexes then act as transcription factors in the nucleus: They bind to certain sites on DNA, where they regulate the transcription of specific genes.

 Web/CD Activity 45C: *Steroid Hormone Action*

THE VERTEBRATE ENDOCRINE SYSTEM

- ■ FIGURE 45.5 illustrates the major endocrine glands of the human body; TABLE 45.1 lists the major vertebrate hormones. Among the numerous hormones regulating the vertebrate body, tropic hormones have other endocrine glands as their targets. (The hormonal functions of the stomach, small intestine, heart, and thymus are discussed in Chapters 41, 42, and 43. The thymus secretes thymosin and other chemical messengers that stimulate the development and differentiation of T lymphocytes.)

- ■ **The hypothalamus and pituitary integrate many functions of the vertebrate endocrine system (pp. 962–964, FIGURES 45.5, 45.6)** Neurosecretory cells of the hypothalamus integrate endocrine and neural function by influencing the pituitary gland. The posterior pituitary is an extension of the hypothalamus that stores and releases two hormones (oxytocin and antidiuretic hormone, ADH) produced by neurosecretory cells in the hypothalamus. Oxytocin induces uterine contractions and milk ejection, and ADH enhances water reabsorption in the kidneys. Under the direction of releasing and inhibiting hormones conveyed by portal vessels from the hypothalamus, the anterior pituitary produces an array of hormones, including thyroid-stimulating hormone (TSH), follicle-stimulating hormone (FSH), luteinizing hormone (LH), growth hormone (GH), prolactin (PRL), adrenocorticotropic hormone (ACTH), melanocyte-stimulating hormone (MSH), and the endorphins. The tropic hormones TSH and the gonadotropins (FSH and LH) are chemically related and stimulate the thyroid gland and the gonads, respectively, to produce their hormones. GH promotes growth directly and stimulates the production of growth factors. Prolactin, named for its stimulation of lactation in mammals, has diverse effects in different vertebrates. ACTH stimulates the adrenal cortex. MSH influences skin pigmentation in some vertebrates and fat metabolism in mammals. Endorphins, the brain's natural opiates, inhibit the perception of pain.

- ■ **The pineal gland is involved in biorhythms (pp. 964–965, FIGURE 45.5, TABLE 45.1)** The pineal gland secretes melatonin, which influences skin pigmentation, biological rhythms, and reproduction in various vertebrates.

- ■ **Thyroid hormones function in development, bioenergetics, and homeostasis (pp. 965–966, FIGURES 45.5, 45.7, 45.8, TABLE 45.1)** The thyroid gland produces iodine-containing hormones (T_3 and T_4) that stimulate metabolism and influence development and maturation. The thyroid also secretes calcitonin, which lowers calcium levels in the blood.

  Web/CD Case Study in the Process of Science: *How Do Thyroxine and TSH Affect Metabolism?*

- ■ **Parathyroid hormone and calcitonin balance blood calcium (p. 966, FIGURES 45.5, 45.9, TABLE 45.1)** The parathyroid glands raise plasma calcium levels by secreting parathyroid hormone (PTH). PTH works with calcitonin to maintain calcium homeostasis by acting directly on bone and kidneys. In addition, PTH stimulates the kidneys to activate vitamin D, which in turn stimulates the intestines to increase uptake of calcium from food.

- ■ **Endocrine tissues of the pancreas secrete insulin and glucagon, antagonistic hormones that regulate blood glucose (pp. 966–969, FIGURES 45.5, 45.10, TABLE 45.1)** The endocrine portion of the pancreas consists of islet cells that secret insulin and glucagon. High blood glucose levels stimulate the release of insulin, which increases the cellular uptake of glucose, promotes the formation and storage of glycogen in the liver, and stimulates protein synthesis and fat storage. Low blood glucose levels trigger glucagon release, which increases blood glucose by stimulating the conversion of glycogen to glucose in the liver and increasing the breakdown of fat and protein. Type I diabetes mellitus is an autoimmune disorder resulting in a lack of insulin. Type II diabetes is usually caused by the loss of responsiveness of target cells to insulin.

- ■ **The adrenal medulla and adrenal cortex help the body manage stress (pp. 969–971, FIGURES 45.5, 45.11–45.14, TABLE 45.1)** The adrenal medulla releases epinephrine and norepinephrine in response to stress-activated impulses from the nervous system. These hormones mediate various fight-or-flight responses. The adrenal cortex releases corticosteroids (including sex hormones), glucocorticoids, and mineralocorticoids. Glucocorticoids influence glucose metabolism and the immune system; mineralocorticoids affect salt and water balance.

- ■ **Gonadal steroids regulate growth, development, reproductive cycles, and sexual behavior (p. 972, FIGURE 45.5, TABLE 45.1)** The gonads—testes and ovaries—produce varying proportions of three kinds of steroid hormones: androgens, estrogens, and progestins.

  Web/CD Activity 45D: *Human Endocrine Glands and Hormones*

Self-Quiz

1. Which of the following is *not* an accurate statement about hormones?
   a. Hormones are chemical messengers that travel to target cells through the circulatory system.
   b. Hormones often regulate homeostasis through antagonistic functions.
   c. Hormones of the same chemical class usually have the same function.
   d. Hormones are secreted by specialized cells usually located in endocrine glands.
   e. Hormones are often regulated through feedback loops.

2. A distinctive feature of the mechanism of action of thyroid hormones and steroid hormones is that
   a. these hormones are regulated by feedback loops.
   b. target cells react more rapidly to these hormones than to local regulators.
   c. these hormones bind with specific receptor proteins on target-cell plasma membranes.
   d. these hormones bind to receptors inside cells.
   e. these hormones affect metabolism.

3. The relationship between the insect hormones ecdysone and brain hormone
   a. is an example of the interaction between the endocrine and nervous systems.
   b. illustrates homeostasis achieved by positive feedback.
   c. demonstrates that peptide-derived hormones have more widespread effects than steroidal hormones.
   d. illustrates homeostasis maintained by antagonistic hormones.
   e. demonstrates competitive inhibition for the hormone receptor.

4. Growth factors are local regulators that
   a. are produced by the anterior pituitary.
   b. are modified fatty acids that stimulate bone and cartilage growth.
   c. are found on the surface of cancer cells and stimulate abnormal cell division.
   d. are proteins that bind to cell surface receptors and stimulate target-cell growth and development.
   e. include histamines and interleukins and are necessary for cellular differentiation.

5. Which of the following hormones is *incorrectly* paired with its action?
   a. oxytocin—stimulates uterine contractions during childbirth
   b. thyroxine—stimulates metabolic processes
   c. insulin—stimulates glycogen breakdown in the liver
   d. ACTH—stimulates the release of glucocorticoids by the adrenal cortex
   e. melatonin—affects biological rhythms, seasonal reproduction

6. An example of antagonistic hormones controlling homeostasis is
   a. thyroxine and parathyroid hormone in calcium balance.
   b. insulin and glucagon in glucose metabolism.
   c. progestins and estrogens in sexual differentiation.
   d. epinephrine and norepinephrine in fight-or-flight responses.
   e. oxytocin and prolactin in milk production.

7. Which of the following is not an example of the close structural and functional relationship between the nervous and endocrine systems?
   a. the secretion of hormones by neurosecretory cells
   b. the multiple functions of norepinephrine
   c. the stimulation of the adrenal medulla in the short-term response to stress
   d. the embryonic development of the posterior pituitary from the hypothalamus
   e. the alteration of gene expression by steroid hormones

8. A portal vessel carries blood from the hypothalamus directly to the
   a. thyroid.
   b. pineal gland.
   c. anterior pituitary.
   d. posterior pituitary.
   e. thymus.

9. Which of the following is the most likely explanation for hypothyroidism in a patient whose iodine level is normal?
   a. a disproportionate production of T_3 to T_4
   b. hyposecretion of TSH
   c. hypersecretion of TSH
   d. hypersecretion of MSH
   e. a decrease in the thyroid secretion of calcitonin

10. The main target organs for tropic hormones are
    a. muscles.
    b. blood vessels.
    c. endocrine glands.
    d. kidneys.
    e. nerves.

11. Where do releasing hormones originate, and what is their general function?

12. Alcohol inhibits secretion of ADH by the anterior pituitary. Predict how this action of alcohol would affect urination.

13. How does thyroxine turn off its own production?

14. How would a deficiency of receptors in the hypothalamus for adrenal steroids affect levels of those hormones in the blood?

15. Estrogens are to _____ as _____ are to testes.

Go to the website or CD-ROM for more quiz questions.

Go to the website or CD-ROM for more quiz questions.

Evolution Connection

The intracellular receptors used by all the steroid and thyroid hormones are similar enough in structure that they are all considered members of one "superfamily" of proteins. Propose a hypothesis for how the genes for these receptors may have evolved. (*Hint:* See FIGURE 19.3.)

The Process of Science

In your response to the previous question, you came up with a hypothesis. How could you test your hypothesis using DNA sequence data?

Examine how thyroxine and the thyroid-stimulating hormome (TSH) affect basal metabolic rate in the Case Study in the Process of Science, available on the CD-ROM and website.

Science, Technology, and Society

Growth hormone (GH) produced by DNA technology has enabled hundreds of children who suffer from pituitary dwarfism to grow normally and reach a stature within the normal range. Now that the hormone is readily available and relatively inexpensive, many parents who are concerned that their children are not growing fast enough want to use GH to make them grow faster and taller. There can be potentially harmful effects, such as a reduction in body fat and an increase in muscle mass. And no one yet knows if GH injections will have seriously harmful long-term effects in individuals who do not have a hypopituitary condition. What criteria do you think should determine the cases in which GH treatments or other hormone therapies are appropriate?

Answers: 1. c; 2. d; 3. a; 4. d; 5. c; 6. b; 7. e; 8. c; 9. b; 10. c; 11. Hypothalamus; they trigger release of certain other hormones from the anterior pituitary. 12. Alcohol increases the volume of urine produced. 13. By negative feedback in which it inhibits TRH secretion by the hypothalamus. 14. The levels of adrenal steroids in the blood would become very high. 15. Ovaries, androgens.

ANIMAL REPRODUCTION

OVERVIEW OF ANIMAL REPRODUCTION

- Both asexual and sexual reproduction occur in the animal kingdom
- Diverse mechanisms of asexual reproduction enable animals to produce identical offspring rapidly
- Reproductive cycles and patterns vary extensively among animals

MECHANISMS OF SEXUAL REPRODUCTION

- Internal and external fertilization both depend on mechanisms ensuring that mature sperm encounter fertile eggs of the same species
- Species with internal fertilization usually produce fewer zygotes but provide more parental protection than species with external fertilization
- Complex reproductive systems have evolved in many animal phyla

MAMMALIAN REPRODUCTION

- Human reproduction involves intricate anatomy and complex behavior
- Spermatogenesis and oogenesis both involve meiosis but differ in three significant ways
- A complex interplay of hormones regulates reproduction
- Embryonic and fetal development occur during pregnancy in humans and other eutherian (placental) mammals
- Modern technology offers solutions for some reproductive problems

T he many aspects of animal form and function we have studied so far can be viewed, in the broadest context, as adaptations contributing to reproductive success. Individuals are transient. A population transcends finite life spans only by reproduction, the creation of new individuals from existing ones. The two earthworms in the photograph above are mating. Unless disturbed, they will remain above the ground and joined like this for several hours. Each worm produces both sperm and eggs, each donates and receives sperm during mating, and each will pro-

duce fertilized eggs. In a few weeks, sexual reproduction will be completed when many new individuals hatch from the eggs.

Animal reproduction is the subject of this chapter. We will first compare the diverse reproductive mechanisms that have evolved in the animal kingdom, and then examine the details of mammalian, particularly human, reproduction.

OVERVIEW OF ANIMAL REPRODUCTION

Both asexual and sexual reproduction occur in the animal kingdom

There are two principal modes of animal reproduction. **Asexual** (from the Greek, "without sex") **reproduction** is the creation of new individuals whose genes all come from one parent without the fusion of egg and sperm. In most cases, asexual reproduction relies entirely on mitotic cell division. **Sexual reproduction** is the creation of offspring by the fusion of haploid **gametes** to form a **zygote** (fertilized egg), which is diploid. Gametes are formed by meiosis (see FIGURE 13.7). The female gamete, the **ovum** (unfertilized egg), is usually a relatively large and nonmotile cell. The male gamete, the **spermatozoon,** is generally a small, motile cell. Sexual reproduction increases genetic variability among offspring by generating unique combinations of genes inherited from two parents. By producing offspring having various phenotypes, sexual reproduction may enhance the reproductive success of parents when pathogens or other environmental factors change relatively rapidly.

FIGURE 46.1 Two from one: asexual reproduction of a sea anemone *(Anthopleura elegantissima)*. The individual in the center of this photograph is undergoing fission, a type of asexual reproduction. Two smaller individuals will soon form as the parent divides approximately in half. The offspring will be genetic copies of the parent.

Diverse mechanisms of asexual reproduction enable animals to produce identical offspring rapidly

Many invertebrates can reproduce asexually by **fission,** the separation of a parent into two or more individuals of approximately equal size (FIGURE 46.1). Also common among invertebrates, **budding** involves new individuals splitting off from existing ones. For example, in certain cnidarians and tunicates, new individuals grow out from the body of a parent (see FIGURE 13.1). The offspring may either detach from the parent or remain joined, eventually forming extensive colonies. Stony corals, which may be more than 1 m across, are cnidarian colonies of several thousand connected individuals. In another form of asexual reproduction, some invertebrates release specialized groups of cells that can grow into new individuals. For example, the **gemmules** of sponges are formed when cells of several types migrate together within the sponge and become surrounded by a protective coat.

Yet another type of asexual reproduction is **fragmentation,** the breaking of the body into several pieces, some or all of which develop into complete adults. For an animal to reproduce this way, fragmentation must be accompanied by **regeneration,** the regrowth of lost body parts. Reproduction by fragmentation and regeneration occurs in many sponges, cnidarians, polychaete annelids, and tunicates. Many animals can also replace lost appendages by regeneration—most sea stars can grow new arms when injured, for example—but this is not reproduction because new individuals are not created. In sea stars of the genus *Linckia,* a whole new individual may develop from an isolated arm. Thus, a single animal with five arms, if broken apart, could asexually give rise to five offspring.

Asexual reproduction has several potential advantages. For instance, it enables animals living in isolation to produce offspring without locating mates. It can also create numerous offspring in a short amount of time, which is ideal for colonizing a habitat rapidly. Theoretically, asexual reproduction is most advantageous in stable, favorable environments because it perpetuates successful genotypes precisely.

Reproductive cycles and patterns vary extensively among animals

Most animals show definite cycles in reproductive activity, often related to changing seasons. The periodic nature of reproduction allows animals to conserve resources and reproduce when more energy is available than is needed for maintenance and when environmental conditions favor the survival of offspring. Ewes (female sheep), for example, have 15-day reproductive cycles and ovulate at the midpoint of each cycle. These cycles generally occur during fall and early winter, so lambs are usually born in the late winter or spring. Even animals that live in apparently stable habitats, such as the tropics or the ocean, generally reproduce only at certain times of the year. Reproductive cycles are controlled by a combination of hormonal and environmental cues, the latter including such factors as seasonal temperature, rainfall, day length, and lunar cycles (see the discussion of the pineal gland in Chapter 45).

Animals may reproduce asexually or sexually exclusively, or they may alternate between the two modes. In aphids, rotifers, and the freshwater crustacean *Daphnia,* each female can produce eggs of two types, depending on environmental conditions such as the time of year. One type of egg is fertilized, but the other type develops by **parthenogenesis,** a process in which the egg develops without being fertilized. The adults produced by parthenogenesis are often haploid, and eggs are formed without meiosis. In the case of *Daphnia,* the switch from sexual to asexual reproduction is often related to season. Asexual reproduction occurs under favorable conditions and sexual reproduction during times of environmental stress.

Parthenogenesis has a role in the social organization of certain species of bees, wasps, and ants. Male honeybees, or drones, are produced parthenogenetically, whereas females, both sterile workers and reproductive females (queens), develop from fertilized eggs.

Among vertebrates, several genera of fishes, amphibians, and lizards reproduce exclusively by a complex form of parthenogenesis that involves the doubling of chromosomes after meiosis to create diploid "zygotes." For example, there are about 15 species of whiptail lizards (genus *Cnemidophorus*) that reproduce exclusively by parthenogenesis. There are no males in these species, but the lizards imitate courtship and mating behavior typical of sexual species of the same genus. During the breeding season, one female of each mating pair

(a) Both lizards in this photograph are *C. uniparens* females. The one on top is playing the role of a male. Every two or three weeks during the breeding season, individuals switch sex roles.

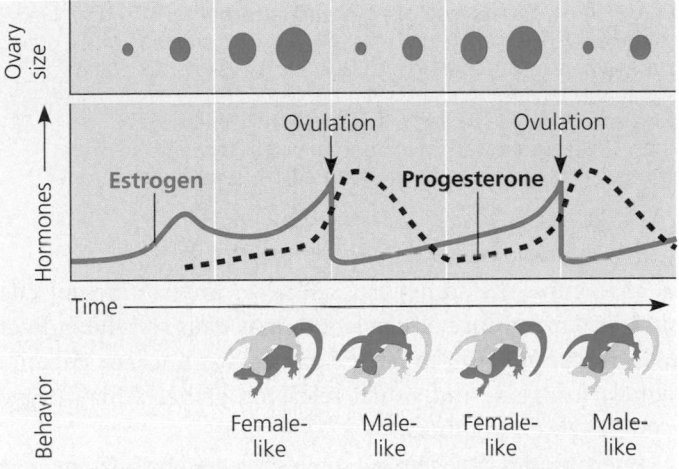

(b) The sexual behavior of *C. uniparens* is correlated with the cycle of ovulation mediated by sex hormones. Before ovulation, an individual behaves like a female. She has relatively large ovaries and her behavior is correlated with high blood levels of estrogen. After ovulation, small ovaries, an abrupt drop in estrogen, and a rapid increase in circulating progesterone are correlated with male behavior.

FIGURE 46.2 Sexual behavior in parthenogenetic lizards. The desert-grassland whiptail lizard *(Cnemidophorus uniparens)* is an all-female species. These reptiles reproduce by parthenogenesis; eggs undergo a chromosome doubling after meiosis and develop into lizards without being fertilized. However, ovulation is enhanced by courtship and mating rituals that imitate the behavior of closely related species that reproduce sexually.

mimics a male (FIGURE 46.2a). The roles change two or three times during the season, female behavior occurring when the level of the female sex hormone estrogen is high prior to ovulation (the release of eggs), and male behavior occurring after ovulation when the level of estrogen drops (FIGURE 46.2b). In fact, ovulation is more likely to occur if one individual is mounted by another during the critical time of the hormone cycle; isolated lizards lay fewer eggs than those that go through the motions of sex. Apparently, these parthenogenetic lizards, which evolved from species having two sexes, still require certain sexual stimuli for maximum reproductive success.

Sexual reproduction presents a special problem for sessile or burrowing animals or for parasites, such as tapeworms, which may have difficulty encountering a member of the opposite sex. One solution to this problem is **hermaphroditism,** in which each individual has both male and female reproductive systems (the term is derived from Hermes and Aphrodite, names of a Greek god and goddess). Although some hermaphrodites fertilize themselves, most must mate with another member of the same species. When this occurs, each animal serves as both male and female, donating and receiving sperm, as we saw for earthworms. Each individual encountered is a potential mate, resulting in twice as many offspring than if only one individual's eggs were fertilized.

Another remarkable reproductive pattern is **sequential hermaphroditism,** in which an individual reverses its sex during its lifetime. In some species, the sequential hermaphrodite is **protogynous** (female first), while other species are **protandrous** (male first). In various species of reef fishes called wrasses, sex reversal is associated with age and size. For example, the Caribbean bluehead wrasse is a protogynous species in which only the largest (usually the oldest) individuals change from female to male (FIGURE 46.3). These fish live in harems consisting of a single male and several females. If the male dies or is removed in experiments, the largest female in the harem changes sex and becomes the new male. Within a week, the transformed individual is producing sperm instead of eggs. In this species, the male defends the harem against intruders, and thus larger size may give a greater reproductive advantage to males than it does to females. In contrast, there are protandrous animals that change from male to female when size increases. In such cases, greater size may

FIGURE 46.3 Sex reversal in a sequential hermaphrodite. In many species of reef fishes called wrasses, sex can change during the animal's life. Sex reversal is often correlated with size. In this scene, a male Caribbean bluehead wrasse and two smaller females are feeding on a sea urchin. All wrasses of this species are born females, but the oldest, largest individuals change sex and complete their lives as males.

increase the reproductive success of females more than it does males. For example, the production of huge numbers of gametes is an important asset for sedentary animals, such as oysters, that release their gametes into the surrounding water. Egg cells are generally much larger than sperm cells, so females produce fewer gametes than males. Larger females tend to produce more eggs than smaller ones, and species of oysters that are sequential hermaphrodites are generally protandrous.

The diverse reproductive cycles and patterns we observe in the animal kingdom are adaptations that have evolved by natural selection. We will see many other examples as we survey the various mechanisms of sexual reproduction.

MECHANISMS OF SEXUAL REPRODUCTION

The mechanisms of **fertilization,** the union of sperm and egg, play an important part in sexual reproduction. Some species have **external fertilization;** eggs are shed by the female and fertilized by the male in the environment (FIGURE 46.4). Other species have **internal fertilization;** sperm are deposited in or near the female reproductive tract, and fertilization occurs within the tract. (The cellular and molecular details of fertilization are discussed in Chapter 47.)

Internal and external fertilization both depend on mechanisms ensuring that mature sperm encounter fertile eggs of the same species

Internal fertilization requires cooperative behavior, leading to copulation. In some cases, uncharacteristic sexual behavior is eliminated by natural selection in a direct manner; for example, female spiders will eat males if specific reproductive signals are not followed during mating. (We will discuss other examples of mating behavior in Chapter 51.) Internal fertilization also requires sophisticated reproductive systems, including copulatory organs that deliver sperm and receptacles for its storage and transport to ripe eggs.

Because external fertilization requires an environment where an egg can develop without desiccation or heat stress, it occurs almost exclusively in moist habitats. Many aquatic invertebrates simply shed their eggs and sperm into the surroundings, and fertilization occurs without the parents actually making physical contact. Timing is still crucial to ensure that mature sperm encounter ripe eggs.

Many fishes and amphibians that use external fertilization exhibit specific mating behaviors, in which one male fertilizes the eggs of one female. Courtship behavior is a mutual trigger for the release of gametes, with two effects: The probability of successful fertilization is increased, and the choice

FIGURE 46.4 The release of eggs and external fertilization. Many amphibians shed gametes into the environment, and fertilization occurs outside the female's body. In most species, behavioral adaptations ensure that a male is present when the female releases eggs. Here, a female frog, clasped by a male (on top), has just released a mass of eggs. The male released sperm (not shown) at the same time, and external fertilization has already occurred in the surrounding water.

of mates may be somewhat selective. Environmental cues such as temperature or day length may cause all the individuals of a population to release gametes at once, or chemical signals from one individual releasing gametes may trigger gamete release in others.

Pheromones are chemical signals released by one organism that influence the behavior of other individuals of the same species. Pheromones are small, volatile or water-soluble molecules that disperse easily into the environment and, like hormones, are active in minute amounts. Many pheromones function as mate attractants. The mate attractants of some female insects can be detected by males as much as a mile away. The pheromone of the female gypsy moth elicits behavioral responses in males at concentrations as low as 1 molecule of pheromone in 10^{17} molecules of other gases in the air. (We will return to pheromones in Chapter 51.)

Species with internal fertilization usually produce fewer zygotes but provide more parental protection than species with external fertilization

All species produce more offspring than survive to reproduce. Species with external fertilization usually produce enormous numbers of zygotes, but the proportion that survive and develop further is often quite small. Internal fertilization usually produces fewer zygotes, but this may be offset by greater protection of the embryos and parental care of the young. Major types of protection include resistant eggshells, development of

FIGURE 46.5 Parental care in an invertebrate. Compared to many insects, giant water bugs *(Belostoma)* produce relatively few offspring, but parental protection enhances the survival of those offspring. Fertilization is internal, and the female glues her fertilized eggs to the back of the male (shown here). Whereas the males of most insect species provide no parental care for their offspring, the male giant water bug carries them for days, frequently fanning water over them. This treatment helps keep the eggs moist, aerated, and free of parasites.

the embryo within the reproductive tract of the female parent, and parental care of the eggs and offspring.

Many species of terrestrial animals produce eggs that can withstand harsh environments. Birds, reptiles, and monotremes have amniote eggs with calcium and protein shells that resist water loss and physical damage. By contrast, the eggs of fishes and amphibians have only a gelatinous coat.

Rather than secreting a protective shell around the egg, many animals retain the embryo, which develops within the female reproductive tract. Among mammals, marsupials such as kangaroos and opossums retain their embryos for a short period in the uterus; the embryos then crawl out and complete fetal development attached to a mammary gland in the mother's pouch. The embryos of eutherian (placental) mammals develop entirely within the uterus, being nourished by the mother's blood supply through a special organ, the placenta (discussed in Chapter 34 and later in this chapter).

When a kangaroo crawls out of its mother's pouch for the first time, or when a human is born, it still is not capable of independent existence. We are familiar with adult birds feeding their young and mammals nursing their offspring, but parental care is much more widespread than we might suspect, and it takes a variety of unusual forms. In one species of tropical frog,

for instance, the male carries the tadpoles in his stomach until they metamorphose and hop out as young frogs. There are also many cases of parental care among invertebrates (FIGURE 46.5).

Complex reproductive systems have evolved in many animal phyla

To reproduce sexually, animals must have systems that produce gametes and deliver them to the gametes of the opposite sex. These reproductive systems are varied. The least complex systems do not even contain distinct **gonads,** the organs that produce gametes in most animals. The most complex reproductive systems contain many sets of accessory tubes and glands that carry and protect the gametes and developing embryos. Many animals whose body plans are otherwise relatively simple possess highly complex reproductive systems. The reproductive systems of parasitic flatworms, for example, are among the most complex in the animal kingdom (FIGURE 46.6).

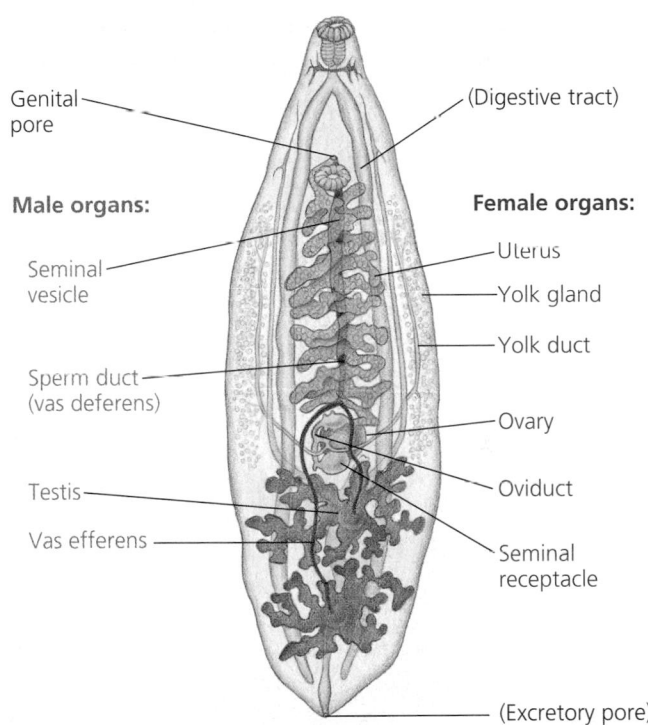

FIGURE 46.6 Reproductive anatomy of a parasitic flatworm. Most flatworms (phylum Platyhelminthes) are hermaphroditic, with complex male and female reproductive systems. Both systems open to the outside through the genital pore. Sperm produced in the testes pass through a pair of ducts (vasa efferentia) into a single sperm duct (vas deferens) and are stored in the seminal vesicle. During copulation, sperm are ejaculated into the female system (usually of another individual) and then move through the uterus to the seminal receptacle. (In some flatworms, sperm are injected into the body tissues through the body wall and then migrate to the female reproductive tract.) Eggs from the ovary pass into the oviduct, where they are fertilized by sperm from the seminal receptacle and coated with yolk and tough shell material secreted by the yolk glands. From the oviduct, the fertilized, shelled eggs pass into the uterus from which they are shed through the genital pore. Usually only a small fraction of the eggs will develop into the next generation of adult worms.

Among the least complex systems are those of polychaete worms (phylum Annelida). Most polychaetes have separate sexes but do not have distinct gonads; rather, the eggs and sperm develop from undifferentiated cells lining the coelom. As the gametes mature, they are released from the body wall and fill the coelom. Depending on the species, mature gametes may be shed through the excretory openings, or the swelling mass of eggs may split the body open, killing the parent and spilling the eggs into the environment.

Most insects have separate sexes with complex reproductive systems (FIGURE 46.7). In the male, sperm develop in a pair of testes and are conveyed along a coiled duct to two seminal vesicles, where they are stored. During mating, sperm are ejaculated into the female reproductive system. In the female, eggs develop in a pair of ovaries and are conveyed through ducts to the vagina, where fertilization occurs. In many species the female reproductive system includes a **spermatheca,** a sac in which sperm may be stored for a year or more.

The basic plan of all vertebrate reproductive systems is quite similar, but there are some important variations. In many non-mammalian vertebrates, the digestive, excretory, and reproductive systems have a common opening to the outside, the **cloaca,** which was probably present in the ancestors of all vertebrates. By contrast, most mammals lack a cloaca and have a separate opening for the digestive tract, and most female mammals have separate openings for the excretory and reproductive systems as well. The uterus of most vertebrates is partly or completely divided into two chambers. However, in humans and other mam-

mals that produce only a few young at a time, as well as in birds and many snakes, the uterus is a single structure. Male reproductive systems differ mainly in the copulatory organs. Many nonmammalian vertebrates do not have a well-developed penis and simply evert the cloaca to ejaculate.

MAMMALIAN REPRODUCTION

Human reproduction involves intricate anatomy and complex behavior

Reproductive Anatomy of the Human Male

In most mammalian species, including humans, the male's external reproductive organs are the scrotum and penis. The internal reproductive organs consist of gonads that produce gametes (sperm cells) and hormones, accessory glands that secrete products essential to sperm movement, and ducts that carry the sperm and glandular secretions (FIGURE 46.8).

The male gonads, or **testes** (singular, **testis**), consist of many highly coiled tubes surrounded by several layers of connective tissue. These tubes are the **seminiferous tubules,** where sperm form. The **Leydig cells** scattered between the seminiferous tubules produce testosterone and other androgens.

Production of normal sperm cannot occur at the body temperatures of most mammals, and the testes of humans and

FIGURE 46.7 Insect reproductive anatomy.

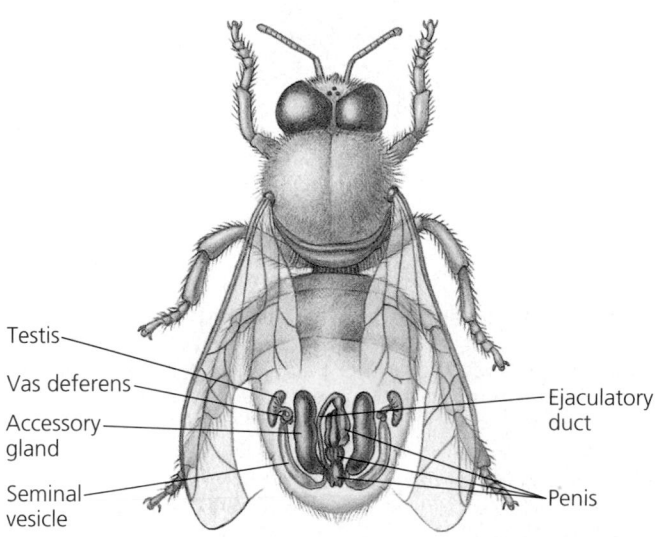

Testis
Vas deferens
Accessory gland
Seminal vesicle
Ejaculatory duct
Penis

(a) Male honeybee. Sperm form in the testes, pass through the sperm duct (vas deferens), and are stored in the seminal vesicle. The male ejaculates sperm along with fluid from the accessory glands. Males of some species of insects and other arthropods have appendages called claspers that grasp the female during copulation.

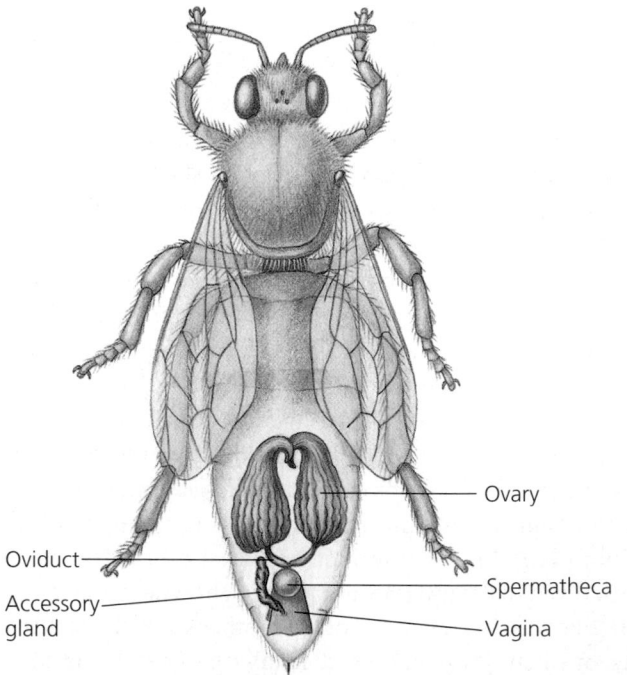

Oviduct
Accessory gland
Ovary
Spermatheca
Vagina

(b) Female honeybee. Eggs develop in the ovaries, pass through the oviducts, and into the vagina. A pair of accessory glands (only one is shown) add protective secretions to the eggs in the vagina. After mating, sperm are stored in the spermatheca, a sac connected to the vagina by a short duct.

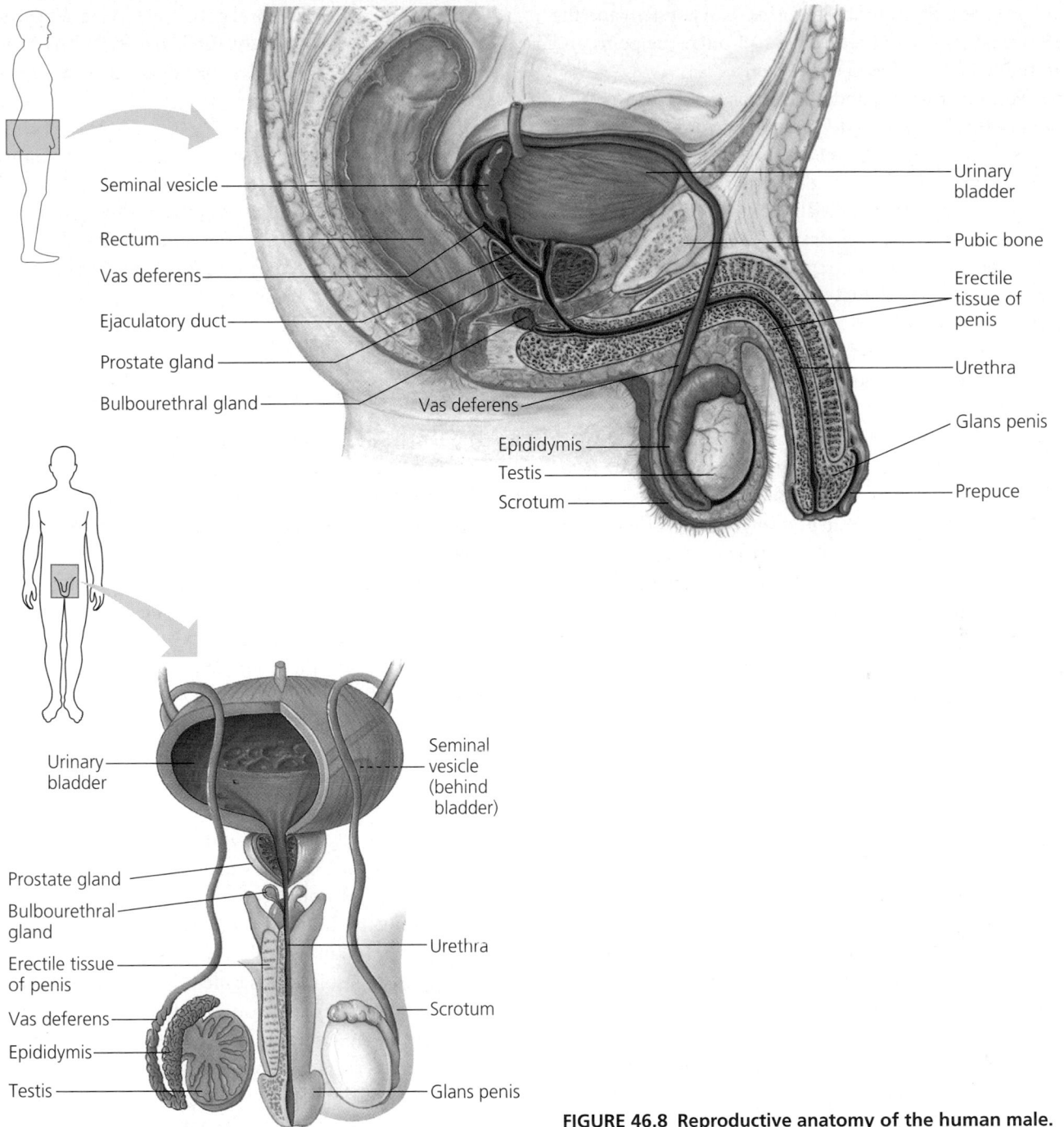

Seminal vesicle
Rectum
Vas deferens
Ejaculatory duct
Prostate gland
Bulbourethral gland
Vas deferens
Epididymis
Testis
Scrotum

Urinary bladder
Pubic bone
Erectile tissue of penis
Urethra
Glans penis
Prepuce

Urinary bladder
Seminal vesicle (behind bladder)
Prostate gland
Bulbourethral gland
Erectile tissue of penis
Vas deferens
Epididymis
Testis
Urethra
Scrotum
Glans penis

FIGURE 46.8 Reproductive anatomy of the human male.

many other mammals are held outside the abdominal cavity in the **scrotum,** which is a fold of the body wall. The temperature in a scrotum is about 2°C below that in the abdominal cavity. The testes develop high in the abdominal cavity and descend into the scrotum just before birth. In many rodents, the testes are drawn back into the abdominal cavity, and sperm maturation is interrupted between breeding seasons. Some mammals whose body temperature is low enough to allow sperm maturation, such as monotremes, whales, and elephants, retain the testes within the abdominal cavity permanently.

From the seminiferous tubules of a testis, the sperm pass into the coiled tubules of the **epididymis.** It takes about 20 days for sperm to pass through the 6-m-long tubules of each epididymis of a human male. During this passage, the sperm become motile and gain the ability to fertilize. During **ejaculation,** the sperm are propelled from the epididymis through the muscular **vas deferens.** These two ducts (one from each epididymis) run from the scrotum around and behind the urinary bladder, where each joins a duct from the seminal vesicle, forming a short **ejaculatory duct.** The ejaculatory ducts open into the

urethra, the tube that drains both the excretory system and the reproductive system. The urethra runs through the penis and opens to the outside at the tip of the penis.

Three sets of accessory glands (the seminal vesicles, prostate, and bulbourethral glands) add secretions to the **semen,** the fluid that is ejaculated. A pair of **seminal vesicles** contributes about 60% of the total volume of the semen. The fluid from the seminal vesicles is thick, yellowish, and alkaline. It contains mucus, the sugar fructose (which provides most of the energy used by the sperm), a coagulating enzyme, ascorbic acid, and prostaglandins, local regulators discussed in Chapter 45.

The **prostate gland** is the largest of the semen-secreting glands. It secretes its products directly into the urethra through several small ducts. Prostatic fluid is thin and milky, contains anticoagulant enzymes and citrate (a sperm nutrient). The prostate gland is the source of some of the most common medical problems of men over age 40. Benign (noncancerous) enlargement of the prostate occurs in more than half of all men in this age group and in virtually all men over 70. Prostate cancer is one of the most common cancers in men. It is treated surgically or with drugs that inhibit gonadotropins, resulting in reduced prostate activity and size.

The **bulbourethral glands** are a pair of small glands along the urethra below the prostate. Before ejaculation they secrete a clear mucus that neutralizes any acidic urine remaining in the urethra. Bulbourethral fluid also carries some sperm released before ejaculation, which is one reason for the high failure rate of the withdrawal method of birth control.

A man usually ejaculates about 2 to 5 mL of semen, and each milliliter may contain about 50 to 130 million sperm. Once in the female reproductive tract, prostaglandins in the semen thin the mucus at the opening of the uterus and stimulate contractions of the uterine muscles, which help move the semen up the uterus. The semen is slightly alkaline, and this helps neutralize the acidic environment of the vagina, protecting the sperm and increasing their motility. When first ejaculated, the semen coagulates, making it easier for uterine contractions to move it along; then anticoagulants liquify the semen, and the sperm begin swimming through the female tract.

The human **penis** is composed of three cylinders of spongy erectile tissue derived from modified veins and capillaries. During sexual arousal, the erectile tissue fills with blood from the arteries. As this tissue fills, the increasing pressure seals off the veins that drain the penis, causing it to engorge with blood. The resulting erection is essential to insertion of the penis into the vagina. Rodents, raccoons, walruses, and several other mammals also possess a **baculum,** a bone that is contained in, and helps stiffen, the penis. Temporary impotence, a reversible inability to achieve an erection, can result from alcohol consumption, certain drugs, and emotional problems. Several drugs and penile implant devices are available for men with nonreversible impotence due to nervous system or circu-

latory problems. A new oral drug called Viagra® promotes the action of the local regulator nitric oxide (NO), enhancing relaxation of smooth muscles in the blood vessels of the penis. This allows blood to enter the erectile tissue and sustain an erection.

The main shaft of the penis is covered by relatively thick skin. The head, or **glans penis,** has a much thinner covering and is consequently more sensitive to stimulation. The human glans is covered by a fold of skin called the foreskin, or **prepuce,** which may be removed by circumcision. Circumcision, which arose from religious traditions, has no verifiable basis in health or hygiene.

Reproductive Anatomy of the Human Female

The female's external reproductive structures are the clitoris and two sets of labia surrounding the clitoris and vaginal opening. The internal reproductive organs consist of a pair of gonads and a system of ducts and chambers to conduct the gametes and house the embryo and fetus (FIGURE 46.9).

The female gonads, the **ovaries,** lie in the abdominal cavity, flanking, and attached by a mesentery to, the uterus. Each ovary is enclosed in a tough protective capsule and contains many follicles. A **follicle** consists of one egg cell surrounded by one or more layers of follicle cells, which nourish and protect the developing egg cell. All of the 400,000 follicles a woman will ever have are formed before her birth. Of these, only several hundred will release egg cells during the woman's reproductive years. Starting at puberty and continuing until menopause, usually one follicle matures and releases its egg cell during each menstrual cycle. The cells of the follicle also produce the primary female sex hormones, the estrogens. The egg cell is expelled from the follicle in the process of **ovulation** (FIGURE 46.10). The remaining follicular tissue then grows within the ovary to form a solid mass called the **corpus luteum.** The corpus luteum secretes additional estrogens and progesterone, the hormone that maintains the uterine lining during pregnancy. If the egg cell is not fertilized, the corpus luteum disintegrates, and a new follicle matures during the next cycle.

The female reproductive system is not completely closed, and the egg cell is released into the abdominal cavity near the opening of the **oviduct,** or fallopian tube. The oviduct has a funnel-like opening, and cilia on the inner epithelium lining the duct help collect the egg cell by drawing fluid from the body cavity into the duct. The cilia also convey the egg cell down the duct to the **uterus,** also known as the womb. The uterus is a thick, muscular organ that can expand during pregnancy to accommodate a 4-kg fetus. The inner lining of the uterus, the **endometrium,** is richly supplied with blood vessels.

The neck of the uterus is the **cervix,** which opens into the vagina. The **vagina** is a thin-walled chamber that forms the birth canal through which the baby is born; it is also the repository for sperm during copulation.

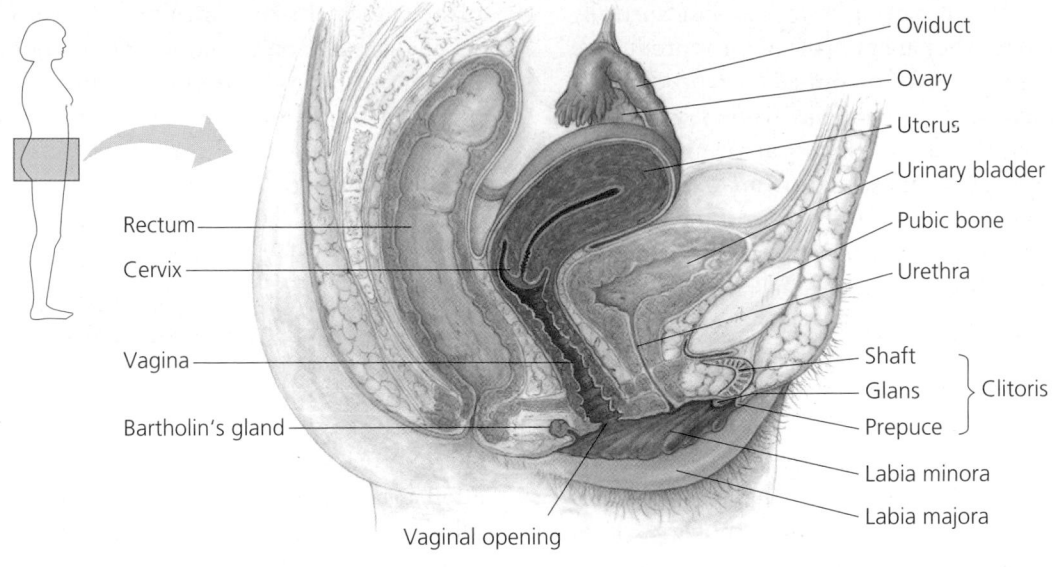

Oviduct
Ovary
Uterus
Urinary bladder
Pubic bone
Urethra

Rectum
Cervix

Vagina

Bartholin's gland

Shaft
Glans } Clitoris
Prepuce

Labia minora
Labia majora

Vaginal opening

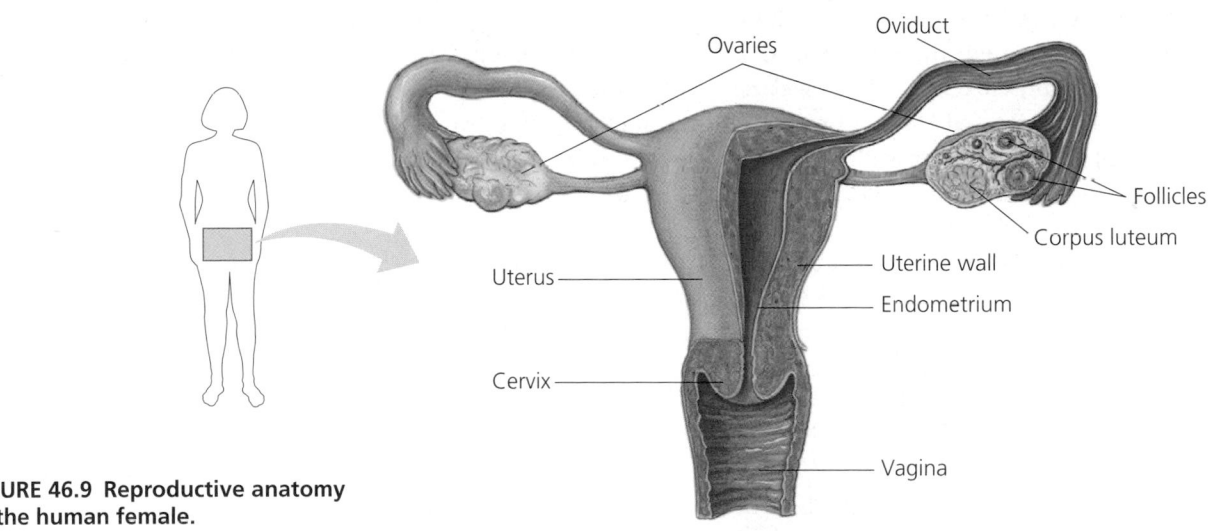

Ovaries
Oviduct

Follicles
Corpus luteum

Uterus
Uterine wall
Endometrium

Cervix

Vagina

FIGURE 46.9 Reproductive anatomy of the human female.

At birth, and usually until sexual intercourse or vigorous physical activity ruptures it, a vascularized membrane called the **hymen** partly covers the vaginal opening in humans. The vaginal opening and the separate urethral opening are located within a region called the **vestibule,** bordered by a pair of slender skin folds, the **labia minora.** A pair of thick, fatty ridges, the **labia majora,** encloses and protects the labia minora and vestibule. At the front edge of the vestibule, the **clitoris** consists of a short shaft supporting a rounded glans, or head, covered by a small hood of skin, the prepuce. During sexual arousal, the clitoris, vagina, and labia minora all engorge with blood and enlarge. The clitoris consists largely of erectile tissue. Richly supplied with nerve endings, it is one of the most sensitive points of sexual stimulation. During sexual arousal, **Bartholin's glands,** located near the vaginal opening, secrete mucus into the vestibule, keeping it lubricated and facilitating intercourse.

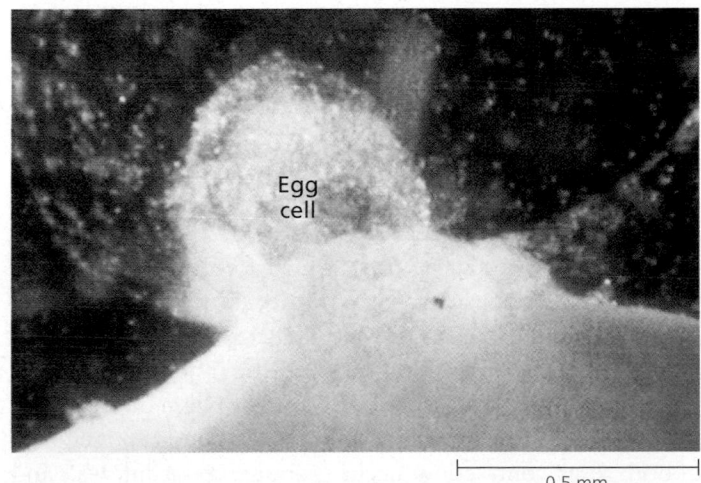

Egg cell

0.5 mm

FIGURE 46.10 Ovulation. An egg cell is released from a follicle at the surface of the ovary. The orangish mass below the ejected egg cell is part of a mammalian ovary.

Mammary glands are present in both sexes but normally function only in women. They are not part of the reproductive system but are important to mammalian reproduction. Within the glands, small sacs of epithelial tissue secrete milk, which drains into a series of ducts opening at the nipple. Fatty (adipose) tissue forms the main mass of the mammary gland of a nonlactating mammal. The low level of estrogen in males prevents the development of both the secretory apparatus and the fat deposits, so male breasts remain small, and the nipples are not connected to the ducts.

Human Sexual Response

Many vertebrates and invertebrates have elaborate and complex mating behavior (see Chapter 51). Human sexuality is characterized by a diversity of stimuli and responses, with a common, underlying physiological pattern.

Two types of physiological reactions predominate in both sexes: **vasocongestion,** the filling of a tissue with blood caused by increased blood flow through the arteries of that tissue, and **myotonia,** increased muscle tension. Both skeletal and smooth muscle may show sustained or rhythmic contractions, including those associated with orgasm.

The sexual response cycle can be divided into four phases: excitement, plateau, orgasm, and resolution. An important function of the excitement phase is preparation of the vagina and penis for **coitus** (sexual intercourse). During this phase, vasocongestion is particularly evident in erection of the penis and clitoris; enlargement of the testes, labia, and breasts; and vaginal lubrication. Myotonia may occur, resulting in nipple erection or tension of the arms and legs.

The plateau phase continues these responses. In females, the outer third of the vagina becomes vasocongested, while the inner two-thirds becomes slightly expanded. This change, coupled with the elevation of the uterus, forms a depression that receives sperm at the back of the vagina. Breathing increases and heart rate rises, sometimes to 150 beats per minute—not in response to the physical effort of sexual activity, but as an involuntary response to stimulation of the autonomic nervous system (see FIGURES 48.17 and 48.18).

Orgasm is characterized by rhythmic, involuntary contractions of the reproductive structures in both sexes. Male orgasm has two stages. Emission is the contraction of the glands and ducts of the reproductive tract, which forces semen into the urethra. Expulsion, or ejaculation, occurs when the urethra contracts and the semen is expelled. During female orgasm, the uterus and outer vagina contract, but the inner two-thirds of the vagina do not. Orgasm is the shortest phase of the sexual response cycle, usually lasting only a few seconds. In members of both sexes, contractions occur at about 0.8-sec intervals and may involve the anal sphincter and several abdominal muscles.

The resolution phase completes the cycle and reverses the responses of the earlier stages. Vasocongested organs return to their normal size and color, and muscles relax. Most of the changes during resolution are completed in 5 minutes. Loss of penile and clitoral erection, however, may take longer. An initial loss of erection is rapid in both sexes, but a return of the organs to their nonaroused size may take as long as an hour.

Spermatogenesis and oogenesis both involve meiosis but differ in three significant ways

Spermatogenesis, the production of mature sperm cells, is a continuous and prolific process in the adult male. Each ejaculation of a human male contains 100 to 650 million sperm cells, and males can ejaculate daily with little loss of fertilizing capacity.

Spermatogenesis occurs in the seminiferous tubules of the testes. FIGURE 46.11 describes the process in some detail. The stem cells that give rise to sperm, **spermatogonia,** are located at the periphery of each seminiferous tubule. The developing sperm cells move toward the central opening (lumen) of the tubule as they undergo meiosis and differentiation. The four cells that results from meiosis all develop into mature sperm cells.

The structure of a sperm cell fits its function (FIGURE 46.12). In most species, a head containing the haploid nucleus is tipped with a special body, the **acrosome,** which contains enzymes that help the sperm penetrate the egg. Behind the head, the sperm cell contains large numbers of mitochondria (or a single large one, in some species) that provide ATP for movement of the tail, which is a flagellum. Mammalian sperm shape varies from species to species, with the head a slender comma shape, an oval form (as in the human sperm), or nearly spherical.

Oogenesis is the development of ova—mature, unfertilized egg cells. FIGURE 46.13, p. 986, shows the process and its location, the ovary. In the developing female embryo, **oogonia,** the stem cells that give rise to ova, multiply and then begin meiosis, but the process stops at prophase I. The cells at this stage, called **primary oocytes,** remain quiescent within small follicles until puberty, when they are reactivated by hormones. Beginning at puberty, FSH (follicle-stimulating hormone) periodically stimulates a follicle to grow and induces its primary oocyte to complete meiosis I and start meiosis II. Meiosis then stops again; the **secondary oocyte,** released during ovulation, does not continue meiosis II right away. In humans, penetration of the egg cell by the sperm triggers the completion of meiosis, and only then is oogenesis actually complete.

Oogenesis differs from spermatogenesis in three major ways. First, during the meiotic divisions of oogenesis, cytokinesis is unequal, with almost all the cytoplasm monopolized

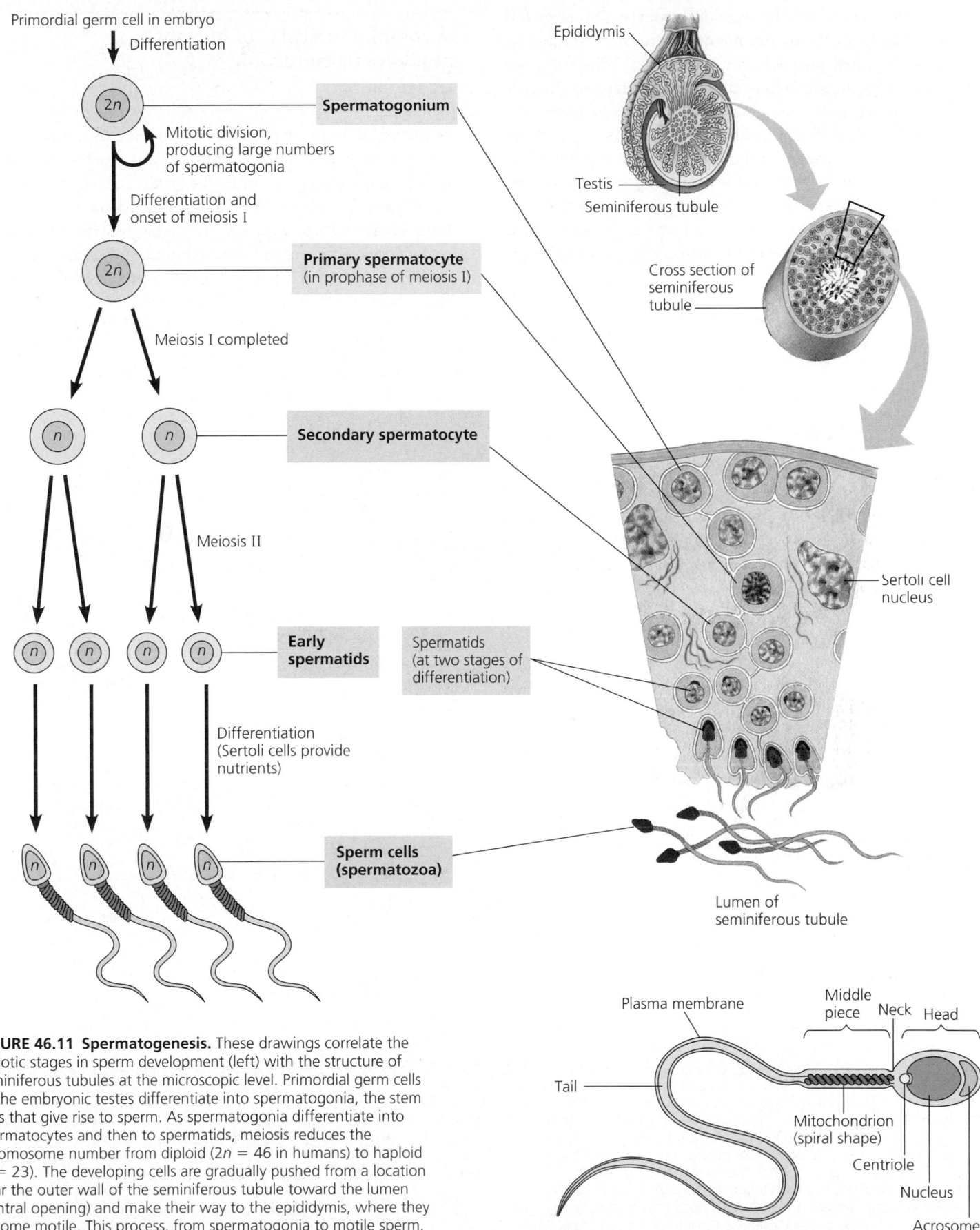

Primordial germ cell in embryo

↓ Differentiation

(2n) ——— **Spermatogonium**

Mitotic division, producing large numbers of spermatogonia

Differentiation and onset of meiosis I

(2n) ——— **Primary spermatocyte** (in prophase of meiosis I)

Meiosis I completed

(n) (n) ——— **Secondary spermatocyte**

Meiosis II

(n) (n) (n) (n) ——— **Early spermatids**

Differentiation (Sertoli cells provide nutrients)

——— **Sperm cells (spermatozoa)**

Epididymis

Testis

Seminiferous tubule

Cross section of seminiferous tubule

Sertoli cell nucleus

Spermatids (at two stages of differentiation)

Lumen of seminiferous tubule

FIGURE 46.11 Spermatogenesis. These drawings correlate the meiotic stages in sperm development (left) with the structure of seminiferous tubules at the microscopic level. Primordial germ cells of the embryonic testes differentiate into spermatogonia, the stem cells that give rise to sperm. As spermatogonia differentiate into spermatocytes and then to spermatids, meiosis reduces the chromosome number from diploid ($2n = 46$ in humans) to haploid ($n = 23$). The developing cells are gradually pushed from a location near the outer wall of the seminiferous tubule toward the lumen (central opening) and make their way to the epididymis, where they become motile. This process, from spermatogonia to motile sperm, takes 65 to 75 days in the human male. In mature males, about 3 million spermatogonia start the process each day.

Plasma membrane

Tail

Middle piece

Neck

Head

Mitochondrion (spiral shape)

Centriole

Nucleus

Acrosome

FIGURE 46.12 Structure of a human sperm cell.

by a single daughter cell, the secondary oocyte. This large cell can go on to form the ovum; the other products of meiosis, smaller cells called polar bodies, degenerate. This contrasts with spermatogenesis, when all four products of meiosis develop into mature sperm (compare FIGURES 46.11 and 46.13). Second, while the cells from which sperm develop continue to divide by mitosis throughout the male's life, this is not the case for oogenesis in the female. At birth, an ovary already contains all the primary oocytes it will ever have. Third, oogenesis has long "resting" periods, in contrast to spermatogenesis, which produces mature sperm from precursor cells in an uninterrupted sequence.

A complex interplay of hormones regulates reproduction

The Male Pattern

In the male, the principal sex hormones are the androgens, of which testosterone is the most important. Androgens, steroid hormones produced mainly by the Leydig cells of the testes, are directly responsible for the primary and secondary sex characteristics of the male. Primary sex characteristics are those associated with the reproductive system: development of the vasa deferentia and other ducts, development of the external

FIGURE 46.13 Oogenesis.

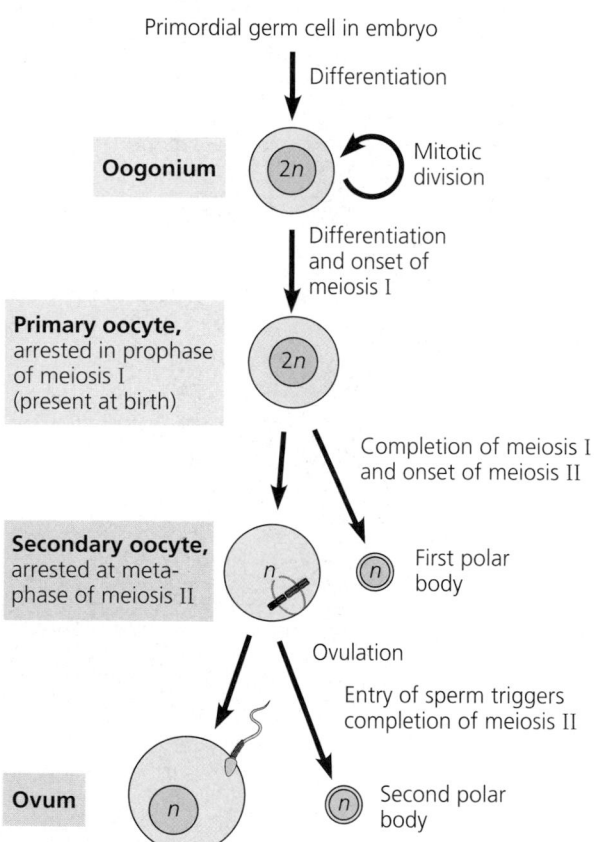

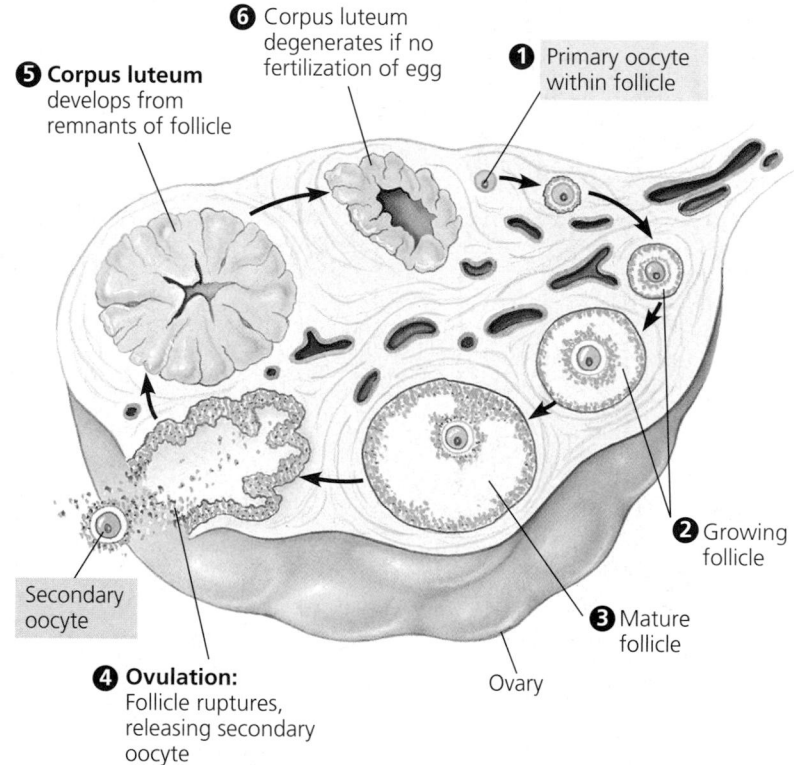

(a) The production of ova begins with differentiation of the primordial germ cells in the female embryo into oogonia, which in turn develop into primary oocytes. By birth, a female's lifetime supply of primary oocytes is present in her ovaries. Each primary oocyte is arrested at prophase of meiosis I. Starting at puberty, a single primary oocyte completes meiosis I each month, developing into a secondary oocyte. The meiotic divisions in oogenesis involve unequal cytokinesis, with the smaller cells becoming polar bodies (the first polar body may or may not divide again). The secondary oocyte completes meiosis II only if a sperm cell enters it. After meiosis II, the haploid nuclei of the sperm and the mature ovum fuse—fertilization.

(b) This cutaway view of an ovary illustrates the developmental stages of an ovarian follicle that accompany oogenesis. In response to FSH, several follicles start to grow, but usually only one matures. For convenience, the stages are presented as a cycle (arrows), although they occur at different times and are never actually present simultaneously within the ovary. In a real ovary, a follicle stays in one place as it goes through the sequential stages.

FIGURE 46.14 Hormonal control of the testes. The anterior pituitary secretes two gonadotropic hormones with different effects on the testes, luteinizing hormone (LH) and follicle-stimulating hormone (FSH). LH and FSH are themselves both regulated by gonadotropin-releasing hormone (GnRH) from the hypothalamus. LH, FSH, and GnRH concentrations in the blood are regulated by negative feedback by androgens. GnRH is also controlled by negative feedback from LH and FSH (not shown). In human males, these feedback loops keep the hormones at relatively constant levels, but in many other mammalian species, seasonal cycles in hormone concentration regulate breeding patterns.

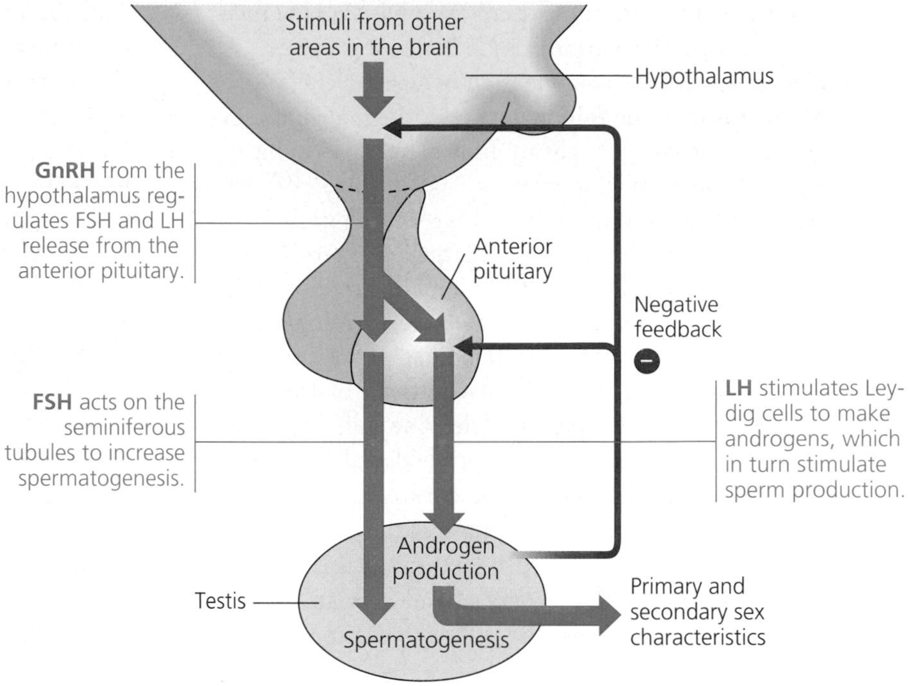

Stimuli from other areas in the brain
Hypothalamus
GnRH from the hypothalamus regulates FSH and LH release from the anterior pituitary.
Anterior pituitary
Negative feedback
FSH acts on the seminiferous tubules to increase spermatogenesis.
LH stimulates Leydig cells to make androgens, which in turn stimulate sperm production.
Androgen production
Testis
Spermatogenesis
Primary and secondary sex characteristics

reproductive structures, and sperm production. Secondary sex characteristics are features that are not directly related to the reproductive system, including deepening of the voice, distribution of facial and pubic hair, and muscle growth (androgens stimulate protein synthesis). Androgens are also potent determinants of behavior in mammals and other vertebrates. In addition to specific sexual behaviors and sex drive, androgens increase general aggressiveness and are responsible for such actions as singing in birds and calling by frogs. Hormones from the anterior pituitary and hypothalamus control both androgen secretion and sperm production by the testes (FIGURE 46.14).

The Female Pattern

In the female, the pattern of hormone secretion and the reproductive events the hormones regulate are cyclic, very different from the male pattern. Whereas males produce sperm continuously, females release only one egg or a few eggs at one time during each cycle. Control of the female cycle is quite complex.

Two different types of cycles occur in female mammals. Humans and many other primates have **menstrual cycles,** whereas other mammals have **estrous cycles.** In both cases, ovulation occurs at a time in the cycle after the endometrium has started to thicken and develop a rich blood supply, which prepares the uterus for the possible implantation of an embryo. One difference between the two types of cycles involves the fate of the uterine lining if pregnancy does not occur. In menstrual cycles, the endometrium is shed from the uterus through the cervix and vagina in a bleeding called **menstruation.** In estrous cycles, the endometrium is reabsorbed by the uterus, and no extensive bleeding occurs.

Other major distinctions include more pronounced behavioral changes during estrous cycles than during menstrual cycles and stronger effects of season and climate on estrous cycles. Whereas human females may be receptive to sexual activity throughout their cycles, most mammals will copulate only during the period surrounding ovulation. This period of sexual activity, called **estrus** (from the Latin *oestrus,* frenzy, passion), is the only time vaginal changes permit mating. Estrus is sometimes called heat, and indeed the female's body temperature increases slightly. The length and frequency of reproductive cycles vary widely among mammals. The human menstrual cycle averages 28 days (although cycles vary, ranging from about 20 to 40 days). In contrast, the estrous cycle of the rat is only 5 days. Bears and dogs have one cycle per year, but elephants cycle several times per year.

Let's examine the reproductive cycle of the human female in more detail as a case study of how a complex function is coordinated by hormones. The term *menstrual cycle* refers specifically to the changes that occur in the uterus (FIGURE 46.15, p. 988). By convention, the first day of a woman's menstrual period, the first day of menstruation, is designated day 1 of the cycle. The **menstrual flow phase** of the cycle, during which menstrual bleeding (loss of most of the functional layer of the endometrium) occurs, usually persists for a few days (FIGURE 46.15d). Then the thin remaining endometrium begins to regenerate and thicken for a week or two, during what is called the **proliferative phase** of the menstrual cycle. During the last phase, the **secretory phase,** usually about two weeks in duration, the endometrium continues to thicken, becomes more vascularized, and develops glands that secrete a fluid rich in glycogen. If an embryo has not implanted in the uterine lining by the end of the secretory phase, a new menstrual flow commences, marking day 1 of the next cycle.

Paralleling the menstrual cycle is an **ovarian cycle** (FIGURE 46.15c). It begins with the **follicular phase,** during which several follicles in the ovary begin to grow. The developing egg cell in each of these follicles enlarges, and its coat of follicle cells thickens. Of the several follicles that start to grow, however, only one usually continues to enlarge and mature while the others disintegrate. The maturing follicle develops an internal fluid-filled cavity and grows very large, forming a bulge near the surface of the ovary. The follicular phase ends with **ovulation,** when the follicle and adjacent wall of the ovary rupture, releasing the secondary oocyte. The follicular tissue that remains in the ovary after ovulation develops into the corpus luteum. Endocrine cells of the corpus luteum secrete female hormones during the **luteal phase** of the ovarian cycle. The next cycle begins with a new growth of follicles.

Hormones coordinate the menstrual and ovarian cycles in such a way that growth of the follicle and ovulation are synchronized with preparation of the uterine lining for possible implantation of an embryo. Five hormones participate in an elaborate scheme involving both positive and negative feedback. These hormones are gonadotropin-releasing hormone (GnRH), secreted by the hypothalamus; follicle-stimulating hormone (FSH) and luteinizing hormone (LH), the two gonadotropins secreted by the anterior pituitary; and estrogens (a family of closely related hormones) and progesterone, the female sex hormones secreted by the ovaries. The relative levels of the pituitary and ovarian hormones in blood plasma are graphed in FIGURE 46.15a and b, and correlated with the phases of the ovarian and menstrual cycles. As you read the following discussion, refer to the figure as a guide to understanding how the hormones regulate the female reproductive system.

During the follicular phase of the ovarian cycle, the pituitary secretes small amounts of FSH and LH in response to stimulation by GnRH from the hypothalamus. At this time, the cells of immature ovarian follicles have receptors for FSH but not for LH. The FSH stimulates follicle growth, and the

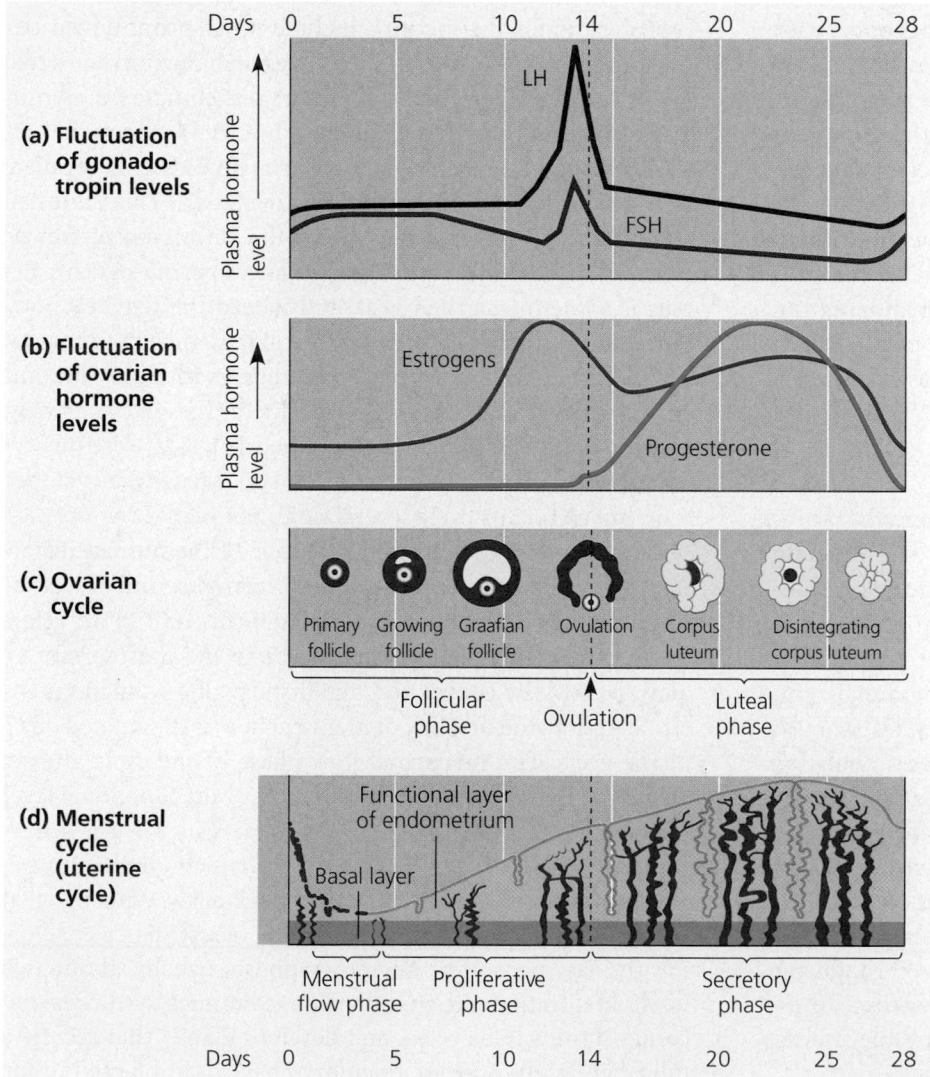

FIGURE 46.15 The reproductive cycle of the human female. Hormones coordinate the ovarian and menstrual cycles, preparing the uterine lining (endometrium) for implantation of an embryo even before ovulation. **(a)** Changes in LH and FSH levels. **(b)** Changes in the levels of estrogens and progesterone. **(c)** The ovarian cycle consists of a follicular phase, during which follicles grow and secrete increasing amounts of estrogens; ovulation; and a luteal phase, during which the corpus luteum secretes estrogens and progesterone. Length of the follicular phase varies; the luteal phase usually lasts 13 to 15 days. **(d)** The menstrual cycle consists of a menstrual flow phase, a proliferative phase, and a secretory phase. Menstruation, the shedding of the endometrium, occurs during the menstrual flow phase. The first day of flow marks day 1 of the menstrual cycle. During the proliferative phase, estrogens from the growing follicle stimulate the endometrium to thicken and become increasingly vascularized. During the secretory phase, the endometrium continues to thicken, its arteries enlarge, and endometrial glands grow. These endometrial changes require estrogens and progesterone, secreted by the corpus luteum after ovulation. Thus, the secretory phase of the menstrual cycle parallels the luteal phase of the ovarian cycle. Disintegration of the corpus luteum at the end of the luteal phase reduces the amount of estrogens and progesterone available to the endometrium, so it is shed. In the event of pregnancy, additional mechanisms maintain high levels of estrogens and progesterone, preventing loss of the endometrium.

cells of these growing follicles secrete estrogens. Notice in FIGURE 46.15b that there is a slow rise in the amount of estrogens secreted during most of the follicular phase. This small increase in estrogens inhibits secretion of the pituitary hormones, keeping the levels of FSH and LH relatively low during most of the follicular phase. These hormonal relationships change radically and rather abruptly when the rate of secretion of estrogens by the growing follicle begins to rise steeply. Whereas a slow rise of estrogens inhibits the secretion of pituitary gonadotropins, a high concentration of estrogens has the opposite effect and *stimulates* the secretion of gonadotropins by acting on the hypothalamus to increase its output of GnRH. You can see this response in FIGURE 46.15a as a steep increase of FSH and LH levels that occurs soon after the increase in the concentration of estrogens. The effect is greater for LH because the high concentration of estrogens, in addition to stimulating GnRH secretion, also increases the sensitivity of LH-releasing mechanisms in the pituitary to the hypothalamic signal (GnRH). By now, the follicles have receptors for LH and can respond to this hormonal cue. In an example of positive feedback, the increase in LH concentration caused by increased secretion of estrogens from the growing follicle induces final maturation of the follicle, and ovulation occurs about a day after the LH surge.

Following ovulation, LH stimulates the transformation of the follicular tissue left behind in the ovary to form the corpus luteum, a glandular structure. (It is for this "luteinizing" function that LH is named.) Under continued stimulation by LH during the luteal phase of the ovarian cycle, the corpus luteum secretes estrogens and a second steroid hormone, progesterone. The corpus luteum usually reaches its maximum development about 8 to 10 days after ovulation. As the levels of progesterone and estrogens rise, the combination of these hormones exerts negative feedback on the hypothalamus and pituitary, inhibiting the secretion of LH and FSH. Near the end of the luteal phase, the corpus luteum disintegrates (possibly a result of prostaglandins secreted by its own cells). Consequently, concentrations of estrogens and progesterone decline sharply. The dropping levels of ovarian hormones liberate the hypothalamus and pituitary from the inhibitory effects of these hormones. The pituitary then begins to secrete enough FSH to stimulate the growth of new follicles in the ovary, initiating the follicular phase of the next ovarian cycle.

How is the ovarian cycle synchronized with the menstrual cycle? Estrogens, secreted in increasing amounts by growing follicles, are a hormonal signal to the uterus, causing the endometrium to thicken. Thus, the follicular phase of the ovarian cycle is coordinated with the proliferative phase of the menstrual cycle. *Before* ovulation, the uterus is already being prepared for a possible embryo. *After* ovulation, estrogens and progesterone secreted by the corpus luteum stimulate continued development and maintenance of the endometrium, including an enlargement of arteries supplying blood to the uterine lining and the growth of endometrial glands that secrete a nutrient fluid that can sustain an early embryo before it actually implants in the uterine lining. Thus, the luteal phase of the ovarian cycle is coordinated with the secretory phase of the menstrual cycle. The rapid drop in the level of ovarian hormones when the corpus luteum disintegrates causes spasms of arteries in the uterine lining that deprive the endometrium of blood. Disintegration of the endometrium results in menstruation and the beginning of a new menstrual cycle. In the meantime, ovarian follicles that will stimulate renewed thickening of the endometrium are just beginning to grow. Cycle after cycle, the maturation and release of egg cells from the ovary is integrated with changes in the uterus, the organ that must accommodate an embryo if the egg cell is fertilized. In the absence of pregnancy, a new cycle begins. We will soon see that there are "override" mechanisms that prevent disintegration of the endometrium in the event of pregnancy.

In addition to their role in coordinating reproductive cycles, estrogens are also responsible for the secondary sex characteristics of the female. The hormones induce deposition of fat in the breasts and hips, increase water retention, affect calcium metabolism, stimulate breast development, and mediate female sexual behavior.

Menopause.   On average, human females undergo menopause, the cessation of ovulation and menstruation, between the ages of 46 and 54. Apparently, during these years the ovaries lose their responsiveness to gonadotropins (the hormones FSH and LH) from the pituitary, and menopause results from a decline in production of estrogens by the ovary. Menopause is an unusual phenomenon; in most species, females as well as males retain their reproductive capacity throughout life. Is there an evolutionary explanation for menopause? Why might natural selection have favored females who stopped reproducing? One intriguing (and highly controversial) hypothesis proposes that during early human evolution, undergoing menopause after having some children actually increased a woman's fitness; losing the ability to reproduce allowed her to provide better care for her children and grandchildren, thereby increasing the survival of individuals bearing her genes.

Embryonic and fetal development occur during pregnancy in humans and other eutherian (placental) mammals

From Conception to Birth

In placental mammals, **pregnancy,** or **gestation,** is the condition of carrying one or more **embryos,** new developing individuals, in the uterus. Pregnancy is preceded by **conception,** the fertilization of the egg by a sperm cell, and continues until the birth of the offspring. Human pregnancy averages 266 days (38 weeks) from conception, or 40 weeks from the start of the last menstrual cycle. Duration of pregnancy in other species correlates with body size and the extent of development of the young at birth. Many rodents (mice and rats) have gestation periods of about 21 days, whereas those of dogs are closer to 60 days. In cows, gestation averages 270 days (almost the same as humans); in giraffes, it is about 420 days; and in elephants, gestation is more than 600 days.

Human gestation can be divided for convenience of study into three **trimesters** of about 3 months each. The first trimester is the time of most radical change for both the mother and the baby. Fertilization occurs in the oviduct (FIGURE 46.16). About 24 hours later, the resulting zygote begins dividing, a process called **cleavage.** Cleavage continues, with the embryo becoming a ball of cells by the time it reaches the uterus about 3 to 4 days after fertilization. By about 1 week after fertilization, cleavage has produced an embryonic stage called the **blastocyst,** a sphere of cells containing a flattened cavity. In a process that takes about 5 more days, the blastocyst implants into the endometrium. Differentiation of body structures now begins in earnest. (Embryonic development will be described in detail in Chapter 47.) During implantation, the blastocyst becomes embedded in the endometrium, which responds by growing over the blastocyst. The embryo obtains nutrients directly from the endometrium during the first 2 to 4 weeks of development. Meanwhile, tissues grow out from the developing embryo and mingle with the endometrium to form the **placenta.** This disk-shaped organ, containing embryonic and maternal blood vessels, grows to about the size of a dinner plate and weighs somewhat less than 1 kg. Diffusion of material between maternal and embryonic circulations provides nutrients, exchanges respiratory gases, and disposes of metabolic wastes for the embryo. Blood from the embryo travels to the placenta through arteries of the umbilical cord and returns via the umbilical vein, passing through the liver of the embryo (FIGURE 46.17).

The first trimester is also the main period of **organogenesis,** the development of the body organs (FIGURE 46.18). The heart begins beating by the fourth week and can be detected with a stethoscope by the end of the first trimester. By the end of the eighth week, all the major structures of the adult are present in rudimentary form. At this point, the embryo is called a **fetus.** Although well differentiated, the fetus is only 5 cm long by the end of the first trimester. Because of its rapid organogenesis, the embryo is most sensitive during the first trimester to such threats as radiation and drugs that can cause birth defects.

The first trimester is also a time of rapid change for the mother. The embryo secretes hormones that signal its presence and control the mother's reproductive system. One embryonic hormone, **human chorionic gonadotropin (HCG),** acts like pituitary LH to maintain secretion of progesterone and estrogens by the corpus luteum through the first trimester. In the absence of this hormonal override, the decline in maternal LH due to inhibition of the pituitary by progesterone would result in menstruation and spontaneous abortion of the embryo. Levels of HCG in the maternal blood are so high that some is excreted in the urine, where it can be detected in pregnancy tests. High levels of progesterone initiate changes in the pregnant woman's reproductive system, including increased mucus in the cervix that forms a protective plug, growth of the maternal part of the placenta, enlargement of the uterus, and (by negative feedback on the hypothalamus and pituitary) cessation of ovulation and menstrual cycling. The breasts also enlarge rapidly and are often quite tender.

During the second trimester, the fetus grows rapidly to about 30 cm and is very active. The mother may feel movements during the early part of the second trimester, and fetal activity may be visible through the abdominal wall by the middle of this

FIGURE 46.16 Formation of the zygote and early postfertilization events.

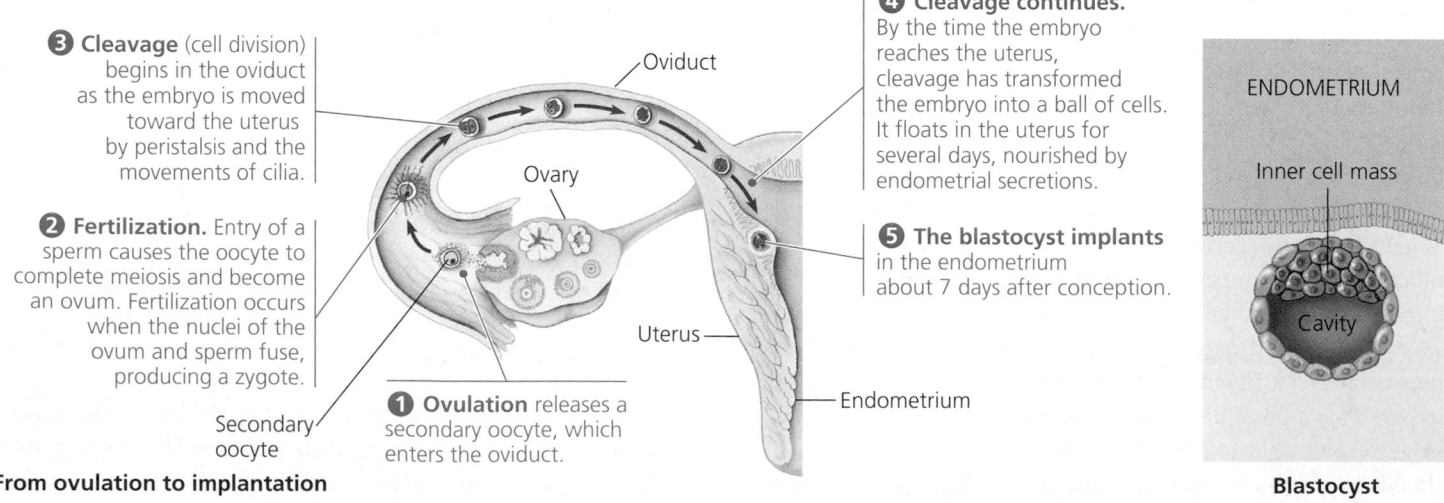

❸ **Cleavage** (cell division) begins in the oviduct as the embryo is moved toward the uterus by peristalsis and the movements of cilia.

❷ **Fertilization.** Entry of a sperm causes the oocyte to complete meiosis and become an ovum. Fertilization occurs when the nuclei of the ovum and sperm fuse, producing a zygote.

Secondary oocyte

❶ **Ovulation** releases a secondary oocyte, which enters the oviduct.

Oviduct

Ovary

Uterus

Endometrium

❹ **Cleavage continues.** By the time the embryo reaches the uterus, cleavage has transformed the embryo into a ball of cells. It floats in the uterus for several days, nourished by endometrial secretions.

❺ **The blastocyst implants** in the endometrium about 7 days after conception.

ENDOMETRIUM

Inner cell mass

Cavity

From ovulation to implantation

Blastocyst

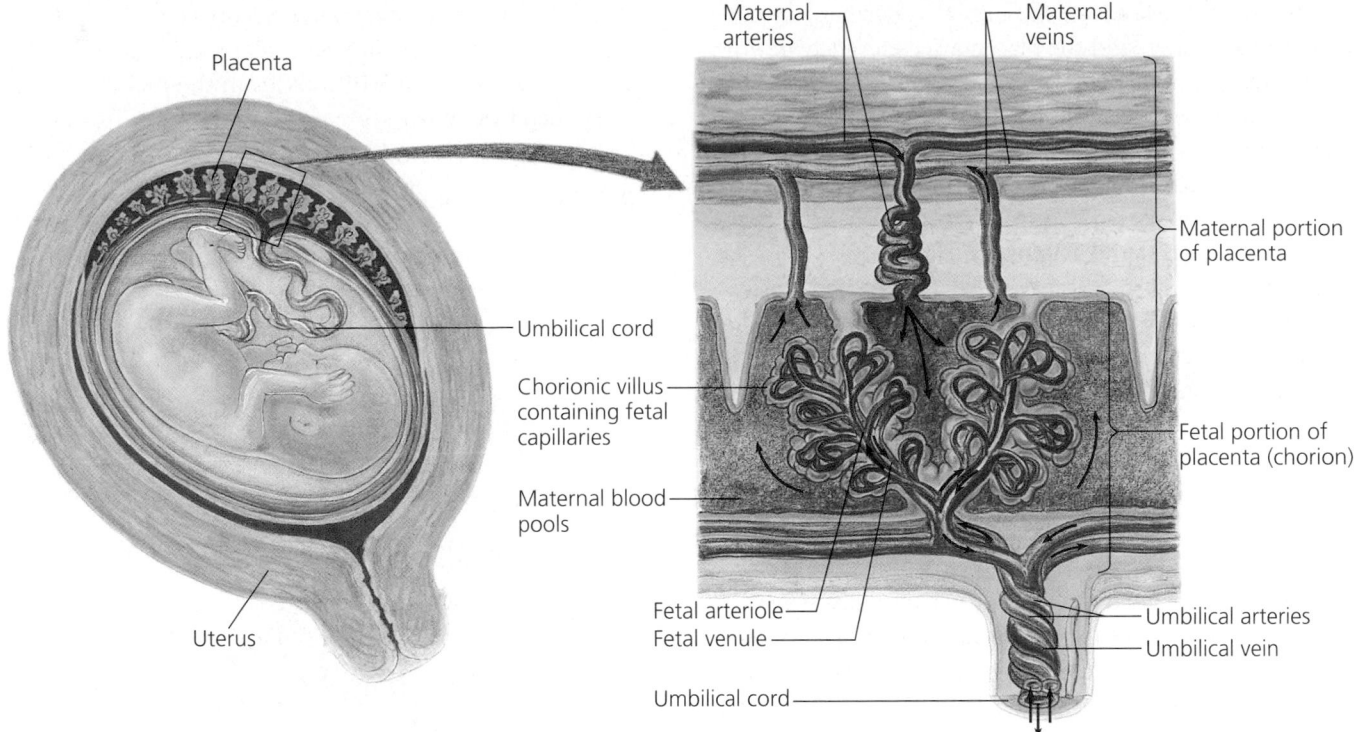

Placenta

Umbilical cord

Uterus

Maternal arteries

Maternal veins

Maternal portion of placenta

Chorionic villus containing fetal capillaries

Maternal blood pools

Fetal portion of placenta (chorion)

Fetal arteriole
Fetal venule

Umbilical arteries
Umbilical vein

Umbilical cord

FIGURE 46.17 Placental circulation. From the fourth week of development until birth, the placenta, a combination of maternal and embryonic tissues, transports nutrients, respiratory gases, and wastes between the embryo or fetus and the mother. Maternal blood enters the placenta in arteries, flows through blood pools in the endometrium, and leaves via veins. Embryonic or fetal blood, which remains in vessels, enters the placenta through arteries and passes through capillaries in fingerlike chorionic villi, where oxygen and nutrients are acquired. As indicated in the drawing, the fetal (or embryonic) capillaries and villi project into the maternal portion of the placenta. Fetal blood leaves the placenta through veins leading back to the fetus. Materials are exchanged by diffusion, active transport, and selective absorption between the fetal capillary bed and the maternal blood pools.

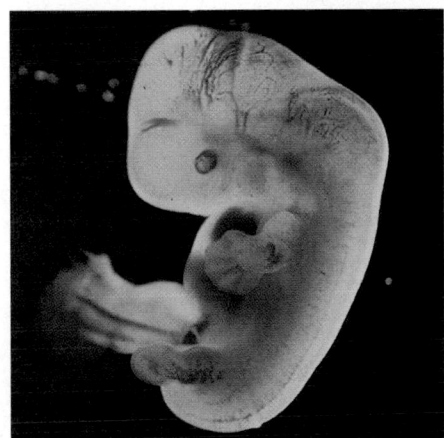

(a) 5 weeks. Limb buds, eyes, the heart, the liver, and rudiments of all other organs have started to develop in the embryo, which is only about 1 cm long.

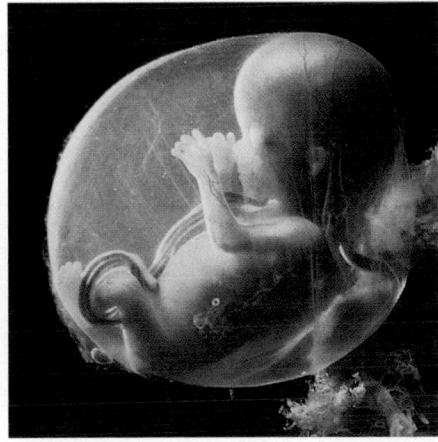

(b) 14 weeks. Growth and development of the offspring, now called a fetus, continue during the second trimester. This fetus is about 6 cm long.

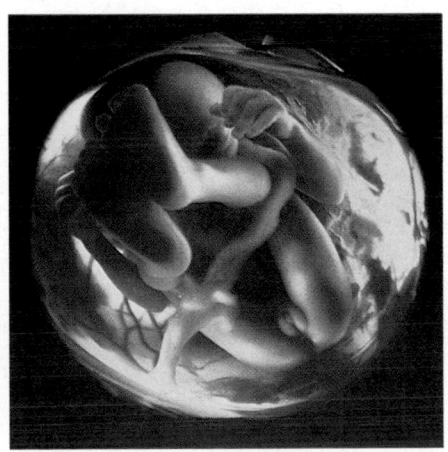

(c) 20 weeks. By the end of the second trimester (at 24 weeks), the fetus grows to about 30 cm in length.

FIGURE 46.18 Human fetal development.

time period. Hormone levels stabilize as HCG declines, the corpus luteum deteriorates, and the placenta secretes its own progesterone, which maintains the pregnancy. During the second trimester, the uterus will grow enough for the pregnancy to become obvious.

The third and final trimester is one of rapid growth of the fetus to about 3–3.5 kg in weight and 50 cm in length. Fetal activity may decrease as the fetus fills the available space within the embryonic membranes. As the fetus grows and the uterus expands around it, the mother's abdominal organs become compressed and displaced, leading to frequent urination, digestive blockages, and strain in the back muscles. A complex interplay of hormones (estrogens and oxytocin) and local regulators (prostaglandins) induces and regulates labor (FIGURE 46.19). Estrogens, which reach their highest level in the mother's blood during the last weeks of pregnancy, trigger the formation of oxytocin receptors on the uterus. Oxytocin, produced by the fetus and the mother's posterior pituitary, stimulates powerful contractions by the smooth muscles of the uterus. Oxytocin also stimulates the placenta to secrete prostaglandins, which enhance the contractions. In turn, the physical and emotional stresses associated with the contractions stimulate the release of more oxytocin and prostaglandins, a positive feedback system that underlies the three stages of labor.

Birth, or **parturition,** occurs through a series of strong, rhythmic uterine contractions, commonly known as **labor** (FIGURE 46.20). The first stage is the opening up and thinning of the cervix, ending with complete dilation. The second stage is expulsion, or delivery, of the baby. Continuous strong contractions force the fetus down and out of the uterus and vagina. The umbilical cord is cut and clamped at this time. The final stage of labor is delivery of the placenta, which normally follows the baby.

Lactation is an aspect of postnatal care unique to mammals. After birth, decreasing levels of progesterone free the anterior pituitary from negative feedback and allow prolactin secretion. Prolactin stimulates milk production after a delay of 2 or 3 days. The release of milk from the mammary glands is controlled by oxytocin (see p. 962 and FIGURE 45.6).

Reproductive Immunology

Pregnancy is an immunological enigma. Half of the embryo's genes are inherited from the father, and thus many of the chemical markers present on the surface of

THE PROCESS OF SCIENCE

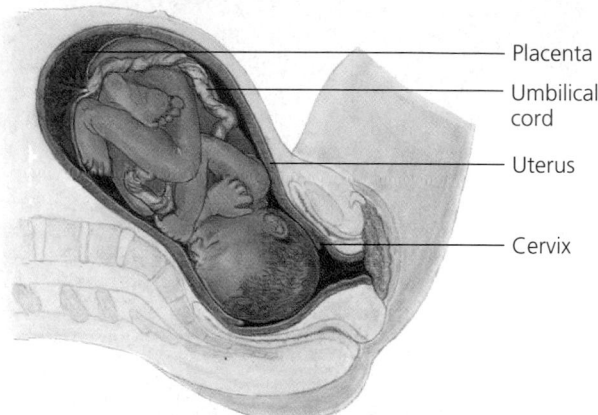

❶ Dilation of the cervix

— Placenta
— Umbilical cord
— Uterus
— Cervix

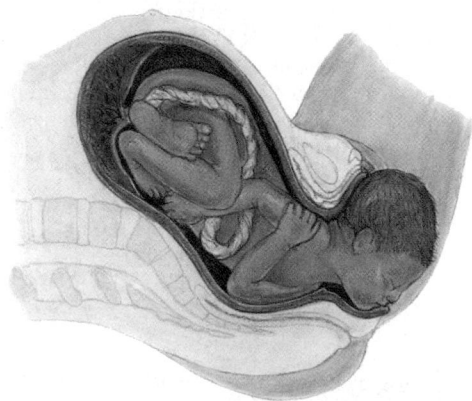

❷ Expulsion: delivery of the infant

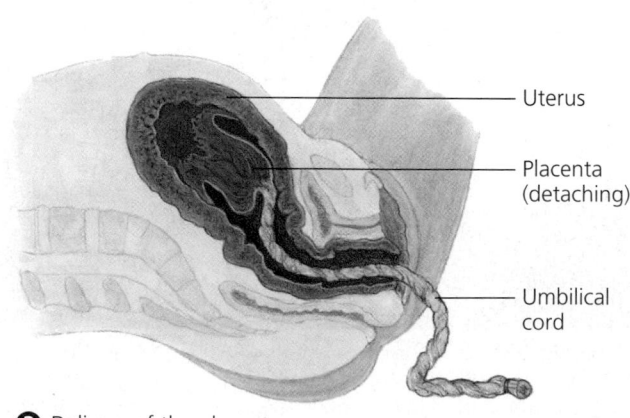

❸ Delivery of the placenta

— Uterus
— Placenta (detaching)
— Umbilical cord

FIGURE 46.20 The three stages of labor.

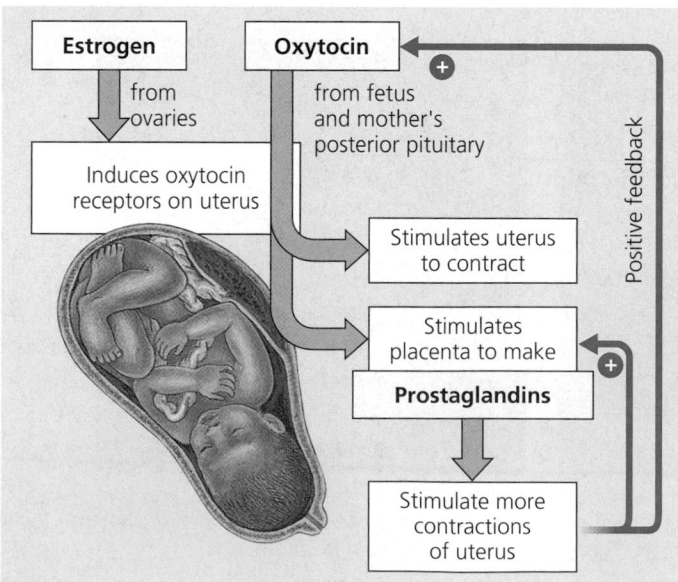

Estrogen → from ovaries → Induces oxytocin receptors on uterus

Oxytocin → from fetus and mother's posterior pituitary → Stimulates uterus to contract

Stimulates placenta to make **Prostaglandins** → Stimulate more contractions of uterus

Positive feedback ⊕

FIGURE 46.19 Hormonal induction of labor.

the embryo will be foreign to the mother. Why, then, does the mother not reject the embryo as a foreign body as she would repel a tissue or organ graft bearing antigens from another person? Reproductive immunologists are working to solve this puzzle.

A major key to the puzzle is a tissue called the *trophoblast* (see FIGURE 47.15). Arising along with the embryo itself from the blastocyst, the trophoblast brings about implantation by growing into the endometrium and later develops into the fetal part of the placenta (see FIGURES 46.16 and 46.17). To some extent, the trophoblast is a barrier separating the embryo from maternal tissue, but, like the embryo, it carries paternal antigens on the surfaces of its cells. So how can it protect the embryo from rejection?

Several lines of research suggest that the trophoblast prevents an immune response against the embryo by somehow interfering with the mother's T lymphocytes, important players in the immune system. For example, there is evidence that the trophoblast produces a chemical signal that induces the development of "suppressor" T cells in the uterus. It is proposed that the suppressor T cells prevent other T cells from attacking the foreign tissue, perhaps by blocking the action of interleukin-2, a cytokine required for a normal immune response (see FIGURES 43.11 and 43.13).

An interesting part of the suppressor-cell hypothesis is the suggestion that the suppressor cells act, paradoxically, only after other white blood cells have identified the trophoblast as foreign tissue and have taken the first steps of the immune response. If this immunological alarm is not intense enough—that is, if the father's cell surface antigens are too similar to the mother's—then no suppressor cells arise, and the embryo *is* attacked by the immune system. Similarity between maternal and paternal antigens may thus account for some cases of women who have multiple miscarriages. In support of this idea, some frequent miscarriers have been successfully treated before pregnancy with sensitizing injections of their mate's antigens.

A very different hypothesis is that the trophoblast secretes an enzyme that rapidly breaks down local supplies of tryptophan, an amino acid necessary for T cell survival and function. At least in mice, this enzyme seems to be essential for maintaining pregnancy. And more recent work on mice has led to yet another hypothesis, which involves a *nonspecific* part of the defense system. A protein found on mouse trophoblast cells protects the embryo's cells from attack by complement (see FIGURE 43.17). Scientists are now searching for comparable proteins made by the human trophoblast.

Contraception

Contraception, the deliberate prevention of pregnancy, can be achieved in a number of ways. Some contraceptive methods prevent the release of mature eggs (secondary oocytes) and sperm

from gonads, others prevent fertilization by keeping sperm and egg apart, and still others prevent implantation of an embryo or abort the embryo (FIGURE 46.21, p. 994). The following brief introduction to the biology of these methods makes no pretense of being a contraception manual. For more complete information, you should consult a physician or health center personnel.

Fertilization can be prevented by abstinence from sexual intercourse or by any of several barriers that keep live sperm from contacting the egg. Temporary abstinence, often called the **rhythm method** of birth control or **natural family planning,** depends on refraining from intercourse when conception is most likely. Because the egg can survive in the oviduct for 24 to 48 hours and sperm for up to 72 hours, a couple practicing temporary abstinence should not engage in intercourse during the few days before and after ovulation. The most effective methods for timing ovulation combine several indicators, including changes in cervical mucus and body temperature during the menstrual cycle. Thus, natural family planning requires that the couple be knowledgeable about these physiological signs. A pregnancy rate of 10% to 20% is typically reported for couples practicing natural family planning. (*Pregnancy rate* is the number of women who become pregnant during a year out of every 100 women using a particular family-planning method, expressed as a percentage.) Some couples use the natural family planning method to *increase* the probability of conception.

The several **barrier methods** of contraception that block the sperm from meeting the egg have pregnancy rates of less than 10%. The **condom,** used by the male, is a thin, latex rubber or natural membrane sheath that fits over the penis to collect the semen. The barrier device most commonly used by females is the **diaphragm,** a dome-shaped rubber cap fitted into the upper portion of the vagina before intercourse. Both of these devices are more effective when used in conjunction with a spermicidal (sperm-killing) foam or jelly. More recently introduced barriers include the cervical cap, which fits tightly around the opening of the cervix and is held in place for a prolonged period by suction, and the vaginal pouch, or female condom.

Intrauterine devices (IUDs), which are small plastic or metal devices that fit into the uterine cavity, prevent implantation of the blastocyst in the uterus. IUDs have a low pregnancy rate but cause harmful effects in a small percentage of women. Problems of persistent vaginal bleeding, uterine infection, perforation of the uterus, tubal pregnancy (implantation of the embryo in the oviduct), and spontaneous expulsion of the devices have been reported and have led to lawsuits and the removal of IUDs from the market. Newer IUD products that prevent implantation by delivering synthetic progesterone locally to the endometrium are now available.

As a method of preventing fertilization, coitus interruptus, or withdrawal (removal of the penis from the vagina before ejaculation), is unreliable. Sperm may be present in secretions that precede ejaculation, and a lapse in timing or willpower can result in late withdrawal.

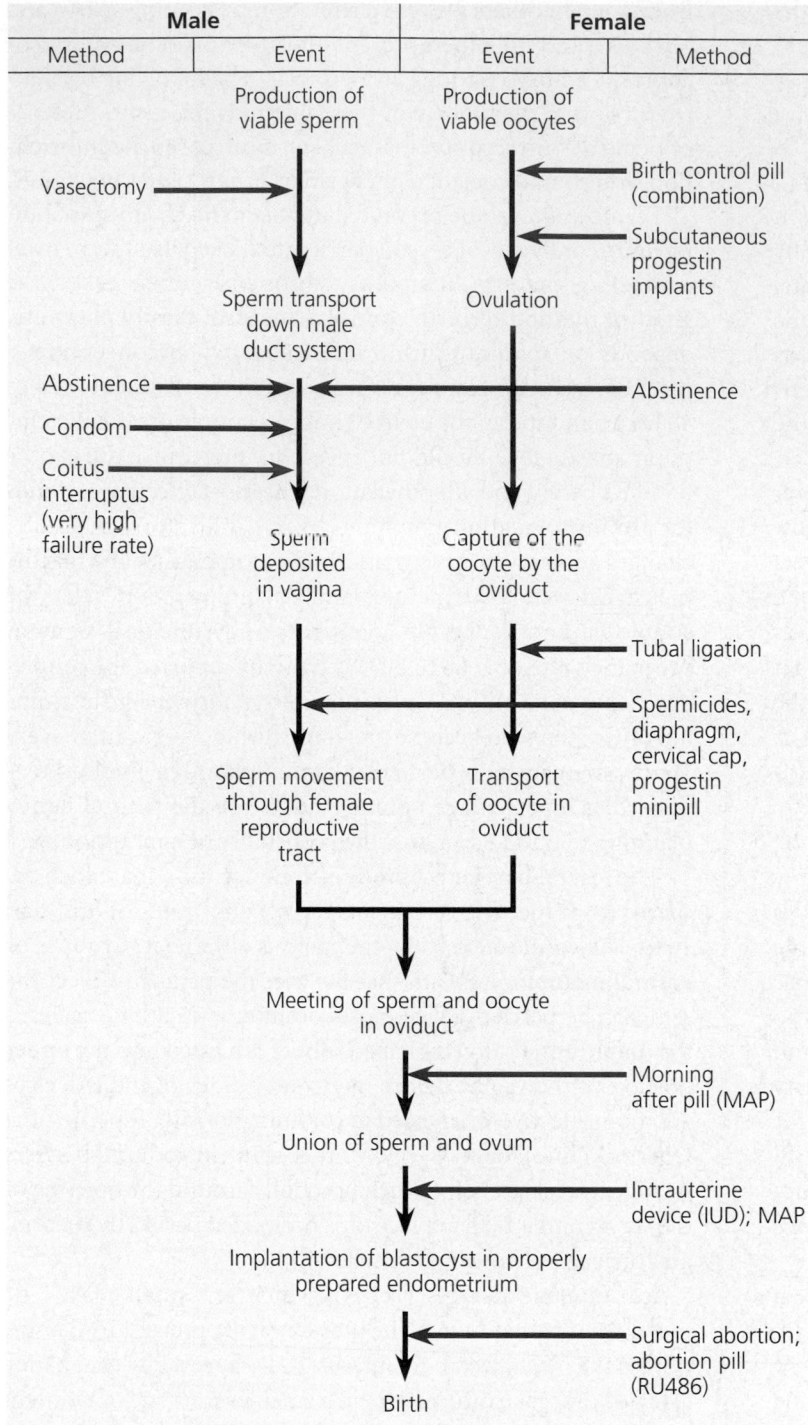

Male		Female	
Method	Event	Event	Method

Production of viable sperm

Production of viable oocytes

Vasectomy →

← Birth control pill (combination)

← Subcutaneous progestin implants

Sperm transport down male duct system

Ovulation

Abstinence →

← Abstinence

Condom →

Coitus interruptus (very high failure rate) →

Sperm deposited in vagina

Capture of the oocyte by the oviduct

← Tubal ligation

← Spermicides, diaphragm, cervical cap, progestin minipill

Sperm movement through female reproductive tract

Transport of oocyte in oviduct

Meeting of sperm and oocyte in oviduct

← Morning after pill (MAP)

Union of sperm and ovum

← Intrauterine device (IUD); MAP

Implantation of blastocyst in properly prepared endometrium

← Surgical abortion; abortion pill (RU486)

Birth

FIGURE 46.21 Mechanisms of some contraceptive methods. Red arrows indicate where these methods, devices, or products interfere with the flow of events from the production of sperm and egg (secondary oocyte) to the birth of a baby.

Except for complete abstinence from sexual intercourse, the methods that prevent the release of gametes are the most effective means of birth control. Chemical contraceptives—**birth control pills**—have pregnancy rates of less than 1%, and sterilization is nearly 100% effective. The most commonly used birth control pills are combinations of synthetic estrogens and a synthetic progestin (progesteronelike hormone).

These two hormones act by negative feedback to stop the release of GnRH by the hypothalamus, and of FSH (an estrogen effect) and LH (a progestin effect) by the pituitary. By blocking LH release, the progestin prevents ovulation. As a backup mechanism, the estrogen inhibits FSH secretion so that no follicles develop. Combination birth control pills can be prescribed in high doses as morning after pills (MAP). Taken within 3 days of unprotected intercourse, they prevent fertilization or implantation, with an effectiveness of about 75%.

A second type of birth control pill, called the minipill, contains only progestin. The minipill prevents fertilization mainly by altering a woman's cervical mucus so that it blocks sperm from entering the uterus. In 1990, the FDA approved a progestin-only capsule (Norplant) that is implanted under the skin. Steadily releasing a tiny amount of progestin into the blood, the implant produces effective birth control for about 5 years. A product called Depo-Provera is a synthetic progestin that is injected every 3 months.

Birth control pills have been the center of much debate, particularly because of long-term harmful effects of the estrogens in combination pills. No solid evidence exists for cancers caused by the pill, but cardiovascular problems are a major concern. Birth control pills have been implicated in abnormal blood clotting, atherosclerosis, and heart attacks. Smoking while using chemical contraception increases the risk of mortality tenfold or more. Although the pill places women at risk for these diseases, it eliminates the dangers of pregnancy; women on birth control pills have mortality rates about one-half those of pregnant women.

Sterilization is the permanent prevention of gamete release. **Tubal ligation** in women usually involves cauterizing or tying off (ligating) a section of the oviducts to prevent eggs from traveling into the uterus. **Vasectomy** in men is the cutting of each vas deferens to prevent sperm from entering the urethra. Both male and female sterilization are relatively safe and free from harmful effects. Both are also difficult to reverse, so the procedures should be considered permanent.

Abortion is the termination of a pregnancy in progress. Spontaneous abortion, or miscarriage, is very common; it occurs in as many as one-third of all pregnancies, often before the woman is even aware she is pregnant. In addition, each year about 1.5 million women in the United States choose abortions performed by physicians. A drug called mifepristone, or RU486, developed in France, enables a woman to terminate pregnancy nonsurgically within the first

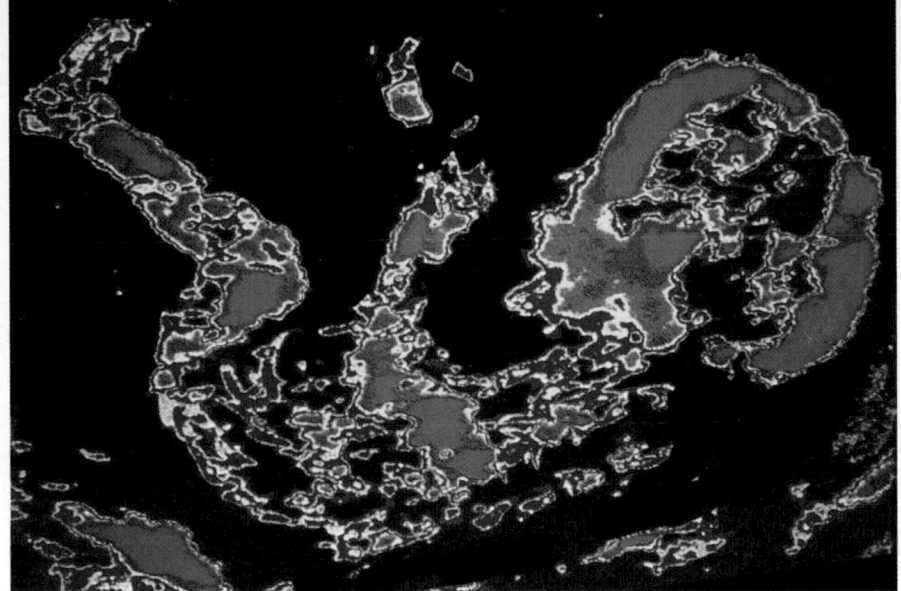

FIGURE 46.22 Ultrasound imaging. This color-enhanced image shows a fetus in the uterus at about 18 weeks. The image is produced on a computer screen when high-frequency sounds from an ultrasound scanner held against a pregnant woman's abdomen bounce off the fetus.

7 weeks. An analog of progesterone, RU486 blocks progesterone receptors in the uterus, thus preventing progesterone from maintaining pregnancy. It is taken with a small amount of prostaglandin to induce uterine contractions.

Of all contraceptives for sexually active individuals, latex condoms are the only ones that offer some protection against sexually transmitted diseases, including AIDS. This protection is, however, not absolute.

Modern technology offers solutions for some reproductive problems

Recent scientific and technological advances have made it possible to deal with problems of reproduction in striking ways. For example, it is now possible to diagnose many genetic diseases and congenital (present at birth) disorders while the fetus is in the uterus. Amniocentesis and chorionic villus sampling are invasive techniques in which amniotic fluid or fetal cells are obtained for genetic analysis (see FIGURE 14.17). Non-invasive procedures usually use high-frequency sound waves, or ultrasound imaging, to detect fetal condition (FIGURE 46.22). An alternate new technique relies on the fact that a few fetal blood cells leak across the placenta into the mother's bloodstream. A blood sample from the mother yields enough fetal cells that can be identified with specific antibodies (which bind to proteins on the surface of fetal cells) and then tested for genetic disorders.

Diagnosing genetic diseases of fetuses poses important ethical questions. To date, essentially all detectable disorders remain untreatable in the uterus, and many cannot be corrected even after birth. Parents may be faced with difficult decisions about whether to terminate a pregnancy or cope with a child who may have profound defects and a short life expectancy. These are complex issues that demand careful, informed thought and competent counseling.

In the past 25 years, several techniques have been developed to assist childless couples who want to have offspring. For cases of male infertility, sperm from anonymous donors are widely available from sperm banks. Sperm are deposited in the vagina or cervix when it is possible for the woman to conceive. Another procedure, called *in vitro* **fertilization,** was developed for women whose oviducts are blocked. Ova are surgically removed following hormonal stimulation of the follicles. The ova are then fertilized in culture dishes in a laboratory. After about 2½ days, when the embryo has reached the eight-cell stage, it is placed in the uterus and allowed to implant. *In vitro* fertilization is costly but has a success rate similar to the pregnancy rate resulting from insemination by sexual intercourse. Embryos can be frozen for later use if the first attempt is unsuccessful. A couple may choose to have oocytes obtained from another woman fertilized with the male partner's sperm, although more serious ethical issues must be resolved when donors are used. Thousands of children have thus far been conceived by *in vitro* fertilization, and there is no evidence of any abnormalities associated with this procedure.

One area of reproductive research that has traditionally received little attention is male contraception. Male chemical contraceptives have proved elusive. Testosterone will block the release of pituitary gonadotropins, but testosterone itself stimulates spermatogenesis. Estrogens are effective, but they inhibit sex drive and can be feminizing. The best prospects so far are for antagonists of GnRH, which are potent inhibitors of spermatogenesis. Several experimental drugs and birth control techniques for both men and women are currently being tested.

■   ■   ■

In this chapter we have considered the structural and physiological bases of animal reproduction. The next chapter focuses on the mechanics of development that transform a zygote into an animal form and on other topics in animal development.

CHAPTER 46 REVIEW

Go to the Campbell Biology website (www.campbellbiology.com) to explore an interactive version of the Chapter Review.

Summary of Key Concepts

OVERVIEW OF ANIMAL REPRODUCTION

■ **Both asexual and sexual reproduction occur in the animal kingdom** (p. 975) Asexual reproduction produces offspring whose genes all come from a single parent. Sexual reproduction requires the fusion of male and female gametes to form a diploid zygote.

■ **Diverse mechanisms of asexual reproduction enable animals to produce identical offspring rapidly** (p. 976, FIGURE 46.1) Fission, budding, and fragmentation with regeneration are mechanisms of asexual reproduction in various invertebrates.

■ **Reproductive cycles and patterns vary extensively among animals** (pp. 976–978, FIGURE 46.2) Animals may reproduce exclusively sexually or asexually, or they may alternate between the two, depending on environmental conditions. Variations on these two modes are made possible through parthenogenesis, hermaphroditism, and sequential hermaphroditism. Reproductive cycles are controlled by hormones and environmental cues, such as changes in temperature, rainfall, day length, and seasonal lunar cycles.

MECHANISMS OF SEXUAL REPRODUCTION

■ In external fertilization, eggs shed by the female are fertilized by sperm in the external environment. In internal fertilization, egg and sperm unite within the female's body.

■ **Internal and external fertilization both depend on mechanisms ensuring that mature sperm encounter fertile eggs of the same species** (p. 978, FIGURE 46.4) External and internal fertilization require critical timing, often mediated by environmental cues, pheromones, and/or courtship behavior. Internal fertilization requires important behavioral interactions between male and female animals, as well as compatible copulatory organs.

■ **Species with internal fertilization usually produce fewer zygotes but provide more parental protection than species with external fertilization** (pp. 978–979, FIGURE 46.5) Greater protection of embryos and parental care of the young usually follow the production of relatively few offspring by internal fertilization.

■ **Complex reproductive systems have evolved in many animal phyla** (pp. 979–980, FIGURES 46.6, 46.7) Reproductive systems range from the production of gametes by undifferentiated cells in the body cavity to complex assemblages of male and female gonads with accessory tubes and glands that carry and protect gametes and developing embryos. The reproductive systems of insects and flatworms are among the most complex in the animal kingdom.

MAMMALIAN REPRODUCTION

■ **Human reproduction involves intricate anatomy and complex behavior** (pp. 980–984, FIGURES 46.8, 46.9) External reproductive structures of the human male are the scrotum and penis. The male gonads, or testes, reside in the cool environment of the scrotum. They possess endocrine Leydig cells surrounding sperm-forming seminiferous tubules that successively lead into the epididymis, vas deferens, ejaculatory duct, and urethra, which exits at the tip of the penis. Externally,

the human female has a vestibule containing separate openings of the vagina and urethra, the labia minora bordering the vestibule, the labia majora, and the clitoris. Internally, the vagina is connected to the uterus, which connects to two oviducts. Two ovaries (female gonads) are stocked with follicles containing diploid primary oocytes formed before the woman's birth. Beginning at puberty, one or more follicles mature during each menstrual cycle. The oocyte contained in a maturing follicle completes the first meiotic division, and a secondary oocyte, which is haploid, is expelled from the surface of the ovary during ovulation. After ovulation, the remnant of the follicle forms a corpus luteum that secretes progesterone and estrogen for a variable duration, depending on whether or not pregnancy occurs. Although separate from the reproductive system, the mammary glands, or breasts, evolved in association with parental care. Both males and females experience the erection of certain body tissues due to vasocongestion and myotonia, which culminate in orgasm.

Web/CD Activity 46A: *Reproductive System of the Human Male*
Web/CD Activity 46B: *Reproductive System of the Human Female*
Web/CD Case Study in The Process of Science: *What Might Obstruct the Male Urethra?*

■ **Spermatogenesis and oogenesis both involve meiosis but differ in three significant ways** (pp. 984–986, FIGURES 46.11, 46.13) Cytokinesis is unequal in oogenesis, producing one large ovum. Production of sperm is continuous; in humans, the number of future egg cells is set at birth. Spermatogenesis is an uninterrupted sequence, but there are long delays in oogenesis.

■ **A complex interplay of hormones regulates reproduction** (pp. 986–989, FIGURES 46.14, 46.15) Androgens from the testes cause the development of primary and secondary sex characteristics in the male. Androgen secretion and sperm production are both controlled by hypothalamic and pituitary hormones. Female hormones are secreted in a rhythmic fashion reflected in the menstrual or estrous cycle. In both types of female cycles, the endometrium thickens in preparation for possible implantation. The menstrual cycle, however, includes endometrial bleeding and lacks the clear-cut period of sexual receptivity limited to the heat period of the estrous cycle. The human menstrual cycle consists of the menstrual flow phase, proliferative phase, and secretory phase. The ovarian cycle includes the follicular and luteal phases. The female reproductive cycle is orchestrated by cyclic secretion of GnRH from the hypothalamus and of FSH and LH from the anterior pituitary. The developing follicle produces estrogens, and the corpus luteum secretes progesterone and estrogens. Positive and negative feedback produce the changing levels of these five hormones, which coordinate the menstrual and ovarian cycles.

■ **Embryonic and fetal development occur during pregnancy in humans and other eutherian (placental) mammals** (pp. 989–995, FIGURES 46.16–46.21) Human pregnancy can be divided into three trimesters. Organogenesis is completed by 8 weeks. Birth, or parturition, results from strong, rhythmic uterine contractions associated with labor. Positive feedback involving the hormones estrogen and oxytocin, and prostaglandins, regulate labor. A pregnant woman's acceptance of her "foreign" fetus may be due to the suppression of the immune response in her uterus. Contraceptive methods include preventing the release of mature gametes from the gonads, preventing gamete union in the female tract, and preventing implantation of the zygote.

■ **Modern technology offers solutions for some reproductive problems** (p. 995, FIGURE 46.22) Current methods of detecting fetal condition include ultrasound imaging, amniocentesis, and chorionic villus sampling. Current technology also allows in vitro fertilization.

1. Which of the following characterizes parthenogenesis?
   a. An individual may change its sex during its lifetime.
   b. Specialized groups of cells may be released and grow into new individuals.
   c. An organism is first a male and then a female.
   d. An egg develops without being fertilized.
   e. Both members of a mating pair have male and female reproductive organs.

2. Which of the following structures is *incorrectly* paired with its function?
   a. gonads—gamete-producing organs
   b. spermatheca—sperm-transferring organ found in male insects
   c. cloaca—common opening for reproductive, excretory, and digestive systems
   d. baculum—bone that stiffens the penis, found in some mammals
   e. endometrium—lining of the uterus; forms the maternal part of the placenta

3. Which of the following male and female structures are *least* alike in function?
   a. seminiferous tubules—vagina
   b. Leydig cells of testes—follicle cells of ovaries
   c. testes—ovaries
   d. spermatogonia—oogonia
   e. vas deferens—oviduct

4. A difference between estrous and menstrual cycles is that
   a. nonmammalian vertebrates have estrous cycles, whereas mammals have menstrual cycles.
   b. the endometrial lining is shed in menstrual cycles but reabsorbed in estrous cycles.
   c. estrous cycles occur more frequently than menstrual cycles.
   d. estrous cycles are not controlled by hormones.
   e. ovulation occurs before the endometrium thickens in estrous cycles.

5. Peaks of LH and FSH production occur during
   a. the flow phase of the menstrual cycle.
   b. the beginning of the follicular phase of the ovarian cycle.
   c. the period surrounding ovulation.
   d. the end of the luteal phase of the ovarian cycle.
   e. the secretory phase of the menstrual cycle.

6. In protandrous hermaphroditism
   a. some individuals may change from male to female.
   b. individuals fertilize themselves.
   c. males rather than females release pheromones.
   d. diploid ova are produced.
   e. the adult gonads are undifferentiated.

7. During human gestation, organogenesis occurs
   a. in the first trimester.
   b. in the second trimester.
   c. in the third trimester.
   d. while the embryo is in the oviduct.
   e. during the blastocyst stage.

8. Which pharmacological strategy is most likely to result in a successful male contraceptive?
   a. block FSH receptors on spermatogonia
   b. maintain high circulating concentrations of androgen
   c. block testosterone receptors on Leydig cells
   d. block androgen receptors within the hypothalamus
   e. maintain high circulating concentrations of FSH

9. Fertilization of human eggs most often takes place in the
   a. vagina.
   b. ovary.
   c. uterus.
   d. oviduct (fallopian tube).
   e. vas deferens.

10. In male mammals, the excretory and reproductive systems share the
   a. testes.
   b. urethra.
   c. ureter.
   d. vas deferens.
   e. prostate.

11. What is the most important difference between the outcome of sexual reproduction and that of asexual reproduction?

12. Arrange the following human organs in the correct sequence for the travel of sperm: epididymis, testis, urethra, vas deferens.

13. What hormonal change triggers the onset of menstruation?

14. _____ is to males as tubal ligation is to _____.

15. How does the timing of mating in humans contrast with that in most other mammals?

Evolution Connection

In animals, hermaphroditism is often found in sessile species (that is, animals that are fixed to a surface, like plants). Mobile species, such as arthropod and vertebrate species, are less often hermaphroditic. Why do you suppose this is the case?

The Process of Science

While parental care is widespread in the animal kingdom, this trait is by no means universal. Imagine you are studying the evolution of parental care in an animal group of interest to you and have mapped the distribution of care behavior on a phylogenetic tree as shown below (see Chapter 25 to review phylogenetic trees). What would be the simplest interpretation of how this behavior has evolved in these animals? If the outgroup exhibited parental care, how would your interpretation change?

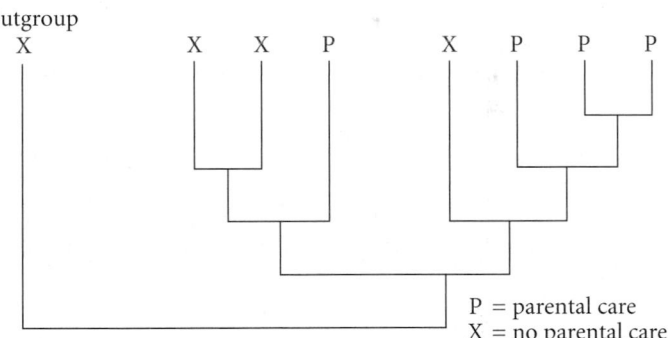

Outgroup

P = parental care
X = no parental care

Analyze evidence of urethra obstruction in the Case Study in the Process of Science, available on the website and CD-ROM.

Science, Technology, and Society

New techniques for sorting sperm, combined with *in vitro* fertilization, make it possible for a couple to choose their baby's sex. Would you want to do this if you had the chance? Why or why not? What potential problems can you foresee if this procedure becomes widely available?

Answers: 1. d; 2. b; 3. a; 4. b; 5. c; 6. a; 7. a; 8. a; 9. d; 10. b; 11. The offspring of sexual reproduction are genetically diverse. 12. Testis, epididymis, vas deferens, urethra 13. A drop in levels of estrogen and progesterone 14. Vasectomy; females 15. Human females are potentially receptive to mating throughout the year, in contrast to the seasonal mating of most other mammals.

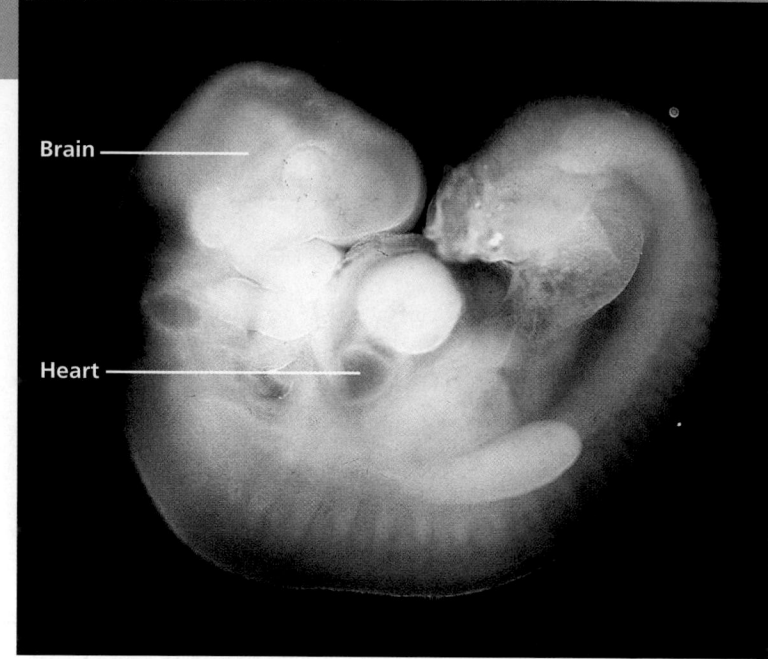

Brain

Heart

CHAPTER 47

ANIMAL
DEVELOPMENT

THE STAGES OF EARLY EMBRYONIC DEVELOPMENT

- From egg to organism, an animal's form develops gradually: *the concept of epigenesis*
- Fertilization activates the egg and brings together the nuclei of sperm and egg
- Cleavage partitions the zygote into many smaller cells
- Gastrulation rearranges the blastula to form a three-layered embryo with a primitive gut
- In organogenesis, the organs of the animal body form from the three embryonic germ layers
- Amniote embryos develop in a fluid-filled sac within a shell or uterus

THE CELLULAR AND MOLECULAR BASIS OF MORPHOGENESIS AND DIFFERENTIATION IN ANIMALS

- Morphogenesis in animals involves specific changes in cell shape, position, and adhesion
- The developmental fate of cells depends on cytoplasmic determinants and cell-cell induction: *a review*
- Fate mapping can reveal cell genealogies in chordate embryos
- The eggs of most vertebrates have cytoplasmic determinants that help establish the body axes and differences among cells of the early embryo
- Inductive signals drive differentiation and pattern formation in vertebrates

I t is difficult to imagine *that each of us began life as a single cell about the size of the period at the end of this sentence. Less than a month after conception, our brains were taking form and our developing hearts had already begun to pulsate (see photograph). It took a total of only about nine months—the length of a school year—to be transfigured from zygote to newborn human, built of billions of differentiated cells organized into specialized tissues and organs.*

By combining molecular genetics with classical approaches to embryology, developmental biologists are now beginning to answer many of the questions about how a single fertilized egg cell gives rise to a specific animal. Chapter 21 described some of the recent research, introducing you to a number of basic genetic and cellular mechanisms involved in development. The animals discussed there were mainly invertebrates. In this chapter we concentrate mainly on the embryonic development of animals with backbones.

Molecular biologist Sidney Brenner has wryly commented that developmental biology is about "how to make a mouse," and indeed, mice are favorite experimental models of researchers studying vertebrate development, in particular the development of mammals. Frogs and chickens are also important model organisms for understanding vertebrate development. From the study of these animals and of certain invertebrates (including the ones you met in Chapter 21), scientists have worked out both the stages of embryonic development and many of the cellular and molecular events that underlie them.

THE STAGES OF EARLY EMBRYONIC DEVELOPMENT

From egg to organism, an animal's form develops gradually: *the concept of epigenesis*

The question of how an egg becomes an animal has been asked for centuries. As recently as the 18th century, the prevailing view was that the egg or sperm contains an embryo that is a preformed, miniature infant (FIGURE 47.1). Development was thought to be simply the enlargement of the embryo. This idea of **preformation** came to include the notion that the embryo must contain all its descendants: a series of successively smaller embryos within embryos, like Russian nesting dolls. One theologian proposed that Eve, in the Garden of Eden, stored all future humanity within her.

The competing theory of embryonic development was **epigenesis,** originally proposed 2,000 years earlier by Aristotle. According to this idea, the form of an animal emerges gradually from a relatively formless egg. As microscopy improved during the 19th century, biologists could see that embryos took shape in a series of progressive steps, and epigenesis displaced preformation as the favored explanation among embryologists.

Modern biology, of course, has completely discarded the idea of a tiny person living in an egg or sperm cell. But when interpreted in broader terms, the concept of preformation may have some merit. Although an embryo's form emerges gradually as it develops from a fertilized egg, *something* is preformed in the zygote. As discussed in Chapter 21, an organism's development is largely determined by the genome of the zygote and the organization of the cytoplasm of the egg cell. Messenger RNA, proteins, and other substances made by the mother are heterogeneously distributed in the unfertilized egg, and these substances have a profound effect on the development of the future embryo in most animal species (mammals may be an exception, as we will discuss later). After fertilization produces a zygote, cell division partitions the cytoplasm in such a way that nuclei of different embryonic cells are exposed to different cytoplasmic environments. This sets the stage for the expression of different genes in different cells. As cell division continues and the embryo develops, inherited traits emerge by mechanisms that selectively control gene expression, leading to the differentiation (specialization) of cells. The timely communication of instructions, telling cells precisely what to do when, is essential. This information transfer occurs by cell signaling among different embryonic cells. Along with cell division and differentiation, development involves morphogenesis, the process by which an animal takes shape. Thus, the overall process of development is one of epigenesis.

In the first half of the chapter we will survey the early stages of embryonic development, when the body plan of an animal emerges from the fertilized egg. In the second half of the chapter we will look at the cellular and molecular mechanisms that play major roles in the developmental process. We begin with the fertilization of an egg cell by a sperm.

Fertilization activates the egg and brings together the nuclei of sperm and egg

The gametes, sperm and egg that unite during fertilization, are both highly specialized cell types produced by a complex series of developmental events in the testes and ovaries of the parents (see Chapter 46). The main function of fertilization is to combine haploid sets of chromosomes from two individuals into a single diploid cell, the zygote. Another key function is activation of the egg: Contact of the sperm with the egg's surface initiates metabolic reactions within the egg that trigger the onset of embryonic development.

Fertilization has been studied most extensively by combining the gametes of sea urchins in the laboratory. Although the details of fertilization vary with different animal groups, sea urchins (phylum Echinodermata) provide a good general model for the important events of fertilization. While sea urchins are not vertebrates or even chordates, they are deuterostomes (see Chapter 32), and their early development is similar to that of vertebrates.

The Acrosomal Reaction

The eggs of sea urchins are fertilized externally after the animals release their gametes into the surrounding seawater. When a sperm cell is exposed to molecules from the slowly dissolving jelly coat that surrounds an egg, a vesicle at the tip of the sperm called the acrosome discharges its contents by exocytosis (FIGURE 47.2, p. 1000). This **acrosomal reaction** releases hydrolytic enzymes that enable an elongating structure called the *acrosomal process* to penetrate the jelly coat of the egg. The tip of the acrosomal process is coated with a protein that adheres to specific receptor molecules located under the jelly coat, on the vitelline layer just external to the plasma membrane of the egg. In sea urchins and many other animals,

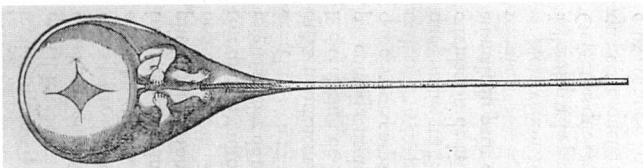

FIGURE 47.1 A "homunculus" inside the head of a human sperm. According to one version of the preformation idea, a sperm contains a preformed, miniature infant, which simply grows in size during embryonic development. This engraving was made in 1694.

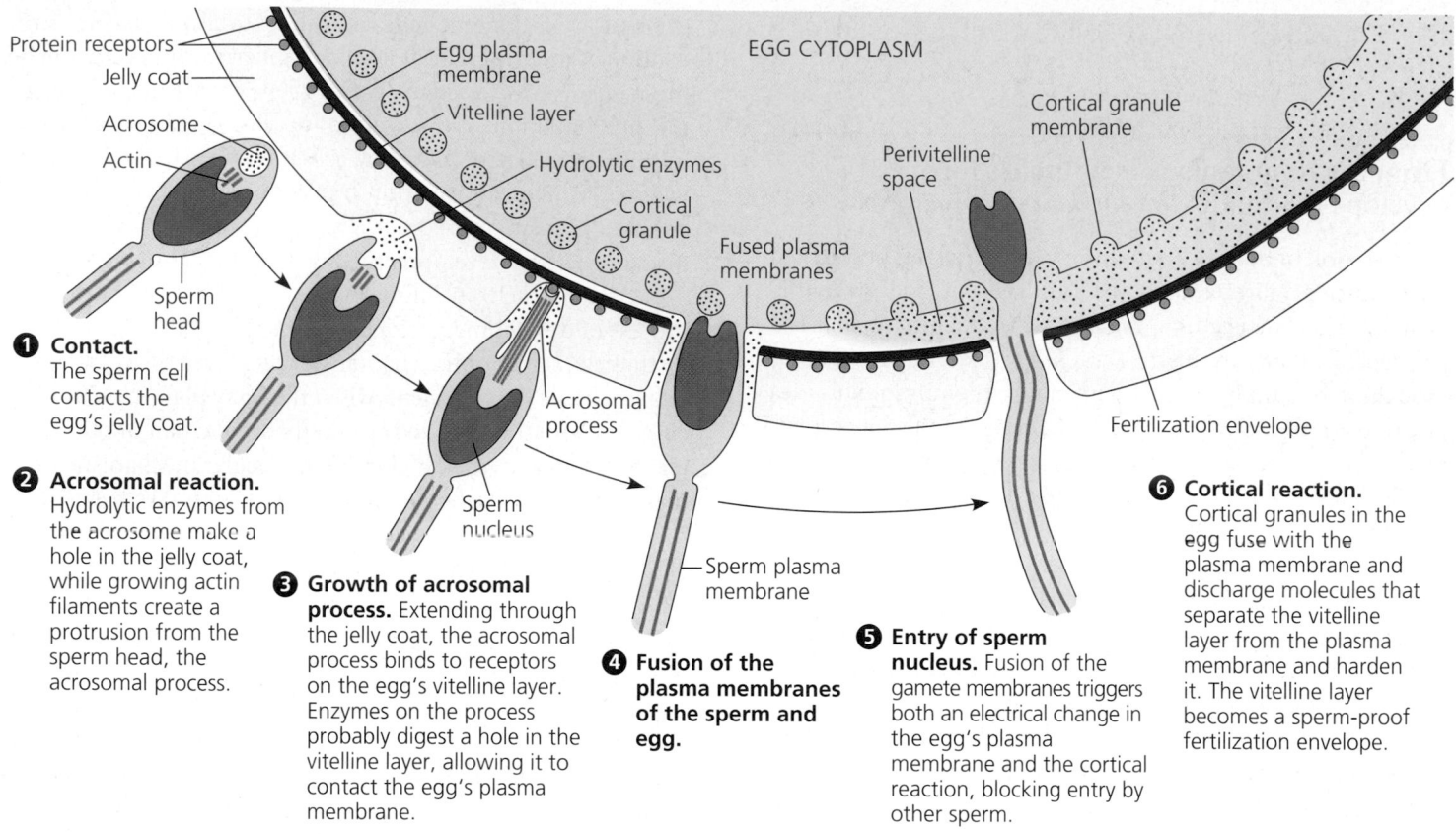

1 Contact.
The sperm cell contacts the egg's jelly coat.

2 Acrosomal reaction.
Hydrolytic enzymes from the acrosome make a hole in the jelly coat, while growing actin filaments create a protrusion from the sperm head, the acrosomal process.

3 Growth of acrosomal process. Extending through the jelly coat, the acrosomal process binds to receptors on the egg's vitelline layer. Enzymes on the process probably digest a hole in the vitelline layer, allowing it to contact the egg's plasma membrane.

4 Fusion of the plasma membranes of the sperm and egg.

5 Entry of sperm nucleus. Fusion of the gamete membranes triggers both an electrical change in the egg's plasma membrane and the cortical reaction, blocking entry by other sperm.

6 Cortical reaction. Cortical granules in the egg fuse with the plasma membrane and discharge molecules that separate the vitelline layer from the plasma membrane and harden it. The vitelline layer becomes a sperm-proof fertilization envelope.

FIGURE 47.2 The acrosomal and cortical reactions during sea urchin fertilization. The events following contact of a single sperm and egg ensure that only one sperm nucleus enters the cytoplasm of the egg.

"lock-and-key" recognition of molecules ensures that eggs will be fertilized only by sperm of the same species. This specificity is especially important when fertilization occurs externally in water, where gametes of other species are likely to be present.

The acrosomal reaction leads to the fusion of sperm and egg plasma membranes and the entry of a single sperm nucleus into the cytoplasm of the egg. Fusion of the membranes causes ion channels to open in the egg cell's plasma membrane, allowing sodium ions to flow into the egg cell and change the membrane potential, the voltage across the membrane (see Chapter 8, pp. 149–150). This membrane depolarization, as such a change is called, is common among animal species. Occurring within about 1 to 3 seconds after a sperm binds to the vitelline layer, the depolarization is also called the **fast block to polyspermy** because it prevents more than one sperm cell from fusing with the egg's plasma membrane. Without this block, multiple sperm could fertilize the egg—resulting in an aberrant number of chromosomes in the zygote.

The Cortical Reaction

Another major effect of the fusion of egg and sperm plasma membranes is the **cortical reaction,** a series of changes in the outer zone (cortex) of the egg cytoplasm (see FIGURE 47.2, step

6). The fusion of sperm and egg triggers a signal-transduction pathway that causes the egg's endoplasmic reticulum to release calcium (Ca^{2+}) into the cytosol. The calcium release from the ER begins at the site of sperm entry and then propagates in a wave across the fertilized egg (FIGURE 47.3). Apparently the signaling pathway leads to production of the "second messengers" IP_3 and DAG (see FIGURE 11.15). The IP_3 opens ligand-gated calcium channels in the ER membrane; the Ca^{2+} released then triggers the opening of other channels, and so on. Within seconds, the high concentration of Ca^{2+} brings about a change in vesicles called **cortical granules,** which lie just under the egg's plasma membrane. Responding to the Ca^{2+} increase, the cortical granules fuse with the plasma membrane and release their contents into the perivitelline space between the plasma membrane and the vitelline layer. Enzymes from the granules separate the vitelline layer from the plasma membrane while mucopolysaccharides produce an osmotic gradient, drawing water into the perivitelline space and swelling it. The swelling pushes the vitelline layer away from the plasma membrane, and other enzymes harden it. The result is that the vitelline layer becomes the **fertilization envelope,** which resists the entry of additional sperm. By this time, usually about a minute after sperm and egg fuse, the voltage across the plasma membrane has returned to normal, and the fast block

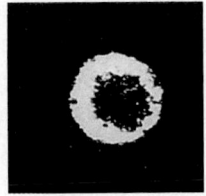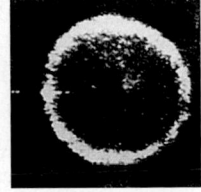

500 μm

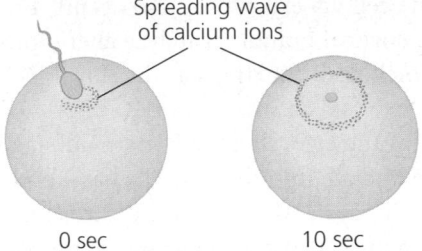

Spreading wave
of calcium ions

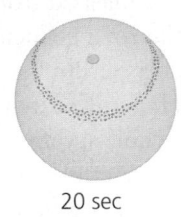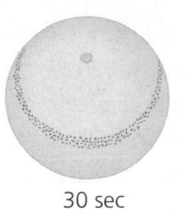

0 sec   10 sec   20 sec   30 sec

FIGURE 47.3 A wave of Ca²⁺ release during the cortical reaction. A fluorescent dye that glows when it binds free Ca^{2+} was used in this experiment to track the cortical reaction from the point of sperm contact (0 sec) during the fertilization of a fish egg (LM). The spreading wave of calcium ions, released into the cytosol from the endoplasmic reticulum, triggers the release of more and more Ca^{2+}. The high cytosolic concentration of Ca^{2+} causes the cortical granules to fuse with the plasma membrane, leading to formation of the fertilization membrane. The Ca^{2+} also helps activate metabolic changes inside the fertilized egg.

to polyspermy no longer functions. But the fertilization envelope, along with other changes in the egg's surface, functions as a **slow block to polyspermy.**

Activation of the Egg

The sharp rise in the egg's cytosolic concentration of Ca^{2+} not only triggers the cortical reaction but also incites metabolic changes within the egg cell. The unfertilized egg has a very slow metabolism, but within a few minutes of fertilization, the rates of cellular respiration and protein synthesis increase substantially. With these rapid changes, the egg cell is said to be activated. In sea urchins and many other species, the DAG produced during the cortical reaction activates a membrane protein that transports H^+ out of the egg, making the egg cytosol slightly alkaline. This pH change seems to be indirectly responsible for the metabolic responses of the egg to fertilization.

Although the binding and fusion of sperm are triggers for egg activation, sperm cells do not contribute any materials required for activation. Indeed, the unfertilized eggs of many species can be artificially activated by the injection of Ca^{2+} or by a variety of mildly injurious treatments, such as temperature shock. This artificial activation switches on the metabolic responses of the egg and causes it to begin developing by parthenogenesis (without fertilization by a sperm). It is even possible to artificially activate an egg that has had its own nucleus removed, showing that the molecules involved in activation reside in the cytoplasm. (Of course, embryonic development of such an egg terminates at a very early stage, because the embryo lacks a genome.) The fact that an egg lacking a nucleus can begin making new kinds of proteins upon activation means that mRNA coding for these proteins must be stockpiled in an inactive form in the cytoplasm of the unfertilized egg.

While the activated egg gears up its metabolism, the nucleus of the sperm cell within the egg starts to swell. After about 20 minutes, the sperm nucleus merges with the egg nucleus, creating the diploid nucleus of the zygote. DNA synthesis begins, and the first cell division occurs in about 90 minutes (in the case of sea urchins and some frogs). The events of fertilization in sea urchins are summarized in FIGURE 47.4.

Fertilization in Mammals

In Chapter 46 you learned how *in vitro* fertilization of human eggs is enabling some couples with fertility problems to have children. The techniques for test-tube fertilization have also made it possible for developmental biologists to study the process of fertilization in mammals. Many of the events turn out to be similar to what has been observed in sea urchins, but there are also important differences.

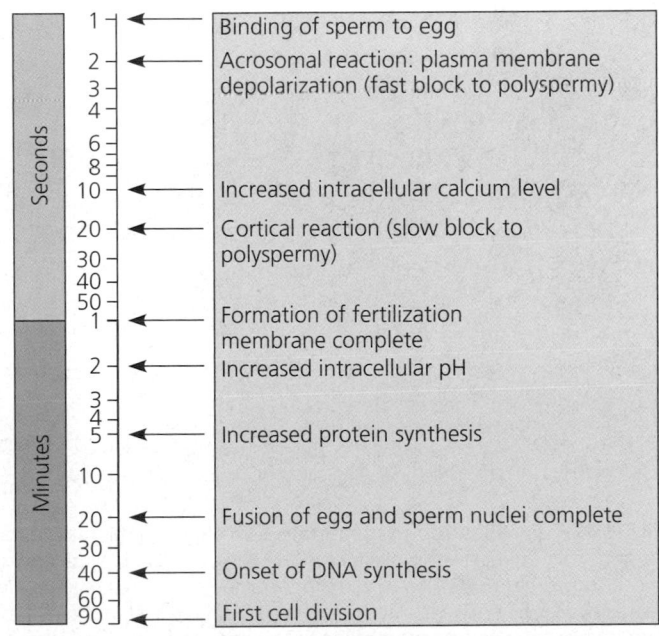

Seconds	
1	Binding of sperm to egg
2	Acrosomal reaction: plasma membrane depolarization (fast block to polyspermy)
3	
4	
6	
8	
10	Increased intracellular calcium level
20	Cortical reaction (slow block to polyspermy)
30	
40	
50	

Minutes	
1	Formation of fertilization membrane complete
2	Increased intracellular pH
3	
4	Increased protein synthesis
5	
10	
20	Fusion of egg and sperm nuclei complete
30	
40	Onset of DNA synthesis
60	
90	First cell division

FIGURE 47.4 Timeline for the fertilization of sea urchin eggs. The process begins when a sperm cell contacts the jelly coat of an egg (top of chart). Notice that the scale is logarithmic.

In contrast to the external fertilization of sea urchins and most other marine invertebrates, fertilization in terrestrial animals, including mammals, is generally internal. Secretions in the mammalian female reproductive tract alter certain molecules on the surface of sperm cells that have been deposited during the male's ejaculation and also increase the motility of the sperm. This enhancement of sperm function in the female reproductive tract, called capacitation, requires about 6 hours in humans.

The mammalian egg (actually a secondary oocyte at this stage; see FIGURE 46.13) is cloaked by follicle cells that were released with the egg during ovulation. A capacitated sperm cell must migrate through this layer of follicle cells before it reaches the **zona pellucida,** the extracellular matrix of the egg (FIGURE 47.5). The zona pellucida consists of three different glycoproteins forming filaments that are cross-linked in a three-dimensional network. One of the glycoproteins, ZP3, also functions as a sperm receptor, binding to a complementary molecule on the surface of the sperm head. The binding of the sperm head to receptor molecules induces the acrosome of the sperm cell to release its contents in an acrosomal reaction similar to that of sea urchin sperm. Protein-digesting enzymes and other hydrolases spilled from the acrosome enable the sperm cell to penetrate the zona pellucida and reach the plasma membrane of the egg. The acrosomal reaction also exposes a protein in the sperm membrane that binds and fuses with the egg membrane.

The binding of a sperm cell to the egg triggers depolarization of the egg membrane, which functions as a fast block to polyspermy, as in sea urchin fertilization. A cortical reaction, in which granules in the cortex of the egg release their contents to the outside of the cell via exocytosis, also occurs. Enzymes released from the cortical granules catalyze alterations of the zona pellucida, which then functions as the slow block to polyspermy.

Fingerlike extensions of the egg cell, called microvilli, take the whole sperm cell, tail and all, into the egg. The basal body of the sperm's flagellum divides and forms two centrosomes (with centrioles) in the zygote. These will generate the mitotic spindle for cell division; unfertilized mammalian eggs have no centrosomes of their own.

In contrast to sea urchin fertilization, the haploid nuclei of sperm and egg do not fuse immediately in mammals. Instead, the envelopes of both nuclei disperse, and the chromosomes from the two gametes share a common spindle apparatus during the first mitotic division of the zygote. Thus, it is not until after this first division, as diploid nuclei form in the two daughter cells, that the chromosomes from the two parents come together in common nuclei to form the genome of the offspring.

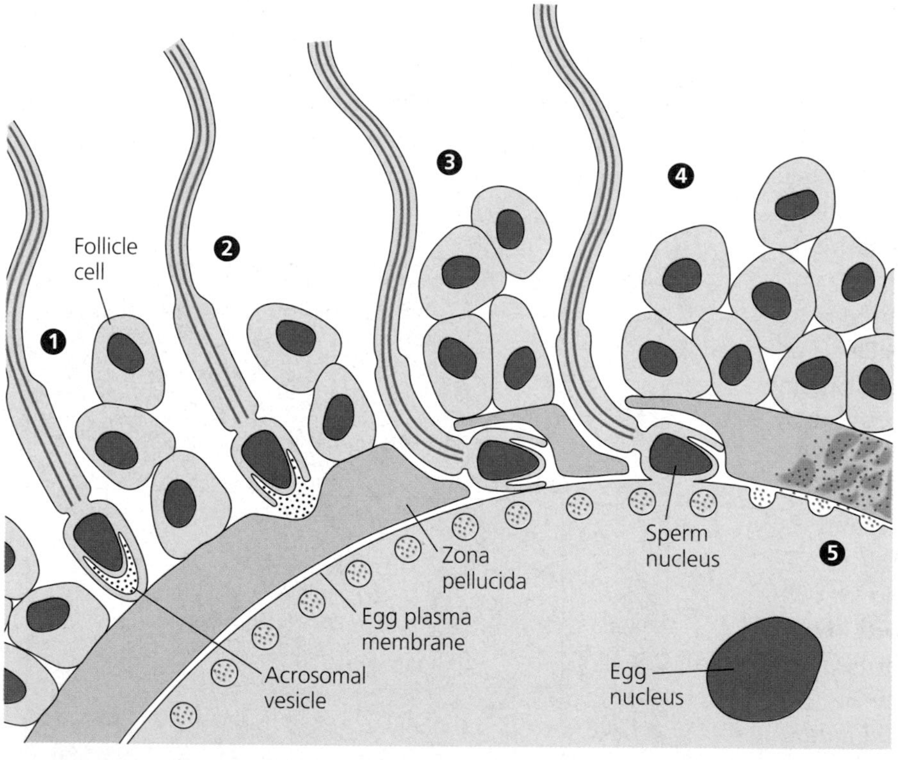

Cleavage partitions the zygote into many smaller cells

Fertilization is followed by three successive stages that begin to build the animal's body. First, a special type of cell division, called cleavage, creates a multicellular embryo, the blastula, from the zygote. The second stage, gastrulation, produces a three-layered embryo called the gastrula. The third stage, called organogenesis, generates rudimentary organs from which adult structures grow.

Cleavage is a succession of rapid cell divisions that follow fertilization (FIGURE 47.6). During cleavage, the cells undergo the S (DNA synthesis) and M (mitosis) phases of the cell cycle but often virtually skip the G_1 and G_2 phases (see FIGURE 12.4). The embryo does not enlarge

FIGURE 47.5 Fertilization in mammals. ❶ The sperm migrates through the coat of follicle cells and binds to receptor molecules in the zona pellucida of the egg. (Receptor molecules are not shown here.) **❷** This binding induces the acrosomal reaction, in which the sperm releases hydrolytic enzymes into the zona pellucida. **❸** With the help of these enzymes, the sperm reaches the plasma membrane of the egg, and membrane proteins of the sperm bind to receptors on the egg membrane. **❹** The plasma membranes fuse, making it possible for the contents of the sperm cell to enter the egg. **❺** Enzymes released during the egg's cortical reaction harden the zona pellucida, which now functions as a block to polyspermy.

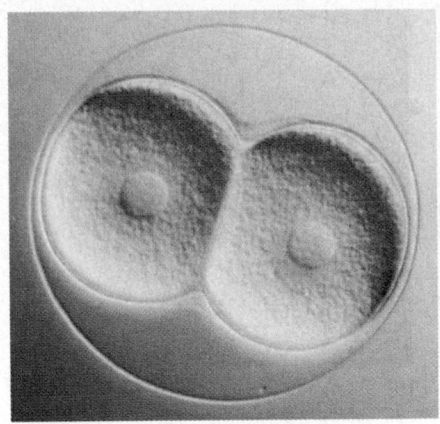

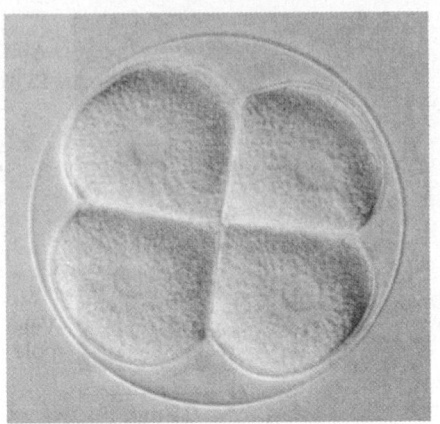

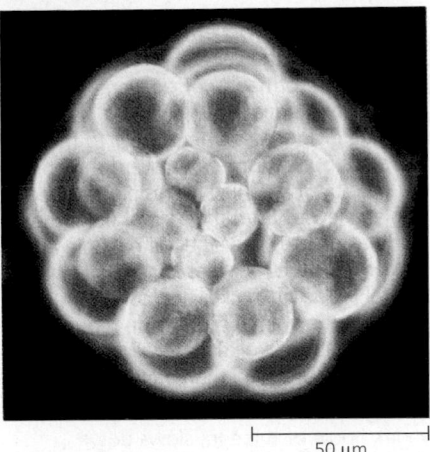

50 μm

(a) The two-cell stage, following the first cleavage division, occurs about 45–90 minutes after fertilization. Notice that the fertilization envelope is still present.

(b) The second cleavage division produces the four-cell stage.

(c) In a few hours, further cleavage divisions have formed a multicellular ball. The embryo is still inside the fertilization envelope, from which the larva will eventually hatch.

FIGURE 47.6 Cleavage in an echinoderm (sea urchin) embryo. Cleavage is a series of cell divisions that transform the zygote into a ball of much smaller cells, called blastomeres (LMs).

during this period of development. Cleavage simply partitions the cytoplasm of one large cell, the zygote, into many smaller cells called **blastomeres,** each with its own nucleus. Thus, different regions of cytoplasm present in the original undivided egg cell end up in separate blastomeres. And because the regions may contain different cytoplasmic components, the partitioning sets the stage for subsequent developmental events.

With the apparent exception of mammals, most animals have both eggs and zygotes with a definite polarity. During cleavage in such organisms, the planes of division follow a specific pattern relative to the poles of the zygote. The polarity is defined by the heterogeneous distribution of substances in the cytoplasm, including specific mRNA and proteins, and **yolk** (stored nutrients). In many frogs and other animals, the distribution of yolk is a key factor in influencing the pattern of cleavage. Yolk is most concentrated at one pole of the egg, called the **vegetal pole,** while the opposite pole, the **animal pole,** has the lowest concentration of yolk. The animal pole is also the site where the polar bodies of oogenesis are budded from the cell (see FIGURE 46.13), and in some animals it marks the point where the anterior (head) end of the embryo will form.

The hemispheres of the zygote are named for their respective poles (FIGURE 47.7). In the eggs of many frogs, the animal and vegetal hemispheres have different coloration due to the heterogeneous distribution of cytoplasmic substances. The animal hemisphere has melanin granules embedded in the outer layer of the cytoplasm (the cortex), giving it a deep gray hue, while the vegetal hemisphere contains the yellow yolk. A rearrangement of the amphibian egg cytoplasm occurs at the time of fertilization. The plasma membrane and associated cortex rotate toward the point of sperm entry, probably because the centrosome brought into the egg by the sperm cell reorganizes the

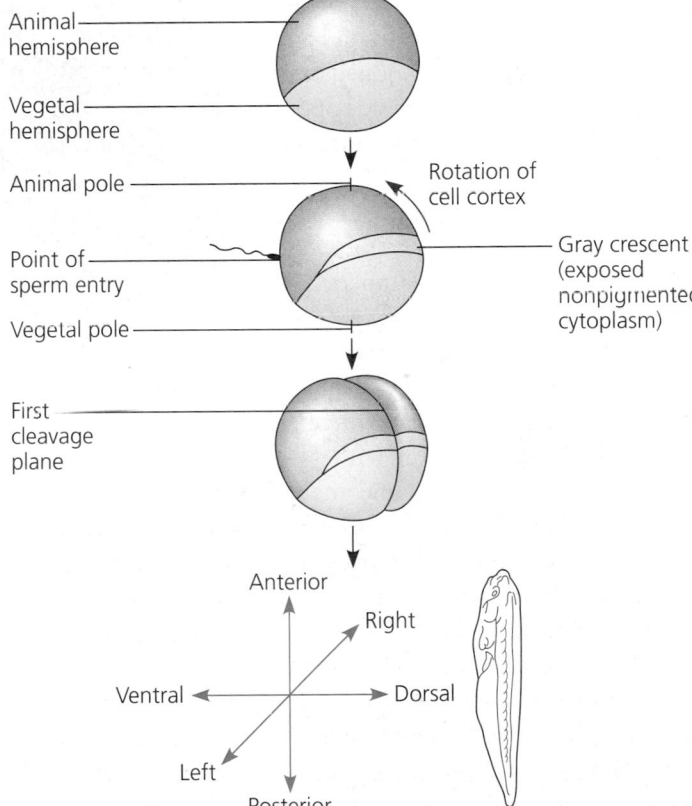

FIGURE 47.7 The establishment of the body axes and the first cleavage plane in an amphibian. The "animal" hemisphere of the egg, which will become the anterior end of the embryo, is dark gray due to melanin in its outer cytoplasm (cortex); the "vegetal" hemisphere is yellow because it contains the yolk. At fertilization, the pigmented cortex slides over the underlying cytoplasm toward the point of sperm entry, exposing a region of lighter-colored cytoplasm. The exposed region, known as the gray crescent, is opposite the point of sperm entry. The first cleavage division bisects the gray crescent. The crescent marks the location of the dorsal side of the future embryo. Thus, all three axes of the embryo are established before the zygote begins cleavage.

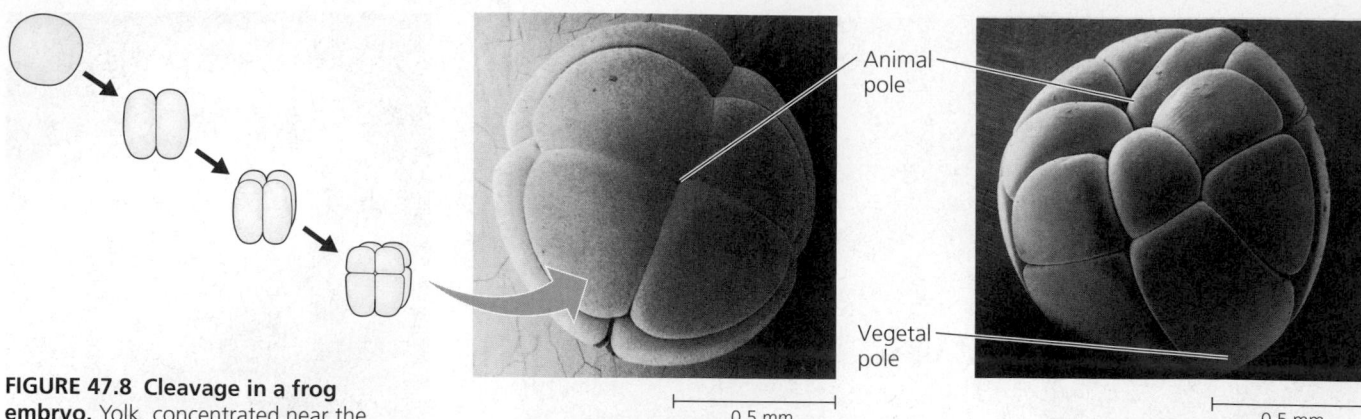

Animal pole

Vegetal pole

FIGURE 47.8 Cleavage in a frog embryo. Yolk, concentrated near the vegetal pole of the egg, slows down the formation of cleavage furrows in the vegetal hemisphere. (a, b, and c are SEMs.)

a, b, and c: Courtesy of Drs. R. G. Kessel and C. Y. Shih, all rights reserved.

0.5 mm

(a) Eight-celled embryo. After two equal divisions through the poles, the third cleavage division is perpendicular to the polar axis but is displaced toward the animal pole by yolk. As a result, the four blastomeres near the animal pole are smaller than the other four.

0.5 mm

(b) Morula (16–64 cells). As cleavage continues, the cells near the animal pole divide more frequently than the yolk-laden cells near the vegetal pole.

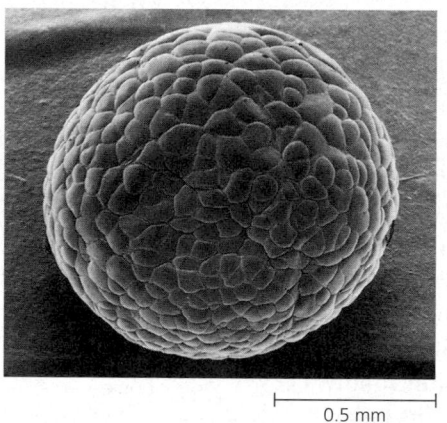

0.5 mm

(c) Blastula (at least 128 cells).

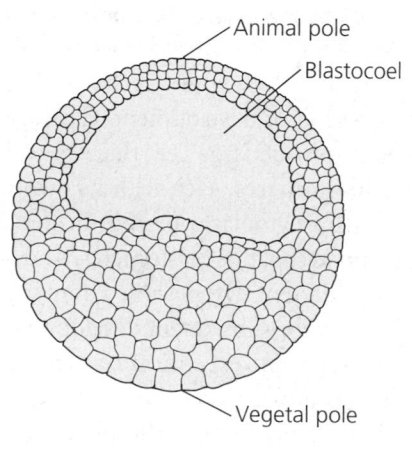

Animal pole

Blastocoel

Vegetal pole

(d) Cross section of blastula. The blastocoel is in the animal hemisphere.

cytoskeleton. The rotation exposes a light-gray region of cytoplasm, a band called the **gray crescent.** Located near the equator of the egg on the side opposite the sperm entry, the gray crescent is an important early marker of the polarity of the amphibian egg: It corresponds to the dorsal side of the later embryo.

Because yolk tends to impede cell division, cleavage of the frog zygote occurs more rapidly in the animal hemisphere than in the vegetal hemisphere, resulting in an embryo with different-sized cells. In comparison with frog eggs, those of sea urchins and many other animals have less yolk but still have an animal-vegetal axis, owing to differential distribution of other substances. Without the restraint imposed by yolk, their cleavage divisions all occur at about the same rate, producing blastomeres of virtually equal size.

In both sea urchins and frogs, the first two cleavage divisions are meridional (vertical), resulting in four cells that each extend from animal to vegetal pole. The third division is equatorial (horizontal), producing an eight-celled embryo with two tiers of four cells each (FIGURE 47.8a). The general pattern up to this point is the same in both types of embryos. In fact, echinoderms, chordates, and the other animal phyla grouped as deuterostomes share many features of early embryonic development. These similarities help distinguish the deuterostomes from the protostomes, such as mollusks, annelids, and arthropods (see FIGURE 32.7).

Continued cleavage produces a solid ball of cells known as the **morula,** Latin for "mulberry," in reference to the lobed surface of the embryo at this stage (FIGURE 47.8b). A fluid-filled cavity called the **blastocoel** forms within the morula, creating a hollow ball called the **blastula** (FIGURE 47.8c). In sea urchins, the blastocoel is centrally located in the blastula. However, in frogs, because of unequal cell division, the blastocoel is located in the animal hemisphere (FIGURE 47.8d).

Yolk is most plentiful and has its most pronounced effect on cleavage in the eggs of birds, reptiles, many fishes, and insects. In birds, for example, the part of the egg we commonly call the yolk is actually the egg cell (ovum), swollen with yolk nutrients. This enormous cell is surrounded by a protein-rich solution (the egg white) that will provide additional nutrients for the growing embryo. Cleavage of the fertilized egg is restricted to a small disc of yolk-free cytoplasm at the animal pole of the egg cell. This incomplete division of a yolk-rich egg is known as **meroblastic cleavage.** It contrasts with **holoblastic cleavage,** the complete division of eggs having little yolk (as in sea urchins) or a moderate amount of yolk (as in frogs).

The yolk-rich eggs of insects, such as *Drosophila,* undergo a unique type of meroblastic cleavage (see FIGURE 21.11). The zygote's nucleus is situated *within* a mass of yolk. Cleavage begins with the nucleus undergoing mitotic divisions that are not accompanied by cytokinesis. These mitotic divisions produce several hundred nuclei, which migrate to the outer edge of the egg. After several more rounds of mitosis, a plasma membrane forms around each nucleus, and the embryo, now a blastula, consists of a single layer of about 6,000 cells surrounding a mass of yolk.

Gastrulation rearranges the blastula to form a three-layered embryo with a primitive gut

The morphogenetic process called **gastrulation** is a dramatic rearrangement of the cells of the blastula. Gastrulation differs in detail from one animal group to another, but a common set of cellular changes drives this spatial rearrangement of an embryo. These general cellular mechanisms are changes in cell motility, changes in cell shape, and changes in cellular adhesion to other cells and to molecules of the extracellular matrix (see FIGURE 7.29). The result of gastrulation is that some of the cells at or near the surface of the blastula move to an interior location, and three cell layers are established. The three-layered embryo is called the **gastrula.** The positioning of the cell layers in the gastrula allows cells to interact with each other in new ways.

The three layers produced by gastrulation are embryonic tissues called **ectoderm, endoderm,** and **mesoderm,** also collectively termed the embryonic germ layers. The ectoderm forms the outer layer of the gastrula; the endoderm lines the embryonic digestive tract; and the mesoderm partly fills the space between the ectoderm and the endoderm. Eventually, these three cell layers develop into all the parts of the adult animal. For instance, our nervous system and the outer layer (epidermis) of our skin come from ectoderm; the innermost lining of our digestive tract and associated organs, such as the liver and pancreas, arise from endoderm; and most other organs and tissues, such as the kidney, heart, muscles, and the inner layer of our skin (dermis), develop largely from mesoderm.

Let's examine gastrulation in a sea urchin embryo (FIGURE 47.9). The wall of the sea urchin blastula consists of a single layer of cells. Gastrulation begins at the vegetal pole where individual cells detach from the blastula wall and enter the blastocoel as migratory cells called *mesenchyme cells.* The remaining cells flatten slightly to form a vegetal plate that buckles inward, a process called **invagination.** The buckled vegetal plate then undergoes extensive rearrangement of its cells, a process that transforms the shallow invagination into a deeper, narrower pouch called the **archenteron,** or primitive gut. The open end of the archenteron, which will become the anus, is called the **blastopore.** A second opening forms at the other end of the archenteron, forming the mouth end of what is now a rudimentary digestive tube. Gastrulation has produced an embryo with a primitive gut and three germ layers: ectoderm (color-coded blue throughout this chapter), endo-

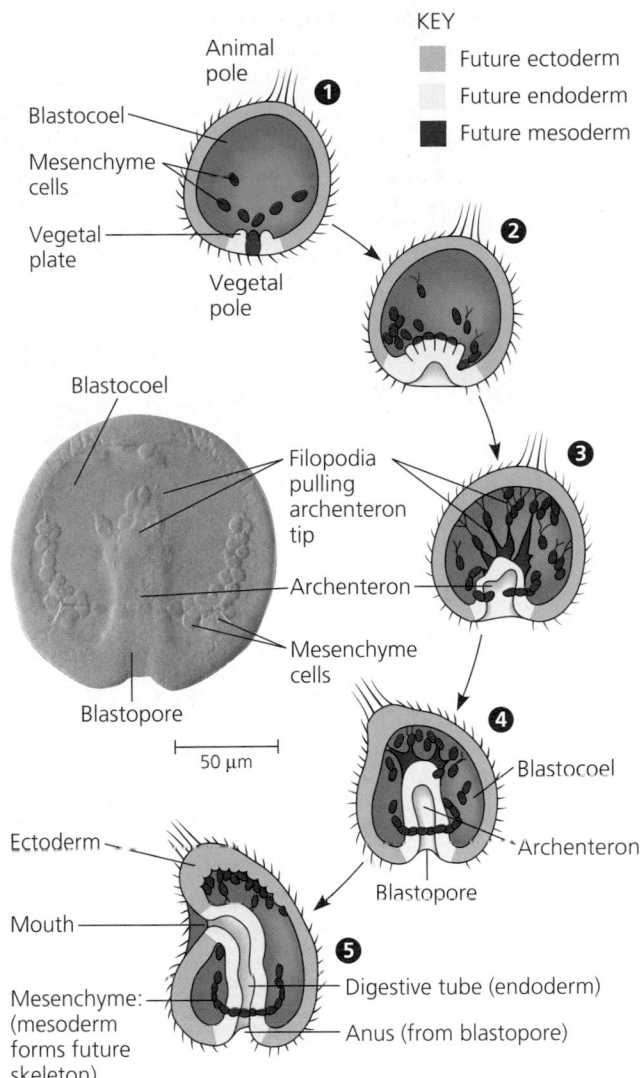

FIGURE 47.9 Sea urchin gastrulation. ❶ Formed by cleavage, the blastula consists of a single layer of ciliated cells surrounding the blastocoel. Gastrulation begins with the migration of mesenchyme cells from the flattened "vegetal plate" into the blastocoel. ❷ The vegetal plate in this early gastrula invaginates (buckles inward). Mesenchyme cells begin to form thin extensions (filopodia). ❸ Endoderm cells form the archenteron (future digestive tube). Mesenchyme cells form filopodial connections between the tip of the archenteron and the ectoderm cells of the blastocoel wall (inset, LM). ❹ Contraction of the filopodia in a late gastrula drags the archenteron the rest of the way across the blastocoel, where the endoderm of the archenteron will fuse with ectoderm of the blastocoel wall. ❺ Gastrulation is complete. The gastrula has a functional digestive tube formed from the endoderm of the archenteron, with a mouth and an anus. Ectoderm forms the embryo's ciliated outer surface. Some of the mesenchyme cells of the mesoderm have secreted calcium carbonate ($CaCO_3$), which will form a simple internal skeleton.

derm (yellow), and mesoderm (red). Thus, the triploblastic (three-layered) body plan characteristic of most animal phyla is established very early in development (see FIGURE 32.6). In the sea urchin, the gastrula eventually develops into a ciliated larva that drifts near the ocean surface as plankton, feeding on bacteria and unicellular algae.

FIGURE 47.10 Gastrulation in a frog embryo.

① The blastocoel of the frog blastula is off-center and surrounded by a wall that is more than one cell thick. At this stage, the colors indicate regions of the blastula that will become the embryo's three germ layers.

KEY

■ Future ectoderm

□ Future endoderm

■ Future mesoderm

② Gastrulation begins when a small tuck, the dorsal lip of the blastopore, appears on one side of the blastula. The tuck is formed by cells changing shape and pushing inward from the surface. Additional cells that will become endoderm and mesoderm then roll inward over the dorsal lip (involution) and move away from the blastopore into the interior of the gastrula. Meanwhile, cells of the animal pole, which will form ectoderm, spread over the embryo's outer surface.

③ Externally, the lip of the blastopore starts becoming circular. Internally, the three germ layers start forming as cells continue migrating inward. The advancing endoderm, mesoderm, and the archenteron, lined by endoderm, are filling the space occupied by the blastocoel.

④ Late in gastrulation, the circular blastopore surrounds a plug of yolk cells (the yolk plug). The three germ layers are now in place, ready for organogenesis.

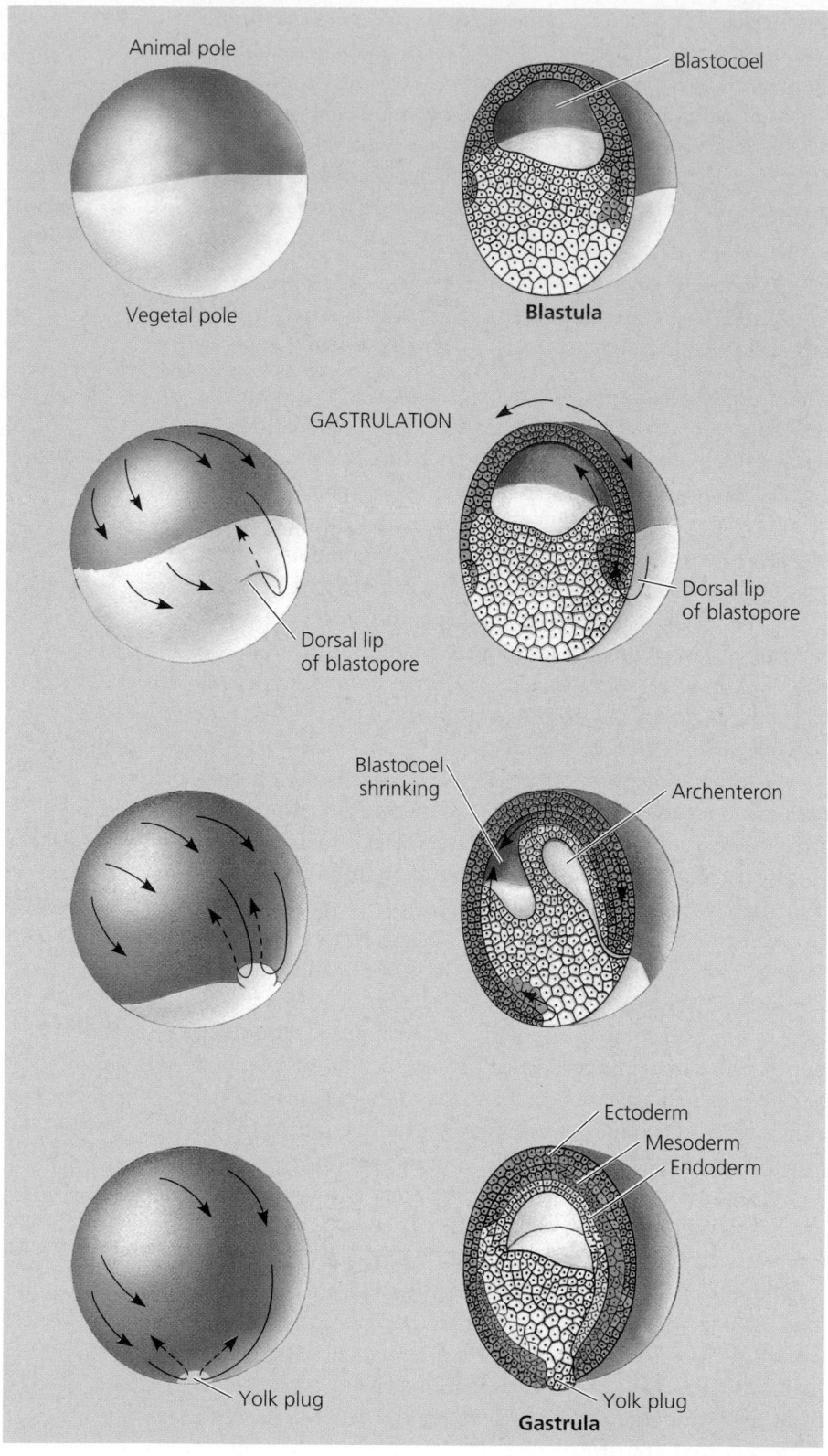

Gastrulation during frog development also produces a three-layered embryo with an archenteron, as shown in FIG-URE 47.10. The mechanics of gastrulation are more compli-cated in a frog, however, because of the large, yolk-laden cells of the vegetal hemisphere, and because the wall of the blastula is more than one cell thick in most species. The first sign of gastrulation is a small crease on one side of the blastula caused by the invagination of a group of cells there. This invagination becomes the dorsal side of the blastopore, called the **dorsal lip** of the blastopore. The dorsal lip of the blastopore forms where the gray crescent was located in the zygote (see FIGURE 47.7). Successive invaginations of groups of cells near the dorsal lip result in the completion of the circular blastopore.

Frog gastrulation continues with cells on the surface of the embryo rolling over the edge of the dorsal lip into the interior of the embryo, a process called **involution.** Once inside the embryo, these cells move away from the blastopore along the roof of the blastocoel. Involution continues, with migrating internal cells becoming organized into layered mesoderm and endoderm, and the archenteron forming within the endoderm. Eventually, the complex cell movements of gastrulation produce a three-layered embryo. As the process is completed, the circular lip of the blastopore encircles a **yolk plug** consisting of large, food-laden cells; these protruding cells will move inward as expansion of the ectoderm causes the blastopore to shrink. At this point, the cells remaining on the surface make up the ectoderm, surrounding the layers of mesoderm and endoderm. With the three germ layers in place, gastrulation is complete, and the embryo's organs begin to form.

In organogenesis, the organs of the animal body form from the three embryonic germ layers

Various regions of the three germ layers develop into the rudiments of organs during the process of **organogenesis** (TABLE 47.1). Three kinds of morphogenetic changes—folds, splits, and dense clustering (condensation) of cells—are the first evidence of organ building. The organs that begin to take shape first in the embryos of frogs and other chordates are the neural tube and notochord, the skeletal rod characteristic of all chordate embryos (see FIGURE 34.6).

Table 47.1 Derivatives of the Three Embryonic Germ Layers in Vertebrates

Germ Layer	Organs and Tissues in the Adult
Ectoderm	Epidermis of skin and its derivatives (e.g., skin glands, nails); epithelial lining of mouth and rectum; sense receptors in epidermis; cornea and lens of eye; nervous system; adrenal medulla; tooth enamel; epithelium of pineal and pituitary glands.
Endoderm	Epithelial lining of digestive tract (except mouth and rectum); epithelial lining of respiratory system; liver; pancreas; thyroid; parathyroids; thymus; lining of urethra, urinary bladder, and reproductive system.
Mesoderm	Notochord; skeletal system; muscular system; circulatory and lymphatic systems; excretory system; reproductive system (except germ cells, which start to differentiate during cleavage); dermis of skin; lining of body cavity; adrenal cortex.

FIGURE 47.11 on page 1008 shows early organogenesis as it occurs in a frog. The **notochord** is formed from dorsal mesoderm that condenses just above the archenteron, and the neural tube originates as a plate of dorsal ectoderm just above the developing notochord. The neural plate soon folds inward, rolling itself into the **neural tube,** which will become the central nervous system—the brain and spinal cord. These organs are hollow in most chordates because of this mechanism of development. In frogs the notochord elongates and stretches the embryo along its anterior-posterior axis. Later, the notochord will function as a core around which mesodermal cells gather and form the vertebrae. Parts of the notochord between the vertebrae persist as the vertebral discs in adults. (These are the discs that can "slip," causing back pain.)

Other condensations occur in strips of mesoderm lateral to the notochord, which separate into blocks called **somites.** The somites are arranged serially on both sides along the length of the notochord (FIGURE 47.11c). Cells from the somites not only give rise to the vertebrae of the backbone, they also form the muscles associated with the axial skeleton. This serial origin of the axial skeleton and muscles reinforces the point made in Chapter 34 that chordates are basically segmented animals, although the segmentation becomes less obvious later in development. (There are signs of segmentation even in the adult, as in the series of vertebrae in a human or the segments of chevron-shaped muscles in a fish.) Lateral to the somites, the mesoderm splits into two layers that form the lining of the body cavity, or coelom.

As organogenesis progresses, morphogenesis and cellular differentiation continue to refine the organs that arise from the three embryonic germ layers. Review TABLE 47.1, which lists the embryonic sources of the major organs and tissues in frogs and other vertebrates. Unique to vertebrate embryos, a band of cells called the *neural crest* develops along the border where the neural tube pinches off from the ectoderm. Cells of the neural crest subsequently migrate to various parts of the embryo, forming pigment cells of the skin, some of the bones and muscles of the skull, the teeth, the medulla of the adrenal glands, and peripheral components of the nervous system, such as sensory and sympathetic ganglia.

Embryonic development of the frog leads to a larval stage, the tadpole, which hatches from the jelly coat that originally cloaked the egg. Later, metamorphosis will transform the frog from the aquatic, herbivorous tadpole to the terrestrial, carnivorous adult.

Amniote embryos develop in a fluid-filled sac within a shell or uterus

All vertebrate embryos require an aqueous environment for development. In the case of fish and amphibians, the egg is laid in the surrounding sea or pond and needs no special

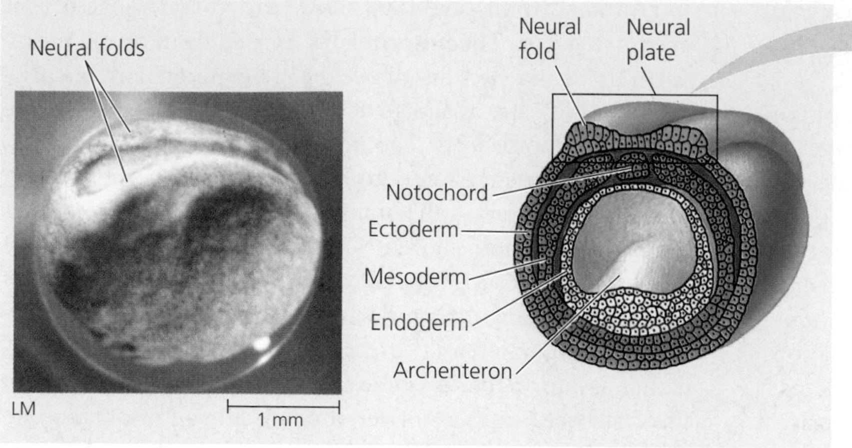

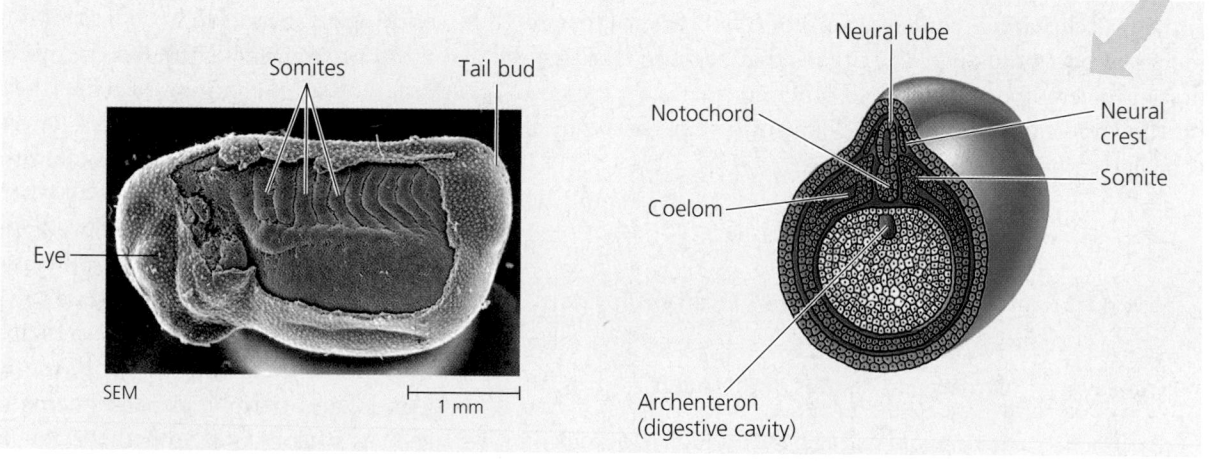

(a) A cross section of a frog embryo at the beginning of organogenesis.
Notice the three germ layers and the rudiments of the notochord and neural plate. The notochord has developed from dorsal mesoderm, and the dorsal ectoderm has thickened, forming the neural plate. Two pronounced ridges, the neural folds, form the lateral edges of the neural plate.

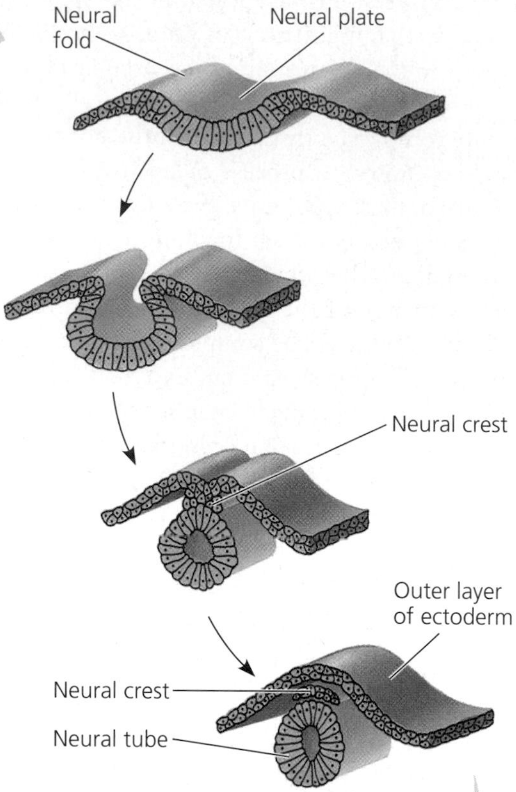

Neural folds

Neural fold / Neural plate

Notochord
Ectoderm
Mesoderm
Endoderm
Archenteron

LM
1 mm

Neural fold / Neural plate

Neural crest

Outer layer of ectoderm

Neural crest
Neural tube

(b) Formation of the neural tube from the neural plate.

FIGURE 47.11 Organogenesis in a frog embryo.

Somites Tail bud

Eye

SEM
1 mm

Neural tube
Notochord
Neural crest
Somite
Coelom
Archenteron (digestive cavity)

(c) Somites. In the diagram on the right, showing a cross section of an embryo with a completed neural tube, somites flank the notochord. Formed from mesoderm, the somites will give rise to segmental structures such as vertebrae and skeletal muscles. The lateral mesoderm has begun to separate into the two tissue layers that line the coelom. The scanning electron micrograph is a side view of a whole embryo at the tail-bud stage; part of the ectoderm has been removed to reveal the somites.

water-filled enclosure. The movement of vertebrate animals onto land required solving the problem of reproduction in dry environments, and two major solutions evolved: the shelled egg of reptiles and birds and the uterus of placental mammals. Within the shell or uterus, the embryos of birds, reptiles, and mammals are surrounded by fluid within a sac formed by a membrane called the amnion. Vertebrates of these three classes are therefore called **amniotes** (see Chapter 34). We have already examined the embryonic development of a vertebrate that lacks an amnion, the frog. For

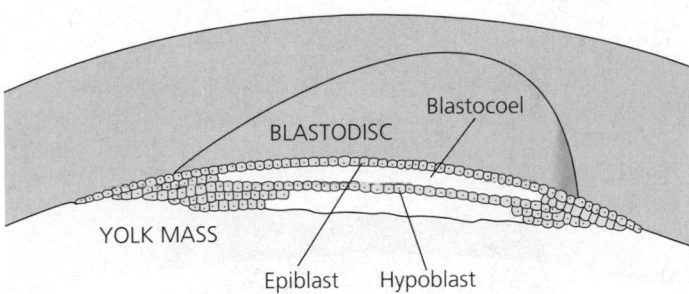

❶ Cleavage. Because of the large amount of yolk, cleavage is meroblastic, or incomplete. Cell division is restricted to a small cap of cytoplasm at the animal pole. Cleavage produces a blastodisc that rests on the large, undivided mass of yolk. The blastodisc becomes arranged into two layers (epiblast and hypoblast) that bound the blastocoel, forming the avian version of a blastula.

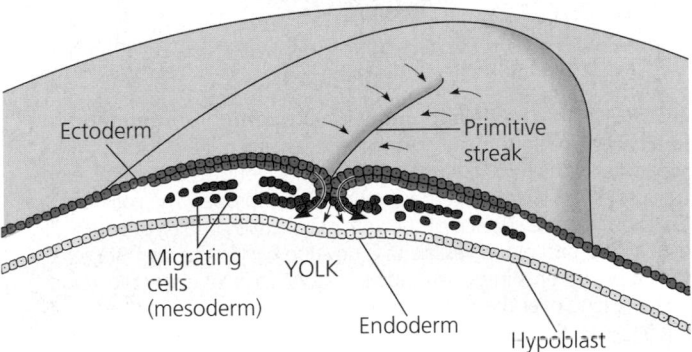

❷ Gastrulation. During gastrulation, some cells of the epiblast migrate (arrows) into the interior of the embryo through the primitive streak, shown here in transverse section. Some of these cells move laterally to form mesoderm, while others migrate downward to form the endoderm.

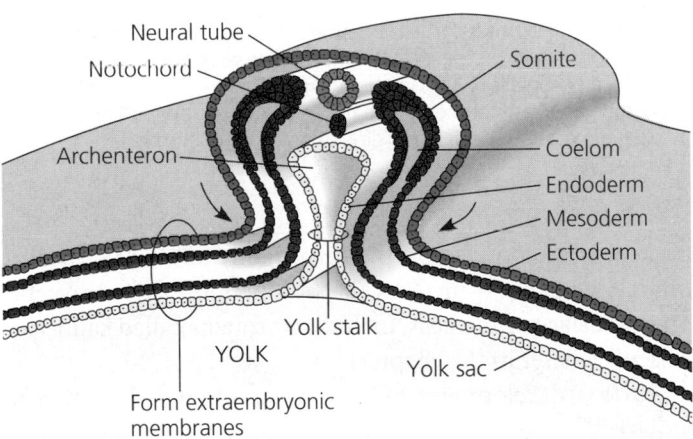

❸ Early organogenesis. The archenteron is formed when lateral folds pinch the embryo away from the yolk. About midway along its length, the embryo will remain attached to the yolk by the yolk stalk, formed mainly by hypoblast cells. The notochord, neural tube, and somites form much as they do in the frog. The three germ layers and hypoblast cells also contribute to the system of extraembryonic membranes that support further development.

FIGURE 47.12 Cleavage, gastrulation, and early organogenesis in a chick embryo.

comparison, we now study the early development of two amniotes, a bird and a mammal.

Avian Development

After fertilization, a bird egg undergoes meroblastic cleavage in which cell division occurs only in a small region of yolk-free cytoplasm atop the large mass of yolk. The early cleavage divisions produce a cap of cells called the **blastodisc,** which rests on the undivided yolk. The blastomeres then sort into upper and lower layers, the epiblast and hypoblast (FIGURE 47.12, step ❶). The cavity between these two layers is the avian version of the blastocoel, and this embryonic stage is the avian equivalent of the blastula, although its form is different from the hollow ball of an early frog embryo.

Gastrulation, as in the frog embryo, involves cells moving from the surface of the embryo to an interior location. In birds, however, the route of this cell migration is very different (FIGURE 47.12, step ❷). Some cells of the upper cell layer (epiblast) move toward the midline of the blastodisc, then detach and move inward toward the yolk. The medial movement on the surface and the inward movement of cells at the blastodisc's midline produce a groove called the **primitive streak.** As the streak lengthens over the surface of the blastodisc, it marks what will become the bird's anterior-posterior axis. The primitive streak is functionally equivalent to the frog blastopore, but it is a linear tuck rather than a ring.

All the cells that will form the embryo come from the epiblast. Some of the epiblast cells that pass through the primitive streak move laterally into the blastocoel, producing the mesoderm. Other epiblast cells, which will produce the endoderm, migrate through the streak and downward, pushing out the cells of the hypoblast. The epiblast cells that remain on the surface give rise to the ectoderm. Although the hypoblast contributes no cells to the embryo, it seems to help direct the formation of the primitive streak before the onset of gastrulation and is required for normal development. The hypoblast cells later segregate from the endoderm and eventually form portions of a sac surrounding the yolk and a stalk connecting the yolk mass to the embryo. After the three germ layers are formed, the borders of the embryonic disc fold downward and come together, pinching the embryo into a three-layered tube joined at midbody to the yolk (FIGURE 47.12, step ❸). Neural tube formation, development of the notochord and somites, and other events in organogenesis occur much as in the frog embryo. FIGURE 47.13 (p. 1010) shows some of the organs in a 2-day-old chick embryo.

Notice in FIGURE 47.12, step 3, that only part of each germ layer contributes to the embryo itself (which is the protrusion that arises from the blastodisc). The tissue layers that are outside the embryo proper develop into four **extraembryonic membranes** that support further embryonic development within the egg. These four "membranes," each a sheet of cells, are the yolk sac, the amnion, the chorion, and the allantois.

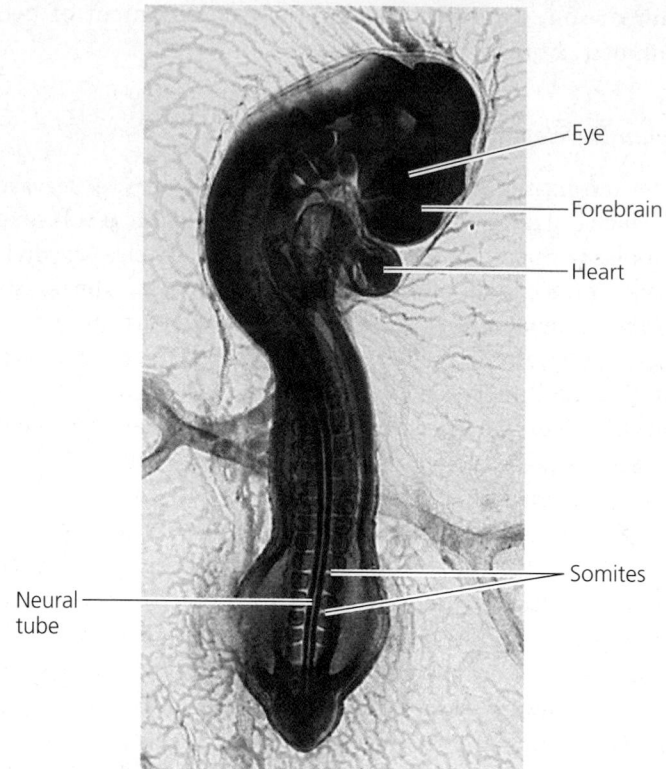

FIGURE 47.13 **Organogenesis in a chick embryo.** Rudiments of most major organs have already formed in this chick, which is about 56 hours old (LM).

The diagram in FIGURE 47.14 shows these membranes for a developing chick, and the legend describes the functions of each membrane.

Mammalian Development

In most mammalian species, fertilization takes place in the oviduct, and the earliest stages of development occur while the embryo completes its journey down the oviduct to the uterus (see Chapter 46). In contrast to the large, yolky eggs of birds and reptiles, the egg of a placental mammal is quite small, storing little in the way of food reserves. As already mentioned, the mammalian egg and zygote do not exhibit any obvious polarity with respect to the contents of the cytoplasm, and cleavage of the yolk-lacking zygote is holoblastic. However, mammalian gastrulation and early organogenesis follow a pattern similar to that of birds and reptiles. (Recall from Chapter 34 that mammals descended from reptilian stock during the early Mesozoic era.)

Cleavage is relatively slow in mammals. In the case of humans, the first division is complete about 36 hours after fertilization, the second division at about 60 hours, and the third division at about 72 hours. The blastomeres are equal in size. An important event during early mammalian development is the process of *compaction,* which occurs at the eight-cell stage. Before compaction, the cells of the early embryo are loosely

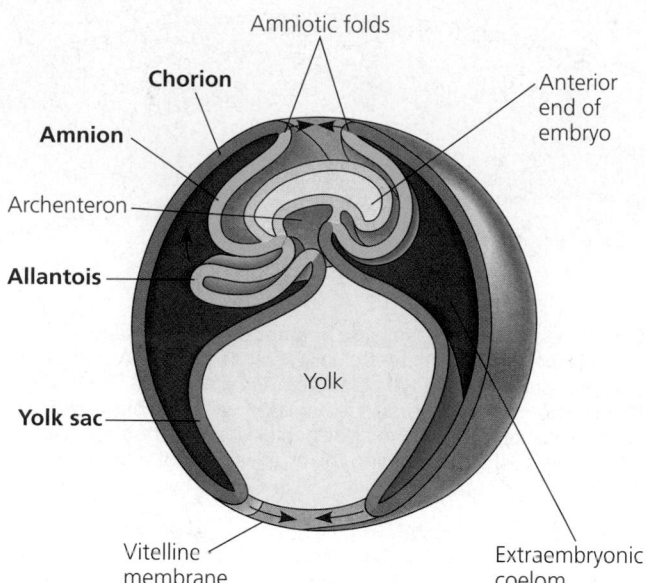

FIGURE 47.14 **The development of extraembryonic membranes in a chick.** The four major membranes (labeled in bold type) provide support services for the embryo. Each is a sheet of cells that develops from epithelial sheets external to the embryo proper. The yolk sac expands over the surface of the yolk mass. Cells of the **yolk sac** will digest yolk, and blood vessels that develop within the membrane will carry nutrients into the embryo. Lateral folds of extraembryonic tissue extend over the top of the embryo and fuse to form two more membranes, the amnion and the chorion, which are separated by an extraembryonic extension of the coelom. The **amnion** encloses the embryo in a fluid-filled amniotic sac, protecting the embryo from drying out, and it, along with the **chorion**, cushions the embryo against mechanical shocks. The fourth membrane, the allantois, originates as an outpocketing of the embryo's hindgut. The **allantois** is a sac that extends into the extraembryonic coelom. It functions as a disposal sac for uric acid, the insoluble nitrogenous waste of the embryo. As the allantois expands, it presses the chorion against the vitelline membrane, the inner lining of the eggshell. Together, the allantois and chorion form a respiratory organ for the embryo. Blood vessels that form in the epithelium of the allantois transport oxygen to the embryo. The extraembryonic membranes of reptiles and birds are adaptations associated with the special problems of development on land.

packed; after compaction the cells tightly adhere to one another. Compaction involves the production of new proteins on the surface of the cells, including proteins called cadherins (discussed later in the chapter).

Further development of the human embryo is shown in FIGURE 47.15. ❶ By about 7 days after fertilization, the embryo has over 100 cells arranged around a central cavity. This is the embryonic stage known as the **blastocyst.** Protruding into one end of the blastocyst cavity is a cluster of cells called the **inner cell mass,** which will subsequently develop into the embryo proper and some of the extraembryonic membranes. The outer epithelium surrounding the cavity is the **trophoblast,** which, along with mesodermal tissue, will form the fetal portion of the placenta. The embryo reaches the uterus by the blastocyst stage and soon begins to implant in the uterine lining (endometrium).

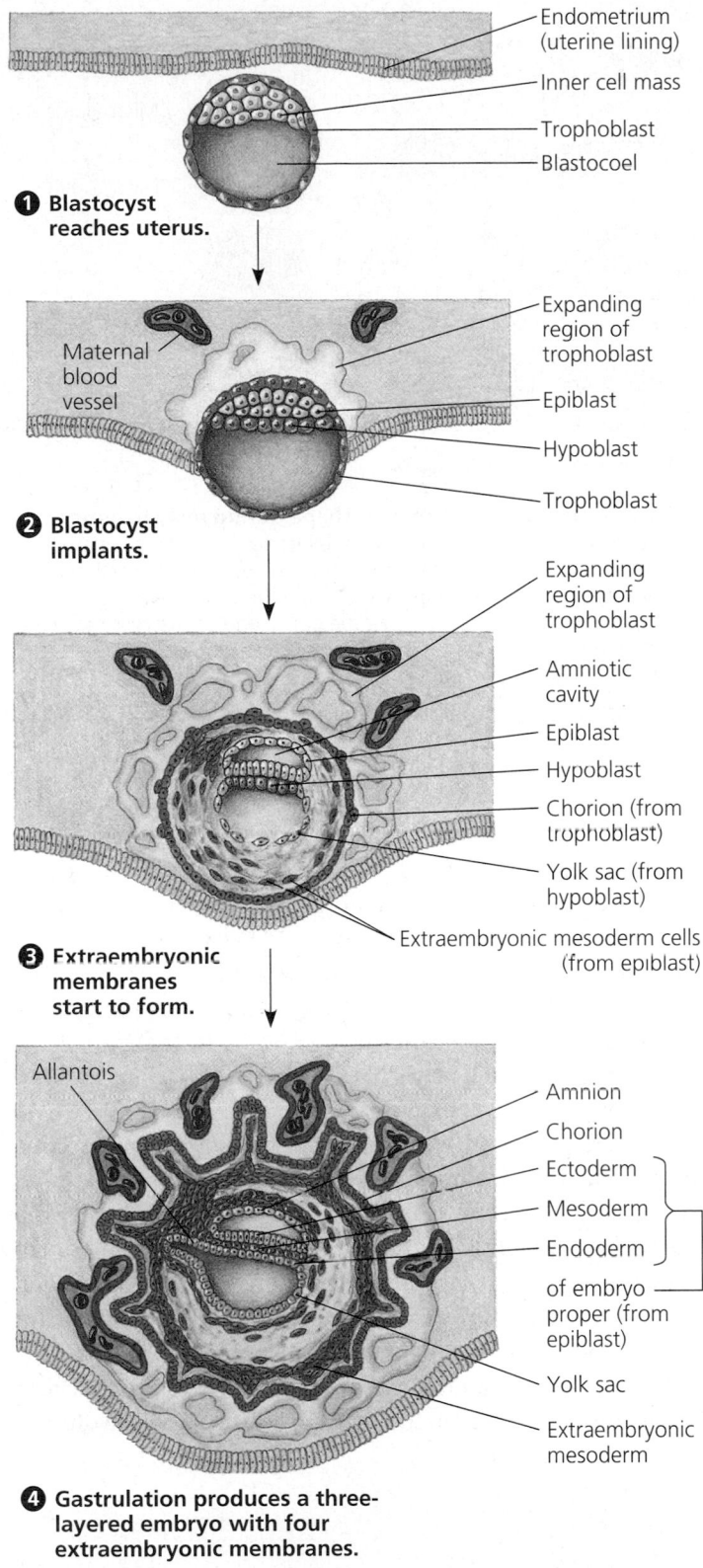

Endometrium (uterine lining)
Inner cell mass
Trophoblast
Blastocoel

❶ Blastocyst reaches uterus.

Expanding region of trophoblast
Maternal blood vessel
Epiblast
Hypoblast
Trophoblast

❷ Blastocyst implants.

Expanding region of trophoblast
Amniotic cavity
Epiblast
Hypoblast
Chorion (from trophoblast)
Yolk sac (from hypoblast)
Extraembryonic mesoderm cells (from epiblast)

❸ Extraembryonic membranes start to form.

Allantois
Amnion
Chorion
Ectoderm
Mesoderm
Endoderm
of embryo proper (from epiblast)
Yolk sac
Extraembryonic mesoderm

❹ Gastrulation produces a three-layered embryo with four extraembryonic membranes.

FIGURE 47.15 Early development of a human embryo and its extraembryonic membranes. This series of drawings illustrates four stages in transverse section. See the text for a description of each stage.

❷ The trophoblast initiates implantation by secreting enzymes that enable the blastocyst to penetrate the endometrium. Bathed in blood spilled from eroded capillaries in the endometrium, the trophoblast thickens and extends fingerlike projections into the surrounding maternal tissue. (The placenta will later form from this proliferated trophoblast and the region of endometrium it invades; see FIGURE 46.17.) Around the time of implantation, the inner cell mass of the blastocyst forms a flat disc with an upper layer of cells, the *epiblast,* and a lower layer, the *hypoblast*. These layers are homologous to those of birds, and, as in birds, the embryo will develop almost entirely from epiblast cells, while the hypoblast cells will form the yolk sac.

❸ Now the extraembryonic membranes begin to develop. The trophoblast is giving rise to the chorion and continues to expand into the endometrium. The epiblast has begun to form the amnion, surrounding a fluid-filled cavity. Mesodermal cells that will become part of the placenta are also derived from the epiblast.

❹ Gastrulation occurs by the inward movement of cells from the epiblast through a primitive streak to form mesoderm and endoderm, just as it does in the chick. We now have a three-layered embryo surrounded by proliferating extraembryonic mesoderm. Four extraembryonic membranes have formed, which are homologous to those of reptiles and birds. The *chorion,* which develops from the trophoblast, completely surrounds the embryo and the other extraembryonic membranes. The *amnion* begins as a dome above the proliferating epiblast and will eventually enclose the embryo in a fluid-filled amniotic cavity. (The fluid from this cavity is the "water" expelled from the vagina of the mother when the amnion breaks just prior to childbirth.) Below the developing embryo proper, the *yolk sac* encloses another fluid-filled cavity. Although this cavity contains no yolk, the membrane that surrounds it is given the same name as the homologous membrane in birds and reptiles. The yolk sac membrane of mammals is a site of early formation of blood cells, which later migrate into the embryo proper. The fourth extraembryonic membrane, the *allantois,* develops as an outpocketing of the embryo's rudimentary gut, as it does in the chick. The allantois is incorporated into the umbilical cord, where it forms blood vessels that transport oxygen and nutrients from the placenta to the embryo and rid the embryo of carbon dioxide and nitrogenous wastes. Thus, the extraembryonic membranes of shelled eggs, where embryos are nourished with yolk, were conserved as mammals diverged from reptiles in the course of evolution, but with modifications adapted to development within the reproductive tract of the mother.

Organogenesis begins with the formation of the neural tube, notochord, and somites. By the end of the first trimester of human development, rudiments of all the major organs have developed from the three germ layers, as was summarized in TABLE 47.1.

THE CELLULAR AND MOLECULAR BASIS OF MORPHOGENESIS AND DIFFERENTIATION IN ANIMALS

Now that we have described the main events of embryonic development in animals, the remainder of the chapter will address the cellular and molecular mechanisms by which they occur. Although biologists are far from a full understanding of these mechanisms, several key principles have emerged as fundamental to the development of all animals.

Morphogenesis in animals involves specific changes in cell shape, position, and adhesion

Morphogenesis is a major aspect of development in both animals and plants, but only in animals does it involve the *movement* of cells. Movement of parts of a cell can bring about changes in cell shape or enable a cell to migrate from one place to another within the embryo. Changes in both cell shape and cell position are involved in cleavage, gastrulation, and organogenesis.

Changes in the shape of a cell usually involve reorganization of the cytoskeleton (see Chapter 7). Consider, for example, how the cells of the neural plate form the neural tube (FIGURE 47.16). First, microtubules oriented parallel to the dorsal-ventral axis of the embryo apparently help lengthen the cells in that direction. At the dorsal end of each cell is a parallel array of actin filaments oriented crosswise. These contract, giving the cells a wedge shape that forces the ectoderm layer to bend inward. Similar cell-shape changes are observed for other invaginations (inpocketings) and evaginations (outpocketings) of tissue layers throughout development.

The cytoskeleton also drives the active movement of cells from one place to another in developing animals. Cells "crawl" within the embryo by using cytoskeletal fibers to extend and retract cellular protrusions. This type of motility is akin to the amoeboid movement described in FIGURE 7.27b, but in contrast to the thick pseudopodia of amoeboid cells, the cellular protrusions of migrating embryonic cells are usually flat sheets (lamellipodia) or spikes (filopodia). In the gastrulation of some organisms, invagination is initiated by wedging of cells on the surface of the blastula, but the penetration of cells

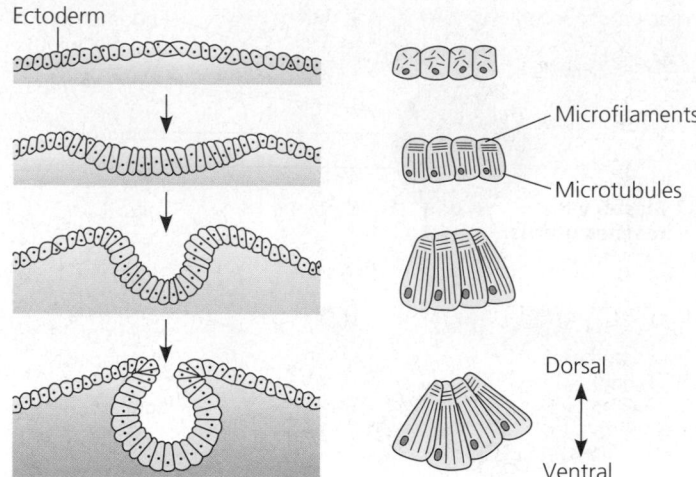

FIGURE 47.16 Change in cellular shape during morphogenesis. Reorganization of the cytoskeleton is associated with morphogenetic changes in embryonic tissues. As indicated here, in the formation of the neural tube of vertebrates, microtubules help elongate the cells of the neural plate. Microfilaments at the dorsal end of the cells may then contract, deforming the cells into wedge shapes and causing the ectoderm to bend inward.

deeper into the embryo involves the extension of filopodia by cells at the leading edge of the migrating tissue. By dragging behind them the cells that follow them through the blastopore, the leading cells help move a sheet of cells from the embryo's surface into the blastocoel, where it forms the endoderm and mesoderm of the embryo (see FIGURE 47.9). There are also many situations in which cells migrate individually, as when the cells of the neural crest disperse to various parts of the embryo.

Cell crawling is also involved in a type of morphogenetic movement called **convergent extension** (FIGURE 47.17). In convergent extension, the cells of a tissue layer rearrange themselves in such a way that the sheet of cells becomes narrower (converges) while it becomes longer (extends). When many cells wedge between one another, the tissue can extend dramatically. Convergent extension is important in early embryonic development. It occurs, for example, as the archenteron elongates in sea urchin embryos and during involution in the frog gastrula.

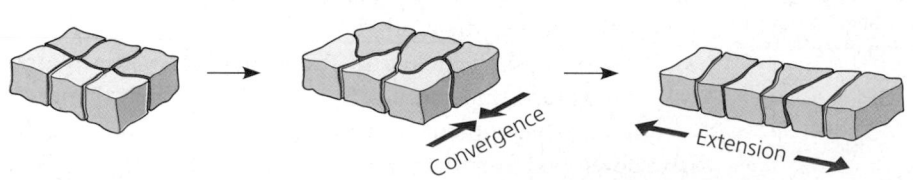

FIGURE 47.17 Convergent extension of a sheet of cells. In this simplified diagram, changes in cell shape and position cause the layer of cells to become narrower and longer. An unknown signal causes the cells to elongate in a particular directon and to crawl between each other. The result of this convergence is the extension of the cell sheet in a direction perpendicular to the convergence.

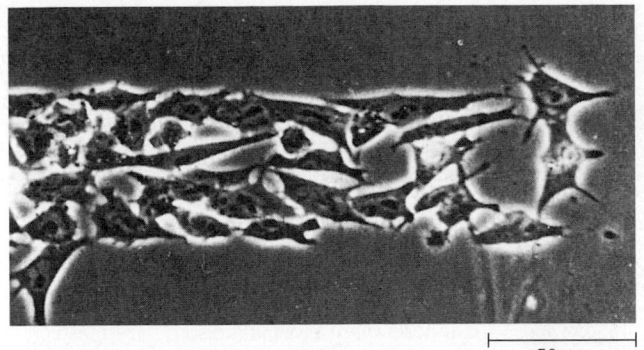

FIGURE 47.18 The extracellular matrix and cell migration. The micrographs here provide evidence that the orientation of the extracellular fibers of the ECM is controlled by the orientation of the cytoskeleton in the cells that secrete the extracellular materials. This is one way that a group of cells can influence the path along which another group of cells migrates during the development of tissues and organs.

50 μm

(a) Cells from the neural crest migrate along a strip of fibronectin fibers placed on an artificial underlayer (LM).

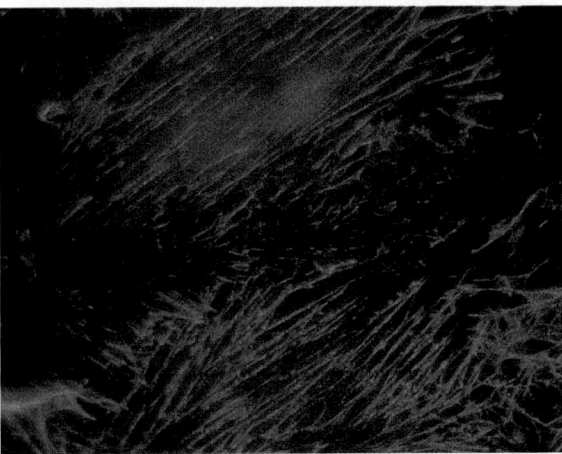 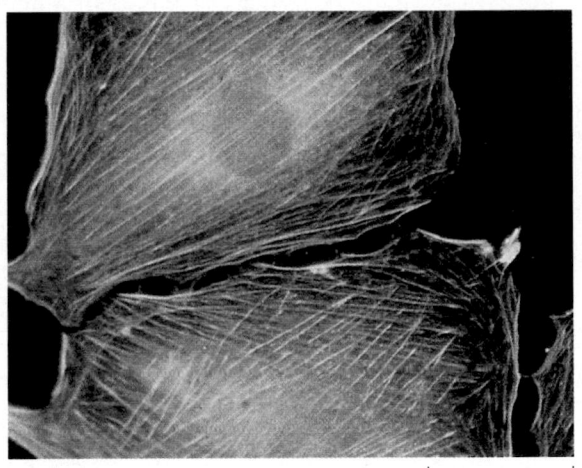

25 μm

(b) Two different fluorescent dyes demonstrate the close relationship between the orientation of fibronectin fibers of the ECM (left) and that of contractile microfilaments of the cytoskeleton (right) for two migrating cells (LMs). Notice that the orientations of the intracellular microfilaments and extracellular fibers correspond.

Scientists do not yet understand exactly what triggers and directs convergent extension. However, the process probably involves the extracellular matrix (ECM), the mixture of secreted glycoproteins lying outside the plasma membranes of cells (see FIGURE 7.29). The ECM is known to help guide cells in many types of morphogenetic movements. ECM fibers may function as tracks, directing migrating cells along particular routes (FIGURE 47.18). Several kinds of extracellular glycoproteins, including various fibronectins, help cells move by providing anchorage for cell crawling. Other substances in the ECM keep cells on the correct paths by *inhibiting* migration in certain directions. Thus, depending on the substances they secrete, nonmigratory cells situated along migration pathways may promote or inhibit movement of other cells. As migrating cells move along specific paths through the embryo, receptor proteins on their surfaces pick up directional cues from the immediate environment. Signals from the receptors direct the cytoskeletal elements to propel the cell in the proper direction.

A good example of the role of the ECM in embryonic cell migration is the movement of mesoderm cells along fibro-

nectin during amphibian gastrulation. Fibronectin fibers line the roof of the blastocoel, and as the mesoderm moves into the interior of the embryo, the cells at the free edge of the mesoderm sheet migrate along the fibers. Researchers can disrupt the ability of cells to attach to fibronectin by, for example, injecting embryos with antibodies against fibronectin. Such a disruption can block the inward movement of the mesoderm.

The glycoproteins that attach migrating cells to an underlying ECM also play a role in holding cells together when migrating cells reach their destinations and tissues and organs take shape. Also contributing to cell migration and stable tissue structure are glycoproteins called **cell adhesion molecules (CAMs),** which are located on the surfaces of cells and bind to CAMs on other cells. CAMs vary in either amount or chemical identity from one type of cell to another, and these differences help regulate morphogenetic movements and tissue building.

One important class of cell-to-cell adhesion molecules is the **cadherins,** which are so named because they require the

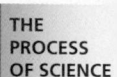

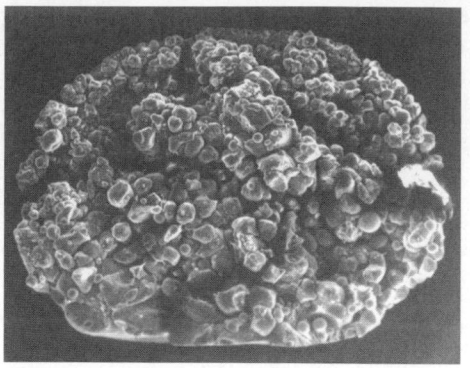

(a) Experimental embryo

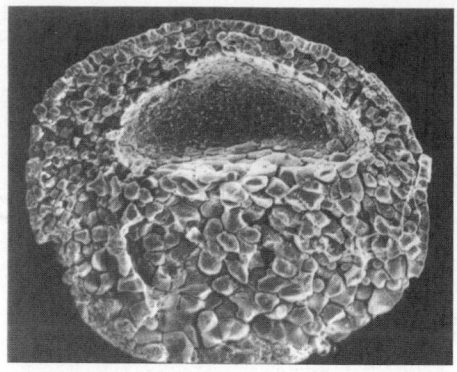

(b) Control embryo

FIGURE 47.19 The role of a cadherin in frog blastula formation. **(a)** Researchers injected *Xenopus* oocytes with nucleic acid complementary to the mRNA encoding a cadherin known as EP cadherin. The "antisense" nucleic acid causes the destruction of the normal mRNA so that no protein is produced (see discussion of RNA interference in Chapter 20). The absence of EP cadherin protein interfered with the development of the blastula. The blastocoel did not form properly, and the cells arranged themselves in a disorganized fashion. **(b)** In control embryos, the blastocoel formed normally.

THE PROCESS OF SCIENCE

presence of calcium ions for proper function. There are many different cadherins, and the gene for each cadherin is expressed in specific locations at specific times during embryonic development. Researchers have vividly demonstrated the importance of one particular cadherin in the formation of the frog blastula (FIGURE 47.19).

The developmental fate of cells depends on cytoplasmic determinants and cell-cell induction: *a review*

Coupled with the morphogenetic changes that give an animal and its parts their characteristic shapes, development also requires the timely differentiation of many kinds of cells in specific locations. As we discussed in Chapter 21, two general principles integrate our current knowledge of the genetic and cellular mechanisms that underlie differentiation during embryonic development:

1. *In many animal species (mammals may be a major exception), the heterogeneous distribution of cytoplasmic determinants in the unfertilized egg leads to regional differences in the early embryo.* By partitioning the heterogeneous cytoplasm of a polarized egg, cleavage parcels out different mRNAs, proteins, and other molecules to different blastomeres. These local differences in cytoplasmic composition help specify the body axes and influence the expression of genes that affect the developmental fate of cells. Thus, cytoplasmic determinants are responsible for the initial differences between cells in the early embryos of many kinds of animals.
2. *Subsequently, in induction, interactions among the embryonic cells themselves induce changes in gene expression. These interactions eventually bring about the differentiation of the many specialized cell types making up a new animal.* Induction may be mediated by diffusible chemical signals, or if the cells are actually in contact, by cell-surface interactions.

(In Chapter 21 we saw how both types of cell signaling are involved in the development of the nematode vulva.)

It will help to keep these two principles in mind when we take a closer look at the molecular and cellular mechanisms of differentiation and pattern formation in vertebrate development. First, however, let's look at some historic experiments that provided early researchers with information about cell fates.

Fate mapping can reveal cell genealogies in chordate embryos

You may recall that in the case of the nematode *Caenorhabditis elegans*, it has been possible to map the developmental history of every cell, beginning with the first cleavage division of the zygote (see FIGURE 21.4). This sort of complete cell lineage has not been determined for other animals, but biologists have been making more general territorial diagrams of embryonic development, called **fate maps,** for over 70 years. Classic studies performed in the 1920s by German embryologist W. Vogt established that in embryos whose axes are defined early in development, it is often possible to ascertain which parts of the embryo will be derived from each region of the zygote or blastula. Following the lead of earlier work on marine worms and mollusks, Vogt charted fate maps for amphibian embryos (FIGURE 47.20a). Using nontoxic dyes, he labeled cells of different regions of the surface of amphibian blastulas with different colors and later sectioned the embryos to see where the colors turned up. Vogt's results were among the earliest indications that the lineage (the "genealogy") of cells making up the three germ layers created by gastrulation is traceable to cells in the blastula (compare FIGURES 47.20a and 47.10). Later researchers developed more sophisticated techniques that allowed them to mark an individual blastomere during cleavage and then follow the marker as it was distributed to all the mitotic descendants of that cell (FIGURE 47.20b).

THE PROCESS OF SCIENCE

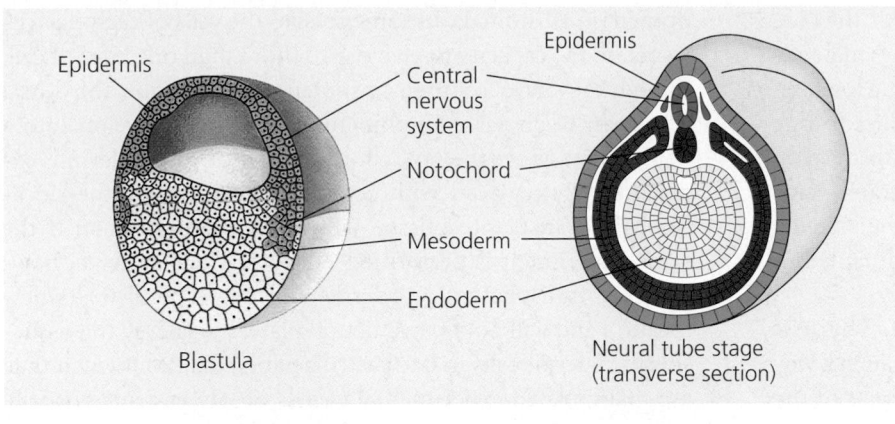

Epidermis

Central
nervous
system

Notochord

Mesoderm

Endoderm

Blastula

Epidermis

Neural tube stage
(transverse section)

(a) Fate map of a frog embryo. The fates of cells of a frog embryo were determined in part by marking different regions of the blastula surface with dyes of various colors and then determining the locations of dyed cells at later stages of development, such as at this neural tube stage.

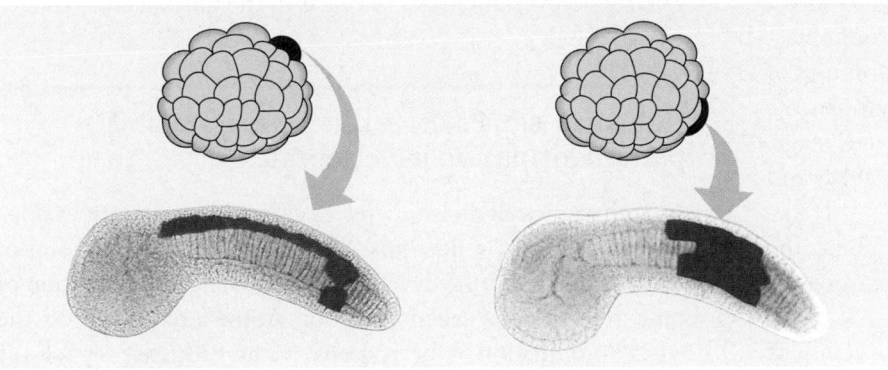

(b) Cell lineage analysis in a tunicate. Shown here are the results of a very specific type of fate map called a cell lineage analysis. The drawings depict 64-cell embryos of a tunicate, an invertebrate chordate. An individual cell can be injected with a dye, enabling a researcher to determine which cells in a later embryo are derived from the marked cells. The two light micrographs of larvae contrast the regions that develop from the two different blastomeres indicated in the drawings.

FIGURE 47.20
Fate maps for two chordates.

Developmental biologists have combined fate-mapping studies with experimental manipulation of parts of embryos in which they study whether a cell's fate can be changed by moving it elsewhere in the embryo. Two important conclusions have emerged. First, in most animals, certain early "founder cells" generate specific tissues of the older embryo. Second, as development proceeds, a cell's *developmental potential*—the range of structures that it can give rise to—becomes restricted. Starting with the normal embryo's fate map, researchers can examine how the differentiation of cells is altered in experimental situations or in mutant embryos.

The eggs of most vertebrates have cytoplasmic determinants that help establish the body axes and differences among cells of the early embryo

In order to understand at the molecular level how embryonic cells acquire their fates, we must return to the question of how the basic axes of the embryo are established. It turns out that much of what you learned in Chapter 21 about the roles of cytoplasmic determinants in the development of fruit flies also applies to vertebrate animals.

Polarity and the Basic Body Plan

As you have learned, a bilaterally symmetrical animal has an anterior-posterior axis, a dorsal-ventral axis, and left and right sides. Establishing this basic body plan is a first step in morphogenesis and is prerequisite to the development of tissues and organs. In mammals, polarity is not obvious until after cleavage, although recent research suggests that the location at which the sperm enters the egg plays a role in determining the axes. However, in most species, basic instructions (where the head end will be, and so on) are set down earlier. For example, in many frogs, as we saw in FIGURE 47.7, the locations of melanin and yolk in the unfertilized egg respectively define the animal and vegetal hemispheres, which in turn determine the anterior-posterior body axis. Fertilization then triggers the formation of the gray crescent, whose position determines the dorsal-ventral axis.

Restriction of Cellular Potency

The asymmetric distribution of cytoplasmic determinants in an egg does not necessarily lead to differences among the earliest blastomeres. The first cleavage may occur along an axis that produces two identical blastomeres, which will have equal developmental potential. This is what happens in amphibians, for instance. The first two blastomeres are similar, and if they are experimentally separated, each can develop into a normal tadpole. In other words, both blastomeres are totipotent. However, in experiments in which the entire gray crescent goes to one blastomere, only that blastomere has the capacity to develop into a normal tadpole when the blastomeres are separated (FIGURE 47.21). Thus, the fates of embryonic cells can be affected not only by the distribution of cytoplasmic determinants but also by the zygote's characteristic pattern of cleavage.

In many species, only the zygote is totipotent. The first cleavage plane divides cytoplasmic determinants in such a way that each blastomere will give rise only to specific parts of the embryo. In contrast, the cells of mammalian embryos remain totipotent until they become arranged into the trophoblast and inner cell mass of the blastocyst. The early blastomeres of mammals seem to receive equivalent amounts of cytoplasmic components from the egg. Indeed, up to the eight-cell stage, the blastomeres of a mammalian embryo all look alike, and each can form a complete embryo if isolated.

However similar or different the cells in an early embryo, the progressive restriction of potency is a general feature of development in all animals. In some species the cells of early gastrulas retain the capacity to give rise to more than one kind of cell, though they have lost their totipotency. If left alone, the dorsal ectoderm of an early amphibian gastrula will develop into a neural plate above the notochord. If the dorsal ectoderm is experimentally replaced with ectoderm from some other location, the transplanted tissue will form a neural plate. If the same experiment is performed on a late-stage gastrula, however, the transplanted ectoderm will not respond to its new location and will not form a neural plate. In general, the tissue-specific fates of cells in late gastrulae are fixed. Even when they are manipulated experimentally, cells of late gastrulae usually give rise to the same types of cell as in the normal embryo.

Inductive signals drive differentiation and pattern formation in vertebrates

As embryonic cell division creates cells with different developmental potential, a new possibility arises. Now one group of cells can influence the development of a neighboring group of cells, in the process called induction. At the molecular level, the effect of induction—the response to an inductive signal—is usually the switching on of a set of genes that make the receiving cells differentiate into a specific tissue. You have already learned about the role of induction in vulval development in *C. elegans* (see FIGURE 21.17). Induction is an essential process in the development of many tissues in other animals as well.

The "Organizer" of Spemann and Mangold

The importance of induction in amphibian development was dramatically demonstrated in the 1920s by German zoologist Hans Spemann and his student Hilde Mangold. In a series of transplantation experiments, they discovered that the dorsal lip of the blastopore in the early gastrula plays a key role in embryonic development, initiating a chain of inductions that results in the formation of the neural tube and other organs. In their most famous experiment, they took a piece of dorsal lip from one embryo and grafted it on the ventral side of a second embryo (FIGURE 47.22). The result was the development of a second notochord and neural tube in the recipient embryo, at the location of the graft. Subsequently, other organs and structures formed, creating a nearly complete second embryo attached to the first. Spemann referred to the dorsal lip of the blastopore as the *primary organizer* of the embryo because of its early and crucial role in development. (We now know of other "organizers" that act even earlier than the dorsal lip of the blastopore.)

Developmental biologists are working intensively to identify the molecular basis of induction by Spemann's organizer. An important clue has come from studies of a growth factor called *bone morphogenetic protein 4* (BMP-4). (Bone morphogenetic

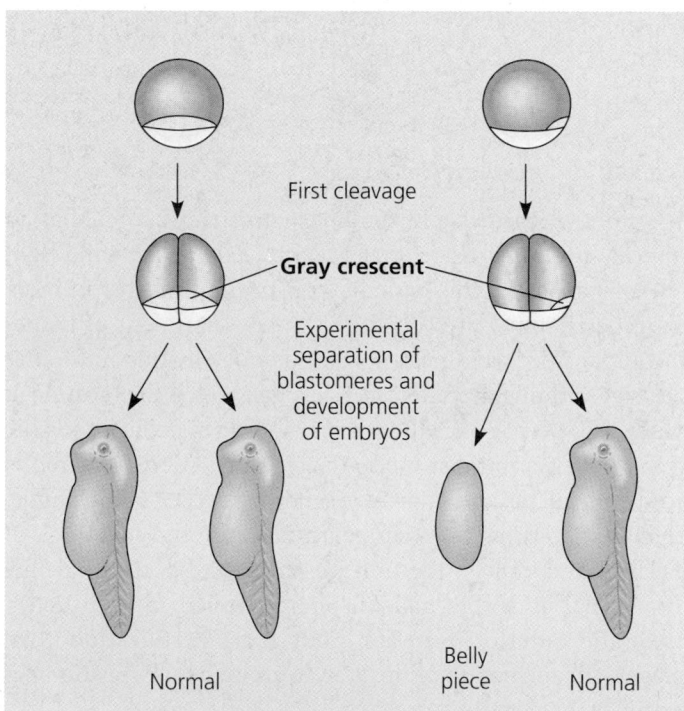

FIGURE 47.21 Experimental demonstration of the importance of cytoplasmic determinants in amphibians. The first cleavage division of an amphibian zygote normally divides the gray crescent evenly between the two blastomeres, which retain their totipotency after the division (left). However, when one blastomere receives the entire gray crescent (right), only that blastomere develops normally. The other blastomere, lacking cytoplasmic determinants present in the crescent, gives rise to an abnormal embryo without dorsal structures.

First cleavage

Gray crescent

Experimental separation of blastomeres and development of embryos

Normal

Belly piece

Normal

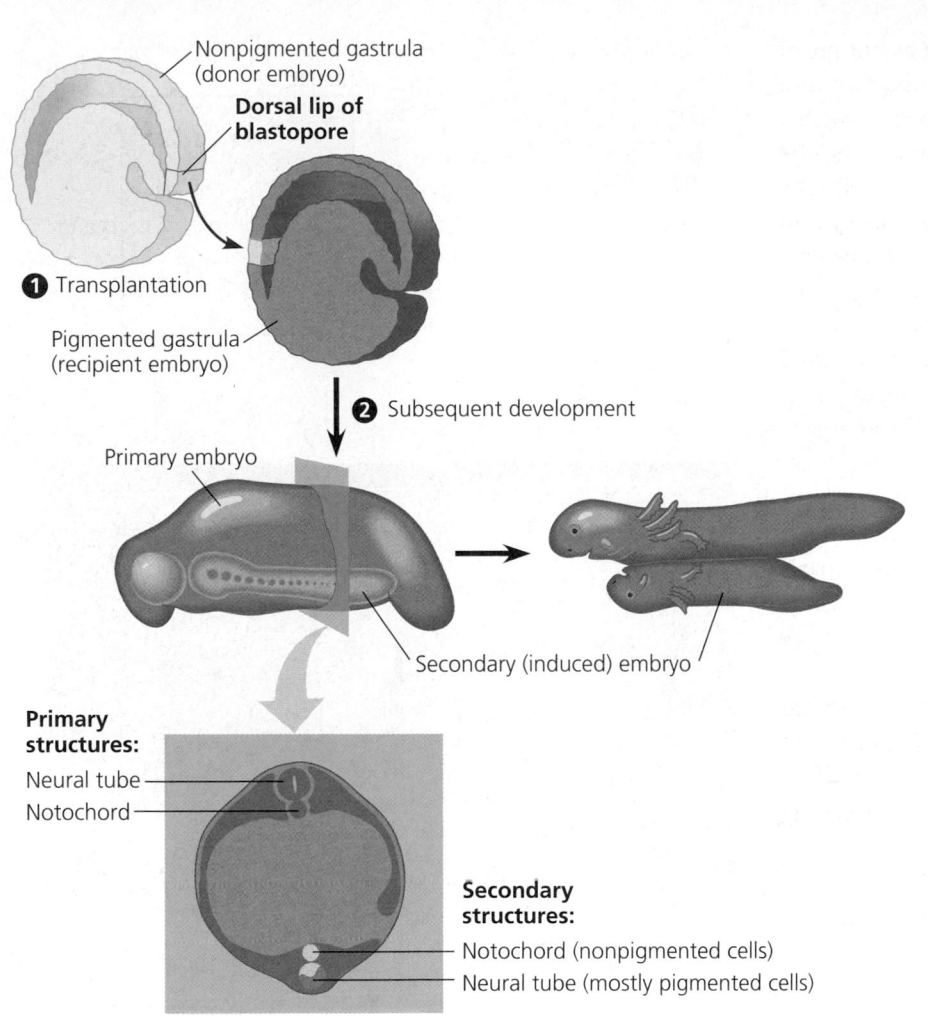

Nonpigmented gastrula
(donor embryo)

**Dorsal lip of
blastopore**

❶ Transplantation

Pigmented gastrula
(recipient embryo)

❷ Subsequent development

Primary embryo

Secondary (induced) embryo

**Primary
structures:**

Neural tube

Notochord

**Secondary
structures:**

Notochord (nonpigmented cells)

Neural tube (mostly pigmented cells)

❶ Spemann and Mangold transplanted a piece of the dorsal lip of a nonpigmented newt gastrula (shown in gray) to the ventral side of the early gastrula of a pigmented newt (red).

❷ The recipient embryo formed a second notochord and neural tube in the region of the transplant, and eventually most of a second embryo. When Mangold and Spemann examined the interior of the double embryo, they found that many cells of the secondary structures came from the (pigmented) recipient, not the (nonpigmented) donor of the graft. This finding indicated that the grafted tissue had "organized" (induced) cells of the recipient to form the extra structures.

FIGURE 47.22 The "organizer" of Spemann and Mangold. The dorsal lip of the blastopore in the early gastrula of an amphibian plays a critical role in inducing the development of other parts of the embryo, as Hans Spemann and Hilde Mangold demonstrated with this experiment in 1924.

THE PROCESS OF SCIENCE

proteins, a family of related proteins with a variety of developmental roles, derive their name from members of the family that are important in bone formation.) Amphibian BMP-4 is active exclusively in cells on the *ventral* side of the gastrula. One major function of organizer cells seems to be to inactivate BMP-4 on the dorsal side of the embryo by producing proteins that bind to BMP-4, rendering it unable to signal. Proteins related to BMP-4 and its inhibitors are also found in other animals, including invertebrates such as the fruit fly. The ubiquity of these molecules suggests that they evolved long ago and may participate in the development of many different organisms.

The induction that causes dorsal ectoderm to develop into the neural tube is only one of many cell-cell interactions that transform the three germ layers into organ systems. Many inductions seem to involve a sequence of inductive steps that progressively determine the fate of cells. For example, in the late gastrula of the frog, ectoderm cells destined to become the lenses of the eyes receive inductive signals from the ectodermal cells that will become the neural plate. Additional inductive signals probably come from endodermal cells and mesodermal cells. Finally, inductive signals from the optic cup, an outgrowth of the developing brain, complete the determination of the lens-forming cells.

Pattern Formation in the Vertebrate Limb

Inductive signals play a major role in **pattern formation**—the development of an animal's spatial organization, the arrangement of organs and tissues in their characteristic places in three-dimensional space. The molecular cues that control pattern formation, called **positional information,** tell a cell where it is with respect to the animal's body axes and help to determine how the cell and its descendants respond to future molecular signals.

In Chapter 21 we discussed pattern formation in the development of the body segments of *Drosophila*. For understanding pattern formation in vertebrates, an important model system has been limb development in the chick. The wings and legs of chicks, like all vertebrate limbs, begin as bumps of tissue called limb buds (FIGURE 47.23). Each component of a chick limb, such as a specific bone or muscle, develops with a precise location and orientation relative to three axes: the proximal-distal axis (the "shoulder-to-fingertip" axis), the anterior-posterior axis (the "thumb-to-little finger" axis), and the dorsal-ventral axis (the "knuckle-to-palm" axis). The embryonic cells within a limb bud respond to positional information indicating location along these three axes.

A limb bud consists of a core of mesodermal tissue covered by a layer of ectoderm. By removing or transplanting different pieces of tissue, researchers have discovered two critical organizer regions in the limb bud, regions with profound effects on the limb's development. These two organizer regions are present in all vertebrate limb buds, including those that will develop into forelimbs (such as wings or arms) and those destined to become hindlimbs. In recent years researchers have established that the cells of these regions secrete proteins that provide key positional information to the other cells of the bud.

One limb-bud organizer region is the **apical ectodermal ridge (AER),** a thickened area of ectoderm at the tip of the bud (see FIGURE 47.23a). The AER is required for the outgrowth of the limb along the proximal-distal axis and for patterning along this axis. The cells of the AER produce and secrete several proteins of the fibroblast growth factor (FGF) family. These proteins appear to be the growth signal that promotes limb bud outgrowth. If the AER is surgically removed and beads soaked with FGF are put in its place, a nearly normal limb will develop. The AER and other ectoderm of the limb bud also appear to guide pattern formation along the limb's dorsal-ventral axis. In experiments where the ectoderm of a limb bud is detached from the mesoderm and then replaced with its orientation rotated 180°, the limb elements that form have reversed dorsal-ventral orientation. (This is equivalent to reversing the palm and back of your hand.)

The second major limb-bud organizer region is the **zone of polarizing activity (ZPA),** located where the posterior side of the bud is attached to the body. The ZPA is necessary for proper pattern formation along the anterior-posterior axis of the limb. Cells nearest the ZPA give rise to posterior structures, such as the digit homologous to our little finger, and cells farthest from the ZPA form anterior structures, such as the avian equivalent of our thumb. A tissue transplantation experiment demonstrating the importance of the ZPA is shown in FIGURE 47.24. Grafting a second ZPA on the anterior side of a limb bud leads to the formation of extra digits, in a mirror-image arrangement.

In the ZPA-transplantation experiment, the extra digits develop from the host limb bud and not from the graft, supporting the hypothesis that the grafted ZPA produces some sort of inductive signal. Indeed, researchers have discovered that the cells of the ZPA secrete an important

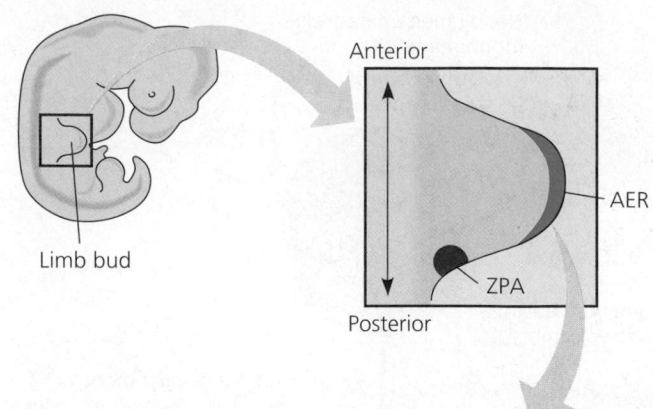

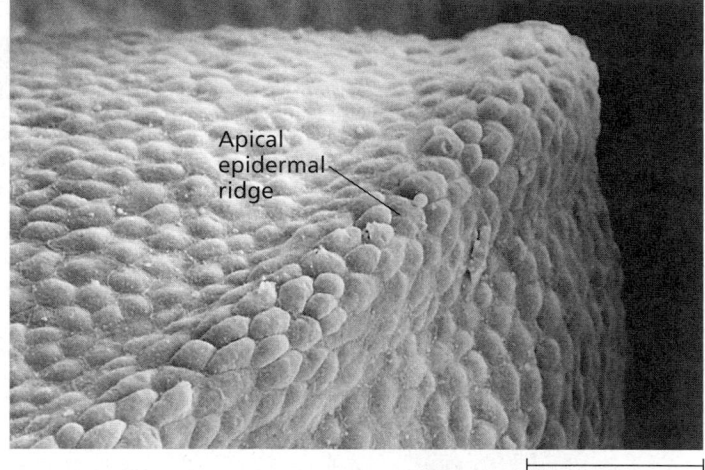

(a) Vertebrate limbs develop from protrusions called limb buds, each consisting of mesoderm cells covered by a layer of ectoderm. Two regions play key "organizer" roles in limb pattern formation: the apical ectodermal ridge (AER) and the zone of polarizing activity (ZPA). (SEM)

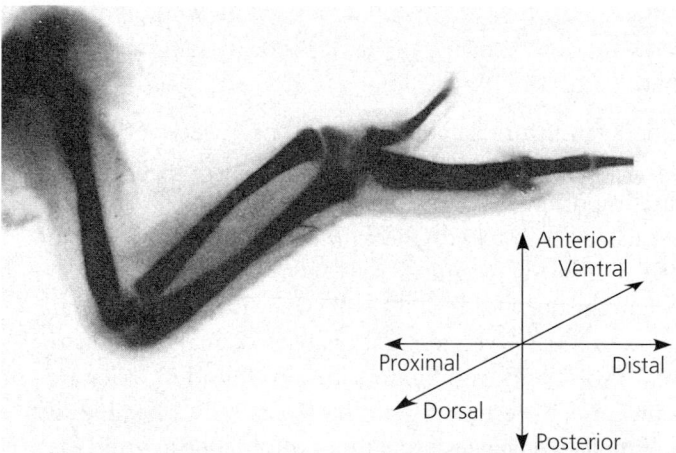

(b) As the bud develops into a limb, such as this wing of a chick embryo, a specific pattern of tissues emerges. Pattern formation requires each embryonic cell to receive some kind of positional information indicating location along the three axes of the limb. The AER and ZPA secrete molecules that help provide this information.

FIGURE 47.23 Organizer regions in vertebrate limb development.

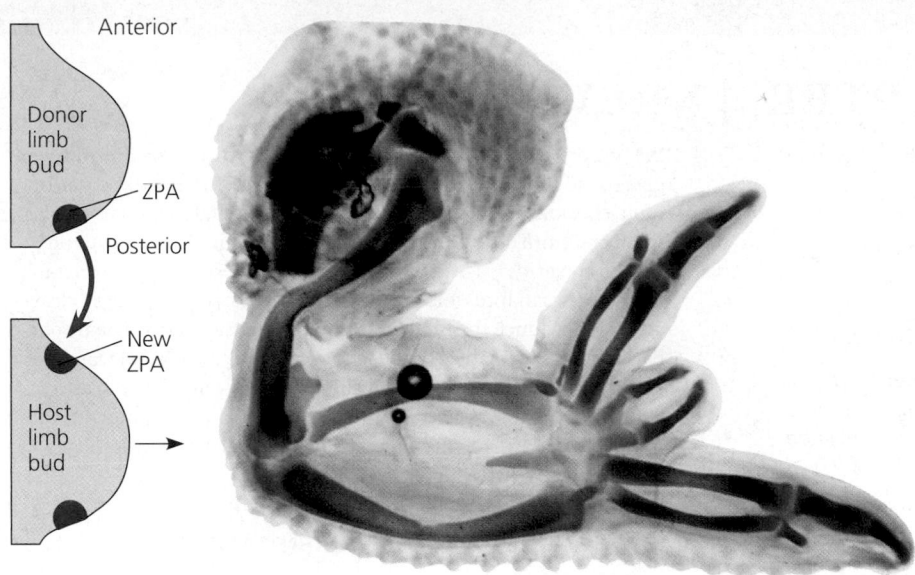

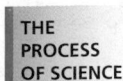
FIGURE 47.24 The experimental manipulation of positional information. The zone of polarizing activity (ZPA) issues molecular signals that establish the anterior-posterior axis of a vertebrate limb. In the experiment shown here, a second ZPA was added to the anterior margin of a chick limb bud by transplanting ZPA tissue from another limb bud. Cells near the grafted ZPA, like cells near the host bud's own ZPA, apparently receive positional information indicating "posterior." The pattern that emerges in the developing limb is a mirror image, with an arrangement of digits equivalent to two human hands joined together at the thumbs. In a subsequent experiment, researchers showed that a protein called Sonic hedgehog could substitute for the ZPA cells. Thus Sonic hedgehog functions as a positional cue.

protein growth factor called *Sonic hedgehog.** If cells genetically engineered to produce large amounts of Sonic hedgehog are implanted in the anterior region of a normal limb bud, a mirror-image limb results—just as if a ZPA had been grafted there. Evidence from studying the mouse version of Sonic hedgehog suggests that extra toes in mice—and perhaps also in humans—can result from the production of the protein in the wrong part of the limb bud. Proteins very similar to Sonic hedgehog have turned out to be important positional cues in a number of developmental situations, including segment formation in the *Drosophila* embryo. (The fruit-fly version, called simply Hedgehog, is the product of a segment-polarity gene; see FIGURE 21.13.)

Is Sonic hedgehog a morphogen? Recall from Chapter 21 that a morphogen is a substance that provides positional information *in the form of a concentration gradient* along an embryonic axis. Although gradients may be involved in the functioning of Sonic hedgehog, the Sonic hedgehog protein itself may not act in a graded manner; instead it may induce the production of something else that does. Researchers are currently investigating this issue.

---

* Sonic hedgehog gets its name from two sources, its similarity to a *Drosophila* protein called Hedgehog, which is involved in segmentation of the fly embryo, and a video game character.

In any case, we can conclude from experiments like the one in FIGURE 47.24 that pattern formation requires cells to receive and interpret environmental cues that vary from one location to another. These cues, acting together along three axes, tell cells where they are in the three-dimensional realm of a developing organ. In vertebrate limb development, we now know that specific proteins serve as some of these cues. In other words, organizer regions such as the AER and the ZPA function as signaling centers.

What determines whether a limb bud develops into a forelimb or a hindlimb? The cells receiving the signals from the AER and ZPA respond according to their developmental histories. Before the AER or ZPA issues its signals, earlier developmental signals have set up patterns of gene expression distinguishing the future forelimbs from the future hindlimbs. These differences cause the cells of the forelimb and hindlimb limb buds to react differently to the same positional cues.

A hierarchy of gene activations eventually affects the expression of homeobox-containing (*Hox*) genes in cells of the developing limb. *Hox* genes seem to be involved in specifying the identity of various regions of the limbs, as well as of the body as a whole (see Chapter 21). Thus, pattern formation is a chain of events involving many steps of signaling and differentiation. Various pathways of pattern formation occur in all the different parts of the developing embryo, eventually producing the final set of differentiated structures in the fully formed animal.

CHAPTER 47 REVIEW

Go to the Campbell Biology website (www.campbellbiology.com) to explore an interactive version of the Chapter Review.

Summary of Key Concepts

THE STAGES OF EARLY EMBRYONIC DEVELOPMENT

■ **From egg to organism, an animal's form develops gradually:** *the concept of epigenesis* (p. 999) An embryo is not preformed in an egg; it develops by epigenesis, the gradual, gene-directed acquisition of form.

■ **Fertilization activates the egg and brings together the nuclei of sperm and egg** (pp. 999–1002, FIGURE 47.2) Fertilization both reinstates diploidy and activates the egg to begin a chain of metabolic reactions that triggers the onset of embryonic development. The acrosomal reaction, which occurs when the sperm meets the egg, releases hydrolytic enzymes that digest through material surrounding the egg. Gamete fusion depolarizes the egg cell membrane and sets up a fast block to polyspermy. Sperm-egg fusion also initiates the cortical reaction, involving a signal-transduction pathway in which calcium ions stimulate cortical granules to erect a fertilization envelope that functions as a slow block to polyspermy. In mammalian fertilization, the cortical reaction hardens the zona pellucida as a slow block to polyspermy.

■ **Cleavage partitions the zygote into many smaller cells** (pp. 1002–1005, FIGURE 47.8) Fertilization is followed by cleavage, a period of rapid cell division without growth, which results in the production of a large number of cells called blastomeres. Holoblastic cleavage, or division of the entire egg, occurs in species whose eggs have little or moderate amounts of yolk. Meroblastic cleavage, incomplete division of the egg, occurs in species with yolk-rich eggs. Cleavage planes usually follow a specific pattern relative to the animal and vegetal poles of the zygote. In many species, cleavage creates a multicellular ball called the blastula, which contains a fluid-filled cavity, the blastocoel.

Web/CD Activity 47A: Sea Urchin Development Video

■ **Gastrulation rearranges the blastula to form a three-layered embryo with a primitive gut** (pp. 1005–1007, FIGURE 47.10) Gastrulation transforms the blastula into a gastrula, which has a rudimentary digestive cavity (the archenteron) and three embryonic germ layers: the ectoderm, endoderm, and mesoderm.

Web/CD Case Study in the Process of Science: What Determines Cell Differentiation in the Sea Urchin?

■ **In organogenesis, the organs of the animal body form from the three embryonic germ layers** (p. 1007, FIGURE 47.11) Early events in organogenesis in vertebrates include formation of the notochord by condensation of dorsal mesoderm, development of the neural tube from folding of the ectodermal neural plate, and formation of the coelom from splitting of lateral mesoderm.

Web/CD Activity 47B: Frog Development Video

■ **Amniote embryos develop in a fluid-filled sac within a shell or uterus** (pp. 1007–1011, FIGURES 47.12, 47.14, 47.15) Meroblastic cleavage in the yolk-rich, shelled eggs of birds and reptiles is restricted to a small disc of cytoplasm at the animal pole. A cap of cells called the blastodisc forms and begins gastrulation with the formation of

the primitive streak. In addition to the embryo, the three germ layers give rise to the four extraembryonic membranes: the yolk sac, amnion, chorion, and allantois. The eggs of placental mammals are small and store little food, exhibiting holoblastic cleavage with no obvious polarity. Gastrulation and organogenesis, however, resemble the processes in birds and reptiles. After fertilization and early cleavage in the oviduct, the blastocyst implants in the uterus. The trophoblast initiates formation of the fetal portion of the placenta, and the embryo proper develops from a single layer of cells, the epiblast, within the blastocyst. Extraembryonic membranes homologous to those of birds and reptiles function in intrauterine development.

THE CELLULAR AND MOLECULAR BASIS OF MORPHOGENESIS AND DIFFERENTIATION IN ANIMALS

■ **Morphogenesis in animals involves specific changes in cell shape, position, and adhesion** (pp. 1012–1014, FIGURES 47.16, 47.17) Cytoskeletal rearrangements are responsible for changes in both shape and position of cells. Both kinds of changes are involved in tissue invaginations, as occurs in gastrulation, for example. The extracellular matrix provides anchorage for cells and also helps guide migrating cells toward their destinations. Cell adhesion molecules on cell surfaces are also important for cell migration and for holding cells together in tissues.

■ **The developmental fate of cells depends on cytoplasmic determinants and cell-cell induction:** *a review* (p. 1014)

■ **Fate mapping can reveal cell genealogies in chordate embryos** (pp. 1014–1015, FIGURE 47.20) Experimentally derived fate maps of embryos have shown that specific regions of the zygote or blastula develop into specific parts of older embryos.

■ **The eggs of most vertebrates have cytoplasmic determinants that help establish the body axes and differences among cells of the early embryo** (pp. 1015–1016, FIGURE 47.21) When cytoplasmic determinants are heterogeneously distributed in an egg, they serve as the basis for setting up differences among parts of the egg and, later, among the blastomeres resulting from cleavage of the zygote. Cells that receive different cytoplasmic determinants undergo different fates.

■ **Inductive signals drive differentiation and pattern formation in vertebrates** (pp. 1016–1019, FIGURES 47.22–47.24) Cells in a developing embryo receive and interpret positional information that varies with location. This information is often in the form of signal molecules secreted by cells in special "organizer" regions of the embryo, such as the dorsal lip of the blastopore in the amphibian gastrula and the apical ectodermal ridge of the vertebrate limb bud. The signal molecules influence gene expression in the cells that receive them, leading to differentiation and the development of particular structures.

Self-Quiz

1. The cortical reaction of sea urchin eggs functions directly in the
   a. formation of a fertilization envelope.
   b. production of a fast block to polyspermy.
   c. release of hydrolytic enzymes from the sperm cell.
   d. generation of an electrical impulse by the egg cell.
   e. fusion of egg and sperm nuclei.

2. Which of the following is common to both avian and mammalian development?
   a. holoblastic cleavage
   b. epiblast and hypoblast
   c. trophoblast
   d. yolk plug
   e. gray crescent

3. The archenteron develops into
   a. the mouth in protostomes.
   b. the blastocoel.
   c. the endoderm.
   d. the lumen of the digestive tract.
   e. the placenta.

4. In a frog embryo, the blastocoel is
   a. completely obliterated by yolk platelets.
   b. lined with endoderm during gastrulation.
   c. located primarily in the animal hemisphere.
   d. the cavity that becomes the coelom.
   e. the cavity that later forms the archenteron.

5. Amphibians, unlike reptiles, generally lay their eggs in water or moist places. This difference is related to the absence (in amphibians) versus the presence (in reptiles) of
   a. extraembryonic membranes.
   b. yolk.
   c. cleavage.
   d. gastrulation.
   e. development of the brain from ectoderm.

6. In an amphibian embryo, a band of cells called the neural crest
   a. rolls up to form the neural tube.
   b. develops into the main sections of the brain.
   c. produces cells that migrate to form teeth, skull bones, and other structures in the embryo.
   d. has been shown by experiments to be the organizer region of the developing embryo.
   e. induces the formation of the notochord.

7. Differences in the development of different cells in the early frog embryo (zygote to blastula) are due to
   a. the differences between meroblastic and holoblastic cleavage.
   b. the heterogeneous distribution of cytoplasmic determinants, such as proteins and mRNA.
   c. inductive interactions occurring between the developing cells.
   d. concentration gradients for regulatory molecules such as BMP-4.
   e. the position of the cells relative to the zone of polarizing activity (ZPA).

8. During convergent extension
   a. cells on the opposite side of the embryo follow converging developmental pathways leading to bilateral symmetry.
   b. the cells of the neural folds adhere to one another to complete the neural tube.
   c. the cells of a tissue layer reorganize forming a narrowed elongated sheet.
   d. the dorsal-ventral axis is established.
   e. cell adhesion molecules are expressed, causing the eight blastomeres to adhere tightly to one another.

9. In the early development of an amphibian embryo, an important "organizer" is the
   a. neural tube.
   b. notochord.
   c. archenteron roof.
   d. dorsal lip of the blastopore.
   e. dorsal ectoderm.

10. The occurrence of genetically identical twins indicates that in humans
    a. only the zygote is totipotent.
    b. the progressive restriction of potency hypothesis does not apply.

c. the first cleavage event must be transverse to the animal-vegetal axis of the zygote.
d. cell divisions producing the earliest blastomeres do not result in an asymmetric distribution of cytoplasmic determinants.
e. the primary organizer continues to function well past gastrulation.

11. Gastrulation in a frog embryo forms a new cavity, the _____, which is lined by _____ and which develops into the animal's _____ tract.

12. What is the embryonic basis for the dorsal, hollow nerve cord that is common to all members of our phylum?

13. What are somites?

14. What usually comes first in the developmental history of a cell in an animal embryo, its migration within the embryo or its differentiation into a specialized cell?

15. Name the four extraembryonic membranes of a developing chick. Which of these are also present in mammals?

Go to the Website or CD-ROM for more quiz questions.

Evolution Connection

Evolution in both insects and vertebrates has involved the repeated duplication of body segments, followed by the fusion of some segments and specialization of their structure and function. What parts of vertebrate anatomy reflect the vertebrate segmentation pattern? Can you guess what vertebrate body part is a product of segment fusion and specialization?

The Process of Science

The "snout" of a frog tadpole bears a sucker. A salamander tadpole has a mustache-shaped structure called a balancer in the same area. You perform an experiment in which you transplant ectoderm from the side of a young salamander embryo to the snout of a frog embryo. You find that the tadpole that develops has a balancer. When you transplant ectoderm from the side of a slightly older salamander embryo to the snout of a frog embryo, the frog tadpole ends up with a patch of salamander skin on its snout. Suggest a hypothesis to explain the results of this experiment in terms of the mechanisms of development. How might you test your hypothesis?

Conduct transplantation experiments of primary mesenchyme cells in the sea urchin embryo to study cell differentiation in the Case Study in the Process of Science, available on the website and CD-ROM.

Science, Technology, and Society

Nerve cells transplanted from aborted fetuses can relieve the symptoms of Parkinson's disease, a brain disorder. Fetal tissue transplants might also be used to treat epilepsy, diabetes, Alzheimer's disease, and spinal cord injuries. Why might tissues from a fetus be particularly useful for replacing diseased or damaged cells? There is controversy over whether the U.S. government should allow fetal tissues from induced abortions to be used in transplant research. Opponents would allow only tissues from miscarriages to be used. Why would most researchers prefer to use tissues from surgically aborted fetuses? What is your position on this issue, and why?

Answers: 1. a; 2. b; 3. d; 4. c; 5. a; 6. c; 7. b; 8. c; 9. d; 10. d; 11. Archenteron; endoderm; digestive 12. The nerve cord, which becomes the brain and spinal cord, develops from a dorsal ectodermal plate that folds to form an interior tube. 13. In a vertebrate embryo, somites are blocks of mesoderm that give rise to segmental structures, such as the vertebrae. 14. Migration 15. Both birds and mammals have an amnion, chorion, allantois, and yolk sac.

CHAPTER 48

NERVOUS SYSTEMS

AN OVERVIEW OF NERVOUS SYSTEMS

- Nervous systems perform the three overlapping functions of sensory input, integration, and motor output
- Networks of neurons with intricate connections form nervous systems

THE NATURE OF NERVE SIGNALS

- Every cell has a voltage, or membrane potential, across its plasma membrane
- Changes in the membrane potential of a neuron give rise to nerve impulses
- Nerve impulses propagate themselves along an axon
- Chemical or electrical communication between cells occurs at synapses
- Neural integration occurs at the cellular level
- The same neurotransmitter can produce different effects on different types of cells

EVOLUTION AND DIVERSITY OF NERVOUS SYSTEMS

- The ability of cells to respond to the environment has evolved over billions of years
- Nervous systems show diverse patterns of organization

VERTEBRATE NERVOUS SYSTEMS

- Vertebrate nervous systems have central and peripheral components
- The divisions of the peripheral nervous system interact in maintaining homeostasis
- Embryonic development of the vertebrate brain reflects its evolution from three anterior bulges of the neural tube
- Evolutionarily older structures of the vertebrate brain regulate essential automatic and integrative functions
- The cerebrum is the most highly evolved structure of the mammalian brain
- Regions of the cerebrum are specialized for different functions
- Research on neuron development and neural stem cells may lead to new approaches for treating CNS injuries and diseases

The scanning electron micrograph on this page shows an unusual juxtaposition of the basic components of animal nervous systems and computers—a single nerve cell (a neuron) on the surface of a microprocessor. Your own nervous system, made of living neurons, is using processes vastly more complex than those of a computer as you read and comprehend these words. Indeed, the human nervous system is probably the most intricately organized aggregate of matter on Earth. A single cubic centimeter of the human brain may contain well over 50 million nerve cells, each of which may communicate with thousands of other neurons in information-processing networks that make the most elaborate computer look primitive. These neural pathways control our every feeling, perception, and movement and enable us to learn, remember, think, and be conscious of ourselves and our surroundings.

The nervous system, endocrine system, and immune system often cooperate and interact in regulating internal body functions and behavior (see Chapters 43 and 45). In the maintenance of homeostasis, certain parts of the brain receive and process data about the body's internal environment and correct imbalances by sending out commands to other organs. Neural and endocrine pathways can also collaborate during responses to stress— enhancing or suppressing immune function, for example.

Despite their structural and functional linkage, the nervous system and endocrine system play somewhat different roles in body coordination. With its incomparable structural complexity, the nervous system can integrate vast amounts of information, such as that required for human thought and speech. Timing can also be a key difference. The endocrine system may take minutes, hours, or even days to act, partly because of the time it takes for hormones to be made and carried in the blood to their target organs. In contrast, the nervous system is a signaling network with branches carrying information directly to and from specific sites. Neurons are specialized for the fast transmission of impulses—as quickly as 150 m/sec (over 330 mph). In humans, information can travel from the brain to the hands (or vice versa) in a few milliseconds.

Animal survival and reproduction depends on rapid and flexible responses to changes in the environment, and a diversity of nervous systems has evolved in various animal phyla. This chapter discusses the form and function of these nervous systems. As we have seen in many other organ systems, animal nervous systems are remarkably similar at the cellular level—in how neurons work, for example—but differ at higher levels of organization, such as the structure and function of their brains.

AN OVERVIEW OF NERVOUS SYSTEMS

Nervous systems perform the three overlapping functions of sensory input, integration, and motor output

In general, a nervous system has three overlapping functions: sensory input, integration, and motor output (FIGURE 48.1). **Sensory receptors,** such as the light-detecting cells in the eyes, collect information about the physical world outside the body as well as processes inside the organism; this **sensory input** is then conveyed to integration centers. Integration is the process by which the input is interpreted and associated with appropriate responses of the body. The circular arrow in FIGURE 48.1 indicates that integration is a continuous background activity. For the most part, integration is carried out in the **central nervous system (CNS),** which consists of the brain and spinal cord in vertebrates. **Motor output** is the conduction of signals from the integration center, the CNS, to **effector cells,** the muscle cells or gland cells that actually carry out the body's responses to stimuli. The signals are conducted by **nerves,** bundles of individual nerve fibers held together by connective tissue. The nerves that communicate motor and sensory signals between the central nervous system and the rest of the body are collectively called the **peripheral nervous system (PNS).** From receptor to effector, information is communicated along a pathway of neurons by a combination of electrical and chemical signals. In this chapter we concentrate on communication within the nervous system. Chapter 49 connects the nervous system to its inputs

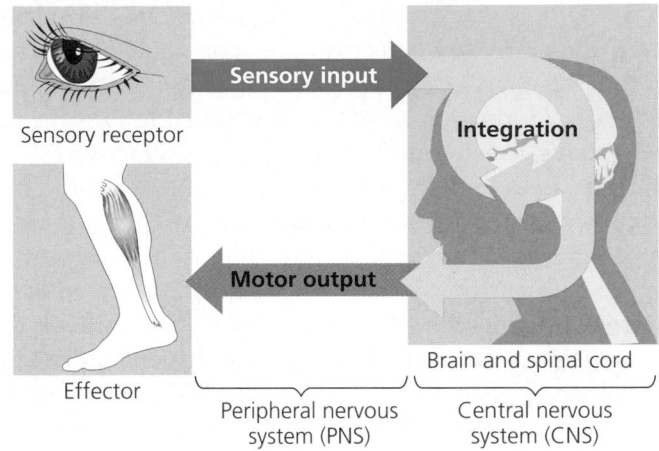

FIGURE 48.1 Overview of a vertebrate nervous system.
Integration of sensory input and motor output is not usually rigid and linear, but involves the continuous background activity symbolized by the brain's cyclical arrow in this diagram.

and outputs by discussing sensory receptors and the physiology of movement.

Networks of neurons with intricate connections form nervous systems

Neuron Structure and Synapses

The structural and functional unit of the nervous system is the **neuron,** or **nerve cell** (FIGURE 48.2a). A neuron has a **cell body,** which contains the nucleus and other organelles, and it has

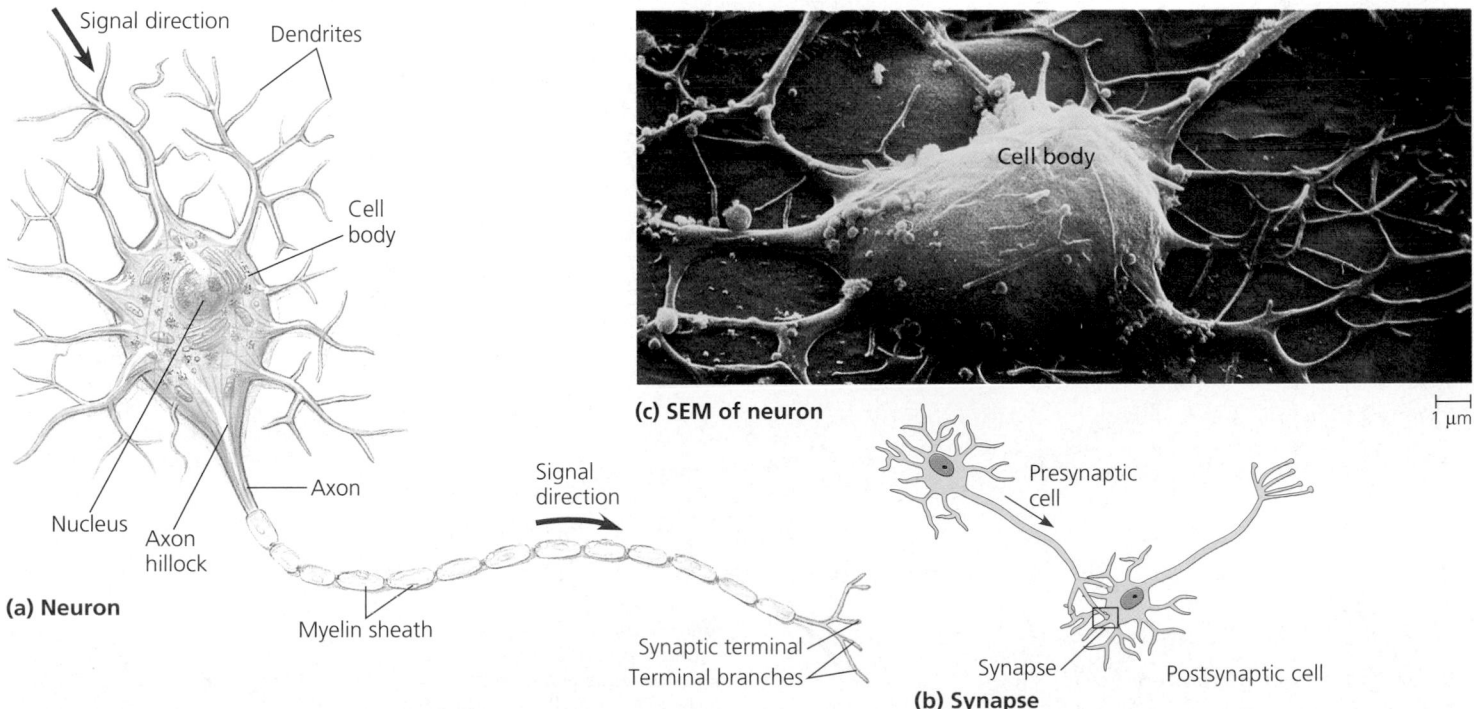

FIGURE 48.2 Structure of a vertebrate neuron.

fiberlike processes (extensions) of two general types. **Dendrites** (from the Greek *dendron,* tree) are short, highly branched processes that receive incoming messages from other cells and carry this information as an electrical signal toward the cell body. **Axons,** usually much longer than dendrites, convey outgoing messages from the neuron to other cells. A neuron may have many dendrites, but it never has more than one axon. Some axons, such as the ones connecting your spinal cord to your foot, can be over a meter long. The conical region of the axon where it joins the cell body is called the **axon hillock;** this region plays an essential role in the transmission and integration of nerve signals. Many axons are enclosed by an insulating layer called the **myelin sheath.** The supporting cells that form this sheath are described later in this section.

Axons have specialized endings called **synaptic terminals,** which relay signals from the neuron to other cells by releasing chemical messengers called **neurotransmitters.** The site of contact between a synaptic terminal and a target cell is called a **synapse** (FIGURE 48.2b). The target cell may be either another neuron or an effector, such as a muscle cell or gland cell. The transmitting cell is called the **presynaptic cell,** and the target cell is called the **postsynaptic cell.** An axon may be branched, and each branch may give rise to hundreds or thousands of synaptic terminals.

A Simple Nerve Circuit—the Reflex Arc

The simplest type of nerve circuit regulates a **reflex,** or automatic response, and is called a **reflex arc.** For example, this is the type of circuit that causes a clam to shut its shell when a shadow (which might be a potential predator) passes overhead.

The simplest reflex arcs require only two kinds of nerve cells. A **sensory neuron** receives information from a sensory receptor about a change in a stimulus such as light, pressure, or sound and passes this information on to a **motor neuron.** In turn, the motor neuron signals an **effector cell,** a muscle cell or gland cell that carries out the response. FIGURE 48.3 illustrates this sequence in the human knee-jerk reflex, the one that makes your leg jerk forward when the doctor hits your kneecap with a small hammer. The sudden, unexpected stretching of the tendon attached to the quadriceps (the front thigh muscle) is sensed by a specialized sensory neuron in the quadriceps. This sensory neuron signals a motor neuron to quickly counter the stretch by commanding the quadriceps to contract, causing your leg to jerk forward slightly.

Actually, our knee-jerk reflex involves more than this simple sensory/motor circuit. Contraction of the front thigh muscle is accompanied by inhibition of the back thigh muscles

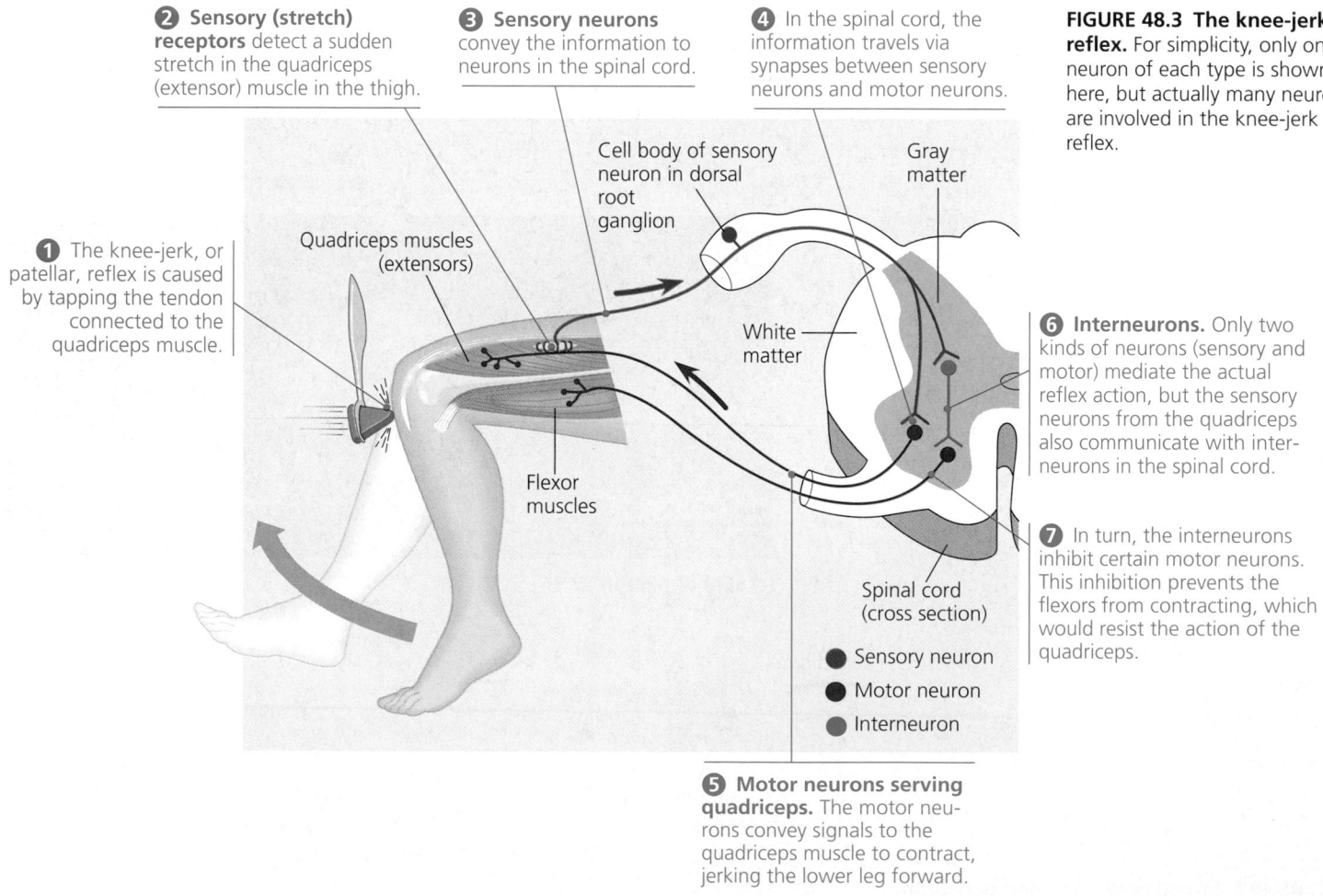

2 **Sensory (stretch) receptors** detect a sudden stretch in the quadriceps (extensor) muscle in the thigh.

3 **Sensory neurons** convey the information to neurons in the spinal cord.

4 In the spinal cord, the information travels via synapses between sensory neurons and motor neurons.

FIGURE 48.3 The knee-jerk reflex. For simplicity, only one neuron of each type is shown here, but actually many neurons are involved in the knee-jerk reflex.

1 The knee-jerk, or patellar, reflex is caused by tapping the tendon connected to the quadriceps muscle.

Quadriceps muscles (extensors)

Cell body of sensory neuron in dorsal root ganglion

Gray matter

White matter

6 **Interneurons.** Only two kinds of neurons (sensory and motor) mediate the actual reflex action, but the sensory neurons from the quadriceps also communicate with interneurons in the spinal cord.

Flexor muscles

Spinal cord (cross section)

● Sensory neuron
● Motor neuron
● Interneuron

7 In turn, the interneurons inhibit certain motor neurons. This inhibition prevents the flexors from contracting, which would resist the action of the quadriceps.

5 **Motor neurons serving quadriceps.** The motor neurons convey signals to the quadriceps muscle to contract, jerking the lower leg forward.

that flex the lower leg (pull it toward the body), and this inhibition involves a second nerve circuit. The sensory neurons from the quadriceps form synapses not only with motor neurons but also with **interneurons** in the spinal cord. In turn, these interneurons inhibit motor neurons to the flexor muscles, preventing them from contracting.

Most nerve circuits include large collections of interneurons that intervene between sensory receptors and effector cells to organize or integrate the most appropriate behaviors. Even in the simplest animals, most interneurons are constantly active, essentially "talking" to each other. This continuous internal "conversation" within the CNS provides context for interpreting sensory input and directing an appropriate action in response. The large collections of interneurons that make up vertebrate brains, particularly human brains, bring an ongoing history to bear on what action is taken. If, for example, you decide that you don't want your knee to jerk forward when a doctor taps it with the hammer, interneurons passing from your brain to the spinal cord segment shown in FIGURE 48.3 can inhibit that reflex.

The cell bodies of motor neurons and interneurons are generally located in the gray matter of the central nervous system; the gray matter of the spinal cord is shown in FIGURE 48.3. The outer white matter in the spinal cord consists of motor and sensory axons. Notice, however, that the cell body of the sensory neuron in FIGURE 48.3 is located *outside* the spinal cord in a structure called the dorsal root ganglion. A **ganglion** (plural, **ganglia**) is a cluster of nerve cell bodies, often of similar function, located in the peripheral nervous system. Similar clusters within vertebrate brains are called **nuclei** (not to be confused with the nuclei of individual cells).

Adapted for different functions, sensory neurons, motor neurons, and interneurons differ markedly in shape, and there is also a variety of shapes within each class. Examples of this structured diversity are shown in FIGURE 48.4. The human nervous system involves the coordinated activity of trillions of nerve cells.

Types of Nerve Circuits

Nerve circuits show three basic patterns of organization. One kind of circuit takes information from a single source, such as an eye, to several parts of the brain; in this case the information from a single presynaptic neuron spreads out to several postsynaptic neurons. In a second kind of circuit, information from several presynaptic neurons converges at a single postsynaptic neuron. The convergent circuits can bring together information from

several sources, such as vision, touch, and hearing, to identify an object in the environment. In a third type of circuit, information flows in a circular path, from one neuron to others and then back to its source. In the human brain, memories may be processed via circular paths and eventually stored.

Supporting Cells (Glia)

Supporting cells, or **glia** (from the Greek, glue) are essential for the structural integrity of the nervous system and for the normal functioning of neurons. Glia outnumber neurons tenfold to fiftyfold. Until recently, researchers assumed that glia play a supportive role without actually participating in nerve signaling. However, recent studies have suggested that some synaptic interactions do occur between glia and neurons.

There are several types of glia in the brain and spinal cord, and as a group these cells do much more than simply glue neurons together. In the developing embryo, supporting cells called radial glia form tracks along which neurons migrate or grow out from the neural tube, the structure that gives rise to the CNS (see FIGURE 47.11). In the mature CNS, glia called **astrocytes** provide structural and metabolic support for neurons. Astrocytes also induce the formation of tight junctions (see FIGURE 7.30) between cells lining the capillaries in the

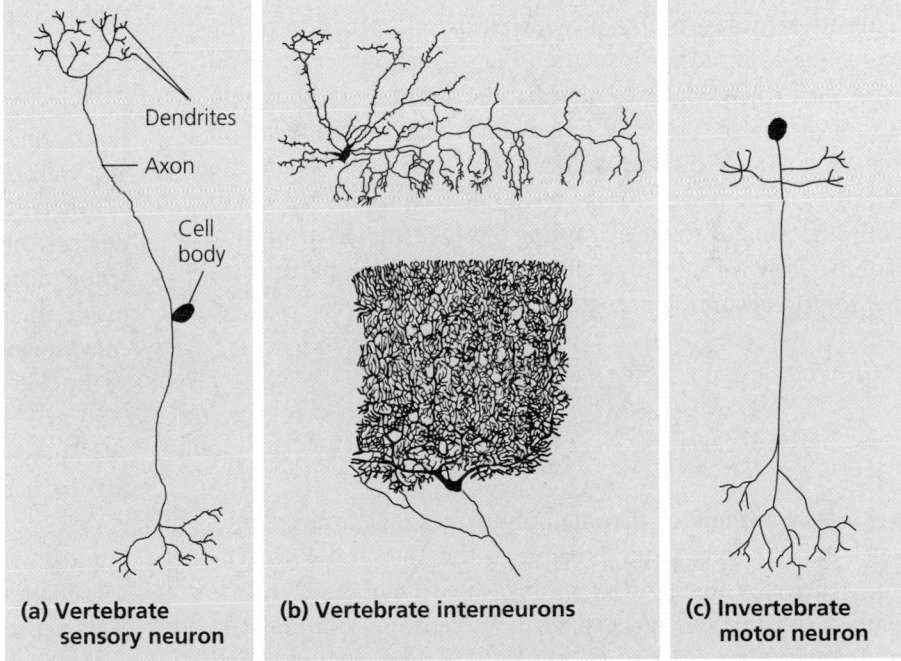

(a) Vertebrate sensory neuron **(b) Vertebrate interneurons** **(c) Invertebrate motor neuron**

Labels: Dendrites, Axon, Cell body

FIGURE 48.4 Structural diversity of neurons. These examples illustrate some of the variety in neuron shape. Cell bodies and dendrites are black; axons are magenta. **(a)** Vertebrate sensory neuron. The short, multibranched dendrites communicate with sensory receptor cells. A single, long axon, which is usually myelinated, conveys signals from the dendrites to synapses with neurons in the CNS. The cell body is connected only to the axon. This configuration is markedly different from that of the neuron in FIGURE 48.2a, which is a vertebrate motor neuron. **(b)** Two types of interneurons found in the mammalian brain. The top one has multiple dendrites and a branched axon; the bottom one has finely branched, meshlike dendrites. **(c)** An invertebrate motor neuron. In contrast to the vertebrate motor neuron in FIGURE 48.2a, the cell body connects only to the dendrites.

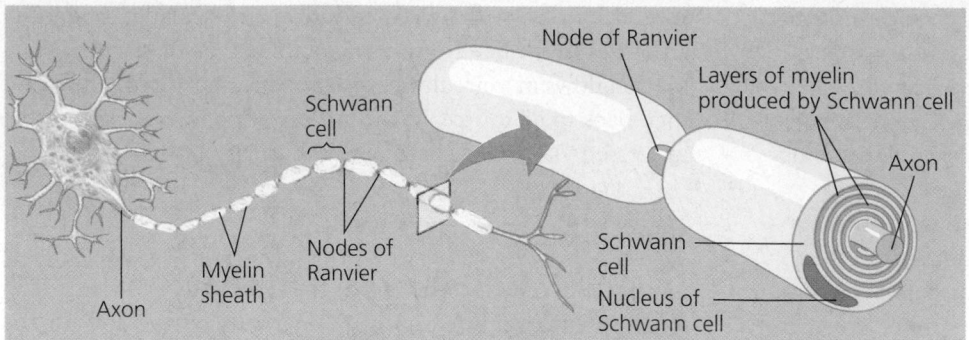

FIGURE 48.5 Schwann cells. In the peripheral nervous system, supporting cells called Schwann cells wrap many axons with an insulating myelin sheath. Gaps between successive Schwann cells are called nodes of Ranvier.

brain. The result is the **blood-brain barrier,** which restricts the passage of most substances into the brain, allowing the extracellular chemical environment of the CNS to be tightly controlled. Recent evidence also suggests that astrocytes communicate with one another and with neurons via chemical signals.

Oligodendrocytes (in the CNS) and **Schwann cells** (in the PNS) are glia that form insulating myelin sheaths around the axons of many neurons. The structure of a myelinated PNS axon is shown in FIGURE 48.5. Neurons become myelinated in a developing nervous system when Schwann cells or oligodendrocytes grow around axons, wrapping them in many layers of membrane, somewhat like a jelly roll. These membranes are mostly lipid, which is a poor conductor of electrical currents. Thus, the myelin sheaths provide electrical insulation of the axon, analogous to the insulation that covers copper electrical wires. We will see later in this chapter that the myelin sheath also increases the speed of propagation of nerve impulses. In the degenerative disease known as multiple sclerosis, myelin sheaths gradually deteriorate, resulting in a progressive loss of coordination due to the disruption of nerve impulse transmission. Clearly, supporting cells are indispensable partners of neurons in a working nervous system.

THE NATURE OF NERVE SIGNALS

THE PROCESS OF SCIENCE

Descriptions of the anatomy of neurons and nervous systems were made possible in the 18th and 19th centuries by increasingly sophisticated compound microscopes, but this did not explain how neurons communicate with each other. Toward the end of the 18th century Luigi Galvani discovered that frog muscle cells produce electricity, and in the 19th century Hermann von Helmholtz and others found that the electrical activity of nerve cells provides a means of carrying signals from one end of a cell to another and also between cells. Work over the past hundred years has shown that virtually all nerve signals are changes in the voltage across the plasma membranes of nerve cells. These voltage changes are caused by the movement of ions across the plasma membrane by way of specialized ion channels. Learning how these cellular processes work is central to an understanding of the nervous system. First we will examine the voltage gradient that exists across a cell's plasma membrane.

Every cell has a voltage, or membrane potential, across its plasma membrane

As we saw in Chapter 8, all living cells have an electrical charge difference (electrical potential, or voltage) across their plasma membranes. The plasma membrane is electrically polarized, meaning that it is more negatively charged on one side than the other. The potential difference across the membrane is called the **membrane potential.**

Measuring Membrane Potentials

Electrophysiologists can measure a cell's membrane potential as a voltage by using microelectrodes connected to a sensitive voltmeter or oscilloscope (FIGURE 48.6a). Precise mechanical devices called micromanipulators (seen next to the microscope in FIGURE 48.6b) are used to position one electrode just inside the cell for comparison with a reference electrode located outside the cell. The voltmeter indicates the magnitude of the charge separation across the membrane, typically about -50 to -100 mV (millivolts) in an animal cell. The minus sign indicates that the inside of the cell is negative in charge with respect to the outside. A neuron in its resting state (that is, not transmitting an electrical signal) usually has a membrane potential of about -70 mV (about 5% of the voltage in a flashlight battery). This membrane potential of an unstimulated neuron is called the **resting potential.**

A number of invertebrates, including squid, lobsters, and earthworms, have some unusually large neurons that make excellent research models for studying nerve impulses. For example, the squid nervous system includes some neurons whose axons have diameters of about 1 mm. These giant axons are relatively easy to impale with microelectrodes. Once the electrodes are in place, they can be used to measure the voltage of the resting potential as well as to record changes in voltage due to ion currents that occur during the transmission of a

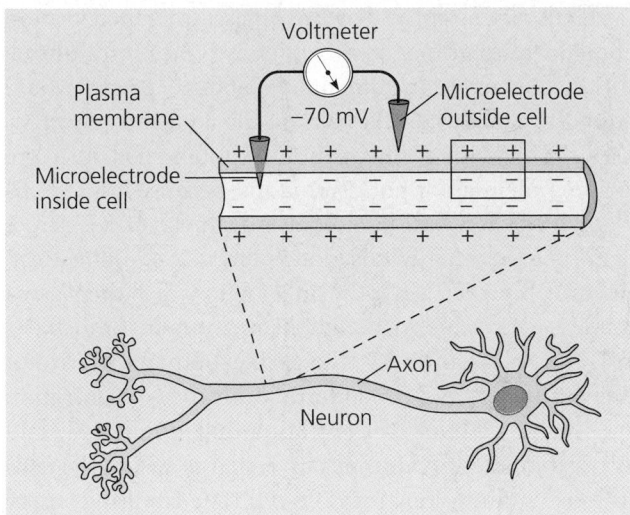

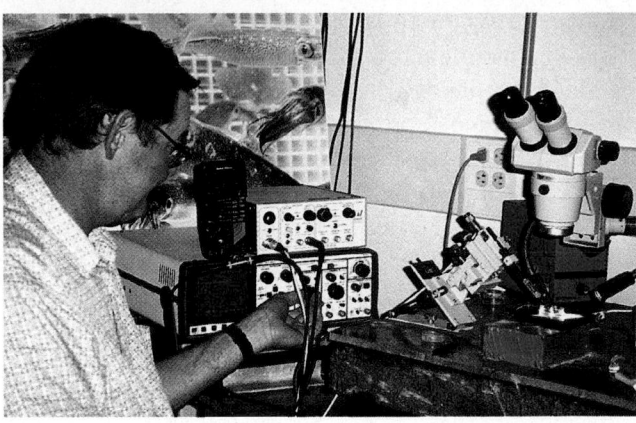

FIGURE 48.6 Measuring membrane potentials. (a) Microelectrodes inside and outside the cell measure the voltage (membrane potential) across a cell's plasma membrane. **(b)** Apparatus for measuring membrane potentials.

(a)

(b)

nerve impulse. Much of the pioneering research on membrane potentials and on the nature of nerve signals was performed using squid giant axons.

How a Cell Maintains a Membrane Potential

The membrane potential exists because of differences in the ionic composition of the intracellular and extracellular fluids (FIGURE 48.7a). The action of the sodium-potassium pump maintains these ionic differences (FIGURE 48.7b). In-

side a cell the principal cation is potassium (K^+), although there is also some sodium (Na^+). Outside a cell the situation is reversed, with Na^+ the principal cation and K^+ having a much lower concentration. Inside a cell the principal anions are proteins, amino acids, sulfate, phosphate, and other negatively charged ions that we can group and symbolize by A^-; chloride (Cl^-) is also present but in a relatively low concentration. Outside a cell, Cl^- is the main anion, with other anions present but less important in the context of membrane potentials.

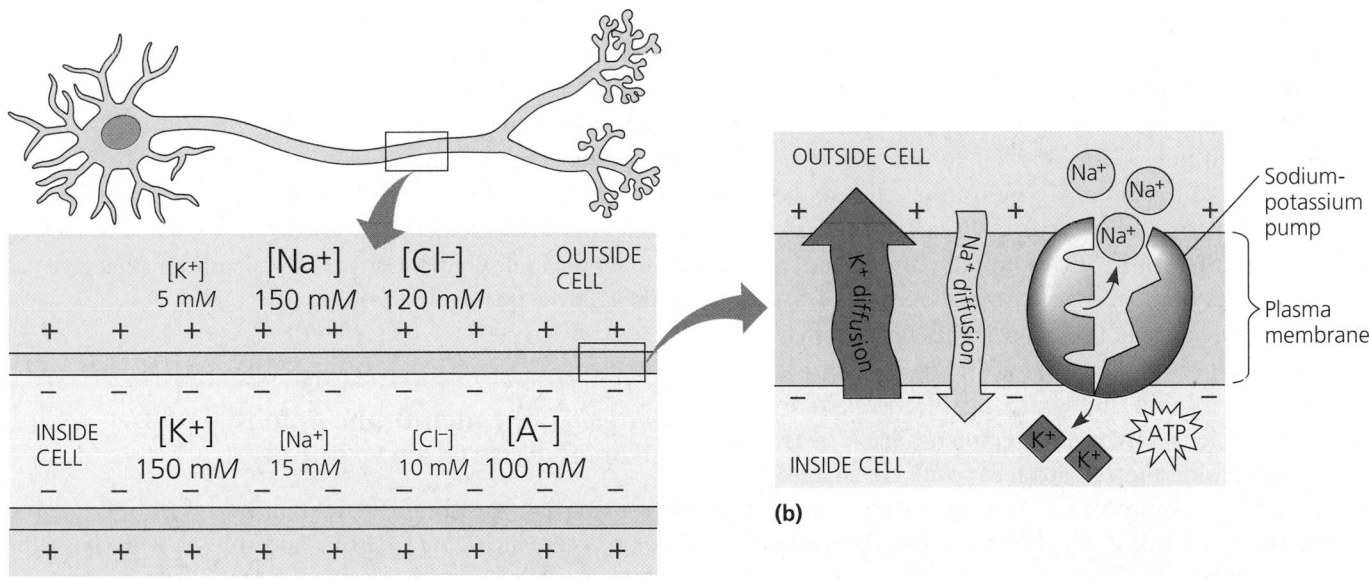

(a)

(b)

FIGURE 48.7 The basis of the membrane potential. (a) Shown here are the approximate concentrations for a mammalian cell (in millimoles per liter, abbreviated mM) of potassium, [K^+]; sodium [Na^+]; chloride, [Cl^-]; and anions that remain inside the cell, [A^-]. K^+ diffuses out of the cell down its concentration gradient, but the A^- anions cannot follow, so the interior of the cell develops a net negative charge. **(b)** There is a steady diffusion of K^+ out of the cell and steady diffusion of Na^+ into the cell; the thickness of the arrows indicates the relative permeability of the membrane to K^+ and Na^+ (the permeability mainly reflects the number of ion-specific channels). Over time, diffusion would cause the ionic gradients shown in part (a) to dissipate. Dissipation is prevented by the sodium-potassium pump, which uses ATP to actively transport Na^+ out of the cell and K^+ into the cell.

Recall from Chapter 8 that the plasma membrane is a phospholipid bilayer with associated membrane proteins. Ions, being electrically charged, cannot dissolve in lipid and therefore cannot directly diffuse across the lipid bilayer of the plasma membrane. In order to cross the membrane, ions must either be pumped by membrane proteins or move passively through ion channels, which are aqueous pores made up of specific transmembrane protein molecules. These channels are selective for specific ions; some allow only Na^+ to cross, others allow only K^+, and still others only Cl^-. Depending on how many ion channels of each kind are present in the plasma membrane of a cell, it is possible for the membrane to have very different permeabilities to different ions. Most cells, including neurons, have much greater permeability to K^+ than to Na^+, suggesting that the membrane has many more potassium channels than sodium channels (FIGURE 48.7b). In a resting neuron, for instance, potassium permeability is about fiftyfold higher than sodium permeability. Because the internal anions (A^-) are primarily large organic molecules such as proteins, they cannot cross the membrane and thus are a pool of internal negative charge that remains in the cell.

Although selective ion channels control the types of ions that diffuse passively into and out of a cell, they do not determine the direction or rate of passage. As you learned in Chapter 8, ions diffuse down an electrochemical gradient made up of two components: the concentrations of the specific ion in different regions (its chemical gradient); and the relative electrical charge in those regions (the electrical gradient). We would expect the ions that pass through ion channels to diffuse along the electrochemical gradient until reaching an equilibrium point when each type of ion enters the cell at the same rate it exits. Given this tendency toward equilibrium, how does the distribution of ions shown in FIGURE 48.7a give rise to a membrane potential?

Consider the case of potassium ions. There is a large concentration gradient for diffusion of K^+ out of the cell, and the membrane has a high permeability to potassium. Thus, there is a net outward flux (efflux) of K^+ driven by the concentration gradient. This transfers positive charge from the inside to the outside of the cell. The inside of the cell becomes progressively more negative with respect to the outside, creating an electrical gradient across the membrane. In essence, this electrical gradient competes with the K^+ concentration gradient: The increasing negative charge inside the cell attracts positively charged potassium, supporting an influx of K^+. When the efflux driven by the concentration gradient is equal to the influx driven by the potential gradient, K^+ ions are in equilibrium.

If K^+ were the only ion that could cross the membrane, the voltage across the membrane would continue to build up until the influx of K^+ down the electrical gradient exactly balanced the efflux of K^+ down its concentration gradient. At that point, there would be no further net transfer of charge across the membrane, and the membrane potential would reach a stable, resting value. For the potassium concentration gradient shown in FIGURE 48.7a, which is typical of a resting neuron, a stable membrane potential of about −85 mV would be required to exactly counterbalance the concentration gradient of K^+. This value of the membrane potential is called the equilibrium potential for potassium ions, because it is the potential at which there will be no net movement of K^+ across the membrane (in other words, potassium is at equilibrium).

Potassium, however, is not the only ion to which the plasma membrane is permeable. Although the membrane is much less permeable to Na^+ than to K^+, the permeability to Na^+ is not zero (see FIGURE 48.7b). For sodium, both the concentration gradient (higher sodium concentration outside the cell) and the electrical gradient (more negative charge inside the cell) tend to move sodium ions into the cell. The resulting steady trickle of positive charge into the cell, carried by Na^+, makes the actual value of the membrane potential somewhat less negative than would be expected if the membrane were permeable only to potassium ions. This explains why the membrane potential of a resting neuron is typically about −70 mV rather than about −85 mV.

Over time, a steady influx of sodium would cause a progressive increase in the sodium concentration inside a cell. Also, because the influx of Na^+ makes the cell interior less negative than the −85 mV required to balance the potassium concentration gradient, there would be a steady efflux of potassium and a progressive decline in the internal K^+ concentration. In other words, if the situation were left unchecked, the concentration gradients for Na^+ and K^+ shown in FIGURE 48.7a would gradually dissipate. This is prevented by the sodium-potassium pump shown in FIGURE 48.7b. This protein, found in abundance in neurons, uses energy from ATP to drive the active transport of sodium back out of the cell, against both the concentration and electrical gradients for sodium. At the same time, the pump moves potassium into the cell, thus restoring the concentration gradient for this ion as well (see FIGURE 8.15 for details). In essence, the cell uses metabolic energy, in the form of ATP, to maintain the ionic gradients across the membrane that give rise to the steady-state membrane potential.

Changes in the membrane potential of a neuron give rise to nerve impulses

All cells have a membrane potential; however, only certain kinds of cells, including neurons and muscle cells, have the ability to generate large changes in their membrane potentials. Collectively these cells are called **excitable cells.** The membrane potential of an excitable cell in a resting (unexcited) state is called the resting potential. As we will see, a change in this voltage may result in an electrical impulse. Changes in the membrane potential are made possible by the specialized ion channels of excitable cells. The ion channels we have discussed so far are called *ungated,* since they are open all of the time. These are the channels that are responsible for the resting potential, as we have just discussed. Excitable cells, including neurons, also have **gated ion channels,** which open or close in response to stimuli. The resulting change in ion concentrations may cause a change in the membrane potential in response to the stimulus. In the case of a sensory neuron, the stimulus may come from the organism's environment—for example, light in the case of photoreceptors in the eye, or vibrations in the air in the case of receptors in the ear. In the case of an interneuron, the stimulus is ordinarily produced by the other neurons that provide chemical or electrical inputs to the cell.

Gated ion channels open or close in response to only one kind of stimulus. **Chemically-gated ion channels** open or close in response to a chemical stimulus, such as a neurotransmitter released from a synaptic terminal. **Voltage-gated ion channels** respond to a change in membrane potential. In addition, a gated ion channel (like other ion channels) usually allows only one kind of ion to pass through it. Thus, there are chemically-gated sodium channels and chemically-gated potassium channels, as well as voltage-gated sodium channels and voltage-gated potassium channels. Each type of channel plays an essential role in the generation and transmission of electrical signals.

Graded Potentials: Hyperpolarization and Depolarization

A change in membrane potential is a localized electrical event at the point of stimulation. Let us consider what happens in a region of a dendrite that is stimulated by a neurotransmitter. The specific effect of this stimulus on membrane polarization depends on the type of chemically-gated ion channel that is opened. FIGURE 48.8 a and b show two types of local responses. Some stimuli trigger a **hyperpolarization,** an increase in the voltage across the membrane (FIGURE 48.8a). One of the ways a stimulus can produce a hyperpolarization is by opening a potassium channel, which increases K^+ outflow and causes the inside of the cell to become more negative. In contrast, a **depolarization** is a reduction in the voltage across the membrane (FIGURE 48.8b). One of the ways this can occur is by a stimulus opening a sodium channel; the increased inflow of Na^+ makes the inside of the cell less negative. These voltage changes are called **graded potentials** because the magnitude of change (either hyperpolarization or depolarization) depends on the strength of the stimulus: A larger stimulus will open more channels, producing a larger change in permeability and thus a larger change in the membrane potential.

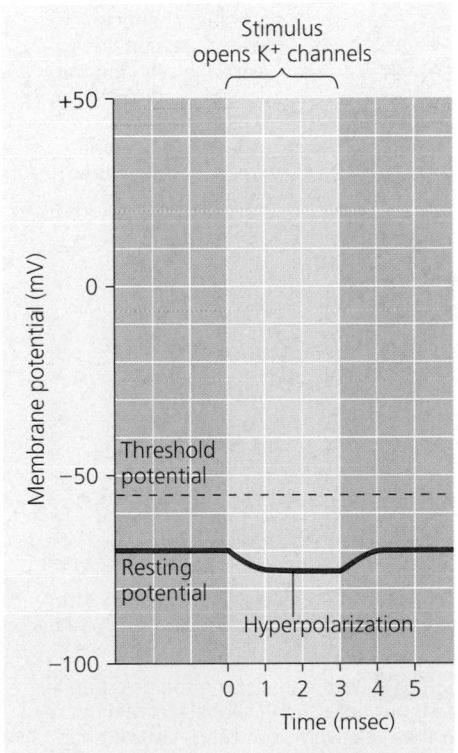

(a) Graded potential: hyperpolarization

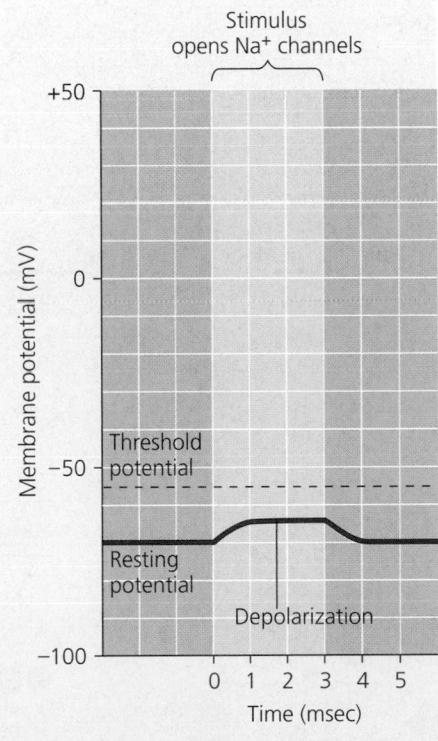

(b) Graded potential: depolarization

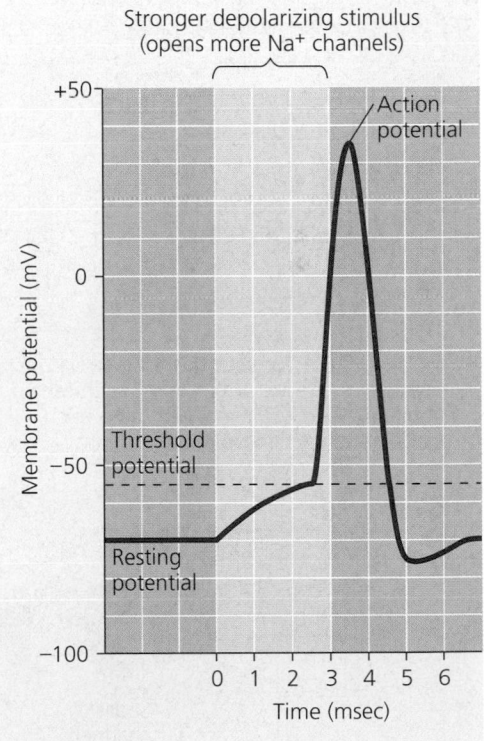

(c) Action potential

FIGURE 48.8 Graded potentials and the action potential in a neuron. Environmental changes can alter the cell's membrane potential. **(a)** One way a neuron can be hyperpolarized is by stimuli that open potassium channels. **(b)** A neuron can be depolarized by stimuli that open sodium channels. **(c)** A depolarizing stimulus of sufficient strength will change the membrane potential to a critical level called the threshold potential. This triggers an action potential, or nerve impulse. Unlike a graded potential, an action potential is an all-or-none event; the size of the action potential is not affected by the strength of the stimulus that triggered it.

The Action Potential: All or Nothing Depolarization

Depolarization of a neuron's plasma membrane is graded with stimulus intensity only up to a particular voltage, called the **threshold potential.** If a sufficiently strong stimulus causes depolarization to reach this threshold, it triggers a different type of response, called an **action potential** (FIGURE 48.8c). In a neuron, an action potential can be generated only in the axon. The action potential is typically triggered by a graded depolarization that originates in a dendrite or in the cell body and spreads along the membrane to the axon. The threshold potential is typically about 15 to 20 mV more positive than the resting potential—that is, about −50 to −55 mV for the plasma membrane of an axon. Hyperpolarizing stimuli do not produce action potentials; in fact, as we will see, hyperpolarization makes it more difficult for a stimulus to raise the membrane potential to threshold.

The action potential of an axon is the nerve impulse. It is a nongraded all-or-none event, meaning that the magnitude of the action potential is independent of the strength of the depolarizing stimulus that triggered it, provided the depolarization reaches threshold. Once an action potential is triggered, the membrane potential goes through a stereotypical sequence of changes, illustrated in FIGURE 48.9. The membrane

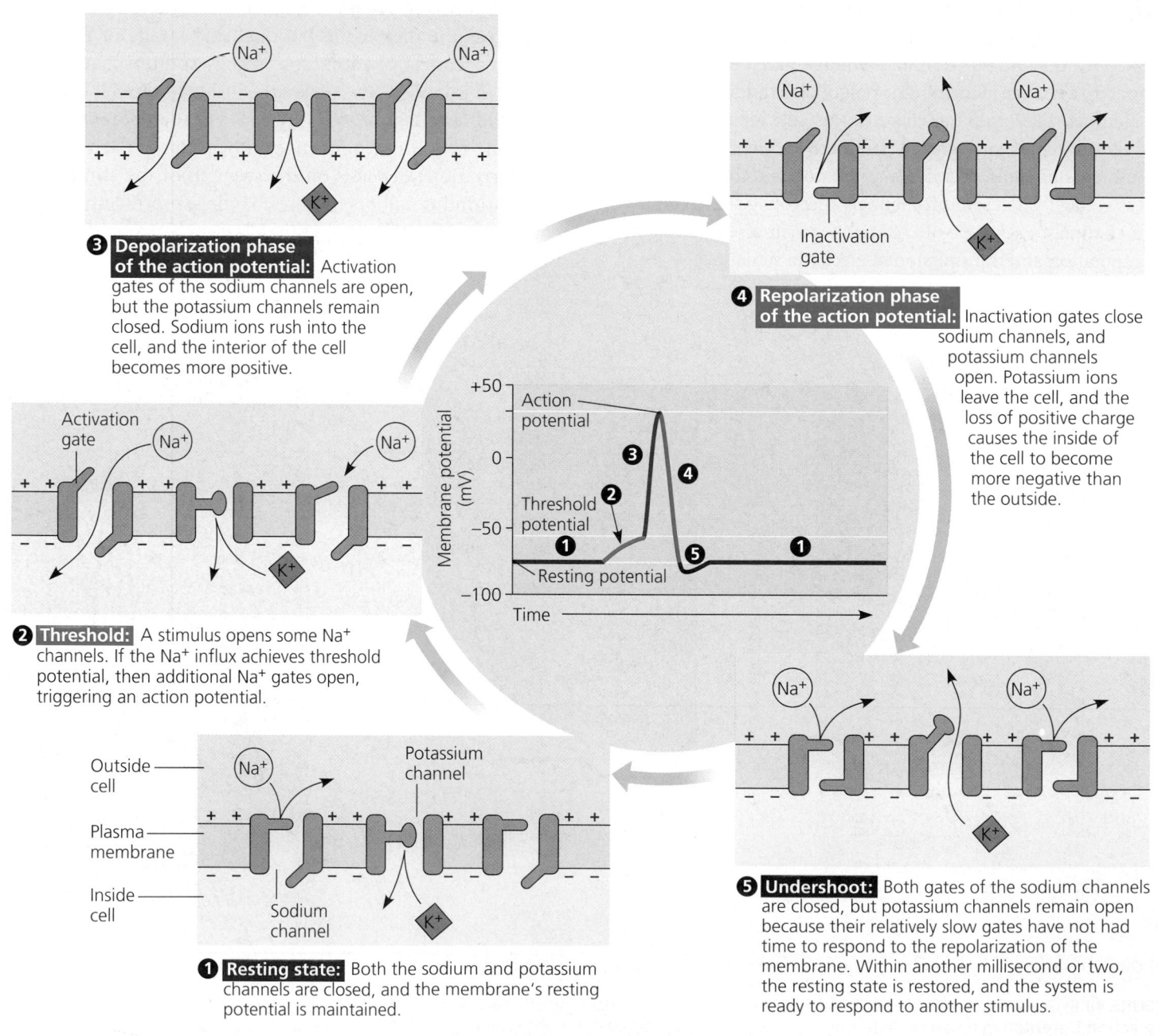

❸ **Depolarization phase of the action potential:** Activation gates of the sodium channels are open, but the potassium channels remain closed. Sodium ions rush into the cell, and the interior of the cell becomes more positive.

❹ **Repolarization phase of the action potential:** Inactivation gates close sodium channels, and potassium channels open. Potassium ions leave the cell, and the loss of positive charge causes the inside of the cell to become more negative than the outside.

❷ **Threshold:** A stimulus opens some Na⁺ channels. If the Na⁺ influx achieves threshold potential, then additional Na⁺ gates open, triggering an action potential.

❶ **Resting state:** Both the sodium and potassium channels are closed, and the membrane's resting potential is maintained.

❺ **Undershoot:** Both gates of the sodium channels are closed, but potassium channels remain open because their relatively slow gates have not had time to respond to the repolarization of the membrane. Within another millisecond or two, the resting state is restored, and the system is ready to respond to another stimulus.

FIGURE 48.9 The role of voltage-gated ion channels in the action potential. The circled numbers on the action potential graph correspond to the five diagrams of voltage-gated sodium and potassium channels in a neuron's plasma membrane.

polarity reverses abruptly, with the interior of the cell becoming positive with respect to the outside. This reversal (the spike in the graph) is followed by a rapid repolarizing phase, during which the membrane potential returns to its resting level. The whole event is typically over within a few milliseconds.

Note that an action potential is an example of positive feedback (see Chapter 1). Depolarization of an axon's membrane to the threshold potential triggers an even larger depolarization, the action potential.

The action potential arises because the plasma membranes of axons have voltage-gated ion channels. The gates of these ion channels open and close in response to changes in membrane potential, controlling rapid influx and efflux of ions. Voltage-gated channels are activated when graded depolarization of an axon's membrane reaches the threshold level. Two types of voltage-gated channels, sodium channels and potassium channels, contribute to the action potential (see FIGURE 48.9). Each voltage-gated potassium channel has a single voltage-sensitive gate that is closed in the resting state and *opens slowly* in response to depolarization. By contrast, each voltage-gated sodium channel has two voltage-sensitive gates: 1) an activation gate that is closed in the resting state and *opens rapidly* in response to depolarization; and 2) an inactivation gate that is open in the resting state and *closes slowly* in response to depolarization.

What is the status of the voltage-gated channels in the instant after a threshold potential is reached? The K^+ gates are opening slowly, and K^+ is beginning to trickle out of the cell. At the same instant, Na^+ is gushing into the cell because the activation gates of the Na^+ channels are wide open, but the inactivation gates are just beginning to close slowly. It is this rapid influx of Na^+ relative to the K^+ efflux that accounts for the voltage change called the action potential.

Note in FIGURE 48.9 that the action potential *spikes,* with the rapid depolarization followed immediately by a *repolarization* that is just as rapid. Two factors underlie this repolarizing phase of the action potential as the membrane potential is returned to its resting state. First, the sodium channel inactivation gate, which is slow to respond to changes in voltage, has time to respond to depolarization by closing, returning sodium permeability to its low resting level. Second, potassium channels, whose voltage-sensitive gates also respond relatively slowly to depolarization, have had time to open. As a result, K^+ flows rapidly out of the cell during repolarization, helping restore the internal negativity of the resting neuron. The potassium channel gates are also the main cause of an *undershoot,* or hyperpolarization, that follows the repolarizing phase (see FIGURE 48.9). Instead of returning immediately to their resting position, these relatively slow-moving gates remain open during the undershoot, allowing potassium to keep flowing out of the neuron. The continued potassium outflow makes the membrane potential more negative than the resting potential—that is, it hyperpolarizes the membrane.

Notice that during the undershoot, both the activation gate and the inactivation gate of the sodium channel are closed. If a second depolarizing stimulus arrives during this period, it will be unable to trigger an action potential because the inactivation gates have not had time to reopen after the preceding action potential. This period when the neuron is insensitive to depolarization is called the **refractory period,** and it sets the limit on the maximum frequency with which action potentials can be generated.

If the action potential is an all-or-none event with amplitude (size) unaffected by the intensity of the stimulus, how can the nervous system distinguish strong stimuli from weaker ones that are still sufficient to trigger action potentials? Strong stimuli result in a greater frequency of action potentials than weaker stimuli; if a stimulus is intense, the neuron will fire repeatedly, producing action potentials as rapidly as the refractory period will allow. Thus, it is the number of action potentials per second, not their amplitude, that codes for stimulus intensity in the nervous system.

Nerve impulses propagate themselves along an axon

For an action potential to function as a long-distance signal, it must somehow "travel" along the axon to the far end of the cell. The action potential does not actually travel but is regenerated repeatedly along the axon. The regeneration mechanism is the same as for graded depolarization: Na^+ entering the cell creates an electrical current that depolarizes the next neighboring region of the membrane. In the case of an action potential, however, the Na^+ influx is extensive enough to assure that the neighboring region will be depolarized above threshold, triggering a new action potential at that position (FIGURE 48.10, p. 1032). Thus, the effect of an action potential is like tipping the first of a row of standing dominoes. Just as the first domino's fall is relayed to the end of the row, the first action potential generates an action potential in the next part of the membrane, and so on to the end of the axon.

What prevents Na^+ entry from reexciting the region *behind* the action potential, which would cause the impulse to spread back along the axon toward the cell body as well as in the normal direction of propagation? Recall that an action potential is followed by a refractory period, when inactivation gates of sodium channels are closed and an action potential cannot be triggered. A wave of depolarization passing a point along the axon cannot induce another action potential behind it, but only in the forward direction. Thus, the axon is normally a one-way avenue for the conduction of nerve impulses.

Several factors affect the speed at which action potentials propagate along an axon. One factor is the diameter of the axon: The larger the axon's diameter, the faster the speed of

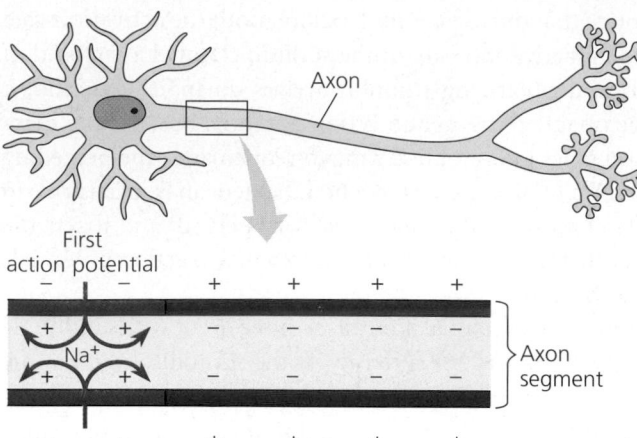

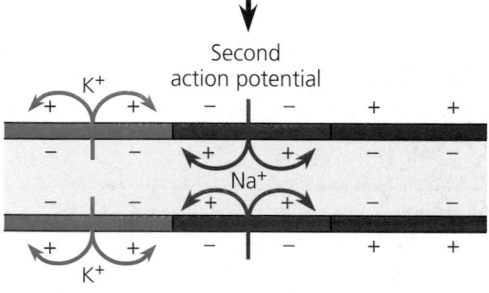

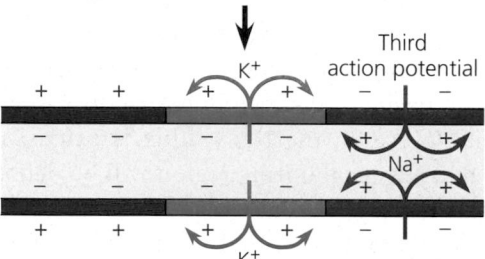

FIGURE 48.10 Propagation of the action potential. The three parts of this figure show the changes that occur in a portion of an axon at three successive times as a nerve signal passes from left to right. At each point along the axon, the voltage-gated channels go through the sequence described in FIGURE 48.9, reproducing the sequence of voltage changes associated with the action potential. (The "status" of the membrane is color-coded to FIGURE 48.9.)

1 An action potential is generated as sodium ions flow inward across the membrane at one location.

2 The depolarization of the first action potential has spread to the neighboring region of the membrane, depolarizing it and initiating a second action potential. At the site of the first action potential, the membrane is repolarizing as K⁺ flows outward.

3 A third action potential follows in sequence, with repolarization in its wake. In this way, local currents of ions *across* the plasma membrane give rise to a nerve impulse that is propagated *along* the axon.

transmission. This is because resistance to the flow of electrical current is inversely proportional to the cross-sectional area of the "wire" that conducts the current. In a thick axon, the depolarization associated with an action potential at a particular location can effectively reach farther along the interior of the axon and set up a new action potential at a greater distance away than in a thin axon. Transmission speed varies from several centimeters per second in very thin axons to about 100 m/sec in the giant axons of certain invertebrates, including squid and lobsters. These giant axons function in behavioral responses requiring great speed, such as the backward tail-flip that enables a threatened lobster or crayfish to escape.

A different means of speeding the propagation of action potentials has evolved in vertebrates. Recall that many axons in vertebrate nervous systems are myelinated, coated with insulating layers of membranes deposited by oligodendrocytes or Schwann cells (see FIGURE 48.5). The voltage-gated ion channels that produce the action potential are concentrated in the nodes of Ranvier, small gaps between successive Schwann cells along the axon. Also, extracellular fluid is in contact with the axon membrane only at the nodes, so that the flow of ions between the inside and outside of the axon can occur only in these regions. For these reasons, the action potential does not propagate itself in the regions between the nodes. Rather, the Na⁺ current generated by the action potential at a node travels all the way to the next node, where it stimulates depolarization and a new action potential (FIGURE 48.11). This mechanism is called **saltatory conduction** (from the Latin *saltare*, to leap) because the action potential appears to "jump" along the axon from node to node. Saltatory conduction can transmit impulses at speeds up to 150 m/sec in myelinated neurons.

We have seen how stimulating the dendrite of a neuron can produce an action potential in the axon that is propagated to the very tip of the cell. Our next step is to find out how the impulse is transmitted from a neuron to another neuron or an effector (muscle or gland cell).

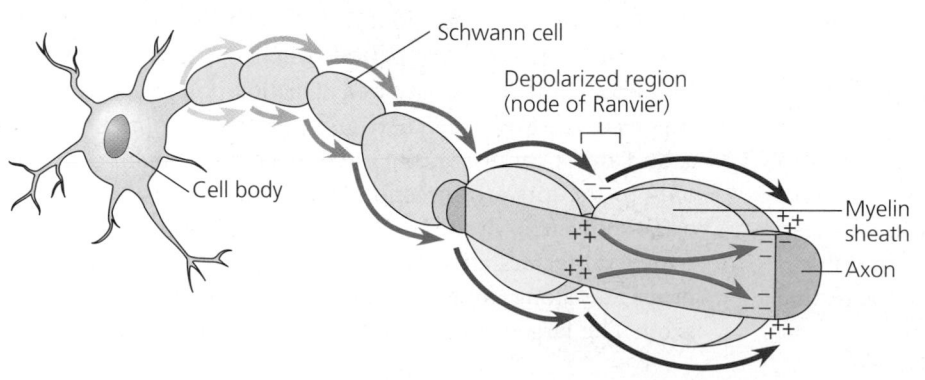

FIGURE 48.11 Saltatory conduction. In a myelinated axon, the ion current during an action potential at one node of Ranvier spreads along the interior of the axon to the next node (blue arrows), where it will trigger an action potential. The action potential thus jumps from node to node as it propagates along the axon (red arrows).

Chemical or electrical communication between cells occurs at synapses

Synapses are unique cell junctions that control communication between a neuron and another cell. Synapses are found between two neurons, between sensory receptors and sensory neurons, between motor neurons and the muscle cells they control, and between neurons and gland cells. Here we focus on synapses between neurons, which usually conduct signals from an axon's synaptic terminals to dendrites or cell bodies of the next cells in a neural pathway. Synapses are of two types: electrical synapses and chemical synapses.

Electrical Synapses

An electrical synapse allows action potentials to spread directly from the presynaptic cell to the postsynaptic cell. The cells are connected by gap junctions (see FIGURE 7.30), intercellular channels that allow the local ion currents of an action potential to flow between neurons. The giant axons of lobsters and other crustaceans are connected end to end and coupled by electrical synapses. These make it possible for impulses to travel from neuron to neuron without delay and with no loss of signal strength. Electrical synapses in the central nervous systems of vertebrates synchronize the activity of neurons responsible for some rapid, stereotypical movements. For example, electrical synapses in the brain enable some fishes to flap their tail very rapidly when escaping from predators. However, chemical synapses are much more common than electrical synapses in vertebrates and most invertebrates.

Chemical Synapses

At a chemical synapse, a narrow gap, or **synaptic cleft,** separates the presynaptic cell from the postsynaptic cell. Because of the cleft, the cells are not electrically coupled, and an action potential occurring in the presynaptic cell cannot be transmitted directly to the membrane of the postsynaptic cell. Instead, a series of events converts the electrical signal of the action potential arriving at the synaptic terminal into a chemical signal that travels across the synapse, where it is converted back into an electrical signal in the postsynaptic cell.

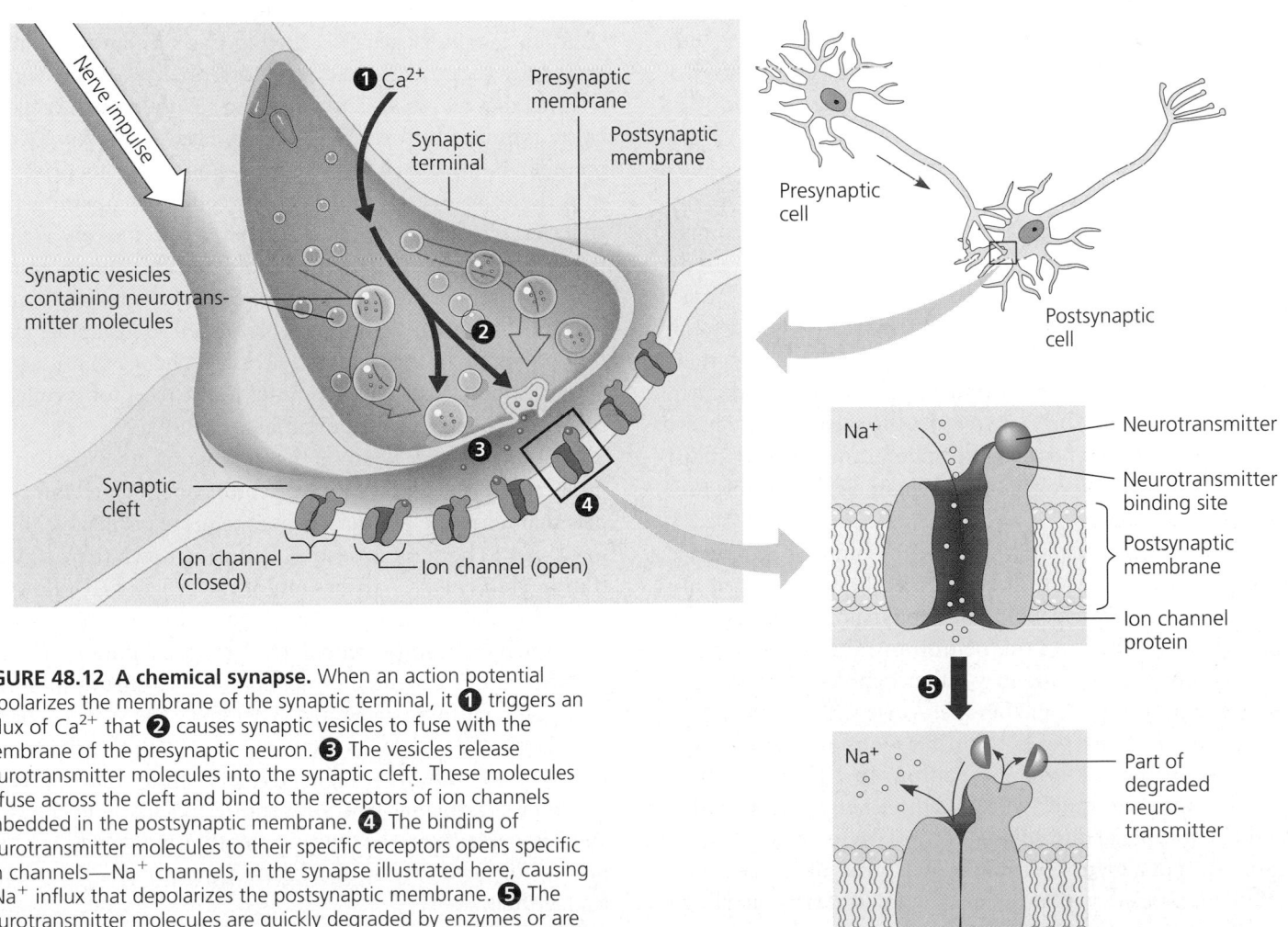

FIGURE 48.12 A chemical synapse. When an action potential depolarizes the membrane of the synaptic terminal, it ❶ triggers an influx of Ca^{2+} that ❷ causes synaptic vesicles to fuse with the membrane of the presynaptic neuron. ❸ The vesicles release neurotransmitter molecules into the synaptic cleft. These molecules diffuse across the cleft and bind to the receptors of ion channels embedded in the postsynaptic membrane. ❹ The binding of neurotransmitter molecules to their specific receptors opens specific ion channels—Na^+ channels, in the synapse illustrated here, causing a Na^+ influx that depolarizes the postsynaptic membrane. ❺ The neurotransmitter molecules are quickly degraded by enzymes or are taken up by another neuron, closing the ion channels and terminating the synaptic response.

The key to understanding the function of a chemical synapse is to examine its structure. The cytoplasm at the tip of the presynaptic axon contains numerous sacs called **synaptic vesicles** (FIGURE 48.12). Each vesicle contains thousands of molecules of a **neurotransmitter,** the substance that is released as an intercellular messenger into the synaptic cleft. Although many different neurotransmitters have been discovered in the nervous systems of animals, most neurons secrete only one kind of neurotransmitter. However, a single postsynaptic neuron may receive chemical signals from a variety of neurons that secrete different neurotransmitters from their synaptic terminals.

A presynaptic neuron dispatches neurotransmitter molecules into the synapse when an action potential arrives at the synaptic terminal and depolarizes the **presynaptic membrane,** the surface of the synaptic terminal that faces the cleft. Calcium ions play a central role in this conversion of the electrical impulse into a chemical signal. Depolarization of the presynaptic membrane causes Ca^{2+} to rush into the neuron through voltage-gated calcium channels. The sudden rise in the cytosolic concentration of Ca^{2+} stimulates the synaptic vesicles to fuse with the presynaptic membrane and spill the neurotransmitter into the synaptic cleft by exocytosis (see Chapter 8). Hundreds of vesicles may respond in unison to a single action potential. The neurotransmitter diffuses the short distance from the presynaptic membrane to the **postsynaptic membrane,** the plasma membrane of the cell body or dendrite on the other side of the synapse.

The postsynaptic membrane is specialized to receive the chemical message. Projecting from the extracellular surface of the membrane are proteins that function as specific receptors for neurotransmitters. The receptors are part of selective ion channels that open and close, controlling movements of ions across the postsynaptic membrane. A receptor is keyed to a particular type of neurotransmitter, and when it binds to this chemical, the gate of the ion channel opens, allowing specific ions, such as Na^+, K^+, or Cl^-, to cross the membrane. Thus, the ion channels of the postsynaptic membrane are *chemically* gated, in contrast to the voltage-gated channels responsible for the action potential.

The ion movements resulting from the binding of neurotransmitter to its receptors alter the membrane potential of the postsynaptic cell. Depending on the type of receptors and the ion channels they control, neurotransmitters binding to the postsynaptic membrane may either depolarize or hyperpolarize the postsynaptic membrane. As we will soon see, depolarization and hyperpolarization have opposite effects on the activity of the postsynaptic neuron. In either case, the neurotransmitter is quickly removed, either by enzymatic breakdown or uptake into adjacent cells. This removal ensures that the effect of a neurotransmitter on a postsynaptic cell will be brief and precise, so that the next action potential arriving at the synapse will be transmitted. For example, the neurotransmitter known as acetylcholine is rapidly degraded by cholinesterase, an enzyme present in both the synaptic cleft and the postsynaptic membrane.

Note that one important function of the synapse is to allow nerve impulses to be transmitted only in a single direction over a neural pathway. Synaptic vesicles are present only in synaptic terminals, and thus only the presynaptic membrane can discharge neurotransmitters. And receptors are restricted to the postsynaptic membrane, ensuring that only this membrane can receive a chemical signal from another neuron.

Neural integration occurs at the cellular level

A single neuron may receive information from numerous neighboring neurons via thousands of synapses, some of them excitatory and some of them inhibitory (FIGURE 48.13). If you reexamine the physiology of the knee-jerk reflex in FIGURE 48.3, you'll see examples of both types of synapses. The synapses between the sensory neurons and the motor neurons that stimulate the quadriceps muscles to contract are excitatory; the synapses between the interneurons and the motor neurons innervating the flexor muscles are inhibitory.

At an excitatory synapse, the binding of neurotransmitter molecules to postsynaptic receptors opens a type of gated channel that allows Na^+ to enter and K^+ to leave the cell. Because the driving force is greater for Na^+ than for K^+ (remember, both voltage and concentration gradients drive Na^+ into the cell—see FIGURE 48.7), the effect of opening these channels is a net flow of positive charge into the cell. This depolarizes the plasma membrane, moving the membrane potential closer to the threshold voltage and making it more likely that the axon of the postsynaptic cell will generate an action potential. In this case, the electrical change that is caused by the binding of neurotransmitter to the receptor is called an **excitatory postsynaptic potential,** or **EPSP.**

At an inhibitory synapse, the binding of neurotransmitter molecules to the postsynaptic receptors opens gated ion channels that make the plasma membrane more permeable to K^+, to Cl^-, or to both of these ions. Diffusing down their concentration gradients, K^+ rushes out of the cell and Cl^- flows in. These ion fluxes hyperpolarize the membrane—that is, they push the membrane potential to a voltage even more negative than the resting potential—making it more difficult for an action potential to be generated. Therefore, the voltage change associated with chemical signaling at an inhibitory synapse is called an **inhibitory postsynaptic potential,** or **IPSP.** Whether a particular neurotransmitter results in an EPSP or an IPSP depends on the type of receptors and gated ion channels on the postsynaptic membrane responding to that neurotransmitter.

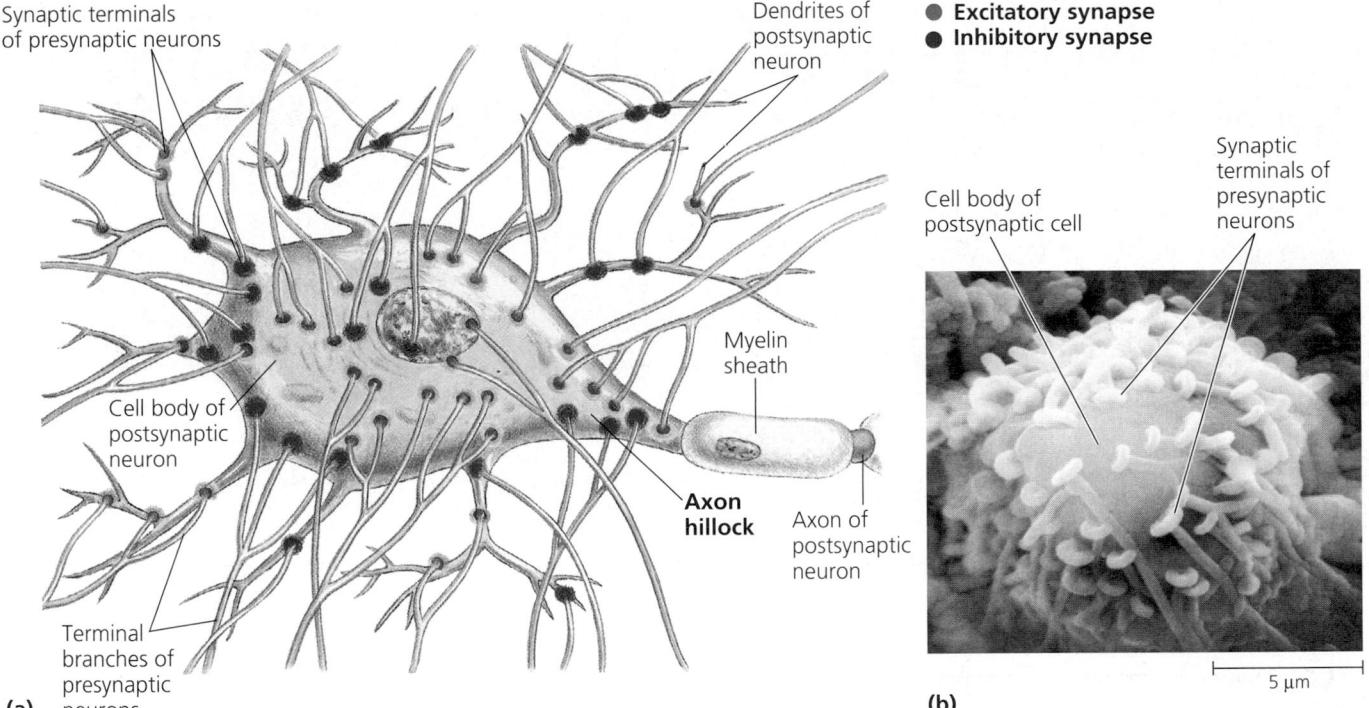

Synaptic terminals
of presynaptic neurons

Dendrites of
postsynaptic
neuron

● **Excitatory synapse**
● **Inhibitory synapse**

Cell body of
postsynaptic cell

Synaptic
terminals of
presynaptic
neurons

Myelin
sheath

Cell body of
postsynaptic
neuron

**Axon
hillock**

Axon of
postsynaptic
neuron

Terminal
branches of
presynaptic
neurons

(a)

(b)

5 μm

FIGURE 48.13 Integration of multiple synaptic inputs. (a) Each neuron, especially in the central nervous system, is on the receiving end of thousands of synapses, some excitatory (green) and others inhibitory (red). At any instant an action potential may be generated at the axon hillock if the combined effect of ion currents induced by excitatory and inhibitory synapses depolarizes the membrane to the threshold potential. Synapses close to the axon hillock generally have a stronger effect on the membrane potential than other synapses. **(b)** This micrograph reveals numerous synaptic terminals of presynaptic neurons that communicate with a single postsynaptic cell (SEM).

Both EPSPs and IPSPs are graded potentials that vary in magnitude with the number of neurotransmitter molecules binding to receptors on the postsynaptic membrane. The local change in membrane voltage, either depolarization or hyperpolarization, lasts only a few milliseconds because the neurotransmitters are removed soon after their release into the synapse. Also, the electrical impact on the postsynaptic cell decreases with distance away from the synapse. For the postsynaptic cell to "fire" (generate an action potential), the local ion currents due to EPSPs must be strong enough to reach and depolarize the membrane at the axon hillock to the threshold potential, usually about -50 mV. The axon hillock (see FIGURE 48.2) is the region where voltage-gated sodium channels open and generate an action potential when some stimulus has depolarized the membrane to the threshold.

A single EPSP at one synapse, even one close to the axon hillock, is not usually strong enough to trigger an action potential (FIGURE 48.14, p. 1036). However, several synaptic terminals acting simultaneously on the same postsynaptic cell, or a smaller number of synaptic terminals discharging neurotransmitters repeatedly in rapid-fire succession, can have a cumulative impact on the membrane potential at the axon hillock, raising it to threshold. This additive effect of postsynaptic potentials is called **summation**. Notice in FIGURE 48.14a that repeated subthreshold EPSPs that do not overlap in time do not depolarize the membrane to threshold.

There are two types of summation: temporal summation and spatial summation. In temporal summation, chemical transmissions from one or more synaptic terminals occur so close together in time that each postsynaptic potential affects the membrane before the voltage has returned to the resting potential after the previous stimulation (FIGURE 48.14b). In spatial summation, several different synaptic terminals, usually belonging to different presynaptic neurons, stimulate a postsynaptic cell at the same time and have an additive effect on the membrane potential (FIGURE 48.14c). By reinforcing one another through temporal or spatial summation, the ion currents associated with several EPSPs can depolarize the membrane at the axon hillock to threshold, causing the neuron to fire. Summation also applies to IPSPs: two or more IPSPs can hyperpolarize the membrane to a more negative voltage than any single release of a neurotransmitter at an inhibitory synapse can achieve. Furthermore, IPSPs and EPSPs counter each other's electrical effects (FIGURE 48.14d).

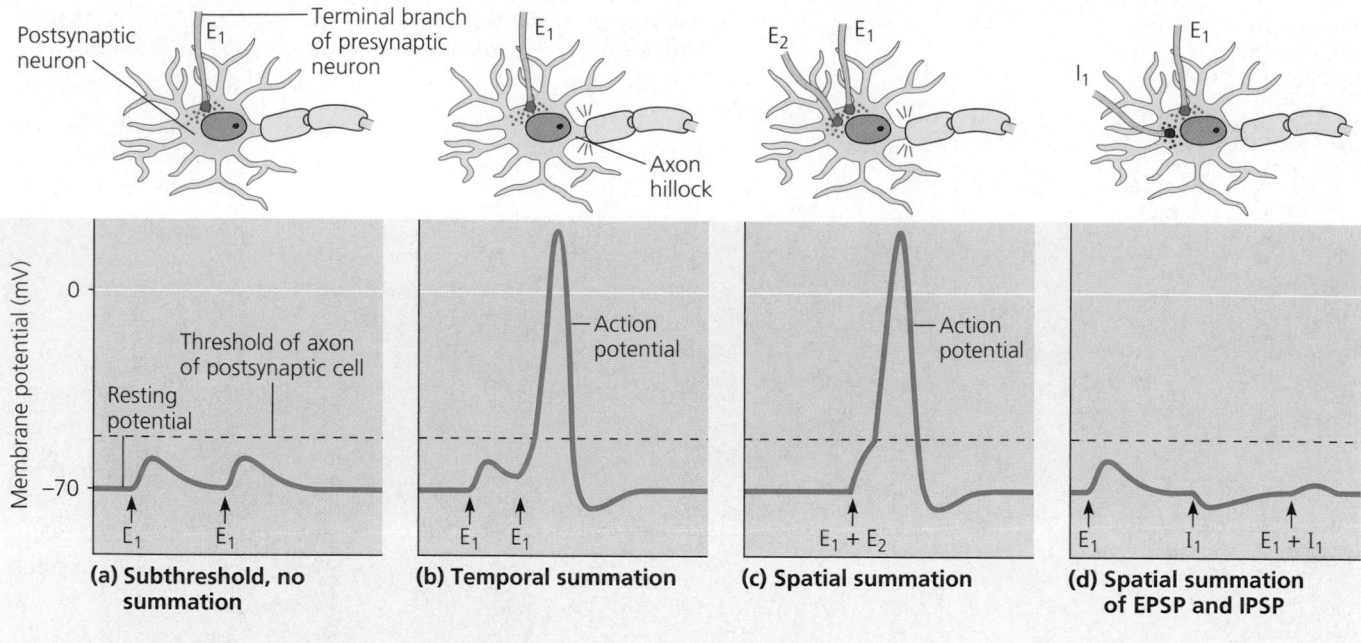

FIGURE 48.14 Summation of postsynaptic potentials. These graphs trace changes in membrane potentials at a postsynaptic neuron's axon hillock. The arrows indicate times when signals trigger changes in membrane potentials at two excitatory synapses (E_1 and E_2, green) and at one inhibitory synapse (I_1, red). Like most single EPSPs, those shown at E_1 and E_2 cannot depolarize the membrane at the axon hillock all the way to the threshold level, and thus without summation do not trigger an action potential.

The axon hillock is the neuron's integrating center, the region where the membrane potential represents the summed effect of all EPSPs and IPSPs. At any instant, the membrane potential at the axon hillock is an average of the depolarization due to summation of all EPSPs and the hyperpolarization due to summation of all IPSPs. Whenever EPSPs overpower IPSPs enough for the membrane potential at the axon hillock to reach threshold, an action potential is generated, and the impulse is transmitted along the axon to the next synapse. A few milliseconds later, after the refractory period, the neuron may fire again if the sum of all synaptic inputs at that moment is still sufficient to depolarize the membrane at the axon hillock to threshold level. On the other hand, by that time the sum of all EPSPs and IPSPs may put the membrane potential at the axon hillock at a voltage more negative than the threshold, or may even hyperpolarize the membrane to a potential more negative than the resting potential, thereby desensitizing the neuron for the moment.

Action potentials, remember, are all-or-none events. But now we see that the occurrence of these nerve impulses depends on the ability of the neuron to integrate quantitative information in the form of multiple excitatory and inhibitory inputs, each involving the specific binding of a neurotransmitter to a receptor on the postsynaptic membrane.

The same neurotransmitter can produce different effects on different types of cells

Dozens of different substances, many of them small, nitrogen-containing organic molecules, are known to function as neurotransmitters, and researchers expect to find many more. TABLE 48.1 summarizes the major known neurotransmitters. Notice that a particular neurotransmitter can trigger different responses in postsynaptic cells. This versatility depends on the receptors present on different postsynaptic cells and on the receptor's mode of action. Many neurotransmitters bind to chemically-gated ion channel proteins, altering the membrane permeability of the postsynaptic cell (see FIGURE 48.12). This type of synaptic communication can take only a few milliseconds, serving the rapid and precise transfer of information at a single synapse. Other neurotransmitters take much longer (up to several minutes) because they communicate via complex signal-transduction pathways in the postsynaptic cell. In some cases neurotransmitters in the brain—such as those regulating mood, attention, and arousal—remain active long enough after their release to diffuse to many synapses and modulate their activity.

Acetylcholine

Acetylcholine is one of the most common neurotransmitters in both invertebrates and vertebrates. In the vertebrate central nervous system, acetylcholine can be inhibitory or excitatory, depending on the type of receptor. At the vertebrate neuromuscular junction, the synapse between a motor neuron and a skeletal muscle cell, acetylcholine is released from the synaptic terminal of the motor neuron. It binds with a receptor that has a direct stimulatory effect on the muscle cell plasma membrane. The effect is excitatory, depolarizing the membrane of the postsynaptic muscle cell. A second type of acetylcholine receptor in heart muscle activates a signal-transduction pathway whose G proteins have two effects: They inhibit adenyl cyclase and open K^+ channels in the muscle cell membrane, making it less able to generate an action potential. Both effects reduce the strength and rate of cardiac muscle cell contraction.

Biogenic Amines

The **biogenic amines** are neurotransmitters derived from amino acids. One group, known as catecholamines, consists of neurotransmitters produced from the amino acid tyrosine. This group includes **epinephrine** and **norepinephrine,** which also function as hormones (see Chapter 45), and a closely related compound called **dopamine.** Another biogenic amine, **serotonin,** is synthesized from the amino acid tryptophan. The biogenic amines often affect biochemical processes within the postsynaptic cell. In many cases they bind to specific receptors on the postsynaptic membrane, triggering signal-transduction pathways that affect the activities of specific enzymes in the postsynaptic cell.

The biogenic amines most commonly function as transmitters within the CNS. However, norepinephrine also functions in a branch of the peripheral nervous system called the autonomic nervous system, which we will examine shortly. Dopamine and serotonin are widespread in the brain and affect sleep, mood, attention, and learning. Imbalances of these neurotransmitters are associated with several disorders. For example, the degenerative illness Parkinson's disease is associated with a lack of dopamine in the brain, and an excess of dopamine is linked to schizophrenia. Some psychoactive drugs, including LSD and mescaline, apparently produce their hallucinatory effects by binding to serotonin and dopamine receptors in the brain.

Table 48.1 The Major Known Neurotransmitters

Neurotransmitter	Structure	Functional Class	Secretion Sites
Acetylcholine		Excitatory to vertebrate skeletal muscles; excitatory or inhibitory at other sites	CNS; PNS; vertebrate neuromuscular junction
Biogenic Amines Norepinephrine		Excitatory or inhibitory	CNS; PNS
Dopamine		Generally excitatory; may be inhibitory at some sites	CNS; PNS
Serotonin		Generally inhibitory	CNS
Amino Acids GABA (gamma aminobutyric acid)	H_2N—CH_2—CH_2—CH_2—COOH	Inhibitory	CNS; invertebrate neuromuscular junction
Glycine	H_2N—CH_2—COOH	Inhibitory	CNS
Glutamate	H_2N—CH—CH_2—CH_2—COOH, COOH	Excitatory	CNS; invertebrate neuromuscular junction
Aspartate	H_2N—CH—CH_2—COOH, COOH	Excitatory	CNS
Neuropeptides (a very diverse group, only two of which are shown) Substance P	Arg—Pro—Lys—Pro—Gln—Gln—Phe—Phe—Gly—Leu—Met	Excitatory	CNS; PNS
Met-enkephalin (an endorphin)	Tyr—Gly—Gly—Phe—Met	Generally inhibitory	CNS

Other Chemical Neurotransmitters

Four amino acids are known to function as CNS neurotransmitters: **gamma aminobutyric acid (GABA), glycine, glutamate,** and **aspartate.** GABA, believed to be the transmitter at most inhibitory synapses in the brain, produces IPSPs by increasing the chloride permeability of the postsynaptic membrane.

Several **neuropeptides,** relatively short chains of amino acids, serve as neurotransmitters. In common with the biogenic amines, neuropeptides often operate via signal-transduction pathways. A neuropeptide called **substance P** is a key excitatory signal that mediates our perception of pain. The **endorphins** are neuropeptides that function as natural analgesics, decreasing the perception of pain by the CNS. Neurochemists first discovered endorphins in the 1970s while studying the mechanism of opium addiction. Candace Pert and Solomon Snyder of Johns Hopkins University found specific receptors for the opiates morphine and heroin on neurons in the brain. It seemed odd that humans would have receptors keyed to chemicals from a plant (the opium poppy). Further research showed that, in fact, the drugs bind to these receptors in the brain by mimicking endorphins, the natural painkillers produced in the brain during times of physical or emotional stress, such as the labor of childbirth (see FIGURE 2.19). In addition to relieving pain, endorphins also decrease urine output (by affecting ADH secretion; see Chapter 45), depress respiration, produce euphoria, and have other emotional effects through specific pathways in the brain. An endorphin is also released from the anterior pituitary gland as a hormone that affects specific regions of the brain. Once again, we see the overlap between endocrine and nervous system control.

Gaseous Signals of the Nervous System

In common with many other types of cells, some neurons of the vertebrate PNS and CNS utilize gas molecules, notably nitric oxide (NO, see Chapter 45) and carbon monoxide (CO), as local regulators. For example, during sexual arousal of human males, certain neurons release NO gas into the erectile tissue of the penis. In response, smooth muscle cells in the blood vessel walls of the erectile tissue dilate and the spongy erectile tissue fills with blood, producing an erection. The male impotence drug, Viagra®, increases the ability to achieve and maintain an erection during sexual arousal by inhibiting an enzyme that slows the muscle-relaxing effects of NO.

Many cells release gas molecules in response to chemical signals. For instance, the neurotransmitter acetylcholine released by neurons into the walls of blood vessels stimulates the endothelial cells of the vessels to synthesize and release NO. In turn, NO signals the neighboring smooth muscle cells to relax, dilating the vessels. The discovery of this mechanism in the late 1980s explained the medicinal action of nitroglycerin, which had been used for a century to treat angina (chest pain associated with reduced blood supply to the heart). Enzymes convert nitroglycerin to NO, which dilates the blood vessels that supply cardiac muscle.

Unlike typical neurotransmitters, NO and other gaseous messengers are not stored in cytoplasmic vesicles; cells synthesize them on demand. They diffuse into neighboring target cells, produce a change, and are broken down—all within a few seconds. In many of its targets, including smooth muscle cells, NO works like many hormones, stimulating a membrane-bound enzyme to synthesize a second messenger that directly affects cellular metabolism.

Now that we have discussed some key concepts about neurons, nerve signals, and neurotransmitters, we can consider how this cellular level of life contributes to the workings of the overall nervous system.

EVOLUTION AND DIVERSITY OF NERVOUS SYSTEMS

The ability of cells to respond to the environment has evolved over billions of years

The fact that you are reading the words on this page illustrates what an extraordinary product of biological evolution our human brains are. At the cellular level, however, the physiology of excitability originated billions of years ago with prokaryotes that could sense changes in their environment and respond in ways that enhanced survival and reproductive success—for example, by chemotaxis steering certain bacteria to food sources (see Chapter 27). Modification of this simple single-celled behavior of sensing and reacting to environmental chemicals provided multicellular organisms with a mechanism for communication between cells of the body. By the time of the Cambrian explosion nearly 600 million years ago (see Chapter 34), systems of nerve cells that allowed animals to sense and move rapidly had evolved in essentially their modern forms.

Nervous systems show diverse patterns of organization

While there is remarkable uniformity in how nerve cells function throughout the animal kingdom, there is great diversity in how nervous systems are organized. What distinguishes the various levels of complexity among animal nervous systems is not so much their basic building blocks, the neurons themselves, but how these cells are networked. These networks are

responsible for the feats of sensing and acting we will examine in Chapter 49 and for the animal behaviors we will describe in Chapter 51.

Some simple multicellular animals lack a nervous system altogether; sponges, for instance, have no cells specialized for impulse conduction (see FIGURE 33.3). The simplest animals with nervous systems, the cnidarians, have bodies organized around radially symmetrical cavities that ingest and expel food (see FIGURE 33.4). In some of these animals, such as the hydra shown in FIGURE 48.15a, the neurons controlling the contractions and expansions of these cavities are arranged in diffuse **nerve nets.** Nerve nets also play a role as components of the nervous system of more complex animals. For example, the sea star shown in FIGURE 48.15b has a nerve ring connected to radial nerves that link to a nerve net in each arm, allowing for more complex movement patterns than the hydra. And a net-like system of neurons in the walls of our intestines controls the smooth muscles that function in peristalsis (see Chapter 41).

Greater complexity of nervous systems and more complex behavior evolved with **cephalization,** which included the clustering of sensory neurons and other nerve cells to form a small brain near the anterior (head) end and mouth region in animals with elongated, bilaterally symmetrical bodies. In relatively simple cephalized animals such as the planarian shown in FIGURE 48.15c, small brains and longitudinal **nerve cords,** which control the animals' directional movements, constitute

the first clearly defined central nervous system (CNS). In more complex invertebrates such as annelids and insects, behavior is regulated by more complicated brains and ventral nerve cords containing segmentally arranged ganglia (FIGURE 48.15d and e). The longitudinal nerve cord of vertebrates, such as the salamander shown in FIGURE 48.15h, runs along the dorsal rather than the ventral surface of the body and does not contain segmental ganglia (though there are segmental ganglia lateral to the nerve cord—see FIGURE 48.2).

Mollusks are good examples of how nervous system organization correlates with how various animals live and interact with their environments. Sessile or slow-moving mollusks such as clams or chitons have little or no cephalization and relatively simple sense organs (FIGURE 48.15f). In contrast, cephalopod mollusks (squid and octopuses) have the most sophisticated nervous systems of any invertebrates, rivaling even those of some vertebrates. The large brain of a squid or octopus, accompanied by large, image-forming eyes and rapid signaling along giant axons, correlates well with the active predatory life of these animals (FIGURE 48.15g). Researchers have demonstrated in laboratory experiments that octopuses can learn to discriminate among visual patterns and perform complex tasks.

Now that we have surveyed nervous systems of varying complexity, let's take a closer look at the structure and function of nervous systems in our own subphylum.

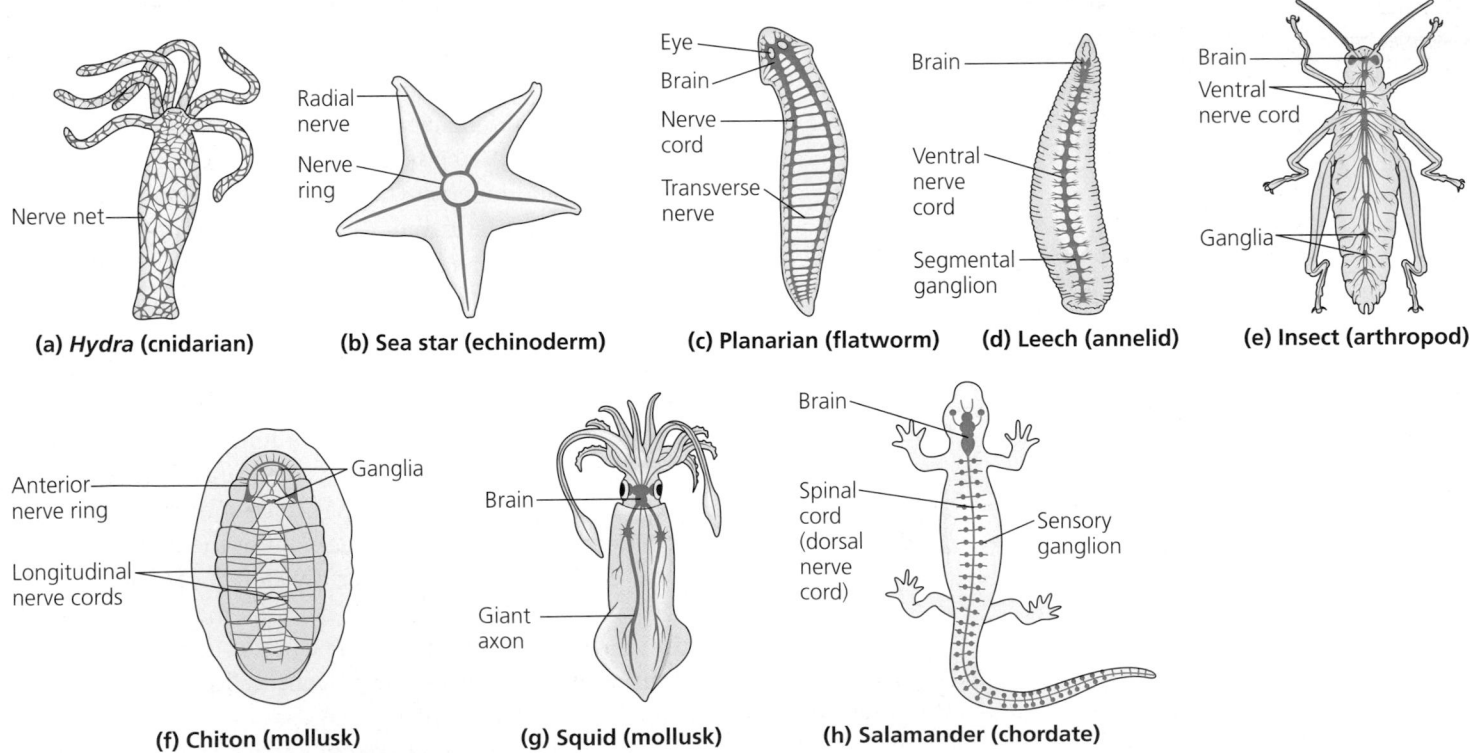

FIGURE 48.15 Diversity in nervous systems.

VERTEBRATE NERVOUS SYSTEMS

Vertebrate nervous systems have central and peripheral components

All vertebrate nervous systems have some fundamental similarities, including distinct central and peripheral elements and a high degree of cephalization. In all vertebrates the brain and spinal cord make up the central nervous system (CNS), and the peripheral nervous system (PNS) is everything outside the CNS (FIGURE 48.16). The brain provides the integrative power that underlies the complex behavior of all vertebrates. The spinal cord, which runs lengthwise inside the vertebral column (spine), integrates simple responses to certain kinds of stimuli (such as the knee-jerk reflex) and conveys information to and from the brain. The vertebrate PNS transmits information to and from the CNS and regulates the internal environment of the organism.

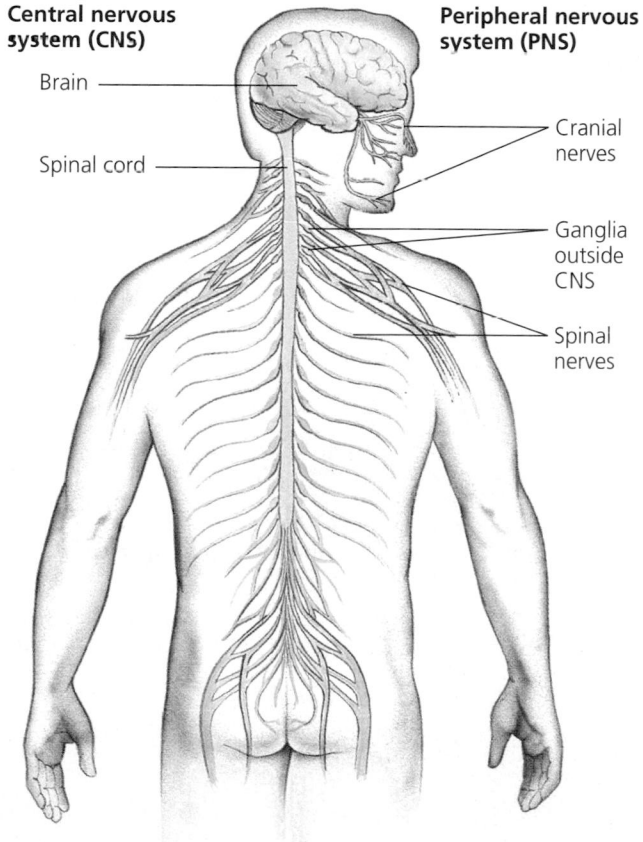

Central nervous system (CNS)

Brain

Spinal cord

Peripheral nervous system (PNS)

Cranial nerves

Ganglia outside CNS

Spinal nerves

FIGURE 48.16 The nervous system of a vertebrate. The components of the central nervous system (brain and spinal cord) develop from the dorsal, hollow nerve cord, a hallmark of chordates. Cranial nerves (originating in the brain), spinal nerves (originating in the spinal cord), and ganglia outside the central nervous system make up the peripheral nervous system.

Neurons of the central nervous system surround a continuous system of fluid-filled cavities. The vertebrate CNS is derived from the dorsal hollow nerve cord of the embryo—one of the phylogenetic hallmarks of chordates (see Chapter 34). This development from a tube explains why both the brain and spinal cord contain fluid-filled spaces. The narrow **central canal** of the spinal cord is continuous with the fluid-filled spaces, called **ventricles,** of the brain. These cavities are filled with **cerebrospinal fluid,** which is formed in the brain by filtration of the blood. Circulating through the central canal and ventricles (and then draining back into the veins), the cerebrospinal fluid conveys nutrients, hormones, and white blood cells across the blood-brain barrier described earlier to different parts of the brain. Among the most important functions of cerebrospinal fluid is to act as a shock absorber, cushioning the brain. Also protecting the brain and spinal cord are layers of connective tissue, called meninges. In mammals, cerebrospinal fluid circulates between two of the meninges, providing an additional cushion for the brain.

Axons within the CNS are located in well-defined bundles, or tracts, whose myelin sheaths give them a whitish appearance. In cross sections of the brain and spinal cord, this **white matter** is clearly distinguishable from **gray matter,** which consists mainly of dendrites, unmyelinated axons, and clusters of nerve-cell bodies, or nuclei.

The divisions of the peripheral nervous system interact in maintaining homeostasis

Structurally, the vertebrate PNS consists of paired cranial and spinal nerves and associated ganglia (see FIGURE 48.16). The **cranial nerves** originate in the brain and innervate organs of the head and upper body. The **spinal nerves** originate in the spinal cord and innervate the entire body. Mammals have 12 pairs of cranial nerves and 31 pairs of spinal nerves. Most of the cranial nerves and all of the spinal nerves contain both sensory and motor neurons; a few of the cranial nerves are sensory only (the olfactory and optic nerves, for example).

Because most nerves contain a diversity of neurons that play different roles, it is convenient to divide the PNS into a hierarchy of components that differ in function (FIGURE 48.17). The **sensory division** of the PNS is made up of the sensory, or afferent (incoming), neurons that convey information to the CNS from sensory receptors that monitor the external and internal environment. The **motor division** is composed of the motor, or efferent (outgoing), neurons that convey signals from the CNS to effector cells. The motor division is divided, in turn, into two functional divisions, called the somatic and autonomic nervous systems.

The **somatic nervous system** carries signals to skeletal muscles, mainly in response to *external* stimuli. The somatic nervous system is often considered voluntary because it is

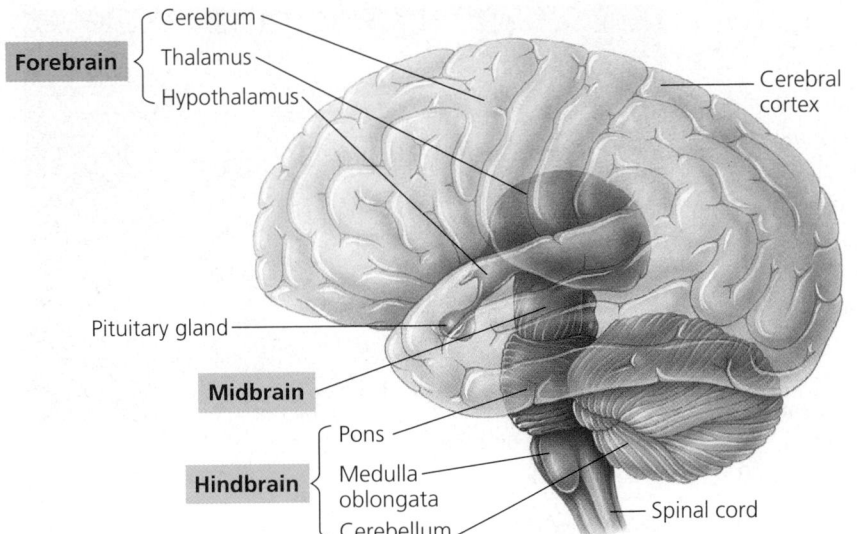

FIGURE 48.20 The main parts of the human brain.

Forebrain
- Cerebrum
- Thalamus
- Hypothalamus

Cerebral cortex

Pituitary gland

Midbrain

Hindbrain
- Pons
- Medulla oblongata
- Cerebellum

Spinal cord

---

cerebral cortex. Major brain centers that develop from the diencephalon—the division of the forebrain that evolved earliest in vertebrate history—are the thalamus, epithalamus, and hypothalamus.

The three regions derived from the midbrain and hindbrain give rise to structures deep within the brain that are collectively called the brainstem. The adult brainstem consists of the midbrain (from the mesencephalon), the pons (from the metencephalon), and the medulla oblongata (from the myelencephalon). The metencephalon also gives rise to another major brain center, the cerebellum, which is not part of the brain stem. FIGURE 48.20 reviews brain anatomy in a three-dimensional view.

Evolutionarily older structures of the vertebrate brain regulate essential automatic and integrative functions

The Brainstem

The brainstem, sometimes called the "lower brain," forms a stalk with caplike swellings at the anterior end of the spinal cord. The three parts of the brainstem—the medulla oblongata, the pons, and the midbrain—are derived from the embryonic hindbrain and midbrain (see FIGURE 48.19). They function in homeostasis, coordination of movement, and conduction of information to higher brain centers.

The Medulla and Pons. Several centers in the brainstem contain nerve cell bodies that send axons to many different areas of the cerebral cortex and cerebellum, releasing neuro-transmitters such as norepinephrine, dopamine, serotonin, and acetylcholine. These signals from the brainstem cause changes in attention, alertness, appetite, and motivation. The **medulla oblongata,** commonly called the **medulla,** contains centers that control several visceral (automatic, homeostatic) functions, including breathing, heart and blood vessel activity, swallowing, vomiting, and digestion. The **pons** also participates in some of these activities, having nuclei that regulate the breathing centers in the medulla, for example.

All the bundles of axons carrying sensory information to and motor instructions from higher brain regions pass through the brainstem, making data conduction one of the most important functions of the medulla and pons. The brainstem also helps coordinate large-scale body movements such as walking. Most of the descending axons carrying instructions about movement to the spinal cord from the midbrain and forebrain cross from one side of the CNS to the other as they pass through the medulla. As a result, the right side of the brain controls much of the movement of the left side of the body, and vice versa.

The Midbrain. The **midbrain** contains centers for the receipt and integration of several types of sensory information. It also serves as a projection center, sending coded sensory information along neurons to specific regions of the forebrain. Prominent nuclei of the midbrain are the inferior and superior colliculi, which are part of the auditory and visual systems, respectively. All fibers involved in hearing either terminate in or pass through the inferior colliculi. In nonmammalian vertebrates, the superior colliculi take the form of prominent optic lobes and may be the only visual centers. In mammals, vision is integrated in the cerebrum, leaving the superior colliculi to coordinate visual reflexes, such as automatically turning your head when your peripheral vision picks up something moving toward you from the side.

The Reticular System, Arousal, and Sleep

As anyone who has sat through a lecture on a warm spring day knows, attentiveness and mental alertness vary from moment to moment. Arousal is a state of awareness of the external world. The counterpart of arousal is sleep, when an individual continues to receive external stimuli but is not conscious of them. Sleep and arousal are controlled by several centers in the brainstem and cerebrum.

A system of neurons called the **reticular formation,** containing over 90 separate nuclei, passes through the core of the brain stem (FIGURE 48.21). Part of the reticular formation, the reticular activating system (RAS), regulates sleep and arousal. Acting as a sensory filter, the RAS selects which information reaches the cerebral cortex, and the more input the cortex receives, the more alert and aware a person is. But arousal is not just a generalized phenomenon; certain stimuli can be ignored while the brain is actively processing other input. Also, specific centers in the brain stem regulate sleep and wakefulness. The pons and medulla contain nuclei that cause sleep when stimulated, and the midbrain has a center that causes arousal. Serotonin may be the neurotransmitter of the sleep-producing centers. Drinking milk before bedtime may induce sleep because milk contains large amounts of tryptophan, the amino acid from which serotonin is synthesized.

Sleep and wakefulness produce different patterns in the electrical activity of the brain, which can be recorded in an **electroencephalogram,** or **EEG** (FIGURE 48.22). As a general

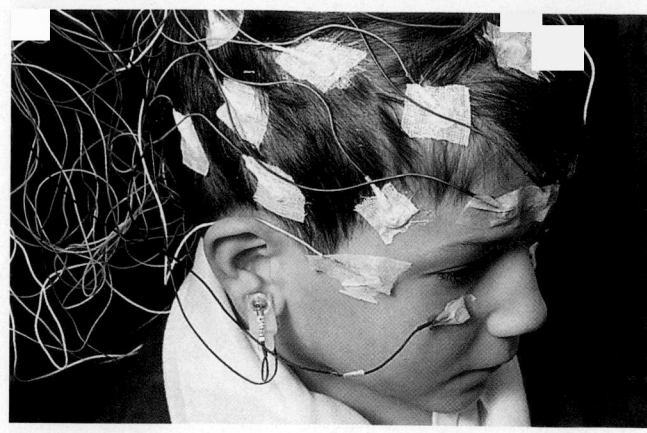

(a) Electrodes on scalp

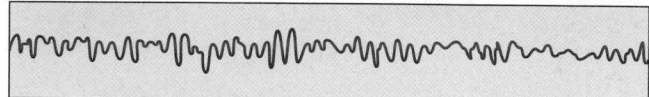

(b) Awake but quiet (alpha waves)

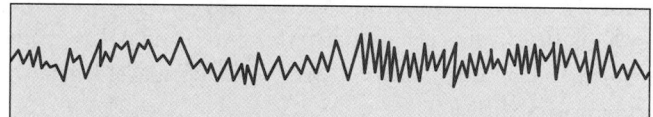

(c) Awake during intense mental activity (beta waves)

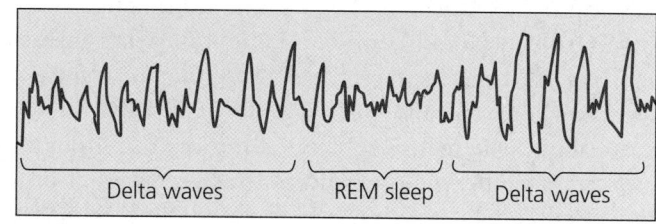

Delta waves · REM sleep · Delta waves

(d) Asleep

FIGURE 48.22 Brain waves recorded by an electroencephalogram (EEG).

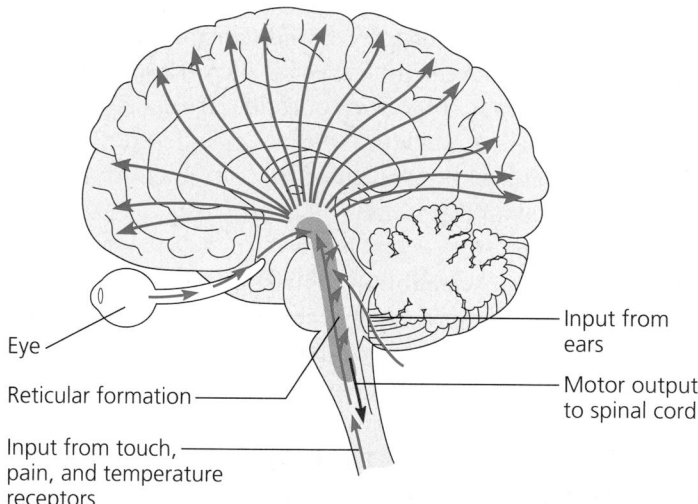

Eye

Reticular formation

Input from touch, pain, and temperature receptors

Input from ears

Motor output to spinal cord

FIGURE 48.21 The reticular formation. This system of over 90 separate nuclei extends through the brainstem. The reticular formation receives input from sensory receptors (blue arrows). Working closely with the cerebral cortex (green arrows), it prevents sensory overload, filtering out familiar and repetitive information that constantly enters the nervous system. Part of this filtering system, the reticular activating system (RAS), is a vital link in determining states of arousal and consciousness.

rule, the less mental activity taking place, the more synchronous the brain waves of the EEG. When a healthy person is lying quietly with closed eyes, slow and synchronous alpha waves predominate. When the eyes are opened or the person solves a complex problem, faster beta waves take over, indicating desynchronization of the parts of the brain.

A sleeping person's EEG reflects the fact that sleep is a dynamic process. In the early stages, theta waves, more irregular than beta waves, often predominate. Deeper sleep produces delta waves, which are quite slow and highly synchronized. Deep sleep also includes periods when a desynchronized EEG reminiscent of wakefulness occurs. During these periods, called REM (rapid eye movement) sleep, the eyes move actively across the visual field behind closed lids. Most dreaming occurs during REM sleep. Like sleep, dreaming has been ascribed magical or prophetic importance, but its true function remains unknown.

The question of why we sleep remains a compelling research problem. All birds and mammals sleep and show a characteristic sleep-wake cycle. One hypothesis is that sleep is involved in the consolidation of learning and memory, and experiments show that regions of the brain activated during a learning task can become active again during sleep.

The Cerebellum

The **cerebellum** develops from part of the metencephalon (see FIGURE 48.19). It functions in coordination and error-checking during motor, perceptual, and cognitive performances. There is strong evidence that the cerebellum is involved in learning and remembering motor responses, because such learning can be blocked by damage to one of its major subdivisions. The cerebellum receives sensory information about the position of the joints and the length of the muscles, as well as information from the auditory and visual systems. It also receives input from the motor pathways, telling it which actions are being commanded from the cerebrum. The cerebellum uses this information to provide automatic coordination of movements and balance. Hand-eye coordination is one example of such a function. If the cerebellum is damaged, the eyes can follow a moving object, but they will not stop at the same place as the object.

The Thalamus and Hypothalamus

The embryonic diencephalon, a division of the forebrain, develops into three adult brain regions: the epithalamus, thalamus, and the hypothalamus (see FIGURE 48.19). The **epithalamus** includes a choroid plexus, one of several clusters of capillaries that produce cerebrospinal fluid. The epithalamus also has a tiny projection, the pineal gland. The endocrine function of the pineal gland was described in Chapter 45. The thalamus and hypothalamus are major integrating centers.

The **thalamus** is the main input center for sensory information going to the cerebrum and the main output center for motor information leaving the cerebrum. It contains many different nuclei, each one dedicated to sensory information of a particular type. Incoming information from all the senses is sorted out in the thalamus and sent on to the appropriate higher brain centers for further interpretation and integration. The thalamus also receives input from the cerebrum and other parts of the brain that regulate emotion and arousal.

Although it weighs only a few grams, the **hypothalamus** is one of the most important brain regions for homeostatic

regulation. We saw in Chapter 45 that the hypothalamus is the source of two sets of hormones, posterior pituitary hormones and releasing hormones that act on the anterior pituitary (see FIGURE 45.6). The hypothalmus contains the body's thermostat, as well as centers for regulating hunger, thirst, and many other basic survival mechanisms. Hypothalamic nuclei also play a role in sexual and mating behaviors, the fight-or-flight response, and pleasure. Stimulation of specific centers can cause what are known as "pure" behaviors. For example, rats placed in an experimental situation where they can press a lever to stimulate a "pleasure" center will do so to the exclusion of eating and drinking. Stimulation of another area can produce rage.

The Hypothalamus and Circadian Rhythms. Animals, including humans, exhibit all kinds of regularly repeated, rhythmic behaviors. What maintains our daily rhythms—when, for example, we sleep, our blood pressure is highest, or our sex drive peaks? Many animals exhibit seasonal rhythms, reproducing or migrating only in the spring or fall, for instance. We have already discussed circadian (daily) and seasonal rhythms in plants (see Chapter 39).

Numerous studies have assessed the relative importance of external cues and internal timekeepers in maintaining rhythmic behavior. These studies show that circadian rhythms usually have a strong internal component, referred to as a **biological clock.** Locating the internal mechanisms responsible for behavioral rhythms has been challenging for researchers. An early hypothesis that the location of these control mechanisms varies across taxonomic boundaries has proved true. For instance, fruit flies (*Drosophila*) appear to have many biological clocks throughout their body and at the outer edges of their wings. In mammals, a pair of structures called the **suprachiasmatic nuclei (SCN)** in the hypothalamus functions as a biological clock. Experiments with rodents have revealed that the cells of the SCN produce specific proteins in response to changing light-dark cycles. The function of this or any other biological clock may be the regulation of a variety of physiological processes, such as hormone release, hunger, and heightened sensitivity to external stimuli that motivate specific rhythmic behaviors.

Researchers have also investigated the role of external cues in circadian rhythms. Usually, a clock's rhythm does not exactly match events in the environment, and external cues are necessary to keep cycles timed to the outside world. Light is a common external cue for circadian rhythms; visual information received by the SCN via sensory neurons from the eyes enables the mammalian clock to remain synchronized with the natural cycles of day length and darkness. For example, activity of the North American flying squirrel normally begins with the onset of darkness and ends at dawn, which suggests that light is an important external cue. If a squirrel is placed in constant light or constant darkness, its rhythmic activity continues, but the duration of its activity cycle (one period of activity plus one period of inactivity) falls a little more out of sync with the outside world each day (FIGURE 48.23). The squirrel's internal clock continues to run without external cues, but on its own cycle, which turns out to be 21 minutes longer than 24 hours. External cues, such as day length and night length, adjust the clock, so that the rhythmic behavior it controls is synchronized with the outside world.

Human circadian rhythms have been studied by placing individuals in comfortable living quarters deep underground, where they could make their own schedules with no external cues of any kind. Under these *free-running* conditions (see Chapter 39), the biological clock of humans seems to have a period of about 25 hours, but with much individual variation. Like other animals, humans use external cues to adjust their rhythms to 24 hours in the real world.

The cerebrum is the most highly evolved structure of the mammalian brain

The cerebrum develops from the embryonic telencephalon, an outgrowth of the forebrain that arose early in vertebrate evolution as a region supporting olfactory reception as well as auditory and visual centers. The cerebrum is divided into right and left **cerebral hemispheres** (FIGURE 48.24a). Each hemisphere consists of an outer covering of gray matter, the cerebral cortex, mentioned earlier; internal white matter; and a cluster of nuclei called **basal nuclei** deep within the white matter. The basal nuclei are important centers for planning and learning movement sequences. Damage in this

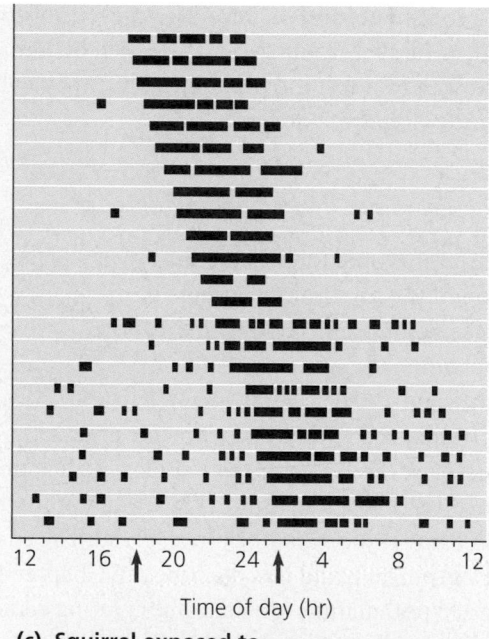

(a)

THE PROCESS OF SCIENCE

FIGURE 48.23 Activity rhythms in a nocturnal mammal. An inhabitant of forests in North America, the northern flying squirrel is active at night and usually sleeps in a nest cavity in a hollow tree from dawn to dusk. **(a)** To study its activity rhythms, researchers have placed flying squirrels in cages containing an exercise wheel in which a squirrel can run. The wheel is connected to a chart recorder, which moves graph paper past a pen at a fixed speed. When a squirrel runs on the wheel, the pen is activated and marks the graph paper. The graphs shown in (b) and (c) trace the activity of two flying squirrels kept under different light conditions for 23 days. The longer black bars indicate periods of extended activity. **(b)** This graph shows the activity pattern of a squirrel exposed to 12 hours of darkness, simulating natural conditions. **(c)** This graph shows the activity pattern of a squirrel held in constant darkness for 23 days. Although the activity of both squirrels remained rhythmic throughout the recording period, with a distinct period of extended activity every day, the high-activity period of the squirrel shifted each day (actually by 21 minutes). After 23 days its period of greatest activity was nearly 8 hours out of synchronization with the actual time of day. (The small magenta arrows indicate when the period of greatest activity began on days 1 and 23.) Experiments with flying squirrels, humans, and other animals show that environmental cues are needed to keep biological clocks tuned to external conditions.

(b) Squirrel exposed to 12 hours of darkness per day

(c) Squirrel exposed to constant darkness

region can cause devastating motor disorders in humans such as Parkinson's disease and Huntington's disease, or even render a person passive and immobile because the nuclei no longer allow motor impulses to be sent to the muscles.

The cerebral cortex ("gray matter") is the largest and most complex part of the mammalian brain, and the part that has changed the most during vertebrate evolution. Some of its components are also found in the brains of reptiles, which share a common ancestor with mammals (see FIGURE 34.20). Unique to mammals, however, is the **neocortex,** an additional outer layer of cortex consisting of six sheets of neurons running tangential to the brain surface. Among mammals, greater cognitive abilities and more sophisticated behavior are associated with the relative size of the cerebral cortex and the presence of convolutions that increase the surface area of the neocortex. Although less than 5 mm thick, the human neocortex has a surface area of about 0.5 m^2 and accounts for about 80% of the total brain mass. Nonhuman primates and cetaceans (whales and porpoises, for example) also have exceptionally large and complex neocortices. In fact, the surface area (relative to body size) of a porpoise's neocortex is second only to that of a human.

Like the rest of the cerebrum, the cerebral cortex is divided into right and left sides, each of which is responsible for the opposite half of the body. The left hemisphere receives information from, and controls the movement of, the right side of the body, and vice versa. A thick band of fibers (cerebral white matter) known as the **corpus callosum** communicates between the right and left cerebral hemispheres (see FIGURE 48.24a).

In the discussion of cerebral functions that follows, the words *cognition* and *cognitive* will come up often. Cognition is defined as the process of knowing, including both awareness and judgment. Thus, the cognitive functions of the brain include learning, decision making, consciousness, and an integrated sensory awareness of the surroundings.

Regions of the cerebrum are specialized for different functions

Each side of the cerebral cortex is customarily described as having four lobes, called the frontal, temporal, occipital, and parietal lobes. Researchers have identified a number of functional areas within each lobe (FIGURE 48.24b). These areas include primary sensory areas for input of different kinds of sensory information, and association areas that integrate the input with information from other parts of the brain.

Two functional cortical areas, the primary motor cortex and the primary somatosensory cortex, form the boundary between the frontal lobe and the parietal lobe. The motor cortex functions mainly in sending commands to the skeletal muscles, signaling appropriate responses to any sensory stimuli. The

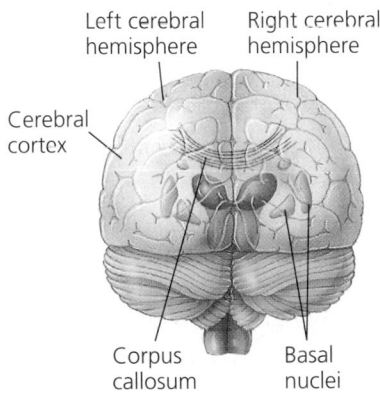

(a) **Back of brain**

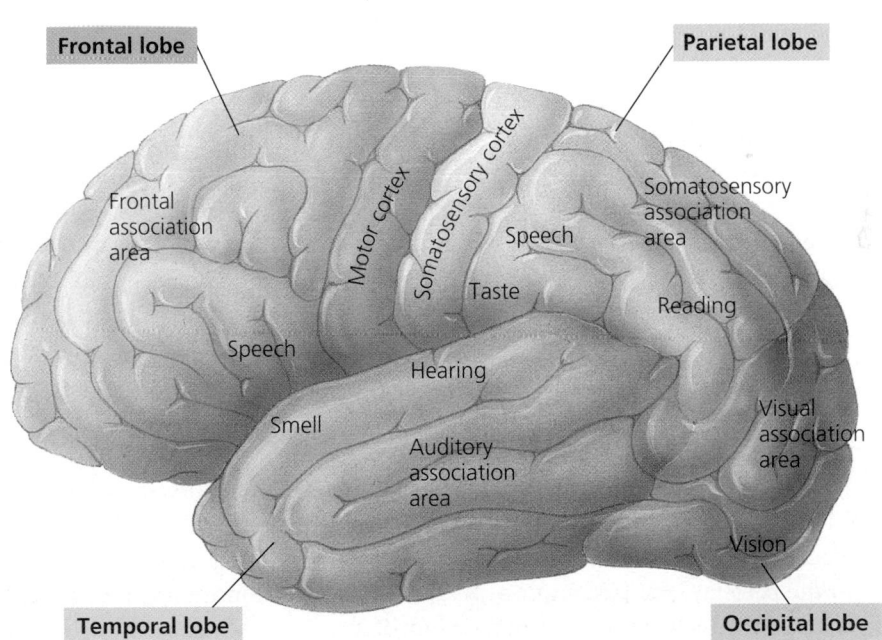

(b) **Left side of brain**

FIGURE 48.24 Structure and functional areas of the cerebrum. (a) This rear view of the human brain shows the bilateral nature of the cerebral hemispheres. The corpus collosum (large fiber tracts connecting the hemispheres) and basal nuclei (ganglia) are completely covered (and not visible from the surface) by the convoluted gray matter of the cerebral cortex. **(b)** The surface of each cerebral hemisphere is divided into four lobes. (A fifth lobe, the insula, located within the fold separating the temporal and parietal lobes, is not shown in this drawing of the left hemisphere.) Specialized functions are localized in each lobe. The association areas of the left hemisphere have different functions than those of the right hemisphere.

FIGURE 48.25 Primary motor and somatosensory areas of the human cerebral cortex. In these cross-sectional maps of the cortex, the surface area devoted to each body part is symbolized by the relative sizes of the body parts in the cartoon.

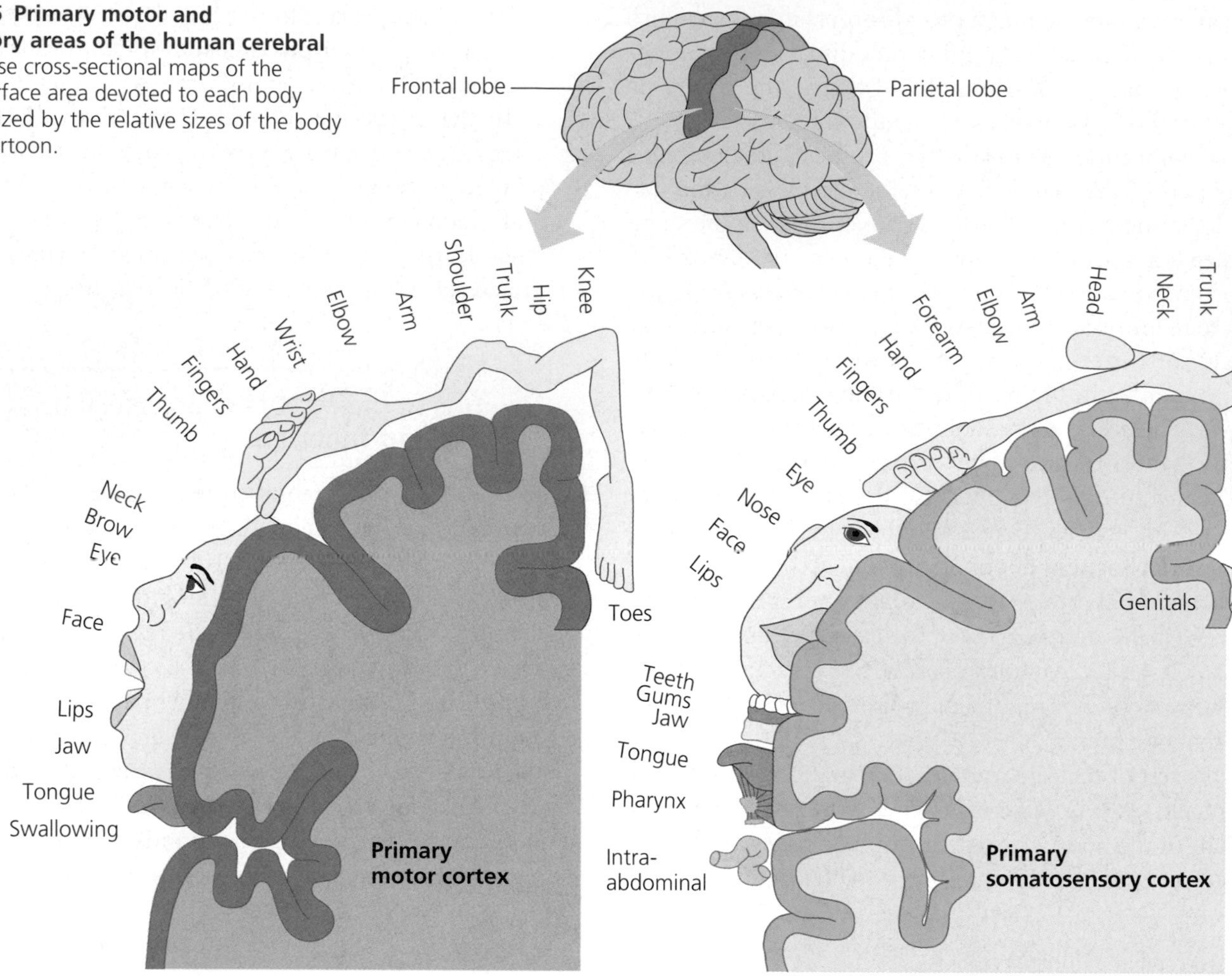

somatosensory cortex receives and partially integrates signals from touch, pain, pressure, and temperature receptors throughout the body. The proportion of motor or somatosensory cortex devoted to a particular part of the body is correlated with the relative importance of sensory or motor information for that part of the body (FIGURE 48.25).

Integrative Function of the Association Areas

Sensory information coming into the cortex, mainly via the thalamus, is first directed to primary sensory areas within the lobes: visual information to the occipital lobe; auditory input to the temporal lobe; and somatosensory information about touch, pain, pressure, temperature, and the position of muscles and limbs to the parietal lobe (see FIGURE 48.24b). Information about taste goes to a separate sensory region of the parietal lobe. Olfactory information is sent first to "primitive" regions of the cortex and then via the thalamus to an interior region of the frontal lobe (by "primitive" regions here, we mean cerebral regions that are similar between mammals and reptiles). All this information is then passed to the adjacent association areas, which integrate ("associate") the input from different types of senses, assess the significance of the overall

sensory input, and then transmit signals to still more association areas, these located in the frontal lobes. The association regions of the frontal lobes then compose an appropriate motor response plan, which is used by the primary motor cortex to direct the movement of skeletal muscles.

The major increase in the size of the neocortex that occurred during mammalian evolution expanded the association areas that integrate higher cognitive functions that make more complex behavior and learning possible. The cortical surface of a rat's brain is relatively smooth and is occupied mainly by primary sensory areas, which receive direct input from sensory receptors. In contrast, the cortical surface of our brains is much more convoluted and consists mainly of association regions. Regional specializations in this association cortex first appear during brain development as infants and children learn routines for sensing and acting. While these regional specializations are roughly the same in different humans, damage to one brain area early in development can frequently cause redirection of its normal functions to other areas. Perhaps the most dramatic example of this is observed in some human infants with intractable epilepsy who are treated by removing the entire right or left cerebral hemisphere. Amazingly, the remaining hemisphere takes over most of the functions normally provided by two hemispheres. Some plasticity is maintained in adult brains, for they can sometimes recover from damage to a portion of the cerebral cortex by developing or using new circuits.

Lateralization of Brain Function

During an infant's or child's brain development, competing functions segregate and displace each other into the cortex of opposite cerebral hemispheres. The left hemisphere becomes most adept at language, math, logic operations, and the processing of serial sequences of information. It has a bias for the detailed, speed-optimized activities required for skeletal motor control and processing of fine visual and auditory details. The right hemisphere is stronger at pattern recognition, face recognition, spatial relations, nonverbal ideation, emotional processing in general, and parallel processing of many kinds of information. Understanding and generating the stress and intonation patterns of speech that convey its emotional content emphasizes right-hemisphere function, as does music. The right hemisphere appears to specialize in perception of the relationship between images and the whole context in which they occur, whereas the left hemisphere is better at focused perception. While working with their hands, most right-handed people use the left hand (right hemisphere) for context or holding and use the right hand (left hemisphere) for fine detailed movement.

Language and Speech

THE PROCESS OF SCIENCE The systematic and detailed mapping of higher cognitive functions to specific brain areas began in the 19th century when physicians performed post-mortem examinations of the brains of patients with language impairments. Those who could understand language but not speak usually had damage to a frontal lobe area, now called Broca's area, just in front of motor cortex controlling face and lips. Studies of brain activity using a modern imaging technology called PET (for positron emission tomography) have confirmed that Broca's area is active during generation of speech (FIGURE 48.26, lower left image). Autopsies during the 19th century also suggested that damage to a posterior portion of the temporal lobe, now called Wernicke's area (upper left image of FIGURE 48.26), could abolish the ability to comprehend speech but leave speech generation intact.

More detailed modern studies, both of brain lesions and imaging of brain activity, have now demonstrated that language is processed by multiple areas in the cortex. An instruction to simply read a printed word out loud activates the visual cortex (upper right image of FIGURE 48.26) and Broca's area. Frontal and temporal areas become active when meaning must be attached to words, as in generating verbs to go with nouns or in grouping together related words or concepts (FIGURE 48.26, lower right image). Researchers can monitor the effect of

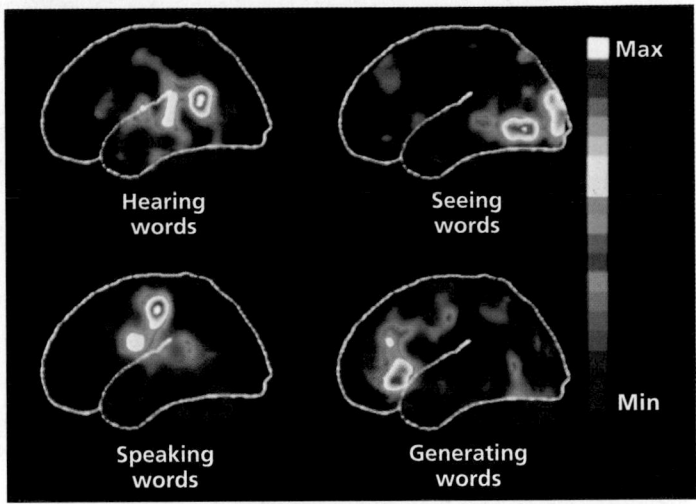

FIGURE 48.26 Mapping language areas of the cerebral cortex. A modern imaging technology called PET scanning maps regions of an organ that are most metabolically active. These four computer-generated brain maps show "hot spots" of activity in the brain of one individual under four different conditions, all related to speech.

practice on this cognitive development. An instruction to associate a verb with each noun projected on a screen initially activates frontal lobes. But after 15 minutes of practice, the activation has contracted mainly to those areas used in simply reading a word out loud. This illustrates an important principle of brain function: novel stimuli or instructions cause the greatest mobilization of brain areas and resources. Once a situation or procedure has become familiar, it is accommodated by a much lower level of brain activity.

Emotions

Two components of the cerebral cortex, the hippocampus and the olfactory cortex, are common to reptiles and mammals. In mammals, these structures, together with some inner portions of the cortex's lobes and sections of the thalamus and hypothalamus, form a ring around the brainstem called the **limbic system** (FIGURE 48.27, p. 1050). By interacting with sensory areas of the neocortex and other higher brain centers, the limbic system generates the feelings we generally call emotions.

The limbic system is central to some of the behaviors, such as the extended nurturing of infants and an emotional bonding to other individuals, that distinguish mammals from most reptiles and amphibians. The limbic system mediates primary emotions that manifest in such behaviors as laughing and crying. But the system also attaches emotional "feelings" to basic survival-related programs of the brainstem, such as feeding, aggression, and sexuality.

Structures of the emotional brain centered in the limbic system form early in development, and provide a foundation for the higher cognitive functions that appear later during the

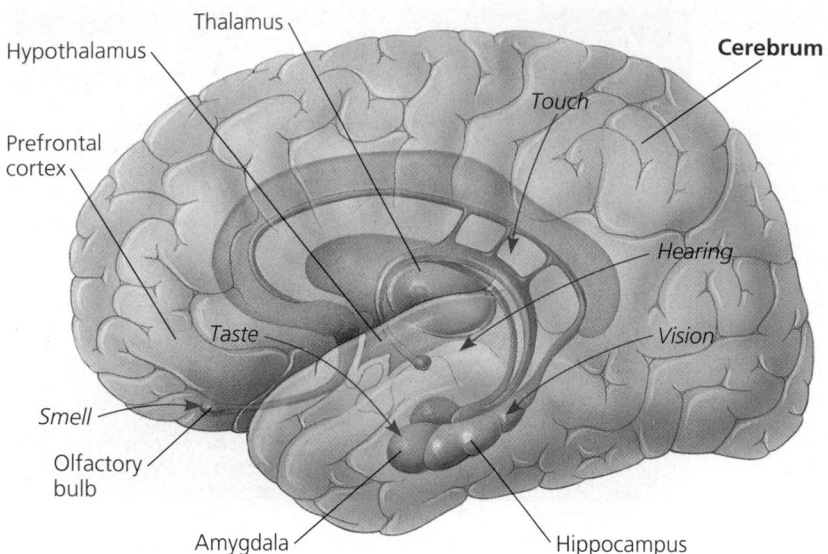

FIGURE 48.27 The limbic system. Portions of the diencephalon (thalamus and hypothalamus) and inner parts of the cerebral cortex, including the amygdala and hippocampus, make up this functional center of human emotions and memory. Signals from the nose enter the brain through the olfactory bulb, which is part of the limbic system. Other sensory information enters the limbic system via other parts of the cerebral cortex (arrows). A "higher" center of integration in the cerebrum, the prefrontal cortex, apparently consults the limbic system and other brain centers in processing and retrieving memories and may use memories to modify behavior.

development of neocortical areas. We, along with other primates, are born with emotional brain circuits prepared to support bonding to a caretaker, to recognize the crude elements of a human face, to have visual and vocal interactions with that caretaker, and to express fear, distress and anger. Learning and memory processes are then set in motion to build a history of what sensing and motor actions "work" in obtaining warmth and food. We also begin very early to distinguish "right" from "wrong"—for example, by causing happy or angry facial or vocal expressions from a caretaker. The amygdala, a nucleus in the temporal lobe (see FIGURE 48.27), is central in recognizing the emotional content of facial expressions and laying down emotional memories.

This emotional memory system seems to appear earlier in development than the system that supports explicit recall of events, which requires another limbic structure, the hippocampus. Adults who learn to avoid an aversive situation, such as a picture that is always followed by a mild electric shock, remember the picture and experience autonomic arousal—as measured by heart rate increase or sweating—if the picture is presented again. Some adults with damage to the hippocampus report that they do not recognize the picture, but the autonomic arousal mediated by the amygdala still occurs, for the emotional memory mediated by this structure is still intact. Conversely, other patients with damage just to the amygdala do not exhibit the autonomic reactions but recall the picture, because the hippocampus which is involved in explicit memory is functioning.

As children develop, primary emotions such as pleasure and fear are associated with different situations in a process that requires portions of the neocortex, especially the frontal lobes. The existence of learning processes by which we accumulate emotional reactions to different situations are revealed by their disruption in some patients who have lesions or tumors in their frontal lobes. Their behavior is superficially normal most of the time—intellect and memory seem intact—but motivation, foresight, goal formation, and decision making are flawed. These patients are subject to bursts of immodest, impolite, or other inappropriate behaviors; they act almost as though their free will had been taken from them. Their emotions and feelings also are diminished. Frontal lobotomies, which disrupt the connection between the frontal lobes and the limbic system, were once performed widely to treat severe emotional disorders. However, the resulting docility was usually accompanied by a loss of the ability to concentrate, plan, and work toward goals, and thus drug therapy has replaced lobotomy for treating such severely ill individuals.

Memory and Learning

Our daily life is a constant checking of what is happening right now against what just happened a few moments ago. We hold information, anticipations, or goals for a time in **short-term memory** locations in the frontal lobes and then release them if they become irrelevant. Should we wish to retain knowledge of a face or a phone number, the mechanisms of **long-term memory** are activated in a process that requires the hippocampus, part of the limbic system (see FIGURE 48.27). If we later need to remember a name or number, we can fetch it from that long-term memory and place it back in our working memory. The transfer of information from short-term to long-term memory is enhanced by rehearsal ("practice makes perfect"), positive or negative emotional states mediated by the amygdala, and the association of new data with data previously learned and stored in long-term memory (it's easier to learn a new card game if you already have "card sense" from playing other games).

Many sensory and motor association areas of the cerebral cortex outside the classical language areas are involved in storing and retrieving words and images from our mental dictionary. Studies of patients with brain lesions and imaging studies on unimpaired patients suggest, for example, that knowledge of unique persons is associated with the anterior portion of the left temporal lobe, knowledge of animals with activity in the middle lower part of this lobe, and knowledge of tools with the lower posterior part.

The memorization of phone numbers, facts, and places—which can be very rapid and may require only one exposure to the relevant item—may rely mainly on rapid changes in the strength of existing nerve connections. In contrast, the slow learning and remembering of skills and procedures, such as a person's experience when trying to improve his or her tennis game, appear to involve cellular mechanisms very similar to those responsible for brain growth and development. Nerve cells actually make new connections. Skill memories usually involve motor activities that are learned by repetition without consciously remembering specific information. You perform learned motor skills, such as walking, tying your shoes, riding a bicycle, or writing without consciously recalling the individual steps required to do these tasks correctly. Once a skill memory is learned, it is difficult to unlearn. For example, a person who has played tennis for years with a self-taught, awkward backhand has a much tougher time learning the correct form than a beginner just learning the game. Bad habits, as we know, are difficult to break.

Many exciting new experiments are documenting changes in gene expression and synaptic molecules that correlate with memory formation. Functional changes at certain synapses in the hippocampus and amygdala seem to be directly related to memory storage and emotional conditioning. One kind of change, called **long-term depression (LTD),** is decreased responsiveness to an action potential by a postsynaptic cell. LTD is induced by repeated, weak stimulation. A second kind of synaptic change, called **long-term potentiation (LTP),** is an enhanced responsiveness to action potentials by a postsynaptic cell. It can result when a presynaptic cell bombards a synapse with a series of brief, repeated action potentials that strongly depolarize the postsynaptic membrane. With LTP established, a single action potential from the presynaptic cell has a much greater effect at the synapse than before. Lasting for a matter of hours, days, or weeks, depending on the number and frequency of repeated action potentials, LTP may be what happens when a memory is being stored or when learning takes place. LTP is associated with the release of the excitatory neurotransmitter glutamate by the presynaptic cell. Glutamate binds with a specific class of receptors in the postsynaptic membrane, opening gated channels that are highly permeable to calcium ions (Ca^{2+}). In turn, the Ca^{2+} triggers a cascade of enzyme activity in the postsynaptic cell, making it more responsive to stimulation.

Human Consciousness

When we use the word *consciousness,* we are usually referring to our own human version of self-consciousness, in which we experience ourselves as not only sensing, acting, and feeling in the present moment but also thinking about the past and the future. How many of these capabilities extend to other animals is a subject of a vigorous debate (see Chapter 51). Until relatively recently the study of human consciousness has been considered outside the province of hard science, more appropriate as a subject for philosophy and religion. This situation is changing as we learn more about the brain activities that underlie our behaviors. Brain imaging studies are showing changes in neuronal activity that correlate with conscious perceptual choices, unconscious versus conscious processing, retrieval of memories, and engaging working memory to plan for the future. There is a growing consensus that consciousness is an emergent property of the brain that recruits activities in many areas of the cortex. Several models suggest a scanning mechanism that repetitively sweeps across the brain to unite many of these disparate processes into a unified conscious moment. Rather than just being a matter of describing the details of how the areas described above are put together, approaching consciousness may also require the understanding of dynamic whole-brain patterns of activity that bear no more direct relationship to individual nerve cells than do hurricanes to their constituent water molecules.

Understanding whole-brain phenomena such as consciousness represents one end of the spectrum of questions about nervous systems. At the other (microscopic) end is a frontier where neuroscientists are studying how neurons develop and organize into networks.

Research on neuron development and neural stem cells may lead to new approaches for treating CNS injuries and diseases

Unlike the peripheral nervous system, the mammalian central nervous system cannot repair itself when damaged or assaulted by disease. The human brain can make new connections between surviving neurons and thus sometimes compensate for damage, as in the remarkable recoveries of some stroke victims. Generally speaking, however, spinal cord injuries, strokes, brain injuries, and diseases that destroy brain cells, such as Parkinson's and Alzheimer's, have devastating effects. Current research on nerve cell development and the discovery of neural stem cells enhance our fundamental knowledge of the nervous system and may one day make it possible for physicians to repair or replace damaged neurons.

THE PROCESS OF SCIENCE

Nerve Cell Development

One of the key questions in neurobiology is how certain cells in a developing animal differentiate into neurons, migrate to their proper location, grow axons to specific places, and make synapses with the proper target cells. How is this accomplished without ending up with a useless tangle of neurons? The labs of Corey Goodman (University of California, Berkeley) and Marc Tessier-Lavigne (University of California, San Francisco) study how neurons find their way during development of the central nervous system. Their work combines elements of cell-to-cell communication (Chapter 11), control of gene expression (Chapter 19), and the genetic basis of development (Chapter 21).

To reach their target cells, axons must grow anywhere from a few microns up to a meter in length (for example, from the human spinal cord to the foot). Axons do not follow a straight path from point A to point B; rather, molecular signposts along the way direct and redirect the growing axon in a series of mid-course corrections that result in a meandering, but not random, trek. The responsive region at the leading edge of the growing axon is called the **growth cone.** Signal molecules released by target cells bind to receptors on the plasma membrane of the growth cone, triggering a signal-transduction pathway (FIGURE 48.28). The response of the axon may be to grow toward the source of the signal molecules (attraction) or grow away from the source (repulsion). Cell adhesion molecules (CAMs) on the axon's growth cone also play a role; they attach to complementary cell adhesion molecules on surrounding cells that provide tracks for the growing axon to follow. Nerve growth factor released by astrocytes and growth-promoting proteins produced by the neurons themselves contribute to the process by stimulating growth of axons.

Goodman, Tessier-Lavigne, and their colleagues have found the genes, gene products, and process of axon guidance to be remarkably similar in nematode worms (*C. elegans*), insects (*Drosophila*), and vertebrates, indicating evolutionary conservation of the genes and basic mechanism for this complex process.

Development occurs in a space and time sequence that would seem difficult to replicate in order to repair or replace neurons in humans with damaged nervous systems. The

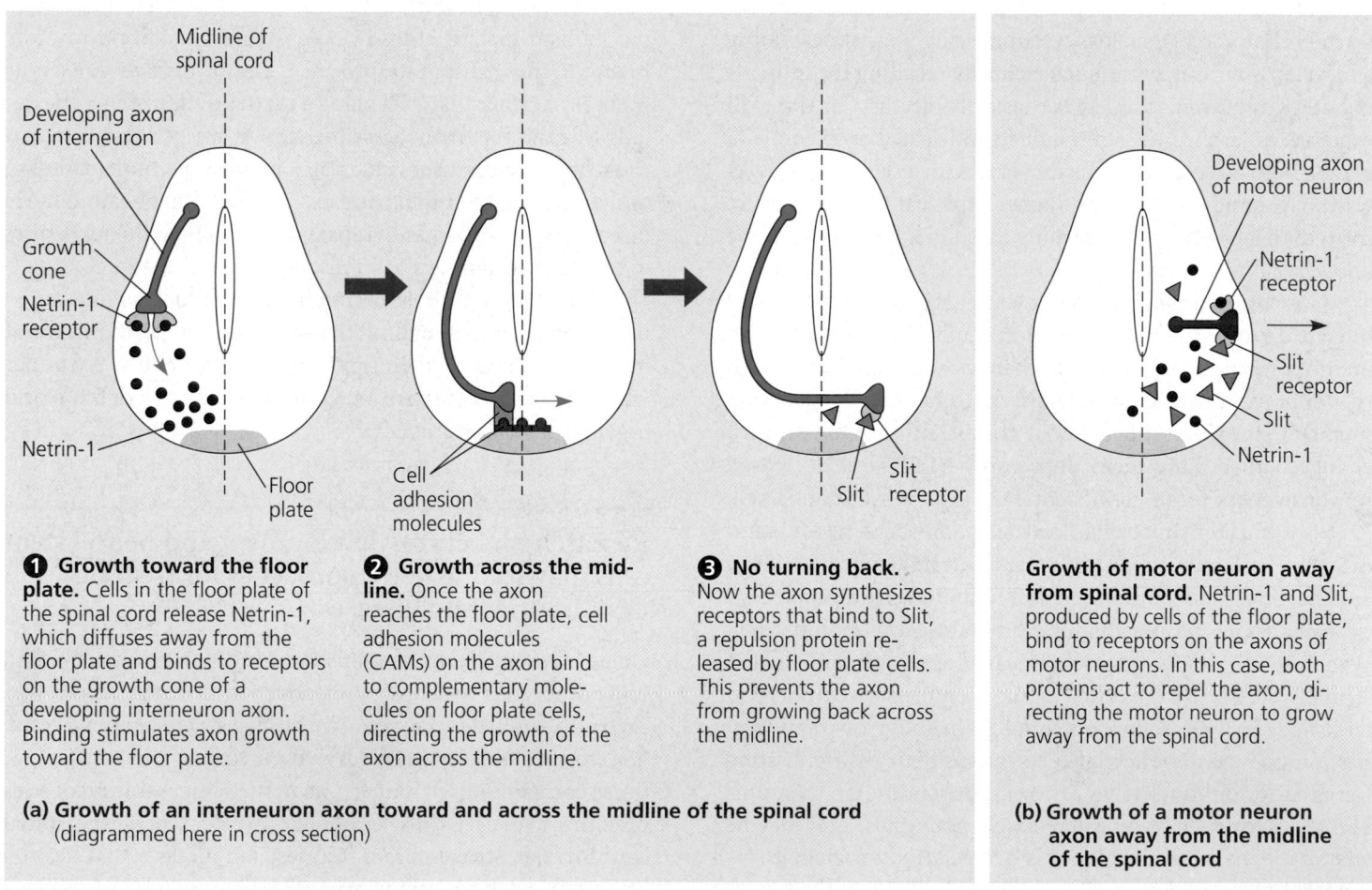

❶ Growth toward the floor plate. Cells in the floor plate of the spinal cord release Netrin-1, which diffuses away from the floor plate and binds to receptors on the growth cone of a developing interneuron axon. Binding stimulates axon growth toward the floor plate.

❷ Growth across the midline. Once the axon reaches the floor plate, cell adhesion molecules (CAMs) on the axon bind to complementary molecules on floor plate cells, directing the growth of the axon across the midline.

❸ No turning back. Now the axon synthesizes receptors that bind to Slit, a repulsion protein released by floor plate cells. This prevents the axon from growing back across the midline.

Growth of motor neuron away from spinal cord. Netrin-1 and Slit, produced by cells of the floor plate, bind to receptors on the axons of motor neurons. In this case, both proteins act to repel the axon, directing the motor neuron to grow away from the spinal cord.

(a) Growth of an interneuron axon toward and across the midline of the spinal cord (diagrammed here in cross section)

(b) Growth of a motor neuron axon away from the midline of the spinal cord

FIGURE 48.28 How do developing axons know which way to go? The region at the leading edge of a developing axon is called the growth cone; this area displays various receptors and cell adhesion molecules that bind to guidance signals from surrounding cells. This figure illustrates a few of the factors involved in axon guidance. In the examples shown here, **(a)** interneurons are induced to cross the midline of the spinal cord whereas **(b)** motor neurons are induced to grow away from the spinal cord, eventually to synapse with muscle or gland cells.

growing axon expresses different genes at different times during its development, and is influenced by surrounding cells that it moves away from. Work continues on deciphering these developmental mechanisms, however, with the ultimate goal of repairing damaged neurological tissue. Using the right combination of attractants, repellants, growth-associated proteins, and growth factors, researchers hope to coax damaged axons to regrow, following the correct pathway and forming connections with the correct targets.

Neural Stem Cells

Until 1998, it was "common knowledge" that you were born with all the brain cells you would ever have. In that year, however, Fred Gage (Salk Institute for Biological Studies, La Jolla, California) and Peter Ericksson (Sahlgrenska University Hospital, Göteborg, Sweden) made a startling announcement: the human brain *does* produce new nerve cells in adulthood. Newly-divided cells were found in the hippocampus, an area of the brain involved with memory and learning (see FIGURE 48.27). It is not clear what function these new cells play in the human brain. However, mice that live in stimulating environments and run on exercise wheels have more new brain cells in their hippocampus and perform better on learning tasks than genetically-identical mice who live in standard cages. Perhaps people can also improve their brain capacity by stimulating their minds and exercising their bodies, resulting in a greater capacity for learning.

The discovery of young brain cells in adult brains came about because of a bit of scientific serendipity and the generosity of some terminally ill cancer patients. Ericksson had been on sabbatical in Gage's lab, where the chemical marker bromodeoxyuridine (BrdU) was injected in mice to label the DNA of dividing cells. In a postmortem study, the brains of the mice were examined for new brain cells, which were identified by the marker. After Ericksson returned to Sweden for clinical work, he was on call with a cancer specialist who happened to mention that a group of terminally ill cancer patients was being given bromodeoxyuridine as part of a study to monitor tumor growth. Ericksson recognized this substance as the marker being used to look for new brain cells in the Gage lab. The patients agreed to donate their brains to science upon their death, and new neurons were found in all the patients (FIGURE 48.29).

Mature brain cells, with their extensive processes and intricate connections to other cells, clearly are not able to undergo cell division. Therefore, the new brain cells must have come from stem cells. Recall from Chapter 21 that stem cells are relatively unspecialized cells that continually divide. Some offspring cells may differentiate into specialized cells under the right conditions, while cell division retains a supply of undifferentiated cells.

One of the difficulties of conducting research on stem cells is finding a source of human stem cells. Various ethical and political issues surround the use of embryonic stem cells. In

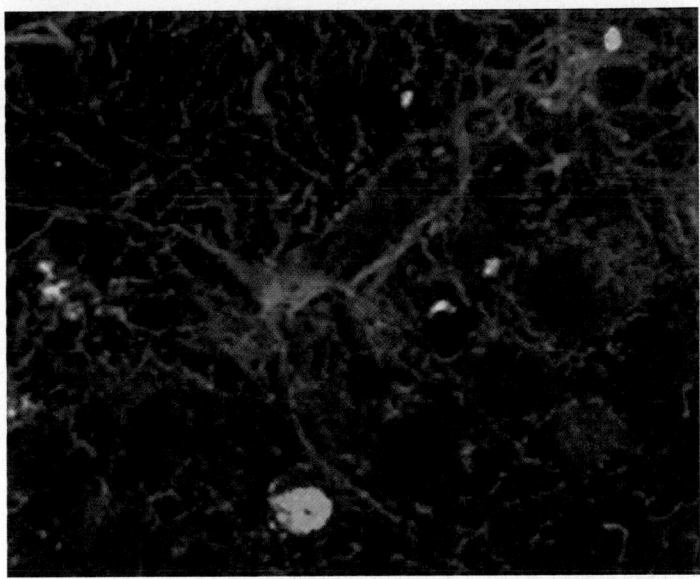

FIGURE 48.29 A newly-born brain cell in the adult human brain. The cell that is colored both red and green is a newly-formed neuron from the hippocampus of a recently-deceased adult human cancer victim. All the red cells are neurons. The green indicates that this particular neuron is the result of a recent cell division. In the process of DNA replication, the DNA incorporated bromodeoxyuridine, a marker that had been injected into the patient to monitor tumor growth. The marker has a green glow when illuminated in a special type of light microscope called a fluorescence microscope.

addition to the stem cells of an early embryo, certain tissues of an adult also have stem cells, though they may be less developmentally versatile than embryonic stem cells. For example, the bone marrow has stem cells that can differentiate into the various kinds of blood cells (see FIGURE 42.15). The existence of stem cells in the brain, however, is an exciting surprise.

In May 2001, Gage and his colleagues announced that they had cultured neural progenitor cells from the brains of recently-deceased individuals and surgical tissue samples. The term "progenitor" refers to the fact that these stem cells are committed to becoming nerve cells; they are not as "plastic" as embryonic stem cells. In the lab, the neural progenitor cells divided 30–70 times and differentiated into neurons and astrocytes. Gage noted, "These results confirm that the adult human brain contains cells that can continue to divide and differentiate."

One of the goals of further research is to find a way to induce the body's own neural progenitor cells to differentiate into glia or specific types of neurons when and where they are needed. Also, cultured neural progenitor cells could be transplanted into damaged neurological tissue.

In addition to the studies we've described here, many other researchers are exploring various approaches to preventing and treating central nervous system disorders and injuries.

■   ■   ■

In the next chapter we examine in more detail the input and output elements of the nervous system—sensory receptors and motor effectors.

CHAPTER 48 REVIEW

Go to the Campbell Biology website (www.campbellbiology.com) to explore an interactive version of the Chapter Review.

Summary of Key Concepts

AN OVERVIEW OF NERVOUS SYSTEMS

■ **Nervous systems perform the three overlapping functions of sensory input, integration, and motor output (p. 1023, FIGURE 48.1)** The nervous system's three main functions are sensory input, integration, and motor output to effector cells. The central nervous system (CNS) integrates information, while the nerves of the peripheral nervous system (PNS) communicate sensory and motor signals between the CNS and the rest of the body.

■ **Networks of neurons with intricate connections form nervous systems (pp. 1023–1026, FIGURES 48.2, 48.3)** Neurons have structural specializations that permit them to interact with large numbers of other neurons. Highly branched dendrites collect information which is directed to the cell body and then to axons which convey signals to further locations. Synaptic terminals at the ends of axons release neurotransmitter molecules into the synapses, thereby relaying neural signals to the dendrites or cell bodies of other neurons or effectors. The simplest nerve circuit is the reflex arc, consisting of a sensory neuron which, on stimulation, signals a motor neuron. The motor neuron then signals an effector cell such as a muscle or gland cell to change its activity. Most nerve circuits, however, interpose large numbers of interneurons between the primary sites of sensing and acting. Animal brains are formed from clusters of the cell bodies of these interneurons. Neurons require the support of surrounding glial cells.

Web/CD Activity 48A: Neuron Structure

THE NATURE OF NERVE SIGNALS

Virtually all nerve signals are changes in the voltage across nerve plasma membranes, and understanding the cellular and molecular basis of these signals is essential to ultimately explaining the functions of brains.

■ **Every cell has a voltage, or membrane potential, across its plasma membrane (pp. 1026–1028, FIGURE 48.7)** The membrane potential of a nontransmitting neuron is due to the unequal distribution of ions, particularly sodium and potassium, across the plasma membrane; the cytosol is more negatively charged than the extracellular fluid. Membrane potential is maintained by differential ion permeabilities and the sodium-potassium pump.

Web/CD Activity 48B: Nerve Signals: Action Potentials

■ **Changes in the membrane potential of a neuron give rise to nerve impulses (pp. 1028–1031, FIGURES 48.8, 48.9)** A stimulus that affects the membrane's permeability to ions can either depolarize or hyperpolarize the membrane relative to the membrane's resting potential. This local voltage change is a graded potential, its magnitude proportional to the strength of the stimulus. An action potential, or nerve impulse, is a rapid, transient, all-or-none depolarization of the neuron's membrane. A local depolarization to the threshold potential opens voltage-gated sodium channels, and the rapid influx of Na^+ brings the membrane potential to a positive value. The membrane potential is restored to its normal resting value by the closing of the Na^+ channels. A refractory period follows an action potential,

corresponding to the period when the voltage-gated Na^+ channels are inactivated. The frequency of action potentials varies with the intensity of the stimulus.

Web/CD Case Study in the Process of Science: What Triggers Nerve Impulses?

■ **Nerve impulses propagate themselves along an axon (pp. 1031–1032, FIGURES 48.10, 48.11)** Once an action potential is initiated in an axon, a wave of depolarization propagates a series of action potentials to the end of the axon. The rate of transmission of a nerve impulse is directly related to the diameter of the axon. Saltatory conduction, a mechanism by which action potentials jump between the nodes of Ranvier of myelinated axons, speeds nervous impulses in vertebrates.

■ **Chemical or electrical communication between cells occurs at synapses (pp. 1033–1034, FIGURE 48.12)** Synapses between neurons conduct signals from the axon of a presynaptic cell to a dendrite or cell body of a postsynaptic cell. Electrical synapses directly pass an action potential between two neurons via gap junctions. In a chemical synapse, a depolarization stimulates the fusion of synaptic vesicles with the presynaptic membrane and the release of neurotransmitter molecules into the synaptic cleft. Neurotransmitters bind to receptor proteins associated with particular ion channels on the postsynaptic membrane. The neurotransmitter is rapidly broken down by enzymes, or taken back up into surrounding cells.

Web/CD Activity 48C: Signal Transmission at a Chemical Synapse

■ **Neural integration occurs at the cellular level (pp. 1034–1036, FIGURES 48.13, 48.14)** A single neuron may receive information via thousands of synapses on its dendrites and cell body. Whether or not it generates an action potential in response depends on temporal and spatial summation of excitatory postsynaptic potentials (EPSPs) and inhibitory postsynaptic potentials (IPSPs) at the axon hillock.

■ **The same neurotransmitter can produce different effects on different types of cells (pp. 1036–1038, TABLE 48.1)** The action of neurotransmitters can be fast and local, or slow and diffuse. One of the most common invertebrate and vertebrate neurotransmitters is acetylcholine. Other transmitters that have been identified include the biogenic amines (epinephrine, norepinephrine, dopamine, and serotonin); several amino acids; and some neuropeptides, such as the analgesic endorphins. Some neurons also release gases, such as nitric oxide, to signal other cells.

EVOLUTION AND DIVERSITY OF NERVOUS SYSTEMS

■ **The ability of cells to respond to the environment has evolved over billions of years (p. 1038)** Our nervous system has a very ancient history, for its basic cellular components and signaling mechanisms were present over 600 million years ago. What has changed most during evolution of the most elaborate nervous systems is the number and complexity of the neural networks making up animal brains.

■ **Nervous systems show diverse patterns of organization (pp. 1038–1039, FIGURE 48.15)** Nervous systems range in complexity from simple nerve nets to highly centralized nervous systems having complex brains and central nerve cords. The nerve cord lies near the ventral body surface in invertebrates, and near the dorsal surface in vertebrates.

VERTEBRATE NERVOUS SYSTEMS

- **Vertebrate nervous systems have central and peripheral components (p. 1040, FIGURE 48.16)** The vertebrate central nervous system (brain and spinal cord) is the integrating link between the information coming from and going to the peripheral nervous system. Derived from the embryonic dorsal hollow nerve cord, the CNS contains contiguous spaces filled with cerebro-spinal fluid. Gray matter (mainly nerve cell bodies, dendrites, and unmyelinated axons) is distinguishable from white matter (mainly myelinated axons) in the brain and spinal cord.

- **The divisions of the peripheral nervous system interact in maintaining homeostasis (pp. 1040–1042, FIGURES 48.17, 48.18)** The vertebrate PNS consists of paired cranial and spinal nerves and associated ganglia. Functionally, the PNS consists of the sensory, or afferent, division, which brings information from sensory receptors to the CNS, and the motor, or efferent, division, which carries signals away from the CNS to effector cells. The motor division consists of the somatic nervous system, which carries signals to skeletal muscles, and the autonomic nervous system, which regulates the primarily automatic, visceral functions of smooth and cardiac muscles. The autonomic nervous system consists of the parasympathetic and sympathetic divisions, which are usually antagonistic in effect on target organs.

- **Embryonic development of the vertebrate brain reflects its evolution from three anterior bulges of the neural tube (pp. 1042–1043, FIGURES 48.19, 48.20)** All vertebrate brains develop and diversify from three embryonic regions: the forebrain, the midbrain, and the hindbrain.

- **Evolutionarily older structures of the vertebrate brain regulate essential automatic and integrative functions (pp. 1043–1046, FIGURES 48.21–48.23)** In the human brain, the medulla oblongata, the pons, and the midbrain make up the brainstem. The medulla oblongata and pons of the hindbrain work together to control homeostatic functions such as breathing rate, and to conduct sensory and motor signals between the spinal cord and higher brain centers. The midbrain receives, integrates, and projects sensory information to the forebrain. The reticular system of the brainstem regulates cycles of sleep and wakefulness. The cerebellum is involved in learning and remembering motor responses, as well as regulating their execution. The thalamus is the main input and output center through which sensory and motor information passes to and from the cortex. The hypothalamus regulates homeostasis and basic survival behaviors such as feeding, fighting, fleeing, and reproducing. Nuclei in the hypothalamus also regulate circadian rhythms.

- **The cerebrum is the most highly evolved structure of the mammalian brain (pp. 1046–1047, FIGURE 48.24)** The cerebrum has two hemispheres, each of which consists of cerebral cortex overlaying internal basal nuclei that are important in learning and controlling movement. A neocortex seen only in mammals forms the highly folded surface of the cortex.

- **Regions of the cerebrum are specialized for different functions (pp. 1047–1051, FIGURES 48.25–48.27)** Each cerebral hemisphere has four lobes—occipital, temporal, parietal, and frontal—that contain centers, respectively, for vision, audition, somatic sensing, and planning and movement. Association areas integrate the information that comes from different sensory areas, and frontal regions then compose a motor response plan. The left hemisphere is normally specialized for focused high speed serial information processing essential to language and logic operations. The right hemisphere emphasizes more global parallel processing involved in pattern recognition, nonverbal ideation, and emotional processing. Areas in the frontal and temporal lobes are essential for the generation and understanding of language. As part of the limbic system, portions of the frontal lobes and the amygdala are involved in building an emotional repertoire. The frontal lobes are a site of short-term memory and can interact with the hippocampus and amygdala to consolidate long-term memory. Neuroscientists are increasingly optimistic that a scientific explanation of human consciousness may be attainable. We are a considerable distance, however, from understanding how the emergent property of consciousness rises from the correlated activities of the nerve cells that make up our brains and nervous systems.

- **Research on neuron development and neural stem cells may lead to new approaches for treating CNS injuries and diseases (pp. 1051–1053, FIGURES 48.28, 48.29)** During neuron development, signal molecules direct an axon's growth until connections are made with the correct target cells. The adult brain contains stem cells that can differentiate into mature nerve cells. To replace neurons lost to trauma or disease, researchers hope to simulate the process of neuron development and induce stem cells to differentiate into nerve cells.

Self-Quiz

1. Which of the following occurs when a stimulus depolarizes a neuron's membrane?
   a. Na^+ diffuses out of the cell.
   b. The action potential approaches zero.
   c. The membrane potential changes from the resting potential to a voltage closer to the threshold potential.
   d. The depolarization is all or none.
   e. The inside of the cell becomes more negative in charge relative to the outside of the cell.

2. Action potentials are usually propagated in only one direction along an axon because
   a. the nodes of Ranvier conduct only in one direction.
   b. the brief refractory period prevents opening of voltage-gated Na^+ channels.
   c. the axon hillock has a higher membrane potential than the tips of the axon.
   d. ions can flow along the axon only in one direction.
   e. both sodium and potassium voltage-gated channels open in one direction.

3. The depolarization of the presynaptic membrane of an axon *directly* causes
   a. voltage-gated calcium channels in the membrane to open.
   b. synaptic vesicles to fuse with the membrane.
   c. an action potential in the postsynaptic cell.
   d. the opening of chemically sensitive gates that allow neurotransmitters to spill into the synaptic cleft.
   e. an EPSP or IPSP in the postsynaptic cell.

4. What is the neocortex?
   a. a primitive brain region common to reptiles, birds, and mammals
   b. a region deep in the cortex that is associated with the formation of emotional memories
   c. a central part of the cortex that receives olfactory information
   d. an additional outer layer of neurons running along the cerebral cortex that is unique to mammals
   e. the association area of the frontal lobe that is involved in higher cognitive functions

5. Which of the following provides evidence that emotional brain circuits form early during human development?
   a. Humans are more likely to be able to recall emotional memories from childhood than factual memories.
   b. Infants are able to understand language before they are able to speak.
   c. Emotional brain circuits involve more "primitive" parts of the brain that evolved before the neocortex develops.
   d. Young infants are able to bond to a caregiver and to express fear, distress, and anger.
   e. Individuals with damage to the amygdala no longer have autonomic responses to stressful stimuli.

6. Which of the following structures or regions is *incorrectly* paired with its function?
   a. limbic system—the motor control of speech
   b. medulla oblongata—homeostatic control center
   c. cerebellum—coordination of movement and balance
   d. corpus callosum—band of fibers connecting left and right cerebral hemispheres
   e. hypothalamus—production of hormones and regulation of temperature, hunger, and thirst

7. Receptor sites for neurotransmitters are located on the
   a. tips of axons.
   b. axon membranes in the regions of the nodes of Ranvier.
   c. postsynaptic membrane.
   d. membranes of synaptic vesicles.
   e. presynaptic membrane.

8. All the following electrical changes of neurons are graded events *except*
   a. EPSPs.
   b. IPSPs.
   c. action potentials.
   d. depolarizations caused by stimuli.
   e. hyperpolarizations caused by stimuli.

9. Of the following components of the nervous system, which is the *most inclusive*?
   a. brain
   b. spinal cord
   c. central nervous system
   d. gray matter
   e. neuron

10. Which of the following best describes what is known about how axons grow toward their proper target cell?
    a. Axons grow in a direct path, attracted by signal molecules that are released from the target cell.
    b. Cells along the growth path release signal molecules that either attract or repel the axon, and the interaction of CAMs on the growth cone and neighboring cells may provide tracks that guide axon growth.
    c. Nerve growth factor released by astrocytes stimulates neural progenitor cells to differentiate into neurons, whose axons then grow toward an increasing concentration of signal molecules.
    d. Axons produce growth-promoting proteins only in the growth cone, causing the axon to grow in an outward direction toward its target cell.
    e. Glia first migrate to the target cell, leaving a trail of CAMs along the path that the growth cone of the axon follows.

11. (a) Arrange the following neurons into the correct sequence for information flow during the knee-jerk reflex: interneuron, sensory neuron, motor neuron. (b) Which of the neuron types is located entirely within the central nervous system?

12. What is the function of the myelin sheath?

13. Contrast excitatory and inhibitory synapses in how they change a receiving cell's membrane potential relative to triggering an action potential.

14. How would a drug that inhibits the parasympathetic nervous system affect a person's pulse?

15. Identify which of these brain structures includes all others in the list: amygdala, limbic system, forebrain, thalamus, cerebrum.

Evolution Connection

Neurons fire in an all-or-none event. This on/off signaling is an evolutionary adaptation of animals that must sense and act in a complex environment. It is possible to imagine a nervous system in which the action potentials were graded, with the amplitude depending on the size of the stimulus. What advantage does on/off signaling have over a graded (continuously variable) kind of signaling?

The Process of Science

From what you know about the action potential and synapses, propose two or three hypotheses for how various anesthetics might prevent pain.

Experiment with a sciatic nerve from a frog in the Case Study in The Process of Science, available on the website and CD-ROM.

Science, Technology, and Society

Alcohol's depressant effects on the nervous system cloud judgment and slow reflexes. Alcohol consumption is a factor in most fatal traffic accidents in the United States. What are some other impacts of alcohol abuse on society? What are some of the responses of people and society to alcohol abuse? Do you think this is primarily an individual or societal problem? Do you think our responses to alcohol abuse are appropriate and proportional to the seriousness of the problem? Defend your position.

Answers: 1. c; 2. b; 3. a; 4. d; 5. d; 6. a; 7. c; 8. c; 9. c; 10. b; 11. (a) sensory neuron ⟶ interneuron ⟶ motor neuron (b) interneuron 12. It speeds up conduction of signals along some dendrites and axons. 13. Release of neurotransmitters from an excitatory synapse moves the receiving cell's membrane potential closer to threshold; release of neurotransmitters from an inhibitory synapse moves the receiving cell's membrane potential farther from threshold. 14. The pulse would probably increase, reflecting an increase in the heart rate. 15. Forebrain.

CHAPTER 49

SENSORY AND MOTOR MECHANISMS

SENSING, ACTING, AND BRAINS

- The brain's processing of sensory input and motor output is cyclical rather than linear

INTRODUCTION TO SENSORY RECEPTION

- Sensory receptors transduce stimulus energy and transmit signals to the nervous system
- Sensory receptors are categorized by the type of energy they transduce

PHOTORECEPTORS AND VISION

- A diversity of photoreceptors has evolved among invertebrates
- Vertebrates have single-lens eyes
- The light-absorbing pigment rhodopsin triggers a signal-transduction pathway
- The retina assists the cerebral cortex in processing visual information

HEARING AND EQUILIBRIUM

- The mammalian hearing organ is within the inner ear
- The inner ear also contains the organs of equilibrium
- A lateral line system and inner ear detect pressure waves in most fishes and aquatic amphibians
- Many invertebrates have gravity sensors and are sound-sensitive

CHEMORECEPTION—TASTE AND SMELL

- Perceptions of taste and smell are usually interrelated

MOVEMENT AND LOCOMOTION

- Locomotion requires energy to overcome friction and gravity
- Skeletons support and protect the animal body and are essential to movement
- Physical support on land depends on adaptations of body proportions and posture
- Muscles move skeletal parts by contracting
- Interactions between myosin and actin generate force during muscle contractions
- Calcium ions and regulatory proteins control muscle contraction
- Diverse body movements require variation in muscle activity

In the gathering dusk *a male moth's antennae detect the chemical attractant of a female moth somewhere upwind. The male takes to the air, following the scent trail toward the female. Suddenly, vibration sensors in the moth's abdomen signal the presence of ultrasonic chirps of a rapidly approaching bat. The bat's sonar enables the mammal to locate moths and other flying insect prey. Reflexively, the moth's nervous system alters the motor output to its wing muscles, sending the insect into an evasive spiral toward the ground. Although it is probably too late for the moth in the photograph on this page, many moths can escape because they can detect a bat's sonar about 30 m away. The bat has to be within 3 m to sense the moth, but since the bat flies faster, it may still have time to detect, home in, and catch its prey.*

The outcome of this interaction depends on the abilities of both predator and prey to sense important environmental stimuli and to produce appropriate coordinated movement. Although not all of an animal's moment-by-moment interactions with its environment are as dramatic as such predator-prey struggles, the detection and processing of sensory information and the generation of motor output provide the physiological basis for all animal behavior.

In Chapter 48 we examined how the nervous system transmits and integrates sensory and motor information, finishing with a description of our human brains. We start this chapter with a brief overview of how sensory and motor information is processed in our brains. Then we will examine in a number of different animal groups both the sensory receptors that receive information from the environment and also the structure and function of muscles, the motor effectors that bring about movement in response to that information. We will also study skeletons in the context of body movement.

SENSING, ACTING, AND BRAINS

The brain's processing of sensory input and motor output is cyclical rather than linear

We can trace the origins of sensing and acting back to the appearance, in simple prokaryotes, of cellular structures able to sense pressure or chemicals in the environment and then direct movement in an appropriate direction. These structures have been transformed during evolution into diverse mechanisms specialized to sense different types of energy and generate many different levels of physical movement in response. Alongside the evolution of the wide variety of sensory receptors that we will describe in the first section of this chapter, the brain mechanisms described in Chapter 48 also evolved. These processes in the brain interpret sensory input and organize motor output to the different kinds of effector organs described in the second portion of the chapter.

Although it is customary to think of sensations causing brain changes which then cause action as a linear process—sensing → brain analysis → action—such a description implies that an animal is rather like a computer passively waiting for instructions before it acts. This is not the case. All animals are in constant motion, probing the environment with that motion, sensing the changes that result, and then using the information to generate the next action. This is a continuous cycle rather than a linear sequence, the brain carrying on background activity which is constantly updated as sensing and acting proceed (FIGURE 49.1).

Sensations begin as different forms of energy such as light, heat, sound, and smells are detected by specialized sensory receptor cells and ultimately converted to action potentials that travel to the brain. In most vertebrates, sensory signals generally go first to the thalamus, the gateway to the cerebral cortex (see FIGURE 48.20). This gateway is influenced by instructions coming back from the cortex selecting which sensory input seems more important at the moment. The information is then sent on to the many parts of the brain that contribute to forming our perceptions, our awareness and interpretation of the stimuli. For example, information relevant to identifying sounds or visual objects streams mainly to regions of the temporal lobes, while information about their motion and location is sent to the parietal lobes. Limbic regions are central in determining the importance of the sensory input to the organism (see FIGURES 48.24b and 48.27). Our memories of similar sounds and objects can strongly influence our final perceptions, so that in some cases we perceive what we expect to hear or see, rather than what is really there. Thus the process of perception starts with a fairly simple input—information about physical sensations carried by action potentials to the brain—and ends with a very complex result that can be biased by our history of sensing.

FIGURE 49.1 Catching a fly ball is no simple feat. From the moment the bat makes contact, a cycle of sensory input, processing in the brain, and responses by muscles eventually place the fielder's glove in the right place at the right time. Brain activity is constantly updated as it senses how the position of the body and glove relative to the ball's flight has been changed by the previous instant's motor output. And, of course, experience (learning and practice) enhance this marvel of coordination.

This description of simple sensations leading to complex perceptions invites a comparison with the subsequent motor actions that drive behaviors. A relatively simple outcome—action potentials in many motor neurons driving the contraction of striated muscles to produce a single coordinated behavior—has an extremely complex beginning. At the same time that information about the location, identity, and meaning of objects in the world streams forward to higher centers in the frontal lobes associated with making behavioral choices, motor-planning regions of the neocortex, basal ganglia, and cerebellum are evaluating a vast array of learned skills—movement programs that are potential responses. The single motor behavior chosen for the next moment is the result of a very complex decision-making mechanism.

We considered the complexity of information processing in the brain in Chapter 48, and we wanted to remind you of that complexity at the beginning of this chapter. However, the main purpose of this chapter is not to deal further with these cognitive processes, but rather to examine, in several different invertebrate and vertebrate groups, the beginning and the end of sensing and acting. We start first with sensory processes—how information about the external and internal environment is collected and conveyed to the brain—and then consider the structure and function of muscles that carry out the movement instructions generated by the brain.

INTRODUCTION TO SENSORY RECEPTION

As you learned in Chapter 48, information is transmitted through the nervous system in the form of nerve impulses, or action potentials, which are all-or-none events (see FIGURE 48.8c). An action potential triggered by light striking the eye is the same as an action potential triggered by air vibrating in the ear. The ability to distinguish any type of stimulus, such as sight or sound, depends on the part of the brain that receives the signal. What matters is where impulses go, not what triggers them.

Action potentials that reach the brain via sensory neurons are called **sensations.** Once the brain is aware of sensations, it interprets them, giving us the **perception** of the stimuli. Perceptions, such as colors, smells, sounds, and tastes, are constructions of the brain and do not exist outside it. So, if a tree falls and no one is present to hear it, is there a sound? The fall certainly produces pressure waves in the air, but if sound is defined as a perception, then there is none unless sensory receptors detect the waves and an animal's brain perceives them.

Sensory receptors transduce stimulus energy and transmit signals to the nervous system

Sensations, and the perceptions they evoke in the brain, begin with **sensory reception,** the detection of the energy of a stimulus by sensory cells. Most **sensory receptors** are specialized neurons or epithelial cells that exist singly or in groups with other cell types within sensory organs, such as the eyes and ears. Sensory receptors called **exteroreceptors** detect stimuli outside the body, such as heat, light, pressure, and chemicals. Other sensory receptors called **interoreceptors** detect stimuli within the body, such as blood pressure and body position.

All stimuli represent forms of energy, and the general function of receptor cells is to convert the energy of stimuli into changes in membrane potentials and then transmit signals to the nervous system. This task consists of four functions: sensory transduction, amplification, transmission, and integration.

Sensory Transduction

The actual detection of a stimulus involves the conversion of stimulus energy into a change in the membrane potential of a receptor cell, a process called **sensory transduction.** The initial response of the sensory receptor to a stimulus is a change in its membrane permeability, resulting in a graded change in membrane potential called a **receptor potential.** (Recall from Chapter 48 that a graded potential is a change in the voltage across the membrane that is proportional to the strength of the stimulus.) In some cases a stimulus such as pressure can stretch the membrane and increase ion flow. In other cases, specific receptor molecules on the membrane of a receptor cell open or close gates to

ion channels when the stimulus is present. FIGURE 49.2 (p. 1060) illustrates the example of sugar triggering a receptor potential that functions in our sense of taste. We will examine other specific examples of sensory transduction later in the chapter.

Amplification

The strengthening of stimulus energy that is otherwise too weak to be carried into the nervous system is called **amplification.** Amplification of the signal may occur in accessory structures of a complex sense organ, as when sound waves are enhanced by a factor of more than 20 before reaching the receptors of the inner ear. Amplification also may be a part of the transduction process itself. An action potential conducted from the eye to the brain has about 100,000 times as much energy as the few photons of light that triggered it. Signal transduction pathways in the receptor cells contribute to this amplification (see FIGURE 49.2).

Transmission

Once the energy in the stimulus has been transduced into a receptor potential, **transmission,** or the conduction of impulses to the CNS (central nervous system), can occur. In some instances, such as in the case of "pain cells," the receptor itself is actually a sensory neuron that conducts action potentials to the CNS. Other receptors are separate cells that must transmit chemical signals (neurotransmitters) across synapses to sensory neurons (see FIGURE 49.2). If the receptor also functions as the sensory neuron, the intensity of the receptor potential will affect the frequency of action potentials that travel as sensations to the CNS. For separate receptor cells, the strength of the stimulus and receptor potential affect the amount of neurotransmitter released by the receptor at its synapse with a sensory neuron, which in turn determines the frequency of action potentials generated by the sensory neuron. Many sensory neurons spontaneously generate signals at a low rate. Therefore, a stimulus does not really switch the production of action potentials on or off; it modulates their frequency (FIGURE 49.2 illustrates this phenomenon). In this way the CNS is sensitive not only to the presence or absence of a stimulus but also to changes in stimulus intensity.

Integration

The processing of information, or **integration,** begins as soon as information is first received. Signals from receptors are integrated through the summation of graded potentials, as are those within the nervous system. One type of integration by receptor cells is **sensory adaptation,** a decrease in responsiveness during continued stimulation (not to be confused with the term adaptation as used in an evolutionary context). Without sensory adaptation, you would feel every beat of your heart and

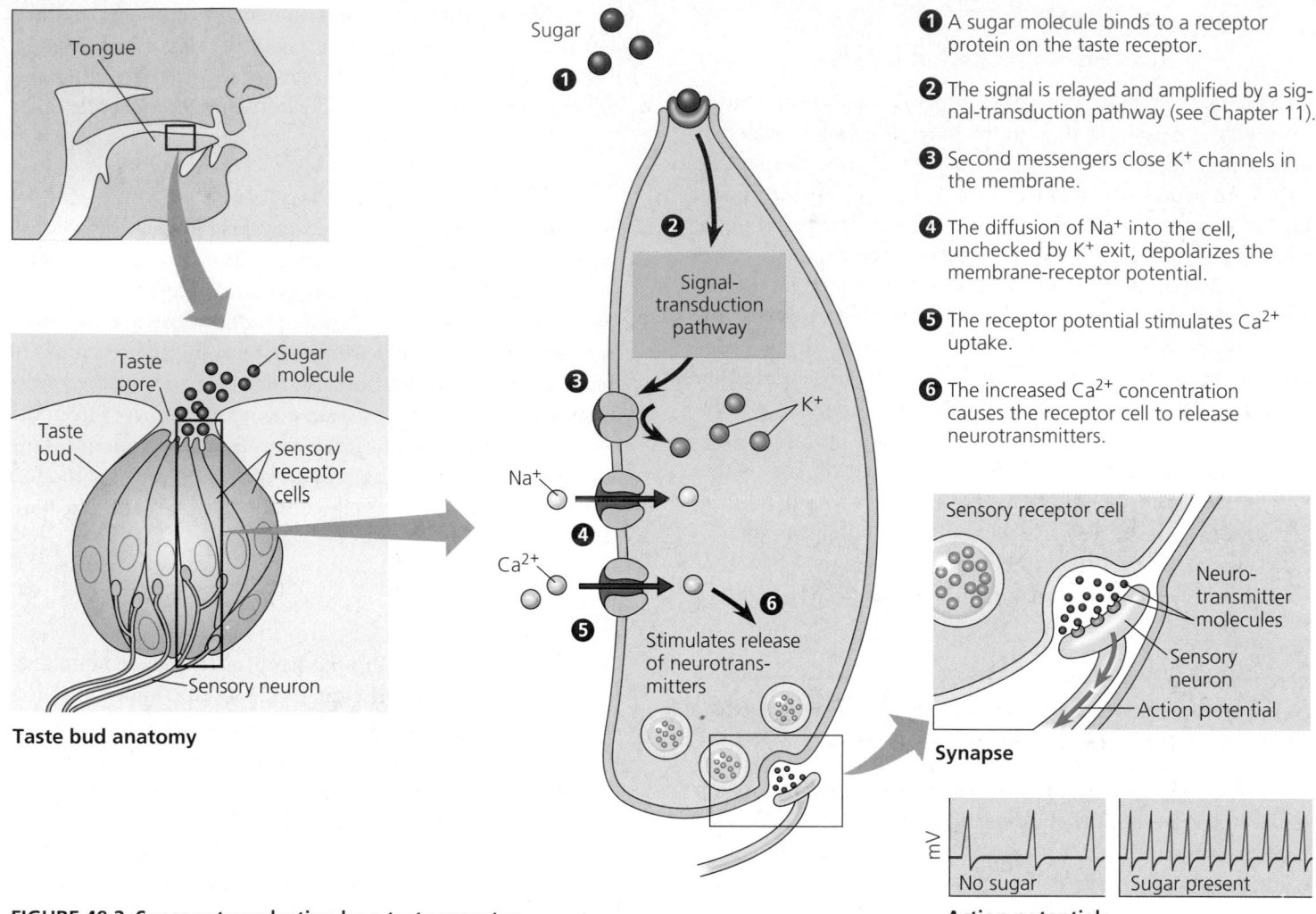

Taste bud anatomy

① A sugar molecule binds to a receptor protein on the taste receptor.

② The signal is relayed and amplified by a signal-transduction pathway (see Chapter 11).

③ Second messengers close K^+ channels in the membrane.

④ The diffusion of Na^+ into the cell, unchecked by K^+ exit, depolarizes the membrane-receptor potential.

⑤ The receptor potential stimulates Ca^{2+} uptake.

⑥ The increased Ca^{2+} concentration causes the receptor cell to release neurotransmitters.

FIGURE 49.2 Sensory transduction by a taste receptor.

Action potentials

every bit of clothing on your body. Receptors are selective in the information they send to the CNS, and adaptation reduces the likelihood that a continuous stimulus will be transmitted.

Another important aspect of sensory integration is the sensitivity of the receptors. The threshold for transduction by receptor cells varies with conditions. For example, the thresholds of glucose receptors in the human mouth can vary over several orders of magnitude of sugar concentration as both the general state of nutrition and the amount of sugar in the diet change.

The integration of sensory information occurs at all levels within the nervous system, and the cellular actions just described are only the first steps. Complex receptors such as the eyes have higher levels of integration as signals converge on sensory nerves, and the CNS further processes all incoming signals.

Sensory receptors are categorized by the type of energy they transduce

Based on the type of energy they detect (transduce), sensory receptors fall into five categories: mechanoreceptors, pain re-

ceptors, thermoreceptors, chemoreceptors, and electromagnetic receptors. One amazing feature of many of these receptors is that they can detect the smallest physical unit of stimulus possible. Most photoreceptors can detect a single quantum (photon) of light, chemoreceptors can detect a single molecule or odorant, and the mechanoreceptor cells of our inner ear can detect a motion of only a few angstroms.

Mechanoreceptors are stimulated by physical deformation caused by such stimuli as pressure, touch, stretch, motion, and sound—all forms of mechanical energy. Bending or stretching of the plasma membrane of a mechanoreceptor cell increases its permeability to both sodium and potassium ions, resulting in a depolarization (receptor potential).

The human sense of touch relies on mechanoreceptors that are actually modified dendrites of sensory neurons (FIGURE 49.3). Receptors that detect light touch are close to the surface of the skin; they transduce very slight inputs of mechanical energy into receptor potentials. Receptors responding to strong pressure and vibrations in the body are in deep skin layers. Other touch receptors detect hair movement—the sensitivity of a cat's whiskers is an example.

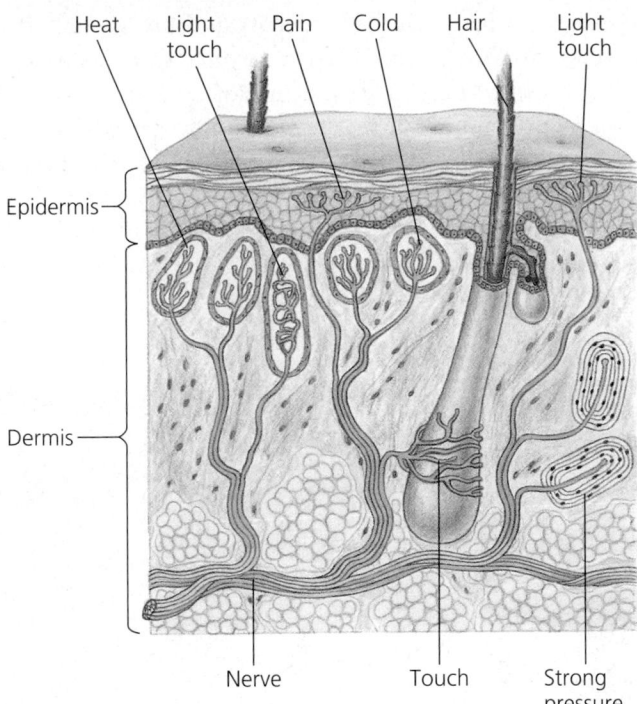

Heat | Light touch | Pain | Cold | Hair | Light touch

Epidermis

Dermis

Nerve | Touch | Strong pressure

FIGURE 49.3 Sensory receptors in human skin. Each mechanoreceptor is a modified dendrite of a sensory neuron. Most receptors in the inner layer of the skin (dermis) are encapsulated by one or more layers of connective tissue. Those in the outer skin layers (epidermis) and touch receptors wound around the base of hairs are naked dendrites. Touch receptors at the base of the stout whiskers of mammals such as cats and many rodents are extremely sensitive and enable the animal to detect nearby objects in the dark.

An example of an interoreceptor stimulated by mechanical distortion is the **muscle spindle,** or stretch receptor (see FIGURE 48.3). This mechanoreceptor monitors the length of skeletal muscles. The muscle spindle contains modified muscle fibers attached to sensory neurons and runs parallel to muscle. When the muscle is stretched, the fibers of the spindle are also stretched, depolarizing the sensory neurons and triggering action potentials that are transmitted back to the spinal cord.

The **hair cell** is a common type of mechanoreceptor that detects motion. Hair cells are found in the vertebrate ear and in the lateral line organs of fishes and amphibians, where they detect movement relative to the environment (see FIGURE 34.12a). The "hairs" are either specialized cilia or microvilli. They project upward from the surface of the hair cell into either an internal compartment, such as the human inner ear, or an external environment, such as a pond. When the cilia or microvilli bend in one direction, they stretch the hair cell membrane and increase its perme-

ability to sodium and potassium ions, leading to an increase in the rate of impulse production in a sensory neuron. When the cilia bend in the opposite direction, ion permeability decreases, reducing the number of action potentials in the sensory neuron. This specificity allows hair cells to respond to the direction of motion as well as to its strength and speed (FIGURE 49.4).

Pain receptors in humans are a class of naked dendrites in the epidermis of the skin called **nociceptors** (see FIGURE 49.3). Most animals probably experience pain, although we cannot say what perceptions other animals actually associate with stimulation of their pain receptors. Pain is one of the most important sensations because the stimulus becomes translated into a defensive reaction, such as withdrawal from danger. Rare individuals who are born without any pain sensation may die from such conditions as a ruptured appendix because they cannot feel the associated pain and are unaware of the danger.

Different groups of pain receptors respond to excess heat, pressure, or specific classes of chemicals released from damaged or inflamed tissues. Some of the chemicals that trigger pain include histamine and acids. Prostaglandins increase pain by sensitizing the receptors—that is, lowering their threshold (see Chapter 45 to review prostaglandins). Aspirin and ibuprofen reduce pain by inhibiting prostaglandin synthesis.

Thermoreceptors, responding to either heat or cold, help regulate body temperature by signaling both surface and body core temperature. There is still debate about the identity of thermoreceptors in the mammalian skin. Possible candidates are two receptors consisting of encapsulated, branched dendrites (see FIGURE 49.3). Many researchers, however, believe that these structures are actually modified pressure receptors and maintain that naked dendrites of certain sensory neurons are the actual thermoreceptors in the skin. There is general agreement that cold and heat receptors in the skin, as well as interothermoreceptors in the anterior hypothalamus of the brain, send information to the body's thermostat, located in the posterior hypothalamus.

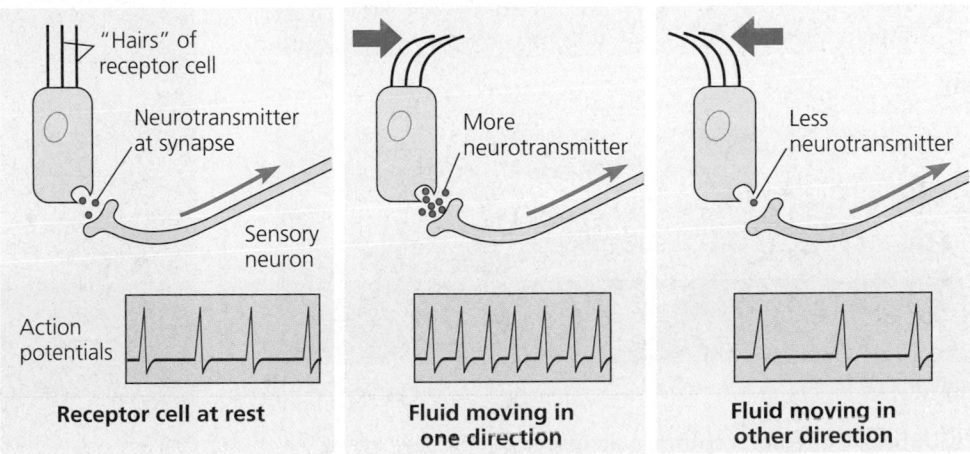

"Hairs" of receptor cell

Neurotransmitter at synapse

Sensory neuron

Action potentials

Receptor cell at rest

More neurotransmitter

Fluid moving in one direction

Less neurotransmitter

Fluid moving in other direction

FIGURE 49.4 Mechanoreception by a hair cell. The "hairs" are either cilia or cellular extensions called microvilli.

Chemoreceptors include both general receptors that transmit information about the total solute concentration in a solution and specific receptors that respond to individual kinds of molecules. Osmoreceptors in the mammalian brain, for example, are general receptors that detect changes in the total solute concentration of the blood and stimulate thirst when osmolarity increases (see Chapter 44). Water receptors in the feet of house flies respond to pure water or to a dilute solution of virtually any substance. Most animals also have receptors specific to important molecules, including glucose, oxygen, carbon dioxide, and amino acids. In all these examples, the stimulus molecule binds to a specific site on the membrane of the receptor cell and initiates changes in membrane permeability. Two other groups of chemoreceptors show intermediate specificity. **Gustatory** (taste) and **olfactory** (smell) **receptors** respond to categories of related chemicals. In taste, humans often classify such general categories as sweet, sour, salty, or bitter. Two of the most sensitive and specific chemoreceptors known are present in the antennae of the male silkworm moth (FIGURE 49.5). They detect the two chemical components of the female moth sex pheromone.

Electromagnetic receptors detect various forms of electromagnetic energy, such as visible light, electricity, and magnetism. **Photoreceptors,** which detect the radiation we know as visible light, are often organized into eyes. Snakes have extremely sensitive infrared receptors that detect the body heat of prey standing out against a colder background (FIGURE 49.6a).

(a)

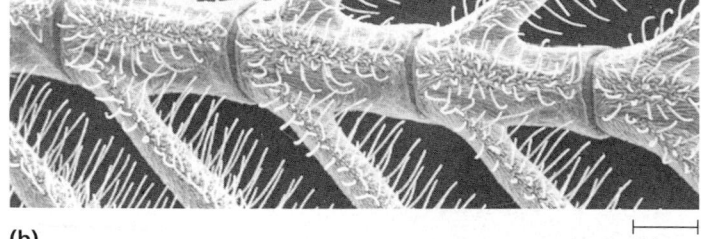

(b)

0.1 mm

FIGURE 49.5 Chemoreceptors in an insect. (a) The antennae of the male silkworm moth *Bombyx mori* are covered with sensory hairs, visible in **(b)** the SEM enlargement. The hairs have chemoreceptors that are highly sensitive to the sex pheromone released by the female.

Some fishes discharge electric currents and use special electroreceptors to locate objects, such as prey, that disturb the electric currents. The platypus, a monotreme mammal, has electroreceptors on its bill that can probably detect electrical fields generated by the muscles of prey, such as crustaceans, frogs, and small fishes. There is also evidence that many animals that home or migrate use the magnetic field lines of Earth to help orient themselves (FIGURE 49.6b). The iron-containing

FIGURE 49.6 Specialized electromagnetic receptors.

Eye

Infrared receptor

(a) This rattlesnake and other pit vipers have a pair of infrared receptors, one between each eye and nostril. The organs are sensitive enough to detect the infrared radiation emitted by a warm mouse a meter away. The snake moves its head from side to side until the radiation is detected equally by the two receptors, indicating that the mouse is straight ahead.

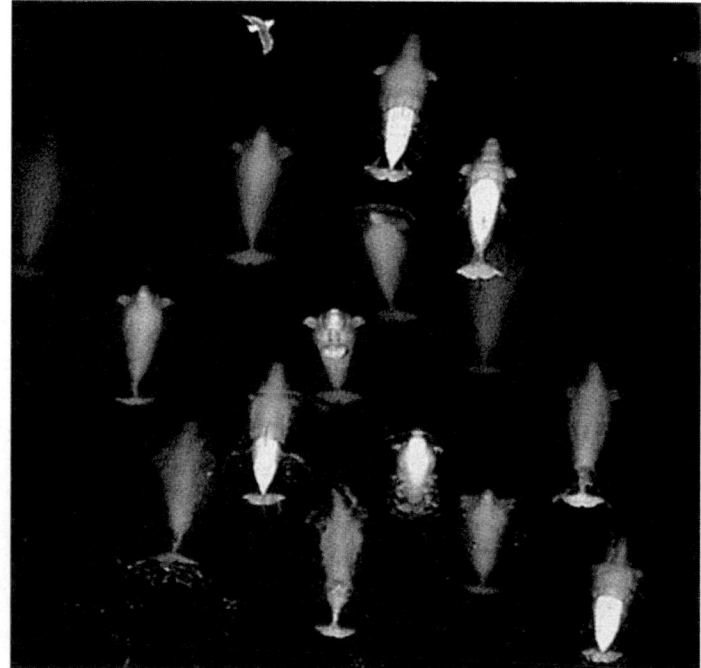

(b) Some migrating animals, such as the beluga whales in this aerial photograph, can apparently sense Earth's magnetic field and use the information, along with other cues, for orientation.

mineral magnetite is found in the skulls of some birds and mammals (including humans), in the abdomens of bees, in the teeth of some mollusks, and in certain protists and prokaryotes that orient with respect to Earth's magnetic field. Once used by sailors as a primitive compass, magnetite may be part of an important orienting mechanism in many animals.

In the next few sections, we'll apply our overview of sensory receptors as we take a closer look at specific senses, such as sight and hearing.

PHOTORECEPTORS AND VISION

A great variety of light detectors has evolved in the animal kingdom, from simple clusters of cells that detect only the direction and intensity of light to complex organs that form images. Despite their diversity, all photoreceptors contain similar pigment molecules that absorb light, and most, if not all, photoreceptors in the animal kingdom may be homologous. Animals as diverse as flatworms, annelids, arthropods, and vertebrates have some of the same, ancient genes associated with the development of photoreceptors in embryos. Thus, the genetic underpinnings of all photoreceptors may have evolved in the earliest bilateral animals (see Chapter 33). The actual eyes that form in an animal depend on developmental patterns regulated by genetic mechanisms that evolved later in the animal's taxonomic group, and whose effects appear to be superimposed on the ancient, homologous mechanism.

A diversity of photoreceptors has evolved among invertebrates

Most invertebrates have photoreceptors, which range from simple clusters of photoreceptor cells to complex image-forming eyes. One of the simplest is the **eye cup** of planarians, which provides information about light intensity and direction without actually forming an image. Photoreceptor cells are located within a cup formed by a layer of cells containing a screening pigment that blocks light. Light can enter the cup and stimulate the photoreceptors only through an opening on one side where there is no screening pigment (FIGURE 49.7). The opening of one eye cup faces left and slightly forward, and the opening of the other cup

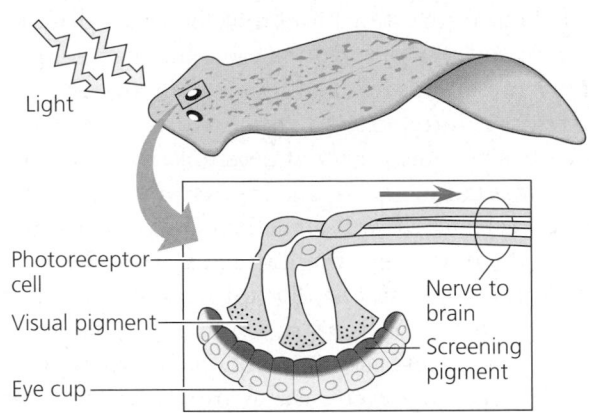

FIGURE 49.7 Eye cups and orientation behavior of a planarian. The brain directs the body to turn until the sensations from the two cups are equal and minimal, causing the animal to move away from light.

faces right-forward. Thus, light shining from one side of the planarian can enter only the eye cup on that side. The brain compares the rate of nerve impulses coming from the two eye cups, and the animal turns until the sensations from the two cups are equal and minimal. The result is that the animal moves directly away from the light source and reaches a shaded location beneath a rock or some other object, a behavioral adaptation that helps hide the planarian from predators.

Image-forming eyes of two major types have evolved in invertebrates: the compound eye and the single-lens eye. **Compound eyes** are found in insects and crustaceans (phylum Arthropoda) and some polychaete worms (phylum Annelida). A compound eye consists of up to several thousand light detectors called **ommatidia** (the "facets" of the eye), each with its own light-focusing lens (FIGURE 49.8). Each ommatidium

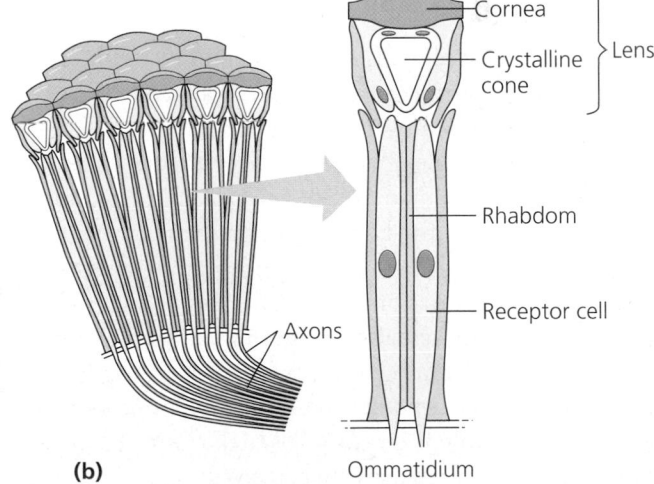

(a)

2 mm

(b)

FIGURE 49.8 Compound eyes. (a) The faceted eyes on the head of a fly, photographed with a stereomicroscope. **(b)** The cornea and crystalline cone of each ommatidium function as a lens that focuses light onto the rhabdom, a stack of pigmented plates on the inside of a circle of receptor cells. The rhabdom traps light and guides it to the receptor cells. The image formed by a compound eye is a mosaic of dots formed by the different intensities of light entering the many ommatidia from different angles.

registers light from a tiny portion of the visual field. Differences in the intensity of light entering the many ommatidia result in a mosaic image. The animal's brain may sharpen the image when it integrates the visual information. The compound eye is extremely acute at detecting movement, an important adaptation for flying insects and small animals constantly threatened with predation. For comparison, consider that the human eye can distinguish light flashes up to about 50 flashes per second. For this reason, the individual images of a movie, which flash at a faster rate, fuse together to create the perception of smooth motion. The compound eyes of some insects, however, recover from excitation rapidly enough to detect the flickering of a light flashing 330 times per second. Such an insect viewing a movie could easily resolve each frame of the film as a separate still image. Insects also have excellent color vision, and some (including bees) can see into the ultraviolet range of the spectrum, which is invisible to us. In studying animal behavior, we cannot extrapolate our sensory world to other species; different animals have different sensitivities and different brain organizations.

Among the invertebrates, **single-lens eyes** are found in some jellies, polychaetes, spiders, and many mollusks. A single-lens eye works on a cameralike principle. The eye of an octopus or squid, for example, has a small opening, the pupil, through which light enters. Analogous to a camera's adjustable aperature (F-stop), the iris of a single-lens eye changes the diameter of the pupil; behind the pupil, a single lens focuses light onto the retina, which consists of light-transducing receptor cells. Also similar to a camera's action, muscles move the lens forward or backward to focus images on the retina.

forms a mucous membrane, the **conjunctiva,** that covers the outer surface of the sclera and helps keep the eye moist. At the front of the eye the sclera becomes the transparent **cornea,** which lets light into the eye and acts as a fixed lens. The conjunctiva does not cover the cornea. The anterior choroid forms the donut-shaped **iris,** which gives the eye its color. By changing size, the iris regulates the amount of light entering the **pupil,** the hole in the center of the iris. Just inside the choroid, the **retina** forms the innermost layer of the eyeball and contains the photoreceptor cells. Information from the photoreceptors leaves the eye at the optic disc, where the optic nerve attaches to the eye. Because there are no photoreceptors in the optic disc, this spot on the lower outside of the retina is a blind spot: Light focused onto that part of the retina is not detected.

The **lens** and **ciliary body** divide the eye into two cavities, one between the lens and the cornea, and a much larger cavity behind the lens within the eyeball itself. The ciliary body constantly produces the clear, watery **aqueous humor** that fills the anterior cavity of the eye. Blockage of the ducts that drain the aqueous humor can produce glaucoma, increased pressure that leads to blindness by compressing the retina. The posterior cavity, filled with the jellylike **vitreous humor,** constitutes most of the volume of the eye. The aqueous and vitreous humors function as liquid lenses that help focus light onto the retina. The lens itself is a transparent protein disc that focuses an image onto the retina. Like squid and octopuses, many fishes focus by moving the lens forward or backward, as in a camera. Humans and other mammals, however, focus by changing the shape of the lens. When viewing a distant object, the lens is flat.

Vertebrates have single-lens eyes

Similar to the single-lens eyes of many invertebrates, the eyes of vertebrates are also cameralike, but they evolved independently in the vertebrate lineage and differ from the single-lens eyes of invertebrates in several details. The human eye, shown in FIGURE 49.9, is capable of detecting an almost countless variety of colors, forming images of objects miles away, and responding to as little as one photon of light. Remember, however, that it is actually the brain that "sees." Thus, to understand vision we must begin by learning how the vertebrate eye generates sensations (action potentials), and then follow these signals to the visual centers of the brain, where images are perceived.

The globe of the vertebrate eye, or eyeball, consists of a tough, white outer layer of connective tissue called the **sclera** and a thin, pigmented inner layer called the **choroid.** A delicate layer of epithelial cells

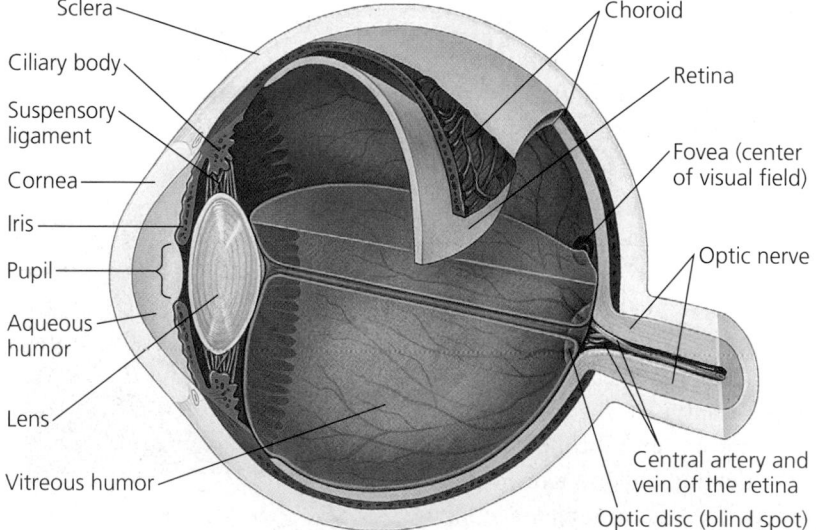

FIGURE 49.9 Structure of the vertebrate eye. In this longitudinal section of the eye, the jellylike vitreous humor is illustrated only in the lower half of the eyeball. The mucous membrane, or conjunctiva, surrounding the sclera (the white of the eye) is not shown.

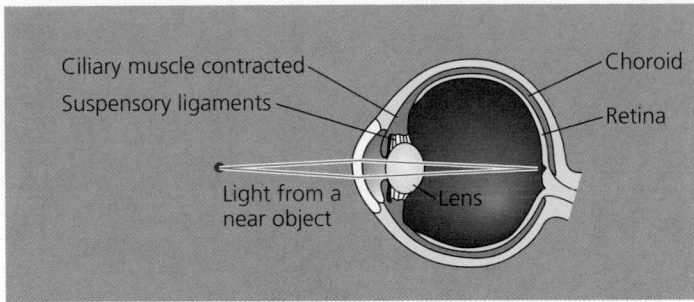

(a) Near vision (accommodation)

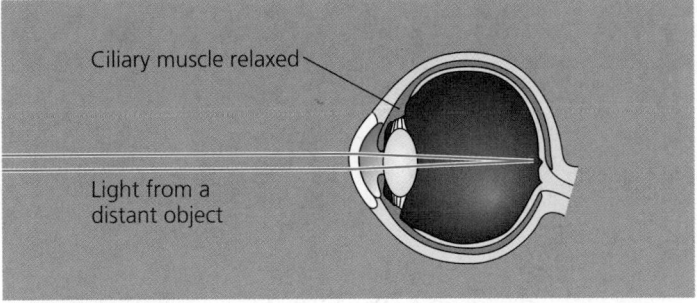

(b) Distance vision

FIGURE 49.10 Focusing in the mammalian eye. The lens bends light and focuses it onto the retina. The thicker the lens, the more sharply the light is bent. The lens is nearly spherical when focusing on near objects and much flatter when focusing on distant objects. Ciliary muscles control the shape of the lens. **(a)** In near vision, the ciliary muscles contract, pulling the border of the choroid layer of the eye toward the lens and causing the suspensory ligaments to relax. With this reduced tension, the elastic lens becomes thicker and rounder. This adjustment of the lens for close vision is known as accommodation. **(b)** In distance vision, the ciliary muscles relax, allowing the choroid to expand and put tension on the suspensory ligaments. The lens is pulled into a flatter shape.

When focusing on a close object, the lens becomes almost spherical, a change called **accommodation** (FIGURE 49.10).

The human retina contains about 125 million **rod cells** and 6 million **cone cells,** two types of photoreceptors named for their shapes. They account for 70% of all sensory receptors in the body, a fact that underscores the importance of the eyes and visual information in how humans perceive their environment.

Rods and cones have different functions in vision, and the relative numbers of these two photoreceptors in the retina are partly correlated with whether an animal is most active during the day or at night. Rods are more sensitive to light but do not distinguish colors; they enable us to see at night, but only in black and white. Because it takes more light to stimulate cones, these receptors contribute very little to night vision. Cones can distinguish colors in daylight. Color vision is found in all vertebrate classes, though not in all species. Most fishes, amphibians, reptiles, and birds have strong color vision, but humans and other primates are among the minority of mammals with this ability. Most mammals are nocturnal, and a maximum number of rods in the retina is an adaptation that gives these animals keen night vision. Cats, usually most active at night, have limited color vision and probably see a pastel world during the day. In the human eye, rods are found in greatest density at the peripheral regions of the retina and are completely absent from the **fovea,** the center of the visual field (see FIGURE 49.9). You cannot see a dim star at night by looking at it directly; if you view it at an angle, however, focusing the starlight onto the regions of the retinas most populated by rods, you will be able to see the star. You achieve your sharpest daylight vision by looking straight at the object of interest because cones are most dense at the fovea, where there are about 150,000 color receptors per mm^2. Some birds have more than a million cones per mm^2, which enables such species as hawks to spot mice and other small prey from high in the sky. In the retina of the eye, as in all biological structures, variations represent evolutionary adaptations.

The light-absorbing pigment rhodopsin triggers a signal-transduction pathway

When the vertebrate lens focuses a light image onto the retina, how do the cells of the retina transduce the stimuli into sensations—action potentials that transmit this information about the environment to the brain? Each rod cell or cone cell has an outer segment with a stack of folded membranes, or discs, in which visual pigments are embedded (FIGURE 49.11a, p. 1066). The visual pigments consist of a light-absorbing pigment molecule called **retinal** (a derivative of vitamin A) bonded to a membrane protein called an **opsin.** Opsins vary in structure from one type of photoreceptor to another, and the light-absorbing ability of retinal is affected by the specific identity of its opsin partner. Rods contain their own type of opsin, which, combined with retinal, makes up the visual pigment **rhodopsin** (FIGURE 49.11b).

When rhodopsin absorbs light, its retinal component changes shape and separates from opsin. This changes the conformation (shape) of the opsin protein. These light-induced changes in retinal plus opsin are referred to as "bleaching" of rhodopsin. In the dark, enzymes convert the retinal back to its original form, and it recombines with opsin to form rhodopsin (FIGURE 49.12, p. 1066). Bright light keeps the rhodopsin bleached and rods become unresponsive; cones take over. When you walk from a bright environment into a dark place, such as walking into a movie theater in the afternoon, you are initially almost blind to faint light. There is not enough light to stimulate the cones, and it takes at least a few minutes for the bleached rods to become functional again.

The light-induced shape-change by opsin triggers a signal-transduction pathway that ultimately results in a receptor potential in the rod cell membrane. First, the altered opsin activates a relay molecule in the signal-transduction pathway,

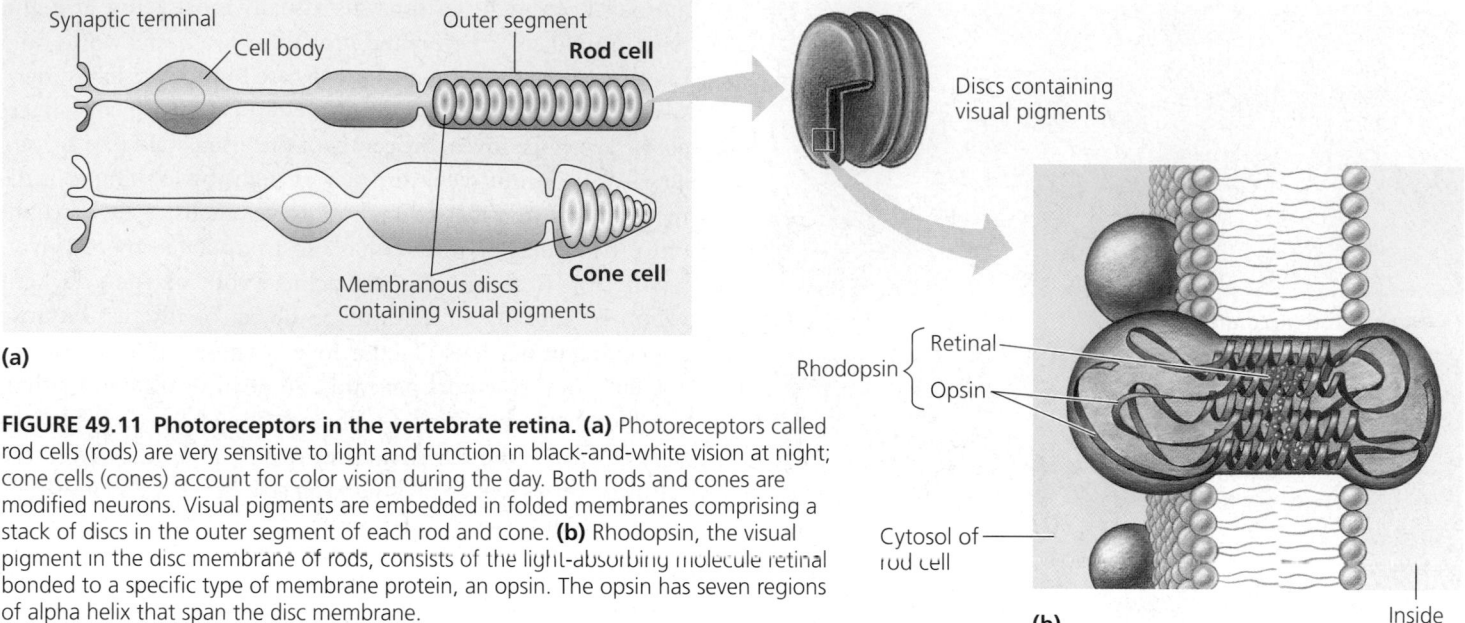

FIGURE 49.11 Photoreceptors in the vertebrate retina. (a) Photoreceptors called rod cells (rods) are very sensitive to light and function in black-and-white vision at night; cone cells (cones) account for color vision during the day. Both rods and cones are modified neurons. Visual pigments are embedded in folded membranes comprising a stack of discs in the outer segment of each rod and cone. **(b)** Rhodopsin, the visual pigment in the disc membrane of rods, consists of the light-absorbing molecule retinal bonded to a specific type of membrane protein, an opsin. The opsin has seven regions of alpha helix that span the disc membrane.

a G protein called transducin, which is also in the disc membrane (FIGURE 49.13). In turn, transducin activates an enzyme that chemically alters the second messenger in the rod cell, a nucleotide called cyclic guanosine monophosphate (cGMP).

In the dark, when rhodopsin is inactive, cGMP is bound to sodium ion channels in the rod cell plasma membrane and keeps those channels open. In this state, the rod cell membrane is actually depolarized and releases the neurotransmitter glutamate (see TABLE 48.1) at its synapses with neighboring cells called **bipolar cells** in the retina (FIGURE 49.14). This steady glutamate release in the dark excites some bipolar cells and inhibits others, depending on what postsynaptic receptor molecules they contain. When light triggers the rhodopsin signal-transduction pathway by altering retinal, an enzyme converts cGMP to GMP, which disengages from the Na^+ channels (see FIGURE 49.13). This closes the channels, decreasing the membrane's permeability to Na^+ and hyperpolarizing the membrane potential. In this case, the cell's receptor potential is a *hyperpolarization* rather than a

depolarization of the membrane. Hyperpolarization slows the rod cell's release of the neurotransmitter glutamate, and this causes either excitation or inhibition of postsynaptic bipolar cells, depending on the bipolar cell's type of glutamate receptor (see FIGURE 49.14).

Color vision involves even more complex signal processing than the rhodopsin mechanism in rods. Color vision results from the presence of three subclasses of cones in the retina, each with its own type of opsin associated with retinal to form visual pigments collectively called **photopsins**. These photoreceptors are known as red cones, green cones, and blue cones, referring to the colors their kind of photopsin is best at absorbing. The absorption spectra for these pigments overlap, and the brain's perception of intermediate hues depends on the differential stimulation of two or more types of cones. For example, when both red and green cones are stimulated, we may see yellow or orange, depending on which of these two populations of cones is most strongly stimulated. Color blindness, more common in males than females because it is generally

FIGURE 49.12 Effect of light on retinal. Retinal exists in two forms, isomers of each other. Absorption of light converts the pigment from the *cis* isomer to the *trans* isomer, which separates from the opsin. This retinal-free opsin triggers the signal-transduction pathway that converts the light signal into an electrochemical signal, a receptor potential, in the rod cell membrane. When the photoreceptor is no longer stimulated by light, enzymes convert the retinal back to the *cis* form, which recombines with opsin to form rhodopsin.

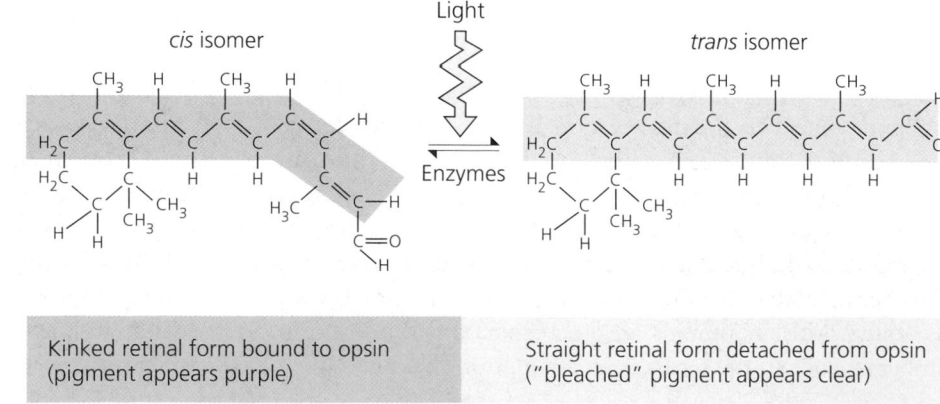

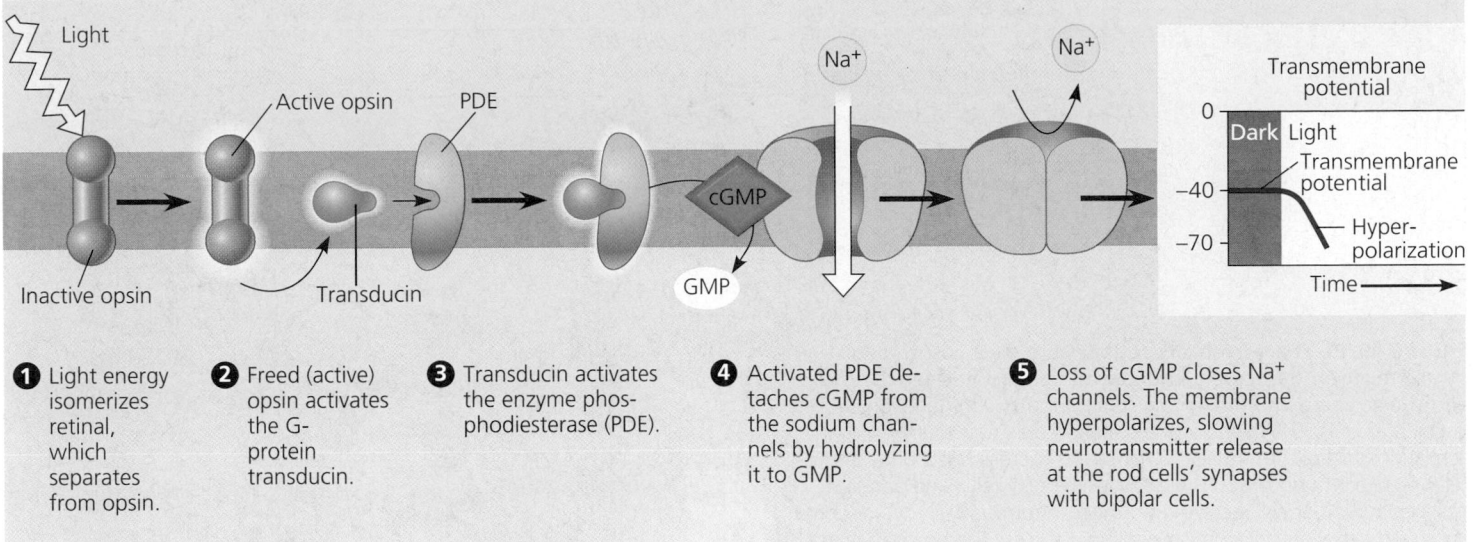

1 Light energy isomerizes retinal, which separates from opsin.

2 Freed (active) opsin activates the G-protein transducin.

3 Transducin activates the enzyme phosphodiesterase (PDE).

4 Activated PDE detaches cGMP from the sodium channels by hydrolyzing it to GMP.

5 Loss of cGMP closes Na⁺ channels. The membrane hyperpolarizes, slowing neurotransmitter release at the rod cells' synapses with bipolar cells.

FIGURE 49.13 From light reception to receptor potential: A rod cell's signal-transduction pathway. Note that the receptor potential, in this case, is a *hyper*polarization of the membrane rather than the more common depolarization.

inherited as a sex-linked trait (see FIGURE 15.9), is due to a deficiency or absence of one or more types of photopsin.

The retina assists the cerebral cortex in processing visual information

Processing of visual information begins in the retina itself. Note again that the axons of rods and cones synapse with neurons called **bipolar cells,** which in turn synapse with **ganglion cells** (FIGURE 49.15, p. 1068). Additional types of neurons in the retina, **horizontal cells** and **amacrine cells,** help integrate the information before it is sent to the brain. The axons of ganglion cells then convey the resulting sensations to the brain as action potentials along the optic nerve.

Signals from the rods and cones may follow either vertical or lateral pathways. In the so-called vertical pathway, information passes directly from the receptor cells to the bipolar cells to the ganglion cells. In the lateral pathway, the horizontal and amacrine cells provide lateral integration of visual signals. Horizontal cells carry signals from one rod or cone to other photoreceptor cells and to several bipolar cells; amacrine cells spread the information from one bipolar cell to several ganglion cells. When a rod or cone stimulates a horizontal cell, the horizontal cell inhibits more distant receptors and bipolar cells that are not illuminated, making the light spot appear lighter and the dark surroundings even darker. This integration, called **lateral inhibition,** sharpens edges and enhances contrast in the image. Lateral inhibition is repeated by the interactions of the amacrine cells with the ganglion cells and occurs at all levels of visual processing.

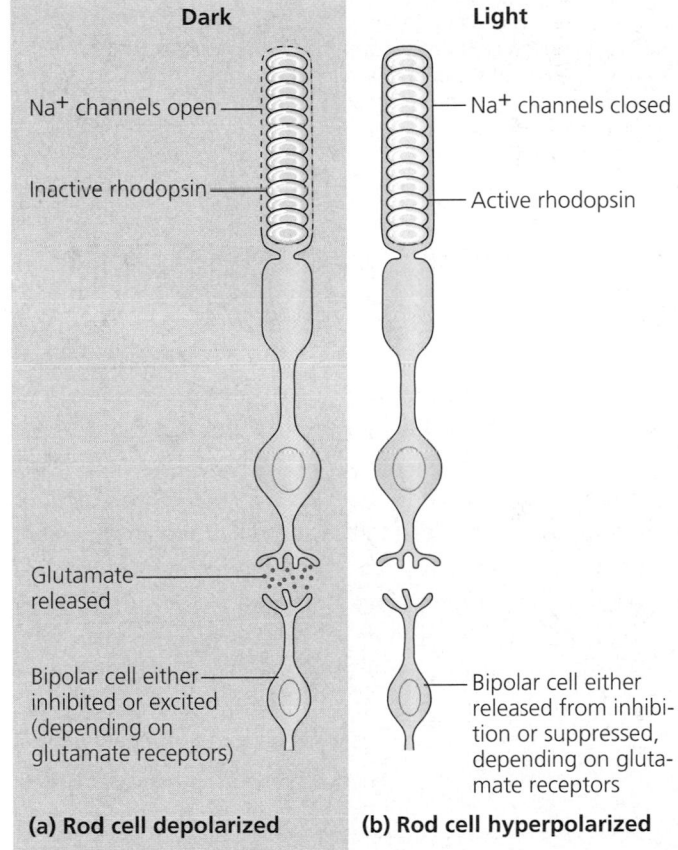

(a) Rod cell depolarized   **(b) Rod cell hyperpolarized**

FIGURE 49.14 The effect of light on synapses between rod cells and bipolar cells. (a) In the dark, rhodopsin is inactive, and the rod cell membrane is highly permeable to sodium and thus depolarized. In this state, the rod cell releases glutamate and regulates the "firing" of two different classes of bipolar cells, which have opposite responses to glutamate. **(b)** In contrast, when light activates rhodopsin, the rod cell membrane becomes less permeable to sodium, and its membrane potential changes (it develops a receptor potential, a hyperpolarization in this case). The synaptic terminals of the rod cell then slow their release of glutamate, enhancing the activity of one class of bipolar cells and suppressing the activity of the other type.

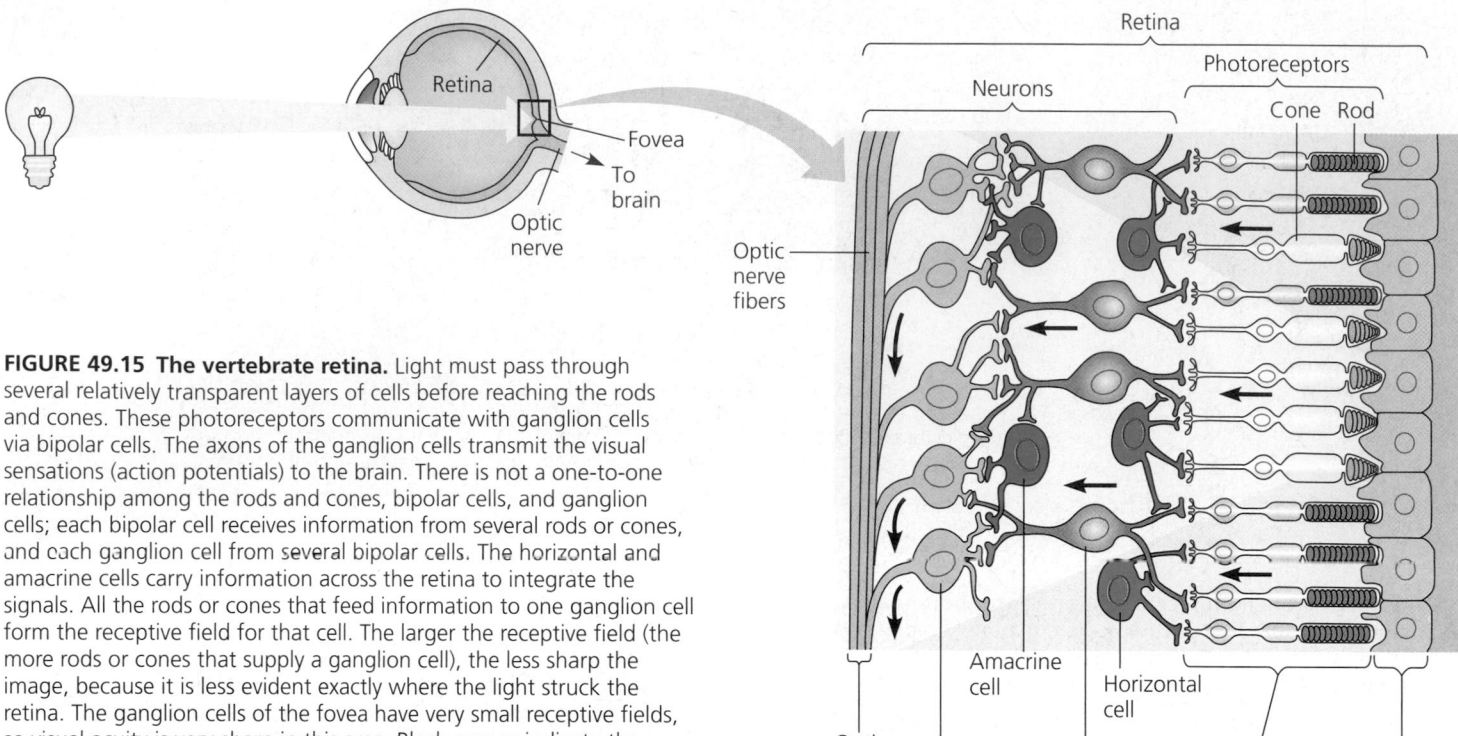

FIGURE 49.15 The vertebrate retina. Light must pass through several relatively transparent layers of cells before reaching the rods and cones. These photoreceptors communicate with ganglion cells via bipolar cells. The axons of the ganglion cells transmit the visual sensations (action potentials) to the brain. There is not a one-to-one relationship among the rods and cones, bipolar cells, and ganglion cells; each bipolar cell receives information from several rods or cones, and each ganglion cell from several bipolar cells. The horizontal and amacrine cells carry information across the retina to integrate the signals. All the rods or cones that feed information to one ganglion cell form the receptive field for that cell. The larger the receptive field (the more rods or cones that supply a ganglion cell), the less sharp the image, because it is less evident exactly where the light struck the retina. The ganglion cells of the fovea have very small receptive fields, so visual acuity is very sharp in this area. Black arrows indicate the pathway of visual information (action potentials) from the retina to the optic nerve.

Axons of ganglion cells form the optic nerves that transmit sensations from the eyes to the brain. The optic nerves from the two eyes meet at the **optic chiasm** near the center of the base of the cerebral cortex (FIGURE 49.16). The nerve tracts of the optic chiasm are arranged in such a way that visual sensations from the left visual field of both eyes are transmitted to the right side of the brain, and visual sensations in the right visual field are transmitted to the left side of the brain. Most of the ganglion cell axons lead to the **lateral geniculate nuclei** of the thalamus. Neurons of the lateral geniculate nuclei continue back to the **primary visual cortex** in the occipital lobe of the cerebrum. Additional interneurons carry the information to other, more sophisticated visual processing and integrating centers elsewhere in the cortex.

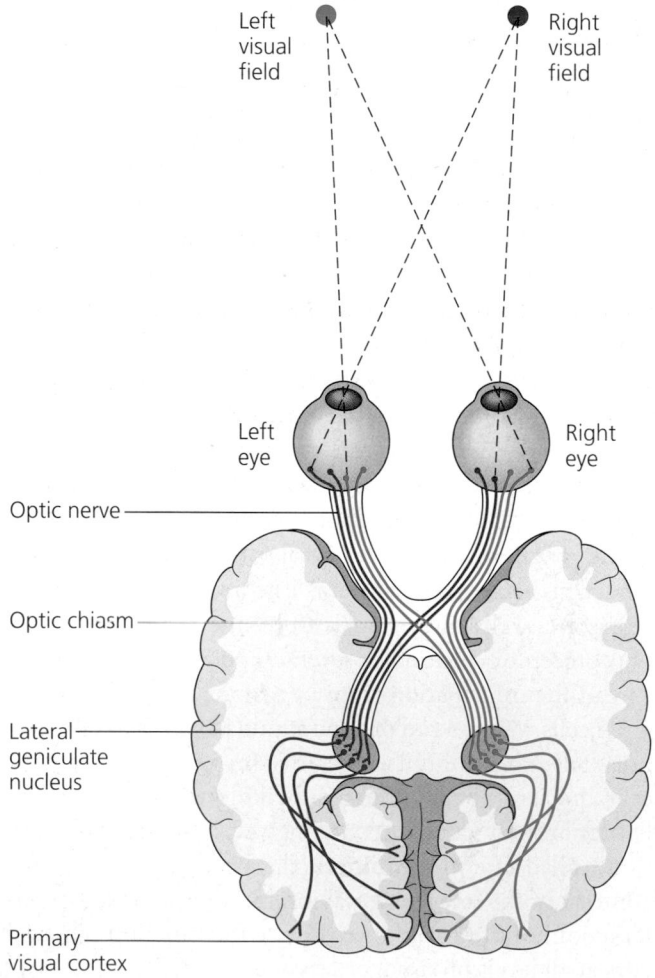

FIGURE 49.16 Neural pathways for vision. Because of the arrangement of neurons in the retinas, optic nerves, and optic chiasm, the right side of the brain receives sensory information about objects in the left visual field (blue), while the left side of the brain receives information from the right visual field (red). Each optic nerve contains about a million axons that synapse with interneurons in the lateral geniculate nuclei. The nuclei relay sensations to the visual cortex, believed to be the first of many brain centers that cooperate in constructing our visual perceptions.

Point-by-point information in the visual field is projected along neurons onto the visual cortex according to its position in the retina, but the information the brain receives is highly distorted. How does the brain convert a complex set of action potentials representing two-dimensional images projected onto our retinas into three-dimensional perceptions of our surroundings? Researchers estimate that at least 30% of the cerebral cortex—hundreds of millions of interneurons in perhaps dozens of integrating centers—take part in formulating what we actually "see." Determining how these centers integrate such components of our vision as color, motion, depth, shape, and detail is the focus of an exciting, fast-moving research effort.

HEARING AND EQUILIBRIUM

Hearing and the perception of body equilibrium, or balance, are related in most animals. Both involve the formation of sensations by mechanoreceptors containing hair cells that produce receptor potentials when the hairs are bent by settling particles or moving fluid. In mammals and most other terrestrial vertebrates, the sensory organs for hearing and equilibrium are closely associated within fluid-filled canals in the ear.

The mammalian hearing organ is within the inner ear

The mammalian ear can be divided into three regions. The **outer ear** consists of the external pinna and the auditory canal, which collect sound waves and channel them to the **tympanic membrane** (eardrum) separating the outer ear from the middle ear. Within the **middle ear,** vibrations are conducted through three ossicles (small bones)—the **malleus** (hammer), **incus** (anvil), and **stapes** (stirrup)—to the inner ear, passing through the **oval window,** a membrane beneath the stapes (FIGURE 49.17a and b, p. 1070). The middle ear also opens into the **Eustachian tube,** which connects with the pharynx and equalizes pressure between the middle ear and the atmosphere, enabling you to "pop" your ears when changing altitude, for example. The **inner ear** consists of a labyrinth of channels within a skull bone (the temporal bone). These channels are lined by a membrane and contain fluid that moves in response to sound or movement of the head.

The part of the inner ear involved in hearing is a coiled organ known as the **cochlea** (Latin, "snail"). The cochlea has two large chambers, an upper vestibular canal and a lower tympanic canal, separated by a smaller cochlear duct (FIGURE 49.17c). The vestibular and tympanic canals contain a fluid called perilymph, and the cochlear duct is filled with a liquid called endolymph. The floor of the cochlear duct, the basilar membrane, bears the **organ of Corti,** which contains the actual receptor cells of the ear, hair cells with hairs projecting

into the cochlear duct (FIGURE 49.17d). Many of the hairs are attached to the tectorial membrane, which hangs over the organ of Corti like a shelf.

How is the anatomy of the ear correlated with the function of hearing? The ear converts the energy of pressure waves traveling through air into nerve impulses that the brain perceives as sound. Vibrating objects, such as the reverberating strings of a guitar or the vocal cords of a speaking person, create percussion waves in the surrounding air. These waves cause the tympanic membrane to vibrate with the same frequency as the sound. The three bones of the middle ear transmit the mechanical movements to the oval window, a membrane on the surface of the cochlea. Vibrations of the oval window produce pressure waves in the fluid within the cochlea.

The cochlea transduces the energy of the vibrating fluid into action potentials. The stapes vibrating against the oval window creates a traveling pressure wave in the fluid of the cochlea that passes into the vestibular canal (FIGURE 49.18a, p. 1071). This wave continues around the tip of the cochlea and through the tympanic canal, dissipating as it strikes the **round window.** The pressure waves in the vestibular canal push downward on the cochlear duct and basilar membrane. The basilar membrane vibrates up and down in response to the pressure waves, and its hair cells alternately brush against and are withdrawn from the tectorial membrane. Deflection of the hairs opens ion channels in the plasma membrane of the hair cells, and positive ions (K^+, in this case) enter. The resulting depolarization increases neurotransmitter release from the hair cell and the frequency of action potentials in the sensory neuron with which the hair cell synapses. This neuron carries the sensations to the brain through the auditory nerve.

Sound is detected by increases in the frequency of impulses in the sensory neuron, but how is the quality of that sound determined? Two important sound variables are volume and pitch. Volume (loudness) is determined by the amplitude, or height, of the sound wave. The greater the amplitude of a sound, the more vigorous the vibrations of fluid in the cochlea, the greater the bending of the hair cells, and the more action potentials generated in the sensory neurons. **Pitch** is a function of a sound wave's frequency, or number of vibrations per second, expressed in hertz (Hz). Short, high-frequency waves produce high-pitched sound, while long, low-frequency waves generate low-pitched sound. Healthy young humans can hear sounds in the range of 20 to 20,000 Hz, dogs can hear sounds as high as 40,000 Hz, and bats can emit and hear clicking sounds of even higher frequency, using this ability to locate objects by sonar.

Pitch can be distinguished by the cochlea because the basilar membrane is not uniform along its length (see FIGURE 49.18b and c). The proximal end near the oval window is relatively narrow and stiff, while the distal end near the tip is wider and more flexible. Each region of the basilar membrane is most affected by a particular vibration frequency. The sensory

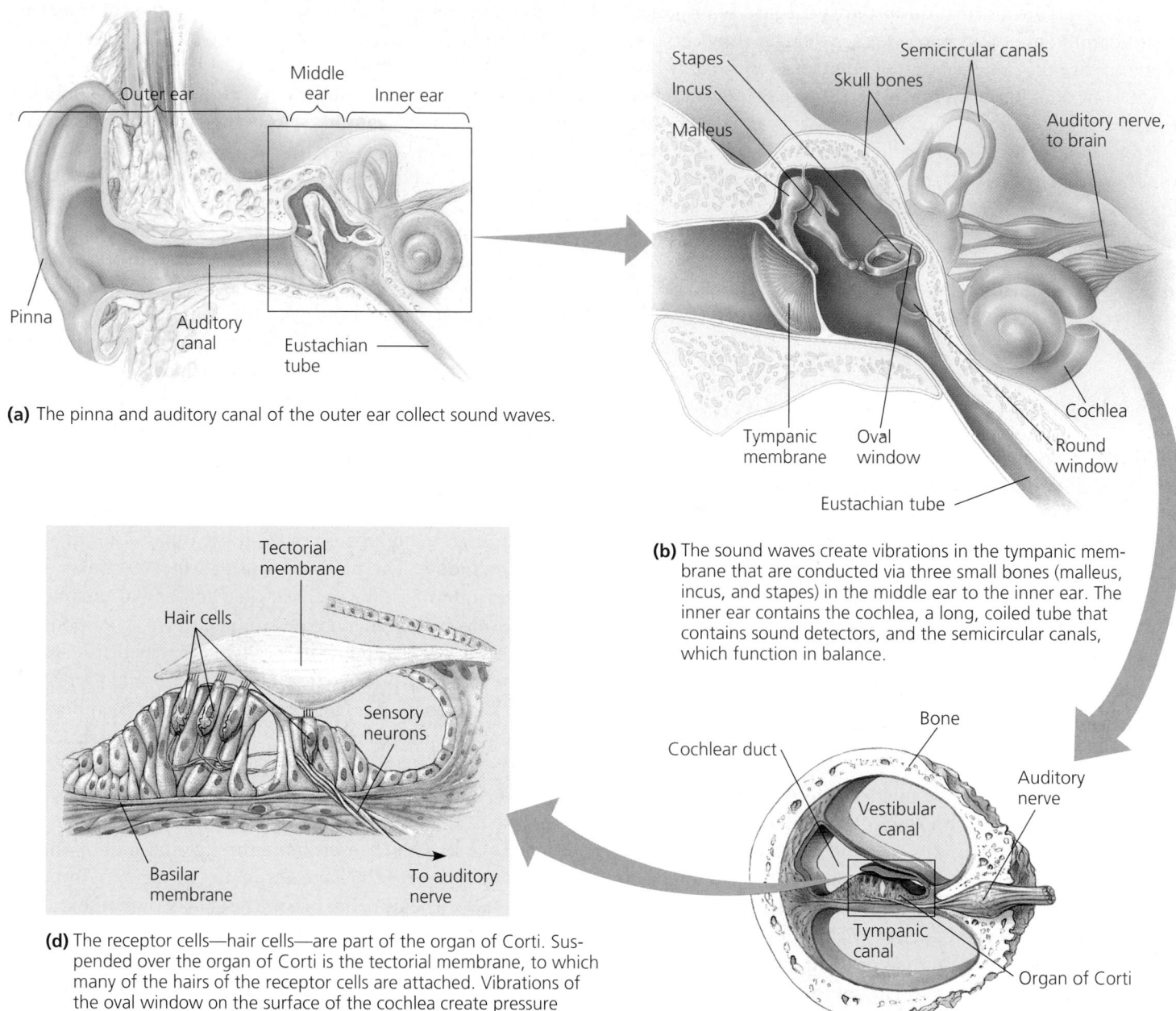

(a) The pinna and auditory canal of the outer ear collect sound waves.

(b) The sound waves create vibrations in the tympanic membrane that are conducted via three small bones (malleus, incus, and stapes) in the middle ear to the inner ear. The inner ear contains the cochlea, a long, coiled tube that contains sound detectors, and the semicircular canals, which function in balance.

(d) The receptor cells—hair cells—are part of the organ of Corti. Suspended over the organ of Corti is the tectorial membrane, to which many of the hairs of the receptor cells are attached. Vibrations of the oval window on the surface of the cochlea create pressure waves in the cochlear fluid. As the basilar membrane vibrates, hair cells repeatedly brush against the tectorial membrane. This stimulus causes the hair cells to depolarize and release neurotransmitter, thereby triggering an action potential in a sensory neuron.

(c) A cross-sectional view of the cochlea shows three canals. The vestibular canal and tympanic canal contain the fluid perilymph. Between these two canals is a smaller cochlear duct filled with endolymph. The organ of Corti sits on the basilar membrane, which forms the floor of the cochlear duct.

FIGURE 49.17 Structure and function of the human ear.

neurons associated with the region vibrating most vigorously at any instant send the most action potentials along the auditory nerve. But the actual perception of pitch depends on neural mapping of the brain. Sensory neurons from the auditory pathway project onto specific auditory areas of the cerebral cortex according to the region of the basilar membrane in which the signal originated. When a particular site of the cortex is stimulated, we perceive a sound of a particular pitch.

The inner ear also contains the organs of equilibrium

Several organs in the inner ear of humans and most other mammals detect body position and balance. Behind the oval window is a vestibule that contains two chambers, the **utricle** and **saccule.** The utricle opens into three **semicircular canals** that complete the apparatus for equilibrium (FIGURE 49.19a).

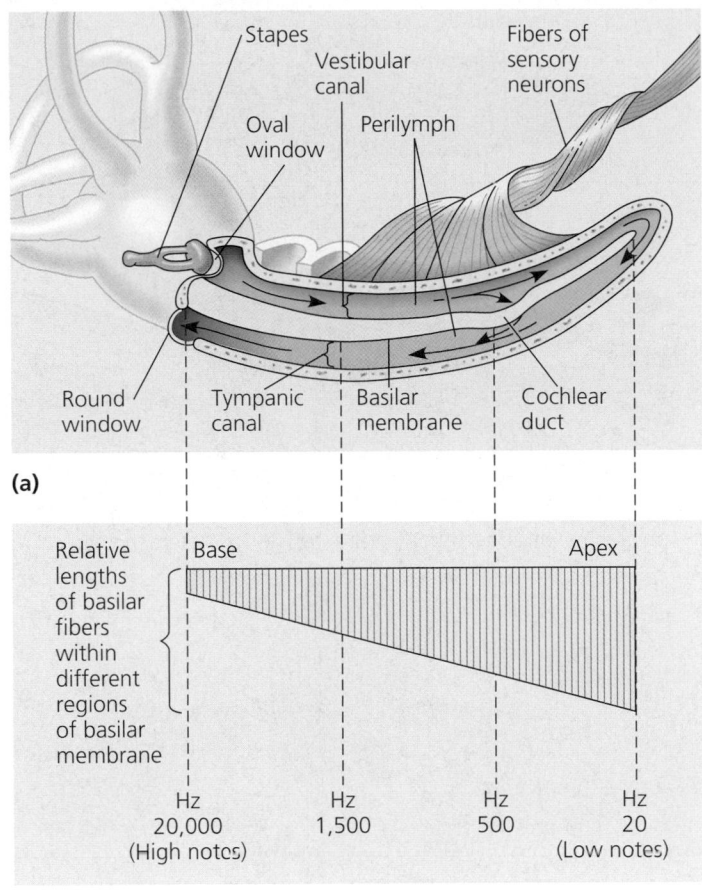

(a)

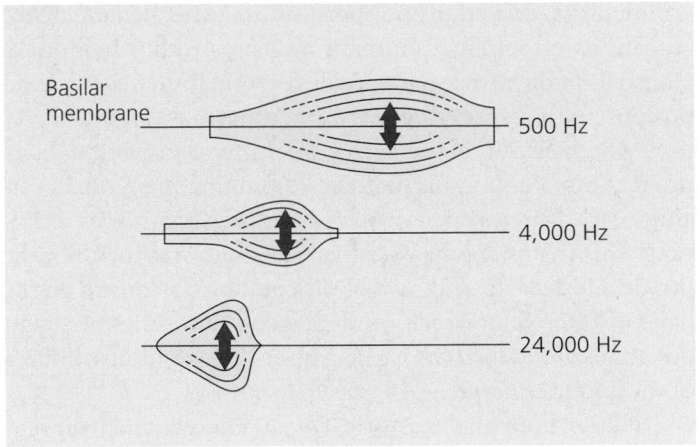

(c)

(b)

FIGURE 49.18 How the cochlea distinguishes pitch. (a) Vibrations of the stapes against the oval window agitate the fluid within the cochlea (uncoiled here), causing pressure waves having a frequency equivalent to the sound waves that entered the ear. The waves (black arrows) pass through the vestibular canal to the apex of the cochlea, then back toward the base of the cochlea via the tympanic canal. The energy causes the cochlear duct, with its basilar membrane and organ of Corti, to vibrate up and down. The bouncing of the basilar membrane stimulates the hair cells within the cochlear duct. **(b)** Fibers span the width of the basilar membrane. Like harp strings, these fibers vary in length, being shorter near the base of the membrane and longer near its apex. The length of the fibers "tunes" specific regions of the basilar membrane to vibrate at specific frequencies. **(c)** Different frequencies of pressure waves in the cochlea cause certain places along the basilar membrane to vibrate, stimulating particular hair cells and sensory neurons. The differential stimulation of hair cells is perceived in the brain as sound of a certain pitch.

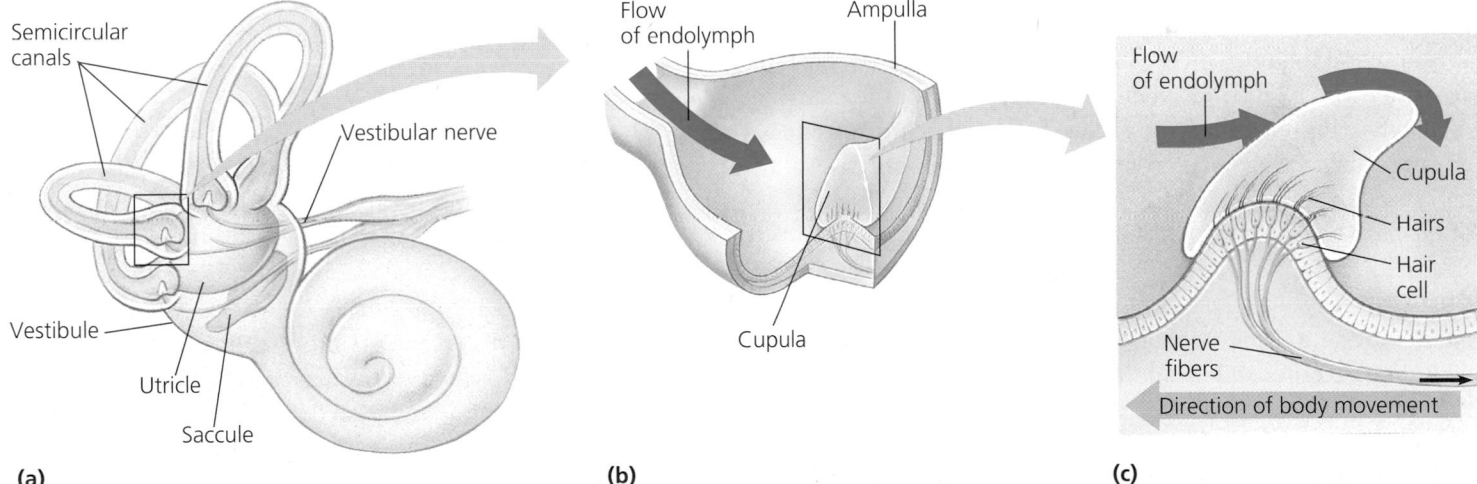

(a)

(b)

(c)

FIGURE 49.19 Organs of balance in the inner ear. (a) Three structures of the inner ear—the utricle and saccule in the vestibule and the semicircular canals—contain hair cells sensitive to balance and body position. The saccule and utricle tell the brain which way is up and also inform it of the body's fixed position in space or of any linear acceleration associated with movement. The semicircular canals are arranged in the three spatial planes.

(b) Each canal has at its base a swelling called an ampulla, which contains a cluster of hair cells with hairs projecting into a gelatinous cap called the cupula. **(c)** When the head changes its rate of rotation, inertia prevents the endolymph in the semicircular canals from moving with the head, so the fluid presses against the cupula, bending the hair cells. The bending increases the frequency of action potentials in the sensory neurons in direct

proportion to the amount of rotational acceleration. The mechanism adjusts quickly if rotation continues at a constant speed: The endolymph begins moving with the head, and the pressure on the cupula is reduced. If rotation stops suddenly, however, the fluid continues to flow through the semicircular canals and again stimulates the hair cells. This new stimulus can cause dizziness.

Sensations related to body position are generated much like sensations of sound in humans and most other mammals. Hair cells in the utricle and saccule respond to changes in head position with respect to gravity and movement in one direction. The hair cells are arranged in clusters, and all the hairs project into a gelatinous material containing many small calcium carbonate particles called otoliths ("ear stones"). Because this material is heavier than the endolymph within the utricle and saccule, gravity is always pulling downward on the hairs of the receptor cells, sending a constant series of action potentials along the sensory neurons of the vestibular branch of the auditory nerve.

Different body angles cause different hair cells and their sensory neurons to be stimulated. When the position of the head changes with respect to gravity (as when the head bends forward), the force on the hair cell changes, and it increases (or decreases—see FIGURE 49.4) its output of neurotransmitter. The brain interprets the resulting changes in impulse production by the sensory neurons to determine the position of the head. By a similar mechanism, the semicircular canals, arranged in the three spatial planes, detect changes in the rate of rotation or angular movements of the head (FIGURE 49.19b and c).

A lateral line system and inner ear detect pressure waves in most fishes and aquatic amphibians

Like other vertebrates, fishes and aquatic amphibians also have inner ears located near the brain. There is no cochlea, but there are a saccule, a utricle, and semicircular canals, structures homologous to the equilibrium sensors of our own ears. Within these chambers in the inner ear of a fish, sensory hairs are stimulated by the movement of otoliths. Unlike the mammalian hearing apparatus, the ear of a fish has no eardrum and does not open to the outside of the body. Vibrations of the water caused by sound waves are conducted through the skeleton of the head to the inner ears, setting the otoliths in motion and stimulating the hair cells. The air-filled swim bladder (see Chapter 34) also vibrates in response to sound and may contribute to the transfer of sound to the inner ear. Some fishes, including catfishes and minnows, have a series of bones called the Weberian apparatus, which conducts vibrations from the swim bladder to the inner ear.

Most fishes and aquatic amphibians also have a **lateral line system** along both sides of the body (FIGURE 49.20). The system contains mechanoreceptors that detect low-frequency waves by a mechanism similar to the function of the inner ear. Water from the animal's surroundings enters the lateral line system through numerous pores and flows along a tube past the mechanoreceptors. The receptor units, called neuromasts, resemble the ampullae in our semicircular canals. Each

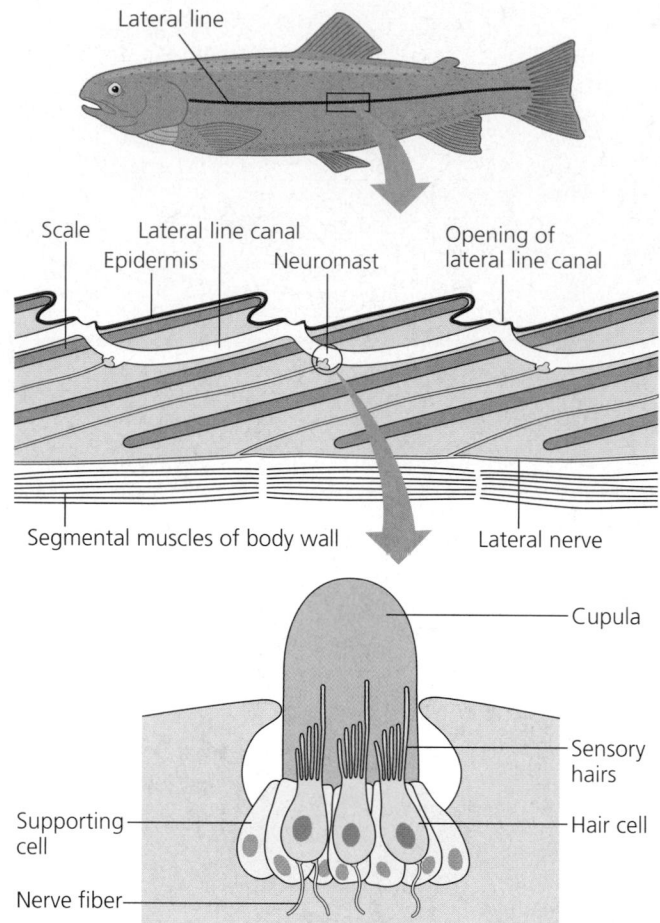

FIGURE 49.20 The lateral line system in a fish. Water flowing through the system bends hair cells. The hair cells transduce the energy into receptor potentials, triggering action potentials, which are conveyed to the brain. The lateral line system enables a fish to monitor water currents, pressure waves produced by moving objects, and low-frequency sounds conducted through water.

neuromast has a cluster of hair cells, with the sensory hairs embedded in a gelatinous cap, the cupula. As the pressure of moving water bends a cupula, the hair cells transduce the energy into receptor potentials and then into action potentials that are transmitted along a nerve to the brain. This information helps the fish perceive its movement through water or the direction and velocity of water currents flowing over its body. The lateral line system also detects water movements or vibrations generated by other moving objects, including prey and predators.

The lateral line system functions only in water. In terrestrial vertebrates, the inner ear has evolved as the main organ of hearing and equilibrium. Some amphibians have a lateral line system as tadpoles, but not as adults living on land. In the ear of a terrestrial frog or toad, sound vibrations traveling in the air are conducted to the inner ear by a tympanic membrane on the body surface and a single middle ear bone. There also is evidence that the lungs of a frog vibrate in response to sound and transmit their vibrations to the eardrum via the auditory tube. A small side pocket of the saccule functions as the main hearing organ of the frog, and it is this outgrowth of the saccule that gave rise to the more elaborate cochlea during the evolution of mammals. Birds also have a cochlea, but like amphibians and reptiles, sound is conducted from the tympanic membrane to the inner ear by a single bone, the stapes.

Many invertebrates have gravity sensors and are sound-sensitive

Most invertebrates have sensory organs called **statocysts** that contain mechanoreceptors and function in their sense of equilibrium (FIGURE 49.21). A common type of statocyst has a layer of hair cells surrounding a chamber containing **statoliths,** which are grains of sand or other dense granules. Gravity causes the statoliths to settle to the low point within the chamber, stimulating hair cells in that location. This is similar to how the saccule and utricle function in vertebrates, and indeed these structures in the vertebrate inner ear are considered to be specialized types of statocysts. The statocysts of invertebrates have various locations. For example, many jellies have statocysts at the fringe of the "bell," giving the animals an indication of body position. Lobsters and crayfish have statocysts near the bases of their antennules. Crayfish have been tricked into swimming upside down in experiments in which the statoliths were replaced with metal shavings that could be pulled to the upper end of the statocysts with magnets.

Many invertebrates demonstrate a general sensitivity to sound, although structures specialized for hearing seem to be

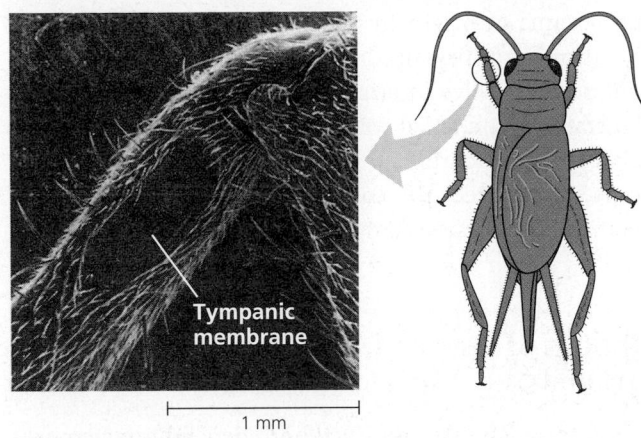

FIGURE 49.22 An insect ear. The tympanic membrane, this one on the front leg of a cricket, vibrates in response to sound waves (SEM). The vibrations stimulate mechanoreceptor cells attached to the inside of the tympanic membrane.

less widespread than gravity sensors. Hearing structures have been most extensively studied in terrestrial insects.

Many (perhaps most) insects have body hairs that vibrate in response to sound waves. Hairs of different stiffness and length vibrate at different frequencies. The hairs are commonly tuned to frequencies of sounds produced by other organisms. A male mosquito locates a mate by means of fine hairs on his antennae. The hairs vibrate in a specific way in response to the hum produced by the beating wings of flying females. A tuning fork that vibrates at the same frequency as a female mosquito's wings will also attract males. Some caterpillars (larval moths and butterflies) have vibrating body hairs that detect the buzzing wings of predatory wasps, warning the caterpillars of danger. Many insects also have localized "ears" (FIGURE 49.22). A tympanic membrane (eardrum) is stretched over an internal air chamber. Sound waves vibrate the tympanic membrane, stimulating receptor cells attached to the inside of the membrane and resulting in nerve impulses that are transmitted to the brain. Some moths can hear notes of such high pitch that they detect the sounds bats produce for sonar, and perception of these sounds triggers the moth's escape maneuver, as mentioned at the beginning of this chapter.

CHEMORECEPTION—TASTE AND SMELL

Many animals use their chemical senses to find mates (as when male silk moths respond to pheromones emitted by females), to recognize territory that has been marked by some chemical substance (as when dogs and cats sniff boundaries that have been staked out by their spraying neighbors), and to help navigate during migration (as when salmon use the unique scent

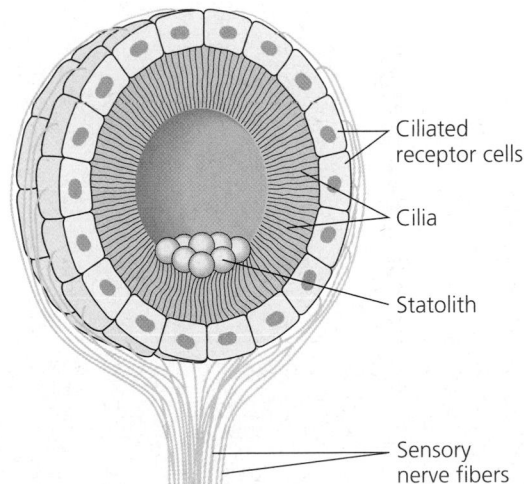

FIGURE 49.21 The statocyst of an invertebrate. The settling of statoliths to the low point within the chamber bends cilia on receptor cells in that location, providing the brain with information about the position of the body.

Ciliated receptor cells

Cilia

Statolith

Sensory nerve fibers

of their streams of origin to return for breeding). Chemical "conversation" is especially important for animals, such as ants and bees, that live in large social groups. In all animals, taste (gustation) and smell (olfaction) are important in feeding behavior. For example, a hydra begins to swallow when chemoreceptors detect the compound glutathione, which is released from prey captured by the hydra's tentacles.

Perceptions of taste and smell are usually interrelated

The perceptions of taste and smell both depend on chemoreceptors that detect specific chemicals in the environment. In the case of terrestrial animals, taste is the detection of certain chemicals that are present in a solution, and smell is the detection of airborne chemicals. However, these chemical senses are usually closely related, and there really is no distinction in aquatic environments.

The taste receptors of insects are located within sensory hairs called sensillae on the feet and mouthparts. The animals use their sense of taste to select food. A tasting hair contains several chemoreceptor cells, each especially responsive to a particular class of chemical stimuli, such as sugar or salt. By integrating sensations (nerve impulses) from these different receptor cells, the insect's brain can apparently distinguish a very large number of tastes (FIGURE 49.23). Insects can also smell airborne chemicals, using olfactory sensillae, usually located on the antennae (see FIGURE 49.5).

In humans and other mammals, the senses of taste and smell are functionally similar and interrelated. In both cases a small molecule must dissolve in liquid to reach the receptor cell and trigger the sensation. That molecule binds to a specific protein in the receptor cell membrane, triggering a depolarization of the membrane and the release of neurotransmitter (see FIGURE 49.2).

The receptor cells for taste are modified epithelial cells organized into **taste buds** scattered in several areas of the tongue and mouth. Most of the taste buds are on the surface of the tongue or are associated with nipplelike projections called papillae on the tongue. Although we cannot distinguish different types of taste receptors from their structures, we recognize four basic taste perceptions—sweet, sour, salty, and bitter—each detected by a chemoreceptor with a certain receptor and signal-transduction pathway. Although each receptor cell is more responsive to a particular type of substance, it can actually be stimulated by a broad range of chemicals. With each taste of food or sip of drink, the brain integrates the differential input from the taste buds, and a complex flavor is perceived.

The olfactory sense of mammals detects certain airborne chemicals. Olfactory receptor cells are neurons that line the upper portion of the nasal cavity and send impulses along their axons directly to the olfactory bulb of the brain (FIGURE 49.24). The receptive ends of the cells contain cilia that extend into the layer of mucus coating the nasal cavity. When an odorous substance diffuses into this region, it binds to specific receptor molecules on the plasma membrane of the olfactory cilia. The binding triggers a signal-transduction pathway in-

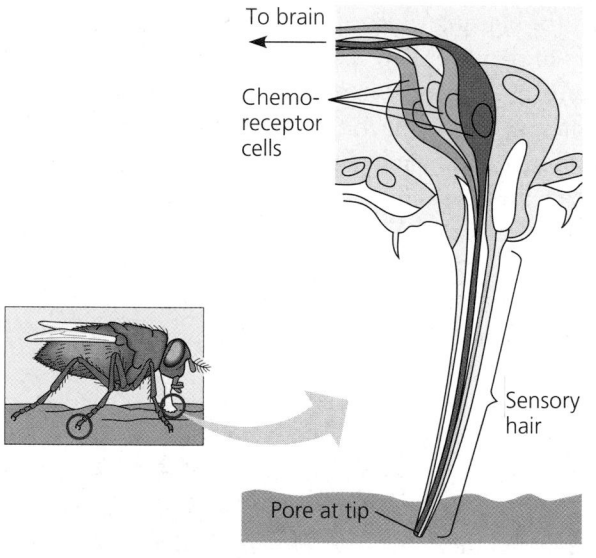

(a)

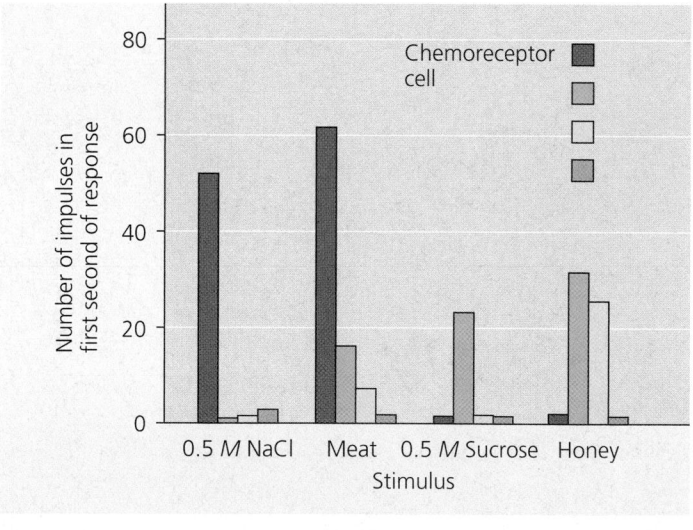

(b)

FIGURE 49.23 The mechanism of taste in a blowfly. (a) Gustatory sensillae (hairs) on the feet and mouthparts each contain four chemoreceptor cells with dendrites that extend to the pore at the tip of the sensory hair. **(b)** Each chemoreceptor (taste) cell is especially sensitive to a particular class of substance; for example, the receptor colored green here is most responsive to sugars. But this specificity is relative; each cell can respond to some extent to a broad range of chemical stimuli. Thus, any natural food probably stimulates two or more of the receptor cells. The brain apparently integrates the frequencies of impulses arriving along the axons of the four classes of receptor cells and distinguishes a great variety of tastes.

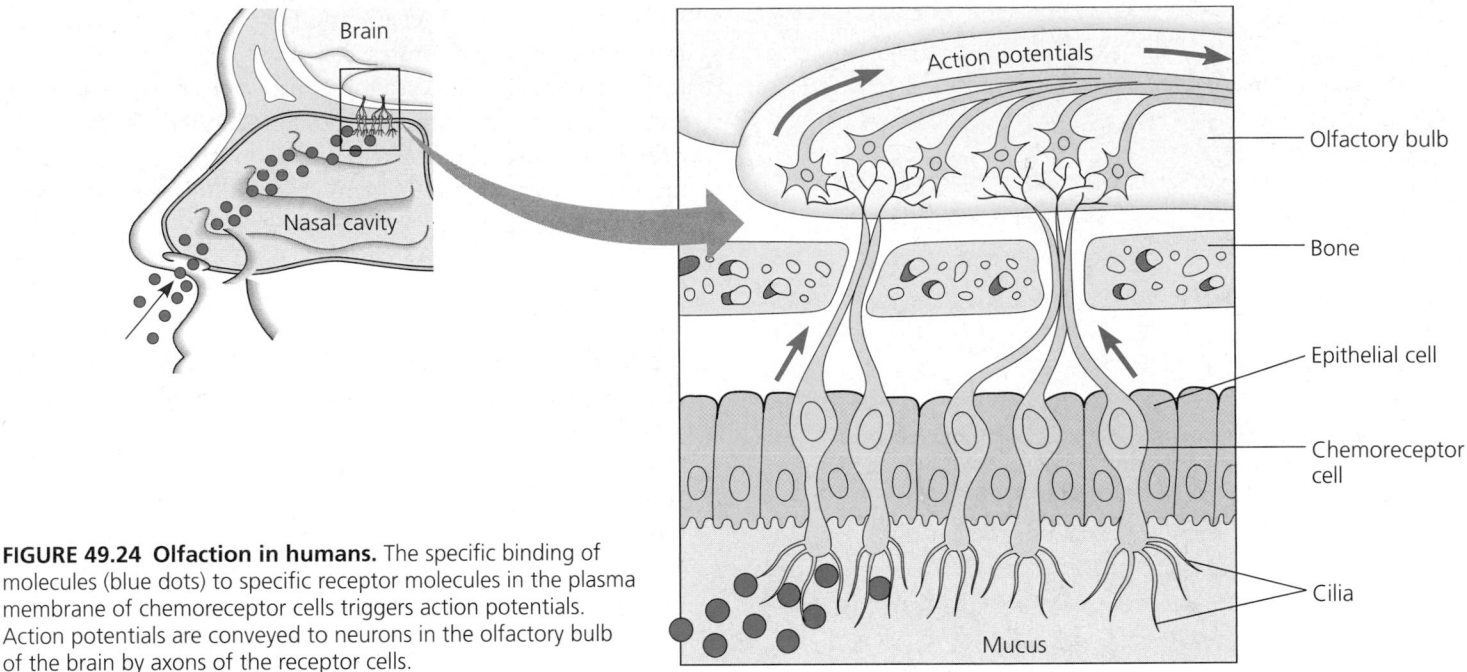

FIGURE 49.24 Olfaction in humans. The specific binding of molecules (blue dots) to specific receptor molecules in the plasma membrane of chemoreceptor cells triggers action potentials. Action potentials are conveyed to neurons in the olfactory bulb of the brain by axons of the receptor cells.

volving a G-protein-signaling pathway and, in many cases, the enzyme adenylyl cyclase and the second messenger cyclic AMP (see FIGURE 11.13). The second messenger opens Na^+ channels in the olfactory receptor cell membrane, depolarizing it and generating action potentials that go to the brain. Humans can distinguish thousands of different odors, but these are probably based on a few primary odors, analogous to the basic tastes of the gustatory system.

Although the receptors and brain pathways for taste and olfaction are independent, the two senses do interact. Indeed, much of what we call taste is really smell. If the olfactory system is blocked, as by a head cold, the perception of taste is sharply reduced.

■   ■   ■

Throughout our discussions of sensory mechanisms we have seen many examples of how sensory inputs to the nervous system result in the specific body movements that we observe as animal behavior. The swimming of planarians away from light, the escape behavior of a moth that hears bat sonar, the "righting" response of a crayfish turned upside-down, and the feeding movements of a hydra when it tastes glutathione—these are just a few cases that we have mentioned so far. Animal behavior flows in a seamless cycle involving continuous brain operations that generate actions, note the consequences of these actions via sensory mechanisms, and then use this information to decide on the next action. The remainder of the chapter focuses on the motor mechanisms that make these animal responses possible: how animals use their muscles and skeletons to move.

MOVEMENT AND LOCOMOTION

Movement is a hallmark of animals. To catch food, an animal must either move through its environment or move the surrounding water or air past itself. Sessile animals such as sponges and many cnidarians stay put, but they wave tentacles that capture prey or use beating cilia to generate water currents that draw and trap small food particles (see Chapter 33). Most animals, however, are mobile and spend a considerable portion of their time and energy actively searching for food, as well as escaping from danger and looking for mates. **Locomotion,** or active travel from place to place, is our focus here.

Locomotion requires energy to overcome friction and gravity

The modes of animal locomotion are diverse. Most animal phyla include species that swim. On land and in the sediments on the floor of the sea and lakes, animals crawl, walk, run, or hop. Active flight (in contrast to gliding downward from a tree or elevated ground) has evolved in only a few animal groups: insects, reptiles, birds, and, among the mammals, bats. A large group of flying reptiles died out millions of years ago, leaving birds and bats as the only flying vertebrates.

In all its forms, locomotion requires that an animal expend energy to overcome two forces that tend to keep it stationary: friction and gravity. Exerting force requires energy-consuming cellular work. Thus, the study of locomotion returns us to the

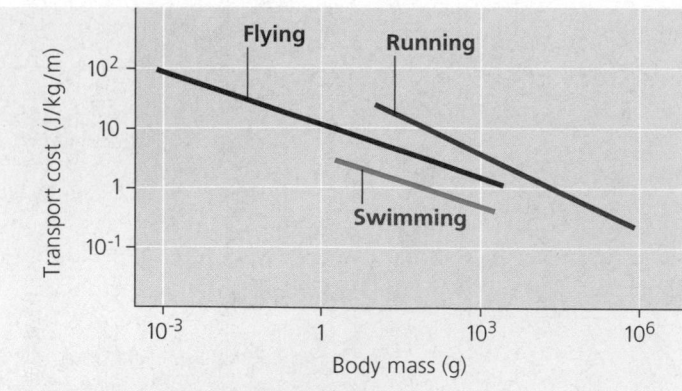

FIGURE 49.25 The cost of transport. This graph compares the transport cost, in joules per kilogram of body weight per meter traveled, for animals specialized for swimming, flying, and running (1 J = 0.24 cal). Notice that both axes are plotted on logarithmic scales.

theme of animal bioenergetics. The energetic cost of transport is different for the various modes of locomotion in different environments. FIGURE 49.25 compares these costs for swimming, running, and flying. Running animals generally consume more energy per meter traveled than equivalently sized animals specialized for swimming, partly because to run (or walk), an animal must expend energy to overcome gravity. Swimming is the most efficient mode of transport (assuming, of course, that an animal is specialized for swimming). If we were to compare energy consumption per minute rather than per meter, we would find that flying animals use more energy than animals swimming or walking for the same amount of time. Each line on the graph in FIGURE 49.25 also shows that a larger animal travels more efficiently than a smaller species specialized for the same mode of transport. For example, a horse consumes less energy per kilogram of body weight than a cat running the same distance. (Of course, total energy consumption is greater for the larger animal.)

Swimming

Because most animals are reasonably buoyant in water, overcoming gravity is less of a problem for swimming animals than for species that move on land or through the air. On the other hand, water is a much denser medium than air, and thus the problem of resistance (friction) is a major one for aquatic animals. A sleek, fusiform (torpedolike) shape is a common adaptation of fast swimmers (see FIGURE 40.6), and swimming tends to be the most energy-efficient means of locomotion.

Animals swim in diverse ways. For instance, many insects and four-legged vertebrates use their legs as oars to push against the water. Squid, scallops, and some cnidarians are jet-propelled, taking in water and squirting it out in bursts. Fishes swim by moving their body and tail from side to side. Whales and other aquatic mammals move by undulating their body and tail up and down.

Locomotion on Land

In general, the problems of locomotion on land are the opposite of those in water. On land, a walking, running, hopping, or crawling animal must be able to support itself and move against gravity, but, at least at moderate speeds, air poses relatively little resistance. When a land animal walks, runs, or hops, its leg muscles expend energy both to propel it and to keep it from falling down. With each step, the animal's leg muscles must also overcome inertia by accelerating a leg from a standing start. For moving on land, powerful muscles and strong skeletal support are more important than a streamlined shape.

Diverse adaptations for traveling on land have evolved in various vertebrates. For example, traveling mainly by hopping, kangaroos have large muscles that generate a lot of power in their hind legs (FIGURE 49.26). When a kangaroo lands, tendons in its hind legs momentarily store energy. The higher the animal hops, the more energy the tendons store. Analogous to the tension of a spring on a pogo stick, the stored energy is available for the next jump and is a cost-free energy boost that

FIGURE 49.26 Energy-efficient locomotion on land. Members of the kangaroo family travel from place to place mainly by leaping forward on their large hind legs. Kinetic energy momentarily stored in tendons after each leap provides a cost-free boost for the next leap. In fact, a large kangaroo hopping along at 30 km/h uses no more energy per minute than it does at 6 km/h. The large tail helps balance the animal's body during leaps and while sitting on the ground.

reduces the total amount of energy the animal must expend to travel. The pogo stick analogy applies to many land animals; the legs of an insect, a dog, or a human, for instance, retain some spring when walking or running, although considerably less than those of a hopping kangaroo.

Maintaining balance is another prerequisite for walking, running, or hopping. A kangaroo's large tail helps balance its body during leaps and also forms a stable tripod with its hind legs when sitting or moving slowly. Illustrating the same principle, a walking cat, dog, or horse keeps three feet on the ground. Bipedal animals, such as humans and birds, keep part of at least one foot on the ground when walking. When running, all four feet (or both feet for bipeds) may be off the ground momentarily, but at running speeds, it is momentum more than foot contact that keeps the body upright.

Crawling poses a very different situation. Because much of its body is in contact with the ground, a crawling animal must exert considerable effort to overcome friction. Earthworms crawl by peristalsis, a type of locomotion dependent on a hydrostatic skeleton, which we examine later. Many snakes crawl by undulating the entire body from side to side. Assisted by large, movable scales on the underside, a snake's body pushes against the ground, driving the animal forward. Boa constrictors and pythons creep straight forward, driven by muscles that lift their belly scales off the ground, tilt the scales forward, and then push them backward against the ground.

Flying

Gravity poses a major problem for a flying animal. For an animal to become airborne, its wings must develop enough lift to overcome the downward force of gravity. The key to flight is the shape of the wings. All types of wings, including those of airplanes, are airfoils—structures whose shape alters air currents in a way that creates lift (see FIGURE 34.26).

Cellular and Skeletal Underpinnings of Locomotion

Underlying the diverse forms of locomotion are fundamental mechanisms common to all animals. At the cellular level, all animal movement is based on one of two basic contractile systems, both of which consume energy to move protein strands against one another. These two systems of cell motility—microtubules and microfilaments—were discussed in Chapter 7. Microtubules are responsible for the beating of cilia and the undulations of flagella. Microfilaments play a major role in amoeboid movement, and they are also the contractile elements of muscle cells. It is the contraction of muscles that concerns us in this chapter, but the work of a muscle in itself cannot translate into movement of the animal. Swimming, crawling, running, hopping, and flying all result from muscles working against some type of skeleton.

Skeletons support and protect the animal body and are essential to movement

The three functions of a skeleton are support, protection, and movement. Most land animals would sag from their own weight if they had no skeleton to support them. Even an animal living in water would be a formless mass with no framework to maintain its shape. Many animals have hard skeletons that protect soft tissues. For example, the vertebrate skull protects the brain, and the ribs form a cage around the heart, lungs, and other internal organs. And skeletons aid in movement by giving muscles something firm to work against. There are three main types of skeletons: hydrostatic skeletons, exoskeletons, and endoskeletons.

Hydrostatic Skeletons

A **hydrostatic skeleton** consists of fluid held under pressure in a closed body compartment. This is the main type of skeleton in most cnidarians, flatworms, nematodes, and annelids (see Chapter 33). These animals control their form and movement by using muscles to change the shape of the fluid-filled compartments. Among the cnidarians, for example, a hydra can elongate by closing its mouth and using contractile cells in the body wall to constrict the central gastrovascular cavity. Because water cannot be compressed very much, decreasing the diameter of the cavity forces it to increase in length. In flatworms (planarians), the interstitial fluid is kept under pressure and functions as the main hydrostatic skeleton. Planarian movement results mainly from muscles in the body wall exerting localized forces against the hydrostatic skeleton. Roundworms (nematodes) hold the fluid in the body cavity (a pseudocoelom; see FIGURE 32.6b) at a high pressure, and contractions of longitudinal muscles result in thrashing movements. In earthworms and other annelids, the coelomic fluid functions as a hydrostatic skeleton. The coelomic cavity is divided by septa between the segments of the worm, and thus the animal can change the shape of each segment individually, using both circular and longitudinal muscles. The hydrostatic skeleton enables earthworms and most other annelids to move by **peristalsis,** a type of locomotion produced by rhythmic waves of muscle contractions passing from head to tail (FIGURE 49.27, p. 1078).

Hydrostatic skeletons are well suited for life in aquatic environments. They may cushion internal organs from shocks and provide support for crawling and burrowing. However, a hydrostatic skeleton cannot support the forms of terrestrial locomotion in which an animal's body is held off the ground, such as walking or running.

Exoskeletons

An **exoskeleton** is a hard encasement deposited on the surface of an animal. For example, most mollusks are enclosed in

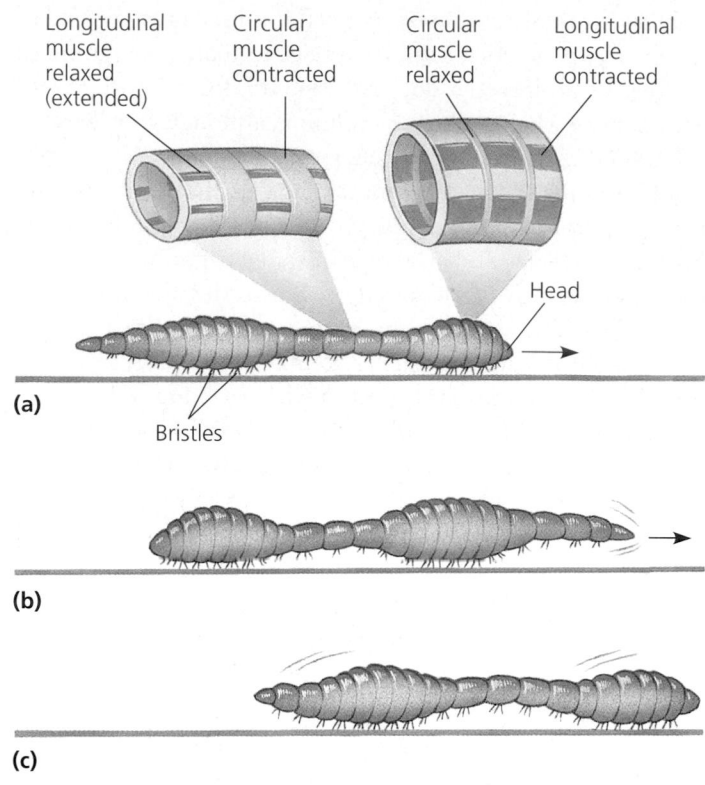

FIGURE 49.27 Peristaltic locomotion in an earthworm. A hydrostatic skeleton, two sets of muscles (one elongating the body, the other shortening it), and bristles holding to the substrate enable an earthworm to crawl over moist ground or burrow through it. Contraction of longitudinal muscles thickens and shortens the worm, while contraction of circular muscles constricts and elongates it. **(a)** As the worm crawls forward, body segments at its head and in front of the tail are short and thick (longitudinal muscles contracted; circular muscles relaxed) and anchored to the ground by bristles. Behind the head and at the tail, segments are thin and elongated (circular muscles contracted; longitudinals relaxed). **(b)** The head has moved forward because circular muscles in the head segments have contracted. Segments behind the head and in front of the tail are now thick and anchored, thus preventing the worm from slipping backward. **(c)** The head segments are thick again and anchored in their new position. The rear segments have released their hold on the ground and have been pulled forward.

calcareous (calcium carbonate) shells secreted by the mantle, a sheetlike extension of the body wall (see FIGURE 33.16). As the animal grows, it enlarges the diameter of the shell by adding to its outer edge. Clams and other bivalves close their hinged shells using muscles attached to the inside of this exoskeleton.

The jointed exoskeleton typical of arthropods is a cuticle, a nonliving coat secreted by the epidermis. Muscles are attached to knobs and plates of the cuticle that extend into the interior of the body. About 30–50% of the cuticle consists of **chitin,** a polysaccharide similar to cellulose. Fibrils of chitin are embedded in a matrix made of protein, forming a composite material that combines strength and flexibility. Where protection is most important, the cuticle is hardened with organic compounds that cross-link the proteins of the exoskeleton. Some crustaceans, such as lobsters, harden portions of their exoskeletons even more by adding calcium salts. In contrast, at the joints of the legs, where the cuticle must be thin and flexible, there is only a small amount of inorganic salts and little cross-linking of proteins. The exoskeleton of an arthropod must periodically be shed (molted) and replaced by a larger case with each spurt of growth by the animal (see FIGURE 5.9).

Endoskeletons

An **endoskeleton** consists of hard supporting elements, such as bones, buried within the soft tissues of an animal. Sponges are reinforced by hard spicules consisting of inorganic material or by softer fibers made of protein (see FIGURE 33.3). Echinoderms have an endoskeleton of hard plates beneath the skin. These ossicles are composed of magnesium carbonate and calcium carbonate crystals, and the separate plates are usually bound together by protein fibers. Sea urchins have a skeleton of tightly bound ossicles, but the ossicles of sea stars are more loosely bound, allowing the animal to change the shape of its arms (see FIGURE 33.38).

Chordates have endoskeletons consisting of cartilage, bone, or some combination of these materials (see FIGURE 40.2). The mammalian skeleton is built from more than 200 bones, some fused together and others connected at joints by ligaments that allow freedom of movement (FIGURE 49.28). Anatomists divide the vertebrate frame into an axial skeleton, consisting of the skull, vertebral column (backbone), and rib cage, and an appendicular skeleton, made up of limb bones and the pectoral and pelvic girdles that anchor the appendages to the axial skeleton. In each appendage, several types of joints provide flexibility for body movement and locomotion.

Physical support on land depends on adaptations of body proportions and posture

An engineer designing a bridge or tall building must take into account the effects of changes in size, or scale. An increase in size from a small-scale model to the real thing has a significant effect on building design. Physical laws dictate that the strength of a building support depends on its cross-sectional area, which increases as the square of its diameter. In sharp contrast, the strain on the supports depends on the building's weight, which increases as the cube of its height or other linear dimension. In common with a bridge or a building, an animal's body "design" must account for the greater demand for support that comes with increasing size. Consistent with phys-

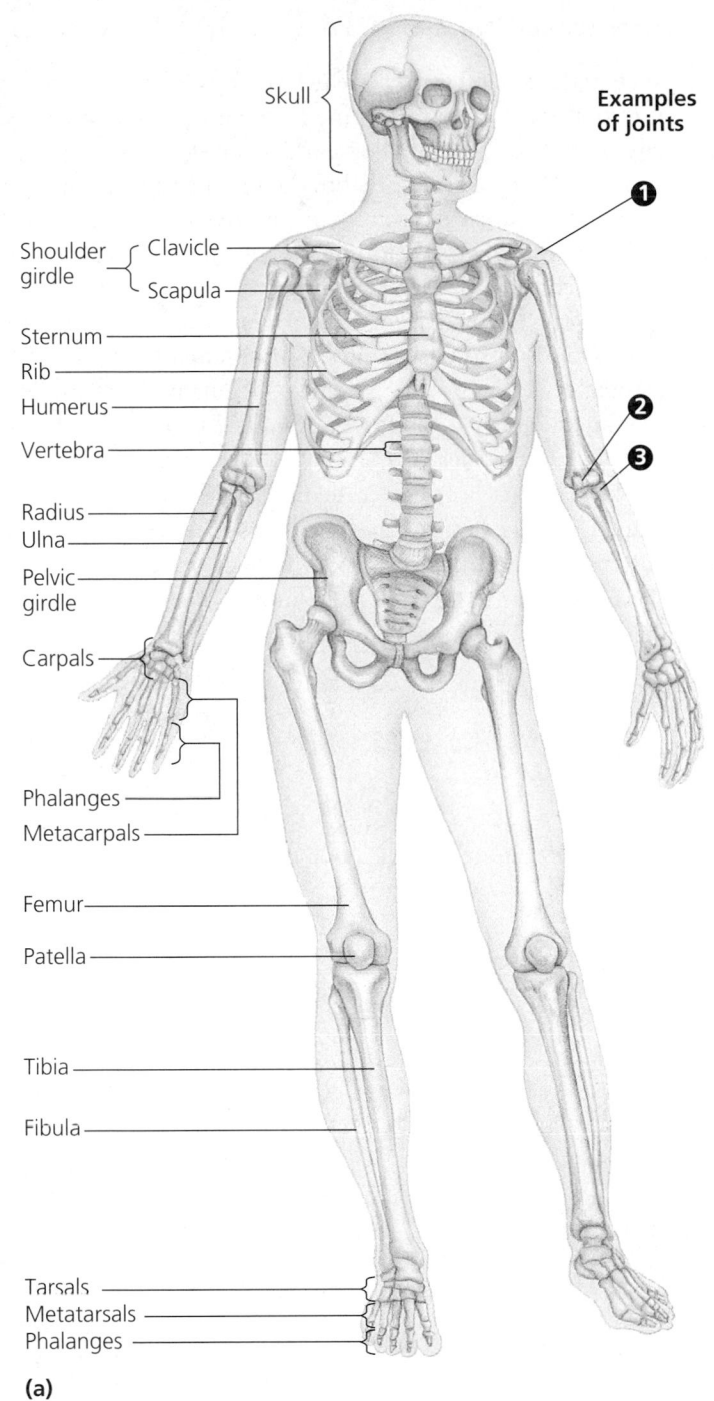

Skull

Shoulder girdle — Clavicle
Scapula

Sternum

Rib

Humerus

Vertebra

Radius
Ulna

Pelvic girdle

Carpals

Phalanges

Metacarpals

Femur

Patella

Tibia

Fibula

Tarsals
Metatarsals
Phalanges

(a)

Examples of joints

❶

❷
❸

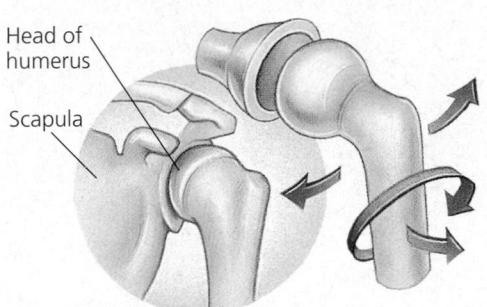

Head of humerus

Scapula

❶ Ball-and-socket joint. Ball-and-socket joints, where the humerus contacts the shoulder girdle and where the femur contacts the pelvic girdle, enable us to rotate our arms and legs and move them in several planes.

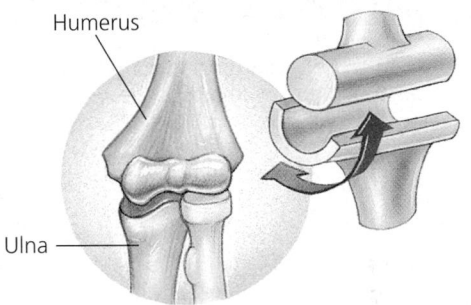

Humerus

Ulna

❷ Hinge joint. Hinge joints, such as between the humerus and the head of the ulna, restrict movement to a single plane.

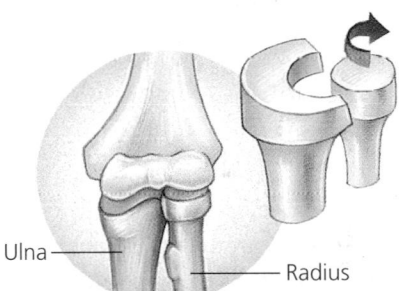

Ulna

Radius

❸ Pivot joint. A pivot joint allows us to rotate the forearm at the elbow.

(b)

FIGURE 49.28 The human skeleton. (a) The axial skeleton (green) provides an axis of support for the upright (bipedal) body and surrounds and protects the brain, spinal cord, lungs, and heart. The appendicular skeleton (gold) supports the arms and legs. Circled numbers indicate locations of different types of joints, described in part b. **(b)** Joints allow great flexibility in body movement, indicated by arrows. Hinge and pivot joints between bones in our wrists and hands enable us to make precise manipulations.

ical laws, a large animal, such as an elephant, has very different body proportions than a small animal, such as a mouse. Imagine a mouse, with its very slender legs, scaled up to elephant size. If the imaginary animal kept its mouselike body proportions, its legs would buckle under its weight.

In simply applying the building analogy, we might predict the size of an elephant's leg bones to be directly proportional to the strain imposed by its body weight. However, our prediction would be inaccurate; an animal's body is complex and nonrigid, and the building analogy only partly explains the relationship between animal body design and support. The size of an animal's legs relative to its body size is only part of the story. It turns out that body posture—the position of the legs relative to the main body—is a more important structural

FIGURE 49.29 Posture helps support large land vertebrates. A coyote stands, walks, and runs with most of its weight supported by muscles and tendons, which hold its legs directly under the body.

Structure and Function of Vertebrate Skeletal Muscle

Vertebrate **skeletal muscle,** which is attached to the bones and is responsible for their movement, is characterized by a hierarchy of smaller and smaller parallel units (FIGURE 49.31). A skeletal muscle consists of a bundle of long fibers running the length of the muscle. Each fiber is a single cell with many nuclei, reflecting its formation by the fusion of many embryonic cells. Each fiber is itself a bundle of smaller **myofibrils** arranged longitudinally. The myofibrils, in turn, are composed of two kinds of **myofilaments. Thin filaments** consist of two strands of actin and one strand of regulatory protein coiled around one another, while **thick filaments** are staggered arrays of myosin molecules.

Skeletal muscle is also called striated muscle because the regular arrangement of the myofilaments creates a repeating pattern of light and dark bands. Each repeating unit is a **sarcomere,** the basic contractile unit of the muscle. The borders of the sarcomere, the **Z lines,** are lined up in adjacent

feature in supporting body weight, at least in mammals and birds (FIGURE 49.29). Muscles and tendons (connective tissue joining muscles to bones), which hold the legs of elephants, coyotes, humans, and other large mammals relatively straight and under the body, bear most of the load.

Muscles move skeletal parts by contracting

As we mentioned earlier, animal movement is based on the contraction of muscles working against some type of skeleton. The action of a muscle is always to contract; muscles can extend only passively.

The ability to move parts of the body in opposite directions requires that muscles be attached to the skeleton in antagonistic pairs, each muscle working against the other (FIGURE 49.30). We flex our arm, for instance, by contracting the biceps, with the hinged joint of the elbow acting as the fulcrum of a lever. To extend the arm, we relax the biceps while the triceps on the opposite side contracts. But how does a muscle actually contract? As always, the key to function is structure. In this section we will examine the structure and mechanism of contraction of vertebrate skeletal muscle and then compare this basic pattern with other types of muscle.

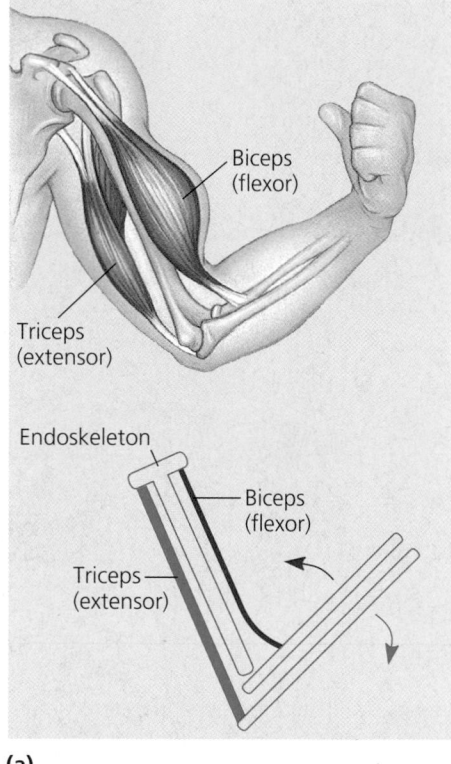

(a)

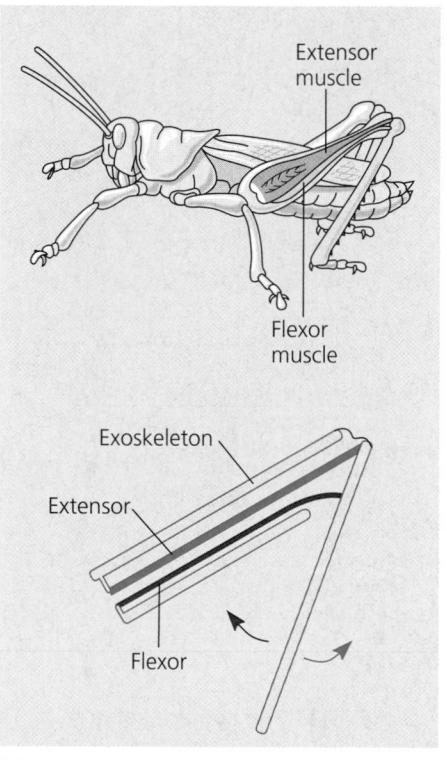

(b)

FIGURE 49.30 The cooperation of muscles and skeletons in movement. Muscles actively contract, but they elongate only when passively stretched. Back-and-forth movement is generally accomplished by antagonistic muscles, each working against the other. This arrangement works with either an endoskeleton or an exoskeleton. **(a)** In humans, contraction of the biceps muscle, represented by red in the bottom diagram, raises (flexes) the forearm. Contraction of the triceps muscle (green) lowers (extends) the forearm. **(b)** Although arthropod muscles are positioned differently and housed within an exoskeleton, the antagonistic action of flexors and extensors is similar to that of a vertebrate. When the flexor muscle (red) in the upper part of a grasshopper's leg contracts, the lower leg is pulled toward the body. In this position the grasshopper is sitting, poised for a jump, as shown here. Alternatively, when the extensor muscle (green) in its upper leg contracts, the leg jerks backward, sending the insect into the air.

in the center of the A band contains only thick filaments. This arrangement of thick and thin filaments is the key to how the sarcomere, and hence the whole muscle, contracts.

Interactions between myosin and actin generate force during muscle contractions

When a muscle contracts, the length of each sarcomere is reduced; that is, the distance from one Z line to the next becomes shorter. In the contracted sarcomere, the A bands do not change in length, but the I bands shorten and the H zone disappears (FIGURE 49.32). These changes can be explained by

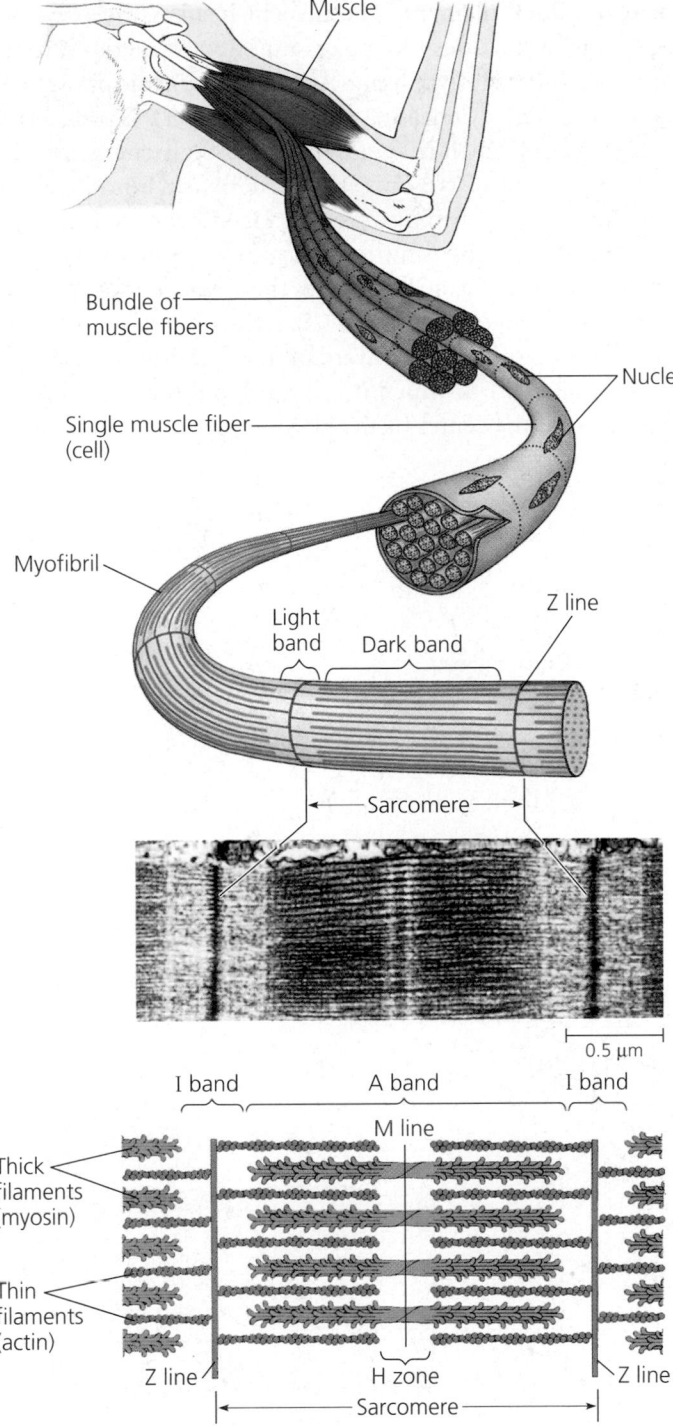

FIGURE 49.31 The structure of skeletal muscle.

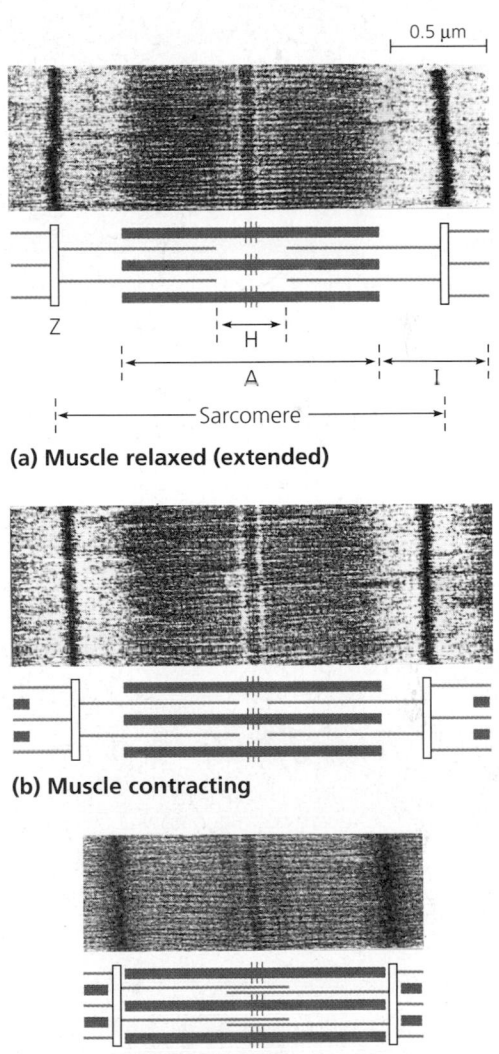

(a) Muscle relaxed (extended)

(b) Muscle contracting

(c) Muscle contracted

FIGURE 49.32 The sliding-filament model of muscle contraction. As seen in these transmission electron micrographs, the lengths of the thick (myosin) filaments (purple) and thin (actin) filaments (orange) remain the same as contraction occurs. (a) In a relaxed muscle, the length of each sarcomere is greater than in a contracting or contracted muscle. (b) During contraction, thick and thin filaments slide past each other, shortening the sarcomere. (c) When the muscle is fully contracted, the sarcomere is markedly shortened; the thin filaments overlap, and there is little or no space between the ends of the thick filaments and the Z lines.

myofibrils and contribute to the striations visible with a light microscope. The thin filaments are attached to the Z lines and project toward the center of the sarcomere, while the thick filaments are centered in the sarcomere. At rest, the thick and thin filaments do not overlap completely, and the area near the edge of the sarcomere where there are only thin filaments is called the **I band.** The **A band** is the broad region that corresponds to the length of the thick filaments. The thin filaments do not extend completely across the sarcomere, so the H zone

the **sliding-filament model** of muscle contraction. According to this model, neither the thin filaments nor the thick filaments change in length when the muscle contracts; rather, the filaments slide past each other longitudinally, so that the degree of overlap of the thin and thick filaments increases. If the region of overlap increases, both the length occupied only by thin filaments (the I band) and the length occupied only by thick filaments (the H zone) must decrease.

The sliding of the filaments is based on the interaction of the actin and myosin molecules that make up the thin and thick filaments. Myosin consists of a long, fibrous "tail" region, with a globular "head" region sticking off to the side. The tail is where the individual myosin molecules cohere to

form the thick filament. The myosin head is the center of bioenergetic reactions that power muscle contractions. It can bind ATP and hydrolyze it into ADP and inorganic phosphate. Some of the energy released by cleaving the ATP is transferred to the myosin, which changes shape to a high-energy configuration (FIGURE 49.33). This energized myosin binds to a specific site on actin, forming a cross-bridge. The stored energy is released, and the myosin head relaxes to its low-energy configuration. This relaxation changes the angle of attachment of the myosin head to the myosin tail. According to this hypothesis, as the myosin bends inward on itself, it exerts tension on the thin filament to which it is bound, pulling the thin filament toward the center of the sarcomere. The bond between

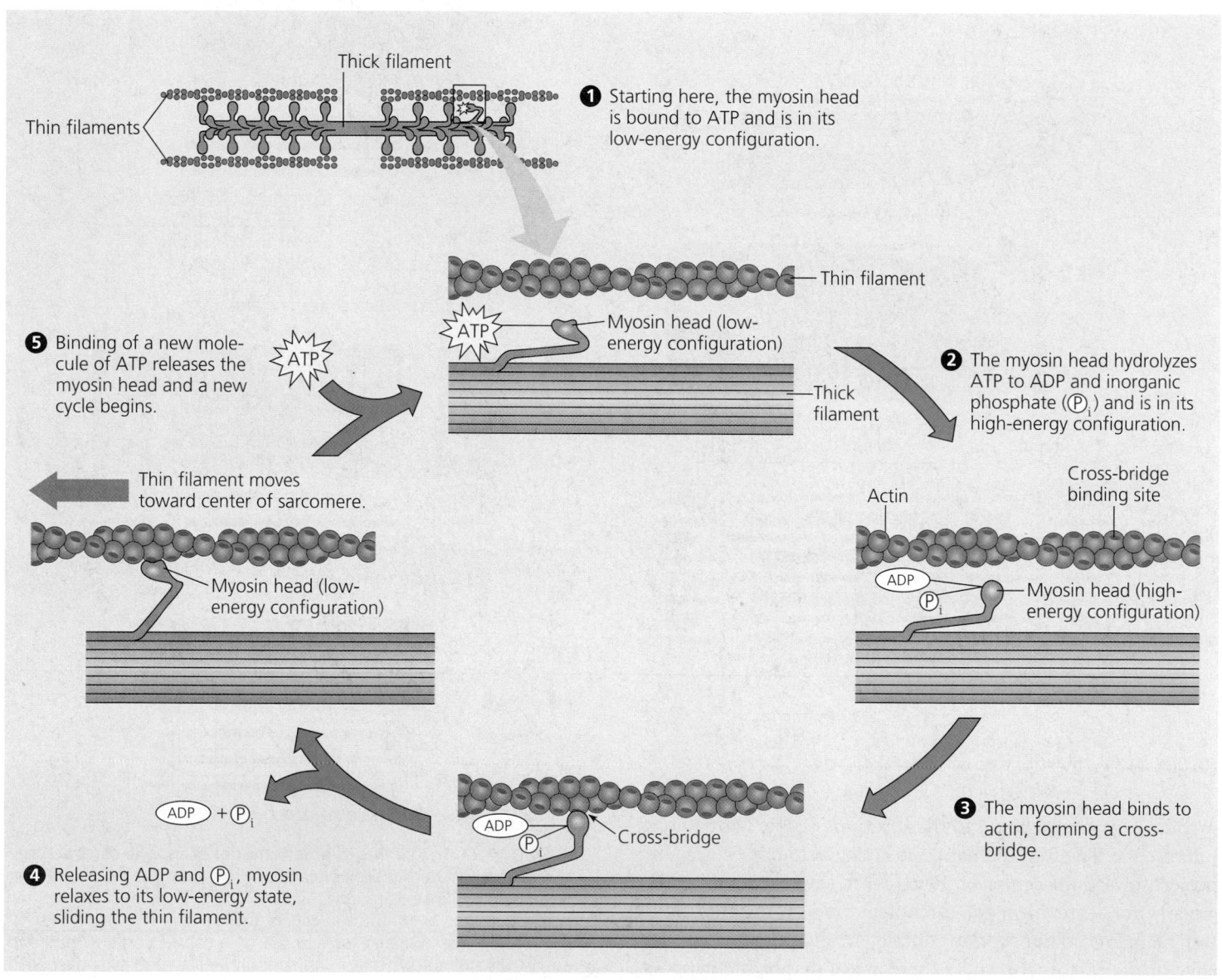

FIGURE 49.33 One hypothesis for how myosin-actin interactions generate the force for muscle contraction.

the low-energy myosin and actin is broken when a new molecule of ATP binds to the myosin head. In a repeating cycle, the free head can then cleave the new ATP to revert to the high-energy configuration and attach to a new binding site on another actin molecule farther along the thin filament. Each of the approximately 350 heads of a thick filament forms and reforms about five cross-bridges per second, driving filaments past each other.

A muscle cell typically stores only enough ATP for a few contractions. Muscle cells also store glycogen between the myofibrils, but most of the energy needed for repetitive muscle contraction is stored in substances called phosphagens. Creatine phosphate, the phosphagen of vertebrates, can supply a phosphate group to ADP to make ATP.

Calcium ions and regulatory proteins control muscle contraction

A skeletal muscle contracts only when stimulated by a motor neuron. When the muscle is at rest, the myosin binding sites on the actin molecules are blocked by the regulatory protein **tropomyosin.** Another set of regulatory proteins, the **troponin complex,** controls the position of tropomyosin on the thin filament (FIGURE 49.34). For a muscle cell to contract, the myosin binding sites on the actin must be uncov-

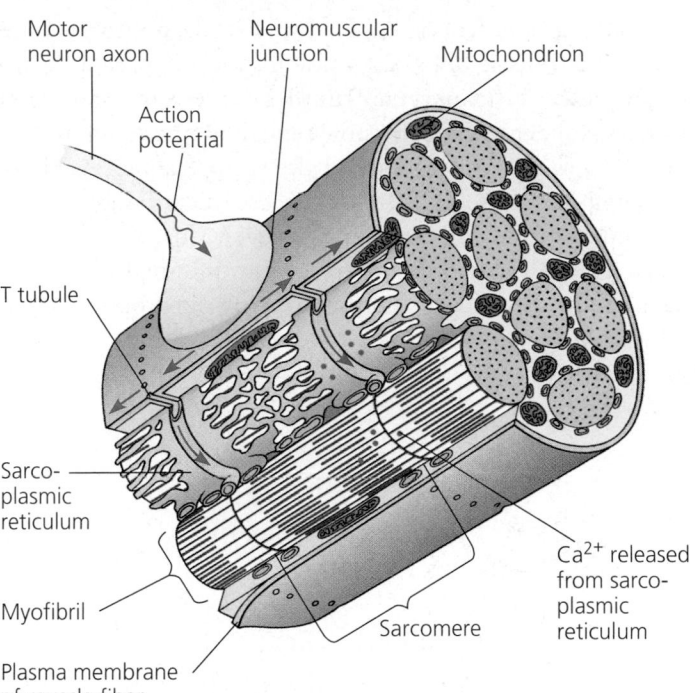

FIGURE 49.35 The roles of the muscle fiber's sarcoplasmic reticulum and T tubules in contraction. Diffusing across the neuromuscular junction, acetylcholine depolarizes the plasma membrane of the muscle fiber, and action potentials (blue arrows) sweep across the fiber and deep into it along T (transverse) tubules. Within the muscle cell the action potentials trigger the release of Ca^{2+} (green dots) from the sarcoplasmic reticulum into the cytoplasm. The Ca^{2+} initiates the sliding of filaments by triggering the binding of myosin to actin.

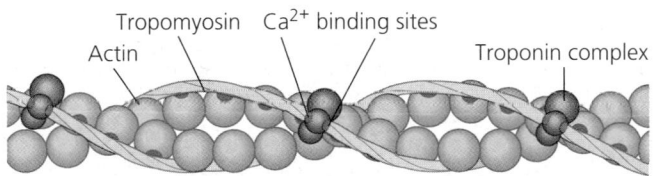

(a) Myosin binding sites blocked; muscle cannot contract

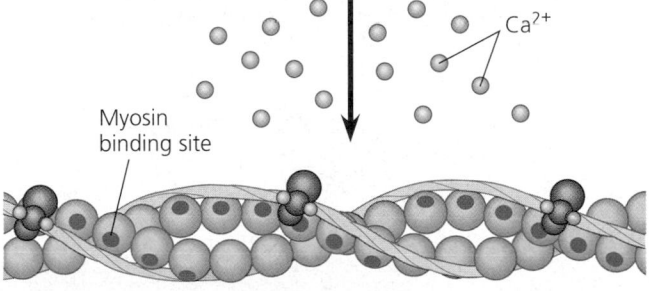

(b) Myosin binding sites exposed; muscle can contract

FIGURE 49.34 Hypothetical mechanism for the control of muscle contraction. The thin filament has two strands of actin twisted into a helix. **(a)** When the muscle is at rest, a long, rodlike tropomyosin molecule blocks the myosin binding sites that are instrumental in forming cross-bridges. **(b)** When another protein complex, troponin, binds calcium ions, the binding sites on actin are exposed, cross-bridges with myosin can form, and the muscle contracts.

ered. This occurs when calcium ions bind to troponin, altering the interaction between troponin and tropomyosin. The Ca^{2+} binding rearranges the tropomyosin-troponin complex, exposing the myosin binding sites on actin. When calcium is present, the sliding of thin and thick filaments can occur, and the muscle contracts. When calcium concentration in the cytosol falls, the binding sites of actin are covered, and contraction stops.

Calcium concentration in the cytosol of the muscle cell is regulated by the **sarcoplasmic reticulum,** a specialized endoplasmic reticulum (FIGURE 49.35). The membrane of the sarcoplasmic reticulum actively transports calcium from the cytosol into the interior of the reticulum, which is thus an intracellular storehouse for calcium.

The stimulus leading to the contraction of a skeletal muscle cell is an action potential in a motor neuron that makes a synaptic connection with the muscle cell. The synaptic terminal of the motor neuron releases the neurotransmitter acetylcholine at the neuromuscular junction, depolarizing the postsynaptic muscle cell and triggering an action potential in the muscle cell. That action potential is the signal for contraction. The action potential spreads deep into the interior of the muscle cell along infoldings of the plasma membrane called **T (transverse) tubules.** Where the transverse tubules contact the sarcoplasmic reticulum, the action potential changes the permeability of the sarcoplasmic reticulum, causing it to release calcium ions. These calcium ions bind to troponin, allowing the muscle to contract. Muscle contraction stops when the sarcoplasmic reticulum pumps the calcium back out of the cytosol, and the tropomyosin-troponin complex again blocks the myosin binding sites as the concentration of calcium falls (FIGURE 49.36).

Diverse body movements require variation in muscle activity

Everyday experience suggests that the action of a whole muscle, such as the biceps, is graded; we can voluntarily alter the extent and strength of contraction. Experimental studies also confirm that whole-muscle contractions are graded. However, at the cellular level, any stimulus that depolarizes the plasma membrane of a single muscle fiber triggers an all-or-none contraction, analogous to the response of neurons to depolarizing stimuli. How does the nervous system produce graded

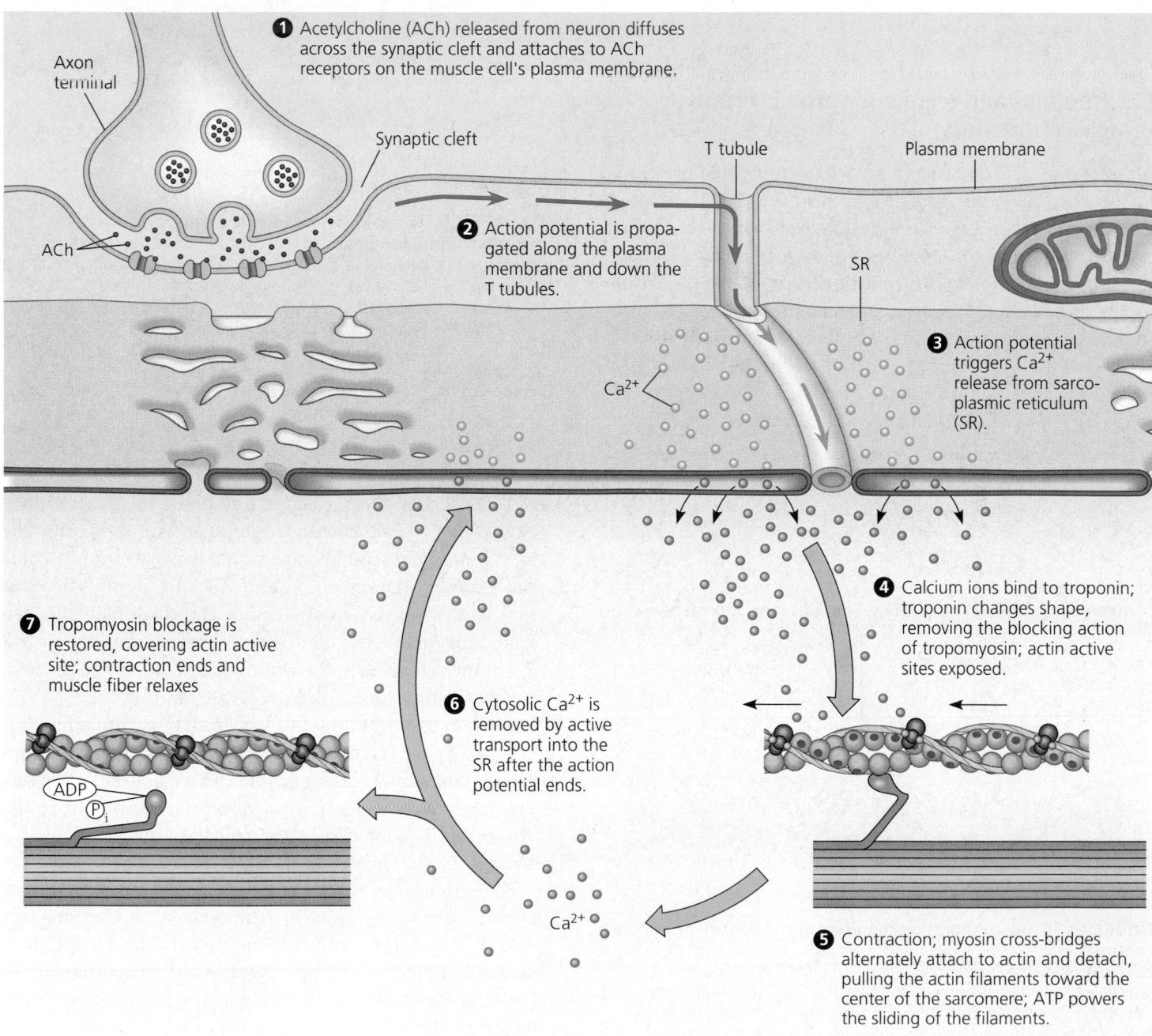

1 Acetylcholine (ACh) released from neuron diffuses across the synaptic cleft and attaches to ACh receptors on the muscle cell's plasma membrane.

Axon terminal

Synaptic cleft

T tubule

Plasma membrane

ACh

2 Action potential is propagated along the plasma membrane and down the T tubules.

SR

Ca²⁺

3 Action potential triggers Ca²⁺ release from sarcoplasmic reticulum (SR).

7 Tropomyosin blockage is restored, covering actin active site; contraction ends and muscle fiber relaxes

ADP
Pᵢ

6 Cytosolic Ca²⁺ is removed by active transport into the SR after the action potential ends.

4 Calcium ions bind to troponin; troponin changes shape, removing the blocking action of tropomyosin; actin active sites exposed.

Ca²⁺

5 Contraction; myosin cross-bridges alternately attach to actin and detach, pulling the actin filaments toward the center of the sarcomere; ATP powers the sliding of the filaments.

FIGURE 49.36 Review of skeletal muscle contraction.

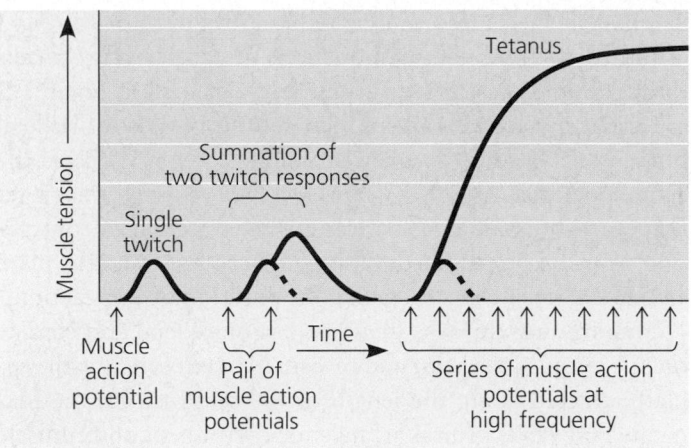

FIGURE 49.37 Temporal summation of muscle cell contractions. This graph compares the tension developed in a muscle in response to a single action potential, a pair of action potentials, and a series of action potentials. The dashed lines show the response that would have resulted if only the first action potential had occurred.

muscle by both determining how many motor units are activated at a given instant and selecting whether large or small motor units are activated. Tension in a muscle can be progressively increased by activating more and more of the motor neurons controlling the muscle, a process called **recruitment** of motor neurons. Depending on the number and size of motor neurons your brain recruits to the task, you can lift a fork, or something much heavier, like your biology textbook.

Some muscles, especially those that hold the body up and maintain posture, are almost always partially contracted. However, prolonged contraction results in muscle fatigue caused by the depletion of ATP, dissipation of the ion gradients required for normal electrical signaling, and the accumulation of lactate (see FIGURE 9.17). In a mechanism that avoids fatigue in postural muscles, the nervous system alternates activation among the various motor units making up the muscle, so that different motor units take turns maintaining the prolonged contraction.

contractions of whole muscles? One way is by varying the frequency of action potentials in the motor neurons controlling the muscle. A single action potential will produce an increase in muscle tension lasting about 100 msec or less, a single twitch (FIGURE 49.37). If a second action potential arrives before the response to the first is over, the tension will sum and produce a greater response. If a muscle receives an overlapping series of action potentials, further summation will occur, with the level of tension depending on the rate of stimulation. And if the rate of stimulation is fast enough, the twitches will blur into one smooth and sustained contraction called **tetanus** (not to be confused with the disease of the same name). Motor neurons usually deliver their action potentials in rapid-fire volleys, and the resulting summation of tension results in smooth contraction typical of tetanus rather than the jerky actions of individual muscle twitches.

The nervous system can also produce graded contraction of a whole muscle by taking advantage of the organization of the muscle cells into motor units. In a vertebrate muscle, each muscle cell is innervated by only one motor neuron, but each branched motor neuron may make synaptic connections with many muscle cells (FIGURE 49.38). There may be hundreds of motor neurons controlling an individual muscle, each with its own pool of muscle fibers scattered throughout the muscle. A **motor unit** consists of a single motor neuron and all the muscle fibers it controls. When the motor neuron fires, all the muscle fibers in the motor unit contract as a group. The strength of the resulting contraction will therefore depend on how many muscle fibers the motor neuron controls. In most muscles there is wide variation in the number of muscle fibers among motor units; some motor neurons may control only a few muscle cells, while others may control hundreds. The nervous system can thus regulate the strength of contraction in the whole

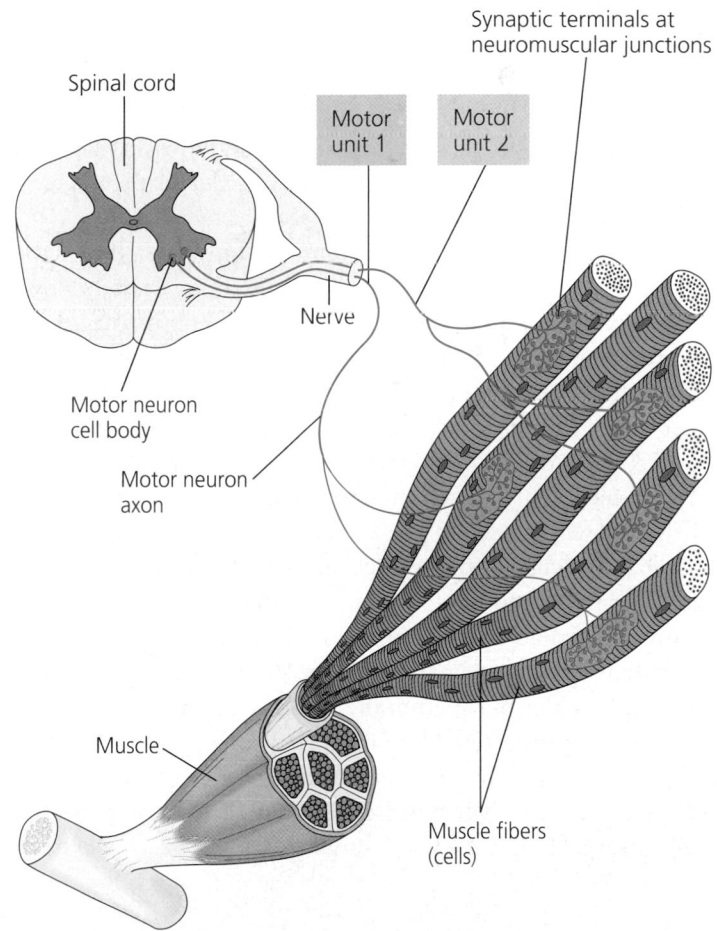

FIGURE 49.38 Motor units in a vertebrate muscle. Each muscle fiber (cell) has a single neuromuscular junction, or synaptic connection, with the motor neuron that controls it. However, each motor neuron typically branches and controls several or many muscle fibers. A motor neuron and all the fibers it controls constitute a contractile apparatus called a motor unit.

Fast and Slow Muscle Fibers

We have seen that the action potential in a skeletal muscle fiber is only a trigger for the contraction; the actual duration of the contraction is controlled by how long the calcium concentration in the cytosol remains elevated. Not all skeletal muscle fibers are identical in this regard. We can identify fast and slow fibers based on the duration of their twitches. **Fast muscle fibers** are used for short, rapid, powerful contractions. In contrast, **slow muscle fibers,** often found in muscles that maintain posture, can sustain long contractions. A slow fiber has less sarcoplasmic reticulum and slower calcium pumps than a fast fiber, so calcium remains in the cytosol longer. This causes a twitch in a slow fiber to last about five times longer than in a fast fiber. Slow fibers are also specialized to make use of a steady supply of energy; they have many mitochondria, a rich blood supply, and an oxygen-storing protein called myoglobin. **Myoglobin,** the brownish-red pigment in the dark meat of poultry and fish, binds oxygen more tightly than hemoglobin, so it can effectively extract oxygen from the blood.

Other Types of Muscle

There are many types of muscles in the animal kingdom, but as noted before, they all share the same fundamental mechanism of contraction: the sliding of actin filaments and myosin filaments past one another. In addition to skeletal muscle, vertebrates have cardiac muscle and smooth muscle (see FIGURE 40.4).

Vertebrate **cardiac muscle** is found in only one place—the heart. Like skeletal muscle, cardiac muscle is striated. However, structural differences between skeletal and cardiac muscle result in differences in their electrical and membrane properties. The junctions between cardiac muscle cells contain specialized regions called **intercalated discs,** where gap junctions (see FIGURE 7.30) provide direct electrical coupling among cells. Thus, an action potential generated in one part of the heart will spread to all the cardiac muscle cells, and the whole heart will contract. Skeletal muscle cells will not contract unless triggered to do so by input from a controlling motor neuron. Cardiac muscle cells, however, can generate action potentials on their own, without any input from the nervous system. The plasma membrane of a cardiac muscle cell has pacemaker properties that cause rhythmic depolarizations, triggering action potentials and causing single cardiac muscle cells to "beat" even when isolated from the heart and placed in cell culture. (But at the whole-organ level, the heart also has a pacemaker, specialized muscle tissue in the wall of the right atrium that coordinates contractions of cardiac muscle cells throughout the heart; see FIGURE 42.7.) The action potentials of cardiac muscle cells are also different from those of skeletal muscle cells, lasting up to twenty times longer. Whereas the action potential of a skeletal muscle cell serves only as a trigger for contraction and does not control the duration of contraction, in a cardiac cell the duration of the action potential plays an important role in controlling the duration of contraction.

Smooth muscle lacks the striations of skeletal and cardiac muscle because the actin and myosin filaments are not all regularly arrayed along the length of the cell. Instead, the filaments may have a spiral arrangement within smooth muscle cells. Smooth muscle also contains less myosin than striated muscle, and the myosin is not associated with specific actin strands. Further, smooth muscle has neither a T tubule system nor a well-developed sarcoplasmic reticulum. Calcium ions must enter the cytosol via the plasma membrane during an action potential, and the amount reaching the filaments is rather small. Contractions are relatively slow, but there is a greater range of control than in striated muscle. Also, smooth muscle can contract over a much greater range of lengths than striated muscle. Smooth muscle is found mainly in the walls of hollow organs such as digestive tract organs and blood vessels.

Invertebrates possess muscle cells similar to vertebrate skeletal and smooth muscle cells. Arthropod skeletal muscles are nearly identical to vertebrate skeletal muscles. However, the flight muscles of insects are capable of independent, rhythmic contraction, so the wings of some insects can actually beat faster than action potentials can arrive from the central nervous system. Another interesting evolutionary adaptation has been discovered in the muscles that hold clam shells closed. The thick filaments of these muscle fibers contain a unique protein called paramyosin that enables the muscles to remain in a fixed state of contraction with a low rate of energy consumption for as long as a month.

■   ■   ■

Although we have discussed sensory receptors and muscles separately in this chapter, they are part of a single integrated system linking together brain, body, and the external world. An animal's behavior, how the animal interacts with its environment, is the product of this system. Behavior is discussed in Unit Eight within the broader context of ecology, the study of interactions between organisms and their environment.

CHAPTER 49 REVIEW

Go to the Campbell Biology website (www.campbellbiology.com) to explore an interactive version of the Chapter Review.

Summary of Key Concepts

SENSING, ACTING, AND BRAINS

- The brain's processing of sensory input and motor output is cyclical rather than linear (p. 1058, FIGURE 49.1) Increasingly complex and varied sensory receptors, effectors, and brains evolved in tandem, linking animal behaviors to the external world. Integration is a constant activity, with sensory receptors reporting to ongoing cycles of brain activity the results of recent movements, and then future movements being appropriately planned.

INTRODUCTION TO SENSORY RECEPTION

- Sensory receptors transduce stimulus energy and transmit signals to the nervous system (pp. 1059–1060, FIGURE 49.2) Sensory receptors are usually modified neurons or epithelial cells that detect environmental stimuli and respond with an electrochemical change in their membrane. Exteroreceptors detect external stimuli; interoreceptors detect internal stimuli. Sensory transduction is the conversion of stimulus energy into a change in a membrane potential called a receptor potential. Signal transduction pathways in receptor cells relay and amplify the signal, leading to neurotransmitter release from the receptor cell.

- Sensory receptors are categorized by the type of energy they transduce (pp. 1060–1063, FIGURES 49.3–49.6) Mechanoreceptors respond to stimuli such as pressure, touch, stretch, motion, and sound. Pain is detected by a group of diverse receptors that respond to excess temperature, pressure, or specific classes of chemicals. Various types of thermoreceptors signal surface and core temperatures of the body. Chemoreceptors respond either to generalized solute concentrations or to specific molecules. Electromagnetic receptors detect energy in the form of different wavelengths of radiation.

PHOTORECEPTORS AND VISION

- A diversity of photoreceptors has evolved among invertebrates (pp. 1063–1064, FIGURES 49.7, 49.8) The light receptors and visual capabilities of invertebrates vary widely, ranging from the simple light-sensitive eye cup of planarians to the image-forming compound eye of insects and crustaceans and the single-lens eye of some jellies, spiders, and many mollusks.

- Vertebrates have single-lens eyes (pp. 1064–1065, FIGURES 49.9, 49.10) The main parts of the vertebrate eye are an outer layer, the sclera, including the transparent cornea; the conjunctiva, a mucous membrane, surrounding all of the sclera except the cornea; the choroid, a pigmented middle layer, which includes the iris, surrounding the pupil; the retina, an inner layer at the back of the eyeball, containing the photoreceptor cells; and the lens, suspended between two chambers in the eye, which focuses light on the retina.

Web/CD Activity 49A: *Structure and Function of the Eye*

- The light-absorbing pigment rhodopsin triggers a signal-transduction pathway (pp. 1065–1067, FIGURES 49.11–49.14) Transduction of the light signal occurs in specialized photoreceptors called rods and cones, which contain light-absorbing retinal bonded to specific membrane proteins, collectively called opsins. When rods and cones absorb light, signal-transduction pathways hyperpolarize their membrane, and they release less neurotransmitter.

- The retina assists the cerebral cortex in processing visual information (pp. 1067–1069, FIGURES 49.15, 49.16) Chemical messages are transmitted from rods and cones to bipolar cells and then to ganglion cells, whose axons in the optic nerve convey action potentials to the brain. Other neurons in the retina integrate information before it is sent to the brain. Most axons of the optic nerves go to the lateral geniculate nuclei of the thalamus, from which neurons lead to the primary visual cortex. Several integrating centers in the cerebral cortex are active in creating visual perceptions.

HEARING AND EQUILIBRIUM

- The mammalian hearing organ is within the inner ear (pp. 1069–1070, FIGURES 49.17, 49.18) The tympanic membrane (eardrum) transmits sound waves to three small bones of the middle ear, which transmit the waves through the oval window to the fluid in the coiled cochlea of the inner ear. Pressure waves vibrate the basilar membrane and the attached organ of Corti, which contains receptor hair cells. The bending of the hairs against the tectorial membrane depolarizes the hair cells, triggering action potentials in the auditory nerve to the brain. Regions of the basilar membrane vibrate more vigorously at different frequencies and transmit to specific auditory areas of the cerebral cortex.

- The inner ear also contains the organs of equilibrium (pp. 1070–1072, FIGURE 49.19) The utricle, saccule, and three semicircular canals in the inner ear function in balance and equilibrium.

- A lateral line system and inner ear detect pressure waves in most fishes and aquatic amphibians (pp. 1072–1073, FIGURE 49.20) The detection of water movement in fishes and aquatic amphibians is accomplished by a lateral line system of clustered hair cell receptors.

- Many invertebrates have gravity sensors and are sound-sensitive (p. 1073, FIGURES 49.21, 49.22) Many arthropods have vibrating exoskeletal hairs and localized "ears," consisting of a tympanic membrane and receptor cells, to sense sounds. Some invertebrates detect position in space by means of statocysts.

CHEMORECEPTION—TASTE AND SMELL

- Perceptions of taste and smell are usually interrelated (pp. 1074–1075, FIGURES 49.23, 49.24) Taste and smell both depend on the stimulation of receptor cells by small, dissolved molecules that bind to proteins on a chemoreceptor membrane. In mammals, taste receptors are organized into taste buds that respond to distinct shapes of molecules. Olfactory receptor cells line the upper part of the nasal cavity and send their axons to the olfactory bulb of the brain.

MOVEMENT AND LOCOMOTION

- **Locomotion requires energy to overcome friction and gravity** (pp. 1075–1077, FIGURES 49.25, 49.26) Overcoming friction is a major problem for swimmers; gravity is less of a problem for swimming animals than for those that move on land or fly. Walking, running, hopping, or crawling on land requires an animal to support itself and move against gravity. Flight requires that wings develop enough lift to overcome the downward force of gravity.

- **Skeletons support and protect the animal body and are essential to movement** (pp. 1077–1078, FIGURES 49.27, 49.28) A hydrostatic skeleton, found in most cnidarians, flatworms, nematodes, and annelids, consists of fluid under pressure in a closed body compartment. Exoskeletons, found in most mollusks and arthropods, are hard coverings deposited on the surface of an animal. Endoskeletons, found in sponges, echinoderms, and chordates, are hard supporting elements embedded within the animal's body.

  Web/CD Activity 49B: *Human Skeleton*

- **Physical support on land depends on adaptations of body proportions and posture** (pp. 1078–1080, FIGURE 49.29) In addition to the skeleton, partially contracted muscles help support large land vertebrates.

- **Muscles move skeletal parts by contracting** (pp. 1080–1081, FIGURES 49.30, 49.31) Muscles, often present in antagonistic pairs, contract and pull against the skeleton to provide movement. Vertebrate skeletal muscle consists of a bundle of muscle cells, each of which contains myofibrils composed of thin filaments of actin and thick filaments of myosin.

  Web/CD Activity 49C: *Skeletal Muscle Structure*

- **Interactions between myosin and actin generate force during muscle contractions** (pp. 1081–1083, FIGURES 49.32, 49.33) The energy to move myosin heads is provided by ATP. Energized myosin heads bind to actin, forming cross-bridges. Bending of the myosin heads pulls the thin filaments (actin) toward the center of the sarcomere, producing a muscle contraction. When ATP binds to the myosin heads, they release, ready to start a new cycle.

  Web/CD Activity 49D: *Muscle Contraction*

- **Calcium ions and regulatory proteins control muscle contraction** (pp. 1083–1084, FIGURES 49.34–49.36) Contraction begins when impulses from a motor neuron are transmitted to the muscle cell membrane through release of acetylcholine at the neuromuscular junction. Action potentials travel to the interior of the cell along the T tubules, stimulating the release of calcium ions from the sarcoplasmic reticulum. The calcium ions bind to the regulatory troponin-tropomyosin complex on the thin filaments, exposing the myosin binding sites on the actin; muscle contraction proceeds.

  Web/CD Case Study in the Process of Science: *How Do Electrical Stimuli Affect Muscle Contraction?*

- **Diverse body movements require variation in muscle activity** (pp. 1084–1086, FIGURES 49.37, 49.38) A muscle twitch results from a single stimulus. More rapidly delivered signals produce a graded contraction by summation. Tetanus is a state of smooth and sustained contraction obtained when motor neurons deliver a volley of action potentials. A motor unit consists of a branched motor neuron and the muscle fibers it innervates. Multiple motor unit recruitment results in stronger contractions. Cardiac muscle, found only in the heart, consists of striated, branching cells that are electrically connected by intercalated discs. Cardiac muscle cells can generate action potentials without neural input. In smooth muscle, contractions are slow but can be sustained over long periods of time.

Self-Quiz

1. Which of the following receptors is incorrectly paired with its category?
   a. hair cell—mechanoreceptor
   b. muscle spindle—mechanoreceptor
   c. taste receptor—chemoreceptor
   d. rod—electromagnetic receptor
   e. gustatory receptor—electromagnetic receptor

2. Some sharks close their eyes just before they bite. Although they cannot see their prey, their bites are on target. Researchers have noted that sharks often misdirect their bites at metal objects, and they can find batteries buried under the sand of an aquarium. This evidence suggests that sharks keep track of their prey during the split second before they bite in the same way that
   a. a rattlesnake finds a mouse in its burrow.
   b. a male silkworm moth locates a mate.
   c. a bat can find moths in the dark.
   d. a platypus locates its prey in a muddy river.
   e. a flatworm avoids light places.

3. Which of the following is an *incorrect* statement about the vertebrate eye?
   a. The vitreous humor regulates the amount of light entering the pupil.
   b. The transparent cornea is an extension of the sclera.
   c. The fovea is the center of the visual field and contains only cones.
   d. The ciliary muscle functions in accommodation.
   e. The retina lies just inside the choroid and contains the photoreceptor cells.

4. The transduction of sound waves to action potentials takes place
   a. within the tectorial membrane as it is stimulated by the hair cells.
   b. when hair cells are bent against the tectorial membrane, causing them to depolarize and release neurotransmitter molecules that stimulate sensory neurons.
   c. as the basilar membrane becomes more permeable to sodium ions and depolarizes, initiating an action potential in a sensory neuron.
   d. as the basilar membrane vibrates at different frequencies in response to the varying volume of sounds.
   e. within the middle ear as the vibrations are amplified by the malleus, incus, and stapes.

5. When light strikes the pigment rhodopsin in a rod cell, retinal dissociates from opsin, initiating a signal-transduction pathway that
   a. depolarizes the neighboring bipolar cells and initiates an action potential in a ganglion cell.
   b. depolarizes the rod cell, causing it to release the neurotransmitter glutamate, which excites bipolar cells.
   c. hyperpolarizes the rod cell, reducing its release of glutamate, which excites some bipolar cells and inhibits others.
   d. hyperpolarizes the rod cell, increasing its release of glutamate, which excites amacrine cells but inhibits horizontal cells.
   e. converts cGMP to GMP, opening sodium channels and hyperpolarizing the membrane, causing the rhodopsin to become bleached.

6. Clams and lobsters both have exoskeletons, but lobsters have much greater mobility. Why?
   a. Clams only have adductor muscles that hold the shell closed, whereas lobsters have both abductor and adductor muscles.
   b. The paramyosin of clam muscles holds them in a low-energy state of contraction, whereas lobster muscles are very similar to vertebrate striated muscles.
   c. Clams can only grow by adding to the outer edge of the shell, whereas lobsters molt and repeatedly replace their exoskeleton with a larger, more flexible one.
   d. The lobster skeleton can actively contract, while the clam skeleton lacks its own contractile mechanism.
   e. Lobsters have a jointed exoskeleton, allowing for the flexible movement of appendages and body parts at the joints.

7. The role of calcium in muscle contraction is
   a. to break the cross-bridges as a cofactor in the hydrolysis of ATP.
   b. to bind with troponin, changing its shape so that the myosin binding sites on the actin filament are exposed.
   c. to transmit the action potential across the neuromuscular junction.
   d. to spread the action potential through the T tubules.
   e. to reestablish the polarization of the plasma membrane following an action potential.

8. Tetanus refers to
   a. the partial sustained contraction of major supporting muscles.
   b. the all-or-none contraction of a single muscle fiber.
   c. a stronger contraction resulting from multiple motor unit summation.
   d. the result of wave summation, which produces a smooth and sustained contraction of a muscle.
   e. the state of muscle fatigue caused by the depletion of ATP and the accumulation of lactate.

9. Which of the following is a true statement about cardiac muscle cells?
   a. They lack an orderly arrangement of actin and myosin filaments.
   b. They have less extensive sarcoplasmic reticulum and thus contract more slowly than smooth muscle cells.
   c. They are connected by intercalated discs, through which action potentials spread to all cells in the heart.
   d. They have a resting potential more positive than an action potential threshold.
   e. They contract only when stimulated by neurons.

10. Which of the following changes occurs when a skeletal muscle contracts?
    a. The A bands shorten.
    b. The I bands shorten.
    c. The Z lines slide farther apart.
    d. The thin filaments contract.
    e. The thick filaments contract.

11. (a) What enables a receptor cell in a taste bud to key on just a certain type of chemical in food? (b) What determines the "taste" that is perceived in response to that particular chemical?

12. For each of the following senses in humans, identify the type of receptor: seeing; tasting; hearing; smelling.

13. Explain why our night vision is mostly in black and white rather than color.

14. How does the ear convert sound waves in the air into pressure waves of the fluid in the cochlea?

15. Contrast swimming with walking in terms of the forces an animal must overcome to move.

16. In exercising your arms or legs to strengthen muscles, why is it important to impose resistance during both flexing and extending of the limbs?

17. How does the sarcoplasmic reticulum help regulate muscle contraction?

Go to the website or CD-ROM for more self-quiz questions.

Evolution Connection

In general, locomotion on land requires more energy than locomotion in water. By integrating what you have learned throughout these chapters on animal functions, discuss some of the evolutionary adaptations of mammals that support the high energy requirements for moving on land.

The Process of Science

Although skeletal muscles generally fatigue fairly rapidly, clam shell muscles have a unique protein called paramyosin that allows them to sustain contraction for up to a month. From your knowledge of the cellular mechanism of contraction, propose a hypothesis to explain how paramyosin might work. How would you test your hypothesis experimentally?

Investigate how electrical stimuli affect muscle contraction in the Case Study in the Process of Science, available on the website and CD-ROM.

Science, Technology, and Society

You may know an older person who has broken a bone (often a hip) partly because of osteoporosis, a loss of bone density that affects many women after menopause. Researchers think that prevention is the best way to avoid osteoporosis. They recommend exercise and maximum calcium intake during the teenage years and the twenties. Is it realistic to expect young people to view themselves as future senior citizens? How would you recommend that they be encouraged to develop good health habits that might not pay off for 40 or 50 years?

Answers: 1. e; 2. d; 3. a; 4. b; 5. c; 6. e; 7. b; 8. d; 9. c; 10. b; 11. (a) The specific receptors on the membrane of the taste receptor cell. (b) The part of the brain stimulated by sensory neurons from the taste bud. 12. Photoreceptor; chemoreceptor; mechanoreceptor; chemoreceptor 13. Rods are more sensitive than cones to light, and thus the low light intensity at night stimulates far more rods than cones. 14. Sound waves in air cause the eardrum to vibrate. The small bones attached to the inside of the eardrum transmit this energy to the oval window on the wall of the inner ear. Vibrations of the oval window set in motion the fluid in the inner ear, which includes the cochlea. 15. Friction (drag) resists an animal moving through water, but gravity has little effect because of the animal's buoyancy. Air poses little resistance to an animal walking on land, but the animal must support itself against the force of gravity and accelerate appendages from standing starts. 16. This exercises both muscles of antagonistic pairs, which only do work when they are contracting. 17. By reversibly storing and releasing Ca^{2+}, the sarcoplasmic reticulum regulates the cytosolic concentration of this ion, which is required in the cytosol for the binding of myosin to actin.

ECOLOGY

In late April 2001, the municipal water supply of North Battleford, Saskatchewan became contaminated with Cryptosporidium, a parasitic protozoan present in the feces of humans and domesticated animals, such as cattle. Dozens of people became ill, and three died from drinking tap water. The probable cause was pollution of the North Saskatchewan River with cattle feces, along with a failure at the municipal water treatment plant that allowed the parasite into the town's water supply. A year earlier, 7 people died and 2,300 others became ill in Walkerton, Ontario, from drinking water contaminated with E. coli, an intestinal bacterium in humans and other mammals. Such water crises are no big surprise to David Schindler, one of the world's best-known freshwater ecologists.

Dr. Schindler is the Killam Memorial Professor of Ecology at the University of Alberta. His research centers on the dynamics of freshwater lakes, especially in boreal (far northern) environments. His studies have illuminated numerous threats to aquatic ecosystems, including phosphate pollution, acid precipitation, climate change, and changes in land use, such as deforestation and intensive livestock operations. His many awards include the Canada Gold Medal for Science and Engineering in 2001 and the first Stockholm Water Prize in 1991, considered the Nobel Prize for water science. Our interview reinforces one of this book's themes: the connection between science, technology, and society.

How did your interests as an ecologist become focused on freshwater lakes and boreal environments?

Well, the freshwater interest was natural. I grew up in lake country. As for the interest in the boreal, that developed fairly early, too. When I was 15, I wanted to go and see Canada. I'd read novels set in northwestern Ontario that were written around the turn of the century. The day I got my driver's license, I headed for Canada with an old German hired man as a chaperone. And I fell in love with the boreal. Never looked back.

Before Alberta, you were a government scientist at The Experimental Lakes Project

**Interview
DAVID SCHINDLER**

in northern Ontario, where your research included classic experiments that led to the banning of phosphates in detergents. Tell us about that research.

When the project started in 1968, our mandate was to test water management issues at the level of whole-lake ecosystems. The main objective was the study of eutrophication, the overfertilization of lakes with mineral nutrients. Eutrophication was causing algal blooms in the St. Lawrence, the Great Lakes, and a lot of European lakes. We started by selecting lakes in northern Ontario for our studies and setting up a main campsite there. The big issue at the time was whether phosphorus was the main culprit in eutrophication. Laboratory experiments implicated phosphorus, but my bosses at the time had a hard time convincing politicians and managers to invest millions or billions of dollars in phosphorus management schemes based solely on these small-scale experiments. Evidence from whole-lake experiments would be more convincing.

Why was there resistance to managing the input of phosphorus into freshwater ecosystems?

One of the big sources of phosphorus in those days was phosphate detergents, and there was a big political lobby defending the use of these detergents. The compa-

nies that produced detergents promoted any type of study that could demonstrate that something other than phosphorus was causing the eutrophication of lakes. These companies were pointing the finger at carbon as the main problem.

So how did you test whether it was carbon, phosphorus, or some other nutrient that was causing the eutrophication and algal blooms?

In our best-known experiment, we divided a lake into two basins. We added just carbon and nitrogen to one basin. We included phosphorus along with carbon and nitrogen in the other basin. We got a tremendous algal bloom within weeks after adding phosphorus, but no change with just the carbon and nitrogen. So it was pretty definite evidence that phosphorus was the culprit.

And that was the end of phosphate detergents?

We published our results in *Science* in 1974, and that pretty well set off a cascade of phosphorus regulations for detergents and sewage effluent. Eastern Canada was looking for a solution to the eutrophication problem, so the response there was quick. But it took 17 years to get all the U.S. states in the Great Lakes region on board. By then, most European countries had also implemented phosphorus restrictions.

That helped solve one problem. What are the most serious current threats to freshwater ecosystems?

There's a whole variety of problems. Acid precipitation is one. Warming of the climate is another. Changes in land use can affect lakes. And easier access to remote lakes is another problem. There are more people with more leisure time. With snowmobiles and four-wheel-drive ATVs, people can easily reach lakes that were inaccessible 20 years ago, dragging along all sorts of power equipment. Even our most remote lakes are being exploited as a result. For example, 80% of the walleye fisheries in Alberta have collapsed in the last ten years.

What impact does agriculture have on the water quality in rivers and lakes?

It can be substantial. For example, if you bulldoze a forest to pasture cows on the land, you increase the runoff of nutrients into the water four- or fivefold at least. And if you plow that land, plant crops, and add fertilizer, you increase the yield of nutrients to the water even more. This is an example of how changes in land use can affect lakes. And now we have these huge, intensive livestock operations—up to 30,000 cattle or 80,000 hogs. The manure has to be dealt with somewhere. Very few of these big livestock operations have sewage treatment. The animals have a lot of the same intestinal microbes that humans have. We would no longer think of discharging raw sewage into a river from a city like Edmonton without treating it. But we do it all the time with intensive livestock operations. We're not handling our fresh water very well. And I think a big problem is that we've tended to look at the water issues, such as land use and acid precipitation, in isolation, when it's usually a combination of factors damaging the ecosystems. And some of the factors affecting one lake may be different from the factors affecting another lake.

Is there much hope that politicians and government agencies responsible for water quality will be patient enough to base their decisions on such detailed evaluation of water resources?

Many of these decisions are made by small municipalities, usually in a science vacuum. The tendency is to look only at the positive side of the balance sheet. For example, if it looks like an intensive livestock operation or a new housing subdivision is going to make money for local businesses, the project is usually approved without much consideration for water issues or other environmental factors. People who make the decision may look upstream, but they never look downstream!

Can you elaborate on your point that many decisions affecting environmental quality are made in a scientific vacuum?

Most industries have big lobbying arms that make sure that their interests are in the politicians' ear all the time. Scientists don't do that. And politicians seldom think to ask scientists what they think about key issues. There's a lot of marvelous research done by the federal government and university scientists that's never used in environmental decisions. In Canada, the most water-rich country in the world, what happened at Walkerton and North Battleford has people yelling about our water crisis. Yet all the science we need to prevent or fix those problems has been established for 20 years. It was just ignored. The science ought to be on the table.

The public pressure on politicians must increase when environmental mishaps such as Walkerton and North Battleford make headline news.

Yes, as horrible as those incidents were, they did make people more aware that they can have big problems with freshwater resources, no matter how water-rich a country is. Maybe people are starting to make the connection that these waters are the sewers for the rest of the landscape we're mucking up. And when you have to drink from the sewer, you want to be pretty careful about what goes into it. I think people are going to start demanding some action. But I'm concerned that it may not be the right action. I see all this demand for higher water standards and better technology without much demand for increased protection of watersheds or better training for the people who operate the technology we do have. But I think scientists could influence how the public views the issue.

So, how *can* ecologists and other scientists have more impact on public policy?

First, I think most scientists should take some sort of course in public communication. I think we should expect scientists to be able to explain to the public any of their research results that bear on critical issues for society. Taxpayers pay most scientists' research budgets and most university salaries. So I think we ought to be giving more back.

It must help a lot when there are some politicians who are actually environmental activists themselves.

Of course. For example, a young, ambitious member of the Canadian Parliament started a series of workshops for politicians called eco-summits. My part is to line up the best science speakers for the topic of each summit. We've had two eco-summits so far, the first on air pollution and the second on water pollution. About a hundred politicians attended the water summit, even though it was held two weeks *before* Walkerton. In addition to the speeches, there are panel discussions that mix politicians with some of the scientists who are most knowledgeable about the summit issues. We kick off the meeting with a prominent keynote speaker who's a big media draw, and of course politicians don't want to miss a media event. For example, Robert F. Kennedy, Jr. talked about the Riverkeepers program as the keynote speaker for the eco-summit on water.

How can undergraduate biology students participate at such interfaces between science and society?

I think a good way for science students to get involved is to join one or two really good environmental groups. Some of these groups, like politicians, don't have enough connection to science. Biology students can help with that, especially if they learn as much as they can about the basic ecology that underlies environmental issues. Working with ecologists in the biology department is a good way to get involved through research. I've had undergraduates publish papers. I myself had two papers accepted for publication before I started graduate school, one in *Science* and one in *Nature*. But I think many students are too shy to see their professors about working with them. They should take the initiative to operate outside their own little peer groups earlier in life.

AN INTRODUCTION TO ECOLOGY AND THE BIOSPHERE

THE SCOPE OF ECOLOGY

- The interactions between organisms and their environments determine the distribution and abundance of organisms
- Ecology and evolutionary biology are closely related sciences
- Ecological research ranges from the adaptations of individual organisms to the dynamics of the biosphere
- Ecology provides a scientific context for evaluating environmental issues

FACTORS AFFECTING THE DISTRIBUTION OF ORGANISMS

- Species dispersal contributes to the distribution of organisms
- Behavior and habitat selection contribute to the distribution of organisms
- Biotic factors affect the distribution of organisms
- Abiotic factors affect the distribution of organisms
- Temperature and water are the major climatic factors determining the distribution of organisms

AQUATIC AND TERRESTRIAL BIOMES

- Aquatic biomes occupy the largest part of the biosphere
- The geographic distribution of terrestrial biomes is based mainly on regional variations in climate

THE SPATIAL SCALE OF DISTRIBUTIONS

- Different factors may determine the distribution of a species on different scales
- Most species have small geographic ranges

O*rganisms are open systems that interact continuously with their environments—a theme that has already surfaced many times in this book. The scientific study of the interactions between organisms and their environments is called* **ecology** *(from the Greek* oikos, *home, and* logos, *to study). It is these interactions that determine both the distribution and abundance of organisms, resulting in the two questions that ecologists so often ask about organisms: Where do they live? And how many are*

there? Ecology is an enormously complex and exciting area of biology that is also of critical practical importance. Photographs of our planet taken by Apollo astronauts, such as the one on this page, remind us that Earth is a finite home in the vastness of space, not an unlimited frontier for human activity. The science of ecology can provide us with the basic understanding that we need to manage our planet's limited resources.

This chapter introduces the breadth of ecology and surveys some of the factors, both living and nonliving, that affect the distribution of organisms.

THE SCOPE OF ECOLOGY

Humans have always had an interest in the distribution and abundance of other organisms. As hunters and gatherers, prehistoric people had to learn where game and edible plants could be found in greatest abundance. Naturalists, from Aristotle to Darwin, made the process of observing and describing organisms in their natural habitats an end in itself, rather than simply a means of survival. Because extraordinary insight can still be gained through this descriptive approach of discovery science (see Chapter 1), natural history remains a fundamental part of ecology.

Although ecology has a long history as a descriptive science, it is becoming increasingly experimental. In spite of the difficulty of conducting field experiments over large areas and over many years, many ecologists are testing hypotheses in the field. For example, in 1993, oceanographers added dissolved iron to a 64-km^2 area of open ocean in the equatorial Pacific to test on a large scale the hypothesis that a shortage of iron as an inorganic nutrient was limiting production in this ecosystem. The experiment was repeated in 1995 on an even larger scale with

THE PROCESS OF SCIENCE

similar results—greatly increased phytoplankton growth when the iron was added. These kinds of field experiments make ecology a rapidly growing and exciting science.

The interactions between organisms and their environments determine the distribution and abundance of organisms

Ecologists use observations and manipulative experiments to test hypotheses aimed at such questions as, Why don't sequoia trees grow in Colorado? Why are there no malaria-carrying mosquitoes in Minnesota? Why are there so many deer in Ohio? What caused the extinction of the passenger pigeon?

FIGURE 50.1 shows the geographic range of the red kangaroo in Australia and graphically illustrates the basis for two major questions ecologists try to answer. What factors limit the geographic range, or *distribution,* of a species? And what factors determine its *abundance*? Given a hypothesis, or suggested explanation, for one of these questions, ecologists make predictions of what should be observed in nature or what the outcome of an experiment should be. In some cases, they can devise mathematical models that enable them to simulate the possible results of large-scale experiments that are impossible

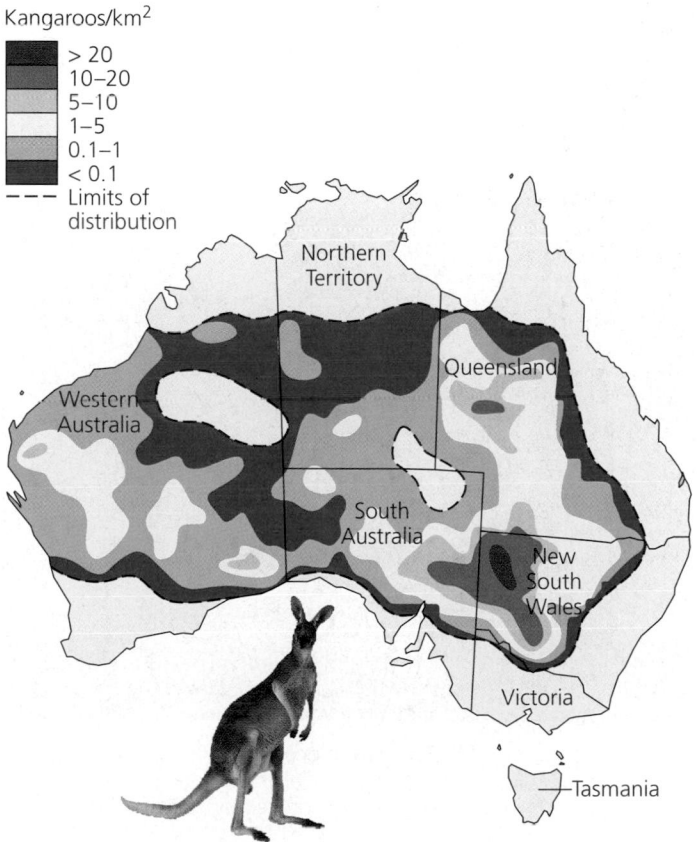

Kangaroos/km²

> 20
10–20
5–10
1–5
0.1–1
< 0.1
--- Limits of distribution

Northern Territory

Western Australia

Queensland

South Australia

New South Wales

Victoria

Tasmania

FIGURE 50.1 Distribution and abundance of the red kangaroo in Australia, based on aerial surveys. This kangaroo species occurs throughout arid regions of the continent.

to conduct in the field. In this approach, important variables and their hypothetical relationships are described through mathematical equations. The potential ways in which the variables interact can then be studied. For example, many ecologists and climatologists use sophisticated computer programs to develop models that predict the effects that human activities will have on climate and how the resulting climatic changes will affect geographic distributions of life-forms during the next century. Of course, such simulations are only as good as the basic information on which the models are based, and obtaining that information requires extensive laboratory work and fieldwork.

The environment of any organism includes **abiotic components** (nonliving chemical and physical factors), such as temperature, light, water, and nutrients, and **biotic** (living) **components**—all the organisms that are part of any individual's environment. Other organisms may compete with an individual for food and resources, prey upon it, or change its physical and chemical environment. As we will see, questions about the relative importance of various environmental components are frequently at the heart of ecological studies—and some accompanying controversies.

Ecology and evolutionary biology are closely related sciences

Many biologists recognize Charles Darwin as an able naturalist whose observations laid the groundwork for the later development of ecology. Indeed, it was the geographic distribution of organisms and their exquisite adaptations to specific environments that provided Darwin with evidence for evolution (see Chapter 22). An important cause of evolutionary change is the interaction of organisms with their environment. Thus, events that occur in the framework of **ecological time** (minutes, months, years) translate into effects over the longer scale of **evolutionary time** (decades, centuries, millennia, and longer). For instance, hawks feeding on field mice have an immediate (ecological) impact on the prey population by killing certain individuals, thereby reducing population size and altering the gene pool. One long-term evolutionary effect of this predator-prey interaction may be selection for mice with fur coloration that camouflages the animals.

Ecological research ranges from the adaptations of individual organisms to the dynamics of the biosphere

Because there are many levels and types of interactions between organisms and their environments, the questions ecologists address are wide-ranging and complex. Ecology can be divided into four increasingly comprehensive levels of study, from the ecology of individual organisms to the dynamics of ecosystems (FIGURE 50.2).

Organismal ecology is concerned with the morphological, physiological, and behavioral ways in which individual organisms meet the challenges posed by their biotic and abiotic environments. The geographic distribution of organisms is often limited by the abiotic conditions they can tolerate.

A **population** is a group of individuals of the same species living in a particular geographic area. **Population ecology** concentrates mainly on factors that affect how many individuals of a particular species live in an area.

A **community** consists of all the organisms of all the species that inhabit a particular area; it is an assemblage of populations of many different species. Thus, **community ecology** deals with the whole array of interacting species in a community. This level of research focuses on the ways in which interactions such as predation, competition, and disease affect community structure and organization.

An **ecosystem** consists of all the abiotic factors in addition to the entire community of species that exist in a certain area. An ecosystem—a lake, for example—may contain many different communities. In **ecosystem ecology,** the emphasis is on energy flow and the cycling of chemicals among the various biotic and abiotic components.

Looking beyond the four basic levels of ecology, we come to **landscape ecology,** which deals with arrays of ecosystems and how they are arranged in a geographic region. A **landscape** or **seascape** consists of several different ecosystems linked by exchanges of energy, materials, and organisms. The landscape level of research focuses on the ways in which interactions among populations, communities, and ecosystems are affected by the juxtaposition of different ecosystems, such as

FIGURE 50.2 Sample questions at different levels of ecology.

(a) Organismal ecology. How do diving whales select their feeding areas?

(b) Population ecology. What factors limit the number of striped mice that can inhabit a particular area?

(c) Community ecology. What factors influence the diversity of tree species that make up a particular forest?

(d) Ecosystem ecology. What processes recycle vital chemical elements, such as nitrogen, within a savanna ecosystem?

streams, lakes, old-growth forests, and forest patches that have had their trees removed by clear-cut logging.

The **biosphere** is the global ecosystem—the sum of all the planet's ecosystems. The most extensive level in ecology, the biosphere includes the atmosphere to an altitude of several kilometers, the land down to and including water-bearing rocks at least 3,000 m below ground, lakes and streams, caves, and the oceans to a depth of several kilometers. An example of research at the biosphere level is the analysis of how changes in atmospheric CO_2 concentration affect global climate.

Ecology provides a scientific context for evaluating environmental issues

The science of ecology should be distinguished from the informal use of the same word to refer to environmental concerns. And yet, we need to understand the often complicated and delicate relationships between organisms and their environments in order to address environmental problems.

Much of our current awareness about the environment had its beginnings in 1962 with Rachel Carson's *Silent Spring* (FIGURE 50.3). In that now-classic book, Carson warned that the widespread use of pesticides such as DDT was causing population declines in many nontarget organisms. Today, acid precipitation, localized famine aggravated by land misuse and population growth, the poisoning of soil and streams with toxic wastes, and the growing list of species extinct or endangered because of habitat destruction are just a few of the problems that threaten the home we share with millions of other forms of life.

Many influential ecologists, including David Schindler, the scientist you met in this unit's interview, recognize their responsibility to educate legislators and the general public about decisions that affect the environment. An important part of this responsibility is communicating the scientific complexity of environmental issues. Politicians and lawyers often want definitive answers to such environmental questions as, How much old-growth forest is needed to save spotted owls? While ecological studies can certainly provide essential information

for making policy decisions on habitat preservation, responses to such questions often include further questions: How many owls must be saved? With what certainty must they be saved? How long can they survive in this amount of forest? Ecologists can help answer these questions so that the public can make informed decisions about environmental concerns.

Although our ecological information is always incomplete, we cannot abstain from making decisions until all the answers are known. A guiding principle here is the **precautionary principle,** which can be expressed simply as "Look before you leap" or "An ounce of protection is worth a pound of cure." Aldo Leopold, the famous wildlife conservationist, expressed the precautionary principle well when he wrote, "To keep every cog and wheel is the first precaution to intelligent tinkering."

FACTORS AFFECTING THE DISTRIBUTION OF ORGANISMS

Ecologists have long recognized striking global and regional patterns in the distribution of organisms within the biosphere. Kangaroos occur in Australia but not in North America, while pronghorn antelope occur in the western United States but not in Europe or Africa. More than a century ago, Darwin, Wallace, and other naturalists began to recognize broad patterns of geographic distribution by naming biogeographic realms (FIGURE 50.4). We now associate these realms with patterns of continental drift that followed the breakup of Pangaea (see FIGURE 25.4).

Biogeography is the study of the past and present distribution of individual species. We discussed biogeography in the context of geologic history in Chapters 22 and 25. The field of

FIGURE 50.3 Rachel Carson. Although her *Silent Spring*, a book seminal to the modern environmental movement, focused on the biosphere's hangover from the pesticide DDT, Carson's message was much broader: "The 'control of nature' is a phrase conceived in arrogance, born of the Neanderthal age of biology and philosophy, when it was supposed that nature exists for the convenience of man."

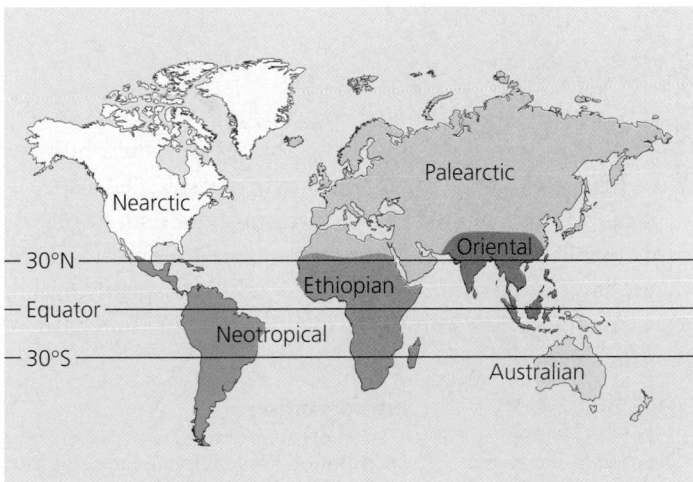

FIGURE 50.4 Biogeographic realms. Continental drift and barriers such as deserts and mountain ranges all contribute to the distinctive floras and faunas found in Earth's major regions. The realms are not sharply delineated but grade together in zones where taxa from adjacent realms coexist.

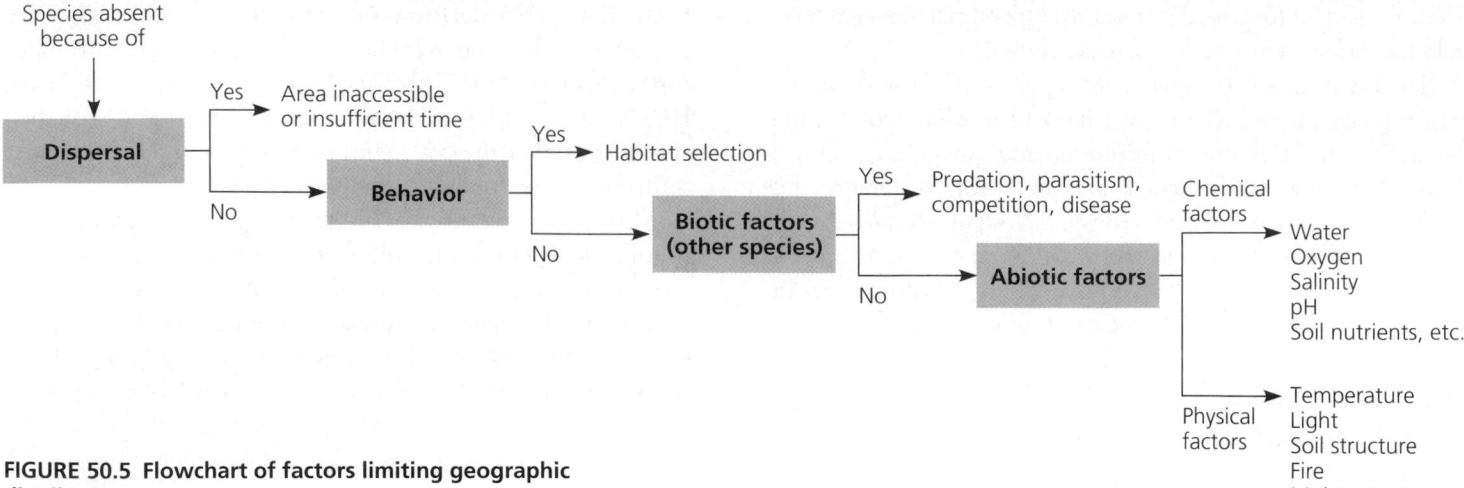

FIGURE 50.5 Flowchart of factors limiting geographic distribution.

biogeography provides a good starting point to understanding what limits geographic distributions. To determine what limits the geographic distribution of any particular species, ecologists ask a series of questions (FIGURE 50.5). Let's work our way through this flowchart of inquiry.

Species dispersal contributes to the distribution of organisms

Why are there no kangaroos in North America? The biogeographer answers with the simplest response: They could not get there because of barriers to dispersal; the area was inaccessible to kangaroos. The **dispersal** of organisms is a critical process for understanding both geographic isolation in evolution (see Chapter 24) and the broad patterns of current geographic distributions.

Species Transplants

THE PROCESS OF SCIENCE One way to determine if dispersal is a key factor limiting distribution is by observing the results when humans have accidentally or intentionally transplanted a species to areas where that species was previously absent. Some organisms can survive in new areas but cannot reproduce there, so we cannot determine the success of a transplant until at least one life cycle is complete. The two possible outcomes of the transplant direct further research.

Outcome	Interpretation
Transplant successful	Distribution limited because the area is inaccessible, time has been too short to reach the area, or the species fails to recognize the area as suitable living space
Transplant unsuccessful	Distribution limited either by other species or by physical and chemical factors

If a transplant is successful, then the *potential* range of the species is larger than its *actual* range, as FIGURE 50.6 illustrates for a hypothetical species. If a species does not occupy all of its potential range, we must determine why. Does the species lack suitable means of dispersal to reach new areas? Some animal species can in fact move into new areas but "choose" not to do so. For these species, we must study their mechanisms of habitat selection.

If the species cannot survive and reproduce in the transplant areas, we must ask whether biotic or abiotic factors exclude it from these areas. Limits imposed by other species (biotic factors) may involve the negative effects of predators, parasites, pathogens, or competitors. Or the transplant area could lack required positive effects of interdependent species, such as pollinators that are present within the actual range of a transplanted angiosperm. If other species do not set limits on the range, we are left with the last possibility that physical or chemical factors (abiotic factors) set the geographic range

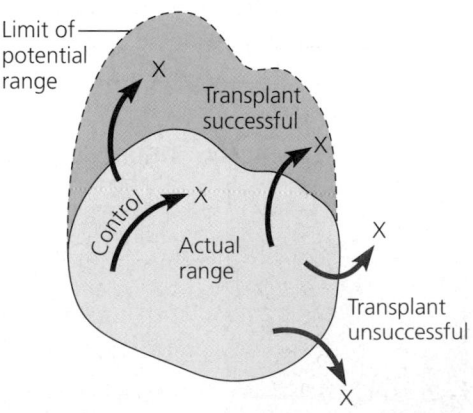

FIGURE 50.6 Set of transplant experiments for a hypothetical species. The results of many separate transplant experiments would be needed to define the limits of the potential range; only four are shown here.

limits. For example, many tropical plant species cannot withstand freezing temperatures, and the frost line effectively limits their distribution.

A proper transplant experiment should have a *control,* transplants done within the existing distribution to provide data on the effects of handling and transplanting the individual organisms. However, ecologists rarely conduct transplant experiments today. Instead, they document the outcomes when species have been transplanted for other purposes, such as to introduce game animals, or when species have been accidentally transplanted.

Problems with Introduced Species

Although many species are naturally restricted to particular biogeographic realms by their dispersal abilities, humans have managed to move species around the globe, particularly during the last 200 years. In fact, the most spectacular examples of dispersal affecting distribution occur when species that have been deliberately or accidentally introduced by humans explode to occupy a new area. Let's examine a couple of examples.

The African Honeybee. The African honeybee is a good example of how the human introduction of a species to new areas can have unpredicted and undesirable consquences. The African honeybee *(Apis mellifera scutellata)* is a very aggressive subspecies of honeybee that was brought to Brazil in 1956 to breed a variety that would produce more honey in the tropics than the standard Italian honeybee *(Apis mellifera ligustica).* The African bees escaped by accident in 1957 and have been spreading throughout the Americas ever since (FIGURE 50.7). Because African bees are aggressive, they may drive out the established colonies of Italian honeybees. In other situations, hybrids form between the African and Italian subspecies. In 1982, the African honeybee crossed the Panama Canal. It reached Mexico in 1985 and southern Texas in 1990. Moving roughly 110 km per year, it crossed the border into California in 1994 and is currently spreading into the southern United States. By December 2000, the African bee had reached northern Texas, southern California, southern Nevada, and all of Arizona. Unfortunately, African bees are aggressive toward humans and domestic animals, and accounts of severe stinging and even deaths have served to map the spread of the African bee. By 2000, ten people had been killed by these bees in the United States, and beekeepers are understandably

worried that the African bee will damage the established honeybee industry. What factors limit the distribution of the African honeybee in North America? Will this species be able to live as far north as northern California and North Carolina? Will cold winters prevent it from moving farther north? Ecologists do not yet know the answers to these questions.

The Zebra Mussel. In 1988, the zebra mussel *(Dreissena polymorpha),* a fingernail-sized mollusk native to the freshwater Caspian Sea of Asia, was discovered in Lake St. Claire near Detroit. No one knows for sure how the species got transplanted there. The best guess is that around 1985, a ship carried larvae of the mussel in its ballast water from a freshwater port in Europe to the Great Lakes, where the ballast water was emptied with no concern about what organisms it might contain.

The zebra mussel quickly became a pest in North America. It reproduces rapidly and forms dense clusters several layers thick on hard surfaces. The mussels were first noticed when

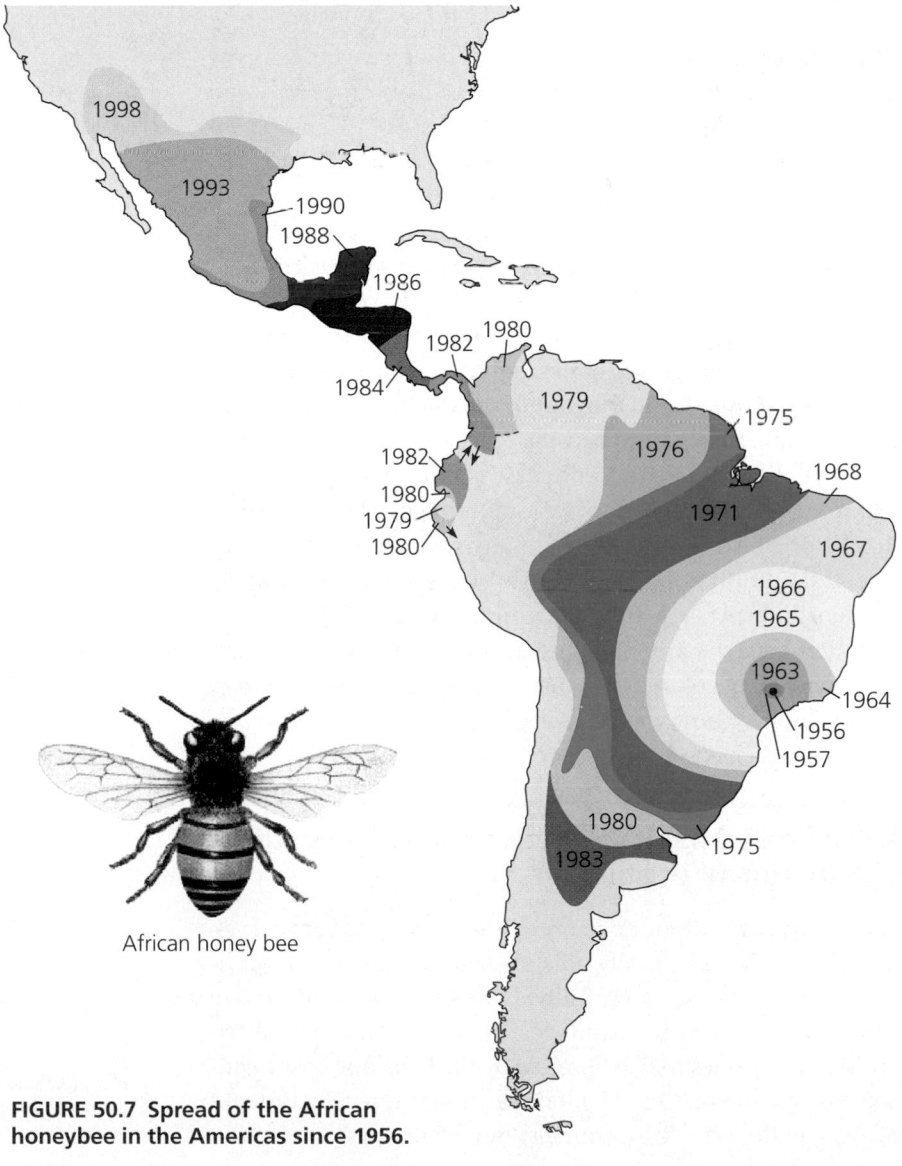

African honey bee

FIGURE 50.7 Spread of the African honeybee in the Americas since 1956.

they reached densities of 750,000 per square meter in water pipes in Lake Erie and clogged the water intakes of city water systems, electrical power stations, and other industrial facilities in the Great Lakes.

Since 1988, zebra mussels have spread rapidly in the river systems of the central United States (FIGURE 50.8). Because they are such efficient suspension feeders, zebra mussels actually make the water much clearer, but they alter the native communities of organisms in the process. By feeding on phytoplankton, the mussels depress populations of zooplankton; and the clearer water admits more sunlight, increasing the growth of rooted aquatic plants in shallow waters. In the Hudson River in New York, phytoplankton biomass was reduced about 85% after zebra mussels invaded; zooplankton, which feed on phytoplankton, declined by more than 70%. Some fish and ducks eat zebra mussels, but these predators are too few to slow down population growth of the mussels. Zebra mussels crowd out native mollusk species by colonizing all hard surfaces, including the shells of freshwater clams. The result can be local extinction of the native species.

The Tens Rule. Not all introduced species thrive in their new homes, and many of the species humans have moved around the globe have failed to colonize new areas. For example, bird introductions into continental areas are usually failures. In North America, for example, only 13 species of introduced birds are common, although 98 species have been introduced. In Europe, only 13 successful establishments of birds are recorded out of 85 species introduced. A rough generalization for the success of introduced species is the *tens rule,* which makes the statistical prediction that an average of one out of ten introduced species become established, and one out of ten established species become common enough to become pests.

The ability of species to disperse is important on a global scale but rarely an important factor limiting the local distribution of organisms. Species have many special adaptations for dispersal and often colonize nearby areas. On a global scale, barriers to dispersal help to determine the biogeographic distribution patterns among continents and islands. The ability of humans to overcome these barriers and move species is just one example of how we are altering the biosphere.

Behavior and habitat selection contribute to the distribution of organisms

Some organisms do not occupy all their potential range, even though they are physically able to disperse into the unoccupied areas. In these cases, individuals seem to avoid certain habitats, even when the habitats are suitable. Thus, the distribution of a species may be limited by the behavior of individuals in selecting habitat. Habitat selection is typically thought of only with respect to animals, but plant species can also

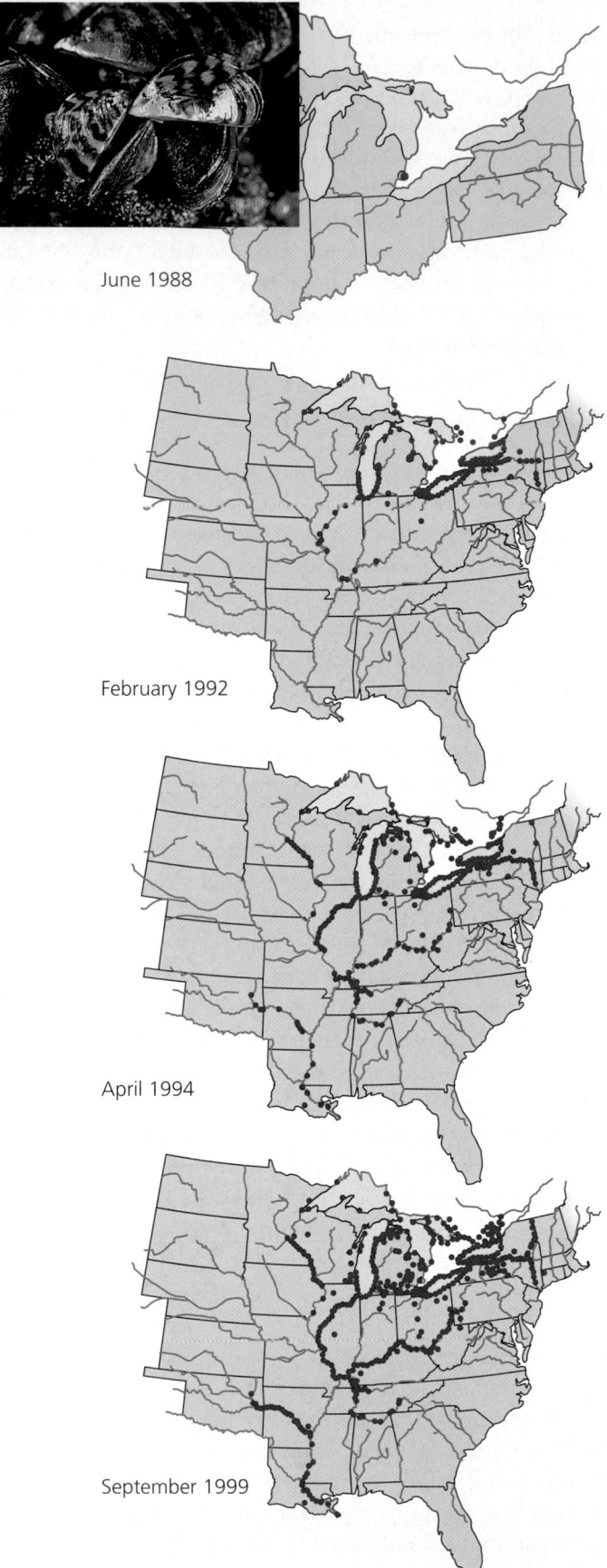

June 1988

February 1992

April 1994

September 1999

FIGURE 50.8 Expansion of the geographic range of the zebra mussel (*Dreissena polymorpha*) since its discovery near Detroit in 1988.

select their habitats, even though individual plants cannot move. How organisms select the type of habitat to occupy is one of the least understood ecological processes.

Insects often have very stereotyped oviposition (egg-depositing) behavior, and this may restrict their local distribution to certain host plants. Consider the European corn borer, for example, whose larvae will feed on a wide variety of plants but occur almost exclusively on corn because the ovipositing females are attracted by volatile odors produced by the corn plant. The complex chemical signaling between plants and herbivorous insects and pollinators is an important research area in the study of plant-herbivore interactions.

Anopheline mosquitoes are important carriers of disease, and their ecology has been studied extensively because of the difficulty of malaria eradication in tropical areas. Each mosquito species is usually associated with a particular type of breeding place, and one of the striking observations that a student of malaria first makes is that large areas of water in the tropics are completely free of dangerous mosquitoes. Habitat selection for oviposition sites by female mosquitoes appears to restrict their distribution. Larvae can develop successfully over a much wider range of conditions than those in which eggs are laid.

The key point in these examples is that evolution does not produce perfect organisms for every suitable habitat. Adaptation can never be exact and instantaneous, and we must be careful not to expect perfection in organisms (see Chapter 23). We may judge a tropical mosquito deficient for not laying eggs in all suitable rice field habitats, but this failing may only reflect the fact that rice fields are a recent habitat in evolutionary time. In addition, not all behavior that has evolved remains adaptive, particularly in ecosystems modified by humans. Environmental conditions may change such that behaviors that were formerly adaptive are now maladaptive. Ground-nesting birds on islands, for example, are threatened if new ground predators (rats, for example) colonize the island. Populations cannot evolve overnight. Even with suitable genetic variation in a population, natural selection may not be able to operate quickly enough to adjust habitat selection behavior to some abrupt environmental change.

Biotic factors affect the distribution of organisms

Frequently, a species cannot complete its full life cycle if transplanted to an area it did not originally occupy. One reason for this inability to survive and reproduce may be negative interactions with other organisms through predation, disease, and competition. Or there may be an absence of other species upon which the transplanted species depends, such as the absence of specific pollinators in the case of a transplanted plant species.

In predation we find some of the clearest cases of biotic limitation on geographic distribution. Carnivorous predators, such as wolves, kill their prey. And herbivores, such as grazing mammals and most insects, also function as predators, eating parts of plants or whole plants.

Let's examine a specific case of a predator limiting the distribution of a prey species. In certain marine ecosystems, there is often an inverse relationship between the abundance of sea urchins and the abundance of seaweeds (large algae, such as kelp). Where sea urchins that graze on kelp are common, kelp cannot become established. Thus, the local distribution of kelp can be limited by sea urchins. This kind of interaction can be tested by "removal and addition" experiments. If the hypothesis that sea urchins are a biotic factor limiting kelp distribution is correct, then kelp should invade an area from which sea urchins have been removed. Conversely, if urchins are added to an area rich in kelp, the kelp should be eliminated. One complication is that there are often several other herbivores in addition to sea urchins; thus, careful manipulative experiments are needed to determine the reasons for the kelp being absent. You can see the results of some careful predator-removal experiments in FIGURE 50.9.

Many organisms are limited in their local distribution by the presence of food resources, predators, diseases, and competitors. Unfortunately, some of the most dramatic cases occur when humans introduce (either accidentally or intentionally) exotic predators or diseases into new areas and wipe out native species. We will see examples of these impacts in Chapter 55 when we discuss conservation ecology.

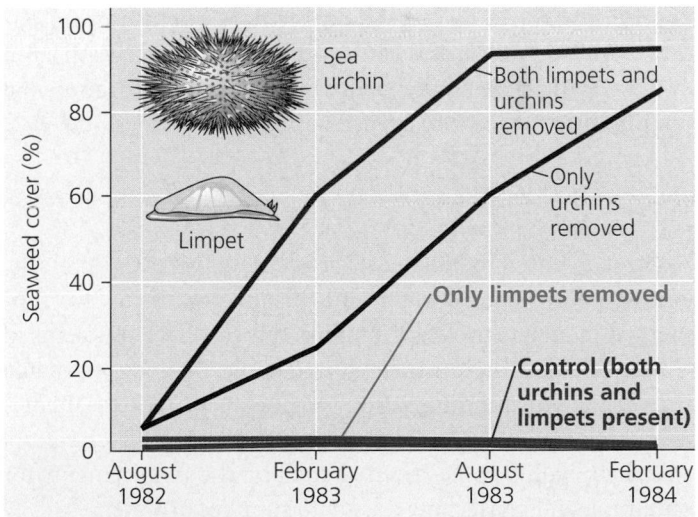

FIGURE 50.9 Predator-removal experiments. Researchers tested the effects of two herbivores, sea urchins and limpets (mollusks), on the abundance of seaweeds in adjacent subtidal areas near Sydney, Australia. In areas where both sea urchins and limpets were present (red line), there was virtually no algal cover present. Predator-removal experiments in areas adjacent to the control site supported the hypothesis that sea urchins are the main herbivores limiting distribution of the seaweeds.

THE PROCESS OF SCIENCE

Abiotic factors affect the distribution of organisms

Patterns of geographic distributions of organisms on a global scale are a broad reflection of the influence of abiotic factors. These patterns mainly mirror regional differences in temperature, rainfall, salinity (saltiness), and light. Throughout this discussion, it is important to remember that the environment varies in both space and time. Although two regions of Earth may experience different conditions at any given time, daily and annual fluctuations of abiotic factors sometimes blur or accentuate the distinctions between those regions.

Temperature

Environmental temperature is an important factor determining the distribution of organisms because of its effect on biological processes and the inability of most organisms to regulate body temperature precisely. Cells may rupture if the water they contain freezes (at temperatures below 0°C), and the proteins of most organisms denature at temperatures above 45°C. In addition, few organisms can maintain a sufficiently active metabolism at very low or very high temperatures. Within a suitable range, however, most biochemical reactions and physiological processes occur more rapidly at higher temperatures. Extraordinary adaptations enable some organisms, including thermophilic prokaryotes (see Chapter 27), to live outside the temperature range habitable by other life.

The internal temperature of an organism is affected by heat exchange with its environment (see Chapter 44), and most organisms cannot maintain tissue temperatures more than a few degrees above or below the ambient temperature. As endotherms, mammals and birds are the major exceptions, but even endotherms function best within certain environmental temperature ranges, which vary with the species.

Water

Water is essential to life, but its availability varies dramatically among habitats. Freshwater and marine organisms live submerged in an aquatic environment, but they face problems of water balance if their intracellular osmolarity does not match that of the surrounding water. Organisms in terrestrial environments encounter a nearly constant threat of desiccation, and their evolution has been shaped by the requirements for obtaining and conserving adequate supplies of water.

Sunlight

Sunlight provides the energy that drives nearly all ecosystems, although only plants and other photosynthetic organisms use this energy source directly. Light intensity is not the most important factor limiting plant growth in many terrestrial environments, but shading by a forest canopy makes competition for light in the understory intense. In aquatic environments, the intensity and quality of light limit the distribution of photosynthetic organisms. Every meter of water depth selectively absorbs about 45% of the red light and about 2% of the blue light passing through it. As a result, most photosynthesis in aquatic environments occurs relatively near the surface. And the photosynthetic organisms themselves absorb some of the light that penetrates, further reducing light levels in the waters below.

Light is also important to the development and behavior of the many organisms that are sensitive to photoperiod, the relative lengths of daytime and nighttime. Photoperiod is a more reliable indicator than temperature for cuing seasonal events, such as flowering by plants or migration by animals.

Wind

Wind amplifies the effects of environmental temperature on organisms by increasing heat loss due to evaporation and convection (the wind-chill factor). It also contributes to water loss in organisms by increasing the rate of evaporation in animals and transpiration in plants. In addition, wind can have a substantial effect on the morphology of plants by inhibiting the growth of limbs on the windward side of trees; limbs on the leeward side grow normally, resulting in a "flagged" appearance.

Rocks and Soil

The physical structure, pH, and mineral composition of rocks and soil limit the distribution of plants and of the animals that feed upon them, thus contributing to the patchiness we see in terrestrial ecosystems. In streams and rivers, the composition of the substrate can affect water chemistry, which in turn influences the resident algae, plants, and animals. In marine environments, the structure of the substrates in the intertidal zone and on seafloors determines the types of organisms that can attach to or burrow in those habitats.

Now that we have surveyed the various factors that affect the distribution of organisms, let's focus on the major role that climate plays in structuring the biosphere.

Temperature and water are the major climatic factors determining the distribution of organisms

Four abiotic factors—temperature, water, light, and wind—are the major components of **climate,** the prevailing weather conditions at a locality. Temperature and water are especially important as factors determining the geographic range of organisms.

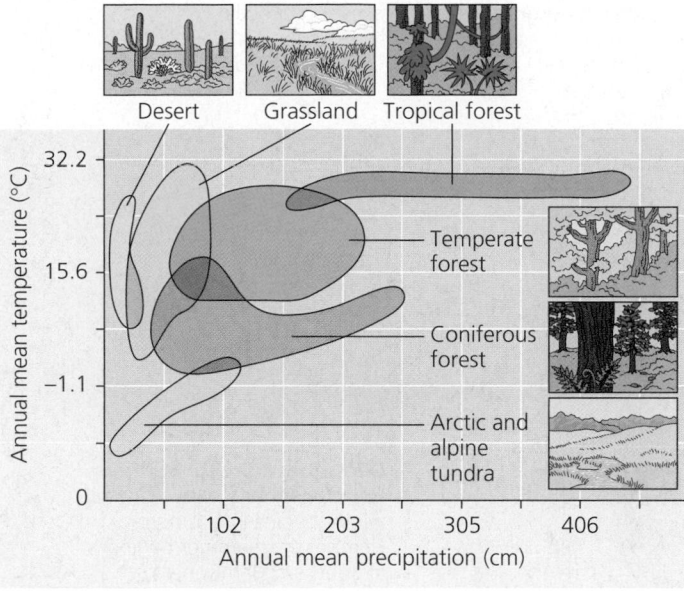

Desert Grassland Tropical forest

FIGURE 50.10 A climograph for some major kinds of ecosystems (biomes) in North America. The areas plotted here encompass the range of annual mean temperature and precipitation occurring in the biomes.

Climate and Biomes

We can see the great impact of climate on the distribution of organisms by constructing a climograph, a plot of the temperature and rainfall in a particular region. For example, FIGURE 50.10 shows a climograph denoting annual mean temperature and precipitation for some of the biomes of North America. **Biomes** are major types of ecosystems, those that occupy broad geographic regions. Examples of biomes are coniferous forests, deserts, and grasslands. Notice in FIGURE 50.10 that the range of rainfall occurring in northern coniferous forests is similar to that of temperate forests, but the temperature ranges are different. Grasslands are generally drier than either kind of forest, and deserts are drier still.

Annual means for temperature and rainfall are reasonably well correlated with the biomes we find in different regions. However, we must always be careful to distinguish *correlation* from *causation*, a cause-and-effect relationship. Although our climograph provides circumstantial evidence that temperature and rainfall are important to the distribution of biomes, it does not prove that these variables govern their geographic location. Only a detailed analysis of the water and temperature tolerances of individual species could establish the controlling effects of these variables.

We can also see in our climograph that factors other than mean temperature and precipitation must play a role in determining which biomes are found where, because there are regions where biomes overlap. For example, certain areas in

North America with a particular temperature and precipitation combination support a temperate forest, but other areas with the same values for these variables support a coniferous forest; still others support a grassland. How do we explain this variation? First, remember that the climograph is based on annual *averages*. Often it is not only the mean or average climate that is important but also the pattern of climatic variation. For example, some areas may get regular precipitation throughout the year, whereas others with the same annual amount of precipitation have distinct wet and dry seasons. A similar phenomenon may occur with respect to temperature. Other factors, such as the bedrock in an area, may greatly affect mineral nutrient availability and soil structure, which in turn affect the kind of vegetation that will develop.

With these complex considerations in mind, let's take a closer look at how climate affects the geographic distribution of organisms.

Global Climate Patterns

Earth's global climate patterns are largely determined by the input of solar energy and the planet's movement in space. The sun's warming effect on the atmosphere, land, and water establishes the temperature variations, cycles of air movement, and evaporation of water that are responsible for dramatic latitudinal variations in climate. Because solar radiation is most intense when the sun is directly overhead, the shape of the Earth causes latitudinal variation in the intensity of sunlight (FIGURE 50.11). However, the planet is also tilted on its axis by 23.5° relative to its plane of orbit around the sun, and this tilt

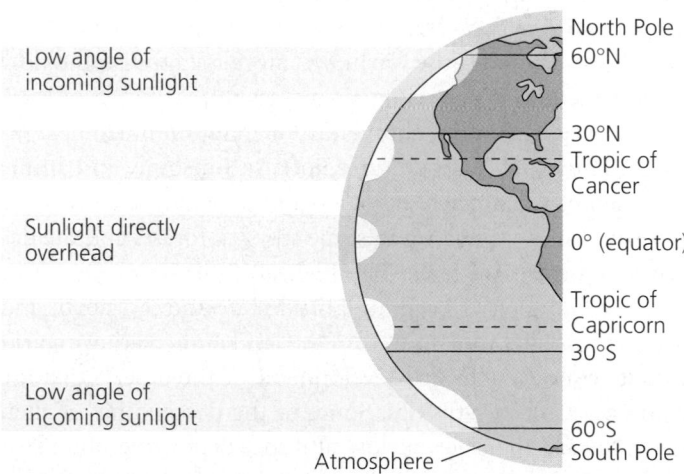

FIGURE 50.11 Solar radiation and latitude. Because sunlight strikes the equator perpendicularly, more heat and light are delivered there per unit of surface area than at higher northern and southern latitudes, where sunlight has a longer path through the atmosphere and strikes the curved surface of Earth at an oblique angle.

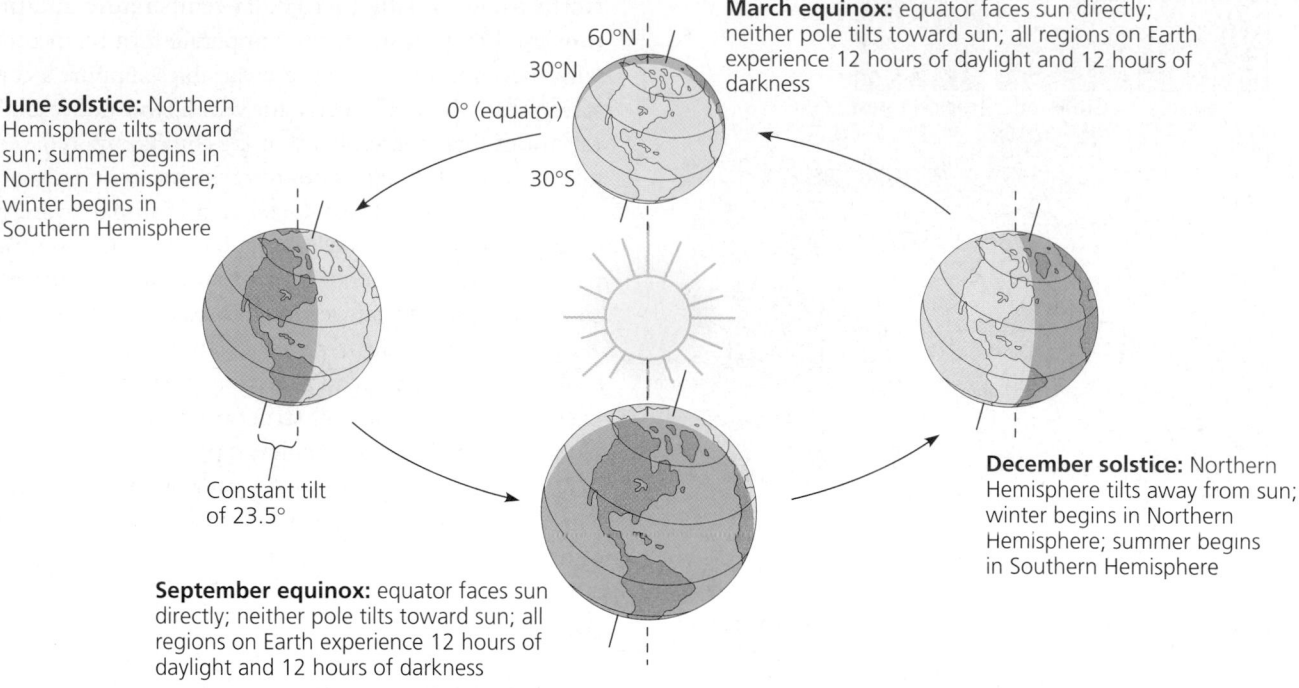

June solstice: Northern Hemisphere tilts toward sun; summer begins in Northern Hemisphere; winter begins in Southern Hemisphere

60°N
30°N
0° (equator)
30°S

March equinox: equator faces sun directly; neither pole tilts toward sun; all regions on Earth experience 12 hours of daylight and 12 hours of darkness

Constant tilt of 23.5°

December solstice: Northern Hemisphere tilts away from sun; winter begins in Northern Hemisphere; summer begins in Southern Hemisphere

September equinox: equator faces sun directly; neither pole tilts toward sun; all regions on Earth experience 12 hours of daylight and 12 hours of darkness

FIGURE 50.12 What causes the seasons? The permanent tilt of the Earth on its axis causes seasonal variation in temperature and light intensity as the planet revolves around the sun.

causes seasonal variation in the intensity of solar radiation (FIGURE 50.12). The **tropics** (those regions that lie between 23.5° north latitude and 23.5° south latitude) experience the greatest annual input and the least seasonal variation in solar radiation of any region on Earth. The seasonality of light and temperature increases steadily toward the poles; polar regions have long, cold winters with periods of continual darkness and short summers with periods of continual light.

Intense solar radiation near the equator initiates a global circulation of air, creating precipitation and winds (FIGURE 50.13). High temperatures in the tropics evaporate water from Earth's surface and cause warm, wet air masses to rise and flow toward the poles. The rising air masses release much of their water content, creating abundant precipitation in tropical regions. Thus, high temperatures, intense sunshine, and ample rainfall are all characteristic of a tropical climate, fostering the growth of lush vegetation in some tropical forests and the development of coral reefs. The high-altitude air masses, now dry, descend toward Earth at latitudes around 30° north and south, absorbing moisture from the land and creating an arid climate conducive to the development of the deserts that are common at these latitudes. Some of the descending air then flows toward the poles at low altitude, depositing abundant precipitation (though less than in the tropics) where the air masses again rise and release moisture in the vicinity of 60° latitude. Broad expanses of coniferous forest dominate the landscape at these fairly wet but generally cool latitudes. Some of the cold, dry rising air then flows to the poles, where it descends and flows back toward the equator, absorbing moisture

and creating the comparatively rainless and bitterly cold climates of the Arctic and Antarctica.

Local and Seasonal Effects on Climate

Proximity to bodies of water and topographic features such as mountain ranges create a climatic patchiness on a regional scale, and smaller features of the landscape contribute to local climatic variation. These regional and local variations in climate contribute to the patchiness of the biosphere.

Ocean currents influence climate along the coasts of continents by heating or cooling overlying air masses, which may then pass across the land. Evaporation from the ocean is also greater than it is over land, and coastal regions are generally moister than inland areas at the same latitude. The cool, misty climate produced by the cold California current that flows southward along the western United States supports a rain forest ecosystem dominated by large coniferous trees in the Pacific Northwest and large redwood groves farther south. Similarly, the warm Gulf Stream flowing northward from the Caribbean Sea and across the North Atlantic tempers the climate on the west coast of the British Isles, making it warmer during winter than the coast of New England, which is actually farther south but is cooled by a current flowing south from the coast of Greenland.

As every vacationer knows, oceans (and large inland bodies of water) generally moderate the climate of nearby terrestrial environments. During a warm summer day, when the land is hotter than a large lake or the ocean, air over the land heats and rises, drawing a cool breeze from the water across the land. At night, by contrast, air over the warmer ocean or lake rises, establishing a circulation that draws cooler air from the land out over the water, replacing it with warmer air from off-shore. Proximity to water does not always moderate climate, however. Several regions (including the coast of central and southern California) have a Mediterranean-like climate; in summer, cool, dry ocean breezes are warmed when they contact the land, absorbing moisture and creating hot, rainless summers just a few miles inland.

Mountains also have a significant effect on solar radiation, local temperature, and rainfall. South-facing slopes in the Northern Hemisphere receive more sunlight than nearby north-facing slopes and are therefore warmer and drier. In many of the mountains of western North America, spruce and other conifers occupy the north-facing slopes, whereas shrubby, drought-resistant vegetation inhabits many south-facing slopes. In addition, at any particular latitude, air temperature declines approximately 6°C with every 1,000-m increase in elevation, paralleling the decline of temperature with increasing latitude. In the north temperate zone, for example, a 1,000-m increase in elevation produces a temperature change equivalent to that over an 880-km increase in latitude. This is one reason mountain communities are similar to those at lower elevation farther from the equator.

When warm, moist air approaches a mountain, it rises and cools, releasing moisture on the windward side of the range. On the leeward side of the mountain, cooler, dry air descends,

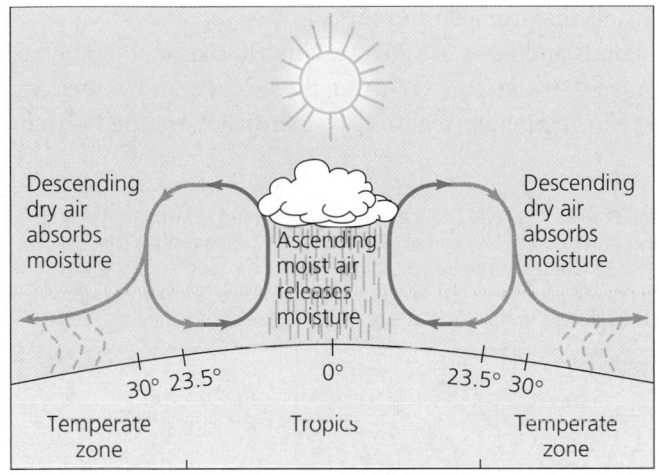

(a) Air circulation and precipitation near the equator. Air masses in the lower atmosphere are warmed by solar radiation and by heat radiating from Earth. Air expands, decreases in density, and rises when it is warmed. Thus, warm air at the equator rises, creating an area of light, shifting winds known as the doldrums. The warm air masses expand as they rise, and because this expansion distributes their internal heat energy over a larger volume, they also cool as they move upward through the atmosphere. Cool air holds less water vapor than warm air, and the rising air masses drop large amounts of rain in the tropics. The now-dry air masses flow toward the two poles at high altitude, cooling further as they spread away from the equator. The density of the air masses increases as they cool, and the air masses descend, absorbing water from the land and creating bands of arid climate around 30° latitude.

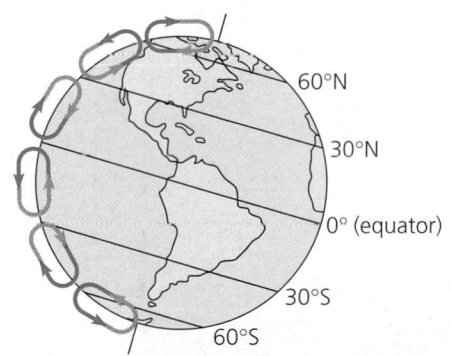

(b) Global air circulation. The movement of heated air creates three major air circulation cells on either side of the equator. Within each circulation cell, rising air (blue) releases moisture as precipitation, and descending air (brown) absorbs moisture, creating arid conditions.

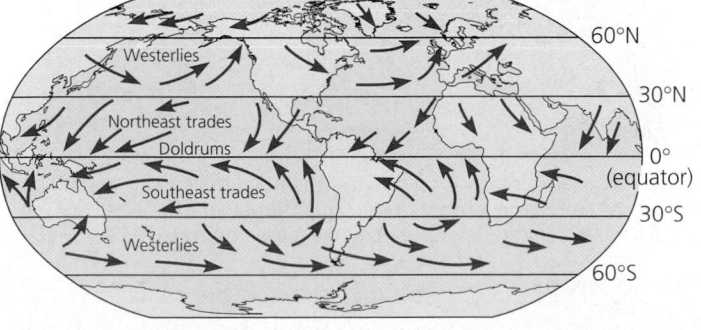

(c) Global wind patterns. Air flowing in the lower levels of the circulation cells, close to Earth's surface, creates predictable global wind patterns. However, as Earth rotates on its axis, land near the equator moves faster than that at the poles, deflecting the winds from the vertical paths illustrated in part (b) and creating more easterly and westerly flows. Cooling trade winds blow from east to west in the tropics and subtropics. In contrast, prevailing westerlies blow from west to east in the temperate zones.

FIGURE 50.13 Global air circulation, precipitation, and winds.

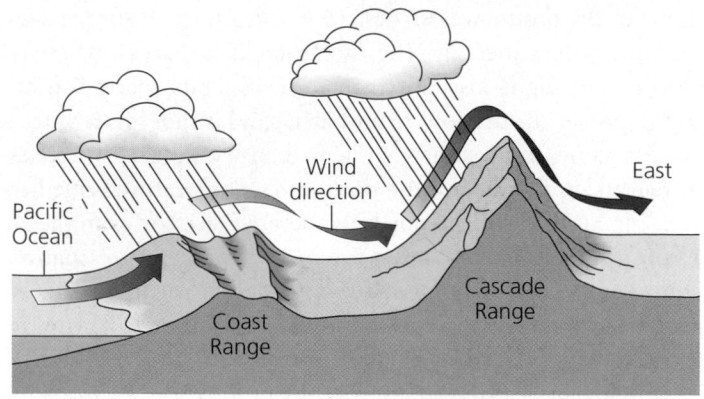

FIGURE 50.14 How mountains affect rainfall. This drawing represents major landforms across the state of Washington. As moist air moves in off the Pacific Ocean and encounters the westernmost mountains (the Coast Range), it flows upward, cools at higher altitudes, and drops a large amount of water. Some of the world's tallest trees, the Douglas firs, thrive here. Farther inland, precipitation increases again as the air moves up and over higher mountains (the Cascade Range). On the eastern side of the Cascades, there is little precipitation. As a result of this rain shadow, much of central Washington is very arid, almost qualifying as desert.

absorbing moisture and producing a rain shadow (FIGURE 50.14). Deserts commonly occur on the leeward side of mountain ranges, a phenomenon evident in the Great Basin and Mojave Desert of western North America, the Gobi Desert of Asia, and in the small deserts that characterize the southwest corners of some Caribbean islands.

Seasonality generates local environmental variation in addition to the global changes in day length, solar radiation, and temperature described earlier. Because of the changing angle of the sun over the course of the year, the belts of wet and dry air on either side of the equator undergo slight seasonal shifts in latitude that produce marked wet and dry seasons around 20° latitude, where tropical deciduous forests grow. In addition, seasonal changes in wind patterns produce variations in ocean currents, sometimes causing the upwelling of nutrient-rich, cold water from deep ocean layers, thus nourishing organisms that live near the surface.

Ponds and lakes are also sensitive to seasonal temperature changes (FIGURE 50.15). During the summer and winter, many lakes in temperate regions are thermally stratified—that is,

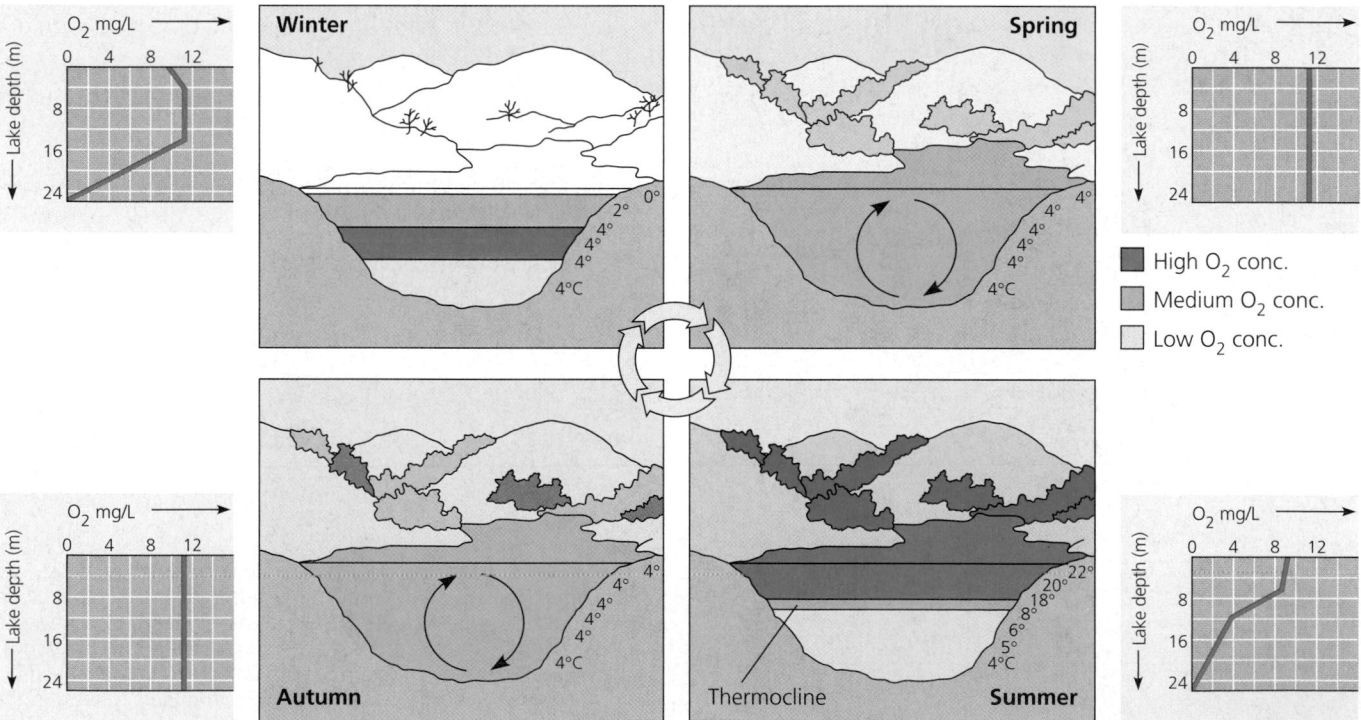

In winter, the coldest water in the lake (0°C) lies just below the surface ice; water is progressively warmer at deeper levels of the lake; typically 4–5°C at the bottom.

In spring, as the sun melts the ice, the surface water warms to 4°C and sinks below the cooler layers immediately below, eliminating the thermal stratification established in winter. In the absence of thermal layering, spring winds mix the water to great depth, bringing oxygen (O_2) to the bottom waters (see graphs) and nutrients to the surface.

In autumn, as surface water cools rapidly, it sinks below the underlying layers, remixing the lake water until the surface begins to freeze and the winter temperature profile is reestablished.

In summer, the lake regains a distinctive thermal profile, with warm water at the surface separated from cold bottom water by a narrow vertical zone of rapid temperature change, called a thermocline.

FIGURE 50.15 Lake stratification and seasonal turnover. Lakes in temperate zones tend to stratify by temperature and density in winter and summer. The biannual mixing of lake waters occurs because water is most dense at 4°C, and water at that temperature sinks below water that is either warmer or colder.

layered vertically according to temperature. Such lakes undergo a biannual mixing of their waters as a result of changing water temperature profiles. This **turnover,** as it is called, brings oxygenated water from the surface of lakes to the bottom and nutrient-rich water from the bottom to the surface in both spring and autumn (see FIGURE 50.15). These cyclic changes in the abiotic properties of lakes are essential for the survival and growth of organisms at all levels within this ecosystem.

Microclimate

Climate also varies on a very fine scale, called **microclimate.** For example, ecologists often refer to the microclimate on a forest floor or under a rock. Many features in the environment influence microclimates by casting shade, affecting evaporation from soil, and changing the patterns of wind. Forest trees frequently moderate the microclimate below. Cleared areas generally experience greater temperature extremes than the forest interior because of greater solar radiation and wind currents that are established by the rapid heating and cooling of open land; evaporation is generally greater in clearings as well. Low-lying ground is usually wetter than high ground and tends to be occupied by different species of trees within the same forest. If you have ever lifted a log or large stone in the woods, you are well aware that there are organisms (such as salamanders, worms, and insects) that live in the shelter of this microenvironment, buffered from the extremes of temperature and moisture. Every environment on Earth is similarly characterized by a mosaic of small-scale differences in the abiotic factors that influence the local distributions of organisms.

Long-Term Climate Change

If temperature and moisture are the master limiting factors for the geographic ranges of plants and animals, the climatic warming that is under way during the 21st century will have profound effects on the biosphere. (The causes and consequences of global warming are discussed in detail in Chapter 54.) One way to get a glimpse of the kinds of changes that may occur is to look back at the changes that have occurred in temperate regions since the end of the last ice age.

The last continental glaciers began retreating in North America and Eurasia about 16,000 years ago. The northward expansion of tree distributions lagged behind the retreat of the ice. A detailed record of these migrations is captured in fossil pollen deposited in lakes and ponds. (It may seem odd to think of trees "migrating," but recall from Chapter 38 that wind and animals can disperse seeds, sometimes over great distances.) In North America, oaks and maples moved rapidly in a northeastward direction from the Mississippi Valley, while hickories advanced more slowly. Hemlocks and white pines moved northwestward from refuges along the Atlantic coast. The important conclusion from this research is that various tree species advanced their range at different rates. If you were sitting in New Hampshire and lived a very long time, you would have seen sugar maple trees arrive 9,000 years ago, hemlock 7,500 years ago, and beech trees 6,500 years ago.

If we can determine the climatic limits of current geographic distributions for organisms, we can make predictions about how distributions will change with climatic warming. A major question when applying this approach to plants is whether seed dispersal is rapid enough to sustain the migration of each species as climate changes. For example, fossils suggest that eastern hemlock was delayed nearly 2,500 years in its movement north at the end of the Ice Age partly because of relatively slow seed dispersal.

Let's look at a specific case of how the fossil record of past tree migrations can inform predictions about the biological impact of the current global warming trend. FIGURE 50.16 shows the current and potential geographic range of the American beech (*Fagus grandifolia*) under two climate-change models. These models predict that the potential northern limit of the beech's range will move 700–900 km north in the next century, and the southern range limit will shrink to the north as well. If

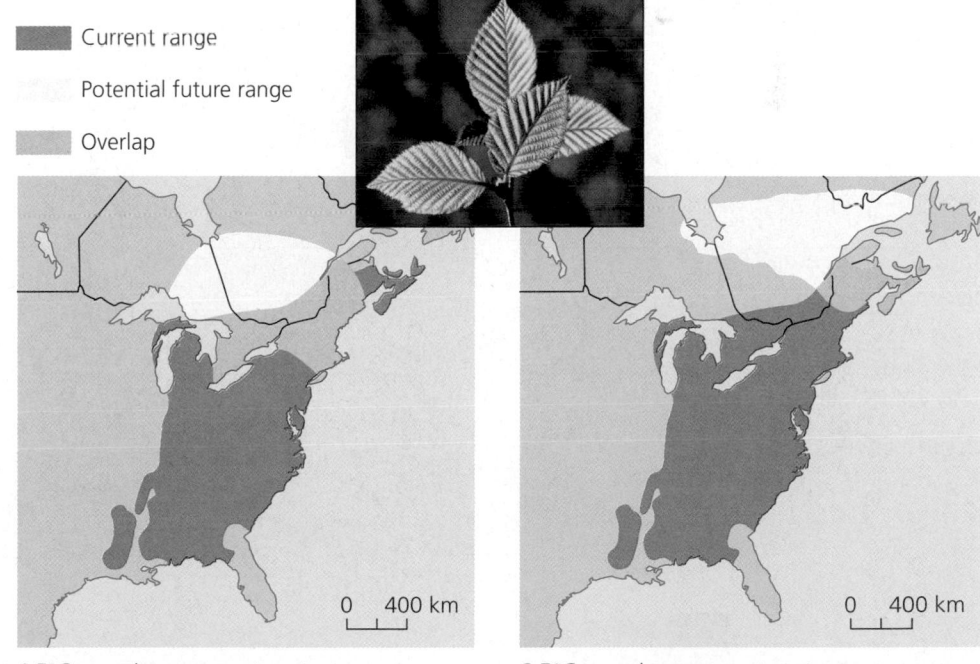

Current range

Potential future range

Overlap

4.5°C warming over next century   6.5°C warming over next century

FIGURE 50.16 Current geographic range and predicted future range for the American beech (*Fagus grandifolia*) under two climate-change scenarios.

left to natural processes, the beech must move 7–9 km per year to the north to keep pace with the warming climate. By contrast, since the end of the Ice Age, the beech migrated into its present range at a rate of only 0.2 km per year. If these predictions are even approximately correct, migrating species such as beech trees will require human assistance to move into new ranges where they can survive as the climate warms. If this does not occur, beech and many other species may become extinct.

AQUATIC AND TERRESTRIAL BIOMES

Having examined some of the factors that determine the distribution of organisms on Earth, we now turn to a brief survey of the major types of ecosystems, the biomes, beginning with the aquatic biomes (FIGURE 50.17).

Aquatic biomes occupy the largest part of the biosphere

Aquatic biomes account for the largest part of the biosphere in terms of area. Ecologists distinguish between freshwater biomes and marine biomes on the basis of physical and chemical differences. For example, marine biomes generally have salt concentrations that average 3%, whereas freshwater biomes are usually characterized by a salt concentration of less than 1%.

Covering about 75% of Earth's surface, oceans have always had an enormous impact on the biosphere. The evaporation of seawater provides most of the planet's rainfall, and ocean temperatures have a major effect on world climate and wind patterns. In addition, marine algae and photosynthetic bacteria supply a substantial portion of the world's oxygen and consume huge amounts of atmospheric carbon dioxide.

Freshwater biomes are closely linked to the soils and biotic components of the terrestrial biomes through which they pass or in which they are situated. The particular characteristics of a freshwater biome are also influenced by the patterns and speed of water flow and the climate to which the biome is exposed.

Vertical Stratification of Aquatic Biomes

Many aquatic biomes exhibit pronounced vertical stratification of physical and chemical variables. Light is absorbed by both the water itself and the microorganisms in it, so that its intensity decreases rapidly with depth. Ecologists distinguish between the upper **photic zone,** where there is sufficient light for photosynthesis, and the lower **aphotic zone,** where little light penetrates. Water temperature also tends to be stratified, especially during summer and winter (see FIGURE 50.15). Heat energy from sunlight warms the surface waters to whatever depth the sunlight penetrates, but the deeper waters remain quite cold. In the ocean and in many temperate-zone lakes, a narrow stratum of

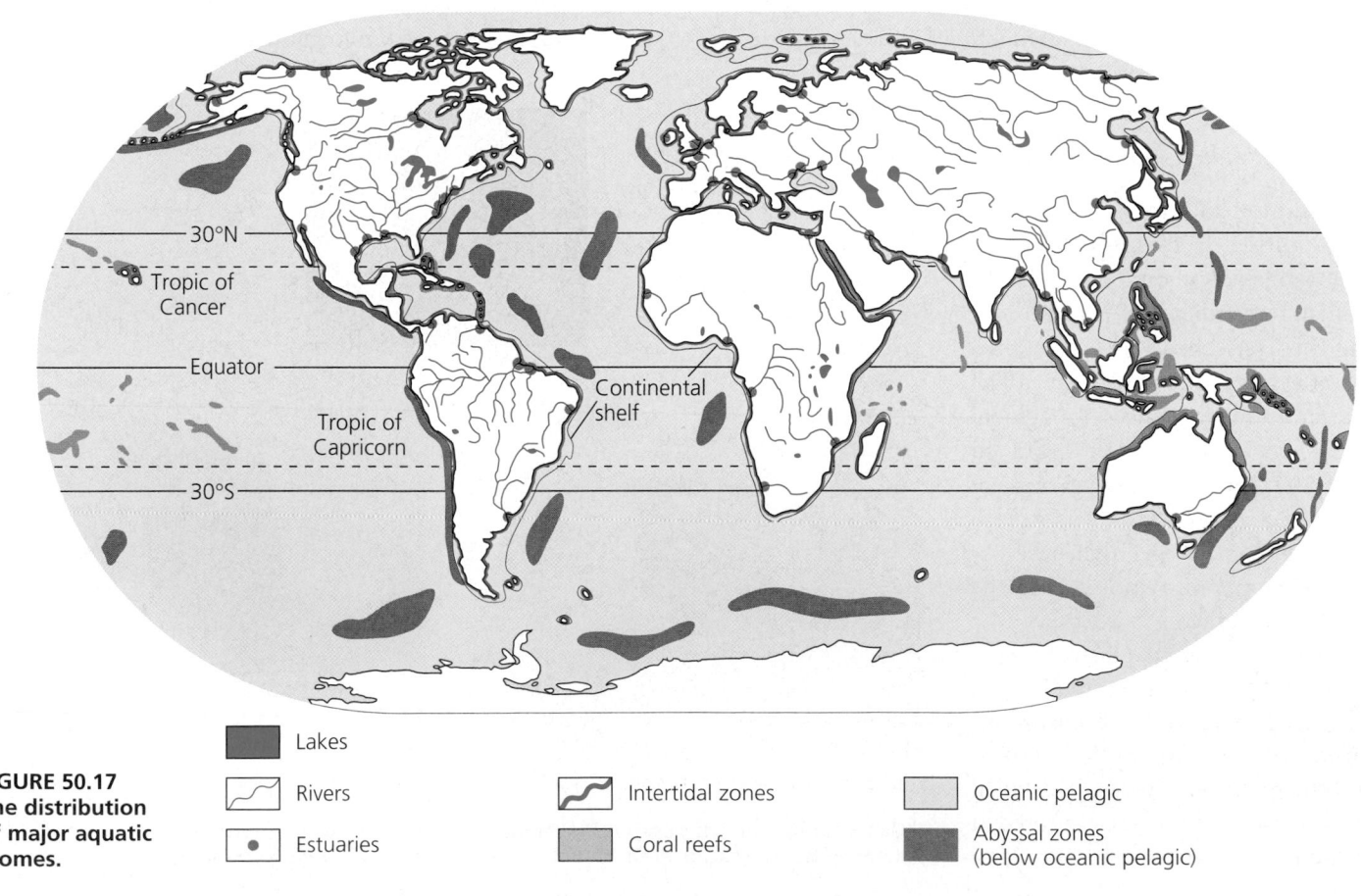

FIGURE 50.17 The distribution of major aquatic biomes.

Lakes
Rivers
Estuaries
Intertidal zones
Coral reefs
Oceanic pelagic
Abyssal zones (below oceanic pelagic)

rapid temperature change called a **thermocline** separates a more uniformly warm upper layer from more uniformly cold deeper waters. At the bottom of all aquatic biomes, the substrate is called the **benthic zone.** Made up of sand and organic and inorganic sediments ("ooze"), the benthic zone is occupied by communities of organisms collectively called **benthos.** A major source of food for the benthos is dead organic matter called **detritus.** In lakes and oceans, detritus "rains" down from the productive surface waters of the photic zone.

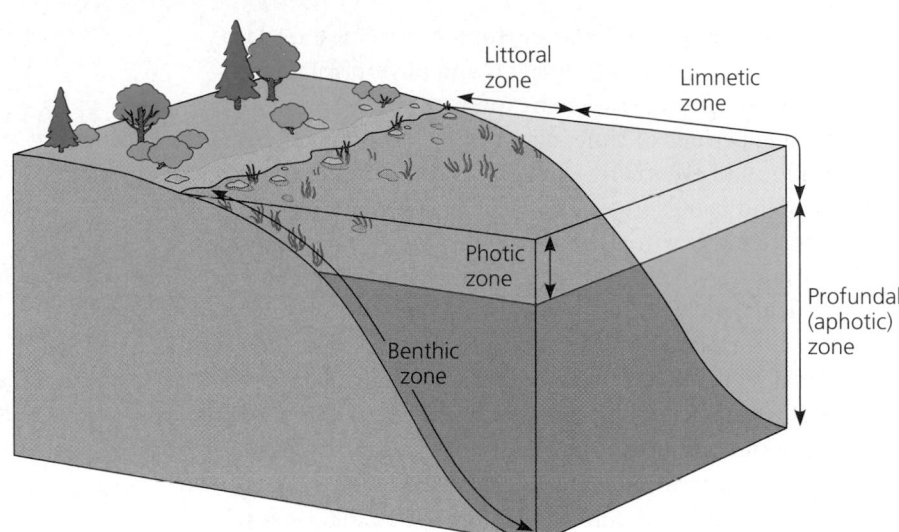

FIGURE 50.18 Zonation in a lake.

Freshwater Biomes

There are two general categories of freshwater biomes: standing bodies of water (such as lakes and ponds) and moving bodies of water (rivers and streams). Standing bodies of water range from small ponds that are a few square meters to large lakes that are thousands of square kilometers in area. In most lakes, communities of organisms are distributed according to the depth of the water and its distance from shore (FIGURE 50.18). Rooted and floating aquatic plants flourish in the **littoral zone,** the shallow, well-lit waters close to shore. In a lake, the well lit, open surface waters farther from shore, called the **limnetic zone,** are occupied by a variety of phytoplankton consisting of algae and cyanobacteria. These organisms photosynthesize and reproduce at a high rate during spring and summer. Zooplankton, mostly rotifers and small crustaceans, graze on the phytoplankton. The zooplankton are consumed by many small

fish, which in turn become food for larger fish, semiaquatic snakes and turtles, and fish-eating birds.

Most of the small organisms of a lake's limnetic zone are short-lived, and their remains continually sink into a deep aphotic region, called the **profundal zone,** and down to the benthic zone. Microbes in the profundal and benthic zones use oxygen for cellular respiration as they decompose this detritus.

Lakes are often classified according to their production of organic matter. **Oligotrophic** lakes are deep and nutrient-poor, and the phytoplankton in the limnetic zone are relatively sparse and not very productive (FIGURE 50.19a). **Eutrophic** lakes, in contrast, are usually shallower, and the nutrient content of their water is high. As a result, the phytoplankton are very productive, and the waters are often quite murky (FIGURE 50.19b).

(a) An oligotrophic lake

FIGURE 50.19 Freshwater biomes.
(a) Oligotrophic lakes, such as Lake Baikal (shown here), in Siberia, are nutrient-poor with a small surface area relative to depth. The bottom sediments are low in decomposable organic matter, limiting bacterial populations in the benthos. The shortage of nutrients in the water limits photosynthesis by plankton in the limnetic zone; as a result, the water is clear and

(b) A eutrophic lake

oxygen-rich and usually supports diverse populations of fish and invertebrates.
(b) Eutrophic lakes, such as this one called Peck's Pond, in the Pocono Mountains of eastern Pennsylvania, are nutrient-rich and generally have a larger surface area relative to depth. The availability of nutrients supports a high rate of photosynthesis, leading to murkier water than that of an oligotrophic lake. High

(c) A stream flowing into a river

organic content in the benthos leads to high decomposition rates and potentially low oxygen supplies in the profundal and benthic zones.
(c) Rivers and streams support substantially different biological communities than those in ponds and lakes. The photograph shows the mouth of a stream, flowing into the much larger Snake River, in Idaho.

Between the oligotrophic and eutrophic extremes are lakes with a moderate amount of nutrients and phytoplankton productivity; these lakes are said to be **mesotrophic.**

Over long periods of time, oligotrophic lakes may become mesotrophic and eventually eutrophic as runoff from the surrounding land brings in additional sediments and nutrients. Unfortunately, human activities often speed this natural process dramatically. Runoff from fertilized lawns and agricultural fields and the dumping of municipal wastes enrich the lakes with excessive amounts of nitrogen and phosphorus; the normally low concentrations of these nutrients limit the growth of algae, including phytoplankton. The result of such pollution is often a population explosion of algae, the production of much detritus, and the eventual depletion of oxygen supplies. Such "cultural eutrophication" makes the water unusable and degrades the lake's aesthetic value (see Chapter 54).

Streams and rivers are bodies of water moving continuously in one direction (FIGURE 50.19c). At the headwaters of a stream (perhaps a spring or snowmelt), the water is often cold and clear, and it carries little sediment and relatively few mineral nutrients. The channel is usually narrow, with a swift current passing over a rocky substrate. Farther downstream, where numerous tributaries may have joined to form a river, the water may be more turbid, carrying substantially more sediment (from the erosion of soil) and nutrients. The channel near the mouth of a river is relatively wide, and the substrate is generally silty from the deposition of sediments over long periods of time.

The nutrient content of flowing water biomes is largely determined by the terrain and vegetation through which the water flows. Fallen leaves from dense, overhanging vegetation can add substantial amounts of organic matter, and the weathering of rocks can increase the concentration of inorganic nutrients. The often turbulent flow of a stream constantly oxygenates the water, whereas the sometimes murkier, warmer waters of a large river may contain relatively little oxygen. Stream- and river-dwelling animals exhibit evolutionary adaptations that enable them to resist being swept away. The smaller ones are typically flat in shape and can attach to rocks temporarily. Many arthropods live on the underside or downstream side of rocks, thereby exploiting a microhabitat that is relatively free of turbulent flow.

Many streams and rivers have been polluted by human activities. For centuries, humans have used streams and rivers as depositories of waste, thinking that these materials would be diluted and carried downstream. While some pollutants are carried far from their source, many settle to the bottom, where they can be taken up by aquatic organisms. Even the pollutants that are carried away contribute to estuary, ocean, and lake pollution. Humans have also changed flow patterns, using stream channelization to speed water flow and dams to restrain it. In many cases, dams have completely changed the downstream ecosystems, altering the intensity and volume of water flow and affecting fish and invertebrate populations (FIGURE 50.20).

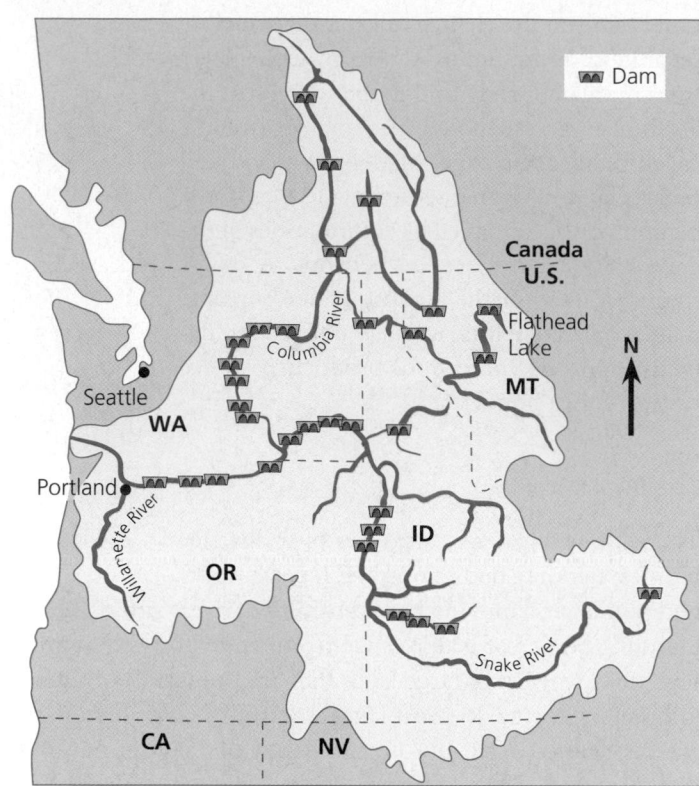

FIGURE 50.20 Damming the Columbia River Basin. If Lewis and Clark lived today, they would have a hard time navigating the Columbia. This map shows only the largest of the 250 dams that have altered freshwater ecosystems throughout the Pacific Northwest. The great concrete obstacles make it difficult for salmon to swim upriver to their breeding streams, though many dams now have "fish ladders" that provide detours.

Wetlands

In the simplest sense, a **wetland** is an area covered with water that supports aquatic plants. In fact, wetlands range from periodically flooded regions to soil that is permanently saturated during the growing season. These conditions favor the growth of specially adapted plants called hydrophytes ("water plants"), which can grow in water or in soil that is periodically anaerobic due to the presence of water. Hydrophytes include floating pond lilies and emergent cattails, many sedges, tamarack, and black spruce. Both the hydrology and the vegetation of an area are important determinants of its classification as a wetland—a classification that can be critical when federal, state, and local governments are making preservation decisions based on rigorous, and often conflicting, definitions.

There are different types of wetlands, ranging from marshes to swamps to bogs. All these varieties, however, generally develop in one of three topographic situations. Basin wetlands develop in shallow basins, ranging from upland depressions to filled-in lakes and ponds (FIGURE 50.21a). Riverine wetlands develop along shallow and periodically flooded banks of rivers and streams. Fringe wetlands occur along the coasts of large lakes and seas, where water flows back and forth because of

(a) **Wetlands.** This marsh in Pennsylvania is an example of a basin wetland. Marshes are usually covered with water year-round. Predominant plants are emergent, with stems and leaves extending above the water surface; they include the pond lilies, reeds, and cattails visible in this photograph. Other kinds of wetlands include swamps (dominated by woody plants), bogs (dominated by sphagnum mosses), and seasonal pools.

FIGURE 50.21 Wetlands and estuaries.

(b) **An estuary.** This view of an estuary that is part of Chesapeake Bay, in Maryland, shows the intimate association of river mouths and the marine environment into which they carry water. Unfortunately, the land surrounding Chesapeake Bay is heavily populated and industrialized, and pollution that enters the bay through four major rivers has made it unsuitable for many plant and animal species. What was once a bountiful natural source of seafood and other resources has been degraded and rendered less productive by human activity.

rising lake levels or tidal action. Thus, fringe wetlands include both freshwater and marine biomes. Marine coastal wetlands are closely linked to estuaries, which we examine shortly.

Ecologically, wetlands are among the richest of biomes. They contain a diverse community of invertebrates, supporting a wide variety of birds. Herbivores from crustaceans to muskrats consume algae, detritus, and plants. In addition to the rich diversity of species that is supported by wetlands, the ecological and economic value of wetlands exceeds that expected from their geographic extent alone; they provide water-storage basins that reduce the intensity of flooding, and they improve water quality by filtering pollutants. In the past, humans have often regarded wetlands as wastelands—as sources of mosquitoes, flies, and bad odors—and have destroyed many wetlands, mostly filling in with earth to provide land for agriculture and development. Both governments and private organizations are now attempting to protect remaining wetlands through acquisition, economic incentives, and regulation. A great deal of research is under way to determine how wetlands can be created or restored.

Estuaries

The area where a freshwater stream or river merges with the ocean is called an **estuary;** it is often bordered by extensive coastal wetlands called mudflats and salt marshes (FIGURE 50.21b). Salinity varies spatially within estuaries, from nearly that of fresh water to that of the ocean; it also varies over the course of a day with the rise and fall of the tides. Nutrients from the river enrich estuarine waters, making estuaries one of the most biologically productive biomes on Earth.

Salt marsh grasses, algae, and phytoplankton are the major producers in estuaries. This environment also supports a variety of worms, oysters, crabs, and many of the fish species that humans consume. Many marine invertebrates and fishes use estuaries as a breeding ground or migrate through them to freshwater habitats upstream. Estuaries are also crucial feeding areas for many semiaquatic vertebrates, particularly waterfowl.

Although estuaries support a wide variety of commercially valuable species, areas around estuaries are also prime locations for commercial and residential developments. In addition, estuaries are unfortunately at the receiving end for pollutants dumped upstream. Very little undisturbed estuarine habitat remains, and a large percentage has been totally eliminated by landfill and development. Many states have now—rather belatedly—taken steps to preserve their remaining estuaries.

Zonation in Marine Communities

Similar to the communities in freshwater lakes, marine communities are distributed according to depth of the water, degree of light penetration, distance from shore, and open water versus bottom (FIGURE 50.22, p. 1110). Marine communities illustrate graphically the limitations on distributions that result from these abiotic factors. There is a photic zone where phytoplankton, zooplankton, and many fish species occur, and an aphotic zone below. Because water absorbs light so well and the ocean is so deep, most of the ocean volume is virtually devoid of light, except for tiny amounts produced by a few luminescent fishes and invertebrates. The zone where land meets water is called the **intertidal zone;** beyond the intertidal zone is the **neritic zone,** the shallow regions over the continental shelves; and past the continental

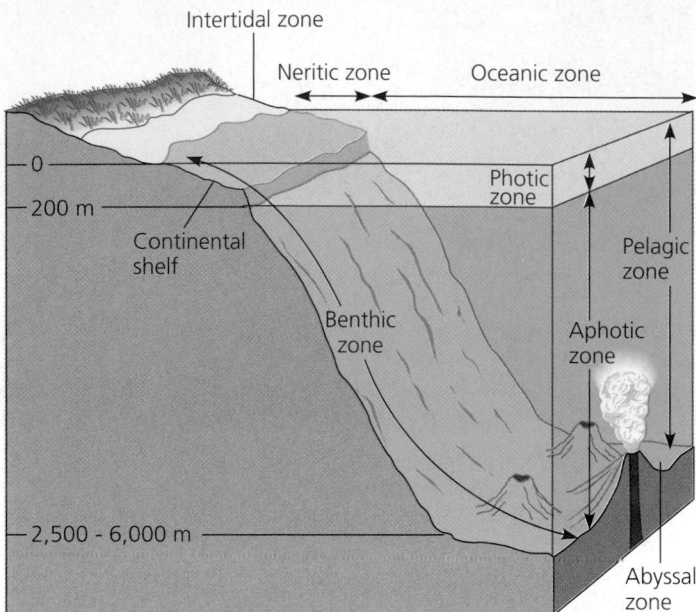

Intertidal zone

Neritic zone　　Oceanic zone

Continental shelf

Benthic zone

Photic zone

Pelagic zone

Aphotic zone

Abyssal zone

0

200 m

2,500 - 6,000 m

FIGURE 50.22 Zonation in the marine environment. The marine environment can be classified on the basis of three physical criteria: light penetration (photic and aphotic zones), distance from shore and water depth (intertidal, neritic, and oceanic zones), and whether it is open water or bottom (pelagic and benthic zones). The abyssal zone is the benthic region in deep oceans. Ecologists often combine two designations, such as the oceanic pelagic zone, to identify the location of a biome.

shelf is the **oceanic zone,** reaching very great depths. Finally, open water of any depth is the **pelagic zone,** at the bottom of which is the seafloor, or **benthic zone.**

Intertidal Zones

An intertidal zone is alternately submerged and exposed by the twice-daily cycle of tides. Intertidal communities are therefore subject to huge daily variations in the availability of seawater (and the nutrients it carries) and in temperature. Perhaps most significant of all, intertidal organisms are subject to the mechanical forces of wave action, which can dislodge them from the habitat.

The rocky intertidal zone is vertically stratified and provides excellent examples of distributional limitations over short distances (FIGURE 50.23a). Most of the organisms have structural adaptations that enable them to attach to the hard substrate in this physically tumultuous environment. On sandy substrates (beaches) or mudflats, the intertidal zone is not as clearly stratified. Wave action and tides constantly move the particles of mud and sand, and few large algae or plants occupy these habitats. Many animals, such as suspension-feeding worms and clams and predatory crustaceans, bury themselves in sand or mud, feeding when the tides bring sources of food. Other surface-dwelling organisms, such as crabs and shorebirds, are scavengers or predators on these organisms.

Partly because of our strong attraction to the seashore, humans have had a long-term impact on intertidal ecosystems. The recreational use of ocean shores has caused a severe decline in the numbers of many beach-nesting birds and sea turtles. Incoming tides carry in polluted water and old fishing lines and plastic debris that can harm wildlife. The most dramatic intertidal pollutant is probably oil, which harms not only birds and marine mammals but also intertidal algae and invertebrates. The ultimate outcome of oil pollution in intertidal zones is reduced species diversity, with increases in the populations of a few oil-resistant species.

Coral Reefs

In warm tropical waters in the neritic zone, **coral reefs** constitute a conspicuous and distinctive biome. Currents and waves constantly renew nutrient supplies to the reefs, and sunlight penetrates to the ocean floor, allowing photosynthesis.

Coral reefs are dominated by the structure of the coral itself, formed by a diverse group of cnidarians that secrete hard external skeletons made of calcium carbonate (see Chapter 33). These skeletons vary in shape, forming a substrate on which other corals, sponges, and algae grow (FIGURE 50.23b). Red algae encrusted with calcium carbonate also add large amounts of limestone to most reefs, as do bryozoans. The coral animals themselves suspension-feed on microscopic organisms and particles of organic debris. They also obtain organic molecules from the photosynthesis of symbiotic dinoflagellate algae that live in their tissues. Coral animals can survive without the dinoflagellates, but their rate of calcium carbonate deposition is much slower without them; thus, reef formation by corals depends on this symbiotic association.

Some coral reefs cover enormous expanses of shallow ocean, but this delicate biome is easily degraded by pollution and development, as well as by souvenir hunters who gather the coral skeletons. High water temperatures (greater than 30°C) cause corals to "bleach"—to expel their symbiotic dinoflagellates and die. In 1998, coral reefs around the world suffered moderate to severe bleaching, and there is great concern that global warming could destroy many coral reefs. Corals are also subject to damage from native and introduced predators, such as the crown-of-thorns sea star, which has undergone a population explosion in many regions and destroyed a number of coral reefs in parts of the western Pacific Ocean. Reef communities are very old and grow very slowly; they may not be able to withstand continued human encroachment and dramatic climate changes.

The Oceanic Pelagic Biome

Most of the ocean's water lies far from shore in the **oceanic pelagic biome,** constantly mixed by ocean currents. Nutrient concentrations are generally lower in the open ocean than in coastal areas because the remains of plankton and other organ-

(a) Intertidal zone

(b) Coral reefs

(c) Benthos: a deep-sea vent community

FIGURE 50.23 Examples of marine biomes.
(a) Intertidal zones. This photograph of rocky intertidal zones on the Oregon coast was taken at low tide to illustrate the vertical zonation of algae and animals. The density of organisms in each of the three major zones is roughly proportional to the percentage of time the zone is submerged. Organisms in the uppermost zone—grazing mollusks, suspension-feeding barnacles, and a few algae—are submerged only during the highest tides and have numerous adaptations that prevent dehydration and overheating. The middle zone, generally submerged at high tide and exposed at low tide, is inhabited by a diverse array of algae,

sponges, sea anemones, mollusks, crustaceans, echinoderms, and small fishes. The bottom intertidal zone is exposed only during the lowest tides. A dense cover of seaweeds in this zone often harbors a diversity of invertebrates and fishes. **(b) Coral reefs.** This coral reef in Fiji illustrates some of the immense variety of microorganisms, invertebrates, and fishes that live among the coral and algae, making coral reefs one of the most diverse and productive biomes on Earth. **(c) Benthos: a deep-sea vent community.** The species composition of benthos varies dramatically with water depth. Pictured here is a vent community, first discovered at a depth of 2,500 m in the late

1970s. These communities are found at spreading centers on the seafloor, where hot magma superheats the water. About a dozen species of prokaryotes identified near the vents are chemoautotrophic producers that obtain energy by oxidizing H_2S formed by a reaction of the hot water with dissolved sulfate (SO_4^{2-}). Among the animals in these communities are giant tube-dwelling worms (pictured here), some more than 1 m long. They are apparently nourished by chemo-synthetic prokaryotes that live as symbionts within the worms. Many other invertebrates, including arthropods and echinoderms, are also abundant around the vents.

isms sink below the photic region into the dark, lower benthic zone. In some tropical areas, surface waters are lower in nutrients than the surface waters of temperate oceans because a year-round thermal stratification prevents an exchange of nutrients between the surface and the deep. Temperate oceans generally are more productive, because, like temperate lakes, they experience a nutrient turnover in the spring and, to a limited extent, in the fall. The springtime recirculation of nutrients from the depths stimulates a surge of photosynthetic plankton growth.

Photosynthetic plankton grow and reproduce rapidly in the photic region of the oceanic biome. Modern sampling methods, which take bacterial photosynthesis into account, show that the plankton's rate of organic food production is higher than formerly thought. Nonetheless, photosynthetic plankton account for less than half the photosynthetic activity on Earth. Zooplankton, including protozoans, worms, copepods, shrimp-like krill, jellies, and the small larvae of invertebrates and fishes, graze on the photosynthetic plankton. Most plankton exhibit morphological structures, such as bubble-trapping

spines, lipid droplets, gelatinous capsules, and air bladders, that help them stay within the photic zone.

The oceanic pelagic biome also includes free-swimming animals, called nekton, that can move against the currents to locate food. Examples of nekton are large squids, fishes, sea turtles, and marine mammals. These animals feed either on plankton or on each other. Although many of these animals feed in the photic region of the pelagic zone, others live at great depths, where fish may have enlarged eyes, enabling them to see in the very dim light, and luminescent organs that attract mates and prey. Many pelagic birds, such as petrels, terns, albatrosses, and boobies, catch fish in the surface waters.

Benthos

The ocean bottom below the neritic and pelagic zones is the benthic zone, as in other aquatic biomes. Nutrients reach the seafloor from the waters above by "raining down" in the form of detritus. Although the benthic zone in shallow, near-coastal waters may receive substantial sunlight, light and temperature decline dramatically with depth.

Neritic benthic communities are extremely productive, consisting of bacteria, fungi, seaweeds and filamentous algae, numerous invertebrates, and fishes. Species composition of these communities varies with distance from the shore, water depth, and composition of the bottom.

Organisms in the very deep benthic communities, or **abyssal zone,** are adapted to continuous cold (about 3°C), extremely high water pressure, near or total absence of light, and low nutrient concentrations. However, oxygen is usually present in abyssal waters, and a fairly diverse community of invertebrates and fishes occupies this region. Marine scientists have also discovered a unique assemblage of organisms associated with **deep-sea hydrothermal vents** of volcanic origin in midocean ridges (FIGURE 50.23c). In this dark, hot, oxygen-deficient environment, the food producers are not photosynthesizing organisms but chemoautotrophic prokaryotes (see Chapter 27). The organic molecules they synthesize support a food chain that includes giant polychaete worms, arthropods, echinoderms, and fishes.

The geographic distribution of terrestrial biomes is based mainly on regional variations in climate

All the abiotic factors we covered earlier in the chapter, but especially climate, are important in determining why a particular terrestrial biome is found in a certain area. Because there are latitudinal patterns of climate over Earth's surface (see FIGURES 50.11–50.13), there are also latitudinal patterns of biome distribution. For example, coniferous forests extend in a broad band across North America, Europe, and Asia (FIGURE 50.24).

Terrestrial biomes are often named for major physical or climatic features and for their predominant vegetation. For example, temperate grasslands are dominated by various grass species and are generally found in middle latitudes, where the climate is more moderate than in the tropics or polar regions. Each biome is also characterized by microorganisms, fungi, and animals adapted to that particular environment. Temperate grasslands, for example, are more likely than forests to be populated by large grazing mammals.

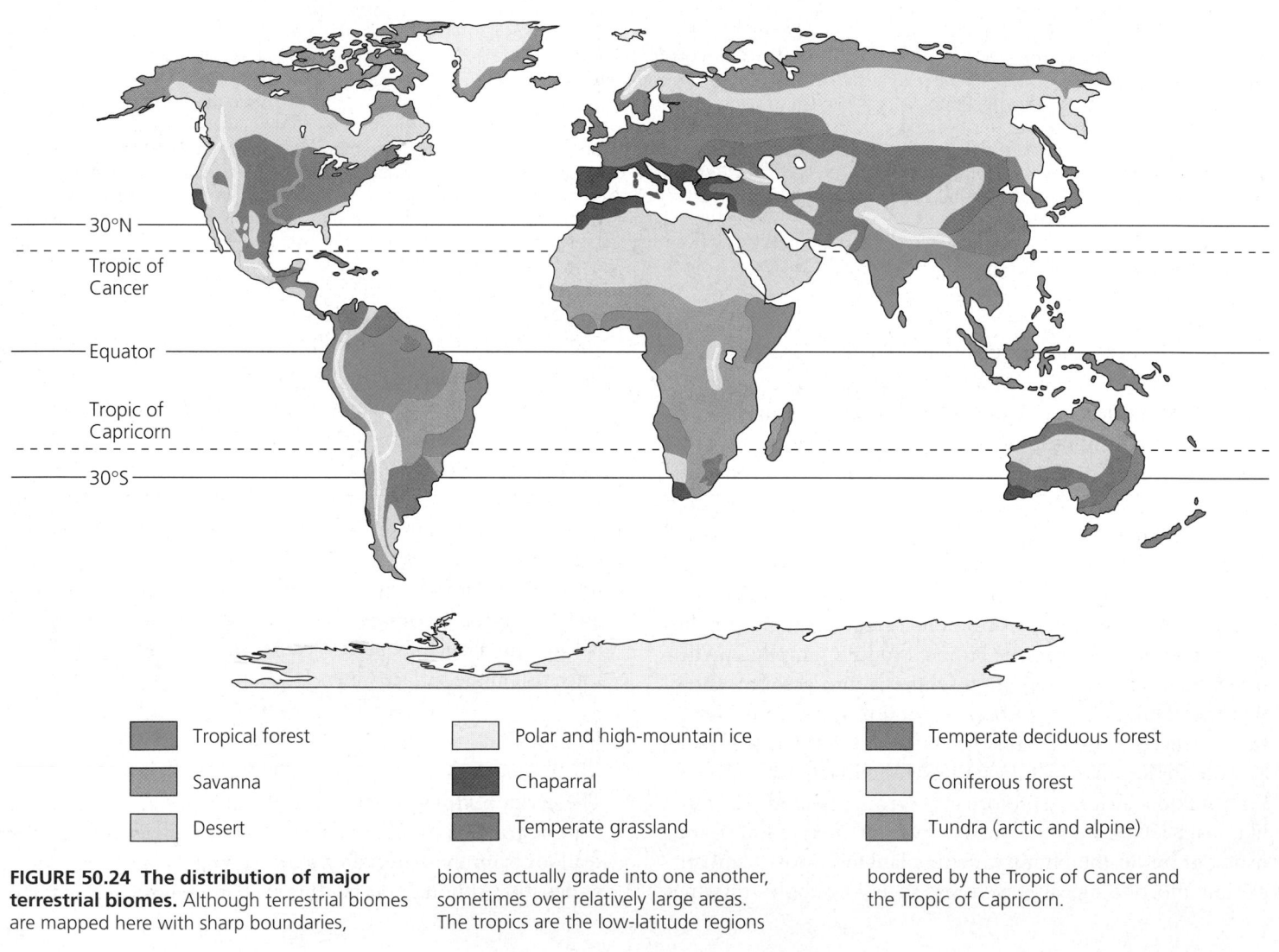

Tropical forest

Savanna

Desert

Polar and high-mountain ice

Chaparral

Temperate grassland

Temperate deciduous forest

Coniferous forest

Tundra (arctic and alpine)

FIGURE 50.24 The distribution of major terrestrial biomes. Although terrestrial biomes are mapped here with sharp boundaries, biomes actually grade into one another, sometimes over relatively large areas. The tropics are the low-latitude regions bordered by the Tropic of Cancer and the Tropic of Capricorn.

Vertical stratification is an important feature of terrestrial biomes, and the shapes and sizes of plants largely define the layering. For example, in many forests, the layers consist of the upper **canopy,** then the low-tree stratum, the shrub understory, the ground layer of herbaceous plants, the forest floor (litter layer), and finally the root layer. Other (nonforest) biomes have similar, though usually less pronounced vertical strata. For instance, grasslands have a canopy formed by an herbaceous layer of grass species, a litter layer, and a root layer. The root layer in arctic tundra is shallower than in most other biomes because a permanently frozen stratum called **permafrost** underlies it.

Vertical stratification of a biome's vegetation provides many different habitats for animals, which often occupy well-defined feeding groups, from the insectivorous and carnivorous birds and bats that feed above canopies to the small mammals, numerous worms, and arthropods that forage the litter and root layers for food.

Terrestrial biomes usually grade into each other, without sharp boundaries. The area of intergradation may be wide or narrow and is called an *ecotone*.

The actual species composition of any one kind of biome varies from one location to another. For instance, in the northern coniferous forest (taiga) of North America, red spruce is common in the east but does not occur in most other areas, where black spruce and white spruce are abundant. Although the vegetation of African deserts superficially resembles that of North American deserts, the plants are in different families. Such "ecological equivalents" can arise because of convergent evolution (see Chapter 25).

Biomes are dynamic, and natural disturbance rather than stability tends to be the rule. As a result of disturbance, biomes usually exhibit extensive patchiness, with several communities represented in any particular area. Hurricanes create openings in tropical and temperate forests. In northern coniferous forests, old trees die and fall over, or snowfall may break branches and small trees, producing openings or gaps that allow deciduous species, such as aspen and birch, to grow. In many biomes, even the dominant plants depend on periodic disturbance. For example, fire is an integral component of grasslands, savannas, chaparral, and many coniferous forests. Before agricultural and urban development, much of the southeastern United States was dominated by a single conifer species, the longleaf pine. Without periodic burning, deciduous trees tended to replace the pines. Forest managers now use fire as a tool to help maintain many coniferous forests.

In many biomes today, extensive human activities have radically altered the natural patterns of periodic physical disturbance. Most of the eastern United States, for example, is classified as temperate deciduous forest, but human activity has eliminated all but a tiny percentage of the original forest. Fires, which used to be part of life on the Great Plains, are now controlled for the sake of agricultural land use. Humans have altered much of Earth's surface, replacing original biomes with urban and agricultural ones.

FIGURE 50.25, on pages 1113–1117, surveys the major terrestrial biomes, beginning near the equator and moving toward the poles.

FIGURE 50.25 Examples of terrestrial biomes.

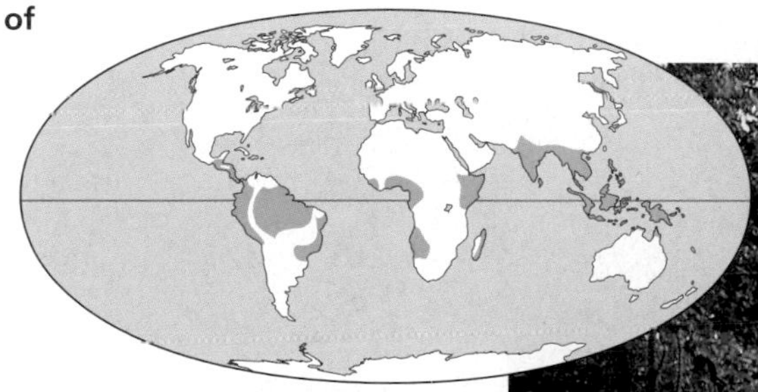

(a) Tropical forest. The photograph shows a tropical rain forest in Costa Rica. Tropical rain forests have pronounced vertical stratification. Trees in the canopy make up the topmost stratum. The canopy is often closed, so that little light reaches the ground below. When an opening does occur, perhaps because of a fallen tree, other trees and large woody vines grow rapidly, competing for light and space as they fill the gap. Many of the trees are covered with epiphytes (plants that grow on other plants rather than in soil), such as orchids and bromeliads. Rainfall, which varies from region to region in the tropics, is the prime determinant of the vegetation growing in an area. In lowland areas that have a prolonged dry season or scarce rainfall at any time, tropical dry forests predominate. The plants found there are a mixture of thorny shrubs and trees and succulents. In regions with distinct wet and dry seasons, tropical deciduous trees are common.

FIGURE 50.25 Examples of terrestrial biomes (continued).

(b) Savanna. This Kenyan savanna is a showcase of large herbivores and their predators. Actually, the dominant herbivores here and in other savannas are insects, especially ants and termites. Grasses and scattered trees are the dominant plants. Fire is an important abiotic component, and the dominant plant species are fire adapted. The luxuriant growth of grasses and forbs (small broadleaf plants) during the rainy season provides a rich food source for animals. However, large grazing mammals must migrate to greener pastures and scattered watering holes during regular periods of seasonal drought.

(c) Desert. Sparse rainfall (less than 30 cm per year) largely determines that an area will be a desert. Some deserts have soil surface temperatures above 60°C during the day. Other deserts, such as those west of the Rocky Mountains and in central Asia, are relatively cold. The Sonoran Desert of southern Arizona (shown here) is characterized by giant saguaro cacti and deeply rooted shrubs. Evolutionary adaptations of desert plants and animals include a remarkable array of mechanisms that store water. The "pleated" structure of saguaro cacti enables the plants to expand when they absorb water during wet periods. Some desert mice never drink, deriving all their water from the metabolic breakdown of the seeds they eat. Many desert plants also rely on CAM photosynthesis, a metabolic adaptation that conserves water in this arid environment (see Chapter 10). Protective adaptations that deter feeding by mammals and insects, such as spines on cacti and poisons in the leaves of shrubs, are also common in desert plants.

FIGURE 50.25 Examples of terrestrial biomes *(continued).*

(d) Chaparral. Dense, spiny, evergreen shrubs dominate chaparral biomes, midlatitude coastal areas with mild, rainy winters and long, hot, dry summers. Plants of the chaparral, such as those in this California scrubland, are adapted to and dependent on periodic fires. The dry, woody shrubs are frequently ignited by lightning and by careless human activities, creating summer and autumn brushfires in the densely populated canyons of southern California and elsewhere. Some of the shrubs produce seeds that will germinate only after a hot fire. Food reserves stored in their fire-resistant roots enable them to resprout quickly and use nutrients released by fires.

(e) Temperate grassland. The veldts of South Africa, the puszta of Hungary, the pampas of Argentina and Uruguay, the steppes of Russia, and the plains and prairies of central North America are all temperate grasslands. The key to the persistence of grasslands is seasonal drought, occasional fires, and grazing by large mammals, all of which prevent establishment of woody shrubs and trees. Temperate grasslands, such as the tallgrass prairie in Kansas (shown here), once covered much of central North America. Because grassland soil is both deep and rich in nutrients, these habitats provide fertile land for agriculture. Most grassland in the United States has been converted to farmland, and very little natural prairie exists today.

FIGURE 50.25 Examples of terrestrial biomes (continued).

(f) Temperate deciduous forest. Dense stands of deciduous trees are the trademark of temperate deciduous forests, such as this one in Great Smoky Mountains National Park in North Carolina. Temperate deciduous forests occur throughout midlatitudes where there is sufficient moisture to support the growth of large trees. More open than rain forests and not as tall, a mature temperate deciduous forest has distinct vertical layers, including one or two strata of trees, an understory of shrubs, and an herbaceous stratum. Deciduous forest trees drop their leaves before winter, when temperatures are too low for effective photosynthesis and water lost through transpiration is not easily replaced from frozen soil. Many temperate deciduous forest mammals also enter a dormant winter state called hibernation, and some bird species migrate to warmer climates. Virtually all the original deciduous forests in North America were destroyed by logging and land clearing for agriculture and urban development. In contrast to drier biomes, these forests tend to recover after disturbance, and today we see deciduous trees dominating undeveloped areas over much of their former range.

(g) Coniferous forest. Cone-bearing trees, such as pine, spruce, fir, and hemlock, dominate coniferous forests. Coastal coniferous forests of the U.S. Pacific Northwest, such as the one shown here in Olympic National Park in western Washington, are actually temperate rain forests. Warm, moist air from the Pacific Ocean supports these unique communities, which like most coniferous forests are dominated by one or a few tree species. Extending in a broad band across northern North America and Eurasia to the southern border of the arctic tundra, the northern coniferous forest, or taiga, is the largest terrestrial biome on Earth (see FIGURE 50.24). Taiga receives heavy snowfall during winter. The conical shape of many conifers prevents too much snow from accumulating on and breaking their branches. Coniferous forests are being logged at an alarming rate, and the old-growth stands of these trees may soon disappear.

FIGURE 50.25 **Examples of terrestrial biomes** *(continued).*

(h) Tundra. Permafrost (permanently frozen subsoil), bitterly cold temperatures, and high winds are responsible for the absence of trees and other tall plants in this arctic tundra in central Alaska (photographed in autumn). Although the arctic tundra receives very little annual rainfall (see FIGURE 50.10), water cannot penetrate the underlying permafrost and accumulates in pools on the shallow topsoil during the short summer. Tundra covers expansive areas of the Arctic, amounting to 20% of Earth's land surface. High winds and cold temperatures create similar plant communities, called alpine tundra, on very high mountaintops at all latitudes, including the tropics.

THE SPATIAL SCALE OF DISTRIBUTIONS

Throughout this chapter, we have examined abiotic factors, such as climate, and biotic factors, such as predation, that contribute to the distribution of the biosphere's diverse species. We have assumed that it is reasonably straightforward to map the geographic range of a species. But this assumption breaks down as we map the detailed distribution of a species in a local area. No species occurs everywhere throughout its range.

Different factors may determine the distribution of a species on different scales

FIGURE 50.26 (p. 1118) illustrates the problem of describing a species' geographic range on different scales. At one extreme, the range of a species is defined by the worldwide extent of occurrence, a line drawn on a map around the outermost points at which the species has been observed. This is the definition of geographic range used in bird field guides and other natural history guidebooks. At the other extreme, we could measure a much smaller area within the larger geographic range and map the location of each individual. If a particular habitat is not occupied by the species, this region would not be included in its geographic range. Ecologists would like to know the actual areas occupied by each species throughout its range, but such detailed data are not available for most organisms. The important point of FIGURE 50.26 is that we can measure geographic ranges on several spatial scales and that even for a single species, there may be several answers to the question: What limits geographic distribution? Abiotic influences, such as climate, are paramount on the global scale. However, at the local level, more subtle biotic interactions, such as symbiosis, may be the major explanation for why a species occurs in one locale and not another.

Most species have small geographic ranges

Most species in all taxonomic groups have small geographic ranges; only a small minority of species are widespread.

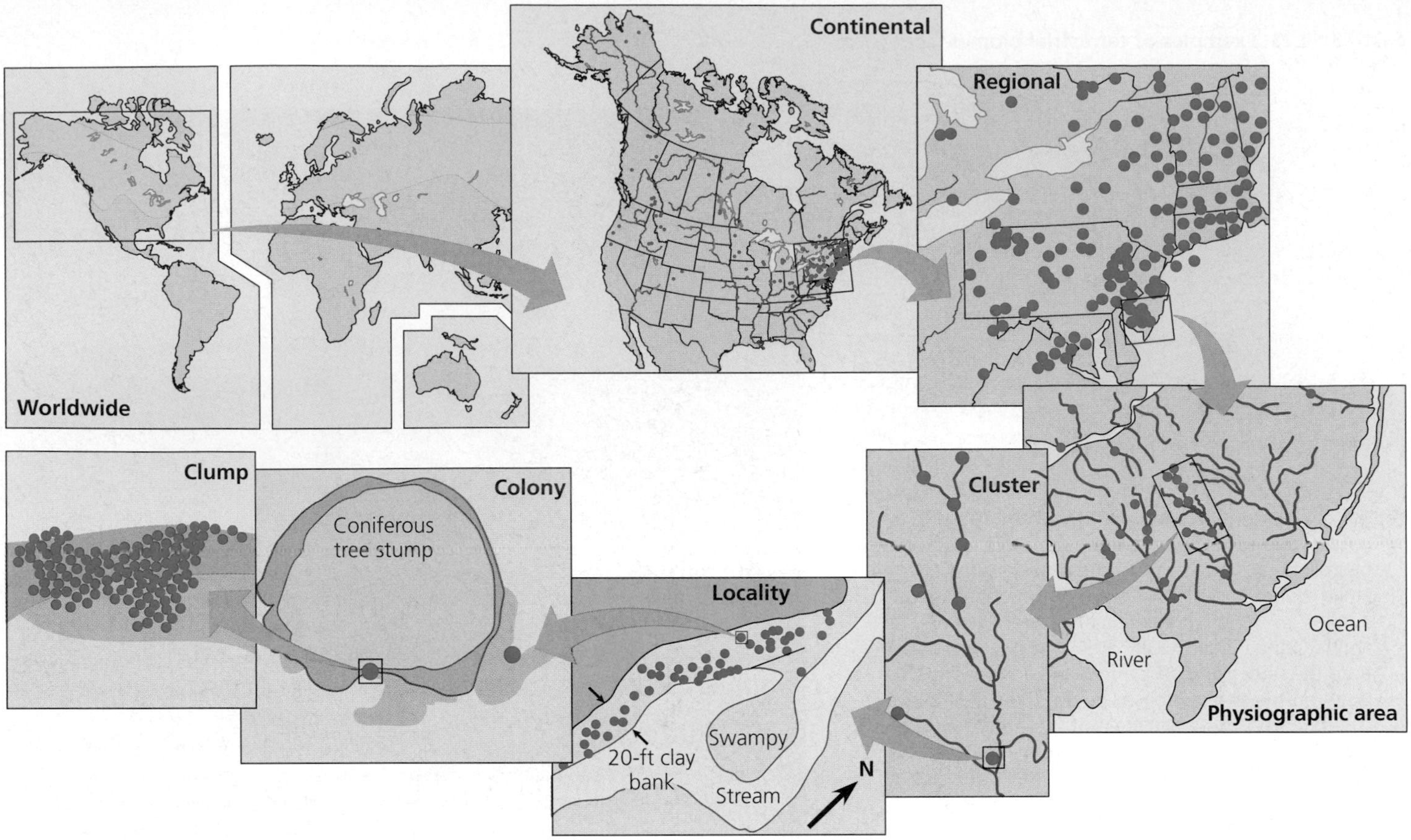

Continental

Regional

Worldwide

Clump

Colony

Coniferous tree stump

Cluster

Locality

Ocean

River

Physiographic area

Swampy

20-ft clay bank

Stream

N

FIGURE 50.26 A hierarchy of scales for analyzing the geographic distribution of the moss *Tetraphis*. The question: What limits geographic distribution? There may be different answers, depending on the scale of our analysis.

FIGURE 50.27 illustrates this for North American birds and British vascular plants. Ecologists do not know the reason for this pattern, even though it occurs in plants and animals, in aquatic and terrestrial groups, in invertebrates and vertebrates. This pattern of small ranges is related to the observation that most species are relatively rare in nature, and only common organisms tend to have widespread geographic ranges. But this observation simply begs the question: Why are some organisms rare and others common? To explain these patterns of life in the biosphere is one of the research challenges ecologists face. In Chapters 52 and 53, we'll examine how ecologists are trying to answer such questions about abundance and distribution in the biosphere. And throughout this ecology unit, we'll see the impact that we humans, by far the most abundant and widely distributed of all large animals, are having on the entire biosphere.

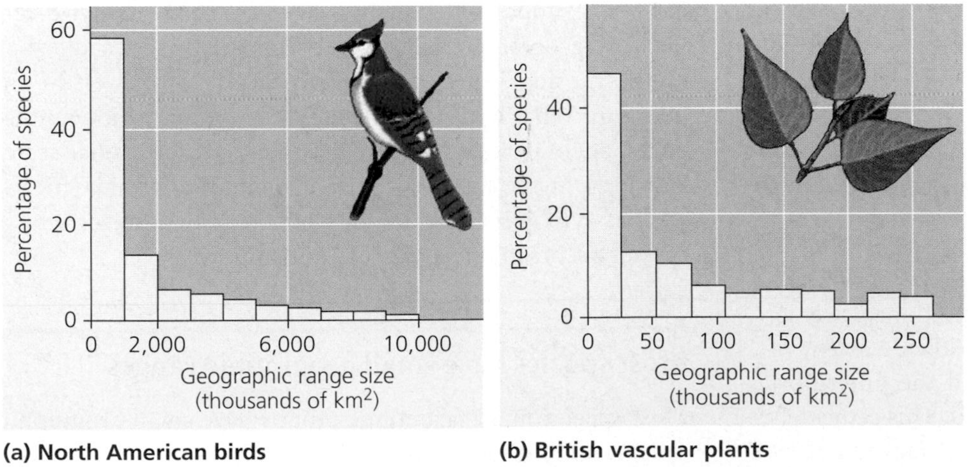

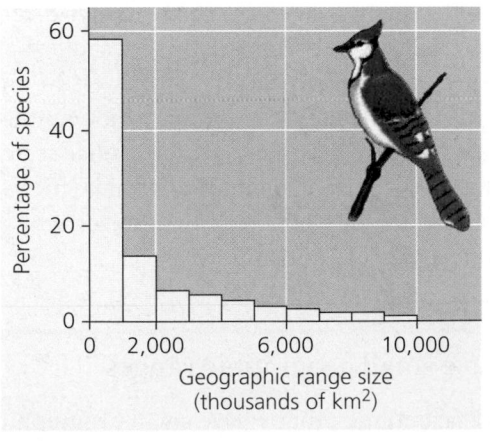

(a) North American birds

(b) British vascular plants

FIGURE 50.27 Most species have small geographic ranges. This concept is illustrated by **(a)** 1,370 species of North American birds and **(b)** 1,499 species of British vascular plants.

CHAPTER 50 REVIEW

Go to the Campbell Biology website (www.campbellbiology.com) to explore an interactive version of the Chapter Review.

Summary of Key Concepts

THE SCOPE OF ECOLOGY

- **The interactions between organisms and their environments determine the distribution and abundance of organisms (p. 1093, FIGURE 50.1)** The central questions of ecology are, Who lives where? and How many are there? Ecologists use observations and experiments to test hypothetical explanations for the environmental limitations of distribution and abundance. The factors of the environment include both abiotic (nonliving) and biotic (living) components.

- **Ecology and evolutionary biology are closely related sciences (p. 1093)** Events that occur in ecological time affect life on the scale of evolutionary time.

- **Ecological research ranges from the adaptations of individual organisms to the dynamics of the biosphere (pp. 1093–1095, FIGURE 50.2)** Ecological research spans increasingly comprehensive levels of organization, from the individual organism through populations, communities, ecosystems, and landscapes to the biosphere (the global ecosystem).

- **Ecology provides a scientific context for evaluating environmental issues (p. 1095, FIGURE 50.3)** Most ecologists favor the precautionary principle of "Look before you leap."

   Web/CD Activity 50A: *Science, Technology, and Society: DDT*

FACTORS AFFECTING THE DISTRIBUTION OF ORGANISMS

- **Species dispersal contributes to the distribution of organisms (pp. 1096–1098, FIGURES 50.6–50.8)** Transplanted species may disrupt the ecosystem at the new site, even causing the extinction of native species.

- **Behavior and habitat selection contribute to the distribution of organisms (pp. 1098–1099)** A species may use only a subset of the habitat in which it could survive.

- **Biotic factors affect the distribution of organisms (p. 1099, FIGURE 50.9)** Biotic factors involve interactions with other species, as in predation and competition.

- **Abiotic factors affect the distribution of organisms (p. 1100)** Among important abiotic factors are temperature, water, sunlight, wind, and rocks and soil.

   Web/CD Activity 50B: *Adaptations to Biotic and Abiotic Factors*
   Web/CD Case Study in the Process of Science: *How Do Abiotic Factors Affect Distribution of Organisms?*

- **Temperature and water are the major climatic factors determining the distribution of organisms (pp. 1100–1106, FIGURES 50.10–50.16)** Global patterns of distribution are set by climate and seasonality, which reflect the input of solar energy and Earth's rotation around the sun.

AQUATIC AND TERRESTRIAL BIOMES

- **Aquatic biomes occupy the largest part of the biosphere (pp. 1106–1112, FIGURES 50.17–50.23)** Aquatic biomes are often stratified vertically with regard to light penetration, temperature, and community structure. Eutrophic lakes are high in nutrients and productivity; oligotrophic lakes are nutrient-poor. Rivers and streams contain freshwater communities that change significantly from the source to the final destination in an ocean or lake. An estuary is the zone where a river or stream enters the ocean; it is marked by fluctuations in salinity.

   Oceanic zones include the intertidal zone, the neritic zone, and the oceanic zone. In the tropics, coral reefs are found in the warm, nutrient-rich waters of the neritic zone. The oceanic pelagic biome includes most of the open ocean. Photosynthetic plankton in the photic region of the pelagic zone are the primary food source for the rest of the community. Benthic, or bottom, communities subsist largely on detritus that rains down from the pelagic zone.

   Web/CD Activity 50C: *Aquatic Biomes*

- **The geographic distribution of terrestrial biomes is based mainly on regional variations in climate (pp. 1112–1117, FIGURES 50.24, 50.25)** Near the equator, where photoperiod and temperature are nearly constant, the amount and pattern of rainfall determines biomes, including tropical rain forest and savanna. Deserts are inhabited by plants and animals adapted to extremely dry conditions. Chaparral is a dry scrubland found where winters are mild and rainy and summers are hot and dry. Temperate grasslands occur on nutrient-rich, deep soils where periodic fires and drought and the grazing of large mammals inhibit the growth of woody plants. Temperate deciduous forests occur in midlatitudes where there is sufficient moisture to support the growth of large, broadleaf deciduous trees. Coniferous forests include coastal temperate rain forests and the northern coniferous forest, or taiga. The largest terrestrial biome, taiga, is characterized by long, cold, snowy winters and short summers. Arctic tundra occurs at the northernmost limits of plant growth, where cold temperatures, wind, and permafrost limit plants to low shrubby or mat-like forms. Alpine tundra occurs at high altitudes.

   Web/CD Activity 50D: *Terrestrial Biomes*

THE SPATIAL SCALE OF DISTRIBUTIONS

- **Different factors may determine the distribution of a species on different scales (p. 1117–1118, FIGURE 50.26)** Climate may limit a species' distribution on a global scale, while local distribution may depend more on biotic factors such as predators.

- **Most species have small geographic ranges (pp. 1117–1118, FIGURE 50.27)** Only a small minority of species have extensive geographic ranges.

Self-Quiz

1. Which of the following levels of study in ecology includes all other levels in the list?
   a. population
   b. organism
   c. landscape
   d. ecosystem
   e. community

2. Which statement about dispersal is *incorrect*?
   a. Dispersal is a common component of the life cycles of plants and animals.
   b. Colonization of devastated areas after floods or volcanic eruptions depends on dispersal.
   c. Dispersal occurs only on an evolutionary time scale.
   d. Seeds are important dispersal stages in the life cycles of most flowering plants.
   e. The ability to disperse can limit the geographic distribution of a species.

3. Which of the following biomes is *correctly* paired with the description of its climate?
   a. savanna—cool temperature, precipitation uniform during the year
   b. tundra—long summers, mild winters
   c. temperate deciduous forest—relatively short growing season, mild winters
   d. temperate grasslands—relatively warm winters, most rainfall in summer
   e. tropical forests—nearly constant photoperiod and temperature

4. The oceans affect the biosphere in all of the following ways *except*
   a. producing a substantial amount of the biosphere's oxygen.
   b. removing carbon dioxide from the atmosphere.
   c. moderating the climate of terrestrial biomes.
   d. regulating the pH of freshwater biomes and terrestrial groundwater.
   e. being the source of most terrestrial rainfall.

5. Which of the following is *correctly* paired with its description?
   a. neritic zone—shallow area over continental shelf
   b. benthic zone—surface water of shallow seas
   c. pelagic zone—seafloor
   d. aphotic zone—zone in which light penetrates
   e. intertidal zone—open water at the edge of the continental shelf

6. Which of the following do all terrestrial biomes have in common?
   a. annual average rainfall in excess of 25 cm
   b. a distribution predicted almost entirely by rock and soil patterns
   c. clear boundaries between adjacent biomes
   d. vegetation demonstrating vertical stratification
   e. cold winter months

7. The tree line on mountains illustrates the altitudinal limitation for the geographic ranges of trees. Which of the following observations or experiments would be *least* helpful for studying the causes for a particular tree line in the Rocky Mountains?
   a. Analyze the growth rates of individual trees as they occur closer and closer to the tree line.
   b. Transplant small seedlings above the tree line.
   c. Determine the position of the tree line on south-facing slopes and north-facing slopes.
   d. Sow seeds of trees in tundra areas above the tree line.
   e. Measure the rate of photosynthesis for a random sample of tree seedlings in a laboratory greenhouse at sea level.

8. The growing season would generally be shortest in which biome?
   a. tropical rain forest
   b. savanna
   c. taiga
   d. temperate deciduous forest
   e. temperate grassland

9. Imagine some cosmic catastrophe that jolts Earth so that it is no longer tilted. Instead, its axis is perpendicular to the line between the sun and Earth. The most predictable effect of this change would be
   a. no more night and day.
   b. a big change in the length of the year.
   c. a cooling of the equator.
   d. a loss of seasonal variations at northern and southern latitudes.
   e. the elimination of ocean currents.

10. While climbing up mountains, one observes transitions in biological communities that are analogous to the changes one encounters
   a. in biomes at different latitudes.
   b. at different depths in the ocean.
   c. in a community through different seasons.
   d. in an ecosystem as it evolves over time.
   e. traveling across the United States from east to west.

11. Why is it more accurate to define the biosphere as the global *ecosystem* rather than the global *community*?

12. What causes summer in the Northern Hemisphere?

13. How do fires help maintain savannas as grassland ecosystems?

14. What two abiotic factors account for the rarity of trees in arctic tundra?

15. Why does sewage cause algal blooms in lakes?

16. What is the energy source for deep-sea vent communities?

17. What is the most likely explanation for a species having a potential range that is much larger than its actual range?

Go to the website on CD-ROM for more quiz questions.

Evolution Connection

Discuss how the concept of time applies to ecological situations and evolutionary changes. Do ecological time and evolutionary time ever correspond? If so, what are some examples?

The Process of Science

Hiking up a mountain one day, you notice a plant species that has one growth form at low elevations and a very different growth form at high elevations. You wonder if these represent two genetically distinct populations of this species, each adapted to the prevailing conditions where they are found, or if this species has simply evolved the capacity for developmental flexibility and can assume either growth form, depending on local conditions. What experiments could you design to distinguish between these two hypotheses?

Study how changes in sunlight, water, wind, and temperature affect organisms in the Case Study in the Process of Science, available on the CD-ROM and website.

Science, Technology, and Society

In pet shops throughout North America, you can purchase a variety of fishes, birds, and reptiles that do not originate from North America. Present scenarios in which such pet trade could endanger native plants and animals. Should governments regulate the pet trade? Are there currently any restrictions on what species a pet shop can sell in your city? How would you balance such regulation against a person's individual rights?

Answers: 1. c; **2.** c; **3.** e; **4.** d; **5.** a; **6.** d; **7.** e; **8.** c; **9.** d; **10.** a; **11.** Because the biosphere includes both abiotic and biotic components of all inhabited spaces. **12.** Because of the fixed angle of Earth's polar axis relative to the orbital plane around the sun, the Northern Hemisphere is tilted toward the sun during a portion of the annual orbit, the part of the orbit that corresponds to the summer months. **13.** By repeatedly preventing the spread of trees and other woody plants. **14.** Long, very cold winters (short growing season) and permafrost. **15.** The sewage adds minerals that stimulate growth of the algae. **16.** Chemicals that spew up from Earth's interior. **17.** Dispersal is a key factor limiting distribution of the species.

BEHAVIORAL BIOLOGY

INTRODUCTION TO BEHAVIOR AND BEHAVIORAL ECOLOGY

- What is behavior?
- Behavior has both proximate and ultimate causes
- Behavior results from both genes and environmental factors
- Innate behavior is developmentally fixed
- Classical ethology presaged an evolutionary approach to behavioral biology
- Behavioral ecology emphasizes evolutionary hypotheses

LEARNING

- Learning is experience-based modification of behavior
- Imprinting is learning limited to a sensitive period
- Bird song provides a model system for understanding the development of behavior
- Many animals can learn to associate one stimulus with another
- Practice and exercise may explain the ultimate bases of play

ANIMAL COGNITION

- The study of cognition connects nervous system function with behavior
- Animals use various cognitive mechanisms during movement through space
- The study of consciousness poses a unique challenge for scientists

SOCIAL BEHAVIOR AND SOCIOBIOLOGY

- Sociobiology places social behavior in an evolutionary context
- Competitive social behaviors often represent contests for resources
- Natural selection favors mating behavior that maximizes the quantity or quality of mating partners
- Social interactions depend on diverse modes of communication
- The concept of inclusive fitness can account for most altruistic behavior
- Sociobiology connects evolutionary theory to human culture

The study of animal behavior *is undoubtedly one of the oldest branches of biology. Tens of thousands of years ago, behavioral knowledge was essential to human survival. By learning the habits of the animals around them, early humans increased their chances of securing a meal and decreased their chances of becoming a meal. Thus, our ancestors' awareness of animal behavior ultimately enhanced their Darwinian fitness. More generally, our own behavior, as well as that of other animals, has its ultimate basis in evolution.*

Consider the male magnolia warbler in the photograph on this page. To our ears, his song is a musical "weete, weete, weetchew," with the last note slurring upward. However, as one experienced birder is fond of telling novices, "It's not music to a bird's ears. Birds sing for practical reasons—to attract mates, to let other males or females know where they are, to hold a territory where they can raise and feed their young. For birds, singing is about survival and passing genes to the next generation."

Bird song provides an excellent introduction to the subject of behavior. For one thing, bird song is amenable to modern experimental methods. As a result, biologists are beginning to understand a great deal about development, functions, and consequences of bird song. Bird song is also attractive to researchers because of several striking parallels between bird song and human speech. It has also become a model system for animal behavior research because it demonstrates a very important generalization: Behavior is influenced by both genetic and environmental factors. The study of bird song has provided guideposts for the study of other complex behaviors that are less understood.

Studying an animal's behavior is essential to understanding the animal's evolution and ecological interactions. This chapter emphasizes the nature of animal behavior, how biologists study it, and the function of behavior in the relationship between an animal and its environment.

INTRODUCTION TO BEHAVIOR AND BEHAVIORAL ECOLOGY

What is behavior?

Most of what we call behavior consists of an animal's muscular activity that is externally visible. In some cases, it is the whole body that moves, as when a predator chases its prey. In other cases, the behavior is a movement of a body part, even though the animal stays in place, as when we signal a direction by extending our arm and pointing a finger. There are also examples of behavior in which muscular activity is less obvious, as when a bird sings by using muscles to force air from its lungs and shape the sounds in its throat. And there are even some nonmuscular activities that count as behavior, as when an animal secretes a sex attractant. Finally, we might include learning as a behavioral process, even when any observable behavior it produces occurs only later. For example, a young bird may memorize a song that it hears an adult of its species singing. But the first observable muscular activity based on this memory may not occur until months later, when the bird begins to sing the song for itself. Thus, in addition to studying observable behaviors, mainly in the form of muscle-powered activities, behavioral biologists also study the mechanisms underlying those behaviors, which may not involve muscles at all. Put another way, we can think of **behavior** as what an animal does and how it does it, a definition broad enough to include nonmotor components of behavior such as learning and memory.

Behavior has both proximate and ultimate causes

When we observe a certain behavior, we are apt to ask both *proximate* and *ultimate* questions. In the study of animal behavior, proximate questions are mechanistic, concerned with the environmental stimuli, if any, that trigger a behavior, as well as the genetic and physiological mechanisms underlying a behavioral act. Ultimate questions address the evolutionary significance of a behavior. To emphasize the distinction (and also the connection) between proximate and ultimate causation, consider the observation that the magnolia warbler, like many animals, breeds in spring and early summer. In terms of proximate causation, a reasonable hypothesis is that breeding is triggered by the effect of increased day length on an animal's photoreceptors. Many animals can be stimulated to begin breeding by experimentally lengthening their daily exposure to light. This stimulus results in neural and hormonal changes that induce behavior associated with reproduction, such as singing and nest building in birds.

In contrast to proximate questions, ultimate questions take such forms as, Why did natural selection favor this behavior and not a different one? Hypotheses addressing "why" questions propose that the behavior maximizes fitness in some

particular way. A reasonable hypothesis for why many animals reproduce in spring and early summer is that this is when breeding is most productive or adaptive. For warblers and many other birds, an abundant supply of insects in the spring provides ample food for rapidly growing offspring. Individuals that attempt to breed at other times would be at a selective disadvantage. Increased day length itself has little adaptive significance, but since it is the most reliable indicator of time of year, there has been selection for a proximate mechanism that depends on increased day length. In brief, proximate mechanisms produce behaviors that ultimately evolved because they reflect Darwinian fitness in some particular way. Behavioral biologists also use the comparative methods of phylogenetic biology (see Chapter 25) to formulate hypotheses about the evolution of behavior. Phylogenetic trees based on molecular, morphological, or behavioral data and illustrating the most likely evolutionary history of a closely related group of species enable researchers to estimate when a particular behavior arose in a lineage, whether it arose once or repeatedly, and which kinds of behavior occurred in ancestors.

Behavior results from both genes and environmental factors

A myth that is still perpetuated by the popular media is that behavior is due *either* to genes (nature) *or* to environmental influences (nurture). In biology, however, the nature-versus-nurture issue is not about either/or; it is about how both the genes and the environment influence the development of phenotypes, including behavioral phenotypes. If we consider the development of any behavioral trait, we find a series of environmental and genetic influences. As we discussed in Chapter 14, phenotype depends on both genes and the environment; behavioral traits have genetic and environmental components, as do all of an animal's anatomical and physiological features.

Thus, we can move from the question of whether behavior is genetic or environmental to ask how different factors influence a particular behavior. One approach to the question is in terms of the norm of reaction (see Chapter 14); we measure, for a particular genotype, what behavioral phenotypes develop in a range of environments. In some cases, much the same behavior develops in all environments; in other cases, the behavior is variable, depending on the environmental experience. It takes thorough research on each case to describe the genetic and environmental influences, as the case study in FIGURE 51.1 illustrates.

In a few cases, researchers have been able to link behaviors to specific genes. For instance, Marla Sokolowski, of Toronto University, studied a genetic polymorphism in a gene called *dg2* in fruit flies (*Drosophila*). The gene influences the level of a protein that functions in signaling within cells. One allele of *dg2* causes a relatively low level of the protein. This results in a

① Nests made with long strips—no tucking behavior.
Several species of brightly colored African parrots, commonly known as lovebirds, build cup-shaped nests inside tree cavities. Females typically make nests with thin strips of vegetation (or, in the laboratory, paper) that they cut with their beaks. In one species, Fischer's lovebird (*Agapornis fischeri*), the bird cuts relatively long strips and carries them back to the nest one at a time in her beak.

Fischer's lovebird

② Nests made with short strips—tucking behavior.
In contrast, the peach-faced lovebird (*A. roseicollis*) cuts shorter strips and usually carries several at a time by tucking them into the feathers of the lower back. Tucking is a fairly complex behavior because the strips must be held just right and pushed in firmly and the feathers then smoothed over.

Peach-faced lovebird

③ Hybrid nests made with intermediate-length strips— in first mating season, unsuccessful tucking behavior.
These two species are closely related and have been experimentally interbred. The resultant hybrid females exhibited an intermediate kind of nest-building behavior. The strips cut by the hybrid birds were intermediate in length; even more interesting was the birds' hybrid manner of handling the strips. They usually made some attempt to tuck them into their rear feathers, but in some cases they did not let go after turning and trying to tuck the strips. In other cases the strips were manipulated or inserted improperly or simply dropped. The result was almost a total failure to transport strips by this method. Eventually, the birds learned to transport the strips in their beaks. Even so, they always made at least token tucking attempts.

Hybrid lovebird

④ In later seasons, only head-turning behavior.
After several years, the birds still turned their heads to the rear before flying off with a strip. These observations demonstrate that the phenotypic differences in the behavior of the two species are based on different genotypes. We also see that the behavior can be modified by experience; the hybrid birds eventually learned to transport the strips.

Hybrid lovebird

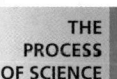
THE
PROCESS
OF SCIENCE

FIGURE 51.1 Genetic and environmental components of behavior: a case study.

behavioral phenotype called "sitter," in which the fly moves less than average. A different allele causes a higher level of the protein and a "rover" behavioral phenotype, in which the fly moves about more than average. In this example, a simple difference in one gene changes the behavioral phenotype. However, a complete account of the development of movement patterns in flies would require a look at many more genes in addition to *dg2*. The *dg2* gene interacts with other genes, and environmental conditions are also influential. Most behavioral traits are polygenic, with environmental variables producing broad norms of reaction.

So what do the media reports of newly discovered genes for complex human behavioral traits, such as depression, violence, or alcoholism, really mean? According to Robert Plomin, director of the Center for Developmental and Health Genetics at Pennsylvania State University, research into the heritability of behavior is the best demonstration of the importance of environment. As Plomin puts it, genes and nongenetic environmental factors "build on each other."

The environmental factors that affect behavior are all conditions in which the genes underlying behavior are expressed. This includes the chemical environment within cells, as well as all the hormonal and other chemical and physical conditions experienced by a developing animal within an egg or womb. It also includes the multiple interactions among components of an animal's nervous system and effectors, as well as the varied chemical, visual, auditory, or tactile interactions with other organisms.

Innate behavior is developmentally fixed

If behavior has both genetic and environmental underpinnings, what do we mean when we say a particular behavior is innate? For example, newly hatched, still-blind birds of many species beg for food by raising their heads, opening their mouths, and cheeping loudly when a parent lands on the side of the nest. This behavior is often attributed to genetic programming without any environmental influence. However, it is imprecise to say that any behavior is due solely to genes. All genes, including those whose expression underlies innate behavior, require an environment (a physical body) to be expressed. The key point about innate behavior is that the range of environmental differences among individuals does not appear to alter the behavior. Although usage of the term *innate* varies, in behavioral biology it refers to behavior that is *developmentally fixed;* all individuals exhibit virtually the same behavior despite the inevitable environmental differences within and outside their bodies during development and throughout life.

How did innate behavior evolve? Performing certain behaviors automatically, without having any specific experience, may have maximized fitness to the point that genes for variant behavior were lost. For example, there are some things that a young animal has to get right on the first try if it is to stay alive. Kittiwakes are gulls that nest on cliff ledges. Uniquely among gull species, kittiwakes show an innate aversion to cliff edges; they turn away from the edge. Kittiwake chicks in earlier generations that did not show the edge-aversion response failed to become ancestors to modern kittiwakes.

Classical ethology presaged an evolutionary approach to behavioral biology

THE PROCESS OF SCIENCE

Modern behavioral biology has its roots in a research field known as **ethology,** which originated in the 1930s with naturalists who tried to understand how a variety of animals behave in their natural habitats (FIGURE 51.2). Foremost among these naturalists were Karl von Frisch, Konrad Lorenz, and Niko Tinbergen, who shared a Nobel Prize in 1973 for their discoveries. How animals can carry out many behaviors without ever having seen them performed was one of the major subjects of early ethological research. Ethologists focused on proximate mechanisms, but with an eye toward the genetic links to behavior and the adaptive nature of behavior, an orientation that helped connect behavioral biology with evolution and ecology.

Early ethologists developed the concept of a **fixed action pattern (FAP),** a sequence of behavioral acts that is essentially unchangeable and usually carried to completion once initiated. A FAP is triggered by an external sensory stimulus known as a

THE PROCESS OF SCIENCE

FIGURE 51.2 Niko Tinbergen's experiments on the digger wasp's nest-locating behavior.

❶ A female digger wasp excavates and cares for four or five separate underground nests. She will fly to each nest daily, bringing food to the single larva in each nest. Biologist Niko Tinbergen designed field experiments to test his hypothesis that the wasp uses visual landmarks to keep track of where her nests are located. First, Tinbergen marked a wasp's nest with a ring of pinecones.

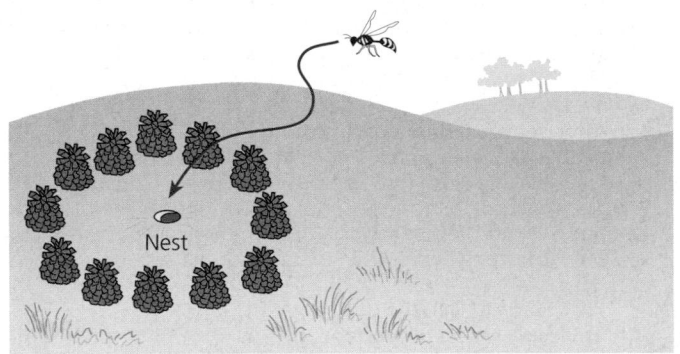

❷ After the mother wasp visited the nest and flew away, Tinbergen moved the pinecones a few feet to one side of the nest. When the wasp returned, she flew to the center of the pinecone circle instead of the nearby nest. The results of such experiments supported the hypothesis that digger wasps use landmarks to keep track of their nests and that they can learn new visual cues.

❸ In a follow-up experiment, Tinbergen restored the pinecones to the actual nest site but arranged them in a triangle instead of a circle. He placed a circle of stones to one side of the nest. The returning wasp flew to the stone ring, a result supporting the hypothesis that the insect was cued by the arrangement of the landmarks rather than the physical objects themselves.

sign stimulus. In many cases, the sign stimulus is some feature of another species. For example, some moths instantly fold their wings and drop to the ground in response to the ultrasonic signals sent out by predatory bats (see the introduction to Chapter 49). The ultrasonic signals are the cue that triggers avoidance behavior in the moths.

A classic case of sign stimuli and FAPs is seen in the male three-spined stickleback fish, which attacks other males that invade his territory. The stimulus for the attack behavior is the red belly of the intruder. The stickleback will not attack an invading fish lacking a red underside, but will readily attack nonfishlike models as long as some red is present (FIGURE 51.3). Tinbergen, who first reported these findings, was inspired to look into the matter by his casual observation that his fish responded aggressively when a red truck passed their tank. As it turns out, red coloration of body parts tends to trigger either aggressive or sexual behavior in many species that have color vision.

Classic experiments on FAPs and sign stimuli have shown that many animals tend to use a relatively limited subset of the sensory information available to them and to behave stereotypically in many situations. In contrast, humans often tend to respond to an entire situation, and we generally base our actions on more diverse information. If a stickleback processed information like a human, it would realize that the models in FIGURE 51.3 are not real fish, despite their red bellies. Actually, relatively simple, stereotypical behaviors seem to occur in all animals, including humans. Human infants grasp strongly with their hands in response to a tactile stimulus. An infant's smile could also be considered a FAP; it is readily induced by simple stimuli, such as a sound or a figure consisting of two dark spots on a white circle, a kind of rudimentary representation of a face.

A sign stimulus is a key feature in an animal's environment, leading the animal to respond quickly and appropriately. However, because the animal responds to the sign rather than to the whole environmental context, the animal can often be tricked into inappropriate behavior, as in the stickleback's response to the models in FIGURE 51.3. In some cases, such misdirected responses can be detrimental. For example, mayflies swarm and mate above water, and females then deposit their eggs on the water surface. The breeding mayflies detect a water surface by the polarization pattern of the light reflected from it. Unfortunately, roads can produce a similar polarization pattern in reflected light, and mayflies now often lay their eggs on roads, where the eggs perish (FIGURE 51.4). Many mayfly species are endangered, and their behavioral mechanism for detecting water may drive them to extinction in environments where roads are becoming more abundant than standing water. Given the potential of sign stimuli to mislead animals into maladaptive behavior, we might ask why animals use them at all. Some ethologists have suggested that FAPs triggered by simple cues prevent an animal from wasting time processing or integrating a wide variety of inputs.

(a) Female mayfly ovipositing (egg-laying) on black plastic sheet used in agriculture

(b) Male mayfly attracted to a dry asphalt road

FIGURE 51.4 Mayflies laying eggs on human-made surfaces.
Mayflies normally lay their eggs on water, but the pattern of polarized light that they use to find water is also produced by reflections from black plastic sheets and roads.

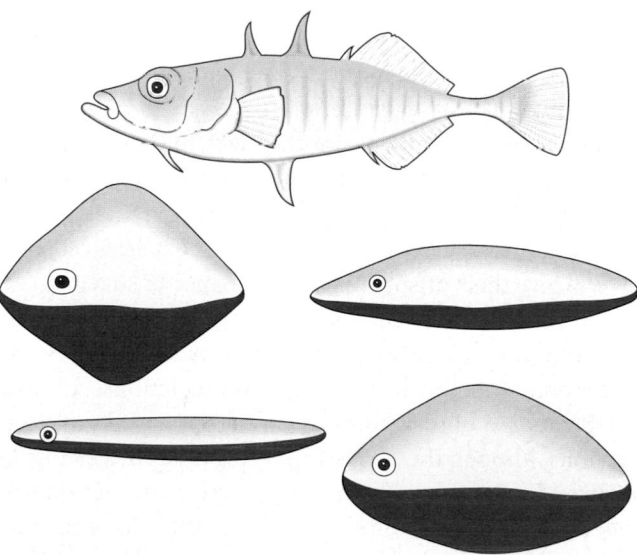

FIGURE 51.3 Classic demonstration of innate behavior.
Aggression in a male three-spined stickleback fish is triggered by a simple visual cue. The realistic model at the top without a red underside produces no response. All the others produce strong responses because they have the required red underside.

Although some simple behavior patterns can be understood in terms of sign stimuli and fixed action patterns, the concept is too simplistic to account for much of animal behavior. Though historically important in the development of behavioral biology, sign stimuli and fixed action patterns are not major focuses of behavioral research today.

Behavioral ecology emphasizes evolutionary hypotheses

Evolution is the core theme of biology, and the study of behavior in an ecological context emphasizes evolutionary (ultimate) explanations. **Behavioral ecology** is the research field that views behavior as an evolutionary adaptation to the natural ecological conditions of animals. Natural selection will favor behavioral patterns that enhance survival and reproductive success. Thus, an animal showing optimal behavior maximizes its reproductive fitness. However, most animal behavior is less than optimal; for instance, environmental change or constraints on an animal's sensory powers may result in suboptimal behavior. Therefore, although we can expect animal behavior to be well adapted, it may not be optimally adapted.

In the remainder of this section, we will explore some examples of research in behavioral ecology.

Songbird Repertoires

THE PROCESS OF SCIENCE

Many songbirds have a repertoire of song types. Some of these songs sound identical to us but can be easily distinguished when analyzed with an instrument called a sound spectrograph (FIGURE 51.5). Why has natural selection favored this multisong behavior over the expert vocalization of a single tune?

Following the approach of behavioral ecology, you could formulate several testable hypotheses, all starting with "A repertoire increases fitness because . . ." You might postulate that a repertoire increases fitness because it makes an older, more experienced male more attractive to females. For this hypothesis to be true, it must be the case that (1) males learn more song types as they get older, making repertoire size a reliable indicator of age, and (2) females prefer to mate with males having large repertoires. Thus, your hypothesis makes two clearly testable predictions.

To test the first prediction, you can determine whether there is a correlation between a male's age and the size of his song repertoire. If there is not, your hypothesis will be invalidated, which would be informative although perhaps disappointing. As it turns out, some songbird species show this correlation, while others do not. Next, you can determine whether females are more sexually stimulated by a large song repertoire than by a small one. This can be done by playing tape-recorded male songs to females that have been made es-

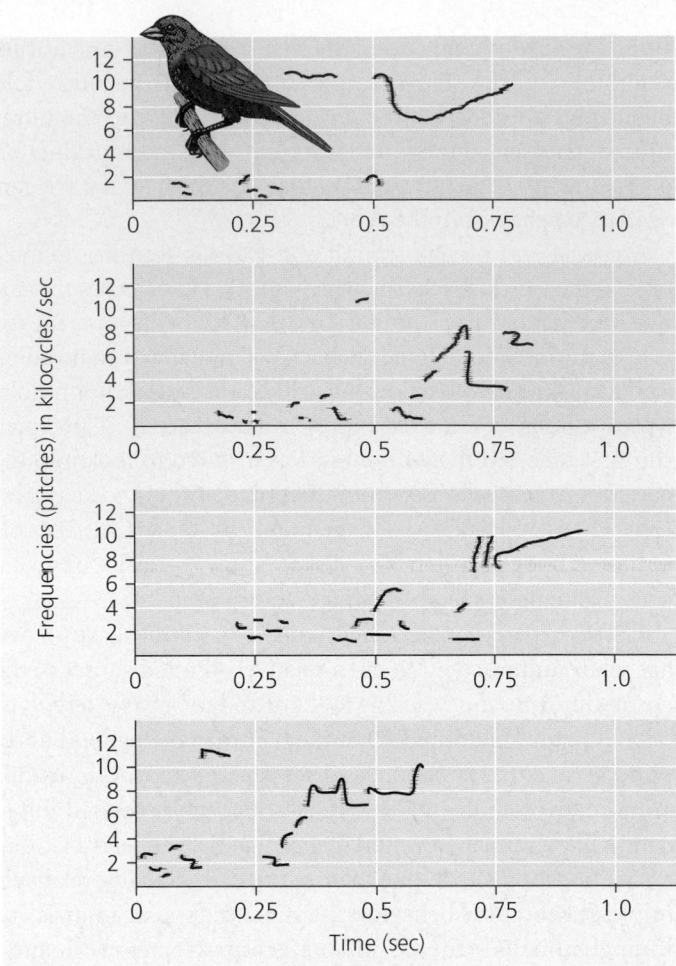

FIGURE 51.5 The repertoire of a songbird. These sonograms, or "voiceprints," show a graph of a sound's frequency (perceived as pitch) versus time. Four distinct song types of one male brown-headed cowbird are illustrated. Individual cowbirds generally have three to six song types, but other species have dozens or even hundreds of types.

pecially receptive by the temporary administration of a female hormone. Such females indicate their song preferences by assuming a copulatory posture, even though there is no male around. FIGURE 51.6 shows another way to assess female responses. All this work may lead to an evolutionary explanation: Large bird-song repertoires result in females mating more often or earlier in the season with experienced males, who will give their offspring a better chance to survive.

Now suppose you did not use evolutionary principles to guide your research on song repertoires. Without an approach that generates testable hypotheses and predictions, you would probably record observations of numerous aspects of singing behavior. Although these efforts may produce interesting data, they would not *explain* the behavior. Alternatively, you might hypothesize that repertoires have nothing to do with fitness, but that male birds simply find variety more pleasurable than singing the same boring song over and over again. The method for testing such a hypothesis would not be clear, emphasizing again how much more productive it is to use evolutionary principles as a guide to behavioral research.

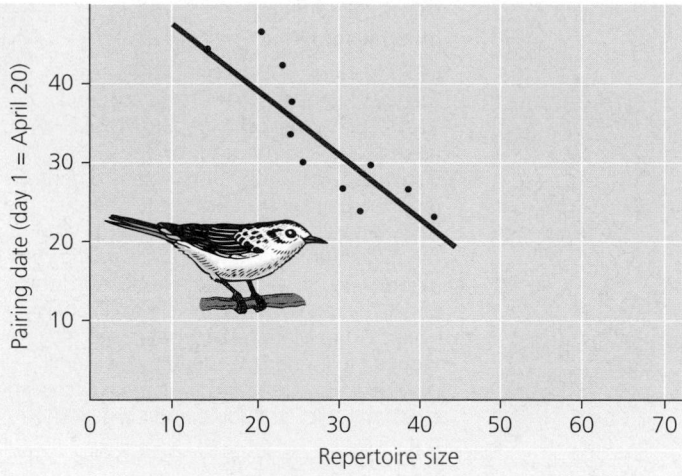

FIGURE 51.6 Female warblers prefer males with large song repertoires. Male sedge warblers with large repertoires attract females to pair with them earlier in the breeding season, as well as more often, than males with small repertoires. Males with large repertoires, as well as the females that prefer large repertoires, benefit from pairing early because breeding early tends to be more successful than breeding late in the season.

Cost-Benefit Analysis of Foraging Behavior

Because adequate nutrition is so essential to an animal's survival and reproductive success, we should expect natural selection to refine behaviors that enhance the efficiency of feeding. Food-obtaining behavior includes not only eating, but also any mechanisms an animal uses to recognize, search for, and capture food items. The term **foraging** encompasses this full set of behaviors. **Optimal foraging theory** views foraging behavior as a compromise between feeding costs and feeding benefits. How do foraging behaviors balance the benefits of nutrition against the costs of obtaining food? Some behavioral ecologists are applying cost-benefit analysis to study the proximate and ultimate causes of diverse foraging strategies.

Reto Zach, of the University of British Columbia, conducted a classic cost-benefit analysis of feeding behavior in crows of the Pacific Northwest. Crows are opportunistic feeders that avail themselves of a variety of food items. On Mandarte Island, off British Columbia, crows search the rocky tide pools for gastropod mollusks called whelks. The bird grasps its prey in its beak, then flies upward and drops the whelk onto the rocks to break the shell. If successful, the crow can dine on the mollusk's soft parts. If the shell doesn't break, the crow flies up and drops the shell again and again until the shell breaks. Of course, the higher the bird flies before dropping a whelk, the fewer the number of drops required to break the shell. But there is an energy cost correlated with the height of the crow's ascent. Zach predicted that crows would, on average, fly to a height that would provide the most food relative to the amount of total energy required to break the whelk shells. To determine the optimal height, Zach erected a 15 m pole and

then dropped shells of relatively uniform size onto rocks from different heights along the pole. He then tabulated the data and calculated the average total effort required to break shells by dropping them from different heights:

Height of Drop (m)	Average Number of Drops Required to Break Shell	Total Flight Height (Number of Drops × Height per Drop)
2	55	110
3	13	39
5	6	30
7	5	35
15	4	60

Note that a height of about 5 m is optimal for breaking shells with the least amount of work. Although dropping the shells from 7 or 15 m resulted in breakage with fewer drops, that advantage was small compared with the effort it would take to fly that high. When Zach measured the average flight height for crows in their whelk-eating behavior, it was 5.23 m, very close to the prediction based on an optimal trade-off between energy gained (food) versus energy expended.

The feeding behavior of the bluegill sunfish provides further support for optimal foraging theory. These animals feed on small crustaceans called *Daphnia*, generally selecting larger individuals, which supply the most energy. However, smaller prey will be selected if larger prey are too far away (FIGURE 51.7a, p. 1128). The optimal foraging approach predicts that the proportion of small to large prey eaten will also vary with the overall density of prey. At very low prey densities, bluegill sunfish should exhibit little size selectivity, because all the prey encountered are needed to meet energy requirements. At higher prey densities, it is more efficient to concentrate on larger crustaceans. In actual experiments, bluegill sunfish did become more selective at higher prey densities, though not to the extent that would theoretically maximize efficiency (FIGURE 51.7b). Young bluegill sunfish forage fairly efficiently, but not as close to the optimum as older individuals, who are apparently able to make more complex distinctions. It may be that younger fish judge size and distance less accurately because their vision is not yet completely developed; learning may also be involved.

Behavioral ecologists recognize that energy costs and benefits are not the only factors affecting foraging behavior. Animals require not only calories, but also essential nutrients, and this, too, affects an animal's choices of food items. Moreover, foraging behavior tends to minimize the risk of the predator becoming the prey while foraging. As the following example shows, there is more to analyzing foraging behavior than weighing energy costs versus energy benefits.

The smallmouth bass readily consumes both minnows and crayfish. The fact that it does not show an overall preference suggests that the trade-offs balance, with minnows being the optimal prey in some situations and crayfish in others. Minnows

(a)

FIGURE 51.7 Feeding by young bluegill sunfish.

(a) In feeding on *Daphnia* (water fleas), the fish do not feed randomly but tend to select prey based on "apparent size," information about both prey size and distance. Confronted with various potential prey, the animal will pursue the one that looks largest. Small prey (low energy yield) at the middle distance in this example may be ignored. But small prey at the closest distance may be taken with a relatively small energy expenditure. More distant but larger prey will require more energy expenditure but will provide a high yield and thus may be chosen over small prey at middle and even close distances.

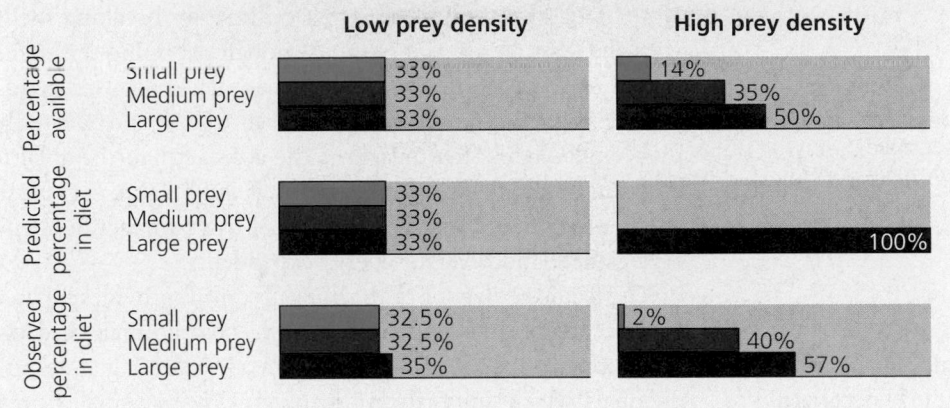

(b)

(b) When prey are at low density, calculations based on optimal foraging theory predict that bluegill sunfish will not be selective, but will eat any size prey as it is encountered. At higher prey densities, the ratio of energy intake to energy expended can be maximized by concentrating only on larger prey. In the experiments described here, we see that bluegills did forage nonselectively at lower prey density. At higher prey density, they favored larger prey, though not to the extent predicted.

contain more usable energy per unit weight (the crayfish has a lot of hard-to-digest exoskeleton), but they may require a greater energy expenditure to pursue. However, even though crayfish may be easier to catch, their large claws and aggressive resistance make them harder to subdue. Trade-offs also include the relative abundance and size of each type of prey. The brain of the small-mouth bass is somehow able to integrate all the relevant variables and formulate responses resulting in a highly efficient foraging behavior that involves switching between minnows and crayfish as conditions change. The proximate mechanisms responsible for this process are not known, but we might predict that they include specific cues that trigger innate as well as mainly experience-based behavior, to which we turn next.

LEARNING

Learning is experience-based modification of behavior

Analyzing the genetic and environmental underpinnings of behavior can help scientists understand the extent to which behavior can vary among individuals of a species. In this section, we examine various forms of **learning,** which is the modifica-tion of behavior resulting from specific experiences. As we have said, even though an animal need not witness a developmentally fixed (innate) behavior to perform it, experience is still involved. Most innate behaviors improve with performance as animals learn to carry them out more efficiently. Conversely, it might seem that some things, such as the different human languages, are completely learned. It is true that the ability to speak a specific language, such as English or Spanish, has little or no genetic basis. However, the ability to learn *a* language is a function of a complex brain that develops in a particular environmental context under the guidance of a human genome.

The alarm calls of vervet monkeys (*Cercopithecus aethiops*) provide an example of how animals improve their performance of a behavior by means of learning. Dorothy Cheney and Richard Seyfarth, of the University of Pennsylvania, studied vervet monkeys in Amboseli National Park, Kenya. The vervets there give distinct alarm calls when they see leopards, eagles, or snakes. When a vervet sees a leopard, it gives a loud barking sound; when it sees an eagle, it gives a short double-syllabled cough; and the snake alarm call is a "chutter." Leopards, eagles, and snakes all eat vervet monkeys, which are about the size of domestic cats. Upon hearing a particular alarm call, the other members of the vervet group behave in an appropriate way: They run up a tree on hearing the alarm call for a leopard (vervets are nimbler than leopards in the trees); look up on hearing the alarm call for an eagle; and look down on hearing the alarm call for a snake (FIGURE 51.8).

Infant vervet monkeys give alarm calls, but in a relatively undiscriminating way. For example, they give the "eagle" alarm call on seeing any bird, including harmless birds such as bee-eaters. With age, the monkeys improve their accuracy. In fact, adult vervet monkeys give the eagle alarm only on seeing an eagle belonging to either of the two species that eat vervets. The mechanism by which infants learn how to give the right call probably includes experiencing the behavior of other members of the group. For instance, if the infant gives the call on the right occasion—an eagle alarm when there is an eagle overhead, for example—another member of the group will also give the eagle call almost immediately. But if the infant gives the call when there is only a bee-eater flying by, the adults in the group are silent. The social confirmation of the infant's call probably helps it to learn when to give each call. Thus, vervet monkeys have an initial, unlearned tendency to give calls on seeing potentially threatening objects in the environment. Learning then fine-tunes the call such that adults only give calls in response to genuine dangers.

Learning Versus Maturation

Learning often affects innate (developmentally fixed) behavior, but changes in innate behaviors are not always due to learning. For instance, behavior may change because of ongoing developmental changes in neuromuscular systems, a process called **maturation.** We commonly speak of birds "learning" to fly, and you may have seen fledgling birds awkwardly fluttering about as if they were practicing. However, young birds have been experimentally reared in restrictive devices so that they could never flap their wings until an age when their normal kin were already flying. Such birds flew immediately and normally when released. Thus, the improvement must have resulted from neuromuscular maturation, not from learning.

In many cases, the distinction between learning and maturation is less obvious. When an adult herring gull brings food to its chick, the adult bends its head down and moves its beak, which has a red spot. The chick pecks the beak, stimulating the adult to regurgitate the food. Experiments have shown that the red spot swung horizontally at the end of a beak is the signal that elicits the chick's pecking. However, newly hatched chicks are indiscriminate and will peck at a wide variety of objects, whereas chicks a week or two old show their strongest response to accurate models of their parents' beaks. Would you predict that this change in the chick's behavior results from maturation or learning? Experiments involving herring gulls and laughing gulls show that maturation is not involved. A laughing gull chick that has been reared by a herring gull will respond more strongly to its foster parent's beak type than to the beak of its own species. The reverse is true for a herring gull chick cross-fostered by laughing gulls. This is an example

FIGURE 51.8 Vervet monkeys learn correct use of alarm calls. Vervet monkeys give different alarm calls on seeing different sources of danger. For example, on seeing a python, vervets give a distinct "snake" alarm call (inset), and the members of the group stand upright and look down. Vervets probably learn when to use appropriate calls from the behavior of other members of their group.

of how learning can modify a behavior that is basically developmentally fixed.

Habituation

Habituation is a very simple type of learning that involves a loss of responsiveness to stimuli that convey little or no information. Examples are widespread. A hydra contracts when disturbed by a slight touch; it stops responding, however, if disturbed repeatedly by such a stimulus. Many mammals and birds recognize alarm calls of members of their species, but they eventually stop responding if these calls are not followed by an actual attack (the "cry-wolf" effect). In terms of ultimate causation, habituation may increase fitness by allowing an animal's nervous system to focus on stimuli that signal food, mates, or real danger instead of wasting time or energy on a vast number of other stimuli that are irrelevant to the animal's survival and reproduction.

Imprinting is learning limited to a sensitive period

Some of the most interesting cases where learning interacts closely with innate behavior involve the phenomenon known as **imprinting,** learning that is limited to a specific time period in an animal's life and that is generally irreversible. You may have seen young ducks or geese following their mother.

Mother-offspring bonding in species with parental care is a critical part of the reproductive cycle. If bonding fails to happen, the parent will not initiate care of the infant. The result is certain death for the offspring and loss of reproductive fitness for the parent. But how do the young know whom—or what—to follow? In his most famous study, Konrad Lorenz divided a clutch of graylag goose eggs, leaving some with the mother and putting the rest in an incubator. The young reared by the mother showed normal behavior, following her about as goslings and eventually growing up to interact and mate with other geese. When the artificially incubated eggs hatched, the geese spent their first few hours with the researcher instead of with their mother. From that day on, they steadfastly followed Lorenz and showed no recognition of their own mother or other adults of their own species (FIGURE 51.9). As adults, the birds continued to prefer the company of Lorenz and other humans over that of their own species, and they sometimes even initiated courtship behavior with humans.

Apparently, graylag geese have no innate sense of "mother" or "I am a goose, you are a goose." Instead, they simply respond to and identify with the first object they encounter that has certain key characteristics. What is innate in these birds is the ability or tendency to respond; the outside world provides the *imprinting stimulus,* something to which the response will be directed. The most important imprinting stimulus in Lorenz's graylag geese was movement of an object away from the young, although the effect was greater if the object emitted some sound. The sound did not have to be that of a goose, however; Lorenz found that a box with a ticking clock in it was readily and permanently accepted as "mother."

FIGURE 51.9 Imprinting. Konrad Lorenz was "mother" to these imprinted geese.

In contrast to other types of learning, imprinting is distinguished by a **sensitive period,** a limited phase in an animal's development when learning of particular behaviors can take place. Lorenz found, for example, that geese totally isolated from any moving objects during the first two days after hatching, which is the sensitive period for imprinting on parents, failed to imprint on anything afterward. Imprinting has commonly been thought of as involving very young animals and rather short sensitive periods. But it is now clear that a similar learning process occurs in older animals and that the sensitive period may be of various durations. For example, just as a young bird requires imprinting to "know" its parents, the adults must also imprint to recognize their young. For a day or two after their young hatch, adult herring gulls will accept and even defend a strange chick introduced into their nesting territory. After imprinting, which is probably based largely on individually variable cues, such as the call notes of chicks, the adults will kill and eat any strange chick.

By imprinting on their parents, young birds first learn who will care for them and subsequently learn species identity and the kind of bird they should mate with later in life. Not only does this sexual imprinting occur later, but the sensitive period lasts longer. For example, in one study involving two closely related species of finches, young males of one species were reared first with members of their own species, then with members of the other species during their several-week-long sensitive period for sexual identity. Later, when exposed to females of their own species, they mated quite reluctantly. They readily mated with females of the other species, however, even when they had not seen any members of that species for as long as eight years. Identification with the second species had been permanently imprinted. Although a sensitive period and irreversibility characterize imprinting, these phenomena are not always rigidly fixed. For example, the cross-fostered finches did eventually mate with females of their own species.

Bird song provides a model system for understanding the development of behavior

A considerable amount of research on how animals learn has focused on bird songs. Some relatively simple bird calls, such as the crowing of a rooster, seem to develop and improve without any type of learning. Thus, researchers interested in learning have focused on the songbirds, such as the sparrows and canaries, especially on the development of elaborate calls by the males. Such research has revealed that the learning techniques for mastering calls vary, depending on the songbird species.

THE PROCESS OF SCIENCE

Some songbirds have a sensitive period for developing their songs. In white-crowned sparrows, for example, the sensitive period occurs during the first 50 days of life. Although the young bird does not sing during this phase, it memorizes the song of its own species by listening to other white-crowned sparrows sing. Behavioral biologists call this memorized song a template. In laboratory experiments, sparrows can form this template by listening to tape recordings during the 50-day sensitive period. But if a bird is isolated throughout this sensitive period, unable to hear either real sparrows or recordings, it fails to develop the adult song typical of its species.

The sensitive period when a white-crowned sparrow forms a song template is followed by a second learning phase, when the juvenile bird sings some tentative notes that researchers call a subsong. With practice, the sparrow's subsong gradually improves until it sounds like the fully developed adult song of the species. During this practice phase, the juvenile bird hears its own singing and compares that vocal output with the template that was memorized during the sensitive learning period. In fact, sparrows that are experimentally deafened at the end of the sensitive phase fail to improve on their subsong, which they continue to sing into adulthood. During normal song development, once a sparrow's song matches the memorized template, it "crystallizes" as the final song, and the bird sings only the adult white-crowned sparrow song for the rest of its life (FIGURE 51.10a).

Additional experiments revealed even more complexity in how birds learn their songs. When isolated white-crowned spar-

rows more than 50 days old were exposed to live singing adults of another species, they learned the song of the other species. Because a young bird can interact socially with a live singing adult, the live model provides a much stronger and more diverse set of stimuli than a taped song. These strong stimuli can overcome the developmentally fixed tendency to acquire only a white-crowned sparrow song. Again we see that an innate tendency can be modified by experience and is not necessarily inflexible. Furthermore, the sensitive period proved to be longer when the stimulus was a live bird than when it was a taped song.

We humans also have a sensitive period for learning vocalizations. It is well known that foreign languages are learned most easily up until the teen years. Of course, this sensitive period is not rigidly fixed. Adults can learn new languages, but adults usually require much more time and effort to become fluent than a child does. Also, adults are much less flexible than children at learning to produce new sounds.

Until recently, researchers assumed that all birds had to listen to themselves to develop a specific song. But eastern phoebes experimentally deafened early in life, well before they began to sing, developed normal songs of their species. Thus, there are important exceptions to the song-learning scenario seen in white-crowned sparrows. In fact, in certain songbirds, including

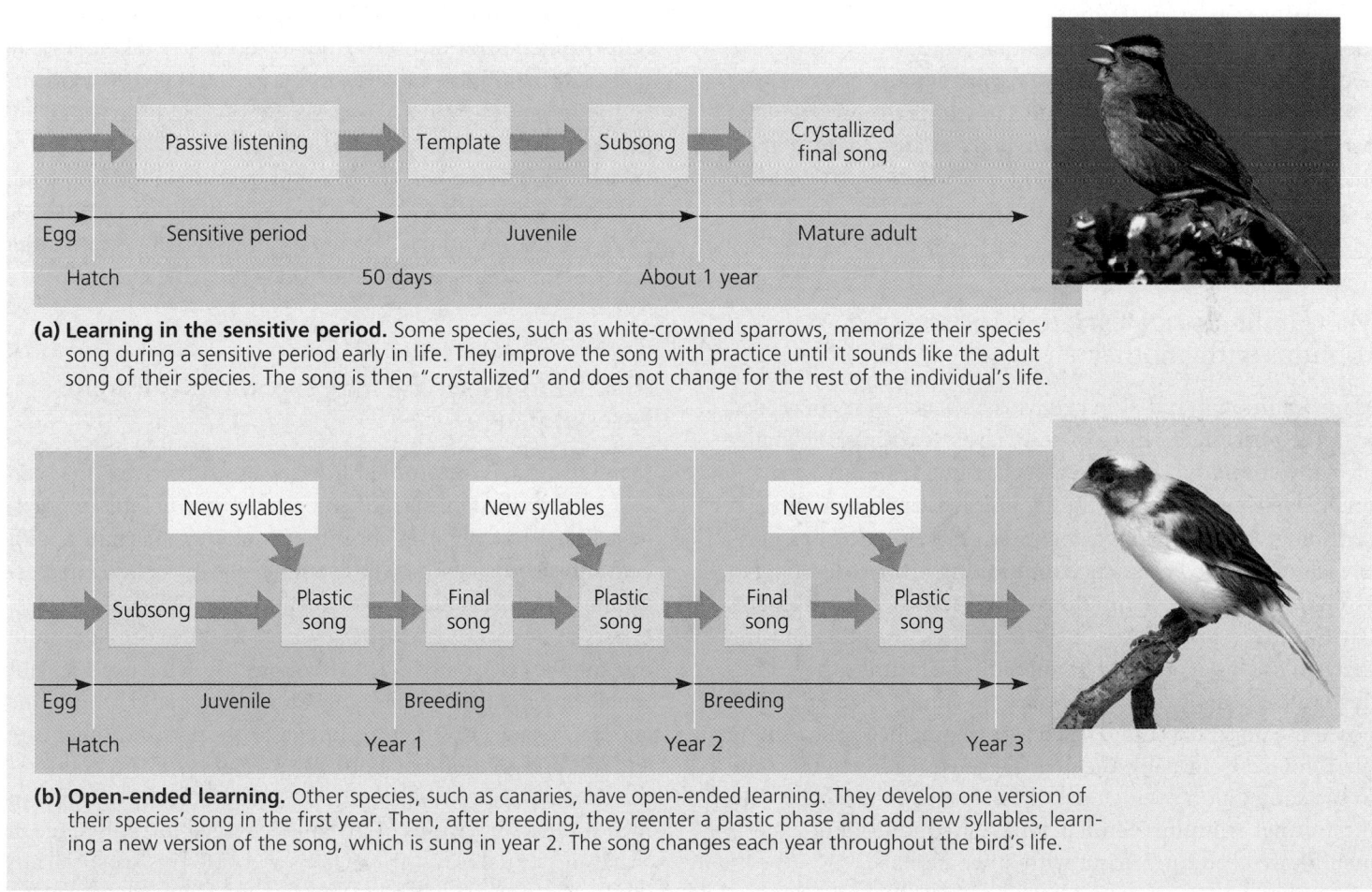

(a) **Learning in the sensitive period.** Some species, such as white-crowned sparrows, memorize their species' song during a sensitive period early in life. They improve the song with practice until it sounds like the adult song of their species. The song is then "crystallized" and does not change for the rest of the individual's life.

(b) **Open-ended learning.** Other species, such as canaries, have open-ended learning. They develop one version of their species' song in the first year. Then, after breeding, they reenter a plastic phase and add new syllables, learning a new version of the song, which is sung in year 2. The song changes each year throughout the bird's life.

FIGURE 51.10 Two kinds of bird-song development.

canaries, a more open-ended development of song occurs (FIG-URE 51.10b). Canaries have no sensitive period for song learning. A young canary begins with a subsong, but the full song it develops is not crystallized, as it is in white-crowned sparrows. Each year, an adult male canary learns a new song. Between breeding seasons, the song becomes flexible again. During this "plastic song" stage, the male creates a new, more elaborate song by adding syllables. The male sings the new song during the next breeding season, after which a plastic song stage leads to still another vocal version. And so on, year after year.

Biologist Fernando Nottebohm identified the region of the forebrain responsible for song learning in canaries. He found that this region in male canaries shows dramatic individual variation in size according to the season and the complexity of an individual male's song. Brains are largest during the breeding season and in males with the most elaborate songs. The shrinking of this brain region at the end of the breeding season may be a mechanism for erasing unneeded songs. Subsequent regeneration of neurons in the brain during the plastic song phase provides a way for new song learning to take place. From an adaptive perspective, the ability of male canaries to learn new songs more than once in their lifetime must be critically important to their fitness, as this adaptation requires a considerable investment of energy. Nottebohm's findings illustrate the important results that can be gleaned from studies melding neurobiology and behavioral ecology.

By analyzing the development of bird song, we can see why a simple division between innate and learned components of behavior has little value. There are both innate and learned influences in bird song development, but the interplay of these factors belies an innate/learned dichotomy.

Many animals can learn to associate one stimulus with another

Many traditional studies in behavioral science were concerned with the proximate causes of **associative learning,** the ability of many animals to learn to associate one stimulus with another. Well known from the laboratory studies of Russian physiologist Ivan Pavlov in the early 1900s, a type of associative learning called **classical conditioning** involves learning to associate an arbitrary stimulus with a reward or punishment. Pavlov sprayed powdered meat into dogs' mouths, causing them to salivate (primarily a physiological rather than a behavioral response). Just before the spraying, however, he exposed the dogs to a sound, such as a ringing bell or a clicking metronome. Eventually, the dogs salivated readily in response to the sound alone, which they had learned to associate with the normal stimulus. Similar types of conditioning experiments have been carried out with other animals.

Another type of associative learning is **operant conditioning,** also called trial-and-error learning. Here, an animal learns to

FIGURE 51.11 Operant conditioning. A young coyote has probably learned from receiving a face full of painful quills to be wary of porcupines.

associate one of its own behaviors with a reward or punishment and then tends to repeat or avoid that behavior. For example, predators quickly learn to associate certain kinds of potential prey with painful experiences and modify their behavior accordingly (FIGURE 51.11).

The best-known laboratory studies involving operant conditioning date from the 1930s work of American psychologist B. F. Skinner. A rat or other animal placed in a "Skinner box" finds and manipulates a lever in the box, usually by accident, and is rewarded by the release of food. The animal quickly learns to associate manipulation of the lever with a food reward. Such learning is the basis for most of the animal training done by humans, in which the trainer typically encourages a behavior by rewarding the animal. Eventually, the animal performs the behavior on command, without always receiving a reward.

Practice and exercise may explain the ultimate bases of play

Many mammals and some birds engage in behavior that can best be described as **play.** Such behavior has no apparent external goal but involves movements closely associated with goal-directed behaviors. For example, playful stalking and attacking of conspecifics (other members of the same species) occurs in many predator species, such as those of the cat and dog families (FIGURE 51.12). Although this behavior does not usually involve painful bites, the animals grab and mouth one another, using movements similar to those used to capture and kill prey. A study of bottlenose dolphins in Australia revealed that young dolphins spend long periods away from their mothers in groups of juveniles engaged in a full range of social and sexual play. African lions and dolphins are social animals, and a social lifestyle is one of the hallmarks of mammalian species that routinely engage in play.

FIGURE 51.12 Play behavior. The roughhousing behavior of these lion cubs evolved in spite of the energy it consumes and the risks it poses. Practicing survival behavior (such as capturing prey), experimenting with social roles, and building a healthy body through exercise are three possible benefits of play.

FIGURE 51.13 Problem solving. Behavioral biologist Bernd Heinrich placed ravens in an experimental situation in which they had to solve the problem of obtaining food hanging from a string. The raven shown here solved the problem by using one foot to pull the string in increments and the other foot to secure the string so the food didn't just drop again. Of interest was the tremendous individual variation in behavior that Heinrich observed. Some ravens never learned to get at the food, while others solved the problem in different ways.

Another common feature of play is that it is potentially dangerous or costly. Baboons sometimes kill and eat juvenile vervet monkeys, and they are most successful in doing so when the monkeys are playing in groups away from adults. In a study of caged ibex, a type of wild goat, play resulted in at least 5 out of 14 kids sustaining injuries that produced limps. "Horsing around" often produces similar injuries in human kids.

Play obviously consumes energy, and the risks to life and limb result in significant additional costs. What could be the ultimate adaptive basis for such seemingly pointless behavior? The "practice hypothesis" suggests that play is a type of learning that allows animals to perfect behaviors needed in functional circumstances. This hypothesis is supported by the observation that play is most common in young animals. However, movements used in play show little improvement after their first few practices. An equally likely ultimate explanation for play is the "exercise hypothesis," which suggests that play is adaptive because it keeps the muscular and cardiovascular systems in top condition. The exercise hypothesis also predicts that play should be especially common in young animals, because they typically do not have to exert themselves in useful activities while under the protection and care of their parents. However, recent studies of beluga whales and several species of dolphins indicate that play, such as creating complex air bubbles, is also common in adults, at least in captivity.

ANIMAL COGNITION

What does an animal's brain do with the information it obtains about the outside world? If a chimpanzee is placed in an area with a banana hung high out of reach and several boxes on the floor, the chimp can "size up" the situation and stack the boxes, enabling it to reach the food. Such novel problem-solving behavior is highly developed in some mammals, especially primates and dolphins, and notable examples have also been observed in some bird species, especially crows, ravens, and jays. Ravens, for example, exhibit marked individual variation in their attempts to solve some of the problems imposed on them by experimenters (FIGURE 51.13).

Watching an animal solve a problem makes us aware that its nervous system has a substantial ability to process information. An expanding research effort on animal cognition seeks to understand information processing at all levels, from the nervous system activities that underlie sophisticated behavior, such as problem solving, to the internal representations animals have about physical objects in their surroundings.

The study of cognition connects nervous system function with behavior

The term *cognition* is variously defined. In a narrow sense, it is synonymous with consciousness, or awareness. In a broad sense—the way we use the term in this book—**cognition** is the ability of an animal's nervous system to perceive, store, process, and use information gathered by sensory receptors.

The study of animal cognition, called **cognitive ethology,** examines the connection between an animal's nervous system and its behavior. Cognitive ethology includes, but is not limited to, the study of animal consciousness, or awareness.

One area of research in cognitive ethology investigates how an animal's brain represents physical stimuli from the environment. For instance, what kind of calculations, if any, does the brain of a dog make about where a Frisbee will land? What is the nature of the dog's internal representation of the spatial relationships between the spinning Frisbee and other objects in the immediate environment? Questions such as these, concerned with internal representations of an animal's physical surroundings, are separate and distinct from questions about consciousness.

Animals use various cognitive mechanisms during movement through space

Directed movement enables animals to avoid predators or poisons, migrate to a more favorable environment, obtain food, and find mates and nest sites. The mechanisms animals use for "finding their way" vary with the spatial scale of the trip and the kind of animal. We will survey three kinds of movement that use mechanisms of increasing cognitive complexity: kinesis and taxis, use of landmarks, and cognitive maps.

Kinesis and Taxis

The simplest mechanisms of movement are kineses and taxes. A **kinesis** is a simple change in activity or turning rate in response to a stimulus. Sow bugs, or wood lice, become more active in dry areas and less active in humid areas, a simple behavior that tends to keep these animals in moist environments. The animals do not move toward or away from specific conditions, but since they slow down in a favorable environment, they tend to stay there. In contrast, a **taxis** is a more or less automatic, oriented movement toward or away from some stimulus. For example, housefly larvae are negatively phototactic after feeding, automatically moving away from light; this simple response presumably ensures that the flies remain in an area where they are harder for predators to detect. Trout exhibit positive rheotaxis (from the Greek *rheos,* current); they automatically swim or orient themselves in an upstream direction, which keeps them from being swept away.

Use of Landmarks Within a Familiar Area

In FIGURE 51.2, we looked at a classic experiment on how digger wasps find their nest entrances. Tinbergen moved a circle of pinecones that had previously surrounded a nest entrance and observed that the wasp landed in the center of the pinecones, even though the nest entrance wasn't there. The wasp was using the pinecones as a **landmark.** The use of landmarks is a more complex cognitive mechanism than a taxis or kinesis. The wasp flies toward a stimulus (the center of the pinecones), as in a taxis, but the pinecone circle is an arbitrary landmark the animal must learn. One nest entrance may have pinecones around it, while another may be next to a pile of stones. Each wasp has to learn the unique landmarks of each individual nest site.

Many animals learn the particular set of landmarks in their area and use those landmarks to find their way within the area. Honeybees, for example, keep track of the nectar supplies from their local flowers and concentrate their foraging on the flowers that are currently most nectar-rich. The bees use landmarks, among other mechanisms, to revisit the most productive flowers. In 2000, a team of biologists reported that they had tracked the foraging movements of honeybees by fitting the insects with tiny transponders (FIGURE 51.14a). Four-day-old honeybees made only short, nonforaging trips out of the hive (FIGURE 51.14b). Slightly older bees, six days old, made longer exploratory trips, but did not yet forage. Only older, fully mature bees foraged, and they moved in a "bee line," back and forth between the hive and nectar-rich flower fields. The earlier exploratory excursions were apparently required to learn a complex set of landmarks throughout the local area.

Cognitive Maps

An animal can move around its environment in a flexible and efficient manner using landmark orientation alone. Honeybees, for instance, might learn ten or so landmarks and locate their hive and flowers in relation to those landmarks. A more powerful mechanism is a cognitive map. A **cognitive map** is an internal representation, or code, of the spatial relationships among objects in an animal's surroundings. It is actually very difficult to distinguish experimentally between an animal that is simply using landmarks and one that is using an internal map. The best evidence for cognitive maps comes from research on birds called jays. A jay stores food in caches, from which the bird can retrieve the food later. An individual jay may store food in thousands of caches. It not only relocates each cache, but also keeps track of food quality, bypassing caches in which the food was relatively perishable and would have decayed. Research by Alan Kamil, of the Universty of Nebraska, suggests that jays use cognitive maps to memorize the locations of their food caches.

Migration Behavior

The most extensive studies of how animal cognition functions in movement have involved animals that exhibit **migration,** regular movement over relatively long distances. Migrating animals generally make one round trip between two regions each year, although there is considerable variation among species.

(a) Researchers fitted honeybees with miniature transponders to follow the movements of individuals of different ages. A transponder is an electronic device that uses the energy of an incoming signal to "reflect" a signal back to a detector.

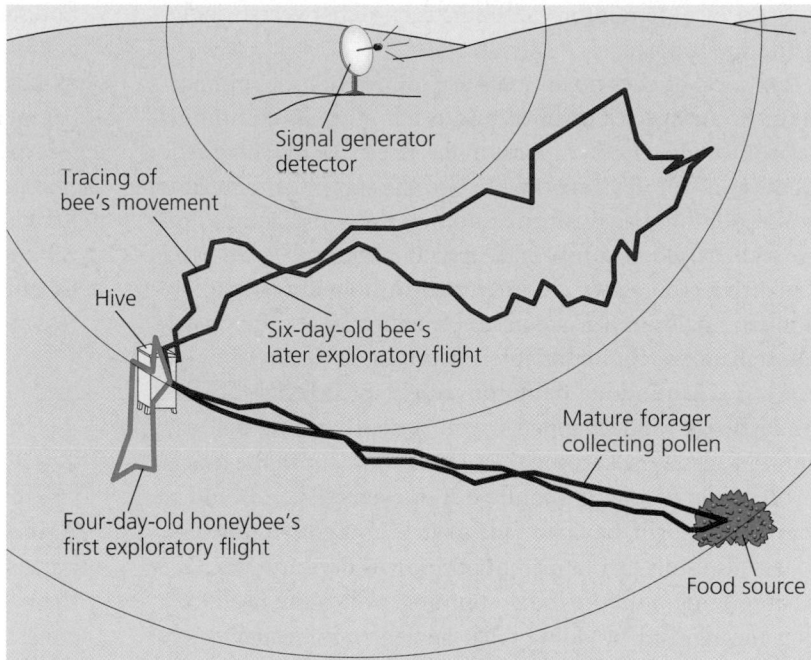

(b) These are tracings of the movements of individual bees. Short exploratory flights led to longer nonforaging flights, during which bees apparently learned local landmarks. By the time a bee is a mature forager, it is very efficient in its "bee line" movements between hive and nectar-rich flowers.

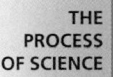
THE PROCESS OF SCIENCE

FIGURE 51.14
Electronic surveillance
of honeybees.

The most notable examples are the migrations of birds, whales, a few butterfly species, and certain oceangoing fish. How is it, for example, that birds called golden plovers find their way over 13,000 km from their arctic breeding grounds to southeastern South America? Even more remarkable, some populations of these birds return to the Hawaiian and Marquesas Islands, small pieces of land in a vast expanse of ocean (FIGURE 51.15).

Migrating animals use one of three mechanisms (or some combination of the three) to find their way. In *piloting,* an animal moves from one familiar landmark to another until it reaches its destination. Piloting is used mostly for short-distance movements. In *orientation,* an animal detects compass directions and travels in a particular straight-line path for a certain distance or until it reaches its destination. The most complex process is *navigation,* in which an animal determines its present location relative to other locations in addition to detecting compass direction (orientation). If you were dropped off at an unfamiliar spot and told that your home was directly to the north, you could use a compass and straight-line travel to get home; that is, you could use orientation. But a compass alone would not be adequate if you were not told which way to go; to choose the right direction, you would need to determine where you were in relation to home. You would need a complex mental picture of your surroundings, a so-called map

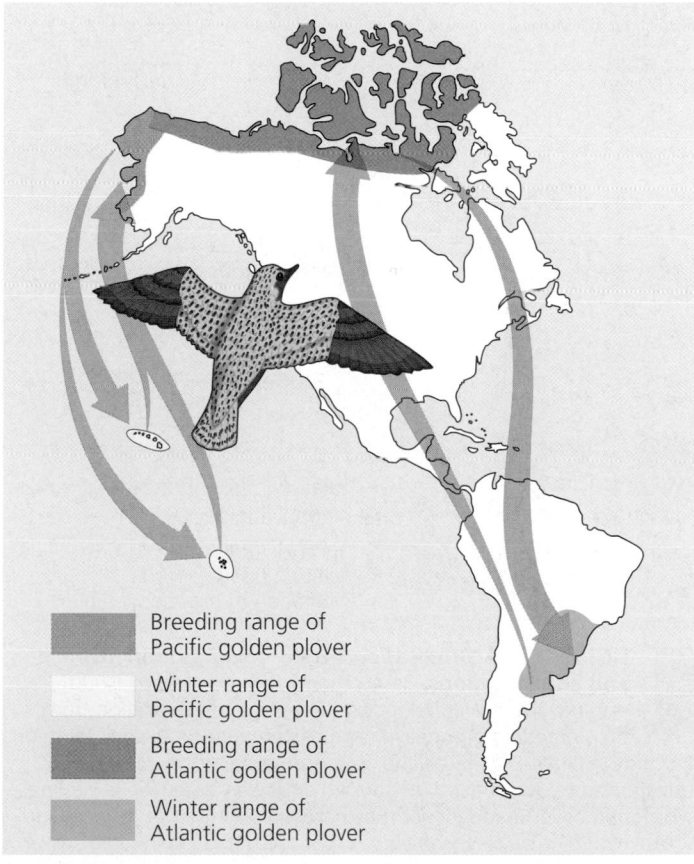

FIGURE 51.15 Migration routes of the golden plover. These birds navigate vast expanses of ocean to the relatively small Hawaiian and Marquesas Islands (yellow). The ground-nesting plovers migrate from warm winter feeding areas to seasonally rich, essentially predator-free Arctic breeding grounds during the short Arctic summer.

Breeding range of
Pacific golden plover

Winter range of
Pacific golden plover

Breeding range of
Atlantic golden plover

Winter range of
Atlantic golden plover

sense. FIGURE 51.16 reinforces this distinction between orientation and navigation.

What sorts of cues do animals use for orientation and navigation? Some species of birds and other animals commonly use a combination of compass references: Earth's magnetic field, the sun (for daytime travel), and the stars (for nighttime travel). Calibrated against one another, these indicators provide excellent, albeit complex, cues to direction.

Cognitive ethologists are interested in how animals utilize cues in moving from place to place. Orienting by the sun or stellar constellations requires an internal timing device to compensate for the continuous daily movement of celestial objects. Consider what would happen if you started walking one day, orienting yourself by keeping the sun on your left. In the morning, you would be heading south; but by evening, you would be heading back north, having made a circle and gotten nowhere. The stars also shift their apparent position as Earth rotates. One night-migrating bird, the indigo bunting, avoids the need for a timing mechanism by doing what ancient human navigators did: fixing on the North Star, which moves little. Many migrants, however, use an internal clock. For example, if an experimental sun is held in a constant position, starlings change their orienta-

tion steadily at a rate of about 15° per hour. This normally compensates for the change in the sun's position as Earth rotates on its axis. The calibration problem is very complex because the apparent location of the sun shifts at a variable rate, being fastest at midday. Furthermore, the apparent position of celestial objects changes as the animal moves over its migration route. Recently, Kenneth and Mary Able, of the State University of New York at Albany, discovered that for one long-distance migrant—the Savannah sparrow—the magnetic and celestial compasses are reset during brief stopovers along the migration route.

The study of consciousness poses a unique challenge for scientists

A simple but profound question is whether nonhuman animals are consciously *aware* of themselves and of the world around them. An equally profound question is whether the study of consciousness (awareness) is within the purview of science.

Many people who have spent a lot of time with pets or wild animals argue that these animals are not behaving like sophisticated robots. But is a dog aware of itself when it is chasing a Frisbee? Do animals "feel" pleasure or sadness as we do? To date, we have no way of answering such questions directly, because conscious awareness is known only to the individual that experiences it, and unlike phenomena that can be studied objectively, it is not associated with any observable behavioral or physiological change.

Donald Griffin, of Princeton University, is a foremost proponent of the view that conscious thinking is an inherent and essential part of the behavior of many nonhuman animals. Griffin argues that if other animals behave in ways we associate with conscious processing in ourselves, perhaps it makes sense to assume that they have the same underlying awareness. In her famous field studies, Jane Goodall has described cognitive decision making in chimpanzees. Griffin suggests that such abilities may extend to many nonprimate branches of the phylogenetic tree as well. He argues, for example, that conscious processes are at the heart of such behaviors as the injury-feigning "strategy" of some species of ground-nesting birds (FIGURE 51.17).

Because of the difficulty of scientifically testing questions about consciousness, some researchers assume the most conservative position—that most animals are not aware. There are, of course, intermediate positions within the argument, and no behavioral biologist would argue that *all* animals behave in ways that indicate consciousness. Moreover, other animals may lack the ability to consciously integrate information (to "think") to the same extent as humans do, but is this a matter of degree—a continuum of abilities—or are humans fundamentally different in some behavioral respect? Ultimately, the answers we find about animal consciousness may profoundly affect how we interact with other animals—and how we view ourselves as well.

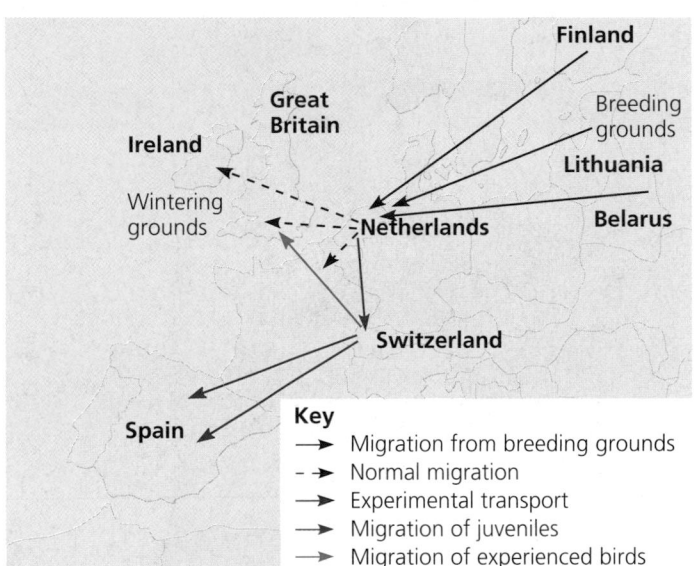

Key
→ Migration from breeding grounds
- → Normal migration
→ Experimental transport
→ Migration of juveniles
→ Migration of experienced birds

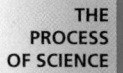
THE PROCESS OF SCIENCE

FIGURE 51.16 Orientation versus navigation in juvenile and adult starlings. Researchers captured about 11,000 starlings in the Netherlands during the birds' migration from their breeding grounds in northeastern Europe to their wintering grounds in Great Britain, Ireland, and northern France. After being transported to Switzerland (red arrow) and released, juvenile starlings, which had never made the journey before, continued to fly west and southwest (blue arrows), which brought them to Spain. Adults, all of whom had made the trip at least once before, flew northwest (green arrow), an atypical direction but one that took them to their usual wintering grounds. Members of both age-groups were able to detect direction, but only the adults exhibited true navigation because they had developed a "map sense" and could determine where their original goal was relative to the site to which they were transported.

FIGURE 51.17 Injury-feigning display. This killdeer uses deception to defend her nest against predators or human disturbance. When danger threatens, she leaves the nest, which is usually concealed, and begins an elaborate display as though she has a broken wing. This behavior has the effect of leading the potential predator away from the nest. As the predator gets close to her, she simply flies away. She returns to the nest only after the danger has passed. Killdeers show extraordinary individual variation in the form and use of this display. In addition, individual killdeers use multiple variations of the display, depending on the type of threat to the nest and whether or not they have experienced the threat before. Some cognitive ethologists point to versatile behavior such as this to support the hypothesis that nonhuman animals are conscious, thinking beings.

SOCIAL BEHAVIOR AND SOCIOBIOLOGY

Sociobiology places social behavior in an evolutionary context

Social behavior, broadly defined, is any kind of interaction between two or more animals, usually of the same species. Although most sexually reproducing species must be social for part of their life cycle in order to reproduce, some species spend most of their lives in close association with conspecifics. Social interactions have long been a research focus for scientists who study behavior. The complexity of behavior increases dramatically when interactions among individuals are considered. Aggression, courtship, cooperation, and even deception are part of the range of social behavior.

Social behavior raises particularly interesting questions about evolutionary adaptation. In morphology and physiology, it is often obvious how some feature of an organism is beneficial. Wings, for instance, are adaptations for flight. But with social behavior, the answers are sometimes less intuitive. As we will see, males and females in some species may have long, complicated courtship periods, which raises the question of why mating could not be done more simply. And some animals behave altruistically (unselfishly), in ways that appear to *reduce* the number of offspring they produce. Thus, social behavior can appear inefficient or even counterproductive to reproductive success. Nevertheless, extensive research is beginning to re-

veal how these behaviors are adaptive and how they could have evolved by natural selection.

The discipline of **sociobiology** applies evolutionary theory to the study and interpretation of social behavior. Much of the evolutionary theory underlying the modern study of social behavior was conceived by the late British biologist William Hamilton, who considered how natural selection acts on the social behavior of individuals. The evolutionary emphasis of Hamilton's work underlies sociobiology and the more general field of behavioral ecology as it is practiced today. The development of sociobiology into a coherent method of analysis and interpretation was further catalyzed in 1975 with the publication of E. O. Wilson's watershed book *Sociobiology: The New Synthesis*.

Competitive social behaviors often represent contests for resources

Because members of a population have a common niche, there is a strong potential for conflict, especially among members of species that normally maintain densities near what the environment can sustain. Sometimes social behavior seems to involve cooperation, as when a group carries out behavior more efficiently than is possible for a single individual (FIGURE 51.18).

FIGURE 51.18 Cooperative prey capture.

(a) African wild dogs. This pack of African wild dogs is killing a wildebeest much larger than themselves.

(b) White pelicans. This line of white pelicans is moving toward a school of fish. This cooperative behavior makes it more difficult for fish to escape by swimming around the birds.

Keep in mind, however, that even when behavior requires some cooperation and seems to be mutually beneficial to interacting individuals, as in mating behavior, each participant usually acts in a way that will maximize its benefits, even if this is at a cost to the other participant. In this section, we examine competitive social interactions, where this "selfish" aspect of behavior is most obvious. Subsequent sections focus on social behaviors involving cooperation.

Agonistic Behavior

In **agonistic behavior,** a contest involving both threatening and submissive behavior determines which competitor gains access to some resource, such as food or a mate. Sometimes the encounter involves tests of strength. More commonly, the contestants engage in threat displays that make them look large or fierce, often with exaggerated posturing and vocalizations. Eventually, one individual stops threatening and presents a submissive or appeasement display, in effect surrendering. Much of this behavior includes **ritual,** the use of symbolic activity, with no serious harm done to either combatant (FIGURE 51.19). Dogs and wolves show aggression by baring their teeth; erecting their ears, tail, and fur; standing upright; and looking directly at their opponent—all of which make the animal appear large and threatening. The eventual loser, on the other hand, sleeks its fur, tucks its tail, and looks away.

The degree to which combat is ritualized depends on the scarcity of the resource and the likelihood that the resource will be available again. For example, male ground squirrels often inflict severe injury upon, or even kill, each other when battling for access to sexually receptive females. In this case, the females for which the ground squirrels are competing are in estrus and receptive to male courtship for only a few hours each year; thus, a male's entire reproductive fitness may depend on his ability to compete against other males that one day.

FIGURE 51.19 Ritual wrestling by rattlesnakes. The rattlesnakes attempt to pin each other to the ground, but they never use their deadly fangs in such combat.

FIGURE 51.20 Reconciliation in chimpanzees. This photo shows two male chimpanzees 10 minutes after a conflict between them. The male at left had threatened the male at right, who had eventually run up a tree. Now the male at left initiates reconciliation by a hand gesture and by making eye contact. Soon after, the two males descended to the ground and groomed each other.

In animals that live in fairly permanent social groups, there are often conflicts in which there is no clear winner and loser; although one of the animals may win a particular face-off, that individual would still benefit from having friendly relations with the "loser." In this case, there is usually some kind of **reconciliation behavior** between the conflicting individuals following the conflict itself. For example, a chimpanzee that has threatened another member of its group may invite reconciliation by a hand gesture, leading to a bout of friendly grooming (FIGURE 51.20). Social primates seem to spend substantial time in reconciliation and pacification-type behavior.

Dominance Hierarchies

Many animals live in social groups maintained by agonistic behavior. Chickens are an example. If several hens who are unfamiliar with one another are put together, they respond by skirmishing and pecking each other. Eventually, the group establishes a clear "pecking order"—a more or less linear **dominance hierarchy.** Within a group, the alpha (top-ranked) hen controls the behavior of all the others, often by mere threats rather than actual pecking. The beta (second-ranked) hen similarly subdues all others except the alpha, and so on down the line to the omega, or lowest-ranking, animal. The advantage to the top-ranked bird is obvious; it is assured access to resources, such as food. And for lower-ranked animals, the system ensures that they do not waste energy or risk harm in futile combat.

Territoriality

A **territory** is an area that an individual defends, usually excluding other members of its own species. Territories are typically used for feeding, mating, and rearing young. A territory is usually fixed in location, its size varying with the species, the territory's function, and the amount of resources available. Song sparrow pairs, for example, may have territories of about 3,000 m², in which they carry out all activities during the several months of their breeding season. Gannets and most other seabirds, in contrast, mate and nest in territories of only a few square meters or less and feed away from their territories (FIGURE 51.21). Bull sea lions defend small territories used only for mating, whereas red squirrels have rather large territories apparently based on feeding patterns. In many species that defend their territories only during the breeding season, individuals may form social groups at other times. Chickadees, for instance, form monogamous breeding pairs that defend small territories in the summer. In the winter, they form larger flocks, enabling the birds to forage more efficiently and benefit from the increased protection from predators that results from membership in a large group.

Note that there is a distinction between a territory and a home range, which is simply the area in which an animal roams about and which is often not defended. In some species, such as breeding song sparrows, territory and home range are the same; but for other species, such as gannets, a territory is considerably smaller than the home range. The distinction is not always clear. Gray squirrels, for example, typically have home ranges that overlap extensively, but one individual may defend part of the range from competitors.

Territories are established and defended through agonistic behavior, and an individual that has gained a territory is often difficult to dislodge. Why do owners usually win? One explanation in behavioral ecology is that a territory is worth more to an owner than to an intruder because the owner is already familiar with it. Thus, because it has more to gain from a territory, an owner is more likely to escalate a battle than is an intruder. In addition, established territory holders are likely to be older and more experienced at engaging in agonistic interactions.

Natural selection does not always favor territoriality, and not all species are territorial. However, for those animals that are, the territory can provide exclusive access to food

FIGURE 51.21 **Territories.** Gannets nest virtually only a peck apart and defend their territories by calling and pecking at each other. This is a population of Australian gannets in New Zealand.

supplies, breeding areas, and places to raise young. Moreover, familiarity with a specific area may help individuals avoid predators. In a territorial species, such benefits outweigh the energy costs of defending the territory and thereby increase fitness.

Ownership of a territory is usually continually proclaimed; this is a primary function of most familiar bird songs, as well as the noisy bellowing of sea lions and the chatter of red squirrels. Other animals may use scent marks or frequent patrols to warn potential invaders (FIGURE 51.22). Gray wolves, which live in packs in huge territories (hundreds of square kilometers), use multiple strategies to advertise territorial boundaries, including scent marking and

(a) **This male cheetah,** a resident of Africa's Serengeti National Park, is spray-urinating on a tree. The odor will serve as a chemical "No Trespassing" sign to other males.

FIGURE 51. 22 **Staking out territory with chemical markers.**

(b) **Another male cheetah** sniffs a marked rock. With their keen olfactory sense, cheetahs can distinguish their own marks from those left by others. These signals help prevent face-to-face meetings that could escalate into violence.

howling. Multiple signals help dispel any ambiguities as to the boundaries of a territory, thereby minimizing the risk that one group will accidentally stray into the territory of a rival pack. This is especially important for wolves because actual face-to-face meetings between groups are often violent.

Defense of territory is usually directed only at conspecifics; a white-crowned sparrow may live within a song sparrow's territory because a different species usually occupies a different niche, or role in the environment, and is less likely to be a direct competitor. Another adaptive reason for concentrating defense on conspecifics is that they are likely to mate with a territory holder's mate.

Although dominance hierarchies and territoriality evolved as a result of their advantages to individuals, such systems have important consequences at the population level because they tend to stabilize density. If resources were allocated evenly among all members of a population, the "fair share" that each individual received might not be enough to sustain anyone, leading to occasional population crashes. With dominance and territoriality, at least some individuals receive an adequate amount of a resource. In fact, if a resource such as food becomes scarce, territories often expand somewhat. In addition, there are usually individuals low in the hierarchy or lacking territories that are ready to move up or step in if one of the successful individuals dies. The result may be relatively stable populations from year to year.

Natural selection favors mating behavior that maximizes the quantity or quality of mating partners

Reproductive behavior includes seeking mates, choosing among potential mates, competing for mates, and, in some species, caring for offspring. Behavioral ecology and its offshoot, sociobiology, seek to explain mating behaviors as outcomes of natural selection reinforcing those variations that enhance reproductive success.

Courtship

Courtship consists of behavior patterns that lead up to copulation (or to gamete release in species with external fertilization; see Chapter 46). In many species, courtship consists of a series of displays and movements by either the male, the female, or both. The elaborate courtship behavior of stickleback fish provides a classic example (FIGURE 51.23). Stickleback courtship lasts only a few minutes, but some animal species court for days or even months. Of what possible benefit to the individual is such elaborate courtship behavior? Put another way, how can natural selection explain the evolution of courtship behavior? Part of the answer is that courtship enables animals to identify potential mates of the same species. This would explain why courtship patterns are often especially elaborate and distinct when two closely related species occupy the same area. Also, courtship can help establish that a poten-

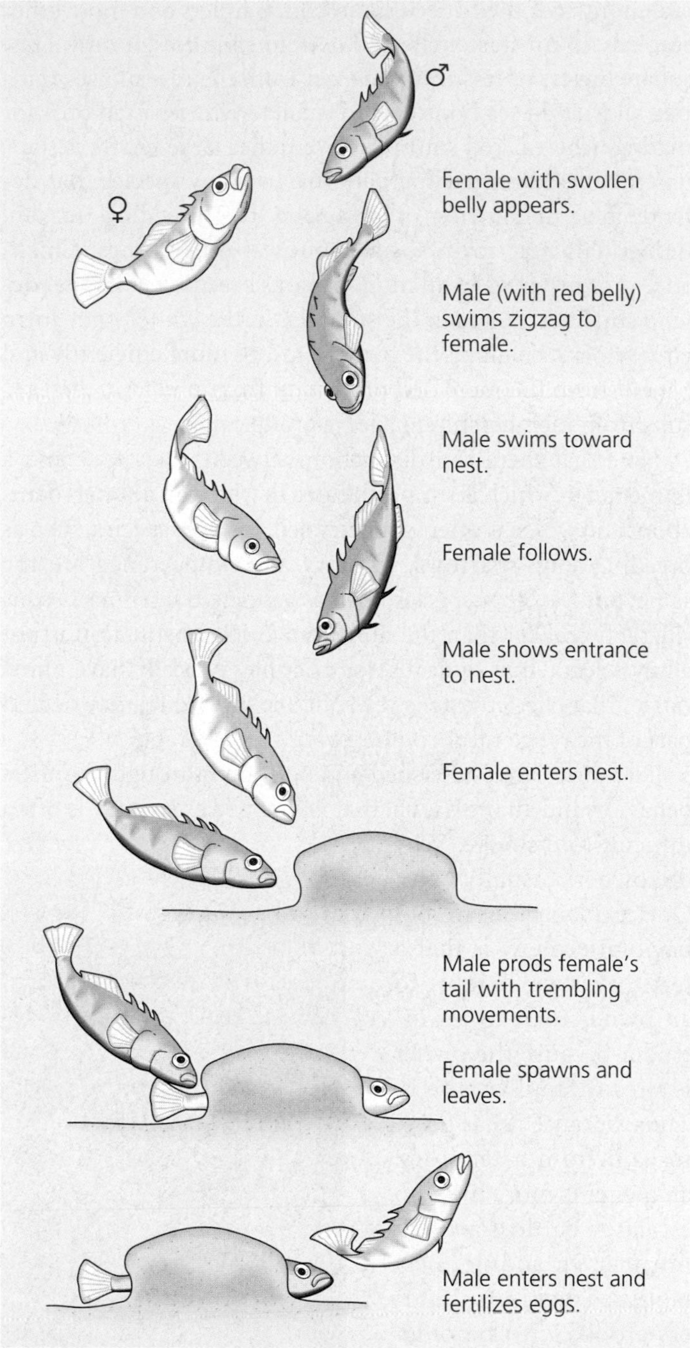
♂
♀
Female with swollen belly appears.
Male (with red belly) swims zigzag to female.
Male swims toward nest.
Female follows.
Male shows entrance to nest.
Female enters nest.
Male prods female's tail with trembling movements.
Female spawns and leaves.
Male enters nest and fertilizes eggs.

FIGURE 51.23 Courtship behavior in the three-spined stickleback. Males are strongly territorial, defending an area in which they have built a nest. If a gravid (egg-carrying) female approaches, her swollen belly inhibits the male's aggressive behavior and elicits zigzag swimming in the male. This entices the female to swim closer, which in turn stimulates the male to swim to the nest and stick his snout inside. This action stimulates the female to wriggle into the tunnel. The male then nuzzles her tail, which stimulates her to spawn, after which she swims out the other end of the nest. The male then enters and deposits sperm on the eggs, after which he immediately and quite aggressively drives the female out of the area, apparently because she lacks the swollen belly that would inhibit his aggression.

tial mate is physiologically ready to reproduce. In sticklebacks, for example, courtship succeeds only if the female displays a swollen belly, which is full of eggs, and the male demonstrates that he has built a nest (see FIGURE 51.23). But if courtship were only a matter of identifying a potential partner that is physiologically competent to reproduce, then such behavior could be much simpler than it is in many species. Much of courtship behavior apparently evolved because of sexual selection, an evolutionary process that we introduced in Chapter 23.

The hypothesis that a courtship behavior is an evolutionary product of sexual selection predicts a basic difference between the mating behaviors of males and females. The difference follows from the potential partners' relative amounts of parental investment in offspring. **Parental investment** refers to the time and resources an individual must spend to produce and nurture offspring. Eggs are generally much larger than sperm and are thus more energetically expensive to produce. Eutherian (placental) mammals produce relatively small eggs (though still much bigger than sperm), but the mother invests considerable time and resources in carrying and nourishing the offspring before their birth. In most species, a male's lower investment per offspring, compared to a female's, means that he can maximize his reproductive output by fertilizing the eggs of many females. Thus, a male's reproductive success is often proportional to his number of partners. This explains why competition among males for mates is common in the animal kingdom. In contrast, the reproductive success of females is less often dependent on the number of partners than it is on the vigor of the limited number of offspring she can produce. This explains why the females of many animal species are so selective in their choices of mates. Healthy mates provide the best opportunity for producing healthy offspring.

This distinction based on parental investment—male competition/female choice—can account for many of the differences between males and females in their morphology and courtship behavior. In some cases, male competition has probably contributed to the evolution of agonistic behaviors and even weapons such as deer antlers, which the males use to compete for mates. But female choice seems to have an even more powerful effect in shaping the secondary sexual characteristics and courtship behaviors of males. For example, the showy displays of peacocks and other male birds during mating season have little to do with direct male-male competition and much to do with advertising robust health to choosy females (see FIGURE 23.16).

As another example of how female choice affects the evolution of males, consider the courtship behavior of stalk-eyed flies (FIGURE 51.24). The eyes of these insects are at the tips of stalks, which are longer in males than in females. During courtship, a male presents himself to a female, front end on. Researchers have documented that females are more likely to mate with those males that have relatively long eyestalks; thus, female choice has been a strong selection factor in the evolution of long eyestalks in males. But why would the females fa-

FIGURE 51.24 Male stalk-eyed fly. These Malaysian insects have eyes at the end of elongated stalks. Females usually select mates with relatively long eyestalks. The stalks are shorter in males with certain genetic disorders.
© Phil Savoie

vor this seemingly arbitrary trait? Behavioral ecologists have correlated certain genetic disorders in the male flies with an inability to develop long eyestalks. Such studies support the hypothesis that females are basing their mate choices on characteristics that are indicators of male quality.

Of course, the distinction between male competition and female choice becomes blurry if you consider that males compete by vying for female choice and not just by engaging in male-male confrontations. And in some species, it is the females who compete directly for mates and the males who are the more selective sex in choosing mates. This is common in species where males provide most of the care for offspring, which raises their parental investment. In the case of sticklebacks, both partners are heavily invested in offspring—the female in her costly production of eggs and the male in his investment in the building and protection of the nest. If you take another look at FIGURE 51.23, you will see that the female choice focuses on courtship signals that advertise a male's ability to provide parental care.

Thus, the specifics of courtship behavior are tied to the natural history of the particular species. But the fascinating variety of dance steps, songs, and display organs are all connected by a comprehensive theory of courtship as an evolutionary product of sexual selection. According to this theory, specific courtship behaviors, like other social behaviors, evolve because they are practiced by the most reproductively successful individuals, which increases the representation of the genes for these behaviors in populations.

Mating Systems

The mating relationship between males and females varies a great deal from species to species. In many species, mating is

promiscuous, with no strong pair-bonds or lasting relationships. In species where the mates remain together for a longer period, the relationship may be **monogamous** (one male mating with one female) or **polygamous** (an individual of one sex mating with several of the other). Polygamous relationships most often involve a single male and many females, called **polygyny,** which can be explained in terms of a difference in parental investment. However, in some species, a single female mates with several males, a relationship called **polyandry.**

The needs of the young are an important ultimate factor in the evolution of mating systems. Most newly hatched birds cannot care for themselves and require a large, continuous food supply that one parent may not be able to provide. In such cases, a male may ultimately leave more viable offspring by helping a single mate than by going off to seek more mates. This may explain why most birds are monogamous. In birds with young that feed and care for themselves almost immediately after hatching, there is less need for parents to stay together. Males of these species can maximize their reproductive success by seeking other mates, and polygyny is relatively common in such birds. In the case of mammals, the lactating female is often the only food source for the young; males usually play no role. And in species where the males protect the females and young, they typically take care of many at once in a harem.

Another factor that influences mating systems and parental care is certainty of paternity. Young born or eggs laid by a female definitely contain the female's genes. But even within a normally monogamous relationship, these young may have been fathered by a male other than the female's usual mate. The certainty of paternity is relatively low in most species with internal fertilization because the acts of mating and birth (or mating and egg laying) are separated over time. This could explain why exclusively male parental care occurs in very few species of birds and mammals. However, certainty of paternity is much higher when egg laying and mating occur together, as in external fertilization. This may explain why parental care in aquatic invertebrates, fishes, and amphibians, when it occurs at all, is at least as likely to be by males as by females (FIGURE 51.25). Male parental care occurs in only 2 out of 28 (7%) fish and amphibian families with internal fertilization, but in 61 out of 89 (69%) families with external fertilization. In fish, even when parental care is given exclusively by males, the mating system is often polygynous, with several females laying eggs in a nest tended by one male.

It is important to point out that when behavioral ecologists use such terms as *certainty of paternity,* they do not mean that animals are aware of those factors when they behave a certain way. Parental behavior correlated with certainty of paternity exists because it has been reinforced over generations by natural selection.

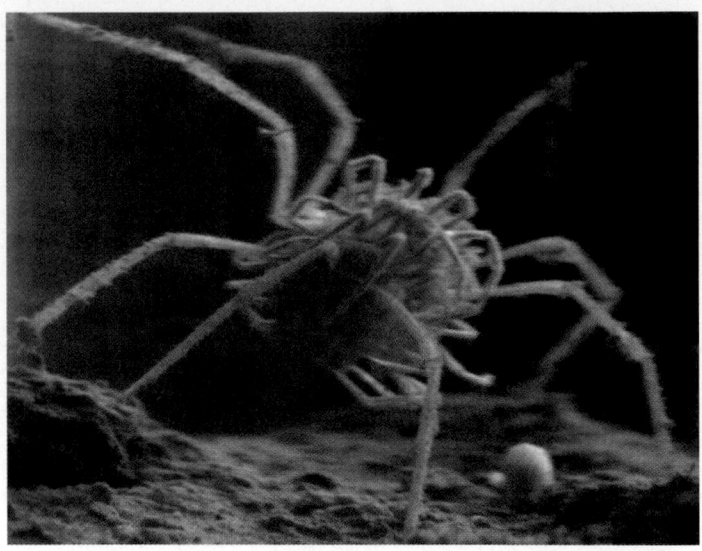

FIGURE 51.25 Paternal care by a sea spider. In many species of marine arthropods called pycnogonids, or sea spiders, a male uses one pair of appendages to carry a ball of eggs that he has fertilized after their release from a mate. The male may mate with a few females, adding the bundle of eggs to those already under his care. The pycnogonid species in this photograph is unusual in that males continue to carry and protect their offspring for some time after hatching.

Social interactions depend on diverse modes of communication

Defining Animal Signals and Communication

Much of what we have discussed under the topics of competitive social interactions and mating behaviors involves animals transmitting information by special behaviors called displays, or signals. In behavioral ecology, a **signal** is a behavior that causes a change in behavior in another animal. The transmission of, reception of, and response to signals make up what we call **communication.** Note that physically forcing another individual to behave in a certain way does not qualify as signaling. For example, if you were to push a sprinter out of the starting blocks at a track meet, the sprinter's behavior would change, but your action would not count as a signal. A signal would be shouting "Ready, set, go!" It is characteristic of signals that they are very efficient in energy costs. It takes less energy to shout "Go!" than it does to push someone down the track.

Singing by male birds is an example of signaling. It transmits the information, "This is my territory. Keep out!" and has the effect that other males are less likely to encroach on the male's territory. This is almost certainly an important message of singing; if we play the tape-recorded songs of another male in a male bird's territory, he becomes highly agitated, approaching and sometimes even attacking the speaker. Another bird has not only ignored his warning but has claimed the territory for himself. This simple experiment is so infallible that some bird-watchers routinely use it to find and see secretive birds that would otherwise stay hidden. This "prerecorded message" procedure makes another important point. We cannot get into an animal's brain to determine whether it has received a message sent by another individual. How, then, do we know when communication has occurred? We usually say that communication has occurred when an act by a *sender* produces a detectable change in the behavior of another individual, the *receiver*. Bird song is communication because it produces a response.

Animals communicate using visual, auditory, chemical (olfactory), tactile, and electrical signals. Which mode is used to transmit information is closely related to an animal's basic lifestyle. Most terrestrial mammals are nocturnal, which makes visual displays relatively ineffective. But olfactory and auditory signals work as well in the dark as in the light, and most mammalian species emphasize these signals. Birds, by contrast, are mostly diurnal and use mostly visual and auditory signals. They almost never use olfactory signals, probably because they can fly faster than chemical signals can travel. (It is hard to imagine a system in which it would be adaptive for a messenger to arrive before its message.) Unlike most mammals, humans are diurnal and in common with birds use mainly visual and auditory communication. Therefore, we can detect the songs and bright colors that birds use to communicate with each other. This may explain why bird-watching is so popular. If humans had the well-developed olfactory abilities of most mammals and could detect their rich world of chemical cues, mammal-sniffing might be as popular as bird-watching.

Pheromones

Animals that communicate by odors emit chemical signals called **pheromones**. These are especially common among mammals and insects and often relate to reproductive behavior. Female silkworm moths, for example, emit a pheromone that can attract males from several kilometers away. Once the moths are together, pheromones also trigger specific courtship behaviors. Another example is the familiar trailing behavior of ants, in which scouts release scents that guide other ants to the food (FIGURE 51.26).

One of the most complex communication systems—certainly among invertebrates—is that of social, or hive, bees. Pheromones produced by a hive's queen and her daughters, the workers, maintain the social order of honeybee colonies. Recent studies indicate that varied blends of two fatty acids, rather than single chemicals, underlie the social behavior and reproduction of honeybees. The context of a chemical signal can be as important as the chemical itself. When male honeybees (drones) are outside the hive (where they can mate with a queen), they are attracted to her pheromone; however, when drones are inside the hive, they are unaffected by the queen's pheromone.

FIGURE 51.26 Fire ants following a pheromone trail. When a worker fire ant of the species *Solenopsis invicta* finds food on a scouting venture, she deposits a scent trail on her trip back to the ant colony. Other workers can then follow this pheromone trail to the food source.

The Dance of the Honeybee

For maximum foraging efficiency, worker bees must convey to one another the location of good food sources, which may change frequently as various flowers bloom or new patches are discovered. How do bees communicate? The study of honeybee communication has a long and rich tradition of experimental research that continues to reveal new elements of the bees' language. The problem was first studied in the 1940s by Karl von Frisch, who carefully watched individual European honeybees (*Apis mellifera carnica*) as they returned to special observation hives. A returning bee would quickly become the center of attention by other bees, called followers (FIGURE 51.27a, p. 1144). The returning bee would go through a repetitive behavior that von Frisch called a dance. If the food source was close to the hive (less than 50 m away), the returning bee moved in tight circles while waggling its abdomen from side to side (FIGURE 51.27b). This dance was usually accompanied by the bee's regurgitating some of the acquired nectar. This behavior, which von Frisch called the "round dance," had the effect of exciting the follower bees and motivating them to leave the hive and search for food that was nearby.

However, bees often forage at great distances from the hive, sometimes in excess of 5 km. In such cases, the round dance is insufficient, lacking both directionality and distance cues necessary for the followers to locate the food source efficiently. A worker returning from a longer distance does a "waggle dance" (FIGURE 51.27c): a half-circle swing in one direction, followed by a straight run and then a half-circle swing in the other direction. This dance seems to indicate both direction and distance. The angle of the straight run in relation to the vertical surface of the open hive is the same as the horizontal angle of the food in relation to the sun. For example, if the bee runs at a 30° angle to the right of vertical, the other workers will fly 30° to the right of the horizontal direction of the sun.

THE PROCESS OF SCIENCE

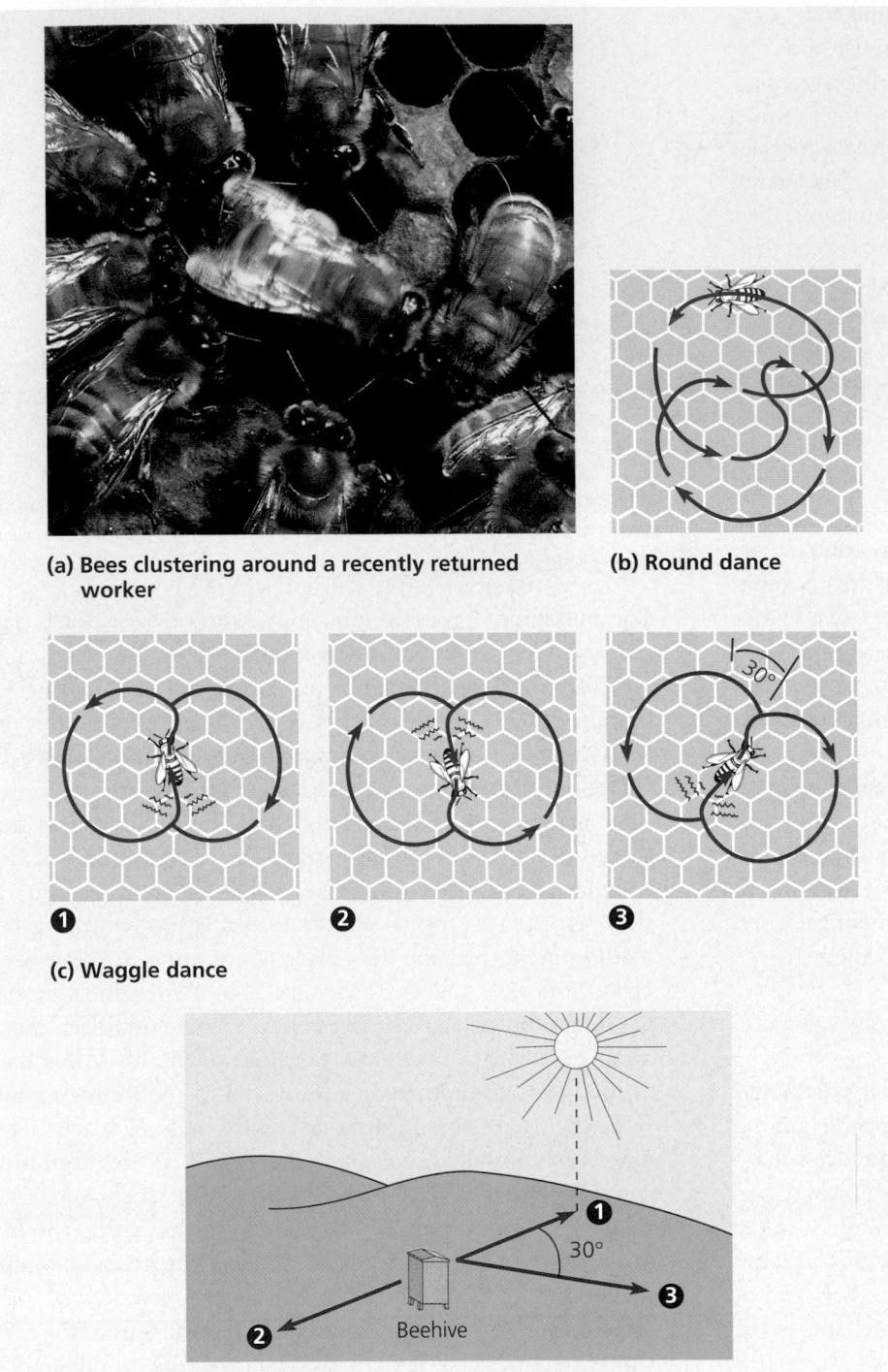

(a) Bees clustering around a recently returned worker

(b) Round dance

(c) Waggle dance

Beehive

FIGURE 51.27 Communication in bees: one hypothesis. (a) Worker bees cluster around one of their sisters, recently returned from a foraging trip. **(b)** The round dance indicates that food is near but may provide no information on directionality or specific distance. **(c)** The waggle dance is performed when food is distant. This dance pattern resembles a figure eight, with a straight run between two semicircular movements. According to von Frisch's hypothesis, the waggle dance indicates both distance and direction. Distance is indicated by the duration of each waggle run or dance and the number of abdominal waggles performed per waggle run. Direction is indicated by the angle (in relation to the vertical surface of the hive) of the straight run that forms part of the dance itself. ❶ For instance, if the straight run is directly upward, this signals that food is in the same direction as the sun. ❷ If the straight run is directly downward, the food is in the direction opposite the sun. ❸ If the angle is 30° to the right of vertical, the food is 30° to the right of the horizontal direction of the sun. And so forth. Odor cues (pheromones) and sound may also convey information about the location and type of food.

Distance to the food is indicated by a variety of elements of the waggle dance. For example, a longer straight run during the dance, and hence an increasing number of abdominal waggles per run, indicates a greater distance to the food source. During waggle dances, the bee also regurgitates nectar; thus, when bees leave to forage, they already "know" the type of food to seek, its distance, and its direction. There is also evidence that the sounds and odors emanating from the dancing bee provide information about the food source.

The concept of inclusive fitness can account for most altruistic behavior

Many social behaviors are selfish, meaning that they benefit the individual at the expense of others, especially competitors. A bird that establishes a territory deprives other individuals of one, and if there is not enough habitat, these other individuals may be unable to breed. Even in species in which individuals do not engage in agonistic behavior, most adaptations that benefit one individual will indirectly harm others. For example, superior foraging ability by one individual may leave less food for others. It is easy to understand the pervasive nature of selfishness if natural selection shapes behavior. Behavior that maximizes an individual's reproductive success will be favored by selection, regardless of how much damage such behavior does to another individual, a local population, or even an entire species.

How, then, can we explain observed examples of what appears to be altruism, or unselfish behavior? On occasion, animals do behave in ways that reduce their individual fitness and increase the fitness of the recipient of the behavior; this is our functional definition of **altruism**. Consider the Belding ground squirrel, which lives in some mountainous regions of the western United States and is vulnerable to predators such as coyotes and hawks (FIGURE 51.28). If a predator approaches, one of the squirrels often gives a high-pitched alarm call. This alerts unaware individuals, who then retreat to their burrows. Careful observations have confirmed that the conspicuous alarm behavior increases the risk of being killed, because it identifies the caller's location.

Another example of altruistic behavior occurs in bee societies, in which the workers are sterile. The workers themselves will never reproduce, but they labor on behalf of a single fertile queen. Furthermore, the workers sting intruders, a behavior that helps defend the hive but results in the death of the worker.

FIGURE 51.28 Altruistic behavior in the Belding ground squirrel. By sounding an alarm call, this Belding ground squirrel warns others of danger, such as an approaching predator. Nearly all alarm calls are given by females.

Still another example of altruism is seen in mole rats, highly social rodents that live in underground chambers and tunnels in southern and northeastern Africa (FIGURE 51.29). The naked mole rat *(Heterocephalus glaber)* is almost hairless and nearly blind and lives in colonies of 75 to 250 or more individuals. The common mole rat *(Cryptomys hottentotus)* has hair and generally lives in smaller colonies. In both species, each colony has only one reproducing female, called the queen, who mates with one to three males, called kings. The

FIGURE 51.29 Two species of colonial mammals.

(a) Naked mole rats

(a) Naked mole rats *(Heterocephalus glaber)* live in underground colonies made up of a single queen, several kings, and often hundreds of nonreproductive individuals. In this photograph, several nonreproductive individuals, who perform all the maintenance activities of the colony, huddle around the queen and her young. Members of each colony are a closely related family unit.

(b) Common mole rat

(b) The common mole rat *(Cryptomys hottentotus)*, widespread in southern Africa, is also colonial with only a queen and several kings reproducing. Compared to naked mole rats, however, common mole rat colonies are smaller and more genetically diverse.

rest of the colony consists of nonreproductive females and males who forage for underground roots and tubers and care for the queen, the kings, and new offspring still dependent on the queen. The nonreproductive individuals may sacrifice their own lives in trying to protect the queen or kings from snakes or other predators that invade the colony.

Inclusive Fitness

How can a naked mole rat, a worker bee, or a Belding ground squirrel enhance its fitness by aiding other members of the population, which are apt to be its closest competitors? How can altruistic behavior be maintained by evolution if it does not enhance—and in fact may even reduce—the reproductive success of the self-sacrificing individuals? Natural selection favors anatomical, physiological, and behavioral traits that increase reproductive success, which in turn propagates the genes responsible for those traits. When parents sacrifice their own personal well-being to produce and aid offspring, this actually increases the fitness of the parents, because it maximizes their genetic representation in the population. But what about helping other close relatives? Like parents and offspring, siblings have half of their genes in common. Therefore, selection might also favor helping one's parents produce more siblings or even helping siblings directly. Evolutionary biologist William Hamilton was the first to realize that selection could result in an animal's increasing its genetic representation in the next generation by "altruistically" helping close relatives other than its own offspring. This realization led to the concept of **inclusive fitness,** which describes the total effect an individual has on proliferating its genes by producing its own offspring *and* by providing aid that enables other close relatives to increase the production of their offspring.

Hamilton's Rule and Kin Selection

Hamilton proposed a quantitative measure for predicting when natural selection would favor altruistic acts among related individuals. The three key variables in an act of altruism are the benefit to the recipient (B), the cost to the altruist (C), and the coefficient of relatedness (r). The benefit and cost measure the change in the average number of offspring produced by the recipient and altruist, respectively, resulting from the altruistic act. Thus, B, the benefit, is the average number of *extra* offspring that the beneficiary of an altruistic act produces; and C, the cost, is how many *fewer* offspring the altruist produces. Suppose, for example, that members of a human population average two children each. Now consider two brothers who are close in age, reproductively mature, equally fertile, but not yet fathers. One of these young men is close to drowning in heavy surf, and his brother risks his own life to swim out and pull his sibling to safety. The benefit to the almost-drowned brother, the recipient of this altruistic act, is

two offspring. Had he drowned, his reproductive output would have been zero. But now, if we use the average, the rescued brother can father two children. The cost to the heroic brother depends on the risk to his own life he took to save his sibling. Let's say that in this kind of surf, an average swimmer has a 5% chance of drowning. Thus, the cost of the altruistic act is 5% the number of offspring we would expect if the altruist had not taken the risky plunge. The cost is 0.05 × 2, or 0.1.

We now know that $B = 2.0$ and $C = 0.1$ for this hypothetical act of altruism, but what about r, the coefficient of relatedness? The **coefficient of relatedness** equals the probability that a particular gene present in one individual will also be inherited from a common parent or ancestor in a second individual. Between two siblings, such as our imaginary brothers at the beach, any gene in one brother has a 50% chance of also being present in the other brother. Thus, for siblings, r is 0.5. One way to see this is to review the segregation of homologous chromosomes that occurs when parents produce gametes by meiosis (FIGURE 51.30; also see Chapter 13).

We can now use values of B, C, and r to evaluate whether natural selection would favor the altruistic act in our imaginary scenario. Natural selection favors altruism if

$$rB > C$$

This inequality is called **Hamilton's rule.** For natural selection to favor an altruistic act, the benefit to the recipient devalued

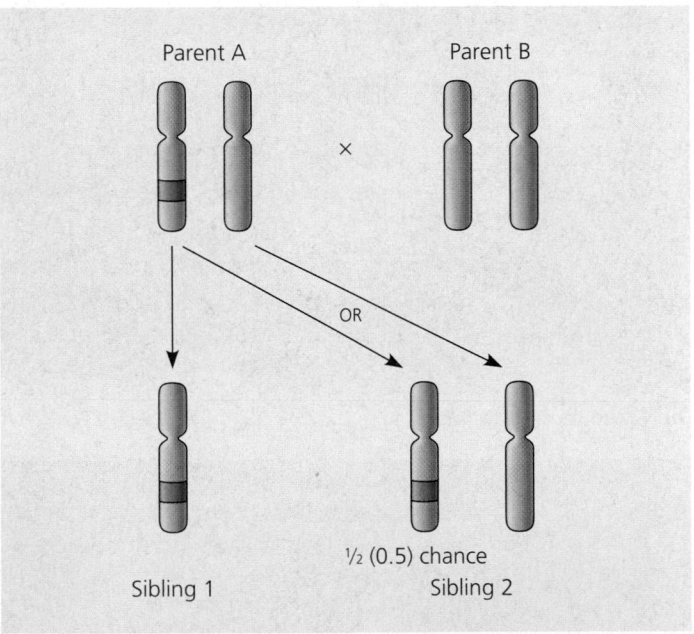

FIGURE 51.30 The coefficient of relatedness between siblings is 0.5. The red band marks a particular gene on a chromosome of a homologous chromosome pair in one parent. As it is the only gene we will follow, the corresponding chromosomes in the other parent are ignored in calculating r, the coefficient of relatedness. Sibling 1 has inherited the gene we are tracing from parent A. There is a probability of 1/2 that sibling 2 will also inherit this gene from parent A. The coefficient of relatedness between the two siblings is 1/2, or 0.5.

(multiplied) by the coefficient of relatedness must exceed the cost to the altruist. For our surfing brothers, $rB = 0.5 \times 2 = 1$ and $C = 0.1$. This satisfies Hamilton's rule; thus, natural selection would favor this altruistic act of one brother saving another. Any particular gene in the altruist will, on average, be passed on to more offspring if that brother risks the rescue than if he does not. (And among those genes may be some that actually contribute to the altruistic behavior, so these genes, too, are propagated.) The natural selection that favors this kind of altruistic behavior by enhancing reproductive success of relatives is specifically called **kin selection.**

Kin selection weakens with hereditary distance. While siblings have an r of 0.5, $r = 0.25$ (1/4) between an aunt and her niece, and $r = 0.125$ (1/8) between first cousins. Notice that as the degree of relatedness decreases, the rB term in the Hamilton inequality also decreases. Would natural selection favor our strong-swimming surfer rescuing his cousin? For this altruistic act, $rB = 0.125 \times 2 = 0.25$, which, luckily for this drowning cousin, is still much smaller than $C = 0.1$, the cost to the altruist. Of course, the degree of risk the altruist takes comes into play, too. If the potential rescuer is a poor swimmer, he may have a 50% chance of drowning instead of the 5% chance for a strong swimmer. In this case, the cost to the altruist would be $0.5 \times 2 = 1$. That's greater than the rB of 0.25 we calculated for the drowning cousin, who better hope a lifeguard is near.

The British geneticist J. B. S. Haldane anticipated the concepts of inclusive fitness and kin selection by jokingly saying that he would lay down his life for two brothers or eight cousins. In today's terms, we would say that he would do this because either two brothers or eight cousins would result in as much representation of Haldane's genes as would two of his own offspring.

If kin selection explains altruism, then the examples of unselfish behavior we observe among diverse animal species should involve close relatives. This expectation is met, but often in complex ways. Like most mammals, female Belding ground squirrels (see FIGURE 51.28) settle close to their site of birth, while males settle at distant sites. Thus, only females are likely to live near close relatives, and nearly all alarm calls are given by females (FIGURE 51.31). However, if all of a female's close relatives are dead, she rarely gives alarm calls. In the case of worker bees, the individuals are sterile, and anything they do to help the entire hive benefits the only permanent member who is reproductively active—the queen, who is their mother.

In the case of naked mole rats, DNA analyses have shown that all the individuals in a colony are closely related. Genetically, the queen appears to be a sibling, daughter, or mother of the kings, and the nonreproductive rats are the queen's direct descendants or her siblings (see FIGURE 51.29a). Hence, when a nonreproductive individual enhances a queen's or king's chances of reproducing, it increases the chances that some genes identical to its own will be passed to the next generation. The scenario for the common mole rat seems to be different

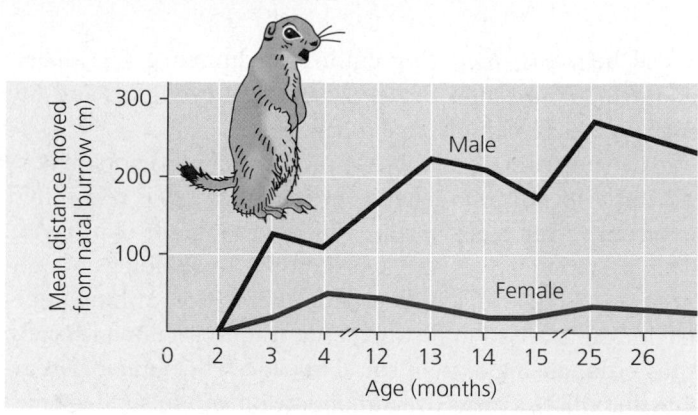

FIGURE 51.31 Kin selection and altruism in the Belding ground squirrel. This graph helps explain the male-female difference in altruistic behavior of ground squirrels. After they are weaned, males disperse much farther from their birthplaces than do females. Therefore, females are more likely to live near close relations, and alarm calls that warn these relatives increase the inclusive fitness of the altruist.

(see FIGURE 51.29b). Some individuals may move from one colony to another, making its colonial groups more genetically diverse and decreasing the opportunities for kin selection.

Some researchers posit that only by living in cooperative groups can mole rats obtain enough food to survive where resources are in short supply. For both the common mole rat and the naked mole rat, cooperative behavior enhances tunneling and underground foraging for roots and tubers, factors that may enable these two species to inhabit arid regions. Three other species of mole rats, which are not colonial, are found only where water and other resources are relatively plentiful. Thus, resource limitation may underlie the evolution of colonial life among these mammals and perhaps the evolution of the kind of altruistic behavior seen in the naked mole rat.

Some animals occasionally behave altruistically toward others who are not relatives. A baboon may help an unrelated companion in a fight, or a wolf may offer food to another wolf even though they share no kinship. Such behavior can be adaptive if the aided individual returns the favor in the future. This sort of exchange of aid is called **reciprocal altruism** and is commonly invoked to explain altruism in humans. Reciprocal altruism is rare in other animals; it is limited largely to species with social groups stable enough that individuals have many chances to exchange aid. It is likely that all behavior that seems altruistic actually has at least the potential to increase fitness in some way.

Sociobiology connects evolutionary theory to human culture

Recall that the main premise of sociobiology is that certain behavioral characteristics exist because they are expressions of genes that have been perpetuated by natural selection. In the last chapter of *Sociobiology: The New Synthesis,* E. O. Wilson speculated about the evolutionary basis of certain kinds of social behavior, including culture, in humans. The debate about the connection between biological evolution and human culture remains heated today.

The spectrum of possible human social behaviors may be circumscribed by our genetic potential, but this is very different from saying that genes are rigid determinants of behavior. This is at the core of the debate about sociobiology. Opponents fear that a sociobiological interpretation of human behavior can be used to justify the status quo in human society, thus rationalizing current social injustices. Sociobiologists argue that this is a gross oversimplification and misunderstanding of what the data tell us about human biology. Sociobiology does not reduce us to robots stamped out of rigid genetic molds. Individuals vary extensively in anatomical features, and we should expect inherent variations in behavior as well. Furthermore, though we are locked into our genotypes, our nervous systems are not "hardwired." Environment intervenes in the pathway from genotype to phenotype for physical traits and even more so for behavioral traits. And because of our capacity for learning and versatility, human behavior is probably more plastic than that of any other animal. Over our recent evolutionary history, we have built up a diversity of structured societies with governments, laws, cultural values, and religions that define what is acceptable behavior and what is not, even when unacceptable behavior might enhance an individual's Darwinian fitness. Perhaps it is our social and cultural institutions that make us truly unique and that provide the only feature in which there is no continuum between humans and other animals (FIGURE 51.32).

■　　■　　■

In this chapter, we have examined the role of behavior in the relationship between animals and their environment. We have focused on the proximate mechanisms underlying behavior

FIGURE 51.32 Both genes and culture build human nature. Teaching of the younger generation by the older is one of the basic ways in which all cultures are transmitted. Sociobiologists see mentoring as an innate tendency with adaptive value that has evolved in the human species.

and on how particular behavior patterns contribute to an animal's survival and reproductive success. In studying an animal's behavior, we are watching it perform on an ecological stage, a setting where evolution by natural selection determines which individuals contribute the greatest number of genes to a population's gene pool. In the next chapter, we take a closer look at populations as ecological and evolutionary units.

CHAPTER 51 REVIEW

Go to the Campbell Biology website (www.campbellbiology.com) to explore an interactive version of the Chapter Review.

Summary of Key Concepts

INTRODUCTION TO BEHAVIOR AND BEHAVIORAL ECOLOGY

■ **What is behavior? (p. 1122)** Behavior mainly consists of observable muscle-powered movements.

■ **Behavior has both proximate and ultimate causes (p. 1122)** Proximate mechanisms include the hormonal, nervous, and environmental stimuli that elicit a particular behavior pattern during the life of an animal. Ultimate causes are the reasons why the behavior pattern evolved over evolutionary time.

■ **Behavior results from both genes and environmental factors (p. 1122–1123, FIGURE 51.1)** An individual's behavior develops under the influence of genes and environment.

■ **Innate behavior is developmentally fixed (pp. 1123–1124)** An innate behavior is one that occurs in all individuals of a population, regardless of individual differences in experience.

■ **Classical ethology presaged an evolutionary approach to behavioral biology (pp. 1124–1126, FIGURES 51.2–51.4)** Early ethologists focused on fixed action patterns (FAPs), essentially unchangeable series of acts usually carried to completion once triggered by an external sensory stimulus (sign stimulus).

- **Behavioral ecology emphasizes evolutionary hypotheses (pp. 1126–1128, FIGURES 51.5–51.7)** Behavioral ecology is based on the theory that animals behave in ways that increase their Darwinian fitness (reproductive success).

  Web/CD Case Study in the Process of Science: *How Can Pillbug Responses to Environments Be Tested?*

LEARNING

- **Learning is experience-based modification of behavior (pp. 1128–1129, FIGURE 51.8)** Learning is modification of behavior resulting from specific experiences. Some apparent learning is due mostly to inherent maturation. Habituation is a simple kind of learning involving loss of sensitivity to unimportant stimuli.

- **Imprinting is learning limited to a sensitive period (pp. 1129–1130, FIGURE 51.9)** Imprinting occurs in various animals and can involve the identity of mating partners as well as of parents.

- **Bird song provides a model system for understanding the development of behavior (pp. 1130–1132, FIGURE 51.10)** Biologists have described two forms of development of bird song: learning during a sensitive period (as in the white-crowned sparrow); and open-ended learning (as in the canary), in which the bird continually adds new components to its song each year.

- **Many animals can learn to associate one stimulus with another (p. 1132, FIGURE 51.11)** Associative learning involves linking one stimulus with another. In operant conditioning, or trial-and-error learning, an animal learns to associate one of its own behaviors with reward or punishment and modifies the behavior accordingly.

- **Practice and exercise may explain the ultimate bases of play (pp. 1132–1133, FIGURE 51.12)** The benefits of play may include the practice of survival behaviors, such as hunting, as well as fulfilling the need for exercise.

ANIMAL COGNITION

- **The study of cognition connects nervous system function with behavior (pp. 1133–1134, FIGURE 51.13)** Cognition is the ability of an animal's nervous system to perceive, store, process, and use information gathered by sensory receptors.

- **Animals use various cognitive mechanisms during movement through space (pp. 1134–1136, FIGURES 51.14–51.16)** Many animals find their way around in space by means of memorized landmarks. A cognitive map, or internal representation of the spatial relationships among objects in their surroundings, is a more powerful navigatory mechanism. Some migrating birds and other animals navigate by calibrating several cues: Earth's magnetic field, the sun, and the stars.

- **The study of consciousness poses a unique challenge for scientists (pp. 1136–1137, FIGURE 51.17)** A debate centers on whether nonhuman animals are conscious, thinking organisms.

SOCIAL BEHAVIOR AND SOCIOBIOLOGY

- **Sociobiology places social behavior in an evolutionary context (p. 1137)** Social behavior encompasses the spectrum of interactions between two or more animals, usually of the same species.

- **Competitive social behaviors often represent contests for resources (pp. 1137–1140, FIGURES 51.18–51.22)** Agonistic behavior involves a contest in which one competitor gains an advantage in obtaining access to a limited resource, such as food or a mate. Some animals have dominance hierarchies, in which high-ranking individuals gain preferential access to resources. Territoriality is a behavior in which an animal defends a specific, fixed portion of its home range against intrusion by other animals of the same species.

- **Natural selection favors mating behavior that maximizes the quantity or quality of mating partners (pp. 1140–1142, FIGURES 51.23–51.25)** Courtship functions to identify that two individuals are of the same species and are ready to breed. The full elaboration of courtship is due to sexual selection, particularly female choice. During courtship, a male may display his genetic quality and (in species with parental care) his readiness for parental care. The mating system of a species consists of the way in which males and females associate for breeding; mating systems may be promiscuous, monogamous, or polygamous, depending partly on the parental investment made by males and females.

- **Social interactions depend on diverse modes of communication (pp. 1142–1144, FIGURES 51.26, 51.27)** Animals communicate by means of signals, in which the behavior of one individual leads to a change in the behavior of another individual.

  Web/CD Activity 51A: *Honeybee Waggle Dance Video*

- **The concept of inclusive fitness can account for most altruistic behavior (pp. 1144–1147, FIGURES 51.28–51.31)** Altruism mainly takes place between genetic relatives and can be explained in terms of inclusive fitness, or kin selection: Genes enhance their proliferation by directing organisms to care for others who share those genes. Hamilton's rule provides a quantitative tool for assessing whether natural selection would favor a particular act of altruism.

- **Sociobiology connects evolutionary theory to human culture (pp. 1147–1148, FIGURE 51.32)** Sociobiology maintains that much of human social behavior can be understood in evolutionary terms.

Self-Quiz

1. Bees can detect wavelengths of light that we cannot see and sense minute amounts of chemicals we cannot smell. But unlike many insects, bees cannot hear very well. Which of the following statements best fits in the perspective of behavioral ecology?
   a. Bees are too small to have functional ears.
   b. Hearing must not contribute much to a bee's fitness.
   c. If a bee could hear, its tiny brain would be swamped with information.
   d. This is an example of a proximate causation.
   e. If bees could hear, the noise of the hive would distract the bees from their work.

2. The nature-versus-nurture controversy centers on
   a. the distinction between proximate and ultimate causes of behavior.
   b. the role of genes in learning.
   c. whether animals have conscious feelings or thoughts.
   d. the extent to which an animal's behavior is innate or learned.
   e. the importance of good parental care.

3. According to the inequality known as Hamilton's rule ($rB > C$),
   a. natural selection could not favor altruism if the altruist loses its life.
   b. natural selection would favor altruistic acts when the benefit to the receiver, reduced by the coefficient of relatedness, exceeds the cost to the altruist.
   c. natural selection is more likely to favor altruistic acts when the beneficiary is an offspring than when it is a sibling.
   d. kin selection is a stronger selection factor than the individual reproductive success favored by natural selection.
   e. Altruism must always be reciprocal.

4. Female spotted sandpipers aggressively court males and then, after mating, leave the clutch for the male to incubate. This sequence may be repeated several times with different males until no available males remain, forcing the female to incubate her last clutch. All of the following terms describe this behavior *except*
   a. polygamy.
   b. polyandry.
   c. polygyny.
   d. promiscuity.
   e. parental investment.

5. Which of the following is *least* likely to involve cognition?
   a. navigation of a sparrow during seasonal migration
   b. being aware of your neighbor's lawn care
   c. territoriality
   d. positive rheotaxis of a fish in a current
   e. optimal foraging

6. Which of the following is *not* true of agonistic behavior?
   a. It is most common among members of the same species.
   b. It may be used to establish and defend territories.
   c. It often involves symbolic conflict and often does not cause serious harm to either the winner or the loser in the encounter.
   d. It is a uniquely male behavior.
   e. It may be used to establish dominance hierarchies.

7. A researcher found that a region of the forebrain in canaries shrinks and regenerates each breeding season. This finding correlates with
   a. the plastic song stage and then the learning of a new, more elaborate song each year.
   b. the crystallization of its adult song from the subsong it developed when it first learned to sing.
   c. the sensitive period in which the male parent bonds with new offspring.
   d. the renewal of nest-building and mating activity each spring.
   e. the sensitive period in which canaries form a template of their species-specific song.

8. The core idea of sociobiology is that
   a. human behavior is rigidly predetermined by inheritance.
   b. humans cannot learn to alter their social behavior.
   c. many aspects of social behavior have an evolutionary basis.
   d. the social behavior of humans is comparable to that of bees.
   e. environment outweighs genes in human behavior.

9. Which of the following provides an example of habituation?
   a. Humpback whales migrating from Hawaii to Alaska are observed singing songs first identified in humpbacks migrating between Alaska and Baja California.
   b. Male sticklebacks attempt to attack any red-colored object near their tank.
   c. Adult brown pelicans are more successful at capturing fish than are juveniles.
   d. Female warblers incubate the eggs cowbirds deposit in their nests.
   e. Aquarium fish are initially startled by tapping on the aquarium glass but eventually ignore it.

10. A honeybee returning to the hive from a food source performs a waggle dance with the run oriented straight to the left on the vertical surface. Most likely, this means that the food is located
    a. 90° left of the hive.
    b. 90° left of the line from the hive to the sun.
    c. in the opposite direction, straight to the right of the hive.
    d. above the hive and slightly to the left.
    e. very close to the hive.

11. When you touch a hot plate, your arm recoils. What are the proximate and ultimate causes of this behavior?

12. What is behavioral ecology?

13. Why is the study of behavior relevant to ecology?

14. Why is a timekeeping mechanism essential for stellar navigation?

15. In terms of ultimate causation, why is "fighting to the death" an unusual form of agonistic behavior among animals?

16. How is a female bird's fitness associated with her ability to choose a mate by keying on displays and adornments that "advertise" the healthiness of the male?

17. What is the ultimate cause for altruistic behavior among kin?

Go to the website or CD-ROM for more quiz questions.

Evolution Connection

In human affairs, we often explain our behavior in terms of subjective feelings or motives or reasons; but evolutionary explanations for behavior work in terms of reproductive fitness. What is the relationship between the two kinds of explanation? For instance, is a human explanation for behavior, such as "falling in love," incompatible with an evolutionary explanation? Does falling in love become more meaningful or less meaningful (or neither) if it has an evolutionary basis?

The Process of Science

Scientists studying scrub jays found that it is common for "helpers" to assist mated pairs of birds in raising their young. The helpers lack territories and mates of their own. Instead, they help the territory owners gather food for their offspring. Propose a hypothesis to explain what advantage there might be for the helpers to engage in this behavior instead of seeking their own territories and mates. How would you test your hypothesis? If your hypothesis is correct, what kind of results would you expect your tests to yield?

Evaluate how to conduct an experiment on pillbug responses to environments in the Case Study in the Process of Science, available on the website and CD-ROM.

Science, Technology, and Society

Researchers are very interested in studying identical twins who were separated at birth and raised apart. So far, the data suggest that these twins are much more alike than researchers would have predicted; they have similar personalities, mannerisms, habits, and interests. What kind of general question do you think researchers hope to answer by studying twins that have been raised apart? Why do identical twins make good subjects for this kind of research? What do the results suggest to you? What are the potential pitfalls of this research? What abuses might occur if the studies are not evaluated critically and if the results are carelessly cited in support of a particular social agenda?

Answers: 1. b; 2. d; 3. b; 4. c; 5. d; 6. d; 7. a; 8. c; 9. e; 10. b; 11. The proximate cause is a simple reflex, a neural pathway linking stimulation of receptors in your finger to motor response by muscles of your arm and hand; the ultimate cause is the natural selection for a behavior that minimizes damage to the body, thereby contributing to survival and reproductive success. 12. The investigation of ultimate causes of behaviors, the evolutionary basis for behaviors as mechanisms that enhance reproductive success. 13. Ecology is the study of organisms' relationships with their environments, and an animal's behavior is part of this organism-environment interface. 14. Because the positions of the stars change with time of night and season. 15. Because ritualized posturing or nonlethal combat can usually produce a winner without injuries that lower reproductive potential for the winner and eliminate it altogether for the loser. 16. She is more likely to have healthy offspring by mating with a healthy male than with a sickly one. 17. Natural selection reinforces altruistic behavior through the reproductive success of closely related individuals that have many genes in common with the altruist, including genes for altruism.

POPULATION ECOLOGY

CHARACTERISTICS OF POPULATIONS

- Two important characteristics of any population are density and the spacing of individuals
- Demography is the study of factors that affect the growth and decline of populations

LIFE HISTORIES

- Life histories are highly diverse, but they exhibit patterns in their variability
- Limited resources mandate trade-offs between investments in reproduction and survival

POPULATION GROWTH

- The exponential model of population growth describes an idealized population in an unlimited environment
- The logistic model of population growth incorporates the concept of carrying capacity

POPULATION-LIMITING FACTORS

- Negative feedback prevents unlimited population growth
- Population dynamics reflect a complex interaction of biotic and abiotic influences
- Some populations have regular boom-and-bust cycles

HUMAN POPULATION GROWTH

- The human population has been growing almost exponentially for three centuries but cannot do so indefinitely
- Estimating Earth's carrying capacity for humans is a complex problem

The size *and activities of the human population are now among Earth's most significant problems. With a population of over 6 billion individuals, our species requires vast amounts of materials and space, including places to live, land to grow our food, and places to dump our waste. Endlessly expanding our presence on Earth, we have devastated the environment for many other species and now threaten to make it unfit for ourselves.*

To understand human population growth, we must consider the general principles of population ecology. It is obvious that no population can continue to grow indefinitely. Species other than humans sometimes exhibit population explosions, but their populations inevitably decline. In contrast to these radical booms and busts, many populations are relatively stable over time, with only minor changes in population size.

In our earlier study of biological populations (see Chapter 23), we emphasized the relationship between population genetics—the structure and dynamics of gene pools—and evolution. Evolution remains our central theme as we now view populations in the context of ecology. Population ecology, the subject of this chapter, is concerned with measuring changes in population size and composition, and with identifying the ecological causes of these fluctuations. Later in this chapter, we will return to our discussion of the human population. Let's first examine some of the structural and dynamic aspects of populations as they apply to any species, such as the monarch butterfly population in the photo on this page.

CHARACTERISTICS OF POPULATIONS

A **population** is a group of individuals of a single species that simultaneously occupy the same general area. They rely on the same resources, are influenced by similar environmental factors, and have a high likelihood of breeding with and interacting with one another. The characteristics of a population are shaped by interactions between individuals and their environments, and natural selection can modify these characteristics.

Two important characteristics of any population are density and the spacing of individuals

At any given moment, every population has geographic boundaries and a population size (the number of individuals it includes). Ecologists begin studying a population by defining boundaries appropriate to the organisms under study and to the questions being posed. A population's boundaries may be natural ones, such as a specific island in Lake Superior where terns nest, or they may be arbitrarily defined by an investigator, such as the oak trees within a specific county in Minnesota. Regardless of such differences, two important characteristics of any population are its density and its dispersion. Population **density** is the number of individuals per unit area or volume—the number of oak trees per square kilometer in the Minnesota county, for example. **Dispersion** is the pattern of spacing among individuals within the geographic boundaries of the population.

Measuring Density

In rare cases, it is possible to determine population size and density by actually counting all individuals within the boundaries of the population. We could count the number of sea stars in a tide pool, for example. Herds of large mammals, such as buffalo or elephants, can sometimes be counted accurately from airplanes (FIGURE 52.1). In most cases, however, it is impractical or impossible to count all individuals in a population. Instead, ecologists use a variety of sampling techniques to estimate densities and total population sizes. For example, they might estimate the number of alligators in the Florida Everglades by counting individuals in a few randomly chosen plots. Or they might count the numbers of oak trees in several randomly placed circular plots of 10-m diameter. Such estimates are more accurate when there are many sample plots and when the habitat is homogeneous.

One sampling technique researchers often use to estimate fish and wildlife populations is the **mark-recapture method.** Traps are placed within the boundaries of the study area, and captured animals are marked with tags, collars, bands, or spots of dye and then immediately released. After a few days or a few weeks—enough time for the marked animals to mix randomly with unmarked members of the population—traps are set again. The proportion of marked (recaptured) animals in the second trapping is assumed equivalent to the proportion of marked animals in the total population:

$$\frac{\text{Number of recaptures in second catch}}{\text{Total number in second catch}}$$

$$= \frac{\text{Number marked in first catch}}{\text{Total population } N}$$

Thus, if there have been no births, deaths, immigration, or emigration, the following proportionality provides an estimate of the population size N:

$$N = \frac{\text{Number marked in first catch} \times \text{Total number in second catch}}{\text{Number of recaptures in second catch}}$$

For example, suppose that 50 snowshoe hares are captured in box traps, marked with ear tags, and released. Two weeks later, 100 hares are captured and checked for ear tags. If 10 hares in this second catch are already marked and thus are recaptures, we would estimate that 10% of the total hare population is marked. Since 50 hares were originally marked, the entire population would be about 500 hares. This method assumes that each marked individual has the same probability of being trapped as each unmarked individual. This is not always a safe assumption, however. An animal that has been caught once may become wary of the traps later on or may learn to return to traps to eat the food used as bait.

In some cases, instead of counting individual organisms, population ecologists estimate density from some index of population size. This usually involves counting signs left by organisms, such as the number of nests, burrows, tracks, or fecal droppings.

FIGURE 52.1 Aerial census for African buffalo (*Syncerus caffer*) in the Serengeti of East Africa. Biologists can count large mammals and birds in open habitats from the air, either directly or from photographs like this one. By repeating these counts over many years, researchers can track population trends.

Patterns of Dispersion

Within a population's geographic range, local densities may vary substantially because the environment is patchy (not all areas provide equally suitable habitat) and because individuals exhibit patterns of spacing in relation to other members of the population.

The most common pattern of dispersion is **clumped,** with the individuals aggregated in patches. Plants may be clumped in certain sites where soil conditions and other environmental factors favor germination and growth. The eastern red cedar is often found clumped on limestone outcrops, where soil is less acidic than in nearby areas. Mushrooms may be clumped on a rotting log. Some animals move in herds (see FIGURE 52.1). Animals often spend much of their time in a particular micro-environment that satisfies their requirements. For example, many forest insects and salamanders are clumped under logs, where the humidity remains high. Herbivorous animals of a particular species are likely to be most abundant where their food plants are concentrated. Clumping of animals may also be associated with mating behavior. For example, mayflies often swarm in great numbers, a behavior that increases mating chances for these insects, which spend only a day or two as reproductive adults. There may also be "safety in numbers"; fish swimming in large schools, for example, are often less likely to be eaten by predators than fish swimming alone or in small groups (FIGURE 52.2a).

In contrast to a clumped distribution of individuals within a population, a **uniform,** or evenly spaced, pattern of dispersion may result from direct interactions between individuals in the population. For example, a tendency toward regular spacing of plants may be due to shading and competition for water and minerals; some plants also secrete chemicals that inhibit the germination and growth of nearby individuals that could compete for resources. Animals often exhibit uniform dispersion as a result of territorial behavior and aggressive social interactions (FIGURE 52.2b). Uniform patterns are not as common in populations as clumped patterns.

Random spacing (unpredictable dispersion) occurs in the absence of strong attractions or repulsions among individuals of a population; the position of each individual is independent of other individuals. For example, trees in a forest are sometimes randomly distributed (FIGURE 52.2c). Random patterns are not as common in nature as one might expect; most populations show at least a tendency toward a clumped distribution.

FIGURE 52.2 Patterns of dispersion within a population's geographic range.

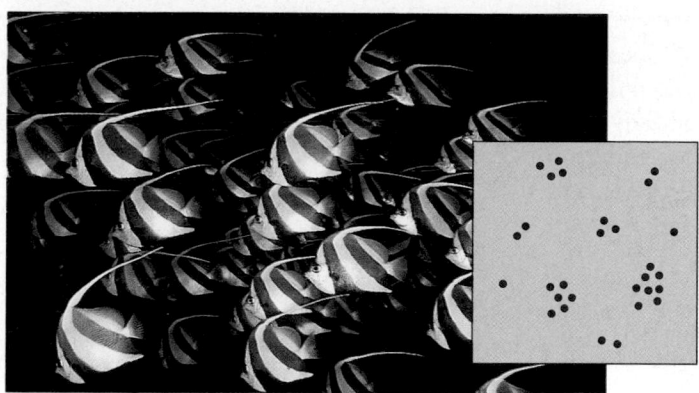

(a) Clumped. Butterfly fish, like many fish, are often found clumped in schools. Schooling may increase the hydrodynamic efficiency of swimming, reduce predation risks, and increase feeding efficiency. Within a school, individuals are more or less evenly spaced.

(b) Uniform. Birds nesting on small islands, such as these king penguins on South Georgia Island in the South Atlantic Ocean, often exhibit uniform spacing.

(c) Random. Trees of the same species are often randomly distributed in tropical rain forests, but this pattern of dispersion is relatively rare in nature.

Demography is the study of factors that affect the growth and decline of populations

Changes in population size reflect the relative rates of processes that add individuals to the population and eliminate individuals from it. Additions occur through births (which we will define here to include all forms of reproduction) and immigration, the influx of new individuals from other areas. Opposing these additions are mortality (death) and emigration, the movement of individuals out of a population. Our focus in this chapter is primarily on factors that influence birth rates and death rates, but you should remember that immigration and emigration may also play a role in population dynamics.

The study of the vital statistics that affect population size is called **demography.** Birth rates vary among individuals (specifically, among females) within a population, depending, in particular, on age; and death rates depend on both age and sex. Let's see how these demographic variables affect population dynamics.

Life Tables and Survivorship Curves

THE PROCESS OF SCIENCE

About a century ago, when life insurance first became available, insurance companies developed an interest in the mathematics of survival. They needed to estimate how long, on average, an individual of a given age could be expected to live. Some of the greatest demographers of the past century worked for life insurance companies. They invented demographic representations called life tables. A **life table** is an age-specific summary of the survival pattern of a population. Population ecologists adapted this approach for nonhuman populations and developed quantitative demography as a branch of biology.

The best way to construct a life table is to follow the fate of a **cohort,** a group of individuals of the same age, from birth until all are dead. The table is constructed from the number of individuals that die in each age-group during the defined time period. Cohort life tables are difficult to collect on wild animals and plants and are available only for a limited number of species.

TABLE 52.1 is a life table for a cohort of Belding ground squirrels (*Spermophilus beldingi*) at Tioga Pass, in California. Much can be learned about a population from a life table. The third column in the table shows the proportion of individuals in a cohort that are still alive at a given age. Notice that the death rates are generally highest among the youngest ground squirrels and among the oldest individuals and that males suffer higher rates of loss than females.

A graphic way of representing the data in a life table is to draw a **survivorship curve,** a plot of the proportion or numbers in a cohort still alive at each age (FIGURE 52.3). Survivorship curves can be classified into three general types. A Type I curve is relatively flat at the start, reflecting low death rates during early and middle life, then drops steeply as death rates increase among older age-groups. Humans and many other large mammals that produce relatively few offspring but provide them with good care often exhibit this kind of curve. In contrast, a Type III curve drops sharply at the left of the graph, reflecting very high death rates for the young, but then flattens out as death rates decline for those few individuals that have survived to a certain critical age. This type of curve is usually associated with organisms that produce very large numbers of offspring but provide little or no care, such as many fishes and marine invertebrates. An oyster, for example, may release millions of eggs, but most offspring die as larvae from predation or other causes. Those few that manage to survive long enough to attach to a suitable substrate and begin growing a hard shell, however, will probably survive for a relatively long time. Type II curves are intermediate, with a constant death

Table 52.1 Life Table for Belding Ground Squirrels (*Spermophilus beldingi*) at Tioga Pass, in the Sierra Nevada Mountains of California*

Age (years)	Females					Males				
	Number Alive at Start of Year	Proportion Alive at Start of Year	Number of Deaths During Year	Death Rate[†]	Average Life Expectancy (years)	Number Alive at Start of Year	Proportion Alive at Start of Year	Number of Deaths During Year	Death Rate[†]	Average Life Expectancy (years)
0–1	337	1.000	207	0.61	1.33	349	1.000	227	0.65	1.07
1–2	252[††]	0.386	125	0.50	1.56	248[††]	0.350	140	0.56	1.12
2–3	127	0.197	60	0.47	1.60	108	0.152	74	0.69	0.93
3–4	67	0.106	32	0.48	1.59	34	0.048	23	0.68	0.89
4–5	35	0.054	16	0.46	1.59	11	0.015	9	0.82	0.68
5–6	19	0.029	10	0.53	1.50	2	0.003	0	1.00	0.50
6–7	9	0.014	4	0.44	1.61	0				
7–8	5	0.008	1	0.20	1.50					
8–9	4	0.006	3	0.75	0.75					
9–10	1	0.002	1	1.00	0.50					

*Males and females have different mortality schedules, so they are tallied separately.
[†]The death rate is the proportion of individuals dying in the specific time interval.
[††]Includes 122 females and 126 males first captured as one-year-olds and therefore not included in the count of squirrels age 0–1.

SOURCE: Data from P. W. Sherman and M. L. Morton, "Demography of Belding's Ground Squirrel," *Ecology* 65(1984): 1617–1628.

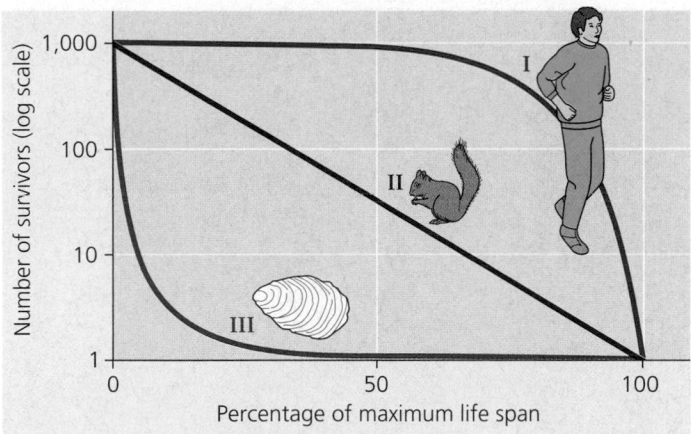

FIGURE 52.3 Idealized survivorship curves. As an example of a Type I curve, humans in developed countries experience high survival rates until old age. At the opposite extreme are Type III curves for organisms such as oysters, which experience very high mortality as larvae but decreased mortality later in life. Type II survivorship curves are intermediate between the other two types and result when a constant proportion of individuals die at each age. Notice that the y axis is logarithmic and that the x axis is on a relative scale, so that species with widely varying life spans can be compared on the same graph.

rate over the life span. This kind of survivorship occurs in some annual plants, various invertebrates such as *Hydra,* some lizard species, and some rodents, such as the gray squirrel.

Many species, of course, fall somewhere between these basic types of survivorship or show more complex patterns. In birds, for example, mortality is often high among the youngest individuals (as in a Type III curve) but fairly constant among adults (as in a Type II curve). Some invertebrates, such as crabs, may show a "stair-stepped" curve, with brief periods of increased mortality during molts (caused by physiological problems or greater vulnerability to predation), followed by periods of lower mortality (when the exoskeleton is hard).

In populations without immigration or emigration, survivorship is one of the two key factors determining changes in population size. Next we consider reproductive output, the other key factor determining population trends.

Reproductive Rates

Demographers who study sexually reproducing species generally ignore males and concentrate on females in the population because only females give birth to offspring. Demographers view populations in terms of females giving rise to new females; males are important only as distributors of genes. How can we describe the reproductive program of a population? The simplest way is to follow the basic approach of the life table and ask how reproductive output varies with age.

A **reproductive table,** or fertility schedule, is an age-specific summary of the reproductive rates in a population. The best way to construct a fertility schedule is to measure the reproductive output of a cohort from birth until death. For sexual species, the reproductive table tallies the number of female offspring produced by each age-group. TABLE 52.2 illustrates a reproductive table for Belding ground squirrels. Reproductive output for sexual species like birds and mammals is a product of the fraction of females of a given age that are breeding and the number of female offspring of those breeding females. By multiplying these together, we can obtain the average output of daughters for each individual in a given age class (the last column in TABLE 52.2). For these ground squirrels, which begin to reproduce at age 1 year, reproductive output rises to a peak at 4 years of age and then falls off in older females.

Table 52.2 Reproductive Table for Belding Ground Squirrels (*Spermophilus beldingi*) at Tioga Pass, in the Sierra Nevada Mountains of California

Age (years)	Proportion of Females Weaning a Litter	Mean Size of Litters (Males + Females)	Mean Number of Females in a Litter	Average Number of Female Offspring*
0–1	0.00	0.00	0.00	0.00
1–2	0.65	3.30	1.65	1.07
2–3	0.92	4.05	2.03	1.87
3–4	0.90	4.90	2.45	2.21
4–5	0.95	5.45	2.73	2.59
5–6	1.00	4.15	2.08	2.08
6–7	1.00	3.40	1.70	1.70
7–8	1.00	3.85	1.93	1.93
8–9	1.00	3.85	1.93	1.93
9–10	1.00	3.15	1.58	1.58

*The average number of female offspring is the proportion weaning a litter multiplied by the mean number of females in a litter.

SOURCE: Data from P. W. Sherman and M. L. Morton, "Demography of Belding's Ground Squirrel," *Ecology* 65 (1984): 1617–1628.

Reproductive tables vary greatly, depending on the species. Squirrels have a litter of two to six young once a year, whereas oak trees drop thousands of acorns each year for tens or hundreds of years. Salmon lay thousands of eggs when they spawn, and mussels and other invertebrates may release hundreds of thousands of eggs in a spawning cycle. Why does one type of life cycle rather than another evolve in a particular population? This is one of the many questions at the interface of population ecology and evolutionary biology.

LIFE HISTORIES

Natural selection will favor traits in organisms that improve their chances of survival and reproductive success. Organisms that survive a long time but do not reproduce are not at all "fit" in the Darwinian sense. In every species, there are trade-offs between survival and traits such as frequency of reproduction, investment in parental care, and the number of offspring produced (seed crops for seed plants and litter size or clutch size for animals). The traits that affect an organism's schedule of reproduction and survival (from birth through reproduction to death) make up its **life history.** Of course, a particular life history, like most characteristics of an organism, is the result of natural selection operating over evolutionary time. Life history traits help determine how populations grow.

FIGURE 52.4 An example of big-bang reproduction. Agaves, or century plants, grow without reproducing for several years and then produce a gigantic flowering stalk and many seeds. After this one-time reproductive effort, the plant dies.

Life histories are highly diverse, but they exhibit patterns in their variability

Because of varying environmental contexts for natural selection, life histories are very diverse. Pacific salmon, for example, hatch in the headwaters of a stream, then migrate to open ocean, where they require one to four years to mature. They eventually return to freshwater streams to spawn, producing thousands of small eggs in a single reproductive opportunity, and then they die. Ecologists call this **big-bang reproduction.** FIGURE 52.4 illustrates big-bang reproduction in agaves. The agave, or century plant, grows in arid climates with sparse and unpredictable rainfall. Agaves grow vegetatively for several years, then send up a large flowering stalk, produce seeds, and die. (We introduced the big-bang reproduction of century plants on the opening page of Chapter 38.) The shallow roots of agaves catch water after rain showers but are dry during droughts. This unpredictable water supply may prevent seed production or seedling establishment for several years at a time. By growing and storing nutrients until an unusually wet year and then putting all its resources into reproduction, the agave's big-bang strategy is a life history adaptation to erratic climate. In another example of big-bang reproduction, annual desert wildflowers generally germinate, grow, produce many small seeds, and then die, all in the span of a month after spring rains. Big-bang (one-time) reproduction is also called **semelparity** (from the Latin *semel,* once, and *parito,* to beget).

In contrast to big-bang reproduction, some lizards produce only a few large eggs during their second year of life, then repeat the reproductive act annually for several years. And some species of oaks do not reproduce until the tree is 20 years old, but then produce vast numbers of large seeds each year for a century or more. Ecologists call this **repeated reproduction** or **iteroparity** (from the Latin *itero,* to repeat).

What factors contribute to the evolution of semelparity versus iteroparity? That is, how much will an individual gain in reproductive success through one strategy versus the other? The key demographic effect of big-bang reproduction is higher reproductive rates. Plants like agaves that reproduce only once typically produce two to five times as many seeds as closely related species that reproduce repeatedly. The critical factor in the evolutionary dilemma of big-bang versus repeated reproduction is the survival rate of the offspring. If their chance of survival is poor or inconsistent, repeated reproduction will be favored.

Limited resources mandate trade-offs between investments in reproduction and survival

Darwinian fitness is measured not by how many offspring are produced but by how many survive to produce their own offspring: Heritable characteristics of life history that result in the most reproductively successful descendants will become more common within the population. If we were to construct a hypothetical life history that would yield the greatest lifetime reproductive output, we might imagine a population of individuals that begin reproducing at an early age, produce many offspring each time they reproduce, and reproduce many times in a lifetime. However, natural selection cannot maximize all these variables simultaneously, because organisms have finite resources, and limited resources mean trade-offs. Ecologists who study the evolution of life histories focus on how these trade-offs operate in specific populations. For example, the production of many offspring with little chance of survival may result in fewer descendants than the production of a few well-cared-for offspring that can compete vigorously for limited resources in an already dense population.

The life histories we observe in organisms represent an evolutionary resolution of several conflicting demands. Time, energy, and nutrients that are used for one thing cannot be used for something else. In the broadest sense, there is a trade-off between reproduction and survival, and this has been demonstrated by several studies. For example, in red deer on the Scottish island of Rhum, females that reproduce in one summer suffer higher mortality over the following winter than do females that did not reproduce (FIGURE 52.5). This cost of reproduction was found even in red deer in the prime of life, but was particularly severe in the older females. And in many insect species, females that lay fewer eggs live longer, suggesting a similar trade-off between investing in current reproduction and survival. There can also be trade-offs between current and future reproduction. When perennial plants produce more seeds in one year, they grow less and have reduced seed production the next year. Moreover, experimental transfers of eggs or nestlings in bird populations have measured the trade-off between reproductive effort and survival. When nestlings of European kestrels were transferred among nests to produce broods of three or four (reduced), five or six (normal), and seven or eight (enlarged), adult kestrels that raised the enlarged broods survived poorly over the following winter (FIGURE 52.6).

As in our red deer and kestrel examples, many life history issues involve balancing the profit of immediate investment in offspring against the cost to future prospects of survival and reproduction. These issues can be summarized by three basic "decisions": when to begin reproducing, how often to breed, and how many offspring to produce during each reproductive episode. The various "choices" are integrated into the life history patterns we see in nature. It is important to clarify our use of the word *choice*. Organisms do not choose consciously when to breed and how many offspring to have. (Humans are an important exception we will consider later in the chapter.) Life history traits are evolutionary outcomes reflected in the development, physiology, and behavior of an organism. Age at maturity and the number of offspring produced during a given reproductive episode are usually maintained within

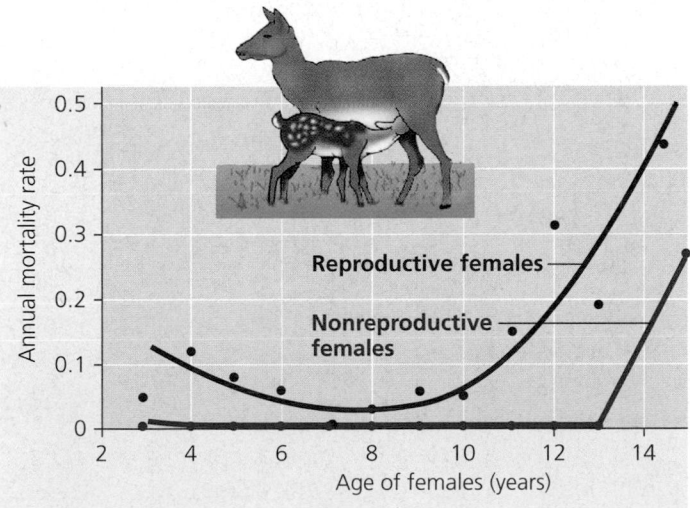

FIGURE 52.5 Cost of reproduction in female red deer on the island of Rhum, in Scotland. Mortality in winter is higher for females that reproduced during the previous summer, no matter what the age of the female.

FIGURE 52.6 Probability of survival over the following year for European kestrels after raising a modified brood. A total of 200 birds were studied from 1985 to 1990 in the Netherlands. Adults with experimentally enlarged broods die more often over the following winter. (Both males and females provide parental care for the nestlings.)

THE PROCESS OF SCIENCE

narrow ranges by stabilizing selection (see Chapter 23). Natural selection molds reproductive patterns in populations; such patterns are not consciously chosen by the organism.

As with all life history adaptations, the number and size of offspring depend on the selective pressures under which the organism evolved. Plants and animals whose young are subject to high mortality rates often produce large numbers of relatively small offspring (FIGURE 52.7a). Thus, plants that colonize disturbed environments usually produce many small seeds, most of which will not reach a suitable environment. Small size might actually benefit such seeds if it enables them to be carried long distances. Birds and mammals that suffer high predation rates also produce large numbers of offspring; examples include quail, rabbits, and mice. In other organisms, extra investment on the part of the parent greatly increases the offspring's chances of survival. Oak, walnut, and coconut trees all have large seeds with a large store of energy and nutrients that the seedlings can use to become established (FIGURE 52.7b). In animals, parental investment in offspring does not always end with incubation or gestation. Primates generally have only one or two offspring at a time. Parental care and an extended period of learning in the first several years of life are very important to offspring fitness in these mammals.

Now that we have analyzed some patterns that underlie diverse life histories, let's examine the effects of these life history traits on the growth of populations.

POPULATION GROWTH

To begin to understand the potential for population increase, consider a single bacterium that can reproduce by fission every 20 minutes under ideal laboratory conditions. At the end of this time, there would be two bacteria, four after 40 minutes, and so on. If this continued for only a day and a half—a mere 36 hours—there would be enough bacteria to form a layer a foot deep over the entire Earth. At the other life history extreme, elephants may produce only six young in a 100-year life span. Still, Darwin calculated that it would take only 750 years for a single pair of elephants to produce a population of 19 million. Obviously, indefinite population increase does not occur for any species, either in the laboratory or in nature. A population that begins at a low level in a favorable environment may increase rapidly for a while, but eventually the numbers must, as a result of limited resources and other factors, stop growing.

As we discussed in Chapter 50, finding the answers to ecological questions depends on a combination of observation and experimentation. The two major forces affecting population growth—birth rates and death rates—can be measured in many populations and used to predict how the populations will change in size over time. Small organisms can be studied in the laboratory to determine how various factors affect their population growth rates, and natural populations

(a)

(b)

FIGURE 52.7 Variation in seed crop size in plants. (a) Most weedy plants, such as this dandelion, grow quickly and produce a large number of seeds. Although most of the seeds will not produce mature plants, their large number and ability to disperse to new habitats ensure that at least some will grow and eventually produce seeds themselves. **(b)** Some plants, such as this coconut palm, produce a moderate number of very large seeds. The large endosperm provides nutrients for the embryo (a plant's version of parental care), an adaptation that helps ensure the success of a relatively large fraction of offspring. Animal species exhibit similar trade-offs between number of offspring and the amount of nutrients provided to each offspring.

can be experimentally manipulated to answer the same questions. Mathematical models can be used for testing hypotheses about the effects of different factors on population growth once we understand how birth and death rates change over time. We can begin to understand population growth by looking at a few simple models of how a population can grow.

The exponential model of population growth describes an idealized population in an unlimited environment

Imagine a hypothetical population consisting of a few individuals living in an ideal, unlimited environment. Under these conditions, there are no restrictions on the abilities of individuals to harvest energy, grow, and reproduce, aside from the inherent physiological limitations that are the result of their life history. The population will increase in size with every birth and with the immigration of individuals from other populations and decrease in size with every death and with the emigration of individuals out of the population. For simplicity, let's ignore the effects of immigration and emigration (a more complex formulation would certainly include these factors). We can define a change in population size during a fixed time interval with the following verbal equation:

$$\text{Change in population size during time interval} = \text{Births during time interval} - \text{Deaths during time interval}$$

Using mathematical notation, we can express this relationship more concisely. If N represents population size and t represents time, then ΔN is the change in population size and Δt is the time interval (appropriate to the life span or generation time of the species) over which we are evaluating population growth. (The Greek letter delta, Δ, indicates change, such as change in time.) We can now rewrite the verbal equation as

$$\frac{\Delta N}{\Delta t} = B - D$$

where B is the number of births in the population during the time interval and D is the number of deaths.

We can now convert the simple model just presented into one in which births and deaths are expressed as the average number of births and deaths per individual during the specified time interval. Let b represent the per capita birth rate, the number of offspring produced per unit time by an average member of the population. If, for example, a population of 1,000 individuals experiences 34 births per year, then the annual per capita birth rate is $^{34}/_{1,000}$, or 0.034. If we know the per capita birth rate and death rate, we can calculate the expected number of births and deaths in a population of any size. For example, if we know that the annual per capita birth rate is 0.034 and the population size is 500, we can use the formula $B = bN$ to calculate the number of births expected in that

population per year:

$$B = bN$$
$$B = 0.034 \times 500$$
$$B = 17 \text{ per year}$$

Similarly, the per capita death rate, symbolized as d, allows us to calculate the expected number of deaths per unit time in a population of any size. If $d = 0.016$ per year, we would expect 16 deaths per year in a population of 1,000 individuals. (Using the formula $D = dN$, how many deaths would you expect per year if $d = 0.010$ annually in populations of 500, 700, and 1,700?) For natural populations or those in the laboratory, the per capita birth rates and death rates can be calculated from estimates of population size and data given in life tables and reproductive tables (see, for example, TABLES 52.1 and 52.2).

We can revise the population growth equation again, this time using per capita birth rates and death rates rather than the numbers of births and deaths:

$$\frac{\Delta N}{\Delta t} = bN - dN$$

One final simplification is in order. Population ecologists are concerned with overall changes in population size, using r to identify the difference in the per capita birth rates and death rates:

$$r = b - d$$

This value, the per capita growth rate, tells whether a population is actually growing (positive value of r) or declining (negative value of r). **Zero population growth (ZPG)** occurs when the per capita birth rates and death rates are equal ($r = 0$). Note that births and deaths still occur in the population, but they balance each other exactly. (Later in this chapter, we will discuss the relevance of ZPG for the human population and the factors preventing the human population from leveling off.)

Using the per capita growth rate, we rewrite the equation for change in population size as

$$\frac{\Delta N}{\Delta t} = rN$$

Finally, most ecologists use the notation of differential calculus to express population growth in terms of instantaneous growth rates:

$$\frac{dN}{dt} = rN$$

If you have not yet studied calculus, don't be intimidated by the form of the last equation; it is essentially the same as the previous one, except that the time intervals Δt are very short and are expressed in the equation as dt. (Do not confuse this use of d to symbolize very small change with our earlier use of d to represent per capita death rate.)

We started this section by describing a population living under ideal conditions, where organisms are constrained only by their life history. In such a situation, the population grows rapidly, because all members have access to abundant food and are free to reproduce at their physiological capacity. Population increase under these ideal conditions is called **exponential population growth,** or geometric population growth. Under these conditions the per capita growth rate may assume the maximum growth rate for the species, called the **intrinsic rate of increase,** denoted as r_{max}. And the equation for exponential population growth is then:

$$\frac{dN}{dt} = r_{max}N$$

The size of a population that is growing exponentially increases rapidly, resulting in a J-shaped growth curve when population size is plotted over time (FIGURE 52.8). Although the intrinsic rate of increase is constant as the population grows, the population actually accumulates more new individuals per unit of time when it is large than when it is small; thus, the curves in FIGURE 52.8 get progressively steeper over time. This occurs because population growth depends on N as well as r_{max}, and larger populations experience more births (and deaths) than small ones growing at the same per capita rate. It is also clear from FIGURE 52.8 that a population with a higher intrinsic rate of increase ($dN/dt = 1.0N$) will grow faster than one with a lower rate of increase ($dN/dt = 0.5N$).

The J-shaped curve of exponential growth is characteristic of some populations that are introduced into a new or unfilled environment or whose numbers have been drastically reduced by a catastrophic event and are rebounding. For example,

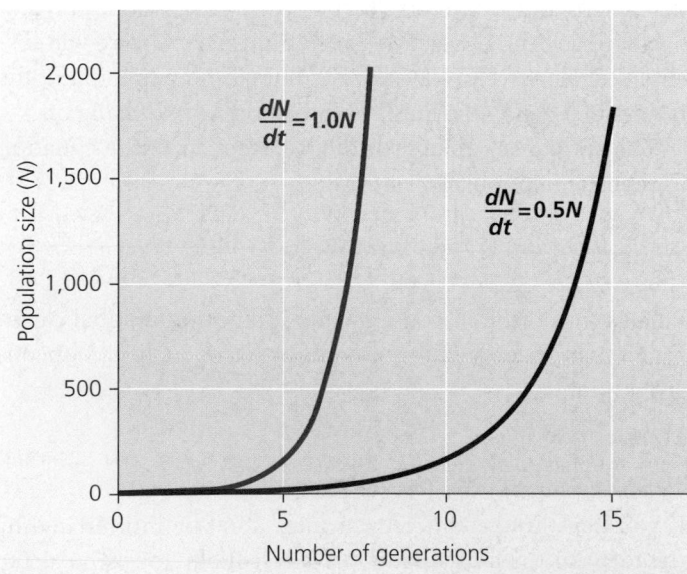

FIGURE 52.8 Population growth predicted by the exponential model. The exponential growth model predicts unlimited population increase under conditions of unlimited resources. This graph compares growth in populations with two different values of r_{max}: 1.0 and 0.5.

FIGURE 52.9 Example of exponential population growth in nature. The whooping crane is an endangered species that has been recovering from near extinction since 1940. Counts of adults are made annually on the wintering grounds at Aransas, Texas. In the year 2000–2001, there were 179 birds in the wintering population in Texas, the population having declined slightly from the preceding year. The overall average rate of increase has been 4% per year since the 1950s.

FIGURE 52.9 illustrates exponential population growth in the whooping crane, an endangered species now recovering from the impact of habitat loss due to agriculture.

The logistic model of population growth incorporates the concept of carrying capacity

The exponential growth model assumes unlimited resources, which is never the case in the real world. No population—neither bacteria nor elephants nor any other organisms—can grow exponentially indefinitely. As any population grows larger in size, its increased density may influence the ability of individuals to harvest sufficient resources for maintenance, growth, and reproduction. Populations subsist on a finite amount of available resources, and as the population becomes more crowded, each individual has access to an increasingly smaller share. Ultimately, there is a limit to the number of individuals that can occupy a habitat. Ecologists define **carrying capacity** as the maximum population size that a particular environment can support at a particular time with no degradation of the habitat. Carrying capacity, symbolized as K, is not fixed, but varies over space and time with the abundance of limiting resources. For example, the carrying capacity for bats may be high in a habitat where flying insects are abundant and there are caves for roosting but lower in a habitat where food is abundant but suitable shelters are less common. Energy limitation is one of the most significant determinants of carrying capacity, although other factors, such as shelters, refuges from predators, soil nutrients, water, and suitable nesting and roosting sites, can be limiting.

Crowding and resource limitation can have a profound effect on the population growth rate. If individuals cannot obtain sufficient resources to reproduce, per capita birth rate will decline. If they cannot find and consume enough energy to maintain themselves, per capita death rates may also increase. A decrease in *b* or an increase in *d* results in a smaller *r* and a lower overall rate of population growth.

The Logistic Growth Equation

We can modify our mathematical model of population growth to incorporate changes in growth rate as the population size nears the carrying capacity (as *N* grows toward *K*). The **logistic population growth** model incorporates the effect of population density on the per capita rate of increase, allowing this rate to vary from a maximum at low population size to zero as carrying capacity is reached. When a population's size is well below the carrying capacity, population growth is rapid, but as *N* approaches *K*, population growth slows down.

Mathematically, we construct the logistic model by starting with the model of exponential population growth and creating an expression that reduces the rate of population increase as *N* increases (FIGURE 52.10). If the maximum sustainable population size is *K*, then *K* − *N* tells us how many additional individuals the environment can accommodate, and $(K - N)/K$ tells us what fraction of *K* is still available for population growth. By multiplying the exponential rate of increase $r_{max}N$ by $(K - N)/K$, we reduce the actual growth rate of the population as *N* increases:

$$\frac{dN}{dt} = r_{max}N\left(\frac{K - N}{K}\right)$$

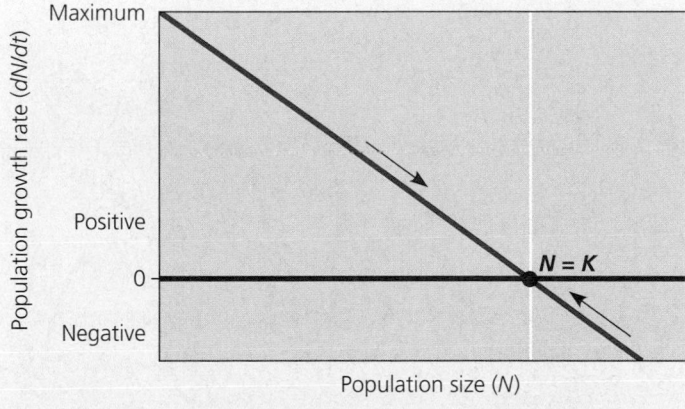

FIGURE 52.10 Reduction of population growth rate with increasing population size (*N*). The logistic model of population growth assumes that the population growth rate *dN/dt* decreases as *N* increases. When *N* is close to 0, the population grows rapidly. However, as *N* approaches *K* (the carrying capacity of the environment), the population growth rate approaches 0, and population growth slows. If *N* is greater than *K*, then the population growth rate is negative, and population size decreases. An equilibrium is reached at the white line when *N* = *K*.

Table 52.3 A Hypothetical Example of Logistic Population Growth, Where $K = 1{,}000$ and $r_{max} = 0.05$ per Individual per Year

Population Size: N	Intrinsic Rate of Increase: r_{max}	$\left(\dfrac{K-N}{K}\right)$	Per Capita Growth Rate: $r_{max}\left(\dfrac{K-N}{K}\right)$	Population Growth Rate:* $r_{max}N\!-\!\left(\dfrac{K-N}{K}\right)$
20	0.05	0.98	0.049	+1
100	0.05	0.90	0.045	+5
250	0.05	0.75	0.038	+9
500	0.05	0.50	0.025	+13
750	0.05	0.25	0.013	+9
1,000	0.05	0.00	0.000	0

*Rounded to the nearest whole number

TABLE 52.3 shows hypothetical calculations for the rate of population increase and changes in *N* at various population sizes for a population growing according to the logistic model. Notice that when *N* is small compared to *K*, the term $(K - N)/K$ is large, and the actual rate of population increase (dN/dt) is close to the intrinsic (maximum) rate of increase. But when *N* is large and resources are limiting, then $(K - N)/K$ is small, and so is the rate of population increase. Zero population growth occurs when the numbers of births and deaths are equal—when *N* equals *K*.

The logistic model of population growth produces a sigmoid (S-shaped) growth curve when *N* is plotted over time (the red line in FIGURE 52.11, page 1162). New individuals are added to the population most rapidly at intermediate population sizes, when there is not only a breeding population of substantial size, but also lots of available space and other resources in the environment. The population growth rate slows dramatically as *N* approaches *K*.

Notice that we haven't said anything about what makes the population growth rate change as *N* approaches *K*. Either the birth rate *b* must decrease, the death rate *d* must increase, or both. Later in the chapter, we will go into some detail about some of the factors affecting *b* and *d*.

How Well Does the Logistic Model Fit the Growth of Real Populations?

The growth of laboratory populations of some small animals, such as beetles or crustaceans, and of microorganisms, such as paramecia, yeasts, and bacteria, fit S-shaped curves fairly well (FIGURE 52.12a, page 1162). These experimental populations are grown in a constant environment lacking predators and other species that may compete for resources, conditions that rarely occur in nature. Even under these laboratory conditions, not all populations show logistic growth patterns. Laboratory

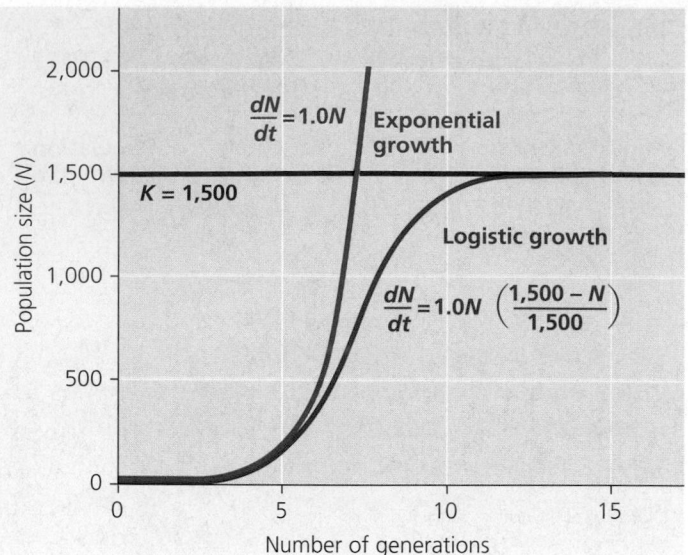

FIGURE 52.11 Population growth predicted by the logistic model. The logistic growth model assumes that there is a maximum population size that the environment can support—the carrying capacity K. The rate of population growth slows as the population approaches the carrying capacity of the environment. The red line shows logistic growth in a population where $r_{max} = 1.0$ and $K = 1,500$ individuals. For comparison, the blue line illustrates a population continuing to grow exponentially with the same r_{max}.

populations of water fleas (*Daphnia*), for example, show exponential growth and overshoot their carrying capacity before settling down to a relatively stable density (FIGURE 52.12b).

Most populations show some deviations from a smooth sigmoid curve. And while many natural populations increase in approximately logistic fashion, the stable carrying capacity is rarely observed. FIGURE 52.12c shows population changes in the song sparrow on a small island in southern British Columbia. The population increases rapidly but suffers periodic catastrophes in winter, so that there is no stable population size.

Some of the basic assumptions built into the logistic model clearly do not apply to all populations. For example, the model incorporates the idea that even at low population levels, each individual added to the population has the same negative effect on population growth rate. Some populations, however, show an *Allee effect* (named after W. C. Allee, of the University of Chicago, who first described it), in which individuals may have a more difficult time surviving or reproducing if the population size is too small. For example, a single plant standing alone may suffer from excessive wind but would be protected in a clump of individuals. And some seabirds that breed in colonies require large numbers at their breeding grounds to provide the necessary social stimulation for reproduction. Moreover, conservation biologists fear that populations of solitary animals, such as rhinoceroses, may become so small that individuals will not be able to locate mates in the breeding season. In all these cases, a greater number of individuals in the population has a positive effect, up to a point, on population growth, rather than a negative effect as assumed by the logistic model.

The logistic model also makes the assumption that populations adjust instantaneously and approach carrying capacity smoothly. In most natural populations, however, there is a lag time before the negative effects of an increasing population are realized. For example, as some important resource, such as food, becomes limiting for a population, reproduction will be reduced, but the birth rate may not be affected immediately because the organisms may use their energy reserves to continue producing eggs for a short time. This may cause the pop-

FIGURE 52.12 How well do these populations fit the logistic population growth model? The dots on these graphs are the actual data points.

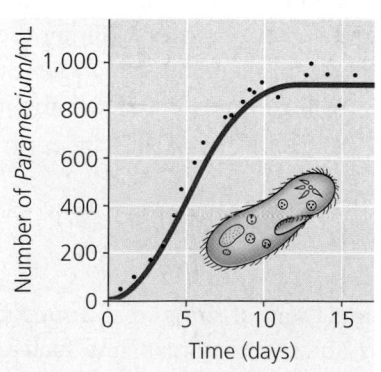

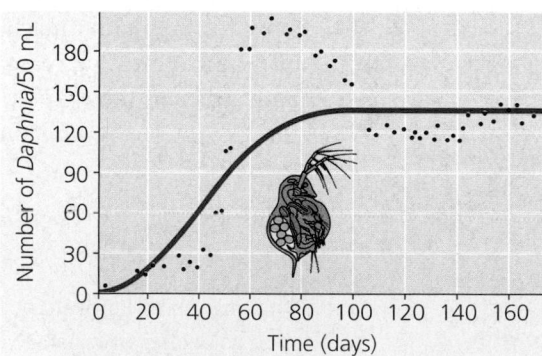

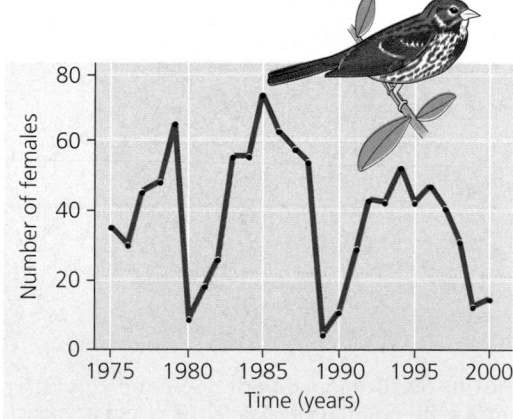

(a) A *Paramecium* population in the lab. The growth of *Paramecium aurelia* in small laboratory cultures closely approximates logistic growth (red curve) if the experimenter maintains a constant environment.

(b) A *Daphnia* population in the lab. The growth of a population of *Daphnia* in a small laboratory culture is not well defined by the logistic model (red curve). This population (black dots) overshoots the carrying capacity of its artificial environment and then settles down to an approximately stable population size.

(c) A song sparrow population in its natural habitat. The population of female song sparrows nesting on Mandarte Island, British Columbia, is periodically reduced by severe winter weather, and population growth is not well described by the logistic model.

ulation to overshoot the carrying capacity. Eventually, deaths will exceed births, and the population may then drop below carrying capacity; even though reproduction begins again as numbers fall, there is a delay until new individuals actually appear. Many populations fluctuate strongly, which makes it difficult to define what is meant by carrying capacity (see FIGURE 52.12c). Others overshoot it at least once before attaining a relatively stable size (see FIGURE 52.12b). We will examine some possible reasons for these fluctuations later in the chapter.

As you will see in the next section, some populations do not necessarily remain at, or even reach, levels where population density is an important factor. In many insects and other small, quickly reproducing organisms that are sensitive to environmental fluctuations, physical variables such as temperature or moisture usually reduce the population well before resources become limiting.

Overall, the logistic model is a useful starting point for thinking about how populations grow and for constructing more complex models. Although it fits few, if any, real populations closely, the logistic model is useful in conservation biology and in pest control to estimate how rapidly a particular population might increase in numbers after it has been reduced to a small size. And like any good starting hypothesis, the logistic model has stimulated much research and many discussions that, whether they support the model or not, lead to a greater understanding of the factors affecting population growth.

The Logistic Population Growth Model and Life Histories

The logistic model predicts different growth rates for low-density populations and high-density populations, relative to the carrying capacity of the environment. At high densities, each individual has few resources available, and the population can grow slowly, if at all. At low densities, the opposite is true: Resources per capita are relatively abundant, and the population can grow rapidly. Different life history features will be favored under these different conditions. At high population density, selection favors adaptations that enable organisms to survive and reproduce with few resources. Thus, competitive ability and maximum efficiency of resource utilization should be favored in populations that are at or near their carrying capacity. At low population density, on the other hand, even in the same species, the "empty" environment should favor adaptations that promote rapid reproduction. Increased fecundity and earlier maturity, for example, would be selected for.

Thus, the life history traits that natural selection favors may vary with population density and environmental conditions. Selection for life history traits that are sensitive to population density can be called **K-selection**, or density-dependent selection. In contrast, selection for life history traits that maximize reproductive success in uncrowded environments (low densities) can be called **r-selection**, or density-independent selection. These names follow from the variables of the logistic equation. K-selection tends to maximize population size and operates in populations living at a density near the limit imposed by their resources (the carrying capacity K). By contrast, r-selection tends to maximize r, the rate of increase, and occurs in variable environments in which population densities fluctuate well below carrying capacity or in open habitats where individuals are likely to face little competition.

In laboratory experiments, researchers have shown that different populations of the same species may show a different balance of K-selected and r-selected traits, depending on conditions. For example, cultures of the fruit fly *Drosophila melanogaster* raised under crowded conditions with minimal food for 200 generations are more productive at high density than populations raised in uncrowded conditions with maximal food. Larvae from cultures selected for living in crowded conditions feed faster than larvae selected for living in uncrowded cultures. The fruit fly genotypes that are most fit at low density do not have high fitness at high density, as predicted by r- and K-selection theory.

POPULATION-LIMITING FACTORS

There are two general questions that we can ask about population growth. First, why do all populations eventually stop increasing? Exponential population growth is rare in nature and always of short duration. What environmental factors stop a population from growing? If we have an introduced weed that is spreading rapidly, what should we do to stop its population growth? Second, why is the population density of a particular species greater in some habitats than in others? Every bird-watcher can tell you what the favorable and unfavorable habitats are for any particular bird species. What determines a favorable habitat, and how do we turn an unfavorable habitat into a good one?

These questions have many practical applications. A conservation biologist might want to turn a declining species population into an increasing one. And in agriculture, the objective may be to get a pest population to decrease. Moreover, agricultural pests may have severe effects in some areas and negligible effects in others. Why? Endangered species, meanwhile, such as humpback whales, require good habitats for survival. What environmental factors create a favorable feeding habitat for humpbacks? All these practical issues involve population-limiting factors. Regulation is one of this book's ten themes (see Chapter 1). In this section, we apply that theme to populations.

The first step in understanding why a population stops growing is to find out how the rates of birth, death, immigration, and emigration change as population density rises. If immigration and emigration offset each other, then a population grows when the birth rate exceeds the death rate and declines

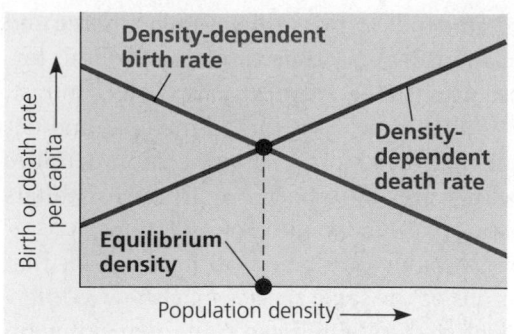

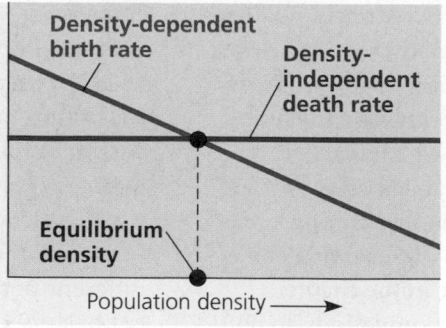

FIGURE 52.13 Graphic model showing how equilibrium may be determined for population density. Population density reaches equilibrium only when the per capita birth rate equals the per capita death rate, and this is possible only if the birth or death rate (or both) changes with density (is a density-dependent rate). In this simple model, immigration and emigration are assumed to be either zero or equal.

when the death rate exceeds the birth rate. FIGURE 52.13 shows a simple graphical model of how a population may stop increasing and reach equilibrium. A death rate that rises as population density rises is said to be **density dependent,** as is a birth rate that falls with rising density. Density-dependent rates are an example of **negative feedback,** a type of regulation you learned about in Chapter 1. In contrast, a birth rate or death rate that does *not* change with population density is said to be **density independent.** With density-independent rates, there is no feedback to slow down population growth.

Negative feedback prevents unlimited population growth

No population stops growing without some type of negative feedback between population density and the vital rates of birth and death. Once we know how birth and death rates change with population density, we need to determine the mechanisms causing these changes. Because populations are affected by a variety of factors that cause negative feedback, it can be a challenge to pinpoint the exact factors at work in a particular population.

Although field studies may eventually shed light on the most important factors producing negative feedback in specific cases, they have not yet provided many generalizations. First, much of the research on populations has been conducted in the temperate zone, and we need many more studies of tropical and polar organisms to complete the picture. Second, birds and mammals have been the subjects of much more research than have other organisms. In particular, insects, which form the dominant group of species on Earth, have not been studied in proportion to their species' richness. Finally, long time periods are required for experimental work on population dynamics, with definitive studies routinely taking 10 to 20 years for completion. With these reservations in mind, let us look at several examples of how birth and death rates change with population density—in some cases where the mechanisms behind these changes are well understood.

Resource limitation in crowded populations can stop population growth by reducing reproduction. For example, crowding can reduce seed production by plants (FIGURE 52.14a). And available food supplies often limit the reproductive output of songbirds; as bird population density increases in a particular habitat, each female lays fewer eggs, a density-dependent response (FIGURE 52.14b). In both of these examples, increasing population density intensifies intraspecific competition for declining nutrients, resulting in a lower birth rate.

Factors other than intraspecific competition for food or nutrients can also cause density-dependent behavior of populations. In many vertebrates and some invertebrates, territoriality,

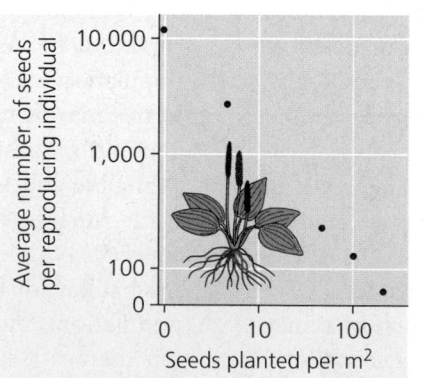

(a) Plantain. The average number of seeds produced by plantain (*Plantago major*), a small herb, decreases with increased sowing density.

FIGURE 52.14 Decreased fecundity at high population densities.

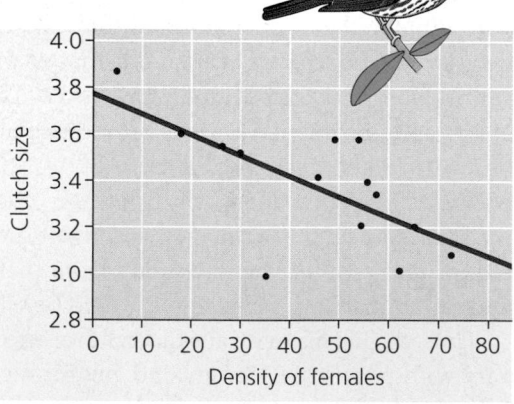

(b) Song sparrow. Clutch size in the song sparrow on Mandarte Island, British Columbia, decreases as density increases. The primary cause for this decline is food shortage. In an experiment in which females were given extra food at high density, they did not suffer a reduction in clutch size.

the defense of a well-bounded physical space, may set a limit on density, so that the space that constitutes the territory becomes the resource for which individuals compete. For oceanic birds such as gannets, which nest on rocky islands where they are relatively safe from predators, the limited number of suitable nesting sites allows only a certain number of pairs to nest and reproduce. Up to a certain population size, most birds can find a suitable nest site, but few birds beyond that threshold breed successfully. Thus, the limiting resource that determines breeding population size for gannets is safe nesting space. As this space fills, birds that cannot obtain a nesting spot do not reproduce. Surplus, or nonbreeding, individuals are a good indication that territoriality is restricting population growth, as it does in many bird populations.

Population density also influences the health and thus the survival of organisms. Plants grown under crowded conditions tend to be smaller and less robust than those grown at lower densities. Small plants are less likely to survive, and those that do survive produce fewer flowers, fruit, and seeds. Gardeners who recognize this phenomenon thin their seedlings to produce the best possible yield. Animals, too, experience increased mortality at high population densities. In laboratory studies of flour beetles, for example, the percentage of eggs that hatch and survive to adulthood decreases steadily as density increases from moderate to high levels (FIGURE 52.15). The main cause of this density-dependent effect is cannibalism of eggs by adult beetles and large larvae.

Predation may also be an important cause of density-dependent mortality for some prey populations if a predator encounters and captures more food as the population density of the prey increases. Many predators, for example, exhibit switching behavior: They begin to concentrate on a particularly common species of prey when it becomes energetically efficient to do so (see the discussion of optimal foraging in Chapter 51). For example, trout may concentrate for a few days on a particular species of insect that is emerging from its aquatic larval stage, then switch as another insect species becomes more abundant. As a prey population builds up, predators may feed preferentially on that species, consuming a higher percentage of individuals; this can cause density-dependent regulation of the prey population.

The accumulation of toxic wastes is another component that can contribute to density-dependent regulation of population size. In laboratory cultures of small microorganisms, for example, metabolic by-products accumulate as the population grows, poisoning the population within this limited, artificial environment. Indeed, ethanol accumulates as a waste product when yeast ferments sugar. The alcohol content of wine is usually less than 13%, the maximum ethanol concentration that most wine-producing yeast cells can tolerate.

The impact of a disease on a population can be density dependent if the transmission rate of the disease depends on a certain level of crowding in the population. For example, tuberculosis strikes a greater percentage of people living in cities than in rural areas.

For some animal species, intrinsic factors, rather than the extrinsic factors just discussed, appear to regulate population size. White-footed mice in a small field enclosure will multiply from a few to a colony of 30 to 40 individuals, but eventually, reproduction will decline until the population ceases to grow. This drop in reproduction is associated with aggressive interactions that increase with population density, and it occurs even when food and shelter are provided in abundance. Although the exact mechanisms by which aggressive behavior affects reproductive rate are not yet understood, we do know that high population densities in mice induce a stress syndrome in which hormonal changes can delay sexual maturation, cause reproductive organs to shrink, and depress the immune system. In this case, high densities cause both an increase in mortality and a decrease in birth rates. Similar effects of crowding occur in wild populations of woodchucks and other rodents.

In these various examples of population regulation by negative feedback, we have seen how increased densities cause population growth rates to decline by affecting reproduction, growth, and survivorship in the individuals that make up the populations. This helps us answer our first question about populations: why all populations eventually stop increasing. Now let's turn to our second question: why certain habitats favor greater population densities.

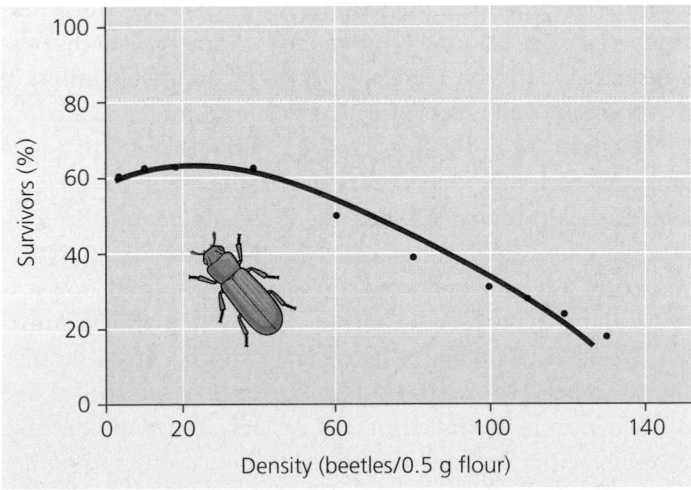

FIGURE 52.15 Decreased survivorship at high population densities. The percentage of flour beetles (*Tribolium confusum*) surviving from egg stage to adult in a laboratory culture decreases as density increases from moderate to high population densities, reducing the numbers of adults in the next generation.

Population dynamics reflect a complex interaction of biotic and abiotic influences

There are good and bad habitats for every species. Carrying capacities can vary in space; some parts of a lake provide better fishing than other parts, for example. Carrying capacities can also vary in time; grasshoppers may be serious pests for farmers in some years and nearly absent in other years. Over the long term, most populations exhibit change. Some remain fairly stable in size, but most populations for which we have long-term data show fluctuations in numbers.

Although we can determine an average population size for many species, the average is often of less interest than the year-to-year or place-to-place trend in numbers. For example, FIGURE 52.16 illustrates the fluctuation in the number of northern pintail ducks from 1955 to 1998. Pintails nest in the prairie regions of the United States and Canada, and their present populations have declined considerably below the levels of the 1950s. Wildlife managers need to find out why these changes occur. Researchers have identified the loss of prairie ponds (either by drying up in droughts or by draining for agriculture) as one key abiotic factor contributing to the pintail duck decline shown in FIGURE 52.16. However, when heavy rains kept prairie ponds full in the 1990s, the pintails did not recover. Pintails nest in the stubble left after grain is harvested. More intensive agriculture in recent years has resulted in early cultivation of stubble fields and the destruction of many pintail nests. As more and more land is taken over by agricultural fields, duck nests also become concentrated in the remaining natural vegetation, enabling predators like foxes and skunks to steal eggs from the nests more efficiently.

Long-term population studies are also challenging the hypothesis that a combination of factors keeps populations of large mammals, such as deer and moose, relatively stable. FIGURE 52.17 shows changes in the moose population on Isle Royale, in Lake Superior, from 1959 to 2000. Moose numbers

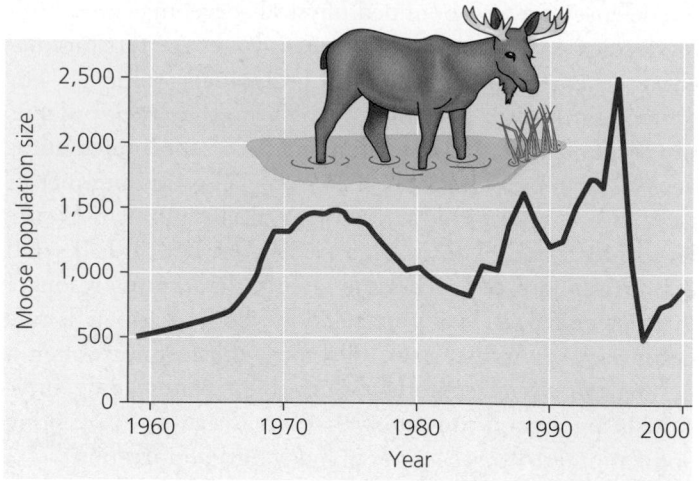

FIGURE 52.17 Long-term study of the moose (*Alces alces*) population of Isle Royale, Michigan. Isle Royale is 544 km² in area and lies in Lake Superior 40 km from shore. Moose colonized the island around 1900 by moving across the ice. However, the lake has not frozen over in recent years, and thus the moose population is isolated from immigration or emigration.

Data courtesy of Rolf O. Peterson, Michigan Technological University, 2001.

have been anything but stable, with two major increases and collapses over the last 40 years. Wolf predation is probably the main cause of the steady decline in numbers from 1973 to 1983. However, the more dramatic collapse in 1995–1996 was caused by severe winter weather and associated food shortage, leading to starvation of more than 75% of the population. The severity of winter loss in large grazers and browsers inhabiting temperate and polar regions is proportional to the harshness of the winter. Colder temperatures increase energy requirements (and therefore the need for food), while deeper snow makes it harder to find food. The result can be widespread death from starvation.

Some populations fluctuate erratically. The Dungeness crab is a classic example (FIGURE 52.18). Only male crabs over a certain size are commercially harvested, on the assumption that there is a vast excess of males for successful reproduction. Females mature at 2 to 3 years of age and release up to 2 million eggs each fall. One key feature of the crabs' population dynamics is cannibalism. Juvenile crabs are cannibalized by older juveniles and by adult crabs. A second key feature is that successful settlement of larval crabs occurs only in shallow waters and depends on ocean currents and water temperature. If winds and currents move larval crabs too far offshore, they cannot reach the ocean bottom to settle successfully. The combination of cannibalism and variable and unpredictable oceanic factors explains the marked fluctuations in populations of the Dungeness crab in the Pacific Northwest. Small changes in environmental variables seem to be magnified by density-dependent cannibalism. These results support the hypothesis that the dynamics of many populations result from a complex interaction of biotic and abiotic factors.

FIGURE 52.16 Decline in the breeding population of the northern pintail (*Anas actua*) from 1955 to 1998. Wildlife managers conduct extensive aerial surveys and ground counts each June throughout the breeding range in Canada and the United States to set hunting regulations for the autumn of each year.

Data from U.S. Fish and Wildlife Service, 2001.

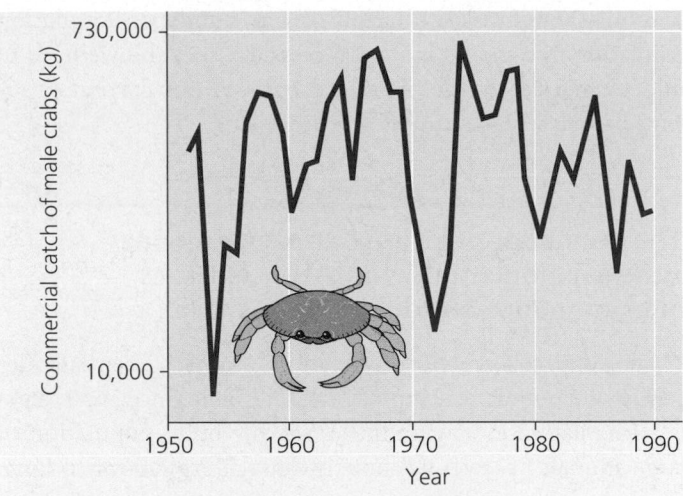

FIGURE 52.18 Extreme population fluctuations. Populations of the Dungeness crab (*Cancer magister*), a commercially important species of the U.S. Pacific Northwest, are well known for their erratic fluctuations. This graph of the commercial catch of male crabs over a 40-year period at Fort Bragg, California, illustrates the general pattern. Researchers developed a mathematical model that simulates a mix of biotic factors (for example, intraspecific competition and cannibalism) and abiotic factors (for example, minor changes in water temperatures caused by alterations in ocean currents). The model accurately predicts the fluctuations in the actual catch.

Some populations have regular boom-and-bust cycles

THE PROCESS OF SCIENCE

Some populations of insects, birds, and mammals fluctuate in density with remarkable regularity that cannot be explained by chance alone. Perhaps the most striking population cycles known are the 10-year cycles of snowshoe hares and lynx in the far northern forests of Canada and Alaska. Lynx are specialist predators of snowshoe hares, so it is not surprising that lynx numbers rise and fall with the numbers of hares (FIGURE 52.19). But why do hare numbers rise and fall in 10-year cycles? There are three main hypotheses.

First, cycles may be caused by food shortage during winter. Hares eat the terminal twigs of small shrubs such as willow and birch in winter and may suffer from malnutrition due to overgrazing. Second, cycles may be due to predator-prey interactions. Many predators other than lynx, such as coyotes, foxes, and great-horned owls, eat hares, and they might overexploit their prey. Third, cycles could be affected by a combination of food resource limitation and excessive predation.

If hare cycles are due to winter food shortage, then they should stop if extra food is added to a field population. Researchers have conducted such experiments for 20 years, over two hare cycles, in the Yukon and report two results. First, hare populations in the areas with extra food increased about threefold in density. The carrying capacity of a habitat for hares can clearly be increased by adding food. Second, the hares with extra food continued to cycle in the same way as the unfed control populations. In particular, cyclic collapses in density occurred in both experimental and control areas, and the decline in numbers could not be stopped by adding food. Thus, food supplies by themselves are not the cause of the hare cycle shown in FIGURE 52.19, so we can discard the first hypothesis.

By putting radio collars on hares, researchers can find individual hares as soon as they die, allowing field ecologists to determine the immediate cause of death. Almost 90% of the hares that died were killed by predators; none appeared to die of starvation. These data support either the second or third hypothesis. By excluding predators from one area with electric fences and by both excluding predators and adding food to another area, ecologists could test which of the remaining two hypotheses is the best explanation for the hare cycle. The results supported the hypothesis that the hare cycle is largely driven by excessive predation but with important impacts from the food supplies available, particularly in winter. Perhaps better-fed hares are more likely to escape from predators. Many different predators contribute to these losses; the cycle is not simply a hare-lynx cycle.

Some small herbivorous mammals, such as voles and lemmings, tend to have 3- to 4-year cycles, and some birds, such as

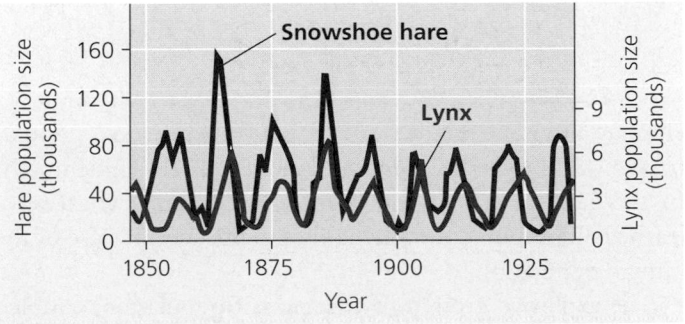

FIGURE 52.19 Population cycles in the snowshoe hare and lynx. Population counts are based on the number of pelts sold by trappers to the Hudson Bay Company. The cycles of lynx populations are probably caused by cyclic fluctuations of the hare, a major food source for lynx. What causes the cycling of the prey population is a challenging research problem. Most patterns of population dynamics are likely caused by multiple interacting factors that are difficult to untangle without direct experimentation.

the ruffed grouse and ptarmigan, have 9- to 11-year cycles. The causes of these cycles undoubtedly vary among species, and perhaps even among populations of the same species. Can we find common patterns in all these cycles? Several ideas have been suggested to explain cycles, but two are now prominent.

One idea is that stress resulting from a high population density may alter hormonal balance, which in turn could reduce fertility and increase aggressiveness. Stress could also be produced by unsuccessful predator attacks or from having to search longer for high-quality food or from parasites that become more common at high population density. However, there are few measurements of stress levels in natural populations, and little is known about how stress affects behavior. Crowding and stress could be a common component in many species of animals that have population cycles, but we do not yet know if it is a major cause of cycles.

Population cycles also result from a time lag in the response of predators to rising prey numbers. Predators reproduce more slowly than their prey, so they always lag behind prey population growth. Some predators, such as birds of prey, can move over large areas and respond very quickly to patches of prey abundance in the environment, eliminating the usual time lag. In other cases, such as the preying on voles and lemmings by weasels, the predators cannot move about as easily and typically show a time lag in their response to prey numbers. Such a time-lag mechanism is a probable factor in the cyclic declines in voles and lemmings, as it is in snowshoe hares.

For the lynx, great-horned owls, weasels, and other predators that depend heavily on a single prey species, the availability of prey is the major factor influencing their population changes. When prey become scarce, predators often turn on one another as well. Coyotes kill foxes and lynx, and great-horned owls kill smaller birds of prey as well as weasels, accelerating the collapse of the predator populations once prey numbers have collapsed in these cyclic systems.

Long-term experimental studies are the key to unraveling the complex causes of population cycles.

HUMAN POPULATION GROWTH

Humans are not exempt from natural processes. No population, including the human population, can grow indefinitely. In this last section of the chapter, we'll apply what we've learned about population dynamics to the specific case of the human population.

The explosive growth of the human population, coupled with massive consumption of the planet's resources by developed nations, is the primary cause of severe environmental degradation and the loss of biological species. Many of the environmental problems that we now confront cannot be solved without thoughtful regulation of our numbers. Population regulation in humans is a controversial issue, and we need to understand clearly the biological basis of our current problems before we can search for solutions.

The human population has been growing almost exponentially for three centuries but cannot do so indefinitely

The exponential growth model in FIGURE 52.8 essentially describes the population explosion of humans since 1650. Ours is a singular case; it is unlikely that any other population of large animals has ever sustained so much growth for so long. The human population increased relatively slowly until about 1650, when approximately 500 million people inhabited Earth (FIGURE 52.20). The population doubled to 1 billion within the next two centuries, doubled again to 2 billion between 1850 and 1930, and doubled still again by 1975 to more than 4 billion. The population now numbers over 6 billion people and increases by about 80 million each year. The world's population increases by 214,000 people per day, equivalent to adding a city the size of Amarillo, Texas, or Madison, Wisconsin, every day. Every week the population increases by the size of San Antonio, Milwaukee, or Indianapolis. It takes only three years for world population growth to add the population equivalent of another United States. If the present growth rate persists, there will be 7.8 billion people on Earth by the year 2025. How might this population increase stop?

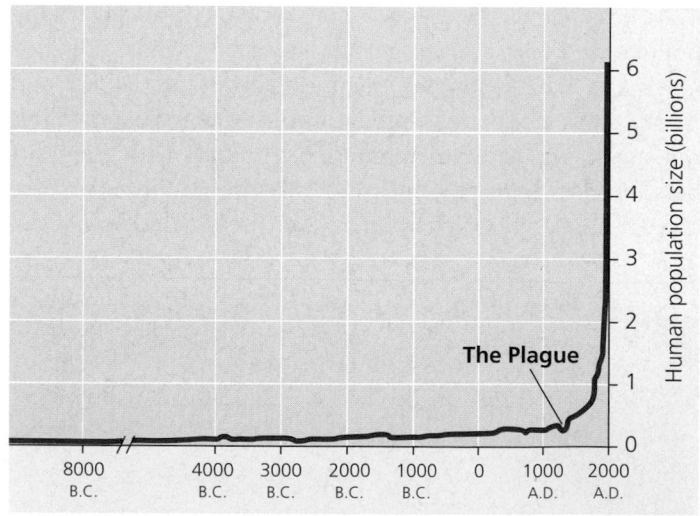

FIGURE 52.20 Human population growth. The human population has grown almost continuously throughout history, but it has skyrocketed since the Industrial Revolution. No other population of large organisms has shown such growth for so long, and the human population must eventually either level off or decline. Whether this reduction in population growth will occur because of decreased birth rates or massive mortality is an open question, one that careful population policies can address.

The Demographic Transition

A regional human population can exist in one of two configurations to maintain population stability:

Zero population growth = High birth rates − High death rates

or

Zero population growth = Low birth rates − Low death rates

The movement from the first toward the second state is called the **demographic transition.** FIGURE 52.21 illustrates the demographic transition for Sweden and Mexico. In Sweden, the demographic transition took about 150 years, but in Mexico, these changes have been quicker.

After 1950, mortality rates declined rapidly in most developing countries, but birth rates have declined in a more variable manner. Birth rate decline has been most dramatic in China. In 1970, the birth rate in China predicted an average family size of 5.9 children; by 1999, the expected family size was 1.85 children. In India, birth rates have fallen more slowly and irregularly. In much of Africa, the transition to lower birth rates is just beginning.

How do such variable birth rates affect the growth of the world's population? Clearly, population dynamics are regional. In the developed nations, populations are near equilibrium (growth rate about 0.1% per year), with reproductive rates near the replacement level (total fertility rate = 2.1 children per female). In many developed countries, including Canada and the United Kingdom, total fertility rates are in fact *below* replacement. These populations will eventually decline if there is no immigration and if the birth rate does not change. About 80% of the world's people now live in the less

developed countries, and most of the current population growth (1.7% per year) is occurring in these nations.

A unique feature of human population growth is our ability to control it with voluntary contraception and family planning. Reduced family size is the key to the demographic transition. However, there is a great deal of disagreement among world leaders as to how much support should be provided for global family planning efforts. Social change and the rising educational and career aspirations of women in many cultures encourage them to delay marriage and postpone reproduction. Delayed reproduction helps to decrease population growth rates and allows us to plan for zero population growth under conditions of low birth rates and low death rates.

Age Structure

One important demographic factor in present and future growth trends is a country's **age structure,** the relative number of individuals of each age (FIGURE 52.22, p. 1170). The relatively uniform age distribution in Italy, for instance, contributes to that country's stable population size; individuals of reproductive age or younger are not disproportionately represented in the population. In contrast, Kenya has an age structure that is bottom-heavy, skewed toward young individuals who will grow up and sustain the explosive growth with their own reproduction. Notice in FIGURE 52.22 that the age structure for the United States is relatively even except for a bulge that corresponds to the "baby boom" that lasted for about two decades after the end of World War II. Even though couples born during those years have had an average of fewer than two children, the nation's overall birth rate still exceeds the death rate because there are still so many "boomers" and their offspring of reproductive age.

Age-structure diagrams not only reveal a population's growth trends, but can also point to future social conditions. Based on the diagrams in FIGURE 52.22, we can predict, for instance, that employment for an increasing number of working-age people will continue to be a significant problem for Kenya in the foreseeable future. In Italy and the United States, a decreasing proportion of working-age people—mostly those of college age today—will soon be supporting an increasing population of retired "boomers." In the United States, it is this demographic feature that has made the future of Social Security and Medicare such a major political issue. Understanding age structures can help us plan for the future.

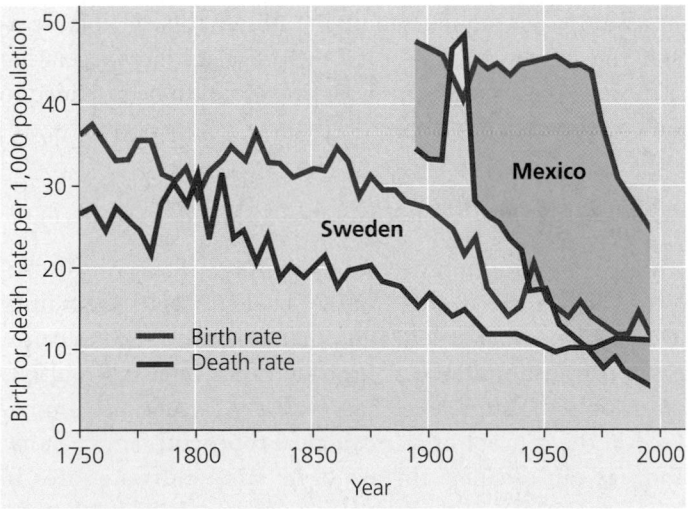

FIGURE 52.21 Demographic transition in Sweden and Mexico, 1750–1997. When births exceed deaths, the population grows (shaded zones). The transition from high birth and high death rates to low birth and low death rates took 150 years in Sweden but has been compressed in Mexico to half that time.

From Population Reference Bureau 2000.

Estimating Earth's carrying capacity for humans is a complex problem

The projected human population of Earth depends on assumptions about future changes in birth and death rates. For 2050, the United Nations projects a population that might

FIGURE 52.22 Age-structure pyramids for the human population of Kenya (growing at 2.1% per year), the United States (growing at 0.6% per year), and Italy (zero growth) for 1995.

range from 7.3 to 10.7 billion people. Even at the low end of this range, without some catastrophe, there will be at least 1.3 billion people added to the population in the next 25 years because of the momentum of population growth. The question that arises from these projections is what size human population the biosphere can support. Is the world already overpopulated? Will it be overpopulated in 2050?

Wide Range of Estimates for Carrying Capacity

What is the carrying capacity of Earth for humans? This question has been asked for more than 300 years by scientists interested in demography. The first known estimate for the carrying capacity of Earth was made by Anton van Leeuwenhoek in 1679. Since then, the estimates of carrying capacity have varied from less than 1 billion to over 1,000 billion (a trillion people). The average of the various estimates is around 10–15 billion. Why should these estimates of carrying capacity be so variable?

Carrying capacity is difficult to estimate, and the scientists who produce these estimates use different methods to get their answers. Some researchers use curves like that produced by the logistic equation (see FIGURE 52.11) to predict the future maximum of the human population. Others generalize from exist-

ing "maximum" population density and multiply this by the area of land that could be inhabited. Still other estimates are based on a single assumed population constraint such as food. Basing carrying capacity on food as the limiting factor is one promising approach. However, such estimates are limited by the assumptions required about the amount of available farmland, the average yield of crops, the prevalent diet (vegetarian or meat eating), and the number of calories to be provided to each person each day.

Ecological Footprint

A more promising approach to estimating the carrying capacity of Earth is to recognize that we have multiple constraints: We need food, fuel, wood, and other amenities such as clothing and transportation. A recent advance in using multiple constraints to estimate human carrying capacity is summarized in the concept of an **ecological footprint**. For each nation, we can calculate the aggregate land and water area in various ecosystem categories that is appropriated by that nation to produce all the resources it consumes and to absorb all the waste it generates. Six types of ecologically productive areas are distinguished in calculating the ecological footprint: arable land (land suitable for crops), pasture, forest, ocean,

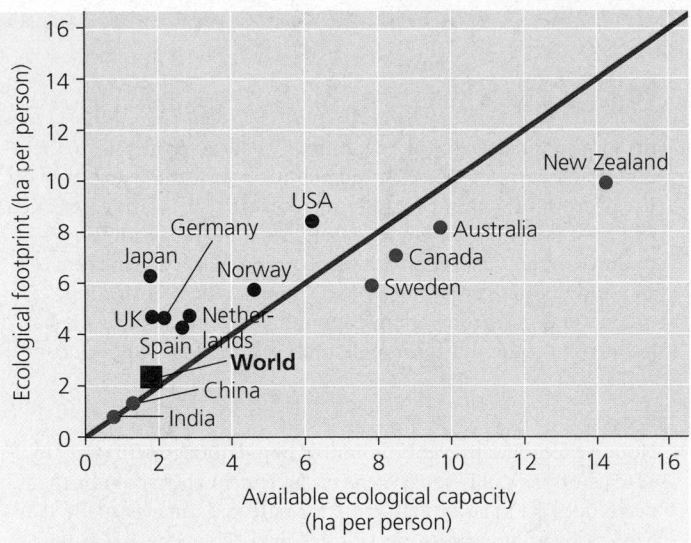

FIGURE 52.23 Ecological footprint in relation to available ecological capacity. The ecological footprint expresses in hectares of land per person the current demand of global resources made by each country. The available ecological capacity measures in land area per person the resource base of each particular country. Countries in black (above the red diagonal) were in an ecological deficit in 1997 when the study was conducted. Countries in blue (below the diagonal) still have resource surpluses.

built-up land, and fossil energy land. Fossil energy land is calculated on the basis of the land required for vegetation to absorb the CO_2 produced by burning fossil fuels. All measures are converted to land area (hectares per person). If we add up all the ecologically productive land on the planet, we find that there is about 2 ha per person (1 ha = 2.47 acres). If we wish to reserve land for parks and conservation, we must reduce this to 1.7 ha per person. This is the benchmark for comparing the ecological footprints of nations.

FIGURE 52.23 graphs the ecological footprints for 13 countries and for the whole world as they stood in 1997. Two inferences are evident from the graph. First, the world in general was *already* in ecological deficit in 1997 when this study was conducted. Second, countries vary greatly in their individual footprint size and in their available ecological capacity (the actual resource base of each country). The United States has an ecological footprint of 8.4 ha per person but has only 6.2 ha per person of available ecological capacity. In other words, the United States has a bigger ecological footprint than its own land and resources can support. By this measure, the U.S. population is already above carrying capacity. By contrast, New Zealand has a larger ecological footprint of 9.8 ha per person but an available capacity of 14.3 ha per person, so it is below its carrying capacity. The overall analysis of human impacts via ecological footprints suggests that the world is already at or slightly above its carrying capacity.

We can only speculate about Earth's ultimate carrying capacity for the human population or about what factors will eventually limit our growth. Perhaps food will be the main factor. Malnutrition and famines are common in some countries, but they result mainly from unequal distribution, rather than inadequate production, of food. So far, technological improvements in agriculture have allowed food supplies to keep up with global population growth. However, we also know, based on principles of energy flow through ecosystems, that environments can support a larger number of herbivores than carnivores (see Chapter 54). If everyone ate as much meat as the wealthiest people in the world, less than half of the present world population could be fed on current food harvests. Nevertheless, it seems unlikely that people in wealthier countries will abandon the consumption of meat.

Perhaps we will eventually be limited by suitable space, like the gannets on ocean islands. Certainly, as our population grows, the conflict over how space will be utilized will intensify, and agricultural land may be developed for housing. There seem to be few limits, however, on how closely humans can be crowded together.

We could also run out of resources other than nutrients and space. Many people are concerned about supplies of nonrenewable resources, such as certain metals and fossil fuels. It is also possible that our population will eventually be limited by the capacity of the environment to absorb the wastes and other insults imposed by humans. For instance, heavy use of commercial fertilizers to produce large food crops already threatens the quality of groundwater in some agricultural areas. In cases such as this, current human occupants could lower Earth's long-term carrying capacity for future generations.

Some technological optimists have suggested that Earth's carrying capacity can be continually increased—that in fact humans have no practical limits to population growth. Technology has undoubtedly increased Earth's carrying capacity for humans, but no population can continue to grow indefinitely. Exactly what the world's human carrying capacity is and under what circumstances we will approach it are topics of great concern and debate. Ideally, human populations would reach carrying capacity smoothly and then level off. Unlike other organisms, we can decide whether zero population growth will be attained through social changes involving individual choice or government intervention or through increased mortality due to resource limitation, plagues, war, and environmental degradation.

■　　　■　　　■

For better or worse, we have the unique responsibility to decide the fate of our species and the rest of the biosphere. We will examine these issues in greater depth when we discuss the concept of sustainability in Chapter 55.

CHAPTER 52 REVIEW

Go to the Campbell Biology website (www.campbellbiology.com) to explore an interactive version of the Chapter Review.

Summary of Key Concepts

CHARACTERISTICS OF POPULATIONS

- **Two important characteristics of any population are density and the spacing of individuals** (pp. 1152–1153, FIGURES 52.1, 52.2) Density is the number of individuals per unit area or volume, and dispersion is the spacing of individuals. Dispersion may range from clumped (most common) to uniform to random, as determined by various environmental and social factors.

  Web/CD Activity 52A: *Techniques for Estimating Population Density and Size*

- **Demography is the study of factors that affect the growth and decline of populations** (pp. 1153–1156, TABLES 52.1, 52.2, FIGURE 52.3) Populations increase from births and immigration and decrease from deaths and emigration. Life tables tabulate the age-specific mortality schedule for cohorts in populations. Survivorship curves plot the numbers in a cohort still alive at each age. A reproductive table gives the age-specific summary of the reproductive rates in a population.

  Web/CD Activity 52B: *Investigating Survivorship Curves*

LIFE HISTORIES

- **Life histories are highly diverse, but they exhibit patterns in their variability** (p. 1156, FIGURE 52.4) Big-bang, or semelparous, organisms reproduce a single time and die. Repeat, or iteroparous, organisms utilize several breeding seasons to reproduce. Life history traits represent trade-offs between conflicting demands for limited time, energy, and nutrients.

- **Limited resources mandate trade-offs between investments in reproduction and survival** (pp. 1157–1158, FIGURES 52.5–52.7) Clutch size and age at first reproduction involve trade-offs between current and future fecundity, fecundity and adult survival, or fecundity and survival of the offspring.

POPULATION GROWTH

- **The exponential model of population growth describes an idealized population in an unlimited environment** (pp. 1159–1160, FIGURE 52.8) If we can ignore immigration and emigration, a population's growth rate is determined by the birth rate minus the death rate. The exponential growth equation $dN/dt = r_{max}N$ represents a population's potential growth in an unlimited environment, where r_{max} is the maximum per capita rate of increase and N is the number of individuals in the population. This model predicts that the larger a population becomes, the faster it grows.

- **The logistic model of population growth incorporates the concept of carrying capacity** (pp. 1160–1163, FIGURES 52.10–52.12, TABLE 52.3) Exponential growth cannot be sustained for long in any population. A more realistic population model limits growth by incorporating carrying capacity (K), the maximum population size that can be sustained by available resources. The logistic equation $dN/dt = r_{max}N$ $(K - N)/K$ describes an S-shaped curve in which population growth levels off as population size approaches carrying capacity. This model

fits some populations well, but for many natural populations, there is no stable carrying capacity, and populations fluctuate regularly or irregularly around some long-term average density. Natural selection will favor traits that allow survival and reproduction with few resources in populations that live at densities near carrying capacity (K), and this density-dependent selection is called K-selection. Adaptations that promote rapid reproduction should be favored at low densities, and this form of natural selection is called r-selection.

POPULATION-LIMITING FACTORS

- **Negative feedback prevents unlimited population growth** (pp. 1164–1165, FIGURES 52.14–52.15) Density-dependent changes in birth and death rates act to curb population increase and can eventually stabilize a population near its carrying capacity. Many density dependent factors produce negative feedback, including intraspecific competition for limited food or space, increased predation, disease, stress due to crowding, or buildup of toxins.

- **Population dynamics reflect a complex interaction of biotic and abiotic influences** (pp. 1165–1167, FIGURES 52.16–52.18) Carrying capacities can vary in space, providing good and poor habitats for a species, or in time, producing population fluctuations. Most natural populations are characterized by instability.

- **Some populations have regular boom-and-bust cycles** (pp. 1167–1168, FIGURE 52.19) Snowshoe hares have 10-year cycles, and field experiments have shown that intensive predation coupled with food shortage in winter drives these cycles. Cyclic fluctuations of herbivores like snowshoe hares or lemmings cause corresponding fluctuations in their predators' populations.

  Biology Labs On-Line: *PopulationEcologyLab*

HUMAN POPULATION GROWTH

- **The human population has been growing almost exponentially for three centuries but cannot do so indefinitely** (pp. 1168–1169, FIGURES 52.20–52.22) Since the Industrial Revolution, human population growth has been sustained by such factors as improved nutrition, medical care, and sanitation, which have lowered death rates rapidly but birth rates more slowly. The age structure of a population has major impacts on societal needs, such as schools and hospitals.

  Web/CD Activity 52C: *Human Population Growth*
  Web/CD Activity 52D: *Analyzing Age-Structure Pyramids*
  Biology Labs On-Line: *DemographyLab*

- **Estimating Earth's carrying capacity for humans is a complex problem** (pp. 1169–1171, FIGURE 52.23) We can estimate the ecological footprints of nations as one measure of how close we are to the carrying capacity of Earth.

Self-Quiz

1. A uniform dispersion pattern for a population may indicate that
   a. the population is spreading out and increasing its range.
   b. resources are heterogeneously distributed.
   c. individuals of the population are competing for some resource, such as water and minerals for plants or nesting sites for animals.
   d. there is an absence of strong attractions or repulsions among individuals.
   e. the density of the population is low.

2. A "cohort" in a human life table consists of
   a. people who are the same age.
   b. people who live in the same city.
   c. people of the same education level.
   d. people who have the same occupation.
   e. people who have the same number of children.

3. The term $(K-N)/K$ influences dN/dt such that
   a. the increase in actual population numbers is greatest when N is small.
   b. as N approaches K, r, the intrinsic rate of increase, becomes smaller.
   c. when N equals K, population growth is zero.
   d. when K is small, the population begins growing exponentially.
   e. as N approaches K, the birth rate approaches zero.

4. A population's carrying capacity is
   a. the number of individuals in that population.
   b. a constant that can be estimated for all populations.
   c. inversely related to r.
   d. the population size that can be supported by available resources for that species within the habitat.
   e. set at 8 billion for the human population.

5. Which life history strategy would be favored by natural selection if survival of offspring is quite low and unpredictable?
   a. big-bang reproduction, or semelparity
   b. production of a large number of large eggs and a great deal of parental care
   c. repeated reproduction, or iteroparity
   d. very early age of first reproduction
   e. relatively late age of first reproduction

6. In a mark-recapture study of a lake trout population, 40 fish were captured, marked, and released. In a second capture, 45 fish were captured; 9 of these were marked. What is the estimated number of individuals in the lake trout population?
   a. 90   d. 800
   b. 200   e. 1,800
   c. 360

7. The population cycle of the snowshoe hare and its predator, the lynx, illustrates that
   a. predators are the only factor controlling the size of prey populations.
   b. the two species must have evolved in close contact because one cannot live without the other.
   c. one should not conclude a cause-and-effect relationship when viewing population patterns without careful observation and experimentation.
   d. both populations are controlled mainly by abiotic factors.
   e. the hare population is r-selected, whereas the lynx population is K-selected.

8. The current size of the human population is closest to
   a. 2 million.   d. 6 billion.
   b. 3 billion.   e. 10 billion.
   c. 4 billion.

9. All these descriptions are characteristic of human populations in industrialized countries *except*
   a. relatively small family size.
   b. several potential reproductions per lifetime.
   c. r-selected life history.
   d. Type I survivorship curve.
   e. relatively even age structure.

10. According to the study of ecological footprints produced in 1997,
    a. the carrying capacity of the world is 10 billion.
    b. the carrying capacity of the world would be higher if all people became vegetarians.
    c. the current demand on global resources by each industrialized country is well below the ecological capacity of those countries.
    d. the United States has a larger ecological footprint than available ecological capacity of its own land.
    e. a technological fix to expand the world's carrying capacity is not ecologically sound.

11. What is the relationship between a population and a species?

12. An aquarium population of guppies has reached a stable population size. We decide to add twice as much guppy food per day to the aquarium, but this turns out to have no effect on population size. What is the most likely explanation for this observation?

13. (a) How does the age structure of the U.S. population explain the current surplus in the Social Security fund? (b) If the system is not changed, why will the surplus give way to a deficit sometime in the next few decades?

14. In the life table for a population with a Type II survivorship curve, what would be the most noticeable feature of the values in the death rate (mortality) column?

15. What trend characterizes a nation that is in demographic transition?

Go to the website or CD-ROM for more quiz questions.

Evolution Connection

Write a paragraph contrasting the conditions that favor semelparous (one-time) versus iteroparous (repeated) reproduction.

The Process of Science

We estimate the size of a population of small mice in a particular field by the mark-recapture method. Our estimate is $N = 350$. Later, we learn from experiments on the behavior of these mice that they can locate a baited trap faster if they have already been rewarded with food by visiting that trap once before. Does this mean that our original estimate of 350 individuals was (a) too low or (b) too high? Explain your answer in terms of the equation for the mark-recapture method.

Link to Biology Labs On-Line for in-depth investigations into predator-prey population dynamics (PopulationEcologyLab) and the human population (DemographyLab).

Science, Technology, and Society

Many people regard the rapid population growth of developing countries as our most serious environmental problem. Others think that the population growth in developed countries, though smaller, is actually a greater threat to the environment. What kinds of problems result from population growth in (a) developing countries and (b) the industrialized world? Which do you think is the greater threat, and why?

Answers: 1. c; 2. a; 3. c; 4. d; 5. c; 6. b; 7. c; 8. d; 9. c; 10. d; 11. A population is a localized subset of a species. 12. The population was already at carrying capacity before we increased food supply, and the key limiting factor was something other than food availability. 13. (a) The largest population segment, the boomers, are currently in the work force in their peak earning years, paying into the system. (b) The boomers will retire over the next few decades and begin drawing from the system at a time when there will be fewer employees paying into Social Security. 14. The mortality is about the same for every age interval of individuals. 15. A transition from high rates of birth and death to low birth and death rates

CHAPTER 53

COMMUNITY ECOLOGY

WHAT IS A COMMUNITY?

- Contrasting views of communities are rooted in the individualistic and interactive hypotheses
- The debate continues with the rivet and redundancy models

INTERSPECIFIC INTERACTIONS AND COMMUNITY STRUCTURE

- Populations may be linked by competition, predation, mutualism, and commensalism
- Trophic structure is a key factor in community dynamics
- Dominant species and keystone species exert strong controls on community structure
- The structure of a community may be controlled bottom-up by nutrients or top-down by predators

DISTURBANCE AND COMMUNITY STRUCTURE

- Most communities are in a state of nonequilibrium owing to disturbances
- Humans are the most widespread agents of disturbance
- Ecological succession is the sequence of community changes after a disturbance

BIOGEOGRAPHIC FACTORS AFFECTING THE BIODIVERSITY OF COMMUNITIES

- Community biodiversity measures the number of species and their relative abundance
- Species richness generally declines along an equatorial-polar gradient
- Species richness is related to a community's geographic size
- Species richness on islands depends on island size and distance from the mainland

On your next walk *through a field or woodland, or even across campus or through a park, try to observe some of the interactions among the species present. You may see birds using trees as nesting sites, bees pollinating flowers, shelf fungi growing on trees, caterpillars feeding on leaves, spiders trapping insects in their webs, ferns growing in shade provided by trees—*

a sample of the many interactions that exist in any ecological theater. In addition to the physical and chemical factors discussed in Chapter 50, an organism's environment includes other individuals of the same species as well as populations of other species living in the same area. Such an assemblage of species living close enough together for potential interaction is called a biological **community.** *In the photograph that opens this chapter, the lion, the zebra, the hyena, the vultures, and the grasses and other plants are all members of a grassland community in East Africa.*

This chapter examines the different kinds of interactions among organisms and addresses the central issue in community ecology: What factors are most significant in structuring a community—in determining its species composition and the relative abundance of species present?

WHAT IS A COMMUNITY?

A community is any assemblage of populations in an area or habitat. Ecologists define the boundaries of a particular community to fit their research questions. We might study the community of decomposers and other organisms in a rotting log, the benthic community in Lake Superior, or the community of trees and shrubs in Shenandoah National Park. A community has a set of properties defined by its species composition, with a structure determined by the interactions between the species. Communities differ dramatically in their **species richness,** the number of species that they contain. But some species are common and some are rare, and so communities also differ in their **relative abundance** of different species. What causes each community to have a certain assemblage of species? Two very different answers to this question have polarized community ecologists for over 50 years.

1174

Contrasting views of communities are rooted in the individualistic and interactive hypotheses

THE PROCESS OF SCIENCE

How can we account for the species found together as members of a community? Two different views on this question emerged among ecologists in the 1920s and 1930s, based primarily on observations of plant distributions. We will briefly describe these historical arguments because they are the forerunners of the closely related rivet and redundancy models, two community models still debated today.

An **individualistic hypothesis** of community structure was first enunciated by H. A. Gleason, of the University of Chicago, in the early 1900s. It depicted the plant community as a chance assemblage of species found in the same area simply because they happen to have similar abiotic requirements—for example, for temperature, rainfall, and soil type. An alternative view, the **interactive hypothesis,** was advocated by F. E. Clements, also in the early 1900s. Clements saw the community as an assemblage of closely linked species, locked into association by mandatory biotic interactions that cause the community to function as an integrated unit—as a superorganism. Evidence for the interactive view was based on the observation that certain species of plants are consistently found together as a group. For example, deciduous forests in the northeastern United States usually include certain species of oak, maple, birch, and beech, along with a specific assemblage of shrubs and vines. These two very different ways of thinking about community structure—individualistic versus interactive— suggest different priorities in studying biological communities. The individualistic hypothesis emphasizes studying single species, while the interactive hypothesis emphasizes entire assemblages of species as the essential units for understanding the interrelationships and distributions of organisms.

These two hypotheses of plant community organization make contrasting predictions about how plant species should be distributed along an environmental gradient, such as a gradual change in moisture or temperature over some distance. The individualistic hypothesis predicts that communities should generally lack discrete geographic boundaries because each species has an independent distribution along the environmental gradient. In other words, each species will be distributed according to its tolerance ranges for abiotic factors that vary along a gradient, and communities should change continuously along the gradient, with the addition or loss of particular species (FIGURE 53.1a). In contrast, the interactive hypothesis predicts that species should be clustered into discrete communities with noticeable boundaries, because the presence or absence of a particular species is largely governed by the presence or absence of other species with which it interacts in its group (FIGURE 53.1b).

In most actual cases, especially where there are broad regions characterized by gradients of environmental variation, the composition of plant communities does seem to change

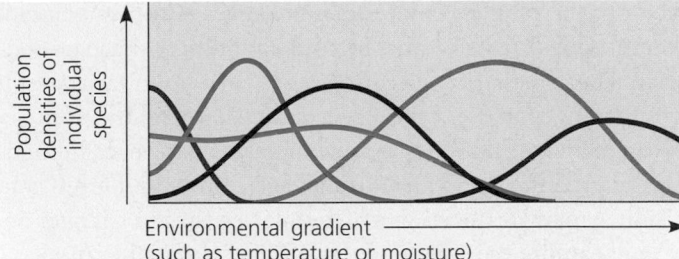

(a) Individualistic hypothesis. The individualistic hypothesis proposes that species are independently distributed along gradients and that a community is simply the assemblage of species that occupy the same area because of similar abiotic needs.

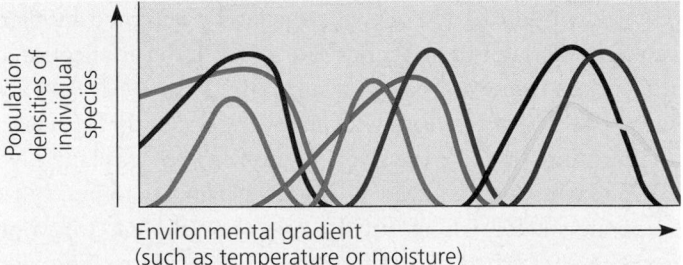

(b) Interactive hypothesis. The interactive hypothesis suggests that communities are discrete groupings of particular species that are closely interdependent and nearly always occur together.

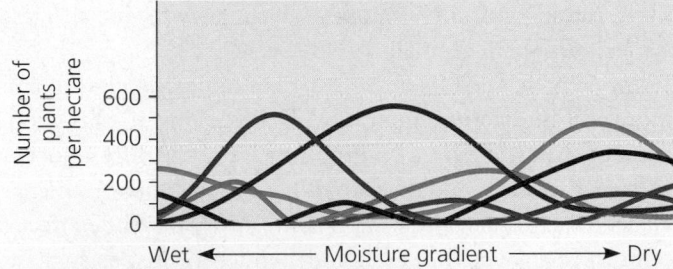

(c) Trees in the Santa Catalina Mountains. The distribution of tree species at one elevation in the Santa Catalina Mountains of Arizona supports the individualistic hypothesis. Each tree species has an independent distribution along the gradient, apparently conforming to its tolerance for moisture, and the species that live together at any point along the gradient have similar physical requirements. Because the vegetation changes continuously along the gradient, it is impossible to delimit sharp boundaries for the communities.

FIGURE 53.1 Testing the individualistic and interactive hypotheses of communities. Ecologist Robert Whittaker tested these two hypotheses by graphing the abundance of different plant species (*y* axis) along environmental gradients of abiotic factors such as temperature or moisture (*x* axis). Each colored curve represents the abundance of one species.

THE PROCESS OF SCIENCE

continuously, with each species more or less independently distributed (FIGURE 53.1c). Such distributions generally support the view of plant communities as relatively loose associations without discrete boundaries. However, where some key factor in the physical environment changes abruptly, adjacent communities are delineated by correspondingly sharp boundaries. The most striking examples of sharp boundaries result from changes in soil type. In general, sharp boundaries between plant communities are rare in nature. However, humans have altered the landscape through agriculture and forestry to produce many artificial sharp boundaries, and in Chapter 55, we will explore the consequences of this habitat alteration.

The debate continues with the rivet and redundancy models

The individualistic hypothesis is generally accepted by plant ecologists today, but further debate arises in the application of these ideas to the animals in a community. In 1981, Paul and Anne Ehrlich, of Stanford University, suggested that species in a community are like the rivets in the wings of airplanes: Not all the rivets are required to hold the wing together, but if someone started taking out the rivets one by one, we would become concerned about flying in that airplane. The **rivet model** of communities is a reincarnation of the interactive model of Clements that was suggested for plant communities. The rivet model suggests that most of the species in a community are associated tightly with other species in a web of life. Thus, reducing or increasing the abundance of one species in a community affects many other species.

In 1992, Brian Walker suggested an opposing view of communities, the **redundancy model.** According to this model, most of the species in a community are not tightly associated with one another, and the web of life is very loose. An increase or decrease in one species in a community has little effect on other species, which operate independently, just as suggested by Gleason 80 years before in his individualistic model of the plant community. Species within a community are redundant. For example, if one predator disappears, another predatory species in the community will usually take its place as a consumer of specific prey. If one pollinator ceases to visit a particular species of flowering plant because the pollinator has disappeared from the area, another pollinator species will do the job.

No matter which (if either) of these two models is correct, it is important to study species relationships in communities. Species in communities do interact, even if some of these interactions are not imperative for the species. It is also important to remember that these two models represent extremes, although most communities actually lie somewhere in the middle. To answer some key questions, we need to determine how species interact in communities and how tight these associations are. For example, what happens to a community when one species is lost or is replaced with a species introduced by humans? Such questions are important because they underlie many of our environmental problems today.

INTERSPECIFIC INTERACTIONS AND COMMUNITY STRUCTURE

There are different **interspecific interactions**—relationships between the species of a community. We will start with the simplest situation: an interaction between just two species.

Populations may be linked by competition, predation, mutualism, and commensalism

The possible interspecific interactions are introduced in TABLE 53.1. In this simplified view, we use a pair of signs, such as $+/-$, to symbolize how an interspecific interaction affects the population densities of the two species. For example, mutual symbiosis (mutualism) is a $+/+$ interaction, meaning that the density of each species is increased in the presence of the other. Predation is an example of a $+/-$ interaction, with a positive effect on the population density of one species (the predator) and a negative effect on the density of the other population (the prey).

Competition

Interspecific competition for resources can occur when resources are in short supply. The weeds growing in a garden are competing with garden plants for soil nutrients and water. Grasshoppers and bison compete in the Great Plains for the grass they eat. Lynx and foxes compete for prey such as snowshoe hares in the northern forests of Alaska and Canada. There is a potential for competition between any two species that need the same limited resource. Some resources, such as oxygen, are not usually in short supply, and although many species use this resource, they do not compete for it. If two populations *do* compete for a resource, the result may be a reduction in the density of one or both species or the local elimination of one of the two competitors.

The Competitive Exclusion Principle.   In 1934, Russian ecologist G. F. Gause studied the effects of interspecific competition in laboratory experiments with two closely related species of protists, *Paramecium aurelia* and *Paramecium caudatum*. Gause

Table 53.1 Interspecific Interactions

Interaction	Effects on Population Density
Competition ($-/-$)	The interaction is detrimental to both species.
Predation ($+/-$) (includes parasitism)	The interaction is beneficial to one species and detrimental to the other.
Mutualism ($+/+$)	The interaction is beneficial to both species.
Commensalism ($+/0$)	One species benefits from the interaction but the other is unaffected.

cultured the protists under stable conditions with a constant amount of food added every day. When he grew the two species in separate cultures, each population grew rapidly and then leveled off at what was apparently the carrying capacity of the culture. But when Gause cultured the two species together, *P. aurelia* apparently had a competitive edge in obtaining food, and *P. caudatum* was driven to extinction in the culture. Gause concluded that two species so similar that they compete for the same limiting resources cannot coexist in the same place. One will use the resources more efficiently and thus reproduce more rapidly. Even a slight reproductive advantage will eventually lead to local elimination of the inferior competitor. Ecologists call Gause's concept the **competitive exclusion principle.**

The Ecological Niche.

The sum total of a species' use of the biotic and abiotic resources in its environment is called the species' **ecological niche.** One way to grasp the concept is through an analogy made by ecologist Eugene Odum: If an organism's habitat is its address, the niche is that habitat plus the organism's occupation. Put another way, an organism's niche is its ecological role—how it "fits into" an ecosystem. The niche of a population of tropical tree lizards, for example, consists of, among many other components, the temperature range it tolerates, the size of branches on which it perches, the time of day in which it is active, and the size and type of insects it eats.

We can now restate the competitive exclusion principle to say that two species cannot coexist in a community if their niches are identical. However, ecologically similar species *can* coexist in a community if there are one or more significant differences in their niches. Numerous tests of competitive exclusion hypotheses include classic field experiments with two species of barnacles that attach to intertidal rocks on the North Atlantic coast (FIGURE 53.2).

Resource Partitioning.

There are two possible outcomes of competition between species having identical niches: Either the less competitive species will be driven to local extinction, or one of the species may evolve enough through natural selection to use a different set of resources. This differentiation of niches that enables similar species to coexist in a community is called **resource partitioning** (FIGURE 53.3, p. 1178). We can think of resource partitioning within a community as "the ghost of competition past"—circumstantial evidence of earlier interspecific competition resolved by the evolution of niche differentiation.

Character Displacement.

As a complement to studies of resource partitioning, a related line of circumstantial evidence for the importance of competition comes from comparisons of closely related species whose populations are sometimes sympatric and sometimes allopatric. Although allopatric populations of such species are similar in structure and use similar resources, sympatric populations often show differences in body structures and in the resources they use. The tendency for characteristics to be more divergent in sympatric populations of two species than in allopatric populations of the same two species is called **character displacement.** The Galápagos finches described in Chapter 25 provide a good example of character displacement in beak sizes and, presumably, in the seeds that they can eat most efficiently. Allopatric populations of *Geospiza fuliginosa* and *G. fortis* have beaks of similar size, but on an island where both species occur, a significant difference in beak depth has evolved (FIGURE 53.4, p. 1178). This difference presumably enables the two species to avoid competition by feeding on seeds of different sizes and probably represents an evolutionary outcome of past competition.

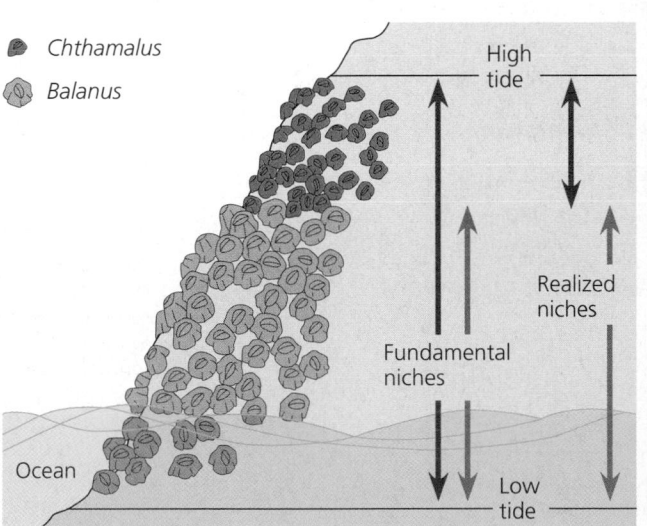

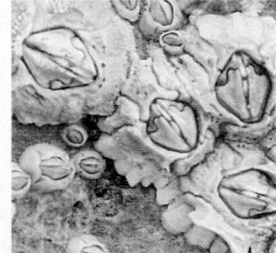

FIGURE 53.2 Testing a competitive exclusion hypothesis in the field. *Balanus balanoides* and *Chthamalus stellatus* are two species of barnacles that grow on rocks exposed during low tide along the Scottish coast. The barnacles have a stratified distribution, with *Balanus* most concentrated on the lower portions of the rocks and *Chthamalus* on the higher portions. The swimming larvae of the barnacles may settle randomly on the rocks and begin to develop into sessile adults, but *Balanus* fails to survive high on the rocks because it is unable to resist desiccation when these areas are exposed to air for several hours during low tides. Its fundamental niche (potential niche) and its realized niche (actual niche) are similar. Even though *Chthamalus* is concentrated primarily on the upper strata of rocks, when ecologist Joseph Connell removed *Balanus* from the lower strata, the *Chthamalus* population spread into that area. Thus, *Chthamalus* could survive lower on the rocks than where it is generally found, were it not for competition from *Balanus*. Its realized niche is only a fraction of its fundamental niche.

Chthamalus

Balanus

High tide

Realized niches

Fundamental niches

Ocean

Low tide

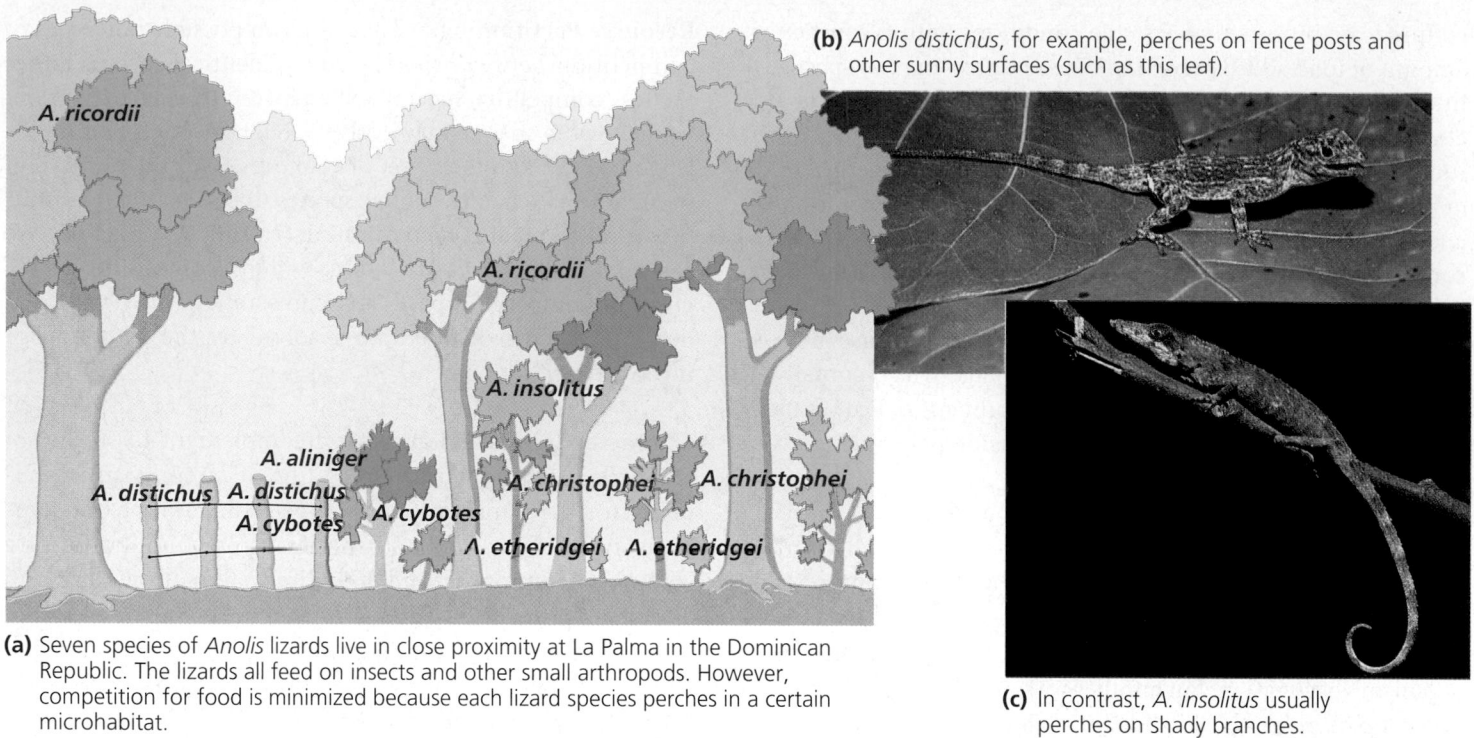

(b) *Anolis distichus,* for example, perches on fence posts and other sunny surfaces (such as this leaf).

(a) Seven species of *Anolis* lizards live in close proximity at La Palma in the Dominican Republic. The lizards all feed on insects and other small arthropods. However, competition for food is minimized because each lizard species perches in a certain microhabitat.

(c) In contrast, *A. insolitus* usually perches on shady branches.

FIGURE 53.3 Resource partitioning in a group of lizards.

Predation

The term **predation** elicits such images as a lion killing and eating an antelope or other prey. But ecologists extend the definition of predation to include **herbivory,** in which an herbivore such as a bison eats part of a plant, and **parasitism,** in which a parasite lives on or in its host organism and depends on the host species for nutrition.

It won't surprise you that predation is a potent factor in adaptive evolution. Eating and avoiding being eaten are prerequisite to reproductive success. Natural selection refines the adaptations of both predators and prey.

Predator Adaptations.   Many important feeding adaptations of predators are both obvious and familiar. Most predators have acute senses that enable them to locate and identify potential prey. In addition, many predators have adaptations such as claws, teeth, fangs, stingers, or poison that help catch and subdue the organisms on which they feed. Rattlesnakes and other pit vipers, for example, locate their prey with special heat-sensing organs located between each eye and nostril, and they kill small birds and mammals by injecting them with toxins through their fangs. Similarly, many herbivorous insects locate appropriate food plants by the using chemical sensors on their feet, and their mouthparts are adapted for shredding tough vegetation. Predators that pursue their prey are gener-

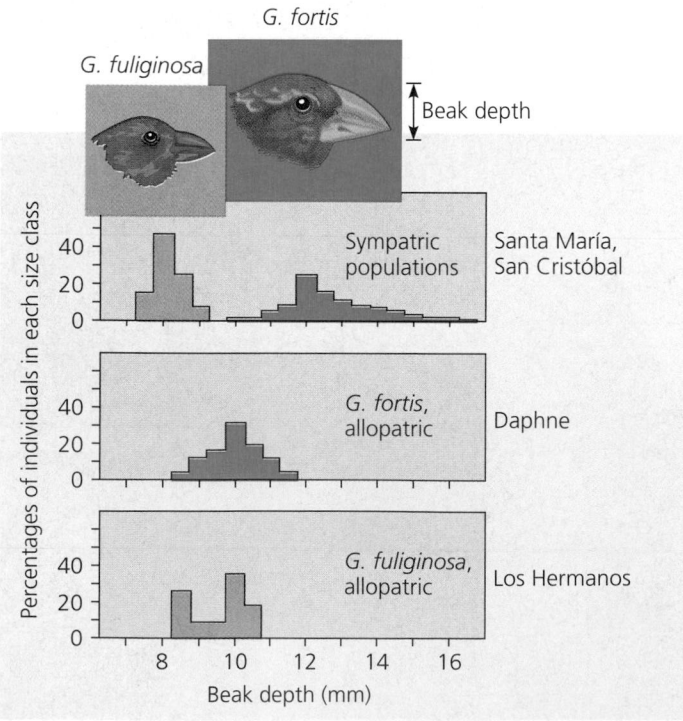

FIGURE 53.4 Character displacement: circumstantial evidence for competition in nature. Although allopatric populations of potential competitors are often similar in morphology and use equivalent resources, sympatric populations may diverge in both characteristics. In this example, two species of Galápagos finches have similar beak morphologies and presumably eat similarly sized seeds where their populations are allopatric on Daphne and Los Hermanos islands. However, where the two species are sympatric on Santa María and San Cristóbal, *Geospiza fuliginosa* has a shallower, smaller beak and *G. fortis* a deeper, larger one. Such evolutionary changes in body structure are thought to reflect resource partitioning. In this case, the two species have adapted to eating different sizes of seeds.

ally fast and agile, whereas those that lie in ambush are often camouflaged in their environments.

Plant Defenses Against Herbivores.   Plants cannot run away from herbivores. Chemical toxins, often in combination with antipredator spines and thorns, are a plant's main arsenal against being eaten to extinction. Among such chemical weapons are the poison strychnine, produced by the tropical vine *Strychnos toxifera;* morphine, from the opium poppy; nicotine, from the tobacco plant; mescaline, from peyote cactus; and tannins, from a variety of plant species. Other defensive compounds that are not toxic to humans but may be distasteful to other herbivores are responsible for the familiar flavors of cinnamon, cloves, and peppermint. Some plants even produce chemicals that imitate insect hormones and cause abnormal development in some insects that eat them.

Animal Defenses Against Predators.   Animals can avoid being eaten by using passive defenses, such as hiding, or active defenses, such as escaping or defending themselves against predators. Fleeing is a common antipredator response, though it can be very costly in terms of energy. Many animals flee into a shelter and avoid being caught without expending the energy required for a prolonged flight. Active self-defense is less common, though some large grazing mammals will vigorously defend their young from predators such as lions. Other behavioral defenses include alarm calls, which often bring in many individuals of the prey species that mob the predator. For example, crows will sometimes gang up and peck on barn owls, which prey on crow eggs.

Many other defenses rely on adaptive coloration, which has evolved repeatedly among animals. Camouflage, or **cryptic coloration,** is a passive defense that makes potential prey difficult to spot against its background (FIGURE 53.5).

Some animals have mechanical or chemical defenses against would-be predators (see the interview with Thomas Eisner on p. 24). Most predators are strongly discouraged by the familiar defenses of skunks and porcupines. Some animals, such as poisonous toads and frogs, can synthesize toxins. Others acquire chemical defense passively by accumulating toxins from the plants they eat. For example, monarch butterflies store poisons

FIGURE 53.6 Aposematic (warning) coloration of a poison-arrow frog. The skin of this tree frog, an inhabitant of rain forests in Costa Rica, produces noxious chemicals. Predators apparently learn to associate the vivid markings of the frog with danger as soon as they touch the frog's skin. In some parts of South America, human hunters in the rain forest tip their arrows with poisons from similar frogs to bring down large mammals.

from the milkweed plants they eat as larvae, making the butterflies distasteful to some potential predators.

Animals with effective chemical defenses are often brightly colored, a warning to predators known as **aposematic coloration** (FIGURE 53.6). This warning coloration seems to be adaptive; there is evidence that predators are more cautious in dealing with bright color patterns in potential prey, perhaps because so many aposematic animals tend to be dangerous prey. In an example of convergent evolution, unpalatable animals in several different taxa have similar patterns of coloration— black with yellow or red stripes characterize unpalatable animals as diverse as yellow-jacket wasps and coral snakes.

A species of prey may gain significant protection through mimicry, a "copycat" adaptation in which one species mimics the appearance of another. In **Batesian mimicry,** a palatable or harmless species mimics an unpalatable or harmful model. In one intriguing example, the larva of the hawkmoth puffs up its head and thorax when disturbed, looking like the head of a small poisonous snake, complete with eyes (FIGURE 53.7). The mimicry even involves behavior; the larva weaves its head back and forth and hisses like a snake. In **Müllerian mimicry,** two or more unpalatable species resemble each other. Presumably,

(a) Hawkmoth larva

(b) Snake

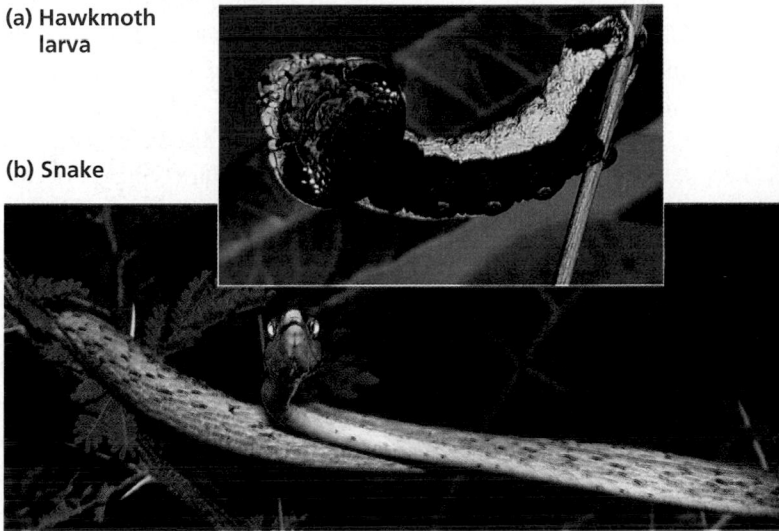

FIGURE 53.7 Batesian mimicry. When disturbed, **(a)** the hawkmoth larva resembles **(b)** a snake.

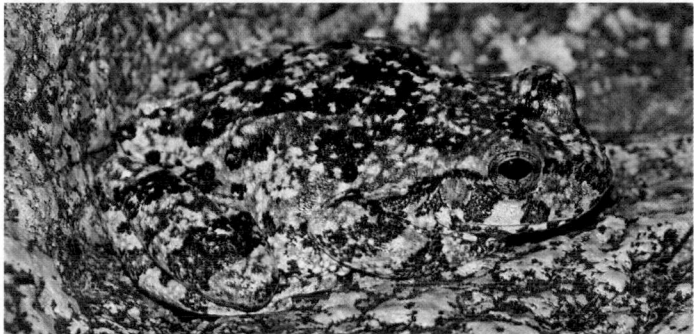

FIGURE 53.5 Camouflage: a canyon tree frog disappearing into a background of granite.

FIGURE 53.8 Müllerian
mimicry. Both **(a)** the cuckoo
bee and **(b)** the yellow jacket
wasp have stingers that release
toxins. The cross-mimicry in
appearance presumably benefits
both species because predators
learn more quickly to avoid any
prey with these distinctive
markings.

(a) Cuckoo bee

(b) Yellow jacket

each species gains an additional advantage because the pooling of numbers causes predators to learn more quickly to avoid any prey with a particular appearance (FIGURE 53.8).

Predators also use mimicry in a variety of ways. For example, some snapping turtles have tongues that resemble a wriggling worm, thus luring small fish; any fish that tries to eat the "bait" is itself quickly consumed as the turtle's strong jaws snap closed.

Parasites and Pathogens as Predators. Parasitism is a symbiotic interaction in which one organism, the **parasite,** derives its nourishment from another organism, its **host,** which is harmed in the process. (Note that this text adopts the most general definition of symbiosis as including parasitism, commensalism, and mutualism, which are all discussed in this chapter. However, some biologists prefer to use the term *symbiosis* more specifically as a synonym for mutualism.)

Parasites that live within their host, such as tapeworms and malarial parasites, are called **endoparasites;** parasites that feed on the external surface of a host, such as mosquitoes and aphids, are called **ectoparasites.** In a special type of parasitism called **parasitoidism,** insects—usually small wasps—lay eggs on living hosts. The larvae then feed on the body of the host, eventually killing it.

In terms of their effects, pathogens, or disease-causing organisms, are similar to parasites. Pathogens are typically bacteria, viruses, or protists, but fungi and prions (protein bodies; see Chapter 18) may also be pathogenic. In contrast to pathogens, which are generally microscopic, many parasites are relatively large, multicellular organisms, such as tapeworms. Also, most parasites inflict nonlethal damage on their hosts—by pilfering nutrients, for example—while many pathogens can inflict lethal harm.

Mutualism

Mutual symbiosis, or **mutualism,** is an interspecific interaction that benefits both species. Mutualistic relationships sometimes require the coevolution of adaptations in both participating species. Changes in either species are likely to affect the survival and reproduction of the other. We have described many examples of mutualism in previous chapters: nitrogen fixation by bacteria in the root nodules of legumes; the digestion of cellulose by microorganisms in the digestive systems of termites and ruminant mammals; photosynthesis by unicellular algae in the tissues of corals; and the exchange of nutrients in mycorrhizae, the association of fungi and the roots of plants. FIGURE 53.9 illustrates another example, a mutualism between ants and acacia trees in Central and South America.

Many mutualistic relationships may have evolved from predator-prey or host-parasite interactions. Most angiosperm plants, for example, have adaptations that attract animals that function in pollination or seed dispersal. Any plants that could

FIGURE 53.9 Mutualism between acacia trees and ants. Certain species of Central and South American acacia trees have hollow thorns that house stinging ants of the genus *Pseudomyrmex*. The ants feed on sugar produced by nectaries on the tree and on protein-rich swellings called Beltian bodies (orange in the photograph) that grow at the tips of leaflets. The acacia benefits from housing and feeding a population of pugnacious ants, which attack anything that touches the tree. The ants sting other insects, remove fungal spores and other debris, and clip surrounding vegetation that happens to grow close to the foliage of the acacia.

derive some benefit by sacrificing such organic materials as nectar rather than pollen or seeds would increase their reproductive success.

Commensalism

Commensalism is an interaction between species that benefits only one of the species involved in the interaction. "Hitchhiking" species, such as algae that grow on the shells of aquatic turtles or barnacles that attach to whales, are sometimes considered commensal. However, the hitchhikers may actually decrease the reproductive success of their hosts slightly by reducing the efficiency of movement in the hosts' search for food or escape from predators.

Commensal associations sometimes involve one species obtaining food that is inadvertently exposed by another. For instance, cowbirds and cattle egrets (also birds) feed on insects flushed out of the grass by grazing bison, cattle, horses, and other herbivores. Because the birds increase their feeding rates when following the herbivores, they clearly benefit from the association. Much of the time, the herbivores may be unaffected by the relationship. However, there are times when they, too, derive some benefit; the birds tend to be opportunistic feeders that occasionally remove and eat ticks and other ectoparasites from the herbivores.

Coevolution and Interspecific Interactions

In describing the adaptations of certain organisms to the presence of other organisms in a community, the term *coevolution* comes up often—perhaps *too* often. There is actually little evidence for coevolution in most cases of interspecific interactions.

Coevolution refers to reciprocal evolutionary adaptations of two interacting species. A change in one species acts as a selective force on another species, in which counteradaptation in turn acts as a selective force on the first species. This linkage of adaptations requires reciprocal genetic change in the interacting populations of the two species. An example of such dual adaptation that probably qualifies as coevolution is the gene-for-gene recognition between a plant species and a species of avirulent pathogen (see FIGURE 39.31). In contrast, the warning coloration of various tree frogs and the aversion reactions of various predators do *not* qualify as coevolution because these are adaptations to a category of other organisms in the community rather than coupled adaptation between just two species.

Despite the current view that coevolution is difficult to demonstrate for most interspecific interactions in communities, biologists agree more generalized adaptation of organisms to other organisms in their environment—to biotic factors— is a fundamental feature of life. Which of the biological interactions we have surveyed is most important in structuring communities? At the present time, most ecologists think that predation and competition are the key processes driving community dynamics. But this conclusion is based mainly on research in temperate communities. Ecologists have far fewer data for interspecific interactions in tropical communities.

Trophic structure is a key factor in community dynamics

The dynamics and structure of a community depend to a large extent on the feeding relationships between organisms—the **trophic structure** of the community. The transfer of food energy from its source in plants and other photosynthetic organisms (primary producers) through herbivores (primary consumers) to carnivores (secondary and tertiary consumers) and eventually to decomposers is referred to as a **food chain.** In the 1920s, Oxford biologist Charles Elton first pointed out that the length of a food chain is usually limited to four or five links, or **trophic levels** (FIGURE 53.10). Elton also recognized that food chains are not isolated units but are hooked together into **food webs.**

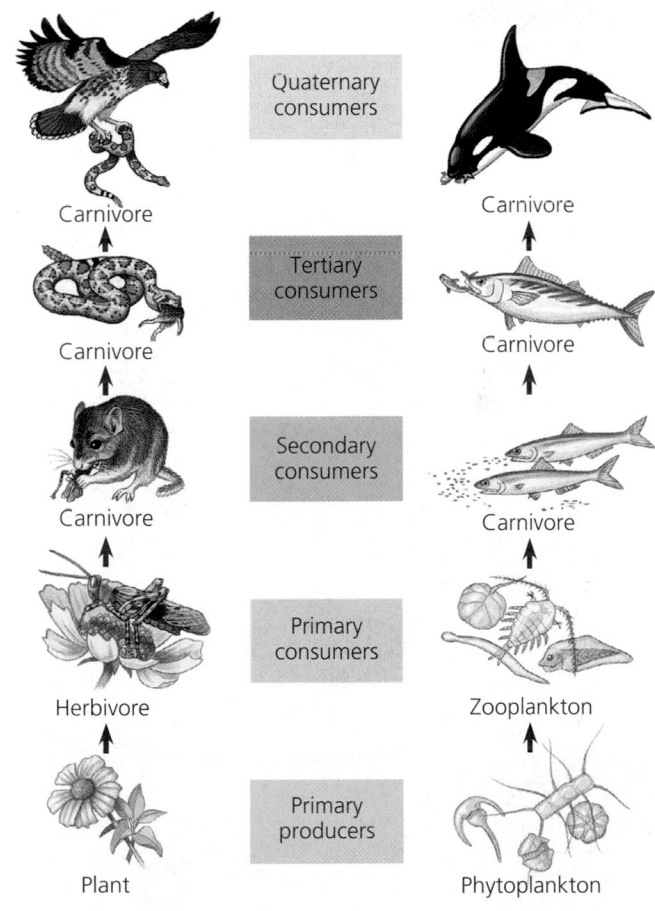

FIGURE 53.10 Examples of terrestrial and marine food chains. The arrows trace energy and nutrients that pass through the trophic levels of a community when organisms feed on one another. Decomposers (detritivores) are not shown here.

Food Webs

Who eats whom in a community? An ecologist can summarize the trophic relationships of a community by diagramming a food web with arrows linking species according to their trophic relationship. For example, FIGURE 53.11 is a simplified food web for an Antarctic pelagic community found in the seasonally productive Southern Ocean. The dominant herbivores in the Antarctic are euphausids (krill, which are crustaceans) and herbivorous plankton such as copepods. These zooplankton species are in turn eaten by various carnivores, including penguins, seals, fish, and baleen whales. Squid, which are carnivores that feed on fish as well as zooplankton, are another important link in these food chains, as they are in turn eaten by seals and toothed whales. During the whaling years, humans became the top predator of this food chain. Having reduced the

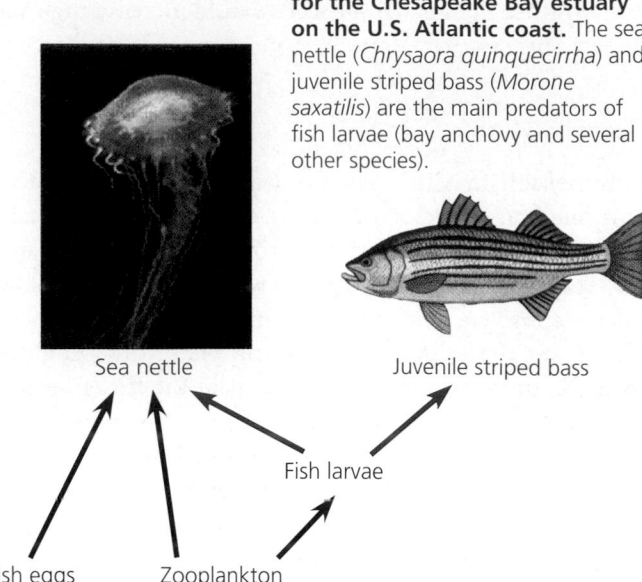

FIGURE 53.12 Partial food web for the Chesapeake Bay estuary on the U.S. Atlantic coast. The sea nettle (*Chrysaora quinquecirrha*) and juvenile striped bass (*Morone saxatilis*) are the main predators of fish larvae (bay anchovy and several other species).

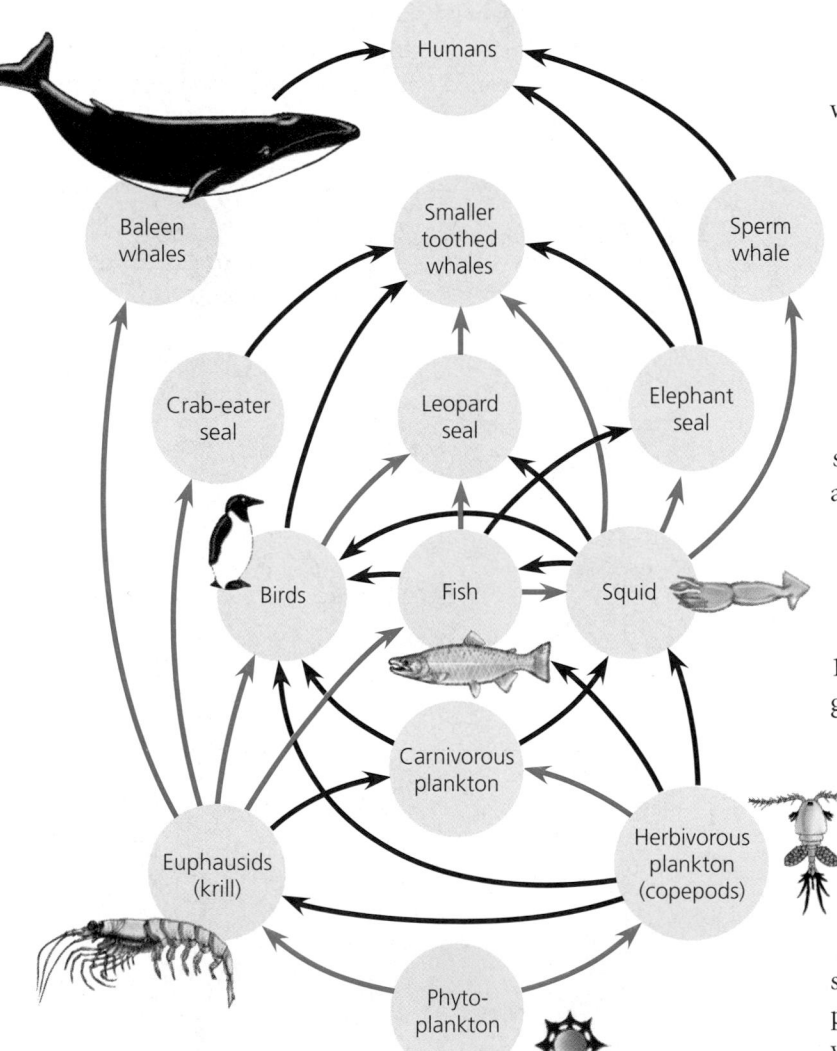

FIGURE 53.11 An antarctic marine food web. Arrows follow the transfer of food from the producers (phytoplankton) through the trophic levels. As in FIGURE 53.10, this diagram omits decomposers, which "feed" at all trophic levels. Blue arrows indicate the major trophic interactions.

whales to low numbers, humans are now harvesting at lower trophic levels, including the harvesting of krill.

What transforms food chains into food webs? First, a given species may weave into the web at more than one trophic level. For example, male horseflies feed as primary consumers on nectar and plant juices, whereas the females are bloodsucking ectoparasites that feed as secondary or even tertiary consumers. Second, most consumers are nonexclusive. For example, foxes are omnivores with diets that include berries and other plant materials, herbivores such as mice, and other predators, such as weasels. Humans are among the most versatile of omnivores.

Food webs can be very complicated, but we can simplify them in two ways. First, we can group species into fairly broad taxonomic groups if their trophic relationships in a community are similar. For example, in FIGURE 53.11, over 100 phytoplankton species are combined in one "functional group," that of primary producers in the food web. A second way to simplify a food web is to isolate a part of the web that interacts very little with the rest of the community.

FIGURE 53.12 illustrates a partial food web for sea nettles (jellies) and juvenile striped bass in Chesapeake Bay. This partial food web can be "pulled out" from a more complex web because these species interact very little with the rest of the community in terms of feeding relationships. Note that sea nettles are both secondary consumers, because they eat zooplankton, and tertiary consumers, because they eat fish larvae, which themselves are secondary consumers of zooplankton.

Animals at successive trophic levels in a food chain tend to be larger with each link (except for parasites). There are upper and lower limits to the size of food a carnivorous animal can eat. The size of an animal and its feeding mechanism puts some upper limit on the size of food it can take into its mouth.

And except in a few cases, large carnivores cannot live on very small food items because they cannot procure enough food in a given time to meet their metabolic needs. Among the exceptions are baleen whales, huge suspension feeders with adaptations that enable them to consume enormous quantities of krill and other small organisms (see FIGURE 41.6).

What Limits the Length of a Food Chain?

Each food chain we can identify in a food web is only a few links long. For example, in the Antarctic web of FIGURE 53.11, there are only four or five links from the producers to any top-level predators. In fact, for all the food webs that ecologists have studied so far, most chains have five or fewer links, although a few are as long as nine links.

Why are food chains relatively short? There are two main hypotheses. The **energetic hypothesis** is the most widely accepted explanation for food chain length. It suggests that the length of a food chain is limited by the inefficiency of energy transfer along the chain. As we will see in the next chapter, only about 10% of the energy stored in the organic matter of any trophic level is converted to organic matter at the next trophic level. If we start with 100 kg of plant material, we can support about 10 kg of herbivore production and 1 kg of carnivore production. If this energetic hypothesis is correct, food chains should be longer in habitats of higher photosynthetic productivity, a prediction that can be tested.

A second hypothesis to explain the limited length of food chains is the **dynamic stability hypothesis.** According to this idea, long food chains are less stable than short chains. Fluctuations at lower trophic levels are magnified at higher levels, potentially causing the extinction of top predators. In a variable environment, top predators must be able to recover from environmental shocks (such as extreme winters) that can reduce the food supply, from producers all the way up the food chain. The longer the food chain, the slower the recovery rate from environmental setbacks for top predators. This hypothesis predicts shorter food chains in unpredictable environments, and this prediction can again be tested by the collection of food chain data.

Most of the data available support the energetic hypothesis as the likely explanation for the shortness of food chains. For example, ecologists have used tree-hole communities in tropical forests as experimental models to test the energetic hypothesis. Many trees have small branch scars that rot to form small holes in the tree trunk. Tree holes hold water and provide habitat for tiny communities sustained by the leaf litter trapped by the water. FIGURE 53.13 shows the results of a set of experiments in which productivity (leaf litter falling into the tree holes) was manipulated. As predicted by the energetic hypothesis, holes with the most leaf litter, and hence the greatest total food supply at the producer level, supported the longest food chains.

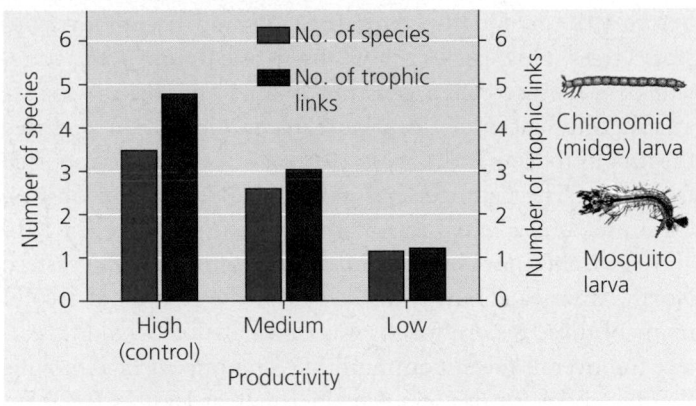

FIGURE 53.13 Test of the energetic hypothesis for the restriction on food chain length. Researchers manipulated the productivity of tree-hole communities in Queensland, Australia, with leaf litter input at three levels: high litter input = natural (control) rate of litter fall; medium = 1/10 natural rate; and low = 1/100 natural rate. Reducing energy input reduced food chain length, a result consistent with the energetic hypothesis. The tree-hole community consists of microbes that break down leaf litter, mosquito larvae that feed on these microbes, predatory midges (chironomids), and other insects that feed directly on leaf litter.

Dominant species and keystone species exert strong controls on community structure

Certain species may have an especially large impact on entire communities, either because of their abundance (dominant species) or because of their pivotal roles in community dynamics (keystone species).

Dominant Species

Dominant species are those species in a community that have the highest abundance or highest **biomass** (the sum weight of all individuals in a population). They exert a powerful control over the occurrence and distribution of other species. For example, the sugar maple is the dominant plant species in many forest communities in eastern North America, and the abundance of maples has a major impact on abiotic factors such as shading and soil, which affect the entire forest community.

Why do some species become dominant in a community? According to one hypothesis, species that are most competitive in exploiting limited resources such as water or nutrients are the most likely species to become dominant. But a competitive edge in resource use may not be the only explanation for why a species becomes dominant. In some communities, a dominant species may owe its abundance to greater success than other potential dominants at avoiding predators.

What happens if we remove a dominant species from a community? Humans have carried out this type of experiment many times by accident. The American chestnut was a dominant tree in the eastern deciduous forests of North America before 1910, making up more than 40% of the canopy (top-story) trees. This species is now absent as a canopy tree as a result of a disease called chestnut blight. This fungal disease attacks only chestnut trees. Humans accidentally introduced the fungus to New York City in 1910 on nursery stock imported from Asia. The pathogen has only a weak impact on Chinese chestnuts, but is lethal to American chestnuts. Between 1910 and 1950, chestnut blight killed all the chestnut trees in eastern North America. Thus, what had been the dominant tree of many of these eastern forests was eliminated. How did this affect the overall forest communities? The impact of removing the dominant species was actually small, at least as far as researchers can tell. The forests have filled in with various species of oaks, hickories, beech, and red maple, and these trees have increased in abundance and replaced the chestnut. Some species of insects, however, were affected. Fifty-six species of moths and butterflies fed on the American chestnut. Of these, 7 species became extinct, but the other 49 species did not rely exclusively on the chestnut for food, and they still survive. No mammals or birds seem to have been seriously affected by this loss of a dominant species. In communities in which dominance is achieved by competitive ability, the loss of a dominant may not have a great effect, because one or more other species with slightly less competitive ability may take over the role of the dominant. This is an example of an outcome predicted by the redundancy model of community structure.

Keystone Species

In contrast to dominant species, most **keystone species** are not especially abundant in a community. They exert strong control on community structure not so much by numerical might as by their ecological roles, or niches. One of the best ways to recognize keystone species is by removal experiments, which is exactly how ecologist Robert Paine first developed the concept of keystone species. The sea star *Pisaster ochraceous* is a keystone predator of mussels in rocky intertidal communities of western North America (FIGURE 53.14a). When Paine removed *Pisaster* manually from rocky intertidal areas, the mussel *Mytilus californianus* was able to monopolize space and exclude other invertebrates and algae from attachment sites (FIGURE 53.14b). *M. californianus* is a dominant species, a superior competitor for open space in the rocky intertidal zone. Predation by *Pisaster* offsets this competitive edge and allows other species to use the space vacated by *Mytilus*. *Pisaster* is not able to eliminate mussels entirely because *Mytilus* can grow too large to be eaten by the sea star. Size-limited predation provides some refuge for this prey species, and these large mussels are able to produce large numbers of fertilized eggs that help recolonize vacant space on the rocks. Paine observed that when sea stars are present, 15 to 20 species of invertebrates and algae occur in the intertidal zone. But after experimental removal of *Pisaster*, species diversity quickly de-

(a) The rocky intertidal zone on the coast of Washington State contains a variety of algal species and invertebrates, including the sea star *Pisaster ochraceous*, which feeds preferentially on mussels but will also consume other invertebrates.

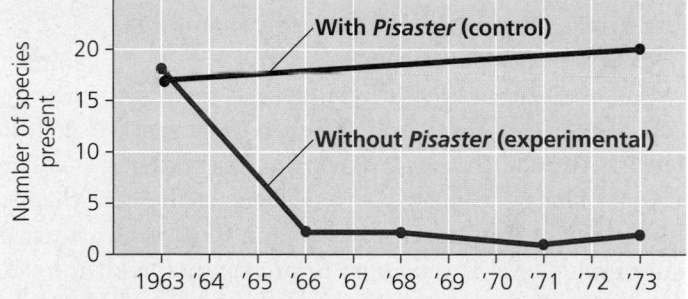

(b) When ecologist Robert Paine experimentally removed *Pisaster* from an intertidal zone in 1963, mussels eventually took over the rock face and eliminated most other invertebrates and algae. In a control area from which *Pisaster* was not removed, there was little change in species diversity.

FIGURE 53.14 Testing a keystone predator hypothesis.

clines to less than 5 species because the mussel, unchecked by the keystone predator, is able to monopolize space. Thus, the important role of *Pisaster* is not reflected in its abundance.

Sea otters, a keystone predator in the North Pacific, offer another example. Once relatively abundant, they were reduced to near extinction by the fur trade during the 19th century. An international treaty provided protection in the 20th century, enabling sea otter populations to recover to very high densities. Sea otters feed on sea urchins, and sea urchins feed mainly on kelp. In areas where sea otters are abundant, sea urchins are rare and kelp forests are well developed. Where sea otters are rare, sea urchins are common and kelp is almost absent. During the last 20 years, sea otters have declined precipitously in large areas off the coast of western Alaska (FIGURE 53.15), sometimes at rates of 25% per year. The loss of this keystone species has allowed sea urchin populations to increase, resulting in the destruction of kelp forests. Ecologists suspect that killer whales are the cause of the sea otter decline. Killer whales have probably been eating sea otters for the past two decades because the previous prey of the whales, mainly seals and sea lions, have declined in density. And the decline of these prey species reflects a decline in the populations of fish species that the seals and sea lions eat. And all of these changes in these Alaskan marine

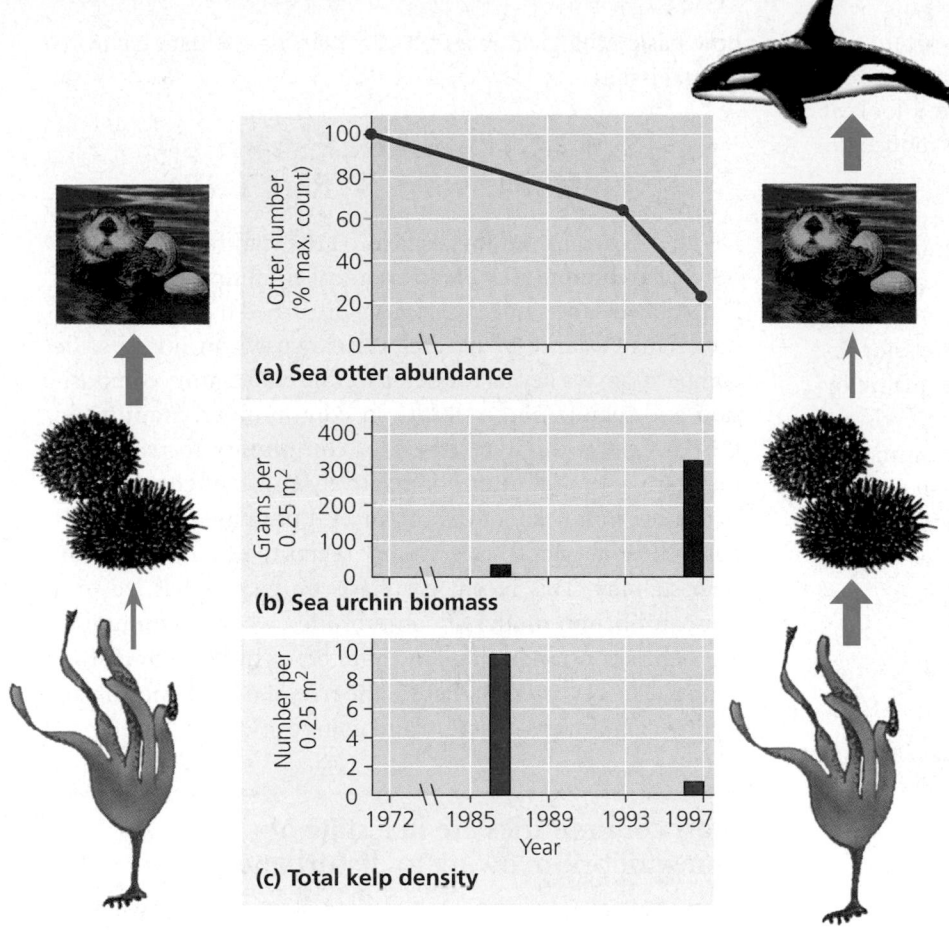

FIGURE 53.15 Sea otters as keystone predators in the North Pacific.
The graphs correlate changes in **(a)** sea otter abundance over time with
changes in two factors; **(b)** sea urchin biomass and **(c)** changes in kelp density
measured from kelp forests at Adak Island (part of the Aleutian Island chain).
The food chains alongside the graphs illustrate a hypothesis for a change in
kelp forest communities due to killer whales preying on sea otters.

(a) Sea otter abundance

(b) Sea urchin biomass

(c) Total kelp density

For example, $V \rightarrow H$ means that an increase
in vegetation will impact (increase) the
numbers or biomass of herbivores, but not
vice versa. In this situation, herbivores are
limited by vegetation, but vegetation is not
limited by herbivory. In contrast, $V \leftarrow H$
means that an increase in herbivore biomass will
have an impact on vegetation (decreasing it),
but not vice versa. A double arrow indicates that
feedback flows in both directions, with each
trophic level sensitive to changes in the biomass
of the other.

Based on these possible interactions, we can
distinguish two models of community organization: the bottom-up model and the top-down
model. The **bottom-up model** postulates $V \rightarrow H$
linkages. In this case, mineral nutrients (N)
control community organization because the
nutrients control plant (V) numbers, which
control herbivore (H) numbers, which in turn
control predator (P) numbers. The simplified
bottom-up model is thus $N \rightarrow V \rightarrow H \rightarrow P$. If
you want to change the community structure of
a bottom-up community, you need to alter biomass at the lower trophic levels. For example, if
you add mineral nutrients to stimulate growth
of vegetation, then all the other trophic levels
should also increase in biomass. But if you add
predators or remove predators from a bottom-
up community, the effect will not extend down
to the lower trophic levels.

In contrast, the **top-down model** postulates that it is
mainly predation that controls community organization because predators control herbivores, which in turn control
plants, which in turn control nutrient levels. The simplified
top-down model is thus $N \leftarrow V \leftarrow H \leftarrow P$. This top-down
control of community structure is also called the *trophic cascade model*. It predicts a cascade of $+/-$ effects down the
trophic levels. Increasing predators will depress herbivore
numbers, the depressed herbivore numbers will have a reduced impact on plant abundance, and the abundant plants
will depress levels of mineral nutrients. For example, in a lake
community with four trophic levels, the trophic cascade
model predicts that removing the top carnivores will increase
the abundance of primary carnivores, decrease herbivores, increase phytoplankton, and eventually decrease concentrations
of mineral nutrients. If there were only three trophic levels in
a lake, removing the primary carnivores would increase the
herbivores and decrease the phytoplankton, causing nutrient
levels to rise. The effects of any manipulation will thus move
down the trophic structure as a series of $+/-$ effects.

Many other models intermediate between the bottom-up
and top-down extremes are possible. For example, all the

communities have probably rippled from human overfishing
in the North Pacific. From case studies such as these, ecologists
are just beginning to outline the key interactions that help
structure aquatic and terrestrial communities.

The structure of a community may be controlled bottom-up by nutrients or top-down by predators

Simplified models based on relationships between adjacent
trophic levels are useful for discussing how communities of
plants, animals, and other organisms might be organized. For
example, consider the three possible relationships between
plants (V for vegetation) and herbivores (H):

$$V \rightarrow H \qquad V \leftarrow H \qquad V \leftrightarrow H$$

The arrows indicate that a change in the biomass (total weight)
of one trophic level causes a change in the other trophic level.

interactions between trophic levels could be reciprocal (↔). The value of these simplified models is that they provide a starting point for the analysis of communities. Let's look at one example where ecologists have applied these models to counter pollution problems in lakes.

Pollution has degraded the freshwater lakes in many countries. Because many freshwater lake communities seem to be structured according to the top-down, or trophic cascade, model, ecologists have a potential method for improving water quality. In lakes with four trophic levels, for example, adding top predators should improve water quality by reducing algal populations. In lakes with three trophic levels, removing fish should improve water quality. We can summarize this strategy for lake restoration, called *biomanipulation,* with this diagram.

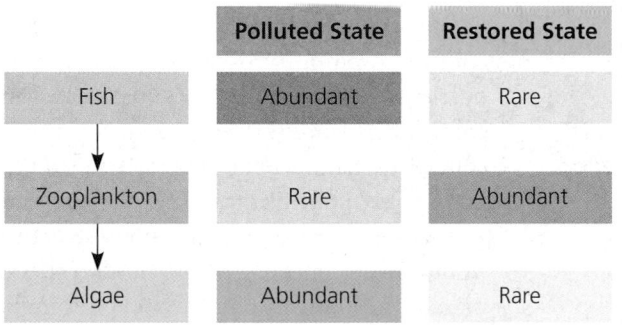

	Polluted State	Restored State
Fish	Abundant	Rare
Zooplankton	Rare	Abundant
Algae	Abundant	Rare

One of the largest biomanipulations of food webs so far was the Lake Vesijärvi experiment in southern Finland. Vesijärvi is a large (110 km²), shallow lake that was heavily polluted with city sewage and industrial wastewater until 1976. Pollution controls then stopped these inputs, and the lake began to recover in water quality. But by 1986, massive blooms of cyanobacteria began to occur. These algal blooms coincided with a very dense population of roach, a plankton-eating fish that had accumulated during the years of mineral nutrient (pollution) input. Roach eat zooplankton, and by reducing zooplankton, they also reduced herbivory on cyanobacteria and other algae, which then increased in abundance. To reverse these changes, ecologists removed 1,018 tons of fish from Lake Vesijärvi between 1989 and 1993, reducing roach to about 20% of their former abundance. At the same time, the ecologists began to stock the lake with pike perch, a predatory fish that eats roach. This added a fourth trophic level to the lake. Biomanipulation was a success in Lake Vesijärvi. The water became clear, and cyanobacterial blooms ended in 1989. The lake continues to remain clear even though the roach removal was stopped in 1993.

Communities will vary in the relative degree of top-down and bottom-up control, and we should not assume that one model will fit all communities. If humans are to manage agricultural landscapes, national parks, reservoirs, and marine fisheries, we need to understand the dynamics of the community we are attempting to manage. This is just one example of how basic ecological research can help us evaluate environmental issues.

DISTURBANCE AND COMMUNITY STRUCTURE

Decades ago, most ecologists favored the traditional view that biological communities are in a state of equilibrium, a more or less stable balance, unless seriously disturbed by human activities. This "balance of nature" view focused on interspecific competition as a key factor determining community composition and maintaining stability in communities. **Stability** in this context is the tendency of a community to reach and maintain an equilibrium, or relatively constant composition of species, in the face of disturbances. But in many communities, at least on a local scale, change seems to be more common than stability. This recent emphasis on change has led to a **nonequilibrium model** of communities, where communities are seen as constantly changing after being buffeted by disturbances. The key question here is the role of disturbances in affecting community structure and composition.

Most communities are in a state of nonequilibrium owing to disturbances

Disturbances are events, such as storms, fire, floods, droughts, overgrazing, or human activities, that damage communities, remove organisms from them, and alter resource availability. The types of disturbances and their frequency and severity vary from community to community. Storms disturb almost all communities, even those in deep oceans. Fire is a significant cause of disturbance in most terrestrial communities; grasslands and chaparral biomes are dependent on regular burning (FIGURE 53.16). Freezing is a frequent occurrence in many rivers, lakes, and ponds, and many streams and ponds are disturbed by spring flooding and seasonal drying. Disturbances often create opportunities for species that have not previously occupied habitats to become established. By gathering data from specific communities over many years, ecologists are beginning to appreciate and understand the impact of disturbances. Let's look at one example.

The marine communities called coral reefs are subject to a variety of physical disturbances associated with tropical storms. At Heron Island Reef, at the southern edge of the Great Barrier Reef, in Queensland, Australia, ecologists used sequential sets of photographs to chronicle changes in coral cover at specific sites over a 30-year period. The scientists measured the area covered by colonies of live coral to estimate the abundance of the coral animals (see Chapter 33). The researchers also measured the "recruitment" of coral animal larvae that settle and found new colonies.

Violent storms were the main type of disturbance to Heron Island Reef, and the amount of damage caused by these cyclones

FIGURE 53.16 Routine disturbance in a grassland community. Historically, prairie grasslands were often swept by fire. These photographs were taken **(a)** before, **(b)** during, and **(c)** after a controlled burn conducted by ecologists studying the long-term effects of fire on a tallgrass prairie in Kansas. (The trees in the photographs were growing along a stream and were not burned during the study.)

(a) Before a controlled burn. A prairie that has not burned for several years has a high proportion of detritus (dead grass).

(b) During. The detritus serves as fuel for fires.

(c) After. Prairies regrow rapidly after a fire burns the detritus. Approximately one month after the controlled burn, virtually all of the biomass in this prairie was living.

depended on the position of the coral colonies on the reef (FIGURE 53.17). Sites exposed more to the open ocean were more subject to wave damage than sites sheltered by the island. Five cyclones passed near Heron Island during the 30 years of study from 1962 to 1992. Of the four study areas graphed in FIGURE 53.17, only the protected area of the inner flat was relatively unaffected by cyclones. Almost every cyclone caused a reduction in coral cover in the exposed pools. The 1972 cyclone completely removed coral cover on the exposed crest ("peak" of the reef), the most severe disturbance observed during the study. Recovery on the exposed crest was slow for the next 25 years.

Recruitment rates of new corals settling on free surfaces were highly variable, and this is typical of many marine invertebrates whose larvae drift in the plankton before settling. During the 30-year Heron Island Reef study, there were no particularly good or bad years for coral recruitment for the area as a whole. Recruitment rate was partly associated with how much free space there was in different areas. Coral larvae need free space to settle because they cannot attach to other living coral or to seaweeds.

What emerges from this case study is a picture of a coral community that changes continually because of the distur-

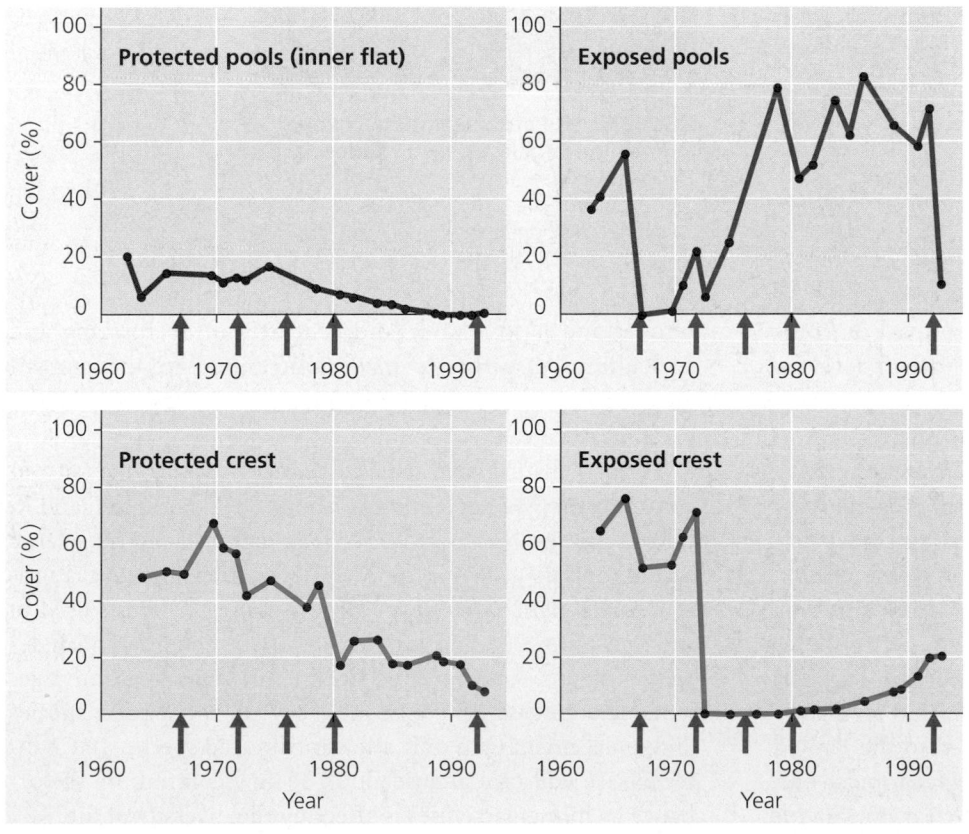

FIGURE 53.17 Storm disturbance to coral reef communities. The graphs compare how five tropical cyclones (red arrows) affected coral cover at four regions of the Heron Island Reef in Australia. Damage from cyclones was highly variable, depending on how well the island protected the four sites. (The gradual decline in coral cover at the protected sites was a normal effect of the coral's upward growth, which increased the exposure of these sites to air over the 30 years of the study.)

bance caused by tropical cyclones and the internal processes of growth and recruitment. Coral reefs are good examples of nonequilibrium communities. And there is mounting evidence that some amount of nonequilibrium resulting from disturbance is the norm for *most* communities. Communities are often in some state of recovery from disturbance.

We tend to think of disturbance as having a negative impact on communities, but this is not always the case. Small-scale disturbances sometimes enhance environmental patchiness, which can be important to the maintenance of species diversity in a community. Frequent small-scale disturbances can also prevent large-scale disturbances. The major fires that occurred in Yellowstone National Park during the summer of 1988 are an example of what can happen in the absence of smaller disturbances. Much of this park was dominated by lodgepole pine, a tree that requires the rejuvenating influence of periodic fires. Lodgepole cones remain closed until exposed to intense heat. When a forest fire destroys the parent trees, the cones open and the seeds are released. The new generation of lodgepole pines can then thrive on nutrients released from the burned trees and on the direct sunlight that was blocked by the old forest. Lodgepole pines that are over 100 years old become increasingly flammable, but for decades, fire suppression by humans had prevented small lightning-induced fires in Yellowstone Park that would have resulted in patches of less flammable trees. As a result, by 1988, about one-third of the Yellowstone trees were 250 to 300 years old. The drought conditions of 1988, combined with the fuel that had accumulated in the forests, resulted in a large-scale fire that destroyed the old lodgepole forests. Demonstrating that communities can often respond very rapidly to even such massive disturbance, burned areas in Yellowstone were largely covered with new vegetation the following year (FIGURE 53.18).

(a) Soon after fire. As this photo taken soon after the fire shows, the burn left a patchy landscape; note the unburned trees in the background.

(b) One year after fire. This photo of the same general area taken the following year indicates how rapidly the community began to recover. A variety of herbaceous plants, different from the species that inhabited the floor of the former forest, cover the ground.

FIGURE 53.18 Patchiness and recovery following a large-scale disturbance. The Yellowstone National Park fire of 1988 destroyed large areas of coniferous forests dominated by lodgepole pines.

Humans are the most widespread agents of disturbance

Of all the animals, humans have the greatest impact on communities worldwide. Logging and clearing for urban development, mining, and farming have reduced large tracts of forests to small patches of disconnected woodlots in many parts of the United States and throughout Europe. Similarly, agricultural development has disrupted what were once the vast grasslands of the North American prairie. After forests are clear-cut and left alone, weedy and shrubby vegetation often colonizes the area and dominates it for many years. This type of vegetation is also found extensively in agricultural fields that are no longer under cultivation and in vacant lots and construction sites that are periodically cleared. Human disturbance of communities is by no means limited to the United States and Europe; nor is it a recent problem. Tropical rain forests are quickly disappearing as a result of clear-cutting for lumber and pastureland. And centuries of overgrazing and agricultural disturbance have contributed to the current famine in parts of Africa by turning seasonal grasslands into great barren areas.

Human disturbance usually reduces species diversity in communities. We currently use about 60% of Earth's land in one way or another, mostly as cropland, forest, and rangeland. Most crops are grown in monocultures, intensive cultivations of a single plant variety over large areas. Even forests used to produce pulpwood and lumber are often replanted in single-species stands. And the effects of intensive grazing on rangelands often include the removal of several native plant species and replacement with only a few introduced species. In Chapter 55, we will take a closer look at how community disturbance by human activities is affecting the diversity of life.

Ecological succession is the sequence of community changes after a disturbance

Changes in community composition and structure are most apparent after some disturbance, such as a large fire or a volcanic eruption, strips away the existing vegetation. The disturbed area may be colonized by a variety of species, which are gradually replaced by a succession of other species. Such transitions in species composition over ecological time represent a process called **ecological succession.** The process is called **primary succession** if it begins in a virtually lifeless area where soil has not yet formed, such as on a new volcanic island or on the rubble (moraine) left behind by a retreating glacier. Often the only life-forms initially present are autotrophic bacteria. Lichens and mosses, which grow from windblown spores, are commonly the first macroscopic photosynthesizers to colonize such areas. Soil develops gradually, as rocks weather and organic matter accumulates from the decomposed remains of the early colonizers. Once soil is present, the lichens and mosses are usually overgrown by other plants, such as grasses, shrubs, and trees that sprout from seeds blown in from nearby areas or carried in by animals. Eventually, an area may be colonized by plants that become the community's prevalent form of vegetation. For primary succession to produce such a community may take hundreds or thousands of years.

Secondary succession occurs where an existing community has been cleared by some disturbance that leaves the soil intact. Succession in Yellowstone following the 1988 fires is an example (see FIGURE 53.18). Often the area begins to return to something like its original state. For instance, forested areas that are cleared for farming will, if abandoned, undergo secondary succession and may eventually return to forest. The earliest plants to recolonize such an area are often herbaceous species that grow from windblown or animal-borne seeds. If the area is not burned or heavily grazed, woody shrubs may in time replace most of the herbaceous species, and forest trees may eventually replace most of the shrubs.

Three key processes may be involved in succession between the early arrivals and the later arriving species. The early arrivals may *facilitate,* or contribute to, the appearance of the later species by making the environment more favorable for the later species. For example, they might make the soil more fertile. Alternatively, the early species may *inhibit* establishment of the later species, so that the later species colonize successfully in spite of, rather than because of, the activities of the early species. Finally, the early species may be completely independent of the later species, so that they *tolerate* the later species but do not help or hinder colonization by the later arrivals. Let's look at how these various processes contribute to primary succession in one specific example.

During the past 300 years, there has been a gradual retreat of glaciers in the Northern Hemisphere. As the glaciers retreat, they leave moraines. Researchers can determine the age of these postglaciation areas of moraine from the age of the new trees growing on the moraine or, in the last 80 years, by direct observation. Ecologists have conducted the most extensive research on moraine succession at Glacier Bay, in southeastern Alaska. Since about 1760, the glaciers there have retreated about 98 km, an extraordinary retreat rate of almost 400 m per year (FIGURE 53.19). Rocky moraines form a nearly pristine but harsh environment for plants to colonize.

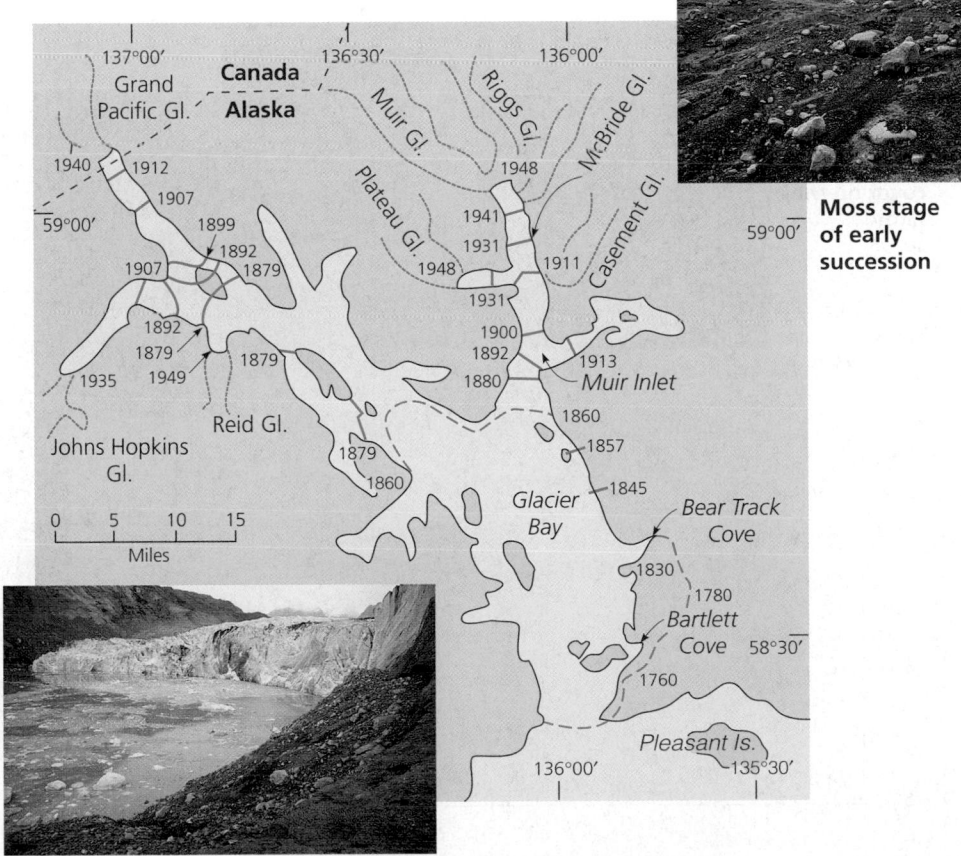

Moss stage of early succession

FIGURE 53.19 A glacial retreat in southeastern Alaska. The dated locations chronicle recession of the glacier since 1760. The broken lines show the approximate edge of the ice in 1760 and 1860, based on historical descriptions. As the ice retreats, it leaves moraines along the edge of the bay (left inset). Primary succession occurs on the moraines (right inset).

Retreating glacier with moraine to right

Table 53.2 The Pattern of Succession on Moraines in Glacier Bay

Years after Deglaciation	Dominant Plant	Other Common Species
0–30	*Dryas*	Fireweed, willows, mosses, cottonwoods
30–80	Alder	Willows
80–200	Sitka spruce	Alder, willows
200–300	Sitka spruce, western hemlock	Mountain hemlock
> 300	Sphagnum moss (in flat areas)	Bog plants

TABLE 53.2 summarizes the pattern of succession on the moraines in Glacier Bay. The exposed moraine is colonized first by pioneering plant species—mosses, fireweed, *Dryas* (a herbaceous angiosperm), willows, and cottonwood. In a few decades, the area is invaded by alder (*Alnus*), which eventually forms dense, pure thickets up to 9 m tall. These alder stands are later invaded by Sitka spruce, which, after another century, forms a dense forest. Western hemlock and mountain hemlock invade the spruce stands, and after another century, the community is a spruce-hemlock forest. This forest, however, will endure on well-drained slopes only. In areas of poor drainage, the forest floor of this spruce-hemlock forest is invaded by sphagnum mosses, which hold large amounts of water and acidify the soil. With the spread of conditions associated with sphagnum, the trees die because the soil is waterlogged and too oxygen-deficient to sustain the trees' roots, and the area becomes a sphagnum bog. Thus, by about 300 years after glacial retreat, the vegetation consists of sphagnum bogs on the poorly drained flat areas and spruce-hemlock forest on the well-drained slopes.

How is succession on glacial moraines related to the environmental impact of the changing vegetation? The bare soil exposed as the glacier retreats is quite basic, with a pH of 8.0–8.4 because of the carbonates contained in the parent rocks. The soil pH falls rapidly with the arrival of vegetation, and the rate of change depends on the vegetation type. The most striking change is caused by alder, which reduces the pH from 8.0 to 5.0 in 30 to 50 years. The leaves of alder are slightly acidic, and as they decompose, they become even more acidic. As the spruce begins to take over from the alder, the pH stabilizes at about 5.0, and it does not change for the next 150 years.

The soil concentrations of mineral nutrients also show marked changes with time. FIGURE 53.20, for example, shows the changes in soil nitrogen levels. One of the characteristic features of the bare soil after glacial retreat is its low nitrogen content. Almost all the pioneer species begin the succession

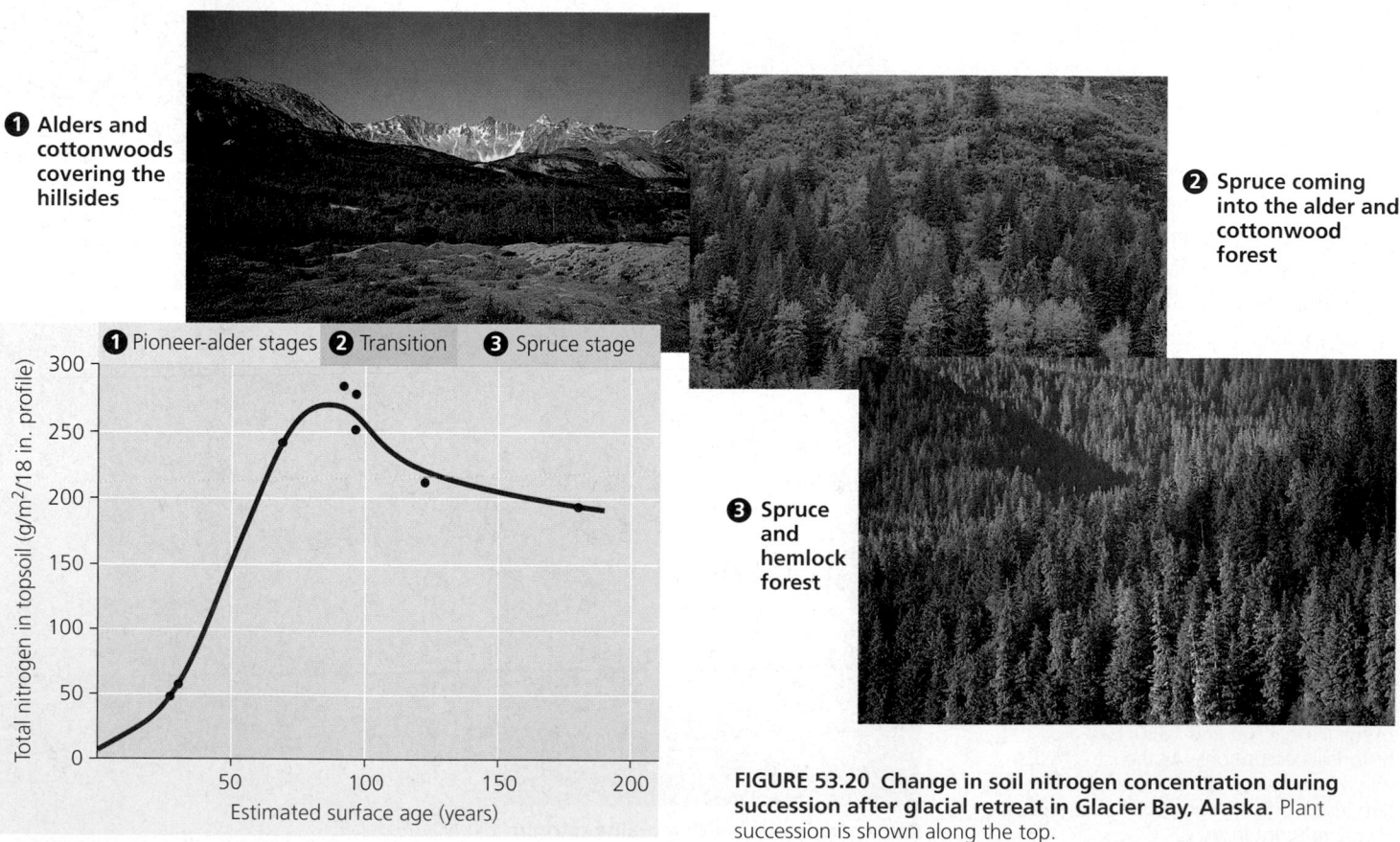

❶ Alders and cottonwoods covering the hillsides

❷ Spruce coming into the alder and cottonwood forest

❸ Spruce and hemlock forest

❶ Pioneer-alder stages ❷ Transition ❸ Spruce stage

FIGURE 53.20 Change in soil nitrogen concentration during succession after glacial retreat in Glacier Bay, Alaska. Plant succession is shown along the top.

with very poor growth and yellow leaves due to inadequate nitrogen supply. The exceptions to this are *Dryas* and, particularly, alder; these species have symbiotic bacteria that fix atmospheric nitrogen (see Chapter 37). Soil nitrogen increases rapidly during the alder stage of succession. The spruce forest develops by spending the capital of soil nitrogen accumulated by the alder. But because spruce trees do not have nitrogen-fixing symbionts, the concentration of soil nitrogen declines again.

Both inhibition and facilitation factor into primary succession on glacial moraines. One way to test experimentally for these processes is to sow seeds of the later successional species into the earlier stages of succession. When seeds of alder and Sitka spruce were sown into the early successional stages at Glacier Bay, they were inhibited in their germination, and the survival of seedlings that did geminate was reduced. Inhibition seems to be the dominant process in the early successional stages. But once alder and *Dryas* have enriched the soil with nitrogen, Sitka spruce uses this nitrogen for growth, so it is facilitated by the pioneer species. The pioneer plants alter the soil properties, which in turn permit new species to grow, and these species in turn alter the environment in different ways, contributing to succession.

BIOGEOGRAPHIC FACTORS AFFECTING THE BIODIVERSITY OF COMMUNITIES

Ecological communities vary extensively in the number of species they contain. Two key factors correlated with a community's **biodiversity,** or species diversity, are its size and geographic location. In the 1850s, both Alfred Wallace and Charles Darwin pointed out that plant and animal life was generally more abundant and varied in the tropics than in other parts of the globe. Darwin and Wallace also noted that small or remote islands have fewer species than large islands or those nearer continents. Such observations imply that biogeographic patterns in biodiversity conform to a set of basic principles rather than being accidents of evolutionary history. How can we explain these large-scale patterns in the diversity of life? We must first learn how to measure the biodiversity of a community.

Community biodiversity measures the number of species and their relative abundance

The biodiversity of a community—the variety of different kinds of organisms that make up the community—has two components. One is **species richness,** or the total number of different species in the community. The other is the **relative abundance** of the different species. For example, imagine two

small forest communities, each with 100 individuals distributed among four different tree species (A, B, C, and D) as follows:

Community 1: 25A, 25B, 25C, 25D

Community 2: 80A, 10B, 5C, 5D

The species richness is the same for both communities because they both contain four species, but the relative abundance is very different (FIGURE 53.21). You would easily notice the four different types of trees in community 1, but without looking carefully, you might see only the abundant species A in the second forest. Most observers would intuitively describe community 1 as the more diverse of the two communities. Indeed, ecologists measure biodiversity as **heterogeneity,** which considers *both* diversity factors: richness and relative abundance.

Censusing the species in a community, a seemingly straightforward activity, is easier said than done, especially for small or highly motile organisms. Various sampling techniques can be employed in such censuses. Due to the fact that most species in a community are relatively rare, sample sizes are

Community 1
A: 25%   B: 25%   C: 25%   D: 25%

Community 2
A: 80%   B: 5%   C: 5%   D: 10%

FIGURE 53.21 Which forest is more diverse? Both forests have the same four tree species (A, B, C, and D). Thus, the two communities are equal in their species richness of trees. But if we factor in the relative abundance of species, then community 1 seems more diverse because of the more equitable representation of the different tree species. Ecologists would say that community 1 has greater heterogeneity, a measure of diversity that includes both species number (richness) and relative abundance.

generally small, resulting in the possibility of large sampling errors. FIGURE 53.22 illustrates this problem for a study of butterfly and moth biodiversity in a community. Measuring biodiversity is hard enough when censusing butterflies, but is much harder when censusing the less visible members of communities, such as mites and nematodes.

Challenging as it is to measure biodiversity in a community—even for just one taxonomic category, such as butterflies—it is very important work. Measuring biodiversity is essential for conservation biology, since we need an inventory of what we hope to protect. Though there is a tendency in biology to focus on individual species, community ecologists tend to lump the species into broader categories, such as the bird species or the tree species of an area. We can use this community based approach to look for large-scale patterns in the biogeography of biodiversity.

Species richness generally declines along an equatorial-polar gradient

Tropical habitats support much larger numbers of species of plants, animals, and other organisms than do temperate and polar regions. A few examples from biodiversity studies will illustrate this global gradient. A 6.6 ha (1 hectare = 10,000 m^2) area in Sarawak (in Malaysia) contained 711 tree species. Contrast this richness with a deciduous forest in Michigan, which contained 10 to 15 species on a plot of 2 ha. And the whole of Europe north of the Alps has only 50 tree species. Ants are also much more diverse in the tropics. There are over 200 species of ants in Brazil, 73 species in Iowa, and 7 species in Alaska. There are 293 species of snakes in Mexico, 126 in the United States, and 22 in Canada. FIGURE 53.23 maps the number of breeding land bird species in different parts of North and Central America. Over 600 bird species occur in Central America, while arctic Canada

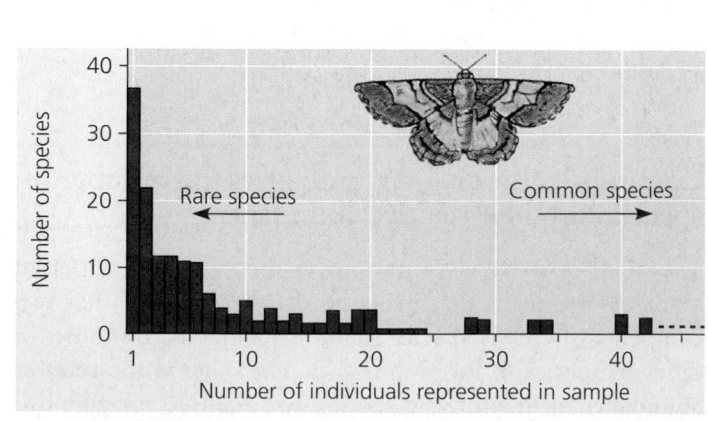

FIGURE 53.23 Geographic pattern of species richness in the land birds of North and Central America. Contour lines give the number of species present.

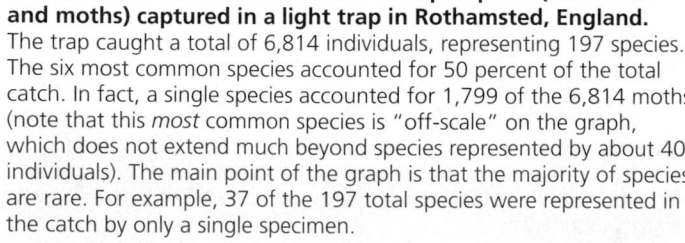

FIGURE 53.22 Relative abundance of Lepidoptera (butterflies and moths) captured in a light trap in Rothamsted, England. The trap caught a total of 6,814 individuals, representing 197 species. The six most common species accounted for 50 percent of the total catch. In fact, a single species accounted for 1,799 of the 6,814 moths (note that this *most* common species is "off-scale" on the graph, which does not extend much beyond species represented by about 40 individuals). The main point of the graph is that the majority of species are rare. For example, 37 of the 197 total species were represented in the catch by only a single specimen.

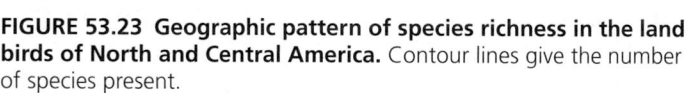

has less than 40 species of birds. Freshwater fishes have their greatest diversity in tropical rivers and lakes. Lakes Victoria, Tanganyika, and Malawi in East Africa contain about 1,450 species of fishes. Over 1,000 species of fishes have been found in the Amazon River in South America, and exploration is still incomplete in this region. In contrast, Central America has 456 fish species, and the Great Lakes of North America have 173 species. Lake Baikal, in Asia, has 39 fish species; and Great Bear Lake, in northwestern Canada, has only 14 species of fishes.

What causes these equatorial-polar gradients in species richness? The two key factors are probably evolutionary history and climate. On the scale of evolutionary time, species diversity may increase in a community because more speciation events have occurred. And tropical communities are generally older than temperate or polar communities. This age difference is partly a consequence of the much longer growing season in the tropics. The growing season in tropical forests is about five times longer than the growing season in the tundra communities of high latitudes. In effect, biological time, and hence time for speciation, runs about five times faster in the tropics than near the poles. And many polar and temperate communities have had to "start over" several times as a result of major disturbance in the form of glaciations.

Most ecologists see climate as the major explanation for the latitudinal gradient in biodiversity. And the two main climatic factors correlated with biodiversity are solar energy input and water availability. Solar energy is the source of both light and heat, two variables that impact the growth of vegetation. The water factor includes both precipitation and humidity.

One way to integrate the energy and water factors of climate is to measure the rate of a community's evapotranspiration, the evaporation of water from soil plus the transpiration of water from plants (see Chapter 36). Evapotranspiration is much higher in hot areas with abundant rainfall than in areas with low temperatures or low precipitation. The species richness of both plants and animals correlates with evapotranspiration (FIGURE 53.24).

Species richness is related to a community's geographic size

One of the earliest biodiversity patterns that scientists recognized is called the **species-area curve,** which quantifies what probably seems obvious: The larger the geographic area of a community we sample, the greater the number of species. Alexander von Humboldt first described this relationship in 1807, based on his explorations. The probable explanation for species-area curves is that larger areas offer a greater diversity of habitats and microhabitats (see Chapter 50). In conservation biology, developing species-area curves for key taxa in a community makes it possible to predict how loss of a certain area of habitat is likely to affect biodiversity.

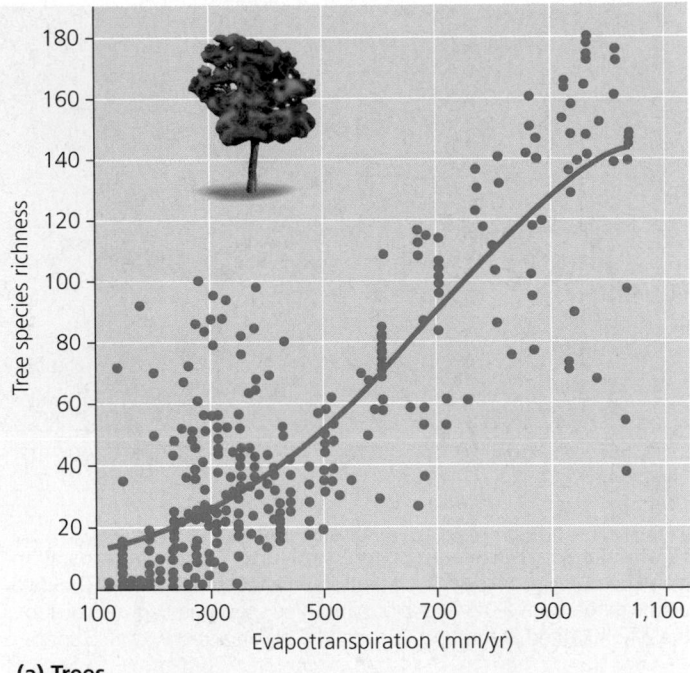

(a) Trees

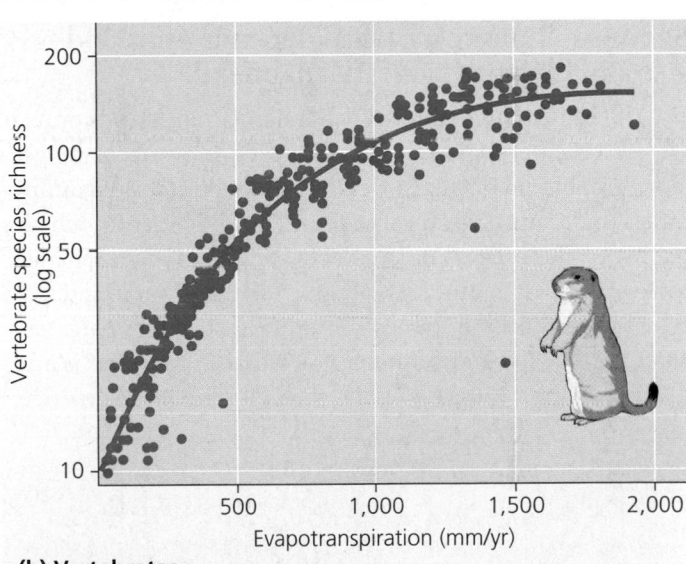

(b) Vertebrates

FIGURE 53.24 Energy and species richness. Species richness of **(a)** trees and **(b)** vertebrates from North America are related to annual available energy at each site, measured by evapotranspiration (which combines solar radiation and temperature). The evapotranspiration values are expressed as their rainfall equivalents in millimeters per year.

FIGURE 53.25 (p. 1194) is a species-area curve for the breeding birds of North America. The slope indicates the extent to which species richness increases with community area. The slopes of different species-area curves vary, depending on the taxon being sampled in the biodiversity survey and the type of community. But the basic concept of increasing diversity with increasing area applies in a variety of situations, from surveys of ant diversity in New Guinea to the number of plant species on islands of different sizes. In fact, island biogeography provides some of the best examples of species-area curves.

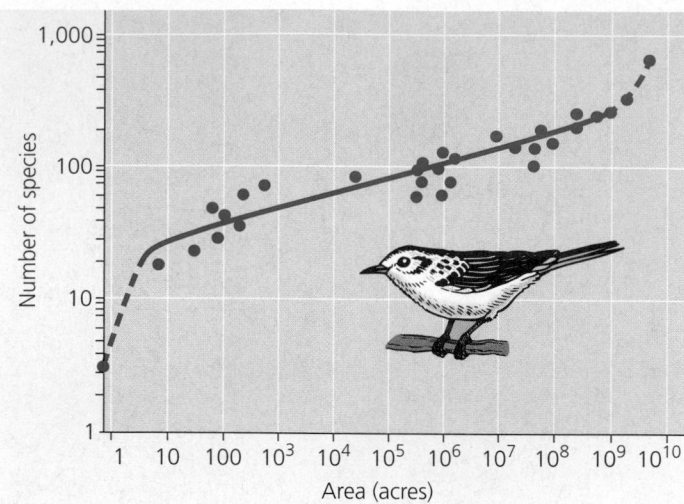

FIGURE 53.25 Species-area curve for North American birds. Both area and number of species are plotted on a logarithmic scale. The data points range from a 0.5-acre plot with three species in Pennsylvania to the whole United States and Canada (4.6 billion acres) with 625 species.

Species richness on islands depends on island size and distance from the mainland

THE PROCESS OF SCIENCE

Because of their limited size and isolation, islands provide excellent opportunities for studying some of the biogeographic factors that affect the species diversity of communities. By "islands," we mean not only oceanic islands, but also habitat islands on land, such as lakes, mountain peaks separated by lowlands, or natural woodland fragments surrounded by areas disturbed by humans—in other words, any patch surrounded by an environment not suitable for the "island" species. In the 1960s, American ecologists Robert MacArthur

and E. O. Wilson developed a general hypothesis of island biogeography to identify the important determinants of species diversity on an island with a given set of physical characteristics.

Imagine that a newly formed oceanic island will receive colonizing species from a distant mainland. Two factors will determine the number of species that eventually inhabit the island: the rate at which new species immigrate to the island and the rate at which species become extinct on the island. And two physical features of the island affect immigration and extinction rates: its size and its distance from the mainland. Small islands will generally have lower immigration rates, because potential colonizers are less likely to reach a small island. For example, birds blown out to sea by a storm are more likely to land by chance on a larger island than on a small one. Small islands will also have higher extinction rates. They generally contain fewer resources and less diverse habitats for colonizing species to partition, increasing the likelihood of local extinctions. Distance from the mainland is also important; for two islands of equal size, a closer island will have a higher immigration rate than one farther away.

The immigration and extinction rates are also affected at any given time by the number of species already present on the island. As the number of species on the island increases, the immigration rate of new species decreases, because any individual reaching the island is less likely to represent a species that is not already present. At the same time, as more species inhabit an island, extinction rates on the island increase because of the greater likelihood of competitive exclusion.

These relationships that make up MacArthur and Wilson's hypothesis of island biogeography are summarized in FIGURE 53.26. Immigration and extinction rates are plotted as a function of the number of species present on the island. The main

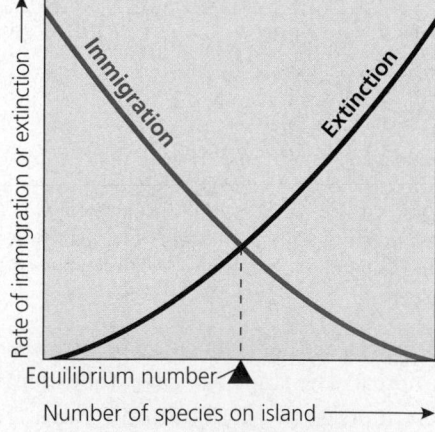

(a) Immigration and extinction rates. The equilibrium number (black triangle) of species on an island represents a balance between the immigration of new species to the island and the extinction of species that are already there.

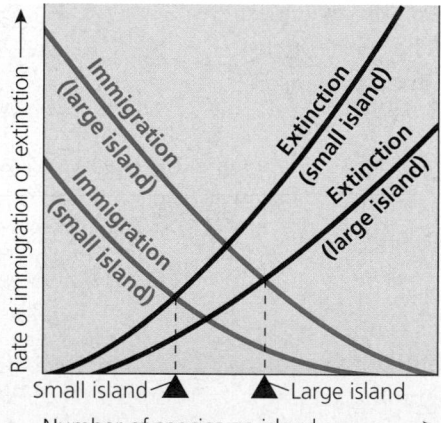

(b) Effect of island size. Large islands may ultimately have a larger equilibrium number of species than small islands because immigration rates tend to be higher and extinction rates tend to be lower on large islands.

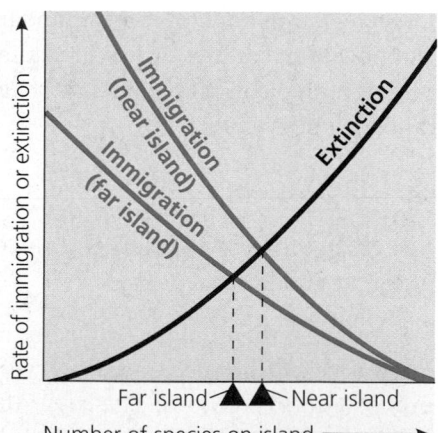

(c) Effect of distance from mainland. Although extinction rates do not differ with an island's distance from a mainland source of species, near islands tend to have larger equilibrium numbers of species than far islands because immigration rates to near islands are higher than those for more distant ones.

FIGURE 53.26 The hypothesis of island biogeography.

prediction of this model is that eventually an equilibrium will be reached where the rate of species immigration matches the rate of species extinction. The number of species at this equilibrium point is correlated with the island's size and distance from the mainland. Any ecological equilibrium, of course, is dynamic; immigration and extinction continue, and the exact species composition may change over time.

MacArthur and Wilson's studies of the diversity of plants and animals on many island chains, such as the Galápagos Islands, support the prediction that species richness increases with island size, in keeping with species-area theory (FIGURE 53.27). Species counts also fit the prediction that the number of species decreases with increasing remoteness of the island.

In the past several decades, the island biogeography hypothesis has come under considerable fire as an oversimplification. Its predictions of equilibria in the species composition of communities may apply in only a limited number of cases and over relatively short time periods, where colonization is the main process affecting species composition. Over a longer period, abiotic disturbances such as storms, adaptive evolutionary changes, and speciation generally alter the species composition and community structure on islands. More important than whether the island biogeography hypothesis has widespread applicability is that it stimulated discussion and research on the effects of habitat size on species diversity, a topic of vital importance for conservation biology, to which we will turn when we discuss habitat fragmentation and biodiversity loss in Chapter 55.

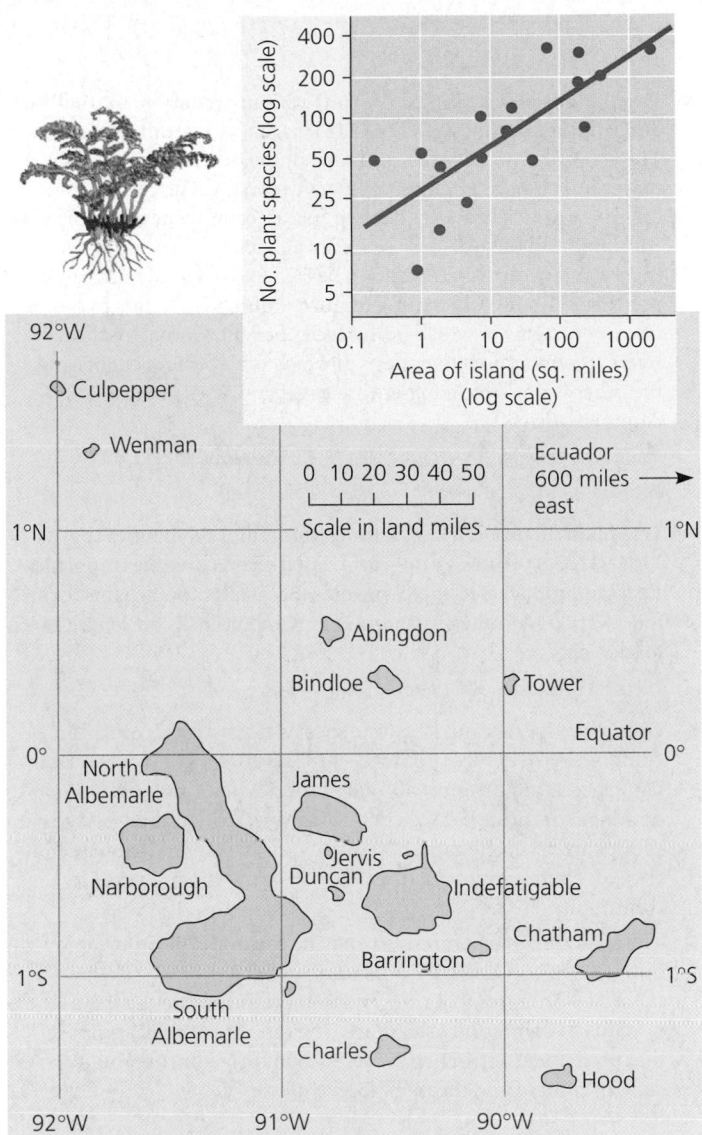

FIGURE 53.27 Number of plant species on the Galápagos Islands in relation to the area of the island.

CHAPTER 53 REVIEW

Go to the Campbell Biology website (www.campbellbiology.com) to explore an interactive version of the Chapter Review.

Summary of Key Concepts

WHAT IS A COMMUNITY?

- **Contrasting views of communities are rooted in the individualistic and interactive hypotheses (pp. 1175–1176, FIGURE 53.1)** The individualistic hypothesis proposes that communities are chance assemblages of independently distributed species with the same abiotic requirements. The interactive hypothesis states that the species within a community are locked into biotic interactions.

- **The debate continues with the rivet and redundancy models (p. 1176)** The rivet model suggests that all the species in a community are linked together in a tight web of interactions, so that the loss of even a single species has strong repercussions for the community. The redundancy model proposes that if a species is lost from a community, other species will fill in the gap.

INTERSPECIFIC INTERACTIONS AND COMMUNITY STRUCTURE

■ **Populations may be linked by competition, predation, mutualism, and commensalism (pp. 1176–1181, TABLE 53.1, FIGURES 53.2–53.9)** The ecological niche is the sum total of the organism's use of the biotic and abiotic resources in its environment. The competitive exclusion principle states that two species cannot coexist in the same community if their niches are identical. Predation includes herbivory and parasitism. Predation has led to diverse adaptations in predators and prey, including mimicry. Mutualism refers to symbiotic interactions in which both species benefit. Commensalism refers to symbiotic interactions in which one species benefits and the other is not affected; there are few if any cases of pure commensalism.

Web/CD Activity 53A: *Interspecific Interactions*
Biology Labs On-Line: *PopulationEcologyLab*

■ **Trophic structure is a key factor in community dynamics (pp. 1181–1183, FIGURES 53.10–53.13)** Food chains link the trophic levels from producers to top carnivores. Branching food chains form food webs. A community's total energy input limits the length of its food chains.

Web/CD Activity 53B: *Food Webs*

■ **Dominant species and keystone species exert strong controls on community structure (pp. 1183–1185, FIGURES 53.14, 53.15)** Dominant species are the most abundant species in a community, and dominance is achieved by having high competitive ability. Keystone species are relatively rare species that exert a disproportionate influence on community structure. They are often top predators in a community.

■ **The structure of a community may be controlled bottom-up by nutrients or top-down by predators (pp. 1185–1186)** The bottom-up model proposes that nutrients and producers are the main determinants of community structure. The top-down model proposes that predators control herbivores who in turn control producers, so control comes from the trophic level above.

DISTURBANCE AND COMMUNITY STRUCTURE

■ **Most communities are in a state of nonequilibrium owing to disturbances (pp. 1186–1188, FIGURES 53.16–53.18)** Increasingly, evidence suggests that disturbance and nonequilibrium instead of stability and equilibrium are the norm for most communities.

■ **Humans are the most widespread agents of disturbance (p. 1188)** Among all animals, humans create the greatest disturbances in communities, usually reducing species diversity. Humans also prevent some naturally occurring disturbances such as fire, which can be important to community structure.

■ **Ecological succession is the sequence of community changes after a disturbance (pp. 1189–1191, FIGURES 53.19, 53.20)** Primary succession occurs where no soil exists when succession begins; secondary succession begins in an area where soil remains after a disturbance.

Web/CD Activity 53C: *Primary Succession*

BIOGEOGRAPHIC FACTORS AFFECTING THE BIODIVERSITY OF COMMUNITIES

■ **Community biodiversity measures the number of species and their relative abundance (pp. 1191–1192, FIGURES 53.21, 53.22)**

The simplest measure of biodiversity is the number of species in a community, or species richness. Species may be rare or common in a community, so relative abundance is another factor in biodiversity.

Web/CD Case Study in the Process of Science: *How Are Impacts on Community Diversity Measured?*

■ **Species richness generally declines along an equatorial-polar gradient (pp. 1192–1193, FIGURES 53.23, 53.24)** Species richness is much greater in the tropics than in temperate and polar regions. Climate is the best explanation for this biodiversity gradient through its impact on energy (heat and light) and water.

■ **Species richness is related to a community's geographic size (p. 1193, FIGURE 53.25)** This ecological principle is formalized as the species-area curve.

■ **Species richness on islands depends on island size and distance from the mainland (pp. 1194–1195, FIGURES 53.26, 53.27)** A hypothesis of island biogeography maintains that species richness on an ecological island levels off at some dynamic equilibrium point, where new immigrations are balanced by extinctions. The hypothesis predicts that species richness is directly proportional to island size and inversely proportional to distance of the island from the source of colonizers.

Web/CD Activity 53D: *Exploring Island Biogeography*

Self-Quiz

1. The concept of trophic structure of a community emphasizes the
   a. prevalent form of vegetation.
   b. keystone predator.
   c. feeding relationships within a community.
   d. effects of coevolution.
   e. species richness of the community.

2. According to the concept of competitive exclusion,
   a. two species cannot coexist in the same habitat.
   b. extinction or emigration are the only possible results of competitive interactions.
   c. competition within a population results in the success of the best-adapted individuals.
   d. two species cannot share the exact same niche in a community.
   e. resource partitioning will allow a species to utilize all the resources of its niche.

3. The effect of a keystone predator within a community may be to
   a. competitively exclude other predators from the community.
   b. maintain species diversity by preying on the prey species that is the dominant competitor.
   c. increase the relative abundance of other predators.
   d. encourage the coevolution of predator and prey adaptations.
   e. create nonequilibrium in species diversity.

4. Food chains are relatively short in communities because
   a. two herbivore species may not feed on the same plant species.
   b. local extinction of one species dooms all the other species in a food web.
   c. energy is lost as it passes from one trophic level to the next higher level.
   d. very few predatory species have evolved.
   e. most plant species are inedible.

5. According to the rivet model of community organization,
   a. two closely related species cannot coexist in the same community.
   b. extinction is rare in well-organized communities.
   c. species can be easily replaced if one should be driven extinct by human actions.
   d. all species in a natural community contribute to its integrity.
   e. communities are loosely structured groups of individualistic species with similar abiotic requirements.

6. An example of cryptic coloration is the
   a. green color of a plant.
   b. bright markings of a poisonous tropical frog.
   c. stripes of a skunk.
   d. mottled coloring of moths that rest on lichens.
   e. bright colors of an insect-pollinated flower.

7. An example of Müllerian mimicry is
   a. a butterfly that resembles a leaf.
   b. two poisonous frogs that resemble each other in coloration.
   c. a minnow with spots that look like large eyes.
   d. a beetle that resembles a scorpion.
   e. a carnivorous fish with a wormlike tongue that lures prey.

8. Predation and parasitism are similar in that both can be characterized as
   a. +/+ interactions.
   b. +/− interactions.
   c. +/0 interactions.
   d. −/− interactions.
   e. symbiotic interactions.

9. Which of the following is the best explanation for the finding that equatorial (tropical) regions have the greatest species richness?
   a. the species-area curve that predicts high richness in large areas
   b. a climate with high levels of solar radiation and water availability
   c. the increased speed of speciation due to higher temperatures in the region
   d. the inverse relationship between evapotranspiration and biodiversity
   e. the greater immigration rate and lower extinction rate found on large tropical islands close to the mainland

10. According to the hypothesis of island biogeography, species richness would be greatest on an island that is
    a. small and remote.
    b. large and remote.
    c. large and close to a mainland.
    d. small and close to a mainland.
    e. environmentally homogeneous.

11. How could a community appear to have relatively little diversity even though it is rich in species?

12. What is a possible advantage to a keystone predator of being specialized to feed mainly on those prey species that are otherwise the most successful among potential prey species?

13. What is the main abiotic factor that distinguishes primary from secondary succession?

14. Explain the distinction between habitat and niche.

15. How is the concept of species-area curves incorporated into the island biogeography hypothesis?

16. What are the trophic levels for a human eating a cheese sandwich?

Go to the website or CD-ROM for more quiz questions.

Evolution Connection

Explain why adaptations of organisms to interspecific interactions do not necessarily represent coevolution. What would a researcher have to demonstrate about an interaction between two species to make a convincing case for coevolution?

The Process of Science

An ecologist studying plants in the desert performed the following experiment. She staked out two identical plots, including a few sagebrush plants and numerous small, annual wildflowers. She found the same five wildflower species in roughly equal numbers on both plots. She then enclosed one of the plots with a fence to keep out kangaroo rats, the most common grain eaters of the area. After two years, four of the wildflower species were no longer present in the fenced plot, but one species had increased drastically. The control plot had not changed in species diversity. Using the principles of community ecology, propose a hypothesis to explain her results. What additional evidence would support your hypothesis?

Measure how human activity affects fly larvae in river environments in the Case Study in the Process of Science, available on the website and CD-ROM.

Science, Technology, and Society

By 1935, hunting and trapping had eliminated wolves from the United States except for Alaska. Because wolves have since been protected as an endangered species, they have moved south from Canada and have become reestablished in the Rocky Mountains and northern Great Lakes. Conservationists who would like to speed up this process have reintroduced wolves into Yellowstone National Park. Local ranchers are opposed to bringing back the wolves because they fear predation on their cattle and sheep. What are some reasons for reestablishing wolves in Yellowstone National Park? What effects might the reintroduction of wolves have on the ecological communities in the region? What might be done to mitigate the conflicts between ranchers and wolves?

Answers: 1. c; 2. d; 3. b; 4. c; 5. d; 6. d; 7. b; 8. b; 9. b; 10. c; 11. This would be the case if one or a few of the species accounted for almost all the organisms in the community, with the other species being rare. 12. The most competitive prey species probably represent the most abundant and dependable food source for the predator. 13. Absence (primary) versus presence (secondary) of soil at the onset of succession. 14. An organism's habitat is where it lives. The organism's niche includes its habitat, but also consists of its trophic relationships, its symbiotic relationships, and other features of its environmental interactions. 15. The island biogeography hypothesis includes the idea that species richness on islands is correlated with their size. 16. Primary consumer of plant products, such as the flour of the bread, and secondary consumer via a dairy product, cheese.

CHAPTER 54

ECOSYSTEMS

THE ECOSYSTEM APPROACH TO ECOLOGY

- Trophic relationships determine the routes of energy flow and chemical cycling in an ecosystem
- Decomposition connects all trophic levels
- The laws of physics and chemistry apply to ecosystems

PRIMARY PRODUCTION IN ECOSYSTEMS

- An ecosystem's energy budget depends on primary production
- In aquatic ecosystems, light and nutrients limit primary production
- In terrestrial ecosystems, temperature, moisture, and nutrients limit primary production

SECONDARY PRODUCTION IN ECOSYSTEMS

- The efficiency of energy transfer between trophic levels is usually less than 20%
- Herbivores consume a small percentage of vegetation: the green world hypothesis

THE CYCLING OF CHEMICAL ELEMENTS IN ECOSYSTEMS

- Biological and geologic processes move nutrients between organic and inorganic compartments
- Decomposition rates largely determine the rates of nutrient cycling
- Nutrient cycling is strongly regulated by vegetation

HUMAN IMPACT ON ECOSYSTEMS AND THE BIOSPHERE

- The human population is disrupting chemical cycles throughout the biosphere
- Combustion of fossil fuels is the main cause of acid precipitation
- Toxins can become concentrated in successive trophic levels of food webs
- Human activities may be causing climate change by increasing carbon dioxide concentration in the atmosphere
- Human activities are depleting atmospheric ozone

An ecosystem *consists of all the organisms living in a community as well as all the abiotic factors with which they interact. As with populations and communities, the boundaries of ecosystems are usually not discrete. Ecosystems can range from a microcosm, such as the terrarium you see here, to lakes and forests. Many ecologists regard the entire biosphere as a global ecosystem, a composite of all the local ecosystems on Earth. The biosphere, or whole-earth ecosystem, is the most inclusive level in the hierarchy of biological organization (see* FIGURE 1.2*).*

The dynamics of an ecosystem involve two processes that cannot be fully described at lower organizational levels: energy flow and chemical cycling. Energy enters most ecosystems in the form of sunlight. It is then converted to chemical energy by autotrophic organisms, passed to heterotrophs in the organic compounds of food, and dissipated in the form of heat. Chemical elements such as carbon and nitrogen are cycled between abiotic and biotic components of the ecosystem. Photosynthetic organisms assimilate these elements in inorganic form from the air, soil, and water and incorporate them into organic molecules, some of which are consumed by animals. The elements are returned in inorganic form to the air, soil, and water by the metabolism of plants and animals and by other organisms, such as bacteria and fungi, that break down organic wastes and dead organisms. The movements of energy and matter through ecosystems are related because both occur by the transfer of substances through photosynthesis and feeding relationships. However, because energy, unlike matter, cannot be recycled, an ecosystem must be powered by a continuous influx of new energy from an external source (the sun). Energy flows through ecosystems, while matter cycles within them.

This chapter describes the dynamics of energy flow and chemical cycling in ecosystems and considers some of the consequences of human intrusions into these processes.

THE ECOSYSTEM APPROACH TO ECOLOGY

Ecosystem ecologists view ecosystems as energy machines and matter processors. By grouping the species in a community into **trophic levels** of feeding relationships, we can follow the transformation of energy in the whole ecosystem and map the movements of chemical elements as they are used by the biotic community. As we discussed in Chapter 53, ecologists assign species to trophic levels on the basis of their main source of nutrition and energy.

Trophic relationships determine the routes of energy flow and chemical cycling in an ecosystem

The trophic level that ultimately supports all others in an ecosystem consists of autotrophs, the **primary producers** of the ecosystem. Most autotrophs are photosynthetic organisms that use light energy to synthesize sugars and other organic compounds, which they then use as fuel for cellular respiration and as building material for growth. Plants, algae, and photosynthetic prokaryotes are the biosphere's main autotrophs, although chemoautotrophic prokaryotes are the primary producers in certain ecosystems, such as deep-sea hydrothermal vents (see FIGURE 50.23).

Organisms in trophic levels above the primary producers are **heterotrophs,** which directly or indirectly depend on the photosynthetic output of primary producers. Herbivores, which eat primary producers (plants or algae), are the **primary consumers.** Carnivores that eat herbivores are **secondary consumers,** and carnivores that eat other carnivores are **tertiary consumers.** Another important group of heterotrophs consists of the detritivores. **Detritivores,** or **decomposers,** are consumers that get their energy from **detritus,** which is nonliving organic material, such as the remains of dead organisms, feces, fallen leaves, and wood. Decomposers play a central role in material cycling (FIGURE 54.1).

Decomposition connects all trophic levels

The fungi, bacteria, invertebrates, and vertebrates that feed as detritivores often form a major link between the primary producers and the consumers in an ecosystem. In streams, for example, much of the organic material that is used by consumers is supplied by terrestrial plants and enters the ecosystem as leaves and other debris that fall into the water or are washed in by runoff. A crayfish might feed on this plant detritus and associated bacteria and fungi at the bottom of a stream and then be eaten by a fish. In a forest, birds might feed on earthworms that have been feeding on leaf litter and its associated bacteria and fungi in the soil. But even more important than this channeling of resources from pro-

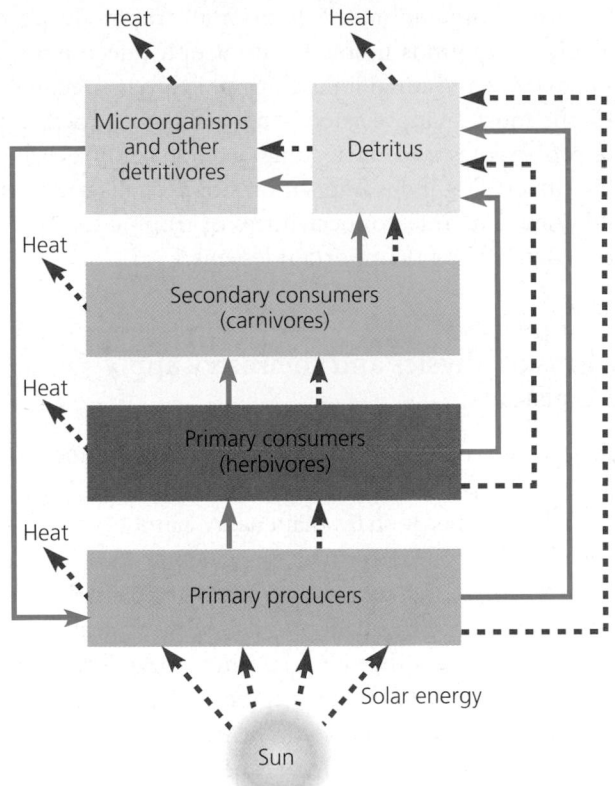

FIGURE 54.1 An overview of ecosystem dynamics. This generalized scheme traces energy flow (broken red lines) and material cycling (solid blue lines) in ecosystems. Energy flow starts from the sun as solar radiation, moves as chemical energy transfers in the food web, and finishes as heat radiated into space. Materials flow through the trophic levels eventually to detritus and then cycle back to the primary producers.

ducers to consumers is the role that detritivores play in making vital chemical elements available to producers.

The organic material that makes up the living organisms in an ecosystem gets recycled. Detritivores break down (decompose) the organic material and recycle the chemical elements in inorganic forms to abiotic reservoirs such as soil, water, and air. Plants and other producers can then reassimilate these elements into organic matter. All organisms perform some decomposition, breaking down organic molecules during cellular respiration, for example. But an ecosystem's main decomposers are prokaryotes and fungi, which secrete enzymes that digest organic material and then absorb the breakdown products (FIGURE 54.2). Accounting for most of the

FIGURE 54.2 Fungi decomposing a log.

conversion of organic materials from all trophic levels into inorganic compounds usable by autotrophs, decomposition by prokaryotes and fungi links all trophic levels. Decomposition is the most unappreciated of all ecological processes because prokaryotes and most fungi are not readily visible to the human eye. Yet if decomposition stopped, all life on Earth would cease. The interconnectedness of trophic levels is one of ecosystem ecology's important lessons.

The laws of physics and chemistry apply to ecosystems

Much of an ecologist's analysis of ecosystem dynamics comes from well-established laws of physics and chemistry. The law of conservation of energy states that energy cannot be created or destroyed but only transformed (see Chapter 6). This means that for all ecosystems, we can potentially trace energy exactly from its input as solar radiation to its release as heat from organisms. Plants and other photosynthetic organisms convert solar energy to chemical energy, but the total amount of energy does not change. The total amount of energy stored in organic molecules plus the amount reflected and dissipated as heat must equal the total incident energy in the form of sunlight. One important objective of ecosystem ecology is to compute such energy budgets and trace the energy flow in particular ecosystems.

The second law of thermodynamics tells us that energy conversions cannot be completely efficient—that some energy will be lost as heat in any conversion process (see Chapter 6). This idea suggests that we can measure the efficiency of ecological energy conversions in the same way we measure the efficiency of light bulbs and car engines.

While energy in ecosystems is dissipated ultimately as heat into outer space, chemical elements are continually recycled. A unit of energy moves through the trophic structure of an ecosystem only once, and without the sun providing continuous energy to Earth, ecosystems would vanish. In contrast, a carbon or nitrogen atom moves from trophic level to trophic level and eventually to the decomposers and then back again in endless cycles. Elements are not lost on a global scale, although they may move out of one ecosystem and into another. The measurement and analysis of these cycles of chemical elements in ecosystems and in the entire biosphere are an important part of ecosystem ecology.

Now that we have considered energy flow and material cycling in the context of trophic structures, let's take a closer look at these dynamics, beginning with the producers.

PRIMARY PRODUCTION IN ECOSYSTEMS

The amount of light energy converted to chemical energy (organic compounds) by an ecosystem's autotrophs during a given time period is called **primary production.** This photosynthetic product is the starting point for studies of ecosystem metabolism and energy flow.

An ecosystem's energy budget depends on primary production

Most primary producers use light energy to synthesize energy-rich organic molecules, which can subsequently be broken down to generate ATP (see Chapter 10). Consumers acquire their organic fuels secondhand (or even third- or fourthhand) through food webs. Therefore, the extent of photosynthetic production sets the spending limit for the energy budget of the entire ecosystem.

The Global Energy Budget

Every day, Earth is bombarded by about 10^{22} joules of solar radiation (1 J = 0.239 cal). This is the energy equivalent of 100 million atomic bombs of the size used at the end of World War II. As described in Chapter 50, the intensity of the solar energy striking Earth and its atmosphere varies with latitude, with the tropics receiving the highest input. Most solar radiation is absorbed, scattered, or reflected by the atmosphere in an asymmetrical pattern determined by variations in cloud cover and the quantity of dust in the air over different regions. The amount of solar radiation reaching the surface of the globe ultimately limits the photosynthetic output of ecosystems.

Much of the solar radiation that reaches the biosphere lands on bare ground and bodies of water that either absorb or reflect the incoming energy. Only a small fraction actually strikes algae, photosynthetic prokaryotes, and plant leaves, and only some of this is of wavelengths suitable for photosynthesis. Of the visible light that does reach photosynthetic organisms, only about 1% is converted to chemical energy by photosynthesis, and this efficiency varies with the type of organism, light level, and other factors. Although the fraction of the total incoming solar radiation that is ultimately trapped by photosynthesis is very small, primary producers on Earth collectively create about 170 billion tons of organic material per year—an impressive quantity.

Gross and Net Primary Production

Total primary production is known as **gross primary production (GPP)**—the amount of light energy that is converted to chemical energy by photosynthesis per unit time. Not all of this production is stored as organic material in the growing plants, because the plants use some of the molecules as fuel in their own cellular respiration. **Net primary production (NPP)** is equal to gross primary production minus the energy used by the primary producers for respiration (R):

$$NPP = GPP - R$$

Net primary production is the measurement of greatest interest to ecologists because it represents the storage of chemical energy that will be available to consumers in an ecosystem. In forests, net production may be as little as one-fourth that of gross production. Trees have larger masses of stems, branches, and roots to support through respiration than do herbs; thus, less energy is lost to respiration in herbaceous and crop communities than in forests.

Primary production can be expressed in terms of energy per unit area per unit time ($J/m^2/yr$) or as **biomass** (weight) of vegetation added to the ecosystem per unit area per unit time ($g/m^2/yr$). Biomass is usually expressed in terms of the dry weight of organic material because water molecules contain no usable energy and because the water content of plants varies over short periods of time. An ecosystem's primary production should not be confused with the *total* biomass of photosynthetic autotrophs present at a given time, called the **standing crop.** Primary production is the amount of *new* biomass added in a given period of time. Although a forest has a very large standing crop biomass, its primary production may actually be less than that of some grasslands, which do not accumulate vegetation because animals consume the plants rapidly and because some of the plants are annuals.

Different ecosystems vary considerably in their production as well as in their contribution to the total production on Earth (FIGURE 54.3). Tropical rain forests are among the most productive terrestrial ecosystems, and because they cover a large portion of Earth, they contribute a large proportion of the planet's overall production. Estuaries and coral reefs also have very high production, but their total contribution to global production is relatively small because these ecosystems are not very extensive. The open ocean contributes more primary production than any other ecosystem, but this is because of its very large size; production per unit area is relatively low. Deserts and tundra also have low production. Satellite images of the globe now provide a means of studying global patterns of primary production (FIGURE 54.4, p. 1202). The most striking impression from these global satellite maps is how unproductive most of the oceans are, in contrast to the high production of tropical forest regions.

In aquatic ecosystems, light and nutrients limit primary production

What controls primary production in ecosystems? To ask this question another way, what factors could we change to increase or decrease primary production for a given ecosystem? We will look first at the factors that limit production in aquatic ecosystems, beginning with marine ecosystems.

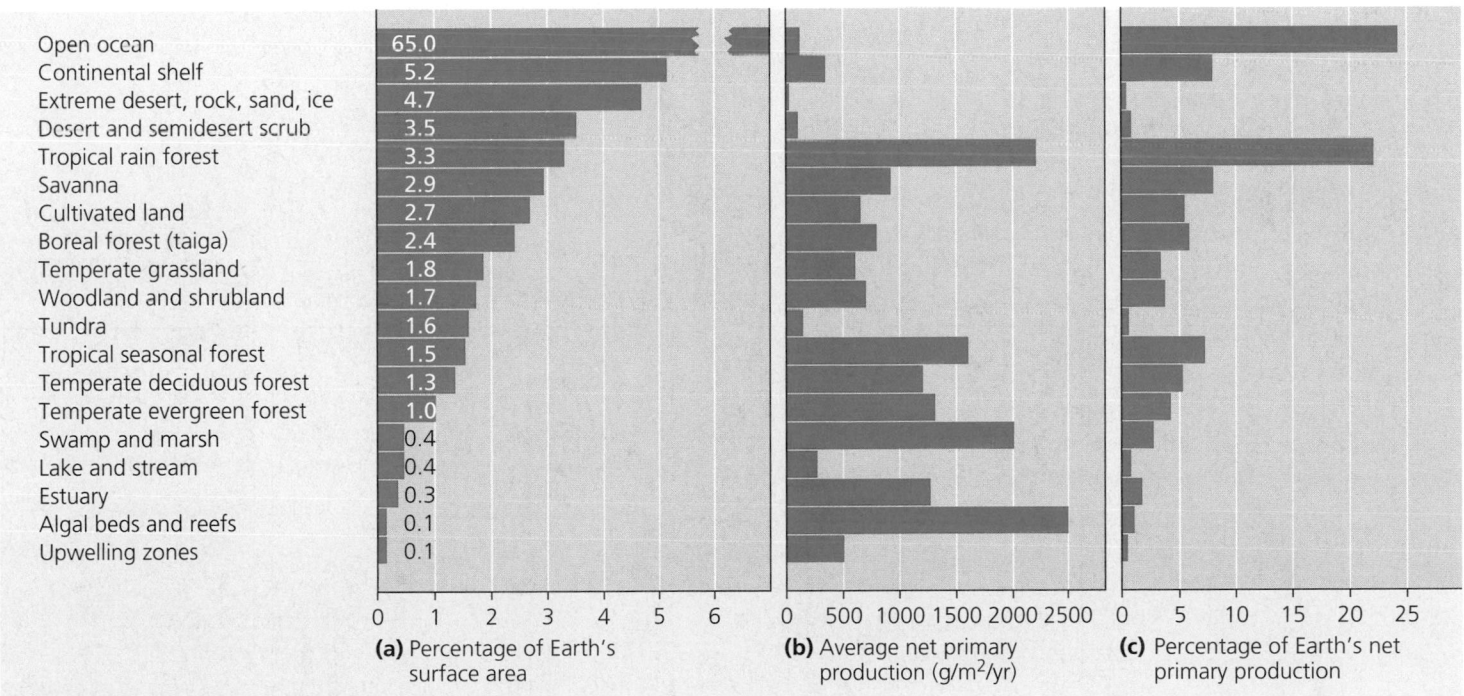

FIGURE 54.3 Primary production of different ecosystems. (a) The geographic extent and **(b)** the production per unit area of different ecosystems determine their total contribution to **(c)** worldwide primary production. Open ocean, for example, contributes a lot to the planet's production despite its low production per unit area because of its large size, whereas tropical rain forest contributes a lot because of its high production. (Aquatic ecosystems are color-coded blue in these histograms; terrestrial ecosystems are green.)

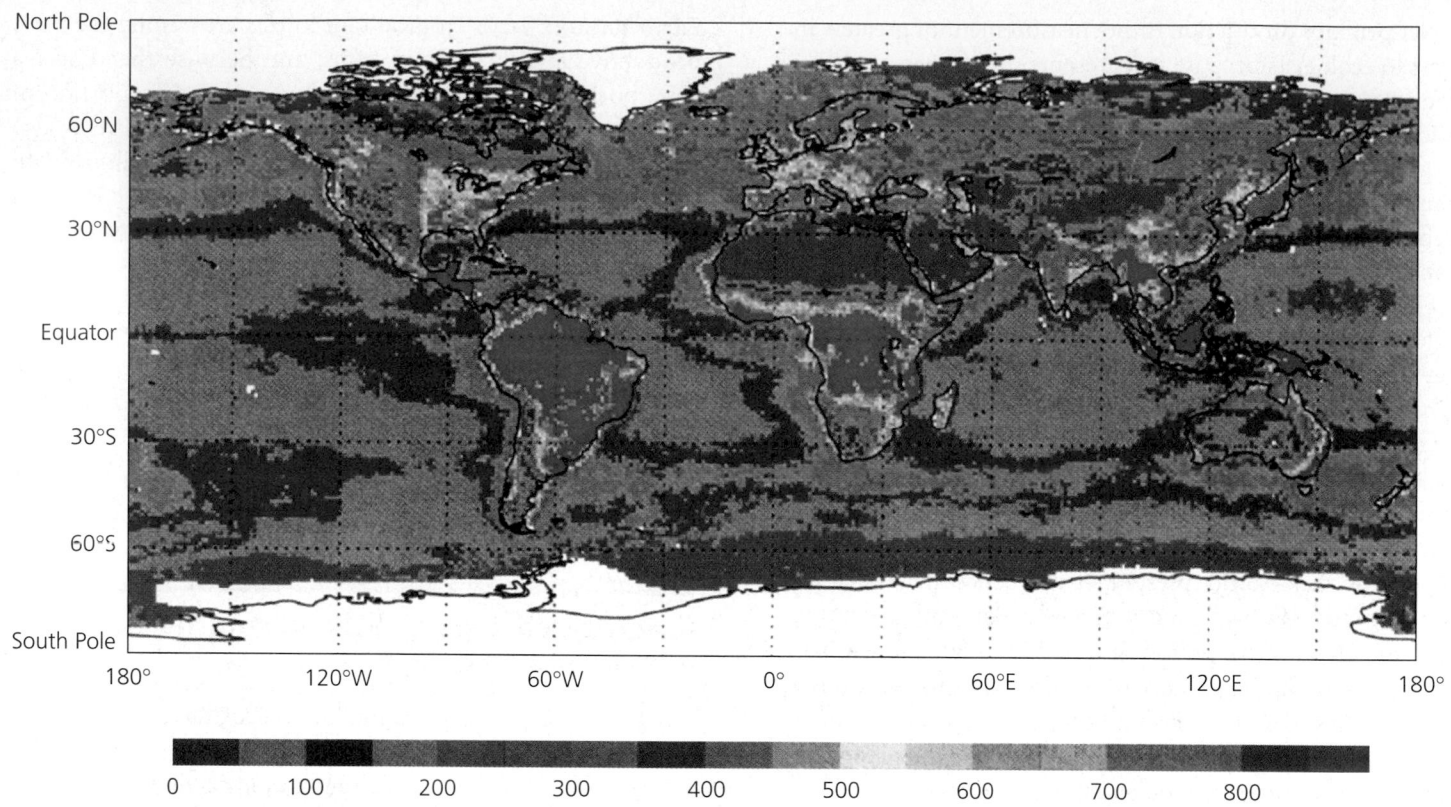

FIGURE 54.4 Regional annual net primary production for Earth. The image is based on data, such as chlorophyll density, collected by satellites. The values on the color key are grams of carbon per square meter per year. Ocean data are averages from 1978 to 1983. Land averages are from 1982 to 1990. Of total global primary production, the ocean contributes 46% and the land 54%.

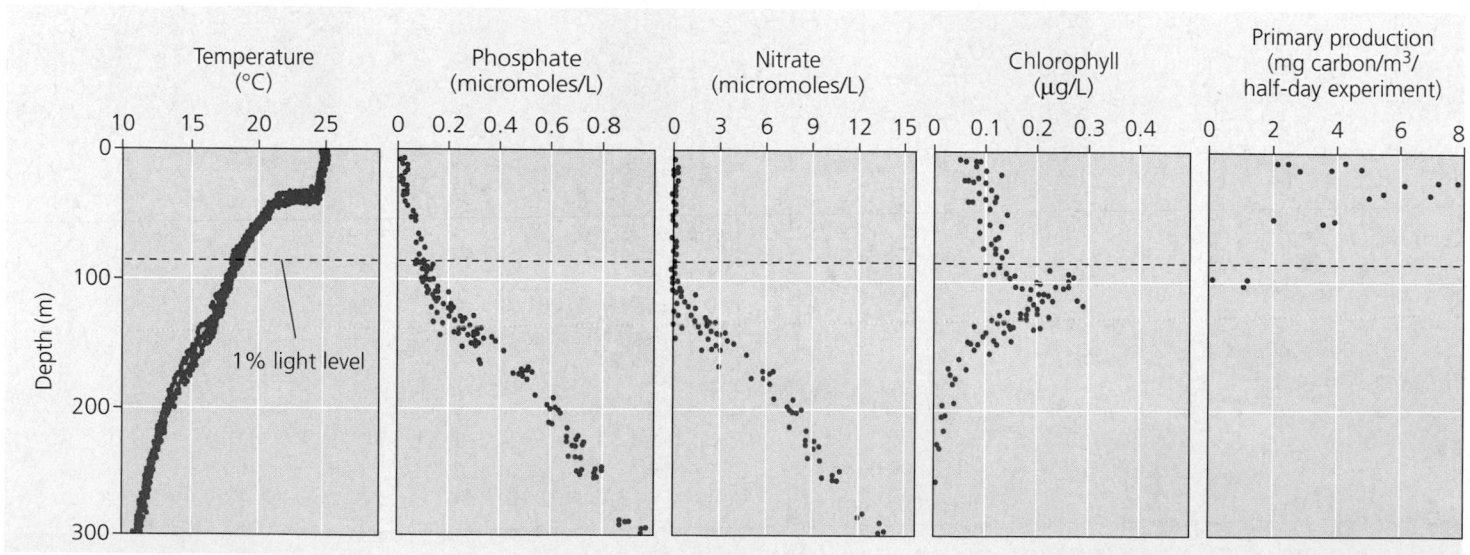

FIGURE 54.5 Vertical distribution of temperature, nutrients, and production in the upper layer of the central North Pacific during summer. The curves are composites of several vertical profiles made over a two-day period at a single location (28°N, 155°W). Chlorophyll is a measure of phytoplankton density. The broken line illustrates the depth of 1% of surface light (the traditional definition of the depth of the photic zone). Note that nitrate has been depleted to undetectable levels in the photic zone, but that most of the primary production (measured from ^{14}C uptake) takes place in that nutrient-depleted zone because of the requirement for light.

Production in Marine Ecosystems

Light is the first variable one might expect to control primary production in oceans, since solar radiation drives photosynthesis. The depth to which light penetrates the ocean affects primary production throughout the photic zone (see FIGURE 50.22). More than half of the solar radiation is absorbed in the first meter of water. Even in "clear" water, only 5–10% of the radiation may reach a depth of 20 m.

If light is the primary variable limiting primary production in the ocean, we would expect production to increase along a gradient from the poles toward the equator, which receives the greatest intensity of light (see FIGURE 50.11). However, you can see that there is no such gradient by reexamining FIGURE 54.4. Some parts of the tropics and subtropics, such as the Sargasso Sea, the Indian Ocean, and the central part of the North Pacific, are very unproductive. In contrast, the North Atlantic, the Gulf of Alaska, and the Southern Ocean off New Zealand are relatively productive areas.

Why are tropical oceans less productive than we would expect from their high year-round intensity of illumination? It is actually *nutrients* more than light that limit primary production in different geographic regions of the ocean. Ecologists use the term **limiting nutrient** for the nutrient that must be added for production to increase. And two elements, nitrogen and phosphorus, are the nutrients that most often limit marine production. In the open ocean, nitrogen and phosphorus concentrations are very low in the photic (upper) zone, where phytoplankton live. Ironically, the nutrient supply is much better in deeper water, where it is too dark for photosynthesis (FIGURE 54.5).

Nitrogen is the one nutrient that limits phytoplankton growth in many parts of the ocean. FIGURE 54.6 illustrates an experimental test of nitrogen versus phosphorus as the limiting factor for production in coastal waters off the south shore of Long Island, New York. Pollution from duck farms along the bays of Long Island adds both nitrogen and phosphorus to the coastal water; but unlike phosphorus, the nitrogen is immediately taken up by algae, and no trace of free nitrogen can be measured in the coastal waters. Nutrient enrichment experiments confirmed that nitrogen was limiting phytoplankton growth. The addition of nitrogen (in the form of ammonium) caused a heavy algal growth in bay water, but the addition of phosphate did not induce algal growth. There are some practical applications of this work, including the prevention of algal blooms caused by pollution that fertilizes the phytoplankton. If nitrogen is the factor now limiting phytoplankton production in the coastal zone, the elimination of phosphates from sewage will not help the problem of coastal pollution unless we first control nitrogen pollution.

Although nitrogen availability generally limits primary production in marine ecosystems, there are some anomalies. Several large areas of the ocean have low phytoplankton densities

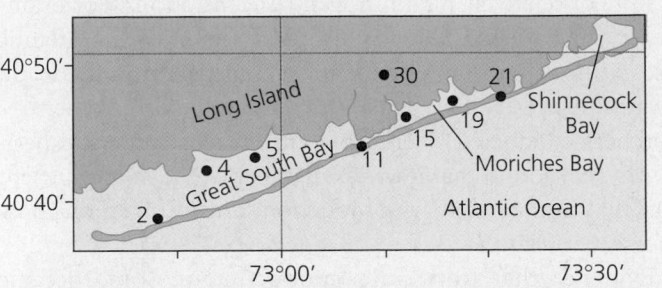

(a) Coast of Long Island, New York. The numbers on the map indicate the data collection locales (stations) and correspond to the numbers along the *x* axes of the two graphs. Duck farms that pollute the coastal waters with phosphate and nitrogenous compounds are concentrated near Moriches Bay.

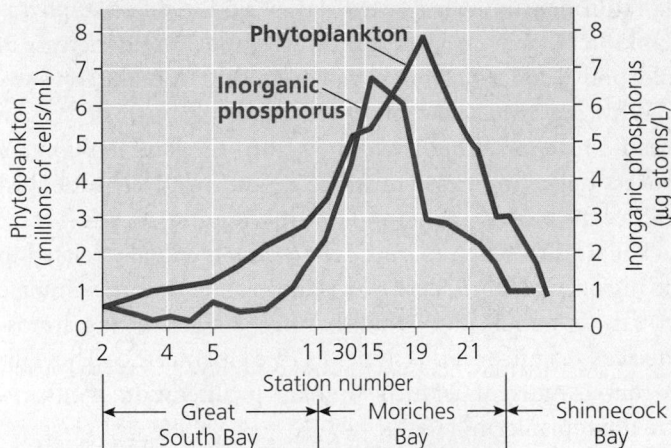

(b) Abundance of phytoplankton and distribution of phosphorus arising from duck farms in Moriches Bay are shown on this graph.

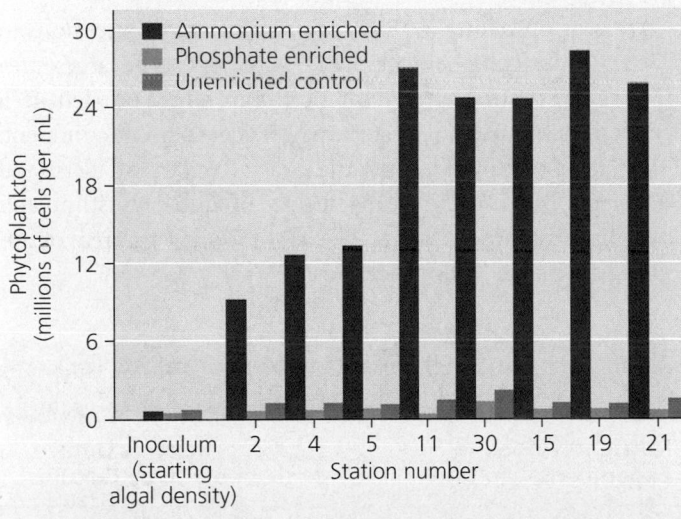

(c) Nutrient enrichment experiments. Researchers cultured the alga *Nannochloris atomus* with water collected from the bays. Adding more phosphorus, which was already in rich supply, had no effect on algal growth. But adding ammonium (a nitrogen source) increased algal density dramatically.

FIGURE 54.6 Experiments on nutrient limitations to phytoplankton production in coastal waters of Long Island.

in spite of relatively high nitrogen concentrations. For example, in the Sargasso Sea, a subtropical region of the Atlantic Ocean, the water is among the most transparent in the world because of a very low density of phytoplankton. When researchers conducted a series of nutrient enrichment experiments, they found that it was availability of the micronutrient iron that limited primary production, not nitrogen or phosphorus (TABLE 54.1).

Evidence that iron limits production in some oceanic ecosystems encouraged marine ecologists to try two large-scale field experiments in the tropical Pacific in 1993 and 1995. The researchers spread low concentrations of dissolved iron over 72 km^2 of ocean and then measured the change in phytoplankton density over a seven-day period. A massive phytoplankton bloom occurred, as indicated by a 27-fold increase in chlorophyll concentration in water samples from the test sites.

Why are iron concentrations naturally low in certain oceanic regions? Windblown dust from the land is the main process delivering iron to the ocean, and relatively little dust reaches the central Pacific and central Atlantic Oceans.

The iron factor in marine ecosystems is actually related to the nitrogen factor. Where iron is limiting, additions stimulate growth of cyanobacteria that fix nitrogen, converting atmospheric N$_2$ to nitrogenous minerals (see Chapter 27). These nitrogenous nutrients in turn stimulate proliferation of eukaryotic phytoplankton:

$$\text{Iron} \xrightarrow{+} \text{Cyanobacteria} \xrightarrow{+} \text{Nitrogen fixation} \xrightarrow{+} \text{Phytoplankton production}$$

Areas of upwelling in the ocean are the main exceptions to primary production being limited by nutrients. The largest area of upwelling occurs in the Antarctic Ocean, where nutrient-rich deep waters circulate to the surface near the Antarctic continent. Other areas of upwelling are the coastal waters off Peru and California. Because the steady supply of nutrients stimulates phytoplankton production at the base of food webs, areas of upwelling are prime fishing locations.

FIGURE 54.7 Remote sensing of primary production in oceans. This 1995 satellite image maps locations of high primary production in the Pacific region around Vancouver Island, British Columbia. A spectral instrument estimates chlorophyll concentrations in surface waters and converts the data to contrasting artificial colors. The areas that are green, red, or white in the color-enhanced photo are the richest in chlorophyll and hence phytoplankton. Note the phytoplankton bloom off the west coast of Vancouver Island. The second big bloom, in the sound between the island and the mainland, is due to a plume of nutrient-rich water from the Fraser River.

Marine ecologists are just beginning to piece together the interplay of factors that determine primary production in the oceans. One of the most important technical breakthroughs is the remote sensing of primary production data collected by satellites (FIGURE 54.7).

Production in Freshwater Ecosystems

Solar radiation limits primary production on a day-to-day basis in lakes, and within any given lake, you can predict the daily primary production from the solar radiation for that day. Temperature is closely linked with light intensity in freshwater systems and is difficult to evaluate as a separate factor. Nutrient limitations are also common in freshwater lakes.

During the 1970s, interest in what controls primary production in freshwater lakes became associated with the environmental issue of increasing water pollution. Sewage and fertilizer runoff from farms and yards added nutrients to lakes. This shifted many lakes from having phytoplankton communities dominated by diatoms or green algae to having phytoplankton communities dominated by cyanobacteria. This process, **eutrophication,** has generally undesirable impacts from a human perspective, including the eventual loss of fish from lakes (see FIGURE 50.19b). Controlling eutrophication requires knowing which polluting nutrient enables the

Table 54.1 Nutrient Enrichment Experiments for Sargasso Sea Samples

Nutrients Added to Experimental Culture*	Relative Uptake of ^{14}C by Cultures†
None (controls)	1.00
N + P only	1.10
N + P + metals (excluding iron)	1.08
N + P + metals (including iron)	12.90
N + P + iron	12.00

*N = nitrogen; P = phosphorus.
†^{14}C uptake by cultures measures primary production.
SOURCE: Data from Menzel and Ryther, *Deep Sea Research* 7(1961):276–281.

cyanobacteria to bloom. Unlike the situation in marine ecosystems, nitrogen is rarely the limiting factor for primary production in lakes. In the 1970s, David Schindler, whom you met in the interview on p. 1190, conducted a series of whole-lake experiments that pointed to phosphorus as the limiting nutrient that stimulated the cyanobacterial blooms (FIGURE 54.8). His research led to the use of phosphate-free detergents and other water quality reforms.

In terrestrial ecosystems, temperature, moisture, and nutrients limit primary production

Of course, water availability varies more among terrestrial ecosystems than among aquatic ones, but terrestrial ecosystems also vary more in temperature. On a large geographic scale, temperature and moisture are the key factors controlling primary production in ecosystems. Note again in FIGURE 54.3b that tropical rain forests, with their warm, wet conditions that promote plant growth, are the most productive of all terrestrial ecosystems.

On a more local scale, mineral nutrients in the soil can play key roles in limiting primary production in terrestrial ecosystems. Primary production removes soil nutrients, sometimes faster than they are replaced. At some point, a nutrient deficiency may cause plant growth to slow or cease. It is unlikely

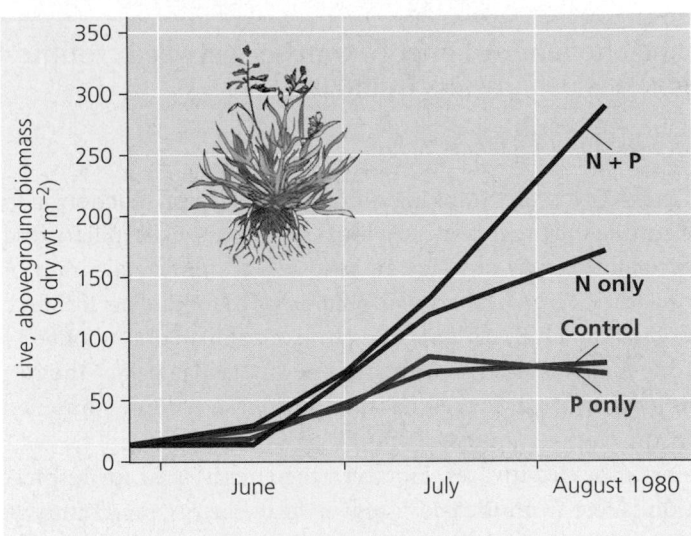

THE PROCESS OF SCIENCE

FIGURE 54.9 Nutrient addition experiments in a Hudson Bay salt marsh. Note that experimental plots receiving phosphorus (P) alone do not outproduce the unfertilized control plots. But adding nitrogen (N) boosts production. Nitrogen is the limiting nutrient, at least until late summer. At this point, phosphorus may become limiting. (What is the basis for this inference?)

that all nutrients will be exhausted simultaneously. If a limiting nutrient controls production, adding a nonlimiting nutrient to the soil will not stimulate production. For example, if nitrogen is limiting, then adding phosphorus, even if it, too, is in relatively short supply, will not stimulate production. But if we add nitrogen, this will stimulate plant growth until some other nutrient—say, phosphorus—becomes limiting. In fact, either nitrogen or phosphorus is the soil nutrient that most commonly limits terrestrial production (FIGURE 54.9).

Scientific studies relating nutrients to production have practical applications in agriculture. Farmers maximize their crop yields by using fertilizers with the right balance of nutrients for the local soil and the type of crop.

SECONDARY PRODUCTION IN ECOSYSTEMS

The amount of chemical energy in consumers' food that is converted to their own new biomass during a given time period is called the **secondary production** of the ecosystem. Consider the transfer of organic matter from producers to herbivores, the primary consumers. In most ecosystems, herbivores manage to eat only a small fraction of the plant material produced. And they cannot digest all the plant material that they *do* eat, as anyone who has walked through a dairy farm will attest. Thus, much of primary production is not used by consumers. Let us analyze this process of energy transfer more closely.

THE PROCESS OF SCIENCE

FIGURE 54.8 The experimental eutrophication of a lake. The far basin of this lake was separated from the near basin by a plastic curtain and fertilized with inorganic sources of carbon, nitrogen, and phosphorus. Within two months, the fertilized basin was covered with a cyanobacterial bloom, which appears white in the photograph. The near basin, which was treated with only carbon and nitrogen, remained unchanged. In this case, phosphorus was the key limiting nutrient, and its addition stimulated the explosive growth of cyanobacteria.

The efficiency of energy transfer between trophic levels is usually less than 20%

Production Efficiency

One way to start thinking about secondary production is to examine the process in individual organisms. Caterpillars feed on plant leaves, and FIGURE 54.10 is a simplified diagram of how the energy a caterpillar obtains from food is partitioned. Out of 200 J (48 cal) consumed by a caterpillar, only about 33 J (one-sixth) is used for growth. The caterpillar passes the rest as feces or uses it for cellular respiration. The energy contained in the feces is not lost from the ecosystem, as it will be consumed by detritivores. However, the energy used for respiration is lost from the ecosystem as heat. Energy pours into an ecosystem as solar radiation and drains away as respiratory heat loss. This is why energy is said to flow through, not cycle within, ecosystems. Only the chemical energy stored as growth (or the production of offspring) by herbivores is available as food to secondary consumers. And our example actually overestimates the conversion of primary production to secondary production because we did not account for all the plant material that herbivores do not even consume. The greenness of terrestrial landscapes due to an abundance of plant material indicates that most net primary production is not converted over the short term to secondary production.

If we view animals as energy transformers, we can ask questions about their relative efficiencies. This is the efficiency measure we'll use:

$$\text{Production efficiency} = \frac{\text{Net secondary production}}{\text{Assimilation of primary production}}$$

Net secondary production is the energy stored in biomass represented by growth and reproduction. Assimilation consists of the total energy taken in and used for growth, reproduction, and respiration. In other words, **production efficiency** is the fraction of food energy that is *not* used for respiration. For the caterpillar in FIGURE 54.10, production efficiency is 33%; 67 J of the 100 J of assimilated energy is used for respiration (note that energy lost as undigestible material in feces does not count as assimilation). Birds and mammals have low production efficiencies, ranging from 1–3%, because they use so much energy to maintain a warm body temperature. Fishes, which are ectotherms (see Chapter 40), have a production efficiency around 10%. Insects are even more efficient, with production efficiencies averaging 40%.

Trophic Efficiency and Ecological Pyramids

Let's scale up now from the production efficiencies of individual consumers to the flow of energy through whole trophic levels.

Trophic efficiency is the percentage of production transferred from one trophic level to the next. Put another way, trophic efficiency is the fraction of net production at one trophic level as a percent of net production at the level below. Trophic efficiencies must always be less than production efficiencies because they take into account not only the energy lost through respiration and materials in feces, but also the energy in organic material in a lower trophic level that is not consumed by the next trophic level. Trophic efficiencies usually range from 5–20%, depending on the type of ecosystem. In other words, 80–95% of the energy available at one trophic level never transfers to the next. And this loss is multiplied over the length of a food chain. If 10% of energy is transferred from primary producers to primary consumers, and 10% of that energy is transferred to secondary consumers, then only 1% of net primary production is available to secondary consumers (10% of 10%).

Pyramids of Production.   This multiplicative loss of energy from a food chain can be represented by a **pyramid of production,** in which the trophic levels are stacked in blocks, with primary producers forming the foundation of the pyramid. The size of each block is proportional to the production of each trophic level (FIGURE 54.11).

Pyramids of Biomass.   One important ecological consequence of low trophic efficiencies can be represented in a **biomass pyramid,** in which each tier represents the standing crop (the total dry weight of all organisms) in a trophic level. Most biomass pyramids narrow sharply from primary producers at the base to top-level carnivores at the apex because energy transfers between trophic levels are so inefficient (FIGURE 54.12a). Some aquatic ecosystems, however, have inverted biomass pyramids, with primary consumers outweighing producers. In the waters of the English Channel, for example, the biomass of zooplankton (consumers) is five times the weight

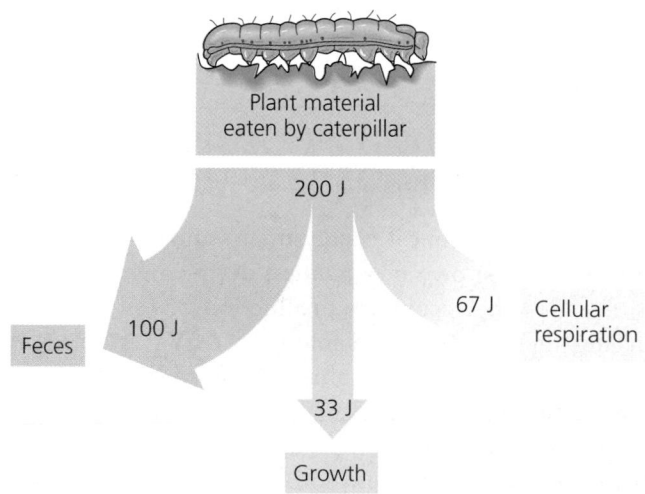

FIGURE 54.10 Energy partitioning within a link of the food chain. Less than 17% of the caterpillar's food is actually converted to caterpillar biomass.

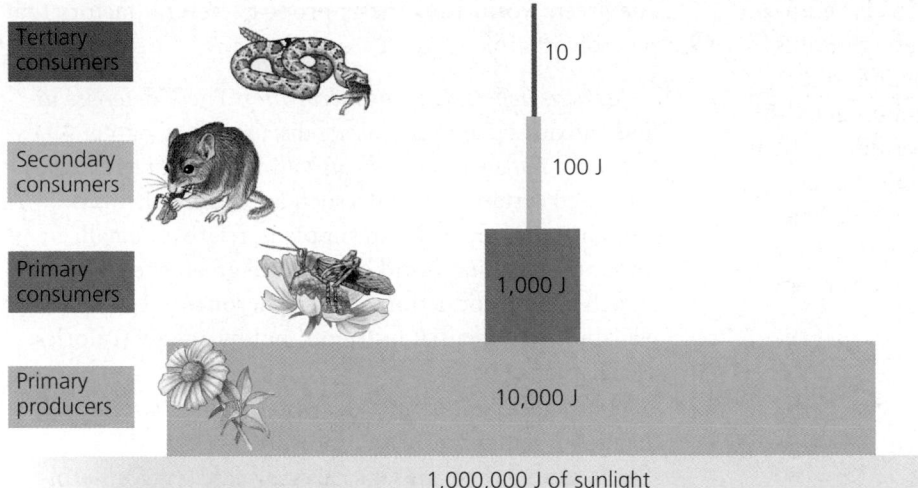

FIGURE 54.11 An idealized pyramid of net production. This example is based on a trophic efficiency of 10% for each link in the food chain. Notice that primary producers convert only about 1% of the energy in the sunlight available to them into net primary production.

Tertiary consumers	10 J
Secondary consumers	100 J
Primary consumers	1,000 J
Primary producers	10,000 J
	1,000,000 J of sunlight

of phytoplankton (producers) (FIGURE 54.12b). Such inverted biomass pyramids occur because the zooplankton consume the phytoplankton so quickly that the producers never develop a large population size, or standing crop. Instead, the phytoplankton grow, reproduce, and are consumed rapidly. Phytoplankton have a short **turnover time,** which means they have a small standing crop biomass compared to their production:

$$\text{Turnover time} = \frac{\text{Standing crop biomass (mg/m}^2)}{\text{Production (mg/m}^2\text{/day)}}$$

Because the phytoplankton continue to replace their biomass at such a rapid rate, they can support a biomass of zooplankton bigger than their own biomass. Nevertheless, the pyramid

of *production* for this ecosystem is upright, like the one in FIGURE 54.11, because phytoplankton have a much higher production than zooplankton.

Pyramids of Numbers. The multiplicative loss of energy from food chains severely limits the overall biomass of top-level carnivores that any ecosystem can support. Only about one one-thousandth of the chemical energy fixed by photosynthesis can flow all the way through a food web to a tertiary consumer, such as a hawk, snake, or shark (see FIGURE 54.11). This explains why food webs usually include only about four or five trophic levels (see Chapter 53).

Because predators are usually larger than the prey they eat, top-level predators tend to be fairly large animals. Thus, the limited biomass at the top of an ecological pyramid is concentrated in a relatively small number of large individuals. This phenomenon is reflected in a **pyramid of numbers,** in which the size of each block is proportional to the number of individual organisms present in each trophic level (FIGURE 54.13). Populations of top predators are typically small, and the animals may be widely spaced within their habitats. As a result, many predators are highly susceptible to extinction (as well as to the evolutionary consequences of small population size discussed in Chapter 23).

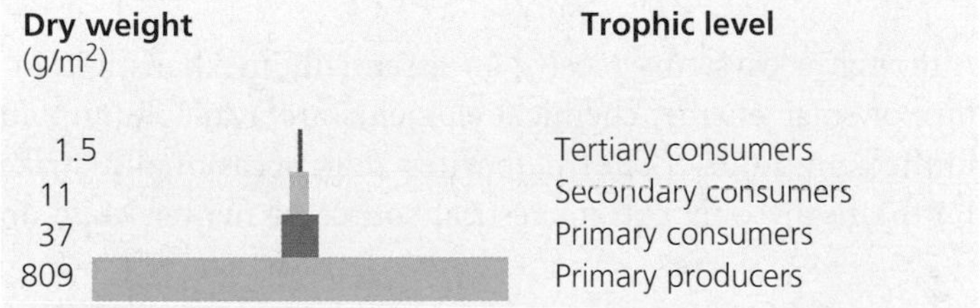

Dry weight (g/m²)	Trophic level
1.5	Tertiary consumers
11	Secondary consumers
37	Primary consumers
809	Primary producers

(a) Florida bog. Most biomass pyramids show a sharp decrease in biomass at successively higher trophic levels, as illustrated by data from a bog at Silver Springs, Florida.

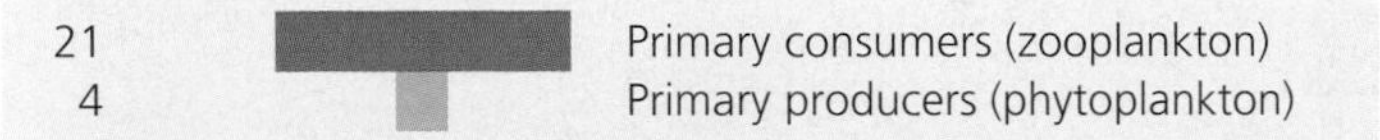

21	Primary consumers (zooplankton)
4	Primary producers (phytoplankton)

(b) English Channel. In some aquatic ecosystems, such as the English Channel, a small standing crop of primary producers (phytoplankton) supports a larger standing crop of primary consumers (zooplankton). This is because the phytoplankton have a short turnover time: The algae reproduce rapidly and are consumed at a high rate.

FIGURE 54.12 Pyramids of biomass (standing crop). Numbers denote the dry weight (g/m²) for all organisms at a trophic level.

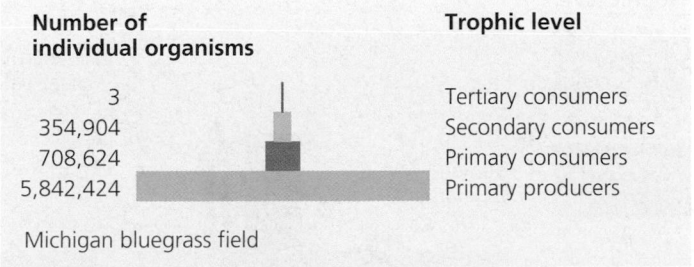

Number of individual organisms	Trophic level
3	Tertiary consumers
354,904	Secondary consumers
708,624	Primary consumers
5,842,424	Primary producers

Michigan bluegrass field

FIGURE 54.13 A pyramid of numbers. In this pyramid of numbers for a bluegrass field in Michigan, only three top carnivores are supported in an ecosystem based on production by nearly 6 million plants.

The dynamics of energy flow through ecosystems have important implications for the human population. Eating meat is a relatively inefficient way of tapping photosynthetic production. A human obtains far more calories by eating grains directly as a primary consumer than by processing that same amount of grain through another trophic level and eating grain-fed animals (beef, chicken, pork, lamb, farm fish). Worldwide agriculture could, in fact, successfully feed many more people than it does today if we all consumed only plant material, feeding more efficiently as primary consumers (FIGURE 54.14).

Herbivores consume a small percentage of vegetation: the green world hypothesis

With so many primary consumers (herbivores) munching away on plants, how can we explain why most terrestrial ecosystems are actually quite green, boosting large standing crops of vegetation? According to the **green world hypothesis,** herbivores consume relatively little plant biomass because they are held in check by a variety of factors, including predators, parasites, and disease.

Let's look at some of the numbers that account for a green world. There is a total of about 83×10^{10} metric tons of carbon stored in the plant biomass of terrestrial ecosystems. And the global rate of terrestrial primary production is about 5×10^{10} metric tons of plant biomass per year. (If you return to the equation on p. 1207, you can now calculate the global turnover rate for Earth's vegetation.) On a global scale, herbivores consume less than 17% of the total net annual production by plants. Thus, in general, herbivores are only a minor nuisance to plants. We do know that some herbivores have the potential to completely strip local vegetation over the short term; an example is the occasional ability of exploding gypsy moth populations to defoliate areas of forest in the northeastern United States. Such exceptions only heighten our curiosity about why Earth is so green.

The green world hypothesis proposes several factors that keep herbivores in check:

- *Plants have defenses against herbivores.* These defenses include noxious chemicals, as we discussed in Chapter 39.
- *Nutrients, not energy supply, usually limit herbivores.* Animals need certain nutrients, such as organic nitrogen (protein), that plants often supply in relatively small amounts. Even in a world of plentiful green energy, the growth and reproduction of many herbivores is limited by availability of essential nutrients, not by energy (calories; see Chapter 41).
- *Abiotic factors limit herbivores.* Unfavorable seasonal changes in temperature and moisture are examples of abiotic factors that can set a carrying capacity for herbivores far below the number that would strip vegetation.
- *Intraspecific competition can limit herbivore numbers.* Territorial behavior and other consequences of competition may maintain herbivores' population densities below what the vegetation could support with food.
- *Interspecific interactions check herbivore densities.* The green world hypothesis postulates that predators, parasites, and disease are the most important factors limiting the growth of herbivore populations. This applies the top-down model of community structure (Chapter 53).

The relationship of herbivores to plants emphasizes the importance of trophic structure for understanding the dynamics of an ecosystem. We now redirect our focus from energy flow to the cycling of nutrients in ecosystems.

THE CYCLING OF CHEMICAL ELEMENTS IN ECOSYSTEMS

Although ecosystems receive an essentially inexhaustible influx of solar energy, chemical elements are available only in limited amounts. (The meteorites that occasionally strike Earth are the only extraterrestrial source of matter.) Life on

Trophic level

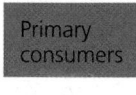
Secondary consumers

Primary consumers

Primary producers

Human vegetarians

Corn

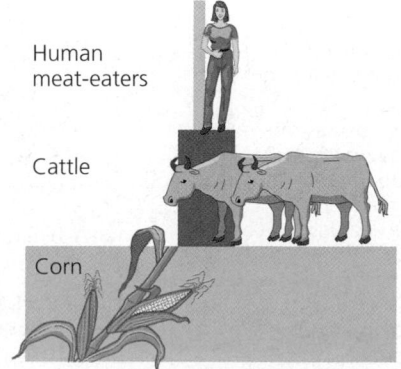
Human meat-eaters

Cattle

Corn

FIGURE 54.14 **Food energy available to the human population at different trophic levels.** Most humans have a diet between these two extremes.

Earth therefore depends on the recycling of essential chemical elements. Even while an individual organism is alive, much of its chemical stock is rotated continuously as nutrients are assimilated and waste products released. Atoms present in the complex molecules of an organism at its time of death are returned as simpler compounds to the atmosphere, water, or soil by the action of decomposers. This decomposition replenishes the pools of inorganic nutrients that plants and other autotrophs use to build new organic matter. Because nutrient circuits involve both biotic and abiotic components of ecosystems, they are also called **biogeochemical cycles.**

Biological and geologic processes move nutrients between organic and inorganic compartments

A chemical's specific route through a biogeochemical cycle varies with the particular element and the trophic structure of the ecosystem. We can, however, recognize two general categories of biogeochemical cycles. Gaseous forms of carbon, oxygen, sulfur, and nitrogen occur in the atmosphere, and cycles of these elements are essentially global. For example, some of the carbon and oxygen atoms a plant acquires from the air as CO_2 may have been released into the atmosphere by the respiration of a plant or animal in some distant locale. Other elements that are less mobile in the environment, including phosphorus, potassium, calcium, and the trace elements, generally cycle on a more localized scale, at least over the short term. Soil is the main abiotic reservoir of these elements, which are absorbed by plant roots and eventually returned to the soil by decomposers, usually in the same general vicinity.

A General Model of Chemical Cycling

Before examining the details of individual cycles, let's look at a general model of nutrient cycling that shows the main reservoirs, or compartments, of elements and the processes that transfer elements between reservoirs (FIGURE 54.15). Most nutrients accumulate in four reservoirs, each of which is defined by two characteristics: whether it contains organic or inorganic materials and whether or not the materials are directly available for use by organisms. One compartment of organic materials is composed of the living organisms themselves and detritus; these nutrients are available to other organisms when consumers feed and when detritivores consume nonliving organic matter. The second organic compartment includes "fossilized" deposits of once-living organisms (coal, oil, and peat), from which nutrients cannot be assimilated directly. Material moved from the living organic compartment to the fossilized organic compartment long ago, when organisms died and were buried by sedimentation over millions of years to become coal, oil, or peat.

Nutrients also occur in two inorganic compartments, one in which they are available for use by organisms and one in

which they are not. The available inorganic compartment includes matter (elements and compounds) that is dissolved in water or present in soil or air. Organisms assimilate materials from this compartment directly and return chemicals to it through the fairly rapid processes of cellular respiration, excretion, and decomposition. Elements in the unavailable inorganic compartment are tied up in rocks. Although organisms cannot tap into this compartment directly, nutrients slowly become available through weathering and erosion. Similarly, unavailable organic materials move into the available compartment of inorganic nutrients through erosion or when fossil fuels are burned and the exhaust enters the atmosphere.

Describing biogeochemical cycles in general is much simpler than actually tracing elements through these cycles, especially since ecosystems exchange elements with other ecosystems. Even in a pond, which has discrete boundaries, several processes add and remove key nutrients. Minerals dissolved in rainwater or runoff from the neighboring land are added to the pond, as are nutrient-rich pollen, fallen leaves, and other airborne material. And of course, carbon, oxygen, and nitrogen cycle between the pond and the atmosphere. Birds may feed on fish or the aquatic larvae of insects, which derived their store of nutrients from the pond, and some of those nutrients may then be excreted or eliminated on land far from the pond's drainage area. Keeping track of the inflow and outflow is even more challenging in less clearly delineated terrestrial ecosystems. Nevertheless, ecologists have worked out the schemes for chemical cycling in several ecosystems, often by adding tiny amounts of radioactive tracers that enable the researchers to follow chemical elements through the various biotic and abiotic components of the ecosystems.

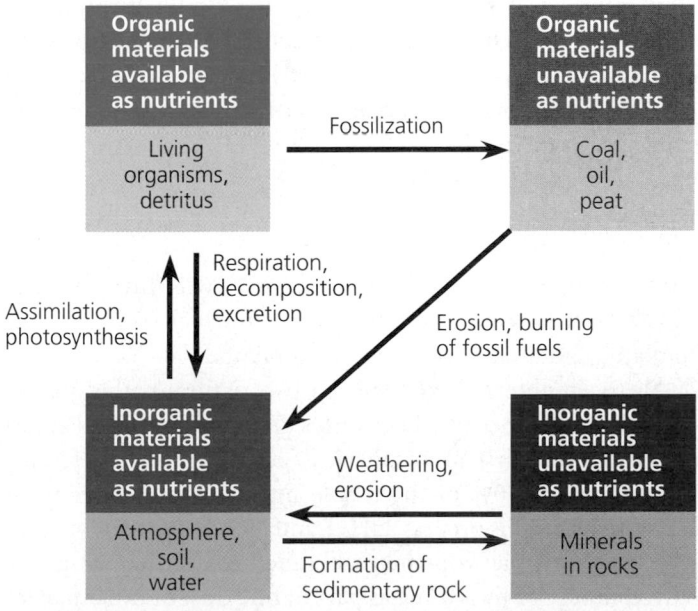

FIGURE 54.15 A general model of nutrient cycling. The biological and geologic processes that move nutrients from one compartment reservoir to another are indicated on the arrows.

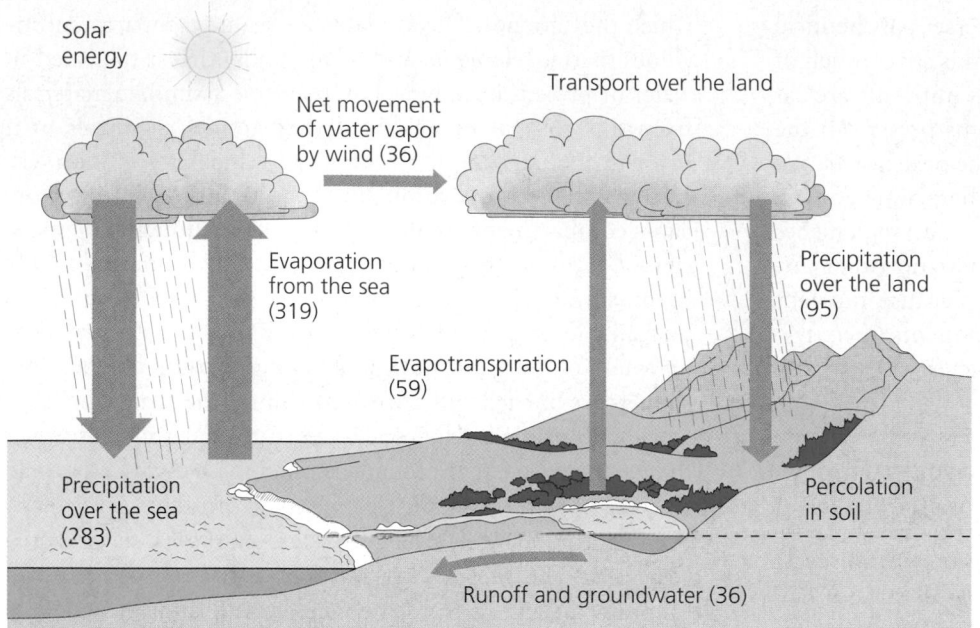

FIGURE 54.16 The water cycle. On a global scale, evaporation exceeds precipitation over the oceans. The result is a net movement of water vapor, carried by the winds, from the ocean to the land. The excess of precipitation over evaporation on land results in the formation of surface and groundwater systems that flow back to the sea, completing the major part of the cycle. Over the sea, evaporation forms most water vapor. On land, however, 90% or more of the vaporization is due to plant transpiration, which together with other types of evaporation is referred to as evapotranspiration. The numbers in this diagram indicate water flow in billion billion (10^{18}) grams per year.

One important cycle, the water cycle, does not fit the generalized scheme of FIGURE 54.15 very well. Although organisms are mostly made of water, very little of the water that cycles through ecosystems is chemically altered by either the biotic or abiotic components. The main exception is the water that is split into hydrogen and oxygen during photosynthesis, but that involves a very tiny fraction of the total water passing through a plant or any other organism. The water cycle is more a physical process than a chemical one; it mainly involves changes between the liquid and gaseous states and the transport of the liquid water and water vapor (FIGURE 54.16). Contrast the water cycle in FIGURE 54.16 with the carbon cycle (FIGURE 54.17), which fits the generalized scheme of biogeochemical cycles much better. Now let's take a closer look at two specific biogeochemical cycles: the nitrogen cycle and the phosphorus cycle.

The Nitrogen Cycle

Although Earth's atmosphere is almost 80% nitrogen, it is mostly in the form of nitrogen gas (N_2), which is unavailable to plants and hence to consumers of plants.

Nitrogen enters ecosystems via two natural pathways, the relative importance of which varies greatly from ecosystem to ecosystem. The first, atmospheric deposition, accounts for approximately 5–10% of the usable nitrogen that enters most ecosystems. In this process, NH_4^+ and NO_3^-, the two forms of nitrogen available to plants, are added to soil by being dissolved in rain or by settling as part of fine dust or other particulates. Some plants, such as the epiphytic bromeliads found in the canopy of tropical rain forests, have aerial roots that can take up NH_4^+ and NO_3^- directly from the moist atmosphere.

The other natural pathway for nitrogen to enter ecosystems is via **nitrogen fixation.** Only certain prokaryotes can fix nitrogen—that is, convert N_2 to minerals that can be used to synthesize nitrogenous organic compounds such as amino acids. Indeed, prokaryotes are vital links at several points in the nitrogen cycle (FIGURE 54.18). Nitrogen is fixed in terrestrial ecosystems by free-living (nonsymbiotic) soil bacteria as well as by symbiotic bacteria (*Rhizobium*) in the root nodules of legumes and certain other plants (see Chapter 37). Some cyanobacteria fix nitrogen in aquatic ecosystems. Organisms that fix nitrogen are, of course, fulfilling their own metabolic requirements, but the excess ammonia (NH_3) they release becomes available to other organisms.

In addition to these natural sources of usable nitrogen, industrial fixation of nitrogen for fertilizer now makes a major contribution to the pool of nitrogenous minerals in terrestrial and aquatic ecosystems.

The direct product of nitrogen fixation is ammonia (NH_3). However, most soils are at least slightly acidic, and NH_3 released into the soil picks up a hydrogen ion (H^+) to form ammonium, NH_4^+, which can be used directly by plants. Because NH_3 is a gas, it can evaporate back to the atmosphere from soils with a pH close to 7 (such as those in the midwestern United States). This NH_3 lost from the soil may then form NH_4^+ in the atmosphere. As a result, NH_4^+ concentrations in rainfall are correlated with soil pH over large regions. This local recycling of nitrogen by atmospheric deposition can be especially pronounced in agricultural areas where both nitrogen fertilizers and lime (a base that decreases soil acidity) are used extensively.

Although plants can use ammonium directly, most of the ammonium in soil is used by certain aerobic bacteria as an

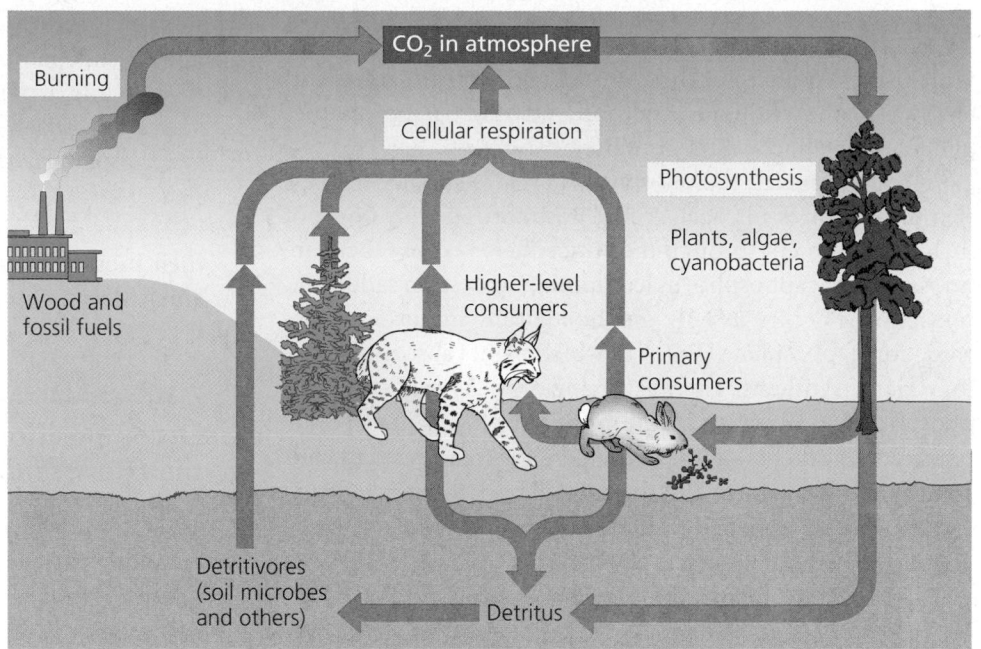

FIGURE 54.17 The carbon cycle. The reciprocal processes of photosynthesis and cellular respiration are responsible for the major transformations and movements of carbon. A seasonal pulse in atmospheric CO_2 is caused by decreased photosynthetic activity during the Northern Hemisphere's winter. On a global scale, the return of CO_2 to the atmosphere by respiration closely balances its removal by photosynthesis. However, the burning of wood and fossil fuels adds more CO_2 to the atmosphere; as a result, the amount of atmospheric CO_2 is steadily increasing.

energy source; their activity oxidizes ammonium to nitrite (NO_2^-) and then to nitrate (NO_3^-), a process called **nitrification.** Nitrate released from these bacteria can then be assimilated by plants and converted to organic forms, such as amino acids and proteins. Animals can assimilate only organic nitrogen, and they do this by eating plants or other animals. Some bacteria can obtain the oxygen they need for metabolism from nitrate rather than from O_2 under anaerobic conditions. As a result of this **denitrification** process, some nitrate is converted back to N_2, returning to the atmosphere. The decomposition of organic nitrogen back to ammonium, a process called **ammonification,** is carried out mainly by bacterial and fungal decomposers. This process recycles large amounts of nitrogen to the soil.

Overall, most of the nitrogen cycling in natural systems involves the nitrogenous compounds in soil and water, not atmospheric N_2. Although nitrogen fixation is important in the buildup of a pool of available nitrogen, it contributes only a tiny fraction of the nitrogen assimilated annually by total vegetation. Nevertheless, many common species of plants depend on their association with nitrogen-fixing bacteria to provide this essential nutrient in a form they can assimilate.

The amount of N_2 returned to the atmosphere by denitrification is also relatively small. The important point is that although nitrogen exchanges between soil and atmosphere are significant over the long term, the majority of nitrogen in most ecosystems is recycled locally by decomposition and reassimilation.

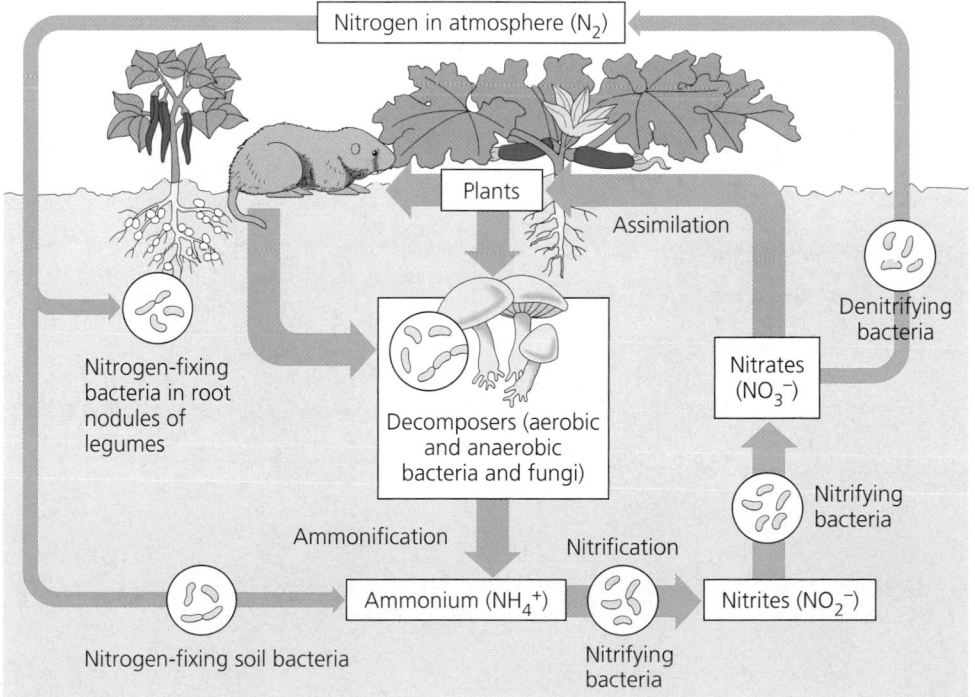

FIGURE 54.18 The nitrogen cycle. The addition of nitrogen from the atmosphere and its return via denitrification involve relatively small amounts compared to the local recycling that occurs in the soil or water. In some ecosystems, NH_4^+ and NO_3^- dissolved in rain add nitrogenous minerals to the soil (a process not included in this diagram).

The Phosphorus Cycle

Organisms require phosphorus as a major constituent of nucleic acids, phospholipids, and ATP and other energy shuttles and as a mineral constituent of bones and teeth.

In some respects, the phosphorus cycle is simpler than either the carbon or nitrogen cycle. Phosphorus cycling does not include movement through the atmosphere because there are no significant phosphorus-containing gases. In addition, phosphorus occurs in only one biologically important inorganic form, phosphate (PO_4^{3-}), which plants absorb and use for organic synthesis. The weathering of rocks gradually adds phosphate to soil (FIGURE 54.19). After producers incorporate phosphorus into biological molecules, it is transferred to consumers in organic form and then added back to the soil by the excretion of phosphate by animals and by the action of bacterial and fungal decomposers on detritus.

Humus and soil particles bind phosphate, so that the recycling of phosphorus tends to be quite localized in ecosystems. However, phosphorus does leach into the water table, gradually draining from terrestrial ecosystems to the sea. Severe erosion can hasten this drain, but the weathering of rocks generally keeps pace with the loss of phosphate. Phosphate that reaches the ocean gradually accumulates in sediments and becomes incorporated into rocks that may later be included in terrestrial ecosystems as a result of geologic processes that raise the seafloor or lower sea level at a particular location. Thus, most phosphate recycles locally among soil, plants, and consumers on the scale of ecological time, while a parallel sedimentary cycle removes and restores terrestrial phosphorus over geologic time. The same general pattern applies to other nutrients that lack atmospheric forms.

FIGURE 54.20 reviews chemical cycling in ecosystems in their most general form. Note again the key role that detritivores (decomposers) play.

Decomposition rates largely determine the rates of nutrient cycling

The rates at which nutrients cycle in different ecosystems are extremely variable, mostly as a result of differences in rates of decomposition. In tropical rain forests, most organic material decomposes in a few months to a few years, while in temperate forests, decomposition takes, on average, four to six years. In the tundra, decomposition can take 50 years, and in aquatic ecosystems, where most decomposition occurs in anaerobic bottom muds, it may occur even more slowly. The temperature and the availability of water and O_2 all affect rates of decomposition and thus nutrient cycling times. Other factors that can influence nutrient cycling are local soil chemistry and the frequency of fires.

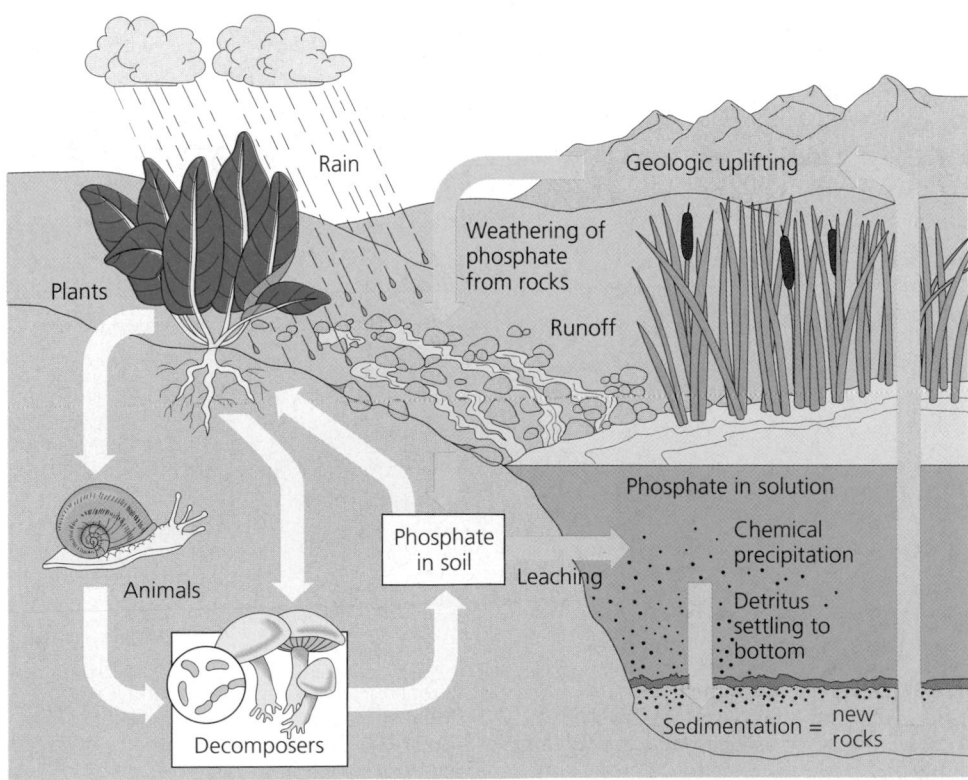

FIGURE 54.19 The phosphorus cycle. Phosphorus, which does not have an atmospheric component, tends to cycle locally (yellow arrows). Generally, small losses from terrestrial systems caused by leaching are balanced by gains from the weathering of rocks. In aquatic systems, as in terrestrial systems, phosphorus is cycled through food webs. Some phosphorus is lost from the ecosystem through chemical precipitation or through settling of detritus to the bottom, where sedimentation may lock away some of the nutrient before biological processes can reclaim it. On a much longer time scale, this phosphorus may become available to ecosystems again through geologic processes such as uplifting (gold arrows).

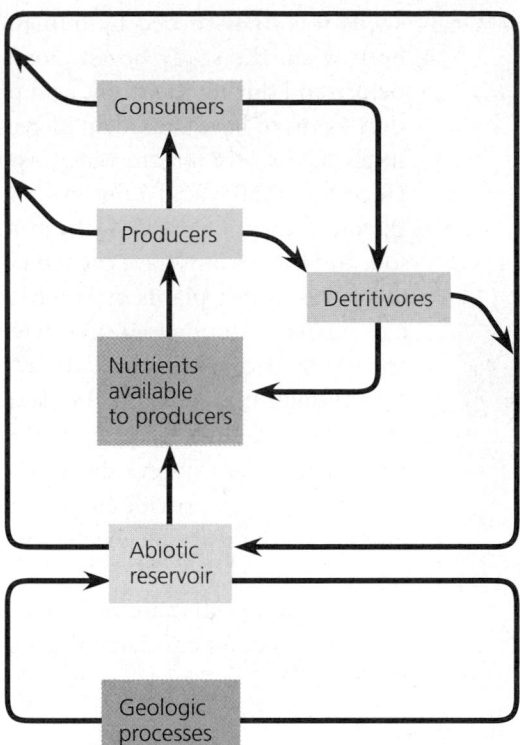

FIGURE 54.20 Review: Generalized scheme for biogeochemical cycles.

In some parts of a tropical rain forest, key nutrients such as phosphorus occur in the soil at levels far below those typical of a temperate forest. At first this might seem paradoxical because tropical forests generally have very high production. The key to this apparent riddle is rapid decomposition in tropical areas because of the warm temperatures and abundant precipitation. In addition, the immense biomass of these forests creates a high demand for nutrients, which are absorbed almost as soon as they become available through the action of decomposers. As a result of rapid decomposition, relatively little organic material accumulates as leaf litter on the floor of tropical rain forests; about 75% of the nutrients in the ecosystem are present in the woody trunks of trees, and about 10% are contained in the soil. Thus, the relatively low concentrations of some nutrients in the soil of tropical rain forests result from a fast cycling time, not an overall scantiness of these elements in the ecosystem.

In temperate forests, where decomposition is much slower, the soil may contain 50% of all the organic material in the ecosystem. The nutrients present in temperate forest detritus and soil may remain there for fairly long periods of time before being assimilated by plants.

In aquatic ecosystems, bottom sediments are comparable to the detritus layer in terrestrial ecosystems, but algae and aquatic plants usually assimilate nutrients directly from the water. Thus, these sediments often constitute a nutrient sink, and aquatic ecosystems can only be very productive if there is

interchange between the bottom layers of water and the surface (see FIGURE 50.16).

Nutrient cycling is strongly regulated by vegetation

Many research groups are conducting **long-term ecological research (LTER)** to monitor the dynamics of natural ecosystems over long periods of time. Let's examine one example of LTER that has helped ecologists understand the key role plants play in regulating nutrient cycles.

THE PROCESS OF SCIENCE

Since 1963, a team of scientists has been conducting a long-term study of nutrient cycling in a forest ecosystem. The study site is the Hubbard Brook Experimental Forest, in the White Mountains of New Hampshire. It is a deciduous forest with several valleys, each drained by a small creek that is a tributary of Hubbard Brook. Bedrock impenetrable to water is close to the surface of the soil, and each valley constitutes a watershed that can drain only through its creek.

The research team first determined the mineral budget for each of six valleys by measuring the inflow and outflow of several key nutrients. They collected rainfall at several sites to measure the amount of water and dissolved minerals added to the ecosystem. To monitor the loss of water and minerals, they constructed a small concrete dam with a V-shaped spillway across the creek at the bottom of each valley (FIGURE 54.21a, p. 1214). About 60% of the water added to the ecosystem as rainfall and snow exits through the stream, and the remaining 40% is lost by transpiration from plants, evaporation from other organisms, and evaporation from the soil.

Preliminary studies confirmed that internal cycling within a terrestrial ecosystem conserves most of the mineral nutrients. Mineral inflow and outflow balanced and were relatively small compared with the quantity of minerals being recycled within the forest ecosystem. For example, only about 0.3% more Ca^{2+} left a valley via its creek than was added by rainwater, and this small net loss was probably replaced by chemical decomposition of the bedrock. During most years, the forest actually registered small net gains of a few mineral nutrients, including nitrogenous ones.

In 1966, one of the valleys, with an area of 15.6 hectares, was completely logged and then sprayed with herbicides for three years to prevent regrowth of plants (FIGURE 54.21b). All the original plant material was left in place to decompose. The inflow and outflow of water and minerals in the experimentally altered watershed were compared with the inflow and outflow in a control watershed for three years. Water runoff from the altered watershed increased by 30–40%, apparently because there were no plants to absorb and transpire water from the soil. Net losses of minerals from the altered watershed were huge. The concentration of Ca^{2+} in the creek increased fourfold, for example, and the concentration of K^+ increased by a

(a) Concrete dams built across streams at the bottom of watersheds enabled researchers to monitor the outflow of water and nutrients from the ecosystem.

(b) Some watersheds were completely logged to study the effects of the loss of vegetation on drainage and nutrient cycling.

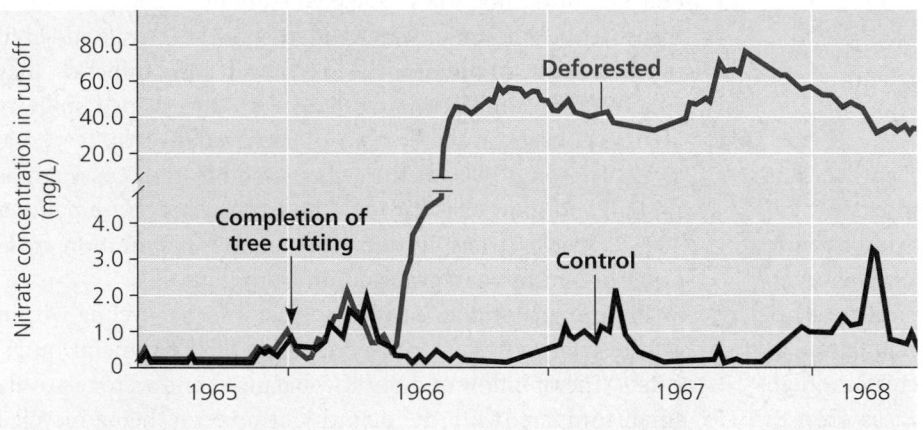

(c) The concentration of nitrate in runoff from the deforested watershed was 60 times greater than in a control (unlogged) watershed.

FIGURE 54.21 Nutrient cycling in the Hubbard Brook Experimental Forest: an example of long-term ecological research.

factor of 15. Most remarkable was the loss of nitrate, which increased in concentration in the creek 60-fold (FIGURE 54.21c). Not only was this vital mineral nutrient drained from the ecosystem, but nitrate in the creek reached a level considered unsafe for drinking water.

This study demonstrated that the amount of nutrients leaving an intact forest ecosystem is controlled by the plants themselves; when plants are not present to retain them, nutrients are lost from the system. These effects are almost immediate, occurring within a few months of logging, and continue as long as plants are absent.

After 35 years, data from Hubbard Brook point to some other long-term trends. It now appears that none of the water-

sheds was undisturbed by human activity even when the study began. For instance, before and during the study, acid precipitation seems to have caused an increase in the levels of Ca^{2+} in stream water. Apparently, since the 1950s, acid rain and snow have dissolved most of the Ca^{2+} from the forest soil, and the streams have carried it away. By the 1990s, forest plants at Hubbard Brook had virtually stopped adding new growth, apparently because of a lack of Ca^{2+}.

Although the Hubbard Brook study and several other LTER projects were designed to assess natural ecosystem dynamics, the results continue to provide important insight into the mechanisms by which human activities affect these processes. In this chapter's last section, we will examine how human intrusions in ecosystem dynamics are altering the biosphere.

HUMAN IMPACT ON ECOSYSTEMS AND THE BIOSPHERE

As the human population has grown in size, our activities and technological capabilities have intruded in one way or another into the dynamics of most ecosystems. Even where we have not completely destroyed a natural system, our actions have disrupted the trophic structure, energy flow, and chemical cycling of ecosystems in most areas of the world. The effects are sometimes local or regional, but the ecological impact of humans can be widespread or even global. For example, acidic gases may be carried by prevailing winds and fall as acid rain hundreds or thousands of miles from the smokestacks emitting the gases.

The human population is disrupting chemical cycles throughout the biosphere

Human activity often intrudes in nutrient cycles by removing nutrients from one part of the biosphere and adding them to another. This may result in the depletion of key nutrients in one area, excesses in another place, and the disruption of the natural chemical cycles in both locations. For example, nutrients in the soil of croplands soon appear in the wastes of humans and livestock. Then they appear in streams and lakes through runoff from fields and sewage discharge. Someone eating a piece of broccoli in Washington, DC, is consuming nutrients that only days before might have been in the soil in California; and a short time later, some of these nutrients will be in the Potomac River on their way to the sea, having passed through an individual's digestive system and the local sewage facilities.

Humans have intruded on nutrient cycles to such an extent that it is no longer possible to understand any cycle without taking human effects into account. In addition to transporting nutrients from one location to another, we have added entirely new materials, many of them toxic, to ecosystems. Let's examine a few specific examples of how humans are impacting the biosphere's chemical dynamics.

Agricultural Effects on Nutrient Cycling

After natural vegetation is cleared from an area, crops may be grown for some time using the existing reserve of nutrients in the soil; nutrient supplementation isn't required. However, in agricultural ecosystems, a substantial fraction of these nutrients is not recycled but is exported from the area in the form of crop biomass (FIGURE 54.22). The "free" period for crop production—when there is no need to add nutrients to the soil—varies greatly. When some of the early North American prairie lands were first tilled, for example, good crops could be produced for many years because the large store of organic materials in the soil continued to decompose and provide nutrients. By contrast, some cleared land in the tropics can be farmed for only one or two years because so few of the ecosystems' nutrients are contained in the soil (see p. 1213). Eventually, in any area under intensive agriculture, the natural store of nutrients becomes exhausted. When this happens, fertilizer must be added. The industrially synthesized fertilizers used extensively today come from oil and are produced at considerable expense of both money and energy.

Agriculture has a great impact on the nitrogen cycle. Cultivation—breaking up and mixing the soil—increases the rate of decomposition of organic matter, releasing usable nitrogen that is then removed from the ecosystem when crops are harvested. As we saw in the case of Hubbard Brook, ecosystems from which living plants are removed lose nitrogen not only because it is removed with the plants themselves, but because without plants to take them up, nitrates continue to be leached from the ecosystem. Industrially synthesized fertilizer is used to make up for the loss of usable nitrogen from agricultural ecosystems.

Recent studies indicate that human activities have approximately doubled the globe's supply of fixed nitrogen available to primary producers. The main cause is industrial nitrogen fixation for fertilizers, but increased cultivation of legumes, with their nitrogen-fixing symbionts, and burning are also

FIGURE 54.22 Agricultural impact on soil nutrients. Transport of harvested plant biomass to market removes mineral nutrients that would otherwise be cycled back to the local soil. To replace the lost nutrients, farmers must apply fertilizers—either organic fertilizers, such as manure or mulch, or manufactured fertilizers.

important. (Fire releases nitrogen compounds stored in soil and vegetation, thereby enhancing the cycling of nitrogen compounds available to photosynthesizers.) In addition to its effects on local soil and water chemistry, the excessive supplements of fixed nitrogen are also associated with a greater release of N_2 and nitrogen oxides into the air by denitrifying bacteria (see FIGURE 54.18). Nitrogen oxides can contribute to atmospheric warming, to the depletion of atmospheric ozone, and in some ecosystems to acid precipitation.

Critical Load and Nutrient Cycles

Excesses of nitrogenous minerals in the soil eventually leach into groundwater or run off directly into freshwater and marine ecosystems. Many rivers contaminated with nitrates and ammonium from agricultural runoff and sewage drain into the North Atlantic Ocean, with the highest nitrogen inputs to the ocean coming from northern Europe. Throughout the Northern Hemisphere, nitrate concentrations in rivers are rising in proportion to human population along the rivers. For example, nitrate levels in the Mississippi River have more than doubled since 1965. Groundwater concentrations of nitrate are also increasing in agricultural areas, sometimes exceeding the maximum level considered safe for drinking water (10 mg of nitrate per liter).

In certain situations, the addition of nitrogen to ecosystems by human activities can actually have positive effects, at least from the human perspective. Adding nitrogenous fertilizers can offset the nitrogen limitation of primary production that is common in terrestrial ecosystems. For example, in Swedish forests, which are nitrogen limited, agricultural runoff of nitrates and ammonia and emissions of nitrogenous compounds from factories are correlated with a 30% boost in the growth rates of trees during the past 50 years. In other words, humans have been inadvertently fertilizing the forests.

The key issue seems to be **critical load,** the amount of added nitrogen that can be absorbed by plants without damaging ecosystem integrity. It is the nitrogen that exceeds critical load that turns up in groundwater and aquatic ecosystems such as lakes, a problem we examine in more detail next.

Accelerated Eutrophication of Lakes

As we discussed in Chapter 50, lakes are classified on a scale of increasing nutrient availability as oligotrophic, mesotrophic, or eutrophic (see FIGURE 50.19). In an oligotrophic lake, primary productivity is relatively low because the mineral nutrients required by phytoplankton are scarce. In other lakes, basin and watershed characteristics result in the addition of more nutrients. These nutrients are captured by the primary producers and then continuously recycled through the lake's food webs. Thus, the overall productivity is higher in mesotrophic lakes and highest in eutrophic (from the Greek, meaning "well nourished") ones.

Human intrusion has disrupted freshwater ecosystems by what is termed **cultural eutrophication.** Sewage and factory wastes; runoff of animal waste from pastures and stockyards; and the leaching of fertilizer from agricultural, recreational, and urban areas has overloaded many streams, rivers, and lakes with inorganic nutrients. This enrichment often results in an explosive increase in the density of photosynthetic organisms (see FIGURE 54.8). Shallower areas become weed choked, making boating and fishing impossible. Large "blooms" of algae and cyanobacteria become common, sometimes resulting in increased oxygen production during the day but reduced oxygen levels at night because of respiration by these large populations of organisms. As the photosynthetic organisms die and organic material accumulates at the lake bottom, detritivores use all the oxygen in the deeper waters. All these effects may make it impossible for some organisms to survive. For example, cultural eutrophication of Lake Erie wiped out commercially important fishes such as blue pike, whitefish, and lake trout by the 1960s. Since then, tighter regulations on the dumping of wastes into the lake have enabled some fish populations to rebound, but many of the native species of fishes and invertebrates have not recovered. Government policies to ensure better water quality should be a high priority in all countries.

Combustion of fossil fuels is the main cause of acid precipitation

The burning of wood and the combustion of coal and other fossil fuels release oxides of sulfur and nitrogen that react with water in the atmosphere to form sulfuric and nitric acid, respectively. The acids eventually fall back to Earth's surface as acid precipitation. **Acid precipitation** is defined as rain, snow, or fog that has a pH less than 5.6 (see FIGURE 3.9). It lowers the pH of aquatic ecosystems and affects the soil chemistry of terrestrial ecosystems.

Although acid precipitation due to fuel combustion has been occurring ever since the Industrial Revolution, emissions have increased during the past hundred years, mainly as a result of ore smelters and electrical generating plants. Ecologists first brought the environmental impact of acid precipitation to public attention in the 1960s, when they began to document damage to forests and lakes in Europe and eastern North America. It is clearly a regional or even a global problem, not a local one. To counter local pollution problems, smelters and generating plants are built with very tall exhaust stacks (over 300 m high). This reduces pollution at ground level, but exports the problem far downwind. The nitrogenous and sulfurous pollutants from fuel combustion can drift hundreds of kilometers before falling as acid precipitation. Ecologists began to realize that lakes in eastern Canada were dying because of air pollution from factories in the midwest. Lakes in southern Norway were losing fish because of acid rain from pollutants generated in Great Britain. By 1980, precipitation in large areas of Europe and North America averaged pH 4.0–4.5, with "record" storms occasionally dropping rain as acidic as pH 3.0 (FIGURE 54.23).

In terrestrial ecosystems, such as the deciduous forests of New England, the change in soil pH due to acid precipitation causes calcium and other nutrients to leach from the soil. The nutrient deficiencies affect the health of plants and limit their growth.

Freshwater ecosystems are particularly sensitive to acid precipitation. The lakes in North America and northern Europe that are most readily damaged by acid precipitation are those underlain by granite bedrock. Such lakes generally have relatively poor buffering capacity because the water is "soft," meaning the concentration of bicarbonate, an important buffer, is low. Fish populations have declined in thousands of such lakes in Norway and Sweden, where the pH of the water has dropped below 5.0. In Canada, newly hatched lake trout die when the pH drops below 5.4. Lake trout are keystone predators in many Canadian lakes, and when they are replaced by acid-tolerant fish such as yellow perch, the dynamics of food webs change dramatically (see FIGURE 53.14 to review keystone predators).

Environmental regulations and new industrial technologies have enabled many developed countries, including the United States, to reduce sulfur dioxide emissions during the past 30 years. In the United States, for example, sulfur dioxide emissions were reduced 25% between 1990 and 1996. The water chemistry in the streams and freshwater lakes of New England is slowly improving after decades of severe acid precipitation.

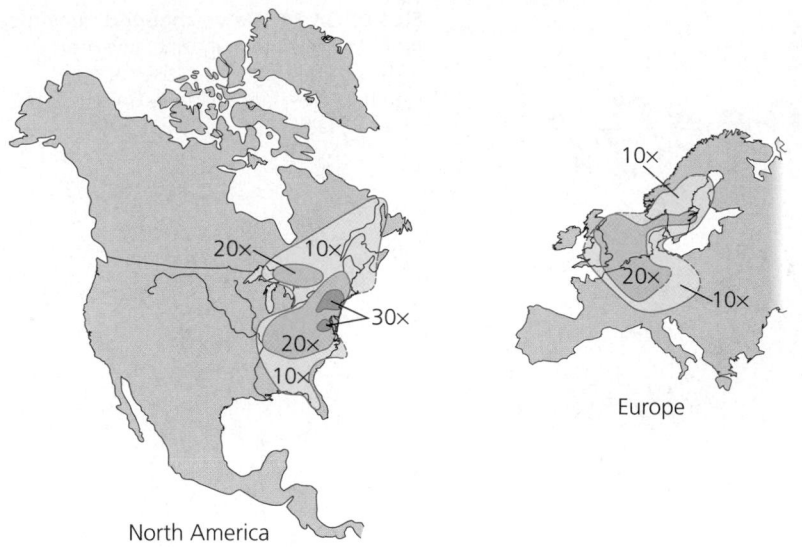

FIGURE 54.23 **Distribution of acid precipitation in North America and Europe. (a)** In these maps of North America and Europe with 1980 data superimposed, the numbers such as 10× identify the acidity of precipitation in those areas compared to a normal rainwater pH of 5.6. In an area marked 10×, precipitation was, on average, ten times more acidic than normal (which means the pH in those areas averaged 4.6). Other areas had precipitation 20× or 30× the normal rainfall acidity. **(b)** This U.S. map profiles pH averages for precipitation in 1999.

(a)

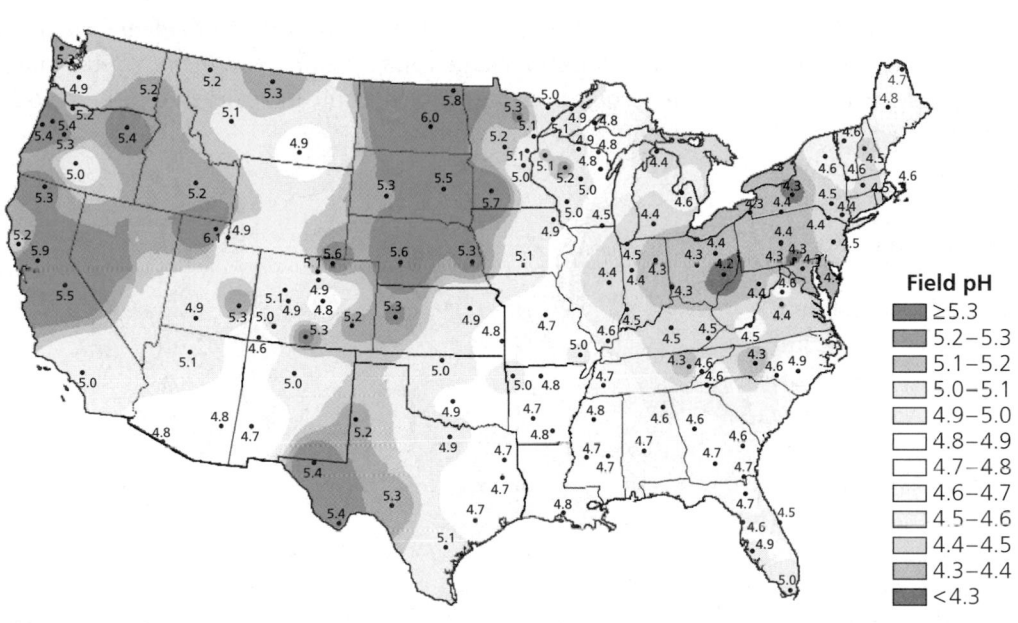

Field pH

■	≥5.3
■	5.2–5.3
■	5.1–5.2
□	5.0–5.1
□	4.9–5.0
□	4.8–4.9
□	4.7–4.8
□	4.6–4.7
□	4.5–4.6
□	4.4–4.5
■	4.3–4.4
■	<4.3

(b)

However, ecologists estimate that it will take another 10 to 20 years for these ecosystems to recover, even if sulfur dioxide emissions continue to decrease.

Toxins can become concentrated in successive trophic levels of food webs

Humans produce an immense variety of toxic chemicals, including thousands of synthetics previously unknown in nature, that are dumped into ecosystems with little regard for the ecological consequences. Many of these poisons cannot be degraded by microorganisms and consequently persist in the environment for years or even decades. In other cases, chemicals released into the environment may be relatively harmless but are converted to more toxic products by reaction with other substances or by the metabolism of microorganisms. For example, mercury, a by-product of plastic production, was once routinely expelled into rivers and the sea in an insoluble form. Bacteria in the bottom mud converted the waste to methyl mercury, an extremely toxic soluble compound that then accumulated in the tissues of organisms, including humans who consumed fish from the contaminated waters.

Organisms acquire toxic substances from the environment along with nutrients and water. Some of the poisons are metabolized and excreted, but others accumulate in specific tissues, especially fat. An example of a class of industrially synthesized compounds that act in this manner are the chlorinated hydrocarbons, which include many pesticides, such as DDT, and the industrial chemicals called PCBs (polychlorinated biphenols). Current research is implicating many of these compounds and others in endocrine system disruption in a large number of animal species, including humans. One of the reasons these toxins are so harmful is that they become more concentrated in successive trophic levels of a food web, a process called **biological magnification.** Magnification occurs because the biomass at any given trophic level is produced from a much larger biomass ingested from the level below. Thus, top-level carnivores tend to be the organisms most severely affected by toxic compounds that have been released into the environment.

A well-known example of biological magnification involves DDT, which has been used to control insects such as mosquitoes and agricultural pests. In the decade after World War II,

FIGURE 54.24 We've changed our tune.
In this 1947 *Time* magazine advertisement, a chorus line sings the praises of a pesticide that was banned in the United States 25 years later.

the chemical industry promoted the benefits of DDT before anybody really understood the ecological consequences (FIG-URE 54.24). By the 1950s, scientists began to understand that DDT persists in the environment and is transported by water to areas far from where it is applied. But by then, the poison had already become a global problem.

Because DDT is soluble in lipids, it collects in the fatty tissues of animals, and its concentration is magnified in higher trophic levels (FIGURE 54.25). Traces of DDT have been found in nearly every organism tested; it has even been found in human breast milk throughout the world. One of the first signs that DDT was a serious environmental problem was a decline in the populations of pelicans, ospreys, and eagles, birds that feed at the top of food chains. The accumulation of DDT (and DDE, a product of its partial breakdown) in the tissues of these birds interfered with the deposition of calcium in their eggshells, a trend that may have already begun because of other environmental contaminants. When these birds tried to incubate their eggs, the weight of the parents broke the shells of affected eggs, resulting in catastrophic declines in their reproduction rates. Rachel Carson's *Silent Spring* helped bring the problem to public attention in the 1960s (see Chapter 50), and DDT was banned in the United States in 1971. A dramatic recovery in populations of the affected bird species followed. The pesticide is still used in many other parts of the world, however.

Human activities may be causing climate change by increasing carbon dioxide concentration in the atmosphere

Many human activities release a variety of gaseous waste products. We once thought that the vastness of the atmosphere could absorb these materials without significant consequences, but we now know that the finite volume of the atmosphere

means that human intrusions can cause fundamental changes in its composition and its interactions with the rest of the biosphere. One pressing problem is the rising level of carbon dioxide in the atmosphere.

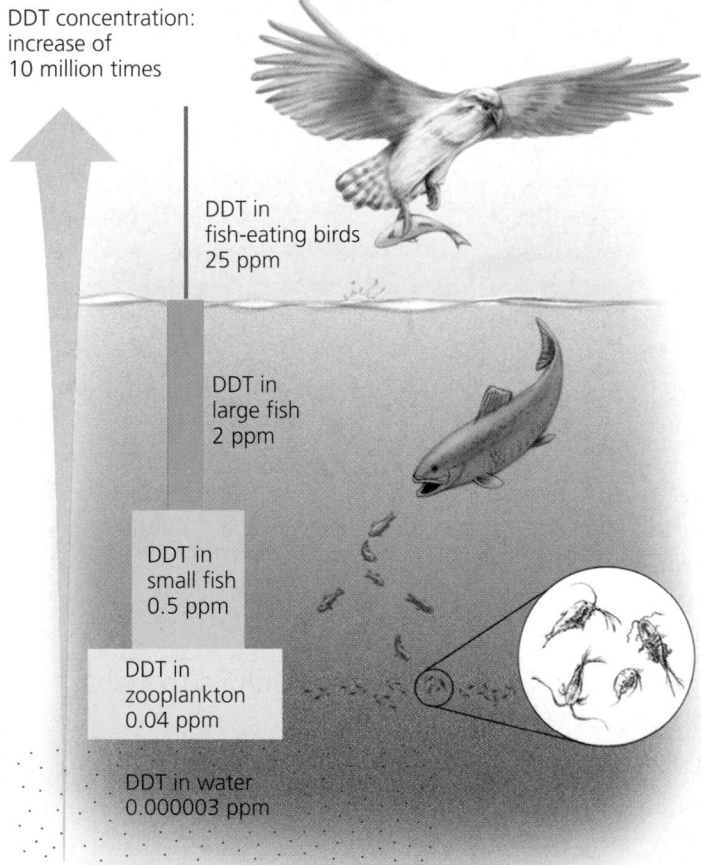

DDT concentration: increase of 10 million times

DDT in fish-eating birds 25 ppm

DDT in large fish 2 ppm

DDT in small fish 0.5 ppm

DDT in zooplankton 0.04 ppm

DDT in water 0.000003 ppm

FIGURE 54.25 Biological magnification of DDT in a food chain.
DDT concentration in a Long Island Sound food chain was magnified by a factor of about 10 million, from just 0.000003 part per million (ppm) as a pollutant in seawater to a concentration of about 25 ppm in the fish-eating osprey, a bird at the top of this food pyramid.

Rising Atmospheric CO₂

Since the Industrial Revolution, the concentration of CO_2 in the atmosphere has been increasing as a result of the combustion of fossil fuels and the burning of enormous quantities of wood removed by deforestation. Various methods of measurement have estimated that the average carbon dioxide concentration in the atmosphere before 1850 was about 274 parts per million (ppm). When a monitoring station on Hawaii's Mauna Loa peak began making very accurate measurements in 1958, the CO_2 concentration was 316 ppm (FIGURE 54.26). Today, the concentration of CO_2 in the atmosphere exceeds 370 ppm, an increase of about 14% since the measurements began. If CO_2 emissions continue to increase at the present rate, by the year 2075, the atmospheric concentration of this gas will be double what it was at the start of the Industrial Revolution.

An increased productivity by vegetation is one predictable consequence of increasing CO_2 levels. In fact, when CO_2 concentrations are raised in experimental chambers such as greenhouses, most plants respond with increased growth. However, because C_3 plants are more limited than C_4 plants by CO_2 availability (see Chapter 10), one effect of increasing CO_2 concentrations on a global scale may be the spread of C_3 species into terrestrial habitats previously favoring C_4 plants. This may have important agricultural implications. For example, corn, a C_4 plant and the most important grain crop in the United States, may be replaced on farms by wheat and soybeans, C_3 crops that will outproduce corn in a CO_2-enriched environment. However, it is not yet possible to predict accurately the gradual and complex effects that rising CO_2 levels will have on species composition in nonagricultural communities.

The Greenhouse Effect

One factor that complicates predictions about the long-term effects of rising atmospheric CO_2 concentration is its possible influence on Earth's heat budget. Much of the solar radiation that strikes the planet is reflected back into space. Although CO_2 and water vapor in the atmosphere are transparent to visible light, they intercept and absorb much of the reflected infrared radiation, re-reflecting some of it back toward Earth. This process retains some of the solar heat. If it were not for this **greenhouse effect,** the average air temperature at Earth's surface would be −18°C, and most life as we know it could not exist. The marked increase in atmospheric CO_2 concentrations during the last 150 years concerns many ecologists and environmentalists because of its potential to increase global temperature.

Global Warming

Scientists continue to construct mathematical models in attempts to predict how increasing levels of atmospheric CO_2 will affect global temperatures. To date, no models are sophisticated enough to include all the biotic and abiotic factors that can influence atmospheric gas concentrations and temperature (for example, cloud cover, CO_2 uptake by photosynthetic organisms, and the effects of particles in the air). However, a number of studies predict a doubling of CO_2 concentration by the end of the twenty-first century and an associated average temperature increase of about 2°C. Supporting these models is a correlation between CO_2 levels and temperatures in prehistoric times. Climatologists can actually measure CO_2 levels in bubbles trapped in glacial ice at different times in Earth's history. Prehistoric temperatures are inferred by several methods, including analysis of past vegetation based on fossils.

An increase of only 1.3°C would make the world warmer than at any time in the past 100,000 years. A worst-case scenario suggests

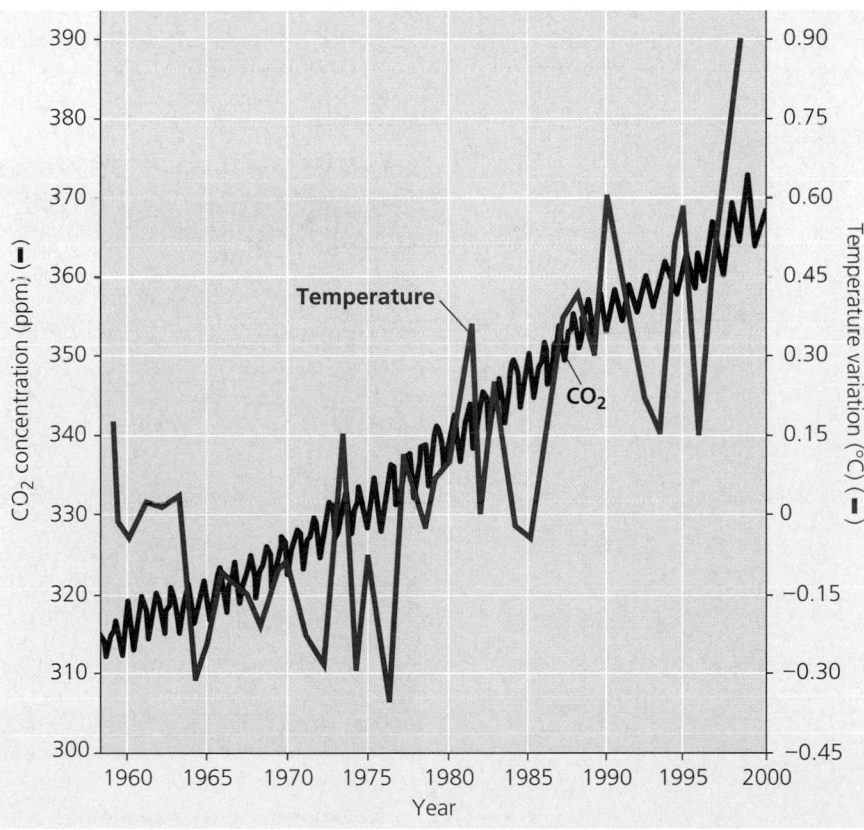

FIGURE 54.26 The increase in atmospheric carbon dioxide and average temperatures from 1958 to 2000. In addition to normal seasonal fluctuations in CO_2 levels, this graph shows a steady increase in the total amount of CO_2 in the atmosphere (black). These measurements are being taken at a relatively remote site in Hawaii, where the air is free from the variable short-term effects that occur near large urban areas. Though average temperatures over the same time period fluctuate a great deal (red), there is a warming trend. Climatologists predict that temperatures could rise about 2°C over the next 100 years if atmospheric CO_2 levels continue to rise at the current rate.

that the warming would be greatest near the poles; the resultant melting of polar ice might raise sea level by an estimated 100 m, gradually flooding coastal areas 150 km (or more) inland from the current coastline. New York, Miami, Los Angeles, and many other cities would then be under water. A warming trend would also alter the geographic distribution of precipitation, making major agricultural areas of the central United States much drier. However, the various mathematical models disagree about the details of how climate in each region will be affected. By studying how *past* periods of global warming and cooling affected plant communities, ecologists are trying to predict the consequences of *future* temperature changes. Analysis of fossilized pollen provides evidence that plant communities change dramatically with changes in temperature. Past climate changes occurred gradually, and plant and animal populations could migrate into areas where abiotic conditions allowed them to survive. A major concern is the projected high rates of climate change—by some estimates, higher than at any time in the past 10,000 years. Many organisms, especially plants that cannot disperse rapidly over long distances, will probably not be able to survive such rapid changes.

Coal, natural gas, gasoline, wood, and other organic fuels central to modern life cannot be burned without releasing CO_2. The apparent warming of the planet that is now under way as a result of the addition of CO_2 to the atmosphere is a problem of uncertain consequences and no simple solutions. Given the importance of combustion to our increasingly industrialized societies, stabilizing CO_2 emissions will require concerted international effort and the acceptance of dramatic changes in both personal lifestyles and industrial processes. Many ecologists believe that this effort suffered a major setback in 2001, when the United States pulled out of the Kyoto Protocol, a 1997 pledge by the industrialized nations to reduce their CO_2 output by about 5% over a ten-year period.

Human activities are depleting atmospheric ozone

Life on Earth is protected from the damaging effects of ultraviolet (UV) radiation by a protective layer of ozone molecules (O_3) that is present in the lower stratosphere between 17 and 25 km above Earth's surface. Ozone absorbs UV radiation, preventing much of it from contacting organisms in the biosphere. Satellite studies of the atmosphere suggest that the ozone layer has been gradually "thinning" since 1975 (FIGURE 54.27).

The destruction of atmospheric ozone probably results mainly from the accumulation of chlorofluorocarbons, chemicals used for refrigeration, as propellants in aerosol cans, and in certain manufacturing processes. When the breakdown products from these chemicals rise to the stratosphere, the chlorine they contain reacts with ozone, reducing it to molec-

FIGURE 54.27 Erosion of Earth's ozone shield.

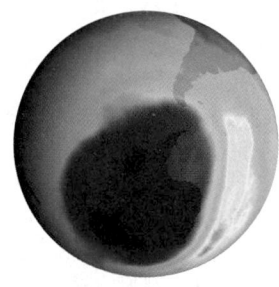

(a) Ozone hole. The ozone hole over the Antarctic is visible as the blue patch in this image based on atmospheric data

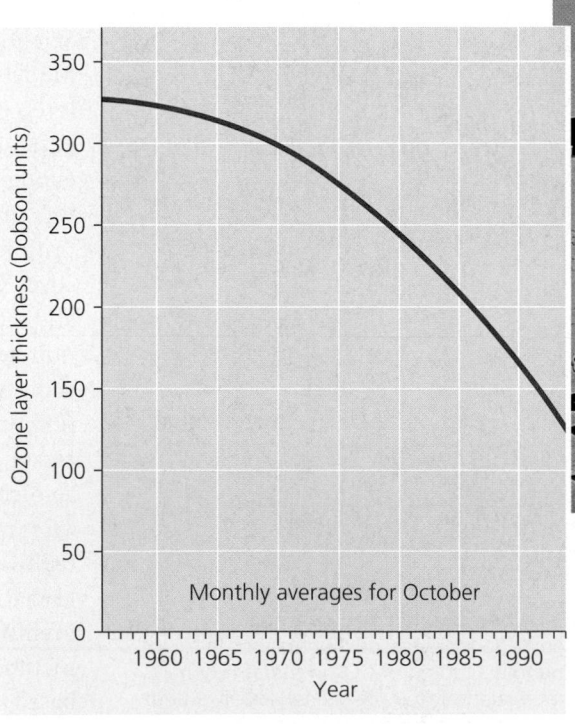

(b) Thickness of the ozone layer. This graph tracks the thickness of the ozone layer over Antarctica in units called Dobsons.

(c) Exposure to UV radiation. Australia, home to these sun lovers, already has the world's highest rate of skin cancer. More intense UV radiation through an eroded ozone shield increases the cancer risk.

ular O_2. Subsequent chemical reactions liberate the chlorine, allowing it to react with other ozone molecules in a catalytic chain reaction. The effect is most apparent over Antarctica, where cold winter temperatures facilitate these atmospheric reactions. Scientists first described the "ozone hole" over Antarctica in 1985 and have since documented that it is a seasonal phenomenon that grows and shrinks in an annual cycle. However, the magnitude of ozone depletion and the size of the ozone hole have generally increased in recent years, and the hole sometimes extends as far as the southernmost portions of Australia, New Zealand, and South America. At the more heavily populated middle latitudes, ozone levels have decreased 2–10% during the past 20 years.

The consequences of ozone depletion for life on Earth may be quite severe. Some scientists expect increases in both lethal and nonlethal forms of skin cancer and in cataracts among humans, as well as unpredictable effects on crops and natural communities, especially the phytoplankton that are responsible for a large proportion of Earth's primary production. The danger posed by ozone depletion is so great that many nations have agreed to end the production of chlorofluorocarbons within a decade. Unfortunately, even if all chlorofluorocarbons were totally banned today, chlorine molecules that are already in the atmosphere will continue to influence stratospheric ozone levels for at least a century.

■   ■   ■

The human impact on Earth's ozone shield is just one more example of how far our technological tentacles reach in disrupting the dynamics of ecosystems and the entire biosphere. How are human activities affecting Earth's biodiversity? In the last chapter of this book, we explore that question and highlight how progress in conservation biology may help slow the loss of species.

CHAPTER 54 REVIEW

Go to the Campbell Biology website (www.campbellbiology.com) to explore an interactive version of the Chapter Review.

Summary of Key Concepts

THE ECOSYSTEM APPROACH TO ECOLOGY

■ **Trophic relationships determine the routes of energy flow and chemical cycling in an ecosystem (p. 1199, FIGURE 54.1)** Energy and nutrients pass from primary producers (autotrophs) to primary consumers (herbivores) and then to secondary consumers (carnivores). Energy *flows* through an ecosystem, entering as light and exiting as heat. Nutrients *cycle* within an ecosystem.

■ **Decomposition connects all trophic levels (pp. 1199–1200, FIGURE 54.2)** Detritivores, mainly bacteria and fungi, recycle essential chemical elements by decomposing organic material and returning elements to inorganic reservoirs.

■ **The laws of physics and chemistry apply to ecosystems (p. 1200)** The laws of physics and chemistry help us understand energy flow through ecosystems. Energy is conserved, but is degraded to heat by ecosystem processes.

PRIMARY PRODUCTION IN ECOSYSTEMS

■ **An ecosystem's energy budget depends on primary production (pp. 1200–1202, FIGURES 54.3, 54.4)** The energy assimilated during photosynthesis is a tiny fraction of the solar radiation reaching Earth; it sets the spending limit for the global energy budget. Gross primary production is the total energy assimilated by an ecosystem in a given time period. Net primary production, the energy accumulated in autotroph biomass, equals gross primary production minus the energy used by the primary producers for respiration. Only net primary production is available to consumers.
Web/CD Case Study in The Process of Science: How Do Temperature and Light Affect Primary Production?

■ **In aquatic ecosystems, light and nutrients limit primary production (pp. 1201–1205, FIGURES 54.5–54.8, TABLE 54.1)** Within the photic zone, the factor that most often limits primary production is a nutrient such as nitrogen.

■ **In terrestrial ecosystems, temperature, moisture, and nutrients limit primary production (p. 1205, FIGURE 54.9)** Climatic factors such as temperature and moisture affect production on a large geographic scale. More locally, a soil nutrient is often the limiting factor in primary production.

SECONDARY PRODUCTION IN ECOSYSTEMS

■ **The efficiency of energy transfer between trophic levels is usually less than 20% (pp. 1206–1208, FIGURES 54.10–54.14)** The amount of energy available to each trophic level is determined by the net primary production and the efficiencies with which food energy is converted to biomass at each link of the food chain. The percentage of energy transferred from one trophic level to the next, called trophic efficiency, is generally 5–20%. Pyramids of production, biomass, and numbers are consequences of the relatively low trophic efficiency.
Web/CD Activity 54A: Pyramids of Production

- **Herbivores consume a small percentage of vegetation: the green world hypothesis (p. 1208)** Predators, disease, competition, nutrient limitations, and other factors keep the populations of herbivores in check.

THE CYCLING OF CHEMICAL ELEMENTS IN ECOSYSTEMS

- **Biological and geologic processes move nutrients between organic and inorganic compartments (pp. 1209–1213, FIGURES 54.15–54.20)** Water moves in a global cycle driven by solar energy. The carbon cycle primarily reflects the reciprocal processes of photosynthesis and cellular respiration. Nitrogen enters ecosystems by atmospheric deposition and nitrogen fixation by prokaryotes, but most of the nitrogen cycling in natural ecosystems involves local cycles between organisms and soil or water. The phosphorus cycle mainly occurs on a more localized scale than the water, carbon, and nitrogen cycles.
  Web/CD Activity 54B: *Energy Flow and Chemical Cycling*
  Web/CD Activity 54C: *The Carbon Cycle*
  Web/CD Activity 54D: *The Nitrogen Cycle*

- **Decomposition rates largely determine the rates of nutrient cycling (pp. 1212–1213)** The proportion of a nutrient in a particular form and its cycling time in that form vary among ecosystems, mostly because of differences in the rate of decomposition.

- **Nutrient cycling is strongly regulated by vegetation (pp. 1213–1214, FIGURE 54.21)** Long-term ecological research projects monitor ecosystem dynamics over relatively long periods of time. Studies in deciduous forests in New Hampshire have shown that logging increases water runoff and can cause huge losses of minerals.

HUMAN IMPACT ON ECOSYSTEMS AND THE BIOSPHERE

- **The human population is disrupting chemical cycles throughout the biosphere (pp. 1214–1216, FIGURE 54.22)** Agricultural practices result in the constant removal of nutrients from ecosystems, so that large supplements are continually required. Considerable amounts of the nutrients in fertilizer move into aquatic ecosystems, where they can stimulate excess algal growth (eutrophication).
  Web/CD Activity 54E: *Water Pollution from Nitrates*

- **Combustion of fossil fuels is the main cause of acid precipitation (pp. 1216–1217, FIGURE 54.23)** North American and European ecosystems downwind from industrial regions have been damaged by rain and snow containing nitric acid and sulfuric acid.

- **Toxins can become concentrated in successive trophic levels of food webs (pp. 1217–1218, FIGURES 54.24, 54.25)** The release of toxic wastes has polluted the environment with harmful substances that often persist for long periods of time. DDT is an example of the many toxic substances that become concentrated along the food chain by biological magnification.

- **Human activities may be causing climate change by increasing carbon dioxide concentration in the atmosphere (pp. 1218–1220, FIGURE 54.26)** Because of the burning of wood and fossil fuels, CO_2 in the atmosphere has been steadily increasing. The ultimate effects may include significant warming and other influences on climate.
  Web/CD Activity 54F: *The Greenhouse Effect*

- **Human activities are depleting atmospheric ozone (pp. 1220–1221, FIGURE 54.27)** The ozone layer reduces the penetration of UV radiation through the atmosphere. Unfortunately, chlorine-containing pollutants are eroding the ozone layer, with dangerous results.

Self-Quiz

1. Which of the following organisms is *incorrectly* paired with its trophic level?
   a. cyanobacteria—primary producer
   b. grasshopper—primary consumer
   c. zooplankton—secondary consumer
   d. eagle—tertiary consumer
   e. fungi—detritivore

2. One of the lessons from a pyramid of production is that
   a. only one-half of the energy in one trophic level is passed on to the next level.
   b. most of the energy from one trophic level is incorporated into the biomass of the next level.
   c. the energy lost as heat or lost in cellular respiration is 10% of the available energy of each trophic level.
   d. production efficiency is highest for primary consumers.
   e. eating grain-fed beef is an inefficient means of obtaining the energy trapped by photosynthesis.

3. The role of decomposers in the nitrogen cycle is to
   a. fix N_2 into ammonia.
   b. release ammonia from organic compounds, thus returning it to the soil.
   c. denitrify ammonia, thus returning N_2 to the atmosphere.
   d. convert ammonia to nitrate, which can then be absorbed by plants.
   e. incorporate nitrogen into amino acids and organic compounds.

4. The Hubbard Brook Experimental Forest study demonstrated all of the following *except* that
   a. most minerals were recycled within a forest ecosystem.
   b. mineral inflow and outflow within a natural watershed were nearly balanced.
   c. deforestation resulted in an increase in water runoff.
   d. the nitrate concentration in waters draining the deforested area became dangerously high.
   e. deforestation caused a large increase in the density of soil bacteria.

5. The recent increase in atmospheric CO_2 concentration is mainly a result of an increase in
   a. primary production.
   b. the biosphere's biomass.
   c. the absorption of infrared radiation escaping from Earth.
   d. the burning of fossil fuels and wood.
   e. cellular respiration by the exploding human population.

6. Which of the following is a result of biological magnification?
   a. Top-level predators may be most harmed by toxic environmental chemicals.
   b. DDT has spread throughout every ecosystem and is found in almost every organism.
   c. The greenhouse effect will be most significant at the poles.
   d. Energy is lost at each trophic level of a food chain.
   e. Many nutrients are being removed from agricultural lands and shunted into aquatic ecosystems.

7. Which of these ecosystems has the *lowest* primary production per square meter?
   a. a salt marsh
   b. an open ocean
   c. a coral reef
   d. a grassland
   e. a tropical rain forest

8. Quantities of mineral nutrients in soils of tropical rain forests are relatively low because
    a. the standing crop is small.
    b. microorganisms that recycle chemicals are not very abundant in tropical soils.
    c. the decomposition of organic refuse and reassimilation of chemicals by plants occur rapidly.
    d. nutrient cycles occur at a relatively slow rate in tropical soils.
    e. the high temperatures destroy the nutrients.

9. Coastal water polluted with phosphate and nitrogenous compounds from duck farms showed detectable levels of phosphates but not nitrogen. In experiments, algae were grown in water samples that were controls or enriched with phosphate or ammonium. The greatest algal growth was observed in the nitrogen-enriched samples; phosphate-enriched and control samples both had similar growth. From these results, one could conclude that
    a. reducing the levels of phosphate in these waters will not help reduce phytoplankton production.
    b. adding nitrogen to these waters will help reduce eutrophication.
    c. the high levels of phosphates in the water is helping to control algal growth.
    d. nitrogen is the limiting nutrient in these waters.
    e. both a and d are reasonable conclusions.

10. Which of the following contributes most to the rate of chemical cycling in an ecosystem?
    a. the rate of primary production
    b. the efficiency of secondary production
    c. the rate of decomposition
    d. the trophic efficiency of the ecosystem
    e. the location of available nutrients in inorganic or organic compartments

11. Why is the transfer of energy in an ecosystem referred to as energy flow, not energy cycling?

12. Why is a pound of bacon so much more expensive than a pound of corn?

13. What is the main abiotic reservoir for carbon?

14. What would most likely happen to the carbon cycle if all the detritivores suddenly "went on strike" and stopped working?

15. Over the short term, why does phosphorus cycling tend to be more localized than carbon, nitrogen, or water cycling?

16. How can clear-cutting a forest (removing all trees) damage the water quality of nearby lakes?

17. How does excessive addition of mineral nutrients to a lake eventually result in the loss of most fish in the lake?

Go to the website or CD-ROM for more quiz questions.

Evolution Connection

Some biologists, struck by the complex interdependence of biotic and abiotic factors that make up ecosystems, have suggested that ecosystems themselves are emergent, "living" systems capable of evolving. One manifestation of this is James Lovelock's Gaia hypothesis, which views Earth itself as a living, homeostatic entity—a kind of superorganism. Critique the idea that ecosystems and the biosphere can evolve by applying the principles of evolution you have learned in this book. If ecosystems are capable of evolving, is this a form of Darwinian evolution? Why or why not?

The Process of Science

With two nearby ponds in a forest as your study site, how would you design a controlled experiment to measure the effect of falling leaves on net primary production in a pond?

Measure effects of temperature and light on oxygen saturation in the Case Study in the Process of Science, available on the website and CD-ROM.

Science, Technology, and Society

The amount of CO_2 in the atmosphere is increasing, and global temperature has increased over the past century; however, scientists do not agree about the extent to which the two phenomena are related. Most say that greenhouse warming is under way and we need to take action now to avoid drastic environmental change. Some say it is too soon to tell and we should gather more data before we act. What are the advantages and disadvantages of doing something now to slow global warming? What are the advantages and disadvantages of waiting until more data are available?

Answers 1. c; **2.** e; **3.** b; **4.** e; **5.** d; **6.** a; **7.** b; **8.** c; **9.** e; **10.** c; **11.** Because energy passes through an ecosystem, entering as sunlight and leaving as heat. It is not recycled within the ecosystem. **12.** Because it took at least 10 pounds of feed corn to produce that pound of bacon **13.** The atmospheric stock of CO_2 **14.** Carbon would accumulate in organic mass, the atmospheric reservoir of carbon would decline, and plants would eventually be starved for CO_2. **15.** Because phosphorus is cycled almost entirely within the soil rather than being transferred over long distance via the atmosphere **16.** Without the growing trees to assimilate minerals from the soil, more of the minerals run off and end up polluting water resources. **17.** The eutrophication (overfertilization) initially causes population explosions of algae and the organisms that feed on them. The respiration of so much life, including the detritivores working on all the organic refuse, consumes most of the lake's oxygen, which the fish require.

CHAPTER 55

CONSERVATION BIOLOGY

THE BIODIVERSITY CRISIS

- The three levels of biodiversity are genetic diversity, species diversity, and ecosystem diversity
- Biodiversity at all three levels is vital to human welfare
- The four major threats to biodiversity are habitat destruction, introduced species, overexploitation, and food chain disruptions

CONSERVATION AT THE POPULATION AND SPECIES LEVELS

- According to the small-population approach, a population's small size can draw it into an extinction vortex
- The declining-population approach is a proactive conservation strategy for detecting, diagnosing, and halting population declines
- Conserving species involves weighing conflicting demands

CONSERVATION AT THE COMMUNITY, ECOSYSTEM, AND LANDSCAPE LEVELS

- Edges and corridors can strongly influence landscape biodiversity
- Conservation biologists face many challenges in setting up protected areas
- Nature reserves must be functional parts of landscapes
- Restoring degraded areas is an increasingly important conservation effort
- The goal of sustainable development is reorienting ecological research and challenging all of us to reassess our values
- The future of the biosphere may depend on our biophilia

Biology is the science of life. *Thus, it is fitting that our final chapter be about preserving life. **Conservation biology** is a goal-oriented science that seeks to counter the **biodiversity crisis,** the current rapid decrease in Earth's great variety of life.*

To date, scientists have described and formally named about 1.5 million species of organisms. We can only estimate how many more currently exist. Some biologists believe that the number is about 10 million, but others estimate it to be between 30 million and 80 million. Some of the greatest concentrations of species are found in the tropics, where the scene in the photograph on this *page is commonplace: tropical forests (such as the one in Ecuador shown here) being destroyed at an alarming rate to make room for and support a burgeoning human population.*

Throughout the biosphere, human activities are altering trophic structures, energy flow, chemical cycling, and natural disturbance—ecosystem processes on which we and other species depend. The amount of human-altered land surface is approaching 50%, and we use over half of all accessible surface fresh water. In the oceans, stocks of many fishes are being depleted by overharvest, and some of the most productive and diverse areas, such as coral reefs and estuaries, are being severely stressed. By some estimates, we are in the process of doing more damage to the biosphere and pushing more species toward extinction than the large asteroid that seems to have triggered the mass extinctions at the close of the Cretaceous period 65 million years ago (see FIGURE 25.6). Globally, the rate of species loss may be as much as 1,000 times higher than at any time in the past 100,000 years.

In this chapter, we take a closer look at the biodiversity crisis and at the science of conservation biology. We will examine some of the research and conservation strategies biologists are using in attempts to slow the rate of species loss. Along the way, we will see that conservation biology relies on research at all levels of ecology, from populations to ecosystems and landscapes.

THE BIODIVERSITY CRISIS

Extinction is a natural phenomenon that has been occurring almost since life first evolved; it is the current *rate* of extinction that underlies the biodiversity crisis. Because we can only estimate the number of species currently existing, we cannot determine the actual rate of species loss or the real magnitude of the biodiversity crisis. We do know for certain that we are experiencing a high rate of species extinction caused by an escalating rate of ecosystem degradation by a single species—*Homo sapiens.* Conservation biology is about trying to understand what is happening to biodiversity, why it is happening, and what we can do about it.

The three levels of biodiversity are genetic diversity, species diversity, and ecosystem diversity

Biodiversity—short for biological diversity—has three main components, or levels (FIGURE 55.1).

Loss of Genetic Diversity

The first level of biodiversity is genetic variation. In addition to the individual variation *within* a population, there is also genetic variation *between* populations, associated with adaptations to local conditions (see Chapter 23). If one local population becomes extinct, then a species has lost some of the genetic diversity that makes adaptation possible. This erosion of genetic diversity is, of course, detrimental to the overall adaptive prospects of the species. But the loss of genetic diversity throughout the biosphere also has implications for human welfare. For example, wild populations of plants closely related to our agricultural species are genetic resources for improving certain crop qualities through plant breeding.

Loss of Species Diversity

The second level of biodiversity is the variety of species in an ecosystem or throughout the entire biosphere—what we called species richness in Chapter 53. Much of the popular and political discussion of the biodiversity crisis centers on species. The U.S. Endangered Species Act (ESA) defines an **endangered species** as one that is "in danger of extinction throughout all or a significant portion of its range." Also defined for protection by the ESA, **threatened species** are those that are likely to become endangered in the foreseeable future throughout all or a significant portion of their range.

Here are just a few examples of why conservation biologists are so concerned about species loss:

- According to the International Union for Conservation of Nature and Natural Resources (IUCN), 13% of the 9,040 known bird species in the world are threatened with extinction. That's 1,183 species! In the past 40 years, population densities of migratory songbirds in the mid-Atlantic United States dropped 50%.
- A recent survey conducted by the Center for Plant Conservation showed that of the approximately 20,000 known plant species in the United States, 200 species have become

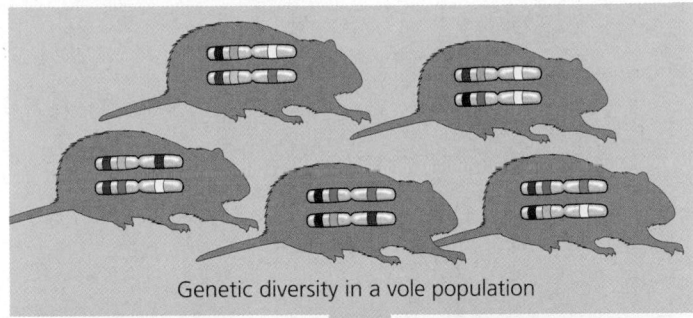

Genetic diversity in a vole population

Species diversity in a coastal redwood ecosystem

Community and ecosystem diversity across the landscape of entire region

FIGURE 55.1 Three levels of biodiversity. (The oversized chromosomes in the voles symbolize the genetic variation within the population.)

extinct since such records have been kept. Another 730 plant species in the United States are endangered or threatened with extinction.
- About 20% of the known freshwater fishes in the world have either become extinct during historical times or are seriously threatened. One of the largest rapid extinction events yet recorded is the ongoing loss of freshwater fishes in Lake Victoria in East Africa. About 200 of the 300 species of cichlids in the lake have been lost, mainly as a result of the recent introduction by Europeans of an exotic predator, the Nile perch.
- Since 1900, 123 freshwater vertebrate and invertebrate species have become extinct in North America, and hundreds more species are threatened. Extinction rates for

(a) Philippine eagle

(b) Chinese river dolphin

(c) Javan rhinoceros

FIGURE 55.2 A hundred heartbeats from extinction. These are just three of the many members of what E. O. Wilson calls the Hundred Heartbeat Club, species with fewer than 100 individuals remaining on Earth.

Extinction of species may be local; for example, a species may be lost in one river system but survive in an adjacent one. Global extinction of a species means that it is lost from *all* its locales. Extinction is often an unseen process. To know for certain that a given species is extinct, we must know its exact distribution. But millions of the world's species have not even been identified. Arthropods (especially insects), nematodes, fungi, protists, and prokaryotes head the list of taxa with great numbers of undiscovered species. But even well-studied taxa, such as birds and mammals, are not completely known. In the past decade, scientists have increased the list of known mammals by about 15%. Without a more complete catalog of species diversity and knowledge of the geographic distribution and ecological roles of Earth's species, our efforts to understand the structure and function of ecosystems on which our survival depends will remain incomplete.

Loss of Ecosystem Diversity

The variety of the biosphere's ecosystems is the third level of biological diversity. Within each ecosystem, the biological community has a network of interactions among populations of different species. The local extinction of one species—say, a keystone predator—can have a negative impact on the overall species richness of the community (see FIGURE 53.14). More broadly, each ecosystem can have an important impact on the whole biosphere. Be it a rain forest, peat bog, or expanse of open ocean, an ecosystem has characteristic patterns of energy flow and chemical cycling. For example, the productive "pastures" of phytoplankton in the oceans may help moderate the greenhouse effect by consuming massive quantities of CO_2 for photosynthesis and for building shells made of bicarbonate.

Some ecosystems are being erased from Earth at an astonishing pace. For example, the cumulative area of all tropical rain forests on the planet is about the size of the 48 contiguous United States, and we lose an area equal to the state of West Virginia each year.

The biodiversity crisis is most often equated to species extinctions, but conservation biologists now realize that the disappearance of a species is often the result of losses in diversity at other levels, including the loss of genetic diversity and ecosystem diversity.

North American freshwater fauna are about five times higher than those for terrestrial animals. About 4% of the known freshwater species will become extinct each decade unless habitat loss and degradation are reversed.

- Harvard biologist Edward O. Wilson has compiled what he grimly calls the Hundred Heartbeat Club. The species that belong are those animals that number fewer than 100 individuals and so are only that many heartbeats away from extinction (FIGURE 55.2).
- Several researchers estimate that at the current rate of destruction, over half of all plant and animal species will be gone by the end of this new century.

Biodiversity at all three levels is vital to human welfare

Why should we care about the loss of biodiversity? Perhaps the purest reason is what E. O. Wilson calls *biophilia*, our sense of connection to nature and other forms of life. The concept that other species are important and should be protected is a pervasive theme of many religions and the basis of the moral argument that we should protect biodiversity. There is also a concern for future human generations. Do we have the right to deprive them of Earth's species richness? Paraphrasing an old Chinese proverb, G. H. Brundtland, former prime minister of Norway, put it this way: "We must consider our planet to be on loan from our children, rather than being a gift from our ancestors."

Benefits of Species Diversity and Genetic Diversity

In addition to the aesthetic and ethical reasons for preserving biodiversity, there are practical reasons as well. Biodiversity is a crucial natural resource, and species that are threatened could provide crops, fibers, and medicines for human use. In the United States, 25% of all prescriptions dispensed from pharmacies contain substances derived from plants. For example, in the 1970s, researchers discovered that the rosy periwinkle from Madagascar contains alkaloids that inhibit cancer cell growth (FIGURE 55.3). The result of this discovery is remission for most victims of two potentially deadly forms of cancer, Hodgkin's disease and a childhood leukemia. There are five other species of periwinkles on Madagascar, and one is approaching extinction.

The loss of species also means the loss of genes. Each species has certain unique genes, and biodiversity represents the sum of all the genomes of all organisms on Earth. Because many millions of species may become extinct before we even know about them, we stand to lose irretrievably the valuable genetic potential held in their unique libraries of genes.

Recently, U.S. National Park Service officials have been negotiating with private industry to sell samples of extremophilic prokaryotes from the numerous hot springs in Yellowstone National Park. The corporations anticipate using DNA extracted from the prokaryotes to mass-produce commercially useful enzymes. Consider the historical example of the polymerase chain reaction (PCR), the gene-cloning technology based on an enzyme extracted from thermophilic prokaryotes from hot springs (see FIGURE 20.7). Many researchers and industry officials are enthusiastic about the potential that such "bioprospecting" holds for the future development of new medicines, foods, petroleum substitutes, industrial chemicals, and other important products.

Ecosystem Services

The benefits that individual species provide to humans are often substantial, but saving individual species is only part of the rationale for saving ecosystems. Humans evolved in Earth's ecosystems, and our bodies are finely adjusted to these systems. While it is possible to survive in a world with considerably less biodiversity, it is important to realize that humans are dependent on ecosystems and on interactions with other

FIGURE 55.3 The rosy periwinkle (*Catharanthus roseus*): a plant that saves lives. Before alkaloids that inhibit cancer cell growth were discovered in the rosy periwinkle over 20 years ago, Hodgkin's disease and acute lymphocytic leukemia were two of the deadliest cancers. Now most victims are cured. This plant is one of hundreds used to treat human diseases.

species. By allowing the extinction of species and the degradation of habitats to continue, we risk our own species' survival.

In the urban and suburban settings in which most of us live today, it is easy to lose sight of the vital ecosystem services on which we depend. ***Ecosystem services*** encompass all the processes through which natural ecosystems and the species they contain help sustain human life on Earth. Here are just a few of these ecosystem services:

- Purification of air and water
- Reduction of the severity of droughts and floods
- Generation and preservation of fertile soils
- Detoxification and decomposition of wastes
- Pollination of crops and natural vegetation
- Dispersal of seeds
- Nutrient cycling
- Control of many agricultural pests by natural enemies
- Protection of coastal shores from erosion
- Protection from ultraviolet rays
- Moderation of weather extremes
- Provision of aesthetic beauty

Human life would cease without these ecosystem services; and yet we generally undervalue them, perhaps because we don't attach a monetary value to them.

In a 1997 article, ecologist Robert Costanza and colleagues attempted to put a dollar figure on ecosystem services. Their bottom-line estimate was $33 trillion per year, nearly twice as much as the gross national product of all the countries of the globe ($18 trillion). It is, of course, difficult to speculate about the dollar value of ecosystem services. Perhaps it is more realistic to do the accounting on a small scale. What, for example, is the true price of building a dam or clear-cutting a patch of forest if we include the dollar loss of ecosystem services in the cost column?

One large-scale experiment illustrates how little we understand about ecosystem services. Biosphere II, in Oracle, Arizona, was an attempt to create a closed ecosystem covering 1.27 ha (3.1 acres). Benefactors curious about the outcome invested over $200 million to build the giant airtight terrarium (FIGURE 55.4). Biosphere II had a forest with soil, a miniature ocean, and several other "ecosystems." In 1991, with much fanfare, eight people entered Biosphere II for what was supposed to be two years of isolated habitation. But the artificial biosphere failed, and the experiment had to be stopped after 15 months. Oxygen concentration dropped to 65% of Earth's atmospheric O_2 concentration, and CO_2 concentration fluctuated wildly. Most of the vertebrate species became extinct in Biosphere II, and all of the pollinators died. There were population explosions of cockroaches and other pests. But in a sense, the experiment was not a failure, for it taught us that no one yet knows how to engineer a system that can provide humans with all the life-support services that natural ecosystems produce for free.

The four major threats to biodiversity are habitat destruction, introduced species, overexploitation, and food chain disruptions

Habitat Destruction

Human alteration of habitat is the single greatest threat to biodiversity throughout the biosphere. Massive destruction of habitats throughout the world has been brought about by agriculture, urban development, forestry, mining, and environmental pollution. The IUCN (see p. 1225) implicates destruction of physical habitat in 73% of the species designated extinct, endangered, vulnerable, or rare.

Though most studies have focused on terrestrial ecosystems, habitat loss also appears to be a major threat to marine biodiversity, especially on continental coasts and coral reefs. About 93% of Earth's coral reefs, among the most species-rich aquatic communities, have been damaged by human activities. At the current rate of destruction, 40–50% of the reefs could be lost in the next 30 to 40 years. About a third of the planet's marine fish species utilize coral reefs, which occupy only about 0.2% of the ocean floor.

In addition to habitat destruction over large regions, many natural landscapes have been fragmented, broken up into small patches (FIGURE 55.5). FIGURE 55.6 illustrates how forest areas in southern Wisconsin were fragmented over a 119-year period. Forest fragmentation is also occurring at a rapid rate in tropical forests. For example, tropical rain forest losses around Veracruz, Mexico, exceeded 85% during the 20-year span from 1967 to 1987. Deforestation continued to proceed up from the lowlands, and by 2000, only 8% of the original forest remained, in the form of an archipelago of small forest islands. Deforestation in this area is due mostly to clearing for cattle ranches. The human population of this region has more than doubled in the last 25 years.

In almost all cases, habitat fragmentation leads to species loss. The prairies of North America are a good ex-

FIGURE 55.4 What scientists learned about ecosystem services from the world's largest terrarium. Biosphere II, in Arizona, covers an area the size of two football fields. The eight biospherians who entered the container in 1991 all had to abandon Biosphere II within 15 months. An investment of over $200 million was not enough to create a system that provides all the ecosystem services required to sustain human life. In fact, a greater appreciation for the pricelessness and complexity of ecosystem services and the biodiversity that provides them was perhaps the most important lesson from Biosphere II.

FIGURE 55.5 Fragmentation of a forest ecosystem. In this aerial photograph of Mt. Hood National Forest in the western United States, you can see the "islands" of coniferous forest that were created when much of the original forest was cut for timber.

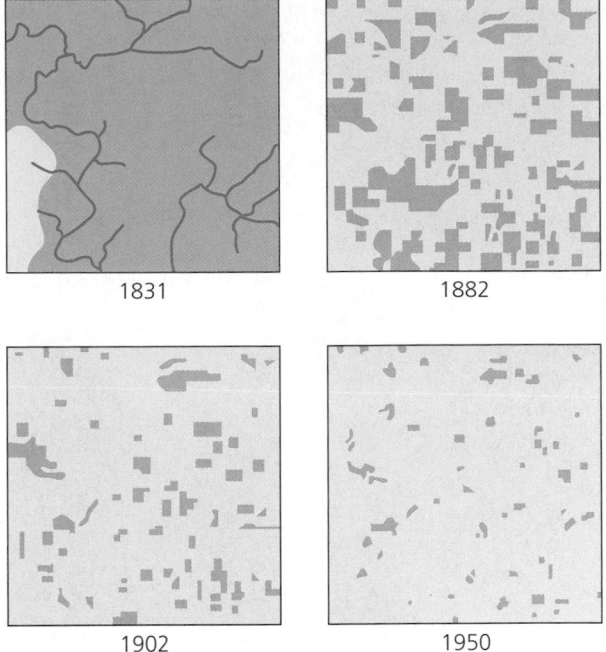

1831

1882

1902

1950

FIGURE 55.6 The history of habitat reduction and fragmentation in a Wisconsin forest. Between 1831 and 1950, more than 95% of the original forest (green) in Cadiz Township was lost, and the remaining 5% consisted of small fragments.

ample. Prairie covered about 800,000 ha of southern Wisconsin when Europeans first arrived, but now occupies less than 0.1% of its original area. Plant diversity surveys of 54 Wisconsin prairie remnants were conducted in 1948–1954 and then repeated in 1987–1988. During the few decades between surveys, the prairie fragments lost between 8% and 60% of their plant species, depending on the fragment.

Introduced Species

Ranking second behind habitat loss as a cause of the biodiversity crisis, introduced species have probably contributed to about 40% of the extinctions recorded since 1750. Sometimes called exotic species, **introduced species** are those that humans move from the species' native locations to new geographic regions. (We discussed zebra mussels and Africanized bees in the United States in FIGURES 50.7 and 50.8.) In some cases, the introductions are intentional. For example, Euro-

pean red foxes were intentionally introduced to Australia in the late 1800s because of an interest in fox hunting. By preying on medium-sized native mammals, foxes have contributed to several of these mammals' becoming extinct. Another example, mentioned earlier, was the disastrous 1960s introduction of the Nile perch to Lake Victoria (FIGURE 55.7a, p. 1230). In other cases, humans transplant species accidentally. For instance, the brown tree snake was accidentally introduced to Guam as a "stowaway" in military cargo after World War II (FIGURE 55.7b). Since then, 12 species of birds and 6 lizard species on which the snakes prey have become extinct on Guam. All 18 of these species continue to live on Guam's small offshore islands, which the snake has not colonized. Intentional or not, introduced species that gain a foothold usually disrupt their adopted community, often by preying on native organisms or outcompeting native species for resources.

Humans have also introduced many species with the best of intentions. For example, the U.S. Department of Agriculture encouraged the import of a Japanese plant called kudzu to the American South in the 1930s to help control erosion, especially along irrigation canals. At first, the government paid farmers to plant kudzu vines. The enthusiasm for the new vines led to kudzu festivals in southern towns, complete with the crowning of kudzu queens. But kudzu celebrations ended decades ago as the invasive plant took over vast expanses of the southern landscape (FIGURE 55.7c). Another introduced plant called purple loosestrife is claiming over 200,000 acres of wetlands per year, crowding out native plants and the animals that

(a) Nile perch, Lake Victoria

(b) Brown tree snake, Guam

(c) Kudzu

(d) European starlings

(e) Argentine ants

(f) The seaweed *Caulerpa*

FIGURE 55.7 A small sample of disastrous species introductions. (a) One of the largest freshwater fishes (up to 2 m long and weighing up to 450 kg), the Nile perch was introduced to Lake Victoria in East Africa to provide high-protein food for the growing human population. Unfortunately, the perch's main effect has been to wipe out about 200 smaller native species, reducing its own food supply to a critical level. **(b)** The brown tree snake, accidentally introduced to Guam at the end of World War II, has probably eliminated 18 species of birds and lizards in its new home. **(c)** Kudzu has invaded much of the U.S. South. **(d)** European starlings have displaced many native songbirds in North America. **(e)** Argentine ants are seen here ganging up on a red ant native to California. **(f)** In this underwater photo of a California lagoon, you can see an aquarium-bred, hyper-vigorous variety of the seaweed *Caulerpa* (in foreground) crowding out the native eelgrass.

feed on the native flora. The story is similar for the introduction to the United States of a bird called the European starling (FIGURE 55.7d). A citizens group intent on introducing all plants and animals mentioned in Shakespeare's plays imported 120 starlings to New York's Central Park in 1890 (the starling is mentioned in just one line of Shakespeare's *Henry IV*). From that foothold, starlings spread rapidly across North America. In less than a century, the population increased to about 100 million, displacing many of the native songbird species in the United States and Canada.

The ease of travel by ships and airplanes has accelerated the transplant of species, especially unintentional introductions. For example, fire ants, which can inflict very painful beelike stings, reached the southeastern United States in the early 1900s from South America, probably in the hold of a produce ship. Fire ants have been extending their range northward and westward ever since. In Texas, for example, fire ants have apparently managed to eliminate about two-thirds of the native ant species. And another accidentally introduced ant species, the Argentine ant, is decimating populations of native ants in California (FIGURE 55.7e).

An even more recent example of introduced species is the appearance in 2000 of an alga called *Caulerpa* in a California lagoon (FIGURE 55.7f). The small seaweed was probably introduced by someone dumping a home saltwater aquarium. Native to Caribbean waters, the California invader is a variety of the alga that has been domesticated and selectively bred as an aquarium alga for its vigor and resistance to disease and herbivores. An earlier invasion of the Mediterranean Sea by this super seaweed is displacing many of the native algae there, and the same thing could happen now all along the Pacific coast of North America.

Introduced species are, of course, an international problem. But in the United States alone, there are at least 50,000 introduced species, with a cost to the economy of over $130 billion in damage and control efforts. And that does not include the priceless loss of native species.

Overexploitation

Overexploitation refers generally to the human harvesting of wild plants or animals at rates exceeding the ability of populations of those species to rebound. It is possible for overexploitation to endanger certain plant species, such as rare trees that produce valuable wood or some other commercial product. But overexploitation more often refers to the overhunting and overfishing of animals. Especially susceptible to overexploitation are large species with low intrinsic reproductive rates, such as elephants, whales, rhinoceroses, and other animals considered valuable by humans. Species on small islands are particularly vulnerable to extinction due to overexploitation. For example, by the 1840s, humans had overhunted the great auk, a large, flightless seabird, to extinction on islands in the Atlantic Ocean because of a demand for feathers, eggs, and meat (FIGURE 55.8).

The decline of the African elephant, the largest extant terrestrial animal, is a classic example of the impact of overhunting. African elephants take 10 to 11 years to reach sexual maturity, and then a fertile female has a single calf every 3 to 9 years. The potential rate of population increase is only about 6% per year, a low growth rate. Elephant populations have been declining in most of Africa during the last 50 years. Only in South Africa have elephant populations been stable or increasing. Illegal hunting for ivory is the major cause of this collapse of elephant populations. When the price of ivory increased during the 1970s, the amount of poaching for ivory grew dramatically. Currently, there is a ban on ivory trade, but this is having little impact in central and eastern Africa, where poaching is rampant.

Overfishing has dramatically reduced the population sizes of many commercially important fish species. Just over a century ago, British biologist T. H. Huxley declared, "Probably all the great sea fisheries are inexhaustible: that is to say that nothing we do seriously affects the number of fish." Huxley and his contemporaries grossly underestimated an increasing demand for protein by an exploding human population coupled with overexploitation made possible by new harvesting

FIGURE 55.8 The great auk (*Pinguinis impennis*). Endemic to islands in the North Atlantic Ocean, the great auk was hunted to extinction by 1844.

FIGURE 55.9 North Atlantic bluefin tuna auctioned in a Japanese fish market. In spite of quotas, the high price that bluefin tuna brings may doom the species to extinction.

technologies, such as long-line fishing and modern trawlers. Many populations of fishes that humans consume have now been reduced to levels that cannot sustain further exploitation. The fate of the North Atlantic bluefin tuna is just one example. Until the past few decades, this big tuna was considered a sport fish of little commercial value—just a few cents per pound for cat food. Then, beginning in the 1980s, wholesalers began airfreighting fresh, iced bluefin to Japan for sushi and sashimi. In that market, the fish now brings up to $100 per pound (FIGURE 55.9). With that kind of demand, the results are predictable. It took just ten years to reduce the North American bluefin population to less than 20% of its 1980 size. The collapse of the northern cod fishery off Newfoundland in the 1990s is a recent example of how it is even possible to overharvest what had been a very common species.

Disruption of Food Chains

Like falling dominoes, the extinction of one species can doom its predators. But this is likely only if the predator feeds exclusively on one species, which is a rare trophic arrangement. Certainly, host-specific parasites can become extinct if their host becomes extinct. But such extinctions have not been the subject of much research.

Most of the evidence for secondary extinctions of larger organisms due to loss of prey is circumstantial. For example, the forest eagle of New Zealand, which preyed on large ground birds, became extinct around 1400 following the extinction of flightless birds called moas. After humans reached New Zealand around A.D. 1000, they probably hunted all 11 species of the large, tame moas to extinction. Although it is a reasonable hypothesis that the disappearance of the forest eagle was

related to the loss of its main prey, we cannot be sure of a cause-and-effect relationship. Similarly, the decline of the black-footed ferret on the Great Plains of North America paralleled the decline of its main prey, prairie dogs. But other factors may have contributed to the decrease in ferret populations. Because most predators are not so specialized in the prey they'll eat, food chain disruption is probably less important as a cause of extinction than habitat destruction, introduced species, and overexploitation.

Now that we have an overview of the biodiversity crisis and its causes, let's examine how conservation biologists hope to apply basic principles of evolutionary biology and ecology to slow the loss of biodiversity at its various levels.

CONSERVATION AT THE POPULATION AND SPECIES LEVELS

Among biologists focusing on conservation at the population and species levels, there are two main approaches, which we will call the small-population approach and the declining-population approach.

According to the small-population approach, a population's small size can draw it into an extinction vortex

A species is designated as endangered when its populations are very small. Conservation biologists who adopt the **small-population approach** study the processes that can cause very small populations to finally become extinct. In other words, it is a population's smallness itself that finally drives it to extinction after such factors as habitat loss have taken their toll on population size. At the center of this concept is the **extinction vortex,** a downward spiral unique to small populations. A small population is prone to positive feedback loops of inbreeding and genetic drift that draw the population down the vortex toward smaller and smaller population size until no individuals exist (FIGURE 55.10).

The key factor driving the extinction vortex is the loss of the genetic variation on which a population depends for adaptive evolution. Both inbreeding and genetic drift can cause a loss of genetic variation, and both of these processes intensify as a population shrinks (see Chapter 23 to review how genetic drift reduces genetic variation in a population).

Not all populations are doomed by low genetic diversity. A number of plant species, such as the lousewort *Pedicularis* and several grasses, seem to have inherently low genetic variability. Furthermore, low genetic variability does not necessarily lead to permanently small populations. For example, overhunting

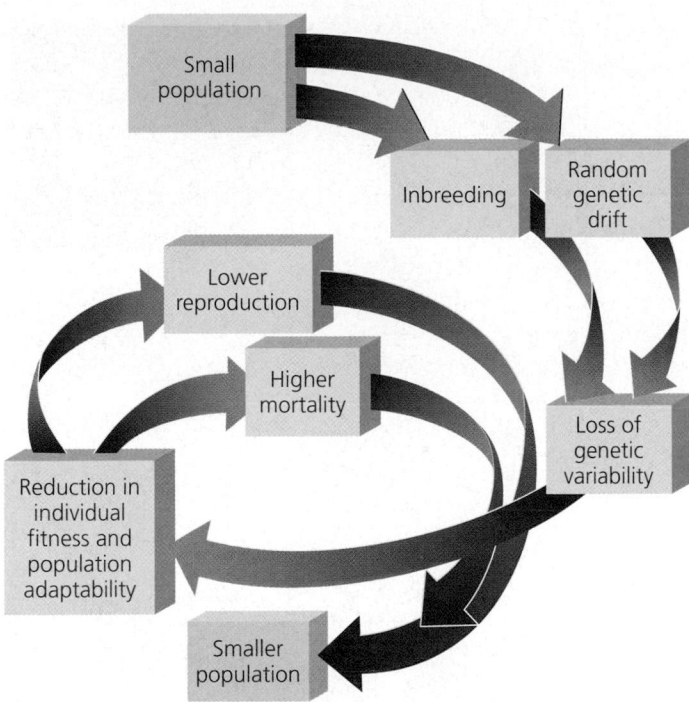

FIGURE 55.10 The extinction vortex of the small-population approach. Small populations can fall into a vortex of positive feedback loops leading to smaller and smaller population size.

of northern elephant seals in the 1890s reduced the species to only 20 individuals—clearly a bottleneck with reduced genetic variation. Since that time, however, the northern seal populations have rebounded to about 150,000 individuals today. Genetic variation in these populations remains relatively low. Among plants, many populations of cord grass (*Spartina anglica*), which thrives in saltmarshes, are genetically uniform at many loci. *S. anglica* arose from a few parent plants only about a century ago by hybridization and allopolyploidy (see FIGURE 24.15). Having spread by cloning, this species now dominates large areas of tidal mudflats in Europe and Asia. Thus, in some cases, low genetic diversity is associated with population expansion rather than decline. But these cases may stand out precisely because they are so unusual. Thus, conservation biologists have good reason to be concerned about very small populations with low genetic variation.

How Small Is Too Small for a Population?

How small does a population have to be before it starts down the extinction vortex? The answer depends on the type of organism and several other factors and must be evaluated case by case. For example, large predators that feed high on the food chain usually require very large individual ranges, resulting in very low population densities. Thus, not all rare species concern conservation biologists. But whatever the number,

most populations presumably require some minimum size to remain viable.

Minimum Viable Population Size (MVP). At some minimal population size, rare species will be able to sustain their numbers and survive. That number is the **minimum viable population size (MVP).** For a given species, MVP is usually estimated using computer models that integrate many factors. For example, the calculation may include an estimate for how many individuals in a small population are likely to be killed by some natural catastrophe such as fire or flood. Once in the extinction vortex, two or three bad weather years in a row could finish off a population that is already below MVP.

Conservation biologists factor a population's MVP into what is called the **population viability analysis (PVA).** The objective of the analysis is to make a reasonable prediction of a population's chances for survival, usually expressed as a certain probability of survival (for example, a 99% chance of survival) over a particular time (for instance, 100 years).

Effective Population Size (N_e). Genetic variation is the key issue in the small-population approach. The *total* size of a population may be misleading because only certain members of the population breed successfully and pass their genetic alleles on to offspring. Therefore, a meaningful estimate of MVP requires the researcher to determine the **effective population size,** which is based on the breeding potential of the population. The following formula incorporates the sex ratio of breeding individuals into the estimate of effective population size, abbreviated N_e:

$$N_e = \frac{4N_f \, N_m}{N_f + N_m}$$

where N_f and N_m are, respectively, the numbers of females and males that successfully breed. Applying this formula to an idealized population whose total size is 1,000 individuals, N_e will also be 1,000 if every individual breeds and the sex ratio is 500 females to 500 males. In this case, $N_e = (4 \times 500 \times 500)/(500 + 500) = 1,000$. Deviation from these conditions (not all individuals breed and/or there is not a 50:50 sex ratio) reduces N_e. For instance, if the total population size is 1,000 but only 400 females breed with 400 males, then $N_e = (4 \times 400 \times 400)/(400 + 400) = 800$, or 80% of the total population size.

In actual study populations, N_e is always some fraction of the total population. Thus, simply censusing a small population—determining the total number of individuals—does not provide a good measure of whether the population is large enough to avoid extinction. Whenever possible, conservation programs are geared to sustain total population sizes that include, at least, the minimum viable number of *reproductively active* individuals. Numerous life history traits can influence N_e, and alternate formulas for estimating N_e take into account family size, maturation age, genetic relatedness among population members, the effects of gene flow between geographically separated populations, and population fluctuations.

Remember that the conservation goal of sustaining effective population size (N_e) above minimum viable population size (MVP) stems from the concern that populations retain enough genetic diversity to be evolutionarily adaptable. Populations with low N_e are prone to inbreeding, reduced heterozygosity, and the random effects of genetic drift and bottlenecking (see Chapter 23). The basic premise of the small-population approach will seem less abstract in light of three case studies.

Case Study: The Greater Prairie Chicken and the Extinction Vortex

When Europeans arrived in North America, the greater prairie chicken (*Tympanuchus cupido*) was common from New England to Virginia and all across the western prairies of the United States and Canada. Agriculture later fragmented the populations of the greater prairie chicken in the central and western states and provinces. For example, in Illinois alone, greater prairie chickens numbered in the millions in the 19th century but declined to 25,000 birds by 1933. And by 1993, the Illinois population of prairie chickens was down to only 50, though large populations remained in Kansas, Minnesota, and Nebraska.

Researchers found that the decline in the Illinois prairie chicken population was associated with a decrease in the hatching rate of eggs. Was this due to low levels of genetic diversity? By comparing DNA samples from the endangered Illinois population with DNA extracted from feathers in museum specimens collected earlier, the biologists confirmed that genetic variation had indeed declined in their study population in Jasper County, Illinois. As a further test of the extinction vortex hypothesis, the scientists imported genetic variation by transplanting birds from the larger populations in Kansas, Minnesota, and Nebraska. Over a five-year period ending in 1997, the researchers moved over 270 greater prairie chickens into their study site in Jasper County (FIGURE 55.11). The viability of eggs rapidly improved, and the population rebounded. The researchers concluded that the Jasper County population of prairie chickens was on its way down the extinction vortex until rescued by a transfusion of genetic variation from other populations.

Case Study: Population Viability Analysis for Two Popular Herbs

For his doctoral dissertation in environmental sciences in 1994 at the University of Quebec, Patrick Nantel presented a population viability analysis of two edible herbaceous plants, American ginseng (*Panax quinquefolius*) and wild leek (*Allium tricoccum*). These perennial herbs are

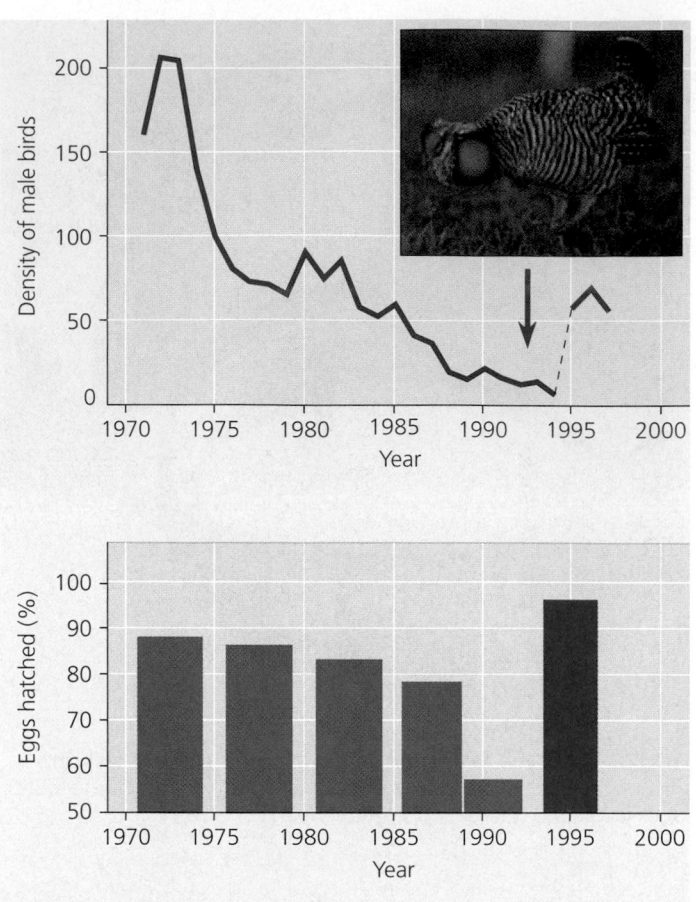

FIGURE 55.11 The decline of the greater prairie chicken (*Tympanuchus cupido*) in central Illinois from 1970 to 1997. The population collapse was mirrored in a reduction in fertility. In 1992, researchers began experimental translocations (blue arrow) of prairie chickens from Minnesota, Kansas, and Nebraska in an attempt to increase genetic variability. The population rebounded strongly.

found in deciduous forest communities in eastern North America (FIGURE 55.12), and both plants are at risk of extinction. The key factors in their decline are destruction and fragmentation of habitat and overharvesting by people who collect the herbs for food. Extinction has already claimed some populations of the two plants. Nantel's PVA for surviving populations in southeastern Canada incorporated data on trends in the numbers of individuals capable of reproducing in two-, three-, and four-year periods. Computer simulations projected the likely effect of environmental influences on these populations. Minimum viable population sizes generated by these computer models were about 170 ginseng plants and between 300 and 1,030 leek plants. There are only about 20 known ginseng populations in Canada having more than 170 individuals, and leek populations of more than a few hundred are rare. Thus, most populations of American ginseng and wild leek in Canada are currently too small to persist unless completely protected from harvesting by humans. Nantel's work is an example of the increasing use of predictive models in planning conservation strategy.

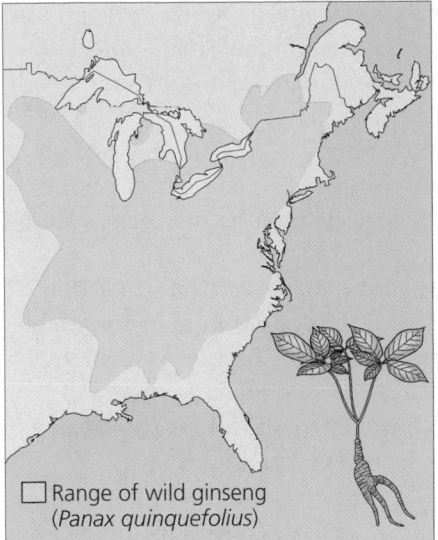

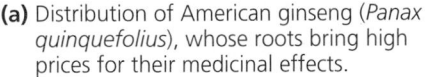

☐ Range of wild ginseng
(*Panax quinquefolius*)

(a) Distribution of American ginseng (*Panax quinquefolius*), whose roots bring high prices for their medicinal effects.

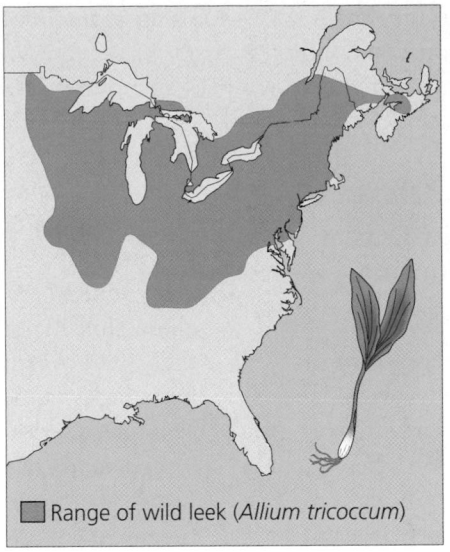

☐ Range of wild leek (*Allium tricoccum*)

(b) Distribution of wild leek (*Allium tricoccum*), valued for its edible bulb.

FIGURE 55.12 Two species of edible plants whose persistence is threatened by habitat loss and overharvesting.

Case Study: Analysis of Grizzly Bear Populations

THE PROCESS OF SCIENCE

Mark Shaffer, currently with the Wilderness Society, performed one of the first population viability analyses as part of a long-term study of grizzly bears (*Ursus artos*) in Yellowstone National Park and its surrounding areas (FIGURE 55.13). Grizzly bears require very large areas of habitat. For instance, estimates of the grizzly's minimum habitat needs in western Canada are about 5 million ha for a population of 50 individuals and about 200 million ha for 1,000 individuals. A threatened species in the United States, the grizzly is currently found in only 4 of the 48 contiguous states. Its populations in those states have been drastically reduced and fragmented: In 1800, an estimated 100,000 grizzlies ranged over about 500 million ha of more or less continuous habitat, while today there are six virtually isolated populations totaling about 1,000 individuals with a total range of less than 5 million ha. The Yellowstone population is the largest, with about 200 bears in an area of about 1 million ha.

Attempting to determine viable sizes for the U.S. grizzly populations, Shaffer used life history data obtained for individual Yellowstone bears over a 12-year period and simulated the effects of environmental factors on survival and reproduction. His models predicted that a total grizzly bear population of 70 to 90 individuals in suitable habitat will have about a 95% chance of surviving for 100 years. Achieving a 99% chance of survival for a century or a 95% chance for 200 years requires enough habitat to support at least 100 bears. Because of habitat limitations, however, recovery targets—the specific goals mandated by the Endangered Species Act—for several of the U.S. populations have been tentatively set at fewer than 100 bears. In such cases, biologists are hopeful that the small populations can be sustained by careful monitoring and special protective measures.

Concerned that policy decisions have been made without information on potential losses of genetic variability in grizzly bear populations, Fred Allendorf and his co-workers at the University of Montana developed a computer model that augmented Shaffer's work. Using detailed life history and kinship data from individual bears for populations in Montana, Wyoming, and British Columbia, Allendorf's model estimated that the effective population size (N_e) of grizzly populations is only about 25% of the total population size. Usually, only a few dominant males breed, and locating females may be difficult, since individuals inhabit such extensive areas. Moreover, females may reproduce only when there is abundant food. Thus, even the relatively large Yellowstone population of 200 bears has an effective population size of only 50, a level that might lead to a loss of genetic variability and possibly fitness. Effective population size could be increased if there was migration between isolated populations of grizzlies. Computer models predict that introducing only two unrelated bears each decade into populations of 100 individuals would reduce the loss of genetic variation by about half. For the grizzly bear, and probably for many other species whose populations are very small, finding ways to promote dispersal among populations may be one of the most urgent conservation needs.

FIGURE 55.13 Long-term monitoring of a grizzly bear population. The ecologist is fitting this tranquilized bear with a radiotransmitter so that its movements can be tracked and compared with other individuals in the Yellowstone National Park population.

The three case studies we have examined bridge small-population theory to practical applications in conservation. Next, we look at an alternative approach to understanding the biology of extinction.

The declining-population approach is a proactive conservation strategy for detecting, diagnosing, and halting population declines

We saw that the small-population approach emphasizes minimum viable population size and the extinction vortex as ways to understand the extinction process. There are, of course, interventions based on small-population theory, including introducing genetic variation from one population to another. But the **declining-population approach** is even more action oriented, focusing on threatened and endangered populations even if they are far greater than minimum viable size. To conservation biologists who lean toward this approach, a downward trend in a species may be cause for alarm and, when possible, corrective action.

The distinction between a declining population (which may be small) and a small population (which may be declining) is less important than the different priorities of the two basic conservation approaches. Practitioners of both the small-population and declining-population approaches recognize that most modern extinctions are due to the human factors of habitat destruction, introduced species, and overexploitation. But the small-population approach emphasizes smallness itself as an ultimate cause of a population's extinction, especially through loss of genetic diversity. In contrast, the declining-population approach emphasizes the environmental factors that caused a population decline in the first place. If, for example, an area is deforested, then species that depend on trees will decline and become locally extinct, whether or not they retain genetic variation.

The declining-population approach requires that population declines be evaluated on a case-by-case basis, with researchers carefully dissecting the causes of a decline before recommending or trying corrective measures. If, for example, the biological magnification of a particular toxic pollutant is causing a decline in some top-level consumer such as a predatory bird, then only reduction or elimination of the poison in the environment can save that particular species. Rarely is the situation so straightforward, but there are procedures to help with even complex cases.

Steps in the Diagnosis and Treatment of Declining Populations

THE PROCESS OF SCIENCE

Like all scientific processes, analyses in conservation biology rarely follow exact formulas for investigation, but we *can* identify a series of logical steps that are common in the declining-population approach:

1. *Confirm that the species is presently in decline or that it was formerly more widely distributed or more abundant.* This step requires assessment of population trends and distribution.

2. *Study the species' natural history to determine its environmental requirements.* Existing research literature on the natural history of this or related species may help with this step.

3. *Determine all the possible causes of the decline.* In listing all possible hypotheses for the decline, human activities that could contribute to losses may become evident, but hypotheses cannot be restricted to human causes. For example, a series of unusually harsh winters could cause local declines in populations of certain species.

4. *List the predictions of each hypothesis for the decline.* Ideally, the investigation would emphasize contrasting predictions based on the different hypotheses (see Chapter 1).

5. *Test the most likely hypothesis first, designing an experiment to determine if this factor is the main cause of the decline.* Many factors may be correlated with the decline without being the direct cause. In the ideal experiment, researchers remove the suspected agent of decline to see if the experimental population rebounds relative to a control population. It may turn out that there are multiple causes of decline.

6. *Apply the results of this diagnosis to the management of the threatened species.* This requires monitoring recovery until the problem of decline is resolved.

As with our discussion of the small-population approach, the declining-population approach will seem less abstract in the context of a case study.

Case Study: Diagnosing and Treating the Decline of the Red-Cockaded Woodpecker

To practice conservation biology, we must understand the often subtle habitat requirements of an endangered species. The red-cockaded woodpecker (*Picoides borealis*) is an endangered, endemic species (found nowhere else) originally found throughout the southeastern United States. This species requires mature pine forests, preferably ones dominated by the longleaf pine. Such habitats have been destroyed or fragmented by logging and agriculture. Most woodpeckers nest in dead trees, but the red-cockaded woodpecker drills its nest holes in mature, living pine trees (FIGURE 55.14a). The heartwood (deep wood) of mature longleaf pines is usually rotted and softened by fungi, allowing the woodpeckers adequate space for nesting once they excavate into the heartwood. Red-cockaded woodpeckers also drill small holes around the entrance to their nest cavity, which causes resin from the tree to ooze down the trunk. The resin seems to repel certain predators, such as corn snakes, that eat bird eggs and nestlings.

(a) **A red-cockaded woodpecker** at the entrance to its nest site in a longleaf pine tree

(b) **Forest that can sustain red-cockaded woodpeckers** has low undergrowth

FIGURE 55.14 Habitat requirements of the red-cockaded woodpecker.

(c) **Forest that cannot sustain red-cockaded woodpeckers** has high, dense undergrowth that impacts the woodpeckers' access to feeding grounds

Another critical habitat factor for this woodpecker is that the understory of plants around the pine trunks must be of low profile (FIGURE 55.14b). Breeding birds tend to abandon nests when vegetation among the pines is thick and higher than about 15 feet (FIGURE 55.14c). Apparently, the birds require a clear flight path between their home trees and the neighboring feeding grounds. Historically, periodic fires swept through longleaf pine forests, keeping the undergrowth low.

The recent recovery of the red-cockaded woodpecker from near extinction to sustainable populations was achieved by recognizing the key habitat factors and protecting some longleaf pine forests that support viable numbers of the birds. The use of controlled fires to reduce forest undergrowth helps maintain mature pine trees as well as the woodpeckers.

Designing a recovery program for the red-cockaded woodpecker was complicated by the social organization of this species. These birds live in groups of a breeding pair and up to four helpers, mostly males. Helpers do not breed but assist in incubating eggs and feeding nestlings. Some young birds disperse to new territories, but most remain behind as helpers to breeders. They may eventually attain breeding status when older birds die, but the wait may take years, and even then, helpers must compete to fill breeding vacancies. Young birds that disperse as members of new groups also have a tough path to reproductive success. New groups usually occupy abandoned territories or start at a new site and must excavate the cavities needed for nesting. This nest building can take several years. Individuals generally have a better chance of reproducing by remaining behind and competing when breeding vacancies open than by dispersing and excavating homes in new

territories. Perhaps this behavioral feature contributed to the decline of the red-cockaded woodpecker.

Ecologists tested the hypothesis that behavior constrains the ability of red-cockaded woodpecker populations to rebound by constructing cavities in pine trees at 20 sites in North Carolina. The results were dramatic: 18 of the 20 sites were colonized by red-cockaded woodpeckers, and new breeding groups formed only in areas where artificial cavities were drilled. The experiment supported the hypothesis that this woodpecker species was leaving much suitable habitat unoccupied because of an absence of breeding cavities. And the research informed a management strategy for reversing the decline of the red-cockaded woodpecker. A combination of controlled burning to clear understory vegetation and excavation of breeding cavities in unoccupied areas that provide good habitat has enabled a once endangered species to rebound. This example of the declining-population approach to conservation biology illustrates the need for case-by-case investigation of the factors contributing to a species' decline.

Conserving species involves weighing conflicting demands

Determining population numbers and habitat needs is only part of the effort to save species. It is also necessary to weigh a species' biological and ecological needs against other conflicting demands. Conservation biology often highlights the relationship between science, technology, and society—one of the themes of this book. For example, an ongoing, sometimes bitter debate in the U.S. Pacific Northwest pits saving habitat for populations of the northern spotted owl, timber wolf, grizzly bear, and bull trout against demands for jobs in the timber, mining, and other resource extraction industries. Programs to restock wolves in Yellowstone and to bolster the populations of grizzly bears and other large carnivores are opposed by some recreationists concerned for their safety and by many ranchers concerned with potential losses of livestock.

Large, high-profile vertebrates are not always the focal point in these conflicts, but habitat use is almost always at issue. Should work proceed on a new highway bridge if it destroys the only remaining habitat of a species of freshwater mussel? If you were the owner of a coffee plantation growing varieties that thrive in bright sunlight, do you think you would be willing to change to shade tolerant coffee varieties that are less productive and less profitable but support large numbers of songbirds?

In addition to questions about human habitat needs, another important factor to weigh is the ecological role of species. Because we will not be able to save every endangered species, we must determine which ones are most important for conserving biodiversity as a whole. Species do not exert equal influence on community and ecosystem processes. Some organisms, called *keystone species,* have disproportionately large impacts relative to their numbers (see Chapter 53). Some keystone species significantly modify habitats, creating diverse patches that support numerous other species. Keystone mutualists provide other species with nutrients, defense against predators and parasites, or, in the case of pollinators, the means to reproduce. Identifying keystone species and finding ways to sustain their populations can ensure the continuance of numerous other species and can be central to the survival of whole communities. Conservation must move beyond its preoccupation with single species like the northern spotted owl and look at the whole community and ecosystem as an important unit of biodiversity.

CONSERVATION AT THE COMMUNITY, ECOSYSTEM, AND LANDSCAPE LEVELS

Most preservation efforts in the past have focused on saving endangered species, but today, conservation biology increasingly aims to sustain the biodiversity of entire communities and ecosystems. On a broader scale yet, the principles of community and ecosystem ecology are being brought to bear on studies of the biodiversity of whole landscapes. In an ecological sense, a **landscape** is a regional assemblage of interacting ecosystems, such as a forest or forest patches, adjacent open fields, wetlands, streams, and streamside (riparian) habitats.

Landscape ecology is the application of ecological principles to the study of human land-use patterns. Understanding landscape dynamics is critically important in conservation because many species use more than one kind of ecosystem, and many live on the borders between ecosystems. The goal of landscape ecology, of which ecosystem management is part, is to understand patterns of landscape use in the past, present, and foreseeable future and to make biodiversity conservation a functional part of the picture. Such a broad view requires understanding community and ecosystem ecology as well as human population dynamics and economics.

Edges and corridors can strongly influence landscape biodiversity

The boundaries, or *edges,* between ecosystems (between a lake and the surrounding forest, for example, or between cropland and suburban housing tracts) and within ecosystems (such as roadsides and rock outcroppings) are defining features of landscapes (FIGURE 55.15). An edge has its own set of physical conditions, such as soil type, topography, and disturbance features, that differ from those on either side. For instance, the soil surface of an edge between a forest patch and a burned area receives more sunlight and is usually hotter and drier than the forest interior but cooler and wetter than the soil surface in the burned area. Blown-down trees are a common disturbance feature of forest edges, which are less protected from strong winds than are forest interiors.

Associated with their specific physical features, edges also have their own communities of organisms. Some organisms thrive in edge communities because they require resources of the two adjacent areas. For instance, a bird called the ruffed grouse (*Bonasa umbellatus*) needs forest habitat for nesting, winter food, and shelter, as well as forest openings with dense shrubs and herbs for summer food. White-tailed deer also thrive in edge habitats, where they can browse on woody shrubs, and deer populations often expand when forests are logged.

The proliferation of edge species can have positive or negative effects on a community's biodiversity. A recent study of edge communities in a tropical rain forest in Cameroon showed that these areas may be important sites of speciation. On the other hand, communities in which edges have proliferated as a result of human alterations often have reduced biodiversity owing to the preponderance of edge-adapted species. For example, populations of the brown-headed cowbird (*Molothrus ater*), an edge-adapted species that lays its eggs in the nests of other birds, are currently expanding in many areas of the western United States. Cowbirds forage in open fields on insects disturbed by or attracted to cattle and other large herbivores, but they need forests where they can parasitize the nests of other birds. Cowbird numbers are burgeoning where forests are being heavily cut and fragmented, creating more edge habitat and open land for cattle, horses, and sheep.

(a) Natural edges between ecosystems. In this landscape in Kakadu National Park in northern Australia, you can see edges of a dry forest, a rocky area with grassy islands, and a flat, grassy lakeshore.

(b) Edges created by human activity. Human activities that degrade and fragment habitats often create edges that are more abrupt than those seen in natural landscapes. Pronounced edges (roads) surround clear-cuts in this photograph of a heavily logged rain forest in Malaysia.

FIGURE 55.15 Edges between ecosystems.

Increasing cowbird parasitism and loss of habitat are correlated with declining populations of several of the cowbird's host species—migratory songbirds such as the yellow warbler, red-eyed vireo, and American redstart.

Another important landscape feature, especially where habitats have been severely fragmented, is a **movement corridor,** a narrow strip or series of small clumps of quality habitat connecting otherwise isolated patches. Streamside habitats often serve as corridors, and government policy in some nations prohibits destruction of these riparian areas. In areas of heavy human use, artificial corridors are sometimes constructed. For example, highways bisect habitat patches required for survival of the few remaining Florida panthers (*Felis concolor coryi*).

FIGURE 55.16 An artificial corridor. This pass under a highway allows movement between protected areas for the few remaining Florida panthers. High fences along the highway reduce road kills of panthers and other species.

The state of Florida has erected high fences to reduce road kills and artificial corridors under highways to allow movement through protected areas for the panthers (FIGURE 55.16).

Movement corridors can promote dispersal and reduce inbreeding in declining populations. Corridors are especially important to species that migrate among different habitats seasonally. However, a corridor can also be harmful—as, for example, in the spread of disease, especially among small populations in closely situated habitat patches. The effects of corridors have not been thoroughly studied, and researchers tend to evaluate their potential impact on a case-by-case basis.

Conservation biologists face many challenges in setting up protected areas

Conservation biologists are applying current ecological research in setting up reserves or protected areas to slow the loss of biodiversity. National parks are examples of such protected places. In choosing locations for protection and designing nature reserves, conservation biologists face many challenges. If a community is subject to fire, grazing, and predation, for example, should the reserve be managed to minimize the risks of these processes to endangered or threatened species? Or should the reserve be left as natural as possible, with such processes as fires ignited by lightning allowed to play out without any human intervention? This is just one of the debates that arise among people who share an interest in the health of national parks and other protected areas.

Governments have set aside about 7% of the world's land in various forms of reserves. How are these protected locations selected? Much of the focus has been on hot spots of biological diversity. A **biodiversity hot spot** is a relatively small area with an exceptional concentration of endemic species and a

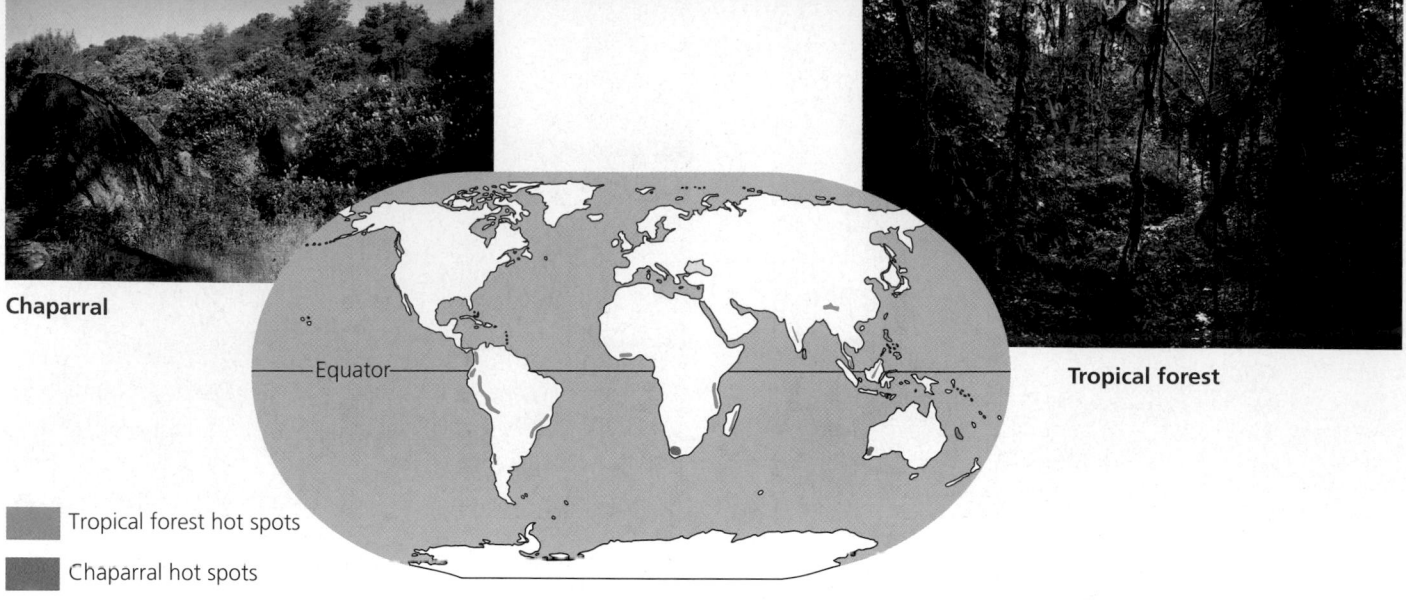

Chaparral

Tropical forest

Equator

■ Tropical forest hot spots

■ Chaparral hot spots

FIGURE 55.17 Some biodiversity hot spots. Only dry shrubland (such as chaparral) and tropical forest hot spots are mapped here.

large number of endangered and threatened species. For example, nearly 30% of all bird species are confined to only about 2% of Earth's land area. And about 50,000 plant species, or 20% of all known plant species, inhabit just 18 hot spots making up a total of only 0.5% of the global land surface. Overall, the "hottest" of the biodiversity hot spots, including rain forests and dry shrublands (such as California's chaparral), total less than 1.5% of Earth's land but are home to a third of all species of plants and vertebrates (FIGURE 55.17). Conservation biologists have also identified aquatic ecosystems, including certain river systems and coral reefs, as biodiversity hot spots.

Biodiversity hot spots are obviously good choices for nature reserves. However, recognizing hot spots is not always straightforward. And even if all hot spots could be protected, that effort would fall woefully short of conserving the planet's biodiversity. One problem is that a hot spot for one taxonomic group, such as birds, may not be a hot spot for some other taxonomic group, such as butterflies. Designating an area as a biodiversity hot spot is often taxonomically biased toward saving vertebrates and plants, with less attention paid to invertebrates and microorganisms. Some biologists are also concerned that the hot-spot strategy focuses so much of the conservation effort on such a small fraction of Earth's land.

As conservation biologists learn more about the requirements for achieving minimum viable population sizes for endangered species, it is becoming clear that most national parks and other reserves are

far too small. For example, FIGURE 55.18 compares the boundaries of Yellowstone and Grand Teton National Parks with the actual area required to prevent extinction of grizzly bears. The *biotic boundary,* the area needed to sustain the grizzly, is more than ten times as large as the *legal boundary,* the actual area of the parks. Given political and economic realities, it is unlikely that many existing parks will be enlarged, and most new reserves will also be too small. Areas of private and public land surrounding reserves will have to contribute to the conservation of biodiversity. In particular, this means integrating how land is used for agriculture and forestry into conservation strategies.

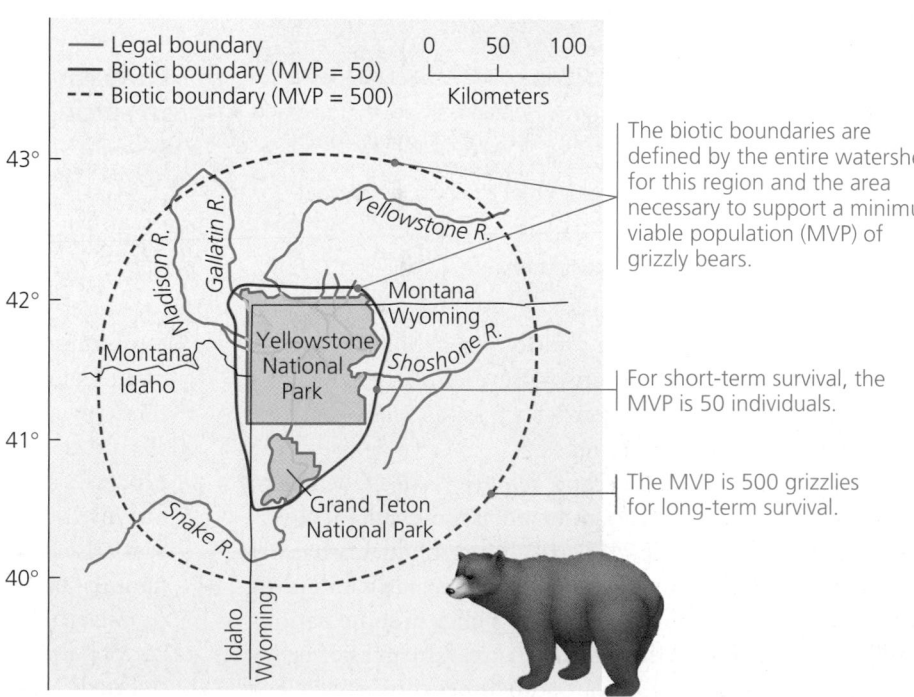

— Legal boundary
— Biotic boundary (MVP = 50)
--- Biotic boundary (MVP = 500)

0   50   100
Kilometers

The biotic boundaries are defined by the entire watershed for this region and the area necessary to support a minimum viable population (MVP) of grizzly bears.

For short-term survival, the MVP is 50 individuals.

The MVP is 500 grizzlies for long-term survival.

FIGURE 55.18 The legal and biotic boundaries for grizzly bears in Yellowstone and Grand Teton National Parks.

Nature reserves must be functional parts of landscapes

Nature reserves are biodiversity islands in a sea of habitat degraded to varying degrees by human activity. It is important to realize, however, that protected "islands" are not isolated from their surroundings and that nonequilibrium ecology applies to nature reserves as well as the landscapes in which they are embedded.

An earlier policy—that protected areas should be set aside to remain unchanged forever—was based on the old concept that ecosystems are balanced, self-regulating units. However, as we have discussed in Chapter 53, disturbance is a functional component of all ecosystems, and management policies that ignore natural disturbances or attempt to prevent them have generally proved to be self-defeating. For instance, setting aside an area of a fire-dependent community, such as a portion of a tallgrass prairie, chaparral, or dry pine forest, with the intention of saving it is unrealistic if periodic burning is excluded. Without the dominant disturbance, the fire-adapted species are usually outcompeted by other species, and biodiversity is reduced.

Because human disturbance and fragmentation are increasingly common landscape features, patch dynamics, population dynamics, edges, and corridors are important in the design and management of protected areas. Unfortunately, there are many more questions than answers. For example, is it better to create one large preserve or a group of smaller preserves? One argument for extensive preserves is that large, far-ranging animals with low-density populations, such as the grizzly bear, require extensive habitats. In addition, more extensive areas have proportionately smaller perimeters than smaller areas and are therefore less affected by edges. An argument favoring smaller, disjunct preserves is that they may slow the spread of disease throughout a population. Often outweighing all other considerations, recent and ongoing land use by humans may largely dictate the size and shape of protected areas. Conservationists typically inherit the land that is useless for exploitation by agriculture or forestry.

Several nations have adopted an approach to landscape management called zoned reserve systems. A **zoned reserve** is an extensive region of land that includes one or more areas undisturbed by humans surrounded by lands that have been changed by human activity and are used for economic gain. The key challenge of the zoned reserve concept is the development of a social and economic climate in the surrounding lands that is compatible with the long-term viability of the protected core area. These surrounding areas continue to be used to support the human population, but with regulations that prevent the types of extensive alterations likely to impact the protected areas. As a result, surrounding tracts of land serve as buffer zones against further intrusion into the undisturbed areas.

The small Central American nation of Costa Rica has become a world leader in establishing zoned reserves. In exchange for reducing its international debt, the Costa Rican government established eight zoned reserves, called "conservation areas" (FIGURE 55.19). Costa Rica is making progress toward managing its zoned reserves, and the buffer zones provide a steady, lasting supply of forest products, water, and

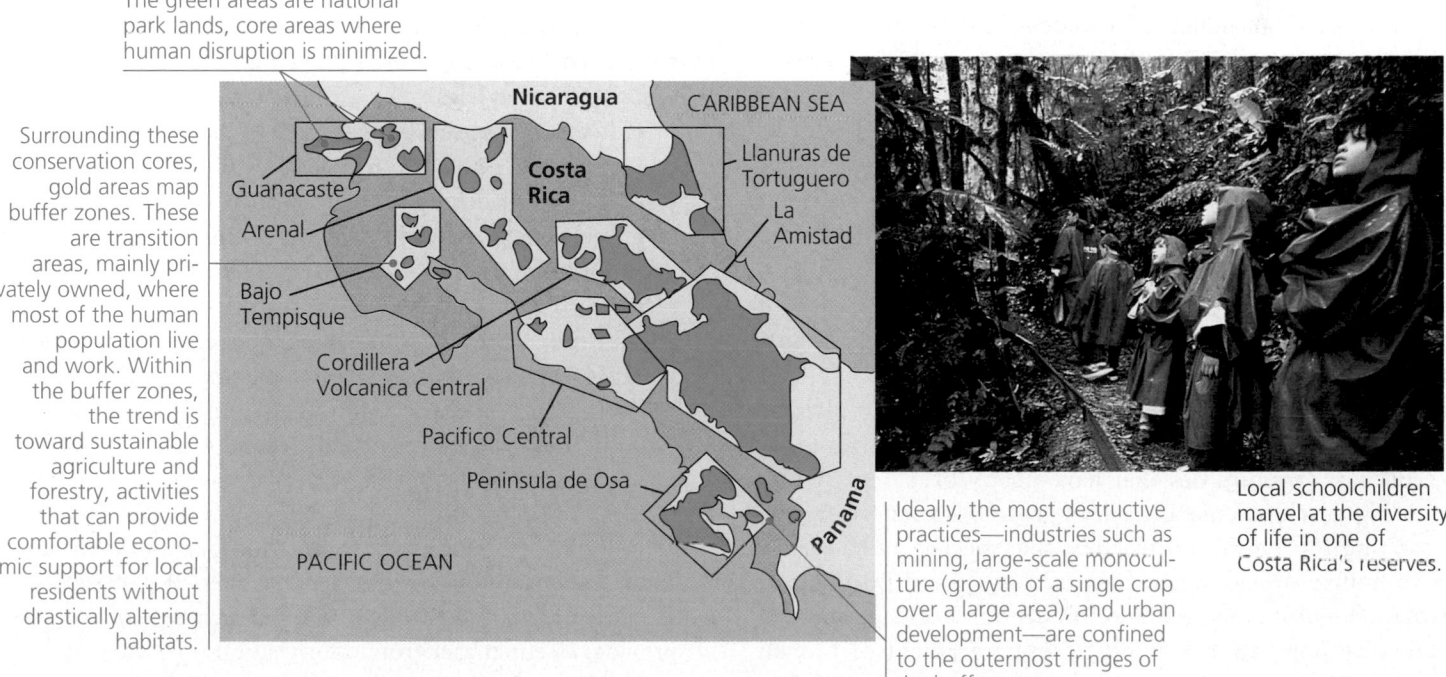

The green areas are national park lands, core areas where human disruption is minimized.

Surrounding these conservation cores, gold areas map buffer zones. These are transition areas, mainly privately owned, where most of the human population live and work. Within the buffer zones, the trend is toward sustainable agriculture and forestry, activities that can provide comfortable economic support for local residents without drastically altering habitats.

Ideally, the most destructive practices—industries such as mining, large-scale monoculture (growth of a single crop over a large area), and urban development—are confined to the outermost fringes of the buffer zone.

Local schoolchildren marvel at the diversity of life in one of Costa Rica's reserves.

FIGURE 55.19 Zoned reserves in Costa Rica.

(a)

(b)

FIGURE 55.20 An endangered, endemic species in its unique habitat. (a) The Florida scrub jay (*Aphelocoma coerulescens*) inhabits desertlike scrub communities in central Florida. **(b)** New housing developments and expanding citrus groves threaten this bird and the remaining fragments of its unique habitat. Ecologists at the Archbold Biological Station, a small reserve with a stable scrub jay population, have found that housing developments do not provide enough food (arthropods) for the jays and are associated with higher mortality of adults. Even if a housing development contains some scrub habitat, it is a difficult place for these birds to rear enough young to offset the increased death rate. Archbold researchers predict that the Florida scrub jay will survive only if reserves of intact oak scrub habitat are maintained and properly managed with prescribed fire.

hydroelectric power and also support sustainable agriculture and tourism. An important goal is providing a stable economic base for people living there. As ecologist Daniel Janzen, a leader in tropical conservation, has said, "The likelihood of long-term survival of a conserved wildland area is directly proportional to the economic health and stability of the society in which that wildland is embedded." Destructive practices that are not compatible with long-term ecosystem conservation and from which there is often little local profit are gradually being discouraged. Such destructive practices include massive logging, large-scale single-crop agriculture, and extensive mining. Costa Rica looks to its zoned reserve system to maintain at least 80% of its native species.

The continued high rate of human exploitation of ecosystems leads to the prediction that considerably less than 10% of the biosphere will ever be protected as nature reserves. Sustaining biodiversity often involves working in landscapes that are almost entirely human dominated. For example, the Florida scrub jay, an endangered endemic species, inhabits dry scrub oak communities that have nearly been replaced by housing developments and citrus groves (FIGURE 55.20). Attempting to understand whether this species could coexist with human development, avian ecologist Reed Bowman, at the Archbold Biological Station in central Florida, examined scrub jay population viability across a gradient of human density. Unfortunately, housing developments, even if they contain some scrub habitats, turn out to be relatively poor

environments for the jay. Bowman is now convinced that long-term survival of this bird depends on reserves of contiguous, intact scrub surrounded by areas where some natural vegetation remains—the zoned reserve concept applied to suburbia.

Restoring degraded areas is an increasingly important conservation effort

Eventually, some areas that are altered by human activity are abandoned. For instance, the soils of many tropical areas become unproductive and are abandoned less than five years after being cleared for farming. Mining activities may last for several decades, but the lands are then abandoned in a degraded state. Many ecosystems are also damaged inadvertently by the dumping of toxic chemicals or such mishaps as oil spills. These degraded habitats and ecosystems are increasing in area because the natural rates of recovery by successional processes are slower than the rate of degradation by human activities.

A new subdiscipline of conservation biology called **restoration ecology** applies ecological principles in an effort to return degraded ecosystems to conditions as similar as possible to their natural, predegraded state. Restoration ecology seeks to reverse population and community declines. One basic assumption of restoration ecology is that most environmental damage is reversible. This optimism must be balanced by a second basic assumption—that communities are not infinitely resilient to damage.

Biological communities can recover naturally from many types of disturbances through a series of restoration mechanisms that occur during the various stages of ecological succession (see Chapter 53). The amount of time required for such natural recovery is more closely related to the spatial scale of the disturbance than the type of disturbance: The larger the area disturbed, the longer the time frame for recovery. Whether the disturbance is natural or caused by humans seems to make little difference in this size-time relationship (FIGURE 55.21). One of the goals of restoration ecology is to identify the processes that most limit the speed of recovery so that those factors can be manipulated to reduce the time it takes for a community to bounce back from the impact of disturbances. Thus, understanding the specific characteristics of succession after each type of disturbance and for each type of ecosystem provides essential background for restoration ecologists.

Two key strategies in restoration ecology are bioremediation and augmentation of ecosystem processes. **Bioremediation** is

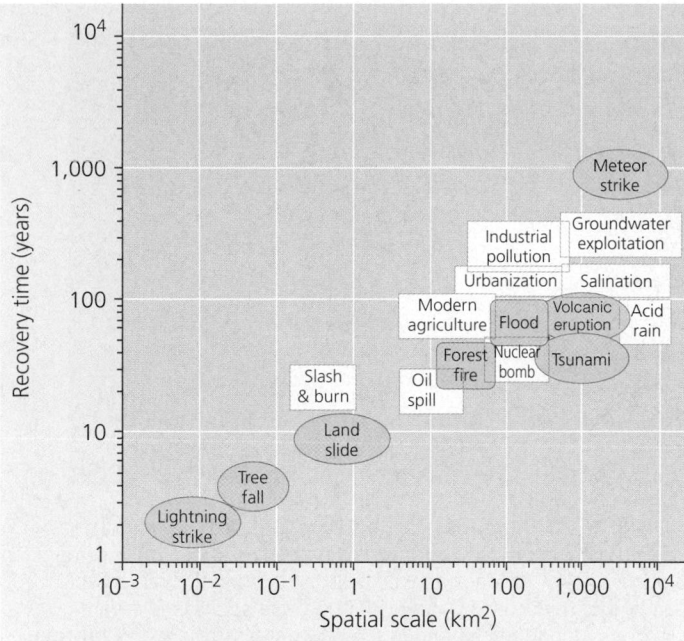

FIGURE 55.21 The size-time relationship for community recovery from natural (salmon-colored ellipses) and human-caused (white rectangles) disasters. Note that the scales are logarithmic. The aim of restoration ecology is to reduce the recovery time by manipulating ecological factors that slow recovery.

FIGURE 55.22 Restoration of degraded roadsides in the tropics. Forest ecologist Ariel Lugo has monitored rapid regrowth of indigenous communities along roadsides in Puerto Rico. An exotic plant, *Albizzia procera* (shown here), which thrives on nitrogen-poor soils, first colonized these sites after the original forest was removed and soils were depleted of nutrients. Apparently, the rapid buildup of organic material from dense stands of *Albizzia* enabled indigenous plants to recolonize the area and overgrow the exotic plant in a relatively brief time.

the use of living organisms, usually prokaryotes, fungi, or plants, to detoxify polluted ecosystems (see Chapter 27). Some plants adapted to soils with heavy metals are capable of accumulating high concentrations of potentially toxic metals such as zinc, nickel, lead, and cadmium. Restoration ecologists can use these plants to revegetate sites degraded by mining and other human activities and then harvest the plants to recover the particular metals. A number of researchers are also focusing on the ability of certain prokaryotes and lichens to concentrate metals. Researchers in the United Kingdom recently discovered a lichen species that grows on soil polluted with uranium dust left over from mining. Useful as a biological monitor of uranium and potentially as a remediator, the lichen concentrates uranium in a dark pigment similar to melanin in human skin. And several extremophilic bacteria and archaea thrive in natural environments similar to industrially polluted sites. Restoration ecologists have achieved some success in using the bacterium *Pseudomonas,* supplied with growth stimulants, to clean up oil spills on beaches. More common still is the use of certain prokaryotes to metabolize toxins in dump sites. Genetic engineering may become increasingly important as a tool for improving the performance of certain species as bioremediators.

In contrast to bioremediation, which is a strategy for *removing* harmful substances, biogiocal *augmentation* uses organisms to *add* essential materials to a degraded ecosystem. Augmenting ecosystem processes requires determining what

factors, such as chemical nutrients, have been removed from an area and are limiting its rate of recovery. Encouraging the growth of plants that thrive in nutrient-poor soils often speeds up the rate of successional changes that can lead to recovery of damaged sites. Ariel Lugo, director of the U.S. Forest Service's Institute of Tropical Forestry in Puerto Rico, has evidence of a positive effect of an introduced plant species on the recovery of native vegetation (FIGURE 55.22). Thriving on nitrogen-poor soils, the leguminous plant *Albizzia procera,* exotic in Puerto Rico, helps set the stage for recolonization by native tropical forest species.

To date, the most extensive and successful restoration projects have been in marginally disturbed wetlands in landscapes where biodiversity has not been greatly depleted. In these projects, restoring the natural water flow patterns and replanting indigenous vegetation have led to recolonization by animal populations. Restoring viable populations of highly sensitive wetland species to heavily degraded wetlands is much more challenging, as are similar restoration efforts in most ecosystems.

Because of the novelty of restoration science, the complexity of ecosystems, and the unique features of each situation, restoration ecologists usually must learn as they go. Many

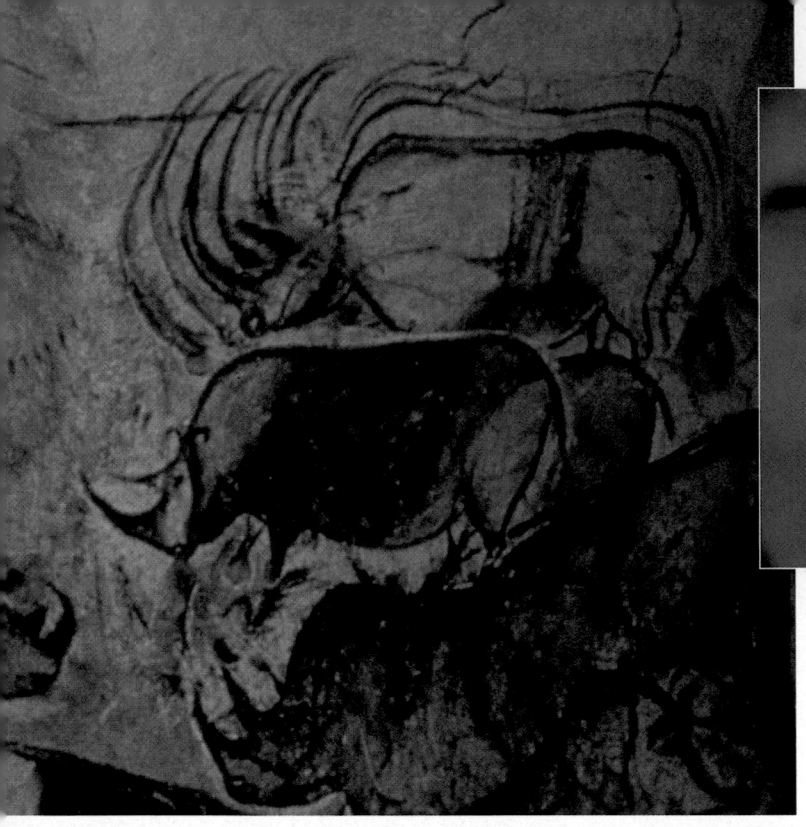

(a)

(b)

FIGURE 55.23 Biophilia, past and present. (a) Art history goes way back—and so does our fascination with and dependence on biodiversity. A Cro-Magnon wildlife artist created this remarkable painting of rhinoceroses about 30,000 years ago. Three cave explorers found the painting in a prehistoric art gallery on Christmas Eve, 1994, when they ventured into a cavern near Vallon-Pont d'Arc, in southern France. **(b)** Biologist Carlos Rivera Gonzales, who is participating in a biodiversity survey in a remote region of Peru, could not resist a closer look at a tiny tree frog.

restoration ecologists advocate **adaptive management,** which is the use of the experimental method in trying several promising types of management to find out what works best. The key to adaptive management and the key to restoration ecology is to consider alternative ways of accomplishing goals and to learn from mistakes. The long-term objective of restoration is to speed the reestablishment of the predisturbance ecosystem. But a pragmatic initial goal is often to approximate the original ecosystem, which can be accomplished much sooner than complete restoration to the original state.

The goal of sustainable development is reorienting ecological research and challenging all of us to reassess our values

Facing increasing loss and fragmentation of habitats, how can we best manage Earth's resources? If we are to conserve most of a nation's species, which habitat patches are most crucial? Among the limited choices, which areas are most practical to protect and manage if we are to save rare species or the greatest number of species?

We must understand the complex interconnections of the biosphere if we are to make sensible decisions about how to conserve these networks. To this end, many nations, scientific societies, and private foundations have embraced the concept of sustainable development, the long-term prosperity of human societies and the ecosystems that support them. The forward-looking Ecological Society of America, the world's largest organization of professional ecologists, endorses a research agenda called the **Sustainable Biosphere Initiative.** The goal of this initiative is to define and acquire the basic ecological information necessary for the intelligent and responsible development, management, and conservation of Earth's resources. The research agenda includes studies of global change, including interactions between climate and ecological processes; biological diversity and its role in maintaining ecological processes; and the ways in which the productivity of natural and artificial ecosystems can be sustained. This initiative requires a strong commitment of human and economic resources.

Sustainable development, of course, is not *just* about science. It depends on most of us reassessing our values. Those of us living in affluent developed nations are responsible for the greatest amount of environmental degradation. Reality demands that we distinguish what we need from what we want, learn to revere the natural processes that sustain us, and reduce our orientation toward short-term personal gain. The current state of the biosphere demonstrates that we are treading precariously on uncharted ecological ground and that the importance of our scientific and personal efforts cannot be overstated.

Conservation science is an intersection of numerous facets of biology, including ecology, evolution, physiology, molecular biology, genetics, and behavior. Efforts to sustain ecosystem processes and stem the loss of biodiversity also connect life science with the social sciences, economics, and humanities. In fact, we'll end our book with a note of optimism based on our humanity.

The future of the biosphere may depend on our biophilia

Despite the uncertainties about the future of the biosphere, now is not the time for gloom and doom but the time to reconnect with the rest of nature. Not many people today live in truly wild environments or even visit such places often. Our modern lives are very different from those of early humans, who hunted and gathered and painted wildlife murals on cave walls. But our behavior reflects remnants of our ancestral attachment to nature and the diversity of life—what Edward O. Wilson calls *biophilia* (FIGURE 55.23). Biophilia includes our sense of connection to diverse organisms and also our attraction to pristine landscapes with clean water and lush vegetation. We evolved in natural environments rich in biodiversity, and we still have an affinity for such settings. Wilson makes the case that our biophilia is innate, an evolutionary product of natural selection acting on a brainy species whose survival depended on a close connection to the environment and a practical appreciation of plants and animals.

It will come as no surprise that most biologists have embraced the concept of biophilia. After all, these are people who have turned their passion for nature into careers. But biophilia strikes a harmonic chord with biologists for another reason. If biophilia is evolutionarily embedded in our genomes, then there is hope that we can become better custodians of the biosphere. If we all pay more attention to our biophilia, a new environmental ethic could catch on among individuals and societies. And that ethic is a resolve never to allow a species to become extinct through human activities or any ecosystem to be destroyed as long as there are reasonable ways to prevent such ecological violence. It is an environmental ethic that balances out another human trait—our tendency to "subdue Earth." Yes, we should be motivated to preserve biodiversity because we depend on it for food, medicine, building materials, fertile soil, flood control, habitable climate, drinkable water, and breathable air. But maybe we can also work harder to prevent the extinction of other forms of life just because it is the ethical thing for us to do as the most thoughtful species in the biosphere. Again, Wilson sounds the call: "Right now, we're pushing the species of the world through a bottleneck. We've got to make it a major moral principle to get as many of them through this as possible. It's the challenge now and for the next century. And there's one good thing about our species: We like a challenge!"

It is appropriate that we end this textbook with biophilia, for biology is a scientific expression of our desire to know nature. We are most likely to save what we appreciate, and we are most likely to appreciate what we understand. By learning about the processes and diversity of life, we also become more aware of ourselves and our place in the biosphere. We hope this book serves you well in this lifelong adventure.

CHAPTER 55 REVIEW

Go to the Campbell Biology website (www.campbellbiology.com) to explore an interactive version of the Chapter Review.

Summary of Key Concepts

THE BIODIVERSITY CRISIS

- The three levels of biodiversity are genetic diversity, species diversity, and ecosystem diversity (pp. 1225–1226, FIGURES 55.1, 55.2) Biodiversity consists of the various kinds of ecosystems, the species richness of communities in those ecosystems, and the genetic variation within and between populations of each species.

- Biodiversity at all three levels is vital to human welfare (pp. 1226–1228, FIGURES 55.3, 55.4) Other species provide humans with food, fiber, and medicines. Estimates by ecologists and economists indicate the enormous economic value of ecosystem services.

- The four major threats to biodiversity are habitat destruction, introduced species, overexploitation, and food chain disruptions (pp. 1228–1232, FIGURES 55.5–55.9) Human alteration of habitat poses the single greatest threat to biodiversity. Competition and predation by introduced species and excessive harvesting for commerce and sport are other significant threats. Extinctions at one trophic level can impact other trophic levels.

  Web/CD Activity 55A: *Madagascar and the Biodiversity Crisis*
  Web/CD Activity 55B: *Introduced Species: Fire Ants*

CONSERVATION AT THE POPULATION AND SPECIES LEVELS

- According to the small-population approach, a population's small size can draw it into an extinction vortex (pp. 1232–1236, FIGURES 55.10–55.13) When a population drops below a minimum viable population size (MVP), its loss of genetic variation due to inbreeding and genetic drift can trap it in a vortex of continued decline leading to extinction. The MVP may be measured as effective population size, the number of breeding individuals.

- **The declining-population approach is a proactive conservation strategy for detecting, diagnosing, and halting population declines** (pp. 1236–1237, FIGURE 55.14) This conservation approach seeks the causes of population declines and addresses those causes in developing ways to stop the declines.

- **Conserving species involves weighing conflicting demands** (pp. 1237–1238) Conservation solutions often require resolving conflicts between the habitat needs of endangered species and human demands for economic development and living space.

CONSERVATION AT THE COMMUNITY, ECOSYSTEM, AND LANDSCAPE LEVELS

- **Edges and corridors can strongly influence landscape biodiversity** (pp. 1238–1239, FIGURES 55.15, 55.16) Boundaries (edges) between ecosystems and along prominent features within ecosystems have unique sets of physical conditions and communities of species. Edges become more extensive as habitat fragmentation increases, and edge-adapted species may become more dominant. Movement corridors may promote dispersal and help sustain populations, or they may promote harmful conditions (such as disease).

- **Conservation biologists face many challenges in setting up protected areas** (pp. 1239–1240, FIGURES 55.17, 55.18) Areas with exceptionally high concentrations of endemic species, called biodiversity hot spots, are also hot spots of extinction, and thus prime candidates for protection. Most national parks and other protected areas are too small to save endangered species without protection in surrounding areas.

- **Nature reserves must be functional parts of landscapes** (pp. 1241–1242, FIGURES 55.19, 55.20) Sustaining biodiversity in reserves requires management to ensure that human activities in the surrounding landscape do not harm the protected habitats. The zoned reserve model recognizes that conservation efforts often involve working in landscapes that are largely human dominated.

- **Restoring degraded areas is an increasingly important conservation effort** (pp. 1242–1244, FIGURES 55.21, 55.22) Restoration ecology often involves bioremediation (the use of organisms to detoxify polluted ecosystems) and augmentation of ecosystem processes such as ecological succession.

  Web/CD Case Study in The Process of Science: *How Are Potential Prairie Restoration Sites Analyzed?*

- **The goal of sustainable development is reorienting ecological research and challenging all of us to reassess our values** (p. 1244) Sustainable development, the long-term prosperity of human societies and the ecosystems supporting them, depends on ecological knowledge and on a commitment to promote ecosystem processes and biodiversity.

  Web/CD Activity 55C: *Conservation Biology Review*

- **The future of the biosphere may depend on our biophilia** (p. 1245, FIGURE 55.23) Our innate sense of connection to nature may eventually motivate a realignment of our environmental priorities.

Self-Quiz

1. Extinction is a natural phenomenon. It is estimated that 99% of all species that ever lived are now extinct. Why, then, do we say that we are now in a biodiversity crisis?
   a. Because of our biophilia, humans feel ethically responsible for protecting endangered species.
   b. Scientists have finally identified most of the species on Earth and are thus able to quantify the number of species becoming extinct.
   c. The current rate of extinction is as much as 1,000 times higher than at any time in the last 100,000 years.
   d. Humans have greater medical needs than at any previous time in history, and many potential medicinal compounds are being lost as plant species become extinct.
   e. Most biodiversity hot spots have been destroyed by recent ecological disasters.

2. One level of the biodiversity crisis is the potential loss of ecosystems. The most likely serious consequence of a loss in ecosystem diversity would be the
   a. increase in global warming and thinning of the ozone layer.
   b. loss of ecosystem services on which humans depend.
   c. increase in the dominance of edge-adapted species.
   d. loss of a source of genetic diversity to preserve endangered species.
   e. loss of species for "bioprospecting."

3. A population of strictly monogamous swans consists of 40 males and ten females. The effective population size (N_e) for this population is
   a. 50.   d. 20.
   b. 40.   e. 10.
   c. 32.

4. Which of the following conditions is the most likely indicator of a population in an extinction vortex?
   a. The population is divided into smaller populations.
   b. The species is rare.
   c. The effective population size of the species is around 500.
   d. Genetic measurements indicate a continuing loss of genetic variation.
   e. All populations are connected by corridors.

5. The application of ecological principles to return a degraded ecosystem to its natural state is specifically characteristic of
   a. population viability analysis.
   b. landscape ecology.
   c. conservation ecology.
   d. restoration ecology.
   e. resource conservation.

6. What is the greatest threat to biodiversity?
   a. overexploitation of commercially important species
   b. introduced species that compete with or prey on native species
   c. the high rate of destruction of tropical rain forests
   d. disruption of trophic relationships as more and more prey species become extinct
   e. human alteration, fragmentation, and destruction of terrestrial and aquatic habitats

7. Which of the following statements about the declining-population approach to conservation is *not* correct?
   a. We need information on whether or not the population in question is in decline.
   b. We need to do something quickly, even if we have no information, because conservation biology is a crisis discipline.
   c. Several hypotheses about why the population is declining should be evaluated.
   d. A proposed reason for the decline should be tested experimentally.
   e. Humans may not be the cause of every population decline.

8. According to the small-population approach, what would be the best strategy for saving a population that is in an extinction vortex?
   a. determining the minimum viable population size by taking into account the effective population size
   b. establishing a nature reserve to protect its habitat
   c. introducing individuals from other populations to increase genetic variation
   d. sterilizing the least fit individuals
   e. reducing the population size of its predators and competitors

9. Which of the following statements about protected areas is *not* correct?
   a. We now protect 25% of the land areas of the planet.
   b. National parks are only one type of protected area.
   c. Most protected areas are small in size.
   d. Protected area management must be coordinated with management of lands outside the protected zone.
   e. Biodiversity hot spots are important areas to protect.

10. What is the Sustainable Biosphere Initiative?
    a. a failed experiment that tried to create an artificial, self-sufficient biosphere
    b. a research agenda to study biodiversity and support sustainable development
    c. a conservation practice that sets up zoned reserves surrounded by buffer zones
    d. the declining-population approach to conservation that seeks to identify and remedy causes of species' declines
    e. a conservation program that uses adaptive management to experiment and learn while working with disturbed ecosystems

11. What is an introduced species?

12. What is a biodiversity hot spot?

13. How is a landscape different from an ecosystem?

14. How can "living on the edge" be a good thing for some species, such as white-tailed deer and cowbirds?

15. As complementary strategies of restoration ecology, contrast the way bioremediation and augmentation use organisms to alter the chemical composition of a degraded ecosystem.

16. Why is a concern for the well-being of future generations essential for progress toward sustainable development?

17. Contrast the small-population approach with the declining-population approach in conservation biology.

Go to the website or CD-ROM for more quiz questions.

Evolution Connection

You learned in this chapter that while extinction is a natural process, the current high rate of extinction caused by human disturbance to world ecosystems is of great concern. What are the implications of this rapid extinction rate for the restoration of biological diversity in the future, as compared with the far slower extinction rates that characterized much of the past history of Earth?

The Process of Science

Suppose that you are in charge of planning a forest reserve, and one of your main goals is to help sustain locally beleaguered populations of woodland birds. Parasitism by the brown-headed cowbird is an escalating problem in the area. Reading research reports, you note that female cowbirds are usually reluctant to penetrate more than about 100 m into a forest and that some woodland birds are known to reduce cowbird nest parasitism by restricting their nesting to the denser, more central regions of forests. The forested area you have to work with is about 1,000 m by 6,000 m. A recent logging operation removed about half of the trees on one of the 6,000-m sides; the other three sides are adjacent to deforested pastureland. Your plan must include space for a small maintenance building, which you estimate to take up about 100 m^2. It will also be necessary to build a road, 10 m by 1,000 m, across the reserve. Where would you construct the road and the building, and why?

Analyze potential sites for a prairie restoration project in the Case Study in The Process of Science, available on the website and CD-ROM.

Science, Technology, and Society

Some organizations are starting to envision a sustainable society—one in which each generation inherits sufficient natural and economic resources and a relatively stable environment. The Worldwatch Institute, an environmental policy organization, estimates that we must reach sustainability by the year 2030 to avoid economic and environmental disaster. To get there, we must begin shaping a sustainable society during the next ten years or so. In what ways is our current system not sustainable? What might we do to work toward sustainability, and what are the major roadblocks to achieving it? How would your life be different in a sustainable society?

Answers: 1. c; 2. b; 3. c; 4. d; 5. d; 6. e; 7. b; 8. c; 9. a; 10. b; 11. A species that has been accidentally or intentionally transferred from one location to another, where it did not occur naturally. 12. A relatively small area with a disproportionate number of endemic species, including endangered species. 13. A landscape is more inclusive in that it consists of several interacting ecosystems in the same region. 14. They use a combination of resources from the two ecosystems on either side of the edge. 15. In bioremediation, certain organisms are used to remove harmful chemicals from the environment; in augmentation, certain organisms are used to add essential chemicals to the environment. 16. Sustainable development is a long-term goal—longer than a human lifetime. Concern only with personal gain in the here and now is an obstacle to sustainable development because it discourages behavior that benefits future generations. 17. The small-population approach focuses on the need to introduce genetic diversity to populations that are below minimum viable size. The declining-population approach concentrates on correcting the factors that contribute to a population's decline.

THE METRIC SYSTEM

Measurement	Unit and Abbreviation	Metric Equivalent	Metric-to-English Conversion Factor	English-to-Metric Conversion Factor
Length	1 kilometer (km)	$= 1000 \ (10^3)$ meters	1 km = 0.62 mile	1 mile = 1.61 km
	1 meter (m)	$= 100 \ (10^2)$ centimeters $= 1000$ millimeters	1 m = 1.09 yards 1 m = 3.28 feet 1 m = 39.37 inches	1 yard = 0.914 m 1 foot = 0.305 m
	1 centimeter (cm)	$= 0.01 \ (10^{-2})$ meter	1 cm = 0.394 inch	1 foot = 30.5 cm 1 inch = 2.54 cm
	1 millimeter (mm)	$= 0.001 \ (10^{-3})$ meter	1 mm = 0.039 inch	
	1 micrometer (μm) (formerly micron, μ)	$= 10^{-6}$ meter $(10^{-3}$mm$)$		
	1 nanometer (nm) (formerly millimicron, mμ)	$= 10^{-9}$ meter $(10^{-3}$μm$)$		
	1 angstrom (Å)	$= 10^{-10}$ meter $(10^{-4}$μm$)$		
Area	1 hectare (ha)	= 10,000 square meters	1 ha = 2.47 acres	1 acre = 0.405 ha
	1 square meter (m²)	= 10,000 square centimeters	1 m² = 1.196 square yards 1 m² = 10.764 square feet	1 square yard = 0.8361 m² 1 square foot = 0.0929 m²
	1 square centimeter (cm²)	= 100 square millimeters	1 cm² = 0.155 square inch	1 square inch = 6.4516 cm²
Mass	1 metric ton (t)	= 1000 kilograms	1 t = 1.103 tons	1 ton = 0.907 t
	1 kilogram (kg)	= 1000 grams	1 kg = 2.205 pounds	1 pound = 0.4536 kg
	1 gram (g)	= 1000 milligrams	1 g = 0.0353 ounce 1 g = 15.432 grains	1 ounce = 28.35 g
	1 milligram (mg)	$= 10^{-3}$ gram	1 mg = approx. 0.015 grain	
	1 microgram (μg)	$= 10^{-6}$ gram		
Volume (solids)	1 cubic meter (m³)	= 1,000,000 cubic centimeters	1 m³ = 1.308 cubic yards 1 m³ = 35.315 cubic feet	1 cubic yard = 0.7646 m³ 1 cubic foot = 0.0283 m³
	1 cubic centimeter (cm³ or cc)	$= 10^{-6}$ cubic meter	1 cm³ = 0.061 cubic inch	1 cubic inch = 16.387 cm³
	1 cubic millimeter (mm³)	$= 10^{-9}$ cubic meter $(10^{-3}$ cubic centimeter$)$		
Volume (Liquids and Gases)	1 kiloliter (kl or kL)	= 1000 liters	1 kL = 264.17 gallons	1 gallon = 3.785 L
	1 liter (L)	= 1000 milliliters	1 L = 0.264 gallons 1 L = 1.057 quarts	1 quart = 0.946 L
	1 milliliter (mL)	$= 10^{-3}$ liter	1 mL = 0.034 fluid ounce	1 quart = 946 mL
		= 1 cubic centimeter	1 mL = approx. $\frac{1}{4}$ teaspoon	1 pint = 473 mL 1 fluid ounce = 29.57 mL
			1 mL = approx. 15–16 drops (gtt.)	1 teaspoon = approx. 5 mL
	1 microliter (μl or μL)	$= 10^{-6}$ liter $(10^{-3}$ milliliters$)$		
Time	1 second (s)	$= \frac{1}{60}$ minute		
	1 millisecond (ms)	$= 10^{-3}$ second		
Temperature	Degrees Celsius (°C) (Absolute zero, when all molecular motion ceases, is −273 °C. The Kelvin (K) scale, which has the same size degrees as Celsius, has its zero point at absolute zero. Thus, 0° K = −273°C.)		$°F = \frac{9}{5}°C + 32$	$°C = \frac{5}{9}(°F - 32)$

A COMPARISON OF THE LIGHT MICROSCOPE AND THE ELECTRON MICROSCOPE

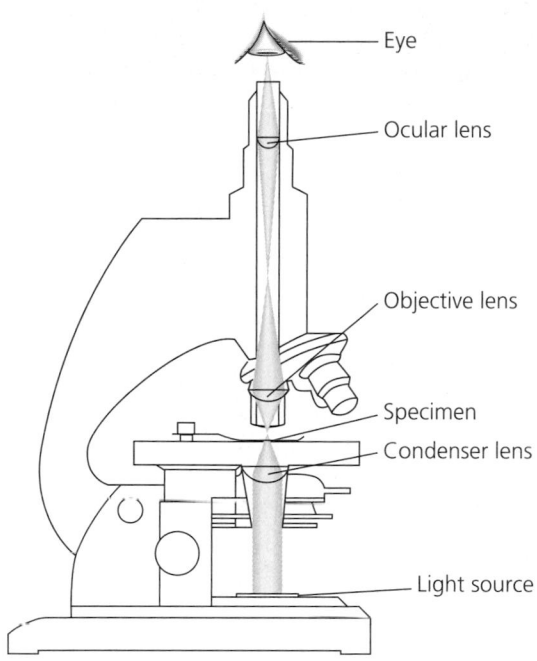

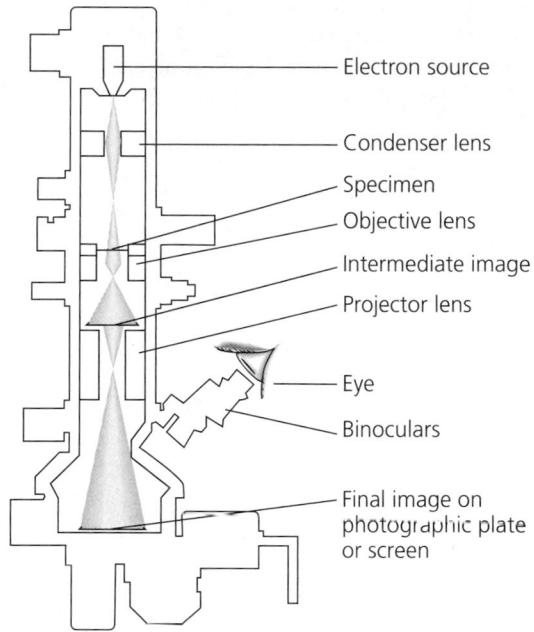

(a) Light Microscope. In light microscopy, light is focused on a specimen by a glass condenser lens; the image is then magnified by an objective lens and an ocular lens, for projection on the eye or on photographic film.

(b) Electron Microscope. In electron microscopy, a beam of electrons (top of the microscope) is used instead of light, and electromagnets are used instead of glass lenses. The electron beam is focused on the specimen by a condenser lens; the image is magnified by an objective lens and a projector lens, for projection on a screen or on photographic film.

CLASSIFICATION OF LIFE

This appendix presents the taxonomic classification used for the major extant groups of organisms discussed in this text; not all phyla are included. The classification reviewed here is based on the three-domain system, which assigns the two major groups of prokaryotes, archaea and bacteria, to separate domains (with eukaryotes making up the third domain). This classification contrasts with the traditional five-kingdom system, which groups all prokaryotes in a single kingdom, Monera. The rationale for alternative classification schemes is discussed in Unit Five of the text. The taxonomic turmoil includes debates about the number and boundaries of kingdoms. In this review, asterisks (*) indicate "candidate kingdoms," major clades of prokaryotes and protists that many systematists have elevated from the phylum level to the kingdom level.

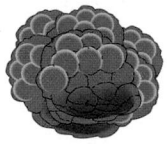

DOMAIN ARCHAEA

*Kingdom Euryarchaeota (methanogens, halophiles, some thermophiles)

*Kingdom Crenarchaeota
(most thermophiles)

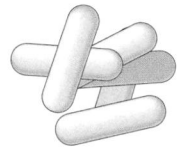

DOMAIN BACTERIA

*Kingdom Proteobacteria

*Kingdom Gram-positive bacteria

*Kingdom Cyanobacteria

*Kingdom Spirochetes

*Kingdom Chlamydia

DOMAIN EUKARYA

The five-kingdom classification scheme unites all the eukaryotes generally called protists in a single kingdom, Protista. This book adopts the cladistic argument for dividing the protists into several "candidate kingdoms" (*). This review also includes some protistan groups of less certain phylogeny.

*Kingdom Parabasala (e.g., trichomonads)

*Kingdom Diplomonadida (diplomonads)

*Kingdom Euglenozoa
Phylum Euglenophyta (euglenoids)
Phylum Kinetoplastida (kinetoplastids, e.g., trypanosomes)

*Kingdom Alveolata
Phylum Dinoflagellata (dinoflagellates)
Phylum Apicomplexa (apicomplexans)
Phylum Ciliophora (ciliates)

*Kingdom Stramenopila
Phylum Phaeophyta (brown algae)
Phylum Oomycota (water molds)
Phylum Chrysophyta (golden algae)
Phylum Bacillariophyta (diatoms)

*Kingdom Rhodophyta (red algae)

*Kingdom Chlorophyta (green algae: chlorophytes and charophyceans, which some biologists now place in the plant kingdom)

*Kingdom Mycetozoa
Phylum Myxogastrida
(plasmodial slime molds)
Phylum Dictyostelida
(cellular slime molds)
Protists of less certain taxonomic affinities
Phylum Rhizopoda (some amoebas)
Phylum Actinopoda
(heliozoans, radiolarians)
Phylum Foraminifera (forams)

Kingdom Plantae
"Bryophytes"
Phylum Hepatophyta (liverworts)
Phylum Anthocerophyta (hornworts)
Phylum Bryophyta (mosses)
"Pteridophytes" (seedless vascular plants)
Phylum Lycophyta (lycophytes)
Phylum Pterophyta (ferns, horsetails, whisk ferns)
"Seed plants"
"Gymnosperms"
Phylum Ginkgophyta (ginkgo)
Phylum Cycadophyta (cycads)
Phylum Gnetophyta (gnetae)
Phylum Coniferophyta (conifers)
"Angiosperms"
Phylum Anthophyta (flowering plants)

Kingdom Fungi

 Phylum Chytridiomycota (chytrids)

 Phylum Zygomycota (zygomycetes)

 Phylum Ascomycota (sac fungi)

 Phylum Basidiomycota (club fungi)

 Deuteromycetes (imperfect fungi)

 Lichens (symbiotic associations of fungi and algae)

Kingdom Animalia

 Phylum Porifera (sponges)

 Phylum Cnidaria (cnidarians)

  Class Hydrozoa (hydrozoans)

  Class Scyphozoa (jellies)

  Class Anthozoa (sea anemones and coral animals)

 Phylum Ctenophora (comb jellies)

 Phylum Platyhelminthes (flatworms)

  Class Turbellaria (free-living flatworms)

  Class Trematoda (flukes)

  Class Monogenea (monogeneans)

  Class Cestoidea (tapeworms)

 Phylum Bryozoa (bryozoans)

 Phylum Phoronida (phoronids)   } "Lophophorate phyla"

 Phylum Brachiopoda (brachiopods)

 Phylum Rotifera (rotifers)

 Phylum Nemertea (proboscis worms)

 Phylum Mollusca (mollusks)

  Class Polyplacophora (chitons)

  Class Gastropoda (gastropods)

  Class Bivalvia (bivalves)

  Class Cephalopoda (cephalopods)

 Phylum Annelida (segmented worms)

  Class Oligochaeta (oligochaetes)

  Class Polychaeta (polychaetes)

  Class Hirudinea (leeches)

 Phylum Nematoda (roundworms)

 Phylum Arthropoda (This review groups arthropods into a single phylum, but some zoologists now split the arthropods into multiple phyla.)

  Class Arachnida (spiders, ticks, scorpions)

  Class Diplopoda (millipedes)

  Class Chilopoda (centipedes)

  Class Crustacea (crustaceans)

  Class Insecta (insects)

 Phylum Echinodermata (echinoderms)

  Class Asteroidea (sea stars)

  Class Ophiuroidea (brittle stars)

  Class Echinoidea (sea urchins and sand dollars)

  Class Crinoidea (sea lilies)

  Class Concentricycloidea (sea daisies)

  Class Holothuroidea (sea cucumbers)

 Phylum Chordata (chordates)

  Subphylum Urochordata (urochordates: tunicates)

  Subphylum Cephalochordata (cephalochordates: lancelets)

  Subphylum Vertebrata (vertebrates)

   Class Myxini (hagfishes)

   Class Cephalaspidomorphi (lampreys)

   Class Chondrichthyes (cartilaginous fishes)

   Class Actinopterygii (ray-finned fishes)

   Class Actinistia (lobe-finned fishes) } "Osteichthyes" (bony fishes)

   Class Dipnoi (lungfishes)

   Class Amphibia (amphibians)

   Class Mammalia (mammals)

   Class Testudines (turtles)

   Class Lepidosauria (lizards, snakes, tuataras) } "Reptiles"

   Class Crocodilia (crocodiles, alligators)

   Class Aves (birds)

CREDITS

This section constitutes a continuation of the copyright page.

PHOTO-PICTURE CREDITS

Page v, photos in screen shot Graham Kent; **p. vi, Unit One** Pui-Shing Ho; **Unit Two** Molecular Probes; **p. vii, Unit Three** Walter Gehring; **Unit Four** Oxford University Museum/Dorling Kindersley; **Unit Five** *Lactarius indigo* by Mary Elizabeth Banning. Courtesy of the New York State Museum, Albany, N.Y; **Unit Six** Ardea; **p. viii, Unit Seven** Darren Bennett/Animals Animals; **Unit Eight** Kevin Schafer/Tom Stack & Associates; **p. xv Thomas Eisner** Robert Barker, Cornell University; **George Langford** Michelle Bosch; **Nancy Hopkins** Anne Dowie/Benjamin Cummings; **Peter and Rosemary Grant** Rita Nannini; **Paul Sereno** Mike Hettwer; **Joanne Chory, Flossie Wong-Staal** James Aronovsky; **David Schindler** Richard Siemens

Chapter 1 opening photo The Brett Weston Archive; **1.1 (a)** Mike Hettwer; **(b, d)** James Aronovsky; **(c)** Michelle Bosch; **1.2 (1)** Benjamin Cummings; **(2)** Biological Photo Service; **(3)** Manfred Kage/Peter Arnold, Inc.; **(4)** Dr. Jeremy Burgess/SPL/Photo Researchers, Inc.; **(5)** Photodisc; **(6)** Carol Fuegi/CORBIS; **1.3 (a)** John Foxx; **(b)** Eyewire; **(c)** Michael Fogden/Bruce Coleman, Inc.; **(d)** Wolfgang Bayer/Bruce Coleman, Inc.; **(e)** Jeff Lepore/Photo Researchers, Inc.; **(f)** Tom and Pat Leeson/Photo Researchers, Inc.; **(g)** CORBIS; **1.5** Photodisc; **1.6 (a)** Photodisc; **(b)** Janice Sheldon; **(c)** T. D. Parsons, D. Kleinfeld, F. Raccuia-Behling and B. Salzberg. *Biophysical Journal*, July 1989. Photo courtesy of Brian Salzberg; **(d)** Nicolae Simionescu; **1.7** Photodisc; **1.9** Charles H. Phillips; **1.11 (a)** A. B. Dowsett/Science Source/Photo Researchers, Inc.; **(b)** Ralph Robinson/Visuals Unlimited; **(c)** D. P. Wilson/Science Source/Photo Researchers, Inc.; **(d)** Photodisc; **(e)** Eyewire; **(f)** CORBIS; **1.12 (a)** Manfred Kage/Peter Arnold, Inc.; **(b)** OMIKRON/Photo Researchers, Inc.; **(c)** W. L. Dentler, University of Kansas/BPS; **1.13** Mike Hettwer; **1.14** Richard Milner; **1.16** P. A. Sutherland/1998, OtherWorld Images, USA; **1.17 (a)** Tom Van Sant/Geosphere Project, Santa Monica/Science Photo Library/Photo Researchers, Inc.; **1.18** Mary DeChirico/Benjamin Cummings; **1.22** Strongin/NYU; **1.23 (a)** Hank Morgan/Photo Researchers, Inc.; **(b)** Peter Menzel; **Table 1.1, p. 22** Photodisc; **p. 23** Hank Morgan/Photo Researchers, Inc.

Unit One interview Robert Barker, Cornell University

Chapter 2 opening photo Thomas Eisner and Daniel Aneshausley; **2.1 (top)** Frank Krahmer/Getty Images/Planet Earth Pictures, Ltd.; **(bottom)** N. L. Max, University of California/BPS; **2.2 (left)** Chip Clark; **(center, right)** Stephen Frisch/Benjamin Cummings; **2.3** Grant Heilman; **2.4** Ivan Polunin/Bruce Coleman, Inc.; **2.6 (a)** Terraphotographics/Biological Photo Service; **(b)** From M. C. Ratazzi et al., *Am J Human Genet* 28:143–154, 1976; **2.7** Simon Fraser/Medical Physics, RVI, Newcastle-Upon-Tyne/Science Photo Library/Photo Researchers, Inc.; **2.8** Stuart Isett/CORBIS/Sygma; **2.15** Stephen Frisch/Benjamin Cummings; **2.20** Runk/Schoenberger/Grant Heilman; **p. 40** Phil Degginger/Color-Pic

Chapter 3 opening photo NASA; **3.2 (left)** Photodisc; **(right)** From *Scanning Electron Microscopy in Biology*, by R. G. Kessel and C. Y. Shih, Springer-Verlag, New York. 1974, p.147; **3.3** George Bernard/Animals Animals; **3.4** DUOMO; **3.6** Flip Nicklin/Minden Pictures; **3.10** Lightwave Photography/Animals Animals

Chapter 4 opening photo Pui-Shing Ho; **4.1** Roger Ressmeyer/CORBIS; **4.5** Manfred Kage/Peter Arnold, Inc.; **4.8 (a, b)** Digital Vision

Chapter 5 opening photo Martin Shields; **5.1 (a)** Estate of Linus Pauling; **(b)** Will & Deni McIntyre, Photo Researchers, Inc.; **5.6 (a)** John N. A. Lott, McMaster University/Biological Photo Service; **(b)** H. Shio and P. B. Lazarow; **5.8** J. Litray/Visuals Unlimited; **5.9 (a)** F. Collet/Photo Researchers, Inc.; **(b)** CORBIS; **5.11 (a)** The American Dairy Association; **(b)** Lara Hartley/Benjamin Cummings; **5.19 (a, b)** Lawrence Berkeley National Laboratory; **5.21 (left)** Martin Shields; **(right)** Vollrath & Edmunds, *Nature* 340:305–317; **5.26** P. B. Sigler from Z. Xu, A. L. Horwich, and P. B. Sigler, *Nature* (1997) 388:741–750. © 1997 Macmillan Magazines, Ltd.; **5.27** Marie Green, University of California, Riverside

Chapter 6 opening photo Jean-Marie Bassot/Photo Researchers, Inc.; **6.2** Alain McGlaughlin/Benjamin Cummings; **6.3 (a)** Anne Dowie/Benjamin Cummings; **(b)** Benjamin Cummings; **6.4** Brian Capon; **6.14 (a, b)** Thomas Steitz, Yale University; **6.21** R. Rodewald, University of Virginia/Biological Photo Service

Unit Two interview Michelle Bosch

Chapter 7 opening photo Molecular Probes; **7.2 (a, b)** William L. Dentler, University of Kansas/Biological Photo Service; **7.4** S. C. Holt, University of Texas Health Center/Biological Photo Service; **7.6** J. David Robertson; **7.9 (top)** From L. Orci and A. Perrelet, *Freeze-Etch Histology*. (Heidelberg: Springer-Verlag, 1975) © 1975 Springer-Verlag; **(center)** From A. C. Faberge, *Cell Tiss. Res.* 151 (1974):403. © 1974 Springer-Verlag; **(bottom)** U. Aebi et al. *Nature* 323 (1996):560–564, figure 1a. Used by permission; **7.10** D. W. Fawcett/Photo Researchers, Inc.; **7.11** R. Bolender, D. Fawcett/Photo Researchers, Inc.; **7.12** Don Fawcett/ Visuals Unlimited; **7.13 (a)** R. Rodewald, University of Virginia/Biological Photo Service; **(b)** Daniel S. Friend, Harvard Medical School; **7.15** E. H. Newcomb; **7.17** Daniel S. Friend, Harvard Medical School; **7.18** W. P. Wergin and E. H. Newcomb, University of Wisconsin, Madison/Biological Photo Service; **7.19** From S. E. Fredrick and E. H. Newcomb, *The Journal of Cell Biology* 43 (1969):343. Provided by E. H. Newcomb; **7.20** John E. Heuser, Washington University School of Medicine, St. Louis, MO; **7.22** Kent McDonald; **7.23 (a)** Richard Kessel/Visuals Unlimited; **(b)** Dennis Kunkel; **7.24 (a)** OMIKRON/Science Source/Photo Researchers, Inc.; **(b)** W. L. Dentler, University of Kansas/Biological Photo Service; **(d)** W.L. Dentler, University of Kansas/Biological Photo Service; **7.26** From Hirokawa Nobutaka *The Journal of Cell Biology* 94 (1982):425 by copyright permission of The Rockefeller University Press; **7.28** G. F. Leedale/Photo Researchers, Inc.; **7.30 (a)** From Douglas J. Kelly, *The Journal of Cell Biology* 28 (1966):51 by copyright permission of The Rockefeller University Press; **(b)** From L. Orci and A. Perrelet, *Freeze-Etch Histology*. (Heidelberg: Springer-Verlag, 1975) © 1975 Springer-Verlag; **(c)** From C. Peracchia and A. F. Dulhunty, *The Journal of Cell Biology* 70 (1976):419 by copyright permission of The Rockefeller University Press; **7.31** Boehringer Ingelheim International GmbH, photo Lennart Nilsson/Albert Bonniers Forlag AB, *The Body Victorious*, Delacorte Press, Dell Publishing Co., Inc. **Table 7.1 (top left)** Biophoto Associates/Photo Researchers, Inc.; **(top and center right, bottom left)** David M. Phillips/ Visuals Unlimited; **(center left)** Ed Reschke; **(bottom right)** Noran Instruments; **Table 7.2 (left)** Mary Osborn, Max Planck Institute; **(center)** Frank Solomon and J. Dinsmore, Massachusetts Institute of Technology; **(right)** Mark S. Ladinsky and J. Richard McIntosh, University of Colorado

Chapter 8 8.3 Philipa Claude; **8.13 (a, b)** Cabisco/Visuals Unlimited; **8.19 (a)** R. N. Band and H. S. Pankratz, Michigan State University/Biological Photo Service; **(b)** D. W. Fawcett/Photo Researchers, Inc.; **(c, both)** M. M. Perry and A. B. Gilbert, *J. Cell Science* 39 (1979) 257. Copyright 1979 by The Company of Biologists Ltd.

Chapter 9 opening photo Photodisc

Chapter 10 opening photo Photodisc; **10.1 (a)** Renee Lynn/Photo Researchers, Inc.; **(b)** Bob Evans/Peter Arnold, Inc.; **(c)** Fred Speigel; **(d)** Sue Barns; **(e)** Paul Johnson/Biological Photo Service; **10.2 (center)** M. Eichelberger/Visuals Unlimited; **(bottom)** Courtesy of W. P. Wergin and E. H. Newcomb, University of Wisconsin/Biological Photo Service; **10.10** Christine L. Case, Skyline College; **10.19 (left)** C. F. Miescke/Biological Photo Service; **(right)** Photodisc

Chapter 11 opening photo Eric Schabtach and Ira Herskowitz; **11.02** Dale Kaiser

Chapter 12 opening photo J. M. Peters; **12.1 (a)** Biophoto Associates/Photo Researchers, Inc.; **(b)** C. R. Wyttenbach, University of Kansas/Biological Photo Service; **(c)** Biophoto/ Science Source/Photo Researchers, Inc.; **12.2** Courtesy of J. M. Murray, University of Pennsylvania; **12.3** Biophoto/Photo Researchers, Inc.; **12.5 (all)** Conly Rieder; **12.6 (top)** Richard Mcintosh; **(bottom)** Matthew Schibler, from *Protoplasma* 137 (1987):29–44; **12.8 (a)** David M. Phillips/Visuals Unlimited; **(b)** Micrograph by B. A. Palevitz. Courtesy of E. H. Newcomb, University of Wisconsin; **12.9** Carolina Biological Supply/ Phototake; **12.15** Gunter Albrecht-Buehler, Northwestern University; **p. 231** Carolina Biological Supply/Phototake

Unit Three interview Anne Dowie/Benjamin Cummings

Chapter 13 opening photo Bill Davilla/Retna, Ltd., **13.1** Roland Birke/OKAPIA/Photo Researchers, Inc.; **13.2** Courtesy of The Inouye and Yaneshiro Families; **13.3 (4)** SIU/Visuals Unlimited; **(5)** CNRI/SPL/Photo Researchers, Inc.

Chapter 14 opening art Bettmann Archive; **14.13** Photodisc; **14.14 (a)** Photodisc; **(b)** Anthony Loveday/Benjamin Cummings; **14.15 (left)** Tony Brain/Science Source/Photo Researchers, Inc.; **(right)** Bill Longcore/Photo Researchers, Inc.; **14.16** Nancy Wexler, Columbia University; **p. 268** Breeder/owner: Patricia Speciale; photographer: Norma JubinVille

Chapter 15 opening photo Reprinted with permission from Peter Lichter and David Ward, *Science* 247 (1990). ©1990 American Association for the Advancement of Science; **15.2 (a)** Jean Claude Levy/Phototake; **(b)** Carolina Biological Supply/Phototake; **15.10** Grant Heilman/Grant Heilman Photography; **15.12** Martin Gallardo, Universidad Austral de Chile; **15.14 (a)** CNRI/Science Photo Library/Photo Researchers, Inc.; **(b)** Greenlar/The Image Works; **15.16** From L. P. Hosticka and M. R. Hanson, "Induction of plastid mutations by nitrosomethylurea," *Journal of Heredity* 75 (1984) 242–246, figure 3

Chapter 16 opening photo National Cancer Institute; **16.2** Oliver Meckes/Photo Researchers Inc.; **16.4 (a, b)** From J. D. Watson, *The Double Helix*, NY: Atheneum Press, 1968, p. 215. © 1968 by J. D. Watson. Courtesy of Cold Spring Harbor Laboratory Archives; **16.5** Richard Wagner, UCSF Graphics; **16.10** From D. J. Burks and P. J. Stambrook, *The Journal of Cell Biology* 77 (1978):762 by the copyright permission of The Rockefeller University Press. Photo provided by P. J. Stambrook; **16.19** Peter Lansdorp

Chapter 17 opening photo Harry Noller, UC Santa Cruz, From *Science* Vol. 291, p. 2526; **17.5** Keith V. Wood; **17.16** Thomas Steitz, Yale University; **17.20** B. Hamkalo and O. L. Miller, Jr.; **17.22** Reprinted with permission from O. L. Miller, Jr., B. A. Hamkalo, and C. A. Thomas, Jr., *Science* 169 (1970):392. Copyright © 1970 American Association for the Advancement of Science

Chapter 18 18.2 (a, b, and d) Robley C. Williams, University of California, Berkeley/BPS; **(c)** John Cardamone Jr., University of Pittsburgh/BPS; **18.7** C. Dauguet/Institute Pasteur/Photo Researchers, Inc.; **18.8 (a)** CDC/Phototake; **(b)** Keith V. Wood/Photo Researchers, Inc.; **18.9 (a, top)** Sherman Thomson/Visuals Unlimited; **(bottom)** Norm Thomas/Photo Researchers, Inc.; **(b)** N. Obalka, N. Yeager, R. Beachy, and C. Fauquet, The Scripps Research Institute; **18.14** Dennis Kunkel/Phototake

Chapter 19 opening photo M. B. Roth and J. Gall; **19.1 (a, top)** S. C. Holt, University of Texas, Health Science Center, San Antonio/BPS; **(bottom)** Courtesy of Victoria Foe; **(b)** Barbara Hamkalo; **(c)** From J. R. Paulsen and U. K. Laemmli, *Cell* 12 (1977):817–828; **(d, top and bottom)** G. F. Bahr/AFIP; **19.2** O. L. Miller Jr., Department of Biology, University of Virginia; **19.4** Evelyne Cudel-Epperson, MSU

Chapter 20 opening photo Department of Energy, Joint Genome Institute. Photograph by Michael Anthony; **20.9** Repligen Corporation; **20.14 (a, b)** Incyte Pharmaceuticals, Inc., Palo

Alto, CA, from R. F. Service, *Science* (1998) 282:396–399, with permission from *Science*; **20.17** Cellmark Diagnostics, Inc., Germantown, Maryland; **20.18** PPL Therapeutics; **20.20** Peter Berger, Institut für Biologie, Freiburg

Chapter 21 opening photo Walter Gehring; **21.1 (a)** Dwight Kuhn; **(b)** Hans Pfletschinger/Peter Arnold, Inc.; **21.3 (a)** Oliver Meckes/Photo Researchers Inc.; **(b)**ArsNatura; **(c)** Stanton Short/Jackson Laboratory; **(d)** Nancy Hopkins, MIT; **(e)** Elliot Meyerowitz; **21.4** J. E. Sulston and H. R. Horvitz, *Dev. Biol.* 56 (1977):110–156; **21.7 (b)** The Roslin Institute, Edinburgh; **21.12 (a, both)** Wolfgang Driever; **(b, both)** Dr. Ruth Lahmann, The Whitehead Institution; **21.13 (a–c)** Jim Langeland, Steve Paddock, Sean Carroll, University of Wisconsin and The Howard Hughes Medical Institute; **21.14 (both)** F. R. Turner, Indiana University; **21.19** Dwight Kuhn; **21.20 (b, all photos)** Elliot Meyerowitz

Unit Four interview Rita Nannini

Chapter 22 opening photo Courtesy Department of Library Services, American Museum of Natural History; **22.1** Down House; **22.2** James L. Amos/The National Audubon Society Collection/Photo Researchers, Inc.; **22.4** CORBIS; **22.5 (iguana)** Joe McDonald/Animals Animals **(painting of HMS *Beagle*)** Courtesy of the National Maritime Museum, London; **22.6 (a, c)** Tui de Roy/Bruce Coleman, Inc.; **(b)** Mike Putland/Ardea London Ltd.; **22.8** Adrian Davies/Bruce Coleman, Inc.; **22.10 (a)** E. S. Ross, California Academy of Sciences; **(b)** Ken G. Preston-Mafhan/Animals Animals; **(c)** P. and W. Ward/Animals Animals; **22.11 (a)** Erich Lessing/Art Resource; **(b)** Anne Dowie/Benjamin Cummings; **(b)** Jack Wilburn/Earth Scenes/Animals Animals; **22.12** Jack Fields/Photo Researchers, Inc.; **22.15** Tom Van Sant/Geosphere Project, Santa Monica/Science Photo Library/Photo Researchers, Inc.; **22.16** Kenneth Kaneshiro; **22.17** Philip Gingerich 1991. Reprinted with permission of *Discover* Magazine; **22.18** Department of Library Services/American Museum of Natural History.

Chapter 23 opening photo David Hillis; **23.1** J. Antonovics/Visuals Unlimited; **23.2 (a)** David Cavagnaro; **(b)** NOAA National Geophysical Data Center; **23.6** 1993 *Time* Magazine; **23.7 (a, b)** Fred Nijhout, Duke University; **23.9 (a)** NASA; **(b, c)** Janice Britton-Davidian, ISEM, UMR 5554 CNRS, Université Montpellier II. Reprinted by permission from *Nature*, Vol. 403, 13 January 2000, p. 158. © 2000 Macmillan Magazines Ltd.; **(d)** Dorling Kindersley; **23.14** Thomas B. Smith, San Francisco State University; **23.16** The Far Side, by Gary Larson © 1981 FarWorks, Inc. All rights reserved. Used with permission.

Chapter 24 opening photo Frans Lanting/Minden Pictures; **24.2 (a, left)** John Shaw/Tom Stack & Associates; **(a, right)** Don & Pat Valenti/Tom Stack & Associates; **(b, all photos)** Photodisc; **24.3** Barbara Gerlach/Tom Stack & Associates; **24.4 (left)** Ralph A. Reinhold/Animals Animals/Earth Scenes; **(center)** Breck P. Kent/Animals Animals/Earth Scenes; **(right)** E. R. Degginger/Animals Animals/Earth Scenes; **24.7 (center)** Richard Sisk/Panoramic Images; **(left)**John Shaw/Bruce Coleman, Inc.; **(right)** Michael Fogden/Bruce Coleman, Inc.; **24.10** Kevin Schafer; **24.14 (all)** University of Amsterdam; **24.16 (all)** Ole Seehausen, Universities of Hull and Leiden; **24.20 (a)** Bob Gibbons, Ardea London Ltd.; **(b)** Tom McHugh, Steinhart Aquarium/Photo Researchers, Inc.; **24.21** Jane Burton/Bruce Coleman, Inc.

Chapter 25 opening photo Oxford University Museum/Dorling Kindersley; **25.1 (a)** Georg Gerster/Photo Researchers, Inc.; **(b)** Dorling Kindersley; **(c)** Tom Till; **(d)** Manfred Kage/Peter Arnold, Inc.; **(e)** Walter H. Hodge/Peter Arnold, Inc.; **(f)** Dr. Martin Lockley, University of Colorado; **(g)** Courtesy Dr. David A. Grimaldi. Photo by Jacklyn Beckett/The American Museum of Natural History, N.Y.; **(h)** F. Latreille/Cerpolex/Cercles Polaires Expeditions; **25.6** Benjamin Cummings; **25.10 (a, b)** Tom McHugh/Photo Researchers, Inc.

Unit Five interview Mike Hettwer

Chapter 26 opening illustration Chip Clark NMNH, artist Peter Sawyer; **26.3 (a)** S. M. Awramik, University of California/BPS; **(b)** Sue Barns; **26.4 (a)** John Stolz; **(b, c)** S. M. Awramik, University of California/BPS; **26.5** Theodore J. Bornhorst/Michigan Technological University; **26.6** Nicholas J. Butterfield, University of Cambridge; **26.7 (a, b)** Andrew H. Knoll, Harvard University; **26.9** Pfizer Inc.; **26.12** F. M. Menger and Kurt Gabrielson, Emory University; **26.14** George Luther, University of Delaware

Chapter 27 opening photo Dr. Tony Brain and David Parker/Science Photo Library/Photo Researchers, Inc.; **27.1** Jack Dykinga; **27.3 (a)** Meckes/Ottawa/Photo Researchers, Inc.; **(b)** Manfred Kage/Peter Arnold Inc.; **(c)** David Chase/CNRI/Phototake; **27.4** Heide Schulz, Max Planck Institute for Marine Microbiology; **27.5 (a, b)** Christine Case; **27.6** Fran Heyl Associates, photo by David Hasty; **27.7** J. Adler; **27.8 (a)** S. W. Watson. © *Journal of Bacteriology*, American Society of Microbiology; **(b)** N. J. Lang/University of California/BPS; **27.9** John Durham/ SPL/Photo Researchers Inc.; **27.10** H. S. Pankratz, T. C. Beaman/Biological Photo Service; **27.11** Sue Barns; **27.14** Helen E. Carr/BPS; **27.15** Ken Lucas/Biological Photo Service; **27.16** Dr. Tony Brain/Science Source/Photo Researchers, Inc.; **27.17 (a)** Larry West/Bruce Coleman, Inc.; **(b)** Centers for Disease Control; **27.18** Martin Bond/The National Audubon Society Collection/Photo Researchers, Inc.; **27.19** Exxon Corporation; **Table 27.3 (*Rhizobium*)** L. Evans Roth/BPS; **(*Chromatium*)** Biological Photo Service; **(Myxobacteria)** Phototake; **(*Bdellovibrio*)** Alfred Pasieka/Peter Arnold, Inc.; **(*Helicobacter pylori*)** Photo Researchers, Inc.; **(chlamydias)** Moredon Animal Health/SPL/Photo Researchers Inc.; **(*Leptospira*)** CNRI/SPL/Photo Researchers, Inc.; **(*Streptomyces*)** Frederick P. Mertz, Visuals Unlimited; **(mycoplasmas)** David M. Phillips, Visuals Unlimited; **(cyanobacteria)** Sue Barns

Chapter 28 opening photo M. I. Walker/Photo Researchers, Inc.; **28.1 (a)** Yuuji Tsukii/Hosei University; **(b)** David Phillips, Visuals Unlimited; **(c)** George Barron; **(d)** Joyce Photographics, Photo Researchers, Inc.; **28.3** Eric V. Grave, Photo Researchers, Inc.; **28.9** Jerome Paulin, Visuals Unlimited; **28.10** David M. Phillips/Visuals Unlimited; **28.11** Oliver Meckes/Science Source/Photo Researchers, Inc.; **28.12** Biophoto Associates/Photo Researchers, Inc.; **28.13** Masamichi Aikawa; **28.14 (a)** Eric Grave/Photo Researchers, Inc.; **(b)** Manfred Kage/Peter Arnold, Inc.; **(c)** M. Abbey/Visuals Unlimited; **28.16** Fred Rhoades/Mycena Consulting; **28.17 (a)** Kent Wood/Science Source/Photo Researchers, Inc.; **(b)** Fred Rhoades/Mycena Consulting; **28.18** Biological Photo Service; **28.19** Anne Wertheim/Animals Animals; **28.20** W. Lewis Trusty/Animals Animals; **28.21** Courtesy of J. R. Waaland, University of Washington/BPS; **28.22 (a)** Courtesy of J. R. Waaland, University of Washington/BPS; **(b)** D. P. Wilson/Eric & David Hosking, Photo Researchers, Inc. **(c)** Gary Robinson/Visuals Unlimited; **28.23 (a)** Manfred Kage/ Peter Arnold, Inc.; **(b)** Estate of J. Metzner/Peter Arnold, Inc.; **(c)** Laurie Campbell/NHPA; **28.24** Courtesy of W. L. Dentler, University of Kansas; **28.26 (all)** Akiro Kihara, Hosei University; **28.27 (a)** and **(b)** Eric Grave/Photo Researchers, Inc.; **28.28** Manfred Kage/Peter Arnold, Inc.; **28.29 (plasmodium)** Ray Simmons/Photo Researchers, Inc.; **(sporangia on plasmodium)** R. Calentine/Visuals Unlimited; **28.30** Robert Kay, MRC Cambridge

Chapter 29 opening photo Otto Rogge; **29.2 (a)** Heather Angel/Natural Visions; **(b)** Linda Graham; **29.3 (a)** Ed Reschke; **(b)** F. A. L. Clowes; **29.4 (a)** Linda Graham; **(b)** Graham Kent; **29.5** Karen Renzaglia, Southern Illinois University; **29.7, 29.8, 29.9 (a) and (b)**, and **29.10** Linda Graham; **29.11** Graham Kent; **29.12 (a, b)** Charles Wellman/Centre for Paleontology; **29.15 (a, both)** Claudia Lipke; **(b)** Courtesy of David Hanson, photographed by Claudia Lipke; **(c)** Laurie Campbell, NHPA; **29.17, 29.18**, and **29.19 (a–c)** Linda Graham; **29.20** Chip Clark; **29.21 (a)** Robert and Linda Mitchell; **(b)** Michael Viard/Peter Arnold, Inc.; **(c)** Laurie Campbell, NHPA; **(d)** E. R. Degginger/Earth Scenes; **29.24 (a)** Glenn Oliver/Visuals Unlimited; **(b)** E. Webber/Visuals Unlimited; **(c)** Jack M. Bostrack/Visuals Unlimited; **29.25** The Field Museum of Natural History

Chapter 30 opening photo National Museum of Natural History, Smithsonian Institution; **30.3** Graham Kent; **30.5 (a)** S. W. Carter/Photo Researchers, Inc.; **(b)** Geoff Bryant/ Photo Researchers, Inc. **(c)** Linda Graham; **30.6 (a)** Loran M. Whitelock; **(b)** Fred Spiegel, University of Arkansas; **30.7 (a)** Michael P. L. Fogden/Bruce Coleman, Inc.; **(b)** William Hahn; **(c)** Doug Sokell/Visuals Unlimited; **30.8 (a)** Stuart Westmorland/Photo Researchers, Inc.; **(b)** Sequoia National Park Service; **(c)** Joseph Sohm; Chromo Sohm Inc./CORBIS; **(d)** William Mullins/Photo Researchers, Inc.; **(e)** Walter H. Hodge/Peter Arnold, Inc.; **(f)** J. Coke, Westvaco Corp.; **(g)** Jaime Plaza, Wildlight; **(g inset)** Jaime Plaza, Royal Botanic Gardens Sydney; **30.10 (a, pollen cones)** Linda Graham; **(a, LM of pollen cone)** John D. Cunningham/Visuals Unlimited; **(a, pollen)** Visuals Unlimited; **(b, ovulate cones)** Linda Graham; **(b, LM of ovulate cone)** Stan W. Elems/Visuals Unlimited; **(b, ovule)** Biophoto Associates/Photo Researchers, Inc.; **30.11 (b)** Stephen McCabe, University of California Santa Cruz Arboretum; **(c)** Heather Angel/Natural Visions; **(d)** Bob & Ann Simpson/Visuals Unlimited; **(e)** John Chellman/Earth Scenes; **(f)** Ed Reschke/Peter Arnold, Inc.; **30.13 (b)** Heather Angel/Natural Visions; **30.16 (a)** Jane Burton/Bruce Coleman, Inc.; **(b)** Scott Camazine/Photo Researchers, Inc.; **(c)** Dwight R. Kuhn; **30.18 (a)** D. Wilder; **(b)** Bob and Clara Calhoun/Bruce Coleman, Inc.; **(c)** Merlin D. Tuttle/Bat Conservation International; **30.19 (a)** Martin Miller/Visuals Unlimited; **(b)** G. Prange/Visuals Unlimited; **Table 30.1 (cherries)** Eyewire; **(raspberry, pineapple)** Photodisc

Chapter 31 opening illustration *Lactarius indigo* by Mary Elizabeth Banning. Courtesy of the New York State Museum, Albany, N.Y.; **31.1 (both)** Fred Rhoades/Mycena Consulting; **31.2** N. Allin and G. L. Barron, University of Guelph/Biological Photo Service; **31.5** Martha J. Powell and Peter Letcher; **(inset)** William Barstow, Department of Botany, University of Georgia, Athens; **31.6** Matt Meadows/Peter Arnold, Inc.; **31.7 (left)** George Barron; **(right)** Ed Reschke/Peter Arnold, Inc.; **31.8** G. L. Barron, University of Guelph/Biological Photo Service; **31.09 (a)** E. R. Degginger/Earth Scenes; **(b)** Jacana/Photo Researchers Inc.; **(c)** J. L. Lepore/Photo Researchers, Inc.; **31.10** Fred Spiegel, University of Arkansas; **31.11 (a)** Kerry T. Givens/Tom Stack & Associates; **(b)** Frans Lanting/Minden Pictures; **(c)** Tom Volk, University of Wisconsin; **31.12** Biophoto Associates/Photo Researchers, Inc.; **31.13** Rob Simpson/Visuals Unlimited; **31.14** Jack Bostrack/Visuals Unlimited; **(inset)** M. F. Brown/Visuals Unlimited; **31.15** Stephen J. Kron, University of Chicago; **31.16** Fred Rhoades/Mycena Consulting; **31.17** V. Ahmadijian/Visuals Unlimited; **31.18** R. L. Peterson, University of Guelhp/BPS; **31.19** R. Ronacordi/Visuals Unlimited; **31.20 (a)** Stuart Bebb/Oxford Scientific Films/Animals Animals/Earth Scenes; **(b)** Holt Studios/Photo Researchers, Inc.; **(c)** David Cavagnaro/Visuals Unlimited; **31.21** Christine Case; **p. 632** Dr. Jeremy Burgess/The National Audubon Society Collection/Photo Researchers, Inc.

Chapter 32 opening photo Charles & Sandra Hood/Bruce Coleman Ltd.; 32.10 Hans Pfletschinger/Peter Arnold, Inc.; 32.11 Carolina Biological/Visuals Unlimited

Chapter 33 opening photo Gary Braasch; **33.2** Andrew J. Martinez/Photo Researchers, Inc.; **33.4 (a)** Ken Lucas/Planet Earth Pictures; **(b)** Claudia Mills/Friday Harbor Labs; **33.6 (a)** Andrew J. Martinez/Photo Researchers Inc.; **(b)** Robert Brons/BPS; **(c)** Kevin McCarthy/Offshoot Stock; **(d)** Chris Huss/The Wildlife Collection; **33.7** Robert Brons/BPS; **33.8** Fred Bavendam/Peter Arnold, Inc.; **33.9** Bill Wood/Bruce Coleman, Inc.; **33.11** Centers for Disease Control; **33.12** Stanley Fleger/Visuals Unlimited; **33.13** W. I. Walker/Photo Researchers, Inc.; **33.14 (a)** Colin Milkins, Oxford Scientific Films/ Animals Animals; **(b)** Fred Bavendam/Peter Arnold, Inc.; **33.15** Bill Wood/NHPA; **33.17** Jeff Foott/Tom Stack & Associates; **33.19 (a)** Tony Craddock/Science Photo Library/Photo Researchers, Inc.; **(b)** CORBIS; **33.20** H. W. Pratt/Biological Photo Service; **33.22 (a)** Tom McHugh/Steinhart Aquarium/Photo Researchers, Inc.; **(b)** Fred Bavendam/Peter Arnold, Inc.; **(c)** Jonathan Blair (CORBIS); **33.24 (a)** A.N.T./NHPA; **(b)** Sea Studios; **(c)** Kjell Sandved; **(d)** Astrid & Hanns-Frieder Michler/Science Photo Library/Photo Researchers, Inc.; **33.25 (a)** Reprinted with permission from A. Eizinger and R. Sommer, Max Planck Institut für Entwicklungsbiologie, Tübingen. © 2000 American Association for the Advancement of Science; **(b)** L. S. Stepanowicz/Photo Researchers, Inc.; **33.27** Chip Clark; **33.28** Milton Tierney, Jr./Visuals Unlimited; **33.29 (a)** Frans Lanting/Minden Pictures; **(b)** David Scharf; **(c)** Diana Sammatoro, Pennsylvania State University; **33.30 (a)** Paul Skelcher/Rainbow; **33.31 (a)** Robert and Linda Mitchell; **(b)** Ed Degginger; **33.32** George Grall/National Geographic Society; **33.34 (a–e)** John Shaw/Tom Stack & Associates; **33.35 (a)** Frans Lanting/Minden Pictures; **(b)** Tom McHugh/Photo Researchers, Inc.; **(c)** C. R. Wyttenbach/ University of Kansas/BPS; **33.37 (a)** Jeffrey L. Rotman/Peter Arnold, Inc.; **(b)** Gary Milburn/Tom Stack & Associates; **(c)** Dave Woodward/Tom Stack & Associates; **(d)** Marty Snyderman; **(e)** Carl Roessler/Animals Animals; **(f)** Fred Bavendam/Peter Arnold, Inc.

Photo Researchers, Inc.; **51.13** Bernd Heinrich/University of Vermont; **51.14** Elizabeth A. Capaldi; **51.17** Jeff Foott/Bruce Coleman, Inc.; **51.18 (a)** Bruce Davidson/Animals Animals; **(b)** G. R. Higbee/Photo Researchers, Inc.; **51.19** Visuals Unlimited; **51.20** Frans B. M. De Waal; **51.21** Doug Wechsler/Animals Animals; **51.22 (a)** Jonathan Scott/Seaphot Ltd./Planet Earth Pictures; **(b)** Michael Dick/Animals Animals; **51.24** Philip Savoie; **51.25** David White, Ocean Engineering Division, Southhampton Oceanography Centre, UK; **51.26** Robert Vander Meer, University of Florida; **51.27** Kenneth Lorenzen, University of California, Davis; **51.28** Stephen Kraseman/ Peter Arnold, Inc.; **51.29 (a)** Jennifer Jarvis, University of Cape Town; **(b)** Tom McHugh/ Photo Researchers, Inc.; **51.32** Ivan Polunin/Bruce Coleman, Inc.

Chapter 52 opening photo Eckart Pott/NHPA; **52.1** A.R.E. Sinclair; **52.2 (a)** Sophie de Wilde/Jacana/Photo Researchers, Inc.; **(b)** Frans Lanting/Minden Pictures; **(c)** Will & Deni McIntyre/Photo Researchers, Inc.; **52.4** E. R. Degginger; **52.6** Hans Reinhard/OKAPIA/Photo Researchers, Inc.; **52.7 (a)** R. Calentine/Visuals Unlimited; **(b)** Max and Bea Hunn/Visuals Unlimited; **52.17** Alan G. Nelson/Animals Animals; **52.19** Alan Carey/Photo Researchers, Inc.

Chapter 53 opening photo Richard D. Estes/Photo Researchers, Inc.; **53.2** Heather Angel/ Natural Visions; **53.3 (b)** Joseph T. Collins/Photo Researchers, Inc.; **(c)** Kevin de Queiroz, National Museum of Natural History; **53.5** C. Allan Morgan/Peter Arnold, Inc.; **53.6** Kevin Schafer/Tom Stack & Associates; **53.7 (a)** Lincoln Brower, Sweet Briar College; **(b)** Peter J. Mayne; **53.8 (a)** Edward S. Ross; **(b)** Runk/Schoenberger/Grant Heilman Photography, Inc.; **53.9** Robert and Linda Mitchell; **53.12** D. L. Breitburg, T. Loher, C. A. Pacey, and A. Gerstein. 1997. Varying effects of low dissolved oxygen on trophic interactions in an estuarine food web. *Ecological Monographs* 67:490. Copyright © 1997 Ecological Society of America. Reprinted with permission; **53.14** Karen Oberhauser; **53.15** J. A. Lubena and S. A. Levin. 1998. The spread of a reinvading species: expansion of the California sea otter. *American Naturalist* 131: fig. 1, p. 529, fig. 2, p. 535. Copyright © 1988 The University of Chicago Press, Chicago. Reprinted with permission; **53.16 (a–c)** Frank Gilliam, Marshall University; **53.17** Paul A. Souders/CORBIS; **53.18 (a)** Grant Heilman/Grant Heilman Photography; **(b)** Jeff Foot/Tom Stack & Associates; **53.19 (all)**, **53.20 (all)** Tom Bean/ Tom & Susan Bean, Inc.

Chapter 54 opening photo E. R. Degginger/Color Pic, Inc.; **54.2** Gregory G. Dimijian/Photo Researchers, Inc.; **54.4** C. B. Field, M. J. Behrenfeld, J. T. Randerson, and P. Falkowski. 1998. Primary production of the biosphere: integrating terrestrial and oceanic components. *Science* 281:237–240; **54.7** Institute of Ocean Sciences, Department of Fisheries and Oceans, Canada; **54.8** Reprinted with permission from D. W. Schindler, *Science* 184 (1974):897, figure 1.49. © 1974 American Association for the Advancement of Science; **54.21 (a)** John D. Cunningham/ Visuals Unlimited; **(b)** Provided by the Northeastern Forest Experiment Station, Forest Service, United States Department of Agriculture; **54.22** Robert Estall/CORBIS; **54.27 (a)** NASA/ Goddard Space Flight Center; **(c)** Bill Bachman/Bill Bachman & Associates Pty., Ltd.

Chapter 55 opening photo Juan Manual Renjifo/Earth Scenes; **55.2 (a)** Daniel Heuclin/NHPA; **(b)** Mark Carwardine/Still Pictures/Peter Arnold, Inc.; **(c)** Dieter & Mary Plage/Bruce Coleman, Inc.; **55.3** Richard Shiell/Animals Animals; **55.4** Roger Ressmeyer/CORBIS; **55.5** Gary Braasch/ Woodfin Camp & Associates; **55.7(a)** Gary Kramer; **(b)** Michael Fogden/Animals Animals; **(c)** David Dennis/Animals Animals; **(d)** Laurie Campbell/NHPA; **(e)** Marc Dantzker; **(f)** Rachel Woodfield, Merkel & Associates; **55.8** The Academy of Natural Sciences of Philadelphia/ CORBIS; **55.9** Richard Vogel/Gamma Liaison; **55.11** R. L. Westemeier, J. D. Brawn, S. A. Simpson, T. L. Esker, R. W. Jansen, J. W. Walk, E. L. Kershner, J. L. Bouzat, and K. N. Paige. 1998. Tracking the long-term decline and recovery of an isolated population. *Science* 282:1696. Copyright © 1998 by the American Association for the Advancement of Science. Reprinted with permission; **55.13** F. & J. Craighead; **55.14 (a)** Rob Curtis/The Early Birder; **(b)** Raymond K. Gehman/National Geographic Society Image Collection; **(c)** Blanche Haning/The Lamplighter; **55.15 (a)** David Hosking/Photo Researchers, Inc.; **(b)** James P. Blair/National Geographic Image Collection; **55.16** Florida Department of Transportation; **55.17 (a)** John D. Cunningham/Visuals Unlimited; **(c)** Peter Ward/Bruce Coleman, Inc.; **55.19** Frans Lanting/Minden Pictures; **55.20 (a)** Joe McDonald/Bruce Coleman, Inc.; **(b)** Reed Brown; **55.22** Ariel Lugo/USDA/Forest Service, photo by Jesus Ayala O'Neill; **55.23 (a)** Jean Clottes/CORBIS/Sygma; **(b)** Frans Lanting/Minden Pictures

ILLUSTRATION CREDITS

Contributing artists to this and previous editions: Nea Bisek, Chris Carothers, Russell Chun, Rachel Ciemma, Barbara Cousins, Pamela Drury-Wattenmaker, Caitlin Duckwall, Cecile Duray-Bito, Janet Hayes, Darwen Hennings, Vally Hennings, Georg Klatt, Sandra McMahon, Linda McVay, Kenneth Miller, Fran Milner, Elizabeth Morales-Denney, Jackie Osborn, Carla Simmons, Carol Verbeeck, John Waller, Judy Waller and the artists of Precision Graphics.

The following figures are adapted from Christopher K. Mathews and K. E. van Holde, *Biochemistry*, 2nd ed., Menlo Park, CA: Benjamin/Cummings, © 1996 The Benjamin/Cummings Publishing Company, Inc.: **4.5, 9.9, 10.11, 17.15, 19.10**.

The following figures are adapted from Wayne M. Becker, Jane B. Reece, and Martin F. Poenie, *The World of the Cell*, 3rd ed., Menlo Park, CA: Benjamin/Cummings, © 1996 The Benjamin/Cummings Publishing Company, Inc.: **4.6, Table 7.2, 7.5, 8.7, 11.7, 11.12, 12.12, 16.13, 16.16, 17.9, 19.5, 19.15, 20.7.**

Figures 7.7 and **7.22** and cell organelle drawings in figures **7.11, 7.12, 7.13,** and **7.20** are adapted from illustrations by Tomo Narashima in Elaine N. Marieb, *Human Anatomy and Physiology*, 5th ed., San Francisco, CA: Benjamin Cummings, © 2001 Benjamin Cummings, an imprint of Addison Wesley Longman, Inc.: **Figures 7.11, 49.11, 49.12,** and **49.13** are also adapted from *Human Anatomy and Physiology*, 5th ed.

Figures 46.17, 48.2(a), 48.5, 48.11, 48.18, 48.21, 49.28(a), 49.32, and **49.36** are adapted from Elaine N. Marieb, *Human Anatomy and Physiology*, 4th ed., Menlo Park, CA: Benjamin/ Cummings, © 1998 Benjamin/Cummings, an imprint of Addison Wesley Longman, Inc.

3.8 Adapted from a figure by Michael Pique, The Scripps Research Institute.

4.7 Adapted from an illustration by Clark Still, Columbia University.

5.12 Adapted from Robert Wallace, Gerald P. Sanders, and Robert J. Ferl. 1991. *Biology: The Science of Life*, 3rd ed. New York: HarperCollins. © 1991 HarperCollins Publishers, Inc. Reprinted by permission of Addison-Wesley Educational Publishers; **5.17, 5.20** Adapted from D. W. Heinz, W. A. Baase, F. W. Dahlquist, B. W. Matthews. 1993. How amino-acid insertions are allowed in an alpha-helix of T4 lysozyme. *Nature* 361:561; **5.23 (b)** © Irving Geis.

6.1 (inset) Adapted from Bruce Alberts et al. 1989. *Molecular Biology of the Cell*, 2nd ed. New York: Garland Publishing. © 1989 Garland Publishing Inc. Reprinted by permission.

9.5 Adapted from Bruce Alberts et al. 1989. *Molecular Biology of the Cell*, 2nd ed., Fig. 7.17, p. 351. New York: Garland Publishing. © 1989 Garland Publishing. Reprinted by permission.

10.13 Adapted from Richard and David Walker. *Energy, Plants and Man*, Fig. 4.1, p. 69. Sheffield: University of Sheffield. © David and Richard Walker. Reprinted by permission.

12.11 Adapted from Bruce Alberts et al. 1989. *Molecular Biology of the Cell*, 2nd ed., Fig. 13.75. New York: Garland Publishing. © 1989 Garland Publishing. Reprinted by permission.

17.11 Adapted from L. J. Kleinsmith and V. M. Kish. 1995. *Principles of Cell and Molecular Biology*, 2nd ed. New York, NY: HarperCollins. Reprinted by permission of Addison Wesley Educational Publishers.

Chapter 18 opening illustration Adapted from James D. Watson et al. 1987. *Molecular Biology of the Gene*, 4th ed., Fig. 7.10. Menlo Park, CA: Benjamin/Cummings. © 1987 James D. Watson.

19.3 (inset) © Irving Geis

21.4 Adapted from Bruce Alberts et al. 1989. *Molecular Biology of the Cell*, 2nd ed., Fig. 16.32, p. 904. New York: Garland Publishing. © 1989 Garland Publishing. Reprinted by permission; **21.15** Adapted from an illustration by William McGinnis; **21.20 (upper right illustration only)** Adapted from E. Dennis et al. 1993. Manipulating floral identity. *Current Biology* 3:90–93. Reprinted by permission.

22.13 Adapted from R. Shurman et al. 1995. *Journal of Infectious Diseases* 171:1411; **22.16** Adapted from H. L. Carson. 1983. *Genetics* 103:465–482.

23.10 Adapted from A. C. Allison. 1961. Abnormal hemoglobin and erythrocyte enzyme-deficiency traits. In *Genetic Variation in Human Populations*, ed. G. A. Harrison. Oxford: Elsevier Science; **23.11** Adapted from Curtis M. Lively and Mark F. Dybdahl. 2000. Parasite adaptation to locally common host genotypes. *Nature* 405:679–680.

24.9 Adapted from D. Futuyma. 1998. *Evolutionary Biology*, 3rd ed., p. 445. Sunderland, MA: Sinauer Associates; **24.12** Adapted from D. M. B. Dodd. *Evolution* 11:1308–1311; **24.18** Adapted from M. Strickberger. 1990. *Evolution*. Boston: Jones & Bartlett, **24.20** Adapted from L. Wolpert. 1998. *Principles of Development*. Oxford University Press; **24.22** Adapted from M. I. Coates. 1995. *Current Biology* 5:844–848.

25.2 Adapted from D. Futuyma. 1998. *Evolutionary Biology*, 3rd ed., p. 128. Sunderland, MA: Sinauer Associates; **25.5** Data from M. J. Benton. 1995. Diversification and extinction in the history of life. *Science* 268.55; **25.17** Adapted from B. Korber et al. June 9, 2000. Timing the ancestor of the HIV-1 pandemic strains. *Science* 288:1789–1796.

26.2 Adapted from David J. Des Marais. September 8, 2000. When did photosynthesis emerge on Earth? *Science* 289:1703–1705; **26.8** Data from Andrew H. Knoll and Sean B. Carroll. June 25, 1999. *Science* 284:2129–2137.

27.7 Adapted from Gerard J. Tortora, Berdell R. Funke, and Christine L. Case. 1998. *Microbiology: An Introduction*, 6th ed. Menlo Park, CA: Benjamin/Cummings. © 1998 Benjamin/Cummings, an imprint of Addison Wesley Longman, Inc.

28.7 Adapted from W. F. Doolittle. Feb. 2000. Uprooting the tree of life. *Scientific American*, p. 95; **28.25** Adapted from C. F. Delwiche. Oct. 1999. *The American Naturalist* 154:5167 (Supplement).

29.22 Adapted from Raven et al. *Biology of Plants*, 6th ed., Fig. 19.7.

32.12 Adapted from Adouette et al. April 25, 2000. *Proceedings of the National Academy of Sciences*, p. 4454; **32.14** Adapted from *Science*, April 15, 1998, p. 392.

33.36 Adapted from G. K. Davis and N. H. Patel. December 1999. The origin and evolution of segmentation. *TCB/TIBS/TIG Joint Millennium Issue.*

34.15 Adapted from C. Zimmer. *At the Water's Edge*. Free Press, p. 99; **34.16** Adapted from C. Zimmer. *At the Water's Edge*. Free Press, p. 90; **34.23** © Donna Braginetz; **34.30** Adapted from Stephen J. Gould et al. *The Book of Life*. London: Ebury Press, p. 96. Reprinted by permission of Random House UK Ltd.; **34.38** Drawn from photos of fossils: *O. tugenensis* photo in Michael Balter, Early hominid sows division, *ScienceNow*, Feb. 22, 2001, © 2001 American Association for the Advancement of Science. *A. ramidus kadabba* photo by Timothy White, 1999/Brill Atlanta. *A. anamensis, A. garhi,* and *H. neanderthalensis* adapted from *The Human Evolution Coloring Book*. *K. platyops* drawn from photo in Meave Leakey et al., New hominid genus from eastern Africa shows diverse middle Pliocene lineages, *Nature*, March 22, 2001, 410:433. *A. boisei* drawn from a photo by David Brill. *H. ergaster* drawn from a photo at www.inhandmuseum.com; **34.41** Adapted from an illustration by Laurie Grace in A. C. Wilson and R. I. Cann. 1992. The recent African genesis of humans. *Scientific American* 73.

35.25 Pie chart adapted from *Nature*, Dec. 14, 2000, 408:799.

39.13 Adapted from T. C. Moore. 1984. *Biochemistry and Physiology of Plant Hormones*, 2nd ed., Fig. 6.13. Springer Verlag; **39.17 (a), 39.18** Adapted from M. Wilkins. 1988. *Plant Watching*. Facts on File Publ.; **39.29** Adapted with permission from Edward Farmer. 1997. *Science* 276:912. Copyright © 1997 American Association for the Advancement of Science.

43.14, 43.15 Adapted from Gerard J. Tortora, Berdell R. Funke, and Christine L. Case. 1998. *Microbiology: An Introduction*, 6th ed. Menlo Park, CA: Benjamin/ Cummings. © 1998 Benjamin/Cummings, an imprint of Addison Wesley Longman, Inc.; **43.17** Adapted from Lennart Nilsson and Jan Lindberg. 1987. *The Body Victorious*, p. 27. New York: Delacorte Press. Illustration by Urban Frank, Studio Frank and Co., Illustrator.

44.4 Adapted from P. T. Marshall and G. M. Hughes. 1980. *Physiology of Mammals and Other Vertebrates*, 2nd ed. Cambridge: Cambridge University Press. Reprinted with the permission of Cambridge University Press; **44.9 (a)** Adapted from B. Heinrich. 1974. *Science* 185:747–756. © 1974 American Association for the Advancement of Science. Reprinted with permission; **44.9 (b)** Adapted from an illustration by Enid Kotschnig in B. Heinrich. 1987. Thermoregulation in a winter moth. *Scientific American* 105; **44.13** Adapted from Lawrence G. Mitchell, John A. Mutchmor, and Warren D. Dolphin. 1988. *Zoology*. Menlo Park, CA: Benjamin/Cummings. © 1988 The Benjamin/Cummings Publishing Company; **44.16 (b)** Kangaroo rat data from Schmidt-Nielsen. 1990. *Animal Physiology: Adaptation and Environment*, 4th ed., p. 339. Cambridge: Cambridge University Press.

45.4 Adapted from Bruce Alberts et al. 1989. *Molecular Biology of the Cell*, 2nd ed., Fig. 15.9. New York: Garland Publishing. © 1989 Garland Publishing. Reprinted by permission; **45.11** Adapted from Gilbert. *Developmental Biology*, 4th ed., Fig. 7.39, p. 282. Sunderland, MA: Sinauer Associates, Inc. Reprinted by permission.

47.17 From Wolpert et al. 1988. *Principles of Development*, Fig. 8.25, p. 251 right. Oxford: Oxford University Press. By permission of Oxford University Press; **47.20** Adapted from Bruce Alberts et al. 1989. *Molecular Biology of the Cell*, 2nd ed., Fig. 16.29a, p. 904. New York: Garland Publishing. © 1989 Garland Publishing. Reprinted by permission; **47.22 Left:** From Wolpert et al. 1988. *Principles of Development*, Fig. 1.10 (right). Oxford: Oxford University Press. By permission of Oxford University Press. **Right:** Adapted from Gilbert. 1997. *Developmental Biology*, 5th Ed., part of Fig. 15.12, p. 604. Sunderland, MA: Sinauer Associates. Reprinted by permission.

48.9 Adapted from G. Matthews. 1986. *Cellular Physiology of Nerve and Muscle*. Cambridge, MA: Blackwell Scientific Publications. Reprinted by permission of Blackwell Science, Inc.; **48.20** Adapted from Alfred Sherwood Romer and Thomas S. Parsons. 1986. *The Vertebrate Body*, 6th ed. Philadelphia, PA: Saunders. © 1986 by Saunders College Publishing. Reproduced by permission; **48.28** Adapted from John G. Nicholls et al. 2001. *From Neuron to Brain*, 4th ed., Fig. 23.24. Sunderland, MA: Sinauer Associates Inc. © 2001 Sinauer Associates.

49.2 (right) Adapted from *Scientific American*, March 2001, p. 36; **49.30** Adapted from Lawrence G. Mitchell, John A. Mutchmor, and Warren D. Dolphin. 1988. *Zoology*. Menlo Park, CA: Benjamin/Cummings. © 1988 The Benjamin/Cummings Publishing Company.

50.1 Adapted from G. Caughly, N. Shepherd, and J. Short. 1987. *Kangaroos: Their Ecology and Management in the Sheep Rangelands of Australia*, Fig. 1.2, p. 12. Cambridge: Cambridge University Press. Copyright © 1987 Cambridge University Press; **50.6** Adapted from Charles J. Krebs. 2001. *Ecology*, 5th Ed., Fig. 3.2. San Francisco, CA: Benjamin Cummings. © 2001 Benjamin Cummings, an imprint of Addison Wesley Longman, Inc.; **50.7** Adapted from Charles J. Krebs. 2001. *Ecology*, 5th Ed., Fig. 3.1. San Francisco, CA: Benjamin Cummings. Data provided by O. R. Taylor in personal communication to Krebs. © 2001 Benjamin Cummings, an imprint of Addison Wesley Longman, Inc.; **50.8** Adapted from the National Zebra Mussel Information Clearinghouse, NOAA, U.S. Department of Commerce; **50.9** Data from W. J. Fletcher. 1987. Interactions among subtidal Australian sea urchins, gastropods and algae: effects of experimental removals. *Ecological Monographs* 57:89–109; **50.16** Adapted from L. Roberts. 1989. How fast can trees migrate? *Science* 243:736, Fig. 2. © 1989 by the American Association for the Advancement of Science; **50.26** Adapted from R. T. T. Forman. 1964. Growth under controlled conditions to explain the hierarchical distributions of a moss, *Tetraphis pellucida*. *Ecological Monographs* 34:1–25; **50.27 (a)** Data from S. Anderson. 1985. The theory of range-size (RS) distributions. *American Museum Novitates* 2833:1–20; **(b)** Data from K. J. Gaston and J. L. Curnutt. The dynamics of abundance—range size relationships. *Oikos* 81:38–44.

51.2 Adapted from Lawrence G. Mitchell, John A. Mutchmor, and Warren D. Dolphin 1988. *Zoology*. Menlo Park, CA: Benjamin/Cummings. © 1988 The Benjamin/Cummings Publishing Company; **51.3** Adapted from N. Tinbergen. 1951. *The Study of Instinct*. Oxford: Oxford University Press; **51.5** Courtesy of Masakazu Konishi; **51.14(b)** Adapted from Elizabeth A. Capaldi et al. February 2000. Ontogeny of orientation flight in the honeybee revealed by armonic radar. *Nature* 403.

52.5 Adapted from T. H. Clutton-Brock, F. E. Guiness, and S. D. Albon. *Red Deer: Behavior and Ecology of Two Sexes*, p. 77. Chicago: University of Chicago Press. © University of Chicago Press; **52.9** Data from Canadian Wildlife Service, 2001; J. R. Cannon, 1996,

Whooping crane recovery: a case study in public and private cooperation in the conservation of endangered species, *Conservation Biology* 10:813–821; C. S. Binkley and R. S. Miller, 1983, Population characteristics of the whooping crane, *Grus americana*, *Canadian Journal of Zoology* 61:2768–2776; **52.12 (c)** Data courtesy of P. Arcese and J. N. M. Smith, 2001; **52.13** Adapted from J. T. Enright. 1976. Climate and population regulation: the biogeographer's dilemma. *Oecologia* 24:295–310; **52.14 (b)** Data from J. N. M. Smith and P. Arcese; **52.16** Data from U.S. Fish and Wildlife Service, 2001; **52.17** Data courtesy of Rolf O. Peterson, Michigan Technological University, 2001; **52.18** Data from Higgins et al. May 30, 1997. Stochastic dynamics and deterministic skeletons: population behavior of Dungeness crab. *Science*; **52.21** Data from Population Reference Bureau, 2000; **52.22** Adapted from J. A. J. McFalls. 1998. Population: a lively introduction. *Population Bulletin* 53; **52.23** Data from J. Wackernagel et al. 1999. National natural capital accounting with the ecological footprint concept. *Ecological Economics* 29:375–390.

53.3 (a) A. S. Rand and E. E. Williams. 1969. The anoles of La Palma: aspects of their ecological relationships. *Breviora* 327. Museum of Comparative Zoology, Harvard University. © Presidents and Fellows of Harvard College; **53.11** Adapted from E. A. Knox. 1970. Antarctic marine ecosystems. In *Antarctic Ecology*, ed. M. W. Holdgate, 69–96. London: Academic Press; **53.12** Adapted from D. L. Breitburg et al. 1997. Varying effects of low dissolved oxygen on trophic interactions in an estuarine food web, *Ecological Monographs* 67:490. Copyright © 1997 Ecological Society of America; **53.13** Adapted from B. Jenkins. 1992. Productivity, disturbance and food web structure at a local spatial scale in experimental container habitats. *Oikos* 65:252. Copyright © 1992 Oikos, Sweden; **53.15** Adapted from J. A. Estes et al. 1998. Killer whale predation on sea otters linking oceanic and nearshore ecosystems. *Science* 282:474. Copyright © 1998 by the American Association for the Advancement of Science; **53.17** Graphs adapted from J. H. Connell et al. A 30-year study of coral abundance, recruitment, and disturbance at several scales in space and time. *Ecological Monographs* 67:461–488; **53.19, 53.20** Adapted from R. L. Crocker and J. Major. 1955. Soil development in relation to vegetation and surface age at Glacier Bay, *Alaska. Journal of Ecology* 43:427–448; **53.22** Adapted from C. B. Williams. 1964. *Patterns in the Balance of Nature*. London: Academic Press; **53.23** Adapted from R. E. Cook. 1969. Variation in species density of North American birds. *Systematic Zoology* 18:63–84; **53.24** Adapted from D. J. Currie. 1991. Energy and large-scale patterns of animal- and plant-species richness. *American Naturalist* 137:27–49; **53.25** Adapted from F. W. Preston. 1960. Time and space and the variation of species. *Ecology* 41:611–627; **53.27** Adapted from F. W. Preston. 1962. The canonical distribution of commonness and rarity. *Ecology* 43:185–215, 410–432.

54.1 Adapted from D. L. DeAngelis. 1992. *Dynamics of Nutrient Cycling and Food Webs*. New York: Chapman & Hal; **54.4** Adapted from C. B. Field et al. 1998. Primary production of the biosphere: integrating terrestrial and oceanic components. *Science* 281:237–240; **54.5** Adapted from T. L. Hayward. 1991. Primary production in the north Pacific central gyre: a controversy with important implications. *Trends in Ecology and Evolution* 6:28–284; **54.6** Adapted from J. H. Ryther and W. M. Dunstan. 1971. Nitrogen, phosphorus, and eutrophication in the coastal marine environment. *Science* 171:1008–1013; **54.9** Adapted from S. M. Cargill and R. L. Jefferies. 1984. Nutrient limitation of primary production in a sub-arctic salt marsh. *Journal of Applied Ecology* 21:657–668; **54.16** Adapted from R. E. Ricklefs. 1997. *The Economy of Nature*, 4th ed. © 1997 by W. H. Freeman and Company. Used with permission; **54.23** Adapted from G. E. Likens et al. 1981. Interactions between major biogeochemical cycles in terrestrial ecosystems. In *Some Perspectives of the Major Biogeochemical Cycles*, ed. G. E. Likens, 93–123. New York: Wiley.

55.6 Adapted from J. T. Curtis. 1959. *The Vegetation of Wisconsin*. Madison, WI: University of Wisconsin Press; **55.10** Adapted from Charles J. Krebs. 2001. *Ecology*, 5th Ed., Fig. 19.1. San Francisco, CA: Benjamin Cummings. © 2001 Benjamin Cummings, an imprint of Addison Wesley Longman, Inc.; **55.11** Adapted from R. L. Westemeier et al. 1998. Tracking the long-term decline and recovery of an isolated population. *Science* 282:1696. © 1998 by the American Association for the Advancement of Science; **55.18** Adapted from W. D. Newmark. 1985. Legal and biotic boundaries of western North American national parks: a problem of congruence. *Biological Conservation* 33:199. © 1985 Elsevier Applied Science Publishers Ltd., Barking, England; **55.21** Adapted from A. P. Dobson et al. 1997. Hopes for the future: restoration ecology and conservation biology. *Science* 277:515. © 1997 by the American Association for the Advancement of Science.

GLOSSARY

5' cap The 5' end of a pre-mRNA molecule modified by the addition of a cap of guanine nucleotide.

A band The broad region that corresponds to the length of the thick filaments.

A site One of three binding sites for tRNA during translation, it holds the tRNA carrying the next amino acid to be added to the polypeptide chain; A stands for aminoacyl-tRNA site.

abdominal cavity The body cavity in mammals that primarily houses parts of the digestive, excretory, and reproductive systems. It is separated from the more cranial thoracic cavity by the diaphragm.

abiotic components (ā´-bī-ot´-ik) Nonliving chemical and physical factors in the environment.

ABO blood groups Genetically determined classes of human blood that are based on the presence or absence of carbohydrates A and B on the surface of red blood cells. The ABO blood group phenotypes, also called blood types, are A, B, AB, and O.

abscisic acid (ABA) (ab-sis´-ik) A plant hormone that generally acts to inhibit growth, promote dormancy, and help the plant tolerate stressful conditions.

absorption The uptake of small nutrient molecules by an organism's own body; the third main stage of food processing, following digestion.

absorption spectrum The range of a pigment's ability to absorb various wavelengths of light.

abyssal zone (uh-bis´-ul) The very deep benthic communities near the bottom of the ocean. This region is characterized by continuous cold, extremely high water pressure, low nutrients, and near or total absence of light.

acanthodians (ak´-an-thō´-dē-un) A group of ancient jawed fishes from the Devonian period.

acclimatization (uh-klı´-muh-tī-zā´-shun) Physiological adjustment to a change in an environmental factor.

accommodation The automatic adjustment of an eye to focus on near objects.

acetyl CoA (acetyl coenzyme A) The entry compound for the Krebs cycle in cellular respiration; formed from a fragment of pyruvate attached to a coenzyme.

acetylcholine (as´-uh-til-kō´-lēn) One of the most common neurotransmitters; functions by binding to receptors and altering the permeability of the post-synaptic membrane to specific ions, either depolarizing or hyperpolarizing the membrane.

acid A substance that increases the hydrogen ion concentration of a solution.

acid chyme (kīm) A mixture of recently swallowed food and gastric juice.

acid precipitation Rain, snow, or fog that is more acidic than pH 5.6.

acoelomate (uh-sē´-lō-māt) A solid-bodied animal lacking a cavity between the gut and outer body wall.

acrosomal reaction The discharge of a sperm's acrosome when the sperm approaches an egg.

acrosome (ak´-ruh-sōm) An organelle at the tip of a sperm cell that helps the sperm penetrate the egg.

actin (ak´-tin) A globular protein that links into chains, two of which twist helically about each other, forming microfilaments in muscle and other contractile elements in cells.

Actinistia (ak´-tuh-nē´-stē-uh) The class of lobe-finned fishes.

Actinopterygii (ak´-tuh-nop´-tuh-rij´-ē-ī) The class of ray-finned fishes.

action potential A rapid change in the membrane potential of an excitable cell, caused by stimulus-triggered, selective opening and closing of voltage-sensitive gates in sodium and potassium ion channels.

action spectrum A profile of the relative performance of different wavelengths of light.

activation energy The amount of energy that reactants must absorb before a chemical reaction will start.

activator A transcription factor that binds to an enhancer and stimulates transcription of a gene.

active immunity Immunity conferred by recovering from an infectious disease.

active site The specific portion of an enzyme that attaches to the substrate by means of weak chemical bonds.

active transport The movement of a substance across a biological membrane against its concentration or electrochemical gradient with the help of energy input and specific transport proteins.

adaptive peak An equilibrium state in a population when the gene pool has allele frequencies that maximize the average fitness of a population's members.

adaptive radiation The emergence of numerous species from a common ancestor introduced into an environment, presenting a diversity of new opportunities and problems.

adenohypophysis (ad´-uh-nō-hī-pof´-uh-sis) Also called the anterior pituitary, it consists of endocrine cells that synthesize and secrete several hormones directly into the blood.

adenylyl cyclase (ad´-en-uh-lil) An enzyme that converts ATP to cyclic AMP in response to a chemical signal.

adhesion The attraction between different kinds of molecules.

adrenal gland (uh-drē´-nul) An endocrine gland located adjacent to the kidney in mammals; composed of two glandular portions: an outer cortex, which responds to endocrine signals in reacting to stress and effecting salt and water balance, and a central medulla, which responds to nervous inputs resulting from stress.

adrenal medulla (uh-drē´-nul muh-dul´-uh) The central portion of an adrenal gland, controlled by nerve signals, that secretes the fight-or-flight hormones epinephrine and norepinephrine.

adrenocorticotropic hormone A peptide hormone released from the anterior pituitary, it stimulates the production and secretion of steroid hormones by the adrenal cortex.

adventitious Roots extending from stems and leaves above ground.

aerobic (ār-ō´-bik) Containing oxygen; referring to an organism, environment, or cellular process that requires oxygen.

afferent arteriole (af´-er-ent) The blood vessel supplying a nephron.

Afrotheria A branch of mammals that includes sloths, anteaters, and armadillos.

age structure The relative number of individuals of each age in a population.

agglutination An antibody-mediated immune response in which bacteria or viruses are clumped together, effectively neutralized, and opsonized.

aggregate fruit A fruit such as a blackberry that develops from a single flower that has several carpels.

agnathan (ag-nā´-thun) A member of a jawless class of vertebrates represented today by the lampreys and hagfishes.

agonistic behavior (a´-gō-nis´-tik) A type of behavior involving a contest of some kind that determines which competitor gains access to some resource, such as food or mates.

AIDS (acquired immunodeficiency syndrome) The name of the late stages of HIV infection; defined by a specified reduction of T cells and the appearance of characteristic secondary infections.

alcohol fermentation The conversion of pyruvate to carbon dioxide and ethyl alcohol.

alcohols Organic compounds containing hydroxyl groups.

aldehyde (al´-duh-hīd) An organic molecule with a carbonyl group located at the end of the carbon skeleton.

aldosterone (al-dos´-tuh-rōn) An adrenal hormone that acts on the distal tubules of the kidney to stimulate the reabsorption of sodium (Na^+) and the passive flow of water from the filtrate.

alga (plural, **algae**) (al´-guh, al´-jē) A photosynthetic, plantlike protist.

alimentary canal (al´-uh-men´-tuh-rē) A digestive tract consisting of a tube running between a mouth and an anus.

allantois (al´-an-tō´-sis) One of four extra-embryonic membranes; serves as a repository for the embryo's nitrogenous waste.

alleles (uh-lē´-ulz) Alternate versions of a gene.

allometric growth (al´-ō-met´-rik) The variation in the relative rates of growth of various parts of the body, which helps shape the organism.

allopatric speciation (al´-ō-pat´-rik) A mode of speciation induced when the ancestral population becomes segregated by a geographic barrier.

allopolyploid (al´-ō-pol´-ē-ployd) A common type of polyploid species resulting from two different species interbreeding and combining their chromosomes.

all-or-none event An action that occurs either completely or not at all, such as the generation of an action potential by a neuron.

allosteric site A specific receptor site on some part of an enzyme molecule remote from the active site.

alpha (α) helix (al´-fuh hē´-liks) A spiral shape constituting one form of the secondary structure of proteins, arising from a specific hydrogen-bonding structure.

alternation of generations A life cycle in which there is both a multicellular diploid form, the sporophyte, and a multicellular haploid form, the gametophyte; characteristic of plants.

alternative RNA splicing A type of regulation at the RNA-processing level in which different mRNA molecules are produced from the same primary transcript depending on which RNA segments are treated as exons and which as introns.

altruism (al´-trū-iz-um) Behavior that reduces an individual's fitness while increasing the fitness of another individual.

altruistic behavior (al´-trū-is´-tik) The aiding of another individual at one's own risk or expense.

Alveolata A protistan clade that includes dinoflagellates, apicomplexans, and the ciliates. Alveolates have small membrane-bounded cavities called alveoli under their cell surfaces. The function of alveoli is unknown.

alveolus (al-vē´-uh-lus) (plural, **alveoli**) (1) One of the deadend, multilobed air sacs that constitute the gas exchange surface of the lungs. (2) One of the milk-secreting sacs of epithelial tissue in the mammary glands.

amacrine cell (am´-uh-krin) Neurons of the retina that help integrate information before it is sent to the brain.

amine (uh-mēn´) An organic compound with one or more amino groups.

amino acid (uh-mēn´-ō) An organic molecule possessing both carboxyl and amino groups. Amino acids serve as the monomers of proteins.

amino group A functional group that consists of a nitrogen atom bonded to two hydrogen atoms; can act as a base in solution, accepting a hydrogen ion and acquiring a charge of +1.

aminoacyl-tRNA synthetase An enzyme that joins each amino acid to the correct tRNA.

aminopeptidase An enzyme found within the small intestine that splits off one amino acid at a time, beginning at the opposite end of the polypeptide containing a free carboxyl group.

ammonia A small and very toxic nitrogenous waste produced by metabolism.

ammonites Shelled cephalopod animals that were the dominant invertebrate predators for millions of years ending with the mass extinctions at the end of the Cretaceous period.

amniocentesis (am´-nē-ō-sen-tē´-sis) A technique for determining genetic abnormalities in a fetus by the presence of certain chemicals or defective fetal cells in the amniotic fluid, obtained by aspiration from a needle inserted into the uterus.

amnion (am´-nē-on) The innermost of four extraembryonic membranes; encloses a fluid-filled sac in which the embryo is suspended.

amniote A vertebrate possessing an amnion surrounding the embryo; reptiles, birds, and mammals are amniotes.

amniotic egg (am´-nē-ot´-ik) A shelled, water-retaining egg that enables reptiles, birds, and egg-laying mammals to complete their life cycles on dry land.

amoeba (uh-mē´-buh) A type of protist characterized by great flexibility and the presence of pseudopodia.

amoebocyte (uh-mē´-buh-sīt) An amoebalike cell that moves by pseudopodia, found in most animals; depending on the species, may digest and distribute food, dispose of wastes, form skeletal fibers, fight infections, and change into other cell types.

Amphibia (am-fib´-ē-uh) The vertebrate class of amphibians, represented by frogs, salamanders, and caecilians.

amphipathic molecule (am´-fē-path´-ik) A molecule that has both a hydrophilic region and a hydrophobic region.

amplification The strengthening of stimulus energy that is otherwise too weak to be carried into the nervous system.

anabolic pathway (an´-uh-bol´-ik) A metabolic pathway that synthesizes a complex molecule from simpler compounds.

anaerobic (an´-ār-ō´-bik) Lacking oxygen; referring to an organism, environment, or cellular process that lacks oxygen and may be poisoned by it.

anaerobic respiration The use of inorganic molecules other than oxygen to accept electrons at the "downhill" end of electron transport chains.

anagenesis (an´-uh-jen´-uh-sis) A pattern of evolutionary change involving the transformation of an entire population, sometimes to a state different enough from the ancestral population to justify renaming it as a separate species; also called phyletic evolution.

analogy The similarity of structure between two species that are not closely related; attributable to convergent evolution.

anaphase The fourth subphase of mitosis, in which the chromatids of each chromosome have separated and the daughter chromosomes are moving to the poles of the cell.

anaphylactic shock (an´-uh-fi-lak-tic) An acute, life-threatening, allergic response.

anapsids One of three groups of amniotes based on key differences between their skulls.

anatomically modern humans Fully modern humans.

anatomy The study of the structure of an organism.

anchorage dependence The requirement that to divide, a cell must be attached to the substratum.

androgens (an´-drō-jenz) The principal male steroid hormones, such as testosterone, which stimulate the development and maintenance of the male reproductive system and secondary sex characteristics.

aneuploidy (an´-yū-ploy-dē) A chromosomal aberration in which certain chromosomes are present in extra copies or are deficient in number.

angiosperm (an´-jē-ō-sperm) A flowering plant, which forms seeds inside a protective chamber called an ovary.

anhydrobiosis (an´-hī´-drō-bī-ō´-sis) The ability to survive in a dormant state when an organism's habitat dries up. Also called cryptobiosis.

animal pole The portion of the egg where the least yolk is concentrated. Opposite of vegetal pole.

anion (an´-ī-on) A negatively charged ion.

annual A plant that completes its entire life cycle in a single year or growing season.

antennae Sensory appendages found in uniramians and crustaceans.

anterior Referring to the head end of a bilaterally symmetrical animal.

anterior pituitary Also called the adenohypophysis, it consists of endocrine cells that synthesize and secrete several hormones directly into the blood.

anther (an´-ther) The terminal pollen sac of a stamen, inside which pollen grains with male gametes form in the flower of an angiosperm.

antheridium (an´-thuh-rid´-ē-um) (plural, **antheridia**) In plants, the male gametangium, a moist chamber in which gametes develop.

Anthocerophyta The phylum of hornworts, small herbaceous (nonwoody) plants.

Anthophyta (an-thof´-uh-duh) The phylum containing all angiosperms.

anthropoid (an´-thruh-poyd) A member of a primate group made up of the apes (gibbon, orangutan, gorilla, chimpanzee, and bonobo), monkeys, and humans.

antibiotic A chemical that kills bacteria or inhibits their growth.

antibody An antigen-binding immunoglobulin, produced by B cells, that functions as the effector in an immune response.

anticodon (an´-tī-kō´-don) A specialized base triplet at one end of a tRNA molecule that recognizes a particular complementary codon on an mRNA molecule.

antidiuretic hormone A hormone that is part of an elaborate feedback scheme that helps regulate the osmolarity of the blood.

antigen (an´-tuh-jen) A foreign macromolecule that does not belong to the host organism and that elicits an immune response.

antigen presentation The process by which an MHC molecule cradles a fragment of an intracellular protein antigen in its hammocklike groove, carries it to the cell surface, and "presents" the protein to an antigen receptor on a nearby T cell.

antigen receptor Transmembrane versions of antibody molecules that B cells and T cells use to recognize specific antigens. Also called membrane antibodies.

antigen-presenting cell (APCs) Cells that ingest bacteria and viruses and then destroy them. Class II MHC molecules in these cells collect peptide remnants of this degradation and present them to helper T cells.

anura (uh-nū´-ra) The order of frogs and toads that includes tailless tetrapod amphibians.

anurans The group of frogs and toads.

aphotic zone (ā´-fō´-tik) The part of the ocean beneath the photic zone, where light does not penetrate sufficiently for photosynthesis to occur.

apical dominance (ā´-pik-ul) Concentration of growth at the tip of a plant shoot, where a terminal bud partially inhibits axillary bud growth.

apical ectodermal ridge A limb-bud organizing region consisting of a thickened area of ectoderm at the tip of a limb bud.

apical meristem (ā´-pik-ul mār´-uh-stem) Embryonic plant tissue in the tips of roots and in the buds of shoots that supplies cells for the plant to grow in length.

apicomplexan (ap´-ē-kom-pleks´-un) One of a group of parasitic protozoans, some of which cause human diseases.

apoda (ap´-uh-duh) The order of caecilians that includes legless amphibians.

apodans The group of caecilians.

apomixis (ap´-uh-mik´-sis) The asexual production of seeds.

apomorphic character (ap´-ō-mōr´-fik) A derived phenotypic character, or homology, that evolved after a branch diverged from a phylogenetic tree.

apoplast (ap´-ō-plast) In plants, the nonliving continuum formed by the extracellular pathway provided by the continuous matrix of cell walls.

apoptosis Programmed cell death brought about by signals that trigger the activation of a cascade of "suicide" proteins in the cells destined to die.

aposematic coloration (ap´-ō-sō-mat´-ik) The bright coloration of animals with effective physical or chemical defenses that acts as a warning to predators.

appendix A small, fingerlike extension of the vertebrate cecum; contains a mass of white blood cells that contribute to immunity.

aquaporin A transport protein in the plasma membrane of a plant or animal cell that specifically facilitates the diffusion of water across the membrane (osmosis).

aqueous humor (ā´-kwē-us hyū´-mer) Plasmalike liquid in the space between the lens and the cornea in the vertebrate eye; helps maintain the shape of the eye, supplies nutrients and oxygen to its tissues, and disposes of its wastes.

aqueous solution (ā´-kwē-us) A solution in which water is the solvent.

Arachida The animal class that includes scorpions, spiders, ticks, and mites.

Archaea (ar´-kē-uh) One of two prokaryotic domains, the other being the Bacteria.

Archaezoa Primitive eukaryotic group that includes diplomonads, such as Giardia; some systematists assign kingdom status to archezoans.

archegonium (ar-ki-gō´-nē-um) (plural, **archegonia**) In plants, the female gametangium, a moist chamber in which gametes develop.

archenteron (ar-ken´-tuh-ron) The endoderm-lined cavity, formed during the gastrulation process, that develops into the digestive tract of an animal.

archosaurs The reptilian group that includes crocodiles, alligators, dinosaurs, and birds.

arteriole (ar-tār´-ē-ōl) A vessel that conveys blood between an artery and a capillary bed.

arteriosclerosis A cardiovascular disease caused by the formation of hard plaques within the arteries.

artery A vessel that carries blood away from the heart to organs throughout the body.

arthropod Segmented coelomates with exoskeletons and jointed appendages.

Arthropoda (ar-throp´-uh-duh) The most diverse phylum in the animal kingdom; includes the horseshoe crab, arachnids (e.g., spiders, ticks, scorpions, and mites), crustaceans (e.g., crayfish, lobsters, crabs, barnacles), millipedes, centipedes, and insects. Arthropods are characterized by a chitinous exoskeleton, molting, jointed appendages, and a body formed of distinct groups of segments.

artificial selection The selective breeding of domesticated plants and animals to encourage the occurrence of desirable traits.

ascocarps Macroscopic fruiting bodies of sac fungi.

ascus (plural, **asci**) A saclike spore capsule located at the tip of the ascocarp in dikaryotic hyphae; defining feature of the Ascomycota division of fungi.

asexual reproduction A type of reproduction involving only one parent that produces genetically identical offspring by budding or by the division of a single cell or the entire organism into two or more parts.

aspartate An amino acid that functions as a CNS neurotransmitter.

associative learning The acquired ability to associate one stimulus with another; also called classical conditioning.

assortative mating A type of nonrandom mating in which mating partners resemble each other in certain phenotypic characters.

astrocytes Glial cells that provide structural and metabolic support for neurons.

asymmetric carbon A carbon atom covalently bonded to four different atoms or groups of atoms.

asymmetric cell division Cell division in which one daughter cell receives more cytoplasm than the other during mitosis.

atherosclerosis A cardiovascular disease in which growths called plaques develop on the inner walls of the arteries, narrowing their inner diameters.

atom The smallest unit of matter that retains the properties of an element.

atomic nucleus An atom's central core, containing protons and neutrons.

atomic number The number of protons in the nucleus of an atom, unique for each element and designated by a subscript to the left of the elemental symbol.

atomic weight The total atomic mass, which is the mass in grams of one mole of the atom.

ATP (adenosine triphosphate) (a-den´-ō-sēn trī-fos´-fāt) An adenine-containing nucleoside triphosphate that releases free energy when its phosphate bonds are hydrolyzed. This energy is used to drive endergonic reactions in cells.

ATP synthase A cluster of several membrane proteins found in the mitochondrial crista (and bacterial plasma membrane) that function in chemiosmosis with adjacent electron transport chains, using the energy of a hydrogen ion concentration gradient to make ATP. ATP synthases provide a port through which hydrogen ions diffuse into the matrix of a mitrochondrion.

atrial natriuretic factor (ā´-trē-al na´-trē-ū-ret´-ik) A peptide hormone that opposes the renin-angiotensin-aldosterone system (RAAS).

atrioventricular (AV) node A region of specialized muscle tissue between the right atrium and right ventricle. It generates electrical impulses that primarily cause the ventricles to contract.

atrioventricular valve A valve in the heart between each atrium and ventricle that prevents a backflow of blood when the ventricles contract.

atrium (ā´-trē-um) (plural, **atria**) A chamber that receives blood returning to the vertebrate heart.

autogenesis model According to this model, eukaryotic cells evolved by the specialization of internal membranes originally derived from prokaryotic plasma membranes.

autoimmune disease An immunological disorder in which the immune system turns against itself.

autonomic nervous system (ot´-ō-nom´-ik) A subdivision of the motor nervous system of vertebrates that regulates the internal environment; consists of the sympathetic and parasympathetic divisions.

autopolyploid (ot´-ō-pol´-ē-ploid) A type of polyploid species resulting from one species doubling its chromosome number to become tetraploid, which may self-fertilize or mate with other tetraploids.

autosome (ot´-ō-sōm) A chromosome that is not directly involved in determining sex, as opposed to a sex chromosome.

autotroph (ot´-ō-trōf) An organism that obtains organic food molecules without eating other organisms or substances derived from other organisms. Autotrophs use energy from the sun or from the oxidation of inorganic substances to make organic molecules from inorganic ones.

auxins (ok´-sinz) A class of plant hormones, including indoleacetic acid (IAA), having a variety of effects, such as phototropic response through the stimulation of cell elongation, stimulation of secondary growth, and the development of leaf traces and fruit.

auxotroph (ok´-sō-trōf) A nutritional mutant that is unable to synthesize and that cannot grow on media lacking certain essential molecules normally synthesized by wild-type strains of the same species.

Aves (ā´-vēz) The vertebrate class of birds, characterized by feathers and other flight adaptations.

axillary bud (ak´-sil-ār-ē) A structure that has the potential to form a vegetative branch. The bud appears in the angle formed between a leaf and a stem.

axon (ak´-son) A typically long extension, or process, from a neuron that carries nerve impulses away from the cell body toward target cells.

B lymphocyte (B cell) A type of lymphocyte that develops in the bone marrow and later produces antibodies, which mediate humoral immunity.

Bacteria One of two prokaryotic domains, the other being the Archaea.

bacterial artificial chromosome (BAC) An artificial version of a bacterial chromosome that can carry inserts of 100,000–500,000 base pairs.

bacteriophage (bak-tēr´-ē-ō-fāj) A virus that infects bacteria; also called a phage. *See* phage.

bacteriorhodopsin A photosynthetic pigment found in halophiles. It is very similar to the visual pigments in the retinas of our eyes.

bacterium (plural, **bacteria**) A prokaryotic microorganism in Domain Bacteria.

bacteroids A form of *Rhizobium* contained within the vesicles formed by the root cells of a root nodule.

baculum (bak´-ū-lum) A bone that is contained in, and helps stiffen, the penis of rodents, raccoons, walruses, and several other mammals.

balanced polymorphism The ability of natural selection to maintain diversity in a population.

bark All tissues external to the vascular cambium in a plant growing in thickness, consisting of phloem, phelloderm, cork cambium, and cork.

Barr body A dense object lying along the inside of the nuclear envelope in female mammalian cells, representing an inactivated X chromosome.

barrier methods Contraception that relies upon a physical barrier to block the passage of sperm. Examples include condoms and diaphragms.

Bartholin's glands (bar´-tō-linz) Glands near the vaginal opening in a human female that secrete lubricating fluid during sexual arousal.

basal body (bā´-sul) A eukaryotic cell organelle consisting of a 9 + 0 arrangement of microtubule triplets; may organize the microtubule assembly of a cilium or flagellum; structurally identical to a centriole.

basal metabolic rate (BMR) The minimal number of kilocalories a resting animal requires to fuel itself for a given time.

basal nuclei A cluster of nuclei deep within the white matter of the cerebrum.

base A substance that reduces the hydrogen ion concentration of a solution.

basement membrane The floor of an epithelial membrane on which the basal cells rest.

base-pair substitution A point mutation; the replacement of one nucleotide and its partner in the complementary DNA strand by another pair of nucleotides.

basidiocarps Elaborate fruiting bodies of a dikaryotic mycelium of a club fungus.

basidium (plural, **basidia**) A reproductive appendage that produces sexual spores on the gills of mushrooms. The fungal division Basidiomycota is named for this structure.

basophil A circulating leukocyte that produces histamine.

Batesian mimicry (bāt´-zē-un mim´-uh-krē) A type of mimicry in which a harmless species looks like a species that is poisonous or otherwise harmful to predators.

behavior What an animal does and how it does it.

behavioral ecology A heuristic approach based on the expectation that Darwinian fitness (reproductive success) is improved by optimal behavior.

benign tumor A mass of abnormal cells that remains at the site of origin.

benthic zone The bottom surfaces of aquatic environments.

benthos (ben´-thōz) The communites of organisms living in the benthic zone of an aquatic biome.

beta cell The source of insulin within the islets of Langerhans, nestled within the pancreas.

beta oxidation A metabolic sequence that breaks fatty acids down to two-carbon fragments which enter the Krebs cycle as acetyl CoA.

beta (β) pleated sheet One form of the secondary structure of proteins in which the polypeptide chain folds back and forth, or where two regions of the chain lie parallel to each other and are held together by hydrogen bonds.

biennial (bī-en´-ē-ul) A plant that requires two years to complete its life cycle.

big-bang reproduction A life history in which adults have but a single reproductive opportunity to produce large numbers of offspring, such as the life history of the Pacific salmon. Also known as semelparity.

bilateral symmetry Characterizing a body form with a central longitudinal plane that divides the body into two equal but opposite halves.

bilateria (bī´-luh-tēr´-ē-uh) Members of the branch of eumetazoans possessing bilateral symmetry.

bile A mixture of substances that is produced in the liver, stored in the gall bladder, and acts as a detergent to aid in the digestion and absorption of fats.

binary fission The type of cell division by which prokaryotes reproduce. Each dividing daughter cell receives a copy of the single parental chromosome.

binomial The two-part latinized name of a species, consisting of genus and specific epithet.

biodiversity crisis The current rapid decline in the variety of life on Earth, largely due to the effects of human culture.

biodiversity hot spot A relatively small area with an exceptional concentration of endemic species.

bioenergetics The study of how organisms manage their energy resources.

biogenesis The principle that all life arises by the reproduction of preexisting life.

biogenic amines Neurotransmitters derived from amino acids.

biogeochemical cycles Any of the various nutrient circuits, which involve both biotic and abiotic components of ecosystems.

biogeography The study of the past and present distribution of species.

biological clock An internal timekeeper that controls an organism's biological rhythms; marks time with or without environmental cues but often requires signals from the environment to remain tuned to an appropriate period. *See also* circadian rhythm.

biological magnification A trophic process in which retained substances become more concentrated with each link in the food chain.

biological species concept The definition of a species as a population or group of populations whose mem-

bers have the potential in nature to interbreed and produce fertile offspring; a biological species is also called a sexual species.

biomass The dry weight of organic matter comprising a group of organisms in a particular habitat.

biome (bī-ōm) One of the world's major ecosystems, classified according to the predominant vegetation and characterized by adaptations of organisms to that particular environment.

bioremediation The use of living organisms to detoxify and restore polluted and degraded ecosystems.

biosphere (bī´-ō-sfēr) The entire portion of Earth inhabited by life; the sum of all the planet's ecosystems.

biotechnology The manipulation of living organisms or their components to produce useful products.

biotic (bī-ot´-tik) Pertaining to the living organisms in the environment.

biotic components All the organisms that are part of the environment.

bipolar cell Neurons that synapse with the axons of rods and cones in the retina of the eye.

birth control pills Chemical contraceptives that inhibit ovulation, retard follicular development, or alter a woman's cervical mucus to prevent sperm from entering the uterus.

bisexual flower A flower equipped with both stamens and carpels.

blade A leaflike structure of a seaweed that provides most of the surface area for photosynthesis.

blastocoel (blas´-tuh-sēl) The fluid-filled cavity that forms in the center of the blastula embryo.

blastocyst An embryonic stage in mammals; a hollow ball of cells produced one week after fertilization in humans.

blastodisc An embryonic cap of dividing cells resting on a large undivided yolk.

blastomeres Small cells of an early embryo.

blastopore (blas´-tō-pōr) The opening of the archenteron in the gastrula that develops into the mouth in protostomes and the anus in deuterostomes.

blastula (blas´-tyū-luh) The hollow ball of cells marking the end stage of cleavage during early embryonic development.

blood A type of connective tissue with a fluid matrix called plasma in which blood cells are suspended.

blood pressure The hydrostatic force that blood exerts against the wall of a vessel.

blood vessel A set of tubes through which the blood moves through the body.

blood-brain barrier A specialized capillary arrangement in the brain that restricts the passage of most substances into the brain, thereby preventing dramatic fluctuations in the brain's environment.

body cavity A fluid-containing space between the digestive tract and the body wall.

bolus A lubricated ball of chewed food.

bond energy The quantity of energy that must be absorbed to break a particular kind of chemical bond; equal to the quantity of energy the bond releases when it forms.

bone A type of connective tissue, consisting of living cells held in a rigid matrix of collagen fibers embedded in calcium salts.

book lungs Organs of gas exchange in spiders, consisting of stacked plates contained in an internal chamber.

bottleneck effect Genetic drift resulting from the reduction of a population, typically by a natural disas-

ter, such that the surviving population is no longer genetically representative of the original population.

bottom-up model A model of community organization in which mineral nutrients control community organization because nutrients control plant numbers, which in turn control herbivore numbers, which in turn control predator numbers.

Bowman's capsule (bō´-munz) A cup-shaped receptacle in the vertebrate kidney that is the initial, expanded segment of the nephron where filtrate enters from the blood.

brachiopod Also called lamp shells, these animals superficially resemble clams and other bivalve mollusks, but the two halves of the brachiopod shell are dorsal and ventral to the animal rather than lateral, as in clams.

brain hormone A hormone produced by neurosecretory cells in the insect brain. It promotes development by stimulating the prothoracic glands to secrete ecdysone.

brainstem The hindbrain and midbrain of the vertebrate central nervous system. In humans, it forms a cap on the anterior end of the spinal cord, extending to about the middle of the brain.

breathing control center A brain center that directs the activity of organs involved in breathing.

bronchioles Fine branches of the bronchus that transport air to alveoli.

bronchus (bron´-kus) (plural, **bronchi**) One of a pair of breathing tubes that branch from the trachea into the lungs.

brown algae One of a group of marine, multicellular, autotrophic protists, the most common type of seaweed. Brown algae include the kelps.

brown fat A special tissue in some mammals, located in the neck and between the shoulders, that is specialized for rapid heat production.

Bryophyta The phylum of mosses. Note that the term "bryophyte" refers instead to the informal group of mosses, liverworts, and hornworts.

bryophyte (brī-uh-fīt) A moss, liverwort, or hornwort; a nonvascular plant that inhabits the land but lacks many of the terrestrial adaptations of vascular plants.

bryozoan Colonial animals that superficially resemble mosses.

budding An asexual means of propagation in which outgrowths from the parent form and pinch off to live independently or else remain attached to eventually form extensive colonies.

budgets Used in reference to the gains and losses of various materials and energy. Most energy and materials budgets are interconnected, with changes in the flux of one component affecting the exchanges of other components.

buffer A substance that consists of acid and base forms in a solution and that minimizes changes in pH when extraneous acids or bases are added to the solution.

bulbourethral gland (bul´-bō-yū-rē´-thrul) One of a pair of glands near the base of the penis in the human male that secrete fluid that lubricates and neutralizes acids in the urethra during sexual arousal.

bulk flow The movement of water due to a difference in pressure between two locations.

bulk-feeder Animals that eat relatively large pieces of food.

bundle-sheath cell A type of photosynthetic cell arranged into tightly packed sheaths around the veins of a leaf.

C₃ plant A plant that uses the Calvin cycle for the initial steps that incorporate CO_2 into organic material, forming a three-carbon compound as the first stable intermediate.

C₄ plant A plant that prefaces the Calvin cycle with reactions that incorporate CO_2 into four-carbon compounds, the end product of which supplies CO_2 for the Calvin cycle.

cadherins An important class of cell-to-cell adhesion molecules.

calcitonin (kal´-si-tō´-nin) A mammalian thyroid hormone that lowers blood calcium levels.

callus A mass of dividing, undifferentiated cells at the cut end of a shoot.

calmodulin (kal-mod´-yū-lin) An intracellular protein to which calcium binds in its function as a second messenger in hormone action.

calorie (cal) The amount of heat energy required to raise the temperature of 1 g of water by 1°C; also the amount of heat energy that 1 g of water releases when it cools by 1°C. The Calorie (with a capital C), usually used to indicate the energy content of food, is a kilocalorie.

Calvin cycle The second of two major stages in photosynthesis (following the light reactions), involving atmospheric CO_2 fixation and reduction of the fixed carbon into carbohydrate.

CAM plant A plant that uses crassulacean acid metabolism, an adaptation for photosynthesis in arid conditions, first discovered in the family Crassulaceae. Carbon dioxide entering open stomata during the night is converted into organic acids, which release CO_2 for the Calvin cycle during the day, when stomata are closed.

Cambrian explosion A burst of evolutionary origins when most of the major body plans of animals appeared in a relatively brief time in geologic history; recorded in the fossil record about 545 to 525 million years ago.

cAMP receptor protein (CRP) A regulatory protein that directly stimulates gene expression.

canopy The uppermost layer of vegetation in a terrestrial biome.

capillary (kap´-il-ār-ē) A microscopic blood vessel that penetrates the tissues and consists of a single layer of endothelial cells that allows exchange between the blood and interstitial fluid.

capillary bed A network of capillaries that infiltrate every organ and tissue in the body.

capsid The protein shell that encloses a viral genome. It may be rod-shaped, polyhedral, or more complete in shape.

capsule A sticky layer that surrounds the cell walls of some bacteria, protecting the cell surface and sometimes helping to glue the cell to surfaces.

carbohydrate (kar´-bō-hī´-drāt) A sugar (monosaccharide) or one of its dimers (disaccharides) or polymers (polysaccharides).

carbon fixation The incorporation of carbon from CO_2 into an organic compound by an autotrophic organism (a plant, another photosynthetic organism, or a chemoautotrophic bacterium).

carbonyl group (kar´-buh-nēl´) A functional group present in aldehydes and ketones and consisting of a carbon atom double-bonded to an oxygen atom.

carboxyl group (kar-bok´-sil) A functional group present in organic acids and consisting of a single carbon atom double-bonded to an oxygen atom and also bonded to a hydroxyl group.

carboxylic acid (kar´-bok-sil´-ik) An organic compound containing a carboxyl group.

carboxypeptidase An enzyme found within the small intestine that splits off one amino acid at a time, beginning at the end of the polypeptide that has a free carboxyl group.

carcinogen (kar-sin´-uh-jin) A chemical agent that causes cancer.

cardiac cycle (kar´-dē-ak) The alternating contractions and relaxations of the heart.

cardiac muscle A type of muscle that forms the contractile wall of the heart; its cells are joined by intercalated discs that relay each heartbeat.

cardiac output The volume of blood pumped per minute by the left ventricle of the heart.

cardiovascular disease (kar´-dē-ō-vas´-kyū-ler) Diseases of the heart and blood vessels.

cardiovascular system A closed circulatory system with a heart and branching network of arteries, capillaries, and veins; the system is characteristic of vertebrates.

carnites The group of birds with a carina, or sternal keel, supporting their large breast muscles.

carnivore An animal, such as a shark, hawk, or spider, that eats other animals.

carotenoid (kuh-rot´-uh-noyd) An accessory pigment, either yellow or orange, in the chloroplasts of plants. By absorbing wavelengths of light that chlorophyll cannot, carotenoids broaden the spectrum of colors that can drive photosynthesis.

carpel (kar´-pul) The female reproductive organ of a flower, consisting of the stigma, style, and ovary.

carrier In human genetics, an individual who is heterozygous at a given genetic locus, with one normal allele and one potentially harmful recessive allele. The heterozygote is phenotypically normal for the character determined by the gene but can pass on the harmful allele to offspring.

carrying capacity The maximum population size that can be supported by the available resources, symbolized as K.

cartilage (kar´-til-ij) A type of flexible connective tissue with an abundance of collagenous fibers embedded in chondrin.

Casparian strip (kas-par´-ē-un) A water-impermeable ring of wax around endodermal cells in plants that blocks the passive flow of water and solutes into the stele by way of cell walls.

catabolic pathway (kat´-uh-bol´-ik) A metabolic pathway that releases energy by breaking down complex molecules to simpler compounds.

catabolite activator protein (CAP) (kuh-tab´-uh-līt) In *E. coli*, a helper protein that stimulates gene expression by binding within the promoter region of an operon and enhancing the promoter's ability to associate with RNA polymerase.

catalyst A chemical agent that changes the rate of a reaction without being consumed by the reaction.

catastrophism The hypothesis by Georges Cuvier that each boundary between strata corresponded in time to a catastrophe, such as a flood or drought, that had destroyed many of the species living there at that time.

catecholamines A class of compounds, including epinephrine and norepinephrine, that are synthesized from the amino acid tyrosine.

cation (kat´-ī-on) An ion with a positive charge, produced by the loss of one or more electrons.

cation exchange A process in which positively charged minerals are made available to a plant when hydrogen ions in the soil displace mineral ions from the clay particles.

CD4 A T cell surface protein, present on most helper T cells, CD4 binds to part of the class II MHC protein.

CD8 A T cell surface protein that enhances the interaction between the antigen-presenting infected cell and a cytotoxic T cell.

cDNA library A limited gene library using complementary DNA. The library includes only the genes that were transcribed in the cells examined.

cecum (sē´-kum) (plural, **ceca**) A blind outpocket of a hollow organ such as an intestine.

cell adhesion molecules Glycoproteins that contribute to cell migration and stable tissue structure.

cell body The part of a cell, such as a neuron, that houses the molecules.

cell center A region in the cytoplasm near the nucleus from which microtubules originate and radiate.

cell cycle An ordered sequence of events in the life of a eukaryotic cell, from its origin in the division of a parent cell until its own division into two; composed of the M, G₁, S, and G₂ phases.

cell cycle control system A cyclically operating set of molecules in the cell that triggers and coordinates key events in the cell cycle.

cell division The reproduction of cells.

cell fractionation The disruption of a cell and separation of its organelles by centrifugation.

cell lineage The ancestry of a cell.

cell plate A double membrane across the midline of a dividing plant cell, between which the new cell wall forms during cytokinesis.

cell wall A protective layer external to the plasma membrane in plant cells, bacteria, fungi, and some protists. In plant cells, the wall is formed of cellulose fibers embedded in a polysaccharide-protein matrix. The primary cell wall is thin and flexible, whereas the secondary cell wall is stronger and more rigid and is the primary constituent of wood.

cell-mediated immunity The type of immunity that functions in defense against fungi, protists, bacteria, and viruses inside host cells and against tissue transplants, with highly specialized cells that circulate in the blood and lymphoid tissue.

cellular differentiation The structural and functional divergence of cells as they become specialized during a multicellular organism's development; dependent on the control of gene expression.

cellular respiration The most prevalent and efficient catabolic pathway for the production of ATP, in which oxygen is consumed as a reactant along with the organic fuel.

cellular slime mold A type of protist that has unicellular amoeboid cells and multicellular reproductive bodies in its life cycle.

cellulose (sel´-yū-lōs) A structural polysaccharide of cell walls, consisting of glucose monomers joined by β-1, 4-glycosidic linkages.

Celsius scale (sel´-sē-us) A temperature scale (°C) equal to 5/9 (°F − 32) that measures the freezing point of water at 0°C and the boiling point of water at 100°C.

central canal The narrow cavity in the center of the spinal cord that is continuous with the fluid-filled ventricles of the brain.

central nervous system (CNS) In vertebrate animals, the brain and spinal cord.

central vacuole A membranous sac in a mature plant cell with diverse roles in reproduction, growth, and development.

centriole (sen´-trē-ōl) A structure in an animal cell composed of cylinders of microtubule triplets arranged in a 9 + 0 pattern. An animal cell usually has a pair of centrioles involved in cell division.

centromere (sen´-trō-mēr) The centralized region joining two sister chromatids.

centrosome (sen´-trō-sōm) Material present in the cytoplasm of all eukaryotic cells, important during cell division; the microtubule-organizing center.

cephalization (sef´-uh-luh-zā´-shun) An evolutionary trend toward the concentration of sensory equipment on the anterior end of the body.

cephalochordate (sef´-uh-lō-kōr´-dāt) A chordate without a backbone, represented by lancelets, tiny marine animals.

cerebellum (sār´-ruh-bel´-um) Part of the vertebrate hindbrain (rhombencephalon) located dorsally; functions in unconscious coordination of movement and balance.

cerebral cortex (suh-rē´-brul) The surface of the cerebrum; the largest and most complex part of the mammalian brain, containing sensory and motor nerve cell bodies of the cerebrum; the part of the vertebrate brain most changed through evolution.

cerebral hemisphere The right or left side of the vertebrate brain.

cerebrospinal fluid (suh-rē´-brō-spī´-nul) Blood-derived fluid that surrounds, protects, against infection, nourishes, and cushions the brain and spinal cord.

cerebrum (suh-rē´-brum) The dorsal portion, composed of right and left hemispheres, of the vertebrate forebrain; the integrating center for memory, learning, emotions, and other highly complex functions of the central nervous system.

cervix (ser´-viks) The neck of the uterus, which opens into the vagina.

chaparral (shap´-uh-ral´) A scrubland biome of dense, spiny evergreen shrubs found at midlatitudes along coasts where cold ocean currents circulate offshore; characterized by mild, rainy winters and long, hot, dry summers.

chaperonin Protein molecules that assist the proper folding of other proteins.

character A heritable feature.

Charophyceans (kār´-uh-fī´-sē-unz) The green algal group that shares two ultrastructural features with land plants. They are considered to be the closest relatives of land plants.

checkpoint A critical control point in the cell cycle where stop and go-ahead signals can regulate the cycle.

chelicerae (kuh-lis´-uh-rē) Clawlike feeding appendages characteristic of the chelicerate group.

Chelicerata (kuh-lis´-uh-rot´-uh) The animal phylum that includes horseshoe crabs, scorpions, ticks, spiders, and an extinct group called the eurypterids.

chelicerates Members of the animal phylum that includes horseshoe crabs, scorpions, ticks, spiders, and an extinct group called the eurypterids.

chemical bond An attraction between two atoms resulting from a sharing of outer-shell electrons or the presence of opposite charges on the atoms; the bonded atoms gain complete outer electron shells.

chemical energy Energy stored in the chemical bonds of molecules; a form of potential energy.

chemical equilibrium In a reversible chemical reaction, the point at which the rate of the forward reaction equals the rate of the reverse reaction.

chemical reaction A process leading to chemical changes in matter; involves the making and/or breaking of chemical bonds.

chemically-gated ion channels Specialized ion channels that open or close in response to a chemical stimulus.

chemiosmosis (kem´-ē-oz-mō´-sis) An energy-coupling mechanism that uses energy stored in the form of a hydrogen ion gradient across a membrane to drive cellular work, such as the synthesis of ATP. Most ATP synthesis in cells occurs by chemiosmosis.

chemoautotroph (kē-mō-ot´-ō-trōf) An organism that needs only carbon dioxide as a carbon source but that obtains energy by oxidizing inorganic substances.

chemoheterotroph (kē-mō-het´-er-ō-trōf) An organism that must consume organic molecules for both energy and carbon.

chemokine A group of about 50 different proteins secreted by blood vessel endothelial cells and monocytes. These molecules bind to receptors on many types of leukocytes and induce numerous changes central to inflammation.

chemoreceptor A receptor that transmits information about the total solute concentration in a solution or about individual kinds of molecules.

chiasma (plural, **chiasmata**) (kī-az´-muh, kī-az´-muh-tuh) The X-shaped, microscopically visible region representing homologous chromatids that have exchanged genetic material through crossing over during meiosis.

Chilopoda (kī-lop´-uh-duh)The animal class that includes centipedes.

chitin (kī-tin) A structural polysaccharide of an amino sugar found in many fungi and in the exoskeletons of all arthropods.

chlorophyll (klōr´-ō-fil) A green pigment located within the chloroplasts of plants. Chlorophyll *a* can participate directly in the light reactions, which convert solar energy to chemical energy.

chlorophyll *a* A type of blue-green photosynthetic pigment that participates directly in the light reactions.

chlorophyll *b* A type of yellow-green accessory photosynthetic pigment that transfers energy to chlorophyll *a*.

chloroplast (klōr´-ō-plast) An organelle found only in plants and photosynthetic protists that absorbs sunlight and uses it to drive the synthesis of organic compounds from carbon dioxide and water.

choanocyte (kō-an´-uh-sīt) A flagellated feeding cell found in sponges. Also called a collar cell, it has a collarlike ring that traps food particles around the base of its flagellum.

cholecystokinin (CCK) (kō´-luh-sis´-tuh-kī´-nin) A hormone released from the walls of the duodenum in response to the presence of amino acids or fatty acids.

cholesterol (kō-les´-tuh-rol) A steroid that forms an essential component of animal cell membranes and acts as a precursor molecule for the synthesis of other biologically important steroids.

Chondrichthyes (kon-drik´-thēz) The vertebrate class of cartilaginous fishes, represented by sharks and their relatives.

chondrin A protein-carbohydrate complex secreted by chondrocytes; chondrin and collagen fibers form cartilage.

chondrocytes Cartilage cells.

chordate (kōr´-dāt) A member of a diverse phylum of animals that possess a notochord; a dorsal, hollow nerve cord; pharyngeal gill slits; and a postanal tail as an embryo.

chorion (kōr´-ē-on) The outermost of four extra-embryonic membranes; contributes to the formation of the mammalian placenta.

chorionic villus sampling (CVS) (kōr´-ē-on´-ik vil´-us) A technique for diagnosing genetic and congenital defects in a fetus by removing and analyzing a small sample of the fetal portion of the placenta.

choroid A thin, pigmented inner layer of the vertebrate eye.

chromatin (krō´-muh-tin) The complex of DNA and proteins that makes up a eukaryotic chromosome. When the cell is not dividing, chromatin exists as a mass of very long, thin fibers that are not visible with a light microscope.

chromista In some classification systems, a kingdom consisting of brown algae, golden algae, and diatoms.

chromosome (krō´-muh-sōm) A threadlike, gene-carrying structure found in the nucleus. Each chromosome consists of one very long DNA molecule and associated proteins. *See* chromatin.

chromosome theory of inheritance A basic principle in biology stating that genes are located on chromosomes and that the behavior of chromosomes during meiosis accounts for inheritance patterns.

chromosome walking A DNA mapping technique that begins with a gene or other sequence that has already been cloned, mapped, and sequenced and "walks" along the chromosomal DNA from that locus, producing a map of overlapping restriction fragments.

chylomicron (kī-lō-mī´-kron) Small intracellular globules composed of fats that are mixed with cholesterol and coated with special proteins.

chymotrypsin (kī´-muh-trip´-sin) An enzyme found in the duodenum. It is specific for peptide bonds adjacent to certain amino acids.

chytrid (kī´-trid) Mainly aquatic primitive fungi that form uniflagellated spores (zoospores). The chytrids and fungi are now thought to form a monophyletic branch of the eukaryotic tree.

ciliary body A portion of the vertebrate eye associated with the lens. It produces the clear, watery aqueous humor that fills the anterior cavity of the eye.

ciliate (sil´-ē-it) A type of protozoan that moves by means of cilia.

cilium (sil´-ē-um) (plural, **cilia**) A short cellular appendage specialized for locomotion, formed from a core of nine outer doublet microtubules and two inner single microtubules ensheathed in an extension of plasma membrane.

circadian rhythm (ser-kā´-dē-un) A physiological cycle of about 24 hours that is present in all eukaryotic organisms and that persists even in the absence of external cues.

clade Each evolutionary branch in a cladogram.

cladistics (kluh-dis´-tiks) A taxonomic approach that classifies organisms according to the order in time at which branches arise along a phylogenetic tree, without considering the degree of morphological divergence.

cladogenesis (klā´-dō-jen´-uh-sis) A pattern of evolutionary change that produces biological diversity by budding one or more new species from a parent species that continues to exist; also called branching evolution.

cladogram A dichotomous phylogenetic tree that branches repeatedly, suggesting a classification of organisms based on the time sequence in which evolutionary branches arise.

class In classification, the taxonomic category above order.

class Arachnida The animal group that includes scorpions, spiders, ticks, and mites.

class Chilopoda The centipede group of animals.

class Diplopoda The millipede group of animals.

class I MHC molecules A collection of cell surface glycoproteins encoded by a family of genes called the major histocompatibility complex. In humans, these glycoproteins are also known as the HLA, human leukocyte antigens. Class I MHC molecules are found on all nucleated cells.

class II MHC molecules A collection of cell surface glycoproteins encoded by a family of genes called the major histocompatibility complex. In humans, these glycoproteins are also known as the HLA, human leukocyte antigens. Class II MHC molecules are restricted to a few specialized cell types.

classical conditioning A type of associative learning; the association of a normally irrelevant stimulus with a fixed behavioral response.

cleavage The process of cytokinesis in animal cells, characterized by pinching of the plasma membrane; specifically, the succession of rapid cell divisions without growth during early embryonic development that converts the zygote into a ball of cells.

cleavage furrow The first sign of cleavage in an animal cell; a shallow groove in the cell surface near the old metaphase plate.

climate The prevailing weather conditions at a locality.

cline Graded variation in some traits of individuals that parallels a gradient in the environment.

clitoris (klit´-uh-ris) An organ in the female that engorges with blood and becomes erect during sexual arousal.

cloaca (klō-ā´-kuh) A common opening for the digestive, urinary, and reproductive tracts in all vertebrates except most mammals.

clonal selection The mechanism that determines specificity and accounts for antigen memory in the immune system; occurs because an antigen introduced into the body selectively activates only a tiny fraction of inactive lymphocytes, which proliferate to form a clone of effector cells specific for the stimulating antigen.

clone (1) A lineage of genetically identical individuals or cells. (2) In popular usage, a single individual organism that is genetically identical to another individual. (3) As a verb, to make one or more genetic replicas of an individual or cell. *See also* gene cloning.

cloning Using a somatic cell from a multicellular organism to make one or more genetically identical individuals.

cloning vector An agent used to transfer DNA in genetic engineering. A plasmid that moves recombinant DNA from a test tube back into a cell is an example of a cloning vector, as is a virus that transfers recombinant DNA by infection.

closed circulatory system Circulatory systems in which blood is confined to vessels and is kept separate from the interstitial fluid.

club fungus The common name for members of the phylum Basidiomycota. The name comes from the clublike shape of the basidium.

clumped Describing a dispersion pattern in which individuals are aggregate in patches.

cnidocyte (nī´-duh-sīt) A specialized cell for which the phylum Cnidaria is named; consists of a capsule containing a fine coiled thread, which, when discharged, functions in defense and prey capture.

cochlea (kok´-lē-uh) The complex, coiled organ of hearing that contains the organ of Corti.

codominance A phenotypic situation in which the two alleles affect the phenotype in separate, distinguishable ways.

codon (kō´-don) A three-nucleotide sequence of DNA or mRNA that specifies a particular amino acid or termination signal; the basic unit of the genetic code.

coefficient of relatedness The probability that a particular gene present in one individual will also be inherited from a common parent or ancestor in a second individual.

coelom (sē´-lōm) A body cavity completely lined with mesoderm.

coelomate (sē´-lō-māt) An animal whose body cavity is completely lined by mesoderm, the layers of which connect dorsally and ventrally to form mesenteries.

coenocytic (sē´-nō-sit´-ik) Referring to a multinucleated condition resulting from the repeated division of nuclei without cytoplasmic division.

coenzyme (kō-en´-zīm) An organic molecule serving as a cofactor. Most vitamins function as coenzymes in important metabolic reactions.

coevolution The mutual influence on the evolution of two different species interacting with each other and reciprocally influencing each other's adaptations.

cofactor Any nonprotein molecule or ion that is required for the proper functioning of an enzyme. Cofactors can be permanently bound to the active site or may bind loosely with the substrate during catalysis.

cognition The ability of an animal's nervous system to perceive, store, process, and use information obtained by its sensory receptors.

cognitive ethology The scientific study of cognition; the study of the connection between data processing by nervous systems and animal behavior.

cognitive map A representation within the nervous system of spatial relations among objects in an animal's environment.

cohesion The binding together of like molecules, often by hydrogen bonds.

cohesion species concept The idea that specific evolutionary adaptations and discrete complexes of genes define species.

cohort A group of individuals of the same age, from birth until all are dead.

coitus (kō´-uh-tus) The insertion of a penis into a vagina, also called sexual intercourse.

coleoptile (kō´-lē-op´-tul) The covering of the young shoot of the embryo of a grass seed.

coleorhiza (kō´-lē-uh-rī´-zuh) The covering of the young root of the embryo of a grass seed.

collagen A glycoprotein in the extracellular matrix of animal cells that forms strong fibers, found extensively in connective tissue and bone; the most abundant protein in the animal kingdom.

collagenous fibers Tough fibers of the extracellular matrix. They are made of collagen that are nonelastic and do not tear easily when pulled lengthwise.

collecting duct The location in the kidney where filtrate from renal tubules is collected; the filtrate is now called urine.

collenchyma cell (kō-len´-kim-uh) A flexible plant cell type that occurs in strands or cylinders that support young parts of the plant without restraining growth.

colloblasts Adhesive structures on the tentacles of ctenophores.

colon (kō´-len) The tubular portion of the vertebrate alimentary tract between the small intestine and the anus; functions mainly in water absorption and the formation of feces.

columnar The column shape of one type of epithelial cell.

commensalism (kuh-men´-suh-lizm) A symbiotic relationship in which the symbiont benefits but the host is neither helped nor harmed.

community All the organisms that inhabit a particular area; an assemblage of populations of different species living close enough together for potential interaction.

community ecology The study of how interactions between species affect community structure and organization.

companion cell A type of plant cell that is connected to a sieve-tube cell by many plasmodesmata and whose nucleus and ribosomes may serve one or more adjacent sieve-tube cells.

competitive exclusion principle The concept that when populations of two similar species compete for the same limited resources, one population will use the resources more efficiently and have a reproductive advantage that will eventually lead to the elimination of the other population.

competitive inhibitor A substance that reduces the activity of an enzyme by entering the active site in place of the substrate whose structure it mimics.

complement A set of about 20 serum proteins that carry out a cascade of steps leading to the lysis of microbes.

complement fixation An immune response in which antigen-antibody complexes activate complement proteins.

complement system A group of at least 20 blood proteins that cooperate with other defense mechanisms; may amplify the inflammatory response, enhance phagocytosis, or directly lyse pathogens; activated by the onset of the immune response or by surface antigens on microorganisms or other foreign cells.

complementary DNA (cDNA) A DNA molecule made *in vitro* using mRNA as a template and the enzyme reverse transcriptase. A cDNA molecule therefore corresponds to a gene, but lacks the introns present in the DNA of the genome.

complete digestive tract A digestive tube that runs between a mouth and an anus; also called alimentary canal. An incomplete digestive tract has only one opening.

complete dominance A type of inheritance in which the phenotypes of the heterozygote and dominant homozygote are indistinguishable.

complete flower A flower that has sepals, petals, stamens, and carpels.

complete metamorphosis The transformation of a larva into an adult that looks very different, and often functions very differently in its environment, than the larva.

compound A substance consisting of two or more elements in a fixed ratio.

compound eye A type of multifaceted eye in insects and crustaceans consisting of up to several thousand light-detecting, focusing ommatidia; especially good at detecting movement.

compressor A small molecule that cooperates with a repressor protein to switch an operon off.

concentration gradient An increase or decrease in the density of a chemical substance in an area. Cells often maintain concentration gradients of ions across their membranes. When a gradient exists, the ions or other chemical substances involved tend to move from where they are more concentrated to where they are less concentrated.

conception The fertilization of the egg by a sperm cell.

condensation reaction A reaction in which two molecules become covalently bonded to each other through the loss of a small molecule, usually water; also called dehydration reaction.

conduction The direct transfer of thermal motion (heat) between molecules of objects in direct contact with each other.

cone cell One of two types of photoreceptors in the vertebrate eye; detects color during the day.

conformer A characterization of an animal in regard to environmental variables. The animal is a conformer if it allows some conditions within its body to vary with certain external changes.

conidium (plural, **conidia**) A naked, asexual spore produced at the ends of hyphae in ascomycetes.

conifer A gymnosperm whose reproductive structure is the cone. Conifers include pines, firs, redwoods, and other large trees.

Coniferophyta (kuh-nif´-er-uh-fī´-tuh) The largest of the four gymnosperm phyla, the reproductive structure is the cone. Conifers include pines, firs, redwoods, and other large trees.

conjugation (kon´-jū-gā-shun) In bacteria, the direct transfer of DNA between two cells that are temporarily joined.

conjunctiva (kon´-junk-tī´-vuh) A mucous membrane that helps keep the eye moist; lines the inner surface of the eyelid and covers the front of the eyeball, except the cornea.

connective tissue Animal tissue that functions mainly to bind and support other tissues, having a sparse population of cells scattered through an extracellular matrix.

conodonts (kō´-nuh-donts) The group of ancient vertebrates that date back as far as 510 million years.

conservation biology A goal-oriented science that seeks to counter the biodiversity crisis, the current rapid decrease in Earth's variety of life.

contraception The prevention of pregnancy.

contractile vacuole A membranous sac that helps move excess water out of the cell.

control elements Segments of noncoding DNA that help regulate transcription of a gene by binding proteins called transcription factors.

convection The mass movement of warmed air or liquid to or from the surface of a body or object.

convergent evolution The independent development of similarity between species as a result of their having similar ecological roles and selection pressures.

convergent extension A mechanism of cell crawling in which the cells of a tissue layer rearrange themselves in such a way that the sheet of cells becomes narrower while it becomes longer.

cooperativity An interaction of the constituent subunits of a protein whereby a conformational change in one subunit is transmitted to all the others.

copepods (kō´-puh-podz) A group of small crustaceans that are important members of marine and freshwater plankton communities.

coral reefs Warm water, tropical, ecosystems dominated by the hard skeletal structures secreted primarily by the resident cnidarians.

cork cambium (kam´-bē-um) A cylinder of meristematic tissue in plants that produces cork cells to replace the epidermis during secondary growth.

cornea (kor´-nē-uh) The transparent frontal portion of the sclera, which admits light into the vertebrate eye.

corpus callosum (kor´-pus kuh-lō-sum) The thick band of nerve fibers that connect the right and left cerebral hemispheres in placental mammals, enabling the hemispheres to process information together.

corpus luteum (kor´-pus lū-tē-um) A secreting tissue in the ovary that forms from the collapsed follicle after ovulation and produces progesterone.

cortex Ground tissue that is between the vascular and dermal tissue in a root or dicot stem.

cortical granules Vesicles that begin just under the egg plasma membrane prior to their involvement in the cortical reaction.

cortical nephrons Nephrons located almost entirely in the renal cortex. These nephrons have a reduced loop of Henle.

cortical reaction A series of changes in the cortex of the egg cytoplasm during fertilization.

corticosteroids A family of steroids synthesized by and released from the adrenal cortex.

cotransport The coupling of the "downhill" diffusion of one substance to the "uphill" transport of another against its own concentration gradient.

cotyledons (kot´-uh-lē´-donz) The one (monocot) or two (dicot) seed leaves of an angiosperm embryo.

countercurrent exchange The opposite flow of adjacent fluids that maximizes transfer rates; for example, blood in the gills flows in the opposite direction in which water passes over the gills, maximizing oxygen uptake and carbon dioxide loss.

countercurrent heat exchanger A special arrangement of blood vessels that helps trap heat in the body core and is important in reducing heat loss in many endotherms.

courtship Behavior patterns that lead up to copulation or gamete release.

covalent bond (kō-vā´-lent) A type of strong chemical bond in which two atoms share one pair of valence electrons.

cranial nerves Nerves that leave the brain and innervate organs of the head and upper body.

craniata The chordate subgroup that possess a cranium.

crassulacean acid metabolism (CAM) A type of metabolism in which carbon dioxide is taken in at night and incorporated into a variety of organic acids.

crista (plural, **cristae**) (kris´-tuh, kris´-tē) An infolding of the inner membrane of a mitochondrion that houses the electron transport chain and the enzyme catalyzing the synthesis of ATP.

crocodilia The reptile group that includes crocodiles and alligators.

crossing over The reciprocal exchange of genetic material between nonsister chromatids during synapsis of meiosis I.

cross-pollination The transfer of pollen from flowers of one plant to flowers of another plant of the same species.

Crustacea (kruh-stā-shuh) The animal phylum that includes mostly aquatic animals such as crabs, lobsters, crayfish, and shrimp.

crustacean A member of a major arthropod phylum that includes lobsters, crayfish, crabs, shrimps, and barnacles.

cryptic coloration Camouflage, making potential prey difficult to spot against its background.

cuboidal The cubic shape of a type of epithelial cell.

cuticle (kyū´-tuh-kul) (1) A waxy covering on the surface of stems and leaves that acts as an adaptation to prevent desiccation in terrestrial plants. (2) The exoskeleton of an arthropod, consisting of layers of protein and chitin that are variously modified for different functions.

cyanobacteria (sī-an´-ō-bak-tēr´-ē-uh) Photosynthetic, oxygen-producing bacteria (formerly known as blue-green algae).

Cycadophyta (sī-kuh-dof´-uh-duh) A phylum of gymnosperms that superficially resemble palms. Cycads bear naked seeds on sporophylls, leaves specialized for reproduction.

cyclic AMP (cAMP) Cyclic adenosine monophosphate, a ring-shaped molecule made from ATP that is a common intracellular signaling molecule (second messenger) in eukaryotic cells (for example, in vertebrate endocrine cells). It is also a regulator of some bacterial operons.

cyclic electron flow A route of electron flow during the light reactions of photosynthesis that involves only photosystem I and that produces ATP but not NADPH or oxygen.

cyclic photophosphorylation The generation of ATP by cyclic electron flow.

cyclin (sī´-klin) A regulatory protein whose concentration fluctuates cyclically.

cyclin-dependent kinase (Cdk) A protein kinase that is active only when attached to a particular cyclin.

cystic fibrosis A genetic disorder that occurs in people with two copies of a certain recessive allele; characterized by an excessive secretion of mucus and consequent vulnerability to infection; fatal if untreated.

cytochrome (sī´-tō-krōm) An iron-containing protein, a component of electron transport chains in mitochondria and chloroplasts.

cytokines In the vertebrate immune system, protein factors secreted by macrophages and helper T cells as regulators of neighboring cells.

cytokinesis (sī´-tō-kuh-nē´-sis) The division of the cytoplasm to form two separate daughter cells immediately after mitosis.

cytokinins (sī´-tō-kī´-nins) A class of related plant hormones that retard aging and act in concert with auxins to stimulate cell division, influence the pathway of differentiation, and control apical dominance.

cytological maps Charts of chromosomes that locate genes with respect to chromosomal features.

cytoplasm (sī´-tō-plaz´-um) The entire contents of the cell, exclusive of the nucleus, and bounded by the plasma membrane.

cytoplasmic determinants The maternal substances in the egg that influences the course of early development by regulating the expression of genes that affect the developmental fate of cells.

cytoplasmic streaming A circular flow of cytoplasm, involving myosin and actin filaments, that speeds the distribution of materials within cells.

cytoskeleton A network of microtubules, microfilaments, and intermediate filaments that branch throughout the cytoplasm and serve a variety of mechanical and transport functions.

cytosol (sī´-tō-sol) The semifluid portion of the cytoplasm.

cytotoxic T cell (T$_C$) A type of lymphocyte that kills infected cells and cancer cells.

daily torpor A daily decrease in metabolic activity and corresponding body temperature during times of inactivity for some small mammals and birds. The physiological changes during resting periods enable these organisms to survive on energy stores in their tissues.

dalton A measure of mass for atoms and subatomic particles.

Darwinian fitness The contribution an individual makes to the gene pool of the next generation, relative to the contributions of other individuals.

day-neutral plant A plant whose flowering is not affected by photoperiod.

decapod A relatively large group of crustaceans that includes lobsters, crayfish, crabs, and shrimp.

decomposers Any of the saprotrophic fungi and bacteria that absorb nutrients from nonliving organic material such as corpses, fallen plant material, and the wastes of living organisms, and convert them into inorganic forms.

deep green An international initiative focusing on the deepest phylogenetic branching within the plant kingdom to identify and name the major plant clades.

deep-sea hydrothermal vents A dark, hot, oxygen-deficient environment associated with volcanic activity. The food producers are chemoautotrophic prokaryotes.

dehydration reaction A chemical reaction in which two molecules covalently bond to each other with the removal of a water molecule.

deletion (1) A deficiency in a chromosome resulting from the loss of a fragment through breakage. (2) A mutational loss of one or more nucleotide pairs from a gene.

demographic transition A shift from zero population growth in which birth rates and death rates are high to zero population growth characterized instead by low birth and death rates.

demography The study of statistics relating to births and deaths in populations.

denaturation (dē-nā´-chur-ā´-shun) For proteins, a process in which a protein unravels and loses its native conformation, thereby becoming biologically inactive. For DNA, the separation of the two strands of the double helix. Denaturation occurs under extreme conditions of pH, salt concentration, and temperature.

dendrite (den´-drīt) One of usually numerous, short, highly branched processes of a neuron that conveys nerve impulses toward the cell body.

density The number of individuals per unit area or volume.

density dependent Any characteristic that varies according to an increase in population density.

density-dependent factor Any factor that has a greater impact on a population as the population density increases.

density-dependent inhibition The phenomenon observed in normal animal cells that causes them to stop dividing when they come into contact with one another.

density-independent factor Any factor that affects a population by the same percentage, regardless of density.

deoxyribonucleic acid (DNA) (dē-ok´-sē-rī´-bō-nū-klā´-ik) A double-stranded, helical nucleic acid molecule capable of replicating and determining the inherited structure of a cell's proteins.

deoxyribose The sugar component of DNA, having one less hydroxyl group than ribose, the sugar component of RNA.

depolarization An electrical state in an excitable cell whereby the inside of the cell is made less negative relative to the outside than at the resting membrane potential. A neuron membrane is depolarized if a stimulus decreases its voltage from the resting potential of -70 mV in the direction of zero voltage.

depolarized The condition of a membrane that is more negatively charged on one side than on the other.

deposit-feeder A heterotroph, such as an earthworm, that eats its way through detritus, salvaging bits and pieces of decaying organic matter.

dermal tissue system The protective covering of plants; generally a single layer of tightly packed epidermal cells covering young plant organs formed by primary growth.

descent with modification Darwin's initial phrase for the general process of evolution.

desmosome (dez´-muh-sōm) A type of intercellular junction in animal cells that functions as an anchor.

determinate cleavage A type of embryonic development in protostomes that rigidly casts the developmental fate of each embryonic cell very early.

determinate growth A type of growth characteristic of animals, in which the organism stops growing after it reaches a certain size.

determination The progressive restriction of developmental potential, causing the possible fate of each cell to become more limited as the embryo develops.

detritivore A consumer that derives its energy from nonliving organic material.

detritus (di-trī´-tus) Dead organic matter.

deuterostomes (dū´-ter-ō-stōmz) One of two distinct evolutionary lines of coelomates, consisting of the echinoderms and chordates and characterized by radial, indeterminate cleavage, enterocoelous formation of the coelom, and development of the anus from the blastopore.

development The sum of all of the changes that progressively elaborate an organism's body.

diacylglycerol (DAG) A second messenger produced by the cleavage of a certain kind of phospholipid in the plasma membrane.

diaphragm (1) A sheet of muscle that forms the bottom wall of the thoracic cavity in mammals; active in ventilating the lungs. (2) A dome-shaped rubber cup fitted into the upper portion of the vagina before sexual intercourse. It serves as a physical barrier to block the passage of sperm.

diapsids One of three groups of amniotes based on key differences between their skulls.

diastole (dī-as´-tō-lē) The stage of the heart cycle in which the heart muscle is relaxed, allowing the chambers to fill with blood.

diastolic pressure Blood pressure that remains between heart contractions.

diatom (dī´-uh-tom) A unicellular photosynthetic alga with a unique, glassy cell wall containing silica.

dicot (dī´-kot) A subdivision of flowering plants whose members possess two embryonic seed leaves, or cotyledons.

differentiation See cellular differentiation.

diffusion The spontaneous tendency of a substance to move down its concentration gradient from a more concentrated to a less concentrated area.

digestion The process of breaking down food into molecules small enough for the body to absorb.

dihybrid (dī´-hī´-brid) An organism that is heterozygous with respect to two genes of interest. A dihybrid results from a cross between parents doubly homozygous for different alleles. For example, parents of genotype *AABB* and *aabb* produce a dihybrid of genotype *AaBb*.

dikaryon (dī-kār´-ē-on) A mycelium of certain septate fungi that possesses two separate haploid nuclei per cell.

dikaryotic A mycelium with two haploid nuclei per cell, one from each parent.

dinoflagellate (dī´-nō-flaj´-uh-let) A unicellular photosynthetic alga with two flagella situated in perpendicular grooves in cellulose plates covering the cell.

dinosaurs An extremely diverse group of ancient reptiles varying in body shape, size, and habitat.

dioecious (dī-ē´-shus) Referring to a plant species that has staminate and carpellate flowers on separate plants.

dipeptidase An enzyme found attached to the intestinal lining. It splits small peptides.

diploblastic Having two germ layers.

diploid cell (dip´-loid) A cell containing two sets of chromosomes ($2n$), one set inherited from each parent.

Diplopoda (duh-plop´-uh-duh) The animal class that includes millipedes.

Dipnoi The class of lungfishes.

directional selection Natural selection that favors individuals at one end of the phenotypic range.

disaccharide (dī-sak´-uh-rīd) A double sugar, consisting of two monosaccharides joined by dehydration synthesis.

dispersal The distribution of individuals within geographic population boundaries.

dispersion The pattern of spacing among individuals within geographic population boundaries.

dissociation curve A chart showing the relative amounts of oxygen bound to hemoglobin when the pigment is exposed to solutions varying in their partial pressure of dissolved oxygen.

distal tubule In the vertebrate kidney, the portion of a nephron that helps refine filtrate and empties it into a collecting duct.

disturbance A force that changes a biological community and usually removes organisms from it. Disturbances, such as fire and storms, play pivotal roles in structuring many biological communities.

disulfide bridge Strong covalent bonds formed when the sulfur of one cysteine monomer bonds to the sulfur of another cysteine monomer.

diversifying selection Natural selection that favors extreme over intermediate phenotypes.

DNA fingerprint An individual's unique collection of DNA restriction fragments, detected by electrophoresis and nucleic acid probes.

DNA ligase (lī´-gās) A linking enzyme essential for DNA replication; catalyzes the covalent bonding of the 3' end of a new DNA fragment to the 5' end of a growing chain.

DNA methylation The addition of methyl groups ($-CH_3$) to bases of DNA after DNA synthesis; may serve as a long-term control of gene expression.

DNA microarray assays A method to detect and measure the expression of thousands of genes at one time. Tiny amounts of a large number of single-stranded DNA fragments representing different genes are fixed to a glass slide. These fragments, ideally representing all the genes of an organism, are tested for hybridization with various samples of cDNA molecules.

DNA polymerase (puh-lim´-er-ās) An enzyme that catalyzes the elongation of new DNA at a replication fork by the addition of nucleotides to the existing chain.

DNA probe A chemically synthesized, radioactively labeled segment of nucleic acid used to find a gene of interest by hydrogen-bonding to a complementary sequence.

DNA-binding domain A part of the three-dimensional structure of a transcription factor that binds to DNA.

domain A taxonomic category above the kingdom level. The three domains are Archaea, Bacteria, and Eukarya.

dominance hierarchy A linear "pecking order" of animals, where position dictates characteristic social behaviors.

dominant allele In a heterozygote, the allele that is fully expressed in the phenotype.

dominant species Those species in a community that have the highest abundance or highest biomass. These species exert a powerful control over the occurrence and distribution of other species.

dopamine A biogenic amine closely related to epinephrine and norepinephrine.

dormancy A condition typified by extremely low metabolic rate and a suspension of growth and development.

dorsal Pertaining to the back of a bilaterally symmetrical animal.

dorsal lip The dorsal side of the blastopore.

double circulation A circulation scheme with separate pulmonary and systemic circuits, which ensures vigorous blood flow to all organs.

double covalent bond A type of covalent bond in which two atoms share two pairs of electrons; symbolized by a pair of lines between the bonded atoms.

double fertilization A mechanism of fertilization in angiosperms, in which two sperm cells unite with two cells in the embryo sac to form the zygote and endosperm.

double helix The form of native DNA, referring to its two adjacent polynucleotide strands wound into a spiral shape.

Down syndrome A human genetic disease resulting from having an extra chromosome 21, characterized by mental retardation and heart and respiratory defects.

downy mildews Members of the group Oomycota, they are heterotrophic stramenopiles that lack chloroplasts, typically have cell walls made of cellulose, and generally live on land as parasites of plants.

Duchenne muscular dystrophy (duh-shen') A human genetic disease caused by a sex-linked recessive allele; characterized by progressive weakening and a loss of muscle tissue.

duodenum (dū-ō-dē'-num) The first section of the small intestine, where acid chyme from the stomach mixes with digestive juices from the pancreas, liver, gallbladder, and gland cells of the intestinal wall.

duplication An aberration in chromosome structure resulting from an error in meiosis or mutagens; duplication of a portion of a chromosome resulting from fusion with a fragment from a homologous chromosome.

dynein (dī'-nin) A large contractile protein forming the side-arms of microtubule doublets in cilia and flagella.

E site One of three binding sites for tRNA during translation, it is the place where discharged tRNAs leave the ribosome; E stands for exit site.

ecdysone (ek'-duh-sōn) A steroid hormone that triggers molting in arthropods.

ecdysozoa One of two distinct clades within the protostomes. It includes the arthropods.

echinoderms (uh-kī'-nō-derm) Sessile or slow-moving animals that include sea stars, sea urchins, brittle stars, crinoids, and basket stars.

ecological efficiency The ratio of net productivity at one trophic level to net productivity at the next lower level.

ecological footprint A method to use multiple constraints to estimate the human carrying capacity of Earth by calculating the aggregate land and water area in various ecosystem categories that is appropriated by a nation to produce all the resources it consumes and to absorb all the waste it generates.

ecological niche (nich) The sum total of a species' use of the biotic and abiotic resources in its environment.

ecological species concept The idea that ecological roles (niches) define species.

ecological succession Transition in the species composition of a biological community, often following ecological disturbance of the community; the establishment of a biological community in an area virtually barren of life.

ecology The study of how organisms interact with their environments.

ecosystem All the organisms in a given area as well as the abiotic factors with which they interact; a community and its physical environment.

ecosystem ecology The study of energy flow and the cycling of chemicals among the various biotic and abiotic factors in an ecosystem.

ectoderm (ek'-tō-derm) The outermost of the three primary germ layers in animal embryos; gives rise to the outer covering and, in some phyla, the nervous system, inner ear, and lens of the eye.

ectomycorrhizae (ek'-tō-mī'-kō-rī'-zē) A type of mycorrhizae in which the mycelium forms a dense sheath, or mantle, over the surface of the root. Hyphae extend from the mantle into the soil, greatly increasing the surface area for water and mineral absorption.

ectoparasites Parasites that feed on the external surface of a host.

ectotherm (ek'-tō-therm) An animal, such as a reptile, fish, or amphibian, that must use environmental energy and behavioral adaptations to regulate its body temperature.

ectothermic Organisms that do not produce enough metabolic heat to have much effect on body temperature.

Ediacaran period The last period of the Precambrian era.

effector cell A muscle cell or gland cell that performs the body's responses to stimuli; responds to signals from the brain or other processing center of the nervous system.

efferent arteriole The blood vessel draining a nephron.

egg-polarity genes Another name for maternal effect genes, these genes control the orientation (polarity) of the egg.

ejaculatory duct The short section of the ejaculatory route in mammals formed by the convergence of the vas deferens and a duct from the seminal vesicle. The ejaculatory duct transports sperm from the vas deferens to the urethra.

elastic fibers Long threads made of the protein elastin. Elastic fibers provide a rubbery quality to the extracellular matrix that complements the nonelastic strength of collagenous fibers.

electrocardiogram (ECG or EKG) A record of the electrical impulses that travel through cardiac muscle during the heart cycle.

electrochemical gradient The diffusion gradient of an ion, representing a type of potential energy that accounts for both the concentration difference of the ion across a membrane and its tendency to move relative to the membrane potential.

electroencephalogram (EEG) A medical test that measures different patterns in the electrical activity of the brain.

electrogenic pump An ion transport protein generating voltage across the membrane.

electromagnetic receptor Receptors of electromagnetic energy, such as visible light, electricity, and magnetism.

electromagnetic spectrum The entire spectrum of radiation ranging in wavelength from less than a nanometer to more than a kilometer.

electron A subatomic particle with a single negative charge; one or more electrons move around the nucleus of an atom.

electron microscope (EM) A microscope that focuses an electron beam through a specimen, resulting in resolving power a thousandfold greater than that of a light microscope. A transmission electron microscope (TEM) is used to study the internal structure of thin sections of cells. A scanning electron microscope (SEM) is used to study the fine details of cell surfaces.

electron shell An energy level representing the distance of an electron from the nucleus of an atom.

electron transport chain A sequence of electron carrier molecules (membrane proteins) that shuttle electrons during the redox reactions that release energy used to make ATP.

electronegativity The attraction of an atom for the electrons of a covalent bond.

electroporation A technique to introduce recombinant DNA into cells by applying a brief electrical pulse to a solution containing cells. The electricity creates temporary holes in the cells' plasma membranes, through which DNA can enter.

element Any substance that cannot be broken down to any other substance.

elimination The passing of undigested material out of the digestive compartment.

embryo New developing individuals.

embryo sac (em'-brē-ō) The female gametophyte of angiosperms, formed from the growth and division of the megaspore into a multicellular structure with eight haploid nuclei.

embryonic lethals Mutations with phenotypes leading to death at the embryo or larval stage.

embryophyte Another name for land plants, recognizing that land plants share the common derived trait of multicellular, dependent embryos.

emulsification The process that keeps tiny fat droplets from coalescing.

enantiomers (en-an'-tē-ō-mer) Molecules that are mirror images of each other.

endangered species A species that is in danger of extinction throughout all or a significant portion of its range.

endemic species (en-dem'-ik) Species that are confined to a specific, relatively small geographic area.

endergonic reaction (en'-der-gon'-ik) A nonspontaneous chemical reaction in which free energy is absorbed from the surroundings.

endocrine gland (en'-dō-krin) A ductless gland that secretes hormones directly into the bloodstream.

endocrine system The internal system of chemical communication involving hormones, the ductless glands that secrete hormones, and the molecular receptors on or in target cells that respond to hormones; functions in concert with the nervous system to effect internal regulation and maintain homeostasis.

endocytosis (en'-dō-sī-tō'-sis) The cellular uptake of macromolecules and particulate substances by localized regions of the plasma membrane that surround the substance and pinch off to form an intracellular vesicle.

endoderm (en´-dō-derm) The innermost of the three primary germ layers in animal embryos; lines the archenteron and gives rise to the liver, pancreas, lungs, and the lining of the digestive tract.

endodermis The innermost layer of the cortex in plant roots; a cylinder one cell thick that forms the boundary between the cortex and the stele.

endomembrane system The collection of membranes inside and around a eukaryotic cell, related either through direct physical contact or by the transfer of membranous vesicles.

endometrium (en´-dō-mē´-trē-um) The inner lining of the uterus, which is richly supplied with blood vessels.

endomycorrhizae (en´-dō-mī´-kō-rī´-zē) A type of mycorrhizae that, unlike ectomycorrhizae, do not have a dense mantle ensheathing the root. Instead, microscopic fungal hyphae extend from the root into the soil.

endoparasites Parasites that live within a host.

endoplasmic reticulum (ER) (en´-dō-plaz´-mik ruh-tik´-yū-lum) An extensive membranous network in eukaryotic cells, continuous with the outer nuclear membrane and composed of ribosome-studded (rough) and ribosome-free (smooth) regions.

endorphin (en-dōr´-fin) A hormone produced in the brain and anterior pituitary that inhibits pain perception.

endoskeleton A hard skeleton buried within the soft tissues of an animal, such as the spicules of sponges, the plates of echinoderms, and the bony skeletons of vertebrates.

endosperm A nutrient-rich tissue formed by the union of a sperm cell with two polar nuclei during double fertilization, which provides nourishment to the developing embryo in angiosperm seeds.

endospore A thick-coated, resistant cell produced within a bacterial cell exposed to harsh conditions.

endosymbiotic theory (en´-dō-sim´-bī-ot´-ik) A hypothesis about the origin of the eukaryotic cell, maintaining that the forerunners of eukaryotic cells were symbiotic associations of prokaryotic cells living inside larger prokaryotes.

endothelium (en´-dō-thē´-lē-um) The innermost, simple squamous layer of cells lining the blood vessels; the only constituent structure of capillaries.

endotherm An animal that uses metabolic energy to maintain a constant body temperature, such as a bird or mammal.

endothermic Organisms with bodies that are warmed by heat generated by metabolism. This heat is usually used to maintain a relatively stable body temperature higher than that of the external environment.

endotoxin A component of the outer membranes of certain gram-negative bacteria responsible for generalized symptoms of fever and ache.

energetic hypothesis The concept that the length of a food chain is limited by the inefficiency of energy transfer along the chain.

energy The capacity to do work (to move matter against an opposing force).

energy coupling In cellular metabolism, the use of energy released from an exergonic reaction to drive an endergonic reaction.

energy level The different states of potential energy for electrons in an atom.

enhancer A DNA sequence that recognizes certain transcription factors that can stimulate transcription of nearby genes.

enterocoelous (en´-ter-ō-sē´-lus) The type of development found in deuterostomes. The coelomic cavities form when mesoderm buds from the wall of the archenteron and hollows out.

enterogastrones A category of hormones secreted by the wall of the duodenum.

enteropeptidase An intestinal enzyme that directly or indirectly triggers activation of other enzymes within the intestinal lumen.

entomology The study of insects.

entropy (en´-truh-pē) A quantitative measure of disorder or randomness, symbolized by *S*.

environmental grain An ecological term for the effect of spatial variation, or patchiness, relative to the size and behavior of an organism.

enzyme (en´-zīm) A protein serving as a catalyst, a chemical agent that changes the rate of a reaction without being consumed by the reaction.

epicotyl (ep´-uh-cot´-ul) The embryonic axis above the point at which the cotyledons are attached.

epidermis (1) The dermal tissue system in plants. (2) The outer covering of animals.

epididymis (ep´-uh-did´-uh-mus) A coiled tubule located adjacent to the testes where sperm are stored.

epigenesis (ep´-uh-jen´-uh-sis) The progressive development of form in an embryo.

epiglottis A cartilaginous flap that blocks the top of the windpipe, the glottis, during swallowing, which prevents the entry of food or fluid into the respiratory system.

epinephrine A hormone produced as a response to stress; also called adrenaline.

epiphyte (ep´-uh-fīt) A plant that nourishes itself but grows on the surface of another plant for support, usually on the branches or trunks of tropical trees.

episome (ep´-uh-sōm) A genetic element that can exist either as a plasmid or as part of the bacterial chromosome.

epistasis A phenomenon in which one gene alters the expression of another gene that is independently inherited.

epithalamus A brain region, derived from the diencephalon, that contains several clusters of capillaries that produce cerebrospinal fluid.

epithelial tissue (ep´-uh-thē´-lē-ul) Sheets of tightly packed cells that line organs and body cavities.

epitope A localized region on the surface of an antigen that is chemically recognized by antibodies; also called antigenic determinant.

erythrocyte (eh-rith´-rō-sīt) A red blood cell; contains hemoglobin, which functions in transporting oxygen in the circulatory system.

erythropoietin (eh-rith´-rō-poy´-uh-tin) A hormone produced in the kidney when tissues of the body do not receive enough oxygen. This hormone stimulates the production of erythrocytes.

esophagus (eh-sof´-uh-gus) A channel that conducts food, by peristalsis, from the pharynx to the stomach.

essential amino acids The amino acids that an animal cannot synthesize itself and must obtain from food. Eight amino acids are essential in the human adult.

essential fatty acids Certain unsaturated fatty acids that animals cannot make.

essential nutrient A chemical element that is required for a plant to grow from a seed and complete the life cycle, producing another generation of seeds.

estivation (es´-tuh-vā´-shun) A physiological state characterized by slow metabolism and inactivity, which permits survival during long periods of elevated temperature and diminished water supplies.

estrogens (es´-trō-jenz) The primary female steroid sex hormones, which are produced in the ovary by the developing follicle during the first half of the cycle and in smaller quantities by the corpus luteum during the second half. Estrogens stimulate the development and maintenance of the female reproductive system and secondary sex characteristics.

estrous cycle (es´-trus) A type of reproductive cycle in all female mammals except higher primates, in which the nonpregnant endometrium is reabsorbed rather than shed, and sexual response occurs only during midcycle at estrus.

estrus A period of sexual activity associated with ovulation.

estuary The area where a freshwater stream or river merges with the ocean.

ethology The study of animal behavior in natural conditions.

ethylene (eth´-uh-lēn) The only gaseous plant hormone, responsible for fruit ripening, growth inhibition, leaf abscission, and aging.

euchromatin (ū-krō´-muh-tin) The more open, unraveled form of eukaryotic chromatin that is available for transcription.

eudicots (ū´-di-kots) A large subgroup of traditionally dicot angiosperms including roses, peas, buttercups, sunflowers, oaks, and maples.

euglenoid A group of protistans, including *Euglena* and its relatives, characterized by an anterior pocket or chamber from which one or two flagella emerge.

eukaryotic cell (ū´-kār-ē-ot´-ik) A type of cell with a membrane-enclosed nucleus and membrane-enclosed organelles, present in protists, plants, fungi, and animals; also called eukaryote.

eumetazoa (ū´-met-uh-zō´-uh) Members of the subkingdom that includes all animals except sponges.

euryhaline Organisms that can tolerate substantial changes in external osmolarity.

eurypterids Mainly marine and freshwater, extinct, chelicerates. These predators, also called water scorpions, ranged up to 3 meters long.

eustachian tube The tube that connects the middle ear to the pharynx.

eutherian mammals (ū-thēr´-ē-um) Placental mammals; those whose young complete their embryonic development within the uterus, joined to the mother by the placenta.

eutrophic Pertaining to a highly productive lake, having a high rate of biological productivity supported by a high rate of nutrient cycling.

evaporation The removal of heat energy from the surface of a liquid that is losing some of its molecules.

evaporative cooling The property of a liquid whereby the surface becomes cooler during evaporation, owing to a loss of highly kinetic molecules to the gaseous state.

evolution All the changes that have transformed life on Earth from its earliest beginnings to the diversity that characterizes it today.

evolutionary species concept The idea that evolutionary lineages and ecological roles can form the basis of species identification.

exaptation (ek´-sap-tā´-shun) A structure that evolves and functions in one environmental context but that can perform additional functions when placed in some new environment.

excitable cells Cells that have the ability to generate changes in their membrane potentials.

excitatory postsynaptic potential (EPSP) An electrical change (depolarization) in the membrane of a postsynaptic neuron caused by the binding of an excitatory neurotransmitter from a presynaptic cell to a postsynaptic receptor; makes it more likely for a postsynaptic neuron to generate an action potential.

excretion The disposal of nitrogen-containing waste products of metabolism.

exergonic reaction (ek´-ser-gon´-ik) A spontaneous chemical reaction in which there is a net release of free energy.

exocytosis (ek´-sō-sī-tō´-sis) The cellular secretion of macromolecules by the fusion of vesicles with the plasma membrane.

exoenzymes Powerful hydrolytic enzymes secreted by a fungus outside its body to digest food.

exon A coding region of a eukaryotic gene. Exons, which are expressed, are separated from each other by introns.

exoskeleton A hard encasement on the surface of an animal, such as the shells of mollusks or the cuticles of arthropods, that provides protection and points of attachment for muscles.

exotoxin (ek´-sō-tok´-sin) A toxic protein secreted by a bacterial cell that produces specific symptoms even in the absence of the bacterium.

exponential population growth The geometric increase of a population as it grows in an ideal, unlimited environment.

expression vector A cloning vector that contains the requisite prokaryotic promoter just upstream of a restriction site where a eukaryotic gene can be inserted.

external fertilization The fusion of gametes that parents have discharged into the environment.

exteroreceptor Sensory receptors that detect stimuli outside the body, such as heat, light, pressure, and chemicals.

extracellular digestion The breakdown of food outside cells.

extracellular matrix (ECM) The substance in which animal tissue cells are embedded consisting of protein and polysaccharides.

extraembryonic membranes Four membranes (yolk sac, amnion, chorion, allantois) that support the developing embryo in reptiles, birds, and mammals.

extreme halophile Microorganisms that live in unusually highly saline environments such as the Great Salt Lake or the Dead Sea.

extreme thermophiles Microorganisms that thrive in hot environments (often 60–80°C).

extremeophile Microorganisms that live in extreme environments. They are further classified as either methanogens, extreme halophiles, or extreme thermophiles.

F factor A fertility factor in bacteria, a DNA segment that confers the ability to form pili for conjugation and associated functions required for the transfer of DNA from donor to recipient. It may exist as a plasmid or integrated into the bacterial chromosome.

F plasmid The plasmid form of the F factor.

F₁ generation The first filial, or hybrid, offspring in a genetic cross-fertilization.

F₂ generation Offspring resulting from interbreeding of the hybrid F₁ generation.

facilitate The positive effect of early species on the appearance of later species in ecological succession.

facilitated diffusion The spontaneous passage of molecules and ions, bound to specific carrier proteins, across a biological membrane down their concentration gradients.

facultative anaerobe (fak´-ul-tā´-tiv an´-uh-rōb) An organism that makes ATP by aerobic respiration if oxygen is present but that switches to fermentation under anaerobic conditions.

family In classification, the taxonomic category above genus.

fast block to polyspermy The depolarization of the egg membrane within 1–3 seconds after sperm binding to the vitelline layer. The reaction prevents additional sperm from fusing with the egg's plasma membrane.

fast muscle fibers Muscle cells used for rapid, powerful contractions.

fat (triacylglycerol) (trī-as´-ul-glis´-uh-rol) A biological compound consisting of three fatty acids linked to one glycerol molecule.

fate maps Territorial diagrams of embryonic development that reveal the future development of individual cells and tissues.

fatty acid A long carbon chain carboxylic acid. Fatty acids vary in length and in the number and location of double bonds; three fatty acids linked to a glycerol molecule form fat.

feces The wastes of the digestive tract.

feedback inhibition A method of metabolic control in which the end product of a metabolic pathway acts as an inhibitor of an enzyme within that pathway.

fermentation A catabolic process that makes a limited amount of ATP from glucose without an electron transport chain and that produces a characteristic end product, such as ethyl alcohol or lactic acid.

fertilization The union of haploid gametes to produce a diploid zygote.

fertilization envelope The swelling of the vitelline layer away from the plasma membrane.

fetus (fē´-tus) A developing human from the ninth week of gestation until birth; has all the major structures of an atult.

fiber A lignified cell type that reinforces the xylem of angiosperms and functions in mechanical support; a slender, tapered sclerenchyma cell that usually occurs in bundles.

fibrin (fī´-brin) The activated form of the blood-clotting protein fibrinogen, which aggregates into threads that form the fabric of the clot.

fibrinogen The inactive form of the plasma protein that is converted to the active form fibrin, which aggregates into threads that form the framework of a blood clot.

fibroblast (fī´-brō-blast) A type of cell in loose connective tissue that secretes the protein ingredients of the extracellular fibers.

fibronectin A glycoprotein that helps cells attach to the extracellular matrix.

fibrous connective tissue A dense tissue with large numbers of collagenous fibers organized into parallel bundles. This is the dominant tissue in tendons and ligaments.

fibrous root systems Root systems common to monocots consisting of a mat of thin roots that spread out below the soil surface.

filament The stalk of a stamen.

filtrate Fluid extracted by the excretory system from the blood or body cavity. The excretory system produces urine from the filtrate after extracting valuable solutes from it and concentrating it.

filtration In the vertebrate kidney, the extraction of water and small solutes, including metabolic wastes, from the blood by the nephrons.

first law of thermodynamics The principle of conservation of energy. Energy can be transferred and transformed, but it cannot be created or destroyed.

fixed action pattern (FAP) A sequence of behavioral acts that is essentially unchangeable and usually carried to completion once initiated.

flaccid (flas´-id) Limp. Walled cells are flaccid in isotonic surroundings, where there is no tendency for water to enter.

flagellum (fluh-jel´-um) (plural, **flagella**) A long cellular appendage specialized for locomotion. The flagella of prokaryotes and eukaryotes differ in both structure and function.

flower In an angiosperm, a short stem with four sets of modified leaves, bearing structures that function in sexual reproduction.

fluid mosaic model The currently accepted model of cell membrane structure, which envisions the membrane as a mosaic of individually inserted protein molecules drifting laterally in a fluid bilayer of phospholipids.

fluid-feeder An animal that lives by sucking nutrient-rich fluids from another living organism.

follicle (fol´-uh-kul) A microscopic structure in the ovary that contains the developing ovum and secretes estrogens.

follicle-stimulating hormone (FSH) A protein hormone secreted by the anterior pituitary that stimulates the production of eggs by the ovaries and sperm by the testes.

follicular phase That portion of the ovarian cycle during which several follicles in the ovary begin to grow.

food chain The pathway along which food is transferred from trophic level to trophic level, beginning with producers.

food vacuole A membranous sac formed by phagocytosis.

food web The elaborate, interconnected feeding relationships in an ecosystem.

foot The portion of a moss sporophyte that gathers sugars, amino acids, water, and minerals from the parent gametophyte via transfer cells.

foraging Behavior necessary to recognize, search for, capture, and consume food.

foram A marine protozoan that secretes a shell and extends pseudopodia through pores in its shell.

forebrain One of three ancestral and embryonic regions of the vertebrate brain; develops into the thalamus, hypothalamus, and cerebrum.

fossil A preserved remnant or impression of an organism that lived in the past.

fossil record The chronicle of evolution over millions of years of geologic time engraved in the order in which fossils appear in rock strata.

founder effect Genetic drift attributable to colonization by a limited number of individuals from a parent population.

fovea (fō´-vē-uh) An eye's center of focus and the place on the retina where photoreceptors are highly concentrated.

fragile X syndrome A hereditary mental disorder, partially explained by genomic imprinting and the addition of nucleotides to a triplet repeat near the end of an X chromosome.

fragmentation A means of asexual reproduction whereby a single parent breaks into parts that regenerate into whole new individuals.

frameshift mutation A mutation occurring when the number of nucleotides inserted or deleted is not a multiple of three, resulting in the improper grouping of the following nucleotides into codons.

free energy The portion of a system's energy that can perform work when temperature is uniform throughout the system. The change in free energy of a system is calculated by the equation $\Delta G = \Delta H - T\Delta S$, where T is absolute temperature.

free energy of activation The initial investment of energy necessary to start a chemical reaction; also called activation energy.

frequency-dependent selection A decline in the reproductive success of a morph resulting from the morph's phenotype becoming too common in a population; a cause of balanced polymorphism in populations.

fruit A mature ovary of a flower that protects dormant seeds and aids in their dispersal.

functional group A specific configuration of atoms commonly attached to the carbon skeletons of organic molecules and usually involved in chemical reactions.

Fungi (fun´-jē) The kingdom that contains the fungi.

fusiform initials The cambium cells within the vascular bundles. The name refers to the tapered ends of these elongated cells.

G protein A GTP-binding protein that relays signals from a plasma membrane signal receptor, known as a G-protein linked receptor, to other signal-transduction proteins inside the cell. When such a receptor is activated, it in turn activates the G protein, causing it to bind a molecule of GTP in place of GDP. Hydrolysis of the bound GTP to GDP inactivates the G protein.

G₀ phase A nondividing state in which a cell has left the cell cycle.

G₁ phase The first growth phase of the cell cycle, consisting of the portion of interphase before DNA synthesis begins.

G₂ phase The second growth phase of the cell cycle, consisting of the portion of interphase after DNA synthesis occurs.

gallbladder An organ that stores bile and releases it as needed into the small intestine.

gametangium (gam´-uh-tan´-jē-um) (plural, **gametangia**) The reproductive organ of bryophytes, consisting of the male antheridium and female archegonium; a multichambered jacket of sterile cells in which gametes are formed.

gamete (gam´-ēt) A haploid cell such as an egg or sperm. Gametes unite during sexual reproduction to produce a diploid zygote.

gametophore The mature gamete-producing structure of a gametophyte body of a moss.

gametophyte (guh-mē´-tō-fīt) The multicellular haploid form in organisms undergoing alternation of generations that mitotically produces haploid gametes that unite and grow into the sporophyte generation.

gamma aminobutyric acid (GABA) An amino acid that functions as a CNS neurotransmitter.

ganglion (gang´-glē-un) (plural, **ganglia**) A cluster (functional group) of nerve cell bodies in a centralized nervous system.

gap genes Mutations in these genes cause "gaps" in *Drosophila* segmentation. The normal gene products map out the basic subdivisions along the anterior-posterior axis of the embryo.

gap junction A type of intercellular junction in animal cells that allows the passage of material or current between cells.

gas exchange The uptake of molecular oxygen from the environment and the discharge of carbon dioxide to the environment.

gastric juice The collection of fluids secreted by the epithelium lining the stomach.

gastrin A digestive hormone, secreted by the stomach, that stimulates the secretion of gastric juice.

gastrovascular cavity An extensive pouch that serves as the site of extracellular digestion and a passageway to disperse materials throughout most of an animal's body.

gastrula (gas´-trū-luh) The two-layered, cup-shaped embryonic stage.

gastrulation (gas´-trū-lā´-shun) The formation of a gastrula from a blastula.

gated channel A protein channel in a cell membrane that opens or closes in response to a particular stimulus.

gated ion channel A gated channel for a specific ion. By opening and closing such channels, a cell alters its membrane potential.

gel electrophoresis (ē-lek´-trō-fōr-ē´-sis) The separation of nucleic acids or proteins, on the basis of their size and electrical charge, by measuring their rate of movement through an electrical field in a gel.

gene A discrete unit of hereditary information consisting of a specific nucleotide sequence in DNA (or RNA, in some viruses).

gene amplification The selective synthesis of DNA, which results in multiple copies of a single gene, thereby enhancing expression.

gene cloning The production of multiple copies of a gene.

gene flow The loss or gain of alleles in a population due to the migration of fertile individuals or gametes between populations.

gene pool The total aggregate of genes in a population at any one time.

gene therapy The alternation of the genes of a person afflicted with a genetic disease.

generalized transduction The random transfer of bacterial genes from one bacterium to another.

genetic drift Changes in the gene pool of a small population due to chance.

genetic engineering The direct manipulation of genes for practical purposes.

genetic map An ordered list of genetic loci (genes or other genetic markers) along a chromosome.

genetic recombination The general term for the production of offspring with new combinations of traits inherited from the two parents.

genetically modified (GM) organism An organism that has acquired one or more genes by artificial means; also known as a transgenic organism.

genetics The scientific study of heredity and hereditary variation.

genome (jē´-nōm) The complete complement of an organism's genes; an organism's genetic material.

genomic imprinting The parental effect on gene expression whereby identical alleles have different effects on offspring, depending on whether they arrive in the zygote via the ovum or via the sperm.

genomic library (juh-nō´-mik) A set of thousands of DNA segments from a genome, each carried by a plasmid, phage or other cloning vector.

genomics The study of whole sets of genes and their interactions.

genotype (jē´-nō-tīp) The genetic makeup of an organism.

genus (jē´-nus) (plural, **genera**) A taxonomic category above the species level, designated by the first word of a species' binomial Latin name.

geographic range The geographic area in which a population lives.

geographic variation Differences in genetic structure between populations.

geologic time scale A time scale established by geologists that reflects a consistent sequence of historical periods, grouped into four eras: Precambrian, Paleozoic, Mesozoic, and Cenozoic.

geometric isomers Compounds that have the same molecular formula but differ in the spatial arrangements of their atoms.

geometric population growth A rapid J-shaped growth curve that typically occurs when members have access to abundant food and are free to reproduce at their physiological capacity.

germ layers Three main layers that form the various tissues and organs of an animal body.

gestation (jes-tā´-shun) Pregnancy; the state of carrying developing young within the female reproductive tract.

gibberellins (jib´-uh-rel´-inz) A class of related plant hormones that stimulate growth in the stem and leaves, trigger the germination of seeds and breaking of bud dormancy, and stimulate fruit development with auxin.

gill A localized extension of the body surface of many aquatic animals, specialized for gas exchange.

gill circulation The flow of blood through gills.

Ginkgophyta A phylum of gymnosperms represented by a single extant species, *Ginkgo biloba*, characterized by fanlike leaves that turn gold and are deciduous in autumn.

glandular epithelia Epithelia that secrete chemical solutions.

glans penis The head end of the penis.

glia Supporting cells that are essential for the structural integrity of the nervous system and for the normal functioning of neurons.

glial cell (glē-ul) A nonconducting cell of the nervous system that provides support, insulation, and protection for the neurons.

glomerulus (glō-mār´-ū-lus) A ball of capillaries surrounded by Bowman's capsule in the nephron and serving as the site of filtration in the vertebrate kidney.

glucagon (glū´-kuh-gon) A peptide hormone secreted by pancreatic endocrine cells that raises blood glucose levels; an antagonistic hormone to insulin.

glucocorticoid A corticosteroid hormone secreted by the adrenal cortex that influences glucose metabolism and immune function.

glutamate An amino acid that functions as a CNS neurotransmitter.

glyceraldehyde-3-phosphate (G3P) (glis´-er-al´-de-hīd) The carbohydrate produced directly from the Calvin cycle.

glycine (glī´-sēn) An amino acid that functions as a CNS neurotransmitter.

glycocalyx (glī´-kō-kā´-liks) A fuzzy coat on the outside of animal cells, made of sticky oligosaccharides.

glycogen (glī´-kō-jen) An extensively branched glucose storage polysaccharide found in the liver and muscle of animals; the animal equivalent of starch.

glycolysis (glī-kol´-uh-sis) The splitting of glucose into pyruvate. Glycolysis is the one metabolic pathway that occurs in all living cells, serving as the starting point for fermentation or aerobic respiration.

glycoprotein A protein covalently attached to a carbohydrate.

glycosidic linkage A covalent bond formed between two monosaccharides by a dehydration reaction.

gnathostomes (nā´-thuh-stōm) The vertebrate subgroup that possess jaws.

Gnetophyta A phylum of gymnosperms consisting of just three extant genera that are very different in appearance.

golden algae Typically unicellular, biflagellated, algae with yellow and brown carotene and xanthophyll accessory pigments.

Golgi apparatus (gol´-jē) An organelle in eukaryotic cells consisting of stacks of flat membranous sacs that modify, store, and route products of the endoplasmic reticulum.

gonadotropins (gon´-uh-dō-trō´-pinz) Hormones that stimulate the activities of the testes and ovaries; a collective term for follicle-stimulating and luteinizing hormones.

gonads (gō´-nadz) The male and female sex organs; the gamete-producing organs in most animals.

G-protein linked receptor A signal receptor protein in the plasma membrane that responds to the binding signal molecule by activating a G protein.

graded potential A local voltage change in a neuron membrane induced by stimulation of a neuron, with strength proportional to the strength of the stimulus and lasting about a millisecond.

gradualism A view of Earth's history that attributes profound change to the cumulative product of slow but continuous processes.

graft versus host reaction An attack against a patient's body cells by lymphocytes received in a bone marrow transplant.

Gram stain A staining method that distinguishes between two different kinds of bacterial cell walls.

gram-negative The group of bacteria with a structurally more complex cell wall made of less peptidoglycan. Gram-negative bacteria are often more toxic than gram-positive bacteria.

gram-positive The group of bacteria with simpler cell walls with a relatively large amount of peptidoglycan. Gram-positive bacteria are usually less toxic than gram-negative bacteria.

granum (gran´-um) (plural, **grana**) A stacked portion of the thylakoid membrane in the chloroplast. Grana function in the light reactions of photosynthesis.

gravitropism (grav´-uh-trō´-pizm) A response of a plant or animal in relation to gravity.

gray crescent A light-gray region of cytoplasm located near the equator of the egg on the side opposite the sperm entry.

gray matter Regions of dendrites and clusters of nerve-cell bodies within the CNS.

green algae Photosynthetic protists that include unicellular, colonial, and multicellular species with grass green chloroplasts; closely related to true plants.

greenhouse effect The warming of planet Earth due to the atmospheric accumulation of carbon dioxide, which absorbs infrared radiation and slows its escape from the irradiated Earth.

gross primary production (GPP) The total primary production of an ecosystem.

ground meristem A primary meristem that gives rise to ground tissue in plants.

ground tissue A tissue of mostly parenchyma cells that makes up the bulk of a young plant and fills the space between the dermal and vascular tissue systems.

growth factor A protein that must be present in the extracellular environment (culture medium or animal body) for the growth and normal development of certain types of cells.

growth hormone A protein of about 200 amino acids that affects a wide variety of target tissues and has both direct effects and tropic effects.

guard cell A specialized epidermal plant cell that forms the boundaries of the stomata.

gustatory receptors Taste receptors.

guttation The exudation of water droplets, caused by root pressure in certain plants.

gymnosperm (jim´-nō-sperm) A vascular plant that bears naked seeds—seeds not enclosed in specialized chambers.

habituation A very simple type of learning that involves a loss of responsiveness to stimuli that convey little or no information.

hair cell A type of mechanoreceptor that detects sound waves and other forms of movement in air or water.

half-life The number of years it takes for 50% of an original sample of an isotope to decay.

Hamilton's Rule The principle that for natural selection to favor an altruistic act, the benefit to the recipient, devalued by the coefficient of relatedness, must exceed the cost to the altruist.

haploid cell (hap´-loyd) A cell containing only one set of chromosomes (n).

Hardy-Weinberg equilibrium The condition describing a nonevolving population (one that is in genetic equilibrium).

Hardy-Weinberg formula A formula for calculating the frequencies of genotypes in a gene pool from the frequencies of alleles, and vice versa.

Hardy-Weinberg theorem An axiom maintaining that the sexual shuffling of genes alone cannot alter the overall genetic makeup of a population.

haustorium (plural, **haustoria**) In parasitic fungi, a nutrient-absorbing hyphal tip that penetrates the tissues of the host but remains outside the host cell membranes.

Haversian system (hā-ver´-shun) One of many structural units of vertebrate bone, consisting of concentric layers of mineralized bone matrix surrounding lacunae, which contain osteocytes, and a central canal, which contains blood vessels and nerves.

heart A muscular pump that uses metabolic energy to elevate hydrostatic pressure of the blood. Blood then flows down a pressure gradient through blood vessels that eventually return blood to the heart.

heart rate The rate of heart contraction.

heat The total amount of kinetic energy due to molecular motion in a body of matter. Heat is energy in its most random form.

heat of vaporization The quantity of heat a liquid must absorb for 1 gram of it to be converted from the liquid to the gaseous state.

heat-shock protein A protein that helps protect other proteins during heat stress, found in plants, animals, and microorganisms.

heavy chains Polypeptide chains that contribute to the structure of an antibody. Two identical heavy chains and two identical light chains, joined by disulfide brides, form a Y-shaped antibody molecule.

helicase An enzyme that untwists the double helix of DNA at the replication forks.

heliozoan Sun animals that live in fresh water. They have skeletons made of siliceous or chitinous unfused plates.

helper T cell (T_H) A type of T cell that is required by some B cells to help them make antibodies or that helps other T cells respond to antigens or secrete lymphokines or interleukins.

hemocyanin (hē´-muh-sī´-uh-nin) A type of respiratory pigment that uses copper as its oxygen-binding component. Hemocyanin is found in the hemolymph of arthropods and many mollusks.

hemoglobin (hē´-mō-glō-bin) An iron-containing protein in red blood cells that reversibly binds oxygen.

hemolymph In invertebrates with an open circulatory system, the body fluid that bathes tissues.

hemophilia A human genetic disease caused by a sex-linked recessive allele, characterized by excessive bleeding following injury.

hepatic portal vessel A large circulatory channel that conveys nutrient-laden blood from the small intestine to the liver, which regulates the blood's nutrient content.

Hepatophyta The phylum of liverworts, small herbaceous (nonwoody) plants.

herbivore A heterotrophic animal that eats plants.

herbivory The consumption of plant material by an herbivore.

heredity The transmission of traits from one generation to the next.

hermaphrodite (her-maf´-rō-dīt) An individual that functions as both male and female in sexual reproduction by producing both sperm and eggs.

hermaphroditism (her-maf´-rō-dī-tizm) A condition in which an individual has both female and male gonads and functions as both a male and female in sexual reproduction by producing both sperm and eggs.

heterochromatin (het´-er-ō-krō´-muh-tin) Nontranscribed eukaryotic chromatin that is so highly compacted that it is visible with a light microscope during interphase.

heterochrony (het´-uh-rok´-ruh-nē) Evolutionary change in the timing or rate of development.

heterocyst (het´-er-ō-sist) A specialized cell that engages in nitrogen fixation on some filamentous cyanobacteria.

heterogeneity A measurement of biological diversity considering richness and relative abundance.

heterokaryon A mycelium formed by the fusion of two hyphae that have genetically different nuclei.

heteromorphic (het´-er-ō-mōr´-fik) Referring to a condition in the life cycle of all modern plants in which the sporophyte and gametophyte generations differ in morphology.

heterosporous (het´-er-os´-pōr-us) Referring to plants in which the sporophyte produces two kinds of spores that develop into unisexual gametophytes, either female or male.

heterotroph (het´-er-ō-trōf) An organism that obtains organic food molecules by eating other organisms or their by-products.

heterozygote advantage Greater reproductive success of heterozygous individuals compared to homozygotes; tends to preserve variation in gene pools.

heterozygous (het´-er-ō-zī´-gus) Having two different alleles for a given genetic character.

hibernation A physiological state that allows survival during long periods of cold temperatures and reduced food supplies, in which metabolism decreases, the heart and respiratory system slow down, and body temperature is maintained at a lower level than normal.

high-density lipoprotein (HDL) A cholesterol-carrying particle in the blood, made up of cholesterol and other lipids surrounded by a single layer of phospholipids in which proteins are embedded. An HDL particle carries less cholesterol than a related lipoprotein, LDL, and may be correlated with a decreased risk of blood vessel blockage.

hindbrain One of three ancestral and embryonic regions of the vertebrate brain; develops into the medulla oblongata, pons, and cerebellum.

histamine (his´-tuh-mēn) A substance released by injured cells that causes blood vessels to dilate during an inflammatory response.

histone (his´-tōn) A small protein with a high proportion of positively charged amino acids that binds to the negatively charged DNA and plays a key role in its chromatin structure.

histone acetylation The attachment of acetyl groups to certain amino acids of histone proteins.

HIV (human immunodeficiency virus) The infectious agent that causes AIDS; HIV is an RNA retrovirus.

holdfast A rootlike structure that anchors a seaweed.

holoblastic cleavage (hō´-lō-blas´-tik) A type of cleavage in which there is complete division of the egg, as in eggs having little yolk (sea urchin) or a moderate amount of yolk (frog).

homeobox (hō´-mē-ō-boks´) A 180-nucleotide sequence within a homeotic gene encoding the part of the protein that binds to the DNA of the genes regulated by the protein.

homeosis (hō´-mē-ō´-sis) Evolutionary alteration in the placement of different body parts.

homeostasis (hō´-mē-ō-stā´-sis) The steady-state physiological condition of the body.

homeotic gene (hō´-mē-ot´-ik) Any of the genes that control the overall body plan of animals by controlling the developmental fate of groups of cells.

Hominid (hah´-mi-nid) A species on the human branch of the evolutionary tree; a member of the family Hominidae, including *Homo sapiens* and our ancestors.

hominid A term that refers to mammals that are more closely related to humans than to any other living species.

hominoid A term that refers to great apes and humans.

homologous chromosomes (hō-mol´-uh-gus) Chromosome pairs of the same length, centromere position, and staining pattern that possess genes for the same characters at corresponding loci. One homologous chromosome is inherited from the organism's father, the other from the mother.

homologous structures Structures in different species that are similar because of common ancestry.

homology (hō-mol´-uh-jē) Similarity in characteristics resulting from a shared ancestry.

homosporous (hō-mos´-pōr-us) Referring to plants in which a single type of spore develops into a bisexual gametophyte having both male and female sex organs.

homozygous (hō´-mō-zī´-gus) Having two identical alleles for a given trait.

horizontal cell Neurons of the retina that help integrate information before it is sent to the brain.

hormone Any one of the many circulating chemical signals found in all multicellular organisms that are formed in specialized cells, travel in body fluids, and coordinate the various parts of the organism by interacting with target cells.

hornworts Members of the phylum Anthocerophyta, they are small herbaceous (nonwoody) plants.

host The larger participant in a symbiotic relationship, serving as home and feeding ground to the symbiont.

host range The limited range of host cells that each type of virus can infect and parasitize.

human chorionic gonadotropin (HCG) (kōr´-ē-on´-ik gon´-uh-dō-trō´-pin) A hormone secreted by the chorion that maintains the corpus luteum of the ovary during the first three months of pregnancy.

Human Genome Project An international collaborative effort to map and sequence the DNA of the entire human genome.

humoral immunity (hyū´-mer-al) The type of immunity that fights bacteria and viruses in body fluids with antibodies that circulate in blood plasma and lymph, fluids formerly called humors.

humus (hyū´-mus) Decomposing organic material found in topsoil.

Huntington's disease A human genetic disease caused by a dominant allele; characterized by uncontrollable body movements and degeneration of the nervous system; usually fatal 10–20 years after the onset of symptoms.

hybrid zone A region where two related populations that diverged after becoming geographically isolated make secondary contact and interbreed where their geographic ranges overlap.

hybridization The mating, or crossing, of two varieties.

hydration shell The sphere of water molecules around each dissolved ion.

hydrocarbon An organic molecule consisting only of carbon and hydrogen.

hydrogen bond A type of weak chemical bond formed when the slightly positive hydrogen atom of a polar covalent bond in one molecule is attracted to the slightly negative atom of a polar covalent bond in another molecule.

hydrogen ion A single proton with a charge of +1. The dissociation of a water molecule (H_2O) leads to the generation of a hydroxide ion (OH^-) and a hydrogen ion (H^+).

hydrolysis (hī-drol´-uh-sis) A chemical process that lyses, or splits, molecules by the addition of water; an essential process in digestion.

hydrophilic (hī´-drō-fil´-ik) Having an affinity for water.

hydrophobic (hī´-drō-fō´-bik) Having an aversion to water; tending to coalesce and form droplets in water.

hydrophobic interaction A type of weak chemical bond formed when molecules that do not mix with water coalesce to exclude the water.

hydrostatic skeleton (hī´-drō-stat´-ik) A skeletal system composed of fluid held under pressure in a closed body compartment; the main skeleton of most cnidarians, flatworms, nematodes, and annelids.

hydroxide ion A water molecule that lost a proton.

hydroxyl group (hī-drok´-sil) A functional group consisting of a hydrogen atom joined to an oxygen atom by a polar covalent bond. Molecules possessing this group are soluble in water and are called alcohols.

hymen A thin membrane that partly covers the vaginal opening in the human female; ruptured by sexual intercourse or other vigorous activity.

hyperpolarization An electrical state whereby the inside of the cell is made more negative relative to the outside than at the resting membrane potential. A neuron membrane is hyperpolarized if a stimulus increases its voltage from the resting potential of

$^-$70 mV, reducing the chance that the neuron will transmit a nerve impulse.

hypertonic In comparing two solutions, referring to the one with a greater solute concentration.

hypha (plural, **hyphae**) (hī´-fuh, hī´-fē) A filament that collectively makes up the body of a fungus.

hypocotyl (hī´-puh-cot´-ul) The embryonic axis below the point at which the cotyledons are attached.

hypothalamus (hī´-pō-thal´-uh-mus) The ventral part of the vertebrate forebrain; functions in maintaining homeostasis, especially in coordinating the endocrine and nervous systems; secretes hormones of the posterior pituitary and releasing factors, which regulate the anterior pituitary.

hypotonic solution In comparing two solutions, the one with a lower solute concentration.

I band The area near the edge of the sarcomere where there are only thin filaments.

ileum The last of three sections of the small intestine primarily involved in the absorption of nutrients and water.

imaginal disk (ih-maj´-in-ul) An island of undifferentiated cells in an insect larva, which are committed (determined) to form a particular organ during metamorphosis to the adult.

immune adherence The collective action of antibodies, complement, and phagocytes. Microbes coated with antibodies and complement proteins adhere to blood vessel walls, making the pathogens easier prey for phagocytic cells circulating in the blood.

immunization Also called vaccination, it is the exposure of an organism to agents that can no longer cause disease but retain the ability to act as antigens, thereby stimulating an immune response and immunological memory.

immunoglobulin (Ig) (im´-ū-nō-glob´-ū-lin) One of the class of proteins comprising the antibodies.

imperfect fungi Molds with no known sexual stages, also called deuteromycetes. This is an informal grouping without phylogenetic basis. If a mycologist discovers a sexual stage of an imperfect fungus, the species is reclassified into a particular phylum, depending upon the type of sexual structures.

imprinting A type of learned behavior with a significant innate component, acquired during a limited critical period.

***in vitro* fertilization** (vē´-trō) Fertilization of ova in laboratory containers followed by artificial implantation of the early embryo in the mother's uterus.

***in vitro* mutagenesis** A technique to discover the function of a gene by introducing specific changes into the sequence of a cloned gene, reinserting the mutated gene into a cell, and studying the phenotype of the mutant.

inclusive fitness The total effect an individual has on proliferating its genes by producing its own offspring and by providing aid that enables other close relatives to increase the production of their offspring.

incomplete dominance A type of inheritance in which F_1 hybrids have an appearance that is intermediate between the phenotypes of the parental varieties.

incomplete flower A flower lacking sepals, petals, stamens, or carpels.

incomplete metamorphosis A type of development in certain insects, such as grasshoppers, in which the larvae resemble adults but are smaller and have different body proportions. The animal goes through a series of molts, each time looking more like an adult, until it reaches full size.

incus The second of the three middle ear bones.

indeterminate cleavage A type of embryonic development in deuterostomes, in which each cell produced by early cleavage divisions retains the capacity to develop into a complete embryo.

indeterminate growth A type of growth characteristic of plants, in which the organism continues to grow as long as it lives.

individualistic hypothesis The concept, put forth by H.A. Gleason, that a plant community is a chance assemblage of species found in the same area simply because they happen to have similar biotic requirements.

induced fit The change in shape of the active site of an enzyme so that it binds more snugly to the substrate, induced by entry of the substrate.

inducer A specific small molecule that inactivates the repressor in an operon.

induction The ability of one group of embryonic cells to influence the development of another.

inflammatory response A line of defense triggered by penetration of the skin or mucous membranes, in which small blood vessels in the vicinity of an injury dilate and become leakier, enhancing the infiltration of leukocytes; may also be widespread in the body.

ingestion A heterotrophic mode of nutrition in which other organisms or detritus are eaten whole or in pieces.

ingroup In a cladistic study of evolutionary relationships among taxa of organisms, the group of taxa that is actually being analyzed.

inhibit The negative effect of early species on the appearance of later species in ecological succession.

inhibiting hormone A kind of hormone released from the hypothalamus that makes the anterior pituitary stop secreting hormone.

inhibitory postsynaptic potential (IPSP) (pōst´-sin-ap´-tik) An electrical charge (hyperpolarization) in the membrane of a postsynaptic neuron caused by the binding of an inhibitory neurotransmitter from a presynaptic cell to a postsynaptic receptor; makes it more difficult for a postsynaptic neuron to generate an action potential.

inner cell mass A cluster of cells in a mammalian blastocyst that protrudes into one end of the cavity and subsequently develops into the embryo proper and some of the extraembryonic membranes.

inner ear One of three main regions of the vertebrate ear; includes the cochlea, organ of Corti, and semicircular canals.

inositol trisphosphate (IP₃) (in-ō´-suh-tol) A second messenger that functions as an intermediate between certain nonsteroid hormones and a third messenger, a rise in cytoplasmic Ca^{2+} concentration.

Insecta The class of arthropods typically having six legs, two pairs of wings, and one pair of antennae. Insect species outnumber all other forms of life combined.

insertion A mutation involving the addition of one or more nucleotide pairs to a gene.

insertion sequence The simplest kind of transposon, consisting of inverted repeats of DNA flanking a gene for transposase, the enzyme that catalyzes transposition.

insight learning The ability of an animal to perform a correct or appropriate behavior on the first attempt in a situation with which it has had no prior experience.

insulin (in´-sū-lin) A vertebrate hormone that lowers blood glucose levels by promoting the uptake of glucose by most body cells and the synthesis and storage of glycogen in the liver; also stimulates protein and fat synthesis; secreted by endocrine cells of the pancreas called islets of Langerhans.

insulinlike growth factors A group of peptides produced by the liver, it circulates in blood plasma and directly stimulates bone and cartilage growth.

integral protein Typically transmembrane proteins with hydrophobic regions that completely span the hydrophobic interior of the membrane.

integration The interpretation of sensory signals within neural processing centers of the central nervous system.

integrin Receptor proteins built into the plasma membrane that interconnect the extracellular matrix and the cytoskeleton.

integuments (in-teg´-ū-ment) Layers of sporophyte tissues that contribute to the structure of an ovule of a seed plant.

interactive hypothesis The concept, put forth by F. E. Clements, that a community is an assemblage of closely linked species, locked into association by mandatory biotic interactions that cause the community to function as an integrated unit, a sort of superorganism.

intercalated discs Specialized junctions between cardiac muscle cells that provide direct electrical coupling among cells.

interferon (in´-ter-fēr´-on) A chemical messenger of the immune system, produced by virus-infected cells and capable of helping other cells resist the virus.

interleukin-1 (IL-1) A cytokine secreted by a macrophage that is in the process of phagocytizing and presenting antigen. IL-1, in combination with the antigen, activates the helper T cell to produce IL-2 and other cytokines.

interleukin-2 (IL-2) A cytokine that helps B cells that have contacted antigen differentiate into antibody-secreting plasma cells.

intermediate filament A component of the cytoskeleton that includes all filaments intermediate in size between microtubules and microfilaments.

internal fertilization Reproduction in which sperm are typically deposited in or near the female reproductive tract and fertilization occurs within the tract.

interneuron (in´-ter-nūr´-on) An association neuron; a nerve cell within the central nervous system that forms synapses with sensory and motor neurons and integrates sensory input and motor output.

internode The segment of a plant stem between the points where leaves are attached.

interoreceptor Sensory receptors that detect stimuli within the body, such as blood pressure and body position.

interphase The period in the cell cycle when the cell is not dividing. During interphase, cellular metabolic activity is high, chromosomes and organelles are duplicated, and cell size may increase. Interphase accounts for 90% of the cell cycle.

intersexual selection Individuals of one sex (usually females) are choosy in selecting their mates from individuals of the other sex, also called mate choice.

interspecific competition Competition for resources between plants, between animals, or between decomposers when resources are in short supply.

interstitial cells (in´-ter-stish´-ul) Cells scattered among the seminiferous tubules of the vertebrate testis that secrete testosterone and other androgens, the male sex hormones.

interstitial fluid The internal environment of vertebrates, consisting of the fluid filling the spaces between cells.

intertidal zone The shallow zone of the ocean where land meets water.

intracellular digestion The joining of food vacuoles and lysosomes to allow chemical digestion to occur within the cytoplasm of a cell.

intrasexual selection A direct competition among individuals of one sex (usually the males in vertebrates) for mates of the opposite sex.

intrinsic rate of increase The difference between the number of births and the number of deaths, symbolized as r_{max}; the maximum population growth rate.

introgression (in´-trō-gresh´-un) The transplantation of genes between species resulting from fertile hybrids mating successfully with one of the parent species.

intron (in´-tron) A noncoding, intervening sequence within a eukaryotic gene.

invagination The infolding of cells.

inversion An aberration in chromosome structure resulting from an error in meiosis or from mutagens; specifically, reattachment of a chromosomal fragment to the chromosome from which the fragment originated, but in a reverse orientation.

invertebrate An animal without a backbone; invertebrates make up 95% of animal species.

involution Cells rolling over the edge of a lip into the interior.

ion (ī´-on) An atom that has gained or lost electrons, thus acquiring a charge.

ionic bond (ī-on´-ik) A chemical bond resulting from the attraction between oppositely charged ions.

ionic compound Compounds resulting from the formation of ionic bonds, also called a salt.

islets of Langerhans Clusters of endocrine cells that secrete glucagon and insulin directly into the bloodstream.

isogamy (ī-sog´-uh-mē) A condition in which male and female gametes are morphologically indistinguishable.

isomer (ī´-sō-mer) One of several organic compounds with the same molecular formula but different structures and therefore different properties. The three types of isomers are structural isomers, geometric isomers, and enantiomers.

isomorphic Referring to alternating generations in which the sporophytes and gametophytes look alike, although they differ in chromosome number.

isopods One of the largest groups of crustaceans, primarily marine, but including pill bugs common under logs and moist vegetation next to the ground.

isotonic (ī´-sō-ton´-ik) Having the same solute concentration as another solution.

isotope (ī´-sō-tōp) One of several atomic forms of an element, each containing a different number of neutrons and thus differing in atomic mass.

iteroparity A life history in which adults produce large numbers of offspring over many years.

jejunum The middle section of the small intestine primarily involved in the absorption of nutrients and water.

joule (J) A unit of energy: 1 J = 0.239 cal; 1 cal = 4.184 J.

juvenile hormone (JH) A hormone in arthropods, secreted by the corpora allata glands, that promotes the retention of larval characteristics.

juxtaglomerular apparatus (juks´-tuh-gluh-mār´-ū-ler) A specialized tissue located near the afferent arteriole that supplies blood to the glomerulus.

juxtamedullary nephrons Nephrons with well-developed loops of Henle that extend deeply into the renal medulla.

karyogamy (kār´-ē-og´-uh-mē) The fusion of nuclei of two cells, as part of syngamy.

karyotype (kār´-ē-ō-tīp) A method of organizing the chromosomes of a cell in relation to number, size, and type.

ketone An organic compound with a carbonyl group of which the carbon atom is bonded to two other carbons.

keystone predator A predatory species that helps maintain species richness in a community by reducing the density of populations of the best competitors so that populations of less competitive species are maintained.

keystone species Species that are not usually abundant in a community yet exert strong control on community structure by the nature of their ecological roles or niches.

kilocalorie (kcal) A thousand calories; the amount of heat energy required to raise the temperature of 1 kg of water by 1°C.

kin selection A phenomenon of inclusive fitness, used to explain altruistic behavior between related individuals.

kinesis (kuh-nē´-sis) A change in activity or turning rate in response to a stimulus.

kinetic energy (kuh-net´-ik) The energy of motion, which is directly related to the speed of that motion. Moving matter does work by imparting motion to other matter.

kinetochore (kuh-net´-uh-kōr) A specialized region on the centromere that links each sister chromatid to the mitotic spindle.

kinetoplastid A group of protistans, including *Trypanosoma*, which have a single large mitochondrion associated with a kinetoplast that houses extranuclear DNA.

kingdom A taxonomic category, the second broadest after domain.

kingdom Plantae The traditional embryophyte definition of the plant kingdom.

kingdom Streptophyta The name given to the group that includes the traditional plant kingdom and the green algae most closely related to plants, the charophyceans and a few related groups.

kingdom Viridiplantae (vī-rid´-ē-plant-ē) The broadest version of the plant kingdom that includes the members of the kingdom Streptophyta plus the chlorophytes (non-charophycean green algae).

Koch's postulates A set of four criteria for determining whether a specific pathogen is the cause of a disease.

Krebs cycle A chemical cycle involving eight steps that completes the metabolic breakdown of glucose molecules to carbon dioxide; occurs within the mitochondrion; the second major stage in cellular respiration.

K-selection The concept that in certain (K-selected) populations, life history is centered around producing relatively few offspring that have a good chance of survival.

labia majora A pair of thick, fatty ridges that enclose and protect the labia minor and vestibule.

labia minora A pair of slender skin folds that enclose and protect the vestibule.

labor A series of strong, rhythmic contractions of the uterus that expel a baby out of the uterus and vagina during childbirth.

lactation The continued production of milk.

lacteal (lak´-tē-al) A tiny lymph vessel extending into the core of an intestinal villus and serving as the destination for absorbed chylomicrons.

lactic acid fermentation The conversion of pyruvate to lactate with no release of carbon dioxide.

lagging strand A discontinuously synthesized DNA strand that elongates in a direction away from the replication fork.

lancelet One of a group of invertebrate chordates.

landmark A point of reference for orientation during navigation.

landscape Several different primarily terrestrial ecosystems linked by exchanges of energy, materials, and organisms.

landscape ecology The application of ecological principles to the study of land-use patterns; the scientific study of the biodiversity of interacting ecosystems.

larva (lar´-vuh) (plural, **larvae**) A free-living, sexually immature form in some animal life cycles that may differ from the adult in morphology, nutrition, and habitat.

larynx (lār´-inks) The voicebox, containing the vocal cords.

lateral geniculate nuclei The destination in the thalamus for most of the ganglion cell axons that form the optic nerves.

lateral inhibition A process that sharpens the edges and enhances the contrast of a perceived image, by inhibiting receptors lateral to those that have responded to light.

lateral line system A mechanoreceptor system consisting of a series of pores and receptor units (neuromasts) along the sides of the body of fishes and aquatic amphibians; detects water movements made by an animal itself and by other moving objects.

lateral meristem (mār´-uh-stem) The vascular and cork cambium, a cylinder of dividing cells that runs most of the length of stems and roots and is responsible for secondary growth.

lateral roots Roots that arise from the outermost layer of the pericycle of an established root.

law of independent assortment Mendel's second law, stating that each allele pair segregates independently during gamete formation; applies when genes for two characteristics are located on different pairs of homologous chromosomes.

law of segregation Mendel's first law, stating that allele pairs separate during gamete formation, and then randomly re-form as pairs during the fusion of gametes at fertilization.

leading strand The new continuous complementary DNA strand synthesized along the template strand in the mandatory 5' ⟶ 3' direction.

learning A behavioral change resulting from experience.

lens The structure in an eye that focuses light rays onto the retina.

lepidosaurs The reptilian group that includes lizards, snakes, and two species of New Zealand animals called tuataras.

leukocyte (lū´-kō-sīt) A white blood cell; typically functions in immunity, such as phagocytosis or antibody production.

Leydig cell Located between the seminiferous tubules of the testes, these cells produce testosterone and other androgens.

lichen (lī´-ken) The mutualistic collective formed by the symbiotic association between a fungus and a photosynthetic alga.

life cycle The generation-to-generation sequence of stages in the reproductive history of an organism.

life history The series of events from birth through reproduction and death.

life table A table of data summarizing mortality in a population.

ligament A type of fibrous connective tissue that joins bones together at joints.

ligand (lig´-und) A molecule that binds specifically to a receptor site of another molecule.

ligand-gated ion channel Protein pores in the plasma membrane that open or close in response to a chemical signal, allowing or blocking the flow of specific ions.

ligand-gated ion channel receptor A signal receptor protein in a cell membrane that can act as a channel for the passage of a specific ion across the membrane. When activated by a signal molecule, the receptor either allows or blocks passage of the ion, resulting in a change in ion concentration that usually affects cell functioning.

light chains Polypeptide chains that contribute to the structure of an antibody. Two identical light chains and two identical heavy chains, joined by disulfide bridges, form a Y-shaped antibody molecule.

light microscope (LM) An optical instrument with lenses that refract (bend) visible light to magnify images of specimens.

light reactions The steps in photosynthesis that occur on the thylakoid membranes of the chloroplast and that convert solar energy to the chemical energy of ATP and NADPH, evolving oxygen in the process.

lignin (lig´-nin) A hard material embedded in the cellulose matrix of vascular plant cell walls that functions as an important adaptation for support in terrestrial species.

limbic system (lim´-bik) A group of nuclei (clusters of nerve cell bodies) in the lower part of the mammalian forebrain that interact with the cerebral cortex in determining emotions; includes the hippocampus and the amygdala.

limnetic zone The well-lit, open surface waters of a lake farther from shore.

linkage map A genetic map based on the frequencies of recombination between markers during crossing over of homologous chromosomes. The greater the frequency of recombination between two genetic markers, the farther apart they are assumed to be. *See also* genetic map.

linked genes Genes that are located on the same chromosome.

lipase An enzyme that hydrolyzes fat molecules in the intestinal lumen.

lipid (lip´-id) One of a family of compounds, including fats, phospholipids, and steroids, that are insoluble in water.

lipoprotein A protein bonded to a lipid; includes the low-density lipoproteins (LDLs) and high-density lipoproteins (HDLs) that transport fats and cholesterol in blood.

littoral zone The shallow, well-lit waters of a lake close to shore.

liver The largest organ in the vertebrate body. The liver performs diverse functions such as producing bile, preparing nitrogenous wastes for disposal, and detoxifying poisonous chemicals in the blood.

liverworts Members of the phylum Hepatophyta, they are small herbaceous (nonwoody) plants.

loams The most fertile of all soils, loams are made up of roughly equal amounts of sand, silt, and clay.

local regulator A chemical messenger that influences cells in the vicinity.

locomotion Active movement from place to place.

locus (lō´-kus) (plural, **loci**) A particular place along the length of a certain chromosome where a given gene is located.

logistic population growth A model describing population growth that levels off as population size approaches carrying capacity.

long-day plant A plant that flowers, usually in late spring or early summer, only when the light period is longer than a critical length.

long-term depression (LTD) A reduced responsiveness to an action potential (nerve signal) by a receiving neuron.

long-term memory The ability to hold, associate, and recall information over one's life.

long-term potentiation (LTP) An enhanced responsiveness to an action potential (nerve signal) by a receiving neuron.

loop of Henle The long hairpin turn, with a descending and ascending limb, of the renal tubule in the vertebrate kidney; functions in water and salt reabsorption.

loose connective tissue The most widespread connective tissue in the vertebrate body. It binds epithelia to underlying tissues and functions as packing material, holding organs in place.

lophophorate animals The group of animals that includes the Bryozoa, Phoronida, and Brachiopoda. Animals in these groups possess a lophophore.

lophophore (lof´-uh-fōr) A horseshoe-shaped or circular fold of the body wall bearing ciliated tentacles that surround the mouth.

Lophotrochozoa One of two distinct clades within the protostomes. It includes annelids and mollusks.

low-density lipoprotein (LDL) A cholesterol-carrying particle in the blood, made up of cholesterol and other lipids surrounded by a single layer of phospholipids in which proteins are embedded. An LDL particle carries the blood correlate with a tendency to develop blocked blood vessels and heart disease.

lungs The invaginated respiratory surfaces of terrestrial vertebrates, land snails, and spiders that connect to the atmosphere by narrow tubes.

luteal phase That portion of the ovarian cycle during which endocrine cells of the corpus luteum secrete female hormones.

luteinizing hormone (LH) (lū´-tē-uh-nī´-zing) A protein hormone secreted by the anterior pituitary that stimulates ovulation in females and androgen production in males.

lymph The colorless fluid, derived from interstitial fluid, in the lymphatic system of vertebrate animals.

lymph node Organs located along lymph vessels. They filter lymph and help attack viruses and bacteria.

lymphatic system A system of vessels and lymph nodes, separate from the circulatory system, that returns fluid and protein to the blood.

lymphocyte A white blood cell. The lymphocytes that complete their development in the bone marrow are called B cells, and those that mature in the thymus are called T cells.

lysogenic cycle (lī´-sō-jen´-ik) A phage replication cycle in which the viral genome becomes incorporated into the bacterial host chromosome as a prophage and does not kill the host.

lysosome (lī´-so-sōm) A membrane-enclosed bag of hydrolytic enzymes found in the cytoplasm of eukaryotic cells.

lysozyme (lī´-sō-zīm) An enzyme in perspiration, tears, and saliva that attacks bacterial cell walls.

lytic cycle (lit´-ik) A type of viral replication cycle resulting in the release of new phages by death or lysis of the host cell.

M phase The mitotic phase of the cell cycle, which includes mitosis and cytokinesis.

macroevolution Evolutionary change on a grand scale, encompassing the origin of new taxonomic groups, evolutionary trends, adaptive radiation, and mass extinction.

macromolecule A giant molecule formed by the joining of smaller molecules, usually by a condensation reaction. Polysaccharides, proteins, and nucleic acids are macromolecules.

macronutrient A chemical substance that an organism must obtain in relatively large amounts. *See also* micronutrient.

macrophage (mak´-rō-fāj) An amoeboid cell that moves through tissue fibers, engulfing bacteria and dead cells by phagocytosis.

major histocompatibility complex (MHC) A large set of cell surface antigens encoded by a family of genes. Foreign MHC markers trigger T-cell responses that may lead to the rejection of transplanted tissues and organs.

malignant tumor A cancerous tumor that is invasive enough to impair functions of one or more organs.

malleus The first of the three middle ear bones.

malnourished An animal whose diet is missing one or more essential nutrients.

malpighian tubule (mal-pig´-ē-un) A unique excretory organ of insects that empties into the digestive tract, removes nitrogenous wastes from the blood, and functions in osmoregulation.

Mammalia The class that includes endothermic vertebrates that possess mammary glands and hair.

mammary glands Exocrine glands that secrete milk to nourish the young. These glands are characteristic of mammals.

mandibles Jawlike structures found in uniramians and crustaceans.

mantle A heavy fold of tissue in mollusks that drapes over the visceral mass and may secrete a shell.

mantle cavity A water-filled chamber that houses the gills, anus, and excretory pores of a mollusk.

map units A measurement of the distance between genes; one map unit is equivalent to a 1% recombination frequency.

mark-recapture method A sampling technique used to estimate wildlife populations.

marsupial (mar-sū´-pē-ul) A mammal, such as a koala, kangaroo, or opossum, whose young complete their embryonic development inside a maternal pouch called the marsupium.

mass number The sum of the number of protons and neutrons in an atom's nucleus.

mast cell A vertebrate body cell that produces histamine and other molecules that trigger the inflammatory response.

maternal effect genes A gene that, when mutant in the mother, results in a mutant phenotype in the offspring, regardless of the genotype.

matrix The nonliving component of connective tissue, consisting of a web of fibers embedded in homogeneous ground substance that may be liquid, jellylike, or solid.

matter Anything that takes up space and has mass.

mechanoreceptor A sensory receptor that detects physical deformations in the body's environment associated with pressure, touch, stretch, motion, and sound.

medulla Also called the medulla oblongata, it is the lowest part of the vertebrate brain; a swelling of the hindbrain dorsal to the anterior spinal cord that controls autonomic, homeostatic functions, including breathing, heart and blood vessel activity, swallowing, digestion, and vomiting.

medusa (muh-dū´-suh) The floating, flattened, mouth-down version of the cnidarian body plan. The alternate form is the polyp.

megapascal (MPa) (meg´-uh-pas-kal´) A unit of pressure equivalent to 10 atmospheres of pressure.

megaphylls The larger leaves of modern vascular plants served by a highly-branched vascular system.

megaspores A spore from a heterosporous plant that develops into a female gametophyte bearing archegonia.

meiosis (mī-ō´-sis) A two-stage type of cell division in sexually reproducing organisms that results in cells with half the chromosome number of the original cell.

meiosis I The first division of a two-stage process of cell division in sexually reproducing organisms that results in cells with half the chromosome number of the original cell.

meiosis II The second division of a two-stage process of cell division in sexually reproducing organisms that results in cells with half the chromosome number of the original cell.

melanocyte-stimulating hormone A hormone that regulates the activity of pigment-containing cells in the skin of some vertebrates.

melatonin A modified amino acid hormone secreted by the pineal gland.

membrane attack complex (MAC) A molecular complex including complement proteins that generates a 7–10-nm diameter pore in a bacterial membrane, causing the cell to die.

membrane potential The charge difference between the cytoplasm and extracellular fluid in all cells, due to the differential distribution of ions. Membrane potential affects the activity of excitable cells and the transmembrane movement of all charged substances.

memory cell A clone of long-lived lymphocytes, formed during the primary immune response, that remains in a lymph node until activated by exposure to the same antigen that triggered its formation. Activated memory cells mount the secondary immune response.

menstrual cycle (men´-strū-ul) A type of reproductive cycle in higher female primates, in which the nonpregnant endometrium is shed as a bloody discharge through the cervix into the vagina.

menstrual flow phase That portion of the menstrual cycle when menstrual bleeding occurs.

menstruation The shedding of portions of the endometrium during a menstrual cycle.

meristem (mār´-uh-stem) Plant tissue that remains embryonic as long as the plant lives, allowing for indeterminate growth.

meristem identity genes Plant genes that promote the switch from vegetative growth to flowering.

meroblastic cleavage (mār´-ō-blas´-tik) A type of cleavage in which there is incomplete division of yolk-rich egg, characteristic of avian development.

mesentery (mez´-en-tār-ē) A membrane that suspends many of the organs of vertebrates inside fluid-filled body cavities.

mesoderm (mez´-ō-derm) The middle primary germ layer of an early embryo that develops into the notochord, the lining of the coelom, muscles, skeleton, gonads, kidneys, and most of the circulatory system.

mesohyl (mes´-uh-hil) A gelatinous region between the two layers of cells of a sponge.

mesophyll (mez´-ō-fil) The ground tissue of a leaf, sandwiched between the upper and lower epidermis and specialized for photosynthesis.

mesophyll cell A loosely arranged photosynthetic cell located between the bundle sheath and the leaf surface.

mesotrophic Lakes with moderate amounts of nutrients and phytoplankton productivity intermediate to oligotrophic and eutrophic systems.

messenger RNA (mRNA) A type of RNA, synthesized from DNA, that attaches to ribosomes in the cytoplasm and specifies the primary structure of a protein.

metabolism (muh-tab´-uh-lizm) The totality of an organism's chemical reactions, consisting of catabolic and anabolic pathways.

metamorphosis (met´-uh-mōr´-fuh-sis) The resurgence of development in an animal larva that transforms it into a sexually mature adult.

metanephridium (met´-uh-nuh-frid´-ē-um) (plural, **metanephridia**) In annelid worms, a type of excretory tubule with internal openings called nephrostomes that collect body fluids and external openings called nephridiopores.

metaphase The third subphase of mitosis, in which the spindle is complete and the chromosomes, attached to microtubules at their kinetochores, are all aligned at the metaphase plate.

metaphase plate An imaginary plane during metaphase in which the centromeres of all the duplicated chromosomes are located midway between the two poles.

metapopulation A subdivided population of a single species.

metastasis (muh-tas´-tuh-sis) The spread of cancer cells to locations distant from their original site.

methanogen Microorganisms that obtain energy by using carbon dioxide to oxidize hydrogen, producing methane as a waste product.

microclimate Very fine scale variations of climate, such as the specific climatic conditions underneath a log.

microevolution A change in the gene pool of a population from generation to generation.

microfilament A solid rod of actin protein in the cytoplasm of almost all eukaryotic cells, making up part of the cytoskeleton and acting alone or with myosin to cause cell contraction.

micronutrient An element that an organism needs in very small amounts and that functions as a component or cofactor of enzymes. *See also* macronutrient.

microphylls The small leaves of lycophytes that have only a single, unbranched vein.

microspores A spore from a heterosporous plant that develops into a male gametophyte with antheridia.

microtubule A hollow rod of tubulin protein in the cytoplasm of all eukaryotic cells and in cilia, flagella, and the cytoskeleton.

microvillus (plural, **microvilli**) One of many fine, fingerlike projections of the epithelial cells in the lumen of the small intestine that increase its surface area.

midbrain One of three ancestral and embryonic regions of the vertebrate brain; develops into sensory integrating and relay centers that send sensory information to the cerebrum.

middle ear One of three main regions of the vertebrate ear; a chamber containing three small bones (the hammer, anvil, and stirrup), that convey vibrations from the eardrum to the oval window.

middle lamella (luh-mel´-uh) A thin layer of adhesive extracellular material, primarily pectins, found between the primary walls of adjacent young plant cells.

mimicry A phenomenon in which one species benefits by a superficial resemblance to an unrelated species. A predator or species of prey may gain a significant advantage through mimicry.

mineral In nutrition, a chemical element other than hydrogen, oxygen, or nitrogen that an organism requires for proper body functioning.

mineral nutrients Essential chemical elements absorbed from the soil in the form of inorganic ions.

mineralocorticoid A corticosteroid hormone secreted by the adrenal cortex that regulates salt and water homeostasis.

minimum dynamic area The amount of suitable habitat needed to sustain a viable population.

minimum viable population size (MVP) The smallest number of individuals needed to perpetuate a population.

mismatch repair The cellular process that uses special enzymes to fix incorrectly paired nucleotides.

missense mutation The most common type of mutation, a base-pair substitution in which the new codon makes sense in that it still codes for an amino acid.

mitochondrial matrix The compartment of the mitochondrion enclosed by the inner membrane and containing enzymes and substrates for the Krebs cycle.

mitochondrion (mī´-tō-kon´-drē-on) (plural, **mitochondria**) An organelle in eukaryotic cells that serves as the site of cellular respiration.

mitosis (mī-tō´-sis) A process of nuclear division in eukaryotic cells conventionally divided into five stages: prophase, prometaphase, metaphase, anaphase, and telophase. Mitosis conserves chromosome number by equally allocating replicated chromosomes to each of the daughter nuclei.

mitotic phase The phase of the cell cycle that includes mitosis and cytokinesis.

mitotic spindle An assemblage of microtubules and associated proteins that is involved in the movements of chromosomes during mitosis.

model organism An organism chosen to study broad biological principles.

modern synthesis A comprehensive theory of evolution emphasizing natural selection, gradualism, and populations as the fundamental units of evolutionary change; also called neo-Darwinism.

molarity A common measure of solute concentration, referring to the number of moles of solute in 1 L of solution.

mold A rapidly growing, asexually reproducing fungus.

mole The number (mol) of grams of a substance that equals its molecular weight in daltons and contains Avogadro's number of molecules.

molecular clocks Evolutionary timing methods based on the observation that at least some regions of genomes evolve at constant rates.

molecular formula A type of molecular notation indicating only the quantity of the constituent atoms.

molecular weight The sum of the weights of all the atoms in a molecule.

molecule Two or more atoms held together by covalent bonds.

molting A process in arthropods in which the exoskeleton is shed at intervals to allow growth by the secretion of a larger exoskeleton.

monoclonal antibody (mon´-ō-klōn´-ul) A defensive protein produced by cells descended from a single cell; an antibody that is secreted by a clone of cells and, consequently, is specific for a single antigenic determinant.

monocot (mon´-ō-kot) A subdivision of flowering plants whose members possess one embryonic seed leaf, or cotyledon.

monoculture Cultivation of large land areas with a single plant variety.

monocyte An agranular leukocyte that is able to migrate into tissues and transform into a macrophage.

monoecious (muh-nē´-shus) Referring to a plant species that has both staminate and carpellate flowers on the same individual.

monogamous A type of relationship in which one male mates with just one female.

monohybrid An organism that is heterozygous with respect to a single gene of interest. A monohybrid results from a cross between parents homozygous for different alleles. For example, parents of genotypes *AA* and *aa* produce a monohybrid genotype of *Aa*.

monomer (mon´-uh-mer) The subunit that serves as the building block of a polymer.

monophyletic (mon´-ō-fī-let´-ik) Pertaining to a taxon derived from a single ancestral species that gave rise to no species in any other taxa.

monosaccharide (mon´-ō-sak´-uh-rīd) The simplest carbohydrate, active alone or serving as a monomer for disaccharides and polysaccharides. Also known as simple sugars, the molecular formulas of monosaccharides are generally some multiple of CH_2O.

monosomic A chromosomal condition in which a particular cell has only one copy of a chromosome, instead of the normal two; the cell is said to be monosomic for that chromosome.

monotreme (mon´-uh-trēm) An egg-laying mammal, represented by the platypus and echidna.

morphogen A substance, such as bicoid protein, that provides positional information in the form of a concentration gradient along an embryonic axis.

morphogenesis (mōr´-fō-jen´-uh-sis) The development of body shape and organization during ontogeny.

morphogens A substance that provides positional information in the form of a concentration gradient along an embryonic axis.

morphological species concept The idea that species are defined by measurable anatomical criteria.

morphospecies A species defined by its anatomical features.

morula (mōr´-yuh-luh) A solid ball of blastomeres formed by early cleavage.

mosaic A pattern of development in which an organism consists of two sets of cells that differ according to which X chromosome is inactivated.

mosaic development A pattern of development, such as that of a mollusk, in which the early blastomeres each give rise to a specific part of the embryo. In some animals, the fate of the blastomeres is established in the zygote.

mosaic evolution The evolution of different features of an organism at different rates.

motor division The efferent neurons that convey information from the CNS to the effector cells.

motor neuron A nerve cell that transmits signals from the brain or spinal cord to muscles or glands.

motor output The conduction of signals from a processing center in a central nervous system to effector cells.

motor unit A single motor neuron and all the muscle fibers it controls.

movement corridor A series of small clumps or a narrow strip of quality habitat (usable by organisms) that connects otherwise isolated patches of quality habitat.

MPF (M-phase-promoting factor) A protein complex required for a cell to progress from late interphase to mitosis; the active form consists of cyclin and *cdc2*, a protein kinase.

mucous membrane (myū´-kus) Smooth moist epithelium that lines the digestive tract and air tubes leading to the lungs.

Müllerian mimicry (myū-lār´-ē-un) A mutual mimicry by two unpalatable species.

multifactorial A type of phenotypic character influenced by genetic and environmental factors.

multigene family A collection of genes with similar or identical sequences, presumably of common origin.

multiple fruit A fruit such as pineapple that develops from an inflorescence, a group of flowers tightly clustered together. When the walls of the many ovaries start to thicken, they fuse together and become incorporated into one fruit.

multiregional hypothesis The idea that modern humans evolved in each region of the Earth from local populations of *Homo erectus*.

muscle spindle An interoreceptor stimulated by mechanical distortion. Also called a stretch receptor.

muscle tissue Tissue consisting of long muscle cells that are capable of contracting when stimulated by nerve impulses.

mutagen (myū´-tuh-jen) A chemical or physical agent that interacts with DNA and causes a mutation.

mutagenesis (myū´-tuh-gen´-uh-sis) The creation of mutations.

mutant phenotypes Traits that are alternatives to the wild type.

mutation (myū-tā´-shun) A rare change in the DNA of a gene ultimately creating genetic diversity.

mutualism (myū´-chū-ul-izm) A symbiotic relationship in which both participants benefit.

mycelium (mī-sē´-lē-um) The densely branched network of hyphae in a fungus.

mycorrhiza (mī´-kō-rī´-zuh) A mutualistic association of plant root and fungus.

mycosis The general term for a fungal infection.

myelin sheath (mī´-uh-lin) In a neuron, an insulating coat of cell membrane from Schwann cells that is interrupted by nodes of Ranvier where saltatory conduction occurs.

myofibril (mī´-ō-fī´-bril) A fibril collectively arranged in longitudinal bundles in muscle cells (fibers); composed of thin filaments of actin and a regulatory protein and thick filaments of myosin.

myofilaments The thick and thin filaments that form the myofibrils.

myoglobin (mī´-uh-glō´-bin) An oxygen-storing, pigmented protein in muscle cells.

myosin (mī´-uh-sin) A type of protein filament that interacts with actin filaments to cause cell contraction.

myotonia Increased muscle tension.

NAD$^+$ Nicotinamide adenine dinucleotide, a coenzyme present in all cells that helps enzymes transfer electrons during the redox reactions of metabolism.

NADP$^+$ An acceptor that temporarily stores energized electrons produced during the light reactions.

natural family planning A form of contraception that relies upon refraining from sexual intercourse when conception is most likely to occur; also called the rhythm method.

natural killer (NK) cell A nonspecific defensive cell that attacks tumor cells and destroys infected body cells, especially those harboring viruses.

natural selection Differential success in the reproduction of different phenotypes resulting from the interaction of organisms with their environment. Evolution occurs when natural selection causes changes in relative frequencies of alleles in the gene pool.

natural theology A philosophy dedicated to discovering the Creator's plan by studying nature. Adaptations of organisms are viewed as evidence that the Creator had designed each and every species for a particular purpose.

negative feedback A primary mechanism of homeostasis, whereby a change in a physiological variable that is being monitored triggers a response that counteracts the initial fluctuation.

negative pressure breathing A breathing system in which air is pulled into the lungs.

nematocysts (nem´-uh-tuh-sists) Stinging components of cnidocytes.

nephron (nef´-ron) The tubular excretory unit of the vertebrate kidney.

neritic zone (nuh-rit´-ik) The shallow regions of the ocean overlying the continental shelves.

nerve A ropelike bundle of neuron fibers (axons and dendrites) tightly wrapped in connective tissue.

nerve net A weblike system of neurons, characteristic of radially symmetrical animals such as Hydra.

nervous tissue Tissue made up of neurons and supportive cells.

net primary production (NPP) The gross primary production of an ecosystem minus the energy used by the producers for respiration.

neural crest A band of cells along the border where the neural tube pinches off from the ectoderm; the cells migrate to various parts of the embryo and form the pigment cells in the skin, bones of the skull, the teeth, the adrenal glands, and parts of the peripheral nervous system.

neural tube A tube of cells running along the dorsal axis of the body, just dorsal to the notochord. It will give rise to the central nervous system.

neurohypophysis (ner´-ō-hī-pof´-uh-sis) Also called the posterior pituitary, it is an extension of the brain. The neurohypophysis stores and secretes oxytocin and antidiuretic hormone, both produced by neurosecretory cells in the hypothalamus.

neuron (ner´-on) A nerve cell; the fundamental unit of the nervous system, having structure and properties that allow it to conduct signals by taking advantage of the electrical charge across its cell membrane.

neuropeptides Relatively short chains of amino acids that serve as neurotransmitters.

neurosecretory cells Hypothalamus cells that receive signals from other nerve cells, but instead of signaling to an adjacent nerve cell or muscle, they release hormones into the bloodstream.

neurotransmitter A chemical messenger released from the synaptic terminal of a neuron at a chemical synapse that diffuses across the synaptic cleft and binds to and stimulates the postsynaptic cell.

neutral variation Genetic diversity that confers no apparent selective advantage.

neutralization An immune response in which an antibody binds to and blocks the activity of an antigen.

neutron An electrically neutral particle (a particle having no electrical charge), found in the nucleus of an atom.

neutrophil The most abundant type of leukocyte. Neutrophils tend to self-destruct as they destroy foreign invaders, limiting their life span to but a few days.

niche *See* ecological niche.

nitric oxide A local regulator gas produced by many types of cells.

nitrogen fixation The assimilation of atmospheric nitrogen by certain prokaryotes into nitrogenous compounds that can be directly used by plants.

nitrogenase (nī-troj´-uh-nāz) An enzyme complex, unique to certain prokaryotes, that reduces N_2 to NH_3.

nitrogen-fixing bacteria Microorganisms that restock nitrogenous minerals in the soil by converting nitrogen to ammonia.

nociceptor A class of naked dendrites in the epidermis of the skin.

node A point along the stem of a plant at which leaves are attached.

nodes of Ranvier (ran-vēr´) The small gaps in the myelin sheath between successive glial cells along the axon of a neuron; also, the site of high concentration of voltage-gated ion channels.

nodules Swellings on the roots of legumes. Nodules are composed of plant cells that contain nitrogen-fixing bacteria of the genus *Rhizobium*.

noncompetitive inhibitor A substance that reduces the activity of an enzyme by binding to a location remote from the active site, changing its conformation so that it no longer binds to the substrate.

noncyclic electron flow A route of electron flow during the light reactions of photosynthesis that involves both photosystems and produces ATP, NADPH, and oxygen. The net electron flow is from water to NADP$^+$.

noncyclic photophosphorylation (fō´-tō-fos-fōr´-uh-lā´-shun) The production of ATP by noncyclic electron flow.

nondisjunction An accident of meiosis or mitosis, in which the members of a pair of homologous chromosomes or sister chromatids fail to move apart properly.

nonequilibrium model The model of communities that emphasizes that they are not stable in time but constantly changing after being buffeted by disturbances.

nonpolar covalent bond A type of covalent bond in which electrons are shared equally between two atoms of similar electronegativity.

nonsense mutation A mutation that changes an amino acid codon to one of the three stop codons, resulting in a shorter and usually nonfunctional protein.

nonshivering thermogenesis The increased production of heat in some mammals by the action of certain hormones that cause mitochondria to increase their metabolic activity and produce heat instead of ATP.

norepinephrine A hormone produced in response to stress.

norm of reaction The range of phenotypic possibilities for a single genotype, as influenced by the environment.

notochord A long flexible rod that runs along the dorsal axis of the body in the future position of the vertebral column.

nuclear envelope The membrane in eukaryotes that encloses the nucleus, separating it from the cytoplasm.

nuclear lamina A netlike array of protein filaments that maintains the shape of the nucleus.

nuclease A team of enzymes that hydrolyze DNA and RNA into their component nucleotides.

nucleic acid A polymer (polynucleotide) consisting of many nucleotide monomers; serves as a blueprint for proteins and, through the actions of proteins, for all cellular activities. The two types are DNA and RNA.

nucleic acid hybridization Base pairing between a gene and a complementary sequence on another nucleic acid molecule.

nucleic acid probe (nū-klā´-ik) In DNA technology, a labeled single-stranded nucleic acid molecule used to tag a specific nucleotide sequence in a nucleic acid sample. Molecules of the probe hydrogen-bond to the complementary sequence wherever it occurs; radioactive or other labeling of the probe allows its location to be detected.

nucleoid (nū´-klē-oid) A dense region of DNA in a prokaryotic cell.

nucleoid region The region in a prokaryotic cell consisting of a concentrated mass of DNA.

nucleolus (nū-klē´-ō-lus) (plural, **nucleoli**) A specialized structure in the nucleus, formed from various chromosomes and active in the synthesis of ribosomes.

nucleoside (nū´-klē-ō-sīd) An organic molecule consisting of a nitrogenous base joined to a five-carbon sugar.

nucleosome (nū´-klē-ō-sōm) The basic, beadlike unit of DNA packaging in eukaryotes, consisting of a segment of DNA wound around a protein core composed of two copies of each of four types of histone.

nucleotide (nu-kle-o-tıd) The building block of a nucleic acid, consisting of a five-carbon sugar covalently bonded to a nitrogenous base and a phosphate group.

nucleotide excision repair The process of removing and then correctly replacing a damaged segment of DNA using the undamaged strand as a guide.

nucleus (1) An atom's central core, containing protons and neutrons. (2) The chromosome-containing organelle of a eukaryotic cell. (3) A cluster of neurons.

obligate aerobe (ob´ lig et ār´ ōb) An organism that requires oxygen for cellular respiration and cannot live without it.

obligate anaerobe (ob´-lig-et an´-uh-rōb) An organism that cannot use oxygen and is poisoned by it.

oceanic pelagic biome Most of the ocean's waters far from shore, constantly mixed by ocean currents.

oceanic zone The region of water lying over deep areas beyond the continental shelf.

olfactory receptors Smell receptors.

oligodendrocytes (ol´-ig-ō-den´-druh-sīt) Glial cells that form insulating myelin sheaths around the axons of neurons in the central nervous system.

oligotrophic lake A nutrient-poor, clear, deep lake with minimum phytoplankton.

ommatidia (ōm´-uh-tid´-ē-uh) The facets of the compound eye of arthropods and some polychaete worms.

omnivore A heterotrophic animal that consumes both meat and plant material.

oncogene (on´-kō-jēn) A gene found in viruses or as part of the normal genome that is involved in triggering cancerous characteristics.

one gene-one polypeptide hypothesis The premise that a gene is a segment of DNA that codes for one polypeptide.

ontogeny (on-toj´-uh-nē) The embryonic development of an organism.

oogamy (ō-og´-uh-mē) A condition in which male and female gametes differ, such that a small, flagellated sperm fertilizes a large, nonmotile egg.

oogenesis (ō´-uh-jen´-uh-sis) The process in the ovary that results in the production of female gametes.

open circulatory system An arrangement of internal transport in which blood bathes the organs directly and there is no distinction between blood and interstitial fluid.

operant conditioning (op´-er-ent) A type of associative learning in which an animal learns to associate one of its own behaviors with a reward or punishment and then tends to repeat or avoid that behavior. Also called trial-and-error learning.

operator In prokaryotic DNA, a sequence of nucleotides near the start of an operon to which an active repressor can attach. The binding of the repressor prevents RNA polymerase from attaching to the promoter and transcribing the genes of the operon.

operculum A protective flap that covers the gills of fishes.

operon (op´-er-on) A unit of genetic function common in bacteria and phages, consisting of coordinately regulated clusters of genes with related functions.

opportunistic Microorganisms that are normal residents of a host but can cause illness when the host's defenses are weakened by such factors as poor nutrition or a recent bout with the flu.

opposable thumb An arrangement of the fingers such that the thumb can touch the ventral surface of the fingertips of all four fingers.

opsin A membrane protein bonded to a light-absorbing pigment molecule.

opsonization An immune response in which the binding of antibodies to the surface of a microbe facilitates phagocytosis of the microbe by a macrophage.

optic chiasm The arrangement of the nerve tracts of the eye such that the visual sensations from the left visual field of both eyes are transmitted to the right side of the brain and the sensations from the right visual field of both eyes are transmitted to the left side of the brain.

optimal foraging theory The basis for analyzing behavior as a compromise of feeding costs versus feeding benefits.

oral cavity The mouth of an animal.

order In classification, the taxonomic category above family.

organ A specialized center of body function composed of several different types of tissues.

organ of Corti The actual hearing organ of the vertebrate ear, located in the floor of the cochlear canal in the inner ear; contains the receptor cells (hair cells) of the ear.

organ system A group of organs that work together in performing vital body functions.

organelle (ōr-guh-nel´) One of several formed bodies with specialized functions, suspended in the cytoplasm of eukaryotic cells.

organic chemistry The study of carbon compounds (organic compounds).

organ identity gene A plant gene in which a mutation causes a floral organ to develop in the wrong location.

organismal ecology The branch of ecology concerned with the morphological, physiological, and behavioral ways in which individual organisms meet the challenges posed by their biotic and abiotic environments.

organogenesis (ōr-gan´-ō-jen´-uh-sis) The development of organ rudiments from the three germ layers.

orgasm Rhythmic, involuntary contractions of certain reproductive structures in both sexes during the human sexual response cycle.

origins of replication Sites where the replication of a DNA molecule begins.

osculum A large opening in a sponge that connects the spongocoel to the environment.

osmoconformer An animal that does not actively adjust its internal osmolarity because it is isotonic with its environment.

osmolarity (oz´-mō-lār´-uh-tē) Solute concentration expressed as molarity.

osmoregulation The control of water balance in organisms living in hypertonic, hypotonic, or terrestrial environments.

osmoregulator An animal whose body fluids have a different osmolarity than the environment, and that must either discharge excess water if it lives in a hypotonic environment or take in water if it inhabits a hypertonic environment.

osmosis (oz-mō´-sis) The diffusion of water across a selectively permeable membrane.

osmotic pressure (oz-mot´-ik) A measure of the tendency of a solution to take up water when separated from pure water by a selectively permeable membrane.

Osteichthyes (os-tē-ik´-thēz) The vertebrate class of bony fishes, characterized by a skeleton reinforced by calcium phosphate; the most abundant and diverse vertebrates.

osteoblasts Bone-forming cells that deposit a matrix of collagen.

osteons The repeating organizational units forming the microscopic structure of hard mammalian bone.

ostracoderm (os-trak´-uh-derm) An extinct agnathan; a fishlike creature encased in an armor of bony plates.

outer ear One of three main regions of the ear in reptiles, birds , and mammals; made up of the auditory canal and, in many birds and mammals, the pinna.

outgroup A species or group of species that is closely related to the group of species being studied, but clearly not as closely related as any study-group members are to each other.

"out of Africa" hypothesis The idea that modern humans evolved from a second migration out of Africa that occurred about 100,000 years ago, replacing all the regional populations of hominids derived from the first migrations of *Homo erectus* out of Africa about 1.5 million years ago.

oval window In the vertebrate ear, a membrane-covered gap in the skull bone, through which sound waves pass from the middle ear to the inner ear.

ovarian cycle (ō-vār´-ē-un) The cyclic recurrence of the follicular phase, ovulation, and the luteal phase in the mammalian ovary, regulated by hormones.

ovary (ō´-vuh-rē) (1) In flowers, the portion of a carpel in which the egg-containing ovules develop. (2) In animals, the structure that produces female gametes and reproductive hormones.

overnourishment A diet that is chronically excessive in calories.

oviduct (ō´-vuh-duct) A tube passing from the ovary to the vagina in invertebrates or to the uterus in vertebrates.

oviparous (ō-vip´-uh-rus) Referring to a type of development in which young hatch from eggs laid outside the mother's body.

ovoviviparous (ō´-vō-vī-vip´-uh-rus) Referring to a type of development in which young hatch from eggs that are retained in the mother's uterus.

ovulation The release of an egg from ovaries. In humans, an ovarian follicle releases an egg during each menstrual cycle.

ovule (ō´-vyūl) A structure that develops in the plant ovary and contains the female gametophyte.

ovum (ō´-vum) The female gamete; the haploid, unfertilized egg, which is usually a relatively large, non-motile cell.

oxidation The loss of electrons from a substance involved in a redox reaction.

oxidative phosphorylation (fos´-for-uh-lā-shun) The production of ATP using energy derived from the redox reactions of an electron transport chain.

oxidizing agent The electron acceptor in a redox reaction.

oxytocin A hormone produced by the hypothalamus and released by the posterior pituitary. It induces contractions of the uterine muscles and causes the mammary glands to eject milk during nursing.

P generation The parent individuals from which offspring are derived in studies of inheritance; P stands for parental.

P site One of three binding sites for tRNA during translation; it holds the tRNA carrying the growing polypeptide chain; P stands for peptidyl-tRNA site).

p53 gene The "guardian angel of the genome," p53 is expressed when a cell's DNA is damaged. Its product, p53 protein, functions as a transcription factor for several genes.

pacemaker A specialized region of the right atrium of the mammalian heart that sets the rate of contraction; also called the sinoatrial (SA) node.

paedogenesis (pē´-dō-jen´-uh-sis) The precocious development of sexual maturity in a larva.

paedomorphosis (pē´-duh-mōr´-fuh-sis) The retention in an adult organism of the juvenile features of its evolutionary ancestors.

pain receptor A category of interoreceptors that detect pain.

pair-rule genes Genes that define the modular patterns in terms of pairs of segments in *Drosophila*. Mutations in these genes result in embryos with half the normal segment number because every other segment fails to develop.

paleoanthropology The study of human origins and evolution.

paleontology (pā´-lē-un-tol´-ō-jē) The scientific study of fossils.

pancreas (pan´-krē-us) A gland with dual functions: The nonendocrine portion secretes digestive enzymes and an alkaline solution into the small intestine via a duct; the endocrine portion secretes the hormones insulin and glucagon into the blood.

Pangaea (pan-jē-uh) The supercontinent formed near the end of the Paleozoic era when plate movements brought all the landmasses of Earth together.

parabasalids A group of protistans, including the trichomonads, that lacks mitochondria.

parabronchi The sites of gas exchange in bird lungs. They allow air to flow past the respiratory surface in just one direction.

paraphyletic (pār´-uh-fī-let´-ik) Pertaining to a taxon that excludes some members that share a common ancestor with members included in the taxon.

parasite (pār´-uh-sīt) An organism that absorbs nutrients from the body fluids of living hosts.

parasitism (pār´-uh-sit-izm) A symbiotic relationship in which the symbiont (parasite) benefits at the expense of the host by living either within the host (as an endoparasite) or outside the host (as an ectoparasite).

parasympathetic division One of two divisions of the autonomic nervous system; generally enhances body activities that gain and conserve energy, such as digestion and reduced heart rate.

parathyroid glands Four endocrine glands, embedded in the surface of the thyroid gland, that secrete parathyroid hormone and raise blood calcium levels.

parathyroid hormone (PTH) A peptide hormone secreted by the parathyroid glands that raise blood calcium level.

parazoa (pār´-uh-zō-uh) Members of the subkingdom of animals consisting of the sponges.

parenchyma cell (puh-ren´-kim-uh) A relatively unspecialized plant cell type that carries most of the metabolism, synthesizes and stores organic products, and develops into a more differentiated cell type.

parental investment The time and resources an individual must spend to produce and nurture offspring.

parental types Offspring with a phenotype that matches one of the parental phenotypes.

parsimony (par´-suh-mō´-nē) In scientific studies, the search for the least complex explanation for an observed phenomenon.

parthenogenesis (par´-thuh-nō-jen´-uh-sis) A type of reproduction in which females produce offspring from unfertilized eggs.

partial pressure The concentration of gases; a fraction of total pressure.

parturition The expulsion of a baby from the mother, also called birth.

passeriformes The order of perching birds.

passive immunity Temporary immunity obtained by acquiring ready-made antibodies or immune cells; lasts only a few weeks or months because the immune system has not been stimulated by antigens.

passive transport The diffusion of a substance across a biological membrane.

pattern formation The ordering of cells into specific three-dimensional structures, an essential part of shaping an organism and its individual parts during development.

peat Extensive deposits of undecayed organic material formed primarily from the wetland moss *Sphagnum*.

peatlands Extensive high latitude boreal wetlands occupied by *Sphagnum*.

pedigree A family tree describing the occurrence of heritable characters in parents and offspring across as many generations as possible.

pelagic zone (puh-laj´-ik) The area of the ocean past the continental shelf, with areas of open water often reaching to very great depths.

penis The copulatory structure of male mammals.

PEP carboxylase An enzyme that adds carbon dioxide to phosphoenolpyruvate (PEP) to form oxaloacetate.

pepsin An enzyme present in gastric juice that begins the hydrolysis of proteins.

pepsinogen The inactive form of pepsin that is first secreted by specialized (chief) cells located in gastric pits of the stomach.

peptide bond The covalent bond between two amino acid units, formed by a dehydration reaction.

peptidoglycan (pep´-tid-ō-glī´-kun) A type of polymer in bacterial cell walls consisting of modified sugars cross-linked by short polypeptides.

perception The interpretation of sensations by the brain.

perennial (puh-ren´-ē-ul) A plant that lives for many years.

perforin (per´-fuh-rin) A protein secreted by a cytotoxic T cell that lyses (ruptures) an infected cell by perforating its membrane.

pericarp The thickened wall of a fruit.

pericycle (pār´-uh-sī-kul) A layer of cells just inside the endodermis of a root that may become meristematic and begin dividing again.

periderm (pār-uh-derm) The protective coat that replaces the epidermis in plants during secondary growth, formed of the cork and cork cambium.

peripheral nervous system (PNS) The sensory and motor neurons that connect to the central nervous system.

peripheral protein Protein appendages loosely bound to the surface of the membrane and not embedded in the lipid bilayer.

peripheral resistance The impedance of blood flow by the arterioles.

peristalsis (pār´-uh-stal´-sis) Rhythmic waves of contraction of smooth muscle that push food along the digestive tract.

peristome The upper part of the moss capsule (sporangium) often specialized for gradual spore discharge.

peritubular capillaries The network of tiny blood vessels that surrounds the proximal and distal tubules in the kidney.

permafrost A permanently frozen stratum below the arctic tundra.

peroxisome (puh-rok´-suh-sōm) A microbody containing enzymes that transfer hydrogen from various substrates to oxygen, producing and then degrading hydrogen peroxide.

petal A modified leaf of a flowering plant. Petals are the often colorful parts of a flower that advertise it to insects and other pollinators.

petiole (pet´-ē-ōl) The stalk of a leaf, which joins the leaf to a node of the stem.

pH A measure of hydrogen ion concentration equal to $-\log [H^+]$ and ranging in value from 0 to 14.

phage (fāj) A virus that infects bacteria; also called a bacteriophage.

phagocytosis (fag´-ō-sī-tō´-sis) A type of endocytosis involving large, particulate substances.

pharynx (fār´-inks) An area in the vertebrate throat where air and food passages cross; in flatworms, the muscular tube that protrudes from the ventral side of the worm and ends in the mouth.

phase change A shift from one developmental phase to another.

phenetics (fuh-net´-iks) An approach to taxonomy based entirely on measurable similarities and differences in phenotypic characters, without consideration of homology, analogy, or phylogeny.

phenotype (fē´-nō-tīp) The physical and physiological traits of an organism.

pheromone (fār´-uh-mōn) A small, volatile chemical signal that functions in communication between animals and acts much like a hormone in influencing physiology and behavior.

phloem (flō´-um) The portion of the vascular system in plants consisting of living cells arranged into elongated tubes that transport sugar and other organic nutrients throughout the plant.

phoronids Tube-dwelling marine worms ranging from 1 mm to 50 cm in length.

phosphate group (fos´-fāt) A functional group important in energy transfer.

phospholipid (fos´-fō-lip´-id) A molecule that is a constituent of the inner bilayer of biological membranes, having a polar, hydrophilic head and a nonpolar, hydrophobic tail.

phosphorylated A molecule that has been the recipient of a phosphate group.

photic zone (fō´-tic) The narrow top slice of the ocean, where light permeates sufficiently for photosynthesis to occur.

photoautotroph (fō´-tō-ot´-ō-trōf) An organism that harnesses light energy to drive the synthesis of organic compounds from carbon dioxide.

photoheterotroph (fō´-tō-het´-uh-rō-trōf) An organism that uses light to generate ATP but that must obtain carbon in organic form.

photon (fō´-ton) A quantum, or discrete amount, of light energy.

photoperiodism (fō´-tō-pēr´-ē-ō-dizm) A physiological response to day length, such as flowering in plants.

photophosphorylation (fō´-tō-fos´-fōr-uh-lā´-shun) The process of generating ATP from ADP and phosphate by means of a proton-motive force generated by the thylakoid membrane of the chloroplast during the light reactions of photosynthesis.

photopsin (fō-top´-sin) One of a family of visual pigments in the cones of the vertebrate eye that absorb bright, colored light.

photoreceptor Receptors of light.

photorespiration A metabolic pathway that consumes oxygen, releases carbon dioxide, generates no ATP, and decreases photosynthetic output; generally occurs on hot, dry, bright days, when stomata close and the oxygen concentration in the leaf exceeds that of carbon dioxide.

photosynthesis The conversion of light energy to chemical energy that is stored in glucose or other organic compounds; occurs in plants, algae, and certain prokaryotes.

photosystem The light-harvesting unit in photosynthesis, located on the thylakoid membrane of the chloroplast and consisting of the antenna complex, the reaction-center chlorophyll *a*, and the primary electron acceptor. There are two types of photosystems, I and II; they absorb light best at different wavelengths.

photosystem I One of two light-harvesting units of a chloroplast's thylakoid membrane; it uses the P700 reaction-center chlorophyll.

photosystem II One of two light-harvesting units of a chloroplast's thylakoid membrane; it uses the P680 reaction-center chlorophyll.

phototropism (fō´-tō-trō´-pizm) Growth of a plant shoot toward or away from light.

phragmoplast An alignment of cytoskeletal elements and Golgi-derived vesicles across the mid-line of a dividing plant cell.

phylogeny (fī-loj´-uh-nē) The evolutionary history of a species or group of related species.

phylum (fī´-lum) A taxonomic category. Phyla are divided into classes.

phylum Anthocerophyta The group of hornworts, small herbaceous (nonwoody) plants.

phylum Bryophyta A formal group of mosses. Note that the term "bryophyte" refers instead to the informal group of mosses, liverworts, and hornworts, nonvascular plants that inhabit the land but lack many of the terrestrial adaptations of vascular plants.

phylum Chelicerata The animal group that includes horseshoe crabs, scorpions, ticks, spiders, and an extinct group called the eurypterids.

phylum Crustacea The animal group that includes mostly aquatic animals such as crabs, lobsters, crayfish, and shrimp.

phylum Hepatophyta The group of liverworts, small herbaceous (nonwoody) plants.

phylum Trilobita An extinct group of arthropods with pronounced segmentation.

phylum Uniramia The animal group that includes centipedes, millipedes, and insects.

physiology The study of the functions of an organism.

phytoalexin (fī´-tō-uh-lek´-sin) An antibiotic, produced by plants, that destroys microorganisms or inhibits their growth.

phytochrome (fī´-tuh-krōm) A pigment involved in many responses of plants to light.

phytoplankton (fī´-tō-plank´-ton) Algae and photosynthetic bacteria that drift passively in the pelagic zone of an aquatic environment.

phytoremediation An emerging, nondestructive technology that seeks to cheaply reclaim contaminated areas by taking advantage of the remarkable ability of some plant species to extract heavy metals and other pollutants from the soil and to concentrate them in easily harvested portions of the plant.

pilus (plural, **pili**) (pī´-lus, pī´-lī) A surface appendage in certain bacteria that functions in adherence and the transfer of DNA during conjugation.

pineal gland (pin´-ē-ul) A small endocrine gland on the dorsal surface of the vertebrate forebrain; secretes the hormone melatonin, which regulates body functions related to seasonal day length.

pinocytosis (pī´-nō-sī-tō´-sis) A type of endocytosis in which the cell ingests extracellular fluid and its dissolved solutes.

pitch A function of a sound wave's frequency, or number of vibrations per second, expressed in hertz.

pith The core of the central vascular cylinder of monocot roots, consisting of parenchyma cells, which are ringed by vascular tissue; ground tissue interior to vascular bundles in dicot stems.

pits Thinner regions in the walls of tracheids and vessels where only primary walls are present.

pituitary gland (puh-tū´-uh-tār-ē) An endocrine gland at the base of the hypothalamus; consists of a posterior lobe (neurohypophysis), which stores and releases two hormones produced by the hypothalamus, and an anterior lobe (adenohypophysis), which produces and secretes many hormones that regulate diverse body functions.

placenta (pluh-sen´-tuh) A structure in the pregnant uterus for nourishing a viviparous fetus with the mother's blood supply; formed from the uterine lining and embryonic membranes.

placental mammal A member of a group of mammals, including humans, whose young complete their embryonic development in the uterus, joined to the mother by a placenta.

placental transfer cells Plant cells that enhance the transfer of nutrients from parent to embryo.

Placoderm (plak´-ō-derm) A member of an extinct class of fishlike vertebrates that had jaws and were enclosed in a tough, outer armor.

planarians Carnivores that prey on smaller animals or feed on dead animals.

plankton Mostly microscopic organisms that drift passively or swim weakly near the surface of oceans, ponds, and lakes.

plantae (plan´-tā) The kingdom that contains the plants.

plasma (plaz´-muh) The liquid matrix of blood in which the cells are suspended.

plasma cell A derivative of B cells that secretes antibodies.

plasma membrane The membrane at the boundary of every cell that acts as a selective barrier, thereby regulating the cell's chemical composition.

plasmid (plaz´-mid) A small ring of DNA that carries accessory genes separate from those of a bacterial chromosome; also found in some eukaryotes, such as yeast.

plasmodesma (plaz´-mō-dez´-muh) (plural, **plasmodesmata**) An open channel in the cell wall of plants through which strands of cytosol connect from adjacent cells.

plasmodial slime mold (plaz-mō´-dē-ul) A type of protist that has amoeboid cells, flagellated cells, and an amoeboid plasmodial feeding stage in its life cycle.

plasmogamy The fusion of the cytoplasm of cells from two individuals; occurs as one stage of syngamy.

plasmolysis (plaz-mol´-uh-sis) A phenomenon in walled cells in which the cytoplasm shrivels and the plasma membrane pulls away from the cell wall when the cell loses water to a hypertonic environment.

plasmolyze The shrinkage of a cell due to water loss.

plastid One of a family of closely related plant organelles, including chloroplasts, chromoplasts, and amyloplasts (leucoplasts).

platelet A small enucleated blood cell important in blood clotting; derived from large cells in the bone marrow.

play Behavior with no apparent external goal but involves movements closely associated with goal-directed behaviors.

pleated sheet One form of the secondary structure of proteins in which the polypeptide chain folds back and forth, or where two regions of the chain lie parallel to each other and are held together by hydrogen bonds.

pleiotropy (plī´-uh-trō-pē) The ability of a single gene to have multiple effects.

plesiomorphic character (plēz´-ē-ō-mōr´-fik) A primitive phenotypic character possessed by a remote ancestor.

pluralistic species concept The idea that there is no universal explanation for the cohesion of individuals that make up species.

pluripotent stem cell A cell within bone marrow that is a progenitor for any kind of blood cell.

podocytes Specialized cells of Bowman's capsule that are permeable to water and small solutes but not to blood cells or large molecules such as plasma proteins.

point mutation A change in a gene at a single nucleotide pair.

polar covalent bond A type of covalent bond between atoms that differ in electronegativity. The shared electrons are pulled closer to the more electronegative atom, making it slightly negative and the other atom slightly positive.

polar molecule A molecule (such as water) with opposite charges on opposite sides.

polarity A lack of symmetry. Structural differences in opposite ends of an organism or structure, such as the root end and shoot end of a plant.

pollen grains The structures that contain the immature male gametophytes.

pollination (pol´-uh-nā´-shun) The placement of pollen onto the stigma of a carpel by wind or animal carriers, a prerequisite to fertilization.

poly(A) tail The modified end of the 3' end of an mRNA molecule consisting of the addition of some 50 to 250 adenine nucleotides.

polyandry (pol´-ē-an´-drē) A polygamous mating system involving one female and many males.

polygamous A type of relationship in which an individual of one sex mates with several of the other.

polygenic inheritance (pol´-ē-jen´-ik) An additive effect of two or more gene loci on a single phenotypic character.

polygyny (puh-lij´-en-ē) A polygamous mating system involving one male and many females.

polymer (pol´-uh-mer) A long molecule consisting of many similar or identical monomers linked together.

polymerase chain reaction (PCR) (puh-lim´-uh-rās) A technique for amplifying DNA *in vitro* by incubating with special primers, DNA polymerase molecules, and nucleotides.

polymorphic (pol´-ē-mōr´-fik) Referring to a population in which two or more physical forms are present in readily noticeable frequencies.

polymorphism (pol´-ē-mōr´-fizm) The coexistence of two or more distinct forms of individuals (polymorphic characters) in the same population.

polynucleotide (pol´-ē-nū´-klē-ō-tīd) A polymer consisting of many nucleotide monomers; serves as a blueprint for proteins and, through the actions of proteins, for all cellular activities. The two types are DNA and RNA.

polyp (pol´-ip) The sessile variant of the cnidarian body plan. The alternate form is the medusa.

polypeptide (pol´-ē-pep´-tīd) A polymer (chain) of many amino acids linked together by peptide bonds.

polyphyletic Pertaining to a taxon whose members were derived from two or more ancestral forms not common to all members.

polyploidy (pol´-ē-ploy´-dē) A chromosomal alteration in which the organism possesses more than two complete chromosome sets.

polyribosome (pol´-ē-rī´-bō-sōm) An aggregation of several ribosomes attached to one messenger RNA molecule.

polysaccharide (pol´-ē-sak´-uh-rīd) A polymer of up to over a thousand monosaccharides, formed by dehydration reactions.

population A group of individuals of one species that live in a particular geographic area.

population ecology The study of how members of a population interact with their environment, focusing on factors that influence population density and growth.

population genetics The study of genetic changes in populations; the science of microevolutionary changes in populations.

population viability analysis (PVA) A method of predicting whether or not a population will persist.

positional information Signals to which genes regulating development respond, indicating a cell's location relative to other cells in an embryonic structure.

positive feedback A physiological control mechanism in which a change in some variable triggers mechanisms that amplify the change.

positive pressure breathing A breathing system in which air is forced into the lungs.

posterior Pertaining to the rear, or tail, of a bilaterally symmetrical animal.

posterior pituitary An extension of the hypothalamus composed of nervous tissue that secretes hormones made in the hypothalamus; a temporary storage site for hypothalamic hormones.

postsynaptic cell The target cell as a synapse.

postsynaptic membrane The plasma membrane of the cell body or dendrite on the other side of the synapse.

postzygotic barrier (pōst´-zī-got´-ik) Any of several species-isolating mechanisms that prevent hybrids produced by two different species from developing into viable, fertile adults.

potential energy The energy stored by matter as a result of its location or spatial arrangement.

Precautionary Principle A guiding principle in making decisions about the environment, it cautions to "look before you leap" or otherwise consider carefully the potential consequences of actions.

predation An interaction between species in which one species, the predator, eats the other, the prey.

preformation The concept that development was simply the enlargement of the embryo.

pregnancy The condition of carrying one or more embryos in the uterus.

preprophase band Microtubules in the cortex (outer cytoplasm) of a cell that are concentrated into a ring.

prepuce (prē´-pyūs) A fold of skin covering the head of the clitoris and penis.

presynaptic cell The transmitting cell at a synapse.

presynaptic membrane The surface of the synaptic terminal that faces the synaptic cleft.

prezygotic barrier (prē´-zī-got´-ik) A reproductive barrier that impedes mating between species or hinders fertilization of ova if interspecific mating is attempted.

primary cell wall A relatively thin and flexible layer first secreted by a young plant cell.

primary consumer An herbivore; an organism in the trophic level of an ecosystem that eats plants or algae.

primary electron acceptor A specialized molecule sharing the reaction center with the chlorophyll *a* molecule; it accepts an electron from the chlorophyll *a* molecule.

primary germ layers The three layers (ectoderm, mesoderm, endoderm) of the late gastrula, which develop into all parts of an animal.

primary growth Growth initiated by the apical meristems of a plant root or shoot.

primary immune response The initial immune response to an antigen, which appears after a lag of several days.

primary oocyte (ō´-uh-sīt) A diploid cell, in prophase I of meiosis, that can be hormonally triggered to develop into an ovum.

primary producer An autotroph, which collectively make up the trophic level of an ecosystem that ultimately supports all other levels; usually a photosynthetic organism.

primary production The amount of light energy converted to chemical energy (organic compounds) by autotrophs in an ecosystem during a given time period.

primary structure The level of protein structure referring to the specific sequence of amino acids.

primary succession A type of ecological succession that occurs in a virtually lifeless area, where there were originally no organisms and where soil has not yet formed.

primary transcript An initial RNA transcript; also called pre-mRNA.

primary visual cortex The destination in the occipital lobe of the cerebrum for most of the axons from the lateral geniculate nuclei.

primase An enzyme that joins RNA nucleotides to make the primer.

primer An already existing RNA chain bound to template DNA to which DNA nucleotides are added during DNA synthesis.

primitive streak A groove on the surface of an early avian embryo along the future long axis of the body.

principle of allocation The concept that each organism has an energy budget, or a limited amount of total energy available for all of its maintenance and reproductive needs.

principle of parsimony The premise that a theory about nature should be the simplest explanation that is consistent with the facts.

prion An infectious form of protein that may increase in number by converting related proteins to more prions.

probe *See* nucleic acid probe.

procambium (prō-kam´-bē-um) A primary meristem of roots and shoots that forms the vascular tissue.

product An ending material in a chemical reaction.

profundal zone The deep aphotic region of a lake.

progestin (prō-jes´-tin) One of a family of steroid hormones, including progesterone, produced by the mammalian ovary; progestins prepare the uterus for pregnancy.

prognathic jaws Longer jaws found in our hominoid ancestors.

progymnosperms An extinct group of plants that is probably ancestral to gymnosperms and angiosperms.

prokaryotic cell (prō-kār´-ē-ot´-ik) A type of cell lacking a membrane-enclosed nucleus and membrane-enclosed organelles; found only in the domains Bacteria and Archaea.

prolactin A hormone produced by the anterior pituitary gland with a great diversity of effects in different vertebrate species.

proliferative phase That portion of the menstrual cycle when the endometrium regenerates and thickens.

prometaphase The second subphase of mitosis, in which discrete chromosomes consisting of identical sister chromatids appear, the nuclear envelope fragments, and the spindle microtubules attach to the kinetochores of the chromosomes.

promiscuous A type of relationship in which mating occurs with no strong pair-bonds or lasting relationships.

promoter A specific nucleotide sequence in DNA that binds RNA polymerase and indicates where to start transcribing RNA.

prophage (prō´-faj) A phage genome that has been inserted into a specific site on the bacterial chromosome.

prophase The first subphase of mitosis, in which the chromatin is condensing and the mitotic spindle begins to form, but the nucleolus and nucleus are still intact.

prosimians A suborder of primates, the premonkeys, that probably resemble early arboreal primates.

prostaglandin (PG) (pros´-tuh-glan´-din) One of a group of modified fatty acids secreted by virtually all tissues and performing a wide variety of functions as messengers.

prostate gland (pros´-tāt) A gland in human males that secretes an acid neutralizing component of semen.

protandrous A form of sequential hermaphroditism in which the male sex occurs first.

proteasome A giant protein complex that recognizes and destroys proteins tagged for elimination by the small protein ubiquitin.

protein (prō´-tēn) A three-dimensional biological polymer constructed from a set of 20 different monomers called amino acids.

protein kinase An enzyme that transfers phosphate groups from ATP to a protein.

protein phosphatase An enzyme that removes phosphate groups from proteins, often functioning to reverse the effect of a protein kinase.

proteoglycan (prō´-tē-ō-glī´-kun) A glycoprotein in the extracellular matrix of animal cells, rich in carbohydrate.

protobionts Aggregates of abiotically produced molecules.

protoderm (prō´-tō-derm) The outermost primary meristem, which gives rise to the epidermis of roots and shoots.

protogynous A form of sequential hermaphroditism in which the female sex occurs first.

proton (prō´-ton) A subatomic particle with a single positive electrical charge, found in the nucleus of an atom.

proton pump An active transport mechanism in cell membranes that consumes ATP to force hydrogen ions out of a cell and, in the process, generates a membrane potential.

protonema A mass of green, branched, one-cell thick filaments produced by germinating moss spores.

protonephridium (prō´-tō-nuh-frid´-ē-um) An excretory system, such as the flame-cell system of flatworms, consisting of a network of closed tubules having external openings called nephridiopores and lacking internal openings.

proton-motive force The potential energy stored in the form of an electrochemical gradient, generated by the pumping of hydrogen ions across biological membranes during chemiosmosis.

proto-oncogene (prō´-tō-on´-kō-jēn) A normal cellular gene corresponding to an oncogene; a gene with a potential to cause cancer but that requires some alteration to become an oncogene.

protoplast The contents of a plant cell exclusive of the cell wall.

protoplast fusion The fusing of two protoplasts from different plant species that would otherwise be reproductively incompatible.

protostome (prō´-tō-stōm) A member of one of two distinct evolutionary lines of coelomates, consisting of the annelids, mollusks, and arthropods, and characterized by spiral, determinate cleavage, schizocoelous formation of the coelom, and development of the mouth from the blastopore.

protozoan (prō´-tō-zō´-un) (plural, **protozoa**) A protist that lives primarily by ingesting food, an animal-like mode of nutrition.

protracheophyte polysporangiophytes A group of Silurian moss-like ancestors that were like bryophytes in lacking lignified vascular tissue but were different in having independent, branched, sporophytes that were not dependent on gametophytes for their growth.

provirus Viral DNA that inserts into a host genome.

proximal tubule In the vertebrate kidney, the portion of a nephron immediately downstream from Bowman's capsule that conveys and helps refine filtrate.

proximate causation The immediate mechanisms underlying an organism's behavioral, physiological, or morphological response, in contrast to ultimate, or evolutionary causation.

pseudocoelom A body cavity that is not completely lined by tissue derived from mesoderm.

pseudocoelomate (sū´-dō-sē´-lō-māt) An animal whose body cavity is not completely lined by mesoderm.

pseudogenes DNA segments very similar to real genes but which do not yield functional products.

pseudopodium (sū´-dō-pō´-dē-um) (plural, **pseudopodia**) A cellular extension of amoeboid cells used in moving and feeding.

pteridophytes Seedless plants with true roots with lignified vascular tissue. The group includes ferns, whisk ferns, and horsetails.

pterosaurs Winged reptiles that lived during the time of dinosaurs.

pulmocutaneous The route of circulation that directs blood to the skin and lungs.

pulmonary circuit The branch of the circulatory system that supplies the lungs.

pulse The rhythmic stretching of the arteries caused by the pressure of blood forced through the arteries by contractions of the ventricles during systole.

punctuated equilibrium A theory of evolution advocating spurts of relatively rapid change followed by long periods of stasis.

Punnett square A diagram used in the study of inheritance to show the results of random fertilization.

pupil The opening in the iris, which admits light into the interior of the vertebrate eye; muscles in the iris regulate its size.

purine (pyū´-rēn) One of two families of nitrogenous bases found in nucleotides. Adenine (A) and guanine (G) are purines.

pyloric sphincter (pī-lōr´-ik sfink´-ter) In the vertebrate digestive tract, a muscular ring that regulates the passage of food out of the stomach and into the small intestine.

pyrimidine (puh-rim´-uh-dēn) One of two families of nitrogenous bases found in nucleotides. Cytosine (C), thymine (T), and uracil (U) are pyrimidines.

pyrogen Molecules that set the body's thermostat to a higher temperature. They are released by certain leukocytes.

Q_{10} effect The phenomenon that the rates of most enzyme-mediated reactions increase by a factor of 2–3 for every 10°C temperature increase.

quantitative character A heritable feature in a population that varies continuously as a result of environmental influences and the additive effect of two or more genes (polygenic inheritance).

quaternary structure (kwot´-er-nār-ē) The particular shape of a complex, aggregate protein, defined by the characteristic three-dimensional arrangement of its constituent subunits, each a polypeptide.

quiescent center A region located within the zone of cell division in plant roots, containing meristematic cells that divide very slowly.

R plasmid A bacterial plasmid carrying genes that confer resistance to certain antibiotics.

radial cleavage A type of embryonic development in deuterostomes in which the planes of cell division that transform the zygote into a ball of cells are either parallel or perpendicular to the polar axis, thereby aligning tiers of cells one above the other.

radial symmetry Characterizing a body shaped like a pie or barrel, with many equal parts radiating outward like the spokes of a wheel; present in cnidarians and echinoderms.

radiata Members of the radially symmetrical animal phyla, including cnidarians.

radiation The emission of electromagnetic waves by all objects warmer than absolute zero.

radicle An embryonic root of a plant.

radioactive dating A method of determining the age of fossils and rocks using half-lives of radioactive isotopes.

radioactive isotope An isotope (an atomic form of a chemical element) that is unstable; the nucleus decays spontaneously, giving off detectable particles and energy.

radiometric dating A method paleontologists use for determining the ages of rocks and fossils on a scale of absolute time, based on the half-life of radioactive isotopes.

radula A straplike rasping organ used by many mollusks during feeding.

random Describing a dispersion pattern in which individuals are spaced in a patternless, unpredictable way.

ras gene This gene codes for Ras protein, a G protein that relays a growth signal from a growth-factor receptor on the plasma membrane to a cascade of protein kinases that ultimately results in the stimulation of the cell cycle. Many ras oncogenes have a point mutation that leads to a hyperactive version of the Ras protein that can lead to excessive cell division.

ratites (rat´-īts) The group of flightless birds.

ray initials Cambium cells that produce radial files of parenchyma cells known as xylem rays and phloem rays.

ray-finned fish A bony fish having fins supported by thin, flexible skeletal rays. All but one living species of bony fishes are rayfins. See lobe-finned fish.

reactant A starting material in a chemical reaction.

reaction center The chlorophyll a molecule and the primary electron acceptor in a photosystem; they trigger the light reactions of photosynthesis. The chlorophyll donates an electron, excited by light energy, to the primary electron acceptor, which passes an electron to an electron transport chain.

reading frame The way a cell's mRNA-translating machinery groups the mRNA nucleotides into codons.

receptacle The site of attachment of the floral organs to the stem.

receptor potential An initial response of a receptor cell to a stimulus, consisting of a change in voltage across the receptor membrane proportional to the stimulus strength. The intensity of the receptor potential determines the frequency of action potentials traveling to the nervous system.

receptor-mediated endocytosis (en´-dō-sī-tō´-sis) The movement of specific molecules into a cell by the inward budding of membranous vesicles containing proteins with receptor sites specific to the molecules being taken in; enables a cell to acquire bulk quantities of specific substances.

recessive allele In a heterozygote, the allele that is completely masked in the phenotype.

reciprocal altruism Altruistic behavior between unrelated individuals, whereby the current altruistic individual benefits in the future when the current beneficiary reciprocates.

recognition species concept The idea that specific mating adaptations become fixed in a population and form the basis of species identification.

recombinant An offspring whose phenotype differs from that of the parents.

recombinant chromosomes Chromosomes created when crossing over combines the DNA from two parents into a single chromosome.

recombinant DNA A DNA molecule made in vitro with segments from different sources.

reconciliation behavior Post-conflict behavior that renews friendly relations.

recruitment The process of progressively increasing the tension of a muscle by activating more and more of the motor neurons controlling the muscle.

rectum The terminal portion of the large intestine where the feces are stored until they are eliminated.

red blood cell A blood cell containing hemoglobin, which transports O_2. Also called erythrocyte.

redox reaction (rē´-doks) A chemical reaction involving the transfer of one or more electrons from one reactant to another; also called oxidation-reduction reaction.

reducing agent The electron donor in a redox reaction.

reduction The addition of electrons to a substance involved in a redox reaction.

redundancy model The concept, put forth by Henry Gleason, that most of the species in a community are not tightly coupled with one another, meaning that the web of life is very loose. An increase or decrease in one species in a community has little effect on other species, which operate independently.

reflex An automatic reaction to a stimulus, mediated by the spinal cord or lower brain.

reflex arc The simplest type of nerve circuit.

refractory period (rē-frak´-tōr-ē) The short time immediately after an action potential in which the neuron cannot respond to another stimulus, owing to an increase in potassium permeability.

regeneration The regrowth of body parts from pieces of an organism.

regulative development A pattern of development, such as that of a mammal, in which the early blastomeres retain the potential to form the entire animal.

regulator A characterization of an animal in regard to a particular environmental variable. The animal is a regulator for that variable if it uses mechanisms of homeostasis to moderate internal change in the face of external fluctuation.

regulatory gene A gene that codes for a protein, such as a repressor, that controls the transcription of another gene or group of genes.

relative abundance Differences in the abundance of species within a community.

relative fitness The contribution of one genotype to the next generation compared to that of alternative genotypes for the same locus.

releaser A signal stimulus that functions as a communication signal between individuals of the same species.

releasing hormone A hormone produced by neurosecretory cells in the hypothalamus of the vertebrate brain that stimulates or inhibits the secretion of hormones by the anterior pituitary.

renal artery The blood vessel bringing blood to the kidney.

renal cortex The outer portion of the vertebrate kidney.

renal medulla The inner portion of the vertebrate kidney, beneath the renal cortex.

renal vein The blood vessel draining the kidney.

renin-angiotensin-aldosterone system (RAAS) A part of a complex feedback circuit that normally partners with antidiuretic hormone in osmoregulation.

repeated reproduction A life history in which adults produce large numbers of offspring over many years. Also known as iteroparity.

repetitive DNA Nucleotide sequences, usually noncoding, that are present in many copies in a eukaryotic genome. The repeated units may be short and arranged tandemly (in series) or long and dispersed in the genome.

replacement hypothesis Another name for the "out of Africa" hypothesis.

replication fork A Y-shaped region on a replicating DNA molecule where new strands are growing.

repressible enzyme An enzyme whose synthesis is inhibited by a specific metabolite.

repressor A protein that suppresses the transcription of a gene.

reproductive table An age-specific summary of the reproductive rates in a population.

Reptilia (rep-til´-ē-uh) The vertebrate class of reptiles, represented by lizards, snakes, turtles, and crocodilians.

residual volume The amount of air that remains in the lungs after forcefully exhaling.

resolving power A measure of the clarity of an image; the minimum distance that two points can be separated and still be distinguished as two separate points.

resource partitioning The division of environmental resources by coexisting species such that the niche of each species differs by one or more significant factors from the niches of all coexisting species.

respiratory medium The source of oxygen. It is typically air for terrestrial animals and water for aquatic organisms.

respiratory pigment Special proteins that transport most of the oxygen in blood.

respiratory surface The part of an animal where gases are exchanged with the environment.

resting potential The membrane potential characteristic of a nonconducting, excitable cell, with the inside of the cell more negative than the outside.

restriction enzyme A degradative enzyme that recognizes and cuts up DNA (including that of certain phages) that is foreign to a bacterium.

restriction fragment length polymorphisms (RFLPs) Differences in DNA sequence on homologous chromosomes that can result in different patterns of restriction fragment lengths (DNA segments resulting from treatment with restriction enzymes); useful as genetic markers for making linkage maps.

restriction site A specific sequence on a DNA strand that is recognized as a "cut site" by a restriction enzyme.

reticular fibers Very thin and branched fibers made of collagen. They form a tightly woven fabric that is continuous with the collagenous fibers of the extracellular matrix.

reticular formation A system of neurons, containing over 90 separate nuclei, that passes through the core of the brain stem.

retina (ret´-uh-nuh) The innermost layer of the vertebrate eye, containing photoreceptor cells (rods and cones) and neurons; transmits images formed by the lens to the brain via the optic nerve.

retinal The light-absorbing pigment in rods and cones of the vertebrate eye.

retrotransposons Transposable elements that move within a genome by means of an RNA intermediate, a transcript of the retrotransposon DNA.

retrovirus (ret´-trō-vī´-rus) An RNA virus that reproduces by transcribing its RNA into DNA and then inserting the DNA into a cellular chromosome; an important class of cancer-causing viruses.

reverse transcriptase (tran-skrip´-tās) An enzyme encoded by some RNA viruses that uses RNA as a template for DNA synthesis.

Rh factor A category of erythrocyte antigen that generates antibodies of the IgG class.

rhizoids Long tubular single cells or filaments of cells that anchor bryophytes to the ground. Rhizoids are not composed of tissues, they lack specialized conducting cells, and they do not play a primary role in water and mineral absorption.

rhodopsin A visual pigment consisting of retinal and opsin. When rhodopsin absorbs light, the retinal changes shape and dissociates from the opsin, after which it is converted back to its original form.

rhythm method A form of contraception that relies upon refraining from sexual intercourse when conception is most likely to occur; also called natural family planning.

ribonucleic acid (RNA) (rī´-bō-nū-klā´-ik) A type of nucleic acid consisting of nucleotide monomers with a ribose sugar and the nitrogenous bases adenine (A), cytosine (C), guanine (G), and uracil (U); usually single-stranded; functions in protein synthesis and as the genome of some viruses.

ribose The sugar component of RNA.

ribosomal RNA (rRNA) (rī´-buh-sōm´-ul) The most abundant type of RNA, which together with proteins, forms the structure of ribosomes. Ribosomes coordinate the sequential coupling of tRNA molecules to mRNA codons.

ribosome A cell organelle constructed in the nucleolus and functioning as the site of protein synthesis in the cytoplasm; consists of rRNA and protein molecules, which make up two subunits.

ribozyme (rī´-bō-zīm) An enzymatic RNA molecule that catalyzes reactions during RNA splicing.

ritual A type of symbolic activity.

rivet model The concept, put forth by Paul and Anne Ehrlich, that many or most of the species in a community are associated tightly with other species in a web of life. An increase or decrease in one species in a community affects many other species.

RNA interference (RNAi) A technique to silence the expression of selected genes in nonmammalian organisms. The method uses synthetic double-stranded RNA molecules matching the sequence of a particular gene to trigger the breakdown of the gene's messenger RNA.

RNA polymerase An enzyme that links together the growing chain of ribonucleotides during transcription.

RNA processing Modification of RNA before it leaves the nucleus, a process unique to eukaryotes.

RNA splicing The removal of noncoding portions (introns) of the RNA molecule after initial synthesis.

rod cell One of two kinds of photoreceptors in the vertebrate retina; sensitive to black and white and enables night vision.

root cap A cone of cells at the tip of a plant root that protects the apical meristem.

root hair A tiny projection growing just behind the root tips of plants, increasing surface area for the absorption of water and minerals.

root pressure The upward push of water within the stele of vascular plants, caused by active pumping of minerals into the xylem by root cells.

root system All of a plant's roots that anchor it in the soil, absorb and transport minerals and water, and store food.

rosette cellulose-synthesizing complexes Rose-shaped array of proteins that synthesize the cellulose microfibrils of the cell walls of charophyceans and land plants.

rough ER That portion of the endoplasmic reticulum studded with ribosomes.

round window The point of contact between the stapes and the cochlea. It is where the vibrations of the stapes create a traveling series of pressure waves in the fluid of the cochlea.

r-selection The concept that in certain (r-selected) populations, a high reproductive rate is the chief determinant of life history.

rubisco Ribulose carboxylase, the enzyme that catalyzes the first step of the Calvin cycle (the addition of CO_2 to RuBP, or ribulose bisphosphate).

ruminant An animal, such as a cow or a sheep, with an elaborate, multicompartmentalized stomach specialized for an herbivorous diet.

S phase The synthesis phase of the cell cycle; the portion of interphase during which DNA is replicated.

SA (sinoatrial) node The pacemaker of the heart, located in the wall of the right atrium. At the base of the wall separating the two atria is another patch of nodal tissue called the atrioventricular node (AV).

sac fungi Members of the phylum Ascomycetes, these range in size and complexity from unicellular yeasts to minute leafspot fungi to elaborate cup fungi and morels. About half of the sac fungi live with algae in the mutualistic associations called lichens.

saccule A chamber in the vestibule behind the oval window that participates in the sense of balance.

salivary amylase A salivary gland enzyme that hydrolyzes starch.

salivary glands Exocrine glands associated with the oral cavity. The secretions of salivary glands contain substances to lubricate food, adhere together chewed pieces into a bolus, and begin the process of chemical digestion.

salt Compounds resulting from the formation of ionic bonds, also called an ionic compound.

saltatory conduction (sol´-tuh-tōr´-ē) Rapid transmission of a nerve impulse along an axon resulting from the action potential jumping from one node of Ranvier to another, skipping the myelin-sheathed regions of membrane.

saprobe An organism that acts as a decomposer by absorbing nutrients from dead organic matter.

sarcomere (sar´-kō-mēr) The fundamental, repeating unit of striated muscle, delimited by the Z lines.

sarcoplasmic reticulum A specialized endoplasmic reticulum that regulates the calcium concentration in the cytosol.

saturated fatty acid A fatty acid in which all carbons in the hydrocarbon tail are connected by single bonds, thus maximizing the number of hydrogen atoms that can attach to the carbon skeleton.

savanna (suh-van´-uh) A tropical grassland biome with scattered individual trees, large herbivores, and three distinct seasons based primarily on rainfall, maintained by occasional fires and drought.

scaffolding protein A type of large relay protein to which several other relay proteins are simultaneously attached to increase the efficiency of signal transduction.

scanning electron microscope (SEM) A microscope that uses an electron beam to scan the surface of a sample to study details of its topography.

schizocoelous The type of development found in protostomes. Initially solid masses of mesoderm split to form coelomic cavities.

Schwann cells Glial cells that form insulating myelin sheaths around the axons of neurons in the peripheral nervous system.

scion (sī´-un) The twig grafted onto the stock when making a graft.

sclera (sklār´-uh) A tough, white outer layer of connective tissue that forms the globe of the vertebrate eye.

sclereid (sklār´-ē-id) A short, irregular sclerenchyma cell in nutshells and seed coats and scattered through the parenchyma of some plants.

sclerenchyma cell (skluh-ren´-kē-muh) A rigid, supportive plant cell type usually lacking protoplasts and possessing thick secondary walls strengthened by lignin at maturity.

scrotum A pouch of skin outside the abdomen that houses a testis; functions in cooling sperm, thereby keeping them viable.

scutellum (skū-tel´-um) A specialized type of cotyledon found in the grass family.

seascape Several different primarily aquatic ecosystems linked by exchanges of energy, materials, and organisms.

second law of thermodynamics The principle whereby every energy transfer or transformation increases the entropy of the universe. Ordered forms of energy are at least partly converted to heat, and in spontaneous reactions, the free energy of the system also decreases.

second messenger A small, nonprotein, water-soluble molecule or ion, such as calcium ion or cyclic AMP, that relays a signal to a cell's interior in response to a signal received by a signal receptor protein.

secondary cell wall A strong and durable matrix often deposited in several laminated layers for plant cell protection and support.

secondary compound A chemical compound synthesized through the diversion of products of major metabolic pathways for use in defense by prey species.

secondary consumer A member of the trophic level of an ecosystem consisting of carnivores that eat herbivores.

secondary growth The increase in girth of the stems and roots of many plants, especially woody, perennial dicots.

secondary immune response The immune response elicited when an animal encounters the same antigen at some later time. The secondary immune response is more rapid, of greater magnitude, and of longer duration than the primary immune response.

secondary oocyte A haploid cell resulting from meiosis I in oogenesis, which will become an ovum after meiosis II.

secondary plant body This is the tissue produced during secondary growth in diameter. Secondary growth produces secondary xylem (wood) and secondary phloem, and the cork cambium, which produces a tough, thick covering for stems and roots that replaces the epidermis.

secondary production The amount of chemical energy in consumers' food that is converted to their own new biomass during a given time period.

secondary structure The localized, repetitive coiling or folding of the polypeptide backbone of a protein due to hydrogen bond formation between peptide linkages.

secondary succession A type of succession that occurs where an existing community has been cleared by some disturbance that leaves the soil intact.

secretion (1) The discharge of molecules synthesized by a cell. (2) In the vertebrate kidney, the discharge of wastes from the blood into the filtrate from the nephron tubules.

secretory phase That portion of the menstrual cycle when the endometrium continues to thicken, becomes more vascularized, and develops glands that secrete a fluid rich in glycogen.

sedimentary rock (sed´-uh-men´-tuh-rē) Rock formed from sand and mud that once settled in layers on the bottom of seas, lakes, and marshes. Sedimentary rocks are often rich in fossils.

seed An adaptation for terrestrial plants consisting of an embryo packaged along with a store of food within a resistant coat.

seed coat A tough outer covering of a seed, formed from the outer coat of an ovule; in a flowering plant, it encloses and protects the embryo and endosperm.

seedless vascular plants The collective name for the phyla Lycophyta (lycophytes) and Pteridophyta (ferns, whisk ferns, and horsetails).

segmentation genes The genes of the embryo that direct the actual formation of segments after the embryo's axes are defined.

segment-polarity genes Genes that set the anterior-posterior axis of each segment in *Drosophila*.

selection coefficient The difference between two fitness values, representing a relative measure of selection against an inferior genotype.

selective channels Selective passageways across a cell membrane formed by specialized membrane transport proteins.

selective permeability A property of biological membranes that allows some substances to cross more easily than others.

self-incompatibility The capability of certain flowers to block fertilization by pollen from the same or a closely related plant.

semelparity A life history in which adults have but a single reproductive opportunity to produce large numbers of offspring, such as the life history of the Pacific salmon. Also known as "big-bang reproduction."

semen (sē´-mun) The fluid that is ejaculated by the male during orgasm; contains sperm and secretions from several glands of the male reproductive tract.

semicircular canals A three-part chamber of the inner ear that functions in maintaining equilibrium.

semiconservative model Type of DNA replication in which the replicated double helix consists of one old strand, derived from the old molecule, and one newly made strand.

semilunar valve A valve located at the two exits of the heart, where the aorta leaves the left ventricle and the pulmonary artery leaves the right ventricle.

seminal vesicle (sem´-uh-nul ves´-uh-kul) A gland in males that secretes a fluid component of semen that lubricates and nourishes sperm.

seminiferous tubules (sem´-uh-nif´-uh-rus) Highly coiled tubes in the testes in which sperm are produced.

sensation An impulse sent to the brain from activated receptors and sensory neurons.

sensitive period A limited phase in an individual animal's development when learning of particular behaviors can take place.

sensory adaptation The tendency of sensory neurons to become less sensitive when they are stimulated repeatedly.

sensory division The afferent neurons that convey information to the CNS from the sensory receptors that monitor the external and internal environment.

sensory input Information about the external and internal environment collected by sensory receptors.

sensory neuron A nerve cell that receives information from the internal and external environments and transmits the signals to the central nervous system.

sensory reception The detection of the energy of a stimulus by sensory cells.

sensory receptor Cellular systems that collect information about the physical world outside the body and inside the organism.

sensory transduction The conversion of a stimulus signal into an electrical signal by a sensory receptor cell.

sepal (sē´-pul) A whorl of modified leaves in angiosperms that encloses and protects the flower bud before it opens.

septa Cross-walls that divide fungal hyphae into cells. Septa generally have pores large enough to allow ribosomes, mitochondria, and even nuclei to flow from cell to cell; singular septum.

sequential hermaphroditism A reproductive pattern in which an individual reverses its sex during its lifetime.

serotonin A biogenic amine synthesized from the amino acid tryptophan.

seta (sē´-tuh) The elongated stalk of a moss sporophyte.

sex chromosomes One of the pair of chromosomes responsible for determining the sex of an individual.

sex-linked gene A gene located on a sex chromosome.

sexual dimorphism (dī-mōr´-fizm) A special case of polymorphism based on the distinction between the secondary sex characteristics of males and females.

sexual reproduction A type of reproduction in which two parents give rise to offspring that have unique combinations of genes inherited from the gametes of the two parents.

sexual selection Selection based on variation in secondary sex characteristics, leading to the enhancement of sexual dimorphism.

shared derived character An evolutionary novelty unique to a particular clade.

shared primitive character A homology common to a taxon more inclusive than the one being defined.

shoot system The aerial portion of a plant body, consisting of stems, leaves, and flowers.

short-day plant A plant that flowers, usually in late summer, fall, or winter, only when the light period is shorter than a critical length.

short-term memory The ability to hold information, anticipations, or goals for a time and then release them if they become irrelevant.

sickle-cell disease A human genetic disease of red blood cells caused by the substitution of a single amino acid in the hemoglobin protein; it is the most common inherited disease among African Americans.

sieve plate In a plant, a pore in the end wall of a sieve-tube member through which phloem sap flows.

sieve-tube member A chain of living cells that form sieve tubes in phloem.

sign stimulus An external sensory stimulus that triggers a fixed action pattern.

signal A behavior that causes a change in behavior in another animal.

signal peptide A stretch of amino acids on a polypeptide that targets the protein to a specific destination in a eukaryotic cell.

signal recognition particle (SRP) A protein-RNA complex that recognizes a signal peptide as it emerges from the ribosome.

signal-transduction pathway A mechanism linking a mechanical or chemical stimulus to a specific cellular response.

signature sequences Regions of small-subunit ribosomal RNA (SSU-rRNA) that have unique nucleotide sequences acquired by the accumulation of mutations in the ancestor of that taxonomic group.

simple epithelium An epithelium consisting of a single layer of cells that all touch the basal lamina.

single nucleotide polymorphisms (SNPs) One base-pair variation in the genome sequence.

single-lens eye The cameralike eye found in some jellies, polychaetes, spiders, and many mollusks.

single-strand binding protein During DNA replication, molecules that line up along the unpaired DNA strands, holding them apart while the DNA strands serve as templates for the synthesis of complimentary strands of DNA.

sink habitat A habitat where mortality exceeds reproduction.

sinuses Spaces surrounding the organs of the body in animals with open circulatory systems.

sister chromatids Replicated forms of a chromosome joined together by the centromere and eventually separated during mitosis or meiosis II.

skeletal muscle Striated muscle generally responsible for the voluntary movements of the body.

sliding-filament model The theory explaining how muscle contracts, based on change within a sarcomere, the basic unit of muscle organization, stating that thin (actin) filaments slide across thick (myosin) filaments, shortening the sarcomere; the shortening of all sarcomeres in a myofibril shortens the entire myofibril.

slow block to polyspermy The formation of the fertilization envelope and other changes in the egg's surface that prevent polyspermy.

slow muscle fibers Muscle cells that can sustain long contractions.

small intestine The longest section of the alimentary canal. It is the principal site of the enzymatic hydrolysis of food macromolecules and the absorption of nutrients.

small nuclear ribonucleoprotein (snRNP) (rī´-bō-nū´-klē-ō-prō´-tēn) One of a variety of small particles in the cell nucleus, composed of RNA and protein molecules; functions are not fully understood, but some form parts of spliceosomes, active in RNA splicing.

smooth ER That portion of the endoplasmic reticulum that is free of ribosomes.

smooth muscle A type of muscle lacking the striations of skeletal and cardiac muscle because of the uniform distribution of myosin filaments in the cell.

snowball Earth The hypothesis that glaciers covered the planet's land masses from pole to pole 750 to 570 million years ago.

social behavior Any kind of interaction between two or more animals, usually of the same species.

sociobiology The study of social behavior based on evolutionary theory.

sodium-potassium pump A special transport protein in the plasma membrane of animal cells that transports sodium out of the cell and potassium into the cell against their concentration gradients.

solute (sol´-ūt) A substance that is dissolved in a solution.

solution A homogeneous, liquid mixture of two or more substances.

solvent The dissolving agent of a solution. Water is the most versatile solvent known.

somatic cell (sō-mat´-ik) Any cell in a multicellular organism except a sperm or egg cell.

somatic nervous system The branch of the motor division of the vertebrate peripheral nervous system composed of motor neurons that carry signals to skeletal muscles in response to external stimuli.

somites Paired blocks of mesoderm just lateral to the notochord of a vertebrate embryo.

soredia Small clusters of lichen hyphae with embedded algae.

sori Clusters of fern sporangia on the backs of green leaves or on special, nongreen leaves (sporophylls). Sori may be arranged in various patterns, such as parallel lines or dots, that are useful in fern identification.

source habitat A habitat where reproduction exceeds mortality and from which excess individuals disperse.

Southern blotting A hybridization technique that enables researchers to determine the presence of certain nucleotide sequences in a sample of DNA.

specialized transduction The transfer of only those genes near the prophage site on the bacterial chromosome.

speciation (spē-sē-ā´-shun) The origin of new species in evolution.

species A group whose members possess similar anatomical characteristics and have the ability to interbreed.

species diversity The number and relative abundance of species in a biological community.

species richness The number of species in a biological community.

species selection A theory maintaining that species living the longest and generating the greatest number of species determine the direction of major evolutionary trends.

species-area curve The biodiversity pattern, first noted by Alexander von Humboldt, noting that the larger the geographic area of a community we sample, the greater the number of species.

specific epithet The second part of a binomial, it refers to one species within a genus.

specific heat The amount of heat that must be absorbed or lost for 1 gram of a substance to change its temperature by 1°C.

spectrophotometer An instrument that measures the proportions of light of different wavelengths absorbed and transmitted by a pigment solution.

spermatheca (sper´-muh-thē´-kuh) A sac in the female reproductive system where sperm are stored.

spermatogenesis The continuous and prolific production of mature sperm cells in the testis.

spermatozoon The male gamete.

sphincter (sfink´-ter) A ringlike valve, consisting of modified muscles in a muscular tube, such as a digestive tract; closes off the tube like a drawstring.

spinal nerve In the vertebrate peripheral nervous system, a nerve that carries signals to or from the spinal cord.

spiral cleavage A type of embryonic development in protostomes, in which the planes of cell division that transform the zygote into a ball of cells occur obliquely to the polar axis, resulting in cells of each tier sitting in the grooves between cells of adjacent tiers.

spiral valve A corkscrew-shaped ridge that increases surface area and prolongs the passage of food along the short digestive tract.

spliceosome (splī´-sē-ō-sōm) A complex assembly that interacts with the ends of an RNA intron in splicing RNA, releasing the intron, and joining the two adjacent exons.

spongocoel (spon´-gō-sēl) The central cavity of a sponge.

spontaneous generation The incorrect notion that life can emerge from inanimate material.

sporangium (plural, **sporangia**) A capsule in fungi and plants in which meiosis occurs and haploid spores develop.

spore In the life cycle of a plant or alga undergoing alternation of generations, a meiotically produced haploid cell that divides mitotically, generating a multicellular individual, the gametophyte, without fusing with another cell.

spore mother cells The cells that undergo meiosis and generate haploid spores within a sporangium.

sporophylls Lycophyte leaves specialized for reproduction.

sporophyte (spōr´-ō-fīt) The multicellular diploid form in organisms undergoing alternation of generations that results from a union of gametes and that meiotically produces haploid spores that grow into the gametophyte generation.

sporopollenin (spōr´-uh-pol´-uh-nin) A secondary product, a polymer synthesized by a side branch of a major metabolic pathway of plants that is resistant to almost all kinds of environmental damage; especially important in the evolutionary move of plants onto land.

sporozoites (spōr´-uh-zō´-īts) Tiny infectious cells of apicomplexans that spread disease.

squamata The group that includes lizards and snakes.

squamous The flat, "fried-egg" shape of one type of epithelial cell.

stability The tendencey of a biological community to resist change and return to its original species composition after being disturbed.

stabilizing selection Natural selection that favors intermediate variants by acting against extreme phenotypes.

stamen (stā´-men) The pollen-producing male reproductive organ of a flower, consisting of an anther and filament.

standard metabolic rate (SMR) The metabolic rate of a resting, fasting, and nonstressed ectotherm.

stapes The third of the three middle ear bones.

starch A storage polysaccharide in plants consisting entirely of glucose.

statocyst (stat´-uh-sist´) A type of mechanoreceptor that functions in equilibrium in invertebrates through the use of statoliths, which stimulate hair cells in relation to gravity.

statolith Sensory organs that contain mechanoreceptors and function in the sense of equilibrium.

stele The central vascular cylinder in roots where xylem and phloem are located.

stem cell In the bone marrow, a type of cell that gives rise to all the types of blood cells.

stenohaline Organisms that cannot tolerate substantial changes in external osmolarity.

stereoisomer A molecule that is a mirror image of another molecule with the same molecular formula.

steroid A type of lipid characterized by a carbon skeleton consisting of four rings with various functional groups attached.

sticky end A single-stranded end of a double-stranded DNA restriction fragment.

stigma (plural, **stigmata**) The sticky part of a flower's carpel, which traps pollen grains.

stipe A stemlike structure of a seaweed.

stock The plant that provides the root system when making a graft.

stoma (stō´-muh) (plural, **stomata**) A microscopic pore surrounded by guard cells in the epidermis of leaves and stems that allows gas exchange between the environment and the interior of the plant.

stomata A microscopic pore surrounded by guard cells in the epidermis of leaves and stems that allows gas exchange.

Stramenopila A diverse protistan clade which includes several heterotrophic groups and a variety of photosynthetic protists.

stratified epithelium An epithelium consisting of more than one layer of cells in which some but not all cells touch the basal lamina.

stress-induced proteins Special molecules, including heat-shock proteins, that are produced within cells in response to exposure to marked increases in temperature and to other forms of severe stress such as toxins, rapid pH changes, and viral infections.

strict aerobe An organism that can survive only in an atmosphere of oxygen, which is used in aerobic respiration.

strict anaerobe An organism that cannot survive in an atmosphere of oxygen. Other substances, such as sulfate or nitrate, are the terminal electron acceptors in the electron transport chains that generate their ATP.

stroke volume The amount of blood pumped by the left ventricle in each contraction.

stroma (strō´-muh) The fluid of the chloroplast surrounding the thylakoid membrane; involved in the synthesis of organic molecules from carbon dioxide and water.

stromatolite Rock made of banded domes of sediment in which are found the most ancient forms of life: prokaryotes dating back as far as 3.5 billion years.

structural formula A type of molecular notation in which the constituent atoms are joined by lines representing covalent bonds.

structural isomers Compounds that have the same molecular formula but differ in the covalent arrangements of their atoms.

style The stalk of a flower's carpel, with the ovary at the base and the stigma at the top.

substance P A neuropeptide that is a key excitatory signal that mediates our perception of pain.

substrate The reactant on which an enzyme works.

substrate-feeders Organisms that live in or on their food source, eating their way through the food.

substrate-level phosphorylation The formation of ATP by directly transferring a phosphate group to ADP from an intermediate substrate in catabolism.

sugar sink A plant organ that is a net consumer or storer of sugar. Growing roots, shoot tips, stems, and fruit are sugar sinks supplied by phloem.

sugar source A plant organ in which sugar is being produced by either photosynthesis or the breakdown of starch. Mature leaves are the primary sugar sources of plants.

sulfhydryl group A functional group consisting of a sulfur atom bonded to a hydrogen atom (—SH).

summation A phenomenon of neural integration in which the membrane potential of the postsynaptic cell in a chemical synapse is determined by the total activity of all excitatory and inhibitory presynaptic impulses acting on it at any one time.

supporting cell In the nervous system, a cell that protects, insulates, and reinforces a neuron.

suppressor T cell (T_S) A type of T cell that causes B cells as well as other cells to ignore antigens.

suprachiasmatic nuclei (SCN) A pair of structures in the hypothalamus of mammals that functions as a biological clock.

surface tension A measure of how difficult it is to stretch or break the surface of a liquid. Water has a high surface tension because of the hydrogen bonding of surface molecules.

survivorship curve A plot of the number of members of a cohort that are still alive at each age; one way to represent age-specific mortality.

suspension-feeder An aquatic animal, such as a clam or a baleen whale, that sifts small food particles from the water.

sustainable agriculture Long-term productive farming methods that are environmentally safe.

sustainable development The long-term prosperity of human societies and the ecosystems that support them.

swim bladder An adaptation, derived from a lung, that enables bony fishes to adjust their density and thereby control their buoyancy.

symbiont (sim´-bē-unt) The smaller participant in a symbiotic relationship, living in or on the host.

symbiosis An ecological relationship between organisms of two different species that live together in direct contact.

sympathetic division One of two divisions of the autonomic nervous system of vertebrates; generally increases energy expenditure and prepares the body for action.

sympatric speciation (sim-pat´-rik) A mode of speciation occurring as a result of a radical change in the genome of a subpopulation, reproductively isolating the subpopulation from the parent population.

symplast In plants, the continuum of cytoplasm connected by plasmodesmata between cells.

synapomorphies Shared derived characters; homologies that evolved in an ancestor common to all species on one branch of a fork in a cladogram, but not common to species on the other branch.

synapse (sin´-aps) The locus where one neuron communicates with another neuron in a neural pathway; a narrow gap between a synaptic terminal of an axon and a signal-receiving portion (dendrite or cell body) of another neuron or effector cell. Neurotransmitter molecules released by synaptic terminals diffuse across the synapse, relaying messages to the dendrite or effector.

synapsids One of three groups of amniotes based on key differences between their skulls.

synapsis The pairing of replicated homologous chromosomes during prophase I of meiosis.

synaptic cleft (sin-ap´-tik) A narrow gap separating the synaptic knob of a transmitting neuron from a receiving neuron or an effector cell.

synaptic terminal A bulb at the end of an axon in which neurotransmitter molecules are stored and released.

synaptic vesicles Membranous sacs containing thousands of neurotransmitters at the tip of the presynaptic axon.

syngamy (sin´-gam-ē) The process of cellular union during fertilization.

systematics The study of biological diversity in an environmental context, encompassing taxonomy and involving the reconstruction of phylogenetic history.

systemic acquired resistance (SAR) A defensive response in infected plants that helps protect healthy tissue from pathogenic invasion.

systemic circulation The branch of the circulatory system that supplies all body organs and then returns oxygen-poor blood to the right atrium via the veins.

systole (sis´-tō-lē) The stage of the heart cycle in which the heart muscle contracts and the chambers pump blood.

T (transverse) tubules Infoldings of the plasma membrane of skeletal muscle cells.

T cell receptor Antigen receptors on a T cell. Unlike antibodies, T cell receptors are never produced in a secreted form.

T lymphocyte (T cell) A type of lymphocyte responsible for cell-mediated immunity that differentiates under the influence of the thymus.

taiga (tī´-guh) The coniferous or boreal forest biome, characterized by considerable snow, harsh winters, short summers, and evergreen trees.

taproot A root system common to eudicots consisting of one large, vertical root (the taproot) that produces many smaller lateral, or branch roots.

target cell A cell that responds to a regulatory signal, such as a hormone.

TATA box A promoter DNA sequence crucial in forming the transcription initiation complex.

taxis (tak´-sis) Movement toward or away from a stimulus.

taxon (plural, **taxa**) The named taxonomic unit at any given level.

taxonomy (tak-son´-uh-mē) The branch of biology concerned with naming and classifying the diverse forms of life.

Tay-Sachs disease A human genetic disease caused by a dysfunctional enzyme that fails to break down brain lipids of a certain class; seizures, blindness, and degeneration of motor and mental performance usually become manifest a few months after birth.

T-dependent antigen Antigens that can stimulate antibody production only with help from T helper cells. Most protein antigens are T-dependent.

telomerase An enzyme that catalyzes the lengthening of telomeres. The enzyme includes a molecule of RNA that serves as a template for new telomere segments.

telomere (tel´-uh-mēr) The protective structure at each end of a eukaryotic chromosome. Specifically, the tandemly repetitive DNA at the end of the chromosome's DNA molecule. *See also* repetitive DNA.

telophase The fifth and final subphase of mitosis, in which daughter nuclei are forming and cytokinesis has typically begun.

temperate deciduous forest A biome located throughout midlatitude regions where there is sufficient moisture to support the growth of large, broadleaf deciduous trees.

temperate phages Phages that are capable of using either the lytic or lysogenic cycle.

temperate virus A virus that can reproduce without killing the host.

temperature A measure of the intensity of heat in degrees, reflecting the average kinetic energy of the molecules.

template strand The DNA strand that provides the template for ordering the sequence of nucleotides in an RNA transcript.

tendon A type of fibrous connective tissue that attaches muscle to bone.

tension Negative pressure on water or solutions. For example, prior to a medical injection, fluid is drawn up into a syringe because of tension created by pulling the plunger upwards.

terminal bud Embryonic tissue at the tip of a shoot, made up of developing leaves and a compact series of nodes and internodes.

terminator A special sequence of nucleotides in DNA that marks the end of a gene. It signals RNA polymerase to release the newly made RNA molecule, which then departs from the gene.

territory An area that an individual or individuals defend and from which other members of the same species are usually excluded.

tertiary consumer A member of a trophic level of an ecosystem consisting of carnivores that eat mainly other carnivores.

tertiary structure (ter-shē-ār-ē) Irregular contortions of a protein molecule due to interactions of side chains involved in hydrophobic interactions, ionic bonds, hydrogen bonds, and disulfide bridges.

testcross Breeding of an organism of unknown genotype with a homozygous recessive individual to determine the unknown genotype. The ratio of phenotypes in the offspring determines the unknown genotype.

testis (plural, **testes**) The male reproductive organ, or gonad, in which sperm and reproductive hormones are produced.

testosterone The most abundant androgen hormone in the male body.

tetanus (tet´-uh-nus) The maximal, sustained contraction of a skeletal muscle, caused by a very fast frequency of action potentials elicited by continual stimulation.

tetrad A paired set of homologous chromosomes, each composed of two sister chromatids. Tetrads form during prophase I of meiosis.

tetrapod A vertebrate possessing two pairs of limbs, such as amphibians, reptiles, birds, and mammals.

thalamus (thal´-uh-mus) One of two integrating centers of the vertebrate forebrain. Neurons with cell bodies in the thalamus relay neural input to specific areas in the cerebral cortex and regulate what information goes to the cerebral cortex.

thallus (plural, **thalli**) A seaweed body that is plantlike but lacks true roots, stems, and leaves.

therapsids The ancestral group of mammals that is part of the synapsid branch of reptilian phylogeny.

thermocline A narrow stratum of rapid temperature change in the ocean and in many temperate-zone lakes.

thermodynamics (ther´-mō-dī-nam´-iks) (1) The study of energy transformations that occur in a collection of matter. *See* first law of thermodynamics and second law of thermodynamics. (2) A phenomenon in which external DNA is taken up by a cell and functions there.

thermoreceptor An interoreceptor stimulated by either heat or cold.

thermoregulation The maintenance of internal temperature within a tolerable range.

theropods A group of relatively small, bipedal, carnivorous dinosaurs.

thick filament A filament composed of staggered arrays of myosin molecules; a component of myofibrils in muscle fibers.

thigmomorphogenesis A response in plants to chronic mechanical stimulation, resulting from increased ethylene production; an example is thickening stems in response to strong winds.

thigmotropism (thig´-mō-trō´-pizm) The directional growth of a plant in relation to touch.

thin filament The smaller of the two myofilaments consisting of two strands of actin and two strands of regulatory protein coiled around one another.

thiol Organic compounds containing sulfhydryl groups.

thoracic cavity The body cavity in mammals that houses the lungs and heart. It is surrounded in part by ribs and separated from the lower abdominal cavity by the diaphragm.

threatened species Species that is likely to become endangered in the foreseeable future throughout all or a significant portion of its range.

three-domain system A system of taxonomic classification based on three basic groups: Bacteria, Archaea, and Eukarya.

threshold potential The potential an excitable cell membrane must reach for an action potential to be initiated.

thrombus A clump of platelets and fibrin that block the flow of blood through a blood vessel.

thylakoid (thī´-luh-koyd) A flattened membrane sac inside the chloroplast, used to convert light energy to chemical energy.

thymus (thī´-mus) An endocrine gland in the neck region of mammals that is active in establishing the immune system; secretes several messengers, including thymosin, that stimulate T cells.

thyroid gland An endocrine gland that secretes iodine-containing hormones (T_3 and T_4), which stimulate metabolism and influence development and maturation in vertebrates, and cacitonin, which lowers blood calcium levels in mammals.

thyroid-stimulating hormone (TSH) A hormone produced by the anterior pituitary that regulates the release of thyroid hormones.

Ti plasmid A plasmid of a tumor-inducing bacterium that integrates a segment of its DNA into the host chromosome of a plant; frequently used as a carrier for genetic engineering in plants.

tidal volume The volume of air an animal inhales and exhales with each breath.

tight junction A type of intercellular junction in animal cells that prevents the leakage of material between cells.

T-independent antigen Antigen that can stimulate antibody production without the help of IL-2.

tissue An integrated group of cells with a common structure and function.

tolerate The neutral effect of early species on the appearance of later species in ecological succession, neither helping nor hindering the colonization of later arrivals.

tonoplast A membrane that encloses the central vacuole in a plant cell, separating the cytosol from the cell sap.

Top-Down Model A model of community organization in which predation controls community organization because predators control herbivores, which in turn control plants, which in turn control nutrient levels; also called the trophic cascade model.

topsoil A mixture of particles derived from rock, living organisms, and humus.

torpor In animals, a physiological state that conserves energy by slowing down the heart and respiratory systems.

torsion A characteristic of gastropods in which the body rotates during development.

totipotency The ability of embryonic cells to retain the potential to form all parts of the animal.

totipotent The ability of a cell to form all parts of the mature organism.

trace element An element indispensable for life but required in extremely minute amounts.

trachea (trā´-kē-uh) The windpipe; that portion of the respiratory tube that has C-shaped cartilagenous rings and passes from the larynx to two bronchi.

tracheae (trā´-kē-ē) Tiny air tubes that branch throughout the insect body for gas exchange.

tracheal system A gas exchange system of branched, chitin-lined tubes that infiltrate the body and carry oxygen directly to cells in insects.

tracheid (trā´-kē-id) A water-conducting and supportive element of xylem composed of long, thin cells with tapered ends and walls hardened with lignin.

transcription The synthesis of RNA on a DNA template.

transcription factor A regulatory protein that binds to DNA and stimulates transcription of specific genes.

transcription initiation complex The completed assembly of transcription factors and RNA polymerase bound to the promoter.

transduction A DNA-transfer process used by phages to carry bacterial genes from one host cell to another.

transfer cells Companion cells with numerous ingrowths of their walls which increase the cells' surface area and enhance the transfer of solutes between apoplast and symplast.

transfer RNA (tRNA) An RNA molecule that functions as an interpreter between nucleic acid and protein language by picking up specific amino acids and recognizing the appropriate codons in the mRNA.

transformation (1) The conversion of a normal animal cell to a cancerous cell. (2) A change in genotype and phenotype due to the assimilation of external DNA by a cell.

translation The synthesis of a polypeptide using the genetic information encoded in an mRNA molecule. There is a change of "language" from nucleotides to amino acids.

translocation (1) An aberration in chromosome structure resulting from an error in meiosis or from mutagens; specifically, attachment of a chromosomal fragment to a nonhomologous chromosome. (2) During protein synthesis, the third stage in the elongation cycle when the RNA carrying the growing polypeptide moves from the A site to the P site on the ribosome. (3) The transport via phloem of food in a plant.

transmission The conduction of impulses to the central nervous system.

transmission electron microscope (TEM) A microscope that passes an electron beam through very thin sections, primarily used to study the internal ultrastructure of cells.

transpiration The evaporative loss of water from a plant.

transpiration-to-photosynthesis ratio The amount of water lost per gram of carbon dioxide assimilated into organic material by photosynthesis.

transport protein A transmembrane protein that helps a certain substance or class of closely related substances to cross the membrane.

transport vesicle A tiny membranous sac in a cell's cytoplasm carrying molecules produced by the cell.

transposon (trans-pō´-zon) A transposable genetic element; a mobile segment of DNA that serves as an agent of genetic change.

triacylglycerol Three fatty acids linked to one glycerol molecule.

triiodothyrodine (trī´-ī-ō´-dō-thī´-rō-dēn) One of two very similar hormones produced by the thyroid gland and derived from the amino acid tyrosine.

Trilobita An extinct phylum of arthropods with pronounced segmentation.

trilobites Members of an extinct phylum of arthropods with pronounced segmentation.

trimester In human development, one of three 3-month long periods of pregnancy.

triplet code A set of three-nucleotide-long words that specify the amino acids for polypeptide chains.

triploblastic Possessing three germ layers: the endoderm, mesoderm, and ectoderm. Most eumetazoa are triploblastic.

trisomic A chromosomal condition in which a particular cell has an extra copy of one chromosome, instead of the normal two; the cell is said to be trisomic for that chromosome.

trochophore A ciliated larva common to the life cycle of many mollusks, it is also characteristic of marine annelids and some other lophotrochozoans.

trophic level (trō´-fik) Any of the several levels of a food chain, whose species are based on their main nutritional source. The trophic level that ultimately supports all others consists of autotrophs, or primary producers.

trophic structure The different feeding relationships in an ecosystem, which determine the route of energy flow and the pattern of chemical cycling.

trophoblast The outer epithelium of the blastocyst, which forms the fetal part of the placenta.

tropic hormone A hormone that has another endocrine gland as a target.

tropical rain forest The most complex of all communities, located near the equator where rainfall is abundant; harbors more species of plants and animals than all other terrestrial biomes combined.

tropics Latitudes between 23.5° north and south.

tropism A growth response that results in the curvature of whole plant organs toward or away from stimuli due to differential rates of cell elongation.

tropomyosin The regulatory protein that blocks the myosin binding sites on the actin molecules.

troponin complex The regulatory proteins that control the position of tropomyosin on the thin filament.

true-breeding Plants that produce offspring of the same variety when they self-pollinate.

trypsin An enzyme found in the duodenum. It is specific for peptide bonds adjacent to certain amino acids.

tubal ligation A means of sterilization in which a woman's two oviducts (Fallopian tubes) are tied closed to prevent eggs from reaching the uterus; a segment of each oviduct is removed.

tube feet Extensions of a network of hydraulic canals that function in locomotion, gas exchange, and feeding.

tumor A mass of abnormal cells within otherwise normal tissue, caused by the uncontrolled growth of a transformed cell.

tumor antigen A foreign macromolecule, associated with a tumor, which does not belong to the host organism and that elicits an immune response.

tumor-suppressor gene A gene whose protein products inhibit cell division, thereby preventing uncontrolled cell growth (cancer).

tundra A biome at the extreme limits of plant growth; at the northernmost limits, it is called arctic tundra, and at high altitudes, where plant forms are limited to low shrubby or matlike vegetation, it is called alpine tundra.

tunicates Members of the subphylum Urochordata.

turgid (ter´-jid) Firm. Walled cells become turgid as a result of the entry of water from a hypotonic environment.

turgor pressure The force directed against a cell wall after the influx of water and the swelling of a walled cell due to osmosis.

turnover The mixing of waters as a result of changing water-temperature profiles in a lake.

tympanic membrane Another name for the eardrum.

type I diabetes mellitus One of the two forms of diabetes, it is an autoimmune disorder usually requiring insulin injections several times a day.

type II diabetes mellitus One of the two forms of diabetes, it is characterized by either a deficiency of insulin or more commonly, by reduced responsiveness in target cells due to some change in insulin receptors.

tyrosine kinase An enzyme that catalyzes the transfer of phosphate groups from ATP to the amino acid tyrosine on a substrate protein.

tyrosine kinase receptor A receptor protein in the plasma membrane that responds to the binding of a signal molecule by catalyzing the transfer of phosphate groups from ATP to tyrosines on the cytoplasmic side of the receptor. The phosphorylated tyrosines activate other signal-transduction proteins within the cell.

ultimate causation The evolutionary explanation for a behavioral, physiological, or morphological response, in contrast to proximate causation; the immediate mechanisms that underlie a response.

ultracentrifuge A machine that spins test tubes at the fastest speeds to separate liquids and particles of different densities.

undernourishment A diet that is chronically deficient in calories.

uniform Describing a dispersion pattern in which individuals are evenly distributed.

uniformitarianism Charles Lyell's idea that geologic processes have not changed throughout Earth's history.

Uniramia The animal phylum that includes centipedes, millipedes, and insects.

uniramians Members of the animal phylum that includes centipedes, millipedes, and insects.

unisexual flower A flower missing either stamens or carpels.

unsaturated fatty acid A fatty acid possessing one or more double bonds between the carbons in the hydrocarbon tail. Such bonding reduces the number of hydrogen atoms attached to the carbon skeleton.

urbilateria The original group of bilateral animals that were relatively complex with true coeloms.

urea A soluble form of nitrogenous waste excreted by mammals and most adult amphibians.

ureter A duct leading from the kidney to the urinary bladder.

urethra A tube that releases urine from the body near the vagina in females or through the penis in males; also serves in males as the exit tube for the reproductive system.

uric acid An insoluble precipitate of nitrogenous waste excreted by land snails, insects, birds, and some reptiles.

urinary bladder The pouch where urine is stored prior to elimination.

urochordate A chordate without a backbone, commonly called a tunicate, a sessile marine animal.

urodela The order of salamanders that includes tetrapod amphibians with tails.

uterus A female organ where eggs are fertilized and/or development of the young occurs.

utricle A chamber behind the oval window that opens into the three semicircular canals.

vaccination A procedure that presents the immune system with a harmless variant or derivative of a pathogen, thereby stimulating the immune system to mount a long-term defense against the pathogen.

vaccine A harmless variant or derivative of a pathogen that stimulates a host's immune system to mount defenses against the pathogen.

vacuole (vak´-ū-ōl) A membrane-enclosed sac taking up most of the interior of a mature plant cell and containing a variety of substances important in plant reproduction, growth, and development.

vagina Part of the female reproductive system between the uterus and the outside opening; the birth canal in mammals; also accommodates the male's penis and receives sperm during copulation.

valence The bonding capacity of an atom generally equal to the number of unpaired electrons in the atom's outermost shell.

valence electron The electrons in the outermost electron shell.

valence shell The outermost energy shell of an atom, containing the valence electrons involved in the chemical reactions of that atom.

van der Waals interactions Weak attractions between molecules or parts of molecules that are brought about by localized charge fluctuations.

variation Differences between members of the same species.

vas deferens The tube in the male reproductive system in which sperm travel from the epididymis to the urethra.

vasa recta The capillary system that serves the loop of Henle.

vascular bundle (vas´-kyū-ler) A strand of vascular tissues (both xylem and phloem) in a plant stem.

vascular cambium A continuous cylinder of meristematic cells surrounding the xylem and pith that produces secondary xylem and phloem.

vascular plant A plant with vascular tissue. Vascular plants include all modern species except the mosses and their relatives.

vascular tissue Plant tissue consisting of cells joined into tubes that transport water and nutrients throughout the plant body.

vascular tissue system A system formed by xylem and phloem throughout the plant, serving as a transport system for water and nutrients, respectively.

vasectomy The cutting of each vas deferens to prevent sperm from entering the urethra.

vasocongestion The filling of a tissue with blood caused by increased blood flow through the arteries of that tissue.

vasoconstriction A decrease in the diameter of superficial blood vessels triggered by nerve signals that contract the muscles of the vessel walls.

vasodilation An increase in the diameter of superficial blood vessels triggered by nerve signals that relax the muscles of the vessel walls.

vegetal pole The portion of the egg where most yolk is concentrated. Opposite of animal pole.

vegetative reproduction Cloning of plants by asexual means.

vein A vessel that returns blood to the heart.

ventilation Any method of increasing contact between the respiratory medium and the respiratory surface.

ventral Pertaining to the underside, or bottom, of a bilaterally symmetrical animal.

ventricle (ven´-truh-kul) (1) A heart chamber that pumps blood out of a heart. (2) A space in the vertebrate brain, filled with cerebrospinal fluid.

venule (ven´-ūl) A vessel that conveys blood between a capillary bed and a vein.

vertebrate A chordate animal with a backbone: the mammals, birds, reptiles, amphibians, and various classes of fishes.

vesicle A sac made of membrane inside of cells.

vessel element A specialized short, wide cell in angiosperms; arranged end to end, they form continuous tubes for water transport.

vestibule The cavity enclosed by the labia minora, it is the space into which the vagina and urethral opening empty.

vestigial organs Structures of marginal, if any, importance to an organism. They are historical remnants of structures that had important functions in ancestors.

villi (singular, **villus**) (1) A fingerlike projection of the inner surface of the small intestine. (2) A fingerlike projection of the chorion of the mammalian placenta. Large numbers of villi increase the surface areas of these organs.

viral envelope A membrane that cloaks the capsid that in turn encloses a viral genome.

viroid (vī´-royd) A plant pathogen composed of molecules of naked circular RNA only several hundred nucleotides long.

virulent virus A virus that reproduces only by a lytic cycle.

visceral mass One of the three main parts of a mollusk, it contains most of the internal organs.

visceral muscle Smooth muscle found in the walls of the digestive tract, bladder, arteries, and other internal organs.

visible light That portion of the electromagnetic spectrum detected as various colors by the human eye, ranging in wavelength from about 380 nm to about 750 nm.

vital capacity The maximum volume of air that a respiratory system can inhale and exhale.

vitalism The belief that natural phenomena are governed by a life force outside the realm of physical and chemical laws.

vitamin An organic molecule required in the diet in very small amounts; vitamins serve primarily as coenzymes or parts of coenzymes.

vitamin D The active form functions as a hormone, acting in concert with parathyroid hormone in bone and promoting the uptake of calcium from food within the intestines.

vitreous humor The jellylike material that fills the posterior cavity of the vertebrate eye.

viviparous (vī-vip´-uh-rus) Referring to a type of development in which the young are born alive after having been nourished in the uterus by blood from the placenta.

vocal cord One of a pair of stringlike tissues in the larynx. Air rushing past the tensed vocal cords makes them vibrate, producing sounds.

voltage-gated ion channels Specialized ion channels that open or close in response to changes in membrane potential.

water molds Members of the group Oomycota, they are heterotrophic stramenopiles that lack chloroplasts. They typically have cell walls made of cellulose.

water potential The physical property predicting the direction in which water will flow, governed by solute concentration and applied pressure.

water vascular system A network of hydraulic canals unique to echinoderms that branches into extensions called tube feet, which function in locomotion, feeding, and gas exchange.

wavelength The distance between crests of waves, such as those of the electromagnetic spectrum.

wetland An ecosystem intermediate between an aquatic one and a terrestrial one. Wetland soil is saturated with water permanently or periodically.

white blood cell A blood cell that functions in defending the body against infections and cancer cells. Also called leukocyte.

white matter Tracts of axons within the CNS.

white rusts Members of the group Oomycota, they are heterotrophic stramenopiles that lack chloroplasts, typically have cell walls made of cellulose, and generally live on land as parasites of plants.

wild type An individual with the normal phenotype.

wobble A violation of the base-pairing rules in that the third nucleotide (5' end) of a tRNA anticodon can form hydrogen bonds with more than one kind of base in the third position (3' end) of a codon.

xerophytes Plants adapted to arid climates.

x-ray crystallography A technique that depends on the diffraction of an X-ray beam by the individual atoms of a molecule to study the three-dimensional structure of a molecule.

xylem (zī´-lum) The tube-shaped, nonliving portion of the vascular system in plants that carries water and minerals from the roots to the rest of the plant.

yeast A unicellular fungus that lives in liquid or moist habitats, primarily reproducing asexually by simple cell division or by budding of a parent cell.

yeast artificial chromosomes (YACs) Yeast artificial chromosomes; vectors that combine the essentials of a eukaryotic chromosome, an origin for DNA replication, a centromere, and two telomeres, with foreign DNA.

yolk Nutrients stored in an egg.

yolk plug Large food-laden endodermal cells surrounded by the blastopore of an amphibian gastrula.

yolk sac One of four extraembryonic membranes that supports embryonic development; the first site of blood cells and circulatory system function.

Z line The borders of a sarcomere.

zero population growth A period of stability in population size when the per capita birth rates and death rates are equal.

zona pellucida The extracellular matrix of a mammalian egg.

zone of elongation This is the region of the root tip adjacent to the zone of cell division. Cells sometimes elongate to more than ten times their original length.

zone of maturation This is the region of the root tip adjacent to the zone of cell elongation. As cells finish elongating, they begin to specialize in structure and function.

zone of polarizing activity A limb-bud organizing region located where the posterior side of the bud is attached to the body.

zoned reserve An extensive region of land that includes one or more areas undisturbed by humans surrounded by lands that have been changed by human activity and are used for economic gain.

zygote The diploid product of the union of haploid gametes in conception; a fertilized egg.

INDEX

A *t* following a page number indicates a table; an *f* following a page number indicates a figure.